Emilio Mutis Duplat
IX 10/007
Austin, Texas, U.S.A.

Special Publication 5
Mineral Deposits Division, Geological Association of Canada

Mineral Deposits of Canada

A Synthesis of Major Deposit-types, District Metallogeny, the Evolution of Geological Provinces, and Exploration Methods

Wayne D. Goodfellow, Editor

Published in Partnership by
Geological Association of Canada, Mineral Deposits Division
and
Geological Survey of Canada

National Library of Canada Cataloguing in Publication Data

Main entry under title:

Mineral Deposits of Canada: A Synthesis of Major Deposit-types, District Metallogeny, the Evolution of Geological Provinces, and Exploration Methods

Geological Association of Canada, Mineral Deposits Division, Special Publication No. 5

ISBN 978-1-897095-24-9

1. Ore deposits--Canada. 2. Metallogeny--Canada. 3. Physiographic provinces--Canada. 4. Mines and mineral resources--Canada. 5. Prospecting--Canada. I. Goodfellow, Wayne D. (Wayne David), 1949- II. Geological Association of Canada III. Series: Special publication (Geological Association of Canada. Mineral Deposits Division); no. 5

QE376.M46 2007 553.0971 C2007-901431-3

Publisher

Geological Association of Canada - Mineral Deposits Division

Department of Earth Sciences

Memorial University of Newfoundland

St. John's, Newfoundland, Canada A1B 3X5

Technical Editing: Elizabeth J. Ambrose, Lee Ewert, Jennifer Owen, Ingrid Kjarsgaard, and Shawn Hood

Technical Editing of French Abstracts: Benoit Couture

Desk Layout: Elizabeth J. Ambrose, Lee Ewert

Printed and bound in Canada by Friesens Corporation

Available From

Geological Association of Canada

Department of Earth Sciences

Memorial University of Newfoundland

St. John's, NL A1B 3X5 CANADA

Tel: (709) 737-7660; Fax: (709) 737-2532

Web: www.gac.ca

E-mail: gac@esd.mun.ca

The Geological Survey of Canada Bookstore

Geological Survey of Canada

601 Booth Street

Ottawa, Ontario K1A 0E8

Tel:1-613-995-4342; Fax:1-613-943-0646

Web: http://gsc.nrcan.gc.ca/bookstore

E-mail: gscbookstore@nrcan.gc.ca

Gold-rich specimen from the Dome Mine, Timmins, Ontario that was presented to GSC in 1923 by the then Board of Directors of Dome Mines. Weight is 136 lbs (61.8kg) of which about 20% by weight is gold. This specimen is a high-grade Timiskaming conglomerate hosting folded carbonate-pyrite-quartz veins with spectacular visible gold. It most likely came from the bonanza East Dome area which was discovered in 1910. It consists of sub-rounded to sub-angular altered and non-altered clasts and folded cross-cutting veins of coarse pyrite, ankerite and minor quartz, shattered and invaded by gold. GSC sample from National Mineral Collection No 10003. Photograph by Igor Bilot, GSC. (see Dubé and Gosselin., this volume).

Mineral Deposits of Canada

A Synthesis of Major Deposit-Types, District Metallogeny, the Evolution of Geological Provinces, and Exploration Methods

Edited by
Wayne D. Goodfellow

DVD Contents

1. Appendices consisting of spreadsheets containing the key attributes (about 40) of deposits in Canada and worldwide.
2. ArcGIS multiparameter, attributed, and co-registered database of mineral deposits of Canada and the world.
3. Ore photo library of major deposits including a) location map, b) geological maps and cross-sectional geology, c) genetic model(s), and d) photographs and photomicrographs of representative rocks, including host lithologies, ore facies, and alteration facies.
4. Digital files of all figures and tables from papers within the volume.
5. An Endnote bibliography of all literature referenced in the volume. A Microsoft WORD file of all references is also included.
6. PDF files of all papers in the volume.

Mineral Deposits of Canada
A Synthesis of Major Deposit-types, District Metallogeny, the Evolution of Geological Provinces, and Exploration Methods
an Introduction

Wayne D. Goodfellow

Geological Survey of Canada, 601 Booth Street, Ottawa, Ontario, K1A 0E8
Corresponding author's email: wgoodfel@nrcan.gc.ca

The *Mineral Deposits of Canada* volume is an outcome of the Consolidation and Synthesis of Mineral Deposits Knowledge (CSMDK) project which was initiated as part of the Geological Survey of Canada's (GSC) program entitled the Consolidation of Canada's Geoscience Knowledge (CCGK). The final stage of producing this volume was supported by the Deep Search project under the GSC's TGI3 program.

The overall objective of the CCGK program was to capture the current knowledge of major mineral deposits in Canada. This objective was addressed by compiling and synthesizing knowledge and data on mineral deposits in Canada and elsewhere, with a focus on information that was acquired over the past 10 years, by updating the geoscience infrastructure for mineral deposits, by improving data and knowledge accessibility using web-based systems, and ultimately by publication of the results in this volume. This research was undertaken by a collaborative and multidisciplinary team of bedrock and Quaternary geologists, mineral deposits geologists and geochemists, geophysicists, surficial geochemists, and GIS specialists.

The initial plan of the CSMDK project was to release all the synthesis products and data online using an interactive and user friendly web-based system. To this end, a web site of preliminary products was established which can be accessed at: http://gsc.nrcan.gc.ca/mindep/. Many of the knowledge and digital data products that appear in this volume were released in draft form at this web site. A measure of the success of this web site is provided by the 50,000 to 150,000 hits per month since the web site was established in February, 2006.

After the Mineral Deposits of Canada web site was established, it was decided that although the web site met the needs of our target audience, it did not meet the needs of those who prefer a final hardcopy version of all synthesis products. Concerns were expressed that the transient nature of web sites may result in the future loss of valuable knowledge and data on Canada's mineral deposits. This led to discussions with the Mineral Deposits Division (MDD) of the Geological Association of Canada (GAC) on a possible volume on the mineral deposits of Canada jointly funded by MDD/GAC and the GSC. After this volume was approved by both the MDD and the GSC, it was decided that the GSC contributions under the CCGK program would be augmented with contributions from the university community, and provincial and territorial geoscience organizations. It was also decided that the volume would be produced in unrestricted colour to take advantage of the wealth of knowledge and data captured by colour geological maps and sections, geophysical and geochemical maps, complex colour plots of geoscience data, 2D and 3D geological and genetic models, and rock photographs and photomicrographs.

With respect to content, the volume is designed to have a global perspective but a Canadian focus, and consists of five major themes: 1) an overview of the economic and geological contexts of mineral deposits in Canada; 2) a current synthesis of the economically important deposit-types, including VMS, SEDEX, MVT, magmatic Ni-Cu-PGE, porphyry deposits, unconformity uranium, kimberlite diamond, IOCG, and several types of lode gold deposits - quartz-carbonate veins, Au-rich VMS, intrusion-related Au, and epithermal Au; 3) the metallogeny of economically important mining camps; 4) the geological evolution and metallogeny of Canadian geological provinces; and 5) the current understanding of geophysical and geochemical exploration methods applied to specific deposit-types.

The *Mineral Deposits of Canada* volume also contains an extensive digital geoscience database which is housed on four high-capacity DVDs. This database includes: 1) spreadsheets with up to 40 key parameters of Canadian and global mineral deposits; 2) ArcGIS Canadian and global mineral deposit databases; 3) an ore photo library of major deposits in Canada; 4) digital files of all figures and tables in papers in the volume; and 5) an Endnote bibliography of all references in individual papers and digital products in the DVD. The objective of the DVDs is to provide a readily accessible and user friendly resource for researchers, teachers and students in the university community, for geoscientists in government agencies, and for explorationists in the mining and exploration industry.

Although the volume has been compared with the *Geology of Mineral Deposit Types* published by the Decade of North American Geology (DNAG), it is in fact very different. The DNAG volume is much more focused and provides a synthesis of most deposit-types regardless of their economic importance to Canada. As such, it represents a valuable source of knowledge on most types of deposits in Canada and elsewehere. The *Mineral Deposits of Canada* (MDC) volume, however, does not attempt to consider all or even most deposit-types, but focuses on 12 major deposit-types, most of which are, have been, or have the potential to be of major economic importance to Canada. Furthermore, the synthesis of major deposit-types represents no more than about 25% of the MDC volume. Other major areas covered, which are not considered specifically by the DNAG volume, include the economic context of Canada's mineral deposits, the metallogeny of mining districts, the geological evolution and metallogeny of geological provinces in Canada, and the current status of geophysical and geochemical exploration methods.

An immense debt of gratitude is owed to a very dedicated group of individuals who helped make this volume a reality. First of all, I would like to thank the many geoscientists across Canada who have contributed up-to-date syntheses and data on complex areas of geoscience in diverse fields and topics that typically encompass a large range of disciplines. Secondly, I wish to thank the 70 reviewers listed below from virtually every continent who provided critical and thorough reviews of all 45 papers in the volume, sometimes with a two-week turnaround, without which the MDC volume would not be possible. Thirdly, I wish thank Elizabeth Ambrose, Jennifer Owen, Benoît Couture, Ingrid Kjarsgaard, and Shawn Hood for technical editing, Elizabeth Ambrose and Lee Ewart for desktop layout of manuscripts, Shawn Hood for building Endnote bibliographies of each paper, John Grant and Jennifer Shaw for building the Mineral Deposits of Canada web site (http://gsc.nrcan.gc.ca/mindep/), and Igor Bilot for cutting and photographing rocks for the Ore Photo Library. I am also indebted to Dirk Templeman-Kluit, Chair of the Publications Committee, Mineral Deposits Division (MDD) of the GAC for stick handling this volume through MDD and providing unwavering support. Finally, I would like to thank my wife, Marianne, for helping complete the final editing of the volume.

Reviewers

Dr. Cam Allen
Dr. Kevin Ansdell
Mr. Stu Averill
Dr. Sarah-Jane Barnes
Dr. Georges Beaudoin
Dr. Steve Beresford
Dr. Eion Cameron
Dr. Ray Coish
Mr. David Crabtree
Dr. Tony Davidson
Dr. Benoît Dubé
Dr. Murray Duke
Dr. Michael R. Easton
Dr. Catharine Farrow
Dr. Sunil Gandhi
Dr. Michel Gauthier
Dr. Richard Goldfarb
Dr. Wayne Goodfellow
Dr. David Gray
Dr. Michael H. Gunning
Dr. Jacob J. Hanley
Dr. Jim Hibbard
Dr. Larry Hulbert
Dr. Christopher Jauer
Dr. Michel Jebrak
Dr. Ian Jonasson
Dr. Karen Kelly
Dr. Andy Kerr
Dr. John Ketchum
Dr. Jon Kim
Dr. Rod Kirkham
Dr. Timothy Kusky
Dr. Benoit Lafrance
Dr. Alain Leclair
Dr. Marc Legault
Dr. David Lentz
Dr. Patrick Mercier-Langevin
Dr. C.M. Lesher
Dr. Peter Lightfoot
Dr. Robert L. Linnen
Dr. John Lydon
Dr. Roger MacQueen
Dr. Steve McCutcheon
Dr. Bernd Milkereit
Dr. Roger Mitchell
Dr. Thomas Monecke
Dr. Bill Morris
Dr. James Mungall
Dr. Tony Naldrett
Dr. Warren Nokleberg
Mr. Tom Nowicki
Dr. Serge Paradis
Dr. Suzanne Paradis
Dr. Graham Parson
Dr. John Percival
Dr. Jan Peter
Dr. Steve Piercey
Dr. Edward Ripley
Dr. Steven Rowins
Dr. Bob Ryan
Dr. Hamish Sandeman
Dr. Don Sangster
Dr. Roger G. Skirrow
Dr. Paul Taufen
Dr. Mike Thomas
Dr. Robert Thompson

An Overview of the Economic and Geological Contexts of Canada's Major Mineral Deposit Types

John W. Lydon

Geological Survey of Canada, 562 Booth Street, Ottawa, Ontario K1A 0E4
Corresponding author's email: jlydon@nrcan.gc.ca

Abstract

Seven mineral deposit types have contributed over 90% of the value of non-ferrous metalliferous mineral production in Canada. Based on average 1996 to 2005 inflation-adjusted metal prices, to the end of 2005 the most productive mineral deposit types have been 1) magmatic Ni-Cu deposits (>$372 billion), mainly from Proterozoic rocks in the Sudbury and Thompson areas; 2) volcanogenic massive sulphide (VMS) deposits ($192 billion), mainly from Archean greenstone belts of Quebec and Ontario, the Proterozoic volcanic belts of Manitoba, and Paleozoic volcanic rocks of New Brunswick; and 3) lode gold deposits ($132 billion), mainly from quartz-carbonate veins of Archean greenstone belts of Quebec and Ontario. Collectively, porphyry, sedimentary exhalative (SEDEX), Mississippi Valley, and uranium deposit types have contributed about $140 billion, and diamonds, a relatively new but growing mineral commodity for Canada, has contributed $8 billion. The dollar equivalent of metal contents per tonne of ore mined over the past five years range from about $130/t to about $350/t for most underground base metal and diamond mines, and $90/t to $300/t for most underground Au mines. Dollar equivalent of metal contents exceed $450/t only in a few metal mines. The average ore dollar equivalent of metal contents range from $10/t to $45/t for open pit metal mines. The most valuable ores are those of U deposits of the Athabasca Basin, where past production has averaged dollar equivalent of metal contents of $540/t, and current reserves are worth $1,000/t to $11,000/t, based on the ten year average value for U, or over three times these values based on the average 2006 uranium price.

About 50% of production and 57% of the $1.57 trillion of the non-ferrous metal and diamond content of total mineral resources are associated with volcanic arcs and back-arcs that were accreted to, or built upon, continental margins during the assembly of supercontinents. Deposit types include VMS, porphyry, komatiitic Ni-Cu deposits, and intrusion-associated Au, as well as orogenic lode gold deposits associated with collisional tectonism. Mineral deposits of mafic-ultramafic magmas, whose emplacement is associated with structures that dislocate or perforate continental crust and penetrate the mantle, and include magmatic Ni-Cu and kimberlite diamond deposits, account for 40% of production and 33% of total mineral resources. Mineral deposits associated with intracontinental or epicontinental sedimentary basins account for the remaining 10% of both production and total mineral resources.

Résumé

Plus de 90 % de la valeur de la production canadienne de minéraux métallifères non ferreux provient de sept types de gîtes minéraux. D'après les prix moyens des métaux de 1996 à 2005, ajustés pour tenir compte de l'inflation, jusqu'à la fin de 2005 les types de gîtes minéraux les plus productifs ont été les suivants: 1) gîtes magmatiques de Ni-Cu (>372 milliards de dollars), 2) gîtes de sulfures massifs volcanogènes (192 milliards de dollars) et 3) gîtes d'or filonien (132 milliards de dollars). Ensemble, les gîtes porphyriques, SEDEX, de type Mississippi Valley et d'uranium ont fourni une production d'une valeur d'environ 140 milliards de dollars et ceux de diamants, un produit relativement nouveau pour le Canada, une valeur de 8 milliards de dollars. L'équivalent en dollars du contenu en métaux par tonne de minerai extrait au cours des cinq dernières années a varié d'environ 130 $/tonne à environ 350 $/tonne pour la plupart des mines souterraines de métaux communs et de diamants et de 90 $/tonne à 300 $/tonne pour la plupart des mines d'or souterraines. Pour les mines de métaux à ciel ouvert, l'équivalent en dollars du contenu en métaux varie de 10 $/tonne à 45 $/tonne. Les plus précieux des minerais ont été ceux des gîtes d'uranium du bassin d'Athabasca, où l'équivalent en dollars du contenu en métal s'est établi à 540 $/tonne et où les réserves courantes valent de 1000 à 11 000 $/tonne, soit plus de trois fois ces valeurs d'après le prix moyen de l'uranium en 2006.

Environ 50 % de la production et 57 % de la valeur de 1,57 trillion de dollars du contenu en métaux non ferreux et en diamants de l'ensemble des ressources minérales sont associés aux arcs volcaniques et aux arrières-arcs qui se sont accolés, ou se sont construits, sur les marges continentales lors de l'assemblage de protocontinents. Les types de gîtes comprennent les gîtes de SMV, les gîtes porphyriques, les gîtes komatiitiques de Ni-Cu et les gîtes d'Au associés à des intrusions ainsi que les gîtes orogéniques d'or filonien associés au tectonisme de collision. Les gîtes magmatiques de kimberlites diamantifères constituées de magmas mafiques-ultramafiques associés à des structures de dislocation ou de perforation de la croûte continentale et pénétrant le manteau fournissent 40 % de la production et 33 % du total des ressources minérales. Les gîtes minéraux associés aux bassins sédimentaires intracontinentaux ou épicontinentaux fournissent les 10 % restants de la production et du total des ressources minérales.

Introduction

One of the objectives of the Geological Survey of Canada's (GSC) Consolidation of Canada's Geoscience Knowledge (CCGK) program was to establish a national Cooperative Geological Mapping Strategy (CGMS) agreed to by the federal and provincial governments. In order to facilitate discussions among non-technical decision-makers in governments to achieve the program's objective, a resource required in the early stages in its mandate was an overview of Canada's mineral resources that emphasized their socio-economic contexts. This overview was delivered to the program in October 2003 and subsequently released to the public as GSC Open File 4668 (Lydon et al., 2004).

The interest shown by the geoscience community in this overview of Canada's mineral resources, especially by those geoscientists whose expertise is other than in mineral

Lydon, J.W., 2007, An overview of the economic and geological contexts of Canada's major mineral deposit types, *in* Goodfellow, W.D., ed., Mineral Deposits of Canada: A Synthesis of Major Deposit-Types, District Metallogeny, the Evolution of Geological Provinces, and Exploration Methods: Geological Association of Canada, Mineral Deposits Division, Special Publication No. 5, p. 3-48.

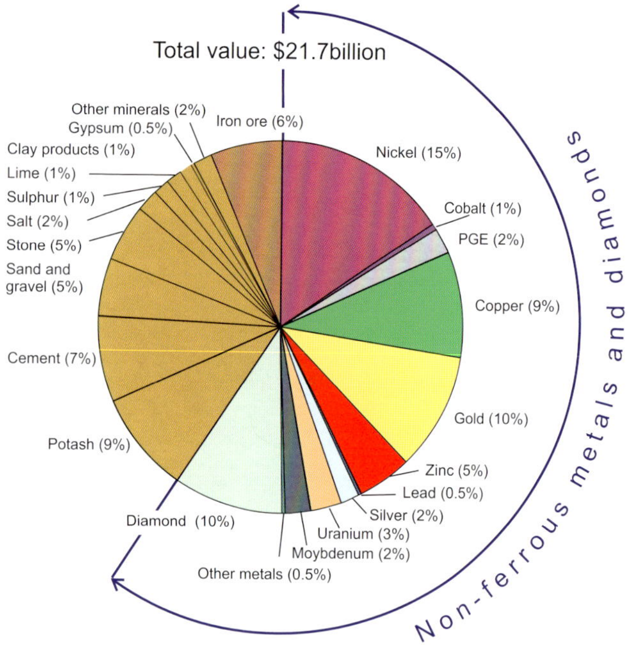

FIGURE 1. Canadian production of non-fuel mineral resources during 2004 by commodity. Only non-ferrous metals and diamonds are discussed in this article. Data from McMullen and Birchfield (2005).

deposits, has prompted the update and expansion of the information contained in GSC Open File 4668 presented here. The main revisions include

1. An update of production and reserve statistics to December 2005.
2. Realignment of categories of mineral resources to more closely conform to CIM standards established by the Canadian Institute of Mining (CIM) Standing Committee on Reserve Definitions and published in the CIM Bulletin, October, 2000. The major impact of this was to reduce the amount of metal contained in the "resources not being mined" category of GSC Open File 4668, especially for those deposits whose resources have been re-estimated since the introduction of National Instrument 43-101 by the Canadian Securities administrators in 2001.
3. Reconciliation of the databases used for GSC Open File 4668 with those expert databases of Appendix 1, which resulted in the addition of deposits, mainly of the lode gold deposit type, to the database.
4. The addition of some other deposit types to the database, namely Ag-rich veins, W skarns, and diagenetic Cu.

Scope of Overview

Objectives

Non-technical decision-makers in government need to know the socio-economic impacts of mining and in particular their geographical distribution, the contribution of mining to the economy, where new development could take place, and how governments can influence the discovery and development of new mineral deposits. Geoscientists who are not expert in mineral deposits need sufficient understanding of the geological attributes and settings of mineral deposits in order to place them in the context of their own fields of expertise, so that their knowledge can be applied to problems of mineral resources. These two sets of information are the essence of economic geology. The one emphasizes that the right economic conditions must be in place before a mineral deposit becomes an economic resource, and the second emphasizes that mineral deposits only occur where the right geological conditions exist, and not where economic or political priorities would like them to be.

In attempting to combine both the economic and geological aspects of mineral resources, this article is divided into three parts:

1. An outline of the range of factors that influence mineral resources statistics.
2. A summary of current and historical contributions by mineral resources to Canada's economy, and a brief description of the relationship between the geographical distribution of mineral resources and the geological architecture of Canada.
3. A brief description of each mineral deposit type of economic importance to Canada, concentrating on its most significant economic and geological characteristics.

Dollar Equivalent of Metal Contents

It is difficult to directly compare the economic value of different resource categories of different mineral deposits using conventionally reported statistics because different mineral deposit types are mined for different commodities and different units of measurement are employed for different commodities and resource categories. Production from metalliferous mineral deposits is usually reported as the quantities of metal produced and/or the values of metal recovered (e.g. Fig. 1). In contrast, mineral reserves and mineral resources are usually reported as tonnages of ores, and the grades of the economic components that those ores contain. Comparisons are further complicated by the use of different weight units: metric tonnes (t), short tons, or pounds for base metals; troy ounces, grams, or kilograms for precious metals; and specialized units such as short ton units or metric tonne units of WO_3 for tungsten (containing 15.86 lbs and 7.93 kg of tungsten, respectively). In order to allow a more direct comparison between different deposits, the unit of measurement of mineral resources used here is the dollar equivalent ($Eq) of the metals contained in the ores. This number is obtained by multiplying the amount of metal contained in a metric tonne of ore by the average inflation-adjusted ten year price of the metal (Table 1). Thus production from, or mineral resources in, a deposit is expressed as a single dollar equivalent number for each resource category, and represents the sum of the dollar equivalents of all economic commodities contained in the category. Grades are expressed as $Eq/t, which again is a sum of the dollar equivalents of all economic commodities contained in a tonne of ore.

Although a 'dollar equivalent value' has the advantage of giving a direct indication of the magnitude of the economic worth of a mineral deposit, it is not a measure of the actual and current economic value of a mineral resource. It does not take into account that only a certain proportion of a mineral deposit can be mined (because of engineering and other factors), and that only a proportion of the metals contained in the ores can be recovered by the mineral beneficiation

TABLE 1. Average metal prices for the 10 year period of 1996 to 2005. Source of data is the United States Geological Survey Annual Commodity Reviews. Prices for Pb are those of the London Metal Exchange and the prices for U are from a compilation by Cameco Limited. Prices have been inflation adjusted to 2005 Canadian dollars.

Year	Cu[1]	Ni[1]	Co[1]	Mo[1]	W[1]	Sn[1]	Zn[1]	Pb[2]	U[3]	Ag[1]	Au[1]	Pt[1]	Pd[1]	Cd[1]	Bi[1]	Sb[1]
Unit	tonne	tonne	tonne	tonne	tonne	tonne	tonne	tonne	tonne	kg	kg	kg	kg	tonne	tonne	tonne
1994	4,399	11,409	97,687	4,492	5,675	14,621	1,953	984	43,766	306	22,265	23,743	9,012	4,476	12,874	7,051
1995	5,366	14,469	113,306	14,604	7,818	16,121	2,071	1,109	52,615	291	21,807	24,042	8,655	7,137	14,934	8,844
1996	4,097	12,775	95,816	8,522	8,323	15,496	1,920	1,316	69,347	284	21,320	21,809	7,124	4,659	13,715	5,524
1997	3,978	11,684	86,847	8,439	8,071	14,195	2,404	1,053	52,967	265	18,036	21,521	9,992	1,898	13,023	3,647
1998	3,098	8,272	84,415	10,363	6,557	14,705	2,025	944	47,864	318	16,954	21,520	16,645	1,103	14,181	3,072
1999	2,925	10,510	65,598	10,315	5,927	14,106	2,062	878	46,684	295	15,738	21,299	20,414	540	14,839	2,621
2000	3,281	14,576	56,400	9,518	5,927	13,765	2,068	765	36,003	271	15,191	29,802	37,535	372	13,021	2,455
2001	2,884	10,111	39,565	8,846	8,071	11,813	1,650	800	39,156	240	14,876	29,166	33,401	863	14,026	2,438
2002	2,852	11,558	25,997	14,113	6,936	10,986	1,452	772	44,000	253	17,063	29,768	18,637	1,091	11,813	3,311
2003	2,800	14,352	34,830	17,363	6,305	11,172	1,334	767	44,926	235	17,490	33,276	9,727	1,939	9,430	3,549
2004	3,959	18,540	70,759	39,795	6,179	16,411	1,552	1,188	65,102	288	17,723	36,600	10,044	1,626	9,521	3,844
2005	4,516	17,619	42,218	87,349	17,654	14,268	1,689	1,184	90,650	279	17,145	34,680	7,404	4,008	10,234	3,874
10 Year AVERAGE METAL PRICE (1994-2005)																
	3439	13000	60245	16667	7995	13692	1816	967	53670	273	17154	27944	17092	1810	12380	3433
20 YEAR AVERAGE METAL PRICE (1984-2005)																
	3944	13957	59416	15920	7730	15359	2058	1033	51262	299	20559	27769	12337	6653	14083	4150

1 Plunkett and Jones, 1999; United States Geological Survey, 2006
2 Natural Resources Canada, 2005
3 Cameco Corporation, 2006
Note: original data has been converted to metric units; prices in $US have been adjusted using the inflation rates of United States Department Labor (2006) and $USD converted to $CAD using contemporaneous exchange rates of Antweiler (2006).

process. Furthermore, the dollar equivalent value of a deposit or its ores does not provide an indication of the profitability of the mining ventures, because costs of mining, beneficiation, transport, smelting, etc. vary widely depending on the mining methods and the metallurgical properties of the ores. Finally, because the prices of metals vary over time, the dollar equivalent value of resources reported here can be taken only as the maximum value that could have been derived from the deposit if it were completely mined out over the period 1995 to 2005.

Production, Reserves, and Resources Statistics

Definitions

The amount of the valuable commodity(ies) that may be potentially mined at a profit are calculated using a mathematical model of the mineral deposit based on a systematic sampling by drill cores. The model contains assumptions on a range of factors that include

i) the validity of the geological model used to interpolate between sampling points;

ii) the choice of cut-off grades that define the boundaries of the mineral resource;

iii) the mining method, which determines what parts of the deposit are feasible to mine and the amount of ore dilution by unmineralized host rocks that will take place;

iv) the mineral beneficiation methods, which determine the proportion of the ore minerals that can be recovered from the ores;

v) various other economic, marketing, legal, environmental, social, and governmental factors that affect a potential mining operation.

The grade and amount of the valuable commodities estimated for a mineral deposit are classified as 'mineral reserves' only if technical studies and tests have shown that it would be feasible and economically viable to mine the mineral deposit and deliver the products to a treatment plant. Otherwise, the estimated grades and tonnage of a mineral deposit are classified as 'mineral resources'. In most cases, mineral resources are considered to be mineral reserves only if they are part of a mineral deposit that is being actively mined, or the mineral deposit is under active development in preparation for mining. Even so, part of the mineral inventory of a mineral deposit that is being actively mined may be classified as mineral resources if, at the time of reporting, that part of the deposit in which they occur has not been included in the current mining plan.

In Canada, the criteria for reporting and classifying mineral resources are stipulated by National Instrument 43-101 (NI 43-101), which was introduced by the Canadian Securities Administrators in 2001. National Instrument 43-101 uses the definitions of different categories of mineral reserves and resources formulated by the Canadian Institute of Mining Standing Committee on Reserve Definitions (CIM Standards, *CIM Bulletin*, October 2000). The CIM Standard subdivides mineral reserves into "proven" and "probable" according to the degree of certainty of whether they will be mined, and mineral resources into "measured", "indicated", and "inferred" according to the degree of certainty of their existence. The criteria for establishing the degree of certainty of existence of mineral resources are not specifically defined by the CIM Standards and it is left to the professional judgment of the "qualified person" calculating the quantities and grades of a mineral resource to decide whether they are classified as measured, indicated, or inferred. Hence, the statistics for mineral resources contain an element of subjectivity.

The grades and tonnages for mineral reserves and mineral resources are very changeable, especially for active mines. As mining progresses, reserves are diminished but these may

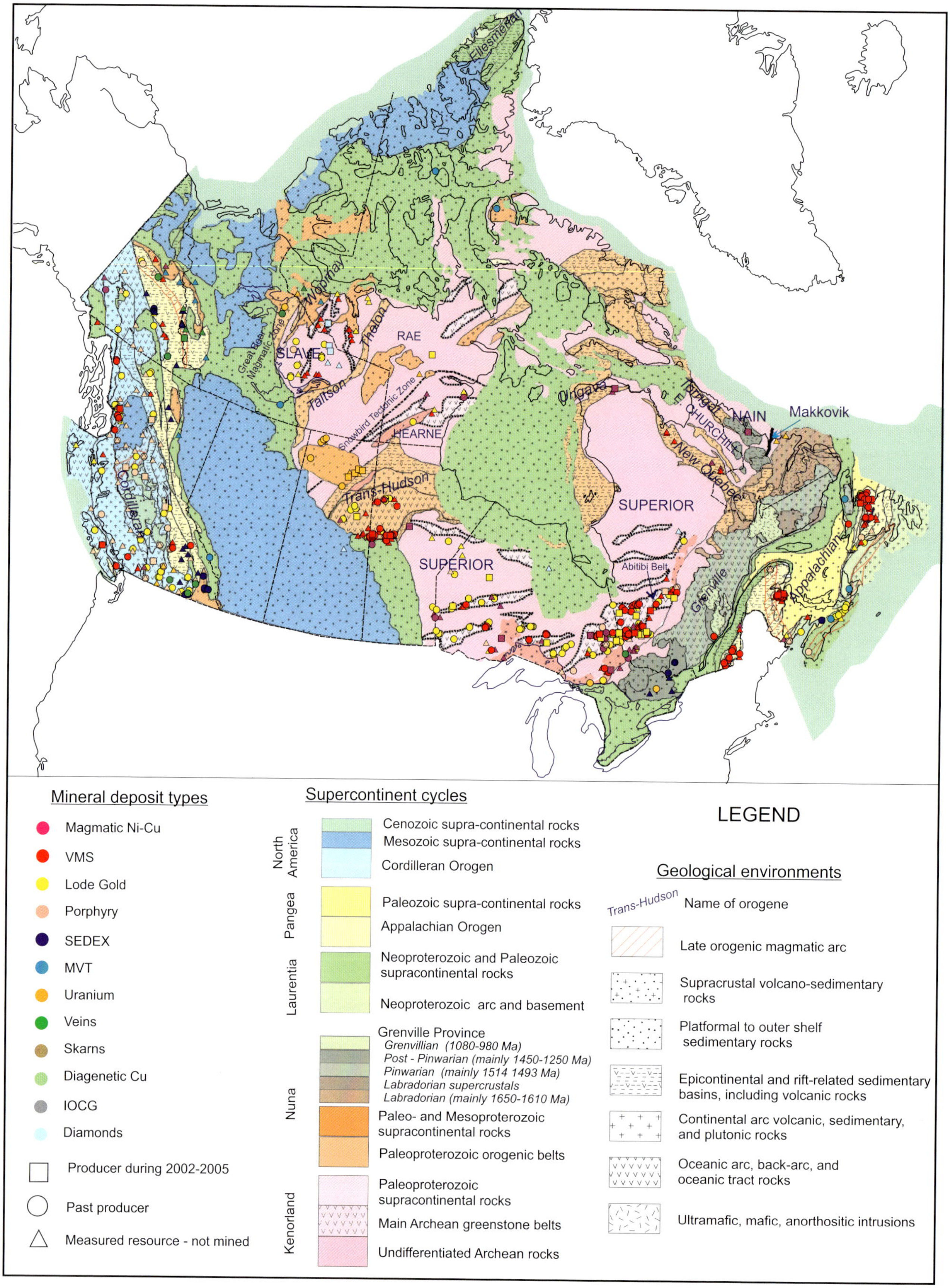

FIGURE 2. Simplified geological map of Canada showing geological domains by supercontinent cycle and the distribution of non-ferrous metalliferous and kimberlite diamond deposits. Deposits are colour-coded by mineral deposit types and shapes of symbol reflect production status. Only deposits for which a mineral resource has been measured are plotted (see Appendix 1, DVD).

be replaced as continued mine development allows mineral resources to be reclassified as mineral reserves and, in turn, continued exploration discovers new mineral resources. Similarly, changing economic, social, or political conditions, or improvements in geological knowledge of the mineral deposit, may change the parameters of the mathematical mine model referred to above. Modifications to the mine model may cause an upward or downward revision to the grades and tonnages of all mineral resource categories. Thus, there is not a single objective set of tonnage and grade statistics for a single mineral deposit because the statistics are specific for a particular time and a specific qualified person.

Production usually refers to the amount of metal in the product that is delivered by a mine to a smelter or refinery. In order to fully reconcile production statistics with statistics for reserves and resources, it is necessary to know the amount and grades of the ore that is mined, that is in stockpiles, and that is milled, and also the recovery rates for each commodity by the beneficiation process. Although a most corporations now report the amount and grade of ore milled and recovery rates, and some also report the amount of ore mined in a reconciliation of reserve data, the lack of ready public availability of all the information that is needed to calculate the amount of metal contained in ores mined has necessitated author's estimates in the compilation of statistics presented here.

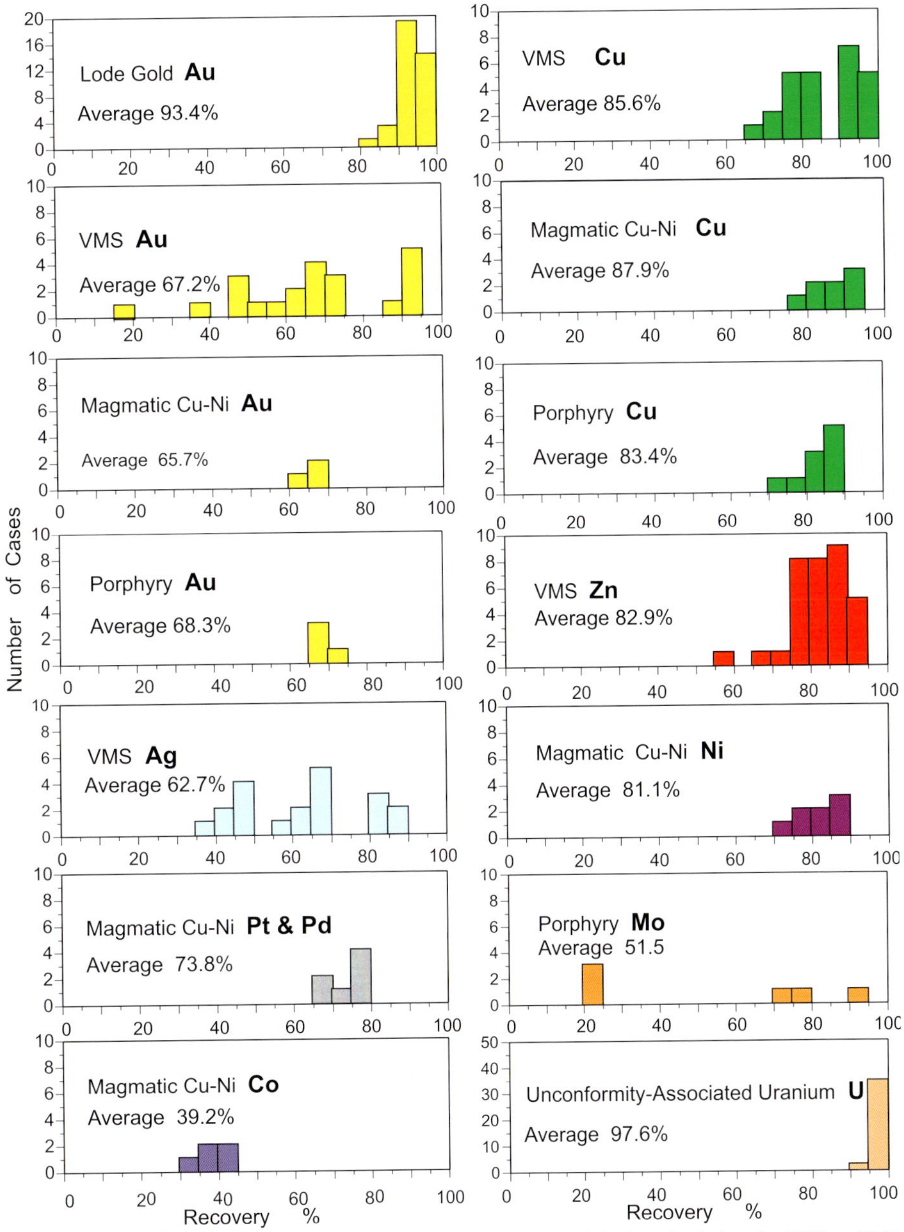

FIGURE 3. Recovery rates of metals for different deposit types mined in Canada during the 2002 to 2005 period. The average recovery rate indicated is the value used to calculate mill-head grades from metal production statistics in cases where ore grades or mill-head grades are not reported in conjunction with tonnes of ore milled. Note that there was no production from SEDEX or Mississippi Valley-type deposits during 2002 to 2005 and therefore recovery rates for these deposit types have not been compiled.

Compilation of Statistics

Geoscientists and mineral explorationists are primarily interested in tonnages and grades of mineral resources, whereas economists and financial specialists are more interested in the amounts and values of metals produced. The two sets of statistics are related by the amounts and grades of mill feed, the recovery efficiency of the concentrating process, and metal prices. Both sets of statistics are given in Appendix 1 (see accompanying DVD), which attempts to list all non-ferrous metal and diamond deposits in Canada for which a resource has been measured (Fig. 2). A summary of these statistics, in terms of quantities of contained metals for each deposit-type is given in Table 2, and a summary of the dollar equivalent metal content is given in Table 3.

Statistics for production were compiled by a variety of methods. For deposits mined since 1970, production statistics were obtained by adding annual production statistics compiled in digital format by provincial and the federal governments and, particularly for deposits mined since 2000, from corporate annual reports or annual information forms filed with the Canadian Securities Commission (www.SEDAR.com). The data used as production are, in most cases, for "ore milled", which is more consistently reported than "ore mined". Where mill head grades were not reported, they were estimated using average recovery rates for the deposit type (Fig. 3). If only the amount of metal recovered is reported, the tonnage and grade of the mill feed is estimated based on average recovery rates and the grade of

TABLE 2. Summary of the metal contents of the major mineral deposit types in the resource categories referred to in the text. For complete details see Appendix 1 (DVD).

PRODUCTION

Mineral deposit type	Cu	Ni	Co	Mo	W	Sn	Zn	Pb	U	Ag	Au	Pt	Pd	Cd	Bi	Sb	Diamonds	Ore
	tonnes	tonnes	tonnes	tonnes	tonnes	tonnes	tonnes	tonnes	tonnes	kg	kg	kg	kg	tonnes	tonnes	tonnes	carats	tonnes
Magmatic Ni-Cu	15,655,055	18,542,859	490,779	-	-	-	-	-	-	7,229,354	248,751	855,091	1,026,718	-	-	-	-	1,367,109,021
VMS	17,620,713	-	-	-	-	-	48,733,065	7,205,972	-	56,794,310	1,212,262	-	-	-	-	-	-	1,102,504,212
Lode Gold	363,489	-	-	-	-	-	81,069	65	-	3,199,358	7,540,758	-	-	-	-	-	-	1,000,730,530
Porphyry	9,827,155	-	-	413,650	1,120	21,244	1,880	-	-	4,593,087	364,744	-	-	-	-	-	-	3,085,567,642
Sedex	8,729	-	-	-	-	11,415	13,546,974	12,395,959	-	13,865,766	26,325	-	-	3,095	22	414	-	238,750,407
MVT	-	-	-	-	-	-	8,946,696	2,646,650	-	739,909	-	-	-	-	-	-	-	109,334,889
U - Athabasca	-	-	-	-	-	-	-	-	199,448	-	-	-	-	-	-	-	-	19,847,461
U - Except Athabasca	5,163	127	226	-	-	-	-	100	206,611	1,042,341	-	-	-	-	-	-	-	174,013,147
Veins	261,020	9,948	20,000	-	-	-	413,131	519,809	-	21,208,897	43,111	-	-	300	-	-	-	33,873,389
Skarns	11,500	-	-	-	63,307	-	88,869	27,290	-	40,586	-	-	-	-	-	-	-	5,299,756
Diagenetic Cu	-	-	-	-	-	-	-	-	-	-	-	-	-	-	-	-	-	-
IOCG	-	-	-	-	-	-	-	-	-	-	-	-	-	-	-	-	-	-
Diamonds	-	-	-	-	-	-	-	-	-	-	-	-	-	-	-	-	50,265,046	30,902,895
TOTALS	43,752,823	18,552,935	511,005	413,650	64,427	32,659	71,811,684	22,795,845	406,058	108,713,609	9,435,952	855,091	1,026,718	3,395	22	414	50,265,046	7,167,933,349

RESERVES

Mineral deposit type	Cu	Ni	Co	Mo	W	Sn	Zn	Pb	U	Ag	Au	Pt	Pd	Cd	Bi	Sb	Diamonds	Tonnes of ore
	tonnes	tonnes	tonnes	tonnes	tonnes	tonnes	tonnes	tonnes	tonnes	kg	kg	kg	kg	tonnes	tonnes	tonnes	carats	tonnes
Magmatic Ni-Cu	3,108,453	4,259,242	127,480	-	-	-	-	-	-	-	56,473	143,273	220,809	-	-	-	-	280,508,000
VMS	1,186,362	-	-	-	-	-	5,392,801	724,791	-	6,183,836	221,173	-	-	-	-	-	-	108,463,433
Lode Gold	-	-	-	-	-	-	-	-	-	-	561,524	-	-	-	-	-	-	138,788,299
Porphyry	3,049,014	-	-	90,344	-	-	-	-	-	394,353	206,660	-	-	-	-	-	-	1,172,967,749
Sedex	-	-	-	-	-	-	-	-	-	-	-	-	-	-	-	-	-	-
MVT	-	-	-	-	-	-	-	-	-	-	-	-	-	-	-	-	-	-
U - Athabasca	-	16,342	1,140	-	-	-	-	-	266,630	-	-	-	-	-	-	-	-	2,809,981
U - Except Athabasca	-	-	-	-	-	-	-	-	-	-	-	-	-	-	-	-	-	-
Veins	-	-	-	-	-	-	-	-	-	-	-	-	-	-	-	-	-	-
Skarns	1,735	-	-	-	9,325	-	-	-	-	-	-	-	-	-	-	-	-	694,000
Diagenetic Cu	-	-	-	-	-	-	-	-	-	-	-	-	-	-	-	-	-	-
IOCG	-	-	-	-	-	-	-	-	-	-	-	-	-	-	-	-	-	-
Diamonds	-	-	-	-	-	-	-	-	-	-	-	-	-	-	-	-	125,604,000	77,600,000
TOTALS	7,345,564	4,275,584	128,620	90,344	9,325	-	5,392,801	724,791	266,630	6,578,189	1,045,830	143,273	220,809	-	-	-	125,604,000	1,781,831,463

MEASURED and INDICATED RESOURCES

Mineral deposit type	Cu	Ni	Co	Mo	W	Sn	Zn	Pb	U	Ag	Au	Pt	Pd	Cd	Bi	Sb	Diamonds	Tonnes of ore
	tonnes	tonnes	tonnes	tonnes	tonnes	tonnes	tonnes	tonnes	tonnes	kg	kg	kg	kg	tonnes	tonnes	tonnes	carats	tonnes
Magmatic Ni-Cu	3,383,118	5,204,674	108,148	-	-	-	-	-	-	88,170	38,330	130,914	280,231	-	-	-	-	701,137,044
VMS	9,263,115	-	-	-	-	-	17,986,579	2,511,518	-	20,131,409	377,104	-	-	-	-	-	-	822,473,918
Lode Gold	2,826	-	-	-	-	-	15,011	8,294	-	138,592	1,622,993	-	-	-	-	-	-	572,305,020
Porphyry	30,400,200	-	-	2,226,465	242,930	82,716	38,420	-	-	4,792,785	2,285,023	-	-	-	-	-	-	11,564,147,534
Sedex	20,689	-	-	-	-	-	13,047,133	6,016,100	-	9,045,602	30,381	-	-	-	-	-	-	206,205,535
MVT	49,592	-	-	-	-	-	4,993,282	1,378,347	-	628,974	-	-	-	-	-	-	-	91,782,538
U - Athabasca	-	-	-	-	-	-	-	-	48,670	-	-	-	-	-	-	-	-	8,926,400
U - Except Athabasca	-	-	-	-	-	-	-	-	38,812	10,772	-	-	-	-	-	-	-	26,401,947
Veins	145,956	-	-	-	-	-	-	-	-	63,396	6,034	-	-	-	-	-	-	4,140,914
Skarns	2,608	-	-	-	495,566	-	227,760	56,940	-	109,500	-	-	-	-	-	-	-	66,983,050
Diagenetic Cu	1,553,080	-	-	-	-	-	-	-	-	418,100	-	-	-	-	-	-	-	40,200,000
IOCG	135,852	-	18,963	2,426	-	-	-	-	-	53,370	21,633	-	-	-	2,217	-	-	37,613,200
Diamonds	-	-	-	-	-	-	-	-	-	-	-	-	-	-	-	-	177,173,870	166,167,000
TOTALS	44,957,035	5,204,674	127,111	2,228,891	738,496	82,716	36,308,185	9,971,199	87,482	35,480,671	4,381,499	130,914	280,231	0	2,217	0	177,173,870	14,308,484,100

INFERRED RESOURCES

Mineral deposit type	Cu	Ni	Co	Mo	W	Sn	Zn	Pb	U	Ag	Au	Pt	Pd	Cd	Bi	Sb	Diamonds	Tonnes of ore
Magmatic Ni-Cu	2,508,564	2,250,194	73,109	-	-	-	-	-	-	-	30,131	92,326	255,100	-	-	-	-	177,616,000
VMS	382,966	-	-	-	-	-	805,031	82,200	-	1,196,017	45,158	-	-	-	-	-	-	21,322,000
Lode Gold	-	-	-	-	-	-	-	-	-	26,688	1,135,794	-	-	-	-	-	-	278,000,236
Porphyry	41,651	-	-	-	-	-	-	-	-	-	17,633	-	-	-	-	-	-	24,205,000
Sedex	-	-	-	-	-	-	6,097,526	2,411,544	-	-	-	-	-	-	-	-	-	112,910,000
MVT	33,112	-	-	-	-	-	1,524,704	1,346,824	-	2,433,388	-	-	-	-	-	-	-	23,578,000
U - Athabasca	-	-	-	-	-	-	-	-	87,975	-	-	-	-	-	-	-	-	1,798,600
U - Except Athabasca	-	-	-	-	-	-	-	-	3,505	-	-	-	-	-	-	-	-	2,300,015
Veins	-	-	-	-	-	-	-	-	-	-	-	-	-	-	-	-	-	-
Skarns	-	-	-	-	-	-	-	-	-	-	-	-	-	-	-	-	-	-
Diagenetic Cu	-	-	-	-	-	-	-	-	-	-	-	-	-	-	-	-	-	-
IOCG	-	-	-	-	-	-	-	-	-	-	-	-	-	-	-	-	-	-
Diamonds	-	-	-	-	-	-	-	-	-	-	-	-	-	-	-	-	90,826,690	66,601,000
TOTALS	2,966,293	2,250,194	73,109	-	-	-	8,427,261	3,840,568	91,480	3,656,092	1,228,715	92,326	255,100	-	-	-	90,826,690	708,330,851

TABLE 3. Summary of the relative proportions of metals, in terms of the dollar equivalent of metal contents contributed by various deposit types to the total metal content of each resource category. All numbers are in millions of 2005 Canadian dollars.

PRODUCTION Mineral deposit type	Cu $Eq	Ni $Eq	Co $Eq	Mo $Eq	W $Eq	Sn $Eq	Zn $Eq	Pb $Eq	U $Eq millions	Ag $Eq	Au $Eq	Pt $Eq	Pd $Eq	Cd $Eq	Bi $Eq	Sb $Eq	Diamonds $Eq	Total $Eq
Magmatic Ni-Cu	53,838.1	241,053.1	29,566.7	0.0	0.0	0.0	0.0	0.0	0.0	1,973.8	4,267.0	23,894.8	17,549.0	0.0	0.0	0.0	0.0	372,142.6
VMS	60,598.0	0.0	0.0	0.0	0.0	0.0	88,482.1	6,965.9	0.0	15,506.6	20,794.9	0.0	0.0	0.0	0.0	0.0	0.0	192,347.4
Lode Gold	1,250.0	0.0	0.0	0.0	0.0	0.0	147.2	0.1	0.0	873.5	129,352.7	0.0	0.0	0.0	0.0	0.0	0.0	131,623.5
Porphyry	33,795.8	0.0	0.0	6,894.5	9.0	290.9	3.4	0.0	0.0	1,254.1	6,256.8	0.0	0.0	0.0	0.0	0.0	0.0	48,504.3
Sedex	30.0	0.0	0.0	0.0	0.0	156.3	24,596.5	11,982.9	0.0	3,785.8	451.6	0.0	0.0	5.6	0.3	1.4	0.0	41,010.4
MVT	0.0	0.0	0.0	0.0	0.0	0.0	16,244.0	2,558.5	0.0	202.0	0.0	0.0	0.0	0.0	0.0	0.0	0.0	19,004.5
U - Athabasca	0.0	0.0	0.0	0.0	0.0	0.0	0.0	0.0	10,704.3	0.0	0.0	0.0	0.0	0.0	0.0	0.0	0.0	10,704.3
U - Except Athab	17.8	1.7	13.6	0.0	0.0	0.0	0.0	0.1	11,088.8	284.6	0.0	0.0	0.0	0.0	0.0	0.0	0.0	11,406.5
Veins	897.7	129.3	1,204.9	0.0	0.0	0.0	750.1	502.5	0.0	5,790.7	739.5	0.0	0.0	0.5	0.0	0.0	0.0	10,015.2
Skarns	39.5	0.0	0.0	0.0	506.1	0.0	161.4	26.4	0.0	11.1	0.0	0.0	0.0	0.0	0.0	0.0	0.0	744.5
Diagenetic Cu	0.0	0.0	0.0	0.0	0.0	0.0	0.0	0.0	0.0	0.0	0.0	0.0	0.0	0.0	0.0	0.0	0.0	0.0
IOCG	0.0	0.0	0.0	0.0	0.0	0.0	0.0	0.0	0.0	0.0	0.0	0.0	0.0	0.0	0.0	0.0	0.0	0.0
Diamonds	0.0	0.0	0.0	0.0	0.0	0.0	0.0	0.0	0.0	0.0	0.0	0.0	0.0	0.0	0.0	0.0	8,252.9	8,252.9
TOTALS	150,466.9	241,184.0	30,785.3	6,894.5	515.1	447.2	130,384.7	22,036.3	21,793.1	29,682.1	161,862.5	23,894.8	17,549.0	6.1	0.3	1.4	8,252.9	845,756.3

RESERVES Mineral deposit type	Cu $Eq	Ni $Eq	Co $Eq	Mo $Eq	W $Eq	Sn $Eq	Zn $Eq	Pb $Eq	U $Eq millions	Ag $Eq	Au $Eq	Pt $Eq	Pd $Eq	Cd $Eq	Bi $Eq	Sb $Eq	Diamonds $Eq	Total $Eq
Magmatic Ni-Cu	10,690.0	55,369.2	7,680.0	0.0	0.0	0.0	0.0	0.0	0.0	0.0	968.7	4,003.7	3,774.1	0.0	0.0	0.0	0.0	82,485.7
VMS	4,079.9	0.0	0.0	0.0	0.0	0.0	9,791.4	700.6	0.0	1,688.4	3,794.0	0.0	0.0	0.0	0.0	0.0	0.0	20,054.3
Lode Gold	0.0	0.0	0.0	0.0	0.0	0.0	0.0	0.0	0.0	0.0	9,632.3	0.0	0.0	0.0	0.0	0.0	0.0	9,632.3
Porphyry	10,485.6	0.0	0.0	1,505.8	0.0	0.0	0.0	0.0	0.0	107.7	3,545.0	0.0	0.0	0.0	0.0	0.0	0.0	15,644.1
Sedex	0.0	0.0	0.0	0.0	0.0	0.0	0.0	0.0	0.0	0.0	0.0	0.0	0.0	0.0	0.0	0.0	0.0	0.0
MVT	0.0	0.0	0.0	0.0	0.0	0.0	0.0	0.0	0.0	0.0	0.0	0.0	0.0	0.0	0.0	0.0	0.0	0.0
U - Athabasca	0.0	212.4	68.7	0.0	0.0	0.0	0.0	0.0	14,310.0	0.0	0.0	0.0	0.0	0.0	0.0	0.0	0.0	14,591.1
U - Except Athab	0.0	0.0	0.0	0.0	0.0	0.0	0.0	0.0	0.0	0.0	0.0	0.0	0.0	0.0	0.0	0.0	0.0	0.0
Veins	0.0	0.0	0.0	0.0	0.0	0.0	0.0	0.0	0.0	0.0	0.0	0.0	0.0	0.0	0.0	0.0	0.0	0.0
Skarns	6.0	0.0	0.0	0.0	74.6	0.0	0.0	0.0	0.0	0.0	0.0	0.0	0.0	0.0	0.0	0.0	0.0	80.5
Diagenetic Cu	0.0	0.0	0.0	0.0	0.0	0.0	0.0	0.0	0.0	0.0	0.0	0.0	0.0	0.0	0.0	0.0	0.0	0.0
IOCG		0.0	0.0	0.0	0.0	0.0	0.0	0.0	0.0	0.0	0.0	0.0	0.0	0.0	0.0	0.0	0.0	0.0
Diamonds	0.0	0.0	0.0	0.0	0.0	0.0	0.0	0.0	0.0	0.0	0.0	0.0	0.0	0.0	0.0	0.0	11,811.1	11,811.1
TOTALS	25,261.6	55,581.7	7,748.7	1,505.8	74.6	0.0	9,791.4	700.6	14,310.0	1,796.0	17,940.0	4,003.7	3,774.1	0.0	0.0	0.0	11,811.1	154,299.2

MEASURED and INDICATED RESOURCES Mineral deposit type	Cu $Eq	Ni $Eq	Co $Eq	Mo $Eq	W $Eq	Sn $Eq	Zn $Eq	Pb $Eq	U $Eq millions	Ag $Eq	Au $Eq	Pt $Eq	Pd $Eq	Cd $Eq	Bi $Eq	Sb $Eq	Diamonds $Eq	Total $Eq
Magmatic Ni-Cu	11,634.6	67,659.6	6,515.3	0.0	0.0	0.0	0.0	0.0	0.0	24.1	657.5	3,658.3	4,789.8	0.0	0.0	0.0	0.0	94,939.2
VMS	31,856.1	0.0	0.0	0.0	0.0	0.0	32,657.3	2,427.8	0.0	5,496.5	6,468.8	0.0	0.0	0.0	0.0	0.0	0.0	78,906.4
Lode Gold	9.7	0.0	0.0	0.0	0.0	0.0	27.3	8.0	0.0	37.8	27,840.5	0.0	0.0	0.0	0.0	0.0	0.0	27,923.3
Porphyry	104,547.0	0.0	0.0	37,109.6	1,942.2	1,132.5	69.8	0.0	0.0	1,308.6	39,196.8	0.0	0.0	0.0	0.0	0.0	0.0	185,306.5
Sedex	71.2	0.0	0.0	0.0	0.0	0.0	23,689.0	5,815.6	0.0	2,469.7	521.1	0.0	0.0	0.0	0.0	0.0	0.0	32,566.7
MVT	170.5	0.0	0.0	0.0	0.0	0.0	9,066.0	1,332.4	0.0	171.7	0.0	0.0	0.0	0.0	0.0	0.0	0.0	10,740.7
U - Athabasca	0.0	0.0	0.0	0.0	0.0	0.0	0.0	0.0	2,612.1	0.0	0.0	0.0	0.0	0.0	0.0	0.0	0.0	2,612.1
U - Except Athab	0.0	0.0	0.0	0.0	0.0	0.0	0.0	0.0	2,083.0	2.9	0.0	0.0	0.0	0.0	0.0	0.0	0.0	2,086.0
Veins	501.9	0.0	0.0	0.0	0.0	0.0	0.0	0.0	0.0	17.3	103.5	0.0	0.0	0.0	0.0	0.0	0.0	622.8
Skarns	9.0	0.0	0.0	0.0	3,962.0	0.0	413.5	55.0	0.0	29.9	0.0	0.0	0.0	0.0	0.0	0.0	0.0	4,469.5
Diagenetic Cu	5,341.1	0.0	0.0	0.0	0.0	0.0	0.0	0.0	0.0	114.2	0.0	0.0	0.0	0.0	0.0	0.0	0.0	5,455.2
IOCG	467.2	0.0	1,142.4	40.4	0.0	0.0	0.0	0.0	0.0	14.6	371.1	0.0	0.0		27.4	0.0	0.0	2,063.1
Diamonds	0.0	0.0	0.0	0.0	0.0	0.0	0.0	0.0	0.0	0.0	0.0	0.0	0.0	0.0	0.0	0.0	15,584.6	15,584.6
TOTALS	154,608.3	67,659.6	7,657.7	37,150.0	5,904.2	1,132.5	65,922.9	9,639.0	4,695.2	9,687.3	75,159.4	3,658.3	4,789.8	0.0	27.4	0.0	15,584.6	463,276.2

INFERRED RESOURCES Mineral deposit type	Cu $Eq	Ni $Eq	Co $Eq	Mo $Eq	W $Eq	Sn $Eq	Zn $Eq	Pb $Eq	U $Eq millions	Ag $Eq	Au $Eq	Pt $Eq	Pd $Eq	Cd $Eq	Bi $Eq	Sb $Eq	Diamonds $Eq	Total $Eq
Magmatic Ni-Cu	8,627.0	29,252.0	4,404.4	0.0	0.0	0.0	0.0	0.0	0.0	0.0	516.9	2,580.0	4,360.3	0.0	0.0	0.0	0.0	49,740.5
VMS	1,317.0	0.0	0.0	0.0	0.0	0.0	1,461.7	79.5	0.0	326.5	774.6	0.0	0.0	0.0	0.0	0.0	0.0	3,959.3
Lode Gold	0.0	0.0	0.0	0.0	0.0	0.0	0.0	0.0	0.0	7.3	19,483.2	0.0	0.0	0.0	0.0	0.0	0.0	19,490.5
Porphyry	143.2	0.0	0.0	0.0	0.0	0.0	0.0	0.0	0.0	0.0	302.5	0.0	0.0	0.0	0.0	0.0	0.0	445.7
Sedex	0.0	0.0	0.0	0.0	0.0	0.0	11,071.0	2,331.2	0.0	0.0	0.0	0.0	0.0	0.0	0.0	0.0	0.0	13,402.2
MVT	113.9	0.0	0.0	0.0	0.0	0.0	2,768.3	1,301.9	0.0	664.4	0.0	0.0	0.0	0.0	0.0	0.0	0.0	4,848.5
U - Athabasca	0.0	0.0	0.0	0.0	0.0	0.0	0.0	0.0	4,721.6	0.0	0.0	0.0	0.0	0.0	0.0	0.0	0.0	4,721.6
U - Except Athab	0.0	0.0	0.0	0.0	0.0	0.0	0.0	0.0	188.1	0.0	0.0	0.0	0.0	0.0	0.0	0.0	0.0	188.1
Veins	0.0	0.0	0.0	0.0	0.0	0.0	0.0	0.0	0.0	0.0	0.0	0.0	0.0	0.0	0.0	0.0	0.0	0.0
Skarns	0.0	0.0	0.0	0.0	0.0	0.0	0.0	0.0	0.0	0.0	0.0	0.0	0.0	0.0	0.0	0.0	0.0	0.0
Diagenetic Cu	0.0	0.0	0.0	0.0	0.0	0.0	0.0	0.0	0.0	0.0	0.0	0.0	0.0	0.0	0.0	0.0	0.0	0.0
IOCG	0.0	0.0	0.0	0.0	0.0	0.0	0.0	0.0	0.0	0.0	0.0	0.0	0.0	0.0	0.0	0.0	0.0	0.0
Diamonds	0	0	0	0	0	0	0	0	0	0	0	0	0	0.0	0.0	0.0	8,378.7	8,378.7
TOTALS	10,201.1	29,252.0	4,404.4	0.0	0.0	0.0	15,300.9	3,712.6	4,909.7	998.2	21,077.1	2,580.0	4,360.3	0.0	0.0	0.0	8,378.7	105,175.1

the previous year-end reserves. Large corporations in particular tend to report production as rounded amounts of major commodities, and include all minor or by-product commodities in an umbrella "other revenues" category. Where there is sufficient reason to suppose recovery of these minor components (e.g. platinum group elements in Sudbury ores), estimates are made based on the ratios of minor components to major components in the previous year-end reserves, or in other cases (e.g. Eskay Creek) on metal ratios of the pre-production estimates of the total mineral resource.

For deposits mined prior to 1970, data were obtained from compilations by provincial governments, the World Minerals Geoscience Database of the Geological Survey of Canada, the compilations of the various deposit types in appendices to this volume, the Canada Minerals Yearbook for the years 1955 to 1970, Cranstone (2002), and Gosselin and Dubé (2005). For these deposits, it has been assumed, unless there is information to the contrary, that the resource reported for the deposit in the general literature was the amount that was mined. For deposits or districts that were mined prior to 1951 and whose cumulative production records are not in provincial databases (such as the Ni-Cu mining of the Sudbury district), estimates of pre-1951 production were based on the annual graphs of Cranstone (2002). Data for reserves are from corporate annual information forms filed with the Canadian Securities Commission (www.SEDAR.com) and corporate annual reports. Data for measured, indicated, and inferred resources for deposits mined since 2002 are from corporate annual information forms.

For deposits not mined since 2002 or that have never been mined, most of the data are historical estimates of resources that do not conform to CIM Standards as required by NI 43-101. These historical estimates of 'reserves' are all included in the measured and indicated resources category used here. Data for these historical estimates of resources were obtained from the compilations mentioned above and Natural Resources Canada (1990). An attempt was made to update estimates for mineral deposits not being mined, but re-assessed since the introduction of NI 43-101. Most of these deposits, reflecting mineral exploration priorities of the time, are of the lode gold type. However, the attempt was not systematic or complete.

For purposes of this article, measured resources and indicated resources are combined into a single 'measured and indicated resources' category, and like the 'production', 'reserves', and 'inferred resources' categories, are used here in the sense that the data incorporates estimates by the author or refers to dollar equivalent values, unless otherwise stated.

IMPORTANT NOTICE: As noted above, the data reported in Appendix 1 (DVD) contain estimates made by the author, and so should be treated only as approximations of the production and mineral resources of the deposits listed. Typographic errors may have occurred in transcription and mistakes made in calculations, and so caution should be exercised in using the data for purposes other than the one here, which is to obtain an overall approximation of the quantity and distribution of Canada's non-ferrous metal and diamond resources.

Metal Prices

Most of the metal prices used here are the 10 year (1996-2005) averages of the average annual prices reported by the United States Geological Survey Annual Commodity Reviews. These reviews compile data from a variety of sources including *Metal Week, Platt's Metals Week, Metal Bulletin*, and *Engineering and Mining Journal*, which collectively report data from a variety of metal exchanges. For base metals, the United States Geological Survey Annual Commodity Reviews uses United States domestic producer prices, which tend to be 5 to 10% higher than prices averaged from the London Metal Exchange (for Pb, the prices are up to 40% higher, and so the London Metal Exchange price for Pb is used in Table 1). These prices have been converted into metric units, inflation-adjusted by using the inflation indices reported by the U.S. Department of Labor, and converted into 2005 Canadian dollars by using the historical annual average exchange rates between the United States' dollar and Canadian dollar as compiled by Werner Antweiler, University of British Columbia, (http://fx.sauder.ubc.ca, last accessed 2007). Prices of U (since 1988) are those compiled by Cameco Corporation (http://www.cameco.com, last accessed 2007).

Economic Contexts and Characteristics of Canada's Mineral Resources

Socio-economic Contexts of Metalliferous Mining in Canada

In 2004, Canada's non-fuel mineral production was worth \$21.7 billion (Fig. 1; McMullen and Birchfield, 2005). This production consists of 23 mineral categories of which 12 are metals, 1 is diamonds, and the rest consist of a range of industrial and construction commodities and Fe ore (Fig. 1). The commodities dealt with in this article are the non-ferrous metals (i.e. all metal categories exclusive of Fe ore in Fig. 1) and diamonds, which together accounted for \$12.9 billion or 59.5% of Canada's non-fuel mineral production in 2004.

In 2005, the value of Canada's non-fuel mineral production rose by 7.7% to \$13.3 billion, due mainly to an increase in the price of metals led by a 65.4% increase in the price of U (Birchfield, 2006). In terms of quantities of metals, production of Zn and Ag actually decreased by about 15% and Au by 8%. The value of diamond production fell by 19.7%, due in part to the increase in the value of the Canadian dollar against the United States dollar and in part due to the lower quality of diamonds mined during the year. The mining industry contributed \$42 billion, or 3.9% of Canada's gross domestic product, in 2005 of which mining contributed 23.7%, primary metal manufacturing (smelting, refining, etc.) 29.2%, metal fabrication 33.7%, and the remaining 13.4% was contributed by non-metallic mineral fabrication. Crude minerals, smelted and refined outputs, and mineral products contributed \$64.2 billion to the value of Canada's exports in 2005. This represented 14.7% of Canada's total domestic exports of \$435.8 billion. Metallic mineral and mineral product exports accounted for 75.8% (\$48.7 billion) of the total value, non-metal exports accounted for 18.8% (\$12.1 billion), and coal and coke accounted for 5.4% (\$3.5 billion) (Birchfield, 2006).

Employment in the metal mining industry in 2005 included 21,519 employees in mining, which spawned an additional 84,000 jobs in primary metal manufacturing and 199,000 jobs in metal fabricating. Most of the people employed in mining and primary metal manufacturing live in northern communities of Canada's provinces and territories, particularly those that historically owe their founding to the mining industry, such as Timmins in Ontario, Noranda in Quebec, Flin Flon in Manitoba, and Yellowknife in Northwest Territories, as well as over 100 other communities. The number of people involved in mining in 2005 was 4.8% lower than in 2004, continuing a 10 year trend of declining employment as the number of mines being operated in Canada continues to decline. In 2005, four metal-producing mines closed and only one (Voisey's Bay Ni-Cu mine in Labrador) opened.

On the brighter side, the potential for new mines in Canada was given a strong vote of confidence by the international mineral exploration industry with an investment of \$1.3 billion in 2005, doubling the amount that was invested in 2000 (Bouchard, 2006). This makes Canada the preferred destination for exploration dollars for the third consecutive year, ahead of South America and Australia. Of this total, $949 million was spent in searching for new mineral deposits, $225 million on appraising known deposits with a view to a production decision, and $130 million in exploration at existing mines. Of the $949 million spent on looking for new mineral deposits, $381.6 million (40%) was for precious metals, particularly Au, $206.8 million (22%) for diamonds, $196.7 million (21%) for base metals, and $86.5 million (9%) for U. About 80% ($761.8 million) of this exploration expenditure was more or less evenly spread across Ontario, Nunavut, British Columbia, Quebec, and Saskatchewan, with each jurisdiction receiving between $126.5 and $168.3 million in exploration investment. Exploration expenditures make substantial contributions to the economies of northern and remote communities, which are used as logistical support and supply centres, and provide employment opportunities for the local residents.

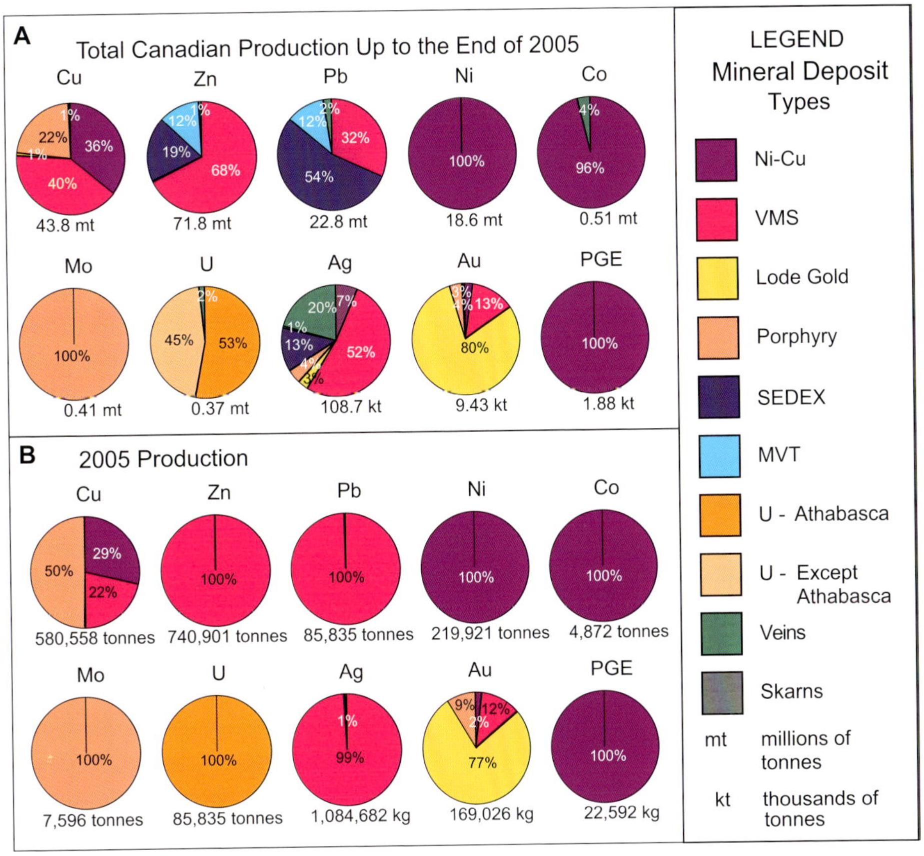

FIGURE 4. (A) Distribution of metals by deposit type for ores mined in Canada up to the end of 2005. (B) Distribution of metals by deposit type for ores mined in Canada during 2005.

Junior mining companies were responsible for attracting 58% of the 2005 mineral exploration investment in Canada, for the second year in a row surpassing the investment made by the major mining companies, a situation not encountered since 1987 (Bouchard, 2006). The success in attracting this investment capital has been helped, since October 2000, by the 15% federal investment tax credit tied to the flow-through-share mechanism, and similar tax credits and other measures in different provinces and territories (Bouchard, 2006).

Canada is the world's leading producer of potash (not dealt with here) and U, and third in the production of Ni, platinum group elements (PGE), and diamonds. Canada is no longer one of the top three producers of Zn, Pb, Cu, or Au, more because Canadian production of these commodities has gradually declined over the past decade rather than increased production by other countries.

Economic Characteristics of Mineral Deposit Types

The total value of non-ferrous metal and diamond ores mined in Canada to date is here estimated to be $Eq845.7 billion, based on the data compilation and processing methodologies described above. This number is a minimum, because it does not include many of the smaller deposits mined in the 19th and the beginning of the 20th centuries, and because the production records for some mines are incomplete. However, insomuch as >95% of the metals produced in Canada has taken place since 1920 (see plots in Cranstone, 2002), this number probably accounts for well over 90% of the metals that have been mined and also captures all diamond production.

Different metals (Fig. 1) tend to occur in different deposit types (Fig. 4). The great majority of non-ferrous metal and diamond production in Canada has been derived from eleven mineral deposit types (Fig. 5), of which only six are currently being mined (Fig. 6). Although most deposit types produce more than one metal, either as co-products or by-products, some deposit types, such as lode gold and uranium deposits, are essentially monometallic with >95% of the value of the ores being supplied by one commodity (Fig. 5). The geological characteristics of these deposit types are described and discussed in detail elsewhere in this volume (Dubé and Gosselin, 2007; Eckstrand and Hulbert; 2007; Galley et al., 2007a; etc.), but are briefly summarized, together with their economic characteristics, in later sections of this article.

Production

The main changes that have contributed to the differences in the relative importance of metal sources between historical (Fig. 4A) and current (Fig. 4B) mineral production has been the decline in the number of volcanogenic massive sulphide (VMS) deposits being mined, and the cessation of mining from sedimentary exhalative (SEDEX), Mississippi Valley-type (MVT), paleoplacer U, and most vein deposits. These changes have mainly impacted the sources for Zn, Pb, Ag, and, to a lesser extent, Cu (Fig. 4). By far the major contributor to the value of Canada's non-ferrous metal and diamond production has been magmatic Ni-Cu deposits (Fig. 5), whose \$Eq372.1 billion metal content represents 44% of Canada's total non-ferrous metal and diamond primary production and 34% of 2005 production with most of this value (65%) being attributable to Ni (Fig. 5). Magmatic Ni-Cu deposits have historically produced (Fig. 4A), and currently account (Fig. 4B) for nearly all of Canada's Ni, Co, and PGE primary output.

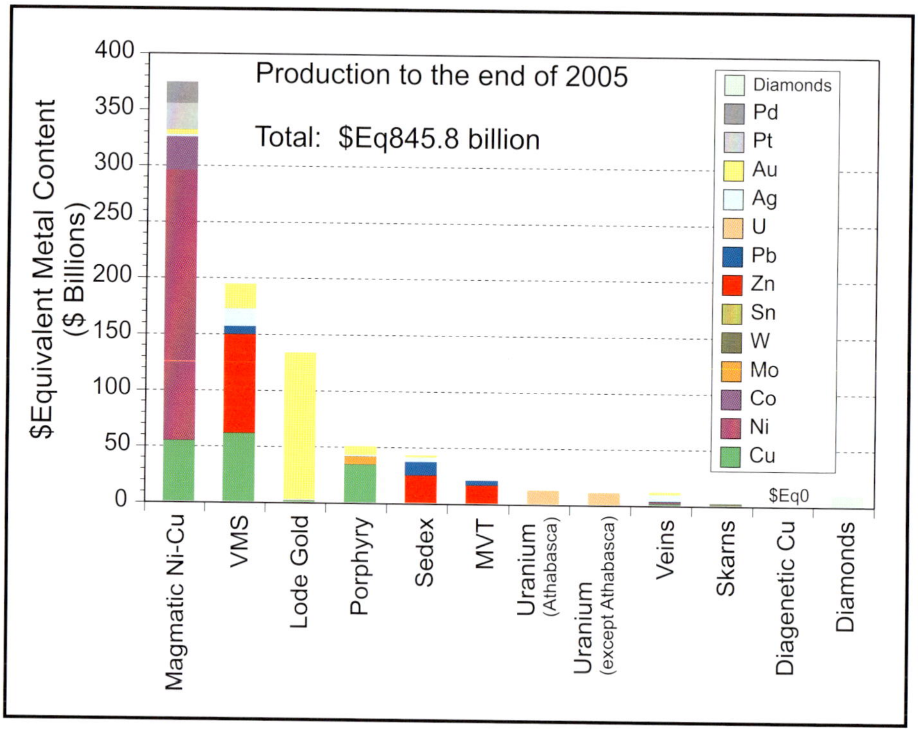

FIGURE 5. Stacked bar graphs showing the dollar equivalent of metal content of Canadian ores that were mined up to the end of 2005.

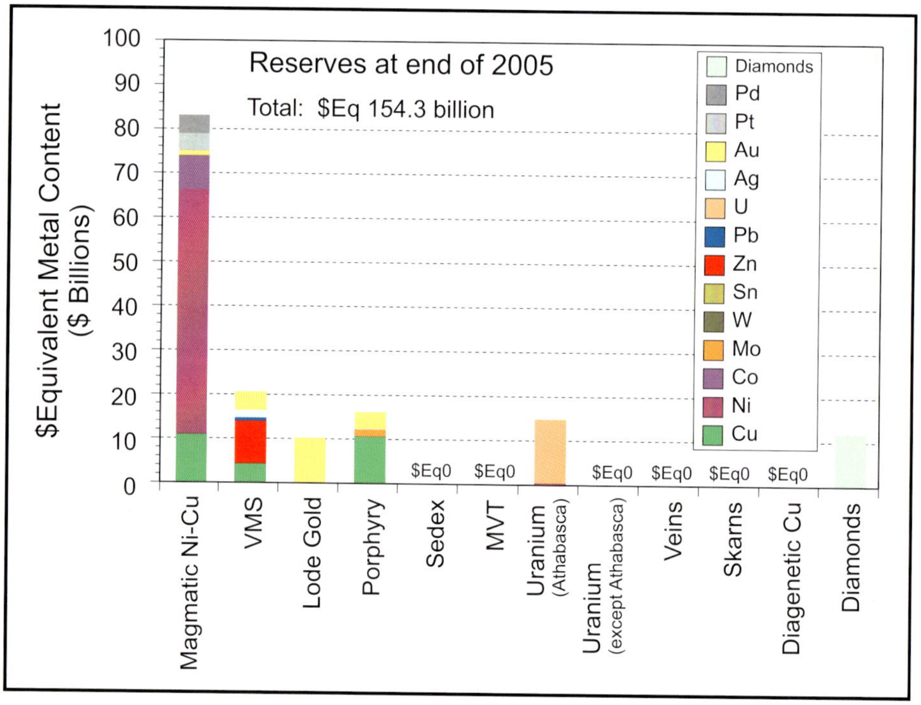

FIGURE 6. Stacked bar graphs showing the dollar equivalent of metal content of Canadian ore reserves by deposit type at the end of 2005.

Volcanogenic massive sulphide and lode gold deposits have been the mainstay of the Canadian metal mining industry, having supported 153 and 194 mines, respectively, across the country from Newfoundland to Vancouver Island. The \$Eq192.3 billion of production makes VMS deposits the second most productive mineral deposit type (Fig. 5), and accounts for 23% of total Canadian historical non-ferrous metal and diamond production and 20% of 2005 production (Fig. 4). This deposit type has produced 68% of Canada's Zn, 52% of its Ag, 40% of its Cu, 32% of its Pb, and 13% of its Au (Fig. 4A), and currently produces virtually all of its Zn, Pb, and Ag (Fig. 4B). Lode gold deposits are Canada's third most valuable mineral deposit type, having produced \$Eq131.6 billion of metal, of which 98% is attributable to Au (Fig. 5). Lode gold deposits account for 80% of Canada's Au production. The amount of Au produced (Table 2) is understated because it does not include small mines, especially those of the 19th and early part of the 20th century for which records are sparse, and the amount of by-product metal is also understated, because most lode gold producers report only the amount of Au.

Porphyry deposits in Canada did not make a significant contribution to Canada's mineral production until the early 1970s, and their \$Eq48.5 billion of production is only 5.7% of total Canadian non-ferrous metal and diamond production (Fig. 5; Table 2) but 22% of all Cu production (Fig. 4A). About 70% of the value of production from porphyry deposits is attributable to Cu, 14% to Mo, 13% to Au, and 2.5% to Ag. Production from SEDEX deposits has totaled \$Eq40.8 billion (Fig. 5), and supplied 19% of the Zn, 54% of the Pb, and 13% of the Ag (Fig. 4A) produced in Canada to the end of 2005. Mississippi Valley-type deposits have contributed \$Eq19.0 billion in production, mainly from the Zn content, and about 12% of both Canada's Zn and Pb production (Fig. 4A).

Total U production from the Athabasca Basin to the end of 2005 was \$Eq10.7 billion, about \$1.0 billion less than the total amount of U that has been produced in all other parts of

Canada (Fig. 5). However, with reserves (Fig. 6) sufficient for another 25 years of production at 2005 rates of mining, production of U from the Athabasca Basin will far surpass production from other areas in the future. Despite the world-leading production and the very high values of the ores, U production from the Athabasca Basin contributed only 11% of Canada's non-ferrous metal and diamond production in 2005, because the quantities of U produced are relatively small. Veins, together with skarn deposits, mined mainly in the early part of the 20th century, have contributed about $Eq10.0 billion in production, about 57% of this value being attributable to Ag (Fig. 5; Table 3). Production of diamonds in Canada began in 1998 and to the end of 2005 had a value of $Eq8.2 billion (Fig. 5). Production of diamonds during 2005 was $1.4 billion or 11% of Canada's total non-ferrous metal and diamond production (Fig. 1).

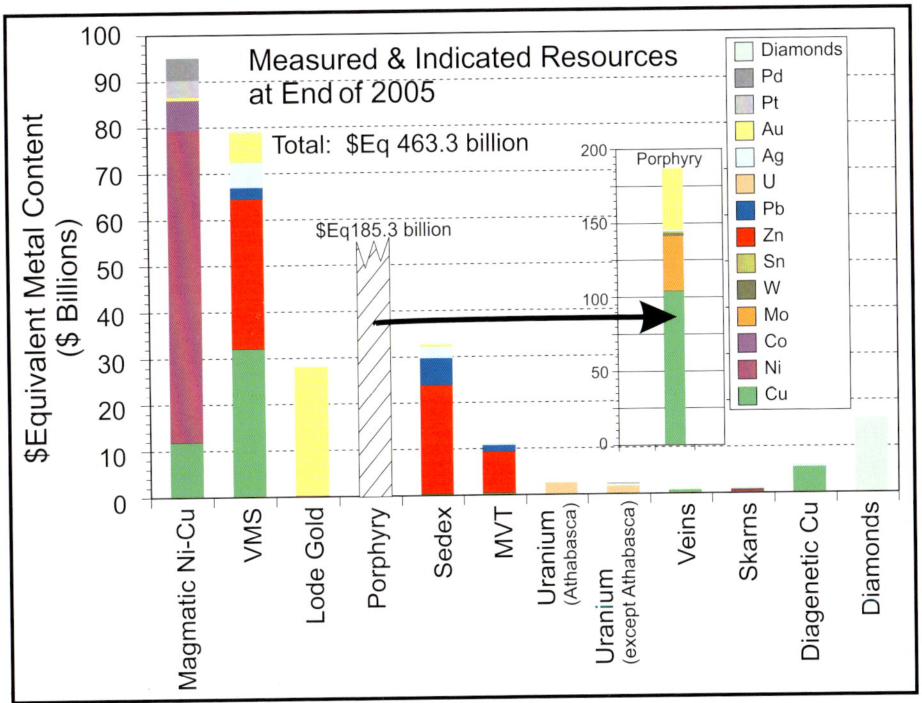

Figure 7. Stacked bar graphs showing the dollar equivalent of metal content of Canadian measured and indicated mineral resources by deposit type at the end of 2005. These statistics include deposits whose historical 'reserves' were reported prior to 2001 and are not compliant with National Instrument 43-101 Standards.

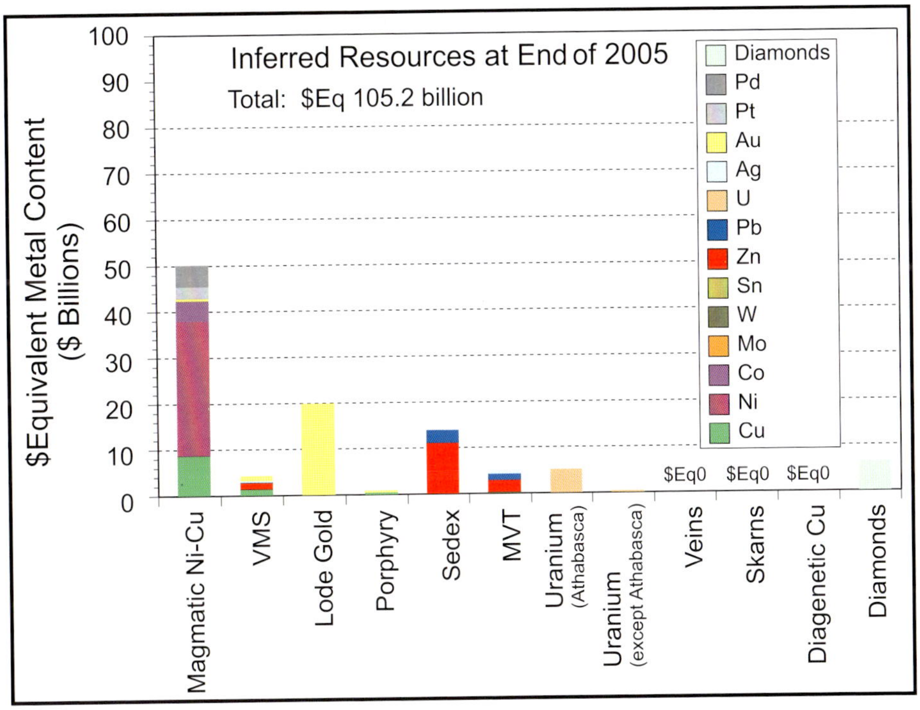

Figure 8. Stacked bar graphs showing the dollar equivalent of metal content of Canadian inferred mineral resources by deposit type at the end of 2005.

Reserves

Magmatic Ni-Cu deposits also dominate Canada's non-ferrous metal and diamond reserves as well as past and current production, containing 53% ($Eq82.5 billion) of Canada's $Eq182.3 billion total (Fig. 6; Table3). About 67% of the value Magmatic Ni-Cu reserves is attributable to the Ni content. VMS deposits, with $Eq20.0 billion of reserves, has the second highest reserve value and represents about 13% of the value of Canada's total non-ferrous metal and diamond reserves. Remaining reserves are distributed between Porphyry deposits ($Eq15.6 billion), U deposits of the Athabasca basin ($14.6 billion), Kimberlite diamonds ($Eq11.8 billion), Lode Gold deposits ($Eq9.6 billion), and a small amount in tungsten skarns (<$Eq0.1 billion). Almost 50% the value of VMS reserves is in their Zn content, and 67 % of the value of Porphyry deposit reserves is in their Cu content (Fig.6; Table 3). Lode Gold deposits contain only 54% of Canadian Au reserves (Fig. 6; Table 3). At 2005 rates of production, the reserves for Lode Gold deposits are sufficient for only four more years of mining, which forecasts impending mine closures and decreased production, unless the concerted exploration efforts ($381.6 million in 2005) to find new deposits meet with significant success.

Measured and Indicated Resources

Measured and indicated resources are the inventory from which the next generation of Canada's mines is likely to be developed. The resources are spread over every mineral deposit type (Fig. 7), but many of these resources are in frontier areas and their future development at this time is uncertain. The statistics for this category include historical estimates of resources, which if re-estimated to NI 43-101 requirements may be reclassified differently, such as to inferred resources or to an informal category such as "potential resources".

Porphyry deposits are the major repository for Canada's un-mined mineral resources, containing $Eq185.3 billion (Fig. 7; Table 3) or 40% of the $Eq463.3 billion of all Canadian mineral resources in this category, including 100% of its Mo, 68% of its Cu, 53% of its Au, and 14% of its Ag, as well as significant quantities of W and Sn. Most of these resources are in British Columbia. Measured and indicated resources of $Eq94.9 billion in Magmatic Ni-Cu deposits and $Eq78.8 billion in VMS deposits represent 20% and 17%, respectively, of Canada's total of this mineral resource category (Fig. 7; Table 3). There are $Eq32.5 billion Zn, Pb and Ag in measured and indicated resources in SEDEX deposits, mainly in theYukon and British Columbia, with most of this value being in their Zn content. Measured and indicated Lode Gold resources of $Eq27.9 billion contain 37% of Canada's Au resources (Fig. 7). Measured and indicated resources of Kimberlite diamonds ($Eq15.6 billion) are mainly in Ekati and in the three deposits currently at the advanced exploration stage (Victor, Snap Lake, Gaché Hue) prior to production decisions which may result in them being reclassified as reserves. Most of Canada's un-mined $Eq15.5 billion mineral resources of MVT deposits are in the western part of Northwest Territories. All other mineral deposit types contribute 3.7 % to the measured and indicated resource category.

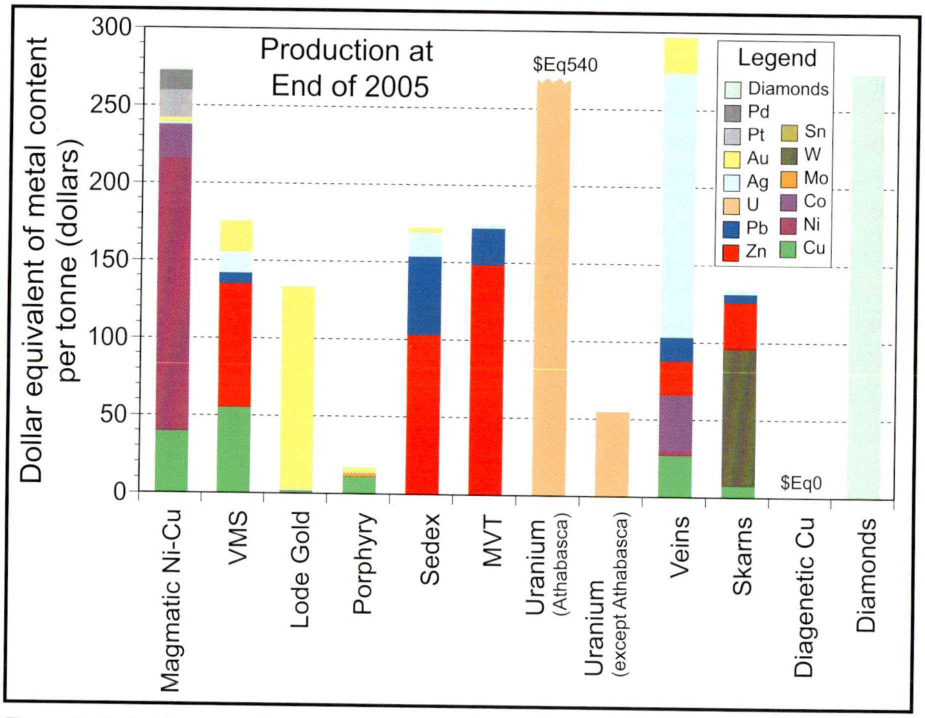

FIGURE 9. Stacked bar graphs showing the dollar equivalent of the average metal content per tonne of ore for Canadian production up to the end of 2005.

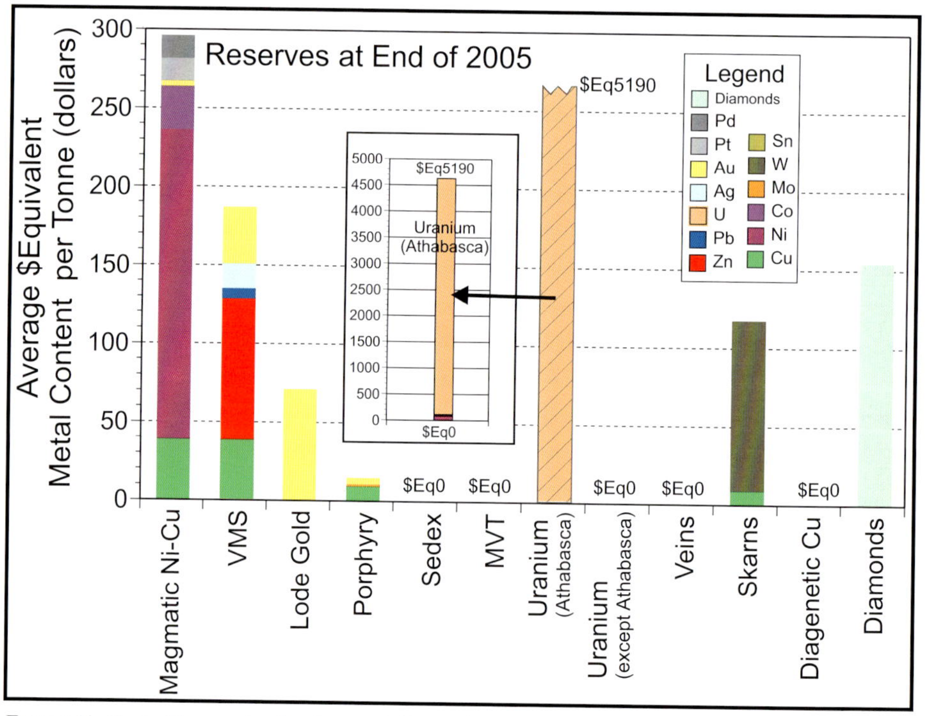

FIGURE 10. Stacked bar graphs showing the dollar equivalent of the average metal content per tonne of ore for Canadian reserves at the end of 2005.

Inferred Resources

The statistics for this resource category applies only to those deposits whose mineral resources have been estimated to NI 43-101 requirements. Insomuch as most exploration effort over the past decade has been directed towards diamonds, Au, Ni and U deposits, the proportion of mineral resources in the inferred resources category is valid on ly for the Magmatic Ni-Cu, Lode Gold, Uranium, and kimberlite diamond mineral deposit types.

Magmatic Ni-Cu deposits dominate the inferred resources category (Fig. 8), constituting 47% of the $Eq105.1 billion total (Fig. 8; Table 3). Lode Gold deposits contain $Eg19.5 billion in inferred resources, SEDEX deposits $13.4 billion (the result of a recent re-estimation of resources at Howards Pass), and kimberlite diamonds $9.7 billion (Fig. 8). Volcanogenic massive sulphide, MVT,and Athabasca basin U, deposits all have inferred resources in the $Eq3.9 billion to $Eq4.7 billion range (Fig. 8; Table 3).

Values per Tonne

The average values per tonne statistics for the different resource categories (Figs. 9, 10, 11, and 12) are proxies for the grades of ores. The Athabasca Basin deposits contain the world's richest U ores, with reserves containing a phenomenal average $Eq5,190/t in sufficient quantities for another 25 years of production at 2005 rates of mining. The average equivalent dollars per tonne for magmatic Ni-Cu ores are $Eq272/t for production (Fig. 9) and $Eq294/t for reserves (Fig. 10), which are the highest amongst all base metal mineral deposits types currently being mined. Average grades of VMS ores that have been mined at $Eq174/t is about the same as other Zn-rich underground base metal mineral deposit types (Fig. 9) but, as discussed later, there is a large range from deposit to deposit, both in dollar equivalents of ore per tonne and the relative proportions of different metals.

The low $Eq15.7/t for production (Fig. 9), $Eq13.4/t for reserves (Fig. 10), $Eq28.1/t for measured and indicated resources (Fig. 11), and $Eq18.4/t for inferred resources (Fig. 12), reflect the low grade of porphyry deposits in general. These large low-grade deposits are currently mined only by the low-cost open-pit method, but on-going research into developing underground mining methods and technology that would rival open-pit costs gives hope that in the future porphyry deposits can be mined without the environmental concerns associated with open-pit mining (Morgan, 2005).

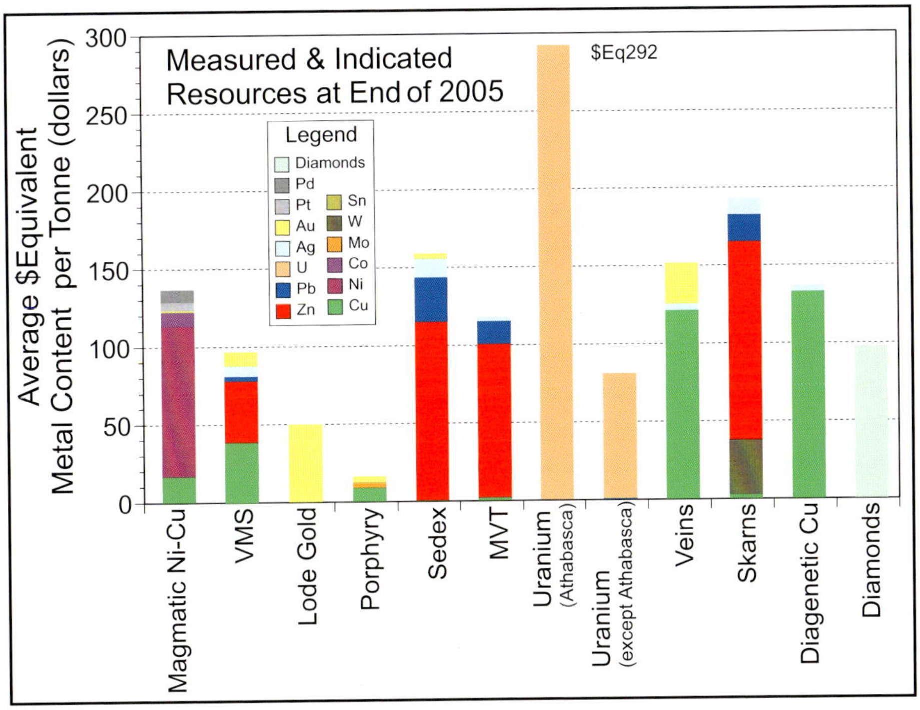

FIGURE 11. Stacked bar graphs showing the dollar equivalent of the average metal content per tonne of ore for Canadian measured and indicated resources at the end of 2005.

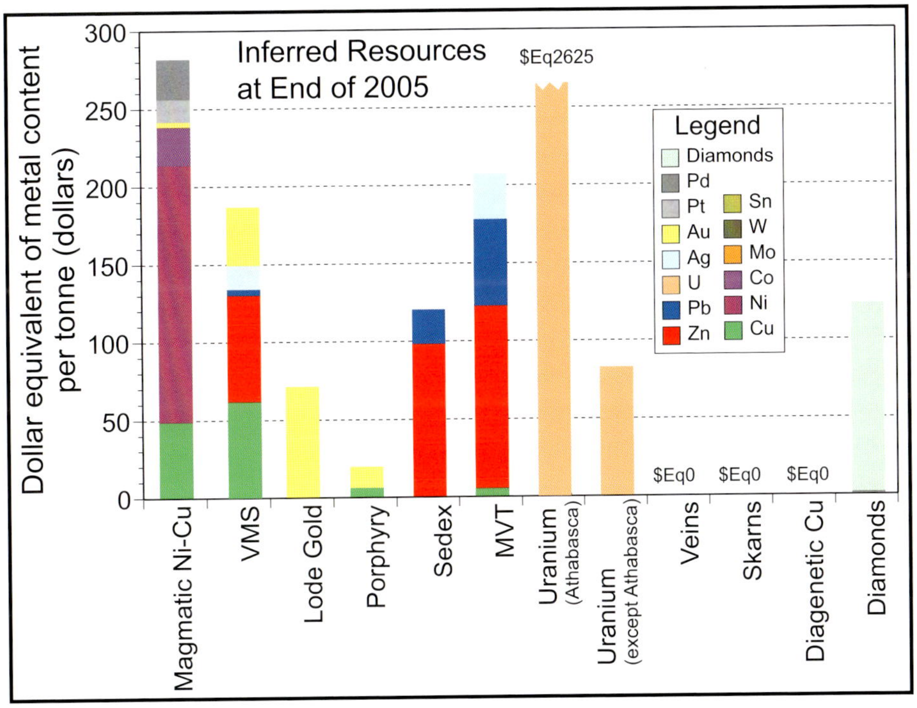

FIGURE 12. Stacked bar graphs showing the dollar equivalent of the average metal content per tonne of ore for Canadian measured and indicated resources at the end of 2005. These statistics include deposits whose historical 'reserves' were reported prior to 2001 and are not compliant with National Instrument 43-101 Standards.

The $Eq157.9/t metal content of measured and indicated resources for SEDEX deposits (Fig. 11) is only slightly lower than the $Eq171.1/t content of ores that have been mined, suggesting that at least part of these resources are marginal to being economically mineable, if a suitable surface transportation and power infrastructure were in place.

The average $Eq173.8/t for production from MVT deposits is about the same for other deposit types mined for Cu, Zn, and/or Pb by underground methods (Fig. 9), but the average $Eq117.0/t for measured and indicated resources (Fig. 11) is lower, largely because of the influence of the large tonnages but lower grades of the Gayna River deposits.

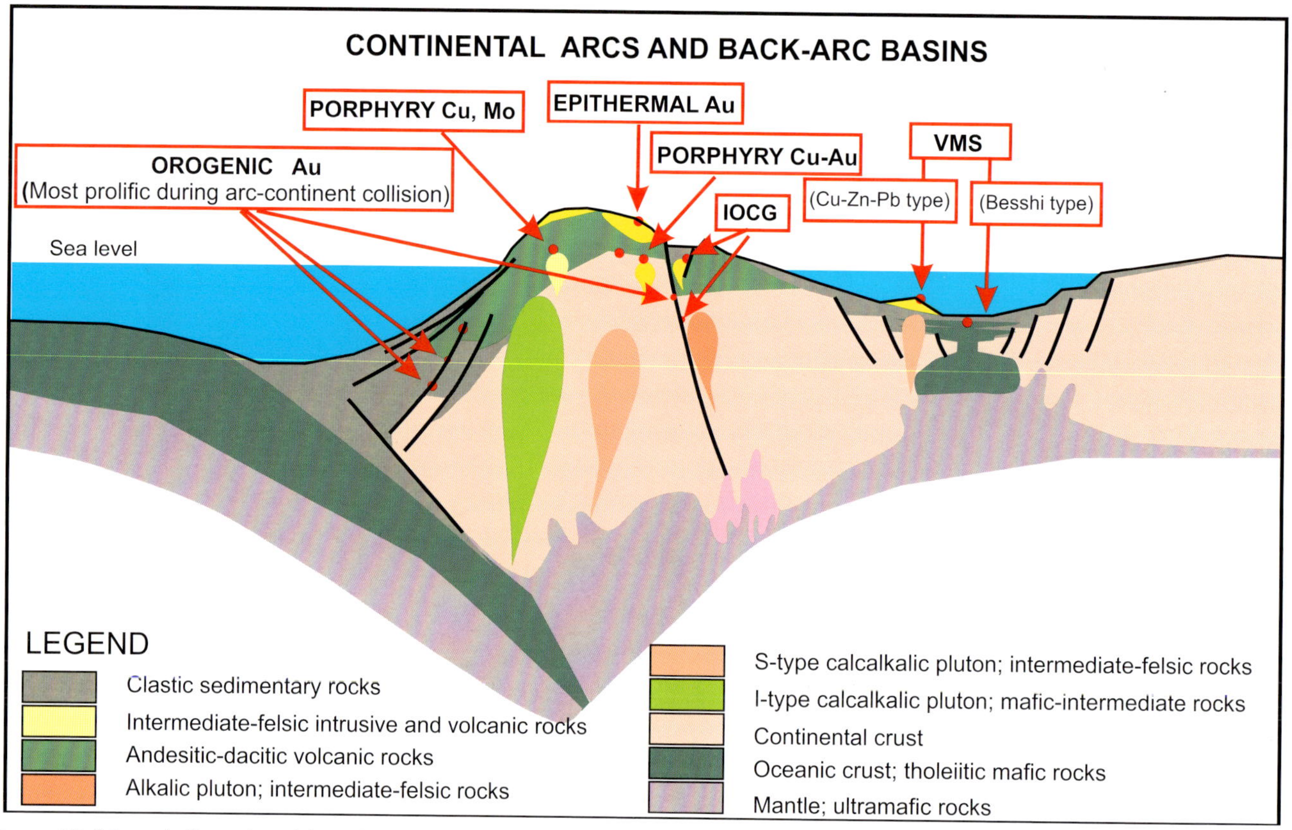

FIGURE 13. Schematic illustration of the major geological characteristics of major mineral deposit types that typically occur in continental arc and back-arc environments.

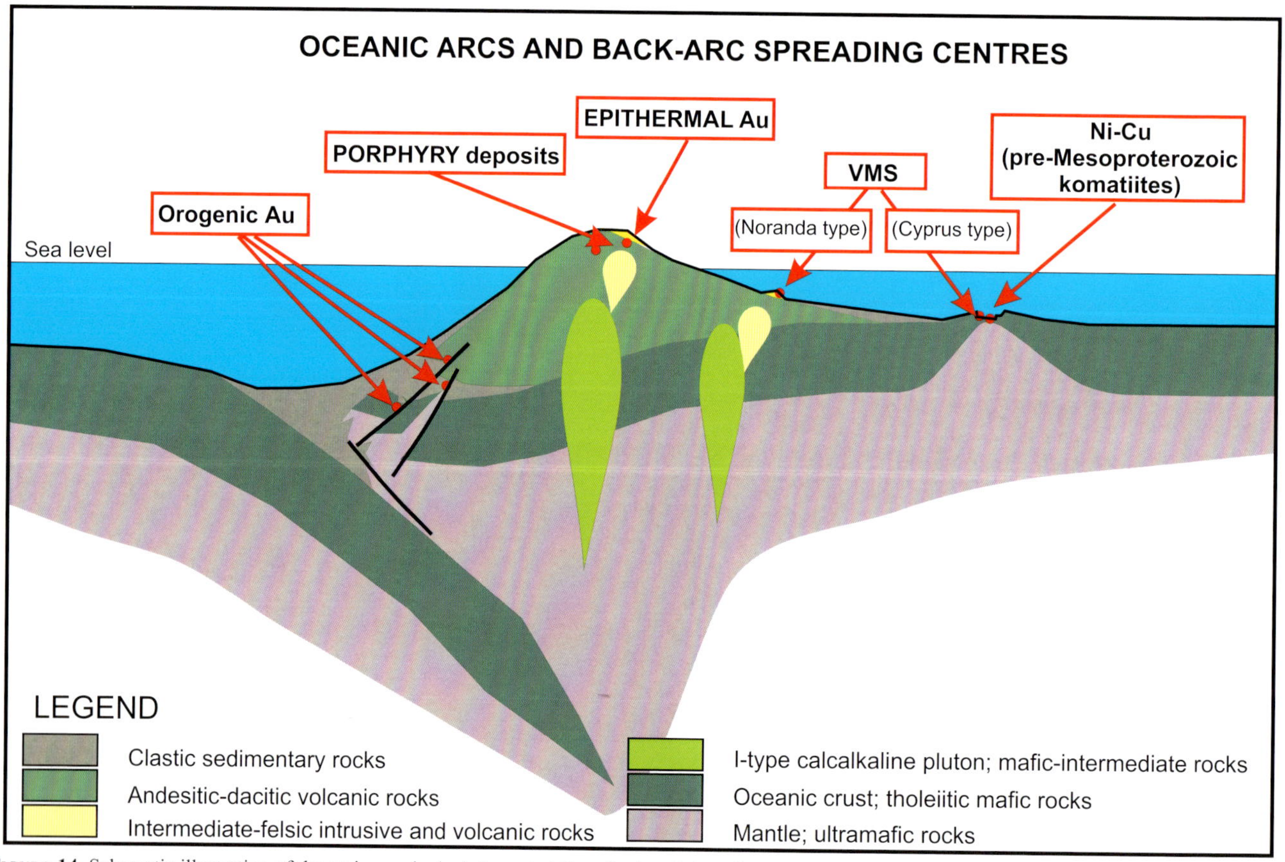

FIGURE 14. Schematic illustration of the major geological characteristics of mineral deposit types that typically occur in oceanic arc environment and back-arc spreading centres.

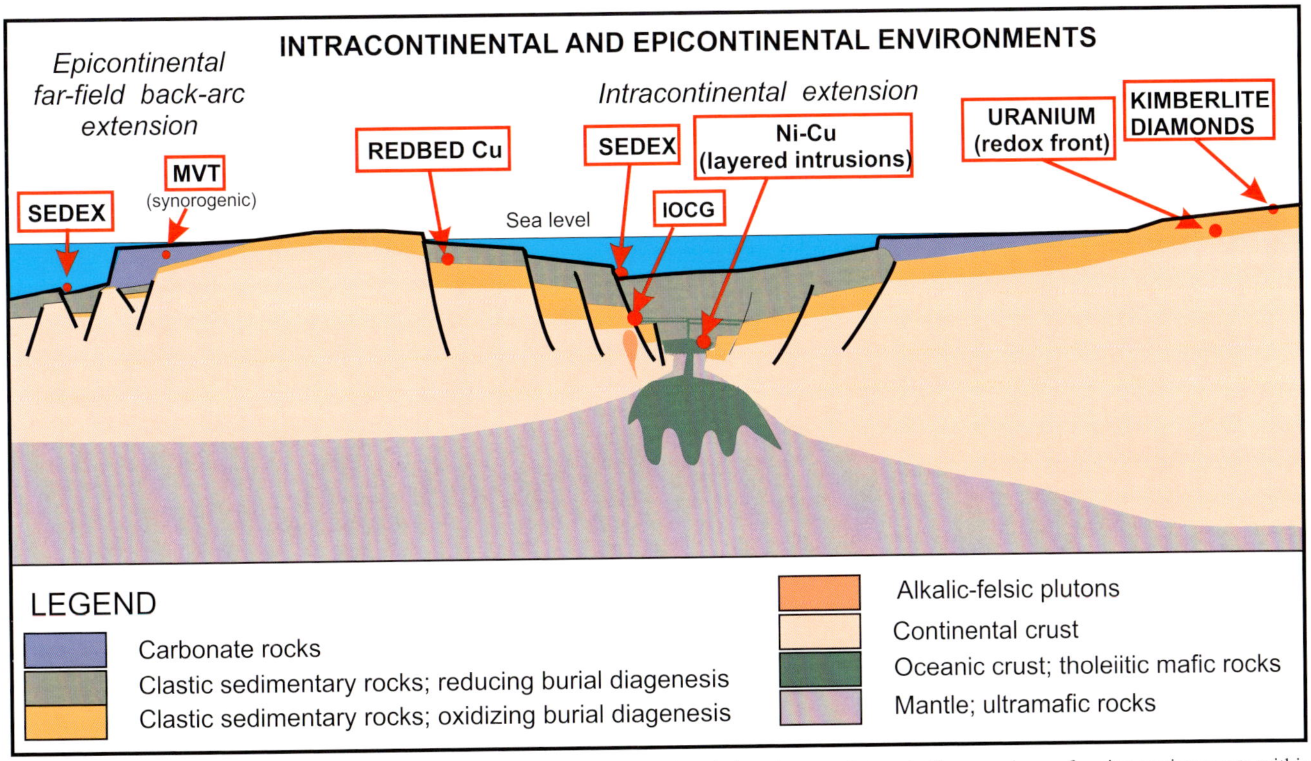

FIGURE 15. Schematic illustration of the major geological characteristics of mineral deposit types that typically occur in ore-forming environments within the interior regions of continents.

Unlike metals, the value of kimberlite diamond ores depends on the quality of the diamonds as well as its grade. The average value of ores produced from the Ekati mine has been \$253/tonne and that of Diavik \$366/tonne. Reserves at Ekati are estimated to have an in situ value of \$91/tonne, at Diavik \$256/tonne and at Jericho \$117/tonne, based on the very tenuous assumptions that the average value of diamonds at the three mines are \$133/ct, \$80/ct, and \$81/ct respectively.

Geological Environments and Distribution of Canada's Mineral Resources

Geological Environments of Mineral Resources

There is increasing acceptance that plate tectonics and the supercontinent cycle operated back into the Archean (e.g. Kerrich and Polat, 2006). The most productive geological environments for the formation of metalliferous mineral deposits are in volcanic arcs and back arcs that were built upon, or accreted to, continental margins during supercontinent assembly (e.g. Barley and Groves, 1992) (Figs. 13, 14). Most of the mineral deposit types, particularly porphyry deposits (Sinclair, 2007) and intrusion-related and epigenetic lode gold subtypes (Dubé and Gosselin, 2007; Hart, 2007) in subaerial arcs, and VMS deposits (Galley et al., 2007a) in submarine arcs and back arcs, are directly related to magmatic and/or associated convective hydrothermal systems (e.g. Lydon, 1996). The age of these deposits (Fig. 16) correspond to periods of subduction, which in the case of deposits formed in oceanic arcs or continental arcs of microcontinents may be up to a few tens of millions of years prior to accretion (e.g. Percival, 2007; van Staal, 2007). The orogenic subtype of lode gold deposits (Dubé and Gosselin, 2007) is related to hydrothermal systems generated by collisional tectono-thermal processes.

Spreading centres in oceanic crust are not highly productive for mineral deposits, even though mid-oceanic spreading centres are the most common site for modern metalliferous hydrothermal deposition. Ophiolites, which generally are oceanic crust generated at back-arc spreading centres (Fig. 14) and obducted onto continents, contain only the relatively small Cyprus-type VMS deposits and podiform chromite deposits (Bédard et al., 2007). However, old oceanic crust, formed during the Paleoproterozoic or earlier, contains Ni-Cu deposits associated with komatiitic lava flows, such as the Archean deposits of the Abitibi Belt and possibly those of the Thompson Belt (Eckstrand and Hulbert, 2007).

Mineral deposit types that form in or on continental crust (Fig. 15) during the long periods of supercontinent cohesion can be divided into two major categories:

1) Mineral deposits of mafic-ultramafic magmas, whose emplacement is along structures that dislocate or perforate continental crust and penetrate the mantle. Most of these structures have their origins in tensile stresses produced by lateral spreading at the head of mantle plumes, which allows the upwelling of mantle magmas into the continental crust (Ernst, 2007). Magmatic Ni-Cu deposits that segregated from these magmas are the major mineral deposit type, but the crustal perforation that produced Canada's most productive deposits of this type was due to a large meteorite impact (Ames and Fallow, 2007; Eckstrand and Hulbert, 2007). Diamondiferous kimberlite diatremes and dykes (Kjarsgaard, 2007) are the other major mineral deposit type in this category.

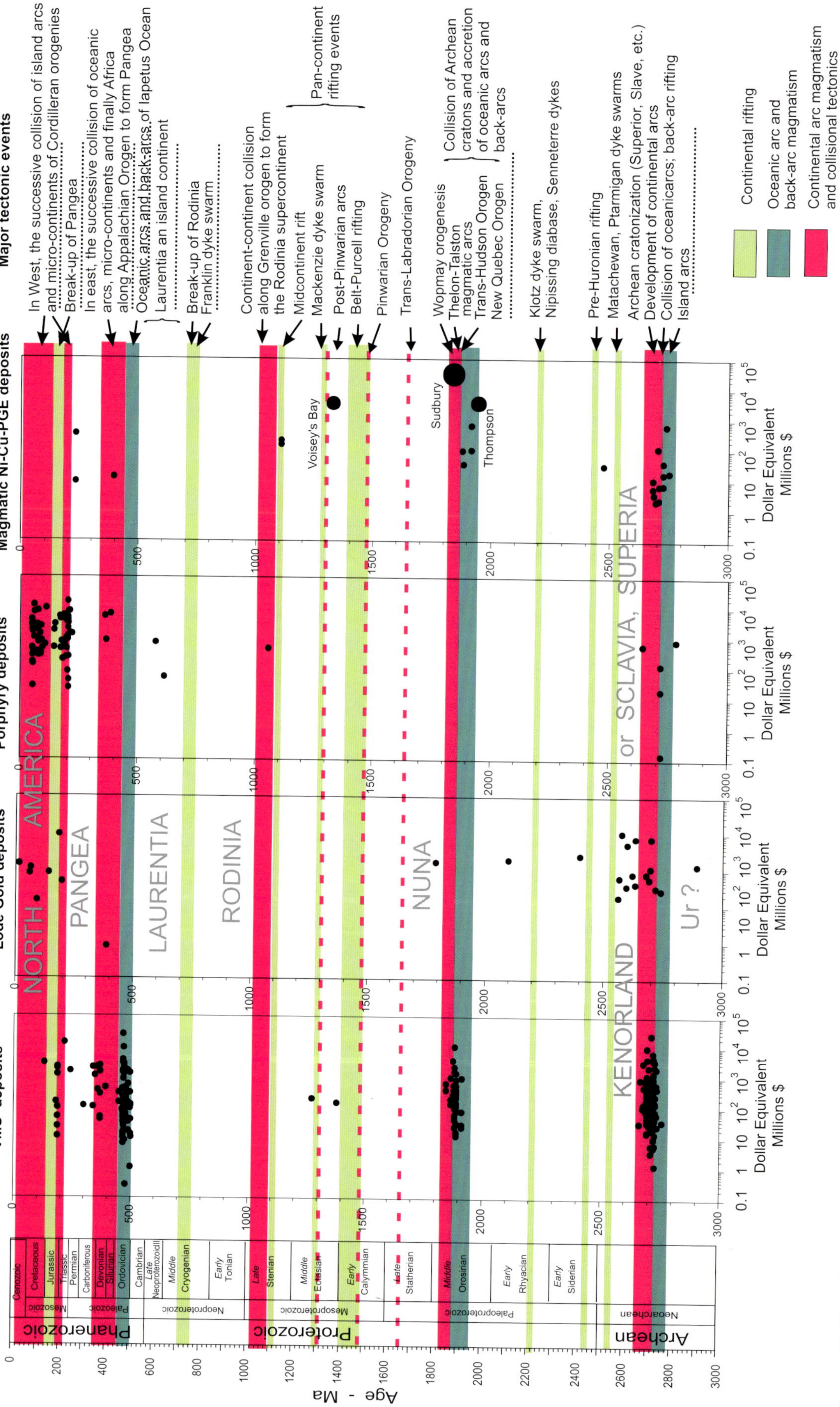

FIGURE 16. Plots of ages versus dollar equivalent of the total metal contents of deposits of orogens (volcanogenic, lode gold, porphyry) and anorogenic mafic magmatism (magmatic Ni-Cu-PGE), showing the relationships between the geochronological distribution of mineral deposits, major tectonic/magmatic events, and the supercontinent cycle. Ages of magmatic host rocks are from the National Geochronological Knowledge Base (www.ims1.ess.nrcan.gc.ca/geochron), and the remainder is from the databases of Appendix 1 (DVD). Deposits or districts referred to in text are labeled.

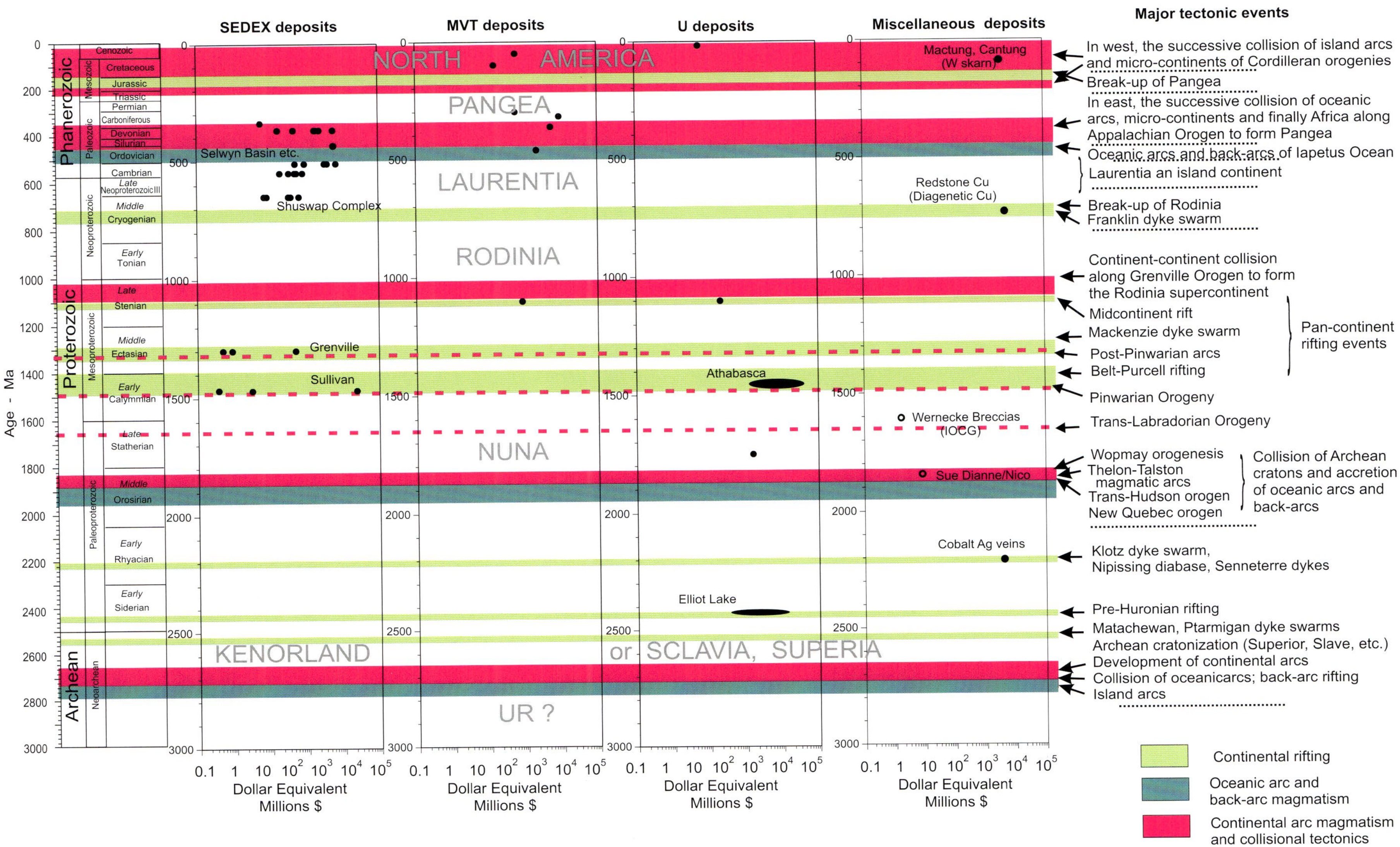

FIGURE 17. Plots of ages versus dollar equivalent of the total metal contents of deposits related to combinations of sedimentary process and tectono-thermal events (sedimentary exhalative, Mississippi Valley-type, uranium) and miscellaneous deposits of orogens and intracontinental environments. See text for discussions. Ages are from the databases of Appendix 1 (DVD).

TABLE 4. Summary of the dollar equivalent of metal and diamond content of Canada's mineral resources by geological environment and supercontinent cycle.

Supercontinent Cycle	Orogens $Eq millions	Sedimentary basins $Eq millions	Anorogenic mafic-ultramafic magmatism $Eq millions	Sudbury $Eq millions	TOTAL $Eq millions	Totals without Sudbury $Eq millions
PRODUCTION						
North America	66,424	181	8,253	-	74,858	74,858
Pangea	58,138	29,500	-	-	87,638	87,638
Rodinia	548	-	-	-	548	548
Nuna	69,584	44,730	267	330,949	445,529	114,580
Kenorland	223,146	7,715	6,322	-	237,183	237,183
TOTAL	417,840	82,126	14,789	330,949	845,703	514,755
RESERVES						
North America	15,454	-	11,811	-	27,265	27,265
Pangea	6,455	-	-	-	6,455	6,455
Rodinia	-	-	-	-	-	-
Nuna	20,507	14,591	16,125	49,215	100,438	51,223
Kenorland	18,881	-	1,259	-	20,140	20,140
TOTAL	61,298	14,591	29,195	49,215	154,299	105,084
MEASURED, INDICATED & INFERRED RESOURCES						
North America	198,748	1,102	18,668	-	218,518	218,518
Pangea	48,625	57,672	5,296	-	111,592	111,592
Rodinia	261	3,492	-	-	3,752	3,752
Nuna	35,147	9,520	35,208	60,778	140,654	79,876
Kenorland	80,584	-	13,352	-	93,935	93,935
TOTAL	363,364	71,786	72,523	60,778	568,451	507,673
TOTAL MINERAL RESOURCES						
North America	280,626	1,283	43,647	-	325,556	325,556
Pangea	113,838	87,172	5,296	-	206,306	206,306
Rodinia	809	3,492	-	-	4,300	4,300
Nuna	125,238	68,841	51,600	440,942	686,621	245,679
Kenorland	322,611	7,715	20,933	-	351,259	351,259
TOTAL	843,122	168,503	121,475	440,942	1,574,042	1,133,100

2) Mineral deposits of intracontinental and epicontinental sedimentary basins. Intracontinental sedimentary basins accumulate in depressions of continental crust caused either by rifting or by crustal warping in the foreland of thrust belts, and epicontinental sedimentary basins are the successors to the rifts along which oceans initially opened. Rifting in these sedimentary basins may be reactivated by far field extensional tectonics related to an approaching subduction zone during the initial stages of the accretionary phase of the supercontinent cycle (Fig. 15). These extensional synsedimentary faults are the conduits for basinal brines to the seafloor to form SEDEX Zn-Pb-Ag deposits (Goodfellow and Lydon, 2007), whereas MVT Zn-Pb deposits (Paradis et al., 2007) are formed by the migrations of basinal brines from epicontinental sedimentary basins into adjacent platformal carbonates in response to tectonic basin inversion during continental accretion. Unconformity-related U deposits (Jefferson et al., 2007) and diagenetic Cu deposits (e.g. Kirkham, 1996a,b) are likewise formed by the migration of basin or basement groundwaters across a groundwater redox boundary. Paleoplacers (e.g. Roscoe, 1996), the most important source for South Africa's world-leading Au production and for Canada's Elliot Lake former U production, accumulate as heavy mineral concentrations at the edges of sedimentary basins in fluviatile or estuarine sediments.

Geochronology of Canada's Mineral Deposits

The age of a mineral deposit is critical to relating ore formation to tectonic and geological processes. The ages of mineralization for deposits that are coeval with magmatic host rocks (VMS, magmatic Ni-Cu-PGE, porphyry deposits) are relatively well known because of a rapidly expanding database of U-Pb ages of magmatic zircon, particularly the National Geochronological Knowledge Base (www.ims1.ess.nrcan.gc.ca/geochron). The ages of synsedimentary deposits (SEDEX) can be estimated from paleontological dating or by bracketing the stratigraphic horizon hosting the deposit by radiometric dating of any intercalated volcanic rocks. The ages of epigenetic deposits (lode gold and MVT) requires specific dating of gangue, alteration, or ore minerals (Dubé and Gosselin, 2007; Paradis et al, 2007), and as a result of the far fewer geochronological studies of this nature, the age distribution of epigenetic deposits is less well documented than mineral deposits that were formed at about the same time as their host rocks (Figs. 16, 17).

With some caveats, the geochronological distribution of mineral deposits in Canada (Figs. 16, 17) conforms to the timing of global supercontinent cycles (e.g. Hoffman, 1988;

McCulloch and Bennett, 1994; Taylor and McLennan, 1995; Rogers and Santosh, 2003; Hawkesworth and Kemp, 2006) and correlate with the age spans of the geological environments described above that these cycles control. The geochronological distribution of VMS deposits (Fig. 16) marks the relatively short periods of supercontinent assembly over the entire span of preserved geological history, but spatially (Fig. 2) are confined to the margins of ancient continents whose growth involved the accretion of oceanic arcs/back-arcs and the construction of submarine continental arcs/back-arcs. Orogens formed by continent-continent collision, such as the Grenville (Figs. 2, 16) (Corriveau et al., 2007) or by intracrustal melting, such as the Taltson-Thelon magmatic zone (Chacko et al., 2000) (Fig. 2) lack a significant VMS resource. Orogenic lode gold deposits, associated with collisional tectonism and its aftermath (Fig. 13, 14) have approximately the same geographic distribution as VMS deposits (Fig. 2), but are generally tens of millions of years younger (Fig. 16) (e.g. Percival, 2007). Porphyry deposits also mark continental arc construction during supercontinent assembly but their incidence, like other high-level deposits (e.g. epithermal Au, Figs. 13, 14), decreases with age (Fig. 16) due to the deepening of the erosion level with time.

The geochronological distribution of some mineral deposit types reflects evolutionary changes to the lithosphere, oceans, or atmosphere that exert an impact on the geological environments in which mineral deposits are formed. Magmatic Ni-Cu deposits in komatiitic lava flows and related sills or dykes, which constitute the majority of Ni-Cu deposits in Figure 16, are confined to rocks older than about 1800 Ma (Eckstrand and Hulbert, 2007), when the Earth's crust was thinner, heat flow higher, and the mantle less depleted (e.g. Taylor and McLennan, 1995). SEDEX deposits (Fig. 17) globally are confined to sedimentary basins younger than 1800 Ma (i.e. on the Nuna or later supercontinents). MVT deposits, which, like SEDEX deposits, are also genetically related to the migration of basinal brines of marine sedimentary basins (Paradis et al., 2007) are nearly all of Phanerozoic age (Fig. 17). The key explanation to this difference in geochronological distribution may be related to the erosion-prone geological setting of MVT deposits in the forelands of thrust and fold belts, which has resulted in the destruction of Proterozoic deposits. Placer U deposits are known only on the Kenorland supercontinent when the atmosphere lacked significant free oxygen.

Distribution of Mineral Deposits in Canada

The geographical distribution of all mineral deposits with respect to orogens and continental interiors of supercontinental cycles are shown in Fig. 2 and a summary of resources by geological environment and supercontinent cycle in Table 4. Mineral deposits of orogens have contributed $Eq417.8 billion (49.4%) of production and account for $Eq843.1 billion (53.6%) of total resources (Table 4). Exclusive of the deposits related to the Sudbury meteorite impact structure (which are not related to the supercontinent cycle) the proportions rise to 81.2% of production and 74.4% of total resources. The distribution of accreted oceanic arcs, continental arcs, and major structural dislocations at continental margins are therefore the main control on the distribution of the bulk of mineral deposits.

Kenorland

Fragments of a single Kenorland (Williams et al., 1991), or more than one smaller Archean- Paleoproterozoic supercontinent(s) such as Sclavia and Superia (Bleeker, 2003), are preserved as the Superior, Slave, Churchill (now Rae, Hearne, Wyoming, and Eastern Churchill subprovinces), Nain, and Makkovik geological provinces and form the central core of Canada's land mass (Fig. 2). These Archean cratons consist of ca. 2700 to 2800 Ma oceanic arcs and back-arc basins that aggregated, together with fragments of ca. 3400 to 2800 Ma continental crust (e.g. David et al., 2003) possibly derived from an earlier Ur supercontinent (Rogers, 1996) via successive collisional events over the ca. 2700 to 2650 Ma time interval (e.g. Bleeker and Hall, 2007; Percival, 2007). Continental arcs and associated back-arc basins formed at the margins of the oceanic arc nuclei both before and during collisional events, and shed detritus to form sedimentary basins both between and within volcanic arcs. Syn- and post-orogenic plutons, dominantly of tonalite, granodiorite, and granite (TGG) composition, invaded the volcano-sedimentary complexes of the newly aggregated crust to become the dominant lithology over large areas. Most tectonic and magmatic activity came to an end by about 2650 Ma (e.g. Bleeker and Hall, 2007; Percival, 2007). The geological architecture of Kenorland is thus very different from later supercontinents in that it may be viewed as an aggregation of newly formed orogens over its entire area, whereas later supercontinents that consist of a cratonic interior and orogens are restricted to linear to arcuate belts along their contemporary margins.

By far the economically most productive geological environments of the Archean provinces are the greenstone belts, which represent the remains of the tholeiitic or calc-alkaline volcanic arcs that have been preserved between tonalite-trondhjemite-granodiorite (TTG) intrusions and not buried by sedimentary basins. These greenstone belts have produced $Eq115.8 billion from lode gold deposits and $Eq102.6 billion from VMS deposits, which together represent 92.1% of production from Kenorland mineral deposits and separately account for 88.0% and 53.4%, respectively, of total Canadian production from these deposit types. The most productive area is the Abitibi greenstone belt in the Superior Province (Card and Poulsen, 1998) of Quebec and Ontario (Fig. 2), but lode gold and VMS deposits occur elsewhere, particularly in the western part of the Superior Province and the Slave Province (Fig. 2). If there is a geological reason for the high metal endowment of the Abitibi belt in comparison to other Archean greenstone belts, it needs to be determined in order to evaluate whether other Archean greenstone belts of Canada have a comparable potential. Magmatic Cu-Ni deposits in komatiite lavas of back-arc basins (Eckstrand and Hulbert, 2007), particularly of the Timmins area, Ontario, and the Val d'Or area, Quebec (Fig. 2), and porphyry deposits (Sinclair, 2007) in the general Chibougamau area, Quebec, and the Timmins and Matachewan areas, Ontario, have produced $Eq2.6 billion and $Eq2.1 billion, respectively, and together account for 1.3% of production from Kenorland mineral deposits.

Deposits associated with anorogenic mafic-ultramafic intrusions include the 2738 Ma Lac des Isles PGE-rich deposit in western Ontario, the low-grade and large-tonnage Dumont sill in western Quebec, and the closed Rankin Inlet mine in Nunavut (Eckstrand and Hulbert, 2007). Perhaps the Ag-rich veins of the highly productive Cobalt district in eastern Ontario (Ruzicka and Thorpe, 1996a), which are associated with the rifting event marked by the ca. 2200 Ma (Andrews et al., 1986) Nipissing diabase sills (Fig. 2), should also be included in this category. Collectively, these deposits have contributed $Eq6.3 billion or 2.7% of Kenorland mineral production. Mineral deposits of Kenorland sedimentary basins consist mainly of the ca. 2450 Ma uraniferous paleoplacers of Elliot Lake (Roscoe, 1996) at the base of the Huronian Supergroup in western Ontario, which have produced $Eq7.7 billion or 3.3% of Kenorland mineral production. The general lack of a platformal cover on Archean cratons led Hoffman (1990) to suggest that the assembly of these cratons produced an anomalously thick continental keel, which may give the Canadian Shield its prospectivity for diamonds (Kjarsgaard, 2007).

Nuna

Break-up of the Archean supercontinent occurred during 2170 to 1920 Ma and assembly of the Nuna supercontinent was completed by about 1830 Ma (Hoffman, 1988, 1989) along the Torngat, New Quebec, Ungava, Trans-Hudson, Taltson-Thelon, and Wopmay orogenic belts (from east to west, Fig. 2) (St-Onge and Lucas, 1996). The most productive of these Nuna orogens is the complex Trans-Hudson Orogen (Corrigan et al., 2007) containing the magmatic Ni-Cu deposits of the Thompson belt of Manitoba hosted by 1883 Ma komatiitic sills of a rifted foreland margin (Eckstrand and Hulbert, 2007; Layton-Matthews et al., 2007) and VMS deposits of the Flin Flon and Snow Lake mining districts in Manitoba and Saskatchewan (Galley et al., 2007b) in the Reindeer Zone, a collage of of Archean crustal fragments, 1920 to 1869 Ma oceanic arcs, and 1870 to 1830 post-accretion continental arcs.

The rifted continental margin of the Ungava Orogen, which represents a ca. 1800 Ma arc-continent collision (St. Onge and Lucas, 1996) contains the 1918 to 1883 Ma magmatic Ni-Cu in komatiitic sills that are being mined in northern Quebec (Lesher, 2007) (Fig. 2). The New Quebec Orogen (Fig. 2), forming the zone between the Superior Province and the southeast Churchill Province, contains small magmatic Ni-Cu-PGE deposits in sills of a ca. 2170 Ma sediment-sill complex associated with rifting of the Superior Craton, and small VMS deposits in a younger ca. 1880 Ma sediment-volcanic sequence (Wardle and Hall, 2002) (Fig. 2). Low-grade U deposits are hosted by felsic volcanics of the 1805 to 1860 Ma Aillik Group (Gandhi and Bell, 1996) that formed during the 1700 to 1900 Ma Makkovikian Orogeny marking the collision of the Makkovik Province with the Nain Province.

The 1910 to 1860 Ma calc-alkaline plutonic rocks of the Torngat Orogen (St-Onge and Lucas, 1996) and the 2000 to 1900 Ma, highly deformed, dioritic to granitic plutons of the Taltson-Thelon magmatic zone (Hoffman, 1988) are both interpreted to be the roots of a magmatic arc and do not preserve the shallow levels of arcs that are the most favourable for the formation of mineral deposits. The Taltson-Thelon magmatic zone may, alternatively, reflect intracrustal melting and not a continental suture between the Rae sub-province in the east and the Slave Province in the west (Chacko et al., 2000; De et al., 2000). The Wopmay Orogen preserves the western rifted Proterozoic passive margin (Coronation Supergroup) of the Slave Province with subsidence beginning about 1970 Ma (Hoffman, 1989), which was translated eastwards over the Slave craton as a fold and thrust belt by the collision of the 1950 to 1910 Ma Hottah arc and was followed by development of the 1880 to 1860 Ma calc-alkaline Great Bear magmatic arc above an east-dipping subduction zone. The Great Bear magmatic zone contains Canada's two principal deposits suggested to be iron oxide copper-gold type (Corriveau, 2007) (Fig. 2).

Several major metallogenic events are associated with episodic rifting of the Nuna supercontinent (Fig. 17). The magmatic Cu-Ni deposits include Voisey's Bay (Naldrett and Li, 2007), hosted by 1290 to 1340 Ma troctolite intrusions associated with the anorogenic Nain plutonic suite, those of the 1108 Ma Crystal Lake gabbro, and those in the Marathon area that are associated with the mid-continental rift. Although there are a number of Paleoproterozoic and Mesoproterozoic sedimentary basins preserved from this time period, only two have produced mineral deposits of economic significance. The ca. 1500 to 1320 Ma Belt-Purcell sedimentary basin of southeastern British Columbia (Lydon, 2007) is a sedimented continental rift that hosts the large Sullivan SEDEX deposit. The ca. 1750 to 1650 Ma Athabasca Basin (Jefferson et al., 2007) has produced the very rich unconformity-associated U deposits of Saskatchewan, which formed by the circulation of basinal and basement fluids at 1500 and 1350 Ma, times which coincide with continental extensional events (Fig. 17).

By far the most significant metallogenetic event in the Nuna supercontinent was the 1850 Ma meteorite impact that produced the Sudbury structure which contains 64.2% of Nuna's $Eq686.6 billion and 28.0% of Canada's $1574 billion total non-ferrous metal and diamond resources. The timing of the impact during accretion of Nuna may have been a factor in its prodigious Ni-Cu endowment.

Although major mineral deposit types of the orogens of Nuna (VMS, lode gold, and komatiite-associated Ni-Cu deposits) are the same as for the greenstone belts of Kenorland, their values (Table 4) and relative proportions are very different. VMS, lode gold, and Cu-Ni deposits constitute 33.2%, 4.8%, and 59.8%, respectively, of Nuna's $Eq125.2 billion total resource in orogens, whereas the comparable statistics for Kenorland are 44.0%, 49.9%, and 2.4%, respectively, of a $322.6 billion total resource in greenstone belts. The large increase in the ratio of komatiitic Ni-Cu to VMS resources but major decrease in ratio of lode gold to VMS resources obviously reflect a dramatic change to the architecture of and/or to the processes of continental accretion between the Archean and the Paleoproterozoic. Another major change from the Archean to Paleoproterozoic that continues into younger supercontinents is the increase in the relative amount of resources contained in supracrustal sedimentary basins (Table 4), presumably mainly due to the increase with decreasing age of the area of supracrustal rocks that have been preserved (Fig. 2).

Rodinia / Laurentia

Accretion continued intermittently along the southeastern margin of the Nuna supercontinent from the end of the Paleoproterozoic to the end of the Mesoproterozoic (e.g. Zhao et al., 2004), culminating with the 1250 to 980 Ma (Rivers et al., 2002) arc-continent and continent-continent collisions of the Grenville Orogeny to form the Rodinia supercontinent. The earliest of three major accretionary events is represented by 1677 to 1646 Ma calc-alkaline to alkaline plutonism and volcanism of the Trans-Labrador arc and related 1650 to 1620 Ma trimodal mafic-anorthositic-monzogranitic magmatism on which were developed 1600 to 1510 Ma successor sedimentary basins (Wakeham Bay Group and correlatives) (Gower and Krogh, 2002). The second major event involved the accretion of a ca. 1500 Ma oceanic arc as the Quebecia terrane (Martin and Dickin, 2005) and a 1514 to 1493 Ma granitic continental arc of the Pinwarian Orogeny (Gower and Krogh, 2002). Post-Pinwarian 1450 to 1250 Ma continental arc volcanics and back arc or intracontinental sedimented rifts with associated carbonate platforms, occurring mainly in the western part of the Canadian portion of the Grenville Orogen, are host to small VMS and SEDEX/MVT deposits (Gauthier and Chartrand, 2005). Important ilmenite deposits (not compiled here) are associated with 1180 to 950 Ma (Davidson, 1998) anorthosites of western Quebec.

Rifting, which led to the break-up of Rodinia and the spawning of Laurentia, started at about 730 Ma in western Canada and 615 Ma in eastern Canada. Neoproterozoic sedimentary basins that filled these rifts are preserved at the western, northern, and eastern margins of what was to become Laurentia. The major mineral deposit of these basins occur in western Canada, notably the diagenetic Cu redstone deposit of western Northwest Territories near the base of the Windemere Supergroup (Figs. 2, 17) and possibly the highly metamorphosed Zn-Pb deposits of presumed Neoproterozoic age in the Shuswap complex of southern British Columbia (Höy, 1982). Final continental separation of Siberia from Laurentia probably did not take place until the Cambrian (Sears and Price, 2000, 2003), but rifting of the western Laurentian continental margin with associated SEDEX deposits continued through the Paleozoic (Fig. 17) until Devonian 390 to 320 Ma arc magmatism and Early Mississippian back-arc spreading (Nelson and Colpron, 2007). This semi-permanent rifting of the western margin of Laurentia illustrates that epicontinental rifting associated with continental separation may be difficult to distinguish from epicontinental rifting caused by far-field extensional tectonics prior to arc accretion, and opens to debate the geotectonic setting of SEDEX deposits in general.

The total non-ferrous metal and diamond resources of Rodinia are very small compared to other supercontinental cycles, mainly because of the extremely small inventory for deposits of its orogens. A major reason for this may be that the high-grade metamorphism generally prevalent throughout the Grenville Orogen may have hindered the recognition of geological environments and hydrothermal alteration patterns favourable for mineralization (Corriveau et al., 2007).

Pangea

The assembly of Pangea, in contrast to the assembly of older supercontinents, did not involve significant generation of new continental lithosphere (e.g. Kemp et al., 2006). The Appalachian Orogen (van Staal, 2007) records the rifting of Rodinia starting at 615 Ma; the closure of an arm of the Iapetus Ocean with the docking of the Avalonia terrane of Gondwanaland origin at ca. 425 Ma; the closure of the Rheic Ocean with the accretion of the Meguma terrane, also of peri-Gondwanaland origin at ca. 395 Ma; and the final collision of Laurentia and Gondwanaland during the Carboniferous-Permian to form the Pangea supercontinent. By far the most important mineral production from the Appalachian Orogen (Table 4) has come from VMS deposits of accreted 485 to 435 Ma arc and back arc, and 510 to 477 Ma ocean tract environments that formed within the realm of the Iapetus Ocean. The most important are those of the Bathurst mining camp of New Brunswick (Goodfellow, 2007a) in the 480 to 455 Ma Tetagouche-Exploits back arc of peri-Gondwanaland (Ganderian) origin (van Staal, 2007) and deposits in the central mineral belt of Newfoundland of peri-Laurentia origin, including the Buchans district in the 481 to 460 Ma Annieopsquotch accretionary tract and the Ducks Pond deposit (currently entering production) in the 513 to 486 Ma Penobscot arc/back arc (van Staal, 2007). A modest production has come from lode gold deposits, notably the Hope Brook mine of Newfoundland in an accreted Neoproterozoic continental fragment, deposits of the Meguma area of Nova Scotia (Sangster and Smith, 2007) associated with ca. 370 Ma orogenic tectonism and magmatism (van Staal, 2007), and smaller deposits associated with ca. 440 to 420 Ma Salinic and ca. 420 to 390 Ma Acadian tectonism and magmatism, such as those of the Baie Verte Peninsula in Newfoundland (Sangster et al., 2007). Porphyry deposits are associated with Middle Devonian to Early Carboniferous Neo-Acadian granites, especially the Gaspé porphyry Cu deposit and associated skarns of Quebec and the Mount Pleasant Sn-W porphyry of New Brunswick, and small magmatic Ni-Cu deposits occur in Early Devonian layered mafic-ultramafic intrusions in New Brunswick. The economically important asbestos deposits (not compiled here) of the eastern townships of Quebec occur in 489 to 477 Ma ophiolites that formed during the separation of the Dashwoods microcontinent from the Laurentia supercontinent and later obducted onto the supercontinent's margin at ca. 425 Ma (van Staal, 2007).

The northern margin of Laurentia was mainly passive from the beginning of its rifting at 723 Ma and accumulated up to 6000 m of Neoproterozoic to Silurian sediments prior to accretion of the Mesoproterozoic-Silurian Pearya microcontinent and the subsequent late Silurian – early Devonian Boothia uplift (Dewing et al., 2007). Rocks of arc or oceanic tract affinity, but without mineral deposits, are known only in the accreted Pearya fragment. Similarly, at the western margin of Laurentia, the evidence for offshore or continent-margin arcs prior to Devonian time is very limited, and it too was the site of continuous sediment accumulation, albeit with semi-continuous rifting, from the Neoproterozoic to the beginning of arc activity of the Cordilleran Orogeny in the Middle to Late Devonian (Nelson and Colpron, 2007). These epicontinental sedimentary basins contain 62.7% of

Canada's total SEDEX-type mineral resource, notably in the Selwyn Basin of Yukon (Goodfellow, 2007b), and the Kechika trough and Kootenay arc of British Columbia (Fig. 2). Their coeval platformal sedimentary sequences, together with successor sedimentary basins on the Appalachian Orogen, contain 92.8% of Canada's total MVT mineral resource, notably in the Polaris deposit of Nunavut, the Pine Pont deposits of Northwest Territories, and deposits along the western margin of the Cordillera (Fig. 2). The ratio of mineral resources in sedimentary basins to mineral resources of orogens for the Pangea continental cycle is three times higher than for the Nuna supercontinent cycle (Table 4), and is probably more a reflection of the relative amounts of preservation of sedimentary basins in the two supercontinent cycles (Fig. 2) than global evolution of geological processes. The mineral resources of Pangean sedimentary basins are dominated by Zn and Pb, in contrast to U for Nuna, which, considering that the world's major U deposits are all of Proterozoic age (Jefferson et al., 2007), is largely a reflection of increasing free-oxygen levels with time. It is debatable whether MVT deposits (Paradis et al, 2007) should be considered to be among the earliest (because they are generally thought to have been formed by the migration of basinal brines from sedimentary basins of 'passive' margins in response to uplift by a developing orogen) or among the latest (because they are products of sedimentary basins formed prior to the beginning of new accretion) deposits of a supercontinent cycle. Insomuch as both the break-up of Pangea with the opening of the Atlantic and the earliest accretion of exotic terranes in the Cordillera occur in the late Permian, MVT deposits formed earlier are here considered to be Pangean deposits (Fig. 17).

North America

The North America cycle (Figs. 16, 17) cannot be regarded to be of the same geotectonic significance as the Nuna or Rodinia supercontinent cycle because it had not yet culminated in the assembly of a supercontinent. Mineral resources are dominated by deposits of the Cordilleran Orogen ($Eq347.2 billion) with kimberlite diamonds ($Eq54.8 billion) being the next most important resource (Table 4). To the end of 2005, porphyry deposits (Sinclair, 2007) have contributed 62.4% of the $Eq66.4 billion of production from the Cordilleran Orogen, followed by lode gold (19.4%), VMS deposits (11.9%), and veins, skarns, etc. (6.4%). The same deposit types account for 83.8%, 5.5%, 7.4%, and 2.5%, respectively, of the $Eq347.2 billion of total non-ferrous metal and diamond resources, the remaining 0.8% by Ni-Cu deposits. The preponderance of porphyry deposits and intrusion-related and epithermal lode gold deposits is a reflection of the youth of the orogen and the preservation of high-level deposits of subaerial arc magmatism. The $Eq26.6 billion of total VMS resources is substantially less than for both Nuna and Pangea, and perhaps reflects the smaller proportion of oceanic arcs compared to continental arcs in the Cordilleran Orogen (Fig. 2).

Volcanogenic massive sulphide deposits of the Findlayson Lake area of southern Yukon (Peter et al., 2007), the Tusequah Chief deposit in northwest British Columbia, and deposits in the Eagle Bay assemblage in southeastern British Columbia (Fig. 2) are associated with arc-related bimodal 360 to 350 Ma magmatism developed on both the continental margin and the attenuated crust of the peri-Laurentia terranes. These peri-Laurentia terranes were formed by the separation of ribbon continents during the period of rifting with which the youngest mid-Devonian SEDEX deposits of the Selwyn Basin and Kechika trough are associated (Nelson and Colpron, 2007). The ca. 370 Ma Myra Falls VMS deposits on Vancouver Island in the Devonian Sicker Group, although of similar age, belong to Wrangellia, a terrane with peri-Siberian linkages, accreted to Intermontane terranes during mid-Jurassic and to North America prior to the Eocene (Nelson and Colpron, 2007) (Fig. 2). Similarly, the very large 217 Ma Windy Craggy VMS deposit of northeastern British Columbia and the high-grade Greens Creek VMS deposit of Alaska are hosted in a Late Triassic rift of the Alexander terrane, also with peri-Siberian linkages (Nelson and Colpron, 2007).

Most of the porphyry and epithermal, intrusion-related, and mesothermal lode gold deposits are related to arc magmatism and collisional tectonics of the Laurentian margin. Closure of marginal basins and accretion of the peri-Laurentia terranes began during the late Permian to early Triassic with the accretion of the innermost pericratonic terranes (Slide Mountain, Quesnel, and Yukon-Tanana), but voluminous arc magmatism, including the Takla and Nicola groups, did not begin until the late Triassic. Early Jurassic arcs were superimposed on Triassic architecture and in Quesnel and Yukon-Tanana terranes, and migrated eastwards towards the continent. In the Stikine terrane, Lower Jurassic volcanogenic strata of the Hazelton Group are widespread. Late Triassic to Early Jurassic, ca. 210 to 185 Ma (Fig. 16) Cu-Au and Cu-Mo porphyry deposits of Stikine and Quesnel terranes are the most important group of deposits in British Columbia (Fig. 2) and include producers such as Highland Valley, Gibraltar, Mt. Polley, and Kemess. This metallogenic province extends into the Yukon-Tanana terrane of Yukon (Fig. 2).

In mid-Jurassic time, the Stikine crustal block collided with the already-accreted Quesnel terrane and western North America, terminating the Quesnel and eastern Stikine (Hazelton) arcs, and trapping the Cache Creek accretionary prism and ocean floor between them (Nelson and Colpron, 2007). Just prior to collision, the narrow Eskay rift developed in the Stikine crustal block and became filled with clastic sediments and a ca. 175 Ma bimodal volcanic suite that hosts the Au-rich Eskay Creek VMS deposit.

Following the mid-Jurassic accretionary event, the accreted terranes and epicontinental miocline were tectonically translated eastwards over the cratonic margin, forming a thickening crust on which later Jurassic through Tertiary arcs were built. Four major metallogenic events are associated with these Jurassic through Tertiary arcs. Late Jurassic-Early Cretaceous plutonism produced the 140 Ma Endako porphyry Mo deposit, which is Canada's main source of Mo. The mid-Cretaceous period of 130 to 90 Ma produced the first plutons making up the Coast Plutonic Complex, and a cluster of younger 97 to 92 Ma plutons that gave rise to porphyry Au deposits such as Brewery Creek in Yukon (Hart, 2007); the tungsten skarn deposits of Cantung and Mactung near the Yukon-Northwest Territories boundary (Fig. 2); the Pb-Zn veins of the Keno Hill district; and the Zn-Pb replacements of the Sa Dena Hes deposit of Yukon. One of the largest lode Au resources in the Cordillera, the Bralorne Pioneer mesothermal vein deposit, is associated with mid Cretaceous thrusting. Late Cretaceous 90 to 65 Ma plutons produced significant porphyry Au deposits in Alaska, the 73 Ma Casino porphyry Cu-Mo-Au deposit, the 70 Ma Ruby Creek porphyry Mo deposits in Yukon, and the currently producing 82 Ma Huckleberry porphyry Cu-Mo deposit in central British Columbia. Porphyry Mo deposits were the most common metallogenic expression during 65 to 37 Ma plutonism and volcanism, and include the past-producing Bell and Granisle deposits, and the coastal Kitsault deposit. The 58 Ma Logtung porphyry W-Mo deposit, located near the British Columbia-Yukon border, also belongs to this metallogenic episode. The youngest Porphyry deposit in the Canadian Cordillera is the 37 Ma Catface Cu-Mo deposit on Vancouver Island (Sinclair, 2007). Epithermal lode gold deposits, such as the Grew Creek deposit in Yukon, formed during this 65 to 37 Ma time period, which coincided with 430 km dextral movement along the Tintina fault system. Dextral fault movement, mainly on the Queen Charlotte fault system, and the formation of the Cinola epithermal lode gold deposit on Graham Island (Fig. 2) were the main tectonic and metallogenic events since 37 Ma in the Canadian Cordillera, after which the magmatic arcs shifted mainly to the Aleutian arc of Alaska (Nelson and Colpron, 2007).

Geological and Economic Characteristics of Canada's Mineral Deposit Types

The distribution of the various mineral deposit types is illustrated here as a series of maps (e.g. Fig. 18) showing their location with respect to the dominant geological environment of supercontinent cycles and equivalent billion dollar values of the different categories of resources. In order to avoid excessive superposition of the pie diagram symbols on these maps, deposits of the same type and approximately the same age occurring in the same area have been grouped into districts. On each map, only the ten or so districts (or in some cases single isolated deposits) with the greatest values are labeled.

Magmatic Ni-Cu-Platinum Group Element Deposits

Geological Contexts and Distribution

Magmatic Ni-Cu-PGE deposits consist of sulphides associated with mafic and ultramafic igneous rocks whose magmas originate in the upper mantle. The orebodies originate as segregations of Fe sulphide melts from the magma that, because of their high density, have gravitationally settled and concentrated near the base of the magma body (Barnes and Lightfoot, 2005; Eckstrand and Hulbert, 2007). Economic deposits are most likely to occur if the magmas have been contaminated with crustal sulphur (Eckstrand and Hulbert, 2007) and, for komatiite-hosted deposits, are in the magnesium-rich part of their compositional range (Barrie et al., 1993). The main economic commodities are Ni, Cu, and platinum group elements, particularly Pd and Pt. The orebodies are classified into two main types, depending on whether Ni-Cu or PGE are the main economic commodities, and into subtypes depending on their geological setting (Eckstrand and Hulbert, 2007).

Astrobleme-impact melt sheet: The basal part of the 1850 Ma meteorite-impact melt sheet of the Sudbury Igneous Complex (Ames and Farrow, 2007) is the only example of this subtype in the world and contains about 74% of Canada's magmatic Ni-Cu total resources. Its original 200 km wide crater was near the boundary of ca. 2711 Ma Neoarchean gneisses to the north and ca. 2450 Ma volcano-sedimentary rocks of the Huronian Supergroup to the south (Fig. 8). This unique event was very fortuitous for Canada's Ni endowment, with Sudbury ores contributing 89% of production, 60% of 2005 reserves, 34% of measured and indicated resources, and 74% of inferred resources of the magmatic Ni-Cu type of mineral deposit in Canada (data in Appendix 1, DVD).

Komatiitic volcanic flows and related intrusions: Paleoproterozoic komatiitic flows and associated intrusions of the 1883 Ma Thompson Nickel Belt of Manitoba (Layton-Matthews et al., 2007) and the 1918 Ma Raglan Horizon of the Cape Smith-Wakeham Bay belt in northern Quebec (Lesher, 2007) (Fig. 18) constitute the second most productive group of magmatic Ni-Cu deposits in Canada, contributing, respectively, 8.3% and 0.8% of the production, 9.1% and 9.0% of the reserves, and 10.6% and 1.5% of the measured and indicated resources. Magmatic Ni-Cu deposits of Archean komatiitic flows are most common in the Abitibi greenstone belt of Ontario and Quebec (Fig. 18), and some, such as the Alexo, Langmuir, Marbridge, Redstone, and Texmont deposits, have been mined. However, production from these deposits has been relatively minor, amounting to about $Eq0.57 billion or 0.15% of the total production from Canadian magmatic Ni-Cu deposits. A large resource of Ni is contained in the Dumont Sill (Fig. 18) but at a very low grade of 0.5% Ni.

Ni-Cu rift-associated mafic sills and dykes: The 1108 Ma Crystal Lakes gabbro (Fig. 18) in western Ontario, which is probably a northern extension of the Duluth Complex of Minnesota (Eckstrand, 1996), is Canada's only example of a magmatic Ni-Cu deposit associated with rifting and flood basalts. The 1270 Ma Muskox intrusion and associated Coppermine flood basalts in the Slave Province of Nunavut, though of the right geological environment, has so far not revealed substantial Ni-Cu sulphide segregations.

Ni-Cu in other mafic/ultramafic intrusions: The Voisey's Bay deposit (Naldrett and Li, 2007) came into production during 2005, and contains 19.5% of the equivalent dollar metal content of reserves and 14.7% of the equivalent dollar value of measured and indicated resources of Canadian magmatic Ni-Cu deposits. The deposit is associated with 1290 to 1340 Ma troctolites of the anorogenic Nain Plutonic Suite in Labrador (Fig. 18) that were emplaced at the 1850 Ma collisional contact between the Archean Nain Province in the east and the Churchill (Rae) Province to the west (Eckstrand and Hulbert, 2007).

Magmatic Cu-Ni deposits associated with tholeiitic intrusions are usually hosted by an ultramafic phase. The deposits range in age from Archean to Mesozoic and have been found across Canada from the 396 Ma St. Stephen deposit in New Brunswick in the east to the 232 Ma Wellgreen and Canask deposits in the west. The Proterozoic Lyn Lake deposit in Manitoba, with about $Eq3.1 billion of production, and the

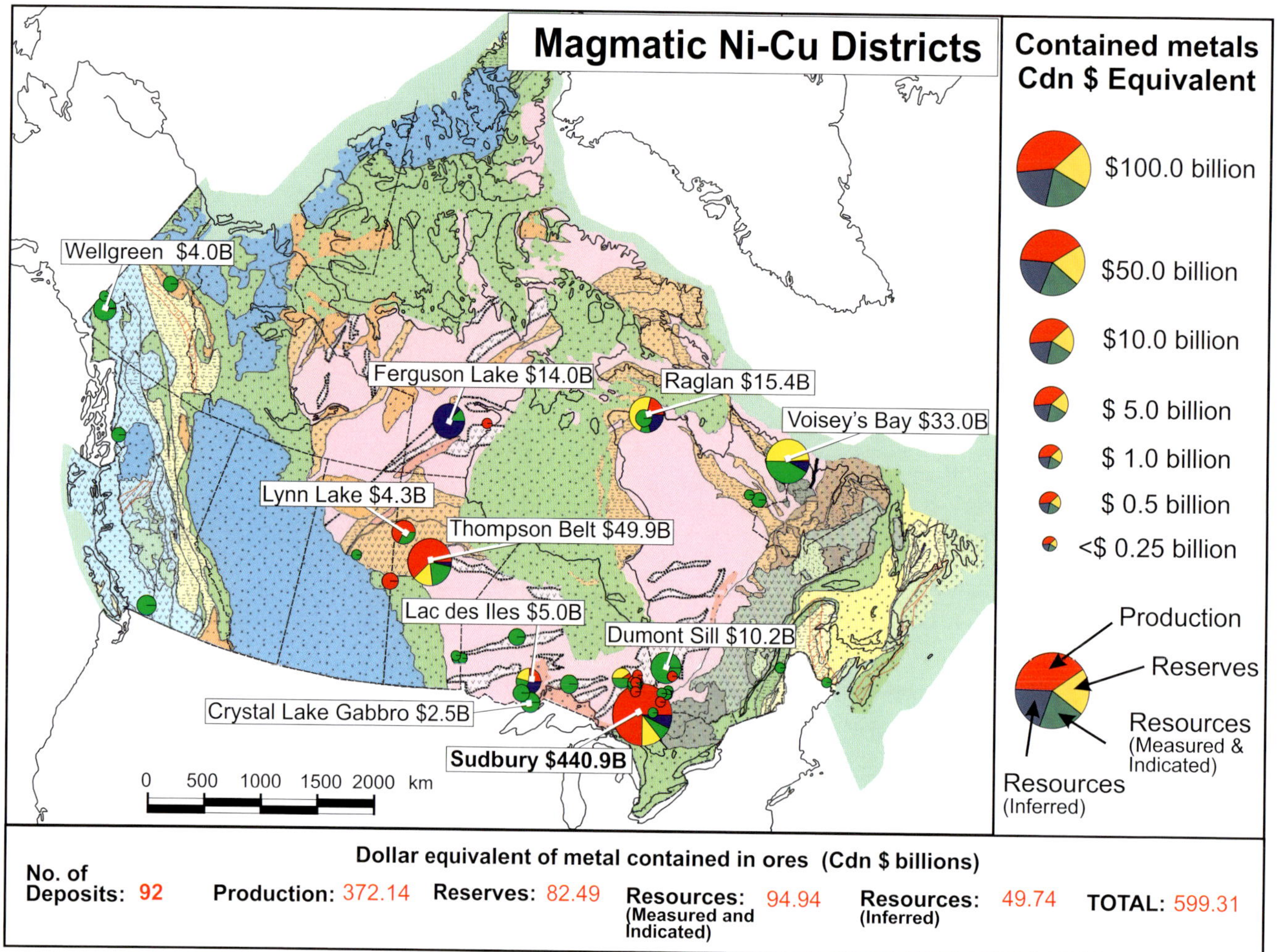

FIGURE 18. Geological map of Canada showing the distribution of magmatic Ni-Cu districts and deposits. Legend for the map is as for Figure 2. The diameter of each pie chart is proportional to the dollar equivalent of total metal contained in the district or deposit, and the subdivisions of the pie charts represent the relative amounts in the different categories of mineral resources. A district represents the combined totals of deposits of similar mineral deposit type and age that occur in the same area. The names and total dollar equivalent of metal content are indicated only for districts or deposits with the highest metal contents.

Archean Montcalm deposit in the Abitibi belt near Timmins (Fig. 18), with about \$Eq0.2 billion in production and \$Eq0.9 billion in reserves, are the most notable examples of this mineral deposit subtype. The Ferguson Lake deposit in Nunavut (Fig. 18), which has the largest single Ni resource in Canada outside of mining areas, appears to be of the tholeiite intrusion type (Carter, 2006).

PGE magmatic breccias of mafic/ultramafic stocks and sills: The Lac des Isles deposit in western Ontario (Fig. 18) is Canada's only deposit that has been mined primarily for its PGE content, consisting mainly of Pd, although subordinate Pt, Ni, and Cu are also produced. The deposit is associated with a 2738 Ma multiphase stock, one of a 30 km diameter ring of similar intrusions in the area.

There are no examples of reef-type or stratiform deposits of well layered, differentiated mafic/ultramafic intrusions in Canada, for which the deposits of the Bushveld Complex of South Africa are the prime examples of this PGE subtype.

Economic Context and Statistics

Magmatic Ni-Cu deposits provide most of the world's Ni supply, though Ni laterite deposits formed by the weathering of ultramafic rocks will probably in time become the main source of Ni (Eckstrand and Hulbert, 2007). The magmatic Ni-Cu mineral deposit type is extremely important to Canada, having produced ores with \$Eq372.1 billion metal content (Figs. 5, 18) or about 44% of all non-ferrous metal and diamond production in Canada. Most of this value (65%) is attributable to Ni (Fig. 5) with 15% coming from Cu, 11% from PGE (Pt+Pd), and 8% from Co. Magmatic Ni-Cu deposits have historically produced (Fig. 4A) and currently account (Fig. 4B) for nearly all of Canada's Ni, Co, and PGE primary output. During 2004, the Ni alone produced from magmatic Ni-Cu deposits accounted for 15% of the value of Canada's total non-fuel mineral production (Fig. 1). Combined with all PGE and Co production, and about 28% of all Cu production, the proportion contributed by magmatic Ni-Cu deposits rises to about 21%. The Ni production is about 12% of the world's 2004 Ni supply, making Canada the third most important Ni producer (McMullen and Birchfield, 2005).

Magmatic Ni-Cu deposits contain a greater value of economic commodities than any other mineral deposit type in Canada in all mineral resource categories (Figs. 5, 6, 8),

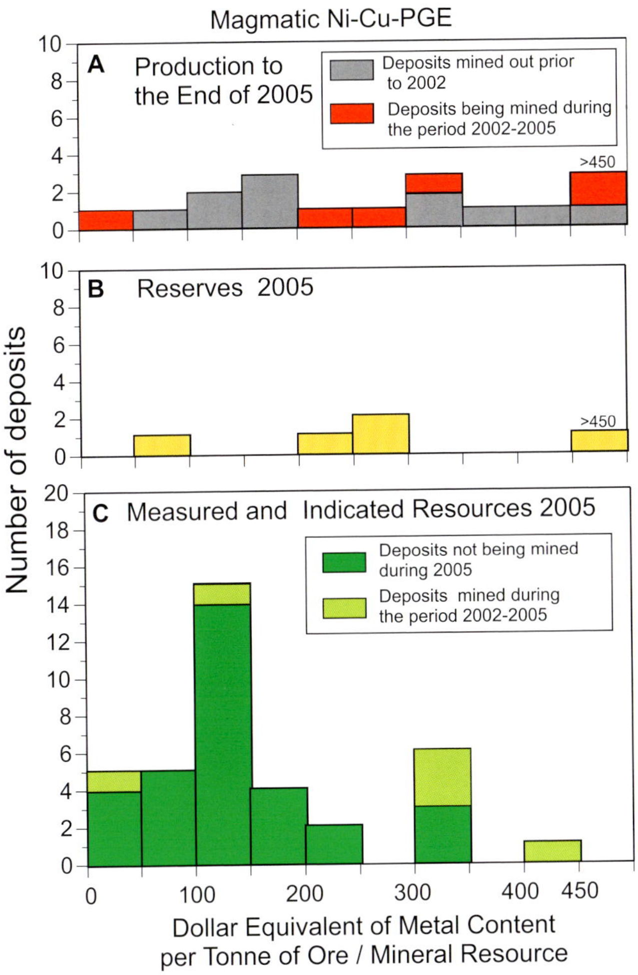

FIGURE 19. Histograms showing the distribution by $50/t increments of the average weighted grades of magmatic Ni-Cu deposits for **(A)** production to the end of 2005; **(B)** reserves at the end of 2005; **(C)** measured and indicated resources at the end of 2005.

except in the measured and indicated resource category which is dominated by porphyry deposits (Fig. 7). As a proportion of Canada's total non-ferrous metal and diamond resources, magmatic Ni-Cu deposits comprise 45% ($Eq82.5 billion) of reserves, 21% ($Eq95.0 billion) of measured and indicated resources, and 51% ($Eq49.7 billion) of inferred resources (Fig. 18).

The main economic attraction of magmatic Ni-Cu deposits is the high average value of the ores, with about 65% of this value, on average, being in the Ni content (Figs. 9, 10, 11). The weighted average values per tonne for magmatic Ni-Cu deposits are $Eq272/t for production (Fig. 9) and $Eq294/t for reserves (Fig. 10). The lower $Eq135/t for measured and indicated resources (Fig. 11) and $Eq280/t for inferred resources (Fig. 12) are due to the lower grade resources of undeveloped deposits such as the Dumont Sill in Quebec. The average value per tonne of ores milled (Fig. 9) or in reserves (Fig. 10) is the highest for all mineral deposit types, except for the high-grade U ores of the Athabasca Basin and some high-grade Ag veins. Production from individual districts range from a low of $Eq41/t for the open pit Lac des Iles Pd-Pt deposit (but note that this value is still much higher than most porphyry open pit mines) to $Eq599/t for the recently opened Voisey's Bay mine (where, as is common practice, the higher grade ores are mined first). Ores from currently producing underground mines have a metal content of more than $Eq220 (Fig. 19A). The dollar equivalent per tonne of metal content of reserves (Fig. 19B) are approximately the same as for production (Fig. 19A), with the exception of the 50% increase in the value for Lac des Isles reserves ($Eq73/t), which includes a higher grade resource to be mined by underground methods. The dollar equivalent of metal content of measured and indicated resources (Fig. 19C) have a very high value of in excess of $Eq300/t for all producing districts with the exception of the Thompson district ($Eq104/t) and the open pit resources at Lac des Isles ($Eq43/t). The measured and indicated resources for most other magmatic Ni-Cu deposits that are not being mined are in the range of $Eq100/t to $Eq150/t (Fig 19C). This is about $Eq50/t higher than for other massive sulphide deposits such as VMS deposits, which are also most commonly mined by underground methods, making the magmatic Ni-Cu deposits a potentially more attractive exploration target than the more common VMS type of deposit.

VMS Deposits

Geological Contexts and Distribution

Volcanogenic massive sulphide deposits are lenses of Fe, Cu, Zn, and sometimes Pb sulphides, usually with valuable amounts of Ag and Au, that were formed on the seafloor by hot springs on and around submarine volcanoes (Galley et al., 2007a). The amount of sulphide in a single deposit, which may consist of several lenses, averages between 5 and 10 Mt depending upon the VMS subtype, but ranges from tens of thousands to hundreds of millions of tonnes (Galley et al., 2007a). VMS deposits are hosted by submarine volcanic rocks, especially lavas or volcaniclastic sediments of felsic compositions. This spatial association of VMS deposits with felsic volcanic centres is illustrated, for example, by the Abitibi Belt (Fig. 2) where felsic volcanics, although forming only about 6% of the total volume of volcanic rocks, host about 90% of the ore tonnage (Barrie et al., 1993). Volcanogenic massive sulphide deposits clustering around these felsic volcanic centres, which average about 30 km in diameter (Sangster, 1980; Galley et al., 2007a), form distinct mining districts containing ten or more deposits, such as at Noranda (Gibson and Galley, 2007), Matagami, and Bathurst (Goodfellow, 2007a) districts (Fig. 2). Volcanogenic massive sulphide districts associated with felsic volcanic centres are likely the submarine equivalent of subaerial porphyry systems with the bulk of the economic metals supplied by magmatic hydrothermal fluids (Lydon, 1996). The leaching of metals from footwall rocks by convection of heated seawater, the prevalent genetic model for VMS deposits (Franklin et al, 2005; Galley et al., 2007a), probably supplied only a minor proportion of metals in these geological settings. However, convection of seawater formed the overprint of hydrothermal alteration typical of VMS deposits (Galley et al., 2007a) and supplied all the metals for convection systems driven by the heat of anhydrous magmas, such as Cyprus-type VMS deposits of back-arc spreading centres (Fig. 14).

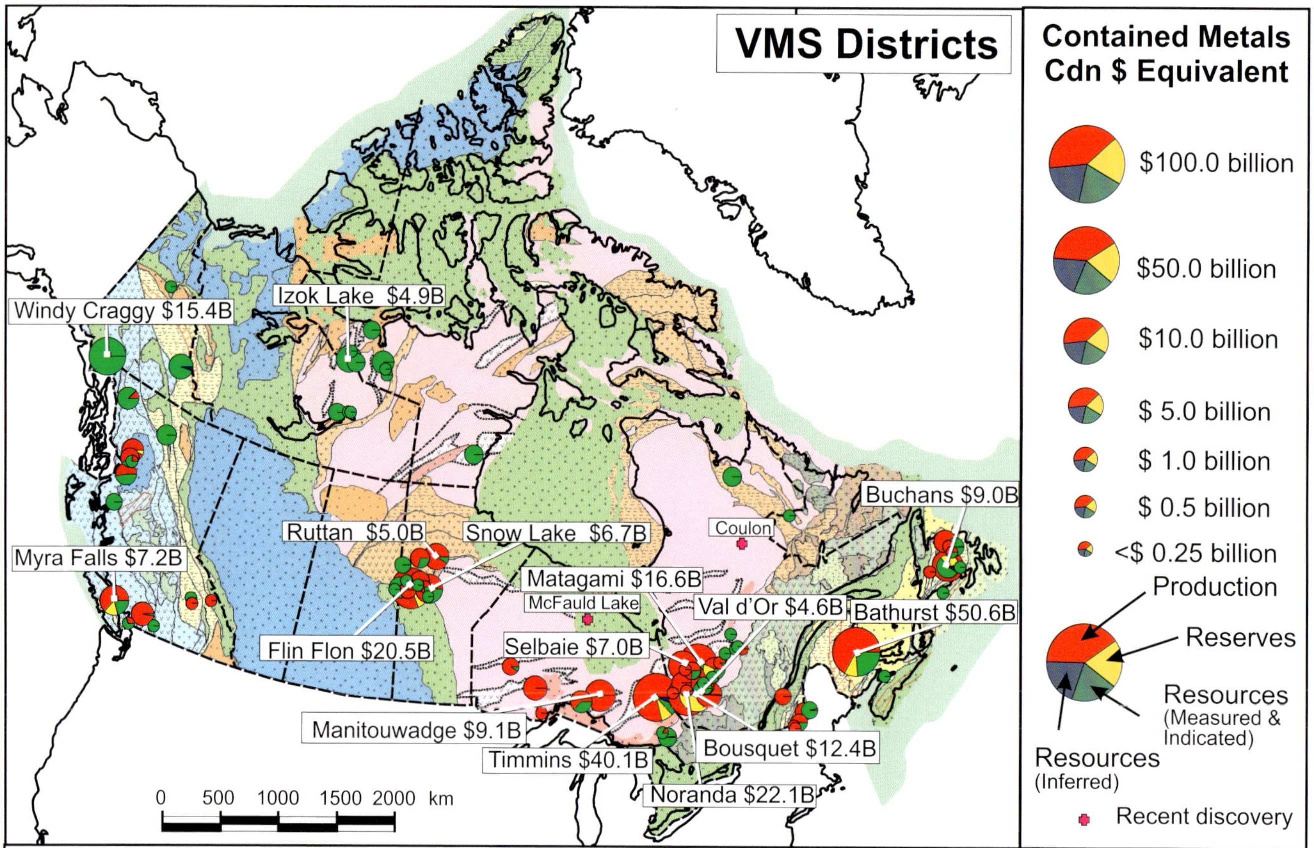

Dollar equivalent of metal contained in ores (Cdn $ billions)

No. of Deposits: 280 | Production: 192.32 | Reserves: 20.05 | Resources: (Measured and Indicated) 78.67 | Resources: (Inferred) 3.96 | TOTAL: 295.00

FIGURE 20. Geological map of Canada showing the distribution of volcanogenic massive sulphide districts and deposits. Legend for the map is as for Figure 2. The diameter of each pie chart is proportional to the dollar equivalent of total metal contained in the district or deposit, and the subdivisions of the pie charts represent the relative amounts in the different categories of mineral resources. A district represents the combined totals of deposits of similar mineral deposit type and age that occur in the same area. The names and total dollar equivalent of metal content are indicated only for districts or deposits with the highest metal contents.

A total of 153 VMS deposits have been mined in all provinces and territories of Canada with the only exceptions being Prince Edward Island and Alberta. Their distribution (Fig. 20) reflects the distribution of ancient orogens (Fig. 20), particularly 2600 to 2800 Ma volcanic belts of the Archean Superior Craton of Quebec and Ontario, and Slave Craton of Nunavut and Northwest Territories; 1800 to 1900 Ma volcanic arcs of the Trans-Hudson Orogen in Manitoba and Saskatchewan; 460 to 500 Ma volcanic belts of the Appalachian Orogen in New Brunswick, Newfoundland, and the eastern townships of Quebec; and 130 to 390 Ma volcanic belts of the Cordilleran Orogen in British Columbia and Yukon (Figs. 16, 20).

Over half of Canadian production from VMS deposits has come from Archean rocks (Fig. 21). The Abitibi Belt has supplied $Eq91.3 billion of the $Eq102.6 billion of metals mined from Archean rocks, of which $Eq33.17 billion has been contributed by the giant Kidd Creek deposit. The Appalachian Orogen has been the second most productive metallogenetic province in Canada, but here again it is a single giant deposit, the Brunswick No. 12, that has contributed $Eq26.7 billion of a total $Eq46.8 billion. The Proterozoic

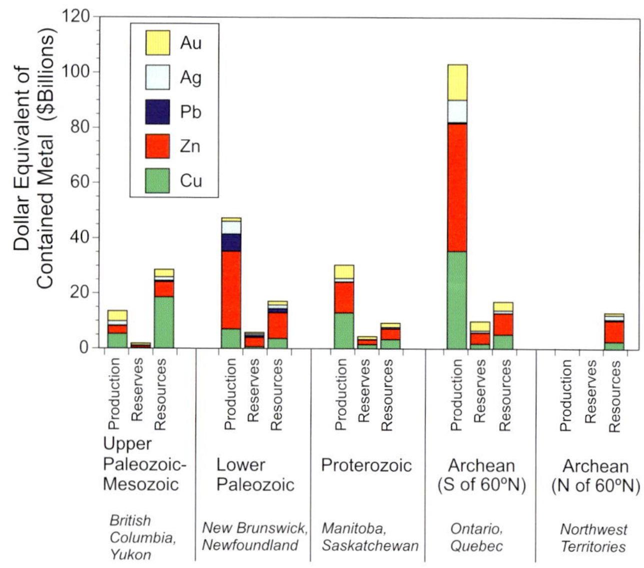

FIGURE 21. Stacked bar graphs showing the dollar equivalent of the metal content of different resource categories for volcanogenic massive sulphide deposits in Canada for the major geological time intervals. The geographical areas in which most of the deposits in each time category occur are indicated in italics.

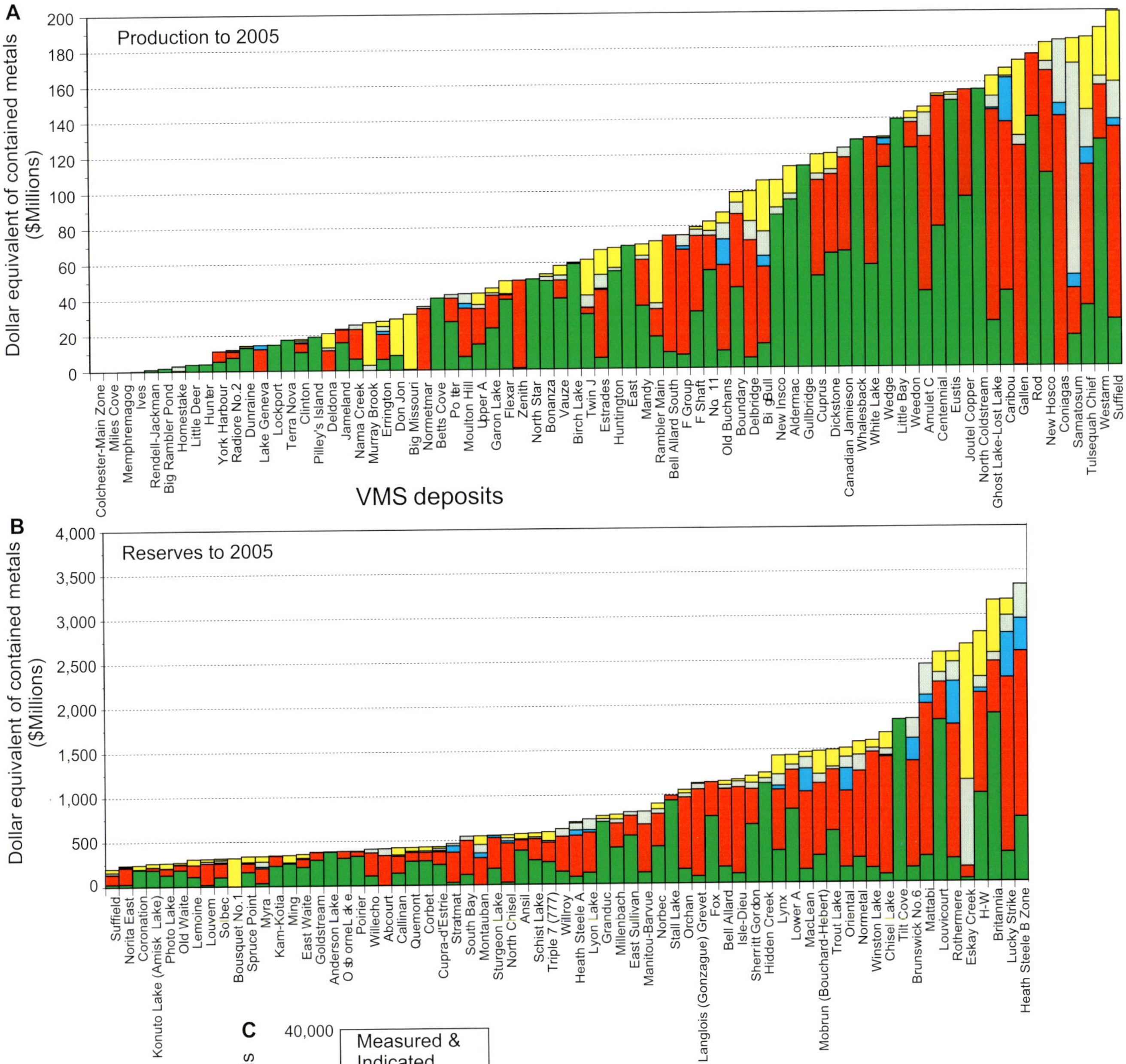

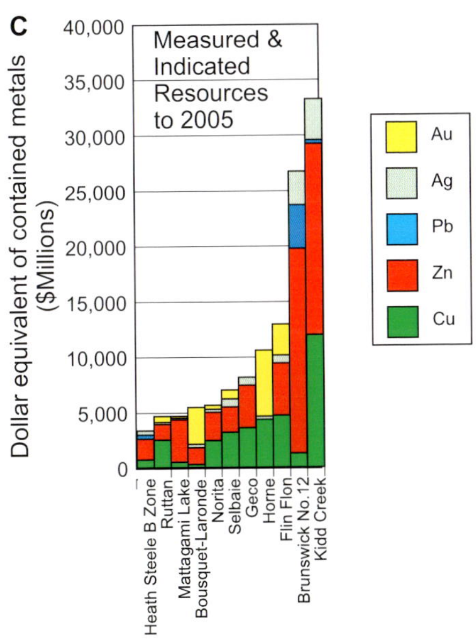

FIGURE 22. Stacked bar charts for all volcanogenic massive sulphide deposits in Appendix 1 (DVD) showing the dollar equivalent of metal content of total metal resources. Note the difference in scale between **(A)**, **(B)**, and **(C)**. The deposits are arranged in order of the dollar-equivalent value. Note the large variation in the relative economic importance of the different metals in different deposits and the lack of correlation between the size (total dollar-equivalent value) and relative metal values.

deposits of Manitoba and Saskatchewan at $Eq29.9 billion rank third (Fig. 21), of which $Eq12.9 billion was contributed by the Flin Flon deposit. The geological age distribution for reserves follows the same pattern as for production (Fig. 21) but measured, indicated, and inferred resources are dominated by the $Eq15.4 billion contained in the Mesozoic Windy Craggy deposit of British Columbia and the $Eq12.6 billion of measured and indicated resources of the Archean Slave Province (Fig. 20).

Economic Context and Statistics

Volcanogenic massive sulphide deposits have been a mainstay of the Canadian metalliferous mining industry, the $Eq192.4 billion of metals that have been mined being second only to magmatic Ni-Cu deposits for a single deposit type (Fig. 5). As a proportion of Canada's total non-ferrous metal and diamond wealth, VMS deposits account for 23% of production, 11% ($Eq20.1 billion) of reserves, 17% ($Eq78.7 billion) of measured and indicated resources, and 4% ($Eq4.0) of inferred resources (Table 3). Volcanogenic massive sulphide deposits have been the source for 68% of Canada's Zn, 52% of its Ag, 40% of its Cu, 32% of its Pb, and 13% of its Au production (Fig. 4A), and currently produces virtually all of its Zn, Pb, and Ag, and 22% of its Cu (Fig. 4B). On average, 46.0% of their value has been contributed by Zn, 31.5% by Cu, 10.8% by Au, 8.1% by Ag, and 3.6% by Pb, but the relative proportion of the different metals within individual deposits varies widely and the amount of metal contained in different deposits varies over four orders of magnitude (Fig. 22). Some VMS deposits are so rich in Au that it is the main economic commodity (Fig. 22) and have been classified as a separate mineral deposit type (Dubé et al., 2007), although some deposits may have been enriched in Au by the superimposition of epigenetic Au-rich mineralization (Mercier-Langevin et al., 2007).

An alarming statistic is that reserves at $Eq20.1.0 billion (Fig. 6) are sufficient only for four more years of mining at the 2005 rate, and all currently producing VMS deposits will be mined out in less than a decade (with the possible exception of the 777/Callinan mine near the Saskatchewan-Manitoba border). Thus, production from this deposit type will continue to rapidly decline unless some of its $Eq78 billion in measured and indicated resources (Fig. 7) are converted into reserves by the opening of new mines. Outside of the Windy Craggy deposit, which is not likely to be mined in the near future because of environmental issues, the largest resources are in the Hackett River-Izok Lake areas of Northwest Territories, which are also unlikely to be mined until a surface transportation infrastructure is established, such as the proposed Bathurst Inlet port and road project (e.g. Nunavut Impact Review Board, ftp.nunavut.ca/nirb/nirb_reviews/current_reviews/03UN114-BIPAR_PROJECT/02-REVIEW, last accessed March 2007). The relatively recent discovery of the McFauld Lake and Coulon deposits (Fig. 20) demonstrates the potential of the Superior Province to the north of the Abitibi Belt. Furthermore, the relatively recent discovery of the Ansil West deposit in the Noranda camp and the Bracemac deposit in the Matagami camp demonstrates that additional deposits may still be found in established mining areas.

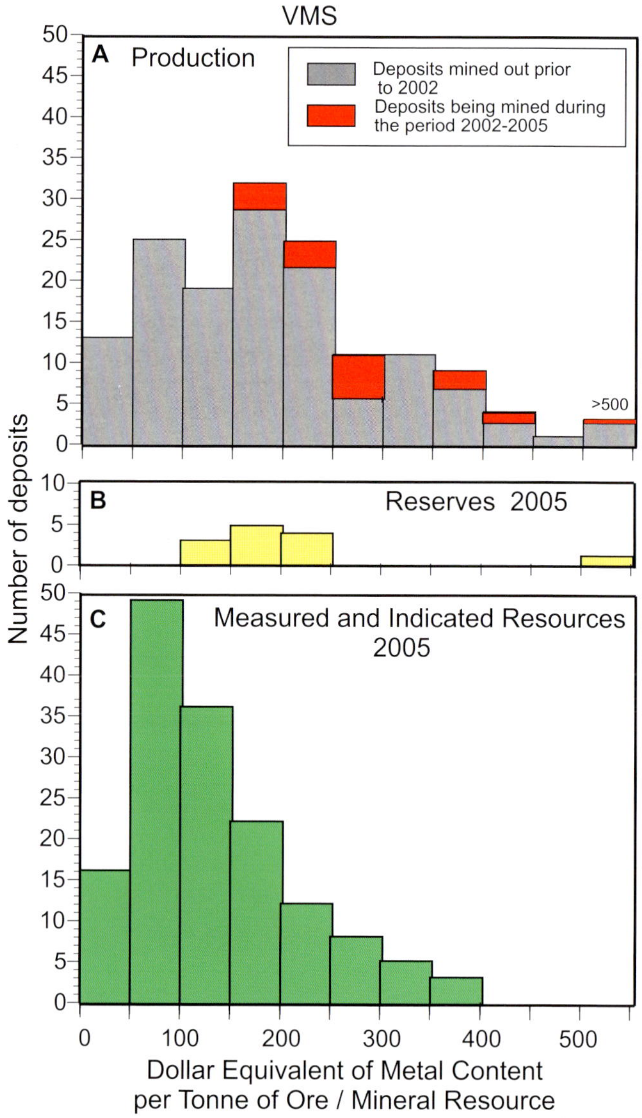

FIGURE 23. Histograms showing the distribution by $50/t increments of the average weighted grades of volcanogenic massive sulphide deposits for **(A)** production to the end of 2005; **(B)** reserves at the end of 2005; and **(C)** measured and indicated resources at the end of 2005.

The main attraction in exploring for VMS deposits is the reasonably high value of their ores (Figs. 5, 23) and the high level of understanding, compared to other deposit types, of geological, geochemical, and geophysical vectors to their ores. Their polymetallic composition provides the added economic benefit of being a buffer against fluctuations in prices for any one metal. The weighted average metal content per tonne of ore is $Eq174/t for production (Fig. 9), $Eq185/t for reserves (Fig. 10), $Eq96/t for measured and indicated resources (Fig. 11), and $Eq186 for inferred resources (Fig. 12). The average value per tonne for production is at the low end of the range of ores mined during the past five years (Fig. 23A), which suggests that most deposits mined in the past with ore values of less than $Eq150/t may not be economically mineable today (except those mined by open pit). Deposits with metal contents per tonne of more than $Eq400/t include deposits unusually rich in precious metals (e.g. Eskay Creek and Homestake) and those poor in

Fe sulphides (deposits of the Buchans district). Values of the metal contents for reserves (Fig. 23B) are generally lower than those for production (Fig. 23A), a reflection of most VMS producing mines coming to the end of their mining lives when lower grade ores are last to be mined. The bulk of VMS measured and indicated resources have metal contents less than $Eq150/t (Fig. 23C) and, like low-value past producers, may also not be candidates for mining under modern economic conditions.

Lode Gold Deposits

Geological Characteristics and Distribution

A lode gold deposit is a hydrothermal deposit whose principal commodity is Au. Sixteen subtypes of lode gold deposits (Poulsen et al., 2000) can be distinguished among three main groups:

1. Those formed at depths of 5 to 10 km, and variably termed orogenic, mesothermal, shear zone-related, or greenstone-related quartz-carbonate vein deposits; these deposits consist of structurally controlled quartz-carbonate veins and typically occur in greenschist-facies metamorphic rocks of orogenic belts (Robert, 1996; Dubé and Gosselin, 2007). The majority of deposits are adjacent to major, deep-seated reverse-oblique faults, particularly dilational zones (Hodgson, 1989) and were probably originally related to collision, obduction, and/or transtensional tectonics, but have a complex and long-lived structural history (Goldfarb et al., 2005). The mineralization generally took place during the later stages of orogenic crustal shortening, post-dating peak metamorphism of the host rocks. The mineralization is generally coeval with felsic to intermediate intrusions of the lode Au district and it is debated whether the genesis of the Au deposits is related to these intrusions or whether both are different expressions of the same deep crustal thermal event (Goldfarb et al., 2005). The Au deposits are associated with large-scale carbonate alteration, particularly along major faults, throughout the district (Dubé and Gosselin, 2007). This group of lode Au deposits accounts for 83.5% of Canada's total lode Au resources.
2. Intrusion-related deposits formed at 1 to 5 km depth, generally in the same way as porphyry deposits, and include the same range of mineral deposit styles, such as sulphide-rich veins, breccia pipes, skarns, etc. (Poulsen et al., 2000). Almost half the Canadian deposits of this type occur in Archean rocks (Appendix 1, DVD) with most of the remainder occurring in Mesozoic rocks of the Cordillera. It should be noted that this definition of 'intrusive-related Au deposits', also used by Dubé and Gosselin (2007, Appendix 1, DVD) is not the same as that used by Goldfarb et al. (2005) and Groves et al. (2005) who restrict the term to generally low-grade deposits associated with reduced granitoids that occur on the foreland side of Phanerozoic continental arcs. Intrusion-related lode Au deposits accounts for 14.6% of Canada's total lode Au resources.
3. Epithermal deposits formed at less than 1 km depth and typically consist of quartz-bearing veins or low-grade precious metal disseminations formed from acid (quartz-alunite-kaolinite subtype) or neutral-alkaline (quartz-carbonate-adularia subtype) hydrothermal fluids generated by the devolatilization of felsic magmas and mixing of the magmatic fluids with meteoric waters in subaerial or shallow-water environments associated with calc-alkaline to alkaline volcanos of continental arcs and back arcs (Taylor, 1996; Simmons et al., 2005). Most preserved epithermal deposits in Canada are Mesozoic or younger in age and occur in the Cordillera because the shallow parts of continental arcs are not commonly preserved in older orogens. Epithermal lode gold deposits account for 2.8% of Canada's total lode gold resources.

Gold-rich VMS deposits (Dubé et al., 2007) and Au-rich porphyry deposits in lode gold deposits, are here included with VMS deposits and porphyry deposits, respectively

Most orogenic lode gold deposits in the world were formed during the intervals 2800 to 2550 Ma, 2100 to 1800 Ma, and 600 to 50 Ma, and most epithermal deposits that have been preserved were formed during the less than 50 Ma period (Goldfarb et al., 2005). Canadian lode gold deposits have a similar chronological distribution, which corresponds to the major periods of continental accretion (Fig. 16). The most productive lode hold districts in Canada are of the orogenic lode gold type, led by the Timmins, Kirkland Lake, and Val d'Or-Noranda mining centres in the Abitibi Belt in Ontario and Quebec (Fig. 24). Although age dating of deposits is sparse, most seem to have formed in the interval 2740 to 2620 Ma, up to 100 Ma later than their host rocks (Dubé and Gosselin, 2007). Like the age of their host rocks, they show a general decrease in age from north to south (Percival, 2007). The largest unmined orogenic Au resources are in the Tundra deposit of the Mackenzie district, the Meadowbank and the Hope Bay deposits, occurring in the Archean Slave Province of Nunavut (Fig. 24). The major Proterozoic orogenic lode gold deposits are in the Lynn Lake and Flin Flon areas of Manitoba and adjacent areas in Saskatchewan, though with a total resource metal content of about $Eq4.5 billion, are small compared to the $Eq143.6 billion content in Archean deposits. Similarly, Phanerozoic orogenic lode gold deposits of the Appalachians ($Eq2.0 billion) in the east and the Cordillera ($Eq5.3 billion) in the west, are similarly small in comparison. The largest intrusion-related deposits are those of the Hemlo district in western Ontario with a total resource of $Eq11.0 billion (Fig. 24).

Economic Context and Statistics

Orogenic lode gold deposits have supplied 80% of Canada's historic lode Au production (Fig. 4A) and in 2004 contributed 77% of the Au produced in Canada, which in turn accounted for 10% of Canada's total non-fuel mineral production (Fig. 1). With over 280 deposits that have been mined in Canada, lode Au deposits have not only been a major employer in the mining industry but the economic base for the establishment of northern communities, such as Timmins in Ontario and Val d'Or in Quebec. Supporting 27 different mining operations during 2003 to 2005,which is more than any other deposit type, and attracting 49.5% if the $1.1 billion invested in mineral exploration during 2005 (Natural Resources Canada, 2006), lode gold deposits continue to underpin a major part of the socio-economic infra-

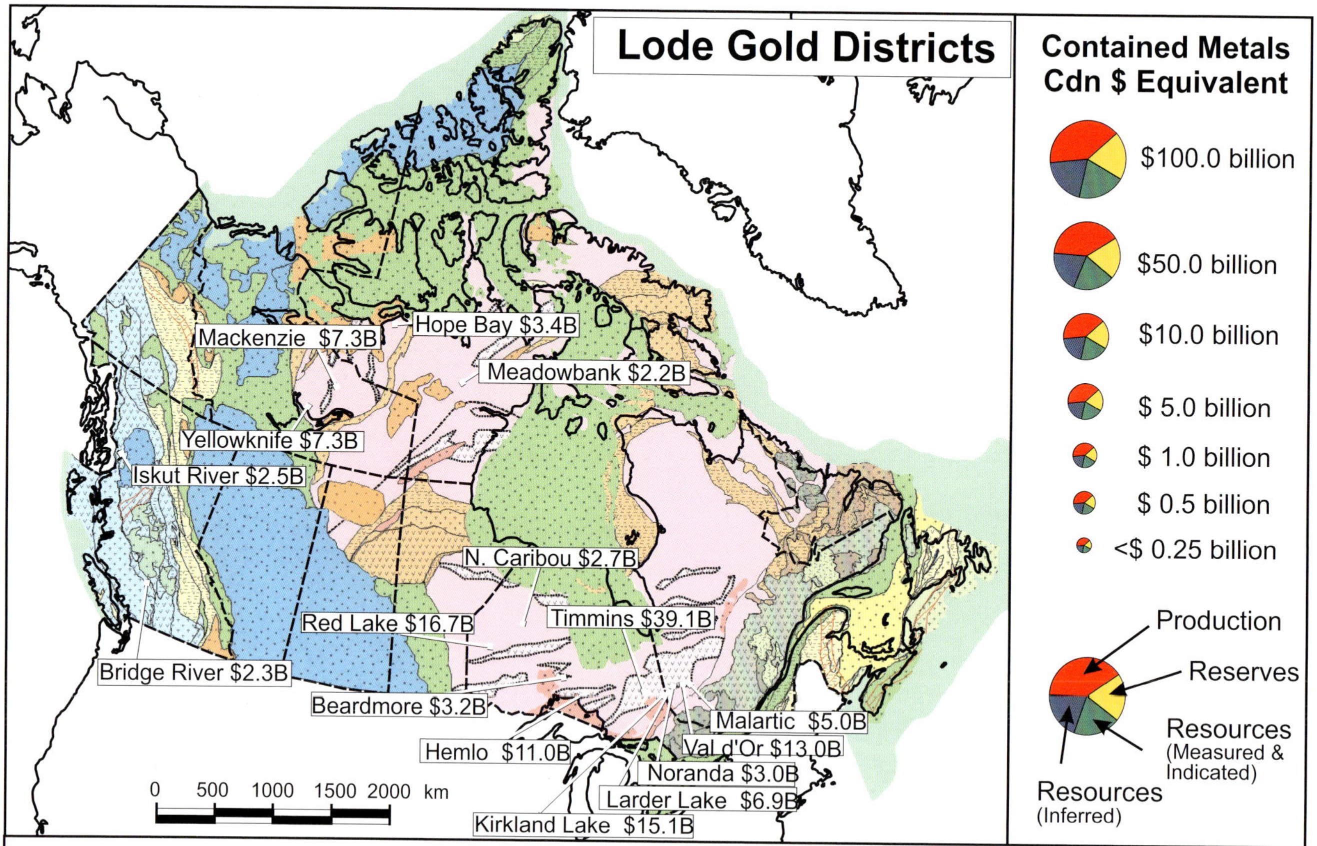

Dollar equivalentof metal contained in ores (Cdn $ billions)

No. of Deposits: 243 **Production:** 131.62 **Reserves:** 9.63 **Resources:** (Measured and Indicated) 27.92 **Resources:** (Inferred) 19.49 **TOTAL:** 188.66

FIGURE 24. Geological map of Canada showing the distribution of lode gold districts and deposits. Legend for the map is as for Figure 2. The diameter of each pie chart is proportional to the dollar equivalent of total metal contained in the district or deposit and the subdivisions of the pie charts represent the relative amounts in the different categories of mineral resources. A district represents the combined totals of deposits of similar mineral deposit type and age that occur in the same area. The names and total dollar equivalent of metal content are indicated only for districts or deposits with the highest metal contents.

structure of the Canadian mining and exploration industries. The development of lode gold deposits has often been the vanguard in the opening of new districts to the exploration for other mineral deposit types, so the planned opening of Meadowbank deposits in Nunavut may very well be the harbinger of increased economic prosperity to that territory.

Canada's total production from lode gold deposits to the end of 2005 is $Eq131.62 billion, representing 15.6% of total non-ferrous production. Lode gold is the third most economically productive deposit type after magmatic Ni-Cu and VMS deposits (Fig. 5, Table 3). On average, almost 95% of the contained Au is recovered (Fig. 3), this high recovery rate compared to base metals recovery rates (Fig. 3) being due to the chemical leaching methods for Au compared to the mechanical differential flotation methods for metal sulphide ore minerals. Nearly all the value (98%) of production from lode gold deposits (Fig. 5) comes from the Au content, but this proportion may be somewhat overstated because Au-producing mining companies usually do not report the amount of byproduct metals, such as Ag or Cu, unless they make a substantial contribution to total revenues. The total amount of Au produced (Table 2) is also understated because it does not include all small mines, especially those of the 19th and early part of the 20th century for which records are sparse. The metal content of reserves in lode gold deposits at $Eq9.63 billion is the lowest of all deposit types currently being mined (Fig. 6) and represents only 5.3% of Canada's total non-ferrous metal and diamond reserves. More lode gold deposits have been re-evaluated than other deposit types since the introduction of the NI 43-101 reporting standard. This has had the effect of reclassifying a higher proportion of historical 'reserve' estimates for lode gold deposits as inferred resources in comparison to the arbitrary reassignment of most historical reserves for other deposit types to measured and indicated resources in Appendix 1 (DVD). It also explains why the $Eq27.9 billion for lode gold (Table 3) represents only 6.1% of Canada's total non-ferrous metal and diamond measured and indicated resources but the $Eq19.49 billion for lode gold (Table 3) is a much higher 20.2% of Canada's total non-ferrous metal and diamond inferred resources.

The weighted average metal content per tonne of ore is $Eq132/t for production, $Eq69.4/t for reserves, $Eq48.8/t for measured and indicated resources, and $Eq70.1 for inferred resources (Figs. 5, 6, 7, and 8, respectively). These are the lowest dollar equivalent of metal contents of all mineral deposit types exploited by underground mining. The relatively low ore values are economic to mine because recovery rates of Au from lode gold deposits average 93.4%, which is much higher than the recovery rates for base metals (Fig. 3), and the beneficiation process produces Au bars at the mine site, which avoids both the transportation costs of bulky base metal sulphide concentrates and smelting charges. In contrast to other deposit types, dollar equivalent per tonne values for reserves (Fig. 10) and resources categories (Figs. 11, 12) are very low compared to production values (Fig. 9). This is because the tonnage of the reserves and resources categories are dominated by those reported for the Porcupine joint venture, aimed at open-pit mining of remaining low-grade (1.2-2.6 g/t Au) resources in the Timmins area, particularly around the Dome mine (Appendix 1, DVD). Though low for lode gold deposits in general, the $ $Eq29/t to $Eq65/t of these open-pit lode gold ores is high compared to the open pit ores of porphyry mineral deposits (Figs. 9, 10, 11, 12).

As with other deposit types, there is a large range in the dollar equivalent of metal content in the mineral resource categories for different deposits ranging from about $Eq20/t to over $Eq450/t (Fig. 25). Most deposits fall in the $Eq50/t to $Eq150/t range (Fig. 25A,C). Deposits mined during the 2002 to 2005 period have much the same range of dollar equivalent per tonne of metal grades as those mined prior to 2005 (Fig. 25A), indicating that there has not been a major shift in mineral deposit economics over the past 50 years, presumably because higher labour costs have been offset by more efficient mining techniques.

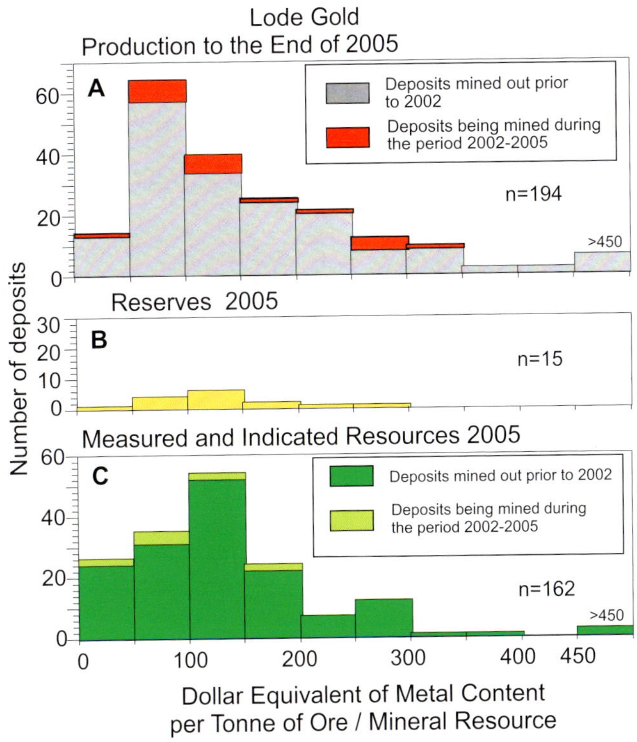

FIGURE 25. Histograms showing the distribution by $50/t increments of the average weighted grades of lode gold deposits for **(A)** production to the end of 2005; **(B)** reserves at the end of 2005; **(C)** measured and indicated resources at the end of 2005.

Porphyry Deposits

Geological Context and Distribution

Porphyry deposits are large, low-grade Cu, Mo, Au, Ag, W, and/or Sn magmatic-hydrothermal deposits spatially associated with felsic to intermediate porphyritic intrusions within 1 to 6 km of the paleosurface that formed in continental- and island-arc geological settings (Kirkham, 1972; Kirkham and Sinclair, 1995; Seedorf et al., 2005; Sinclair, 2007). The deposits typically contain hundreds of millions of tonnes of ore but the grades are generally <1% Cu and <0.1% Mo so that they can only be economically mined by open pit methods. Porphyry Cu deposits generally occur above magma cupolas in the root zones of subaerial andesitic to dacitic stratovolcanoes (Seedorf et al., 2005). Porphyry Mo deposits are typically associated with A-type granites of back-arc continental extensional geological environments, and porphyry Au deposits have an affinity with alkaline intrusions (Sinclair, 2007).

The volume of rocks mineralized by ore-bearing veins and replacements is up to 4 km^3, with the deposit generally zoned from a barren core, through an interior shell of veins rich in metal sulphides, to upper and outer zones containing Pb-, Zn-, Ag-, and/or Au-rich skarns, mantos, and epithermal veins (Einaudi, 1982; Jones, 1992; Kirkham and Sinclair, 1995; Seedorf et al., 2005; Sinclair, 2007). The shallow part of the porphyry system may be capped by a zone of siliceous quartz-alunite-pyrite alteration formed by ultra-acidic hot spring and fumarolic activity.

The great bulk of Canadian porphyry mineral resources are in British Columbia (Fig. 26), which accounts for about 85% of production and 87% of total porphyry mineral resources. This is because they are most abundantly preserved in the younger orogens where the depth of erosion has not reached the lower depth limits of porphyry formation. Yukon accounts for only 0.5% of production and 5% of total resources. Outside of the Cordillera, porphyry deposits occur in the Archean of the Superior Province and in the Appalachian Orogen (Fig. 26). Of these, the Appalachian Gaspé copper and the Archean Troilus deposits in Quebec, have been the most important producers with production estimated as $Eq11.8 billion and $Eq2.4billion, respectively (Fig. 26).

Economic Context and Statistics

Porphyry deposits provide about 60% of the world's primary Cu and virtually all the world's Mo and Re, as well as significant proportions of primary Au, Ag, and Sn (Sinclair, 2007). Most of this production comes from circum-Pacific Mesozoic to Cenozoic orogenic belts as well as scattered deposits from the Mesozoic Alpine-Tethyan Orogen of Europe and Asia, and other orogens as old as Archean. Increasing interest and exploration efforts are being directed towards central Asia following the discovery during 1998 to 2003 of the 2.5 billion tonne Oyu Tolgoi deposit in the Paleozoic Tuva Mongol arc, Mongolia (Perelló et al., 2001).

The main economic attraction of porphyry deposits is their very large tonnages that provide the long mine life preferred for corporate planning and stability purposes. However, the capital investment required for the development of such large-scale operations is correspondingly large. For example, estimates for preproduction development costs are over $1.1 billion for the Galore Creek deposit in British Columbia (NovaGold Resources Inc., 2006) and over $1.4

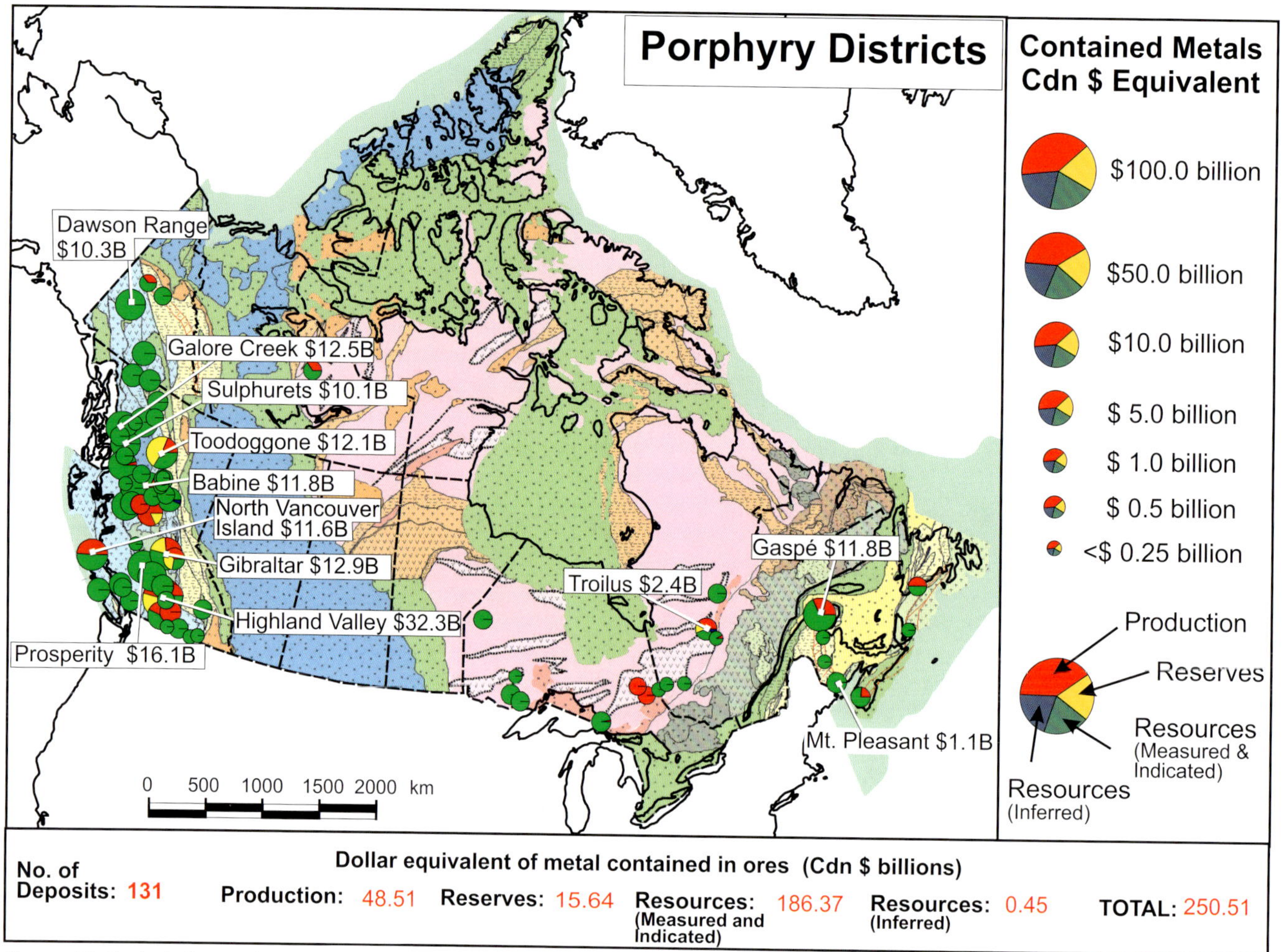

FIGURE 26. Geological map of Canada showing the distribution of porphyry districts and deposits. Legend for the map is as for Figure 2. The diameter of each pie chart is proportional to the dollar equivalent of total metal contained in the district or deposit, and the subdivisions of the pie charts represent the relative amounts in the different categories of mineral resources. A district represents the combined totals of deposits of similar mineral deposit type and age that occur in the same area. The names and total dollar equivalent of metal content are indicated only for districts or deposits with the highest metal contents.

billion for the Oyu Tolgoi deposit in Mongolia (Ivanhoe Mines Ltd., 2006). An important factor for the development of porphyry deposits is the presence of a higher grade (2-4% Cu) supergene zone of secondary Cu enrichment, which allows the accelerated repayment of the large preproduction investment. Mitigating against the development of porphyry deposits, is the social factor of a widespread public perception of the negative environmental impacts of open-pit mining, and this has prompted some major mining companies to research methods of bulk underground mining methods that could compete with the low mining costs of the open-pit method (Morgan, 2005)

The value of metals mined from Canadian porphyry deposits is about $Eq48.5billion with about 70% of this value being attributable to Cu, 14% to Mo, 13% to Au, and 2.5% to Ag. In 2005, porphyry deposits accounted for 100% of the Mo, 50% of the Cu, and 9% of the Au produced in Canada (Fig. 4B). Other metals, particularly W, Sn, and In from the East Kemptville and Mount Pleasant deposits, have contributed <1% of total production value. The contribution of Re, currently worth about $1200/kg and recovered from Mo concentrates, is not reported. On average, 83.4% of the Cu, 68% of the Au, and 51% of the Mo are recovered, so overall 75% of the metal value of porphyry ores is converted into economic wealth. The majority of Canadian porphyry resources contain less than $Eq20/t (Fig. 27), which is less than 10% the value of ores for other base metal mineral deposit types, and so is dependent on the economies of large-scale mining to be economically viable.

Porphyry reserves contain about $Eq15.6 billion in metal value (Fig. 6), the main contrast to past production (Fig. 5) being the larger proportion (21%) contributed by Au. Measured and indicated resources (Fig. 7) are estimated to contain about $Eq185.3 billion of metal based mainly on historical estimates, the great bulk of which are in British Columbia (Fig. 26), which represents 40% of Canada's total in this resource category.

Sedimentary Exhalative Deposits

Geological Contexts and Distribution

SEDEX deposits are stratiform to stratabound layers or lenses of Fe, Zn, and Pb sulphides, usually with a significant Ag content, that occur in sedimentary basins and were formed by the submarine venting of hydrothermal fluids (Goodfellow and Lydon, 2007). The sequence of sedimentary rocks that hosts the deposits are typically the sag phase

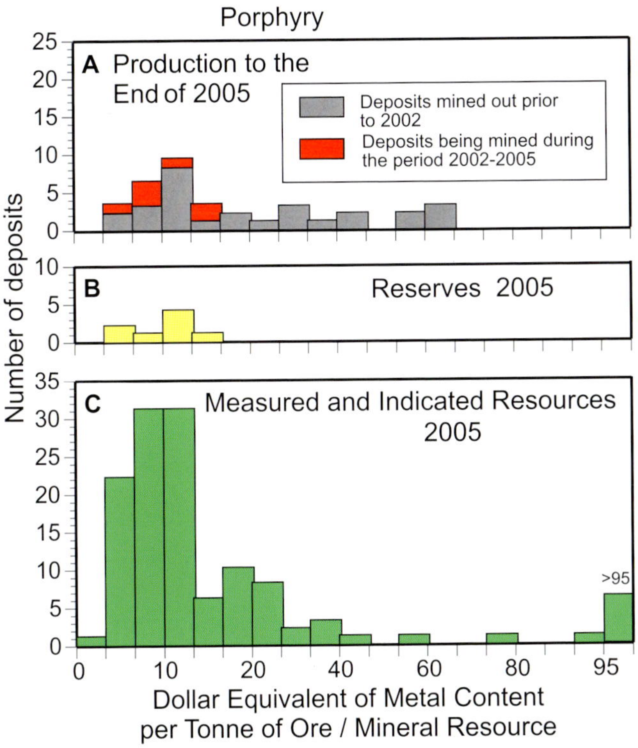

FIGURE 27. Histograms showing the distribution by \$50/t increments of the average weighted grades of porphyry deposits for **(A)** production to the end of 2005; **(B)** reserves at the end of 2005; **(C)** measured and indicated resources at the end of 2005.

of an intra- or epicontinental rift though some deposits (e.g. Sullivan, British Columbia) occur in the rift-fill part of the sequence. The question as to whether the lithospheric extension causing the rifting is related to mantle plume-head spreading or far-field back-arc extension has not as yet been addressed. Sedimentary exhalative deposits are typically much larger than VMS deposits and those that have been mined generally contain about 50 to more than 100 Mt of ore (Goodfellow and Lydon, 2007).

Most of Canada's SEDEX deposits are in British Columbia and Yukon (Fig. 28), where they occur in the Mesoproterozoic Belt-Purcell basin (Lydon, 2007) and in basinal facies of the Paleozoic miogeocline, notably the Selwyn Basin (Goodfellow, 2007b). A number of small Zn-Pb deposits in Cambrian carbonate rocks of the Kootenay arc and eastern Shuswap metamorphic complex in southern British Columbia may be of the SEDEX type. However, the high degree of deformation in these areas, which has transposed both compositional layering and sulphide mineralization into the foliation, leaves doubt as to whether these carbonate-hosted deposits are Irish-type SEDEX deposits or Mississippi Valley type. The same enigma applies to Zn-Pb occurrences in marble-rich sequences of Mesoproterozoic-Neoproterozoic supracrustal sedimentary basins of the Grenville Province in Ontario and Quebec (Corriveau et al., 2007), which have geological and metallogenetic affinities to the large Zn-Pb deposits of the Balmat-Edwards area of northern New York State. In the eastern part of Canada, the Walton deposit and MVT deposits (Paradis et al., 2007) of the Mississippian Windsor Basin may represent the transatlantic equivalent of the highly productive Irish ore-forming environment that produced deposits with both seafloor (SEDEX) and epigenetic (MVT) characteristics.

Economic Context and Statistics

Sedimentary exhalative deposits are an important source for Zn, Pb, and Ag. Judging from global production statistics for Zn (Jorgenson, 2004) and Pb (Smith, 2003), and with the caveat of the uncertainty of the mineral deposit types from which China obtains its world-leading 23.5% supply of primary Zn and 22.4% of primary Pb supply, it is estimated that SEDEX deposits provide more than 30% of the world's Zn and more than 40% of the world's Pb primary metal supply. The Mt. Isa, George Fisher, Cannington, McArthur River, and Broken Hill mines of Australia and the Red Dog mine of Alaska are the current major SEDEX producers. The main economic attractions of SEDEX deposits are the potential for very large tonnages of ore and that a high proportion of deposits of this type are strongly zoned with a high-grade core passing laterally into lower grade mineralization. Mining of these high-grade ores early in the mining operation allows for a shorter pay-back period.

The major production in Canada from SEDEX deposits came from the Sullivan deposit in British Columbia, which closed in 2001, and the Anvil Range deposits of Yukon, which last produced in 1996. Total production from SEDEX deposits has totaled \$Eq41 billion (Fig. 5), and supplied 19% of the Zn, 54% of the Pb, and 13% of the Ag (Fig. 4A) produced in Canada to the end of 2005. There are \$Eq32.5 billion Zn, Pb, and Ag in measured and indicated resources (Fig. 7) and \$Eg13.4 billion in inferred resources (Fig. 8) of this deposit type, mainly in Yukon and British Columbia, notably the deposits of the Howards Pass, MacMillan Pass, and the Gataga districts (Fig. 28). The \$Eq157.9/t metal content of measured and indicated resources is only slightly lower than the \$Eq171.1/t content of ores that have been mined (Fig. 29), suggesting that at least part of these resources are marginal to being economically mineable, if a suitable surface transportation and power infrastructure were in place. The value of the average Canadian SEDEX mineral resource (\$Eq155/t) is higher than the value of the average global SEDEX mineral resource (\$Eq135/t), which gives Canadian deposits a competitive edge in attracting development investment when the high-grade deposits currently being mined in other countries become exhausted. Recent exploration drilling in the Howards Pass district has shown that mineralization occurs in the area between the XY and Anniv deposits, which will likely lead to an increase in the amount of mineral resources currently estimated for the district.

The \$Eq171/t average value for Canadian production from SEDEX deposits is about 30% less than the global production average value (Fig. 29). One reason for this is that mining of large foreign deposits has largely been restricted to the higher grade cores that do not include the lower grade surrounding zone(s). For example, the approximately 124 Mt of production from the Mt. Isa mine has an average grade of 6.7% Zn, 6.3% Pb, and 150 g/t Ag (Perkins, 1990), or about \$Eq223/t, but the average grade of the 367 Mt of the remaining mineral resources (that is potentially mineable at the Black Star and Mount Isa open pits) is about 4.2 % Zn, 3.2% Pb, and 68 g/t Ag (Xstrata Report, 2006), or about

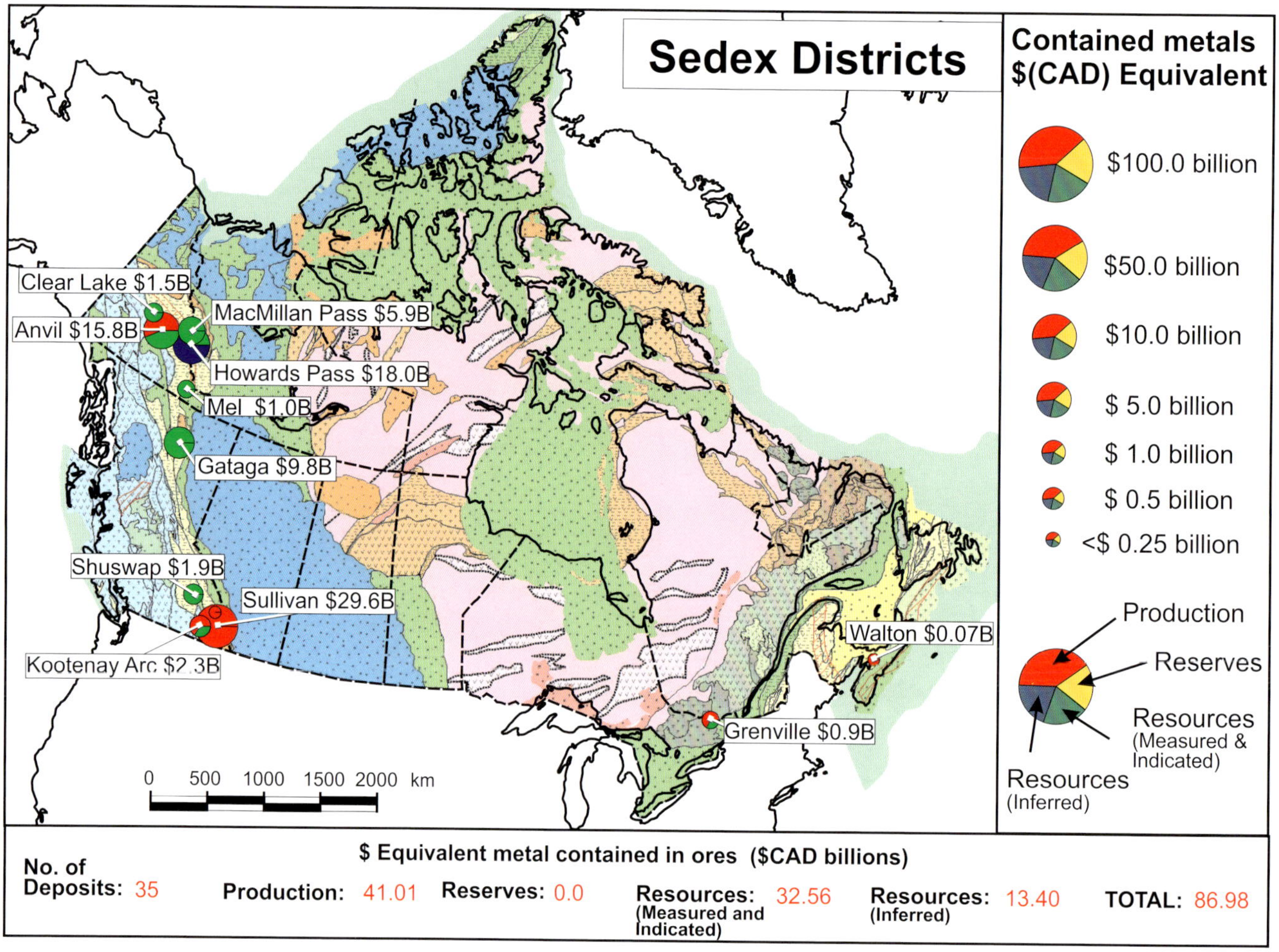

FIGURE 28. Geological map of Canada showing the distribution of SEDEX districts and deposits. Legend for the map is as for Figure 2. The diameter of each pie chart is proportional to the dollar equivalent of total metal contained in the district or deposit, and the subdivisions of the pie charts represent the relative amounts in the different categories of mineral resources. A district represents the combined totals of deposits of similar mineral deposit type and age that occur in the same area. The names and total dollar equivalent of metal content are indicated only for districts or deposits with the highest metal contents.

$Eq125/t. Another reason is that the global production average includes deposits with large tonnages of exceptionally high-grade ore, such as the 150 Mt of $Eq256/t that has been mined at Broken Hill, Australia (Mackenzie and Davies, 1990) and the 72 Mt of $Eq327/t being mined at Red Dog, Alaska (TeckCominco Annual Report for 2005). Examples of large deposits with exceptionally high grades have not yet been found in Canada.

Mississippi Valley-Type Deposits

Geological Context and Distribution

Mississippi Valley-type (MVT) deposits are irregular stratabound to discordant epigenetic replacements and/or open-space fillings of carbonate sedimentary rocks by Zn and Pb sulphides (Leach et al., 2005; Paradis et al., 2007). Individual bodies are generally scattered over a district that may be hundreds of square kilometres in area and contain tens to hundreds of individual deposits (Leach et al., 2005) and range from a few hundred thousand of tonnes to tens of million of tonnes (Paradis et al., 2007). Mississippi Valley-type deposits tend to occur in platformal carbonate sequences situated in the foreland regions of orogens, a geo-

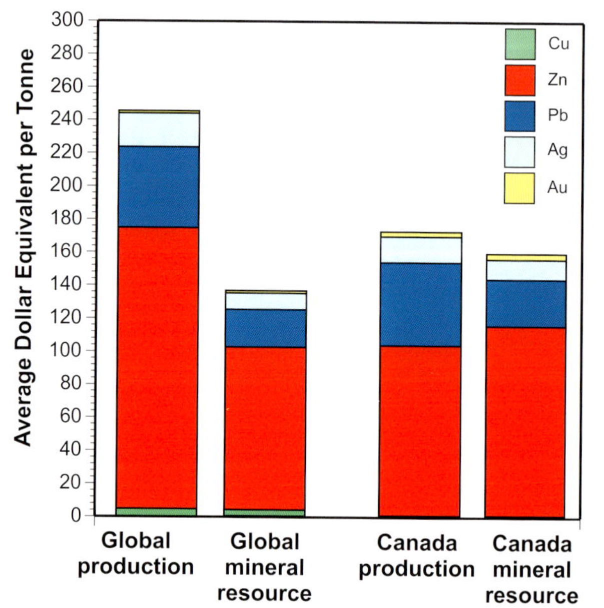

FIGURE 29. Bar charts comparing the average weighted grades, expressed as dollar equivalent per tonne, of Canadian SEDEX deposits with the world-wide average.

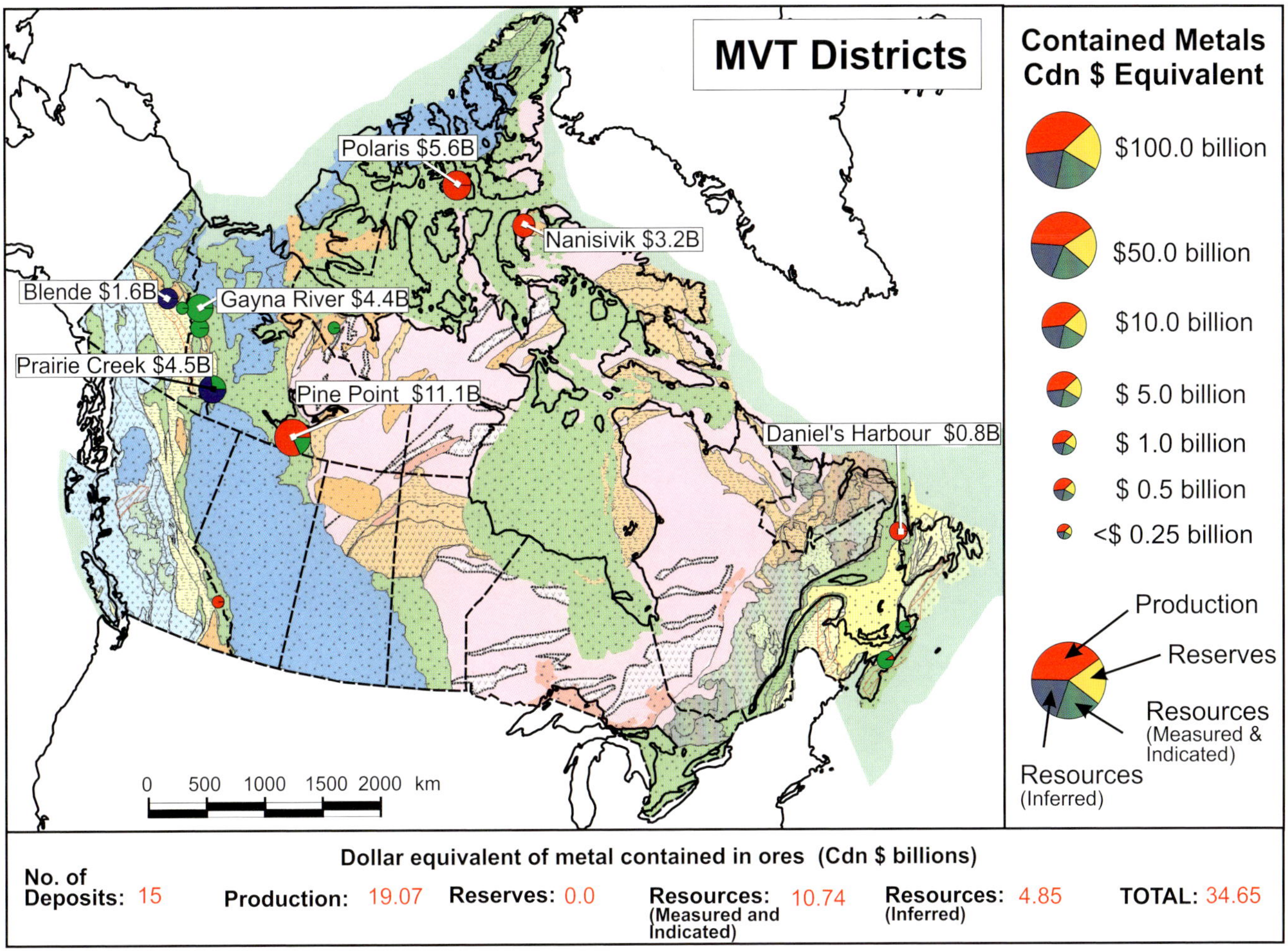

FIGURE 30. Geological map of Canada showing the distribution of Mississippi Valley-type districts and deposits. Legend for the map is as for Figure 2. The diameter of each pie chart is proportional to the dollar equivalent of total metal contained in the district or deposit, and the subdivisions of the pie charts represent the relative amounts in the different categories of mineral resources. A district represents the combined totals of deposits of similar mineral deposit-type and age that occur in the same area. The names and total $Equivalent metal content are indicated only for districts or deposits with the highest metal contents.

tectonic setting thought to reflect the zone of tectonically induced migration of metalliferous brines from basins of the continental margin into carbonate aquifers of the platformal sequences (e.g. Garven, 1985; Leach et al., 2001, 2005; Paradis et al., 2007). Most MVT deposits in Canada (Fig. 30) are foreland to the Cordilleran Orogen of Yukon and Northwest Territories where they occur in Neoproterozoic to Devonian carbonates. Those of the Maritimes occur in Ordovician to Mississippian carbonates adjacent to the Appalachian Orogen. Polaris, in Ordovician carbonates, is related to the Paleozoic Ellesmerian tectonism (Symons and Sangster, 1992) as may Nanisivik in Neoproterozoic carbonates, though there is debate that it may be related to Grenville-age compressional tectonics (Dewing et al., 2007). The Gays River mineralization was formed by the ca. 297 Ma (Kontak et al., 1994) Pennsylvanian-Permian migration of basinal brines, presumably under the influence of Alleghanian tectonism.

Economic Context and Statistics

The mining of MVT deposits was especially important to Canada during the 1980s when production from the deposits at Pine Point in Northwest Territories, Polaris on Little Cornwallis Island, Nanisivik on Baffin Island, and Newfoundland Zinc (Daniel's Harbour) in western Newfoundland, contributed about 27% of the Zn and 15% of the Pb to Canada's primary production of these metals. This production declined to 20% of the Zn and 8% of the Pb during the 1990s until the closure of the Polaris and Nanisivik mines in 2002, when all production in Canada from this deposit type came to an end. Although the value of the metal content of the ores are comparable to that of both the VMS and the SEDEX deposit types (Fig. 9), MVT ores tend to be coarser grained and the carbonate host rocks relatively soft, both of which decrease beneficiation costs. The carbonate host rocks, while advantageous in averting acid mine drainage, tend to be highly productive aquifers increasing water-control costs, which can be a major factor, as at the Pine Point and Newfoundland Zinc deposits.

Canadian production from MVT has totaled about $Eq19.0 billion, 95% of it coming from just three mines: Pine Point at $Eq9.1 billion, Polaris at $Eq5.6 billion, and Nanisivik at $Eq3.2 billion. The average value of MVT ores that have been mined is $Eq173.8/t, about the same for other deposit types mined for Cu, Zn, and/or Pb by underground methods (Fig. 9), but ranges from $Eq81/t for the 741,000

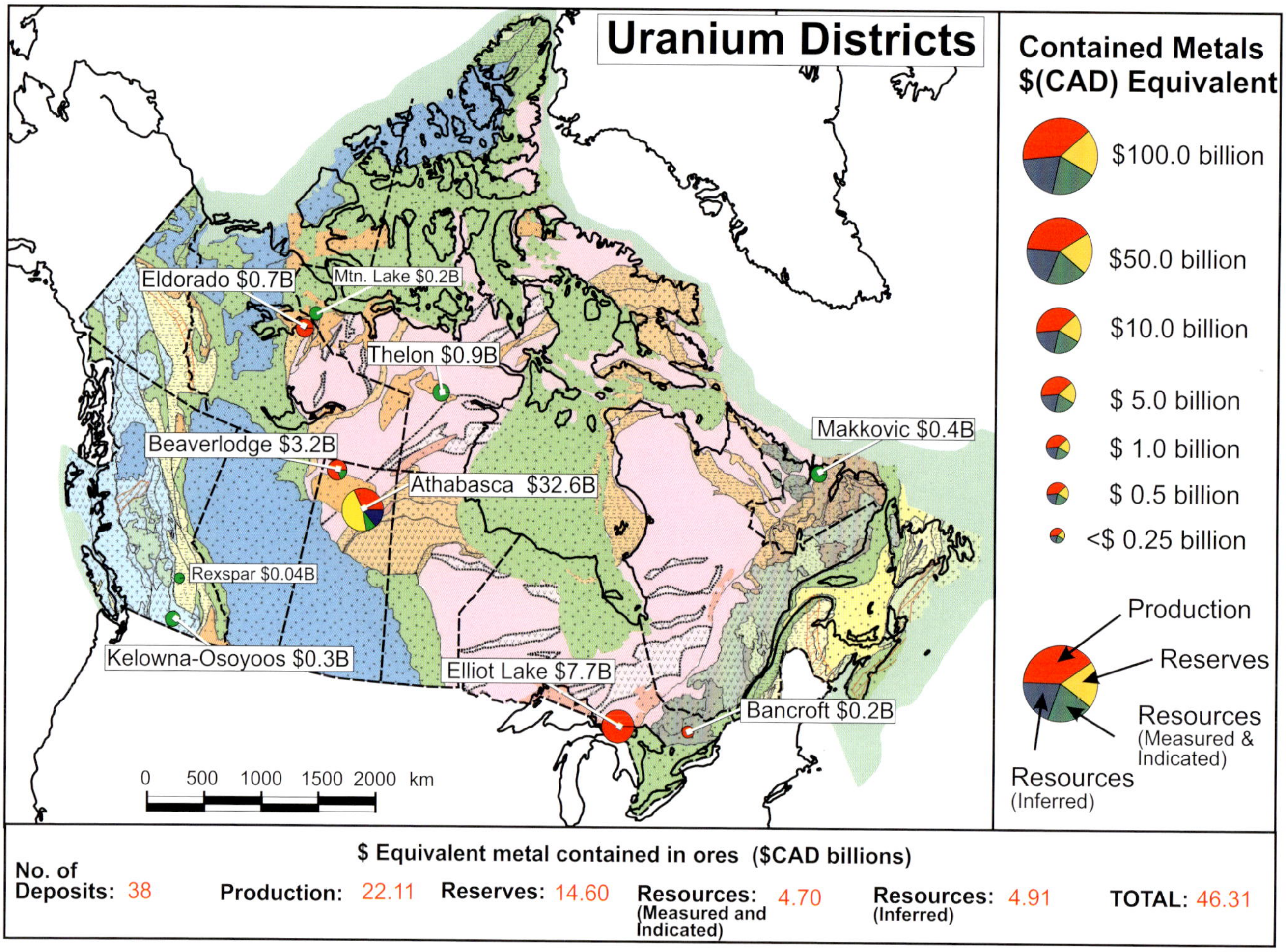

FIGURE 31. Geological map of Canada showing the distribution of uranium districts and deposits. Legend for the map is as for Figure 2. The diameter of each pie chart is proportional to the dollar equivalent of total metal contained in the district or deposit, and the subdivisions of the pie charts represent the relative amounts in the different categories of mineral resources. A district represents the combined totals of deposits of similar mineral deposit type and age that occur in the same area. The names and total dollar equivalent of metal content are indicated only for districts or deposits with the highest metal contents.

tonnes of ore mined at Gays River to $Eq278/t for the ores from Pine Point. For all deposits, most of this value is attributable to the Zn content and the remainder mainly to Pb (Fig. 5). In contrast to other deposit types from which galena is mined (VMS, SEDEX, and some vein types), MVT deposits on average do not seem to have a significant Ag content. The low Ag value in ores for MVT deposits (Fig. 9) is because Ag grades for production are reported for only two of the fourteen deposits contributing to these statistics. Similarly, the compilation of Leach et al. (2005) reports Ag grades for only 22 of 88 deposits with reported Zn and Pb grades.

Mississippi Valley-type measured and indicated resources are estimated at $Eq10.7 billion (Fig. 7), and like past production, most of this value is from the Zn content. Measured and indicated resources have an average value of $Eq117/t (Fig. 11) and is much lower than for other base metal deposits mined by underground methods because about 40% of the resources are contributed by the low-grade ($Eq88/t) deposits of the Gayna River district of Northwest Territories (Fig. 30). The highest grade measured and indicated resources are in the Prairie Creek deposit ($Eq353/t), which has unusually high Ag and Cu contents. The Prairie Creek deposit, with its estimated minimum total $Eq4.5 billion of measured, indicated, and inferred resources (Fig. 30) has already undergone most of its preproduction development, but actual production is awaiting resolution of land-use issues. Ground water management is likely to be a major negative factor for development of the $Eq2 billion of measured and indicated resources remaining in the Pine Point district (Fig. 30) that are largely on the Great Slave Reef property (Hannigan, 2007), and the low grades and remote location are likewise negative factors for the development of the Gayna River resources.

Uranium Deposits

Geological Context and Distribution

Most of Canada's U resources are related to the global phenomenon of U concentration during the Proterozoic, starting with the U paleoplacers of Elliot Lake at the beginning of the Paleoproterozoic and culminating in the hydrodynamic events during the Mesoproterozoic that formed the epigenetic deposits of the Athabasca sedimentary basins. Other types of U deposits include veins, pegmatites, and sandstone-hosted deposits.

Unconformity-related uranium deposits: The unconformity-related deposits of the Athabasca Basin (Jefferson et al., 2007) (Fig. 31) contain more than ten times the amount mined from other deposit types and currently contribute all of Canada's substantial uranium production. The deposits tend to be small (0.1 to 5.0 Mt), but with grades ranging up to 20% U contain very valuable ores (Figs. 9, 10). The ores are thought to have been formed during the exchange of groundwaters between the hydrochemical environment of basement rocks and the different hydrochemical environment of the overlying Athabasca Group sedimentary rocks, with U deposition taking place close to the unconformity surface presumable due to redox reactions. There is strong evidence of a spatial relationship between U deposits and graphite-bearing lithological units in the basement (Jefferson et al., 2007). Polymetallic deposits contain significant Ni, Co, Cu, Pb, Zn, Mo, and in some cases Au, Ag, and PGE in addition to U. These polymetallic deposits typically occur in basal sandstones and conglomerates of the Athabasca Group where the unconformity surface has been displaced by local faults and where groundwater from the Paleoproterozoic basement flowed upwards into the Athabasca Group (Jefferson et al., 2007). Monometallic deposits, containing only U in significant concentrations, typically occur along fracture, shear, or breccia zones for up to 400 m into the basement below the unconformity and are thought to have formed by the ingress of formational waters of the Athabasca Group downward into the basement (Jefferson et al., 2007). Sedimentation in the Athabasca Basin began at 1730 to 1750 Ma and continued to at least 1650 Ma. Formation of the U deposits occurred during one or both of two major hydrothermal events at about 1500 and 1350 Ma (Fig. 17) with remobilization events at 1176, 900, and 300 Ma (Jefferson et al., 2007). Other examples of unconformity-related uranium deposits are known at the base of the Thelon Basin in Nunavut (Fig. 31).

Paleoplacer uranium deposits: Pyritic quartz pebble conglomerates in the Matinenda Formation near the base of the Proterozoic Huronian Supergroup are host to the paleoplacer uraninite deposits mined in the Elliot Lake district (Roscoe, 1996) during the period of 1957 to 1992. The age of the Huronian paleoplacers (Fig. 17) of the Elliot Lake district are approximately that of the felsic volcanic rocks of the Coppercliff Formation near the base of the Huronian Supergroup, which have been dated at 2450 Ma (Roscoe, 1996).

Other uranium deposits: The veins of the Beaverlodge district (Fig. 31) are mostly hosted by the Paleoproterozoic Tazin Group and are close to the contact with the overlying Mesoproterozoic Martin Formation of the Athabasca Group. The ore mineralogy of the veins in the Beaverlodge district (Fig. 31) is generally monometallic, consisting of pitchblende and lesser brannerite with carbonate gangue (Ruzicka, 1996). Classified by Ruzicka (1996) as "veins in shear zones" they may be exhumed representatives of the monometallic basement vein variety of unconformity-associated uranium deposits (Mazimhaka and Hendry, 1989; Jefferson et al., 2007). Similarly, the veins of the Eldorado area (Fig. 31), hosted by 1860 to 1875 Ma volcanic rocks of the Great Bear magmatic zone but with a mineralization occurring over the interval 1775 to 1665Ma with Pb isotope resetting at 1500 and 1420 Ma (Ruzicka and Thorpe, 1996b), may represent exhumed veins that formed below the unconformity at the base of the Hornby Bay Basin. The disseminated pitchblende, breccia fillings, and veins of the Makkovic area (Fig. 31) are hosted by aerially extensive flows, tuffs, and volcaniclastic sequences formed during the 1700 to 1900 Ma Makkovikian Orogeny (Gandhi and Bell, 1996). The mineralization occurs in felsic volcanic rocks that dominate the bimodal upper part of the 1805 to 1860 Ma Aillik Group. Isotopic ages of the deposits are between 1750 and 1800 Ma (Ghandi and Bell, 1996).

Pegmatite-associated and disseminated uraninite and uranothorite in Grenvillian rocks of the Bancroft area, Ontario, are associated with granitic to syenitic pegmatites but also occur in siliceous sulphidic marbles and calc-silicate skarns (Carter and Colvine, 1985). Low-grade uranium phosphate mineralization occurs in Miocene and surficial sediments near Kelowna, British Columbia, and in sandstones of the Proterozoic Hornby Basin in Nunavut.

Economic Context and Statistics

Most of Canada's uranium production prior to 1984 came from the low-grade paleoplacer ores of the Elliot Lake district in Ontario, with a smaller amount produced from veins of the Beaverlodge district of Saskatchewan. Over their total production history, these deposits produced $Eq10.8 billion, almost totally as uranium (Fig. 5). However, the average $Eq59.5/t of these ores (Fig. 9) could not compete on the opened market with the much higher grade $Eq539.3/t grades of the unconformity-related deposits of the Athabasca Basin of Saskatchewan, resulting in the closure of the Elliot Lake mines in 1996 (Leadbeater, 1998) after producing about 144,000 t of U ($Eq7.7 billion) in 160 Mt of ore at a grade of about 0.9%U. Production of U from fault-associated veins of the Beaverlodge area amounted to about $Eq3.2 billion (Fig. 31) and about 6000 t of uranium ($Eq0.5 billion) was recovered at the Eldorado Ag-rich polymetallic veins (Ruzicka and Thorpe, 1996b). The amount of ore mined from individual mineral deposits of the Athabasca Basin is generally small, so that their overall contribution to the value of national mineral resource inventory is also relatively small, contributing only 3% of Canada's total non-fuel mineral production in 2004 (Fig. 1) and only 1.27% of total historic non-ferrous metal and diamond production (Table 3), but obviously will far exceed the combined production from other areas when currently known reserves and resources have been mined. The $Eq14.6 billion of reserves in the Athabasca Basin deposits constitute 8% of Canada's non-ferrous metal and diamond reserves (Table 3). Unlike other deposit types, most of the unmined mineral resources of the Athabasca Basin are classified as reserves, leaving only $Eq2.6 billion as measured and indicated resources, which constitute only 0.6% of Canada's mineral resources in this category. Known reserves and resources in the Athabasca Basin are sufficient for another 25 years of mining (Saskatchewan Interactive, 2006). Uranium in paleoplacers, veins, and pegmatites is generally considered to be of too low a grade or too low a tonnage to economically compete with the unconformity-associated type of U deposit. However, this outlook may change if the price of U continues its upward trend of the past two years.

Uranium has a much higher value than other non-precious metals (Table 1), which accounts for the unconformity-related ores of the Athabasca Basin having the highest weighted average value per tonne of all mineral deposit types at $Eq540/t for production (Fig. 9) and $Eq5190/t for reserves (Fig. 10). Due to the rapid rise in the price of U during 2004 and 2005 (and continuing through 2006), the dollar equivalent value for U ores would be more than double the figures given here if the current U price were used rather than its 10 year average price. As with other deposit types, there is a great deal of variation between deposits, ranging from a high of over $Eq10,000/t for production (and reserves) at the high-grade McArthur River deposit, down to about $Eq300/t at the Rabbit Lake and Cluff Lake deposits. The very high value per tonne for reserves of $Eq5190/t (Fig. 10) is because nearly half the tonnage of current reserves is in the very high-grade McArthur River and Cigar Lake deposits (Appendix 1, DVD). The relatively low value of $Eq293/t for measured and indicated resources of the Athabasca Basin U deposits (Fig. 11) is because the bulk of the resources are in the low-grade (0.12% U) deposits of the Hidden Bay area.

Outside of the Athabasca Basin, significant (>$Eq0.1 billion) U resources are restricted to the Kiggavik deposits of the Thelon Basin, deposits in the Makkovic area of Labrador, and deposits in Miocene sediments near Kelowna, British Columbia (Fig. 31). The Kiggavik Main deposit, with nearly 2.5 Mt of ore valued at $Eq264/t, seems to have better economic prospects than the 1.9 Mt Blizzard deposit in British Columbia, valued at $Eq112/t, or the 6.2 Mt Michelin deposit in Labrador, valued at $Eq54/t.

Miscellaneous Deposits

Geological Contexts and Distribution

Various mineral deposit types, which occur in Canada and are economically important elsewhere in the world, are either included in the seven deposit types outlined above nor, except for iron oxide copper-gold deposits (IOCG; Corriveau, 2007), described or compiled elsewhere in this volume. They are grouped here as veins, IOCG, skarns, and diagenetic copper.

Veins are infillings of fractures by ore minerals, and although veins form at least part of the mineralization of various deposit types, particularly lode gold and porphyry deposits, some are wide enough and rich enough to have been mined for metals other than Au. The economically most important of these vein deposits are Ag-rich and represented in Canada by the Ag-Co-Sb veins of the Cobalt area of Ontario, the Ag-Pb-Zn veins of the Elsa (Keno Hill) area of Yukon, and the Ag-U veins of the eastern Great Slave Lake area, Northwest Territories. The Ag veins of the Cobalt area (Fig. 32) generally occur within a few hundred metres of the Nipissing diabase sills and the unconformity between the Huronian Supergroup and metavolcanics of the Archean basement (Marshall and Watkinson, 2000). U-Pb ages of baddelyite from the Nipissing diabase and vein-related rutile have similar ages of 2219 and 2217 Ma, respectively (Andrews et al., 1986), and led Marshall and Watkinson (2000) to suggest that the two events are genetically related. Most deposits of the Elsa area occur in Mississippian Keno Hill quartzite and are zoned with respect to the Mayo Lakes pluton. Sinclair et al. (1980) obtained a 90 Ma K-Ar age for the mineralized veins, which probably represents distal veins of intrusion-related lode Au systems of the Cretaceous Tintina Au province (Lynch, 1989; Lynch et al., 1990; Groves et al., 2003).

Iron oxide copper-gold deposits (Corriveau, 2007) are a relatively new mineral deposit type (Hitzman et al., 1992) inspired by the 1975 discovery of the immense Cu-Au-U-Ag-REE Olympic Dam deposit in the Gawler Craton of South Australia (Roberts and Hudson, 1983). The debate continues as to whether hematite/magnetite-rich deposits with Cu, Au, and other metal enrichments are a distinct genetic class or whether they are local variants of other deposit types, particularly porphyry deposits (Williams et al., 2005; Corriveau, 2007) formed by magmas or magmatic-hydrothermal systems interacting with pre-existing highly saline sulphate-rich groundwaters (e.g. Barton and Johnson, 2000). The only IOCG deposits in Canada with a measured resource are the Sue-Diane and NICO deposits of the 1850 to 1880 Ma Great Bear Magmatic Zone of the Northwest Territories (Fig. 32). The 1600 Ma Wernecke breccias in the Wernecke and Ogilvie mountains, Yukon, in places, are mineralized with hematite/magnetite and associated Cu-Co-Au-Ag-U enrichments (Thorkelson et al., 2001; Hunt et al., 2005)

Skarn deposits are characterized by Fe-enriched pervasive alteration of carbonate-rich rocks by magmatic hydrothermal fluids at the margins of felsic plutons that contain metals in economic concentrations. Although skarns are usually a part of the mineralization associated with porphyry Cu and Mo deposits, some skarn deposits seem to have been formed in the absence of porphyry-type mineralization. Of particular interest are the W-rich skarns (Dawson, 1996) associated with Cretaceous stocks of Yukon and Northwest Territories and the Cu-rich skarns of the Whitehorse copper belt, Yukon (Dawson and Kirkham, 1996).

Diagenetic copper deposits include the Kuperschiefer-type, redbed copper and volcanic redbed copper deposit types of Kirkham (1996a,b). Although there has been no production from this type of deposit in Canada, elsewhere in the world they have been prolific producers, notably in the central African copper belt and in the Kuperschiefer of Poland. Deposits commonly contain >1 Mt of Cu. They are formed when oxidized fluids carrying soluble cupric compounds enter a reducing, H_2S-rich environment and precipitate as Cu sulphides or in some cases as native Cu. They are usually thin but widespread sheets that mark the interface between oxidizing and reducing conditions during basinal dewatering stages of diagenesis. The two examples of diagenetic copper deposits, the Redstone deposit in Neoproterozoic sediments of the western Northwest Territories and the Zone (or Dot) 47 deposit in shear zones within the Coppermine basalts of Nunavut (Fig. 32), have not been mined, but remain potentially large but remote Cu resources.

Economic Context and Statistics

Veins: The mining of veins was ideally suited to the mining methods of the 19th and early 20th century, and therefore was important to Canada's mining history. Mining of Ag-rich veins in the Cobalt area of Ontario (Fig. 32) produced $Eq4.7 billion and during its peak production in 1908 supported 34 different mines (Udd, 2000). Mining at Cobalt attracted the investments to develop a railroads transporta-

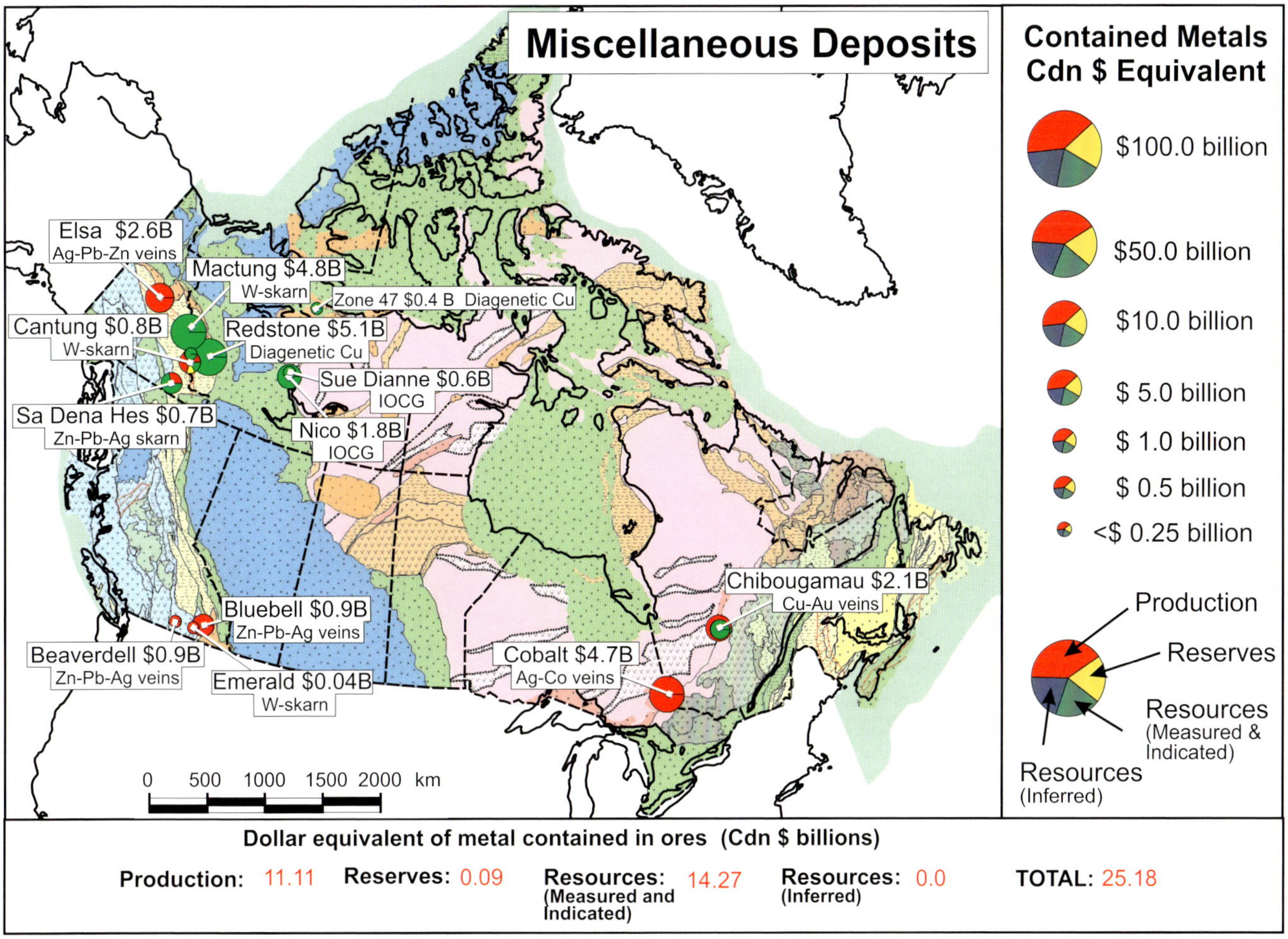

FIGURE 32. Geological map of Canada showing the distribution of selected veins, skarns, iron oxide copper-gold and diagenetic copper deposits. Legend for the map is as for Figure 2. The diameter of each pie chart is proportional to the dollar equivalent of total metal contained in the district or deposit, and the subdivisions of the pie charts represent the relative amounts in the different categories of mineral resources. A district represents the combined totals of deposits of similar mineral deposit type and age that occur in the same area. The names and total dollar equivalent of metal content are indicated only for districts or deposits with the highest metal contents.

tion infrastructure northward from the Great Lakes – St. Lawrence corridor, which in turn provided the means for settlement in these areas and helped set the scene for the discovery of the mineral deposits of the prolific Abitibi belt to the north. The Keno Hill deposits of the Elsa area were discovered in 1906 in the wake of the Klondike Au rush and 16 deposits from the area were intermittently mined during the period of 1921 to 1988, producing ores with Ag, Pb, and Zn contents of about $Eq2.4 billion. The opening of the mines provided a much needed economic boost at a time when the population of Yukon had declined to 4,157 people in 1921 with the decline of Klondike placer Au mining. The Echo Bay and Camsell River area Ag-rich polymetallic (Ag-U-Cu-Co-Ni-Bi-Sb-Pb-Zn) veins at the east end of Great Bear Lake together produced metals with a value of about $Eq0.85 billion, with about half the metal value derived from Ag. However, the area has historical importance, not because of its Ag production but for its U production from the Eldorado mine. First opened in 1933 as a source of Ra, it closed in 1940 only to be re-opened in 1942 as a source of U for the Manhattan Project that led to the development of nuclear arms. Copper-Au veins are currently being mined in the Chibougamau district of Quebec.

Iron oxide copper-gold deposits: The NICO and Sue-Diane deposits together have a metal content of $Eq2.3 billion, which is a negligible proportion (0.15%) of Canada's total non-ferrous metal and diamond resources. Most of the value in the $Eq1.8 billion NICO deposit lies in its Co content (74%), with the remainder contributed by Au (24%) and Bi (2%). At the $Eq0.6 billion Sue-Dianne deposit, most of the value is contributed by Cu (89%) and the remainder by Mo (8%) and Ag (3%).

Skarns: Although skarns are commonly a part of the mineralization styles included in porphyry deposits, the W-bearing skarns along the Yukon-Northwest Territories boundary (Fig. 2) are not part of recognized porphyry systems. They are economically and strategically important because they are the repository of the western world's largest tungsten resource. Only the $Eq0.8 billion Cantung deposit, with a $Eq111.6/t resource, has been mined but the much larger $Eq4.8 billion Mactung deposit has not, perhaps mainly because of its lower grade $Eq60.2/t resource. The mining of small amounts of W in Canada was part of the war effort both during 1914 to 18 (Burnt Hill, Nova Scotia) and during 1939 to 1945 (Indian Path, Nova Scotia; Outpost Island, Northwest Territories; Emerald, British Columbia) (Udd,

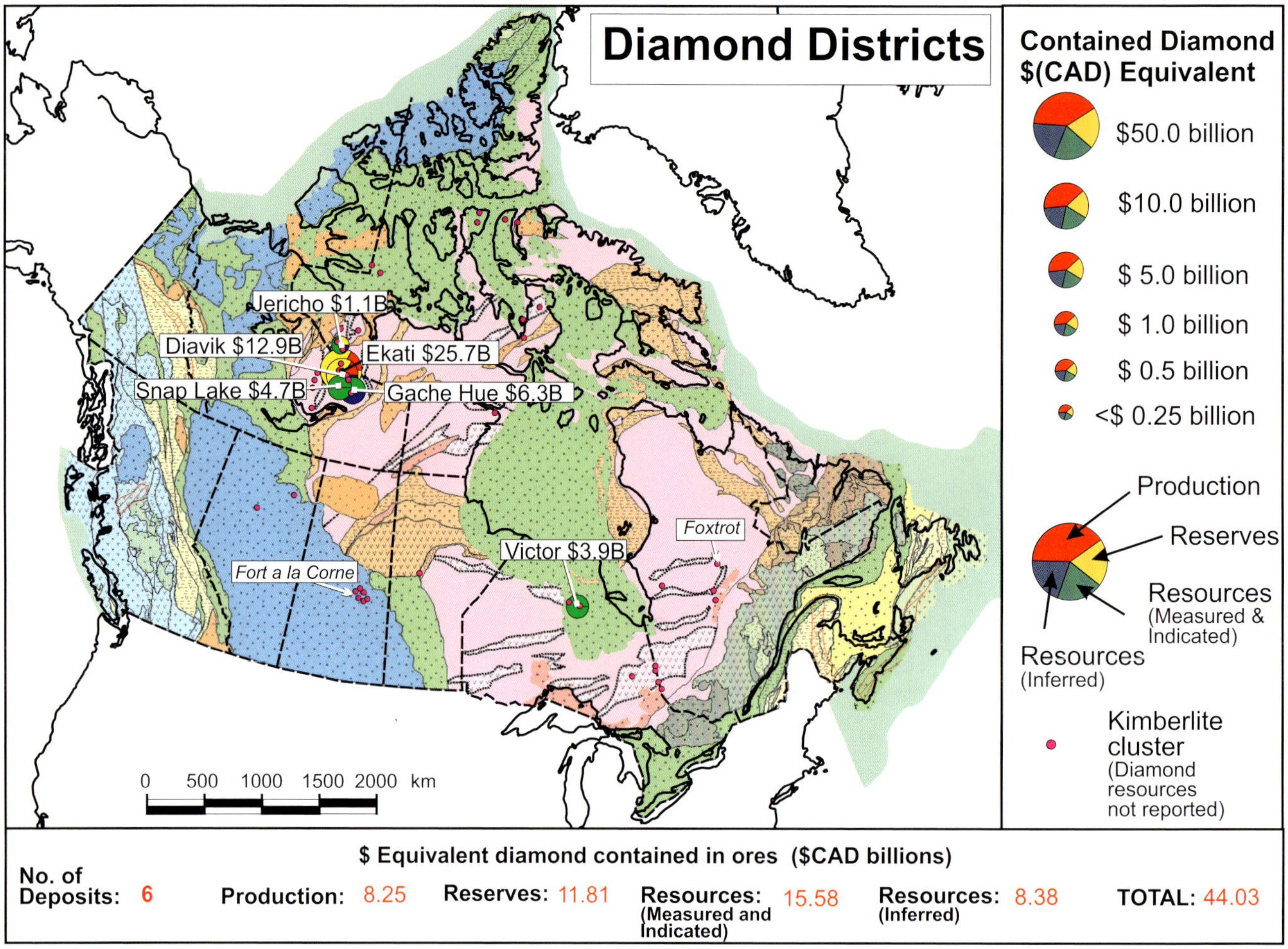

FIGURE 33. Geological map of Canada showing the distribution of kimberlite diamond deposits and other kimberlite pipes. Legend for the map is as for Figure 2. The diameter of each pie chart is proportional to the dollar equivalent of total metal contained in the district or deposit, and the subdivisions of the pie charts represent the relative amounts in the different categories of mineral resources. A district represents the combined totals of deposits of similar mineral deposit type and age that occur in the same area.

2000). During the 1950s, the Emerald (Fig. 33), and nearby Dodger and Feeney skarn deposits were important sources of W in the free world (Udd, 2000). Canada's position as a western world leader in W production seemed to have been consolidated by the discovery in 1954 and start of production in 1962 of the Cantung skarn deposit in Northwest Territories near the Yukon-Northwest Territories border (Fig. 32) and by the 1962 discovery of the large Mactung deposit and the 1984 start of W production from the Mount Pleasant porphyry deposit in New Brunswick (Fig. 26). Until 1986, Canada was producing about 8% of the world's W supply, but in 1986 low-priced W exports from the People's Republic of China forced the closure of Canadian W mines (Natural Resources Canada, 2004).

Diagenetic copper deposits are an important source of Cu in other parts of the world. In Canada, the Redstone deposit in the western part of Northwest Territories, with a $Eq5.1 billion resource, is Canada's prime example, and there is a high potential for increasing the $Eq0.4 billion known resources of the Coppermine River area of Nunavut.

Kimberlite Diamond Deposits

Geological Context and Distribution

Kimberlite diamond deposits (Kjarsgaard, 2007) are diatremes and craters of kimberlite that contain diamond xenocrysts of sufficient grade and stone quality to be economic. The great majority of kimberlite pipes do not contain diamonds. Kimberlites (Skinner and Clement, 1979; Mitchell, 1986) are thought to have formed by the partial melting of carbonated garnet lherzolite in the asthenosphere at pressures of more than 5 GPa (~150 km depth). Kimberlites acquire their diamond content as their magmas rise through the subcontinental lithospheric mantle where diamonds may have previously formed in peridotite (mainly harzburgite) or eclogite of old lithospheric mantle under conditions of high pressure (>150 km depth) but relatively low temperature (<1200ºC) (Boyd et al., 1985). The formation of diamonds appears to have been episodic over Earth's history, with major diamond-forming events occurring at 3.2 to 3.6 Ga, 2.7 to 2.9 Ga, 1.6 to 2.0 Ga, and 1.0 to 1.2 Ga (data in Gurney et. al., 2005), which broadly coincides with the time periods of major supercontinent assembly and generation of new continental crust (e.g. Kemp et al., 2006). Melting of the asthenosphere is thought to be triggered by tensile tectonic stresses and kimberlites reach the surface along upward propagating, fluid-filled fractures (Gurney et al., 2005). Lamproites, which may contain diamonds but are only of minor economic importance, are generated by partial melting of lithospheric mantle under conditions of greater hydrous content than kimberlites (Foley et al., 1987).

Diamond-bearing kimberlite (and lamproite) diatreme events also appear to have been episodic, occurring in the intervals 1140 to 1200 Ma, 456 to 542 Ma, 340 to 260 Ma, 147 to 173 Ma, 53 to 124 Ma, and 20 to 22 Ma (data in Gurney et. al., 2005). Most of these events are restricted to specific regions, whereas others occurred contemporaneously on different continents (Kjarsgaard, 2007).

The pressure-temperature stability field of diamonds mentioned above is met only in the environment of low geothermal gradients in the mantle root zones of thick old continental lithosphere of continental nuclei, giving rise to "Clifford's Rule" that diamond deposits are found only in terranes older than 1.5 Ga (Clifford, 1966), particularly Archean cratons (Janse, 1984). Most of Canada's diamond resources are in the Archean Slave Province of Northwest Territories and Nunavut (Fig. 33), with kimberlite eruptive events occurring during the Cambrian (Snap Lake at 523-535 Ma; Gacho Kue at 542 Ma), the Jurassic (Jericho at 172 Ma), and Paleogene (Ekati at 53-56 Ma) (Kjarsgaard, 2007). The Archean Superior Province hosts the Victor kimberlite cluster of Ontario, probably of Jurassic age, and the Foxtrot (Renard), Quebec, of Neoproterozoic age (ca. 630 Ma) (Kjarsgaard, 2007). The Archean Sask Craton underlies the Fort à la Corne cluster of Upper Cretaceous age (95-105 Ma). Kimberlites have been found in the Arctic areas of the Archean/Paleoproterozoic Churchill Province and the Paleoproterozoic (1.8-2.4 Ga) of Alberta, as well as in probable Paleoproterozoic terranes of southeastern British Columbia and Nunavut (Kjarsgaard, 2007) but their diamond potential has yet to be determined.

Economic Context and Statistics

Although diamonds had been discovered in glacial drift south of the Great Lakes in the 19th century and in the Peterborough (prior to 1920) and Porcupine (1971) areas of Ontario, and the first in situ diamonds were questionably recovered from the Pain de Sucre kimberlite dyke just west of Montreal in 1967 by de Beers, it was the discovery of the Ekati diamond-bearing cluster of kimberlite pipes in the Lac de Gras area that started the largest staking rush in Canadian history (Brummer, 1978). The attraction in exploring for diamonds is their very high value and the fact mines can be developed in areas without a permanent surface transportation infrastructure because the mine product has very little bulk and can be transported by air. World-wide, the value of diamondiferous kimberlites ranges up to about \$800/t though the majority of kimberlite ores are in the \$50/t to \$100/t range (Kjarsgaard, 2007). Canadian mines rank amongst the world's most valuable with production to date averaging \$267/t (Fig. 9), and reserves averaging about \$150/t (Fig. 10). In 2005, Canadian production accounted for approximately 13.5% of world output on a value basis, making Canada the world's third largest producer by value. With the scheduled openings of the Snap Lake mine in 2007, the Victor mine in 2008, and the Gacho Kué mine in 2011, Canada's share of world diamond production is expected to increase to over 20% (Perron, 2006).

Diamond mines are the largest private employers in the Northwest Territories, with the creation to 2005 of about 6700 direct and indirect jobs, and have also resulted in the creation of several hundred companies by Aboriginals (Perron, 2006). About \$250.7 million was spent on diamond exploration during 2005 (Bouchard, 2006), accounting for 19.3% of total mineral exploration investment attracted to Canada. The bulk of this was directed towards Northwest Territories and Nunavut, again bringing more economic activity to northern communities.

The value of diamond resources is more difficult to calculate than metal resources because the value of the contained commodity is dependent not only on grade and recovery rate but also on quality. The value of recovered diamonds per carat can vary widely between different kimberlite pipes and even within a single pipe. For example, production at Ekati has averaged about \$Eq133/t whereas at Diavik it has averaged \$Eq366/t. The average value per tonne of ore at the Ekati mine varies from US\$40 in the Pigeon pipe to US\$186 in the Panda pipe (Kjarsgaard, 2007). To the end of 2005, Canada's diamond production has been \$Eq8.2 billion (Fig. 5), with reserves at \$Eq11.8 billion (Fig. 6), and measured and indicated resources are estimated at an additional \$Eq15.6 billion (Fig. 7). With additional promising kimberlite pipe clusters under current evaluation, such as at the Fort a la Corne, Saskatchewan and the Foxtrot (Renard), Quebec properties, these resource estimates are likely to substantially increase in the future.

Conclusions

In terms of the value of economic commodities of their ores, magmatic Ni-Cu is the most important deposit type to Canada, thanks mainly to the giant 1850 Ma meteorite impact that produced the unique Sudbury structure. Based on the inflation-adjusted average 1996 to 2005 metal prices, magmatic Ni-Cu deposits have contributed \$Eq372.1 billion or 44% of Canada's total non-ferrous metal and diamond production and \$Eq82.5 billion or 45% of 2005 reserves. However, porphyry deposits, mainly of British Columbia, contain the bulk of the measured and indicated resources categories, which amount to \$186.4 billion or 44% of Canada's total. Volcanogenic massive sulphide and lode gold deposits have been the mainstay of Canada's mining industry and together have supported at least 347 significant mines located in all provinces of Canada except Alberta and Prince Edward Island. Production from VMS and lode gold deposits has amounted to \$Eq192.3 billion and \$Eq131.6 billion, respectively. Kimberlite diamond deposits, a relatively new mineral deposit type to be mined in Canada, contributed 11% of Canada's total non-ferrous metal and diamond production in 2005. With several deposits at an advanced stage of evaluation and the deposit type attracting 22% of the \$949 million dollars spent on exploration for new deposits in 2005, mining of kimberlite diamonds will be of increasing economic importance in the future. The most valuable ores are the unconformity-related U deposits of the Athabasca Basin, with average production at \$Eq540/t and reserves at

$Eq4,590/t, compared to the average $Eq130-$Eq180/t for production from underground base metal and gold mines.

The most productive geological environments for mineral deposits are in orogens formed by the collision and accretion of volcanic arcs during the assembly of supercontinents, and this setting hosts nearly all VMS, lode gold, porphyry, and komatiite-associated Ni-Cu deposits. Exclusive of the Sudbury impact structure, mineral deposits of orogens have accounted for 81% of production with 51% coming from Archean greenstone belts. Deposits of epicontinental and intracontinental sedimentary basins account for 16% of production, mainly from Proterozoic U and SEDEX deposits, and Phanerozoic SEDEX and MVT deposits. Deposits associated with anorogenic mafic-ultramafic magmatism, including kimberlite intrusion, account for the remaining 3%.

There has been no production from SEDEX and MVT since 2001, when the Sullivan SEDEX mine closed. Over the past decade the number of VMS and lode gold mine closures has greatly outstripped mine openings for both deposit types, so that at current rates of production reserves for both types will be depleted within ten years. A single mineral deposit is a finite mining resource and it may take decades to unravel the geological complexity of an area to discover new resources to replace them. Providing the geoscience infrastructure necessary for the discovery of new mineral deposits in such a large country as Canada is the key to narrowing the search areas and to attracting the exploration investment to bring about a replenishment of mineral resources.

Acknowledgements

Many colleagues of the Consolidation of Canada's Geoscience Knowledge Minerals Synthesis Project contributed either directly or indirectly to this overview of Canada's mineral resources by freely sharing their expert knowledge and compilations of the various mineral deposit types. The list would include nearly all authors of this volume. Judy Goodwin is thanked for her technical editing and M.D. Thomas and W.D. Goodfellow for their reviews of an early version of the manuscript.

References

Ames, D.E., and Farrow, C.E.G., 2007, Metallogeny of the Sudbury mining camp, Ontario, *in* Goodfellow, W.D., ed., Mineral Deposits of Canada: A Synthesis of Major Deposit-Types, District Metallogeny, the Evolution of Geological Provinces, and Exploration Methods: Geological Association of Canada, Mineral Deposits Division, Special Publication No. 5, p. 329-350.

Andrews, A.J., Masliwec, A., Morris, W.A., Owsiacki, L.R., and York, D., 1986, The silver deposits at Cobalt and Gowganda, Ontario. II: An experiment in age determinations employing radiometric and paleomagnetic measurements: Canadian Journal of Earth Sciences, v. 23, p. 1507-1518.

Antweiler, W., 2006, Pacific Exchange Rate Service, University of British Columbia, Sauder School of Business, http://fx.sauder.ubc.ca/ (accessed August 2006)

Barley, M.E., and. Groves, D.I., 1992, Supercontinent cycles and the distribution of metal deposits through time: Geology, v. 20, p. 291-294.

Barnes, S-J., and Lightfoot, P.C., 2005, Formation of magmatic nickel sulfide deposits and processes affecting their copper and platinum group element contents, *in* Hedenquist, J.W., Thompson, J.F.H., Goldfarb, R.J., and Richards, J.P., eds., 100th Anniversary Volume: Economic Geology, p. 179-213.

Barrie, C.T., Ludden, J.N., and Green, T.H., 1993, Geochemistry of volcanic rocks associated with Cu-Zn and Ni-Cu deposits in the Abitibi Subprovince: Economic Geology, v. 88, p. 1341-1358.

Barton, M.D., and Johnson, D.A., 2000, Alternative brine sources for Fe oxide (-Cu-Au) systems: Implications for hydrothermal alteration and metals, *in* Porter, T.M., ed., Hydrothermal Iron Oxide Copper-Gold and Related Deposits: A Global Perspective: Australian Mineral Foundation, Adelaide, v. 2, p. 43-60.

Bédard, J.H., Pagé, P., Bécu, V., Schroetter, J.-M., and Tremblay, A., 2007, Overview of the geology and Cr-PGE potential of the Southern Québec Ophiolite Belt, *in* Goodfellow, W.D., ed., Mineral Deposits of Canada: A Synthesis of Major Deposit-Types, District Metallogeny, the Evolution of Geological Provinces, and Exploration Methods: Geological Association of Canada, Mineral Deposits Division, Special Publication 5, p. 433-448.

Birchfield, G., 2006, Canadian overview, *in* Canadian Minerals Yearbook, Review and Outlook: Natural Resources Canada, Mining and Metals Sector, p. 1-6.

Bleeker, W., 2003, The late Archean record: a puzzle in ca. 35 pieces: Lithos, v. 71, p. 99-134.

Bleeker, W., and Hall, B., 2007, The Slave Craton: Geology and metallogenic evolution, *in* Goodfellow, W.D., ed., Mineral Deposits of Canada: A Synthesis of Major Deposit-Types, District Metallogeny, the Evolution of Geological Provinces, and Exploration Methods: Geological Association of Canada, Mineral Deposits Division, Special Publication No. 5, p. 849-879.

Bouchard, G., 2006, Canadian exploration scene, *in* Canadian Minerals Yearbook, Review and Outlook: Natural Resources Canada, Mining and Metals Sector, p. 7-10.

Boyd, F.R, Gurney, J.J., and Richardson, S.H., 1985, Evidence for a 150-200 km thick Archaean lithosphere from diamond inclusion thermobarometry: Nature, v. 315, p. 387-389.

Brummer, J.J., 1978, Diamonds in Canada: Transactions of the Canadian Institute of Mining and Metallurgy, v. 81, p. 144-159.

Card, K.D., and Poulsen, K.H., 1998, Geology and mineral deposits of the Superior Province of the Canadian Shield: Chapter 2, *in* Lucas, S.B., and St-Onge, M.R., co-ordinators, Geology of the Precambrian Superior and Grenville Provinces and Precambrian Fossils in North America: Geological Survey of Canada, Geology of Canada, No. 7, p. 13-194 (also Geological Society of America, The Geology of North America, v. C-1).

Carter, N.C., 2006, Report on Revised Estimates of Mineral Resources Ferguson Lake Nickel-Copper-Cobalt-PGE Property, Ferguson Lake Area, Kivalliq Region, Nunavut Territory: prepared for Starfield Resources Inc., SEDAR.COM filing, 145 p.

Carter, T.R., and Colvine, A.C., 1985, Metallic mineral deposits of the Grenville province, southeastern Ontario: Canadian Institute of Mining and Metallurgy Bulletin, v. 78, p. 95-106.

Chacko, T., De, S.K., Creaser, R.A., and Muehlenbachs, F., 2000, Tectonic setting of the Taltson magmatic zone at 1.9-2.0 Ga: A granitoid-based perspective: Canadian Journal of Earth Sciences, v. 37, p. 1597-1609.

Clifford, T.N., 1966, Tectono-metallogenic units and metallogenic provinces of Africa: Earth and Planetary Science Letters, v. 1, p. 421-434.

Corrigan, D., Galley, A.G., and Pehrsson, S., 2007, Tectonic evolution and metallogeny of the southwestern Trans-Hudson Orogen, *in* Goodfellow, W.D., ed., Mineral Deposits of Canada: A Synthesis of Major Deposit-types, District Metallogeny, the Evolution of Geological Provinces, and Exploration Methods: Geological Association of Canada, Mineral Deposits Division, Special Publication 5, p. 881-902.

Corriveau, L., 2007, Iron oxide copper-gold deposits: a Canadian perspective, *in* Goodfellow, W.D., ed., Mineral Deposits of Canada: A Synthesis of Major Deposit-Types, District Metallogeny, the Evolution of Geological Provinces, and Exploration Methods: Geological Association of Canada, Mineral Deposits Division, Special Publication 5, p. 307-328.

Corrigan, D., Galley, A.G., and Pehrsson, S., 2007, Tectonic evolution and metallogeny of the southwestern Trans-Hudson Orogen, *in* Goodfellow, W.D., ed., Mineral Deposits of Canada: A Synthesis of Major Deposit-Types, District Metallogeny, the Evolution of Geological Provinces, and Exploration Methods: Geological Association of Canada, Mineral Deposits Division, Special Publication 5, p. 881-902.

Cranstone, D.A., 2002, A History of Mining and Mineral Exploration in Canada and an Outlook for the Future: Natural Resources Canada, Metals and Mining Sector, 47 p.

David, J., Parent, M., Stevenson, R., Nadeau, P., and Godin, L., 2003, The Porpoise Cove supracrustal sequence, Inukjuak area: A unique example of Paleoarchean crust (ca. 3.8 Ga) in the Superior Province: Geological

Association of Canada, Program with Abstracts, 28 p. (CD-ROM).

Davidson, A., 1998, An overview of Grenville Province geology, Canadian Shield; Chapter 3, *in* Lucas, S.B., and St-Onge, M.R., co-ordinators, Geology of the Precambrian Superior and Grenville provinces and Precambrian Fossils in North America: Geological Survey of Canada, Geology of Canada, No. 7, p. 205-270 (also Geological Society of America, The Geology of North America, v. C-1).

Dawson, K.M., 1996, Skarn tungsten, *in* Eckstrand, O.R., Sinclair, W.D., and Thorpe, R.I., eds., Geology of Canadian Mineral Deposit Types: Geological Survey of Canada, Geology of Canada, No. 8, p. 495-502 (also Geological Society of America, The Geology of North America, v. P-1).

Dawson, K.M., and Kirkham, R.V., 1996, Skarn copper in Eckstrand, O.R., Sinclair, W.D., and Thorpe, R.I., eds., Geology of Canadian Mineral Deposit Types: Geological Survey of Canada, Geology of Canada, No. 8: p. 460-476 (also Geological Society of America, The Geology of North America, v. P-1).

De, S.K, Chacko, T, Creaser, R.A., and Muehlenbachs, F., 2000, Geochemical and Nd-Pb-O isotope systematics of granites from the Taltson magmatic zone, NE Alberta: Implications for early Proterozoic tectonics in western Laurentia: Precambrian Research, v. 102, p. 221-249.

Dewing, K., Turner, E., and Harrison, J.C., 2007, Geological history, mineral occurrences and mineral potential of the sedimentary rocks of the Canadian Arctic Archipelago, *in* Goodfellow, W.D., ed., Mineral Deposits of Canada: A Synthesis of Major Deposit-Types, District Metallogeny, the Evolution of Geological Provinces, and Exploration Methods: Geological Association of Canada, Mineral Deposits Division, Special Publication No. 5, p. 733-753.

Dubé, B., and Gosselin, P., 2007, Greenstone-hosted quartz-carbonate vein deposits, *in* Goodfellow, W.D., ed., Mineral Deposits of Canada: A Synthesis of Major Deposit-Types, District Metallogeny, the Evolution of Geological Provinces, and Exploration Methods: Geological Association of Canada, Mineral Deposits Division, Special Publication No. 5, p. 49-73.

Dubé, B., Gosselin, P. Mercier-Langevin, P., Hannington, M., and Galley, A., 2007, Gold-rich volcanogenic massive sulphide deposits, *in* Goodfellow, W.D., ed., Mineral Deposits of Canada: A Synthesis of Major Deposit-Types, District Metallogeny, the Evolution of Geological Provinces, and Exploration Methods: Geological Association of Canada, Mineral Deposits Division, Special Publication No. 5, p. 75-94.

Eckstrand, O.R., 1996, Nickel-copper sulphide, *in* Eckstrand, O.R., Sinclair, W.D., and Thorpe, R.I., eds., Geology of Canadian Mineral Deposit Types: Geological Survey of Canada, Geology of Canada, No. 8, p. 584-605 (also Geological Society of America, The Geology of North America, v. P-1).

Eckstrand, O.R., and Hulbert, L., 2007, Magmatic Nickel-Copper-Platinum Group Element Deposits, *in* Goodfellow, W.D., ed., Mineral Deposits of Canada: A Synthesis of Major Deposit-Types, District Metallogeny, the Evolution of Geological Provinces, and Exploration Methods: Geological Association of Canada, Mineral Deposits Division, Special Publication No. 5, p. 205-222.

Einaudi, M.T., 1982, Description of skarns associated with porphyry copper plutons, *in* Titley, S.R., ed., Advances in Geology of the Porphyry Copper Deposits: University of Arizona Press, Tuscon, Arizona, p. 139-183.

Ernst, R.E., 2007, Large igneous provinces in Canada through time and their metallogenic potential, *in* Goodfellow, W.D., ed., Mineral Deposits of Canada: A Synthesis of Major Deposit-Types, District Metallogeny, the Evolution of Geological Provinces, and Exploration Methods: Geological Association of Canada, Mineral Deposits Division, Special Publication 5, p. 929-937.

Foley, S.S., Venurelli, G., Green, D.H., and Toscani, L., 1987, The ultrapotassic rocks: Characteristics, classification and constraints for petrogenetic models: Earth Science Reviews, v. 24, p. 81.

Franklin, J.M., Gibson, H.L., Jonasson, I.R., and Galley, A.G., 2005, Volcanogenic massive sulfide deposits, *in* Hedenquist, J.W., Thompson, J.F.H., Goldfarb, R.J., and Richards, J.P., eds., 100th Anniversary Volume: Economic Geology, p. 523-560.

Galley, A.G., Hannington, M.D., and Jonasson, I.R., 2007a, Volcanogenic massive sulphide deposits, *in* Goodfellow, W.D., ed., Mineral Deposits of Canada: A Synthesis of Major Deposit-Types, District Metallogeny, the Evolution of Geological Provinces, and Exploration Methods: Geological Association of Canada, Mineral Deposits Division, Special Publication No. 5, p. 141-161.

Galley, A.G., Syme, R., and Bailes, A. 2007b, Metallogeny of the Paleoproterozoic Flin Flon Belt, Manitoba and Saskatchewan, *in* Goodfellow, W.D., ed., Mineral Deposits of Canada: A Synthesis of Major Deposit-Types, District Metallogeny, the Evolution of Geological Provinces, and Exploration Methods: Geological Association of Canada, Mineral Deposits Division, Special Publication No. 5, p. 509-531.

Gandhi, S.S., and Bell, R.T., 1996, Volcanic-associated uranium, *in* Eckstrand, O.R., Sinclair, W.D., and Thorpe, R.I., eds., Geology of Canadian Mineral Deposit Types: Geological Survey of Canada, Geology of Canada, No. 8: p. 269-276 (also Geological Society of America, The Geology of North America, v. P-1).

Garven, G., 1985, The role of regional flow in the genesis of the Pine Point deposit, Western Canada sedimentary basin: Economic Geology, v. 80, p. 307-324.

Gauthier, M., and Chartrand, F., 2005, Metallogeny of the Grenville Province revisited: Canadian Journal of Earth Sciences, v. 42, p. 1719-1734.

Gibson, H., and Galley, A.G., 2007, Volcanogenic massive sulphide deposits of the Archean, Noranda District, Quebec, *in* Goodfellow, W.D., ed., Mineral Deposits of Canada: A Synthesis of Major Deposit-Types, District Metallogeny, the Evolution of Geological Provinces, and Exploration Methods: Geological Association of Canada, Mineral Deposits Division, Special Publication No. 5, p. 533-552.

Goldfarb, R.J., Baker, T., Dubé, B., Groves, D.I., Hart, C.J.R., and Gosselin, P., 2005, Distribution, character, and genesis of gold deposits in metamorphic terranes, *in* Hedenquist, J.W., Thompson, J.F.H., Goldfarb, R.J., and Richards, J.P., eds., 100th Anniversary Volume: Economic Geology, p. 405-450.

Goodfellow, W.D., 2007a, Metallogeny of the Bathurst mining camp, northern New Brunswick, *in* Goodfellow, W.D., ed., Mineral Deposits of Canada: A Synthesis of Major Deposit-Types, District Metallogeny, the Evolution of Geological Provinces, and Exploration Methods: Geological Association of Canada, Mineral Deposits Division, Special Publication No. 5, p. 449-469.

——— 2007b, Base metal metallogeny of the Selwyn Basin, Canada, *in* Goodfellow, W.D., ed., Mineral Resources of Canada: A Synthesis of Major Deposit-Types, District Metallogeny, the Evolution of Geological Provinces, and Exploration Methods: Geological Association of Canada, Mineral Deposits Division, Special Publication 5, p. 553-579.

Goodfellow, W.D., and Lydon, J.W., 2007, Sedimentary exhalative deposits, *in* Goodfellow, W.D., ed., Mineral Deposits of Canada: A Synthesis of Major Deposit-Types, District Metallogeny, the Evolution of Geological Provinces, and Exploration Methods: Geological Association of Canada, Mineral Deposits Division, Special Publication No. 5, p. 163-183.

Gosselin, P., and Dubé, B., 2005, Gold Deposits of Canada: Distribution, Geological Parameters and Gold Content: Geological Survey of Canada, Open File 4896, 105 p.

Gower, C.F., and Krogh, T.E., 2002, A U-Pb geochronological review of the Proterozoic history of the eastern Grenville Province: Canadian Journal of Earth Sciences, v. 39, p. 795-829.

Groves, D.I., Goldfarb, R.J., Robert, F., and Hart, C.J., 2003, Gold deposits in metamorphic belts: Overview of current understanding, outstanding problems, future research, and exploration significance: Economic Geology, v. 98, p. 1-29.

Groves, D.I., Condie, K.C., Goldfarb, R.J., Hronsky, J.M.A., and Veilreicher, R.M., 2005, Secular changes in global tectonic processes and their influence on the temporal distribution of gold-bearing mineral deposits: Economic Geology, v. 100, p. 203-224.

Gurney, J.J., Helmstaedt, H.H., Le Roex, A.P., Nowicki, T.E., Richardson, S.H., and Westerlund, K.J., 2005, Diamonds: Crustal distribution and formation processes in time and space and an integrated deposit model, *in* Hedenquist, J.W., Thompson, J.F.H., Goldfarb, R.J., and Richards, J.P., eds., 100th Anniversary Volume: Economic Geology, p. 143-177.

Hannigan, P., 2007, Metallogeny of the Pine Point Mississippi Valley-type zinc-lead district, southern Northwest Territories, *in* Goodfellow, W.D., ed., Mineral Deposits of Canada: A Synthesis of Major Deposit-Types, District Metallogeny, the Evolution of Geological Provinces, and Exploration Methods: Geological Association of Canada, Mineral Deposits Division, Special Publication No. 5, p. 609-632.

Hart, C., 2007, Reduced Intrusion-related gold systems, *in* Goodfellow, W.D., ed., Mineral Deposits of Canada: A Synthesis of Major Deposit-Types, District Metallogeny, the Evolution of Geological Provinces, and Exploration Methods: Geological Association of Canada, Mineral Deposits Division, Special Publication 5, p. 95-112.

Hawkesworth, C.J., and Kemp, I.S., 2006, Evolution of the continental crust: Nature, v. 443, p. 811-817.

Helmstaedt, H.H., and Gurney, J.J., 1995, Geotectonic controls of primary diamond deposits: Journal of Geochemical Exploration, v. 53, p. 125-144.

Hitzman, M.W., Oreskes, N., and Einaudi, M.T., 1992, Geological characteristics and tectonic setting of Proterozoic iron oxide (Cu-U-Au-REE) deposits: Precambrian Research, v. 58, p. 241-287.

Hodgson, C.J., 1989, The structure of shear-related vein-type gold deposits: A review: Ore Geology Reviews, v. 4, p. 231-273.

Hoffman, P.F., 1988, United Plates of America, the birth of a craton: Early Proterozoic assembly and growth of Laurentia: Annual Reviews of Earth and Planetary Sciences, v. 16, p. 543-603.

——— 1989, Precambrian geology and tectonic history of North America, *in* Bally, A.W., and Palmer, A.R., eds., The Geology of North America – An Overview: Geological Society of America, Decade of North American Geology, v. A, p. 447-511.

——— 1990, Subdivision of the Churchill Province and extent of the Trans-Hudson Orogen of North America, *in* Lewry, J.F., and Stauffer, M.R., eds., The Early Proterozoic Trans-Hudson Orogen of North America: Geological Association of Canada, Special Paper 37, p. 15-39.

Höy, T., 1982, Stratigraphic and structural setting of stratabound lead-zinc deposits in southeastern B.C.: Canadian Institute of Mining and Metallurgy Bulletin, v. 75, p. 114-134.

Hunt, J., Baker, T., and Thorkelson, D., 2005, Regional-scale Proterozoic IOCG-mineralized breccia systems: Examples from the Wernecke Mountains, Yukon, Canada: Mineralium Deposita, v. 40, p. 492-514.

Ivanhoe Mines Ltd., 2006, Annual Information Form for 2005: Canadian Securities Administrators filing, SEDAR.com, 80 p.

Jones, B.K., 1992, Application of metal zoning to gold exploration in porphyry copper systems: Journal of Geochemical Exploration, v. 43, p. 127-155.

Janse, A.J.A., 1984, Kimberlites – where and when, *in* Glover, J.E., and Harris, P.G., eds., Kimberlite Occurrence and Origin: A Basis for Conceptual Models in Exploration: University of Western Australia, Geology Department and Extension Services, Publication No.8, p. 19-62.

Jorgenson, J.D., 2004, Zinc, *in* Minerals Yearbook: Volume I – Metals and Minerals: United States Geological Survey, p. 84.1-84.6.

Jefferson, C.W., Thomas, D.J., Gandhi, S.S., Ramaekers, P., Delaney, G., Brisbin, D., Cutts, C., Quirt, D., Portella, P., and Olson, R.A., 2007, Unconformity-associated uranium deposits of the Athabasca Basin, Saskatchewan and Alberta, *in* Goodfellow, W.D., ed., Mineral Deposits of Canada: A Synthesis of Major Deposit-Types, District Metallogeny, the Evolution of Geological Provinces, and Exploration Methods: Geological Association of Canada, Mineral Deposits Division, Special Publication 5, p. 273-305.

Kemp, A.I.S., Hawkesworth, C.J., Paterson, B.A., and Kinny, P.D., 2006, Episodic growth of the Gondwana supercontinent from from hafnium and oxygen isotope ratios: Nature, v. 439, p. 580-583.

Kerrich, R., and Polat, A., 2006, Archean greenstone-tonalite duality: Thermochemical mantle convection models or plate tectonics in the early Earth global dynamics?: Tectonophysics, v. 415, p. 141-165.

Kirkham, R.V., 1972, Porphyry deposits, *in* Report of Activities: Geological Survey of Canada, Paper 72-1, Part B, p. 62-64.

——— 1996a, Sediment-hosted stratiform copper, *in* Eckstrand, O.R, Sinclair, W.D., and Thorpe, R.I., eds., Geology of Canadian Mineral Deposit Types: Geological Survey of Canada, Geology of Canada, No. 8, p. 223-240 (also Geological Society of America, The Geology of North America, v. P-1).

——— 1996b, Volcanic redbed copper, *in* Eckstrand, O.R., Sinclair, W.D., and Thorpe, R.I., eds., Geology of Canadian Mineral Deposit Types: Geological Survey of Canada, Geology of Canada, No. 8, p. 241-252 (also Geological Society of America, The Geology of North America, v. P-1).

Kirkham, R.V., and Sinclair, W.D., 1995, Porphyry copper, gold, molybdenum, tungsten, tin, silver, *in* Eckstrand, O.R., Sinclair, W.D., and Thorpe, R.I., eds., Geology of Canadian Mineral Deposit Types: Geological Survey of Canada, Geology of Canada, No. 8, p. 421-446 (also Geological Society of America, The Geology of North America, v. P-1).

Kjarsgaard, B.A., 2007, Kimberlite diamond deposits, *in* Goodfellow, W.D., ed., Mineral Deposits of Canada: A Synthesis of Major Deposit-Types, District Metallogeny, the Evolution of Geological Provinces, and Exploration Methods: Geological Association of Canada, Mineral Deposits Division, Special Publication 5, p. 245-271.

Kontak, D.J., McBride, S., and Farrar, E., 1994, $^{40}Ar/^{39}Ar$ dating of fluid migration in a Mississippi Valley-type deposit: Gays River Zn-Pb deposit, Nova Scotia: Economic Geology, v. 89, p. 1501-1517.

Layton-Matthews, D., Lesher, C.M., Burnham, O.M., Liwanag, J., Halden, N.M., Hulbert, L., and Peck, D.C., 2007, Magmatic Ni-Cu-platinum-group element deposits of the Thompson Nickel Belt, *in* Goodfellow, W.D., ed., Mineral Deposits of Canada: A Synthesis of Major Deposit-Types, District Metallogeny, the Evolution of Geological Provinces, and Exploration Methods: Geological Association of Canada, Mineral Deposits Division, Special Publication 5, p. 409-432.

Leach, D.L., Bradley, D., Lewchuk, M., Symons, D.T.A., Premo, W., de Marsily, G., and Brannon, J., 2001, Mississippi Valley-type lead-zinc deposits through geological time: Implications from recent age-dating research: Mineralium Deposita, v. 36, p. 7111-740.

Leach, D.L., Sangster, D.F., Kelly, K.D., Large, R.R., Garven, G., Allen, C.R., Gutzmer, J., and Walters, S., 2005, Sediment-hosted lead-zinc: a global perspective, *in* Hedenquist, J.W., Thompson, J.F.H., Goldfarb, R.J., and Richards, J.P., eds., 100th Anniversary Volume: Economic Geology, p. 561-607.

Leadbeater, D., 1998, The development of Elliot Lake,"Uranium capital of the world": A background to the layoffs of 1990–1996: Elliot Lake tracking study: Laurentian University, Institute for Northern Ontario Research and Development, ELTAS Analysis Series, No.1A19, 50 p.

Lesher, C.M., 2007, Ni-Cu-(PGE) Deposits in the Raglan Area, Cape Smith Belt, New Québec, *in* Goodfellow, W.D., ed., Mineral Deposits of Canada: A Synthesis of Major Deposit-Types, District Metallogeny, the Evolution of Geological Provinces, and Exploration Methods: Geological Association of Canada, Mineral Deposits Division, Special Publication 5, p. 351-386.

Lydon, J.W., 1996, Characteristics of volcanogenic massive sulphides: Interpretations in terms of hydrothermal convection systems and magmatic hydrothermal systems: Boletin Geologico y Minero, v. 107, p. 215-264.

——— 2007, Geology and metallogeny of the Belt-Purcell basin, *in* Goodfellow, W.D., ed., Mineral Deposits of Canada: A Synthesis of Major Deposit-Types, District Metallogeny, the Evolution of Geological Provinces, and Exploration Methods: Geological Association of Canada, Mineral Deposits Division, Special Publication 5, p. 581-607.

Lydon, J.W., Goodfellow, W.D., Dubé, B., Sinclair, W.D., Paradis, S., Corriveau, L.C., and Gosselin, P., 2004, A Preliminary Overview of Canada's Mineral Resources: Geological Survey of Canada, Open File 4668, 27 p.

Lynch, J.V.G., 1989, Large-scale hydrothermal zoning reflected in the tetrahedrite-freibergite solid solution, Keno Hill Ag-Pb-Zn District, Yukon: Canadian Mineralogist, v. 27, p. 384-400.

Lynch, J.V.G., Longstaffe, F.J., and Nesbitt, B.E. 1990. Stable isotope and fluid inclusion indications of large-scale hydrothermal paleoflow, boiling, and fluid mixing in the Keno Hill Ag–Pb–Zn district, Yukon Territory, Canada: Geochimica et Cosmochimica Acta, v. 54, p. 1045-1059.

Mackenzie, D.H., and Davies, R.H., 1990. Broken Hill lead-silver-zinc deposit at Z.C. mines, *in* Hughes, F.E., ed., Geology of the Mineral Deposits of Australia and Papua New Guinea: Australasian Institute of Mining and Metallurgy, Monograph Series No. 14, p. 1079-1084.

Marshall, D., and Watkinson, D.H., 2000, The Cobalt mining district: Silver sources, transport and deposition: Exploration and Mining Geology, v. 9, p. 81-92.

Martin, C., and Dickin, A.P., 2005, Styles of Proterozoic crustal growth on the southeast margin of Laurentia: Evidence from the central Grenville Province northwest of Lac St.-Jean, Quebec: Canadian Journal of Earth Sciences, v. 42, p. 1643-1652.

Mazimhaka, P., and Hendry, H., 1989, Uranium deposits in the Beaverlodge area, northern Saskatchewan: Their relationship to the Martin Group (Proterozoic) and the underlying basement, *in* Uranium Resources and Geology of North America: IAEA Technical Document 500, Vienna, p. 297-315.

McCulloch, M.T., and Bennett, V.C, 1994, Progressive growth of the Earth's continental crust and depleted mantle: Geochemical constraints: Geochimica et Cosmochimica Acta, v. 58, p. 4717-4738.

McMullen, M., and Birchfield, G., 2005, General review, *in* Canada Minerals Yearbook: Natural Resources Canada, p. 1.1-1.20.

Mercier-Langevin, P., Dubé, B., Lafrance, B., Hannington, M., Galley, A., Moorhead, J., and Gosselin, P., 2007, Metallogeny of the Doyon-Bousquet-LaRonde mining camp, Abitibi Greenstone Belt, Quebec, *in* Goodfellow, W.D., ed., Mineral Deposits of Canada: A Synthesis of Major Deposit-Types, District Metallogeny, the Evolution of Geological Provinces, and Exploration Methods: Geological Association of Canada, Mineral Deposits Division, Special Publication 5, p. 673-701.

Morgan, R., 2005, Going underground: *Mining Journal*, January 14th, p. 2.

Mitchell, R.H., 1986, Kimberlites: Mineralogy, Geochemistry and Petrology: Plenum Press, New York, 422 p.

Naldrett, A.J., and Li, C., 2007, The Voisey's Bay deposit, Labrador, Canada, *in* Goodfellow, W.D., ed., Mineral Deposits of Canada: A Synthesis of Major Deposit-Types, District Metallogeny, the Evolution of Geological Provinces, and Exploration Methods: Geological Association of Canada, Mineral Deposits Division, Special Publication 5, p. 387-407.

Natural Resources Canada, 1990, Canadian Mineral Deposits Not Being Mined in 1989: Mining and Metals Sector, Publication MR 223, unpaginated.

——— 2004, Tungsten, *in* Canada Minerals Year Book: Natural Resources Canada, Minerals and Metals Sector, p. 57.1-57.5.

——— 2005, Lead, *in* Mineral and Metal Commodity Reviews: Natural Resources Canada, Minerals and Metals Sector, p. 29.1-29.11.

——— 2006, Canada's Minerals and Metals Industry: An Economic Overview 2005: Natural Resources Canada, Minerals and Metals Sector, www.nrcan.gc.ca/mms/pdf/econo05_e.pdf, 15 p.

Nelson, J., and Colpron, M., 2007, Tectonics and metallogeny of the British Columbia, Yukon and Alaskan Cordillera, 1.8 Ga to the present, *in* Goodfellow, W.D., ed., Mineral Deposits of Canada: A Synthesis of Major Deposit-Types, District Metallogeny, the Evolution of Geological Provinces, and Exploration Methods: Geological Association of Canada, Mineral Deposits Division, Special Publication 5, p. 755-791.

NovaGold Resources Inc., 2006, Annual Information Form for 2005: Canadian Securities Administrators filing, SEDAR.com, 50 p.

Paradis, S., Hannigan, P., and Dewing, K., 2007, Mississippi Valley-Types lead-zinc deposits, *in* Goodfellow, W.D., ed., Mineral Deposits of Canada: A Synthesis of Major Deposit-Types, District Metallogeny, the Evolution of Geological Provinces, and Exploration Methods: Geological Association of Canada, Mineral Deposits Division, Special Publication 5, p. 185-203.

Percival, J.A., 2007, Geology and metallogeny of the Superior Province, Canada, *in* Goodfellow, W.D., ed., Mineral Deposits of Canada: A Synthesis of Major Deposit-Types, District Metallogeny, the Evolution of Geological Provinces, and Exploration Methods: Geological Association of Canada, Mineral Deposits Division, Special Publication 5, p. 903-928.

Perelló, J., Cox, D., Garamjav, D., Sanjdorj, S., Diakov, S., Schissel, D., Munkhbat, T-O., and Oyun, G., 2001, Oyu Tolgoi, Mongolia: Siluro-Devonian porphyry Cu-Au-(Mo) and high-sulfidation Cu mineralization with a Cretaceous chalcocite blanket: Economic Geology, v. 96, p. 1407-1428.

Perkins, W.G., 1990, Mount Isa copper orebodies, *in* Hughes, F.E., ed., Geology of the Mineral Deposits of Australia and Papua New Guinea: Australasian Institute of Mining and Metallurgy, Monograph Series No. 14, p. 935-941.

Perron, L., 2006, Diamonds, *in* Mineral Year Book Annual Review 2005: Natural Resources Canada, Mines and Minerals Sector, p. 17-20.

Peter, J.M., Layton-Matthews, D., Piercey, S., Bradshaw, G., Paradis, S., and Boulton, A., 2007, Volcanic-hosted massive sulphide deposits of the Finlayson Lake District, Yukon *in* Goodfellow, W.D., ed., Mineral Deposits of Canada: A Synthesis of Major Deposit-Types, District Metallogeny, the Evolution of Geological Provinces, and Exploration Methods: Geological Association of Canada, Mineral Deposits Division, Special Publication 5, p. 471-508.

Plunkett, P.A., and Jones, T.S., 1999, Metal Prices in the United States through 1998: United States Geological Survey, 183 p.

Poulsen, K.H., Robert, F., and Dubé, B., 2000, Geological Classification of Canadian Gold Deposits: Geological Survey of Canada, Bulletin 540, 106 p.

Rivers, T., Ketchum, J., Indares, A., and Hynes, A., 2002, The high pressure belt in the Grenville Province: Architecture, timing, and exhumation: Canadian Journal of Earth Sciences, v. 39, p. 867-893.

Robert, F., 1996, Quartz-carbonate vein gold, *in* Eckstrand, O.R., Sinclair, W.D., and Thorpe, R.I., eds., Geology of Canadian Mineral Deposit Types: Geological Survey of Canada, Geology of Canada, No. 8, p. 350-366 (also Geological Society of America, The Geology of North America, v. P-1).

Roberts, D.E., and Hudson, G.R.T., 1983, The Olympic Dam copper-uranium-gold deposit, Roxby Downs, South Australia: Economic Geology, v. 78, p. 799-822.

Rogers, J.J.W., 1996, A history of continents in the past three billion years: Journal of Geology, v. 104, p. 91-107.

Rogers, J.J.W., and Santosh, M., 2003, Supercontinents in Earth history: Gondwana Research, v. 6, p. 357-368.

Roscoe, S.M., 1996, Paleoplacer uranium, gold, *in* Eckstrand, O.R., Sinclair, W.D., and Thorpe, R.I., eds., Geology of Canadian Mineral Deposit Types: Geology of Canada, No. 8, p. 10-23 (also Geological Society of America, The Geology of North America, v. P-1).

Ruzicka, V., 1996, Veins in shear zones, *in* Eckstrand, O.R., Sinclair, W.D., and Thorpe, R.I., eds., Geology of Canadian Mineral Deposit Types: Geology of Canada, No. 8, p. 278-283 (also Geological Society of America, The Geology of North America, v. P-1).

Ruzicka, V., and Thorpe, R.I., 1996a, Arsenide silver-cobalt veins, *in* Eckstrand, O.R., Sinclair, W.D., and Thorpe, R.I., eds., Geology of Canadian Mineral Deposit Types: Geology of Canada, No. 8, p. 288-296 (also Geological Society of America, The Geology of North America, v. P-1).

——— 1996b, Arsenide vein uranium-silver, *in* Eckstrand, O.R., Sinclair, W.D., and Thorpe, R.I., eds., Geology of Canadian Mineral Deposit Types: Geology of Canada, No. 8, p. 296-306 (also Geological Society of America, The Geology of North America, v. P-1).

Sangster, D.F., 1980, Quantitative characteristics of volcanogenic massive sulphide deposits, Part 1: Metal content and size distribution of massive sulphide deposits in volcanic centres: Canadian Institute of Mining and Metallurgy Bulletin, v. 73, p. 74-81.

Publication 5, p. 971-982.

Sangster, A.L., and Smith, P.K., 2007, Metallogenic summary of the Meguma gold deposits, Nova Scotia, *in* Goodfellow, W.D., ed., Mineral Deposits of Canada: A Synthesis of Major Deposit-Types, District Metallogeny, the Evolution of Geological Provinces, and Exploration Methods: Geological Association of Canada, Mineral Deposits Division, Special Publication 5, p. 723-732.

Sangster, A.L., Douma, S.L., and Lavigne, J., 2007, Base metal and gold deposits of the Betts Cove Complex, Baie Verte Peninsula, Newfoundland, *in* Goodfellow, W.D., ed., Mineral Deposits of Canada: A Synthesis of Major Deposit-Types, District Metallogeny, the Evolution of Geological Provinces, and Exploration Methods: Geological Association of Canada, Mineral Deposits Division, Special Publication 5, p. 703-721.

Saskatchewan Interactive, 2006, Uranium History by Decade: http://interactive.usask.ca/ski/mining/search/mineral_types/ energy/uranium/history_uranium.html.

Sears, J.W., and Price, R.A., 2000, New look at the Siberian connection; no SWEAT: Geology, v. 28, p. 423-426.

——— 2003, Tightening the Siberian connection to western Laurentia: Bulletin of the Geological Society of America, v. 115, p. 943-953.

Seedorf, E., Dilles, J.H., Profett Jr., J.M., Einaudi, M.T., Zurcher, L., Stavast, W.J.A., Johnson, D.A., and Barton, M.D., 2005, Porphyry deposits: Characteristics and origin of hypogene features, *in* Hedenquist, J.W., Thompson, J.F.H., Goldfarb, R.J., and Richards, J.P., eds., 100th Anniversary Volume: Economic Geology, p. 251-298.

Simmons, S.F., White, N.C., and John, D.A., 2005, Geological characteristics of epithermal precious and base metal deposits, *in* Hedenquist, J.W., Thompson, J.F.H., Goldfarb, R.J., and Richards, J.P., eds., 100th Anniversary Volume: Economic Geology, p. 485-522.

Sinclair, W.D., 2007, Porphyry deposits, *in* Goodfellow, W.D., ed., Mineral Deposits of Canada: A Synthesis of Major Deposit-Types, District Metallogeny, the Evolution of Geological Provinces, and Exploration

Methods: Geological Association of Canada, Mineral Deposits Division, Special Publication 5, p. 223-243.

Sinclair, A.J., Tessari, D.J., and Harakal, J.E., 1980, Age of Ag-Pb-Zn mineralization, Keno Hill-Galena Hill area, Yukon Territory: Canadian Journal of Earth Sciences, v. 17, p. 1100-1103.

Skinner, E.M.W., and Clement, C.R., 1979, Mineralogical classification of southern African kimberlites, *in* Boyd, F.R., and Meyer, H.O.A., eds., Kimberlites Diatremes and Diamonds: Their Geology, Petrology and Geochemistry: American Geophysical Union, Washington, p. 129-139.

Smith, G.R., 2003, Lead, *in* Minerals Yearbook: Volume I – Metals and Minerals: United States Geological Survey, p. 43.1-43.9.

St-Onge, M.R., and Lucas, S.B., 1996, Paleoproterozoic orogenic belts, *in* Lecheminant, A.N., Richardson, D.G., Dilabio, R.N.W., and Richardson, K.A., eds., Searching for Diamonds in Canada: Geological Survey of Canada, Open File 3228, p. 17-24.

Symons, D.T.A., and Sangster, S.F., 1992, Late Devonian paleomagnetic age for the Polaris Mississippi Valley-type Zn-Pb deposit, Canadian Arctic Archipelago: Canadian Journal of Earth Sciences, v. 29, p. 15-25.

Taylor, B.E., 1996, Epithermal gold deposits, *in* Eckstrand, O.R., Sinclair, W.D., and Thorpe, R.I., eds., Geology of Canadian Mineral Deposit Types: Geology of Canada, No. 8, p. 329-350 (also Geological Society of America, The Geology of North America, v. P-1).

Taylor, S.R., and McLennan, S.M., 1995, The geochemical evolution of the continental crust: Reviews of Geophysics, v. 33, p. 241-265.

Thorkelson, D.J., Mortensen, J.K., Davidson, G.J., Creaser, R.A., Perez, W.A., and Abbott, J.G., 2001, Mesoproterozoic intrusive breccias in Yukon, Canada: The role of hydrothermal systems in reconstructions of North America and Australia: Precambrian Research, v. 111, p. 31-56.

Udd, J.E., 2000, A Century of Achievement: The Development of Canada's Mineral Industries: Canadian Institute of Mining, Metallurgy and Petroleum, Special Volume 52, 206 p. (Updated appendix listing chronology of events available http://www.nrcan.gc.ca/mms/stude-etudi/chro_e.htm).

United States Geological Survey, 2006, Mineral Commodity Summaries 2006: 199 p.

van Staal, C.R., 2007, Pre-Carboniferous tectonic evolution and metallogeny of the Canadian Appalachians, *in* Goodfellow, W.D., ed., Mineral Deposits of Canada: A Synthesis of Major Deposit-Types, District Metallogeny, the Evolution of Geological Provinces, and Exploration Methods: Geological Association of Canada, Mineral Deposits Division, Special Publication 5, p. 793-818.

Wardle, R.J., and Hall, J., 2002, Proterozoic evolution of the northeastern Canadian Shield: Lithoprobe Eastern Canadian Shield Onshore-Offshore Transect (ESCOOT), introduction and summary: Canadian Journal of Earth Sciences, v. 39, p. 563-567.

Williams, H., Hoffman, P.F., Lewry, J.F., Monger, J.W.H., and Rivers, T., 1991, Anatomy of North America: Thematic geologic portrayals of the continent: Tectonophysics, v. 187, p. 117-134.

Williams, P.J., Barton, M.D., Johnson, D.A., Fontboté, L, de Haller, A., Mark, G., Oliver, N.H.S., and Marschik, R., 2005, Iron oxide copper-gold deposits: Geology, space-time distribution, and possible modes of origin, *in* Hedenquist, J.W., Thompson, J.F.H., Goldfarb, R.J., and Richards, J.P., eds., 100th Anniversary Volume: Economic Geology, p. 5371-405.

Xstrata Report, 2006, Ore Reserves and Mineral Resources: Xstrata public release, 1 March 2006.

Zhao, G., Sun, M., Wilde, S.A., and Li, S., 2004, A Paleo-Mesoproterozoic supercontinent: Assembly, growth and breakup: Earth-Science Reviews, v. 67, p. 91-123.

Greenstone-Hosted Quartz-Carbonate Vein Deposits

Benoît Dubé and Patrice Gosselin

Geological Survey of Canada, 490 de la Couronne, Quebec, Quebec G1K 9A9
Corresponding author's email: bdube@nrcan-rncan.gc.ca

Abstract

Greenstone-hosted quartz-carbonate vein deposits typically occur in deformed greenstone belts of all ages, especially those with variolitic tholeiitic basalts and ultramafic komatiitic flows intruded by intermediate to felsic porphyry intrusions, and sometimes with swarms of albitite or lamprophyre dyke. They are distributed along major compressional to transtensional crustal-scale fault zones in deformed greenstone terranes commonly marking the convergent margins between major lithological boundaries, such as volcano-plutonic and sedimentary domains. The large greenstone-hosted quartz-carbonate vein deposits are commonly spatially associated with fluvio-alluvial conglomerate (e.g. Timiskaming conglomerate) distributed along major crustal fault zones (e.g. Destor Porcupine Fault). This association suggests an empirical time and space relationship between large-scale deposits and regional unconformities.

These types of deposits are most abundant and significant, in terms of total gold content, in Archean terranes. However, a significant number of world-class deposits are also found in Proterozoic and Paleozoic terranes. In Canada, they represent the main source of gold and are mainly located in the Archean greenstone belts of the Superior and Slave provinces. They also occur in the Paleozoic greenstone terranes of the Appalachian orogen and in the oceanic terranes of the Cordillera.

The greenstone-hosted quartz-carbonate vein deposits correspond to structurally controlled complex epigenetic deposits characterized by simple to complex networks of gold-bearing, laminated quartz-carbonate fault-fill veins. These veins are hosted by moderately to steeply dipping, compressional brittle-ductile shear zones and faults with locally associated shallow-dipping extensional veins and hydrothermal breccias. The deposits are hosted by greenschist to locally amphibolite-facies metamorphic rocks of dominantly mafic composition and formed at intermediate depth (5-10 km). The mineralization is syn- to late-deformation and typically post-peak greenschist -facies or syn-peak amphibolite-facies metamorphism. They are typically associated with iron-carbonate alteration. Gold is largely confined to the quartz-carbonate vein network but may also be present in significant amounts within iron-rich sulphidized wall-rock selvages or within silicified and arsenopyrite-rich replacement zones.

There is a general consensus that the greenstone-hosted quartz-carbonate vein deposits are related to metamorphic fluids from accretionary processes and generated by prograde metamorphism and thermal re-equilibration of subducted volcano-sedimentary terranes. The deep-seated, Au-transporting metamorphic fluid has been channelled to higher crustal levels through major crustal faults or deformation zones. Along its pathway, the fluid has dissolved various components - notably gold - from the volcano-sedimentary packages, including a potential gold-rich precursor. The fluid then precipitated as vein material or wall-rock replacement in second and third order structures at higher crustal levels through fluid-pressure cycling processes and temperature, pH and other physico-chemical variations.

Résumé

Les gîtes de filoniens à veines de quartz-carbonates dans des roches vertes reposent généralement au sein de ceintures de roches vertes de tout âge, mais tout particulièrement dans celles qui présentent des basaltes tholéiitiques à texture variolaire et des coulées ultramafiques komatiitiques dans lesquels se sont mis en place des intrusions porphyriques de composition intermédiaire à felsique et, parfois, des essaims de dykes d'albitite ou de lamprophyre. Ces gîtes sont répartis le long d'importantes zones de failles d'échelle crustale formées dans un régime allant de la compression à la transtension, au sein de terrains de roches vertes déformés, où elles coïncident habituellement avec d'importantes limites lithologiques qui témoignent d'une marge convergence, comme celles qui séparent des domaines sédimentaires de domaines volcano-plutoniques. Les plus gros gisements du genre sont souvent associés, sur le plan spatial, à des conglomérats fluvio-alluvionnaires (p. ex. le conglomérat de Timiskaming) répartis le long d'importantes zones de failles d'échelle crustale (p. ex. la faille de Destor-Porcupine). Cette association suppose un lien empirique aussi bien temporel que spatial entre les gros gisements et les discordances régionales.

Les gîtes de ce type sont plus abondants et importants, quant au contenu total en or, dans les terrains archéens. Cependant, de nombreux gisements de calibre mondial reposent aussi dans des terrains protérozoïques et paléozoïques. Au Canada, ils constituent la principale source d'or et sont concentrés dans les ceintures de roches vertes archéennes des provinces du lac Supérieur et des Esclaves, mais on en a aussi découvert dans le terrains de roches vertes paléozoïque de l'orogène des Appalaches et dans les terrains océaniques de la Cordillère.

Ces gîtes constituent des minéralisations épigénétiques à contrôle structural complexe caractérisées par des réseaux simples à complexes de filons de quartz carbonates laminés porteurs d'or produits par le remplissage de failles. Ces filons sont logés dans des failles et des zones de cisaillement à comportement fragile-ductile formées en régime compressif, qui présentent un pendage moyen à fort, auxquels sont associés, par endroits, des brèches hydrothermales et des veines d'extension à faible pendage. Les gîtes, qui se sont formés à des profondeurs intermédiaires (de 5 à 10 km), sont encaissés dans des roches métamorphiques, de composition principalement mafique, du faciès des schistes verts et, par endroits, du faciès des amphibolites. La mise en place de la minéralisation est contemporaine des phases intermédiaires et tardives de la déformation et s'est déroulée après l'atteinte des conditions maximales du métamorphisme au faciès des schistes verts ou lors de l'atteinte des conditions maximales du métamorphisme au faciès des amphibolites. La minéralisation est généralement associée à une altération à carbonates de fer. L'or est en grande partie piégé dans un réseau de filons de quartz-carbonates, mais il est aussi présent en quantités importantes dans les épontes de roches encaissantes sulfurées riches en fer ou de zones des remplacement silicifiées et riches en arsénopyrite.

Dubé, B., and Gosselin, P., 2007, Greenstone-hosted quartz-carbonate vein deposits, *in* Goodfellow, W.D., ed., Mineral Deposits of Canada: A Synthesis of Major Deposit-Types, District Metallogeny, the Evolution of Geological Provinces, and Exploration Methods: Geological Association of Canada, Mineral Deposits Division, Special Publication No. 5, p. 49-73.

On croit que l'existence des gîtes de filons de quartz-carbonates dans des roches vertes est liée à celle de fluides métamorphiques issus de processus d'accrétion, et qu'ils sont le produit d'un métamorphisme prograde et d'une remise en équilibre thermique de terrains volcano-sédimentaires subductés. Les fluides métamorphiques de grande profondeur qui ont transporté l'or se sont élevés dans la croûte en empruntant d'importantes failles ou zones de déformation d'échelle crustale. Le long de leur parcours, ils ont dissous divers éléments, dont l'or, dans les assemblages volcano-sédimentaires, qui pouvaient comprendre un précurseur riche en or. Les fluides ont ensuite précipité sous forme de veines ou ont remplacé les roches encaissantes dans des structures de deuxième et de troisième ordres, à des niveaux crustaux supérieurs, selon une succession de cycles liés à des variations de la pression hydrostatique, de la température, du pH et d'autres paramètres physico-chimiques.

Definition

Simplified Definition

Greenstone-hosted quartz-carbonate vein deposits occur as quartz and quartz-carbonate veins, with valuable amounts of gold and silver, in faults and shear zones located within deformed terranes of ancient to recent greenstone belts commonly metamorphosed at greenschist facies.

Scientific Definition

Greenstone-hosted quartz-carbonate vein deposits are a subtype of lode gold deposits (Poulsen et al., 2000) (Fig. 1). They are also known as mesothermal, orogenic (mesozonal and hypozonal - the near surface orogenic epizonal Au-Sb-Hg deposits described by Groves et al. (1998) are not included in this synthesis), lode gold, shear-zone-related quartz-carbonate or gold-only deposits (Hodgson and MacGeehan, 1982; Roberts, 1987; Colvine, 1989; Kerrich and Wyman, 1990; Robert, 1990; Kerrich and Feng, 1992; Hodgson, 1993; Kerrich and Cassidy, 1994; Robert, 1995; Groves et al., 1998; Hagemann and Cassidy, 2000; Kerrich et al., 2000; Poulsen et al., 2000; Goldfarb et al., 2001; Robert and Poulsen, 2001; Groves et al., 2003; Goldfarb et al., 2005; Robert et al., 2005). The focus of the following text is mainly on Canadian examples and particularly those deposits found in the Abitibi Archean greenstone belt. For a complete global perspective, readers are referred to the above list of selected key references.

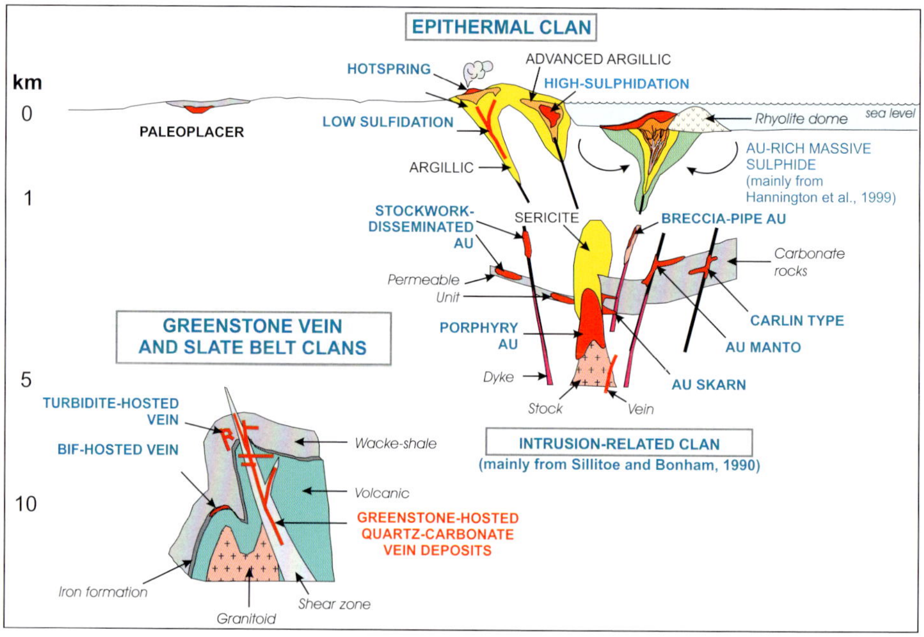

FIGURE 1. Inferred crustal levels of gold deposition showing the different types of gold deposits and the inferred deposit clan (from Dubé et al., 2001; modified from Poulsen et al., 2000).

Greenstone-hosted quartz-carbonate vein deposits are structurally controlled, complex epigenetic deposits that are hosted in deformed and metamorphosed terranes. They consist of simple to complex networks of gold-bearing, laminated quartz-carbonate fault-fill veins in moderately to steeply dipping, compressional brittle-ductile shear zones and faults, with locally associated extensional veins and hydrothermal breccias. They are dominantly hosted by mafic metamorphic rocks of greenschist to locally lower amphibolite facies and formed at intermediate depths (5-10 km). Greenstone-hosted quartz-carbonate vein deposits are typically associated with iron-carbonate alteration. The relative timing of mineralization is syn- to late-deformation and typically post-peak greenschist-facies or syn-peak amphibolite-facies metamorphism. They are formed from low salinity, H_2O-CO_2-rich hydrothermal fluids with typically anomalous concentrations of CH_4, N_2, K, and S. Gold is mainly confined to the quartz-carbonate vein networks but may also be present in significant amounts within iron-rich sulphidized wall rock. Greenstone-hosted quartz-carbonate vein deposits are distributed along major compressional to transpressional crustal-scale fault zones in deformed greenstone terranes of all ages, but are more abundant and significant, in terms of total gold content, in Archean terranes. However, a significant number of world-class deposits ($\geq$100 t Au) are also found in Proterozoic and Paleozoic terranes. International examples of this subtype of gold deposits include Mt. Charlotte, Norseman, and Victory (Australia), Bulyanhulu (Tanzania), and Kolar (India) (Fig. 2). Canadian examples include Sigma-Lamaque (Québec), Dome and Pamour (Ontario), Giant and Con (Northwest Territories), San Antonio (Manitoba), Hammer Down (Newfoundland), and Bralorne-Pioneer (British Columbia). Detailed characteristics and references are found in the text below. The reader may refer to Appendix 1 for a list of geographical, geological, and economical characteristics of Canadian gold deposits with more than 250 000 oz Au in combined production and reserves (data from Gosselin and Dubé, 2005b).

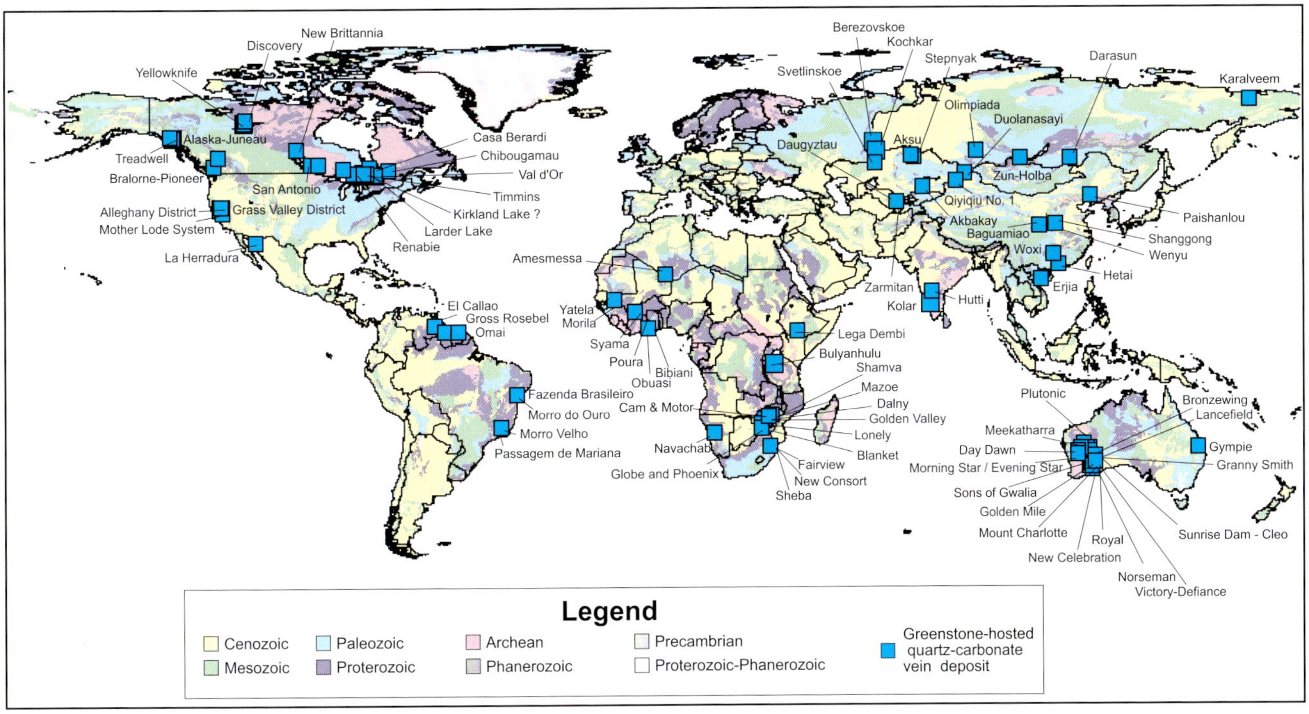

FIGURE 2. World distribution of greenstone-hosted quartz-carbonate vein deposits containing at least 30 tonnes of Au.

Economic Characteristics of Greenstone-Hosted Quartz-Carbonate Vein Deposits

Summary of Economic Characteristics

The total world production and reserves of gold, including the Witwatersrand paleoplacer deposits, stands at more than 126 420 metric tonnes Au (Gosselin and Dubé, 2005a). World production and reserves for the greenstone-hosted quartz-carbonate vein deposit subtype is 15 920 metric tonnes Au (Gosselin and Dubé, 2005a), which is equivalent to 13% of the total world production and puts them in second place for world productivity behind paleoplacers. Total Canadian production and reserves, at 9 280 metric tonnes Au, represent 7% of the world total. However, Canadian production and reserves for the greenstone-hosted quartz-carbonate vein subtype are 5 510 metric tonnes, which constitutes 35% of the world production for this deposit subtype, and 59% of the total Canadian production and reserves of gold. The Superior province contains 86% (4 760 metric tonnes) of Canadian gold production and reserves for greenstone-hosted quartz-carbonate vein deposits (Gosselin and Dubé, 2005a,b). The Abitibi sub-province is the main source and represents 81% (4 470 metric tonnes) of the total Canadian gold.

There are 103 known greenstone-hosted quartz-carbonate vein deposits world-wide containing at least 30 tonnes (~1 M oz) Au (production and reserves), including 31 Canadian deposits, whereas 33 other deposits in Canada, and several hundred worldwide, contain more than 7.5 tonnes (~250 000 oz) but less than 30 tonnes (Gosselin and Dubé, 2005b). A select group of 41 world-class deposits contains more than 100 tonnes Au, including 11 giant deposits with more than 250 tonnes. In this group of world-class deposits, six are from the Abitibi greenstone belt of the Canadian Archean Superior Province (Fig. 3). The Superior Province is the largest and best preserved Archean craton in terms of greenstone-hosted gold endowment, followed by the Yilgarn craton of Australia.

The temporal and geographic distribution of greenstone-hosted quartz-carbonate vein deposits is shown on Figure 2. Greenstone-hosted quartz-carbonate vein deposits occur in greenstone terranes of all ages. Although they are present in Paleozoic to Tertiary terranes, they are mainly concentrated in Precambrian terranes, and particularly in those of late Archean age. In Canada, all the world-class deposits but one (Bralorne-Pioneer) are of late Archean age. Their concentration in the Archean is thought to be related to 1) continental growth and the related higher number of large-scale collisions between continental fragments (and/or arc complex), and 2) the associated development of major faults and large-scale hydrothermal fluid flow during the supercontinent cycle and mantle plume activity (cf. Barley and Groves, 1992; Condie, 1998; Kerrich et al., 2000; Goldfarb et al., 2001).

Grade and Tonnage Characteristics

Greenstone-hosted quartz-carbonate vein deposits are second on total tonnage of gold only to the Witwatersrand paleoplacers of South Africa. The largest greenstone-hosted quartz-carbonate vein deposit in terms of total gold content is the Golden Mile complex in Kalgoorlie, Australia, with more than 1800 tonnes Au (Gosselin and Dubé, 2005a). The Hollinger-McIntyre deposit in Timmins, Ontario, is the second largest deposit of such type ever found with 987 tonnes Au (Gosselin and Dubé, 2005a). In contrast to the Golden Mile complex, open pit mining of the Hollinger-McIntyre deposit is now impossible due to housing, which leaves a significant part of the total gold content of the deposit inaccessible.

The average grade of greenstone-hosted quartz-carbonate deposits is fairly consistent, ranging from 5 to 15 g/t Au, whereas the tonnage is highly variable and ranges from a few thousand tonnes to over 100 million tonnes of ore, although more typically these deposits contain only a few million tonnes of ore (Fig. 4).

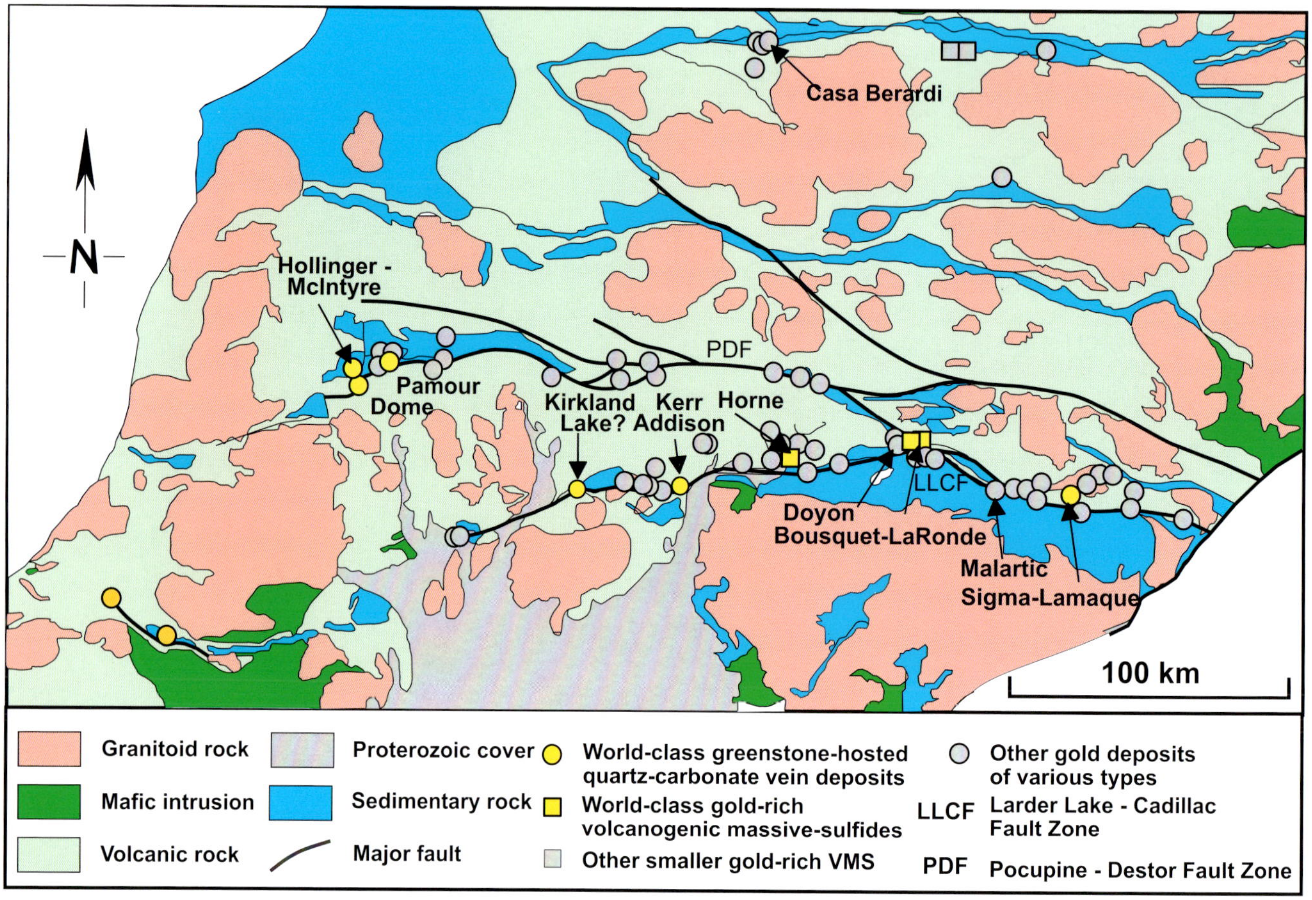

FIGURE 3. Simplified geological map of the Abitibi greenstone belt showing the distribution of major fault zones and gold deposits. Modified from Poulsen et al. (2000). See Appendix 1 for deposit details.

Comparison of Grade and Tonnage Characteristics with the Global Range

In Canada, this type of gold deposit is widely distributed from the Paleozoic greenstone terranes of the Appalachian Orogen on the east coast (e.g. Hammer Down and Deer Cove Newfoundland, Dubé et al., 1993; Gaboury et al., 1996), through the Archean greenstone belts of the Superior (Dome and Sigma-Lamaque) and Slave provinces (Con and Giant) in central Canada, to the oceanic terranes of the Cordillera (Bralorne-Pioneer).

The average gold grade of world-class Canadian deposits is 10 g/t, which is slightly higher than the average for this type of deposit worldwide (7.6 g/t, Fig. 5). World-class deposits in Canada have on average lower tonnage (20.91 Mt of ore) than those worldwide (39.91 Mt). Perhaps this is in part because mining in Canada has traditionally taken place underground, whereas in other countries open pits have also been developed.

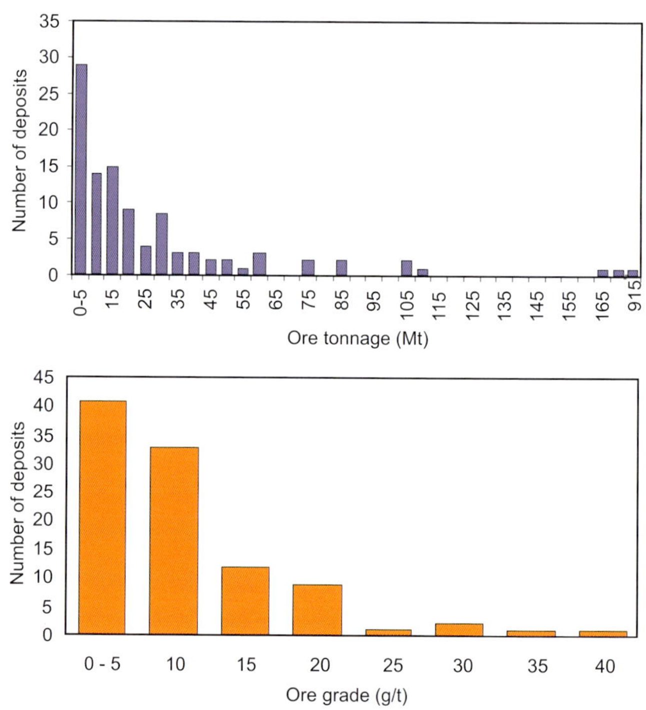

FIGURE 4. Tonnage and grade repartition for gold deposits in the world containing at least 30 tonnes of Au in combined production and reserves.

Geological Characteristics of Greenstone-Hosted Quartz-Carbonate Vein Deposits

Physical Properties

Mineralogy

The main gangue minerals in greenstone-hosted quartz-carbonate vein deposits are quartz and carbonate (calcite, dolomite, ankerite, and siderite), with variable amounts of

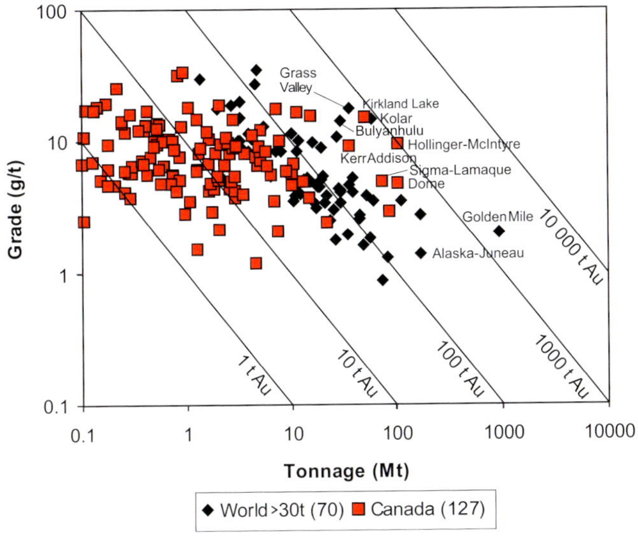

FIGURE 5. Tonnage versus grade relationship of Canadian and world Au deposits containing at least 30 tonnes of Au in combined production and reserves.

white micas, chlorite, tourmaline, and sometimes scheelite. The sulphide minerals typically constitute less than 5 to 10% of the volume of the orebodies. The main ore minerals are native gold with, in decreasing amounts, pyrite, pyrrhotite, and chalcopyrite and occur without any significant vertical mineral zoning. Arsenopyrite commonly represents the main sulphide in amphibolite-facies rocks (e.g. Con and Giant) and in deposits hosted by clastic sediments. Trace amounts of molybdenite and tellurides are also present in some deposits, such as those hosted by syenite in Kirkland Lake (Thompson et al., 1950; Fig. 6A, B).

Textures

This type of gold deposit is characterized by moderately to steeply dipping, laminated fault-fill quartz-carbonate veins (Fig. 7A, B, C) in brittle-ductile shear zones and faults, with or without fringing shallow-dipping extensional veins and breccias (Fig. 7D, E). Quartz vein textures vary according to the nature of the host structure (extensional vs. compressional). Extensional veins typically display quartz and carbonate fibres at a high angle to the vein walls and with multiple stages of mineral growth (Fig. 7E), whereas the laminated veins are composed of massive, fine-grained quartz. When present in laminated veins, fibres are subparallel to the vein walls (Robert et al., 1994; Robert and Poulsen, 2001).

Dimensions

Individual vein thickness varies from a few centimetres up to 5 metres, and their length varies from 10 up to 1000 m. The vertical extent of the orebodies is commonly greater than 1 km and reaches 2.5 km in a few cases (e.g. the Kirkland Lake deposit, Charlewood, 1964).

Morphology

The gold-bearing shear zones and faults associated with this deposit type are mainly compressional and they commonly display a complex geometry with anastomosing and/or conjugate arrays (Daigneault and Archambault, 1990; Hodgson, 1993; Robert et al., 1994; Robert and Poulsen, 2001). The laminated quartz-carbonate veins typically infill the central part of, and are subparallel to slightly oblique to, the host structures (Hodgson, 1989; Robert et al., 1994; Robert and Poulsen, 2001) (Fig. 8). The shallow-dipping extensional veins are either confined within shear zones, in which case they are relatively small and sigmoidal in shape, or they extend outside the shear zone and are planar and laterally much more extensive (Robert et al., 1994).

FIGURE 6. (A) Quartz-breccia vein, Main Break, Kirkland Lake. (B) High-grade quartz veinlets, hosted by syenite with visible gold, disseminated pyrite, and traces of tellurides, Main Break, Kirkland Lake.

Stockworks and hydrothermal breccias may represent the main mineralization styles when developed in competent units such as the granophyric facies of differentiated gabbroic sills (e.g. San Antonio deposit, Robert et al., 1994; Robert and Poulsen, 2001), especially when developed at shallower crustal levels. Ore-grade mineralization also occurs as disseminated sulphides in altered (carbonatized) rocks along vein selvages. Due to the complexity of the geological and structural setting and the influence of strength anisotropy and competency contrasts, the geometry of vein networks varies from simple (e.g. Silidor deposit), to fairly complex with multiple orientations of anastomosing and/or conjugate sets of veins, breccias, stockworks, and associated structures (Dubé et al., 1989; Hodgson, 1989, Belkabir et al., 1993; Robert et al., 1994; Robert and Poulsen, 2001). Layer anisotropy induced by stiff differentiated gabbroic sills

FIGURE 7. (A) Laminated fault-fill veins, Pamour mine, Timmins. (B) Close-up of photo A showing a laminated fault-fill vein with iron-carbonatized wall-rock clasts. (C) Boudinaged fault-fill vein, section view, Dome mine. (D) Arrays of extensional quartz veins, Pamour mine. (E) Extensional quartz-tourmaline "flat vein" showing multiple stages of mineral growth perpendicular to the vein walls, Sigma mine (from Poulsen et al., 2000). (F) Tourmaline-quartz vein, Clearwater deposit, James Bay area.

within a matrix of softer rocks, or, alternatively, by the presence of soft mafic dykes within a highly competent felsic intrusive host, could control the orientation and slip directions in shear zones developed within the sills; consequently, it may have a major impact on the distribution and geometry of the associated quartz-carbonate vein network (Dubé et al., 1989; Belkabir et al., 1993). As a consequence, the geometry of the veins in settings with large competence contrasts will be strongly controlled by the orientation of the hosting bodies and less by external stress. The anisotropy of the stiff layer and its orientation may induce an internal strain different from the regional one and may strongly influence the success of predicting the geometry of the gold-bearing vein network being targeted in an exploration program (Dubé et al., 1989; Robert et al., 1994).

Host Rocks

The veins in greenstone-hosted quartz-carbonate vein deposits are hosted by a wide variety of host rock types; mafic and ultramafic volcanic rocks and competent iron-rich differentiated tholeiitic gabbroic sills and granitoid intrusions are common hosts. However, there are commonly district-specific lithological associations acting as chemical and/or structural traps for the mineralizing fluids as illustrated by tholeiitic basalts and flow contacts within the Tisdale Assemblage in Timmins (cf. Hodgson and MacGeehan, 1982; Brisbin, 1997). A large number of deposits in the Archean Yilgarn craton are hosted by gabbroic ("dolerite") sills and dykes (Solomon et al., 2000) as illustrated by the Golden Mile dolerite sill in Kalgoorlie (Bartram and McGall, 1971; Travis et al., 1971; Groves et al., 1984), whereas in the Superior Province, many deposits are associated with porphyry stocks and dykes (Hodgson and McGeehan, 1982). Some deposits are also hosted by and/or along the margins of intrusive complexes (e.g. Perron-Beaufort/North Pascalis deposit hosted by the Bourlamaque batholith in Val d'Or (Belkabir et al., 1993; Robert, 1994)). Other deposits are hosted by clastic sedimentary rocks (e.g. Pamour, Timmins).

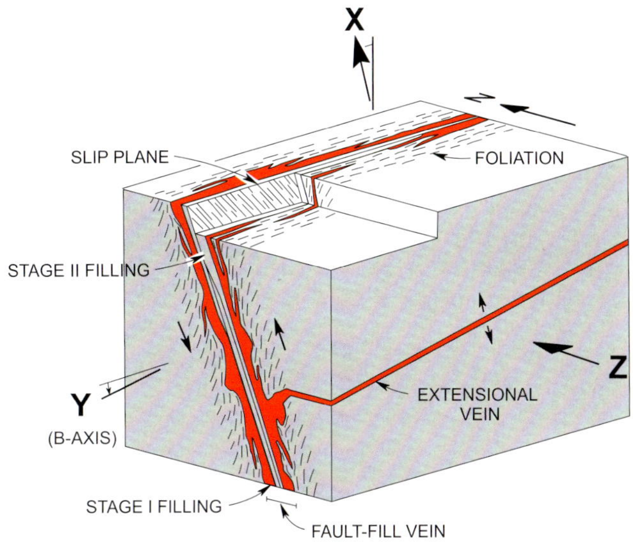

FIGURE 8. Schematic diagram illustrating the geometric relationships between the structural element of veins and shear zones and the deposit-scale strain axes (from Robert, 1990).

Chemical Properties

Ore Chemistry

The metallic geochemical signature of greenstone-hosted quartz-carbonate vein orebodies is Au, Ag, As, W, B, Sb, Te, and Mo, typically with background or only slightly anomalous concentrations of base metals (Cu, Pb, and Zn). The Au/Ag ratio typically varies from 5 to 10. Contrary to epithermal deposits, there is no vertical metal zoning. Palladium is locally present as illustrated by the syndefor-

FIGURE 9. (A) Large boudinaged iron-carbonate vein, Red Lake district. (B) Iron carbonate pervasive replacement of an iron-rich gabbroic sill, Tadd prospect, Chibougamau. (C) Green carbonate rock showing fuchsite-rich replacement and iron-carbonate veining in a highly deformed ultramafic rock, Larder Lake. (D) Green carbonate alteration showing abundant green micas replacing chromite-rich ultramafics, Baie Verte, Newfoundland.

FIGURE 10. **(A)** Diopside vein in biotite-actinolite-microcline-rich gold-bearing alteration, Madsen mine, Red Lake. **(B)** Auriferous metasomatic hydrothermal layering with actinolite-rich and biotite-microcline-rich bands, Madsen mine, Red Lake. **(C)** Gold-rich No. 8 vein with visible gold in a quartz carbonate-actinolite-diopside-rich vein, Madsen mine, Red Lake.

mation auriferous quartz or hematite-quartz veins hosted by Proterozoic iron formation in Brazil (Olivo et al., 1995).

Alteration Mineralogy and Chemistry

At a district scale, greenstone-hosted quartz-carbonate vein deposits are associated with large-scale carbonate alteration (Fig. 9A, B) commonly distributed along major fault zones and associated subsidiary structures. At a deposit scale, the nature, distribution, and intensity of the wall-rock alteration is controlled mainly by the composition and competence of the host rocks and their metamorphic grade. Typically, the proximal alteration haloes are zoned and characterized – in rocks at greenschist facies – by iron-carbonatization and sericitization, with sulphidation of the immediate vein selvages (mainly pyrite, less commonly arsenopyrite). Altered rocks show enrichments in CO_2, K_2O, and S, and leaching of Na_2O. Further away from the vein, the alteration is characterized by various amounts of chlorite and calcite, and locally magnetite (Phillips and Groves, 1984; Dubé et al., 1987; Roberts, 1987). The dimensions of the alteration haloes vary with the composition of the host rocks and may envelope entire deposits hosted by mafic and ultramafic rocks. Pervasive chromium- or vanadium-rich green micas (fuchsite and roscoelite) and ankerite with zones of quartz-carbonate stockworks are common in sheared ultramafic rocks (Fig. 9C, D). Common hydrothermal alteration assemblages that are associated with gold mineralization in amphibolite-facies rocks include biotite, amphibole, pyrite, pyrrhotite, and arsenopyrite, and, at higher grades, biotite/phlogopite, diopside, garnet, pyrrhotite and/or arsenopyrite (cf. Mueller and Groves, 1991; Witt, 1991; Hagemann and Cassidy, 2000; Ridley et al., 2000, and references therein), with variable proportions of feldspar, calcite, and clinozoisite (Fig. 10). The variations in alteration styles have been interpreted as a direct reflection of the depth of formation of the deposits (Colvine, 1989; Groves, 1993). The alteration mineralogy of the deposits hosted by amphibolite-facies rocks, in particular the presence of diopside, biotite, K-feldspar, garnet, staurolite, andalusite, and actinolite, suggests that they share analogies with gold skarns, especially when they (1) are hosted by sedimentary or mafic volcanic rocks, (2) contain a calc-silicate alteration assemblage related to gold mineralization with an Au-As-Bi-Te metallic signature, and (3) are associated with granodiorite-diorite intrusions (cf. Meinert, 1998; Ray, 1998). Canadian examples of deposits hosted in amphibolite-facies rocks include the replacement-style Madsen deposit in Red Lake (Dubé et al., 2000) and the quartz-tourmaline vein (Fig. 7F) and replacement-style Eau Claire deposit in the James Bay area (Cadieux, 2000; Tremblay, 2006).

Geological Properties

Continental Scale

Greenstone-hosted quartz-carbonate-vein deposits are typically distributed along crustal-scale fault zones (cf. Kerrich et al., 2000, and references therein) characterized by several increments of strain (e.g. Cadillac-Larder Lake fault) (Figs. 3, 11A, B, 12A, B), and, consequently multiple generations of steeply dipping foliations and folds resulting in a complex deformational history. These crustal-scale fault zones are the main hydrothermal pathways towards higher crustal levels. However, the deposits are spatially and genetically associated with second- and third-order compressional reverse-oblique to oblique brittle-ductile high-angle shear and high-strain zones (Fig. 12C), which are commonly located within 5 km of the first order fault and are best developed in its hanging wall (Robert, 1990). Brittle faults may also be the main host to gold mineralization as illustrated by the Kirkland Lake Main Break, a brittle structure hosting the giant Kirkland Lake deposit exploited by seven mines that have collectively produced more than 760 metric tonnes of gold (Fig. 13) (Thomson, 1950; Kerrich and Watson, 1984;

FIGURE 11. (A) Mylonitic foliation, Cadillac-Larder Lake Break, Val d'Or. **(B)** Close-up showing mylonitic foliation within Cadillac-Larder Lake Break, Val d'Or.

Ayer et al., 2005; Ispolatov et al., 2005 and references therein). Greenstone-hosted quartz-carbonate vein deposits typically formed late in the tectonic-metamorphic history (Groves et al., 2000; Robert et al., 2005) and the mineralization is syn- to late-deformation and typically post-peak greenschist-facies and syn-peak amphibolite-facies metamorphism (cf. Kerrich and Cassidy, 1994; Hagemann and Cassidy, 2000). Most world-class greenstone-hosted quartz-carbonate vein deposits are hosted by greenschist-facies rocks. Important exceptions include Kolar (India), which formed at amphibolite facies.

Greenstone-hosted quartz-carbonate vein deposits are also commonly spatially associated with Timiskaming-like regional unconformities (Fig. 14A, B, C). Several deposits are hosted by, or located next to, such unconformities (e.g. the Pamour and Dome deposits), suggesting an empirical temporal and spatial relationship between large gold deposits and regional unconformities (Poulsen et al., 1992; Hodgson, 1993; Robert, 2000; Dubé et al., 2003; Robert et al., 2005).

District Scale

In this section, some of the key geological characteristics of prolific gold districts are presented with a special emphasis on Archean deposits. Only a brief overview is presented here, and the reader is referred to key papers by Hodgson and MacGeehan (1982), Hodgson (1993), Robert and Poulsen (1997), Hagemann and Cassidy (2000), Poulsen et al. (2000), Groves et al. (2003), and Robert et al. (2005), among others, for more information.

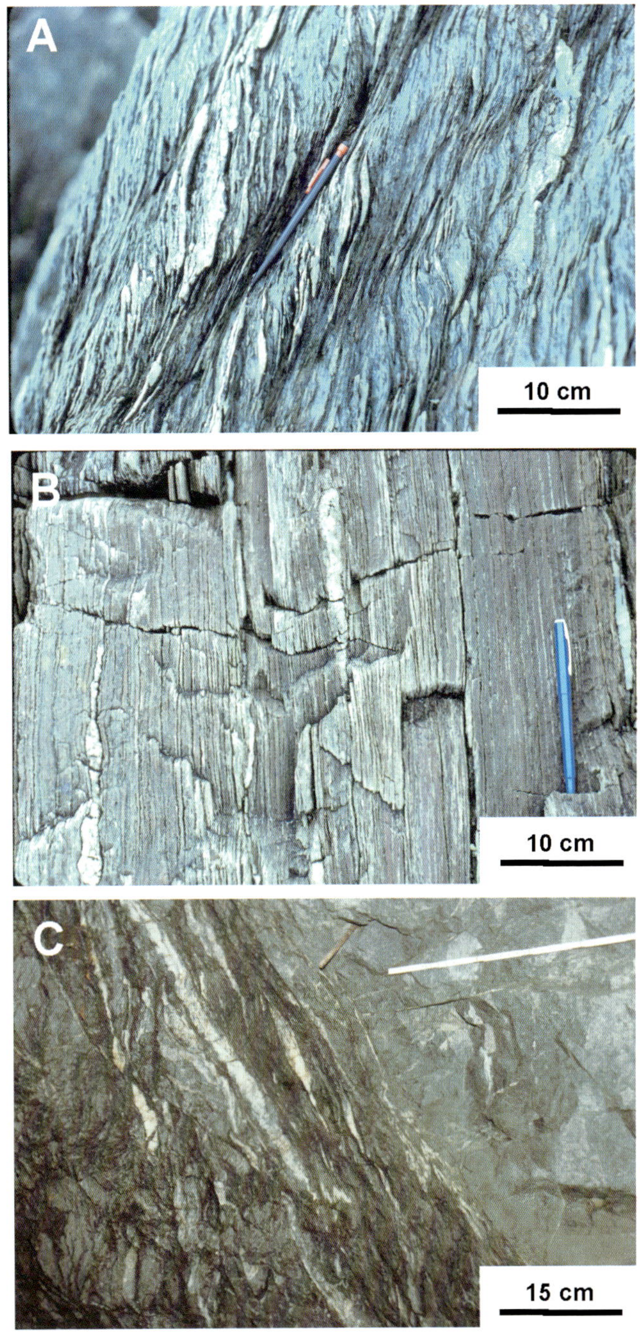

FIGURE 12. (A) Vertical section of shear bands indicating a reverse-oblique sense of motion recorded by the gold-bearing Cape Ray fault zone, Newfoundland (from Dubé et al., 1996). **(B)** Section view showing reverse-oblique mylonite, Cape Ray fault zone, Newfoundland. **(C)** Section view showing auriferous quartz vein hosted by a second-order reverse shear zone, Cooke mine, Chapais, Quebec (from Dubé and Guha, 1992).

Large gold camps are commonly associated with curvatures, flexures, and dilational jogs along major compressional fault zones, such as the Porcupine-Destor fault in Timmins or the Larder Lake-Cadillac fault in Kirkland Lake (Fig. 3), which have created dilational zones that allowed migration of hydrothermal fluids (cf. Colvine et al., 1988; Sibson, 1990; Phillips et al., 1996; McCuaig and Kerrich,

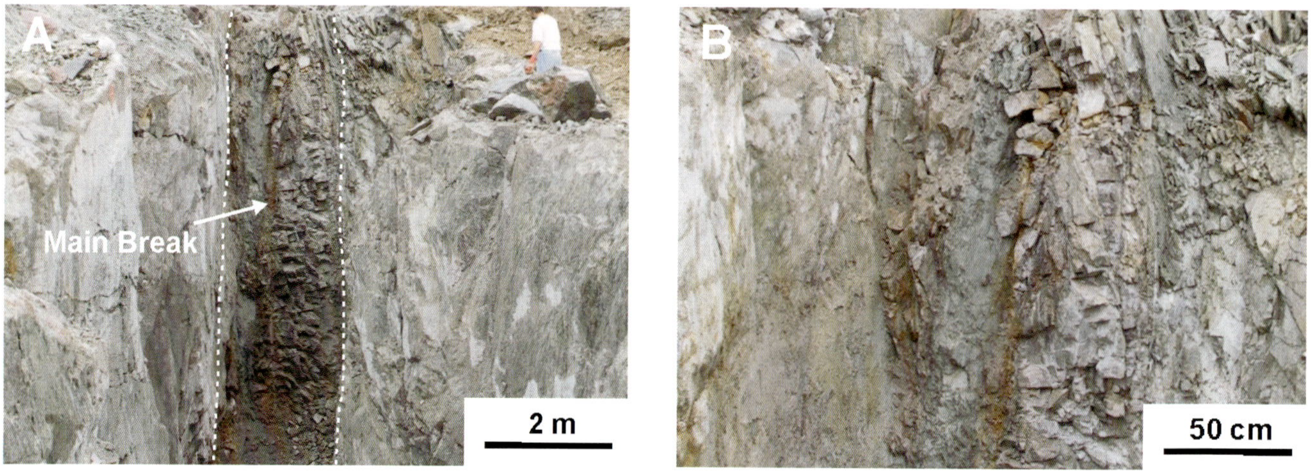

FIGURE 13. (A) Section view showing the 25 M oz Kirkland Lake Main Break. (B) Close-up of photo (A) showing the Kirkland Lake Main Break in section view; note the brittle nature of the structure with gouges.

1998; Hagemann and Cassidy, 2000; Kerrich et al., 2000; Groves et al., 2003; Goldfarb et al., 2005; Ispolatov et al., 2005; Robert et al., 2005). In terms of geological setting, large gold districts, such as Timmins, are mainly underlain by tholeiitic basalts (commonly variolitic) (Fig. 14D) and ultramafic komatiitic flows that are intruded by intermediate to felsic porphyries, and locally swarms of albitite and/or lamprophyre dykes (cf. Hodgson and MacGeehan, 1982). Irrelevant to their age, Timiskaming-like regional unconformities, distributed along major faults or stratigraphical dis-

FIGURE 14. (A) Timiskaming conglomerate, Kirkland Lake. (B) Mineralized quartz veins hosted by a carbonatized Timiskaming conglomerate, Pamour mine, Timmins. (C) Mineralized quartz vein hosted in a discrete brittle-ductile high-strained zone hosted by weakly deformed Timiskaming conglomerate, Kirkland Lake. (D) Variolitic basalt, Vipond Formation, Tisdale Assemblage, Timmins.

continuities, are also typical of large gold camps. In terms of hydrothermal alteration, the main characteristic at the district scale is the presence of large-scale iron-carbonate alteration, the width of which gives some indication as to the size of the hydrothermal system(s) (e.g. Timmins). Protracted magmatic activity with synvolcanic and syn- to late tectonic intrusions emplaced along structural discontinuities (e.g. Destor-Porcupine Fault) may also be highly significant. In many cases, U-Pb dating of intrusive rocks indicates that they are older than gold mineralization, in which case these rocks may have provided a competent structural trap or induced anisotropy in the layered stratigraphy that influenced and partitioned the strain. In other cases, the intrusive rocks are post-mineralization. However, the possibility remains that the thermal energy provided by some intrusions contributed to large-scale and long-lived hydrothermal fluid circulation (cf. Wall, 1989).

The presence of other deposit types in a district, such as volcanogenic massive sulphide (VMS) or Ni-Cu deposits, is also commonly thought to be a favourable factor (cf. Hodgson, 1993; Huston, 2000). The provinciality of the high Au content of a district may be related to specific fundamental geological characteristics in terms of favourable source-rock environments or gold reservoirs (Hodgson, 1993). The local geological "heritage" of the district, in addition to ore-forming processes, may thus be a major factor to take into account.

Knowledge Gaps at District Scale: One of the main remaining knowledge gaps at district scale is the structural evolution, and in some cases, the tectonic significance of the large-scale faults that control the distribution of the greenstone-hosted quartz-carbonate-vein deposits. The nature and significance of the early stage(s) of deformation (e.g. D0-D1) of major fault zones to the circulation of gold-bearing fluids and the formation of large gold deposits remain obscure. For example, despite decades of work in the Timmins' district, the structural evolution of the Porcupine-Destor Fault, a poorly exposed, regionally extensive, steeply dipping, long-lived fault (active between ca. 2680-2600 Ma), and its definite relationship to gold mineralization, remain controversial (cf. Hurst, 1936; Pyke, 1982; Bleeker, 1995; 1997; Hodgson and Hamilton, 1989; Hodgson et al., 1990; Brisbin, 1997; Ayer et al., 2005; Bateman et al., 2005, and references therein). The processes controlling the distribution of the large gold districts along such crustal-scale structures are poorly understood and therefore remain an avenue for future research (Robert et al., 2005). Key questions remain, such as the reason(s) why the Timmins district contains a large number of world-class gold deposits, why some large-scale Archean fault zones in greenstone belts are devoid of significant gold deposits, and why the gold grade in some districts is significantly higher.

Deposit Scale

The location of higher grade mineralization (ore shoots) within a deposit has been the subject of investigation since the early works of Newhouse (1942) and McKinstry (1948). Ore shoots represent a critical element to take into account when defining and following the richest part of an orebody. Two broad categories of ore shoots are recognized: 1) geometric, and 2) kinematic (Poulsen and Robert, 1989; Robert et al., 1994; Robert and Poulsen, 2001). As outlined by Poulsen and Robert (1989), geometric ore shoots are controlled by the intersection of a given structure (i.e., a fault, a shear zone, or a vein) with a favourable lithological unit, such as a competent gabbroic sill, a dyke, an iron formation, or a particularly reactive rock. The geometric ore shoot will be parallel to the line of intersection. The kinematic ore shoots are syndeformation and syn-formation of the veins, and are defined by the intersection between different sets of veins or contemporaneous structures. The plunge of kinematic ore shoots is commonly at a high angle to the slip direction.

Structural traps, such as fold hinges or dilational jogs along faults or shear zones, are also key elements in locating the richest part of an orebody. However, multiple factors are commonly involved, as mentioned by Groves et al. (2003), and world-class and giant-size deposits commonly exhibit complex geometries. This complexity is mainly due to the longevity of the hydrothermal system and/or multistage, barren and/or gold-bearing hydrothermal, structural, and magmatic events (Dubé et al., 2003; Groves et al., 2003; Ayer et al., 2005). This is especially well illustrated at the Dome mine, where low-grade colloform-crustiform ankerite veins cut the 2690 ± 2 Ma Paymaster porphyry (Corfu et al., 1989) (Fig. 15A). These ankerite veins have been deformed; they are typically boudinaged and are cut by extensional, en echelon, auriferous quartz veins (Fig. 15B, C). The <2673.9 ± 1.8 Ma Timiskaming conglomerate (Ayer et al., 2003, 2005) contains clasts of the ankerite veins in the Dome open pit (Fig. 15D, E), whereas the Timiskaming conglomerate is itself carbonatized, cut by auriferous quartz veins and locally contains spectacular visible gold (Fig. 15F). Argillite and sandstone above the Timiskaming conglomerate are themselves folded and cut by auriferous quartz veins (Dubé et al., 2003). These chronological relationships illustrate the superimposed hydrothermal and structural events involved in the formation of the giant deposit with post-magmatic carbonate veining predating the deposition of the Timiskaming conglomerate, which in turn precedes formation of the bulk of the gold mineralization.

Distribution of Canadian Greenstone-Hosted Quartz-Carbonate Vein Districts

The most productive Canadian metallogenic districts for greenstone-hosted quartz-carbonate vein deposits occur in (Late) Archean greenstone belts of the Superior, Churchill, and Slave provinces (Table 1). The Abitibi greenstone belt contains the majority of the productive districts, including the very large Timmins, Kirkland Lake, Larder Lake, Rouyn-Noranda, and Val d'Or districts. The Kirkland Lake gold deposit is included here as a greenstone-hosted quartz-carbonate deposit, however, the structural timing of gold deposition and its origin is still the subject of debate (Kerrich and Watson, 1984; Cameron and Hattori, 1987; Robert and Poulsen, 1997; Ayer et al., 2005; Ispolatov et al., 2005) as the deposit shares strong analogies with tellurium-rich syndeformation gold deposits associated with alkaline magmatism as defined by Jensen and Barton (2000). Other younger greenstone belts of the Appalachian and Cordilleran orogens are also favourable terranes for quartz-carbonate vein-type gold deposits (Fig. 16). Districts listed in Table 1 also include

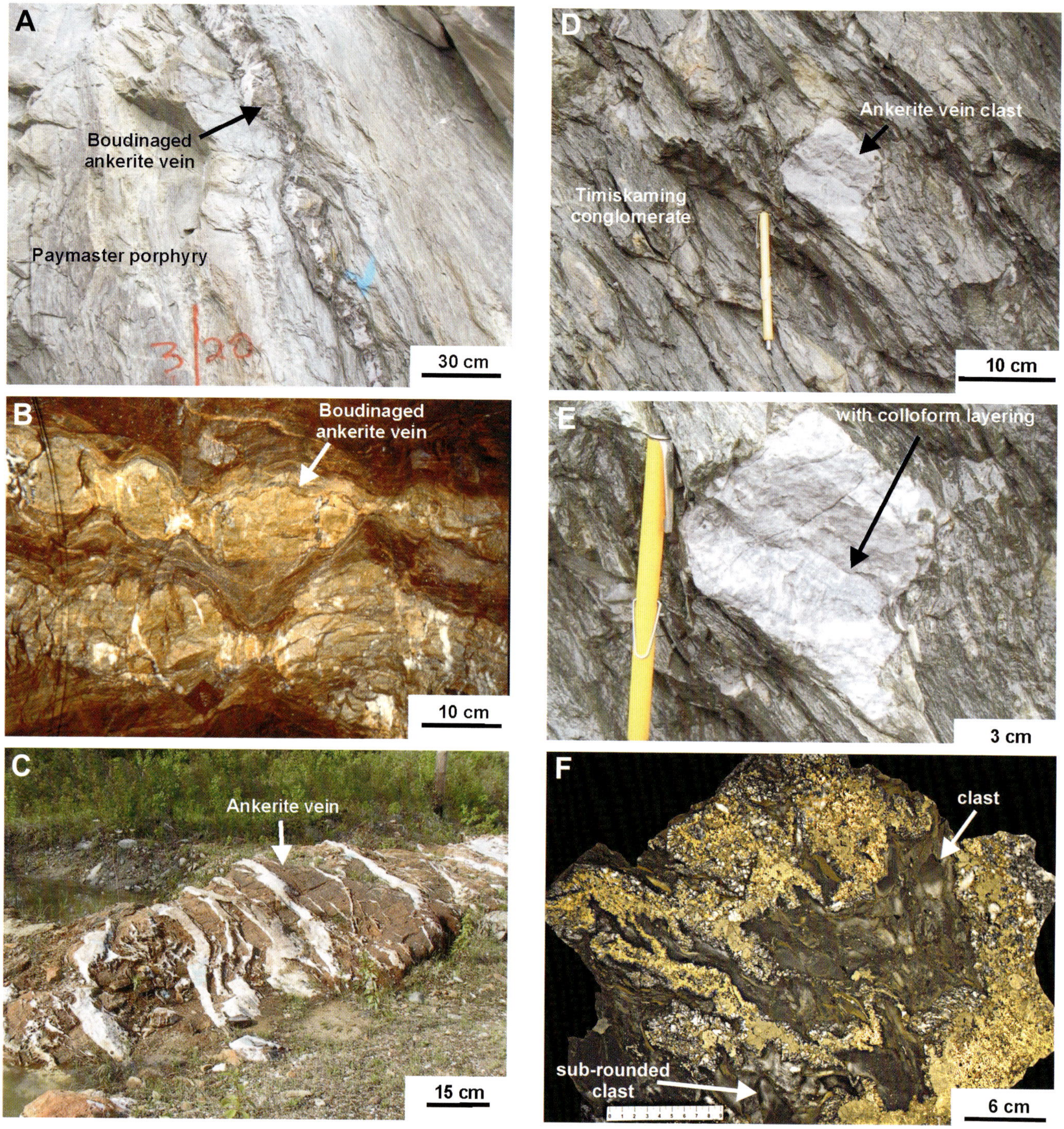

Figure 15. **(A)** Boudinaged ankerite vein with late quartz veins cutting the Paymaster porphyry, Dome mine. **(B)** Boudinaged ankerite veins with syndeformation late extensional quartz veins, Dome mine (from Poulsen et al., 2000). **(C)** Massive ankerite Kurst vein cut by late gold-bearing extensional quartz vein, Dome mine area. **(D)** Ankerite vein clast within Timiskaming conglomerate, Dome mine (from Dubé et al., 2003). **(E)** Close-up of photograph (D) (from Dubé et al., 2003). **(F)** High-grade Timiskaming conglomerate hosting folded carbonate-pyrite veins with spectacular visible gold. The specimen was presented to the Geological Survey of Canada in 1923 by the then Board of Directors of Dome Mines. Weight is 136 lbs (61.8 kg) of which about 20% by weight is gold. It most likely came from the bonanza East Dome area, which was discovered in 1910. It consists of subrounded to subangular altered and nonaltered clasts and folded crosscutting veins of coarse pyrite, ankerite, and minor quartz shattered and invaded by gold. Geological Survey of Canada National Mineral collection Sample No. 1003. Photograph by Igor Bilot, Geological Survey of Canada.

deposits hosted by iron formation (BIF-hosted vein or Homestake-type; Poulsen et al., 2000).

The geographical and temporal distribution of greenstone-hosted quartz-carbonate vein deposits containing at least 30 t Au is included in Figure 2. The greatest concentration of deposits is found in the Archean, particularly in the Late Archean in Canada (Fig. 16). Proterozoic gold deposits occur in the United States as exemplified by the Homestake deposit, a giant iron-formation-hosted vein and disseminated Au-Ag deposit, as well as in greenstone belts of Brazil and western Africa. However, Canadian deposits of Proterozoic age are rare; they include the New Britannia deposit in the

TABLE 1. Most productive Canadian districts for greenstone-hosted quartz-carbonate vein deposits.

District	Geological Province	Production & Reserves (tonnes Au)*	Resources (tonnes Au)*
Timmins	Superior/Abitibi	2,072.9	78.5
Kirkland Lake	Superior/Abitibi	794.8	72.6
Val d'Or	Superior/Abitibi	638.9	171.6
Rouyn-Noranda	Superior/Abitibi	519.6	66.5
Larder Lake	Superior/Abitibi	378.7	14.5
Malartic	Superior/Abitibi	278.7	23.2
Red Lake**	Superior/Uchi	128.0	17.2
Joutel	Superior/Abitibi	61.4	27.5
Matheson	Superior/Abitibi	60.4	9.7
Cadillac	Superior/Abitibi	22.1	25.1
Pickle Lake	Superior/Uchi	90.4	8.1
Rice Lake	Superior/Uchi	51.6	25.2
Beardmore-Geraldton	Superior/Wabigoon	123.5	35.1
Michipicoten	Superior/Wawa	41.1	2.8
Mishibishu	Superior/Wawa	26.7	16.8
Goudreau-Lolshcach	Superior/Wawa	8.8	19.6
Flin Flon	Churchill	62.2	12.7
Lynn Lake	Churchill	19.5	14.6
La Ronge	Churchill	3.4	5.6
Keewatin	Churchill-Hearne	7.2	252.4
Yellowknife	Slave	432.8	16.6
MacKenzie	Slave	38.1	286.6
Cassiar	Cordillera	14.9	55.4
Baie Verte	Appalachian/Dunnage	10.3	8.9

*as of December 31, 2002
**does not include the Campbell-Red Lake, Cochenour, and MacKenzie Red Lake deposits as they are not considered typical greenstone-hosted quartz-carbonate deposits

Flin Flon district (Manitoba) and other smaller deposits of the Churchill Province, as well as gold-bearing quartz-carbonate veins in the central metasedimentary belt of the Grenville Province (Carter, 1984; Jourdain et al., 1990; Easton and Fyon, 1992). Mesozoic and Cenozoic deposits are less common, but are important within Circum-Pacific collisional orogenic belts (e.g. the Mesozoic Mother Lode and Alleghany districts, and the Cenozoic Alaska-Juneau and Treadwell deposits, USA). The only world-class Mesozoic Canadian deposit (Fig. 16) is the Bralorne-Pioneer deposit (British Columbia). Other smaller deposits (not represented in Fig.16) were also formed in the Cordilleran during the Mesozoic, and in the Appalachians during Paleozoic times.

Additionally, three important unexploited deposits (as of December 31, 2004) are noted on Figure 16:

1) Hope Bay (Hope Bay district, Northwest Territories, 210 t Au in unmined reserves and resources),
2) Moss Lake (Shebandowan district, Ontario, 69 t Au, resources),
3) Box (Athabaska district, Saskatchewan, 29 t Au, resources, as of December 1998).

The following deposits, which are located inside districts represented on Figure 16, also contain important unmined resources (as of December 31, 2004, unless otherwise indicated):

1) Tundra (Mackenzie district, Northwest Territories, 262 t Au),
2) Goldex (Val d'Or district, Quebec, 56 t Au),

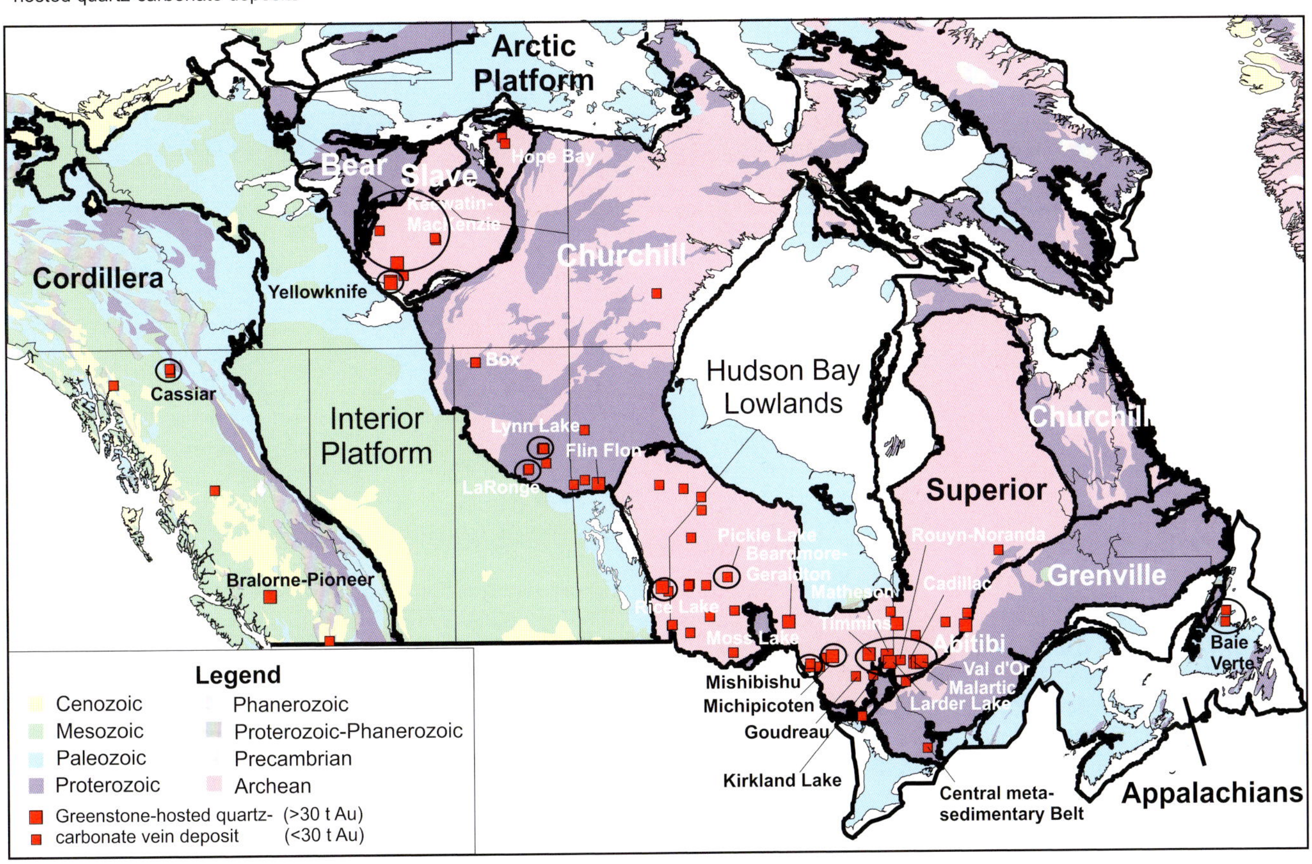

FIGURE 16. Location of Canadian greenstone-hosted quartz-carbonate vein districts. See Appendix 1 for deposit details.

FIGURE 17. Fine-grained chloritized albitite dyke on the 4175 foot level of the McIntyre mine, intruding sericitized Pearl Lake porphyry. Both the albitite dyke and the altered porphyry are cut by quartz-ankerite-albite veins (from Brisbin, 1997; photograph by Nadia Melnik-Proud, caption after Melnik-Proud, 1992; photo obtained by B. Dubé from D. Brisbin).

3) Taurus (Cassiar district, British Columbia, 50 t Au, as of December 1999),

4) Lapa-Pandora-Tonawanda (Cadillac district, Quebec, 54 t Au including 36 t Au as reserves).

Associated Mineral Deposit Types

Greenstone-hosted quartz-carbonate vein deposits are thought to represent the main component of the greenstone deposit clan (Fig. 1) (Poulsen et al., 2000). However, in metamorphosed terranes, other types of gold deposits formed in different tectonic settings and/or crustal levels, such as Au-rich VMS or intrusion-related gold deposits, may have been juxtaposed against greenstone-hosted quartz-carbonate vein deposits during the various increments of strain that characterize Archean greenstone belts (Poulsen et al., 2000). Although these different gold deposits were formed at different times, they now coexist along major faults. Examples include the Bousquet 2 - Dumagami and LaRonde Penna Au-rich VMS deposits that are distributed a few kilometres north of the Cadillac-Larder Lake fault east of Noranda (Fig. 3), where the fault zone hosts the former O'Brien and Thompson Cadillac greenstone-hosted quartz-carbonate vein deposits. Intrusion-related syenite-associated disseminated gold deposits, such as the Holt-McDermott and Holloway mines in the Abitibi greenstone belt of Ontario, occur mainly along major fault zones, in association with preserved slivers of Timiskaming-type sediments and consequently are spatially associated with greenstone-hosted quartz-carbonate vein deposits (Robert, 2001).

Genetic and Exploration Models

Poulsen et al. (2000) has indicated that one of the main problems in deformed and metamorphosed terranes, such as those underlain by greenstone belts, is that many primary characteristics may have been obscured by overprinting deformation and metamorphism to the extent that they are difficult to recognize. This is particularly the case with gold-rich VMS or intrusion-related deposits. But since greenstone-hosted quartz-carbonate vein deposits are syn- to late main phase of deformation, their primary features are, in most cases, relatively well preserved (Groves et al., 2000). Consequently, once a deposit is appropriately classified, exploration models are relatively well defined (cf. Hodgson, 1990, 1993; Groves et al., 2000, 2003). Since the early 1980s, several different genetic models have been proposed to explain the formation of greenstone-hosted quartz-carbonate vein deposits and this has resulted in significant controversy. Some of this controversy is caused by the difficulty in metamorphosed greenstone terranes to classify certain key deposits, such as Hemlo (Lin, 2001; Muir, 2002; Davis and Lin, 2003), due to the poor preservation of primary characteristics largely obscured by post-mineralization deformation and metamorphism. Thus, adequate classification of gold deposits is a key to formulating successful exploration models (Poulsen et al., 2000). An excellent review of the various proposed genetic models, and the pros and cons of each of these, has been presented by Kerrich and Cassidy (1994). Since then, Hagemann and Cassidy (2000), Kerrich et al. (2000), Ridley and Diamond (2000), Groves et al. (2003), and Goldfarb et al. (2005), among others, have also revisited the subject. Only a brief summary is presented here.

Several genetic models have been proposed during the last two decades without attaining a definite consensus. One of the main controversies is related to the source of the fluids. The ore-forming fluid is typically a 1.5 ± 0.5 kb, 350 ± 50°C, low-salinity H_2O-CO_2 ± CH_4 ± N_2 fluid that transported gold as a reduced sulphur complex (Groves et al., 2003). Several authors have emphasized a deep source for gold, with fluids related to metamorphic devolatilization, and deposition of gold over a continuum of crustal levels (cf. Colvine, 1989; Powell et al., 1991; Groves et al., 1995). Others have proposed a magmatic source of fluids (cf. Spooner, 1991), a mantle-related model (Rock and Groves, 1988), drifting of a crustal plate over a mantle plume (Kontak and Archibald, 2002), anomalous thermal conditions associated to upwelling asthenosphere (Kerrich et al., 2000), or deep convection of meteoric fluids (Nesbitt et al., 1986). Hutchinson (1993) has proposed a multi-stage, multi-process genetic model in which gold is recycled from pre-enriched source rocks and early formed, typically subeconomic gold concentrations. Hodgson (1993) also proposed a multi-stage model in which the gold was, at least in part, recycled from gold-rich district-scale reservoirs that resulted from earlier increments of gold enrichment.

The debate on gold genesis was, at least in part, based upon interpretations of stable isotope data, and after more than two decades, it is still impossible to unequivocally distinguish between a fluid of metamorphic, magmatic, or mantle origin (Goldfarb et al., 2005). The significant input of meteoric waters in the formation of quartz-carbonate greenstone-hosted gold deposits is now, however, considered unlikely (Goldfarb et al., 2005). The magmatic and mantle-related models mainly based on spatial relationships between the deposits and intrusive rocks, are challenged by crosscutting field relationships combined with precise U-Pb zircon dating. These show that, in most cases, the proposed magmatic source for the ore-forming fluid is significantly older than the quartz-carbonate veins. For example, in the Timmins area, the quartz-carbonate veins hosting the gold mineralization at the Hollinger-McIntyre deposit cut an albitite dyke intruding the Pearl Lake porphyry (Fig. 17). One such albitite dyke was dated at 2673 +6/-2 Ma

(Marmont and Corfu, 1989) and more recently at 2672.8 ± 1.1 Ma (Ayer et al., 2005). Thus the albitite dyke is ca.15 Ma younger than the 2689 ± 1 Ma Pearl Lake porphyry and various porphyries in the regions ranging in age from 2691 to 2687 Ma (Corfu et al., 1989; Ayer et al., 2003). These chronological relationships rule out the possibility that the ore fluids could be related to known intrusions. An alternative to the magmatic fluid source model is one in which intrusions have provided the thermal energy responsible, at least in part, for fluid circulation (cf. Wall, 1989). The mantle-related model was mainly based on the close spatial relationship between lamprophyre dykes and gold deposits (Rock and Groves, 1988). Key arguments against such a model have been presented by Wyman and Kerrich (1988, 1989). Recently, Dubé et al. (2004) have demonstrated that the lamprophyre dykes spatially associated with gold mineralization at the Campbell-Red Lake deposit, although different than the typical greenstone-hosted quartz-carbonate vein deposit, are at least 10 Ma younger than the main stage of gold mineralization.

Each of these models has merit, and various aspects of all or some of them are potentially involved in the formation of quartz-carbonate greenstone-hosted gold deposits in metamorphic terranes. However, the overall geological settings and characteristics suggest that the greenstone-hosted quartz-carbonate vein deposits are related to prograde metamorphism and thermal re-equilibration of subducted volcano-sedimentary terranes during accretionary or collisional tectonics (cf. Kerrich et al., 2000, and references therein). The deep-seated, Au-transporting fluid has been channelled to higher crustal levels through major crustal faults or deformation zones (Figs. 1, 18). Along its pathway, the fluid has dissolved various components, notably gold, from the volcano-sedimentary packages, which may include a potential gold-rich precursor. The fluid will then precipitate sulphides, gold, and gangue minerals as vein material or wall-rock replacement in second- and third-order structures at higher crustal levels through fluid-pressure cycling processes (Sibson et al., 1988) and temperature, pH, and other physicochemical variations.

Nevertheless, the source of the ore fluid, and hence of gold in greenstone-hosted quartz-carbonate vein deposits, remains unresolved (Groves et al. 2003). According to Ridley and Diamond (2000), a model based on either metamorphic devolatilization or granitoid magmatism best fits most of the geological parameters. These authors indicated that the magmatic model could not be ruled out simply on the basis of a lack of exposed granite in proximity of a deposit with a similar age, because the full subsurface architecture of the crust is unknown. Ridley and Diamond (2000) also indicated that the fluid composition should not be expected to reflect the source. The fluid travels great distances and its measured composition now reflects the fluid-rock interactions along its pathway, or a mixed signature of the source and the wall rocks (Ridley and Diamond, 2000).

In terms of exploration, at the geological province or terrane scale, geological parameters that are common in highly auriferous volcano-sedimentary belts include 1) reactivated crustal-scale faults that controlled emplacement of porphyry-lamprophyre dyke swarms; 2) complex regional-scale

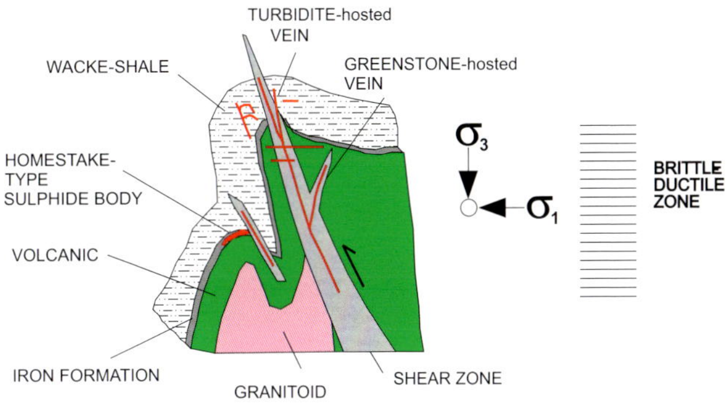

FIGURE 18. Schematic diagram illustrating the setting of greenstone-hosted quartz-carbonate vein deposits (from Poulsen et al., 2000).

geometry of mixed lithostratigraphic packages; and 3) evidence for multiple mineralization or remobilization events (Groves et al., 2003). The empirical spatial and potentially genetic (?) relationship between large gold deposits and a Timiskaming-like regional unconformity represents a key first-order exploration target irrelevant to the deposit type or the mineralization style, as illustrated by large gold districts such as Timmins, Kirkland Lake, and Red Lake (Poulsen et al., 1992; Hodgson, 1993; Robert, 2000; Dubé et al., 2000, 2003, 2004; Robert et al., 2005).

Knowledge Gaps

Several outstanding problems remain for greenstone-hosted quartz-carbonate vein deposits. As mentioned above, the sources of fluid and gold remain unresolved (Ridley and Diamond, 2000). Other critical elements are listed in Hagemann and Cassidy (2000) and Groves et al. (2003). In practical terms, the three most outstanding knowledge gaps to be addressed are 1) better definition of the key geological parameters controlling the formation of giant gold deposits; 2) controls on the high-grade content of deposits or parts of deposits; 3) controls on the distribution of large gold districts, such as Timmins or Val d'Or; and 4) the influence of the early stage structural history of crustal scale faults on their gold endowment. The classification of gold deposit types remains a problem, which is more than an academic exercise as it has a major impact on exploration strategies (e.g. what type of deposit to look for, where, and how?) (Poulsen et al., 2000). However, the reasons why geological provinces, such as the Superior province and the Yilgarn craton are so richly endowed are now much better understood (Robert et al., 2005). It is also believed that integrated research programs, such as the Geological Survey of Canada EXTECH, Natmap, or Targeted Geoscience Initiative, where various aspects of the geology of a gold mining district or camp are addressed, remain an excellent approach for developing additional understanding of these deposits. The most fundamental elements to take into account to successfully establish the complex evolution and relationships between mineralizing event(s), geological setting, and deformation/metamorphism phase(s) are 1) basic chronological field relationships, combined with 2) accurate U-Pb geochronology.

Acknowledgements

This synthesis has been made possible by the kind cooperation of numerous company, government, and university geologists who shared their knowledge and who have allowed surface and underground visits to many gold deposits. We benefited from numerous discussions with colleagues from the provincial surveys and from the Geological Survey of Canada. The first author would like to extend his deepest appreciation to F. Robert and H.K. Poulsen for constructive suggestions, collaboration, and discussions on gold deposits during the last twenty years. W. Goodfellow and I. Kjarsgaard are thanked for their editorial contribution. Careful constructive reviews by R. Goldfarb, M. Gauthier, and S. Castonguay have led to substantial improvements.

References

Akande, S.O., 1985, Coexisting precious metals, sulfosalts and sulfide minerals in the Ross gold mine, Holtyre, Ontario: Canadian Mineralogist, v. 23, p. 95-98.

Alldrick, D.J., 1983, The Mosquito Creek Mine, Cariboo Gold Belt (93H/4), *in* Geological Fieldwork 1982: British Columbia Ministry of Energy, Mines and Petroleum Resources, Paper 1983-1, p. 99-112.

Ames, H.G., 1948, Perron mine, *in* Structural Geology of Canadian Ore Deposits - A Symposium: Canadian Institute of Mining and Metallurgy, Special Volume 1, p. 893-898.

Ames, D.E., Franklin, J.M., and Froese, E., 1991, Zonation of hydrothermal alteration at the San Antonio gold mine, Bisset, Manitoba, Canada: Economic Geology, v. 86, p. 600-619.

Anglin, C.D., and Franklin, J.M., 1985, Gold mineralization in the Beardmore-Geraldton area of northwestern Ontario: Structural considerations and the role of iron formation: Geological Survey of Canada, Current Research Paper 85-1B, p. 193-201.

Anglin, C.D., Jonasson, I.R., and Franklin, J.M., 1996, Sm-Nd dating of scheelite and tourmaline: Implications for the genesis of Archean gold deposits, Val d'Or, Canada: Economic Geology, v. 91, p. 1372-1382.

Archambault, G., Guha, J., Tremblay, A., and Kanwar, R., 1984, Implications of the geomechanical interpretation of the Copper Rand deposit on the Dore Lake shear belt, *in* Guha, J., And Chown, E.H., eds., Chibougamau - Stratigraphy and Mineralization: Proceedings of the Chibougamau Symposium and Field Trip: Canadian Institute of Mining, Special Volume 34, p. 300-318.

Armitage, A.E., Tella, S., and Miller, A.R., 1993, Iron-formation-hosted gold mineralization and its geological setting, Meliadine Lake area, District of Keewatin, Northwest Territories: Geological Survey of Canada, Current Research 1993-C, p. 187-195.

Armitage, A.E., James, R.S., and Goff, S.P., 1996, Gold mineralization in Archean banded iron formation, Third Portage Lake area, Northwest Territories, Canada: Exploration and Mining Geology, v. 5, p. 1-15.

Armstrong, H.S., 1943, Gold ores of the Little Long Lac area, Ontario: Economic Geology, v. 38, p. 204-252.

Ayer, J.A., Barr, E., Bleeker, W., Creaser, R.A., Hall, G., Ketchum, J.W.F., Powers, D., Salier, B., Still, A., and Trowell, N.F., 2003, New geochronological results from the Timmins area: Implications for the timing of late-tectonic stratigraphy, magmatism and gold mineralization: *in* Summary of Field work and Other Activities: Ontario Geological Survey, Open File Report 6120, p. 33-1 to 33-11.

Ayer, J.A., Thurston, P.C., Bateman, R., Dubé, B., Gibson, H.L., Hamilton, M.A., Hathway, B., Hocker, S.M., Houlé, M.G., Hudak, G., Ispolatov, V.O., Lafrance, B., Lesher, C.M., MacDonald, P.J., Péloquin, A.S., Piercey, S.J., Reed, L.E., and Thompson, P.H., 2005, Overview of results from the greenstone architecture project: Discover Abitibi Initiative: Ontario Geological Survey, Open File Report 6154, 146 p.

Backman, O.L., 1936, The geology of the Siscoe gold mine: Canadian Mining Journal, v. 57, p. 467-475.

Bacon, W.R., 1978, Lode gold deposits in Western Canada: Canadian Institute of Mining and Metallurgy Bulletin, v. 71, p. 96-104.

Barley, M.E., and Groves, D.I., 1992, Supercontinent cycles and the distribution of metal deposits through times: Geology, v. 20, p. 291-294.

Barnett, E.S., Hutchinson, R.W., Adamcik, A., and Barnett, R., 1982, Geology of the Agnico-Eagle deposit, Quebec, *in* Hutchinson, R.W., Spence, C.D., and Franklin, J.M., eds., Precambrian Sulphide Deposits, H.S. Robinson Memorial Volume: Geological Association of Canada, Special Paper 25, p. 403-426.

Barrett, R.E., and Johnston, A.W., 1948, Central Patricia Mine, *in* Structural Geology of Canadian Ore Deposits - A Symposium: Canadian Institutde of Mining and Metallurgy, Special Volume 1, p. 368-372.

Barret, T.L., and Sherlock, R.L., 1996, Geology, lithogeochemistry and volcanic setting of the Eskay Creek Au-Ag-Cu-Zn deposit, Northwestern British Columbia: Exploration and Mining Geology, v. 5, p. 339-368.

Barron, K.M., Duke, N,A., and Hodder, R.W., 1989, Petrology of the Springpole Lake alkalic volcanic complex, *in* Geoscience Research Grant Program; Summary of Research 1988-1989: Ontario Geological Survey, Miscellaneous Paper 143, p. 133-145.

Bartram, G.D., and McCall, G.J.H., 1971, Wall-rock alteration associated with auriferous lodes in the Golden Mile, Kalgoorlie, *in* Glover, J.E., ed., Symposium on Archaean Rocks: Geological Society of Australia, Special Publication 3, p. 191-199.

Basnett, R., 1999, Seabee Mine, *in* Ashton, K.E., and Harper, C.T., eds., MINExpo '96 Symposium: Advances in Saskatchewan Geology and Mineral Exploration: Saskatchewan Geological Society, Special Publication No. 14, p. 72-79.

Bateman, R.J., Ayer, J.A., Dubé, B., and Hamilton, M.A., 2005, The Timmins - Porcupine Gold Camp, Northern Ontario: The Anatomy of an Archaean Greenstone and Its Gold Mineralization. Discover Abitibi Initiative: Ontario Geological Survey, Open File Report 6158, 90 p.

Beaudoin, G., Hubert, C., Trudel, P., and Perreault, G., 1987, Géologie du projet Callahan, cantons de Vassan et de Dubuisson - District de Val-d'Or: Ministère de l'Éenergie et des Ressources, Gouvernement du Québec, MB 87-48.

Belkabir, A., and Hubert, C., 1995, Geology and structure of a sulfide-rich gold deposit: an example from the Mouska gold mine, Bousquet district, Canada. Economic Geology, v. 90, p. 1064-1079.

Belkabir, A., Robert, F., Vu, L., and Hubert, C., 1993, The influence of dikes on auriferous shear zone development within granitoid intrusions: The Bourlamaque pluton, Val d'Or district, Abitibi greenstone belt: Canadian Journal of Earth Sciences, v. 30, p. 1924-1933.

Bergmann, A., 1947, Development and ore possibilities of the area between McKenzie Island and East Bay in the Red Lake mining district of Ontario: Precambrian, v. 20, p. 4-7.

Bevier, M.L., and Gebert, J.S., 1991, U-Pb geochronology of the Hope Bay - Elu Inlet area, Bathurst Block, northeastern Slave Structural Province, Northwest Territories: Canadian Journal of Earth Sciences, v. 28, p. 1925-1930.

Blackburn, C.E., Johns, G.W., Ayer, J., and Davis, D.W., 1991, Wabigoon Subprovince, Chapter 9, *in* Thurston, P.C., Williams, H.R., Sutcliffe, R.H., and Stott, G.M., eds., Geology of Ontario: Ontario Geological Survey, Special Volume 4, p. 303-381.

Blais, A., 1991, Copper Rand mine, *in* Guha, J., Chown, E.H., and Daigneault, R., eds., Litho-tectonic Framework and Associated Mineralization of the Eastern Extremity of the Abitibi Greenstone Belt (8th Symposium of the International Association on the Genesis of Ore Deposits, Field Trip 3): Geological Survey of Canada, Open File 2158, p. 63-67.

Blais, M., 1955, L'altération hydrothermale en bordure des filons aurifères de la mine O'Brien, comté d'Abitibi-Est: Le Naturaliste Canadien, v. 132, p. 77-98.

Bleeker, W., 1995, Surface geology of the Porcupine camp, *in* Heather, K.B., ed., Precambrian '95 - Tectonics and Metallogeny of Archean Crust in the Abitibi-Kapuskasing-Wawa Region, Field Trip Guidebook: Geological Survey of Canada, Open File Report 3141, p. 13-47.

—— 1997, Geology and mineral deposits of the Porcupine camp: Geological Association of Canada, Field Trip Guide Book B6, 10-37 p.

Boldy, J., Drouin, M., Hilgendorf, C., Davidson, D., Boniwell, J.B., and Gingerich, J., 1984, Case history of a gold discovery, Eastmain River area, Quebec, *in* Guha, J., and Chown, E.H., eds., Chibougamau - Stratigraphy and Mineralization: Proceedings of the Chibougamau Symposium, Chibougamau, Quebec, Sept. 2-6, 1984, CIM Special Volume 34, p. 441-456.

Brisbin, D.I., 1997, Geological Setting of Gold Deposits in the Porcupine Gold Camp, Timmins, Ontario: Ph.D. thesis, Queen's University, Kingston, Ontario, 523 p.

Brisson, H., 1998, Caractéristiques, chronologie et typologie des minéralisations aurifères du la région du lac Shortt (Québec), sous-province archéenne de l'Abitibi: Université du Québec à Chicoutimi, Ph.D. thesis, 277 p.

Brodie Hicks, H., 1945, The geology of the Central Patricia mine, Ontario: Precambrian, v. 18, p. 7-9.

Brown, D.A., 1987, Geological Setting of Volcanic-Hosted Silbak Premier Mine, Northwestern British Columbia (104 A/4, B/1): M.Sc. thesis, University of British Columbia, Vancouver, British Columbia, 216 p.

Brown, R.A., 1948, O'Brien mine, *in* Structural Geology of Canadian Ore Deposits: A Symposium: Canadian Institute of Mining and Metallurgy, Special Volume 1, p. 809-816.

Bruce, D.E., and Miller, J.H.L., 1990, Equity Silver mines; A success story, *in* Hollister, V.F., ed., Discoveries of Valuable Minerals and Precious Metals Deposits Related to Intrusions and Faults: Society of Mining, Metalllurgy and Exploration Case Histories of Mineral Discoveries, v. 2, p. 144-161.

Bruce, E.L., and Samuel, W., 1937, Geology of the Little Long Lac Mine: Economic Geology, v. 32, p. 318-334.

Buffam, S.W., and Allen, R.B., 1948, Chesterville mine, *in* Structural Geology of Canadian Ore Deposits - A Symposium: Canadian Institute of Mining and Metallurgy, Special Volume 1, p. 662-671.

Bullis, H.R., Pratico, W., and Webb, D.R., 1987, The Con Mine, *in* Padgham, W.A., ed., Yellowknife Guide Book: A Guide to the Geology of the Yellowknife Volcanic Belt and Its Bordering Rocks: Geological Associaton of Canada, Mineral Deposits Division, p. 175-181.

Bullis, H.R., Hureau, R.A., and Penner, B.D., 1994, Distribution of gold and sulfides at Lupin, Northwest Territories: Economic Geology, v. 89, p. 1217-1227.

Burrows, D.R., and Spooner, E.T.C., 1986, McIntyre Cu-Au deposit, Timmins, Ontario, Canada, *in* Macdonald, A.J., Downes, M., Pirie, J., and Chater, A.M., eds., Proceeding of Gold '86, An International Symposium on the Geology of Gold: p. 23-39.

Burrows, D.R., Spooner, E.T.C., Wood, P.C., and Jemielita, R.A., 1993, Structural controls on formation of the Hollinger-McIntyre Au quartz vein system in the Hollinger Shear Zone, Timmins, Southern Abitibi Greenstone Belt, Ontario: Economic Geology, v. 88, p. 1643-1663.

Byrne, N.W., 1950, The Discovery Yellowknife gold mine, Giauque Lake, Yellowknife mining area, N.W.T.: Precambrian, v. 23, p. 8-12.

Byron, M., 1994, Anomalous haloes of pathfinder elements for gold, Upper Canada deposit, Kirkland Lake, Ontario: Exploration and Mining Geology, v. 3, p. 161-179.

Cadieux, A.-M., 2000, Géologie du gîte aurifère Eau Claire, propriété Clearwater, Baie James, Québec: M.Sc. thesis, Université Laval, Québec, 242 p.

Callan, N.J., and Spooner, E.T.C., 1998, Repetitive hydraulic fracturing and shear zone inflation in an Archean granitoid-hosted, ribbon banded, Au-quartz vein system, Renabie area, Ontario, Canada: Ore Geology Reviews, v. 12, p. 237-266.

Cameron, E.M., and Hattori, K., 1987, Archean gold mineralization and oxidized hydrothermal fluids: Economic Geology, v. 82, p. 1177-1191.

Cann, R.M., and Godwin, C.I., 1980, Geology and age of the Kemess porphyry copper-molybdenum deposit, north-central British Columbia: Canadian Institute of Mining and Metallurgy Bulletin, v. 73, p. 94-99.

Carignan, J., and Gariépy, C., 1993, Pb isotope geochemistry of the Silidor and Launay gold deposits: Implications for the source of Archean Au in the Abitibi subprovince: Economic Geology, v. 88, p. 1722-1730.

Carrier, A., 1994, Évolution structurale et métallogénique du gisement aurifère Silidor, Abitibi, Québec: M.Sc. thesis, University of Quebec at Montreal, Montreal, Quebec, 272 p.

Carrier, A., Jébrak, M., Angelier, J., and Holyland, P., 2000, The Silidor deposit, Rouyn-Noranda district, Abitibi belt: Geology, structural evolution and paleostress modeling of an Au quartz vein-type deposit in an Archean trondhjemite: Economic Geology, v. 95, p. 1049-1065.

Carter, T.R., 1984, Metallogeny of the Grenville Province, Southeastern Ontario: Ontario Geological Survey, Open File Report 5515, 422 p.

Cattalani, S., Barrett, T.J., Maclean, W.H., Hoy, L., Hubert, C., and Fox, J.S., 1993, Métallogénèse des gisements Horne et Quémont (région de Rouyn-Noranda): Ministère de l'Énergie et des Ressources, Québec, ET 90-07, 121 p.

Chabot, F., 1998, Minéralisation aurifère dans le pluton de Bourlamaque: la mine Beaufor, Section 3A, *in* Pilote, P., Moorhead, J., and Mueller, W., eds., Développement d'un arc volcanique, la région de Val d'Or, ceinture de l'Abitibi - Volcanologie physique et évolution métallogénique: Geological Association of Canada, Mineralogical Association of Canada, Guidebook 1998 A2, p. 59-64.

Champigny, N., and Sinclair, A.J., 1982, The Cinola gold deposit, Queen Charlotte Islands, British Columbia: *in* Hodder, R.W., and Petruk, W., Geology of Canadian Gold Deposits: Canadian Institute of Mining and Metallurgy, Special Volume 24, p. 243 - 254

Charlewood, G.H., 1964, Geology of Deep Developments on the Main Ore Zone at Kirkland Lake: Ontario Department of Mines, Geological Circular No. 11, 49 p.

Chi, G., Dubé, B., and Williamson, K., 2002, Preliminary fluid-inclusion microthermometry study of fluid evolution and temperature-pressure conditions in the Goldcorp High-Grade zone, Red Lake mine, Ontario, *in* Current Research 2002: Geological Survey of Canada, p. 1-12.

Childe, F., 1996, U-Pb geochronology and Nd and Pb isotopic characteristics of the Au-Ag-rich Eskay Creek VMS deposit, northwestern British Columbia: Economic Geology, v. 91, p. 1209-1224.

Church, B.N., 1985, Geology of the Mount Attwood-Phoenix area, Greenwood (82E/2): *in* Geological Fieldwork 1984: British Columbia Ministry of Energy, Mines and Resources, Paper 1985-2, p. 17-21.

Church, B.N., and Jones, L.D., 1997, Metallogeny of the Greenwood mining camp (082E02), British Columbia Ministry of Employment and Investment; Energy and Minerals Division, Online Report, http://www.em.gov.bc.ca/Mining/Geolsurv/Minfile/mapareas/greenwd.htm, 20 p.

Coker, W.B., Fox, J.S., and Sopuck, V.J., 1982, Organic centre-lake sediments: Application in the geochemical exploration for gold in the Canadian Shield of Saskatchewan: *in* Hodder, R.W., and Petruk, W., Geology of Canadian Gold Deposits: Canadian Institute of Mining and Metallurgy, Special Volume 24, p. 267-280.

Colvine, A.C., 1989, An empirical model for the formation of Archean gold deposits: Products of final cratonization of the Superior Province, Canada, *in* Keys, R.R., Ramsay, W.R.H., and Groves, D.I., eds., The Geology of Gold Deposits: The Perspective in 1988: Economic Geology, Monograph 6, p. 37-53.

Colvine, A.C., Fyon, J.A., Heather, K.B., Marmont, S., Smith, P.M., and Troop, D.G., 1988, Archean Lode Gold Deposits in Ontario: Ontario Geological Survey, Miscellaneous Paper 139, 136 p.

Condie, K., 1998, Episodic continental growth and supercontinents; A mantle avalanche connection? Earth and Planetary Science Letters, v. 163, p. 97-108.

Coombe, W., Lewry, J.F., and Macdonald, R., 1986, Regional geological setting of gold in the La Ronge Domain, Saskatchewan, *in* Clark, L.A., ed., Gold in the Western Shield: Canadian Institute of Mining and Metallurgy, Special Volume 38, p. 26-56.

Corfu, F., Krogh, T.E., Kwok, Y.Y., and Jensen, L.S., 1989, U-Pb zircon geochronology in the southwestern Abitibi greenstone belt, Superior Province: Canadian Journal of Earth Sciences, v. 26, p. 1747-1763.

Cormie, J.M., 1936, Geology and ore deposits of the Central Patricia gold mine, Ontario: Economic Geology, v. 31, p. 93-103.

Couture, J.F., 1990, Carte géologique des gîtes métallifères des districts de Rouyn-Noranda et de Val-d'Or (partie sud des feuillets SNRC 32C et 32D ouest): Ministère de l'Énergie et des Ressources du Québec, #2109, DV-9011, scale 250000.

Couture, J.F., and Guha, J., 1990, Relative timing of emplacement of an Archean lode-gold deposit in an amphibolite terrane: The Eastmain River deposit, northern Quebec: Canadian Journal of Earth Sciences, v. 27, p. 1621-1636.

Couture, J.F., and Pilote, P., 1993, The geology and alteration patterns of a disseminated, shear zone-hosted mesothermal gold deposit: The Francoeur 3 deposit, Rouyn-Noranda, Quebec: Economic Geology, v. 88, p. 1664-1684.

Cyr, J.B., Pease, R.B., and Schroeter, T.G., 1984, Geology and mineralization at Equity Silver mine: Economic Geology, v. 79, p. 947-968.

Daigneault, R., and Allard, G.O., 1990, Le Complexe du lac Doré et son environnement géologique, Région de Chibougamau - Sous-Province de l'Abitibi: Ministère de l'Énergie et des Ressources, Québec, MM-89-03, 275 p.

Daigneault, R., and Archambault, G., 1990, Les grands couloirs de déformation de la Sous-Province de l'Abitibi, *in* Rive, M., Verpaelst, P., Gagnon, Y., Lulin, J.M., Riverin, G., and Simard, A., eds., The Northwestern Quebec Polymetallic Belt: A Summary of 60 Years of Mining and Exploration: Canadian Institute of Mining and Metallurgy, Special Volume 43, p. 43-64.

Darling, R., Vu, L., Dussault, C., Waitzenegger, B., and Popov, V., 1986, New gold target in Quebec; Geology of the Ferderber gold deposit, Val d'Or: Canadian Mining Journal, v. 107, p. 25-29.

Davidson, S.C., and Banfield, A.F., 1944, Geology of the Beattie gold mine, Duparquet, Quebec: Economic Geology, v. 39, p. 535-556.

Davis, D.W., and Lin, S., 2003, Unraveling the geologic history of the Hemlo Archean gold deposit, Superior Province, Canada; A U-Pb geochronological study: Economic Geology, v. 98, p. 51-67.

Davis, D.W., and Smith, P.M., 1991, Archean gold mineralization in the Wabigoon Subprovince, a product of crustal accretion: Evidence from U-Pb geochronology in the Lake of the Woods area, Superior Province, Canada: Journal of Geology, v. 99, p. 337-353.

Davis, W.J., and Zaleski, E., 1998, Geochronological investigations of the Woodburn Lake group, western Churchill Province, Northwest Territories: preliminary results: *in* Radiogenic Age and Isotopic Studies, Report 11: Geological Survey of Canada, Current Research 1998-F, p. 89-97.

Derry, D.R., Hopper, C.H., and Mcgowan, H.S., 1948, Matachewan Consolidated mine, *in* Structural Geology of Canadian Ore Deposits: A Symposium: Canadian Institute of Mining and Metallurgy, Special Volume 1, p. 638-643.

Diakow, L.J., and Metcalfe, P., 1997, Geology of the Swannell ranges in the vicinity of the Kemess copper-gold porphyry deposit, Attycelley Creek (NTS94E/2), Toodoggone River map area: *in* Lefebure, D.V., McMillan, W.J., and McArthur, J.G., eds., Geological Fieldwork 1996: British Columbia Ministry of Energy, Mines and Petroleum Resources, Paper 1997-1, p. 101-115.

Dion, C., and Maltais, G., 1998, La mine d'or Joe Mann, *in* Pilote, P., ed., Géologie et métallogénie du district minier de Chapais-Chibougamau: Ministère des Ressources Naturelles du Québec, DV 98-03, p. 125-131.

Dougherty, E.Y., 1934, Mining geology of the Vipond gold mine, Porcupine district, Ontario: Transactions of the Canadian Institute of Mining and Metallurgy and of the Mining Society of Nova Scotia, v. 37, p. 260-284.

Downes, M.J., Hodges, D.J., and Derweduwen, J., 1984, A free carbon- and carbonate-bearing alteration zone associated with the Hoyle Pond gold occurrence, Ontario, Canada, *in* Foster, R.P., ed., Gold '82: The Geology, Geochemistry and Genesis of Gold Deposits: Proceedings of the International Symposium: Balkema, Rotterdam, p. 435-448.

Dubé, B., and Guha, J., 1989, Étude métallogénique (aurifère) du filon-couche de Bourbeau (région de Chibougamau): Ministère de l'Énergie et des Ressources du Québec, MM 87-03, 104 p.

—— 1992, Relationship between NE trending regional faults and Archean mesothermal gold-copper mineralization: Cooke mine, Abitibi greenstone belt, Quebec, Canada: Economic Geology, v. 87, p. 1525-1540.

Dubé, B., Guha, J., and Rocheleau, M., 1987, Alteration patterns related to gold mineralization and their relation to CO_2/H_2O ratios: Mineralogy and Petrology, v. 37, p. 267-291.

Dubé, B., Poulsen K.H., and Guha, J., 1989, The effects of layer anisotropy on auriferous shear zones: The Norbeau mine, Quebec: Economic Geology, v. 84, p. 871-878.

Dubé, B., Lauzière, K., and Poulsen, H.K., 1993, The Deer Cove deposit: An example of thrust-related breccia-vein type gold mineralization in the Baie Verte Peninsula, *in* Current Research: Newfoundland and Labrador Geological Survey, Report 93-1D, p. 481.

Dubé, B., Dunning, G.R., Lauziere, K., and Roddick, J.C., 1996, New insights into the Appalachian Orogen from geology and geochronology along the Cape Ray fault zone, southwest Newfoundland: Geological Society of America Bulletin, v. 108, p. 101-116.

Dubé, B., Dunning, G.R., nd Lauzière, K., 1998a, Geology of the Hope Brook Mine, Newfoundland, Canada: A preserved Late Proterozoic high-sulfidation epithermal gold deposit and its implications for exploration: Economic Geology, v. 93, p. 405-436.

Dubé, B., Lauzière, K., and Boisvert, 1998b, Lithological map and hydrothermal alteration map of the high-sulfidation Hope Brook gold deposit: Geological Survey of Canada, Open File 3606, scale 1:2500.

Dubé, B., Balmer, W., Sanborn-Barrie, M., Skulski, T., and Parker, J., 2000, A Preliminary Report on Amphibolite-Facies, Disseminated-Replacement-Style Mineralization at the Madsen Gold Mine, Red Lake, Ontario: Geological Survey of Canada, Current Research 2000-C17, 12 p.

Dubé, B., O'Brien, S., and Dunning, G.R., 2001a, Gold deposits in deformed terranes: Examples of epithermal and quartz-carbonate shear-zone-related gold systems in the Newfoundland Appalachians and their implications for exploration: North Atlantic Mineral Symposium, St. John's, Newfoundland, May 27-30, 2001, p. 31-35.

Dubé, B., Williamson, K., and Malo, M., 2001b, Preliminary report on the geology and controlling parameters of the Goldcorp Inc. High Grade zone, Red Lake mine, Ontario, *in* Current Research 2001: Geological Survey of Canada, p. 1-13.

—— 2002, Geology of the Goldcorp Inc. High-Grade Zone, Red Lake Mine, Ontario: An Update: Geological Survey of Canada, Current Research 2001-C26, 13 p.

—— 2003a, Gold Mineralization within the Red Lake Mine Trend: Example from the Cochenour-Willans Mine Area, Red Lake, Ontario, with New Key Information from the Red Lake Mine and Potential Analogy with the Timmins Camp: Geological Survey of Canada, Current Research 2003-C21, 15 p.

Dubé, B., Mercier-Langevin, P., Lafrance, B., Hannington, M.D., Moorhead, J., Davis, D., Galley, A., and Pilote, P., 2003b, The Doyon-Bousquet-Laronde Archean Au-rich VMS gold camp: The example of the world-class LaRonde deposit, Abitibi, Quebec: CAnadian Institue of Mining and Metalurgy, Timmins, Extended Abstract, p. 3-10.

Dubé, B., Williamson, K., McNicoll, V., Malo, M., Skulski, T., Twomey, T., and Sanborn-Barrie, M., 2004, Timing of gold mineralization in the Red Lake gold camp, northwestern Ontario, Canada: New constraints from U-Pb geochronology at the Goldcorp high-grade zone, Red Lake mine and at the Madsen Mine: Economic Geology, v. 99, p. 1611-1641.

Dubé, P.L., Hubert, C., Brown, A.C., and Simard, J.-M., 1991, The Telbel orebody of the Agnico-Eagle mine in the Joutel area of the Abitibi greenstone belt, Quebec, Canada: A stratabound, gold-bearing massive siderite deposit with early diagenetic pyritization, *in* Ladeira, E.A., ed., Brazil Gold '91: The Economics, Geology, Geochemistry and Genesis of Gold Deposits: Proceedings of a Symposium, Belo Horizonte, May 1991, Balkena, Rotterdam, p. 493-498.

Durocher, M.E., 1983, The nature of hydrothermal alteration associated with the Madsen and Starratt-Olsen gold deposits, Red Lake area: Ontario Geological Survey, Miscellaneous Paper 110, p. 111-140.

Durocher, M.E., and Hugon, H., 1983, Structural geology and hydrothermal alteration in the Flat Lake-Howery Bay Deformation Zone, Red Lake area, *in* Wood, J., White, O.L., Barlow, R.B., and Colvine, A.C., eds., Summary of Field Work, 1982: Ministry of Natural Resources, Ontario Geological Survey, Miscellaneous Paper 116, p. 216-219.

Dugas, F., 1963, Prospecting possibilities in the Belleterre area, Western Quebec: Canadian Mining Journal, v. 84, p. 85-89.

Easton, R.M., and Fyon, J.A., 1992, Metallogeny of the Grenville Province, *in* Thurston, P.C., Williams, H.R., Sutcliffe, R.H., and Stott, G.M., eds., Geology of Ontario: Ontario Geological Survey, Special Volume 4, p. 1217-1254.

Ettlinger, A.D., Meinert, L.D., and Ray, G.E., 1992, Gold skarn mineralization and fluid evolution in the Nickel Plate deposit, British Columbia: Economic Geology, v. 87, p. 1541-1565.

Fahrni, K.C., 1966, Geological relations at Copper Mountain, Phoenix and Granisle mines, *in* Tectonic History and Mineral Deposits of the Western Cordillera: Canadian Institute of Mining and Metallurgy, Special Volume 8, p. 315-320.

Ferguson, S.A., Buffman, B.S.W., Carter, O.F., Griffis, A.T., Holmes, T.C., Hurst, M.E., Jones, W.A., Lane, H.C., and Longley, C.S., 1968, Geology and Ore Deposits of Tisdale Township, District of Cochrane: Ontario Department of Mines, Geological Report 58, 117 p.

Ferguson, S.A., Groon, H.A., and Haynes, R., 1971, Gold Deposits of Ontario. Part 1: Districts of Algoma, Cochrane, Kenora, Rainy River, and Thunder Bay: Ontario Department of Mines and Northern Affairs, Ontario Division of Mines, Mineral Resources Circular 13, 315 p.

Fraser, R.J., 1993, The Lac Troilus Gold-Copper deposit, Northwestern Quebec: A possible archean porphyry system: Economic Geology, v. 88, p. 1685-1699.

Fyles, J.T., 1984, Geological Setting of the Rossland Mining Camp: British Columbia Ministry of Energy, Mines and Petroleum Resources, Bulletin 74, 61 p.

Fyon, A., Breaks, T.W., Heather, K.B., Jackson, W.L., Muir, T.L., Stott, G.M., and Thurston, P.C., 1992, Metallogeny of metallic mineral deposits in the Superior Province of Ontario, Chapter 22 *in* Thurston, P.C., Williams, H.R., Sutcliffe, R.H., and Stott, G.M., eds., Geology of Ontario: Ministry of Northern Development and Mines, Special Volume No. 4, Part 2, p. 1091-1176.

Gaboury, D., and Daigneault, R., 1999, Evolution from sea floor-related to sulfide-rich quartz veint-type gold mineralization during deep submarine volcanic construction: the Géant Dormant gld mine, Archean Abitibi belt, Canada: Economic Geology, v. 94, p. 3-22.

—— 2000, Flat vein formation in a transitional crustal setting by self-induced fluid pressure quilibrium - An example from the Géant Dormant gold mine, Canada: Ore Geology Reviews, v. 17, p. 155-178.

Gaboury, D., Daigneault, R., Tourigny, G., and Gobeil, C., 1996a, An Archean volcanic-related gold-sulfide-quartz vein orebody: The Géant Dormant mine, Abitibi Subprovince, Québec, Canada: Exploration and Mining Geology, v. 5, p. 197-213.

Gaboury, D., Dubé, B., Laflèche, M., and Lauzière, K., 1996b, Geology of the mesothermal Hammer Down Gold Deposit, Newfoundland: Canadian Journal of Earth Sciences, v. 33, p. 335-350.

Gaboury, D., Daigneault, R., and Beaudoin, G., 2000, Volcanogenic-related origin of sulfide-rich quartz veins: evidence from O and S isotopes at the Géant Dormant gold mine, Abitibi belt, Canada: Mineralium Deposita, v. 35, p. 21-36.

Galley, A.G., Ames, D.E., and Franklin, J.M., 1989, Results of studies on gold metallogeny of the Flin Flon belt, *in* Galley, A.G., Investigations by the Geological Survey of Canada in Manitoba and Saskatchewan during the 1984 - 1989 Mineral Development Agreements: Geological Survey of Canada, Open File 2133, p. 25-32.

Gaulin, H., and Trudel, P., 1990, Caractéristiques pétrographiques et géochimiques de la minéralisation aurifère à la mine Elder, Abitibi, Québec: Canadian Journal of Earth Sciences, v. 27, p. 1637-1650.

Gauthier, N., Rocheleau, M., Kelly, D., and Gagnon, Y., 1990, Controls on the distribution of gold mineralization within the Cadillac tectonic zone, Rouyn-Beauchastel segment, Abitibi belt, Quebec, *in* Rive, M., Verpaelst, P., Gagnon, Y., Lulin, J.M., Riverin, G., and Simard, A., eds., The Northwestern Quebec Polymetallic Belt: A Summary of 60 Years of Mining and Exploration: Canadian Institute of Mining and Metallurgy, Special Volume 43, p. 185-198.

Gill, J.E., and Byers, A.R., 1948, Surf Inlet and Pugsley mines, *in* Structural Geology of Canadian Ore Deposits: A Symposium: Canadian Institute of Mining and Metallurgy, Special Volume 1, p. 99-104.

Goldfarb, R.J., Groves, D.I., and Gardoll, D., 2001, Orogenic gold and geologic time: A global synthesis: Ore Geology Reviews, v. 18, p. 1-75.

Goldfarb, R.J., Baker, T., Dubé, B., Groves, D.I., Hart, C.J.R., Robert, F., and Gosselin, P., 2005, World distribution, productivity, character, and genesis of gold deposits in metamorphic terranes, *in* Hedenquist, J.W., Thompson, J.F.H., Goldfarb, R.J., and Richards, J.P., eds., Economic Geology One Hundredth Anniversary Volume: 1905 - 2005: Society of Economic Geologists, p. 407-450.

Gordon, J.B., Lovel, H.L., and de Grijs, J., 1971, Gold Deposits of Ontario: Part 2, Districts of Muskoka, Nipissing, Parry Sound, Sudbury, Timiskaming and Counties Southern Ontario: Ontario Ministry of Natural Resources, Ontario Geological Survey, Mineral Deposits Circular 18, 253 p.

Gosselin, P., and Dubé, B., 2005a, Gold Deposits of the World: Distribution, Geological Parameters and Gold Content: Geological Survey of Canada, Open File 4895, 1 CD-ROM.

—— 2005b, Gold deposits of Canada: Distribution, geological parameters and gold content. Geological Survey of Canada, Open File 4896, 1 CD-ROM.

Gray, M.D., and Hutchinson, R.W., 2001, New evidence for multiple periods of gold emplacement in the Porcupine mining district, Timmins area, Ontario, Canada: Economic Geology, v. 96, p. 453-475.

Greig, C.U., Anderson, R.G., Daubeny, P.H., Bull, K.F., and Hinderman, T.K., 1994, Geology of the Cambria Icefield: Regional setting and Red Mountain gold deposit, northwestern British Columbia: Geological Survey of Canada, Current Research 1994-A, p. 45-56.

Groves, D.I., Goldfarb, R.J., Robert, F., and Hart, C.J.R., 2003, Gold deposits in metamorphic belts: Overview of current understanding, outstanding problems, future research, and exploration significance: Economic Geology, v. 98, p. 1-29.

Groves, D.I., Phillips, G.N., Ho, S.E., Henderson, M.E., Clark, M.E., and Woad, G.M., 1984, Controls on distribution of Archean hydrothermal gold deposits in Western Australia, *in* Foster, R.P., ed., Gold '82: The Geology, Geochemistry and Genesis of Gold Deposits: Proceedings of the International Symposium: Balkema, Rotterdam, p. 689-712.

Groves, D.I., Goldfarb, R.J., Know-Robinson, C.M., Ojala, J., Gardoll, S., Yun, G., and Holyland, P., 2000, Late-kinematic timing of orogenic gold deposits and significance for computer-based exploration techniques with emphasis on the Yilgarn block, Western Australia: Ore Geology Reviews, v. 17, p. 1-38.

Groves, D.J., 1993, The crustal continuum model for late Archean lode-gold deposits of the Yilgarn Block, Western Australia: Mineralium Deposita, v. 28, p. 366-374.

Groves, D.J., Ridley, J.R., Bloem, E.J.M., Gebre-Mariam, M., Hronsky, J.M.A., Knight, J.T., McCuaig, T.C., McNaughton, N.J., and Ojala, J., 1995, Lode-gold deposits of the Yilgarn Block: products of late-Archaean crustal-scale over-pressured hydrothermal systems, *in* Coward, R.P., and Ries, A.C., eds., Early Precambrian Processes: Geological Society of London, Special Publication 95, p. 155-172.

Groves, D.J., Goldfarb, R.J., Gebre-Mariam, M., Hagemann, S.G., and Robert, F., 1998, Orogenic gold deposits: A proposed classification in the context of their crustal distribution and relationships to other gold deposit types: Ore Geology Reviews, v. 13, p. 7-27.

Gummer, P.K., 1986, Geology of the Seabee gold deposit, Tabbernor Lake greenstone belt, Saskatchewan, *in* Clark, L.A., ed., Gold in the Western Shield: Canadian Institute of Mining and Metallurgy, Special Volume 38, p. 272-284.

Hagemann, S.G., and Cassidy, K.F., 2000, Archean orogenic lode gold deposits, *in* Hagemann, S.G., and Brown, P.E., eds., Gold in 2000: Society of Economic Geologists, Reviews in Economic Geology, v. 13, p. 9-68.

Hall, R.S., and Rigg, D.M., 1986, Geology of the west anticline zone, Musselwhite prospect, Opapimiskan Lake, Ontario, Canada, *in* Macdonald, A.,J., Downes, M., Pirie, J., and Chater, A.M., eds., Proceeding of Gold '86, An International Symposium on the Geology of Gold: p. 124-136.

Hannington, M.D., Poulsen, K.H., Thompson, J.F.H., and Sillitoe, R.H., 1999, Volcanogenic gold in the massive sulfide environment, Chapter 14 *in* Barrie, C.T., and Hannington, M.D., Volcanic-Associated Massive Sulfide Deposits: Processes and Examples in Modern and Ancient Settings: Reviews in Economic Geology, v. 8, p. 319-350.

Harding, W.D., 1936, Geology of the Birch-Springpole Lakes area: Ontario Department of Mines; Annual Report 1936, v. 45, part 4, p. 1-33.

Hardy, R.V., 1993, A light stable isotope and fluid inclusion study of the Sheep Creek gold camp, Salmo, British Columbia, *in* Programs with Abstracts: Geological Association of Canada, Mineralogical Association of Canada Joint Annual Meeting, Edmonton, Alberta, p. A40.

Heather, K.B., 1991, Geological and structural setting of gold mineralization, Renabie portion of the Missanabie-Renabie gold district, *in* Sage, R.P., and Heather K.B., eds., The Structure, Stratigraphy and Mineral Deposits of the Wawa Area: Geological Association of Canada, Guidebook 1991 A6, p. 81-111.

Hitchins, A.C., and Orssich, C.N., 1995, The Eagle zone gold-tungsten sheeted vein porphyry deposit and related mineralization, Dublin Gulch, Yukon Territory, Section Gold ($\pm$Ag $\pm WO_3$ $\pm$Cu) deposits associated with porphyry-type systems, *in* Schroeter, T.G., ed., Porphyry Deposits of the Northwestern Cordillera of North America: Canadian Institute of Mining, Metallurgy and Petroleum, Special Volume 46, p. 803-810.

Hodgson, C.J., 1989, The structure of shear-related, vein-type gold deposits: A review: Ore Geology Reviews, v. 4, p. 635-678.

—— 1990, An overview of the geological characteristics of gold deposits in the Abitibi Subprovince, *in* Ho, S.E., Robert, F., and Groves, D.I., eds., Gold and Base Metal Mineralization in the Abitibi Subprovince, Canada, with Emphasis on the Quebec Segment: Geology Department and University Extension, University of Western Australia, Perth, Publication 24, p. 63-100.

—— 1993, Mesothermal lode-gold deposits, *in* Kirkham, R.V., Sinclair, W.D., Thorpe, R.I., and Duke, J.M., eds., Mineral Deposits Modelling: Geological Association of Canada, Special Paper 40, p. 635-678.

Hodgson, C.J., and Hamilton, J.V., 1989, Gold mineralization in the Abitibi greenstone belt: end-stage results of Archean collisional tectonics? *in* Keays, R.R., Ramsay, W.R.H., and Groves, D.I., eds., The Geology of Gold deposits; The Perspective in 1988: Economic Geology, Monograph 6, p. 86-100.

Hodgson, C.J., and MacGeehan, P.J., 1982, Geological characteristics of gold deposits in the Superior Province of the Canadian Shield, *in* Hodder, R.W., and Petruks, W., eds., Geology of Canadian Gold Deposits: Canadian Institute of Mining and Metallurgy, Special Volume 24, p. 211-229.

Hodgson, C.J., Hamilton, J.V., and Piroshco, D.W., 1990, Structural setting of gold deposits and the tectonic evolution of the Timmins-Kirkland Lake area, southwestern Abitibi greenstone belt, *in* Ho, S.E., Robert, F., and Groves, D.I., eds., Gold and Base Metal Mineralization in the Abitibi Subprovince, Canada, with Emphasis on the Quebec Segment: Geology Department and University Extension, University of Western Australia, Perth, Publication 24, p. 101-120.

Horwood, H.C., 1948, Howey and Hasaga mines *in* Structural Geology of Canadian Ore Deposits: A Symposium: Canadian Institute of Mining and Metallurgy, Special Volume 1, p. 340-345.

Horwood, H.C., and Pye, E.G., 1955, Geology of Ashmore Township: Annual report of the Department of Mines, v. 60, pt 5, Ontario Department of Mines, 105 p.

Höy, T., and Dunne, K.P.E., 2001, Metallogeny and Mineral Deposits of the Nelson-Rossland Map-Area; Part II, The Early Jurassic Rossland Group, Southeastern British Columbia: British Columbia Ministry of Energy and Mines, Bulletin 109, 195 p.

Hurst, M.E., 1936, Recent studies in the Porcupine area: Canadian Institute of Mining and Metallurgy, v. 39, p. 448-458.

Huston, D.L., 2000, Gold in volcanic-hosted massive sulfide deposits: Distribution, genesis and exploration, *in* Hagemann, S.G., and Brown, P.E., eds., Gold in 2000: Society of Economic Geologists, Reviews in Economic Geology, v. 13, p. 401-426.

Hutchison, R.W., 1993, A multi-stage, multi-process genetic hypothesis for greenstone-hosted gold lodes: Ore Geology Reviews, v. 7, p. 349-382.

Isachsen, C.E., Bowring, S.A., and Padgham, W.A., 1991, U-Pb zircon geochronology of the Yellowknife volcanic belt, N.W.T., Canada: New constraints of the timing and duration of greenstone belt magmatism: Journal of Geology, v. 99, p. 55-67.

Ispolatov, V., Lafrance, B., Dubé, B., Hamilton, M., and Creaser, R., 2005, Geology, structure, and gold mineralization, Kirkland Lake and Larder Lake areas (Gauthier and Teck townships): Discover Abitibi Initiative: Ontario Geological Survey, Open File Report 6159, 170 p.

Issigonis, M.J., 1980, Occurrence of disseminated gold deposits in porphyries in Archean Abitibi belt, northwest Quebec, Canada: Transactions of the Institution of Mining and Metallurgy, v. 89, p. 157-158.

James, 1948, Siscoe mine, *in* Structural Geology of Canadian Ore Deposits: A Symposium: Canadian Institute of Mining and Metallurgy, Special Volume 1, p. 876-882.

Jefferson, C.W., Lustwerk, R.L., and Lambert, M.B., 1992a, Stratigraphy, facies changes and structure in auriferous, iron-rich, Archean sedimentary sequences around the Back River volcanic complex, northeastern Slave province, NWT, *in* Richardson, D.G., and Irving, M., eds., Project Summaries: Canada-Northwest Territories Mineral Development Subsidiary Agreement 1987-1991: Geological Survey of Canada, Open File 2484, p. 185-188.

Jefferson, C.W., Dufresne, R.A., Olson, R.A., and Rice, R., 1992b, Stratigraphy and sedimentology of auriferous Archean iron formations in the vicinity of George Lake, Eastern Slave Province, *in* Richardson, D.G., and Irving, M., eds., Project Summaries: Canada-Northwest Territories Mineral Development Subsidiary Agreement 1987-1991: Geological Survey of Canada, Open File 2484, p. 199-203.

Jenkins, C.L., Trudel, P., and Perreault, G., 1989, Progressive hydrothermal alteration associated with gold mineralization of the Zone 1 intrusion of the Callahan property, Val-d'Or region, Quebec: Canadian Journal of Earth Sciences, v. 26, p. 2495-2506.

Jenkins, C.L., Vincent, R., Garson, D.F., Robert, F., and Poulsen, K.J., 1997, Index-level Database for Lode Gold Deposits of the World: Geological Survey of Canada, Open File 3490, CD-ROM.

Jenney, C.P., 1941, Geology of the Omega mine, Larder Lake, Ontario: Economic Geology, v. 36, p. 424-447.

Jensen, E.P., and Barton, M.D., 2000, Gold deposits related to alkaline magmatism, *in* Hagemann, S.G., and Brown, P.E., eds., Gold in 2000: Society of Economic Geologists, Reviews in Economic Geology, v. 13, p. 279-314.

Jourdain, V., Gauthier, M., and Guha, J., 1990, Métallogénie de l'or dans le sud-est de l'Ontario: Geological Survey of Canada, Open File 2287, 52 p.

Kempthorne, R.H., 1957, Bevcon mine, *in* Structural Geology of Canadian Ore Deposits: A Symposium: Canadian Institute of Mining and Metallurgy, Special Volume 2, p. 416-419.

Kerr, D.J., and Gibson, H.L., 1993, A comparison of the Horne volcanogenic massive sulfide deposit and intracauldron deposits of the Mine sequence, Noranda, Quebec: Economic Geology, v. 88, p. 1419-1442.

Kerrich, R., 1983, Geochemistry of Gold Deposits of the Abitibi Greenstone Belt: Canadian Institute of Mining and Metallurgy, Special Volume 27, 75 p.

Kerrich, R., and Cassidy, K.F., 1994, Temporal relationships of lode gold mineralization to accretion, magmatism, metamorphism and deformation – Archean to present: A review: Ore Geology Reviews, v. 9, p. 263-310.

Kerrich, R., and Feng, R., 1992, Archean geodynamics and the Abitibi-Pontiac collision: Implications for advection of fluids at transpressive collisional boundaries and the origin of giant quartz vein systems: Earth Sciences Reviews, v. 32, p. 33-60.

Kerrich, R., and Watson, G.P., 1984, The Macassa Mine Archean lode gold deposit, Kirkland Lake, Ontario: geology, patterns of alteration, and hydrothermal regimes: Economic Geology, v. 79, p. 1104-1130.

Kerrich, R., and Wyman, D.A., 1990, Geodynamic setting of mesothermal gold deposits: An association with accretionary tectonic regimes: Geology, v. 18, p. 882-885.

Kerrich, R., Goldfarb, R., Groves, D., and Garwin, S., 2000, The geodynamic of world-class gold deposits: characteristics, space-time distribution and origins, *in* Hagemann, S.G., and Brown, P.E., eds., Gold in 2000: Society of Economic Geologists, Reviews in Economic Geology, v. 13, p. 501-551.

Kerswill, J.A., 1993, Models for iron-formation-hosted gold deposits: *in* Kirkham, R.V., Sinclair, W.D., Thorpe, R.I., and Duke, J.M., eds., Mineral Deposit Modeling: Geological Association of Canada, Special Paper 40, p. 171-199.

—— 1996, Iron-formation-hosted stratabound gold, Chapter 15, *in* Eckstrand, O.R., Sinclair, W.D., and Thorpe, R.I., eds., Geology of Canadian Mineral Deposit Types: Geological Survey of Canada, Geology of Canada, No. 8, p. 367-382.

King, J.E., Davis, W.J., Van Nostrand, T., and Relf, C., 1989, Archean to Proterozoic deformation and plutonism of the western Contwoyto Lake map area, central Slave Province, District of Mackenzie, N.W.T: Geological Survey of Canada, Current Research, part C, Paper 89-1C, M44-89/1C, 412 p.

Kinkel, A.R., 1948, Buffalo Ankerite mine, *in* Structural Geology of Canadian Ore Deposits - A Symposium: Canadian Institute of Mining and Metallurgy, Special Volume 1, p. 515-519.

Kishida, A., and Kerrich, R., 1987, Hydrothermal alteration zoning and gold concentration at the Kerr-Addison Archean lode gold deposit, Kirkland Lake, Ontario: Economic Geology, v. 82, p. 649-690.

Koulomzine, 1948, Consolidated Central Cadillac mine, *in* Structural Geology of Canadian Ore Deposits - A Symposium: Canadian Institute of Mining and Metallurgy, Special Volume 1, p. 816-821.

Kontak, D.L., and Archibald, D.A., 2002, $^{40}Ar/^{39}Ar$ dating of hydrothermal biotite from high-grade gold ore, Tangier gold deposit, Nova Scotia; Further evidence for 370 Ma gold metallogeny in the Meguma Terrane: Economic Geology, v. 97, p. 619-628.

Kreczmer, M.J., and Deveaux, P.J., 1985, Exploration and geology of Tartan Lake and Puffy Lake deposits, Flin Flon area, Manitoba, *in* Clark, L.A., ed., Gold in the Western Shield: Canadian Institute of Mining and Metallurgy, Special Volume 38, p. 359-360.

Krupka, K.M., Ohmoto, H., and Wickman, F.E., 1977, A new technique in neutron activation analysis of Na/K ratios of fluid inclusions and its application to the gold-quartz veins at the O'Brien mine, Quebec, Canada: Canadian Journal of Earth Sciences, v. 14, p. 2760-2770.

Kuhns, R.J., Sawkins, F.J., and Ito, E., 1994, Magmatism, metamorphism, and deformation at Hemlo, Ontario, and the timing of Au-Mo mineralization in the Golden Giant mine: Economic Geology, v. 89, p. 720-756.

Labine, R.J., 1991, Hoyle Pond Mine, *in* Fyon, J.A., and Green, A.H., eds., Geology and Ore Deposits of the Timmins District, Ontario (Guidebook to Field Trip 6): Geological Survey of Canada, Open File 2161, p. 114-123.

Lane, H.C., 1968, Preston East Dome Mine, *in* Ferguson, S.A., Buffam, B.S.W., and Carter, O.F., eds., Geology and Ore Deposits of Tisdale Township: Ontario Geological Survey, Report 58, p. 143-151.

Lau, M.H.S., 1988, Structural geology of the vein system in the San Antonio gold mine Bissett, Manitoba, Canada: M.Sc. thesis, University of Manitoba, Winnipeg, Manitoba, 154 p.

Laurus, K.A., and Fletcher, W.K., 1999, Gold distribution in glacial sediments and soils at Boston Property, Nunavut, Canada: Journal of Geochemical Exploration, v. 67, p. 271-285.

Leclair, 1992, The Ida Point gold showing, Hope Bay greenstone belt, Northwest Territories, *in* Richardson, D.G., and Irving, M., eds., Project

Summaries; Canada-Northwest Territories Mineral Development Subsidiary Agreement 1987-1991: Geological Survey of Canada, Open File 2484, p. 61-63.

Leitch, C.H.B., 1990, Bralorne: A Mesothermal, shield-type vein gold deposit of Cretaceous age in southwestern British Columbia: Canadian Institute of Mining and Metallurgy Bulletin, v. 83, p. 53-80.

Lennan, W.B., 1986, Ray Gulch tungsten skarn deposit, Dublin Gulch area, central Yukon, *in* Morin, J.A., ed., Mineral deposits of Northern Cordillera, Proceedings of the Mineral Deposits of Northern Cordillera Symposium, December 1983: Canadian Institute of Mining and Metallurgy, Special Volume 37, p. 245-254.

Leroy, O.E., 1912, The Geology and Ore Deposits of Phoenix, Boundary District, British Columbia: Geological Survey of Canada, Memoir No. 21, 110 p.

Lewis, T.D., 1998, Musselwhite: An exploration success: Canadian Institute of Mining and Metallurgy Bulletin, v. 91, p. 51-55.

Lin, S., 2001, Stratigraphic and structural setting of the Hemlo gold deposit, Ontario, Canada: Economic Geology, v. 96, p. 477-507.

Longley, C.S., and Lazier, T.A., 1948, Paymaster Mine, *in* Structural Geology of Canadian Ore Deposits - A Symposium: Canadian Institute of Mining and Metallurgy, Special Volume 1, p. 520-528.

Lustwerk, R.L., 1992, Geochemistry and petrography of iron formation and other sedimentary rocks associated with the Back River volcanic and sedimentary complex, NWT, *in* Richardson, D.G., and Irving, M., eds., Project Summaries: Canada-Northwest Territories Mineral Development Subsidiary Agreement 1987-199: Geological Survey of Canada, Open File 2484, p. 189-192.

Macdonald, A.J., Lewis, P.D., Thompson, J.F.H., Nadaraju, G., Bartsch, R., Bridge, D.J., Rhys, A., Roth, T., Kaip, A., Godwin, C.I., and Sinclair, A.J., 1996, Metallogeny of an Early to Middle Jurassic arc, Iskut River area, Northwestern British Columbia: Economic Geology, v. 91, p. 1098-1114.

Macdonald, J.M., 1982, The McLeod-Cockshutt and Hard Rock mines, Geraldton: Examples of an iron-formation related gold deposit, Section Mineral Deposit Programs, *in* Wood, J., White, O.L., Barlow, R.B., and Colvine, A.C., eds., Summary of Field Work, 1982: Ontario Geological Survey, Miscellaneous Paper 116, p. 188-191.

Mcdonald, J.M., Duke, J.M., and Hauser, R.L., 1993, Geological setting of the NERCO Con mine and the relationships of gold mineralization to metamorphism, Yellowknife, N.W.T.: Exploration and Mining Geology, v. 2, p. 139-154.

Magnan, M., and Blais, A., 1995, The Copper Rand Mine (Au-Cu-Ag), Section Day 4: Porphyry and epithermal type mineralization in the Doré Lake Complex and the Roy Group, *in* Pilote, P., ed., Precambrian '95 - Metallogeny and Geologic Evolution of the Chibougamau Area - from Porphyry Cu-Au-Mo to Mesothermal Lode Gold Deposits: Geological Survey of Canada, Open File 3143, p. 87-94.

Marmont, S., 1986, The geological setting of the Detour Lake gold mine, Ontario, Canada, *in* Macdonald, A.J., Downes, M., Pirie, J., and Chater, A.M., eds., Proceeding of Gold '86, An International Symposium on the Geology of Gold: p. 81-96.

Marmont, S., and Corfu, F., 1989, Timing of gold introduction in the Late Archean tectonic framework of the Canadian Shield: Evidence from U-Pb zircon geochronology of the Abitibi subprovince, *in* Keays, R.R., Ramsay, W.R.H., and Groves, D.I., eds., The Geology of Gold Deposits; The Perspective in 1988: Economic Geology, Monograph 6, p. 101-111.

Marquis, P., Brown, A.C., Scherkus, E., Trudel, P., and Hoy, L.D., 1992, Géologie de la mine Donald J. LaRonde (Abitibi): Ministère des Ressources Naturelles du Québec, ET 89-06, 106 p.

Mather, W.B., 1937, Geology and paragenesis of the gold ores of the Howey mine, Red Lake, Ontario: Economic Geology, v. 32, p. 131-153.

Mathews, W.H., 1953, Geology of the Sheep Creek Camp: British Columbia Department of Mines, Bulletin No. 31, 94 p.

Mckay, G.A., Cormie, A.M., and Coulson, C.J., 1948, Leitch mine, *in* Structural Geology of Canadian Ore Deposits - A Symposium: Canadian Institute of Mining and Metallurgy, Special Volume 1, p. 385-389.

McLaren, D.C., 1946, The Omega gold mine: Canadian Mining Journal, v. 67, p. 941-950.

—— 1947, Hasaga Gold Mines Limited: Canadian Mining Journal, v. 68, p. 168-174.

Mcmurchy, R.C., 1941, Geology of the Powell mine: Engineering and Mining Journal, v. 142, p. 47.

—— 1948, Powell Rouyn mine, *in* Structural Geology of Canadian Ore Deposits - A Symposium: Canadian Institute of Mining and Metallurgy, Special Volume 1, p. 739-747.

Meinert, L.D., 1998, A review of skarns that contain gold, *in* Lentz, D.R., ed., Mineralized Intrusion-Related Skarn Systems: Mineralogical Association of Canada, Short Course, v. 26, p. 359-414.

Melling, D.R., Watkinson, D., Poulsen, K.H., Chorlton, L.B., and Hunter, 1986, The Cameron Lake gold deposit, Northwestern Ontario, Canada: Geological setting, structure, and alteration, *in* Macdonald, A.J., Downes, M., Pirie, J., and Chater, A.M., eds., Proceeding of Gold '86, An International Symposium on the Geology of Gold, p. 149-169.

Melnik-Proud, N., 1992, The Geology and Ore Controls In and Around the McIntyre-Hollinger Complex, Timmins, Ontario: Ph.D. thesis, Queen's University, Kingston, Ontario, 353 p.

Metcalfe, P., and Moors, J.G., 1993, Lithostratigraphy and mineralization of the Bronson Corridor, Iskut River area, northwestern British Columbia: Geological Association of Canada, Annual Meeting, Edmonton, AB, Abstracts 16, p. 70.

Méthot, Y., and Trudel, P., 1987, Géologie de la mine Marban, région de Malartic: Ministère de l'Éenergie et des Ressources, Gouvernement du Québec, MB 87-53, 71 p.

Michibayashi, K., 1995, Two-phase syntectonic gold mineralization and barite remobilization within the main ore body of the Golden Giant mine, Hemlo, Ontario, Canada., Ore Geology Reviews, v. 10, p. 31-50.

Mihalynuk, M.G., and Marriott, C.C., 1991, The Polaris-Taku deposit: Geologic setting and recent mineral exploration results *in* Exploration in British Columbia 1991: British Columbia Ministry of Energy, Mines and Petroleum Resources, p. 127-131.

Miller, A.R., Balog, M.J., and Tella, S., 1995, Oxide iron-formation-hosted lode gold, Meliadine Trend, Rankin Inlet Group, Churchill Province, Northwest Territories: Geological Survey of Canada, Current Research 1995-C, p. 163-174.

Mills, J.W., 1950, Structural control of orebodies as illustrated by the use of vein contours at the O'Brien gold mine, Cadillac, Quebec: Economic Geology, v. 45, p. 786-807.

Morasse, S., Hodgson, C.J., Guha, J., and Coulombe, A., 1986, Preliminary report on the geology of the Lac Shortt gold deposit, Demaraisville area, Quebec, Canada, *in* Macdonald, A.J., Downes, M., Pirie, J., and Chater, A.M., eds., Proceeding of Gold '86, An International Symposium on the Geology of Gold: p. 191-196.

Morasse, S., Wasteneys, H.A., Cormier, M., Helmstaedt, H., and Mason, R., 1995, A pre-2686 Ma intrusion-related gold deposit at the Kiena mine, Val d'Or, Québec, southern Abitibi Subprovince: Economic Geology, v. 90, p. 1310-1321.

Moritz, R.P., and Crocket, J.H., 1990, Mechanics of formation of the gold-bearing quartz-fuchsite vein at the Dome mine, Timmins area, Ontario: Canadian Journal of Earth Sciences, v. 27, p. 1609-1620.

Morrow, H.F., 1949, The geology of Hard Rock gold mine's quartz stringer ore zones: Precambrian, p. 11-39.

Mortensen, J.K., 1993, U-Pb geochronology of the eastern Abitibi Subprovince. Part 2: Noranda - Kirkland Lake area: Canadian Journal of Earth Sciences, v. 30, p. 29-41.

Mueller, A.G., and Groves, D.I., 1991, The classification of Western Australia greenstone-hosted gold deposits according to wallrock-alteration mineral assemblages: Ore Geology Reviews, v. 6, p. 291-331.

Muir, T.L., 1997, Hemlo Gold Deposit Area, Precambrian Geology: Ontario Geological Survey, Report 289, 219 p.

—— 2002, The Hemlo gold deposit, Ontario, Canada: Principal deposit characteristics and constraints on mineralization: Ore Geology Reviews, v. 21, p. 1-66.

Nesbitt, B.E., Murowchick, J.B., and Muehlenbachs, K., 1986, Dual origins of lode gold deposits in the Canadian Cordillera: Geology, v. 14, p. 506-509.

Newhouse, W.H., 1942, Structural features associated with the ore deposits described in this volume, *in* Newhouse, W.H., ed., Ore Deposits as Related to Structural Features: Princeton University Press, Princeton, N.J., p. 9-53.

Olivo, G.R., and Williams-Jones, A.E., 2002, Genesis of the auriferous C quartz-tourmaline vein of the Siscoe mine, Val d'Ord district, Abitibi subprovince, Canada: Structural, mineralogical and fluid inclusion constraints: Economic Geology, v. 97, p. 929-947.

Olivo, G.R., Gauthier, M., Bardoux, M., Leao de Sa, E., Fonseca, J.T.F., and Carbonari, F., 1995, Palladium-bearing gold deposit hosted by

Proterozoic Lake Superior-type iron-formation at Cauê iron mine, Itabira district, southern Sao Francisco craton, Brazil: Geologic and structural controls: Economic Geology, v. 90, p. 118-134.

Ostry, G., and Halden, N.M., 1995, Geology of the Puffy Lake Au deposit, Sherridon district, Manitoba: Exploration and Mining Geology, v. 4, p. 51-63.

Pan, Y., and Fleet, M.E., 1992, Calc-silicate alteration in the Hemlo gold deposit, Ontario: Mineral assemblages, P-T-X constraints and significance: Economic Geology, v. 87, p. 1104-1120.

Penczak, R.S., and Mason, R., 1997, Metamorphosed Archean epithermal As-Au-Sb-Zn-(Hg) vein mineralization at the Campell Mine, Northwestern Ontario: Economic Geology, v. 92, p. 696-719.

—— 1999, Characteristics and origin of Archean premetamorphic hydrothermal alteration at the Campbell gold mine, northwestern Ontario, Canada: Economic Geology, v. 94, p. 507-528.

Phillips, G.N., and Groves, D.I., 1984, Fluid access and fluid-wall rock interaction in the genesis of the Archean gold-quartz vein deposit at Hunt mine, Kambalda, Western Australia, *in* Foster, R.P., ed., Gold '82: The Geology, Geochemistry and Genesis of Gold Deposits: Proceeding of the Symposium Gold' 82, Balkema, Rotterdam, p. 389-416.

Phillips, G.N., Groves, D.I., and Kerrich, R., 1996, Factors in the formation of the giant Kalgoorlie gold deposit: Ore Geology Reviews, v. 10, p. 295-317.

Picard, S., 1990, Le gisement Silidor, *in* Rive, M., Verpaelst, P., Gagnon, Y., Lulin, J.M., Riverin, G., and Simard, A., eds., The Northwestern Quebec Polymetallic Belt: A Summary of 60 Years of Mining and Exploration: Canadian Institute of Mining and Metallurgy, Special Volume 43, p. 175-183.

Pilote, P., and Guha, J., 1998, Part B - Metallogeny of the eastern extremity of the Abitibi belt, *in* Pilote, P., Dion, C., and Morin, R., eds., Metallogeny of the Chibougamau District: Geological Evolution and Development of Distinct Mineralized Systems Through Time: Geological Association of Canada, Mineralogical Association of Canada, Guidebook 1998 B3, p. 29-40.

Pilote, P., Guha, J., and Daigneault, R., 1990a, Contexte structural et minéralisations aurifères des gîtes Casa-Berardi, Abitibi, Québec: Canadian Journal of Earth Sciences, v. 27, p. 1672-1685.

Pilote, P., Guha, J., Daigneault, R., Robert, F., Cloutier, J.Y., and Golightly, J.P., 1990b, The Structural Evolution of the Casa-Berardi East Gold Deposits, Casa-Berardi Township, Quebec,*in* Rive, M., Verpaelst, P., Gagnon, Y., Lulin, J.M., Riverin, G., and Simard, A., eds., The Northwestern Quebec Polymetallic Belt: A Summary of 60 Years of Mining and Exploration: Canadian Institute of Mining and Metallurgy, Special Volume 43, p. 337-348.

Poulsen, 1995, Disseminated and replacement gold, *in* Eckstrand, O.R., Sinclair, W.D.T., and Thorpe, R.I., eds., Geology of Canadian Mineral Deposit Types: Geological Survey of Canada, Geology of Canada No. 8, p. 383-392.

—— 1996, Carlin-type gold deposits and their potential occurrence in the Canadian Cordillera, *in* Current Research 1996-A: Geological Survey of Canada, p. 1-9.

Poulsen, K.H., and Robert, F., 1989, Shear zones and gold: Practical examples from the southern Canadian Shield, *in* Bursnall, J.T., ed., Mineralization and Shear Zones: Geological Association of Canada, Short Course Notes 6, p. 239-266.

Poulsen, K.H., Ames, D.E., Lau, M.H.S., and Brisbin, D.I., 1986, Preliminary report on the structural setting of gold in the Rice Lake area, Uch subprovince, southeastern Manitoba: Geological Survey of Canada, Current Research 1986, p. 213-221.

Poulsen, K.H., Card, K.D., and Franklin, J.M., 1992, Archean tectonic and metallogenic evolution of the Superior Province of the Canadian Shield: Precambrian Research, v. 58, p. 25-54.

Poulsen, K.H., Mortensen, J.K., and Murphy, D.C., 1997, Styles of intrusion-related gold mineralization in the Dawson-Mayo area, Yukon Territory, *in* Current Research 1997-A: Geological Survey of Canada, p. 1-10.

Poulsen, K.H., Robert, F., and Dubé, B., 2000, Geological Classification of Canadian Gold Deposits: Geological Survey of Canada, Bulletin 540, 106 p.

Powell, R., Will, T.M., and Phillips, G.N., 1991, Metamorphism in Archean greenstone belts: Calculated fluid compositions and implications for gold mineralization: Journal of Metamorphic Geology, v. 9, p. 141-150.

Pressacco, R., ed., 1999, Special Project: Timmins Ore Deposit Descriptions: Ontario Geological Survey, Open File 5985, 222 p.

Proudlove, D.C., Hutchinson, R.W., and Rogers, D.S., 1989, Multiphase Mineralization in Concordant and Discordant Gold Veins, Dome Mine, South Porcupine, Ontario, Canada, *in* Keays, R.R., Ramsay, W.R.H., and Groves, D.I., eds., The Geology of Gold Deposits: The Perspective in 1988: Economic Geology Monograph 6, p. 112-12.

Pyke, D.R., 1982, Geology of the Timmins Area, District of Cochrane: Ontario Department of Mines, Geological Report, v. 219, 141 p.

Quirion, 1991, Geologie de la mine d'or Lac Shortt, *in* Guha, J., Chown, E.H., and Daigneault R., eds., Litho-Tectonic Framework and Associated Mineralization of the Eastern Extremity of the Abitibi greenstone belt (Field Trip 3): Geological Survey of Canada, Open File 2158, p. 116-131.

Ray, G.E., 1998, Skarns and skarn deposits in the Canadian Cordillera, *in* Lefebure, D.V., ed., Metallogeny of Volcanic Arcs: British Columbia Geological Survey, Short Course Notes, Open File 1998-5, Section B, p. 45.

Ray, G.E., Shearer, J.T., and Niels, R.J.E., 1986, The geology and geochemistry of the Carolin gold deposit, southwestern British Columbia, Canada, *in* Macdonald, A.J., Downes, M., Pirie, J., and Chater, A.M., eds., Proceeding of Gold '86, An International Symposium on the Geology of Gold: p. 470-487.

Ray, G.E., Dawson, G.L., and Webster, I.C.L., 1996, The stratigraphy of the Nicola Group in the Hedley district, British Columbia, and the chemistry of its intrusions and Au skarns: Canadian Journal of Earth Sciences, v. 33, p. 1105-1126.

Rebagliati, C.M., Bowen, B.K., Copeland, D.J., and Niosi, D.W.A., 1995, Kemess South and Kemess North porphyry gold-copper deposits, northern British Columbia, Section Porphyry copper (±Au ±Mo) deposits of the calc-alkalic suite, *in* Schroeter, T.G., ed., Porphyry Deposits of the Northwestern Cordillera of North America: Canadian Institute of Mining, Metallurgy and Petroleum, Special Volume 46, p. 377-396.

Richard, M., Hubert, C., Brown, A.C., and Sirois, R., 1990, The Pierre Beauchemin gold mine: A structurally controlled deposit within a subhorizontal layered composite granitoid, *in* Rive, M., ed., The Northwestern Quebec Polymetallic Belt: A Summary of 60 Years of Mining Exploration. Proceedings of the Rouyn-Noranda 1990 Symposium May 28 - June 1, 1990: Canadian Institute of Mining and Metallurgy, Special Volume 42, p. 211-219.

—— 1991, The Pierre Beauchemin gold mine, *in* Tourigny, G., and Verpaelst, P., eds., Control on Base Metal and Gold Mineralization, Bousquet-Rouyn-Noranda Area: Society of Economic Geologists, Guidebook v. 10, p. 98-106.

Richardson, D.J., and Ostry, G., 1996, Gold Deposits of Manitoba: Manitoba Energy and Mines, Economic Geology Report ER86-1 (2nd edition), 114 p.

Ridler, J.R., 1970, Relationship of Mineralization to Volcanic Stratigraphy in the Kirkland-Larder Lakes Area, Ontario: Geological Association of Canada Proceedings, v. 21, p. 33-42.

Ridley, J.R., and Diamond, L.W., 2000, Fluid chemistry of orogenic lode gold deposits and implications for genetic models, *in* Hagemann, S.G., and Brown, P.E., eds., Gold in 2000: Society of Economic Geologists, Reviews in Economic Geology, v. 13, p. 141-162.

Ridley, J.R., Groves, D., I., and Knight, J.T., 2000, Gold deposits in amphibolite and granulite facies terranes of the Archean Yilgarn craton, Western Australia: Evidence and implications for synmetamorphic mineralization, *in* Spry, P.G., Marshall, B., and Vokes, F.M., eds., Metamorphosed and Metamorphogenic Ore Deposits: Society of Economic Geologists, Reviews in Economic Geology, v. 11, p. 265-290.

Robert, F., 1990, Structural setting and control of gold-quartz veins of the Val d'Or area, southeastern Abitibi subprovince, *in* Ho, S.E., Robert, F., and Groves, D.I., eds., Gold and Base-Metal Mineralization in the Abitibi Subprovince, Canada, with Emphasis on the Quebec Segment: University of Western Australia, Short Course Notes, v. 24, p. 167-210.

—— 1994, Vein fields in gold districts: The example of Val d'Or, southeastern Abitibi subprovince, Quebec, *in* Canadian Shield: Geological Survey of Canada, Current Research 1994-C, p. 295-302.

—— 1995, Quartz-carbonate vein gold, *in* Eckstrand, O.R., Sinclair, W.D.T., and Thorpe, R.I., eds., Geology of Canadian Mineral Deposit Types: Geological Survey of Canada, Geology of Canada No. 8, p. 350-366.

—— 1997, A preliminary geological model for syenite-associated disseminated gold deposits in the Abitibi belt, Ontario and Quebec: Geological Survey of Canada, Current Research 1997, p. 201-210.

—— 2000, World-class greenstone gold deposits and their exploration: 31st International Geological Congress, Rio de Janeiro, Brasil, August, 2000, V. De Presentacionnes, CD-ROM, doc. SG304e, p. 4.

—— 2001, Syenite-associated disseminated gold deposits in the Abitibi-greenstone belt, Canada: Mineralium Deposita, v. 36, p. 503-516.

Robert, F., and Brown, A.C., 1986a, Archean gold-bearing quartz veins at the Sigma mine, Abitibi greenstone belt, Quebec. Part I: Economic Geology, v. 81, p. 578-592.

—— 1986b, Archean gold-bearing quartz veins at the Sigma mine, Abitibi greenstone belt, Quebec. Part II: Economic Geology, v. 81, p. 593-616.

Robert, F., and Poulsen, K.H., 1997, World-class Archaean gold deposits in Canada: An overview: Australian Journal of Earth Sciences, v. 44, p. 329-351.

—— 2001, Vein formation and deformation in greenstone gold deposits, *in* Richards, J.P., and Tosdal, R.M., eds., Structural Controls on Ore Genesis: Society of Economic Geologists, Reviews in Economic Geology, v. 14, p. 111-155.

Robert, F., Poulsen, K.H., and Dubé, B., 1994, Structural analysis of lode gold deposits in deformed terranes and its application: Geological Survey of Canada, Short course notes, Open File Report 2850, 140 p.

Robert, F., Poulsen, K.H., Cassidy, K.F., and Hodgson, C.J., 2005, Gold metallogeny of the Yilgarn and Superior cratons, *in* Goldfarb, R.J., and Richards, J.P., eds., Economic Geology One Hundredth Anniversary Volume: 1905 - 2005: Society of Economic Geologists, p. 1001-1033.

Roberts, R.G., 1987, Archean lode gold deposits: Geoscience Canada, v. 14, p. 1-19.

Robertson, R., 2001, Hope Bay partners evaluate project's economics: The Northern Miner, November 5, 2001, p. B1.

Rock, N.M.S., and Groves, D.I., 1988, Can lamprophyres resolve the genetic controversy over mesothermal gold deposits? Geology, v. 16, p. 538-541.

Rogers, C., and Houle, J., 1999, Geological setting of the Kemess South Au-Cu porphyry deposit and local geology between Kemess Creek and Bicknell Lake (NTS 94E/2) *in* Geological Fieldwork, 1998: British Columbia Ministry of Energy, Mines and Petroleum Resources, Paper 1999-1, p. 103-114.

Rogers, J.A., 1982, The geology and ore deposits of the No. 8 shaft area, Dome mine, *in* Hodder, R.W., and Petruck, W., eds., Geology of Canadian Gold Deposits: Canadian Institute of Mining and Metallurgy, Special Volume 24, p. 161-168.

—— 1992, The Arthur W. White mine, Red Lake area, Ontario: Detailed structural interpretation the key to successful grade control and exploration: Canadian Institute of Mining and Metallurgy Bulletin, v. 85, p. 37-44.

Roozendaal, Hawley, Siddle, and Webb, 1997, GMD Resource Corp; Rediscovering the Discovery mine, *in* Igboji, I.E., Goff, S.P., and Beales, P., Exploration Overview, 1996: Mining, Exploration and Geological Investigations: Department of Indian and Northern Affairs, p. 3-3 3-32.

Ropchan, J.R., Luinstra, B., Fowler, A.D., Benn, K., Ayer, J., Berger, B., Dahn, R., Labine, R., and Amelin, Y., 2002, Host-rock and structural controls on the nature and timing of gold mineralization at the Holloway mine, Abitibi subprovince, Ontario: Economic Geology, v. 97, p. 291-309.

Roussy, J., 2003, Relations etnre la distribution de l'or, la structure, la composition des veines et de l'altération hydrothermale à la mine Beaufor, Val-d'Or, Abitibi, Québec: M.Sc. thesis, Université Laval, Quebec, Canada, 316 p.

Rhys, D.A., 1996, The Snip and Johnny Mountain gold mines: Early Jurassic intrusive-related vein deposits, Iskut River area, northwestern British Columbia: *in* Exploration in British Columbia 1995: British Columbia Ministry of Employment and Investment, Geological Survey Branch, p. 163.

Rhys, D.A., and Godwin, C.I., 1992, Preliminary structural interpretation of the Snip mine (104B/11), *in* Grant, B., and Newell, J.M., eds., Geological Fieldwork 1991: British Columbia Ministry of Energy, Mines and Petroleum Resources, Paper 1992-1, p. 549-554.

Sanborn, M., 1987, The Role of Brittle-Ductile Shear in the Formation of Gold-bearing Quartz-Carbonate Veins in the West Carbonate Zone of the Cochenour Willans Gold Mine, Red Lake, Ontario: M.Sc. thesis, University of Toronto, Toronto, Ontario, 148 p.

Sanborn-Barrie, M., Skulski, T., and Parker, J., 2001, Three hundred million years of tectonic history recorded by the Red Lake greenstone belt, Ontario: Geological Survey of Canada, Current Research 2001, p. 1-14.

Sansfaçon, R., and Hubert, C., 1990, The Malartic gold district, Abitibi greenstone belt, Quebec; geological setting, structure and timing of gold emplacement at Malartic Gold Fields, Barnat, East-Malartic, Canadian Malartic and Sladen mines, Section Val d'Or and Malartic mining camps, *in* Rive, M., Verpaelst, P., Gagnon, Y., Lulin, J.M., Riverin, G., and Simard, A., eds., The Northwestern Quebec Polymetallic Belt: A Summary of 60 Years of Mining and Exploration: Canadian Institute of Mining and Metallurgy, Special Volume 43, p. 221-236.

Sansfaçon, and Trudel, 1988, Géologie de la mine Malartic Gold Fields-Région de Malartic: Service Géologique du Nord-Ouest, Ministère de l'Énergie et des Ressources, MB 88-24, 72 p.

Sauvé, P., 1985, Géologie de la mine Bevcon: Ministère de l'Énergie et des Ressources, service de la Géologie, MB 85-04.

Sauvé, P., Imreh, L., and Trudel, P., 1993, Description des gîtes d'or de la région de Val d'Or: Ministère des Ressources Naturelles du Québec, MM 91-03, 178 p.

Savoie, A., Sauvé, P., Trudel, P., and Perrault, 1990, Geologie de la mine Doyon, Cadillac, Quebec, *in* Rive, M., Verpaelst, P., Gagnon, Y., Lulin, J.M., Riverin, G., and Simard, A., eds., The Northwestern Quebec Polymetallic Belt: A Summary of 60 Years of Mining and Exploration: Canadian Institute of Mining and Metallurgy, Special Volume 43, p. 401-412.

Sawyer, E.W., and Barnes, S.-J., 1994, Thrusting, magmatic intraplating, and metamorphic core complex development in the Archaean Belleterre-Angliers Greenstone Belt, Superior Province, Quebec, Canada: Precambrian Research, v. 68, p. 183-200.

Scales, M., 1997, Glimmer shines: Canadian Mining Journal, v. 118, p. 20-21.

Schmitt, H.R., Cameron, E.M., Hall, G.E.M., and Vaive, J., 1993, Mobilization of gold into lake sediments from acid and alkaline mineralized environments in the southern Canadian Shield: Gold in lake sediments and natural waters: Journal of Geochemical Exploration, v. 48, p. 329-358.

Schroeter, T.G., 1985, Muddy Lake prospect (104K/1W), *in* Geological Fieldwork 1984: British Columbia Ministry of Energy, Mines and Petroleum Resources, Paper 1985-1, p. 352-358.

Schroeter, T.G., Lane, B., and Bray, A., 1992, Geologic setting and mineralization of the Red Mountain mesothermal gold deposit, *in* Exploration in British Columbia 1991: British Columbia Ministry of Energy, Mines and Petroleum Resources, Paper 1992-1, p. 117-125.

Schultz, D.J., and Kerrich, R., 1991, Structural controls and geochemical character of the Seabee gold mine, Laonil Lake, Glennie Domain, *in* Summary of Investigations 1991: Saskatchewan Geological Survey, Miscellaneous Report 91-4, p. 101-108.

Shelton, K.,L. Costello, C.S., and van Hees, E.H.P., 2000, Contrasting styles of Archaean greenstone gold deposition: Colomac gold mine, Canadian Northwest Territories: Journal of Geochemical Exploration, v. 69-70, p. 303-307.

Sherlock, R.L., Alexander, R.B., March, R., Kellner, J., and Barclay, W.A., 2001, Geological Setting of the Meadowbank Iron-Formation-Hosted Gold Deposits, Nunavut: Geological Survey of Canada, Current Research 2001-C11, 16 p.

Sibbald, T.I.I., 1986, Bedrock geological mapping, Sulphide Lake area, *in* Macdonald, R., and Sibbald, T.I.I., eds., Summary of Investigations, 1986: Saskatchewan Geological Survey, Miscellaneous Report 86-4, p. 63-64.

Sibbick, S.J., Rebagliati, C.M., Copeland, D.J., and Lett, R.E., 1992, Soil geochemistry of the Kemess South porphyry gold-copper deposit (94E/2E): *in* Grant, B., and Newell, J.M., eds., Geological Fieldwork 1991: British Columbia Ministry of Energy, Mines and Petroleum Resources, Paper 1992-1, p. 349-361.

Sibson, R.H., 1990, Faulting and fluid flow, *in* Nesbitt, B.E., ed., Short Course on Fluids in Tectonically Active Regimes of the Continental Crust: Mineralogical Association of Canada, Short Course Handbook, v. 18, p. 93-132.

Sibson, R.H., Robert, F., and Poulsen, K.H., 1988, High-angle reverse faults, fluid-pressure cycling, and mesothermal gold-quartz deposits: Geology, v. 16, p. 551-555.

Simard, J.-M., and Genest, R., 1990, Géologie de la mine Agnico-Eagle, Joutel (Québec), Section Matagami, Joutel, and Casa-Berardi mining camps, *in* Rive, M., Verpaelst, P., Gagnon, Y., Lulin, J.M., Riverin, G., and Simard, A., eds., The Northwestern Quebec Polymetallic Belt: A Summary of 60 Years of Mining and Exploration: Canadian Institute of Mining and Metallurgy, Special Volume 43, p. 373-382.

Sinclair, W.D., 1982, Gold deposits of the Matachewan area, Ontario, *in* Hodder, R.W., and Petruck, W., eds., Geology of Canadian Gold Deposits: Canadian Institute of Mining and Metallurgy, Special Volume 24, p. 83-93.

Sinclair, W.D., Pilote, P., Kirkham, R.V., Robert, F., and Daigneault, R., 1994, A preliminary report of porphyry Cu-Mo-Au and shear-zone hosted Cu-Au deposits in the Chibougamau area, Quebec: *in* Current Research 1994-C: Geological Survey of Canada, p. 303-309.

Smit, H., Sieb, M., and Swanson, C., 1996, The Dublin Gulch intrusive-hosted gold deposit, *in* Lefebvre, D.V., ed., New Mineral Deposit Models of the Cordillera: Cordilleran Roundup Short Course, British Columbia Geological Survey, Ministry of Energy, Mines and Petroleum Resources, p. 157-158.

Smith, P.M., 1986, Duport, a structurally controlled gold deposit in northwestern Ontario, Canada: *in* Macdonald, A.J., Downes, M., Pirie, J., and Chater, A.M., eds., Proceeding of Gold '86, An International Symposium on the Geology of Gold: p. 197-212.

Smith, J.P., Spooner, E.T.C. Broughton, D.W., and Ploeger, F.R., 1993, Archean Au-Ag-(W) quartz vein/disseminated mineralisation within the Larder Lake-Cadillac Break, Kerr Addison-Chesterville system, North East Ontario, Canada: Ontario Geoscience Research Grant Program, Grant No. 364, Ontario Geological Survey, Open File Report 5831, 310 p.

Solomon, M., Groves, D.I., and Jaques, A.L., 2000, The lode gold deposits of the Western Australian Shield, Chapter 5, *in* Solomon, M., and Groves, D.I., eds., The Geology and Origin of Australia's Mineral Deposits: Hobart: Centre for Ore Deposit Research, University of Tasmania, Oxford Monographs on Geology and Geophysics, No. 24, p. 54-84.

Spooner, E.T.C., 1991, The magmatic model for the origin of Archean Au-quartz vein ore systems: Assessment of the evidence, *in* Ladeira, E.A., ed., Brazil Gold '91: The Economics, Geology, Geochemistry and Genesis of Gold Deposits: Rotterdam, Brookfield, p. 313-318.

Stewart, P.W., 1992, The Origin of the Hope Brook Mine, Newfoundland: A Shear Zone-Hosted Acid Sulphate Gold Deposit: Ph.D. thesis, University of Western Ontario, London, Ontario, 398 p.

Stott, G.M., and Corfu, F., 1992, Uchi Subprovince, Chapter 6 *in* Thurston, P.C., Williams, H.R., Sutcliffe, R.H., and Stott, G.M., eds., Geology of Ontario: Ontario Geological Survey, Special Volume 4, p. 145-236.

Studemeister, P.A., and Kilias, S., 1987, Alteration pattern and fluid inclusions of gold-bearing quartz veins in Archean trondhjemite near Wawa, Ontario, Canada: Economic Geology, v. 82, p. 429-439.

Taylor, R.B., 1948, Delnite mine, *in* Structural Geology of Canadian Ore Deposits - A Symposium: Canadian Institute of Mining and Metallurgy, Special Volume 1, p. 504-507.

Tella, S., Schau, M., Armitage, A.E., Seemayer, B.E., and Lemkow, D., 1992, Precambrian geology and economic potential of the Meliadine Lake-Barbour Bay region, District of Keewatin, Northwest Territories, *in* Current Research Part C: Geological Survey of Canada, Paper 92-1C, p. 1-11.

Teasdale, N., Brown, A.C., and Tourigny, G., 1996, Gîtologie de la mine Bousquet 2: Ministère des Ressources Naturelles, Secteur des Mines, Gouvernement du Québec, MB 96-37, 43 p.

Tessier, A.C., 1990 Structural evolution and host rock dilation during emplacement of gold-quartz vein at the Perron deposit, Val d'Or Quebec: M.Sc. thesis, Queen's University, Kingston, Ontario, 242 p.

Théberge, L., 1997, Géologie structurale et métallogénie de la mine Casa-Berardi Ouest, Abitibi, Québec: M.Sc. thesis, Université du Québec à Montréal, 74 p.

Thiboutot, H., 1986, Eastmain River gold deposit, Quebec, *in* Macdonald, A.,J., Downes, M., Pirie, J., and Chater, A.M., eds., Proceeding of Gold '86, An International Symposium on the Geology of Gold: p. 155-157.

Thompson, J.F.H., and Newberry, R.J., 2000, Gold deposits related to reduced granitic intrusions, *in* Hagemann, S.G., and Brown, P.E., eds., Gold in 2000: Society of Economic Geologists, Reviews in Economic Geology, v. 13, p. 377-400.

Thomson, J.E., 1948, Omega mine, *in* Structural Geology of Canadian Ore Deposits - A Symposium: Canadian Institute of Mining and Metallurgy, Special Volume 1, p. 658-662.

—— 1950, Geology of Teck Township and the Kenogami Lake Area, Kirkland Lake Gold Belt: Annual Report 1948, Ontario Department of Mines, p. 1-53.

Thomson, J.E., Charlewood, G.H., Griffin, K., Hawley, J.E., Hopkins, H., MacIntosh, C.G., Orgizio, S.P., Perry, O.S., and Ward, W., 1950, Geology of the Main Ore Zone at Kirkland Lake: Annual Report 1948, Ontario Department of Mines, p. 54-196.

Thrall, B., 1999, Heap leaching in extreme northern climates; overview of Brewery Creek mine, Yukon, Canada, *in* Udd, J.E., and Keen, A.J., eds., Mining in the Arctic: Proceedings of the 5th International Symposium: Yellowknife, Northwest Territories, 14-17 June 1998, p. 161-164.

Tompkins, J.R., 1988, The Puffy Lake gold project, *in* Brawner, C.O., ed., Gold mining '88: Second International Conference on Gold Mining: Society of Mining Engineers, p. 482-493.

Tourigny, G., 1995, Structural setting and style of gold-bearing shear zones in the Belleterre district, Témiscamingue, Québec: Exploration and Mining Geology, v. 4, p. 1-14.

Tourigny, G., and Tremblay, A., 1997, Origin and incremental evolution of brittle/ductile shear zones in granitic rocks: natural examples from the southern Abitibi Belt, Canada: Journal of Structural Geology, v. 19, p. 15-27.

Tourigny, G., Hubert, C., Brown, A.C., Crépeau, R., Trudel, P., Hoy, L., and Kheang, L., 1992, Géologie de la mine Bousquet: Ministère des Resources Naturelles du Québec, ET 89-09, 99 p.

Tourigny, G., Doucet, D., and Bourget, A., 1993, Geology of the Bousquet 2 mine: An example of a deformed, gold bearing polymetallic sulfide deposit: Economic Geology, v. 88, p. 1578-1597.

Travis, G.A., Woodall, R., and Bartram, G.D., 1971, The geology of the Kalgoorlie Goldfield, *in* Glover, J.E., ed., Symposium on Archean Rocks: Geological Society of Australia, Special Publication 3, p. 175-190.

Tremblay, A., 2001, Postmineralization faults in the Beaufor gold deposit, Abitibi greenstone belt, Canada: geometry, origin, and tectonic implications for the Val-d'Or mining district: Economic Geology, v. 96, p. 509-524.

Tremblay, L.P., 1950, Northeast Part of Giauque Lake Map Area Northwest Territories: Geological Survey of Canada, Paper 50-18, 37 p.

Tremblay, M., 2006, Minéralisation et déformation du gîte aurifère de la zone Eau Claire, propriété Clearwater, Baie James: M.Sc. thesis, Université du Québec à Chicoutimi, Québec, 179 p.

Trenholme, 1948, Belleterre mine, *in* Structural Geology of Canadian Ore Deposits - A Symposium: Canadian Institute of Mining and Metallurgy, Special Volume 1, p. 796-803.

Troop, D.G., 1985, Preliminary report on geology and metasomatism at the Ross Mine and vicinity, District of Cochrane, *in* Wood, J., White, O.L., Barlow R.B., and Colvine, A.C., eds., Summary of Field Work and Other Activities 1985: Ontario Geological Survey, Miscellaneous Report 126, p. 320-325.

—— 1986, Multiple orebody types and vein morphologies, Ross mine, district of Cochrane, *in* Thurston, P.C., White, O.L., Barlow, R.B., Cherry, M.E., and Colvine, A.C., eds., Summary of Field Work and Other Activities, 1986: Ontario Geological Survey, Miscellaneous Report 132, p. 413-420.

Trudel, P., 1985, Géologie de la mine Beaufor, région de Val d'Or: Ministère de l'Énergie et des Ressources du Québec, MB 85-42, 33 p.

Trudel, P., and Sauvé, P., 1992, Synthese des caracteristiques gitologiques des gisements d'or du district de Malartic: Ministere de l'Energie et Ressources, Quebec, MM 98-04.

Trudel, P., Methot, and Perrault, G., 1989, Géochimie de la minéralisation aurifère à la mine Eldrich, region de Rouyn-Noranda, Québec, Canada: Journal of Geochemical Exploration, v. 32, p. 415-428.

Tully, D.W., 1963, The geology of the Upper Canada mine: Canadian Institute of Mining and Metallurgy Bulletin, v. 56, p. 24-34.

Turek, A., Heather, K.B., Sage, R.P., and Van Schmus, W.R., 1996, U-Pb zircon ages for the Missanabie-Renabie area and their relation to the rest of the Michipicoten greenstone belt, Superior Province, Ontario, Canada: Precambrian Research, v. 76, p. 191-211.

Tyson, 1945, Report on gold belts in the Little Long Lac-Sturgeon River district: Canadian Mining Journal, v. 66, p. 839-850.

Van Breemen, O., Jefferson, C.W., and Bursey, R.L., 1992, Precise 2683 Ma age of turbidite-hosted auriferous iron formations in the vicinity of George Lake, eastern Slave structural province, NWT., *in* Richardson, D.G., and Irving, M., eds., Project Summaries: Canada-Northwest Territories Mineral Development Subsidiary Agreement 1987-1991: Geological Survey of Canada, Open File 2484, p. 205-207.

van Hees, E.H.P., 1979, Auriferous Ankerite Vein Genesis in the Aunor Mine, Timmins, Ontario: M.Sc. thesis, University of Western Ontario, London, Ontario, 138 p.

van Hees, E.H.P., Shelton,K., McMenamy, T., Ross, Jr., L., Cousens, B., Falck, H., Robb, N,m and Canam, T., 1999, Metasedimentary influence on metavolcanics-rock-hosted greenstone gold deposits: Geochemistry of the Giant mine, Yellowknife, N.W.T.: Geology, v. 27, p. 71-74.

Vu, L., 1990, Geology of the Ferderber gold deposit and gold potential of the Bourlamaque Batholith, Belmoral Mines Ltd., Val d'Or, Quebec, *in* Rive, M., Verpaelst, P., Gagnon, Y., Lulin, J.M., Riverin, G., and Simard, A., eds., The Northwestern Quebec Polymetallic Belt: A Summary of 60 Years of Mining and Exploration: Canadian Institute of Mining and Metallurgy, Special Volume 43, p. 237-244.

—— 1991, Geology of the Belmoral (Ferderber) gold mine, Val d'Or, Quebec, *in* Chartrand, F., ed., Geology and Gold, Rare Element, and Base Metal Mineralization of the Val d'Or Area, Quebec: Society of Economic Geologists, Guidebook 9, p. 37-44.

Vu, L., Darling, R., Beland, J., and Popov, V., 1987, Structure of the Ferderber gold deposit, Belmoral Mines Ltd., Val d'Or, Quebec: Canadian Institute of Mining and Metallurgy Bulletin, v. 89, p. 68-77.

Walker, 1969, Gold-quartz veins of the Sheep Creek camp, British Columbia, *in* Newhouse, W.H., ed., Ore Deposits as Related to Structural Features: United States National Research Council, p. 177-178.

Wall, V.J., 1989, Fluids and Metamorphism: Ph.D. thesis, Monash University, Melbourne, Australia, 234 p.

Walsh, J.F., Kesler, S.E., Duff, D., and Cloke, P.L., 1988, Fluid inclusion geochemistry of high-grade, vein-hosted gold ore at the Pamour Mine, Porcupine Camp, Ontario: Economic Geology, v. 83, p. 1347-1367.

Webb, S.A., 1994, The Discovery mine, rediscovered, *in* Kusick, R., and Goff, S.P., eds., Exploration Overview 1994; Northwest Territories; mining, exploration and geological investigations: Northern Affairs Program, Northwest Territories Geology Division, p. 64-65.

Williams, H.R., Stott, G.M., Heather, K.B., Muir, T.L., and Sage, R.P., 1992, Wawa Subprovince, Chapter 12, *in* Thurston, P.C., Williams, H.R., Sutcliffe, R.H., and Stott, G.M., eds., Geology of Ontario: Ontario Geological Survey, Special Volume 4, p. 485-539.

Wilson, M.E., and Lee, A.C., 1948, Senator-Rouyn mine, *in* Structural Geology of Canadian Ore Deposits - A Symposium: Canadian Institutde of Mining and Metallurgy, Special Volume 1, p. 735-739.

Wilson, M.E., and Mcquarry, W.R., 1948, Stadacona mine, *in* Structural Geology of Canadian Ore Deposits - A Symposium: Canadian Institutde of Mining and Metallurgy, Special Volume 1, p. 776-782.

Witt, W.K., 1991, Regional metamorphic controls on alteration associated with gold mineralization in the Eastern Goldfields province, Western Australia: Implications for the timing and origin of Archean lode-gold deposits: Geology, v. 19, p. 982-985.

Wiwchar, M.B., 1957, Consolidated Discovery Yellowknife Mine, *in* Structural Geology of Canadian Ore Deposits - A Symposium: Canadian Institutde of Mining and Metallurgy, Special Volume 2, p. 201-209.

Wojdak, P.J., and Sinclair, A.J., 1984, Equity Silver silver-copper-gold deposit: alteration and fluid inclusion studies: Economic Geology, v. 79, p. 969-990.

Workman, A.W., 1986, Geology of the McDermott gold deposit, Kirkland Lake area, Northeastern Ontario, Canada: *in* Macdonald, A.J., Downes, M., Pirie, J., and Chater, A.M., eds., Proceeding of Gold '86, An International Symposium on the Geology of Gold: p. 184-190.

Wood, P.C., Burrows, D.R., Thomas, A.V., and Spooner, E.T.C., 1986, The Hollinger-McIntyre Au-quartz vein system, Timmins, Ontario, Canada; Geologic characteristics, fluid properties and light stable isotope geochemistry, *in* Macdonald, A.J., Downes, M., Pirie, J., and Chater, A.M., eds., Proceeding of Gold '86, An International Symposium on the Geology of Gold: p. 56-80.

Wyman, D., and Kerrich, R., 1988, Alkaline magmatism, major structures, and gold deposits: Implications for greenstone belt gold metallogeny: Economic Geology, v. 83, p. 454-461.

—— 1989, Archean shoshonitic lamprophyres associated with Superior Province gold deposits: Distribution, tectonic setting, noble metal abundances, and significance for gold mineralization, *in* Keays, R.R., Ramsay, W.R.H., and Groves, D.I., eds., The Geology of Gold Deposits: The Perspective in 1988: Economic Geology Monograph 6, p. 651-657.

Wyman, D., Kerrich, R., and Fryer, B.J., 1987, Gold mineralization overprinting iron formation at the Agnico-Eagle deposit, Quebec, Canada: Mineralogical, microstructural and geochemical evidence, *in* Macdonald, A.J., Downes, M., Pirie, J., and Chater, A.M., eds., Proceeding of Gold '86, An International Symposium on the Geology of Gold: p. 108-123.

Yeager, D.A., and Metcalfe, P., 1990, Geology of the Stonehouse gold deposit, Iskut River area, B.C., *in* Vancouver '90: Geological Association of Canada/Mineralogical Association of Canada Joint Annual Meeting, Vancouver, British Columbia, Program with Abstracts 15, p. 143.

Young, M., and Helmstaedt, H.H., 2001, Tectonic Evolution of the Northern Pickle Lake Greenstone Belt, Northwestern Superior Province, Ontario: Geological Survey of Canada, Current Research 2001-C20, 22 p.

Yule, A., Mckenzie, C., and Zentilli, M., 1990, Hope Brook: A new Appalachian gold deposit in Newfoundland, Canada, and the significance of hydrothermally altered mafic dikes: Chronique de la Recherche Miniere, v. 498, p. 29-42.

Zaleski, E., L'Heureux, R., Duke, N., Wilkinson, L., and Davis, W.J., 1999, Komatiitic and felsic volcanic rocks overlain by quartzite, Woodburn Lake group, Meadowbank River area, western Churchill Province, Northwest Territories (Nunavut): Geological Survey of Canada, Current Research 1999-C, p. 9-18.

Ziehlke, D.V., 1983, The Nor-Acme gold deposit, Manitoba: Canadian Institute of Mining and Metallurgy Bulletin, v. 76, p. 80.

Zweng, P.L., Mortensen, J.K., and Dalrymple, G.B., 1993, Thermochronology of the Camflo gold deposit, Malartic, Quebec: Implications for magmatic underplating and the formation of gold-bearing quartz veins: Economic Geology, v. 88, p. 1700-1721.

Gold-rich Volcanogenic Massive Sulphide Deposits

Benoît Dubé[1], Patrice Gosselin[1], Patrick Mercier-Langevin[1],
Mark Hannington[2], and Alan Galley[3]

1. Geological Survey of Canada, 490 rue de la Couronne, Québec, Québec G1K 9A9
2. Department of Earth Sciences, University of Ottawa, Marion Hall, 140 Louis Pasteur,Ottawa, Ontario K1N 6N5
3. Geological Survey of Canada, 601 Booth Street, Ottawa, Ontario K1A 0E8
Corresponding author's email: bdube@nrcan.gc.ca

Abstract

Gold-rich volcanogenic massive sulphide deposits (Au-rich VMS) form a subtype of both volcanogenic massive sulphide and lode-gold deposits. Their diagnostic features are stratabound to discordant, volcanic-hosted massive sulphide lenses with associated discordant stockwork feeder zones in which average Au grades (in g/t) exceed associated combined Cu, Pb, and Zn grades (in weight percent); Au is thus the main commodity. The Au-VMS deposits are present in both recent seafloor and deformed and metamorphosed submarine volcanic settings. They occur in a variety of volcanic terranes from mafic bimodal through felsic bimodal to bimodal siliciclastic in greenstone belts of all ages, typically metamorphosed to greenschist and lower amphibolite facies, and intruded by subvolcanic intrusions and dyke-sill complexes. The deposits are commonly located in proximity of intermediate to felsic volcanic centres, at or close to the interface between intermediate to felsic volcanic domes and basalt-andesite or clastic sediments. Several of the largest Au-VMS deposits are located in Canada: Horne, Bousquet 2-Dumagami, LaRonde Penna, and Eskay Creek. The first three deposits are hosted within the Archean Blake River Group, which is therefore an important geological assemblage for this style of Au deposit.

In metamorphosed greenstone terranes, the gangue minerals may include quartz, sericite, aluminous silicates, such as andalusite, kyanite, pyrophyllite, and Zn-rich staurolite, and Mn-rich garnet. Sulphide minerals are mainly pyrite and base-metal sulphides with a complex assemblage of minor phases including bornite, tennantite, sulphosalts, arsenopyrite, mawsonite, and tellurides. The Au has most commonly an uneven distribution within the deposit due to both primary depositional controls and subsequent tectonic remobilization. The chemical signature of the ore is diverse, and dominated by Au, Ag, and Cu or Zn with locally high concentrations of As, Sb, Bi, Pb, Se, Te, and Hg. The Eskay Creek deposit is a low-temperature Au-rich VMS deposit characterized by a mineralogical assemblage of stibnite, realgar, cinnabar, and arsenopyrite with various proportions of barite. It is clearly distinct from other Au-rich VMS deposits. Potassic alteration characterized by K-feldspar, typical of deposits with an Au-Zn-Pb-Ag association, occurs at Eskay Creek, especially in the footwall alteration zone. The advanced argillic alteration found at LaRonde Penna and Bousquet 2-Dumagami is thought to be typical of deposits from the Au-Cu association. Where present, the metamorphosed advanced argillic and more discrete massive silicic alteration assemblages indicate high-sulphidation conditions similar to those encountered in some epithermal environments.

There are two genetic models for Au-rich VMS: 1) conventional syngenetic volcanic-hosted Au-poor VMS mineralization overprinted during regional deformation by Au mineralization; and 2) syngenetic VMS deposits characterized by an anomalous fluid chemistry (with magmatic input) and/or deposition within a shallow-water to subaerial volcanic setting equivalent to epithermal conditions, in which boiling may have had a major impact on the fluid chemistry. The deformation and metamorphism that commonly overprint the mineralization in ancient terranes have obscured the original relationships and led to considerable debate about the syntectonic versus synvolcanic origin of Au-rich VMS.

Résumé

Les gîtes de sulfures massifs volcanogènes riches en or (SMV riches en Au) constituent un sous-type des gîtes de sulfures massifs volcanogènes (SMV) ainsi que des gîtes d'or primaires. De manière caractéristique, ils consistent en lentilles stratoïdes à discordantes de sulfures massifs encaissées dans des roches volcaniques, lesquelles surmontent des zones nourricières discordantes formées d'un enchevêtrement de filonnets minéralisés (stockwerk). Les valeurs de la teneur moyenne en or (en g/t) de ces minéralisations excèdent celles des teneurs combinées en Cu, Pb et Zn (en pourcentage); l'or constitue donc la principale substance utile de ces gîtes. Les gîtes de SMV riches en Au sont présents aussi bien dans des environnements volcaniques des fonds marins actuels que dans des environnements volcaniques plus anciens où les roches sont déformées et métamorphisées. Ils se retrouvent dans toute une gamme de terrains volcaniques, des successions mafiques aux successions bimodales à dominante felsique et jusqu'aux empilements de roches volcaniques bimodales et de roches silicoclastiques des ceintures de roches vertes de tous âges, métamorphisées de manière caractéristique au faciès des schistes verts ou au faciès des amphibolites inférieur, dans lesquelles se sont mis en place des intrusions subvolcaniques et des complexes de dykes et de filons-couches. Les gîtes se trouvent couramment à proximité de centres volcaniques de composition intermédiaire à felsique, à l'interface, ou à proximité de l'interface, entre des dômes de roches volcaniques intermédiaires à felsiques et des successions de basalte-andésite ou de roches sédimentaires clastiques. Plusieurs des plus grands gisements de SMV riches en Au se trouvent au Canada : Horne, Bousquet 2-Dumagami, LaRonde Penna et Eskay Creek. Parmi ceux-ci, les trois premiers sont encaissés dans le Groupe de Blake River d'âge Archéen, lequel constitue par conséquent un important assemblage géologique pour ce style de gîte aurifère.

Dans les terrains métamorphiques de roches vertes, les minéraux de gangue peuvent comprendre le quartz, la séricite, des silicates alumineux comme l'andalousite, la kyanite, la pyrophyllite et la staurotide riche en Zn, ainsi qu'un grenat riche en Mn. Les minéraux sulfurés sont principalement la pyrite et des sulfures de métaux communs avec une association complexe de phases mineures incluant la bornite, la tennantite, des sulfosels, l'arsénopyrite, la mawsonite et des tellurures. L'or présente en général une distribution inégale dans le gîte en raison à la fois de contrôles primaires

Dubé, B., Gosselin, P. Mercier-Langevin, P., Hannington, M., and Galley, A., 2007, Gold-rich volcanogenic massive sulphide deposits, *in* Goodfellow, W.D., ed., Mineral Deposits of Canada: A Synthesis of Major Deposit-Types, District Metallogeny, the Evolution of Geological Provinces, and Exploration Methods: Geological Association of Canada, Mineral Deposits Division, Special Publication No. 5, p. 75-94.

sur la formation de la minéralisation et de la remobilisation tectonique subséquente. La signature chimique du minerai peut prendre diverses formes et elle est dominée par Au, Ag et Cu ou Zn avec par endroits des concentrations élevées de As, Sb, Bi, Pb, Se, Te et Hg. Le gisement d'Eskay Creek est un gîte de SMV riches en Au de basse température, caractérisé par une paragenèse de stibine, réalgar, cinabre et arsénopyrite avec diverses proportions de barytine. Il se distingue nettement des autres gîtes de SMV riches en Au. Au gisement d'Eskay Creek, on peut observer une altération potassique, caractérisée par la présence de feldspath potassique et typique des gîtes présentant une association Au-Zn-Pb-Ag, surtout dans la zone d'altération de l'éponte inférieure. L'altération argileuse acide aux gisements de LaRonde Penna et de Bousquet 2-Dumagami serait caractéristique des gîtes présentant l'association Au-Cu. Lorsque présentes, les associations de minéraux témoignant du métamorphisme de zones d'altération argileuse acide ou de zones de silicification massive plus discrètes révèlent des conditions indicatrices d'une forte activité en soufre similaires à celles observées dans certains milieux épithermaux.

Il existe deux modèles génétiques pour les gîtes de SMV riches en Au : 1) minéralisation syngénétique classique de SMV pauvres en or dans des roches volcaniques à laquelle s'est superposée une minéralisation en Au lors de la déformation régionale; et 2) gîtes syngénétiques de SMV caractérisés par une composition chimique anomale des fluides (avec apport magmatique) ou par le dépôt dans un cadre volcanique en eau peu profonde ou subaérien où les conditions sont équivalentes à celles des gîtes épithermaux, dans lesquelles l'ébullition peut avoir une incidence majeure sur la composition chimique des fluides. La déformation et le métamorphisme, dont les effets se superposent à la minéralisation dans les terrains anciens, ont obscurci les relations originales et suscité un débat considérable sur l'origine syntectonique ou synvolcanique des gîtes de SMV riches en Au.

Definition

Simplified Definition

A lens of Fe-, Cu-, Zn-, and Pb-sulphides, with significant amounts of Au and Ag, formed on or below the seafloor from fluids vented on and around submarine volcanoes.

Scientific Definition

Gold-rich volcanogenic massive sulphide (Au-rich VMS) deposits form a subtype of both volcanogenic massive sulphide (VMS) and lode-Au deposits (Poulsen and Hannington, 1996; Hannington et al., 1999; Huston, 2000; Poulsen et al., 2000) (Fig.1). Like most VMS deposits, they consist of semimassive to massive, stratabound to locally discordant sulphide lenses underlain by discordant stockwork feeder zones. The main difference between Au-rich VMS and other VMS deposits is their average Au content (in g/t), which exceeds the associated combined Cu, Pb, and Zn grades (in weight percent) (Poulsen et al., 2000). Gold is thus the main commodity; however, the polymetallic nature of this deposit subtype makes it more resistant to fluctuating metal prices, resulting in a very attractive exploration target. Several Au-rich VMS deposits discussed by Hannington et al. (1999) and Huston (2000) are not included within this subtype of Au deposits as their Au to base metal ratio is too low. Furthermore, the focus of the following text is mainly on Canadian examples and particularly those deposits found in the Archean Abitibi greenstone belt. For a complete global perspective the readers are referred to Large et al. (1989), Poulsen and Hannington (1996), Sillitoe et al. (1996), Hannington et al. (1999), Huston (2000), and Franklin et al. (2005), among others.

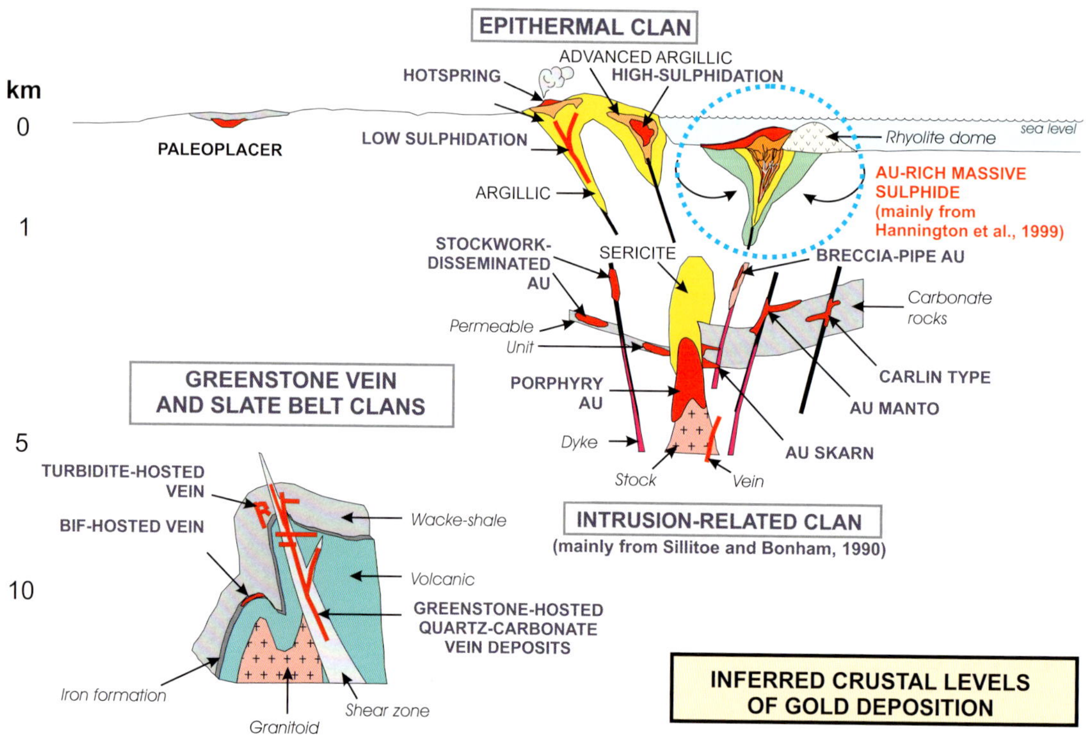

FIGURE 1. Schematic illustration of the various types of gold deposits shown at their inferred crustal levels of formation (Dubé et al., 2001, modified from Poulsen et al., 2000). BIF=banded iron formation.

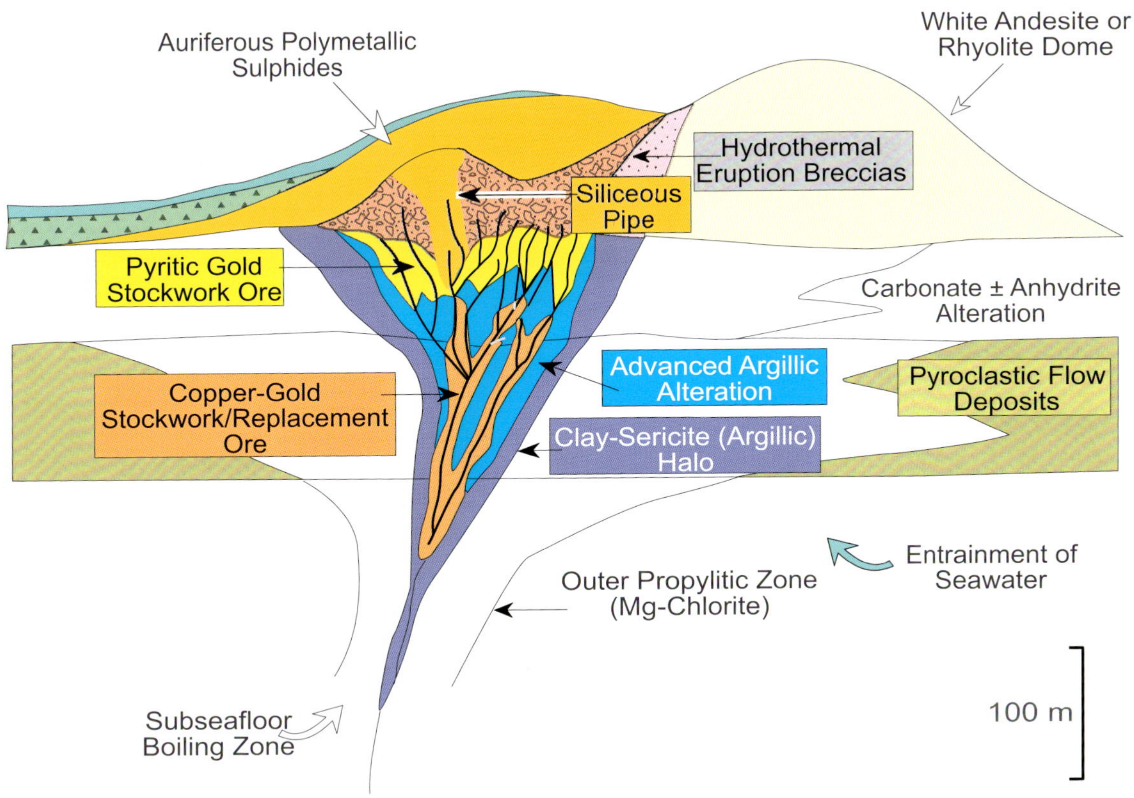

FIGURE 2. Schematic illustration of geological setting and hydrothermal alteration associated with Au-rich high-sulphidation volcanogenic massive sulphide hydrothermal systems (inspired by Hannington et al., 1999).

Gold-rich VMS deposits occur in both recent seafloor and in deformed and metamorphosed submarine volcanic settings within greenstone belts of various ages. In the latter, they may contain local syntectonic quartz-sulphide or, more rarely, quartz-tourmaline veins, which add to their complexity. They occur in a variety of submarine volcanic terranes, from mafic bimodal through felsic bimodal to bimodal siliciclastic. Their host strata are commonly underlain by coeval subvolcanic intrusions and sill-dyke complexes, and are typically metamorphosed to greenschist and lower amphibolite facies. The Au has most commonly an uneven distribution within the deposit due to both primary depositional controls and subsequent tectonic modification and remobilization. Some Au-rich VMS deposits are characterized by metamorphosed advanced argillic and massive silicic alteration indicative of an oxidized low-pH hydrothermal fluid that differs significantly from the mainly reduced, near neutral to weakly acidic fluids (of low-sulphidation conditions) typical of most ancient and modern VMS deposits (Sillitoe et al., 1996). Where present, the metamorphosed advanced argillic and massive silicic alteration assemblages are thought to indicate high-sulphidation conditions similar to those encountered in some epithermal environments (Fig. 2). In such cases, the Au-rich VMS deposits are commonly interpreted as shallow-water submarine equivalents to subaerial epithermal deposits (Hannington, 1993; Sillitoe et al., 1996; Hannington et al., 1999; Fig. 2).

Three types of Au-rich VMS deposits have been proposed based on common metallic associations (Huston and Large, 1989; Hannington et al., 1999): 1) an Au-Zn-Pb-Ag association in which Au is concentrated towards the top or along the margins of the massive sulphide lens; 2) an Au-Cu association where Au is concentrated at the base of the massive sulphide lens or within the underlying stringer zone; and 3) a pyritic Au group where Au is concentrated within massive pyrite zones with low base metals content.

Key characteristics of Au-rich VMS deposits and references are found in the text below.

Economic Characteristics of Gold-Rich Volcanogenic Massive Sulphide Deposits

Total world production and reserves of Au, including the Witwatersrand placer deposit, is 126,420 tonnes (t) (Gosselin and Dubé, 2005a). The Canadian production and reserves of Au stand at 9,280 t, representing 7% of the world total (Gosselin and Dubé, 2005a). Globally, Au-rich VMS deposit production and reserves account for 1370 t Au equivalent to 1% of the world production and reserves, whereas the Canadian total for Au-rich VMS is 870 t representing close to 10% of the Canadian production and reserves (Gosselin and Dubé, 2005a,b). The Blake River Group of the Abitibi subprovince, with 755 tonnes (t) of Au from Au-rich VMS, represents 13% of the known Au in the Abitibi greenstone belt and 85% of the production and reserves for that subtype of Au deposits in Canada. The Blake River Group, in particular within the Doyon-Bousquet-LaRonde district, is therefore an important geological assemblage for this type of Au deposit (Dubé et al., 2003).

Grade and Tonnage Characteristics

Au-rich VMS deposits range in size from small sulphide lenses with less than 3 t of Au, to giant-sized lenses and

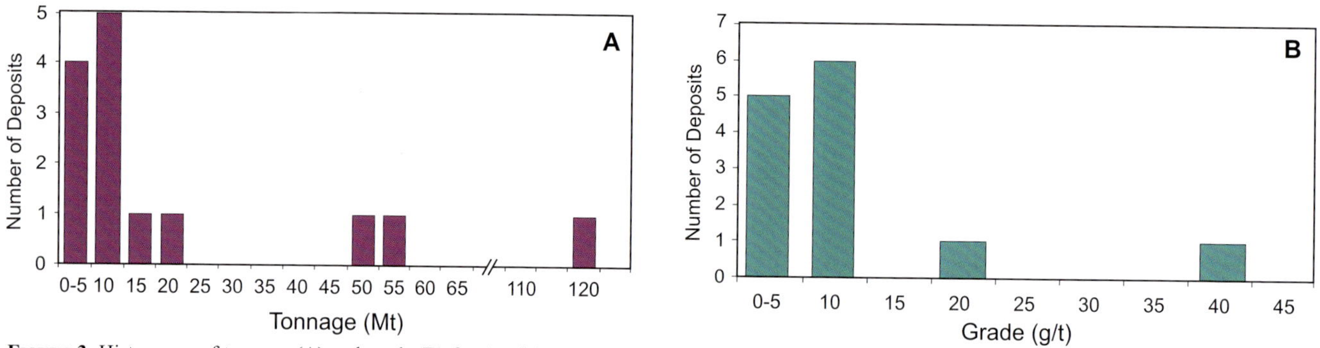

FIGURE 3. Histograms of tonnage **(A)** and grade **(B)** for Au-rich volcanogenic massive sulphide deposits worldwide (numbers include production, reserves and resources) (data from Gosselin and Dubé, 2005a).

stockwork-stringer zones of more than 50 million tonnes (Mt) of ore containing over 300 t of Au (Fig. 3A). Gold grade is typically greater than 4 g/t, with one deposit (Eskay Creek) reaching as high as 38 g/t (Figs. 3B, 4). Average Au grade for Canadian Au-rich VMS deposits is 5.9 g/t, however it may vary from 2.9 g/t up to 38 g/t. There are presently only eleven Au-rich VMS deposits in the world containing at least 30 t Au (approximately 1 Moz) in production, reserves, and resources (Fig. 5, Table 1). World-class deposits (≥100 t Au) form a select group of six deposits that includes the Paleoproterozoic Boliden deposit in Sweden (125 t Au produced) (Grip and Wirstam, 1970; Bergman-Weihed et al., 1996), one of the best known international examples, and Mount Morgan (Australia, 321 t Au in production, reserves, and resources). Some of the largest Au-rich VMS deposits are Canadian: Horne in the Noranda district (Cu-Au, 331 t of Au produced from 54.3 Mt of ore at 6.1 g/t Au, Kerr and Gibson, 1993), LaRonde Penna (Au-Zn-Ag-Cu) and Bousquet 2-Dumagami (Au-Ag-Cu-Zn, 112 t of Au produced) in the Doyon-Bousquet-LaRonde district, and Eskay Creek in British Columbia (Au-Ag-Cu-Zn-As-Sb-Hg, 81 t of Au produced and 37 t in reserves and resources). LaRonde Penna is the second largest Au-rich VMS deposit in Canada; it is also the largest Au deposit presently being mined in Canada. About 12.3 Mt of ore and 43.4 t of Au (1.4 Moz) have been extracted from the Penna shaft since the beginning of its production to the end of 2005 (Mercier-Langevin et al., 2007c). Reserves and resources at December 31, 2005 were evaluated at 6.74 Moz Au from 46.5 Mt at an average grade of 4.51 g/t Au, 2.04% Zn, 0.34% Cu, and 42.67 g/t Ag (Agnico-Eagle Mines, 2005 annual report).

Geological Setting of Gold-Rich Volcanogenic Massive Sulphide Deposits

Global-Scale Setting

Gold-rich VMS deposits occur in a variety of submarine volcanic terranes from mafic bimodal through felsic bimodal to bimodal siliciclastic in greenstone belts of all ages, typically metamorphosed to greenschist or lower amphibolite facies, and intruded by subvolcanic intrusions and dyke-sill complexes. The tectonic setting is commonly inferred to be island arcs, rifted arcs, back-arc basins, or back-arc rifts (Hannington et al., 1999; Huston, 2000). In modern volcanic settings, Au-rich VMS deposits are commonly associated with rifted-arc and incipient back-arc environments rather than mature back-arc spreading centres (Hannington et al., 1999). According to these authors, an association with rifted continental crust and continental margin arc environments may be particularly important for some districts (e.g. Boliden, Eskay Creek).

District-Scale Setting

Gold-rich VMS deposits coexist regionally with Au-poor VMS deposits, so that districts endowed with such deposits may well contain undiscovered Au-rich mineralization (Hannington et al., 1999). Their host strata are commonly underlain by coeval subvolcanic intrusions and sills or dykes. Consequently, large volumes of effusive rhyolite and associated felsic pyroclastic rocks (lithic tuffs, crystal tuffs, etc.) and the occurrence of subvolcanic intrusions or dyke swarms of tonalitic to granitic composition (Hannington et al., 1999) are important features of Au-rich VMS deposits. Apophyses developed on oxidized granitoids may be particularly important (Huston, 2000). Areas of transitional subaerial to shallow submarine volcanism are potentially very prospective (Hannington et al., 1999). Shallow-water volcanic complexes can be traced most readily through detailed mapping of volcanic and sedimentary facies (Hannington et al., 1999), and textures indicative of boiling, a process potentially responsible for the elevated Au content (see below), may also be a useful exploration guide (Huston, 2000).

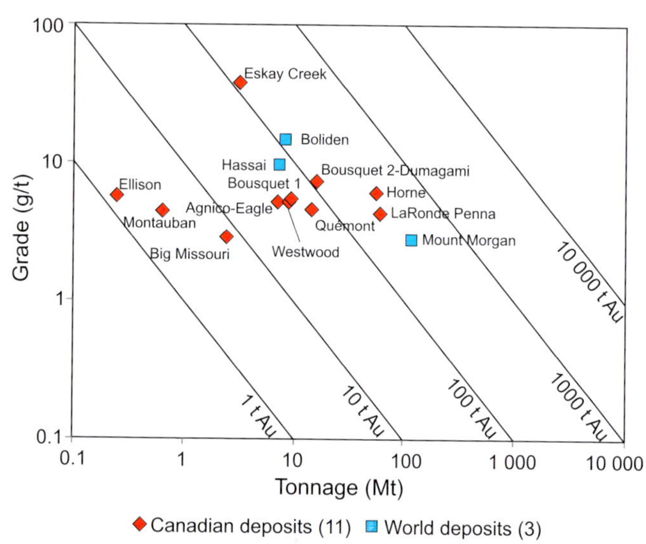

FIGURE 4. Tonnage versus grade of all Canadian Au-rich volcanogenic massive sulphide deposits and three international (>30 t Au) deposits (numbers include production, reserves, and resources) (data from Gosselin and Dubé, 2005a,b).

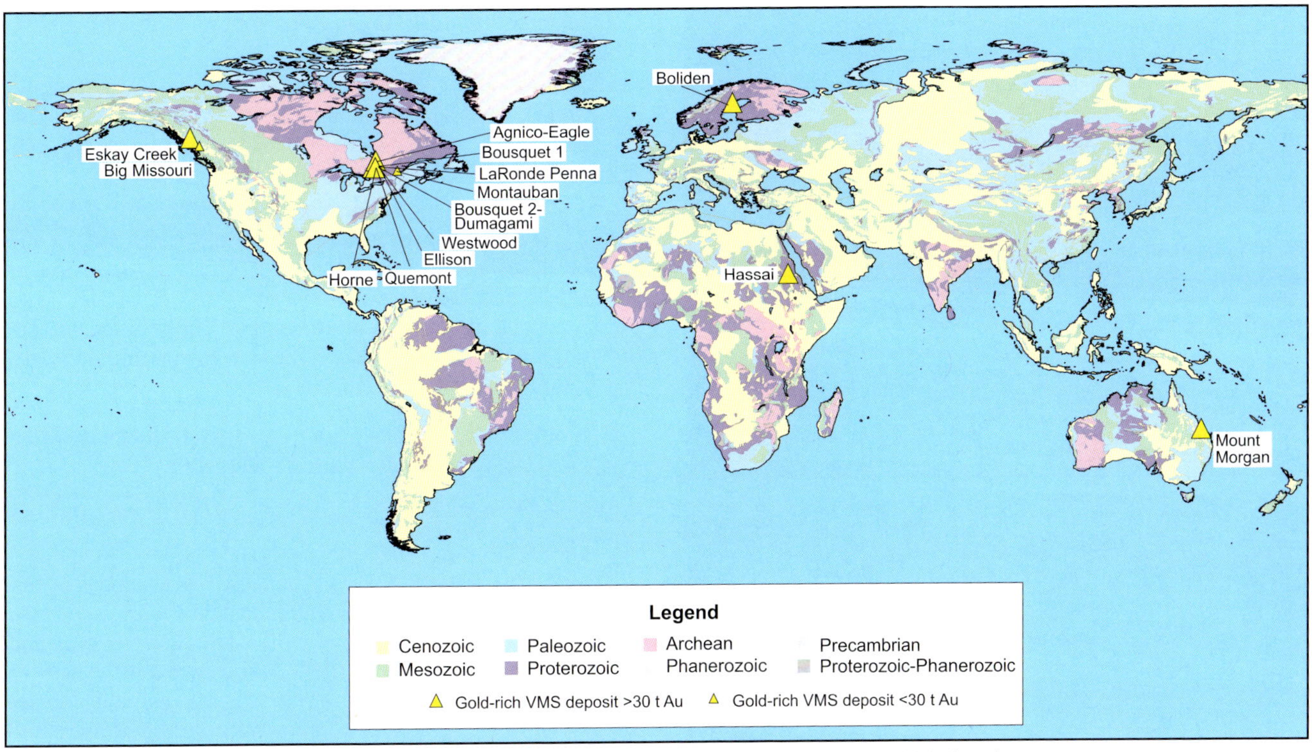

FIGURE 5. Map showing locations of Au-rich volcanogenic massive sulphide deposits.

Most deposits occur near synvolcanic fault-bounded blocks and horsts near extensional faults that commonly served as conduits for vent fluid. Post-volcanic deformation and metamorphism often caused remobilization of Au into faults and shear zones adjacent to the original sulphide lenses (Tourigny et al., 1993; Hannington et al., 1999). Pyritic quartz-sericite schists are a common host rock. Semiconformable zones of hydrothermal alteration are characterized by large aureoles of distal chloritic/propylitic (with sericite) alteration, surrounding more proximal zones of intense alteration near the deposit. These proximal zones may be characterized by sericitic to silicic/advanced argillic (aluminous) alteration indicative of low pH (acid) alteration (see below).

Base-metal enrichment and sodium-depletion anomalies may be important district-scale features (Huston, 2000). Barite commonly occurs near and within the younger deposits; barite-rich caps, barren pyrite lenses or exhalites are also found stratigraphically below or above deposits (Hannington et al., 1999).

The two largest Canadian Au deposits of this type (Horne and LaRonde Penna) are located a few kilometres from the Cadillac-Larder Lake fault, suggesting a possible empirical relationship with such large crustal faults.

Deposit-Scale Settings

The association of Au-rich VMS deposits with felsic volcanic rocks and tonalitic to granitic subvolcanic intrusions make these rock types important targets for exploration.

When deposits are chemically or mineralogically zoned, Au may have been leached from the base and centre of massive sulphide lenses and reprecipitated at the top due to gradual chemical refining of the ore (Ohmoto, 1996). Thus, variations in metal zonation, metal ratios, and ore thickness can be used to find the richest part of an orebody (Huston, 2000) even in deformed and metamorphosed terranes where these primary characteristics are commonly preserved despite

TABLE 1. Grade and tonnage of Au-rich volcanogenic massive sulphide deposits with at least 30 tonnes Au in production, reserves and resources.

Deposit Name	Country	Tonnage (Mt)	Au (g/t)	Ag (g/t)	Cu (%)	Zn (%)
Bousquet 1*	Canada	9.26	5.44			
Agnico Eagle**	Canada	6.93	5.18			
Dumagami*	Canada	15.48	7.25	5.54	0.52	0.03
Horne**	Canada	54.3	6.1	13	2.22	
LaRonde Penna*	Canada	58.76	4.31	44.96	0.33	2.17
Westwood*	Canada	8.61	5.19			
Quemont**	Canada	13.92	4.74	19.53	1.21	1.82
Eskay Creek**	Canada	3.12	37.99	1705.36		
Mt. Morgan**	Australia	116.04	2.77	0.51	0.5	
Hassai**	Sudan	7.18	9.74			
Boliden**	Sweden	8.3	15.09	48.31	1.42	

* Data from Mercier-Langevin et al. (2007).
** Data from Gosselin and Dubé (2005a).

FIGURE 6. **(A)** Visible gold coating a late north-south-trending brittle fault plane, Dumagami mine. **(B)** Pyrite-rich stringer zone underneath lens 20 South, LaRonde Penna. **(C)** Quartz-andalusite-muscovite-pyrite schist, Bousquet 2 mine.

regional deformation and metamorphism (Mercier-Langevin, 2005).

Joints, faults, shears, and veins, either within or adjacent to sulphide lenses, may contain remobilized Au or Au reintroduced during deformation/metamorphism yielding attractive areas of high Au grades, as documented at the Bousquet 2-Dumagami deposit (Fig. 6A) (Marquis et al., 1990a; Tourigny et al., 1993).

Recognition of systematic mineralogical variations in alteration zones and distinctive proximal alteration assemblages, such as advanced argillic alteration and/or Mn-rich garnet-biotite assemblage, are important guides.

Possible geochemical anomalies are dependent on deposit Au-Cu versus Au-Zn-Pb-Ag associations. Deposits of the Au-Cu association show significant proximal depletion of Na and K, and enrichment of Cu, As, Mo, Bi, and Te, with distal enrichment of Zn, Ca, Mn, and C. Those of the Au-Zn-Pb-Ag association are characterized by an enrichment of As, Sb, and particularly Hg (Huston, 2000).

Geological Characteristics of Gold-Rich Volcanogenic Massive Sulphide Deposits

Physical Properties

Morphology

The typical morphology of Au-rich VMS deposits consists of a lenticular massive sulphide body with associated underlying discordant stockwork-stringer feeders and replacement zones (Fig. 6B). Some deposits, such as LaRonde Penna, contain stacked massive sulphide lenses (Fig. 7). The orebodies are commonly tabular and stratabound to discordant (e.g. LaRonde Penna 20 South lens). In most cases they have been deformed and tilted, and have a foliation-parallel pipe-like geometry due to their strong transposition along the main foliation and stretching lineation. In these cases, the stockwork-stringer zones may have been transformed to foliation-parallel sulphide veinlets in schistose, altered rocks with quartz, white mica, and sometimes aluminous silicates (Fig. 6C). At Horne, zones of auriferous sulphide veinlets with Fe-chlorite selvages account for some of the Au-rich ore (Kerr and Mason, 1990), however, the deposit lacks a well defined stringer zone (Poulsen et al., 2000). Early VMS mineralization at the Doyon deposit (Quebec) is overprinted by a large, telescoped epithermal or intrusion-related gold deposit associated with high-level emplacement of subvolcanic intrusions (Galley et al., 2003).

Dimensions

The vertical extent of the stockwork is typically larger than its lateral extension. In some cases where the deposits are overturned, the orebody has more than 2 km of known vertical extension (Horne and LaRonde Penna deposits). The lateral extension of the deposit is typically a few hundred metres. The thickness of the massive sulphide lenses is highly variable, especially when submitted to deformation (shortening), but commonly in the order of a few tens of metres.

Host Rocks

The mineralization is typically hosted by felsic volcanic flows and volcaniclastic rocks (or their metamorphosed equivalents) near or at the interface with basaltic andesite, andesite or clastic sedimentary strata (e.g. LaRonde Penna, Eskay Creek, and Boliden) (Fig. 8A,B).

The Horne deposit is contained within a fault-bounded block of tholeiitic rhyolite flows and pyroclastic breccias and tuffs in contact with andesite flows to the east (Fig. 9). It is

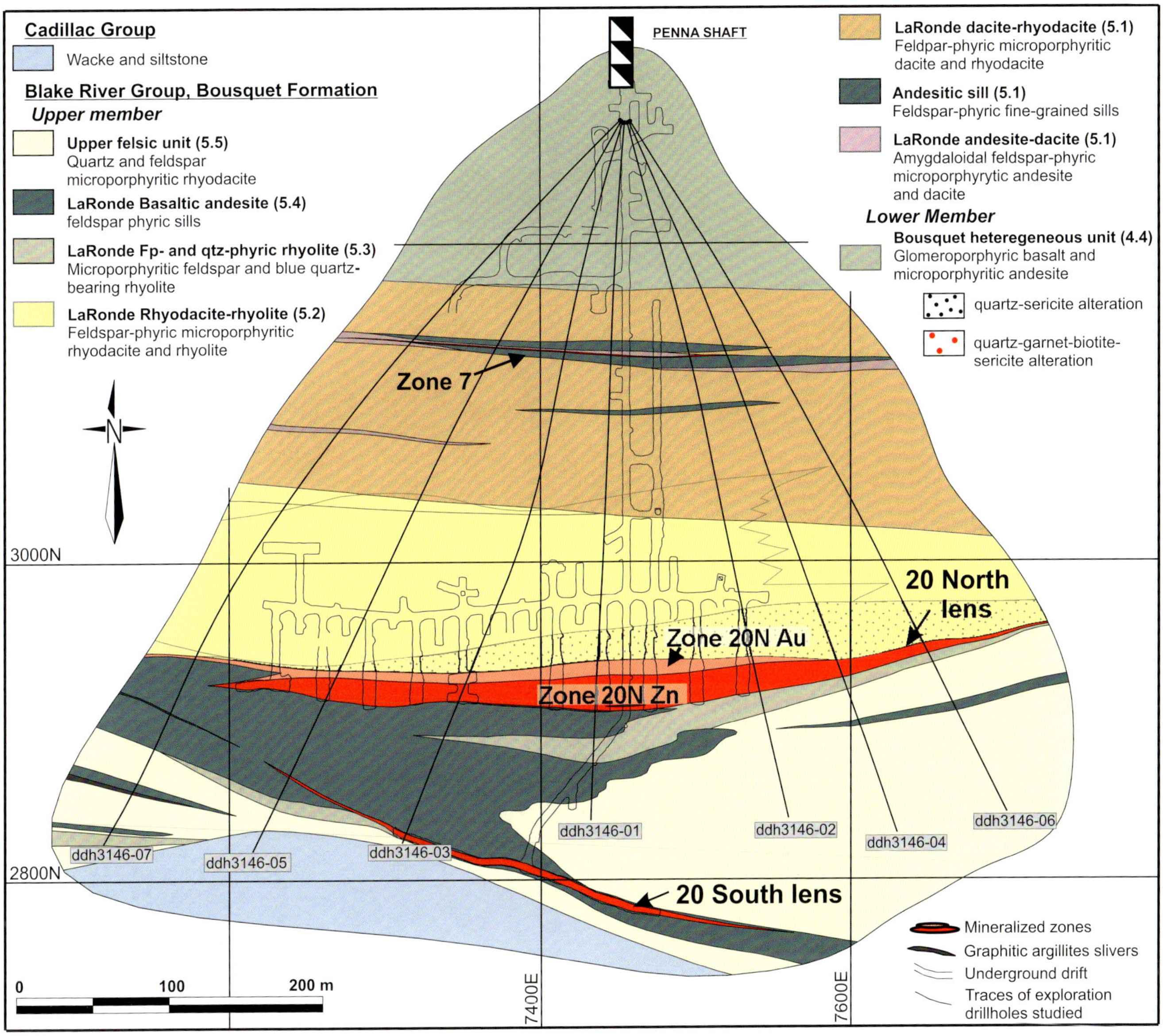

FIGURE 7. Geological plan of level 146 (1460 metres below surface) from drill core logging and drift mapping of the upper to middle portion of the LaRonde Penna deposit showing stratigraphy and sulphide lens stacking (modified from Dubé et al., 2004).

juxtaposed against andesite flows and a diorite intrusion to the south, and rhyolites to the north that contain the Quemont deposit, another auriferous massive sulphide deposit (Poulsen et al., 2000) potentially related to the same giant hydrothermal system responsible for the formation of the Horne deposit.

Textures

Banded and stratiform massive sulphide lenses and adjacent stockworks are commonly transposed by the main foliation in deformed deposits. In such cases, syntectonic sulphide veins may have developed, adding to the complexity and controversy of the deposits. Well preserved primary sulphide layering is rare to absent.

Mineralogy

The sulphide mineralogy of the Au-bearing ores is commonly more complex than in traditional Au-poor VMS deposits (Hannington et al., 1999). Sulphide minerals are mainly pyrite (Fig. 10), chalcopyrite, sphalerite, pyrrhotite, and galena with a complex assemblage of minor phases including locally significant amounts of bornite (Bousquet 2-Dumagami; Fig. 11A,B), tennantite, sulphosalts, arsenopyrite, mawsonite, and tellurides (Table 2). The strong association of tellurides with Au suggests a possible magmatic input in the hydrothermal fluid. The Boliden deposit contains nearly fifty different ore minerals (Ödman, 1941; Grip and Wirstam, 1970; Bergman Weihed et al., 1996), whereas more than twenty-five major and trace minerals have been identified in the ores at LaRonde Penna including arsenopyrite, tetrahedrite, tennantite, bornite, Pb-Sb and Ag-Sb sulphosalts, Cu-Sn-sulphides, native Bi, Bi-tellurides, Ag (-Au) tellurides, electrum and rare selenides (Fig. 12A,B; Dubé et al., 2004).

The Eskay Creek deposit is a low-temperature Au-rich VMS deposit characterized by a mineralogical assemblage of stibnite, realgar, cinnabar, and arsenopyrite with variable proportions of barite (Roth et al., 1999; Sherlock et al.,

Figure 8. (A) Rhyodacitic talus breccia in the footwall of the 20 North massive sulphide lens, LaRonde Penna. **(B)** Rhyodacitic lapilli tuff (unit 5.5) in the footwall of the 20 South massive sulphide lens, upper member of the Bousquet Formation, LaRonde Penna.

1999). The 21A zone consists of stratabound to stratiform lenses of semimassive to massive stibnite and realgar, whereas the 21B zone is a stratiform sulphide-sulphosalt Zn-Pb-Au-Ag zone. The sedimentary textures of the stratiform 21B zone are consistent with its detrital origin (Fig. 13A,B) (Poulsen et al., 2000); it is thus clearly distinct from other Au-rich VMS deposits.

As indicated by Hannington et al. (1999), Au occurs mainly as native metal and Au-tellurides in Cu-Au VMS deposits, whereas auriferous, polymetallic (Au-Zn-Pb-Ag) VMS typically contain electrum, which is often Ag- or Hg-rich (cf. Huston et al., 1992). In some deposits, Au is mainly hosted in commonly refractory arsenic-rich pyrite and arsenopyrite and present as submicroscopic inclusions or structurally bound to the crystal lattice (Huston et al., 1992; Larocque et al., 1993, and references therein). In metamorphosed deposits such as LaRonde Penna, metamorphic remobilization and segregation has had an impact on the local distribution of Au in the ores and has played an important role in generating non-refractory Au minerals (Dubé et

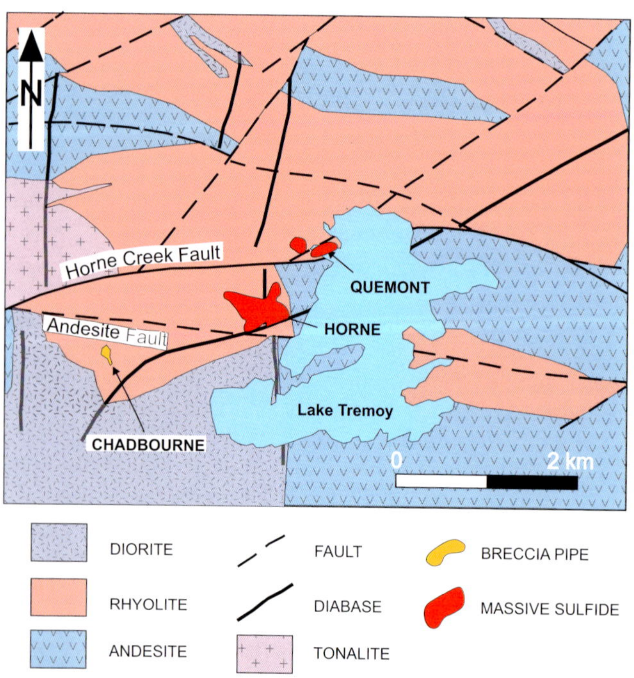

Figure 9. Simplified map of the geology around the Horne deposit, Rouyn-Noranda district, modified from Cattalani et al. (1993); massive sulphide deposits are projected to surface (from Poulsen et al., 2000).

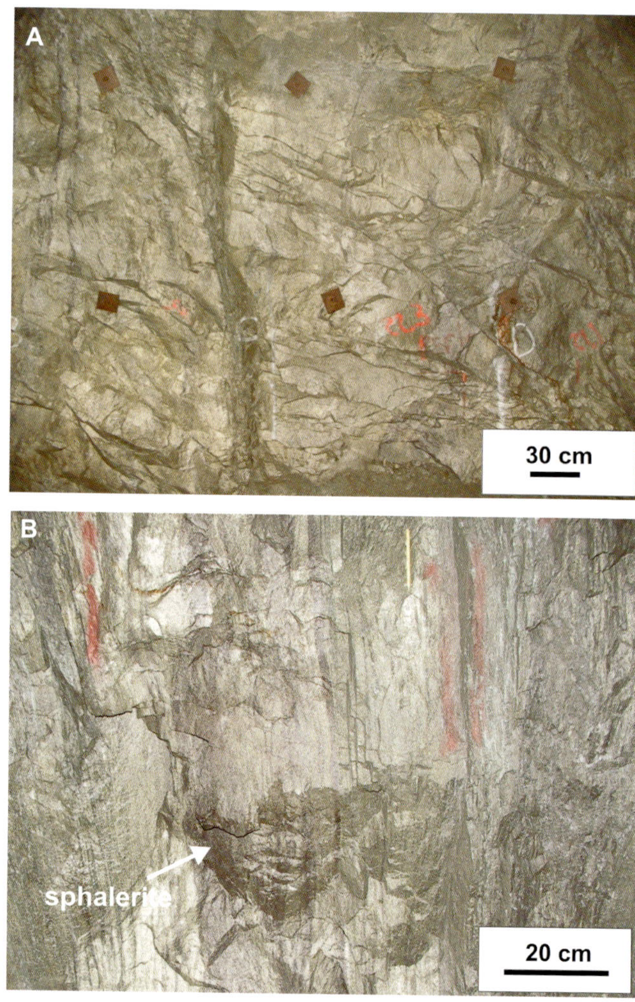

Figure 10. (A) Massive pyrite zone, lens 20 North, LaRonde Penna. **(B)** Massive pyrite with layers of sphalerite, lens 20 South, grading 34 g/t Au and 8% Zn over 10 metres, LaRonde Penna.

TABLE 2. Mineralogy of the LaRonde and selected Au-VMS deposits in the Doyon-Bousquet-LaRonde district (from Dubé et al., 2004).

	LaRonde[2]		Bousquet 2		Dumagami		Bousquet 1	
Major Minerals	Pyite	FeS_2	Pyite	FeS_2	Pyite	FeS_2	Pyite	FeS_2
	Pyrrhotite	$Fe_{1-x}S$	Chalcopyite	$CuFeS_2$	Chalcopyite	$CuFeS_2$	Chalcopyite	$CuFeS_2$
	Chalcopyite	$CuFeS_2$	Bornite	Cu_5FeS_4	Sphalerite	ZnS		
	Sphalerite	ZnS						
	Galena	PbS						
Minor Minerals	Arsenopyrite	FeAsS	Sphalerite	ZnS	Galena	PbS	Sphalerite	ZnS
	Tetrahedrite (Ag)	$Cu_{12}(Sb,Ag)_4S_{13}$	Galena	PbS	Arsenopyrite	FeAsS	Pyrrhotite	$Fe_{1-x}S$
	Stannite	Cu_2FeSnS_4	Tennantite	$(Te)(Cu,Zn,Fe)_{13}(As,Sb,Te)_4S_{12}$	Bornite	Cu_5FeS_4	Magnetite	$FeFe_2O_4$
	Magnetite	$FeFe_2O_4$	Chalcocite	Cu_2S	Chalcocite	Cu_2S	Arsenopyrite	FeAsS
			Pyrrhotite	$Fe_{1-x}S$	Covellite	CuS	Galena	PbS
			Magnetite	$FeFe_2O_4$	Digenite	Cu_9S_5	Bornite	Cu_5FeS_4
Trace Minerals	Electrum	AuAg	Electrum	AuAg	Electrum	AuAg	Electrum	AuAg
	Hessite	Ag_2Te	Hessite	Ag_2Te	Altaite	PbTe	Au-Telluride	Ag_3AuTe_2; $AuTe_2$
	Native Bi		Petzite	Ag_3AuTe_2	Cu-Ag-sulphide		Altaite	PbTe
	Bi-telluride[3]		Altaite	PbTe	Stannite	Cu_2FeSnS_4	Stannite	Cu_2FeSnS_4
	Ag-Bi sulphide Mathildite	$AgBiS_2$	Tellurobismuthite	Bi_2Te_3	Alabandite	MnS	Gudmundite	FeSbS
	Cu-Pb-Bi-sulphide[4]		Mawsonite	$Cu_6Fe_2SnS_8$				
	Pb-Sb sulphosalts[5]		Colusite	$Cu_{26}V_2(As,Sn,Zn,Ge)_6S_{32}$				
	Ag-Sb sulphosalts[6]		Renierite	Cu_3GeFeS_4				
	Gudmundite	FeSbS						
	Clausthalite	PbSe						
	Colusite	$Cu_{26}V_2(As,Sn,Zn,Ge)_6S_{32}$						
	Scheelite	$CaWO_4$						
	Cassiterite	SnO_2						

[1]LaRonde: Dubé et al. (2004); Bousquet 2: Tourigny et al. (1993); Dumagami: Marquis et al. (1990a); Bousquet 1: Tourigny et al. (1992).

[2]All ore zones, excluding Zone 5

[3]Bi-telluride = joseite-like compositions Bi_4Te_2S;

[4]Cu-Pb-Bi-sulphide = wittichenite; Cu_3BiS_3 and unidentified Mn-Bi-Pb-Sb sulphosalt;

[5]Pb-Sb sulphosalts = boulangerite $Pb_5Sb_4S_{11}$ and bournonite $PbCuSbS_3$;

[6]Ag-Sb sulphosalts = dyscrasite Ag_3Sb.

al., 2004). At LaRonde Penna, free Au (as electrum) accounts for the majority (>90%) of the Au in the ore (Dubé et al., 2004). The Au grains are typically very fine (1-5 microns) and occur mainly as inclusions in recrystallized pyrite and chalcopyrite, and within microfractures in recrystallized pyrite. The electrum typically occurs intimately intergrown with other remobilized trace minerals (Fig. 12B).

Chemical Properties

Ore Chemistry

The chemical signature of the ore is dominated by Au, Ag, and Cu or Zn with locally high concentrations of As, Sb, Bi, Pb, Se, Te, and Hg. At Eskay Creek, elevated Sb, As, Hg, and Ba are characteristic of the high-grade ore (Roth et al., 1999). Where associated with Cu, Au is commonly concentrated within the stockwork-stringer zone in the immediate footwall of the massive sulphide lens (e.g. LaRonde Penna, 20 North Au lens below the 20 North Zn massive sulphide lens, Dubé et al., 2004) (Huston, 2000). Where associated with Zn, Au is located toward the upper part (Huston, 2000) or throughout the massive sulphide lens (e.g. 20 South lens at LaRonde Penna, Dubé et al., 2004; Fig 10B). Silver is commonly more abundant than Au and the Ag/Au ratios typically vary from 1:2 to 10:1.

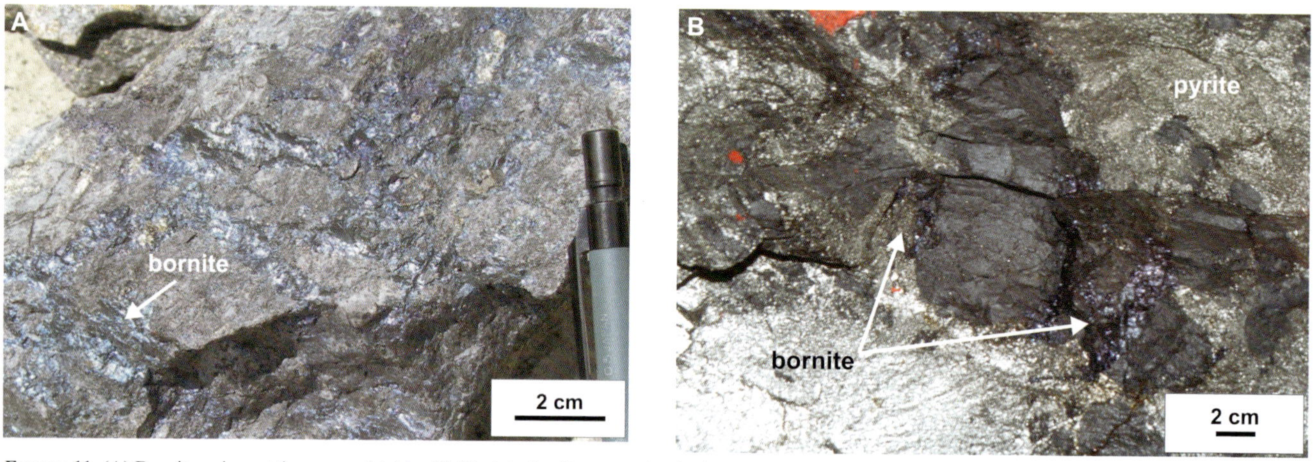

Figure 11. (A) Bornite veins cutting across highly silicified dacite (Bousquet 2 mine). **(B)** Bornite veinlets cutting highly silicified clast within massive pyrite lens (Bousquet 2 mine).

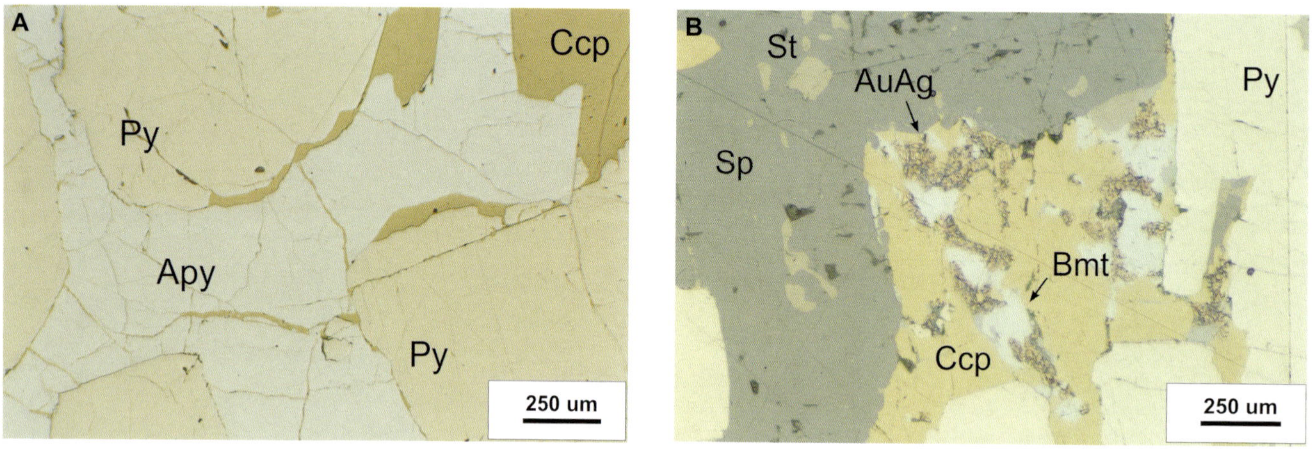

Figure 12. Photomicrographs showing typical ore textures and mineral assemblages of the different ore zones in the LaRonde mine (Apy: arsenopyrite; AuAg: electrum; Bmt: bismuthinite; Ccy: chalcopyrite; Py: pyrite; Sp: sphalerite; St: stannite (from Dubé et al., 2004).

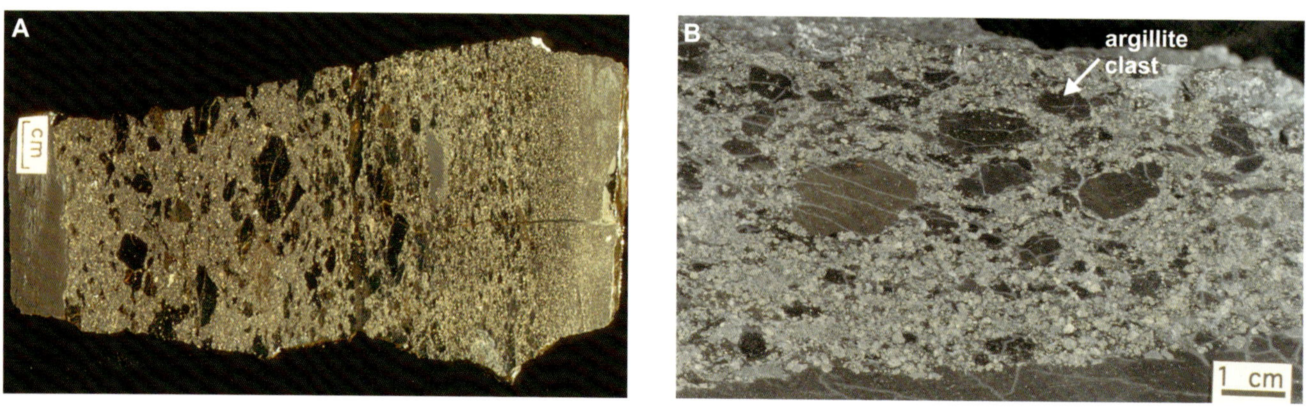

Figure 13. (A) Graded sulphide-silicate turbidite bed. **(B)** Tetrahedrite-rich (light grey) sedimentary debris flow containing siliceous argillite clasts (dark grey to black). Both photographs from the 21B zone, Eskay Creek (from Poulsen et al., 2000).

Alteration Mineralogy

In the Doyon-Bousquet-LaRonde district, the alteration assemblages proximal to or hosting the ore are commonly characterized by semiconformable to discordant zones of metamorphosed advanced argillic (aluminous) alteration with quartz, sericite, andalusite and/or kyanite, pyrophyllite and by local Zn-rich staurolite (Figs. 14A-D, 15A) and massive silicic alteration with strong to complete leaching of Na_2O, CaO, MgO, and K_2O (Fig. 15A,C; Valliant and Barnett, 1982; Dubé et al., 2004, in press). SiO_2, Al_2O_3, and TiO_2 have commonly been affected by residual enrichment due to the removal of the other oxides, although SiO_2 could have also been added through silicification. The advanced argillic alteration Index (AAAI=100 (SiO_2/(SiO_2+10MgO+10CaO+10Na_2O)) has been proposed recently to quantify such intense acid leaching with SiO_2 enrichment and to help in mapping the various alteration zones (Williams and Davidson, 2004). Andalusite and/or kyanite are commonly

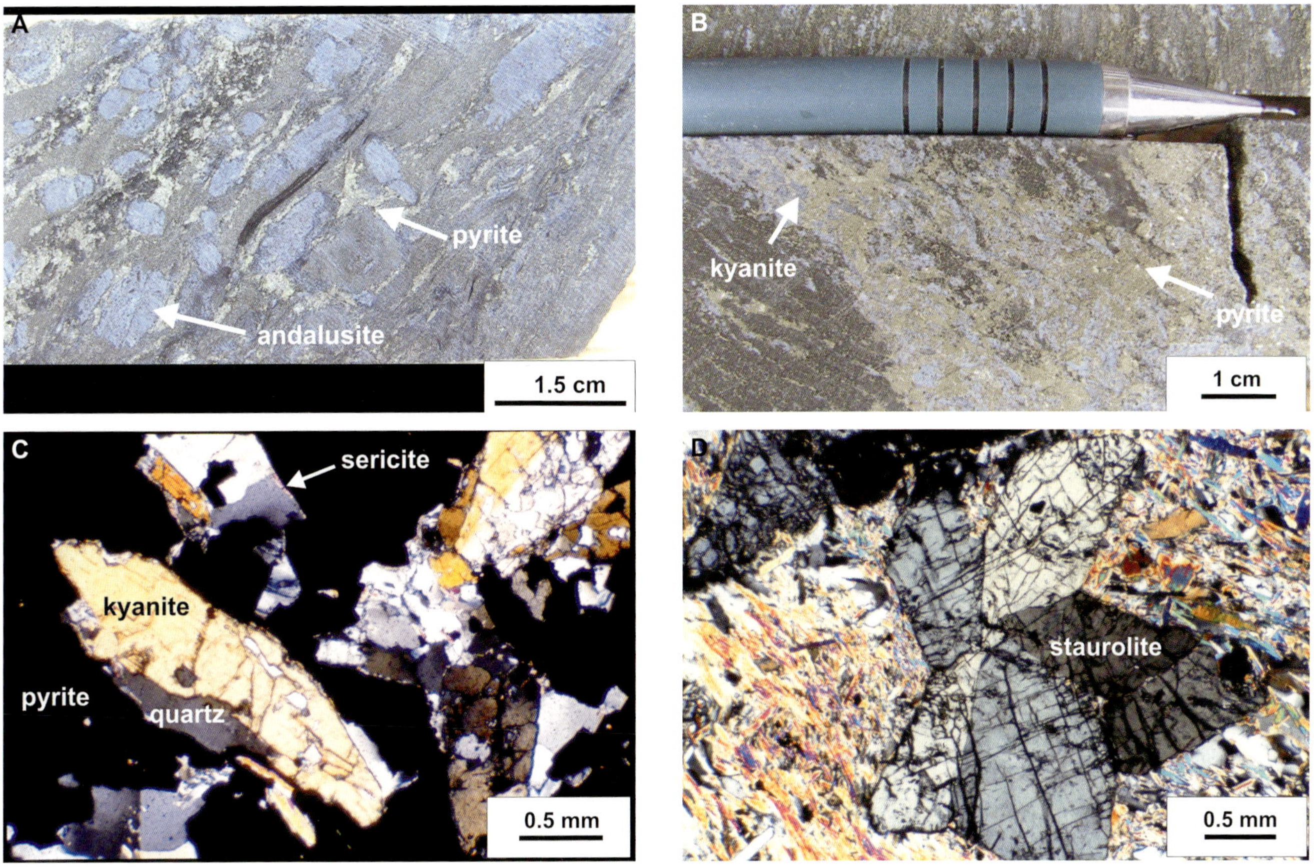

FIGURE 14. **(A)** Andalusite, kyanite, and pyrite replacement and stockwork ore zone, LaRonde Penna. **(B)** Kyanite and pyrite-rich replacement-style ore zone, LaRonde Penna. **(C)** Kyanite crystals (yellow, brown) in pyrite (black) and quartz (white, grey), LaRonde Penna (polarized light). **(D)** Zn-rich staurolite (blue, grey, and beige) in sericitic matrix, LaRonde Penna (polarized light).

retrograded into pyrophyllite (e.g. LaRonde Penna, Bousquet 2-Dumagami) (Marquis et al., 1990c; Tourigny et al., 1993; Dubé et al., 2004, in press).A proximal quartz-biotite-Mn-rich garnet assemblage or an outer quartz-manganiferous garnet-Zn-rich staurolite-chloritoid-biotite-muscovite-chlorite assemblage may be present, especially in the footwall of the mineralization (Fig. 15B; Valliant and Barnett, 1982, Dubé et al., 2004). Green chromium mica may also be locally present, as illustrated by the presence of chromium-rich phengite in both the immediate footwall and hanging wall of the 20 South lens at LaRonde Penna (Dubé et al., 2004, in press; Fig. 15D) and in the footwall of the Rambler deposit in Newfoundland. The North and South ore zones at Montauban are associated with disseminated pyrite, sphalerite, and chalcopyrite, with cordierite-anthophyllite and quartz-biotite garnet assemblages within quartz-biotite and quartz-sillimanite gneisses (Morin, 1987). Potassic alteration, characterized by K-feldspar, occurs at Eskay Creek, especially in the footwall alteration zone (Roth et al., 1999). Huston (2000) proposed that the advanced argillic alteration is more typical of the Au-Cu subclass of Au-rich VMS deposits, whereas potassic feldspar is more common of those characterized by the Au-Zn-Pb-Ag association. Tourmaline is present at Boliden as lens-shaped auriferous tourmaline ore located beneath the massive sulphide zone within the sericitic alteration, as well as minor high-grade quartz-tourmaline veins (Bergman Weihed et al., 1996). Traces to minor amounts of tourmaline are also present at LaRonde Penna (Dubé et al., in press).

At the Horne deposit, most rhyolitic rocks within the fault-bounded block have been affected by weak sericitization and silicification that become more intense near the orebodies, where alteration is characterized by a quartz-sericite±pyrite assemblage (Poulsen et al., 2000). Chlorite alteration, which locally contains elevated Cu and Au values, is largely restricted to the immediate footwall and sidewall of the deposit, except for local discordant zones in the footwall (Barrett et al., 1991).

Distribution of Canadian Gold-Rich Volcanogenic Massive Sulphide Districts

Due to the low number of Au-rich VMS deposits, there are relatively few districts containing Au-rich VMS deposits in Canada and around the world. Districts listed in Table 3 contain Au-rich VMS deposits as well as other Au-poor VMS, epithermal, intrusion-related (Au-Cu sulphide-rich veins), or greenstone-hosted quartz-carbonate vein deposits. However, production, reserves, and resources figures listed are for deposits of the Au-rich VMS subtype only.

Gold-rich VMS deposits are distributed in areas where hydrothermal activity took place in tectonically active, arc- or back-arc-related seafloor environments representing volcanic terranes located in orogenic belts and greenstone belts, frequently near major crustal-scale faults and large subvol-

FIGURE 15. **(A)** Quartz-muscovite-andalusite schist in the footwall of the 20 North lens, LaRonde Penna. **(B)** Deformed Mn-rich garnet-biotite alteration zone at the base of the 20 North lens, LaRonde Penna. **(C)** Massive silicic alteration (97% SiO_2) cut by pyrite stringers at the base of the massive sulphide lens, Bousquet 2 mine. **(D)** Banded green mica alteration hosting the 20 South lens, LaRonde Penna.

canic intrusions. In Canada, such settings are found in Archean greenstone belts within the Superior Province (Rouyn-Noranda and Doyon-Bousquet-LaRonde districts), in the Paleozoic Dunnage zone of the Appalachians (e.g. Baie Verte district), and the Mesozoic Intermontane Belt of the Canadian Cordillera (Iskut River district). High metamorphic grade, Proterozoic volcanic rocks hosting the Montauban deposit (Montauban district), Grenville Province, are thought to be related to island-arc or back-arc environments (MacLean et al., 1982; Nadeau et al., 1999).

Canadian Au-rich VMS deposits with a total Au content of at least 30 t Au are located mainly in the Abitibi greenstone belt and are Late Archean in age (Fig. 16). Figure 16 also shows that the total Au content of the Canadian Archean Au-VMS deposits is larger than anywhere else. They have an average grade of 5.35 g/t Au and their tonnage ranges from 6.93 Mt (Agnico-Eagle Telbel) to more than 58 Mt (LaRonde Penna). The Horne and Quemont deposits are located in the Rouyn-Noranda district (see Fig. 17 for location of districts). This district also contains a large number of greenstone-hosted quartz-carbonate vein and VMS deposits that are not sufficiently enriched in Au to fall into the Au-rich VMS category.

The Doyon-Bousquet-LaRonde district hosts three producing mines (LaRonde Penna, Doyon, and and Mouska) and several past producers (Bousquet 1, Bousquet 2-Dumagami, Mooshla A and B, and MicMac), as well as two significant undeveloped deposits (Westwood and Ellison, Fig. 18; Mercier-Langevin et al., 2007c). It is the second largest Au camp in Quebec and one of the most prolific Archean Au-rich VMS districts with more than 462 t Au (past production, reserves, and resources). The LaRonde Penna deposit represents one of the largest complexes of Au-rich VMS lenses in the world. The VMS deposits in the camp are hosted by the upper portion of the Blake River Group

TABLE 3. List of Canadian Au-rich volcanogenic massive sulphide districts.

District	Geological Province	Production (tonnes Au)	Resources +Reserves*
Doyon-Bousquet-LaRonde	Superior/Abitibi	195	267
Rouyn-Noranda	Superior/Abitibi	397	
Joutel	Superior/Abitibi	36	
Iskut River	Cordilleran	83	43
Montauban	Grenville	3	

*(tonnes Au)

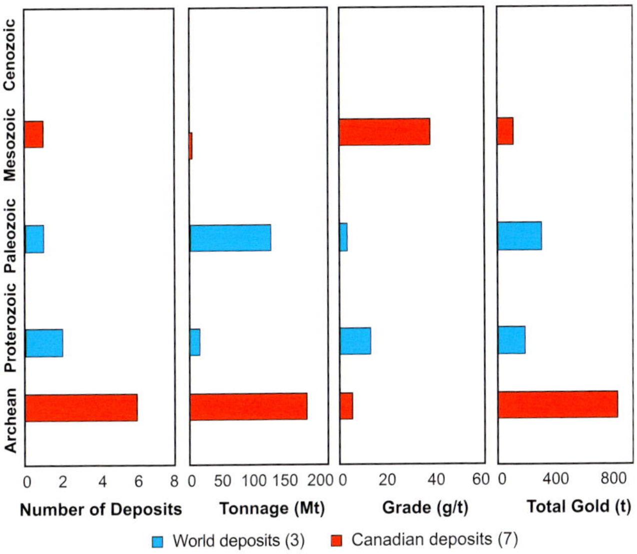

FIGURE 16. Comparison of Au-rich volcanogenic massive sulphide deposits (≥30 t Au) from different geological periods (data from Gosselin and Dubé, 2005a,b).

(Bousquet Formation), which also hosts most of the VMS deposits located in the nearby world-class Noranda district. The Doyon mine is the second largest deposit in the district with 206 t of Au in production, reserves, and resources. Early VMS mineralization at Doyon is overprinted by a telescoped sulphide-rich epithermal or intrusion-related, pre-to early deformation vein-type deposit associated with high-level emplacement of subvolcanic intrusions (Galley and Lafrance, in press). The only other operating mine in this district is the greenstone-hosted quartz-carbonate vein Mouska deposit (Belkabir and Hubert, 1995).

The Joutel district contains the Agnico-Eagle (Telbel) Au-rich VMS deposit, where disseminated to semimassive sulphide lenses and stringers are hosted by carbonate-altered volcanogenic sedimentary rocks and felsic tuffs. The origin of the deposit has been the subject of debate between syngenetic exhalative (Barnett et al., 1982; Dubé et al., 1991) and syndeformation epigenetic (Wyman et al., 1986), but colloform textures, breccias, and a ribboned sulphide facies have led Gauthier et al. (2000) to infer a shallow-water setting close to the Mattabi style of VMS deposits. Two Cu-Zn VMS deposits exploited in the early 1960s (the Joutel Copper and Poirier mines) occur less than 5 km from the Agnico-Eagle deposit, and within the area are also other VMS deposits (such as the Selbaie Cu-Zn-Au-Ag deposit) and greenstone-hosted quartz-carbonate veins Au deposits (e.g. Casa-Berardi).

The only Canadian deposit of Proterozoic age is the Montauban deposit (Grenville Province, Quebec), which has produced some 2.8 t Au from ore grading 4.5 g/t. There is no other known Au deposit in the Montauban area. Proterozoic examples outside Canada include Boliden (Sweden) and Hassai (Sudan). Boliden is a controversial Paleoproterozoic deposit located within the upper part of 1.89 to 1.87 Ga dacite-rhyodacite calc-alkaline volcanic rocks near the interface with a sedimentary sequence (Bergman Weihed et al., 1996). The deposit is characterized by predeformation massive pyrite±arsenopyrite ore, overprinted by less abundant auriferous quartz-chalcopyrite-sulphosalt and quartz-tourmaline veins, with associated andalusite-sericite-quartz

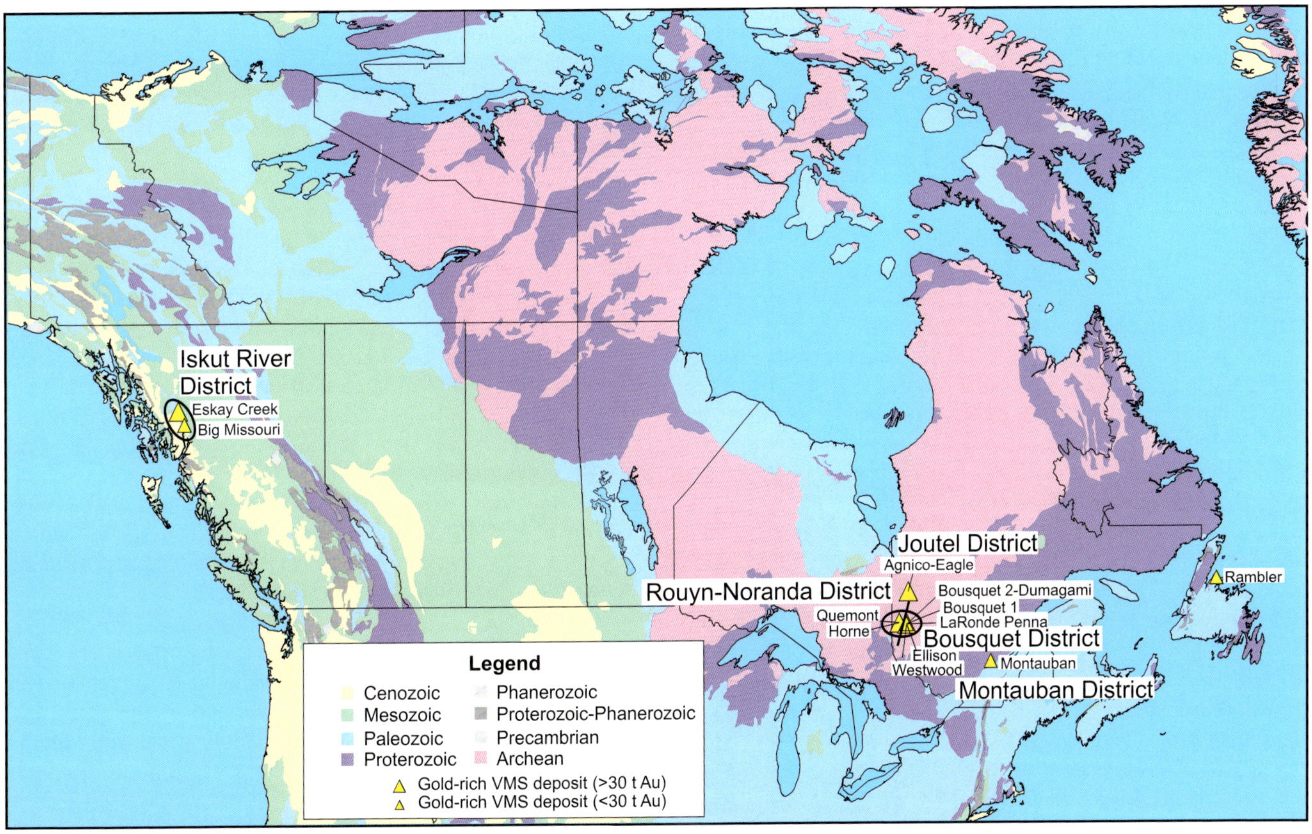

FIGURE 17. Locations of Au-rich volcanogenic massive sulphide districts in Canada.

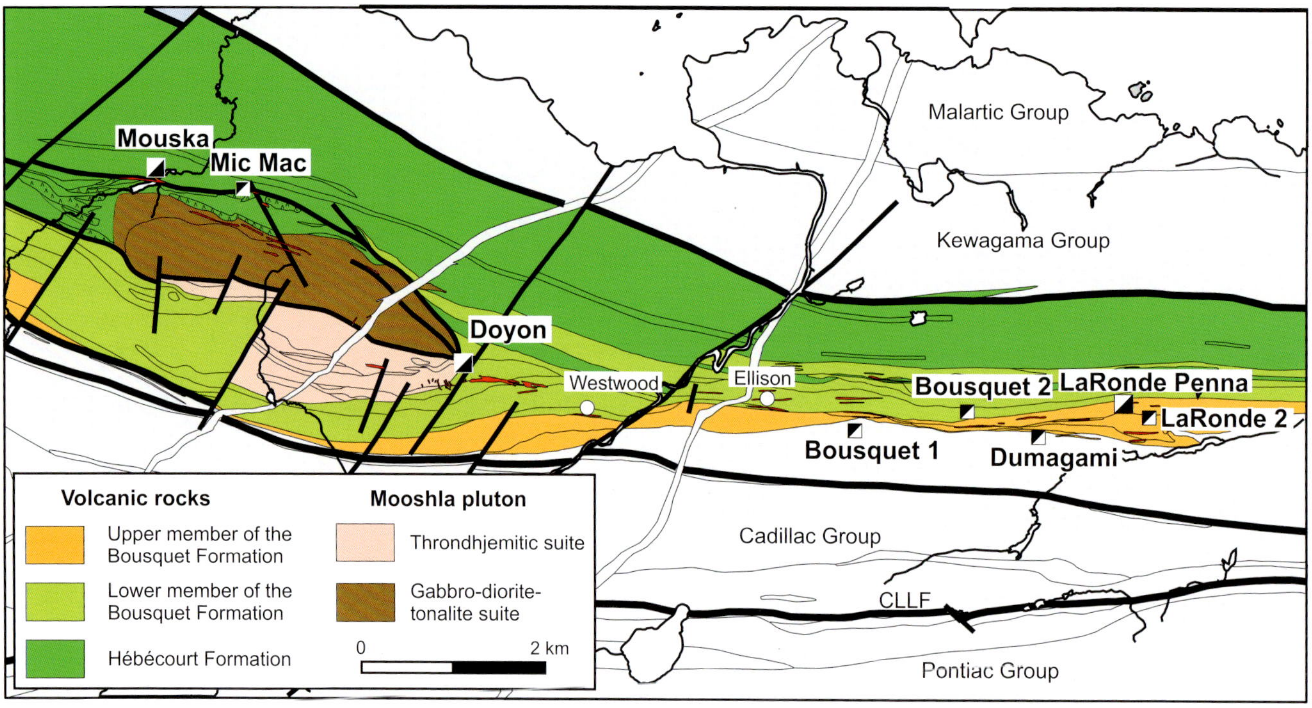

FIGURE 18. Simplified geological map of the Doyon-Bousquet-LaRonde district showing the main geological features and mine locations (from Dubé et al., 2003). CLLF = Cadillac-Larder Lake Fault.

assemblages (Nilsson, 1968; Bergman Weihed et al., 1996). The deposit has been interpreted as a replacement metamorphic deposit (Ödman, 1941), a synvolcanic exhalative deposit (Rickard, 1986), a subseafloor replacement deposit (Allen et al., 1996), a high-sulphidation epithermal deposit (Bergman Weihed et al, 1996), and a subseafloor replacement deposit transitional between a VMS and a high-sulphidation deposit (Doyle and Allen, 2003).

A small Paleozoic deposit, the Rambler Cu-Zn-Au-Ag deposit in Newfoundland (6.9 t Au from ore grading 1.54 g/t), is located in a greenstone belt of the Dunnage zone (Appalachians) (Hibbard, 1983; Coates, 1990). Although not a gold-rich VMS due to average Cu grades exceeding Au grades, the Rambler Main mine (399,093 t at 1.3% Cu, 2.16% Zn, 5.14 g/t Au, and 29.14 g/t Ag, Coates, 1990) was a small but significantly gold-rich portion of the deposit. Rambler occurs along the flank of a felsic dome and is hosted by felsic volcaniclastic rocks (dacitic tuffs, lapilli tuffs, and agglomerates) overlain by mafic to intermediate flows and volcaniclastic rocks (Coates, 1990). Similarly to the LaRonde Penna deposit, a portion of the footwall of the Main mine orebody at Rambler is characterized by Au-rich quartz-sericite-green mica schists with up to 4.8 g/t Au. The Baie Verte district has a number of other deposits of the greenstone-hosted quartz-carbonate vein subtype as well (e.g. Stog'er Tight, 3.4 t Au). The largest Paleozoic deposit is Mount Morgan (116 Mt at 2.77 g/t Au for 321 t of Au, (Gosselin and Dubé, 2005a) in Australia.

The Intermontane belt of the Canadian Cordillera hosts the Mesozoic Iskut River district with two deposits: one is the world-class, high-grade Eskay Creek deposit (3.1 Mt of ore at 38 g/t Au, Gosselin and Dubé, 2005b); the other is the smaller Big Missouri deposit (2.5 Mt of ore at 2.9 g/t Au) (British Columbia MINFILE database #104B 046). Other significant deposits of this district are Au-Cu sulphide-rich veins (Snip deposit, 32 t Au) (British Columbia MINFILE database #104B 250) and (mainly) low-sulphidation epithermal deposits (Silbak Premier, 65 t Au) (Gosselin and Dubé, 2005b). The Eskay Creek deposit is hosted by Lower Jurassic rocks of the Hazelton Group in the Stikine Terrane in a stratigraphic sequence consisting of felsic flow-banded volcanic rocks and breccias of a flow dome complex overlain by marine argillites and pillow basalts (MacDonald et al., 1996). The deposit consists of stratabound to stratiform lenses of semimassive to massive stibnite and realgar (21A Zone) and a stratiform sulphide-sulphosalt Zn-Pb-Au-Ag zone composed of stibnite, arsenopyrite, pyrite, sphalerite, galena, and tetrahedrite (21B Zone) (Poulsen et al., 2000). The sedimentary textures of the stratiform ore in the 21B Zone are consistent with a debris flow, whereas the 21A Zone, with its underlying stockwork and disseminated ore zones, shares some analogies with epithermal mineralization (Poulsen et al., 2000). Alteration of the underlying volcanic rocks includes chloritization, silicification, and sericitization, and is comparable to that of other VMS deposits. Most workers regard Eskay Creek as having characteristics of both epithermal and VMS deposit-types (MacDonald et al., 1996; Sillitoe et al., 1996). The comparatively low tonnage, which is significantly smaller than that of Archean deposits, is compensated by the much higher ore grade.

Associated Mineral Deposit Types

Gold-rich VMS deposits are thought to have formed under a variety of conditions. At one end of the spectrum, they are thought to represent the shallow water equivalents of subaerial epithermal Au deposits (Hannington, 1993; Poulsen and Hannington, 1996, Sillitoe et al., 1996) (Fig. 19). These may coexist regionally with Au-poor VMS deposits (e.g. Horne), and could also possibly coexist with deposits of the epithermal clan (including deposits of the high-sulphidation type) and intrusion-related Au deposits formed lower in the stratigraphic sequence (Sillitoe et al., 1996; Hannington et al., 1999) (e.g. Doyon mine in the Doyon-Bousquet-LaRonde district) (Fig. 1). As there is an empirical relationship between large Au-rich VMS deposits and major fault zones (e.g. LaRonde Penna and Horne with the Cadillac-Larder Lake fault), they may coexist with syn-deformation Au deposits of various types, including greenstone-hosted quartz-carbonate veins.

Genetic and Exploration Models

Conventional Models

Gold-rich VMS deposits have been the subject of much controversy during the last 20 to 30 years, particularly the timing of Au deposition (synvolcanic or syntectonic) relative to the formation of the massive sulphide orebody. Consequently, their origin remains questioned (Arnold and Sillitoe, 1989; Poulsen and Hannington, 1996; Sillitoe et al., 1996; Huston, 2000, and references therein), especially in metamorphosed Precambrian terranes (Valliant and Hutchinson, 1982; Stone, 1988; Stone et al., 1988; Tourigny et al., 1989; Marquis et al., 1990a,b,c; Bergman Weihed et al., 1996). As discussed by Poulsen and Hannington (1996), there are basically two genetic models: 1) conventional syngenetic volcanic-hosted Au-poor base metal mineralization overprinted during regional-scale deformation and metamorphism by syndeformation Au mineralization (Tourigny et al., 1989, 1990; Marquis et al., 1990a,b,c; Yeats and Groves, 1998); and 2) syngenetic VMS deposits that differ from the conventional massive sulphide deposits by an anomalous fluid chemistry and/or deposition within a shallow-water to subaerial volcanic setting (Poulsen and Hannington, 1996; Hannington et al., 1996; Sillitoe et al., 1996; Huston, 2000). In such a shallow-water environment, boiling may have had a major impact on the fluid chemistry, including its high Au enrichment relative to base metals, and its precipitation due to rapid change in pH and decrease in temperature (Poulsen and Hannington, 1996). In some cases, Au-rich VMS deposits are thought to represent a transition between abyssal base metal-rich seafloor sulphide deposits and subaerial high-sulphidation epithermal Au deposits; as such, they are considered the submarine equivalent to subaerial high-sulphidation epithermal deposits (Poulsen and Hannington, 1996; Sillitoe et al., 1996; Hannington et al., 1999; Huston, 2000).

In the case of a syndeformational Au mineralization overprinting conventional VMS deposits, the key geological features in support of such a model include location of deposits in highly deformed and metamorphosed sequences close to regional-scale faults, the presence of sulphide veins that are discordant to regional foliations, and ore zones that locally are parallel to foliation (Doyon-Bousquet-LaRonde district) (cf. Tourigny et al., 1989; Marquis et al., 1990a,b,c) or transposed along a discordant post-ore shear zone (or high-strain zone), as illustrated at Boliden.

In the synvolcanic Au-rich VMS deposits model, regional metamorphism and deformation have served only to modify the geometry of the deposit, and its mineralogy by locally

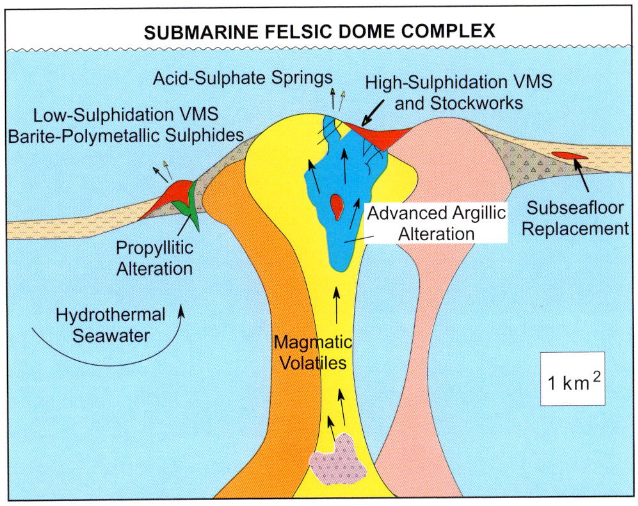

FIGURE 19. Schematic section showing the position of high- and low-sulphidation volcanogenic massive sulphide environments in relationship to the submarine volcanic setting (from Sillitoe et al., 1996).

redistributing some of the constituents into structurally controlled sites (Valliant and Barnett, 1982; Tourigny et al., 1993; Dubé et al., 2004, in press). The world-class Eskay Creek deposit contains relatively undeformed Au-Ag orebodies that undoubtedly have exhalative affinities (Poulsen and Hannington, 1996). Furthermore, in a re-examination of the geology of the Horne deposit, Kerr and Mason (1990) offered evidence against a late-tectonic superposition of Au. Foremost among these are the observations that, locally, an unmineralized debris flow has unconformably cut down into the Au-Cu mineralization, and that otherwise barren pyroclastic tuffs in the stratigraphic hanging-wall contain Au-rich blocks of underlying massive sulphides. The Horne orebodies also contain elevated Au concentrations (e.g. >3 g/t Au) over intervals in excess of several tens of metres (Barrett et al., 1991). In some districts, the Au-rich VMS deposits occur within the same stratigraphic sequences and adjacent to Au-poor VMS deposits. For example, the Au-rich orebody at Montauban contains substantial quantities of base metals and occurs along strike from the Montauban Zn-Pb orebody (Poulsen and Hannington, 1996). The base metal deposits are considered by most geologists to be of VMS-type. If one accepts the exhalative nature of the deposits, the most relevant inquiry is to the nature of the parameters that distinguish base metal massive sulphide deposits that contain little Au from auriferous base metal massive sulphide deposits that have a relatively high base metal to Au ratio (Poulsen and Hannington, 1996).

Recent work in the Doyon-Bousquet-LaRonde district demonstrates that the Bousquet Formation (2699-2697 Ma) of the Blake River Group has played a key role in the formation of the Au-rich VMS lenses (Dubé et al., 2004, in press; Mercier-Langevin et al., 2004; Lafrance et al., 2005; Mercier-Langevin, 2005). Although almost all volcanic contacts (hiatus) host massive sulphide lenses or transposed stringers, the largest VMS lenses are located in the upper transitional to calc-alkaline portion of the Bousquet Formation (Dubé et al., 2004; Mercier-Langevin et al., 2004; Lafrance et al., 2005). The large LaRonde Penna deposit is thought to be, at least in part, genetically related to rhyo-

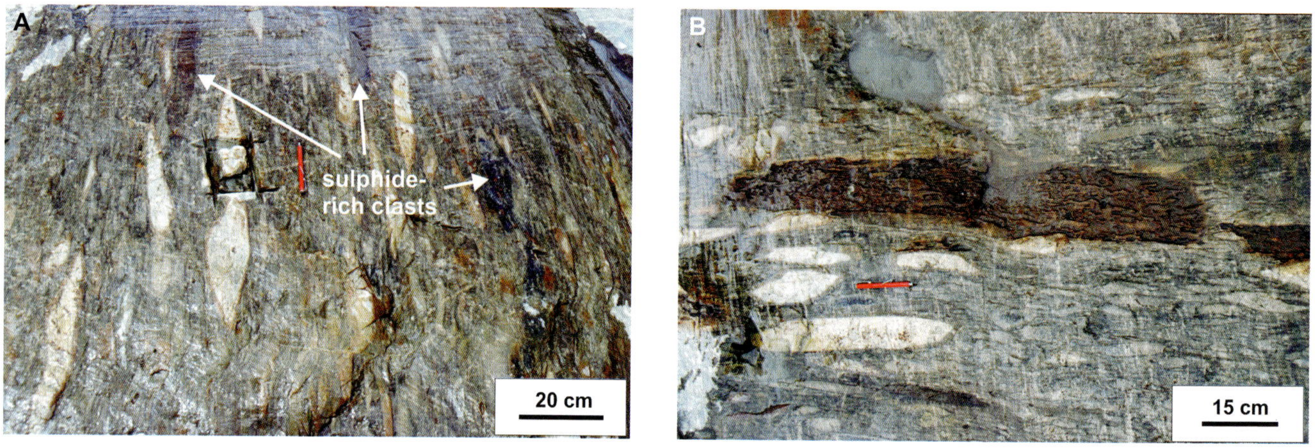

Figure 20. **(A)** Sulphide-rich clasts within a debris-flow stratigraphically above the massive sulphides lens number 6 mined at LaRonde No. 2 shaft. **(B)** Close-up of a sulphide-rich clast.

dacitic to rhyolitic calc-alkaline domes, and to the presence of gabbroic to dioritic high-level sills-dykes (known as andesite) and blue quartz-bearing rhyolites in its hanging wall, which are believed to have acted as a less permeable cap. The presence of Au-bearing sulphide-rich clasts (up to 6 g/t Au) in a debris flow stratigraphically above a massive sulphide lens (lens number 6) mined at LaRonde shaft No. 2 is definitive evidence of pre-deformation Au-rich VMS mineralization (Mercier-Langevin, 2005; Fig. 20A,B).

The impact of regional structures on the primary synvolcanic control on the distribution of the mineralization is well illustrated in the Doyon-Bousquet-LaRonde district, where a number of large orebodies of varying size, type, and preservation are located about 1 km apart (Mercier-Langevin et al., 2007c). They plunge steeply towards the southwest, parallel to the main stretching lineation (L2). The spacing of the deposits in the camp is thought to be controlled by synvolcanic faults, whereas their plunge result from a combination of factors including the superimposed main stage of D2 deformation (Dubé et al., 2003). Detailed mineralogy and trace element geochemistry of the ore and alteration at LaRonde Penna demonstrated that synvolcanic Au and base metals have been locally remobilized and concentrated during superimposed deformation-metamorphism events (Dubé et al., 2004). Despite extensive metamorphic recrystallization and deformation, a crude metal zoning of Cu, Zn, Pb, Ag, and Au, which mimics primary metal zonation in less deformed massive sulphides, is still evident in the massive sulphide lenses (Dubé et al., 2004; Mercier-Langevin, 2005). The intimate association of Au with the base metals, both at LaRonde Penna and in the nearby Bousquet 2-Dumagami Au-Cu-Zn orebody, is evidence for the existence of an auriferous polymetallic protore from which strongly remobilized Au-bearing sulphides were derived (Teasdale et al., 1996; Mercier-Langevin, 2005). Remobilized, auriferous chalcopyrite veins, which have previously been considered to be products of syntectonic Cu-Au mineralization (Marquis et al., 1990a), contain locally abundant sphalerite, pyrrhotite, and galena and merge with foliation-parallel massive sulphides of similar mineralogy, suggesting that the structurally controlled Au-rich ores are mainly products of local metamorphic remobilization. Partial replacement of pyrite by remobilized chalcopyrite has been suggested as evidence for the late-stage introduction of Cu and Au (Marquis et al., 1990a). However, pressure-solution effects such as these are a common feature of metamorphosed massive sulphide ores and do not require that new Cu was added during metamorphism (Dubé et al., 2004, and references therein). Other trace elements that are intimately associated with Au (e.g. Bi, As, and Te) are enriched in the ores, but no more than in many other Archean Cu-Zn VMS deposits. There is clear textural evidence for remobilization of Ag, Pb, Bi, and Sb minerals together with Au in the chalcopyrite veinlets surrounding large pyrite porphyroblasts. However, remobilized constituents of the ores do not appear to have moved beyond the margins of the orebodies and the primary metal zonation of Pb, Cu, and other elements has been preserved (Dubé et al., 2004). The implication is that these elements were likely present in the original ores at their current levels and have simply been concentrated in metamorphically remobilized sulphides during deformation (Dubé et al., 2004). Thus, the Au-base metal association at the LaRonde Penna strongly suggests the formation of a polymetallic Au-VMS system (Dubé et al., 2003). Synvolcanic Au and base metals have been locally remobilized and concentrated during superimposed deformation-metamorphism events as proposed by Tourigny et al. (1993) and Teasdale et al. (1996) for the nearby Bousquet 2-Dumagami deposit.

Two end-member explanations for primary Au enrichments in VMS deposits are proposed (Poulsen and Hannington, 1996). One is where the hydrothermal fluid is dominated by evolved seawater from which local boiling has caused the deposition of significant Au. In this case, the resulting auriferous base metal massive sulphide deposit should have few distinguishing field parameters compared to Au-poor massive sulphides. Another scenario is one where corrosive leaching by acidic, relatively oxidized, and sometimes sulphur-rich fluids (from disproportionation of magmatic SO_2?) that may be of magmatic origin (Hedenquist and Lowenstern, 1994; Sillitoe et al., 1996), mixed with and diluted by seawater, contributes to high-Au enrichment relative to base metals. Hydrothermal alteration types (i.e. sericitic and advanced argillic) are therefore more typical of subaerial magmatic hydrothermal systems (high-sulphidation epithermal). Both conditions may be encountered in an emerging (or submerging) volcanic arc (Fig. 21). This model is attractive in that it allows for regional coexistence of Au-poor with Au-rich massive sulphide deposits (Poulsen and Hannington, 1996; Sillitoe et al., 1996; Hannington et al., 1999).

Advances in Genetic/Exploration Models of the Last Decades

Gold-rich VMS deposits, like other VMS deposits, occur at major lithological contacts that mark a hiatus or a distinctive change in volcanic and sedimentary facies (e.g. the rhyolite-andesite contact at Horne, and the dacite/rhyodacite-andesite contact at LaRonde Penna and Bousquet 2-Dumagami). In the Doyon-Bousquet-LaRonde district, the emplacement of intermediate to felsic domes could have provided the magmatic heat source in the immediate environment and may also have contributed part of the Au to the system (Mercier-Langevin et al., 2007a,b). Proximity to transitional to calc-alkaline volcanic centres is a key element in terms of exploration in this district (Dubé et al., 2004, in press; Mercier-Langevin et al., 2004). Other Au-VMS deposits may be located in districts that host VMS deposits that are not particularly Au-rich (e.g. Horne in the Noranda district, and Boliden in the Skellefte district). Depending on shape and depth, the sulphide contents of many of these deposits are sufficient to produce geophysical responses. Airborne and ground electromagnetic methods are effective at detecting massive sulphide mineralization but only where the sulphide grains are electrically connected. If they are not, they are, from an electromagnetic point of view, considered as disseminated sulphides; in which case they should be detectable by induced polarization surveys (Poulsen and Hannington, 1996).

At a local scale, the presence of aluminous alteration (originally proposed by Valliant et al., 1983), with a mineral assemblage characterized by andalusite, kyanite, and staurolite with Mn-rich garnet, is a useful exploration guide to some Au-VMS deposits, such as Bousquet 1 and Bousquet 2-Dumagami, LaRonde Penna, Boliden, and Au-rich VMS deposits in ancient metamorphosed terranes (Dubé et al., 2004, in press). Because of their ductility, these alteration zones have accommodated most of the post-ore strain and are commonly transformed into schists and transposed subparallel to high-strain zones. Consequently, the alteration will coincide with deformation zones regardless of genetic and age relationships. Aluminous schists with anomalous Au and/or zinc values in intermediate to felsic, transitional to calc-alkaline volcanic or volcaniclastic products located underneath a sedimentary cover, represent excellent exploration targets (Dubé et al., in press). However, exploration strategies for Au-rich VMS must consider the diverse metal content and zonation of alteration minerals even within a single hydrothermal system. All the Au-rich VMS lenses are not associated exclusively with aluminous alteration (Hannington et al., 1999; Huston, 2000; Dubé et al., 2004; Mercier-Langevin, 2005). The quartz-biotite-Mn-rich garnet alteration zone in transitional to calc-alkaline dacites-rhyodacites represents a first order alteration footprint especially well developed in the footwall of some large VMS lenses, such as in the upper and intermediate parts of 20 North zone at LaRonde Penna. However, the andalusite- and kyanite-rich aluminous alteration is a key ore host or vector towards the ore at depth at LaRonde Penna and at Bousquet 2-Dumagami (Marquis et al., 1990c; Tourigny et al., 1990, 1992; Dubé et al., 2004, in press). Thus, there are potentially significant 3-D variations in the hydrothermal alteration and mineralization styles even within a single deposit; exploration strategies for Au-rich VMS deposits should take this diversity into account (Dubé et al., 2004, in press).

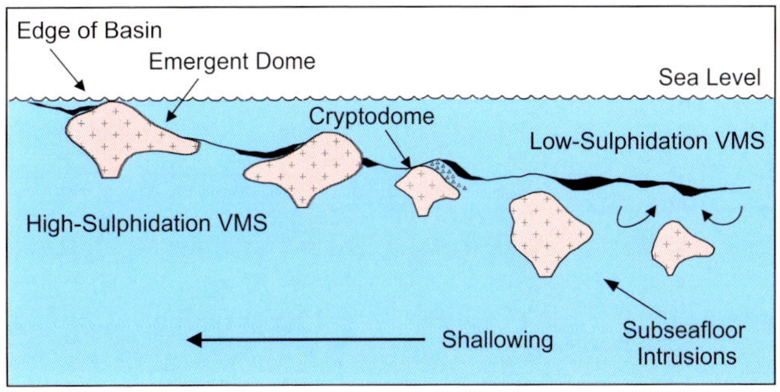

FIGURE 21. Development of high-sulphidation versus low-sulphidation conditions in a submarine setting in relation to the depth of emplacement of associated subvolcanic intrusions (from Hannington et al., 1999).

Knowledge Gaps

The reason(s) why some districts or specific rock formations are so prolific and contain major Au-rich VMS deposits adjacent to Au-poor VMS deposits remains one of the most critical questions to be addressed. The best example is the Archean Blake River Group (2701-2696 Ma; Lafrance et al., 2005) of the Abitibi subprovince: it yielded slightly more than 755 t of Au, which represents 85% of the production and reserves for this subtype of Au deposits in Canada. The Blake River Group is, therefore, a particularly important exploration target for this style of Au deposit (Dubé et al., 2003). It also hosts many conventional VMS deposits (Gibson and Watkinson, 1990). The geochemical composition and volcanic facies of the Au-rich VMS-hosting Bousquet Formation of the Blake River Group, as well as its geological setting, are key elements responsible, at least in part, for the formation of the district (Lafrance et al., 2003; Dubé et al., 2004, in press; Mercier-Langevin et al., 2004; Mercier-Langevin, 2005). However, more work remains to be done to fully understand the unique Au endowment of this particular volcanic group (e.g. Mercier-Langevin et al., 2007c). As proposed by Hodgson (1993) and Huston (2000), it is also possible that the "provinciality" of the VMS deposits high-Au grade may be related to specific fundamental geological characteristics in terms of favourable source-rock environments or Au reservoirs. The local geological "heritage" (endowment) of the district, in addition to ore-forming processes, may indeed be a major factor to take into account.

The role played by synvolcanic and subvolcanic intrusions, such as the 2700 +3/-2 Ma Flavrian-Powell subvolcanic intrusive complex (Galley and Van Breemen, 2002) near Horne, or the multistage Mooshla intrusion in the Doyon-Bousquet-LaRonde district, remains to be better defined. The synvolcanic Mooshla intrusion (2696.9 ±1 Ma) (Lafrance et al., 2005) is contemporaneous to the upper member of Bousquet Formation and hosts intrusion-related pre-main stage deformation quartz- and sulphide-rich veins at the Doyon mine (Gosselin, 1998; Galley et al., 2003; Galley and Lafrance, in press). Despite its relatively small size, it has been proposed that the Mooshla intrusion may have provided metals and/or the thermal effect needed to initiate or maintain the hydrothermal fluid circulation (Valliant and Hutchinson, 1982).

At the district scale, the two largest Canadian Au deposits of this type (Horne and LaRonde Penna) are located a few kilometres from the Cadillac-Larder Lake fault. This empirical relationship between such a crustal-scale fault and the synvolcanic VMS mineralization remains intriguing.

The presence in some Au-rich VMS deposits of metamorphosed, advanced argillic (aluminous) and massive silicic alteration, indicative of an oxidized, low-pH hydrothermal fluid, has been used to suggest an analogy between some Au-VMS and high-sulphidation epithermal deposits (Hannington, 1993; Sillitoe et al., 1996; Hannington et al., 1999, and references therein). In such cases, the Au-rich VMS deposits are interpreted as shallow-water submarine equivalents to subaerial epithermal deposits. However, direct magmatic input into the hydrothermal fluid(s) responsible for the formation of Au-VMS deposits with high-sulphidation affinity remains to be demonstrated, as well as the source of the low-pH characteristic of the fluid. Also, as pointed out by Huston (2000), the shallow-water setting proposed for Au-VMS deposits is often difficult to establish clearly and field criteria to better define the water depth in ancient metamorphosed settings need to be improved. Deformation and metamorphism that commonly overprint mineralization in ancient terranes have induced transposition and local remobilization of Au, and sometimes base metals, into structures adjacent to the original massive sulphide lenses. This has obscured the original relationships and led to considerable debate about syntectonic versus synvolcanic origin for the formation of the deposits and/or their Au content (e.g. Boliden). In some cases, deformation is so severe that it is extremely difficult to define the exact chronology of events, the timing of Au precipitation, and whether the deposit is the end product of superimposed hydrothermal systems. A definitive conclusion about the timing of mineralization with respect to deformation and metamorphism is often impossible. The use of new geochronological techniques, such as Re-Os dating of arsenopyrite, may help resolve the debate (Huston, 2000). This new technique may provide the age of a sulphide phase that is directly associated with the mineralization and which sometimes hosts the Au. Combined with field relationships and U-Pb dating of the host sequence, Re-Os age determination of pyrite and/or arsenopyrite (Arne et al., 2001) could consequently determine the exact timing of Au precipitation within a VMS deposit and contribute to the debate between the syntectonic versus synvolcanic model.

Acknowledgements

This synthesis has been made possible by the kind cooperation of numerous companies, governments, and university geologists who shared their knowledge and who have allowed surface and underground tours of many Au deposits. We benefited from numerous discussions with colleagues from the provincial surveys, in particular Géologie Québec, and from the Geological Survey of Canada. The first author would like to extend his deepest appreciation to F. Robert and H.K. Poulsen for constructive suggestions, collaboration, and discussions on Au deposits during the last twenty years as well as to Agnico-Eagle Mines Ltd., Barrick Gold Corp., and Cambior Inc. for their support and great collaboration. W. Goodfellow and I. Kjarsgaard are thanked for their editorial contribution. Careful constructive reviews by M. Jébrak, W. Goodfellow, G. Beaudoin, and S. Castonguay have led to substantial improvements.

References

Allen, R.A., Weihed, P., and Svenson, S.A., 1996, Setting of Zn-Cu-Au-Ag massive sulfide deposits in the evolution and facies architecture of the 1.9 Ga marine volcanic arc, Skellefte district, Sweden: Economic Geology, v. 91, p. 1022-1053.

Arne, D.C., Bierlein, F.P., Morgan, J.W., and Stein, H.J., 2001, Re-Os dating of sulfides associated with gold mineralization in Central Victoria, Australia: Economic Geology, v. 96, p. 1455-1459.

Arnold, G.O., and Sillitoe, R.H., 1989, Mount Morgan gold-copper deposit, Queensland, Australia: Evidence for an intrusion-related replacement origin: Economic Geology, v. 84, p. 1805-1816.

Barnett, E.S., Hutchinson, R.W., Adamcik, A., and Barnett, R., 1982, Geology of the Agnico-Eagle deposit, Quebec: in Hutchinson, R.W., Spence, C.D., Franklin, J.M., eds., Precambrian Sulphide Deposits; Geological Association of Canada: Special Paper 25, p. 423-426.

Barrett, T.J., Cattalani, S., and MacLean, W.H., 1991, Massive sulphide deposits of the Noranda area, Quebec. I. The Horne Mine: Canadian Journal of Earth Sciences, v. 28, p. 465-488.

Belkabir, A., and Hubert, C., 1995, Geology and structure of a sulphide-rich gold deposit: an example from the Mouska gold mine, Bousquet district, Canada: Economic Geology, v. 90, p. 1064-1079.

Bergman Weihed, J., Bergström, U., Billström, K., and Weihed, P., 1996, Geology, tectonic setting, and origin of the paleoproterozoic Boliden Au-Cu-As deposit, Skellefte District, Northern Sweden: Economic Geology, v. 91, p. 1073-1097.

Cattalani, S., Barrett, T.J., MacLean, W.H., Höy, L., Hubert, C., and Fox, J., 1993, Métallogénèse des gisements Horne et Quémont (region de Rouyn-Noranda): Ministère de l'Énergie et des Ressources, Québec, ET-90-07, 121 p.

Coates, H., 1990, Geology and mineral deposits of the Rambler property in metallogenic framework of base and precious metal deposits, Central and Western Newfoundland (Field trip 1), *in* Field Trip Guidebook 8th IAGOD Symposium: Geological Survey of Canada, Open File 2156. p. 184-193.

Doyle, M.G., and Allen, R.L., 2003, Subseafloor replacement in volcanic-hosted massive sulfide deposit: Ore Geology Reviews, v. 23, p. 183-222.

Dubé, L.-M., Hubert, C., Brown, A.C., Simard, J.M., 1991, The Telbel orebody of the Agnico-Eagle mine in the Joutel area of the Abitibi greenstone belt, Quebec, Canada: A stratabound, gold-bearing massive siderite deposit with early diagenetic pyritization, *in* Ladeira, E.A., ed., Brazil Gold '91 : The Economics, Geology, Geochemistry and Genesis of Gold Deposits. Proceedings of a Symposium: Belo Horizonte, May, 1991, p. 493-498.

Dubé, B., O'Brien, S., and Dunning, G.R., 2001, Gold deposits in deformed terranes: examples of epithermal and quartz-carbonate shear-zone-related gold systems in the Newfoundland Appalachians and their implications for exploration: North Atlantic Mineral Symposium, St. John's, Newfoundland, Canada, Extended abstracts volume, May 27-30, 2001, p. 31-35.

Dubé, B., Mercier-Langevin, P., Lafrance, B., Hannington, M., Moorhead, J., Davis, D., and Pilote, P., 2003, The Doyon-Bousquet-LaRonde Archean Au-rich VMS gold camp: The example of the world-class LaRonde deposit, Abitibi and its implications for exploration. Keynote talk at the CIM Timmins, 2003. Field conference. Ore deposits at depth: Challenges and opportunities; September 23-26, 2003. Technical Sessions Abstract Volume, p. 3-10.

Dubé, B., Mercier-Langevin, P., Hannington, M., Davis, D., et Lafrance, B., 2004, Le gisement de sulfures massifs volcanogènes aurifères LaRonde, Abitibi, Québec: altération, minéralisations, genèse et implications pour l'exploration. Ministères des Resources naturelles de la faune et des parcs. MB 2004-03. 112 p.

Dubé, B., Mercier-Langevin, P., Hannington, M., Lafrance, G., Gosselin, G., and Gosselin, P., in press, The LaRonde-Penna giant Au-rich volcanogenic massive sulphide deposit, Abitibi, Quebec: Mineralogy and geochemistry of alteration and implications for genesis and exploration: Economic Geology.

Franklin, J.M., Gibson, H.L., Jonasson, I.R., and Galley, A.G., 2005, Volcanogenic massive sulfide deposits, *in* Hedenquist, J.W., Thompson, J.F.H., Goldfarb, R.J., and Richards, J.P., eds., Economic Geology, 100th Anniversary Volume, p. 523-560.

Galley, A., and Lafrance, B., in press, Évolution et métallogénie du Pluton de Mooshla: Ministère des Ressources Naturelles et de la Faune du Québec.

Galley, A., and Van Breemen, O., 2002, Timing of Synvolcanic Magmatism in Relation to Base-Metal Mineralization, Rouyn-Noranda, Abitibi Volcanic Belt, Québec; Radiogenic Age and Isotopic Studies. Report 15: Geological Survey of Canada, Current Research 2002-F8, 9 p.

Galley, A., Pilote, P., and Davis, D., 2003, Gold-related subvolcanic Mooshla intrusive complex, Bousquet Mining District, P.Q., *in* Ore Deposits at Depth, CIM 2003 Field Conference, Timmins, Ontario, Abstract volume, p. 16.

Gauthier, M., Baillargeon, F., and Legault, M., 2000, Un gisement d'or caméléon: Eagle-Telbel, Joutel, Abitibi: Conference given at the Ministère des Ressources Naturelles du Québec.

Gibson, H.L., and Watkinson, D.H., 1990, Volcanogenic massive sulphide deposits of the Noranda cauldron and shield volcano, Quebec, *in* The Northwestern Quebec Polymetallic Belt: The Canadian Institute of Mining and Metallurgy, Special Volume 43, p. 119-132.

Gosselin, G., 1998, Veines de quartz aurifères précoces à la zone Ouest de la Mine Doyon, Canton de Bousquet, Preissac, Abitibi. Thèse de maîtrise, Université du Québec à Chicoutimi, 128 p.

Gosselin, P., and Dubé, B., 2005a, Gold Deposits of the World: Distribution, Geological Parameters and Gold Content: Geological Survey of Canada, Open File 4895, 1 CD-ROM.

Gosselin, P., and Dubé, B., 2005b, Gold deposits of Canada: Distribution, Geological Parameters and Gold Content: Geological Survey of Canada, Open File 4896, 1 CD-ROM.

Grip, E., and Wirstam, A., 1970, The Boliden Sulphide Deposit: Sveriges Geologiska Undersökning, v. C51, 68 p.

Hannington, M.D., 1993, Shallow submarine hydrothermal systems in modern island arc settings: The Gangue, no. 43, p. 6-8.

Hannington, M.D., Poulsen, K.H., Thompson, J.F.H., and Sillitoe, R.H., 1999, Volcanogenic gold in the massive sulfide environment, *in* Barrie, C.T., and Hannington, M.D., eds., Volcanic-Associated Massive Sulfide Deposits: Processes and Examples in Modern and Ancient Settings: Reviews in Economic Geology, Vol. 8, p. 325-356.

Hedenquist, J.W., and Lowenstern, J.B., 1994, The role of magmas in the formation of hydrothermal ore deposits: Nature, v. 370, p. 519-527.

Hibbard, J., 1983, Geology of the Baie Verte Peninsula, Newfoundland: Government of Newfoundland and Labrador, Department of Mines and Energy, Mineral Development Division, Memoir 2, 279 p.

Hodgson, C.J., 1993, Mesothermal lode-gold deposits, *in* Kirkham, R.V., ed., Mineral Deposit Modeling: Geological Association of Canada, Special Paper 40, p. 635-678.

Huston, D.L., 2000, Gold in volcanic-hosted massive sulfide deposits; distribution, genesis, and exploration, *in* Hagemann, S.G. ed., Gold in 2000: Reviews in Economic Geology, v. 13, p. 401-426.

Huston, D.L., and Large, R.R., 1989, A chemical model for the concentration of gold in volcanogenic massive sulphide deposits: Ore Geology Reviews, v. 4, p. 171-200.

Huston, D.L., Bottril, R.S., Creelman, R.A., Zaw, K., Ramsden, T.R., Rand, S.W., Gemmel, J.B., Jablonski, W., Sie, S.H., and Large, R.R., 1992, Geologic and geochemical controls on the mineralogy and grain size of gold-bearing phases in eastern Australian volcanic-hosted massive sulfide deposits: Economic Geology, v. 87, p. 542-563.

Kerr, D.J., and Mason, R., 1990, A re-appraisal of the geology and ore deposits of the Horne mine complex at Rouyn-Noranda, Quebec, *in* Rive, M., ed., The Northwestern Quebec Polymetallic Belt: A Summary of 60 Years of Mining Exploration, 1990: Canadian Institute of Mining and Metallurgy, Special Volume 43, p. 153-165.

Kerr, D.J., and Gibson, H.L., 1993, A comparison of the Horne volcanogenic massive sulfide deposit and intracauldron deposits of the Mine sequence, Noranda, Quebec: Economic Geology, v. 88, p. 1419-1442.

Lafrance, B., Moorhead, J., et Davis, D., 2003, Cadre géologique du camp minier de Doyon-Bousquet-LaRonde. Ministère des Ressources Naturelles, Québec: ET 2002-07, 45 p.

Lafrance, B., Davis, D.W., Goutier, J., Moorhead, J., Pilote, P., Mercier-Langevin, P., Dubé, B., Galley, A.G., and Mueller, W.U., 2005, Nouvelles datations isotopiques dans la portion québécoise du Groupe de Blake River et des unités adjacentes. Ressources Naturelles et Faune Québec. RP 2005-01. 15 p.

Large, R.R., Huston, D.L., McGoldrick, P.J., Ruxton, P.A., and McArthur, G., 1989, Gold distribution and genesis in Australian volcanogenic massive sulphide deposits and their significance for gold transport models, *in* Keays, R.R., Ramsay, W.R.H., and Groves, D.I., eds., The Geology of Gold Deposits: The Perspective in 1988: Economic Geology, Monograph 6, p. 520-536.

Larocque, A.C., Hodgson, C.J., and Lafleur, P.-J., 1993, Gold distribution in the Mobrun volcanic-associated massive sulfide deposit, Noranda, Quebec: A preliminary evaluation of the role of metamorphic remobilization: Economic Geology, v. 88, p. 1443-1459.

MacDonald, A.J., Lewis, P.D., Thompson, J.F.H., Nadaraju, G., Bartsch, R.D., Bridge, D.J., Rhys, D.A., Roth, T., Kaip, A., Godwin, C.J., and Sinclair, A.J., 1996, Metallogeny of an Early to Middle Jurassic arc, Iskut River area, northwestern British Columbia: Economic Geology, v. 91, p. 1098-1114.

MacLean, W.H., Stamatelopoulou-Seymour, K., and Prabhu, M.K., 1982, Sr, Y, Zr, Nb, Ti and REE in Grenville amphibolites at Montauban-les-Mines, Quebec: Canadian Journal of Earth Sciences, v. 19, p. 633-644.

Marquis, P., Hubert, C., Brown, A.C., and Rigg, D.M., 1990a, Overprinting of early, redistributed Fe and Pb-Zn mineralization by late-stage Au-Ag-Cu deposition at the Dumagami mine, Bousquet district, Abitibi, Quebec: Canadian Journal of Earth Sciences, v. 27, p. 1651-1671.

——— 1990b, An evaluation of genetic models for gold deposits of the Bousquet district, Quebec, based on their mineralogic, geochemical, and structural characteristics: Canadian Institute of Mining and Metallurgy, Special Volume 43, p. 383-399.

Marquis, P., Brown, A.C., Hubert, C., and Rigg, D.M., 1990c, Progressive alteration associated with auriferous massive sulfide bodies at the Dumagami Mine, Abitibi greenstone belt, Quebec: Economic Geology, v. 85, p. 746-764.

Mercier-Langevin, P., 2005, Géologie du gisement de sulfures massifs volcanogènes aurifères LaRonde, Abitibi, Québec. Ph.D. Thesis, Institut national de la recherche scientifique, Eau, Terre et Environnement, 694 p.

Mercier-Langevin, P., Dubé, B., Hannington, M.D., Davis, D., et Lafrance, B., 2004, Contexte géologique et structural des sulfures massifs volcanogènes aurifères du gisement LaRonde, Abitibi. Ministères des Resources naturelles de la faune et des parcs. ET 2003-03. 60 p.

Mercier-Langevin, P., Dubé, B., Hannington, M.D., Davis, D.W., Lafrance, B., and Gosselin, G., 2007a, Geology and geochronology of the LaRonde Penna deposit, Abitibi greenstone belt, Quebec: Controls on the genesis of a world-class gold-rich volcanogenic massive sulfide complex: Economic Geology, submitted.

Mercier-Langevin, P., Dubé, B., Hannington, M.D., Richer-Laflèche, M., and Gosselin, G., 2007b, Lithogeochemistry and paleotectonic setting of the LaRonde Penna world-class Au-rich volcanogenic massive sulfide deposit, Abitibi greenstone belt, Quebec: Economic Geology, submitted.

Mercier-Langevin, P., Dubé, B., Lafrance, B., Hannington, M., Galley, A., Moorhead, J., and Gosselin, P., 2007c, Metallogeny of the Doyon-Bousquet-LaRonde mining camp, Abitibi Greenstone Belt, Quebec, *in* Goodfellow, W.D., ed., Mineral Deposits of Canada: A Synthesis of Major Deposit-Types, District Metallogeny, the Evolution of Geological Provinces, and Exploration Methods: Geological Association of Canada, Mineral Deposits Division, Special Publication 5, p. 673-701.

Morin, G., 1987, Gîtologie de la région Montauban. Ministère de L'Énergie et des Ressources Québec, Report MM-86-02, 59 p.

Nadeau, L., Brouillette, P., and Hébert, C., 1999, New observations on relict volcanic features in medium-grade gneiss of the Montauban group, Grenville province, Quebec, *in* Current Research 1999-E: Geological Survey of Canada, p. 149-160.

Nilsson, C.A., 1968, Wall rock alteration at the Boliden deposit, Sweden: Economic Geology, v. 63, p. 472-494.

Ödman, O.H., 1941, Geology and Ores of the Boliden Deposit, Sweden: Sveriges Geologiska Undersökning, Series C, v. 438, 190 p.

Ohmoto, H., 1996, Formation of volcanogenic massive sulphide deposits: The Kuroko perspective: Ore Geology Reviews, v. 10, p. 135-177.

Poulsen, K.H., and Hannington, M.D., 1996, Volcanic-associated massive sulphide gold, *in* Eckstrand, O.R., Sinclair, W.D., and Thorpe, R.I., eds., Geology of Canadian Mineral Deposit Types: Geology of Canada, v. 8, p. 183-196.

Poulsen, K.H., Robert, F., and Dubé, B., 2000, Geological classification of Canadian gold deposits: Geological Survey of Canada, Bulletin 540, 106 p.

Rickard, D., ed., 1986, The Skellefte Field: Sveriges Geologiska Undersökning, Series Ca, v. 62, 54 p.

Roth, T., Thompson, J.F.H., and Barrett, T.J., 1999, The precious metal-rich Eskay Creek deposit, north western British Columbia, *in* Tucker, B.C., and Hannington, M.D., eds., Volcanic-Associated Massive Sulfide Deposits; Processes and Examples in Modern and Ancient Settings: Reviews in Economic Geology, v. 8, p.357-373.

Sherlock, R.L., Roth, T., Spooner, E.T.C., and Bray, C.J., 1999, Origin of the Eskay Creek precious metal-rich volcanogenic massive sulfide deposit; fluid inclusion and stable isotope evidence: Economic Geology, v. 94, p. 803-824.

Sillitoe, R.H., and Bonham, H.F., 1990, Sediment-hosted gold deposits; distal products of magmatic-hydrothermal systems: Geology, v. 18, p. 157-161.

Sillitoe, R.H., Hannington, M.D., and Thompson, J.F.H., 1996, High-sulfidation deposits in the volcanogenic massive sulfide environment: Economic Geology, v. 91, p. 204-212.

Stone, W.E., 1988, Nature and Significance of Metamorphism in Gold Concentration, Bousquet Township, Abitibi Greenstone Belt, Northwest Quebec: Ph.D. thesis, University of Western Ontario, London, Ontario, 441 p.

Stone, W.E., Valliant, R.I., and Bateman, P.W., 1988, Wall rock alteration, regional metamorphism and gold concentration in the Bousquet gold district, Abitibi greenstone belt, Quebec, Canada, *in* Bicentennial Gold'88 extended abstracts oral programme: Geological Society of Australia. Abstracts No 22, p. 51-55.

Teasdale, N., Brown, A.C., and Tourigny, G., 1996, Gîtologie de la mine Bousquet 2. Ministère des Ressources naturelles, Québec; MB 96-37, 43 p.

Tourigny, G., Brown, A.C., Hubert, C., and Crépeau, R., 1989, Synvolcanic and syntectonic gold mineralization at the Bousquet Mine, Abitibi greenstone belt: Economic Geology, v. 84, p. 1875-1890.

——— 1990, Syn-volcanic and syn-tectonic gold mineralization at the Bousquet Mine, Abitibi greenstone belt, Quebec: Economic Geology, v. 85, p. 1875-1890.

Tourigny, G., Hubert, C., Brown, A.C., Crépeau, R., Trudel, P., Hoy, L., and Kheang, L., 1992, Géologie de la mine Bousquet: ET 89-09. MInistère de l'Énergie et des Ressources du Québec. 99 p.

Tourigny, G., Doucet, D., and Bourget, A., 1993, Geology of the Bousquet 2 mine: An example of a deformed, gold-bearing, polymetallic sulfide deposit: Economic Geology, v. 88, p. 1578-1597.

Valliant, R.I., and Barnett, R.L., 1982, Manganiferous garnet underlying the Bousquet gold orebody, Quebec: Metamorphosed manganese sediment as a guide to gold ore: Canadian Journal of Earth Sciences, v. 19, p. 993-1010.

Valliant, R.I., and Hutchinson, R.W., 1982, Stratigraphic distribution and genesis of gold deposits, Bousquet Region, North-western Quebec, *in* Hodder, R.W., and Petruk, W., eds., Geology of Canadian Gold Deposits: Canadian Institute of Mining and Metallurgy, Special Volume 24, p. 27-40.

Valliant, R.I., Barnett, R.L., and Hodder, R.W., 1983, Aluminum silicate-bearing rock and its relation to gold mineralization; Bousquet Mine, Bousquet township, Quebec: Canadian Institute of Mining and Metallurgy, Bulletin, v. 76, p. 81-90.

Williams, N.C., and Davidson, G.J., 2004, Possible submarine advanced argillic alteration at the Basin Lake prospect, Western Tasmania, Australia: Economic Geology, v. 99, pp. 987-1002.

Wyman, D.A., Kerrick R., and Fryer, B.J., 1986, Gold mineralization overprinting iron formation at the Agnico-Eagle deposit, Quebec, Canada: Mineralogical, microstructural and geochemical evidence, *in* MacDonald, A.J., ed., Proceedings of Gold '86, an international symposium on the geology of gold: Toronto, p. 108-123.

Yeats, C.J., and Groves, D.I., 1998, The Archean Mount Gibson gold deposits, Yilgarn craton, Western Australia: Products of combined synvolcanic and syntectonic alteration and mineralisation: Ore Geology Reviews, v. 13, p. 103-129.

Reduced Intrusion-Related Gold Systems

Craig J.R. Hart

Yukon Geological Survey, Box 2703 (K-10), Whitehorse, Yukon, Canada Y1A 2C6

and

Centre for Exploration Targeting (M006), The University of Western Australia, Crawley, WA, Australia, 6009

Corresponding author's email: craig.hart@uwa.edu.au

Abstract

Reduced intrusion-related gold systems (RIRGS) are characterized by widespread arrays of sheeted auriferous quartz veins that preferentially form in the brittle carapace at the top of small plutons, where they form bulk-tonnage, low-grade Au deposits characterized by a Au-Bi-Te-W metal assemblage, such as the Fort Knox and Dublin Gulch deposits. RIRGS also include a wide range of intrusion-related mineral deposit styles (skarns, replacements, veins) that form within the region of hydrothermal influence surrounding the causative pluton, and are characterized by proximal Au-W-As and distal Ag-Pb-Zn metal associations, thereby generating a zoned mineral system. Plutons that generate RIRGS form in tectonic settings characterized by weak post-collisional extension behind a thickened continental margin. Such settings are also conducive to the formation of W deposits, and thereby generate a regional Au-W metallogenic association, but individual plutons can generate both W and Au deposits. Associated magmas are diverse and have characteristics of I-, S-, and A-type granitoids. The most prolific Au systems comprise metaluminous, moderately reduced, moderately fractionated, biotite>>hornblende>pyroxene quartz monzonites that have mixed with volatile-rich lamprophyric melts. The magmas have a reduced primary oxidation state that form ilmenite-series plutons. This reduced state causes associated sulphide assemblages to be characterized by pyrrhotite, and quartz veins that host methane-rich inclusions. RIRGS mostly form at a depth of 5 to 7 km and generate mineralizing fluids that are low salinity, aqueous carbonic in composition and are, therefore, unlike typical porphyry Cu deposits. The RIRGS class was developed on well-studied examples in Yukon and Alaska. Other suggested Canadian examples are in southeastern British Columbia and New Brunswick; numerous global examples have been suggested, but many are controversial.

Résumé

Les systèmes aurifères associés à des intrusions réduites (SAAIR) sont caractérisés par des réseaux étendus de filons de quartz aurifère dans des zones feuilletées, qui se forment de manière préférentielle dans la carapace à déformation fragile développée au sommet de petits plutons. Ces systèmes filoniens aurifères forment des gîtes d'or à faible teneur caractérisés par une association métallique à Au-Bi-Te-W comme les gisements de Fort Knox et de Dublin Gulch. Les SAAIR englobent également une gamme étendue de styles de minéralisations associés à des intrusions (skarns, gîtes de remplacement, gîtes filoniens) qui se forment dans la zone d'influence hydrothermale entourant le pluton auquel ils doivent leur formation et qui sont caractérisés par des associations métalliques de nature proximale à Au-W-As et de nature distale à Ag-Pb-Zn , produisant ainsi des systèmes à zonalité minérale. Les plutons engendrant des SAAIR se forment dans des cadres tectoniques caractérisés par une faible extension consécutive à une collision à l'arrière d'une marge continentale épaissie. De tels cadres sont également favorables à la formation de gîtes de tungstène, engendrant ainsi une association métallogénique régionale à Au-W, mais des plutons individuels peuvent produire des gîtes de W et des gîtes de Au. Les magmas associés sont divers et présentent des caractéristiques des granitoïdes des types I, S, et A. Les systèmes aurifères les plus prolifiques renferment des monzonites quartziques à biotite>>hornblende>pyroxène modérément réduites, modérément fractionnées et de caractère métalumineux, qui se sont mélangées à des bains lamprophyriques riches en matières volatiles. Les magmas présentent un état d'oxydation initial réduit qui produit des plutons de la série de l'ilménite. Cet état réduit fait en sorte que les paragenèses sulfurées associées sont caractérisées par la pyrrhotine et que des filons de quartz renferment des inclusions riches en méthane. Les SAAIR se forment principalement à des profondeurs de 5 à 7 km et engendrent des fluides minéralisateurs aquo-carboniques de faible salinité, qui sont ainsi différents de ceux qui caractérisent les gîtes porphyriques de cuivre. La classe des SAAIR a été élaborée d'après des exemples bien étudiés au Yukon et en Alaska; d'autres exemples sont proposés au Canada dans le sud-est de la Colombie-Britannique et au Nouveau-Brunswick alors que de nombreux exemples ont été proposés ailleurs dans le monde, mais plusieurs d'entre eux restent controversés.

Definition

Reduced intrusion-related Au systems (RIRGS) include a wide range of Au-only mineral deposit styles that are considered to have had a direct genetic link with a cooling felsic intrusion during their formation. Associated deposit styles may be as varied as skarns, veins, disseminations, stockworks, replacements, and breccias (Hart et al., 2000a) and, therefore, define ores that are broader in classification than simply a deposit and were thus identified as a system (Lang et al., 2000). The most diagnostic deposit style within the RIRGS classification is intrusion-hosted, sheeted arrays of thin, low-sulphide quartz veins with a Au-Bi-Te-W signature, which typically comprise bulk tonnage, low-grade Au resources. The host or associated intrusions characteristically have moderately low primary oxidation states, making them reduced, ilmenite-series (Ishihara, 1981) granitoids. The best

Hart, C.J.R., 2007, Reduced intrusion-related gold systems, *in* Goodfellow, W.D., ed., Mineral deposits of Canada: A Synthesis of Major Deposit Types, District Metallogeny, the Evolution of Geological Provinces, and Exploration Methods: Geological Association of Canada, Mineral Deposits Division, Special Publication No. 5, p. 95-112.

examples of RIRGS include Fort Knox (Alaska) and Dublin Gulch (Yukon).

These gold systems have only been recognized as a new deposit class since 1999 and, as such, are in a juvenile state of understanding with still rapidly evolving data collection, interpretation, and nomenclature (see Hart, 2005 for details; see Mair et al., 2006a as a current example). As an example of nomenclature evolution, many of these dominantly intrusion-hosted systems were originally considered to be Au porphyry deposits (Hollister, 1992; Sinclair, 2007), whereas country rock-hosted disseminated and stockwork systems dominantly distal to plutons were considered to be "Carlin-like" deposits (Poulsen 1996). RIRGS are distinct from intrusion-related Au deposits as defined by Sillitoe (1991, 1995), because the deposits are characteristically associated with porphyry Cu systems that are related to highly oxidized and more mafic intrusions. Therefore, within the intrusion-related clan, two different types of Au mineralizing systems can be identified using the prefixes "reduced" and "oxidized". The RIRGS are a distinct class that lacks anomalous Cu, have associated W, low sulphide volumes, and a reduced sulphide mineral assemblage, and that are associated with felsic, moderately reduced (ilmenite-series) plutons, whereas oxidized intrusion-related Au deposits are mostly Au-rich (or relatively Cu-poor) variants of the porphyry Cu deposit model associated with more mafic, oxidized, magnetite-series plutons.

Intrusion-hosted, sheeted-vein styles of mineralization typically occur in all associated RIRGS plutons to some degree, but the classification can include skarns, replacements, disseminations, veins, and stockworks that may develop within, beyond, or above the pluton's thermal aureole. However, except for the skarn deposit model which has long recognized a "reduced" variant (Einaudi et al., 1981; Meinert, 1998), specific deposit models for the other Au mineralization styles are lacking because such styles are common to many deposit types. Shallow-level equivalents of RIRGS such as Brewery Creek occur locally and are termed "epizonal". Intrusion-related Au-bearing vein deposits certainly occur, but their characteristics are so varied that establishing a set of defining characteristics to construct a model has proven difficult (e.g., Sillitoe and Thompson 1998). This has, in part, resulted in considerable confusion in distinguishing intrusion-related Au and orogenic Au vein deposits (variably discussed in Goldfarb et al., 2000, 2005; Hart et al., 2002; Groves et al., 2003; Hart and Goldfarb, 2005). Herein, the characteristics of RIRGS are defined and distinguished from other deposit types.

Distribution

Canada

The RIRGS classification was developed in response to exploration and Au deposit discoveries in the 1990s in Alaska (USA) and Yukon (Canada) in the northern North American Cordillera (Fig.1). The most significant economic mineralization is in the Fairbanks area of central Alaska, where the actively mined Fort Knox deposit (Bakke, 1995) serves as a type example of a RIRGS. Significant but unmined systems occur at Dublin Gulch (Hitchins and Orssich, 1995; Maloof et al., 2001), Scheelite Dome (Mair et al., 2000, 2006a) and Clear Creek (Marsh et al., 2003), where they form the core of the Tombstone Gold Belt in the central Yukon. All of these Tombstone Gold Belt systems are hosted mainly in, and directly formed from, reduced mid-Cretaceous (95–91 Ma) plutons that define a discontinuous belt that spans 1000 km across central Yukon and Alaska (Hart et al., 2004a).

This RIRGS model was adopted in the late 1990s to classify many Au deposits and districts throughout interior Alaska and Yukon, and assembled into the vast Tintina Gold Province (TGP; Tucker and Smith, 2000). However, many of the identified deposits and districts in the TGP lack Au ores with RIRGS characteristics or are dominated by placer Au deposits with uncertain lode sources (Hart et al., 2002). Although the TGP contains many important RIRGS, it also includes other types of lode Au deposits. Thus, the >800 km-long Tombstone Gold Belt best describes the distribution of the unequivocal RIRGS, whereas TGP describes a large geographical area with a significant Au endowment (Fig. 1).

Elsewhere in the Cordillera, "plutonic-related" systems with some similar characteristics to RIRGS are associated with Cretaceous Bayonne suite intrusions in southeastern British Columbia (Fig. 2), where they form a 350 km-long belt (Logan et al., 2000). Outside of the Cordillera, the best known Canadian examples may be those in the Acadian orogeny of southern New Brunswick (McLeod and McCutcheon, 2000), such as Clarence Stream and Lake George, although these are better known for their W-Sn-Mo-Sb mineralization. Potential Archean examples in Canada have been suggested to occur in the southern Superior Province of Ontario and Quebec (Robert, 2001; Malartic, Fig. 2), but these deposits may be distinct because they are unlike typical RIRGS.

Global

Although the RIRGS classification is still developing, numerous Phanerozoic global examples suggested in early compilations, such as those of Thompson et al. (1999), Thompson and Newberry (2000), and Lang et al. (2000), have defined a preliminary distribution that indicates potential RIRGS in Paleozoic and Mesozoic orogenic belts, although supporting documentation for these classifications is mostly lacking. Suggested examples include the Bolivian Polymetallic Belt, Yanshanian orogen of the North China craton, Tien Shan of central Asia, New England and Lachlan provinces in Australia, and the Bohemian Massif and the Iberian Peninsula in Europe (Fig. 2). Specific deposits include Timbarra (Mustard, 2001), Kidston (Baker and Tullemans, 1990), and potentially other Au deposits in eastern Australia (Blevin, 2004); Penedona and Jales (Portugal), Salave and Solomon (Spain), Mokrsko and Petrackova hora (Czech Republic), Vasilkovskoe (Kazakstan), Niuxinshan (China), Sn-rich Kori Kollo (Bolivia), and Petza River and Miller Mountain (USA; Thompson et al., 1999; Lang and Baker, 2001). Some Paleozoic Au giants in central Asia have been interpreted by some workers (e.g., Cole et al., 2000; Wall et al., 2004) to show features that resemble RIRGS (i.e., Jilau, Muruntau, Sukhoi Log, Kumtor), but such associations are highly controversial and many workers argue that these

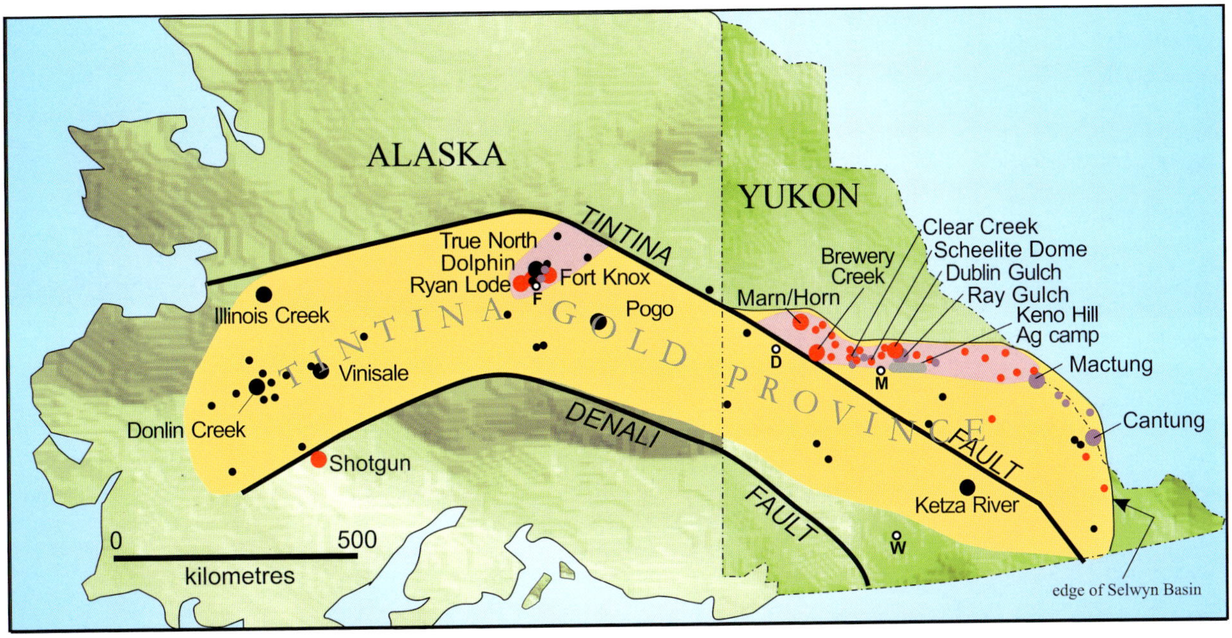

FIGURE 1. Distribution of the Tintina Gold Province (TGP) across Alaska and Yukon as shown in yellow. Individual Au deposits (large circles), notable occurrences, and those mentioned in the text (small circles) are shown. Not all Au mineralization in the TGP represents RIRGS. Gold deposits and occurrences considered to be RIRGS are shown in red; those that are ambiguous or controversial in origin are highlighted in black; and those dominated by W, but associated with the same plutonic event as RIRGS, are shown in purple. The RIRGS that define the Tombstone Gold Belt (shown in pink) underlie central Yukon; its western extent has been offset along the Tintina Fault and displaced to its present location, where it forms part of the Fairbanks district. Notably, the TGP is composed of numerous different Au districts with varying deposit types and ages of mineralization, but reduced intrusion-related Au systems are mostly limited to the Tombstone Gold Belt. F=Fairbanks, D=Dawson, M=Mayo, W=Whitehorse.

deposits are likely orogenic Au deposits (Groves et al., 2003; Goldfarb et al., 2005).

Global Archean and Proterozoic RIRGS examples have also been suggested, but Proterozoic examples such as Tennant Creek and Telfer (Rowins, 2000) in northern Australia are Cu-rich and lack most RIRGS characteristics. Similarly, Boddington (McCuaig et al., 2001) and Tower Hill (Witt, 2001), both of Western Australia, have been suggested as Archean examples, but mostly lack critical RIRGS features, and indicate the unlikelihood of formation or preservation of good Archean examples.

Time

The mid-Cretaceous was the most favourable time for formation of RIRGS along Canada's western margin, in adjacent eastern Alaska, and in southeastern British Columbia. Some southwestern Alaskan suggested examples, such as Shotgun, are latest Cretaceous. The less well understood occurrences in New Brunswick are Devonian. Globally, although there is still much controversy as to what deposits truly are RIRGS, the Phanerozoic appears to clearly dominate, particularly the Cretaceous and the mid-Paleozoic (Devonian to Carboniferous), which were the most favorable times for formation of RIRGS. Proterozoic and Archean examples are controversial and too few to assess in terms of timing, but the lack of W deposits of these ages emphasizes the unlikelihood of this as a significant era.

Grade and Tonnage

The grades and tonnages of deposits classified as RIRGS are wide-ranging due to the variation of deposit styles within the classification. The variation further broadens with the inclusion of controversial Alaskan deposits, such as the high-grade Pogo and large tonnage Donlin Creek orebodies. Figure 3 shows the grade-tonnage relationships for most of the suggested Canadian and Alaskan examples, as well as Timbarra in Australia.

The most characteristic deposit style, intrusion-hosted sheeted vein deposits, is best represented by mineralization at Fort Knox and Dublin Gulch. Both deposits have minable reserve grades of approximately 0.9 g/t Au and cutoffs of 0.4 to 0.5 g/t, but at Fort Knox, material as low as 300 ppb Au may be stockpiled. The grades of individual veins are 5 to 50 g/t Au, but most ore blocks have an average of 3 to 5 veins per metre within otherwise barren host rocks, thus yielding ~1 g/t ores. Gold grade is, therefore, mainly controlled by vein density. Whereas Fort Knox and Dublin Gulch have similar overall grades, Fort Knox's lower-grade ores are enriched by higher-grade and overprinting, late-stage quartz shear veins. Sheeted vein arrays also occur at deposits such as Brewery Creek (Classic Zone), Dolphin, Shotgun, and Gil, but are not the main ore hosts because each deposit has other features that control grade distribution. Tonnages in the RIRGS are likely to range from 10 to 300 Mt, with grades likely to be from 0.7 to 1.5 g/t Au, hosting approximately 10 to 300 t (0.3–10 Moz) of contained Au. Fort Knox has a resource of about 210 t (7 Moz) of contained Au.

Few of the skarn, replacement, and disseminated deposits within the RIRGS have defined grades and tonnages. The Marn and Horn skarn orebodies in Yukon indicate high

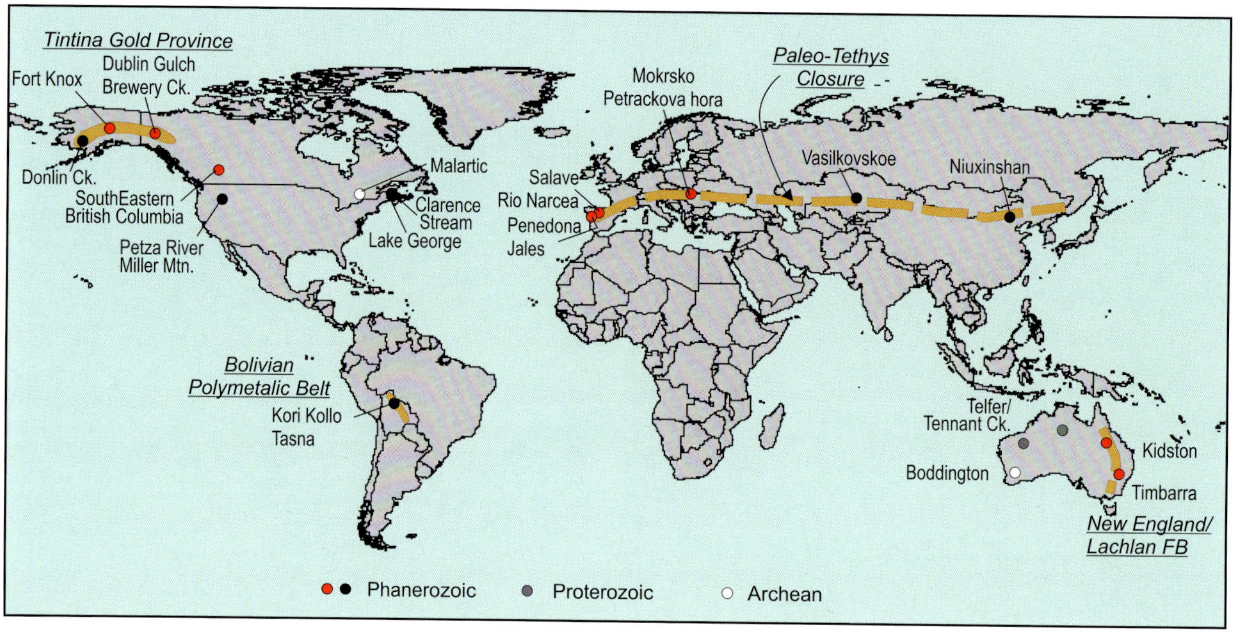

FIGURE 2. Global distribution of Au deposits suggested to be RIRGS. The Phanerozoic examples, for which there is a high degree of confidence, are shown in red. Ambiguous examples are shown in black, grey, and white. Modified from Thompson et al. (1999), Lang et al. (2000), and Lang and Baker (2001).

grades (8–10 g/t Au), but small tonnages (~50 000–300 000 t). Ryan Lode in central Alaska is the best potential example of an intrusion-related vein, with an average grade of 3 g/t over 3.6 Mt of ore (Bakke et al., 2000) along the margin of a small, unroofed stock. Scheelite Dome represents a huge, although undefined, country rock-hosted RIRGS, with skarn, vein, disseminated, and replacement deposit styles (Hulstein et al., 1999; Mair et al., 2000). Tungsten ores are associated with RIRGS, most significantly is the Ray Gulch deposit (7.26 Mt of 0.87% WO_3), which forms in contact with the Dublin Gulch pluton (Lennan, 1983; Brown et al., 2002).

The epizonal Brewery Creek deposit had an overall grade of 1.44 g/t Au and 17.2 Mt of ore, divided between about eight orebodies, prior to mining and extraction of 9 t (285 000 oz) Au from 1996 to 2002. Parts of some individual orebodies had considerably higher average grades of 4 g/t Au. The overall tonnage of Brewery Creek is misleading because it was explored only for its easily recoverable oxide resource. The underlying sulphide orebody remains undefined, but is the focus of current exploration efforts. The Donlin Creek deposit, at 323 Mt of 2.7 g/t Au and with a Au-As-Sb-Hg signature similar to that of Brewery Creek, had also been suggested to be an epizonal RIRGS (Ebert et al., 2000; Thompson and Newberry 2000), but recent studies (Ebert et al., 2003; Goldfarb et al., 2004) indicate otherwise.

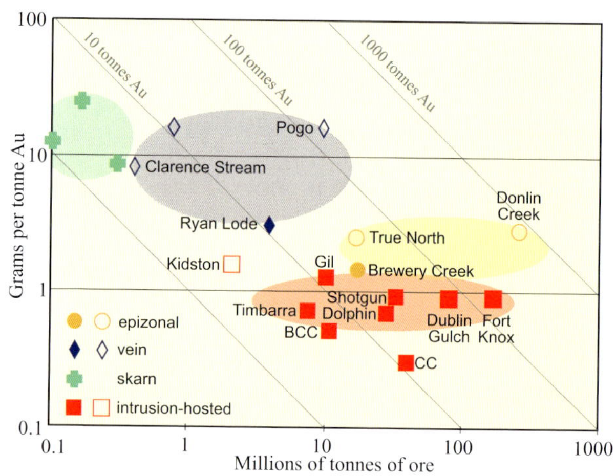

FIGURE 3. Grade and tonnages of deposits considered to be RIRGS, with emphasis on the intrusion-hosted mineralization style. Other deposit types and controversial deposits are shown for comparison. Open symbols are used for deposits that are controversial. BCC = Brewery Creek Classic zone, CC = Clear Creek.

Geological Attributes

Tectonic Setting and Magmatic Associations

Due to the evolving understanding of this deposit class, and the injudicious incorporation of numerous international examples, the associated tectonic settings are poorly constrained, with back-arc, foreland fold belts, collisional, post-collisional, and magmatic arc settings in orogenic belts having all been proposed (e.g., Thompson et al., 1999; Goldfarb et al., 2000). Examples of widely varying global and generally poorly understood settings for major RIRGS include the Paleo-Tethys sutures, the margins of the North China Craton, the Tasman Orogen, the northern North American Cordilleran orogen, and the Andes (Thompson et al., 1999; Lang et al., 2000; Thompson and Newberry 2000). However, the setting of the best-studied examples in the Tombstone Gold Belt are well understood (Fig. 4; Mair et al., 2006b). Associated RIRGS deposits and occurrences in Yukon are directly related to

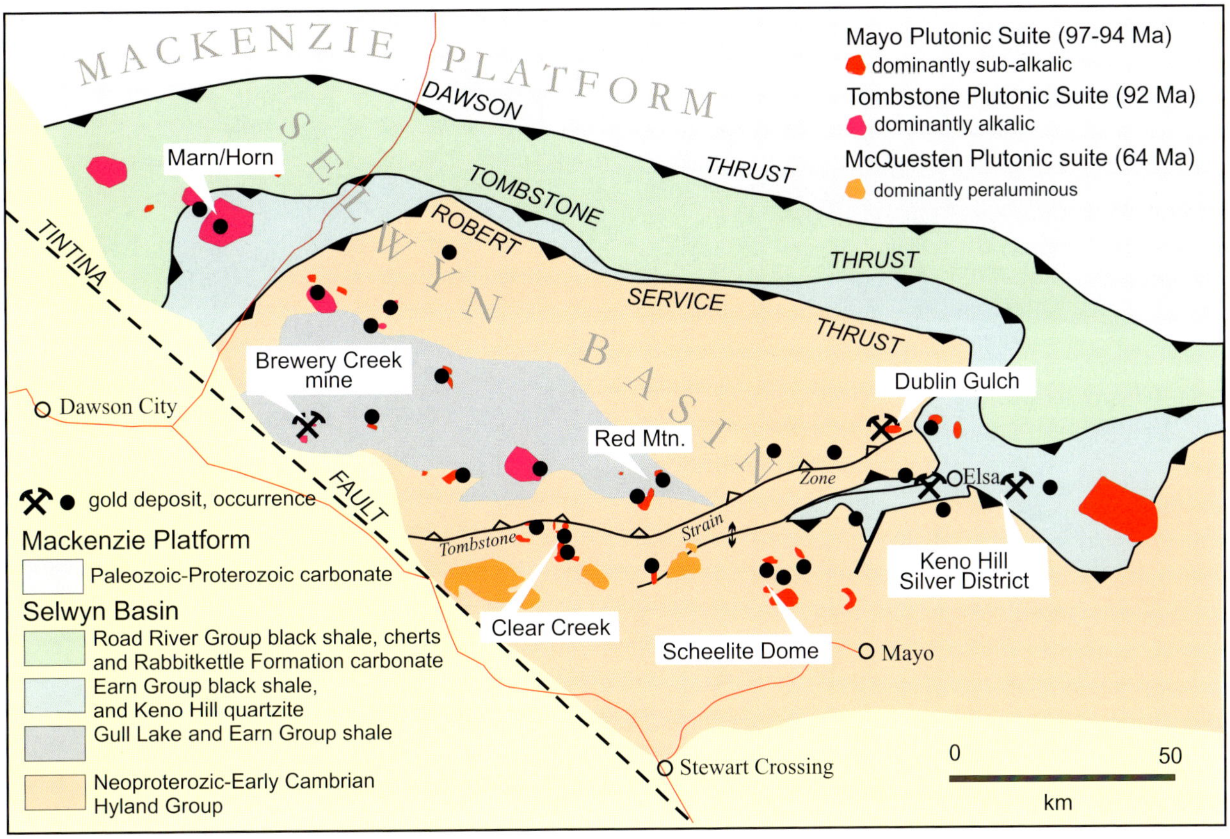

FIGURE 4. Regional geological setting of the western Tombstone Gold Belt in west-central Yukon. The region is dominated by tectonically thickened Selwyn Basin clastic strata; Mackenzie Platform carbonates dominate north of the Dawson Thrust. Tectonism peaked at ca 105 Ma (Mair et al., 2006b) and was followed by the emplacement of the Mayo Plutonic Suite, then finally by the Tombstone Plutonic Suite, during a period of weak, crustal extension at the end of the mid-Cretaceous. This tectonic and magmatic setting, within variably calcareous and carbonaceous miogeoclinal sedimentary rocks on the cratonic margin far inboard from the subduction-related, continental margin arc, is considered ideal for the formation of RIRGS as well as for W deposits. Modified from Murphy (1997).

specific plutonic suites that were emplaced into tectonically thickened Paleozoic and older miogeoclinal strata that overlie the ancient cratonic margin. The plutons are within the dominantly clastic and locally carbonaceous strata of the Selwyn Basin (Murphy, 1997; Poulsen et al., 1997), and do not occur within the presumably thicker crust of the adjacent carbonate platforms (Fig. 4).

Magmatism has resulted from melts generated and emplaced following Jurassic-Cretaceous subduction, collision, and obduction of outboard terranes that resulted in thickening of the ancient continental margin (Mair et al., 2006a). The associated and resultant plutonic episode that spans 15 to 20 million years formed hundreds of plutons, stocks, dikes, and sills that intruded strata throughout the Selwyn Basin, and related tectonic elements in Alaska. However, those plutons that generated and commonly host the Au mineralization (as well as W mineralization) preferentially belong to the most inboard and youngest of the plutonic suites that together comprise the Tombstone-Tungsten belt (Mortensen et al., 2000; Hart et al., 2004a). These plutons were emplaced during a brief (~5 m.y.) period of weak extension at circa 93 Ma that followed foreland-directed thrusting and crustal thickening, and were the last magmatic response of the mid-Cretaceous Cordilleran orogeny.

Despite suggestions indicating otherwise, the plutons of the Tombstone-Tungsten belt, those associated with RIRGS mineralization, are neither reduced continental "arc" or I-type (Newberry et al., 1995; McCoy et al., 1997; Thompson et al., 1999), nor typical crustal melt-derived or S-type granitoids (Anderson, 1983, 1988; Gordey and Anderson, 1993). They do, however, locally have characteristics of each type. Most have general granitic to monzonitic to granodioritic compositions, are mainly metaluminous, but locally peraluminous, and have some geochemical features similar to calc-alkaline granites. However, many plutons or phases are highly felsic, and have peraluminous compositions with accessory muscovite, tourmaline, and garnet. Other associated plutons are alkaline, mainly potassic, locally with silica undersaturated compositions that are variably enriched in LILEs and HFSEs (Anderson, 1987; Hart et al., 2005) and characteristic of A-type granitoids. The plutons with dominantly metaluminous, peraluminous, and alkalic characteristics define the Mayo, Tungsten, and Tombstone plutonic suites, respectively, and notably, all have associated Au mineralization. The diversity and contrasting features of the various lithologies and the geochemistry of the plutons defies conventional plutonic categorization. Detailed work at Scheelite Dome led Mair (2005) to the construction of a model whereby highly volatile, enriched lithosphere-derived lamprophyric melts interact and eventually mix with ascending and fractionating felsic melts extracted from the lower crust. The result is a wide range of lithological and geochemical characteristics in the associated plutons. The Mayo Plutonic Suite, which has the strongest Au association and affinities with the RIRGS model and includes the Fort Knox and Dublin Gulch plutons, are metaluminous, moderately reduced, moderately fractionated, biotite>>hornblende>pyroxene quartz monzonites (Hart et al., 2005; Mair, 2005)

Therefore, the RIRGS preferentially formed in association with the youngest, furthest inboard, moderately reduced (ilmenite-series) plutonic suite that developed during weak post-collisional extension behind a thickened continental margin. All Yukon, Alaskan, and British Columbia examples are associated with plutons that intruded the ancient continental margin or pericratonic terranes that had been already regionally metamorphosed. A potentially recognizable feature of such plutons may be their paradoxical metaluminous or alkalic character, despite having a highly radiogenic isotopic ancestry that suggests an ancient crustal derivation. The RIRGS-associated plutons in the Yukon and Alaska, even the most metaluminous examples, have high radiogenic initial strontium ratios >0.71 and epsilon Nd values from –7 to –15 (Farmer et al., 2000; Lang, 2001; Mair, 2005; compiled in Hart et al., 2005 and references therein).

Geological Settings and Ore Controls

The RIRGS are best developed in and surrounding the apices of small, cylindrical-shaped plutons that intruded sedimentary or metasedimentary country rocks (Figs. 5 and 6). Intrusion-hosted mineralization is preferentially sited in tensional zones that develop in the pluton's brittle carapace near the country rock contact.

Pluton size is important because batholiths are unlikely to develop into mineralizing systems. The RIRGS are generally well developed, surrounding small (<2 km^2) isolated plutons with mineralization in the intrusion and in the hornfelsed thermal aureole. Larger plutons (2–10 km^2) may have apophyses or later phases that are preferentially mineralized. Roof zones immediately above plutons may also be mineralized, in particular where there is large surface area of contact between the pluton and reactive country rocks.

Pluton geometry is also important. Elongate plutons reflect structural controls on pluton emplacement and indicate a dominant extensional direction that may be important for localizing later mineralization. Cylinder-shaped plutons with steep sides and domed or cupola-like roofs are preferred geometries because these features enhance fluid focusing (Fig. 5). Sharp shoulders also provide regions of structural and rheological contrast that may enhance development of fluid focusing structures (Stephens et al., 2004).

Depth of pluton emplacement may be a feature critical to RIRGS formation. These systems generally lack multidirectional, interconnected vein stockworks that are characteristic of porphyry deposits. This is likely due to their deeper levels of emplacement (5–9 km; Baker and Lang, 2001; Mair et al., 2006a), whereby the increased confining pressure prevents

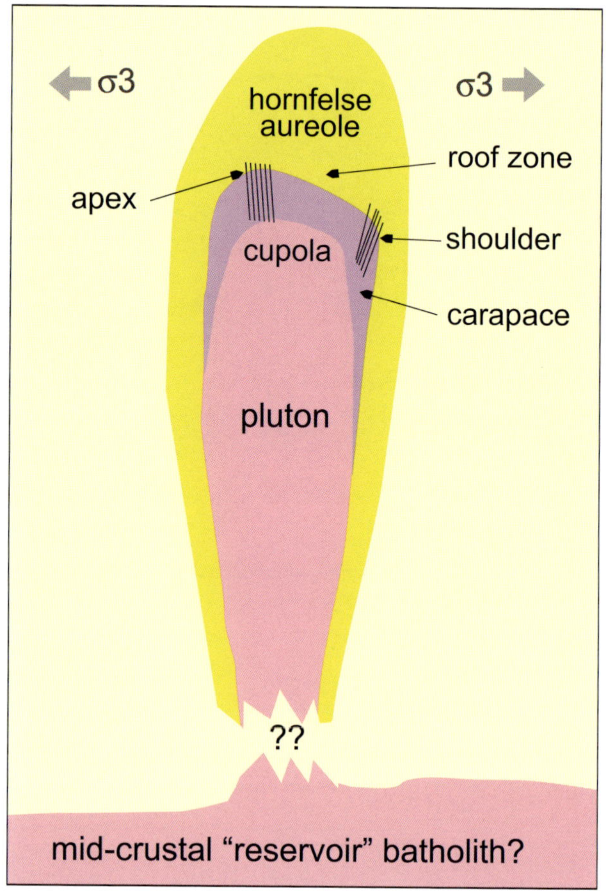

FIGURE 5. Hypothetical cross section of a small (100 m–5 km across) pluton, probably derived from a larger magmatic reservoir, and intruding into extensional regimes at higher crustal levels. Of note is the asymmetric hornfels aureole and the early-chilled and more brittle marginal carapace. Preferred sites of intrusion-hosted Au mineralization are above the cupola, where exsolved fluids will accumulate, and mineralized fractures developed in the pluton's apex and shoulders. Epizonal styles of mineralization are associated with dike and sill complexes that would be hosted near the top of the hornfels aureole.

rapid fluid exsolution and explosive pressure release, and the development of high permeability stockworks and breccias. As well, the depth precludes the entraining of significant volumes of meteoric water and the formation of broad alteration haloes. As a result, fluid flow and mineralization in most RIRGS systems is largely controlled by structural features that impinge on the thermally driven hydrothermal system (Hart et al., 2000b; Stephens et al., 2000, 2004; Mair, 2005).

The dominant structural control on RIRGS is weak extension that forms arrays of parallel fractures in the brittle carapace that are filled with thin (0.1–5 cm), auriferous, low-sulphide quartz veins that form extensive, intrusion-hosted sheeted arrays (Fig. 7). Hornfels quartzite forms a brittle host lithology for mineralized quartz veins that range from shattered "stockworky" fractures to veins several m in width (O'Dea et al., 2000). Solitary fracture, fissure, and shear-hosted veins occur in the pluton, in the hornfels, and as far as several km from the pluton, and may fill structures that were active while creating space during pluton emplacement (Stephens et al., 2004). At Fort Knox, an economically im-

FIGURE 6. Looking west into the Fort Knox (Alaska) open pit (July 2004). The light colored rocks are the ca. 92 Ma Vogt pluton that hosts the ore. The deposit is entirely within the exposed part of the pluton, whose top is only barely exposed by erosion as shown by the thin remnant of overlying dark, hornfels country rocks.

portant set of late, through-going quartz-rich shear veins cut through the main part of the intrusion-hosted orebody (Bakke et al., 2000), but controls on their formation are uncertain. Disseminated mineralization may also form near pre-existing structures, which may be older thrust faults. The disseminated deposit style preferentially develops, however, in more epizonal settings where such brittle fractures are more common, hornfelsing is more apparent, and fluid flow is more diffuse.

A final feature controlling mineralization in RIRGS is chemical reactivity of the country rocks (Hart et al., 2000a). Limestone units within the thermal aureole are obvious sites for skarn formation, and the presence of reduced skarns may indicate plutons that are prospective for intrusion-hosted sheeted vein deposits within the larger system. Most associated skarns are scheelite dominant, but they may be overprinted by a lower-temperature Au mineralizing event (Mair, 2005). Hornfelsed, variably calcareous, clastic host rocks are host to auriferous arsenopyrite-rich mineralization, either as replacements, or with diopside ± chlorite ± actinolite skarns that can be widespread, occurring several kilometres away from the causative pluton (Hart et al., 2000a; Mair et al., 2006a).

Deposit Size, Morphology, and Architecture

Areas influenced by fluid interactions from the causative pluton in RIRGS are generally restricted to the limits of the hornfelsed zones, which themselves may extend for as far as 3 km from the pluton's margins. Deposit size and geometry are also dependent on the style of mineralization, and because intrusion-hosted sheeted vein arrays are the most economically significant style, they are preferentially discussed. The Fort Knox deposit is contained within the 1000 m × 600 m surface exposure of the Vogt stock (Bakke et al., 2000). Geometry of the top part of the orebody is controlled by the limits of intrusive rocks, and at depth by cut-off grade that is affected by the density of late, higher-grade shear veins. At Dublin Gulch, there are four zones of sheeted mineralization within the 6 km × 2 km pluton (Fig. 8a). The mineralized zones are on the pluton's shoulders and the largest zone, the Eagle zone, is in the barely unroofed pluton's western margin and in the adjacent country rocks. The Eagle zone covers a surface area of approximately 1000 m × 500 m, has been intersected by drilling at depths of 300 m, and remains open at depth (Smit et al., 1996; Hitchins and Orssich, 1995). The best grades form a shell within a few hundred metres of the pluton's margins, which likely represents the carapace; grades then decrease with depth (Fig. 8b). Similarly, lower vein density and lower grades are indicated at the deeper levels of the Fort Knox orebody.

Elsewhere, sheeted vein arrays are commonly tens of metres wide and hundreds of metres long, and usually entirely within the intrusion, but less commonly in well-developed hornfels zones above the pluton. Within sheeted veins zones, ore grade intersections of at least 3 to 5 veins per metre are best developed near the wallrock contacts, likely due to preferential fracturing of the more brittle, early-chilled pluton carapace, as well as the obvious rheological differences (Stephens et al., 2004).

Intrusion-hosted orebodies may be aligned with the strike direction of extensional fractures that likely result from the same far-field stresses that controlled pluton emplacement and, therefore, may parallel the elongate direction of the pluton. These mineralized corridors are commonly filled with pegmatite, aplite, and lamprophyre dikes.

The epizonal mineralization at Brewery Creek is distal to the causative pluton and occurs in ten orebodies that are distributed over a 12 km-long corridor known as the "reserve trend" (Diment and Craig, 1999; Lindsay et al., 2000). Thick monzonitic sills along this trend are preferentially fractured compared to surrounding carbonaceous shales, and they host most of the ore. Individual orebodies are generally a few hundred metres long. The thickness of most of the orebodies is uncertain, because most were only explored to their limits of oxidation (~30 m); however, recent drilling indicates locally high-grade sulphide ores at greater depths (e.g., 6.8 g/t Au over 15.3 m at 70 m depth; Alexco Resources Corp., 2006).

Ore Paragenesis, Mineralogy, and Zonation

The ore paragenesis, mineralogy, and related zonation is controlled by the temperature of fluids during mineral deposition and fluid-wallrock interactions. As such, the ore stages define various metal assemblages that vary in time and space, are well developed with their distance away from the causative pluton, and change according to the nature of the country rocks. Within the paragenesis, the earliest ore stages as characterized by Brown et al. (2002), and Mair et al. (2006b) are within skarns developed on the pluton's margins. Features of vein paragenesis are illustrated schematically in Figure 9 and are summarized below from the work of

FIGURE 7. **(A)** Outcrop-scale exposure of an array of intrusion-hosted, sheeted quartz veins in the apex of the Rhosgobel pluton at Clear Creek, Yukon. One- to three-cm-wide veins of simple, single-stage quartz with minor sulphides and scheelite fill parallel, extensional fractures that may extend for hundreds of metres. Individual veins may contain 50 g/t Au, but host rocks are barren such that vein density is the grade-controlling feature. Wide intersections generally average 1 g/t Au. Note 14 cm-long marker pen (circled in red) for scale. **(B)** Ore from Fort Knox i s dominated by a series of subparallel, white and grey, single-stage quartz-sulphide veins that are parallel with the east–west elongate direction of the pluton, and other structural elements such as dikes and shears (in this photo, the veins are offset by a small fracture). It should be noted that there are not interconnected, multidirectional quartz vein stockworks that are typical of porphyry deposits. Alteration, mainly sericite-calcite, is limited to the immediate vein selvages.

many authors, including Maloof et al. (2001), Marsh et al. (2003), and Mair et al. (2006b).

Early ore stages are dominated by high-temperature (650°C) anhydrous diopsidic pyroxene-plagioclase skarns that locally contain scheelite. Lower-temperature (420°C), and sometimes overprinting, hydrous, either biotite-, zoisite-, or actinolite-dominant skarn assemblages may be Au-bearing where they contain significant amounts of sulphide minerals. High-temperature sulphide assemblages are dominated by pyrrhotite>>chalcopyrite, but lower-temperature assemblages are dominated by arsenopyrite with various Bi-Te-Sb-Pb-Au minerals and alloys (Fig. 10). All silicate assemblages lack garnet, which is a feature of reduced skarns (Meinert, 1998).

Early veins are dominantly intrusion-hosted and are characterized by alkali feldspar-, mica-, and scheelite-bearing quartz veins that may host only sparse sulphide minerals and may lack Au. Slightly lower-temperature, sheeted vein arrays are similar but may host up to a few percent pyrite or arsenopyrite, but with blebs of various Au-Bi-Te alloys, and are the key hosts to Au mineralization. At the system scale, and more commonly outside of the intrusion, are more sulphide-rich arsenopyrite and antimony veins. The last gasps of the hydrothermal event generate Ag-Pb-Zn-bearing quartz veins which occur in the most distal locations, often beyond the limits of the hornfelsed aureole. These distal Ag-rich galena-sphalerite quartz-carbonate veins also form the 30 km-long Keno Hill Ag district and may indicate a regional zoning within the Tombstone Gold Belt.

The mineralogy of intrusion-hosted sheeted veins, such as those at Dublin Gulch and Fort Knox, comprise a single stage of massive, translucent grey or white quartz, locally with subordinate coarse-grained, white K-feldspar or mica that are most commonly on the vein margins. Some veins are "dry", lacking quartz, and consist of sulphide mineral-coated fractures with biotite, which form a limonitic gouge where oxidized. Low sulphide contents (0.1%–2%) typify intrusion-hosted assemblages, which are dominated by pyrite > pyrrhotite > arsenopyrite with accessory scheelite and bismuthinite. Arsenopyrite is much more abundant in country rock-hosted veins (up to 10 vol.%), and is common with pyrrhotite in replacement-style mineralization. Accessory molybdenite occurs locally in thin quartz veins; chalcopyrite, sphalerite, and galena are sparse. Cassiterite has been reported near some occurrences (e.g., Tin Dome at Dublin Gulch) but is rare and does not occur with Au.

Native Bi and Bi-bearing sulphosalt minerals are commonly described from various mineralization styles in RIRGS. They occur late in the paragenesis and are typically alloyed with Te, Pb, Sb, or Au. Tellurobismutite, maldonite, tetradymite, native Bi, boulangerite, and Bi-Pb sulphosalts have been recognized at Fort Knox (McCoy et al., 1997, 2002). At Dublin Gulch, approximately 40% of the Au occurs as complex intergrowths with native Bi (Hitchins and Orssich, 1995), but Au also occurs in Bi-Pb sulphosalts with galena and molybdenite (Maloof et al., 2001), and with bismuthinite, tetradymite, tellurobismuthite (Hitchins and Orssich, 1995). The common association of very high fineness free Au within or adjacent to native Bi or bismuth-bearing minerals suggests that Au may have exsolved from earlier, high-temperature Bi-alloys, upon cooling (Fig. 10).

Ore mineralogy at the epizonal Brewery Creek deposit is distinctly different but relatively simple. It is dominated by early pyrite, with associated arsenian pyrite and arsenopyrite (Diment and Craig, 1999; Lindsay et al., 2000). Oxidized mineralization appears disseminated with limonite, but hypogene ore shows sulphide minerals in, and adjacent to, spaced, sulphidized fractures lacking quartz. Gold is refractory, being associated with acicular arsenopyrite and the arsenian pyrite overgrowths on early pyrite. Although late stibnite veins are also a common feature, their association with Au is equivocal.

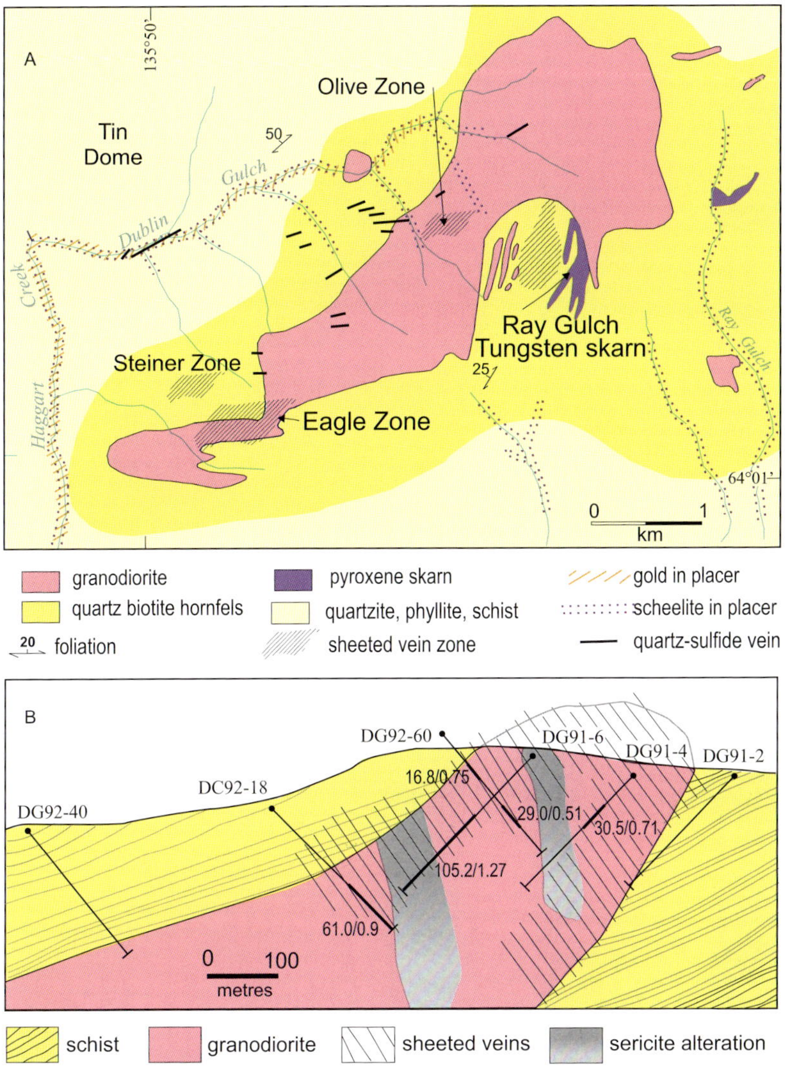

FIGURE 8. (A) The Dublin Gulch pluton is a good example of a RIRGS with many different styles of mineralization, including intrusion-hosted sheeted vein arrays, scheelite-bearing skarns, tin greisens, Au- and arsenopyrite-bearing quartz veins, and creeks with placer Au and scheelite. Bismuthinite has also been found in the placer concentrates. Beyond the limits of this figure, and the hornfels zone, Ag-Pb-Zn-bearing quartz veins also occur. **(B)** Cross section of the Eagle zone at the west end of the Dublin Gulch pluton. Sheeted vein arrays are shown schematically but are mostly concentrated in the margins of the apical region of the pluton and do not appreciably occur in the adjacent, less-competent wall rocks. Modified from Smit et al. (1996), with drill intersections (given as metres/grams per tonne Au) from Hitchins and Orssich (1995).

Geochemical Features

The RIRGS are geochemically distinguished from oxidized intrusion-related mineralizing systems by their dominance in Au, associated W, and lack of anomalous Cu. Tungsten enrichments and deposits occur within the RIRGS, such as the Ray Gulch W skarn at Dublin Gulch, but Au and W enrichments may be spatially distinct (Brown et al., 2002), or later Au may overprint early W veins or skarns (Mair et al., 2006a). Bismuth and Te are common elements but are also characteristic features of many other deposit models.

Sulphide-rich (10%) sheeted veins at Clear Creek contain up to 30 g/t Au, as much as 1 wt.% As, and 100 to 1000 ppm Bi and W (Marsh et al., 1999). At Fort Knox, Au strongly correlates with Bi and Te, and but weak enrichments of W, Mo, Sb, and As do not correlate with elevated Au (Bakke, 1995; Flanigan et al., 2000). High Au grade intersections at Dublin Gulch have elevated Bi, As, Sb, Cu, and Zn concentrations, although again Au grades solely correlate with Bi (Maloof et al., 2001) and probably with Te, but such data are unavailable. Factor analysis on the Clear Creek veins indicated a dominant Au-rich As-Au-Bi±Sb,Te element association, and a base metal suite of Ag-Bi-Pb±As, Au that represents a cooler/later assemblage (Marsh et al., 1999). Tungsten displays little correlation with either element suite, but had values of >300 ppm in many Au-rich samples. Similarly, mineralization at Scheelite Dome was characterized by both a Au-Te-Bi±W±As and a Ag-Pb-Zn-Cd-Sb±Cu±Au elemental association, with the latter characteristic of ores in the nearby Keno Hill Ag district (Mair et al., 2000).

Epizonal Au at Brewery Creek lacks the typical W, Bi, and Te association. Instead, Au ores occur with arsenopyrite and are, therefore, dominated by As. Late fractures host stibnite enrichments but Au-Sb correlations are erratic. Additionally, Hg was recovered during mining, thus giving Brewery Creek an As-Sb-Hg association.

Geochemical zoning occurs at the pluton scale, with elemental zonation reflecting the cooling trend of the hydrothermal fluids, with a component of country-rock buffering. The pluton's geochemical influence is typically on the order of 1 to 3 km, but can be larger in roof zones above the pluton, for example >10 km at Scheelite Dome. Intrusion-hosted ores are dominated by a Au-W-Bi-Te signature with Au correlating well with Bi and Te, but not at all with W (Fig. 11). Geochemical signatures of high-temperature skarns adjacent to the pluton may be similar, but in some systems, As and W enrichments may be more significant than Bi-Te signatures. Percent-level Cu and Bi, with 8.6 g/t Au at the Marn deposit (Brown and Nesbitt, 1987), occur only in skarns formed from alkalic or more oxidized plutons. Hornfels-hosted ores, which may be veins, disseminations, or replacements, are more sulphide-rich than intrusion-hosted ores, and are characterized by elevated As that correlates with Au (Flannigan et al., 2000). Distal mineralization, which forms at or beyond the limits of hornfels, is dominated by a Pb-Ag-Zn or Sb-rich geochemical association, and typically occurs as fault-filled veins.

Vertical zonation patterns within a system may mimic the lateral zonation, but may be less evident due to the much broader thermal gradients that occur above the pluton. Mineralization or entire orebodies (e.g., Gil in the Fairbanks

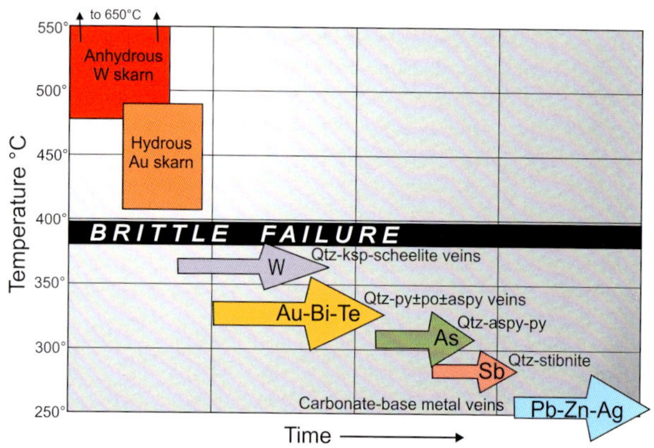

FIGURE 9. Schematic paragenesis of evolving types and metal associations of mineralization in typical cooling RIRGS. As well as time, the lower axis could also represents distance away from the fluid source such that As-, Sb- and Ag-P-Zn veins are almost exclusively beyond the causative pluton.

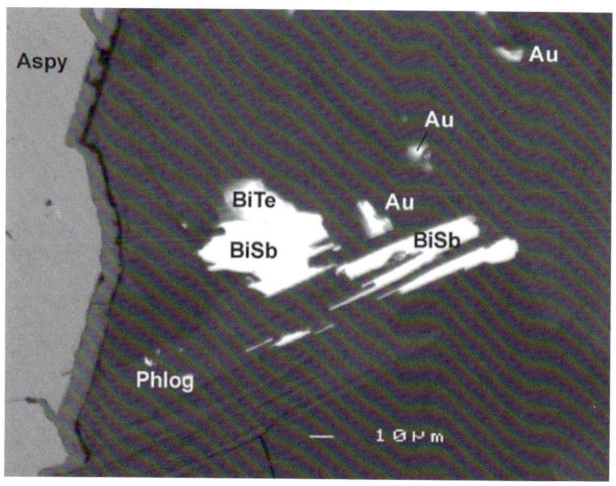

FIGURE 10. SEM backscatter image of Scheelite Dome Au-W-As-Bi skarn (Tom zone), emphasizing the occurrence of Bi-Te-Sb alloys and Au. Gold occurs as small, 5 to 20 micron particles near larger Bi-Te-Sb alloys that occur as intergrain infillings. Gold may be deposited in high-temperature alloys and exsolved out upon cooling. This sample contains 17 g/t Au, 12 ppm Te, 426 ppm Bi, 161 ppm W, and 3300 ppm As. Gold in vein-hosted ores has a similar character and metal tenor. Figure and data courtesy of John Mair.

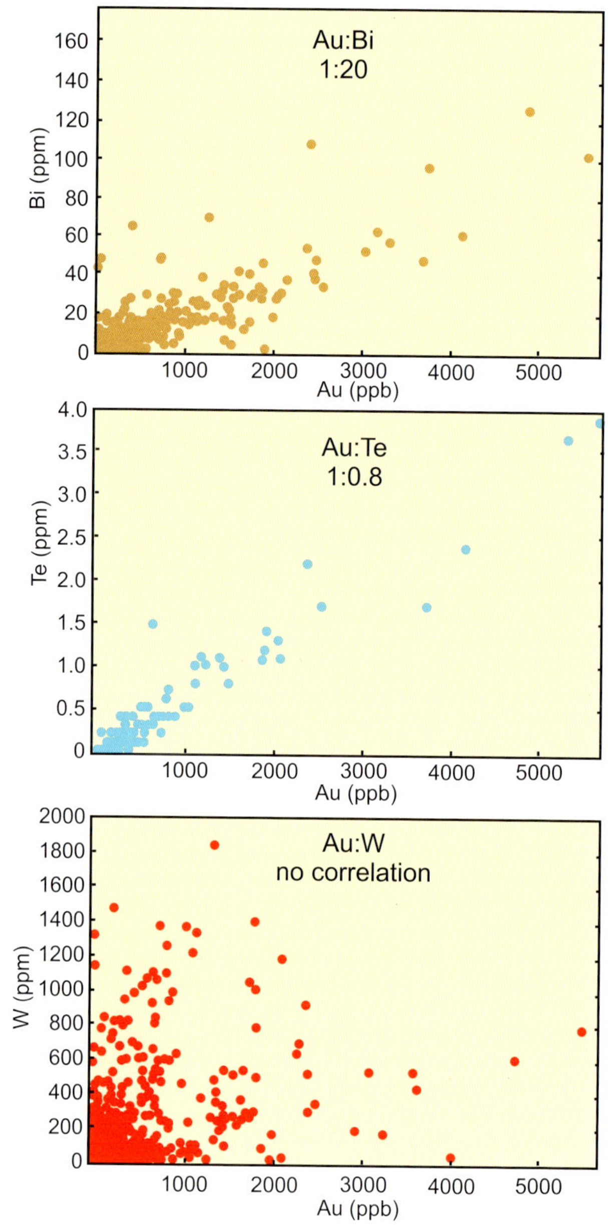

FIGURE 11. Binary element plots of intrusion-hosted ores dominated by sheeted veins that demonstrate a Au-Bi-Te-W association. Gold displays a high positive correlation with Bi and Te, but not with W. These data are from the Rhosgobel zone at Clear Creek (Yukon); each point represents the analysis of 1.5 m of drill core. Note that there are fewer analyses for Te.

area) may be within a roof zone altered to hornsfel above unexposed plutons, which may be several kilometres below and fail to show apparent zoning trends. Within the pluton, suggestions of W or Mo enrichments at depth (Bakke, 1995) have not been supported by data, but decreases in Au grades with depth are evident at Fort Knox. A key aspect of vertical zonation within RIRGS are the more apparent variations within shallowly emplaced systems. The Brewery Creek deposit and nearby occurrences (Ida, Oro) emphasize that the fluid systems associated with these more shallowly emplaced systems (~3 km) take on epizonal characteristics and metal signatures in response to the lower temperatures.

Alteration Mineralogy and Fluid Geochemistry

Alteration in intrusion-hosted ores is neither extensive nor intensive, and is typically limited to 0.5 to 3 cm-wide selvages adjacent to the veins with intervening, apparently fresh, barren rock (Fig. 7B). Alteration proximal to veins most commonly consists of either texture-destructive K-feldspar replacement (Maloof et al., 2001) or pervasive carbonate replacement of mafic minerals. An adjacent sericite-dominant ± pyrite ± carbonate assemblage overprinting plagioclase and mafic minerals is common. In particular, this alteration distinguishes Au-bearing from (higher-temperature?) non-Au-bearing veins at Fort Knox (P. Jensen, pers. commun., 2002). Chlorite alteration may be more distal, but is not pervasive throughout the host rocks. Vein-proximal alteration may be cryptic and more easily observed on weathered surfaces. Alternatively, alteration of country rocks surrounding a causative pluton may be both intensive and extensive. Although typically dominated by prograde alteration to biotite-quartz ± pyrrhotite hornfels, later retrograde fluid-dominated alteration may be widespread. For example, sericite after biotite may occupy large regions of the hornfels aureole, such as at Scheelite Dome (Mair et al., 2000), and is also a more focused feature of distal Ag-Pb-Zn veins in the Keno Hill district (Lynch, 1989). At the epizonal Brewery Creek deposit, PIMA (Portable Infrared Mineral Analyser) studies indicate that hypogene ores contain illite surrounded by a more extensive ankerite alteration halo (R. Diment, pers. commun., 2001).

The fluid chemistry of the intrusion-hosted sheeted veins of Yukon deposits has been well documented by Baker and Lang (2001), Baker (2002), Marsh et al. (2003), and Mair et al. (2006). Most Au-W-Bi-Te veins consist of early high-temperature (380°–300°C), CO_2-rich (5%–14%), low-salinity (2–6 wt.% NaCl equiv.) aqueous carbonic fluids with CH_4 and N_2. These fluids cooled and locally unmixed to yield lower temperature (mostly 280°–250°C, but as low as 160°C), immiscible, low-salinity (0.2 wt.% NaCl equiv.) and high-salinity (6–15 wt.% NaCl equiv.) aqueous fluids lacking significant CO_2, which formed the As-, Sb-, and Ag-Pb-Zn veins. In alkalic magmatic systems that have associated Cu mineralization, such as Mike Lake and Emerald Lake, the fluids evolved to become highly saline with 30 to 55 wt.% NaCl equivalent. Skarns and veins interacting with country rocks have elevated CH_4, which likely results from interactions with the sedimentary rocks causing more evolved fluids to become increasingly reduced (Mair et al., 2006a). In epizonal systems such as Brewery Creek, the fluids are aqueous and lack CO_2, but are characterized locally by high salinities. Fluid inclusion barometry indicates Au deposition at most sites (Dublin Gulch, Clear Creek, Scheelite Dome) at depths of 3 to 9 km (mostly 5–7 km).

Comprehensive light stable isotopic studies of oxygen and sulphur have been undertaken at the Clear Creek (Marsh et al., 2003) and Scheelite Dome (Mair et al., 2006a) systems. The $\delta^{18}O_{quartz}$ of the Au-W-Bi-Te sheeted quartz veins range from 14 to 16 per mil, slightly heavier than the host granitic rocks at 11 to 13 per mil, and similar to values of 13 to 16 per mil for host sedimentary rocks. Antimony and Ag-Pb-Zn veins hosted by country rocks have $\delta^{18}O$ values of 17 to 20 per mil reflecting extensive wall-rock interaction. Sulphur isotope data display values of 0 to –3 per mil for intrusion-hosted quartz Au-W-Bi-Te veins, 2 to –7 per mil for various skarn mineralization, –7 to –10 per mil for country rock-hosted arsenopyrite veins, and –9 to –11 per mil for Ag-Pb-Zn veins. Like the oxygen isotope and fluid inclusion data, the $\delta^{34}S$ values indicate a trend that reflects progressive interaction with sedimentary country rocks as the hydrothermal system evolves and cools. Therefore, in addition to phase separation, progressive cooling and reduction of the fluid were likely mechanisms of Au deposition (Mair et al., 2006a).

Empirical and Genetic Exploration Models

Empirical Model

A schematic empirical model for RIRGS (Fig. 12) incorporates observable features from the "type" deposits and occurrences in the Yukon and Alaska. The key empirical features have been described above, but those features that are considered as critically distinguishing attributes are listed below.

System

Mineralization extends beyond the limits of the intrusion, and locally beyond the thermal aureole yielding a broad mineralizing system (Fig. 8A). The size of the system is generally dictated by the limits of the thermal aureole, commonly several kilometres across, but can be dependent on the depth of erosion with the broadest and best developed mineralization at the top of or above the pluton (Fig. 5).

Diverse Mineralization

Differing styles of mineralization emphasize not only the extent of the mineralizing system, but also the involvement of the country rock and its role in creating mineral system diversity. Chemically reactive and/or physically brittle sedimentary strata result in a diversity of mineralization styles, whereas the causative pluton is typically dominated by solely sheeted vein sets.

Zoned Deposit Types

RIRGS typically deposit metals in intrusion-hosted, contact, pluton-proximal, and pluton-distal settings, and thus exhibit a predictable zonation of differing deposit styles outward from the central, mineralizing pluton (Fig. 12). Skarns and replacements are generally pluton proximal, with an increase in structural control on more distal mineralization. There is also crustal-scale vertical zonation, with epizonal occurrences forming at shallower levels.

Concentric Metal Zoning

Predictable metal signatures develop broad-scale zoning surrounding and above a central causative pluton, due primarily to the effects of steep thermal gradients on fluid chemistry (Figs. 9, 12). Gradients and metal zones are steeper on the sides of the pluton and broadly developed above it (Fig. 5). Zoning is somewhat analogous to that identified in porphyry systems (e.g., Jones, 1992).

Metal Associations

Gold, as well as W, may form ore, but Au does not directly correlate with W (Fig. 11). Bismuth and Te are enriched in intrusion-hosted Au ores and correlate with Au. Arsenic enrichments characterize hornfels-hosted mineralization and form regional-scale geochemical anomalies.

Pluton Features

Associated plutons are generally small and solitary, with "smoking gun" characteristics that indicate they were the source of the hydrothermal fluids. Features that provide evidence of high volatile contents, fractionation, and fluid ex-

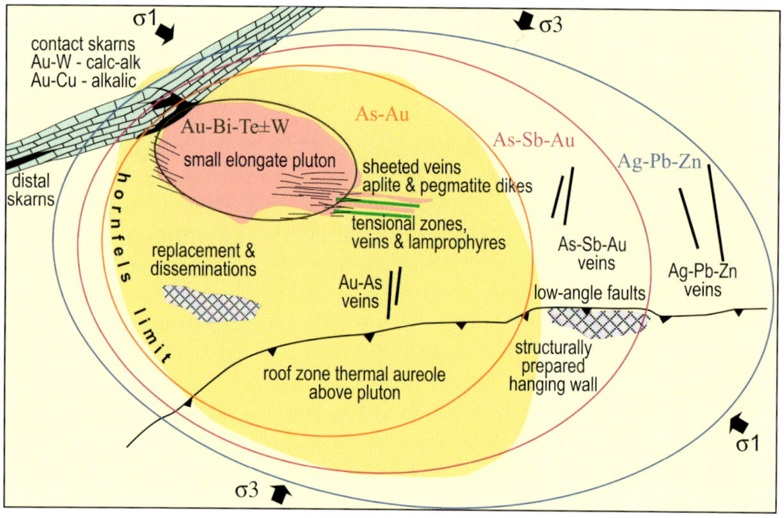

Figure 12. General plan model of RIRGS from the Tintina Gold Province. Of note are the wide range of mineralization styles and geochemical variations that vary predictably outward from a central pluton. Scale is dependent on the size of the exposed pluton, which is likely to range from 100 m to 5 km in diameter. Modified from Hart et al. (2002).

solution include the presence of hornblende in biotite granitoids, textural and grain-size variations, aplite and pegmatite dikes, tourmaline veins, miarolitic cavities and vugs, greisen-style alteration, unidirectional solidification textures, and cupola-hosted mineralization (Fig. 13).

Redox State

The RIRGS are associated with felsic, ilmenite-series plutons that lack magnetite and, as such, have low magnetic susceptibilities, low ferric:ferrous ratios (<0.3), and flat aeromagnetic responses. Associated mineralization has a reduced character, with pyrrhotite commonly the dominant sulphide mineral and fluids that may locally contain methane.

Genetic Model

The RIRGS genetic model requires that the ore-generating cooling pluton reaches volatile saturation and that a fluid exsolves from the melt. Metals and volatiles such as sulphur and halogens presumably preferentially partition from the melt into an exsolving aqueous-carbonic ore fluid phase. Pressure, or depth of emplacement, exerts the greatest control on volatile saturation, particularly because volatiles are easily dissolved in felsic melts under higher pressures (Burnham and Ohmoto, 1980). However, volatile saturation is also induced by magmatic processes such as fractional crystallization, magma mixing, or simple cooling. Pluton emplacement depth appears, therefore, to be critical and explains why RIRGS are typically associated with a specific suite of plutons distributed over a broad area; such plutons likely represent melt crystallization at the same general crustal level.

At the pluton scale, mineralization is limited to regions above and outward from the site of volatile saturation. Being less dense than the melt, fluids will migrate to the uppermost parts of the less viscous portion of the magma chamber, which is usually the volatile-rich magmatic cupola immediately under the earlier-formed carapace (Candela and Blevin, 1995). Fluids will invade fractures in the carapace and opportunistically leak into and react with adjacent country rocks. Mineral occurrences are, therefore, most commonly sited at the pluton's apex, in the igneous carapace, or in hornfelsed country rocks adjacent to and above the pluton. The host plutons to many RIRGS likely have magma volumes that are too small to provide the large amount of metals and volatiles contained in these deposits, thereby suggesting the participation of larger volumes of primary magmatic fluids and metals (Candela and Piccoli, 2005). These could include deeper unexposed batholiths or mafic lamprophyric melts.

The metallogeny of mineral deposits associated with intrusive rocks is mainly controlled by the associated magma's primary oxidation state (Ishihara, 1981) and the degree of magma fractionation (Thompson et al., 1999; Fig. 14). Highly oxidized magmas in arc environments that are relatively unfractionated will have a metal tenor dominated by Cu and potentially Au. In reduced magmatic systems, Cu is likely stripped out by the early precipitation of sulphide melt blebs, whereas W behaves incompatibly and, therefore, is enriched during fractionation. Tungsten is relatively easy to mobilize in reduced, CO_2-rich hydrothermal systems. Controls on the partitioning of Au in evolving magmatic systems is poorly understood, but empirically it is associated with and enriched in both highly oxidized and moderately reduced systems (e.g., Leveille et al., 1988; Rowins, 2000; Mungall, 2002); in the latter case, this may lead to formation of a RIRGS.

Knowledge of magmatic fluid physical and chemical properties for most intrusion-related mineral deposits was originally based on comparison with the porphyry model that emphasized the role of highly saline aqueous fluids in metal transport under oxidized magmatic conditions (e.g., Burnham, 1979). The RIRGS are, however, dominated by aqueous carbonic fluids that have high volumes of CO_2 and low salinities, are largely reduced, and only locally contain highly saline fluids (Baker and Lang, 2001; Marsh et al., 2003; Mair et al., 2006b). These fluids are unlike those in porphyry Cu deposits.

The concentration of metals through magma fractionation have been suggested by McCoy et al. (1997) for the Alaskan RIRGS and was shown to occur in the more siliceous Timbarra system of eastern Australia (Mustard et al., 2006). Conversely, Mair (2005), working in Yukon, has suggested that input of volatile-rich mafic magmas were likely responsible for volatile enrichment that assisted the early, less-volatile, siliceous magmas in reaching fluid saturation. It is uncertain but probable that this mixing also played a role in metal enrichment as well.

Because an underlying feature of the RIRGS classification is the direct genetic association of mineralization with a causative magma, geochronology is of fundamental importance in processes studies. Many lines of evidence suggest that

FIGURE 13. Vug in the Fort Knox pluton filled with coarse-grained quartz, feldspar, and (chloritized) amphibole. Features such as this indicate that the pluton reached fluid saturation and exsolved a hydrothermal fluid. A pluton's response to high volatile contents is the formation of pegmatitic and aplitic dikes as well as a mineralizing fluid.

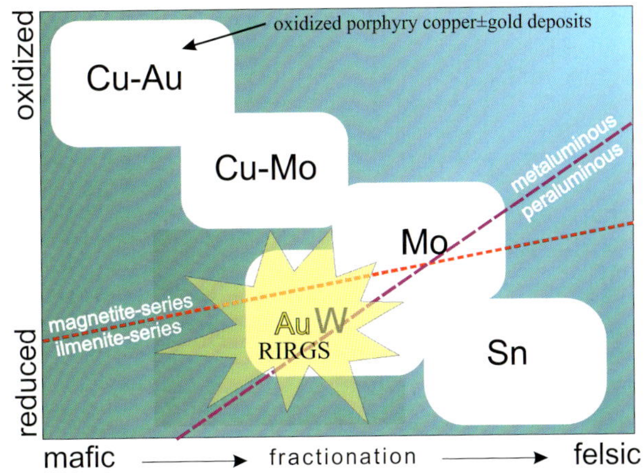

FIGURE 14. Schematic plot emphasizing the variations in metal association as a function of the primary magmatic oxidation state and the lithologic character of the associated plutonic rocks. Gold, associated with RIRGS plot in the field occupied by W systems but notably is far removed from the more characteristic Au-Cu field that is associated with highly oxidized and more mafic magmas. The result is that Au can be enriched in both oxidized and reduced magmas, but that a reduced oxidation state may be necessary for Au enrichment in fractionated systems. The corollary is that fractionated oxidized systems are likely to be depleted in Au. Modified from Thompson et al. (1999).

these magmatic systems cool quickly, indicating that magmatic and hydrothermal processes must be essentially coeval. Comparative geochronology on magmatic and hydrothermal phases using different analytical methods indicates that an entire magmatic-hydrothermal system can be discerned to within a 2 to 3 million year window (Selby et al., 2001; Hart et al., 2004b).

Key Exploration Criteria

At the regional scale, exploration should focus on the foreland parts of orogenic belts where felsic plutons have intruded ancient continental margins, inland of accreted terranes or collisional zones. These regions may be historically better recognized for their W or Sn metal tenor, and may also host Ag-rich veins or Au placers that are associated with the plutons. All mineralizing plutons that belong to the same suite or time interval are potential targets for RIRGS. Prospective plutons were preferably intruded deeper than ~5 km to keep these low-volume hydrothermal systems contained in the melts and subsequently focused during exsolution. RIRGS associated with shallower plutons are characterized by more diffuse epizonal styles of mineralization and a Au-As-Sb-Hg signature. Associated plutons will have low primary oxidation states and are, therefore, easy to differentiate from magnetite-series plutons of true continental margin arcs that have associated Cu-Mo porphyry deposits.

At the deposit scale, targeting the pluton's carapace is critical such that those plutons that are barely unroofed are considered the best locations for RIRGS. Roof zones above plutons are also highly prospective, but may be difficult to target as they are rarely noted on geological maps. Deeply eroded plutons, recognized by their large circular-shaped surface areas, are unlikely to yield large-tonnage intrusion-hosted sheeted vein deposits, but may nevertheless have hornfels with Au-bearing skarns or veins. Understanding the structural controls on pluton emplacement may be key to developing targets and preferred deposit orientations within a magmatic-hydrothermal system (Stephens et al., 2004). Additional exploration techniques and exploration pathways to discovery are described in Hart et al. (2002).

Geochemical

Regional geochemical surveys are very good at identifying mineralizing plutons, particularly where characterized by broad As aureoles, such as those of the Tombstone Gold Belt (Fig. 15). Placer Au may occur in related drainages in significant amounts (>100 000 oz; e.g., Allen et al., 1999). Placer scheelite is also a feature of many occurrences. Soil geochemistry can be extremely effective locally at delineating potential mineralization within the area of a causative pluton, and recognizing mineralized portions of its hornfelsed zone (e.g., Diment and Craig, 1999; Hulstein et al., 1999). Soil lines should cross the extensional direction that may mimic a pluton's elongation direction. Gold grades can be up to several grams per tonne in some soils, but low anomaly thresholds (25 ppb Au) may be required for surveys with low geochemical response (Diment and Craig, 1999). Anomalous Bi, Te, or W values, or multi-element analyses using metal ratios or factor analysis can assist in interpretation of vein types or predicting more proximal (i.e., intrusion-hosted) ores in areas with poor rock exposure (Marsh et al., 1999; Mair et al., 2000).

Geophysical

Geophysical methods that identify Au mineralization in RIRGS are still elusive, but potential-field methods are ideal at assisting interpretations of geological settings where ores could be found. Regional aeromagnetic surveys are effective at identifying unmapped or unexposed plutons or locating roof zones (Fig. 16). Associated plutons have low magnetic responses; however, pyrrhotite concentrations in hornfelsed aureoles may yield doughnut-shaped signatures for exposed plutons and simple bulls eyes for roof zones of unexposed plutons (Hart et al., 2002). This response is pronounced in reducing sedimentary rocks where pyrrhotite formation is

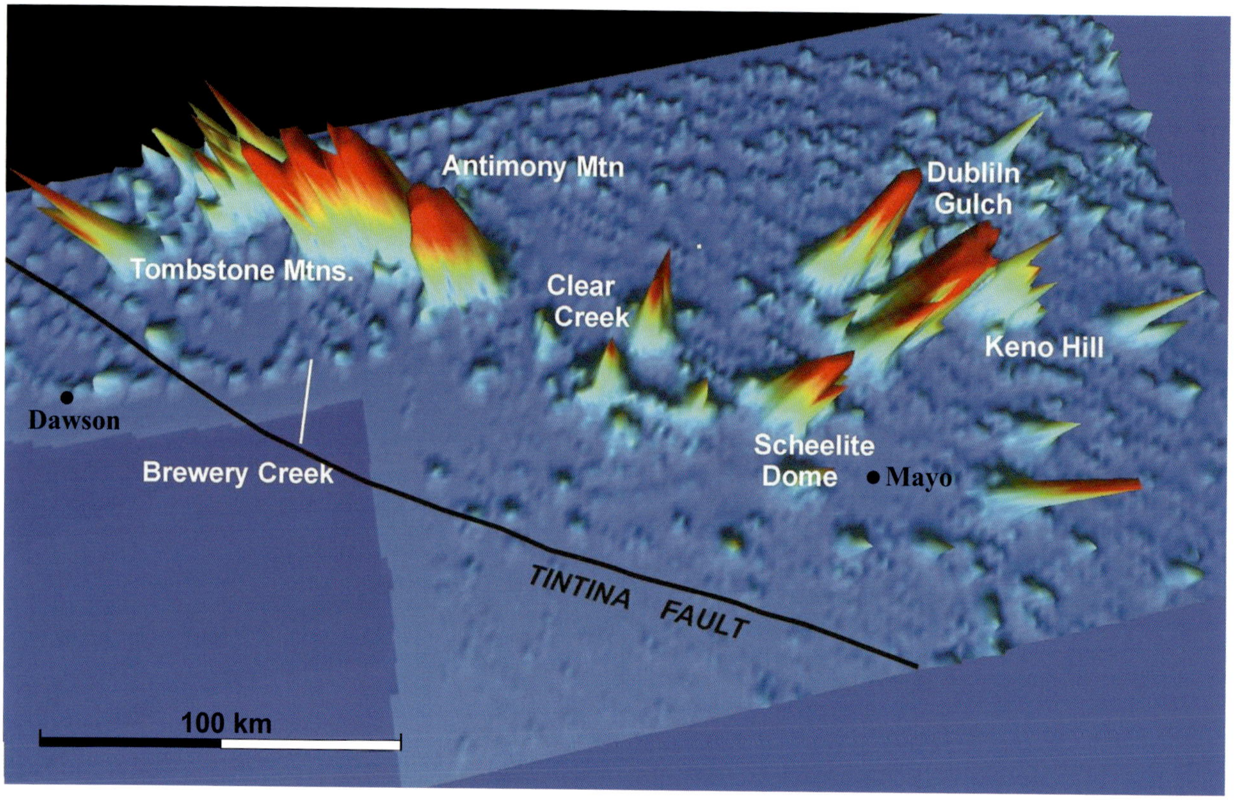

FIGURE 15. Gridded and contoured regional stream silt geochemical data for arsenic for the west-central Yukon. Sample density is one per 13 km^2. Background values are <50 ppm As, and regions in red have values >500 ppm As. Data are compiled from Geological Survey of Canada open file reports (Geological Survey of Canada, 1978, 1990, 1991). Despite the dominance of arsenopyrite in the Brewery Creek orebodies, note the absence of a significant As anomaly for this poorly exposed property.

more likely but otherwise may be lacking.

Within intrusion-hosted systems, geophysical exploration methods have largely yielded ambiguous to poor results (Bakke, 1995). Within the hornfelsed aureoles, however, magnetic methods allow identification of major structures as lows produced by alteration of pyrrhotite. Induced polarization methods are useful for identifying reactive and sulphidized zones within the hornfels when targeting disseminated-, replacement-, or skarn-type ores (M. Powers and G. Carlson, pers. commun. 1999).

Knowledge Gaps

- This evolving RIRGS classification suffers from nomenclature issues and there is a need for a framework for all intrusion-related deposit models to emphasize distinguishing characteristics (Hart, 2005). In particular, there is uncertainty about the inclusion of many of the variants of intrusion-related mineralization styles (e.g., epizonal, veins), as well as other global or Archean examples, which result in considerable uncertainty in this deposit classification.

- Knowledge gaps identified by Thompson and Newberry (2000), and Lang and Baker (2001) emphasize the need for improved basic deposit descriptions and focused research on individual systems. Despite a good geological overview of the most significant recognized RIRGS (Bakke, 1995; Bakke et al., 2000), Fort Knox remains among the most poorly studied and understood example. In contrast, Dublin Gulch (Hitchins and Orssich, 1995; Maloof et al., 2001), Scheelite Dome (Mair, 2004; Mair et al., 2006b), and Clear Creek (Marsh et al., 2003; Stephens et al., 2004) have all seen more focused studies.

- Many critical questions still need to be answered. What features of a causative pluton's magmatic character are important in the formation of RIRGS? Are the Tombstone Gold Belt plutons unique? Why are these deposits not more common in similar tectonic settings elsewhere, or in Precambrian rocks?

- The controlling features of most of the ore-forming processes, such as metal enrichment and volatile saturation remain poorly understood, and knowledge gaps are nearly identical to those identified by Sinclair (2007) for porphyry deposits associated with oxidized magmas. More specific to RIRGS, however, is the role of oxidation state on controlling Au concentration in magmatic systems.

- There is controversy about the most appropriate model for many Au deposits; in particular, the Pogo and Donlin Creek deposits of interior Alaska have been interpreted as both RIRGS (Smith et al., 1999; Ebert et al., 2000; Rhys

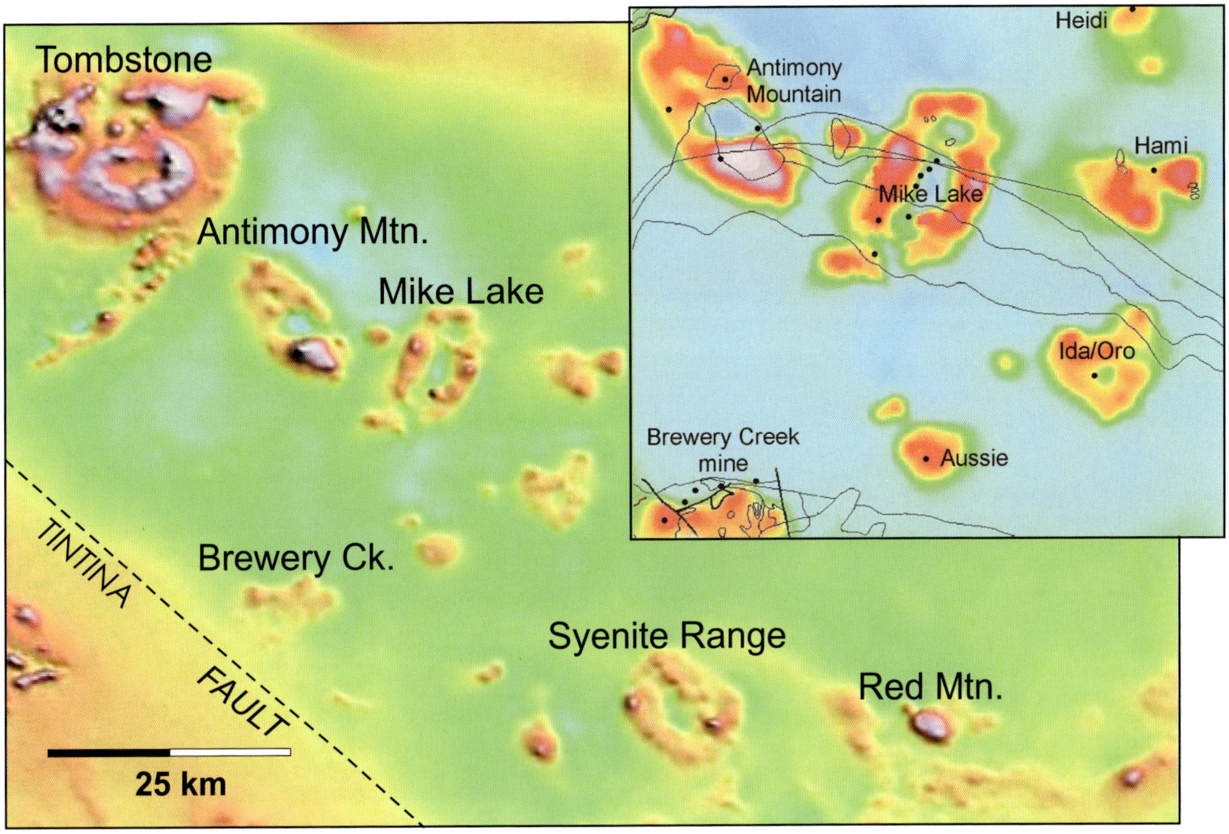

FIGURE 16. Residual total magnetic intensity from airborne aeromagnetic data indicate that, despite their low magnetic response, many plutons can be identified aeromagnetically by the positive signatures generated by magnetic hornfels zones that likely develop from contact metamorphic pyrrhotite aureoles. The results are doughnut-shaped features with lows generated by the ilmenite-series plutons forming the "holes". The inset shows detailed magnetic features in the Mike Lake area, with the black dots indicating mineral occurrences and the lines indicating geological contacts. Notably, mineralization at the Heidi, Hami, Ida/Oro, and Aussie occurrences are associated with large magnetic expressions, but little or no outcrop of plutonic rocks. These occurrences, as well as the Red Mountain occurrence, likely represent roof zones above the tops of the mostly unexposed plutons.

et al., 2003) and orogenic (Goldfarb et al., 2000, 2004, 2005; Groves et al., 2003). Even Central Asian gold deposits, such as Muruntau (Sillitoe, 1991; Wall et al., 2004) and Jilau (Cole et al., 2000), have been considered to be RIRGS. It is important, therefore, that a clear set of characteristics that distinguish RIRGS from orogenic gold deposits be established.

Areas of High Potential in Canada

Good potential still remains in central Yukon, where numerous plutonic systems intruding the continental margin were originally (ca. 1994) assessed for only their intrusion-hosted ores with little effort directed towards hornfels-hosted ores. Although the RIRGS classification still requires improvement, significant advances in understanding were made during a time of limited exploration activity (1998–2003) and the fruits of that research may not have been effectively applied. As well, little effort has been directed toward evaluating roof zones or toward the search for unroofed plutons.

Many RIRGS prospects still have considerable exploration potential, each with numerous untested drill targets. Scheelite Dome, in particular, has a 10 km-long, >100 ppb Au soil anomaly that has not been fully explored. Clear Creek's Bear Paw zone, with drill intersections of 2.3 g/t Au over 31.8 m, remains open in many directions. Dublin Gulch's Eagle zone, currently at 72 t (2.3 million oz) of contained Au, remains open and three other zones of sheeted veins are yet to be fully drill tested.

In southeastern British Columbia, most effort has been directed towards large solitary veins, but potential also exists in targeting bulk tonnage deposits in pluton cupolas. The greatest potential in BC exists with the Bayonne plutonic suites, but plutons tend to be large, so smaller stocks or apophyses should be considered.

Exploration for RIRGS in Archean and Proterozoic terranes has been limited, and perhaps rightfully so considering the lack of appropriate tectonic environments for their formation or preservation during these time. However, the value of Bi-Te signatures as Au pathfinders for intrusion-related systems in these environments appear to have been broadly considered.

Acknowledgements

The author is indebted to R. Goldfarb, J. Mair, D. Groves, T. Baker, J. Stephens, J. Thompson, L. Lewis, M. Burke, J. Lang, A. Bakke, M. Lindsay, R. Hulstein, A. Doherty, G. Abbott, D. Murphy, G. Carlson, E. Marsh, C. Freeman, D. McCoy, and R. Newberry for discussions and insight, and to the numerous exploration geologists who shared their knowledge. Much of the work presented herein is based on these discussions, because citable material is not available. Reviews by R. Goldfarb and S.M. Rowins are appreciated. Comments and the invitation to participate in this volume from W. Goodfellow are also appreciated. The Yukon Geological Survey and The University of Western Australia are thanked for their continuing support.

References

Alexco Resources Corp., 2006, Alexco Announces Results from Brewery Creek: Yukon Diamond Drilling, News Release. http://www.alexcoresource.com/s/NewsReleases.asp.

Allen, T.L., Hart, C.J.R., and Marsh, E.E., 1999, Placer gold and associated heavy minerals of the Clear Creek drainage, central Yukon: Past to present, *in* Roots, C.F., and Emond, D.S., ed., Yukon exploration and geology 1998: Exploration and Geological Services Division, Yukon Region, Indian and Northern Affairs Canada, p. 197–214.

Anderson, R.G., 1983, Selwyn Plutonic Suite and its relationship to tungsten mineralization, southeastern Yukon and District of Mackenzie: Geological Survey of Canada, Current Research, Paper 83-1B, p. 151–163.

—— 1987, Plutonic rocks in the Dawson map area, Yukon Territory: Geological Survey of Canada, Current Research, Paper 87-1A, p. 689–697.

—— 1988, An overview of some Mesozoic and Tertiary plutonic suites and their associated mineralization in the northern Canadian Cordillera: Canadian Institute of Mining and Metallurgy, Special Volume 39, p. 96–113.

Baker, E.M., and Tullemans, F.J., 1990, Kidston gold deposit: Australasian Institute of Mining and Metallurgy, Monograph 14, p. 1461–1465.

Baker, T., 2002, Emplacement depth and carbon dioxide-rich fluid inclusions in intrusion-related gold deposits: Economic Geology, v. 97, p. 1111–1117.

Baker, T., and Lang, J.R., 2001, Fluid inclusion characteristics of intrusion-related gold mineralization, tombstone-Tungsten magmatic belt, Yukon Territory, Canada: Mineralium Deposita, v. 36, p. 563–582.

Bakke, A.A., 1995, The Fort Knox “porphyry” gold deposit: Structurally controlled stockwork and shear quartz vein, sulphide-poor mineralization hosted by Late Cretaceous pluton, east-central Alaska: Canadian Institute of Mining and Metallurgy, Special Volume 46, p. 795–802.

Bakke, A., Morrel, B., Odden, J., Bergstron, T., and Woodman, J., 2000, Kinross Gold USA’s activities in the Fairbanks mining district, K2K: British Columbia and Yukon Chamber of Mines, Special Volume 2, p. 89–98.

Blevin, P.L., 2004, Redox and compositional parameters of interpreting the granitoids metallogeny of eastern Australia: Implications for gold-rich ore systems: Resource Geology, v. 54, p. 241–252.

Brown, I.J., and Nesbitt, B.E., 1987, Gold-copper-bismuth mineralization in hedenbergitic skarn, Tombstone Mountains, Yukon: Canadian Journal of Earth Sciences, v. 24, p. 2362–2372.

Brown, V.S., Baker, T., and Stephens, J.R., 2002, Ray Gulch tungsten skarn, Dublin Gulch, central Yukon: Gold-tungsten relationships in intrusion-related ore systems and implications for gold exploration, *in* Emond, D.S., Weston, L.H., and Lewis, L.L., ed., Yukon exploration and geology 2001: Exploration and Geological Services Division, Yukon Region, Indian and Northern Affairs Canada, p. 259–268.

Burnham, C.W., 1979, Magma and hydrothermal fluids, in Barnes, H. L., ed., Geochemistry of hydrothermal ore deposits, 2nd edition: New York, Wiley, p. 71–136.

Burnham, C.W., and Ohmoto, H., 1980, Late-stage processes of felsic magmatism: Kozan Chishitsu (Mining Geology), v. 8, p. 1–11.

Candela, P.A., and Blevin, L.P., 1995, Do some miarolitic cavities preserve evidence of magmatic volatile phase permeability?: Economic Geology, v. 90, p. 2310–2316.

Candela, P.A., and Piccoli, P.M., 2005, Magmatic processes in the development of porphyry-type ore systems: Economic Geology 100th Anniversary Volume, p. 25–38.

Cole, A., Wilkinson, J.J., Halls, C., and Serenko, T.J., 2000, Geological characteristics, tectonic setting, and preliminary interpretations of the Jilau gold-quartz vein deposit, Tajikistan: Mineralium Deposita, v. 35, p. 600–618.

Diment, R., and Craig, S., 1999, Brewery Creek gold deposit, central Yukon, *in* Roots, C.F., and Emond, D.S., ed., Yukon exploration and geology 1998: Exploration and Geological Services Division, Yukon Region, Indian and Northern Affairs Canada, p. 225–230.

Ebert, S., Miller, L., Petsel, S., Dodd, S., Kowalczyk, P., Tucker, T.L., and Smith, M.T., 2000, Geology, mineralization, and exploration at the Donlin Creek project, southwestern Alaska: British Columbia and Yukon Chamber of Mines, Special Volume 2, p. 99–114.

Ebert, S.W., Baker, T., and Spencer, R.J., 2003, Fluid inclusion studies at the Donlin Creek gold deposit, Alaska: Possible evidence for reduced porphyry-Au to sub-epithermal transition: Society for Geology Applied to Mineral Deposits, Proceedings of the Seventh Biennial Meeting, Athens, August 24–28, p. 263–266.

Einaudi, M.T., Meinert, L.D., and Newberry, R.J., 1981, Skarn Deposits: Economic Geology, 75th Anniversary Volume, p. 317–391.

Farmer, G.L., Mueller, S., Marsh, E., Goldfarb, R.J., and Hart, C.J.R., 2000, Isotopic evidence on sources of Au-related mid-Cretaceous Tombstone Plutonic Suite granitic rocks, Clear Creek district, Yukon [abs.]: Geological Society of America, Cordilleran Section Abstracts with Programs, v. 32, p. A-13.

Flanigan, B., Freeman, C., Newberry, R., McCoy, D., Hart, C., 2000, Exploration models for mid and Late Cretaceous intrusion-related gold deposits in Alaska and the Yukon Territory, Canada, *in* Cluer, J.K., Price, J.G., Struhsacker, E.M., and others, eds., Geology and ore deposits 2000: The Great basin & beyond, Geological Society of Nevada, Symposium Proceedings, p. 591–614.

Geological Survey of Canada, 1978, Regional stream sediment and water geochemical reconnaissance data, central Yukon Territory, NTS 106D and parts of 106C, E and F; 116A and part of 116H; and parts of 116B, C, F and G: Geological Survey of Canada, Open File 518, 519, and 520, 42 maps, scale: 1:250,000.

—— 1990, Regional stream sediment and water geochemical reconnaissance data, Yukon Territory (106D; parts of 106C, 106E, 106F): 42 maps, Geological Survey of Canada, Open File 2175, 42 maps, scale: 1:250,000, 210 p.

—— 1991, Regional stream sediment and water geochemical reconnaissance data, Yukon Territory (116 B, parts of 116 C, 116 F & 116 G): 41 maps, Geological Survey of Canada, Open File 2365, 41 maps, scale: 1:250,000, p. 198.

Goldfarb, R., Hart, C., Miller, M, Miller, L., Farmer, G.L., and Groves, D., 2000, The Tintina Gold Belt: A global perspective: British Columbia and Yukon Chamber of Mines, Special Volume 2, p. 5–34.

Goldfarb, R.J., Ayuso, R., Miller, M.L., Ebert, S.W., Marsh, E.E., Petsel, S.A., Miller, L.D., Bradley, D., Johnson, C., and McClelland, W., 2004, The Late Cretaceous Donlin Creek deposit, southwestern Alaska-controls on epizonal formation: Economic Geology, v. 99, p. 643–671.

Goldfarb, R.J., Bake, T., Dubé, B., Groves, D.I., Hart, C.J.R., and Gosselin, P., 2005, Distribution, character, and genesis of gold deposits in metamorphic terranes: Economic Geology, 100th Anniversary Volume, p. 407–450.

Gordey, S.P., and Anderson, R.G., 1993, Evolution of the northern Cordilleran miogeocline, Nahanni map area (105I), Yukon and Northwest Territories: Geological Survey of Canada, Memoir 248, 214 p.

Groves, D.I., Goldfarb, R.J., Robert, F., and Hart, C.J.R., 2003, Gold deposits in metamorphic belts: Overview of current understanding, outstanding problems, future research, and exploration significance: Economic Geology, v. 98, p. 1–29.

Hart, C.J.R., 2005, Classifying, distinguishing and exploring for intrusion-related gold systems: The Gangue: Newsletter of the Geological Association of Canada Mineral Deposits Division, v. 87, p. 1, 4–9.

Hart, C.J.R., and Goldfarb, R.J., 2005, Distinguishing intrusion-related from orogenic gold systems: Proceedings of the 2005 New Zealand Minerals Conference, Auckland, November 13–16, p. 125–133.

Hart, C.J.R., Baker, T., and Burke, M., 2000a, New exploration concepts for country-rock hosted, intrusion-related gold systems, Tintina Gold Belt:

British Columbia and Yukon Chamber of Mines, Special Volume 2, p. 145–172.

Hart, C.J.R., Baker, T., Lindsay, M.J., Oliver, N.H.S., Stephens, J.R., and Mair, J.L., 2000b, Structural controls on Tombstone Plutonic Suite gold deposits, Tintina Gold Belt, Yukon [abs.]: Geological Society of America Abstracts with Programs, Cordilleran Section, v. 32, p. 6.

Hart, C.J.R., McCoy, D., Goldfarb, R.J., Smith, M., Roberts, P., Hulstein, R., Bakke, A.A., and Bundtzen, T.K., 2002, Geology, exploration and discovery in the Tintina gold province, Alaska and Yukon: Society of Economic Geologists, Special Publication 9, p. 241–274.

Hart, C.J.R., Goldfarb, R.J., Lewis, L.L., and Mair, J.L., 2004a, The Northern Cordillera Mid-Cretaceous Plutonic Province: Ilmenite/magnetite-series granitoids and intrusion-related mineralisation: Resource Geology, v. 54, n. 3, p. 253–280.

Hart, C.J.R., Villeneuve, M.E., Mair, J.L., Goldfarb, R.J., Selby, D., Creaser R.A., and Wijns, C., 2004b, Comparative U-Pb, Re-Os, and Ar-Ar geochronology of mineralizing plutons in Yukon and Alaska [abs.]: The University of Western Australia Centre for Global Exploration, Publication 33, p. 347–349.

Hart, C.J.R., Mair, J.L., Goldfarb, R.J., and Groves, D.I., 2005, Source and redox controls of intrusion-related metallogeny, Tombstone-Tungsten Belt, Yukon, Canada: Transactions of the Royal Society of Edinburgh: Earth Science, v. 95, p. 339–356.

Hitchins, A.C., and Orssich, C.N., 1995, The Eagle zone gold-tungsten sheeted vein porphyry deposit and related mineralization, Dublin Gulch, Yukon Territory: Canadian Institute of Mining and Metallurgy, Special Volume 46, p. 803–810.

Hollister, V.F., 1992, On a proposed plutonic porphyry gold deposit model: Nonrenewable Resources, v. 1, p. 293–302.

Hulstein, R., Zuran, R., Carlson, C.G., and Fields, M., 1999, The Scheelite dome gold project, central Yukon, *in* Roots, C.F., and Emond, D.S., ed., Yukon exploration and geology 1998: Exploration and Geological Services Division, Yukon Region, Indian and Northern Affairs Canada, p. 243–248.

Ishihara, S., 1981, The granitoid series and mineralization: Economic Geology 75th Anniversary Volume, p. 458–484.

Jones, B.K., 1992, Application of metal zoning to gold exploration in porphyry copper systems: Journal of Geochemical Exploration, v. 43, n. 2, p. 127–155.

Lang, J.R., 2001, Regional and system-scale controls on the formation of copper and or gold magmatic-hydrothermal mineralization: University of British Columbia Mineral Deposit Research Unit, Special Publication 2, 115 p.

Lang, J.R., and Baker, T., 2001, Intrusion-related gold systems: the present level of understanding: Mineralium Deposita, v. 36, p. 477–489.

Lang, J.R., Baker, T., Hart, C.J.R., and Mortensen, J.K., 2000, An exploration model for intrusion-related gold systems: Society of Economic Geology Newsletter, v. 40, p. 1, 6–15.

Lennan, W.B., 1983, Ray Gulch tungsten skarn deposit, Dublin Gulch area, central Yukon: Canadian Institute of Mining, Metallurgy and Petroleum, Special Volume 37, p. 245–254.

Leveille, R.A., Newberry, R.J., and Bull, K.F., 1988, An oxidation state-alkalinity diagram for discriminating some gold-favorable plutons; an empirical and phenomenalogical approach [abs.]: Geological Society of America Abstracts with Programs, v. 20, p. 142.

Lindsay, M.J., Baker, T., Oliver, N.H.S., Diment, R., and Hart, C.J.R., 2000, The magmatic and structural setting of the Brewery Creek gold mine, central Yukon, *in* Emond, D., and Weston, L., ed., Yukon exploration and geology 1999: Exploration and Geological Services Division, Yukon Region, Indian and Northern Affairs Canada, p. 219–227.

Logan, J., Lefebure, D., and Cathro, M., 2000, Plutonic related gold-quartz veins and their potential in British Columbia: British Columbia and Yukon Chamber of Mines, Special Volume 2, p. 197–225.

Lynch, J.V.G., 1989, Large scale hydrothermal zoning reflecting in tetrahedrite-freibergite solid solution, Keno Hill Ag-P-Zn district, Yukon: Canadian Mineralogist, v. 27, p. 383–400.

Mair, J.L., 2004, Tectonic setting, magmatism and magmatic-hydrothermal systems at Scheelite Dome, Tombstone Gold Belt, Yukon: Critical constraints on intrusion-related gold systems: Unpublished Ph.D. thesis, Perth, University of Western Australia, 197 p.

Mair, J.L., Hart, C.J.R., Goldfarb, R.J., O'Dea, M., and Harris, S., 2000, Geology and metallogenic signature of gold occurrences at Scheelite Dome, Tombstone gold belt, Yukon, in Emond, D., and Weston, L., ed., Yukon exploration and geology 1999: Exploration and Geological Services Division, Yukon Region, Indian and Northern Affairs Canada, p. 165–176.

Mair, J.L., Goldfarb, R.J., Johnson, C.A., Hart, C.J.R., and Marsh, E.E., 2006a, Geochemical constraints on the genesis of the Scheelite Dome intrusion-related gold deposit, Tombstone Gold Belt, Yukon, Canada: Economic Geology, v. 101, p. 523–553.

Mair, J.L., Hart, C.J.R., and Stephens, J., 2006b, Deformation history of the northwestern Selwyn Basin, Yukon, Canada: Implications for orogen evolution and mid-Cretaceous magmatism: Geological Society of America Bulletin, v. 118, p. 304–323.

Maloof, T.L., Baker, T., and Thompson, J.F.H., 2001, The Dublin Gulch intrusion-hosted deposit, Tombstone Plutonic Suite, Yukon Territory, Canada: Mineralium Deposita, v. 36, p. 583–593.

Marsh, E.E., Hart, C.J.R., Goldfarb, R.J., and Allen, T.L., 1999, Geology and geochemistry of the Clear Creek gold occurrences, Tombstone gold belt, central Yukon Territory,*in* Roots, C.F., and Emond, D.S., ed., Yukon exploration and geology 1998: Exploration and Geological Services Division, Yukon Region, Indian and Northern Affairs Canada, p. 185–196.

Marsh, E.E., Goldfarb, R.J., Hart, C.J.R., and Johnson, C.A., 2003, Geology and geochemistry of the Clear Creek intrusion-related gold occurrences, Tintina gold province, Yukon, Canada: Canadian Journal of Earth Sciences, v. 40, n. 5, p. 681–699.

McCoy, D., Newberry, R.J., Layer, P.W., DiMarchi, J.J., Bakke, A.A., Masterman, J.S., and Minehane, D.L., 1997, Plutonic-related gold deposits of interior Alaska: Economic Geology, Monograph 9, p. 191–241.

McCoy, D., Newberry, R.J., Severin, K., Marion, P., Flanigan, B., and Freeman, C., 2002, Paragenesis and metal associations in interior Alaska gold deposits: An example from the Fairbanks district: Mining Engineering, v. 54, 1, p. 33–38.

McCuaig, T.C., Behn, M., Stein, H.J., Hagemann, S.G., McNaughton, N.J., Cassidy, K.F., Champion, D.C., and Wyborn, L., 2001, The Boddington gold mine: A new style of Archaean Au-Cu deposit [abs.]: Australian Geological Survey Organisation, 4th International Archaean Symposium, 2001/37, p. 453–455.

McLeod, M.J., and McCutcheon, S.R., 2000, Gold environments in New Brunswick: Map plate 2000-8: New Brunswick Dept. of Natural Resources and Energy, Minerals and Energy Division: scale: 1:500,000.

Meinert, L.D., 1998, A review of skarns that contain gold, *in* Lentz, D.R., ed., Mineralized porphyry/skarn systems: Mineralogical Association of Canada, Short Course Series, v. 26, p. 359–414.

Mortensen, J.K., Hart, C.J.R., Murphy, D.C., Heffernan, S., Tucker, T.L., and Smith, M.T., 2000, Temporal evolution of Early and Mid-Cretaceous magmatism in the Tintina gold belt: British Columbia and Yukon Chamber of Mines, Special Volume 2, p. 49–58.

Mungall, J.E., 2002, Roasting the mantle: Slab melting and the genesis of major Au and Au-rich Cu deposits: Geology, v. 30, p. 915–918.

Murphy, D.C., 1997, Geology of the McQuesten River region, northern McQuesten and Mayo map areas, Yukon Territory: Exploration and Geological Services Division, Yukon Region, Indian and Northern Affairs Canada, Bulletin 6, 95 p.

Mustard, R., 2001, Granite-hosted gold mineralization at Timbarra, northern New South Wales: Mineralium Deposita, v. 36, p. 542–562.

Mustard, R., Ulrich, T., Kamenetsky, V.S., and Mernagh, T., 2006, Gold and metal enrichment in natural granitic melts during fractional crystallization: Geology, v. 34, p. 85–88.

Newberry, R.J., McCoy, D.T., and Brew, D.A., 1995, Plutonic-hosted gold ores in Alaska: Igneous versus metamorphic origins: Resource Geology, Special Issue 18, p. 57–100.

O'Dea, M., Carlson, G., Harris, S., Fields, M., Tucker, T.L., and Smith, M.T., 2000, Structural and metallogenic framework for the Scheelite Dome Deposit, Yukon Territory: British Columbia and Yukon Chamber of Mines, Special Volume 2, p. 115–129.

Poulsen, K.H., 1996, Carlin-type gold deposits and their potential occurrence in the Canadian Cordillera, Geological Survey of Canada, Current Research, Report 1996-A, p. 1–9.

Poulsen, K.H., Mortensen, J.K., and Murphy, D.C., 1997, Styles of intrusion-related gold mineralization in the Dawson-Mayo area, Yukon Territory: Geological Survey of Canada, Current Research, Report: 1997-A/B, p. 1–10.

Rhys, D., DiMarchi, J., Smith, M., Friesen, R., and Rombach, C., 2003, Structural setting, style and timing of vein-hosted gold mineralization at the Pogo Deposit, east central Alaska: Mineralium Deposita, v. 38, p. 863–875.

Robert, F., 2001, Syenite-associated disseminated gold deposits in the Abitibi greenstone belt, Canada: Mineralium Deposita, v. 36, p. 503–516.

Rowins, S.M., 2000, Reduced porphyry copper-gold deposits: A new variation on an old theme: Geology, v. 28, p. 491–494.

Selby, D., Creaser, R.A., and Hart, C.J.R., 2001, Timing relationship between plutonism and gold mineralization: Re-Os molybdenite study of the reduced intrusion-related gold deposits of the Tombstone Plutonic Suite, Yukon and Alaska [abs.]: Joint Annual Geological Association of canada—Mineralogical Association of Canada, Abstracts and Program, St. Johns, Newfoundland, May 27-30, p. 134.

Sillitoe, R.H., 1991, Intrusion-related gold deposits, *in* Foster, R.P., ed., Gold metallogeny and exploration: Glasgow, Blackie, p. 165–209.

—— 1995, Gold-rich porphyry copper deposits: Geological model and exploration implications: Geological Association of Canada, Special Paper 40, p. 465– 478.

Sillitoe, R.H., and Thompson, J.F.H., 1998, Intrusion-related vein gold deposits: Types, tectono-magmatic settings, and difficulties of distinction from orogenic gold deposits: Resource Geology, v. 48, p. 237–250.

Sinclair, W.D., 2007, Porphyry deposits, *in* Goodfellow, W.D., ed., Mineral deposits of Canada: A synthesis of major deposit types, district metallogeny, the evolution of geological provinces, and exploration methods: Geological Association of Canada, Mineral Deposits Division, Special Publication 5, p. 223-243.

Smit, H., Sieb, M., and Swanson, C., 1996, The Dublin Gulch intrusive-hosted gold deposit: British Columbia Geological Survey, Cordilleran Round Up 1996, Short Course Notes, p. F.3–F.13.

Smith, M., Thompson, J., Bressler, J., Layer, P., Mortensen, J., Abe, I., and Takaoka, H., 1999, Geology of the Liese zone, Pogo property, East-Central Alaska: Society of Economic Geology Newsletter, v. 38, p. 1, 12–21.

Stephens, J.R., Oliver, N.H.S., Baker, T., and Hart, C.J.R., 2000, Structural evolution and controls on gold mineralization at Clear Creek, Yukon, *in* Emond, D., and Weston, L., ed., Yukon exploration and geology 1999: Exploration and Geological Services Division, Yukon Region, Indian and Northern Affairs Canada, p. 151–163.

Stephens, J.R., Mair, J.L., Oliver, N.H.S., Hart, C.J.R., Baker, T., Blenkinsop, T.G., Vearncombe, J.R., and Reddy, S.M., 2004, Structural and mechanical controls on intrusion-related deposits of the Tombstone gold belt, Yukon, Canada, with comparisons to other vein-hosted ore-deposit types: Journal of Structural Geology, v. 26, p. 1025–1041.

Thompson, J.F.H., and Newberry, R.J., 2000, Gold deposits related to reduced granitic intrusions: Society of Economic Geologists, Reviews 13, p. 377–400.

Thompson, J.F.H., Sillitoe, R.H., Baker, T., Lang, J.R., and Mortensen, J.K., 1999, Intrusion-related gold deposits associated with tungsten-tin provinces: Mineralium Deposita, v. 34, p. 323–334.

Tucker, T.L., and Smith, M.T., 2000, The Tintina Gold Belt: Concepts, exploration, and discoveries: British Columbia and Yukon Chamber of Mines, Special Volume 2, p. 225

Wall, V.J., Graupner, T., Yantsen, V., Seltmann, R., and Hall, G.C., 2004, Muruntau, Uzbekistan: A giant thermal aureole gold (TAG) system [abs.]: Perth, University of Western Australia, Centre for Global Metallogeny, Extended Abstracts, v. 33, p. 199–203.

Witt, W.K., 2001, Tower Hill gold deposits, Western Australia: An atypical, multiply deformed Archaean gold-quartz deposit: Australian Journal of Earth Sciences, v. 48, p. 81–99.

EPITHERMAL GOLD DEPOSITS

BRUCE E. TAYLOR

Geological Survey of Canada, 601 Booth Street, Ottawa, Ontario K1A 0E8
Corresponding author's email: btaylor@nrcan.gc.ca

Abstract

Epithermal Au (±Ag) deposits form in the near-surface environment, from hydrothermal systems typically within 1.5 km of the Earth's surface. They are commonly found associated with centres of magmatism and volcanism, but form also in shallow marine settings. Hot-spring deposits and both liquid- and vapour-dominated geothermal systems are commonly associated with epithermal deposits. Epithermal Au deposits are commonly consider to comprise one of three subtypes: high sulphidation, intermediate sulphidation, and low sulphidation, each denoted by characteristic alteration mineral assemblages, occurrences, textures, and, in some cases, characteristic suites of associated geochemical elements (e.g. Hg, Sb, As, and Tl). Base metal (Cu, Pb, and Zn) and sulphide minerals may also occur in addition to pyrite and native Au or electrum. In some epithermal deposits, notably those of the intermediate-sulphidation subtype, base metal sulphides may comprise a significant ore constituent.

Canadian Au production from epithermal deposits has been minor (<5%), compared to that from transitional and intrusion-related Au deposits, or to other lode Au deposits. The shallow origin of epithermal Au deposits makes them more susceptible to erosion, and, accordingly, epithermal Au deposits have represented a high-grade, readily mineable, exploration target largely in Tertiary and younger volcanic centres, including the Cordillera. However, a number of older epithermal Au deposits have also been discovered, including several Proterozoic examples in Canada. Thus, older terranes need not be excluded entirely from exploration.

Modern geothermal and volcanic systems provide natural laboratories for the study of epithermal deposits, guiding theoretical models and laboratory experiments, and expanding our understanding of potential environments and vectors to mineralized systems. Yet, a principal, unanswered question still remains: do rich Au deposits form from Au-rich sources, or from exceptionally efficient mechanisms or processes of Au precipitation?

Résumé

Les gîtes d'or (+/- argent) épithermaux se forment dans le milieu subsuperficiel, de manière caractéristique à moins de 1,5 km de la surface de la Terre. Ils sont couramment présents dans des centres magmatiques ou volcaniques à la surface du globe, mais se forment également dans des cadres marins de faible profondeur. Des dépôts de sources thermales ainsi que des systèmes géothermiques tant à dominante liquide qu'à dominante vapeur sont couramment associés aux gîtes épithermaux. Deux sous-types de gîtes sont définis d'après les minéraux d'altération communs qu'ils renferment: les gîtes à quartz-(kaolinite)-alunite (QAL) formés à partir de fluides acides avec un important apport magmatique et les gîtes à adulaire-séricite (ADS) formés à partir de fluides quasi neutres en grande partie composés d'eau météorique. Des métaux communs peuvent être présents dans l'un ou l'autre sous-type.

La production canadienne d'or tirée de gîtes épithermaux est mineure (< 5%), mais ailleurs dans le monde, cette classe de gîtes à teneur élevée et facilement exploitables constitue une cible d'exploration. Les gîtes épithermaux profonds ou associés à des intrusions ont fourni une plus importante contribution à la production canadienne d'or. La faible profondeur à laquelle se forment les gîtes d'or épithermaux fait en sorte qu'ils sont davantage susceptibles d'érosion et ils ont par conséquent tendance à être découverts dans des terrains plus récents. De toute évidence, un enfouissement rapide a préservé un certain nombre de gîtes d'or épithermaux dans des terrains plus anciens, dont on trouve des exemples dans des roches du Protérozoïque. Ces occurrences permettent d'afficher un certain optimisme quant à l'exploration de terrains plus anciens au Canada.

Les systèmes géothermiques contemporains constituent des laboratoires naturels d'étude des gîtes épithermaux ainsi que des guides pour l'élaboration de modèles théoriques et d'expériences à mener en laboratoire. De telles études permettent continuellement d'améliorer la compréhension et fournissent des guides pour l'exploration; chacune contribuant à répondre à l'importante question qui reste sans réponse : les riches gîtes se forment-ils à partir de sources riches ou sont-ils attribuables à des facteurs déterminés par un processus de précipitation ou de concentration?

Definition

Simplified Definition

Epithermal deposits of Au (± Ag) comprise veins and disseminations near the Earth's surface (≤1.5 km), in volcanic and sedimentary rocks, sediments, and, in some cases, also in metamorphic rocks. The deposits may be found in association with hot springs and frequently occur at centres of young volcanism. The ores are dominated primarily by precious metals (Au, Ag), but some deposits may also contain variable amounts base metals such as Cu, Pb, and Zn.

Scientific Definition

Epithermal Au deposits are a type of lode deposit (e.g. Poulsen, 1996; Poulsen et al., 2000) consisting of economic concentrations of Au (± Ag and base metals). These deposits form in a variety of host rocks from hydrothermal fluids, primarily by replacement (i.e. by solution and reprecipitation), or by open-space filling (e.g. veins, breccias, pore spaces). The form of deposits originating by open-space filling typically reflects that of the structural control of the hydrothermal fluids (planar vs. irregular fractures, etc). The deposits are commonly young, generally Tertiary or Quaternary. They may be of similar age as their host rocks when these are volcanic in origin, or (typically) younger than their host.

Early workers (e.g. Lindgren, 1922, 1933; Emmons, 1924) emphasized a broad depth-zoning classification of hydrothermal metal deposits in which epithermal deposits were interpreted to have formed in a 'shallow' regime, qual-

Taylor, B.E., 2007, Epithermal gold deposits, *in* Goodfellow, W.D., ed., Mineral Deposits of Canada: A Synthesis of Major Deposit-Types, District Metallogeny, the Evolution of Geological Provinces, and Exploration Methods: Geological Association of Canada, Mineral Deposits Division, Special Publication No. 5, p. 113-139.

itatively on the order of ≤1500 m. Depth is, however, challenging to quantify, except where a datum (e.g. the Earth's surface) may be recognized, or a stratigraphic reconstruction made. Depth may be inferred, for example, from a depth-to-boiling curve, plus evidence of boiling in fluid inclusions and an estimated formation temperature based on a geothermometer (e.g. trace element or stable isotope distributions), or, simply, from textures indicative of boiling-induced supersaturation (e.g. quartz pseudomorphs of bladed calcite). In any case, the generally accepted shallow origin of epithermal Au deposits is a central (and genetically important) characteristic. This environment is marked by rapid changes in temperature and pressure of the hydrothermal fluids that may be accompanied by boiling and mixing with other fluids, causing changes in pH and oxidation state and, consequently, precipitation of Au (and other metals). Epithermal deposits may also be characterized by the presence of other volatile accessory elements (Hg, Sb, Tl, etc.).

Many geologists today regard 'epithermal' Au deposits to be primarily associated with continental volcanism or magmatism, although similar processes of ore deposition occur in other, near-surface environments (e.g. seafloor volcanogenic massive sulphide (VMS) deposits, submarine volcanic arc systems, and non-volcanic vein deposits; see Gosselin and Dubé, 2005a-d; Dubé et al., 2007). These other deposits, especially those with similar characteristics from marine environments, may also be considered epithermal deposits in a broad sense. Many extensive reviews of epithermal deposits, among them Buchanan (1981), Hayba et al. (1985), Berger and Bethke (1986), Heald et al. (1987), White and Hedenquist (1990), Arribas (1995), Richards (1995), Hedenquist et al. (1996, 2000), and Cooke and Simmons (2000) have provided a wealth of background information. This chapter will largely focus on the continental environment, emphasizing young deposits in the Canadian Cordillera, as well as examples of epithermal Au deposits found in older terranes elsewhere in Canada.

Epithermal Au deposits are distinguished on the basis of the sulphidation state of the sulphide mineralogy as belonging to one of three subtypes: (1) high sulphidation (previously called quartz-(kaolinite)-alunite, alunite-kaolinite, enargite-Au, or high sulfur: Ashley, 1982; Hedenquist, 1987; Bonham, 1988); (2) intermediate sulphidation (Hedenquist et al., 2000); or (3) low sulphidation (previously called adularia-sericite). High-sulphidation subtype deposits usually occur close to magmatic sources of heat and volatiles, and form from acidic hydrothermal fluids containing magmatic S, C, and Cl. Low-sulphidation subtype fluids are thought to be near-neutral, dominated by meteoric waters, but containing some magmatic C and S. In addition, some geologists also refer to 'hot-spring' deposits as an additional subtype of epithermal deposit that may form as surface expressions of hydrothermal systems, typically of the low-sulphidation subtype sometimes associated with acidic, steam-heated alteration zones. See Henley et al. (1984), Hayba et al. (1985), Heald et al. (1987), Hedenquist (1987), Bonham (1988), Berger and Henley (1989), and Panteleyev (1996a-c) for discussions and original definition of these terms. Hedenquist et al. (2000) is recommended for a more recent and very comprehensive summary of current usage, classification, and deposit characteristics.

The Blackdome and most of the deposits in the Toodoggone River camp in British Columbia, and the Mt. Skukum deposit, Yukon, are among the best Canadian examples of volcanic-hosted, low-sulphidation subtype epithermal Au deposits (Table 1). The Cinola deposit, British Columbia (Champigny and Sinclair, 1982), hosted by sedimentary rocks, is an example of a low-sulphidation hot-spring deposit (in particular, its upper part). High-sulphidation deposits are less well represented in Canada, but include the volcanic-hosted Al deposit, Toodoggone River camp, British Columbia and the metamorphosed Hope Brook deposit, Nova Scotia (Table 1). Numerous areas of advanced argillic alteration mineral assemblages and associated Au prospects formed by high-sulphidation systems of Neoproterozoic age are also known in the Burin Peninsula, Newfoundland (e.g. Hickey's Pond: O'Brien et al., 1999). The locations of these and other Canadian epithermal deposits or districts, and selected deposits elsewhere in the world, are shown in Figures 1 and 2, respectively, and include, especially, those deposits or districts noted in the text.

Numerous examples of both low-sulphidation and high-sulphidation deposits in volcanic and sedimentary host rocks exist world wide, especially in younger volcanic terranes. Classic examples of the high-sulphidation subtype include Summitville, Colorado (Bethke et al., 2005) and Nansatsu, Japan (Hedenquist et al., 1994). The Creede district, Colorado (e.g. Heald et al., 1987) and Hishikari, Japan (Izawa et al., 1990; Hayashi et al., 2001, and references therein) are good examples of volcanic-hosted low-sulphidation subtype deposits. Others are noted below.

Diagnostic Features of Epithermal Gold Deposits

Geological, mineralogical, and geochemical features of epithermal Au deposits are listed for each of three deposit subtypes in Table 2. Distinctive features typically present include key alteration mineral assemblages (low sulphidation: sericite, adularia, kaolinite, calcite, rhodochrosite, Fe-chlorite, quartz; high sulphidation: alunite, kaolinite, pyrophyllite, sericite, adularia (illite), chlorite, barite; Table 2), ore mineral assemblages (low sulphidation: electrum, Hg-Sb-As sulphides, base metal sulphides; high sulphidation: native Au, electrum, tellurides, base metal sulphides; Table 2), geological evidence for shallow emplacement: sinter deposits, fluid inclusion or textural evidence (e.g. lamellar calcite, or their quartz pseudomorphs) for boiling, hydrothermal breccias and eruption deposits, open-space crustiform veins, and marked ^{18}O depletion of wall rocks. Vertical zoning of alteration minerals, lower Au:Ag ratios in electrum with depth, and spatial and temporal separation of Au and abundant base metals are also characteristic (though not universal) of epithermal Au deposits.

Although most known epithermal Au deposits are Tertiary in age, the mineralogical and geological characteristics noted above have led to the recognition of much older epithermal deposits, including recrystallized and deformed examples in metamorphic terranes. Although high-sulphidation-related alteration is distinctive, corroboration of an epithermal setting by low $^{18}O/^{16}O$ ratios can, particularly for the low-sulphidation subtype, provide a unique record of alteration by meteoric waters, one that survives metamorphism and requires a shallow origin. For example, the Carolina Slate

TABLE 1. Comparative mineralogical, geological, and production data for selected epithermal Au deposits in Canada and several non-Canadian type examples.*

District and/or deposit	Age[1] Host [Min.]	Size[2] Ore[5]	(R+P) Au[6]	Grade[3]	Ag/Au	Base Metal	%S	Ad	Al	Cpy	En	Ss	Ags	Sp	Gn	Ba	Fl	Rc	Cc	Ank	Host rock	Alt'n.[7] vn → w.r.	Form[8]	Selected Refs.
								Mineralogy[4]									Carbonates							
HIGH-SULPHIDATION TYPE:[9] *Volcanic host rocks*																								
Toodoggone River, B.C.	189-198;182																							
Al (Bonanza; Thesis)	[196]	0.348	3.21	9.6					X	X		X	X	X	X	X					and./dacite	Si/A	vn	10,11,29,30,34
BV	[190-197]	0.053	0.55	10.4		x										X					andesite		vn,bx	29
Equity Silver, B.C.[14]	57.2 [>48;57.2]	31.42	24.41	4.2	128.2	X				X		X	X	X							dacite/tuff/congl.	A	vn,st,diss.	17,18,19
Summitville, Colorado	20.2-22.0 [22.3]	83.51	3.5		1.2	X	5?		XX	XX	X	XX	X	X	X	X					qtz. Latite	Vgy-Si/Qtz-Al/A	repl., vn	28,31
Nansatsu, Japan	3.4-7.6 [2.7-5.5]		>18	3-6	0.1-1.0	x	≤10		X		X	x	X	X	X	x			X		andesite	Vgy-Si/Al-A/Ph/P	repl.	32
El Indio, Chile[19]	13.7 [8.6]	8.7	108	1.7-218	0.5-10	XX	≤30[11]		X	[X]	[XX]	X	x	[x]	[x]	X					rhyodacite	Si/Ph-A (Al-A)	vn,bx	35,36
INTERMEDIATE-SULPHIDATION TYPE *(possible example; variant of low-sulphidation subtype)*																								
Stewart-Iskut region, B.C.	210																		X					
Silbak-Premier	[194.8?]	9.622	66.24	7.0	22.6	XX	≤5[12]	X[13]		X		X	X	X	X	X		X	X	X	and./dacite	Si/K/Ph/P	vn,st,bx	21
LOW-SULPHIDATION TYPE: *Volcanic and plutonic host rocks*																								
Mt. Skukum, Y.T.	53.2 [50.7]	0.200	2.49	25.0	0.9		<1	X	x							x	X	x	XX	x	andesite	Si/±K/Ph/A/P	vn,bx,st	7,8,9
Mt. Nansen, Y.T.	Tertiary	0.288	3.15	11.1	39.0					X				X	X									14
Laforma, Y.T.	>140 [78?]	0.191	2.13	11.2		X				X		x		X	x	X			X	X	granodiorite	Ph	vn	33
Venus, Y.T.	L. Jur.	>0.07	>0.66	9.3	26.5	XX	15-60						x		X	X					and./dacite	Si/A	vn	5,6,16
Toodoggone River, B.C.	189-198;182																							
Lawyers	[180]	0.880	6.73	7.4	46.7	X		X	X	X		X	X	X	X	X					andesite	Si/A/P	st,bx	1,2,3,24,30,34
Baker (Chapelle)		0.055	1.05	19.5	9.1		3-15			X		X	X	X	X				X		and./basalt	Si/Ph/A/P	vn	13
Blackdome, B.C.	Eocene [>24,<51.5]	0.368	7.35	20.6	3.1	x	≤5	X		X			x	x	x				X		and./dacite	Si/K/A/P	vn,bx	4,23,26,34
Stewart-Iskut region, B.C.	210																		X					
Sulphurets (Snowfield)	[192.7]	25	0.78	2.4	0.6	X				X		X		X	X	X	X		X		bslt.-and./andesite	K/Ph/A/P	st,vn,diss.	40
Creede, Colorado	Tertiary	1.4	21.0	1.5	400	X		X	x			x	X	X	X	X	X	X						22,28
LOW-SULPHIDATION TYPE: *Sedimentary and/or mixed host rocks*																								
Cinola, B.C.	Tert./Cret. [14]	23.80	58.31	2.45	2	x	≤10		x	x				x	x				X		congl./s.s./shale	Si/A	diss.,bx,vn	25,27,34
Dusty Mac, B.C.	Eocene	0.093	0.60	7.2	21.5	X	≤15			x			x	x	x		X		X		s.s./sh./and. pyroclastic	Si/A	bx,st	34,39
Hishikari, Japan	0.51-1.78/Cret. [0.8-1.0]		121.7	70	1.27	x		XX		X		x	x	X	x				X	x	shale-s.s./and./dacite	Si/A	vn	37,38

* Principal deposits plus several others selected to represent part of the spectrum of variation in type and setting; modified from Taylor (1996).

1. Based on reported mineral ages; Ma., exclusive of uncertainty limits, Host=age of host rocks; [Min.] = age of mineralization. 2. P = cumulative production; R = reserves; 3. Average grade in g/t; 4. characteristic; in addition to quartz (+pyrite±sericite±clays); 5. tonnes of ore x10^6; 6. grams of gold (Au) x10^6; 7. Alteration facies, vein (vn) to wall rock (w.r.): Vgy-Si; vuggy silica; Qtz, quartz; Al, alunite (advanced argillic); Si, silicification; K, potassic; Ph, phyllic (sercitic); A, argillic/advanced argillic; P, propylitic (sequence from vein to wall rock); 8. Form of deposit (in order of importance) vn, vein; bx, breccia; st, stockwork; diss., disseminated, repl., replacement; 9. Classification is based on available data; uncertain in some cases; 10. main gold deposition probably not from alunite-kaolinite type system, *sensu stricto*, see text; 11. older, alunite-associated (Cu) veins contain 30-90% sulphide; 12. base metal-rich veins and breccias contain 20-45% suphide; 13. potassium feldspar; species not confirmed; 14. contains metamorphosed advanced argillic mineral as semblage; low-pH conditions approached those of magmatic-hydrothermal QAL subtype deposits; 15. in oxidized ore.

Abbreviations: %S*, per cent sulphide; Ad, adularia; Al, alunite; Cpy, chalcopyrite; En, enargite; Ss, sulphosalts (e.g., tennantite-tetrahedrite); Ags, silver sulphides; Sp, sphalerite; Gn, galena; Ba, barite; Fl, fluorite; CO_3^*, carbonate; Rc rhodochrosite; Cc, calcite; Ank, ankerite; XX = abundant; X = present; x = minor to rare; blank = absent or unknown; and. = andesite; bslt. = basalt; congl. = conglomerate; s.s. = sandstone; lms. = limestone; sh. = shale; Tert. = Tertiary; Cret. = Cretaceous; L. Jur. = Lower Jurassic; *NB: [ ] = not in paragenetic association with Au*.

References: 1) Schroeter, 1986; 2) Schroeter, 1985; 3) Schroeter, 1982; 4) D. Rennie, 1986, written comm.; 5) Walton, 1986; 6) Walton and Nesbitt, 1986; 7) McDonald et al., 1986; 8) McDonald and Godwin, 1986; 9) Pride and Clark, 1985; 10) Clark and Williams-Jones, 1986; 11) Schroeter, 1986; 12) Andrew et al., 1986; 13) Barr et al., 1986; 14) Morin and Downing, 1984; 15) Duke and Godwin, 1986; 16) McFall, 1981; 17) Shen and Sinclair, 1982; 18) Cyr et al., 1984; 19) Wojdak and Sinclair, 1984; 20) Love, 1989; 21) McDonald, 1990; 22) Mosier et al., 1985; 23) Faulkner, 1986; 24) Vuli miri et al., 1986; 25) Champigny and Sinclair, 1982; 26) Vivian et al., 1987; 27) Christie, 1989; 28) Heald et al., 1987; 29) Diakow et al., 1993; 30) Clark and Williams-Jones, 1991; 31) Stoffregen, 1987; 32) Hedenquist et al., 1994; 33) McInnes et al., 1990; 34) B.C. Geol. Survey (MINFILE/pc), 1992; 35) Jannas et al., 1990; 36) Siddeley and Araneda, 1986; 37) Izawa et al., 1990; 38) Bakken and Einaudi, 1986; 39) Church, 1973; 40) Margolis, 1993.

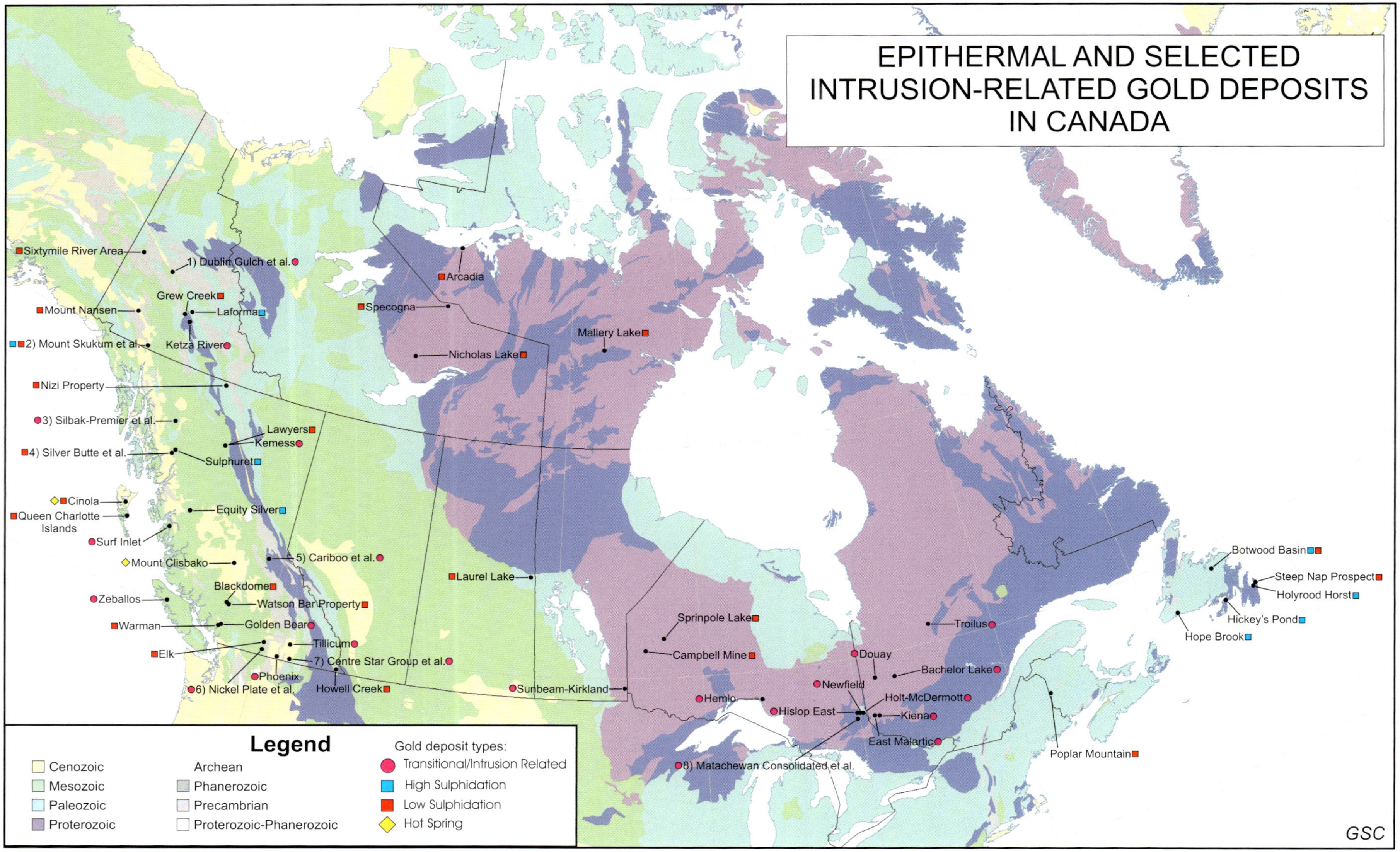

All Mines listed are largest deposits for area; in cases where more than one deposit is located, "et al." has been indicated as follows:
1) Dublin Gulch et al., includes: Dublin Gulch, Eagle Zone, Brewery Creek; **2)** Mount Skukum et al., includes: Mount Skukum, Skukum Creek, Mount Reid, Berney; **3)** Silbak-Premier et al., includes: Silbak-Premier, Spectrum,Banks, Banker, Tel, Yellow Giant, Johnny Mountain, Stonehouse, Snip, Twin Zone, Scottie, Salmon Gold, Premier, Bush, Silbak, Premier Gold; **4)** Silver Butte et al., includes: Silver Butte, SIB, Goldwedge, ; **5)** Cariboo et al., includes: Cariboo, Aurum, QR, Dome, Quesnel River; **6)** Nickel Plate et al., includes: Nickel Plate, Hedley; **7)** Centre Star Group et al., includes: Centre Star Group, Josie, Le Roi No. 2; **8)** Matachewan Consolidated et al., includes: Matachewan Consolidated, Young-Davidson, Ryan Lake.

References: Brown and Cameron, 1999; Dubé et al., 1998; Gosselin and Dubé, 2005b,d; Panteleyev, 1996a,b,c, 2005a,b; Poulsen, 1996; Poulson et al., 2000; Taylor, 1996, this paper; Turner et al., 2003.

FIGURE 1. Location of selected Canadian epithermal Au deposits and prominent examples elsewhere in the world, classified by subtype as referred to in the text. Names and locations of deposits from sources are listed.

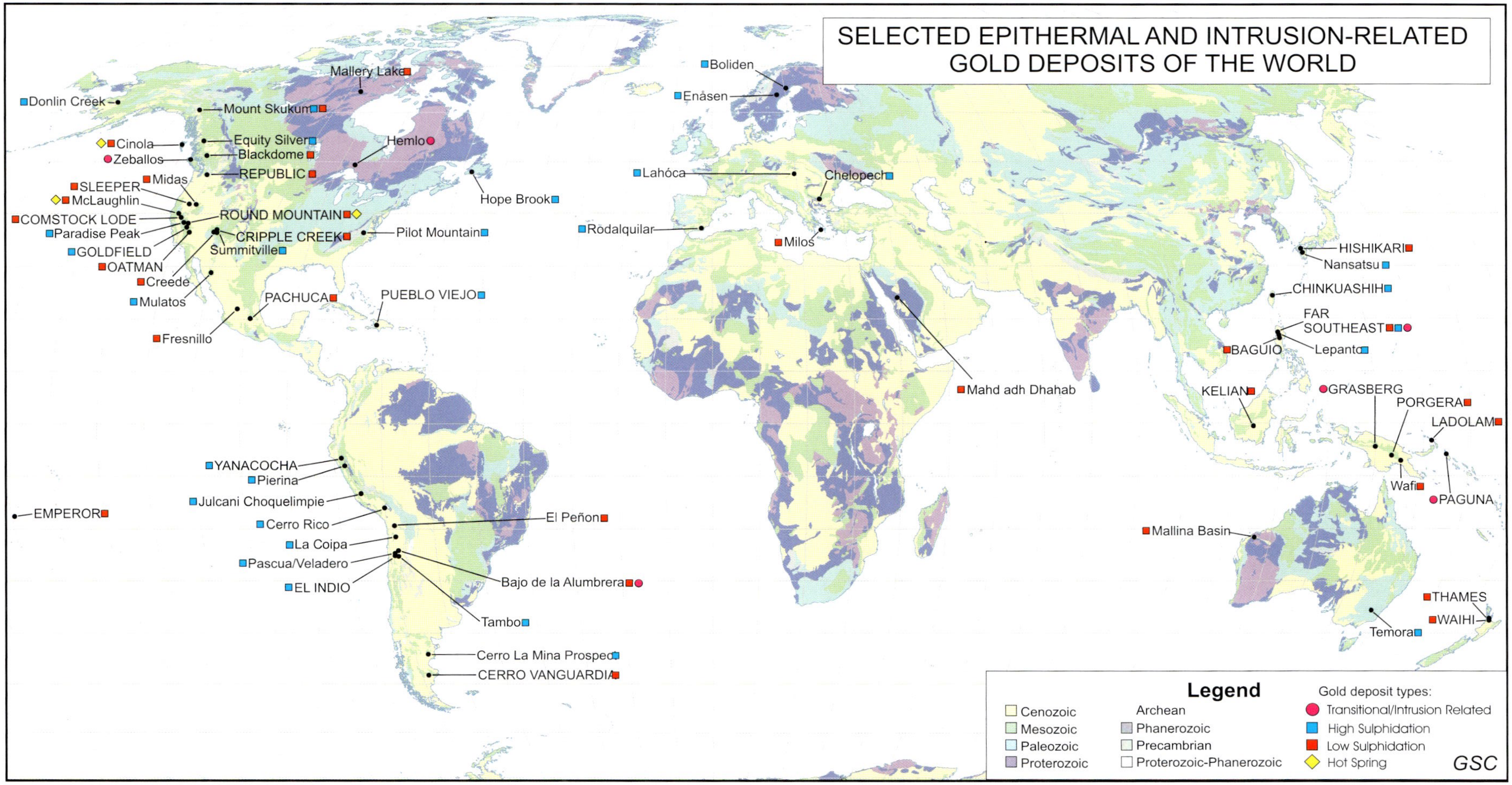

References:
Arancibia et al., 2006; Bethke et al., 2005; Carman, 2003; Deyell et al., 2005; Dubé et al., 1998; Fifarek and Rye, 2005; Goldfarb et al., 2004; Gosselin and Dubé, 2005a,c; Hedenquist et al., 2000; Huston et al., 2002; Klein and Criss, 1988; Naden et al., 2005; Panteleyev, 1996a,b,c, 2005a,b; Poulsen, 1996, 2000; Sillitoe, 1992, 1997; Taylor, 1996, this paper; Turner et al., 2003.

N.B.: Giant and Bonanza Gold deposits indicated by capitalization of deposit name, e.g., EL INDIO.

FIGURE 2. Global distribution of selected Canadian and non-Canadian epithermal and intrusion-related Au deposits of the world. The association of many (young) deposits with the circum-Pacific Belt emphasizes their genetic link to magmatic centres. Giant or Bonanza deposits (Sillitoe, 1992) are labeled in uppercase font. Deposits shown include those noted in the text or represented in accompanying figures. Names and locations of deposits from sources are listed.

TABLE 2. Summary of geological setting, definitive characteristics[1] and several examples of typical epithermal Au deposit subtypes.

	HIGH-SULPHIDATION subtype	LOW-SULPHIDATION subtype	
	Hosted in volcanic rocks	Hosted in volcanic and plutonic rocks	Hosted in sedimentary and mixed host rocks
Geological setting	volcanic terrane, often in caldera-filling volcaniclastic rocks; hot spring deposits and acid lakes may be associated	Spatially related to instrusive centre; veins in major faults, locally ring fracture type faults; hot springs may be present	In calcareous to clastic sedimentary rocks; may be at depth by magma; can form at variety of depths
Ore mineralogy	native gold, electrum, tellurides; magmatic-hydrothermal: (+bn), en, tennantite, cv, sp, gn; Cu typically > Zn, Pb; Au-stage may be distinct, base-metal poor; steam-heated: base-metal poor; gangue: quartz (vuggy silica), barite	electrum (lower Au/Ag with depth), gold; sulphides include: sp, gn, cpy, ss); sulphosalts; gangue: quartz, adularia, calcite, chlorite; ± barite, anhydrite in deeper deposits variable metal content, high sulphide veins closer to intrusions	gold (micrometre): within or on sulphides (e.g. pyrite unoxidized ore), native (in oxidized ore), electrum, Hg-Sb-sulphides, pyrite, minor base metals; gangue: quartz, calcite
Alteration mineralogy	advanced argillic + alunite, kaolintie, pyrophyllite (deeper); ± sericite (illite); adularia, carbonate absent; chlorite and Mn-minerals rare; no selenides; barite with Au; steam-heated: vertical zoning	sericitic replaces argillic facies (adularia ± sericite ± kaolinite); Fe-chlorite, Mn-minerals, selenides present; carbonate and/or rhodochrosite) may be abundant, lamellar if boiling occurred; quartz-kaolinite-alunite-subtype minerals possible steam-heated zone; clays	silicification, decalcification, sericitization, sulphidation; alteration zones may be controlled by stratigraphic permeability rather than by faults and fractures; quartz (may be chalcedonic)-sericite (illite)-montmorillonite
Host rocks	silicic to intermediate (andesite)	intermediate to silicic intrusive/extrusive rocks	felsic intrusions; most sedimentary rocks except massive carbonates (hosts to mantos and skarns)
$^{18}O/^{16}O$ - shift in wall rocks	may be less pronounced, or superposed on earlier high-^{18}O alteration	moderate to large; pronounced in and immediately adjacent to veins	very limited ^{18}O-shift of altered rocks, if present at all
C-H-S isotopes	magmatic fluids indicated ($\delta^{13}C_{CO_2} \cong -5\pm2$; $\delta D_{H_2O} \cong -35\pm10$; $\delta^{18}O_{H_2O} \cong +7\pm2$; $\delta^{34}S_{\Sigma S} \cong 0$); magmatic-hydrothermal alunite $\delta^{34}S$>sulphide minerals; $\delta D \cong -35\pm10$; steam-heated alunite $\delta^{34}S \cong$ sulphides, $\delta^{18}O$ data indicate hydrothermal origin	magmatic water (H_2O) may be obscured by mixing; surface waters dominate; C, S typically indicate a magmatic source, but mixtures with wall rock derived C, S possible	hydrogen isotope data (sericite, clays, fluid inclusions) in some cases indicate presence of evolved surface waters; organic carbon ($\delta^{13}C \cong -26\pm2$) may be derived from wall rocks
Ore fluids (examples from fluid inclusion studies)	160-240°C; ≤1 wt.% NaCl (late fluids); possibly to 30 wt.% NaCl in early fluids; boiling common; (Nansatsu district, Japan; Hedenquist et al., 1994)	sulphide-poor: 180-31°C, ≤1 wt.% NaCl, about 1.0 molal CO_2 (Mt. Skukum: McDonald, 1987) sulphide-rich: ave. 25°C, <1 to 4 wt.% NaCl (Silbak-Premier: McDonald, 1990)	bimodal: 150-160 (most); 270-280°C, ≤15 wt.% NaCl; nonboiling: (Cinola: Shen et al., 1982); 230-250°C, ≤1 wt.% NaCl; nonboiling (Dusty Mac: Zhang et al., 1989)
Age of mineralization and host rocks	host rocks and mineralization of similar age	mineralization variably younger (>1 Ma) than host rocks	mineralization variably younger (>1 Ma) than host rocks
Deposit size	small areal extent (e.g. 1 km^2) and size (e.g. 2500-3500 kg Au)	may occur over large area (e.g. several tens of km^2); may be large (e.g. 100 000 kg Au).	may have large areal extent (e.g. >>1 km^2), large size (e.g. 58 000 kg Au), low grades (e.g. 2.5 g/t)
Examples Canadian	Equity Silver, B.C.; Mt. Skukum, Yukon (only: alunite 'cap') Al deposit, Toodoggone River, B.C.	Blackdome, B.C.; Mt. Skukum, Yukon (Cirque vein) Silbak-Premier, B.C. (intermediate sulphidation)	Cinola, B.C.
Foreign	Summitville, Colorado Kasuga, Japan	Creede, Colorado (intermediate sulphidation)	Hishikari, Japan
Modern analogues:	Matsukawa, Japan[2]	Broadlands, New Zealand[3]	Salton Sea geothermal field, California[4]

1) based, in part, on Heald et al., 1987; Taylor, 1987; Berger and Henley, 1989; Panteleyev, 1991; Rye et al., 1992; Sillitoe, 1993; Hedenquist et al., 2000; Izawa et al. 1990, 1993; and data reported for Canadaian deposits and other examples cited in the text; 2) Nakamura et al., 1970; 3) Browne in Henley and Hedenquist, 1986; 4) Williams and McKibben, 1989, but analogy not complete.
Abbreviations: bn = bornite; cpy = chalcopyrite; cv = covellite; en = enargite; gn = galena; py = pyrite; sp = sphalerite; ss = sulphosalts.

Belt is host to Au deposits containing pyrophyllite, andalusite, topaz, and traces of diaspore (metamorphosed advanced argillic assemblage) marked by oxygen isotope depletion from meteoric-recharged hydrothermal systems (Klein and Criss, 1988), confirming an epithermal origin (see also Feiss et al., 1993).

Among Canadian examples, oxygen and hydrogen isotope measurements of vein quartz and of fluid inclusions, respectively, from the unmetamorphosed Au-Ag mineralized, low-sulphidation altered stockwork associated with Middle Proterozoic rhyolite dykes in the Mallery Lake area, Nunavut, confirm a meteoric origin and epithermal setting (Turner et al., 2001). Oxygen isotope data on rocks from the Hope Brook zone (B. Taylor and P. Stewart, unpub. data, 1990) suggest a magmatically dominated (high-temperature) origin rather than a shallow, meteoric (e.g. steam-heated) alteration system. Detailed geologic and mineralogical studies (Dubé et al., 1998) have corroborated this conclusion. Similarly, the association of Hg-Sb-As-Tl (e.g. Harris, 1989) and range in $\delta^{34}S$ of pyrite (Cameron and Hattori, 1985) at the controversial Hemlo mine (Ontario), a disseminated Au deposit in the Precambrian Shield, might suggest a metamorphosed epithermal deposit. However, mineralized host rocks are not depleted in ^{18}O (Kuhns, 1988), suggesting deeper level, magmatically dominant fluids. This is also supported by S isotope fractionations between sulphate-sulphide mineral pairs (Hattori and Cameron, 1986).

Associated Mineral Deposit Types

Other deposit types that may be found broadly associated with epithermal deposits (i.e. within epithermal districts or camps) are those that share a common genetic link to magmatic centres (e.g. veins, skarns, and mantos; Sillitoe, 1993; Sillitoe and Thompson, 1998; Lang and Baker, 2001). Some vein and/or replacement deposits, typically of the intermediate sulphidation subtype (e.g. Silbak Premier, British Columbia, Table 1; Hedenquist et al., 2000), might be generally referred to as 'deep epithermal' or 'transitional' according to Panteleyev (1986, 1991). Because of the possibility of association, the locations of selected 'intrusion-related' or 'transitional' Au deposits in Canada have been included in Figure 1.

Often a 'barren gap' intervenes between the epithermal and 'deep epithermal' portions of magmatically heated geothermal systems. Close juxtaposition of epithermal vein and porphyry- or intrusion-related deposits may imply induced changes in the relative positions of meteoric geothermal systems and magmatic heat sources. Superposition of Au-bearing epithermal veins in the Coromandel Peninsula, New Zealand on slightly older Cu-Mo-Au porphyry deposits (e.g. Waihi deposit: Brathwaite and Faure, 2002), and superposition of a low-sulphidation subtype deposit on a porphyry Cu-type deposit in the Philippines (Acupan: Cooke and Bloom, 1990) are two examples. The extent of such superposition, or 'telescoping', may be tectonically and/or climactically controlled by rapid rates of uplift and high erosion, and volcanic sector collapse (Sillitoe, 1993; Müller et al., 2002a,b).

Deep epithermal veins ("transitional" deposits of Panteleyev, 1986) or replacement deposits associated with conventional epithermal deposits may comprise a component of the genetic link between a degassing high-level magma and an overlying mineralized epithermal systems. The quartz-monzonite porphyry intrusion known beneath the Summitville, Colorado high-sulphidation magmatic-hydrothermal system is essentially coeval with Au mineralization and hydrothermal alteration (Bethke et al., 2005) and provides an example of the porphyry-epithermal linkage for high-sulphidation deposits. Intrusion-related vein deposits in the Sulphurets, Mt. Washington, and Zeballos camps, British Columbai, are possible Canadian examples (British Columbia Ministry of Energy, Mines, and Petroleum Resources, 1992; Margolis, 1993). Other, related hydrothermal deposits that may be associated with epithermal vein deposits, and represent the mesothermal counter part, include Au-bearing skarns (high-temperature, silica-replacement deposits; e.g., Hedley, British Columbia) and manto deposits (sulphide rich replacement; e.g., Ketza River, British Columbia) in carbonate rocks, and intrusion-adjacent deposits sometimes referred to collectively as such 'intrusion-related' deposits (e.g. Thompson et al., 1999; Lang and Baker, 2001).

Disseminated and vein Au deposits associated with alkalic intrusions (e.g. Howell Creek, Fernie, British Columbia: Brown and Cameron, 1999) have gained attention as a distinct type of intrusion-related Au deposit (e.g. Richards, 1995; Jensen and Barton, 2000; Robert, 2001). The Au-Te epithermal deposit(s) at Cripple Creek, Colorado, represent a classic association with alkalic (diatreme) magmatic rocks; no extrusive rocks are present at this level of erosion (e.g. Kelley et al., 1998).

Hot-spring deposits, including siliceous sinters, steam-heated alteration zones, and brecciated root zones (e.g. Cinola, British Columbia: Christie, 1989; see also Izawa et al., 1993), cap many modern geothermal systems, and may be associated with either low- or high-sulphidation epithermal deposits. Their formation at the Earth's surface makes them very susceptible to erosion, however. Modern and recent deposits are most common, and examples are found worldwide, including Japan, Indonesia, New Zealand (e.g. Champagne Pool, Wairakai), Nevada (Round Mountain: Sander and Einaudi, 1990), and California (McLaughlin: Sherlock, 2005). Older, Jurassic, sinters, and related Au-Ag epithermal deposits are also known from Patagonia, Argentina (Schalamuk et al., 1997). Gold grades are typically variable and subeconomic, but the presence of Au, along with Hg, Sb, As, S, and Tl, may suggest a mineralized root zone, or deeper epithermal deposit.

Volcanogenic massive sulphide (VMS) deposits that form at or near the seafloor from submarine hot springs and sub-seafloor geothermal systems are epithermal deposits in the broad sense (e.g. Sillitoe et al., 1996). Gold-bearing VMS deposits (e.g. Horne mine, Noranda: Dubé et al., 2007; Eskay Creek 21B, British Columbia: Roth et al., 1999) are recognized as a type of VMS deposit (Dubé et al., 2007), and both high-sulphidation and low-sulphidation variants have been recognized (Sillitoe et al., 1996). Similarly, island arc settings may also host submarine calderas and epithermal deposits (e.g. Pueblo Viejo high-sulphidation subtype: cf. Kesler et al., 2005; Sillitoe et al., 2006; Milos, Greece: Naden et al., 2005). Indeed, many of the same processes regarding origins and causes of metal deposition and enrichment associated with continental epithermal deposits apply to these shallow marine settings. Comparison of S isotope geochemistry between adjacent subaerial and submarine mineralized systems in Papua New Guinea (Gemmell et al., 2004) offers further supportive evidence. Thus, the submarine environment should not be ignored with regard to epithermal deposits, despite the fact that some features (steam-heated alteration caps, dominance of meteoric waters, etc) familiar among subaerial epithermal and hot-spring deposits are necessarily absent.

Economic Characteristics of Epithermal Gold Deposits

Summary of Economic Characteristics

Gold (±Ag) is the principal commodity of epithermal Au deposits, occurring usually as native Au, or in electrum alloyed with Ag. It may also occur in tellurides, or as inclusions in sulphides. Copper and the other base metals, Pb and Zn, may also occur with Au, especially in transitional epithermal deposits with high Ag grades. Indeed, the common presence of enargite has led to the term enargite-Au deposits for some high-sulphidation subtype deposits (Ashley, 1982).

Epithermal (vein-) deposits, compared to the low-grade, bulk-tonnage porphyry deposits or the 'Carlin-type' deposits, are typically small in size (e.g. 10^6 to 10^8 tonnes of ore; high-sulphidation deposits tend to be smaller than low-sulphidation deposits) and, consequently, have a short mining life. However, epithermal Au deposits can reach high grades, a few to several tens of g/t, or more in exceptional cases (e.g. 70 g/t, Hishikari, Japan; to ~200 g/t, El Indio, Chile; Table 1).

Epithermal Au deposits represent a minor proportion, typically a few percent, of the total Au (reserves + production) in Canada. For example, epithermal deposits yielded an average annual production of 2725 kg/year, or about 2.7% of the total annual Au produced in Canada from 1985 to 1987. Owing largely to occurrence tending to favour geologically younger terrane, epithermal Au contributed relatively more (as much as 24%) of the total Au produced in British Columbia and the Yukon during the same period.

Grade and Tonnage Characteristics

The sizes (in 10^6 tonnes) of the principal Canadian epithermal Au vein deposits and selected 'type' deposits elsewhere are listed in Table 1, and plotted versus Au grade (g/t) by class of deposit in Figure 3. The mean grade and tonnage of several classic examples of non-Canadian deposits (low-sulphidation: Creede, Colorado, and Hishikari, Japan; high-sulphidation: Summitville, Colorado; average Au-bearing porphyry deposit; and average Carlin-type deposit, Nevada) are plotted for comparison. The estimated sizes (ca. 0.05 to 42 Mt of ore) give an order of magnitude basis for comparison; definition of size depends on cut-off grades and economics. Ore comprises disseminated Au in silicified and/or finely veined rocks in the Cinola deposit, British Columbia, and in areas of the Sulphurets district, British Columbia. Here, grades are typically lower, but tonnages larger, than in other, vein-type, epithermal deposits (Table 1). Based on a reported grade of 2.45 g/t and 23.80 Mt of ore,

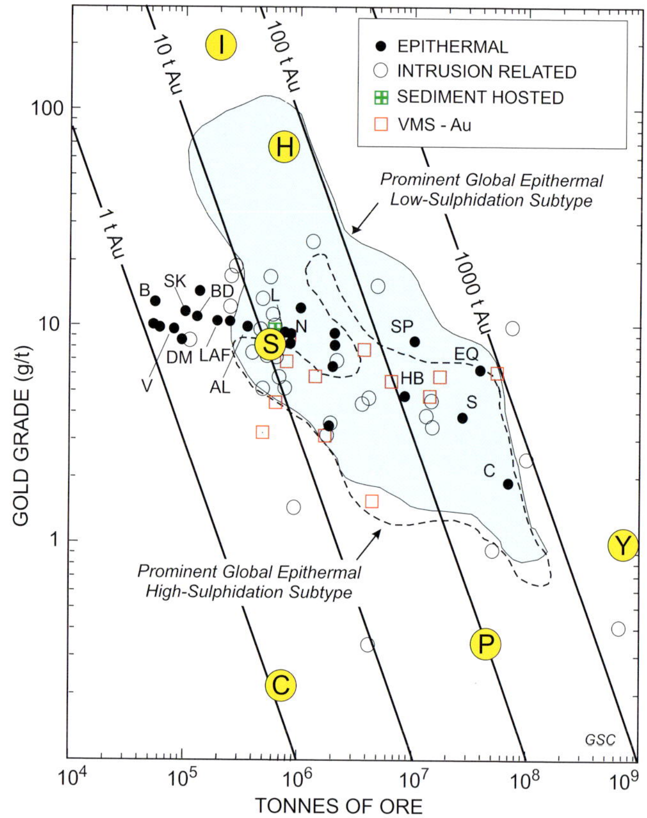

FIGURE 3. Plot of Au grade (g/t) versus tonnage (economic, or reserves+production) for selected Canadian epithermal Au deposits and prominent examples elsewhere in the world, classified by subtype as referred to in the text. Canadian epithermal deposits (filled circles; see Table 1) include AL = Al; B = Baker; BD = Blackdome; C = Cinola; DM = Dusty Mac; EQ = Equity Silver; HB = Hope Brook; L = Lawyers; LAF = Laforma; N = Mt. Nansen; SK = Mt. Skukum; SP = Silbak Premier; S = Sulphurets camp (Brucejack Lake, Sulphuret, West Zone deposits); and V = Venus. Hydrothermal vein deposits of a possible 'transitional' or 'deep epithermal' deposits are represented by open circles, sediment-hosted deposits by a green square with cross, and Au-bearing VMS deposits ('marine epithermal') by open red squares (see Appendix 1 in Dubé et al., 2007). The median grades and tonnages for several comparable types of deposits (yellow-filled circles) from Cox and Singer (1986) include porphyry Cu-Au [**P**]; low-sulphidation Creede-type [**C**]; intermediate sulphidation: polymetallic vein deposits associated with felsic intrusions [**M**]; and high-sulphidation: Summitville deposit [**S**]; and Lawyers deposit, Toodoggone River district, British Columbia [**L** is similar to the 'Comstock-type', Nevada (not plotted) of Cox and Singer, 1986]. Median values for the low-sulphidation Hishikari, Japan vein deposit [**H**], and for the high-sulphidation El Indio, Chile, deposit [**I**] are from Hedenquist et al. (2000). Fields for prominent low-sulphidation (blue shading) and high-sulphidation (dashed line) epithermal Au deposits worldwide (global) are based on data in Hedenquist et al. (1996; 2000).

the Cinola deposit is potentially the second largest epithermal Au deposit in Canada. For vein-style epithermal deposits, the Au grades of Canadian examples (most ~2.5-25 g/t) are similar to those of a majority of 'mesothermal' quartz-carbonate deposits (see Gosselin and Dubé, 2005a-d), but of generally smaller size, and are distinguished from the latter by higher Ag:Au ratios (>1:1).

Canadian epithermal Au deposits are comparable in size and grade to many deposits found in the major epithermal terranes of the world, as illustrated in Figure 3. The largest epithermal deposits (in tonnes of ore) and the richest deposits (in g/t) are found outside of Canada. Fields for prominent low-sulphidation subtype epithermal Au deposits and for high-sulphidation subtype epithermal Au deposits (from data in Hedenquist et al., 1996, 2000) overlap, but suggest that high-sulphidation subtype deposits tend to be comparable in size and grade to the smaller of the low-sulphidation subtype deposits. Several non-Canadian epithermal deposits selected by subtype to reflect the global range of grade and tonnage are plotted together with Canadian deposits in Figure 3.

Exploration Properties of Epithermal Gold Deposits

Physical Properties

The mineralogy, textural features, host rocks, morphology, and selected chemical properties found typically in epithermal Au deposits are summarized in Table 2. Key features are emphasized below.

Mineralogy

Quartz is the predominant gangue mineral in all epithermal Au deposits, whereas distinctive ore and gangue minerals characterize high-sulphidation and low- sulphidation deposit subtypes. Mineralogical zoning around veins or replacement zones may be present in both subtypes, recording chemical and/or thermal gradients. Both subtypes of deposits can contain very fine-grained Au and gangue mineral assemblages, especially in hot-spring and steam-heated environments that form above boiling hydrothermal systems (Henley and Ellis, 1983).

In high-sulphidation subtype deposits, native Au and electrum are typically associated with pyrite+enargite±covellite±bornite±chalcocite. In addition to sulphosalts and base metal sulphides, tellurides and bismuthinite are present in some deposits. Total sulphide contents are generally higher in high-sulphidation than low-sulphidation subtype deposits, but high sulphide contents may also characterize transitional polymetallic low- sulphidation deposits (e.g. Silbak Premier, British Columbia). Where base metals are present in high-sulphidation deposits, the Cu abundance can vary significantly («0.1-5%: Sillitoe, 1993), and typically dominate that of Zn. Principal gangue minerals include quartz ('vuggy silica'), alunite, barite (especially associated with Au), and, in some deposits, S; manganese minerals and fluorite are rare. Calcite is not characteristic of high-sulphidation subtype deposits due to the high acidity of the hydrothermal fluids.

Native Au and electrum occur in low-sulphidation subtype vein deposits that often contain only a few percent or less of sulphides (usually pyrite; e.g., Blackdome, British Columbia). In deposits in which sulphide minerals are abundant (e.g. Venus; Silbak-Premier: sulphide-rich stage), sulphides commonly include chalcopyrite, tetrahedrite, galena, sphalerite, and arsenopyrite in addition to pyrite. The principal gangue minerals include calcite, chlorite, adularia, barite, rhodochrosite, fluorite, and sericite.

In sediment-hosted low-sulphidation deposits, the characteristic assemblage of gangue minerals commonly includes cinnabar, orpiment-realgar, and stibnite, in addition to jasperoid, quartz, dolomite, and calcite. Chalcedonic quartz veins and jasperoid are typically associated with ore, whereas calcite veins are often more common further from ore, or are paragenetically late. In siliceous sinter associated with hot-spring deposits, sulphate minerals, clays, and minor pyrite constitute the typical gangue assemblage; vertical zoning of alteration mineral assemblages is characteristic.

In some deposits hosted by volcaniclastic rocks (e.g. McDermitt, Nevada), micrometre-size Au grains are typical, although visible (recrystallized) native Au may occur in oxidized portions of some deposits. Gold can occur coating sulphides and/or encapsulated in quartz in silicified rocks, accompanied by Hg-, Sb-, and As-bearing sulphides. At Cinola, British Columbia, a rare example in Canada (Christie, 1989; see also Poulsen, 1996), Au is most abundant in the subsurface silicified sediments and hydrothermal breccias. Inclusion of sediment-derived hydrocarbons may occur during vein formation in sedimentary rocks, or deposits within hydrothermal systems encompassing sedimentary rocks (e.g. Owen Lake, British Columbia: Thomson et al., 1992).

Textures

Vuggy silica has a porous texture formed by removal of minerals, particularly feldspars during reaction with very acidic fluids and concentration of residual silica (e.g. Summitville, Colorado: Stoffregen, 1987). Massive, quartz-rich zones may result from further silicification (i.e. by addition of silica). Examples include alteration zones at Mt. Skukum, Yukon (alunite cap zone: Love, 1989) and at the Al deposit (Toodoggone River, British Columbia: Diakow et al., 1993). In high-sulphidation subtype deposits, coarse-grained alunite is characteristic, whereas alunite from steam-heated zones (high-sulphidation subtype caps to epithermal systems), and from supergene weathering of sulphide deposits, is typically very fine grained to microcrystalline.

Lamellar or platy ('angel wing') calcite, in some cases pseudomorphically replaced by silica (e.g. Mt. Skukum, Yukon), is of particular significance because it forms in boiling zones in low-sulphidation subtype systems (e.g. de Ronde and Blattner, 1988; Simmons and Christenson, 1994). Rhombic adularia has been similarly associated with boiling (Keith and Muffler, 1978; Dong and Morrison, 1995).

Unique to (unmetamorphosed) hot-spring deposits, are non-horizontal laminated or 'bedded' lenses that may contain textures formed by silica fossilization of plant matter (root casts, etc.), and vertical crystallization textures and structures.

Dimensions

High-sulphidation deposits of magmatic hydrothermal origin (e.g. Rye et al., 1992) are typically of smaller dimension than low-sulphidation subtype deposits, and are found in close proximity to, and often topographically above, a related source of magmatic heat and volatiles. Altered rocks of the Summitville, Colorado, deposit, for example, crop out over an area of 1.5 by 1.0 km (Heald et al., 1987). Shallow, steam-heated environments, in contrast, may produce widespread altered areas, typically (but not always) barren; bulk-tonnage mining of these zones may be possible if they are mineralized. For example, mineralized areas altered to quartz+clay+alunite (+barite+dickite) at the Al deposit, Toodoggone River area, British Columbia, measure about 250 m by as much as 1.5 km (Diakow et al., 1993). Fault-controlled, quartz-(kaolinite)-alunite alteration zones meas-

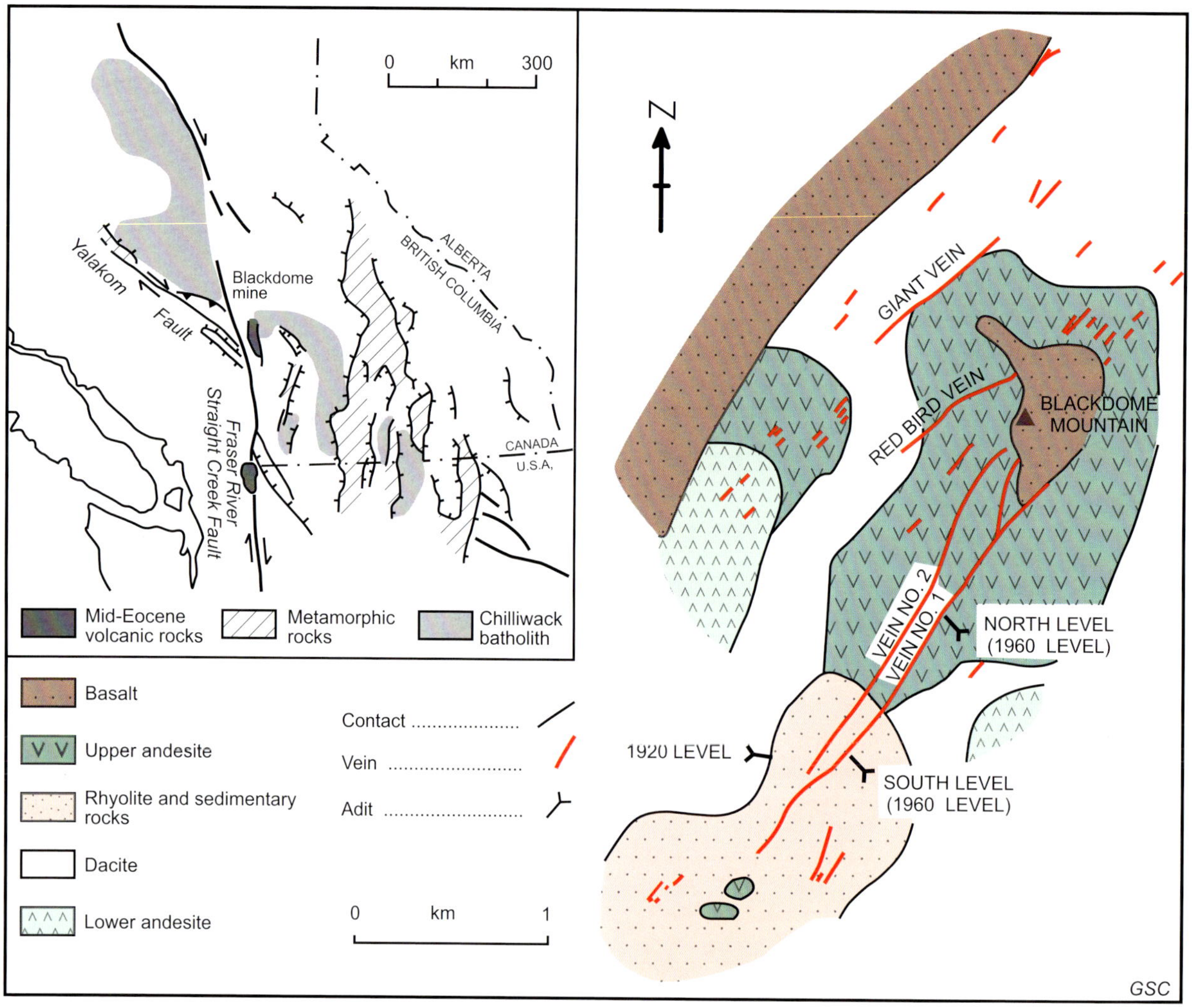

FIGURE 4. Geological map illustrating setting of the epithermal Au vein deposit at the Blackdome mine, British Columbia (from Taylor, 1996; data from D. Rennie, unpub. rep., 1987). Inset map illustrates the regional tectonic setting of the Blackdome mine area (red square), between the Yalakom (55-45 Ma) and younger Fraser River Straight Creek (40-35 Ma) strike-slip fault (Coleman and Parrish, 1990; R.R. Parrish, pers. comm., 1991).

uring roughly 200 by 250 m occur topographically above the Mt. Skukum deposit, in an area partially removed by erosion (McDonald, 1987). Similarly altered prospective areas occur in folded Neoproterozoic (Avalonian) rocks in the Burin Peninsula, Newfoundland affected by advanced argillic alteration (pyrophyllite-alunite-specularite; e.g., O'Brien et al., 1999). Two of these measure approximately 125 by 225 m (Hickey's Pond prospect) and 4700 m x 4 km (Stewart prospect), although evidence of similar alteration is present over a length of approximately 100 km. The Avalonian Carolina slate belt hosts high-sulphidation subtype gold deposits (abandon mines) of similar age, alteration style, and dimension (c.f. Klein and Criss, 1988; O'Brien et al., 1999).

Low-sulphidation subtype deposits in some cases cover larger areas than typical of high-sulphidation deposits, even though alteration mineral assemblages are restricted to generally narrow zones enclosing veins and breccias. At the Blackdome mine, British Columbia (Fig. 4), quartz veins as much as 0.7 m thick and 2200 m long, are contained within an area approximately 2 by 5 km. Veins comprising the Lawyers deposit and the Baker mine in the Toodoggone district, British Columbia, are commonly 2 to 7 m wide and as much as several hundred metres in length. Veins and breccia zones as wide as 40 m and as long as 1200 m comprise the Main zone of the Silbak-Premier deposit in British Columbia (McDonald, 1990). Elsewhere, mineralized veins in low-sulphidation subtype epithermal deposits have been mined for a strike length of more than 5 km at Creede, Colorado (Heald et al., 1987), and occur for a distance of about 2 km at the high-grade (e.g. 70 g/t) Hishikari mine, Japan (Izawa et al., 1990). Alteration zones around the veins of the Hishikari deposit have been mapped in an area measuring as much as 2 km wide by more than 3 km long (Izawa et al., 1990).

Hot-spring deposits comprising surface lenses or aprons of silica (siliceous sinter), may be several hundred metres in diameter, but only metres to tens of metres in thickness. Discordant hydrothermal conduits beneath these deposits may extend over a hundred or more metres in the vertical dimension, and resemble funnel-like forms in section, decreasing from perhaps many tens of metres to a few metres with depth (e.g. Christie, 1989).

Morphology

The morphology of epithermal vein-style deposits can be quite variable. Deposits may consist of roughly tabular lodes controlled by the geometry of the principal faults they occupy (e.g. Cirque vein, Mt. Skukum; Fig. 5; Table 1), or comprise a host of interrelated fracture fillings in stockwork, breccia, lesser fractures, or, when formed by replacement of rock or void space, they may take on the morphology of the lithologic unit or body of porous rock (e.g. irregular breccia pipes and lenses) replaced. Volumes of rock mineralized by replacement may be discordant and irregular, or concordant and tabular, depending on the nature of porosity, permeability, and water-rock interaction. In deposits of very near-surface origin (e.g. Cinola), an upward enlargement of the volume of altered and mineralized rocks may be found centred about the hydrothermal conduits. Hot-spring deposits tend to comprise subhorizontal aprons or lenses of sinter about their upflow zones and subhorizontal replacement zones in the shallow subsurface. Phreatic eruptions produce discordant zones of breccia-like deposits; clasts may be partially rounded (e.g. Izawa et al., 1990).

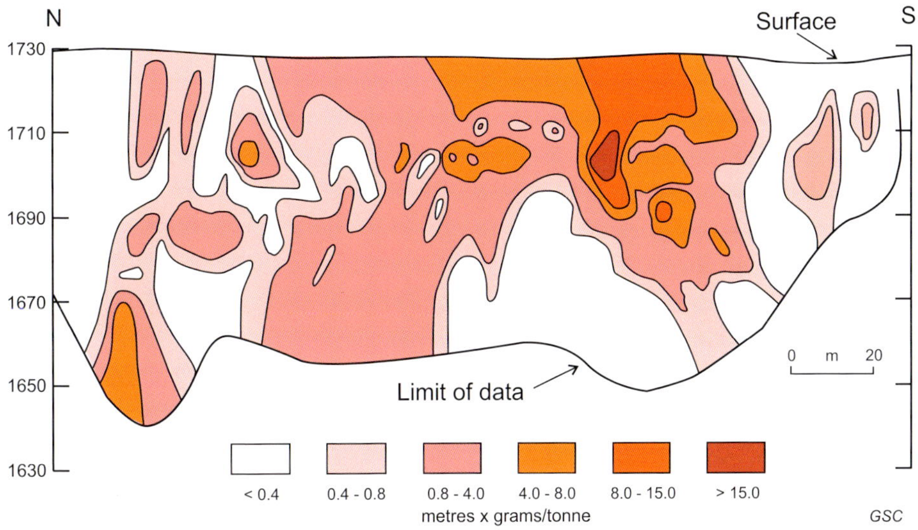

Figure 5. Longitudinal section of the Cirque vein, Mt. Skukum (from Taylor, 1996, modified after McDonald, 1987) illustrating distribution of Au (thickness x grade).

Brecciation of previously emplaced veins (e.g. Mt. Skukum, Yukon) can form permeable zones along irregularities in fault planes: vertically plunging ore zones in faults with strike-slip motion and horizontal ore zones in dip-slip faults. Topographic (i.e. paleosurface) control of boiling by hydrostatic pressure can also result in horizontal or subhorizontal mineralized zones, limiting the vertical distribution of ore (as suggested in Fig. 5; Cirque vein, Mt. Skukum, Yukon). The distribution of high-sulphidation alteration in steam-heated settings (possibly in the Toodoggone River camp, British Columbia) may also reflect a topographic control of the paleo-water table.

Host Rocks

Nearly any rock type, even metamorphic rocks, may host epithermal Au deposits, although volcanic, volcaniclastic, and sedimentary rocks tend to be more common. Typically, epithermal deposits are younger than their enclosing rocks, except in the cases where deposits form in active volcanic settings and hot springs. Here, the host rocks and epithermal deposits can be essentially synchronous with spatially associated intrusive or extrusive rocks, within the uncertainty of the determined ages in some cases (e.g. high-sulphidation Summitville deposit, Colorado: Bethke et al., 2005; low-sulphidation El Peñon deposit, Northern Chile: Arancibia et al., 2006).

Chemical Properties

Ore Chemistry

Gold:silver ratios of epithermal Au deposits may vary widely both between and within deposits from 0.5 for the high-sulphidation type Kasuga deposit, Japan (Hedenquist et al., 1994), for example, to >>500 in the Cerro Rico de Potosi deposit, Peru (Erickson and Cunningham, 1993). Differing magmatic metal budgets (Sillitoe, 1993) and depths of formation (Hayba et al., 1985) have been suggested to influence this ratio. At the Lawyers deposit, Toodoggone River district, British Columbia, the Ag:Au ratio varies northward in the deposit, from less than 20 to more than 80 (average = 46.7; Table 1), and higher ratios are also found at deeper levels of the deposit (Vulimiri et al., 1986). Typically, Ag:Au ratios for epithermal deposits, though variable, tend to be higher in low-sulphidation subtype deposits than in high-sulphidation subtype deposits (Table 1). The deep epithermal (mesothermal) Equity Silver deposit, British Columbia (e.g. Cyr et al., 1984; Wojdak and Sinclair, 1984) has the highest Ag:Au ratio (approximately 128; Table 1) among Canadian epithermal deposits.

High precious metal/base metal ratios in hot-spring deposits (and steam-heated zones in general) are thought to be characteristic. Buchanan (1981) suggested that base metals precipitate in deeper, more saline, liquid-dominated portions of the system, whereas deposition of Au occurs in an upper, gas-rich, or boiling portion of the geothermal system, resulting in the observed metal separation.

Whereas base metals may accompany Au (±Ag) in variable amounts in intrusion-related or transitional deposits, more volatile elements commonly occur with Au (±Ag) in shallower epithermal and hot-spring environments. These elements characteristically include Hg, Sb, Tl, As, and native S.

Alteration Mineralogy and Chemistry

Examples of alteration mineral zoning and its relationship to lithology are illustrated for portions of three Canadian deposits in Figure 6A-C (sediment-hosted low-sulphidation subtype: Cinola, British Columbia; volcanic-hosted low- to intermediate-sulphidation subtype: Silbak-Premier, British Columbia; Fig. 5C), and rhyolite/andesite-hosted high-sulphidation alteration topographically above the low-sulphidation subtype (Mt. Skukum deposit, Yukon). Alteration was, in each case, structurally controlled, cross-cutting the host rocks. Symmetrical zoning developed about some veins (e.g.

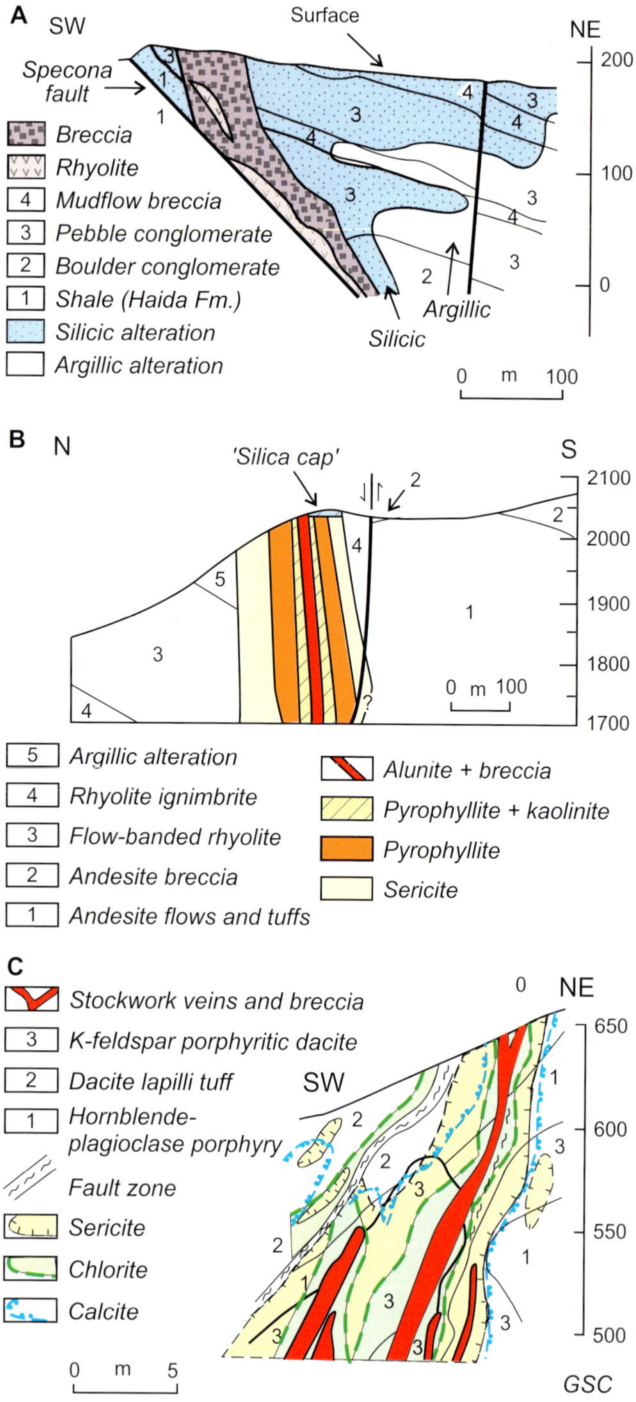

FIGURE 6. Geological cross-sections of representative Canadian epithermal deposits illustrating alteration mineral zoning and selected features (from Taylor, 1996). **(A)** Cinola sedimentary-hosted Au deposit (after Christie, 1989; low-sulphidation subtype), also illustrating localization of both magma (interpreted from faulted dyke) and hydrothermal fluids by the Specona fault, and the control exerted by primary lithological permeability on the distribution of zones of silicification and argillic alteration. **(B)** Cross-section through a portion of a principal normal fault in the Mt. Skukum area (after Love, 1989; low-sulphidation subtype, volcanic host) illustrating the distribution of associated alunite+silica, and advanced argillic alteration mineral assemblages. Supergene alteration has been superposed on quartz-(kaolinite)-alunite alteration zones. **(C)** Cross-section through a portion of the Silbak-Premier deposit (intermediate sulphidation; after McDonald, 1990) illustrates hydrothermal propylitic, sericitic, and potassic alteration mineral assemblages in relation to fault-controlled vein stockwork and breccia, and to porphyritic dacite.

Mt. Skukum, Yukon), but was markedly asymmetrical in other cases (e.g. Cinola, British Columbia).

In both high-sulphidation and low-sulphidation deposit subtypes, hydrothermal alteration mineral assemblages are commonly regularly zoned about vein- or breccia-filled fluid conduits, but may be less regularly zoned in near-surface environments, or where permeable rocks have been replaced. Characteristic alteration mineral assemblages in both deposit subtypes can give way to propylitically altered rocks containing quartz+chlorite+albite+carbonate±sericite, epidote, and pyrite. The distribution and formation of the earlier formed propylitic mineral assemblages generally bears no obvious direct relationship to ore-related alteration mineral assemblages.

Altered rocks in low-sulphidation deposits generally comprise two mineralogical zones: (1) inner zone of silicification (replacement of wall rocks by quartz or chalcedonic silica); and (2) outer zone of potassic-sericitic (phyllic) alteration (quartz+K-feldspar and/or sericite, or sericite and illite-smectite). Adularia is the typical K-feldspar, but its prominence varies greatly; it may be absent altogether. Chlorite and carbonate are present in many deposits, especially in wall rocks of intermediate composition, and in some cases (e.g. Shasta deposit, Toodoggone River, British Columbia: Thiersch and Williams-Jones, 1990; Silbak-Premier: McDonald, 1990) chloritic alteration accompanied the potassic alteration and silicification. Argillic alteration (kaolinite and smectite) occurs still farther from the vein. In some deposits (e.g. Cinola: Christie, 1989), argillic alteration predates silicification, giving evidence of the waxing and waning of hydrothermal systems. Argillic mineral assemblages are commonly superposed on the above, or form in higher level alteration zones (e.g. Toodoggone River area, British Columbia: Diakow et al., 1993), where adularia is replaced by kaolinite; smectite may occur furthest from the veins.

Silicified rocks are common in epithermal deposits, as is quartz gangue in veins. For example, in both the volcanic-hosted Blackdome deposit, British Columbia, and sedimentary hosted Cinola deposit, irregularly silicified and mineralized wall rocks are common adjacent to faults and fractures. Silicified and decarbonated host rocks characterize Carlin-type Au deposits in Nevada (e.g. Bagby and Berger, 1986). The silicification of wall rocks (and the distribution of ore) was apparently controlled by available primary permeability of bedding planes or rock fabric. Secondary permeability can also be produced by physical and chemical processes involving the hydrothermal fluids themselves. The sudden release of pressure on hydrothermal fluid (e.g. by faulting) can cause brecciation, creating pore space permeability (e.g. Cinola, British Columbia, breccia zone in Fig. 6A). This can occur in geothermal systems within several hundred metres of the Earth's surface (e.g. Hedenquist and Henley, 1985). Dissolution of carbonate upon reaction between hydrothermal fluids and wall rocks also can produce secondary permeability.

Based on the nature of silicification, Bagby and Berger (1986) distinguished two types of sediment-hosted deposits: jasperoidal- and Carlin-type; the latter is no longer considered a subtype of epithermal Au deposits. Jasperoidal deposits may occur in clastic sedimentary rocks, where silicification is characterized by quartz veins (commonly) and/or replacement (e.g. Cinola, British Columbia). Other effects of alteration are otherwise similar, and include decarbonation (where rocks originally contained carbonate) and argillization. Alteration minerals include alunite, quartz, calcite, illite, cinnabar, orpiment, realgar, stilbite, pyrite, pyrrhotite, marcasite, and arsenopyrite.

Advanced argillic alteration mineral assemblages that characterize high-sulphidation deposits include quartz+kaolinite+alunite+dickite+pyrite in and adjacent to veins or zones of replacement in the magmatic-hydrothermal environment. Pyrophyllite occurs in place of kaolinite at the higher temperatures and pressures of deeper deposits. In some outer zones (e.g. alunite 'cap'; Mt. Skukum), argillic (smectite)±sericite mineral assemblages may occur (Fig. 6B). These alteration minerals indicate a very low pH hydrothermal environment (possibly below even that for alunite stability; Stoffregen, 1987) of high oxidation state (hematite and sulphate are stable). Zones of silica replacement and 'vuggy silica' are characteristic, and carbonates are absent. Topaz and tourmaline in high-temperature zones indicate the presence of F and B in the acidic hydrothermal fluids.

Acid-sulphate (high-sulphidation) type alteration fluids form by the dissolution of large amounts of magmatic SO_2 in high-temperature hydrothermal systems, and also by reaction of host rocks with steam-heated meteoric waters acidified by oxidation of H_2S (probably of magmatic origin: e.g., Rye et al., 1992; Bethke et al., 2005), or by dissolution of CO_2. Two contemporaneous fluids are typically found to have been significant in epithermal Au deposits, and particularly in the high-sulphidation subtype (e.g. Summitville, Colorado: Bethke et al., 2005; Pierina, Peru: Fifarek and Rye, 2005). These are a saline fluid (typically ~10-40 wt.% NaCleq) found often in the deeper portion of the hydrothermal system, associated with mineralized zones, and a low-density (<10% NaCleq) fluid more commonly found in the upper part of the hydrothermal system. The more saline fluids are often interpreted to have evolved from magmatic fluids and to have transported the metals to meteoric-water dominated geothermal system. In rare situations where recharge of surficial waters is limited (e.g. <10 Ma, Andes) magmatic waters may dominate throughout (e.g. Pascua-Lama, Chile: Deyell et al., 2005). Lower acidity, highly saline fluids are thought responsible for intermediate sulphidation deposits typically rich in base metal and Fe sulphide minerals (Hedenquist et al., 2000).

Fluids attributed to low-sulphidation hydrothermal systems are typically less saline than those in high-sulphidation systems, although fluids of two different salinities are also common: a deeper fluid of perhaps approximately 10 to 20 wt.% NaCleq is responsible for metal transport, along with a shallower, dilute fluid (<1 wt.% NaCleq). Stable isotope data, along with other geochemical attributes, indicate that the two fluids evolved from a single, low to intermediate salinity fluid via boiling (e.g. Waihi, New Zealand: Brathwaite and Faure, 2002). The primary fluids in low-sulphidation subtype deposits are commonly inferred to have largely evolved from meteoric rather than magmatic water, or comprise some mixture of the two (e.g. Hishikari, Japan: Faure et al., 2002).

Distinguishing steam-heated environments, localized above deeper, boiling hydrothermal systems (Henley and Ellis, 1983), whether of low-sulphidation or high-sulphidation subtype, from shallow magmatic hydrothermal environments is not always straightforward. Steam-heated, high-sulphidation alteration zones may occur as 'blankets' above low-sulphidation deposits (e.g. Sulphurets, B.C.; Margolis, 1993) and also at the top of high-sulphidation systems, and may or may not directly overlie mineral deposits (Henley, 1985). Alunite is a characteristic mineral of hypogene high-sulphidation alteration, but may also occur in steam-heated environments, above either low- or high-sulphidation systems. In addition, alunite may form during supergene weathering of sulphide deposits. Distinguishing the alunite formed in different environments may be aided by texture: hypogene alunite is typically coarse-grained; supergene alunite is typically fine-grained and poorly crystallized; the presence of halloysite, iron oxides, jarosite, possible supergene enrichment, and subhorizontal mineral zoning characterize the supergene nature of the alteration. In metamorphosed deposits, alunite texture is a less reliable criterion, and low-temperature minerals usually no longer exist. Stable isotope studies, however, can distinguish supergene alunite, steam-heated alunite, and magmatic-hydrothermal hypogene alunite (Rye et al., 1992; Rye, 2005).

Stable isotope techniques offer a particularly powerful tool to map and decipher paleo-hydrothermal systems represented by altered and mineralized rocks, and to guide exploration, even in highly metamorphosed terrane. Oxygen and hydrogen isotope, and fluid inclusion studies have thus far indicated that Au precipitating hydrothermal fluids in epithermal deposits typically comprise mixtures of low-salinity, meteoric waters and more saline waters. The saline waters can be either magmatic water or evolved waters formed by reaction of hydrothermal fluids with host rocks, by boiling and/or by mixing of magmatic volatiles and other evolved fluids. Saline fluids have been shown to be especially important in the transport of base metals. Volatile components (e.g. metals, F, Cl, CO_2, SO_2, etc.) may be added to geothermal systems by subsurface magmatic degassing, forming fluids with magmatic isotopic characteristics that are common to high-sulphidation systems. Magmatic water and volatiles tend to constitute a minor, if even detectable, component of fluids in low-sulphidation subtype deposits (except for C and S in some cases).

The hydrothermal fluids responsible for alteration in selected Canadian and non-Canadian deposits, including fields for several groups of deposits, are plotted in a δD versus $\delta^{18}O$ diagram in Figure 7. The fluids responsible for alteration and mineralization largely represent altered or 'evolved' meteoric waters whose isotopic compositions have been shifted to higher $^{18}O/^{16}O$ and D/H (deuterium-to-hydrogen) ratios than those of pure local meteoric waters (compare with present day meteoric water, 'PDMW', Fig. 7). Such isotopic alteration or evolution of the fluids occurs during chemical, isotopic, and mineralogical hydrothermal alteration of the host rocks, and is denoted by curved reaction paths in Figure 7 (e.g. field of fluids for the Summitville

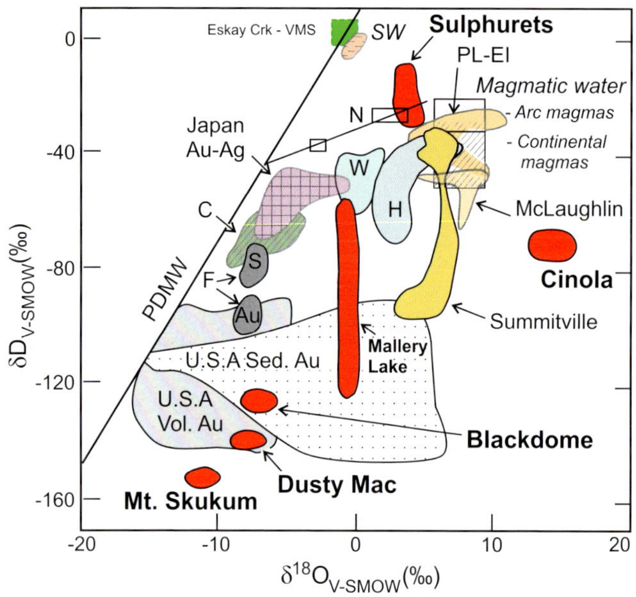

FIGURE 7. Plot of $\delta^{18}O$ versus δD for present day meteoric waters and for waters in equilibrium with gangue minerals in selected epithermal Au deposits; Canadian examples are shown in red (modified from Taylor, 1996, with additions). This diagram illustrates the origin and oxygen isotope enrichment of meteoric waters in many epithermal vein systems. Abbreviations are for the Finlandia vein [**F**], Colqui district, Peru (**S** = sulphide stage; **Au** = precious metal stage); **PDMW** = present day meteoric water; **SW** = seawater (note similar composition for fluids in the Eskay Creek Au deposit: Sherlock et al., 1999); U.S.A. SED. Au = sedimentary rock-hosted Au in the United States; U.S.A. Vol. Au = volcanic rock-hosted Au in the United States. Magmatic water composition defined by Taylor (1987, 1992). Sources of data for Canadian deposits: Taylor (1987); Blackdome: Vivian et al. (1987); Cinola: B. Taylor and A. Christie (unpub. data, 1991); Dusty Mac: Zhang et al. (1989); Mallery Lake: Turner et al. (2001); Sulphurets: Margolis (1993), and Mt. Skukum: McDonald (1987) and B. Taylor (unpub. data, 1991);. Sources for non-Canadian deposits: [**H**] Hishikari, Japan: Faure et al. (2002); McLaughlin, California: Sherlock et al. (1995); [**N**] Nansatsu, Japan (two fluid compositions identified on plausible mixing line): Hedenquist et al. (1994); [**PL-EI**] Pascua-Lama - El Indio, Chile (data for Pascua-Lama): Deyell et al. (2005); Summitville, Colorado: Bethke et al. (2005); [**W**] Waihi, New Zealand: Braithwaite and Faure (2002); others, including Creede, Colorado [**C**], Finlandi vein Colqui, Peru [**F**: **Au**- and **S**-rich stages] as cited in Taylor (1987). Mixing with meteoric waters that evolved by water/rock reaction is indicated by the Summitville field of waters. Mixing of magmatic fluids and local meteoric water in the Nansatsu deposit, Japan [**N**] is shown by the straight dashed mixing line.

deposit). The mixing of meteoric waters (PDMW) with either modified fluids or magmatic fluids is indicated by straight lines (e.g. such indicated for the Nansatsu deposit, 'N'). Variations of δD or $\delta^{18}O$ of the fluids may be accompanied by variations in wt.% NaCl, such as by mixing of a (deeper) saline fluid and a (shallower) dilute fluid (e.g. Taylor, 1987). Evidence of the mixing of distinct fluids with distinct isotopic ratios and salinities has been reported for some vein deposits of the deep epithermal or 'transitional' category (e.g. Finlandia vein, Peru: Kamilli and Ohmoto, 1977).

Altered sedimentary wall rocks are generally less depleted in ^{18}O and the hydrothermal fluids are more enriched in ^{18}O than in volcanic terranes. For example, the markedly higher $^{18}O/^{16}O$ ratios of hydrothermal fluids at Cinola, compared to those at Mt. Skukum and Blackdome (see Fig. 7), can be attributed to higher initial $^{18}O/^{16}O$ ratios of the local waters and to their reaction with sedimentary wall rocks. Involvement of seawater or low-latitude meteoric water is indicated for the Sulphurets area (Margolis, 1993). Meteoric waters formed the major component of the ore-forming fluids at the Blackdome (Vivian et al., 1987), Dusty Mac (Zhang et al., 1989), and Mt. Skukum (McDonald, 1987) deposits. Data reported in Diakow et al. (1991, 1993) indicate a broadly similar scenario for low-sulphidation deposits of the Toodoggone River area, British Columbia An unusual range in δD (-151 to -54 per mil) and $\delta^{18}O$ (recalculated: -7.6 to -2.6 per mil) for vein-depositing fluids in the Laforma vein (Table 1; McInnes et al., 1990) are consistent, along with S isotope data, with a mixing scenario involving magmatic volatiles and meteoric waters. Margolis (1993) inferred progressive mixing of magmatic water and seawater during potassic, sericitic, and advanced argillic alteration at Sulphurets, British Columbia, on the basis of isotopic data and water-rock reaction modeling. In some cases (e.g. Creede, Colorado), incorporation of dilute (fresh meteoric) fluids occurred abruptly, and late, in the paragenesis (e.g. Foley et al., 1989).

Recognition of the source of S by means of its isotopic composition depends on the relative mass balance for the contributing sources that are isotopically distinct. Host rock S, or biogenically precipitated S (e.g. Eskay Creek: Roth et al., 1999; Roth and Taylor, 2000, submitted) may comprise a significant component in some low-sulphidation deposits, whereas in high-sulphidation deposits, magmatic S (as S02; $\delta^{34}S$ about 0-4 per mil for felsic magmas: Taylor, 1987) dominates. Where magmatic sources of wall rock S (e.g. sulphide minerals) dominate, magmatic S isotope values may characterize low-sulphidation deposits also.

Carbon isotope data for calcite or fluid inclusion CO_2 typically reveals its magmatic (ultimately mantle) origin, even in systems dominated by meteoric water (e.g. Laforma, Yukon: McInnes et al., 1990; Mt. Skukum, Yukon: McDonald, 1987). Admixture with terrestrial C sources may also occur (e.g. organic carbon in the Owen Lake deposit: Thomson et al., 1992).

Fluid inclusions typically have been shown to contain predominantly fluids of low salinity (less than approximately 5 wt.% NaCleq) and have filling temperatures of 150 to 300ºC, with maxima in the range of approximately 260 to 280°C (e.g. Equity Silver: Shen and Sinclair, 1982; Blackdome: Vivian et al., 1987). Vapour-dominated systems at or near a boiling water table tend to evolve toward a rather uniform temperature of about 240°C due to the limitation imposed by a maximum in the enthalpy of steam+liquid (e.g. White et al., 1971). Some deep epithermal (transitional) environments close to genetically related intrusions are characterized by higher temperatures, salinities, and CO_2 contents (e.g. Baker, 2002). The occurrence of fluid inclusions formed at different times in a dynamic system complicates interpretation of the evolution of the system. Temporal changes in the Creede hydrothermal system, identified by abrupt changes in the chemical and isotopic compositions of fluid inclusions between different growth zones, or in fracture planes through crystals, demonstrate that the identity of ore-transporting fluids can be obscured by inappropriate sampling and analysis (Foley et al., 1989).

Geological Properties

Continental Scale

Exploitation of deposits in active volcanic systems (e.g. Ladolam, Lihir Island, Papua New Guinea: Müller et al., 2002a,b; Carman, 2003), the lure of shallow, easily extracted high-grade deposits, and even the sensational exposure of assay sample adulteration at the Busang epithermal Au prospect, Borneo (e.g. Hutchinson, 1997), have resulted in increased general public awareness of epithermal deposits and mining activities. Recent increase in scientific/technical documentation of deposits, review papers, exploitation of geothermal systems, and laboratory/theoretical studies have helped to clarify geological settings, epithermal Au deposit characteristics worldwide, and clarified processes of transport and precipitation of Au. Large deposits appear to require a sustained (magmatic) heat source, and efficient, localized processes (e.g. cooling, degassing/boiling, fluid mixing, and wall-rock reaction) leading to supersaturation and precipitation of ore minerals. Whether an Au-rich source, especially efficient Au precipitation, or a particular setting or climatic influence is necessary to produce very large deposits remains unanswered (cf. Sillitoe, 1992).

Epithermal Au deposits may be found in association with volcanic activity in numerous tectonic settings, including island-arc volcanoes (e.g. Papua New Guinea: Sillitoe, 1989), and continental-based arcs and volcanic centres (e.g. Silverton caldera, Colorado). The shallow formation of epithermal Au deposits suggests a higher probability of erosion, especially the high-sulphidation deposits that frequently occur in active arc environments. Accordingly, epithermal Au deposits, especially in volcanic terranes, are commonly Tertiary in age, although numerous examples are also known of pre-Tertiary deposits, including the Lower Paleozoic high-sulphidation subtype Gidginbung Au deposit, Lachlan fold belt, New South Wales, Australia (Lindhorst and Cook, 1990). Early Devonian hot spring sinter deposits in Scotland (Nicholson, 1989), and other examples of Paleozoic age are known in Australia, from Queensland (Wood et al., 1990) and the Pilabara Craton (Huston et al., 2002). Even much older, Late Proterozoic epithermal Au deposits are also known: the Hope Brook mine, Newfoundland, (Dubé et al., 1998) and the Mahd adh Dhahab deposit, Arabian Shield (Huckerby et al., 1983). Some deposits of even greater antiquity have survived erosion, deformation, and metamorphism (e.g. Proterozoic Mallery Lake deposit, Nunavut: Turner et al., 2001, 2003), whereas many others were subsequently metamorphosed and deformed (e.g. Paleoproterozic Enåsen deposit, Sweden: Hallberg, 1994), which inhibits recognition of their epithermal (especially low-sulphidation) origins.

The tectonic setting of epithermal Au deposits is characterized by extension, at least at the district scale or larger, localizing and facilitating emplacement of magma and, at higher levels, hydrothermal fluids. Regional strike-slip fault systems may bind rhomb-shaped extensional zones or pull-apart basins. Fault jogs or transitions from one fault to another create local environments of extension (see also Goldfarb et al., 2004, for similar control on location of intrusion-related Donlin Creek deposit, Alaska). Regionally extensive rift zones can also provide the extensional framework (e.g. Northern Nevada rift: John et al., 2003). The deposits of the Toodoggone River area, British Columbia, for example, are thought to have formed in an elongate, tectonically controlled graben in the medial portion of an island arc (Diakow et al., 1991, 1993). The preservation of these Early Jurassic epithermal deposits may have to do with the fact that the Toodoggone River area was one of active deposition of younger rocks, rather than one of constructional volcanism and uplift, in a climate providing high erosion rates such as found today in Melanesia (e.g. Chivas et al., 1984).

Small volcanic- and volcaniclastic-hosted deposits in Canada are also found in other structural-tectonic settings of a more local nature. These include the Dusty Mac deposit, British Columbia (e.g. Church, 1973; Zhang et al., 1989, Table 15.1-1), located in breccia and stockwork zones along reverse faults at the margin of the White Lake basin. Gold-mineralized zones of silicification and argillic alteration along faults in the Tintina Trench with Eocene rhyolitic dyke, are characterized by superimposed steam-heated or supergene high-sulphidation alteration mineral assemblages (cf. Duke and Godwin, 1986). Sediment-hosted Au (±Ag) deposits occur in a variety of settings in which sedimentary sequences have been intruded by magmas and also in sedimentary rocks not obviously closely associated with intrusions. In some cases, the deposits are located in the outer zones of paleo-hydrothermal systems associated with intrusions (e.g. Cinola; Equity Silver, British Columbia).

District Scale

Epithermal Au deposits are, in many cases, structurally controlled; the same features that served as the conduits for hydrothermal fluids may have facilitated processes leading to Au deposition (e.g. rapid cooling, boiling, fluid mixing, water-rock reaction, decompression, to name a few). The deposits may be of similar age to their host rocks where these are volcanic, or they may be much younger. A magmatic heat source is commonly inferred. The deposits comprise veins and/or related mineralized breccia and wall rock (e.g. Mt. Skukum), or replacement bodies associated with zones of silicification (e.g. Cinola). Principal geological and other characteristics of each subtype of epithermal Au (±Ag) deposit are listed in Table 2 (see Table 1 for data on individual examples). Both high-sulphidation and low-sulphidation deposit subtypes (distinguished by alteration characteristics) share many features in common. Modern geothermal systems have many features in common with epithermal deposits.

Caldera ring fractures (e.g. Summitville, Colorado: Lipman, 1975), radial fractures (e.g. Lake City, Colorado: Slack, 1980), extensional faulting due to tension above resurgent domes (e.g. Creede, Colorado) may create favourable vein-hosting environments in volcanic terranes. Extensional, pull-apart basins formed between regional strike-slip faults, or at transitions between these faults, provide favourable sites for intrusions and epithermal deposits. Northeast-trending, regional Eocene strike-slip faulting was related to extensional synvolcanic faults at Blackdome, British Columbia, for example, that controlled the emplacement of dykes and Au-bearing quartz veins (Fig. 4). Synchronous tectonic and hydrothermal activity is indicated in some deposits by the fact that many of the vein-bearing faults were active during and after vein filling (e.g. Blackdome, Mt. Skukum, and Toodoggone River deposits); tectonic vein breccias and displaced mineralized and altered rocks resulted. Similar orientations of normal faults southeast of the Blackdome area and east of the Fraser River-Straight Creek fault have been attributed to northwest extension along the Yalakom fault (e.g. Ewing, 1980; Coleman and Parrish, 1990).

Knowledge Gaps

Tertiary terrane was once thought to be virtually the only fertile ground for the occurrence of epithermal Au deposits. And, certainly, a greater number of important epithermal deposits are known in association with young centres of magmatic activity. Thus, the focus of much of the exploration in Canada for epithermal deposits has been in the Cordillera. Within the last twenty-odd years, however, an increasing number of epithermal Au deposits have been recognized in pre-Tertiary terranes. The apparent metamorphic stability of alunite, recognition of abundant aluminosilicate minerals as potential indicators of pre-metamorphic argillic alteration, and the association of zones of replacement by massive quartz have led to the recognition of high-sulphidation subtype deposits in older terrane, even when the deposits have been extensively deformed (e.g. Hope Brook: Dubé et al., 1998). These discoveries emphasize the need to recognize the preservation of near-surface crust in ancient terranes, and to better understand the tectonic environments and conditions that hold higher potential for such preservation.

Low-sulphidation subtype epithermal Au deposits are harder to recognize in ancient terranes, owing to the facts that their commonly found alteration mineral assemblages are not unique, especially in regional metamorphic terranes, or may no longer be present, depending on the grade of subsequent metamorphism, and that these deposits are often not as intimately associated with igneous rocks as is the tendency for the high-sulphidation subtype deposits. In this case, oxygen isotope techniques can be used to support geological evidence for an epithermal environment by providing a measurable, unique, and robust criterion of near-surface origin of the paleo-geothermal system. Moreover, the distribution of the geothermal system can potentially be mapped using oxygen isotope data, even in deformed rocks. A wider application of this approach could enhance recognition of potentially fertile terrane.

Modern geothermal systems, both subaerial and submarine, commonly associated with centres of active volcanism provide excellent analogs to mineralized epithermal systems (e.g. Cooke and Simmons, 2000). These systems, as well as epithermal districts themselves, should be examined to establish to what extent high-sulphidation and low-sulphidation subtype deposits represent a spectrum of characteristics. The essential contribution of magmatic volatiles to form high-temperature, high-sulphidation alteration, and simply the need of a viable heat source to sustain low-sulphidation systems may explain the apparent lack of a common overlap in space and time.

Where do intermediate sulphidation systems 'fit' in time and space relative to low- and high-sulphidation with respect to the development and evolution of high-level magmatism and their related geothermal systems? The spatial, geochemical, and chronological links need to be strengthened between deposit subtypes in the classic epithermal Au environment. Notwithstanding the utility of general relationships depicted by schematic cross sections in common usage (e.g. Fig. 15-2 in Poulsen, 1996), a stronger understanding of any such connections between these environments could facilitate the task of the explorationist.

Comparative studies of both poorly and highly mineralized hydrothermal systems need to be undertaken in order to understand and better define characteristics or specific geologic features of a 'regional' nature (e.g. magmatic-hydrothermal evolution with respect to tectonic setting and climate) as possible predictors of better mineralized terrane.

Deposit Scale

The geological settings of low-, intermediate- and high-sulphidation subtype epithermal deposits are summarized for comparison in Table 2. With respect to a high-level magmatic intrusive centre, the geological properties of these deposit subtypes are broadly those of a 'distal' versus 'proximal' settings, but in both time (relative to magmatic emplacement and active versus passive degassing; e.g., Taylor 1987, 1992). These environments and selected geological properties are illustrated schematically in Figure 8.

The locations of epithermal Au deposits are typically determined by those features that define the hydrothermal system 'plumbing', i.e., provide the hydrological control and control on magmatic emplacement (e.g. structural controls on fluid flow and magmatic emplacement; topographical/paleosurface control of hydrology, boiling elevation, hydrothermal eruption). Extensional faults are especially important, whether due to local, volcanic-related features (e.g. resurgent doming: Creede, Colorado), or to regional tectonism (e.g. rifting zones, or pull-apart basins associated with strike-slip faults: Mt. Skukum, Yukon: Love, 1989; Love et al., 1998; Blackdome, British Columbia: Coleman and Parrish, 1990; R.R. Parrish, pers. comm., 1991; El Peñon, Chile: Arancibia et al., 2006). Fault intersections and fault plane inflections provide zones for vein thickening and zones of brecciation during synchronous movement and vein growth.

High-sulphidation deposits are typically associated with andesitic to rhyolitic rocks and with geologic features associated with sites of active volcanic venting and doming, including among others ring fractures, caldera fill breccias, hot springs, and acidic crater lakes. It is the dominance of directly derived or evolved magmatic fluids that buffer the hydrothermal fluids to low pH and result in the distinct character of the high-sulphidation subtype. Orebodies primarily consist of zones of silica-rich replacement. Bodies of massive 'vuggy silica' and marked advanced argillic alteration mineral assemblages are typical.

Low-sulphidation deposits that occur further removed from active magmatic vents may be more apparently controlled by structural components, zones of fluid mixing, and emplacement of smaller magmatic bodies (e.g. dykes). Meteoric waters dominate the hydrothermal systems, which are more nearly pH neutral in character. Low-sulphidation related geothermal systems are more closely linked to pas-

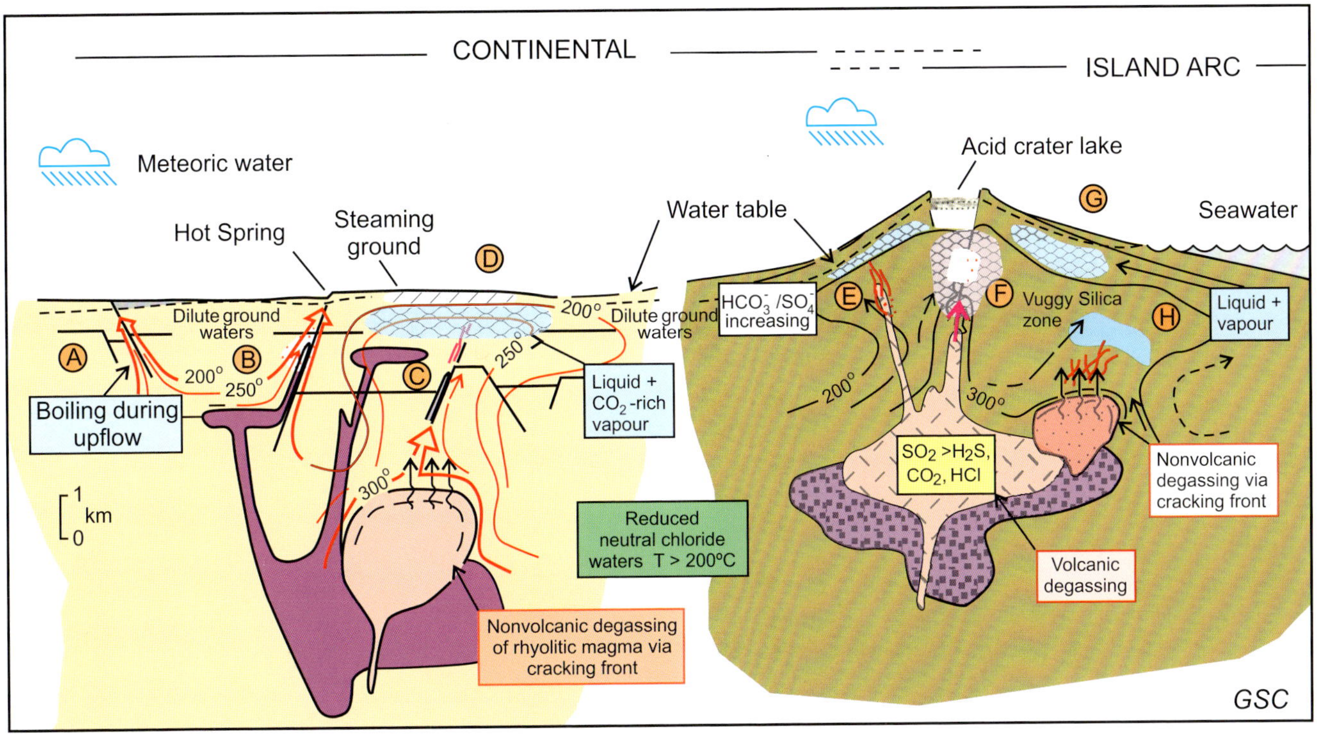

FIGURE 8. Schematic cross-section illustrating the general geological and hydrological settings of quartz-(kaolinite)-alunite and adularia-sericite deposits (from Taylor, 1996; partially adapted from Henley and Ellis, 1983, and Rye et al., 1992). Characteristics shown evolve with time; all features illustrated are not implied to be synchronous. Interpreted settings are indicated for several Canadian deposits discussed in the text; see also Table 1. Local environments and examples of low-sulphidation deposits include: (A) basin margin faults: Dusty Mac; (B) disseminated ore in sedimentary rocks: Cinola; (C) veins in degassing, CO_2-rich, low sulphide content, low-sulphidation systems: Blackdome, Mt. Skukum; (E) porphyry-associated vein-stockwork, sulphide-rich (intermediate sulphidation) and sulphide-poor stages: Silbak-Premier; and (H) disseminated replacement associated with porphyry-type and stockwork deposits, involving seawater: Sulphurets. Examples of high-sulphidation environments include: (D and G) steam-heated advanced argillic alteration (quartz-kaolinite-alunite) zone: Toodoggone River district, British Columbia; (F) magmatic-hydrothermal, high-sulphidation vuggy quartz zone (± aluminosilicates, corundum, alunite) Summitville, Colorado, or Nansatsu district, Japan. Fluid flow parallels isotherms. Upflow zones are shown schematically by arrowhead-shaped isotherms. Volcanic degassing refers to magmatic degassing driven by depressurization during emplacement ('first boiling'). Nonvolcanic degassing refers to vapour exsolution during crystallization ('second boiling'). The SO_2 disproportionates to H_2S and H_2SO_4 during ascent beneath environment (F). Note that free circulation occurs only in crust above about 400°C. All shown temperatures are in Celsius degrees.

sive rather than to active magmatic degassing (if at all), and sustained by the energy provided by cooling, subvolcanic intrusions or deeper subvolcanic magma chambers.

Some deposits with mostly low-sulphidation characteristics with respect to their alteration mineral assemblages have sulphide ore mineral assemblages that represent a sulphidation state between that of high-sulphidation and low-sulphidation deposits. Such deposits tend to be more closely spatially associated with intrusions, and Hedenquist et al. (2000) suggest the term 'intermediate sulphidation' for these deposits.

Zones of boiling, as indicated by mineral textures (bladed carbonate, rhombohedral adularia), are potential sites of Au deposition, especially in low-sulphidation subtype deposits, and that may be related to (and predicted from) paleo-topography. A stationary zone of boiling increases the potential for a high-grade deposit. Similarly, structural control may influence sites of fluid mixing, which can also lead to metal precipitation.

Distribution of Canadian Epithermal Districts

Epithermal, deep epithermal/transitional, or intrusion-related deposits in Canada, as illustrated in Figure 1, are primarily found in the Cordillera in close association with centres of magmatism. Although dominantly young (Tertiary) as a deposit class, older epithermal Au deposits also occur in Canada (Fig. 1, Table 1), or may be suspected in older (even metamorphosed) terranes based on evidence of epithermal alteration. Examples include the low- and high-sulphidation type, Jurassic epithermal deposits of the Toodoggone River area, British Columbia, the low-sulphidation Proterozoic Mallery Lake deposit, Nunavut, and the high-sulphidation Hope Brook deposit, Newfoundland. The Avalonian terrane is notably prospective for both high- and low-sulphidation deposits from the abundant evidence of advanced argillic alteration (e.g. Mills et al., 1999; O'Brien et al., 1999, 2001), and this type of evidence suggests Ordovician terrane in Newfoundland may also be prospective (e.g. O'Driscoll and Wilton, 2005). Whereas advanced argillic alteration can be recognized in metamorphosed terranes, recognition of the origins of sericitic and argillic alteration that formed in high- and low-sulphidation hydrothermal systems may be problematic in metamorphosed terrane on the basis of textural and mineralogical grounds alone. Yet, the existence of ancient unmetamorphosed examples (e.g. Mallery Lake, Nunavut), plus occurrences of deeper, transitional or intrusion-related deposits (Fig. 1), suggest that older deposits remain to be discovered. Exploration of preserved continental volcanic centres and associated epithermal and transitional or intrusion-related deposits in rocks at least as old as the Late Proterozoic should be considered.

Genetic and Exploration Models

Stable isotope and fluid inclusion studies have contributed significantly to our knowledge of the origins, pressures, temperatures, and chemical compositions of hydrothermal fluids responsible for epithermal Au deposits. Studies of modern geothermal systems, hot springs, and volcanic gases have greatly increased our understanding of epithermal deposits because active geothermal systems offer modern-day analogs for physical and chemical parameters that can be directly measured, and compared to inferences made from ancient water-rock interaction and alteration zones (Hedenquist et al., 2000). For example, comparisons between active, steam-dominated geothermal systems like the Geysers-Clear Lake system, California, and hot-spring deposits such as Cinola, British Columbia or the McLaughlin deposit, California (Sherlock, 2005) aid the interpretation of features of hot-spring deposits in the geologic record, and add quantitative constraints on parameters of formation, metal segregation, and concentration.

Determination of mineral solubility, metal volatility and transport, and phase relations, as well as numerical water-rock reaction simulations, have contributed to our quantization of the chemical and physical nature of mineralizing hydrothermal fluids, and also to our understanding of the processes that lead to the transport and deposition of Au, Ag, and base metals.

Lindgren (1922, 1933) suggested that degassing magmas are sources of many ore-forming constituents in epithermal Au deposits, and this supposition appears to be essentially correct for magmatic-hydrothermal high-sulphidation deposits (Stoffregen, 1987; Rye et al., 1992). However, for many deposits (e.g. the majority of low-sulphidation subtypes) O- and H-isotope data permit only a very small fraction (i.e. <10%) of the hydrothermal water to be of magmatic origin, despite the close association of some deposits with cooling magmatic rocks, whereas C and S isotope studies indicate a significant magmatic contribution in many cases. Thus, a mineralizing fluid can have a complex origin, one involving links to degassing magmas as well as the dominance of local recharge waters to fuel the hydrothermal system.

Schematic cross-sections illustrating the principal physical environments of low-sulphidation and high-sulphidation epithermal vein and hot-spring deposits and their related geothermal systems are shown in Figure 8. This figure emphasizes features of genetic significance found in at least some of the more prominent Canadian deposits.

Slow-cooling epizonal plutons, such as those shown in Figure 8, may undergo 'subvolcanic' (passive) degassing as they crystalllize (i.e. (second boling') rather than 'volcanic' (active) degassing, and can provide mineralizing constituents to overlying or adjacent meteoric hydrothermal systems via protracted leakage of magmatic volatiles across cracking fronts at the margins of the crystallizing magma. Variations on this theme derive also from differences in the S content of rhyolitic (lower) to andesitic (higher) magmatic volatiles, from differences in crustal level at which magmatic degassing occurs, and from the relative proportions of magmatic and meteoric fluids involved through time. Additional discussion can be found, for example, in Henley (1985), Stoffregen (1987), White and Hedenquist (1990), Giggenbach (1992), Rye et al. (1992), Sillitoe (1993), and Heinrich et al. (2005).

The two principal (end-member) geochemical environments of epithermal mineralization and alteration are determined largely by the dominance in each case of two different fluids. On the one hand, magmatic-hydrothermal environments that are dominated (buffered) by acidic, magmatic fluids containing CO_2-, HCl-, and SO_2-rich vapour produce high-sulphidation mineral assemblages characterized by oxidized forms of iron (e.g. hematite) and S (e.g. alunite), and by base leaching of wall rocks leaving marked (residual) silica enrichment. This environment may overlie porphyry systems (Sillitoe and Bonham, 1984). On the other hand, near neutral, more reduced, meteoric-dominated waters containing Cl, H_2S, and CO_2, yield low-sulphidation (adularia-sericite) mineral assemblages through hydrolysis reactions involving feldspar in the wall rocks. The chemical state of these fluids becomes largely wall-rock buffered.

Two fundamentally different hypotheses regarding the source of Au in epithermal deposits are (1) metals are supplied directly by actively or passively degassing magma (e.g. Taylor, 1987; 1988) that also provides heat to the paleo-hydrothermal system, and (2) the metals are leached from the rocks that host the geothermal system. On the one hand, isotopic confirmation of the importance of meteoric waters has encouraged proponents of the second hypothesis. On the other hand, isotopic data also indicate that S and C are of magmatic origin in certain deposits (e.g. Summitville, Colorado: Rye et al., 1992) and in active geothermal systems (e.g. Taylor, 1987). Symonds et al. (1987) demonstrated the transport and variable degrees of volatility of precious and base metals, and their association with magmatic Cl, C, and S, by sampling and analyzing very high-temperature volcanic gases. Metals and other constituents were shown to separate upon cooling, illustrating the potential for direct magmatic contribution to geothermal systems and for temperature-related metal zoning.

At magmatic temperatures, Au solubility is high in Cu-Fe sulphides, which may represent a significant Au source depending on the passive degassing history of a porphyry-forming magma (e.g. Kesler et. al., 2002). The oxidation state or sulphide activity of magma at initial vapour saturation, for example, might influence the amount of extractable Au that could be contributed to associated epithermal deposits. From the phase relations of NaCl brines and calculated fluid speciation from 150 to 450ºC involving brine, S, and Au, Heinrich et al. (2004) proposed that an excess of sulphide over Fe in an exsolved, one-phase Au-rich magmatic fluid is essential for the transport of Au to the hydrothermal system (see also Heinrich, 2005). The presence of Au in vapour-dominated portions of hydrothermal systems, the sublimation of metals from volcanic vapours (e.g. Symonds et al., 1987), the separation of base and precious metals in epithermal deposits (Buchanan, 1981), and phase-relations and experimental data (Williams-Jones and Heinrich, 2005), suggest that significant amounts of Au can be transported by a low-density (subcritical) fluid upon phase separation from the initially exsolved magmatic fluid.

Alteration mineral assemblages are characteristic of two end-member chemical environments of alteration and mineralization: low to very low pH, oxidized fluids (high-sulphidation subtype) and near neutral, more reduced fluids (low- and intermediate-sulphidation subtypes). These two environments are contrasted in Figure 9, which also represents stability fields for selected mineral and isopleths of Au solubility (after Giggenbach, 1992). Field (G) in Figure 9 illustrates the temperature and oxidation state of the meteoric geothermal fluids discharged from the Geysers geothermal field, California (Lowenstern and Janik, 2003) as an example of typical low-sulphidation systems. Such fluids may further evolve by water/rock reaction or by mixing with fluids chemically influenced by water/rock interaction (e.g. Mt. Skukum, field SK, in Fig. 9). Magmatic fluids of higher temperature and a relatively more oxidized (i.e. more negative R_H) nature dominate hydrothermal systems hosting high-sulphidation subtype deposits. These fluids may mix with surface waters and/or with geothermal waters similar to those from the Geysers, as shown by the field for Summitville fluids (field S in Fig. 9).

The upwardly welling, highly acidic, magmatic-hydrothermal plume may produce a high-sulphidation mineralization event that is likely to be short-lived, limited by the shallow degassing of the magma in response to depressurization during its ascent (so-called 'first boiling'), and by the eventual neutralization of the fluids due to reaction with wall rocks and/or dilution by meteoric fluids. In contrast, meteoric fluids heated by cooling magmatic rocks can provide potential fluids for mineralization and alteration over somewhat longer periods of time, and at sites further removed from the magmatic heat source. With time, the meteoric water dominated environment may encroach upon the earlier, hotter, hydrothermal-magmatic environment.

Active geothermal systems provide instructive analogues to low-sulphidation hydrothermal systems. Geochemical studies of dominantly volcanic-hosted geothermal systems in the Taupo Volcanic Zone, New Zealand (see Henley and Hedenquist, 1986) have demonstrated the existence of two principal types of fluids: (1) a deep chloride water, generally 200 to about 300°C, and (2) a shallower, less than 100 to 200°C steam-heated, low-chlorinity, acidic water. The interface between waters of markedly different salinity has been described in the Salton Sea geothermal system by Williams and McKibben (1989). These deep chloride waters produce low-sulphidation subtype alteration (e.g. Henley, 1985), and where they are rapidly depressurized, degas CO_2 and H_2S, cool, and precipitate precious and base metals (Clark and Williams-Jones, 1990). The well scales studied by Clark and Williams-Jones (1990) revealed a vertical separation of precious metals (higher) and base metals (lower) analogous to that described by Ewers and Keays (1977) for the Broadlands geothermal field (New Zealand), and by Buchanan (1981) for a number of deposits.

Sub-millimetre-scale variations in $\delta^{18}O$, as much as 6 per mil in vein quartz from Hishikari at times of Au precipitation, indicate that intermittent vein opening permitted introduction of deep, metal-bearing fluids to the veins. The deep-sourced fluids mixed with meteoric water, boiled (indicated by bladed quartz), cooled, and precipitated Au (Hayashi, et al., 2001). The bladed quartz analyzed by Hayashi, et al. (2001) formed subsequent to initial boiling by replacement

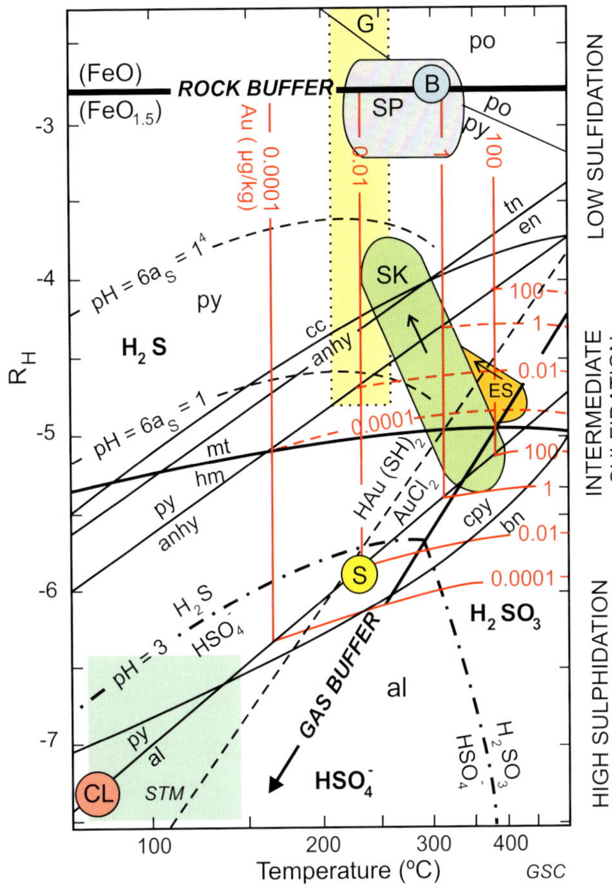

FIGURE 9. Diagram of redox potential (RH = log f_{H_2}/f_{H_2O}) versus temperature (from Taylor, 1996; modified after Giggenbach, 1992; Hedenquist et al., 1994). Calculated isopleths of Au in µg/kg solubility (as the dissolved species HAu(SHh; (Giggenbach, 1992) are shown in red. An equimolar isopleth for $HAu(SH)_2$ and $AuCl_2$ (after Hedenquist et al., 1994) is shown for pH=3 and 1.0 wt.% Cl (or pH=5 and 10 wt.% Cl). The thiogold complex $HAu(SH)_2$ probably dominates as the Au-transporting agent in much of the epithermal environment at pH<5 (Giggenbach, 1992). Redox conditions for mineral deposition in the Broadlands-Ohaaki geothermal system (B), Summitville deposit (S), and in Crater Lakes (CL) (e.g. Ruapehu, New Zealand) are from Giggenbach (1992). Fields showing approximate conditions of formation, and their variation with time shown by arrows, for Mt. Skukum (SK), Equity Silver (ES), and Silbak-Premier (SP; also Blackdome and others) are based on data in references cited in Table 1. Field (G) represents the redox state for total discharge (meteoric origin) from the Geysers geothermal field, California (Lowenstern and Janik, 2003). Also shown are approximate conditions for steam-heated deposits (STM) of high-sulphidation subtype. The diagram shows stability limits and reactions for several minerals discussed in the text. Diagram represents a large variation in system composition, and not all mineral reactions are implied to occur in all deposits. The 'gas buffer' curve represents redox control by a magmatic SO_2-H_2S gas mixture; the Fe^{+2}/Fe^{+3} couple provides redox control in 'rock dominated' systems (rock buffer). High-sulphidation deposits form under oxidizing, acidic conditions of the lower one-third of the diagram. Conditions of formation of low-sulphidation deposits are represented by the upper half of diagram, with most forming near the rock buffer curve. Isopleths of Au solubility (in µg/kg) are shown in red for two sets of buffered conditions, one set buffered by the coexistence of pyrite and alunite, the other by pyrite and anhydrite. Dashed isopleths apply to the pyrite-anhydrite buffered conditions. Abbreviations are: aS = a_{H_2S}/a_{SO_4}; ai = alunite; anhy = anhydrite; bn = bornite; cc = calcite; cpy = chalcopyrite; en = enargite; hm = hematite; mt = magnetite; po = pyrrhotite; py = pyrite; tn = tennantite. For alternative representations of mineral/gas phase stability at different conditions see Cooke and Simmons (2000).

of calcite. Any inheritance of ^{18}O from the isotopically heavier calcite would indicate a lower estimate for the $\delta^{18}O$ of the deeper, evolved fluid. Nevertheless, the mixing of metalliferous and dilute fluids and consequent metal precipitation (via cooling, dilution, and oxidation) are clearly demonstrated.

Simple, conductive cooling of Au-bearing fluids is sufficient to cause Au precipitation (see Fig. 9). Boiling can also cause cooling, chemical fractionation, and an increase in pH associated with acid vapour loss that leads to saturation and precipitation of chloride-complexed metals (e.g. Cu, Pb, Zn: Drummond and Ohmoto, 1985; Spycher and Reed, 1989; Williams-Jones and Heinrich, 2005). Also, degassing of initially CO_2-rich fluids in gas-rich systems depletes the liquid in H_2S that is carried off in a CO_2-rich vapour. The loss of H_2S eventually leads to precipitation of sulphide-complexed metals (e.g. Au; Drummond and Ohmoto, 1985; Henley, 1985; Hayashi and Ohmoto, 1991). Carbon dioxide and hydrogen sulphide are well correlated in some geothermal fluids (Fig. 11 in Taylor, 1987). Boiling and chemical fractionation of the hydrothermal fluid provides an explanation for the separation of precious and base metals. This separation results in a vertical zoning where fluids are upwardly flowing (Clark and Williams-Jones, 1990), or in relative temporal stages, such as at Silbak-Premier, British Columbia, and El Indio, Chile. As a corollary, larger vein deposits require the movement of larger amounts of fluid through localized zones of boiling, and thus the importance of structural analysis in exploration is obvious. Neutralization and cooling of ore fluids may also occur (1) by mixing with dilute groundwaters, and (2) by water-rock reaction (e.g. sulphidation of ferrous iron-bearing minerals), especially during formation of disseminated and replacement-type orebodies.

Steam-heated acid waters formed by the oxidation and condensation of H_2S (boiled off deeper geothermal reservoirs) in groundwater produce high-sulphidation subtype alteration of the volcanic rocks (Henley and Hedenquist, 1986). The Champagne Pool, in the CO_2-rich Waiotapu geothermal field (steam-heated, high-sulphidation subtype alteration), New Zealand, is a hydrothermal eruption feature below which Au and Ag are being deposited in response to boiling and loss of H_2S over the approximate temperature interval 250 to 175°C (Hedenquist, 1986). Ore-grade, Au-bearing amorphous sulphides precipitate in the pool at 75°C, and base metal sulphides occur below the zone of boiling. Acidic waters produce advanced argillic alteration and, with variation in P_{CO_2}, evolve to cause the replacement of adularia and albite by sericite. Thus, by chemical evolution, a geothermal field, initially boiling and producing high-sulphidation subtype alteration, may eventually produce minerals characteristic of low-sulphidation subtype alteration.

The precious metal content of steam-heated alteration zones may also be related to the rate of fluid ascent versus the extent of boiling and H_2S loss: faster moving fluids and/or those less depleted in H_2S may produce higher grades of precious metals in steam-heated alteration zones. This might apply to the ascension of boiling magmatic hydrothermal plumes as well as to boiling meteoric and marine geothermal fluids.

Exploration for epithermal Au deposits entails, for a comprehensive approach, judicial application of methodologies to assess the geological characteristics (e.g. tectonic/structural setting, petrological association, mode of occurrence, geochronology), mineralogical and geochemical characteristics (mineral assemblage, mineral/rock chemical compositions, isotopic composition, exploration geochemical techniques), and geophysical characteristics (e.g. electrical and magnetic properties). The results of the application of these techniques are compared to one or more 'models' that represent empirically determined associations of characteristics. Hedenquist et al. (2000) is a useful and comprehensive reference in this regard.

Assessment of Geological Characteristics

Volcanic arcs and belts with abundant intermediate to felsic rocks and associated rift systems host epithermal Au deposits of many ages. Evidence of high-level magmatism in more deeply eroded terrane may still offer possibilities for transitional or intrusions-related deposits. Geologic mapping, including alteration of mineral assemblages, and attention to structural control(s) provides a fundamental means of assessment.

Assessment of Mineralogical and Geochemical Characteristics

Mapping and recognition of alteration mineral assemblages are reasonably straightforward in unmetamorphosed terrane. New instrumental technologies, such as Short-Wave Infrared Spectroscopy (SWIR; e.g., PIMA®, Portable Infrared Mineral Analyzer; Ducart et al., 2006), in addition to portable (to field offices) XRF analyzers and X-Ray diffractometers, are finding increased application for mineralogical and elemental identification in the field. In deformed and metamorphosed terranes, however, interpretive mineralogical or alteration mapping may be problematic. In particular, distinction of high-sulphidation alteration formed in steam-heated zones (which may form above either high- or low-sulphidation systems), from high-temperature alteration may affect interpretation of the deposit subtype and exploration strategy. The nature and origin of highly silicic zones should also be determined, particularly in deformed terranes where the usual textural criteria may no longer be applied. Oxygen isotope mapping, using whole-rock analysis, can be used to map paleo-hydrothermal systems, even in highly metamorphosed and deformed terranes. In particular, oxygen isotope techniques can be especially useful to decipher the origins of chlorite-sericite-bearing mineral assemblages, and assist in interpreting the origins (e.g. residual, vuggy silica zone from silicic zones from near-surface, lower temperature silicification).

Two applications of oxygen isotope techniques for the exploration of epithermal Au deposits are shown in Figures 10 and 11 representing examples of high-sulphidation and low-sulphidation systems, respectively. The Pilot Mountain area in the Carolina Slate Belt contains a number of previously mined high-sulphidation epithermal Au deposits. The greenschist-metamorphosed terrane bears mineralogical evidence of argillic and advanced argillic alteration shown on the basis of oxygen isotope characteristics (Klein and Criss, 1988) to have formed in a meteoric-water recharged, high-sulphidation system. Isotopic zoning above the associated high-level stock documents upflow of magmatic-hydrothermal fluid that was most intense in the area of vuggy silica

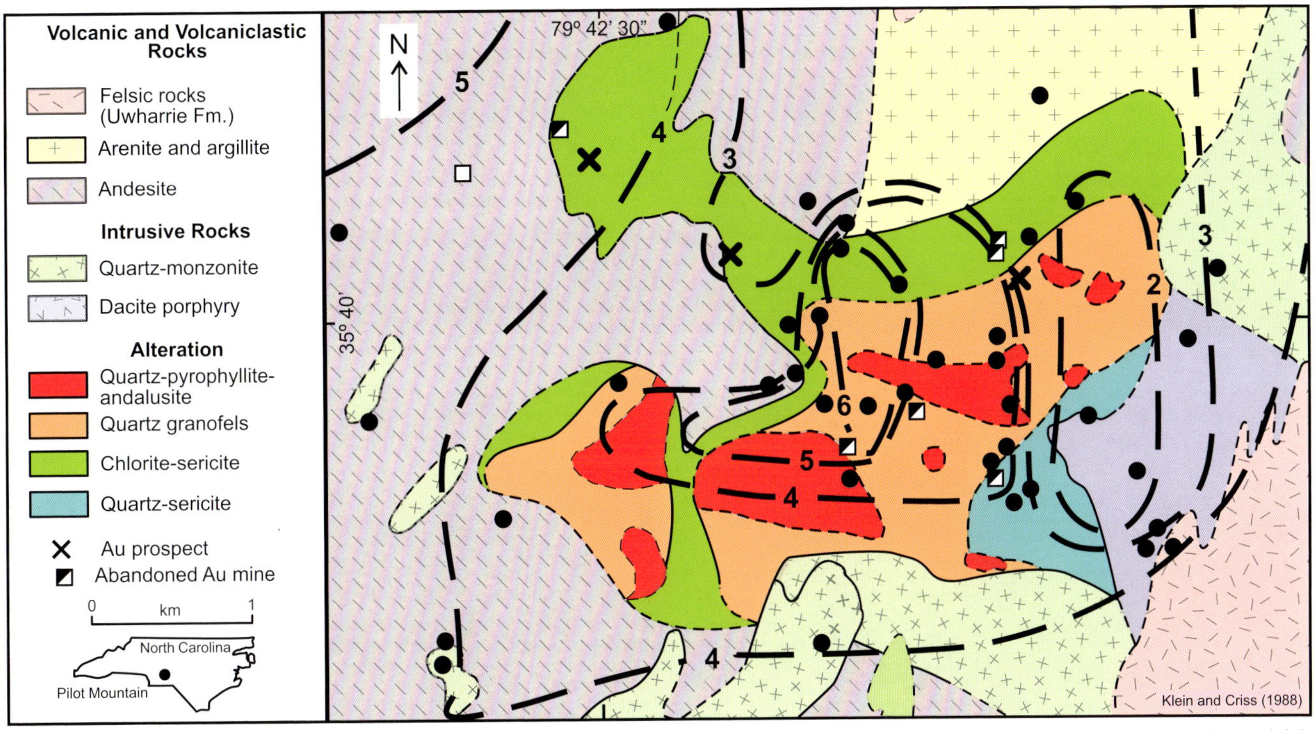

FIGURE 10. A $\delta^{18}O_{WHOLE\text{-}ROCK}$ isopleth map of the high-sulphidation epithermal Au district of Pilot Mountain in the Carolina-Avalon slate belt, Randolph County, North Carolina (after Klein and Criss, 1988). Klein and Criss (1988) infer the greenschist metamorphosed terrane to be westward tilted, exposing a vertical section through part of a high-level quartz monzonite stock, dacite porphyry, and argillic and advanced argillic alteration zones at the apex of the stock. Areas mapped as 'quartz granofels' are interpreted to represent a metamorphosed 'vuggy silica' zone associated with Au (note greater number of Au mines). Alteration and mineralization at Pilot Mountain are analogous to that at the Hope Brook mine, Newfoundland (Dubé et al., 1998), and similar to areas of high-sulphidation alteration of broadly similar age in the Burin and Avalon Peninsulas, Newfoundland (see discussion in text). Fifty-three samples, virtually all of which yielded low $\delta^{18}O_{WHOLE\text{-}ROCK}$ values, record a near-surface, meteoric water-recharged geothermal system over an area of more than 30 km^2 (larger than the 4 x 6 km area shown). The relative increase in $\delta^{18}O_{WHOLE\text{-}ROCK}$ in altered rocks above the stock probably reflects the upflow of isotopically heavier, magmatic-hydrothermal fluids. Mapped anomalies of Au, Mo, Sn, Cu, and B in soil samples agree well with isotopically mapped zones of magmatic-hydrothermal influence (Klein and Criss, 1988). Continued hydrothermal upflow is indicated by the northwest-trending zone of chlorite-sericite; a higher density of samples would permit a more detailed isotopic definition of hydrothermal flow patterns.

alteration and Au mineralization (Fig. 10). The isotopically mappable affects of related meteoric hydrothermal alteration cover at least 30 km^2, well beyond definitive mineralogical zones, and distinguish this area from other, non-mineralized sites of plutonism in the slate belt (Klein and Criss, 1988). The application of this technique to mapping meteoric high-sulphidation systems in undeformed terrane is straightforward (e.g. Tonopah, Nevada: Taylor, 1973), whereas the study at Pilot Mountain demonstrates the potential of this technique for exploration in older terranes.

Oxygen isotope zoning about veins in the low-sulphidation epithermal Au deposit at Hishikari, Japan (Fig. 11) shows definitive effects of hydrothermal water/rock interaction in a surface zone as much as 200 m or more in length above blind vein deposits. Whereas clay mineral alteration can also be recognized and mapped at the surface of this young unmetamorphosed deposit (Izawa et al., 1990), whole-rock oxygen isotope anomalies as telltale indicators of epithermal fluid flow can survive even high-grade regional metamorphism.

Soil and rock geochemical analyses may prove fruitful. The Pilot Mountain, North Carolina, district (Keith and Criss, 1988; see also caption to Fig. 10) provides a particularly good example, in a greenschist-facies metamorphic terrane, of the correspondence between mineralogical alteration, oxygen isotope zoning, and geochemical soil anomalies. In Canada, geochemical mapping of the typical epithermal pathfinder elements (Hg, Sb, As, Tl, in addition to Au and Ag), plus 'intrusion-related' elements (e.g. Mo, Cu, Sn, B) may also be tested in both soil and till, as well as rocks. Aluminosilicates, corundum, sulphides, specular hematite, and alunite may, among other minerals, also prove useful in till analysis.

Assessment of Geophysical Parameters

In contrast to their application in the exploration of other types of ore deposits, geophysical techniques have been less useful in the discovery of epithermal deposits (Sillitoe, 1995; Hedenquist et al., 2000). Except for the use of aeromagnetic surveys as a very powerful aid in regional geologic mapping, the application of other geophysical techniques for epithermal Au deposits in Canada appears less fruitful.

Knowledge Gaps

Upon comparison of many features, both regional and local, of 16 bonanza (>30 tonnes Au) and giant (>200 tonnes Au) epithermal Au deposits, Sillitoe (1992) concluded that, although complex arc environments and unusual igneous rock types seemed more prospective, no single feature could be isolated as an apparent cause or explanation. Either an unusually rich source of Au or an unusually effective depositional process was necessary to effect such concentrations

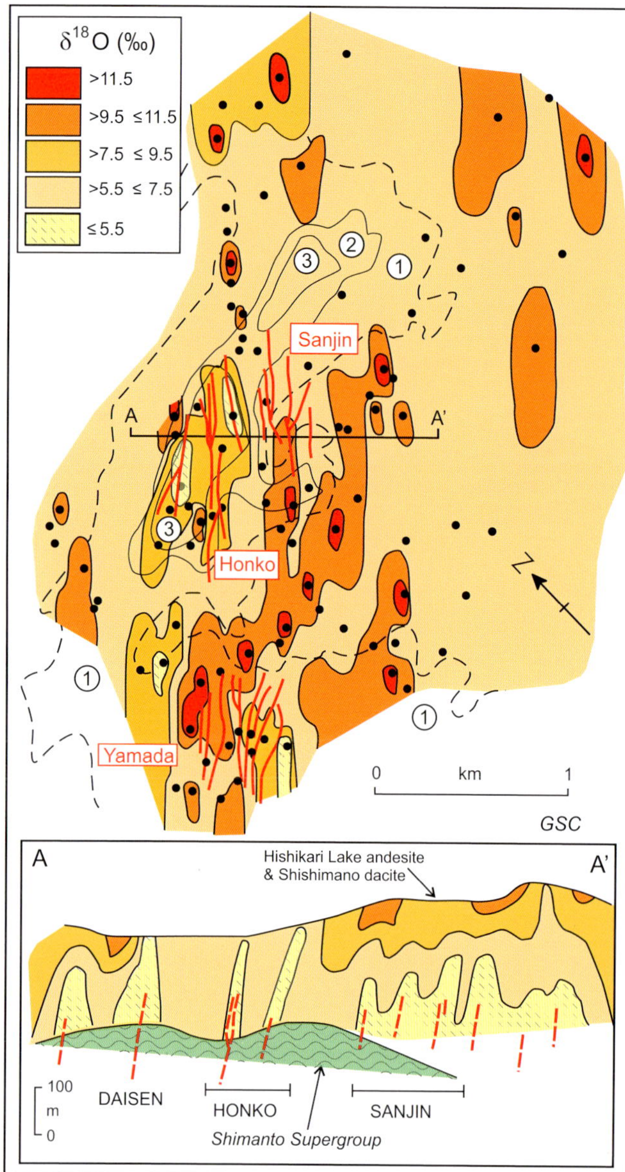

Figure 11. The $\delta^{18}O$ isopleth map and cross-section of the low-sulphidation vein Au deposit at Hishikari area, Japan, from Taylor (1996; based on data in Naito et al., 1993) show oxygen isotope zoning of altered wall rocks associated with vein emplacement and mineralization (see also Faure et al., 2002). Positions of Daisen quartz vein, Honko vein system, and Sanjin vein system (shown in red) are projected to surface and into the plane of section A-A'. Oxygen isotope data of Naito et al. (1993) were recontoured to emphasize the structural control of fluid flow indicated by oxygen isotope zoning. Fresh andesite and dacite are presumed to have a whole-rock $\delta^{18}O$ value of 8.5‰ (Naito et al., 1993). Locations of whole-rock oxygen isotope samples are shown by filled circles, but the locations of subsurface samples, which provide the basis for the isotopic zoning pattern in the cross-section, are omitted for simplicity. Alteration mineral zones (after Izawa et al., 1990) are (1) chlorite-smectite and/or sericite-smectite; (2) quartz-smectite and/or kaolin minerals; and (3) cristobalite-smectite+kaolinite or halloysite. Chlorite-sericite alteration dominates at depth. Note that prospective is otopic zones extended to the surface, some 200 m in many cases, above blind vein deposits.

of Au. This 'chicken or egg' conclusion remains as a principal enigma, a key question in the knowledge gap. A complete spectrum in Au endowments in epithermal deposits worldwide might be expected if Au endowment depended only on the Au content of the source materials. Sillitoe (1992) emphasized the necessity for the "geologically unexpected" in the environments of these rich deposits and the likelihood that that this factor resulted in usually effective Au precipitation. Despite many detailed studies since Sillitoe's paper (1992), universal agreement on this central question appears to elude.

A firmer understanding of links between porphyries and epithermal systems is evolving, and an understanding of the temporal differences in magmatic and hydrothermal evolution that explains the lack of direct linkages (e.g. low-sulphidation and porphyry Cu-Au deposits). Efficient trapping of hydrothermally transported Au is certainly required for an economic deposit, and processes such as mixing, boiling, cooling, and oxidation are known to have occurred at the time of gold precipitation (e.g. Hayashi et al., 2001). As a corollary, further studies of the processes of magmatic degassing (both active and passive), associated metal migration, and the influence of the oxidation state of the magma on metal availability and migration would seem to be helpful.

A sufficient number of ancient epithermal Au deposits, both low- and high-sulphidation subtypes, are now known to raise the level of understanding needed regarding the likelihood of preservation and rates of destruction of the epithermal regime of the crust. Clearly very old examples have survived.

Acknowledgements

This chapter was, in part, extracted and updated from Taylor (1996). Lillian Munro assisted greatly by data, map, and literature compilation. Benoit Dubé, Wayne Goodfellow, and Ian Jonasson provided helpful review comments on short notice.

References

Andrew, K.P.E., Gordon, C.I., and Cann, R.M., 1986, Wolf epithermal precious metal vein prospect central British Columbia (93F/3), *in* Geological Fieldwork 1985: British Columbia Ministry of Energy, Mines, and Petroleum Resources, Paper 1986-1, p. 317-320.

Arancibia, G., Matthews, S.J., Cornejo, P., Perez de Arce, C., Zuluaga, J. I., and Kasanevan, S., 2006, $^{40}Ar/^{39}Ar$ and K-Ar geochronology of magmatic and hydrothermal events in a classic low-sulphidation epithermal bonanza deposit; El Peñon, Northern Chile: Mineralium Deposita, v. 41, p. 505-516.

Arribas, A., Jr., 1995, Characteristics of high-sulphidation epithermal deposits, and their relation to magmatic fluid: Mineralogical Association of Canada, Short Course Handbook, v. 23, p. 419-454.

Ashley, R.P., 1982, Occurrence model for enargite-gold deposits, *in* Erickson, R.L., ed., Characteristics of Mineral Deposit Occurrences: United States Geological Survey, Open File Report 82-795, p. 144-147.

Bagby, W.C., and Berger, B.R., 1986, Geologic characteristics of sediment-hosted disseminated precious-metal deposits in the western United States: Reviews in Economic Geology, v. 2, p. 169-199.

Baker, T., 2002, Emplacement depth and carbon dioxide-rich fluid inclusions in intrusion-related gold deposits: Economic Geology, v. 97, p. 1111-1117.

Bakken. B. M., and Einaudi, M. T., 1986, Spatial and temporal relations between wall rock alteration and gold mineralization, Main Pit, Carlin gold mine, Nevada, U.S.A., *in* Macdonald, A.J., ed., Proceedings of Gold '86: An International Symposium on the Geology of Gold: Konsult International, Toronto, Ontario, p. 388-403.

Barr, D.A., Drown, T.J., Law, T.W., McCreary, G.A., Muir, W.W., Paxton, J.M., and Roscoe, R.L., 1986, The Baker mine operation: The Canadian Institute of Mining and Metallurgy Bulletin, v. 79, p. 61-68.

Berger, B. R., and Bethke, P. M., (eds.) 1986, Geology and Geochemistry of Epithermal Systems: Reviews in Economic Geology, v. 2, Society of Economic Geologists, 298 p.

Berger, B.R., and Henley, R.W., 1989, Advances in the understanding of epithermal gold-silver deposits, with special reference to the western United States, *in* Keays, R.R., Ramsay, W.R.H., and Groves, D.I. eds., The Geology of Gold Deposits: The Perspective in 1988: Economic Geology, Monograph 6, p. 405-423.

Bethke, P.M., Rye, R.O., Stoffregen, R.E., and Vikre, P.G., 2005, Evolution of the magmatic-hydrothermal acid-sulfate system at Summitville, Colorado: Integration of geological, stable isotope, and fluid inclusion evidence: Chemical Geology, v. 215, p. 281-315.

Bonham, J.F., Jr., 1988, Models for volcanic-hosted precious metal deposits; a review, *in* Schafer, R.W., Cooper, J.J., and Vikre, P.G., eds., Bulk Mineable Precious Metal Deposits of the Western United States: Geological Society of Nevada, p. 259-271.

Braithwaite, R.L., and Faure, K., 2002, The Waihi epithermal gold-silver-base metal sulfide-quartz vein system, New Zealand: Temperature and salinity controls on electrum and sulfide deposition: Economic Geology, v. 97, p. 269-290.

British Columbia Ministry of Energy, Mines and Petroleum Resources, 1992, Minfile/pc V. 3.0: Information Circular 1992-28.

Brown, D.A., and Cameron, R., 1999, Sediment-hosted, disseminated gold deposits related to alkalic intrusions in the Howell Creek structure, southeastern British Columbia (82G/2, 7), *in* Geological Fieldwork 1998: British Columbia Ministry of Energy, Mines and Petroleum Resources, Paper 1999-1, p. 179-192.

Buchanan, L.J., 1981, Precious metal deposits associated with volcanic environments in the southwest, *in* Dickenson, W.R., and Payne, W.D., eds., Relations of Tectonics to Ore Deposits in the South Cordillera, Arizona: Geological Society Digest, v. XIV, p. 237-261.

Cameron, E.M., and Hattori, K., 1985, The Hemlo gold deposit, Ontario: A geochemical and isotopic study: Geochimica et Cosmochimica Acta, v. 49, p. 2041-2050.

Carman, G.D., 2003, Geology, mineralization, and hydrothermal evolution of the Ladolam Gold Deposit, Lihir Island, Papua New Guinea, *in* Simmons, S., and Graham, I., eds., Volcanic, Geothermal, and Ore-Forming Fluids: Rulers and Witnesses of Processes within the Earth: Society of Economic Geologists, Special Publication 10, p. 247-284.

Champigny, N., and Sinclair, A.J., 1982, The Cinola gold deposit, Queen Charlotte Islands British Columbia, *in* Hodder, R.W., and Petruk, W., eds., Proceedings of the Gold Symposium: The Canadian Institute of Mining and Metallurgy, Special Volume 24, p. 243-254.

Chivas, A.R., O'Neil, J.R., and Katchan, G., 1984, Uplift and submarine formation of some Melanesian porphyry copper deposits: Stable isotope evidence: Earth and Planetary Science Letters, v. 68, p. 326-334.

Christie, A.B., 1989, Cinola deposit, Queen Charlotte Islands, *in* Geological Fieldwork 1988: British Columbia Ministry of Energy, Mines, and Petroleum Resources, Paper 1989-1, p. 423-428.

Church, N.B., 1973, Geology of the White Lake Basin: British Columbia Department of Mines and Petroleum Resources, Bulletin 61, 120 p.

Clark, J.R., and Williams-Jones, A.E., 1986, Geology and genesis of epithermal gold-barite mineralization, Verrenas deposit, Toodoggone district, B.C. (abstract): Geological Association Canada-Mineralogical Association Canada, Annual Meeting, Ottawa, 1986, Program with Abstracts, v. 11, p. 57.

——— 1990, Analogues of epithermal gold-silver deposition in geothermal well scales: Nature, v. 346, p. 644-645.

——— 1991, New K-Ar isotopic ages of epithermal alteration from the Toodoggone River area, British Columbia (94E) in Geological Fieldwork 1990: British Columbia Ministry of Energy, Mines and Petroleum Resources, Paper 1991-1, p. 207-216.

Coleman, M.E., and Parrish, R.R., 1990, Eocene deformation in the Bridge River terrane, B.C.: Constraints on dextral movement on the Fraser fault (abstract): Geological Association Canada, Program with Abstracts, v. 15, p. A26.

Cooke, D.R., and Bloom, M.S., 1990, Epithermal and subjacent porphyry mineralization, Acupan, Baguio district, Philippines: A fluid-inclusion and paragenetic study, *in* Hedenquist, J., White, N.C., and Siddeley, G., eds., Epithermal Gold Mineralization of the Circum-Pacific; Geology, Geochemistry, Origin and Exploration, II: Journal of Geochemical Exploration, v. 35, p. 297-340.

Cooke, D.R., and Simmons, S.F., 2000, Characteristics and genesis of epithermal gold deposits: Society of Economic Geologists Reviews, v. 13, p. 221-244.

Cox, D.P., and Singer, D.A., eds., 1986, Mineral Deposit Models: United States Geological Survey, Bulletin 1693, United States Government Printing Office, Washington, D.C., 379 p.

Cunningham, C.G., 1985, Relationship between disseminated gold deposits and a regional paleo-thermal anomaly in Nevada: United States Geological Survey, Open File Report 85-722, 20 p.

Cyr, J.B., Pease, R.B., and Schroeter, R.G., 1984, Geology and mineralization at Equity Silver mine: Economic Geology, v. 79, p. 947-968.

de Ronde, C.E.J., and Blattner, P., 1988, Hydrothermal alteration, stable isotopes, and fluid inclusions of the Golden Cross epithermal gold-silver deposit, Waihi, New Zealand: Economic Geology, v. 83, p. 895-917.

Deyell, C.L., Leonardson, R., Rye, R.O., Thompson, J.F.H., Bissig, T., and Cooke, D.R., 2005, Alunite in the Pascua-Lama high-sulfidation deposit: constraints on alteration and ore deposition using stable isotope geochemistry: Economic Geology, v. 100, p. 131-148.

Diakow, L.J., Panteleyev, A., and Schroeter, T.G., 1991, Jurassic epithermal deposits in the Toodoggone River area, Northern British Columbia: Examples of well preserved volcanic-hosted, precious metal mineralization: Economic Geology, v. 86, p. 529-554.

——— 1993, Geology of the Early Jurassic Toodoggone Formation and gold-silver deposits in the Toodoggone River map area, northern British Columbia: British Columbia Geological Survey, Bulletin 86, 72 p.

Dong, G., and Morrison, G.W., 1995, Adularia in epithermal veins, Queensland: Morphology, structural state and origin: Mineralium Deposita, v. 30, p. 11-19.

Drummond, S.E., and Ohmoto, H., 1985, Chemical evolution and mineral deposition in boiling hydrothermal systems: Economic Geology, v. 80, p. 126-147.

Dubé, G., Dunning, G., and Lauziere, K., 1998, Geology of the Hope Brook Mine, Newfoundland, Canada: A preserved Late Proterozoic high-sulfidation epithermal gold deposit and its implications for exploration: Economic Geology, v. 93, p. 405-436.

Dubé, B., Gosselin, P., Mercier-Langevin, P., Hannington, M., and Galley, A., 2007, Gold-rich volcanogenic massive sulphide deposits, *in* Goodfellow, W.D., ed., Mineral Deposits of Canada: A Synthesis of Major Deposit-Types, District Metallogeny, the Evolution of Geological Provinces, and Exploration Methods: Geological Association of Canada, Mineral Deposits Division, Special Publication 5, p. 75-94.

Ducart, D.F., Crósta, A. P., Filho, C.R.S., and Coniglio, J., 2006, Alteration mineralogy at the Cerro La Mina epithermal prospect, Patagonia, Argentina: Field mapping, short-wave infrared spectroscopy, and ASTER images: Economic Geology, v. 101, p. 981-996.

Duke, J.L., and Godwin, C.I., 1986, Geology and alteration of the Grew Creek epithermal gold-silver prospect, south-central Yukon, *in* Yukon Geology: Exploration and Geological Services Division, Yukon, Indian and Northern Affairs Canada, v. 1, p. 72-82.

Emmons, W.H., 1924, Primary downward changes in ore deposits (with discussion by J. E. Spurr): Transactions of the American Institute of Mining, Metallurgical and Petroleum Engineers Inc. (AIME), v. 70, p. 964-997.

Erickson, G.E., and Cunningham, C.G., 1993, Epithermal precious-metal deposits hosted by the Neogene and Quaternary volcanic complex of the central Andes, *in* Kirkham, R.V., Sinclair, W.D., Thorpe, R.I., and Duke, J.M., eds., Mineral Deposit Modeling: Geological Association of Canada, Special Paper 40, p. 419-431.

Ewers, G.R., and Keays, R.R., 1977, Volatile and precious metal zoning in the Broadlands geothermal field, New Zealand: Economic Geology, v. 72, p. 1337-1354.

Ewing, T.E., 1980, Paleocene tectonic evolution of the Pacific Northwest: Journal of Geology, v. 88, p. 619-638.

Faulkner, E.L., 1986, Blackdome deposit (920/7E, 8W), *in* Geological Fieldwork 1985: British Columbia Ministry of Energy, Mines and Petroleum Resources, Paper 1986-1, p. 106-109.

Faure, K., Matsuhisa, Y., Metsugi, H., Mizota, C., and Hayashi, S., 2002, The Hishikari Au-Ag epithermal deposit, Japan: oxygen and hydrogen isotope evidence in determining the source of paleohydrothermal fluids: Economic Geology, v. 97, p. 481-498.

Feiss, P.G., Vance, R.K., and Wesolowski, D.J., 1993, Volcanic-rock hosted, gold and base-metal mineralization associated with Neoproterozoic-

Early Paleozoic back-arc extension in the Carolina terrane, southern Appalachian Piedmont: Geology, v. 17, p. 548-551.

Fifarek, R.H., and Rye, R.O., 2005, Stable isotope geochemistry of the Pierina high-sulfidation Au-Ag deposit, Peru: influence of hydrodynamics on SO_2-^{4}-H_2S sulfur isotopic exchange in magmatic-steam and steam-heated environments: Geochemistry of sulfate minerals in high and low temperature environments: Chemical Geology, v. 215, p. 253-279.

Foley, N.K., Bethke, P.M., and Rye, R.O., 1989, A reinterpretation of the δD_{H_2O} of inclusion fluids in contemporaneous quartz, sphalerite, Creede mining district, Colorado: A generic problem for shallow ore bodies?: Economic Geology, v. 84, p. 1966-1977.

Gemmell, J.B., Sharpe, R., Jonasson, I.R., and Herzig, P.M., 2004, Sulfur isotope evidence for magmatic contributions to submarine and subaerial gold mineralization: Conical seamount and the Ladolam gold deposit, Papua New Guinea: Economic Geology, v. 99, p. 1711-1725.

Giggenbach, W., 1992, Magma degassing and mineral deposition in hydrothermal systems along convergent plate boundaries: Economic Geology, v. 87, p. 1927-1944.

Goldfarb, R.J., Ayuso, R.A., Miller, M.L., Ebert, S., Marsh, E.E., Petsel, S.A., Miller, L.D., Bradley, D., Johnson, C., and McClelland, W., 2004, The Late Cretaceous Donlin Creek gold deposit, Southwestern Alaska: Controls on epizonal ore formation: Economic Geology, v. 99, p. 643-671.

Gosselin, P., and Dubé, B., 2005a, Gold deposits and gold districts of the world: Geological Survey of Canada, Open File 4893, 1 colour poster.

——— 2005b, Gold deposits and gold districts of Canada: Geological Survey of Canada, Open File 4894, 1 Colour poster.

——— 2005c, Gold deposits of the world: distribution, geological parameters, and gold content: Geological Survey of Canada, Open File 4895, 1 CD-ROM.

——— 2005d, Gold deposits of Canada: distribution, geological parameters, and gold content: Geological Survey of Canada, Open File 4896, 1 CD-ROM.

Hallberg, A., 1994, The Enåsen gold deposit, central Sweden: 1. A Paleoproterozoic high-sulphidation epithermal gold mineralization: Mineralium Deposita, v. 29, p. 150-162.

Harris, D.C., 1989, The Mineralogy and Geochemistry of the Hemlo Gold Deposit, Ontario: Geological Survey of Canada, Economic Geology Report 38, 88 p.

Hattori, K., and Cameron, E.M., 1986, Archean magmatic sulphate: Nature, v. 319, p. 45-47.

Hayashi, K.-I., and Ohmoto, H., 1991, Solubility of gold in NaCl- and H_2S-bearing aqueous solution at 259-350°C: Geochimica et Cosmochimica Acta, v. 55, p. 2111-2126.

Hayashi, K.-I., Maruyama, T., and Satoh, H., 2001, Precipitation of gold in a low-sulfidation epithermal gold deposit: Insights from submillimeter-scale oxygen isotope analysis of vein quartz: Economic Geology, v. 96, p. 211-216.

Hayba, D.O., Bethke, P.M., Heald, P.M., and Foley, N.K., 1985, Geologic, mineralogic, and geochemical characteristics of volcanic-hosted epithermal precious-metal deposits: Reviews in Economic Geology, v. 2, p. 129-167.

Heald, P., Foley, N.K., and Hayba, D.O., 1987, Comparative anatomy of volcanic-hosted epithermal deposits: Acid-sulphate and adularia-sericite types: Economic Geology, v. 82, p.1-26.

Hedenquist, J.W., 1986, Waiotapu geothermal field, Chapter 6 *in* Henley, R.W., Hedenquist, J.W., and Roberts, P.J., eds., Guide to Active Epithermal (Geothermal) Systems and Precious Metal Deposits: Mineral Deposits, Monograph 26, 211 p.

——— 1987, Mineralization associated with volcanic-related hydrothermal systems in the circum-Pacific Basin, *in* Horn, M.K. ed., Transactions of the Fourth Circum-Pacific Energy and Mineral Resources Conference, Singapore: American Association of Petroleum Geologists, p. 513-524.

Hedenquist, J.W., and Henley, R.W., 1985, Hydrothermal eruptions in the Waiotapu geothermal system, New Zealand: Their origin, associated breccias, and relation to precious metal mineralization: Economic Geology, v. 80, p. 1640-1668.

Hedenquist, J.W., Matsuhisa, Y., Izawa, E., White, N.C., Giggenbach, W.F., and Aoki, M., 1994, Geology, geochemistry, and origin of high sulfidation Cu-Au mineralization in the Nansatsu district, Japan: Economic Geology, v. 89, p. 1-30.

Hedenquist, J.W, Izawa, E., Arribas, A.R., and White, N.C., 1996, Epithermal Gold Deposits: Styles, Characteristics, and Exploration: The Society of Resource Geology: Resource Geology Special Publication number 1, 70 p.

Hedenquist, J.W., Arribas, A.R., and Gonzalez-Urien, E., 2000, Exploration for epithermal gold deposits, Chapter 7 *in* Hagemann, S.G., and Brown, P.E., eds., Gold in 2000: Society of Economic Geologists, Reviews in Economic Geology, v. 13, p. 245-277.

Heinrich, C.A., 2005, The physical and chemical evolution of low-salinity magmatic fluids at the porphyry to epithermal transition: A thermodynamic study: Mineralium Deposita, v. 39, p. 864-889.

Heinrich, C.A., Driesner, T., Stefansson, A., and Seward, T.M., 2004, Magmatic vapor contraction and the transport of gold from the porphyry environment to epithermal ore deposits: Geology, v. 32, p. 761-764.

Heinrich, C.A., Bierlein, F.P., Foster, D.A., Gray, D.R., and Davidson, G.J., 2005, The physical and chemical evolution of low-salinity magmatic fluids at the porphyry to epithermal transition: a thermodynamic study: Mineralium Deposita, v. 39, p. 864-889.

Henley, R.W., 1985, The geothermal framework of epithermal deposits, *in* Berger, B.R., and Bethke, P.M., eds., Geology and Geochemistry of Epithermal Systems: Society of Economic Geologists, Reviews in Economic Geology, Volume 2, p. 1-24.

Henley, R.W., and Ellis, A.J., 1983, Geothermal systems, ancient and modern: Earth Science Reviews, v. 19, p. 1-50.

Henley, R.W., and Hedenquist, J.W., 1986, Introduction to the geochemistry of active and fossil geothermal systems, Chapter 1 *in* Henley, R.W., Hedenquist, J.W., and Roberts, P.J., eds., 1986, Guide to Active Epithermal (Geothermal) Systems and Precious Metal Deposits: Mineral Deposits, Monograph 26, 211 p.

Henley, R.W., Truesdell, A.H., and Barton, P.B., Jr., 1984, Fluid-Mineral Equilibria in Hydrothermal Systems: Society of Economic Geologists, Reviews in Economic Geology, Volume 1, 267 p.

Huston, D.L., Blewett, R. S., Keillor, B., Standing, J., Smithies, R.H., Marshall, A., Mernagh, T.P., and Kamprad, J., 2002, Lode gold and epithermal deposits of the Mallina Basin, North Pilbara Terrain, Western Australia: Economic Geology, v. 97, p. 801-818.

Huckerby, J.A., Moore, J.M., and Davis, G.R., 1983, Tectonic control of mineralization at Mahd adh Dhahab gold mine, western Saudia Arabia: The Institute of Mining and Metallurgy Transactions, v. 92, p. Bl71-B182.

Hutchinson, B., 1997, Fools' Gold: the Marketing of a Global Market Fraud: House of Collectibles, 266 p.

Izawa, E., Naito, K., Ibaraki, K., and Suzuki, R., 1993, Mudstones in a hydrothermal eruption crater above the gold-bearing vein system of the Yamada deposit at Hishikari, Japan: Resource Geology, Special Issue No. 14, p. 85-92.

Izawa, E., Urashima, Y., Ibaraki, K., Suzuki, R., Yokoyama, T., Kawasaki, K., Koga, A., and Taguchi, S., 1990, The Hishikari gold deposit: High-grade epithermal veins in Quaternary volcanics of southern Kyushu, Japan: Journal of Geochemical Exploration, v. 36, p. 1-56.

Jannas, R.R., Beane, R.E., Ahler, B.A., and Brosnahan, D.R., 1990, Gold and copper mineralization at the El Indio deposit, Chile, *in* Hedenquist, J., White, N.C., and Siddeley, G., eds., Epithermal Gold Mineralization of the Circum-Pacific: Geology, Geochemistry, Origin and Exploration, II: Journal of Geochemical Exploration, v. 36, p. 233-266.

Jensen, E.P., and Barton, M.D., 2000, Gold deposits related to alkaline magmatism, *in* Hagemann, S.G., and Brown, P.E., eds., Gold in 2000: Reviews in Economic Geology, v. 13, p. 279-314.

John, D.A., Hofstra, A.H., Fleck, R.J., Brummer, J.E., and Saderholm, E.C., 2003, Geologic setting and genesis of the Mule Canyon low-sulfidation epithermal gold-silver deposit, North-Central Nevada: Economic Geology, v. 98, p. 425-463.

Kamilli, R.J., and Ohmoto, H., 1977, Paragenesis, zoning, fluid inclusion, and isotopic studies of the Finlandia vein, Colqui District, central Peru: Economic Geology, v. 72, p. 950-982.

Keith, T.E.C., and Muffler, L.P.J., 1978, Minerals produced during cooling and hydrothermal alteration of ash flow tuff from Yellowstone drill hole Y-5: Journal of Volcanology and Geothermal Research, v. 3, p. 373-402.

Kelley, K.D., Romberger, S.B., Beaty, D.W., Pontius, J.A., Snee, L.W., Stein, H.J., and Thompson, T. B., 1998, Geochemical and geochronological constraints on the genesis of the Au-Te deposits at Cripple Creek, Colorado: Economic Geology, v. 93, p. 981-1012.

Kesler, S.E., Chryssoulis, S.L., and Simon, G., 2002, Gold in porphyry copper deposits: Its abundance and fate: Ore Geology Reviews, v. 21, p. 103-124.

Kesler, S.E., Campbell, I.H., Smith, C.N., Hall, C.M., and Allen, C.M., 2005, Age of the Pueblo Viejo gold-silver deposit and its significance to models for high-sulfidation epithermal mineralization: Economic Geology, v. 100, p. 253-272.

Klein, T.L., and Criss, R.E., 1988, An oxygen isotope and geochemical study of meteoric-hydrothermal systems at Pilot Mountain and selected other localities, Carolina Slate Belt: Economic Geology, v. 83, p. 801-821.

Kuhns, R.J., 1988, The Golden Giant Deposit, Hemlo, Ontario: Geologic and Geochemical Relationships between Mineralization, Alteration, Metamorphism, Magmatism and Tectonism: Ph.D.thesis, University of Minnesota, Minneapolis, Minnesota, 254 p.

Lang, J. R., and Baker, T., 2001, Intrusion-related gold systems: the present level of understanding: Mineralium Deposita, v. 36, p. 477-489.

Lindgren, W., 1922, A suggestion for the terminology of certain minerals deposits: Economic Geology, v. 17, p. 202-294.

——— 1933, Mineral Deposits: McGraw-Hill, New York, 4th edition, 930 p.

Lindhorst, J.W., and Cook, W.G., 1990, Gidginbung gold-silver deposit, Temora, *in* Hughes, F.E., ed., Geology of the Mineral Deposits of Australia and Papua New Guinea: The Australasian Institute of Mining and Metallurgy; Monograph No. 14, p. 1365-1370.

Lipman, P.W., 1975, Evolution of the Platoro Caldera Complex and Related Volcanic Rocks, southeastern San Juan Mountains, Colorado: United States Geological Survey, Professional Paper 852, 128 p.

Love, D. S., 1989, Geology of the Epithermal Mount Skukum Gold Deposit, Yukon Territory: Geological Survey of Canada, Open File 2123, 46 p.

Love, D.S., Clark, A.H., Hodgson, C.J., Mortensen, J.K., Archibald, D.A., and Farrar, E., 1998, The timing of adularia-sericite-type mineralization and alunite-kaolinite-type alteration, Mount Skukum epithermal gold deposit, Yukon Territory, Canada: $^{40}Ar/^{39}Ar$ and U-Pb geochronology: Economic Geology, v. 93, p. 437-462.

Lowenstern, J.B., and Janik, C.J., 2003, The origins of reservoir liquids and vapors from The Geyser's Geothermal Field, California, *in* Simmons, S.D., and Graham, I. eds., Volcanic, Geothermal, and Ore-Forming Fluids: Rulers and Witnesses of Processes within the Earth: Society of Economic Geologists, Special Publication 10, p. 181-195.

Margolis, J., 1993, Geology and Intrusion-Related Copper-Gold Mineralization, Sulphurets, British Columbia: Ph.D. thesis, University of Oregon, Eugene, Oregon, 289 p.

McDonald, B.W.R., 1987, Geology and Genesis of the Mount Skukum Tertiary Epithermal Gold-Silver Vein Deposit, southwestern Yukon Territory (NTS 105D SW): M.Sc. thesis, University of British Columbia, Vancouver, British Columbia, 177 p.

McDonald, B.W.R., and Godwin, C.I., 1986, Geology of main zone at Mt. Skukum, Wheaton River area, southern Yukon, *in* Yukon Geology: Exploration and Geological Services Division, Yukon, Indian and Northern Affairs Canada, v. 1, p. 6-10.

McDonald, R.W.R., Stewart, E.B., and Godwin, C.1., 1986, Exploration geology of the Mt. Skukum epithermal gold deposit, southwestern Yukon, *in* Yukon Geology: Exploration and Geological Services Division, Yukon, Indian and Northern Affairs Canada, v. 1, p. 11-18.

McDonald, D.W.A., 1990, The Silbak Premier Silver-Gold Deposit: A Structurally Controlled, Base Metal-Rich Cordilleran Epithermal Deposit, Stewart, B.C.: Ph.D. thesis, University of Western Ontario, London, Ontario, 411 p.

McFall, J., 1981, The geology and mineralization of the Venus Pb-Zn-Ag mine Yukon Territory (abstract): The Canadian Institute of Mining and Metallurgy Bulletin, v. 74, p. 64.

McInnes, B.I.A, Crocket, J.H., and Goodfellow, W.D., 1990, The Laforma deposit, an atypical epithermal-Au system at Freegold Mountain, Yukon Territory, Canada: Journal of Geochemical Exploration, v. 36, p. 73-102.

Mills, J., O'Brien, S.J., Dubé, B., Mason, R., and O'Driscoll, C.F., 1999, The Steep Nap Prospect: A low-sulphidation, gold-bearing epithermal vein system of late Neoproterozoic age, Avalon Zone, Newfoundland Appalachian, *in* Current Research: Newfoundland Geological Survey, Newfoundland Department of Mines and Energy, Report 99-1, p. 255-274.

Morin, J., and Downing, D.A., 1984, Gold-silver deposits and occurrences in Yukon Territory: Exploration and Geological Services Division, Yukon, Indian and Northern Affairs Canada, Open File 1984-2, scale 1:2 000 00.

Mosier, D.L., Sato, T., and Singer, D.A., 1985, Grade and tonnage model of Creede epithermal veins, *in* Tooker, E.W. ed., Geologic Characteristics of Sediment- and Volcanic-Hosted Disseminated Gold Deposits - Search for an Occurrence Model: United States Geological Survey, Bulletin 1646, p.146-149.

Müller, D., Herzig, P.M., Scholten, J.C., and Hunt, S., 2002a, Ladolam gold deposit, Lihir Island, Papua New Guinea: Gold mineralization hosted by alkaline rocks, *in* Goldfarb, R.J., and Nielsen, R.L., eds., Integrated Methods for Discory: Global Exploration in the Twenty-First Century: Society of Economic Geologists, Special Publication 9, p. 367-382.

Müller, D., Kaminiski, K., Uhlig, S., Graupner, T., Herzig, P.M., and Hunt, S., 2002b, The transition from porphyry- to epithermal-style gold mineralization at Ladolam, Lihir Island, Papua New Guinea: A reconnaissance study: Mineralium Deposita, v. 37, p. 61-74.

Naden, J., Kilias, S.P., and Darbyshire, D.P.F., 2005, Active geothermal systems with entrained seawater as modern analogs for transitional volcanic-hosted massive sulfide and continental magmato-hydrothermal mineralization: the example of Milos Island, Greece: Geology, v. 33, p. 541-544.

Naito, K., Matsuhisa, Y., Izawa, E., and Takaoka, H., 1993, Oxygen isotopic zonation of hydrothermally altered rocks in the Hishikari gold deposit, southern Kyushu, Japan: Its implications for mineral prospecting: Resource Geology, Special Issue No. 14, p.71-84.

Nakamura, H., Sumi, K., Karagiri, K., and Iwata, T., 1970, The geological environment of the Matsukawa geothermal area, Japan: Geothermics, Special Issue 2, p. 221-231.

Nicholson, K., 1989. Early Devonian geothermal systems in northeast Scotland: Exploration targets for epithermal gold: Geology, v. 17, p. 568-571.

O'Brien, S.J., Dubé, B., and O'Driscoll, C.F., 1999, High-sulphidation, epithermal-style hydrothermal systems in late Neoproterozoic Avalonian rocks on the Burin Peninsula, Newfoundland: Implications for gold exploration, *in* Current Research: Newfoundland Geological Survey, Newfoundland Department of Mines and Energy, Report 99-1, p. 275-296.

O'Brien, S.J., Dunning, G.R., Dubé, B., O'Driscoll, C.F., Sparkes, B., Israel, and Ketchum, J., 2001, New insights into the Neoproterozoic geology of the central Avalon Peninsula (Parts of NTS map areas 1N/6, 1N/7 and 1N/3), eastern Newfoundland, *in* Current Research: Newfoundland Geological Survey, Newfoundland Department of Mines and Energy, Report 2001-1, p. 169-189.

O'Driscoll, J.M., and Wilton, D.H.C., 2005, Preliminary geochronological, geochemical and isotopic studies of auriferous systems in the Botwood Basin and environs, central Newfoundland, *in* Current Research: Newfoundland Geological Survey, Newfoundland Department of Mines and Energy, Report 05-1, p. 207-222.

Panteleyev, A., 1986, A Canadian Cordilleran model for epithermal gold-silver deposits: Geoscience Canada, v. 13, p. 101-111.

——— 1991, Gold in the Canadian Cordillera: A focus on epithermal and deeper environments, *in* McMillan, W.J., Höy, T., MacIntyre, D.G., Nelson, J.L., Nixon, G.T., Hammack, J.L., Panteleyev, A., Ray, G.E., and Webster, I.C.L., eds., Ore Deposits, Tectonics and Metallogeny of the Canadian Cordillera: British Columbia Ministry of Energy, Mines and Petroleum Resources, Paper 1991-4, p. 163-212.

——— 1996a, Hot-Spring Au-Ag ($H0_3$), *in* Lefebure, D.V., and Höy, T., eds., Selected British Columbia Mineral Deposit Profiles, v. 2 – Metallic Deposits: Ministry of Employment and Investment, Open File 1996-13, p. 33-36.

——— 1996b, Epithermal Au-Ag-Cu high sulphidation (H04), *in* Lefebure, D.V., and Höy, T., eds., Selected British Columbia Mineral Deposit Profiles, v. 2 – Metallic Deposits, Ministry of Employment and Investment, Open File 1996-13, p. 37-39.

——— 1996c, Epithermal Au-Ag-Cu low sulphidation ($H0_5$), *in* Lefebure, D.V., and Höy, T., eds., Selected British Columbia Mineral Deposit Profiles, v. 2 – Metallic Deposits, Ministry of Employment and Investment, Open File 1996-13, p. 41-44.

——— 2005a, Epithermal Au-Ag-Cu: High sulphidation: British Columbia Mineral Deposit Profiles, British Columbia Geological Survey, Report H04, modified for Yukon by A. Fonseca, Yukon Geological Survey

——— 2005b, Epithermal Au-Ag-Cu: Low sulphidation: British Columbia Mineral Deposit Profiles, British Columbia Geological Survey, Report H05, modified for Yukon by A. Fonseca, Yukon Geological Survey

Poulsen, K.H., 1996, Lode gold, *in* Eckstrand, O.R., Sinclair, W.D., and Thorpe, R.I., eds., Geology of Canadian Mineral Deposit Types: Geological Survey of Canada, Geology of Canada, No. 8, p. 323-328 (also Geological Society of America, The Geology of North America, v. P-1).

Poulsen, K.H., Robert, F., and Dubé, B., 2000, Geological Classification of Canadian Gold Deposits: Geological Survey of Canada, Bulletin 540, 106 p.

Pride, M.J., and Clark, G.S., 1985, An Eocene Rb-Sr isochron for rhyolite plugs, Mt. Skukum area, Yukon Territory: Canadian Journal of Earth Sciences, v. 22, p.1747-1753.

Richards, J.P., 1995, Alkalic-type epithermal gold deposits, a review: Mineralogical Association of Canada, Short Course Handbook, v. 23, p. 367-400.

Robert, F., 2001, Syenite-associated disseminated gold deposits in the Abitibi greenstone belt, Canada: Mineralium Deposita, v. 36, p. 503-516.

Roth, T., and Taylor, B.E., 2000, Sulfur isotope and textural zoning of pyrite in mudstone about the polymetallic Eskay Creek Deposit, northwestern British Columbia, Canada, *in* Volcanic Environments and Massive Sulfide Deposits: International Conference and Field Meeting, Special Publication - Centre for Ore Deposit and Exploration Studies, Program and Abstracts, v.3, p. 177-180.

Roth, T., and Taylor, B.E., (submitted), Variations in sulphur and carbon isotopes in the Eskay Creek deposit and related argillite host rocks: Economic Geology.

Roth, T., Thompson, J.F.H., and Barrett, T.J., 1999, The precious metal-rich Eskay Creek Deposit, north western British Columbia, *in* Tucker, B.C., and Hannington, M.D., eds., Volcanic-Associated Massive Sulfide Deposits: Processes and Examples in Modern and Ancient Settings: Reviews in Economic Geology, v. 8, p. 357-373.

Rye, R. O., 2005, A review of the stable-isotope geochemistry of sulfate minerals in selected igneous environments and related hydrothermal systems: Geochemistry of sulfate minerals in high and low temperature environments: Chemical Geology, v. 215, p. 5-36.

Rye, R.O., Bethke, P.M., and Wasserman, M.D., 1992, The stable isotope geochemistry of acid sulfate alteration: Economic Geology, v. 87, p. 225-262.

Sander, M.V., and Einaudi, M., 1990, Epithermal deposition of gold during transition from propylitic to potassic alteration at Round Mountain: Economic Geology, v. 85, p. 285-311.

Schalamuk, I.B., Zubia, M., Genini, A., and Fernandez, R.R., 1997, Jurassic epithermal Au-Ag deposits of Patagonia, Argentina: Ore Geology Reviews, v. 12, p. 173-186.

Schroeter, T.G., 1982, Toodoggone River (94E), *in* Geological Fieldwork 1981: British Columbia Ministry of Energy, Mines and Petroleum Resources, Paper 1982-1, p. 122-133.

_____ 1985, Toodoggone River area, *in* Geological Fieldwork 1984: British Columbia Ministry of Energy, Mines and Petroleum Resources, Paper 1985-1, p. 291-298.

——— 1986, Brief studies of selected gold deposits in southern British Columbia, *in* Geological Fieldwork 1985: British Columbia Ministry of Energy, Mines and Petroleum Resources, Paper 1986-1, p. 15-22.

Shaw, D.R., Ashley, R.P., and Carter, L.M.H. (eds.), 1990, Epithermal Gold deposits – Part 1: United States Geological Survey, Bulletin 1857-H, 31 p.

——— 1991, Epithermal Gold deposits – Part 2: United States Geological Survey, Bulletin 1857-I, 148 p.

Shen, K., and Sinclair, A.J., 1982, Preliminary results of a fluid inclusion study of Sam Goosly deposit, Equity Mines Ltd., Houston, *in* Geological Fieldwork 1981: British Columbia Ministry of Energy, Mines and Petroleum Resources, Paper 1982-1, p. 229-233.

Shen, K., Champigny, N., and Sinclair, A.J., 1982, Fluid inclusion and sulphur isotope data in relation to genesis of the Cinola Gold deposit, Queen Charlotte Islands, B.C., *in* Hodder, R.W., and Petruk, W., eds., Geology of Canadian Gold Deposits, Proceedings of the CIM Gold Symposium, September 1980: The Canadian Institute of Mining and Metallurgy, Special Volume 24, p. 255-257.

Sherlock, R.L., 2005, The relationship between the McLaughlin gold-mercury deposit and active hydrothermal systems in the Geysers-Clear Lake area, Northern Coast Ranges, California: Ore Geology Reviews, v. 26, p. 349-382.

Sherlock, R. L., Tosdal, R. M., Leherman, N. J., Graney, J. R., Losh, S., Jowett, E. C., Kesler, S. E., 1995, Origin of the McLaughlin mine sheeted vein complex: Metal zoning, fluid inclusion and isotopic evidence: Economic Geology, v. 90, p. 2156-2181.

Sherlock, R.L., Roth, T., Spooner, E.T.C., and Bray, C.J., 1999, Origin of the Eskay Creek precious metal-rich volcanogenic massive sulphide deposit: Fluid inclusion and stable isotope evidence: Economic Geology, v. 94, p. 803-824.

Siddeley, G., and Araneda, R., 1986. The El Indio-Tambio gold deposits, Chile, *in* Macdonald, A.J. ed., Proceedings Gold '86, an International Symposium on the Geology of Gold: Toronto, Ontario, p. 445-456.

Sillitoe, R.H., 1989, Gold deposits in western Pacific island arcs: the magmatic connection, *in* Keays, R.R., Ramsay, W.R.H., and Groves, D.I., eds., The Geology of Gold Deposits: The Perspective in 1988: Economic Geology, Monograph 6, p. 274-291.

——— 1992, Giant and bonanza gold deposits, *in* Whiting, B.H., Mason, R., and Hodgson, C.J. eds., Epithermal and Sub-volcanic Settings, Giant Ore Deposits, p. 231-38.

——— 1993, Epithermal models: genetic types, geometrical controls and shallow features, *in* Kirkham, R.V., Sinclair, W.D., Thorpe, R.I., and Duke, J.M. eds., Mineral Deposit Modeling: Geological Association of Canada, Special Paper 40, p. 403-417.

——— 1995, Exploration and Discovery of Ase- and Precious-Metal Deposits in the Circum-Pacific Region During the Last 25 years: Resource Geology, Special Issue 19, 119 p.

——— 1997, Characteristics and controls of the largest porphyry copper-gold and epithermal gold deposits in the circum-Pacific region: Australian Journal of Earth Sciences, v. 44, p. 373-388.

Sillitoe, R.H., and Bonham, H.F., Jr., 1984, Volcanic landforms and ore deposits: Economic Geology, v. 79, p. 1286-1298.

Sillitoe, R. H., and Thomson, J.F.H., 1998, Intrusion-related vein gold deposits: types, tectono-magmatic settings and difficulties of distinction from orogenic gold deposits: Resource Geology, v. 48, p. 237-250.

Sillitoe, R.H., Hannington, M.D., and Thompson, J.F.H., 1996, High-sulfidation deposits in the volcanogenic massive sulfide environment: Economic Geology, v. 91, p. 204-212.

Sillitoe, R.H., Hall, D.J., Redwood, S.D., and Waddell, A.H., 2006, Pueblo Viejo high-sulfidation epithermal gold-silver deposit, Dominican Republic: A new model of formation beneath barren limestone cover: Economic Geology, v. 101, p. 1427-1435.

Simmons, S.F., and Christenson, B.W., 1994, Origins of calcite in a boiling geothermal system: American Journal of Science, v. 294, p. 361-400.

Slack, J.F., 1980, Multistage vein ores of the Lake City district, western San Juan mountains, Colorado: Economic Geology, v. 75, p. 963-991.

Spycher, N.F, and Reed, M.H., 1989, Evolution of a Broadlands-type epithermal ore fluid along alternative P-T paths: Implications for the transport and deposition of base, precious, and volatile metals: Economic Geology, v. 84, p. 328-359.

Stoffregen, R.E., 1987, Genesis of acid-sulfate alteration and Au-Cu-Ag mineralization at Summitville, Colorado: Economic Geology, v. 82, p. 1575-1591.

Symonds, R.P., Rose, W.I., Reed, M.H., Lichte, F.E., and Finnegan, D.L., 1987, Volatilization, transport and sublimation of metallic and non-metallic elements in high temperature gases at Merapi Volcano, Indonesia: Geochimica et Cosmochimica Acta, v. 51, p. 2083-2101.

Taylor, B.E., 1987, Stable isotope geochemistry of ore-forming fluids, *in* Kyser, T.K. ed., Stable Isotope Geochemistry of Low Temperature Fluids: Mineralogical Association of Canada, Short Course Handbook, v. 13, p.337-445.

——— 1988, Degassing of rhyolitic magmas: Hydrogen isotope evidence and implications for magmatic-hydrothermal ore deposits, *in* Taylor, R.P., and Strong, D.F., eds., Recent Advances in the Geology of Granites-Related Mineral Deposits: Canadian Institute of Mining and Metallurgy, Special Publication, v. 39, p. 33-49.

——— 1992, Degassing of H_2O from rhyolite magma during eruption and shallow intrusion, and the isotopic composition of magmatic water in hydrothermal systems: Geological Survey of Japan, Report No. 279, p. 190-194.

——— 1996, Epithermal gold deposits, *in* Eckstrand, O.R., Sinclair, W.D., and Thorpe, R.I., eds., Geology of Canadian Mineral Deposit Types: Geological Survey of Canada, Geology Canada, no. 8, p. 329-350. (also Geological Society of America, The Geology of North America, V. P-1).

Taylor, H.P., Jr., 1973, O18/O16 evidence for meteoric-hydrothermal alteration and ore deposition in the Tonopah, Comstock Lode, and Goldfield Mining Districts, Nevada: Economic Geology, v. 68, p. 747-764.

Thiersch, P., and Williams-Jones, A.E., 1990, Paragenesis and ore controls of the Shasta Ag-Au deposit, Toodoggone River area, British Columbia, *in* Geological Fieldwork 1989: British Columbia Ministry of Energy, Mines and Petroleum Resources, Paper 1990-1, p. 315-321.

Thomson, M.L., Mastalerz, M., Sinclair, A.J., and Bustin, R.M., 1992, Fluid source and thermal history of an epithermal vein deposit, Owen Lake, central British Columbia: Evidence from bitumen and fluid inclusions: Mineralium Deposita, v. 27, p. 219-225.

Thompson, J.F.H., and Newberry, R.J., 2000, Gold deposits related to reduced granitic intrusions: Ch. 11, *in* Hagemann, S.G., and Brown, P.E. eds., Gold in 2000, Reviews in Economic Geology v. 13, Society of Economic Geologists, p. 377-400.

Thompson, J.F.H., Sillitoe, R.H., Baker, T., Lang, J.R., and Mortensen, J. K., 1999, Intrusion-related gold deposits associated with tungsten-tin provinces: Mineralium Deposita, v. 34, p. 323-334.

Tooker, E.W. (ed.), 1985, Geologic Characteristics of Sediment- and Volcanic-Hosted Disseminated Gold Deposits - Search for an Occurrence Model: United States Geological Survey, Bulletin 1646, 150 p.

Turner, W.A., Richards, J.P., Nesbitt, B.E., Muehlenbachs, K., and Biczok, J.L., 2001, Proterozoic low-sulfidation epithermal Au-Ag mineralization in the Mallery Lake area, Nunavut, Canada: Mineralium Deposita, v. 36, p. 442-457.

Turner, W.A., Heaman, L.M., and Creaser, R.A., 2003, Sm-Nd fluorite dating of Proterozoic low-sulfidation epithermal Au-Ag deposits and U-Pb zircon dating of host rocks at Mallery Lake, Nunavut, Canada: Canadian Journal of Earth Sciences, v. 40, p. 1789-1804.

Vivian, G., Morton, R.D., Changkakoti, A., and Gray, J., 1987, Blackdome Eocene epithermal Ag-Au deposit, British Columbia, Canada - nature of ore fluids: Transactions of the Institution of Mining and Metallurgy (Section B; Applied Earth Science), v. 96, p. B9-B14.

Vulimiri, M.R., Tegart, P., and Stammers, M.A., 1986, Lawyers gold-silver deposits, British Columbia, *in* Morin, J.A., Mineral Deposits of Northern Cordillera: Proceedings of the Mineral Deposits of Northern Cordillera Symposium: Canadian Institute of Mining and Metallurgy, Special Volume 37, p. 191-201.

Walton, L., 1986, Textural characteristics of the Venus vein and implications for ore shoot distribution, *in* Yukon Geology: Exploration and Geological Services Division, Yukon, Indian and Northern Affairs Canada, v. 1-1984, p. 67-71.

Walton, L., and Nesbitt, B.E., 1986, Evidence for late stage Au-galena mineralization in the Venus arsenopyrite-pyrite-quartz vein, southwest Yukon (abstract): Geological Association of Canada Mineralogical Association of Canada, Annual Meeting, Ottawa, 1986, Program with Abstracts, v. 11, p. 141.

White, D.E., Muffler, L.J.P., and Truesdell, A.H., 1971, Vapor-dominated hydrothermal systems compared with hot-water systems: Economic Geology, v. 66, p. 75-97.

White, N.C., and Hedenquist, J.W., 1990, Epithermal environments and styles of mineralization: variations and their causes, and guidelines for exploration, *in* Hedenquist, J., White, N.C., and Siddeley, G. eds., Epithermal Gold Mineralization of the Circum-Pacific, Geology, Geochemistry, Origin and Exploration, II: Journal of Geochemical Exploration, v. 36, p. 445-474.

Williams, A.E., and McKibben, M.A., 1989, A brine interface in the Salton Sea geothermal system, California: Fluid geochemical and isotopic characteristics: Geochimica et Cosmochimica Acta, v. 53, p. 1905-1920.

Williams-Jones, A.E., and Heinrich, C.A., 2005, Vapor transport of metals and the formation of magmatic-hydrothermal ore deposits: Economic Geology v. 100, p. 1287-1312.

Wojdak, P.J., and Sinclair, A.J., 1984, Equity Silver silver-copper-gold deposit; alteration and fluid inclusion studies: Economic Geology, v. 79, p. 969-990.

Wood, D.G., Porter, R.G., and White, N.C., 1990, Geological features of some Paleozoic epithermal gold occurrences in northeastern Queensland, Australia, *in* Hedenquist, J., White, N.C., and Siddeley, G., eds., Epithermal Gold Mineralization of the Circum-Pacific, Geology, Geochemistry, Origin and Exploration, II: Journal of Geochemical Exploration, v. 36, p. 413-443.

Zhang, X., Nesbitt, B.E., and Muehlenbachs, K., 1989, Gold mineralization in the Okanagen Valley, southern British Columbia: Fluid inclusion and stable isotope studies: Economic Geology, v. 84, p. 410-424.

Volcanogenic Massive Sulphide Deposits

Alan G. Galley[1], Mark D. Hannington[2], and Ian R. Jonasson[1]

1. Geological Survey of Canada, 601 Booth Street, Ottawa, Ontario K1A 0E8

2. Department of Earth Sciences, University of Ottawa, Marion Hall, 140 Louis Pasteur,Ottawa, Ontario K1N 6N5

Corresponding author's email: agalley@nrcan.gc.ca

Abstract

Volcanogenic massive sulphide (VMS) deposits, also known as volcanic-associated, volcanic-hosted, and volcano-sedimentary-hosted massive sulphide deposits, are major sources of Zn, Cu, Pb, Ag, and Au, and significant sources for Co, Sn, Se, Mn, Cd, In, Bi, Te, Ga, and Ge. They typically occur as lenses of polymetallic massive sulphide that form at or near the seafloor in submarine volcanic environments, and are classified according to base metal content, gold content, or host-rock lithology. There are close to 350 known VMS deposits in Canada and over 800 known worldwide. Historically, they account for 27% of Canada's Cu production, 49% of its Zn, 20% of its Pb, 40% of its Ag, and 3% of its Au. They are discovered in submarine volcanic terranes that range in age from 3.4 Ga to actively forming deposits in modern seafloor environments. The most common feature among all types of VMS deposits is that they are formed in extensional tectonic settings, including both oceanic seafloor spreading and arc environments. Most ancient VMS deposits that are still preserved in the geological record formed mainly in oceanic and continental nascent-arc, rifted-arc, and back-arc settings. Primitive bimodal mafic volcanic-dominated oceanic rifted arc and bimodal felsic-dominated siliciclastic continental back-arc terranes contain some of the world's most economically important VMS districts. Most, but not all, significant VMS mining districts are defined by deposit clusters formed within rifts or calderas. Their clustering is further attributed to a common heat source that triggers large-scale subseafloor fluid convection systems. These subvolcanic intrusions may also supply metals to the VMS hydrothermal systems through magmatic devolatilization. As a result of large-scale fluid flow, VMS mining districts are commonly characterized by extensive semi-conformable zones of hydrothermal alteration that intensifies into zones of discordant alteration in the immediate footwall and hanging wall of individual deposits. VMS camps can be further characterized by the presence of thin, but areally extensive, units of ferruginous chemical sediment formed from exhalation of fluids and distribution of hydrothermal particulates.

Résumé

Les gîtes de sulfures massifs volcanogènes (SMV) sont connus sous diverses appellations parmi lesquelles on peut mentionner les gîtes de sulfures massifs associés à des roches volcaniques, encaissés dans des roches volcaniques ou logés dans des assemblages volcano-sédimentaires. Ils constituent des sources considérables de Zn, Cu, Pb, Ag et Au, ainsi que des sources importantes de Co, Sn, Se, Mn, Cd, In, Bi, Te, Ga et Ge. Ils consistent généralement en lentilles de sulfures massifs polymétalliques formées dans des milieux volcaniques sous-marins, au sein ou à proximité du fond océanique, et sont classés d'après leur contenu en métaux communs ou en or ou selon la lithologie des roches encaissantes. Près de 350 gîtes SMV ont été découverts au Canada et plus de 800, de par le monde. Dans l'histoire de la production minière du Canada, 27 % du cuivre, 49 % du zinc, 20 % du plomb, 40 % de l'argent et 3 % de l'or ont été extraits de gisements SMV. On trouve de tels gîtes aussi bien dans des terrains volcaniques sous-marins datant de 3,4 Ga que dans les fonds océaniques actuels où de nouveaux gîtes sont en cours de formation. La caractéristique la plus commune à tous les gîtes de SMV tient à leur formation dans des milieux tectoniques de distension, parmi lesquels on peut mentionner les fonds océaniques en expansion et les arcs. La plupart des anciens gîtes SMV conservés dans les archives géologiques se sont formés dans des milieux océaniques et continentaux d'arc naissant, d'arc de divergence et d'arrière-arc. Quelques-uns des districts à gisements SMV les plus importants dans le monde sur le plan économique se trouvent dans des terrains océaniques primitifs d'arc de divergence caractérisés par un volcanisme bimodal à dominante mafique, de même que dans des terrains continentaux d'arrière arc caractérisés par un volcanisme bimodal à dominante felsique et la présence de matériaux silicoclastiques. La plupart des principaux districts miniers à gisements SMV consistent en amas de gisements formés dans des rifts ou des caldeiras. Leur regroupement est attribuable à l'existence d'une source de chaleur commune qui donne naissance à de vastes réseaux de convection de fluides sous le plancher océanique. Les intrusions subvolcaniques qui produisent cette chaleur peuvent aussi fournir des métaux aux réseaux hydrothermaux des gîtes SMV par le biais d'un dégagement magmatique de matières volatiles. En raison de l'écoulement de fluides sur une grande étendue, les districts miniers à gisements SMV se caractérisent souvent par la présence de vastes zones semi-concordantes d'altération hydrothermale, qui gagnent en intensité pour devenir des zones d'altération discordantes, dans l'éponte inférieure et l'éponte supérieure immédiates des gisements. Ces districts se distinguent aussi par la présence d'unités minces mais étendues de sédiments chimiques ferrugineux qui résultent de l'exhalaison et de la diffusion de particules hydrothermales.

Definition

Volcanogenic massive sulphide (VMS) deposits are also known as volcanic-associated, volcanic-hosted, and volcano-sedimentary-hosted massive sulphide deposits. They typically occur as lenses of polymetallic massive sulphide that form at or near the seafloor in submarine volcanic environments. They form from metal-enriched fluids associated with seafloor hydrothermal convection. Their immediate host rocks can be either volcanic or sedimentary. VMS deposits are major sources of Zn, Cu, Pb, Ag, and Au, and significant sources for Co, Sn, Se, Mn, Cd, In, Bi, Te, Ga, and Ge. Some also contain significant amounts of As, Sb, and Hg. Historically, they account for 27% of Canada's Cu production, 49% of its Zn, 20% of its Pb, 40% of its Ag, and

Galley, A.G., Hannington, M.D., and Jonasson, I.R., 2007, Volcanogenic massive sulphide deposits, *in* Goodfellow, W.D., ed., Mineral Deposits of Canada: A Synthesis of Major Deposit-Types, District Metallogeny, the Evolution of Geological Provinces, and Exploration Methods: Geological Association of Canada, Mineral Deposits Division, Special Publication No. 5, p. 141-161.

3% of its Au. Because of their polymetallic content, VMS deposits continue to be one of the most desirable deposit types for security against fluctuating prices of different metals.

VMS deposits form at, or near, the seafloor through the focused discharge of hot, metal-rich hydrothermal fluids. For this reason, VMS deposits are classified under the general heading of "exhalative" deposits, which includes sedimentary exhalative (SEDEX) and sedimentary nickel deposits (Eckstrand et al., 1995). Most VMS deposits have two components (Fig. 1). There is typically a mound-shaped to tabular, stratabound body composed principally of massive (>40%) sulphide, quartz and subordinate phyllosilicates, and iron oxide minerals and altered silicate wall-rock. These stratabound bodies are typically underlain by discordant to semi-concordant stockwork veins and disseminated sulphides. The stockwork vein systems, or "pipes", are enveloped in distinctive alteration halos, which may extend into the hanging-wall strata above the VMS deposit.

VMS deposits are grouped according to base metal content, gold content, and host-rock lithology (Figs. 2, 3, 4). The base metal classification used by Franklin et al. (1981) and refined by Large (1992) and Franklin et al. (2005) is perhaps the most common. VMS deposits are divided into Cu-Zn, Zn-Cu, and Zn-Pb-Cu groups according to their contained ratios of these three metals (Fig. 2). The Cu-Zn and Zn-Cu categories for Canadian deposits were further refined by Morton and Franklin (1987) into Noranda and Mattabi types, respectively, by including the character of their host rocks (mafic vs. felsic, effusive vs. volcaniclastic) and characteristic alteration mineral assemblages (chlorite-sericite dominated vs. sericite-quartz ± carbonate-rich). The Zn-Pb-Cu category was added by Large (1992) in order to more fully represent the VMS deposits of Australia (Fig. 2). Poulsen and Hannington (1995) created a sim-

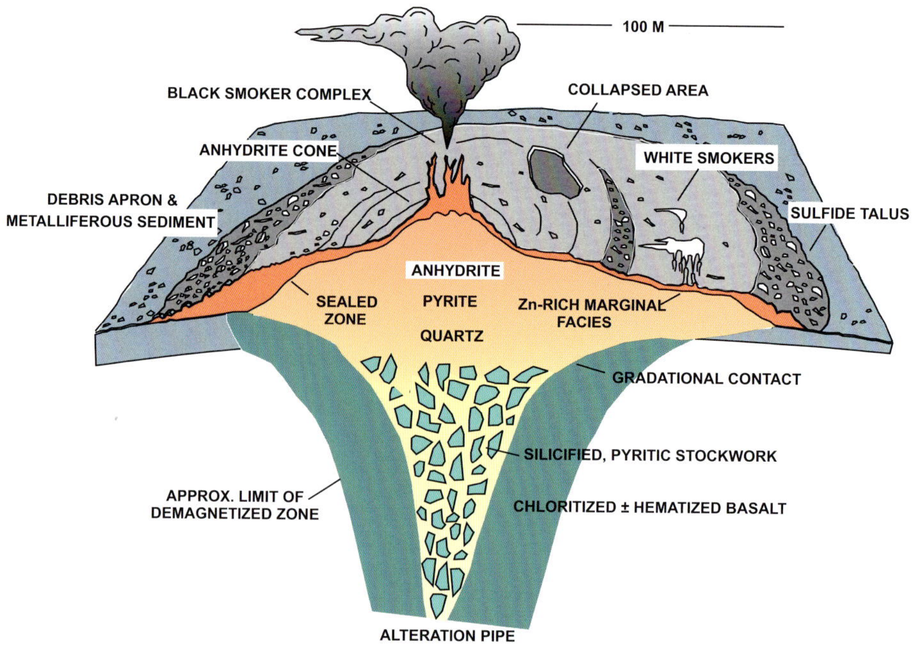

FIGURE 1. Schematic diagram of the modern TAG sulphide deposit on the Mid-Atlantic Ridge. This represents a classic cross-section of a VMS deposit, with concordant semi-massive to massive sulphide lens underlain by a discordant stockwork vein system and associated alteration halo, or "pipe". From Hannington et al. (1998).

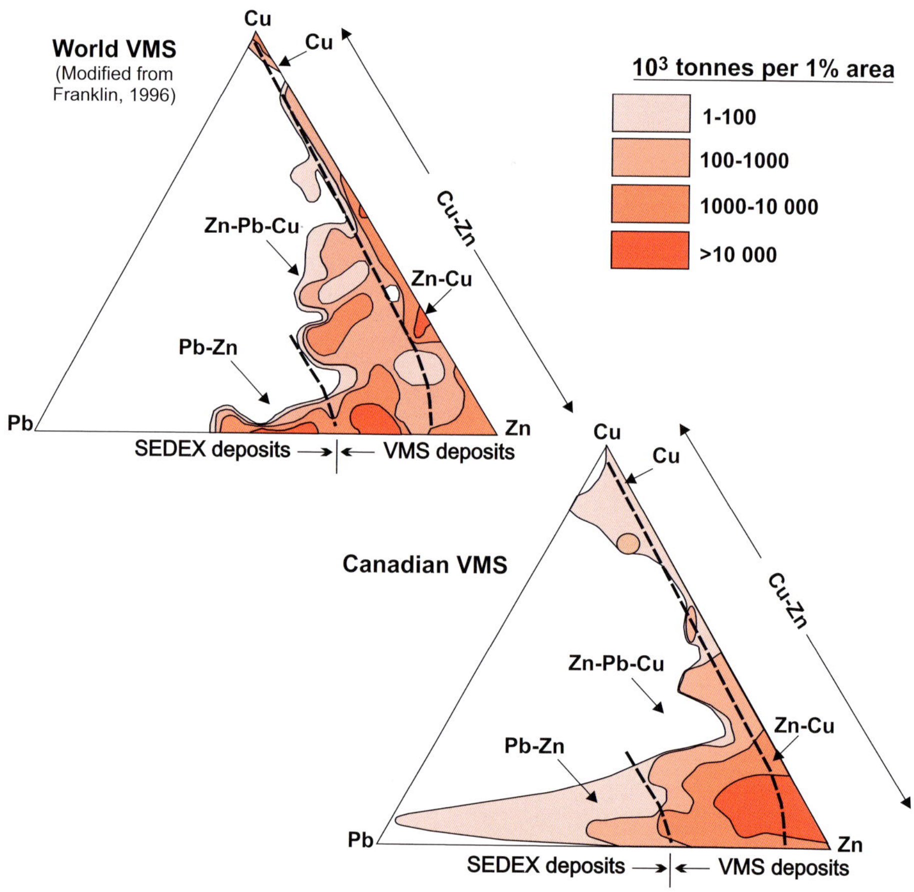

FIGURE 2. Base metal classification scheme of worldwide and Canadian VMS deposits as defined by Franklin et al. (1981) and modified by Large (1992) to include the Zn-Pb-Cu class. The preponderance of Cu-Zn and Zn-Cu VMS deposits in Canada is due to the abundance of Precambrian primitive oceanic arc settings. Worldwide, there is a larger proportion of felsic-hosted, more Pb-rich continental rift and continent margin arc settings.

ple bimodal definition of "normal" versus "Au-rich" VMS deposits (Fig. 3). This originally was intended to identify deposits that are transitional between VMS and epithermal deposits (e.g. Sillitoe et al., 1996) (Fig. 4). Further research has indicated a more complex spectrum of conditions for the generation of Au-rich VMS related to water depth, oxidation state, the temperature of the metal-depositing fluids, and possible magmatic contributions (e.g. Hannington et al., 1999a). In the classification of Poulsen and Hannington (1995) Au-rich VMS deposits are arbitrarily defined as those in which the abundance of Au in ppm is numerically greater than the combined base metals (Zn+Cu+Pb in wt.%, Fig. 3). A third classification system that is gaining acceptance is a five-fold grouping first suggested by Barrie and Hannington (1999), and later modified by Franklin et al. (2005). This system classifies VMS deposits by their host lithologies (Fig. 4), which includes all strata within a host succession defining a distinctive time-stratigraphic event (Franklin et al., 2005). These five different groups are bimodal-mafic, mafic-backarc, pelitic-mafic, bimodal-felsic, and felsic-siliciclastic. To this is added a sixth group of hybrid bimodal felsic, which represent a cross between VMS and shallow-water epithermal mineralization (Fig. 4). These lithologic groupings generally correlate with different submarine tectonic settings. There order here reflects a change from the most primitive VMS environments, represented by ophiolite settings, through oceanic rifted arc, evolved rifted arcs, continental back-arc to sedimented back-arc.

Geographical Distribution

There are close to 850 known VMS deposits worldwide with geological reserves of over 200,000 t. They are located in submarine volcanic terranes that range in age from the 3.4 Ga Archean Pilbara Block, Australia, to actively forming deposits in modern seafloor spreading and oceanic arc terranes (Fig. 5, Table 1). VMS-epithermal hybrids are also forming today in volcanically active shallow submarine (Manus Basin) and lacustrine environments. VMS deposits are recognized on every major continent except Antarctica, although Zn-Pb-Cu deposits are forming in the Bransfield Strait adjacent to the Antarctic Peninsula (Peterson et al., 2004). Cu and Au have been produced from Tertiary-age deposits hosted in ophiolites around the eastern Mediterranean and Oman for over 5000 years. Prior to 2002, VMS deposits are estimated to have supplied over 5 billion tonnes of sulphide ore (Franklin and Hannington, 2002). This includes at least 22% of the world's Zn production, 6% of the world's Cu, 9.7% of the world's Pb, 8.7% of its Ag, and 2.2% of its Au (Singer, 1995).

Over 350 deposits and major VMS occurrences containing geological reserves of more than 200,000 tonnes are known in Canada, of which only 13 are producing mines as of 2006 (Fig. 6, Table 2). Of these, Louvicourt, Bouchard-Hébert, Selbaie, and Konuto have been closed. VMS deposits are known to occur in every province and territory except Alberta and Prince Edward Island. The largest number of deposits is in Quebec (33%), followed in descending order by Manitoba (15%), Newfoundland (12%), British Columbia (10%), Ontario (9%), and New Brunswick (9%). The deposits in New Brunswick have had the highest aggre-

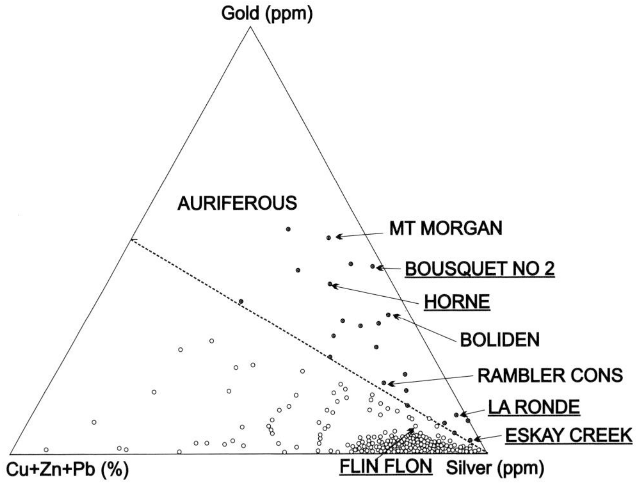

FIGURE 3. Classification of VMS deposits based on their relative base metal (Cu+Zn+Pb) versus precious metal (Au, Ag) contents. Some of Canada's better known auriferous deposits (underlined) are compared to international examples. Despite having produced 170 t of Au, the Flin Flon deposit is not considered an auriferous VMS deposit under this classification. Modified from Hannington et al. (1999c).

gate metal value (Cu+Zn+Pb), followed by Quebec and then Ontario (Fig. 7).

Grade and Tonnage

The over 800 VMS deposits worldwide range in size from 200,000 tonnes to supergiant deposits containing more than 150 million tonnes (Franklin et al., 2005) (Table 3). Among the largest is Rio Tino, Spain's portion of the Iberian Pyrite Belt (IPB), with contained ore in excess of 1.535 Bt. The richest supergiant produced to date is Neves Corvo on the Portuguese side of the IPB, with ore in excess of 270 Mt, with 8.8 Mt of contained metal. At the average metal prices to date for 2006 (Cu=$1.75/lb, Zn=$1.25/lb, Ag=$6.00/oz), this orebody was originally worth in the order of 26 billion dollars (US). Other large districts are the Urals and Rudny Altai of Russia and Kazakhstan with over 70 Mt of contained metals each (Fig. 5). Canada has two supergiant VMS deposits (Windy Craggy and Brunswick No. 12) and two giant VMS deposits (Kidd Creek, and Horne), which are defined as being in the upper 1% of the world's VMS deposits with respect to total original reserves (Fig. 10A). In Canada, the largest VMS mining district is Bathurst, New Brunswick, which contained over 320 Mt of geological resource of massive sulphide containing 30 Mt of combined Zn, Cu, and Pb (Figs. 6, 10A). The 128 Mt Brunswick No. 12 deposit alone contained 16.4 Mt of metal (Table 2). This is followed by the 138.7 Mt Kidd Creek deposit containing 12.6 Mt of metal. The largest known Canadian VMS deposit is the 297 Mt Windy Craggy deposit, but it contains only 4.1 Mt of Cu, Co, and Au. The 50 Mt Horne deposit contains 2.2 Mt of Zn+Cu+Pb, along with over 330 t of Au, making it also a world-class gold deposit (Fig. 10B). The 98 Mt LaRonde VMS deposit contains 258 Mt of gold, and because of its high Au/base metal ratio (Au ppm/Zn+Cu+Pb% = 1.9) it is classified by its owner as a gold deposit rather than a VMS deposit.

Determining the mean and median metal concentrations for Canadian VMS deposits is difficult due to missing or

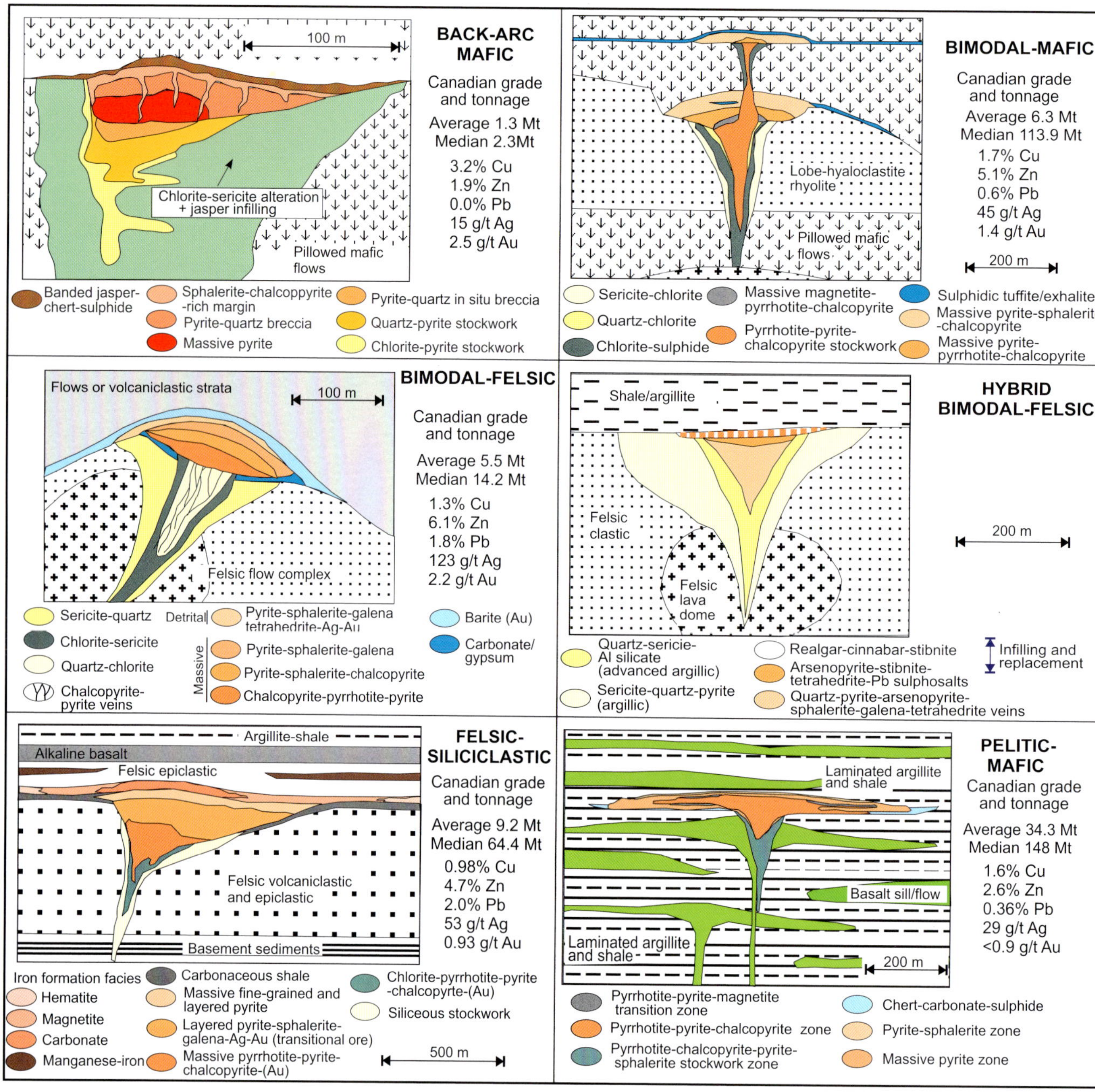

FIGURE 4. Graphic representation of the lithological classifications modified from Barrie and Hannington (1999) by Franklin et al. (2005), with the addition of the hybrid bimodal felsic as a VMS-epithermal subtype of bimodal-felsic. Average and median sizes for each type for representative Canadian deposits shown, along with average grade.

incomplete data for a large number of deposits. Pb grades are known for 34% of Canadian deposits, whereas 55% have known Au grades and 75% have known Ag grades. From the available production data, the mean and median (in brackets) size and grades for past and present producing Canadian deposits are 7 306 521 t grading 4.88% (4.12) Zn, 1.62% (0.70) Cu, 1.639% (1.00) Pb, 63 g/t (37) Ag, and 1.65 g/t (0.88) Au. Figure 9B shows the more meaningful breakdown of tonnage and grade for each of the five Canadian VMS types as defined by host lithology. Bimodal mafic deposits account for the greatest number and, therefore, the largest aggregate tonnage of the five deposit types, with both siliciclastic types accounting for the largest average tonnage. The mafic-siliciclastic deposit types have the highest average tonnage, with the number highly skewed by Windy Craggy. As expected, the three deposit types dominated by mafic volcanic and volcaniclastic rocks have the highest Cu grades, whereas the two felsic-dominated deposit types contain the highest Pb and Ag contents. The bimodal felsic deposit group contains the highest average gold. Mafic-ultramafic-dominated systems can also contain Se, Co, and Ni. The presence of immature sediments (i.e. black shale) within the footwall stratigraphy can also influence hydrothermal fluid composition, as is postulated for he Se-rich Wolverine and KZK deposits in the Finlayson Lake camp (Bradshaw et al., 2003). Possible contributions from devolatilizing subvol-

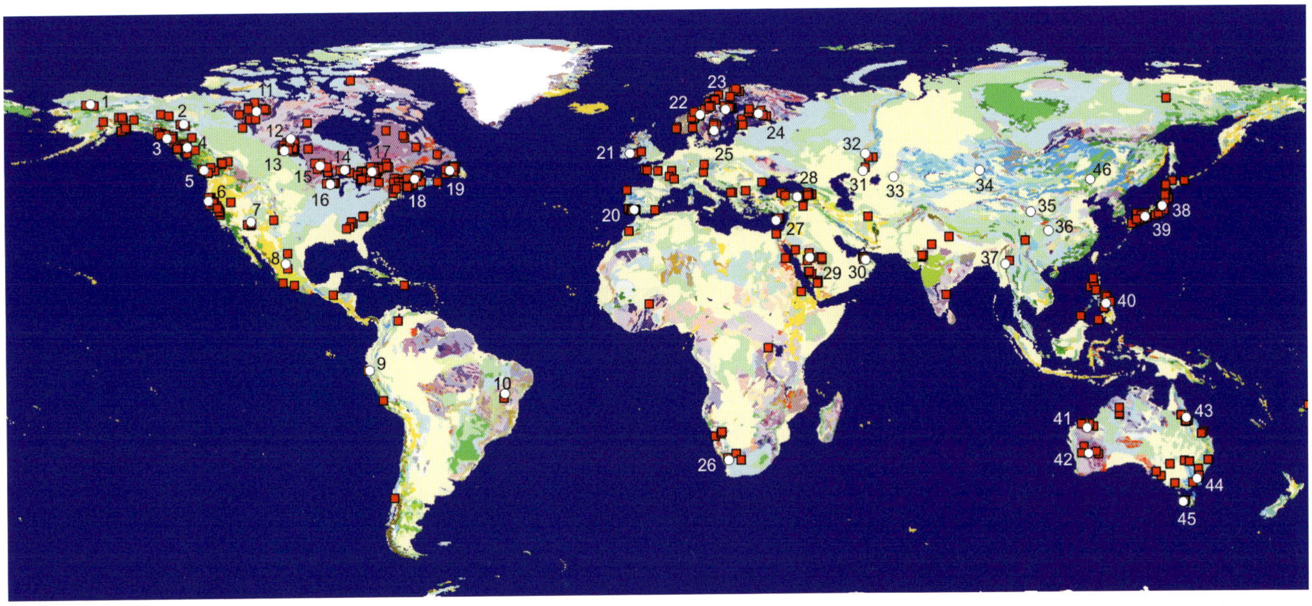

FIGURE 5. Geographical distribution of ancient VMS deposits, with major districts highlighted with respect to known aggregate geological reserves (see Table 1). Modified from Sinclair et al. (1999) and Franklin et al. (2005).

canic intrusions may also account for anomalous Se, Sn, In, Bi, Te, and possibly Au and Sb contents (Hannington et al., 1999c; Yang and Scott, 2003; Dubé et al., 2004).

Geological Attributes

Tectonic Environment

The most common feature among all types of VMS deposits is that they are formed in extensional tectonic settings, including both oceanic seafloor spreading and arc environments (Fig. 11). Modern seafloor VMS deposits are recognized in both oceanic spreading ridge and arc environments (Herzig and Hannington, 1995), but deposits that are preserved in the geological record formed mainly in oceanic and continental nascent-arc, rifted arc, and back-arc settings (Franklin et al. 1998; Allen et al., 2002) (Fig. 11). This is because during subduction-driven tectonic activity much of the ancient ocean-floor is subducted, leaving only a few ophiolite suites as remnant obducted ocean-floor. Examples of these include the Ordovician Bay of Islands ophiolite in Newfoundland and the Late Triassic Cache Creek terrane in British Columbia (Bédard and Hébert, 1996; Nelson and Mihalynuk, 2004).

Nascent, or early arc rifting, results from the initial foundering of older thickened oceanic crust, commonly

TABLE 1. Major world volcanogenic massive sulphide deposits and districts.

No.*	Deposit/District, Country	Tonnage (Mt)	No.*	Deposit/District, Country	Tonnage (Mt)
1	Brooks Range, Alaska	35	23	Skellefte, Sweden	70
2	Finlayson Lake, Yukon	20	24	Outokumpu-Pyhasalmi, Finland	90
3	Windy Craggy, BC & Green's Creek, Alaska	300	25	Bergslagen-Orijarvi, Sweden & Finland	110
4	Northern Cordillera, British Columbia	100	26	Preiska, South Africa	45
5	Myra Falls, British Columbia	35	27	Troodos, Cyprus	35
6	Shasta, California	35	28	Black Sea, Turkey	200
7	Jerome, Arizona	40	29	Saudi Arabia	70
8	Central Mexico	120	30	Semail, Oman	30
9	Tambo Grande, Peru	200	31	Southern Urals, Russia / Kazahkstan	400
10	Amazonian craton, Brazil	35	32	Central Urals, Russia	100
11	Slave Province, Northwest Territories, Nunavut	30	33	Rudny Altai, Kazahkstan / Russia	400
12	Ruttan, Manitoba	85	34	Altai Shan, Mongolia	40
13	Flin Flon-Snow Lake, Manitoba	150	35	North Qilian, China	100
14	Geco, Manitouwadge, Ontario	60	36	Sanjiang, China	50
15	Sturgeon Lake, Ontario	35	37	Bawdwin-Laochang, Burma /	40
16	Ladysmith-Rhinelander, Wisconsin/Michigan	80	38	Hokuroku, Japan	80
17	Abitibi, Ontario-Quebec	600	39	Besshi, Japan	230
18	Bathurst, New Brunswick	495	40	Phillipines arc	65
19	Dunnage Zone, Newfoundland	75	41-42	Pilbara, Yilgarn Western Australia	75
20	Iberian Pyrite Belt, Spain & Portugal	1575	43	Central Queensland, Australia	80
21	Avoca, Ireland	37	44	Lachlan Fold Belt, Australia	100
22	Trondhjeim, Norway	100	45	Mt. Read, Tasmania	200
			46	Sino-Korean Platform	40

* numbers refer to Figure 5; tonnage is approximate

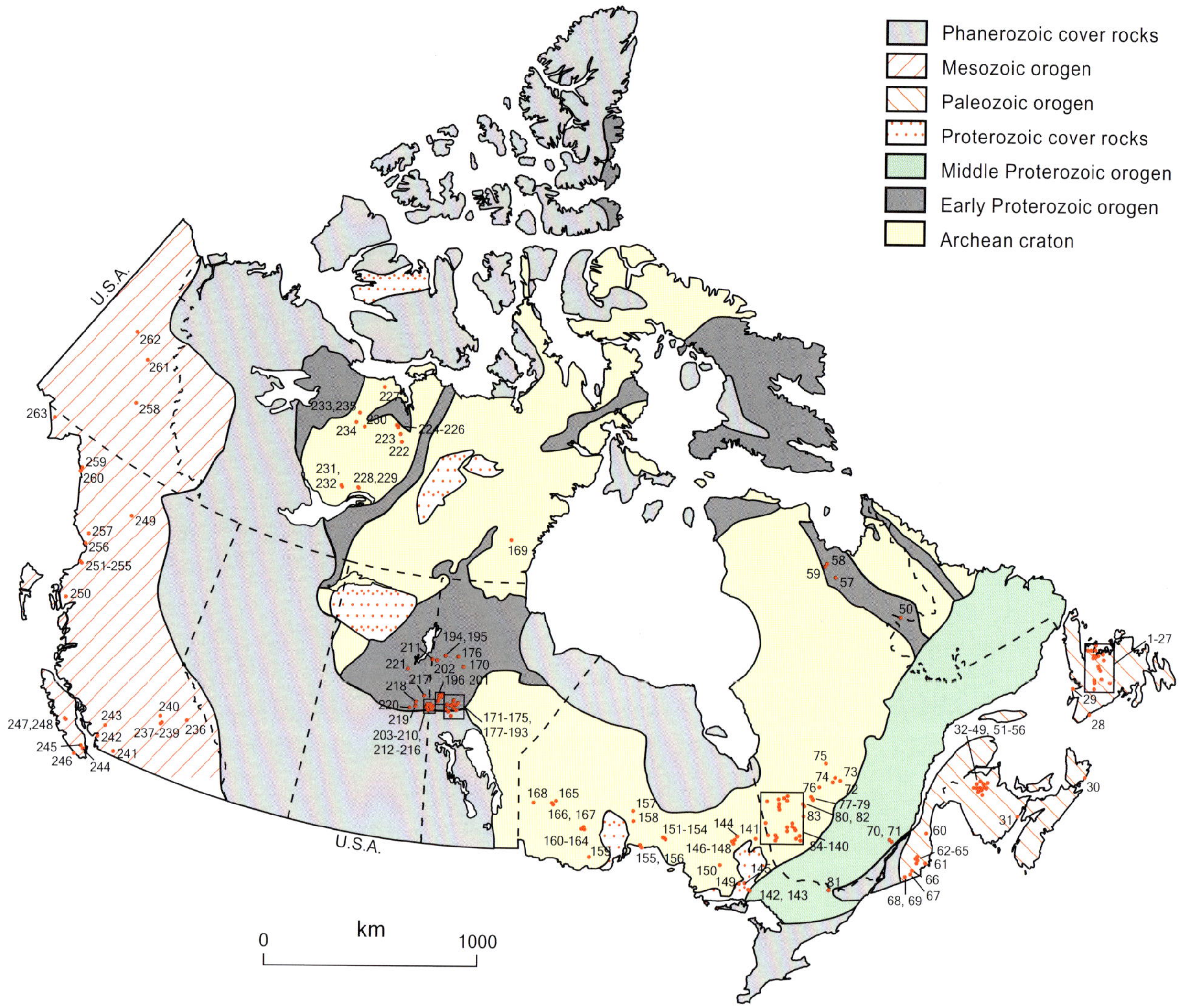

FIGURE 6. Distribution of VMS deposits in Canada by geologic province. Numbers correspond to deposits listed in the national VMS database (Appendix 1).

TABLE 2. Canadian volanogenic massive sulphide deposits presently in production (2005).

Deposit	Location	(1) Mt	Cu wt.%	Zn wt.%	Pb wt.%	Ag g/t	Au g/t	Age
Brunswick No. 12	Bathurst, New Bruswick	229.8	0.46	7.66	3.01	91	0.46	Ordovician
Kidd Creek	Abitibi, Ontario	149.3 (181 mined + all resources)	2.89	6.36	0.22	92	0.05	Archean
LaRonde (incl. LaRonde II)	Abitibi, Quebec	88.1	0.32	1.71		40.9	5.07	Archean
Selbaie	Abitibi, Quebec	47.3	0.98	1.98		20	0.9	Archean
Myra Falls Gp.,Buttle Lake	Wrangellia, British Columbia	29.3	1.83	6.25	0.55	49	2.01	Devonian
Trout Lake	Trans-Hudson Orogen, Manitoba	20	1.83	5.59		17.4	1.73	Paleoproterozic
Louvicourt	Abitibi, Quebec	19.3	3.1	1.71		28.7	0.83	Archean
Triple 7	Trans-Hudson Orogen, Manitoba	14.5	3.32	5.78		37.7	2.71	Paleoproterozic
Bouchard-Hébert	Abitibi, Quebec	10.2	2.11	4.79		15	1.4	Archean
Callinan	Trans-Hudson Orogen, Manitoba	9.16	1.41	3.59		23.5	2.08	Paleoproterozic
Duck Pond*	Central Volcanic Belt,	5.2	3.24	5.97	1.1	61.5	0.88	Ordovician
Perseverence Group *	Abitibi , Quebec	5.1	1.24	15.82		29.4	0.38	Archean
Eskay Creek	Stikine, British Columbia	4	0.33	5.4	2.2	998	26.4	Jurassic
Bell Allard	Abitibi, Quebec	3.2	1.5	13.77		43.5	0.76	Archean
Chisel North	Trans-Hudson Orogen, Manitoba	2.8	0.15	9.36	0.4	22	0.4	Paleoproterozic
Konuto	Trans-Hudson Orogen,	1.28	5.27	1.44		10.6	2.09	Paleoproterozoic

* In preproduction (2006)

(1) Includes production and estimated reserves where applicable.

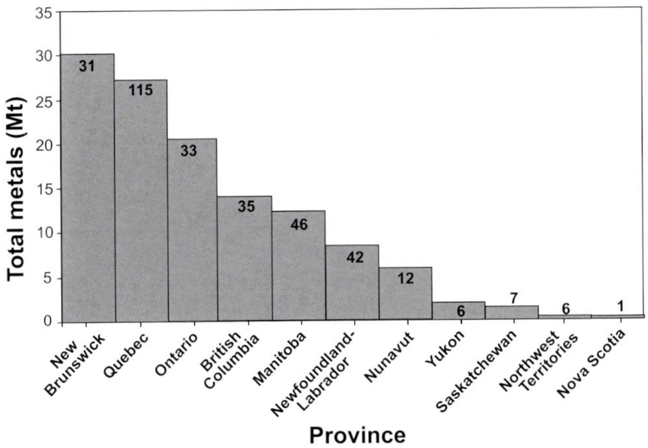

FIGURE 7. Histogram of the total tonnage of base metals from known VMS deposits per province; also shown are the number of deposits. The aggregate tonnage was calculated by total metals represent divided by geological reserves (proven, possible, and probable; non 43-101 compliant).

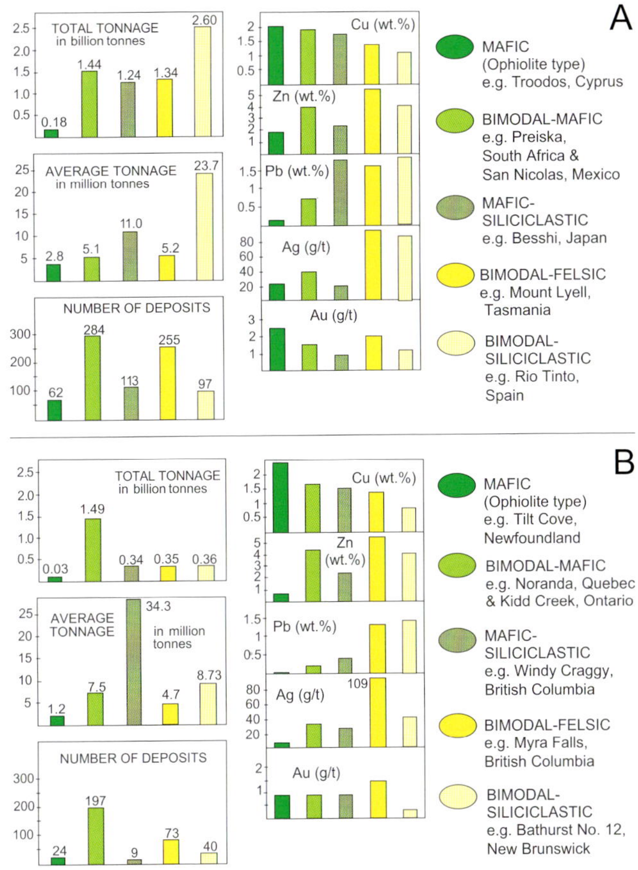

FIGURE 9. Statistics for VMS deposits grouped by lithologic class (Barrie and Hannington, 1999): (A) worldwide deposits; (B) Canadian deposits.

along transform fault sutures (Bloomer et al., 1995). These early suprasubduction terranes are most commonly observed in the ancient rock record at the base of oceanic arc assemblages in which VMS deposits are spatially associated with isolated extrusive rhyolite complexes near the top of thick basalt and basaltic andesite successions. The best Canadian example of these bimodal mafic-dominated caldera settings is the Paleoproterozoic host succession to the Anderson, Stall, and Rod VMS deposits in the Snow Lake camp, Manitoba (Bailes and Galley, 1999). The komatiite-basalt-rhyolite setting for the Archean Kidd Creek deposit is interpreted to be an early primitive arc setting possibly linked to an underlying mantle plume (Wyman et al., 1999), or a rare example of a non-arc VMS setting associated with partial lithospheric melting above a mantle plume (cf. Iceland).

In the idealized evolutionary stages of arc terrane formation, extension of the principal arc assemblage is another common period of VMS formation (Fig. 11). This results in

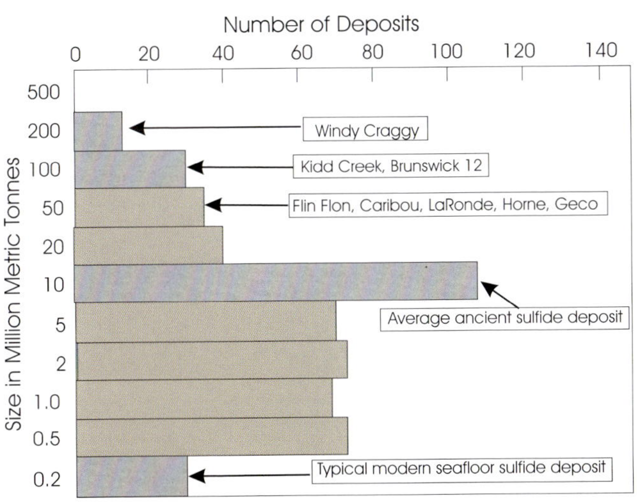

FIGURE 8. Global (proven, possible, and probable; non 43-101 compliant) size distribution for VMS deposits, with deposits over 50 Mt considered "very large" (Table 1), those over 100 Mt considered "giant", and those over 150 Mt defined as "supergiant". Atlantis II Deep, Red Sea, is considered the largest modern example of a seafloor massive sulphide deposit. Best known examples of Canadian very large, giant, and supergiant deposits are shown. Modified from Hannington et al. (1995).

the formation of calderas in which bimodal mafic extrusive successions predominate. This is perhaps the most common arc environment for VMS formation in oceanic arc settings. Bimodal mafic-dominated VMS-hosting calderas include the Archean Noranda and the Paleoproterozoic Flin Flon mining camps (Gibson and Watkinson, 1990; Syme and Bailes, 1993). Rifting of continental margin arcs, in contrast, results in the development of more volcaniclastic-rich bimodal felsic extensional settings. Examples include the Sturgeon Lake camp in the Archean Wabigoon terrane of Ontario (Morton et al., 1990; Whalen et al., 2004) and the Devonian Buttle Lake VMS camp in the Wrangellia Terrane of British Columbia (Barrett and Sherlock, 1996a). Outside Canada, the Paleoproterozoic Skellefte mining district in Sweden (Allen et al., 1996a) and the Cambrian Mount Read VMS district in Tasmania (Corbett, 1992) are other examples of rifted continental margin arc settings. Continued extension in both oceanic and continental margin arc settings results in the development of back-arc basins. In oceanic arc settings, mature back-arc ophiolites also can host VMS deposits. Canadian examples include the Paleoproterozoic Birch-Flexar-Coronation camp on the Saskatchewan side of the Flin Flon mining district (Wyman et al., 1999) and Betts Cove, Newfoundland (Swinden et al., 1988; Bédard et al., 1998). Well known examples outside Canada include the Tethyan ophiolites in Cyprus (Troodos), Oman (Semail), and Turkey (Ergani) (Galley and Koski, 1999, and references therein).

TABLE 3. Examples of large-tonnage volcanogenic massive sulphide deposits of the World (Canadian deposits in red).

NAME	COUNTRY	Orogen	Mtonnes Ore (Geol.)	CU %	PB %	ZN %	AU (g/t)	AG (g/t)	Orebody Age (est. Ma)
SUPERGIANT									
Rio Tinto (Stockwork)	Spain	Hercynian	1200.00	0.15		0.15		7.00	320
Rio Tinto (Massive)	Spain	Hercynian	335.00	0.39	0.12	0.34	0.36	22.00	320
Kholodnina	Russia	Baikal-Vitim	300.00	0.04	0.79	5.2			750
Windy Craggy (Cu,Co)	Canada	N.Cordilleran	297.40	1.38		0.25	0.22	3.83	220
Neves Corvo Group	Portugal	Hercynian	270.00	1.59	0.15	1.41		9.87	320
Gai East	Russia	Uralides (Hercynian)	269.00	1.2		0.7	1.10	7.70	395
Aljustrel Group (total)	Portugal	Hercynian	250.00	1.2	1.2	3.2	1.00	38.00	320
Brunswick #12	Canada	Appalachian	229.80	0.46	3.01	7.66	0.46	91.00	465
Gai	Russia	Uralides (Hercynian)	205.00	1.4	0.06	0.5	1.10	7.90	395
La Zarza	Spain	Hercynian	164.00	1.2	1.1	2.5	1.80	47.00	320
Ducktown	USA	Grenvillean? (Oecee)	163.34	1		0.9	0.30	3.00	1000
GIANT									
Kidd Creek	Canada	Abitibi (Kenoran)	147.88	2.31	0.22	6.18	0.01	87.00	2714
Horne - No. 5 Zone	Canada	Abitibi (Kenoran)	144.00	1		0.9	1.40		2698
Ozernoe	Russia	Baikal-Vitim	130.00	0.01	1.2	6.2			500
Ridder-Sokol	Kazakhstan	Altaides (Hercynian)	125.00	0.3	2	4	2.50	10.00	400
Zyryanov	Kazakhstan	Altaides (Hercynian)	125.00	0.4	2.7	4.5	0.13	20.00	395
Gacun	China	Yidun, Indosinian (Tethyan)	124.00	0.72	4.62	6.66	0.46	157.00	200
Masa Valverde	Spain	Hercynian	120.00	0.5	0.6	1.3	0.80	38.00	320
Sibai	Russia	Uralides (Hercynian)	115.00	1	0.04	1.56	0.60	16.00	392
Tharsis	Spain	Hercynian	110.00	0.5	0.6	2.7	0.70	22.00	320
Yubileinoe	Russia	Uralides (Hercynian)	107.00	1.9	0.1	1.2	2.50	16.00	392
Uchaly	Russia	Uralides (Hercynian)	106.00	1.1		3.8	1.10	15.50	392
Madneuli	Georgia	Caucasian (Tethyan)	102.60	1.29		1.8	0.73	4.31	70
VERY LARGE									
Mount Lyell	Australia	Tasman	98.57	1.17	0.01	0.04	0.39	7.20	495
Rouez	France	Caledonian	90.74	0.6		1.5	1.50	21.00	600
Aznalcollar	Spain	Hercynian	90.00	0.51	0.85	1.8	0.48	37.00	320
LaRonde (incl.LaRonde-II)	Canada	Abitibi (Kenoran)	88.00	0.3		1.7	5.07	40.90	2710
Skorpion	Namibia	Gariep	85.00		0.71	8.05			752
Podolsk	Russia	Uralides (Hercynian)	84.10	2.01	0.13	1.3	1.49	27.60	392
Murgul	Turkey	Pontides (Tethyan)	83.14	0.76	0.05	0.03	0.05	3.70	175
Ruttan	Canada	Trans-Hudson	82.80	1.37	0.08	1.63	0.49	13.11	1900
Tambo Grande 3	Peru	S.Cordilleran	82.00	1	0.3	1.4	0.80	25.00	104
San Nicolas	Mexico	C.Cordilleran	79.90	1.34		2.27	0.53	30.00	136
Pyhasalmi	Finland	Svecokarelian	75.70	0.9	0.06	1.9	0.20	14.00	1921
Sotiel	Spain	Hercynian	75.20	0.56	1.34	3.16	0.21	24.00	320
Los Frailes	Spain	Hercynian	70.00	0.34	2.25	3.92		62.00	320
Heath Steele	Canada	Appalachian	69.90	0.98	0.89	2.69	0.54	47.00	465
Ulaan	Mongolia	Kazakh-Mongol(Hercyn.)	68.00		1.2	2	0.21	53.00	380
Caribou	Canada	Appalachian	64.69	0.51	1.6	4.29	1.89	51.00	465
Crandon	USA	Trans-Hudson	63.50	1		6.5			1870
Flin Flon	Canada	Trans-Hudson	62.93	2.2		4.1	2.85	43.20	1875
Zincgruvan(+Knalla)	Sweden	Svecokarelian	60.00		3.2	10.4		69.00	1890
Tishin	Kazakhstan	Altaides (Hercynian)	60.00	0.5	0.9	5.3	0.90	15.00	395
Geco	Canada	West.Superior (Kenoran)	58.40	1.86	0.15	3.45		50.06	2720
Tambo Grande 1	Peru	S.Cordilleran	56.20	1.6	0.3	1	0.50	26.00	104
Deerni (Cu-Co)	China	Indosinian (Tethyan)	54.00	1.23		1.57	0.42	4.73	260
Horne-H&G Orebodies	Canada	Abitibi (Kenoran)	53.70	2.2			6.10	13.00	2700
Mount Morgan	Australia	Tasman	50.00	0.7	0.05	0.1	4.70	0.60	385
Outokumpu(Cu,Zn,Co)	Finland	Svecokarelian	50.00	3.3	0.005	1.07	0.07	9.00	1970
Artem'yev	Kazakhstan	Altaides (Hercynian)	50.00	1.4	1.6	2.2	1.20	143.00	375
Lousal	Portugal	Hercynian	50.00	0.7	0.8	1.4	0.70	21.00	300
LARGE									
Britannia	Canada	N.Cordilleran	49.31	1.08	0.033	0.26	0.34	4.03	150
Novo-Leninogorsk	Kazakhstan	Altaides (Hercynian)	48.00	0.16	1.43	4.04	1.54	32.80	395
Preiska	South Africa	Namaqua	47.00	1.7		3.8	0.00		1300
Anyox-Hidden Creek	Canada	N.Cordilleran	45.95	1.37			0.17	9.92	195
Hanaoka Mine (total)	Japan	Japan arcs(Tethyan)	43.50	1.2	1.5	4.7	0.40	68.00	15
Aguas Tenidas	Spain	Hercynian	41.00	1.3	0.91	3.1	0.50	37.00	320
Hongtoushan	China	Sino-Korean Platform	40.00	1.75		2.4	0.77	32.40	3000
Maleev	Kazakhstan	Altaides (Hercynian)	40.00	2.3	1.3	7.5	0.75	75.00	390
Orlovskoye	Kazakhstan	Altaides (Hercynian)	40.00	2.4	0.5	2.1	0.80	47.00	392
Ashele (#1)	China	Altayshan (Hercynides)	34.00	2.51		2.98	0.57	104.03	375
Xiaotieshan	China	Tarim-NorthQilian (Caled.)	34.00	1.26	3.39	5.33	2.28	126.20	440
Arctic (Brooks Range,Ak)	USA	N.Cordilleran	32.93	4	0.8	5.5	0.70	55.00	365
Rosebery	Australia	Tasman	32.70	0.58	4.4	14.5	2.70	145.00	495
Liwu	China	Yidun, Indosinian (Tethyan)	31.00	2.5		0.62			430
Belousov	Kazakhstan	Altaides (Hercynian)	30.00	2.6	2.4	9.2	2.00	119.00	395
Lokken (Hoydal)	Norway	Caledonian	30.00	2.3	0.02	1.8	0.29	19.00	450
Jerome- United Verde	USA	Yavapai	30.00	4.8		0.2	1.37	49.70	1800
Bald Mountain	USA	Appalachian	29.98	1.03	<0.05	1.12	0.51	14.40	430
Besshi	Japan	Japan arcs(Tethyan)	29.95	2.6		0.3	0.70	21.00	210
Selbaie	Canada	Abitibi (Kenoran)	29.90	1.21		1.91	0.63	37.02	2730
Myra Falls Group	Canada	N.Cordilleran	29.32	1.83	0.55	6.25	2.00	49.00	365
Garpenberg (+Lappberget)	Sweden	Svecokarelian	29.00	0.3	3.3	5.3	0.65	98.00	1890
Bisha	Eritrea	Pan African	28.60	1.52		4.63	1.68	46.80	850
Vihanti	Finland	Svecokarelian	28.10	0.48	0.36	5.12	0.49	25.00	1910
Falun	Sweden	Svecokarelian	28.10	3	1.5	4	3.00	20.00	1875
Safyanovka	Russia	Uralides (Hercynian)	27.50	3.04		1.4	1.32	25.00	392
McIlvenna Bay	Canada	Trans-Hudson	27.23	0.9	0.1	3.27	0.34	16.43	1900
Mattagami Lake	Canada	Abitibi (Kenoran)	25.60	0.42		5.1	0.30	21.60	2725
Las Cruces (primary)	Spain	Hercynian	25.20	1.25	1.69	3.63	0.38	38.00	320
Granduc	Canada	N.Cordilleran	25.06	1.79	0.021	0.1	0.17	10.63	190
Korbalikhinsk	Russia	Altaides (Hercynian)	25.00	1.46	2.01	9.81			375
Greens Creek	USA	N.Cordilleran	25.00	0.32	5.1	13.9	5.61	706.00	220

* Modified from Franklin et al., 2005

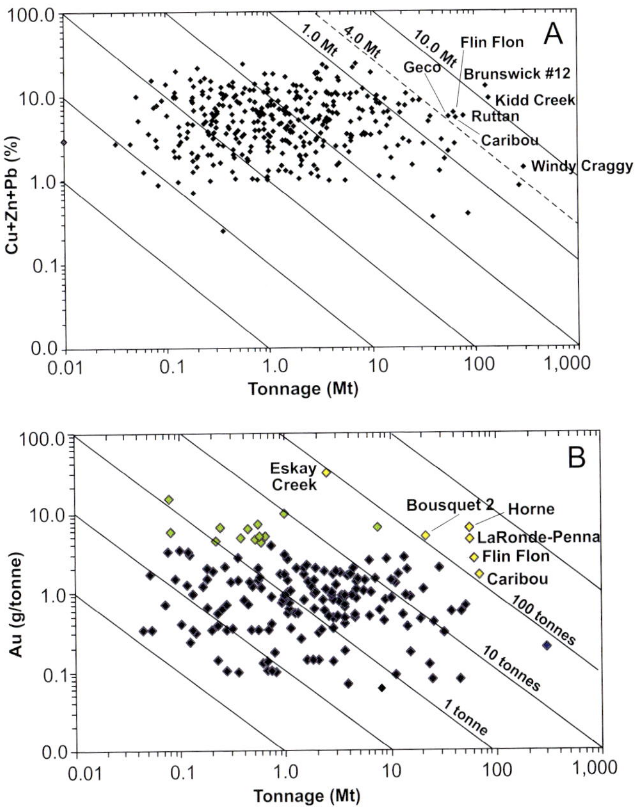

FIGURE 10. Distribution of Canadian VMS deposits with respect to **(A)** aggregate base metal grade versus tonnes and **(B)** contained Au versus tonnes; most auriferous Au deposits contain >4 g/t Au (green diamonds). Those containing over 1000 tonnes of Au (yellow diamonds) include both auriferous VMS deposits and those with moderate Au grades but large tonnages. Giant and supergiant VMS deposits are identified by name. From National VMS database (Appendix 1).

Continental back-arc settings contain some of the world's most economically important VMS districts. These environments are dominated by bimodal siliciclastic rocks ± iron formation and include the Ordovician Bathurst camp of New Brunswick (van Staal et al., 2003) and Finlayson Lake (Piercey et al., 2001). Examples outside Canada include the Archean Golden Grove camp in Western Australia (Sharpe and Gemmell, 2002), the Paleoproterozoic Bergslagen district of Sweden (Allen et al., 1996b), the Cambro-Ordovician Mount Windsor district of Queensland (Doyle and McPhie, 2000), the Devono-Mississippian Iberian Pyrite Belt (Carvalho et al., 1999), and parts of the Devonian Southern Urals VMS districts of Russia and Kazakhstan (Herrington et al., 2002; Franklin et al., 2005).

Other extensional environments may form in post-accretion and/or successor arc settings. Crustal thickening of an accreted ocean-floor-arc assemblage can result in modification of the angle of descent of the subducting slab, cessation of subduction along a section of plate boundary, or a change in the direction of approach of the colliding plates (Ziegler, 1992; Hamilton, 1995). This process results in the generation of strike-slip basins in the older arc assemblages. Magmatism associated with these successor arc basins may be associated with mineralized porphyry systems (Richards, 2003), and the basins may be infilled with both subaqueous and subaerial bimodal volcanic rocks. This can result in the formation of multiple mineral deposit types, including epithermal and VMS deposits. A good example of this is the Lower Jurassic Hazelton Group in British Columbia, which hosts the Eskay Creek Au-rich VMS deposit (Barrett and Sherlock, 1996b; Nelson and Mihalynuk, 2004). When these strike-slip fault systems propagate into a continental margin setting, such as in the modern day Guaymas Basin, Gulf of California, the strike-slip basins begin to infill with terrigenous sediment. They can host mafic siliciclastic-hosted VMS deposits, such as the Triassic Windy Craggy and Green's Creek deposits in British Columbia and Alaska, respectively (Peter and Scott, 1999). These are known as Besshi-type deposits from the type locality in the fore-deep accretionary wedge outboard of the Miocene Japanese islands. Other mafic siliciclastic-hosted VMS deposits occur along modern sedimented seafloor spreading systems such as Middle Valley, on the Juan de Fuca Ridge off the British Columbia coast (Goodfellow et al., 1999).

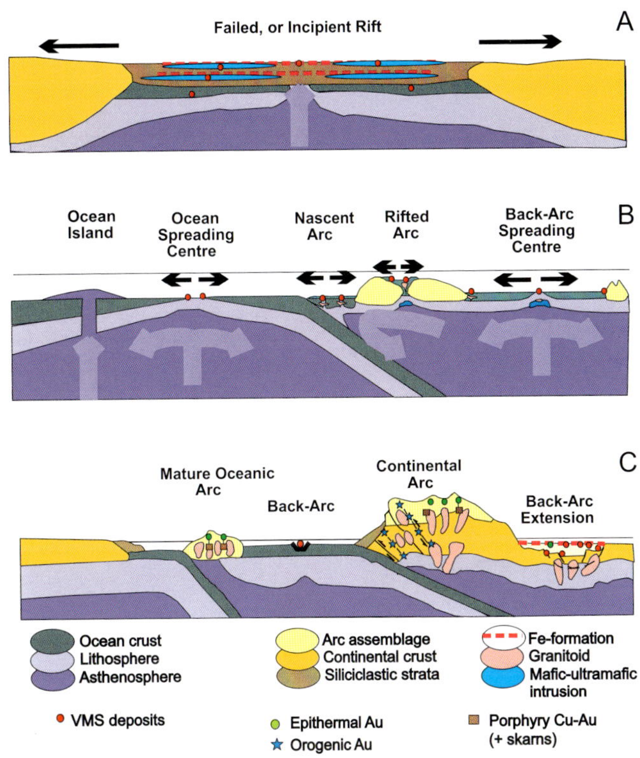

FIGURE 11. There are three principal tectonic environments in which VMS deposits form, each representing a stage in the formation of the Earth's crust. **(A)** Early Earth evolution was dominated by mantle plume activity, during which numerous incipient rift events formed basins characterized by early ocean crust in the form of primitive basalts and/or komatiites, followed by siliciclastic infill and associated Fe-formation and mafic-ultramafic sills. In the Phanerozoic, similar types of incipient rifts formed during transpressional, back-arc rifting (Windy Craggy). **(B)** The formation of ocean basins was associated with the development of ocean spreading centers along which mafic-dominated VMS deposits formed. The development of subduction zones resulted in oceanic arc formation with associated extensional domains in which bimodal mafic, bimodal felsic, and mafic-dominated VMS deposits formed. **(C)** The formation of mature arc and ocean-continent subduction fronts resulted in successor arc and continental volcanic arc assemblages that host most of the felsic-dominated and bimodal siliciclastic deposits. Thin black arrows represent direction of extension and thick, pale arrows represent mantle movement. Modified from Groves et al., 1998

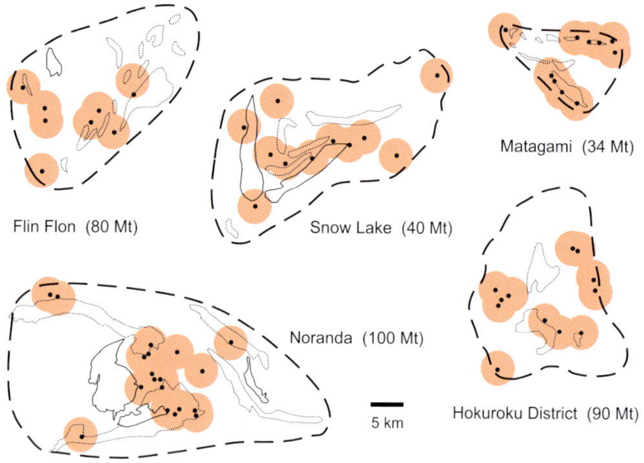

FIGURE 12. A same-scale comparison of selected VMS districts. A 5 km diameter circle around each deposit shows the hypothetical area of influence of proximal-scale alteration about each deposit, all encircled by a dashed line defining the proposed extent of a regional-scale alteration system for each camp based on the presence of known felsic volcanic and synvolcanic formations/intrusions (thin black lines). The Noranda example corresponds closely to the observed alteration. Modified from Sangster (1980).

District-Scale Environments

Most, but not all VMS deposits, occur in clusters that define major mining camps. Sangster (1980) used the distribution of VMS deposits within well known mining districts in Canada to indicate that there was a first-order regional control on their distribution (Fig. 12). In general, the deposit clusters are restricted to either linear rifts or calderas. These features are generated by regional thinning of the basement, decompression melting of the underlying mantle, and generation of mafic magmas (Fig. 13). In ocean spreading-ridge settings, these magmas rise to within a few thousand metres of the seafloor to form elongate gabbroic sills that parallel the seafloor-spreading axes (Stinton and Detrick, 1992). Where pre-existing ocean-floor or arc lithosphere is present, these mafic magmas, at temperatures of 1000 to 1400°C, may underplate the crust, producing intermediate to felsic partial melts and bimodal mafic intrusive/extrusive assemblages. The associated gabbro-diorite-tonalite-trondhjemite intrusive complexes may rise to within 2 to 3 km of the seafloor (Galley, 2003, and references therein). Where extension is taking place in thicker (20-30 km) crust, such as in a continental back-arc setting, magmas may form mid-crustal intrusions. These melts may not intrude into their comagmatic volcanic assemblages but may remain in the underlying basement rocks. These different scenarios result in multiple forms of district-scale alteration and deposit characteristics for a VMS district.

The presence of either mafic or composite high-level subvolcanic intrusions within a rift or caldera setting will drive a subseafloor hydrothermal-fluid convection system (Galley, 1993; Alt, 1995) (Fig. 14). Connate seawater in the porous crust is first heated, causing it to become buoyant. As this heated water rises up synvolcanic fault structures, cold seawater is drawn in above the cooling intrusion. These originally cold, near neutral pH fluids are progressively heated during their downward migration, interacting with the surrounding rocks at progressively higher temperatures. The isotherms above cooling sill complexes are generally horizontal, resulting in the formation of a stratified, district-scale semi-conformable alteration zone controlled in extent by the strike length of the underlying intrusion (Spooner and Fyfe, 1973; Munha and Kerrich, 1980; Lagerblad and Gorbatchev; 1985; Gibson and Watkinson, 1990; Galley, 1993; Alt, 1995; Brauhart et al., 1998; Bailes and Galley, 1999) (Fig. 14). The distribution of the resulting alteration mineral assemblages mimics that of regional metamorphic facies (Spooner and Fyfe, 1973; Alt, 1995; Hannington et al., 2003) (Fig. 15). Hydrothermal fluid reaction zones immediately overlying the intrusions can be altered to amphibolite-facies assemblages, including Fe-Ca-rich amphibole, clinozoisite, Ca-plagioclase, and magnetite (Figs. 15, 16A,B,C). Above this are Na-Ca-rich greenschist-facies assemblages characterized by albite, quartz, chlorite, actinolite, and epidote. Closer to

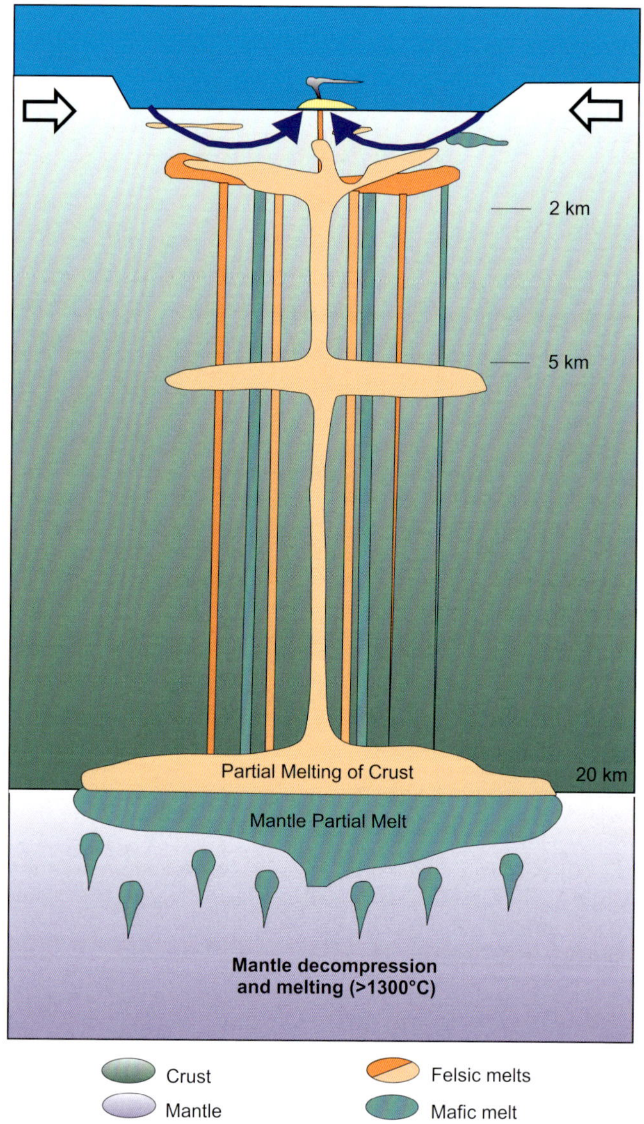

FIGURE 13. VMS environments are characterized by tectonic extension at various scales (open arrows). Extension resulted in crustal thinning, mantle depressurization, and the generation of basaltic melts. Depending on crustal thickness and density, these mafic melts ponded at the base of the crust, resulting in partial melting and generation of granitoid melts. These anhydrous, high-temperature melts quickly rose to a subseafloor environment (<3 km below seafloor), where their heat initiated and sustained convective hydrothermal cells that formed VMS deposits (black arrows).

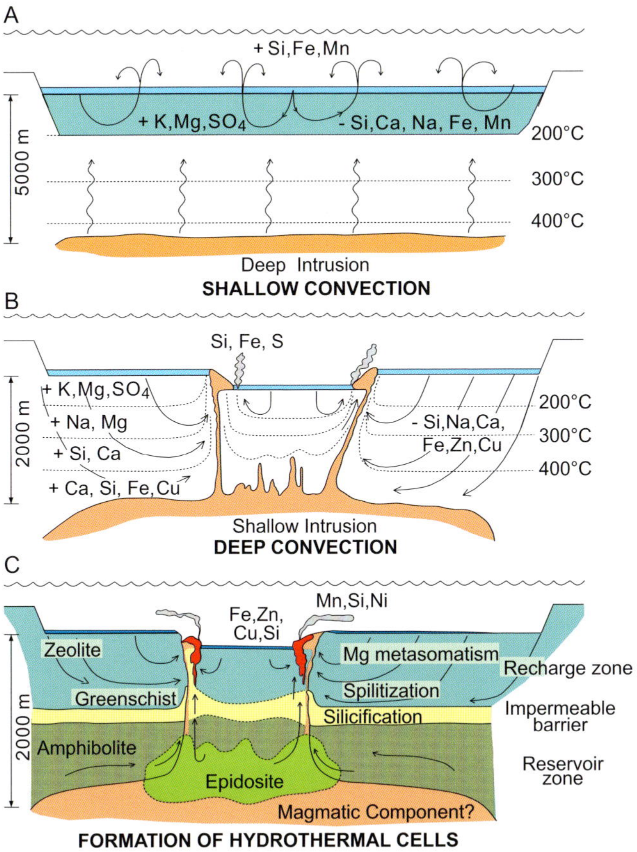

FIGURE 14. The development and maturation of a generic subseafloor hydrothermal system involves three stages. **(A)** The relatively deep emplacement of a subvolcanic intrusion below a rift/caldera and the establishment of a shallow circulating, low-temperature seawater convection system. This results in shallow subseafloor alteration and associated formation of hydrothermal exhalative sediments. **(B)** Higher level intrusion of subvolcanic magmas and resultant generation of a deep-seated subseafloor seawater convection system in which net gains and losses of elements are dictated by subhorizontal isotherms. **(C)** Development of a mature, large-scale hydrothermal system in which subhorizontal isotherms control the formation of semiconformable hydrothermal alteration assemblages. The high-temperature reaction zone next to the cooling intrusion is periodically breached due to seismic activity or dyke emplacement, allowing focused upflow of metal-rich fluids to the seafloor and formation of VMS deposits. From Galley (1993).

the seafloor are zeolite-clay and related subgreenschist mineral assemblages characterized by K-Mg-rich smectites, mixed-layer chlorites, and K-feldspar. In some regional hydrothermal systems, the low-temperature alteration assemblages closest to the seafloor are dominated by carbonate species due to precipitation from shallowly circulating seawater (Fig. 16D). These chemical and mineralogical changes in the ancient rock record can be further revealed by mapping shifts in bulk rock oxygen and hydrogen isotope compositions of the different zones (Green et al., 1983; Taylor and South, 1985; Aggarwal and Longstaffe, 1987; Cathles, 1993; Paradis et al., 1993). These stratified alteration zones can have a strike length of 5 to 50 km and a thickness of 1 to 3 km in caldera settings (Fig. 15). The size and areal morphology of the alteration system is a reflection of the size and areal morphology of the VMS deposit cluster (Fig. 12). The distribution of VMS deposits within this cluster depends on synvolcanic fault distribution relative to the underlying intrusions (Eastoe et al., 1987; Gibson and Watkinson, 1990; Brauhart et al., 1998; Galley, 2003). Faults that acted as conduits for volcanic feeder systems were the focal point for ascent of high-temperature, acidic metal-laden hydrothermal fluids that formed VMS deposits. These fault systems may have remained active through several cycles of volcanic and hydrothermal activity. This may have resulted in several periods of VMS formation at different stratigraphic levels within the rift or caldera structure.

Mafic-dominated, bimodal mafic, and bimodal felsic host rocks are dominated by effusive volcanic successions and accompanying, large-scale hypabyssal intrusions (Fig. 17). This high-temperature subseafloor environment supported high-temperature (>350°C) hydrothermal systems, from which may have precipitated Cu, Cu-Zn, and Zn-Cu- (Pb) VMS deposits with variable Au and Ag contents. Areally extensive, 1 to 5 m thick, Fe-rich "exhalites" (iron formations) may mark the most prospective VMS horizons (Spry et al., 2000; Peter, 2003) (Fig. 18A). These exhalite deposits consist of a combination of fine volcaniclastic material, chert, and carbonates. They formed during the immature and/or waning stages of regional hydrothermal activity when shallowly circulating seawater stripped Fe, Si, and some base metals at <250°C and precipitated them on the seafloor through extensive, but diffuse, low-temperature hydrothermal venting. Formation of exhalites on a basalt-dominated substrate was commonly accompanied by silicification and/or chloritization of the underlying 200 to 500 m of strata (Fig. 18B). Examples of this are observed in the Noranda, Matagami Lake, and Snow Lake VMS camps (Kalogeropoulos and Scott, 1989; Liaghat and MacLean, 1992; Bailes and Galley, 1999). In felsic volcaniclastic-dominated terranes, the generation of Fe-formation exhalites was accompanied by extensive K-Mg alteration of the felsic substrate, as recorded in the Bergslagen district of Sweden (Lagerblad and Gorbatschev, 1985) and in the Iberian Pyrite Belt (Munha and Kerrich, 1980).

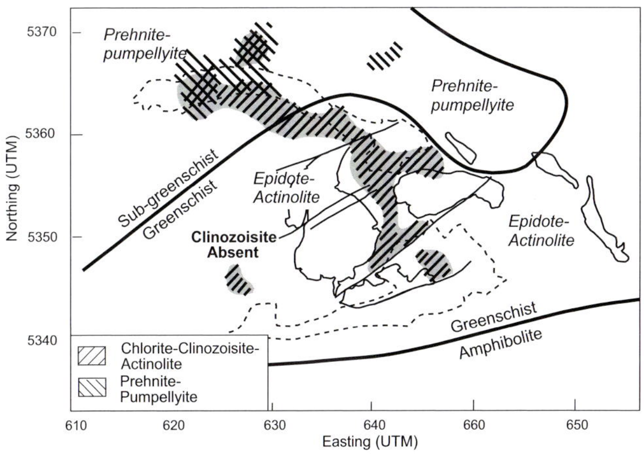

FIGURE 15. Comparison of regional greenschist-facies hydrothermal alteration in the Noranda Volcanic Complex with previously mapped metamorphic isograds (solid lines from Dimroth et al., 1983; Powell et al., 1993). The distribution of greenschist-facies hydrothermal alteration (shaded) suggests that interpreted metamorphic zonation is at least partly a product of early synvolcanic hydrothermal processes. Note that epidote and chlorite in the pre-cauldron sequence are distinct from those of the mine sequence volcanic rocks, even though they are well within the epidote-actinolite subfacies and have been metamorphosed at the same pressure and temperature. Modified from Hannington et al. (2003).

FIGURE 16. **(A)** High-temperature hydrothermally altered mafic volcaniclastic turbidite (left) overlain by a strongly silicified mafic debris flow 1200 m below the Chisel-Lost-Ghost VMS horizon, Snow Lake. This is a regional-scale reaction zone overlain by a high-temperature zone of silica precipitation. **(B)** Strongly silicified pillows with pipe vesicles infilled with actinolite, epidote, and magnetite, and interpillow hyaloclastite completely replaced by the same assemblage. This alteration facies directly overlies the subvolcanic Mooshla intrusion, Bousquet VMS camp, Quebec. **(C)** Epidosite typical of the root zones of VMS hydrothermal upflow zones in which high fluid/rock ratios have resulted in leaching of lithophile, chalcophile, and low field strength elements from the strata. **(D)** Chloritoid-rich zone below the Mattabi deposit, Sturgeon Lake, where Fe-rich hydrothermal fluids overprinted a previously formed carbonate-rich regional alteration zone.

Mafic, felsic, and bimodal siliciclastic volcanic assemblages tend to host volumetrically smaller mafic and/or felsic sill-dyke complexes, and generally contain Zn-Cu-Co and Zn-Pb-Cu-Ag VMS deposits, respectively. More Cu-rich deposits, such as Neves Corvo in the Iberian Pyrite Belt, may also be present in settings proximal to discrete extrusive complexes. The district-scale semiconformable hydrothermal systems consist of low-temperature mineral assemblages, with Mg-K smectite and K-feldspar alteration overlain by extensive units of low-temperature Fe-Si-Mn deposits. Other types of iron formation in VMS districts are interpreted to be products of plume fallout from high-temperature hydrothermal venting, or collection of hypersaline brines within fault-controlled depressions on the seafloor (Peter, 2003). Iron formation horizons can extend for tens of kilometres, as in the Bathurst VMS camp in New Brunswick (Peter and Goodfellow, 1996) (Fig. 18C), the Paleoproterozoic Bergslagen district (Allen et al., 1996b), the Devono-Mississippian Iberian Pyrite Belt in Spain and Portugal (Carvalho et al., 1999), and the Mississippian Finlayson Lake camp, Yukon (Peter, 2003). Mineralogical variations within these regionally extensive iron formations, from oxide through carbonate to sulphide, are indicative of proximity to more focused, higher temperature hydrothermal vent complexes and also reflect stratification of the water column in the basin. The mineralogical variations are accompanied by changes in element ratios such as Fe, Mn, B, P, and Zn (exhalative component) versus Al and Ti (detrital clastic component) (Peter and Goodfellow, 1996).

Deposit-Scale Environments

VMS deposits consist of a massive to semimassive stratabound sulphide lens, and most are underlain by a sulphide-silicate stockwork vein system (Figs. 1, 4). Within this broad framework, there is a spectrum of deposit sizes, morphologies, and compositions, depending on the nature of the synvolcanic faulting, footwall and host-rock lithology, water depth, size and duration of the hydrothermal system, temperature gradients, and degree of preservation. Individual massive sulphide lenses can be over 100 m thick, tens of metres wide, and hundreds of metres in strike length. The 135 Mt Kidd Creek deposit begins at the present erosion surface and extends for over 2000 m downplunge (original strike length), with the five composite orebodies over 500 m wide and individual lenses up to 100 m thick. The stratabound sulphide mound component of a VMS deposit may have a number of morphologies and variable internal structure (e.g. Fig. 1). Observations of modern seafloor hydrothermal vent complexes in effusive, flow-dominated terranes indicate that the deposits begin to form as a series of

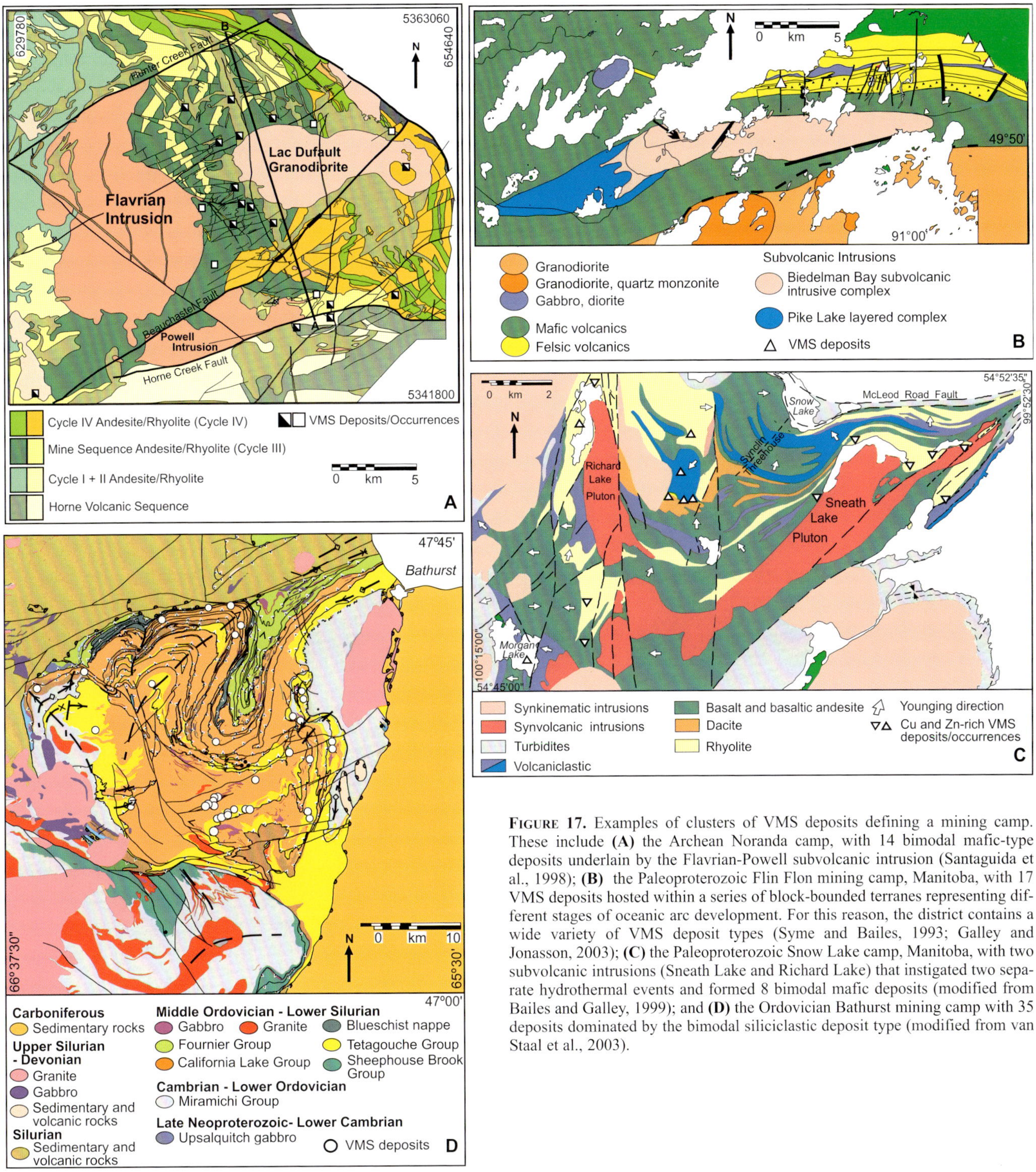

FIGURE 17. Examples of clusters of VMS deposits defining a mining camp. These include **(A)** the Archean Noranda camp, with 14 bimodal mafic-type deposits underlain by the Flavrian-Powell subvolcanic intrusion (Santaguida et al., 1998); **(B)** the Paleoproterozoic Flin Flon mining camp, Manitoba, with 17 VMS deposits hosted within a series of block-bounded terranes representing different stages of oceanic arc development. For this reason, the district contains a wide variety of VMS deposit types (Syme and Bailes, 1993; Galley and Jonasson, 2003); **(C)** the Paleoproterozoic Snow Lake camp, Manitoba, with two subvolcanic intrusions (Sneath Lake and Richard Lake) that instigated two separate hydrothermal events and formed 8 bimodal mafic deposits (modified from Bailes and Galley, 1999); and **(D)** the Ordovician Bathurst mining camp with 35 deposits dominated by the bimodal siliciclastic deposit type (modified from van Staal et al., 2003).

sulphide-silicate-sulphate chimneys (Fig. 19A). These become structurally unstable with continued growth and collapse, and coalesce to form a breccia mound (Fig. 19A,C). Continued circulation of hydrothermal fluids within this breccia mound results in sealing from seawater by a silica, clay, and/or sulphate cap. Progressive deposition of metal sulphides within the mound results in the formation of a complexly textured, semimassive to massive sulphide mound. The flow of hydrothermal fluid through the mound structure commonly results in remobilization of previously deposited metals along a chemical and temperature gradient perpendicular to the seawater interface. This process is referred to as zone refining (Eldridge et al., 1983) and results in a chalcopyrite±pyrrhotite-rich core and a sphalerite±pyrite±galena-rich outer zone (Fig. 20). In extreme cases, much of the base and precious metals can be remobilized out of the sulphide mound and carried into the seawater column by venting hydrothermal fluids and spent fluids (hot seawater). Massive pyritic cores and thin, base- and precious-metal-enriched outer margins are a characteristic of VMS

FIGURE 18. **(A)** Mine contact tuff exhalite horizon (between white lines) that overlies the silicified andesites of the Waite Formation, Noranda, Quebec. **(B)** Silicified basaltic andesite of the Upper Amulet Formation., Noranda, Quebec, as an example of pervasive silica precipitation that occurred in mafic flows directly underlying tuffaceous exhalite units in many Precambrian VMS camps. **(C)** Banded magnetite-chert Fe-formation overlying the Austin Brook massive sulphide deposit, Bathurst camp (photo by J.M. Peter).

deposits that have had a protracted thermal history (e.g. Hannington et al., 1998; Petersen et al., 2000).

Although many VMS deposits have a clastic component, this is usually subordinate to the massive sulphide facies. In many cases, such as the hanging-wall orebody at Buttle Lake, British Columbia (Barrett and Sherlock, 1996a), Kidd Creek, Ontario (Hannington et al., 1999b), and Louvicourt, Quebec, these subordinate clastic facies contain a mixture of sulphide and host-rock fragments (Fig. 19D). Interbedded sulphide and silicate-rich layers form from erosion and periodic collapse of a sulphide mound to form sand- to breccia-sized deposits. Examples where these clastic sulphide components are a dominant part of the deposit include Eskay Creek and Tulsequah Chief, British Columbia (Barrett and Sherlock, 1996a; Sebert and Barrett, 1996), and Buchans, Newfoundland (Walker and Barbour, 1981). In other cases, finely bedded ore lenses may result from high-temperature plume fallout of sulphide particles intermixing with hydrothermal silica, talc, and Mg-smectites, plus ambient background pelagic sedimentation (Peter, 2003, and references therein). Similar finely banded ores can also be a product of dynamic recrystallization of sulphides during regional deformation events. VMS deposits readily accommodate strain during regional deformation because of the ductile nature of massive sulphide bodies, and can therefore display much higher degrees of recrystallization and remobilization than the surrounding volcanic and sedimentary strata.

In some cases, VMS deposits do not form on the seafloor but develop as a result of shallow subseafloor replacement. This occurs when hydrothermal fluids infill primary pore space in either extrusive, autoclastic, volcaniclastic, or epiclastic successions below an impermeable cap. At the Ansil deposit in the Archean Noranda VMS camp, a succession of laminated felsic ash flows/turbidites infilled a small fault-bounded rift on the felsic flow complex. Hydrothermal fluid seepage up the rift margins resulted in unit-by-unit replacement of the laminated volcaniclastic layers by pyrite, sphalerite, and silica (Fig. 21A). This sulphide-impregnated unit was in turn replaced by massive pyrrhotite-chalcopyrite during a second stage of subseafloor replacement (Galley et al., 1996) (Fig. 21B). Some exceptionally large massive sulphide deposits have formed within volcanic depressions infilled with autoclastic and heterolithologic debris flow and talus deposits. These include the Horne No. 5 lens (Kerr and Gibson, 1993) Kidd Creek (Hannington et al., 1999b), and several orebodies at Buttle Lake (Barrett and Sherlock, 1996a) (Fig. 21C).

Most Canadian VMS deposits are characterized by discordant stockwork vein systems that commonly underlie the massive sulphide lenses, but may also be present in the immediate stratigraphic hanging-wall strata (Fig. 21D). These stockwork vein systems occur at the centre of more extensive, discordant alteration zones. They form by interaction between rising hydrothermal fluids, circulating seawater, and subseafloor rocks. The alteration zones and attendant stockwork vein systems may extend vertically below a deposit for several hundred metres. Proximal hanging-wall alteration can manifest itself as a semi-conformable halo up to tens of metres thick (Brunswick No 12, Bathurst) or may continue above the deposit for tens to hundreds of metres as a discordant alteration zone (Ansil, Noranda). In some cases, the proximal alteration zone and attendant stockwork vein mineralization connects a series of stacked massive sulphide lenses (Amulet, Noranda; LaRonde, Bousquet) representing synchronous and/or sequential phases of ore formation during successive breaks in volcanic activity.

In plan view, proximal alteration zones may form a halo up to twice the diameter of the massive sulphide lens (Fig. 22), but with deposits such as Chisel Lake, Snow Lake

FIGURE 19. **(A)** Example of a zoned sulphide chimney from the Endeavour Ridge vent field (I.R. Jonasson). **(B)** Typical textures from a massive sulphide mound, Main vent field, Juan de Fuca Ridge. Mineralogical banding is due to incremental chimney growth, with ovoids representing worm casts. Fragment cemented by later sulphide growth during mound collapse and subsequent invasion by hydrothermal fluid. **(C)** Clastic sandy sulphide ore from Cretaceous Aarja deposit, Semail ophiolite, Oman. This common texture is created by repeated mound collapse resulting from anhydrite dissolution and recementing with later sulphide (photo by I.R. Jonasson). **(D)** Pyrite-sphalerite clast as part of a proximal debris flow, Louvicourt, Val d'Or. 15 cm metal grid for scale.

camp, or Eskay Creek, British Columbia, footwall alteration can be volumetrically extensive and many times the diameter of the massive sulphide lens (Galley et al., 1993). The morphology of proximal alteration zones can vary widely, but generally they tend to widen in proximity to the paleo-seafloor surface, suggesting more intensive interaction between shallowly circulating, or connate, seawater and an ascending hydrothermal fluid. The internal mineralogical zonation of the alteration zones is indicative of these mixing phenomena. A Fe-chlorite-quartz-sulphide±sericite± talc mineral assemblage is commonly associated with the core of stockwork vein mineralization, which becomes increasingly quartz- and sulphide-rich towards the lower contact of the massive sulphide lens. In some cases, talc and/or magnetite occur at the base of the massive sulphide lens and the top of the alteration pipe, as at several of the Matagami district VMS deposits, the Ansil deposit in the Noranda camp, and the Late Triassic Chu Chua deposit in the Slide Mountain terrane of British Columbia. The core zone is cloaked in a wider zone of Fe-Mg-chlorite-sericite, including phengite in the part of this zone that encompasses the immediate hanging wall to the massive sulphide lens. Outboard from this is a zone rich in sericite, phengite, Mg-chlorite, ±albite, ±carbonate, and ±barite. This outer zone may also encompass a portion of the hanging-wall strata above, and lateral to, the massive sulphide lens.

In shallow-water environments (i.e. <1,500 m water depth), boiling may have occurred either in the upflow zone or in the immediate subseafloor. Depending on the extent of boiling, this can result in vertically extensive pyritic stock-

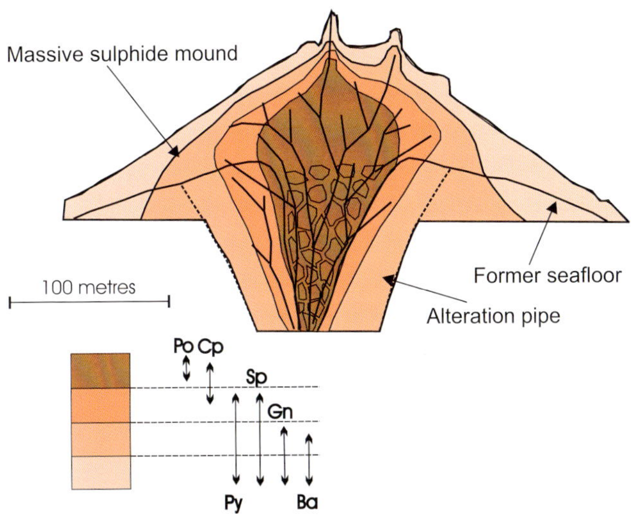

FIGURE 20. Mineral zonation commonly observed within VMS deposits; this zoning is largely a function of hydrothermal fluid temperature and composition. Temperature gradient results in the zoning of sulphide minerals within both the discordant stockwork zone and the conformable sulphide mound. From Lydon (1984).

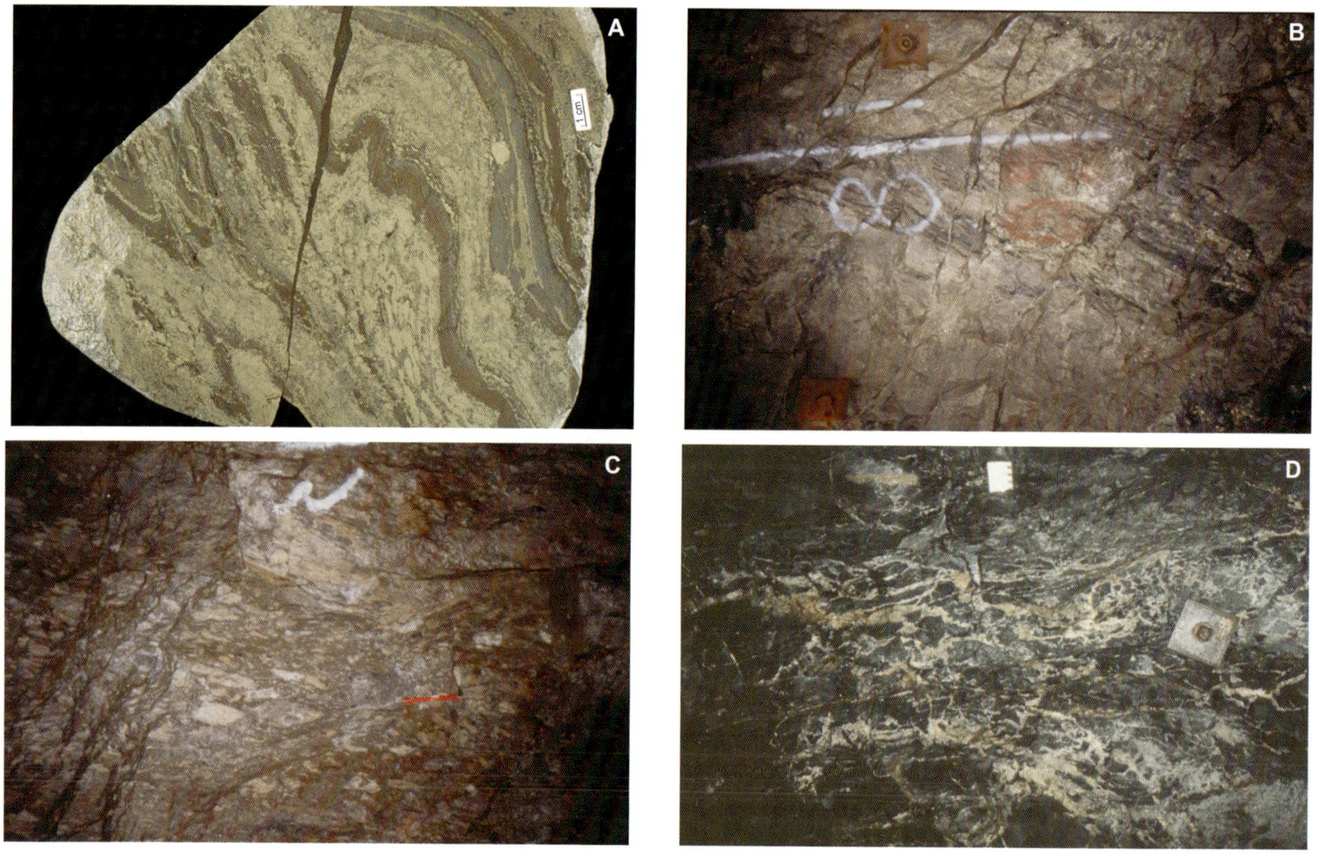

FIGURE 21. (A) Finely bedded tuff partially replaced by massive pyrrhotite-chalcopyrite at the Ansil deposit, Noranda. 15 cm metal plates for scale. (B) Cranston tuff unit with lit-par-lit replacement and in-filling by firstly by pyrite-sphalerite, followed by pyrrhotite-chalcopyrite, Ansil deposit, Noranda. (C) Rhyolite clasts cemented by pyrite-sphalerite rich sulphide groundmass, Louvicourt deposit, Val d'Or. 12 cm red magnet for scale. (D) Well developed pyrrhotite-chalcopyrite vein stockwork zone with intense chlorite alteration of the rhyolite wallrocks, Ansil deposit, Noranda.

work zones, possibly with widespread and intense sericite-quartz-pyrite alteration. The extensive sericite-rich alteration system that underlies the Eskay Creek auriferous VMS deposit may be a product of extensive subsurface boiling of hydrothermal fluids, which resulted in the formation of low-temperature (<200°C) Sb-Hg-As-Pb sulphosalt-rich ore lenses (Sherlock et al., 1999). More advanced argillic alteration may be produced by acidic magmatic volatiles, and this alteration can lead to distinctive aluminosilicate-rich mineral assemblages when metamorphosed to greenschist grade. In the case of the LaRonde deposit, Quebec, "classic" mound-type Zn-Cu-Au massive sulphide lenses are associated with extensive zones of metamorphosed argillic alteration containing pyrite-chalcopyrite-bornite-gold stockwork systems. This may be the result of shallow subsurface boiling and separation of a volatile-rich fluid or focused input of oxidized magmatic fluids (Dubé et al., 2004).

In less extreme cases, distal, low-temperature hydrothermal alteration assemblages associated with VMS deposits may be difficult to distinguish from regional greenschist-facies metamorphic mineral assemblages. When both proximal and regional semiconformable alteration zones are affected by amphibolite-grade regional or contact metamorphism, the originally strongly hydrated alteration mineral assemblages change into a coarse-grained quartz-phyllosilicate-aluminosilicate assemblages that are very distinct from the surrounding unaltered strata (Fig. 23). It then becomes possible to use the systematic variations in these coarse-grained metamorphic mineral assemblages as vectors towards the core of the proximal alteration system or upsection towards the paleo-seafloor (Hodges and Manojlovic, 1993).

Genetic/Exploration Models

Exploration models for VMS systems have several common themes despite the large variety of submarine environments in which the deposits can form. The generation of a VMS-hosting volcanic complex is a response to focused heat flow caused by tectonic extension, mantle depressurization, and the resultant formation of high-temperature mantle melts, crustal partial melts, and common bimodal volcanic succession. The large majority of VMS deposits in Canada form in either bimodal mafic or bimodal felsic volcanic terranes dominated by basalt-basaltic andesite and rhyolite-rhyodacite. Prospective VMS-hosting arc terranes are characterized by bimodal volcanic successions that have a tholeiitic to transitional tholeiitic-calc alkaline composition. The felsic volcanics are characterized by low Zr/Y (<7) and low $(La/Yb)_N$ (<6) ratios, with elevated high field strength element contents (Zr >200 ppm, Y >30 ppm, and elevated LREE and HREE) typical of high-temperature, reduced magmas derived from partially hydrated crust (Barrie et al., 1993; Barrie, 1995; Lentz, 1998). The lower viscosities of the high-temperature felsic magmas result in rapid ascent

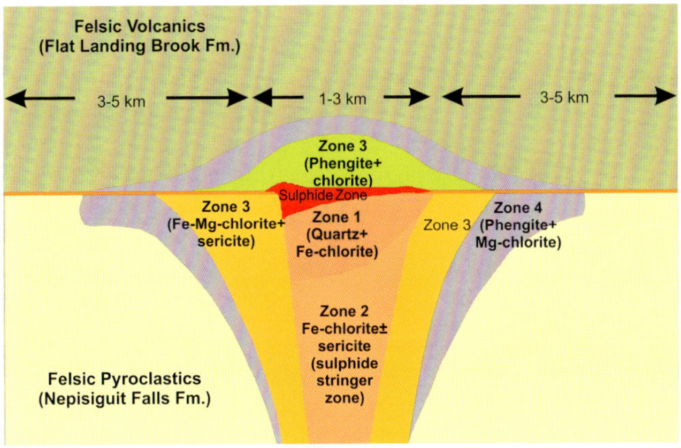

FIGURE 22. A schematic composite section through a VMS alteration system in the Bathurst mining camp as an example of a VMS proximal alteration zone metamorphosed to greenschist-grade mineral assemblages. From Goodfellow et al. (2003).

with minimal heat loss into subseafloor settings where hydrothermal convection can be initiated. For this reason, most prospective VMS environments are characterized by high-level sill-dyke swarms, discrete felsic extrusive centres, and large (>15 km long and 2000 m thick) subvolcanic composite intrusions. The absence of substantial subvolcanic intrusions in some camps may be due to poor preservation as a result of folding and faulting.

The interaction of large volumes of volcanic strata with seawater within these high-heat extensional environments results in the formation of district-scale alteration zones that extend over the strike length of the VMS-hosting extensional feature (spreading ridge, rift, or caldera). Stacked alteration zones can have an aggregate thickness of 2000 to 3000 m, and may be intruded by resurgent phases of the underlying subvolcanic intrusion. Subvolcanic intrusions themselves can display textural features indicating high-level devolatilization and high-temperature magmatic hydrothermal alteration (quartz-epidote-magnetite-ferroactinolite-sulphides). In some cases, this devolatilization may contribute metals to the overlying convective hydrothermal system (Large et al., 1996; Lydon, 1996; Galley, 2003, and references therein). Regional semiconformable alteration systems resemble regional metamorphic zones (zeolite, greenschist, amphibolite), with increasing grade towards the heat source. Most Canadian VMS districts have been affected by regional metamorphism, which has resulted in recrystallization of the original alteration minerals to greenschist and/or amphibolite assemblages. In camps such as Noranda, Bousquet, Sturgeon Lake, Manitouwadge, Snow Lake, Leaf Rapids, and the western Stikine (Tulsequah Chief), regional metamorphism or local contact metamorphism of alteration minerals has produced distinctive coarse-grained mineral assemblages characterized by such minerals as phlogopite, cordierite, anthophyllite, muscovite, staurolite, garnet, andalusite, and kyanite. The metamorphosed alteration can be distinguished from essentially isochemical regional metamorphic mineral assemblages by the losses and gains of various elements during fluid-rock interactions (Fig. 15).

Submarine volcanic stratigraphy that is prospective for VMS mineralization commonly contains ferruginous exhalative horizons as an indication of subseafloor hydrothermal activity. Precambrian VMS-related exhalites are commonly composed of finely bedded, sulphide-rich tuffaceous material. More extensive Algoma-type oxide facies Fe-formations are also common in VMS-prospective back-arc environments of all ages. Both types of exhalite may form proximal to massive sulphide deposits or extend for strike lengths of several kilometres to tens of kilometres (Spry et al., 2000; Peter, 2003). Proximity to a hydrothermal source in these formations is indicated by positive inter-element correlation between hydrothermal components (Eu, Fe, Mn, Pb, Zn, Cd, Au, Ca, Sr, Ba, P, and CO_2) versus clastic components (Si, Ti, Al, Mg, K, and Zr), increasing chondrite normalized EuEu* (hydrothermal input), and decreasing Ce/Ce* (seawater input) towards the source (Peter and Goodfellow, 1996; Peter, 2003). Vertical and horizontal facies vary from oxide through silicate to carbonate, which in some cases, also may reflect proximity to focused hydrothermal activity (Peter, 2003).

Key Exploration Criteria

The following are the major exploration criteria for Canadian VMS deposits and key attributes of VMS-hosting volcanic complexes.

The deposits occur in volcanic belts from Late Archean to Eocene in which extension is indicated by relatively primitive (tholeiitic to transitional) bimodal volcanism in nascent-arc, rifted-arc, and back-arc environments. Some obducted seafloor-spreading centres and rifted continental margins are also prospective.

VMS formation occurs during periods of major ocean-closing and terrane accretion. These include the Late Archean (2.8-2.69 Ga), Paleoproterozoic (1.92-1.87 Ga), Cambro-Ordovician (500-450 Ma), Devono-Mississippian (370-340 Ma), and Early Jurassic (200-180 Ma).

In effusive flow-dominated settings in oceanic arc and continental margin arcs, VMS deposits can be associated with 15 to 25 km long, mafic to composite synvolcanic intrusions. These intrusions are Na-rich and depleted in low field strength elements and have low airborne radiometric responses but commonly show magnetic halos due to surrounding zones of high-temperature fluid interaction. Exploration should be focused up to 3000 m upsection in the comagmatic volcanic suites in the hanging wall of the intrusions. Rhyolites with high Zr (>300 ppm), negative chondrite-normalized Eu anomalies, $(La/Yb)_N$ values of less than 7, $(Gd/Yb)_N$ values of less than 2, and Y/Zr ratios of less

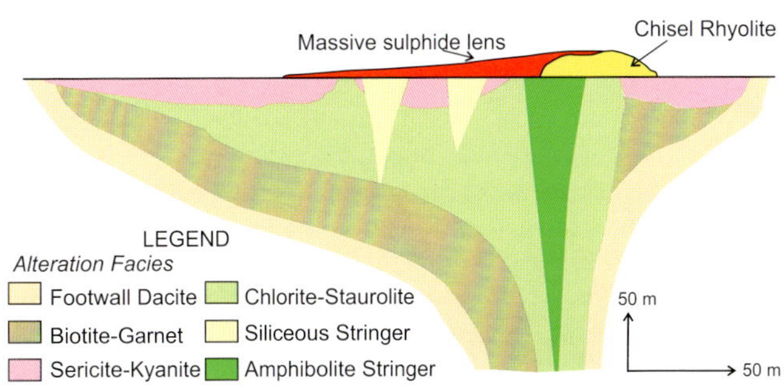

FIGURE 23. A stylized cross-section through the proximal alteration zone at the Chisel deposit, Snow Lake mining camp, illustrating the changes in mineral assemblages that occur when the terrane undergoes lower to middle amphibolite-grade regional metamorphism. From Galley et al. (1993).

than 7 define high-temperature (>900°C) felsic volcanic environments favourable for VMS formation. The presence of synvolcanic dyke swarms and exhalite horizons are indicative of areas of high paleo-heat flow.

In continental back-arc, bimodal siliciclastic-dominated settings aeromagnetic surveys can be used to identify aerially extensive iron formations to target hydrothermally active paleo-seafloor horizons. Variations in the mineralogy of the iron formations and varying element ratios can serve as vectors toward high-temperature hydrothermal centres. Volumetrically minor sill-dyke complexes also may identify higher temperature hydrothermal centres.

In upper greenschist-amphibolite metamorphic terranes, distinctive, coarse-grained mineral suites commonly define VMS alteration zones. These include chloritoid, garnet, staurolite, kyanite, andalusite, phlogopite, and gahnite. More aluminous mineral assemblages commonly occur closer to a high-temperature alteration pipe. Metamorphic mineral chemistry, such as Fe/Zn ratio of staurolite, is also a vector to ore. These largely refractory minerals have a high survival rate in surficial sediments, and can be used through heavy mineral separation as further exploration guides in till-covered areas.

Mineralogy and chemistry can be used to identify large-scale hydrothermal alteration systems in which clusters of VMS deposits may form. Broad zones of semiconformable alteration will show increases in Ca-Si (epidotization-silicification), Ca-Si-Fe (actinolite-clinozoisite-magnetite), Na (spilitization), or K-Mg (mixed chlorite-sericite±K-feldspar). Proximal alteration associated with discordant sulphide-silicate stockwork vein systems includes chlorite-quartz-sulphide- or sericite-quartz-pyrite±aluminosilicate-rich assemblages and is typically strongly depleted in Na and Ca due to high-temperature feldspar destruction. In addition to geochemical analysis, X-ray diffraction, PIMA, and oxygen isotope analysis can assist in vectoring towards higher temperature, proximal alteration zones and associated VMS mineralization. Although PIMA has been used most effectively on alteration systems that contain minerals with a high reflective index, there has been some success in identifying greenschist-facies minerals within Precambrian VMS hydrothermal systems (Thompson et al., 1999)

Knowledge Gaps

Researchers have gathered an impressive amount of knowledge over the last ten years with respect to how, and where, VMS deposits form within various geodynamic regimes. This is due to a combination of studies of modern seafloor environments and detailed and regional-scale studies of ancient VMS environments. These studies have allowed us to place VMS depositional environments within the context of diverse supra-subduction settings that can be identified in deformed and metamorphosed terranes through lithostratigraphic facies evaluation and lithogeochemical analyses. Prospective settings for subseafloor hydrothermal systems can now be determined through identification of synvolcanic intrusions that trigger the systems, geochemical variations in altered rocks and chemical sedimentary horizons, and the use of mineralogy, geochemistry, and isotope geology. The fundamental ingredient for the efficient use of these tools is an appropriate level of understanding of the architecture of the volcanic terranes. Mapping at 1:20 thousand scale and complementary geochronological studies of the Flin Flon, Snow Lake, Leaf Rapids, and Bathurst mining camps were key to understanding the evolution of the various VMS-hosting arc assemblages and at what period of time in this evolution the deposits formed. Detailed lithostratigraphic mapping was essential in unraveling deformation histories and understanding the structural repetitions of prospective ore horizons. At larger scales, we still need a better understanding of the longevity of hydrothermal systems and the character and scale of fluid flow into both volcanic and sedimentary hanging-wall strata. We also need a better understanding of how to prospect for VMS environments through thick drift cover using novel heavy mineral analysis and selective leach methods. Successful exploration under cover requires improved understanding of the processes of secondary and tertiary remobilization of metals and trace elements from a VMS deposit and its associated alteration system.

Some Areas of High Mineral Potential in Canada

The recognition of new classes of high-sulphidation and shallow-water VMS deposits and their genetic association with differentiated magmatic suites in both calc-alkaline and alkaline volcanic arcs opens up new terranes and volcanic environments to exploration that were previously considered non-prospective for VMS. These environments include arc fronts and successor magmatic arcs in addition to primitive rifted-arc and back-arc terranes. Calc-alkaline to alkaline terranes, such as the Triassic Nicola Group and the Lower Jurassic Hazelton Group in British Columbia, should be revisited for atypical VMS deposits. Evolved parts of Archean greenstone terranes, in particular >2.8 Ga terranes in which there was involvement of early sialic crust, should also be considered in this context, e.g. Frotet-Troilus Domain, Grand Nord, North Caribou, and western Slave subprovinces.

Incipient rift environments of the Paleoproterozoic Trans-Hudson Orogen: The presence of large volumes of iron formation and associated VMS mineralization in the Labrador Trough is evidence of extensive hydrothermal systems generated in these 2.1 to 2.0 Ga rift systems on both margins of the orogen. Why did these not develop large VMS deposits as in other Fe-formation-rich environments (e.g. Manitouwadge)?

Intrusions associated with Ni-Cu-PGE mineralization represent large volumes of magma, commonly emplaced at shallow crustal levels as part of volcano-plutonic complexes. If emplaced in a subaqueous environment, these terranes should be highly prospective for mafic siliciclastic or mafic-dominated VMS deposits. These may include the submarine volcanic stratigraphy above the Fox River and Bird River sills in Manitoba and possibly the Bad Vermilion anorthositic complex in southwestern Ontario.

Intra-continental back-arc environments have been recognized as highly prospective for VMS deposits. Where are the continental back-arc environments in the Superior, Slave, and Grenville provinces? Have we explored enough in the >2.8 or <1.5 Ga terranes?

Terranes affected by thin-skinned fold-thrust tectonics present special challenges for exploration but are also highly prospective for VMS deposits. The potential for new exploration targets in areas such as the Central Volcanic Belt of Newfoundland is high, and the lessons learned in the Iberian Pyrite Belt with respect to exploring in such terranes can be applied in these and other similar terranes in Canada.

The so-called oceanic terranes of British Columbia, such as the Triassic Slide Mountain and Cache Creek terrane, should be re-evaluated for their VMS potential in light of the possibility that they represent back-arc and not ocean-basin environments. The presence of boninite and subvolcanic tonalite-trondhjemite intrusions ± rhyolites in these terranes would be key indicators of possible arc-back-arc systems. Boninite, in particular, is an indication of a depleted mantle source typical of nascent to back-arc regimes (Crawford et al., 1989; Stern et al., 1995; Kerrich at al., 1998; Piercey et al., 2001).

Acknowledgements

The authors would like to thank many of our colleagues whose ideas and discussions helped summarize the characteristics of this fascinating deposit type. They include Jim Franklin, Wayne Goodfellow, Steve Piercey, Jan Peter, Benoit Dubé, and Harold Gibson. Jan Peter, and Steve Piercey are thanked for their insightful reviews of the manuscript.

References

Aggarwal, P.K., and Longstaffe, F.J., 1987, Oxygen isotope geochemistry of metamorphosed massive sulfide deposits of the Flin Flon-Snow Lake Belt, Manitoba: Contributions to Mineralogy and Petrology, v. 96, p. 314-325.

Allen, R.L., Lundström, I., Ripa, M., Simeonov, A., Christofferson, H., 1996a, Facies analysis of a 1.9 Ga, continental margin, back arc, felsic caldera province with diverse Zn-Pb-Ag (Cu-Au) sulfide and Fe oxide deposits, Bergslagen region, Sweden: Economic Geology, v. 91, p. 979-1008.

Allen, R. L., Weihed, P., and Svesen, S-A, 1996b, Setting of Zn-Cu-Au-Ag massive sulfide deposits in the evolution and facies architecture of a 1.9 Ga marine volcanic arc, Skellefte district, Sweden: Economic Geology, v. 91, p. 1022-1053.

Allen, R.L., Weihed, P., and Global VMS Research Project Team, 2002, Global comparisons of volcanic-associated massive sulphide districts, *in* Blundell, D.J., Neubauer, F., and Von Quadt, A., eds., The Timing and Location of Major Ore Deposits in an Evolving Orogen: Geological Society of London Special Publication 204, p. 13-37.

Alt, J.C., 1995, Subseafloor processes in mid-ocean ridge hydrothermal systems, *in* Humphris, S., ed., Seafloor Hydrothermal Systems: Physical, Chemical, Biological, and Geological Interactions: American Geophysical Union Monograph 91, p. 85-114.

Bailes, A.H., and Galley, A.G., 1999, Evolution of the Paleoproterozoic Snow Lake arc assemblage and geodynamic setting for associated volcanic-hosted massive sulfide deposits, Flin Flon Belt, Manitoba, Canada: Canadian Journal of Earth Science, v. 36, p. 1789-1805.

Barrett, T.J., and Sherlock, R.L, 1996a, Volcanic stratigraphy, lithogeochemistry, and seafloor setting of the H-W massive sulfide deposit, Myra Falls, Vancouver Island, British Columbia: Exploration and Mining Geology, v. 5, p. 421-458.

——— 1996b, Geology, lithogeochemistry, and volcanic setting of the Eskay Creek Au-Ag-Cu-Zn deposit, northern British Columbia: Exploration and Mining Geology, v. 5, p.339-368.

Barrie, C.T., 1995, Zircon thermometry of high-temperature rhyolites near volcanic-associated sulfide deposits: Geology, v. 23, p. 169-172.

Barrie, C.T., and Hannington, M.D., 1999, Introduction: Classification of VMS deposits based on host rock composition, *in* Barrie, C.T., and Hannington, M.D., eds., Volcanic-Associated Massive Sulfide Deposits: Processes and Examples in Modern and Ancient Settings: Reviews in Economic Geology, v. 8, p. 2-10.

Barrie, C.T., Ludden, J.N., and Green, A.H., 1993, Geochemistry of volcanic rocks associated with Cu-Zn and Ni-Cu deposits in the Abitibi Subprovince: Economic Geology, v. 88, p. 1341-1358.

Bédard, J.H., and Hébert, R., 1996, The lower crust of the Bay of Islands ophiolite, Canada: Petrology, mineralogy, and the importance of syntexis in magmatic differentiation in ophiolites and at ocean ridges: Journal of Geophysical Research, v. 101, p. 25105-25124.

Bédard, J.H., Lauzière, K., Tremblay, A., and Sangster, A., 1998, Evidence from Betts Cove ophiolite boninites for forearc seafloor-spreading: Tectonophysics, v. 284, p. 233-245.

Bloomer, S.H., Taylor, B., MacLeod, C.J., Stern, R.J., Fryer, P., Hawkins J.W., Johnson, L., 1995, Early arc volcanism and the ophiolite problem: A perspective from drilling in the Western Pacific: Geophysical Monograph 88, p. 1-30.

Bradshaw, G.D., Rowins, S.M., Peter, J.M., and Taylor, B.E., 2003, Genesis of the Wolverine deposit, Finlayson Lake district, Yukon: A transitional style of polymetallic massive sulfide mineralization in an ancient continental margin setting: The Gangue, p. 1-7.

Brauhart, C.W., Groves, D.I., and Morant, P., 1998, Regional alteration systems associated with volcanogenic massive sulfide mineralization at Panorama, Pilbara, Western Australia: Economic Geology, v. 93, p. 292-302.

Carvalho, D., Barriga, F.J.A.S., and Munha, J., 1999, Bimodal-siliciclastic systems - The case of the Iberian Pyrite Belt, *in* Barrie, C.T., and Hannington, M.D., eds., Volcanic-Associated Massive Sulfide Deposits: Processes and Examples in Modern and Ancient Settings: Reviews in Economic Geology, v. 8, p. 375-402.

Cathles, L.M., 1993, Oxygen isotope alteration in the Noranda Mining District, Abitibi Greenstone Belt, Quebec: Economic Geology, v. 88, p. 1483-1511.

Corbett, K.D., 1992, Stratigraphic-volcanic setting of massive sulfide deposits in the Cambrian Mount Read volcanics, western Tasmania: Economic Geology, v. 87, p. 564-586.

Crawford, A.J., Falloon, T.J., and Green D.H., 1989, Classification, petrogenesis and tectonic setting of boninite, *in* Crawford, A.J., ed., Boninites and Related Rocks: Unwin Hyman, London, p. 2-48.

Dimroth, E., Imreh, L., Goulet, N., and Rocheleau, M., 1983, Evolution of the south-central segment of the Archean Abitibi Belt, Quebec; Part II, Tectonic evolution and geomechanical model: Canadian Journal of Earth Sciences 20, p. 1355-1373.

Doyle, M.G., and McPhie, J, 2000, Facies architecture of a silicic intrusion-dominated volcanic centre at Highway-Reward, Queensland, Australia: Journal of Volcanology and Geothermal Research, v. 99, p. 79-96.

Dubé, B., Mercier-Langevin, P., Hannington, M.D., Davis, D.W., and LaFrance, B., 2004, Le gisement de sulfures massifs volcanogènes aurifères LaRonde, Abitibi, Québec: alteration, mineralisation genèse et implications pour l'exploration: Ministère de Resources Naturelles, Québec, Open File Report MB 2004-03, 112 p.

Eastoe, C.J., Solomon, M., and Walshe, J.L., 1987, District-scale alteration associated with massive sulfide deposits in the Mount Read volcanics, Tasmania: Economic Geology, v. 82, p. 1204-1238.

Eckstrand, O.R., Sinclair, W.D., and Thorpe, R.I., eds. 1995, Geology of Canadian Mineral Deposit Types, Geology of Canada, No. 8, Decade of North American Geology (DNAG): Geological Society of America, Part 1, p. 183-196.

Eldridge, C. W., Barton, P.B., and Ohmoto, H., 1983, Mineral textures and their bearing on formation of the Kuroko orebodies: Economic Geology Monograph, v. 5, p. 241-281.

Epp, M., and Crocket, J.H., 1999, Geology and geochemistry of the Potterdoal Cu-Zn deposit, Munro township, Ontario, *in* Hannington, M.D., and Barrie, C.T., eds., The Giant Kidd Creek Volcanogenic Massive Sulfide Deposit, Western Abitibi Subprovince, Canada: Economic Geology Monograph 10, p. 593-612.

Franklin, J.M., 1996, Volcanic-associated massive sulphide base metals, *in* Eckstrand, O.R., Sinclair, W.D., and Thorpe, R.I., eds., Geology of Canadian Mineral Deposit Types: Geological Survey of Canada, Geology of Canada, no. 8, p. 158-183.

Franklin, J.M., and Hannington, M.D., 2002, Volcanogenic massive sulfides through time: Geological Society of America, 2002 Annual Meeting, Abstracts with Programs, v. 34, p. 283.

Franklin, J.M., Lydon, J.W., and Sangster, D.F., 1981, Volcanic-associated massive sulfide deposits; in Skinner, B.J., ed., Economic Geology 75th Anniversary Volume: Society of Economic Geologists, p. 485-627.

Franklin, J.M., Hannington, M.D., Jonasson, I.R., and Barrie, C.T., 1998, Arc-related volcanogenic massive sulphide deposits: Proceedings of Short Course on Metallogeny of Volcanic Arcs, January 24-25, Vancouver: British Columbia Geological Survey Open-File 1998-8, p. N1-N32.

Franklin, J.M., Gibson, H.L., Jonasson, I.R., and Galley, A.G., 2005, Volcanogenic Massive Sulfide Deposits, *in* Hedenquist, J.W., Thompson, J.F.H., Goldfarb, R.J., and Richards, J.P., eds., Economic Geology 100th Anniversary Volume: The Economic Geology Publishing Company, p. 523-560.

Galley, A.G., 1993, Semi-conformable alteration zones in volcanogenic massive sulphide districts: Journal of Geochemical Exploration, v. 48, p. 175-200.

——— 2003, Composite synvolcanic intrusions associated with Precambrian VMS- related hydrothermal systems: Mineralium Deposita, v. 38, p. 443-473.

Galley, A.G., and Jonasson, I.R., 2003, Classification and tectonic environments of VMS deposits in the Flin Flon mining camp, Manitoba: Geological Association of Canada/Mineralogical Association of Canada Annual General Meeting, Program with Abstracts (CD-ROM).

Galley, A.G., and Koski, R.A., 1999, Setting and characteristics of ophiolite-hosted volcanogenic massive sulfide deposits, *in* Barrie, C.T., and Hannington, M.D., eds., Volcanic-Associated Massive Sulfide Deposits: Processes and Examples in Modern and Ancient Settings: Reviews in Economic Geology, v. 8, p. 215-236.

Galley, A.G., Bailes, A.H., and Kitzler, G., 1993, Geological setting and hydrothermal evolution of the Chisel Lake and North Chisel Zn-Pb-Ag-Au massive sulphide deposit, Snow Lake, Manitoba: Exploration and Mining Geology, v. 2, p. 271-295.

Gibson, H.L., and Watkinson, D.H., 1990, Volcanogenic massive sulphide deposits of the Noranda Cauldron and Shield Volcano, Quebec, *in* Rive, M., Verpaelst, P., Gagnon, Y., Lulin, J.M., Riverin, G., and Simard, A., eds., The Northwestern Quebec Polymetallic Belt: Canadian Institute for Mining and Metallurgy, Special Volume 43, p. 119-132.

Goodfellow, W.D., Zierenberg, R.A., and ODP 196 Shipboards Science Party, 1999, Genesis of massive sulfide deposits at sediment-covered spreading centers, *in* Barrie, C.T., and Hannington, M.D., eds., Volcanic-Associated Massive Sulfide Deposits: Processes and Examples in Modern and Ancient Settings: Reviews in Economic Geology, v. 8, Society of Economic Geologists, p. 297-324.

Goodfellow, W.D., McCutcheon, S.R., and Peter, J.M., 2003, Massive sulfide deposits of the Bathurst Mining Camp, New Brunswick and Northern Maine: Introduction and summary of findings, *in* Goodfellow, W.D., McCutcheon, S.R., and Peter, J.M., eds., Massive Sulfide Deposits of the Bathurst Mining Camp, New Brunswick, and northern Maine: Economic Geology Monograph 11, Society of Economic Geologists, p. 1-16.

Green, G.R., Ohmoto, H., Date, J., and Takahashi, T., 1983, Whole-rock oxygen isotope distribution in the Fukazawa-Kosaka area, Hokuroko district, Japan, and its potential application to mineral exploration, *in* Ohnoto, H., and Skinner, B.J., ed., The Kuroko and Related Volcanogenic Massive Sulfide Deposits: Economic Geology Monograph 5, Society of Economic Geologists, p. 395-511.

Groves, D.I., Goldfarb, R.J., Gebre-Mariam, H., Hagemann, S.G., and Robert, F., 1998, Orogenic gold deposits - a proposed classification in the context of their crustal distribution and relationship to other gold deposit types: Ore Geology Reviews, v. 13, p. 7-27.

Hamilton, W.B., 1995, Subduction systems and magmatism: Geological Society of London Special Publication, v. 81, p. 3-28.

Hannington, M.D., Jonasson, I.R., Herzig, P.M., and Petersen, S., 1995, Physical and chemical processes of seafloor mineralization at mid-ocean ridges, *in* Humphris, S., ed., Seafloor Hydrothermal Systems: Physical, Chemical, Biological, and Geological Interactions: American Geophysical Union, Geophysical Monograph v. 91, p. 115-157.

Hannington, M.D., Galley, A.G., Herzig, P.M., and Petersen, S., 1998, Comparison of the TAG mound and stockwork complex with Cyprus-type massive sulfide deposits; Proceedings of the Ocean Drilling Program, Scientific Results Volume 158, College Station, TX, p. 389-415.

Hannington, M.D., Barrie, C.T., and Bleeker, W., 1999a, The giant Kidd Creek volcanogenic massive sulfide deposit, western Abitibi Subprovince, Canada, *in* Hannington, M.D., and Barrie, C.T., eds., The Giant Kidd Creek Volcanogenic Massive Sulfide Deposit, Western Abitibi Subprovince, Canada: Economic Geology Monograph 10, p. 1-30.

Hannington, M.D., Bleeker, W., and Kjarsgaard, I., 1999b, Sulfide mineralogy, geochemistry and ore genesis of the Kidd Creek deposit: Part II. The bornite zone, *in* Hannington, M.D., and Barrie, C.T., eds., The Giant Kidd Creek Volcanogenic Massive Sulfide Deposit, Western Abitibi Subprovince, Canada: Economic Geology Monograph 10, p. 225-266.

Hannington, M.D., Poulsen, K.H., Thompson, J.F.H., and Sillitoe, R.H., 1999c, Volcanogenic gold in the massive sulfide environment, *in* Barrie, C.T., and Hannington, M.D., eds., Volcanic-Associated Massive Sulfide Deposits: Processes and Examples in Modern and Ancient Settings: Reviews in Economic Geology 8, p. 325-356.

Hannington, M.D., Santaguida, F., Kjarsgaard, I.M., and Cathles, L.M., 2003, Regional greenschist facies hydrothermal alteration in the central Blake River Group, western Abitibi subprovince, Canada: Mineralium Deposita, v. 38, p. 393-422.

Herrington, R.J., Armstrong, R.N., Zaykov, V.V., Maslennikov, V.V., Tessalina, S.G., Orgeval, J.J., and Taylor, R.N.A., 2002, Massive sulfide deposits in the South Urals; geological setting within the framework of the Uralide Orogen, *in* Brown, D., Juhlin, C., and Puchkov, V., eds., Mountain Building in the Uralides; Pangea to the Present: Geophysical Monograph. 132, p. 155-182.

Herzig, P.M., and Hannington, M.D., 1995, Polymetallic massive sulfides at the modern seafloor: A review: Ore Geology Reviews, v. 10, p. 95-115.

Hodges, D.J., and Manojlovic, P.M., 1993, Application of lithogeochemistry to exploration for deep VMS deposits in high grade metamorphic rocks: Journal of Geochemical Exploration, v. 48, p. 201-224.

Kalogeropoulos, S.I., and Scott, S.D., 1983, Mineralogy and geochemistry of tuffaceous exhalites (Tetsusekiei) of the Fukazawa mine, Hokuroko District, Japan: *in* Ohnoto, H., and Skinner, B.J., eds., The Kuroko and Related Volcanogenic Massive Sulfide Deposits: Economic Geology Monograph 5, p. 412-432.

Kalogeropoulos, S.I., and Scott, S.D., 1989, Mineralogy and geochemistry of an Archean tuffaceous exhalite: the Main Contact Tuff, Millenbach mine area, Noranda, Quebec: Canadian Journal of Earth Sciences, v. 26, p. 88-105.

Kerr, D.J., and Gibson, H.L., 1993, A comparison between the volcanology and geochemistry of volcanic successions hosting the Horne mine deposit and smaller intra-cauldron deposits of the mine sequence: Economic Geology, v. 88, p. 1419-1443.

Kerrich, R., Wyman, D., Fan, J., and Bleeker, W., 1998, Boninite series - low Ti-tholeiite associations from the 2.7 Ga Abitibi greenstone belt: Earth and Planetary Science Letters, v. 164, p. 303-316.

Lagerblad, B., and Gorbatschev, R., 1985, Hydrothermal alteration as a control of regional geochemistry and ore formation in the central Baltic Shield: Geologische Rundschau, v. 74, p. 33-49.

Large, R.R., 1992, Australian volcanic-hosted massive sulphide deposits: features, styles and genetic models: Economic Geology, v. 87, p. 471-510.

Large, R.R., Doyle, M., Raymond, O., Cooke, D., Jones, A., and Heaman, L., 1996, Evaluation of the role of Cambrian granites in the genesis of world class VHMS deposits in Tasmania: Ore Geology Reviews, v. 10, p. 215-230.

Lentz, D.R., 1998, Petrogenetic evolution of felsic volcanic sequences associated with Phanerozoic volcanic-hosted massive sulfide systems: the role of extensional geodynamics: Ore Geology Reviews, v. 12, p. 289-327.

Liaghat, S., and MacLean, W.H., 1992, The Key Tuffite, Matagami mining district: Origin of the tuff components and mass changes: Exploration and Mining Geology, v. 1, p. 197-207.

Lydon, J.W., 1984. Some observations on the morphology and ore textures of volcanogenic sulfide deposits of Cyprus: Geological Survey of Canada, Current Research, Paper 84-01A, p. 601-610.

——— 1996, Characteristics of volcanogenic massive sulfide deposits: Interpretations in terms of hydrothermal convection systems and magmatic hydrothermal systems: Instituto Tecnologico Geominero de Espana, Boletin geologico y minero, v. 107, p. 15-64.

Morton, J.L., and Franklin, J.M., 1987, Two-fold classification of Archean volcanic-associated massive sulfide deposits: Economic Geology, v. 82, p. 1057-1063.

Morton, R.L., Hudak, G.J., Walker, J.S., and Franklin, J.M., 1990, Physical volcanology and hydrothermal alteration of the Sturgeon Lake caldera complex, *in* J.M. Franklin, Schneiders, B.R., and Koopman, E.R., eds., Mineral Deposits in the Western Superior Province, Ontario: Open File 2164, Geological Survey of Canada, p. 74-94.

Munha, J., and Kerrich, R., 1980, Sea water-basalt interaction in spilites from the Iberian Pyrite Belt: Contributions to Mineralogy and Petrology, v. 73, p. 191-200.

Nelson, J., and Mihalynuk, M., 2004, Mega-terranes and deep structures: tectonics and the potential for major new mineral deposits in British Columbia: Mineral Exploration Roundup, Program with Abstracts, p. 26.

Paradis, S., Taylor, B.E., Watkinson, D.H., and Jonasson, I.J., 1993, Oxygen isotope zonation and alteration in the Noranda mining district, Abitibi greenstone belt, Quebec: Economic Geology, v. 88, p. 1512-1525.

Peter, J.M., 2003, Ancient iron-rich metalliferous sediments (iron formations): Their genesis and use in the exploration for stratiform base metal sulphide deposits, with examples from the Bathurst Mining Camp, *in* Lentz, D.R., ed., Geochemistry of Sediments and Sedimentary Rocks: Evolutionary Considerations to Mineral Deposit-Forming Environments, GEOtext 4: Geological Association of Canada, p. 145-173.

Peter, J.M., and Goodfellow, W.D., 1996, Mineralogy, bulk and rare earth element geochemistry of massive sulfide-associated hydrothermal sediments of the Brunswick Horizon, Bathurst Mining Camp, New Brunswick: Canadian Journal of Earth Sciences, v. 33, p. 252-283.

Peter, J.M., and Scott, S.D., 1999, Windy Craggy, northwestern British Columbia: The world's largest Besshi-type deposit, *in* Barrie, C.T., and Hannington, M.D., eds., Volcanic-Associated Massive Sulfide Deposits: Processes and Examples in Modern and Ancient Settings: Reviews in Economic Geology, v. 8, p. 261-295.

Petersen, S., Herzig, P.M., and Hannington, M.D., 2000, Third dimension of a presently forming VMS deposit: TAG Hydrothermal Mound, Mid-Atlantic Ridge, 26°N: Mineralium Deposita, v. 35, p. 233-259.

Petersen, S., Schwarz-Schampera, U., Herzig, P., Hannington, M., and Jonasson, I., 2004, Hydrothermal precipitates associated with bimodal volcanism in the Central Bransfield Strait, Antarctica: Mineralium Deposita, v. 39, p. 358-379.

Piercey, S.J., Murphy, D.C., Mortensen, J.K., and Paradis, S., 2001, Boninitic magmatism in a continental margin setting, Yukon-Tanana terrane, southeastern Yukon, Canada: Geology, v. 29, p. 731-734.

Poulsen, H., and Hannington, M., 1995, Auriferous Volcanogenic Sulfide Deposits, *in* Eckstrand, O.R., Sinclair, W.D., and Thorpe, R.I., eds. Geology of Canadian Mineral Deposit Types, Geology of Canada, no. 8, Decade of North American Geology (DNAG): Geological Society of America, Part 1. p. 183-196.

Powell, W.G., Carmichael, D.M., and Hodgeson, C.J., 1993, Relative timing of metamorphism and tectonism during the evolution of the southern Abitibi greenstone belt. *in* Caty, J.L., ed., L'Exceptionnel potentiel mineral du Quebec, une realite a decouvrir; resume des conferences; seminaire d'information 1993 (The exceptional mineral potential of Quebec, a reality yet to be discovered; information seminar 1993) Conference Abstracts. DV – Direction generale de l'Exploration Geologique et Minerale. p. 49-50.

Richards, J.P., 2003, Tectono-magmatic precursors for porphyry Cu-(Mo-Au) deposit formation: Economic Geology, v. 98, p. 1515-1534.

Sangster, D., 1980, Quantitative characteristics of volcanogenic massive sulphide deposits I. Metal content and size distribution of massive sulphide deposits in volcanic centers: Canadian Institute of Mining and Metallurgy Bulletin, v. 73, p. 74-81.

Santaguida, F., Gibson, H.L., Watkinson, D.H., and Hannington, M.D., 1998, Semi-conformable epidote-quartz hydrothermal alteration in the Central Noranda Volcanic Complex: Relationship to volcanic activity and VMS mineralization: Canadian Minerals Research Organization Project 94E07, Annual Report, The Use of Regional-Scale Alteration and Subvolcanic Intrusions in the Exploration for Volcanic-Associated Massive Sulphide Deposits, p. 139-180.

Sebert, C., and Barrett, T.J., 1996, Stratigraphy, alteration, and mineralization at the Tulsequah chief massive sulfide deposit, northwestern British Columbia; Exploration and Mining Geology 5, 4, p. 281-308.

Sharpe, R., and Gemmell, J.B., 2002, The Archean Cu-Zn magnetite-rich Gossan Hill volcanic-hosted massive sulfide deposit, Western Australia; genesis of a multistage hydrothermal system: Economic Geology, v. 97, p. 517-539.

Sherlock, R.L., Roth, T., Spooner, E.T.C., and Bray, C.J., 1999, The origin of the Eskay Creek precious metal-rich volcanogenic massive sulfide deposit: fluid inclusion and stable isotope data: Economic Geology, v. 94, p. 803-824.

Sillitoe, R.H., Hannington, M.D., and Thompson, J.F.H., 1996, High sulfidation deposits in the volcanogenic massive sulfide environment: Economic Geology, v. 91, p. 204-212.

Sinclair, W.D., Chorlton, L.B., Laramée, R.M., and Eckstrand, O.R., 1999, World Minerals Geoscience Database Project: Digital databases of generalized world geology and mineral deposits for mineral exploration and research; http://gdr.nrcan.gc.ca/minres/index_e.php.

Singer, D.A., 1995, World-class base and precious metal deposits - a quantitative analysis: Economic Geology, v. 90, p. 88-104.

Spooner, E.T.C., and Fyfe, W.S., 1973, Subseafloor metamorphism, heat and mass transfer: Contributions to Mineralogy and Petrology, v. 42, p. 287-304.

Spry, P.G., Peter, J.M., and Slack, J.F., 2000, Meta-exhalites as exploration guides to ore, *in* Spry, P.G., Marshall, B., and Vokes, F.M., eds., Metamorphosed and Metamorphogenic Ore Deposits: Reviews in Economic Geology, v. 11, p. 163-201.

Stern, R.A., Syme, E.C., Bailes, A.H., and Lucas, S.B., 1995, Paleoproterozoic (1.90-1.86 Ga) arc volcanism in the Flin Flon Belt, Trans-Hudson Orogen, Canada: Contributions to Mineralogy and Petrology, v. 119, p. 117-141.

Stinton, J.M., and Detrick, R.S., 1992, Mid-ocean ridge magma chambers: Journal of Geophysical Research, v. 97, p. 197-216.

Swinden, H.S., Kean, B.F., and Dunning, G.R., 1988, Geological and paleotectonic settings of volcanogenic massive sulfide mineralization in Central Newfoundland, *in* Swinden, H.S., and Kean, B.F., eds., The Volcanogenic Sulphide Districts of Newfoundland: A Guidebook and Reference Manual for Volcanogenic Sulphide Deposits in the Early Paleozoic Oceanic Volcanic Terranes of Central Newfoundland: Geological Association of Canada, Mineral Deposits Division, p. 2-27.

Syme, E.C., and Bailes, A.H., 1993, Stratigraphy and tectonic setting of Early Proterozoic volcanogenic massive sulphide deposits, Flin Flon, Manitoba: Economic Geology, v. 88, p. 566-589.

Taylor, B.E., and South B.C., 1985, Regional stable isotope systematics of hydrothermal alteration and massive sulfide deposition in the West Shasta District, California: Economic Geology, v. 80, p. 2149-2163.

Thompson, A. J., Hauff, B., and Robitaille, P.L., 1999, Alteration mapping in exploration: Application of short-wave infrared (SWIR) spectroscopy: Society of Economic Geology Newsletter, v. 39.

van Staal, C.R., Wilson, R.A., Rogers, N., Fyffe, L.R., Langton, J.P., McCutcheon, S.R., McNicoll, V., and Ravenhurst, C.E., 2003, Geology and tectonic history of the Bathurst Supergroup, Bathurst Mining Camp, and its relationships to coeval rocks in southwestern New Brunswick and adjacent Maine - a synthesis, *in* Goodfellow, W.D., McCutcheon, S.R., and Peter, J.M., eds., Massive Sulfide Deposits of the Bathurst Mining Camp, New Brunswick and Northern Maine: Economic Geology Monograph 11, p. 37-60.

Walker, P.N., and Barbour, D.M., 1981,Geology of the Buchans ore horizon breccias, *in* Swanson, E.A., Strong, D.F., and Thurlow, J.G., eds., The Buchans Ore Body; Fifty Years of Geology and Mining: Geological Association of Canada, Special Paper 22, p. 161-185.

Whalen, J.B., McNicoll, V.J., Galley, A.G., and Longstaffe, F.J., 2004, Tectonic and metallogenic importance of an Archean composite high-and low-Al tonalite suite, western Superior Province, Canada: Precambrian Research, v. 132, p. 275-301.

Wyman, D.A., Bleeker, W., and Kerrich, R., 1999, A 2.7 Ga komatiite, low Ti tholeiite, arc tholeiite transition, and inferred proto-arc geodynamic setting of the Kidd Creek deposit: Evidence from precise trace element data, *in* Hannington, M.D., and Barrie, C.T., eds., The Giant Kidd Creek Volcanogenic Massive Sulfide Deposit, Western Abitibi Subprovince, Canada: Economic Geology Monograph 10, p. 511-528.

Yang, K., and Scott, S.D., 2003, Geochemical relationships of felsic magmas to ore metals in massive sulfide deposits of the Bathurst mining camp, Iberian Pyrite Belt, Hokuroko district and the Abitibi Belt, *in* Goodfellow, W.D., McCutcheon, S.R., and Peter, J.M., eds., Massive Sulfide Deposits of the Bathurst Mining Camp, New Brunswick and Northern Maine: Economic Geology, Monograph 11, p. 457-478.

Ziegler, P.L., 1992, Plate tectonics, plate moving mechanisms and rifting: Tectonophysics, v. 215, p. 9-34.

Sedimentary Exhalative (SEDEX) Deposits

Wayne D. Goodfellow and John W. Lydon

Geological Survey of Canada, 601 Booth Street, Ottawa, Ontario, K1A 0E8
Corresponding author's email: wgoodfel@nrcan.gc.ca

Abstract

There are 132 SEDEX (including Irish and BHT subtypes) deposits worldwide with known grade and tonnage, and of these, 50 have geological resources equal to or greater than 20 Mt. In Canada, there are 35 deposits, seven with measured geological resources of more than 20 Mt, including the Sullivan deposit with 162 Mt. Twelve Canadian deposits are past producers, the largest of which is the Sullivan deposit, which produced 149 Mt of 5.33 percent Zn and 5.64 percent Pb.

The morphology of SEDEX deposits is highly variable and includes mounds, lenses, and tabular or sheet-like bodies. Their internal architecture is controlled by the proximity of seafloor sulphides to fluid discharge vents. Vent-proximal deposits typically formed from buoyant hydrothermal fluids, whereas vent-distal deposits formed from fluids that are denser than seawater and pooled in bathymetric depressions that may be remote from seafloor vents.

Most SEDEX deposits are hosted by organic-rich sedimentary rocks that were deposited in basins during periods in the Earth's history when the oceans were stratified with a lower anoxic and H_2S-rich water column. In the Paleozoic Selwyn Basin, for example, there is a close temporal relationship between upward-increasing $\delta^{34}S$ secular trends in sedimentary pyrite, anoxic laminated carbonaceous shales and cherts, and three major SEDEX forming events in the Late Cambrian, Early Silurian, and Late Devonian.

The typical basinal architecture of most SEDEX deposits is a continental rift basin with at least 2 to 5 km of syn-rift, coarse-grained, permeable clastics and related volcanic rocks and/or volcaniclastics overlain by post-rift relatively impermeable basinal shales or carbonates. Hydrothermal discharge to the seafloor was commonly focused at the intersection of extensional and transform faults. There is close temporal and, in many cases, spatial association of SEDEX deposits with basaltic volcanic rocks, dykes, and sills. The low rigidity, permeability, and thermal conductivity of host sediments served to focus and prolong hydrothermal discharge at a restricted number of vent sites, thereby generating deposits that are an order of magnitude larger on average than VMS deposits.

SEDEX deposits most likely formed from oxidized and therefore H_2S-poor fluids generated in geopressured hydrothermal reservoirs within syn-rift clastic (and evaporitic) sediments sealed by fine-grained marine sediments. The large variability in the temperature, salinity, metal content, and redox conditions of SEDEX fluids was controlled by a number of parameters including the local thermal regime, the redox state of the reservoir sediments, and the presence or absence of evaporates. Because most of the fluids that formed SEDEX deposits were probably depleted in reduced S, an essential requirement for this deposit type is a sufficient supply of reduced S at the site of deposition. In the case of well-bedded deposits that formed at the seafloor, the most likely S source is bacteriogenic H_2S generated in an ambient anoxic water column.

Résumé

À l'échelle du monde, il existe 132 gîtes SEDEX, y compris les gîtes irlandais et de type Broken Hill avec teneurs et tonnages connus, et 50 d'entre eux renferment des ressources géologiques supérieures à 20 millions de tonnes. Le Canada possède 35 gîtes de ressources géologiques mesurées, dont sept renferment plus de 20 millions de tonnes, y compris le gisement de Sullivan avec 162 millions de tonnes. Douze gisements canadiens sont des producteurs passés et, le plus important d'entre eux, le gisement Sullivan, a produit 149 millions de tonnes avec 5,33 % de Zn et 5,64 % de Pb. D'une grande diversité, la géomorphologie des gîtes SEDEX comprend buttes, lentilles et corps tabulaires ou feuilletés.

La structure interne de ces gîtes est régie par la proximité des sulfures du plancher océanique par rapport aux évents relâchant des fluides. En général, les gîtes à proximité des évents se forment par l'intensité des fluides hydrothermaux; en revanche, ceux formés à distance des évents sont formés par des fluides plus denses que l'eau de mer et se concentrent dans les dépressions du plancher océanique qui peuvent être à l'écart des évents.

La plupart des gîtes SEDEX sont encaissés dans des sédiments carbonés déposés durant une période de l'histoire de la Terre au cours de laquelle les océans furent stratifiés avec une colonne d'eau anoxique plus basse et riche en H_2S. Dans le bassin de Selwyn du Paléozoïque, notamment, une relation étroite existe entre les tendances séculaires d'augmentation à la hausse de $\delta^{34}S$ dans la pyrite sédimentaire, les shales et les cherts laminés anoxiques et trois événements d'importance dans la formation des gîtes SEDEX au cours du Cambrien supérieur, du Silurien inférieur et du Dévonien supérieur.

La structure type des bassins de la plupart des gîtes SEDEX se caractérise par un bassin d'effondrement continental avec au moins 2 à 5 km de dépôts clastiques, volcaniques ou volcano-détritiques à gros grain, contemporains au rift, recouverts de shales ou de carbonates de bassins imperméables, lesquels sont postérieurs au rift. Les jaillissements hydrothermaux sur le plancher marin se produisaient habituellement à l'intersection des failles d'extension et de transformation. Il existe une association temporelle étroite et, en maints cas, spatiale des gîtes SEDEX avec les roches volcaniques basaltiques, les dykes et les filons-couches. La faible rigidité, la perméabilité et la conductivité thermale des sédiments hôtes ont servi à concentrer et à prolonger le jaillissement thermal dans un nombre restreint de sites d'évents, ce qui a créé des gîtes plus importants que les gîtes de sulfures massifs volcanogènes.

Selon toute probabilité, les gîtes SEDEX ont été formés de fluides oxydés, donc pauvres en H_2S produits par des réservoirs hydrothermaux géopressurés dans les sédiments clastiques (et évaporitiques), contemporains au rift, enveloppés

Goodfellow, W.D., and Lydon, J.W., 2007, Sedimentary exhalative (SEDEX) deposits, *in* Goodfellow, W.D., ed., Mineral Deposits of Canada: A Synthesis of Major Deposit Types, District Metallogeny, the Evolution of Geological Provinces, and Exploration Methods: Geological Association of Canada, Mineral Deposits Division, Special Publication No. 5, p. 163-183.

par du carbonate ou des sédiments marins à grain fin. Un certain nombre de paramètres régissaient les grandes variations de température, de salinité, du contenu en métaux, des conditions de redox des fluides de SEDEX, soit le régime thermique, l'état de redox des sédiments du réservoir et la présence ou l'absence d'évaporite. Vu que la plupart des fluides qui ont formé les gîtes SEDEX étaient appauvris en souffre réduit, un élément essentiel de ce type de gîte est un approvisionnement suffisant de soufre réduit sur l'emplacement du dépôt. Dans le cas des gîtes bien lités qui ont formé le plancher océanique, la source de soufre la plus probable est le H_2S qui est habituellement enrichi dans les colonnes d'eau anoxiques.

Introduction

This synthesis of SEDEX deposits is intended as a synopsis of the major economic, geological, geochemical and genetic attributes of SEDEX deposits, with an emphasis on those that are considered important in mineral exploration. A great deal of the key attributes are based on research undertaken in the Selwyn Basin, Canada (e.g., Goodfellow, 2004; Goodfellow, 2007) and the Sullivan district, southeast B.C. (Höy et al., 2000; Lydon et al., 2000a; Lydon et al., 2007). This was augmented by major research programs on SEDEX deposits of the Northern Territories, Australia, by the Centre for Ore Deposits Research (CODES; e.g., Large et al., 1998; Bull and Rawlings, 1999) and on the Red Dog deposits, Brooks Range, Alaska by the U.S. Geological Survey (e.g., Kelley and Jennings, 2004; Young, 2004).

For a comprehensive analysis of this economically important deposit type, the reader is referred to reviews by Large (1981b), Goodfellow et al. (1993) and Leach et al. (2005).

Definition

Mineral Deposit Type

Sedimentary exhalative (SEDEX) deposits are typically tabular bodies composed predominantly of Zn, Pb, and Ag bound in sphalerite and galena that occur interbedded with iron sulphides and basinal sedimentary rocks, and that were deposited on the seafloor and in associated sub-seafloor vent complexes from hydrothermal fluids vented into mostly reduced sedimentary basins in continental rifts.

Mineral Deposit Subtypes

Subtypes of SEDEX deposits include the Broken Hill type (BHT) and those that formed below but near the seafloor (e.g., Irish-type deposits). The Irish type of SEDEX deposits (e.g., Hitzman and Beaty, 1997) is hosted predominantly by carbonate rocks and these deposits, either individually or collectively (district-wide), may show characteristics of both seafloor deposition and epigenetic features typical of Mississippi Valley-type (MVT) deposits (e.g., Sangster, 1990; Paradis et al., 2007). Irish-type deposits are considered to have formed by ore-forming processes similar to those of SEDEX deposits, but because carbonate platforms are highly soluble in mildly acidic ore fluids, the ores were also deposited in open space that commonly formed by hydrothermal karsting (e.g., dissolution voids, collapse breccias). BHT deposits (Beeson, 1991; Parr and Plimer, 1993; Walters, 1998) are characterized by high metamorphic grade, elevated base metal to S ratios, a spatial and temporal association with Fe-Si-Mn oxide exhalites, and bimodal felsic-mafic volcanic/intrusive and fine- to coarse-grained, mostly clastic sedimentary host rocks.

It is generally recognized that, especially at the deposit scale, no matter the criteria used to define a SEDEX deposit, there is a continuum of characteristics between SEDEX and VMS (volcanogenic massive sulphide) deposits on the one hand, and SEDEX and MVT deposits on the other hand. The distinction between the three classes of deposits is based not only on the typical physical, chemical, and geological attributes of member deposits and their respective geological environments, but also on genetic models for the class. Both SEDEX and MVT deposits occur within marine platforms of thick sedimentary basins and are thought to result from the migration of basinal metalliferous saline fluids, whereas VMS deposits occur in submarine volcanic-sedimentary sequences and are formed from convective hydrothermal fluids driven by, and/or magmatic fluids from, a subvolcanic magma body. Both SEDEX and VMS deposits were formed by hydrothermal systems that vented fluids onto the sea floor, so that the age difference between the ores and the immediate host rocks is by definition small. In contrast, MVT deposits formed in the subsurface and so the age difference between ores and host rocks can be much larger than for SEDEX deposits (Leach et al., 2001; Leach et al., 2005).

Associated Mineral Deposit Types

SEDEX-associated deposits include MVT (Paradis et al., 2007) and stratiform barite deposits (Goodfellow, 2004). The MVT deposits occur in platformal carbonate sequences that are the shallow-water facies equivalent of basinal, typically fine-grained sediments that host SEDEX deposits. Although there is debate regarding the genetic link between SEDEX and MVT deposits, the overlap in timing and the compositional similarities suggests that these deposits formed from similar basinal metalliferous fluids (Goodfellow, 1987; Sangster, 1990). The only major difference is that ore-forming fluids vented at the seafloor in the case of SEDEX deposits whereas most of the sulphide precipitation took place subsurface in the case of MVT deposits. The source of most of the reduced S in both deposit types is most likely bacterial reduced seawater sulphate (Goodfellow, 1987).

Economic Characteristics

Summary

SEDEX deposits are an important resource for Zn and Pb (Table 1). In 2004, SEDEX deposits accounted for 38.8% and 65.1% of the western world's Zn reserves and resources, respectively (Fig. 1; Hunt, 2006). By comparison, VMS and MVT deposits accounted for 30.2% and 15.3%, and 4.2% and 1.4%, of Zn reserves and resources, respectively. In addition, the percentage of SEDEX Zn reserves has shown a consistent increase over the past two decades. In 2004, SEDEX deposits also accounted for 32% of Zn production that was followed by VMS at 27%, Zn oxide at 12%, and MVT at 6%. The size (in tonnes of Pb + Zn metal) of SEDEX deposits is on average about an order of magnitude greater than that of VMS deposits (Goodfellow et al., 1993).

The bulk of the mineralization in most SEDEX deposits resides in the bedded ore facies. The ore minerals in this fa-

TABLE 1. Grade and tonnage of major global SEDEX deposits.

Deposit Name	Deposit Status	Location	Lat. (°)	Long. (°)	Age	Geological Resources (maximum size)						
						Cu (%)	**Zn (%)**	**Pb (%)**	**Ag (g/t)**	**Au (g/t)**	**Ore (Mt)**	**Zn+Pb (Mt)**
HYC (McArthur River)	producer	Australia	-16.43	136.10	Late Paleoproterozoic	0.20	9.20	4.10	41.00		237	31.52
Talvivaara	deposit	Finland	63.98	-28.05	Paleoproterozoic	0.14	0.53		2.60		221	1.17
Mehdiabad	deposit	Iran	36.63	59.18	Cretaceous		7.20	2.30	51.00		218	20.71
Broken Hill	producer	Australia	-31.97	141.47	Paleoproterozoic	0.10	11.00	10.00	180.00	0.10	205	43.05
Ozernoe	deposit	Russia	52.50	112.50	Early Cambrian	1.20	6.20		37.00		180	11.16
Red Dog	producer	U.S.	68.07	-162.80	Mississippian		16.60	4.60	83.00		165	34.98
Sullivan	past producer	Canada	49.71	-116.01	Mesoproterozoic		5.86	6.08	67.36		162	19.33
Gamsberg	deposit	South Africa	-29.25	18.97	Mesoproterozoic		7.10	0.55			150	11.47
Mount Isa	producer	Australia	-20.73	139.48	Late Paleoproterozoic		6.80	5.90	148.00		124	15.75
Arditurri	deposit	Spain	43.17	-1.49	Late Carboniferous	1.00	8.00		50.00		120	9.60
Howards Pass (total)	deposit	Canada	62.56	-129.53	Early Silurian		5.00	2.00	17.00		120	8.40
Century	producer	Australia	-18.75	138.63	Mesoproterozoic		10.20	1.50	36.00		118	13.81
Saladipura	past producer	India	27.65	75.53	Paleoproterozoic		1.25				115	1.44
Big Syncline	deposit	South Africa	-29.20	18.83	Middle Paleo-proterozoic	0.09	2.45	1.01	12.90		101	3.49
Jiashengpan	deposit	China	41.00	109.32	Paleoproterozoic		3.80	1.30			100	100.0
Filizchai	producer	Azerbaijan	41.79	46.47	Early Jurassic	0.64	4.50	2.00		54.00	100	100.00
Broken Hill	producer	South Africa	-29.23	18.78	Middle Paleo-proterozoic	0.34	1.77	3.57	48.10		85	85.0
Black Mountain	deposit	South Africa	-29.23	18.73	Middle Paleo-proterozoic	0.75	0.59	2.67	30.00		82	85.0
Limonitovoye	deposit	Russia	58.24	93.19	Neoproterozoic	0.70	1.90				80	85.0
Navan	current producer	Ireland	53.62	6.82	Mississippian		8.04	2.68			78	8.34
Sindesar Kalan East	deposit	India	25.00	74.17	Paleoproterozoic		2.13	0.51			70	1.85
Changba-Lijiagou	producer	China	34.00	105.50	Middle Devonian		10.99	1.92			68	8.79
Rampura-Agucha	deposit	India	25.83	74.73	Paleoproterozoic		13.60	1.90	45.00		64	9.87
Howards Pass (Anniv)	deposit	Canada	62.47	-129.21	Early Silurian		5.40	2.10	25.00		61	4.58
Meggen	past producer	Germany	51.13	8.08	Middle Devonian	0.17	5.83	0.83			60	4.00
Howards Pass (XY)	deposit	Canada	62.56	-129.53	Early Silurian		5.40	2.10	9.00		59	4.43
Faro	past producer	Canada	62.36	-133.37	Late Cambrian		5.70	3.40	36.00		58	5.24
Rajpura-Dariba	producer	India	24.98	74.13	Paleoproterozoic	0.15	2.04	2.79	113.88	0.28	56	2.68
Tekeli	deposit	Kazakhstan	44.80	78.95	Neoproterozoic	1.00	6.00	5.00			50	5.50
Rosh Pinah	producer	Namibia	-27.95	16.77	Neoproterozoic	0.10	7.00	2.00			50	4.50
Hilton	producer	Australia	-20.57	139.47	Late Mesoproterozoic		9.30	6.50	151.00		49	7.74
Cannington	producer	Australia	-21.87	140.92	Paleoproterozoic		4.30	10.90	493.00		47	7.19
Dugald River	deposit	Australia	-20.25	140.15	Mesoproterozoic		13.28	2.09	42.00		43	6.59
Mokanpura North	deposit	India	25.00	74.13	Paleoproterozoic		2.40	0.60			40	1.20
Zinkgruvan	deposit	Sweden	58.82	15.10	Paleoproterozoic		10.00	1.50		45.00	40	4.60
Cirque	deposit	Canada	57.51	-125.15	Late Devonian		8.00	2.20	47.20		39	3.93
Su-Lik	deposit	U.S.	68.17	-163.20	Mississippian		8.00	2.00	30.00		34	3.40
Aguilar	producer	Argentina	-23.20	-65.70	Early Ordovician	0.05	8.50	6.50	150.00		32	4.83
Elura	past producer	Australia	-31.17	145.65	Devonian		8.62	5.58	108.98		32	4.57
Grum	past producer	Canada	62.27	-133.22	Late Cambrian		4.90	3.10	49.00		31	2.46
Rosh Pinah	producer	South Africa	-27.95	16.77	Neoproterozoic	0.10	7.25	2.10	11.00		30	2.76
Mochia	producer	India	24.36	73.72	Paleoproterozoic		3.79	1.69			27	1.47
Citronen Fjord	deposit	Denmark	83.08	-28.25	Ordovician		7.50				25	1.88
Rammelsberg	past producer	Germany	51.88	10.42	Middle Devonian	1.07	18.06	8.59			25	6.58
Hilton North (George Fisher)	deposit	Australia	-20.53	139.48	Late Mesoproterozoic		12.10	6.40	110.00		23	4.26
Lisheen	producer	Ireland	52.73	7.68	Mississippian		11.50	1.90	26.00		22	2.97
Balmat	producer	U.S.	44.25	-75.40	Mesoproterozoic		9.00	0.50			21	1.99
Franklin	past producer	U.S.	41.12	-74.57	Mesoproterozoic		19.60				20	4.00
Qingchengzi	deposit	China	40.73	123.59	Midfdle Proterozoic		2.00	3.00		75.00	20	1.00
Zhairem	deposit	Kazakhstan	48.33	70.50	Late Devonian	0.50	5.00	2.00			20	1.40

Notes: Data from Appendix 2 (DVD), Goodfellow and Lydon, 2007. Geological resources = production + economic reserves + sub-economic reserves; Mt = million tonnes; Zn+Pb Mt = Mt of combined Zn and Pb metal.

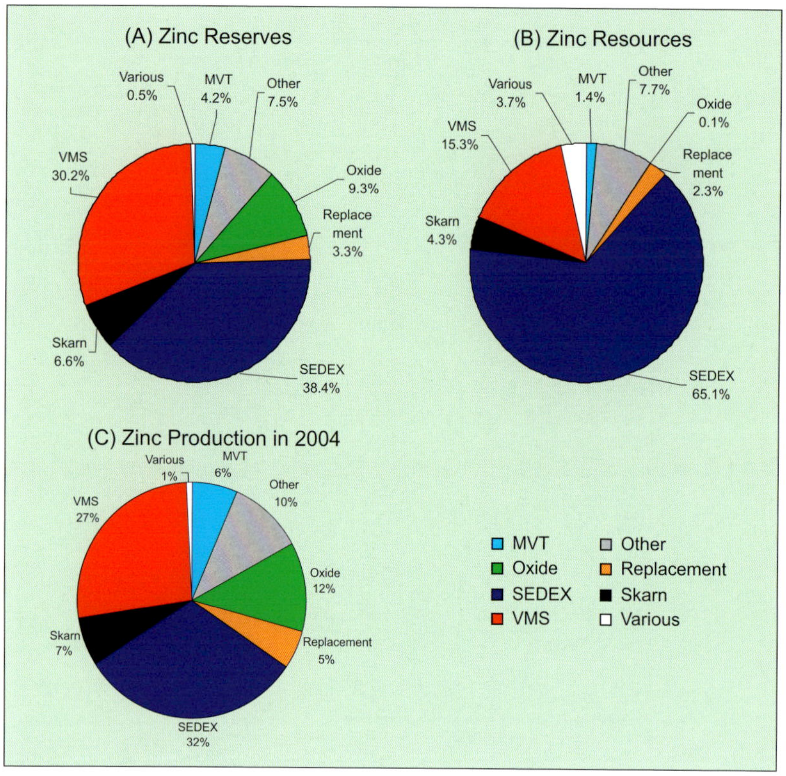

FIGURE 1. Zinc reserves **(A)**, resources **(B)**, and production **(C)** in 2004, broken down by deposit type. From Hunt, 2004.

cies are commonly fine-grained and intergrown, which leads to low metal recoveries during ore beneficiation. Although recrystallization of fine-grained sedimentary sulphides by metamorphism and/or by hydrothermal reworking in the vent complex produces coarser-grained ores from which higher recovery rates are obtained, these rates for SEDEX deposits are, on average, much lower than for VMS, MVT, and BHT deposits, the other major sources of Zn and Pb (Goodfellow et al., 1993).

Grade and Tonnage Characteristics

There are 132 SEDEX including Irish and BHT deposits world wide with grade and tonnage figures (Fig,. 2 and Appendix 2, DVD), and of these, 50 have geological resources greater than 20 Mt (Table 1). In Canada, there are 35 deposits with known geological resources (Appendix 1, DVD), of which seven contain more than 20 Mt, including the Sullivan deposit with 161 Mt and the Howards Pass deposits estimated at 120 Mt (Table 2 and Fig. 2). Of the 35 Canadian deposits, 12 are past producers, the largest of which is the Sullivan deposit that produced 149 Mt of 5.33% Zn and 5.64% Pb (Table 3). The Howards Pass deposit is currently not economic, although it is currently undergoing major exploration and evaluation. Most SEDEX deposits in Canada occur in the Anvil, MacMillan Pass, and Howards Pass districts, Selwyn Basin; the Gataga district, Kechika Trough; the Sullivan district, Belt-Purcell; and the Duncan district, southern British Columbia (Fig. 3).

In order of endowment of SEDEX deposits globally, the major basins are: Mt. Isa-McArthur Basin (seven deposits totaling 112 Mt of Zn + Pb metal), Curnamona Craton (one deposit of 75 Mt Zn+Pb), Selwyn Basin (17 deposits totaling 55 Mt Zn+Pb), Brooks Range (three deposits totaling 40 Mt Zn+Pb), Namaqualand (four deposits totaling 30 Mt Zn+Pb), Rajasthan (five deposits totaling 20 Mt Zn+Pb), Belt-Purcell Basin (one deposit totaling 19 Mt Zn+Pb), Irish Midlands (five deposits totaling 15 Mt Zn+Pb), and Rhenish Basin (two deposits totaling 11 Mt Zn+Pb; Sangster and Hillary, 2000).

Singer (1995) defined giant deposits as the top 10 percentile and super-giants as the top one percentile in terms of metal content. Using these criteria, giant deposits contain more than 8 Mt of Pb+Zn metal. If the criteria are extended to total geological size, giant and super-giant SEDEX deposits contain more than 100 and 300 Mt of ore, respectively. Some obvious conclusions from Table 1 and Figure 4 are:

- Broken Hill is a super-giant deposit based on both metal tonnes and grade; HYC and Mehdiabad are super-giant deposits based on geological resources.
- Nine of the 15 giant or super-giant deposits, based on total geological resources, are Meso-Palaeoproterozoic in age. Six of these - Red Dog, Ozernoe, Arditurri, Mehdiabad, Filizchai and Howards Pass - are Phanerozoic giant or super-giant deposits.
- The Broken Hill and Red Dog deposits stand out as producers in terms of both size and metal grade.

Metal grades are highly variable, with a mean of 0.97 wt. % Cu, 3.28 wt. % Pb, 6.76 wt. % Zn and 63 g/t Ag. Average geological resources are 34.78 Mt of sulphide and 3.165 Mt of Zn + Pb metal Fig. 5. Most of the production from SEDEX deposits in Canada came from the world-class Sullivan deposit in the Belt-Purcell basin, southeastern British Columbia, the Faro, Grum, and Vangorda deposits in the Selwyn Basin, Yukon, and the H.B. and Reeves-MacDonald deposits, southern British Columbia (Table 3).

Geological Attributes of SEDEX deposits

Deposit Morphology

The bulk of the ore is contained in a stratiform sulphide body that typically has a high aspect ratio (i.e., the ratio of the lateral extent of the body to its maximum stratigraphic thickness). Because most SEDEX deposits have an aspect ratio of 20 or more, the most common morphology is represented by sheets or tabular lenses of stratiform sulphides up to a few tens of metres in thickness and more than a km in length (Large, 1983). The Sullivan deposit in southern B.C.(Fig. 6), and the Tom and Jason deposits at MacMillan Pass, Yukon, are wedge-shaped and have a moderately high aspect ratio. The Howards Pass (Fig. 7) and Anniv deposits in the Howards Pass district, on the other hand, are saucer-shaped and taper gradually laterally for distances of several km. Unlike the

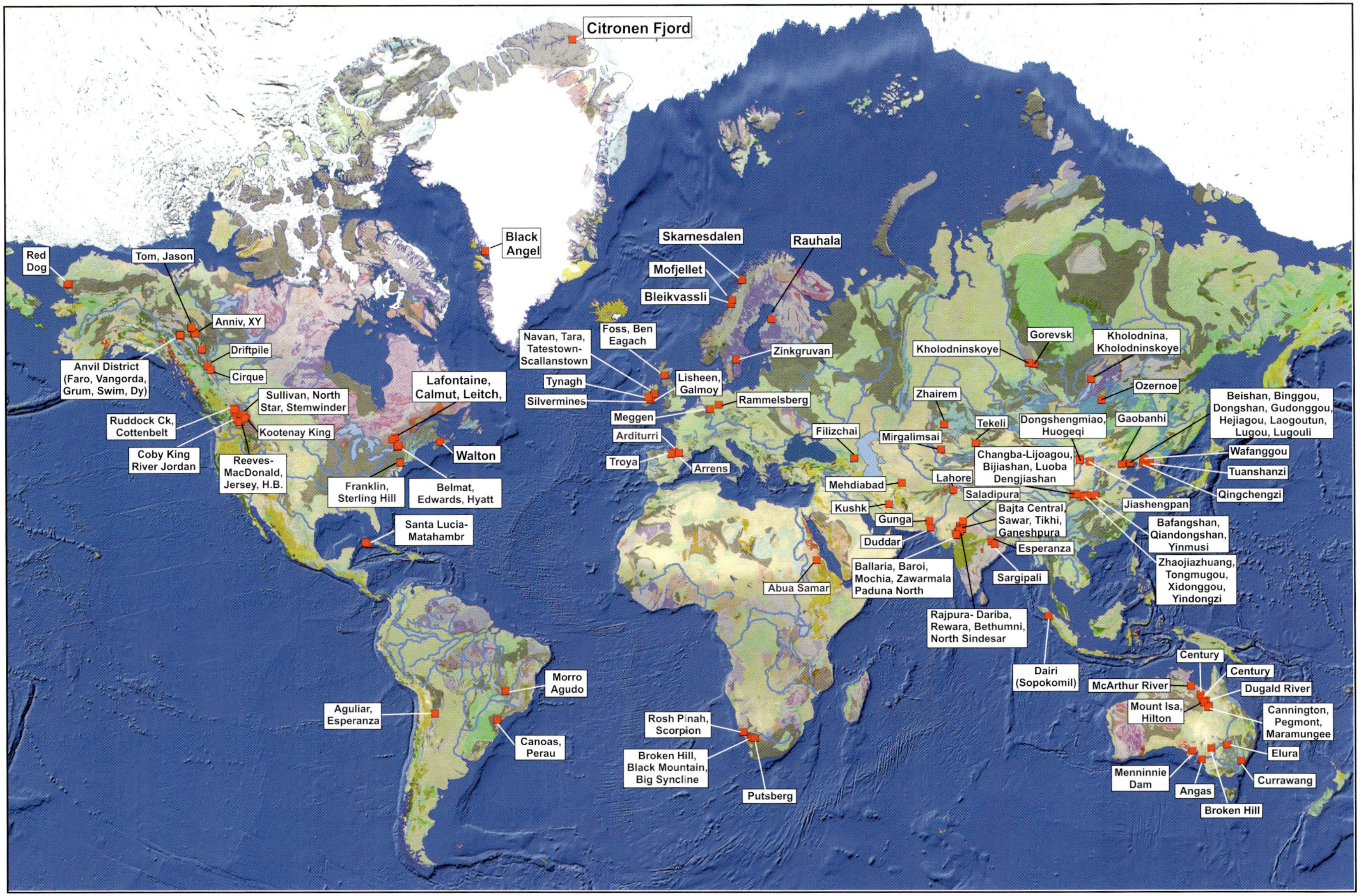

FIGURE 2. Global distribution of SEDEX deposits (includes Irish-type and BHP deposits) plotted on a simplified geological map of the world. Geology from Kirkham and Rafer (2003). Data on SEDEX deposits from Appendix 2 (DVD).

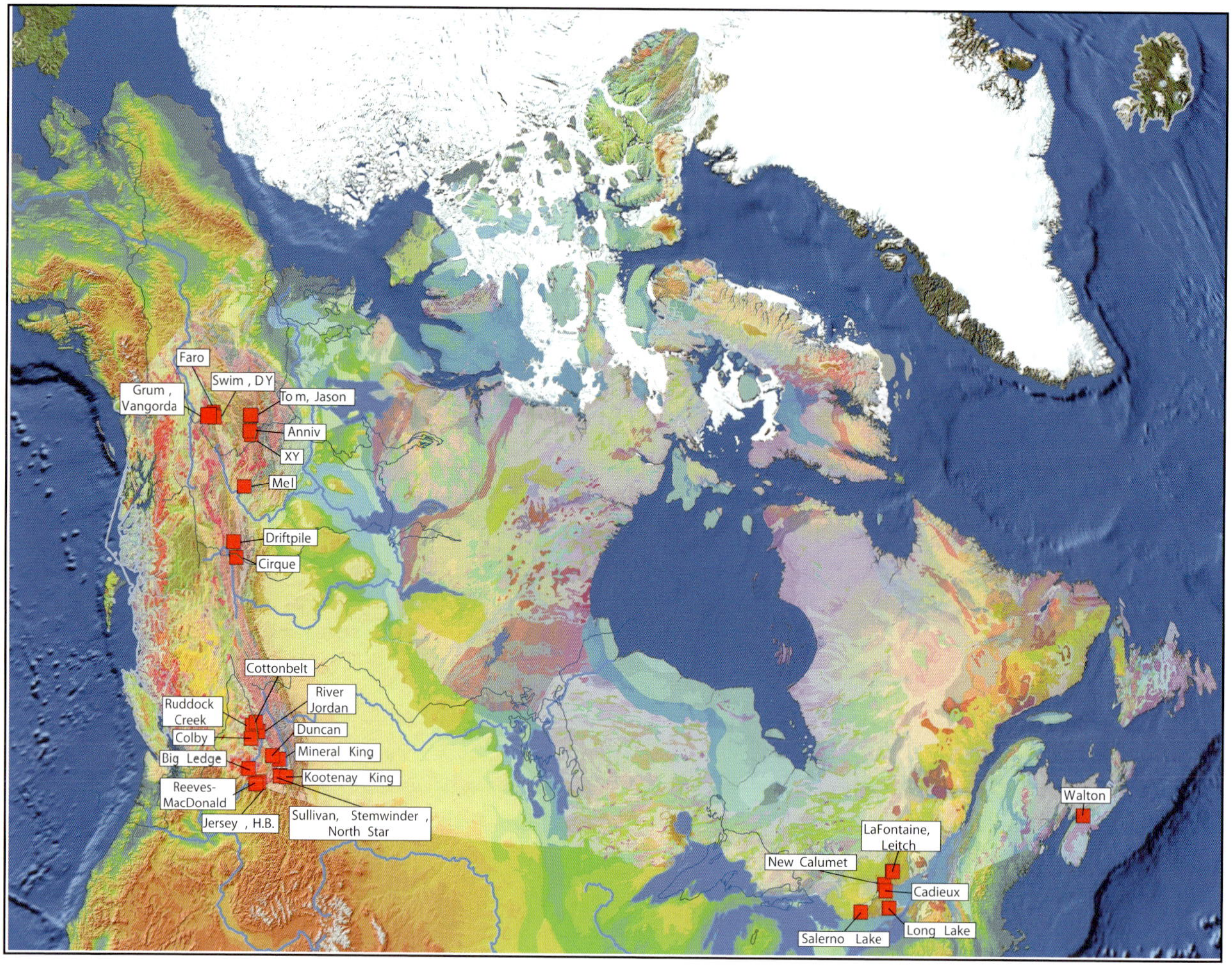

FIGURE 3. Canadian SEDEX deposits plotted on a 1:2 000 000 geological map of Canada by Wheeler et al. (1996). Data on SEDEX deposit from Appendix 1 (DVD).

vent-proximal Tom and Jason deposits (e.g.,Fig. 8A), the Howards Pass deposits (XY, Anniv, and OP) are probably not spatially associated with a hydrothermal seafloor vent (e.g., Fig. 8B). The geometry of distal deposits is controlled by the basin morphology within which hydrothermal metalliferous fluids formed brine pools similar to Atlantis II and Discovery deeps in the Red Sea (Zierenberg, 1990).

Deposit Architecture

Vent-proximal deposits are characterized by four distinct facies: 1. bedded sulphides, 2. vent complex, 3. sulphide stringer zone, and 4. distal hydrothermal sediments. Near the center of fluid up-flow represented by the stringer zone, the bedded sulphides are characteristically infilled, veined, and variable replaced by a higher-temperature mineral assemblage, producing the vent complex (Goodfellow et al., 1993). The distal hydrothermal sediments probably represent plume fallout that has been dispersed by bottom currents or alternatively clastic sulphides shed from sulphide mounds. Examples of deposits with zone-refined vent complexes include the Sullivan, Tom, Jason, and Rammelsberg deposits.

Vent-distal deposits, however, are typically weakly zoned, well bedded, and conform to the basin morphology (e.g., Fig. 8B). There is no evidence of the type of zone refining that accompanies veining, infilling, and replacement of bedded sulphides by a typically higher-temperature assemblage and that characterizes vent-proximal deposits.

Textures and Mineralogy

The bedded facies in both distal and proximal deposits is composed of sulphide minerals, other hydrothermal products such as carbonates, chert, barite, and apatite, and non-hydrothermal clastic, chemical, and biogenic sedimentary rocks. The dominant sulphide mineral in most deposits is pyrite, although in some deposits (e.g., Sullivan and Mt. Isa), pyrrhotite is predominant. The main economic minerals are sphalerite and galena (e.g., Fig. 9A, Howards Pass deposit), although chalcopyrite is an economically important mineral in a few deposits (e.g., Rammelsberg; Hannak, 1981). The ratio of iron sulphides to base metal sulphides ranges from less than 1:1 (e.g., Red Dog; Moore et al., 1986) to greater than 5:1 (e.g., Sullivan; Hamilton et al., 1982). The relative proportion of non-sulphide hydrothermal components is similarly variable. Barite, when present, occurs in major amounts (i.e., more than 25% of the hydrothermal product) and is present in about 25% of Proterozoic and about 75% of Phanerozoic SEDEX deposits (Goodfellow et al., 1993). Silica, usually as chert, is ubiquitous in most stratiform ores and is in part hydrother-

TABLE 2. Grade and tonnage of Canadian SEDEXdeposits.

Deposit Name	Province/ Territory	Lat. (°)	Long. (°)	Age	Age (Ma)	Deposit Status	Geological Resource (size of deposit) Cu (%)	Zn (%)	Pb (%)	Ag (g/t)	Au (g/t)	Ore (Mt)
Sullivan	British Columbia	49.71	-116.01	Mesoprotero-zoic	1470	past producer		5.86	6.08	67.36		161.97
Howards Pass (total)	Yukon	62.56	-129.53	Early Silurian	435	deposit		5.40	2.10	17.00		120.00
Howards Pass (Anniv)	Yukon	62.47	-129.21	Early Silurian	435	deposit		5.40	2.10	25.00		61.00
Howards Pass (XY)	Yukon	62.56	-129.53	Early Silurian	435	deposit		5.40	2.10	9.00		59.00
Faro	Yukon	62.36	-133.37	Late Cambrian	514	past producer		5.70	3.40	36.00		57.60
Cirque	British Columbia	57.51	-125.15	Late Devonian	367	deposit		8.00	2.20	47.20		38.50
Grum	Yukon	62.27	-133.22	Late Cambrian	514	past producer		4.90	3.10	49.00		30.80
Dy	Yukon	62.23	-133.14	Late Cambrian	514	deposit	0.12	6.70	5.50	84.00	0.95	21.10
Tom	Yukon	63.17	-130.14	Late Devonian	367	deposit		7.00	4.61	49.10		15.72
South Cirque	British Columbia	57.51	-125.15	Late Devonian	367	deposit		6.90	1.40	32.00		15.50
Jason	Yukon	63.15	-130.26	Late Devonian	367	deposit		7.40	6.50	65.00		10.10
Jersey	British Columbia	49.10	-117.22	Early Cambrian	525	past producer		3.49	1.65	3.08		7.68
Vangorda	Yukon	62.25	-133.18	Late Cambrian	514	past producer		4.90	3.80	54.00	0.79	7.50
Mel	Yukon	60.35	-127.40	Cambrian	525	deposit		7.10	2.05			6.78
Big Ledge	British Columbia	49.50	-118.15	Neoproterozoic	700	deposit		4.00				6.50
H.B.	British Columbia	49.15	-117.20	Early Cambrian	525	past producer		4.10	3.30	4.80		6.45
Reeves-Mac-Donald	British Columbia	49.02	-117.37	Early Cambrian	525	past producer		3.42	0.98	3.40		5.80
Clear Lake	Yukon	62.78	-135.14	Mississippian	340	deposit		11.40	2.00	38.01		5.57
Ruddock Creek	British Columbia	51.27	-118.98	Late Pro-terozoic	650	deposit		7.50	2.50			5.00
Walton	Nova Scotia	45.21	-64.04	Mississipian	340	past producer	0.04	0.11	0.32	27.70		4.90
Swim	Yukon	62.21	-133.03	Late Cambrian	514	deposit		4.70	3.80	42.00		4.75
New Calumet	Quebec	45.70	-76.68	Mesoprotero-zoic	1300	past producer		5.95	1.66	121.33	0.64	3.65
Duncan	British Columbia	50.37	-116.95	Early Cambrian	525	deposit		3.10	3.30			2.76
River Jordan	British Columbia	51.13	-118.41	Late Pro-terozoic	650	deposit		5.60	5.10	35.00		2.60
Driftpile	British Columbia	58.05	-125.95	Late Devonian	367	deposit		11.90	2.00			2.40
Mineral King	British Columbia	50.35	-116.43	Neoproterozoic	370	past producer		4.12	1.76	27.50		2.10
Upton	Quebec	45.68	-72.67	Ordovician	472	past producer		2.21		15.00		1.00
Colby (Kingfisher)	British Columbia	50.73	-118.73	Proterozoic	1700	deposit		2.60	0.58			1.67
Cottonbelt	British Columbia	51.45	-118.82	Proterozoic	1700	deposit		2.00	6.00	50.00		1.00
Cadieux (Renfrew Zinc)	Ontario	45.41	-76.71	Mesoprotero-zoic	1300	deposit		10.00				1.00
Salerno Lake	Ontario	44.85	-78.47	Mesoprotero-zoic	1300	deposit		5.00				0.75
Long Lake	Ontario	44.69	-76.77	Mesoprotero-zoic	1300	past producer		10.00				0.10
Leitch	Quebec	46.13	-75.96	Mesoprotero-zoic	1300	deposit		8.00				0.08
North Star	British Columbia	49.68	-116.03	Mesoprotero-zoic	1470	past producer		6.12	35.50	673.00		0.06
Lafontaine	Quebec	46.12	-75.96	Mesoprotero-zoic	1300	deposit		12.03				0.03
Stemwinder	British Columbia	49.69	-116.02	Mesoprotero-zoic	1470	past producer		15.60	3.70	76.30		0.03
Kootenay King	British Columbia	49.74	-115.61	Mesoprotero-zoic	1468	past producer		15.60	5.35	66.50		0.01

Notes: Data from Appendix 1 (DVD), Goodfellow and Lydon, 2007

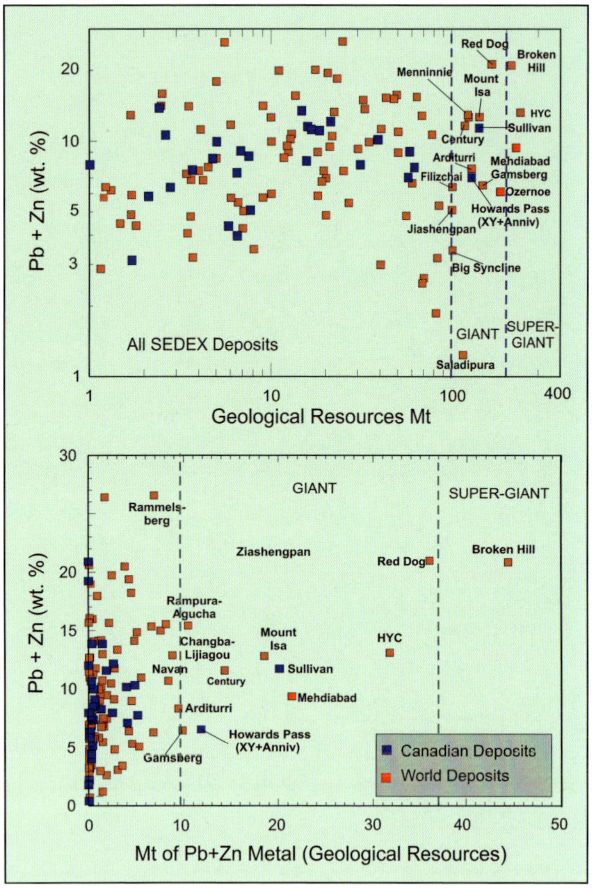

FIGURE 4. Grade vs. tonnage plots for SEDEX (including Irish and BHT) deposits world wide. Top: Pb + Zn (wt. %) vs. geological resources (Mt); Bottom: Pb + Zn (wt. %) vs. Pb + Zn metal (Mt).

mally derived.

The outer margins of the stratiform body, referred to as the distal facies, usually consist of bedded hydrothermal products that have no economic value, due to a decrease in the relative proportion of base metal sulphides with distance from the center of hydrothermal discharge (e.g., Sullivan horizon, Fig. 9C). The distal facies includes laminated pyrite and pyrrhotite, manganese, iron and calcium carbonates, iron oxides, barite, and phosphate. The contact between the bedded ore facies and the distal sedimentary facies is in most cases gradational and economically defined.

In contrast to the regularly layered appearance of the bedded ore facies, the vent complex is heterogeneous in nature and is typically composed of massive zones, replacement patches, irregular veins and/or disseminations of sulphides, carbonates, and silicates (mostly quartz; e.g., Fig. 9B, Tom vent complex). The mineral assemblage is dominated by pyrite, pyrrhotite, galena, sphalerite, ferroan carbonate, dolomite, quartz, tourmaline, and lesser muscovite, chlorite, chalcopyrite, arsenopyrite, and sulphosalt minerals. In deposits in which the vent complex has been well documented (e.g., Tom, Jason, and Sullivan), the contact between the stratabound vent complex and adjacent sedimentary-hydrothermal facies is a discordant replacement contact. The vent complex is formed by the reaction of upflowing hydrothermal fluids with hydrothermal and host sediment, which causes the replacement of the lower-temperature sedimentary minerals by higher-temperature vent assemblages (Turner, 1990; Kelley et al., 2004).

The feeder zone underlying vent complexes is discordant and composed of sulphide, carbonate, and silica veins, impregnations, and replacements that intersects the footwall sedimentary sequence (e.g., Fig. 9D, Sullivan deposit; Fig. 9E, Red Dog deposit). The feeder zone at many deposits appears to be rooted in a synsedimentary fault zone, and fault-scarp breccias, debris flows,

TABLE 3. Grade and tonnage of Canadian SEDEX deposits brought into production.

						Total Production					
Deposit Name	Province/ Territory	Lat. (°)	Long. (°)	Age	Date (Ma)	Cu (%)	Zn (%)	Pb (%)	Ag (g/t)	Au (g/t)	Mt
Sullivan	British Columbia	49.71	-116.01	Mesoproterozoic	1470	0.00	5.33	5.64	62.10	0.00	149.17
Faro	Yukon	62.36	-133.37	Cambrian	514		5.00	3.40	33.00	0.30	53.18
Vangorda	Yukon	62.25	-133.18	Cambrian	514		4.30	3.40	48.00	1.00	5.20
Grum	Yukon	62.27	-133.22	Cambrian	514		5.80	3.40	50.00	0.60	4.60
Jersey	British Columbia	49.10	-117.22	Early Cambrian	525		3.24	1.41	2.64		8.13
H.B.	British Columbia	49.15	-117.20	Early Cambrian	525		4.10	0.77	4.74	0.00	6.66
Reeves-MacDonald	British Columbia	49.02	-117.37	Early Cambrian	525	0.00	3.48	0.99	3.39		5.85
Walton	Nova Scotia	45.21	-64.04	Mississippian	340	0.56	1.44	4.33	374.00		0.36
New Calmut	Quebec	45.70	-76.68	Mesoproterozoic	1300		6.06	1.70	124.43	0.65	3.39
North Star	British Columbia	49.68	-116.03	Mesoproterozoic	1470		6.12	35.50	673.00		0.06
Mineral King	British Columbia	50.35	-116.43	Neoproterozoic	370	0.03	4.30	1.78	27.45		2.10
Kootenay King	British Columbia	49.74	-115.61	Mesoproterozoic	1468		15.60	5.35	66.50		0.01

Notes: Data from Appendix 1 (DVD), Goodfellow and Lydon, 2007.

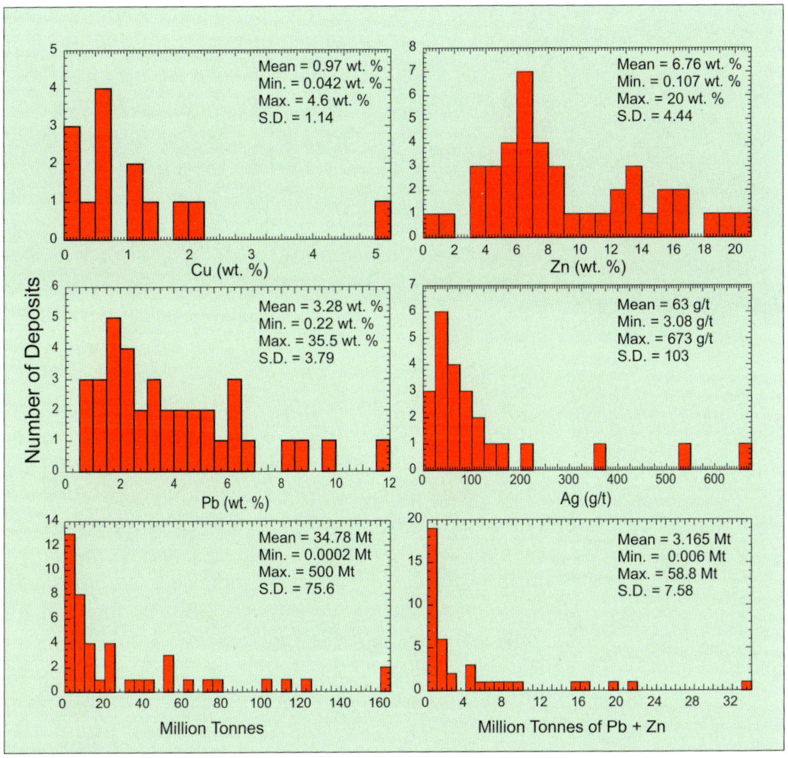

FIGURE 5. Histograms of metal grades and tonnage for global SEDEX deposits.

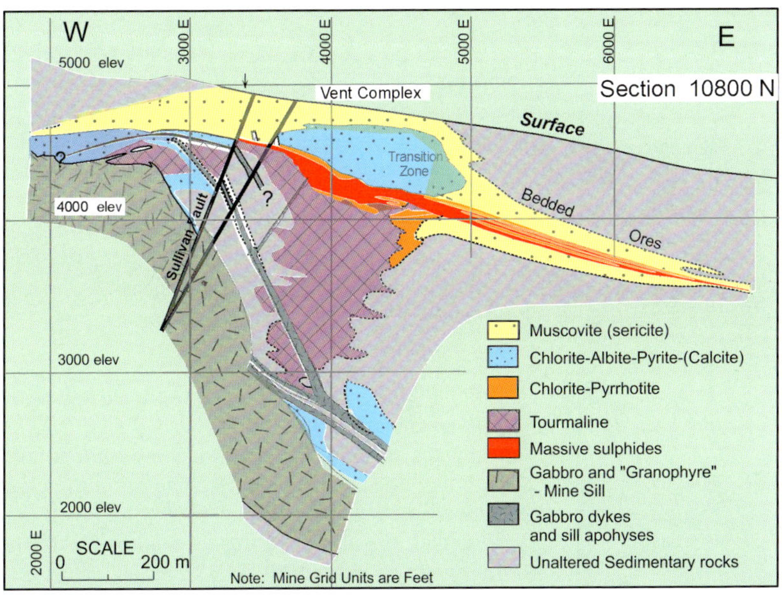

FIGURE 6. Cross section of the Sullivan deposit showing the local stratigraphy, major sulphide types, and core alteration facies (from Lydon et al., 2000a).

and abrupt facies changes associated with SEDEX deposits indicate that the fault was active before, during, and/or after sulphide formation (e.g., Jason and HYC deposits; Turner, 1990; Large and McGoldrick, 1998).

Host Rocks

Most SEDEX deposits are hosted by basinal marine, reduced facies, fine-grained sedimentary rocks that consist mostly of carbonaceous chert and shale. These sedimentary rocks generally represent pelagic and hemipelagic sediments. In some deposits (e.g., Sullivan), pelagic and/or hemipelagic sedimentary rocks are interbedded with basin-wide turbiditic siltstones and sandstones, and locally derived coarse-grained clastic sediments shed off uplifted blocks during extensional tectonism. In the case of the Irish-type deposits, the dominant lithologies are limestone and dolomite. BHT deposits are hosted by a bimodal volcanic and clastic sedimentary sequence that is commonly metamorphosed to amphibolite-granulite facies, as at Broken Hill, Australia.

Chemical Attributes of SEDEX Deposits

Ore Composition

The main economic constituents of SEDEX ores are Zn, Pb, and Ag that occur primarily in sphalerite and galena in both the bedded ores and vent complex. Grades of Pb + Zn are commonly highest near the transition between the zone-refined vent complex and the laterally extensive bedded ore facies (e.g., Tom and Sullivan, Goodfellow and Rhodes, 1990; Lydon et al., 2000a). This increase in base metals and ore-associated elements (e.g., Hg, As, Sb) is caused by the leaching of these elements from the vent complex and reprecipitation in bedded facies adjacent to the vent complex. In a few deposits, such as Rammelsberg in Germany and Mount Isa in Australia, Cu is an important economic resource and is concentrated in what has been interpreted as the vent facies (Goodfellow et al., 1993). Other economically recoverable elements include Sn in the Sullivan deposit (Hauser and Hutchinson, 1983) and Au in deposits of the Anvil district (Jennings and Jilson, 1986) and Sullivan (Conly et al., 2000), although the Au content of SEDEX deposits is typically low (Emsbo, 2000).

In addition to the ore-forming elements, SEDEX deposits are enriched in a large suite of ore-associate elements that may include Fe, Mn, P, Ba, Ca, Mg, Hg, Cd, As, Sb, Se, Sn, In, Ga, Bi, Co, Ni, and Tl (e.g., Goodfellow, 1984; Goodfellow and Rhodes, 1990; Slack et al., 2004).

Textural, Mineralogical, and Chemical Zonation

One of the most characteristic features of many SEDEX deposits is the radial zonation of hydrothermal textures, minerals, and elements about centers of hydrothermal fluid discharge. This lateral zonation away from the center of fluid discharge is controlled mostly by zone refining in the vent complex and is typically accompanied by a decrease in the

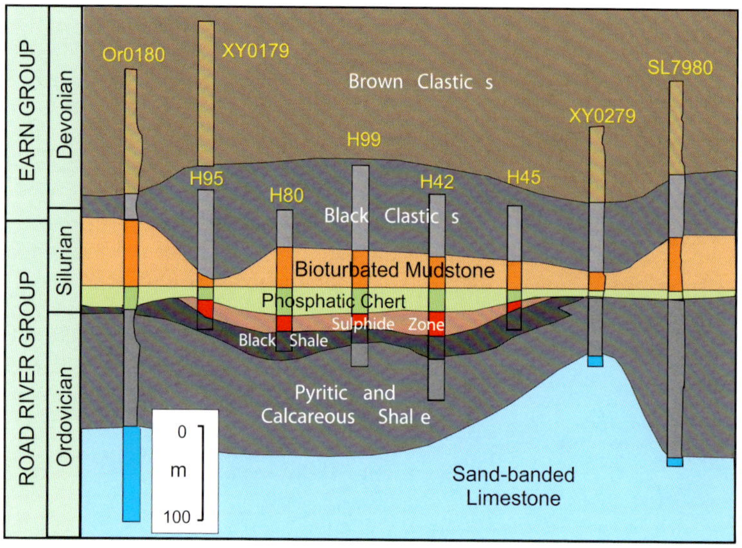

FIGURE. 7. Cross section of the Howards Pass (XY) deposit showing the local stratigraphy in the area of mineralization (from Goodfellow, 2004).

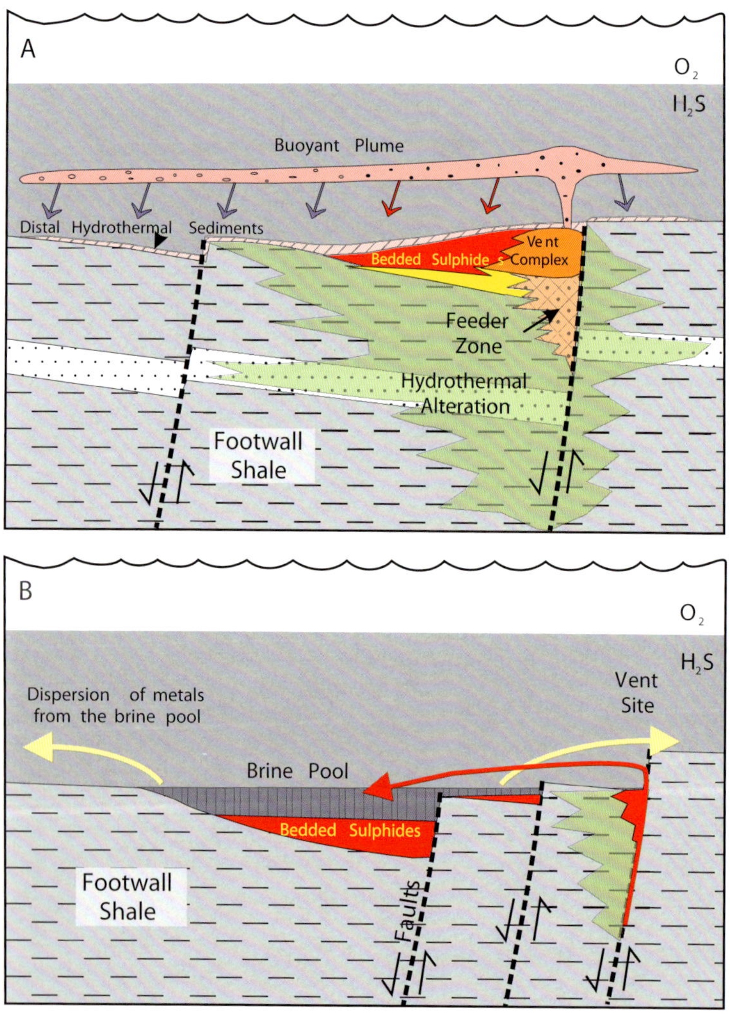

FIGURE. 8. Genetic models for SEDEX deposits. **(A)** Vent-proximal deposits formed from buoyant hydrothermal plume (e.g., Sullivan, B.C.; Tom and Jason, Yukon; Rammelsberg, Germany); **(B)** Vent-distal deposit formed from a bottom-hugging brine (e.g., Howards Pass and Anniv, Yukon; HYC deposit, Australia).

thickness of the stratiform body and the number and/or thickness of individual beds of hydrothermal minerals (Goodfellow et al., 1993).

An increase in the Zn:Pb ratio away from the vent complex is the most pronounced and consistent feature of SEDEX deposits, and has been well documented for the Tom (Goodfellow and Rhodes, 1990), Jason (Turner, 1990), Cirque (Jefferson et al., 1983), Sullivan (Hamilton et al., 1982), and Red Dog (Moore et al., 1986) deposits. Other chemical zonation patterns include increases in Pb:Ag, Cu:Zn+Pb, Fe:Zn, Ba:Zn, and SiO_2:Zn ratios away from the vent complex (e.g., Sullivan and Tom, Hamilton et al., 1982; Goodfellow, 2004).

The Tom deposit is a good example of a deposit that is highly zoned. The bedded ore facies sediments that overlie the fluid discharge conduit are brecciated, veined, and variably replaced by silica, ferroan carbonate, and sulphides. The subsurface replacement processes have produced a marked textural, mineralogical, and chemical zonation from the core of the vent complex to the distal fringes (Goodfellow and Rhodes, 1990). At the margins of the vent complex, fine-grained low-temperature phases of the bedded ore facies (i.e., Fe-poor, cream sphalerite, galena, framboidal pyrite, barite, and chert) are replaced by coarser-grained, higher-temperature assemblages (i.e., coarsely crystalline, Fe-rich black sphalerite, and Hg-rich pink sphalerite, galena, siderite, quartz, pyrrhotite, chalcopyrite, arsenopyrite, and tetrahedrite). These mineralogical changes are reflected by the systematic zonation of Zn, Pb, Cu, Cd, Hg, As, Sb, Ba, Fe, Mn, Ca, and CO_2 within and about the vent complex, and an increase in Pb/Pb+Zn ratios towards the center of hydrothermal fluid discharge (Goodfellow and Rhodes, 1990). Near the core of the vent complex, sedimentary barite has been completely replaced by ferroan carbonate minerals.

Alteration Textures, Mineralogy, and Chemistry

Although many SEDEX deposits are closely associated with an underlying hydrothermal feeder zone, hydrothermal alteration has not been well documented and mapped at most deposits. Alteration minerals that have been reported for SEDEX deposits include quartz, muscovite, chlorite, ankerite, siderite, tourmaline, and sulphides. The sulphide content of alteration zones is typically low, but pyrite, pyrrhotite, galena, sphalerite, chalcopyrite, tetrahedrite, and arsenopyrite may be present.

Hydrothermal alteration associated with SEDEX deposits is commonly widespread and extends for hundreds of metres into the pre- and post-ore sedimentary sequence and up to several kilometres laterally from the deposit. For example, sericite alteration at Sullivan extends more than 200 m below the ore, approximately 4 km in an east–west direction along the Kimberley Fault, and about 6 km south along the Sullivan-North Star corridor (Fig. 10 from

FIGURE. 9. Photographs of SEDEX and barite deposits. **(A)** Bedded facies: sphalerite and galena interlaminated with pyrite, hydrothermal carbonate, and carbonaceous chert, Howards Pass (XY) deposit, Yukon and Northwest Territories; **(B)** Vent complex: pyrite, sphalerite, galena, and ferroan carbonates replacing bedded sulphide facies, Tom Deposit, Selwyn Basin, Yukon; **(C)** Distal hydrothermal sediments: pyrrhotite and pyrite containing disseminated sphalerite interlaminated with fine-grained turbiditic sedimentary rocks, Concentrator Hill, Sullivan deposit, B.C.; **(D)** Vent complex: tourmalinized breccia infilled with pyrrhotite, sphalerite, and chalcopyrite, Sullivan Deposit, B.C.; **(E)** Sulphide stringer zone: black silicified shale cut by a network of brown sphalerite veins, Red Dog deposit, Alaska; **(F)** Well-bedded barren barite deposit, Gataga District, northeastern B.C.

Turner et al., 2000). Large et al. (2000) have described a widespread Zn, Pb, and Tl halo that extends laterally from the McArthur River deposit along the pyritic black shale facies of the Barney Creek Formation for at least 15 km to the west, and hundreds of metres above and below the mineralized zone.

Hydrothermal alteration of the hanging-wall sedimentary sequence has not been documented for most SEDEX deposits. Although post-ore hydrothermal sediments clearly show that hydrothermal fluids continued to vent well after SEDEX formation, the paucity of post-ore hydrothermal alteration described in the literature may indicate that it is too subtle in most cases to be readily recognized. Perhaps the best documented example of post-ore hydrothermal alteration is the albite-chlorite-pyrite alteration of turbiditic sedimentary rocks overlying the Sullivan deposit (Hamilton et al., 1982; Turner et al., 2000). The late albitization of the vent complex and feeder zone is clear evidence that post-ore hydrothermal fluids utilized the same conduits as the ore-forming fluids.

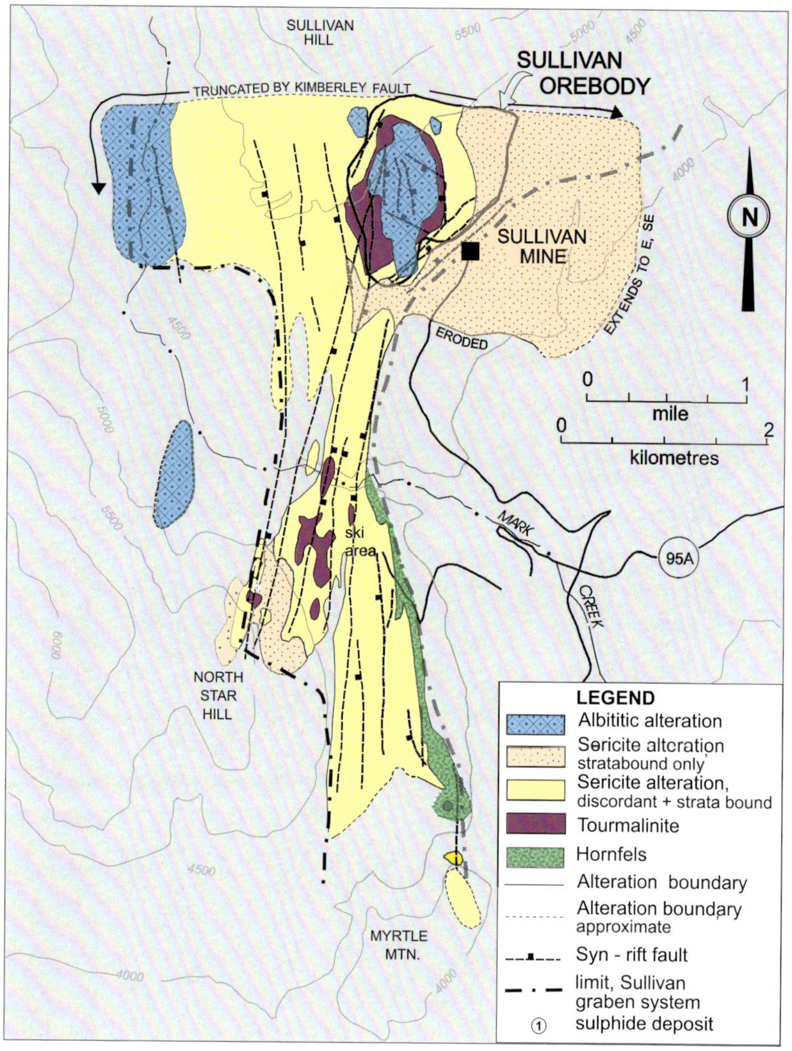

FIGURE 10. Map showing the distribution of major alteration facies within the Sullivan graben system (from Turner et al., 2000).

The nature and extent of the hydrothermal alteration and mineralization in the feeder zone depends to a large degree on the mineralogy and physical properties of footwall sediments, the temperature and chemical composition of the hydrothermal fluids, and hydrostatic pressure or water depth (Goodfellow et al., 1993). As a result, compared to VMS deposits, feeder pipes and associated alteration underlying SEDEX deposits are relatively inconspicuous due to low sulphide contents, less reactive siliciclastic sediments compared to glassy volcanic rocks, and limited seawater recharge through the low-permeability hemipelagic mud that hosts most deposits. Furthermore, subtle but diagnostic hydrothermal assemblages such as clay minerals are obscured by later metamorphism.

Distal and Post-ore Hydrothermal Sediments

Hydrothermal systems that form SEDEX deposits are typically long-lived and highly focused at discrete vent sites. Furthermore, hydrothermal activity commonly continues after sulphide formation, forming post-ore hydrothermal sediments and alteration that extend for hundreds of metres into the stratigraphic hanging-wall sequence. In most SEDEX deposits, the basal contact between the stratiform sulphides and underlying sedimentary rocks is sharp, whereas the top of the sulphide zone grades into an overlying sequence of host sedimentary pelagic and/or clastic sedimentary rocks.

Hydrothermal products that extend for hundreds of meters into the post-ore sequence and provide an important target for mineral exploration include:

1. Laminated and disseminated barite and pyrite (e.g., Tom, Jason, Rammelsberg, Meggen)
2. Manganese and iron carbonates (e.g., Meggen, Silvermines, McArthur River (HYC))
3. Phosphatic and pyritic chert (e.g., Howards Pass, Anniv)
4. Metal-rich laminated pyrite (e.g., HYC, Mt. Isa, Tom, Jason)
5. Albite-chlorite-pyrite alteration (e.g., Sullivan)

Most SEDEX deposits are also surrounded by hydrothermal sediments that extend for up to several km from the sulphide zone. In deposits containing barite (e.g., Tom and Meggen), these distal sediments typically consist of interlayered barite, chert, carbonate and host lithologies. In other deposits, this facies is represented by iron sulphides and/or chert interbedded with host lithologies (e.g., Sullivan, Hamilton et al., 1982). At Howards Pass in the Yukon, the distal fringe of the ore zone consists of a laminated phosphatic chert and pyritic shale (Goodfellow, 2004). A manganese-rich halo has been noted to surround several SEDEX deposits, including the Tom (Large, 1981a) and Meggen (Gwosdz and Krebs,

1977). Widespread lithogeochemical halos have also been documented at the Lady Loretta (Large and McGoldrick, 1998) and McArthur River (HYC; Large et al., 2000) deposits.

Geology—Continental and Global Scale

Tectonic Settings

SEDEX, Irish, and BHT deposits occur in intra-cratonic and epicratonic sedimentary basins (Large, 1980). The tectonic settings include intra-cratonic rifts driven by mantle plumes (e.g., Belt-Purcell basin; Anderson and Goodfellow, 2000), reactivated rifted margins (e.g., Howards Pass; Goodfellow et al., 1986), and far-field back-arc rifting (e.g., MacMillan Pass-Gataga (Fig. 11) and Mount Isa-McArthur districts; Nelson et al., 2002; Betts et al., 2003). Although there is considerable debate on BHT deposits, our interpretation is that the association of the Broken Hill deposit with coeval bimodal igneous rocks in a sedimentary basinal sequence suggests that this deposit type is similar to less-deformed and metamorphosed VSHMS (volcanic-sediment-hosted massive sulphide) deposits in sedimented back-arc continental rifts (e.g., Bathurst Mining Camp, Finlayson Lake, Iberian Pyrite Belt).

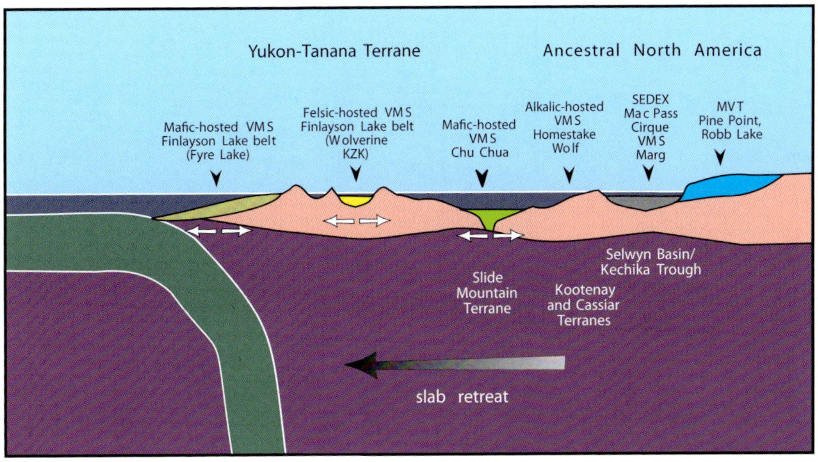

FIGURE 11. Tectonic model for the Devono-Mississipian margin of western North America, showing a complex history of ocean plate subduction and slab rollback, back-arc rifting and the formation of SEDEX and MVT deposits in the northern Canadian Cordillera (from Nelson et al., 2002).

Most SEDEX deposits formed during periods of tectonism which is typically manifested by fault reactivation, intrabasin clastic sedimentation, and in many cases magmatism represented by volcanism and/or sill emplacement. Most deposits occur in reduced marine basins that formed during the sag phase of basin history, adjacent to deeply penetrating faults.

Secular Variations and the Sulphur Cycle

SEDEX, Irish, and BHT deposits formed during several discrete episodes from Paleoproterozoic (about 2000 Ma) to Cretaceous (Fig. 12). The number and size of SEDEX per 50 million years are greatest in the Middle Paleoproterozoic and Paleoproterozoic, respectively. Lead and Ag grades are highest in the Early Mesoproterozoic, due to the influence of the giant Sullivan deposit, and Au grades reach a maximum in Paleoproterozoic and Early Mesoproterozoic deposits although the database of Au grades is very limited (see Table 1 and Appendix 1, DVD).

Most SEDEX deposits coincide with periods in Earth history when the oceans were episodically stratified with a lower anoxic and H_2S-rich water column (Fig. 13A). In the Paleozoic Selwyn Basin, for example, there is a close relationship between upward increasing $\delta^{34}S$ secular trends in sedimentary pyrite, anoxic laminated carbonaceous shales and cherts, and three major SEDEX forming events in the Late Cambrian, Early Silurian, and Late Devonian (Fig. 13B). In the Selwyn Basin, episodes of ocean stratification with reduced bottom waters coincide with global cycles of black shale deposition during the Paleozoic (Arthur and Sageman, 1994), indicating that the Selwyn Basin anoxic episodes reflect global processes. The secular variation of pyrite $\delta^{34}S$ values has been attributed to Rayleigh fractionation of S isotopes under closed or partly closed anoxic conditions, and accompanying increased rates of isotopically light S deposition as sedimentary sulphides causing an increase in $\delta^{34}S$ values for H_2S in the reduced water column (Goodfellow, 1987).

The transition from a ferrous Fe-dominated Archean ocean to a reduced S-dominated stratified Proterozoic and Phanerozoic oceans followed the oxidation and precipitation of Fe in Superior-type iron formations (Fig. 13A). This major change in ocean chemistry was accompanied by the build-up of sulphate, which was essential for bacterial sulphate reduction, and the establishment of H_2S-rich anoxic water columns. The absence of Archean and Early Paleoproterozoic SEDEX deposits is probably controlled, therefore, by the limiting effect of high contents of reduced Fe on the activity of H_2S in anoxic oceans. Under these conditions, metals in S-depleted saline hydrothermal fluids vented into seawater would be expected to be lost to the water column because of a lack of reduced S to precipitate them (Goodfellow et al., 1993).

Basin Architecture

The ideal basinal architecture for the formation of SEDEX deposits (Fig. 14) is a continental rift basin with at least 2 to 5 km of coarse-grained permeable clastics and related volcanics and/or volcaniclastics that form the syn-rift phase overlain by an impermeable cap or seal of basinal shales and/or carbonates (Lydon, 1983; Large, 1986). The source for Zn and Pb is syn-rift dominantly clastic sedimentary rocks (Lydon et al., 2000b). The ideal traps for the metals are reduced subbasins with an adequate supply of bacteriogenic H_2S in the ambient water column (Fig. 14). Although an association with magmatism is not indicated for all SEDEX deposits, mafic volcanic rocks and sills are spatially and temporally associated with many deposits (e.g., Selwyn Basin and Belt-Purcell basin) and bimodal volcanic rocks host most BHT deposits (e.g., Broken Hill; Parr and Plimer, 1993)

Two of the best understood continental basins hosting

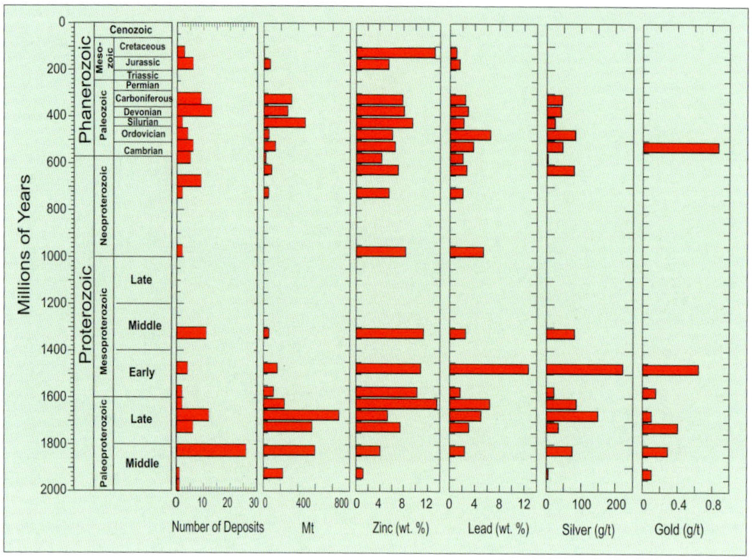

FIGURE 12. Secular variations of tonnage and metal grades for SEDEX deposits world wide. Plotted from data in Appendix 2 (DVD).

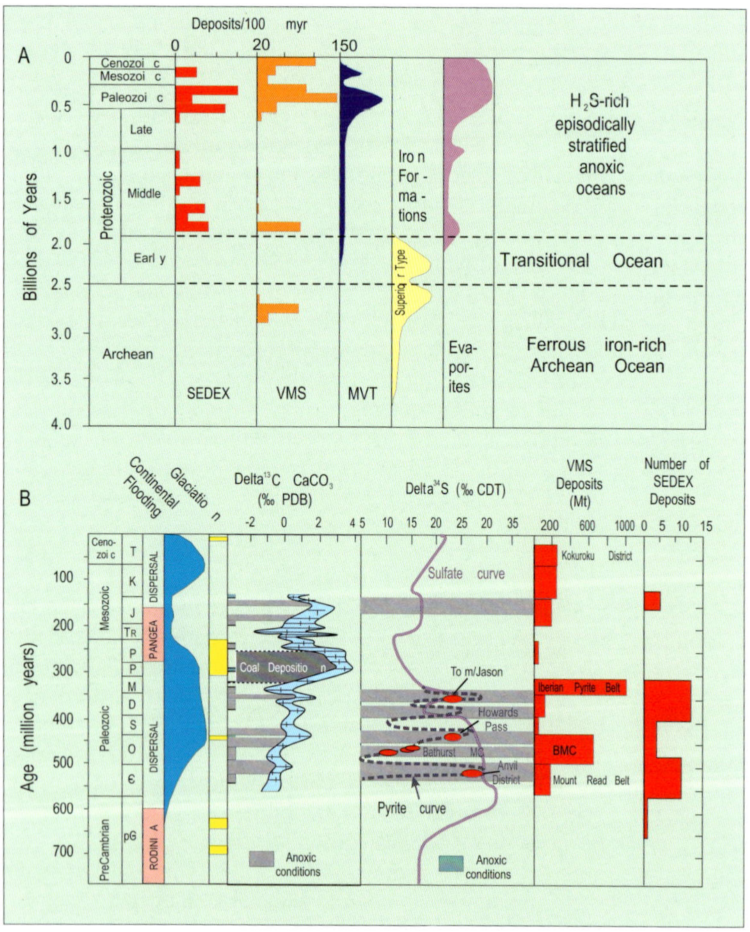

FIGURE 13. **(A)** Secular distribution of SEDEX, VMS, and MVT deposits, iron formation, evaporites, and H_2S-rich anoxic periods (Goodfellow, 2004). **(B)** Late Proterozoic and Phanerozoic tectonic (continent accretion and dispersion; Fischer, 1984) and glacial (Baker, 1997) events, continental flooding (Vail et al., 1977), and sulfur isotope secular curves for evaporites (Claypool et al., 1980), carbonates (Arthur and Sageman, 1994), and pyrite (Goodfellow, 1987). Also shown are global (Arthur and Sageman, 1994) and Selwyn Basin (Goodfellow, 1987) anoxic periods, and the temporal distribution of volcanic-associated massive sulphide (Mosier et al., 1983) and SEDEX (Goodfellow et al., 1993) deposits. BMC = Bathurst Mining Camp.

SEDEX deposits in Canada are the Paleozoic Selwyn Basin that extends from Alaska border through Canada into the U.S. south of the British Columbia (Fig. 15 from Goodfellow et al., 1995), and the Middle Proterozoic Belt-Purcell basin that extends from southeastern British Columbia to Montana and Idaho in the U.S. The Bathurst Mining Camp in northern New Brunswick (Goodfellow and McCutcheon, 2003) and the Finlayson district in the Yukon are two possible examples of BHT deposits (Piercey et al., 2001; Peter et al., 2007).

District Scale (Canada)

At the district scale, SEDEX deposits commonly occur in clusters and formed during several hydrothermal events within local second-order basins adjacent to extensional structures (e.g., Fig. 16, Sullivan deposit) separating basinal facies fine-grained sediments from platformal carbonates sequences. Some deposits are spatially and temporally associated with volcanic and intrusive rocks. Most deposits are hosted by marine basinal sedimentary rocks that include black shales and cherts, carbonates, and intrabasinal clastic sedimentary rocks.

Deposit Scale

SEDEX, Irish, and BHT deposits typically occur in local third-order sedimentary basins. Vent-proximal deposits are associated with active faults that commonly define the margins of local basins (Fig. 8A), whereas vent-distal deposits occupy bathymetric depressions on the seafloor (Fig. 8B). The most recognized control on the siting of vents and associated ores is cross-stratal faults that served as conduits for transporting metalliferous fluids from the hydrothermal reaction zone to the seafloor. Other factors that influenced the location of vents are: 1. basement highs where the thickness of soft sediment is at a minimum, 2. the hydrology of the reaction zone which is partly controlled by the thermal structure, 3. the basement geometry, and 4. the position of recharge zones. In the case of vent-distal deposits, the predominant control other than proximity to vents on the location of deposits is fluid density and local seafloor bathymetry.

Distribution of Canadian Metallogenetic Districts

Major metallogenetic districts that host SEDEX deposits (Fig. 3) are:

- Mississippian Walton District, Carboniferous basin, Nova Scotia
- Late Devonian MacMillan Pass District, Selwyn Basin, northeastern Yukon
- Late Devonian Gataga District, Kechika Trough, northeastern British Columbia

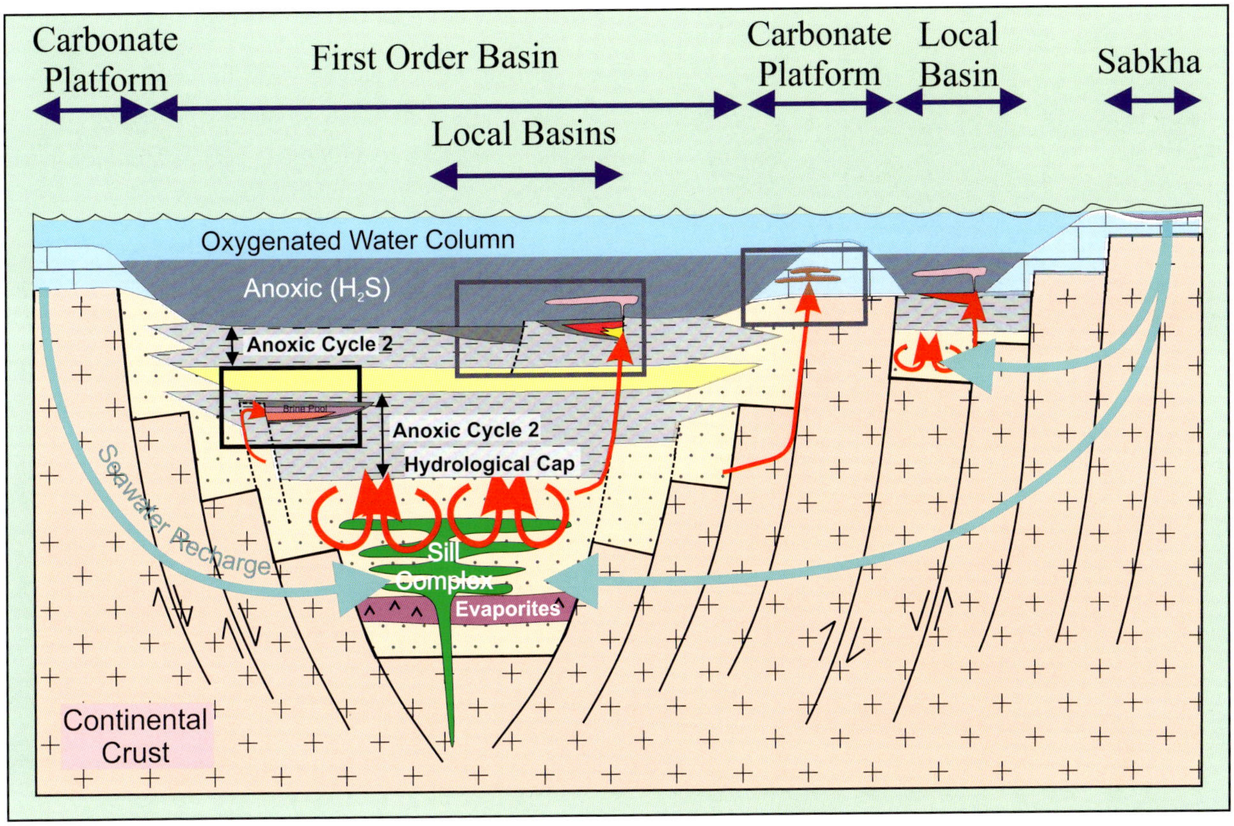

FIGURE 14. Model of the sedimentary basin architecture of productive basins hosting SEDEX deposits.

- Early Silurian Howards Pass District, Selwyn Basin, northeastern Yukon
- Late Cambrian Anvil District, Selwyn Basin, Yukon
- Late Cambrian Duncan District, southern British Columbia
- Neoproterozoic Shuswap District, southern British Columbia
- Mesoproterozoic Sullivan District, Belt-Purcell basin, southeastern British Columbia
- Mesoproterozoic Maniwaki-Gracefield District, Grenville Province, Quebec

The Selwyn Basin occupies the Paleozoic margin of western North America and has been subdivided into five spatially distinct mining districts representing three ages of mineralization: 1. Late Cambrian Anvil, 2. Late Cambrian Duncan, 3. Early Silurian Howards Pass, 4. Late Devonian MacMillan Pass, and 5. Late Devonian Gataga districts. The Middle Proterozoic Belt-Purcell is an intracratonic basin (Price and Sears, 2000) that hosts the Sullivan district (Höy et al., 2000). Other SEDEX districts include the highly metamorphosed Neoproterozoic Shuswap district in southern British Columbia, and the Mesoproterozoic Maniwaki-Gracefield district in the Grenville Province of Quebec. In the Appalachian orogen, the Carboniferous Walton district of Nova Scotia hosts the past-producing Walton deposit.

Genetic and Exploration Models

Conventional Models

SEDEX deposits most likely formed from oxidized and therefore H_2S-poor fluids generated in geopressured hydrothermal reservoirs within syn-rift clastic (and evaporitic) sediments sealed by fine-grained marine sediment or carbonates (Goodfellow et al., 1993; Cooke et al., 2000). During the post-rift or sag phase of basin formation, which is typically long-lived and spans over 200 million years in the case of the Selwyn Basin, Zn and Pb were probably leached from iron oxides coating detrital minerals and homogenized by convecting fluids in the hydrothermal reservoir. The large variability in the temperature, salinity, metal content, and redox conditions of SEDEX fluids was controlled by a number of parameters including the local thermal regime, the redox state of the reservoir sediments, and the presence or absence of evaporates. Ancient SEDEX fluids are similar to metalliferous fluids in modern sedimentary basins, which are characterized by temperatures up to 305°C (e.g., Salton Sea brines; McKibben et al., 1988), high salinities, and very low total S(reduced):total metal ratios.

A mineralizing episode is triggered by tectonic events that re-activate major rift faults that generate rapidly subsiding grabens or half-grabens. The reactivated rift faults can breach the sediment cap, causing the discharge of metal-rich saline fluids (enriched in Fe, Zn, Pb as chloride complexes, and variable SO_4^{2-}) along the faults to the basin floor. For most SEDEX deposits, metal sulphides are precipitated at or

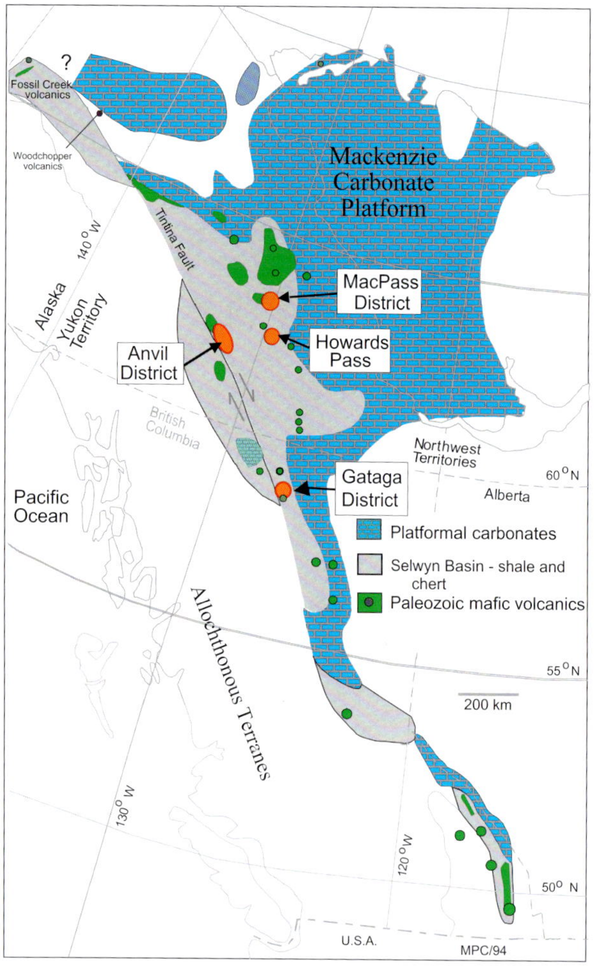

FIGURE 15. Paleofacies distribution and SEDEX deposits of the Selwyn Basin, Yukon and British Columbia (modified from Goodfellow et al., 1995; Cecile et al., 1997).

above the seafloor by reaction with H_2S in the overlying reduced water column (Goodfellow, 1987). Continued fluid discharge from the reservoir is driven by free convection, as cold seawater descends into the basin sedimentary sequence along recharge faults (Yang et al., 2004), where it is heated and leaches Zn and Pb from the reservoir sediments.

Because most fluids that formed SEDEX deposits were probably depleted in reduced S, an important requirement for this deposit-type to form is a sufficient supply of reduced S at the site of deposition (Goodfellow, 1987; Cooke et al., 1998). In the case of well-bedded deposits that formed at the seafloor, the most likely S source is biogenic H_2S that is typically enriched in anoxic water columns (Goodfellow, 2000). The presence or absence of barite is controlled by two major factors: 1. the activity of sulphate in the ore-forming fluids, which limits barium solubility, and 2. the activity of sulphate in the water column, which controls the precipitation of barite. In extreme anoxic basins where all the sulphate has been reduced to sulphide, barium carried by hydrothermal fluids will be lost to the water column because of a lack of sulphate to precipitate barite.

Recent Advances in Genetic/Exploration Models

Major advances in the genetic and exploration models are listed and discussed briefly below:

- The tectonic settings for sedimentary basins that host SEDEX deposits, particularly far-field back-arc extensional environments and intracratonic rifts driven by mantle plumes are better understood.
- The architecture of productive sedimentary basins, particularly as it relates to syn-rift clastic sedimentation, the development of hydrothermal reservoirs, and the onset of subsequent sagging related to conductive cooling of the lithosphere has been documented in detail for several basins.
- The temporal distribution of SEDEX deposits (post-1.8 billion years) is probably controlled by the oxidation of Archean oceans, the buildup of sulphate and associated bacterial sulphate reduction, the development of Proterozoic and Phanerozoic oceans episodically stratified with lower anoxic water columns, and the precipitation of sulphides by the mixing of S-poor metalliferous fluids with ambient H_2S-rich anoxic oceans.
- Detailed documentation of different aspects of SEDEX deposits:
    - Geology and sedimentology of local third-order basins hosting deposits.

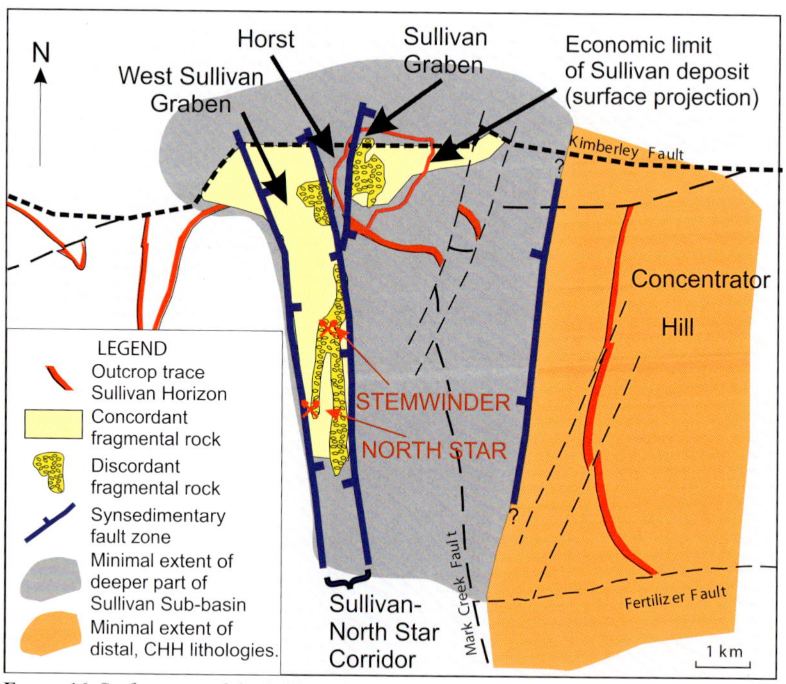

FIGURE 16. Surface map of the Sullivan deposits showing the Sullivan and West Sullivan grabens or third-order basins, and basin bounding faults (from Lydon et al., 2000a).

- Morphology and architecture of mineralized zones.
- Mineralogy and chemical composition of hydrothermal sediments and the development of exploration vectors.
- Mapping of hydrothermal alteration surrounding SEDEX deposits.

- New thermal chemical models which indicate that SEDEX deposits formed from high f_{O2} and therefore H_2S-depleted basinal metalliferous fluids (Cooke et al., 2000).

- The recognition that SEDEX deposits in the Selwyn Basin are genetically related to coeval MVT deposits on the adjacent Mackenzie Carbonate Platform (Goodfellow, 1987).

Key Exploration Criteria for Canadian SEDEX Deposits

Continental Scale

At the continental or basin-wide scale, criteria indicating a high potential for SEDEX and BHT Zn-Pb deposits include:

- Age: post-1800 Ma because of the S cycle.

- Tectonic setting: sedimented intra-cratonic and epicontinental rifts.

- Sedimentary basin architecture: syn-rift clastic sequence overlain by post-rift impervious marine basinal sediments; the recharge of saline brines on adjacent carbonate platforms may be an important source of chloride to complex metals in basinal hydrothermal fluids.

- Anoxic episodes: interglacial, warm oceanic conditions represented by global episodes of anoxic sedimentation (e.g., carbonaceous non-bioturbated shales, laminated pyrite, the absence of benthic fauna) and evaporate formation.

- Magmatic events: syn- and/or post-rift mafic alkali volcanic rocks, and tholeiitic sills and dykes.

- Coeval MVT deposits in adjacent carbonate platforms.

- Geochemical stream anomalies: sediment and water samples that are anomalous in ore-forming and ore-associated elements over a large area of the basin. An example is widespread Zn anomalies that extend for hundreds of kilometres in the Selwyn Basin (e.g., Fig. 17, Selwyn Basin).

District and Deposit Scale

At the district scale, the criteria for evaluating the potential to host SEDEX and BHT Zn-Pb deposits are:

- Second- and third-order basins within sedimented continental rifts.

- Local tectonism: reactivation of extensional faults as evidenced by sedimentary facies and thickness changes, fault scarp breccias, local unconformities, major fault offsets and fault intersections.

- Magmatic centres (e.g., volcanic and sill complexes).

- Reduced seafloor environments: anoxic cycles in the evolution of a given basin.

- Paleoclimate: evaporites in the hydrothermal reaction zone and/or saline recharge fluids on the adjacent platforms that provide a source of chloride to complex base metals.

- Distal hydrothermal sediments: e.g., barite, apatite, pyrite, and Mn-Fe-Ca-Mg carbonates.

- Hydrothermal alteration: pre- and post-ore quartz, muscovite, chlorite, ankerite, siderite, and tourmaline alteration with associated disseminated and vein sulphides of ore-forming and ore-associated elements.

- Geochemically anomalous shales: enriched in base metals and ore-associated elements (e.g., As, Sb, Cd, Mn, P, Ba, Hg, Tl).

Surficial geochemical anomalies: sediment and water samples that are locally anomalous in deposit-forming and ore-associated elements.

Knowledge Gaps

Key knowledge gaps are summarized below.

1. Radiometric and paleontological ages of ore-forming events is required to relate the causes of ore formation to other geological events such as tectonism, magmatism, and ocean stratification. This can be carried out by direct dating of the various ore components (e.g., Nd/Sm in carbonate, sulphides; Rb/Sr in carbonates; U/Pb in xenotime); detailed micropaleontology of Phanerozoic rocks; or by bracketing the age of the ore horizon by U/Pb dating of intercalated volcanic ash horizons.
2. Development of exploration criteria for discovering SEDEX in sedimentary basins. This would involve comparative studies of the geological and hydrothermal evolution of productive and non-productive basins. Essential information would include:
    - History of tectonic and magmatic events within basins.

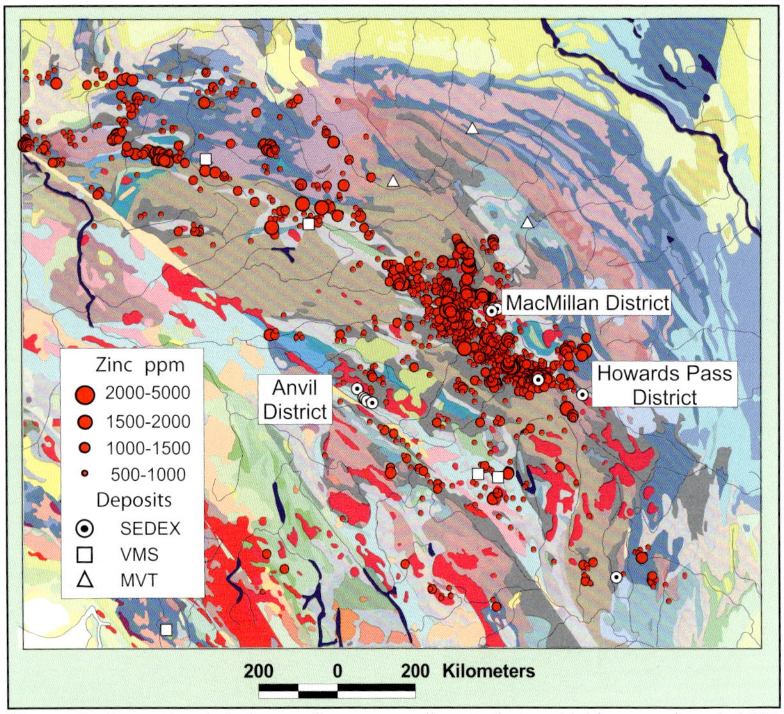

FIGURE 17. Proportional circle map for stream sediments that are highly anomalous (>1,000 ppm) in Zn, Selwyn Basin, Yukon.

- Sediment accumulation rates (needs radiometric dating control).

- Evolutionary pathways of mineralogical and lithogeochemical changes during diagenesis and burial metamorphism. This requires transects of the basin supported by careful stratigraphy, sedimentology, lithogeochemistry and radiometric dating of diagenetic and burial metamorphic minerals.

- Evolutionary paths of pore fluids in the basin (would have to rely on studies of fluid inclusions in cements and diagenetic minerals), with emphasis on determining the stage or circumstances under which pore fluids become metalliferous.

- Paleoclimatology and the history of ocean stratification. This would require geochemical and isotopic (e.g., S, C, Sr) profiles of basinal stratigraphy and the systematic documentation of lithological and sedimentological indicators.

3. Development of prognostic criteria for stratigraphic interval(s) with a high mineral potential and the most favorable spatial position(s) for SEDEX deposits in a sedimentary basin. This requires a fundamental understanding of all the key processes controlling focused expulsion of metalliferous hydrothermal fluids in a sedimentary basin. This could be accomplished by compilation of key data (especially data generated by the oil industry), mathematical modeling of fluid flow using constrained rock property parameters, and targeted research on key basins where the critical components are well preserved.

4. Refine models of SEDEX formation. There is still much debate as to the ore-forming environment required for the deposition and preservation of sedimentary Zn-Pb deposits. It is still not decided whether SEDEX deposits require a brine pool, whether they can form as a sulphide mound, or whether they are predominantly subsurface replacements of fine-grained sediments. These and other unresolved questions on the genesis of SEDEX deposits require further studies of more pristine examples of SEDEX deposits such as those of the Tethyan Belt and Cuba.

5. Determine the impact of ocean evolution on the temporal distribution of SEDEX deposits. The empirical observation that there are no SEDEX deposits older than about 1.8 billion years and the proposed ocean chemical evolution model as it relates to reduced biogenic S supply needs to be tested in sedimentary basins that span the geological time scale. This question is of particular interest to Canada because of the large number of Early Proterozoic sedimentary basins especially in Northern Canada (e.g., Labrador Trough, Labrador; Foxe-Rinkian Basin, western Nunavut; Wopmay Orogen, northern NWT; Athapuscow Aulacogen, NWT; Kilohigok Basin, Bathurst Inlet-Victoria Island, NWT).

Areas of High Mineral Potential in Canada

Using the criteria outlined above for identifying productive basins, major sedimentary basins in Canada are rated for their potential to host SEDEX, Irish-type and BHT sulphide deposits. It should be noted that the confidence level varies with the quality and comprehensiveness of geoscientific data, and the level of exploration by the mining industry.

Basins Hosting SEDEX Deposits

The following basins host SEDEX and VSHMS deposits and are therefore highly prospective for additional new deposits within existing districts and new districts.

- Selwyn Basin, Yukon, Northwest Territories, and British Columbia (Cambrian to Mississippian). Although most deposits are located in the Yukon and British Columbia, the potential for SEDEX deposits is considered high in Selwyn Basin facies that extends the full length of the Canadian Cordillera.

- The Mesoproterozoic Belt-Purcell basin, southeastern British Columbia and northwestern U.S. This continental rift hosts several SEDEX deposits including the world-class Sullivan deposit.

- Finlayson Lake district, Yukon (Middle-Late Devonian).

This sedimented back-arc continental rift is host to several major deposits.

- Bathurst Mining Camp, northern New Brunswick (Middle Ordovician). This back-arc sediment-covered continental rift hosts over 20 deposits greater than 1 Mt, and several past and current producers including the world-class Brunswick No. 12 deposit.

Highly Prospective Sedimentary Basins

The following basins do not contain SEDEX deposits but exhibit all or most of the key attributes of productive basins.

- Borden Basin, Nunavut (Middle Proterozoic)
- Franklinian Basin, northern Nunavut (Paleozoic)
- Wernecke Supergroup, central Yukon (Mesoproterozoic)

Moderately Prospective Sedimentary Basins

The following basins do not contain SEDEX deposits and exhibit most of the key attributes of productive basins but are deficient in at least one fundamental attribute. The Richardson Trough in not underlain by a syn-rift clastic sequence and therefore does not exhibit the classic architecture of productive sedimentary basins. All the remaining sedimentary basins listed below are older than the oldest SEDEX deposits and may be deficient in bacteriogenic S that is considered essential for precipitating base metals vented by H_2S-poor hydrothermal fluids into ambient seawater.

- Richardson Trough, northern Yukon and NWT (Paleozoic)
- Labrador Trough, Labrador (Paleoproterozoic)
- Foxe-Rinkian Basin, western Nunavut (Paleoproterozoic)
- Wopmay Orogen, northern Nunavut (Paleoproterozoic)

Acknowledgements

This synthesis is based largely on previous reviews of SEDEX deposits and research emanating from Geological Survey of Canada research projects on SEDEX deposits of the Selwyn Basin, Yukon, and Sullivan mining district, southern British Columbia. It is intended to have a global perspective but a national focus on SEDEX deposits and sedimentary basins considered to have SEDEX potential. We thank C. Allen and S. Paradis for reviewing the manuscript and offering many excellent suggestions, and I. Kjarsgaard for checking and editing an earlier version of the manuscript and accompanying Canadian and global SEDEX databases. P. Smith of Brook Hunt is thanked for making recent data on global Zn and Pb production, reserves and resources available for plotting in Figure 1.

References

Anderson, H.E., and Goodfellow, W.D., 2000, Geochemistry and isotope chemistry of the Moyie sills: Implications for the early tectonic setting of the Mesoproterozoic Purcell basin, in Lydon, J.W., Höy, T., Knapp, M., and Slack, J., eds., The geological environment of the Sullivan deposit, British Columbia: Geological Association of Canada, Mineral Deposits Division, Special Publication 1, p. 302–321.

Arthur, M., and Sageman, B.B., 1994, Marine shales: Depositional mechanisms and environments of ancient deposits: Annual Review of Earth and Planetary Sciences, v. 22, p. 499–551.

Baker, V.R., 1997, Megafloods and glaciation, in Martini, I.P., ed., Late glacial and postglacial environmental changes; Quaternary, Carboniferous-Permian, and Proterozoic: New York, Oxford University Press, p. 98–108.

Beeson, R., 1991, Broken Hill-type lead-zinc deposits—An overview of their occurrence and geological setting: Institution of Mining and Metallurgy, Transactions, Section B: Applied Earth Science, v. 99, p. 163–175.

Betts, P.G., Giles, D., and Lister, G.S., 2003, Tectonic environment of shale-hosted massive sulfide Pb-Zn-Ag deposits of Proterozoic northeastern Australia: Economic Geology, v. 98, p. 557–576.

Bull, S.W., and Rawlings, D.J., 1999, A tectono-stratigraphic review of the northern Australian Proterozoic zinc belt: Centre for Ore Deposit and Exploration Studies (CODES), Special Publication, v. 2, p. 1–13.

Cecile, M.P., Morrow, D.W., and Williams, G.K., 1997, Early Paleozoic (Cambrian to Early Devonian) tectonic framework, Canadian Cordillera: Bulletin of Canadian Petroleum Geology, v. 45, p. 54–74.

Claypool, G.E., Holser, W.T., Kaplan, I.R., Sakai, H., and Zak, I., 1980, The age curves of sulfur and oxygen isotopes in marine sulfate and their mutual interpretation: Chemical Geology, v. 28, p. 3–4.

Conly, A.G., Goodfellow, W.D., Taylor, R.P., and Lydon, J.W., 2000, Geology, geochemistry, and sulphur isotope geochemistry of the hanging wall sulphide zones and their related hydrothermal alteration, Sullivan Zn-Pb-Ag deposit, in Lydon, J.W., Höy, T., Knapp, M., and Slack, J., eds., The geological environment of the Sullivan deposit, British Columbia: Geological Association of Canada, Mineral Deposits Division, Special Publication 1, p. 541–573.

Cooke, D.R., Bull, S.W., Large, R.R., and McGoldrick, P.J., 1998, Oxidised (sulfate-bearing) brines; their importance for the formation of Australian Proterozoic Pb-Zn (SEDEX) deposits [abs.]: Geoscience for the new Millennium, 14th Australian Geological Convention, Townsville, Queensland, Geological Society of Australia, Abstracts, p. 91.

Cooke, D.R., Bull, S.W., Large, R.R., and McGoldrick, P.J., 2000, The importance of oxidized brines for the formation of Australian Proterozoic stratiform sediment-hosted Pb-Zn (SEDEX) deposits: Economic Geology, v. 95, p. 1–18.

Emsbo, P., 2000, Gold in SEDEX deposits: Reviews in Economic Geology, v. 13, p. 427–437.

Fischer, A.G., 1984, Biological innovations and the sedimentary record, in Holland, H.D., and Trendall, A.F., eds., Patterns of change in Earth evolution: Berlin, Springer-Verlag, p. 145–157.

Goodfellow, W.D., 1984, Geochemistry of rocks hosting the Howards Pass (XY) strata-bound Zn-Pb deposit, Selwyn Basin, Yukon Territory, Canada: Proceeding of the Sixth Quadrennial IAGOD Symposium: Stuttgart, Germany, E. Schweizerbart'sche Verlagsbuchhandlung (Nägele u. Obermiller), p. 91–112.

—— 1987, Anoxic stratified oceans as a source of sulphur in sediment-hosted stratiform Zn-Pb deposits (Selwyn Basin, Yukon, Canada): Chemical Geology, v. 65, p. 359–382.

—— 2000, Anoxic conditions in the Aldridge Basin during formation of the Sullivan Zn-Pb deposit: Implications for the genesis of massive sulphides and distal hydrothermal sediments, in Lydon, J.W., Höy, T., Knapp, and M., Slack, J., eds., The geological environment of the Sullivan deposit, British Columbia: Geological Association of Canada, Mineral Deposits Division, Special Publication 1, p. 218–250.

—— 2004, Geology, genesis, and exploration of SEDEX deposits, with emphasis on the Selwyn Basin, Canada, in Deb, M., and Goodfellow, W.D., eds., Sediment-hosted lead-zinc sulphide deposits: Attributes and models of some major deposits of India, Australia, and Canada: Delhi, India, Narosa Publishing House, p. 24–99.

—— 2007, Base metal metallogeny of the Selwyn Basin, Canada, *in* Goodfellow, W.D., ed., Mineral deposits of Canada: A synthesis of major deposit types, district metallogeny, the evolution of geological provinces, and exploration methods: Geological Association of Canada, Mineral Deposits Division, Special Publication 5, p. 553–579.

Goodfellow, W.D., and McCutcheon, S.R., 2003, Geological and genetic attributes of volcanic-associated massive sulfide deposits of the Bathurst Mining Camp, northern New Brunswick—A synthesis: Economic Geology Monograph 11, p. 245–301.

Goodfellow, W.D., and Rhodes, D., 1990, Geological setting, geochemistry and origin of the Tom stratiform Zn-Pb-Ag-barite deposits, in Abbott, J.G., and Turner, R.J.W., eds., Mineral deposits of the northern Canadian Cordillera: Ottawa, International Association on the Genesis of Ore Deposits, Eighth Symposium, Field Trip 14 Guidebook, p. 177–244.

Goodfellow, W.D., Jonasson, I.R., and Morin, J.A., 1986, Environment of formation of the Howards Pass (XY) Zn-Pb deposit, Selwyn Basin, Yukon: Canadian Institute of Mining and Metallurgy, Special Volume, v. 37, p. 19–50.

Goodfellow, W.D., Lydon, J.W., and Turner, R.W., 1993, Geology and genesis of stratiform sediment-hosted (SEDEX) Zn-Pb-Ag sulphide deposits: Geological Association of Canada, Special Paper 40, p. 201–251.

Goodfellow, W.D., Cecile, M.P., and Leybourne, M.I., 1995, Geochemistry, petrogenesis and tectonic setting of lower Paleozoic alkalic and potassic volcanic rocks, northern Canadian Cordilleran Miogeocline: Canadian Journal of Earth Sciences, v. 32, p. 1236–1254.

Gwosdz, W., and Krebs, W., 1977, Manganese halo surrounding Meggen ore deposit, Germany: Institution of Mining and Metallurgy, Transactions, Section B: Applied Earth Science, v. 86, p. 73–77.

Hamilton, J.M., Bishop, D.T., Morris, H.C., and Owens, O.E., 1982, Geology of the Sullivan orebody, Kimberley, B.C., Canada: Geological Association of Canada, Special Paper 25, p. 597–665.

Hannak, W.W., 1981, Genesis of the Rammelsberg ore deposit near Goslar/Upper Hartz, Federal Republic of Germany, in Wolfe, K.H., ed., Handbook of strata-bound and stratiform ore deposits, 9: Amsterdam, Elsevier, p. 551–642.

Hauser, R.L., and Hutchinson, R.W., 1983, Tin in the Sullivan deposit, Kimberley, B.C.: Geological Association of Canada; Mineralogical Association of Canada, Canadian Geophysical Union; Joint Annual Meeting, Program with Abstracts, v. 8, p. A31.

Hitzman, M.W., and Beaty, D.W., 1997, The Irish Zn-Pb-(Ba) orefield: Society of Economic Geologists, Special Publication 4, p. 112–143.

Höy, T., Anderson, D., Turner, R.J.W., and Leitch, C.H.B., 2000, Tectonic, magmatic and metallogenic evolution of the early synrift phase of the Purcell basin, southeastern British Columbia, in Lydon, J.W., Höy, T., Knapp, M., and Slack, J., eds., The geological environment of the Sullivan deposit, British Columbia: Geological Association of Canada, Mineral Deposits Division, Special Publication 1, p. 32–60.

Hunt, B., 2006, Lead and zinc costs: Mines and projects, 2006 Edition: Surrey, England, Brook Hunt and Associates Ltd.

Jefferson, C.W., Kilby, D.B., Pigage, L.C., and Roberts, W.J., 1983, The Cirque barite-zinc-lead deposits, northeastern British Columbia, in Sangster, D.F., ed., Short course in sediment-hosted stratiform lead-zinc deposits: Mineralogical Association of Canada, Short Course Handbook, v. 8, p. 121–139.

Jenkins, C.L., and Lydon, J.W., 2002, A mineral deposit data base structure and a data base of VMS and SEDEX deposits: Geological Survey of Canada, Open File 4165, (CD-ROM).

Jennings, D.S., and Jilson, G.A., 1986, Geology and sulphide deposits of the Anvil Range, Yukon: Canadian Institute Mining and Metallurgy, Special Volume 37, p. 339–361.

Kelley, K.D., and Jennings, S., 2004, Barite and Zn-Pb-Ag deposits in the Red Dog District, western Brooks Range, Northern Alaska; preface: Economic Geology, v. 99, p. 1267–1280.

Kelley, K.D., Dumoulin, J.A., and Jennings, S., 2004, The Anarraaq Zn-Pb-Ag and barite deposit, Northern Alaska; Evidence for replacement of carbonate by barite and sulfides: Economic Geology, v. 99, p. 1577–1591.

Kirkham, R.V., and Rafer, A.B., 2003, Selected world mineral deposits database. Geological Survey of Canada, Open File 1801 (CD-ROM).

Large, D.E., 1980, Geological parameters associated with sediment-hosted, submarine exhalative Pb-Zn deposits: An empirical model for mineral exploration: Geologisches Jahrbuch, v. D40, p. 59–129.

—— 1981a, Geochemistry of sedimentary rocks in the vicinity of the Tom Pb-Zn-Ag deposit, Yukon Territory, Canada: 8th International Geochemical Exploration Symposium, Hanover, April 10-15, 1980, Transactions, p. 203–217.

—— 1981b, Sediment-hosted submarine exhalative lead-zinc deposits—A review of their geological characteristics and genesis, in Wolf, K.H., ed., Handbook of strata-bound and stratiform ore deposits, 9: New York, Elsevier, p. 469–507.

—— 1983, Sediment-hosted massive sulphide lead-zinc deposits: An em pirical model, in Sangster, D.F., ed., Sediment-hosted stratiform lead-zinc deposits: Mineralogical Association of Canada, Short Course Handbook 8, p. 1–29.

—— 1986, The paleotectonic setting of Rammelsberg and Meggen, Germany—A basin analysis study, in Turner, R.J.W., and Einaudi, M.T., eds., The genesis of stratiform sediment-hosted lead and zinc deposits: Stanford University Publications, Geological Sciences, Conference proceedings, 20, p. 109–112.

Large, R.R., and McGoldrick, P.J., 1998, Lithogeochemical halos and geochemical vectors to stratiform sediment hosted Zn-Pb-Ag deposits; 1, Lady Loretta deposit, Queensland: Journal of Geochemical Exploration, v. 63, p. 37–56.

Large, R.R., Bull, S.W., Cooke, D.R., McGoldrick, P.J., and Williams, P.J., 1998, A genetic model for the H.Y.C. Deposit, Australia; Based on regional sedimentology, geochemistry, and sulfide-sediment relationships: Economic Geology, v. 93, p. 1345–1368.

Large, R.R., Bull, S.W., and McGoldrick, P.J., 2000, Lithogeochemical halos and geochemical vectors to stratiform sediment hosted Zn-Pb-Ag deposits; Part 2, HYC Deposit, McArthur River, Northern Territory: Journal of Geochemical Exploration, v. 68, p. 105–126.

Leach, D.L., Bradley, D., Lewchuk, M.T., Symons, D.T.A., de, M.G., and Brannon, J., 2001, Mississippi Valley-type lead-zinc deposits through geological time; Implications from recent age-dating research: Mineralium Deposita, v 36, p. 711 740.

Leach, D.L., Sangster, D.L., Kelly, K.D., Large, R.R., Garven, G., Allen, C.R., Gutzmer, J., and Walter, S., 2005, Sediment-hosted lead-zinc deposits: A global perspective: Society of Economic Geologists, One Hundredth Anniversary Volume 1905–2005, p. 561–607.

Lydon, J.W., 1983, Chemical parameters controlling the origin and deposition of sediment-hosted stratiform lead-zinc deposits, in Sangster, D.F., ed., Sediment-hosted stratiform lead-zinc deposits: Mineralogical Association of Canada, Short Course Handbook, v. 8, p. 175–250.

Lydon, J.W., Paakki, J.J., Anderson, H.E., and Reardon, N.C., 2000a, An overview of the geology and geochemistry of the Sullivan Deposit, in Lydon, J.W., Höy, T., Knapp, M., and Slack, J., eds., The geological environment of the Sullivan deposit, British Columbia: Geological Association of Canada, Mineral Deposits Division, Special Publication 1, p. 505–522.

Lydon, J.W., Walker, R., and Anderson, E.H., 2000b, Lithogeochemistry of the Aldridge formation and the chemical effects of burial diagenesis, in Lydon, J.W., Höy, T., Knapp, M., and Slack, J., eds., The geological environment of the Sullivan deposit, British Columbia: Geological Association of Canada, Mineral Deposits Division, Special Publication 1, p. 137–179.

Lydon, J.W., 2007, An overview of the economic and geological contexts of Canada's major mineral deposit types, in Goodfellow, W. D., ed., Mineral Deposits of Canada: A Synthesis of Major Deposit-types, District Metallogeny, the Evolution of Geological Provinces, and Exploration Methods, Special Publication 5, Mineral Deposits Division, Geological Association of Canada, p. 3–48.

McKibben, M.A., Eldridge, C.S., and Williams, A.E., 1988, Sulfur and base metal transport in the Salton Sea geothermal system, in Tucker, R., ed., New horizons, 12. Davis, CA, United States, Geothermal Resources Council, Transactions, p. 121–125.

Mosier, D.L., Singer, D.A., and Salem, B.B., 1983, Geologic and grade-tonnage information on volcanic-hosted copper-zinc-lead massive sulphide deposits: USGS Open File Report, 83-89, 77p.

Moore, D.W., Young, L.E., Modene, J.S., and Plahuta, J.T., 1986, Geological setting and genesis of the Red Dog zinc-lead-silver deposit, Western Brooks Range, Alaska: Economic Geology, v. 81, p. 1696–1727.

Nelson, J., Paradis, S., Christensen, J., and Gabites, J., 2002, Canadian Cordilleran mississippi valley-type deposits; A case for Devonian-Mississippi back-arc hydrothermal origin: Economic Geology, v. 97, p. 1013–1036.

Paradis, S., Hannigan, P., and Dewing, K., 2007, Mississippi Valley-type lead-zinc deposits, *in* Goodfellow, W.D., ed., Mineral deposits of Canada: A synthesis of major deposit types, district metallogeny, the evolution of geological provinces, and exploration methods: Geological Association of Canada, Mineral Deposits Division, Special Publication 5, p. 185-203.

Parr, J.M., and Plimer, I.R., 1993, Models for Broken Hill-type lead-zinc-silver deposits: Geological Association of Canada, Special Paper 40, p. 253–288.

Peter, J.M., Layton-Matthews, D., Piercey, S., Bradshaw, G., Paradis, S., and Boulton, A., 2007, Volcanic-hosted massive sulphide deposits of the Finlayson Lake District, Yukon in Goodfellow, W. D., ed., Mineral Deposits of Canada: A Synthesis of Major Deposit-types, District Metallogeny, the Evolution of Geological Provinces, and Exploration Methods, Special Publication 5, Mineral Deposits Division, Geological Association of Canada, p. 471–508.

Piercey, S.J., Paradis, S., Murphy, D.C., and Mortensen, J.K., 2001, Geochemistry and paleotectonic setting of felsic volcanic rocks in the Finlayson Lake volcanic-hosted massive sulfide district, Yukon, Canada: Economic Geology, v. 96, p. 1877–1905.

Price, R.A., and Sears, J.W., 2000, A preliminary palinspastic map of the Mesoproterozoic Belt-Purcell Supergroup, Canada and USA; Implications for the tectonic setting and structural evolution of the Purcell anticlinorium and the Sullivan deposit, in Lydon, J.W., Höy, T., Knapp, M., and Slack, J., eds., The geological environment of the Sullivan deposit, British Columbia: Geological Association of Canada, Mineral Deposits Division, Special Publication 1, p. 61–81.

Sangster, D.F., 1990, Mississippi Valley-type and SEDEX lead-zinc deposits: A comparative examination: Institution of Mining and Metallurgy, Transactions, Section B: Applied Earth Science, v. 99, p. 21–42.

Sangster, D.F., and Hillary, E.M., 2000, SEDEX lead-zinc deposits—Proposed subtypes and their characteristics: Exploration and Mining Geology, v. 7, p. 341–357.

Singer, D.A., 1995, World class base and precious metal deposits—A quantitative analysis: Economic Geology, v. 90, p. 88–104.

Slack, J.F., Kelley, K.D., Anderson, V.M., Clark, J.L., and Ayuso, R.A., 2004, Multistage hydrothermal silicification and Fe-Tl-As-Sb-Ge-REE enrichment in the Red Dog Zn-Pb-Ag district, Northern Alaska: Geochemistry, origin, and exploration applications: Economic Geology, v. 99, p. 1481–1508.

Turner, R.J.W., 1990, Jason stratiform Zn-Pb-barite deposit, Selwyn Basin, Canada (NTS 105-O-1): Geological setting, hydrothermal facies and genesis, in Abbott, J.G., and Turner, R.J.W., eds., Mineral deposits of the northern Canadian Cordillera: International Association on the Genesis of Ore Deposits, Field Trip 14: Guidebook, p. 137–175.

Turner, R.J.W., Leitch, C.H.B., and Ross, K.V., 2000, District-scale alteration associated with the Sullivan deposit, British Columbia, Canada, in Lydon, J.W., Höy, T., Knapp, M., and Slack, J., eds., The geological environment of the Sullivan deposit, British Columbia: Geological Association of Canada, Mineral Deposits Division, Special Publication 1, p. 408–439.

Vail, P.R., Mitchum, R.M., Jr., and Thompson, S., III, 1977, Seismic stratigraphy and global changes of sea level; Part 4: Global cycles of relative changes of sea level: American Association of Petroleum Geologists, Memoir 26, p. 83–97.

Walters, S.G., 1998, Broken hill-type deposits: AGSO Journal of Australian Geology and Geophysics, v. 17, p. 229–237.

Wheeler, J.O., Hoffman, P.F., Card, K.D., Davidson, A., Sanford, B.V., Okulitch, A.V., and Roest, W.R., 1996, Geological map of Canada, Map 1860A, Geological Survey of Canada, CD-ROM.

Yang, J., Large, R.R., and Bull, W., 2004, Factors controlling free thermal convection in faults in sedimentary basins: Implications for the formation of zinc-lead mineral deposits: Geofluids, v. 4, p. 237–247.

Young, L.E., 2004, A geologic framework for mineralization in the western Brooks Range, Alaska: Economic Geology, v. 99, p. 1281–1306.

Zierenberg, R.A., 1990, Deposition of metalliferous sediment beneath a brine pool in the Atlantis II deep, Red Sea, in McMurray, G.R., ed., Gorda Ridge: A seafloor spreading center in the United States exclusive economic zone: New York, Springer-Verlag, p. 131–142.

Mississippi Valley-Type Lead-Zinc Deposits

Suzanne Paradis[1], Peter Hannigan[2], and Keith Dewing[2]

1. Geological Survey of Canada Pacific, 9860 West Saanich Road, Sidney, British Columbia V8L 4B2
2. Geological Survey of Canada Calgary, 3303 - 33 Street North West, Calgary, Alberta T2L 2A7
Corresponding author's email: Suzanne.Paradis@nrcan.gc.ca

Abstract

Mississippi Valley-type (MVT) deposits are epigenetic, stratabound, carbonate-hosted sulphide bodies composed predominantly of sphalerite, galena, iron oxides (pyrite, marcasite), and carbonates (calcite, dolomite). MVT deposits are important Zn and Pb reserves and resources in the world. Silver, barite, gypsum, and fluorite may also be economically recovered from these deposits. Major world MVT deposits are found in Canada (Pine Point, Polaris, Nanisivik, Gays River, and Daniel's Harbour), mid-United States districts (Upper Mississippi Valley, Missouri, Tri-State, and Tennessee), Australia (Lennard Shelf and Coxco), and Europe (Silesia, Alpine, Reocin and Cévennes).

The mineralization occurs as open-space fillings of breccias and fractures, and/or as replacement of the host dolostone. Less commonly, sulphide and gangue minerals occupy primary carbonate porosity.

Most MVT deposits are found in carbonate platforms adjacent to cratonic sedimentary basins; they occur in limestone less frequently. They are also restricted to rocks younger than two billion years and formed during short time intervals, primarily within the Phanerozoic. MVT mineral districts are the product of regional- or subcontinental-scale fluid migration. Deposits are formed by the migration of warm saline aqueous solutions, similar to oilfield brines, through aquifers within platform-carbonate sequences toward the basin periphery. One of the most popular models relates ore-fluid migration to compressive tectonic regimes associated with continental accretion. This model is not universally applicable, however, as some of the MVT deposits most likely formed under an extensional tectonic regime.

Résumé

Les gisements du type Mississippi Valley consistent en des corps sulfurés stratoïdes épigénétiques encaissés dans des carbonates et se composent principalement de sphalérite, de galène, d'oxydes de fer (pyrite et marcassite) et de carbonates (calcite et dolomite). Ils constituent d'importantes réserves et ressources mondiales de Zn et Pb, et il est parfois possible d'extraire de manière rentable de l'argent, de la baryte, du gypse et de la fluorine de ces gisements. Les plus importants gisements de ce genre reposent au Canada (Pine Point, Polaris, Nanisivik, Gays River et Daniel's Harbour), dans les districts du milieu des É.-U. (Upper Mississippi Valley, Missouri, Tri-State, et Tennessee), en Australie (plate-forme de Lennard et Coxco) et en Europe (Silésie, Alpes, Reocin, et Cévennes).

Les minéralisations se présentent comme des remplissages dans des brèches et des fractures ou comme un remplacement de la dolomie encaissante. Moins fréquemment, les minéraux sulfurés et de gangue occupent les pores primaires des carbonates.

La plupart des gisements du type Mississippi Valley reposent dans des plates-formes carbonatées contiguës à des bassins sédimentaires cratoniques, mais parfois, ils sont logés dans du calcaire. Par ailleurs, ils sont encaissés dans des roches de moins de 2 Ga qui se sont formées rapidement, principalement pendant le Phanérozoïque. Les districts minéraux du type Mississippi Valley résultent d'une migration de fluides régionale ou sous continentale. Les gisements sont issus de solutions aqueuses salines chaudes qui sont similaires à des saumures de champ de pétrole qui ont circulé par des aquifères, dans des séquences de roches carbonatées néritiques, jusqu'aux environs d'un bassin. L'un des modèles les plus populaires établit un lien entre la migration du minerai et les régimes tectoniques compressifs rattachés à une accrétion continentale. Ce modèle n'est pas toujours applicable, car certains gisements du type Mississippi Valley se sont fort probablement formés dans un régime tectonique de distension.

Definition

Mississippi Valley-type (MVT) deposits are epigenetic, stratabound, carbonate-hosted bodies composed predominantly of sphalerite, galena, iron sulphides, and carbonates. The deposits account for approximately 27 percent of the world's current lead and zinc resources[1] (Tikkanen, 1986). They are so-named because several classic MVT deposits are located in carbonate rocks within the drainage basin of the Mississippi River in the central United States (US). Important Canadian deposits include Pine Point, Polaris, Nanisivik, Daniel's Harbour, Gays River, Monarch-Kicking Horse, and Robb Lake.

The deposits occur mainly in dolostone as open-space fillings, collapse breccias, and/or as replacement of the carbonate host rock. Less commonly, sulphide and gangue minerals occupy primary carbonate porosity. The deposits are epigenetic, having been emplaced after lithification of the host rocks.

MVT deposits originate from saline basinal metalliferous fluids at temperatures in the range of 75 to 200°C (Leach and Sangster, 1993). They are located in carbonate platform settings, typically in relatively undeformed orogenic foreland rocks, commonly in foreland thrust belts, and rarely in rift zones (Leach and Sangster, 1993).

Individual deposits are generally less than 2 million tonnes, are zinc-dominant, and possess grades that rarely exceed 10% (Pb+Zn). The deposits do, however, characteristically occur in clusters, referred to as "districts". A metallogenic district is defined as a continuous area that contains the expressions of the geological environment and tectonic events that controlled the formation of MVT deposits. For example, the Cornwallis district in Nunavut hosts one deposit[2], the Polaris mine, and approximately 80 showings[3] (Dewing et al., 2007). Another example is the Pine Point district in the Northwest Territories, which hosts the Pine Point

Paradis, S., Hannigan, P., and Dewing, K., 2007, Mississippi Valley-type lead-zinc deposits, *in* Goodfellow, W.D., ed., Mineral Deposits of Canada: A Synthesis of Major Deposit-Types, District Metallogeny, the Evolution of Geological Provinces, and Exploration Methods: Geological Association of Canada, Mineral Deposits Division, Special Publication No. 5, p. 185-203.

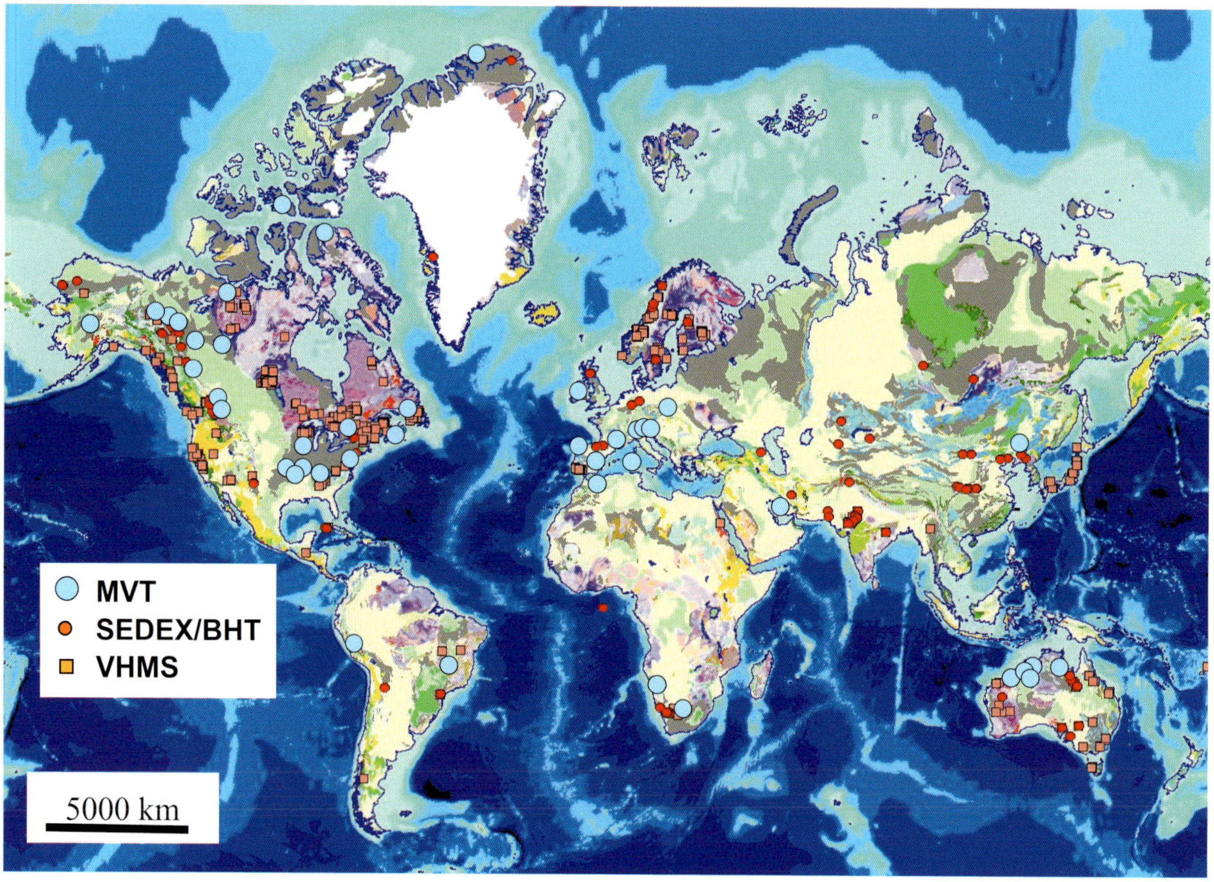

FIGURE 1. Distribution of Mississippi Valley-type deposits and districts worldwide. BHT = Broken Hill-type, MVT = Mississippi Valley-type, SEDEX = sedimentary exhalative, and VHMS = volcanic-hosted massive sulphides.

deposit containing approximately 100 orebodies[4] (Hannigan, 2007). Other districts may have a half-dozen to more than 300 orebodies, which can contain up to several hundred million tonnes of ore scattered over hundreds to thousands of square kilometres (Sangster, 2002).

Mineral Deposit Subtypes

MVT deposits are part of a larger family of carbonate-hosted deposits, all of which are epigenetic and contain zinc. Sub-types of MVT deposits include the carbonate-hosted F (±Ba, ±Zn, ±Pb) deposits (Dunham, 1983; Rowan et al., 1996) and some "Irish-type" deposits (Hitzman and Beaty, 1996). The "Irish-type" deposits are stratabound, structurally controlled, carbonate-hosted, Pb-Zn deposits that have sedimentary exhalative (SEDEX) and/or MVT characteristics. Those that show epigenetic features are considered subtypes of MVT deposits.

Associated Mineral Deposit Types

MVT-associated deposits include a variety of deposits that belong to the spectrum of sediment-hosted base-metal deposits. These include SEDEX deposits, Broken Hill-type Pb-Zn deposits, sandstone-hosted Pb deposits, carbonate-hosted Cu-Pb-Zn deposits (or Kipushi type), fracture-controlled carbonate-hosted F (±Ba) deposits, carbonate-hosted manto-type Ag-Pb-Zn deposits, high-temperature carbonate replacement Pb-Zn (±Fe, ±Cu, ±Ag, ±Au) deposits (Megaw et al., 1996; Smith, 1996; Titley, 1996), and the diapir-related Zn-Pb (±Ag, ±Cd, ±Cu, ±Hg) deposits (Sheppard et al., 1996).

Economic Characteristics

Summary of Economic Characteristics

MVT deposits account for approximately 27% of the world's lead and zinc resources, and they are dispersed throughout the world (Fig. 1). The proportion of the world's primary production of Zn and Pb from MVT deposits is 25% and 26%, respectively (Tikkanen, 1986). A large proportion of Pb and Zn production comes from several classic MVT districts located in the drainage basin of the Mississippi River in the central US and in the East Tennessee district in the eastern United States. MVT deposits also occur in Canada, Australia, Europe (Belgium, Poland, France, Ireland, Spain, Austria, Italy, Bulgaria, Macedonia, and

[1] Resources: Maximum estimate of geological resources from a systematic evaluation of a deposit that has been mined (i.e. production + proven reserves + probable reserves) or that hasn't been mined (i.e. measured resources + indicated resources).
[2] Deposit: A mineral deposit is any natural, but locally restricted, concentration of minerals in the Earth's crust. It carries no necessary profitability implications.
[3] Showing: Occurrences hosting minor in situ mineralization.
[4] Orebody: Concentration of metallic and non-metallic minerals of sufficient ore content to make extraction economically feasible.

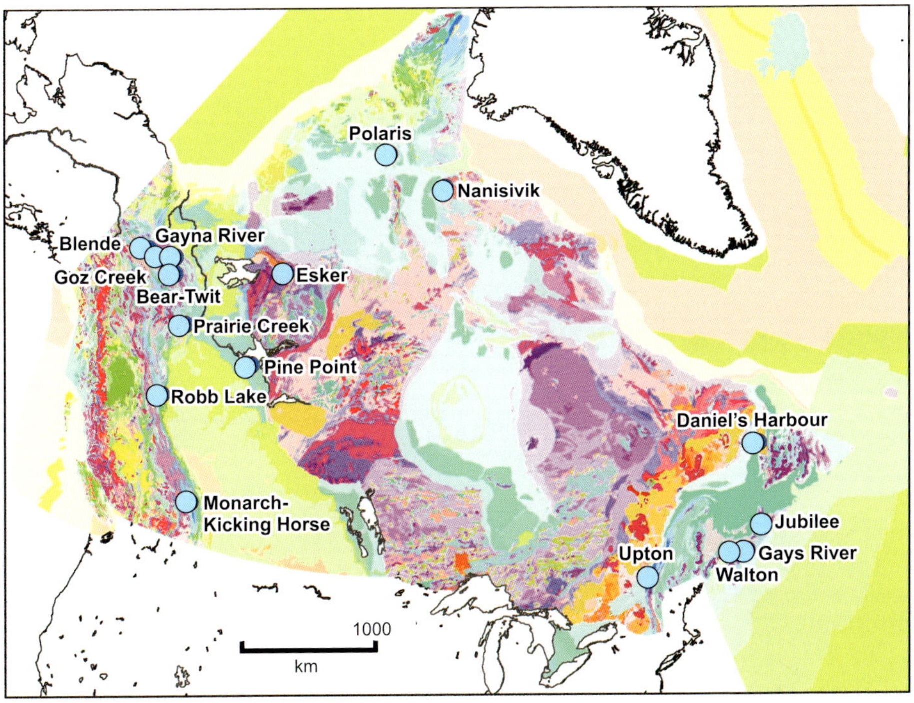

FIGURE 2. Distribution of Mississippi Valley-type deposits plotted on a simplified geological map of Canada (Map D1860A). Major attributes of these deposits are listed in Appendix 1.

Serbia), Africa (Egypt, Tunisia, Algeria, Morocco, and South Africa), Asia (Russia, China, India, Iran, Saudi Arabia, and Turkey), and South America (Brazil and Peru). In Canada, there are 16 major deposits (Fig. 2) that in themselves form districts. Each district contains one to more than 100 sulphide bodies. The Pine Point district, for example, contains 100 orebodies distributed over 1,600 square kilometres.

MVT deposits contributed only negligibly to Canadian Zn and Pb production prior to the opening of the Pine Point mine in 1964. Between 1964 and 2002, about 30% of the annual lead and zinc production in Canada was derived from MVT deposits (Sangster, 2002). With the closure of Polaris and Nanisivik in 2002, however, the proportion of Zn and Pb derived from MVT deposits in Canada dropped to nil.

Grade and Tonnage Characteristics

The size, grade, and metal ratio parameters of individual MVT deposits are difficult to compare. As mentioned by Sangster (1990, 1995) and Leach and Sangster (1993), several deposits were mined before accurate data were recorded. Furthermore, the production and reserve data are usually published as district totals rather than for each individual deposit or orebody. An exception to this is Pine Point, where grade and tonnage are recorded for individual orebodies that range in size from less than 100,000 tonnes to 17.5 million tonnes with average grades of 2% Pb and 6.2% Zn (Hannigan, 2007). Orebodies average 1.32 Mt of Zn-Pb ore grading near 7% Zn and 3% Pb. The total geological resource (produced resource and remaining proven reserves) for the Pine Point district is estimated to be around 80 Mt (Appendix 1).

There are over 80 MVT deposits (with grade and tonnage data) worldwide (Fig. 1), 16 of them in Canada (Fig. 2; Appendix 1). Seven out of 16 Canadian MVT deposits have been mined for a total of 112.5 Mt of ore. Pine Point was the largest deposit with a production of approximately 64,300,000 tonnes of 6.95% Zn and 3.0% Pb (Appendix 1). There are currently no MVT deposits in production in Canada.

Geological resource estimates for Canadian deposits range approximately from 800,000 to 79.3 million tonnes (average 16 Mt) with 0.4 to 17.6% combined Pb and Zn (average 7.69%; Fig. 3). Geological resources tabulated in Appendix 1 for Canadian deposits that have not been in production are generally not in compliance with the National Instrument 43-101 standards. Most non-Canadian individual deposits yield less than 10 million tonnes of Pb+Zn metals, with combined Pb+Zn grades seldom exceeding 15%. These figures are similar to Canadian deposits (Fig. 3). The size (in tonnes of Pb+Zn metals) of non-Canadian MVT districts (e.g. Tri-States, United States; Silesia, Poland) is approximately an order of magnitude greater than the size of individual deposits with combined Pb+Zn grades of between 2 and 5% (Fig. 3).

The metal ratios in Canadian deposits, expressed as Zn/(Zn+Pb) values (Fig. 4), show a bimodal distribution with the majority of deposits having values between 0.6 and 0.8 (median = 0.74), and a smaller group between 0.9 and 1.0 (median = 0.97). The majority of Canadian deposits are Zn-rich relative to Pb, with Zn grades between 4 and 6 wt.% (median = 5.42) and Pb grades between 0 and 2 wt.% (median = 1.4) (Fig. 5). Daniel's Harbour and many deposits in the East Tennessee district (US) are essentially free of lead and have Zn/(Zn+Pb) ratios close to 1.0. Silver and copper content in MVT deposits is low and often not reported. When reported, silver grades vary from 10 to 161 g/t Ag. The

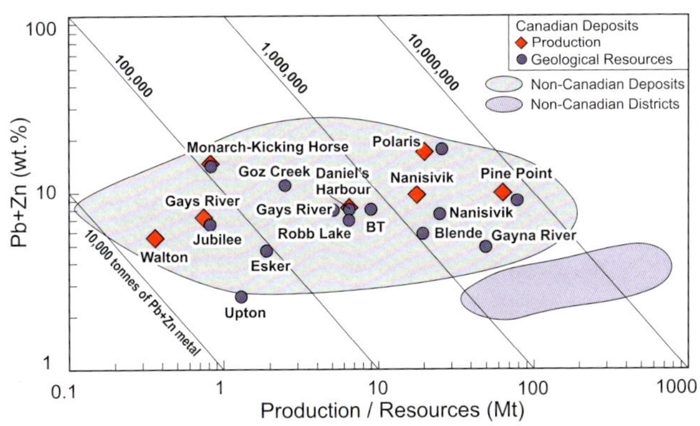

FIGURE 3. Grade-tonnage for Canadian Mississippi Valley-type deposits with geological and production resources. Diagonal lines represent total tonnage of contained lead and zinc metal. The fields for non-Canadian deposits and districts are approximate and include geological resource and grade data from the World's Minerals Project. BT = Bear-Twit.

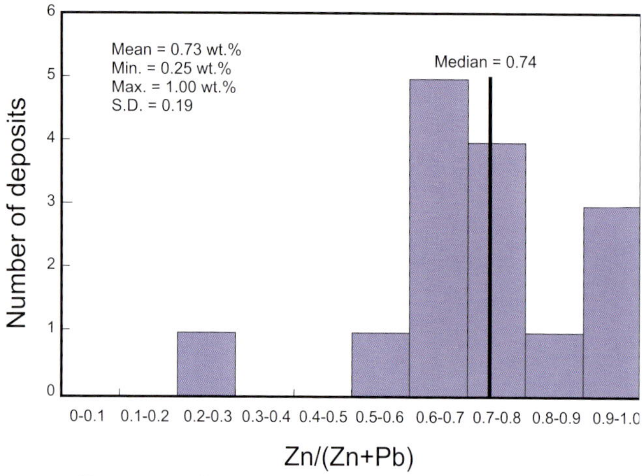

FIGURE 4. Zn/(Zn+Pb) ratios in Canadian MVT deposits.

Walton deposit, with an average value of 350 g/t Ag, has the highest silver grades of any MVT deposit in Canada (Patterson, 1988a,b, 1989). It is also the only Canadian deposit where production reported Cu grade values (Appendix 1).

Geological Attributes of MVT Deposits

Nature of Sulphide Bodies

MVT deposits occur in clusters of a few to hundreds of individual sulphide bodies that vary in character and shape and are often interconnected (Fig. 6). Deposits range from massive replacement zones to open-space fillings of breccias and fractures, to disseminated clusters of crystals that occupy intergranular pore spaces (Leach and Sangster, 1993). Sulphide-hosting structures are most commonly zones of highly brecciated dolomite; in some instances (e.g. Pine Point, Robb Lake, Daniel's Harbour, Nanisivik) these zones are arranged in linear patterns that suggest a tectonic control (Figs. 6, 7, 8). These breccia zones may range from more or less concordant tabular structures, controlled by individual strata, to discordant cylindrical structures within tens of metres of sedimentary sequences. At Pine Point, the orebodies are either tabular or prismatic structures in interconnected paleokarst networks (Fig. 8). The karst networks are directly related to distinct facies and lithofacies features in the Presqu'ile barrier complex. The MVT sulphide bodies are, therefore, discordant on the deposit scale, but stratabound on a district scale.

Dimensions

The dimensions of sulphide bodies can be difficult to measure because of their irregular and variable shape. At Pine Point, the L-36 orebody has dimensions of 1450 m in length, 50 to 400 m in width, and 2.5 to 10 m in thickness and the X-15 orebody is 800 m in length, 400 m in width, and 20 to 30 m in thickness (Hannigan, 2007). At Robb Lake, several bodies extend for more than 300 m along bedding and crosscut more than 50 m of stratigraphic section; others are thin and narrow bodies and pods parallel to bedding (Paradis et al., 1999). At Polaris, the main orebody had dimensions of 800 m in length, 300 m in width, and 150 m in thickness (Dewing et al., 2007).

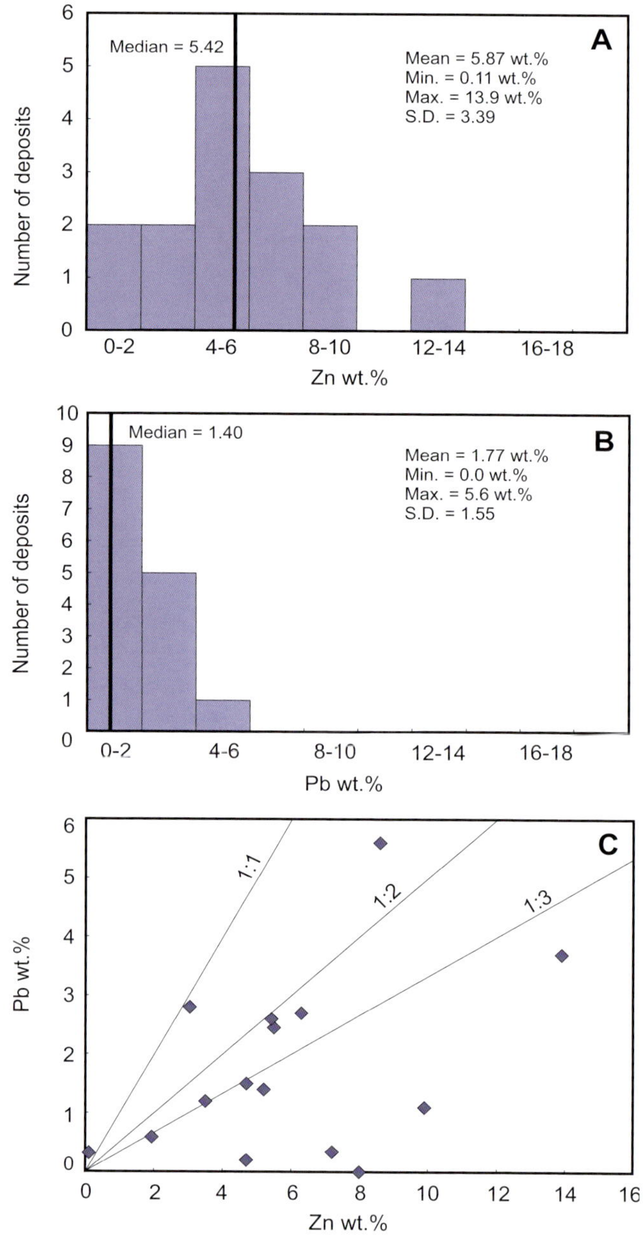

FIGURE 5. **(A)** and **(B)** Histograms of the lead and zinc grades of Canadian deposits. **(C)** Pb wt.% versus Zn wt.% for Canadian MVT deposits.

Host Rocks

The deposits are hosted in carbonate rocks, usually dolostone and less frequently limestone. The dolostone consists of medium- to coarse-grained white sparry dolomite that has replaced a fine-grained dolostone host, which itself has replaced a limestone host. The Pine Point orebodies, for example, are enclosed in large, discordant zones of secondary coarse-grained vuggy dolostone with white saddle dolomite and calcite gangue. The deposit host rocks are mostly fine-grained crystalline dolostone and local limestone (Rhodes et al., 1984). Gays River is an example of a deposit in early diagenetic dolostone without secondary sparry or saddle dolomite (Ravenhurst et al., 1989). In the East Tennessee (US) and Alpine (Europe) districts, the secondary dolomite is only locally developed (Sangster, 1990), whereas at the Jubilee deposit in Nova Scotia, the deposit is

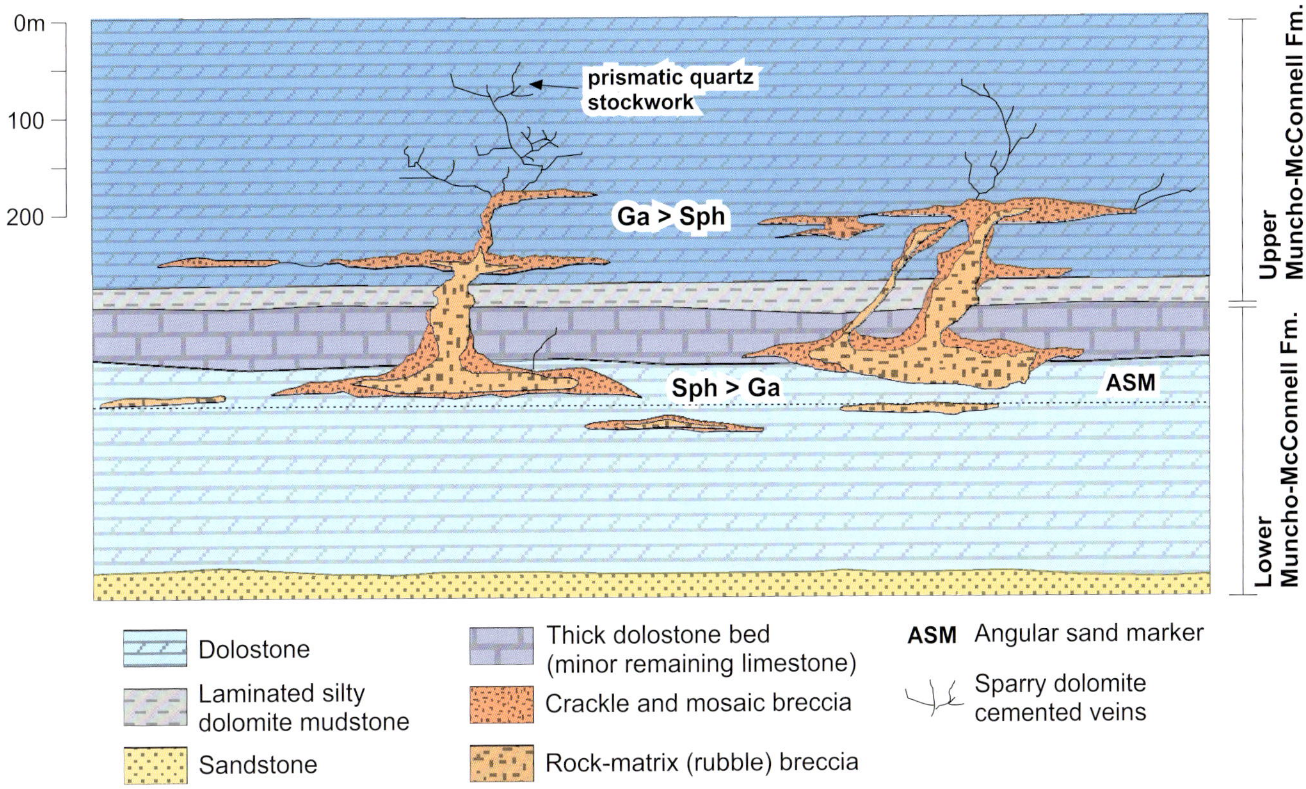

FIGURE 6. Schematic representation of the Robb Lake breccia-hosted Zn-Pb sulphide bodies, showing textural and mineralogical zoning and stratigraphic controls (from Nelson et al., 1999; from Paradis et al., 1999; Paradis and Nelson, 2007).

exclusively hosted in limestone (Paradis et al., 1993; Fallara et al., 1998).

Mineralogy

MVT deposits have simple mineral assemblages that consist of sphalerite, galena, pyrite, marcasite, dolomite, calcite, quartz, and occasionally barite, fluorite, celestite, gypsum, anhydrite, native sulphur, and pyrrhotite. Chalcopyrite, bornite, and other copper minerals are normally not constituents of MVT deposits and are only abundant in some districts,

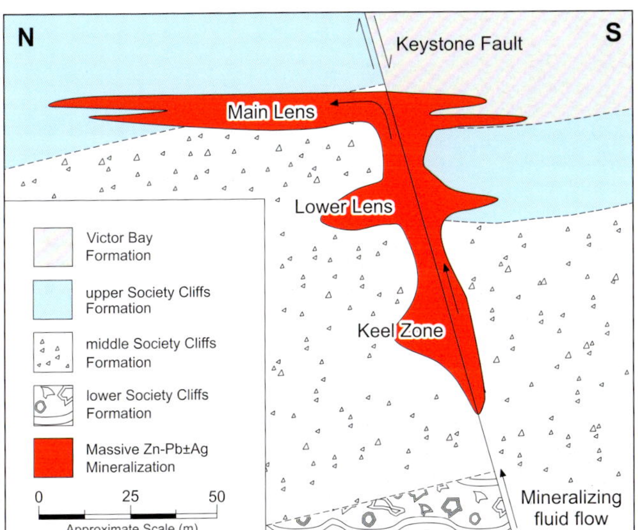

FIGURE 7. Schematic cross-section of the Nanisivik main lens and keel zone (modified from Patterson and Powis, 2002).

FIGURE 8. View eastward of the mined-out karst trend at Pine Point (courtesy of P. Hannigan).

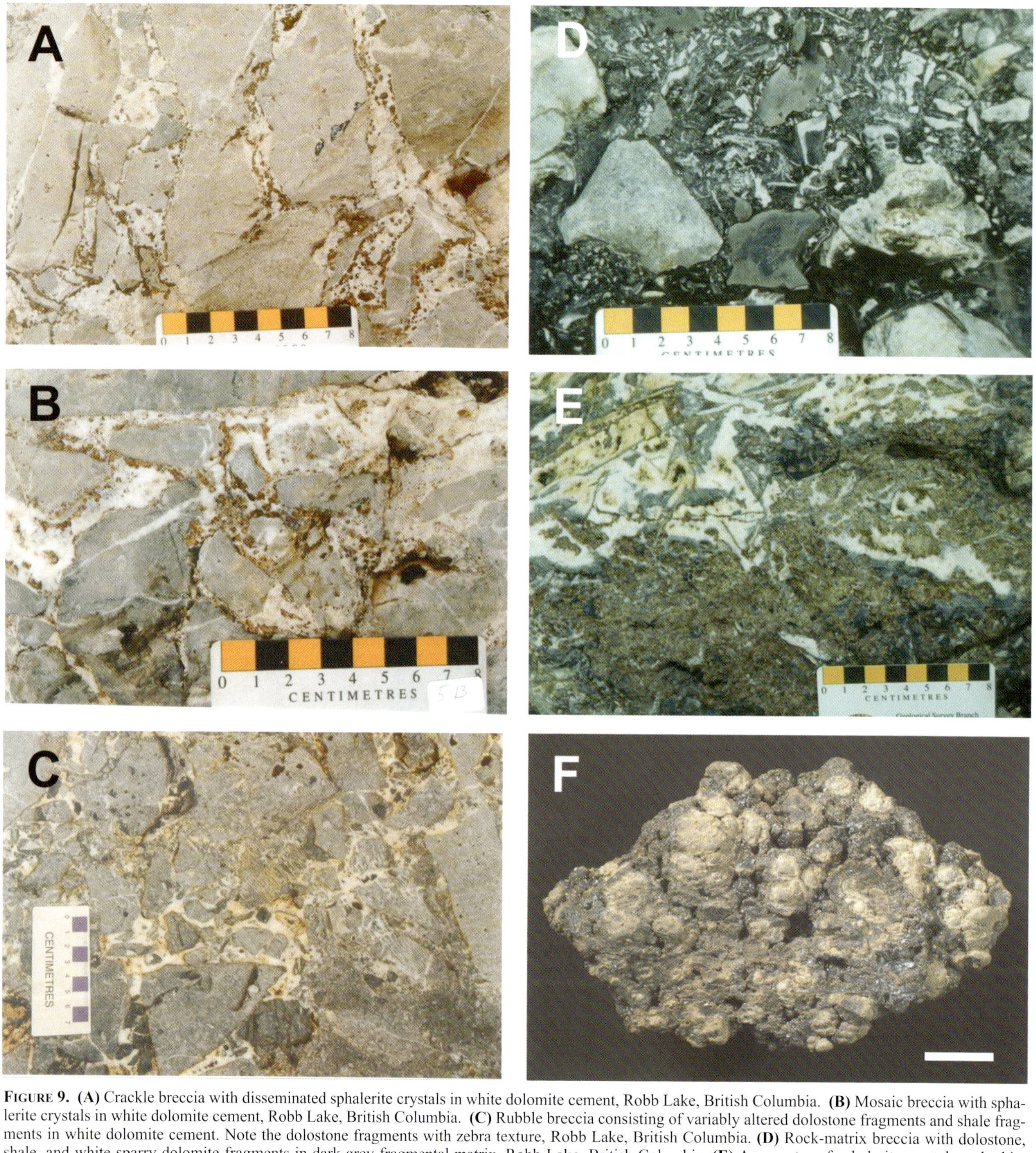

FIGURE 9. (A) Crackle breccia with disseminated sphalerite crystals in white dolomite cement, Robb Lake, British Columbia. (B) Mosaic breccia with sphalerite crystals in white dolomite cement, Robb Lake, British Columbia. (C) Rubble breccia consisting of variably altered dolostone fragments and shale fragments in white dolomite cement. Note the dolostone fragments with zebra texture, Robb Lake, British Columbia. (D) Rock-matrix breccia with dolostone, shale, and white sparry dolomite fragments in dark grey fragmental matrix, Robb Lake, British Columbia. (E) Aggregates of sphalerite crystals and white sparry dolomite along small fractures, Pine Point, Northwest Territories. (F) Aggregates of colloform sphalerite and skeletal galena completely replacing the carbonate, Polaris, Nunavut; scale bar is 1 cm.

such as the Cornwallis district in Canada (Randell and Anderson, 1996; Dewing et al., 2007) and the Viburnum Trend of the Southeast Missouri district in the US (Sangster, 1990; Leach et al., 1995; Foley, 2002). Deposits of the Viburnum Trend have a unique and complex mineralogy that is not typical of most MVT deposits, and includes siegenite, bornite, tennantite, bravoite, digenite, covellite, arsenopyrite, fletcherite, adularia, pyrrhotite, magnetite, millerite, polydymite, vaesite, djurleite, chalcocite, anilite, and enargite (Leach et al., 1995). The carbonate-hosted, Cu-rich showings of the Cornwallis district are associated with Zn-Pb mineralization and are structurally and stratigraphically controlled. They consist of chalcocite and bornite, and lesser chalcopyrite, cuprite, covellite, and native copper as fracture-, vein-, and breccia-filling of the Silurian dolostone (Dewing et al., 2007).

The abundance of iron sulphides relative to other sulphide minerals in MVT deposits ranges from dominant to nil (Leach et al., 1995; St. Marie et al., 2001). For example, at Nanisivik, iron sulphides are abundant, whereas in some Appalachian deposits (e.g. Daniel's Harbour), only trace to minor amounts of iron sulphides are present. Deposits that contain significant iron sulphides can be detected by induced polarization (IP) surveys and ground electromagnetic methods (EM), whereas those that contain only sphalerite and minor galena are generally poor conductors and have variable resistivity (Summer, 1976). At Pine Point, the sphalerite is non-polarizable, however the high concentrations of conductive minerals in the orebodies (i.e. galena, pyrite, and marcasite) made the IP method very successful for locating orebodies (Lajoie and Klein, 1979). Sphalerite and gangue minerals have widely varying densities, so gravity surveys could prove useful in exploring for sulphide bodies containing mainly sphalerite. In the Cornwallis district, the Polaris deposit was drilled based on a gravity anomaly (Dewing et al., 2007).

Organic material, such as hydrocarbon, is common in some MVT deposits (e.g. Polaris, Pine Point, Robb Lake, Jubilee, Walton) but is not present in significant amounts in others (e.g. Monarch-Kicking Horse, Gays River). At Pine Point and Polaris, pyrobitumen is a common, though volumetrically small, component of the deposits that occurs in vugs as crusts or blobs (Dewing et al., 2007; Hannigan, 2007). Other evidence for hydrocarbon generation at Polaris is found in abundant bitumen, hydrocarbon fluid inclusions, thermal maturity values indicating passage into the oil window, and petrographic evidence for localized, direct transformation of alginite into bitumen (Randell, 1994; L.V. Stasiuk, pers. comm., 1992).

Textures

Sulphide textures are mostly related to open-space filling of breccias, fractures, and vugs. Replacement of carbonate host rocks and internal sediments, and sulphide disseminations are also observed (e.g. Polaris, Pine Point, Robb Lake deposits). Internal sediments are defined as well stratified materials that have partly or entirely filled the space between the breccia fragments (Leach and Sangster, 1993). The materials consist of dolomite grains, sand-sized sulphide crystals, and sulphide-carbonate fragments (Kendall, 1960; Sass-Gustkiewicz et al., 1982; Rhodes et al., 1984; Randell and Anderson, 1990; Leach and Sangster, 1993). The mineralized breccias are of several textural types: crackle, mosaic, rubble, and rock-matrix ("trash") breccias (Fig. 9A, B, C, D). Sulphides and white sparry and saddle dolomite constitute the cement between the fragments. Descriptions of the breccias can be found in Ohle (1959; 1985), Sangster (1988; 1995), Leach and Sangster (1993), Paradis et al. (1999), and Nelson et al. (2002). In these open-space features, mineral textures are varied (see Figs. 9 and 10). The sulphides are disseminated, massive, and banded. Disseminated sulphides occur as fine to coarse crystals of sphalerite and galena overlain by, or intergrown with, white, coarse, crystalline, sparry dolomite cement (Fig. 9A, B). Coarse sphalerite crystals occasionally coat the tops of fragments or line the bottoms of cavities creating a texture known as "snow on roof" (Leach and Sangster, 1993; Sangster, 1995). Sphalerite also forms massive aggregates of coarse-grained colloform and botryoidal crystals (Figs. 9E, F, and 10A) and laminae of fine-grained crystals. At Nanisivik, replacement of the dolostone is the main mechanism of massive sulphide emplacement. The deposit consists of massive pyrite, sphalerite, and galena that replace the dolostone along high-angle normal faults and form mantos that shallowly crosscut bedding (Patterson and Powis, 2002). At Polaris, massive carbonate replacement, breccia-fill and vein sulphide ore form a 10 to 30 m thick, high-grade Zn-Pb-Fe, tabular unit hosted in the upper part of the deposit. Elsewhere, the replacement is selective and fol-

FIGURE 10. Various ore textures from Polaris, Nunavut; scale bar is 1 cm. **(A)** Sphalerite and skeletal galena are replacing the carbonate clasts. **(B)** Crystalline sphalerite in solution collapse breccia. **(C)** Massive galena and sparry dolomite in pseudobrecciated carbonate host rock.

lows stylolites, organic-rich layers, fossil-rich bands, and carbonate sand matrix (Randell, 1994). At Daniel's Harbour, selective replacement of the bioturbated limestone by the hydrothermal dolomite produced a pseudobreccia (Lane, 1984). At Monarch-Kicking Horse, Pine Point, and Robb Lake, selective replacement of a variety of primary rock fabrics by the hydrothermal dolomite formed a zebra texture (Fig. 11A, B).

In terms of geophysical parameters, the open-space filling nature of the MVT sulphide mineralization makes it a poor conducting medium for electrical geophysical surveys such as self-potential (SP) and electromagnetic (EM) (Lajoie and Klein, 1979). In most deposits, there is sufficient concentration of conducting minerals such as galena, pyrite, and marcasite, but the electrical continuity among these minerals is poor. This is due to the fact that these conductive minerals are often surrounded by non-conductive and non-polarizable sphalerite and abundant carbonate minerals, which interrupt the conducting paths (Lajoie and Klein, 1979).

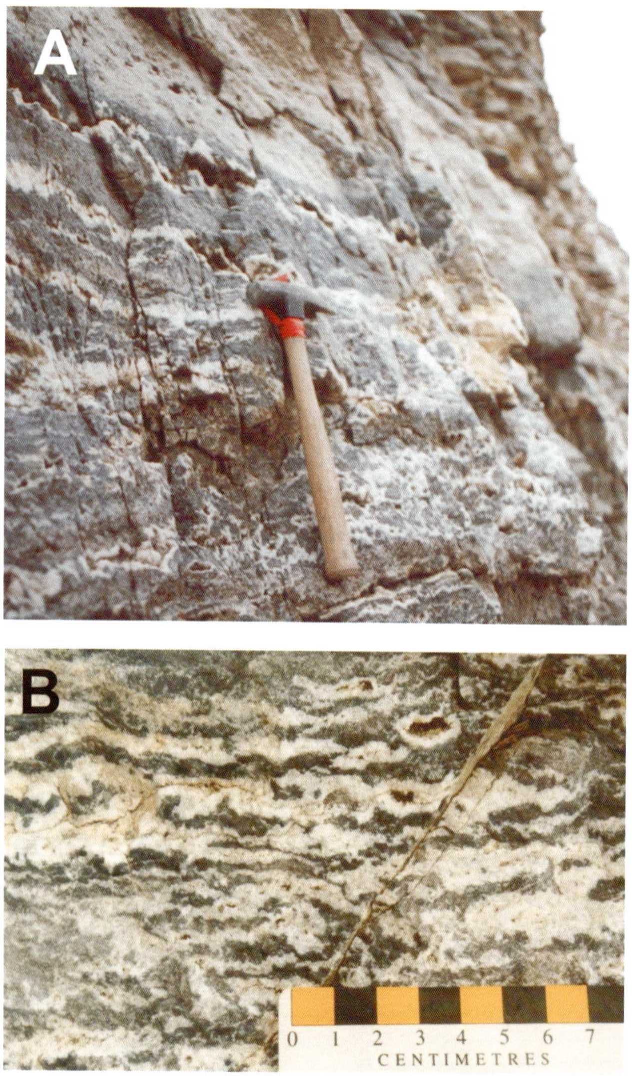

FIGURE 11. **(A)** Hydrothermal dolomite in bedding-parallel bands forming zebra texture (wall of pit N-42, Pine Point mine site, Northwest Territories). **(B)** Close-up of selective replacement of bedding-parallel bands by coarse-crystalline dolomite (Robb Lake, British Columbia).

Chemical Attributes of MVT Deposits

Ore Composition

Lead and zinc are the primary commodities recovered from MVT deposits. Silver, cadmium, germanium, copper, barite, and fluorite, although generally absent in most deposits, are by-products in some deposits. A complex suite of trace minerals may be present in some deposits, and may include some or all of the following minerals: arsenopyrite, bravoite, bornite, chalcopyrite, carrollite, celestite, chalcocite, covellite, digenite, djurleite, enargite, gallite, germanite, greenockite, linnaeite, marcasite, millerite, molybdenite, pyrrhotite, renierite, siegenite, tennantite, tungstenite, and vaesite (Foley, 2002). Elements associated with these minerals are As, Cu, Co, Ni, Cd, Ag, In, Ge, Ga, Sb, Bi, As, Mo, Sn, and Au. Cobalt and Ni are diagnostic accessory elements in deposits of the southeast Missouri and Upper Mississippi Valley districts in the US (Foley, 2002). Thallium and arsenic are enriched in sphalerite of the Polish deposits (Viets et al., 1996; Foley, 2002).

The majority of MVT deposits have essentially no geochemical signature because of limited primary dispersion of elements bound in sphalerite and galena within the carbonate rocks (Lavery et al., 1994). When weathering of the sulphides occurs and minerals such as limonite, cerussite, anglesite, smithsonite, hemimorphite, and pyromorphite are formed (e.g. Austinville-Ivanhoe district, US, and Silesia district, Poland), the soil and stream sediments of the regions surrounding the deposits may contain anomalous concentrations of Pb, Zn, Fe, and trace elements Sb, As, Bi, Ag, Tl, Cd, Mn, and Cu. In the East Tennessee district, detectable Zn, Fe, and Pb anomalies are found in residual soil and stream sediments (Leach et al., 1995).

In the Pine Point district, Pb, Zn, and Fe, which are anomalous in lake sediments, soils, and tills, are used for geochemical dispersion surveys in the exploration for orebodies. Zinc gives larger and more contrasting anomalies in lake sediments and soils than Pb; however, not all Zn anomalies are associated with an orebody. Shale units in the Western Canadian Sedimentary Basin, especially organic-rich ones, give high background values in Zn and Pb, similar in magnitude to those associated with orebodies. Relatively poorly defined Pb anomalies are often present near sulphide bodies, due to low Pb mobility and the generally low Pb/Zn ratios in MVT orebodies.

Alteration Mineralogy and Chemistry

Most MVT deposits show features of hydrothermal brecciation, recrystallization, dissolution, dolomitization, and silicification. The hydrothermal breccias known as collapse breccias result from the dissolution of underlying carbonate beds and are interpreted as meteoric karst or hydrothermal karst (Kyle, 1981; Sass-Gustkiewicz et al., 1982; Leach and Sangster, 1993).

Extensive hydrothermal dolomitization forms an envelope around most deposits, which extends tens to hundreds of metres beyond the sulphide bodies. According to Leach and Sangster (1993), the dolomitic halos can be pre-, syn-, or post-sulphides. This hydrothermal dolomitization consists of coarse crystalline white sparry dolomite and saddle dolomite cement. Dolomitized limestone forms a halo of minimum

TABLE 1. Summary of age dates (Ma) for Mississippi Valley-type deposits in Canada.

Deposit	Age of Host Rocks	Age of Mineralization Period/Epoch	Epoch/Age	Date (Ma)	Method	Reference	Orogeny
Polaris	Late - Middle Ordovician	Late Devonian	Famennian	367±7	Paleomagnetism	Symons and Sangster (1992)	Ellesmerian
		Late Devonian	Famennian	366±15	Sphalerite Rb-Sr	Christensen et al. (1995)	Ellesmerian
Nanisivik	Middle Proterozoic (Neohelikian)	Middle Ordovician	Chazyan	461	Ar-Ar on adularia[1]	Sherlock et al. (2003)	?
		Proterozoic	Middle Proterozoic	1095±10	Paleomagnetism[2]	Symons et al. (2000)	Rifting
Gays River	Early Mississippi (Late Tournaisian-Early Visean)	Pennsylvanian-Permian		297±27	Ar-Ar on biotite[3]	Kontak et al. (1994)	Alleghenian
		Pennsylvanian		303±17	Paleomagnetism	Pan et al. (1993)	Alleghenian
Daniel's Harbour	Early Ordovician	Middle Devonian	Givetian	374±10	Paleomagnetism[4]	Pan and Symons (1993)	Acadian
		Late Devonian-Early Mississippian	Famennian	360±10	Ar-Ar on authigenic feldspar	Hall et al. (1989)	Acadian
Monarch-Kicking Horse	Middle Cambrian	Late Cretaceous		100±12	Paleomagnetism	Symons et al. (1998)	Laramide[5]
Pine Point	Middle Devonian (Late Eifelian-Late Givetian)	Late Devonian-Early Carboniferous		361±13	Sphalerite Rb-Sr[6]	Nakai et al. (1993)	Antler
		Late Devonian-Early Carboniferous		374±21	Rb-Sr isochron	Brannon et al. (1995)	Antler
		Late Devonian-Early Carboniferous		362±9	Rb-Sr isochron on sphalerite + leachate	Nakai et al. (1993)	Antler
		Late Cretaceous-Paleocene		71±13	Paleomagnetism[7]	Symons et al. (1993)	Laramide
Robb Lake	Late Silurian - Middle Devonian	Paleogene/Eocene	Lutetian	47±10	Paleomagnetism[8]	Smethurst et al. (1999)	Laramide
		Paleozoic		No good isochron; but similar to Pine Point at 360	Sphalerite Rb-Sr[9]	Nelson et al. (2002)	Antler
Upton	Early Ordovician	Middle to Late Ordovician			Sr-Sr on barite	Paradis and Lavoie (1996)	Taconian

1 Ar-Ar on adularia was done on a 3 m-wide selvage of the diabase dyke referred to as the Mine dyke; Sherlock et al. (2003) argued that this alteration resulted from the hydrothermal system that emplaced the sulphides and that the Mine dyke (723 Ma) was emplaced before mineralization and therefore gives a maximum age on mineralization. Other geological evidence for this age include 1) sulphides emplaced post-tilting and folding of host strata; 2) sulphides show no displacement across Keystone fault; 3) fault displaced the host stratigraphy by ~150 m, and diabase dyke by 60 m, and 4) Rb-Sr dates on sphalerite similar to Ar-Ar on adularia (Sherlock, pers. comm., 2004).

2 A paleomagnetic study of the dolostone around the Nanisivik Mine yielded an age of 1095±10 Ma (Symons et al., 2000). This age is interpreted to be coeval with the hydrothermal system that formed the sulphide deposit. Symons et al. (2000) argued that the Mine dyke gives a minimum age of mineralization because the main sulphide lens is bisected by the mine dyke.

3 Ar-Ar date is from biotite-muscovite alteration on the surface of breccia clasts under an encrustation of galena cement. This age overlaps the paleomagnetism age of 303±17 Ma (Pan et al., 1993). Other methods have been used to date material from the underlying footwall greywackes of the Meguma Group, which provide indirect evidence. Ravenhurst et al. (1987) obtained a maximum age of ≤330 Ma from a reset Ar-Ar spectrum on muscovite and a minimun age of ≥189±15 Ma from apatite fission tracks. Ravenhurst et al. (1987; 1989) obtained maximum ages between 308±37 and 342±33 Ma based on zircon fission tracks and a maximum age of ≤352±7 Ma based on K-Ar illite dating. Arne et al. (1990) reported a minimum age of 203 to 241±10 Ma from apatite fission track dates of post-ore cooling.

4 Paleomagnetic data indicate that the sphalerite ore acquired its characteristic remanent magnetization during ore genesis and its Early Ordovician carbonate host rocks were remagnetized in the same thermochemical event. Nakai et al. (1993) determined a poor Rb-Sr sphalerite isochron age of 375±75 Ma. Ar-Ar spectra analyses on potassium feldspars from the mine and other showings define thermochemical events at 360±10 Ma and at 210 Ma (Hall et al., 1989). Geological evidence listed by Lane (1990) suggested that ore genesis corresponded to the Acadian Orogeny at ~374±14 Ma.

5 Westervelt (1979), Nesbitt and Muehlenbachs (1991, 1994, 1995), and Yao and Demicco (1995) consider that dolomitization and mineralization predate the Late Jurassic initiation of the Laramide Orogeny because the strata appear folded. Nesbitt and Muehlenbachs (1991; 1994; 1995) obtained a late Paleozoic isotopic signature for MVT mineralization, which led them to support a Late Devonian to Carboniferous fluid flow event during the Antler Orogeny.

6 Rb-Sr sphalerite isochron gives Middle Devonian ages of 361±13 Ma (Nakai et al., 1993) and 374±21 Ma (Brannon et al., 1995). Combined sphalerite and fluid inclusion analyses form a well defined isochron with an age of 362±9 Ma (Nakai et al., 1993).

7 Paleomagnetism indicates that the deposits formed between the mid-Late Cretaceous and Paleocene (71±13 Ma, Symons et al., 1993). Arne (1991) found a maximum age of ~98 Ma and Ravenhurst et al. (1989) an age of ~50-60 Ma for ore formation from apatite fission tracks.

8 The Eocene paleomagnetism age of 47±10 (Smethurst et al., 1999) correponds to the waning stages of the Laramide Orogeny and post-dates the main Laramide thrust-related deformation (i.e. ~185-60 Ma).

9 The combined sphalerite and fluid inclusion analyses do not form a single isochron with a well defined age on a Rb-Sr diagram. A Model 3 age calculation using Rb-Sr data for the sphalerite residues alone gives an apparent age of 348±130 Ma. A Model 3 age calculation using sphalerites and fluid inclusions gives an apparent age of 312±67 Ma. Three of the Robb Lake sphalerite residues fall on the Pine Point isochron. The calculated apparent age for these three samples is 350±100 Ma (Model 3 calculation). Although showing considerable scatter, the Rb-Sr data are permissive of a Paleozoic age for Robb Lake mineralization (Nelson et al., 2002).

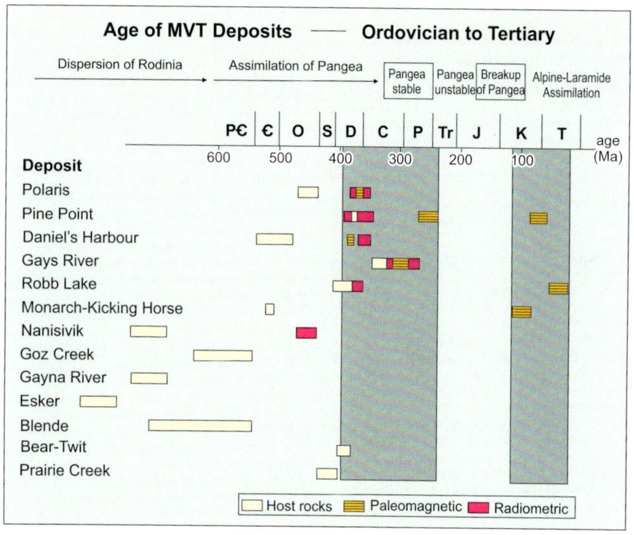

FIGURE 12. Distribution of radiometric and paleomagnetic ages of Canadian MVT deposits and their host rocks (modified from Leach et al., 2001a). The darker green shading indicates the two most important geologic periods for MVT genesis.

strike length of 1 km around and well below the orebody at Polaris (Randell and Anderson, 1996).

At Pine Point, calcite flooding forms halos around the orebodies, giving a coarsely granular appearance to the carbonate host rocks (Hannigan, 2007). Iron, Zn, and Pb display pronounced concentric distribution patterns in the Pine Point and Sulphur Point formations. Iron is the most widely distributed element, Zn is intermediate, and Pb occurs near the centre of the orebodies (Hannigan, 2007). These anomalous patterns decrease gradationally from maximum density and high-grade prismatic cores to barren country rocks. They are widespread in the Pine Point and Sulphur Point formations, negligible in Watt Mountain Formation, and confined to major solution collapse features in the Slave Point Formation (Turner et al., 2002). Iron dispersion highs tend to be displaced north of the deposit (Hannigan, 2007).

Silicification is not widespread in the Canadian MVT deposits, but when present, it forms small discontinuous zones of microcrystalline quartz peripheral to or enveloping the sulphides (e.g. Robb Lake, Esker, Prairie Creek, Goz Creek). Silicification is characteristic of the Tri-State district in the south-central US (Brockie et al., 1968; Sangster, 1995), and the northern Arkansas district (McKnight, 1935; Leach and Sangster, 1993), where it occurs as microcrystalline quartz that forms halos around or mantles the sulphide zones. According to Leach et al. (2001b), ore-related silicification of carbonate host rocks limits the buffering capacity of the rocks near the deposits and therefore influences the extent of metal-bearing water dispersion in some districts.

Formation of authigenic clay and feldspar minerals, as well as alteration of organic matter, is recognized at Polaris (Héroux et al., 1999), Jubilee, and Gays River (Héroux et al., 1996; Bertrand et al., 1998; Chagnon et al., 1998). Alteration of the organic matter and clays at Polaris, Pine Point, Gays River, and Jubilee forms halos of variable sizes around the deposits. The size and intensity of the alteration aureoles correlate with the temperatures of the ore formation and/or with the volume of ore-fluids responsible for ore deposition. These anomalies can be useful exploration tools when integrated with geological data (Héroux et al., 1996; Randell et al., 1996; Sass-Gustkiewicz and Kwiecinska, 1999).

Geology – Continental, District, and Deposit Scale

Continental Scale

MVT deposits are found in shallow water carbonate rocks of platformal settings. Most are found in orogenic forelands where platformal carbonates have some hydrological connection to orogenic belts. A few, however, are found in, or are adjacent to, extensional environments, and fewer still are contained within intracratonic basins (i.e. basins lying entirely within and bounded by a craton). The orogenic foreland-type deposits occur in platform carbonate sequences of an orogenic forebulge, and in foreland thrust belts. MVT deposits of the Pine Point, Ozark, and central Tennessee districts are good examples of deposits in relatively undeformed carbonate sequences of orogenic forelands. Those of the Appalachians, Rocky Mountains MVT belt, and Cornwallis fold belt are examples of deposits in foreland thrust belts. Other deposits, such as those of the Nanisivik, Gays River, Lennard Shelf (Australia), and Alpine (Europe) districts, occur in or adjacent to rift zones. Deposits such as those of the Upper Mississippi Valley district in the US occur in intracratonic basins where numerous small orebodies occur in vertical veins, fractures, and faults along a broad northwest-trending and -plunging half-dome of the Wisconsin arch (Heyl, 1983; Arnold et al., 1996). In the Mackenzie Mountains of northwestern Canada, hundreds of small Zn-Pb showings occur as veins, breccias, and vugs in dolostone or limestone. These are grouped within the Gayna River, Goz Creek, Bear Twit, and Blende deposits.

Many of the MVT deposits of the world possibly formed during large contractional tectonic events at specific times in the Earth's history (Leach et al., 2001a). The most important periods for MVT genesis were during the Devonian-Permian and the Cretaceous-Tertiary (Fig. 12). The Devonian to Permian period saw a series of continental collisions that culminated in the formation of the supercontinent Pangea (Leach et al., 2001a). Over 70% of the total MVT Zn-Pb metals produced so far worldwide were formed at that time. Deposits such as Daniel's Harbour, Polaris, and those of the Lennard Shelf district in Australia are Devonian-Mississippian in age. The Cretaceous–Tertiary period saw the breakup of Pangea, punctuated by the Alpine and Laramide orogenies affecting the western margin of North America and Africa-Eurasia (Leach et al., 2001a). Deposits such as Robb Lake, Pine Point, and Monarch-Kicking Horse in North America may have formed during the Laramide Orogeny in Cretaceous-Tertiary time or during the Devonian-Mississippian Antler Orogeny. The controversy regarding the age of these deposits resides in the disagreement between the paleomagnetic and Rb-Sr dating methods (Table 1). Paleomagnetism at Monarch-Kicking Horse, Robb Lake, and Pine Point indicates that the sulphide mineralization formed between 47±10 and 100±12 Ma (Symons et al., 1993; 1998; Smethurst et al., 1999), whereas radiometric dating by Rb-Sr sphalerite isochron yields Late Devonian-Early Mississippian ages for Pine Point (Nakai et al., 1993; Brannon et al., 1995). Rb-Sr data on sphalerite from the

Robb Lake deposit, although showing considerable scatter due to heterogeneous $^{87}Sr/^{86}Sr$ values in primary fluids, suggest that the mineralization is also of Paleozoic age (Nelson et al., 2002). Radiometric dating has not been done at Monarch-Kicking Horse. The Silesian deposits in Poland are believed to have formed during the Outer Carpathian orogeny in the Tertiary period (Symons et al., 1995). Paleomagnetic, U-Pb, Th-Pb, and Sm-Nd dating of deposits in the Cévennes region of France yielded Early to Middle Eocene ages that correspond to the uplift of the Pyrenees during the closing stages of the Pyrenean orogeny (Rouvier et al., 1995; Lewchuk et al., 1998a; Lewchuk et al., 1998b; Lewchuk et al., 1998c; Henry et al., 2001; Leach et al., 2001a; Rouvier et al., 2001). Furthermore, preliminary paleomagnetic results from the Reocin deposit in Spain are consistent with a Tertiary age for mineralization (Lewchuk et al., 1998b; 1998c).

Criteria Indicating Good Potential for MVT Deposits

1) Tectonic setting: Deposits are hosted in platform carbonate successions developed on the flanks of sedimentary basins; most are found in orogenic foreland thrust belts, few are found in, or adjacent to, extensional environments, and fewer in intracratonic basins.
2) Tectonic events and ages: Deposits formed mainly during large contractional tectonic events at specific times in the history of the Earth, i.e. Devonian-Permian and Cretaceous-Tertiary; a few known deposits are associated with extensional tectonic events that occurred during the Ordovician (e.g. Nanisivik deposit) and Late Devonian-Early Mississippian (e.g. Lennard Shelf deposits in Australia).
3) Coeval SEDEX deposits may be present in adjacent continental rift basins. There is a geographic and temporal linkage between Phanerozoic MVT and SEDEX deposits, particularly in western Canada (Goodfellow et al., 1993; Nelson et al., 2002).

Knowledge Gaps

Given the great diversity of geological features and ore-forming processes for MVT deposits, the most significant obstacle in our understanding of MVT genesis has been the lack of information on the age of many MVT deposits in the world. Once the ages are known, formation of MVT deposits can be connected to large-scale tectonic events.

District Scale

MVT deposits characteristically occur in districts that cover hundreds of square kilometres, such as the Pine Point district (>1600 km^2), Tri-State district (?1800 km^2), southeast Missouri district (~2300 km^2), Upper Mississippi Valley district (~10,000 km^2) and east Alpine district (~10,000 km^2). Within each district, deposits display similar geological features, such as mineralogy, textures, isotope signatures, and tectonic controls, but these characteristics change from district to district. This led Ohle (1959) to suggest that mineralization must be related to regional-scale events rather than to local and isolated events.

At the district scale, MVT deposits principally occur in dolostone (rarely in limestone or sandstone) of platform carbonate sequences, often below major unconformities. Ore controls such as barrier reef complexes, breccias, paleokarsts, depositional margins near carbonate/shale edges, facies tracts, faults, and basement highs, are typically district-specific (see Leach and Sangster, 1993, for further information).

Criteria Indicating Good Potential for MVT Deposits

1) *Tectonic setting:* The presence of platform carbonate sequences that commonly overlie deformed and metamorphosed continental crustal rocks, and have some hydrologic connection to sedimentary basins affected by orogenic events. Deposits usually occur in platform carbonates at shallow depths on flanks of sedimentary basins.
2) *District controls:* The deposits are localized by geologic features that permit upward migration of fluids, such as faults and basement highs. Other structures such as barrier reef complexes, breccias, paleokarsts, depositional margins near carbonate/shale edges, and facies tracts could also be important in trapping mineralized fluids.
3) *Paleolatitude:* Appropriate paleoclimatic conditions that lead to the formation and preservation of carbonate platforms and evaporites.
4) *Geophysical signature:* Induced polarization (IP) methods have proved successful in discovering orebodies in some MVT districts (e.g. Pine Point, Polaris), but only if sufficient conductive minerals such as galena, pyrite, and marcasite were present, and the chargeability of the host carbonates was low. Ground electromagnetic (EM) methods may work for deposits with sufficient concentrations of iron sulphide minerals (e.g. Nanisivik). Seismic, magnetic, and gravity methods can be used to identify some geological features associated with MVT deposits, such as basement highs beneath sedimentary successions, faults, sinkholes, paleokarsts and carbonate/shale facies. Airborne magnetic surveys were successfully used in delineating lineaments and buried Precambrian topography in the southeast Missouri district (Leach et al., 1995). Since sphalerite and gangue minerals have different densities, gravity surveys could prove useful in exploring for sulphide bodies containing mainly sphalerite (e.g. Daniel's Harbour). In the Cornwallis district, for example, the Polaris deposit was drilled based on a 2.2 milligal residual gravity anomaly over the deposit (Dewing et al., 2007). Airborne hyperspectral surveys also located numerous showings in the Cornwallis district.
5) *Geochemical anomalies:* Zinc, lead, iron, silver, and manganese can be detected in residual soil samples and stream sediments. Alteration of clay minerals and organic matter are useful exploration tools when integrated with geological data.

Knowledge Gaps

District scale questions that address key knowledge gaps relating to MVT deposits are listed below.

- Why are MVT deposits mostly associated with dolostone rather than limestone? Is it due to evidence that many dolomites are formed in evaporitic environments and thus provide sulphates that can be reduced to sulphides? Or is it simply a physical relationship where dolomites having greater porosity provide an increased probability of deposition of open-space filling ore minerals?

- Are evaporites critical for the generation of metalliferous fluids?
- What is the contribution of organic matter to MVT ore deposition? Are hydrocarbons critical for mineralization to occur?
- What causes metal zoning in some MVT districts, as well as in individual sulphide bodies?
- What type of ground preparation process is needed for sulphide deposition, and also what governs the location of orebodies in a district?
- What function do regional tectonic processes, such as orogenies, plate-margin interactions, or eustasy, have in the mineralization process?
- How do the local and regional hydrology and paleohydrology relate to dolomitization as well as mineralization? What flow paths are involved and what is the duration of their operation?
- What is the regional extent of dolomitizing and mineralizing fluids and what pathways did they take?
- What were the interactions between fluids and rocks through which they flowed that led to sulphide localization?

Deposit Scale

MVT deposits typically occur in carbonate rocks associated with interconnected paleokarst networks, pre-existing solution collapse-breccias, and related carbonate dissolution features that are located beneath impermeable rock formation barriers such as shales, finely crystalline dolostones, and unconformities, and along faults and fracture zones. Thick porous beds and areas of fractured rocks are required to form mineable thicknesses of sulphides, so at the deposit scale, the ore-controlling structures are commonly zones of solution collapse-breccias. In some instances (e.g. Pine Point), these zones form linear solution channels (called karsts) at the base of a reef and barrier complex (e.g. the Presqu'ile dolomite). They have remarkable continuity along the strike of the barrier and they form tabular orebodies. At intervals along these tabular solution channels, more intensive karstification produces either chimney-like karsts, called prismatic structures, or broader, more anastomosing areas of thicker tabular karsts (see Rhodes et al., 1984). These areas of enhanced solution activity, which are often associated with structural highs, frequently host orebodies. They are also loci for regional faults (extensional or strike-slip systems) that act as conduits for hydrothermal fluids. Major basement faults influence the alignment of orebodies within districts, however minor subsidiary faults, which may be splays off major faults, seem to be most efficient with respect to ground preparation by creating zones of weakness through which dolomitizing and mineralizing solutions can move freely. Porosity in these zones of weakness is enhanced by dolomitization and karsting.

Criteria Indicating Good Potential for MVT Deposits

1. *Local tectonic setting:* The presence of platform carbonate rocks.
2. *Deposit controls:* Barrier reef complexes, paleokarsts, solution-collapse breccias, faults, or fractures, and impermeable rock formation barriers, such as shales overlying porous carbonate rocks, are important ore-controls.
3. *Structures:* Identification of collapse depressions above hydrothermal breccia systems localized in dilatational fault arrays. Optimum fracturing, dilatational space creation, and mineralization may occur where these faults crosscut and intersect carbonates in outer shelf/platform settings. Localization of the hydrofracture system may be controlled by, or require the presence of, an overlying seal (e.g. shale) above the host carbonate; this maintains a "confined" pressure system in the underlying carbonate host (Davies, 1996).
4. *Sulphides:* Disseminated sulphides in the carbonate rocks may indicate proximity to sulphide deposits.
5. *Hydrothermal alteration:* Extensive hydrothermal dolomitization of precursor carbonate rocks is common and mostly appears as cement in sulphide-bearing breccias; this hydrothermal dolomite consists of coarse-grained white sparry dolomite and saddle dolomite. Silicification is also locally present, but is not as extensive as dolomitization. The alteration of clay minerals and organic matter is recognized in some deposits. Calcite flooding forms halos around orebodies.
6. *Geochemical anomalies:* Bitumen and pyrobitumen often occur in carbonate successions hosting MVT deposits and within the sulphide mineralization (e.g. Pine Point, Polaris, Robb Lake). Soft sticky bitumen masses, heavy oil, and pyrobitumen are present locally in the Pine Point district. Many levels of the Presqu'ile barrier complex show an impregnation by heavy oil and bitumen where small pores are filled by droplets (Krebs and Macqueen, 1984). Near some orebodies, these liquid hydrocarbons have been altered to pyrobitumen, forming irregular concentrations in large vugs and glossy globules as well as thin coatings on carbonate crystals (Krebs and Macqueen, 1984). The mineral paragenesis suggests that bitumen was deposited during the ore-stage event at Pine Point. At Polaris, dating of pyrobitumen by Re-Os gives 368±15 Ma, indicating that it is part of the mineralizing event (Selby et al., 2005).

Distribution of Canadian MVT Districts

Geological Distribution

For the purpose of this synthesis, a metallogenic district is defined as a continuous area that contains the expressions of the geological environment and tectonic events that controlled the formation of MVT deposits. In Canada, several MVT districts were identified. Most of these districts consist of several deposits, which comprise one to more than 100 sulphide bodies. For example, the Gayna River district contains more than 100 sulphide occurrences, and the Cornwallis district in Nunavut hosts one ore deposit, the Polaris mine, and approximately 80 showings (Dewing et al., 2007).

The Canadian deposits show a strong concentration along an arcuate circum-continental trend (Fig. 2). This trend is due to the fact that, by definition, MVT deposits are hosted in platformal carbonate rocks, which are located around the periphery of the craton. The largest number of deposits is located in the Mackenzie Mountains of the Yukon and the Northwest Territories, where hundreds of small deposits and a few larger ones (Gayna River, Blende, Bear Twit, Goz Creek, and Prairie Creek; Appendix 1) occur in Proterozoic to Devonian dolostone and limestone. Further south, a linear series of MVT deposits occurs in Cambrian and Silurian-Devonian carbonate rocks of the Canadian Rocky Mountains within the Robb Lake and Monarch-Kicking Horse districts.

Most Canadian deposits are found in deformed carbonate rocks of foreland thrust belts. Districts such as Gayna River, Blende, Goz Creek, Bear Twit, Robb Lake, and Monarch-Kicking Horse are hosted in deformed and thrust-faulted carbonates adjacent to the shelf front in the northern and southern Canadian Cordillera. Only deposits of the Pine Point district occur in weakly deformed carbonate rocks of the orogenic forelands. Daniel's Harbour, Gays River, and Upton in eastern Canada are hosted in deformed carbonate rocks of the Appalachian foreland thrust belts. Nanisivik in northern Canada formed in an extensional environment associated with east-west trending normal faults that divided the area into a series of horsts and grabens (Sherlock et al., 2003).

Temporal Distribution

Canadian MVT deposits are found in rocks ranging in age from Early Proterozoic to Early Mississippian with the majority of deposits in rocks of Paleozoic age (Fig. 12). The absolute age of mineralization is not known for all Canadian deposits. Table 1 summarizes the ages for some deposits. Deposits such as Nanisivik, Pine Point and Robb Lake show contrasting results between radiometric and paleomagnetic methods; reasons for the discrepancy are unclear. Radiometric and paleomagnetic data show that mineralization is Paleozoic in age and coincides mainly with periods of orogenic uplift that occurred in regions adjacent to the respective deposits. These MVT deposits, and those of the Ozark district in the US, have been attributed to large-scale migration of fluids during convergent orogenic processes (Leach et al., 2001a). The association of mineralization with orogenic uplift in a convergent regime supports the topographically-driven fluid flow model of ore fluid migration in compressive tectonic regimes (see below). Only one Canadian deposit so far, Nanisivik, is not associated with a contractional tectonic event but is attributed to mid-Ordovician extensional tectonism (Sherlock et al., 2003).

Genetic and Exploration Models

Conventional Models for Fluid Transport

Recent advances in understanding large-scale fluid flow in the crust, coupled with new geochemical and geological studies of MVT districts, have established that most MVT mineral districts are the products of regional or subcontinental-scale hydrological processes. Deposits formed from warm, saline, aqueous solutions (similar to oil-field brines) that migrated out of sedimentary basins, through aquifers, to the basin periphery and into the platform carbonate sequences. To affect this movement of sulphide-bearing brines, at least three different fluid flow processes have been proposed.

1. The topographic or gravity-driven fluid flow model (Garven and Freeze, 1984; Garven, 1985; Bethke and Marshak, 1990; Garven and Raffensperger, 1997).
2. The sedimentary-diagenetic model (Jackson and Beales, 1967; Sharp, 1978; Cathles and Smith, 1983; Oliver, 1986).
3. The hydrothermal convection model (Morrow, 1998).

The first model involves flushing of subsurface brines out of a sedimentary basin by groundwater flow from recharge areas in elevated regions of a foreland basin to discharge areas in lower elevated regions (Garven and Freeze, 1984; Garven, 1985; Bethke and Marshak, 1990; Garven and Raffensperger, 1997). In this model, subsurface flow is driven away from an uplifted orogen by the hydraulic head produced by tectonic uplift and tends to be concentrated in permeable units of a foreland succession. Considerable geologic evidence listed by Leach and Sangster (1993) supports this model. The model has been proposed for several MVT districts in the world, particularly those of the US mid-continent and the Pine Point district. At Pine Point and the Western Canada sedimentary basin (WCSB), Garven (1985) carried out some hydrogeological simulations and demonstrated that Pine Point formed in less than a million years from circulation of groundwater (rich in Pb and Zn) eastward from the elevated thrust belt of the Laramide Orogen through the Middle Devonian carbonates of the Keg River barrier.

The second model considers that compaction of sediments in a subsiding basin drives a continuous outward flow of pore fluids laterally along aquifers (Jackson and Beales, 1967). The basin-derived fluids or brines acquire heat, metals, and other solutes during migration and deposit sulphides in the host carbonates at favourable sites and under favourable physico-chemical conditions. Maintaining high initial fluid temperatures during transport of up to hundreds of kilometres from basin source to platform depositional site could be a problem. A variation of this model, episodic outward flow, was therefore proposed by Cathes and Smith (1983) and Bethke (1985). Their model involves overpressuring of subsurface aquifers by rapid sedimentation, followed by rapid and episodic release of basinal fluids. Another variation of the second model involves tectonic loading and compression of sediments during the development of orogenic thrust belts, which may have caused the rapid expulsion of formational fluids outward into the foreland basins with the thrust belts behaving like giant squeegees (Oliver, 1986).

The third model involves deep convection circulation of hydrothermal brines due to buoyancy forces related to temperature and salinity variations (Morrow, 1998). This model supports long-lived flow systems that are capable of recycling subsurface solutions many times through the rock mass, and has been invoked to explain regional hydrothermal dolomitization in the Western Canada sedimentary basin (Morrow, 1998), the Manetoe facies of the Northwest Territories (Aulstead et al., 1988; Morrow et al., 1990), the Ordovician gas-producing carbonates of the Michigan Basin (Coniglio et al., 1994), and MVT deposits of the northern

Canadian Rocky Mountains (Nelson et al., 2002). In the latter, Nelson et al. (2002) speculated that mineralization and dolomitization in the WCSB occurred in a far-field back-arc continental rift in response to the Late Devonian-Early Mississippian Antler Orogeny.

Conventional Models for Deposition of Sulphides

Three models involving 1) mixing, 2) sulphate reduction, and 3) reduced sulphur have been proposed for the chemical transport and deposition of sulphides.

The mixing model proposes transport of base metals in fluids of low sulphur content. Mixing of the metal-rich brines with fluids containing hydrogen sulphides at the depositional site triggers sulphide precipitation (Beales and Jackson, 1966; Anderson, 1975; Anderson and Macqueen, 1982; Sverjensky, 1984; Adams et al., 2000). Mixing of the ore fluids with a dilute or cool fluid, or reactions with host rocks to change the pH, are other variants on mixing.

The sulphate reduction model involves transport of base metals and sulphate in the same solution. Precipitation occurs at depositional sites when sulphate is reduced upon reaction with organic matter or methane (Anderson, 1975; Beales, 1975).

The reduced sulphur model requires that the base metals and the reduced sulphur be transported together in the same solution. Precipitation occurs either through cooling, mixing with diluted fluids, changes in pH, or loss of volatiles (Anderson, 1975; Anderson and Macqueen, 1982; Sverjensky, 1984, 1986).

Advances in Genetic and Exploration Models of the Last Decade

In the past, MVT deposits were considered to have few connections to global tectonic processes. Remarkable advances in age dating of MVT deposits in the last 10 years has resulted in improved understanding of the origin of MVT deposits, their links to global Earth tectonic events, and deposit modeling. However, there is still a paucity of information on the ages of MVT formation, and in some cases paleomagnetic and radiometric age dates show contradictory results.

MVT deposits that have been dated successfully show a relationship to large-scale tectonic events. Most MVT deposits formed during contractional tectonic events associated with the assimilation of Pangea in the Devonian to Permian, or during microplate assimilation on the western margin of North America and Africa-Eurasia in the Cretaceous to Tertiary. Few deposits correspond to extensional tectonic events in the Ordovician and early Mississippian time (Leach et al., 2001a). The latter are rare and poorly understood relative to global tectonic events, and more research needs to be done on the subject. Many important questions remain regarding the genesis of MVT deposits; some of them are listed below.

Knowledge Gaps

The current knowledge gaps in our understanding of MVT deposits concern mainly the following domains: 1) the role of tectonic processes, 2) the age of MVT ore-forming events, and 3) the chemical processes.

Tectonic Processes

Leach et al. (2001a) and Bradley and Leach (2003) stressed the genetic links between MVT mineralization and regional- and global-scale tectonic processes. It is now clear that MVT deposits are products of enormous hydrothermal systems that left trace mineralization over a wide area, and that the nearly ubiquitous occurrence of MVT deposits on the flanks of cratonic sedimentary basins reflects focused migration of deep-basin brines into shelf-carbonate sequences. Thus, the regional hydrogeologic framework is of paramount importance in the evaluation of large areas for their potential to contain MVT deposits.

MVT deposits in North America have been attributed to large-scale migration of fluids mainly during convergent orogenic processes. The topographically-driven fluid flow model associated with ore-fluid migration in compressive tectonic regimes best describes MVT mineralization in North America. Other deposits, such as Nanisivik and those of the Lennard Shelf and Alpine districts, have been attributed to continental extension and may require other fluid-driving mechanisms (Leach and Sangster, 1993).

Other questions that address key knowledge gaps also become important in understanding formation of MVT deposits, and could indirectly guide exploration for undiscovered MVT deposits. They are summarized below:

- What is the role of continental extension in the genesis of MVT deposits?
- What are the ages for ore formation in extensional regimes? The new ages will provide information on MVT genesis in the context of global crustal tectonic models and fluid migration.
- Why do MVT deposits form in some, but not all, carbonate platforms in collisional forelands? Certain carbonate platforms are fertile for MVT deposits while others are barren.
- What is it about the late stage of some collisions that induces regional-scale fluid migrations?
- Why are MVT deposits mostly associated with dolostones rather than limestones? Is it due to evidence that many dolomites are formed in evaporitic environments and thus provide sulphates that can be reduced to sulphides? Or is it simply a physical relationship where dolomites having greater porosity provide an increased probability of deposition of open-space filling sulphide minerals?
- What type of ground-preparation process is needed for sulphide deposition and also what governs the location of orebodies in a district?
- What function do regional tectonic processes such as orogenies, plate-margin interactions or eustasy have in the mineralization process?
- How do the local and regional hydrology and paleohydrology relate to dolomitization as well as mineralization? What flow paths are involved and what is the duration of their operation?
- What controls the hydrology of basins and carbonate platforms? Is it related to the distribution of fractures and

faults in the basement rocks and overlying sediments? Do some faults serve to recharge or discharge fluids?

Age of Mississippi Valley-Type Ore-Forming Events

The need to know the absolute age of MVT mineralization remains one of the most critical aspects of research on MVT deposits, and an obstacle to our understanding of MVT genesis. Recent advances in age dating of MVT deposits provide new evidence that there are important genetic relationships between convergent and divergent orogenic events and the formation of MVT deposits (Leach et al., 2001a). The most important periods for MVT genesis are the Devonian-Permian and the Cretaceous-Tertiary when large-scale contractional tectonic events occurred. Other deposits such as Nanisivik and those of the Lennard Shelf district are associated with extensional events in the Ordovician and Early Mississippian, respectively. These may prove to be more abundant, but so far little is known about the role of continental extension and the formation of MVT deposits. There is a paucity of MVT deposits of Precambrian, Early Paleozoic, and Mesozoic ages despite abundant host rocks and seemingly suitable tectonic settings.

Even if important advances in dating have been achieved in the last decade, much still remains to be done, as many deposits and districts have not been successfully dated and there are still many questions to be answered:

- Why are there few known Proterozoic MVT deposits when there is an abundance of SEDEX deposits in Proterozoic time (Goodfellow et al., 1993)?
- What are the temporal and genetic links between MVT and SEDEX deposits? Is there a genetic/temporal link? If yes, could they have formed from the same hydrothermal system?
- What are the temporal and genetic links between MVT and oil/gas deposits?
- What is the age of deposits for which there is controversy regarding timing (e.g. Robb Lake, Pine Point, East Tennessee, Cévennes) due to conflicting results between paleomagnetic and radiometric dating techniques?

Chemical Processes

Chemical processes that localized the deposition of sulphides are critical to the development of models for MVT deposits, yet the specific chemical reactions that led to sulphide deposition remain one of the most controversial aspects of MVT deposits. Several questions remain as to:

- Why do these deposits contain mostly lead and zinc in economic quantities?
- What causes metal zoning in some MVT districts as well as in individual deposits?
- What is the contribution of organic matter to MVT sulphide deposition? Are hydrocarbons critical for mineralization to occur?
- What controls the chemical composition of sulphide-forming fluids? What are the chemical attributes of hydrothermal reaction zones that generate sulphide-forming fluids?
- What alteration vectors are most effective in exploring for MVT deposits?

Key Exploration Criteria For Canadian MVT Deposits

The major exploration criteria for Canadian MVT deposits are summarized below:

- *Geologic settings:* Platformal carbonate rocks peripheral to cratonic sedimentary basins. Deposits are commonly located close to a carbonate shelf margin, at the transition into slope and basinal shales or argillaceous facies.
- *Ages:* Carbonate host rocks that range from Middle Proterozoic to Carboniferous with a majority of deposits in the Mid to Late Paleozoic period.
- *District-scale controls:* The deposits are localized by geologic features that permit upward migration of fluids, such as faults and basement highs. There could be a strong regional fault control system on the localization and emplacement of MVT deposits. Optimum fracturing, dilational space creation, and mineralization may occur where faults intersect barren carbonate rocks in outer shelf/carbonate margin (reef margin) settings. Basement highs are zones where fluid flow is laterally restricted and thus represent possible depositional sites. Basement highs are also sites of reef growth, which are potential host rocks for MVT deposits.
- *Deposit-scale controls:* Proximity of growth faults and intersection of faults, regional and local dolomitization and possibly laterally equivalent iron-formations (as Irish-type). Identification of potential conduits, traps, and prospective stratigraphy provides a means of predicting the potential locations of undiscovered massive sulphide bodies.
- *Paleolatitude:* Appropriate paleoclimatic conditions that lead to the formation and preservation of carbonate platforms and evaporites.
- *Hydrothermal event:* Evidence of hydrothermal fluid discharge may include 1) disseminated sulphides in the carbonate host rocks peripheral to the deposits; 2) SEDEX deposits in sedimentary basins adjacent to the carbonate platform; 3) distal hydrothermal sediments, such as Mn-Fe-Ca-Mg carbonates; 4) stream sediment and water that are enriched in MVT-forming and -associated elements.

Acknowledgements

We are grateful for the knowledgeable assistance of Geological Survey of Canada colleagues R. M. Laramée, L. B. Chorlton, and R.L. Sherlock. Their patience and understanding is hereby recognized with gratitude and thanks. The manuscript benefited from a preliminary review by Ingrid Kjarsgaard, and a final review by Don Sangster and Ian Jonasson.

References

Adams, J.J., Rostron, B.J., and Mendoza, C.A., 2000, Evidence for two-fluid mixing at Pine Point, Northwest Territories: Journal of Geochemical Exploration, v. 69-70, p. 103-108.

Anderson, G.M., 1975, Precipitation of Mississippi Valley-type ores: Economic Geology, v. 70, p. 937-942.

Anderson, G.M., and Macqueen, R.W., 1982, Ore deposit models - 6. Mississippi Valley-type lead-zinc deposits: Geoscience Canada, v. 9, p. 108-117.

Arne, D.C., 1991, Regional thermal history of the Pine Point area, Northwest Territories, Canada, from fission track analysis: Geology, v. 86, p. 428-435.

Arne, D.C., Duddy, I.R., and Sangster, D.F., 1990, Thermochronological constraints on ore formation at the Gays River Pb-Zn deposit, Nova Scotia, Canada, from apatite fission track analysis: Canadian Journal of Earth Sciences, v. 27, p. 1013-1022.

Arnold, B.W., Bahr, J.M., and Fantucci, R., 1996, Paleohydrogeology of the Upper Mississippi Valley Zinc-Lead District: Society of Economic Geologists, Special Publication No. 4, p. 378-389.

Aulstead, K.L., Spencer, R.J., and Krouse, H.R., 1988, Fluid inclusion and isotopic evidence on dolomitization, Devonian of Western Canada: Geochimica et Cosmochimica Acta, v. 52, p. 1027-1035.

Beales, F.W., 1975, Precipitation mechanisms for Mississippi Valley-type ore deposits: Economic Geology, v. 70, p. 943-948.

Beales, F.W., and Jackson, S.A., 1966, Precipitation of lead-zinc ores in carbonate reservoirs as illustrated by Pine Point ore field, Canada: Institution of Mining and Metallurgy Transactions, section B, p. B8278-8285.

Beales, P., Woodward, R., and Campbell, C., 2002, Great Slave Reef, in A Guide to Mineral Deposits of the Northwest Territories: Minerals, Oil and Gas Division, Resources, Wildlife and Economic Development, Government of Northwest Territories, p. 82.

Bell, W.A., 1944, Carboniferous Rocks and Fossil Floras of Northern Nova Scotia: Geological Survey of Canada, Memoir 238, 277 p.

Bertrand, R., Chagnon, A., Héroux, Y., and Savard, M.M., 1998, Hydrothermal alteration of clay minerals and organic matter within and outside the Jubilee carbonate-hosted Zn-Pb deposit, Cape Breton Island, Nova Scotia, Canada: Economic Geology, v. 93, p. 746-756.

Bethke, C.M., 1985, A numerical model of compaction-driven groundwater flow and heat transfer and its application to the paleohydrology of intracratonic sedimentary basins: Journal of Geophysical Research, v. 80, p. 6817-6828.

Bethke, C.M., and Marshak, S., 1990, Brine migration across North America - the plate tectonics of groundwater: Earth and Planetary Science Letters, Annual Review, v. 18, p. 287-315.

Boyles, R.W., 1972, The Geology, Geochemistry and Origin of the Barite, Manganese and Lead-Zinc-Copper-Silver Deposits of the Walton-Cheverie area, Nova Scotia: Geological Survey of Canada, Bulletin 166, 181 p.

Bradley, D.C., and Leach, D.L., 2003, Tectonic controls of Mississippi Valley-Type lead-zinc mineralization in orogenic forelands: Mineralium Deposita, v. 38, p. 652-667.

Brannon, J.C., Cole, S.C., Podosek, F.A., and Misra, K.C., 1995, Radiometric dating of ancient calcite, Th-Pb and U-Pb isochrons for ore-stage and late-stage calcite from central Tennessee zinc district, an Appalachian-Ouachita age MVT deposit: Geological Society of America, Abstracts with Programs, 27, p. 118.

Brock, J., 1973, Notes for Delivery about Welcome North's Interest in Godlin Lake, Northwest Territories (unpublished): Mineral Exploration Group, 13 p.

Brockie, D.C., Hare, E.H.J., and Dingess, P.R., 1968, The geology and ore deposits of the Tri-State district of Missouri, Kansas and Oklahoma, *in* Ridge, J. D., ed., Ore Deposits of the United States, 1933-1967: American Institute of Mining, Metallurgical, and Petroleum Engineers, Inc., New York, NY, p. 400-430.

Brophy, J.A., 1987, Pine Point Mine, *in* Ellis, C. E., ed., Operating Mines, 1984-85: Mineral Industry Report 1984-85, Northwest Territories: Indian and Northern Affairs Canada, p. 15-31.

Carrière, J.J., and Sangster, D.F., 1999, A multidisciplinary study of carbonate-hosted zinc-lead mineralization in the MacKenzie Platform (a.k.a. Blackwater and Lac de Bois platforms) Yukon and Northwest Territories, Canada: Geological Survey of Canada, Open File 3700, 145 p.

Cathles, L.M., and Smith, A.T., 1983, Thermal constraints on the formation of Mississippi Valley-type lead-zinc deposits and their implications for episodic basin dewatering and deposit genesis: Economic Geology, v. 78, p. 983-1002.

Chagnon, A., St.-Antoine, P., Savard, M.M., and Héroux, Y., 1998, Impact of Pb-Zn sulphide precipitation on the clay mineral assemblage in the Gays River Deposit, Nova Scotia, Canada: Economic Geology, v. 93, p. 779-792.

Christensen, J.N., Halliday, A.N., Leigh, K.H., Randell, R.N., and Kesler, S.E., 1995, Direct dating of sulphides - a critical test using Polaris Mississippi Valley-type Zn-Pb deposits: Geochimica et Cosmochimica Acta, v. 59, p. 5191-5197.

Coniglio, M., Sherlock, R., Williams-Jones, A.E., Middleton, D., and Frape, S.K., 1994, Burial and hydrothermal diagenesis of Ordovician carbonates from the Michigan Basin, Ontario, Canada, *in* Purser, B., Tucker, M., and Zenger, D., eds., A Volume in Honor of Dolomieu: International Association of Sedimentologists, Special Publication 21, p. 231-254.

Cumming, G.L., Kyle, J.R., and Sangster, D.F., 1990, Pine Point: A case history of lead isotopic homogeneity in a Mississippi Valley-type district: Economic Geology, v. 85, p. 133-144.

Davies, G.R., 1996, Hydrothermal Dolomite (HTD) Reservoir Facies, Global Perspectives on Tectonic-Structural and Temporal Linkage between MVT and SEDEX Pb-Zn Ore Bodies, and Subsurface HTD Reservoir Facies: Graham Davies Geological Consultants, Canadian Society of Petroleum Geologists, Short Course Notes, 167 p.

Dewing, K., Turner, E., and Harrison, J.C., 2007, Geological history, mineral occurrences and mineral potential of the sedimentary rocks of the Canadian Arctic Archipelago, *in* Goodfellow, W.D., ed., Mineral Deposits of Canada: A Synthesis of Major Deposit-Types, District Metallogeny, the Evolution of Geological Provinces, and Exploration Methods: Geological Association of Canada, Mineral Deposits Division, Special Publication 5, p. 733-753.

Dunham, K., 1983, Ore genesis in the English Pennines, A fluoritic subtype: International Conference on Mississippi Valley-type Lead-Zinc Deposits, Proceedings Volume, p. 271-278.

Ellis, C.E., 1995, Operating Mines, Chapter 2, *in* Beales, P., ed., Mineral Industry Report 1988-89: Northwest Territories, Indian and Northern Affairs Canada, p. 13-24.

Ellis, C.E., and Hearn, K., 1990, Operating Mines, Chapter 2, *in* Ellis, C. E., ed., Mineral Industry Report 1986-87: Northwest Territories, Indian and Northern Affairs Canada, p. 11-32.

Fallara, F., Savard, M.M., and Paradis, S., 1998, A structural, petrographic, and geochemical study of the Jubilee Zn-Pb deposit, Nova Scotia, Canada and a new metallogenic model: Economic Geology, v. 93, p. 757-778.

Foley, N.K., 2002, Chapter E - Environmental geochemistry of platform carbonate-hosted sulphide deposits, http://pubs.usgs.gov/of/2002/of02-195/ChE.txt.

Fraser, S.C., 1996, Geology and Geochemistry of the Prairie Creek Zn, Pb, Ag Deposits, Southern Mackenzie Mountains, Northwest Territories: M.Sc. thesis, University of Alberta, Edmonton, Alberta, 146 p.

Fritz, W.H., 1990, In defense of the escarpment near the Burgess Shale fossil locality: Geoscience Canada, v. 17, p. 106-110.

Fritz, W.H., Cecile, M.P., Norford, B.S., Morrow, D., and Geldsetzer, H.H.J., 1992, Cambrian to Middle Devonian Assemblages, *in* Gabrielse, H., and Yorath, C.J., eds., Geology of the Cordilleran Orogen in Canada: Geology of Canada, no.4, Geological Survey of Canada, p. 153-218.

Garven, G., 1985, The role of regional fluid flow in the genesis of the Pine Point deposit, Western Canada Sedimentary Basin: Economic Geology, v. 80, p. 307-324.

Garven, G., and Freeze, R.A., 1984, Theoretical analysis of the role of groundwater flow in the genesis of stratabound ore deposits. 2. Quantitative results: American Journal of Science, v. 284, p. 1125-1174.

Garven, G., and Raffensperger, J.P., 1997, Hydrogeology and geochemistry of ore genesis in sedimentary basins, in Barnes, H. L., ed., Geochemistry of hydrothermal ore deposits: Wiley, New York, p. 125-189.

Goodfellow, W.D., Lydon, J.W., and Turner, R.J.W., 1993, Geology and genesis of stratiform sediment-hosted (SEDEX) zinc-lead-silver sulphide deposits, *in* Kirkham, R.V., Sinclair, W.D., Thorpe, R.I., and Duke, J.M., eds., Mineral Deposit Modeling: Geological Association of Canada, Special Paper 40, p. 201-251.

Gummer, P.K., Plint, H.E., and Rainbird, R.H., 1996, The Esker Lake prospect: stratabound Pb-Zn-Cu-Ag in emergent inner shelf carbonates, Rocknest Formation, Coronation Supergroup, Northwest Territories: 3, *in* 1996 Exploration Overview: Department of Indian and Northern Development Canada, p. 18-19.

Hall, C.M., York, D., Saunders, C.M., and Strong, D.F., 1989, Laser ^{40}Ar/^{39}Ar dating of Mississippi Valley type mineralization from Western Newfoundland: International Geological Congress, Abstract v. 28, p. 2.10-2.11.

Hannigan, P., 2007, Metallogeny of the Pine Point Mississippi Valley-Types zinc-lead district, Southern Northwest Territories, *in* Goodfellow, W.D., ed., Mineral Deposits of Canada: A Synthesis of Major Deposit-Types, District Metallogeny, the Evolution of Geological Provinces, and Exploration Methods: Geological Association of Canada, Mineral Deposits Division, Special Publication 5, p. 609-632.

Henry, B., Rouvier, H., Le Goff, M., Leach, D., Macquar, J.-C., Thibieroz, J., and Lewchuk, M.T., 2001, Paleomagnetic dating of widespread remagnetization on the southern border of the French Massif Central and implications for fluid-flow and Mississippi Valley-type mineralization: Geophysical Journal International, v. 145, p. 368-380.

Héroux, Y., Chagnon, A., and Savard, M., 1996, Organic matter and clay anomalies associated with base-metal sulphide deposits: Ore Geology Reviews, v. 11, p. 157-173.

Héroux, Y., Chagnon, A., Dewing, K., and Rose, H.R., 1999, The carbonate-hosted base-metal sulphide Polaris deposit in the Canadian Arctic, organic matter alteration and clay diagenesis, *in* Glikson, M., and Mastalerz, M., eds., Organic matter and mineralization, thermal alteration, hydrocarbon generation and role in metallogenesis: Kluwer Academic Publisher, p. 260-295.

Hewton, R.S., 1982, Gayna River: A Proterozoic Mississippi Valley-type Zinc-Lead deposit, *in* Hutchinson, R. W., Spence, C. D., and Franklin, J. M., eds., Precambrian Sulphide Deposits, H. S. Robinson Memorial Volume: Geological Association of Canada, 25, p. 668-700.

Heyl, A.V., 1983, Geologic characteristics of three Mississippi Valley-type districts: International Conference on Mississippi Valley-Type Lead-Zinc Deposits, Proceedings, p. 27-30.

Hitzman, M.W., and Beaty, D.W., 1996, The Irish Zn-Pb(-Ba) orefield, *in* Sangster, D. F., ed., Carbonate-Hosted Lead-Zinc Deposits: Society of Economic Geologists, Special Publication Number 4, p. 112-143.

Jackson, S.A., and Beales, F.W., 1967, An aspect of sedimentary basin evolution: The concentration of Mississippi Valley-type ores during the late stages of diagenesis: Bulletin of Canadian Petroleum Geology, v. 15, p. 393-433.

Jakubek, S.S., Lepinis, Y.J., Macrobbie, D.C., and Sozanski, A.G., 1989, Canadian Mineral Deposits not being Mined in 1989: Energy, Mines and Resources Canada, Mineral Bulletin M38-2, 223 p.

Kendall, D.L., 1960, Ore deposits and sedimentary features, Jefferson City mine, Tennessee: Economic Geology, v. 55, p. 985-1003.

Kontak, D.J., 1992, A Preliminary Report on Geological, Geochemical, Fluid Inclusion and Isotopic Studies of the Gays River Zn-Pb Deposit, Nova Scotia: Nova Scotia Department Natural Resources, Open File Report 92-014, 223 p.

Kontak, D.J., and Sangster, D.F., 1989, Aqueous and liquid petroleum inclusions in barite from the Walton Deposit, Nova Scotia, Canada: A carboniferous, carbonate-hosted Ba-Pb-Zn-Cu-Ag deposit: Economic Geology, v. 93, p. 845-868.

Kontak, D.J., McBride, S., and Farrer, E., 1994, $^{40}Ar/^{39}Ar$ dating of fluid migration in a Mississippi Valley-type deposit: Gays River Zn-Pb deposit, Nova Scotia: Economic Geology, v. 89, p. 1501-1517.

Krebs, W., and Macqueen, R., 1984, Sequence of diagenetic and mineralization events, Pine Point lead-zinc property, Northwest Territories, Canada: Bulletin of Canadian Petroleum Geology, v. 32, p. 434-464.

Kyle, J.R., 1981, Geology of the Pine Point lead-zinc district, *in* Wolf, K. H., ed., Handbook of Strata-Bound and Stratiform Ore Deposits: Elsevier Publishing Company, New York, 9, p. 643-741.

Lajoie, J.J., and Klein, J., 1979, Geophysical exploration at the Pine Point Mines Ltd. zinc-lead property, Northwest Territories, Canada, *in* Hood, P. J., ed., Geophysics and Geochemistry in the Search for Metallic Ores: Geological Survey of Canada, Report 31, p. 653-664.

Lane, T.E., 1984, Preliminary classification of carbonate breccias, Newfoundland Zinc Mines, Daniel's Harbour, Newfoundland: Geological Survey of Canada, Paper 84-1A, p. 505-512.

———1990, Dolomitization, Brecciation and Zinc Mineralization and their Paragenetic, Stratigraphic and Structural Relationships in the Upper St. George Group (Ordovician) at Daniel's Harbour, Western Newfoundland: Ph.D. thesis, Memorial University of Newfoundland, St. John's, Newfoundland, 262 p.

Lavery, N.C., Leach, D.L., and Saunders, J.A., 1994, Lithogeochemical investigations applied to exploration for sediment-hosted Zn-Pb ore deposits, *in* Fontebté, L., and Boni, M., eds., Sediment Hosted Zn-Pb Ores: Springer-Verlag, p. 393-428.

Leach, D.L., and Sangster, D.F., 1993, Mississippi Valley-type lead-zinc deposits, *in* Kirkham, R. V., Sinclair, W.D., Thorpe, R.I., and Duke, J.M., eds., Mineral Deposit Modeling: Geological Association of Canada, Special Paper 40, p. 289-314.

Leach, D.L., Viets, J.B., Foley-Ayuso, N., and Klein, D.P., 1995, Mississippi Valley-Type Pb-Zn deposits (Models 32a, b; Briskey, 1986 a, b), *in* Du Bray, E. A., ed., Preliminary Compilation of Descriptive Geoenvironmental Mineral Deposit Models: U.S. Government Consulting Group, Open File Report 95-831, p. 234-243.

Leach, D.L., Bradley, D., Lewchuk, M.T., Symons, D.T.A., de Marsily, G., and Brannon, J., 2001a, Mississippi Valley-type lead-zinc deposits through geological time: implications from recent age-dating: Mineralium Deposita, v. 36, p. 711-740.

Leach, D.L., Premo, W., Lewchuk, M., Henry, B., LeGoff, M., Rouvier, H., Macquar, J.C., and Thibieroz, J., 2001b, Evidence for Mississippi Valley-type lead-zinc mineralization in the Cévenne region, southern France, during Pyrénées orogeny, Mineral Deposits at the Beginning of the 21st Century: Balkema, Rotterdam, p. 157-160.

Lewchuk, M.T., Henry, B., Rouvier, H., Macquar, J.C., and Leach, D., 1998a, Paleomagnetism of Mississippi Valley-Type mineralization in southern France and Cenozoic orogenesis: Annali di Geophysica, v. 16, p. 53.

Lewchuk, M.T., Rouvier, H., Henry, B., LeGoff, M., and Leach, D., 1998b, Paleomagnetism of carbonates and lead-zinc mineralization in Southern France: IUGG, Birmingham, UK, Abstract.

Lewchuk, M.T., Rouvier, H., Henry, B., Macquar, J.-C., and Leach, D., 1998c, Paleomagnetism of Mississippi Valley-Type mineralization in southern France and Cenozoic orogenesis: European Geophysical Society XXIII general assembly, Nice, France, 20-24 April.

Macqueen, R.W., and Thompson, R. I., 1978, Carbonate-hosted lead zinc occurrences in northeastern British Columbia with emphasis on the Robb Lake deposit: Canadian Journal of Earth Sciences, v. 15, p. 1737-1762.

McKnight, E.T., 1935, Zinc and lead deposits of Northern Arkansas: United States Geological Survey, Bulletin 853, 311 p.

Megaw, P.K.M., Barton, M.D., and Falce, J.I., 1996, Carbonate-hosted lead-zinc (Ag, Cu, Au) deposits of northern Chihuaha, Mexico, *in* Sangster, D. F., ed., Carbonate-hosted lead-zinc deposits: Society of Economic Geologists, Special Publication Number 4, p. 277-289, 664.

Morrow, D., 1998, Regional subsurface dolomitization; Models and constraints: Geoscience Canada, v. 25, p. 57-70.

Morrow, D.W., and Cook, D.G., 1987, The Prairie Creek Embayment and lower Paleozoic Strata of the Southern Mackenzie Mountains: Geological Survey of Canada, Memoir 412, 195 p.

Morrow, D.W., Cumming, G.L., and Aulstead, K.L., 1990, The Gas-Bearing Devonian Manetoe Facies, Yukon and Northwest Territories: Geological Survey of Canada, Bulletin 400, 54 p.

Murphy, J.D., 1973, Mineral Showings on Welcome North Mines Ltd.'s Property in the Godlin Lakes Area 105 P/14, 15 and 106 A/3 (63 50'N, 129 20'W): Indian Affairs and Northern Development Canada, Resident Geologist Report, 12 p.

Nakai, S., Halliday, A.N., Kesler, S.E., Jones, H.D., Kyle, J.R., and Lane, T.E., 1993, Rb-Sr dating of sphalerites from Mississippi Valley-type (MVT) ore deposits: Geochimica et Cosmochimica Acta, v. 57, p. 417-427.

Nawajka, M., Wachowiak, N., Gummer, P. K., Pope, M., and Grotzinger, J. P., 1997, Stratabound Zn-Pb Mineralization of the Esker Horizon Paleoproterozoic Rocknest Formation, Wopmay Orogen, Canada: Geoscience Forum, Yellowknife, Programs and Abstracts of talks and posters.

Nelson, J., Paradis, S., Christensen, J., and Gabites, J., 2002, Canadian Cordilleran Mississippi Valley-type deposits: A case for Devonian-Mississippian back-arc hydrothermal origin: Economic Geology, v. 97, p. 1013-1036.

Nelson, J.L., Paradis, S., and Zantvoort, W., 1999, The Robb Lake carbonate-hosted lead-zinc deposit, northeastern British Columbia: A Cordilleran MVT deposit, *in* Geological and Current Fieldwork 1998 - A Summary of Field Activities: British Columbia Ministry of Energy and Mines, Geological Survey Branch, Paper 1999-1, p. 89-101.

Nesbitt, B.E., and Muehlenbachs, K., 1991, Stable isotopic constraints on the nature of the syntectonic fluid regime of the Canadian Cordillera: Geophysical Research Letters, v. 18, p. 963-966.

Nesbitt, B.E., and Muehlenbachs, K., 1994, Paleohydrology of the Canadian Rockies and origins of brines, Pb-Zn deposits and dolomitization in the Western Canada Sedimentary Basin: Geology, v. 22, p. 243-246.

——— 1995, Geochemistry of syntectonic, crustal fluid regimes along the lithoprobe Southern Canadian Cordillera Transect: Canadian Journal of Earth Sciences, v. 32, p. 1699-1719.

Norris, A.W., and Uyeno, T.T., 1998, Middle Devonian brachiopods, conodonts, stratigraphy, and transgressive-regressive cycles, Pine Point area, south of Great Slave Lake, District of Mackenzie, Northwest Territories: Geological Survey of Canada, Bulletin 522, 191 p.

Ohle, E.L., 1959, Some considerations in determining the origin of ore deposits of the Mississippi Valley-type. Part 1: Economic Geology, v. 54, p. 769-789.

——— 1985, Breccias in Mississippi Valley-type deposits: Economic Geology, v. 75, p. 161-172.

Oliver, J., 1986, Fluids expelled tectonically from orogenic belts, their role in hydrocarbon migration and other geological phenomena: Geology, v. 14, p. 99-102.

Olson, R.A., 1984, Genesis of paleokarst and strata-bound zinc-lead sulfide deposits in a Proterozoic dolostone, Northern Baffin Island, Canada: Economic Geology, v. 79, p. 1056-1103.

Pan, H., and Symons, D.T.A., 1993, Paleomagnetism of the Mississippi Valley-type Newfoundland Zinc deposits: evidence for Devonian mineralization in the northern Appalachians: Geophysical Reference Letters, v. 98, p. 22415-22427.

Pan, H., Symons, D.T.A., and Sangster, D.F., 1993, Paleomagnetism of the Gays River zinc-lead deposit, Nova Scotia: Pennsylvanian ore genesis: Geophysical Reference Letters, v. 20, p. 1159-1162.

Paradis, S., and Lavoie, D., 1996, Multiple-stage diagenetic alteration and fluid history of Ordovician carbonate-hosted barite mineralization, southern Quebec Appalachians: Sedimentary Geology, v. 107, p. 121-139.

Paradis, S., and Nelson, J.L., 2007, Metallogeny of the Robb Lake carbonate-hosted zinc-lead district, northeastern British Columbia, *in* Goodfellow, W.D., ed., Mineral Deposits of Canada: A Synthesis of Major Deposit-Types, District Metallogeny, the Evolution of Geological Provinces, and Exploration Methods: Geological Association of Canada, Mineral Deposits Division, Special Publication 5, p. 633-654.

Paradis, S., Savard, M., and Fallara, F., 1993, Preliminary study on diagenesis and mineralization of the Jubillee Pb-Zn deposit, Nova Scotia: Geological Survey of Canada, Current Research Paper 93-1D, 111-119 p.

Paradis, S., Nelson, J., and Zantvoort, W., 1999, A new look at the Robb Lake carbonate-hosted lead-zinc deposit, northeastern British Columbia: Geological Survey of Canada, Current Research 1999-A, p. 61-70

Paradis, S., Chi, G., and Lavoie, D., 2004, Fluid inclusion and isotopic constraints on the genesis of the Upton carbonate-hosted Ba-Zn-Pb deposit, Canada: Economic Geology, v. 99, p. 807-817.

Patterson, J.M., 1988a, Exploration potential for argentiferous base metals at the Walton deposit, Hants County, Nova Scotia: Nova Scotia Department of Mines and Energy, Report 88-1, p. 129-134.

——— 1988b, Walton Project: Final Report: Nova Scotia Department of Mines and Energy, Open File Report 88-045, 15 p.

——— 1989, Exploration potential for argentiferous base metals at the Walton deposit, Hants County, Nova Scotia: Nova Scotia Department of Mines and Energy, Open File Report 89-022, 5 p.

Patterson, K.M., and Powis, K., 2002, Structural and stratigraphic controls on Zn-Pb-Ag mineralization at the Nanisivik Mississippi Valley-type deposit, northern Baffin Island, Nunavut: Geological Survey of Canada, Current Research 2002-C22, 12 p.

Rainbird, R.H., 1997, Report on the Stratigraphic and Paleoenvironmental Setting of Stratabound Pb-Zn-Cu in the Rocknest Formation, Epworth Group, Asiak Fold-Thrust Belt, NT: Internal report for Rhonda Mining Corporation, 18 p.

Randell, R.N., 1994, Geology of the Polaris Zn-Pb Mississippi Valley-Type Deposit, Canadian Arctic Archipelago: Ph.D. thesis, University of Toronto, Toronto, Ontario, 13 maps, 736 p.

Randell, R.N., and Anderson, G.M., 1990, The Geology of the Polaris Carbonate-Hosted Zn-Pb Deposit, Canadian Arctic Archipelago: Geological Survey of Canada, Current Research Paper 90-1D, 7 p.

———1996, Geology of the Polaris Zn-Pb deposit and surrounding area, Canadian Arctic Archipelago, in Sangster, D.F., ed., Carbonate-hosted lead-zinc deposits: Society of Economic Geologists, Special Publication 4, p. 307-319.

Randell, R.N., Héroux, Y., Chagon, A., and Anderson, G.M., 1996, Organic matter and clay minerals at the Polaris Zn-Pb deposit, Canadian Arctic Archipelago: Society of Economic Geologists International Field Conference on Carbonate-Hosted Lead-Zinc Deposits, St. Louis, Missouri, Extended Abstract, p. 247-248.

Ravenhurst, C., Reynolds, P.H., and Zentilli, M., 1987, Isotopic constraints on the genesis of Zn-Pb mineralization at Gays River, Nova Scotia, Canada: Economic Geology, v. 82, p. 1294-1308.

Ravenhurst, C., Reynolds, P.H., Zentilli, M., Krueger, H.W., and Blenkinsop, J., 1989, Formation of carboniferous Pb-Zn and barite mineralization from basin-derived fluids, Nova Scotia, Canada: Economic Geology, v. 84, p. 1471-1488.

Rhodes, D., Lantos, E.A., Lantos, J.A., Webb, R.J., and Owens, D.C., 1984, Pine Point orebodies and their relationship to the stratigraphy, structure, dolomitization, and karstification of the Middle Devonian barrier complex: Economic Geology, v. 79, p. 991-1055.

Robinson, M., and Godwin, C., 1995, Genesis of the Blende Carbonate-hosted Zn-Pb-Ag deposit, North-Central Yukon Territory: geological inclusion and isotopic constraints: Economic Geology, v. 90, p. 369-384.

Rouvier, H., Henry, B., and Le Goff, M., 1995, Regional remagnetization and Mesozoic levels containing "Mississippi Valley-Type" deposits: The southern border of the French Massif Central: XXI UGGI General Assembly, Boulder, Abstract.

Rouvier, H., Henry, B., Macquar, J.-C., Leach, D.L., Le Goff, M., Thibiéroz, J., and Lewchuk, M., 2001, Réaimantation régionale Eocène, migration de fluides et minéralisations sur la bordure cévenole (France): Bulletin de la Société Géologique de France, v. 4, p. 503-516.

Rowan, E.L., Thibiéroz, J., Bethke, C.M., and de Marsily, G., 1996, Geochemical and hydrologic conditions for fluorite mineralization in regions of continental extension: An example from the Albigeois district, France, in Sangster, D. F., ed., Carbonate-hosted lead-zinc deposits: Society of Economic Geologists, Special Publication Number 4, p. 448-464.

Sangster, D.F., 1988, Breccia-hosted lead-zinc deposits in carbonate rocks, *in* James, N. P., and Choquette, P. W., eds., Paleokarst: Springer-Verlag, New York, NY, p. 102-116.

——— 1990, Mississippi Valley-type and SEDEX lead-zinc deposits: a comparative examination: Institution of Mining and Metallurgy, Transactions, Section B: Applied Earth Science, v. 99, p. 21-42.

——— 1995, Mississippi Valley-Type lead-zinc, *in* Eckstrand, O.R., Sinclair, W.D., and Thorpe, R.I., eds., Geology of Canadian Mineral Deposit Types: Geological Survey of Canada, Geology of Canada, no.8, p. 253-261.

——— 2002, MVT deposits of the world; database documentation of the World Minerals Geoscience Database Project, http://www.nrcan.gc.ca/gsc/mrd/wmgdb/index_e.php.

Sass-Gustkiewicz, M., and Kwiecinska, B., 1999, Organic matter in the Upper Silesian (Mississippi Valley-type) Zn-Pb deposits, Poland: Economic Geology, v. 94, p. 981-992.

Sass-Gustkiewicz, M., Dzulynski, S., and Ridge, J.D., 1982, The emplacement of zinc-lead sulphide ores in the upper Silesian ore district: A contribution to understanding Mississippi Valley-type deposits: Economic Geology, v. 77, p. 392-412.

Selby, D., Creaser, R.A., Dewing, K., and Fowler, M., 2005, Evaluation of bitumen as a ^{187}Re-^{187}Os geochronometer for hydrocarbon maturation and migration: A case study from the Polaris MVT deposit, Canada: Earth and Planetary Science Letters, v. 235, p. 1-15.

Sharp, J.M.J., 1978, Energy and momentum transport model of the Ouachita basin and its possible impact on formation of economic mineral deposits: Economic Geology, v. 73, p. 1057-1068.

Sharp, R.J., Ste-Marie, C.P., and Lorenzini, C., 1995a, Field study of Polaris Mine area, Northwest Territories, Canada, in Misra, K.C., ed., Carbonate-hosted lead-zinc-fluorite-barite deposits of North America. Guidebook Series: Geological Association of Canada, 22, p. 38-41.

Sharp, R.J., Ste-Marie, C.P., Lorenzini, C., Leigh, K.E., Dewing, K., Héroux, Y., and Chagnon, A., 1995b, A field guide to the geology of the Polaris Mine, Little Cornwallis Island, Northwest Territories, Canada, *in* Misra, K.C., ed., Carbonate Hosted Lead-Zinc-Fluorite-Barite Deposits of North America. Guidebook Series: Geological Association of Canada, 22, p. 19-37.

Sheppard, S.M.F., Charef, A., and Bouhel, S., 1996, Diapirs and Zn-Pb mineralization: A general model based on Tunisian (N. Africa) and Gulf Coast (U.S.A.) deposits, *in* Sangster, D.F., ed., Carbonate-Hosted Lead-Zinc Deposits: Society of Economic Geologists, Special Publication Number 4, p. 230-243.

Sherlock, R.L., Lee, J.K.E., and Cousens, B.L., 2003, Geological and geochronological constraints on the timing of mineralization at the Nanisivik zinc-lead Mississippi Valley-type deposit, Northern Baffin Island, Nunavut, Canada: Geological Association of Canada Annual Meeting, Vancouver, Abstract.

——— 2004, Geologic and geochronologic constraints on the timing of mineralization at the Nanisivik Zn-Pb Mississippi Valley-Type deposit, Northern Baffin Island, Nunavut, Canada: Economic Geology, v. 99, p. 279-293.

Smethurst, M.T., Symons, D.T.A., Sangster, D.F., and Lewchuk, M.T., 1999, Paleo-magnetic age for Zn-Pb mineralization at Robb Lake, northeastern British Columbia: Bulletin of Canadian Petroleum Geology, v. 47, p. 548-555.

Smith, D.M., Jr., 1996, Sedimentary basins and origin of intrusion-related carbonate-hosted Zn-Pb-Ag deposits, *in* Sangster, D.F., ed., Carbonate-Hosted Lead-Zinc Deposits: Society of Economic Geologists, Special Publication Number 4, p. 255-263.

St. Marie, J., Kesler, S.E., and Allen, C.R., 2001, Origin of iron-rich Mississippi-Valley-type deposits: Geology, v. 29, p. 59-62.

Summer, J.S., 1976, Principles of Induced Polarization for Geophysical Exploration: Developments in Economic Geology: Elsevier Scientific Publishing Co., New York, v. 5, 165 p.

Sverjensky, D.A., 1984, Oil field brines as ore-forming solutions: Economic Geology, v. 79, p. 23-27.

——— 1986, Genesis of Mississippi Valley-type lead-zinc deposits: Annual Review of Earth and Planetary Sciences, v. 14, p. 177-199.

Symons, D.T.A., and Sangster, S.F., 1992, Late Devonian paleomagnetic age for the Polaris Mississippi Valley-type Zn-Pb deposit, Canadian Arctic Archipelago: Canadian Journal of Earth Sciences, v. 29, p. 15-25.

Symons, D.T.A., Pan, H., Sangster, D.F., and Jowett, E.C., 1993, Paleomagnetism of the Pine Point Zn-Pb deposits: Canadian Journal of Earth Sciences, v. 30, p. 1028-1036.

Symons, D.T. A., Sangster, D.F., and Leach, D.L., 1995, A Tertiary age from paleomagnetism for Mississippi Valley-type zinc-lead mineralization of Upper Silesia, Poland: Economic Geology, v. 90, p. 782-794.

Symons, D.T.A., Lewchuk, M.T., and Sangster, D.F., 1998, Laramide orogenic fluid flow into the Western Canada Sedimentary Basin, evidence from paleomagnetic dating of the Kicking Horse Mississippi Valley-type ore deposit: Economic Geology, v. 93, p. 68-83.

Symons, D.T.A., Symons, T.B., and Sangster, D.F., 2000, Paleomagnetism of the Society Cliffs dolostone and the age of the Nanisivik zinc deposits, Baffin Island, Canada: Mineralium Deposita, v. 35, p. 672-682.

Thompson, R.I., 1989, Stratigraphy, Tectonic Evolution and Structural Analysis of the Halfway River Map Area (94B), Northern Rocky Mountains, British Columbia: Geological Survey of Canada, Memoir 425, 119 p.

Thorkelson, D.J., Mortensen, J.K., Creaser, R.A., Davidson, G.J., and Abbott, J.G., 2001, Early Proterozoic magmatism in Yukon, Canada: constraints on the evolution of northwestern Laurentia: Canadian Journal of Earth Sciences, v. 38, p. 1479-1494.

Thorsteinsson, R., 1959, Cornwallis and Little Cornwallis Islands, District of Franklin, Northwest Territories: Geological Survey of Canada, Memoir 294, 134 p.

Tikkanen, G.D., 1986, World resources and supply of lead and zinc, *in* Bush, W.R., ed., Economics of Internationally Traded Minerals: Society of Mining Engineers, Inc., p. 242-250.

Titley, S.R., 1996, Characteristics of high temperature, carbonate-hosted replacement ores and some comparisons with Mississippi Valley-type ores, *in* Sangster, D.F., ed., Carbonate-Hosted Lead-Zinc Deposits: Society of Economic Geologists, Special Publication Number 4, p. 244-254.

Turner, W.A., Pierce, K.L., and Cairns, K.A., 2002, Great Slave Reef (GSR) project drillhole database. A compilation of the drillhole locations, drill logs, and associated geochemical data for the Great Slave Reef Joint Venture Project; Interior Platform, Northwest Territories, Canada (85B11 to 14): C.S. Lord Northern Geoscience Centre, Open Report 2002-001, CD-ROM.

Utting, J., Keppie, J.D., and Giles, P.S., 1989, Palynology and stratigraphy of the Lower Carboniferous Horton Group, Nova Scotia: Contributions to Canadian Paleontology; Bulletin - Geological Survey of Canada, p. 117-143.

Viets, J.G., Hofstra, A.H., and Emsbo, P., 1996, Solute compositions of fluid inclusions in sphalerite from North America and European Mississippian valley-type ore deposits and ore fluids derived from evaporated seawater, *in* Sangster, D. F., ed., Carbonate-hosted lead-zinc deposits: Society of Economic Geologists, Special Publication Number 4, p. 465-482.

Wachowiak, N., Gummer, P.K., Pope, M., and Grotzinger, J.P., 1997, Stratabound Zn-Pb mineralization of the Esker Horizon, Paleoproterozoic Rocknest Formation, Wopmay Orogen, Canada: Northwest Territories Geoscience Forum 25th Anniversary, Yellowknife, Northwest Territories, Canada, Program and Abstracts of Talks and Posters, p. 96-98.

Westervelt, T., 1979, Structural superposition in the Lake O'hara region, Yoho and Kootenay National Parks, British Columbia, Canada: Ph. D. thesis, University of Wyoming, Laramide, Wyoming, 290 p.

Yao, Q., and Demicco, R.V., 1995, Paleoflow patterns of dolomitizing fluids and paleohydrogeology of the southern Canadian Rocky Mountains: Evidence from dolomite geometry and numerical modeling: Geology, v. 23, p. 791-794.

Magmatic Nickel-Copper-Platinum Group Element Deposits

O. Roger Eckstrand and Larry J. Hulbert

Geological Survey of Canada, 601 Booth Street, Ottawa, Ontario, K1A 0E8
Corresponding author's email: lhulbert@nrcan.gc.ca

Abstract

Magmatic deposits containing exploitable quantities of nickel, copper, and platinum group elements (PGE) are associated with variable quantities of localized sulphide concentrations in mafic and ultramafic rocks. Ni-Cu deposits, nickel being the main economic commodity, are associated with high concentrations of sulphides, and the host bodies are classified based on the nature of the confining magmatic environment: (1) meteorite-impact, (2) rift and continental flood basalt, (3) komatiitic, and (4) other related mafic/ultramafic bodies. Platinum group element deposits are also confined to mafic/ultramafic bodies, but are associated with low quantities of sulphides. Reef-type or stratiform PGE deposits form in large, well-layered mafic/ultramafic intrusions, whereas magmatic breccia-type deposits occurs in stock-like or layered bodies.

The economics and rarity of such deposits with respect to number, grade, tonnage, and mining districts are outlined. In addition, the geological attributes of the various deposit types and subtypes are documented.

Exploration models based on district and local scales are discussed, as well as recent advances and knowledge gaps in this field.

Résumé

Les gîtes magmatiques renfermant des quantités exploitables de nickel, de cuivre et d'éléments du groupe du platine (ÉGP) sont associés à des concentrations localisées de sulfures, en quantités plus ou moins importantes, dans les roches mafiques et ultramafiques. Les gîtes de Ni-Cu, où le nickel est la principale substance utile, sont associés à de fortes concentrations de sulfures et les corps hôtes sont classés d'après la nature des milieux magmatiques qui les renferment : (1) impact météoritique, (2) basaltes de rift et de plateaux continentaux, (3) unités komatiitiques et (4) autres corps mafiques/ultramafiques connexes. Les gîtes d'éléments du groupe du platine sont également restreints aux corps mafiques et ultramafiques, mais sont associés à de faibles quantités de sulfures. Les gîtes d'ÉGP de type horizon minéralisé ou minéralisation stratiforme sont formés dans de grandes intrusions mafiques/ultramafiques bien stratifiées, alors que les gîtes de type brèche magmatique se forment dans des corps s'apparentant à des stocks ou dans des massifs stratifiés.

La valeur et la rareté de ces gîtes sont soulignées en termes de nombres, de teneurs de tonnages et de districts miniers. Les attributs géologiques des divers types et sous-types de gîtes sont en outre documentés.

Des modèles d'exploration à l'échelle du district et à l'échelle locale sont discutés et les progrès récents dans ce domaine ainsi que les lacunes dans nos connaissances sont soulignés.

Definition

A broad group of deposits containing nickel, copper, and platinum group elements (PGE) occur as sulphide concentrations associated with a variety of mafic and ultramafic magmatic rocks (Eckstrand et al., 2004; Naldrett, 2004). The magmas originate in the upper mantle and contain small amounts of nickel, copper, PGE, and variable but minor amounts of S (the one exception to this source of magma is the Sudbury Igneous Complex, or SIC, which will be discussed separately). The magmas ascend through the crust and cool as they encounter cooler crustal rocks. If the original S content of the magma is sufficient, or if S is added from crustal wall rocks, a separate sulphide liquid forms as droplets dispersed throughout the magma. Because the partition coefficients of nickel, copper, and PGE as well as iron favour sulphide liquid over silicate liquid, these elements preferentially transfer into the sulphide droplets from the surrounding magma. The sulphide droplets tend to sink toward the base of the magma because of their greater density, and form sulphide concentrations. On further cooling, the sulphide liquid crystallizes to form the ore deposits that contain these metals.

Among such deposits, two main types are distinguishable. In the first, Ni-Cu sulphide, Ni and Cu are the main economic commodities. These occur as sulphide-rich ores that are associated with differentiated mafic and/or ultramafic sills and stocks, and ultramafic (komatiitic) volcanic flows and sills. The second type is exploited principally for PGE, which are associated with sparsely dispersed sulphides in very large to medium-sized, typically mafic/ultramafic layered intrusions.

In Ni-Cu sulphide deposits (the first type), Ni constitutes the main economic commodity, generally at grades of about 1 to 3 percent. Copper may be either a coproduct or by-product, and Co, PGE, and Au are the usual by-products. However, in some cases, such as Noril'sk-Talnakh, PGE may constitute highly significant coproducts. Other commodities recovered in some cases include Ag, S, Se, and Te. These metals are all associated with the sulphides, which generally make up more than 10 percent of the ore.

The mafic and ultramafic magmatic bodies that host the Ni-Cu sulphide ores are diverse in form and composition, and can be subdivided into the following four subtypes:

1. A meteorite-impact mafic melt sheet that contains basal sulphide ores (Sudbury, Ontario is the only known example).
2. Rift and continental flood basalt-associated mafic sills and dyke-like bodies (Noril'sk-Talnakh, Russia; Jinchuan,

Eckstrand, O.R., and Hulbert, L.J., 2007, Magmatic nickel-copper-platinum group element deposits, *in* Goodfellow, W.D., ed., Mineral Deposits of Canada: A Synthesis of Major Deposit Types, District Metallogeny, the Evolution of Geological Provinces, and Exploration Methods: Geological Association of Canada, Mineral Deposits Division, Special Publication No. 5, p. 205-222.

China; Duluth Complex, Minnesota; Muskox, Nunavut; and Crystal Lake intrusion, Ontario).

3. Komatiitic (magnesium-rich) volcanic flows and related sill-like intrusions (Thompson, Manitoba; Raglan and Marbridge, Quebec; Langmuir, Ontario; Kambalda and Agnew, Australia; Pechenga, Russia; Shangani, Trojan, and Hunter's Road, Zimbabwe).

4. Other mafic/ultramafic intrusions (Voisey's Bay, Labrador; Lynn Lake, Manitoba; Giant Mascot, British Columbia; Kotalahti, Finland; Råna, Norway; and Selebi-Phikwe, Botswana).

The PGE of the second type of deposit include Os, Ir, Ru, Rh, Pt, and Pd. Platinum and Pd are the most abundant of these and determine the economic value of these ores, although Rh, Ni, Cu, and Au are commonly recovered as well.

1. PGE-dominant magmatic sulphide ores are associated with mafic/ultramafic intrusions. There are two principal subtypes of deposits:

2. Reef-type or stratiform PGE deposits, which occur in well layered mafic/ultramafic intrusions (Merensky Reef and UG-2 chromitite layer of the Bushveld Complex, South Africa; J-M Reef of the Stillwater Complex, Montana; Main sulphide zone in the Great Dyke, Zimbabwe).

3. Magmatic breccia type, which occurs in stock-like or layered mafic/ultramafic intrusions (Platreef deposits of the northern Bushveld Complex, South Africa; Lac des Iles deposit and Marathon deposit, Ontario).

Mafic/ultramafic rocks host other types of mineralization as well. These include lateritic nickel deposits, placer Pt deposits, chromite deposits, and titaniferous magnetite deposits. None of these are discussed further.

Economic Characteristics

Magmatic Ni-Cu sulphide deposits provide most of the Ni produced in the world and continue to have substantial reserves. However, lateritic Ni deposits, formed from the weathering of ultramafic rocks, are also substantial sources of Ni, and have global reserves greater than those of Ni-Cu sulphide deposits. Lateritic Ni deposits do not occur in Canada, but will probably in time become the main source of nickel.

Magmatic PGE deposits and Ni-Cu sulphide deposits are the source of essentially all of the world's platinum group elements. Placer deposits have also been mined for Pt in many parts of the world, but are of little significance in Canada and appear to have little potential elsewhere.

Some Ni-Cu-PGE deposits occur as individual sulphide bodies associated with magmatic mafic and/or ultramafic bodies. Others occur as groups of sulphide bodies associated with one or more related magmatic bodies in areas or belts up to tens, even hundreds of kilometres long. Such groups of deposits are known as districts (e.g. Sudbury, Thompson, Noril'sk-Talnakh, Kambalda, Raglan). In total there are 142 Ni-Cu-PGE deposits and districts in the world for which grade and ore tonnage data have been reported that contain more than 100 000 tonnes of resources and/or production, as shown in Figure 1. These include deposits that are economic or possibly economic. The distribution of these deposits in Canada is shown in Figure 2. Among the global deposits/districts

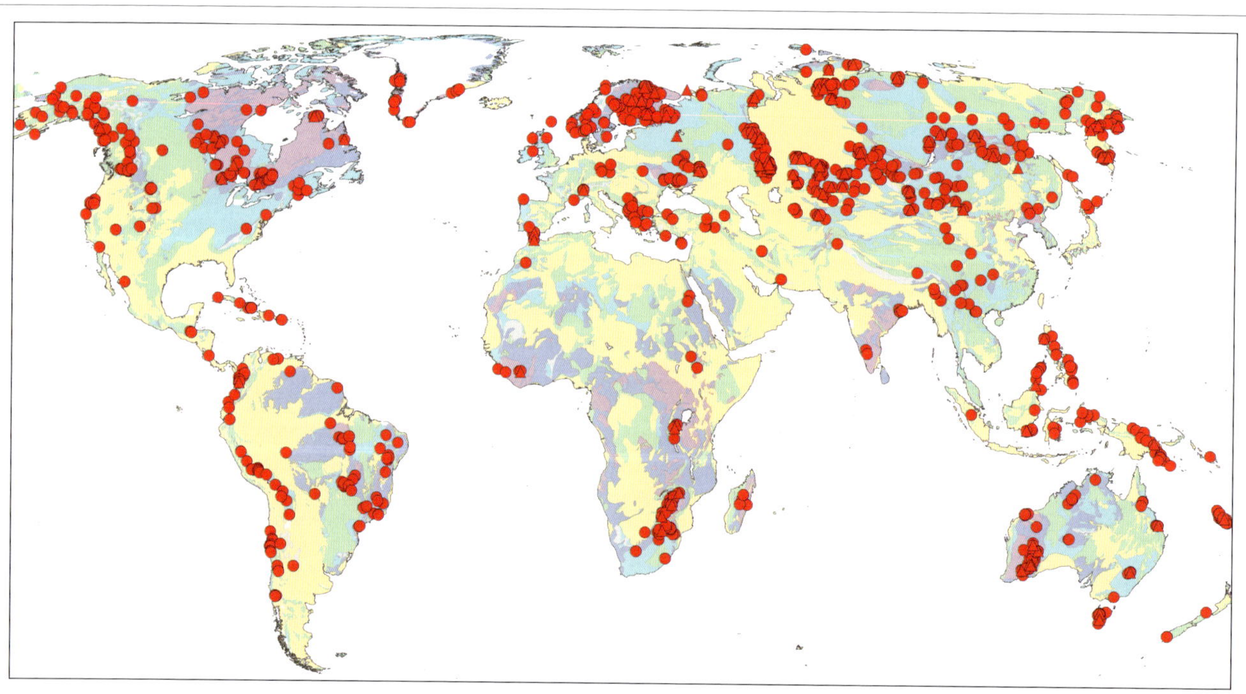

FIGURE 1. World map (after Chorlton, 2003) showing magmatic Ni-Cu-PGE sulphide deposits having resources and/or production greater than 100,000 tonnes of ore.

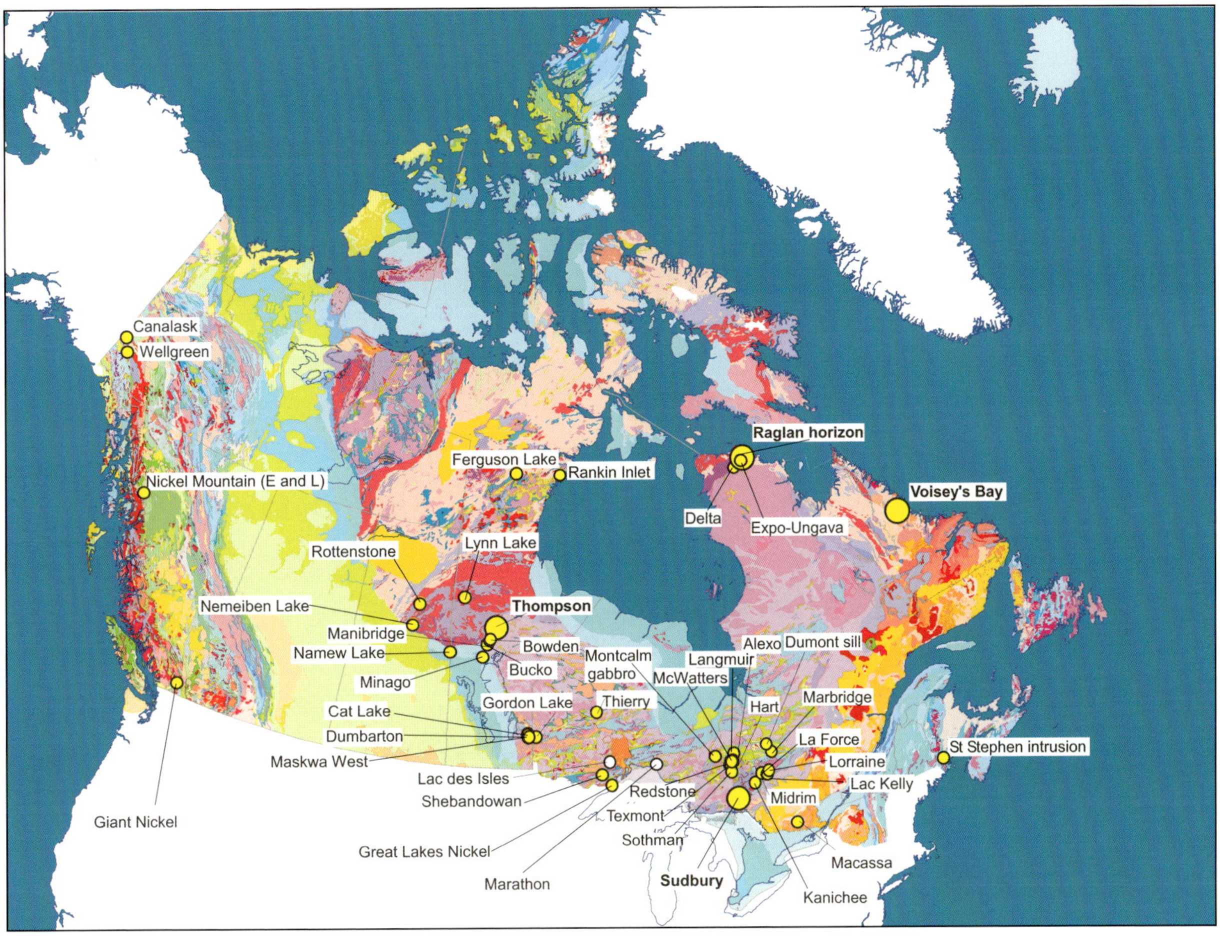

FIGURE 2. Geological map of Canada (after Wheeler et al., 1996), showing the distribution of magmatic Ni-Cu-PGE sulphide deposits with resources and/or production greater than 100 000 tonnes of ore. Ni-Cu deposits are shown in yellow, with PGE deposits shown in white.

there are 51 Ni-Cu deposits/districts and 5 PGE deposits/districts with greater than 10 million metric tonnes (MT), and 13 Ni-Cu deposits/districts and 2 PGE deposits/districts with greater than 100 Mt.

Grade and Tonnage Characteristics

Among Ni-Cu deposits, Ni grades are typically between 0.7 and 3 percent, and Cu grades are between 0.2 and 2 percent (Fig. 3). Ore tonnages of individual deposits range from a few hundred thousands to a few tens of millions (Fig. 3A). Two giant Ni-Cu districts stand out above all the rest in the world: Sudbury, Ontario, and Noril'sk-Talnakh, Russia, with ore tonnages of 1645 and 1903 Mt respectively (Fig. 4). Other major Ni-Cu districts include the Thompson, Voisey's Bay, and Raglan districts in Canada, and Jinchuan (China), Kambalda (Australia), and Pechenga (Russia).

The most important platinum-rich PGE district in the world is the Bushveld Complex, South Africa (Pt/Pd = 1.35), which contains two major types of PGE deposits. The next in importance is the Noril'sk-Talnakh district, which is exceptionally Pd-rich (Pd/Pt = 3.5) as a by-product of its Ni-Cu ores. Stillwater, U.S. is also a significant producer of unusually rich PGE ores (Pd/Pt = 3.6). Canada's only primary producing deposit is the Lac des Iles Pd deposit (Pd/Pt = 9.2). The Sudbury district, because of its size, also produces significant amounts of PGE, although PGE tenors are comparatively low.

Grades and tonnages of global magmatic Ni-Cu deposits (Fig. 4) show that Sudbury and Noril'sk-Talnakh are the only districts that contain in excess of 10 million tons of contained Ni. The other important districts tend to have Ni contents of about 1 to 6 million tonnes.

Geological Attributes

Magmatic Ni-Cu-PGE deposits are consistently found in association with mafic and/or ultramafic magmatic bodies, but these parent bodies occur in diverse geological settings. Their ages are predominantly Archean and Paleoproterozoic (Fig. 3E). In the following account, the two main types, (1) Ni-Cu and (2) PGE, and the four subtypes of Ni-Cu will be treated separately. Each account will begin with regional settings and proceed with progressively more detailed characterization of the deposits, including local geological setting, associated bounding rocks, the magmatic host rocks, and the ores themselves.

Nickel-Copper Deposits

As noted above, these ores are characterized by an abundance of sulphide. Much of the S in the sulphides was de-

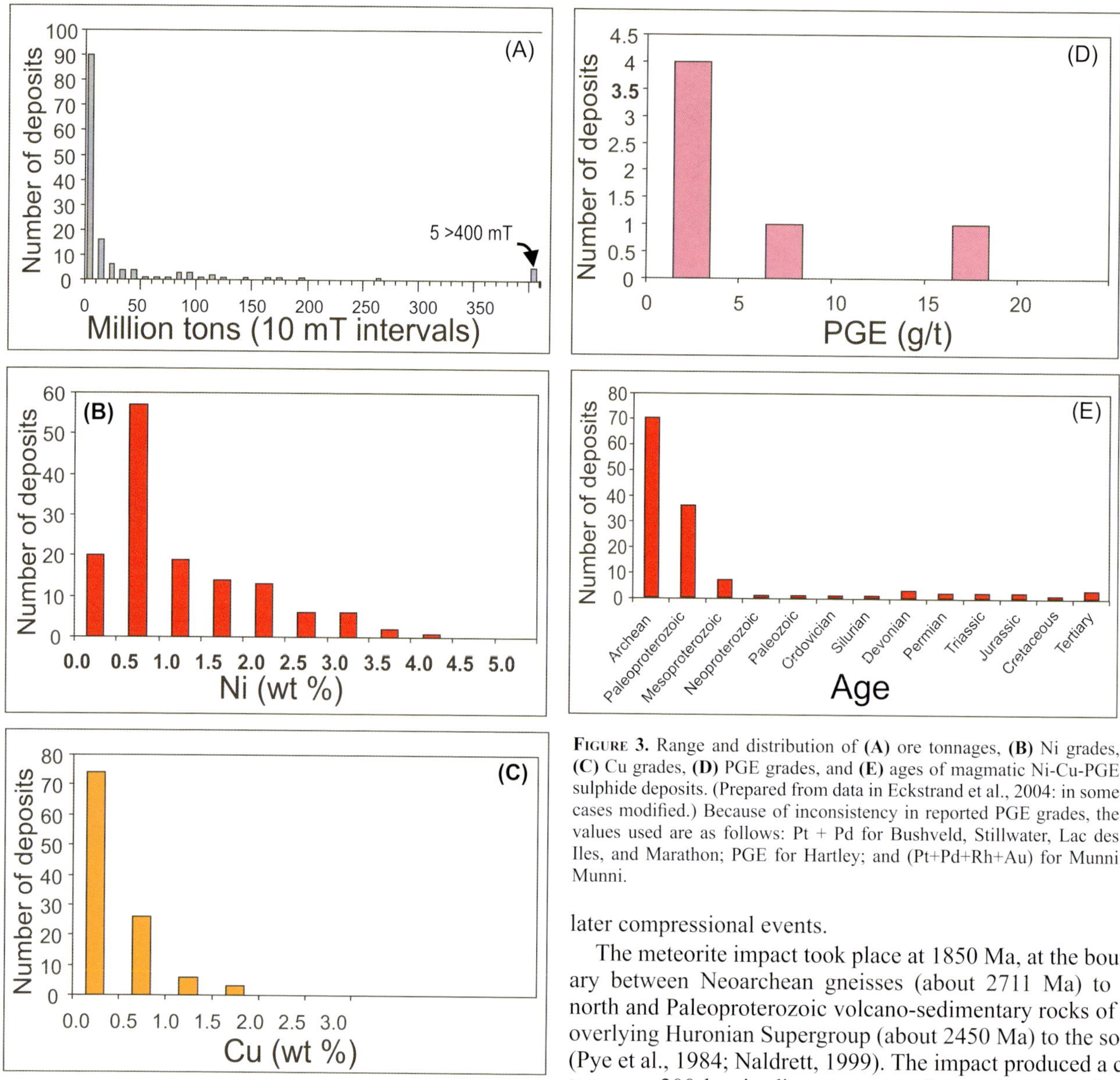

FIGURE 3. Range and distribution of **(A)** ore tonnages, **(B)** Ni grades, **(C)** Cu grades, **(D)** PGE grades, and **(E)** ages of magmatic Ni-Cu-PGE sulphide deposits. (Prepared from data in Eckstrand et al., 2004: in some cases modified.) Because of inconsistency in reported PGE grades, the values used are as follows: Pt + Pd for Bushveld, Stillwater, Lac des Iles, and Marathon; PGE for Hartley; and (Pt+Pd+Rh+Au) for Munni Munni.

rived by assimilation (e.g., Grinenko, 1985). It is likely that the high content of S in the magma caused over saturation of S in the magma, thus producing large quantities of sulphide liquid. As stated above, Ni, Cu, and PGE partition preferentially into sulphide liquid relative to silicate liquid. On cooling, the liquid sulphide crystallizes over a large temperature range to eventually form the common mineral assemblage dominated by pyrrhotite-pentlandite-chalcopyrite.

Meteorite-Impact Subtype

Sudbury is the only known representative of this type of Ni-Cu deposit. Because meteorite impacts are random events on the earth's surface, there is no possible regional geological control on their distribution, with the exception that subsequent geological events could obscure or obliterate their traces. In the case of the Sudbury Igneous Complex (Fig. 5), it is well preserved although strongly deformed by later compressional events.

The meteorite impact took place at 1850 Ma, at the boundary between Neoarchean gneisses (about 2711 Ma) to the north and Paleoproterozoic volcano-sedimentary rocks of the overlying Huronian Supergroup (about 2450 Ma) to the south (Pye et al., 1984; Naldrett, 1999). The impact produced a crater some 200 km in diametre, as well as radiating and concentric fracture/breccia zones that penetrated the surrounding wall rocks for distances of tens of kilometres. The impact generated a high-temperature melt layer that occupied the floor of the impact crater. On cooling, the melt differentiated into a lower norite unit and an overlying granophyre, separated by a thinner gabbro layer. Contacts between these units are gradational, and finer-scale layering is absent. A discontinuous, more mafic basal unit termed the sublayer contains most of the Ni-Cu ores and abundant xenolithic clasts (Souch et al., 1969; Pattison, 1979; Naldrett et al., 1984). The melt also intruded some of the radiating breccia zones, forming many kilometres long quartz diorite dykes (offsets) extending outward from the SIC, and these also contain Ni-Cu ores (Cochrane, 1984).

Subsequent regional overthrusting from the south compressed the southern half of the SIC and produced the presently exposed elongate basin 65 km long and 27 km across (Shanks and Schwerdtner, 1991). The inward dip of the com-

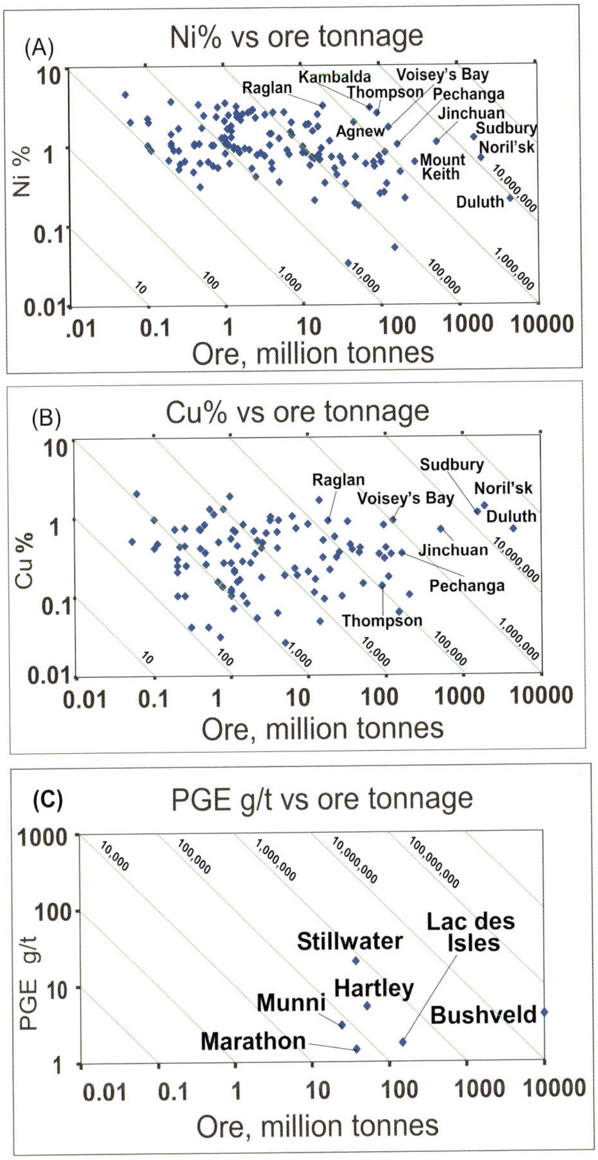

FIGURE 4. Grade and tonnage plots of global magmatic Ni-Cu sulphide deposits. **(A)** Tonnages vs. Ni grades; **(B)** Tonnages vs. Cu grades; **(C)** Tonnages vs. PGE grades. (Prepared from data in Eckstrand et al., 2004: in some cases modified.) Inclined contours show quantities of contained metals in each figure; tonnes for Ni and Cu, and kg for PGE.

plex averages about 30° along the less-deformed north range, and 45° to 60° along the strongly deformed south range. The total thickness of the complex is about 2.5 km.

The impacted country rocks contained significant amounts of S in the form of sulphides. These were incorporated in the initial super-liquidus melt as dissolved S, but with cooling, the melt became saturated with respect to S. Sulphide liquid was thus produced, which extracted Ni, Cu, and PGE from the silicate melt. Another factor contributing to formation of sulphide was the reduced solubility of sulphide in the melt caused by the mixing of mafic and felsic target rocks. The liquid sulphide, along with abundant fragmental material, segregated into a basal mafic noritic unit (sublayer) and collected in depressions (embayments) along the base of the melt sheet. The Murray mine is in such an embayment (Fig. 5; Souch et al., 1969). Sulphide liquid also accompanied melt into the offsets. On cooling, the sulphide liquid crystallized to form Ni-Cu-PGE ores. In some of the embayments, sulphide melt remaining after partial crystallization migrated downward from the SIC into breccia zones in the footwall rocks to produce particularly Cu and PGE-rich sulphide ore veins and masses up to 400 m below the sublayer.

The resulting orebodies associated with the sublayer at the base of the intrusion form irregular lenticular sulphide-rich masses, with the longest dimension plunging steeply as at the Murray mine on the South Range (Fig. 6A), and the Strathcona, McCreedy East, and Fraser mines on the North Range (Fig. 6B; Coats and Snajdr, 1984). Clusters of such orebodies, similarly oriented, lie in the embayments and persist to great depths as at the Creighton mine. The orebodies in the offsets form discontinuous sulphide-rich sheets or lenses with steep dips subparallel to the associated quartz diorite offset. An example is the orebody in the Copper Cliff mine shown in Figure 6C (Cochrane, 1984).

A different kind of ore zone occurs at the Falconbridge East mine, where the ore is irregularly strung out as discontinuous sheets along the Main fault, which separates the felsic norite of the SIC from the Stobie volcanics (Fig. 6D; Owen and Coats, 1984). The deep Cu-PGE-rich ores in the footwall below the SIC form sets of subparallel stringers and veins of massive sulphides (Fig. 6B; Coats and Snajdr, 1984).

The sulphide ores consist of the typical magmatic sulphide minerals. In general order of abundance, they include pyrrhotite, pentlandite, chalcopyrite, and pyrite. Bornite is present in copper-rich ores, and South Range ores typically contain arsenic minerals, including niccolite, maucherite, gersdorfite, and cobaltite. The platinum group elements occur as microscopic grains of numerous minerals, the most abundant of which are michenerite (PdBiTe), moncheite ($PtTe_2$), and sperrylite ($PtAs_2$).

Sudbury ores have many of the same textural features as other magmatic Ni-Cu sulphide ores. Massive ores (Fig. 7C) consist mainly of an annealed mosaic of subequant pyrrhotite grains with shreddy interstitial pentlandite. Breccia ores (Fig. 7D) contain rock clasts and silicate grains suspended in a matrix of sulphide (mostly pyrrhotite with patchy grains of pentlandite; chalcopyrite often penetrates the rock clasts). A distinctive feature of Sudbury sulphide-rich ores and the hosting sublayer is the presence of clasts of ultramafic rock, not exposed elsewhere but likely unmelted residue of one of the rocks impacted by the meteorite.

Rift and Continental Flood Basalt-Associated Subtype

Ni-Cu deposits of the rift and continental flood basalt-associated subtype are the products of the magmatism that accompanies intracrustal rifting events. They include the largest deposit, Noril'sk-Talnakh, (12.6 MT of contained Ni), and several other large deposits, for example, Jinchuan (Chai and Naldrett, 1992) and Duluth. The features that these deposits tend to have in common are that they are associated with large magma systems, and that within these systems the Ni-Cu sulphide ores tend to be associated with conduits

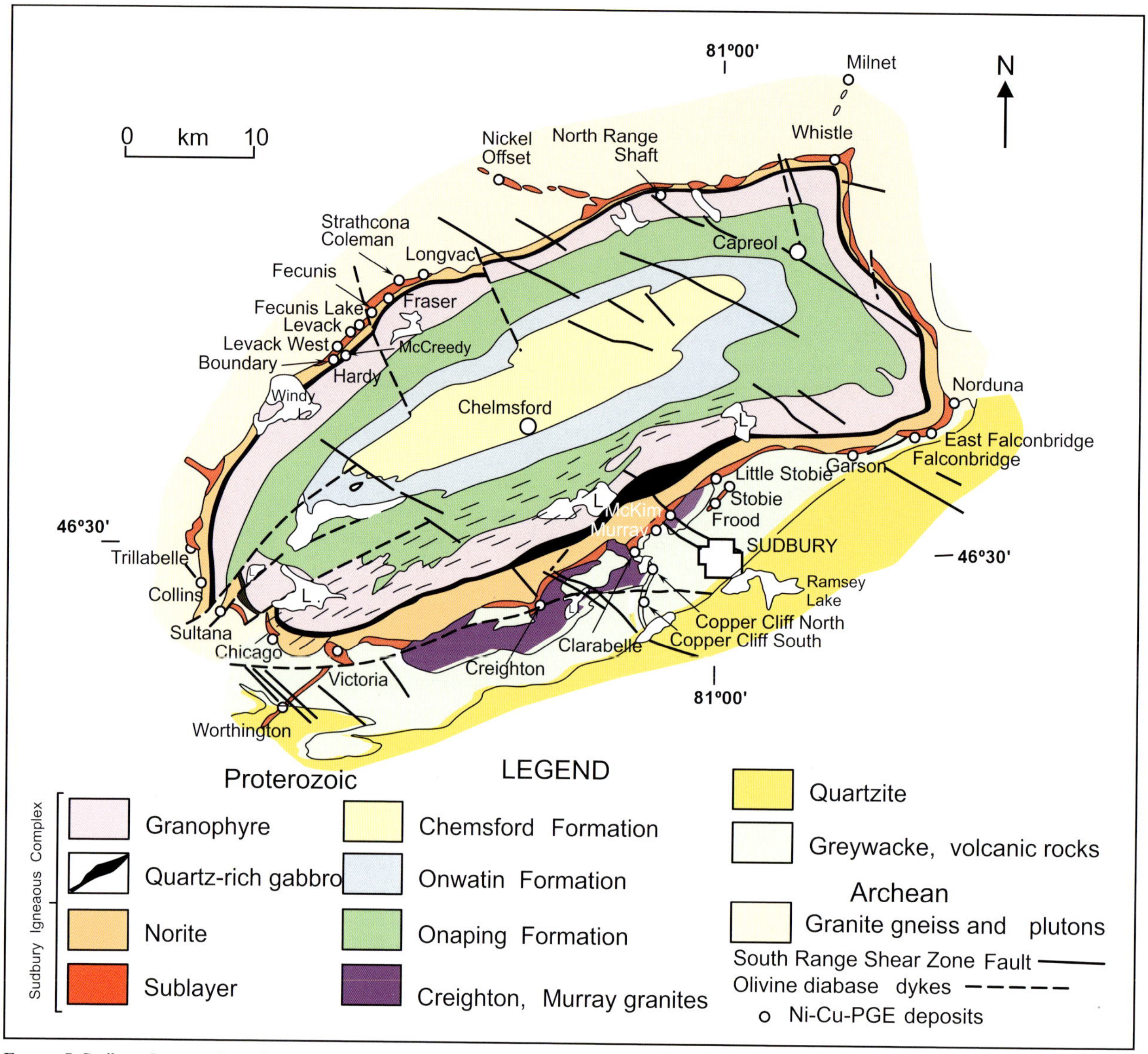

FIGURE 5. Sudbury Igneous Complex: geological map (assembled from Pattison, 1979; Naldrett et al., 1984; Naldrett, 1989; Shanks and Schwerdtner, 1991).

or feeders to the larger igneous masses (in this last respect, Duluth is an exception in which the low-grade Ni-Cu sulphides have not yet proven to be economic). Much of the sulphide has been derived by contamination of the magma through incorporation of S from adjoining wall rocks. Once formed, and if in sufficient quantity, the sulphides tend to settle gravitationally within the moving magma, and collect in the conduits at points where magma velocity is reduced. The sulphides have probably experienced progressive enrichment by repeated extraction of additional metals from successive pulses of magma moving through the conduits (Maier et al., 2001).

Noril'sk-Talnakh: The Ni-Cu-PGE ores of the Noril'sk-Talnakh district (Duzhikov et al., 1992; Naldrett and Lightfoot, 1992) are spatially associated with the huge Siberian flood basalt magmatic suite. In the Noril'sk-Talnakh area, the sedimentary strata form a gentle north–south-trending syncline. Intruded into this sequence are elongate, gently dipping sill-like mafic bodies that underlie the 3.5 km thick lava sequence. These are the units with which the ores are associated (Fig. 8), and that are considered to be feeders to the overlying volcanic rocks. All the ore-bearing sills lie within 7 km of the NNE trending Karayelakh fault, which is thought to be part of the conduit system. The sills have thicknesses of a few tens of metres, lateral extents of a few hundred metres, and lengths of a few kilometres. They consist of a variety of layer-like gabbro-dolerite units (Fig. 9; Distler, 1994). The lowermost unit consists of an olivine-free gabbro-dolerite contact facies overlain by coarser-grained taxitic olivine gabbro-dolerite, which passes upwards into picritic gabbro-dolerite. Olivine-free gabbro-dolerite and anorthosite units make up the upper portions of these bodies. The sills are enveloped by metamorphic aureoles of exceptional thickness (up to 200 metres) and, hence, are considered to have been conduits for the passage of very large volumes of magma.

Three distinct types of Ni-Cu-PGE ore occur in specific

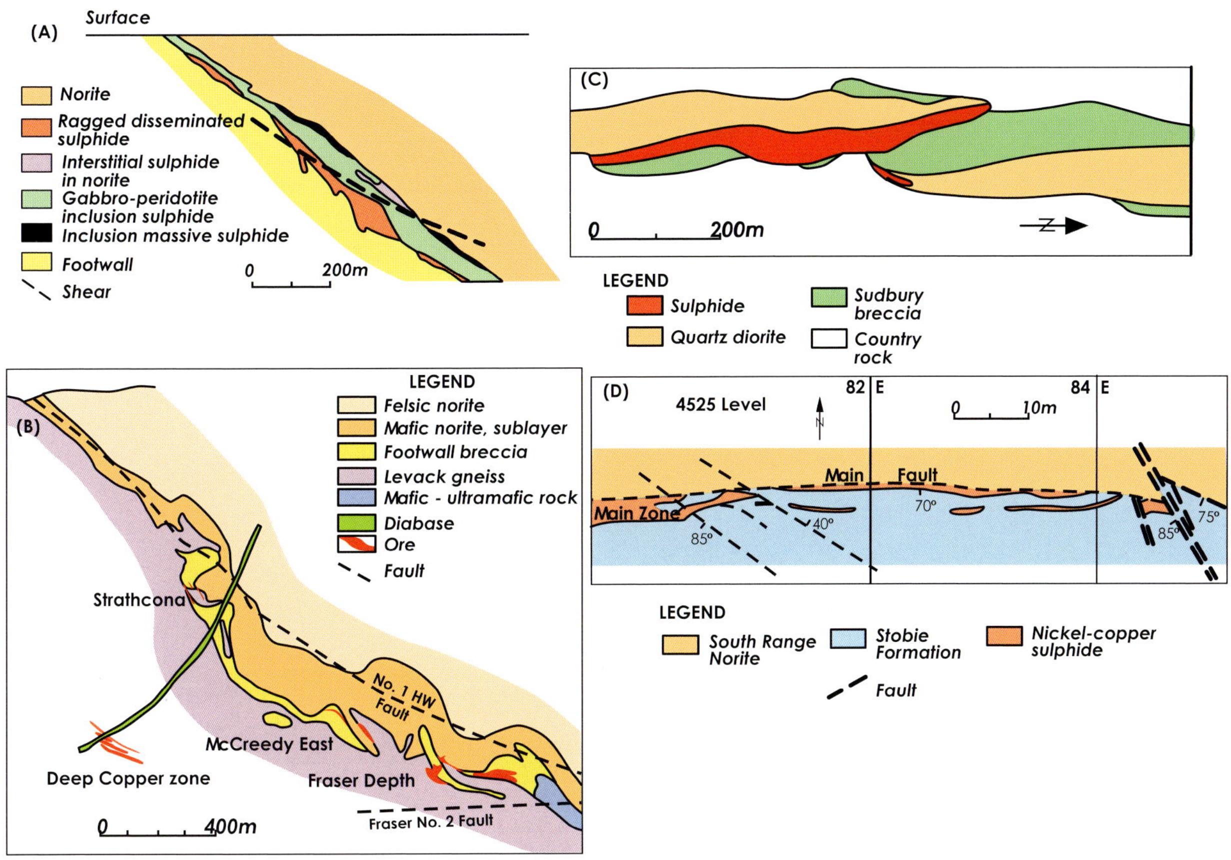

FIGURE 6. Sudbury ore deposits: geological maps and sections. **(A)** Murray mine (after Souch et al., 1969); **(B)** Strathcona, McCreedy East and Fraser mines (after Coats and Snajdr, 1984); **(C)** Copper Cliff South mine (after Cochrane, 1984); **(D)** Falconbridge East mine (after Owen and Coats, 1984).

associations with the mineralized sills, and contribute to the total resources of the Noril'sk-Talnakh ore field (Table 1).

1. Massive sulphide ores occur as flat-lying sheets and lenses at the base of the sills, in some cases protruding downward into the footwall rocks (Figs. 8, 9). One such massive sulphide orebody attains a thickness of over 50 m and lateral dimensions of hundreds of metres. Some of the larger orebodies display remarkable sulphide zonation, ranging from pyrrhotite dominated chalcopyrite-pentlandite assemblages in the outermost and lower parts, through progressively more copper-rich zones, to mainly Cu sulphides, chalcopyrite, cubanite, and mooihoekite together with pentlandite in the central upper parts (Stekhin, 1994). The latter can have up to 25 to 30 percent Cu, 3 to 6 percent Ni, 50 to 60 ppm Pt, and 60 to 200 ppm Pd. This zonation of sulphides is believed to result from fractionation in situ. The mechanism of early cumulate separation and basal segregation of a pyrrhotite-like iron sulphide leaves a Cu-PGE-rich supernatant liquid to crystallize last. These Ni and Cu-rich massive sulphide ores have been the mainstay of Noril'sk production for much of the district's history.

TABLE 1. Noril'sk-Talnakh Ore Field—Measured, Indicated, and Inferred Resources (2003)

Ore type	Ore (Mt)	Ni %	Cu %	PGE (g/t)
Rich (massive)	88.7	3.42	5.38	5–100
Cuprous (Cu breccia)	108.4	0.8	2.64	5–50
Disseminated	1706.3	0.51	1.02	2–10
Total	1903.4	0.66	1.31	

2. Copper breccia ores as semiconformable sheet-like zones occupy the upper contacts of the sills with the overlying rocks (stringer-disseminated ores in Fig. 9). The breccia comprises fragments of both the intrusion and wall rocks in a matrix of mainly massive sulphide. Sulphide stringers and disseminations accompany the breccias.

3. Disseminated sulphide ores form lenticular to tabular layers in picritic gabbro-dolerite units within the sills. The sulphides generally take the form of centimetre-size spheres of chalcopyrite, pentlandite, and pyrrhotite dispersed through the host gabbro-dolerite. This was the first ore type mined at Noril'sk; later it declined in importance with the discovery of massive sulphide ores. However, it is presently an important component in mining reserves again due to the high price of platinum.

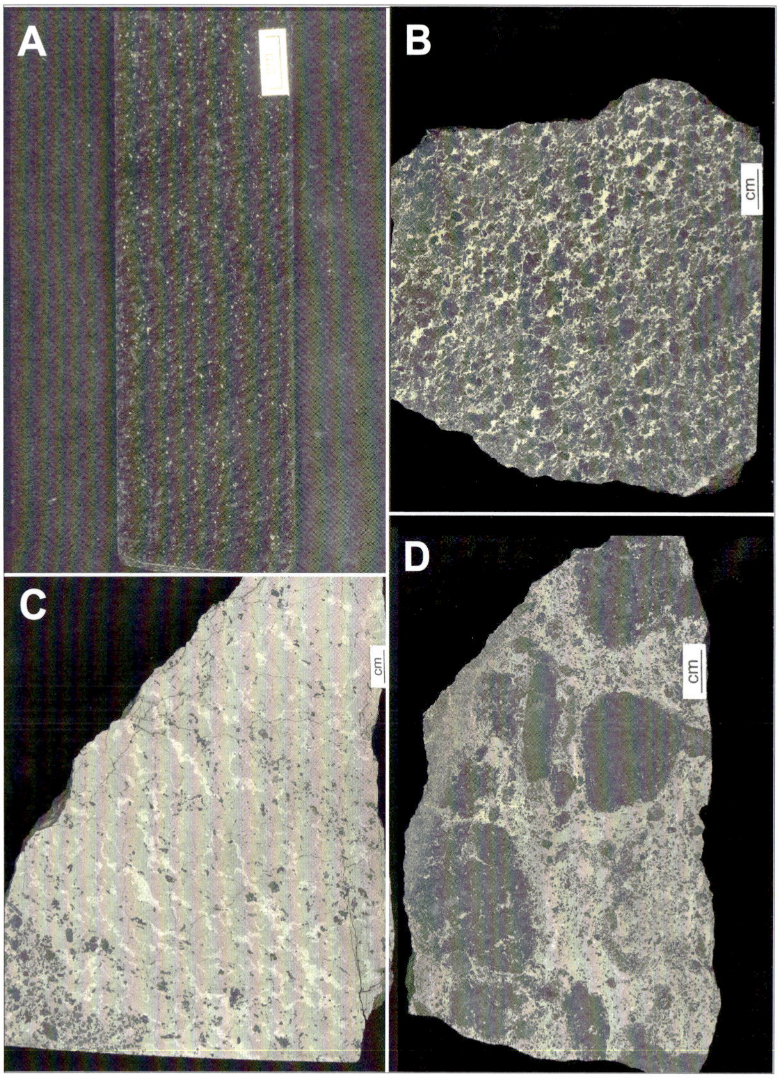

FIGURE 7. Typical magmatic Ni-Cu sulphide ore textures: **(A)** disseminated sulphides, Thompson; **(B)** matrix-textured sulphides, Lynn Lake; **(C)** massive sulphides, Lynn Lake; **(D)** sulphide breccia, Lynn Lake. Pyrrhotite=medium gray, pentlandite=light gray, chalcopyrite=yellow, silicate gangue=dark gray to black. Photographs courtesy of L.J. Hulbert.

Noril'sk-Talnakh ores are exceptionally rich in PGE, to the degree that the precious metals currently have approximately the same value in the ores as the base metals. Noril'sk is the world's leading producer of Pd, and supplies about 20 percent of the world's Pt, second only to the Bushveld.

Komatiitic Volcanic Flow and Sill-Associated Subtype

Komatiitic Ni-Cu deposits are widely distributed in the world, mainly in Neoarchean and Paleoproterozoic terranes (Lesher, 1989). Major Ni-Cu producing districts and other prominent deposits are found in Australia, Canada, Brazil, Zimbabwe, Finland, and Karelia (Russia).

The komatiitic subtype of Ni-Cu sulphide deposits occurs for the most part in two different settings. One setting is as komatiitic volcanic flows and sills in mostly Neoarchean greenstone belts. Greenstone belts are typical terranes found in many Archean cratons, and may represent intra-cratonic rift zones or rifted arcs. They are generally composed of strongly folded, basaltic/andesitic volcanic rocks and related sills, siliciclastic sediments, and granitoid intrusions. They have been metamorphosed to greenschist and amphibolite facies, and typically adjoin tonalitic gneiss terranes. Komatiitic rocks form an integral part of some of these greenstone belts. Examples are the Kambalda district and the Mt. Keith deposit, respectively, from two greenstone belts in Western Australia. The second setting is as Paleoproterozoic komatiitic sills associated with rifting at cratonic margins. Prime examples are the Raglan horizon in the Cape Smith-Wkeham Bay Belt of Ungava, Quebec, and the Thompson district of the Thompson Nickel Belt, in northern Manitoba. The komatiitic rocks are set in a sequence of volcano-sedimentary strata unconformably resting on Archean basement and are weakly (Raglan) to intensely (Thompson) folded and deformed. An additional Paleoproterozoic example is the Pechenga Belt of Ni-Cu sulphide deposits in the Russian Kola Peninsula (Melezhik et al., 1994).

The liquid-equivalent portions of ultramafic komatiitic rocks are magnesium-rich (18%–32% MgO), and therefore the precursor magmas are very hot and fluid. Because of their primitive (high Mg, Ni) composition, the Ni:Cu ratio of the associated sulphide ores is high, in many cases 10:1 or more. The S in the sulphide ores has been derived in significant proportion by contamination from sulphidic wall rocks. The commonly observed close spatial association of these deposits and their hosts with sulphidic sedimentary footwall rocks, and the similarity of S isotopes and other chemical parametres of the magmatic and sedimentary sulphides strongly suggests that the S in these deposits was derived locally from the sediments. This contrasts to some degree with deposits like Noril'sk where, although it is clear that S came from an extraneous source, that source was not likely so near at hand.

Two types of Ni-Cu sulphide ores characterize these deposits. Sulphide-rich ores comprising massive, breccia, and matrix-textured ores (Fig. 7C, 7D, and 7B, respectively) consisting of pyrrhotite, pentlandite, and chalcopyrite occur at the basal contact of the hosting ultramafic flows and sills. These deposits are generally small, in the order of a few million tonnes, and the grades are in the 1.5 to 4 percent range. The second type, sulphide-poor disseminated ore (Fig. 7A), forms internal lens-like zones of sparsely dispersed sulphide blebs that consist mainly of pyrrhotite-pentlandite. Deposits of this type also occur in both sills and flows, but the largest deposits are in sills, with ore tonnages of 10s to 100s of millions, although grades are a modest 0.6 to 0.9 percent Ni. The rich sulphide concentrations of the first type appear to result from significant contamination by S from host rocks, whereas the lower-grade sulphides of the second type may not have

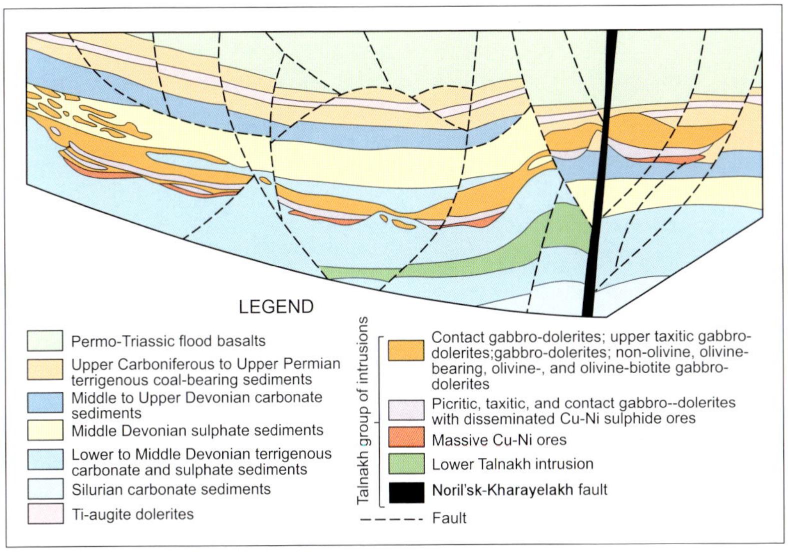

FIGURE 8. Noril'sk-Talnakh: west-east geological section (after Duzhikov et al., 1992).

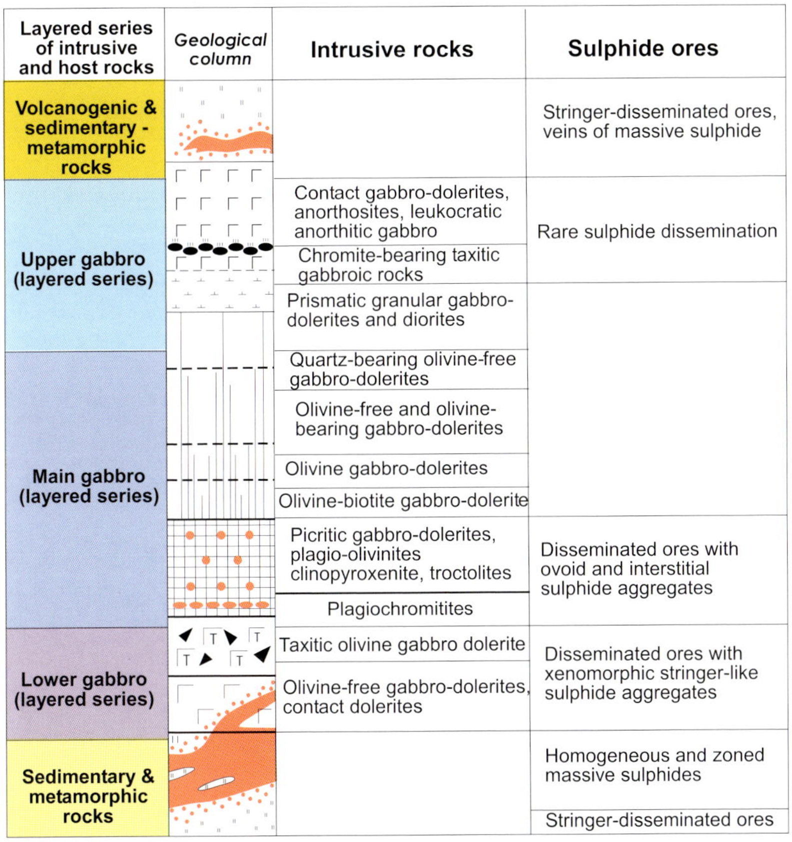

FIGURE 9. Noril'sk-Talnakh: typical stratigraphic profile of an ore-bearing sill (after Distler, 1994).

had an external source of S.

Komatiitic Ores in Greenstone Belt Setting

Canadian examples of this kind of Ni-Cu deposit are best developed in the Abitibi Greenstone Belt. The Alexo, Langmuir, Redstone, and Texmont mines in the Timmins, Ontario area and the Marbridge mine in the Val d'Or area have been minor producers. The deposits in Western Australia are much larger and more economically significant.

Kambalda, Western Australia: Ni sulphide ores of the Kambalda district are typical of the basal contact deposits associated with ultramafic flows in greenstone belts (Gresham and Loftus-Hills, 1981; Gresham, 1986). They occur in the Kambalda komatiite, which is a package of ultramafic flows (2710 Ma) that has been folded into an elongate, doubly plunging anticlinal dome structure about 8 km by 3 km (Fig. 10). The underlying member of this succession is the Lunnon basalt, and the overlying units are a sequence of basalts, slates, and greywackes (2710–2670 Ma). The core of the dome is intruded by a granitoid stock (2662 Ma), whose dykes crosscut the komatiitic hosts and ores.

The Kambalda komatiite is made up of a pile of thinner, more extensive sheet flows and thicker channel flows (Perring et al., 1994). The flows that contain ore are channel flows in the lower part of the pile, and may be up to 15 km long and 100 m thick. These flows are commonly interspersed with sulphidic interflow sediment, from which the S that formed the ores was probably derived (Lesher, 1989).

Most of the orebodies are at the basal contact of the lowermost channel flows (accounting for 80% of reserves), although some do occur in overlying flows in the lower part of the flow sequence (Fig. 11). The orebodies typically form long tabular or lenticular bodies up to 3 km long and 5 m thick. The ores generally consist of massive and breccia sulphides (Fig. 7C,D) at the base, overlain successively by matrix-textured sulphides (Fig. 7B), and disseminated sulphides (Fig. 7A). The sediment that underlies the flow sequence is generally absent beneath the lowermost ore-bearing channel flow, due to thermal erosion by the flow. Structural deformation renders the shape and continuity of ores more complicated in many instances. Because of their weaker competency compared to their wall rocks, sulphide zones are in many cases strung out along, or cut off by, faults and shear zones.

Komatiitic Ores in Rifted Cratonic Margin Setting

There are two major Canadian nickel belts in rifted cratonic settings, both being segments of the Circum-Superior Belt that encircles a large part of the northern Superior province. One is the Raglan horizon in the Cape Smith-Wakeham Bay Belt in the Ungava peninsula of northern Quebec, and the other is the Thompson Nickel Belt in northern Manitoba.

Raglan Horizon, Cape Smith-Wakeham Bay Belt: The Raglan horizon is a series of Ni-Cu ore-bearing komatiitic sills emplaced along the northern contact of the Povungnituk Group, at the base of the overlying Chukotat Group

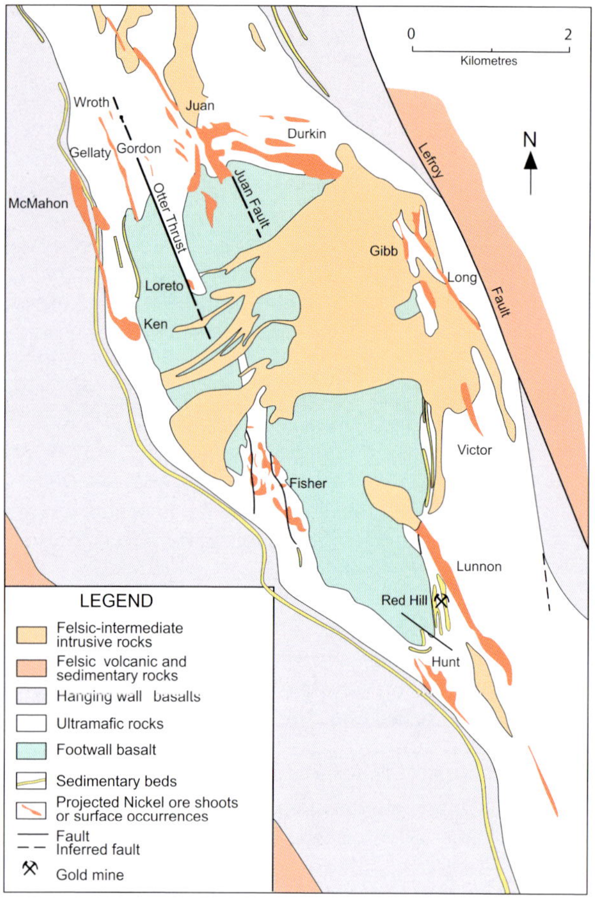

FIGURE 10. Kambalda district: geological map (after Gresham and Loftus-Hills, 1981).

(Fig. 12). Together, these form the southerly leading edge of the Cape Smith-Wakeham Bay Belt, northern Quebec, a thin-skinned thrust belt which overrides the Archean craton. The Povungnituk Group consists of basaltic and rhyolitic volcanic and clastic sedimentary rocks, the products of continental rifting. The Chukotat Group comprises massive and pillowed basalts and related mafic/ultramafic sills.

In addition to the Raglan Horizon of komatiitic sills along the Chukotat contact, there is another wide zone of komatiitic differentiated mafic/ultramafic sills in the interior of the Povungnituk Group. These Paleo-proterozoic suites of komatiitic magmatic rocks (1918 Ma) differ from the greenstone type of komatiites in their lower liquid-equivalent MgO content (up to only 16%–18%) and consequently Ni:Cu ratios of the ores are lower, averaging about 3:1. There are a number of economic Ni-Cu deposits in the Raglan horizon, and as well there are many Ni-Cu occurrences elsewhere in this horizon and in the ultramafic units lower in the Povungnituk Group. The Raglan sills appear to have richer, more abundant sulphide ore, likely because the clastic sediments they intrude are sulphide-rich, and have provided much of the S that contributed to formation of the ores.

The Ni-Cu sulphide deposits of the Raglan horizon have much the same development of ore types as the komatiitic greenstone deposits. The Raglan deposits are basal contact deposits consisting of massive and breccia sulphides at the basal contact, overlain in turn by matrix-textured ores and disseminated sulphides. Tectonic deformation has disrupted and mobilized some of the orebodies. Because of their remoteness and accompanying higher production costs, only the richer deposits can profitably be mined.

Thompson Nickel Belt: The Thompson Nickel Belt (TNB) is a portion of the Paleoproterozoic Circum-Superior Belt (Fig. 13), the rifted cratonic margin of the Archean Superior province (Bleeker, 1991). The Ni sulphide ores that characterize the TNB are associated with ultramafic komatiitic sills (1880 Ma; Hulbert et al., 2005) that intrude a sequence of Paleoproterozoic sedimentary cover rocks (Ospwagan Group). The latter consists of conglomerates, greywackes, iron formation, and pelitic and calcareous sediments capped by mafic to ultramafic volcanics. Most rocks have suffered several periods of intense deformation, and amphibolite to granulite facies metamorphism (about 1820 Ma). Paleoproterozoic strata are tightly infolded with the Archean basement gneisses. Original relationships are strongly deformed and obscured. The TNB on the northwest side abuts against the Paleoproterozoic Churchill province along the relatively late Churchill-Superior Boundary fault.

The ultramafic sills with which the ore is associated intrude the Pipe Formation of the Ospwagan Group. The Pipe Formation consists of pelitic schists and iron formations. All the known deposits in the Moak Lake-Thompson area are associated with sulphide iron formations of the Pipe Formation. The Pipe 2 and Birchtree ultramafic sills intersect a sulphide iron formation near the base of the Pipe Formation, whereas the Thompson ultramafic sill intersects another sulphide iron formation that is higher in the same pelitic unit.

Intense deformation has produced unusual modifications of some of the nickel deposits. Some of the deformational

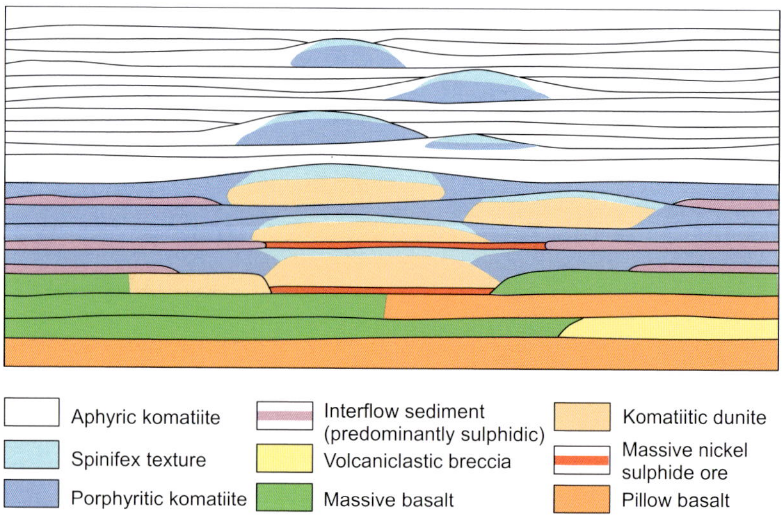

FIGURE 11. Generalized section of komatiitic flows and related nickel deposits (after Lesher, 1989).

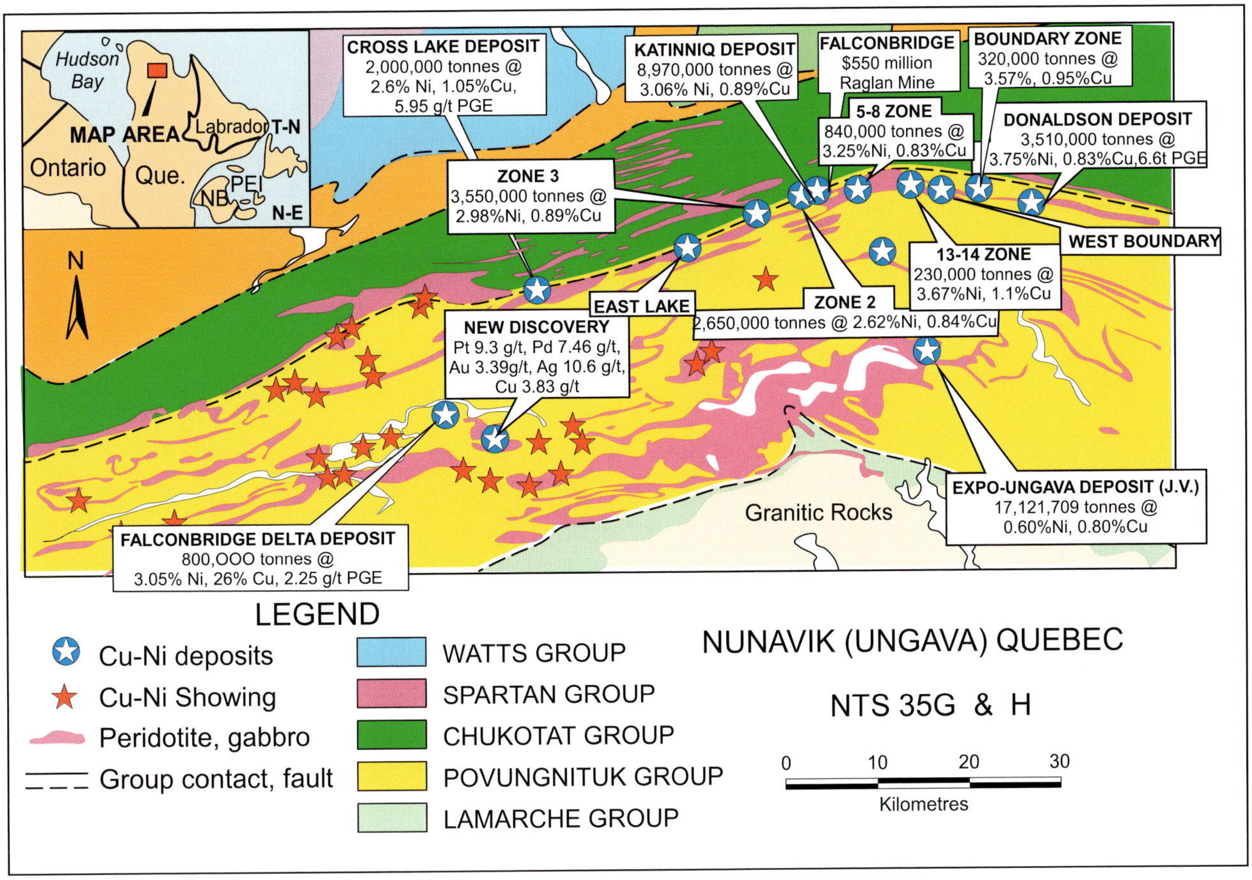

FIGURE 12. Cape Smith Ungava district: geological map (from Canadian Royalties Inc. Web site).

features are due to the weak competency of massive sulphide relative to its wall rocks. The following descriptions are arranged in order of increasing deformational effects experienced by the various deposits. The Pipe 2 nickel deposit consists of massive and stringer sulphide concentrations forming a U shape around the nose of the folded ultramafic sill, and representing the original basal contact sulphide. The Manibridge mineralized ultramafic is laced with pegmatitic dykes that were mobilized out of the surrounding gneisses, and present problems for mining. The Birchtree mine has one ore zone that is an extensive sheet-like shear zone of massive and breccia sulphide. The Soab North mine consists of a partly mineralized ellipsoidal boudin of ultramafic rock with a nearly complete enclosing sheath of massive and breccia sulphide. Ore in the Thompson mine, the principal deposit in the belt, is associated with a highly fragmented ultramafic sill, now dispersed as a zone of ultramafic boudins of all sizes, aligned in a horizon within the pelitic schist unit. The ore consists of nickeliferous sulphides (pyrrhotite-pentlandite) as impregnations in the pelitic schist in a conformable zone that is coextensive with the ultramafic boudins. Massive sulphides are commonly coarsely recrystallized; pentlandite "eyes" up to several cm are not unusual.

Other Mafic/Ultramafic Intrusion-Associated Subtypes

The host mafic/ultramafic intrusions associated with these Ni-Cu sulphide deposit include a variety of types: multiphase stocks (Lynn Lake, Proterozoic; Råna, Silurian), multiphase chonoliths (Kotalahti, 1885 Ma), multiphase sills (Kanichee and Carr Boyd Rocks, Archean), and highly deformed sills (Selebi-Phikwe, Archean). The styles of mineralization are also varied, including massive sulphides, breccia sulphides, stringers and veins, and disseminated sulphides. Voisey's Bay is the most important example.

Voisey's Bay: The Ni-Cu sulphide ores at Voisey's Bay are associated with the troctolitic Voisey's Bay Intrusion, a part of the anorogenic Nain Plutonic Suite in Labrador. These deposits have similarities to those at Noril'sk in that the role of a feeder system appears crucial to the accumulation of sulphides (Li et al., 2001).

The troctolitic intrusions (1290–1340 Ma) straddle the collisional suture (~1850 Ma) between the Archean Nain province gneisses (2843 Ma) to the east and the Paleoproterozoic Churchill (Rae) province gneisses to the west (Ryan et al., 1995; Naldrett et al., 1996; Fig. 14). These intrusions constitute a large magmatic system that includes granites, anorthosite, ferro-diorite, and troctolite. The Voisey's Bay Intrusion intrudes sulphide-bearing Tasiuyak gneiss of the Churchill province, which appear to have been the source of much of the S essential for forming the magmatic sulphides.

The Voisey's Bay intrusion (Fig. 15) consists of a deep western subchamber of troctolite-olivine gabbro that is connected by a subvertical mineralized feeder dyke of ferrodiorite, olivine gabbro, and troctolite. This dyke extends and

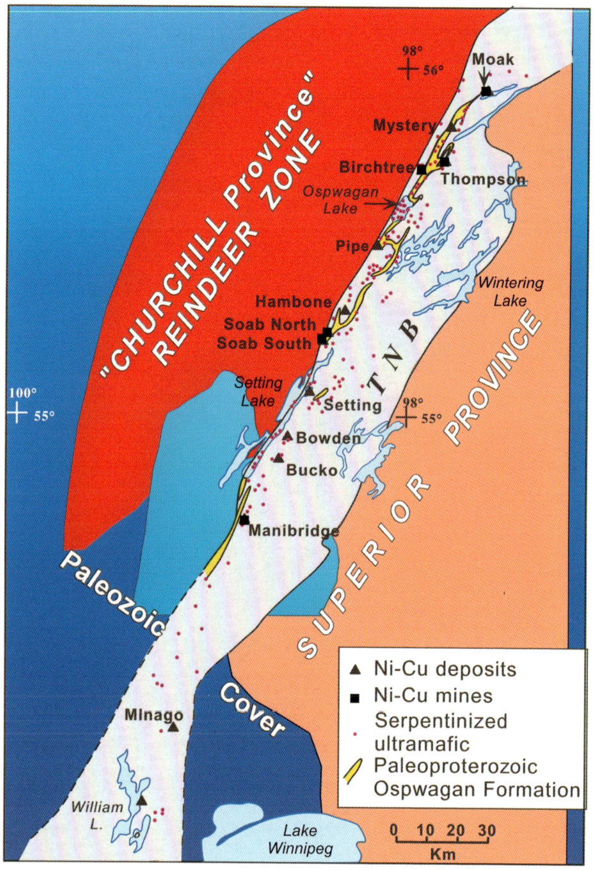

FIGURE 13. Thompson Nickel Belt: regional geology (after Hulbert et al., 2005).

flattens generally eastward for about 3 km to the Eastern Deeps troctolitic chamber, the largest exposed part of the intrusion. Along this strike length, three main Ni-Cu sulphide zones constitute integral widened parts of the feeder dyke. The Reid Brook mineralized zone (Fig. 15B) in the west is a near-vertical, thickened part of the feeder dyke with a central mineralized Leopard Troctolite (augite oikocrysts), sheathed in a mineralized breccia and transected by steep massive sulphide veins. The Ovoid deposit (Fig. 15C) is the richest ore zone. It is a flat-lying spoon-shaped lens of massive sulphide enveloped in mineralized Leopard and variable-textured troctolite and breccia, representing a widened part of the feeder dyke. The Eastern Deeps zone (Fig. 15D) is located where the feeder dyke widens out into the base of the Eastern Deeps troctolite chamber. At the core of this junction is a massive sulphide lens that expands and extends into the Eastern Deeps chamber. The massive sulphide is enclosed in a complex mineralized sheath of variable textured troctolite, Leopard troctolite, and breccia, similar to the assemblages accompanying the Reid Brook and Ovoid mineralized ores.

The feeder system and the Eastern Deeps zone are extensively mineralized in addition to the three zones mentioned above. However, these ores represent sulphide-enriched locations in the feeder system, where it widened and slowed the through-going flow of magma. As a result, the suspended droplets of liquid sulphide settled gravitationally out of the flowing magma and produced accumulations of ponded liquid sulphides that crystallized to form massive Ni-Cu sulphide. Each of the main ore zones includes veins of cross-cutting massive sulphide that transect the other rock units, indicating the later mobility of liquid sulphide.

Sulphide assemblages consist of the usual pyrrhotite-pentlandite-chalcopyrite, with additional troilite and magnetite. Pyrrhotite grain size is exceptionally coarse, up to 20 cm in the massive sulphide ore, whereas pentlandite forms finer exsolution grains and lamellae. The Ni, Cu, and Co resources for the Voisey's Bay deposits are given in Table 2.

Platinum Group Element (PGE) Deposits

Economic Platinum Group Element deposits are extremely rare. Two districts, Bushveld and Noril'sk-Talnakh, supply the majority of the world's PGE, although Noril'sk-Talnakh has not been considered primarily a PGE deposit (Cawthorn, 1999; Cawthorn et al., 2002). Stillwater (Zientek et al., 2002) is the only other significant PGE producer of this type. Lac des Iles (Hinchey and Lavigne, 2005), small by comparison, is Canada's only producer of this type of deposit.

An obvious feature of the few economic PGE deposits in the world is the large size of their host intrusions. An apparent exception is the smaller Lac des Iles intrusion, but it is just one of a number of comagmatic plutons in the area, which together constitute a significant magma system. Mafic magmas have very low contents of PGE. Despite the high R factor of PGE (e.g., the high partition coefficients of PGE), the sulphide has apparently equilibrated with large proportions of magma to form economic PGE deposits.

Another feature shared by most known examples is the small amount of sulphide (less than 3%) with which the PGE are associated. The sparsely disseminated sulphide is mainly chalcopyrite, but also includes pentlandite and pyrrhotite. The PGE minerals occur in very minute quantities that have apparently exsolved from the iron and base metal sulphides during cooling (Cabri, 2002). They include a host of known as well as unnamed minerals. Pentlandite is the only common sulphide mineral that contains a significant amount of any PGE, in this case Pd.

The small amount of sulphide appears due to the fact that the only S involved is the original mantle S, with little or no addition from the intruded wall rocks. Because the solubility of S in mafic magmas is quite low, the amount of sulphide produced when the magma reaches saturation is very small, resulting in small, sparsely dispersed sulphides. This is in distinct contrast with Ni-Cu sulphide deposits in which the ore consists of rich concentrations of sulphide.

Two distinct modes of PGE deposits are (1) the reef type, and (2) the magmatic breccia type. Of the two, only the reef type has proved to be a major producer.

TABLE 2. Voisey's Bay Resources*

Ore type	Ore (Mt)	Ni %	Cu %	Co %
Ovoid deposit (proven)	31	2.88	1.69	0.14
Additional, indicated	97	1.29	0.61	0.08
Additional, inferred	14	1	0.7	0.06
Total	142	1.61	0.85	0.09

*From Inco 2001 Annual Report

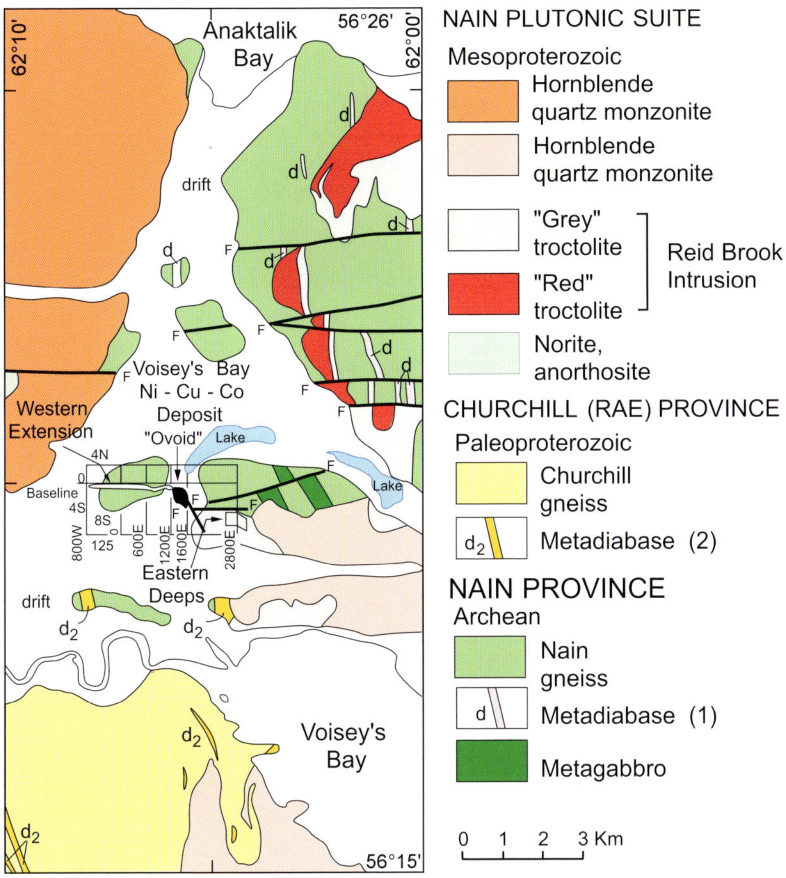

FIGURE 14. Voisey's Bay district: geological map (after Naldrett, 1997)

Reef Subtype

The reef or stratiform subtype of PGE deposits invariably occurs in large, well-layered mafic/ultramafic intrusions (Naldrett, 1989). The most important examples include the Merensky Reef and UG-2 chromitite reef of the Western and Eastern Bushveld, the J-M Reef of the Stillwater Complex, and the Main Sulphide zone of the Great Dyke (Prendergast and Wilson, 1989; Oberthuer, 2002). Other examples include the PGE zones in the Penikat (Finland; Alapieti and Lahtinen, 2002), Munni Munni (Australia; Barnes et al., 1992), and the Rincon del Tigre (Bolivia; Prendergast, 2000) layered intrusions. All PGE reefs are typically more or less conformable, relatively thin layers (from less than one to a few metres) within the well-layered sequence of the intrusions. No significant examples are known in Canada.

The genesis of the Merensky and J-M reefs remains controversial. Because of their great lateral extent (virtually a single layer within the whole of each large intrusion) and the thinness of the reefs, it is appealing to call on a magmatic process operating during the course of formation of the layered intrusions. The most generally accepted model involves the mixing of the residual magma remaining after partial crystallization with a new pulse of magma emplaced above it (Campbell et al., 1983). It has been demonstrated experimentally that this mixing mechanism can induce sulphide saturation. The newly formed sulphide droplets, thus produced then scavenge PGE from the silicate magma and settle to form a sparse sulphide concentration with a rich PGE content as a thin layer on the floor of the overlying magma. An alternative model proposes PGE carried upward by rising fluids (Boudreau and McCallum, 1992).

Bushveld Complex: The Bushveld Complex is a mafic/ultramafic layered intrusion (2060 Ma) that extends over an area of 240 by 350 km in the Kapvaal craton, South Africa (Fig. 16A). It is noted not only for its large size, but also for the remarkable lateral extent of the Merensky Reef and the UG-2 chromitite, the two producing PGE layers (Cawthorn et al., 2002). The Complex's total thickness of over 7 km is made up of four stratigraphic zones: (1) the Lower zone of bronzitites, harzburgites, and dunites; (2) the Critical zone of chromitite, pyroxenite, norite, and anorthosite, which includes the Merensky Reef and UG-2 chromitite as well as numerous additional chromitites; (3) the Main zone of norite and gabbronorite with minor anorthosite and pyroxenite; and (4) the Upper zone of anorthosite, leucogabbro, and diorite, notable for numerous magnetitite layers up to 6 m thick.

The whole of the sequence represents a simple progression of cumulus minerals (Fig. 16B), but actual succession of layered units is complex. Much of the Critical zone is made up of cyclic units, each consisting of all or part of an upward sequence of chromitite, pyroxenite, norite, and anorthosite.

The Merensky Reef occurs near the top of the upper part of the Critical zone, and the UG-2 chromitite at varying depths below the Merensky: about 30 m below at Union, 0 m below at Rustenburg, and 350 m below near Lebowa. The Merensky Reef lies at the base of the Merensky cyclic unit, below the basal pyroxenite (Fig. 16C). It generally comprises a thin pegmatoidal feldspathic pyroxenite layer about 1 m in thickness, bounded above and below by very thin chromitite layers, and containing sparsely disseminated Cu-Ni sulphides (up to 3%). The UG-2 chromitite occurs at the base of the UG-2 cyclic unit. It ranges from 70 to 130 cm in thickness, and has the same lateral extent as the Merensky Reef (see Fig. 16A). Estimated resources contained in the two reefs and the Platreef (discussed below) are shown in Table 3 (Cawthorn, 1999).

The PGE grade of the Merensky Reef is surprisingly uniform throughout the lateral extent of the unit, ranging between 4.9 and 7.3 g/t. This is despite considerable variation along strike in the platinum group mineral assemblages, which include alloys, sulphides, tellurides, and arsenides.

A feature common to sulphide reef-type deposits in layered intrusions is that they tend to occur at, or some distance above, the contact between the lower ultramafic zone and the upper mafic zone. The Bushveld and Stillwater reefs occur some distance above the contact, and the Hartley and Munni

TABLE 3. Bushveld Complex PGE Resources

	Pt g/t	Pd g/t	Mt
Eastern Bushveld			
Merensky	3.2	1.4	1320
UG2	2.4	2	2035
Western Bushveld (N.)			
Merensky	3.2	1.4	435
UG2	2.4	2	675
Western Bushveld (S.)			
Merensky	3.2	1.4	760
UG2	2.4	2	1530
Northern Bushveld			
Platreef	1.3	1.4	3060
Total Bushveld:	2.3	1.7	9815

Tonnages and total average grades are calculated from the grade and total ounces of Pt and Pd estimated by von Gruenewaldt, as cited in Cawthorn, 1999.

Munni reefs (Barnes et al., 1992) lie immediately below this contact.

Magmatic Breccia Subtype

The magmatic breccia subtype of PGE mineralization is characterized by a large zone of sparsely disseminated sulphide in a mafic magmatic host that has a high proportion of breccia clasts, both cognate and exotic. The most important example of this subtype is the Platreef district in the Northern Bushveld Complex, South Africa. Two similar Canadian deposits are in the River Valley intrusion (Tardif, 2000) and the Marathon deposit in the Coldwell Complex (Barrie et al., 2002). These deposits all comprise semiconformable zones of PGE mineralization in a basal breccia unit of a layered mafic/ultramafic intrusion. The Lac des Iles PGE deposit in Canada is different from the preceding examples in that the intrusion is a multiphase stock-like body rather than a layered intrusion. Nevertheless, the deposit comprises disseminated sulphide in a mafic magmatic breccia (Fig. 17), and on this basis, is grouped in this subtype.

Lac des Iles: The Lac des Iles intrusion (2738 Ma) intrudes a Neoarchean gneissic tonalitic terrane. It is one of a 30 km-diametre ring of similar intrusions, and on a larger scale, part of an ENE-trending zone of mafic plutons (Lavigne and Michaud, 2002; Hinchey and Lavigne, 2005). The intrusion consists essentially of a gabbronorite elliptical core, enveloped by a border unit of varitextured gabbro. The Roby Ore zone lies between these two units at the west end of the intrusion and is made up of a combination of varitextured gabbro, which is matrix to a heterolithic gabbro breccia. The varitextured gabbro contains abundant coarse-grained and pegmatitic patches, and the clasts in the heterolithic breccia are mostly cognate mafic rock types. A 20 m-wide north-trending dyke-

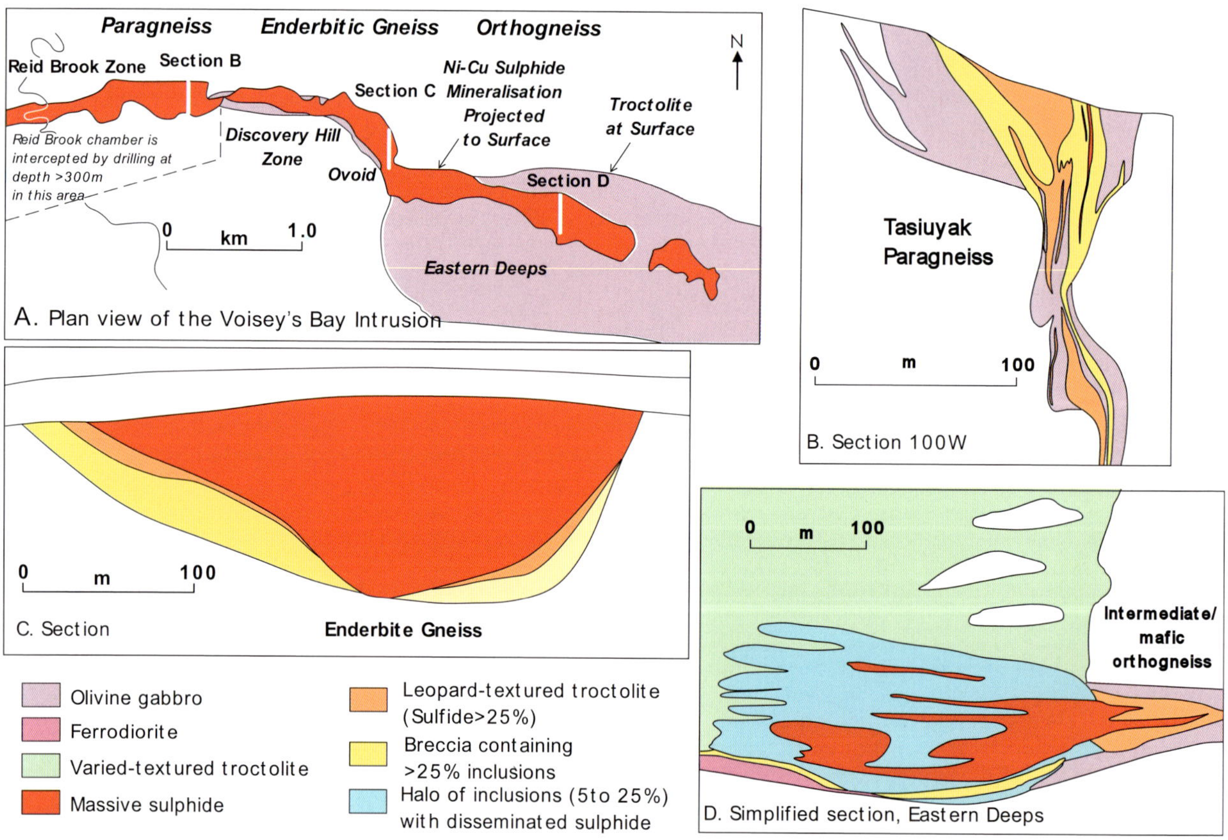

FIGURE 15. Voisey's Bay ore deposits: **(A)** Plan of the Voisey's Bay intrusion feeder and associated ore zones (after Li et al., 2001). The ores are projected to surface. **(B)** Reid Brook zone (after Li and Naldrett, 1999). **(C)** Ovoid orebody (after Li and Naldrett, 1999). **(D)** Eastern Deeps (after Li et al., 2001).

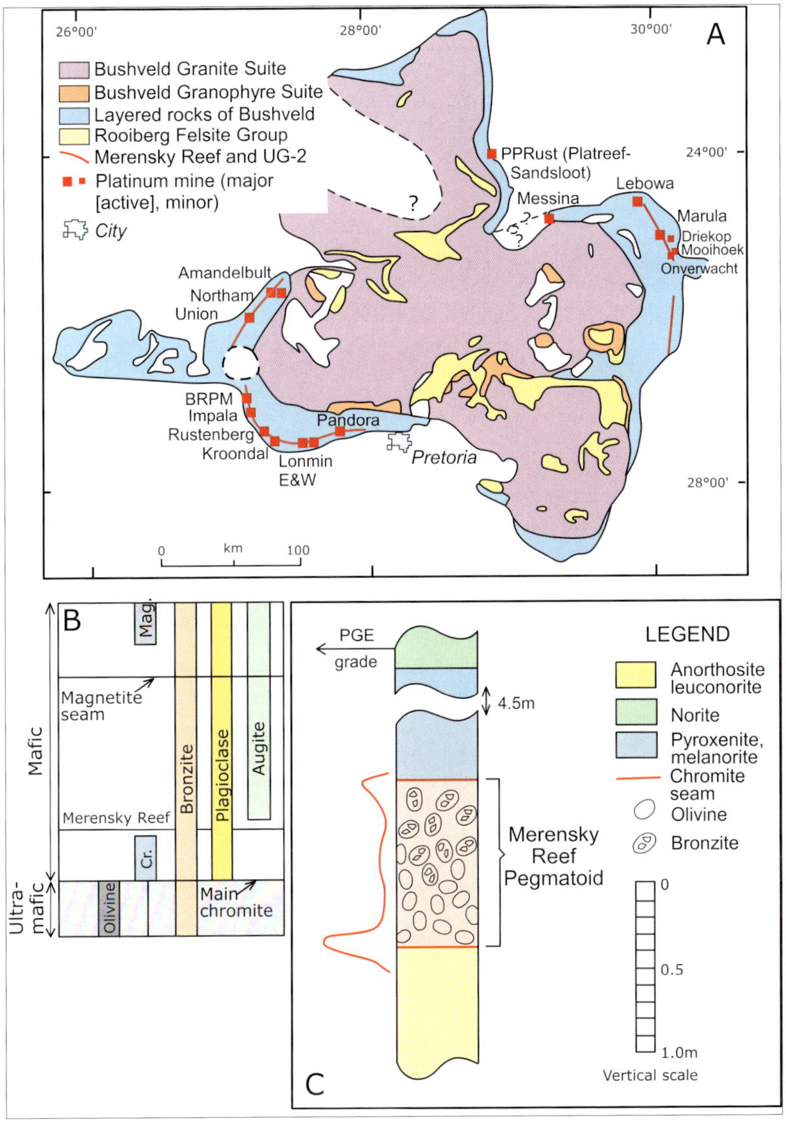

FIGURE 16. Bushveld Complex: **(A)** Geological map showing the trace of the Merensky Reef and platinum mines (modified after Campbell et al., 1983); **(B)** Stratigraphic range of cumulus minerals over the 4 zones of the complex (after Campbell et al., 1983); **(C)** Typical local stratigraphy of the Merensky Reef and profile of PGE grade (after Naldrett, 1989).

like pyroxenite lies between the Roby Ore Zone and the barren gabbronorite to the east, and effectively marks the eastern boundary of mineralization.

The PGE mineralized Roby Ore zone is 950 m long by 8 m wide and is distinguished by the presence of up to 3 percent irregularly disseminated sulphides. These include chalcopyrite, pyrrhotite, pentlandite, and pyrite as grains and patches of submillimetre to a few centimetres size. Sulphide mineralization is coextensive with the varitextured gabbro breccia. PGE mineralization is Pd-rich (Pd:Pt = 9:1) and is locally erratically distributed, but on a mine scale is more or less uniform (Fig. 18). A higher grade zone (about 5 g/t) is localized on a 400 m-long portion of the western part of the pyroxenite dyke and a parallel portion of the adjoining varitextured gabbro/heterolithic breccia. Within this higher-grade zone, the silicates are hydrothermally altered to amphibole, chlorite, and saussuritized feldspar. The PGE minerals are mainly braggite, merenskyite, and kotulskite.

The stock-like Lac des Iles PGE deposit may represent a conduit for mineralized magmatic breccia. If intruded to a higher level in the crust, such a magmatic breccia could have been emplaced as the stratiform basal PGE-mineralized breccia unit of a layered intrusion such as the Platreef, the River Valley intrusion, or the Marathon deposit.

Exploration Models

Because magmatic Ni-Cu-PGE sulphide deposits are invariably associated with mafic and/or ultramafic magmatic bodies, such bodies constitute the first-order target for exploration. From the preceding accounts, it is clear that the different types of deposits are associated with different suites of mafic and/or ultramafic rocks, each of which have somewhat different but typical attributes.

District Scale

The Voisey's Bay discovery has emphasized, as is also the case at Noril'sk-Talnakh, the importance of relatively small intrusions as parts of large magmatic systems. Their role as conduits for large volumes of magma provides sites for accumulations of settled sulphide out of the passing magma. At Voisey's Bay, a dyke-like conduit that led from one magma chamber to a higher one contains the ores. At Noril'sk-Talnakh, sills are the conduits that appear to have fed the flood basalts, and in which the sulphide ores formed. Although of different geometries, the conduits record the passage of differing magmas by exhibiting significant differentiation: well-layered at Noril'sk-Talnakh (Fig. 9), distinct dyke facies at Voisey's Bay (Fig. 15B,C). In the case of the Jinchuan deposits, the exposed ore-laden intrusion itself may be a feeder to a much larger layered magmatic complex, now largely removed by erosion. If this interpretation is correct, the target within a large mafic magmatic province would be smaller differentiated cognate intrusions that may represent magma conduits.

Komatiitic deposits occur in small to medium-sized sills and flows that invariably include ultramafic rocks, either alone or with mafic differentiates, usually gabbros. Those in greenstone belts tend to occupy a limited range of stratigraphy at the district or regional scale. Thus, they form clusters of ultramafic lenses along strike of formations as at the Langmuir and Redstone mines near Timmins, Ontario, or whole formations as at Kambalda (Fig. 10). Similarly, the komatiitic deposits in cratonic margin rift settings occur in lenticular ultramafic sills strung out along strike in long linear belts as at Thompson (Fig. 13) and Raglan horizon (Fig. 12). These groupings of target rocks focus exploration at a district scale.

Ultramafic rocks associated with any of the deposit types

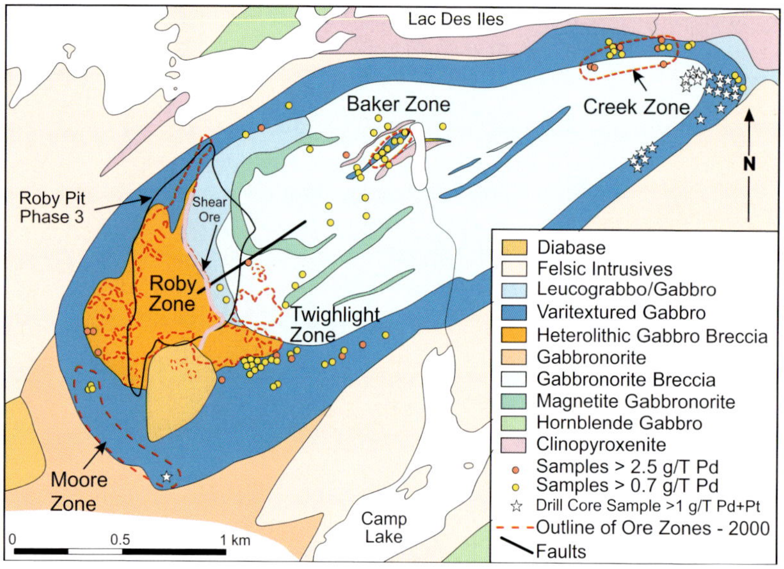

FIGURE 17. Lac des Isle: geological map of intrusion (after Lavigne and Michaud, 2002).

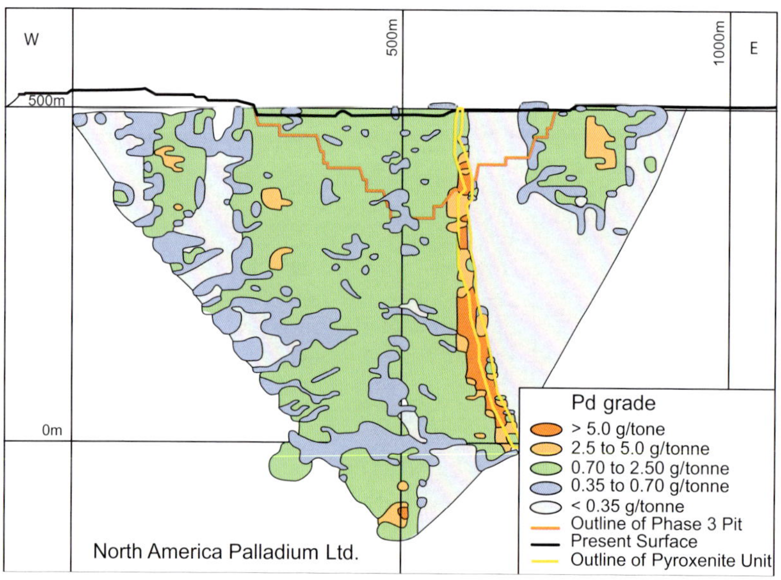

FIGURE 18. Lac des Isle: west-east section showing grade distribution (after Lavigne and Michaud, 2002).

have, in most terranes (especially greenschist facies metamorphism), undergone serpentinization with the accompanying generation of magnetite. Consequently these bodies typically have a well-defined magnetic response. Low-level aeromagnetic surveys thus are indispensable at early exploration stages, especially in poorly exposed areas.

Large layered intrusions are the prime targets in exploration for PGE deposits, and have been recognized in many regions. However, there may still be unidentified bodies in some poorly exposed or poorly mapped areas. Magnetic and gravity surveys could be of use in these areas.

Local Scale

Sulphide-rich Ni-Cu deposits achieve their concentrations mostly through the settling effects of gravity. Consequently, in virtually all magmatic bodies (sills, flows, and dykes), the sulphide-rich ores are most likely to be found at the base of those bodies. Determination of the base of a given body is, thus, an important part of exploration targeting. Within the komatiitic greenstone belt type, the ores are generally located in the lowest flow, which is also generally the most primitive in the pile of flows. Some ores may lie at a somewhat higher level.

In areas that have been intensely deformed and/or faulted, the distribution pattern of sulphide-rich zones may be more complex. For instance, in the Thompson Nickel Belt, some of the sulphide ores are extended far beyond the parent ultramafic bodies.

The exploration of large layered mafic/ultramafic intrusions for PGE deposits should be focused from just below to some distance above the main ultramafic-mafic contact. This is the stratigraphic range of most of the PGE-rich layers in the Bushveld, Stillwater, Great Dyke, Munni Munni, and Rincon del Tigre deposits.

Because chromite is commonly a mineral associated with PGE deposits (e.g., the UG-2 reef in the Bushveld Complex), geochemical surveys should include Cr as well as the obvious suite consisting of Ni, Cu, Co, Pt, and Pd.

Electromagnetic surveys designed to detect conductors should be effective in locating the sulphide-rich (i.e., massive, breccia, and matrix-textured sulphide) deposits. IP methods may identify disseminated sulphides, but the presence of serpentinization in the ultramafic host may render the technique ineffective.

Recent Advances

A much better appreciation of the role of magma dynamics in the concentration and enrichment of magmatic Ni-Cu-PGE sulphide deposits has developed in the last decade or two. The importance of changes in fluid flow, particularly decreases in the rate of flow of magmas, has become clearer. The location of sulphide concentrations in conduits at Talnakh-Noril'sk and Voisey's Bay, and near conduits in certain of the komatiitic deposits, suggests that sulphides accumulate where the flow rate of magma was reduced and the entrained sulphides were able to settle gravitationally to form rich basal concentrations.

Nickel depletion of mafic magmatic rocks in connection with the existence of Ni sulphide deposits has become better documented. It was anticipated that the formation of nickeliferous liquid sulphide in a magma resulted by extraction of nickel from the magma, thereby leaving the magma depleted in nickel. Documentation has supported this theory, and it now plays a part in exploration strategy.

Knowledge Gaps

One of the gaps in our knowledge of Ni-Cu sulphide deposits is knowing the most important factor in triggering sulphide saturation in a given magma. Certain things are clear.

The magma must have a sufficient dissolved content of Ni, Cu, and PGE. Once a liquid sulphide is formed, it will tend to equilibrate with the magma, and this means acquiring the Ni, Cu, and PGE from the magma according to the partition coefficients for those elements. It also is clear that much of the S in magmatic Ni-Cu sulphide deposits has been derived from sulphidic wall rocks, commonly pyritic sediments. Thus, addition of S to the magma by incorporation of such material leads to sulphide saturation. However, it is also known that by increasing the silica content of the magma through incorporation of siliceous wall rock, the solubility of sulphide in the magma is decreased, thereby producing sulphide saturation. It remains unclear which of the two mechanisms is the more critical in producing sulphide saturation. The significance for exploration is whether it is essential to have wall rock rich in sulphide as a source of S in order to better evaluate a priori the nickel potential of a given mafic/ultramafic body. Existing evidence tends to favor the sulphidic wall rock theory, but more investigation of the settings of known nickel sulphide deposits is needed in order to evaluate the importance of the alternative theory.

In the case of PGE reef type deposits, there is still ongoing controversy over the main mechanism of concentration of PGE in the thin extensive "reefs" that are hosted in very large layered mafic/ultramafic intrusions. As noted above, the magmatic theory emphasizing magma mixing is the more favored, but a "fluids from below" theory has some persuasive arguments. This controversy will undoubtedly continue; it is unclear whether there are important exploration ramifications contingent on this question.

Acknowledgements

The authors are grateful for the helpful reviews of M. Lesher and M. Duke. Their comments have led to much improvement of this manuscript. The editorial guidance of W. Goodfellow is also appreciated.

References

Alapieti, T.T., and Lahtinen, J.J., 2002, Platinum-group element mineralization in layered intrusions of northern Finland and the Kola Peninsula, Russia: Canadian Institute of Mining and Metallurgy Special Volume 54, p. 507 546.

Barnes, S.J., Keays, R.R., and Hoatson, D.M., 1992, Distribution of sulphides and PGE within the porphyritic websterite zone of the Munni Munni Complex, Western Australia: Australian Journal of Earth Sciences, v. 39, p. 289–302.

Barrie, C.T., MacTavish, A.D., Walford, P.C., Chataway, R., and Middaugh, R., 2002, Contact-type and magnetitite reef-type Pd-Cu mineralization in ferroan olivine gabbros of the Coldwell Complex, Ontario: Canadian Institute of Mining, Metallurgy and Petroleum, Special Volume 54, p. 321–337.

Bleeker, W., 1991, Thompson Area—General geology and ore deposits, *in* Galley, A.G., Bailes, A.H., Syme, E.C., Bleeker, W., Macek, J.J., and Gordon, T.S., eds., Geology and mineral deposits of the Flin Flon and Thompson belts, Manitoba: Geological Survey of Canada, International Association on the Genesis of Ore Deposits, Guide Book 10, Open File 2165, p. 93–136.

Boudreau, A.E., and McCallum, I.S., 1992, Concentration of platinum-group elements by magmatic fluids in layered intrusions: Economic Geology, v. 87, p. 1830–1848.

Cabri, L.J., ed., 2002, The geology, geochemistry, mineralogy, and mineral beneficiation of platinum-group elements: Canadian Institute of Mining, Metallurgy and Petroleum, Special Volume 54, 852 p.

Campbell, I.H., Naldrett, A.J., and Barnes, S.J., 1983, A model for the origin of the platinum-rich sulfide horizons in the Bushveld and Stillwater complexes: Journal of Petrology, v. 24, p. 133–165.

Cawthorn, R.G., 1999, The platinum and palladium resources of the Bushveld Complex: South African Journal of Science, v. 95, p. 481–489.

Cawthorn, R.G., Merkle, R.K.W., and Viljoen, M.J., 2002, Platinum-group element deposits in the Bushveld Complex, South Africa: Canadian Institute of Mining, Metallurgy and Petroleum, Special Volume 54, p. 389–429.

Chai, G., and Naldrett, A.J., 1992, Characteristics of Ni-Cu-PGE mineralization and genesis of the Jinchuan deposit, northwest China: Economic Geology, v. 87, p. 1475–1495.

Chorlton, L.B., comp., 2003, Generalized geology of the world, age and rock type domains: Geological Survey of Canada, Open File 5529, CD, in prep.

Coats, C.J.A., and Snajdr, P., 1984, Ore deposits of the North Range, Onaping-Levack area, Sudbury: Ontario Geological Survey, Special Volume 1, p. 327–346.

Cochrane, L.B., 1984, Ore deposits of the Copper Cliff offset: Ontario Geological Survey, Special Volume 1, p. 347–359.

Distler, V.V., 1994, Platinum mineralization of the Noril'sk deposits: Ontario Geological Survey, Special Publication 5, p. 243–260.

Duzhikov, O.A., Distler, V.V., Strunin, B.M., Mkrtychyan, A.K., Sherman, M.L., Sluzhenikin, S.S., and Lurye, A.M., 1992, Geology and metallogeny of sulfide deposits Noril'sk region, USSR: Society of Economic Geologists, Special Publication 1, p. 242.

Eckstrand, O.R., Good, D.J., Yakubchuk, A., and Gall, Q., comp., 2004, World dstribution of Ni, Cu, PGE, and Cr deposits and camps: Geological Survey of Canada, unpublished update of Open File 3791a.

Gresham, J.J., 1986, Depositional environment of volcanic peridotite-associated nickel-copper sulfide deposits with special reference to the Kambalda dome: Society for Geology Applied to Mineral Deposits, Special Publication 4, p. 63–90.

Gresham, J.J., and Loftus-Hills, G.D., 1981, The geology of the Kambalda nickel field, Western Australia: Economic Geology, v. 76, p. 1373–1416.

Grinenko, L.N., 1985, Sources of sulfur of the nickeliferous and barren gabbro-dolerite intrusions of the northwest Siberian platform: International Geology Review, v. 27, p. 695–708.

Hinchey, J.G., and Lavigne, M.J., 2005, Geology, petrology, and controls on PGE mineralization of the southern Roby and Twilight zones, Lac des Iles Mine, Canada: Economic Geology, v. 100, p. 43–61.

Hulbert, L.J., Hamilton, M.A., Horan, M.F., and Scoates, R.F.J., 2005, U-Pb Zircon and Re-Os isotope geochronology of mineralized ultramafic intrusions and associated nickel ores from the Thompson Nickel Belt, Manitoba, Canada: Economic Geology, v. 100, p. 29–41.

Lavigne, M.J., and Michaud, M.J., 2002, Geology of North American Palladium Ltd's Roby Zone deposit, Lac des Iles: Exploration and Mining Geology, v. 10, p. 1–17.

Lesher, C.M., 1989, Komatiite-associated nickel sulfide deposits: Reviews in Economic Geology, v. 4, p. 45–102.

Li, C., and Naldrett, A.J., 1999, Geology and petrology of the Voisey's Bay intrusion: Reaction of olivine with sulfide and silicate liquids: Lithos, v. 47, p. 1–32.

Li, C., Naldrett, A.J., and Ripley, E.M., 2001, Critical factors for the formation of a nickel-copper deposit in an evolved magma system, lessons from a comparison of the Pants Lake and Voisey's Bay sulfide occurrences in Labrador, Canada: Mineralium Deposita, v. 36, p. 85–92.

Maier, W.D., Li, C., and de Waal, S.A., 2001, Why are there no major Ni-Cu sulfide deposits in large layered mafic-ultramafic intrusions?: The Canadian Mineralogist, v. 39, Part 2, p. 547–556.

Melezhik, V.A., Hudson-Edwards, K.A., Green, A.H., and Grinenko, L.N., 1994, Pechenga area, Russia; Part 2, Nickel-copper deposits and related rocks: Institution of Mining and Metallurgy, Transactions, Section B: Applied Earth Science, v. 103, p. B146–B161.

Naldrett, A.J., 1989, Stratiform PGE deposits in layered intrusions: Reviews in Economic Geology, vol. 4, p. 135–165.

——1999, Summary, Development of ideas on Sudbury geology, 1992–1998: Geological Society of America, Special Paper 339, p. 431–442.

——2004, Magmatic sulfide deposits; Geology, geochemistry and exploration, Heidelberg, Springer Verlag, 728 p.

Naldrett, A.J., and Lightfoot, P.C., 1992, The Ni-Cu-PGE ores of the Nori'sk region of Siberia: A model for giant magmatic sulfide deposits associated with flood basalts: Society of Economic Geologists, Special Volume 2, p. 81–123.

Naldrett, A.J., Hewins, R.H., Dressler, B.O., and Rao, B.V., 1984, The contact sublayer of the Sudbury igneous complex: Ontario Geological Survey, Special Volume 1, p. 253–274.

Naldrett, A.J., Keats, H., Sparkes, K., and Moore, R., 1996, Geology of the Voisey's Bay Ni-Cu-Co deposit, Labrador, Canada: Exploration and Mining Geology, v. 5, p. 169–179.

Oberthuer, T., 2002, Platinum-group element mineralization of the Great Dyke, Zimbabwe: Canadian Institute of Mining, Metallurgy and Petroleum, Special Volume 54, p. 483–506.

Owen, D.L., and Coats, C.J.A., 1984, Falconbridge and East mines: Ontario Geological Survey, Special Volume 1, p. 371–378.

Pattison, E.F., 1979, The Sudbury sublayer: Its characteristics and relationships with the main mass of the Sudbury Irruptive: Canadian Mineralogist, v. 17, p. 257–274.

Perring, C., Barnes, S., and Hill, R., 1994, Direct evidence for thermal erosion and related nickel-sulfide mineralisation at the base of a komatiite lava channel: Australia, Exploration and Mining Research News, Commonwealth Scientific and Industrial Research Organization, Australia, no. 2, p. 7–11.

Prendergast, M.D., 2000, Layering and precious metals mineralization in the Rincon del Tigre complex, Eastern Bolivia: Economic Geology, v. 95, p. 113–130.

Prendergast, M.D., and Wilson, A.H., 1989, The Great Dyke of Zimbabwe—II, Mineralization and mineral deposits, *in* Prendergast, M.D., and Jones, J., eds., Magmatic sulphides—The Zimbabwe volume: London, The Institution of Mining and Metallurgy, p. 21–42.

Pye, E.G., Naldrett, A.J., and Giblin, P.E., eds., 1984, The geology and ore deposits of the Sudbury Structure: Ontario Geological Survey, Special Volume 1, 603 p.

Ryan, B., Wardle, R.J., Gower, C.F., and Nunn, G.A.G., 1995, Nickel-copper-sulphide mineralization in Labrador: The Voisey Bay discovery and its exploration implications: Newfoundland Department of Natural Resources, Geological Survey Branch, Current Research, Report 95-1, p. 177–204.

Shanks, W.S., and Schwerdtner, W.M., 1991, Crude quantitative estimates of the original northwest–southeast dimension of the Sudbury structure, south central Canadian shield: Canadian Journal of Earth Sciences, v. 28, p. 1677–1686.

Souch, B.E., Podolsky, T., and the Inco Ltd. geological staff, 1969, The sulfide ores of Sudbury: Their particular relationship to a distinctive inclusion-bearing facies of the Nickel Irruptive: Economic Geology Monograph 4, p. 252–261.

Stekhin, A.I., 1994, Mineralogical and geochemical characteristics of the Cu-Ni ores of the Oktyabrsky and Talnakh deposits, *in* Naldrett, A.J., Lightfoot, P.C., and Sheahan, P., eds., The Sudbury-Norilsk Symposium: Ontario Geological Survey, Special Publication 5, p.217–230.

Tardif, N.P., 2000, Regional distribution of platinum, palladium, gold, kimberlite indicator minerals and base metals in surficial sediments, River Valley area, northeastern Ontario: Ontario Geological Survey, Open File 6010, 106 p.

Wheeler, J.O., Hoffman, P.F., Card, K.D., Davidson, A., Sanford, B.V., Okulitch, A.V., and Roest, W.R., 1996, Geological map of Canada: Geological Survey of Canada, A series, 1860A, two sheets, scale: 1:5 000 000.

Zientek, M.L., Cooper, R.W., Corson, S.R., and Geraghty, E.P., 2002, Platinum-group element mineralization in the Stillwater Complex, Montana: Canadian Institute of Mining, Metallurgy and Petroleum, Special Volume 54, p. 459–481.

Porphyry Deposits

W.D. Sinclair

Geological Survey of Canada, 601 Booth Street, Ottawa, Ontario K1A 0E8
Corresponding author's email: sinclair@nrcan.gc.ca

Abstract

Porphyry deposits are the world's most important source of Cu and Mo, and are major sources of Au, Ag, and Sn; significant byproduct metals include Re, W, In, Pt, Pd, and Se. They account for about 50 to 60% of world Cu production and more than 95% of world Mo production. In Canada, they account for more than 40% of Cu production, virtually all Mo production, and about 10% of Au production. Porphyry deposits are large, low- to medium-grade deposits in which primary (hypogene) ore minerals are dominantly structurally controlled and which are spatially and genetically related to felsic to intermediate porphyritic intrusions. They are distinguished from other granite-related deposits such as skarns and mantos by their large size and structural control, mainly stockworks, veins, vein sets, fractures, and breccias. Porphyry deposits typically contain hundreds of millions of tonnes of ore, although they range in size from tens of millions to billions of tonnes; grades for the different metals vary considerably but generally average less than 1%. In porphyry Cu deposits, for example, Cu grades range from 0.2% to more than 1% Cu; in porphyry Mo deposits, Mo grades range from 0.07% to nearly 0.3% Mo. In porphyry Au and Cu-Au deposits, Au grades range from 0.2 to 2 g/t Au. Associated igneous rocks vary in composition from diorite-granodiorite to high-silica granite; they are typically porphyritic epizonal and mesozonal intrusions, commonly subvolcanic. A close temporal and genetic relationship between magmatic activity and hydrothermal mineralization in porphyry deposits is indicated by the presence of intermineral intrusions and breccias that were emplaced between or during periods of mineralization. Porphyry deposits range in age from Archean to Recent, although most economic deposits are Jurassic or younger.

Résumé

Les gîtes porphyriques constituent, à l'échelle mondiale, la plus importante source de cuivre et de molybdène, une grande source d'or, d'argent et d'étain, de même qu'une source considérable de métaux récupérés comme sous-produits, dont le rhénium, le tungstène, l'indium, le platine, le palladium et le sélénium. En outre, ils comptent pour 50 à 60 % environ de la production mondiale de cuivre et pour plus de 95 % de celle de molybdène. Au Canada, ces proportions atteignent plus de 40 %, dans le cas du cuivre, presque 100 %, dans celui du molybdène et quelque 10 %, dans celui de l'or. Les gisements porphyriques consistent en vastes accumulations de minerai à teneur faible ou moyenne où la répartition des minéraux métalliques primaires (hypogènes) est régie principalement par des contrôles structuraux et liée, tant sur le plan spatial que génétique, à des intrusions porphyriques de composition felsique à intermédiaire. Ils se distinguent des autres gisements apparentés à des granites, comme les skarns et les mantos, par leur grande taille et leurs contrôles structuraux, la minéralisation se présentant surtout dans des stockwerks, des filons, des réseaux de filons, des fractures et des brèches. Ils renferment généralement des centaines de millions de tonnes de minerai, quoique leur volume peut varier de quelques dizaines de millions de tonnes à des milliards de tonnes. Leurs teneurs en métaux varient considérablement mais se situent habituellement en moyenne à moins de 1 %. Par exemple, la teneur en Cu des gisements porphyriques de cuivre va de 0,2 % à plus de 1 %, celle en Mo des gisements porphyriques de molybdène, de 0,07 % à presque 0,3 % et celle en Au des gisements porphyriques d'or et de cuivre-or, de 0,2 à 2 g/t. Les roches ignées qui leur sont associés montrent une composition variant de la diorite-granodiorite au granite fortement siliceux et se présentent en général sous forme d'intrusions à texture porphyrique épizonales et mésozonales, habituellement subvolcaniques. Dans les gîtes porphyriques, la présence de brèches et d'intrusions mises en place entre les zones minéralisées pendant ou entre les périodes de minéralisation témoigne d'un étroit lien temporel et génétique entre le magmatisme et la minéralisation hydrothermale. L'âge des gîtes porphyriques s'étend de l'Archéen à l'Holocène, quoique la plupart des gisements rentables datent du Jurassique ou d'une époque plus récente.

Definition

Porphyry deposits are large, low- to medium-grade deposits in which primary (hypogene) ore minerals are dominantly structurally controlled and which are spatially and genetically related to felsic to intermediate porphyritic intrusions (Kirkham, 1972). The large size and structural control (e.g. veins, vein sets, stockworks, fractures, 'crackled zones', and breccias) serve to distinguish porphyry deposits from a variety of deposits that may be peripherally associated, including skarns, high-temperature mantos, peripheral mesothermal veins, and epithermal precious-metal deposits. Secondary minerals may be developed in supergene-enriched zones in porphyry Cu deposits by weathering of primary sulphides. Such zones typically have significantly higher Cu grades, thereby enhancing the potential for economic exploitation.

The metal content of porphyry deposits is diverse. The following subtypes of porphyry deposits were defined by Kirkham and Sinclair (1995) according to the metals that are essential to the economics of the deposit (metals that are byproducts or potential byproducts are listed in brackets):

- Cu (±Au, Mo, Ag, Re, PGE)
- Cu-Mo (±Au, Ag)
- Cu-Mo-Au (±Ag)
- Cu-Au (±Ag, PGE)
- Au (±Ag, Cu, Mo)
- Mo (±W, Sn)
- W-Mo (±Bi, Sn)
- Sn (±W, Mo, Ag, Bi, Cu, Zn, In)
- Sn-Ag (±W, Cu, Zn, Mo, Bi)
- Ag (±Au, Zn, Pb)

Sinclair, W.D., 2007, Porphyry deposits, *in* Goodfellow, W.D., ed., Mineral Deposits of Canada: A Synthesis of Major Deposit-Types, District Metallogeny, the Evolution of Geological Provinces, and Exploration Methods: Geological Association of Canada, Mineral Deposits Division, Special Publication No. 5, p. 223-243.

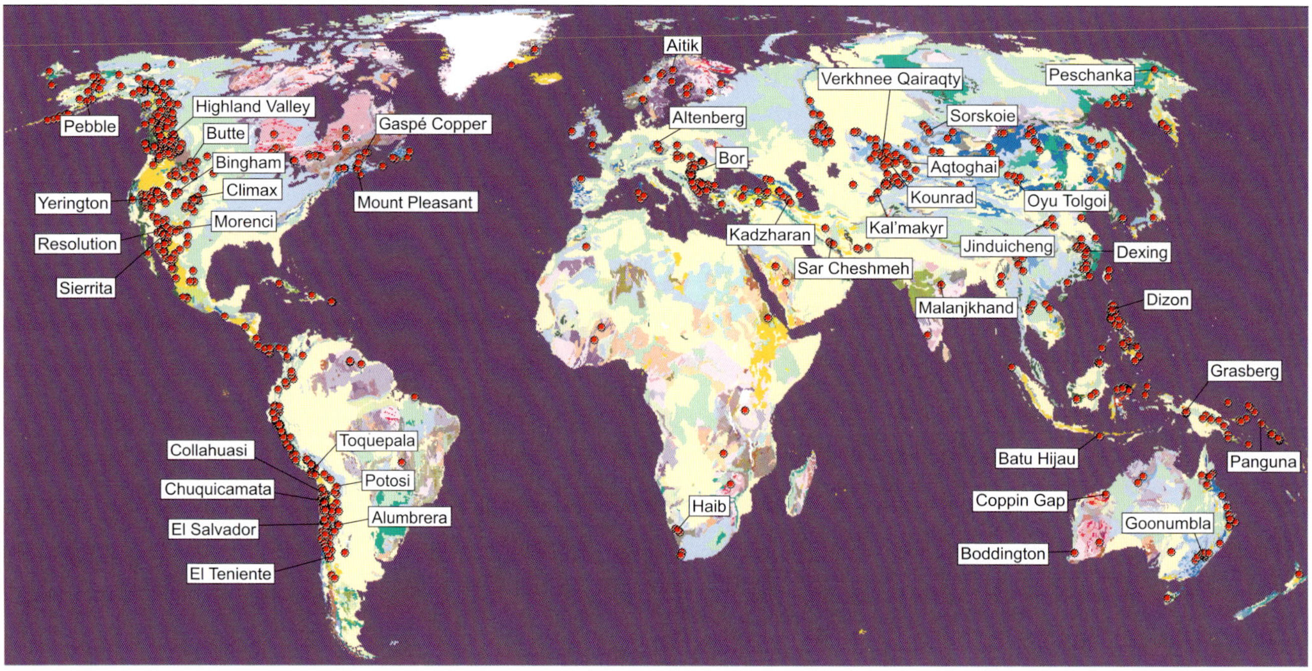

FIGURE 1. Global distribution of porphyry deposits (from Kirkham and Dunne, 2000). World geology is from Chorlton (2004). See Appendix 2 (World) for deposit details.

For deposits with currently subeconomic grades and tonnages, subtypes are based on probable coproduct and byproduct metals, assuming that the deposits were economic.

Geographical Distribution

Porphyry deposits occur throughout the world in a series of extensive, relatively narrow, linear metallogenic provinces (Fig. 1). They are predominantly associated with Mesozoic to Cenozoic orogenic belts in western North and South America, around the western margin of the Pacific Basin, and in the Tethyan orogenic belt in eastern Europe and southern Asia. However, major deposits also occur within Paleozoic orogens in Central Asia and eastern North America and, to a lesser extent, within Precambrian terranes. The distribution of porphyry deposits in Canada is shown in Figure 2.

Importance

Porphyry deposits are the world's most important source of Cu and Mo; they account for about 60 to 70% of world Cu production and more than 95% of world Mo production. Porphyry deposits are also major sources of Au, Ag, and Sn; significant byproduct metals include Re, W, In, Pt, Pd, and Se.

Historical production of Cu and Au from porphyry deposits in Canada relative to total Canadian production is shown in Figure 3. Canadian production of Cu from porphyry deposits began in the 1920s with the exploitation of high-grade ores in the Copper Mountain district, British Columbia, although the first significant large-scale production from a recognized porphyry deposit did not begin until 1955, from the Copper Mountain deposit at Mines Gaspé, Quebec. Copper production from porphyry deposits increased rapidly in the 1960s and early 1970s with the development of deposits in British Columbia such as Bethlehem and Lornex in the Highland Valley district, Granisle and Bell Copper in the Babine district, Ingerbelle in the Copper Mountain district, Brenda, Island Copper, and Gibraltar. In 2000, production of Cu from Canadian porphyry deposits amounted to 267,000 t, or about 2% of total world Cu production and approximately 43% of total Canadian Cu production; the major part of Canadian Cu production was from magmatic Cu-Ni deposits at Sudbury and from numerous volcanogenic massive sulphide (VMS) deposits scattered across the country. About 60% of Canadian Cu reserves are in porphyry deposits, largely in the Cordillera, but they include a considerable amount of low-grade Cu resources that are currently subeconomic.

Since 1970, porphyry deposits have become increasingly important sources of Au in Canada (Fig. 3B). In 2000, porphyry Cu-Au deposits (e.g. Kemess and Mount Polley) and porphyry Au deposits (e.g. Troilus) collectively produced about 15 t Au, representing nearly 10% of total Canadian Au production.

Canadian Mo production in 2000 was entirely from porphyry deposits and totalled 6980 t, of which the Endako porphyry Mo deposit produced about 71%, the Highland Valley porphyry Cu deposit 20%, and the Huckleberry porphyry Cu-Mo deposit the remaining 9%.

At present (2006), no porphyry W-Mo or Sn deposits are in production in Canada. From 1983 to 1985, approximately 1120 t W were produced from the porphyry W-Mo deposit at Mount Pleasant in New Brunswick, and significant W resources remain at Mount Pleasant and in the Logtung W-Mo porphyry deposit in Yukon Territory. To date, most Canadian W production has come from smaller, but higher grade, skarn W deposits such as Cantung, Northwest Territories, which has produced more than 50,000 t W. From 1985 to 1992, more than 21,000 t Sn were produced from the granite-related, porphyry-type Sn deposit at East Kemptville,

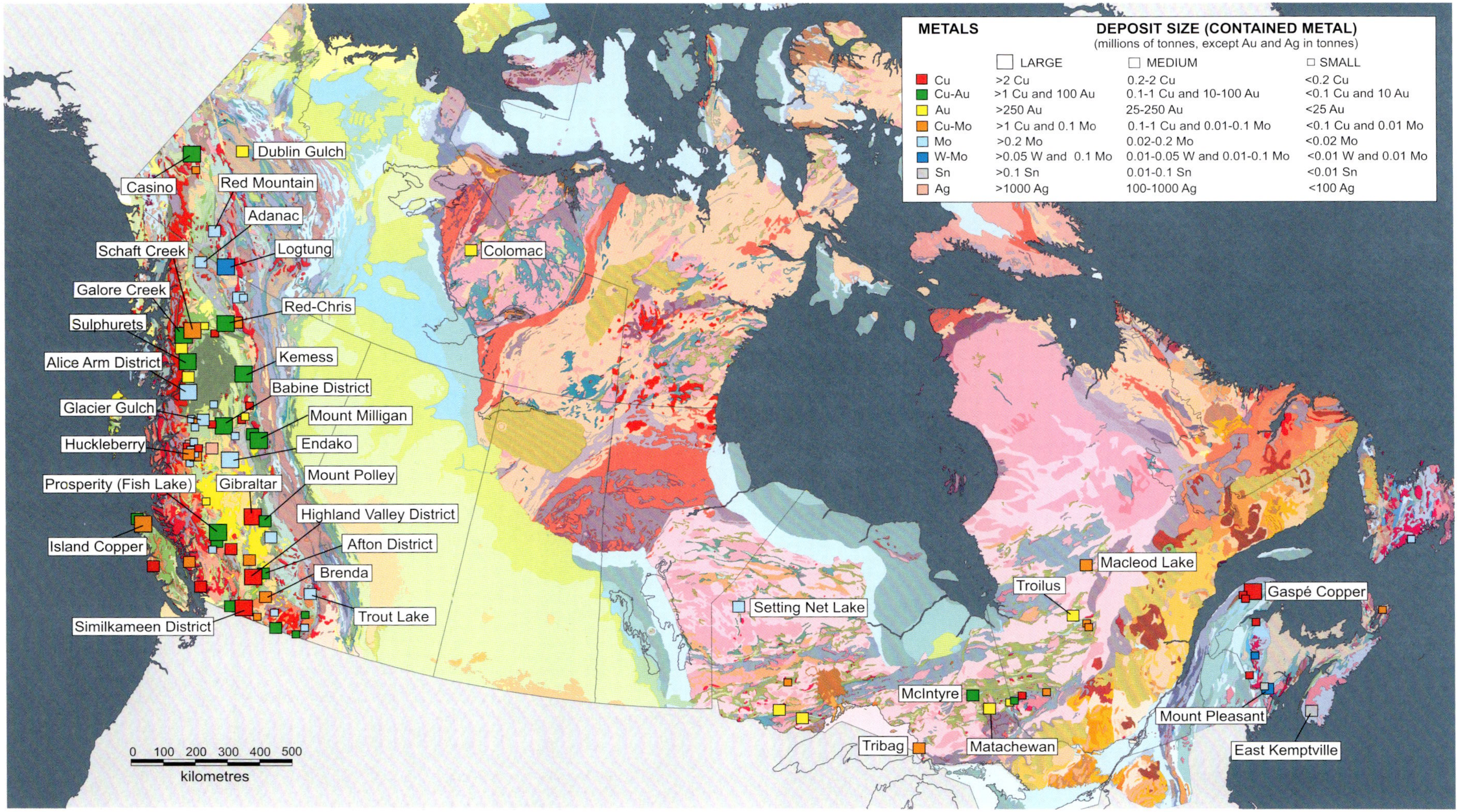

Figure 2. Distribution of selected porphyry deposits in Canada (Kirkham and Dunne, 2000). Canadian geology is from Wheeler et al. (1996). See Appendix 1 (Canada) for deposit details.

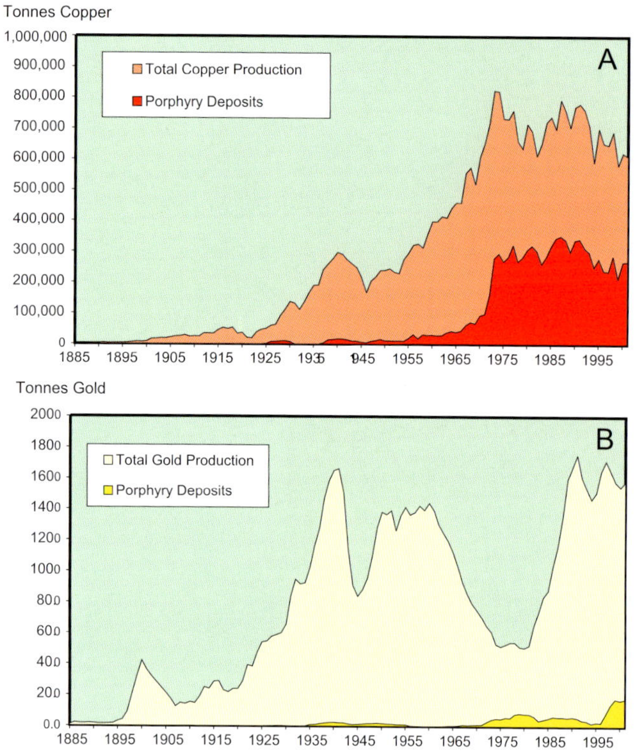

FIGURE 3. Canadian production of copper (A) and gold (B) between 1885 and 2000. Data are from Natural Resources Canada and Statistics Canada.

Nova Scotia. Prior to production from East Kemptville, the only significant production of Sn was from the Sullivan sedimentary exhalative (SEDEX) Pb-Zn-Ag deposit, which produced about 10,000 t Sn from 1941 to 1985 (BC MINFILE: http://www.em.gov.bc.ca/Mining/Geolsurv/Minfile).

Minor metals present in significant amounts in some Canadian porphyry deposits include Re, In, and PGE. During its operations from 1971 to 1995, the Island Copper porphyry Cu-Mo deposit produced about 28 t Re (Perelló et al., 1995). The Mount Pleasant porphyry Sn-polymetallic deposits contain more than 600 t In (Sinclair et al., 2006). Some porphyry Cu-Au deposits contain significant amounts of PGE; the Afton deposit, for example, has a measured and indicated resource of 68.7 Mt grading 1.1% Cu, 0.9 g/t Au, 2.6 g/t Ag, and 0.1 g/t Pd (Bradbrook, 2006).

Grade and Tonnage

Porphyry deposits are large and typically contain hundreds of millions of tonnes of ore, although they range in size from tens of millions to billions of tonnes; grades for the different metals vary considerably but generally average less than 1% (Appendix 1 (Canada) and Appendix 2 (World)).

In **porphyry Cu** deposits, Cu grades range from 0.2% to more than 1% (Fig. 4); Mo content ranges from approximately 0.005 to about 0.03% (Fig. 5); and Au contents range from 0.004 to 0.35 g/t (Fig. 6). Ag content ranges from 0.2 to 5 g/t. Re is also a significant byproduct from some porphyry Cu deposits; at Island Copper, for example, Re was extracted from molybdenite concentrates that contained as much as 1200 ppm Re (unpublished data). Some Au-rich porphyry Cu deposits have relatively high contents of Pt-group elements (PGE) (Mutschler and Mooney, 1993; Tarkian and Stribrny, 1999).

Cu grades in **porphyry Cu-Au** deposits are comparable to those of the porphyry Cu subtype (Fig. 4), but Au contents tend to be consistently higher (0.2-2.0 g/t, Fig. 6). Sillitoe (1993b) suggested that porphyry Cu deposits should contain >0.4 g/t Au to be called Au-rich. However, Au is an important coproduct at grades as low as 0.2 g/t Au (Fig. 7). Kirkham and Sinclair (1995) suggested that a 1:1 ratio (i.e. 1 g/t Au to 1% Cu) could be useful in defining Au-rich porphyry Cu deposits, particularly at higher copper grades; Kesler et al. (2002) suggested a slightly lower ratio of about 0.8:1 should be the division between Au-rich and Au-poor deposits. Although the number of deposits in this class is limited, deposits such as Grasberg in Indonesia, with a resource greater than 2.5 billion tonnes grading 1.1% Cu and 1.04 g/t Au (Freeport-McMoRan Copper and Gold Inc., Annual Report 2000), indicate that porphyry Cu-Au deposits can contain major Au as well as Cu resources. In comparison, the Kemess South deposit in British Columbia contained a geological resource of about 179 Mt grading 0.23% Cu and 0.67 g/t Au prior to mining (Appendix 1 (Canada)). **Porphyry Au** deposits contain 0.8 to 2.0 g/t Au in deposits that range in size from about 30 to greater than 200 Mt of ore

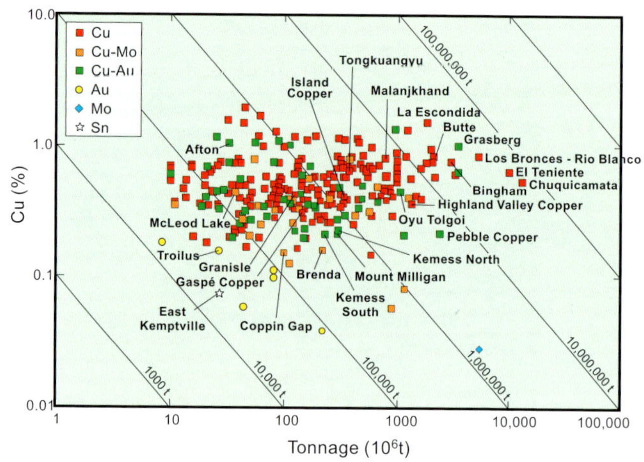

FIGURE 4. Cu grades versus tonnage for Canadian and foreign porphyry deposits (modified from Kirkham and Sinclair, 1995). Diagonal lines indicate tonnes of contained Cu.

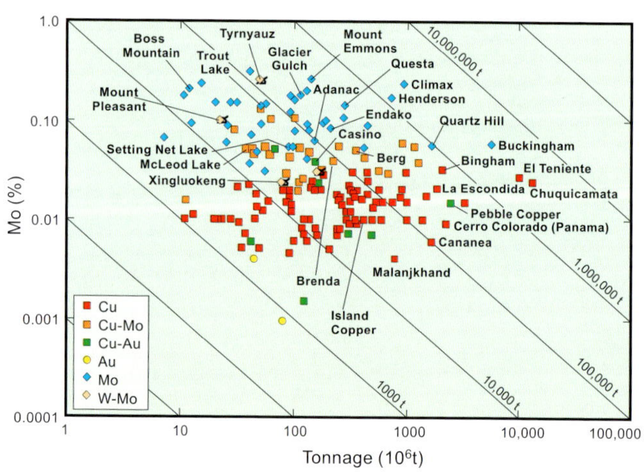

FIGURE 5. Mo grades versus tonnage for Canadian and foreign porphyry deposits (modified from Kirkham and Sinclair, 1995). Diagonal lines indicate tonnes of contained Mo.

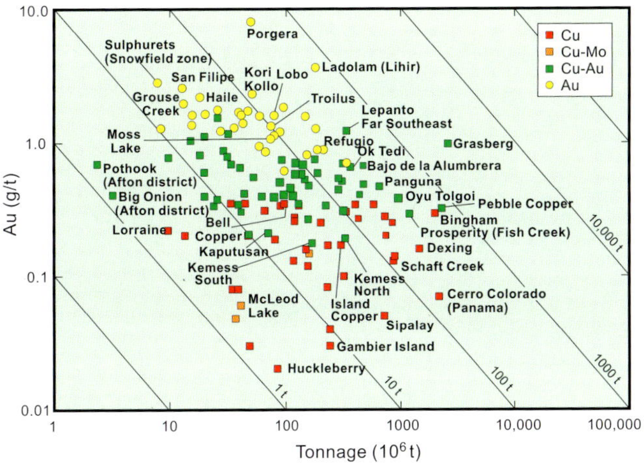

Figure 6. Au grades versus tonnage for Canadian and foreign porphyry deposits (modified from Kirkham and Sinclair, 1995). Diagonal lines indicate tonnes of contained Au.

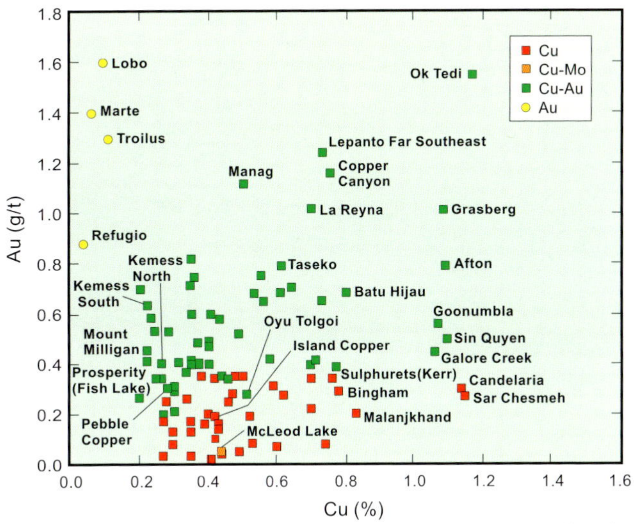

Figure 7. Au versus Cu grades in Canadian and foreign porphyry deposits (modified from Kirkham and Sinclair, 1995).

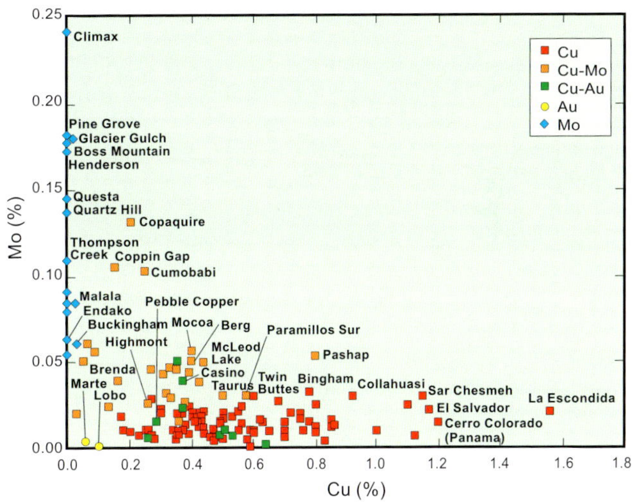

Figure 8. Mo versus Cu grades in Canadian and foreign porphyry deposits (modified from Kirkham and Sinclair, 1995).

(Fig. 6). At the Troilus deposit, Quebec, production and reserves to the end of 2002 totalled 71 Mt grading 0.93 g/t Au and 0.1% Cu (Appendix 1 (Canada)).

Porphyry Mo deposits are divided herein into two subtypes; Climax-type and Endako-type, which correspond generally to Sillitoe's (1980) classification of rift-related and subduction-related porphyry Mo deposits. Climax-type deposits are large and rich, as exemplified by the Climax and Henderson deposits in Colorado, with resources of 907 Mt grading 0.24% Mo and 727 Mt grading 0.17% Mo respectively. These deposits represent one end of a spectrum of Mo-bearing deposits, many of which have lower Mo grades and/or tonnages (Fig. 5). The opposite end of the spectrum in terms of grade is represented by the Endako deposit, for which combined production and reserves to the end of 2002 totalled 341 Mt with an average grade of 0.066% Mo (Appendix 1 (Canada)). The dividing line between rift-related (Climax-type) and subduction-related (Endako-type) deposits is 0.15% Mo (Sillitoe, 1980), although some Endako-type deposits may contain high-grade zones in excess of 0.15% Mo (e.g. Glacier Gulch and Trout Lake, British Columbia and Red Mountain, Yukon Territory). Limited data are available for W and Sn grades in most porphyry Mo deposits, but some, such as Climax, have produced significant amounts of W and Sn.

Porphyry W and W-Mo deposits contain tens of millions of tonnes to more than 100 Mt at grades of 0.1 to 0.3% W (Sinclair, 1995a); **porphyry Sn** deposits contain tens to hundreds of millions of tonnes grading 0.2 to 0.5% Sn (Sillitoe et al., 1975; Sinclair, 1995b).

Porphyry Ag deposits are not well represented; the large, low-grade, vein-stockwork Real de Angeles deposit in Mexico, with reserves of 85 Mt grading 75 g/t Ag, 1.0% Pb, and 0.92% Zn (Pearson et al., 1988), is a possible example.

Cu and Mo contents indicate that a continuum exists between porphyry Cu and porphyry Mo deposits (Fig. 8). End-member deposits are abundant and important economically but deposits with intermediate Cu and Mo contents indicate that porphyry Cu deposits, with minor or no Mo, grade to porphyry Mo deposits with negligible Cu contents (Westra and Keith, 1981; Kirkham and Sinclair, 1995). A continuum may also exist between porphyry Mo and porphyry W-Mo and W deposits, although more data are required to substantiate such a relationship. These examples illustrate some of the difficulties in making sharp distinctions between different porphyry deposit subtypes and one reason for viewing porphyry deposits as a single large class of deposits characterized by diverse metal contents with gradational boundaries between metal subtypes.

Geological Attributes

Temporal Distribution

Porphyry deposits range in age from Archean to Recent, although most are Jurassic or younger (Fig. 9). On a global basis, the peak periods for development of porphyry deposits are Jurassic, Cretaceous, Eocene, and Miocene in age. These ages also correspond to peak periods of porphyry mineralization in Canada, except for Miocene, of which there are no significant deposits in Canada.

Porphyry-type deposits of Precambrian age are not as well represented and, in some cases, their classification is contentious. For example, Goodman et al. (2005) consider Troilus to be an "orogenic" deposit related to metamorphism,

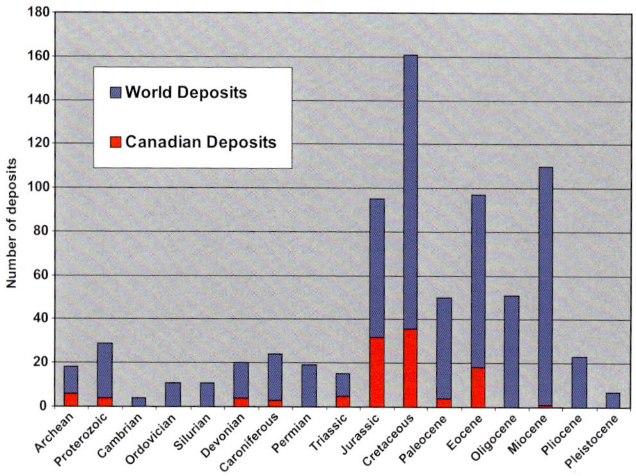

FIGURE 9. Age distribution of porphyry deposits (based on data from Kirkham and Dunne, 2000).

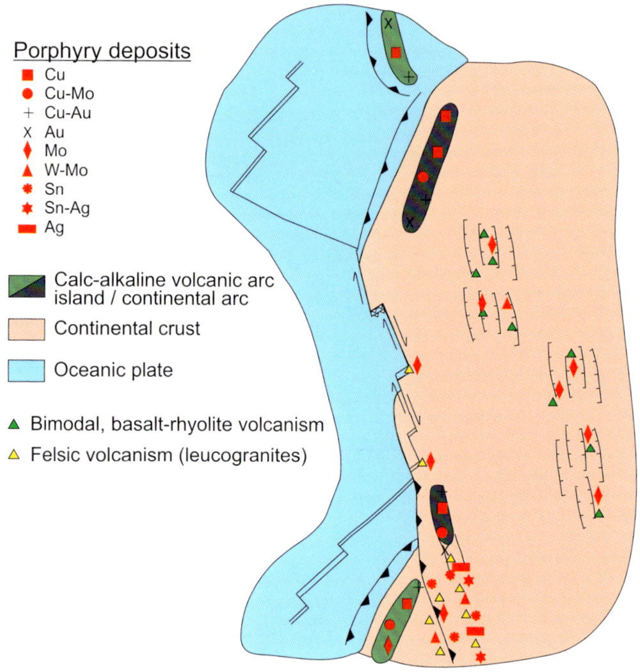

FIGURE 10. Schematic diagram showing the tectonic settings of porphyry deposits (Kirkham and Sinclair, 1995).

rather than a premetamorphic porphyry deposit as described by Fraser (1993). In the Chibougamau district, Cu-Au vein deposits in strongly foliated rocks have been considered to be shear related (e.g. Archambault et al., 1984) but Pilote et al. (1995) demonstrated that they are premetamorphic in origin and likely related to the Clark Lake porphyry deposit. Other examples of Precambrian porphyry-type deposits include Aitik, Sweden (Wanhainen et al., 2003), Malanjkhand, India (Sikka and Nehru, 1997); Tongkuangyu, China (Weixing and Dazhong, 1987); Boddington, Australia (Roth et al., 1991); Haib, Namibia (Minnitt, 1986); Setting Net Lake, Ontario (Ayres et al., 1982), and McIntyre, Ontario (Mason and Melnyk, 1986). The Coppin Gap Cu-Mo deposit in Australia, dated at approximately 3.3 Ga (de Laeter and Martyn, 1986; Williams and Collins, 1990), is the oldest known porphyry deposit in the world.

Continental Scale (Geotectonic Environment)

Tectonic Setting

Porphyry deposits occur in a variety of tectonic settings. Porphyry Cu deposits typically occur in the root zones of andesitic stratovolcanoes in subduction-related, continental- and island-arc settings (Mitchell and Garson, 1972; Sillitoe, 1973, 1988a; Sillitoe and Bonham, 1984; Fig. 10). In southern Arizona, porphyry Cu deposits are associated with granitic rocks emplaced in a continental setting, within or along the margins of calderas that are now largely eroded (Lipman and Sawyer, 1985). Porphyry Cu-Au deposits, such as those associated with Triassic and Lower Jurassic silica-saturated, alkaline intrusions in British Columbia, formed in an island-arc setting, although possibly during periods of extension. Grasberg and Porgera formed in a continental-island-arc collisional zone during or immediately following subduction (Richards and Kerrich, 1993; MacDonald and Arnold, 1994). Porphyry Cu deposits in the Yulong belt of Tibet are related to pull-apart basins in a large, post-subduction strike-slip fault (Hou et al., 2003). Porphyry Au deposits of Tertiary age in the Maricunga belt in Chile appear to have formed in a continental-arc setting along strike to the north from major porphyry Cu deposits of the same general age (Sillitoe, 1992, 1993b; Camus, 2002).

Climax-type porphyry Mo deposits are typically associated with anorogenic granites that have been emplaced in continental settings, particularly rift or extensional environments (Fig. 10). The Climax and Henderson deposits, for example, are genetically related to small stocks connected at depth to a regional batholith emplaced during active extension in the Rio Grande rift (Bookstrom, 1981; Carten et al., 1988b, 1993). Endako, on the other hand, is related to a composite batholith that formed in a continental subduction-related environment (Whalen et al., 2001). Some porphyry Mo deposits, such as Kitsault and Glacier Gulch, occur in areas of continental subduction that are underlain by slab gaps or windows (Kirkham, 1998). Others appear to have formed during extension in areas adjacent to strike-slip faults (e.g. northern Cordillera - Quartz Hill, Adanac, Casmo and Mount Haskins; Fig. 10). The Mount Pleasant W-Mo deposit in southern New Brunswick is associated with F-rich granitic rocks that have intruded along the Gander-Avalon boundary zone during a period of extension possibly related to crustal delamination (Whalen et al., 1994, 1996). A few deposits, such as Questa, New Mexico, are associated with high-silica rhyolites and granites that formed in continental calderas (Lipman, 1988). For most porphyry deposits, however, the depth of erosion is such that caldera settings are conjectural (e.g. Lipman, 1984).

Some porphyry Mo deposits, along with porphyry W-Mo and porphyry Sn deposits, formed in areas of great continental thickness related to collisional tectonic settings, although the deposits generally postdate the collision event. Porphyry Sn deposits in Bolivia, for example, are related to subvolcanic intrusions that were emplaced in thick continental crust above a relatively shallow-dipping subduction zone (Sillitoe, 1976; Grant et al., 1980).

Details of each setting and related controls on magma generation, composition, and emplacement conceivably had

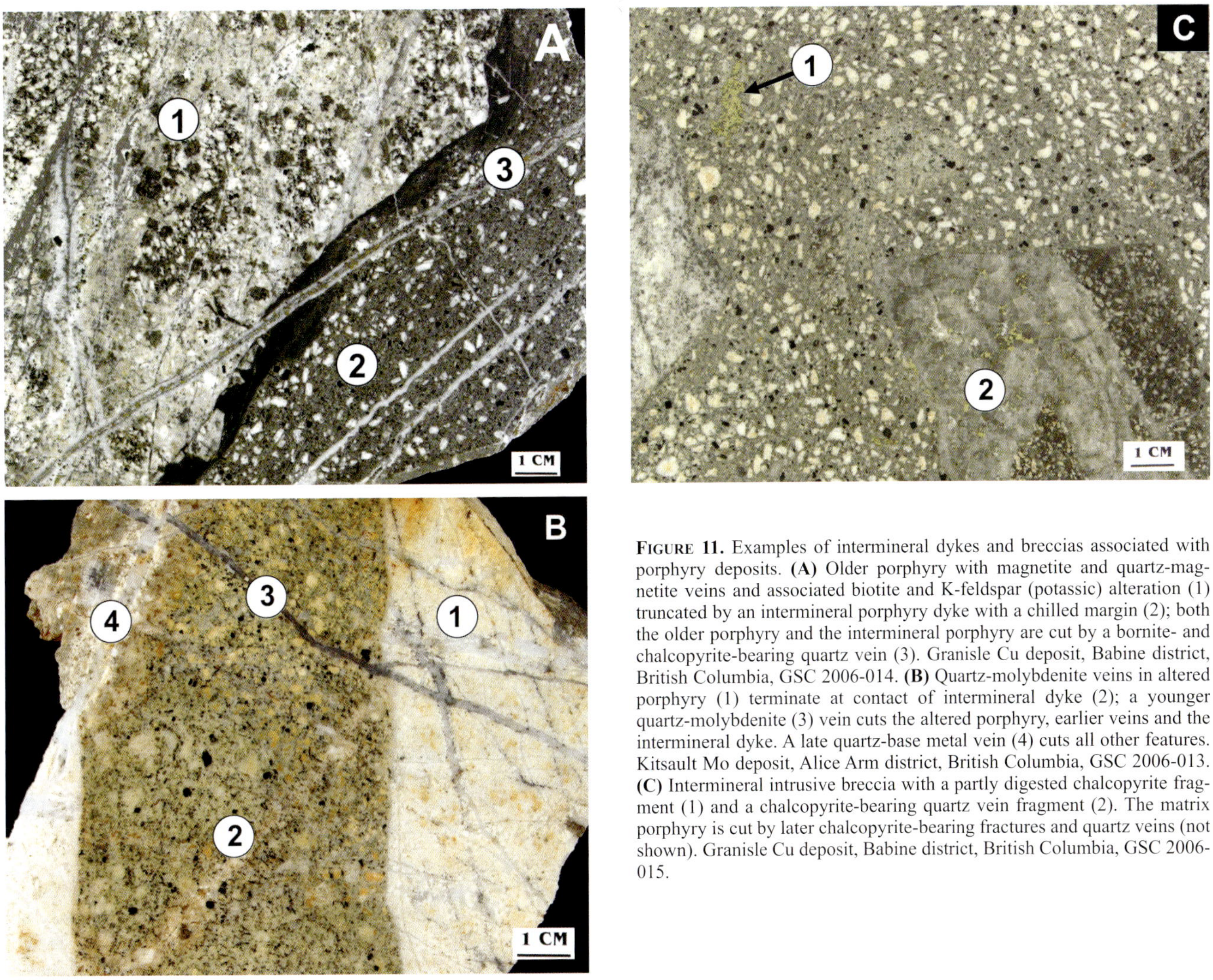

FIGURE 11. Examples of intermineral dykes and breccias associated with porphyry deposits. **(A)** Older porphyry with magnetite and quartz-magnetite veins and associated biotite and K-feldspar (potassic) alteration (1) truncated by an intermineral porphyry dyke with a chilled margin (2); both the older porphyry and the intermineral porphyry are cut by a bornite- and chalcopyrite-bearing quartz vein (3). Granisle Cu deposit, Babine district, British Columbia, GSC 2006-014. **(B)** Quartz-molybdenite veins in altered porphyry (1) terminate at contact of intermineral dyke (2); a younger quartz-molybdenite (3) vein cuts the altered porphyry, earlier veins and the intermineral dyke. A late quartz-base metal vein (4) cuts all other features. Kitsault Mo deposit, Alice Arm district, British Columbia, GSC 2006-013. **(C)** Intermineral intrusive breccia with a partly digested chalcopyrite fragment (1) and a chalcopyrite-bearing quartz vein fragment (2). The matrix porphyry is cut by later chalcopyrite-bearing fractures and quartz veins (not shown). Granisle Cu deposit, Babine district, British Columbia, GSC 2006-015.

a major influence on the size, metal contents, and nature of individual deposits. However, exceptions to typical settings, such as the Tribag and Jogran porphyry Cu-Mo deposits in Ontario that apparently are related to a continental rift environment (Kirkham, 1973; Norman and Sawkins, 1985), and the Malmbjerg porphyry Mo deposit in East Greenland, which is related to the Iceland mantle plume (Schønwandt, 1988; Brooks et al., 2004), indicate that individual porphyry deposits can occur in diverse and unique settings.

Regional Structures

In some cases, the distribution of porphyry deposits can be related to regional structures. The Rio Grande rift system in the western United States, for example, is the locus for porphyry Mo deposits (Bookstrom, 1981). The West Fissure zone along strike of the Eocene porphyry Cu belt in northern Chile, from El Salvador in the south to past Collahausi in the north, was active both during and following porphyry emplacement and hydrothermal activity (Baker and Guilbert, 1987). Also within this belt, cross structures apparently controlled the distribution of individual deposits such as Quebrada Blanca, Collahausi and Escondida (Sillitoe, 1992; Richards et al., 2001). The major Philippine strike-slip fault system in the northern part of the island arc system, similar to the West Fissure zone in northern Chile, was probably also a control on the location of major magmatic and hydrothermal centres, which might be localized in areas that are pull-apart structures at dilational bends. Porphyry Cu and Cu-Au deposits in the Babine district of British Columbia are associated with porphyritic intrusions emplaced within a zone of extension related to the development of pull-apart basins situated between dextral strike-slip faults (McIntyre and Villeneuve, 2001). In many districts, however, perhaps because of intense alteration and multiple intrusions, regional structural control is obscure.

Deposit Scale

Geological Setting and Related Magmatic Rocks

Porphyry deposits occur in close association with porphyritic epizonal and mesozonal intrusions. A close temporal relationship between magmatic activity and hydrothermal mineralization in porphyry deposits is indicated by the presence of intermineral intrusions and breccias that were emplaced between or during periods of mineralization (e.g. Kirkham, 1971; Fig. 11).

The composition of intrusions associated with porphyry deposits varies widely and appears to exert a fundamental

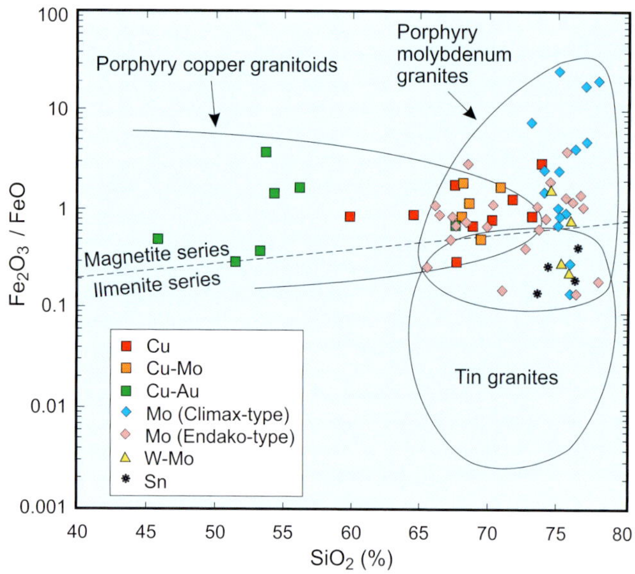

FIGURE 12. SiO_2-(Fe_2O_3/FeO) variation diagram for granitic rocks related to porphyry deposits of Cu, Cu-Mo, Cu-Au, Mo, W-Mo, and Sn (modified from Kirkham and Sinclair, 1995). The dividing line between the magnetite and ilmenite series is from Ishihara (1981) and the field of tin granites is from Lehmann (1990).

control on the metal content of the deposits (Fig. 12). Intrusive rocks associated with porphyry Cu-Au and porphyry Au deposits tend to be low-silica (45-65 wt.% SiO_2), mafic and relatively primitive in composition, ranging from calc-alkaline dioritic and granodioritic plutons to alkalic monzonitic rocks (e.g. Richards, 1990; Ross et al., 1995; Snyder and Russell, 1995; Holliday et al., 2002; Wilson et al., 2003; Cooke et al., 2004). Porphyry Cu and Cu-Mo deposits are associated with intermediate to felsic, calc-alkaline intrusive rocks that range from granodiorite to granite in composition (60-72 wt.% SiO_2) (e.g. Kesler et al., 1975; Titley and Beane, 1981).

Porphyry deposits of Mo (Climax-type), W-Mo, W, and Sn, in comparison, are typically associated with felsic, high-silica (72-77 wt.% SiO_2) and, in many cases, strongly differentiated granitic plutons (Mutschler et al., 1981; White et al., 1981; Kooiman et al., 1986; Guan et al., 1988). Intrusions associated with Endako-type porphyry Mo deposits range more widely in composition, from granodiorite to granite (65-77 wt.% SiO_2, Fig. 12). They also typically have low fluorine contents (<0.1% F) compared to intrusions associated with Climax-type deposits (Mutschler et al., 1981) and Endako-type deposits are distinguished as being fluorine-deficient or fluorine-poor (Theodore and Menzie, 1984; Linnen et al., 1995; Selby et al., 2000).

Oxidation state of granitic rocks, reflected by accessory minerals such as magnetite, ilmenite, pyrite, pyrrhotite, and anhydrite, also influences metal contents of related deposits (Ishihara, 1981). For example, porphyry deposits of Cu, Cu-Mo, Cu-Au, Au, Mo (mainly Climax-type), and W are generally associated with oxidized, magnetite-series plutons, whereas porphyry Sn and some Endako-type Mo deposits are related to reduced, ilmenite-series plutons (Fig. 12).

Small amounts of alkaline magmatic rocks associated with calc-alkaline intrusions may be significant for the formation of related porphyry deposits. For example, lamprophyre dykes are contemporaneous with felsic igneous rocks at Climax, Henderson, and other porphyry Mo deposits and are likely part of a bimodal suite (Bookstrom, 1981; Bookstrom et al., 1988). In addition to melting lower crustal rocks, underplated or injected mafic magmas represented by the lamprophyre dykes may have supplied some volatiles such as CO_2 and SO_2 (Keith et al., 1993). At Bingham, Utah, minette dykes that are coeval with ore-related monzonite represent mafic alkaline magmas that appear to have mixed with the more felsic monzonitic magmas, transferring substantial amounts of sulphur, copper, and other metals to the felsic magma and hydrothermal fluids in the process (Keith et al., 1997). Repeated injections of hot mafic melt into colder felsic magmas also would have triggered periodic eruptions and magmatic degassing resulting in extensive brecciation and mineralization of surrounding rocks (Hattori and Keith, 2001).

Felsic intrusive rocks that are closely associated with some porphyry deposits are characterized by distinct textural features such as comb-quartz layers (Fig. 13) and other unidirectional solidification textures (Shannon et al., 1982; Kirkham and Sinclair, 1988). Comb-quartz layers occur in places along the margins and in the upper parts of small intrusions or cupolas. They generally range from less than one millimetre to several centimetres or more in thickness and are separated by interlayers of fine-grained, porphyritic aplite (Fig. 13A,B). Prismatic quartz crystals in the layers are oriented roughly perpendicular to the planes of layering and appear to have grown on a crystallized aplitic substrate inward toward the centre of the intrusion (Fig. 13C). Closely related features include vein dykes and parting veins (Fig. 13D; Kirkham and Sinclair, 1988; Seedorf et al., 2005). Comb-quartz layers are significant because individual layers likely crystallized from pockets of exsolved magmatic-hydrothermal fluid and the development of multiple layers reflects a continuous supply of magmatic fluid from subjacent magma (Lowenstern and Sinclair, 1996). Small stocks with abundant comb-quartz layers, therefore, represent conduits for large volumes of degassed magmatic-hydrothermal fluids that are capable of producing large porphyry orebodies. For example, at the Henderson Mo deposit, two of the nine associated plutons are characterized by significantly greater textural variation, including well developed comb-quartz layers (termed "brain rock"); these two plutons are responsible for about 80% of the Mo mineralization (Carten et al., 1988a).

Morphology and Architecture

The overall form of individual porphyry deposits is highly varied and includes irregular, oval, solid, or "hollow" cylindrical and inverted cup shapes (e.g. Sutherland Brown, 1969; Lowell and Guilbert, 1970; James, 1971; McMillan and Panteleyev, 1980). Orebodies may occur separately or overlap and, in some cases, are stacked one on top of the other (Wallace et al., 1968; White et al., 1981; Titley et al., 1986; Carten et al., 1988a). Individual orebodies measure hundreds to thousands of metres in three dimensions. Orebodies are characteristically zoned, with barren cores and crudely concentric metal zones that are surrounded by barren pyritic halos with or without peripheral veins, skarns, replacement manto zones and epithermal precious-metal deposits (e.g.

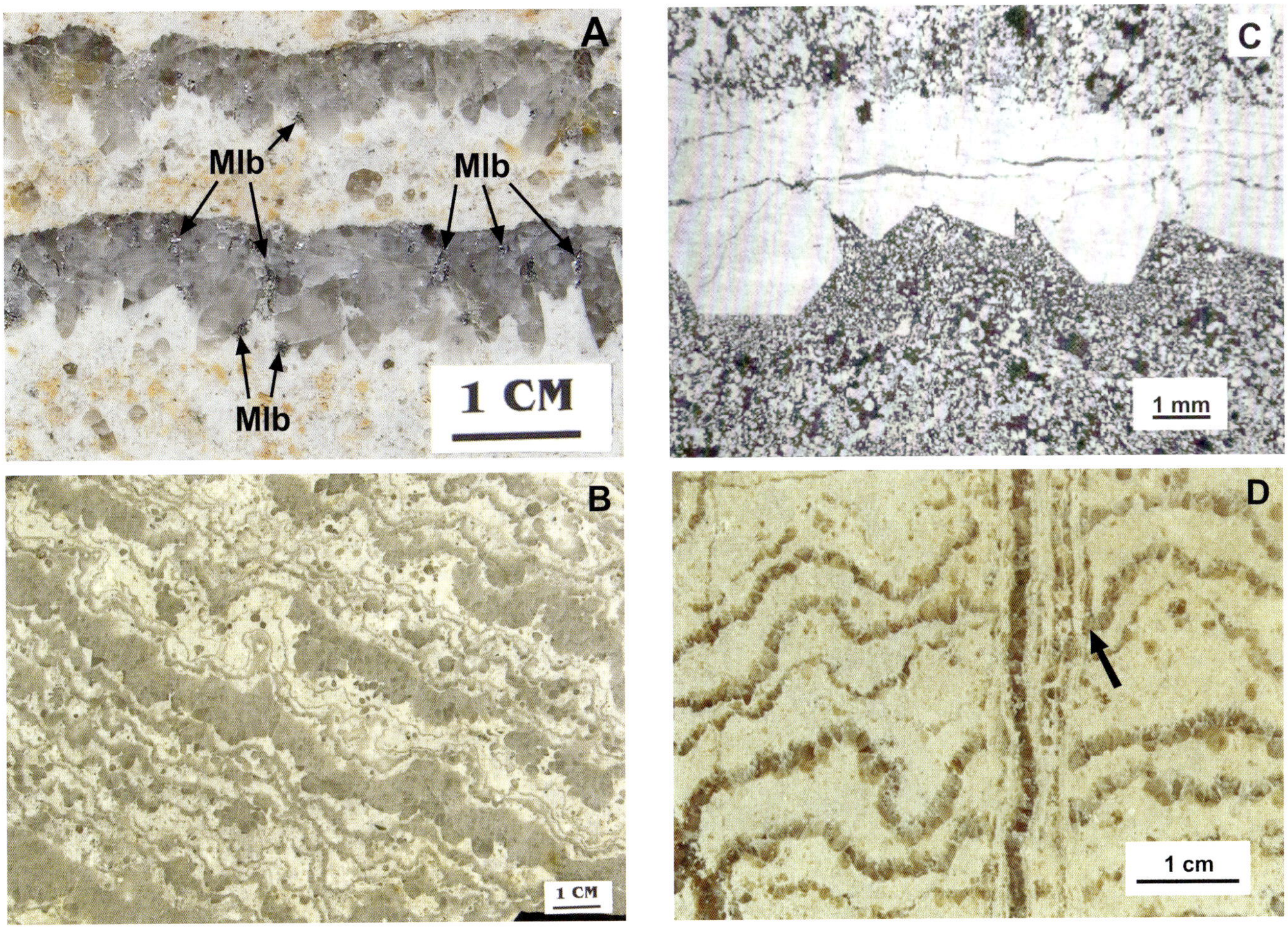

FIGURE 13. Examples of comb-quartz layers in felsic intrusions associated with porphyry deposits. **(A)** Comb-quartz layers containing molybdenite (Mlb) separated by aplite interlayers; growth direction from the top to the bottom. Anticlimax Mo deposit, British Columbia, GSC 2006-018. **(B)** Multiple thin to thick comb-quartz layers separated by aplite interlayers; growth direction of the quartz crystals in the layers was from the upper right to the bottom left. Logtung W-Mo deposit, Yukon Territory, GSC 2006-003. **(C)** Photomicrograph of a comb-quartz layer in aplitic granite showing euhedral termination of quartz crystals approximately perpendicular to the layer; the very fine-grained texture of the granite adjacent to the crystal faces resulted from pressure quenching and consequent rapid cooling related to sudden (catastrophic) release of the fluid phase in which the quartz crystals were growing. North zone, Mount Pleasant Sn deposit, New Brunswick, GSC 204152-T. **(D)** Contorted comb-quartz layers (growth direction towards the bottom of the photograph) cut by a parting vein, which consists of quartz veinlets with numerous septa or partings of aplite; one of the quartz veinlets in the parting vein appears to be rooted at the termination of a comb-quartz layer (arrow). Logtung W-Mo deposit, Yukon Territory, GSC 2006-006.

Einaudi, 1982; Sillitoe, 1988a,b; Jones, 1992). Complex, irregular ore and alteration patterns are due, in part, to the superposition and spatial separation of mineral and alteration zones of different ages.

Associated Structures and Mineralization Styles

At the scale of ore deposits, associated structures can result in a variety of mineralization styles, including veins, vein sets, stockworks, fractures, 'crackled zones', and breccia pipes (Fig. 14). In large, complex, economic porphyry deposits, mineralized veins and fractures typically have a very high density. Orientations of mineralized structures can be related to local stress environments around the tops of plutons or can reflect regional stress conditions (Rehrig and Heidrick, 1972; Heidrick and Titley, 1982; Linnen and Williams-Jones, 1987; Carten et al., 1988a). Where they are superimposed on each other in a large volume of rock, the combination of individual mineralized structures results in higher grade zones and the characteristic large size of porphyry deposits.

Mineralogy

The mineralogy of porphyry deposits is highly varied, although pyrite is typically the dominant sulphide mineral in porphyry Cu, Cu Mo, Cu Au, Au, and Ag deposits, reflecting the fact that these deposits are huge sulphur anomalies (Hunt, 1991). In porphyry deposits of the more lithophile elements, i.e. Sn, W, and Mo, the overall sulphur and sulphide mineral contents are lower. Principal ore and associated minerals of the different porphyry deposit subtypes are as follows.

Porphyry Cu, Cu-Mo, and Cu-Mo-Au deposits: Principal ore minerals are chalcopyrite, bornite, chalcocite, tennantite, enargite, other Cu sulphides and sulphosalts, molybdenite, and electrum; associated minerals include pyrite, magnetite, quartz, biotite, K-feldspar, anhydrite, muscovite, clay minerals, epidote and chlorite.

Porphyry Cu-Au deposits: Principal ore minerals are chalcopyrite, bornite, chalcocite, tennantite, other Cu minerals, native Au, electrum, and tellurides; associated minerals include pyrite, arsenopyrite, magnetite, quartz, biotite, K-

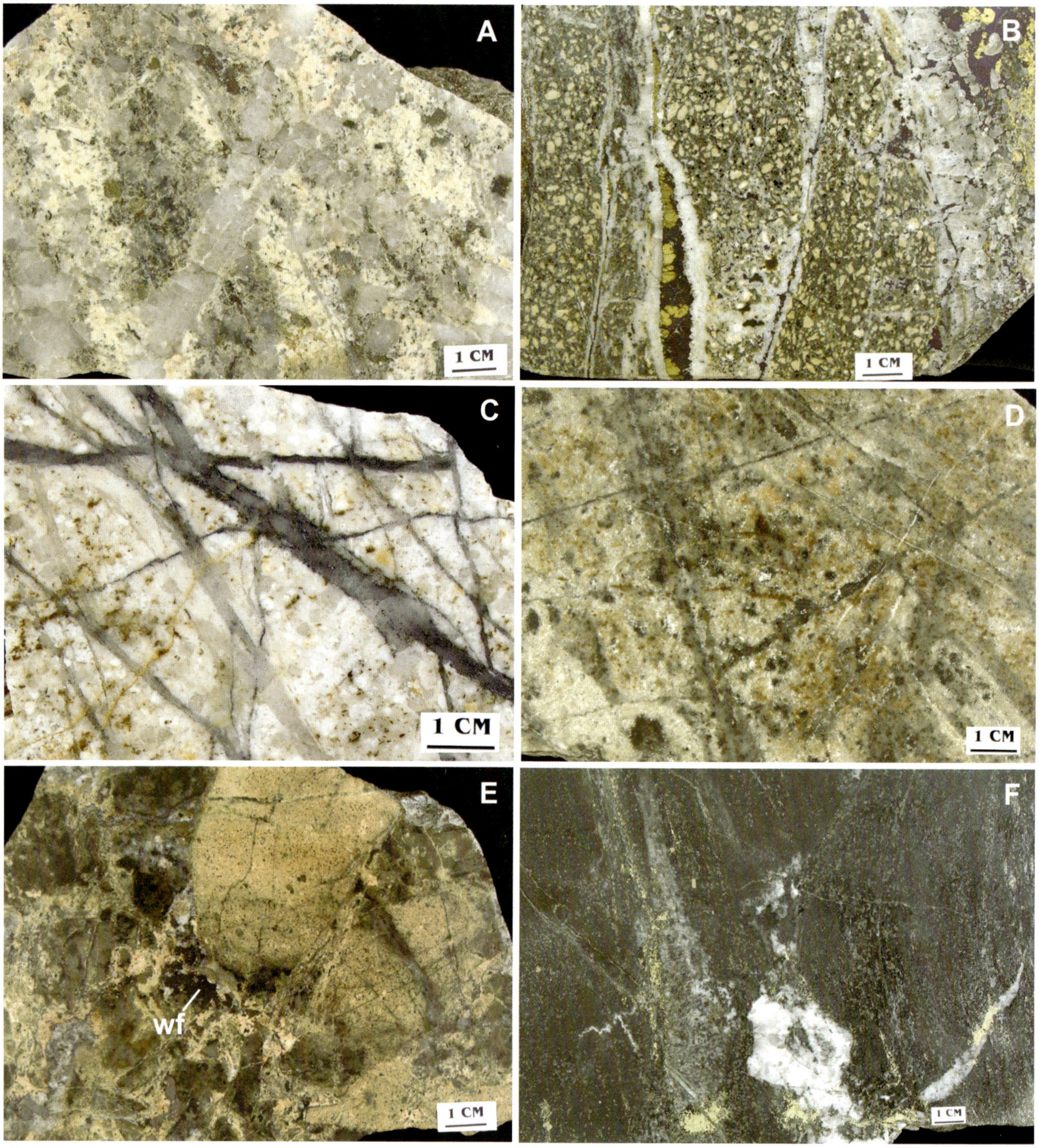

FIGURE 14. Examples of different mineralization styles associated with porphyry deposits. **(A)** Bornite-bearing quartz veins cutting highly sericitized Bethsaida granodiorite. Valley Cu deposit, Highland Valley district, British Columbia, GSC 2006-012. **(B)** Chalcopyrite- and bornite-rich quartz-apatite veins and veinlets cutting biotite-feldspar porphyry. High-grade ore, Granisle Cu deposit, Babine district, British Columbia, GSC 2006-016. **(C)** Quartz-molybdenite stockwork in sericitized granodiorite porphyry; cross-cutting and offset relationships of molybdenite-bearing fractures and quartz veinlets indicate multiple stages of mineralization. Kitsault Mo deposit, Alice Arm district, British Columbia, GSC 2006-007. **(D)** Stockwork of wolframite-bearing fractures cutting intensely altered breccia. Fire Tower zone, Mount Pleasant W-Mo deposit, New Brunswick, GSC 2006-005. **(E)** Mineralized breccia containing granite clasts with wolframite-bearing fractures that are truncated at the margins of the clasts, indicating that the granite was mineralized prior to the incorporation of the clasts in the breccia; wolframite (wf) also occurs as disseminated grains in breccia matrix. Fire Tower zone, Mount Pleasant W-Mo deposit, New Brunswick, GSC 2006-004. **(F)** Chalcopyrite disseminated along foliation planes and in a crosscutting quartz vein in deformed biotite-rich mafic breccia. Troilus Au deposit, Quebec, GSC 2006-010.

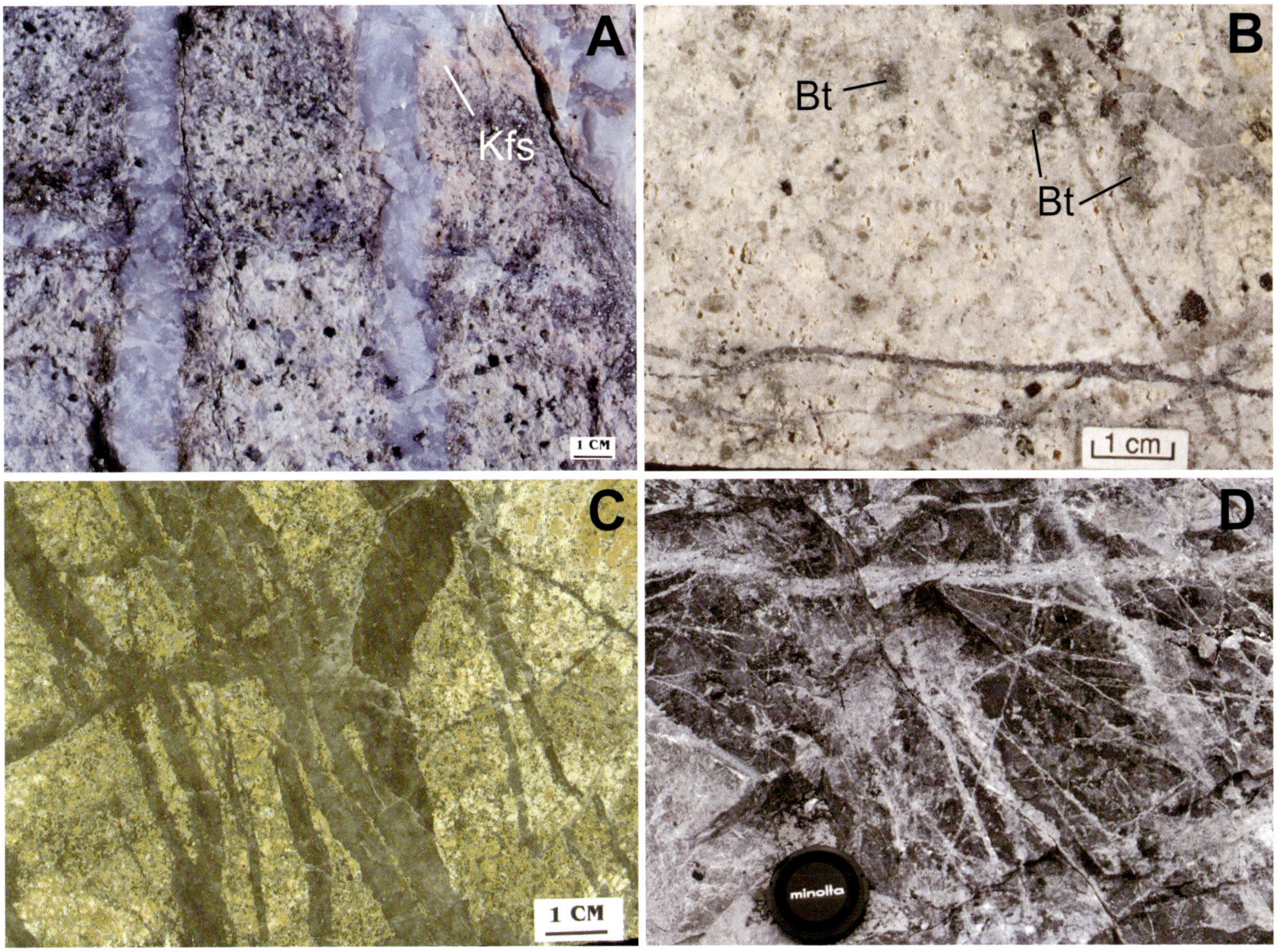

FIGURE 15. Examples of different types of alteration associated with porphyry deposits. **(A)** Potassic K-feldspar (Kfs) alteration around mineralized quartz veins in Bethsaida granodiorite. Valley Cu deposit, Highland Valley district, British Columbia, GSC 2006-017. **(B)** Potassic alteration of granodiorite porphyry consisting of pervasive pink K-feldspar and patches of fine-grained hydrothermal biotite (Bt) associated with quartz-molybdenite veinlets. Red Mountain Mo deposit, Yukon Territory, GSC 2006-008. **(C)** Quartz-pyrite-chalcopyrite vein stockworks in feldspar porphyry heavily overprinted by sericitic (phyllic) alteration. Bell Cu-Au deposit, Babine district, British Columbia, GSC 2006-011. **(D)** Stockwork of wolframite- and molybdenite-bearing fractures with white selvages of quartz-topaz-fluorite-sericite alteration cutting chloritized granite. Fire Tower zone, Mount Pleasant W-Mo deposit, New Brunswick, GSC 2006-009.

feldspar, anhydrite, epidote, chlorite, scapolite, albite, calcite, fluorite, and garnet.

Porphyry Au deposits: Principal ore minerals are native Au and electrum; minor amounts of chalcopyrite, bornite, and molybdenite may also be present. Associated minerals include pyrite, magnetite, quartz, biotite, K-feldspar, muscovite, clay minerals, epidote, and chlorite.

Porphyry Mo deposits: Principal ore minerals are molybdenite, scheelite, wolframite, cassiterite, bismuthinite, and native bismuth; minor amounts of chalcopyrite may also be present. Associated minerals include pyrite, magnetite, quartz, K-feldspar, biotite, muscovite, clay minerals, fluorite, and topaz.

Porphyry W-Mo deposits: Principal ore minerals are scheelite, wolframite, molybdenite, cassiterite, stannite, bismuthinite, and native bismuth; other minerals include pyrite, arsenopyrite, loellingite, quartz, K-feldspar, biotite, muscovite, clay minerals, fluorite, and topaz.

Porphyry Sn and Sn-Ag deposits: Principal ore minerals are cassiterite, tetrahedrite, argentite, stannite, wolframite, chalcopyrite, sphalerite, franckeite, cylindrite, teallite, molybdenite, bismuthinite, other sulphides and sulphosalts, native Ag, and native Bi; associated minerals include pyrite, arsenopyrite, loellingite, quartz, K-feldspar, biotite, muscovite, clay minerals, fluorite, and topaz.

Porphyry Ag deposits: Principal ore minerals are freibergite, stephanite, acanthite, sphalerite, and galena; associated minerals include arsenopyrite, pyrrhotite, pyrite, adularia, quartz, fluorite, and calcite.

Alteration

Hydrothermal alteration is extensive and typically zoned on a deposit scale (Lowell and Guilbert, 1970) as well as around individual veins and fractures (Fig. 15). In many porphyry deposits, alteration zones on a deposit scale consist of an inner potassic zone characterized by K-feldspar and/or biotite (± amphibole ± magnetite ± anhydrite; Fig. 15A,B) and an outer zone of propylitic alteration that consists of quartz, chlorite, epidote, calcite and, locally, albite associated with pyrite. Zones of phyllic alteration (quartz + sericite + pyrite, Fig. 15C) and argillic alteration (quartz + illite +

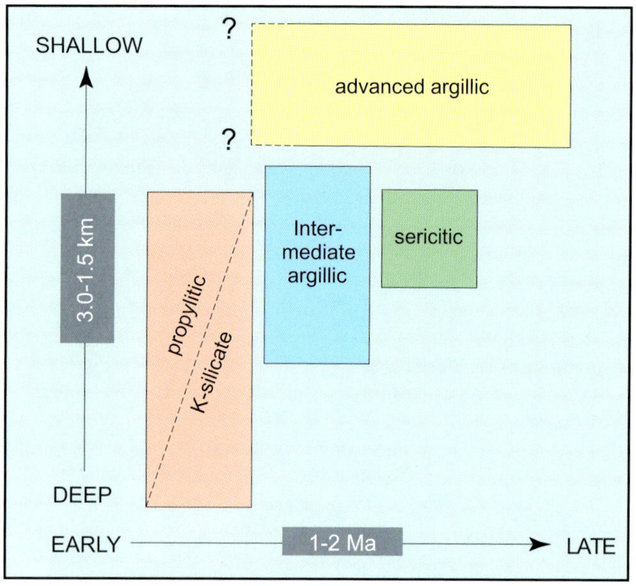

FIGURE 16. Schematic time-depth relations of principal alteration types in Au-rich porphyry Cu systems and other types of porphyry deposits (after Sillitoe, 1993b).

pyrite ± kaolinite ± smectite ± montmorillonite ± calcite) may be part of the zonal pattern between the potassic and propylitic zones, or can be irregular or tabular, younger zones superimposed on older alteration and sulphide assemblages (e.g. Moyle et al., 1990; Sillitoe, 1993b). The spatial and temporal relationships between different types of alteration are shown schematically in Figure 16.

Economic sulphide zones are most closely associated with potassic alteration, as demonstrated by Carson and Jambor (1974) for several porphyry Cu and Cu Mo deposits. Sodic alteration (mainly as secondary albite) is associated with potassic alteration in some porphyry Cu-Au deposits, such as Copper Mountain and Ajax, British Columbia (Preto, 1972; Barr et al., 1976; Ross et al., 1995). Albitic alteration partly overlaps potassic alteration and Cu on the north side of the Ingerbelle deposit at Copper Mountain; at the Ajax deposit, highest Cu grades occur near, but not in, the most intensely altered albitic rocks. Eaton and Setterfield (1993) indicated that the subeconomic Nasivi 3 porphyry Cu deposit in the centre of the shoshonitic Tavua caldera, adjacent to the epithermal Emperor Au mine in Fiji, contains an albitic, Cu bearing core surrounded by peripheral propylitic alteration and overprinted by younger phyllic alteration. Sodic-calcic alteration (oligoclase + quartz + sphene + apatite ± actinolite ± epidote) has been documented in the deep root zones beneath potassically altered porphyry Cu deposits at Yerington and Ann-Mason, Nevada (Carten, 1986; Dilles and Einaudi, 1992).

Alteration mineralogy is controlled in part by the composition of the host rocks. In mafic host rocks with significant iron and magnesium, biotite (± minor hornblende) is the dominant alteration mineral in the potassic alteration zone, whereas K-feldspar dominates in more felsic rocks. In carbonate-bearing host rocks, calc-silicate minerals such as garnet and diopside are abundant.

Alteration mineralogy is also controlled by the composition of the mineralizing system. In more oxidized environments, minerals such as pyrite, magnetite (± hematite), and anhydrite are common, whereas pyrrhotite is present in more reduced environments (Rowins, 2000). Fluorine-rich systems, such as those related to many porphyry Sn and W Mo deposits, and Climax-type porphyry Mo deposits, commonly contain fluorine-bearing minerals as part of the alteration assemblages. At Mount Pleasant, for example, potassic alteration is rare and the principal alteration associated with the W-Mo deposit consists of quartz, topaz, fluorite, and sericite (Fig. 15D), which is surrounded by propylitic alteration consisting of chlorite + sericite (Kooiman et al., 1986). Similarly, alteration in some low grade Sn deposits in Australia (e.g. Ardlethan) grades out from a central zone of quartz + topaz to zones of sericite and chlorite ± carbonate (Scott, 1981). Siems (1989) suggested that lithium silicate alteration (e.g. lithium-rich mica and tourmaline), which accompanies Sn, W, and Mo in some granite-related deposits, is analogous to potassic alteration in porphyry Cu and Mo deposits.

Phyllic alteration zones are not present in all porphyry deposits. In many deposits in which they are present, however, phyllic alteration is superimposed on earlier potassic alteration assemblages (Carson and Jambor, 1979). At Chuquicamata in Chile, for example, a zone of intense phyllic alteration with associated pyrite that extends to depth in the core of the deposit is superimposed on earlier potassic alteration and small amounts of associated Cu sulphides with low Cu grades. This phyllic zone contains higher than average Cu grades and, in the weathered zone, oxidation of the pyrite produced acid leaching that resulted in an extensive deep supergene-enriched Cu zone (Ossandón et al., 2001).

Advanced argillic (high sulphidation) and adularia-type (low sulphidation) epithermal alteration zones with associated precious-metal deposits occur above or near several porphyry Cu and Cu-Mo deposits (e.g. Far Southeast-Lepanto, Hedenquist et al., 1998). These alteration zones, in places, show a marked telescoping of older potassic and younger epithermal alteration (Moyle et al., 1990; Sillitoe, 1990, 1993a,b; Setterfield et al., 1991; Vila and Sillitoe, 1991; Eaton and Setterfield, 1993; Richards and Kerrich, 1993). The advanced argillic assemblages include illite, quartz, alunite, natroalunite, pyrophyllite, diaspore, and high pyrite content. Adularia assemblages, with quartz, sericite, and clay minerals, have lower pyrite contents. Sillitoe (1993a) suggested that advanced argillic or high-sulphidation-type epithermal systems can occur in spatial association with porphyry Cu, Cu-Mo, Cu-Au, and Au deposits, but not with porphyry Mo deposits. Adularia- or low-sulphidation-type epithermal systems probably form from more dilute ore fluids and may or may not occur on the peripheries of porphyry systems. Sillitoe (1993a) also suggested that base-metal-rich epithermal deposits form from more concentrated NaCl brines and, similar to porphyry deposits, are parts of magmatic-hydrothermal systems.

Genetic and Exploration Models

A generalized empirical model for porphyry deposits is illustrated schematically in Figure 17, which shows a porphyry Cu deposit associated with a small subvolcanic porphyritic intrusion and surrounded by a more extensive pyritic zone. The larger scale of the hydrothermal system is

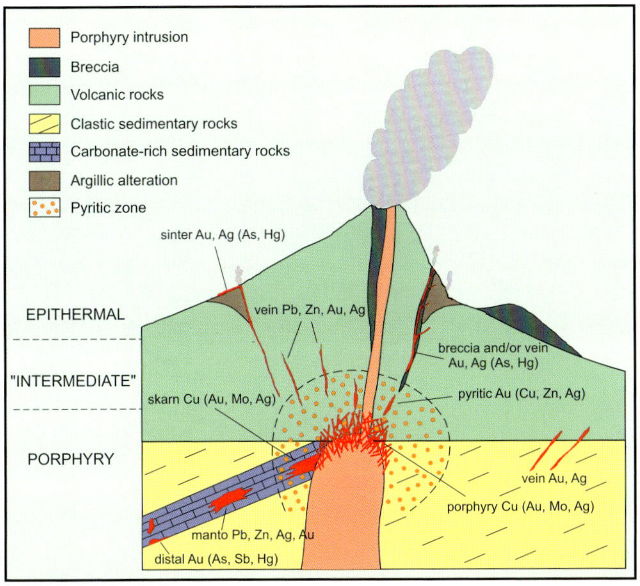

FIGURE 17. Schematic diagram of a porphyry Cu system in the root zone of an andesitic stratovolcano showing mineral zonation and possible relationship to skarn, manto, "mesothermal" or "intermediate" precious-metal and base-metal vein and replacement, and epithermal precious-metal deposits (Kirkham and Sinclair, 1995).

reflected by related, peripheral types of deposits, including skarn Cu, replacement (manto) Zn, Pb, Ag, Au and various types of base- and precious-metal veins and breccia-hosted deposits.

Magmatic-Hydrothermal Model

The most applicable genetic model for porphyry deposits is a magmatic-hydrothermal one, or variations thereupon, in which the ore metals were derived from temporally and genetically related intrusions. Large polyphase hydrothermal systems developed within and above genetically related intrusions and commonly interacted with meteoric fluids (and possibly seawater) on their tops and peripheries. During the waning stages of hydrothermal activity, the magmatic-hydrothermal systems collapsed inward and were replaced by waters of dominantly meteoric origin. Redistribution, and possibly further concentration of metals, occurred in some deposits during these waning stages (Brimhall, 1980; Brimhall and Ghiorso, 1983).

Variations of the magmatic-hydrothermal model for porphyry deposits, commonly referred to as the "orthomagmatic" model, have been presented by such authors as Burnham (1967, 1979), Phillips (1973), and Whitney (1975, 1984). These authors envisaged that felsic and intermediate magmas were emplaced at high levels in the crust and underwent border zone crystallization along the walls and roof of the magma chamber. As a consequence of this crystallization, supersaturation of volatile phases occurred within the magma, resulting in separation of volatiles due to resurgent, or second, boiling. Ore metals and many other components were strongly partitioned into these volatile phases, which became concentrated in the carapace of the magma chamber (Christiansen et al., 1983; Candela and Holland, 1986; Manning and Pichavant, 1988; Candela, 1989; Cline and Bodnar, 1991; Heinrich et al., 1992). When increasing fluid pressures exceeded lithostatic pressures and the tensile strength of the overlying rocks, fracturing of these rocks occurred, permitting rapid escape of hydrothermal fluids into newly created open space. A fundamental control on ore deposition was the pronounced adiabatic cooling of the ore fluids due to their sudden expansion into the fracture and/or breccia systems, thus the importance of structural control on ore deposition in porphyry deposits. Aplitic and micrographic textures in granitic rocks associated with porphyry deposits are the result of pressure-quench crystallization related to the rapid escape of the ore fluids (Shannon et al., 1982; Kirkham and Sinclair, 1988).

Modification of the above orthomagmatic model is required for at least some, if not most, porphyry deposits, in view of studies by Shannon et al. (1982), Carten et al. (1988a), Kirkham and Sinclair (1988), Shinohara et al. (1995) and Cloos (2002). These studies indicate that the small porphyritic stocks associated with porphyry deposits were largely liquid until ore formation was essentially complete and were conduits for enormous volumes of ore-forming fluids produced by degassing of large subjacent magma bodies. In the context of porphyry Mo deposits, Shinohara et al. (1995) examined various mechanisms of fluid transport, including gravitational ascent of bubbles in a static magma, and concluded that convection of the magma itself provided the most favourable model for the transport of fluids to the

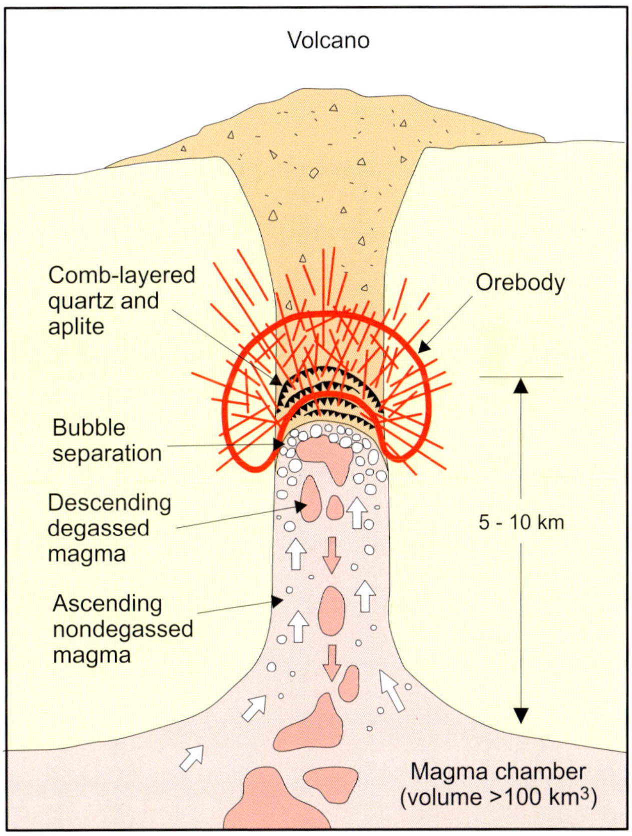

FIGURE 18. Schematic diagram of a convecting magma that is feeding a small subvolcanic intrusion below a porphyry deposit (modified from Shinohara et al., 1995). Fluid separation from the degassing magma occurs near the top of the magma column, forming pockets of magmatic-hydrothermal fluid in which comb-quartz layers grow inward from intrusion margins. Mineralized vein and fracture stockworks form when the fluid pressure exceeds lithostatic pressure and tensile strength of the surrounding rocks.

site of deposit formation. This model is presented schematically in Figure 18. According to this model, fluid-charged, nondegassed magma rises by convection from a deep magma reservoir in a column that is about 300 m in diameter at the top. Fluid separation from the magma occurs at a shallow, subvolcanic depth, focusing volatiles near the top of the magma column and producing degassed magma that descends through the nondegassed magma because of its lower volatile content and consequently greater density. Areas where ore-forming fluids accumulated in cupolas of intrusions associated with porphyry Mo and W-Mo deposits are characterized by abundant comb-quartz layers (Shannon et al., 1982; Carten et al., 1988a; Kirkham and Sinclair, 1988; Lowenstern and Sinclair, 1996). The occurrence of comb-quartz layers in intrusions associated with some porphyry Cu, Cu-Au, and Au deposits, as well as porphyry Mo deposits (Kirkham and Sinclair, 1988; Atkinson and Hunter, 2002; Kirwin, 2005), suggests that the Shinohara et al. (1995) model may be applicable to porphyry deposits in general.

Role of Mafic Magmas

Another modification to the orthomagmatic model involves mixing of mafic magmas with felsic to intermediate, ore-related calc-alkaline magmas prior to the onset of mineralization. Hot mafic melt injected into colder felsic magma can trigger vigorous convection leading to catastrophic magmatic degassing and explosive venting in subvolcanic plutons (Sparks et al., 1977), and extensive brecciation and mineralization of surrounding rocks (Hattori and Keith, 2001). For example, Dietrich et al. (2000) proposed that magma mixing was responsible for the formation of the Bolivian porphyry tin deposits, which are associated with small subvolcanic rhyodacite bodies. They suggested that influx of primitive melt into the lower levels of an upper crustal felsic magma system and subsequent mixing produced hybrid, rhyodacitic melt and triggered volcanic activity. Consequent decompression of the felsic magma system resulted in the release of Sn-W-B-As-bearing magmatic vapour phases from the felsic magma that were focused in subvolcanic vents to form the deposits.

Mafic magmas may also contribute volatiles and metals to the ore-forming fluids. Carten et al. (1993) suggested that, for high-grade porphyry Mo deposits, volatiles (F, Cl, S, CO_2) released from underplated and injected saturated mafic magmas were responsible for stripping metals from the overlying felsic magmas. At Bingham, Utah, mafic alkaline dykes and volcanic rocks that are coeval with ore-related monzonite contain minute magmatic sulphide globules 1 to 500 μm in size and composed chiefly of pyrite, pyrrhotite and chalcopyrite (Keith et al., 1997). Mafic magmas represented by these alkaline rocks appear to have mixed with the more felsic monzonitic magmas prior to mineralization and may have contributed more than half of the S and significant amounts of the Cu, Au, and PGE in the Bingham deposit (Maughan et al., 2002). Halter et al. (2002) further suggested that sulphide melts may control Cu:Au ratios in porphyry deposits by preconcentrating Cu and Au during the evolution of the magmatic system before volatile saturation. Destabilization of the sulphide melts during the later exsolution of chloride-rich volatiles would result in bulk transfer of metals and sulphur to hydrothermal ore-forming fluids.

Source and Evolution of Ore-Forming Fluids

The origin and evolution of metal-bearing fluids responsible for porphyry deposits are complex, depending on their initial compositions and cooling paths. Although some of the metal in the deposits may be derived from upper crustal rocks by meteoric fluids (Sheets et al., 1996), wall rocks of the intrusions and deposits are generally not considered to be viable sources for the metals in porphyry deposits (Kirkham and Sinclair, 1995). The bulk of the metal content of porphyry deposits appears to have been deposited from highly saline fluids (up to 60% NaCl equiv.) of magmatic derivation (Hedenquist and Lowenstern, 1994). In some cases, such hypersaline fluids may have separated directly from felsic magmas (e.g. Cline and Vanko, 1995; Ulrich et al., 1999, 2001; Harris et al., 2003; Webster, 2004). In other cases, the most primitive fluids exsolved from felsic melts are of low to moderate salinity, containing 2 to 12% NaCl equiv. (Redmond et al., 2004; Rusk et al., 2004). On ascending from their igneous source and undergoing decompression and cooling, such fluids commonly separate into a low-density vapour phase and a dense, hypersaline liquid. Although vapour-rich phases can contain significant amounts of metals such as Cu and Au (Rusk et al., 2004), the bulk of the metal content is partitioned into the hypersaline brines (Ulrich et al., 2001; Redmond et al., 2004).

Ore minerals may precipitate from the metalliferous fluids in response to adiabatic expansion (decompression) and fluid cooling, chemical effects of phase separation, interaction with wall rocks, or mixing with meteoric water. In some porphyry copper deposits, such as Bingham and Bajo de la Alumbrera, sulphide deposition was primarily due to cooling of metal-rich, saline brines below 400°C (Ulrich et al., 2001; Redmond et al., 2004). In other porphyry copper deposits, such as Butte and El Salvador, interaction of the magmatic fluid system with meteoric water had a significant impact on metal deposition (Gustafson and Hunt, 1975; Hemley and Hunt, 1992). In tin-rich systems, acid neutralization of the magmatic ore-bearing fluids by feldspar alteration to muscovite-quartz ± topaz ± fluorite (phyllic alteration) or by mixing with meteoric water are effective mechanisms for cassiterite deposition (Heinrich, 1990).

Exploration Methods

Geological Methods

Several features of porphyry deposits conducive to exploration are related to their large size. Metal, mineral, and alteration patterns tend to be large, concentric, and zoned, thus yielding useful clues to areas with exploration potential. Large pyritic halos, for example, may be used to delineate the extent of the deposits, and also the intensity and complexity of the hydrothermal system. In weathered zones, hypogene sulphide minerals are commonly absent due to oxidization and leaching but associated silicate and other gangue minerals that remain may serve as guides to ore. Careful study and interpretation of leached cappings have also been used to differentiate between barren and mineralized deposits, some with major supergene enriched ores (Blanchard, 1968; Anderson, 1982; Perelló et al., 2001).

On a regional scale, the presence of epizonal to mesozonal felsic to intermediate porphyritic intrusions, especially if accompanied by large pyritic alteration zones, indicate that the area could be prospective for porphyry deposits. Porphyry Cu and Cu-Mo deposits are relatively abundant in island- and continental-arc volcanic terranes; porphyry Mo and W-Mo deposits occur mainly in areas of continental crust characterized by extensional tectonic settings with bimodal mafic and felsic magmatism. For Climax-type porphyry Mo deposits in particular, one of the key exploration criteria is a tectonic setting that reflects a changeover from compressional to extensional tectonics (Keith et al., 1993).

Geochemical Methods

Porphyry deposits tend to have large geochemical dispersion halos and reconnaissance stream sediment and soil geochemical surveys have been effective exploration tools in many parts of the world. For deposits that are deeply buried by post-mineralization rocks or sediments, detection of associated geochemical anomalies may require the application of partial extraction techniques (e.g. Kelley et al., 2003). Biogeochemical methods have also been developed for detecting porphyry deposits (Dunn et al., 1996; Dunn, 2000), particularly in arid regions (e.g. Chaffee, 1982).

In most parts of Canada, interpretation of geochemical data must take into account the effects of glaciation, which can be complex. For example, in the Babine district of British Columbia, the use of geochemistry for exploration has been problematic because of the widespread surficial cover, the variable distribution of glacial, glaciolacustrine, and glaciofluvial sediments, and an incomplete understanding of the glacial history of the region (Carter et al., 1995). However, Levson (2001) demonstrated that ice flow and related glacial dispersion were dominantly southeasterly, although they have been influenced locally by topographic control and by late-glacial, westerly ice flow, and that copper anomalies in basal till can be traced northwesterly in the up-ice direction to locate buried mineralization.

Geophysical Methods

The dispersed nature of sulphide distribution in porphyry deposits is particularly amenable for the application of the induced polarization (IP) method, which was originally developed specifically for porphyry Cu exploration (Brant, 1966). Controlled source audio-frequency magnetotelluric (CASMT) surveys have proven to be an extremely effective technique for defining the horizontal limits and depth extent of porphyry systems (Corry et al., 1988). Magnetic surveys can be used to outline porphyry Cu and Cu-Au deposits with abundant hydrothermal magnetite, and pyrrhotite- and/or magnetite-bearing skarn and hornfels zones around porphyry-related intrusive rocks. Conversely, some deposits are characterized by magnetic lows due to the destruction of magnetite in phyllic alteration zones. Gamma ray spectrometry surveys have been used to outline potassic alteration zones closely related to mineralized zones in the Mount Milligan deposit in central British Columbia and the Casino deposit in west-central Yukon Territory (Shives et al., 2000). Recent developments in remote sensing have included the application of hyperspectal imaging to outline hydrothermal alteration around porphyry deposits (e.g. Cudahy et al., 2001; Berger et al., 2003).

Knowledge Gaps

One of the most fundamental problems in exploration for porphyry deposits is to identify factors that can distinguish the relatively few fertile igneous complexes from the many that are not. Despite the close association of porphyry deposits with intermediate to felsic intrusive rocks and general agreement that their formation involves the separation of metalliferous fluids from ore-related magmas, the processes that form porphyry deposits are diverse and significant knowledge gaps remain.

Role of Tectonic Setting

The view that porphyry Cu (±Mo±Au) deposits form in subduction-related island arcs associated with converging plates is overly simplistic (Richards, 2003). The large Andean porphyry Cu systems formed during discrete periods of regional uplift, tectonic shortening, and crustal thickening that have been attributed to increased convergence velocity and convergence angle (Camus, 2002). Giant deposits such as Bingham and Grasberg, although ultimately related to plate convergence, are not located in areas of steady-state subduction and arc volcanism. Climax-type porphyry Mo deposits occur in tectonic settings that are transitional between compressional and extensional tectonics (Keith et al., 1993). Changes or variations in tectonic setting such as arc reversals, low-angle subduction, and slab window and slab gap environments are other factors that must also be considered (e.g. Soloman, 1990; Kirkham, 1998).

Sources of Metals and Sulphur in Porphyry Deposits

Ore-related magmas can be generated in various ways, ranging from mantle melting to melting and/or assimilation of lower or upper crustal rocks, or a subducted plate. Consequently, the metals in porphyry deposits may have diverse origins. For example, although Cu, Au, and PGE are likely derived primarily from mafic, mantle-derived magmas (e.g. McInnes et al., 1999; Skewes et al., 2002), other sources for these metals may include subducted plates (Noll et al., 1996) and lower crustal rocks (e.g. Lang and Titley, 1998; Barra et al., 2002). There is general consensus that Mo, W, and Sn are derived from crustal rocks (Candela and Piccoli, 2005; Seedorf et al., 2005); however, it is debated whether the source regions need to be enriched in specific metals such as Sn and W (e.g. Keqin and Jinchu, 1988) or whether magmatic processes such as fractionation are more important than the primary metal content of magmas (e.g. Lehmann, 1982). The source of abundant sulphur in porphyry deposits is problematic because of the relatively low solubility of sulphur in felsic magmas compared to mafic magmas. To account for this "excess sulphur", Hattori and Keith (2001) have emphasized the role of mafic melt in supplying sulphur to felsic magma and magmatic-hydrothermal fluid. Streck and Dilles (1998), on the other hand, have argued that hydrous magmas may contain abundant oxidized sulphur dissolved as sulphate or anhydrite, and that porphyry ore fluids may be enriched in SO_2 gas via the breakdown of anhydrite.

Controls on Metal Ratios in Porphyry Deposits

Although there is a general relationship between magma composition and metal ratios, other factors are likely involved. For example, preconcentration of Cu and Au in sulphide melts associated with mafic primitive magmas may exert a fundamental control on Cu:Au ratios in porphyry Cu deposits (Keith et al., 1997; Halter et al., 2002). In contrast, porphyry Au deposits in the northern Cordillera are associated with reduced, ilmenite-series granitic rocks in a continental arc setting and have an affiliation toward porphyry W-Mo rather than porphyry Cu systems (McMillan et al., 1995; Hart et al., 2004); the Au in these deposits is likely of crustal origin and may have been concentrated by magmatic fractionation under reducing conditions (Thompson et al., 1999). In porphyry Sn and W deposits, metal ratios are affected by the relative concentrations of Cl and F in the ore-bearing magmatic-hydrothermal fluids, which are in turn controlled by fractional crystallization and the resulting chemistry of the residual melts from which the fluids have exsolved (Audétat et al., 2000).

Relationship between Porphyry Deposits and Other Types of Intrusion-Related Hydrothermal Deposits, such as Epithermal Deposits

It is increasingly clear that high-sulphidation epithermal deposits may be part of porphyry systems (e.g. Hedenquist et al., 1998; Heinrich et al., 2004). Intermediate-sulphidation deposits may be related to porphyry deposits, but do not show such close connections as high-sulphidation deposits, and many low-sulphidation precious metal deposits do not appear to be linked to porphyry deposits (Sillitoe and Hedenquist, 2003).

Critical Factors for the Formation of Giant Porphyry Deposits

Giant-sized porphyry Cu deposits do not appear to have significantly different anatomical characteristics compared to smaller deposits; the primary controls on deposit size are instead likely related to large-scale lithotectonic conditions, the favourable criteria for which are not yet well defined (Clark, 1993). Some evidence suggests that magmas associated with the generation of giant porphyry Cu deposits were highly enriched in Cu, in which case only small amounts of magma would have been required to produce large deposits (e.g. Core et al., 2006). In other cases, intrusions that are not unusually enriched in Cu may have generated porphyry deposits by the efficient collection of Cu from large volumes of magma (Skewes and Stern, 1995; Cloos, 2002). In the latter case, to produce a giant deposit would require significant longevity of the ore-forming process and of the favourable lithotectonic conditions necessary to sustain it.

Areas of High Potential in Canada

Many parts of the Canadian Cordillera are favourable for various types of porphyry deposits. Areas with some of the highest potential for porphyry Cu-Au deposits are those underlain by Late Triassic to Early Jurassic alkaline volcanic and intrusive rocks, which occur mainly in the Quesnel and Stikine terranes. Some of the more favourable areas include the Afton and Cariboo districts, the Mt. Milligan area and the Galore Creek area. In the Toodoggone region, calc-alkaline intrusions of Late Triassic to Early Jurassic age are associated with the Kemess North and South porphyry Cu-Au deposits and the potential for additional deposits is high. Eocene calc-alkaline intrusions in the Babine district, Stikine terrane, have high potential for porphyry Cu-Au deposits and Late Cretaceous to Eocene calc-alkaline intrusions in the Tatsa Ranges, Stikine Terrane are favourable for porphyry Cu-Mo deposits. Mid-Cretaceous calc-alkaline intrusions of the Mayo plutonic suite in the North American terrane have potential for porphyry Au deposits (Hart et al., 2004). Areas with potential for porphyry Mo deposits include the Alice Arm district, Smithers area and the Tahtsa Ranges, where deposits that are transitional between Climax- and Endako-type deposits, such as Kitsault and Glacier Gulch, are associated with Late Cretaceous to Eocene intrusions. These areas may also have potential for low-grade, Endako-type deposits associated with Oligocene intrusions, comparable to the Quartz Hill deposit in nearby Alaska (Wolfe, 1995), or with Late Jurassic intrusions as at Endako. Porphyry Mo and W-Mo deposits in northern British Columbia and southern Yukon Territory are associated with intrusions that range from mid-Cretaceous to Eocene in age (Sinclair, 1995c).

In the Canadian Appalachians, subduction-related mid-Devonian intrusions in New Brunswick and Quebec (Gaspé) have potential for porphyry Cu and Cu-Mo deposits. Mid to Late Devonian granitic plutons that appear to have been emplaced in an extensional environment along the Gander-Avalon boundary in southern New Brunswick have potential for associated porphyry deposits of W, Mo, and/or Sn, such as at Sisson Brook and Mount Pleasant (Sinclair, 2005).

Although exploration for porphyry deposits in Precambrian terranes has been limited, large, low-grade, porphyry-type deposits of Cu, Au, and Mo associated with Archean and Proterozoic intrusions have been recognized and the potential for porphyry deposits in Precambrian terranes is likely underexplored. Numerous porphyry-type deposits occur in the Abitibi and Opatica greenstone belts, including one that is currently in production (Troilus). These and other areas where felsic to intermediate, subvolcanic plutons have intruded shallow marine or subaerial volcanic and sedimentary rocks should be the most favourable areas for exploration.

Acknowledgements

The author has benefited greatly from discussions with many colleagues over many years and from numerous informative visits to porphyry deposits. Much of the data used in this paper was taken from the porphyry and related deposits database compiled by Rod Kirkham and Kathryn Dunne for the GSC's World Minerals Geoscience Database Project. Diagrams were prepared with the assistance of Beth Hillary, Ingrid Kjarsgaard, and Kim Nguyen. Lesley Chorlton provided the generalized world geology database from the World Minerals Geoscience Database Project. Rod Kirkham and Bob Linnen reviewed the manuscript and provided many constructive suggestions.

References

Anderson, J.A., 1982, Characteristics of leached capping and appraisal, *in* Titley, S.R., ed., Advances in Geology of the Porphyry Copper Deposits; Southwestern North America: University of Arizona Press, Tucson, Arizona, p. 275-296.

Archambault, A., Guha, J., Tremblay, A., and Kanwar, R., 1984, Implications of the geomechanical interpretation of the Copper Rand deposit on the Doré Lake shear belt, *in* Guha, J., and Chown, E.H., eds.,

Chibougamau - Stratigraphy and Mineralization: Canadian Institute of Mining and Metallurgy, Special Volume 34, p. 300-318.

Atkinson, W.W., and Hunter, W., 2002, Comb quartz layers in the porphyry copper deposit at Yerington, Nevada: Geological Society of America, 2002, Abstracts with Programs, p. 15.

Audétat, A., Günther, D., and Heinrich, C.A., 2000, Magmatic-hydrothermal evolution in a fractionating granite: A microchemical study of the Sn-W-F-mineralized Mole Granite (Australia): Geochimica et Cosmochimica Acta, v. 64, p. 3373-3393.

Ayres, L.D., Averill, S.A., and Wolfe, W.J., 1982, An Archean molybdenite occurrence of possible porphyry type at Setting Net Lake, northwestern Ontario, Canada: Economic Geology, v. 77, p. 1105-1119.

Baker, R.C., and Guilbert, J.M., 1987, Regional structural control of porphyry copper deposits in northern Chile: Geological Society of America, 1987, Abstracts with Programs, p. 578.

Barr, D.A., Fox, P.E., Northcote, K.E., and Preto, V.A., 1976, The alkaline suite porphyry deposits: a summary, *in* Sutherland Brown, A., ed., Porphyry Deposits of the Canadian Cordillera: Canadian Institute of Mining and Metallurgy, Special Volume 15, p. 359-367.

Barra, F., Ruiz, J., Ryan, M., and Titley, S.R., 2002, A Re-Os study of sulfide minerals from the Bagdad porphyry Cu-Mo deposit, northern Arizona, USA: Mineralium Deposita, v. 38, p. 585-596.

Berger, B.R., King, T.V.V., Morath, L.C., and Phillips, J.D., 2003, Utility of high-altitude infrared spectral data in mineral exploration: application to northern Patagonia Mountains, Arizona: Economic Geology, v. 98, p. 1003-1018.

Blanchard, R., 1968, Interpretation of Leached Outcrops: Nevada Bureau of Mines, Bulletin 166, 196 p.

Bookstrom, A.A., 1981, Tectonic setting and generation of Rocky Mountain porphyry molybdenum deposits, *in* Dickinson, W.D., and Payne, W.D., eds., Relations of Tectonics to Ore Deposits in the Southern Cordillera: Arizona Geological Society Digest, v. 14, p. 251-226.

Bookstrom, A.A., Carten, R.B., Shannon, J.R., and Smith, R.P., 1988, Origins of bimodal leucogranite-lamprophyre suites, Climax and Red Mountain porphyry molybdenum systems, Colorado: Petrologic and strontium isotope evidence: Colorado School of Mines Quarterly, v. 83, p. 1-22.

Bradbrook, C., 2006, Advancing the New Afton copper-gold project - world class grades next to the Trans-Canada Highway: Roundup06, Association for Mineral Exploration BC, 2006, Abstract Volume, p. 38-39.

Brant, A.A., 1966, Geophysics in the exploration for Arizona porphyry deposits, *in* Titley, S.R., and Hicks, C.L., eds., Geology of the Porphyry Copper Deposits, Southwestern North America: University of Arizona Press, Tucson, Arizona, p. 87-110.

Brimhall, G.H., Jr., 1980, Deep hypogene oxidation of porphyry copper potassium-silicate proto-ore at Butte, Montana: a theoretical evaluation of the copper remobilization hypothesis: Economic Geology, v. 75, p. 384-409.

Brimhall, G.H., Jr., and Ghiorso, M.S., 1983, Origin and ore-forming consequences of the advanced argillic alteration process in hypogene environments by magmatic gas contamination of meteoric fluids: Economic Geology, v. 78, p. 73-90.

Brooks, C.K., Tegner, C., Stein, H., and Thomassen, B., 2004, Re-Os and $^{40}Ar/^{39}Ar$ ages of porphyry molybdenum deposits in the East Greenland volcanic-rifted margin: Economic Geology, v. 99, p. 1215-1222.

Burnham, C.W., 1967, Hydrothermal fluids at the magmatic stage, *in* Barnes, H.L., ed., Geochemistry of Hydrothermal Ore Deposits: Holt, Rinehart and Winston Inc., New York, p. 34-76.

——— 1979, Magma and hydrothermal fluids, *in* Barnes, H.L., ed., Geochemistry of Hydrothermal Ore Deposits, 2nd edition: Wiley Interscience, New York, p. 71-136.

Camus, F., 2002, The Andean porphyry systems, *in* Cooke, D.R., and Pongratz, J., eds., Giant Ore Deposits: Characteristics, Genesis and Exploration: CODES Special Publication, v. 4, p. 5-21.

Candela, P.A., 1989, Calculation of magmatic fluid contributions to porphyry-type ore systems: Predicting fluid inclusion chemistries: Geochemical Journal, v. 23, p. 295-305.

Candela, P.A., and Holland, H.D., 1986, A mass transfer model for copper and molybdenum in magmatic hydrothermal systems: the origin of porphyry-type ore deposits: Economic Geology, v. 81, p. 1-19.

Candela, P.A., and Piccoli, P., 2005, Magmatic processes in the development of porphyry-type ore systems, *in* Hedenquist, J.W., Thompson, J.F.H., Goldfarb, R.J., and Richards, J.R., eds., Economic Geology 100th Anniversary Volume: Society of Economic Geologists, Littleton, Colorado, p. 25-37.

Carson, D.J.T., and Jambor, J.L., 1974, Mineralogy, zonal relationships and economic significance of hydrothermal alteration at porphyry copper deposits, Babine Lake area, British Columbia: Canadian Institute of Mining and Metallurgy Bulletin, v. 76, p. 24.

——— 1979, The occurrence and significance of phyllic overprinting at porphyry copper-molybdenum deposits (abstract): Canadian Institute of Mining and Metallurgy, v. 72, p. 78.

Carten, R.B., 1986, Sodium-calcium metasomatism: chemical, temporal, and spatial relationships at the Yerington, Nevada, porphyry copper deposit: Economic Geology, v. 81, p. 1495-1519.

Carten, R.B., Geraghty, E.P., and Walker, B.M., 1988a, Cyclic development of igneous features and their relationship to high-temperature hydrothermal features in the Henderson porphyry molybdenum deposit, Colorado: Economic Geology, v. 83, p. 266-296.

Carten, R.B., Walker, B.M., Geraghty, E.P., and Gunow, A.J., 1988b, Comparison of field-based studies of the Henderson porphyry molybdenum deposit, Colorado with experimental and theoretical models of porphyry systems, *in* Taylor, R.P., and Strong, D.F., eds., Recent Advances in the Geology of Granite-Related Mineral Deposits: The Canadian Institute of Mining and Metallurgy, Special Volume 39, p. 351-366.

Carten, R.B., White, W.H., and Stein, H.J., 1993, High-grade, granite-related molybdenum systems: classification and origin, *in* Kirkham, R.V., Sinclair, W.D., Thorpe, R.I., and Duke, J.M., eds., Mineral Deposit Modeling: Geological Association of Canada, Special Paper 40, p. 521-554.

Carter, N.C., Dirom, G.E., and Ogryzlo, P.L., 1995, Porphyry copper-gold deposits, Babine Lake area, west-central British Columbia, *in* Schroeter, T.G., ed., Porphyry Deposits of the Northwestern Cordillera of North America: Canadian Institute of Mining, Metallurgy and Petroleum, Special Publication 46, p. 247-255.

Chaffee, M.A., 1982, Geochemical prospecting techniques for porphyry deposits, *in* Titley, S.R., ed., Advances in Geology of the Porphyry Copper Deposits; Southwestern North America: University of Arizona Press, Tucson, Arizona, p. 297-307.

Chorlton, L.B., 2004, Generalized geology of the world: age and rock type domains, Geological Survey of Canada; http://gdr.nrcan.gc.ca/minres/data_e.php.

Christiansen, E.H., Burt, D.M., Sheridan, M.F., and Wilson, R.T., 1983, The petrogenesis of topaz rhyolites from the western United States: Contributions to Mineralogy and Petrology, v. 83, p. 16-30.

Clark, A.H., 1993, Are outsize porphyry copper deposits either anatomically or environmentally distinctive? *in* Whiting, B.H., Hodgson, C.J., and Mason, R., eds., Giant Deposits: Society of Economic Geologists, Special Publication 2, p. 213-284.

Cline, J.S., and Bodnar, R.J., 1991, Can economic porphyry copper mineralization be generated by a typical calc-alkaline melt?: Journal of Geophysical Research, v. 96, p. 8113-8126.

Cline, J.S., and Vanko, D.A., 1995, Magmatically generated saline brines related to molybdenum at Questa, New Mexico, USA, *in* Thompson, J.F.H., ed., Magmas, Fluids and Ore Deposits: Mineralogical Association of Canada, Short Course Volume 23, p. 153-174.

Cloos, M., 2002, Bubbling magma chambers, cupolas and porphyry copper deposits, *in* Ernst, G., ed., Frontiers in Geochemistry; Organic, Solution, and Ore Deposit Geochemistry, International Book Series, v. 6, p. 191-217.

Cooke, D.R., Wilson, A.J., and Davies, A.G.S., 2004, Characteristics and genesis of porphyry copper-gold deposits: 24 ct Au Workshop, 2004, CODES Special Publication 5, p. 17-34.

Core, D.P., Kesler, S.E., and Essene, E.J., 2006, Unusually Cu-rich magmas associated with giant porphyry copper deposits: Evidence from Bingham, Utah: Geology, v. 34, p. 41-44.

Corry, C.E., Emer, D.F., and Zonge, K.L., 1988, Controlled source audio-frequency magnetotelluric surveys of porphyry sulfide deposits and prospects in the cordillera of the United States: North American Conference on Tectonic Control of Ore Deposits and the Vertical and Horizontal Extent of Ore Systems, University of Missouri, Rolla, 1988, Proceedings Volume, p. 204-213.

Cudahy, T.J., Wilson, J., Hewson, R., Linton, P., Harris, P., Sears, M., Okada, K., and Hackwell, J.A., 2001, Mapping porphyry-skarn alteration at Yerington, Nevada, using hyperspectral VNIR-SWIR-TIR imaging data: Geoscience and Remote Sensing Symposium, 2001, p. 631-633.

de Laeter, J.R., and Martyn, J.E., 1986, Age of molybdenum-copper mineralization at Coppin Gap, Western Australia: Australian Journal of Earth Sciences, v. 33, p. 65-71.

Dietrich, A., Lehmann, B., and Wallianos, A., 2000, Bulk rock and melt inclusion geochemistry of Bolivian tin porphyry systems: Economic Geology, v. 95, p. 313-326.

Dilles, J.H., and Einaudi, M.T., 1992, Wall-rock alteration and hydrothermal flow paths about the Ann-Mason porphyry copper deposit, Nevada - A 6-km vertical reconstruction: Economic Geology, v. 87, p. 1963-2001.

Dunn, C.E., 2000, Pine bark biogeochemical signatures of the Endako molybdenum camp and surrounding areas, central British Columbia; implications for mineral exploration and environmental baselines: Geological Society of America, 2000, Abstracts with Programs, p. 11-12.

Dunn, C.E., Balma, G.R., and Sibbick, S.J., 1996, Biogeochemical survey using Lodgepole Pine bark: Mount Milligan, central British Columbia (parts of NTS 93N/1 and 93O/4): Geological Survey of Canada, Open File 3290, p. 113.

Eaton, P.C., and Setterfield, T.N., 1993, The relationship between epithermal and porphyry hydrothermal systems within the Tavua Caldera, Fiji: Economic Geology, v. 88, p. 1053-1083.

Einaudi, M.T., 1982, Description of skarns associated with porphyry copper plutons, *in* Titley, S.R., ed., Advances in Geology of the Porphyry Copper Deposits: University of Arizona Press, Tuscon, p. 139-183.

Fraser, R.J., 1993, The Lac Troilus gold-copper deposit, northwestern Quebec: a possible Archean porphyry system: Economic Geology, v. 88, p. 1685-1699.

Goodman, S., Williams-Jones, A.E., and Carles, P., 2005, Structural controls on the Archean Troilus gold-copper deposit, Quebec, Canada: Economic Geology, v. 100, p. 577-582.

Grant, J.N., Halls, C., Sheppard, S.M.F., and Avila, W., 1980, Evolution of the porphyry tin deposits of Bolivia: Mining Geology Special Issue, v. 8, p. 151-173.

Guan, X., Shou, Y., Xiao, J., Lian, S., and Li, J., 1988, A new type of tin deposit - the Yinyan porphyry tin deposit in China, *in* Hutchison, C.S., ed., Geology of Tin Deposits in Asia and the Pacific: Springer Verlag, Berlin, New York, p. 487-494.

Gustafson, L.B., and Hunt, J.P., 1975, The porphyry copper deposit at El Salvador, Chile: Economic Geology, v. 70, p. 857-912.

Halter, W.E., Pettke, T., and Heinrich, C.A., 2002, The origin of Cu/Au ratios in porphyry-type deposits: Science, v. 296, p. 1844-1846.

Harris, A.C., Kamenetsky, V.S., White, N.C., van Achterbergh, E., and Ryan, C.G., 2003, Melt inclusions in veins: Linking magmas and porphyry Cu deposits: Science, v. 302, p. 2109-2111.

Hart, C.J.R., Goldfarb, R.J., Lewis, L.L., and Mair, J.L., 2004, The northern Cordilleran mid-Cretaceous Plutonic Province: ilmenite/magnetite-series granitoids and intrusion-related mineralization: Resource Geology, v. 54, p. 253-280.

Hattori, K., and Keith, J.D., 2001, Contribution of mafic melt to porphyry copper mineralization: evidence from Mount Pinatubo, Philippines, and Bingham Canyon, Utah, USA: Mineralium Deposita, v. 36, p. 799-806.

Hedenquist, J.W., and Lowenstern, J.B., 1994, The role of magmas in the formation of hydrothermal ore deposits: Nature, v. 370, p. 519-527.

Hedenquist, J.W., Arribas, A., Jr., and Reynolds, T.J., 1998, Evolution of an intrusion-centered hydrothermal system; Far Southeast-Lepanto porphyry and epithermal Cu-Au deposits, Philippines: Economic Geology, v. 93, p. 373-404.

Heidrick, T.L., and Titley, S.R., 1982, Fracture and dike patterns in Laramide plutons and their structural and tectonic implication: American Southwest, *in* Titley, S.R., ed., Advances in Geology of the Porphyry Copper Deposits, Southwestern North America: University of Arizona Press, Tucson, Arizona, p. 73-91.

Heinrich, C.A., 1990, The chemistry of hydrothermal tin (-tungsten) ore deposition: Economic Geology, v. 85, p. 457-481.

Heinrich, C.A., Ryan, C.G., Mernach, T.P., and Eadington, P.J., 1992, Segregation of ore metals between magmatic brine and vapor: A fluid inclusion study using PIXE microanalysis: Economic Geology, v. 87, p. 1566-1583.

Heinrich, C.A., Dreisner, T., Stefánsson, A., and Seward, T.M., 2004, Magmatic vapor contraction and the transport of gold form the porphyry environment to epithermal ore deposits: Geology, v. 32, p. 761-764.

Hemley, J.J., and Hunt, J.P., 1992, Hydrothermal ore-forming processes in the light of studies in rock-buffered systems: II. Some general geologic applications: Economic Geology, v. 87, p. 23-43.

Holliday, J.R., Wilson, A.J., Blevin, P.L., Tedder, I.J., Dunham, P.D., and Pitzner, M., 2002, Porphyry gold-copper mineralisation in the Cadia district, eastern Lachlan Fold Belt, New South Wales, and its relationship to shoshonitic magmatism: Mineralium Deposita, v. 37, p. 100-116.

Hou, Z., Ma, H., Khin, Z., Zhang, Y., Wang, M., Wang, Z., and Pan, G., 2003, The Himalayan Yulong porphyry copper belt; product of large-scale strike-slip faulting in eastern Tibet: Economic Geology, v. 98, p. 125-145.

Hunt, J.P., 1991, Porphyry copper deposits; *in* Hutchinson, R.W., and Grauch, R.I., eds., Historical Perspectives of Genetic Concepts and Case Histories of Famous Discoveries: Economic Geology Monograph 8, p. 192-206.

Ishihara, S., 1981, The granitoid series and mineralization, *in* Skinner, B.J., ed., Economic Geology Seventy-fifth Anniversary Volume, 1905-1980: Economic Geology Publishing Co., p. 458-484.

James, A.H., 1971, Hypothetical diagrams of several porphyry copper deposits: Economic Geology, v. 66, p. 43-47.

Jones, B.K., 1992, Application of metal zoning to gold exploration in porphyry copper systems: Journal of Geochemical Exploration, v. 43, p. 127-155.

Keith, J.D., Christiansen, E.H., and Carten, R.B., 1993, The genesis of giant porphyry molybdenum deposits, *in* Whiting, B.H., Hodgson, C.J., and Mason, R., eds., Giant Ore Deposits: The Society of Economic Geologists, Special Publication Number 2, p. 285-317.

Keith, J.D., Whitney, J.A., Hattori, K., Ballantyne, G.H., Christiansen, E.H., Barr, D.L., Cannan, T.M., and Hook, C.J., 1997, The role of magmatic sulphides and mafic alkaline magmas in the Bingham and Tintic mining districts, Utah: Journal of Petrology, v. 38, p. 1679-1690.

Kelley, D.L., Hall, G.E.M., Closs, G.L., Hamilton, I.C., and McEwen, R.M., 2003, The use of partial extraction geochemistry of copper exploration in northern Chile: Geochemistry - Exploration, Environment, Analysis, v. 3, p. 85-104.

Keqin, X., and Jinchu, Z., 1988, Time-space distribution of tin/tungsten deposits in South China and controlling factors of mineralization, *in* Hutshison, C.S., ed., Geology of Tin Deposits in Asia and the Pacific: Springer-Verlag, New York, p. 265-277.

Kesler, S.E., Jones, L.M., and Walker, R.L., 1975, Intrusive rocks associated with porphyry copper mineralization in island arc areas: Economic Geology, v. 70, p. 515-526.

Kesler, S.E., Chryssoulis, S.L., and Simon, G., 2002, Gold in porphyry copper deposits: its abundance and fate: Ore Geology Reviews, v. 21, p. 103-124.

Kirkham, R.V., 1971, Intermineral intrusions and their bearing on the origin of porphyry copper and molybdenum deposits: Economic Geology, v. 66, p. 1244-1250.

——— 1972, Porphyry deposits, in Blackadar, R.G., ed., Report of Activities Part B, November 1971 to March 1972: Geological Survey of Canada, Paper 72-1b, p. 62-64.

——— 1973, Tectonism, volcanism and copper deposits, in Volcanism and Volcanic Rocks: Geological Survey of Canada, Open File 164, p. 129-151.

——— 1998, Tectonic and structural features of arc deposits, *in* Lefebure, D.V., co-ordinator, ed., Metallogeny of Volcanic Arcs: British Columbia Geological Survey, Short Course Notes, Open File 1998-8, Section B, p. 44.

Kirkham, R.V., and Dunne, K.P.E., 2000, World distribution of porphyry, porphyry-associated skarn, and bulk-tonnage epithermal deposits and occurrences: Geological Survey of Canada, Open File 3792a, 26 p.

Kirkham, R.V., and Sinclair, W.D., 1988, Comb quartz layers in felsic intrusions and their relationship to the origin of porphyry deposits, *in* Taylor, R.P., and Strong, D.F., eds., Recent Advances in the Geology of Granite-Related Mineral Deposits: The Canadian Institute of Mining and Metallurgy, Special Volume 39, p. 50-71.

——— 1995, Porphyry copper, gold, molybdenum, tungsten, tin, silver, *in* Eckstrand, O.R., Sinclair, W.D., and Thorpe, R.I., eds., Geology of Canadian Mineral Deposit Types: Geological Survey of Canada, Geology of Canada, no. 8, p. 421-446.

Kirwin, D.J., 2005, Unidirectional solidification textures associated with intrusion-related Mongolian mineral deposits, *in* Seltmann, R., Gerel, O., and Kirwin, D.J., eds., Geodynamics and Metallogeny of Mongolia with a Special Emphasis on Copper and Gold Deposits: International Association on the Genesis of Ore Deposits, Guidebook Series 11, CERCAMS/NHM London, p. 63-84.

Kooiman, G.J.A., McLeod, M.J., and Sinclair, W.D., 1986, Porphyry tungsten-molybdenum orebodies, polymetallic veins and replacement bodies, and tin-bearing greisen zones in the Fire Tower Zone, Mount Pleasant, New Brunswick: Economic Geology, v. 81, p. 1356-1373.

Lang, J.R., and Titley, S.R., 1998, Isotopic and geochemical characteristics of Laramide magmatic systems in Arizona and implications for the genesis of porphyry copper deposits: Economic Geology, v. 93, p. 138-170.

Lehmann, B., 1982, Metallogeny of tin: magmatic differentiation versus geochemical heritage: Economic Geology, v. 77, p. 50-59.

——— 1990, Metallogeny of Tin: Lecture Notes in Earth Sciences: Springer Verlag, Berlin, 211 p.

Levson, V.M., 2001, Quaternary geology of the Babine porphyry copper district: Implications for geochemical exploration: Canadian Journal of Earth Sciences, v. 38, p. 733-749.

Linnen, R.L., and Williams-Jones, A.E., 1987, Tectonic control of quartz vein orientations at the Trout Lake stockwork molybdenum deposit, southeastern British Columbia: Implications for metallogeny in the Kootenay Arc: Economic Geology, v. 82, p. 1283-1293.

Linnen, R.L., Williams-Jones, A.E., Leitch, C.H.B., and Macauley, T.N., 1995, Molybdenum mineralization in a fluorine-poor system: The Trout Lake stockwork deposit, southeastern British Columbia, *in* Schroeter, T.G., ed., Porphyry Deposits of the Northwestern Cordillera of North America: Canadian Institute of Mining, Metallurgy and Petroleum, Special Publication 46, p. 771-780.

Lipman, P.W., 1984, The roots of ash flow calderas in western North America: windows into the tops of granitic batholiths: Journal of Geophysical Research, v. 89, p. 8801-8841.

——— 1988, Evolution of silicic magma in the upper crust: the mid-Tertiary Latir volcanic field and its cogenetic granitic batholith, northern New Mexico, U.S.A.: Transactions of the Royal Society of Edinburgh; Earth Sciences, v. 79, p. 265-288.

Lipman, P.W., and Sawyer, D.A., 1985, Mesozoic ash-flow caldera fragments in southeastern Arizona and their relation to porphyry copper deposits: Geology, v. 13, p. 652-656.

Lowell, J.D., and Guilbert, J.M., 1970, Lateral and vertical alteration-mineralization zoning in porphyry ore deposits: Economic Geology, v. 65, p. 373-408.

Lowenstern, J.B., and Sinclair, W.D., 1996, Exsolved magmatic fluid and its role in the formation of comb-layered quartz at the Cretaceous Logtung W-Mo deposit, Yukon Territory, Canada: Transactions of the Royal Society of Edinburgh; Earth Sciences, v. 87, p. 291-303.

MacDonald, G.D., and Arnold, L.C., 1994, Geological and geochemical zoning of the Grasberg Igneous Complex, Irian Jaya, Indonesia: Journal of Geochemical Exploration, v. 50, p. 143-178.

Manning, D.A.C., and Pichavant, M., 1988, Volatiles and their bearing on the behaviour of metals in granitic systems, *in* Taylor, R.P., and Strong, D.F., eds., Recent Advances in the Geology of Granite-Related Mineral Deposits: The Canadian Institute of Mining and Metallurgy, Special Volume 39, p. 13-24.

Mason, R., and Melnyk, N., 1986, The anatomy of an Archean gold system; the McIntyre-Hollinger Complex at Timmins, Ontario, Canada, *in* Macdonald, A.J., Downes, M., Pirie, J., and Chater, A.M., eds., Gold '86; An International Symposium on the Geology of Gold Deposits, Toronto, 1986, p. 40-55.

Maughan, D.T., Keith, J.D., Christiansen, E.H., Pulsipher, T., Hattori, K., and Evans, N.J., 2002, Contributions from mafic alkaline magmas to the Bingham porphyry Cu-Au-Mo deposit, Utah, USA: Mineralium Deposita, v. 37, p. 14-37.

McInnes, B.I.A., McBride, J.S., Evans, N.J., D.D., L., and Andrew, A.S., 1999, Ore metal recycling in subduction zones: Science, v. 286, p. 512-516.

McIntyre, D.G., and Villeneuve, M.E., 2001, Geochronology of mid-Cretaceous to Eocene magmatism, Babine porphyry copper district, central British Columbia: Canadian Journal of Earth Sciences, v. 38, p. 639-655.

McMillan, W.J., and Panteleyev, A., 1980, Ore deposit models - 1. Porphyry copper deposits: Geoscience Canada, v. 7, p. 52-63.

McMillan, W.J., Thompson, J.F.H., Hart, C.J.R., and Johnston, S.T., 1995, Regional geological and tectonic setting of porphyry deposits in British Columbia and Yukon Territory, *in* Schroeter, T.G., ed., Porphyry Deposits of the Northwestern Cordillera of North America: Canadian Institute of Mining and Metallurgy, Special Volume 46, p. 40-57.

Minnitt, R.C.A., 1986, Porphyry copper mineralization, Haib River, Southwest Africa, Namibia, *in* Anhaeusser, C.R., and Maske, S., eds., Mineral Deposits of Southern Africa: The Geological Society of South Africa, v. II, p. 1567-1585.

Mitchell, A.H., and Garson, M.S., 1972, Relationship of porphyry copper and circum-Pacific tin deposits to palaeo-Benioff zones: Transactions of the Institute of Mining and Metallurgy, v. 81, p. B10-25.

Moyle, A.J., Doyle, B.J., Hoogvliet, H., and Ware, A.R., 1990, Ladolam gold deposit, Lihir Island, *in* Hughes, F.E., ed., Geology of the Mineral Deposits of Australia and Papua New Guinea: The Australasian Institute of Mining and Metallurgy, Melbourne, p. 1793-1805.

Mutschler, F.E., and Mooney, T.C., 1993, Precious-metal deposits related to alkaline igneous rocks- provisional classification, grade-tonnage data and exploration frontiers, *in* Kirkham, R.V., Sinclair, W.D., Thorpe, R.I., and Duke, J.M., eds., Mineral Deposit Modeling: Geological Association of Canada, Special Paper 4, p. 479-520.

Mutschler, F.E., Wright, E.G. Ludington, S., and Abbott, J.T., 1981, Granite molybdenite systems: Economic Geology, v. 76, p. 874-897.

Noll, P.D., Jr., Newsom, H.E., Leeman, W.P., and Ryan, J.G., 1996, The role of hydrothermal fluids in the production of subduction zone magmas: Evidence from siderophile and chalcophile trace elements and boron: Geochimica et Cosmochimica Acta, v. 60, p. 587-611.

Norman, D.I., and Sawkins, F.J., 1985, The Tribag breccia pipes: Precambrian Cu-Mo deposits, Batchawana Bay, Ontario: Economic Geology, v. 80, p. 1593-1621.

Ossandón, C.G., Fréraut, C.R., Gustafson, L.B., Lindsay, D.D., and Zentilli, M., 2001, Geology of the Chuquicamata mine: a progress report: Economic Geology, v. 96, p. 249-270.

Pearson, M.F., Clark, K.F., and Porter, E.W., 1988, Mineralogy, fluid characteristics, and silver distribution at Real de Angeles, Zacatecas, Mexico: Economic Geology, v. 83, p. 1737-1759.

Perelló, J.A., Fleming, J.A., O'Kane, K.P., Burt, P.D., Clarke, G.A., Himes, M.D., and Reeves, A.T., 1995, Porphyry copper-gold-molybdenum deposits in the Island Copper Cluster, northern Vancouver Island, British Columbia, *in* Schroeter, T.G., ed., Porphyry Deposits of the Northwestern Cordillera of North America: The Canadian Institute of Mining and Metallurgy, Special Volume 46, p. 214-238.

Perelló, J., Cox, D., Garamjav, D., Sanjdorj, S., Diakov, S., Schissel, D., Munkhbat, T.-U., and Oyun, G., 2001, Oyu Tolgoi, Mongolia: Siluro-Devonian porphyry Cu-Au-(Mo) and high-sulfidation Cu mineralization with a Cretaceous chalcocite blanket: Economic Geology, v. 96, p. 1407-1428.

Phillips, W.J., 1973, Mechanical effects of retrograde boiling and its probable importance in the formation of some porphyry ore deposits: Institute of Mining and Metallurgy Transactions, v. B82, p. 90-98.

Pilote, P., Robert, F., Sinclair, W.D., Kirkham, R.V., and Daigneault, R., 1995, Porphyry-type mineralization in the Doré Lake Complex: Clark Lake and Merrill Island areas, *in* Pilote, P., ed., Metallogenic Evolution and Geology of the Chibougamau Area - From Porphyry Cu-Au-Mo to Mesothermal Lode Gold Deposits: Geological Survey of Canada, Precambrian '95 Field Trip Guidebook, Open File 3143, p. 65-86.

Preto, V.A., 1972, Geology of Copper Mountain: British Columbia Department of Mines and Petroleum Resources, Bulletin 59, p. 87.

Redmond, P.B., Einaudi, M.T., Inan, E.E., Landtwing, M.R., and Heinrich, C.A., 2004, Copper deposition by fluid cooling in intrusion-centered systems: New insights from the Bingham porphyry ore deposit, Utah: Geology, v. 32, p. 217-220.

Rehrig, W.A., and Heidrick, T.L., 1972, Regional fracturing in Laramide stocks of Arizona and its relationship to porphyry copper mineralization: Economic Geology, v. 67, p. 198-213.

Richards, J.P., 1990, Petrology and geochemistry of alkalic intrusives at the Porgera gold deposit, Papua New Guinea: Journal of Geochemical Exploration, v. 35, p. 141-200.

——— 2003, Tectono-magmatic precursors for porphyry Cu-(Mo-Au) deposit formation: Economic Geology, v. 98, p. 1515-1534.

Richards, J.P., and Kerrich, R., 1993, The Porgera Gold Mine, Papua New Guinea: magmatic hydrothermal to epithermal evolution of an alkalic-type precious metal deposit: Economic Geology, v. 88, p. 1017-1052.

Richards, J.P., Boyce, A.J., and Pringle, M.S., 2001, Geologic evolution of the Escondida area, northern Chile: A model for spatial and temporal localization of porphyry Cu mineralization: Economic Geology, v. 96, p. 271-305.

Ross, K.V., Godwin, C.I., Bond, L., and Dawson, K.M., 1995, Geology, alteration and mineralization of the Ajax East and Ajax West deposits, southern Iron Mask Batholith, Kamloops, British Columbia, *in* Schroeter, T.G., ed., Porphyry Deposits of the Northwestern Cordillera of North America: The Canadian Institute of Mining and Metallurgy, Special Volume 46, p. 565-580.

Roth, E., Groves, D.E., Anderson, G., Daley, L., and Staley, R., 1991, Primary mineralization at the Boddington mine, Western Australia: An Archean porphyry Cu-Au-Mo deposit, *in* Ladeira, E.A., ed., Brazil Gold '91: The Economic Geology, Geochemistry and Genesis of Gold Deposits: p. 481-488.

Rowins, S.M., 2000, Reduced porphyry copper-gold deposits: a new variation on an old theme: Geology, v. 28, p. 491-494.

Rusk, B.G., Reed, M.H., Dilles, J.H., Klemm, L.M., and Heinrich, C.A., 2004, Compositions of magmatic hydrothermal fluids determined LA-ICP-MS of fluid inclusions from the porphyry copper-molybdenum deposit at Butte, MT: Chemical Geology, v. 210, p. 173-199.

Schønwandt, H.K., 1988, Geology and tectonic setting of cratonic porphyry molybdenum deposits in the North Atlantic region, *in* Boissonnas, J., and Omenetto, P., eds., Mineral Deposits within the European Community: Society for Geology Applied to Mineral Deposits, Special Publication No. 6, Springer-Verlag, Berlin-Heidelberg, p. 210-239.

Scott, K.M., 1981, Wall-rock alteration in disseminated tin deposits, southeastern Australia: Proceedings of the Australasian Institute of Mining and Metallurgy, v. 280, p. 17-28.

Seedorf, E., Dilles, J.D., Proffett, J.M., Jr., Einaudi, M.T., Zurcher, L., Stavast, W.J.A., Johnson, D.A., and Barton, M.D., 2005, Porphyry deposits: characteristics and origin of hypogene features, *in* Hedenquist, J.W., Thompson, J.F.H., Goldfarb, R.J., and Richards, J.R., eds., Economic Geology 100th Anniversary Volume: Society of Economic Geologists, Littleton, Colorado, p. 251-298.

Selby, D., Nesbitt, D.E., Muehlenbachs, K., and Prohaska, W., 2000, Hydrothermal alteration and fluid chemistry of the Endako porphyry molybdenum deposit, British Columbia: Economic Geology, v. 95, p. 183-202.

Setterfield, T.N., Eaton, P.C., W.J., R., and Sparks, R.S.J., 1991, The Tavua Caldera, Fiji: a complex shoshonitic caldera formed by concurrent faulting and downsagging: Journal of the Geological Society, v. 148, p. 115-127.

Shannon, J.R., Walker, B.M., Carten, R.B., and Geraghty, E.P., 1982, Unidirectional solidification textures and their significance in determining relative ages of intrusions at the Henderson mine, Colorado: Geology, v. 19, p. 293-297.

Sheets, R.W., Nesbitt, B.E., and Muehlenbachs, K., 1996, Meteoric water component in magmatic fluids from porphyry copper mineralization, Babine Lake area, British Columbia: Geology, v. 24, p. 1091-1094.

Shinohara, H., Kazahaya, K., and Lowenstern, J.B., 1995, Volatile transport in a convecting magma column: Implications for porphyry Mo mineralization: Geology, v. 23, p. 1091-1094.

Shives, R.B.K., Charbonneau, B.W., and Ford, K.L., 2000, The detection of potassic alteration by gamma-ray spectrometry; recognition of alteration related to mineralization: Geophysics, v. 65, p. 2001-2011.

Siems, P.L., 1989, Lithium silicate alteration of tin granites: an analog of potassium silicate alteration in porphyry copper and molybdenite deposits: The Geological Society of America, 1989, Abstracts with Programs, p. 143.

Sikka, D.G., and Nehru, C.E., 1997, Review of Precambrian porphyry Cu±Mo±Au deposits with special reference to Malanjkhand porphyry copper deposit, Madhya Pradesh, India: Journal of the Geological Society of India, v. 49, p. 239-288.

Sillitoe, R.H., 1973, The tops and bottoms of porphyry copper deposits: Economic Geology, v. 68, p. 700-815.

——— 1976, Andean mineralization: a model for the metallogeny of convergent plate margins, *in* Strong, D.F., ed., Metallogeny and Plate Tectonics: Geological Association of Canada, Special Paper 14, p. 59-100.

——— 1980, Types of porphyry molybdenum deposits: Mining Magazine, v. 142, p. 550-551, 553.

——— 1988a, Ores in volcanoes: Seventh Quadrennial IAGOD Symposium, Stuttgart, 1988, Proceedings, p. 1-10.

——— 1988b, Gold and silver deposits in porphyry systems, *in* Schafer, R.W., Cooper, J.J., and Vikre, P.G., eds., Bulk Mineable Precious Metal Deposits of the Western United States, Symposium Proceedings: Geological Society of Nevada, p. 233- 257.

——— 1990, Gold-rich porphyry copper deposits of the circum-Pacific region - An updated overview: The Australasian Institute of Mining and Metallurgy Pacific Rim 90 Congress, 1990, p. 8.

——— 1992, Gold and copper metallogeny of the central Andes - past, present, and future exploration objectives: Economic Geology, v. 87, p. 2205-2216.

——— 1993a, Epithermal models: genetic types, geometrical controls and shallow features, *in* Kirkham, R.V., Sinclair, W.D., Thorpe, R.I., and Duke, J.M., eds., Mineral Deposit Modeling: Geological Association of Canada, Special Paper 40, p. 403-417.

——— 1993b, Gold-rich porphyry copper deposits: geological model and exploration implications, *in* Kirkham, R.V., Sinclair, W.D., Thorpe, R.I., and Duke, J.M., eds., Mineral Deposit Modeling: Geological Association of Canada, Special Paper 40, p. 465-478.

Sillitoe, R.H., and Bonham, H.F., Jr., 1984, Volcanic landforms and ore deposits: Economic Geology, v. 79, p. 1286-1298.

Sillitoe, R.H., and Hedenquist, J.W., 2003, Linkages between volcanotectonic settings, ore-fluid compositions, and epithermal precious metal deposits, *in* Simmons, S.F., and Graham, I., eds., Volcanic, Geothermal, and Ore-Forming Fluids; Rulers and Witnesses of Processes within the Earth: Economic Geology, Special Publication 10, p. 315-343.

Sillitoe, R.H., Halls, C., and Grant, J.N., 1975, Porphyry tin deposits in Bolivia: Economic Geology, v. 70, p. 913-927.

Sinclair, W.D., 1995a, Porphyry Sn, *in* Lefebure, D.V., and Ray, G.E., eds., Selected British Columbia Mineral Deposit Profiles, Volume 1 - Metallics and Coal: British Columbia Ministry of Energy, Mines and Petroleum Resources, Open File 1995-20, p. 97-99.

——— 1995b, Porphyry W, *in* Lefebure, D.V., and Ray, G.E., eds., Selected British Columbia Mineral Deposit Profiles, Volume 1 - Metallics and Coal: British Columbia Ministry of Energy, Mines and Petroleum Resources, v. Open File 1995-20, p. 101-104.

——— 1995c, Molybdenum, tungsten and tin deposits and associated granitoid intrusions in the northern Canadian Cordillera and adjacent parts of Alaska, *in* Schroeter, T.G., ed., Porphyry deposits of the Northwestern Cordillera of North America: The Canadian Institute of Mining and Metallurgy, Special Volume 46, p. 58-76.

——— 2005, Tin, tungsten, molybdenum and indium in New Brunswick, *in* Merlini, S.A.A., ed., Exploration and Mining New Brunswick: New Brunswick Department of Natural Resources, Abstracts, Information Circular 2005-1, p. 91.

Sinclair, W.D., Kooiman, G.J.A., Martin, D.A., and Kjarsgaard, I.M., 2006, Geology, geochemistry and mineralogy of indium resources at Mount Pleasant, New Brunswick, Canada: Ore Geology Reviews, v. 28, p. 123-145.

Skewes, M.A., and Stern, C.R., 1995, The genesis of giant late Miocene to Pliocene copper deposits of central Chile in the context of Andean magmatic and tectonic evolution: International Geology Review, v. 37, p. 893-909.

Skewes, M.A., Arévalo, A., Floody, R., Zuñiga, P.H., and Stern, C.R., 2002, The giant El Teniente breccia deposit: Hypogene copper distribution and emplacement, *in* Goldfarb, R.J., and Nielsen, R.L., eds., Integrated Methods for Discovery: Global Exploration in the Twenty-First Century: Society of Economic Geologists, Special Publication 9, p. 299-332.

Snyder, L.D., and Russell, J.K., 1995, Petrogenetic relationships and assimilation processes in the alkalic Iron Mask batholith, south-central British Columbia, *in* Schroeter, T.G., ed., Porphyry Deposits of the Northwestern Cordillera of North America: The Canadian Institute of Mining and Metallurgy, Special Volume 46, p. 593-608.

Soloman, M., 1990, Subduction, arc reversal, and the origin of porphyry copper-gold deposits in island arcs: Geology, v. 18, p. 630-633.

Sparks, S.J.R., Sigurdsson, H., and Wilson, L., 1977, Magma mixing: a mechanism for triggering acid explosive eruptions: Nature, v. 267, p. 315-318.

Streck, M.J., and Dilles, J.H., 1998, Sulfur content of oxidized arc magmas as recorded in apatite from a porphyry copper batholith: Geology, v. 26, p. 523-526.

Sutherland Brown, A., 1969, Mineralization in British Columbia and the copper and molybdenum deposits: Canadian Institute of Mining and Metallurgy, v. 72, p. 1-15.

Tarkian, M., and Stribrny, B., 1999, Platinum-group elements in porphyry copper deposits: a reconnaissance study: Mineralogy and Petrology, v. 65, p. 161-183.

Theodore, T.G., and Menzie, W.D., 1984, Fluorine-deficient porphyry molybdenum deposits in the western North American Cordillera: Sixth Quadrennial IAGOD Symposium, 1984, Proceedings, p. 463-470.

Thompson, J.F.H., Sillitoe, R.H., Baker, T., Lang, J.R., and Mortenson, J.K., 1999, Intrusion-related gold deposits associated with tungsten-tin provinces: Mineralium Deposita, v. 34, p. 323-334.

Titley, S.R., and Beane, R.E., 1981, Porphyry copper deposits, *in* Skinner, B.J., ed., Economic Geology Seventy-fifth Anniversary Volume, 1905-1980: Economic Geology Publishing Co., p. 214-269.

Titley, S.R., Thompson, R.C., Haynes, F.M., Manske, S.L., Robison, L.C., and White, J.L., 1986, Evolution of fractures and alteration in the Sierrita-Esperanza hydrothermal system, Pima County, Arizona: Economic Geology, v. 81, p. 343-370.

Ulrich, T., Günther, D., and Heinrich, C.A., 1999, Gold concentrations of magmatic brines and the metal budget of porphyry copper deposits: Nature, v. 399, p. 676-679.

Ulrich, T., Günther, D., and Heinrich, C.A., 2001, The evolution of a porphyry Cu-Au deposit, based on LA-ICP-MS analysis of fluid inclusions: Bajo de la Alumbrera, Argentina: Economic Geology, v. 96, p. 1743-1774.

Vila, T., and Sillitoe, R.H., 1991, Gold-rich porphyry systems in the Maricunga gold-silver belt, northern Chile: Economic Geology, v. 86, p. 1238-1260.

Wallace, S.R., Muncaster, N.K., Jonson, D.C., Mackenzie, W.B., Bookstrom, A.A., and Surface, V.A., 1968, Multiple intrusion and mineralization at Climax, Colorado, *in* Ridge, J.D., ed., Ore Deposits of the United States, 1933-1967 (Graton-Sales Volume): American Institute of Mining, Metallurgical, and Petroleum Engineers, New York, p. 605-640.

Wanhainen, C., Broman, C., and Martinsson, O., 2003, The Aitik Cu-Au-Ag deposit in northern Sweden: A product of high salinity fluids: Mineralium Deposita, v. 38, p. 715-726.

Webster, J.D., 2004, The exsolution of magmatic hydrosaline chloride liquids: Chemical Geology, v. 210, p. 33-48.

Weixing, H., and Dazhong, S., 1987, Mineralization and evolution of the Early Proterozoic copper deposits in the Zhongtiao Mountains: Acta Geologica Sinica, v. 61, p. 152-165.

Westra, G., and Keith, S.B., 1981, Classification and genesis of stockwork molybdenum deposits: Economic Geology, v. 76, p. 844-873.

Whalen, J.B., Jenner, G.A., Currie, K.L., Barr, S.A., Longstaffe, F.J., and Hegner, E., 1994, Geochemical and isotopic characteristics of granitoids of the Avalon Zone, southern New Brunswick: possible evidence for repeated delamination events: The Journal of Geology, v. 102, p. 269-282.

Whalen, J.B., Fyffe, L.R., Longstaffe, F.J., and Jenner, G.A., 1996, The position and nature of the Gander-Avalon boundary, southern New Brunswick, based on geochemical and isotopic data from granitoid rocks: Canadian Journal of Earth Sciences, v. 33, p. 129-139.

Whalen, J.B., Anderson, R.G., Struik, L.C., and Villeneuve, M.E., 2001, Geochemistry and Nd isotopes of the François Lake plutonic suite, Endako batholith: host and progenitor to the Endako molybdenum camp, central British Columbia: Canadian Journal of Earth Sciences, v. 38, p. 603-618.

Wheeler, J.O., Hoffman, P.F., Card, K.D., Davidson, A., Sanford, B.V., Okulitch, A.V., and Roest, W.R., 1996, Geological map of Canada: Geological Survey of Canada, Map 1860A, scale: 1:5 000 000.

White, W.H., Bookstrom, A.A., Kamilli, R.J., Ganster, M.W., Smith, R.P., Ranta, D.E., and Steininger, R.C., 1981, Character and origin of Climax-type molybdenum deposits, *in* Skinner, B.J., ed., Economic Geology Seventy-Fifth Anniversary Volume, 1905-1980: Economic Geology Publishing Co., p. 270-316.

Whitney, J.A., 1975, Vapour generation in a quartz monzonite magma: a synthetic model with application to porphyry copper deposits: Economic Geology, v. 70, p. 346-358.

——— 1984, Volatiles in magmatic systems; *in* Whitney, J.A., Henley, R.W., Truesdell, A.H., and Barton, P.B., eds., Fluid-Mineral Equilibria in Hydrothermal Systems: Society of Economic Geologists, Reviews in Economic Geology, v. 1, p. 155-175.

Williams, I.S., and Collins, W.J., 1990, Granite-greenstone terranes in the Pilbara Block, Australia, as coeval volcano-plutonic complexes: evidence from U-Pb zircon dating of the Mount Edgar Batholith: Earth and Planetary Science Letters, v. 97, p. 41-53.

Wilson, A.J., Cooke, D.R., and Harper, B.L., 2003, The Ridgeway gold-copper deposit: a high-grade alkalic porphyry deposit in the Lachlan Fold Belt, New South Wales, Australia: Economic Geology, v. 98, p. 1637-1666.

Wolfe, W.J., 1995, Exploration and geology of the Quartz Hill molybdenum deposit, southeast Alaska, *in* Schroeter, T.G., ed., Porphyry Deposits of the Northwestern Cordillera of North America: The Canadian Institute of Mining and Metallurgy, Special Volume 46, p. 764-770.

Kimberlite Diamond Deposits

B.A. Kjarsgaard

Geological Survey of Canada, 601 Booth Street, Ottawa, Ontario, K1A 0E8
Corresponding author's email: bruce.kjarsgarrd@nrcan.gc.ca

Abstract

Diamonds have formed over a significant period of the Earth's history, from ca. 3.57 Ga to 88 Ma, and probably to present day. Macrodiamonds are interpreted to crystallize from low-density fluids, or carbon- and water-rich melts at pressures >~4.0 GPa and temperatures <~1350°C. These P–T conditions are met within thick, old lithospheric mantle roots that have low paleogeothermal gradients, and these roots lie under ancient continental nuclei. Kimberlite-hosted diamond mines occur in these cratonic shield regions that are older than 2.5 Ga. Macrodiamonds are transported as xenocrysts from the mantle to the surface by kimberlite magmas. The initiation of kimberlite magmatism is at depth in the asthenospheric mantle (>150 km), although the initiation and generation of kimberlite magma is poorly understood. Kimberlites magmas generate a range of rocks that form a wide variety of landforms and intrusions, in many aspects similar to that generated by small-volume alkali basaltic volcanic systems. Kimberlite bodies typically form from multiple intrusive and/or extrusive events; these discrete events form distinct kimberlite phases. These individual kimberlite phases are characterized by differing textures, mineralogy and geochemistry, and diamond grade, size populations and morphology, and value.

Résumé

Des diamants se sont formés pendant une longue période de l'histoire de la Terre, de 3,57 Ga à 88 Ma environ, et se forment probablement encore de nos jours. Selon diverses interprétations, les macrodiamants se formeraient par cristallisation à partir de fluides de faible densité ou encore de bains magmatiques riches en carbone et en eau à des pressions supérieures à environ 4,0 GPa et à des températures inférieures à environ 1350 °C. Ces conditions de pression et de température sont atteintes à l'intérieur d'anciennes et épaisses racines de manteau lithosphérique caractérisées par de faibles paléogradients géothermiques, qui se situent sous d'anciens noyaux continentaux. Les mines de diamants dans des kimberlites gisent dans ces régions de boucliers cratoniques datant de plus de 2,5 Ga. Les macrodiamants sont transportés sous forme de xénocristaux depuis le manteau jusqu'à la surface par des magmas kimberlitiques. L'amorce du magmatisme kimberlitique se produit en profondeur dans le manteau asthénosphérique (>150 km), bien que le début de la formation et la production du magma kimberlitique soient des phénomènes encore mal compris. Les magmas kimberlitiques engendrent toute une gamme de roches donnant lieu à des formes de terrain et à des intrusions très variées, à de nombreux égards similaires à celles engendrées en de moindres volumes par les systèmes volcaniques de basalte alcalin. Les corps kimberlitiques se forment de manière caractéristique à partir de multiples événements intrusifs ou extrusifs; ces événements discrets forment des phases kimberlitiques distinctes, caractérisées par des différences de texture, de minéralogie et de géochimie ainsi que de la teneur en diamants et de caractéristiques de ceux-ci telles que la taille (populations granulométriques), la morphologie et la valeur des pierres.

Definition

Simplified Definition of Deposit Type

Primary deposits of diamonds occur in kimberlite as a sparsely dispersed xenocrystal mineral in pyroclastic, volcaniclastic, resedimented volcaniclastic and sub-volcanic (hypabyssal) rocks of mantle origin (kimberlite-hosted deposits). Secondary deposits formed by the weathering of primary deposits occur in unconsolidated and consolidated sediments (placers, paleoplacers, and volcanogenic sedimentary deposits).

Scientific Definition of Deposit Type

Kimberlite is formed from alkali-poor, H_2O- and CO_2-rich ultrabasic magma that has an enriched incompatible (Ba, Zr, Hf, Ta, Nb, REE) and compatible (Ni, Co, Cr) trace element signature. These magmas gave rise to rocks that form a wide variety of landforms and intrusions, similar to those exhibited by small-volume alkali basalt volcanic systems. Diamond xenocrysts are variably distributed throughout the host rocks at concentration levels of <0.01 to 2.0 ppm.

Deposit Subtypes

Lamproite and orangeite (also called Group II kimberlite) host primary magmatic diamond deposits. Placer and paleoplacer deposits can also be viewed as a subtype.

Distributions

Global Distribution of Kimberlite Diamond Deposits

Major producing or past-producing kimberlite-hosted diamond mines are known from southern Africa (South Africa, Botswana, Zimbabwe), south-central Africa (Tanzania, Democratic Republic of Congo, Angola), western Africa (Sierra Leone), Russia (Yakutia), China, Canada, and the United States (Fig. 1). These deposits are all found in Precambrian terranes and specifically in Archean continental blocks. The absence of kimberlite-hosted diamond mines in countries or continents with significant Archean crustal blocks such as Australia, India, and South America is notable, especially given the formerly significant paleoplacer and placer deposits in India and Brazil.

Kjarsgaard, B.A., 2007, Kimberlite diamond deposits, *in* Goodfellow, W.D., ed., Mineral Deposits of Canada: A Synthesis of Major Deposit Types, District Metallogeny, the Evolution of Geological Provinces, and Exploration Methods: Geological Association of Canada, Mineral Deposits Division, Special Publication No. 5, p. 245-272.

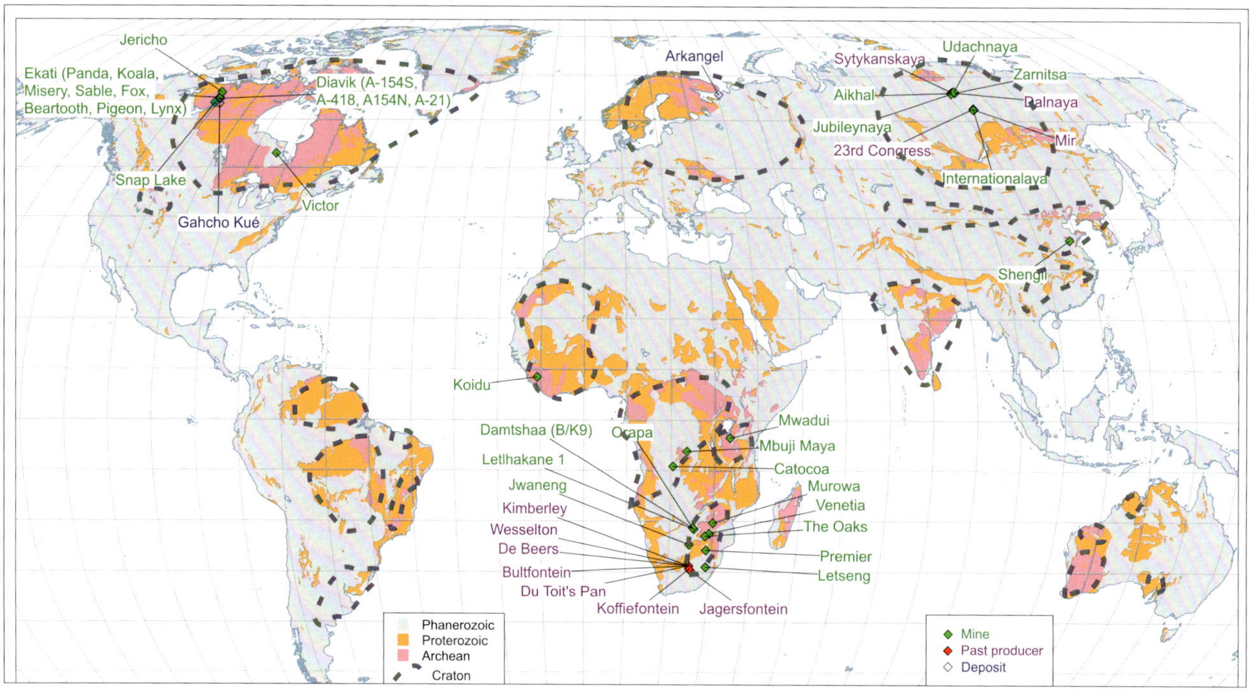

FIGURE 1. Global distribution of kimberlite-hosted diamond mines (active mines and past producers) and significant deposits. Also shown are the major Archean cratons.

Ages of Deposits

A significant observation is that there are specific time periods in the Earth's history when kimberlites with economically viable contents of diamonds erupted contemporaneously on different continents, for example, in Tanzania and Canada at ca. 52 to 56 Ma (Davis and Kjarsgaard, 1997) and in South Africa and Canada at ca. 535 to 542 Ma, whereas at other times kimberlite events are restricted to a specific region, e.g., Yakutia at ca. 360 Ma (Fig. 2; Heaman et al., 2003, 2004). The oldest kimberlite-hosted diamond mine is the Cullinan (or Premier) Mine in South Africa, which has been dated by radiometric methods at ~1200 Ma (Allsopp et al., 1989). The Venetia kimberlite in South Africa is of similar age to the Gahcho Kué #5034 pipe and the Snap Lake sill in the Slave Province, Northwest Territories (NWT), Canada (all ca. 542–535 Ma; Heaman et al., 2003). Kimberlites that host diamond mines in China are dated at ca. 475 to 462 Ma (Dobbs et al., 1994). The Yakutian kimberlites that host the economically important diamond mines have been dated by radiometric methods at ca. 376 to 344 Ma (Davis, 1977, 1978; Kinny et al., 1997), whereas the Jwaneng kimberlite in Botswana is ca. 235 Ma in age (Kinny et al., 1989). Numerous kimberlites hosting diamond mines in southern Africa, for example, Letseng, Lesotho; Jagersfontein, Koffiefontein, Du Toit's Pan, Bultfontein, De Beers, Kimberley and Wesselton in South Africa and Orapa in Botswana, are of similar age (ca. 95–84 Ma; Davis, 1977). Slightly younger in age are kimberlites that host the Mbuji Maya Mine in the Democratic Republic of Congo (71 Ma; Davis, 1977). The youngest kimberlites known to host diamond mines are found at the Ekati and Diavik Mines in the Slave province, NWT (56–53 Ma;

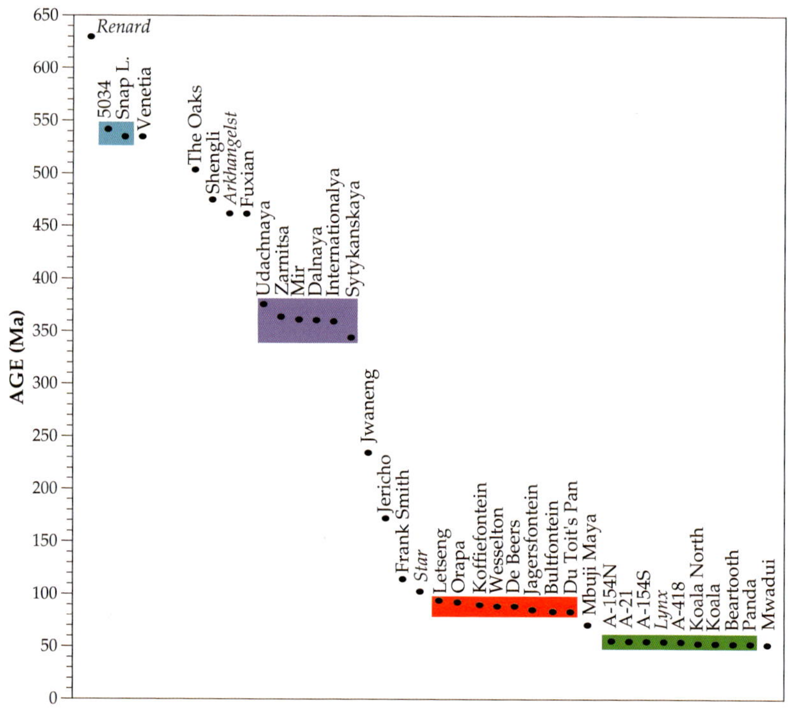

FIGURE 2. Emplacement ages of kimberlite-hosted diamond mines (past producers and active mines) and significant deposits. Data from Heaman et al. (2003, 2004).

Graham et al., 1999; Kjarsgaard et al., 2002; Creaser et al., 2004) and in Tanzania at the Mwadui Mine (52 Ma; Gobba, 1989).

Distribution and Age of Canadian Deposits

The kimberlites that comprise the Ekati, Diavik, Jericho, and Snap Lake mines all lie within the Contwoyto terrain of the Archean Slave province, as does the Gahcho Kué deposit (Fig. 3). The Jericho kimberlite in the north Slave is Jurassic in age (172 Ma). The central Slave kimberlites that host the Ekati (Panda, Beartooth, Koala) and Diavik (A-154N, A-154S, A-418, A-21) mines are Paleogene in age (56.0–53.0 Ma). The southeast Slave Snap Lake and Gahcho Kué (5034) kimberlites are of Cambrian age (523–535 Ma and 542 Ma, respectively), being significantly older (Fig. 2). The Victor diamond mine lies in the Archean Superior craton within the Sachigo sub-province (Fig. 3). Although the age of the Victor kimberlite has not been determined, most other kimberlites in the field have Jurassic emplacement ages in the range of 175 to 180 Ma (Heaman and Kjarsgaard, 2000). The Slave craton is estimated to contain ~US$27B (see Table 1) of diamonds in known economically viable deposits (Fig. 4a). The Superior province, in contrast, contains only one known mine (Victor), which is estimated to contain ~US$3.5B (see Table 1) of diamonds (Fig. 4a). The estimated value (see Table 1) of Canadian diamond deposits by age is:

- Middle Paleogene Ekati and Diavik mines at Lac de Gras contains ~US$19B of economically viable diamonds;
- Middle Jurassic Victor and Jericho deposits contain a ~US$4.1B of diamonds;
- Early Cambrian Snap Lake and Gahcho Kué deposits contains ~US$7.2B of diamonds (Fig. 4b).

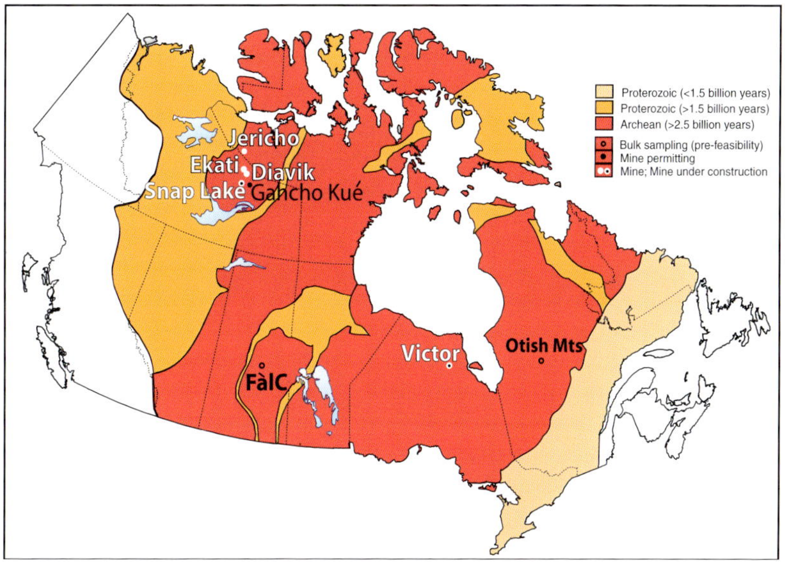

FIGURE 3. Geographic distribution of Canadian diamond mines, diamond deposits, and kimberlites that being bulk sampled as part of advanced exploration projects. Also shown are Proterozoic and Archean terranes (younger sedimentary cover sequences removed).

Pre-feasibility bulk sampling is currently being undertaken on the Cretaceous age ca. 103 Ma Star and Orion kimberlites in the Fort à la Corne area of central Saskatchewan

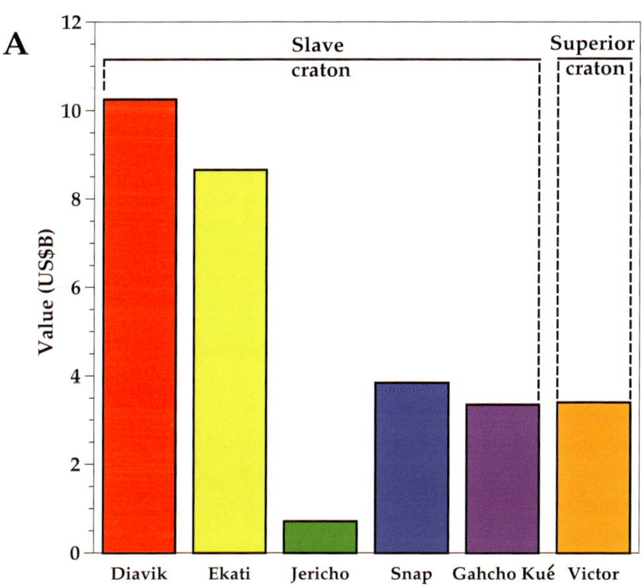

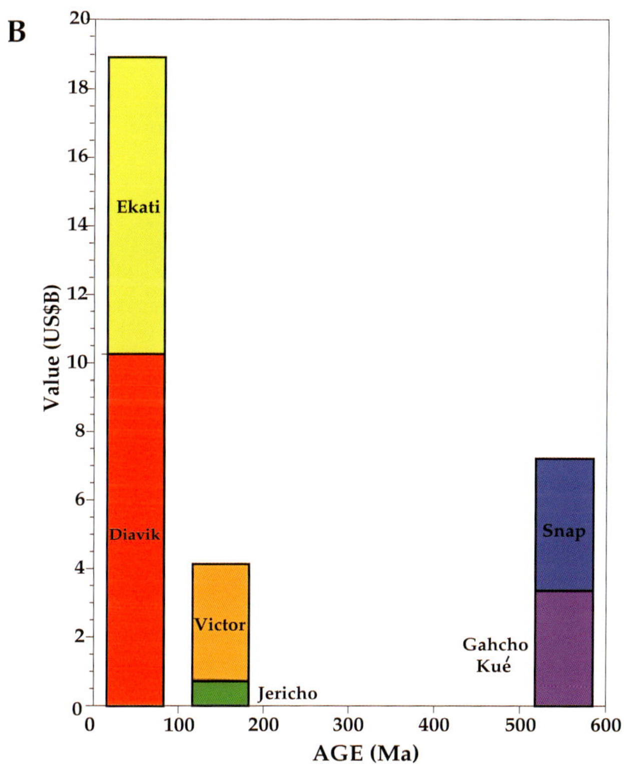

FIGURE 4. **(A)** In situ value of Canadian diamond mines and deposits subdivided by location with respect to Archean cratonic blocks. **(B)** In situ value of Canadian diamond mines and deposits subdivided by emplacement age.

TABLE 1. Canadian kimberlite mine diamond grade, stone value, ore value and tonnage data

PIPE	Grade (c/t) reserve	stone value US$/c reserve	Avg ore value US$/T reserve + resource	Area (ha) of pipe	Mt (tot) reserve	Grade ind/inf resource	MT ind/inf resource	Total tonnes (MT) reserve + resource	Avg grade (c/t) reserve + resource
A-154S	4.8	96	459	2.64	10.5	4.4	0.6	11.1	4.78
A-418	3.4	56	192	1.39	8.7	3.8	0.6	9.3	3.43
A-154N	3.1	82	248	1.23	10.6	2.5	1.6	12.2	3.02
A-21		28	76	1		2.7	4.8	4.8	2.70
Diavik total								**37.4**	
Panda	1.1	168	186	3.3	10.9			15.7	1.11
Panda u/g	1.4				1.4	1	3.4		
Koala	1	138	174	4.5	10	1	6.2	23.9	1.26
Koala u/g						1.2	13.9		
Koala North	0.5	143	72		0.1			1.7	0.50
Koala North u/g						0.5	1.6		
Misery	4.5	34	142	1.1	5.2	3.4	2.2	7.4	4.17
Sable	0.9	82	74	2.4	12.8	0.9	3.8	16.6	0.9
Fox	0.4	129	52	11	5.6	0.4	21.1	26.7	0.4
Beartooth	1.1	100	110	0.5	2	1.1	0.5	2.5	1.1
Pigeon	0.4	100	40	1	6.8	0.4	1.5	8.3	0.4
Lynx	0.8	108	86	0.6	1.4			1.4	0.8
Ekati total								**102.8**	
Jericho	1.2	85	102	3	2.6	0.63	4.5	7.1	0.84
Victor	0.23	458	105	5	32.4			32.4	0.23
Snap Lake	1.46	144	210	n/a	18.3			18.3	1.46
5034	1.6	82	134	1.95	8.7	1.7	4.9	13.6	1.64
Hearne	1.7	70	117	1.5	5.7	1.53	1.5	7.2	1.66
Tuzo		57	66	1.4		1.15	10.6	10.6	1.15
Gahcho Kue total								**31.4**	

Notes:
Diavik Mine grade and tonnage data as per end December 2004 from Aber Resources as reported by Aber Diamond Corporation (2005; see www.Sedar.com) with rough diamond value from Kjarsgaard et al. (2002) and The Northern Miner (2006b, v. 92 #7). Ekati Mine data as per end June 2003 for grade and tonnage from BHP-Billiton Annual Reports (J. Carlson, pers. comm, 2004); with rough diamond value from Kjarsgaard et al. (2002); Lynx data from The Northern Miner (2004, v. 90 #4). For the BHP Billiton data (other than Panda), the stone value and grade figures are exploration data and not run of mine and are probably low, and the stone values in Table 1 were determined over an approximate ten year period, during which time there was significant changes in the 'book' value for diamonds. Ekati Mine proven and probable ore reserves at June 2006 include: open cast, 31 Mt at 0.4 c/t; stockpile 1.2 Mt at 2.5 c/t; underground 11.8 MT at 0.9 c/t (from BHP-Billiton Annual Report 2007; note no stone values are given). Jericho Mine grade, tonnage and rough diamond value from Tahera Corporation (2003; see www.Sedar.com). Victor Mine and Snap Lake Mine grade, tonnage and rough diamond value from De Beers Canada factsheets (www.debeesrcanada.com) and The Northern Miner (2006a, v. 92 #2). Gahcho Kué deposit grade, tonnage and rough diamond value from De Beers Canada factsheets (www.debeesrcanada.com), The Northern Miner (2006a, v. 92 #2) and from Mountain Province Diamonds (2006; see www.Sedar.com). (2006a, v. 92, n. 2), and from Mountain Province Diamonds (2006; see www.Sedar.com).

and the Neoproterozoic age ca. 630 Ma Renard kimberlite in the northern Otish Mountains area of Quebec (Fig. 3).

Grade, Tonnage, and Value Statistics

Economic Characteristics of Global Diamond Deposits

Within a kimberlite diamond deposit, there is significant grade variation (typically greater than an order of magnitude) between individual intrusive phases and/or extrusive phases, and/or resedimented volcaniclastic phases (Fig. 5a). Furthermore, within an individual kimberlite phase in a diamond mine, there can also be significant grade variation (more than an order of magnitude; Fig. 5b). Hence, the compiled diamond grade data as shown in Figure 6a is based on interpreted average grades for mined out deposits, or 'head frame' grades for active mines. Also note that many diamond grades are inferred (e.g., Russian deposits) due to a lack of public domain data, or are based on data from exploration or feasibility studies, which may not be a true representation of actual mine grades due to the small size of the preliminary samples. With these caveats in mind, diamond grade for economic diamond deposits ranges from ~0.02 to ~10 carats per tonne (carat/tonne), more than two orders of magnitude.

In contrast to the vast majority of commodities (e.g., metals), there is not a single value for diamond. The value of an individual rough diamond can be exceptionally variable (<US$1/carat to >US$1,000,000/carat), depending upon size, colour, and quality. For this reason, the average rough diamond value stated in US$/carat is a very important economic parameter. The known range for current and past producing mines is quite variable, ranging from ~US$15/carat to ~US$800/carat (Fig. 6b). Again, note that the caveats stated regarding diamond grade are also applicable to the average rough diamond value; for most active mines, the average rough diamond value is considered proprietary data. In this regard, some of the values shown in Figure 6b are probably low, due to use of exploration or feasibility sampling data. For example, the average value of rough diamonds from the Panda kimberlite from the feasibility study was US$130/carat, whereas the Panda run-of-mine rough diamond value increased ~30% to US$168/carat. This increase can be due to many factors, including lower stone breakage during mining as compared to reverse circulation drilling during exploration sampling, but perhaps more importantly mining typically recovers larger diamonds, which can strongly influence average rough diamond values (this is due to a statistical under sampling of the larger sieve classes of diamond in exploration or feasibility sampling, even when 3,000 to 10,000 carat parcels of diamond are recovered for evaluation purposes). An additional significant problem in comparing average rough diamond values is the fluctuation in book price over time (years to decades) due to supply and demand, and general economic conditions.

Kimberlite diamond deposits are quite variable in area (Fig. 6c) and geometry (dikes, sills, pipes, tephra cones). Due to a lack of public domain data, there is a paucity of ore tonnage reserve or resource data and even if the area of the kimberlite is known, it is difficult to calculate ore tonnage due to the variable geometry of these deposits. This is further complicated by the observation that not all of a kimberlite body may be economic with respect to diamond recovery. For example, the Letseng (Lesotho) kimberlite is 16 hectares (ha) in area, which would place it in the upper quartile for active or past producing kimberlite diamond mines. However, only the K6 phase at Letseng with an area of 4 ha was economic (Fig. 6c). Janse (1993) provided a novel approach to determining ore resource data worldwide by calculating tonnage to 120 m depth. Using this criterion, there are certainly very high tonnage (e.g., Mwadui, Orapa, Jwaneng, Premier) and very low tonnage (e.g., Koidu, The Oaks, Misery, A-154N) deposits.

Due to the wide range of diamond grades and rough diamond values for kimberlite diamond deposits, ore value expressed as US$/tonne (= grade [carat/tonne] × rough diamond value [US$/carat]) is one of the main parameters utilized in assessing the economic potential of a deposit (Figs. 6d, 7). Ore values range from ~US$13/tonne to ~US$1000/tonne.

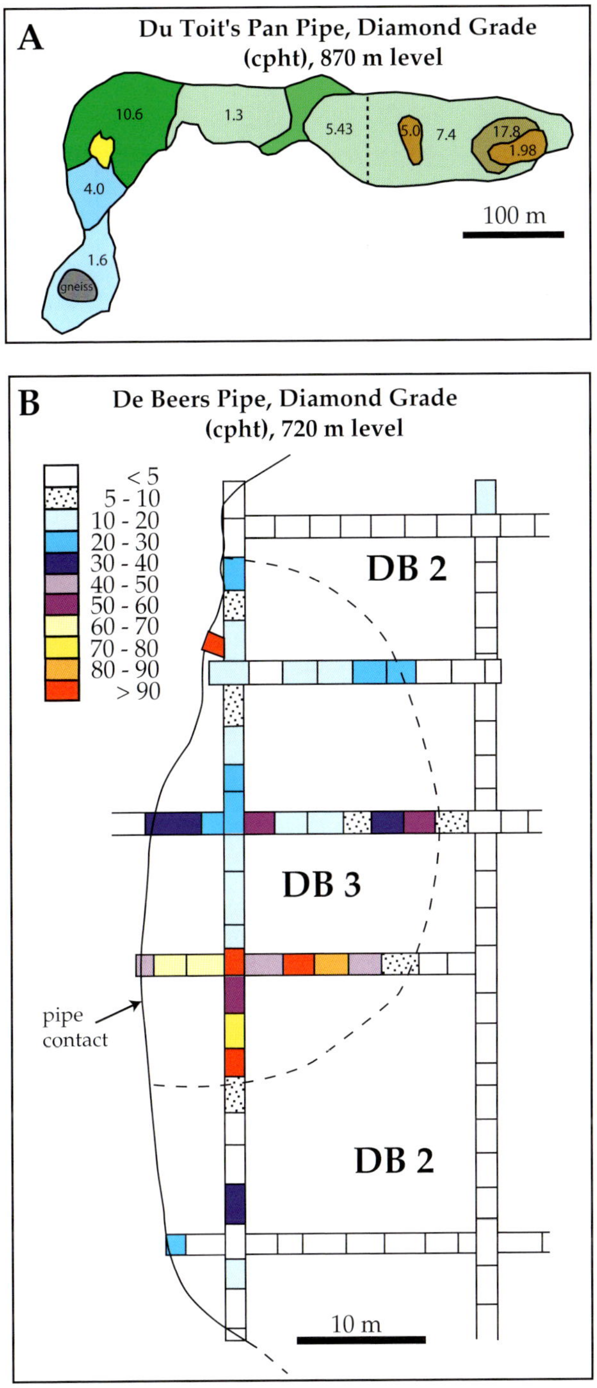

FIGURE 5. **(A)** Du Toit's Pan kimberlite, South Africa, 870 m level, illustrating the variation of diamond grade in cpht in relation to the different geological phases of kimberlite (after Clement, 1982). **(B)** De Beers kimberlite, South Africa, 720 m level, illustrating the variation of diamond grade in cpht within the DB2 and DB3 kimberlite intrusions (adapted from Clement, 1982). Note: cpht = carats per hundred tonnes.

Economic Characteristics of Canadian Deposits

The diamond grade for individual Canadian kimberlite pipes is highly variable (Kjarsgaard and Levinson, 2002). Five of the Lac de Gras kimberlite bodies (Ekati Mine Misery pipe and Diavik Mine A-154S, A-154N, A-418, A-21 pipes) have exceptional diamond grades (2.7–4.8 carats/tonne; Table 1) compared to active and past-producing diamond mines worldwide (Fig. 6a). These diamond grades, while remarkable for a kimberlite, are comparable to or low-

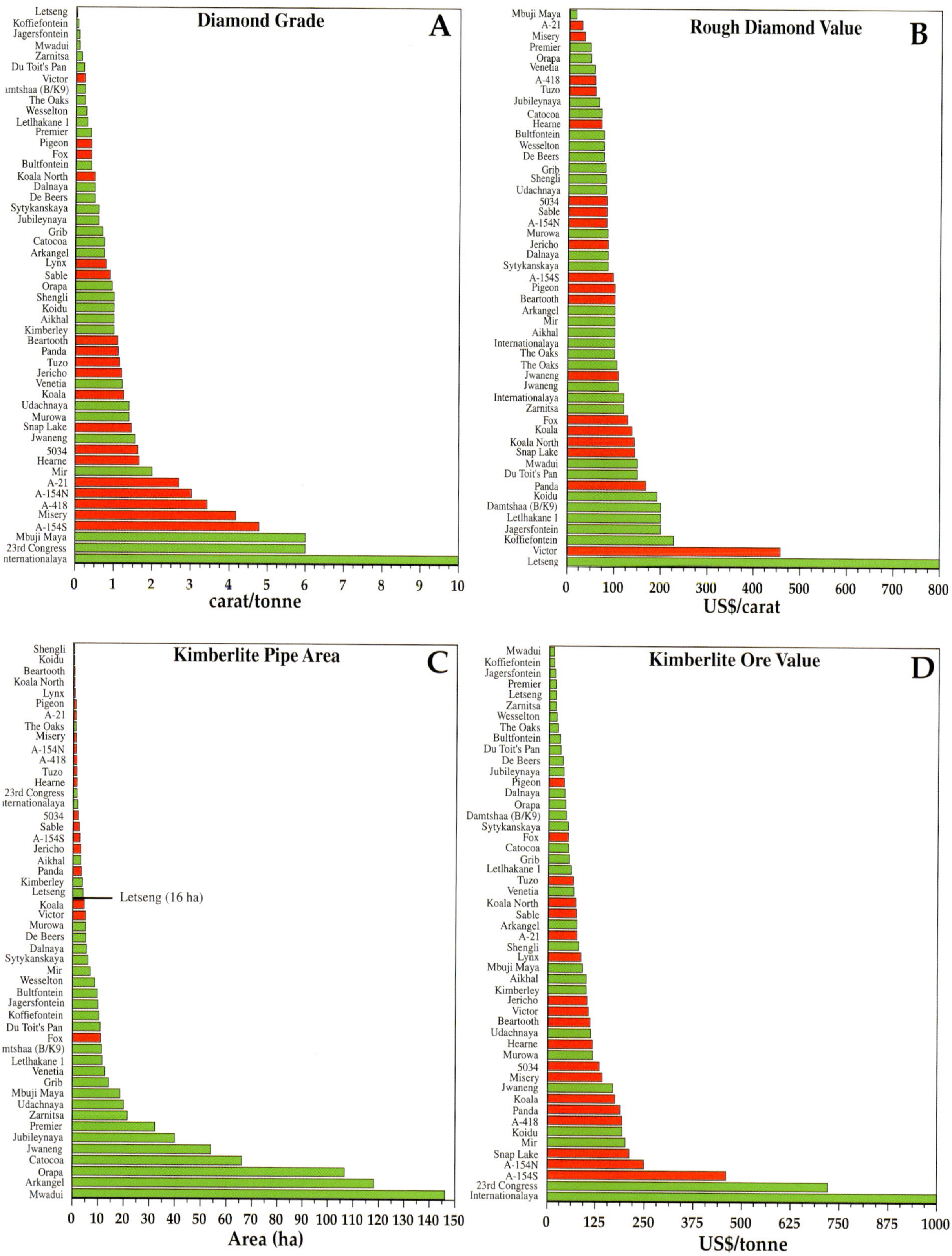

FIGURE 6. **(A)** Histogram illustrating the variation in diamond grade (in carats per tonne) for Canadian (red) and global (green) kimberlite diamond mines (active and past-producers). **(B)** Histogram illustrating the variation in rough diamond value (in US$ per carat) for Canadian (red) and global (green) kimberlite diamond mines (active and past-producers). **(C)** Histogram illustrating the range in hectares (ha) for Canadian (red) and worldwide (green) kimberlite diamond mines (active and past-producers). Note that the Letseng kimberlite total area is 16 ha, but the mineable part of the pipe was 4 ha. **(D)** Histogram illustrating the variation in ore value (in US$ per tonne) for Canadian (red) and global (green) kimberlite diamond mines (active and past-producers).

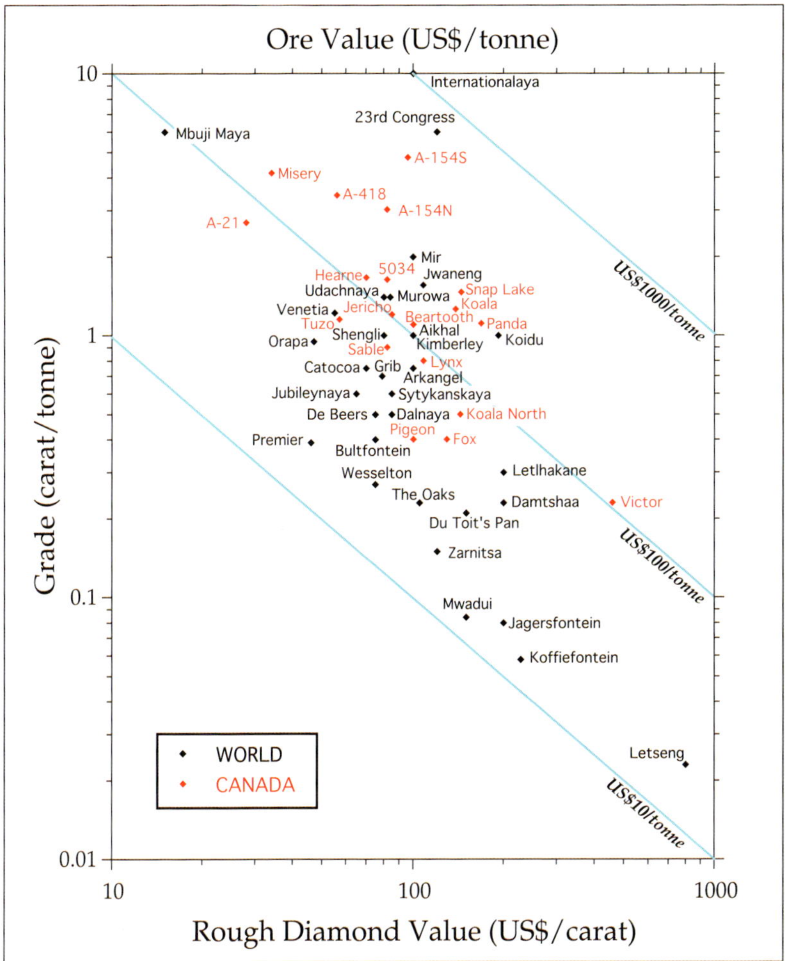

FIGURE 7. Log–log plot of diamond grade (carats per tonne) vs. rough diamond value (US$ per carat) for Canadian (red) and global (black) kimberlite diamond mines (active and past producers). This figure illustrates how kimberlites with quite disparate diamond grades and rough diamond values can have very similar ore values per tonne.

er than reported grades for Mbuji-Mayi (6 carats/tonne), 23rd Congress (6 carats/tonne), and Internationalaya (10 carats/tonne) kimberlites. The Hearne (1.67 carats/tonne), 5034 (1.64 carats/tonne), and Snap Lake (1.46 carats/tonne) kimberlites are high-grade diamond deposits (Table 1, Fig. 6a) and are in the upper echelon for grade as compared to active and past-producing diamond mines worldwide, for example, Mir, Jwaneng, and Murowa. Seven Canadian kimberlites, including the Koala, Jericho, Tuzo, Panda, Beartooth, Sable, and Lynx kimberlites, have average diamond grades ranging from 1.2 to 0.8 carats/tonne as compared to active and past-producing diamond mines worldwide (Table 1, Fig. 6a). The Koala North, Pigeon, Fox, and Victor kimberlites are of lower diamond grade (0.5–0.23 carats/tonne; Table 1, Fig. 6a).

There is a large variation in the average rough diamond value (quality) for individual Canadian kimberlite pipes. The Victor kimberlite has exceptionally valuable rough diamonds, averaging US$458/carat, and is only exceeded in value by diamonds from the Letseng (Lesotho) kimberlite (US$800/carat). The Panda (US$168/carat), Snap Lake (US$144/carat), Koala North (US$143/carat), Koala (US$138/carat), and Fox (US$129/carat) kimberlites all contain high-value diamonds (Table 1, Fig. 6b) and are in the upper echelon for rough diamond value as compared to active or past-producing diamond mines worldwide, for example, Koffiefontein, Letlhakane, and Damstshaa. The Lynx, Beartooth, Pigeon, A-154S, Jericho, A-154N, Sable, and 5034 kimberlites have rough diamond values ranging from US$82/carat to US$108/carat (Table 1, Fig. 6b); these diamond values are all higher than the world average rough diamond value of ~US$68/carat (based on 2004 world production data from Janse, 2005). Canadian kimberlites with low diamond values, for example, Misery (US$34/carat) or A-21 (US$28/carat), have average stone values approximately twice that of Mbuji Maya (US$15/carat).

The average to very small area (< 5 ha) of the known viable economic Canadian kimberlite pipes (with the exception of the 11 ha Fox pipe) is evident from Figure 6c. To date, economically viable Canadian kimberlites are significantly smaller than active diamond mines worldwide, the majority of which are >11 ha in area, although this observation would change if any of the very large (to 200 ha) Fort à la Corne kimberlites are mined.

There is a large variation in the value of ore for individual Canadian kimberlite pipes. The Diavik Mine A-154S kimberlite has an exceptional ore value of US$459/tonne, exceeded worldwide only by the Internationalaya (US$1000/tonne) and 23rd Congress (US$720/tonne) kimberlites in Russia (Figs. 6d, 7). The A-154N, Snap Lake, A-418, Panda, and Koala kimberlites all have very high ore values (US$248/tonne to US$170/tonne; Table 1, Figs. 6d, 7) that are in the upper echelon for ore values as compared to active or past-producing diamond mines worldwide, for example, Mir, Koidu, and Jwaneng. An additional six Canadian kimberlites (Misery, 5034, Hearne, Beartooth, Victor, and Jericho) have high ore values (>US$100/tonne; Table 1, Figs. 6d, 7). Ore values for the Lynx, A-21, Sable, Koala North, Tuzo, Fox, and Pigeon kimberlites are lower at 86, 76, 74, 72, 66, 52, and 40 US$/tonne, respectively (Table 1, Figs. 6d, 7).

In summary, economic Canadian kimberlite pipes tend to be small in area (low tonnage) bodies, that have variable (typically moderate to very high) diamond grades coupled with quite variable (low to very high) rough diamond values. These kimberlites have moderate to very high ore values, with the majority having ore values >US$100/tonne (Figs. 6d, 7). The development of the Ekati (eight orebodies) and Diavik (four orebodies) mines in the Lac de Gras area, Northwest Territories (NWT), required multiple economic pipes within a defined area so as to provide sufficient tonnage for a 15 to 20 year mine life. This is also true of the proposed Gahcho Kué diamond mine (three orebodies) in the NWT. An additional important factor regarding the economics of potential diamond mines in Canada is the operating cost. This will vary with mining method, for example, open pit versus underground mining, and the availability of infrastructure, for example, remote areas of Canada's north (the NWT, Nunavut, northern Ontario, and Québec) versus non-remote (southern to central Ontario, Quebec, and the prairie provinces). These two factors combined can result in potential operating costs varying by an order of magnitude (~US$10/tonne to US$100/tonne).

Contribution of Canadian Rough Diamonds to the World Supply

In 2004, combined diamond production from the Ekati and Diavik mines totaled 12.6 million carats, valued at US$2.1 billion (Perron, 2004). Canadian diamond production in 2004 comprised ~8% of world production by weight, but more importantly, ~16% of world production by value (2004 world production data from Janse, 2005). Compared to other producing countries, Canada ranked third by value (after Botswana and Russia) and sixth by number of carats produced. In 2004, Canadian diamonds from the combined output of the Ekati and Diavik Mines averaged ~US$135/carat, double the world average carat value of ~US$68/carat (based on 2004 world production data from Janse, 2005).

In 2006, Tahera's recently commissioned Jericho diamond mine started production. This mine will contribute an additional ~0.4 million carats/year for the next eight years, once full production is reached in 2007. By the end of this decade, additional rough diamond supplies will be coming on stream from the De Beers Victor (0.6 million carats/year) and Snap Lake (1.5 million carats/year) diamond mines. De Beers is also in the process of permitting the Gahcho Kué diamond deposit, which would have an output of 3 million carats/year. In summary, by 2010 Canada will likely be producing in excess of 15 million carats of rough diamond per year.

Geological Attributes

Continental Scale

Geotectonic Environment

Kimberlite diamond deposits are found within ancient Precambrian terrains older than 1.5 Ga (Clifford, 1966; "Clifford's Rule"). Diamond (i.e., macrodiamonds as opposed to microdiamonds) requires specific pressure and temperature conditions to form and remain stable with respect to graphite. These P–T conditions (typically > 4.0 GPa and < 1350°C) are met within thick, old lithospheric mantle roots with low paleogeothermal gradients that typically lie under ancient continental nuclei. Modern geochronology of Precambrian terrains has led to better temporal understanding of the formation of continental nuclei, which Janse (1984; "Janse's Rule") utilized to suggest kimberlite diamond deposits are in fact found within Archean continental blocks (Fig. 1).

Recent radiometric age determinations of mineral inclusions in macrodiamonds (see summaries in Pearson and Shirey, 2002; Gurney et al., 2005) indicate that the oldest diamonds (peridotite paragenesis, or P-type) formed within the Earth's mantle no earlier than 3.57 Ga. A corollary is that Eoarchean (>3.6 Ga) continental blocks (with Eoarchean mantle roots) are not good targets for diamond exploration, perhaps because the early Earth was too hot and/or the lithospheric mantle to thin for diamonds to form and/or remain stable within the mantle. Pearson and Shirey (2002) and Gurney et al. (2005) have noted that the youngest known diamond formation event is Cretaceous, and that many kimberlite diamond mines contain diamonds that formed at different times in the Earth's history, concluding that diamond formation is episodic. It should be noted that a very limited amount of data is available regarding diamond formation age, and it is highly likely that diamonds are currently forming in the mantle.

The known ages of diamond formation can be utilized to suggest that the most promising exploration targets will be Paleoarchean continental blocks followed by Mesoarchean and Neoarchean blocks, and then Paleoproterozoic continental blocks. The rationale is simply that Paleoarchean mantle roots have the possibility of experiencing the highest number of potential diamond forming events, followed by Mesoarchean mantle roots, Neoarchean mantle roots, and Paleoproterozoic mantle roots. This idea is only valid if the crust and the mantle are coupled, for example, Mesoarchean continental crust overlies Mesoarchean mantle lithospheric roots. In general, this is usually true although there are certainly exceptions (Pearson, 1999a,b). For example, at the Fort à la Corne kimberlite field in central Saskatchewan, Paleoproterozoic rocks of the Trans Hudson Orogen overlie Archean rocks of the Sask craton (Ashton et al., 2005). The lamproite-hosted Argyle diamond mine (Australia), emplaced into the Paleoproterozoic Halls Creek mobile belt in a seemingly "off craton" environment, is however, underlain by Archean age lithospheric mantle (Luguet et al., 2005).

Eclogite paragenesis diamonds (E-type), on the basis of radiometric age determinations on inclusions in diamond, formed from 2.9 to 1.0 Ga. Detailed major-, trace-element and isotopic studies on kimberlite-derived eclogite xenoliths (Jacob, 2004; Jacob et al., 2005) and stable isotopic data on eclogite paragenesis diamonds (e.g., Gurney et al., 2005) indicate a subduction origin. Hence, a second important criteria at continental scale is Archean crust with Archean mantle roots that have experienced (distal) Archean to Mesoproterozoic subduction processes.

It has been previously suggested that certain geological events destroyed diamond-bearing lithospheric mantle roots. Helmstaedt (2003) argued that plume heads associated with basic igneous events, magmatic underplating associated with voluminous basic igneous events, and large scale plate collisions are all processes that destroy lithospheric mantle roots and their contained diamonds. The study of Read et al. (2004) in the Minais Gerais State, Brazil definitively illustrates that over a short time period (<10 m.y.) approximately 75 km of diamond-bearing lithospheric mantle root was removed or destroyed, related to the eruption of voluminous alkaline basic kamafugitic magma. However, the Premier diamond mine lies within the Bushveld complex (and has sampled mantle xenoliths that represent underplated basic magma; Hoal, 2003), which would suggest that voluminous basic magmatism and underplating is not always diamond unfriendly. Furthermore, the observation that the Venetia and The Oaks diamond mines lie in the Limpopo collision belt between the Kaapvaal and Zimbabwe cratons does not support the notion that large-scale plate collisions destroys lithospheric mantle roots and their contained diamonds.

Distribution and Structural Control of Kimberlite Fields within Cratons

The distribution of kimberlite fields within a prospective (cratonic) region of old crust/mantle typically follows a broad linear pattern. This is observed, for example, in Yakutia (Bardet, 1965; Kaminsky et al., 1995). In the Slave craton the kimberlite fields follow a northwest trend from Kennedy Lake through Lac de Gras to northern Napatulik Lake (Fig. 8). Similarly, kimberlite fields in the Superior craton of Ontario follow a broad northwest trend (Fig. 8). Kaminsky et al. (1995) suggest these broad linear zones are areas of long-lived deep-seated major faults. This hypothesis may be applicable to kimberlite fields in Ontario (Fig. 8) with respect to the Lake Timiskaming structural zone (often incorrectly referred to as a rift). In contrast, in the Slave craton the kimberlite fields may be located above or proximal to the leading edge of the Mesoarchean central Slave basement complex (as defined by Bleeker et al., 1999).

The idea of pronounced structural control on the distribution of kimberlite fields, clusters, and pipes is not new (see Marsh, 1973; Nixon 1973; Haggerty, 1982; Mitchell, 1986, Coopersmith, 1993; White et al., 1995; and Jelsma et al., 2004). Structures such as deep fractures or shear zones (cryptic lineaments) are often apparent on aeromagnetic maps and digital elevation models according to Kaminsky et al. (1995) and Jelsma et al. (2004), with most kimberlite fields occurring at the intersections of two or more of these major structures. Jelsma et al. (2004) suggest these fractures or zones extend into the mantle and are re-activated over time. In Ontario, the Timiskaming, Kirkland Lake, and Attawapiskat fields are all at the intersection of major easterly trending structures with the northwest-trending Lake Timiskaming structural zone. Within a kimberlite field, some kimberlite clusters of small aerial extent contain multiple kimberlite-hosted diamond mines. This is certainly true for the five Kimberley area mines within the Kimberley cluster of South Africa and the twelve kimberlites (in two distinct clusters)

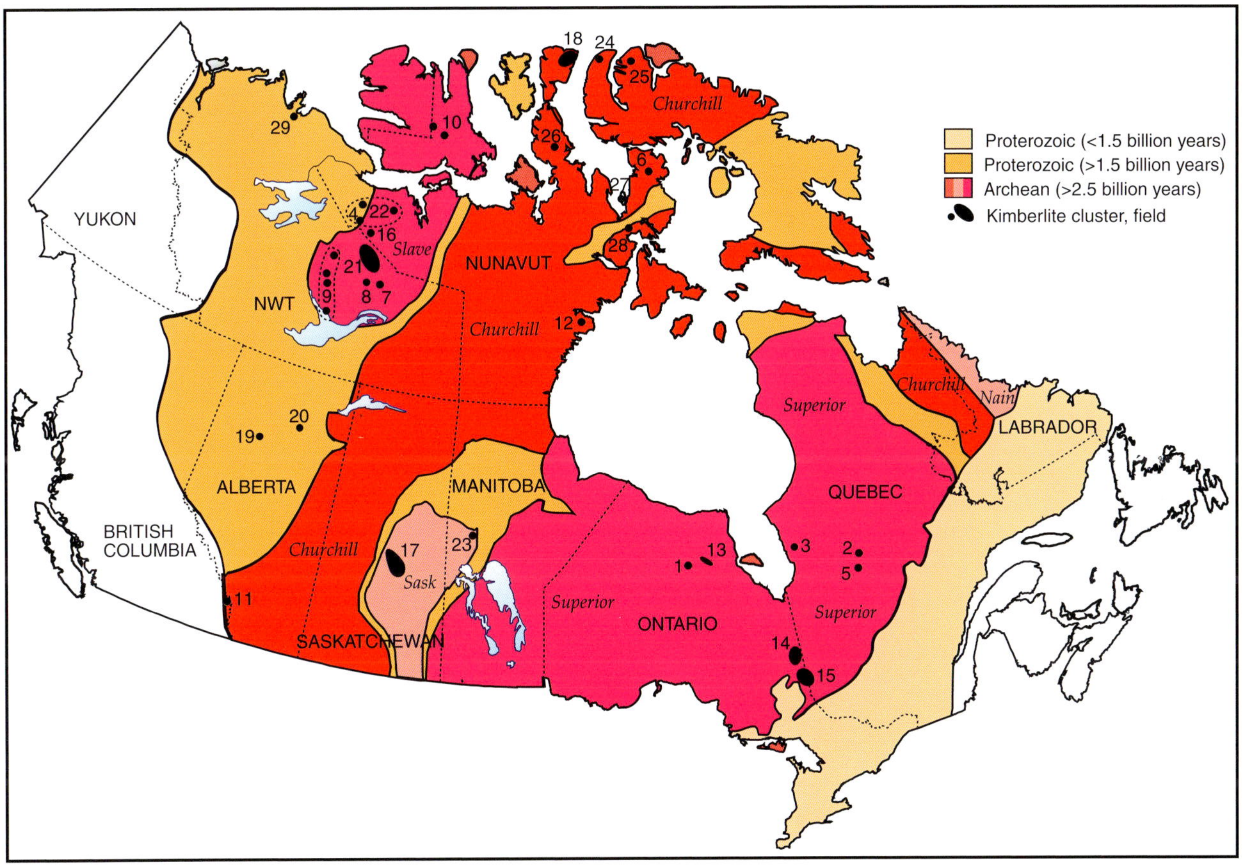

Figure 8. Distribution of Canadian kimberlites fields and clusters with respect to Proterozoic and Archean terranes (younger sedimentary cover sequences removed). The Archean Slave, Superior, Sask, Nain and Churchill cratonic blocks are shown as separate entities and labelled in italics. Localities as follows: (1) Kyle Lake cluster, Ontario; (2) Renard cluster - north Otish Mountains, Quebec; (3) Wemindji sills, Quebec; (4) Anuri, Nunavut; (5) Lac Beaver - south Otish Mountains, Quebec; (6) Aviat cluster, Melville Peninsula, Nunavut; (7) southeast Slave field, Gahcho Kué cluster, NWT; (8) southeast Slave field, Snap Lake area, NWT; (9) southwest Slave field, including the Drybones Bay and Upper Carp Lake clusters, NWT; (10) Victoria Island field (with four distinct clusters), Nunavut and NWT; (11) Crossing Creek cluster, southeast British Columbia; (12) Rankin Inlet field, Nunavut; (13) Attawapiskat field, Ontario; (14) Kirkland Lake field, Ontario; (15) Lake Timiskaming field, Ontario and Quebec; (16) Jericho cluster, Nunavut; (17) Fort à la Corne field (with six distinct clusters), Saskatchewan; (18) Somerset Island field, Nunavut; (19) Buffalo Head Hills field, Alberta; (20) Birch Mountains cluster, Alberta; (21) Lac de Gras field, NWT; (22) Coronation Gulf field, Nunavut; (23) Snow Lake - Wekusko, Manitoba; (24) Brodeur Peninsula cluster, Nunavut; (25) Baffin Island, Nunavut; (26) Boothia Peninsula, Nunavut; (27) Wales Island, Nunavut; (28) Repulse Bay cluster, Nunavut; (29) Darnley Bay cluster, NWT. Adapted from Kjarsgaard and Levinson (2002).

within the Lac de Gras field associated with the Ekati and Diavik mines.

Kimberlite Fields of Canada

Age and Geographic Distribution

Mitchell (1986) defined kimberlites as belonging to clusters, fields, and provinces. Individual kimberlite bodies grouped together are termed clusters (e.g., the Jericho cluster, Nunavut). A number of clusters within an area constitute a kimberlite field (e.g., the Lac de Gras field, Northwest Territories). Several kimberlite fields comprise a kimberlite province (e.g., the Slave kimberlite province). Kimberlites clusters usually form over time intervals of ~1 to 5 m.y., whereas kimberlites fields usually form over time intervals of ~10 to 30 m.y. (Fig. 9). Kimberlites within a kimberlite province may have a wide span of emplacement ages. In the Slave kimberlite province there are six distinct kimberlite emplacement events (Neoproterozoic, Cambrian, Siluro-Ordovician, Permian, Jurassic, Cretaceous-Paleogene; Fig. 9) with the oldest kimberlite ~613 Ma in age and the youngest kimberlite ~45 Ma in age.

The oldest known bona fide kimberlites in Canada are the Mesoproterozoic ca. 1100 Ma occurrences in the Kyle Lake cluster of northern Ontario (Fig. 8, #1). Neoproterozoic age ca. 630 Ma age kimberlites are known in central Quebec from the northern Otish Mountains area (Renard cluster; Fig. 8, #2) and eastern James Bay (Wemindji sills; Fig. 8, #3). The Renard cluster is currently at the bulk sampling stage of exploration. In the northern Slave province (in Nunavut) there is at least one Neoproterozoic age kimberlite (Anuri, ca. 613 Ma; Fig. 8, #4). Cambrian kimberlites are also observed in central Quebec in the southern Otish Mountains area (Lac Beaver, ca. 550 Ma; Fig. 8, #5). On Melville Peninsula within the Churchill province is the Cambrian age ca. 546 Ma Aviat kimberlite (Fig. 8, #6). The southeast Slave field (NWT) contains two economically important clusters of Cambrian age kimberlites, the Gahcho Kué pipes (5034 pipe, 542 Ma; Fig. 8, #7) and the Snap Lake sill (ca. 535 Ma; Fig. 8, #8). The southwest Slave field (in the NWT) contains a number of Siluro-Ordovician kimberlites dated at 463 to 435 Ma (Fig. 8, #9). Permian age kimberlites are known from Victoria Island (Fig. 8, #10).

Consistent with worldwide kimberlite emplacement age data, the majority of Canadian kimberlites are Mesozoic in age (Fig. 9). Triassic age kimberlites are rare in Canada, restricted to the Crossing Creek cluster in south-eastern British Columbia (Fig. 8, #11) and the Rankin Inlet area of Nunavut (Fig. 8, #12). There are Jurassic age kimberlites in four fields

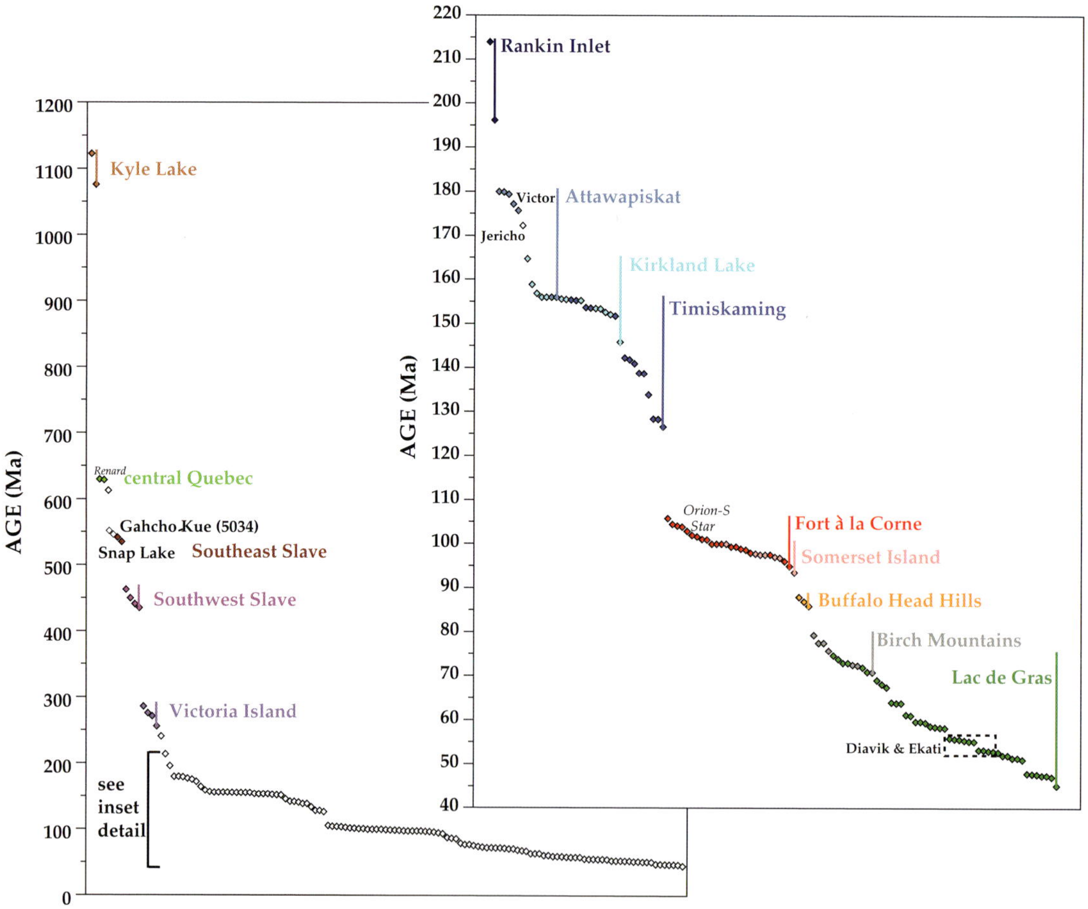

FIGURE 9. Detailed emplacement ages of Canadian kimberlites. Individual fields are colour coded; vertical bars denote the time span of volcanism within that kimberlite field. Data sources: Heaman and Kjarsgaard (2000); Heaman et al. (2003, 2004) and references therein; Kjarsgaard (unpublished data).

in east-central Canada, including the Rankin Inlet, Nunavut, and Attawapiskat, Kirkland Lake, and Lake Timiskaming fields, Ontario (Fig. 8, #12, 13, 14, 15, respectively). The Attawapiskat field hosts the Victor diamond mine. In the northern Slave province, Nunavut, there are also Jurassic age kimberlites, including the 172 Ma Jericho diamond mine (Fig. 8, #16). Cretaceous age kimberlites occur in east-central, central, and western Canada. The youngest kimberlites in the Lake Timiskaming field (ca. 142–126 Ma) are of Early Cretaceous age (Fig. 8 #15). Mid-Cretaceous age kimberlites occur at the Fort à la Corne, Somerset Island, and Buffalo Head Hills kimberlite fields (Fig. 8, #17, 18, 19, respectively). Late Cretaceous age kimberlites are known from the Birch Mountains (Alberta; Fig. 8, #20) and the Lac de Gras (NWT; Fig. 8, #21) kimberlite fields. The youngest kimberlites in Canada are of Paleogene age and occur in the Lac de Gras field where all the kimberlites of economic interest (Diavik and Ekati mine kimberlites) were emplaced within a narrow timespan of 56 to 53 Ma (Kjarsgaard et al., 2002).

Distribution with Respect to Geological Terranes

The majority of kimberlites in Canada (>340) are located in the Archean Slave craton, with kimberlite fields in the southeast (e.g., Snap Lake, Gahcho Kué), southwest (e.g., Carp Lake, Dry Bones Bay), central (Lac de Gras), and north Slave (Coronation Gulf; Fig. 8, #22), and on Victoria Island. In the Archean Superior province there are approximately 75 known kimberlites from the Kyle Lake, Attawapiskat, Kirkland Lake, Lake Timiskaming, Wemindji, and Otish Mountains areas (Fig. 8). The Archean Sask craton contains approximately 75 kimberlites, located in the Fort à la Corne and Candle Lake areas of Saskatchewan, and at Wekusko-Snow Lake in Manitoba (Fig. 8, #23). The Churchill province (Archean-Paleoproterozoic) hosts approximately 95 kimberlites in clusters or fields at Somerset Island, Brodeur Peninsula (Fig. 8, #24), Baffin Island (Fig. 8, #25), Boothia Peninsula (Fig. 8, #26), Melville Peninsula, Wales Island (Fig 8, #27), Repulse Bay (Fig. 8, #28), and Rankin Inlet.

More than 45 kimberlites have been discovered in northern Alberta, the majority in the Buffalo Head Hills field which is underlain by the Paleoproterozoic (2.0–2.4 Ga) Buffalo Head terrane. An additional 8 kimberlites are known from the Birch Mountains field which is underlain by the Paleoproterozoic (2.0–1.8 Ga) Taltson magmatic arc. The five kimberlites in the Crossing Creek area of southeastern B.C. are rootless, that is, they are within a thrust slice and their relationship to basement is unclear. However geological interpretations of Ross et al. (1991) suggest this area is underlain by the Hearn domain of the Churchill province. The basement terrain to the kimberlites at Darnley Bay (NT; Fig. 8, #29) is unknown but assumed to be Proterozoic.

Local Geological Setting

Kimberlites occur in clusters and fields; it is rare to find an isolated kimberlite. Although kimberlite fields are typically associated with the intersections of major structures, in detail individual kimberlites are often located on second- or third-order splays off one of the major structures (Coopersmith, 1993). In Lac de Gras some kimberlites are found at weak points in the upper crust, such as geological contacts, for example, at granite-greywacke contacts or at the margins of diabase dyke-host rock contacts (Kjarsgaard et al., 2002; Lockhart et al., 2004). In the Lac de Gras field, kimberlites of a limited age range are aligned along specific structural trends (Kjarsgaard et al., 2002). For example the economically viable (53–56 Ma) kimberlites are all within a corridor coincident with the 010° trending Lac de Gras diabase dyke swarm (Wilkinson et al., 2001; Stubley, 2004). In more detail, Lockhart et al. (2004) noted a rotation of structural alignment trends of kimberlites in the Lac de Gras field with time, starting with an ~015° trend at 55 Ma, then ~025° at 53 Ma, and ~045° trend at 48 Ma. These data are consistent with different second- or third-order crustal structures being re-activated over a 7 m.y. time period (related to far-field Cordilleran tectonics).

Deposit Scale Attributes of Kimberlites

Kimberlite Mineralogy

Kimberlite exhibits a distinctive inequigranular texture due to the presence of macrocrysts (large 0.5–10 mm rounded to anhedral crystals, typically olivine) set in a fine-grained matrix (Mitchell, 1986). Macrocryst is a non-genetic term for large crystals that could be xenocrysts, or megacrysts. Xenocrysts observed in kimberlite are composed of the following minerals: forsterite, Cr-pyrope garnet, chromian spinels, Cr-diopside, bronzite, amphibole and phlogopite from disaggregated mantle peridotite; almandine-pyrope garnet and omphacite from disaggregated mantle eclogite. Megacrysts observed in kimberlite include Ti-Cr-pyrope, Mg-ilmenite, Cr-diopside, phlogopite, enstatite, and zircon, and are thought to be high-pressure cognate phases of the kimberlite (Mitchell, 1986; Nowell et al., 2004). Primary kimberlite phenocryst/microphenocryst phases include forsterite, spinel, and more rarely phlogopite. The kimberlite groundmass consists of variable proportions of forsterite, spinels, perovskite, monticellite, apatite, phlogopite-kinoshitalite mica, carbonates, and ilmenite, although not all of these groundmass phases occur in any kimberlite. Auto-deuteric serpentine ± diopside ± calcite ± magnetite alteration of macrocrysts and primary mineral phases is common in these rocks, and typically occurs at subsolidus (<800°C) temperatures (Wilson et al., 2007). Deuteric microcrystalline diopside with serpentine is common in crustally contaminated massive volcaniclastic kimberlite in pipes. Stripp et al. (2006) utilized thermodynamic criteria to calculate that diopside + lizardite formed at <380°C and low pCO_2, that is, subsolidus conditions.

Kimberlite Geochemistry

Kimberlites have a characteristic geochemical signature, being rich in the incompatible elements Sr, Ba, LREE (La, Ce, Sm, Nd), Nb, Ta, Zr, P, Th, and U ('alkaline signature'), as well as having high concentrations of the first-order transition elements Mg, Ni, Cr, and Co ('ultramafic signature'). There are essentially no other rocks that have this distinctive alkaline and ultramafic geochemical signature, with the exception of a few rare magnesio-carbonatites. Some ultramafic lamprophyres (e.g., alnöite, aillikite) may also display a 'similar' geochemistry but typically these rocks have higher contents of TiO_2, Al_2O_3, Na_2O and K_2O, and can also be identified petrographically (e.g., see Tappe et al., 2005). It has been recently demonstrated by Grunsky and Kjarsgaard (2007) that it is feasible to use whole rock major-and trace-element geochemistry to distinguish and separate individual eruptive phases within a kimberlite body.

Alteration Mineralogy

At the time of intrusion or eruption, some of the kimberlite macrocryst and primary groundmass minerals may be partially or completely replaced by deuteric serpentine ± diopside ± calcite ± magnetite (see section above on kimberlite mineralogy). Furthermore, the porous nature of volcaniclastic kimberlite makes these rocks highly susceptible to post-emplacement alteration by weathering processes that cause the development clay-rich kimberlite, commonly referred to in older South African literature as "yellow" or "blue" ground.

Kimberlite Nomenclature

Kimberlites worldwide typically form small (almost always <1000 m diameter; ~80 ha) bodies (e.g., Dawson, 1980; Mitchell, 1986). The variety of shapes and sizes of kimberlites in Canada is exceptional and many of the Canadian examples do not resemble the classic South African model of a kimberlite pipe (Hawthorne, 1975). This model (Fig. 10) was constructed by utilizing the upper portion of a kimberlite pipe from Orapa, Botswana (to which a tuff ring was added) and the lower portion of a kimberlite pipe from the Kimberley area of South Africa (where 1400 m of erosion was interpreted to have occurred; Hawthorne, 1975), and is considered to represent a composite model (Kjarsgaard, 2003; Gurney et al. 2005). More recently, the erosion at Kimberley has been estimated at <850 m (Hansom et al., 2006).

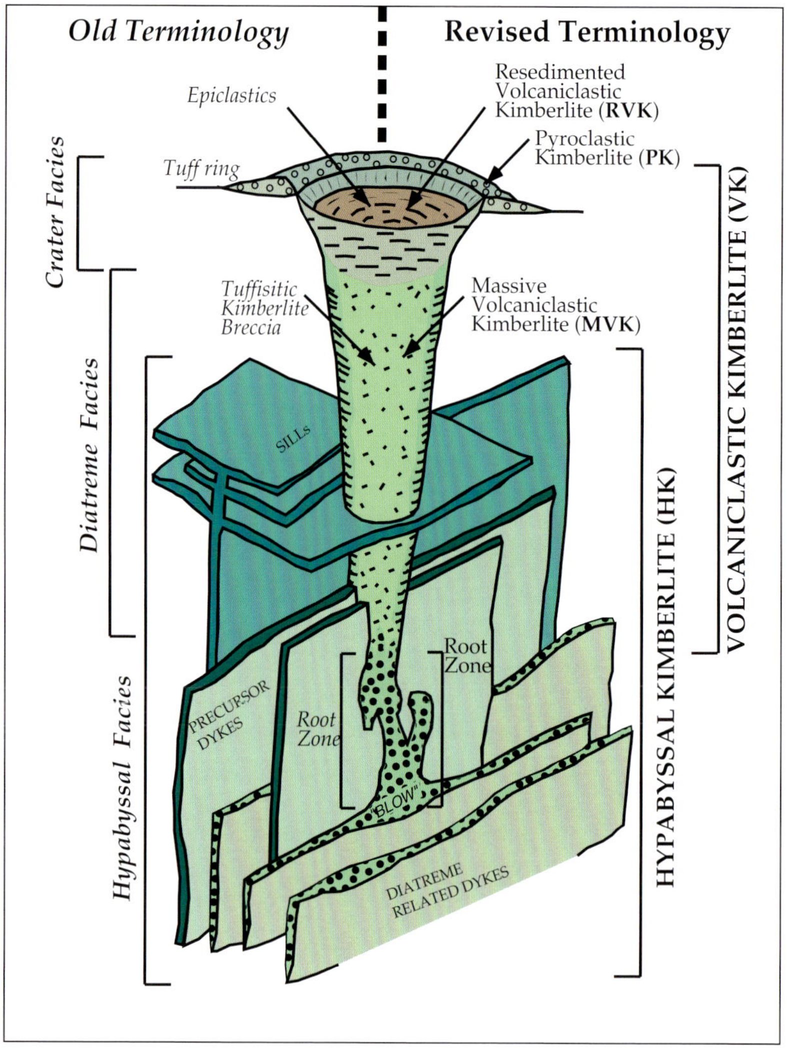

FIGURE 10. The classic South African model of a kimberlite pipe with old nomenclature (left side of figure) and a simpler, revised two-fold nomenclature system (right side of figure) to describe rocks from kimberlite magmatic systems (Mitchell, 1995; Kjarsgaard, 2003; Sparks et al., 2006). PK = pyroclastic kimberlite; RVK = resedimented volcaniclastic kimberlite; MVK = massive volcaniclastic kimberlite; HK = hypabyssal kimberlite. Figure modified after Kjarsgaard, 2003).

Rock types in the classic South African model were assigned to different zones or facies (e.g., Clement, 1975; Dawson, 1980; Clement and Skinner, 1985; Mitchell, 1986, 1995) which included crater facies, diatreme facies, and hypabyssal facies (Fig. 10). The use of the term facies (*the aspect, appearance, and characteristics of a rock, usually reflecting the* ***conditions*** *of its origin*), however, is incorrectly applied because the conditions of origin of diatreme facies and crater facies rocks is highly controversial (compare Mitchell, 1986 with Clement and Reid, 1989, with Lorenz et al., 1999). A simpler, non-genetic, two-fold nomenclature system (Mitchell, 1995; Kjarsgaard, 2003; Sparks et al., 2006) to describe rocks from kimberlite magmatic systems is preferred: volcaniclastic kimberlite (VK), or fragmental rocks, and; hypabyssal kimberlite (HK), or, non-fragmental rocks. The volcaniclastic kimberlites can be subdivided further into pyroclastic kimberlite (PK), re-sedimented volcaniclastic kimberlite (RVK), and massive volcaniclastic kimberlite (MVK). Note also the existence of different sub-types of PK, RVK, and MVK; different RVK subtypes include grain flow deposits and mass flow slumps. The revised terminology is shown in Figure 10 (right side), where it can be compared to the old terminology

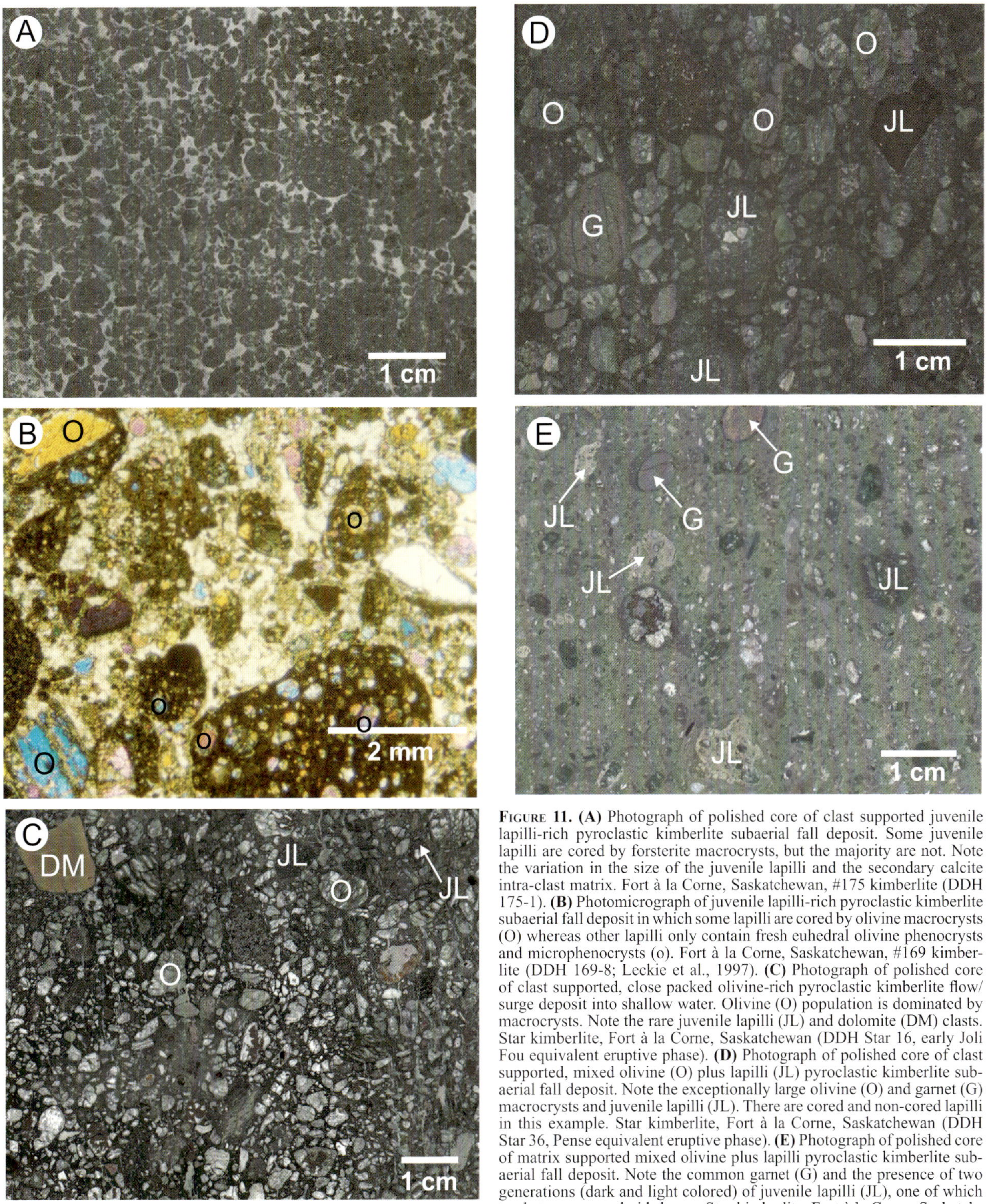

FIGURE 11. **(A)** Photograph of polished core of clast supported juvenile lapilli-rich pyroclastic kimberlite subaerial fall deposit. Some juvenile lapilli are cored by forsterite macrocrysts, but the majority are not. Note the variation in the size of the juvenile lapilli and the secondary calcite intra-clast matrix. Fort à la Corne, Saskatchewan, #175 kimberlite (DDH 175-1). **(B)** Photomicrograph of juvenile lapilli-rich pyroclastic kimberlite subaerial fall deposit in which some lapilli are cored by olivine macrocrysts (O) whereas other lapilli only contain fresh euhedral olivine phenocrysts and microphenocrysts (o). Fort à la Corne, Saskatchewan, #169 kimberlite (DDH 169-8; Leckie et al., 1997). **(C)** Photograph of polished core of clast supported, close packed olivine-rich pyroclastic kimberlite flow/surge deposit into shallow water. Olivine (O) population is dominated by macrocrysts. Note the rare juvenile lapilli (JL) and dolomite (DM) clasts. Star kimberlite, Fort à la Corne, Saskatchewan (DDH Star 16, early Joli Fou equivalent eruptive phase). **(D)** Photograph of polished core of clast supported, mixed olivine (O) plus lapilli (JL) pyroclastic kimberlite subaerial fall deposit. Note the exceptionally large olivine (O) and garnet (G) macrocrysts and juvenile lapilli (JL). There are cored and non-cored lapilli in this example. Star kimberlite, Fort à la Corne, Saskatchewan (DDH Star 36, Pense equivalent eruptive phase). **(E)** Photograph of polished core of matrix supported mixed olivine plus lapilli pyroclastic kimberlite subaerial fall deposit. Note the common garnet (G) and the presence of two generations (dark and light colored) of juvenile lapilli (JL), one of which tends to more ameboid shapes. Star kimberlite, Fort à la Corne, Saskatchewan (DDH Star 39, Cantuar 'north' equivalent eruptive phase).

(left side of Fig. 10). The revised terminology allows for a better understanding of kimberlites that do not resemble classic South African model kimberlites as more recently described by Field et al. (1997), for example, many kimberlite pipes in Canada and in other parts of the world.

Kimberlite Textures

Kimberlites are quite variable in terms of texture, although most textural features that are observed are common to other sub-volcanic systems and volcanic or volcano-sedimentary sequences. The description of these rocks, therefore, uses

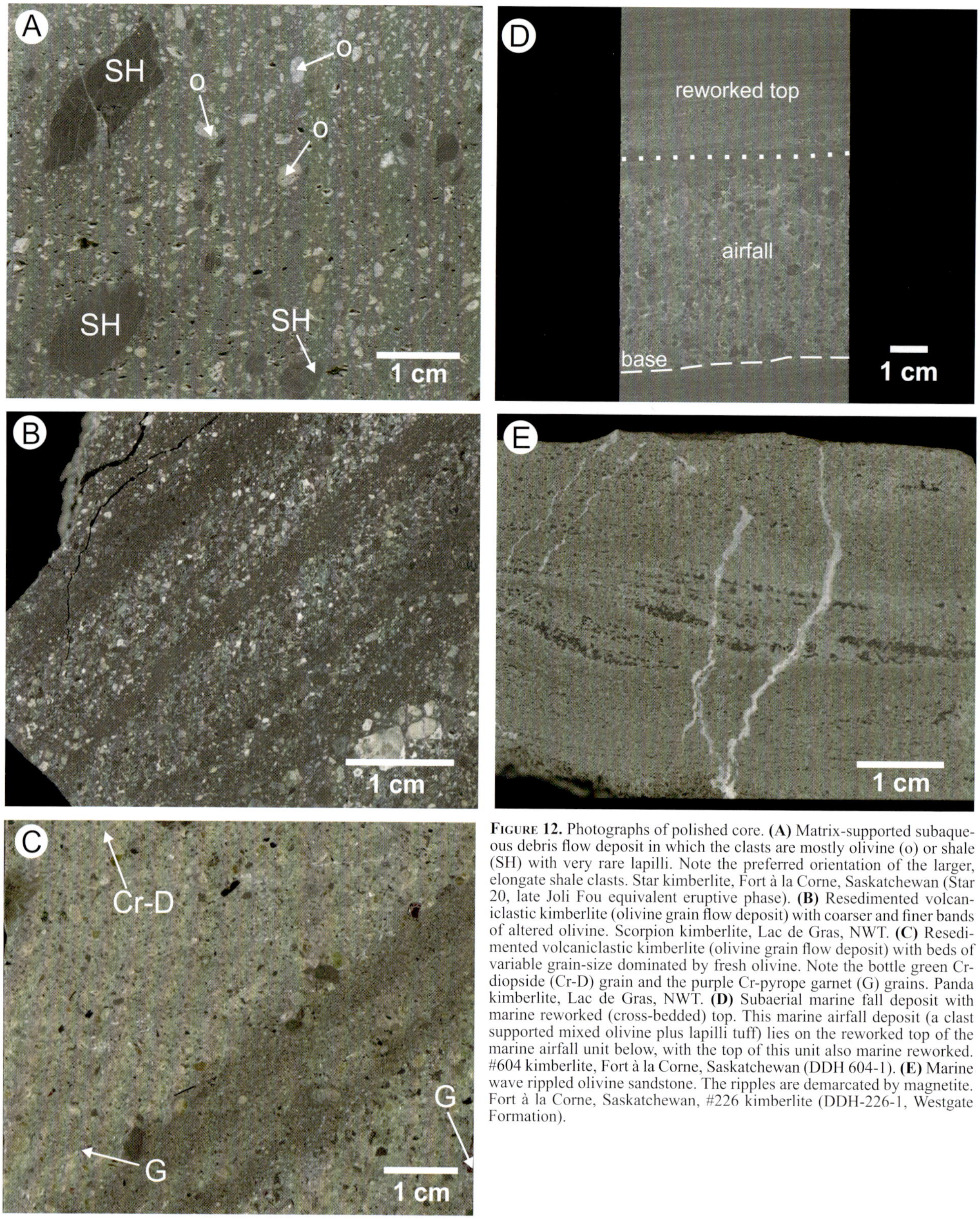

FIGURE 12. Photographs of polished core. **(A)** Matrix-supported subaqueous debris flow deposit in which the clasts are mostly olivine (o) or shale (SH) with very rare lapilli. Note the preferred orientation of the larger, elongate shale clasts. Star kimberlite, Fort à la Corne, Saskatchewan (Star 20, late Joli Fou equivalent eruptive phase). **(B)** Resedimented volcaniclastic kimberlite (olivine grain flow deposit) with coarser and finer bands of altered olivine. Scorpion kimberlite, Lac de Gras, NWT. **(C)** Resedimented volcaniclastic kimberlite (olivine grain flow deposit) with beds of variable grain-size dominated by fresh olivine. Note the bottle green Cr-diopside (Cr-D) grain and the purple Cr-pyrope garnet (G) grains. Panda kimberlite, Lac de Gras, NWT. **(D)** Subaerial marine fall deposit with marine reworked (cross-bedded) top. This marine airfall deposit (a clast supported mixed olivine plus lapilli tuff) lies on the reworked top of the marine airfall unit below, with the top of this unit also marine reworked. #604 kimberlite, Fort à la Corne, Saskatchewan (DDH 604-1). **(E)** Marine wave rippled olivine sandstone. The ripples are demarcated by magnetite. Fort à la Corne, Saskatchewan, #226 kimberlite (DDH-226-1, Westgate Formation).

standard volcanic or sedimentary terminology. However hypabyssal (sub-volcanic) kimberlite lithologies with numerous large forsterite crystals are not termed porphyritic, but macrocrystic as many of the large forsterite crystals are xenocrysts and not phenocrysts.

Volcaniclastic (i.e., fragmental) kimberlite (VK) is exceptionally variable in texture and is subdivided into pyroclastic kimberlite (PK), massive volcaniclastic kimberlite (MVK), and resedimented volcaniclastic kimberlite (RVK). Pyroclastic kimberlite is variable in constituent clast components, grain size, and the degree of clast support. These rocks can be clast-supported and dominated by kimberlite juvenile

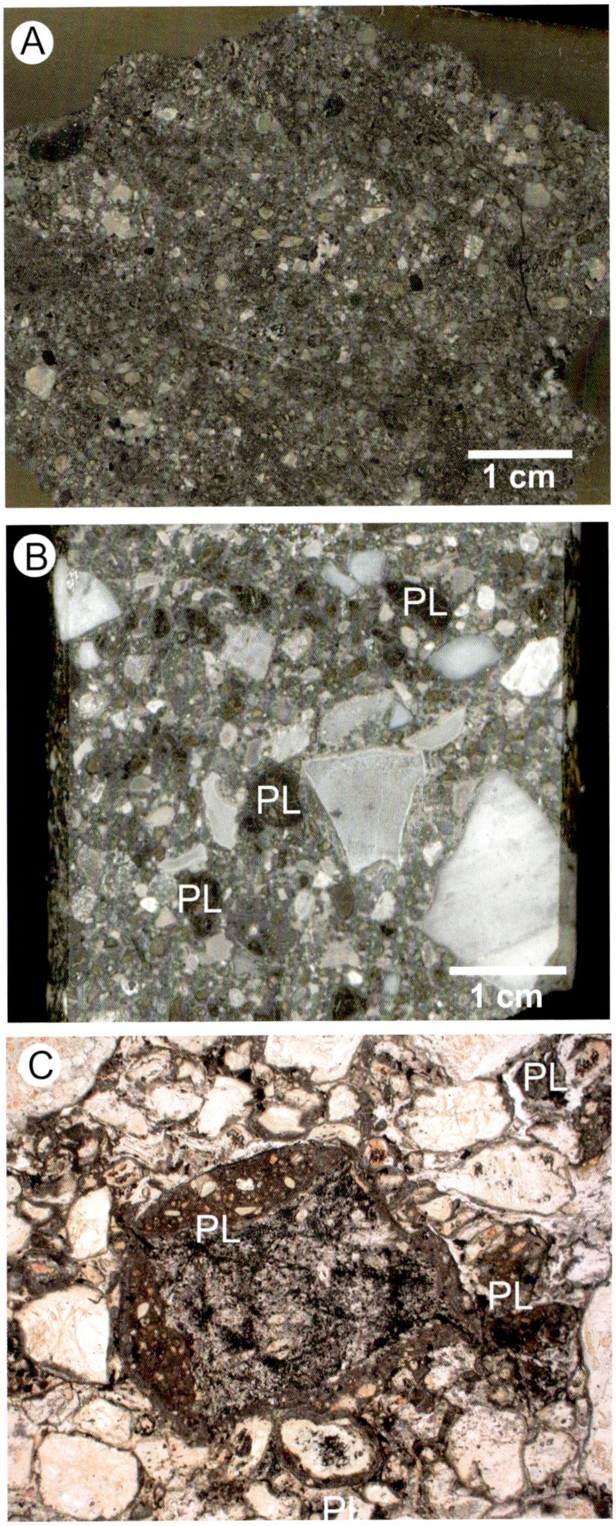

Figure 13. (A) Photograph of polished slab of massive volcaniclastic kimberlite which is relatively poor in pelletal lapilli. Fox kimberlite, Lac de Gras, NWT. **(B)** Photograph of polished core of massive volcaniclastic kimberlite breccia. Pelletal lapilli (PL) of variable size are common. Eastern "TKB". #95-2 kimberlite, New Liskeard, Ontario. **(C)** Photomicrograph of pelletal lapilli (PL) in massive volcaniclastic kimberlite breccia from the Eastern "TKB". #95-2 kimberlite, New Liskeard, Ontario.

lapilli (Fig. 11a,b), or forsterite crystals (Fig. 11c), or mixed clast assemblages of kimberlite juvenile lapilli and forsterite crystals (Fig. 11d). Matrix-supported pyroclastic kimberlite air fall deposits commonly have sub-equal clast proportions of juvenile lapilli and olivine crystals (Fig. 11e), although finer-grained matrix-supported olivine crystal tuffs are known. Base surge deposits and a variety of deposit types from primary subaerial and subaqueous (Fig. 12a) debris flows are associated with kimberlite volcanic sequences (e.g., Zonneveld et al., 2004, 2006; Kjarsgaard et al., 2005, 2006; Harvey et al., 2006).

The resedimentation of volcaniclastic kimberlite tephra results in the formation of volcano-sedimentary deposits (RVK). In the Lac de Gras kimberlite field, many of the kimberlite pipes are in-filled with forsterite-rich grain-flow (talus fan) deposits (Fig. 12b,c) and slumps. Slump deposits in the Fort à la Corne field occur as both extra-crater and, more rarely, intra-crater kimberlite deposits (Zonneveld et al., 2004, 2006; Kjarsgaard et al., 2006). Marine (Fig. 12d, e) and fluvial reworked kimberlite deposits are also common in Fort à la Corne (Leckie et al., 1997, Nixon and Leahy, 1997, Zonneveld et al., 2004, 2006).

Massive volcaniclastic kimberlite (MVK), although heterogeneous at the mm to cm scale (Fig. 13a), is often relatively homogenous at the 10 to 100 cm scale. These massive kimberlite rocks, which can occupy much of the pipe zone in the classic South African model are commonly breccias (Fig. 13b). A hallmark of these MVK rocks is the occurrence of pelletal lapilli (Fig. 13c; compare with Fig. 11b) which are discrete, spherical to elliptical lapilli-sized (2–64 mm) clasts consisting of fine-grained primary kimberlite magma (Clement, 1982; Mitchell, 1986, 1995). The centres of the lapilli are typically cored by a forsterite grain that is mantled by fine-grained to cryptocrystalline kimberlite groundmass minerals (e.g., spinel, perovskite, calcite, apatite), which are commonly oriented around the core, giving a weak to well-developed concentric structure. The replacement of the MVK groundmass by serpentine, calcite and microcrystalline diopside at subsolidus temperatures (Sparks et al., 2006, Stripp et al., 2006) leads to the development of the distinctive (but inappropriately named) tuffisitic kimberlite breccia (TKB) first described at numerous southern African kimberlite occurrences.

Hypabyssal kimberlite (HK) commonly exhibits a distinctive inequigranular texture due to the presence of large, rounded, anhedral macrocrysts (i.e., megacrysts and xenocrysts) plus euhedral to subhedral phenocrysts set in a finer-grained groundmass (Fig. 14a,b). HK breccias contain abundant country rock clasts (Fig. 14c). The HK groundmass is either uniform in appearance or contains segregations of auto-deuteric calcite +/- serpentine. Mitchell (1986, 1997) provides numerous excellent photomicrographs of uniform and segregation textured HK. Flow banding (or flowage differentiation) is a well-documented igneous process in basic sills and dykes (Bhattacharji and Smith, 1964; Drever and Johnston, 1966) that is also observed in kimberlites (Figs. 14d, 15a). Crystal settling appears to be a rare process (Fig. 15b) in kimberlite magmatic systems, but has been observed at the Benfontein sill in South Africa (Dawson and Hawthorne,

FIGURE 14. **(A)** Photograph of polished slab of hypabyssal macrocrystic kimberlite. Note the fresh olivine macrocrysts (O) and the secondary calcite veinlets (white, centre field of view). K4 kimberlite, Somerset Island, Nunavut. **(B)** Photomicrograph of hypabyssal macrocrystic kimberlite with fresh anhedral to subhedral olivine macrocrysts (O), subhedral to euhedral olivine phenocrysts (o), rare phlogopite (p) and spinel phenocrysts. Groundmass is dominated by calcite (tabular and subhedral). Brown turbid areas are groundmass (spinel + serpentine + carbonate + apatite) not resolvable at the scale of the photomicrograph. Jos kimberlite dyke, Somerset Island, Nunavut. **(C)** Photograph of hypabyssal kimberlite breccia outcrop near the margin of the intrusion. Country rock clasts are dominated by the local host rocks (limestone and dolostone). K-20 kimberlite, Somerset Island, Nunavut. **(D)** Photograph of polished slab of flow banded hypabyssal macrocrystic kimberlite, with alternating bands of coarser and finer olivine grains. Peddie kimberlite, New Liskeard, Ontario. **(E)** Photomicrograph of aphanitic hypabyssal kimberlite. Note that with the exception of rare, larger country rock clasts, the sample could easily be confused as a basalt from a purely macroscopic perspective, due to the absence of any olivine macrocrysts and the abundant olivine (o) phenocrysts and microphenocrysts. McLean kimberlite, New Liskeard, Ontario. **(F)** Photomicrograph of aphanitic hypabyssal kimberlite. Note the sample consists of essentially sub-equal amounts of euhedral to subhedral olivine (o) plus minor spinel (s) phenocrysts and groundmass (spinel + serpentine + carbonate + apatite) not resolvable at the scale of the photomicrograph. Peuyuk kimberlite, sub-phase of intrusion "A", Somerset Island, Nunavut.

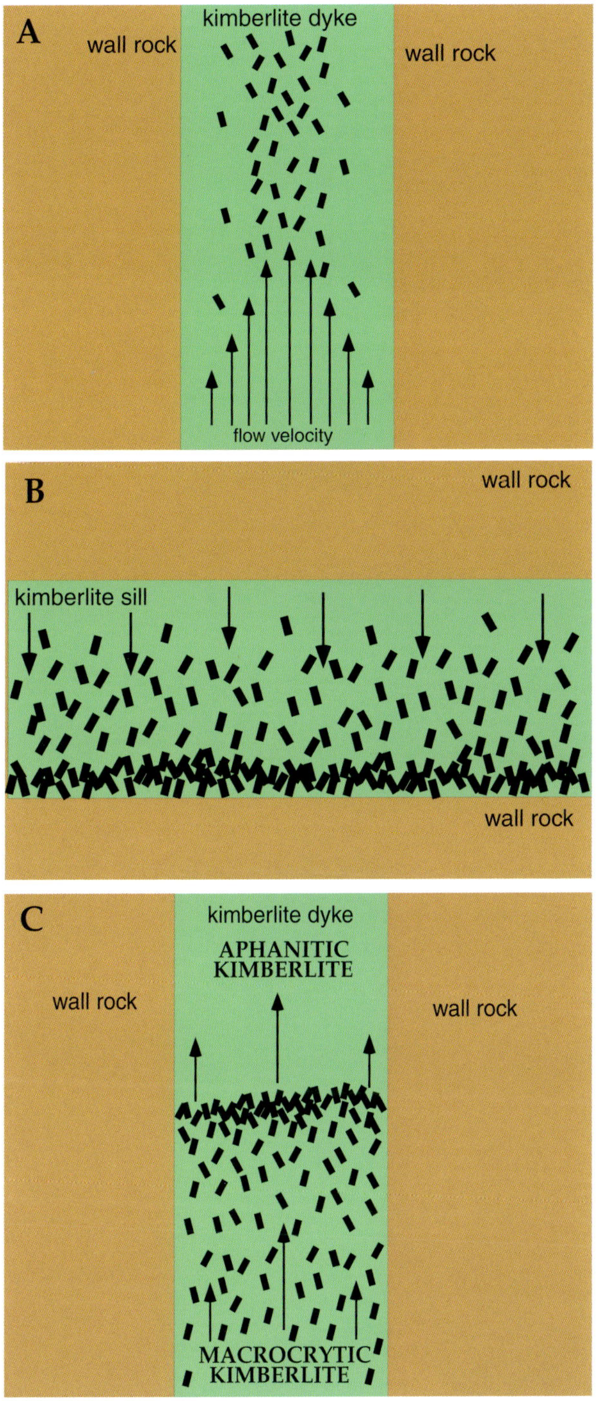

FIGURE 15. **(A)** Schematic representation of flow banding in a kimberlite dyke, with increased flow rates at the centre due to dispersive shear pressure. **(B)** Schematic representation of gravitational crystal settling (magmatic sedimentation) in a kimberlite sill. **(C)** Schematic representation of filter pressing in a kimberlite dyke (all figures adapted from Kjarsgaard, 2003).

1973). Filter pressing in dykes, sills, or plugs (Fig. 15c) and flowage differentiation processes can lead to the formation of aphanitic kimberlite as observed at the McLean and Peuyuk kimberlites (Fig. 14e, f) and the Jericho kimberlite (Price et al., 2000).

Architecture and Morphology of Kimberlites

Shape and area of kimberlite bodies: Kimberlite pyroclastic rocks and volcaniclastic deposits that form volcanic edifices (tephra cones, tuff cones, tuff rings) are not common due to their low preservation (typically <1 m.y.), but do occur in Mali, Tanzania, and Canada (Kjarsgaard, 2007). The best-preserved volcanic edifices in the world are found at the Fort à la Corne kimberlite field, Saskatchewan, and at the Buffalo Head Hills kimberlite field, Alberta. In Saskatchewan, individual volcanic phases have up to ~100 m of preserved relief above the paleo-eruptive surface and are up to ~900 m in diameter (Kjarsgaard, 2007). Fort à la Corne kimberlites with multiple feeder vents have coalesced tephra cones and/or rings, and can form much larger bodies up to 2.5 km diameter (Fig. 16). These feeder vents are quite variable in their morphology, ranging from narrow pipe-like bodies to wider flared vents (Fig. 16), and can be infilled by MVK, VK, and PK (Kjarsgaard et al., 2004, 2006; Harvey et al, 2006). Kimberlites from the Buffalo Head Hills field (Carlson et al., 1999) have some basic similarities to those of Fort à la Corne with regard to kimberlite body geometry, but in detail some slightly different pyroclastic rock types are present (e.g., accretionary lapilli tuffs; Boyer, 2005). Although no kimberlite lava lakes have been identified, large HK bodies in the Lac de Gras area (Nowicki et al., 2007) have been tentatively interpreted as welded spatter deposits. If true, these deposits must have formed from Hawaiian-style fire fountaining at very high effusion rates. The lava flow at the Igwisi Hills, Tanzania could have a similar origin (S. Kurszlaukis, pers. commun., 2007).

Massive volcaniclastic kimberlite (MVK)-dominated pipes are exceptionally variable in area, ranging from ~50 to 500 m, rarely to 800 m diameter. Although they tend to be generally circular in plan view, the pipes are typically highly irregular in shape. This effect is pronounced if local faulting, fracture sets, or host-rock geology has strongly influenced the growth of the pipe during the eruptions (Barnett and Lorig, 2007). Kimberlite pipes of similar geometry to the classic South African model form carrot-shaped pipes (Fig. 10) typically infilled by HK, MVK, and RVK (e.g., Fig. 17 a–c). Examples of this type of pipe in Canada include several kimberlites in the Kirkland Lake and Lake Timiskaming fields, Ontario, and at some kimberlite pipes in the North Slave field (NWT), plus the Fox and Nicholas Bay kimberlites in the Lac de Gras area. The best described example of a pipe dominated by MVK and RVK is at Orapa, Botswana (Field et al., 1997).

The kimberlite pipes in the Lac de Gras field tend to be quite small (50–150 m diameter; <5 ha), steep sided (Fig. 18), and highly irregular (Graham et al., 1999; Nowicki et al., 2003). The pipe infill is in some cases dominated by kimberlite slumps and grain flows resedimented from a pyroclastic tephra cone (Fig. 19; Kjarsgaard, 2003, 2007) that formed during the pipe excavation stage. At Lac de Gras, consequent or subsequent input of pyroclastic kimberlite also contributes to the pipe infill (e.g., Kirkley et al. 1998; Graham et al., 1999; Moss and Russell, 2006). Canadian examples of pipes infilled with resedimented volcaniclastic kimberlite include

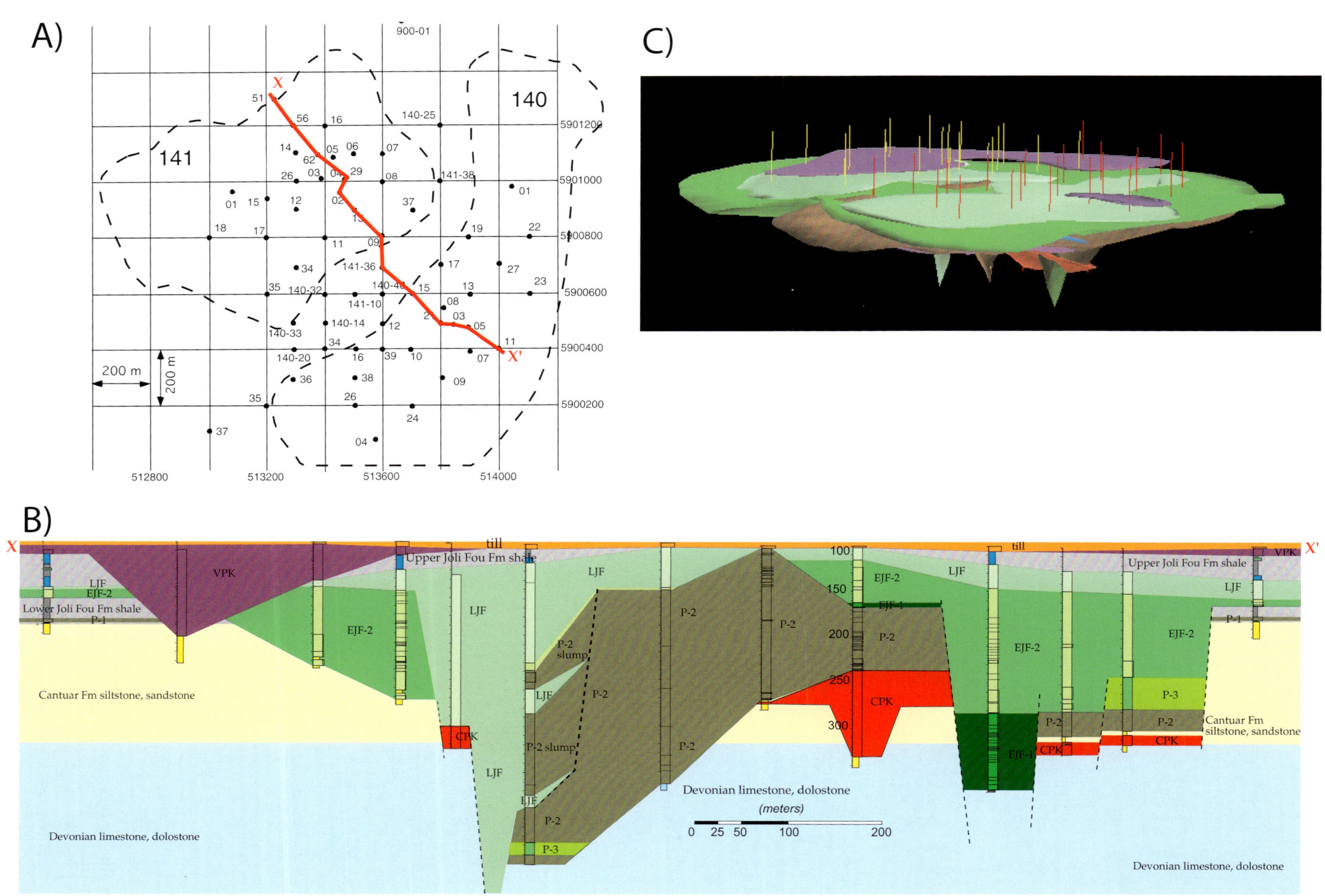

FIGURE 16. **(A)** Plan view of Orion South (140/141) kimberlite, Fort à la Corne field, Saskatchewan. Line of cross section X–X' shown in red. **(B)** Northwest–southest cross section of the Orion South (140/141) kimberlite, 1:1 scale, no vertical exaggeration. Seven distinct eruptive phases are identified in this model: purple = VPK; light green = LJF; green = EJF-1; dark green = EJF-1; brown = P-2; lime green = P-3; red = CPK. One minor phase also shown in brown = P-1. Note that four distinct feeder vents are identified (CPK, EJF-1, P-2, LJF). **(C)** 3-D solids model of the Orion South kimberlite illustrating relationships between the major eruptive phases and their feeder vents. Modified after Harvey et al., (2004).

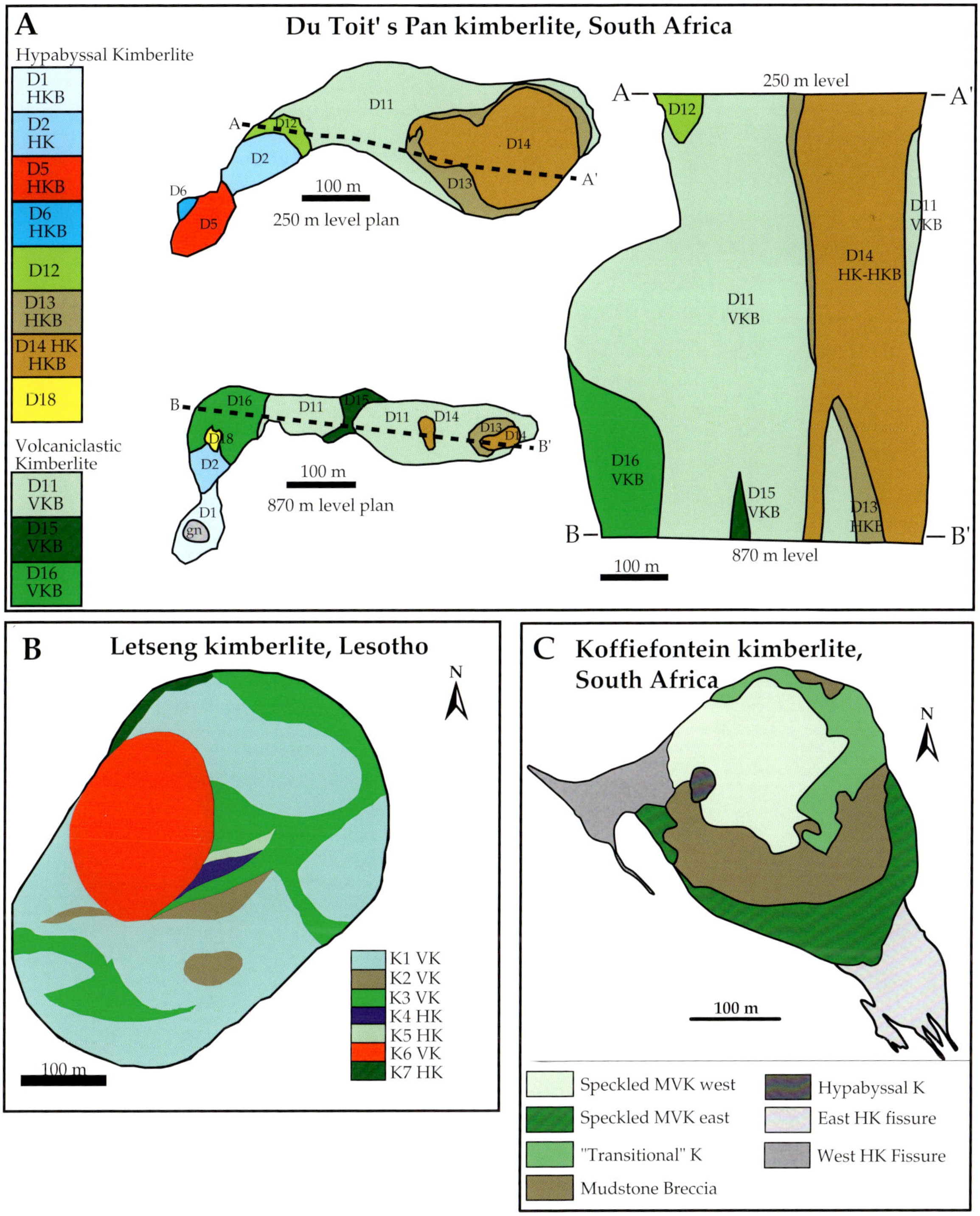

FIGURE 17. **(A)** Plan views of the 250 m and 870 m levels of the Du Toit's Pan, South Africa kimberlite and section view from the 250 m to 870 m level (after Clement, 1982). **(B)** Plan view of the Letseng, Lesotho kimberlite (after Bloomer and Nixon, 1973). **(C)** Plan view of the Koffiefontein, South Africa kimberlite (after Naidoo et al., 2004). Note that for each kimberlite the multiple varieties of distinct kimberlite identified.

the Panda, Koala, Misery, A-154N, and A-154S pipes.

Hypabyssal kimberlite dykes and sills are typically from 0.1 to 3 m in width (or thickness), and rarely to 5 to 10 m. Dyke enlargements (blows) and plugs of hypabyssal kimberlite range up to 100 m in diameter, and individual dykes or sills may be traced discontinuously (bifurcating with en echelon offsets being typical, as is common for basaltic dyke or sill systems) from a few hundred metres up to a few kilometres. The Snap Lake (NWT) kimberlite sill (Kirkley et al., 2003), the Somerset Island K24 kimberlite dyke (Kjarsgaard, 1996), and the Lynx dyke in the Otish mountains, Quebec, are good examples of spatially extensive (>1 km) kimberlite bodies.

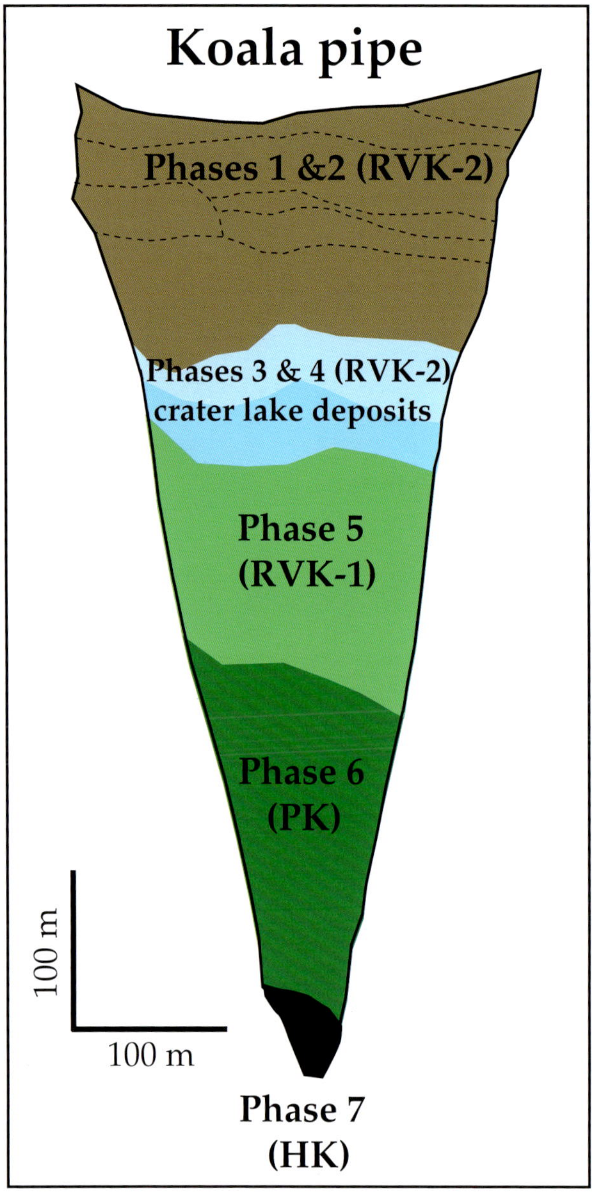

FIGURE 18. Geological cross section of the steep-sided, inverted cone shaped Koala kimberlite body, Ekati Mine, Lac de Gras field. Phase 7 is hypabyssal kimberlite (HK) and Phase 6 is pyroclastic kimberlite (PK) (Crawford et al., 2006). Phase 5 is interpreted here as syn-eruption resedimented volcaniclastic kimberlite (RVK-1); phase 4 and 3 are interpreted here as crater lake sediment;, and phases 2 and 1 are interpreted as post-eruption resedimented volcaniclastic kimberlite (RVK-2). Internal phases within phase 1 and 2 demarcated by dashed lines. Modified after Nowicki et al. (2003) and Crawford et al., (2006).

Irregularly shaped, multiple-lobed root zone complexes consisting dominantly of hypabyssal kimberlite have been observed at the Renard kimberlites, Québec, and the Batty Bay complex, Somerset Island, Nunavut. Massive, plug-like bodies of hypabyssal kimberlite have also been observed at the Notre Dame du Nord (NDN) and Peddie kimberlites in the Lake Timiskaming fields, Ontario.

The Multiphase Intrusive–Extrusive Nature of Kimberlite Bodies: Of crucial importance in understanding the diamond potential of a kimberlite is the recognition that multiple intrusive (HK, MVK) and/or extrusive (MVK, VK, PK) kimberlite phases occur within a single kimberlite body (Figs. 5, 16, 17, 18). Previous (e.g., 1970s, 1980s, 1990s) and also some current (i.e., 2006) geologic and genetic models for the formation of kimberlite pipes are oversimplified. For example, the conceptual classic South African model (Figs. 10, 20c) shows only a simple stratification of rock types, with PK and RVK at the top of the pipe, the pipe infilled by MVK, and HK in the root zone at the base of the pipe. This is not in accord with geological observations (e.g., Moulle, 1885, Wagner, 1914; Nixon, 1973; Clement, 1982). For example, at the Du Toit's Pan (South Africa) kimberlite (Fig. 17a), the upper part of the pipe (the 250 m level) is occupied by one MVK phase and six different HK phases. At the Letseng (Lesotho) kimberlite (Fig. 17b), the upper part of the pipe is occupied by four different MVK phases and three different HK phases. The Koffiefontein (South Africa) kimberlite (Fig. 17c) pipe on the 470 m level is occupied by two different MVK phases, two RVK phases, and three different HK phases. Taken together, the geological observations from three different classic South African model kimberlite pipes are consistent with the idea of multiple kimberlite phases in a single pipe as originally suggested by Moulle (1885) and Wagner (1914). Although Clement and Reid (1989) did note that HK and MVK are observed at the same structural level within a pipe, the models illustrated in Figures 10 and 20 show a simple stratification of different kimberlite rock types with depth in contrast to what is actually observed.

Simplified geologic models constructed for kimberlites from the Fort à la Corne and Lac de Gras area (Field and Scott-Smith, 1999; Scott-Smith, 2006) are illustrated in Figures 20a,b. These models serve little, if any purpose, because they are so oversimplified that they are misleading with respect to the known variation in architecture and morphology, internal geology (i.e., the multiple intrusive/eruptive nature of kimberlites), and hence diamond economics. For example, see Harvey et al. (2006), Kjarsgaard et al. (2006), or Kjarsgaard (2007) for Fort à la Corne kimberlites and Nowicki et al. (2003), Crawford et al. (2006), or Moss and Russell (2006) for Lac de Gras kimberlites.

Genetic and Exploration Models

Conventional Models

Diamonds have formed over a significant period of Earth history, from ~3.57 Ga (Gurney et al., 2005) to 88 Ma (Pearson and Shirey, 2002), and are represented by many diamond forming events (Shirey et al., 2004; Gurney et al., 2005). Diamonds are interpreted to crystallize from low-density fluids, or carbon- and water-rich melts (Bulanova, 1995; Navon, 1999). Macro-diamonds are transported as xenocrysts from the mantle to the surface by kimberlite magmas. Kimberlite-hosted diamond mines occur in a restricted tectonic setting and are observed only in ancient continental shield regions (Clifford, 1966) older than 2.5 Ga (Janse, 1984). The most favourable tectonic environment for economic kimberlites is a thick, old (2.5–3.6 Ga) craton with low heat flow. The kimberlite magmatism is initiated at depth in the mantle (>150 km),

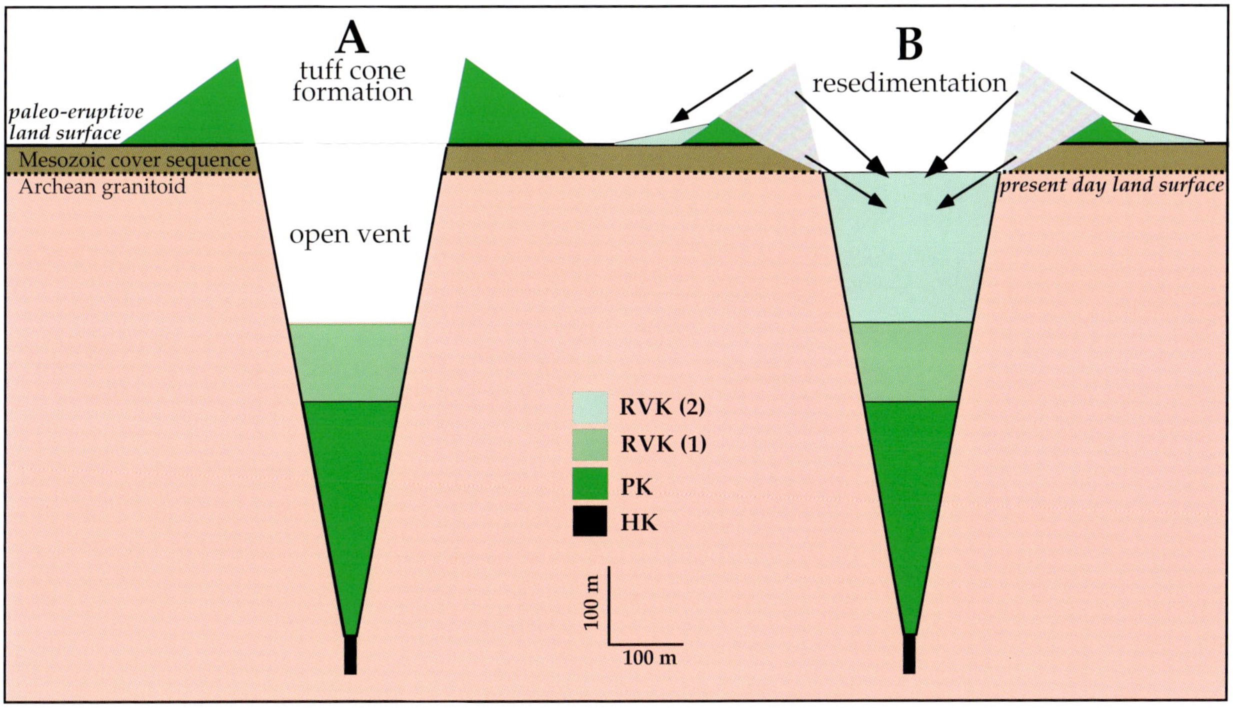

FIGURE 19. (A) Scaled model for the reconstruction of a kimberlite tephra cone at the Koala pipe, NWT. The tephra cone is 600 m in diameter and 108 m high with a crater rim diameter of 300 m an angle of repose of ~36°, and an internal crater wall angle of ~78°. The tephra cone sits on a 32 m thick Mesozoic cover sequence, on top of Archean granitoids. **(B)** Resedimentation of kimberlite (from the tephra cone) and cover sequence sediments by grain flow and slumping processes, into the open excavated pipe. The model assumes it is not feasible for the entire tephra cone to be resedimented into the open pipe, i.e., tephra will also be displaced away from the open pipe. 1:1 scale, no vertical exaggeration. Adapted and modified after Kjarsgaard (2003, 2007)

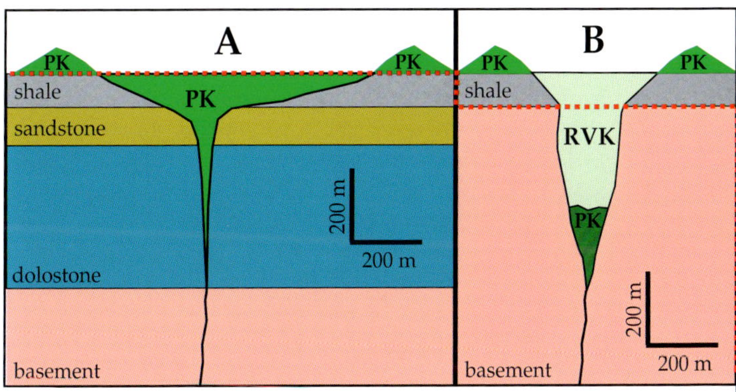

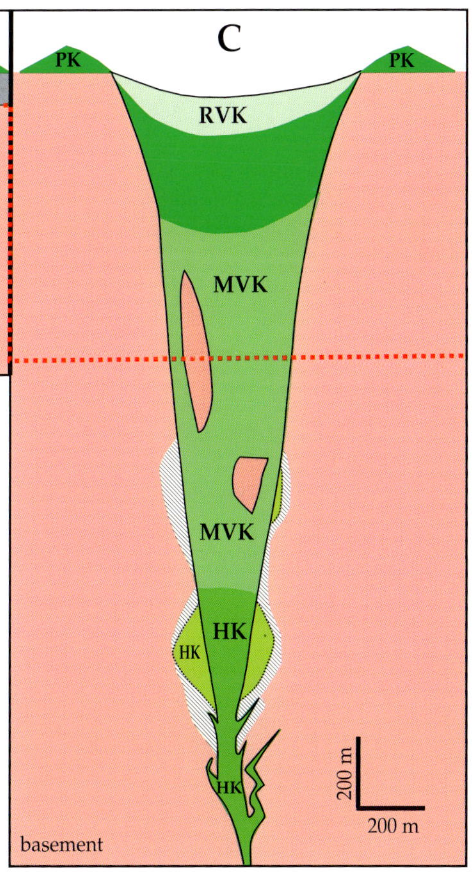

FIGURE 20. Highly oversimplified (non-relevant) models of kimberlite pipes; adapted from Scott-Smith, 2006 (**A**, **B**) and Field and Scott-Smith, 1999 (**C**). Note the original figures are to the correct scale in this diagram. Compare **(A)** with Fig. 16 and note there is no relationship. See the text and also Kjarsgaard et al. (2007) for further discussion of the geometry and architecture of Fort à la Corne kimberlites. Contrast the additional complexity of a Lac de Gras kimberlite (Fig. 19) with that shown in **(B)**. The 'classic South African model' kimberlite pipe shown in **(C)** exhibits a regular change in geologic units from PK and RVK (top) to MVK (middle) to HK (bottom) which is a severe oversimplification of the morphology of this style of kimberlite pipe (compare with Figs. 17 A–C).

although magma generation is poorly understood. Ultra-high pressure majorite garnets that occur as inclusions in diamonds (Moore and Gurney, 1985) and in mantle xenoliths (Haggerty and Sautter, 1990) confirm that some kimberlite magmas formed at depths of at least 300 km, which is consistent with ultra-deep (sub-lithospheric) diamond paragenesis. Hf–Nd isotopic studies on kimberlite (Nowell et al., 2004) indicate that the source area for kimberlite magmas resides at depth in the convecting mantle but taps an isotopic reservoir of ancient, deeply subducted basalt.

Thermobarometric calculations on mineral assemblages from diamondiferous mantle xenoliths (Pearson et al., 1994) and polymineralic diamond inclusions are consistent with diamonds forming in the Earth's mantle at depths greater than ~150 km. On the basis of petrological studies on eclogites and C isotopic studies on E-type diamonds, it is suggested that the C for many of these diamonds originated at or near the Earth's surface and was transported into the mantle via subduction processes (Jacob, 2004; Gurney et al., 2005; Jacob et al., 2005). In contrast, P-type diamonds have a very restricted range of C isotopic compositions, similar to and consistent with a juvenile (mantle) C source. The range in diamond contents of kimberlites (<.001 to >10 carat/tonne) is dependent upon the amount of diamond-bearing eclogite and peridotite lithospheric mantle material plus ultra-deep diamonds sampled, and the degree to which resorption and mechanical sorting of this entrained material occurs during transport to the surface. Kimberlites probably ascend through the mantle at velocities of 10 to 30 km/hr by crack propagation processes (Eggler, 1989). High level, rapid H_2O and CO_2 degassing coupled with groundwater interaction may result in highly explosive near-surface magmatism and the formation of kimberlite pipes, tephra cones, tuff cones, and tuff rings of variable geometry. Multiple intrusive and/or extrusive kimberlite phases are the norm, and inter- and intra-phase diamond variation may exceed an order of magnitude.

Advances of the Last Decade

There have been a number of significant advances in exploration for kimberlite-hosted diamond deposits in the past decade. The geochronology of kimberlites on a local and global scale has illustrated that eruption age is potentially important with respect to diamond prospectivity (Heaman et al., 2003, 2004). This enhanced understanding is due mainly to the development of the U–Pb perovskite radiometric technique by L. Heaman, and the Ar/Ar technique on groundmass phlogopite. A major increase in the understanding of the Earth's lithosphere has arisen due to deep seismic studies in Canada (Musacchio et al., 2004) and southern Africa (Fouch et al., 2004). These studies have been complimented by direct age dating of mantle xenoliths (e.g., Pearson et al., 1995, 2002; Carlson et al., 2000; Irvine et al., 2003). There is, however, significant scope for additional studies to be undertaken in these areas.

The development of single mineral grain thermobarometers (i.e., clinopyroxene method for peridotite derived grains; Nimis and Taylor, 2000) to understand fossil paleogeotherms and thus diamond stability is a major advance in diamond exploration, although these methods are not as precise in comparison to the more time consuming methods of traditional four- or five-phase thermobarometry. Furthermore, a thermobarometer has recently been developed for eclogite xenoliths (Simakov, 1999). Previously, it was only possible to determine eclogite paleotemperature and determine equilibration pressure by assuming a fossil geotherm (based on peridotite data) to understand if the eclogite equilibrated in the diamond or graphite stability field.

The use of garnet xenocryst data to reconstruct the lithospheric mantle in terms of the proportions of eclogite and various types of peridotite by Schulze (1989, 1994) was an important conceptual breakthrough. More recently, the concept of mantle stratigraphy has been attempted by trace element analysis of peridotite garnets (e.g., Griffin et al., 2004), combined with single grain Cr-pyrope thermobarometry on the garnets (Ryan et al., 1996). Unfortunately, these mantle stratigraphic methods do not take into account the amount of chromite harzburgite, chromite dunite, or eclogite in the mantle sample. Importantly, it has been shown by a number of authors (e.g., Gurney et al., 2005) that E-type diamonds can be an exceptionally important and sometimes the dominant component of the total (P-type, E-type, ultra-deep) diamond population within a kimberlite hosted diamond deposit.

Understanding the major element geochemistry of kimberlite indicator minerals Cr-pyrope garnet, Cr-spinel, Mg-ilmenite, and Cr-diopside has undergone a number of significant advances in the past decade. Wyatt et al. (2004) determined a new MgO versus TiO_2 bivariate plot to distinguish kimberlitic from non-kimberlitic ilmenite. Ramsay and Tompkins (1994) provided new insights on Cr-diopside parageneses applicable to diamond exploration. Grütter et al. (2004) published a simple, easy to use garnet discrimination/classification scheme, based on a large data set, using sound petrological principles. This classification allows for discrimination of wehrlite garnet (G12); high-Ti peridotite garnet (G11); harzburgite/dunite (subcalcic) garnet (G10); lherzolite (Ca-saturated) garnet (G9); pyroxenite–websterite–eclogite garnet (G4); eclogite garnet (G3); and megacryst garnet (G1).

The application of kimberlite indicator minerals as diamond indicator minerals has also significantly advanced in the past decade through the use of major- and trace-element data. In the recent Grütter et al. (2004) classification system, diamond harzburgite/dunite (subcalcic) garnet can now be discriminated on the CaO versus Cr_2O_3 bivariate plot. Furthermore, pressure is estimated from the Cr content of these peridotite-derived garnet xenocrysts (Grütter et al., 2006). For eclogite garnets, discrimination of diamond eclogite garnet is accomplished by use of a TiO2 versus Na_2O bivariate diagram (Fipke et al. 1995; Schulze, 1997). New and revised diamond Cr-spinel discrimination plots utilizing Cr_2O_3 with MgO, TiO_2, Al_2O_3 and Fe_2O_3 (e.g., Grütter and Apter, 1998; Kjarsgaard, 1998a) update the original discrimination plots of Sobolev (1971, 1977). These new and revised kimberlite indicator mineral and diamond indicator mineral plots are illustrated in McClenaghan and Kjarsgaard (2007).

A number of significant advances have occurred with regard to the application of airborne magnetic and EM surveys for kimberlite exploration. The application of GPS-controlled flight lines has certainly improved line spacing and the geographic positioning of the survey, which is particularly important for detailed surveys. The application of airborne gravity systems developed over the past decade has not (as yet) found any kimberlites of economic significance that were not previously recognized by magnetic and/or EM surveys. The development and application of seismic and MT methods to kimberlite exploration has been limited (Ford et al., 2007), but the results thus far are encouraging (e.g., Hammer et al., 2004).

There has been a renewed recognition over the past decade that kimberlite diamond deposits are complex, multiple intrusive/extrusive bodies. Although this has been known for over 100 years, over-simplified kimberlite pipe models developed in the 1970s, 80s, and 90s (see Fig. 10), are still (2006), in part, being perpetuated (see Fig. 20), which has led to non-optimum diamond sampling strategies in some cases.

Key Exploration Criteria

Surficial Indicator Mineral and Geochemical Sampling

Kimberlites contain a distinctive suite of resistant minerals of varying abundance known as kimberlite indicator minerals (Gurney and Zweistra, 1995). These include garnets, spinels, ilmenite, clinopyroxene, forsterite, orthopyroxene, and zircon. The indicator minerals are derived from the megacryst suite, disaggregated mantle peridotite, and eclogite, and from the kimberlite itself (see previous section on kimberlite mineralogy). The use of these indicator minerals in regional and detailed glacial till, stream sediment, and soil sampling programs to explore for kimberlite is well established worldwide. McClenaghan and Kjarsgaard (2007) summarize the salient visual, mineralogical, and geochemical characteristics of these minerals as applied to kimberlite and diamond exploration.

The distinctive geochemical signature (see previous section on kimberlite geochemistry) of kimberlite forms the basis for the exploration of these deposits using the chemical composition of glacial tills, stream sediments, and soils. Note, however, that any geochemical anomalies may be of limited areal extent due to dilution of the diagnostic kimberlite signature during dispersion in till and stream sediments (see further discussion in McClenaghan and Kjarsgaard, 2007).

Geophysical Prospecting Methods

The physical properties of kimberlite are quite variable, depending upon the type of kimberlite (HK, MVK, VK, PK, RVK) and style(s) of alteration. Kimberlite bulk densities range from ~2.5 to 3.1 g/ml for non-fragmental HK, to much lower values of 1.6 to 2.5 g/ml for fragmental VK (Katsube and Kjarsgaard, 1996). Kimberlite electrical resistivity measurements range from ~3,000 to 60,000 Ωm for non-fragmental HK to much lower values of ~10 to 3,000 Ωm for fragmental VK (Katsube and Kjarsgaard, 1996). Kimberlite magnetic susceptibility ranges from ~1 to 100 (10^{-3} S.I. units) for both HK and VK (Katsube and Kjarsgaard, 1996). Because kimberlite bodies can be hosted by just about any type of rock, the physical properties of host rocks are widely variable in terms of their magnetic response, electrical resistivity, and density. Hence, kimberlites can be represented by either positive or negative anomalies on airborne or ground magnetic, electromagnetic, and gravity surveys. Furthermore, because of the protracted emplacement of kimberlites over a timespan of up 30 m.y. (Heaman and Kjarsgaard, 2000; Heaman et al., 2003, 2004), they may be represented as both positive and negative magnetic features on airborne or ground magnetic surveys due to reversal(s) of the Earth's magnetic field.

A limited number of seismic (reflection and refraction) surveys have been undertaken over kimberlites with varying results. Recent surveys in Canada have shown that this technique can be effective in delineating sills in granite greenstone terrains (e.g., Snap Lake; Hammer et al., 2004) and sediment-hosted kimberlite tephra cones/feeder vents (e.g., Buffalo Head Hills; Fort à la Corne, Gendzwill and Matieshin, 1996; Carlson et al., 1999). In Canada, the application of airborne multi-spectral (Th, U, K) radiometric surveys has not been shown thus far to be an effective kimberlite exploration tool (Richardson, 1996). However, Macnae (1995) indicated that anomalous responses have been reported for kimberlites in India and Siberia, and noted that, due to limited depth penetration of this method, it would only be useful for outcropping kimberlites. In glaciated terrains such as Canada, this technique may be effective in identifying very kimberlite-rich tills. Airborne and ground hyperspectral imaging techniques to identify alteration minerals has been shown to be effective in non-glaciated terrains (e.g., Australia), however in glaciated terrains, dilution masks or lowers the response of diagnostic target minerals (Kerr et al., 2002). Macnae (1995), Walker (2003) and Ford et al. (2007) provide excellent overviews on the applications of geophysics to kimberlite exploration.

Sampling Strategies

Because there may only be one economically viable kimberlite body within a field or cluster, every potential target needs to be identified and drilled or sampled. To delineate the number of kimberlites within a cluster or field, typically a combination of geophysical techniques (airborne and ground) coupled with indicator mineral (and boulder) sampling and/or geochemical sampling in till, stream sediments, or soil is undertaken. In areas where the kimberlite is exposed or shallowly buried, a preliminary assessment of the diamond potential can be obtained economically by sampling in pits or trenches. Coopersmith (1993) provides a good synopsis of sampling techniques for pits and trenches on exposed or shallowly buried pipes. Deeply buried deposits need to be systematically drilled to be evaluated.

After sampling drill core or pits and trenches, it is crucial to discriminate distinct kimberlite phases within samples. This is accomplished by detailed macroscopic observations coupled with whole-rock geochemistry and geophysical (magnetic, density, multi-spectral gamma) techniques on core or hand samples, or by down-hole geophysical methods. Additional detailed petrographic analysis by microscopy or microbeam methods may also be required. Sampling for diamond indicator minerals or diamonds is undertaken on the basis that discrete kimberlite phases within a kimberlite body are sampled separately. Analysis of microdiamond-size populations, with the diamonds recovered using caustic or acid dissolution techniques, utilizes a variety of statistical techniques (e.g., Rombouts, 1995; Chapman and Boxer, 2004). The analysis of diamond indicator minerals is undertaken utilizing a variety of geochemical screening criteria (see summary in Advances of the Last Decade).

Knowledge Gaps

A significant problem for diamond explorationists is recognizing and understanding the significant variability in the architecture and morphology of kimberlite bodies worldwide. Perhaps even more important is the recognition and understanding of discrete phases within an individual kimberlite body. The various factors (e.g., phreatomagmatic versus primary magmatic volatile driven) involved in the different styles of kimberlite magmatism is poorly understood at present due to an almost complete paucity of supporting data sets. Hence, basic genetic models for kimberlite pipe formation are poorly understood. Detailed geological, mineralogical, geochemical and volcanological models of individual kimberlite bodies are rare. Comparative studies of kimberlites within a kimberlite field are also rare or lacking. Note that these suggestions for future work are certainly not new ideas (see Mitchell, 1986, 1995).

A second major fundamental problem for the modern diamond explorer is knowledge of the thickness and age of lithospheric mantle roots within continental cratonic regions. The thickness aspect can be determined by seismic tomography studies (e.g., van der Lee and Frederickson, 2005). However, it is important to recognize that the present day thickness may be less than the thickness at the time of kimberlite eruption (see Geotectonic Environment).

Determining the age of the mantle lithosphere is much more problematic, as physical samples of mantle peridotite must be obtained from igneous rocks of deep seated origin (i.e., kimberlite, lamproite, ultramafic lamprophyre, alkali basalt). However, whole rock Re–Os + PGE age determination technique for mantle peridotites is quite advanced (Pearson et al., 1995, 2002; Irvine et al., 2003) and coupled with thermobarometric estimates for xenoliths, age versus depth profiles can be constructed. Alternately, less refined lithospheric age data can be construed from laser ablation Re-Os data for sulphide inclusions in forsterite (e.g., Aulbach et al., 2002; Griffin et al., 2002). Certainly, a more refined understanding of the age and formation of lithospheric mantle and its relationship to diamond formation and stability would be invaluable.

More problematic are the following questions:

1. What mantle event(s) cause diamonds to form? Are there related and identifiable events in the crust?

2. What mantle event(s) triggers the partial melting that generates kimberlite magma in the mantle? At what depth (or depths) does this occur and what are the source characteristics? Are there crustal events which also record the events that trigger kimberlite formation?

Areas of High Diamond Potential in Canada

With the largest Precambrian Shield on the Earth and diamond mines in two different Archean blocks (Slave and Superior cratons), where next should one explore in Canada? Quite a number of potential areas with high diamond potential were outlined last decade by Kjarsgaard (1998b), specifically within the Superior craton and in the enigmatic Churchill province. Since that time, quite a number of new discoveries have been made in both the Superior craton (e.g., Wemindji, Renard kimberlites) and Churchill province (e.g., Aviat, Repulse Bay, Rankin Inlet, Boothia Peninsula kimberlites). Both of these cratonic regions are quite large and are still underexplored. The lack of a comprehensive bedrock geological framework for the Rae, the south, central, and northern Hearne domains, and Queen Maude block within the Churchill province, coupled with a paucity of lithospheric mantle information makes it difficult at present to refine the search area. The northwestern (Ontario and Manitoba) and northeastern (Quebec) parts of the Superior province are also perhaps under explored, although this is due in part to thick glacial or glaciolacustrine cover. The true extent and nature of the Archean Sask craton is imperfectly known and understood, and the geometry of the Superior –Sask–Churchill terrain boundaries in the prairie provinces needs refining in order to better understand the potential for kimberlite diamond deposits in this region.

Acknowledgements

Formal manuscript reviews by Prof. D. G. Pearson and an anonymous reviewer as well as an informal review by G. Read were quite helpful. Discussions with numerous industry (BHP Billiton, De Beers, Kennecott, Shore Gold, Ashton, MPH Consulting, etc.) and various consultants and academic colleagues the past 17 years have been extremely useful. They may not agree with all the contents of this chapter! C. Jefferson (GSC) is thanked for continuing moral support. Wayne Goodfellow is thanked for editorial comments and patience. B. Hillary drafted Figure 1.

References

Allsopp, H.L., Bristow, J.W., Smith, C.B., Brown, R., Gleadow, A.J.W., Kramers, J.D., and Garvie, O.G., 1989, A summary of radiometric dating methods applicable to kimberlite and related rocks: Geological Society of Australia, Special Publication 14, p. 343–357.

Ashton, K.E., Lewry, J.F., Heaman, L.M., Hartlaub, R.P., Stauffer, M.R. ,and Tran, H.T., 2005, The Pelican thrust zone: Basal detachment between the Archean Sask Craton and Paleoproterozoic Flin Flon—Glennie Complex, western Trans-Hudson Orogen: Canadian Journal of Earth Sciences, v. 42, 685–706.

Aulbach, S., Griffin, W.L., O'Reilly, S.Y., and Kivi, K., 2002, Sulfides from the lower mantle?: Geochimica et Cosmochimica Acta, v. 66, no. 15A, p.38.

Bardet, M.G., 1965, Les gisements de diamant d'U.R.S.S.: Chron. Mines Rech. Minere, v. 65, p. 1–40.

Barnett, W.P., and Lorig, L. 2007, A model for stress controlled pipe growth: Journal of Volcanology and Geothermal Research, v. 159, p. 108–125.

Bhattacharji, S., and Smith, C.H., 1964, Flowage differentiation: Science, v. 145, p. 150–153.

Bleeker, W., Ketchum, J.W.F., and Davis, W.J., 1999, The Central Slave basement complex: Part II. Age and tectonic significance of high strain zones along the basement-cover contact: Canadian Journal of Earth Sciences, v. 36, p. 1111–1130.

Bloomer, A.G., and Nixon, P.H., 1973, The geology of the Letseng-la-terae kimberlite pipes, *in* Nixon, P.H. (ed.), Lesotho kimberlites: Maseru, Lesotho National Development Corporation, p. 20–32.

Boyer, L., 2005, Volcanology of kimberlites in the Buffalo Head Hills, Alberta: Unpublished M.Sc. thesis, Vancouver, University of British Columbia, 188 p.

Bulanova, G.P., 1995, The formation of diamond: Journal of Geochemical Exploration, v. 53, p. 1–24.

Carlson, R.W., Boyd, F.R., Shirey, S.B., Janney, P.E., Grove, T.L., Bowring, S.A., Schmitz, M.D., Dann, J.C., Bell, D.R., Gurney, J.J., Richardson,

S.H. Tredouz, M., Menzies, A.H., Pearson, D.G., Hart, J.R., Wilson, A.H., and Moser, D., 2000, Continental growth, preservation, and modification in southern Africa: GSA Today, v. 10, p. 1–7.

Carlson, S.M., Hillier, W.D., Hood, C.T., Pryde, R.P., and Skelton, D.N. 1999, The Buffalo Head Hills kimberlites: A newly discovered diamondiferous kimberlite province north-central Alberta, Canada: Red Roof Design, VIIth International Kimberlite Conference, Proceedings: The J.B. Dawson Volume, Cape Town, p. 109–116.

Chapman, J.G., and Boxer, G.L., 2004, Size distribution and analyses for estimating diamond grade and value: Lithos, v. 76, p. 369–376.

Clement, C.R., 1975, The emplacement of some diatreme-facies kimberlites: First International Kimberlite Conference, Physics and Chemistry of the Earth, v. 9, p. 51–59.

Clement, C.R., 1982, A comparative geological study of some major kimberlite pipes in the Northern Cape and Orange Free state: Unpublished Ph.D. thesis, University of Cape Town, 432 p.

Clement, C.R., and Skinner, E.M.W., 1985, A textural-genetic classification of kimberlites: Transactions of the Geological Society of South Africa, v. 88, p. 403–409.

Clement, C.R., and Reid, A.M., 1989, The origin of kimberlite pipes: An interpretation based on a synthesis of geological features displayed by Southern African occurrences: Geological Society of Australia, Special Publication 14.

Clifford, T.N., 1966, Tectono-metallogenic units and metallogenic provinces of Africa: Earth and Planetary Science Letters, v. 1, p. 421–434.

Coopersmith, H.G., 1993, *in* Sheahan, P., and Chater, A., eds., Diamonds: Exploration, sampling, and evaluation: Toronto, Prospectors and Developers Association of Canada, Short Course Proceedings, p. 173-184.

Crawford, B.B., Porritt, L., Nowicki, T., and Carlson, J.A., 2006, Key geological characteristics of the Koala kimberlite, Ekati diamond mine, Canada: Kimberlite Emplacement Workshop, Saskatoon, Saskatchewan, September 7–12, 2006, CD-ROM.

Creaser, R.A., Grütter, H.S., Carlson, J.A., and Crawford, B., 2004, Macrocrystal phlogopite Rb–Sr dates for the Ekati property kimberlites, Slave Province, Canada: Evidence for multiple intrusive episodes in the Paleocene and Eocene: Lithos, v. 76, p. 399–414.

Davis, G.L., 1977, The ages and uranium contents of zircons from kimberlites and associated rocks: Carnegie Institute of Washington. Yearbook, v. 76, p. 631–635.

Davis, G.L., 1978, Zircons from the mantle: Carnegie Institute of Washington, Yearbook v. 77, p. 805–807.

Davis, W.J., and Kjarsgaard, B.A., 1997, A Rb–Sr isochron for a kimberlite from the recently discovered Lac de Gras field, Slave Province, Northwest Canada: Journal of Geology, v. 105, p. 503–509.

Dawson, J.B., 1980, Kimberlites and their xenoliths: Berlin, Springer, 252 p.

Dawson J.B., and Hawthorne, J.B., 1973, Magmatic sedimentation and carbonatitic differentiation in kimberlite sills at Benfontein, South Africa: Journal of the Geological Society of London, v. 129, p. 61–85.

Dobbs, P.N., Duncan, D.J., Hu, S., Shee, S.R., Colgan, E.A., Brown, M.A., Smith, C.B., and Allsop, H.L., 1994, The geology of the Mengyin kimberlites, Shandong, China: Brazil, Companhia de Pesquisa de Recursos Minerais, CPRM Special Publication 1B, p. 329–345.

Drever, H.I., and Johnston, R., 1966, Picritic minor intrusions, *in* Wyllie, P.J., ed., Ultramafic and related rocks: New York, Wiley, p. 71–82.

Eggler, D.H., 1989, Kimberlites: How do they form?: Geological Society of Australia, Special Publication 14, p. 323–342.

Field, M., and Scott-Smith, B.H., 1999, Contrasting geology and near-surface emplacement of kimberlite pipes in southern Africa and Canada: Red Roof Design, VIIth International Kimberlite Conference, Proceedings: The J.B. Dawson Volume, Cape Town, p. 214–237.

Field, M., Gibson, J.G., Wilkes, T.A., Gababotse, J., and Khutjwe, P., 1997. The geology of the Orapa A/K1 kimberlite Botswana: Further insight into the emplacement of kimberlite pipes: Russian Geology Geophysics, v. 38, p. 24–39.

Fipke, C.E., Gurney, J.J., and Moore, R.O., 1995. Diamond exploration techniques emphasizing indicator mineral geochemistry and Canadian examples: Geological Survey of Canada, Bulletin 423, 86 p.

Ford, K., Keating, P., and Thomas, M.D., 2007, Overview of geophysical signatures associated with Canadian ore deposits, in Mineral deposits of Canada: A synthesis of major deposit types, district metallogeny, the evolution of geological provinces, and exploration methods, Goodfellow, W.D., ed.: Geological Association of Canada, Mineral Deposits Division, Special Publication 5, p. 939–970.

Fouch, M.J., James, D.E., Gao, S.S., Van Decar, J.C., and van der Lee, S., 2004, Mantle seismic structure beneath the Kaapvaal and Zimbabwe cratons: Transactions of the Geological Society of South Africa, v. 107, p. 33–44.

Gendzwill, D.J., and Matieshin, S.D., 1996. Seismic reflection survey of a kimberlite intrusion in the Fort à la Corne district, Saskatchewan: Geological Survey of Canada, Open File 3228, p. 241–242.

Gobba, J.M., 1989, Kimberlite exploration in Tanzania: Journal of African Earth Sciences, v. 9, p. 565–578.

Graham, I., Burgess, J.L., Bryan, D., Ravenscroft, P.J., Thomas, E., Doyle, B.J., Hopkins, R., and Armstrong, K.A.m, 1999, Exploration history and geology of the Diavik kimberlites, Lac de Gras, Northwest Territories, Canada: Red Roof Design, VIIth International Kimberlite Conference, Proceedings: The J.B. Dawson Volume, Cape Town, p. 262–279.

Griffin, W.L., Spetsius, Z.V., Pearson, N.J., and O'Reilly, S.Y., 2002, In situ Re–Os analysis of sulfide inclusions in kimberlitic forsterite: New constraints on depletion events in the Siberian lithospheric: Geochemistry, Geophysics, Geosystems—G3, v. 3, no. 11, 25 p.

Griffin, W.L., O'Reilly, S.Y., Doyle, B.J., Pearson, N.J., Coopersmith, H., Kivi, K., Malkovets, V., and Pokhilenko, N., 2004, Lithosphere mapping below the North American plate, Lithos, v. 77, p. 873–922.

Grunsky, E., and Kjarsgaard, B.A., 2007, Classification of eruptive phases of the Star kimberlite, Saskatchewan, Canada based on statistical treatment of whole-rock geochemical analyses: Applied Geochemistry, in press.

Grütter, H.S., and Apter, D.B., 1998, Kimberlite- and lamproite-borne chromite phenocrysts with "diamond-inclusion"-type chemistries [abs.]: University of Cape Town, 7th International Kimberlite Conference, Cape Town, April 13–17, 1998, p. 280–281.

Grütter, H.S., Gurney, J.J., Menzies, A.H., and Winter, F., 2004, An updated classification scheme for mantle-derived garnet, for use by diamond explorers: Lithos, v. 77, p. 841–857.

Grütter, H., Lattia, D., and Menzies, A., 2006, Cr-Saturation arrays in concentrate garnet compositions from kimberlite and their use in mantle barometry: Journal of Petrology, v. 47, p. 801–820.

Gurney, J.J., and Zweistra, P., 1995, The interpretation of the major element compositions of mantle minerals in diamond exploration: Journal of Geochemical Exploration, v. 53, p. 293–310.

Gurney, J.J, Helmstaedt, H.H., Le Roex, A.P., Nowicki, T.E., Richardson, S.H., and Westerlund, K.J., 2005, Diamonds: Crustal distribution and formation processes in time and space and an integrated deposit model: Economic Geology, 100th Anniversary Volume, p. 143–177.

Haggerty, S.E., 1982, Kimberlites in western Liberia: An overview of the geological setting in a plate tectonic framework: Journal of Geophysical research, v. 87, p. 10811–10826.

Haggerty, S.E., and Sautter, V., 1990, Ultradeep (greater than 300 km), ultramafic upper mantle xenoliths: Science v. 248, p. 993–996.

Hammer, P.T.C., Clowes, R.M., and Ramachandran, K., 2004, Seismic reflection imaging of thin kimberlite dykes and sills: Exploration and deposit characterization of the Snap Lake dyke, Canada: Lithos, v. 76, p. 359–368.

Hansom, E.K., Moore, J.M., Robey, J., Bordy, E.M., and Marsh, J.S., 2006, Re-estimation of erosion levels in Group I and II kimberlites between Lesotho, Kimberley, and Victoria West, South Africa: Kimberlite Emplacement Workshop, Saskatoon, Saskatchewan, September 7–12, 2006, CD-ROM.

Harvey, S.E., Kjarsgaard, B.A., Zonneveld, J.P., and McNeil, D., 2004, Update on kimberlite eruptive sequences, volcanic stratigraphy, and 3-D modelling of the 140/141 kimberlite complex, Fort à la Corne field, Saskatchewan: Saskatchewan Geological Survey, Summary of Investigations 2004, 2, Miscellaneous Report 2004-4.2, CD-ROM.

Harvey, S.E., Kjarsgaard, B.A., Zonneveld, J.P., Heaman, L.M., and MacNeil, D, 2006, Volcanology and sedimentology of distinct eruptive phases at the Star kimberlite, Fort à la Corne field: Kimberlite Emplacement Workshop, Saskatoon, Saskatchewan, September 7–12, 2006, CD-ROM.

Hawthorne, J.B., 1975, Model of a kimberlite pipe: Physics and Chemistry of the Earth, v. 9, p. 1–16.

Heaman, L.M., and Kjarsgaard, B.A., 2000, Timing of eastern North American kimberlite magmatism; Continental extension of the great meteor hotspot track: Earth and Planetary Science Letters, v. 178, p. 253–268.

Heaman, L.M., Kjarsgaard, B.A., and Creaser, R., 2003, The timing of kimberlite magmatism in North America: Implications for global kimberlite genesis and diamond exploration: Lithos, v. 71, p. 153–184.

Heaman, L.M., Kjarsgaard, B.A., and Creaser, R., 2004, The temporal evolution of North American kimberlites: Lithos, v. 76, p. 377–398.

Helmstaedt, H.H., 2003, Tectonics and structural considerations in diamond exploration: Diamonds Short Course Notes, Cordilleran Round-Up, Vancouver, British Columbia, January, 29–30, 2003, 8 p.

Hoal, K.O., 2003, Samples of Proterozoic iron-enriched mantle from the premier kimberlite: Lithos, v. 71, p. 259–272.

Irvine, G.J., Pearson, D.G., Kjarsgaard, B.A., Carlson, R.W., Kopylova, M.G., and Dreibus, G., 2003, Evolution of the lithospheric mantle beneath northern Canada: A Re-Os isotope and PGE study of kimberlite-derived xenoliths from Somerset Island and a comparison to the Slave and Kaapvaal cratons: Lithos, v. 71, p. 461–488.

Jacob, D.E., 2004, Nature and origin of eclogite xenoliths from kimberlites: Lithos, v. 77, p. 295–316.

Jacob, D.E., Bizmis, M., and Salters, V.J.M., 2005, Lu-Hf and geochemical systematics of recycled ancient oceanic crust: Evidence from Roberts Victor eclogites: Contributions to Mineralogy and Petrology, v. 148, p. 707–720.

Janse, A.J.A., 1984, Kimberlites—Where and when, *in* Glover, J.E., and Harris, P.G., eds., Kimberlite occurrence and origin: Perth, University of Western Australia, Geology Department/Extension Services, Publication 8, p. 19–62.

Janse, A.J.A., 1993, The aims and economic parameters of diamond exploration, *in* Sheahan, P., and Chater, A., eds., Diamonds: Exploration, sampling, and evaluation: Toronto, Prospectors and Developers Association of Canada, Short Course Proceedings, p. 173–184.

—— 2005, The search for diamonds: Mining Journal, London, 20 p.

Jelsma, H.A., deWit, M.J., Thiart, C., Dirks, P.H.G.M., Viola, G., Basson, I.J., and Anckar, E., 2004, Preferential distribution along transcontinental corridors of kimberlites and related rocks of southern Africa: Transactions of the Geological Society of South Africa, v. 107, p. 301–324.

Kaminsky, F.V., Feldman, A.A., Varlamov, V.A., Boyko, A.N., Olofinsky, L.N., Shofman, I.L., and Vaganov, V.I., 1995, Prognostication of primary diamond deposits: Journal of Geochemical Exploration, v. 53, p. 167–182.

Katsube, T.J., and Kjarsgaard, B.A., 1996, Physical property characteristics of Canadian kimberlites: Geological Survey of Canada, Open File 3228, p. 241–242.

Kerr, D., Budkewitsch, P., Bryan, D., Knight, R., and Kjarsgaard, B.A., 2002, Surficial geology, spectral reflectance characteristics, and their influence on hyperspectral imaging as a drift prospecting technique for kimberlites in the DIAVIK Diamond Mine area, NWT: Geological Survey of Canada, Current Research, no. 2002-C4, 10 p.

Kinny, P.D., Compston, W., Bristow, J.W., and Williams, I.S., 1989, Archean mantle xenocrysts in a Permian kimberlite: Two generations of kimberlitic zircon in Jwaneng DK2, southern Botswana: Geological Society of Australia, Special Publication 14, p. 833–842.

Kinny, P.D., Griffin, B.J., Heaman, L.M., Brakhfogel, F.F., and Spetsius, Z.V., 1997, SHRIMP U–Pb ages of perovskite from Yakutian kimberlites: Russian Geology and Geophysics, v. 38, p. 97–105.

Kirkley, M.B., Kolebaba, M.R., Carlson, J.A., Gonzales, A.M., Dyck, D.R., and Dierker, C., 1998, Kimberlite emplacement processes interpreted from Lac de Gras examples [abs.]: University of Cape Town, 7th International Kimberlite Conference, Cape Town, April 13–17, 1998, 429–431.

Kirkley, M.B., Mogg, T., and McBean, D., 2003, Snap Lake field trip guide: Geological Survey of Canada, 8th International Kimberlite Conference Field Trip Guidebooks; Miscellaneous G 293, CD-ROM.

Kjarsgaard, B.A., 1996, Somerset Island kimberlite field, District of Franklin, Northwest Territories: Geological Survey of Canada, Open File 3228, p. 61–66.

——1998a, Compositional trends of spinel and mica in alkali minettes, southern Alberta, Canada: University of Cape Town, 7th International Kimberlite Conference, Cape Town, April 13–17, 1998, p. 435–437.

——1998b, Occurrence and diamond potential of kimberlites in Canada, [abs.]: Canadian Institute of Mining and Metallurgy, 100th Anniversary Meeting, Montreal, Quebec, May 4–9, p. 77.

——2003, Volcanology of kimberlite: Diamonds Short Course Notes, Cordilleran Round-Up, Vancouver, British Columbia, January, 29–30, 2003, 20 p.

——2007, On the recognition and significance of kimberlite tephra cones, tuff cones, and tuff rings: Journal Volcanology Geothermal Research, in press.

Kjarsgaard, B.A., and Levinson, A.A., 2002, Diamonds in Canada: Gems and Gemology, v. 38, no. 3, p. 208–238.

Kjarsgaard, B.A., Wilkinson, L., and Armstrong, J.A., 2002, Geology, Lac de Gras kimberlite field, central Slave province, Northwest Territories—Nunavut (NTS 76C, D, parts of E, F): Geological Survey of Canada, Open File 3238, colour map with descriptive notes, 1:250 000 scale.

Kjarsgaard, B.A., Harvey, S., Zonneveld, J.P, McNeil, D.H., Heaman, L.M., and White, D., 2004, Geology of the Star kimberlite, Saskatchewan: Saskatchewan Geological Survey, Summary of Investigations 2004, v. 2, Miscellaneous Report 2004-4.2, CD-ROM.

Kjarsgaard, B.A., Zonneveld, J.P., Grunsky, E., Heaman, L.M., McNeil, D., and Du Plessis, P., 2005, Recent advances in the geology of the Star kimberlite, Saskatchewan: Saskatchewan Geological Survey, Summary of Investigations 2005, v. 2, Miscellaneous Report 2005-4.2, CD-ROM.

Kjarsgaard, B.A., Harvey, S., Heaman, L.M., White, D., Grunsky, E., McNeil, D.H., and Zonneveld, J.P, 2006, Kimberlite eruptive sequences, volcanic stratigraphy, and emplacement of the 140/141 kimberlite, Fort à la Corne field, Saskatchewan, Canada Kimberlite Emplacement Workshop, Saskatoon, Saskatchewan, September 7–12, 2006, CD-ROM.

Kjarsgaard, B.A., Leckie, D.A. and Zonneveld, J.P., 2007. Discussion of "Geology and diamond distribution of the 140/141 kimberlite, Fort à la Corne, central Saskatchewan, Canada", by A. Berryman, B.H. Scott-Smith and B.C. Jellicoe (Lithos v. 76, p. 99 - 114). Lithos, in press.

Leckie, D.A., Kjarsgaard, B.A., Bloch, J., McIntyre, D., McNeil, D., Stasiuk, L., and Heaman L., 1997, Emplacement and reworking of Cretaceous, diamond-bearing, crater facies kimberlite of central Saskatchewan, Canada: Geological Society of America, Bulletin 109, p. 1000–1020.

Lockhart, G, Grutter, H., and Carlson, J., 2004, Temporal, geomagnetic ,and related attributes of kimberlite magmatism at Ekati, Northwest Territories, Canada: Lithos, v. 77, p. 665–682.

Lorenz, V., Zimanowski, B., Büttner, R., and Kurszlaukis, S. 1999, Formation of kimberlite diatremes by explosive interaction of kimberlite magma with groundwater: Filed and experimental aspects:. Red Roof Design, VIIth International Kimberlite Conference, Proceedings: The P.H. Nixon Volume, p. 522–528.

Luguet, A., Pearson, D.G., Jacques, A.L., Bulanova, G.P., Smith, C.B., Roffey, S., and Rayner, M.J., 2005, Archean mantle beneath the Halls Creek mobile zone, West Australia revealed by Re-Os isotopes: Geochemica Cosmochemica Acta, v. 69/10S, p. A285.

Macnae, J., 1995, Applications of geophysics for the detection and exploration of kimberlites and lamproites: Journal of Geochemical Exploration, v. 53, p. 213–244.

Marsh, J.S., 1973, Relationships between transform directions and alkaline igneous rock lineaments in Africa and South America: Earth and Planetary Science Letters, v. 18, p. 317–323.

McClenaghan, M.B., and Kjarsgaard, B.A. 2007: Indicator mineral and geochemical exploration methods for kimberlite in glaciated terrain, examples from Canada, *in* Goodfellow, W.D., ed., Mineral deposits of Canada: A synthesis of major deposit types, district metallogeny, the evolution of geological provinces, and exploration methods: Geological Association of Canada, Mineral Deposits Division, Special Publication 5, p. 983-1006.

Mitchell, R.H., 1986. Kimberlites: Mineralogy, geochemistry, and petrology: New York, Plenum Press, 442 p.

——1995, Kimberlites, orangeites, and related rocks: New York, Plenum Press, 410 p.

——1997, Kimberlites, orangeites, lamproites, melilitites, and minettes: A petrographic atlas: Thunder Bay, Ontario, Almaz Press, 243 p.

Moore, R.O., and Gurney J.J., 1985, Pyroxene solid solutions in garnets included in diamond: Nature v. 318 p. 553–555.

Moss, S., and Russell, J.K., 2006, Pyroclastic origins of the mega-graded bed at Diavik, Kimberlite Emplacement Workshop, Saskatoon, Saskatchewan, September 7–12, 2006, CD-ROM.

Moulle, M.A., 1885, Sur la géologie générale et sur le mines de diamants de L'Afrique Du Sud: Annales des Mines, Series 8, v. 7, p. 93–344.

Musacchio, G., White, D.J., Asudeh, I., and Thomson, C.J., 2004, Lithospheric structure and composition of the Archean western Superior Province from seismic refraction/wide-angle reflection and gravity modelling: Journal of Geophysical Research, B, Solid Earth and Planets, v. 109, no. 3, p. 1–28.

Naidoo, P., Stiefenhofer, J., Field, M., and Dobbe, R., 2004, Recent advances in the geology of the Koffiefontein Mine, Orange Free State, South Africa: Lithos, v. 76, p. 161–182.

Navon, O., 1999, Diamond formation in the Earth's mantle: VIIth International Kimberlite Conference, Cape Town, Proceedings: The P.H. Nixon Volume, p. 584–2604.

Nimis, P., and Taylor, W.R., 2000, Single clinopyroxene thermobarometry for garnet peridotites: Part 1: Calibration and testing of a Cr-in-cpx barometer and enstatite-in-cpx thermometer: Contributions to Mineralogy and Petrology, v. 139, p. 541–554.

Nixon, P.H., 1973, Lesotho kimberlites: Maseru, Lesotho National Development Corporation, 350 p.

Nixon, P.H., and Leahy, K., 1997, Diamond-bearing volcaniclastic kimberlites in Cretaceous marine sediments, Saskatchewan, Canada: Russian Geology and Geophysics, v. 38, p. 17–23.

Nowell, G.M., Pearson, D.G., Bell, D.R., Carlson, R.W., Smith, C.B., Kempton, P., and Noble, S.R., 2004, Hf isotope systematics of kimberlites and their megacrysts: New constraints on their source regions: Journal of Petrology, v. 45 p. 1583–1612.

Nowicki, T., Carlson, J.A., Crawford, B.B., Lockhart, G.D., Oshurst, P.A., and Dyck, D.R., 2003, Field guide to the Ekati diamond mine: Geological Survey of Canada, 8th International Kimberlite Conference Field Trip Guidebooks, Miscellaneous G 293, CD-ROM.

Nowicki, T., Porritt, L., Crawford B., and Kjarsgaard, B.A., 2007, Geochemical trends in kimberlites of the Ekati property, Northwest Territories, Canada: Insights on volcanic and resedimentation processes: Journal of Volcanology and Geothermal Research, in press.

Pearson, D.G., 1999a, The age of continental roots: Lithos, v. 48, p.171–194.

——1999b, Evolution of cratonic lithospheric mantle: An isotopic perspective: Geochemical Society, Special Publication, v. 6, p.57–78.

Pearson D.G., and Shirey, S.B., 2002, Isotopic dating of diamonds: Reviews in Economic Geology, v. 12, p. 143–172.

Pearson D.G., Boyd, F.R., Haggerty, S.E., Pasteris, J.D., Field, S.W., Nixon, P.H., and Pokhilenko, N.P., 1994, The characterization of graphite in cratonic lithospheric mantle: A petrological carbon isotope and Raman spectroscopic study: Contributions to Mineralogy and Petrology, v. 115, p. 449–466.

Pearson D.G., Carlson, R.W., Shirey, S.B., Boyd, F.R., and Nixon, P.H., 1995, The stabilization of Archean lithospheric mantle: A Re-Os study of peridotite xenoliths from the Kaapvaal craton: Earth and Planetary Science Letters, v. 134, p. 341–357.

Pearson, D.G., Irvine, G.J., Carlson, R.W., Kopylova, M.G., and Ionov, D.A., 2002, The development of lithospheric keels beneath the earliest continents: Time constraints using PGE and Re-Os isotope systematics: Geological Society of London, Special Publication, v. 199, p. 65–90.

Perron, L., 2004, Diamonds: Ottawa, Natural Resources Canada, Canadian Minerals Yearbook, 2004, 14 p.

Price, S.E., Russell, J.K., and Kopylova, M.G., 2000, Primitive magma from the Jericho Pipe, NWT, Canada: Constraints on primary kimberlite melt chemistry: Journal of Petrology, v. 41, p. 789–808

Ramsay, R.R., and Tompkins, L.A., 1994, The geology, heavy mineral concentrate mineralogy, and diamond prospectivity of the Boa Esperanca and Cana Verde pipes, Corrego D'Anta, Minas Gerais, Brazil: Brazil, Companhia de Pesquisa de Recursos Minerais, CPRM Special Publication 1B, 5th International Kimberlite Conference, Proceedings, p. 329–345.

Read, G., Grutter, H., Winter, S., Luckman, N., Gaunt, F., and Thomsen, F., 2004, Stratigraphic relations, kimberlite emplacement, and lithospheric thermal evolution, Quirico Basin, Minais Gerais State, Brazil: Lithos, v. 77, p. 803–818.

Richardson, K.A., 1996, Geophysical exploration and GIS applications: Introduction: Geological Survey of Canada, Open File 3228, p. 225–229.

Rombouts, L., 1995, Sampling and statistical evaluation of diamond deposits: Journal of Geochemical Exploration, v. 53, p. 351–367.

Ross, G.M., Parrish, R.R., Villeneuve, M.E., and Bowring, S.A., 1991, Geophysics and geochronology of the crystalline basement of the Alberta Basin, western Canada: Canadian Journal of Earth Sciences, v. 28, p. 512–522.

Ryan, C.G., Griffin, W.L., and Pearson, N.J., 1996, Garnet geotherms: A technique for derivation of P–T data from Cr-pyrope garnets: Journal of Geophysical Research, v. 101, p. 5611–5625.

Schulze, D.J., 1989, Constraints on the abundance of eclogite in the upper mantle: Journal of Geophysical Research, v. 94, p. 4205–4212.

——1994, Abundance and distribution of Low-Ca-garnet harzburgite in the sub-cratonic lithosphere of southern Africa: Brazil, Companhia de Pesquisa de Recursos Minerais, CPRM Special Publication 1A, 5th International Kimberlite Conference, p. 327–335.

——1997, The significance of eclogite and Cr-poor megacryst garnets in diamond exploration: Exploration and Mining Geology, v. 6, p. 349–366.

Scott-Smith, B.H., 2006, Canadian kimberlites: geological characteristics relevant to emplacement, Emplacement Workshop, Saskatoon, Saskatchewan, September 7–12, 2006, CD-ROM.

Shirey, S.B., Richardson, S.H., and Harris, J.W., 2004, Age, paragenesis, and composition of diamonds and evolution of the Precambrian mantle lithosphere of southern Africa: Transactions of the Geological Society of South Africa, v. 107, p. 91–106.

Simakov, S.K., 1999, Garnet—Clinopyroxene geobarometry of deep mantle eclogites and eclogite paleogeotherms: University of Cape Town, 7th International Kimberlite Conference, Cape Town, April 13–17, 1998, Proceedings, p. 783–787.

Sobolev, N.V., 1971. On mineralogical criteria of a diamond potential of kimberlites: Geologiya i Geofizika, v. 12 (3), p. 70–78 (in Russian).

——1977, Deep-seated inclusions in kimberlites and the problem of the composition of the upper mantle: Washington, American Geophysical Union, 279 p.

Sparks, R.S.J., Baker, L., Brown, R.J., Field, M., Schumacher, J., Stripp, G., and Walters, A.L., 2006, Dynamics of kimberlite volcanism: Journal of Volcanology and Geothermal Research, v. 155, p.18–48.

Stripp, G.R., Field, M., Schumacher, J.C., Sparks, R.J.S., and Cressey, G., 2006, Post emplacement serpentinization and related hydrothermal metamorphism in a kimberlite from Venetia, South Africa: Journal of Metamorphic Petrology, v. 24, p. 515–534.

Stubley, M., 2004, Spatial distribution of kimberlite in the Slave Province: A geometrical approach: Lithos, v. 77, p. 683–694.

Tappe, S., Foley, S.F., Jenner, G.A., and Kjarsgaard, B.A., 2005, Integrating ultramafic lamprophyre into the IUGS classification of igneous rocks: Rationale and implications: Journal of Petrology, v. 46, no. 9, p. 1893–1900.

van der Lee, S., and Frederiksen, A., 2005, Surface wave tomography applied to the North American upper mantle, *in* Levander, A. and Nolet, G., eds., Seismic earth: Array analysis of broadband seismograms: Washington, American Geophysical Union, p. 67–80.

Wagner, P.A., 1914, The diamond fields of South Africa: Johannesburg, Transvaal Leader, 347 p.

Walker, S., 2003, Potential field geophysics in diamond exploration:Diamonds Short Course Notes, Cordilleran Round-Up, Vancouver, British Columbia, January, 29–30, 2003, 24 p.

White, S.H., de Boorder, H., and Smith, C.B., 1995, Structural controls of kimberlite and lamproite emplacement: Journal of Geochemical Exploration, v. 53, p. 245–264.

Wilkinson, L., Kjarsgaard, B.A., LeCheminant, A.N., and Harris, J., 2001, Diabase dyke swarms in the Lac de Gras area: Significance to kimberlite exploration—initial results: Geological Survey of Canada, Current Research 2001-C8, 17 p.

Wilson, M.R., Kjarsgaard, B.A., and Taylor, B.E., 2007, Stable isotopic composition of magmatic and deuteric carbonate phases in hypabyssal kimberlite, Lac de Gras field, Northwest Territories, Canada: Chemical Geology, in press.

Wyatt, B.A., Baumgartner, M., Anckar, E., and Grütter, H., 2004, Compositional classification of kimberlitic and non-kimberlitic ilmenite: Lithos, v. 77, p. 819–840.

Zonneveld, J.-P., Kjarsgaard, B.A., Harvey, S.E., Heaman, L.M., McNeil, D.H., and Marcia, K.Y., 2004, Sedimentologic and stratigraphic constraints on emplacement of the Star Kimberlite, east-central Saskatchewan: Lithos, v. 76, p. 115–138.

Zonneveld, J.-P., Kjarsgaard, B.A., Harvey, S.E., and McNeil, D.H., 2006, Accommodation space and kimberlite edifice preservation: Implications for volcanological models of Fort à la Corne kimberlites, Kimberlite Emplacement Workshop, Saskatoon, Saskatchewan, September 7–12, 2006, CD-ROM.

Unconformity-Associated Uranium Deposits of the Athabasca Basin, Saskatchewan and Alberta

C.W. Jefferson[1], D.J. Thomas[2], S.S. Gandhi[1], P. Ramaekers[3], G. Delaney[4], D. Brisbin[2], C. Cutts[5], D. Quirt[5], P. Portella[5], and R.A. Olson[6]

1. Geological Survey of Canada, 601 Booth Street, Ottawa, Ontario K1A 0E8
2. Cameco Corporation, 2121 - 11th Street West, Saskatoon, Saskatchewan S7M 1J3
3. MF Resources, 832 Parkwood Drive SE, Calgary, Alberta T2J 2W7
4. Saskatchewan Industry and Resources, 2101 Scarth Street, Regina, Saskatchewan S4P 3V7
5. AREVA Resources Canada Inc., P.O. Box 9204, 817 - 45th Street W., Saskatoon, Saskatchewan S7K 3X5
6. Alberta Geological Survey, Energy Utilities Board, 4th Floor, Twin Atria, 4999 - 98 Avenue, Edmonton, Alberta T6B 2X3
Corresponding author's email: cjefferson@nrcan.gc.ca

Abstract

This review of the geology, geophysics, and origin of the unconformity-associated uranium deposit type is focused on the Athabasca Basin. Pods, veins, and semimassive replacements of uraninite (var. pitchblende) are located close to unconformities between late Paleo- to Mesoproterozoic conglomeratic sandstone basins and metamorphosed basement rocks. The thin, overall flat-lying, and apparently unmetamorphosed but pervasively altered, mainly fluvial strata include red to pale tan quartzose conglomerate, sandstone, and mudstone. Beneath the basal unconformity, red hematitic and bleached clay-altered regolith grades down through chloritic altered to fresh basement gneiss. The highly metamorphosed interleaved Archean to Paleoproterozoic granitoid and supracrustal basement gneiss includes graphitic metapelitic that preferentially hosts reactivated shear zones and many deposits. A broad variety of deposit shapes, sizes and compositions ranges from monometallic and generally basement-hosted veins to polymetallic lenses located just above or straddling the unconformity, with variable Ni, Co, As, Pb and traces of Au, Pt, Cu, REEs, and Fe.

Résumé

Cet examen de la géologie, de la géophysique et de l'origine des gîtes d'uranium associés à des discordances est focalisé sur le bassin d'Athabasca. Les minéralisations d'uraninite (variété pechblende) qui prennent la forme de lentilles, de filons ou de corps semi-massifs de remplacement se situent près de la discordance entre des grès conglomératiques de bassin du Paléoprotérozoïque tardif au Mésoprotérozoïque et les roches du socle métamorphisées. La succession sédimentaire de bassin est mince, repose dans l'ensemble à plat, est apparemment non métamorphisée, mais profondément altérée, et se compose principalement d'unités fluviatiles constituées de conglomérats quartzeux, de grès et de mudstones de couleur rouge à chamois pâle. Sous la discordance marquant la base de la succession sédimentaire, un régolite hématitique rouge, décoloré par endroits par une altération argileuse, passe progressivement vers les profondeurs du socle à des gneiss chloritisés puis à des gneiss non altérés. Les roches très métamorphisées du socle, formées d'une intercalation de gneiss granitoïdes et de gneiss supracrustaux de l'Archéen au Paléoprotérozoïque, incluent des métapélites graphitiques qui renferment de manière préférentielle des zones de cisaillement réactivées et un grand nombre de gîtes. Les gîtes sont de formes, de dimensions et de compositions très variées, passant de minéralisations monométalliques, généralement sous forme de filons encaissés dans le socle, à des lentilles polymétalliques présentant des concentrations variables de Ni, Co, As, Pb et des traces de Au, Pt, Cu, de terres rares et de Fe, qui se situent à cheval sur la discordance ou à peu de distance au-dessus.

Introduction

This synopsis of unconformity-associated uranium (also unconformity-related and -type) deposits emphasizes the Athabasca Basin. The empirical term 'associated' is chosen because some genetic aspects are still under debate and the deposits occupy a wide range of spatial positions and shapes with respect to the unconformity. An expanded version of this paper introduces the final volume for EXTECH IV, Athabasca Uranium Multidisciplinary Study (Jefferson et al., 2007). Citations therefore include the most recent publications of EXTECH IV in addition to classic references.

After a concise definition, the grade, tonnage, and value statistics of unconformity-associated uranium deposits are provided in global and Canadian context. Geological attributes are summarized on continental, district, and deposit scales: favourable expressions of deposits; their size, morphology, and architecture; ore mineralogy and composition; and alteration mineralogy, geochemistry, and zonation. Key exploration criteria are summarized for geology, geochemistry, and geophysics. Genetic and exploration models are reviewed in terms of conventional knowledge and recent advances, with reference to uranium sources, transport and focus of deposition. Conceptual and applied knowledge gaps are evaluated at the district and deposit scales. New lines of research and areas of uranium potential are proposed.

Definition

Unconformity-associated uranium deposits are pods, veins, and semimassive replacements consisting of mainly uraninite dated mostly 1600 to 1350 Ma, and located close to basal unconformities between Proterozoic redbed basins and metamorphosed basement rocks, especially supracrustal gneiss with graphitic metapelite. Prospective basins in Canada (Figs. 1, 2) are 1 to 3 kilometres thick, relatively flat-lying, unmetamorphosed but pervasively altered, Proterozoic (ca. 1.8 to <1.55 Ga), mainly fluvial conglomeratic sandstone. The basement gneiss is paleoweathered with

Jefferson, C.W., Thomas, D.J., Gandhi, S.S., Ramaekers, P., Delaney, G., Brisbin, D., Cutts, C., Quirt, D., Portella, P., and Olson, R.A., 2007, Unconformity-associated uranium deposits of the Athabasca Basin, Saskatchewan and Alberta, *in* Goodfellow, W.D., ed., Mineral Deposits of Canada: A Synthesis of Major Deposit-Types, District Metallogeny, the Evolution of Geological Provinces, and Exploration Methods: Geological Association of Canada, Mineral Deposits Division, Special Publication No. 5, p. 273-305.

variably preserved thicknesses of reddened, clay-altered hematitic regolith grading down through a green chloritic zone into fresh rock. Monometallic, generally basement-hosted ore pods, veins, and breccia in reactivated fault zones. Polymetallic, commonly subhorizontal ore lenses straddle the unconformity, replacing sandstone and altered basement rock with variable amounts of U, Ni, Co, and As, and traces of Au, PGEs, Cu, REEs, and Fe.

Grade, Tonnage, and Value Statistics

Global Unconformity Resources

World uranium resources are contained in some fourteen different types of deposits (Organization for Economic Co-operation and Development, Nuclear Energy Agency, and the International Atomic Energy, 2004), with Proterozoic unconformity deposits constituting more than 33%, mainly in Australia and Canada. Uranium resource data for Canadian and comparative Australian unconformity deposits are compiled with original references in the digital Canada Minerals Database by Gandhi (2007). Appendix 1 (CD-ROM) and Figure 3 summarize individual grades and tonnages of 42 Canadian and Australian deposits, illustrating their current relative importance. Table 1 provides totals for the Athabasca and Thelon basins. The Hornby Bay and Elu basins are less well explored and no unconformity resources have been outlined (Roscoe, 1984; Gandhi, 2007).

Unconformity deposits of the Athabasca Basin are the world's largest storehouse of high-grade U resources and are the sole producers of Canada's primary U. The most spectacular grades and tonnages (Appendix 1, Fig. 3) are those of Cigar Lake (east and west zones combined=875 kilotonnes of ore grading ~15% and containing 131,400 tonnes U) and McArthur River (1017 kilotonnes of ore grading ~22.28% and containing 192,085 tonnes U). The average

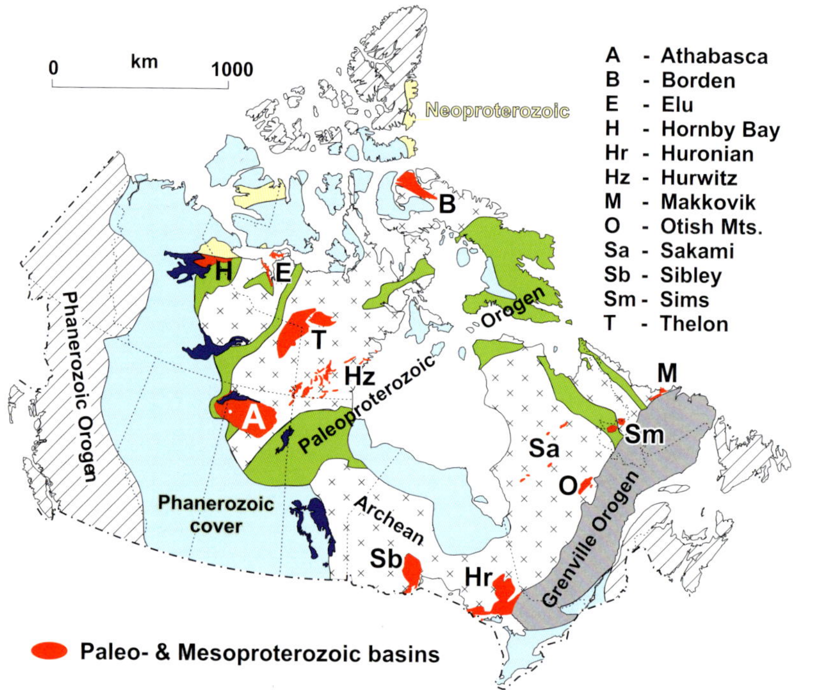

FIGURE 1. Paleo- to Mesoproterozoic basins within the Canadian Shield that contain unconformity-associated uranium deposits (e.g. Athabasca and Thelon) or are considered to have potential for them.

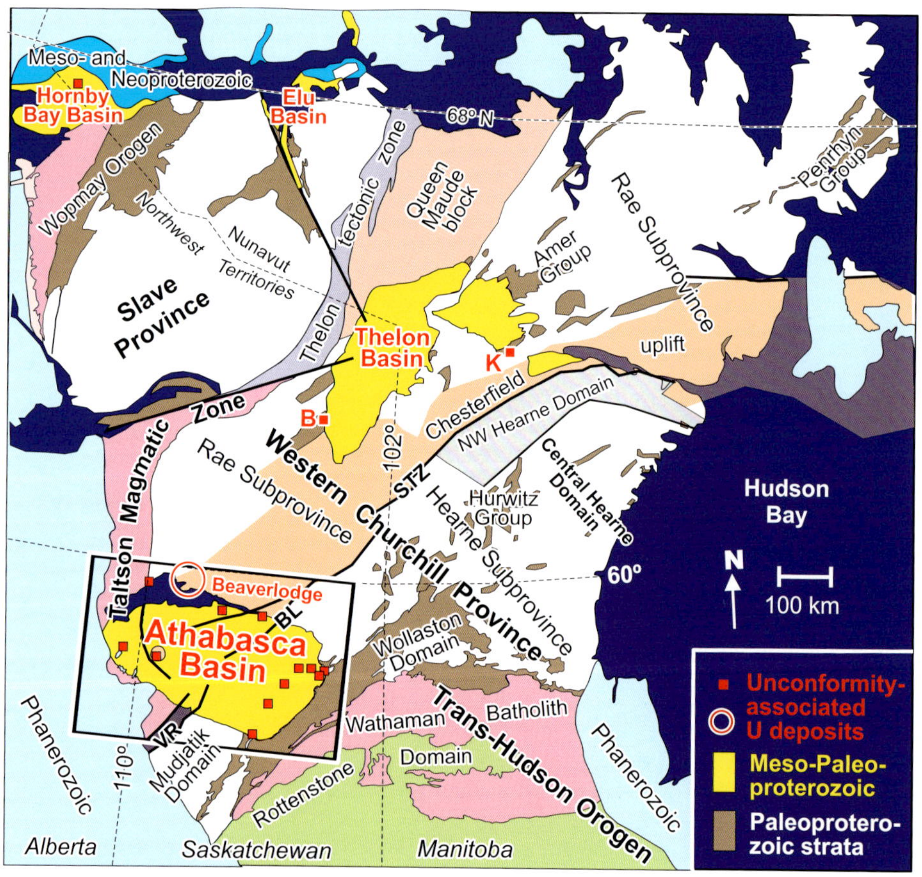

FIGURE 2. Location of unconformity-associated occurrences and the Athabasca Basin relative to major tectonic elements of the northwestern Canadian Shield, after Thomas et al. (2000) and Pehrsson (pers. comm., 2005). Occurrences are listed in the Appendix 1, known conventional resources are plotted in Figure 3, and Athabasca Basin occurrences are located in more detail on Figure 4A. Unconformity–associated prospects of the Thelon Basin are Boomerang Lake (B) and Kiggavik (K). Hornby Bay Basin (HBB) hosts the sandstone-hosted PEC-YUK prospect.

grade for some 30 unconformity deposits in the Athabasca Basin, including these two high-grade examples, drops to 1.97% U, still four times the average grade (0.44% U) of Australian unconformity deposits (Table 1) and more than ten times that of the Beaverlodge district (Smith, 1986, p. 99).

Reasonably assured (terminology of Organization for Economic Co-operation and Development, Nuclear Energy Agency, and the International Atomic Energy, 2004) U resources of the Kombolgie Basin in northern Australia are slightly more than 50% of those in the Athabasca Basin. Aside from the above noted lower grade and the lack of sandstone-hosted deposits, these large-tonnage basement-hosted deposits are geologically similar to the Athabasca resources. They are confined to a relatively small portion of the basin, about 7,500 km^2, known as the Alligator Rivers Uranium field. The Kintyre deposit is a large tonnage and low-grade unconformity deposit in Western Australia, comparable with those in the Alligator Rivers Uranium field. Much of the Kintyre deposit is basement-hosted like those of the Alligator Rivers Uranium field.

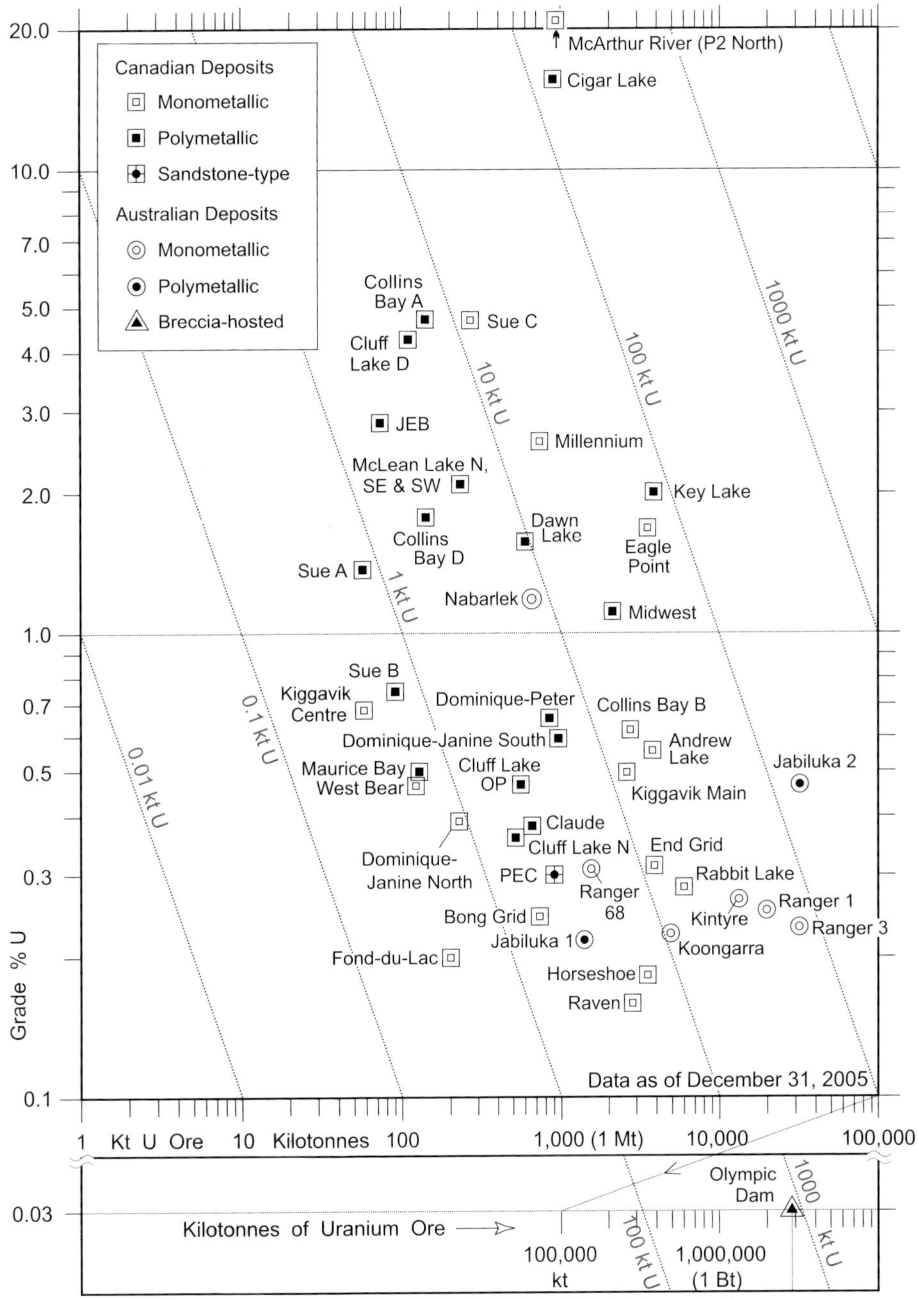

FIGURE 3. Grade versus tonnage plot of the unconformity and selected other types of uranium deposits in Canada and Australia (after Ruzicka, 1996a and Gandhi, 1995). Data linked by name in Appendix 1 (CD-ROM).

Canadian Unconformity-Associated Uranium Resources – Current and Past Producers

In 1997, Canadian U production was entirely from unconformity deposits and represented approximately 34% of the world's total. At that time Canadian U sales were 11,274 tonnes U (29.3 M lbs U_3O_8) reportedly valued at $402.25 million (US), entirely from the Saskatchewan portion of the Athabasca Basin. Canada's production gradually declined to 28% of the world's primary U by 2003. Canadian production may nevertheless reach 50% of world requirements given production from Cigar Lake (Robertson, 2006a) that will be milled at McClean Lake and Rabbit Lake, with continuation of production from McArthur River milled at Key Lake (Uranium Information Centre, Melbourne, Australia, 2006, 2007). Midwest (2,200 tonnes U) is another likely new producer in the Athabasca Basin.

The Athabasca Basin (Fig. 4) is by far the most significant U metallogenic district in Canada, in terms of known deposits and being the only current producer. It covers more than 85,000 km^2, but 96% of its known U resources underlie a limited zone near the eastern margin of the basin. Past producers and new discoveries demonstrate high potential at a number of other places across the basin. Intense new exploration is reevaluating existing prospects and developing new prospects (e.g. Millennium (Robertson, 2006b) and Shea Creek).

The Martin Basin and its closely underlying basement rocks north of Lake Athabasca, known as the Beaverlodge district, produced significant U from the 1950s to 1980s, (Fig. 2; Beck, 1969) with a fascinating history of discovery and development (Reeves and Beck, 1982). Past-producing 'classic vein U' (Ruzicka, 1996b) deposits in the Beaverlodge have long been known to be spatially associated with the unconformity beneath the Martin Group.

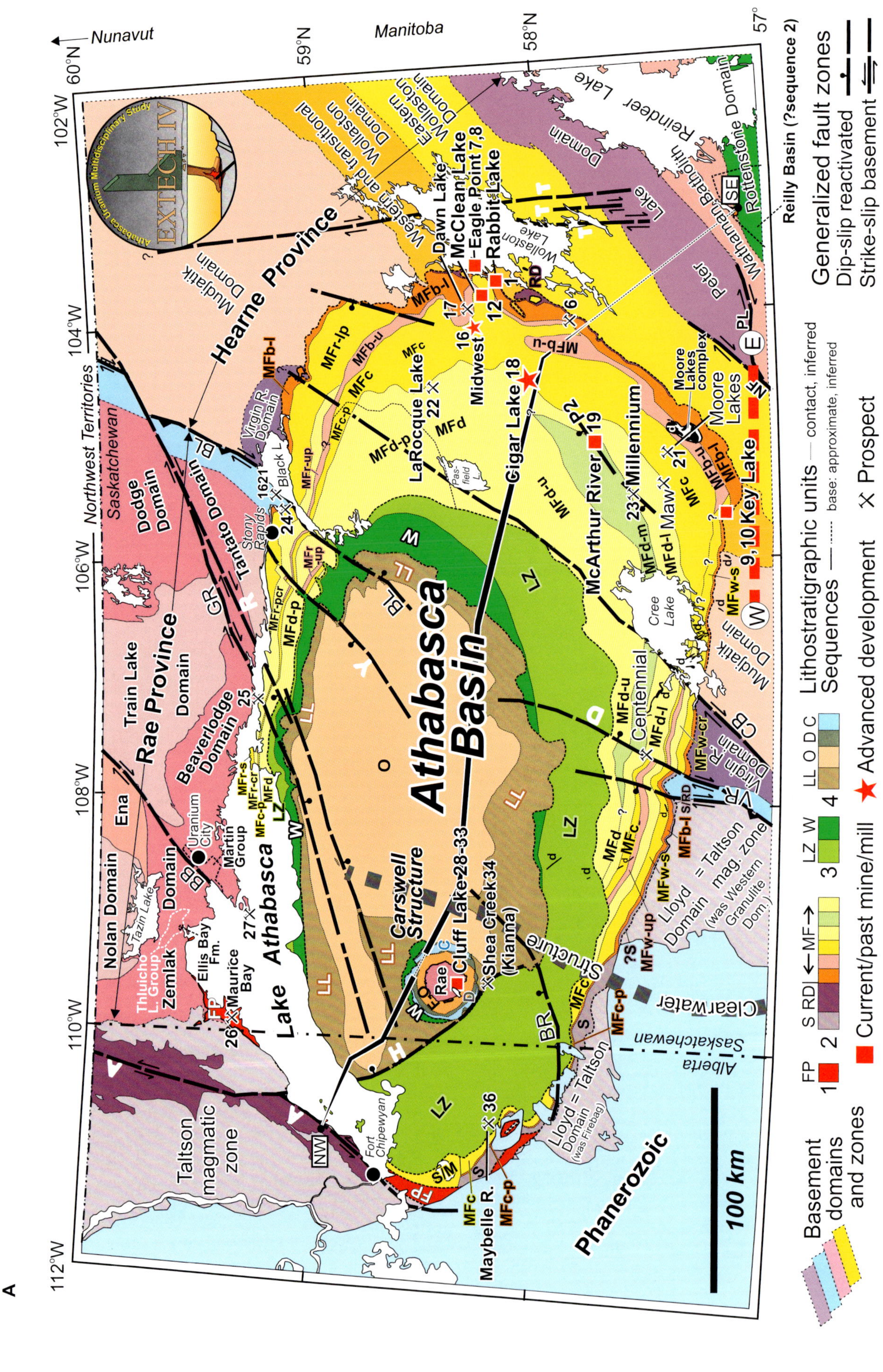
A
Nunavut
Manitoba
Northwest Territories
Saskatchewan
Alberta
60°N
59°N
58°N
57°
112°W
110°W
108°W
106°W
104°W
102°W
Athabasca Basin
Hearne Province
Rae Province
Lake Athabasca
Carswell Structure
Clearwater Structure
Phanerozoic
Mudjatik Domain
Western and transitional Wollaston Domain
Eastern Wollaston Domain
Peter Lake Domain
Reindeer Lake
Wathaman Batholith
Rottenstone Domain
Reilly Basin (?sequence 2)
Wollaston Lake
Dawn Lake
McClean Lake
Eagle Point 7,8
Rabbit Lake
Midwest
Cigar Lake 18
LaRocque Lake
McArthur River 19
Millennium
Moore Lakes complex
Moore Lakes
Key Lake
Cree Lake
Centennial
Maw
Pasfield L.
Black L.
Stony Rapids
Virgin R. Domain
Dodge Domain
Tantato Domain
Train Lake Domain
Beaverlodge Domain
Uranium City
Martin Group
Nolan Domain
Ena Domain
Tazin Lake
Zemlak Domain
Thluicho L. Group
Ellis Bay Fm.
Maurice Bay
Cluff Lake 28-33
Shea Creek 34
(Kianna)
Taltson magmatic zone
Fort Chipewyan
Maybelle R.
Lloyd Domain = Taltson (was Firebag)
Lloyd Domain = Taltson mag. zone (was Western Granulite Dom.)
100 km
EXTECH IV
Athabasca Uranium Multidisciplinary Study
Basement domains and zones
Lithostratigraphic units — contact, inferred
Sequences — base: approximate, inferred
Generalized fault zones
Dip-slip reactivated
Strike-slip basement
Current/past mine/mill
Advanced development
Prospect
FP 1
S RD 2
MF
LZ W 3
LL O D C 4

B

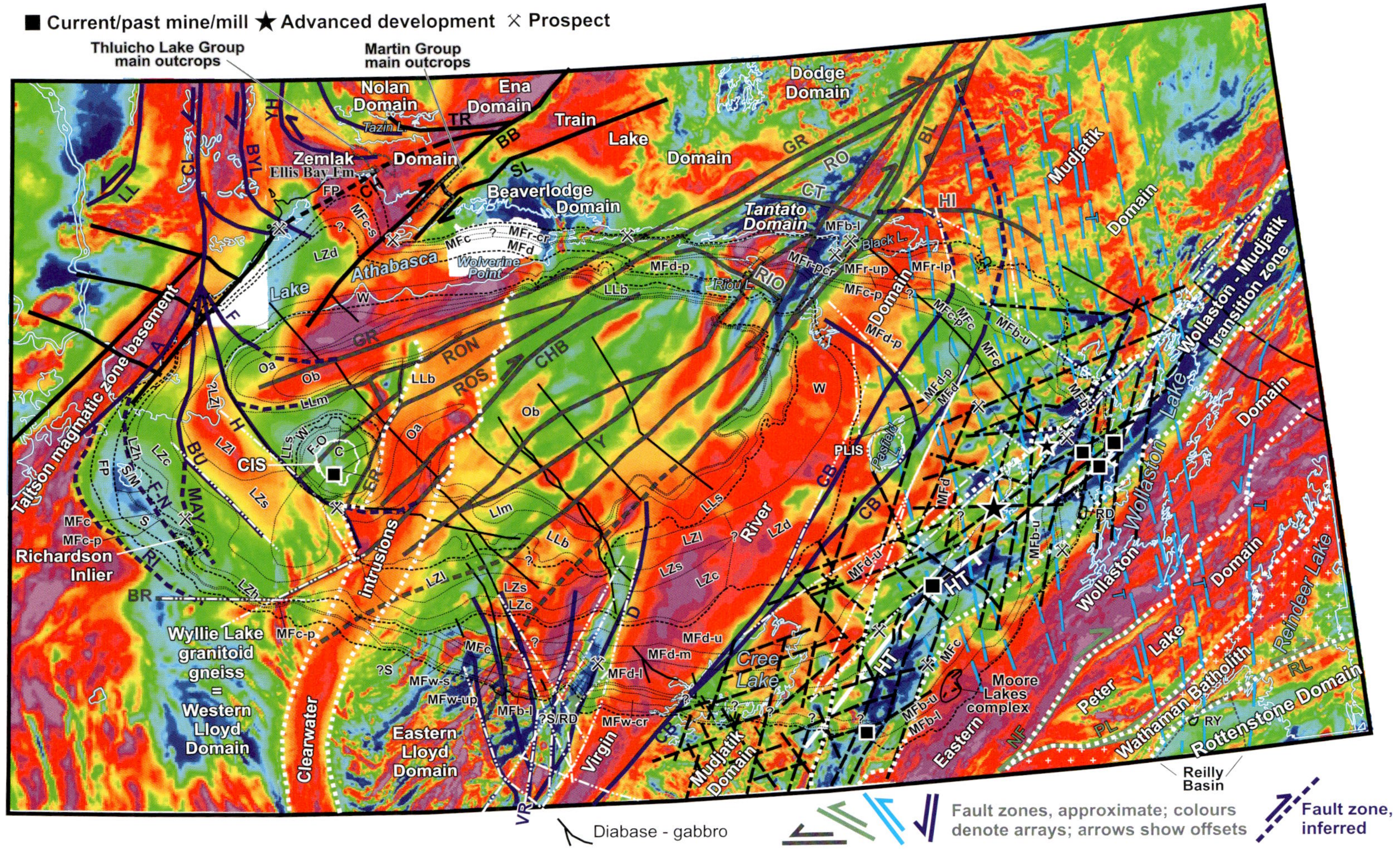

FIGURE 4. Geological setting and unconformity-associated uranium occurrences (numbered as in the Appendix on CD-ROM) of the Athabasca Basin region in northern Saskatchewan and Alberta, where Rae and Hearne are termed 'Province' rather than Subprovince (see Fig. 2). Symbols and fonts are slightly larger for more significant occurrences. Basement geology is after Portella and Annesley (2000a), Thomas et al. (2002), Card et al. (2003, 2007a, b), and Card (2006). Athabasca Basin geology and stratigraphic units (Table 3) are from Ramaekers et al. (2007). C=Carswell, D=Douglas, FP=Fair Point, LL=Locker Lake, LZ=Lazenby Lake, MF=Manitou Falls (members: b=Bird (l=lower, u=upper) c=Collins, d=Dunlop, r=Raibl (up=upper pebbly), w=Warnes (up=upper pebbly)), O=Otherside, RD=Read, S=Smart, W=Wolverine Point, d=diabase. Members of LZ, LL, and O are indicated by lines and labels but only one shade is used per formation. The western Wollaston Domain and the Wollaston – Mudjatik transition of Portella and Annesley (2000a,b) are combined here as "Wollaston-Mudjatik transition zone". CIS=Carswell Structure. Generalized fault zones after Ramaekers (2004) include multiple ductile movements before deposition of Athabasca Group and brittle transcurrent and dip-slip movements during and after deposition; they are named as: A=Allan, BB=Black Bay, BL=Black Lake, BR=Beatty River, BU=Bustard, CB=Cable Bay, CH=Charlot, CHB=Charbonneau, CL=Charles Lake, CT=Clut, D=Dufferin, ER=East Rim, F=Fidler, FN=Fowler–Net Lake, GR=Grease River, H=Harrison, HT=Hudsonian thrusts (general trajectory), LL=Leland Lakes, MAY=Maybelle, NF=Needle Falls, PL=Parker Lake, P2=P2 fault at McArthur River, R=Richardson, RI=Riou, RL=Reilly Lake, RO=Robillard, RON=Robillard north, ROS=Robillard south, SL=St. Louis, T=Tabbernor, VR=Virgin River array (Dufferin is one named fault of many in VR), Y=Yaworski, YH=Yatsore-Hill Island. Arrays of faults with similar orientation and offset are indicated by colour groups. **A)** Simplified bedrock geology. Cross-sections of Figure 5A and B are located along dotted lines labelled NW-SE (along the basin axis) and E-W (south of Key Lake). **B)** Outlines of stratigraphic units of the Athabasca Group (black), basement domains (white outlines) and major reactivated fault systems (heavy coloured lines) on total magnetic field (Geological Survey of Canada, 1987; Pilkington, 1989). Faults are after Portella and Annesley (2000a), Ramaekers (2004), Card et al. (2007a), and Thomas and McHardy (2007). Many late faults have limited offsets that cannot be shown at this scale (see Card, 2006).

TABLE 1. Summary of uranium resources in major Paleo- and Mesoproterozoic districts of northwestern Canada (shaded) and Australia; data from Appendix 1.

District	Kt Ore[1]	% U[2]	Tonnes U
Athabasca Basin	29,811	1.97	587,063
Beaverlodge District[3]	15,717	0.165	25,939
Thelon Basin	11,989	0.405	48,510
Hornby Bay Basin	900	0.3	2,700
Kombolgie Basin	87,815	0.323	283,304
Paterson Terrane	12,200	0.25	30.5
Olympic Dam[4]	2,877,610	0.03	863,283

1. Includes past production.
2. Calculated from Kt ore and tonnes uranium, rounded to significant digits.
3. Past production from two "classic vein-type" (Eldorado and Lorado Mills) and one episyenite-type (Gunnar) deposits.
4. Genetically linked with the 1850 Ma Gawler Range volcano-plutonic complex. Olympic Dam is breccia hosted, not unconformity-associated, but is included here for comparison because it is such a vast individual resource of uranium, of approximately the same age as the unconformity-associated deposits listed here (references in Gandhi, 2007).

Uranium ore veins are present only in the basement near its basal unconformity and are absent where the Martin Group pinches out, despite continuity of favourable basement lithologic units and structures (e.g. Robinson, 1955, p. 73-74; Tremblay, 1968; Mazimhaka and Hendry, 1989; T. Trueman, oral presentation in Saskatoon, November 2005). Exploration is now reevaluating this district.

Canadian Unconformity-Associated Uranium Resources – Potential Producers

The Thelon Basin is very similar in size, general age ,and geological attributes to the Athabasca Basin (Figs. 1, 2; Table 2), yet its reasonably assured U resources are 9% of the Athabasca Basin. These are concentrated in basement-hosted deposits of the Kiggavik trend, an area less than 500 km2 with an average grade similar to that of the Kombolgie Basin. Exploration is active throughout the Thelon Basin, except in the Thelon Game Sanctuary.

The term 'Hornby Bay Basin' refers to the region outlined by exposures of the Hornby Bay Group. This restricted usage excludes the 'Amundsen Basin' (regional stratigraphy defined by Rainbird et al., 1994) that, although used by Kerans et al. (1981) as including Hornby Bay Group, is too broad to provide metallogenic discrimination (see below). The only deposit in Hornby Bay Basin with measured resources (PEC prospect, Appendix 1, Fig. 3) is classified as sandstone type (Bell, 1996) and is the only significant example of this deposit type in the Proterozoic of Canada. Nonetheless, the Hornby Bay Basin is viewed as correlative with the Athabasca and Thelon basins, and is being intensively explored for unconformity deposits. The Elu Basin, extending from Bathurst Inlet north to Hadley Bay on Victoria Island (Figs. 1, 2), is broadly correlative with the Hornby Bay Basin (Campbell, 1979). Additional prospective basins, such as Huronian, upper Hurwitz, Otish, Sibley, and Sims, are discussed under *Ages of Known and Prospective Districts* below.

Geological Attributes

Continental-Scale Geological Attributes

The continental-scale geotectonic environment of significant unconformity-associated U deposits is at the base of flat-lying, fluvial redbed strata on peneplaned tectonometamorphic complexes in the interiors of large cratons. The Athabasca and Thelon basins are located on the western Churchill Province between the eroded remnants of two major orogenic belts: the ca. 1.9 Ga Taltson magmatic zone to Thelon tectonic zone and the ca. 1.8 Ga Trans-Hudson Orogen (Fig. 2). These belts accommodated ductile transpression during convergence of the Slave and Superior provinces (e.g. Hoffman, 1988) and form the western and eastern portions of the Rae and Hearne provinces, respectively. High heat flow could have resulted from the voluminous radiogenic granitoid intrusions within these orogens.

TABLE 2. Comparison of Athabasca and Thelon basins (after Miller and LeCheminant, 1985; Gandhi, 1989; Kyser et al., 2000).

Attribute	Athabasca	Thelon
Graphitic metasedimentary rocks beneath ore	Distinct	Locally
Paleoweathering profile below basal unconformity	Shallow to deep	Shallow to deep
Subbasins developed via reactivated faults	Yes	Yes
Maximum age of sedimentation (Ma)	ca. 1720-1750	ca.1720
Fluorapatite	Yes	Yes
Aeolian sandstone	Possible	Yes
Arkosic sandstone regionally clay altered	Minor	Yes
Quartz overgrowths preserve hematite rims	Yes	Yes
Early detrital kaolin in matrix	Yes	No?
Peak diagenetic clay minerals	Dickite + illite	Illite
Peak diagenetic / hydrothermal temperatures	~240°	~200°
Illite incorporates Mg and Fe	in regolith only	Variable
Corroded zircon grains near ore zones	Local	No?
Regional fresh zircon	Yes	Yes
Extensive aluminum phosphate ± sulphate	Yes	Yes
potassium-feldspar + chlorite at 1 Ga	No	Yes
Late vein carbonate from meteoric water	Yes	Yes
Bleaching and clay alteration halos	Yes	Yes
Sandstone / unconformity-hosted uranium	Yes	One example
Basement-hosted uranium	Yes	Yes
Significant deposits	Yes	One

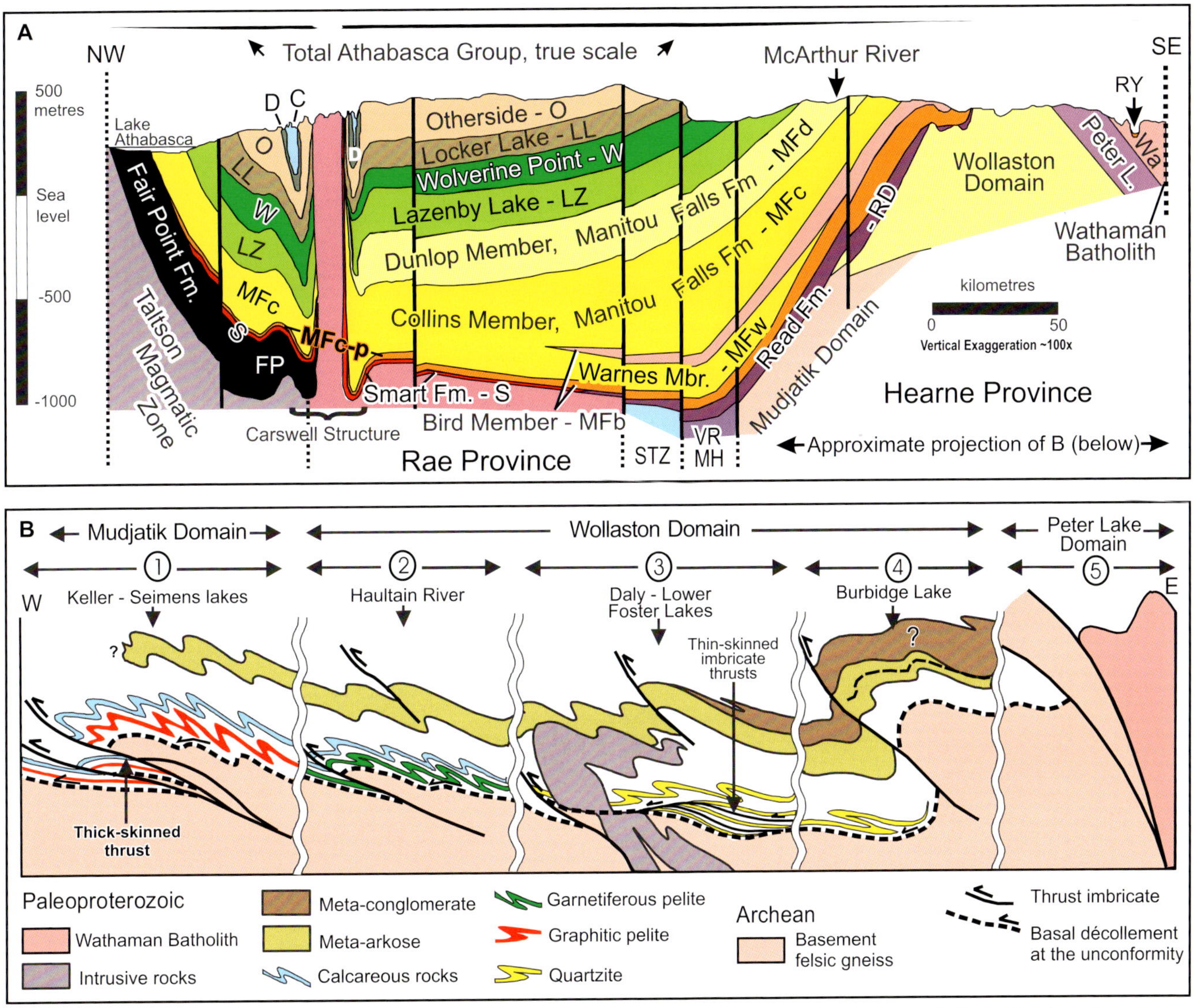

FIGURE 5. **A)** Lithostratigraphic cross-section of the Athabasca Basin, after Ramaekers (1990) and Ramaekers et al. (2007). True 1:1 scale is shown at top. Stratigraphic units in the vertically exaggerated section are formations except those starting with MF, which are members of Manitou Falls Formation; MFw is Warnes Member. Basement domains are diagrammatic. Line of section is shown as NW-SE in Figure 4A. **B)** Diagrammatic structural cross-section south of Key Lake, adapted from Tran (2001) and Györfi et al. (2007), illustrates structural geometry of the Wollaston-Mudjatik transition zone that underlies the most economically productive area of the eastern Athabasca Basin. Line of section is shown as W-E in Figure 4A. The circled numbers 1 through 5 refer to discrete areas mapped by Tran (2001) from which the components of this diagram were obtained.

There is no suggestion that the Athabasca Basin region was a site of enhanced mafic magmatism (Buchan and Ernst, 2004).

Original thicknesses and lateral extents of the most prospective basins were somewhat greater than what is preserved, but far less than foreland or continent-margin basins. The Canadian basins are interpreted as discrete 'lakes of gravel and sand' separated by large areas of limited accommodation (Ramaekers et al., 2007). The maximum cored thickness of the Athabasca Group in any one place is about 1500 m – the four major depositional sequences separated by basin-wide unconformities (Table 3) record repeated deposition and erosion during about 200 Ma. Shallow marine strata are minor or cap the redbed sequences (Figs. 4, 5).

The Thelon Basin developed in an interior position, very much like the Athabasca Basin (Rainbird et al., 2003a), with some 1800 m (Overton, 1977) comprising mainly the Thelon Formation with three depositional fluvial sequences (Hiatt et al., 2003) capped by the thin, volcanic Kuungmi and carbonate Lookout Point formations (Cecile, 1973; Rainbird et al. 2003a). In the Hornby Basin, paleocurrents from the west for basal units (Kerans et al., 1981) indicate the Hornby Bay Basin also formed within an intracratonic setting, but thicknesses are greater and more variable, with evidence for syndepositional compression (MacLean and Cook, 2004).

The older, past-producing Martin Basin north of Lake Athabasca (Mazimhaka and Hendry, 1989) and the prospective Baker Lake Basin east of Thelon Basin (Miller, 1980) are distinct pull-apart structures filled by relatively thick but laterally restricted sequences and host smaller, lower grade, unconformity-associated deposits and occurrences. The younger, Amundsen Embayment, located northwest of Hornby Bay and Elu basins (Young et al., 1979), had a marginal marine setting with no coarse siliciclastic rocks, layercake stratigraphy, and a metallogeny dominated by sedimen-

Table 3. Lithostratigraphic units and unconformity-bounded sequences of the Athabasca Group (after Ramaekers et al., 2007). The framework mineral for all textural types from conglomerate to mudstone is 99% quartz. Every unit contains crossbedding and ripple crosslamination, and most contain single-layer thick quartz pebble or granule beds. Only diagnostic stratigraphic parameters, such as grain-size and desiccation cracks, are summarized here. Aggregate maximum thickness of each formation is summarized in metres in the left-hand column. Inferred minimum age of basement and U-Pb age of volcanic zircon from Wolverine Point are by Rainbird et al. (2007). Douglas Formation Re-Os age is by Creaser and Stasiuk (2007). [S] = present only in Saskatchewan.

Formation [code]	**Member**; [code] (textural lithology) MTG = maximum transported grain size in mm	**Sequence**
Carswell [C] 500 m	Upper and lower carbonate (dololutite, dolorudite, stromatolite, oolite, dolarenite) [S]	4
Douglas [D] 300 m	(dark grey carbonaceous mudstone with desiccation or synaeresis cracks and interbeds of fine to very fine quartzarenite, MTG <2). Organic matter is 1541 ± 13 Ma by Re-Os isochron[S]	
	Now recognized as up-faulted Locker Lake Formation[S]	
Otherside [O] 183 m	Birkbeck [Ob] (quartzarenite with minor thin interbeds of dark mudstone near the top; MTG <2 except for pebbly unit near base)	
	Archibald [Oa] (quartz-pebbly quartzarenite, quartzarenite; MTG<8)	
Locker Lake [LL] 288 m	Marsin [LLm] (quartz-pebbly quartzarenite; MTG 8-16)	
	Brudell [LLb] (thin conglomerate beds in quartzarenite; MTG >16)	
	Snare [LLs] (quartz-pebbly quartzarenite; MTG 2-16, sparse mudstone interbeds <50 cm thick)	
	Unconformity	Unconformity
Wolverine Point [W] 186 m	Claussen [Wc] (interstitial-clay-rich quartzarenite, sparse mudstone interbeds <1 m thick; MTG <2)	3
	Brule [Wb] (interbedded mudstone >50 cm and tuffaceous quartzarenite, common thin intraclast conglomerate; MTG <2 except for local basal lag. Zircon in tuff intraclasts is 1644 ± 13 Ma by U-Pb.	
Lazenby Lake [LZ] 460 m (aggregate thickness excludes laterally equivalent Dowler Member)	Dowler [LZd] (quartzarenite, minor siltstone and quartz-pebbly quartzarenite; MTG < 8)	
	Larter [LZl] (quartz-pebbly quartzarenite, minor mudstone intraclasts; MTG <8)	
	Shiels [LZs] (quartz-pebbly quartzarenite with pebbly layers, rare mudstone beds and intraclasts; MTG >8)	
	Clampitt [LZc] (pebbly base, quartz-pebbly quartzarenite, minor laminated siltstone and mudstone; MTG <8)	
	Hodge [LZh] (5-30 cm basal conglomerate, quartz-pebbly quartzarenite and conglomerate, sandstone intraclasts; MTG >8)	
	Basal unconformity to Mirror Basin	Unconformity
Manitou Falls [MF] 991 m (thickness excludes Warnes and Raibl members laterally equivalent to Bird	Dunlop [MFd] (>1% clay-intraclasts in quartzarenite, mudstone interbeds; MTG <2)[S]	2
	Collins [MFc] (quartzarenite with minor quartz pebbly beds, mudstone interbeds, <1% clay intraclasts, <2% conglomerate interbeds)	
	Warnes [MFw] (quartzarenite and clay-intraclast-rich quartzarenite in Karras Deposystem, from Virgin River area to Alberta)	
	Raibl [MFr] (pebbly quartzarenite in Moosonees Deposystem, northeastern Athabasca Basin; minor clay intraclasts, <2% conglomerate; MTG >2)[S]	
	Bird [MFb] (interbedded >2% quartz-pebble conglomerate, quartz-pebbly quartzarenite, thin mudstone and siltstone interbeds; MTG >2)	
F-O: Undivided Fair Point to Otherside formations in Carswell Structure[S]		
	Local unconformity separates Manitou Falls and Read formations[S]	
S/M: undivided Smart or Manitou Falls formations (only in Alberta)		
Smart [S] 153 m	(quartzarenite with local red mudstone and oncoid interbeds at base). May be a distal equivalent of Read Formation	
Read [RD] 156 m	(basal quartz»lithic pebble conglomerate, interbedded low-angle bedded quartzarenite, quartz-pebbly quartzarenite and quartz pebble conglomerate, common but local red quartz siltstone to mudstone intraclasts and interbeds with desiccation cracks; MTG >2)[S]	
Reilly [RY]	RYcg (conglomeratic quartzarenite)[S]	
	Basal unconformity to Reilly Basin[S]	Unconformity
Fair Point [FP] >380 m	Beartooth [FPb] (0-10% quartz»lithic-pebbly quartzarenite with abundant matrix clay; MTG generally <64)	1
	Lobstick [FPl] (interbedded >2% quartz»lithic conglomerate, quartz»lithic-pebbly quartzarenite and local basal quartz-pebbly red mudstone with minor desiccation cracks; MTG commonly >64	
No formal designation	BL (basal lag, pebbles to boulders)	Unconformity
Basement: interleaved Archean granitoid gneiss, Paleoproterozoic paragneiss, late intrusions and metamorphism (titanite ca. 1750 Ma)		

tary and volcanic copper, rather than U (Delaney et al., 1982).

The influences of plate or plume tectonics on the origins, diagenesis, and mineralization of intracontinental Proterozoic basins have been considered enigmatic (Ross, 2000). Peneplaned and deeply paleoweathered basement gneiss and the relatively thin cover recorded by flat-lying basins of continental strata imply a relatively stable tectonic environment that persisted from before to long after ore deposition. Despite the internal continental setting, complexities within these basins tell a story of subtle but highly influential transpressional and extensional tectonics that reactivated faults vital to ore formation (see the following section and Ramaekers (2004)).

District-Scale Geological Attributes

Unconformities

The first-order favourable attribute is the unconformity at the base of a relatively flat-lying and intracontinental, unmetamorphosed, late Paleoproterozoic to Mesoproterozoic, fluvial, conglomeratic sandstone, diagenetic redbed sequence. Prospective Proterozoic basins in Canada and Australia are typically underlain by extensive red hematitic paleoregoliths (Fraser et al., 1970; Cecile, 1973; Macdonald, 1980, 1985; Miller et al., 1989; Ramaekers, 1990; Gall, 1994). Variable thicknesses of regolith grade down through green chloritic altered rock into fresh basement (Hoeve and Quirt, 1984; Kyser et al., 2000). The regolith is interpreted as a result of regional paleoweathering (see review by Gall and Donaldson, 2006) that has been overprinted by diagenetic bleaching and additional hematite alteration (Macdonald, 1985).

The basement immediately below the Athabasca Group has a vertical paleoweathered profile ranging from a few centimetres up to 220 metres thick, with some deeper pockets and slivers developed along fault zones (Macdonald, 1980). Regionally, the upper portion of the regolith profile exhibits strong red hematitic alteration that grades downward through greenish chloritic alteration into fresh basement rock. Extensively within mineralized districts, a bleached zone variably overprints the top of the red zone of the paleoweathered basement and basal units of the Athabasca Group. The bleached zone comprises buff-coloured clay and quartz. It crosscuts and therefore postdates the red zone beneath. In profiles developed on the basement meta-arkose, a zone of white clay replacement of feldspar and mafic minerals separates the red and green zones. A downward progression from kaolinite to illite and chlorite is common through the regolith profile. Earthy bright red to nearly black crystalline hematite alteration in turn overprints the white alteration in mineralized areas.

In the western part of the Athabasca Basin, the unconformity between the Fair Point Formation and overlying Smart Formation is also marked by local red hematitic alteration (Ramaekers et al., 2007). Therefore, the regolith beneath the eastern part of the Athabasca Basin would have resulted from paleoweathering lasting from before deposition of the Fair Point Formation to before deposition of the Read and Manitou Falls formations, although much of the earlier paleoweathered material was probably transported into the Jackfish Basin as a component of the Fair Point detritus. This is consistent with Macdonald's (1980) suggestion that the upper, soil-textured portion of the paleosoil was mostly eroded before deposition of the Athabasca Group. Alternative views and additional details are presented by Jefferson et al. (2007).

Uranium-Rich Basement Complexes

Many workers have emphasized the favourability of basement domains that have U-enriched granite and pegmatite that were generated during regional high-grade metamorphism and anatexis of metasedimentary rocks (e.g. Thomas, 1983, Annesley et al., 1997; Madore et al., 2000; Cuney et al., 2003; Freiberger and Cuney, 2003; Hecht and Cuney, 2003). Abundant U-bearing minerals in such domains include monazite, zircon, and uraninite. The latter also forms numerous small prospects in pegmatite (Thomas, 1983) that are recorded in the Saskatchewan mineral database, west of the Needle Falls shear zone (Saskatchewan Industry and Resources, 2005). The radiogenic domains, also rich in K, Th, and rare earth elements (REE), may have selectively contributed U to the overlying basins through a variety of mechanisms (see *Genetic Models* below).

In the eastern part of the Athabasca Basin, the great majority of mines and prospects are located where the Athabasca Group unconformably overlies the transition between the western Wollaston and eastern Mudjatik basement domains (Fig. 2; Thomas, 1983; Annesley et al., 2005). This transition contains high proportions of pelitic, quartzose, and arkosic paragneiss that are isoclinally folded and interleaved with Archean orthogneiss and intruded by abundant pegmatite. Many significant deposits in this eastern area are also located at the metamorphosed unconformable contact between the Archean granitoid gneiss and late Paleoproterozoic basal Wollaston Supergroup (Yeo and Delaney, 2007), where it contains graphitic metapelitic gneiss (Annesley et al., 2005). The metapelite constitutes a weak zone that focused deformation, including a major D1 décollement (Fig. 5B), during folding and thrusting (Tran, 2001).

Significant but fewer mined deposits and prospects are also located in the basement complex of the Cluff Lake area that was exposed by the central uplift of the Carswell Structure (Lainé et al. 1985), most again associated with graphitic units and close to the overturned basal unconformity of the Athabasca Group. High-grade intersections have been reported from other western localities in the Athabasca Basin, also associated with graphitic shear zones in supracrustal belts in the underlying basement. These are summarized by Rippert et al. (2000), Brouand et al., (2003), Card (2006), Wheatley and Cutts, (2006), Card et al. (2007), Kupsch and Catuneanu (2007), and Pan et al. (2007).

Reactivated Faults

The close relationship between unconformity-associated U deposits and faults has been known since the classic reports on the Rabbit Lake deposit by Hoeve and Sibbald (1978) and Hoeve et al. (1980), and are targets of exploration in all Canadian basins. Faults in the Athabasca Basin constitute a number of arrays (Figs. 4, 5) with different attributes, such as dextral or sinistral, extensional or transpressional, and ductile or brittle, within which subsidiary splays may be invisible at the district scale but critical at the deposit scale. A number of originally ductile faults underwent repeated brittle reactivation, with offset on the order of tens to hundreds of metres, and were important for focusing mineralizing fluids. The largest of the reactivated faults offset the unconformity by hundreds of metres, e.g., the Dufferin Fault, whereas primary ductile offsets of basement units were at least tens of kilometres. For example, the P2 reverse fault offsets the unconformity by 20 to 40 m and at times was extensional (Bernier, 2004). Regional (Hajnal et al., 2007) to detailed (Györfi et al., 2007) analysis of seismic data shows the deep listric nature of the P2 fault zone and its geometric relationship to folds and thrusts of the Hudsonian Orogeny. Regional aeromagnetic data (Portella and Annesley 2000a, b; Ramaekers et al., 2007) show that many other prospective structures affect the Athabasca Basin.

These reactivated fault arrays are spatially linked with thickness, facies, and paleocurrent changes in the Athabasca Group (Ramaekers et al., 2007; Yeo et al., 2007). The fault zones served as hinge lines during some 200 million years of alternating sedimentation and erosion. Changes in basin polarity and fault valve activity would have influenced basinal fluid flow, possibly up-dip toward basin margins (Hiatt and Kyser, 2007), along paleochannel conglomerate units (Collier and Yeo, 2001; Long, 2007), but mainly vertically along the fault conduits causing basement alteration and mineralization (e.g. Hoeve and Quirt, 1984).

District-Scale Graphitic Metapelite Gneiss

Graphitic basement units are key empirical exploration parameters for U deposits in the Athabasca Basin, and underlie the northeastern and southwestern Thelon Basin (e.g. Boomerang Lake, Fig. 2) (Davidson and Gandhi, 1989) and Kombolgie Basin, Australia (Dahlkamp, 1993). Graphitic metapelite is absent beneath the eastern part of the Thelon Basin along the Kiggavik Trend (Fig. 2) although the basement-hosted deposits are located mainly in supracrustal rocks (Fuchs and Hilger, 1989). Underlying the eastern Athabasca Basin, graphitic units are stratigraphically low in the metasedimentary components of the western Wollaston and eastern Mudjatik basement domains (Fig. 5B); these are members of the Karin Lake Formation, Daly Lake Group, Wollaston Supergroup (Yeo and Delaney, 2007). The protolith of the graphitic metapelitic gneiss appears to unconformably overlie older granitoid gneiss and forms the basal interface of the overlying Wollaston Supergroup. In a tectonostratigraphic reconstruction by Tran (2001), it is interpreted to be part of an east-west facies change in basal Wollaston Supergroup units from metaquartzitic gneiss, through garnetiferous silicate-facies iron formation to weakly sulphidic graphitic metapelite (e.g. Fig. 5B). Graphitic units underlying the western Thelon Basin are identified as part of the Amer Group (Miller and LeCheminant, 1985).

Graphitic metapelitic gneiss units in the Wollaston and Mudjatik domains constitute weak zones between competent units and were foci for local deformation during regional folding, thrusting, and later brittle deformation. Similar graphitic units underlie deposits in the Maybelle River (Pană et al., 2007) and Shea Creek areas of the western Athabasca Basin where they are detectable as conductors and considered as components of supracrustal belts in the Taltson Magmatic Zone (Rippert et al., 2000; Brouand et al., 2003; Card et al., 2007a). These supracrustal components can be traced further east and northeast under the Athabasca Basin toward the Virgin River shear zone (Stern et al, 2003; Card, 2006; Card et al., 2007b). In the Cluff Lake area of the Carswell Structure, the graphitic metapelite and host metasedimentary units were interpreted as part of the Rae Province, such as those exposed north of the Athabasca Basin (Harper, 1982). Recent geochronology shows that magmatism coeval with the Taltson magmatic zone also affected both of these regions (S. Pehrsson, pers. comm., 2006). The graphitic metapelite units in these belts perhaps conducted deep crustal heat upward to the base of the Athabasca Basin where their thermal anomalies drove convection of ore-forming hydrothermal fluids (Hoeve and Quirt, 1984).

Quartz-Dominated, Uranium-Depleted Strata

The conglomeratic sandstone bodies that overlie the regolith and host parts of the polymetallic U deposits are thoroughly oxidized terrestrial redbed sequences with very long and complex diagenetic/hydrothermal-alteration histories (Table 2, Fig. 6). The preserved detrital framework in the Athabasca Basin is more than 99% quartz but much of the sand-grade material has only moderate textural maturity (Hoeve and Quirt, 1984, Ramaekers, 1990; Bernier, 2004; and Collier, 2004). Such compositional maturity in the absence of textural maturity, particularly in conglomeratic units such as the Fair Point, Read, and lower Manitou Falls formations, is typical of primary quartz-dominated detrital mineralogy that developed under a warm, tropical climate and did not require second- or third-cycle reworking (Dal Cin, 1968; Long, 2007).

Whatever the primary depositional mineralogy, the approximately 200 Ma duration of high temperature diagenesis in the Athabasca Basin ensured that any primary feldspar (Sibbald et al. 1976; Quirt, 1985) was altered to clay. Hiatt and Kyser (2000) summarized well established examples showing destruction of feldspar and other unstable minerals during burial and later diagenesis (Millikan, 1988; McBride, 1989). Nevertheless, in the bulk of the Athabasca Group, clay minerals form less than 3% of the rock, therefore if there was primary feldspar it was a very minor component (Macdonald, 1980; Kyser et al., 2000; and Wasyliuk, 2002).

In contrast, variable primary feldspar and lithic fragments are preserved in quartz- and phosphate-cemented units of the Thelon Formation (Kyser et al., 2000) whose coarser grained units have been described as lithic arkose and subarkose (Cecile, 1973; Miller, 1983). Much of the Hornby Bay Group is feldspathic (Kerans et al., 1981). Nevertheless, the sediments filling these basins are also depleted in U as indicated by airborne gamma-ray surveys.

Evidence of preexisting unstable detrital minerals in the Athabasca basin is discussed in detail by Jefferson et al. (2007). Particular attention is paid to altered mafic heavy mineral laminae that now consist of zircon and quartz framework grains surrounded by an abundant irregular matrix of secondary hematite, clay minerals, and Th-rich aluminum phosphate ± sulphate (AP) minerals, one being identified as florencite (Percival, 1989; Mwenifumbo and Bernius, 2007). These AP minerals are regionally abundant and relatively radioactive due to their Th content but essentially lack U.

<u>District-Scale Alteration</u>

In addition to paleoweathering and hydrothermal alteration of basement gneiss below the unconformity, two types of regional-scale alteration have been recognized: basin-wide pre-ore diagenetic sandstone alteration, and subbasin-scale alteration halos that outline trends and clusters of U deposits. One of the earliest recognizable regional diagenetic events in the Athabasca Basin is a pre-ore quartz overgrowth (Q1 event) that encapsulates hematite-coated detrital quartz grains. Extensive pressure solution is recorded by abundant stylolites, particularly in the Fair Point Formation (Sequence 1).

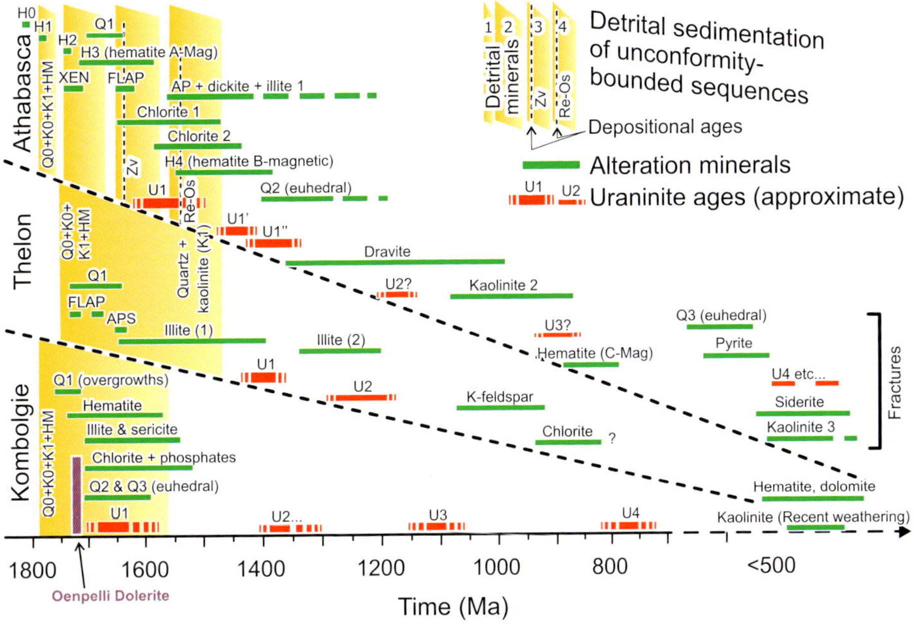

FIGURE 6. Simplified mineral paragenesis of the Paleo- to Mesoproterozoic Athabasca, Thelon, and Kombolgie basins, modified after Kyser et al. (2000, their Fig. 10.21) and Polito et al. (2004, 2005). The ages of 'primary' uraninite (the first deposition of uranium in each deposit) may be different for different deposits, hence U1 (ca. 1500-1600 Ma) and U1'-U1'' ages (e.g. ca. 1350-1400 Ma) might both be primary in the Athabasca Basin; younger ages may be a result of remobilization and addition of uranium over time. Depositional ages in the Athabasca Group are Zv: U-Pb on volcanic zircon in Wolverine Point Formation (Rainbird et al., 2007), and Os-Re: primary organic matter in Douglas Formation (Creaser and Stasiuk, 2007). The ages of depositional sequences 1 through 4 for the Athabasca Group and of uranium deposits are summarized from discussions in text. The 1723 ± 6 Ma age of the Oenpelli Dolerite is by $^{40}Ar/^{39}Ar$ (Kyser et al., 2000) and U-Pb (Edgecombe et al., 2002). AP=aluminum phosphate; APS=aluminum sulphate phosphate; FLAP=fluorapatite; H0 is primary hematite in the paleoweathered regolith; H1 and H2 are very early diagenetic hematite in basal red mudstone beds; XEN=xenotime.

The Q1 event was followed by a complex diagenetic sequence that differs in the Athabasca, Thelon, and Kombolgie basins (Fig. 6, after Kyser et al., 2000), but is independent of stratigraphy. The original mixed clay matrix of the Athabasca Basin was shown by Hoeve and Quirt (1984) to have been dominantly kaolinitic with small amounts of montmorillonite, a range of chlorite minerals, and a low-Mg-Fe illite (after Nickel and Nichols, 1991). Regional diagenesis converted this to a mixture dominated by dickite, a higher crystallinity polymorph of kaolinite, with minor amounts of illite and chlorite (Quirt and Wasyliuk, 1997; Earle et al., 1999; Quirt, 2001; Wasyliuk, 2002). Hoeve and Quirt (1984) used the illite crystallinity to confirm the 200°C burial diagenetic temperature determined from fluid inclusions (Pagel, 1975), as well as to show increased diagenesis with depth in the Rumpel Lake borehole that cores the lower Locker Lake and complete Wolverine Point, Lazenby Lake, and Manitou Falls formations in the east-central part of the basin. The dominant clay mineral analysed by Kyser et al. (2000) and Renac et al. (2002) in the Thelon and Kombolgie basins is illite. Applying the calculations of Hoeve and Quirt (1984) to these basins suggests that sufficient K was present in detrital feldspar and mica to support complete conversion of detrital kaolinite to illite during their diagenesis.

A variation in the regional background dickite of the Athabasca Basin was noted by Earle and Sopuck (1989) in its southeastern part where a large illite anomaly forms a corridor, 10 to 20 km wide, that extends for 100 km northeast from Key Lake to Cigar Lake (Fig. 7). Earle et al. (1999) described the illitic alteration at Key Lake in more detail. The axis of this regional illite anomaly also contains subparallel linear zones of anomalous chlorite and dravite. The dravite is clay sized, blue-green, and concentrated along fractures and disseminated in altered zones, overprinting illite, chlorite, and kaolinite, recording late hydrothermal boron metasomatism. The illite anomaly encompasses all known U deposits and prospects in the southeastern part of the basin (Fig. 4A), notably Key Lake, P-Patch (4 km east of Key Lake), McArthur River, BJ (7.5 km southeast of McArthur River), the Millennium prospect, and is discontinuous around the Cigar Lake mine and the Dawn Lake – Rabbit Lake areas.

Basement rock compositions and structures (Fig. 4B) likely influenced the alteration mineral chemistry of the overlying sandstone. One apparent spatial association is the above-described regional illite (+chlorite+dravite) anomaly that overlies a 5 to 20 km wide aeromagnetic low, where underlying Wollaston Supergroup gneiss includes abundant metaquartzite and metapelite units. The illite anomaly is expressed as local K anomalies in ternary K-U-Th ground spectral gamma-ray surveys (e.g. Shives et al., 2000) but is not evident in regional airborne gamma-ray data (Carson et al., 2002b). Chlorite alteration dominates in the eastern part of this illite region, possibly spatially associated with quartzite ridges, although not all known basement ridges are overlain by this alteration.

A third district-scale alteration effect is subtly evident in airborne radiometric, borehole geophysical, stratigraphic, and mineralogical data. Thorium anomalies correlate directly with increased detrital grain size in the Manitou Falls Formation, but U and K appear to be systematically depleted with respect to Th and other elements irrespective of grain size (Mwenifumbo et al., 2007). These correlations are subtle in the western part of the Athabasca Basin at Shea Creek (Mwenifumbo et al., 2000) but are strongly developed in conglomerate and pebbly sandstone beds of the Manitou Falls Formation in the eastern part of the Basin. Thorium anomalies in coarse-grained beds are evident in all drill logs in transects across exploration sites (Mwenifumbo and Bernius 2007, Yeo et al., 2007b). Thorium is also anomalous in regional airborne spectral gamma-ray data that map Th anomalies in ternary K-U-Th plots (Carson et al., 2002b), coinciding with the outcrop distribution of the Bird and lower Collins members of Manitou Falls Formation (Campbell et al., 2002, 2007). Detailed correlation of Th anomalies in drill core indicates aluminum phosphate host minerals (Mwenifumbo and Bernius, 2007). These are interpreted as alteration products of heavy minerals, such as monazite, that were preferentially altered in situ to aluminum phosphate minerals and released U (Jefferson et al., 2007; Mwenifumbo et al., 2007).

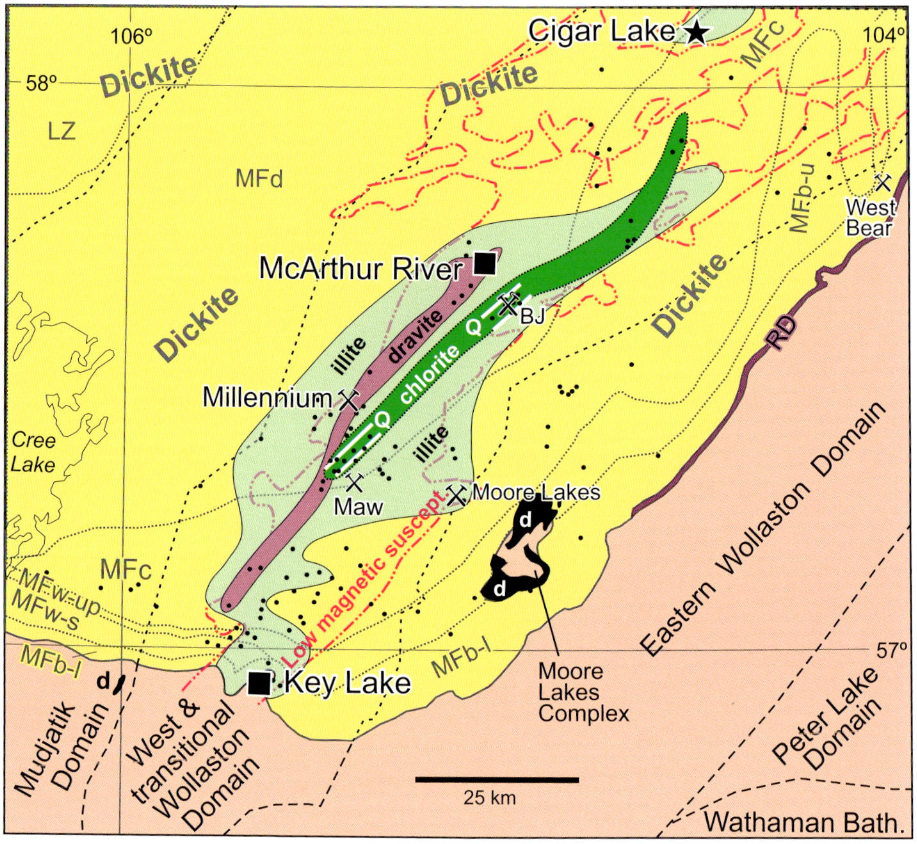

FIGURE 7. Lithogeochemical map of the southeastern part of the Athabasca Basin, showing regional illite, chlorite, and dravite anomalies in the surficial material and outcrops of the Athabasca Group. Also projected to surface are areas of low magnetic susceptibility in basement rocks, and basement quartzite ridges (white bars labelled Q; after Earle and Sopuck (1989)). Basement units (long dash) are projected (short dash) beneath the Athabasca Group (grey) based on aeromagnetic (Fig. 4B) and drill data (dots). RD=Read Formation. Stratigraphic units in the Manitou Falls Formation are MFb-l and MFb-u, lower and upper subunits of Bird Member; MFw-s and MFw-up, sandy and upper pebbly subunits of Warnes Member; MFc, Collins Member; and MFd, Dunlop Member. LZ=Lazenby Lake Formation. d=diabase; square=mine or mill; star=planned mine; crossed hammer=prospect.

Ages of Known and Prospective Districts

Sedimentation began in the eastern Athabasca Basin at about 1740 to 1730 Ma (Orrell et al., 1999; Rainbird et al., 2007) and slightly earlier in the west while the Wollaston fold and thrust belt was still uplifted (Ramaekers et al., 2007). The Barrensland Group of Thelon Basin also has a minimum age of 1750 to 1720 Ma (Miller et al., 1989; Rainbird et al., 2003a). The uppermost ages of the Athabasca and Barrensland groups are weakly constrained. The Wolverine Point Formation in sequence 3 of the Athabasca Group was deposited at 1644 ± 13 Ma (Rainbird et al., 2007). Carbonaceous marine shale of the Douglas Formation was deposited approximately 100 Ma later at 1541 ± 13 Ma (Creaser and Stasiuk, 2007), but there are no age constraints for the dolomitic Carswell Formation at the top of sequence 4, and no other ages in the Thelon or Hornby Bay basins.

Fluorapatite ages of 1640 to 1620 Ma (Rainbird et al., 2003b) in the Athabasca Basin suggest a regional hydrothermal event at about the same time as localized pre-ore alteration minerals developed (1670-1620 Ma; Alexandre et al., 2003). Athabasca Basin U deposits record two primary hydrothermal ore-related events: 1600 to 1500 Ma (Alexandre et al., 2003) and 1460 to 1350 Ma (McGill et al., 1993; Fayek et al., 2002a). These were overprinted by further alteration and U remobilization events at approximately 1176, 900, and 300 Ma (Hoeve and Quirt 1984; Cumming and Krstic 1992; Kyser et al., 2000; Fayek et al., 2002a). Thus, U deposits began to form at the base of the Athabasca Group after the Wolverine Point was deposited 1000 m above, but before the Douglas Formation was deposited an additional 500 m above that. Mineralization therefore took place beneath 1000 to 1500 m of strata, after early diagenesis and during late, high-temperature diagenesis within a time span of at least 100 Ma. Ages of U and associated alteration minerals in northern Australia have similar punctuated histories following primary uraninite deposition at about 1680 Ma (Fig. 6) temporally linked to hydrothermal events recorded in the overlying Kombolgie Subgroup (Polito et al., 2005).

Unconformity-associated U deposits may have formed beneath the Thelon Basin at about the same time as the Athabasca Basin (Kyser et al., 2000), although the oldest date is 1400 Ma (Kiggavik deposit, Fuchs and Hilger, 1989). The Hornby Bay and Elu basins began forming at about the same time as the Thelon and Athabasca basins, also experiencing volcanism at about 1670 Ma (Narakay volcanic complex, Bowring and Ross, 1985), but no unconformity U occurrences have been dated in them.

Pitchblende veins in the Beaverlodge district yield uraninite ages of about 1780 Ma (Koeppel, 1967) near the unconformity between gneiss and the Martin Group. Mafic flows in the upper Martin Group (Gillies Channel Formation) are interpreted to be ca. 1820 Ma based on the age of a diabase dyke with similar geochemistry (Ashton et al., 2004). The Baker Lake Group (1833-1785 Ma, Rainbird et al., 2007) is correlative with the Martin Group (Donaldson, 1968; Fraser et al., 1970; Rainbird et al., 2003a; Ashton et al., 2004) but it differs in having significant intercalated ultrapotassic volcanic rocks (Peterson et al., 2002). Its U prospects appear to be spatially associated with the basal unconformity (Miller, 1979, 1980), although only a few attributes support the unconformity model (Gandhi, 2007).

In the Paleoproterozoic Otish Group of Quebec (Chown and Caty, 1973), unconformity-associated U prospects (Johan et al., 1987) are of unknown age (Ruzicka, 1996a). The Sims Formation, a sequence of conglomerate, arkose, and quartzite of ca 1.8 to 1.45 Ga age located in Labrador (Ware and Hiscott, 1985; Wardle, 2005), unconformably overlies deformed metasedimentary rocks of the Paleoproterozoic Labrador Trough and is being explored for unconformity-associated U (e.g. Consolidated Abaddon Resources, Inc., news release, Feb. 23, 2006). Supracrustal units in the Central Mineral Belt of Labrador may include redbeds and unconformities, however in this region the U deposits with reasonably assured resources, such as Michelin, have been classified as volcanic (Gandhi, 1996).

Older Paleoproterozoic redbed sequences, such as the <1.96 Ga upper Hurwitz Group (Davis et al., 2005), the upper Wollaston Supergroup (Yeo and Delaney, 2007), and upper Huronian Cobalt Group, experienced different atmospheric and tectonic environments. Their dark, locally phosphatic metapelitic units contain regionally elevated U contents; their red quartzite units accumulated in more tectonically active paralic and foreland basin environments and most were stable for only tens of Ma. The Wollaston Supergroup and Hurwitz Group have been variably tectonized and dismembered, forming parts of the basement assemblages beneath the Athabasca and Thelon basins. The Cobalt Group is flat-lying and well enough preserved that unconformity-associated U attributes, such as clay alteration, should be recognizable.

Toward the upper part of the Mesoproterozoic era, the Sibley Group filled the intracontinental Nipigon Basin between 1537 +10/-2 Ma, (Davis and Sutcliffe, 1985) and 1339 ± 33 Ma (Franklin, 1978). It is 900 m thick (Hollings et al., 2004), relatively flat lying, fluvial–lacustrine (Cheadle, 1986), and aeolian (Rogala 2003), and is a target of exploration (e.g. Rampart Ventures Ltd., news release, September 30, 2004), despite being arkosic, carbonate-cemented, and lacking alteration in the areas sampled (Hanley et al., 2003). It does record synsedimentary faulting (Craven et al. 2007b), is underlain by regolith, and some of the underlying basement rocks have high background U contents, graphitic metasedimentary units, and local pitchblende veins.

Deposit-Scale Geological Attributes

Local Geological Settings

The relationship between reactivated faults and U deposits was illustrated at the deposit scale by Ruhrmann (1987) for the Gaertner orebody at Key Lake, McGill et al. (1993) for the McArthur River area, Baudemont and Federowich (1996) for the Dominique-Peter deposit, Baudemont and Pacquet (1996) for the McClean Lake area, Rippert et al., (2000) for Shea Creek, Harvey and Bethune (2007) for the Deilmann orebody at Key Lake, and Tourigny et al. (2007) for the Sue C mine.

Drill core of the McClean Lake area preserves evidence of repeated episodic brittle fault reactivation (Baudemont and Paquet, 1996). The geometries of ore zones and reactivation structures in the Sue C Pit at McClean Lake (Tourigny et al. 2007) demonstrate spatial and geometric relationships between transpressional reactivation of ductile Hudsonian basement fault zones and deposition of uraninite in dilatant jogs, consistent with the overall south-plunging deposit geometry.

In the McArthur River area, structural and stratigraphic analysis of drill core (McGill et al. 1993) documented 40 to 80 m of reverse offset along the P2 fault that hosts the ore pods. The ore pods are localized where cross-faults intersect the P2 fault (Györfi et al., 2007). At Key Lake, McClean Lake, and McArthur River faults that were active during sedimentation (Bernier, 2004; Long, 2007, Yeo et al., 2007) appear to have been reactivated after lithification and early silicification, and were conduits for hydrothermal fluid flow in the vicinity of ore deposits (e.g. Hoeve and Quirt, 1984). The spatial linkage between synsedimentary faults and ore-related fault reactivation is not universal, for example at Cigar Lake there is little evidence of offset along graphitic basement units that underlie the ore zone.

Pre- and syn-Athabasca Group paleotopographic features, such as paleovalleys up to a few dekametres deep and minor fault scarps in the order of metres high, have been documented by detailed sedimentological, isopach, and stratigraphic analysis of the Deilmann Pit (Collier and Yeo, 2001; Harvey and Bethune, 2007; Long, 2007) and of the Sue C Pit (Long, 2001, 2007). The concept of paleovalleys spatially associated with Athabasca ore deposits was introduced by Wallis et al. (1985) or possibly earlier, and Macdonald (1980) showed that metapelite and meta-semipelite units were prone to deeper regolith development, especially along graphitic zones. However, recent EXTECH IV work has clarified long-term temporal and spatial relationships between sedimentation, basement paleotopographic lows, and reactivated fault zones (Ramaekers, 2004, Ramaekers et al., 2007) . These are related specifically to the faults that host the U deposits and are the focus of alteration halos that can be mapped by various methods, such as mineralogy, seismic reflection, magnetotellurics, and gravity.

Subtle basement uplifts developed before and during sedimentation, and grew to hundreds of metres after sedimentation (Jefferson et al., 2001; Györfi et al., 2002, 2007; Bernier, 2004; Ramaekers et al., 2007; Yeo et al., 2007b) in the McArthur River camp, the BJ zone east of the P2 fault, and in a parallel transect across the Wheeler River area to the south, just east of the Millennium deposit. These commonly termed 'quartzite ridges' (Earle and Sopuck, 1989; Fig. 5 of Marlatt et al., 1992) are mainly compressional pop-up structures bounded by outward-divergent faults (Györfi, 2006; Györfi et al., 2007) such as the hanging wall of the P2 fault that hosts the McArthur River U deposit. Faults exposed

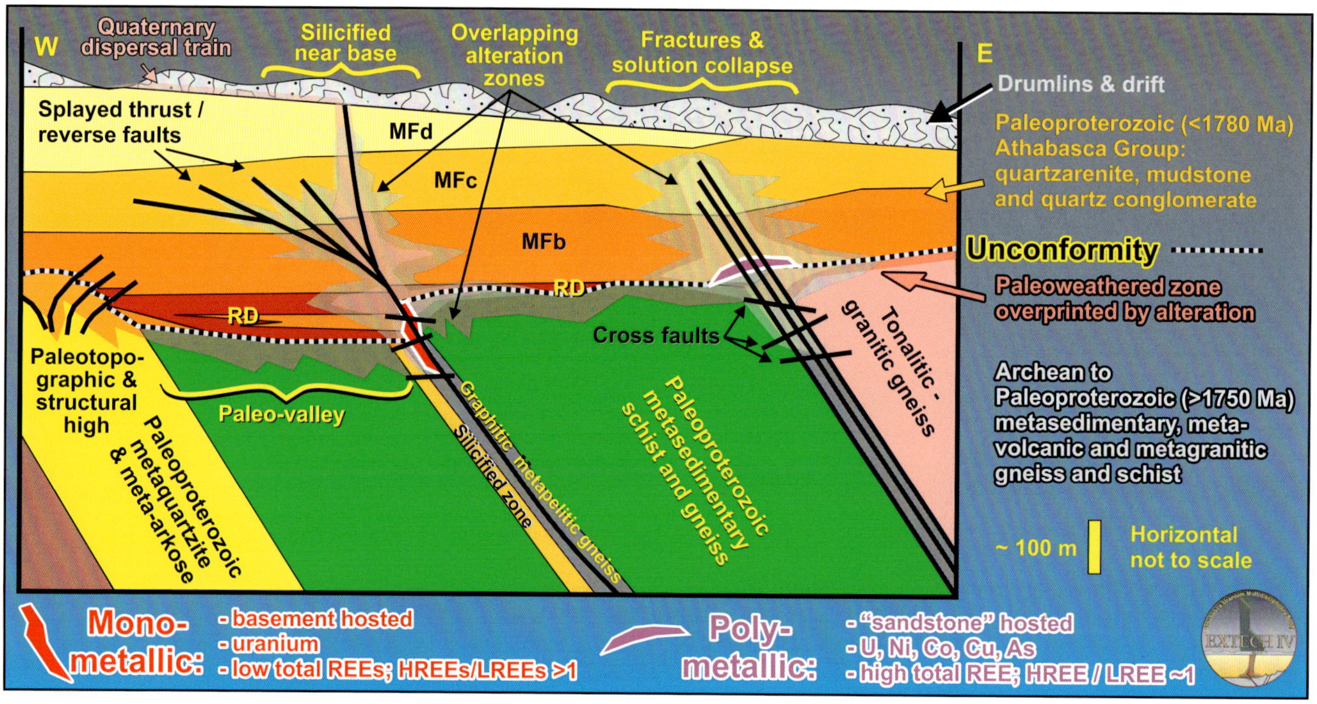

FIGURE 8. Generalized geological elements of mono- and polymetallic unconformity-associated uranium deposits in the eastern part of the Paleoproterozoic Athabasca Basin (after Sibbald et al. (1976), Hoeve and Sibbald (1978), Hoeve and Quirt (1984), McGill et al. (1993), Ruzicka (1996a),Thomas et al. (2000), and Tourigny et al. (2007)). This is an empirical geological framework model illustrating two end-member styles of ore exemplified by the McArthur River (simple) and Cigar Lake (complex) deposits, between which a complete spectrum of styles are known, even within single deposits and deposit groups.

near the unconformity (e.g. Sue C Pit, Tourigny, 2007) splay out into kink folds and bedding-parallel shears in the Manitou Falls Formation. Basal paleotalus deposits, up to 20 m thick, demonstrate that some of the basement uplifts developed before and during deposition, but structural draping of thick tabular units, such as the Read Formation over such basement highs, also records post-depositional uplift.

The graphitic metapelitic gneiss units are also strong conductors, serving as excellent exploration targets for electromagnetic methods in the Athabasca and Thelon basins of Canada and the Kombolgie Basin of Australia. They are also the continuing topic of genetic debate (discussed below under *Focus of Uranium Deposition*).

Deposit Size, Morphology, and Architecture

Deposit tonnages and grades summarized in Appendix 1 and Figure 3 are aggregates for ore zones, ranging from a series of lenses to an individual open-pit mine, such as the Sue C, or several large pods, such as the underground McArthur River mine. The details of individual deposit morphology are highly varied, ranging between end-member styles that reflect both stratigraphic and structural control (Hoeve and Quirt 1984; Sibbald 1985; updated in Thomas et al., 2000) (Figs. 8, 9, 10, 11):

1. fracture-controlled and breccia-hosted replacement, dominantly basement-hosted (Fig. 9C; e.g. McArthur River, Rabbit Lake, Eagle Point, McClean-Sue C, Dominique-Peter, Raven and Horseshoe), and
2. clay-bounded, massive ore developed along the unconformity and just above it in the overlying conglomerate and sandstone of the Athabasca Group (Fig. 9A; e.g. Cigar Lake, Key Lake (Deilmann and Gaertner zones), Collins Bay A, B, and D zones, other McClean deposits, Midwest deposit, and Cluff Lake D zone).

The fracture-controlled basement ore typically occupies steeply to moderately dipping brittle shear, fracture, and breccia zones hundreds of metres in strike length that extend down-dip for tens to 400 m into basement rocks below the unconformity (e.g. Eagle Point, Fig. 9C). Disseminated and massive uraninite/pitchblende occupies fractures and breccia matrix. The high-grade ore lenses are bounded by sheared and brecciated graphitic schist that contains smaller lenses of similar material, forming an envelope of lower grade ore. Typical mining grades for these deposits are on the order of 0.5 to 2% U. Individual lenses of high-grade ore range from 1 to 2 m thick and 3 to 5 m in vertical dimension (e.g. Sue Pit, Tourigny et al., 2002) to massive pods 100 m or more in vertical extent, 90 m in length and 50 m in width, with mining grades in the order of 20 to 25% U at the world-class McArthur River deposit (Jamieson and Spross, 2000).

In contrast, clay-bound ore is developed along the basement-sandstone unconformity and forms flattened elongate pods to flattened linear orebodies typically characterized by a high-grade core (1-15% U_3O_8) surrounded by a lower grade halo (<1% U_3O_8). The largest of these, Cigar Lake, is about 1.9 km long and 50 to 100 m wide with an upward-convex lens-shaped cross-section up to 20 m in thickness (Andrade, 2002). Most of the clay-bound orebodies have root-like extensions into the basement, resulting in elongated, skewed, T-shaped cross-sections (Fig. 9B). In places, U also extends up into the overlying conglomeratic sandstone, along cataclastic breccia and fracture zones. Isolated, small, 'perched' occurrences of disseminated pitchblende are rarely of ore grade but are good indicators of potential ore at depth. These typically are speculated to represent 'young'

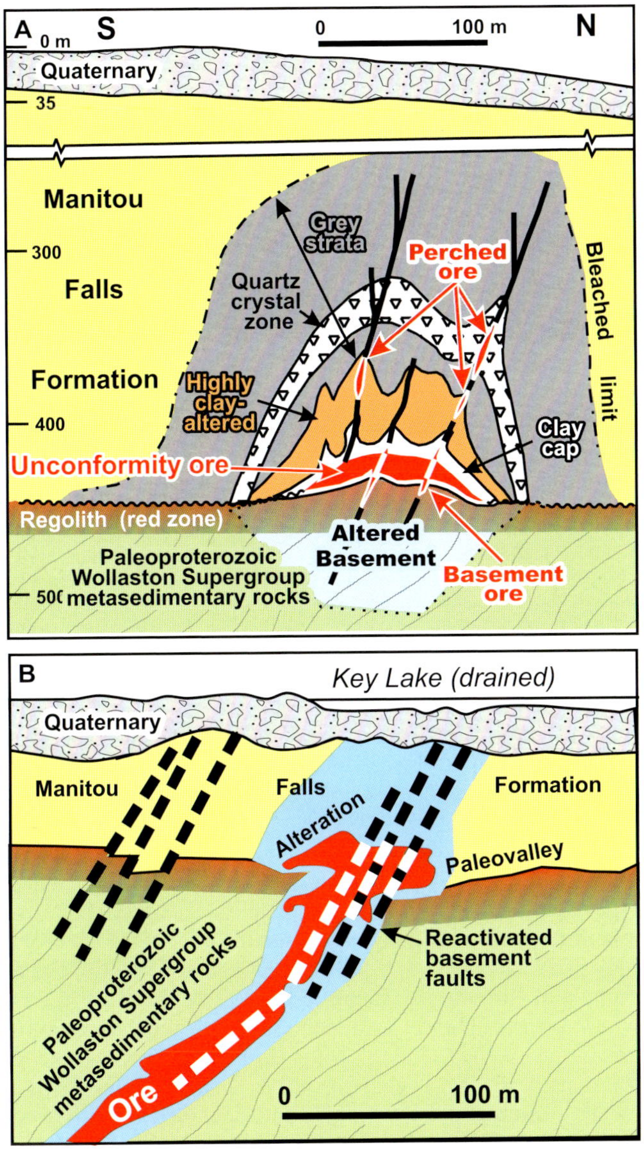

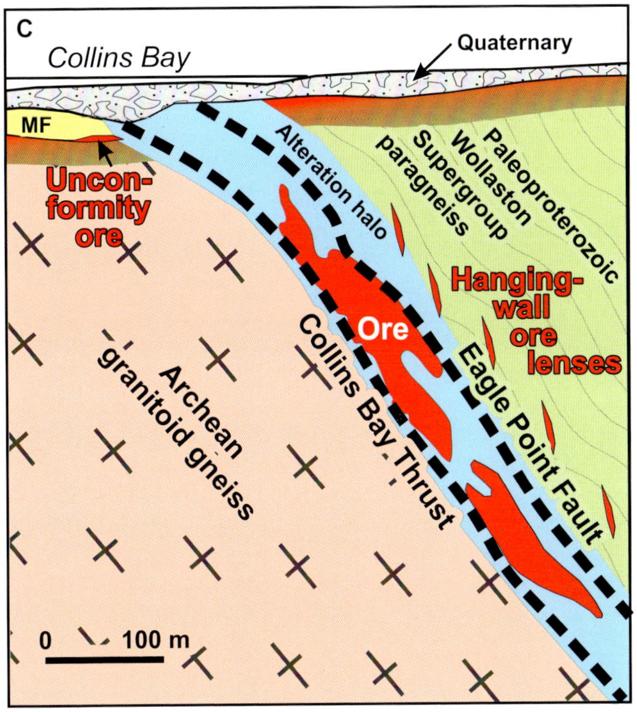

FIGURE 9. Examples of three end-point shapes and positions of unconformity-associated uranium deposits, after Thomas et al. (2000) and Andrade (2002) located in the southeastern part of Athabasca Basin (Fig. 4). **(A)** Cigar Lake (underground, production expected to start about August 2007) is dominantly unconformity ore with minor basement-hosted lenses and perched ore in the overlying Manitou Falls Formation. **(B)** Deilmann (open pit, mined out) at Key Lake included both basement-hosted and unconformity ore. **(C)** Eagle Point is mostly basement hosted (originally mined by open pit and underground; hanging-wall lenses still being developed and mined underground (LeMaitre and Belyk, oral presentation, 2005). Vertical scale=horizontal scale in (B) and (C).

remobilized primary ore. Recent significant intersections in the Shea Creek area (e.g. 27.4% U_3O_8 over 8.8 metres, *Canadian Mining Journal*, July 13, 2005) show that perched ore is a viable exploration target in its own right.

Unconformity-hosted ore deposits show similar ranges in size to fault-hosted ores but in the horizontal dimension. The Cigar Lake deposit overall contains the same order of magnitude of U as McArthur River deposit (Appendix 1), and both deposits comprise major lenses, but those of McArthur River are much more distinct. Only the eastern two lenses at Cigar Lake, with a combined strike length of about 600 m, are scheduled for Phase 1 production, which is estimated to be 496,780 tonnes at an average grade of 20.7% U_3O_8 (Andrade, 2002).

Ore Mineralogy, Chemistry, and Zonation

Unconformity-associated U deposits are dominated by massive to disseminated uraninite (Ruzicka, 1996a). Associated, paragenetically younger, minor coffinite, variable quantities of secondary U minerals, trace to minor sulphide minerals such as galena, pyrite, arsenopyrite, pentlandite, and chalcopyrite, and native gold characterize the varied metal endowment of these deposits (see below). The field term 'pitchblende' is used to refer to the commonly sooty, cryptocrystalline, botryoidal form of uraninite. The sooty appearance of pitchblende is, in part, due to crushing, milling, recrystallization, hydrothermal alteration, and remobilization associated with multistage syn- and post-ore deformation. Much of the ore preserves coarsely crystalline forms of uraninite, and systematic petrographic study of deposits across the Athabasca Basin has revealed paragenetic sequences (e.g. Wilson et al., 2007) that are consistent with classic studies of the Deilmann orebody at Key Lake (Ruhrmann, 1987).

The compositional spectrum of unconformity-associated U deposits can be described in terms of monometallic (also known as simple) and polymetallic (complex) end-members on the basis of associated metals (Everhart and Wright, 1953; Beck, 1969; Ruzicka 1989, 1996a; Thomas et al., 2000; Fig. 8). Polymetallic deposits are typically hosted by sandstone and conglomerate, situated within 25 to 50 m of the basement unconformity. At Cigar Lake, nearly all of the ore is located at the contact with or just above an assemblage consisting of hydrothermally altered paleoregolith and basal sandstone-conglomerate (Andrade, 2000). Polymetallic ores are characterized by anomalous concentrations of sulphide and arsenide minerals containing significant amounts of Ni, Co, Cu, Pb, Zn, and Mo. Some deposits also contain elevated Au, Ag, Se, and platinum-group elements.

Monometallic ores contain just traces of metals other than U and Cu, and are also termed 'simple'. These generally are hosted in basement fractures and faults, typically more than 50 m below the unconformity, with some lenses perched in sandstone well above the unconformity. Eagle Point is an end-member example of a monometallic, entirely basement-hosted U deposit (Thomas et al., 2000). McArthur River exemplifies a super high-grade monometallic deposit that extends from about 20 m above the unconformity (~500 m below surface) to more than 90 metres below the unconformity (McGill et al., 1993; Thomas et al., 2000; Jefferson et al., 2002b). It also contains minor galena, pyrite, chalcopyrite, Ni-Co sulpharsenides, and gold (Gandhi, 2007). A number of deposits have both monometallic and polymetallic components, e.g., Dielmann and Gaertner orebodies at Key Lake (Thomas et al., 2000) that combine to form skewed T-shaped profiles.

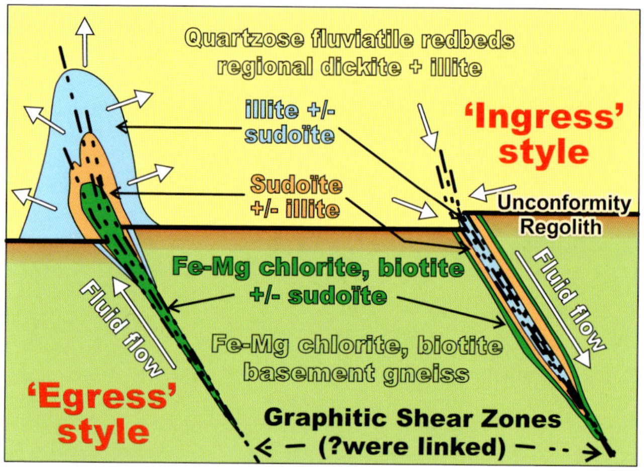

FIGURE 10. Diagrammatic explanation of 'egress'- versus 'ingress'-style alteration zones for unconformity-associated uranium deposits, after Hoeve and Quirt (1984), Sibbald (1985), Fayek and Kyser (1997), and Quirt (2003). Fluids flowing from the basement were reducing; fluids from the redbeds were oxidizing and contained uranium. Models invoke mixing of such fluids at the unconformity to precipitate uranium. The shear zones and fluid flow along them may have been linked.

Alteration Mineralogy and Geochemistry

Alteration mineralogy and geochemistry of unconformity deposits and their host rocks are among the most important exploration criteria in the Athabasca and Thelon basins of Canada and Kombolgie Basin of Australia. The parageneses of these basins (Fig. 6) have been compared by Hoeve and Quirt (1984), Miller and LeCheminant (1985), Kotzer and Kyser (1995), Kyser et al. (2000), and Cuney et al. (2003). Early work on alteration mineralogy in the Athabasca Basin is exemplified by Hoeve et al. (1981a, b), and Hoeve and Quirt (1984). Wasyliuk (2002) set the modern template for exploration using clay mineralogy. Intense clay alteration zones surrounding deposits such as Cigar Lake also constitute natural geological barriers to U migration in ground waters (Percival et al., 1993) and are important geotechnical factors in mining and ore processing (Andrade, 2002).

Comparing the mineralogical and fluid paragenesis of the various host basins helps to assess which parameters might be critical for exploration programs. Differences in local to extensive alteration in different basins may suggest different prospectivity; nevertheless, each alteration system must be understood to design the appropriate exploration strategy. District- and corridor-scale high-temperature diagenesis and hydrothermal alteration involving dickite, illite, dravite, and chlorite are described above for the Athabasca Basin. Each alteration product has detailed, deposit-specific distributions that need to be mapped in three dimensions for each exploration target. The original kaolinitic (detrital) clay is locally preserved in early pre-ore (Q1) silicification, such as at McArthur River, along with dickite, its alteration product (Wasyliuk, 2002; Mwenifumbo et al., 2007). Such relict minerals need to be recognized in order to determine what is anomalous.

Phosphate minerals preserve a rich record of regional to local, low- to high-temperature saline fluid diagenesis involving precipitation, destruction, and re-precipitation of phosphate as xenotime, apatite, and Ca-Sr-LREE-Al-phosphate minerals (AP). Xenotime in the Athabasca Basin typically forms 1 to 10 micron euhedral overgrowths on detrital zircon, lacks U, and is overgrown by quartz and fluorapatite (Rainbird et al., 2003b). This xenotime is significantly post-dated by hydrothermal uraniferous xenotime described by Quirt et al. (1991) for the Maw Zone (MacDougall, 1990). Diagenetic, variably uraniferous, patchy to stratabound fluorapatite cement was dated at ca. 1640 to 1620 Ma by U-Pb SHRIMP geochronology from the Wolverine Point to basal Fair Point formations (Rainbird et al., 2003b). This age is within error of the depositional age of the Wolverine Point Formation (1644 ± 13 Ma, Rainbird et al., 2007). Fluorapatite in the Thelon Basin has a similar paragenesis, locally forming patchy pinkish red cement with approximately 20 to 500 ppm U (Miller, 1983), dated as 1750 to 1720 Ma (Miller et al., 1989), and spatially associated with fracture and breccia zones from the bottom to top of the Thelon Formation.

The pseudocubic AP minerals recognized in trace to minor amounts throughout the Athabasca Group are interpreted as mid- to late-diagenetic to early hydrothermal. Goyazite, intergrown with illite and dravite (Hoeve and Quirt, 1984) was estimated at 1500 to 1250 Ma in age by Kotzer and Kyser (1995). Crandallite, in the same family as goyazite, has been interpreted as the youngest phosphate generation, intermixed with late-formed kaolinite (Hoeve and Quirt, 1984, Wilson, 1985). The AP minerals referred to as 'crandallite group' by Mwenifumbo and Bernius (2007) were specifically identified as florencite, form abundant 3 to 6 micron grains intimately intergrown with illite, dickite, anatase, and hematite, and are most abundant in the lower Manitou Falls Formation around the eastern Athabasca Basin. They contain elevated Th, form clusters resembling detrital grains, and are the matrix to hematite-rich pebbly laminae with relatively abundant detrital zircon and rare xenotime. This assemblage is interpreted as altered from detrital heavy minerals during moderately high-temperature (150-170°C) diagenesis that post-dated Q1 quartz overgrowths. In the Thelon, Hornby Bay, and Elu basins, aluminum phosphate sulphate (APS) minerals are described as concentrated at the base and in the regolith (Gall and Donaldson, 2006). These APS are also pseudocubic but slightly larger than the Athabasca AP minerals, and here too appear to post-date uraniferous fluorapatite.

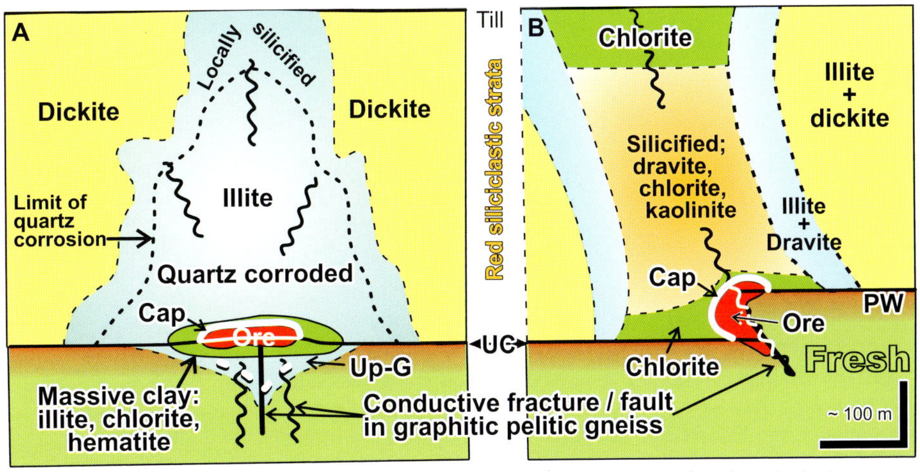

FIGURE 11. Two end-member 'egress-type' sandstone alteration models for uranium deposits in the eastern Athabasca Basin (modified from Matthews et al. (1997), Wasyliuk (2002), and Quirt (2003)). **(A)** Quartz dissolution egress style (e.g. Cigar Lake, Midwest). **(B)** Silicification egress style (e.g. McArthur River, Key Lake). Cap=secondary black-red earthy hematite capping ore bodies, Reg=regolith profile grading from red hematitic saprolite down through green chloritic alteration to fresh basement gneiss, UC=unconformity, Up-G=upper limit of graphite preserved in footwall alteration zone. Fresh=unaltered metamorphic basement: Fe-Mg-chlorite biotite paragneiss. Note that the silicification is a very early component of the right-hand model, and that quartz dissolution is also present there, albeit a minor localized feature.

Other alteration is spatially associated with ore deposits located inside and outside the above-described regional zones. Anomalously high proportions of illite are observed in the 1 to 3% clay matrix of the Athabasca Group strata and in altered basement rock in the vicinity of U deposits. This results in anomalous K_2O/Al_2O_3 ratios in the sandstone (Earle and Sopuck, 1989) that are locally recorded by ground and airborne spectral K-Th-U gamma-ray and till geochemical data (Shives et al., 2000; Campbell et al., 2007). Sudoite, an Al-Mg-rich ditrioctahedral chlorite, is present in the alteration of both sandstone and basement at Cigar Lake (Percival and Kodama, 1989) and McArthur River (Wasyliuk, 2002). Up to five types of chlorite have been documented in the basement (Quirt and Wasyliuk, 1997) – the challenge for chlorite is to distinguish ore-related alteration from that associated with retrograde metamorphism and paleoweathering.

Quartz cement predating ore formation (Q1) is informally termed 'tombstone' where drill core has a polished appearance due to a high density of cement in Athabasca Group strata above or proximal to basement quartzite highs above and to the west of the McArthur River area and in the Millennium area (Yeo et al., 2001b; Mwenifumbo et al., 2004, 2007). The tombstone silicification of the Athabasca Group preserves some of the early detrital kaolinitic clay mineralogy, detrital to early diagenetic hematite and the early regional diagenetic dickite (Wasyliuk, 2002). Even some microbial laminae defined by very finely crystalline hematite are preserved by silicification (Yeo et al., 2007a) very close to the McArthur River deposit.

Quartz dissolution is a major alteration effect above the Cigar Lake deposit, resulting in significant volume reduction and collapse of the Athabasca Group strata. It is superimposed in places on Q1 at McArthur River. Later silicification fronts comprising drusy disseminated to fracture-filling quartz (Q2) are also present in both larger quartz-dissolution alteration systems, for example at Cigar Lake (Andrade 2002) and in the McArthur River area (McGill et al. 1993). Drusy quartz (quartz crystals filling void space) is mostly developed at the periphery of the ore deposits, is related to quartz dissolution in the deposit area by mass balance analysis (Percival, 1989), and was probably synchronous with quartz dissolution during deposit formation (Hoeve and Quirt 1984). Later drusy quartz (Q3) was also described locally within the previous quartz-dissolution zones (Thomas et al., 2000; Andrade, 2002).

The alteration types described above are organized into two broad geometric shapes interpreted as 'egress' and 'ingress' (Fig. 10). The egress and ingress alteration zones are spatially but not necessarily temporally related to ore deposits. Egress-type alteration halos are developed mainly in the conglomeratic sandstone overlying unconformity-associated U deposits (Hoeve and Quirt, 1984). Deposits with egress halos include both basement-hosted and sandstone-hosted types, and the alteration ranges between two distinctive end-member types as illustrated in Figure 11: i) quartz dissolution + illite, and ii) silicified (Q1 +Q2) + later illite-kaolinite-chlorite + dravite. Strata overlying deposits in the northern part of the eastern Athabasca Basin characteristically underwent quartz corrosion with volume losses locally exceeding 90% (Percival, 1989). In contrast, alteration in the McArthur River to Millennium corridor is dominantly represented by the early silicification end-member with local, intense, pre-ore quartz corrosion and little apparent volume loss (Matthews et al., 1997). Around the Deilmann orebody at Key Lake, silicification is minor but late kaolinite and dravite (tourmaline) are superimposed on earlier dickite alteration.

Illite-kaolinite-chlorite alteration halos are up to 400 m wide at the base of the Athabasca Group (Figs. 10, 11), thousands of metres in strike length, and extend several hundred metres above major deposits (e.g. Cigar Lake, Bruneton, 1993; McArthur River , Thomas et al., 2000; Shea Creek, Kister et al., 2003). This alteration typically envelops the main ore-controlling structures, forming plume-shaped or flattened elongate bell-shaped halos that taper gradually upward from the base of the sandstone and narrow sharply downward into the basement. Illite-dominated halos have K_2O/Al_2O_3 ratios >0.18 and MgO/Al_2O_3 ratios <0.15; kaolinite-dominated halos have K_2O/Al_2O_3 and MgO/Al_2O_3 ratios <0.04; and chlorite-rich haloes have MgO/Al_2O_3

ratios >0.125 and K_2O/Al_2O_3 ratios <~0.04 (Sopuck et al., 1983). Compared to background ratios of 0.1 to 0.16 K_2O/Al_2O_3 in the Athabasca Group (Ibrahim and Wu, 1985), Percival (1989) measured K_2O/Al_2O_3 ratios >0.27% for most of the alteration zone at Cigar Lake (n=150).

A bleached and white clay (illite) replacement zone at the top of the unconformity is interpreted as ore-related hydrothermal alteration superimposed on the lateritic-weathered, red-green profile developed in basement rocks (Macdonald, 1980). The bleached white to pale green colour also locally overprints very fine-grained early diagenetic hematite in red to pink mudstone beds and intraclasts of the Read Formation. Intervening sandstone and conglomerate beds in the alteration zones are bleached to very pale green or white. Early diagenetic hematite is also intimately interlayered with clay minerals to form micro-laminae in oncoidal beds that mark the base of the Smart Formation (Yeo et al., 2007a). The matrix of these oncoidal beds is also intensely hematitic, deep maroon in colour, and very fine grained. These oncoidal beds are suspected to have resulted from pedogenic and/or shallow-water processes in ephemeral ponds. Early microbial laminae outlined by microcrystalline hematite are preserved by Q1 silicification within a bleached zone near McArthur River.

Iron pigmentation takes a number of additional forms in the Athabasca Basin. Away from ore zones, late diagenetic, broadly developed hematite transects sandstone and conglomerate beds, is commonly purple to maroon in colour, and forms spots or liesegang bands with roll shapes that tend to follow bedding planes. Bleached zones transecting this late diagenetic hematite are purplish to reddish brown, through very pale purple to nearly white in colour. Bright to very dark, 'brick red', coarse-grained hematite forms caps over ore deposits (Fig. 11). Brick red through deep reddish brown to nearly black hematite also forms dense cement within parts of the Read Formation and Bird Member of Manitou Falls Formation (Table 3), particularly in the lower conglomeratic subunits that overlie the basal unconformity and are near U deposits. This very intense, crystalline hematite alteration is interpreted as hydrothermal in origin. Recent oxidation processes are documented by limonitic alteration along fault zones, and by local limonitic liesegang banding in outcrop and drill core.

Less alteration is evident above basement-hosted deposits with ingress-type alteration zoning (Hoeve and Quirt, 1984). Such deposits are essentially 'blind' exploration targets, except for geophysical methods, although broader geochemical and mineralogical halos above them may provide clues to their existence. They are monomineralic and have very narrow, inverted alteration halos along the sides of the basement structure, grading from illite±sudoïte on the inside, through sudoïte±illite, to Fe-Mg chlorite±sudoïte on the outside against fresh basement rock (Fig. 10; Quirt 2003). Targeting these metallurgically attractive deposits requires exploration geologists to understand the geometry of fault systems. Some alteration zones have both ingress and egress characteristics (e.g. McArthur River), suggesting complex hydrothermal systems involving both processes very close to one another in both time and space. Such fluid-flow complexities have previously been interpreted for Canadian and Australian deposits (e.g. Hoeve and Quirt, 1984; Wilde and Wall, 1987). A good understanding of the basement geology and structural features is required for basement-hosted deposit exploration.

Key Exploration Criteria

Geological Exploration Criteria

The main first-order exploration criterion is Paleo- to Mesoproterozoic redbed basins as described above under continental- and district-scale geological attributes. The sedimentary sequences in Canadian basins with known U resources are depleted in U. On the other hand, a number of such basins in Canada and around the world have yet to reveal such deposits, although prospects have been discovered. Another first-order criterion is basement complexes characterized by relatively high U, well above the Clarke value of about 5 ppm (Thomas, 1983; Annesley et al., 2005). These are deformed and metamorphosed, tectonically interleaved Archean and Paleoproterozoic orthogneiss and paragneiss, intruded by granitoid plutons and pegmatite bodies.

Second-order empirical parameters associated with unconformity mineralization include graphitic metapelite, ductile faults, and other pre-existing complexities (e.g. extensional or compressional flexures, bifurcations, splays, duplex structures, and cross structures) within the basement complex. Repeated brittle reactivations of the ductile structures offset the basal unconformity, and were foci for fluid flow and ore deposition. Reactivation structures in sandstone can be traced into the primary basement fault zone and provide the local structural framework of a prospect. Reactivated fault zones may be localized at hinge lines that separate different depositional subbasins associated with different sequences. Those flexures that developed before and during sedimentation would have provided the most intensive 'ground preparation' for mineralization. In the case of the Athabasca Basin, the first U ore was emplaced during and after deposition of upper sequence 3 (Fig. 7; see *Ages of Known and Prospective Districts* above).

Paleovalleys and post-depositional offsets of the basal unconformity are manifestations of the heterogeneous nature of the basement rocks described above. Paleovalleys are not a pre-requisite for world-class orebodies; for example, the Cigar Lake deposit rests on a small basement high (Andrade, 2002). Intersections of different arrays of steeply dipping faults are especially significant, such as between the P2 fault and cross faults at McArthur River (e.g. Fig. 4 in McGill et al., 1993; Györfi, 2006). A variety of structural sites favourable for U deposition have been documented at intersections between the Rabbit Lake and a number of other fault trends in the Rabbit Lake – Eagle Point area (Rhys, 2002; LeMaitre and Belyk, oral presentation, Targeted Geoscience Initiative Saskatchewan Open House, 2004; Thomas et al., oral presentation, Targeted Geoscience Initiative Saskatchewan Open House, 2005).

Geochemical Exploration Criteria

Geochemistry in various media reflects the pervasive and local mineralogical alteration. The regional background of U is 1 to 2 ppm in lake sediments (mainly glacial till reflecting local bedrock sources; Maurice et al. 1985) and approximately 1 ppm in the Athabasca Group (Quirt, 1985; Wallis et al., 1985; Percival, 1989; Andrade 2002, Table 4.1). Uranium anomalies in lake sediments reach values of 1500 ppm in the Key Lake area (Maurice et al., 1985). Anomalous U (>2.5 ppm) in the Athabasca Group was discovered in the above described clay alteration halos, extending in places to the top of the sandstone, even in sections more than 500 m thick (e.g. Clark, 1987; Thomas et al., 2000). Percival (1989) measured common values of 13 ppm in 'unaltered' sandstone above the clay alteration halo at Cigar Lake, with highly altered sandstone in the clay zone yielding up to 235 ppm U and altered basement giving approximately 95 ppm. The trace elements U, Ni, As, and Co have greater than background concentrations in halos above some deposits and prospects. The dispersion of Ni, As, and Co in such geochemical anomalies is restricted in some cases to tens of metres (Sopuck et al., 1983), thus limiting their usefulness as pathfinder elements.

Lake water and sediment geochemistry (e.g. Coker and Dunn, 1983; Maurice et al., 1985) and radiometric prospecting were significant tools in early regional exploration. Measuring and contouring radon gas emission as an expression of radioactive decay related to underlying U ore deposits has been used with mixed results on reconnaissance to detailed scales (e.g. Dyck, 1969; Scott, 1983), but is still employed today. Analysis of spruce twigs showed that the McClean Lake – Rabbit Lake area is situated in the middle of an immense biogeochemical anomaly that was interpreted as a result of tree roots extracting anomalous U from ground water (Dunn, 1983). Groundwater geochemistry has been employed successfully in the past (e.g. Toulhoat and Beaucaire, 1993), however, given the long history of fluid flow and the still-active but variably constrained groundwater systems in the broadly permeable Athabasca Group, this technique should be re-evaluated.

As exploration advanced to deeper targets, focus shifted to alteration mineralogy reflected by surficial geochemistry. Regional-scale alteration halos of potassic clay minerals (e.g. illite), boron minerals (e.g. dravite), quartz cement, and dissolution are intersected in various places at the present surface where they are incorporated into Quaternary till. These in situ to slightly transported anomalies can be measured in till and rock samples (Earle and Sopuck, 1989; Campbell et al., 2007) and by gamma-ray spectrometry as outlined below.

Favourable basins show geochemical evidence of regional to focused fluid flow resulting in mineralogical expressions such as clay alteration and redox boundaries. Illite, chlorite, dravite, quartz cement and dissolution are the main local vectors in ingress-type or expanded egress-type zonation. This mineralogy can be analysed in the field by portable short-wave infrared (SWIR) spectrometers such as PIMA II© (Integrated Spectronics Ltd.) and FieldSpec Pro. Calibrated software algorithms for semiquantitative analyses (Earle et al., 1999) enhance the usefulness of these spectrometers. Spectrometric methods have the potential to be fully quantitative, given calibration of peak resolution with appropriate mineral standards, and the use of artificial mixtures to develop best-fit algorithms (Zhang et al., 2001; Percival et al., 2002). Infrared spectrometry is particularly useful in distinguishing between the kaolinite-group polytypes of kaolinite and dickite (Wasyliuk, 2002). Normative calculations based on lithogeochemical data further refine the mineralogical identifications.

Airborne gamma-ray spectrometry is here treated as a geochemical tool, because it directly measures U, K, and Th in surficial material. Interpretation of results from such surveys requires knowledge of paleoice-flow directions and till stratigraphy. Campbell et al. (2007) have provided calibration data that document relationships between gamma-ray and surficial geochemical data. This provides a quantitative basis for the use of ground (Shives et al., 2000) and airborne gamma-ray multiparameter geophysical surveys (e.g. Richardson, 1983; Campbell et al., 2002) as geochemical prospecting and lithologic mapping tools. The extensive illite alteration corridor between McArthur River and Key Lake (Fig. 7) does not correlate with K in published reconnaissance gamma-ray data (Carson et al., 2002a, b), although detailed ground gamma-ray spectrometry by Shives et al. (2000) suggests that K does correlate with illite alteration in the McArthur River area.

The mineralogy and chemical composition of Quaternary deposits are strongly related to local bedrock. A variety of ice-flow directions must be considered in tracing surficial materials back to their sources. Campbell (2007) provided an overview of Quaternary geology east of the Snowbird Tectonic Zone. Fenton and Pawlowicz (in press - a, b) reviewed surficial geology in map areas NTS 74 E and 74 L that cover most of the Alberta portion of the Athabasca Basin, compiled regional drift thickness and draped the Quaternary geology on principal 3-component imagery of the RADARSAT-1 data. Campbell et al. (2007) mapped detailed relationships between till composition and airborne gamma-ray data that depend on till stratigraphy and the nature of local bedrock. These reviews highlight morphological features such as sand dunes, drumlins, and eskers, and other indicators of the prevailing southwestward regional ice-flow and complex local ice-flow histories.

In the Athabasca Basin region, the bedrock is broadly the basement gneiss or the Athabasca Group conglomeratic sandstone with their varying degrees of alteration. Transport of gneissic material onto the edges of the basin from the northeast (prevailing ice flow) may be the cause of some anomalous linear features (Campbell et al., 2007). Also, the Athabasca Group material has been transported onto the gneissic basement and Paleozoic strata to the southwest, hence anomalies found there could tend to represent a source somewhere up-ice within the Athabasca Basin. This does not rule out the possibility of anomalies derived from outlying basement-hosted U deposits, above which Athabasca Group cover has been totally eroded.

Geophysical Exploration Criteria

Initial exploration in the Athabasca and similar basins focused on surface expressions of radioactivity associated with near-surface deposits located around the margins of the unconformities. In the Athabasca Basin this included the rim and the uplifted basement pillar of the Carswell meteorite impact structure. Thus initial discoveries included the Rabbit Lake, Key Lake, and Cluff Lake camps (Appendix 1). Detailed follow-up exploration traditionally focused on airborne and ground electromagnetic methods based on recognition of an association between graphitic faults and U at Key Lake (e.g. Matthews et al., 1997). These methods have been and remain the most effective tool to identify the precise location, depth, and characteristics of basement conductors.

Electromagnetic methods also detect ore-related alteration features. High-resolution airborne electromagnetic surveys in Australia have detected shallow but hidden low-resistivity alteration zones and crudely mapped fault offsets of the unconformity (Bisset, 2003). Improved audiomagnetotelluric methods in the McArthur River area of Athabasca Basin have detected deep conductors and shallow alteration zones (Craven et al., 2007; Tuncer et al., in press). Highly altered, clay-rich, quartz-corroded quartzarenite has relatively low resistivity, whereas quartz-rich silicified zones are characterized by high resistivity. Powell et al. (2005) have shown that high resistivity also maps zones of high porosity related to quartz dissolution and fracturing in the Virgin River exploration area (Centennial prospect, Fig. 4A). Detailed multiparameter borehole geophysics has been used to calibrate audiomagnetotelluric data and link them to detailed lithostratigraphic and mineralogical data, especially the resistivity contrasts (Mwenifumbo et al., 2004, 2007).

Airborne magnetic surveys provide the means to extrapolate maps of basement geology from the margins of these Proterozoic basins to their centres (e.g. Pilkington, 1989) with the aid of magnetic susceptibility and related data from outcrop and drill cores that intersect the basement. Card (2006) and Thomas and McHardy (2007) provide modern reviews of this technology and demonstrate its application to the central and eastern Athabasca Basin, respectively. They point out first-order exploration targets, such as faults and favourable basement lithologic units, as mapped by magnetic gradients between Archean gneiss domes and the Wollaston Supergroup (Fig. 4A).

Seismic reflection is a relatively new exploration tool from the mineral industry perspective, although much of our knowledge about the overall depth and shape of the Athabasca and Thelon basins has come from early seismic studies (e.g. Hobson and MacAulay, 1969; Overton, 1977; Suryam, 1981, 1984). Modern seismic reflection provides a continuous structural framework in 2-D and 3-D (White et al., 2007), from near surface to a few kilometres below the unconformity (Györfi et al., 2007) or deep in the crust to Moho (Hajnal et al., 2007) by varying source frequency, shot and geophone spacing and data processing, calibrated with the aid of borehole geophysics (Mwenifumbo et al., 2004). Complete structural sections can be interpreted using local and generic structural analogues (Fig. 5B) to determine fundamental exploration parameters such as the position of and irregularities in the unconformity, and shallow to deep faults.

Ground and airborne gravity can detect alteration zones as negative gravity anomalies (dissolution zones) or positive anomalies (silicified zones), but direct detection of ore deposits is a challenge due to their small dimensions that limit the magnitude of gravity anomalies (Thomas and Wood, 2007). Gravity also provides insights into the geological framework for exploration on both regional and district scales. It is best used in conjunction with multiple other data sets that can help to resolve ambiguities related to factors such as overburden thickness, and bulk densities and dips of deep basement units.

Genetic/Exploration Models

Conventional Models

The first refereed publication of a geological model for a new class of U deposit called 'unconformity type' was by Hoeve and Sibbald (1978), which built on the work and ideas of many geologists following the initial discoveries at Rabbit Lake in 1968, Cluff Lake in 1969, and Key Lake in 1975, by which time the importance of the unconformity had become evident. Most conventional models employed today in the Athabasca Basin are a combination of empirical, spatially associated attributes that invoke late diagenetic to hydrothermal processes with ore formation being spatially and temporally focused by the reactivation of pre-Athabasca Group structures (e.g. Hoeve et al. 1980; Kotzer and Kyser 1995; Baudemont and Paquet, 1996; Fayek and Kyser, 1997; Thomas et al., 2000). These models suggest that oxidizing, U-bearing, basin fluids heated by geothermal gradient eventually attained 200°C (burial depths of ~5-6 km) at the unconformity and reacted with reducing fluids coming out of reactivated basement shear zones. Uranium precipitated as uraninite in fault zones where reduced and oxidized fluids were mixed. Uraninite filled tension gashes and other structural traps during active faulting, and was repeatedly brecciated while new uraninite precipitated. Ore deposits accumulated where these conditions were focused for very long periods of time (Hoeve and Quirt, 1987), perhaps hundreds of millions of years (Kyser et al., 2000). Zones of inferred fluid mixing are characterized by alteration halos that contain illite, kaolinite, dravite, chlorite, euhedral quartz, and locally, Ni-Co-As-Cu sulphide minerals (Hoeve and Quirt, 1984; Wallis et al., 1985; Kotzer and Kyser, 1995). The latter described the chlorite as Mg-chlorite (=clinochlore). At Cigar Lake, most of the chlorite is a less common Al-Mg variety termed sudoïte (Hoeve and Quirt, 1984; Percival and Kodama, 1989). More than one variety of chlorite likely is present in this deposit type (Quirt, 1989, 2003).

Pre-ore to post-ore alteration halos developed around sites of ore deposition where reduced basement fluids circulated upward into the overlying oxidized basin-fluid environment ('egress type' of Fayek and Kyser, 1997). Ingress of basinal fluids downward into the basement developed inverted and condensed alteration zones, mainly in host basement rocks, with a more subtle and/or complex expression in overlying conglomeratic sandstone (Fayek and Kyser, 1997). There are many variations on the ingress and egress alteration themes in the unconformity-associated U deposit model (Quirt and Ramaekers, 2002; Quirt, 2003). Both flow paths may have developed nearby, along the same fault zone or on intersecting faults, especially in structurally complex areas. Hoeve and Quirt (1984, p. 110-114), Wilde and Wall (1987, p. 1167), and Wilde et al. (1989) discussed similar fluid flow concepts for unconformity-associated U deposits in Australia.

Early models by Knipping (1974), revisited by Dahlkamp (1978) and Langford (1978), introduced the role of supergene processes related to pre-Athabasca Group weathering of basement rocks, transport by surface- and groundwater, and deposition within basement rocks under reducing conditions. In the 1970s, deeply buried deposits (i.e. beneath hundreds of metres of the Athabasca Group) were not known, but it is now clear that even the deposits close to surface near the edge of the Athabasca Basin were formed after at least the lower Athabasca Group was deposited because of geochronology of the U deposits and the Athabasca Group. Even for deposits more than 500 m down, alteration effects reach the surface (e.g. Fig. 8).

A magmatic hydrothermal origin was briefly considered (summarized in Hoeve et al., 1980), but there is no local evidence of magmatism that is coeval with U deposition. While work on the unconformity model progressed, discussion continued as to the source of U being directly from basement rock (e.g. Tremblay, 1982) or from secondary sources including the Athabasca Group (Ruzicka, 1996a).

Genetic Models: Advances of the Last Decade

Significant advances have been made since the discovery of Rabbit Lake more than 35 years ago, but many new questions have arisen and some of the fundamental enigmas of Hoeve and Sibbald (1978) remain. A wide variety of U deposit models was developed more than a decade ago (Dahlkamp, 1993) and these are still in use today (Organization for Economic Co-operation and Development, Nuclear Energy Agency, and the International Atomic Energy, 2004). A combined efficiency of source, transport, and deposition of U are required to form a world-class deposit, and this combination still needs to be better understood (Cuney et al., 2003). Uranium deposits may appear at each step of the geological cycle, from magmatic and fluid fractionation in the deep continental crust (e.g. the Tranomaro pyroxenite, Madagascar and Rössing alaskite, Namibia) to evapotranspiration at the surface (e.g. the Yeleerie calcrete, Australia). However, very high-grade, large-tonnage U deposits have only been discovered in the vicinity of unconformities of Mesoproterozoic age. In the following reviews of source, transport, and deposition of U, there is sufficient diversity in unconformity deposits to also require multiple variants on the main model.

Uranium Sources

Suggested primary sources of U for the Athabasca and Thelon basins include radiogenic S-type granites and pegmatites (e.g. Thomas, 1983; Madore et al., 2000), metasedimentary terrains with abundant pelite whose U endowment is well above the 5 ppm Clarke value (Thomas, 1983; Miller and LeCheminant, 1985), and pre-existing U concentrations such as those in the Wollaston Supergroup (Delaney, 1993; Yeo and Delaney, 2007), pegmatites (Thomas, 1983) that intrude the Hearne Province (formerly known as the Cree Lake Zone; Lewry and Sibbald, 1979) and the many deposits in the Beaverlodge (Koeppel, 1967, Tremblay, 1968; Ruzicka, 1996b). As in copper provinces of the world, regions relatively well endowed with U due to a high proportion of radiogenic granitoid intrusions (such as in the Wollaston, Mudjatik, and Wathaman domains of the Trans Hudson Orogen and in the Taltson magmatic zone), have a much better chance of generating world-class deposits given favourable subsequent conditions. Thus a particular set of tectonic conditions was responsible for creating the U-rich western Churchill structural province. The McArthur Basin of Australia also overlies U-rich basement terrains and has a similar metallogenic history (Kyser et al., 2000). The U may have been removed from these primary sources either directly, from the Athabasca Group via detrital heavy minerals, or from the Athabasca Group via detrital clay and hydroxide minerals. Mass balance calculations illustrating the viability of the first two hypotheses are summarized by Jefferson et al. (2007).

How could U have been derived directly from underlying U-rich basement rock (Annesley et al. 1997; Hecht and Cuney, 2000; Madore et al., 2000)? Low-U AP products of altered uraniferous monazite have been documented in basement rocks underlying the Athabasca Group (Hecht and Cuney, 2000; Madore et al., 2000; Cuney et al., 2003), in basement alteration zones proximal to U deposits beneath the Kombolgie Basin of Australia (Gaboreau et al., 2003). However, drill core does not indicate sufficiently large volume and permeability of altered basement immediately beneath the Athabasca Basin to generate the amount of U needed to form major deposits (Hoeve and Quirt, 1984). Transitions to unaltered, very tight basement rocks are sharp outside of shear zones (Fig. 10). Deeper alteration along some fault zones may be partly attributed to paleoweathering. On the other hand, deep seismic profiles (Hajnal et al., 2007) suggest that a large volume of basement rock has been disrupted along the P2 fault to considerable depth.

How could U have been derived from primary detrital minerals in the Athabasca Group? Such minerals are essentially absent except for zircon and rare tourmaline; the other heavy minerals could have been incorporated in the original sediment but destroyed by diagenetic to hydrothermal alteration. Regionally, the group contains ≤1 ppm U despite its proximal detrital source terrains containing 5 to 20 ppm U. Possible original detrital carriers of U include rock fragments and heavy minerals, such as zircon, monazite, and uraninite, that should have been eroded from source terranes but are now absent, except zircon. Detrital U-oxide must have been rare because of the highly oxidizing conditions and lack of organic matter. Detrital ilmenite and magnetite are consistent with the preservation of iron-titanium oxide minerals in the Manitou Falls Formation (Mwenifumbo and Bernius, 2007), however, ilmenite and magnetite do not typically carry significant U. Feldspar is ruled out as a source because it was such a minor component of the Athabasca Group (see *Quartz-Dominated, Uranium-Depleted Strata* above). Zircon (suggested by Kyser et al., 2000) is ruled out because regionally it is fresh with normal U contents (Rayner et al., 2003; Rainbird et al., 2007), and any altered zircon across the Athabasca Basin shows evidence of U uptake, not leaching (Cuney et al., 2003), a characteristic of altered zircon in general (e.g. Rayner et al., 2005).

If there was a detrital mineral to yield U, monazite is favoured by the presence of Th- and REE- rich AP minerals in coarse-grained beds with black bands outlining cross beds and laminae draped over pebbles (e.g. Fig. 5 of Yeo et al., 2000). In situ alteration of monazite has been shown to produce U-poor but Th- and REE-rich AP minerals in the basement (above), in sandstone of the Franceville Basin, and around the Oklo deposits in Gabon (Mathieu et al., 2000; Cuney and Mathieu, 2001), and is suggested by Mwenifumbo et al. (2007) for the lower Manitou Falls Formation of the Athabasca Basin. The former abundance of detrital monazite is indicated by Th contents averaging 18 ppm (Quirt 1985) in the eastern Athabasca Basin associated with the lower Manitou Falls Formation, an average of 40 ppm around the Midwest Deposit (Ibrahim and Wu, 1985), abundant peaks between 20 and 50 ppm Th in borehole gamma-ray logs both near and away from deposits (geochemical analyses reach 730 ppm, Mwenifumbo and Bernius, 2007), and a broad Th anomaly on airborne gamma-ray maps (Carson et al., 2002a, b) that coincides with the entire distribution of the lower Manitou Falls Formation.

Could U have been carried adsorbed on Fe-Ti-oxides-hydroxides, hematite, altered zircon, and clay minerals (Macdonald, 1980; Hoeve and Quirt, 1984)? Such 'chemical sponges' have modern analogues in tropical soils and rivers. The lack of base metals in the unconformity-associated U ores is consistent with fluids leaching only the Athabasca Group because these metals are typically derived from feldspar that is absent from the Athabasca Group but still present in the basement. Inflowing surficial and ground waters could also have carried U (Hoeve and Sibbald, 1978; Macdonald, 1980; Hoeve and Quirt, 1984; Kyser et al., 2000). Unlike the basement, the clastic basin fill had very high permeability, huge volume, and abundant surface area on clastic grains of all types for chemical reaction. The degree of alteration of the Athabasca Basin overall (Kyser et al., 2000), and the infiltration of Cretaceous oil through the Fair Point Formation (Wilson et al., 2007) demonstrate that pervasive fluid flow affected a vast volume of the Athabasca Group.

Transport of Uranium

When, under what conditions, and how far was the U transported from its basement and/or detrital sedimentary sources? The timing of U movement in diagenetic fluids is illuminated by its content in phosphate minerals. Early diagenetic xenotime that overgrows zircon in the Athabasca Group contains virtually no U. Later diagenetic fluorapatite contains U in both Athabasca (Rainbird et al., 2003b) and Thelon (Miller, 1983) basins, yet predates the oldest uraninite ages. The apparently latest diagenetic AP minerals also lack U (Mwenifumbo and Bernius, 2007) but are interpreted to be about the same age as the ore deposits, recording conditions when U was released from precursor(s) and carried in solution elsewhere, presumably toward the ore deposits. Geochronology of U deposits and stratigraphy constrains ore formation to a period after deposition of sequence 3 to before deposition of sequence 4 of the Athabasca Group.

Geochemical factors constraining the development, movement, and mineral chemical changes accompanying fluids in sedimentary basins have been treated extensively in the literature (e.g. Hoeve and Quirt, 1984; Hiatt and Kyser, 2000; Kyser et al., 2000; Hiatt et al., 2003; Polito et al., 2004, 2005) for the three best-known Paleo- and Mesoproterozoic sedimentary basins that host unconformity-associated U deposits: Athabasca, Thelon, and Kombolgie (Fig. 6). The histories of fluid movement in these basins involved multiple low- to high-temperature events over hundreds of millions of years. Low-temperature uraniferous fluids are still in circulation. Diagenetic contrasts between the Athabasca, Thelon, and Kombolgie basins (Fig. 6) resulted in different mineral parageneses that record different equilibrium fluids (Kyser et al., 2000). However, in all three cases, oxidized (fO_2 > -45, in the hematite field), saline (chlorinity up to 6 molal) basinal brines transported the U (Ruzicka, 1996a; Cuney et al., 2003). High fO_2 is based on the lack of organic matter and the pervasive hematite in these basins.

The passage of later diagenetic reducing fluids is recorded by the bleached zone that invades the red regolith and red mudstone in the Read Formation, the drab grey and tan mudstone of the Manitou Falls Formation, and the presence of hydrocarbon and bitumen surrounding uraninite in many of the ore deposits. Proposed origins of the hydrocarbons range from the migration of hydrocarbons through the basin at least twice (Wilson et al., 2007) to abiotic synthesis (e.g. McCready et al., 1999; Sangély et al., 2003).

Acidity was controlled by the kaolinite-illite buffer to a pH of about 4.5 at 200°C (Cuney et al., 2003). Feldspar was either lacking or altered during diagenesis to form the sparse regional illite in the quartzarenite. Early diagenetic brines preserved as inclusions in quartz overgrowths on detrital quartz grains are NaCl-rich and inferred by Cuney et al. (2003) to have been derived from evaporitic layers that once existed in upper strata of the basin. Derome et al. (2002, 2003a, b) have determined that the brines trapped later in pervasively silicified zones and drusy quartz, close to the mineralized zones, became enriched in Ca, and inferred this to have resulted from their earlier interaction with Ca-rich basement rocks. High Ca in the mineralizing fluid is of major importance for accessory mineral alteration and for U mobilization from basement source rocks as shown by (i) incongruent dissolution of monazite with U-P-LREE leaching and new formation of a Th-U silicate with lower Th/U ratios, (ii) new formation of U-poor Ca-Sr-REE hydrated Al-phosphates, and (iii) Ca-REE-U-Al-P enrichment of zircon altered zones (Hecht and Cuney, 2000, 2003; Cuney et al., 2003). On the other hand, removal of significant Ca from basement rocks should have caused albitization, which has not been observed. Derome et al. (2003a) preferred a shallow source, above the preserved Athabasca Group, given the inferred 140°C temperature of the calcic fluid inclusions. Condition (i) above would have applied to alteration of detrital monazite in the Manitou Falls Formation and the Franceville Basin.

The effect of fluid compositions on U-solubility has not been quantified experimentally (Cuney et al., 2003). Uranium solubilities of 30 ppm were calculated by Raffensperger and Garven (1995) for five-molal Na-Ca-Cl solutions at 200°C for a fO_2 of -20, well within the hematite field, and the concentrations of other possible strong U-ligands (e.g. F and P) is only limited by the solubility product of fluorite and apatite (Cuney et al., 2003).

Temperatures during primary mineralization are interpreted in various ways. Pagel et al. (1980), Kyser et al. (2000), and Cuney et al. (2003) interpreted that ore was deposited during peak diagenesis at 180 to 250°C, suggesting a geothermal gradient in the order of 35°C/km. Ramaekers (2004) suggested that either the geothermal gradient beneath the Athabasca Basin was anomalously high (40-50°C/km for a 5 km thick basin-fill) or that the basin-fill was much thicker before erosion. In contrast, fluid inclusion studies by Derome et al. (2003a) indicate that temperature and pressure close to the unconformity decreased from the 'early diagenetic' 160 to 220°C and 1 to 1.25 kbar respectively from Rabbit Lake and Carswell deposits to the mineralization stage of 140 to 160°C and 0.6 kbar. Derome et al. (2003b) found that a late, low-saline, CH_4-bearing, higher temperature fluid (200°C) was derived from the basement, and was commonly mixed with basinal NaCl brines in the Kombolgie Basin but is rarely recorded by fluid inclusions in the Athabasca Basin. Other fluid inclusion results have been obtained from widely distributed unconformity-associated U deposits such as in the Kombolgie Basin of Australia (Derome et al., 2003b) and Shea Creek (Derome et al., 2002), and work is in progress on fluid inclusions and other micro-analytical techniques for samples from Rabbit Lake and McArthur River (M. Cuney, pers. comm., 2005).

Uranium sourcing and transport were essentially independent of the local stratigraphy above the deposits, as shown by differences between the basal Read Formation and the overlying lower Manitou Falls Formation in the eastern Athabasca Basin. Whereas both formations are conglomeratic, the Read Formation has much lower Th, abundant red mudstone with desiccation cracks (both lacking in Manitou Falls Formation), few black laminae (locally abundant in Manitou Falls Formation), and this is the unit that directly overlies the deepest red regolith. Such attributes of the Read Formation record highly oxidizing conditions and subaerial exposure before and during sedimentation. The unconformably overlying Manitou Falls Formation records little evidence of oxidizing conditions or subaerial exposure during sedimentation, and its lower two members have high Th contents. Mwenifumbo et al. (2007) and Yeo et al. (2007) infer from this that the amount of monazite and other labile minerals that contributed to primary sedimentation was small in the Read Formation. Given that the Read and Manitou Falls formations have very similar diagenetic histories, it is highly unlikely that Th was selectively removed from the Read and upper Manitou Falls formations during diagenesis and/or hydrothermal alteration, or Th selectively added to the lower Manitou Falls Formation. The amount of U present as U-oxide in the original sediments of the Read Formation must have been very low owing to its lack of organic matter and inferred highly oxidizing conditions (Cuney et al., 2003). Thus the Read Formation is not considered to have been a good source of U, in contrast to the Fair Point and lower Manitou Falls formations. Large pathways for mineralizing fluid flow are necessitated by these compositional differences and highlighted by the fact that the barren Read Formation has a limited distribution but significant deposits are independent of that. The Read Formation directly overlies the McArthur River ore pods, but is absent over the Cigar Lake deposit.

Focus of Uranium Deposition

Aquifers along the unconformity, brittle reactivated faults (including seismic pumping), crosscutting local structures and alteration (e.g. silicification, clay minerals, and dissolution) were the main controls on fluid flow at the deposit sites. The graphitic metapelitic gneiss units are not only conductive targets and the sites of reactivated faults, but are also widely regarded as a key source of reductant in geochemical process models for unconformity-associated U, albeit without consensus (e.g. Hoeve and Sibbald, 1978; McCready et al., 1999; Wilson et al., 2007). Graphitic units may also have focused U precipitation from hydrothermal fluids by conducting deep heat sources to drive convection (Hoeve and Quirt, 1984) or by serving as anodes of natural electrical systems. Significant U deposits can form in the absence of graphitic units (e.g. Kiggavik, Fuchs and Hilger, 1989; and some of the deposits at Cluff Lake), however these are in the minority and it is not known whether super high-grade deposits, such as McArthur River, can form without graphite.

Intersections of reactivated basement shear zones with offsets of the unconformity and intersections between different fault arrays enhanced fluid flow to focus U deposition, and now guide mine-scale exploration and development. Studies relating fault intersections, inferred fluid flow, and ore locations include Baudemont and Paquet (1996) at McClean Lake; Baudemont and Federovich (1996) at Cluff Lake; and Rhys (2002), D. Brisbin (oral presentation, 32nd International Geological Congress, Florence, Italy, August 20-28, 2004), R. LeMaitre and C. Belyk (oral presentation, Saskatchewan Geological Survey, Open House 2004, November 30, 2004, Saskatoon, Saskatchewan), and D. Thomas (oral presentation, Saskatchewan Geological Society, Uranium Short Course, November 29, 2005, Saskatoon, Saskatchewan) regarding new discoveries at Eagle Point. Similar work during active mining at Sue C Pit, McClean Lake (Tourigny et al., 2002, 2007) showed that the south-raking geometry of elongate ore lenses and pods, together with structural elements in the enclosing shear zone, can predict the overall south-raking geometry of the deposit. They have suggested that en echelon arrays of uraninite veins at Sue C Pit may represent mineralized hybrid extensional shear fractures.

The fault intersections related to at least two ore deposits coincide with paleovalleys that are reflected in thickness, facies changes, and paleocurrents in the Read and lower Manitou Falls formations (Harvey and Bethune, 2007; Long, 2007). Recognition of such valleys provides an additional focus for exploration. Such paleovalleys, the linear geometry of their river channels, and basement ridges may also have influenced fluid flow related to ore formation (Collier and Yeo, 2001). Faults and fracture systems are still open in mine districts and form important present-day aquifers that must be accounted for in mine development and environmental monitoring of groundwater. The fracture systems affect present day thermal conductivity (Mwenifumbo et al., 2004) and host recent limonitic alteration products. Thermal anomalies caused by the conductive properties of graphitic metapelite would have focused upward fluid flow in their vicinities.

Knowledge Gaps

Why would such high-grade, large-tonnage U deposits be found only at the basal unconformities of shallow, late Paleo- to Mesoproterozoic conglomeratic sandstone basins? And why does the Athabasca Basin host deposits that are one or two orders of magnitude larger than in similar basins elsewhere? Further mineralogical and lithogeochemical analysis is needed to test, expand, and focus the three-basin comparison of Kyser et al. (2000), which assigns higher overall U potential to basins with more intense alteration histories. An alternative perspective should also be pursued, i.e., that these different basins may have similar U potential but require different exploration paradigms adapted to their different alteration characteristics.

Whereas it might seem that exploration is at a mature stage in the Athabasca Basin, this single basin is larger than some Canadian provinces and many countries of the world, and only a small part of it has been touched by intensive exploration. Almost every year for the past 30 years, a significant discovery or advancement has been made there. Is the entire basal unconformity surface prospective where it intersects favourable basement domains and reactivated graphitic shear zones? Will additional knowledge provide tools to expand production from the existing clusters of deposits and significant prospects?

Origins of Intracontinental Proterozoic Basins

The triggers and drivers for the development of intracontinental Proterozoic basins and their fluid histories have long been considered enigmatic (Ross 2000). It is becoming clear that the Athabasca Basin developed by late-stage transpressive tectonic processes (Ramaekers, 2004; Ramaekers et al., 2007). Ruzicka (1996a) used terms such as "rapid subsidences" and "rifting" to describe events that triggered hydrologic systems, however such events were neither as "rapid" nor "rifting" as dramatic as in continental rift basins or strike-slip basins – these intra-continental events involved subtle, gentle subsidence, uplift just sufficient to generate cobble and pebble conglomerate, and the rifting led to accommodation space for just slivers of sediment accumulation compared to passive margin basins. Nonetheless, it was indeed tectonism, subtle tilts in the basin floor and reactivation of bounding faults that must have both driven and focused hydrothermal circulation to form the unconformity ore deposits. Much work remains to document the relationships between the different orders of fault systems and their orientations, to determine which faults are most prospective and when they focused ore-forming fluids. The generally accepted protracted fluid history in the Athabasca Basin, the wide range in uraninite ages, and the older regional phosphatic alteration of the Athabasca Basin challenge researchers to tackle regional background samples (e.g. Pagel, 1975) to help place the various alteration and putative ore forming fluid events into a basin-development framework.

Paleoweathering

Is the red-green basement alteration below the unconformity due to post-sedimentary alteration (Cuney, 2003) or a superimposition of such alteration on a primary paleoweathering profile (McDonald, 1980; Hoeve and Quirt, 1984)? Textural evidence of paleosol should include geopetal features and intraclasts of paleosol with red-green zonation preserved in the basal conglomerate. Paleosols are present within the lower Thelon and Manitou Falls formations (e.g. Hiatt et al., 2003; Hiatt and Kyser, 2007), and diaspore has been recognized within the regolith (Hoeve and Quirt, 1984). However in mineralized areas where most data have been collected, the 'white clay' (illite) alteration, where present along the unconformity, extends up into the basal conglomeratic sandstone units and obliterates most paleosol features that might have been preserved in lithic fragments. Well constrained fieldwork and geochemistry comparing the unconformity assemblage proximal and distal to ore would help answer these questions and develop additional exploration vectors.

Mineralogical Anomalies

Regional illite appears to be developed only in certain corridors where basement structures crosscut the basin, such as the McArthur River - Key Lake corridor (Fig. 7) and at Shea Creek (Rippert et al., 2000). Where no basement structure is observed, as in the Erica 1 or Rumpel Lake drillholes, most of the primary kaolinite, or dickite - its regional alteration product - are preserved, and illite is poorly developed.

Regional illite+kaolinite assemblages represent terrestrial strata (most of the Athabasca Group) whereas illite+chlorite (sudoïte), with minor expandable layers in illite, represent marine strata (Wolverine Point and Douglas formations) (Hoeve and Quirt (1984, p. 38-44). Potassium was conserved in the regional diagenetic process and would not have been moved in quantity to form the illite-dominated deposit-related alteration halos. Potassium for illite, as well as Mg and Fe for chlorite in these alteration halos, must have come from basement-derived fluids. Separate linear tourmaline and chlorite alteration zones (see Fig. 7) suggest discrete basement sources for B versus Mg+Fe. How are these related to U potential?

How critical is the degree of quartz domination in the Athabasca Basin compared to correlative basins? A corollary is the importance of possible detrital U carriers such as clay, iron hydroxide, and heavy minerals such as monazite. Do these result from primary sedimentary, diagenetic-alteration-fluid flow, and/or incomplete sampling histories? Do such differences mean different overall U potential or just different exploration strategies?

Geochronology

U-Pb ages on U oxides have errors of several to tens of millions of years, attributed to continuous or/and episodic diffusion of radiogenic Pb out of U oxides (Cuney et al., 2003). It is unknown whether the common ore ages of about 1350, 1000, and 300 Ma are all reset from older primary ages of 1600 to 1500 Ma that have been determined recently on the McArthur, Cigar Lake, and Sue deposits (e.g. Cumming and Krstic, 1992; Fayek et al., 2002a, b; Alexandre et al., 2003).

Calculation of the time necessary to form these massive uraninite orebodies requires numerous parameters, most of which are poorly constrained. Assuming that the mineralizing fluid contained 5 to 10 ppm U, percolated at rates of approximately 0.1 m/year, and had a volume of several tens of cubic kilometres, the formation of Cigar Lake would have required a few million years. In contrast, the Sue C and McArthur River deposits developed as lenses and pods within low-pressure dilatant jogs of transpressive faults by active processes such as seismic pumping (Sibson, 2001; Tourigny et al., 2007), also known as fault valve behaviour (Nguyen et al., 1998). The duration of ore formation was constrained by the duration of fault activity and the effectiveness of U transport.

Typical second-order basin-filling sequences are thought to require 20 to 45 million years for deposition (Krapez 1996); nevertheless, the duration of Precambrian examples is difficult to determine with such precision. The 1740 to 1760 Ma maximum age of the Athabasca Group and the 1644 Ma age of the tuff in sequence 3 allow about 100 Ma for deposition and development of the unconformities at the tops of sequences 1 and 2, and basin-wide phosphate diagenesis. Deposition of upper sequence 3, erosion and deposition of lower sequence 4 (Douglas Formation), pre- and post-ore alteration and mineralization took place during the next 100 Ma, with repeated U remobilization taking place over hundreds of millenia. Will it be possible to temporally correlate these sequences and mineralization events with those of the Thelon, Hornby Bay, and Elu basins? Will the older uraninite ages of ca. 1680 Ma for the Kombolgie Basin (Polito et al., 2005) be discovered in Canadian basins?

Fluid Flow

Is it possible to quantify what different fluid-flow events took place, what were their paths, and which events were responsible for ore deposition over the remarkable 100s of millions of years of alteration in Proterozoic basins? Unpublished modern fluid-flow modeling (Ord, 2003) has been built on a rich base of ideas and data (e.g. Hoeve and Quirt, 1984; Sibbald, 1985; Hoeve and Quirt, 1987; Wilde and Wall, 1987; Quirt, 1989; Bruneton, 1993; Fayek and Kyser, 1997; Kyser et al., 2000; Jefferson et al., 2001; Cuney et al., 2003; Kister et al., 2003; Polito et al., 2004; Ramaekers, 2004; Hiatt and Kyser, 2007; Ramaekers et al., 2007).

How were fluid flows balanced during downward (ingress) and upward and outward (egress) to and from basement fault zones? Was it rectilinear with long horizontal flow paths? Both modern and ancient flow paths reach more than 1000 m depth within the basement, are significant along the basal unconformity and in the regolith, and extend throughout the full preserved thickness of the Athabasca Group up to the Wolverine Point Formation aquitard. The Douglas Formation mudstone aquitard may also have constrained convection. Modern fluid flow around Cigar Lake and McArthur River is along subhorizontal aquifers parallel to bedding with shorter flow paths in near-vertical to 45° fault zones (e.g. Cramer and Smellie, 1992). Ore deposits formed where stable ingress or egress zones fostered prolonged mixing of U-bearing oxidized basinal fluids with reducing basement fluids. An ongoing challenge is to distinguish sites of ore-related focused flow from sites of diffuse fluid flow.

Fluid Chemistry: Causes of Quartz Dissolution and Uranium Precipitation

A basement-derived, reduced fluid was proposed by Hoeve and Sibbald (1978) in the mixing model to explain U deposition, the source of Mg (dravite and sudoïte alteration), B (dravite) in the sandstone, and Ni, Co, Cu, Zn, and Au in the polymetallic deposits. What reducing agents could have come from graphitic metapelite? Hydrocarbons (McCready et al., 1999; Annesley et al., 2001) are unlikely because pure graphite and water do not react below 400°C, such a reaction is thermodynamically unlikely, and bitumen paragenesis is overwhelmingly post-uraninite (Leventhal et al., 1987; Wilson et al., 2007). Could disorganized graphite grains from the basement have reacted with Na-Ca-Cl brines at low temperature to form the CH_4 and N_2 that have been found in fluid inclusions (e.g. Hoeve and Quirt, 1987; Landais et al., 1993; Cuney et al., 2003; Sangély et al., 2003)?

Some deposits are not associated with graphitic metapelite, e.g., the Kiggavik and several Cluff Lake deposits are not associated with graphitic basement rocks. Consideration should be given to the sulphide and mafic mineralogy of basement rock, the activity of Fe^{2+} (Cramer, 1986; Wilde and Wall, 1987; Quirt, 1989 and earlier workers cited therein; Wilde et al., 1989) and decrease in pH of an alkaline fluid, with iron being the electron acceptor to convert U^{6+} to U^{4+} (S. Romberger, Uranium short course, Cordilleran Roundup Vancouver, British Columbia, January 20, 2006), heat flow (Hoeve and Quirt, 1984), and electrochemical potential.

The basement fluid must also have been alkaline and under saturated in silica to cause the ubiquitous quartz dissolution above ore deposits. Slightly warmer basement fluid in equilibrium with rock such as pelitic gneiss would, upon introduction into the sandstone, be under saturated in silica (Hoeve and Quirt; 1984, 1987). Is such a fluid capable of destroying zircon as witnessed at Cigar Lake?

Triggers for Uranium Deposition

For every unconformity-hosted deposit underlying an egress-type alteration zone, how many corresponding basement-hosted deposits might exist with ingress-type alteration? Basement-hosted, monomineralic deposits are difficult to find but are attractive targets because they are suitable for open pit mining if close to surface and more importantly, because they are hosted in relatively competent basement rocks, there is much less need for freeze-wall technology to control water, even if deep; in addition, they are metallurgically attractive. Some deposits may remain to be discovered in areas of shallow Athabasca Group cover. They are unlikely to be preserved more than a few kilometres outside the eastern Athabasca Basin given the depth of erosion (Harper and Yeo, 2005), however, extensive areas of thin regolith around the Thelon Basin remain prospective, with the Kiggavik deposit being one example. Nisto, a small past producer northeast of Black Lake (Fig. 4; Appendix 1; Macdonald et al., 2000), is one possible example outside the Athabasca Basin.

If both ingress- and egress-type fluid flow developed along reactivated basement fault systems, and if hydrothermal convection is integral to the genesis of unconformity-associated U deposits, each fault system that generated a basal sediment- and/or regolith-hosted ore deposit had the potential to generate basement-hosted deposits. Geophysical, geochemical, and mineralogical tools are being continually improved and reapplied in this regard. Reevaluation of historical exploration, particularly drilling, is being undertaken with this difficult model in mind. Process models involving seismic pumping may be tested by further structural analysis, and this may provide structural geological tools to locate sites of ingress.

Acknowledgments

This paper was instigated by Wayne Goodfellow and John Lydon as part of an initiative on Canadian mineral deposit types, Project X-15, under the Earth Sciences Sector Program CCGK (Consolidating Canada's Geoscience Knowledge) of Natural Resources Canada. An early version of this paper was presented at the Prospectors and Developers Association of Canada Annual Meeting in March 2005. A more extensive version of this paper integrates key results of EXTECH IV into its final volume (Jefferson et al., 2007). The EXTECH IV Athabasca Uranium Multidisciplinary Study had a three-year data-acquisition partnership from April 2000 to March 2003, led by the Geological Survey of Canada (GSC), Saskatchewan and Alberta geological Surveys, Cameco Corporation ,and AREVA Resources of Canada Inc. Subsidiary industry-NSERC (Natural Sciences and Engineering Research Council of Canada) partnerships involved the universities of Alberta, Laurentian, and Saskatchewan. Team leaders, mentors, and facilitators include the co-authors of this paper, the cited papers from the EXTECH IV volume (Jefferson and Delaney, 2007), as well as Suraj Ahuja, Nick Andrade, Irvine Annesley, Guy Breton, Mark Fenton, Rod Klassen, Brian McGill, Scott McHardy, Dave Quirt, Charles Roy, and Richard Stern. In April 2003, write up of EXTECH IV became part of the Western Churchill Metallogeny Project, Northern Resources Development Program of the GSC, led by Sally Pehrsson, into the consolidated models for unconformity-associated U deposits. Completion of this paper has been under the Uranium Resources of Canada Project, Secure Canadian Energy Supply Program.

Many of the concepts updated in this paper are ongoing issues for exploration geologists. Colleagues and co-authors have provided insights and guidance in setting priorities and guiding the project through to completion, but retain diverse views regarding some of the interpretations made here. Reviews by Michel Cuney, Wayne Goodfellow, Mike Gunning, Evelyn Inglis, Ingrid Kjarsgaard, Darrel Long, Jeanne Percival, Dave Quirt, and Dirk Tempelman-Kluit substantially improved the manuscript. Rob Rainbird is thanked for providing editorial oversight as well as a thorough scientific review.

References

Alexandre, P., Kyser, K., and Polito, P., 2003, Geochronology of the Paleoproterozoic basement-hosted unconformity-type uranium deposits in northern Saskatchewan, Canada: Uranium Geochemistry 2003, International Conference, Université Henri Poincaré, Nancy, France, April 13-16, 2003, Proceedings, p. 37-40.

Andrade, N., 2002, Geology of the Cigar Lake uranium deposit, *in* Andrade, N., Breton, G., Jefferson, C.W., Thomas, D.J., Tourigny, G., Wilson, W., and Yeo, G.M., eds., The Eastern Athabasca Basin and Its Uranium Deposits: Field Trip A-1 Guidebook: Saskatoon: Geological Association of Canada and Mineralogical Association of Canada, p. 53-72.

Annesley, I.R., Madore, C., Shi, R., and Krogh, T.E., 1997, U-Pb geochronology and thermotectonic events in the Wollaston Lake area, Wollaston Domain: A summary 1994-1996, *in* Summary of Investigations 1997: Saskatchewan Geological Survey, Miscellaneous Report 97-4, p. 162-173.

Annesley, I.R., Madore, C., and Cutler, J., 2001, Synchrotron X-ray analyses of graphitic pelitic gneisses in the vicinity of unconformity-type uranium mineralization, *in* Summary of Investigations, Volume 2: Saskatchewan Geological Survey, Miscellaneous Report 2001-4.2, p. 132-140.

Annesley, I.R., Madore, C., and Portella, P., 2005, Geology and thermotectonic evolution of the western margin of the Trans-Hudson Orogen: Evidence from the eastern sub-Athabasca basement, Saskatchewan: Canadian Journal of Earth Sciences, v. 42, p. 573-597.

Ashton, K.E., Hartlaub, R.P., Heaman, L.M., Morelli, R., Bethune, K.M., and Hunter, R.C., 2004, Paleoproterozoic sedimentary successions of the southern Rae Province: Ages, origins, and correlations: Geological Association of Canada-Mineralogical Association of Canada, Proceedings of the Joint Annual Meeting, St. Catharines, May 12 to 14, 2004, p. 434.

Baudemont, D., and Federowich, J., 1996, Structural control of uranium mineralization at the Dominique-Peter deposit, Saskatchewan, Canada: Economic Geology, v. 91, p. 855-874.

Baudemont, D., and Pacquet, A., 1996, The Sue D and E uranium deposits, northern Saskatchewan: Evidence for structurally controlled fluid circulation in the Athabasca Basin, *in* Ashton, K.E., and Harper, C.T., eds., MinExpo'96 Symposium-Advances in Saskatchewan Geology and Mineral Exploration: Saskatchewan Geological Society, Special Publication No. 14, p. 85-94.

Beck, L.S., 1969, Uranium Deposits of the Athabasca Region, Saskatchewan: Department of Mineral Resources, Geological Sciences Branch, Precambrian Geology Division, Province of Saskatchewan, Report No. 126 with maps 126A, 126B, 126C and 126D, 140 p.

Bell, R.T., 1996, Sandstone uranium, *in* Eckstrand, O.R., Sinclair, W.D., and Thorpe, R.I., eds., Geology of Canadian Mineral Deposit Types, v. 8: Geological Survey of Canada, Geology of Canada, p. 212-219.

Bernier, S., 2004, Stratigraphy of the Late Paleoproterozoic Manitou Falls Formation, in the Vicinity of the McArthur River Uranium Deposit, Athabasca Basin, Saskatchewan, Canada: M.Sc. thesis, Laurentian University, Sudbury, Ontario, 184 p.

Bisset, A., 2003, Detecting unconformity related alteration using Tempest - an example from Arnhem Land, Australia: Uranium Geochemistry 2003, International Conference, Université Henri Poincaré, Nancy, France, April 13-16, 2003, Proceedings, p. 75-78.

Bowring, S.A., and Ross, M.G., 1985, Geochronology of the Narakay volcanic complex: Iimplications for the age of the Coppermine Homocline and Mackenzie igneous events: Canadian Journal of Earth Sciences, v. 22, p. 774-781.

Brouand, M., Cuney, M., and Deloule, E., 2003, Eastern extension of the Taltson orogenic belt and eastern provenance of Athabasca sandstone: IMS 1270 ion microprobe U/Pb dating of zircon from concealed basement plutonic rocks and from overlying sandstone (Canada): Uranium Geochemistry 2003, International Conference, Université Henri Poincaré, Nancy, France, April 13-16, 2003, Proceedings, p. 91-94.

Bruneton, P., 1993, Geological environment of the Cigar Lake uranium deposit: Canadian Journal of Earth Sciences, v. 30, p. 653-673.

Buchan, K.L., and Ernst, R. E., 2004, Diabase dyke swarms and related units in Canada and adjacent regions: Geological Survey of Canada, Map 2022A, scale: 1:5 000 000.

Campbell, F.H.A., 1979, Stratigraphy and sedimentation in the Helikian Elu Basin and Hiukitak Platform, Bathurst Inlet-Melville Sound, N.W.T.: Geological Survey of Canada, Paper 79-8, p. 18.

Campbell, J.E., 2007, Quaternary geology of the eastern Athabasca Basin, Saskatchewan, *in* Jefferson, C.W., and Delaney, G., eds., EXTECH IV: Geology and Uranium EXploration TECHnology of the Proterozoic Athabasca Basin, Saskatchewan and Alberta: Geological Survey of Canada, Bulletin 588, (also Geological Association of Canada, Mineral Deposits Division, Special Publication 4; Saskatchewan Geological Society, Special Publication 18), p. 211-228.

Campbell, J.E., Shives, R.B.K., and Klassen, R.A., 2002, Integrated field investigations of airborne radiometric spectral domains, NEA-IAEA Test Area, eastern Athabasca Basin: A preliminary report, *in* Summary of Investigations, 2002, Volume 2: Saskatchewan Geological Survey, Saskatchewan Industry and Resources, Miscellaneous Report 2002-4.2, 22 p., (CD-ROM).

Campbell, J.E., Klassen, R.A., and Shives, R.B.K., 2007, Integrated field investigations of airborne radiometric data and drift composition, Nuclear Energy Agency-International Atomic Energy Agency Athabasca Test Area, Saskatchewan, *in* Jefferson, C.W., and Delaney, G., eds., EXTECH IV: Geology and Uranium EXploration TECHnology of the Proterozoic Athabasca Basin, Saskatchewan and Alberta: Geological Survey of Canada, Bulletin 588, (also Geological Association of Canada, Mineral Deposits Division, Special Publication 4; Saskatchewan Geological Society, Special Publication 18), p. 533-554.

Card, C.D., 2006, Remote predictive map for the basement to the western Athabasca Basin: Saskatchewan Industry and Resources, Preliminary Geological Map, Open File 2006-45, scale: 1:500 000.

Card, C.D., Campbell, J.E., and Slimmon, W.L., 2003, Basement lithologic framework and structural features of the western Athabasca Basin, *in* Summary of Investigations 2003, Volume 2: Saskatchewan Geological Survey, Saskatchewan Industry Resources,. Miscellaneous Report 2003-4.2, Paper D-3, 17 p., CD-ROM.

Card, C., Pană, D., Stern, R., and Rayner, N., 2007a, New insights into the geological history of the basement rocks to the southwestern Athabasca Basin, Saskatchewan and Alberta, *in* Jefferson, C.W., and Delaney, G., eds., EXTECH IV: Geology and Uranium EXploration TECHnology of the Proterozoic Athabasca Basin, Saskatchewan and Alberta: Geological Survey of Canada, Bulletin 588, (also Geological Association of Canada, Mineral Deposits Division, Special Publication 4; Saskatchewan Geological Society, Special Publication 18), p. 119-134.

Card, C., Portella, P., Annesley, I., and Pană, D., 2007b, Basement rocks to the Athabasca Basin, Saskatchewan and Alberta, *in* Jefferson, C.W., and Delaney, G., eds., EXTECH IV: Geology and Uranium EXploration TECHnology of the Proterozoic Athabasca Basin, Saskatchewan and Alberta: Geological Survey of Canada, Bulletin 588, (also Geological Association of Canada, Mineral Deposits Division, Special Publication 4; Saskatchewan Geological Society, Special Publication 18), p. 69-88.

Carson, J.M., Holman, P.B., Ford, K.L., Grant, J.A., and Shives, R.B.K., 2002a, Airborne Gamma Ray Spectrometry Compilation Series, Lake Athabasca, Alberta-Saskatchewan: Geological Survey of Canada, Open File 4252, scale: 1:1 000 000.

——— 2002b, Airborne Gamma Ray Spectrometry Compilation Series, Wollaston Lake, Saskatchewan: Geological Survey of Canada, Open File 4253, scale: 1:1 000 000.

Cecile, M.P., 1973, Lithofacies Analysis of the Proterozoic Thelon Formation, Northwest Territories (including computer analysis of field data): M.Sc. thesis, Carleton University, Ottawa, Ontario, 119 p.

Cheadle, B.A., 1986, Alluvial-playa sedimentation in the Lower Keweenawan Sibley Group, Thunder Bay district, Ontario: Canadian Journal of Earth Sciences, v. 23, p. 527-542.

Chown, E.H., and Caty, J.L., 1973, Stratigraphy, petrography and paleocurrent analysis of the Aphebian clastic formations of the Mistassini-Otish basin, *in* Young, G.M., ed., Huronian Stratigraphy and Sedimentation: Geological Association of Canada, Special Publication 12, p. 49-71.

Clark, L., 1987, Near surface lithogeochemical halo as an aid to discovery of deeply buried unconformity-type uranium deposits: Journal of Geochemical Exploration, v. 28, p. 71-84.

Coker, W.B., and Dunn, C.E., 1983, Lake water and lake sediment geochemistry, NEA/IAEA Athabasca Test Area, *in* Cameron, E.M., ed., Uranium Exploration in Athabasca Basin, Saskatchewan, Canada: Geological Survey of Canada, Paper 82-11, p. 117-125.

Collier, B., 2003: Detailed Stratigraphy of the Paleoproterozoic Athabasca Group along the Shea Creek - Douglas River Trend and Its Relationship to the Regional Stratigraphy of the Western Athabasca Basin, Canada: M.Sc. thesis, Laurentian University, Sudbury, Ontario, 166 p.

Collier, B., and Yeo, G.M., 2001, Stratigraphy of the Paleoproterozoic Manitou Falls B member at the Deilmann Pit, Key Lake, Saskatchewan, *in* Summary of Investigations, 2001, Volume 2: Saskatchewan Geological Survey, Saskatchewan Energy and Mines, Report 2001-4.2, p. 297-305.

Cramer, J.J., 1986, Sandstone-hosted uranium deposits in northern Saskatchewan as natural analogs to nuclear fuel waste disposal vaults: Chemical Geology, v. 55, p. 269-279.

Cramer, J., and Smellie, J., 1992, Summary report on the meetings in Forsmark, Sweden, December 2-4, 1991; Cigar Lake Project Natural Analog Studies, *in* Cramer, J., and Smellie, J., eds., Cigar Lake Mining Corporation Report CLR-92-1: Atomic Energy of Canada Limited (AECL), Swedish Nuclear Fuel and Waste Management Company (SKB) and United States Department of the Environment (US DOE), 51 p.

Craven, J.A., McNeice, G., Powell, B., Koch, R., Annesley, I.R., Wood, G., Mwenifumbo, C.J., Unsworth, M., and Xiao, W., 2007a, Audio magnetotelluric studies at the McArthur River mining camp and Shea Creek area, northern Saskatchewan, *in* Jefferson, C.W., and Delaney, G., eds., EXTECH IV: Geology and Uranium EXploration TECHnology of the Proterozoic Athabasca Basin, Saskatchewan and Alberta: Geological Survey of Canada, Bulletin 588, (also Geological Association of Canada, Mineral Deposits Division, Special Publication 4; Saskatchewan Geological Society, Special Publication 18), p. 413-424.

Craven, J.A., Shareef, S., Hart, T.R., Martí, A., and Farquharson, C., 2007b, Preliminary Results of the Audiomagnetotelluric Study of the Nipigon Embayment: Lake Nipigon Region Geoscience Initiative: Ontario Geological Survey, Miscellaneous Data Release 192, 3-D movie and report, CD ROM.

Creaser, R.A., and Stasiuk, L.D., 2007, Depositional age of the Douglas Formation, northern Saskatchewan, determined by Re-Os geochronology, *in* Jefferson, C.W., and Delaney, G., eds., EXTECH IV: Geology and Uranium EXploration TECHnology of the Proterozoic Athabasca Basin, Saskatchewan and Alberta: Geological Survey of Canada, Bulletin 588, (also Geological Association of Canada, Mineral Deposits Division, Special Publication 4; Saskatchewan Geological Society, Special Publication 18), p. 341-346.

Cumming, G.L., and Krstic, D., 1992: The age of unconformity-associated uranium mineralization in the Athabasca Basin, northern Saskatchewan: Canadian Journal of Earth Sciences, v. 29, p. 1623-1639.

Cuney, M., and Mathieu, R., 2001, Extreme light rare earth element mobilization by diagenetic fluids in the geological environment of the Oklo natural reactor zones, Franceville Basin, Gabon: Geology, v. 28, p. 743-746.

Cuney, M., Brouand, M., Cathelineau, M., Derome, D., Freiberger, R., Hecht, L., Kister, P., Lobaev, V., Lorilleux, G., Peiffert, C., and Bastoul, A.M., 2003, What parameters control the high grade - large tonnage of the Proterozoic unconformity related uranium deposits?: Uranium Geochemistry 2003, International Conference, Nancy, France, April 13-16 2003, Proceedings, p. 123-126.

Dahlkamp, F.J., 1978, Geological appraisal of the Key Lake U-Ni deposits, northern Saskatchewan: Economic Geology, v. 73, p. 1430-1449.

——— 1993, Uranium Ore Deposits: Springer-Verlag, Berlin, 460 p.

Dal Cin, R., 1968, Climatic significance of roundness and percentage of quartz in conglomerates: Sedimentary Petrology, v. 38, p. 1094-1099.

Davidson, G.I., and Gandhi, S.S., 1989, Unconformity-related U-Au mineralization in the middle Proterozoic Thelon sandstone, Boomerang Lake Prospect, Northwest Territories, Canada: Economic Geology, v. 84, p. 143-157.

Davis, D.W., and Sutcliffe, R.H., 1985, U-Pb ages from the Nipigon Plate and Northern Lake Superior: Geological Society of America Bulletin, v. 96, p. 1572-1579.

Davis, W.J., Rainbird, R.H., Aspler, L.B., and Chiarenzelli, J.R., 2005, Detrital zircon geochronology of the Paleoproterozoic Hurwitz and Kiyuk groups, western Churchill Province, Nunavut: Geological Survey of Canada, 2005-F1, Current Research, 13 p.

Delaney, G., 1993, A re-examination of the context of U-Cu, Cu, and uranium mineralization, Dudderidge Lake, Wollaston Domain, *in* Summary of Investigations 1993: Saskatchewan Geological Survey, Saskatchewan Energy and Mines, Miscellaneous Report 93-4, p. 73-85.

Delaney, G.D., Jefferson, C.W., Yeo, G.M., McLennan, S.M., Bell, R.T., and Aitken, J.D., 1982, Some Proterozoic sediment-hosted metal occurrences of the northeastern Canadian Cordillera: Society of Economic Geologists Couer d'Alene Field Conference, Idaho, Proceedings, p. 97-116

Derome, D., Cuney M., Cathelineau, M., Fabre, C., and Lhomme, T., 2002, Reconstitution of individual fluid inclusion composition using microthermometry, Raman and Laser Induced Breakdown Spectroscopy: Application to the Shea Creek uranium deposit (Saskatchewan, Canada), *in* Abstracts Volume, SS18: The Athabasca Basin and Its Uranium Deposits: Geological Association of Canada - Mineralogical Association of Canada, Annual Meeting, Saskatoon, 1 p.

Derome, D., Cathelineau, M., Cuney M., and Fabre. C., 2003a, Reconstitution of the P, T, X properties of paleo-fluids in the McArthur River unconformity-type uranium deposit (Saskatchewan, Canada): Uranium Geochemistry 2003, International Conference, Nancy, France, April 13-16 2003, Proceedings, p. 141-144.

Derome, D., Cuney M., Cathelineau, M., Dubessy, J., and Bruneton, P., 2003b, A detailed fluid inclusion study in silicified breccias from the Kombolgie sandstones (Northern Territory, Australia): Application to the genesis of Middle-Proterozoic unconformity-type uranium deposits: Journal of Geochemical Exploration, v. 80, p. 259-275.

Donaldson, J.A., 1968, Proterozoic sedimentary rocks of northern Saskatchewan, Report of Activities: Part A: Geological Survey of Canada, Paper 68-1, 131 p.

Dunn, C.E., 1983, Uranium biogeochemistry of the NEA/IAEA Athabasca Test Area, *in* Cameron, E.M., ed., Uranium Exploration in Athabasca Basin, Saskatchewan, Canada: Geological Survey of Canada, Paper 82-11, p. 127-132.

Dyck, W., 1969, Development of Uranium Exploration Methods Using Radon: Geological Survey of Canada, Paper 69-46, 26 p.

Earle, S., and Sopuck, V., 1989, Regional lithogeochemistry of the eastern part of the Athabasca Basin uranium province, Saskatchewan, *in* Muller-Kahle, E., ed., Uranium Resources and Geology of North America: International Atomic Energy Agency, Technical Document TECDOC-500, p. 263-269.

Earle, S., Wheatley, K., and Wasyliuk, K., 1999, Application of reflectance spectroscopy to assessment of alteration mineralogy in the Key Lake area: MinExpo '96 Symposium - Advances in Saskatchewan geology and mineral exploration, Saskatoon, November 21, 22, 1996, Proceedings, p. 109-123.

Edgecombe, S.M., Hazell, M., Page, R.W., Black, L.P., and Sun, S-S., 2002, OZCHRON National Geochronology Database [Digital Datasets]: Canberra, Geoscience Australia, http://www.ga.gov.au/databases/20010926_27.jsp.

Everhart, D.L., and Wright, R., 1953, The geologic character of pitchblende veins: Economic Geology, v. 48, p. 77-96.

Fayek, M., and Kyser, K.K., 1997, Characterization of multiple fluid events and rare-earth-element mobility associated with formation of unconformity-type uranium deposits in the Athabasca Basin, Saskatchewan: The Canadian Mineralogist, v. 35, p. 627-658.

Fayek, M., Harrison, T.M., Ewing, R.C., Grove, M., and Coath, C.D., 2002a, O and Pb isotope analyses of uranium minerals by ion microprobe and U-Pb ages from the Cigar Lake deposit: Chemical Geology, v. 185, p. 205-225.

Fayek, M., Kyser, T.K., and Riciputi, L.R., 2002b, U and Pb isotope analysis of uranium minerals by ion microprobe and the geochronology of the McArthur River and Sue Zone uranium deposits, Saskatchewan, Canada: The Canadian Mineralogist, v. 40, p. 1553-1569.

Fenton, M.M., and Pawlowicz, J.G., in press-a, Preliminary drift thickness and bedrock topography western Athabasca Basin, Alberta: Alberta Energy and Utilities Board, Alberta Geological Survey, Geo-Note.

——— in press-b, Surficial Geology of the western Athabasca Basin, Alberta: update utilizing RADARSAT -1 data: Alberta Energy and Utilities Board, Alberta Geological Survey, Geo-Note.

Franklin, J.M., 1978, The Sibley Group, Ontario, *in* Wanless, R.K., and Loveridge, W.D., eds., Rubidium-Strontium Isotopic Age Studies, Report 2: Geological Survey of Canada, Paper 77-14, p. 31-34.

Fraser, J.A., Donaldson, J.A., Fahrig, W.F., and Tremblay, L.P., 1970, Helikian basins and geosynclines of the northwestern Canadian Shield, *in* Baer, A.J., ed., Basins and Geosynclines of the Canadian Shield: Geological Survey of Canada, Paper 70-40, p. 213-238.

Freiberger, R., and Cuney, M., 2003, New evidence for extensive fluid circulation within the basement in relation to unconformity type uranium deposits genesis in the Athabasca Basin, Saskatchewan, Canada: Uranium Geochemistry 2003, International Conference, Nancy, France, April 13-16 2003, Proceedings, p. 151-154.

Fuchs, H., and Hilger, W., 1989, Kiggavik (Lone Gull): An unconformity related uranium deposit in the Thelon Basin, Northwest Territories, Canada: Uranium Resources and Geology of North America, Technical Committee Meeting, Saskatoon, Vienna, 1-3 September, 1987, Proceedings, p. 429-454.

Gaboreau, S., Vieillard, P., Bruneton, P., and Beaufort, D., 2003, Aluminium phosphates sulphates minerals associated to unconformity-associated uranium deposits of the Kombolgie basin (North Territories, Australia): Uranium Geochemistry 2003, International Conference, Nancy, France, April 13-16 2003, Proceedings, p. 157-160.

Gall, Q., 1994, The Proterozoic Thelon paleosol, N.W.T., Canada: Precambrian Research, v. 68, p. 115-137.

Gall, Q., and Donaldson, J.A., 2006, Diagenetic fluorapatite and aluminum phosphate-sulphate in the Paleoproterozoic Thelon Formation and Hornby Bay Group, northwestern Canadian Shield: Canadian Journal of Earth Sciences, v. 43, p. 617-628.

Gandhi, S.S., 1989, Geology and uranium potential of the Thelon Basin and adjacent basement in comparison with the Athabasca Basin region: Uranium Resources and Geology of North America, Technical Committee Meeting, Saskatoon, Vienna, 1-3 September, 1987, Proceedings, p. 411-428.

——— 1995, An overview of the exploration history and genesis of Proterozoic uranium deposits in the Canadian Shield: Exploration and Research for Atomic Minerals, v. 8, p. 1-47.

——— 1996, Volcanic-associated uranium, Chapter 12 *in* Eckstrand, O.R., Sinclair, W.D., and Thorpe, R.I., eds., Geology of Canadian Mineral Deposit Types: Geological Survey of Canada, Geology of Canada, v. 8, p. 269-276.

——— 2007, Significant Unconformity Associated Uranium Deposits of the Athabasca Basin, Saskatchewan and Alberta, and Selected Related Deposits of Canada and the World: Geological Survey of Canada, Open File 5005, Saskatchewan Industry and Resources, Open File 2007- 11, CD-ROM.

Geological Survey of Canada, 1987, Magnetic Anomaly Map of Canada: Geological Survey of Canada, Map 1255A, scale 1:5 000 000.

Győrfi, I., 2006, Seismic Constraints on the Geological Evolution of the McArthur River Region in View of the Tectonics of the eastern Athabasca Basin, northern Saskatchewan: Ph.D. thesis, University of Saskatchewan, Saskatoon, Saskatchewan, 230 p.

Győrfi, I., Hajnal, Z., Bernier, S., Takács, E., Reilkoff, B., White, D., Powell, B., and Koch R., 2002, Initial seismic images along line 14 of the McArthur River high-resolution survey, *in* Summary of Investigations 2002, Volume 2: Saskatchewan Geological Survey, Saskatchewan Industry and Resources, Miscellaneous Report 2002-4.2, CD-ROM.

Győrfi, I., Hajnal, Z., White, D.J., Takács, E., Reilkoff, B., Annesley, I.R., Powell, B., and Koch, R., 2007, High-resolution seismic survey from the McArthur River region: contributions to mapping the complex P2 uranium ore zone, Athabasca Basin, Saskatchewan, *in* Jefferson, C.W., and Delaney, G., eds., EXTECH IV: Geology and Uranium EXploration TECHnology of the Proterozoic Athabasca Basin, Saskatchewan and Alberta: Geological Survey of Canada, Bulletin 588, (also Geological Association of Canada, Mineral Deposits Division, Special Publication 4; Saskatchewan Geological Society, Special Publication 18), p. 397-412.

Hajnal, Z., Takács, E., White, D.J., Győrfi, I., Powell, B., and Koch, R., 2007, Regional seismic signature of the basement and crust beneath the McArthur River mine district, Athabasca Basin, Saskatchewan, *in* Jefferson, C.W., and Delaney, G., eds., EXTECH IV: Geology and Uranium EXploration TECHnology of the Proterozoic Athabasca Basin, Saskatchewan and Alberta: Geological Survey of Canada, Bulletin 588, (also Geological Association of Canada, Mineral Deposits Division, Special Publication 4; Saskatchewan Geological Society, Special Publication 18), p. 389-396.

Hanly, A., Geboy, N. J., Hiatt, E. E., and Kyser, T. K., 2003, Diagenesis, stratigraphy, paleohydrology, and lead isotope study of the Proterozoic Sibley Group, western Ontario: Geological Society of America, Annual Meeting, Proceedings, p. 234.

Harper, C.T., 1982, Geology of the Carswell Structure, Central part (parts of NTS areas 74K-5, -6, -11, -12): Saskatchewan Energy and Mines, Map 214a and Report 214, 6 p, scale: 1:50 000.

Harper, C.T., Ebel, C., Yeo, G., Card, C., and Nelson, C., 2005, Wollaston Lake Project: geology of the Wollaston Supergroup in the Rabbabou Bay-Wellbelove Bay area, northeast Wollaston Lake, Saskatchewan (parts of NTS 64L-6, -7, -10, and -11), Paper A-6 *in* Summary of Investigations 2005, Volume 2: Saskatchewan Geological Survey, Saskatchewan Industry Resources, Miscellaneous Report 2005-4.2, 25 p. (CD-ROM).

Harvey, S.E., and Bethune, K., 2007, Context of the Deilmann orebody, Key Lake mine, Saskatchewan, *in* Jefferson, C.W., and Delaney, G., eds., EXTECH IV: Geology and Uranium EXploration TECHnology of the Proterozoic Athabasca Basin, Saskatchewan and Alberta: Geological Survey of Canada, Bulletin 588, (also Geological Association of Canada, Mineral Deposits Division, Special Publication 4; Saskatchewan Geological Society, Special Publication 18), p. 249-266.

Hecht, L., and Cuney, M., 2000, Hydrothermal alteration of monazite in the Precambrian basement of the Athabasca Basin: implications for the genesis of unconformity related deposits: Mineralium Deposita, v. 35, p. 791-795.

——— 2003, Tracing the sources of unconformity-type uranium deposits: Uranium Geochemistry 2003, International Conference, Nancy, France, April 13-16 2003, Proceedings, p. 177-180.

Hiatt, E., and Kyser, K., 2000, Links between depositional and diagenetic processes in basin analysis: porosity and permeability evolution in sedimentary rocks, Chapter 4 *in* Kyser, K., ed., Fluids and Basin Evolution: Short Course Series Volume 28 (Series editor Robert Raeside): Mineralogical Association of Canada, p. 63-92.

——— 2007, Sequence stratigraphy, hydrostratigraphy, and mineralizing fluid flow in the Proterozoic Manitou Falls Formation, eastern

Athabasca Basin, Saskatchewan, *in* Jefferson, C.W., and Delaney, G., eds., EXTECH IV: Geology and Uranium EXploration TECHnology of the Proterozoic Athabasca Basin, Saskatchewan and Alberta: Geological Survey of Canada, Bulletin 588, (also Geological Association of Canada, Mineral Deposits Division, Special Publication 4; Saskatchewan Geological Society, Special Publication 18), p. 489-506.

Hiatt, E.E., Kyser, T.K., and Dalrymple, R.W., 2003, Relationships among sedimentology, stratigraphy and diagenesis in the Proterozoic Thelon Basin, Nunavut, Canada: Implications for paleo-aquifers and sedimentary-hosted mineral deposits: The Journal of Geochemical Exploration, v. 80, p. 221-240.

Hobson, G.D., and MacAulay, H.A., 1969, A Seismic Reconnaissance Survey of the Athabasca Formation, Alberta and Saskatchewan (Part of 74) (A co-operative venture with the Saskatchewan Department of Mineral Resources): Geological Survey of Canada, Paper 69-18, 23 p.

Hoeve, J., and Quirt, D.H., 1984, Mineralization and Host Rock Alteration in Relation to Clay Mineral Diagenesis and Evolution of the Middle-Proterozoic, Athabasca Basin, northern Saskatchewan, Canada: Saskatchewan Research Council, SRC Technical Report 187, 187 p.

——— 1987, A stationary redox front as a critical factor in the formation of high-grade, unconformity-type uranium ores in the Athabasca Basin, Saskatchewan, Canada: Bulletin de Minéralogie, v. 110, p. 157-171.

Hoeve, J., and Sibbald, T., 1978, On the genesis of Rabbit Lake and other unconformity-type uranium deposits in northern Saskatchewan, Canada: Economic Geology, v. 73, p. 1450-1473.

Hoeve, J., Sibbald, T.I.I., Ramaekers, P., and Lewry, J.F., 1980, Athabasca Basin unconformity-type uranium deposits: A special class of sandstone-type deposits?, *in* Ferguson, J., and Goleby, A.B., eds., Uranium in the Pine Creek Geosyncline: Vienna, International Atomic Energy Agency, p. 575-594.

Hoeve, J., Rawsthorn, K., and Quirt, D., 1981a: Uranium metallogenetic studies: Collins Bay B Zone, II. Clay mineralogy, *in* Summary of Investigations 1981: Saskatchewan Geological Survey, Miscellaneous Report 81-4, p. 73-75.

——— 1981b, Uranium metallogenetic studies: clay mineral stratigraphy and diagenesis in the Athabasca Group, *in* Summary of Investigations 1981: Saskatchewan Geological Survey, Miscellaneous Report 81-4, p. 76-89.

Hoffman, P., 1988, United plates of America, the birth of a craton: Early Proterozoic assembly and growth of Laurentia: Annual Reviews of Earth and Planetary Sciences 16, p. 543-603.

Hollings, P., Fralick, P., and Kissin, S., 2004, Geochemistry and geodynamic implications of the Mesoproterozoic English Bay granite-rhyolite complex, northwestern Ontario: Canadian Journal of Earth Sciences, v. 41, p. 1329-1338.

Ibrahim, H.J., and Wu, H.F., 1985, Geochemical and mineralogical alteration at the Midwest uranium-nickel deposits, *in* Sibbald, T.I., and Petruk, W., eds., Geology of Uranium Deposits: Canadian Institute of Mining and Metallurgy, Special Volume 32, p. 73-84.

Jamieson, B.W., and Spross, J., 2000, The Exploration and Development of the High Grade McArthur River Uranium Orebody, Fachaufsatze, Erzmetall: Gesellschaft für Bergbau, Metallurgie, Rohstoff- und Umwelttechnik (GDMB), H14503, No. 7/8, p. 457-469.

Jefferson, C.W., and Delaney, G., eds., 2007, EXTECH IV: Geology and Uranium EXploration TECHnology of the Proterozoic Athabasca Basin, Saskatchewan and Alberta: Geological Survey of Canada, Bulletin 588 (also Geological Association of Canada, Mineral Deposits Division, Special Publication 4; Saskatchewan Geological Society, Special Publication 18), 643 p.

Jefferson, C.W., Percival, J.B., Bernier, S., Cutts, C., Drever, G., Jiricka, D., Long, D., McHardy, S., Quirt, D., Ramaekers, P., Wasyliuk, K., and Yeo, G.M., 2001, Lithostratigraphy and mineralogy in the Eastern Athabasca Basin, northern Saskatchewan - progress in year 2 of EXTECH IV, *in* Summary of Investigations 2001, Volume 2, Part B, EXTECH IV Athabasca Uranium Multidisciplinary Study: Saskatchewan Geological Survey, Saskatchewan Energy and Mines, Miscellaneous Report 2001-4.2b (CD-ROM), p. 272-290.

Jefferson, C.W., McGill, B.D., Thomas, D.J., Marlatt, J.M., McHardy, S., Drever, G.L., and Sopuck, V.J., 2002, McArthur River Mine, Field Trip A1. The Eastern Athabasca Basin and its Uranium Deposits, Chapter 5 *in* Andrade, N., Jefferson, C.W., Thomas, D.J., Tourigny, G., Wilson S., and Yeo, G.M., eds., Annual Meeting, Field Trip Guidebook A1: Geological Association of Canada - Mineralogical Association of Canada, p. 73-88.

Jefferson, C.W., Thomas, D.J., Gandhi, S.S., Ramaekers, P., Delaney, G., Brisbin, D., Cutts, C., Portella, P., and Olson, R.A. 2007, Unconformity-associated uranium deposits of the Athabasca Basin, Saskatchewan and Alberta, *in* Jefferson, C.W., and Delaney, G., eds., EXTECH IV: Geology and Uranium EXploration TECHnology of the Proterozoic Athabasca Basin, Saskatchewan and Alberta: Geological Survey of Canada, Bulletin 588, (also Geological Association of Canada, Mineral Deposits Division, Special Publication 4; Saskatchewan Geological Society, Special Publication 18), p. 23-68.

Johan, Z., Picot, P., and Ruhlmann, F., 1987, The ore mineralogy of the Otish Mountains uranium deposit, Quebec: skippenite, Bi,Se,Te, and watkinsonite, $Cu,PbBi_4(Se,S)_8$, two new mineral species: Canadian Mineralogist, v. 25, p. 625-638.

Kerans, C., Ross, G.M., Donaldson, J.A., and Geldsetzer, H.J., 1981, Tectonism and depositional history of the Helikian Hornby Bay and Dismal Lakes Groups, District of Mackenzie, *in* Campbell, F.H.A., ed., Proterozoic Basins of Canada: Geological Survey of Canada, Paper 81-10, p. 157-182.

Kister, P., Laverret, E., Cuney, M., Vieillard, P., and Quirt, D., 2003, 3-D distribution of alteration minerals associated with unconformity-type mineralization: Anne Zone, Shea Creek deposit (Saskatchewan, Canada): Uranium Geochemistry 2003, International Conference, Nancy, France, April 13-16 2003, Proceedings, p. 197-200.

Knipping, H.D., 1974, The concepts of supergene versus hypogene emplacement of uranium at Rabbit Lake, Saskatchewan, Canada, Formation of Uranium Deposits: Vienna, International Atomic Energy Agency, p. 531-548.

Koeppel, V., 1967, Age and History of Uranium Mineralization of the Beaverlodge Area, Saskatchewan: Geological Survey of Canada, Paper 67-31, 111 p.

Kotzer, T., and Kyser, T.K., 1995, Fluid history of the Athabasca Basin and its relation to diagenesis, uranium mineralization and paleohydrology: Chemical Geology, v. 120, p. 45-89.

Krapez, B., 1996, Sequence stratigraphic concepts applied to the identification of basin-filling rhythms in Precambrian successions: Australian Journal of Earth Sciences, v. 43, p. 355-380.

Kupsch, B.G., and Catuneanu, O., 2007, Alteration features and geochemical signatures of the Maybelle River uranium zone, Athabasca Basin, Alberta, *in* Jefferson, C.W., and Delaney, G., eds., EXTECH IV: Geology and Uranium EXploration TECHnology of the Proterozoic Athabasca Basin, Saskatchewan and Alberta: Geological Survey of Canada, Bulletin 588, (also Geological Association of Canada, Mineral Deposits Division, Special Publication 4; Saskatchewan Geological Society, Special Publication 18) p. 347-362.

Kyser, K., Hiatt, E., Renac, C., Durocher, K., Holk, G., and Deckart, K., 2000, Diagenetic fluids in Paleo- and Mesoproterozoic sedimentary basins and their implications for long protracted fluid histories, Chapter 10 *in* Kyser, K., ed., Fluids and Basin Evolution, Short Course Series Volume 28 (Series editor Robert Raeside): Mineralogical Association of Canada, p. 225-262.

Lainé, R., Alonso, D., and Svab, M., 1985, The Carswell Structure Uranium Deposits, Saskatchewan: Geological Association of Canada, Special Paper 29, 230 p.

Landais, P., Dubessy, J., Dereppe, J.M., and Philp, R.P., 1993, Characterization of graphite alteration and bitumen genesis in the Cigar Lake deposit (Saskatchewan, Canada): Canadian Journal of Earth Sciences, v. 30, p. 743-753.

Langford, F.F., 1978, Origin of unconformity-type pitchblende deposits in the Athabasca Basin of Saskatchewan, *in* Kimberly, M.M., ed., Uranium Deposits: Their Mineralogy and Origin, Short Course: University of Toronto, Mineralogical Association of Canada, p. 485-499.

Leventhal, J.S., Grauch R., Threlkeld, C., Lichte, F., and Harper, C., 1987, Unusual organic matter associated with uranium from the Claude deposit, Cluff lake, Canada: Economic Geology, v. 82, p. 1169-1176.

Long, D.G.F., 2001, Architecture of Late Paleoproterozoic fluvial strata in the Manitou Falls Formation at McClean Lake and Key Lake, eastern Athabasca Basin, northern Saskatchewan, *in* Summary of Investigations 2001, Volume 2: Saskatchewan Geological Survey, Saskatchewan Energy and Mines, Miscellaneous Report, 2001-4.2b (CD-ROM B), p. 306-313.

——— 2007, Topographic influences on the sedimentology of the Manitou Falls Formation, eastern Athabasca Basin, Saskatchewan, *in* Jefferson, C.W., and Delaney, G., eds., EXTECH IV: Geology and Uranium EXploration TECHnology of the Proterozoic Athabasca Basin,

Saskatchewan and Alberta: Geological Survey of Canada, Bulletin 588, (also Geological Association of Canada, Mineral Deposits Division, Special Publication 4; Saskatchewan Geological Society, Special Publication 18), p. 267-280.

Macdonald, C., 1980, Mineralogy and Geochemistry of a Precambrian Regolith in the Athabasca Basin: M.Sc. thesis, University of Saskatchewan, Saskatoon, Saskatchewan, 151 p.

——— 1985, Mineralogy and geochemistry of the sub-Athabasca regolith near Wollaston Lake: Geology of Uranium Deposits: Canadian Institute of Mining-Society of Economic Geologists, Uranium Symposium, September 1981, Proceedings, p. 155-158.

Macdonald, J., Stomp, G., and Rediron, S., 2000, Environmental Quality Committees and Northern Mines Monitoring Secretariat Report to Communities 1999-2000: Saskatchewan Northern Affairs, 28 p. http://www.northern.gov.sk.ca/NMMS/eqcannualreports/annualreport99_00.pdf.

Macdougall, D.G., 1990, Rare earth mineralization in the Athabasca Group - Maw Zone, *in* Summary of Investigations 1990: Saskatchewan Geological Survey, Saskatchewan Energy and Mines, Miscellaneous Report 90-4, p. 103-105.

MacLean, B.C., and Cook, D.G., 2004, Revisions to the Paleoproterozoic Sequence A, based on reflection seismic data across the western plains of the Northwest Territories, Canada: Precambrian Research, v. 129, p. 271-289.

Madore, C., Annesley, I., and Wheatley, K., 2000, Petrogenesis, age, and uranium fertility of peraluminous leucogranites and pegmatites of the McClean Lake / Sue and Key Lake / P-Patch deposit areas, Saskatchewan, GeoCanada: The Millennium Geoscience Summit; Joint meeting of the Canadian Geophysical Union, Canadian Society of Exploration Geophysicists, Canadian Society of Petroleum Geologists, Canadian Well Logging Society, Geological Association of Canada, and the Mineralogical Association of Canada, May 29-June 2, 2000, Calgary, Alberta, p. 4.

Marlatt, J., McGill, B., Matthews, R., Sopuck, V., and Pollock, G., 1992, The discovery of the McArthur River uranium deposit, Saskatchewan, Canada, new developments in uranium exploration, resources, production and demand: International Atomic Energy Agency - Nuclear Energy Agency of the OECD, Vienna, 26-29 August, 1991, Technical Committee Meeting, Proceedings, IAEA TECDOC-650, p. 118-127.

Mathieu, R., Cuney, M., and Cathelineau, M., 2000, Geochemistry of paleofluids circulation in the Franceville Basin and around Oklo natural reaction zones (Gabon): Journal of Geochemical Exploration, v. 69-70, p. 245-249.

Matthews, R., Koch, R., Leppin, M., Powell, B., and Sopuck, V., 1997, Advances in integrated exploration for unconformity-associated uranium deposits in western Canada, *in* Gubins, A.G., ed., Geophysics and Geochemistry at the Millennium, Proceedings of Exploration '97. Fourth Decennial International Conference on Mineral Exploration, September 15-18, 1997: Toronto, Prospectors and Developers Association of Canada, p. 993-1002.

Maurice, Y.T., Dyck, W., and Strand, J.G., 1985, Secondary dispersion around the uranium-nickel deposit at Key Lake, northern Saskatchewan, *in* Sibbald, T.I., and Petruk, W., eds., Geology of Uranium Deposits: Canadian Institute of Mining and Metallurgy, Special Volume 32, p. 38-47.

Mazimhaka, P., and Hendry, H., 1989, Uranium deposits in the Beaverlodge area, northern Saskatchewan: their relationship to the Martin Group (Proterozoic) and the underlying basement, Uranium resources and geology of North America: Technical Committee Meeting, Saskatoon, 1-3 September, 1987, Proceedings, Vienna, International Atomic Energy Agency, TECDOC-500, p. 297-315.

McBride, E.F., 1989, Diagenesis of the Maxon sandstone (Early Cretaceous), Marathon Region, Texas: A diagenetic quartzarenite: Journal of Sedimentary Petrology, v. 57, p. 98-107.

McCready, A.J., Annesley, I.R., Parnell, J., and Richardson, L.C., 1999, Uranium-bearing Carbonaceous Matter, McArthur River Uranium Deposit, Saskatchewan, *in* Summary of Investigations 1999: Saskatchewan Geological Survey, Saskatchewan Energy and Mines, p. 110-120.

McGill, B., Marlatt, J., Matthews, R., Sopuck, V., Homeniuk, L., and Hubregtse, J., 1993, The P2 North uranium deposit Saskatchewan, Canada: Exploration Mining Geology, v. 2, 4, p. 321-331.

McKay, A.D., and Miezitis, Y., 2001, Australia's Uranium Resources, Geology and Development of Deposits: Australian Geological Survey Organization - Geoscience Australia, Australia Mineral Resource Report 1 (Internet).

Miller, A.R., 1979, Uranium geology in the central Baker Lake Basin, District of Keewatin, *in* Current Research, Part A: Geological Survey of Canada, Paper 79-1A, p. 57-59.

——— 1980, Uranium Geology of the eastern Baker Lake Basin, District of Keewatin, Northwest Territories: Geological Survey of Canada, Bulletin 330, 63 p.

——— 1983, A progress report: Uranium-phosphorous association in the Helikian Thelon Formation and sub-Thelon saprolite, central District of Keewatin, *in* Current Research, Part A: Geological Survey of Canada, Paper 83-1A, p. 449-456.

Miller, A.R., and LeCheminant, A.N., 1985, Geology and uranium metallogeny of Proterozoic supracrustal successions, central District of Keewatin, N.W.T. with comparisons to northern Saskatchewan, *in* Sibbald, T.I., and Petruk, W., eds., Geology of Uranium Deposits: Canadian Institute of Mining and Metallurgy, Special Volume 32, p. 167-185.

Miller, A.R., Cumming, G., and Krstic, D., 1989, U-Pb, Pb-Pb and K-Ar isotopic study and petrography of uraniferous phosphate bearing rocks in the Thelon Formation, Dubawnt Group, Northwest Territories, Canada: Canadian Journal of Earth Sciences, v. 26, p. 867-880.

Millikan, K.L., 1988, Loss of provenance information through subsurface diagenesis in Plio-Pleistocene sandstones, northern Gulf of Mexico: Journal of Sedimentary Petrology, v. 58, p. 992-1002.

Mwenifumbo, C.J., and Bernius, G., 2007, A Th rich crandallite group mineral: a source of thorium enrichment in the Athabasca Group, Saskatchewan, *in* Jefferson, C.W., and Delaney, G., eds., EXTECH IV: Geology and Uranium EXploration TECHnology of the Proterozoic Athabasca Basin, Saskatchewan and Alberta: Geological Survey of Canada, Bulletin 588, (also Geological Association of Canada, Mineral Deposits Division, Special Publication 4; Saskatchewan Geological Society, Special Publication 18) p. 521-532.

Mwenifumbo, C.J., Pflug, K.A., Elliott, B.E., Jefferson, C.W., Koch, R., Robbins, J., and Matthews, R., 2000, Multiparameter borehole geophysical logging at the Shea Creek and McArthur River Projects: Parameters for exploration, stratigraphy and high-resolution seismic studies, *in* Summary of Investigations 2000, Volume 2: Saskatchewan Geological Survey, Saskatchewan Energy and Mines, Miscellaneous Report 2000-4.2, p. 110-122.

Mwenifumbo, C.J., Elliott, B.E., Jefferson, C.W., Bernius, G. R., and Pflug, K.A., 2004, Physical rock properties from the Athabasca Group: designing geophysical exploration models for unconformity-associated uranium deposits: Journal of Applied Geophysics, v. 55, p. 117-135.

Mwenifumbo, C.J., Percival, J.B., Bernius, G., Elliott, B., Jefferson, C.W., Wasyliuk, K., and Drever, G., 2007, Comparison of geophysical, mineralogical and stratigraphic attributes in drill holes MAC 218 and RL 88, McArthur River uranium camp, Athabasca Basin, Saskatchewan, *in* Jefferson, C.W., and Delaney, G., eds., EXTECH IV: Geology and Uranium EXploration TECHnology of the Proterozoic Athabasca Basin, Saskatchewan and Alberta: Geological Survey of Canada, Bulletin 588, (also Geological Association of Canada, Mineral Deposits Division, Special Publication 4; Saskatchewan Geological Society, Special Publication 18) p. 507-520.

Nguyen, P.T., Cox, S.F., Harris, L.B., and Powell, C.M., 1998, Fault-valve behavior in optimally oriented shear zones: an example at the Revenge gold mine, Kambalda, Western Australia: Journal of Structural Geology, v. 20, p. 1625-1640.

Nickel, E.H., and Nichols, M.C., 1991, Mineral Reference Manual: Van Nostrand Reinhold, New York, 250 p.

Ord, A., 2003, The Athabasca and Witwatersrand mineralizing systems: Uranium Geochemistry 2003, International Conference, Nancy, France, April 13-16 2003, Proceedings, p. 271.

Organization for Economic Co-operation and Development, Nuclear Energy Agency, and the International Atomic Energy, 2004, Uranium 2003: Resources, Production and Demand: Nuclear Energy Agency, Organization for Economic Co-operation and Development, 288 p.

Orrell, S.E., Bickford, M.E., and Lewry, J.F., 1999, Crustal evolution and age of thermotectonic reworking in the western hinterland of Trans-Hudson Orogen, northern Saskatchewan: Precambrian Research, v. 95, p. 187-223.

Overton, A., 1977, Seismic determination of basement depths, *in* Report of Activities Part C: Geological Survey of Canada, Paper 77-1C, p. 19-25.

Pagel, M., 1975, Détermination des conditions physico-chimiques de la silicification diagénétique des grès Athabasca (Canada) au moyen des inclusions fluides: Académie des Sciences Paris, v. 280, Comptes Rendus, Série D, p. 2301-2304.

Pagel, M., Poty, B., and Sheppard, S.M.F., 1980, Contributions to some Saskatchewan uranium deposits mainly from fluid inclusion and isotopic data, *in* Ferguson, S., and Goleby, A., eds., Uranium in the Pine Creek Geosyncline: International Atomic Energy Agency (IAEA), p. 639-654.

Pană, D., Creaser, R.A., Muehlenbachs, K., and Wheatley, K., 2007, Basement geology in the Alberta portion of the Athabasca Basin: context for the Maybelle River area, *in* Jefferson, C.W., and Delaney, G., eds., EXTECH IV: Geology and Uranium EXploration TECHnology of the Proterozoic Athabasca Basin, Saskatchewan and Alberta: Geological Survey of Canada, Bulletin 588, (also Geological Association of Canada, Mineral Deposits Division, Special Publication 4; Saskatchewan Geological Society, Special Publication 18) p. 135-154.

Percival, J.B., 1989, Clay mineralogy, Geochemistry and Partitioning of Uranium Within the Alteration Halo of the Cigar Lake Uranium Deposit, Saskatchewan, Canada:. Ph.D. thesis, Carleton University, Ottawa, Ontario, 343 p.

Percival, J.B., Bell, K., and Torrance, J.K., 1993, Clay mineralogy and isotope geochemistry of the alteration halo at the Cigar Lake uranium deposit: Canadian Journal of Earth Sciences, v. 30, p. 689-704.

Percival, J.B., Wasyliuk, K., Reif, T., Bernier, S., Drever, G., and Perkins, C.T., 2002, Mineralogical aspects of three drill cores along the McArthur River transect using a portable infrared spectrometer, Paper D-14 *in* Summary of Investigations 2002, Volume 2: Saskatchewan Geological Survey, Saskatchewan Industry and Resources, Miscellaneous Report 2002-4.2, (CD-ROM), p. 15.

Peterson, T. D., Van Breemen, O., Sandeman, H., and Cousens, B., 2002: Proterozoic (1.85–1.75 Ga) igneous suites of the western Churchill province: granitoid and ultrapotassic magmatism in a reworked Archean hinterland: Precambrian Research, v. 119, p. 73-100.

Pilkington, M., 1989, Variable-depth magnetization mapping: application to the Athabasca Basin, northern Alberta and Saskatchewan, Canada: Geophysics, v. 54, p. 1164-1173.

Polito, P.A., Kyser, T.K., Marlatt, J., Alexandre, P., Bajwah, Z., and Drever, G., 2004, Significance of alteration assemblages for the origin and evolution of the Proterozoic Nabarlek unconformity-related uranium deposit, Northern Territory, Australia: Economic Geology, v. 99, p. 113-139.

Polito, P.A., Kyser, T.K., Thomas, D., Marlatt, J., and Drever, G., 2005, Re-evaluation of the Petrogenesis of the Proterozoic Jabiluka unconformity-related uranium deposit, Northern Territory, Australia: Mineralium Deposita, v. 40, p. 257-288.

Portella, P., and Annesley, I.R., 2000a, Paleoproterozoic tectonic evolution of the eastern sub-Athabasca basement, northern Saskatchewan: Integrated magnetic, gravity, and geological data, extended abstract *in* GeoCanada: The Millennium Geoscience Summit: Joint meeting of the Canadian Geophysical Union, Canadian Society of Exploration Geophysicists, Canadian Society of Petroleum Geologists, Canadian Well Logging Society, Geological Association of Canada, and the Mineralogical Association of Canada: Calgary, Alberta, Canada, 4 p.

——— 2000b, Paleoproterozoic thermotectonic evolution of the eastern sub-Athabasca basement, northern Saskatchewan: Integrated geophysical and geological data, *in* Summary of Investigations 2000, Volume 2: Saskatchewan Geological Survey, Saskatchewan Energy and Mines, Miscellaneous Report 2000-4.2, p. 191-200.

Powell, B., Leppin, M., Wood, G., O'Dowd, C., and Brisbin, D., 2005, Recent innovative applications of geophysics to new uranium discoveries in the Athabasca Basin (extended synopsis), *in* Symposium on Uranium Production and Raw Materials for the Nuclear Fuel Cycle - Supply and Demand, Economics, the Environment and Energy Security: Vienna, Austria, 20-24 June, 2005, Proceedings, p. 112-114.

Quirt, D.H., 1985, Lithogeochemistry of the Athabasca Group: Summary of sandstone data, *in* Summary of Investigations 1985: Saskatchewan Geological Survey, Saskatchewan Energy and Mines, Miscellaneous Report 85-4, p. 128-132.

——— 1989, Host-Rock Alteration at Eagle Point South: Saskatchewan Research Council, Publication No. R-855-1-E-89, 95 p.

——— 2001, Kaolinite and Dickite in the Athabasca Sandstone, Northern Saskatchewan, Canada: Saskatchewan Research Council, Publication No. 10400-16D01, 27 p.

——— 2003, Athabasca unconformity-type uranium deposits: one deposit type with many variations: Uranium Geochemistry 2003, International Conference, Nancy, France, April 13-16 2003, Proceedings, p. 309-312.

Quirt, D.Q., and Ramaekers, P., 2002, The Athabasca Basin: Sedimentary background to unconformity-type uranium deposits: Geological Association of Canada - Mineralogical Association of Canada Joint Meeting, Abstracts Volume 27, p. 94.

Quirt, D.H., and Wasyliuk, K., 1997, Kaolinite, dickite, and other clay minerals in the Athabasca Group, Canada, and the Kombolgie Formation, Australia: 11th International Clay Conference, Ottawa, Ontario, June 1997, Proceedings, p. A61.

Quirt, D., Kotzer, T., and Kyser, T.K., 1991, Tourmaline, phosphate minerals, zircon and pitchblende in the Athabasca Group: Maw Zone and McArthur River areas, *in* Summary of Investigations 1991: Saskatchewan Geological Survey, Saskatchewan Energy and Mines, Report 91-4, p. 181-191.

Raffensperger, J.P., and Garven, G., 1995, The formation of unconformity-type uranium deposits. 2. Coupled hydrochemical modeling: American Journal of Science, v. 295, p. 639-696.

Rainbird, R.H., Jefferson, C.W., Hildebrand, R.S., and Worth, J.K., 1994, The Shaler Supergroup and revision of Neoproterozoic stratigraphy in the Amundsen Basin, Northwest Territories, *in* Current Research, Part A: Geological Survey of Canada, Paper 94-1A, p. 61-70.

Rainbird, R.H., Hadlari, T., Aspler, L.B, Donaldson, J.A., LeCheminant, A.N., and Peterson, T. D., 2003a, Sequence stratigraphy and evolution of the Paleoproterozoic intracontinental Baker Lake and Thelon basins, western Churchill Province, Nunavut, Canada: Precambrian Research, v. 125, p. 21-53.

Rainbird, R.H., Rayner, N., and Stern, R.A., 2003b, SHRIMP U-Pb geochronology of apatite cements and zircon bearing tuff clasts in sandstones from the Athabasca Group, Athabasca Basin, northern Saskatchewan and Alberta: Saskatchewan Industry Resources, Open House, December 1-3, 2003, Saskatoon, Proceedings, p. 6

Rainbird, R.H., Stern, R.A., Rayner, N., and Jefferson, C.W., 2007, Age, provenance, and regional correlation of the Athabasca Group, Saskatchewan and Alberta, constrained by igneous and detrital zircon geochronology, *in* Jefferson, C.W., and Delaney, G., eds., EXTECH IV: Geology and Uranium EXploration TECHnology of the Proterozoic Athabasca Basin, Saskatchewan and Alberta: Geological Survey of Canada, Bulletin 588, (also Geological Association of Canada, Mineral Deposits Division, Special Publication 4; Saskatchewan Geological Society, Special Publication 18) p. 193-210.

Ramaekers, P., 1990, Geology of the Athabasca Group (Helikian) in northern Saskatchewan: Saskatchewan Energy and Mines, Saskatchewan Geological Survey, Report 195, 49 p.

——— 2004, Development, Stratigraphy and Summary Diagenetic History of the Athabasca Basin, Early Proterozoic of Alberta and Its Relation to Uranium Potential: Alberta Geological Survey, Alberta Energy and Utilities Board, Special Report 62 (PDF), 85 p.

Ramaekers, P., Jefferson, C.W., Yeo, G.M., Collier, B., Long, D.G.F., Catuneanu, O., Bernier, S., Kupsch, B., Post, R., Drever, G., McHardy, S., Jiricka, D., Cutts, C., and Wheatley, K., 2007, Revised geological map and stratigraphy of the Athabasca Group, Saskatchewan and Alberta, *in* Jefferson, C.W., and Delaney, G., eds., EXTECH IV: Geology and Uranium EXploration TECHnology of the Proterozoic Athabasca Basin, Saskatchewan and Alberta: Geological Survey of Canada, Bulletin 588, (also Geological Association of Canada, Mineral Deposits Division, Special Publication 4; Saskatchewan Geological Society, Special Publication 18) p. 155-192.

Rayner, N.M., Stern, R.A., and Rainbird, R.H., 2003, SHRIMP U-Pb detrital zircon geochronology of Athabasca Group sandstones, northern Saskatchewan and Alberta, *in* Current Research 2003: Geological Survey of Canada, 2003-F2, 20 p.

Rayner, N.M., Stern, R.A., and Carr, S.D., 2005, Grain-scale variations in trace element composition of fluid-altered zircon, Acasta Gneiss Complex, northwestern Canada: Contributions to Mineralogy and Petrology, v. 148, p. 721-734.

Reeves, P., and Beck, L., 1982, A history of uranium exploration in the Athabasca Basin, *in* Hanula, M.R., ed., The Discoverers, a 50-year History of the Prospectors and Developers Association, some Famous Prospectors and their Discoveries: Pitt Publishing Company Limited, Toronto, p. 153-161.

Renac, C., Kyser, T.K., Durocher, K., Drever, G., and O'Connor, T., 2002, Comparison of diagenetic fluids in the Proterozoic Thelon and

Athabasca Basins, Canada: Implications for protracted fluid histories in stable intracratonic basins: Canadian Journal of Earth Sciences, v. 39, p. 113-132.

Rhys, D.A., 2002, Geological Report on the Hidden Bay property, Wollaston Lake area, northern Saskatchewan: UEX Corporation, Vancouver, B.C.: 7770K PDF, System for Electronic Document Analysis and Retrieval (SEDAR), 89 p.

Richardson, K.A., 1983, Airborne gamma-ray spectrometric survey of the NEA/IAEA Athabasca Test Area, *in* Cameron, E.M., ed., Uranium Exploration in Athabasca Basin, Saskatchewan, Canada: Geological Survey of Canada, Paper 82-11, p. 1-14.

Rippert, J.C., Koning, E., Robbins, J., Koch, R., and Baudemont, D., 2000, The Shea Creek Uranium Project, West Athabasca Basin, Saskatchewan, Canada, Geological Association of Canada – Mineralogical Association of Canada, Joint Annual Meeting, Calgary, Abstract, v. 26, p. 65.

Robertson, R., 2006a, Cigar Lake: Saskatchewan's next uranium producer: The Northern Miner, v. 91, p. 1-6.

——— 2006b, Millennium: Cameco's newest uranium discovery: The Northern Miner, v. 91, p. 1-14.

Robinson, S.C., 1955, Mineralogy of Uranium Deposits, Goldfields, Saskatchewan: Geological Survey of Canada Bulletin 31. 128 p.

Rogala, B., 2003, The Sibley Group: A Lithostratigraphic, Geochemical and Paleomagnetic Study: M.Sc. thesis, Lakehead University, 150 p.

Roscoe, S.M., 1984, Assessment of Mineral Resource Potential in the Bathurst Inlet Area, NTS 76J, K, N, O Including the Proposed Bathurst Inlet National Park: Geological Survey of Canada, Open File 788, 75 p., scale: 1:3 000 000.

Ross, G.M., 2000, Tectonics of sedimentary basins: A prelude to fluid evolution, Chapter 3 *in* Kyser, K., ed., Fluids and Basin Evolution: Mineralogical Association of Canada, Short Course Series, v. 28, p. 39-62.

Ruhrmann, G., 1987, The Gaertner uranium orebody at Key Lake (northern Saskatchewan, Canada) after three years of mining: An update of the geology, *in* Gilboy, C.F., and Vigrass, L.W., eds., Economic Minerals of Saskatchewan: Saskatchewan Geological Society, Special Publication 8, p. 120-137.

Ruzicka, V.R., 1989, Monometallic and polymetallic deposits associated with the sub-Athabasca unconformity in Saskatchewan, *in* Current Research: Geological Survey of Canada, Paper 89-1C, p. 67-79.

Ruzicka, V.R., 1996a, Unconformity-associated uranium, Chapter 7 *in* Eckstrand, O.R., Sinclair, W.D., and Thorpe, R.I., eds., Geology of Canadian Mineral Deposit Types: Geological Survey of Canada, Geology of Canada, v, 8, p. 197-210.

——— 1996b, Veins in shear zones, Vein Uranium, Section 13.1 *in* Eckstrand, O.R., Sinclair, W.D., and Thorpe, R.I., eds., Geology of Canadian Mineral Deposit Types: Geological Survey of Canada, Geology of Canada, v. 8, p. 278-283.

Sangély, L., Michels, R., Chaussidon, M., Brouand, M., Cuney, M., Richard, L., and Huault, V., 2003, Origin of carbonaceous matter in discordance-type uranium ore-deposit of Saskatchewan (Canada): Clues to an biogenic synthesis: Uranium Geochemistry 2003, International Conference, Nancy, France, April 13-16 2003, Proceedings, p. 323-326.

Saskatchewan Industry and Resources, 2005, Saskatchewan Mineral Deposit Index (SMDI): Government of Saskatchewan, Saskatchewan Industry and Resources.

Scott, F., 1983, Midwest Lake uranium discovery, *in* Cameron, E.M., ed., Uranium Exploration in Athabasca Basin, Saskatchewan, Canada: Geological Survey of Canada, Paper 82-11, p. 41-50.

Shives, R.B.K., Wasyliuk, K., and Zaluski, G., 2000, Detection of K-enrichment, illite chimneys using ground gamma ray spectrometry, McArthur River area, northern Saskatchewan, *in* Summary of Investigations 2000, Volume 2: Saskatchewan Geological Survey, Saskatchewan Energy and Mines, Miscellaneous. Report 2000-4.2, p. 160-169.

Sibbald, T.I.I., 1985, Geology and genesis of the Athabasca Basin uranium deposits, *in* Summary of Investigations 1985: Saskatchewan Geological Survey, Saskatchewan Energy and Mines, Miscellaneous Report 84-4, p. 133-156.

Sibbald, T.I.I., Munday, R.J.C., and Lewry, J.F., 1976, The geological setting of uranium mineralization in northern Saskatchewan, *in* Dunn, C.E., ed., Uranium in Saskatchewan: Saskatchewan Geological Society, Special Publication No. 3, p. 51-98.

Sibson, R.H., 2001, Seismogenic framework for hydrothermal transport and ore deposition: Reviews in Economic Geology, v. 14, p. 25-50.

Smith, E.E.N., 1986, Geology of the Beaverlodge operation Eldorado Nuclear Limited SA, *in* Evans, E.L., ed., Uranium Deposits of Canada: Canadian Institute of Mining and Metallurgy, Special Volume 33, p. 95-119.

Sopuck, V.J., de Carle, E.M., and Cooper, B., 1983, The application of lithogeochemistry in the search for unconformity-type uranium deposits, northern Saskatchewan, Canada, *in* Parslow, G.R., ed., Geochemical Exploration 1982: Journal of Geochemical Exploration, Special Issue 19, p. 77-99.

Stern, R.A.C., Pană, D., and Rayner, N., 2003, SHRIMP U-Pb ages of granitoid basement rocks of the southwestern part of the Athabasca Basin, Saskatchewan and Alberta, *in* Current Research 2003: Geological Survey of Canada, 2003-F3, 20 p.

Suryam, J.V., 1981, Regional seismic profile, Athabasca Basin: Saskatchewan Energy and Mines, Miscellaneous Report 81-4, 115 p.

——— 1984, Seismic investigations on the perimeter of the Athabasca Basin in Northern Saskatchewan (Wollaston Lake-Black Lake areas): Saskatchewan Energy and Mines, Report No. 216B, 38 p.

Thomas, D.J., 1983, Distribution, Geological Controls and Genesis of Uraniferous Pegmatites in the Cree Lake Zone of northern Saskatchewan: M.Sc. thesis, University of Regina, Regina, Saskatchewan, 213 p.

Thomas, M.D., and McHardy, S., 2007, Magnetic insights into basement geology in the area of McArthur River uranium deposit, Athabasca Basin, Saskatchewan, *in* Jefferson, C.W., and Delaney, G., eds., EXTECH IV: Geology and Uranium EXploration TECHnology of the Proterozoic Athabasca Basin, Saskatchewan and Alberta: Geological Survey of Canada, Bulletin 588, (also Geological Association of Canada, Mineral Deposits Division, Special Publication 4; Saskatchewan Geological Society, Special Publication 18) p. 425-440.

Thomas, M.D., and Wood, G., 2007, Geological significance of gravity anomalies in the area of McArthur River uranium deposit, Athabasca Basin, Saskatchewan, *in* Jefferson, C.W., and Delaney, G., eds., EXTECH IV: Geology and Uranium EXploration TECHnology of the Proterozoic Athabasca Basin, Saskatchewan and Alberta: Geological Survey of Canada, Bulletin 588, (also Geological Association of Canada, Mineral Deposits Division, Special Publication 4; Saskatchewan Geological Society, Special Publication 18) p. 441-464.

Thomas, D.J., Matthews, R.B., and Sopuck, V., 2000, Athabasca Basin (Canada) unconformity - type uranium deposits: Exploration model, current mine developments and exploration directions, *in* Geology and Ore Deposits 2000: The Great Basin and Beyond: Geological Society of Nevada Symposium, Reno, Nevada, May 15-18 2000, Proceedings, v. 1, p. 103-126.

Thomas, D.J., Jefferson, C.W., Yeo, G.M, Card, C., and. Sopuck, V.J., 2002, I. Introduction, *in* Andrade, N., Breton, G., Jefferson, C.W., Thomas, D.J., Tourigny, G., Wilson, S., Yeo, G.M., ed., Trip A1: The Eastern Athabasca Basin and Its Uranium Deposits. Field Trip Guidebook: Geological Association of Canada - Mineralogical Association of Canada, p. 1-22.

Toulhoat, P., and Beaucaire, C., 1993, Geochemistry of ground water in the Cigar Lake uranium deposit (Saskatchewan, Canada) and application of isotopes of uranium and lead as exploration guides: Canadian Journal of Earth Sciences, v. 30, p. 754-763.

Tourigny, G., Wilson, G., Breton, G., and Portella, P., 2002, Geology of the Sue C uranium deposit, McClean Lake area, northern Saskatchewan, *in* N. Andrade, G.B., C.W. Jefferson, D.J. Thomas, G. Tourigny, S. Wilson and G.M. Yeo, ed., Trip A1: The Eastern Athabasca Basin and Its Uranium Deposits. Field Trip Guidebook, Geological Association of Canada - Mineralogical Association of Canada, p. 35-51.

Tourigny, G., Quirt, D.H., Wilson, N., Wilson, S., Breton, G., and Portella, P., 2007, Basement geology of the Sue C uranium deposit, McClean Lake area, Saskatchewan, *in* Jefferson, C.W., and Delaney, G., eds., EXTECH IV: Geology and Uranium EXploration TECHnology of the Proterozoic Athabasca Basin, Saskatchewan and Alberta: Geological Survey of Canada, Bulletin 588, (also Geological Association of Canada, Mineral Deposits Division, Special Publication 4; Saskatchewan Geological Society, Special Publication 18) p. 229-248.

Tran, H.T., 2001, Tectonic Evolution of the Paleoproterozoic Wollaston Group in the Cree Lake Zone, northern Saskatchewan, Canada: Ph.D. thesis, University of Regina, Regina, Saskatchewan, 458 p.

Tremblay, L.P., 1968, Geology of the Beaverlodge Mining Area, Saskatchewan: Geological Survey of Canada, Memoir 367, 265 p.

——— 1982, Geology of the Uranium Deposits Related to the Sub-Athabasca Unconformity, Saskatchewan: Geological Survey of Canada, Paper 81-20, 56 p.

Tuncer, V., Unsworth, M., Siripunvaraporn, W., and Craven, J., in press, Exploration for unconformity type uranium deposits with audio-magnetotelluric data: A case study from the McArthur River Mine, Saskatchewan (Canada): Geophysics.

Uranium Information Centre, 2006, World Uranium Mining, Nuclear Issues Briefing Paper 41, July 2006: Uranium Information Centre, Melbourne Australia, http://www.uic.com.au/nip41.htm.

——— 2007, World Uranium Mining, Nuclear Issues Briefing Paper 3, January 2007: Uranium Information Centre, Melbourne Australia, www.uic.com.au/nip03.htm.

Wallis, R.H., Saracoglu, N., Brummer, J.J., and Golightly, J.P., 1985, Geology of the McClean uranium deposits, *in* Geology of Uranium Deposits: Proceedings of the Canadian Institute of Mining – Society of Economic Geologists, Uranium Symposium, September 1981, Proceedings, p. 101-13.

Wardle, R.J., 2005, Uranium in Labrador: Geological Survey, Newfoundland and Labrador Department of Natural Resources, Preliminary Open File Report, 8 p.

Ware, M.J., and Hiscott, R.N., 1985, Sedimentology of Proterozoic cratonic sheet sandstones of the eastern Canadian Shield: Sims Formation, Labrador, Canada: Precambrian Research, v. 30, p. 1-26.

Wasyliuk, K., 2002, Petrogenesis of the Kaolinite-Group Minerals in the eastern Athabasca Basin of northern Saskatchewan: Applications to Uranium Mineralization: M.Sc. thesis, University of Saskatchewan, Saskatoon, Saskatchewan, 140 p.

Wheatley, K., and Cutts, C., 2006, Overview of the Maybelle River uranium mineralization, Alberta, Canada, *in* Uranium production and raw materials for the Nuclear Fuel Cycle - supply and demand, economics, the environment and energy security: International Atomic Energy Agency, Symposium Proceedings, Vienna, Austria, 20-24 June, 2005, STI/PUG/1259 on companion CD to printed volume, 12 p.

White, D.J., Hajnal, Z., Györfi, I., Takács, E., Roberts, B., Mueller, C., Reilkoff, B., Jefferson, C.W., Koch, R., Powell, B., Annesley, I.R., and Brisbin, D.. 2007, Seismic methods for uranium exploration: an overview of EXTECH IV seismic studies at the McArthur River mining camp, Athabasca Basin, Saskatchewan, *in* Jefferson, C.W., and Delaney, G., eds., EXTECH IV: Geology and Uranium EXploration TECHnology of the Proterozoic Athabasca Basin, Saskatchewan and Alberta: Geological Survey of Canada, Bulletin 588, (also Geological Association of Canada, Mineral Deposits Division, Special Publication 4; Saskatchewan Geological Society, Special Publication 18) p. 363-388.

Wilde, A.R., and Wall, V.J., 1987, Geology of the Nabarlek uranium deposit, Northern Territory, Australia: Economic Geology, v. 82, p. 1152-1168.

Wilde, A.R.O., Mernach, T.P., Bloom, M.S., and Hoffman, C.F., 1989, Fluid inclusion evidence on the origin of some Australian unconformity-associated uranium deposits: Economic Geology, v. 84, p. 1627-1642.

Wilson, J.A., 1985, Geology of the Athabasca Group in Alberta: Geological Survey Department, Alberta Research Council, Bulletin No. 49. 78 p.

Wilson, N.S.F., Stasiuk, L.D., and Fowler, M.G., 2007, Origin of organic matter in the Proterozoic Athabasca Basin of Saskatchewan and Alberta, and significance to unconformity-associated uranium deposits, *in* Jefferson, C.W., and Delaney, G., eds., EXTECH IV: Geology and Uranium EXploration TECHnology of the Proterozoic Athabasca Basin, Saskatchewan and Alberta: Geological Survey of Canada, Bulletin 588, (also Geological Association of Canada, Mineral Deposits Division, Special Publication 4; Saskatchewan Geological Society, Special Publication 18) p. 325-340.

Yeo, G.M., and Delaney, G., 2007, The Wollaston Supergroup, stratigraphy and metallogeny of a Paleoproterozoic Wilson cycle in the Trans-Hudson Orogen, Saskatchewan, *in* Jefferson, C.W., and Delaney, G., eds., EXTECH IV: Geology and Uranium EXploration TECHnology of the Proterozoic Athabasca Basin, Saskatchewan and Alberta: Geological Survey of Canada, Bulletin 588, (also Geological Association of Canada, Mineral Deposits Division, Special Publication 4; Saskatchewan Geological Society, Special Publication 18) p. 89-118.

Yeo, G.M., and Jiricka, D., 2002, Dawn Lake deposits, Chapter 2 *in* Andrade, N. Jefferson, C.W., Thomas, D.J., Tourigny, G., Wilson, W., and Yeo, G.M., eds., The eastern Athabasca Basin and Its Uranium Deposits, Field Trip A-1 Guidebook: Geological Association of Canada and Mineralogical Association of Canada Meeting, Saskatoon, p. 23-32.

Yeo, G., Collier, B., Ramaekers, P., Koning, E., Robbins, J., and Jiricka, D., 2001, Stratigraphy of the Athabasca Group in the southwestern Athabasca Basin, Saskatchewan (NTS 74F and 74K), *in* Summary of Investigations 2001, Volume 2: Saskatchewan Geological Survey, Saskatchewan Energy and Mines, Miscellaneous Report 2001-4.2, p. 252-265.

Yeo, G., Percival, J.B., Jefferson, C.W., Ickert, R., and Hunt, P., 2007a, Environmental significance of oncoids and crypto-microbial laminates from the Late Paleoproterozoic Athabasca Group, Saskatchewan and Alberta, *in* Jefferson, C.W., and Delaney, G., eds., EXTECH IV: Geology and Uranium EXploration TECHnology of the Proterozoic Athabasca Basin, Saskatchewan and Alberta: Geological Survey of Canada, Bulletin 588, (also Geological Association of Canada, Mineral Deposits Division, Special Publication 4; Saskatchewan Geological Society, Special Publication 18) p. 315-324.

Yeo, G.M., Jefferson, C.W., and Ramaekers, P., 2007b, Comparison of lower Athabasca Group stratigraphy among deposystems, Saskatchewan and Alberta, *in* Jefferson, C.W., and Delaney, G., eds., EXTECH IV: Geology and Uranium EXploration TECHnology of the Proterozoic Athabasca Basin, Saskatchewan and Alberta: Geological Survey of Canada, Bulletin 588, (also Geological Association of Canada, Mineral Deposits Division, Special Publication 4; Saskatchewan Geological Society, Special Publication 18) p. 465-488.

Young, G.M., Jefferson, C.W., Delaney, G.D., and Yeo, G.M., 1979, Middle and late Proterozoic evolution of the northern Canadian Cordillera and Shield: Geology, v. 7, p. 125-128.

Zhang, G., Wasyliuk, K., and Pan, Y., 2001, The characterization and quantitative analysis of clay minerals in the Athabasca Basin, Saskatchewan: Application of short-wave infrared reflectance spectroscopy: Canadian Mineralogist, v. 39, p. 1347-1363.

IRON OXIDE COPPER-GOLD DEPOSITS: A CANADIAN PERSPECTIVE

LOUISE CORRIVEAU

Geological Survey of Canada, 490 de la Couronne, Québec, Québec G1K 9A9
Corresponding author's email: lcorrive@nrcan.gc.ca

Abstract

Iron oxide copper-gold (±Ag±Nb±P±REE±U) deposits form an attractive target for mineral exploration in Canada as they can host significant resources in base, precious, and strategic metals, as well as energy, and are under-explored across the country. This new deposit type includes a large spectrum of sulphide-deficient, mono- to polymetallic breccias, veins, disseminations, and massive lenses with more than 20% low-Ti magnetite and/or hematite. These hydrothermal deposits are associated with large-scale continental A- to I-type granitic suites with intermediate and mafic facies, alkaline-carbonatite stocks, crustal-scale fault zones, regional sodic-calcic alteration, focused potassic and iron oxide alteration, and coincident aeromagnetic and gravity highs. Shallow- to mid-crustal intracratonic, intra-arc or back-arc continental extensional settings, and late- or post-orogenic settings are currently considered prospective for iron oxide copper-gold (IOCG) deposits. Though there is no producing IOCG mine in Canada, three IOCG deposits are known and exploration is accelerating in prospective Proterozoic granitic, gneissic, and sedimentary belts of the Canadian Shield and the Cordillera, as well as Phanerozoic settings in the Appalachian and Cordilleran orogens.

Résumé

Les gîtes d'oxydes de fer-Cu-Au (±Ag±Nb±P±ÉTR±U) représentent une cible d'exploration attrayante pour le Canada, puisqu'ils peuvent contenir des ressources significatives en métaux usuels, précieux et stratégiques et en énergie et qu'ils n'ont pas suscité un niveau approprié d'exploration à la grandeur du pays. Ce nouveau type de gîte inclut un large éventail de minerais monométalliques à polymétalliques, déficients en sulfures et contenant plus de 20% de magnétite pauvre en titane et/ou d'hématite, qui se présentent sous la forme de brèches, de veines, de disséminations et de lentilles massives. Ces gîtes d'origine hydrothermale sont associés à des suites intrusives granitiques continentales d'envergure, de type A ou I qui présentent des faciès mafiques et intermédiaires, à des stocks de roches alcalines et de carbonatite, à des failles d'échelle crustale, à des zones d'altération sodique-calcique régionales, à des zones d'altération potassique et à oxydes de fer de superficie plus restreinte et à la coïncidence d'anomalies aéromagnétiques et gravimétriques positives. Des milieux continentaux en extension soit intracratoniques, intra-arc ou d'arrière-arc, et des contextes tardi-orogéniques ou postorogéniques sont actuellement considérés comme potentiellement fertiles en gîtes d'oxydes de fer-Cu-Au. Bien qu'il n'y ait aucune mine de ce type de gîte au Canada, trois gîtes ont été découverts et l'exploration s'accélère dans des terrains granitiques, gneissiques ou sédimentaires du Protérozoïque, dans le Bouclier canadien et la Cordillère, ou du Phanérozoïque, dans les Appalaches et la Cordillère.

Definition

Iron oxide copper-gold (IOCG) deposits encompass a wide spectrum of sulphide-deficient low-Ti magnetite and/or hematite orebodies of hydrothermal origin where breccias, veins, disseminations, and massive lenses with polymetallic enrichments (Cu, Au, Ag, U, REE, Bi, Co, Nb, P) are genetically associated with, but either proximal or distal to, large-scale continental, A- to I-type magmatism, alkaline-carbonatite stocks, and crustal-scale fault zones and splays. The deposits are characterized by more than 20% iron oxides. Their lithological hosts and ages are non-diagnostic whereas their alteration zones are distinctive, with sodic-calcic or potassic regional alteration superimposed by focused potassic and iron oxide alteration. The deposits occur at shallow- to mid-crustal levels in anorogenic to orogenic, extensional to compressional, continental settings such as intracratonic and intra-arc rifts, continental magmatic arcs and back-arc basins, and collisional orogens. Currently, known IOCG deposit districts (Fig. 1) occur in Precambrian shields worldwide as well as in circum-Pacific regions (e.g. Porter, 2000, 2002a and papers therein; Gandhi, 2004b; Williams et al., 2005).

Because of the diversity of IOCG deposits, there is debate whether they form a single deposit type or whether they are iron oxide-rich variants of other deposit types. In this synthesis, IOCG deposits are considered to define a bona fide deposit type. Polymetallic deposits lacking significant iron oxides are not considered herein as IOCG deposits, while hydrothermal monometallic, low-Ti magnetite and/or hematite Fe deposits (i.e. Fe the as sole economic metal) with alteration zones typical of IOCG deposits and polymetallic Fe oxide deposits with Nb and REE as important economic commodities are considered end-members of the IOCG deposits spectrum. This broad approach to IOCG deposits is taken because the monometallic hydrothermal Fe oxide deposits with low economic potential and sodic-calcic alteration may be the easiest component of a polymetallic system to identify and may serve as important pathfinders for grass-roots exploration. Understanding what processes make the polymetallic Cu-Au, Bi-Co, or U rich iron oxide deposits different from the monometallic iron oxide deposits will be key to increased exploration effectiveness in the future. However, a current priority is to identify prospective IOCG settings among the frontier plutonic and metamorphic terranes of Canada in support of emerging exploration plays. This target colours not only the definition chosen for IOCG deposits in this synthesis, but also the entire synthesis of their settings where grouping of elements is commonly favoured over the specificity of individual deposit and deposit subtypes.

Iron Oxide Copper-Gold Deposit Subtypes

Opinions diverge about what constitutes the IOCG deposit type. Reasons for this confusion include 1) the brief

Corriveau, L., 2007, Iron oxide copper-gold deposits: A Canadian perspective, *in* Goodfellow, W.D., ed., Mineral Deposits of Canada: A Synthesis of Major Deposit-Types, District Metallogeny, the Evolution of Geological Provinces, and Exploration Methods: Geological Association of Canada, Mineral Deposits Division, Special Publication No. 5, p. 307-328.

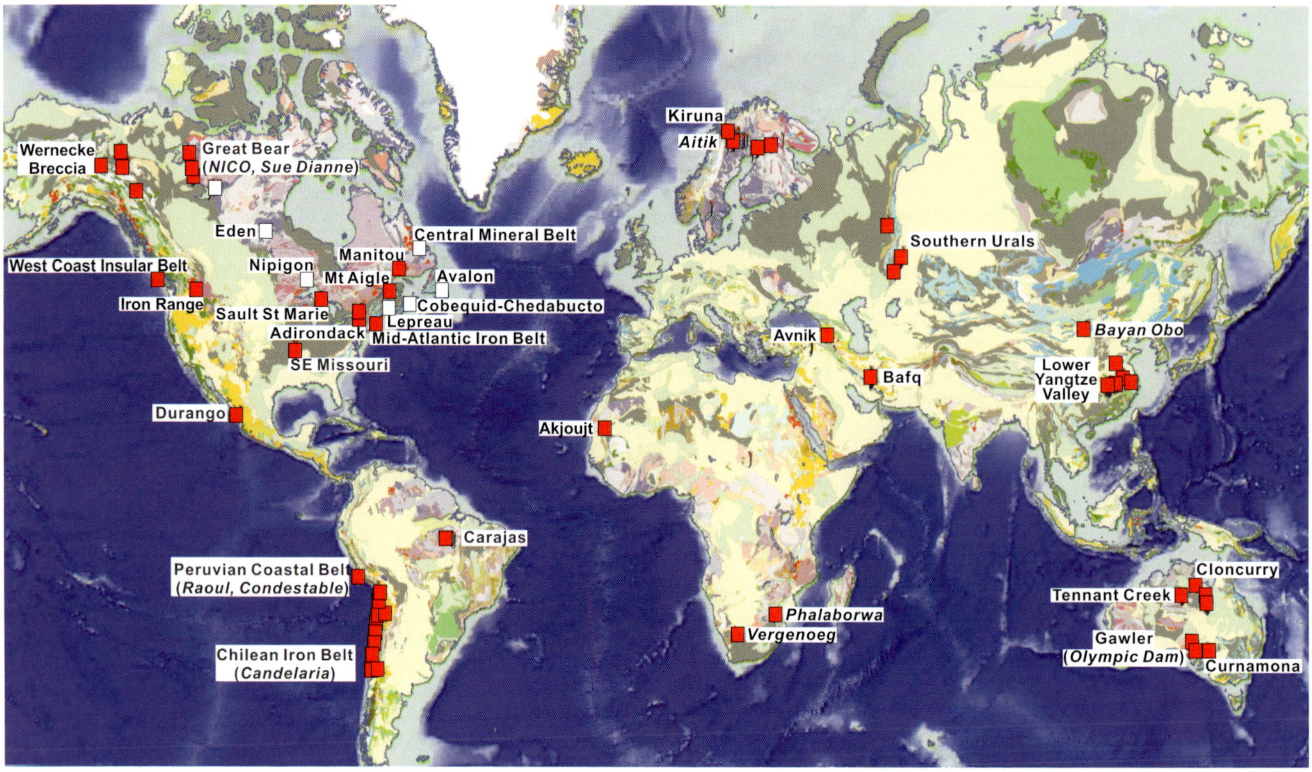

FIGURE 1. Distribution of IOCG districts and important deposits worldwide (red squares) and prospective Canadian settings (white squares) (this work; Gandhi 2004b; Williams et al., 2005). Australia: Gawler (Olympic Dam, Acropolis, Oak Dam, Prominent Hill and Wirrda Well deposits), Cloncurry (Ernest Henry, Eloise, Mount Elliot, Osborne and Starra deposits), Curnamona (North Portia and Cu Blow deposits) and Tennant Creek (Gecko, Peko/Juno and Warrego deposits) districts; Brazil: Carajás district (Cristalino, Alemão/Igarapé Bahia, Salobo, and Sossego deposits); Canada: Great Bear Magmatic Zone (Sue-Dianne and NICO deposits), Wernecke Breccias, West Coast Insular Belt, Iron Range, Manitou (Kwyjibo deposit), Eden, Nipigon, Sault Ste Marie, Mt de l'Aigle, Cobequid-Chedabucto, Pocologan, Avalon and Central Mineral Belt districts; Chile: Chilean Iron Belt (Candelaria, Carmen, El Algarrobo, El Romeral, Manto Verde, and Punta del Cobre deposits); China: Bayan Obo deposit, Lower Yangtze Valley district (Meishan and Daye deposits); Iran: Bafq district (Chogust, Chadoo Malu, Seh Chahoon deposits); Mauritania: Akjoujt deposit; Mexico: Durango district (Cerro de Mercado); Peru: Peruvian Coastal Belt (Raul, Condestable, Eliana, Monterrosas and Marcona deposits); Sweden: Kiruna district (Kiirunavaara, Loussavaara), Aitik deposit; South Africa: Phalaborwa and Vergenoeg deposits; Turkey: Avnik deposit; USA: Southeast Missouri (Pea Ridge and Pilot Knob deposits); Adirondack, and Dover, Edison and Rittenhouse Gap (Mid-Atlantic Iron Belt) districts.

time span since the recognition of this deposit type, 2) the extreme diversity of iron oxide Cu-Au, U, Ag, REE, Bi, Co deposits, hence their many potential subtypes, and 3) the uncertainties surrounding their genesis. The giant Olympic Dam Cu-U-Au deposit in Australia was discovered in 1975, Sue-Dianne in the 1970s, Starra in 1980, La Candelaria in 1987, Osborne in 1988, Ernest Henry in 1991, NICO in 1995, Alemão in 1996, and Prominent Hill in 2001. The early discoveries served to define this group of deposits as the IOCG deposit type in the 1990s (Hitzman et al., 1992). Since then, new discoveries, reclassification of existing deposits and worldwide research led to the recognition of the oxide-mineralizing systems documented by a number of landmark papers and volumes (Porter, 2000, 2002a; Williams et al., 2005 and references therein).

Considering the current state of knowledge, the classification systems are necessarily descriptive and oversimplified. The classification elaborated by Gandhi (2004a) for the World Minerals Geoscience Database Project is used herein. It comprises six subtypes named after world-class deposits or the mineral districts whose characteristics best exemplify the spectrum currently observed (Table 1). Four of them are spatially, and arguably genetically, related to calc-alkaline magmatism and the other two are related to alkaline-carbonatite magmatism.

The Olympic Dam subtype consists of breccia-hosted deposits where polymetallic ore is spatially and temporally associated with iron oxide alteration and as such share similarities with the Olympic Dam Cu-Au-U-Ag-REE deposit at the eastern margin of the Gawler Craton of South Australia (Roberts and Hudson, 1983; Reeve et al., 1990; Oreskes and Einaudi, 1992; Hitzman, 2000; Skirrow et al., 2002). Such deposits have a strong spatial association with regional-scale granitic suites but are rarely hosted within them, Olympic Dam being a notable exception. Among the giant IOCG deposits, most belong to this subtype (e.g. Manto Verde and La Candelaria, Table 2), while fewer belong to the other subtypes (e.g. Salobo deposit).

The Olympic Dam deposit is hosted in a 7 by 5 km (in plan), funnel-shaped, breccia complex with a core of barren hematite-quartz breccia that includes volcaniclastic and sedimentary clasts, peripheral mineralized hematite-granite breccias, and a halo of weakly altered and brecciated granite (Olympic Dam Breccia Complex; Reeve et al., 1990). The complex was formed close to the paleosurface through progressive, polyphase hydrothermal, phreatomagmatic and tectonic brecciation and alteration of the Roxby Downs granite of the Hiltaba intrusive suite slightly after emplacement of the granite itself (Reeve et al., 1990; Cross et al., 1993; Haynes et al., 1995; Johnson and Cross, 1995; Reynolds,

2000). The breccias are not of sedimentary origin as initially interpreted from limited drilling information (Roberts and Hudson, 1983). Alteration dominated by hematite, sericite, chlorite, carbonate ± Fe-Cu sulphides (pyrite, chalcopyrite, bornite, chalcocite) ± uraninite, pitchblende, and REE minerals prevails, and is locally superimposed on a magnetite-siderite assemblage (Haynes et al., 1995; Reynolds, 2000). Similar alteration assemblages are observed in other IOCG deposit and prospects that define the Olympic Cu-Au province of the eastern Gawler Craton, including in the volcanic-hosted Prominent Hill deposit (Skirrow et al., 2002; Belperio and Freeman, 2004).

The felsic to mafic Hiltaba intrusive suite and comagmatic Gawler Range volcanics represent a very fertile ca. 1.59 Ga volcano-plutonic setting with two types of I-type granites (sensus largo) (Wyborn, 2002). The magnetite to hematite-stable Roxby Downs type (granite to syenite, monzonite, quartz monzodiorite, and leucotonalite) is the most oxidized and fractionated variety, has common coarse-grained, porphyritic and megacrystic facies, is comagmatic with the Lower Gawler Range Volcanics, and is associated with the Cu-Au mineralization of the Olympic Cu-Au province. Members may have A-type granite characteristics, a result of strong fractionation and high-temperature magmatism. The more reduced, magnetite to ilmenite Kokatha type is associated with vein Au (±Sn±Ag) deposits and occurs to the west of the Olympic Cu-Au province (Wyborn, 2002).

TABLE 1. Classification of magmatic-hydrothermal iron oxide deposits and related Cu-Au deposits (after Gandhi, 2004a).

Source → *Proximal* → *Distal*

Calc-alkaline magma

Iron Skarn-type	Kiruna-type	Olympic Dam-type	Cloncurry-type
Massive magnetite-garnet-pyroxene	Massive magnetite-apatite-actinolite	Breccia (one or more stages), magnetite-hematite matrix	Hydrothermal veins & disseminations in older 'ironstones' or FeOx mineralization
Stratabound lensoid & irregular bodies at intrusive contact	Tabular, pipe-like & irregular bodies, dykes & veins	Pipe-like & irregular bodies, vent or fault controlled	Stratabound, breccia or fault controlled
Monometallic Fe and related FeOx-Cu-Au deposits	Monometallic Fe & related Cu-FeOx porphyry deposits	Polymetallic: Fe, Cu, Au, Ag, REE	Polymetallic: Cu, Au, Ag, Bi, Co, W
Alteration: Sodic	Alteration: Sodic	Alteration: Potassic	Alteration: Potassic
Magnitogorsk deposit, Russia	Kiirunavaara deposit, Sweden	Olympic Dam deposit, Australia	Osborne & Starra deposits, Australia

Source → *Proximal* → *Distal*

Alkaline-carbonatite magma

Phalaborwa-type	Bayan Obo-type
Within or marginal to intrusion	Hosted by country rock
Veins, layers, disseminations and aggregates; late intrusive phase	Veins, layers, disseminations and aggregates, stratabound lenses
Low Ti magnetite, apatite, olivine, phlogopite, carbonate, fluorite, Cu sulphides, pyrite, PGE, Au, Ag, uranothorianite, baddeleyite	Magnetite (replacive and/or pre-existing), hematite, bastnaesite, phlogopite, Fe-Ti-Cr-Nb oxides, fluorite, monazite, carbonate
Zoning in ore; Na & K alteration	Zoning in ore; Na & K alteration
Phalaborwa deposit, South Africa	Bayan Obo deposit, China

Recent seismic transects across the eastern Gawler Craton highlight that the Olympic Dam deposit occurs within a Palaeoproterozoic orogenic belt along the margin of an Archean core. It is located directly above 1) the intersection of a crustal-scale ramp with the Moho, 2) a striking and seismically anomalous non-reflective lower crustal layer that extends to the Moho, forming a major window in an otherwise reflective Moho, and 3) a highly reflective horizontal sill-like body in the mid crust (Lyons et al., 2004). These results demonstrate that Olympic Dam was not formed in an anorogenic environment. However, the paleotectonic setting of the eastern Gawler Craton during formation of the deposit continues to be debated. One interpretation is that it was an intracontinental back-arc (Ferris et al., 2002) (see section on continental-scale properties, below). Like many other IOCG deposits and prospects, Olympic Dam is a blind deposit discovered under 300 to 400 m of Neoproterozoic and Cambrian sedimentary rocks by drilling of coincident positive magnetic and gravity anomalies during grass-roots exploration (Roberts and Hudson, 1983). Some 75 deposits and prospects are now entered under this subtype in the World Minerals Geoscience Database Project (Gandhi, 2004b).

The Cloncurry subtype, with 55 examples in the World Minerals Geoscience Database Project, is named after the Cloncurry district in the Mount Isa Inlier of northwest Queensland, Australia. It comprises deposits where hydrothermal Cu±Au mineralization overprinted pre-existing 'ironstones' or iron formations (Starra, Salobo) or earlier magnetite-dominated hydrothermal iron oxides (Ernest Henry, Peko) (Mark et al., 2000; Requia and Fontboté, 2000; Skirrow and Walshe, 2002; Gandhi, 2004a). Local remobilization of iron oxides in the host rocks and/or addition of iron from external sources and overprinting of regional-scale potassic and sodic-calcic alteration zones by highly potassic alteration that carried sulphides may have occurred in some IOCG deposits and districts of the Cloncurry subtype (Partington and Williams, 2000; Wyborn, 2002). Fault and shear zones and ductility contrasts at the local and regional scale also play important roles (Partington and Williams, 2000; Marshall, 2003). For example, in the ca. 1.50 Ga Ernest Henry deposit, mineralization postdates peak metamorphism of a ca. 1.74 Ga volcanic host and is broadly synchronous with emplacement of the Williams and Naraku batholiths and widespread sodic-calcic alteration (Mark et al., 2000; Oliver et al., 2004). Early magnetite-apatite mineralization and associated Na-Ca alteration were followed by brecciation in a zone between two parallel shears, and then overprinted by Cu-Au mineralization. Mineralization may postdate deformation and metamorphism of the host (up to

TABLE 2. Resources of selected iron oxide copper-gold deposits.

Deposit	Country	Resources[1]	Grade	Key References
Olympic Dam	Australia	3810 Mt	1.1% Cu, 0.4 kg/t U_3O_8, 0.5 g/t Au	Western Mining press release, 2004
Phalaborwa	South Africa	850 Mt	0.5% Cu (+ Au, Ag, PGE, U, Zr, REE, Ni, Se, Te,	Leroy, 1992
Salobo	Brazil	789 Mt	0.96% Cu, 0.52 g/t Au	Souza and Vieira, 2000
Manto Verde	Chile	600 Mt	0.5% Cu, 0.1 g/t Ag	Sillitoe, 2003
Aitik	Sweden	380 Mt[2] 226 Mt[3]	0.4% Cu, 0.2 g/t Au, 4 g/t Ag[2] 0.37% Cu, 0.2 g/t Au, 3 g/t Ag[3]	Wanhainen et al., 2003
Cristalino	Brazil	500 Mt	1% Cu, 0.30 g/t Au	Tallarico et al., 2004
Candelaria	Chile	470 Mt	0.95% Cu, 0.22 g/t Au, 3.1 g/t Ag	Marschik et al., 2000
Sossego	Brazil	355 Mt	1.1% Cu, 0.28 g/t Au	Haynes, 2000
Igarapé Bahia	Brazil	170 Mt	1.5% Cu, 0.8 g/t Au	Ronze et al., 2000
Ernest Henry	Australia	167 Mt	1.1% Cu, 0.5 g/t Au	Williams and Skirrow, 2000
Prominent	Australia	101 Mt	1.5% Cu, 0.55 g/t Au (+21 Mt at 1.2 g/t Au)	Oxiana Limited, 2005
Bayan Obo	China	48-100 Mt	6% REE_2O_3 (+1 Mt at 0.13% Nb)	Smith and Chengyu, 2000
NICO	Canada	22 Mt[3]	1.08 g/t Au, 0.13% Co, 0.16% Bi	Fortune Minerals, 2007
Osborne	Australia	15.5 Mt	3.0% Cu, 1.05 g/t Au	Gauthier et al., 2001
Sue Dianne	Canada	17 Mt	0.72% Cu, 2.7 g/t Ag	Goad et al., 2000
Starra	Australia	7.4 Mt	1.9% Cu, 3.8 g/tAu	Rotherham et al., 1998
Eloise	Australia	3 Mt	5.5% Cu, 1.4 g/t Au (+Fe, Ni)	Williams and Skirrow, 2000
Peko	Australia	3 Mt	4% Cu, 3.5 g/t Au, 14 g/t Ag, 0.2% Bi	Skirrow and Walshe, 2002
Monakoff	Australia	1 Mt	1.5% Cu, 0.5 g/t Au (Pb, Zn, U)	Williams and Skirrow, 2000
Kiruna district	Sweden	3400 Mt	60% Fe (400 Mt produced)	Frietsch et al., 1979

1: million tonnes of ore calculated; 2: produced; 3: reserve.

upper-amphibolite facies), or be metamorphogenic. The Osborne deposit, formerly considered to be coeval with 1.5 Ga IOCG deposits of the Cloncurry district, appears to contain older components (1595-1600 Ma) that formed syn-peak metamorphism (Gauthier et al., 2001; Williams et al., 2005).

The Kiruna subtype, with more than 355 deposits and prospects in the World Minerals Geoscience Database Project, comprises monometallic Fe, low Ti, and V magnetite-apatite deposits with little or no Cu and Au (Hitzman, 2000). It is named after the classic Kiruna district in northern Sweden and includes the world-class Kiirunavaara massive magnetite deposit in Sweden, the Bafq district in Iran, and several deposits in the Andes (Frietsch et al., 1979; Daliran, 2002; Sillitoe, 2003). These deposits are generally coeval with, and genetically related to, their host volcanic and plutonic rocks. Their massive veins, tabular, pipe-like, or irregular bodies and associated sodic and sodic-calcic alteration in some IOCG deposit districts represent a precursor to significant Cu-Au mineralization both in intracratonic and in continental-arc settings (Hitzman, 2000; Mark and Foster, 2000; Skirrow et al., 2002; Sillitoe, 2003). Large-scale intensive sodic metasomatism results in the formation of very diagnostic albitites.

The Phalaborwa (Palabora) subtype consists of magnetite-rich deposits formed coevally with and proximal to alkaline-carbonatite intrusions (Groves and Vielreicher, 2001; Goff et al., 2004). Primary characteristics are the presence of apatite and strong fenitization, the strong enrichment in REE, F and P, and the extremely high LREE to HREE ratios. The REEs are contained in apatite and in a variety of other REE-bearing phases. Zirconium content may be high, residing in baddeleyite. Titanium content of magnetite in the host intrusions is variable (<1 to 4 wt.% TiO_2) and is higher than in magnetite of most IOCG deposits. Among the alkaline-carbonatite intrusions, the 2.06 Ga Phalaborwa Complex in South Africa is exceptional in its economic Cu grade. It consists of a magnetite-rich, pipe-like orebody hosted in a carbonatite phase of an alkaline pyroxenite intrusion. The orebody is zoned, with Cu sulphides concentrated in the core, and magnetite toward the margin (Harmer, 2000). Copper sulphides (chalcopyrite or bornite and minor chalcocite) postdate the iron-oxide mineralization. Its setting at the margin of an Archean craton is viewed by Vielreicher et al. (2000) as a key element in the development of the ore deposit.

The Bayan Obo subtype consists of magnetite-rich, REE (+Nb) deposits lacking economic Cu-Au and are distal to inferred or known alkaline-carbonatite plutonic sources but display their diagnostic mineral assemblages and metal content (Smith and Chengyu, 2000). At the Bayan Obo deposit in China, the ore occurs as massive, banded, and disseminated forms in marble deposited in grabens on the margin of the Archean North China craton. Deposit mineralogy, with some 70 minerals, is dominated by magnetite, bastnaesite, fluorite, alkali amphiboles, and pyroxenes in association with apatite, phlogopite, and barite (Smith et al., 2000). In the vicinity of the deposit, a link with an alkaline-carbonatite magmatic source for the mineralizing fluids is exposed as carbonatite dykes. The presence of deep-seated faults, active in the area since the Proterozoic, illustrate that significant iron oxide ore may form even if large volumes of coeval magmatism are not exposed at the same level of erosion, provided that channel ways for fluids are available.

The iron-skarn subtype shares some features with the IOCG Kiruna subtype, and includes major deposits in the southern Urals region and in the Peruvian Coastal Belt (Ray and Lefebure, 2000; Herrington et al., 2002; Injoque, 2002; Sillitoe, 2003; Gandhi, 2004a). It was included in the IOCG deposit clan for the World Minerals Geoscience Database.

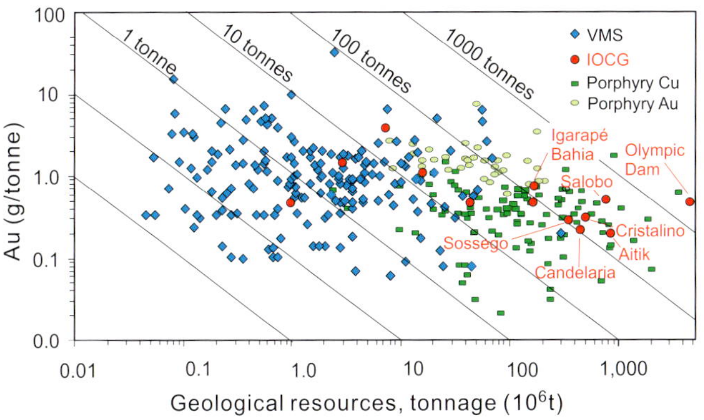

FIGURE 2. Au grade versus tonnage plot of iron oxide copper-gold (IOCG), volcanogenic massive sulphide (VMS), and porphyry Cu deposits from Kirkham and Sinclair (1996), Galley et al. (2007), and Table 2.

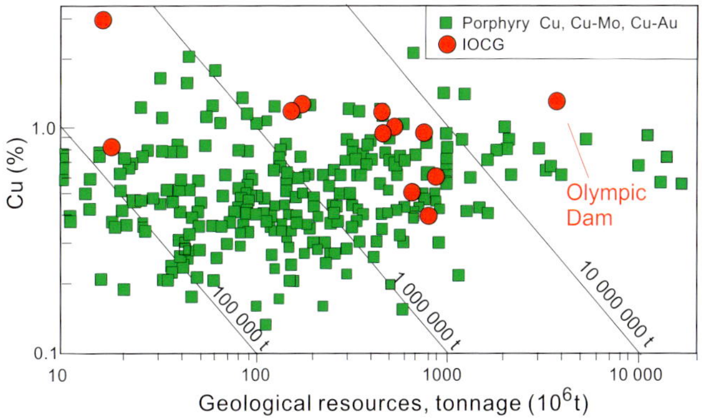

FIGURE 3. Cu grade versus tonnage plot of the geological resources of iron oxide copper-gold (IOCG) deposits and porphyry Cu based on Kirkham and Sinclair (1996).

Associated Mineral Deposit Types

On a district scale, IOCG deposits are known to occur in the vicinity of alkaline and calc-alkaline porphyry Cu-Mo or Cu-Au deposits, Cu-Ag manto deposits, volcanic-hosted uranium ore bodies, hematite-rich massive ironstones, sediment-hosted Au-PGE, polymetallic Ag-Pb-Zn±Au veins, lode Au, and SEDEX deposits (Pollard, 2000; Ferris and Schwarz, 2003; Sillitoe, 2003; British Columbia Geological Survey, 2005). In contrast, they are rarely found in the same setting as a volcanic-hosted massive sulphide deposit. Near surface supergene U, Cu, and/or Au blankets or veins occur locally (Carajás district; Tazava and de Oliveira, 2000).

Economic Characteristics

IOCG deposits can have enormous geological resources (Table 2; Figs. 2, 3) with significant reserves of base, precious, and strategic metals, as well as nuclear energy. They are major sources of Cu, Au, U, REE (LREE), F, and vermiculite; significant sources of Ag, Nb, P, Bi, and Co; and sources of various by-products including PGE, Ni, Se, Te, and Zr. They also contain a number of associated elements, notably As, B, Ba, Cl, Co, Mo, Mn, W, (Pb, Zn).

The 3810 Mt Olympic Dam deposit (Table 2) contains currently the world's largest uranium resource (1.4 Mt), the fourth Cu resource (42.7 Mt), and the fourth Au resource (55.1 M ounces) (Western Mining Corporation, 2004). Production started in 1978 and reached 9 Mt of ore in 2002 (178,120 t Cu and 2890 t U_3O_8). Ore reserves (proven and probable) in 2003 were 730 Mt at 1.6% Cu, 0.5% U, and 0.5 g/t Au (Western Mining Corporation, 2003). Rare earth element resources amount to 2000 Mt at 0.3285% rare-earth oxides, but their recovery is currently not economical (Reynolds, 2000; Orris and Grauch, 2002).

Resources of IOCG deposits for individual commodities can be as high as some of the best volcanogenic massive sulphide (VMS) and porphyry Cu deposits (e.g. Figs. 2, 3). Although gold grades are low in most of the large tonnage IOCG deposits, total Au resources may be very large (Table 2). The Phalaborwa deposit (Table 2) is presently the world's second largest Cu mine and has by-products of Au, Ag, platinum group metals, magnetite, P, U, Zr, REE, Ni, Se, Te, and Bi (Leroy, 1992). The Bayan Obo deposit hosts the world's largest resource of rare earths, and is presently the principal producer of rare earth elements (Orris and Grauch, 2002). In Canada, the largest IOCG deposit is the 22 Mt NICO deposit (Lou Lake), followed by the 17 Mt Sue-Dianne deposit (Table 2).

Exploration Guides

This section focuses on the regional and local geological features of IOCG deposits and their hosts that may help unveil targets in green-field exploration terranes of Canada. Fundamental topics, such as the origin of fluids and the deposits themselves, are not addressed in detail (cf. Hitzman, 2000; Partington and Williams, 2000; Porter, 2000, 2002a; Requia et al., 2003; Sillitoe, 2003; Williams et al., 2005).

Physical Properties

Mineralogy

Ore mineralogy varies considerably among deposits and from one IOCG deposit subtype to another. The principal minerals are bornite, chalcopyrite, and chalcocite. Subordinate minerals include Ag-, Cu-, Ni-, Co-, U-arsenides, autunite, bastnaesite, bismuthinite, brannerite, britholite, carrolite, cobaltite, coffinite, covellite, digenite, electrum, florencite, loellingite, malachite, molybdenite, pitchblende, sulphosalts, uraninite, xenotime, native bismuth, copper, silver and gold, Ag-, Bi-, Co-telluride, and vermiculite (see Table 3 for mineral formulae; Ray and Lefebure, 2000). Gangue mineralogy consists principally of hematite, magnetite, pyrite, pyrrhotite, albite, K-feldspar, sericite, carbonate, chlorite, quartz (crypto-crystalline in some cases), amphibole, pyroxene (aegerine-augite), biotite, apatite (F- or REE-rich), and vonsenite with accessory allanite, barite, epidote, fayalite, fluorite, ilvaite, garnet (andradite, Fe-rich garnet), monazite, perovskite, phlogopite, rutile, scapolite, titanite, and tourmaline (Table 3). The amphibole includes Fe-, Cl-, Na-, or Al-rich hornblende (edenite), actinolite, grunerite, hastingsite, and tschermakitic or alkali amphibole. Carbonates include calcite, ankerite, siderite, and dolomite. Late-stage veins contain fluorite, barite, siderite, hematite, and sulphides.

Textures and Morphology

Iron oxide copper-gold deposit mineralization can be hosted in subvertical to subhorizontal, single or polyphase breccia zones or in mantos, veins, stockwork, volcanic pipes, diatremes, lenses, massive concordant to crosscutting tabular

TABLE 3. Iron oxide copper-gold deposit mineralogy.

Mineral	Formula	Mineral	Formula
albite	$NaAlSi_3O_8$	kainosite	$Ca_2(Y,Ce)_2Si_4O_{12}(CO_3)H_2O$
allanite	$Ca(La,Ce)(Fe^{2+},Mn^{2+})(Al,Fe^{3+})_2[SiO_4][Si_2O_7]O(OH)$	K-feldspar	$(K,Na)AlSi_3O_8$
amphibole	$(Na,K)_{0\text{-}1}Ca_2(Mg, Fe^{2+}, Fe^{3+}, Al)_5[Si_{6\text{-}7}Al_{2\text{-}1}O_{22}](OH,F)_2$	loellingite	$Fe^{2+}As_2$
apatite	$Ca_5(PO_4)_3(OH,F,Cl)$	iron oxide	$FeO - Fe_2O_3$
autunite	$Ca(UO_2)_2(PO_4)_2\cdot 12(H_2O)$	malachite	$Cu_2(CO_3)(OH)_2$
baddeleyite	ZrO_2	molybdenite	MoS_2
barite	$Ba(SO_4)$	monazite	$(Ce,La,Nd,Th)PO_4$
bastnaesite	$(La,Ce,Y)(CO_3)F$	muscovite	$KAl_2(Si_3Al)O_{10}(OH,F)_2$
biotite	$K(Mg,Fe^{2+})_3[AlSi_3O_{10}](OH,F)_2$	olivine	$(Mg,Fe)_2SiO_4$
bismuthinite	Bi_2S_3	pitchblende	$UO_2 - UO_3$
bornite	Cu_5FeS_4	phlogopite	$KMg_3(Si_3Al)O_{10}(F,OH)_2$
brannerite	$(U,Ca,Ce)(Ti,Fe)_2O_6$	pyrite	$Fe^{2+}S_2$
britholite	$Ca_{2.8}Ce_{0.9}Th_{0.6}La_{0.4}Nd_{0.2}Si_{2.7}P_{0.5}O_{12}(OH)_{0.8}F_{0.2}$	pyrrhotite	Fe(1-x)S
carrolite	$CuCo_{1.5}Ni_{0.5}S_4$	quartz	SiO_2
chalcocite	Cu_2S	rutile	TiO_2
chalcopyrite	$CuFeS_2$	scapolite	$(Na,Ca)_4[Al_3Si_9O_{24}]Cl$
chlorite	$(Mg,Al,Fe)_{12}[(Si,Al)_8O_{20}](OH)_{16}$	sericite	$KAl_2(OH)_2(AlSi_3O_{10})$
clinopyroxene	$(Na,Ca)(Fe^{2+},Fe^{3+},Mg)[Si_2O_6]$	sillimanite	Al_2SiO_5
cobaltite	CoAsS	telluride	Bi_2Te_3 + other Ag, Bi, or Co telluride
coffinite	$U(SiO_4)_{1\text{-}x}(OH)_{4x}$	thorite	$(Th,U)SiO_4$
covellite	CuS	titanite	$CaTiSiO_5$
digenite	Cu_9S_5	tourmaline	$Na(Mg,Fe,Mn,Li,Al)_3Al_6[Si_6O_{18}](BO_3)_3(OH,F)_4$
epidote	$Ca_2Fe^{3+}{}_{2.25}Al_{0.75}(SiO_4)_3(OH)$	uraninite	UO_2
florencite	$(Ce,La,Nd)Al_3(PO_4)_2(OH)_6$	vermiculite	$(Fe^{2+}, Mg)_2(Fe^{3+},Al)BO_5\cdot$
fluorite	CaF_2	vonsenite	$(Mg,Fe^{2+},Al)_3(Al,Si)_4O_{10}(OH)_2\cdot 4(H_2O)$
garnet	$Ca_3Fe_3{}^{2+}(SiO_4)_3 - Fe_3{}^{2+}Al_2(SiO_4)_3$	xenotime	$(Yb,Y)(PO_4)$
ilvaite	$CaFe^{2+}{}_3(SiO_4)_2(OH)$	zircon	$ZrSiO_4$

bodies, and mineralized clasts. It can also occur as dissemination through host rock or surrounding single or multiple lenses of massive ore. Alteration associated with IOCG deposits (described in the geological properties section) occurs in the form of vein, veinlet, stockwork, breccia matrix, open-space filling, vein-wall metasomatic front, and texturally preservative or destructive, centimetre- to kilometre-scale penetrative replacement. Superimposition of alteration is common and later stage alteration may strongly obliterate evidence of early phases.

Breccias, where present, form the most striking elements of IOCG deposits. They are generally aligned along fault zones and splays or, in some cases, subparallel to strata or intrusive contacts. They may also crosscut or be superimposed concordantly on host lithology or alteration zones. The breccia zones vary in extent from outcrop to mountain scale (1 m^2 to 10 km^2), such as in the Wernecke Mountains in Canada (Thorkelson, 2000), range in colour from grey, to black, greenish or mottled red and pink, and vary in clast size from <1 cm up to hundreds of metres in length or thickness. Breccias are commonly heterolithic and composed of distinct angular to subangular or, more rarely, rounded lithic and oxide clasts or fine-grained massive material where fragments may be difficult to recognize. Fragments can be derived in situ or may have moved from above or below. Core zones of breccias may contain up to 100% iron ore and grade into crackle breccia, where host rocks display incipient brecciation, and to a weakly fractured margin with iron oxide or carbonate veins. As such, the boundary of breccias tends to be gradational over the scale of centimetres to several metres. They may also be sharp where they abut against faults or where breccia material is intruded into the country rock. Diffuse, wavy layered textures of red and black hematite and partial to complete replacement and pseudomorphs of early magnetite or host rock textures (e.g. lapilli) and filling of microcavities by hematite are common (discussed in detail in Ray and Lefebure, 2000).

Dimensions and Depth

Iron-rich veins, breccias and tabular zones, and alteration halos may reach hundreds of metres in width and many kilometres in length (e.g. the Selwin Line in Australia; Ray and Lefebure, 2000; Sleigh, 2002). As such, IOCG deposits are extremely good targets for regional-scale geophysical and geochemical surveys and integrated expert system studies (as discussed below). The deposits and their breccias where present can be formed at shallow (Olympic Dam type), moderate (Sossego, Ernest-Henry type), or deep (Salobo) levels within the first 10 km of the crust (Kerrich et al., 2000; Davidson et al., 2002). Metamorphogenic IOCG deposits, such as the Osborne deposit in the Cloncurry district and the Jayville deposit in the Adirondack district, occur in settings up to granulite facies (Fig. 1; Gauthier et al., 2001; Johnson and Selleck, 2005), and preservation of metamorphosed IOCG deposits could be expected up to this metamorphic grade as well. The latter is exemplified by the granulite facies Edison iron mines of the New Jersey Highlands (Fig. 1; Puffer and Gorring, 2005).

Architecture

Deposits are associated with crustal-scale to local fault zones and zones of competency contrasts, including dilational jogs (Olympic Dam), duplexes, splays on faults, thrusts and shears (Carajás district and Monakoff deposit), folds (Tennant Creek district), complex intercalation of high- and low-permeability units, and intersections of highly permeable units with fault zones (Candelaria) (Cross et al., 1993; Adshead-Bell, 1998; Marschik et al., 2000; Partington and Williams, 2000; Skirrow, 2000, 2004; Davidson et al., 2002; Sleigh, 2002; Sillitoe, 2003). Lineaments and lineament intersections commonly serve as exploration targets. For example, the location of the Olympic Dam orebody corresponds well to the original target area defined on the basis of regional lineament analysis (O'Driscoll, 1986; Woodall, 1994).

Host Rocks

The host rocks of IOCG deposits encompass coeval or pre-existing sedimentary (iron formation or iron-rich sediments), mafic to felsic volcanic or plutonic rocks, schists, and gneisses that may serve as lithological (e.g. due to competency contrast, permeability, porosity, or reactivity) or chemical traps (e.g. Mark et al., 2000; Skirrow, 2000; Knight et al., 2002; Weihed et al., 2005). As such, rock types do not constitute a diagnostic feature per se. An important unifying link is that IOCG deposit host rocks formed in regionally oxidized settings through which fluids could flow and/or react (Haynes, 2000).

Chemical Characteristics

The ore/gangue/alteration minerals of IOCG deposits (listed in Table 3), notably the light REE-, Bi-, Co- and U-bearing minerals and rutile, have distinctive chemical fingerprints and their mineral chemistry is used as process-forming and source-rock tracers (Morton and Hallsworth, 1999; Daliran, 2002; Zack et al., 2002). The main ore minerals are iron oxides, Cu-Fe sulphides, U oxides, and/or REE-enriched minerals combined in diverse metal associations (e.g. Fe-P; Fe-REE-Nb, Fe-Cu-Au, and Fe-Cu-Au-U-Ag-REE). Anomalously high values for Fe, Cu, Au, U, Ag, ±REEs, especially the LREE (Ce, La, Nd, Pr, Sm, Gd), ±Nb ±P ±Co ±F ±B ±Mo ±Y ±As ±Bi ±Te ±Mn ±Se ±Ba ±Pb ±Ni ±Zn provide diagnostic element suites for geochemical exploration. Ore assemblages may appear highly varied but in fact comprise a series of elements that share certain chemical affinities that facilitate their transportation and/or precipitation together. Iron oxide is either co-transported or pre-existing and acts as a catalyst in the precipitation of other metals (Porter, 2002b).

Some of the ore, gangue, and alteration minerals are resistate minerals able to survive weathering and mechanical dispersal (Morton and Hallsworth, 1999). They commonly have a high mode (up to 60% apatite and 20% Fe oxide in some ore at Bayan Obo) and are unusual (Table 3), providing distinctive and propitious pathfinders in exploration. Techniques that detect chemical dispersion and in situ anomalies are commonly used for IOCG exploration, notably soil, lake, and stream sediment surveys, till geochemistry, litho-geochemistry, and geophysical (aeromagnetic, gravity, and radiometric) surveys.

Geological Settings

Continental-Scale Settings

Key continent- to orogen-scale knowledge in support of IOCG exploration includes information on geotectonic environments, tectonomagmatic nature, and evolution of intrusive and volcanic suites including batholiths and alkaline intrusions, orogenic events, and crust architecture in terms of faults, domain boundaries, or sutures, Moho discontinuities, crust-scale ramps, zones of unusually low reflectivity, and subhorizontal reflectors in the mid-crust. IOCG deposits are regionally and temporally associated with, but can be spatially distinct from, large-scale magmatism that includes voluminous felsic magmatism and mafic to intermediate magmas either within bimodal or more continuous suites. The Hiltaba plutonic suite, that hosts and is broadly coeval with the giant Olympic Dam deposit, is granitic to monzodioritic, A- to I-type, cogenetic with the dominantly felsic Gawler Range volcanic suite, and associated with ultramafic and mafic dykes temporally related to the hydrothermal activity (Johnson and McCulloch, 1995; Creaser, 1996). Strong subhorizontal, mid-crustal reflectors below Olympic Dam point to the presence of large mafic sills (cf. Mandler and Clowes, 1997) and underscore the critical role that mafic magmatism may play in the development of prospective IOCG settings (Lyons et al., 2004; Tornos and Casquet, 2005). The Ernest-Henry, Starra, and Lightning Creek deposits in the Cloncurry district are also related to I-type, magnetite-series diorites to granites (Williams and Naraku batholiths; Wyborn, 1998, 2002; Marshall, 2003). In contrast, the 2576 ± 8 Ma Salobo deposit is coeval with 2573 ±2 Ma A-type granitic magmatism (Requia et al., 2003). Magmatic-hydrothermal activities associated with the differentiation of alkaline-carbonatite magmas can also generate magnetite-rich deposits (e.g. Phalaborwa and Bayan Obo deposits). On this basis, Gandhi (2004a) proposed two main series for IOCG deposit-related magmatism: a calc-alkaline (Kiruna, Olympic Dam, and Cloncurry subtypes) and an alkaline series (Phalaborwa and Bayan Obo subtypes) (Table 1). His approach guides the subtype assignments of deposits in the World Minerals Geoscience Database and is in line with the work of Creaser et al. (1991), which documents the evolution of A-type granites as part of the I-type granite series and the key association of magnetite-bearing I-type granites with IOCG deposits (Partington and Williams, 2000; Wyborn, 2002; Sillitoe, 2003; Gandhi, 2004b).

Discussions of prospective settings for IOCG deposits still focus on intracratonic rift environments (e.g. Kerrich et al., 2000; Faure, 2003), a consequence of former interpretations of the Olympic Dam deposit as plume-related, intracratonic, and anorogenic. Such interpretations hinged on the A-type geochemical signature of coeval Hiltaba suite granitoids (Giles, 1988; Creaser, 1989). Though entrenched in the literature as Anorogenic (Anderson, 1983; Pitcher, 1993), A-type granites are not necessarily anorogenic or alkaline. The term A-type simply refers to granites with high SiO_2, Na_2O+K_2O, Fe/Mg ratios, F, Ga, Sn, Y, REE, and high field strength elements (HFSE), such as Zr and Nb (Collins et al.,

Table 4. Age distribution of selected iron oxide copper-gold deposits.

Age (Ma)	Ages of deposits or districts	Ref.
Phanerozoic	Candelaria, Punta del Cobre, Raul Condestable 0.115 Ga, Mantoverde 0.12 Ga, Carmen 0.13 Ga	1
Phanerozoic	Bayan Obo 0.5 Ga ?	2
Neoproterozoic	Kwyjibo 0.98 Ga	3
Mesoproterozoic	Lyon Mt. 1.04 Ga	4
Mesoproterozoic	Bayan Obo 1.3-1.2 Ga ?	5
Mesoproterozoic	Ernest Henry >1.51 Ga	6
Mesoproterozoic	Olympic Dam 1.59 Ga	
Paleoproterozoic	Osborne, Wernecke 1.60 Ga	7
Paleoproterozoic	Curnamona ~1.61Ga	8
Paleoproterozoic	Tennant Creek ~1.83Ga	9
Paleoproterozoic	Aitik, Sue Dianne, NICO ~1.87Ga	10
Paleoproterozoic	Kiruna 1.89-1.88 Ga	11
Paleoproterozoic	Phalaborwa 2.06 Ga	12
Archean	Igarapé Bahia 2.57 Ga	13
Archean	Salobo 2.58 Ga	14

1 - Mathur et al. (2002), Sillitoe (2003), Gelcich et al. (2005);
2 - Chao et al. (1997), Smith et al. (2000);
3 - Gauthier et al. (2004); 4 - Selleck et al. (2004);
5 - Yang et al. (2003); 6 - Mark et al. (2000);
7 - Gauthier et al. (2001), Thorkelson et al. (2001);
8 - Williams and Skirrow (2000); 9 - Skirrow (2000);
10 - Gandhi et al. (2001), Wanhainen et al. (2003);
11 - Romer et al. (1994); 12 - Harmer (2000);
13 - Tallarico et al. (2004); 14 - Requia et al. (2003).

1982). Such granites can be distinguished from other granite types, from which they can be derived, with the geochemical discriminant diagrams of Whalen et al. (1987). A-type granitoids have high crystallization temperatures (e.g. 900-1000°C) sustained by the emplacement of mantle-derived mafic magmas at depth (Creaser et al., 1991). They form through reworking of fertile crustal settings, hence a common spatial association with the margins of Archean cratons (see discussion in Creaser, 1996). Although extension is needed for A-type granites to form, the tectonic setting does not have to be intracratonic as it can also be within or inboard of active continental margins or in collisional to post-collisional environments (Collins et al., 1982; Whalen et al., 1987; Creaser, 1996; Tornos and Casquet, 2005). The same is true for the tectonic setting of alkaline magmas, including carbonatite complexes, and associated mineral deposits (e.g. Wang et al., 2001; Sillitoe, 2002; Mumin and Corriveau, 2004).

Orogenic activities were taking place at the time of A-type intrusions and formation of the Olympic Dam deposit in and at the margin of the Gawler Craton (e.g. Creaser, 1996; Partington and Williams, 2000). With the recognition of juvenile material (Johnson and McCulloch, 1995) and syntectonic plutonic phases among the Hiltaba suite (Ferris et al., 2002) and the orogenic-type crustal architecture displayed in the seismic transects centred on Olympic Dam, alternative paleotectonic interpretations emerge for this deposit. One interpretation is that the deposit is related to a continental, extensional back-arc setting along which basaltic under-plating took place during crustal thinning (Ferris et al., 2002; Giles et al., 2002). Magmatic arcs can be very fertile (e.g. the Andean polymetallic IOCG deposits, massive magnetite deposits, manto-type and small porphyry Cu deposits, and calcic skarn deposits; Sillitoe, 2003), and a new paradigm is emerging in IOCG exploration with the targeting of active and paleocontinental margins (Gandhi et al., 2001; Sillitoe, 2003; Mumin and Perrin, 2005).

The presence of arcs and successor arcs at the margin of Archean cratons, a demonstrated fertility of the crust in Cu, Au, and Ni, an interconnectivity between crust and mantle in terms of magmas and fault zones with crustal-scale ramp intercepting the mantle, and the development of extensional/dilational settings within what may have been in overall (i.e. orogen-scale) compressional or transpressional tectonic environment are other key continental-scale features that may guide targeting of prospective settings (Hitzman, 2000; Groves and Vielreicher, 2001; Sillitoe, 2003; Tornos and Casquet, 2005). Timing with respect to the Earth's evolution does not appear to be critical in the development of IOCG deposits (Nisbet et al., 2000). Deposits are now known to occur from the Archean to the Mesozoic (Table 4) with the 1.9 to 1.5 Ga interval being the most fertile period currently documented. The occurrence of Archean deposits is significant as it opens Archean terranes to IOCG exploration.

District-Scale Settings

The significant IOCG districts enclose several mineral deposit types in association with large-scale, high-temperature magmatic suites (as discussed above) and crustal-scale fault zones. For example, an orogenic gold province is now recognized to the west of the Olympic Cu-Au province in the Gawler Craton. Gold mineralization is associated with the Kokatha-type granitoids of the Hiltaba suite, whereas IOCG mineralization is associated with Roxby Downs-type granite of the same suite (Wyborn, 2002; Ferris and Schwarz, 2003). The close association between monometallic iron and polymetallic IOCG deposits is also common and represents an empiric vector to target prospective IOCG settings. Such iron ores were not systematically analyzed for the entire spectrum of IOCG-associated metals in the past. Today, they form a first-order target for IOCG exploration and provide a means to define prospective settings. This approach is used in emerging exploration plays in Canada (e.g. Downes, 2003; Marshall et al., 2003). Iron oxide-rich deposits and prospects can be distributed in belts more than 100 km long and more than 10 km wide, with a 10 to 30 km spacing along or in the vicinity of structures, either exposed or cryptic, that may control mineralization, e.g., crust-scale fault zones, long-lived brittle-ductile faults, shear zones and narrow grabens or rifts, splay-offs, dilational jogs, subsidiary brittle fractures, and stratabound zones of high permeability (Ray and Lefebure, 2000; Sleigh, 2002; Belperio and Freeman, 2004). The presence of mono- or polyphase iron oxide(s) matrix breccia at a local to regional scale leads to spectacular outcrop exposures and/or to significant magnetic and gravity anomalies, keys to mineral exploration at the district scale (e.g. Reeve et al., 1990; Hitzman, 2000; Smith, 2002).

Mineralization is broadly coeval and regionally spatially associated with large-scale volcanic and plutonic suites (Johnson and Cross, 1995; Marshall, 2003; Belperio and Freeman, 2004). Crustal-scale fault zones or splay faults and competency contrasts play key roles in channelling fluids and source magmas and influence fluid flow patterns and arrangements of alteration zones, breccias, and/or ore deposition. Fluids that can account for oxide-rich but sulphide-deficient ore need to be saline and relatively oxidized. Their primary source(s) can be magmatic, metamorphic, evaporitic, or brines (Barton and Johnson, 2000, 2004; Haynes, 2000; Skirrow et al., 2002). Though contentious in many districts, the role of magmatic fluid sources in the formation of regional Na-(Ca) alteration, magnetite alteration, and Cu-Au mineralization is well documented in the Cloncurry district (e.g. Ernest Henry and Mt Elliot deposits; Marshall, 2003) whereas, in the same district, pre-existing ironstones (e.g. at Starra Cu-Au mineralization) and possibly some magnetite alteration (e.g. Osborne) had non-magmatic sources (Marshall, 2003). Magmas play a significant role in providing the thermal regime necessary for very efficient fluid flow, mixing, and metal recharge. By doing so, magmatic fluids will be mixed with other fluids if available, making the source of fluids complex to decipher.

Extensional and transtensional movements along faults, including those in overall compressional regimes, can channel hot deep-seated fluids upward and cold surface-waters downward, allowing for fluid mixing, large-scale alteration, and mineralization (Skirrow, 2004). The depth at which fluid and metal recharge and discharge take place will influence the resulting alteration and mineralization patterns, as reviewed in Kerrich et al. (2000). Sodic-calcic alteration zones are commonly zones of fluid recharge in the base and ferrous metals. Focussed potassic alteration and iron oxide(s) matrix breccia are zones of upflow and discharge of metals, whereas laterally extensive, K-feldspar-dominated alteration in oxidized (hematite stable) continental and transitional marine settings may represent recharge zones (Barton and Johnson, 2004). The role of evaporite for recharge has been debated and the reader is referred to Barton and Johnson (2004) for a presentation of the various alternatives and impacts of fluid types on footprints of IOCG deposits. Late- to post-orogenic IOCG mineralization may be associated with extensional shear zones developed during orogenic collapse (Weihed et al., 2005) or with thrusts developed during terrane assembly, in particular if mid- to lower-crustal mafic-ultramafic layered sills are present (e.g. SW Iberia: Tornos and Casquet, 2005; Cloncurry district, Mt Isa region: MacCready et al., 1998; Partington and Williams, 2000).

Deposit-Scale Settings

A striking characteristic of IOCG deposits is their distinctive alteration zones and overprinting relationships. Though IOCG subtypes may have distinct alteration, three main types emerge: sodic-calcic, iron, and potassic alteration. The sodic-calcic (Na-Ca) alteration zones are regional in scale (>1 km wide) and range from intense and pervasive albitization (± clinopyroxene, amphibole, titanite), such as in the Cloncurry and Kiruna districts (Frietsch et al., 1979; Mark and Foster, 2000), to a magnetite-bearing calc-silicate (clinopyroxene, amphibole, garnet) - alkali feldspar (K-feldspar, albite) ± Fe-Cu sulphides assemblage (CAM assemblage of Skirrow et al., 2002). Various alteration assemblages, including albite, actinolite, magnetite, apatite, and late epidote, may also represent Na-Ca alteration (Ray and Lefebure, 2000). If Ca-rich host rocks are present, Fe-rich garnet-clinopyroxene±scapolite skarn assemblages may form (e.g. Candelaria, Chile; Kiruna, Sweden; Shimyoka and Kantonga prospects, Zambia; Ray and Lefebure, 2000). Such alteration types tend to form early in paragenetic sequences and to be barren in terms of significant polymetallic mineralization; others are well mineralized (e.g. Candelaria, Williams et al., 2005). Regional-scale, texturally preservative, K-feldspar-chlorite potassic alteration is observed locally such as at Manto Verde.

A second, iron-rich assemblage is present in some districts and commonly consists of low-Ti magnetite±biotite with minor Fe-Cu sulphides, as is found in parts of the eastern Gawler Craton and in the Cloncurry district (MB assemblage of Skirrow et al., 2002; Williams et al., 2005). Iron can also be in the form of oxides other than magnetite (hematite), carbonate (siderite and ankerite), silicate (chlorite and amphibole, in particular grunerite and hastingsite) or borate (vonsenite) (Downes, 2003; Johnson and Selleck, 2005). Magnetite is generally considered to have formed at greater depth and/or at higher temperature than the subsequent, more focussed and commonly crosscutting hematite alteration.

The main ore stage is associated with potassium-iron alteration zones in which sericite (at Olympic Dam and Prominent Hill) or K-feldspar (most common) prevails in the rock (Skirrow et al., 2002; Belperio and Freeman, 2004) and hematite (specularite, botryoidal hematite, and martite) dominates as the iron phase. K-feldspar-hematite veins (Wyborn, 1998) or an alteration assemblage of hematite-sericite-chlorite-carbonate±Fe-Cu sulphides±U, REE minerals (HSCC of

Skirrow et al., 2002) are observed. Intense chloritization may result in almost total destruction of hydrothermal biotite. Other minerals common in the alteration assemblages are quartz, fluorite, barite, orthoclase, epidote, and tourmaline. Titanite, monazite, perovskite, rutile, xenotime, apatite, and magnetite, where they occur, provide a means of establishing directly the age of alteration. In Table 4, the ages are derived by U-Pb dating of some of these minerals found in alteration zones and ore assemblages (e.g. Teale and Fanning, 2000; Clark et al., 2005; Gelcich et al., 2005) or of zircons in genetically related intrusive and volcanic rocks (Johnson and Cross, 1995; Thorkelson et al., 2001). $^{40}Ar/^{39}Ar$ and Re-Os data on gangue and ore minerals, respectively, have also been used to constrain the ages of mineralization (Marschik and Fontboté, 2001; Mathur et al., 2002; Williams et al., 2005).

Tracers of IOCG mineralization extend commonly for several kilometres (1-6 km) over a width of several hundred metres while the deposit itself may range between 0.5 and 1 km in length and a few hundred metres in width. IOCG deposits are thus detectable by basic to sophisticated geological, geochemical, remote sensing, and geophysical techniques. At the local to district scale, the diagnostic character of alteration and overprinting relationship provides an efficient vector to mineralization while mapping. Because mineralization tends to be deficient in pyrite, it is not systematically associated with intense gossans and key mineralized outcrop in potassic-hematite alteration may look simply like red or striking pink syenitic rocks. With their common coincident gravity and magnetic anomalies, exploration relies heavily on ground and airborne magnetic surveys complemented by existing gravity data or acquisition of target specific data (Gow et al., 1994; Smith, 2002). The gravity response is a consequence of the high density and abundance of the Fe oxides while the magnetic response reflects the proportion of magnetite in the Fe oxides and less commonly of pyrrhotite in the deposit. Late-stage fluid fluxes may induce magnetite formation or destruction through non-redox transformations (Ohmoto, 2003), and consequently may affect the geophysical response of orebodies and their magnetite versus hematite contents and distribution.

The presence of U-rich mineralization with km-scale potassic alteration halos in some IOCG sub-types leads to radiometric anomalies that may be detectable in exposed systems by detailed airborne surveys. U/Th and K/Th ratios are particularly useful as Th varies within a small range so the addition of U and K is readily detected (Gandhi et al., 1996; Smith, 2002). The deposits, through their large disseminated mineralization zones, may produce strong induced polarization and resistivity responses, while the breccia core tends to be electrically connected and conductive, such as in the Candelaria deposit (Ryan et al., 1995; Ray and Lefebure, 2000). Induced polarization and electromagnetic surveys are thus useful exploration tools to complement the magnetic and radiometric surveys (see Smith, 2002 and references therein). Recent case studies have also shown the usefulness of hyperspectral and Aster imaging for mineral exploration in exposed systems. Bands in Aster images are very sensitive to hematite alteration. Structural studies on regional and local scales are also important considering that IOCG deposits are structurally controlled (as discussed above).

In metamorphic terranes, mineralization may postdate metamorphism (e.g. at Ernest Henry), coincide with the last stages of orogenesis (e.g. at Monakof), be coeval with peak metamorphism and anatexis (e.g. at Osborne, Gauthier et al., 2001; Curnamona province, Williams and Skirrow, 2000), or predate metamorphism altogether (suspected IOCG systems in the Grenville Province, Corriveau et al., 2007). Metamorphosed IOCG deposits could occur at any crustal level following orogenesis, hence prognostication for IOCG prospective districts should also include metamorphosed terranes, even those metamorphosed at granulite facies. Following metamorphism, the textures and the mineral assemblages of alteration zones and ores may change dramatically. Mineral assemblage and modes may be unusual, providing a tool for protolith assessment and a vector for exploration (Bonnet and Corriveau, 2007). In such terranes, exploration therefore needs to take into account the metamorphic products of IOCG-related alteration and mineralization. For example, non-metamorphosed K-Fe-rich alteration commonly consists of biotite, K-feldspar, sericite, magnetite, carbonates, actinolite, and/or chlorite. The metamorphosed equivalents at the upper-amphibolite and granulite facies will largely retain the same bulk compositions, though more dehydrated, and will correspond to a variety of quartzofeldspathic gneisses with biotite, magnetite±sillimanite±garnet and/or cordierite±hornblende and/or orthopyroxene (Selleck et al., 2004). The hydrothermal origin of the mineralized gneiss protolith may be difficult to recognize in the field if it is associated closely with gneissic rocks derived from magnetite- and/or hornblende-bearing granitoids. If sericite is very widespread, the rock may even be mapped as 'metapelite' as have been many sericitic and argillic alteration zones in gneissic terranes (Corriveau and Bonnet, 2005). The calcic alteration might produce a variety of calc-silicate rocks that end up being mapped as metasediments, especially if 'metapelites' are found in the vicinity. It can also produce amphibolite and hornblendite with Na-Al-rich hornblende (edenite), magnetite, and plagioclase (likely oligoclase), with or without orthopyroxene. Albitites and K-feldspar-rich rocks will largely retain similar assemblages if metamorphosed at high grade.

Prospective Canadian Iron Oxide Copper-Gold Deposit Settings

Proterozoic Settings

From a Canadian perspective, the demonstration that continental magmatic arc and orogenic settings can be fertile for IOCG deposits is critical as it significantly furthers analogies between the Proterozoic (1.9-1.5 Ga) continental arc and orogenic settings of the Canadian Shield and ancestral North America in the Cordilleran Orogen and the Olympic Cu-Au province of the Gawler Craton. Two IOCG deposits with calculated resources are known in Canada: NICO and Sue-Dianne (Table 2). Both occur in the Great Bear magmatic zone of the Wopmay Orogen (Fig. 1). A third deposit is Kwyjibo in the Manitou district of the Grenville Province (Fig. 1). Beyond these deposits, significant Cu-Au (U, REE) prospects are also known from the Mesoproterozoic

Wernecke Breccias and the Iron Range in the ancestral North America components of the Cordillera and the Sault Ste. Marie area in the Mid-Continent Rift System. Other Proterozoic areas of interest include the Reindeer Zone of the Trans-Hudson Orogen, the Central Mineral Belt that extends across arc terranes of the Makkovick, Churchill, and Grenville provinces, the Nipigon Embayment (e.g. English Bay prospect, Greenwich Lake occurrence; Schneiders et al., 2003), and the Sudbury-Wanapitei area of the Southern Province. Potential IOCG targets also include the Vernot prospects in the Rivière aux Feuilles district, northern Superior Province, the Romanet horst, Labrador Trough, New Quebec Orogen, Quebec (MRNF, 2005), and a series of settings in the Grenville Province (Corriveau et al., 2007), such as the 1.74 to 1.70 Ga Killarney Magmatic Belt, the Disappointment Lake unit in the Wilson Lake terrane, and the 1.40 to 1.35 Ga Bondy Gneiss Complex in the western Grenville Province.

Great Bear Magmatic Zone, Northwest Territories

The Great Bear magmatic zone hosts the Co-Au-Bi NICO and the Cu-Ag Sue-Dianne deposits as well as IOCG prospects, showings, occurrences, and alteration zones in or adjacent to Andean-type, 1.88-1.85 Ga, calc-alkaline, basaltic to rhyolitic caldera-fill complexes and stratovolcanoes, diatremes, and coeval felsic to intermediate epizonal plutons (Hildebrand, 1986; Gandhi, 1994; Gandhi et al., 1996, 2001). This continental magmatic arc formed at the western margin of the Wopmay Orogen following the collision of an arc terrane with the Archean Slave craton (Fig. 4; Gandhi et al., 2001). It is bounded to the east by the domain-bounding Wopmay Fault Zone and to the northwest by the Leith fault and is dissected by brittle faults, some of which are mineralized (Hildebrand et al., 1987). Historical mining and exploration activities of this magmatic belt focussed on vein-type U, Ag, Co, Cu, and/or Au mineralization (e.g. Eldorado, Echo Bay, Contact Lake, Rayrock mines; Byrne, 1969; Reardon, 1990; Gandhi, 1994). In the late 1980s, the IOCG deposit potential of the altered hosts of such veins was recognized following regional mapping by the Geological Survey of Canada of intense calcic-sodic and iron oxide alterations and their linkage to the Kiruna-type iron oxide deposits in the northern Great Bear magmatic zone (Hildebrand, 1986) and to Kiruna- and Olympic Dam-types in the southern Great Bear magmatic zone (Gandhi, 1994) and its extension in the Great Slave Lake area (Badham, 1978). Airborne radiometric surveys, conducted by the Geological Survey of Canada during the U exploration boom in the 1970s, led to the discovery of the Sue-Dianne deposit (Charbonneau, 1988; Gandhi et al., 1996). Subsequent discoveries, such as the NICO deposit, were anchored on decades of exploration and detailed mapping (see review in Gandhi and Lentz, 1990; Goad et al., 2000) and results of the Canada-Northwest Territories Mineral Development Agreement and the follow-up Mineral Initiatives Program between 1987 and 1996 (e.g. Reardon, 1990; Gandhi, 1994; Gandhi et al., 1996).

Mineralization encompasses a large spectrum of sulphide-bearing mono- to polymetallic breccias, veins, disseminations, and massive to layered lenses, many with more than 20% low-Ti magnetite and/or hematite, as well as of sulphide-rich, U, Co, and/or Ag veins. Mineralization can be linked to three IOCG deposit subtypes: Olympic Dam, Kiruna, and Cloncurry. The Olympic Dam-type Cu-Ag Sue-Dianne deposit and the Mar, Nod, Fab, and Damp prospects are hosted by dacite-rhyolite ignimbrites and flows. They contain low-grade U mineralization, as well as base and precious metals. The Sue-Dianne deposit is hosted within a structurally controlled diatreme breccia complex above an unconformity with the metamorphic basement (Goad et al., 2000). The NICO deposit is a Co-Bi deposit with significant by-product Au and is either viewed as a Cloncurry- or Olympic Dam-type deposit (cf. Goad et al., 2000; Gandhi, 2004b). It is hosted in amphibole-magnetite-biotite schists and ironstones/iron oxide alteration unconformably underlying rhyolites of the Great Bear magmatic zone (Goad et al., 2000; Gandhi et al., 2001). Ore mineralogy is varied and includes Fe- and Cu-sulphides, native Au, and Bi, Bi telluride and sulphosalts, bismuthinite, cobaltite, and loellingite (Table 3). Current reserve estimates were obtained following the positive results of recent drilling programs and a positive bankable feasibility study (Fortune Minerals Limited, 2007). Where such data are available (e.g. southern Great Bear), coincident aeromagnetic and gravity highs are very diagnostic of the prospects and deposits and as such furthers linkage with IOCG settings (see Smith, 2002).

Kiruna-type magnetite-apatite-actinolite veins and alteration zones are numerous and associated with monzonitic plutons, such as those of the Mystery Island Suite (Hildebrand, 1986; Gandhi, 1994). They are hosted within 1.87 to 1.86 Ga volcano-sedimentary sequences associated with the Great Bear magmatic activity or in older metasediments of the 1.88 Ga Treasure Lake Group (Gandhi and van Breemen, 2005). At Contact Lake (Fig. 4), Cu-Au-Ag-Co mineralization overprints focussed potassic and iron alteration (potential zones of upflow and discharge) in association with laterally extensive calcic-sodic (Kiruna-type) and potassic alteration (potential zones of fluid recharge). Such alteration types and overprint relationships, combined with the discovery of hydrothermal breccias and diatremes and significant mineralization, point to a high polymetallic resource potential that goes beyond the historical mining of vein-type mineralization and the Kiruna-type classification (Webb, 2001; Corriveau et al., 2006; Mumin and Stewart, 2006). Mineralizing systems that overprint ironstones in Treasure Lake-type metasedimentary rocks comprise pre-, syn-, and post-deformation alteration and share affinities with the Cloncurry sub-type (Corriveau et al., 2006). Those that overprint younger volcanic rocks that were originally devoid of significant Fe oxide share more similarities with the Olympic Dam subtype.

Resemblance of this Paleoproterozoic continental arc with the Andes suggests that it may have as strong a mineral potential in as varied deposit types as the Andes. Moreover, the similarities between the timing and setting of the Great Bear magmatic zone and those of the Wathaman-Chipewyan batholiths of the Trans-Hudson Orogen strongly support revisiting this orogen as well (Mumin and Corriveau, 2004).

Manitou Lake District, Quebec

Kwyjibo, in the Manitou Lake district, is a replacive, Cloncurry subtype, Fe oxide Cu-REE-Mo-F-U-Au deposit

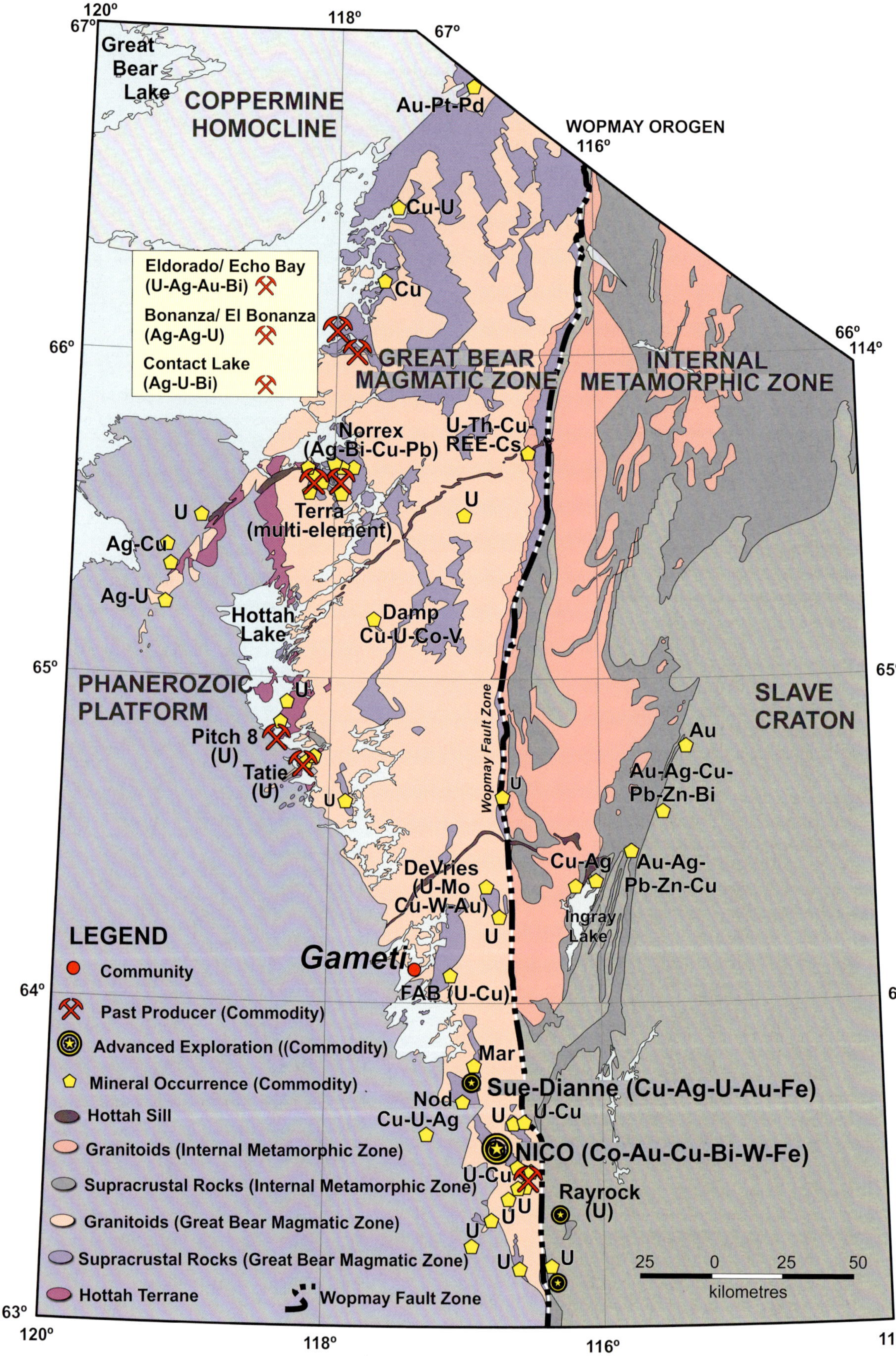

FIGURE. 4. Great Bear Magmatic Zone geology and mineral showings, prospects and past-producing mines (Hoffman and Hall, 1993; NORMIN.db at www.nwtgeoscience.ca/normin).

hosted in a deformed, 1.17 Ga, A-type granite of the Canatiche orthogneiss-paragneiss-granite complex in the eastern Grenville Province of Quebec (Clark et al., 2005). Seven main prospects form the deposit zone: the Josette, Malachite, Andradite, Fluorine, Grabuge, Lingis, and Rodrigue prospects. Other prospects and showings, including the IOCG-type Marmont prospect and the Cu-Zn Manitou showing, occur to the south and west, either in the Canatiche Complex or within the adjacent supracrustal-laden Manitou Complex. Most prospects are rich in magnetite, have polymetallic enrichments in Cu, REE, Y, P, F, and Ag, and are anomalous in Th, U, Mo, W, Zr, and Au. Other Fe oxide bodies are monometallic and akin to magnetite±apatite, Kiruna-type mineralization (Gauthier et al., 2004; Clark et al., 2005). The main polymetallic mineralization event, dated at 0.97 to 0.95 Ga, comprises chalcopyrite, pyrite, fluorite, molybdenite, and REE- and Y-bearing minerals, such as andradite garnet, apatite, bastnaesite, britholite, kainosite, monazite, perovskite, pyrochlore, thorite, uraninite, and xenotime (Table 3), in association with pargasitic hornblende, clinopyroxene, and quartz (Gauthier et al., 2004). The resulting polymetallic hydrothermal veins and stockworks replace an earlier, locally folded, magnetite-rich alteration that forms disseminations, stockworks, breccia (Fig. 5), semimassive, and massive bodies in the host and potentially coeval 1175±4 Ma leucogranite (Clark et al., 2005). Later stages of mineralization likely used the magnetite-rich bodies as chemical traps. Focussed potassic, sodic-calcic, calc-silicate, fluoritic, phosphatic, silicic, hematitic, and sodic alterations are observed. Evidence of bimodal magmatism is observed in the form of a composite mafic-felsic dyke that crosscuts a magnetite breccia and bears cuspate boundaries indicative of magma mingling (see Fig. 3f in Clark et al., 2005). Similarities of this dyke with those of the host magmatic suite at a regional scale suggest that part of the mineralization may be coeval with the Canatiche granite itself, however the age of this dyke is not known and the role of mafic magmatism in the formation of this IOCG setting remains uncertain. The field aspect of this dyke is very distinct from that of the younger late- to post-tectonic, highly discordant 1.03 and 0.99 Ga granites and felsic-mafic plutons and dykes that intrude the Manitou and Canatiche complexes (Wodicka et al., 2003). A distinct period of monazite growth between 1063 and 1047 Ma in the Manitou Complex is attributed to fluid-assisted deformation during tightening of fold structures (Wodicka et al., 2003), an event that may be coeval with the tightening and shearing of the magnetite stockwork described by Clark et al. (2005). Titanites dated at 0.98 Ga from the 1.17 to 1.18 Ga Canatiche Complex granitoids signal a regional thermal event associated with the late-stage intrusions (Wodicka et al., 2003) while oxygen isotope data from magnetite-rich bodies point to the presence of magmatic and/or metamorphic fluids and their local mixing with cooler meteoric water (Clark et al., 2005). The best drill intersections and surface data include the following: 1.83% Cu, 0.96% La+Ce+Sm, 654 ppm Th, 435 ppm U, and 164 ppb Au over 9.5 m (channel sample); 0.36% Cu over 16.5 m and 0.88% La+Ce+Sm over 29.9 m (drill core) (Gauthier et al., 2004).

FIGURE 5. Magnetite breccia, Kwyjibo deposit, Quebec.

Wernecke Breccias, Yukon

"Wernecke Breccias" (Figs. 1, 6) is a collective term for a curvilinear array of 1.60 Ga breccia bodies extending over a 50,000 km^2 area in a west-to-east direction from the Wernecke to the Ogilvie Mountains across the Coal Creek and Wernecke inliers of the Cordillera as well as extending northward and reaching the Hart River Inlier (Bell, 1986; Thorkelson, 2000). The breccias cut greenschist-facies Paleoproterozoic Wernecke Supergroup sedimentary rocks, and 1.71 Ga Bonnet Plume River diorite intrusions enclose fragments of undeformed volcanic rocks apparently not preserved in the exposed tectonostratigraphic record (Delaney, 1981; Thorkelson, 2000; Laughton et al., 2005). A 1595 ± 5 Ma U-Pb age obtained from a hydrothermal titanite within the Slab Mountain breccia matrix provides an age constraint on the regional-scale surges of hydrothermal fluids that formed the breccias. A ca. 1.38 Ga sedimentary pile unconformably overlies the breccias, indicating that most of the hydrothermal activity took place prior to 1.38 Ga. However, evidence for subsequent minor brecciation and hydrothermal alteration exists and is associated with 1.38 and 1.27 Ga intrusive events (Thorkelson et al., 2001).

At a regional scale, the breccias are spatially associated with deep-seated fault zones such as the nearly 600 km long Richardson Fault array, while at the local scale the discharge of fluids used pre-existing discontinuities such as the core of fold structures, intrusive contacts, sedimentary layering, faults, and shear zones (Hunt et al., 2005). Normal faults are very commonly associated with the breccia zones (Fig. 6) but a cogenetic relationship is uncertain, as in many cases faults can be shown to postdate brecciation and breccias are found outside of fault zones (Thorkelson, 2000, p. 45).

The breccias vary in size from outcrop to mountain scale (0.1 to 10 km^2), in colour from grey (sodic) to mottled red and pink (potassic), and in fragment size from <5 cm up to hundreds of metres (Thorkelson, 2000, p. 17, 31-40). Contacts tend to be mostly gradational but sharp intrusions of breccia material into unaltered country rocks also occur. Contacts are also sharp where fault bounded. At least 65 breccia bodies are known, many of which are associated with Fe oxide Cu (±Co, Au, Ag, U, and locally Mo) prospects and showings (Yukon Geological Survey, Minfile 116B (84, 99, 102,

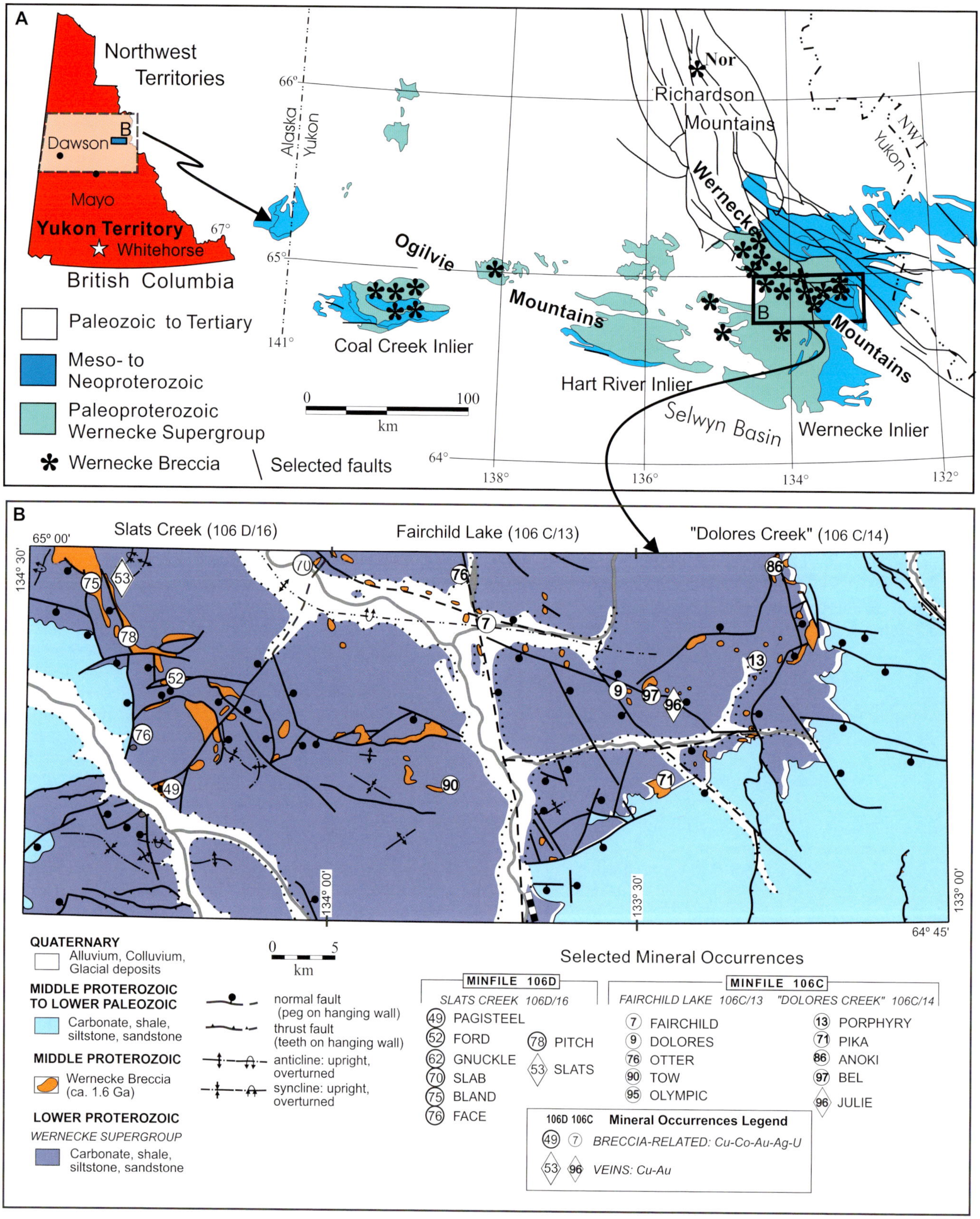

FIGURE 6. Distribution of Wernecke Supergroup and Wernecke Breccias in Yukon **(A)** and example of the spatial relation of breccia to fault zones **(B)** (after Thorkelson, 2000).

103), 106C (6, 7, 13, 15-17, 44, 71, 86), 106D (49, 52, 68, 75-77, 79, 87, 96), 106E (2, 3, 5, 9, 11, 22, 25, 30, 31, 40), 106L (61); Deklerk, 2003). Mineralization occurs as disseminations and veins in Wernecke Supergroup metasediments and in Wernecke Breccias, as clasts and within the matrix of heterolithic breccias, as well as in carbonate veins cutting breccias (Hunt et al., 2005). Brecciation occurred either close to the surface, leading to near-surface vents with volcanic fragments and sodic, low temperature alteration facies, or more commonly, well below the surface leading to potassic alteration facies with locally derived fragments (Laughton et al., 2005; cf. Hunt et al., 2005 for a different interpretation). Hematite is the main Fe oxide but magnetite can be locally abundant. At the scale of mineral properties (e.g. Yukon Olympic property; Copper Ridge Exploration website), kilometre-scale Bouguer gravity anomalies can be associated with slightly offset magnetic anomalies. In areas overlain by Paleozoic sedimentary rocks, blind targets can only be detected by such geophysical anomalies.

The physical and mineralogical characteristics and the age of the Wernecke Breccias share some similarities with those of the Olympic Dam deposit. Their settings could have been contiguous based on the SWEAT continental reconstruction and both were part of a mature Proterozoic orogen. In contrast to the Olympic Dam, the Wernecke Breccias are not associated with exposed magmatism (Hitzman et al., 1992; Thorkelson, 2000; Thorkelson et al., 2001, 2003).

Iron Range District, British Columbia

The Iron Range district forms a belt of iron oxide mineralization that extends intermittently for 20 km along the Iron Range fault system in the Canadian Cordillera (Stinson and Brown, 1995; Ray and Webster, 2000; British Columbia Geological Survey, 2005). The fault cuts across folded clastic sedimentary rocks and gabbro sills of the 1.47 Ga Mesoproterozoic intracratonic basin that hosts the Sullivan Pb-Zn-Ag deposit (e.g. Anderson and Davis, 1995; Schandl and Davis, 2000). The fault zone is interpreted as an intracratonic rift-related normal fault that was active during sedimentation. It was episodically reactivated, and acted as a channel for sill emplacement as well as for the fluids that generated polyphase hydrothermal alteration and associated Fe iron oxide mineralization (Stinson and Brown, 1995; Marshall and Downie, 2002).

The main Iron Range prospects and showings include, in the northern segment of the fault, the Golden Cap, Union Jack, American Flag, O-Ray, Maple Leaf, Keepsake, Rhodesia, La Grande, Cracker Jack, Dakota, Idaho, Pacific, Agnes, Niagara, and Constellation occurrences and, to the south, the Great War showings (British Columbia Geological Survey, 2005). The mineral occurrences display evidence of polyphase hydrothermal activity including quartz veining, brecciation with hematite infill, albitization, and hematite veining. Magnetite also occurs and is pseudomorphed by hematite; hence magnetite may have prevailed at an early stage of the hydrothermal activity (Marshall and Downie, 2002). Pyrite ranges between 2 and 20% while bornite and chalcopyrite are very minor. Intense albite and chlorite alteration is limited mostly to the fault zone itself and to some local infiltration within their hosts, the Middle Aldridge Formation units and the Moyie gabbro sills (e.g. Fig. 3 in Marshall and Downie, 2002). Such a limited alteration contrasts with the regional-scale alteration typical of most IOCG districts.

An intense linear magnetic anomaly overlaps the belt of Fe oxide occurrences, with peak magnetic susceptibility values coinciding with massive lenses of magnetite. A gamma-ray spectrometric survey detected elevated eTh/K values along the Iron Range fault that are associated with albite-rich alteration and breccia zones (Marshall and Downie, 2002).

The Reindeer Zone, Manitoba

The Trans-Hudson Orogen, like many other orogenic belts of Canada, is a green-field exploration setting in terms of IOCG deposits. In an effort to diversify and enhance mineral exploration on its territory, the Manitoba Geological Survey sponsored, under the leadership of H. Mumin at Brandon University, an IOCG targeted reconnaissance assessment of the Reindeer Zone in the Trans-Hudson Orogen of Manitoba. The study resulted in the identification and prioritization of 142 regionally prospective target sites (Mumin and Perrin, 2005). Follow-up assessment of 13 targets, to test the adequacy of the selection protocol, led to the following significant discoveries that were identified and delineated: 1) the Eden Lake REE-phosphate-rich carbonatite complex and 2) the Eden deformation corridor and polymetallic mineral belt (Mumin, 2002; Mumin and Corriveau, 2004; Mumin and Perrin, 2005). The Eden Lake carbonatite complex and its network of high-grade but narrow REE-Y-U-Th veins intrude early 1.87 Ga A-type granitoids and mafic rocks (Halden and Fryer, 1999), in intimate association with mafic, monzonite, syenite and alkali granite phases, and 1.80 Ga syenite (Mumin, 2002; Mumin and Perrin, 2005). Fenitic alteration is widespread, either pervasive or in vein form, and includes clinopyroxene-feldspar, REE-apatite-clinopyroxene, and carbonate-K-feldspar assemblages. High-grade REE veins consist of allanite, britholite, garnet, albite, clinopyroxene, apatite, titanite, and fluorite and reach up to 169,000 ppm total REEs (Mumin and Perrin, 2005). The A-type granitoids and alkaline rocks are not anorogenic as originally proposed (Halden and Fryer, 1999) but were emplaced at the end of deformation in the polyphase Eden deformation corridor (Mumin and Corriveau, 2004).

Though the carbonatite complex does not host significant iron oxides and does not constitute a bona fide IOCG prospect, the REE-rich character of its veining system and carbonatitic host constitute geological properties of the REE-rich iron oxide mineralization end-member and point to a strong connectivity between mantle and crust in the area. The carbonatite complex occurs above a 'piercing point' of the Moho where a crustal-scale ramp, exposed at surface as the Granville Lake structural zone, intercepts the mantle (White et al., 2000; cf. White, 2005 for a different interpretation of crust architecture). The A-type granitoid it intrudes is part of the Andean-type continental arc granitic to mafic magmatism that formed the 1.86 to 1.85 Ga Wathaman-Chipewyan batholith and other intrusions at the margin of the Archean Hearne craton following accretion of Paleoproterozoic oceanic island-arcs, back-arcs, oceanic islands, and Archean fragments (Ansdell, 2005; Corrigan et al., 2005). This complex crust has a demonstrated fertility in Cu (VMS deposits), Au (shear-zone hosted) and Ni (magmatic sulphides) (Beaumont-Smith and Bohm, 2003; Gale, 2003), and the potential IOCG targets are distributed across most components, including the batholith itself. In terms of crustal architecture and magmatism, the Eden Lake carbonatite complex setting is strikingly similar to that of the Olympic Dam deposit, whereas the geological make-up of the zone is comparable to what is known of the Gawler Craton. These traits are major incentives to foster knowledge acquisition in the Reindeer Zone.

Central Mineral Belt, Labrador

The Central Mineral Belt district of Labrador (Fig. 1) hosts numerous Cu, U, and REE showings and prospects including the Michelin uranium deposit (Gandhi, 1986; Swinden et al., 1991; Geological Survey of Newfoundland and Labrador, 2005). The belt encompasses a variety of mineral deposit types and exceeds 250 km in length across several geological entities, namely the Churchill, Nain, Makkovik, and Grenville provinces. The mineral belt occurs adjacent to amalgamated Archean cratons (Superior and Nain) and displays a spatial association with the crustal-scale and domain-bounding Kaipokok Bay shear zone that separates the 1.86 Ga bimodal A-type Aillik Group (Schärrer et al., 1988; Sinclair et al., 2002) from the Moran Lake and the 2.18 Ga Post Hill (formerly Lower Aillik Group) groups (Ketchum et al., 2002). Calc-alkaline continental-arc magmatism of 1.89 to 1.87 Ga and successor rifted or back-arc 1.86 to 1.85 Ga volcanism and sedimentation, including the voluminous felsic A-type melts of the Aillik Group, took place along the continental margin and was followed by polyphase transpression and metamorphism (1.81-1.78 Ga and 1.74-1.71 Ga) and intrusion of A-type granitoids (Ketchum et al., 2002; Sinclair et al., 2002).

Deformation that affected the Moran Lake, Post Hill, and Aillik groups was accompanied by regional-scale Na-(Ca) alteration, characterized by albite-(actinolite-epidote-clinopyroxene-garnet-calcite-quartz)-rich assemblages and Fe-(K-U-Cu-Mo) mineralization (magnetite, hematite, amphibole, pyroxene, pyrite, chalcopyrite, and molybdenite; Post Hill, Michelin North, Michelin South, Emben, and Burnt Lake properties; Marshall et al., 2003; Geological Survey of Newfoundland and Labrador, 2005). Sericite-hematite alteration zones with disseminated sulphides and abundant REE mineralization and hematite- and carbonate-rich breccias also occur locally. Values of 1.09% Cu, 20.5 g/t Ag, 0.72% Cu, and 26.5 g/t Ag have been reported from the Michelin property (Marshall et al., 2003). The Makkovik Province extends eastward to the Baltic Shield and may have similar IOCG potential (Gower et al., 1995; Weihed et al., 2005).

Sault Ste. Marie and Sudbury Districts, Ontario

In the Sault Ste. Marie district, Ontario, the presence of a significant volume of basalts and high-level felsic intrusion associated with the 1 Ga Mid-Continent Rift system, and of magnetite and sulphide veins, fluorite, fault-related iron oxide breccias, Cu mineralization in association with Au and Ag (e.g. Island Copper and Coppercorp prospects) and splays off major crustal faults, such as the Mamainse Point Fault, are deemed positive indicators for IOCG deposits (Atkinson et al., 2004; Tortosa and Moss, 2004). Extending eastward for over 400 km broadly along the Great Lake tectonic zone and the Murray Fault system are a series of Cu-Au-REE (U) showings (Rathbun (Fe), Skead (Au), Wanapitei (Au, Ag), Glad (Au, Cu, Ni), Ashigami (Au Ag), Skadding (Au)) in the Sudbury-Wanapitei area of the Southern Province. The mineralization is associated with 1700 Ma Na-metasomatic zones, brecciation, Ca-Mg-Fe carbonate, chlorite sulphides, and magnetite-bearing alteration (Gates, 1991; Schandl et al., 1994; Rogers et al., 1995).

Cordilleran and Appalachian Prospective Settings

Phanerozoic settings of interest include the Mont de l'Aigle prospect, the Avalon zone, the Cobequid-Chedabucto fault zone, and the Lepreau Iron mine in the Appalachian orogen, and the Insular Range skarn deposits and the Heff deposit in the Cordillera.

The Heff Cu±Au±REE±P-bearing magnetite skarn deposit comprises a magnetite-bearing garnet-pyroxene skarn alteration that formed by the infiltration of hydrothermal fluids from the Heffley Creek Pluton, a pyroxene and magnetite-bearing mafic-ultramafic body in host limestone (Ray and Webster, 2000). Pods and massive lenses of magnetite, up to 10 m thick, host disseminations and veinlets of sulphides. Samples of mineralized skarn share similarities with many IOCG deposits by having more than 25% Fe (magnetite), anomalous Au and Cu, and light REE enrichment (BC Minfile 092INE096, in British Columbia Geological Survey, 2005).

Polymetallic iron oxide Cu, iron oxide Cu-Au-Ag, Cu-Mo, and Pb-Zn-Ag mineralization is found in the vicinity of the Shickshock-Sud fault zone in the Appalachian Orogen of Quebec (e.g. Mont de l'Aigle and adjacent prospects; Simard et al., 2006). Mineralized subsidiary faults crosscut the Lemieux Dome, a subcircular structure with faulted sedimentary rocks, marginal felsic volcanic and pyroclastic rocks, and a mafic to felsic dyke swarm emplaced along fault zones. The prospects comprise a series of quartz-chalcopyrite-dolomite veins and polyphase hematite-quartz-dolomite vein and breccia complexes that may extend for hundreds of metres. As such, the setting shares affinities with the Olympic Dam subtype but the presence of Fe-skarn subtype showings is also significant.

In the Appalachian Orogen, the other extensively explored areas for IOCG deposits are the Avalon zone (Cross Hill and Net Point prospects; Newfoundland; GSNL National Mineral Inventory Number, 001M/10/Cu 005 and 001M/12/Cu 006 in Geological Survey of Newfoundland and Labrador, 2005), the Cobequid-Chedabucto fault zone separating the Avalon and Meguma terranes (Mt. Thom and Bass River prospects; Nova Scotia; O'Reilly, 1996, 2002), and the Lepreau Iron mine in the Pocologan metamorphic suite (New Brunswick; Barr et al., 2002; NB deposit database, URN 590 in New Brunswick Department of Mines, 2005).

Iron Oxide Copper-Gold-Directed Strategies: A Canadian Perspective

Major IOCG provinces are commonly associated with crustal-scale structures and extensive magmatic events that drive large-scale fluid flow into mid- to upper crustal levels along fault zones, other discontinuities, and permeable units. Mixing of magmatic fluids with near-surface meteoritic fluids or brines is commonly invoked. Also invoked are evaporites as the source of NaCl in the mineralizing solutions to carry the metals and to generate regional-scale sodic alteration. The presence of evaporite is, however, not an essential prerequisite for the formation of IOCG deposits as many have magmatic fluid signatures. A-type granites or alkaline intrusions, however, are considered critical, though in some cases the source may be distal from the deposit, as is the case in the Bayan Obo deposit. Consequently, a lack of surface expression of magmatism does not necessarily translate into a lack of IOCG deposit potential.

Advances in genetic/exploration models of the last decade are reviewed in Hitzman (2000), Porter (2000, 2002a), and Williams et al. (2005). It is apparent from these papers that there are currently a lot more knowledge gaps to fill regarding the processes and types of fluids that lead to IOCG deposits, than models that are not in dispute. Nevertheless, consistency in geological properties allows IOCG deposits to be detected by geological, geochemical, remote sensing, and geophysical techniques (Nisbet et al., 2000) but Canadian successes were soundly anchored in field work using alteration, geochemical, and mineral-indicator vectoring on a deposit or district scale (e.g. Goad et al., 2000; Mumin and Perrin, 2005). New developments in exploration include hyperspectral data and Aster satellite image analysis, predictive, soft (qualitative to semi-quantitative) and hard (numerical) modeling, and geographic information system (GIS)-based mineral prospectivity analysis (weights of evidence, fuzzy logic, and neural networks) (Venkataraman et al., 2000; Harris et al., 2001; Lamothe and Beaumier, 2001). Such expert systems are at the frontier of our current knowledge but their ability to circumscribe target areas is constrained to be as good as the data that is entered in the information system.

Geological indicators of IOCG mineralization have significant lateral extent and include diagnostic alteration zones, overprinting relationships, and geochemical fingerprints. Hence alteration mapping and vectoring are important components of IOCG exploration strategies. A key problem in diversifying exploration toward IOCG in Canada is that it requires targeting gneissic and granitic terranes where modern geoscience knowledge may be rare or only at reconnaissance scale. Field geology may be perceived as a traditional tool but, as pointed out by Corriveau and Clark (2005), if supported by modern geosciences, state-of-the-art laboratory facilities, and creative minds, it provides the leading edge in gaining new perspectives on frontier geological settings in terms of mineral potential. This is particularly true for IOCG exploration, as mapping of prospective settings was, in most cases, conducted prior to the recognition of IOCG deposits; hence vectors to ore as significant as the ore-bearing hematite-K-feldspar alteration are commonly missing in available records increasing risk to exploration and expenditures.

This synthesis has highlighted two Canadian IOCG settings in which alteration and breccia are so well developed and exposed that field-anchored research across them could lead to a rapid gain in scientific knowledge and expertise: (1) the Great Bear Magmatic Zone mineralizing systems in the Northwest Territory, and (2) the Wernecke Breccias in the Yukon. As metamorphosed IOCG deposits exist, their search should not be neglected. Two suspected IOCG settings at granulite facies, the Bondy Gneiss Complex and the Disappointment Lake gneiss in the Grenville Province of Quebec and Labrador, provide examples of potential IOCG systems metamorphosed at high grade in Canada. Provision of field examples of IOCG settings, practical means of finding IOCG alteration halos in the field, and field vectors to potential mineralization, combined with assessment of knowledge gaps on key Canadian IOCG systems, and transfer of knowledge to stakeholders will pave the way to discoveries of ore deposits. But first and foremost it will lower a hesitance to explore in such non-classical terranes and decrease the risk and expenditures to do so. In the years to come, Canadians are likely to see, like the Australians have (see Table 2), an increase in discovery of IOCG deposits while significantly enlarging the realm of commodities to be mined at the same time.

Acknowledgments

Colleagues from provincial and federal surveys, industry, and academia have shared with the author internal reports, public-domain information, and personal knowledge of IOCG deposits and host settings. They are thanked for making this synthesis possible. They include S. Gandhi, L. Chorlton, D. Corrigan, B. Dubé, P. Gosselin, B. Hillary, W. Goodfellow, and J. Lydon (Geological Survey of Canada), D. Lefebure and D. Terry (British Columbia), C. Beaumont-Smith (Manitoba), M. McLeod (New Brunswick), C. Gower and S. O'Brien (Newfoundland and Labrador), J. Mason and M. Easton (Ontario), T. Clark, D. Lamothe and S. Perreault (Québec), S. Goff, V. Jackson and L. Ootes (Northwest Territory), and G. Abbot and J. Hunt (Yukon) from government surveys; M. Downe and T. Setterfield (Monster Copper Corporation), M. O'Dea and L. Marshall (Fronteer), F. Chartrand and Y. Trudeau (SOQUEM), R. Goad and K. Neale (Fortune Minerals) and D. Bubar (Avalon Ventures) from industry; G. Beaudoin (University of Laval), M. Jébrak (Université du Québec à Montréal), H. Mumin (Brandon University), and D. Thorkelson (Simon Fraser University) from academia. Many thanks also to J. Lydon and W. Goodfellow, Geological Survey of Canada Mineral Synthesis Project co-leader and leader, respectively. Reviews by S. Gandhi, R. Skirrow (Geoscience Australia), H. Mumin, G. Osborne (Western Mining Corporation), and I. Kjarsgaard (Geological Survey of Canada) on earlier versions of this manuscript are gratefully acknowledged.

References

Adshead-Bell, N.A., 1998, Evolution of the Starra and Selwyn high strain zones, Eastern Fold Belt, Mount Isa Inlier: Implications for Au-Cu mineralization: Economic Geology, v. 93, p. 1450-1462.

Anderson, H.E., and Davis, D.W., 1995, U-Pb geochronology of the Moyie sills, Purcell Supergroup, southeastern British Columbia: Implications for the Mesoproterozoic geological history of the Purcell (Belt) basin: Canadian Journal of Earth Sciences, v. 32, p. 1180-1193.

Anderson, J.L., 1983, Proterozoic anorogenic plutonism of North America: Geological Society of America, Memoir 161, p. 133-152.

Ansdell, K.M., 2005, Tectonic evolution of the Manitoba-Saskatchewan segment of the Paleoproterozoic Trans-Hudson Orogen, Canada: Canadian Journal of Earth Sciences, v. 42, p. 741-759.

Atkinson, B.T., Hailstone, M., Seim, G.W., Wilson, A.C., Draper, D.M., Butorac, S., and Cooper, G.R., 2004, Report of Activities 2003, Resident Geologist Program, Timmins Regional Resident Geologist Report: Timmins and Sault Ste. Marie Districts: Ontario Geological Survey, Open File Report 6130, 87 p.

Badham, J.P.N., 1978, Magnetite-apatite-amphibole-uranium and silver-arsenide mineralizations in lower Proterozoic igneous rocks, East Arm, Great Slave Lake, Canada: Economic Geology, v. 73, p. 1474-1491.

Barr, S.M., White, C.E., and Miller, B.V., 2002, The Kingston terrane, southern New Brunswick, Canada: Evidence for an Early Silurian volcanic arc: Canadian Journal of Earth Sciences, v. 114, p. 964-982.

Barton, M.D., and Johnson, D.A., 2000, Alternative brine sources for Fe-oxide (-Cu-Au) systems: Implications for hydrothermal alteration and metals, *in* Porter, T.M., ed., Hydrothermal Iron Oxide Copper-Gold and Related Deposits: A Global Perspective: PGC Publishing, Adelaide, v. 1, p. 43-60.

Barton, M.D., and Johnson, D.A., 2004, Footprints of Fe-oxide (-Cu-Au) systems, *in* Muhling, J., Goldfarb, R., Vielreicher, N., Bierlein, F., Stumpfl, E., Groves D.I., and Kenworthy, S., eds., Society of Economic Geology 2004 Conference, Predictive Mineral Discovery Under Cover, Extended Abstracts: Centre for Global Metallogeny, The University of Western Australia, no. 33, p. 112-116.

Beaumont-Smith, C.J., and Bohm, C.O., 2003, Tectonic evolution and gold metallogeny of the Lynn Lake greenstone belt, Manitoba (NTS 64C10, 11, 12, 14, 15 and 16), *in* Report of Activities 2003, Manitoba Industry, Trade and Mines: Manitoba Geological Survey, p. 159-170.

Bell, R.T., 1986, Megabreccias in northeastern Wernecke Mountains, Yukon Territory, *in* Current Research, Part A: Geological Survey of Canada, Paper 86-1A, p. 375-384.

Belperio, A., and Freeman, H., 2004, Common geological characteristics of Prominent Hill and Olympic Dam - Implications for iron oxide copper-gold exploration models: Australian Institute of Mining Bulletin, Nov-Dec. Issue, p. 67-75.

Bonnet, A.-L., and Corriveau, L., 2007, Alteration vectors to metamorphosed hydrothermal systems *in* gneissic terranes, *in* Goodfellow, W.D., ed., Mineral Deposits of Canada: A Synthesis of Major Deposit-Types, District Metallogeny, the Evolution of Geological Provinces, and Exploration Methods: Geological Association of Canada, Mineral Deposits Division, Special Publication 5, p. 1035-1049.

British Columbia Geological Survey, 2005, British Columbia Minfile 082FSE014-28, 92, 082FNE132, 092INE096): British Columbia Ministry of Energy and Mines, website.

Byrne, N.W., 1969, Report on the Contact Lake Mine, Ulster Petroleums Limited, Great Bear Lake Area, Mackenzie Mining District, N.W.T.: Department of Indian and Northern Affairs, Assessment Report, No. 060371, 33 p.

Chao, E.C.T., Back, J.M., Minkin, J.A., Tatsumoto, M., Wang Junwen, Conrad, J.E., McKee, E.H., Zonglin, H., Qingrun, M., and Shengguang, H., 1997, The sedimentary carbonate-hosted giant Bayan Obo REE-Fe-Nb ore deposit of Inner Mongolia, China: A cornerstone example for giant polymetallic ore deposits of hydrothermal origin: U.S. Geological Survey, Bulletin 2143, 65 p.

Charbonneau, B.W., 1988, Gamma spectrometric and magnetic anomalies associated with Cu-U mineralization, Faber Lake volcanic belt, District of Mackenzie, N.W.T., *in* Current Research, Part C: Geological Survey of Canada, Paper 88-1C, p. 255-258.

Clark, T., Gobeil, A., and David, J., 2005, Iron oxide-Cu-Au-type and related mineralization in the Manitou Lake area, Grenville Province, Quebec: Variations in composition and alteration style: Canadian Journal of Earth Sciences, v. 42, p. 1829-1847.

Collins, W.J., Beams, S.D., White, A.J.R., and Chappell, B.W., 1982, Nature and origin of A-type granites with particular reference to south-eastern Australia: Contributions to Mineralogy and Petrology, v. 80, p. 189-200.

Corrigan, D., Hajnal, Z., Németh, B., and Lucas, S.B., 2005, Tectonic framework of a Paleoproterozoic arc-continent to continent-continent collisional zone, Trans-Hudson Orogen, from geological and seismic reflection studies: Canadian Journal of Earth Sciences, v. 42, p. 421-434.

Corriveau, L., and Bonnet, A.-L., 2005, Pinwarian (1.5 Ga) volcanism and hydrothermal activity at the eastern margin of the Wakeham Group, Grenville Province, Quebec: Canadian Journal of Earth Sciences, v. 42, p. 1749-1782.

Corriveau, L., and Clark, T., 2005, The Grenville Province: a geological and mineral resources perspective derived from government and academic research initiatives: Canadian Journal of Earth Sciences, v. 42, p. 1637-1642.

Corriveau, L., Mumin, H., Ootes, L., and Landry, J.-Y., 2006, Alteration mapping as vectors towards iron oxide Cu-Au (Ag-U-Co-Bi) mineralization in green-field exploration settings: Case examples from the Great Bear Magmatic Zone, NWT: Geological Association of Canada-Mineralogical Association of Canada, Abstracts with Program,v. 31, p. 30.

Corriveau, L., Perreault, S., and Davidson, A., 2007, Prospective metallogenic settings of the Grenville Province, *in* Goodfellow, W.D., ed., Mineral Deposits of Canada: A Synthesis of Major Deposit-Types, District Metallogeny, the Evolution of Geological Provinces, and Exploration Methods: Geological Association of Canada, Mineral Deposits Division, Special Publication 5, p. 819-847.

Creaser, R.A., 1989, The Geology and Petrology of Middle Proterozoic Felsic Magmatism of the Stuart Shelf, South Australia: Ph.D. thesis, La Trobe University, Melbourne, Australia, 434 p.

——— 1996, Petrogenesis of a Mesoproterozoic quartz latite-granitoid suite from the Roxby Downs area, South Australia: Precambrian Research, v. 79, p. 371-394.

Creaser, R.A., Price, R.C., and Wormald, R.J., 1991, A-type granites revisited: assessment of a residual source model: Geology, v. 19, p. 163-166.

Cross, K.C., Daly S.J., and Flint, R.B., 1993, Olympic Dam deposit, *in* Drexel, J.F., Preiss, W.V., and Parker, A.J., eds., The Geology of South Australia: South Australia Department of Mines and Energy, Bulletin 54, v. 1, p. 132-138.

Daliran, F., 2002, Kiruna-type iron oxide-apatite ores and "apatitites" of the Bafq district, Iran, with an emphasis on the REE geochemistry of their apatites, *in* Porter, T.M., ed., Hydrothermal Iron Oxide Copper-Gold and Related Deposits: A Global Perspective: PGC Publishing, Adelaide, v. 2, p. 303-320.

Davidson, G.J., Davis, B., and Garner, A., 2002, Structural and geomechanical constraints on the emplacement of the Monakoff oxide Cu-Au (Co-U-REE-Ag-Zn-Pb) deposit, Mt Isa Inlier, Australia, *in* Porter, T.M., ed., Hydrothermal Iron Oxide Copper-Gold and Related Deposits: A Global Perspective: PGC Publishing, Adelaide, v. 2, p. 49-75.

Deklerk, R., 2003, Yukon MINFILE 2003 - A Database of Mineral Occurrences: Yukon Geological Survey (CD-ROM).

Delaney, G.D., 1981, The mid-Proterozoic Wernecke Supergroup, Wernecke Mountains, Yukon Territory, *in* Campbell, F.H.A., ed., Proterozoic Basins in Canada: Geological Survey of Canada, Paper 81-10, p. 1-23.

Downes, M., 2003, Exploration for iron oxide copper-gold deposits and an emerging play in Nova Scotia: Monster Copper Corporation website, http://www.monstercopper.com.

Faure, S., 2003, Les gisements protérozoïques de fer polymétalliques: cibles d'exploration au Québec: Consorem, 61 p.

Ferris, G.M., and Schwarz, M.P., 2003, Proterozoic gold province of the central Gawler Craton: MESA Journal, v. 30, p. 4-12.

Ferris, G.M., Schwarz, M.P., and Heithersay, P., 2002, The geological framework, distribution and controls of Fe-oxide Cu-Au mineralisation in the Gawler Craton, South Australia. Part I - Geological and tectonic framework, *in* Porter, T.M., ed., Hydrothermal Iron Oxide Copper-Gold and Related Deposits: A Global Perspective: PGC Publishing, Adelaide, v. 2, p. 9-31.

Fortune Minerals Limited, 2007, New release, January 16, 2007, www.fortuneminerals.com.

Frietsch, R., Rapunen, H., and Vokes, F.M., 1979, The ore deposits in Finland, Norway and Sweden - A review: Economic Geology, v. 74, p. 975-1001.

Gale, G.H., 2003, Vectoring volcanogenic massive sulphide deposits using rare earth elements and other pathfinder elements at the Ruttan mine, Manitoba (NTS 63B5), *in* Report of Activities 2003, Manitoba Industry, Trade and Mines: Manitoba Geological Survey, p. 54-73.

Galley, A.G., Hannington, M.D., and Jonasson, I.R., 2007, Volcanogenic massive sulphide deposits, *in* Goodfellow, W.D., ed., Mineral Deposits of Canada: A Synthesis of Major Deposit-Types, District Metallogeny, the Evolution of Geological Provinces, and Exploration Methods: Geological Association of Canada, Mineral Deposits Division, Special Publication 5, p. 141-161.

Gandhi, S.S., 1986, Uranium in Early Proterozoic Aillik Group, Labrador, *in* Evans, E.L., ed., Uranium Deposits of Canada: Canadian Institute of Mining and Metallurgy, Special Volume 33, p. 70-82.

——— 1994, Geological setting and genetic aspects of mineral occurrences in the southern Great Bear magmatic zone, Northwest Territories, *in* Sinclair, W.D., and Richardson, D.G., eds, Studies of Rare-Metal Deposits in the Northwest Territories: Geological Survey of Canada, Bulletin 475, p. 63-96.

——— 2004a, Magmatic-hydrothermal Fe Oxide±Cu±Au deposits: Classification for a digital database and an overview of selected districts: IAVCEI General Assembly 2004, Pucón, Chile, CD-ROM, Abstracts 01a_pt_169, http://iavcei2004.free.cl/abstracts/Gandhi1_ibxobhffgsnpoxphrspm.doc.

Gandhi, S.S. (comp.), 2004b, World distribution of iron oxide ±Cu ±Au ±U (IOCG) deposits: Geological Survey of Canada, Geoscience Data Repository, Mineral Resources Metadata, http://gdr.nrcan.gc.ca/minres/metadata_e.php?id=8.

Gandhi, S.S., and Lentz, D.R., 1990, Bi-Co-Cu-Au-As and U occurrences in the Snare Group metasediments and felsic volcanics of the southern Great Bear magmatic zone, Lou Lake, Northwest Territories, *in* Current Research, Part C, Canadian Shield: Geological Survey of Canada, Paper 90-1C, p. 239-253.

Gandhi, S.S., and van Breemen, O., 2005, SHRIMP U-Pb geochronology of detrital zircons from the Treasure Lake Group — new evidence for Paleoproterozoic collisional tectonics in the southern Hottah terrane, northwestern Canadian Shield: Canadian Journal of Earth Sciences, v. 42, p. 833-845.

Gandhi, S.S., Prasad, N., and Charbonneau, B.W., 1996, Geological and geophysical signatures of a large polymetallic exploration target at Lou Lake, southern Great Bear magmatic zone, Northwest Territories, *in* Current Research, Part E: Geological Survey of Canada, Current Research 1996-E, p. 147-158.

Gandhi, S.S., Mortensen, J.K., Prasad, N., and van Breemen, O., 2001, Magmatic evolution of the southern Great Bear continental arc, northwestern Canadian Shield: Geochronological constraints: Canadian Journal of Earth Sciences, v. 38, p. 767-785.

Gates, B.I., 1991, Sudbury Mineral Occurrence Study: Ontario Geological Survey, Open File Report 5771, 235 p.

Gauthier, L., Hall, G., Stein, H., and Schaltegger, U., 2001, The Osborne deposit, Cloncurry district: A 1595 Ma Cu-Au skarn deposit, *in* Williams, P.J., ed., A Hydrothermal Odyssey: Extended Conference Abstracts, Economic Geology Research Unit Contribution 59, p. 58-59.

Gauthier, M., Chartrand, F., Cayer, A., and David, J., 2004, The Kwyjibo Cu-REE-U-Au-Mo-F property, Quebec: A Mesoproterozoic polymetallic iron oxide deposit in the Northeastern Grenville Province: Economic Geology, v. 99, p. 1177-1196.

Gelcich, S., Davis, D.W., and Spooner, E.T.C., 2005, Testing the apatite-magnetite geochronometer: U-Pb and $^{40}Ar/^{39}Ar$ geochronology of plutonic rocks, massive magnetite-apatite tabular bodies, and IOCG mineralization in Northern Chile: Geochimica et Cosmochimica Acta, v. 69, p. 3367-3384.

Geological Survey of Newfoundland and Labrador, 2005, National Mineral Inventory Number 013J/12/U 001, 001M/10/Cu 005, 001M/12/Cu 006: Geological Survey of Newfoundland and Labrador, National Mineral Inventory website, http://www.nr.gov.nl.ca/mines&en/geosurvey/mods/helptext/#Intro.

Giles, C.W., 1988, Petrogenesis of the Proterozoic Gawler Range volcanics, South Australia: Precambrian Research, v. 40-41, p. 407-427.

Giles, D., Betts, P., and Lister, G., 2002, Far-field continental backarc setting for the 1.80-1.67 Ga basins of northeastern Australia: Geology, v. 30, p. 823-826.

Goad, R.E., Mumin, A.H., Duke, N.A., Neale, K.L., and Mulligan, D.L., 2000, Geology of the Proterozoic iron oxide-hosted NICO cobalt-gold-bismuth, and Sue-Dianne copper-silver deposits, southern Great Bear Magmatic Zone, Northwest Territories, Canada, *in* Porter, T.M., ed., Hydrothermal Iron Oxide Copper-Gold and Related Deposits: A Global Perspective: PGC Publishing, Adelaide, v. 1, p. 249-267.

Goff, B.H., Weinberg, R., Groves, D.I., Vielreicher, N.M., and Fourie, P.J., 2004, The giant Vergenoeg fluorite deposit in a magnetite-fluorite-fayalite-REE pipe: A hydrothermally altered carbonatite-related pegmatoid?: Mineralogy and Petrology, v. 80, p. 173-199.

Gow, P.A., Wall, V.J., Oliver, N.H.S., and Valenta, R.K., 1994, Proterozoic iron oxide (Cu-U-Au-REE) deposits: Further evidence of hydrothermal origins: Geology, v. 22, p. 633-636.

Gower, C.F., McConnell, J.W., and van Nostrand, T., 1995, New mineral-exploration targets in the Pinware Terrane, Grenville Province, southwest Labrador: Newfoundland and Labrador Department of Mines and Energy, Geological Survey Branch, Report No. 95-1, p. 15-24.

Groves, D.I., and Vielreicher, N.M., 2001, The Phalaborwa (Palabora) carbonatite-hosted magnetite-copper sulfide deposit, South Africa: An end-member of the iron-oxide copper-gold-rare earth element deposit group?: Mineralium Deposita, v. 36, p 189-194.

Halden, N.M., and Fryer, B.J., 1999, Geochemical characteristics of the Eden Lake Complex: Evidence for anorogenic magmatism in the Trans-Hudson Orogen: Canadian Journal of Earth Sciences, v. 36, p. 91-103.

Harmer, R.E., 2000, Mineralisation of the Phalaborwa Complex and the carbonatite connection, *in* Porter, T.M., ed., Hydrothermal Iron Oxide Copper-Gold and Related Deposits: A Global Perspective: PGC Publishing, Adelaide, v. 1, p. 331-340.

Harris, J.R., Wilkinson, L., Heather, K., Fumerton, S., Bernier, M.A., Ayer, J., and Dahn, R., 2001, Application of GIS processing techniques for producing mineral prospectivity maps-A case study: Mesothermal Au in the Swayze Greenstone Belt, Ontario, Canada: Natural Resources Research, v. 10, p. 91-124.

Haynes, D.W., 2000, Iron oxide copper (-gold) deposits: Their position in the ore deposit spectrum and modes of origin, *in* Porter, T.M., ed., Hydrothermal Iron Oxide Copper-Gold and Related Deposits: A Global Perspective: PGC Publishing, Adelaide, v. 1, p. 71-90.

Haynes, D.W., Cross, K.C., Bills, R.T., and Reed, M.H., 1995, Olympic Dam ore genesis: A fluid-mixing model: Economic Geology, v. 90, p. 281-307.

Herrington, R., Smith, M., Maslennikov, V., Belogub, E., and Armstrong, R., 2002, A short review of Palaeozoic hydrothermal magnetite iron-oxide deposits of the south and central Urals and their geological setting, *in* Porter, T.M., ed., Hydrothermal Iron Oxide Copper-Gold and Related Deposits: A Global Perspective: PGC Publishing, Adelaide, v. 2, p. 343-353.

Hildebrand, R.S., 1986, Kiruna-type deposits: their origin and relationship to intermediate subvolcanic plutons in the Great Bear magmatic zone, northwest Canada: Economic Geology, v. 81, p. 640-659.

Hildebrand, R.S., Hoffman, P.F., and Bowring, S.A., 1987, Tectonomagmatic evolution of the 1.9-Ga Great Bear magmatic zone, Wopmay orogen, northwestern Canada: Journal of Volcanology and Geothermal Research, v. 32, p. 99-118.

Hitzman, M.C., 2000, Iron oxide-Cu-Au deposits: What, where, when, and why?, *in* Porter, T.M., ed., Hydrothermal Iron Oxide Copper-Gold and Related Deposits: A Global Perspective: PGC Publishing, Adelaide, v. 1, p. 9-25.

Hitzman, M.W., Oreskes, N., and Einaudi, M.T., 1992, Geological characteristics and tectonic setting of Proterozoic iron oxide (Cu-U-Au-LREE) deposits: Precambrian Research, v. 58, p. 241-287.

Hoffman, P., and Hall, L., 1993, Geology, Slave Craton and Environs, District of Mackenzie, Northwest Territories: Geological Survey of Canada, Open File 2559, scale 1:1 000 000.

Hunt, J.A., Baker, T., and Thorkelson, D., 2005, Regional-scale Proterozoic IOCG-mineralized breccia systems: Examples from the Wernecke Mountains, Yukon, Canada: Mineralium Deposita, v. 40, p. 492-514.

Injoque, J.E., 2002, Fe oxide-Cu-Au deposits in Peru, an integrated view, *in* Porter, T.M., ed., Hydrothermal Iron Oxide Copper-Gold and Related Deposits: A Global Perspective: PGC Publishing, Adelaide, v. 2, p. 97-113.

Johnson, E., and Selleck, B., 2005, The Nature and Significance of the Carthage-Colton Shear Zone and Related Late-to-Post Tectonic Granites and Ore Deposits; Adirondack Mountains, New York: Friends of the Grenville, Field Trip Guidebook, 46 p.

Johnson, J.P., and Cross, K.C., 1995, U-Pb geochronological constraints on the genesis of the Olympic Dam Cu-U-Au-Ag deposit, South Australia: Economic Geology, v. 90, p. 1046-1063.

Johnson, J.P., and McCulloch, M.T., 1995, Sources of mineralizing fluids for the Olympic Dam deposit (South Australia): Sm-Nd isotopic constraints: Chemical Geology, v. 121, p. 177-199.

Kerrich, R., Goldfarb, R., Groves, D., and Garwin, S., 2000, The geodynamics of world-class gold deposits: Characteristics, space-time distribution and origins: Reviews of Society of Economic Geologists, v. 13, p. 501-551.

Ketchum, J.W.F., Culshaw, N.G., and Barr, S.M., 2002, Anatomy and orogenic history of a Paleoproterozoic accretionary belt: The Makkovick Province, Labrador, Canada: Canadian Journal of Earth Sciences, v. 39, p. 711-730.

Kirkham, R.V., and Sinclair, W.D., 1996, Gîtes porphyriques de cuivre, de molybdène, d'or, de tungstène, d'étain et d'argent, *in* Eckstrand, O.R., Sinclair, W.D., and Thorpe, R.I., eds., Geology of Canada Mineral Deposit Types: Geological Survey of Canada, Geology of Canada, No. 8, p. 468-495.

Knight, J., Lowe, J., Sojen, J., Cameron, J., Merrillees, J., Nag, S., Shah, N., Dua G., and Jhala, K., 2002, The Khetri copper belt, Rajasthan: Iron oxide copper-gold terrane in the Proterozoic of NW India, *in* Porter, T.M., ed., Hydrothermal Iron Oxide Copper-Gold and Related Deposits: A Global Perspective: PGC Publishing, Adelaide, v. 2, p. 321-341.

Lamothe, D., and Beaumier, M., 2001, Évaluation du potentiel en minéralisations de type Olympic Dam-Kiruna de la région du lac Manitou (SNRC 22I): Ministère des Ressources naturelles, Québec, EP 2001-01, including annotated map SI-22I-PPO-01I at 1:250 000.

Laughton, J.R., Thorkelson, D.J., Brideau, M-A., Hunt, J.A., and Marshall, D.D., 2005, Early Proterozoic orogeny and exhumation of Wernecke Supergroup revealed by vent facies of Wernecke Breccia, Yukon, Canada: Canadian Journal of Earth Sciences, v. 42, p. 1033-1044.

Leroy, A.J., 1992, Palabora - not just another copper mine: Mineral Industry International, v. 1005, p. 14-19.

Lyons, P., Goleby, B., Jones, L., Drummond, B., and Korsch, R., 2004, Imaging the tectonic environment of the giant Olympic Dam iron oxide-copper-gold deposit, South Australia: 11th Deep Seismic Conference, Program with Abstracts, p. 72.

MacCready, T., Goleby, B.R., Goncharov, A., Drummond, B.J., and Lister, G.S., 1998, A framework of overprinting orogens based on interpretation of the Mount Isa deep seismic transect: Economic Geology, v. 93, p. 1422-1434.

Mandler, H.A.F., and Clowes, R.M., 1997, Evidence for extensive tabular intrusions in the Precambrian shield of Western Canada; a 160-km-long sequence of bright reflections: Geology, v. 25, p. 271-274.

Mark, G., and Foster, D.R.W., 2000, Magmatic-hydrothermal albite-actinolite-apatite-rich rocks from the Cloncurry district, NW Queensland, Australia: Lithos, v. 51, p. 223-245.

Mark, G., Oliver, N.H.S., Williams, P.J., Valenta, R.K., and Crookes, R.A., 2000, The evolution of the Ernest Henry Fe-oxide-(Cu-Au) hydrothermal system, *in* Porter, T.M., ed., Hydrothermal Iron Oxide Copper-Gold and Related Deposits: A Global Perspective: PGC Publishing, Adelaide, v. 1, 123-136.

Marschik, R., and Fontboté, L., 2001, The Candelaria-Punta del Cobre iron oxide Cu-Au (-Zn-Ag) deposits, Chile: Economic Geology, v. 96, p. 1799-1826.

Marschik, R., Leveille, R.A., and Martin, W., 2000, La Candelaria and the Punta del Cobre district, Chile: Early Cretaceous iron-oxide Cu-Au (-Zn-Ag) mineralization, *in* Porter, T.M., ed., Hydrothermal Iron Oxide Copper-Gold and Related Deposits: A Global Perspective: PGC Publishing, Adelaide, v. 1, p. 163-175.

Marshall, L.J., 2003, Brecciation within the Mary Kathleen Group of the Eastern Succession, Mt Isa Block, Australia. Implications of District-Scale Structural and Metasomatic Processes for Fe-Oxide-Cu-Au Mineralisation: Ph.D. thesis, James Cook University, Townsville, Australia, 323 p.

Marshall, L.J., and Downie, C.C., 2002, Geological Report for the Iron Range Project, Deli 1-8, FeO 1-30, HC 1-10, IOx 1-12, Ir 1-36, Luke 1-8, and Tck 1-8 Properties: Eagle Plains Resources Ltd., Unpublished Report, 79 p.

Marshall, L.J., Smith, R., Wilton, D., and Butler, R., 2003, Geology and Preliminary Exploration Results, Central Mineral Belt, Labrador, Canada: Altius Minerals Corporation and Fronteer Development Group, Unpublished Report, 52 p.

Mathur, R., Marschik, R., Ruiz, J., Munizaga, F., Leveille, R.A., and Martin, W., 2002, Age of mineralization of the Candelaria Fe oxide Cu-Au deposit and the origin of the Chilean iron belt, based on Re-Os isotopes: Economic Geology, v. 97, p. 59-71.

Morton, A.C., and Hallsworth, C.R., 1999, Processes controlling the composition of heavy mineral assemblages in sandstones: Sedimentary Geology, v. 124, p. 3-29.

MRNF, 2005, Fiches de gîtes: SIGEOM databank, Ministère des Ressources naturelles et de la Faune, Québec. (http://www.mrnf.gouv.qc.ca/produits-services/mines.jsp).

Mumin, A.H., 2002, Discovery of a carbonatite complex at Eden Lake (NTS 64C9), Manitoba: Manitoba Geological Survey, Report of Activities 2002, p. 187-197.

Mumin, A.H., and Corriveau, L., 2004, The Eden deformation corridor and polymetallic mineral belt: Trans Hudson Orogen, Leaf Rapids District, Manitoba: Manitoba Geological Survey, Report of Activities 2004, p. 69-91.

Mumin, A.H., and Perrin, J.A., 2005, Scoping study for the hydrothermal iron-oxide copper-gold (IOCG) deposit type in Manitoba: Summary of regional investigations: Manitoba Geological Survey, Manitoba Industry, Economic Development and Mines, Geoscientific Paper GP2005-2, 38 p.

Mumin, A.H., and Stewart, P.W., 2006, Mineral potential of the Contact Lake belt: Report on 2005 exploration in the Contact Lake Belt, Great Bear Magmatic Zone IOCG project, District of Mackenzie, Northwest Territories, Canada, NTS 86F/13 & 86K/04, 66° N, 117° 45' E: Alberta Star Development Corporation, Unpublished Report, 233 p.

New Brunswick Department of Mines, 2005, New Brunswick deposit database, URN 590, Lepreau Iron: New Brunswick Department of Mines web site. http://dnre-mrne.gnb.ca/MineralOccurrence.

Nisbet, B., Cooke, J., Richards, M., and Williams, C., 2000, Exploration for iron oxide copper gold deposits in Zambia and Sweden; Comparison with the Australian experience, *in* Porter, T.M., ed., Hydrothermal Iron Oxide Copper-Gold and Related Deposits: A Global Perspective: PGC Publishing, Adelaide, v. 1, p. 297-308.

O'Driscoll, E.S.T., 1986, Observations of the lineament-ore relation: Philosophical Transactions of the Royal Society of London, v. A317, p. 195-218.

Ohmoto, H., 2003, Nonredox transformations of magnetite-hematite in hydrothermal systems: Economic Geology, v. 98, p. 157-161.

Oliver, N.H.S., Cleverley, J.S., Mark, G., Pollard, P.J., Fu, B., Marshall, L.J., Rubenach, M.J., Williams, P.J., and Baker, T., 2004, The role of sodic alteration in the genesis of iron oxide-copper-gold deposits; eastern Mt Isa Block, Australia: Economic Geology, v. 99, p. 1145-1176.

O'Reilly, G.A., 1996, Vein and Breccia Deposits of Fe-Cu-Co-Ni-Ba-Au Associated with the Cobequid-Chedabucto Fault Zone, Nova Scotia: Nova Scotia Information Circular ME 47, 2 p.

——— 2002, From the Nova Scotia mineral inventory files, the Cobalt-bearing Bass River magnetite deposits: Nova Scotia Information Circular, v. 19, No 1, p. 5.

Oreskes, N., and Einaudi, M.T., 1992, Origin of hydrothermal fluids at Olympic Dam: Preliminary results from fluid inclusions and stable isotopes: Economic Geology, v. 87, p. 64-90.

Orris, G.J., and Grauch, R.I., 2002, Rare Earth Element Mines, Deposits, and Occurrences: US Geological Survey, Open File Report 02-189, 174 p.

Oxiana Limited, 2005, Prominent Hill resource upgraded, Australian stock exchange and media release, 19 July 2005, website www.oxiana.com.au.

Partington, G.A., and Williams, P.J., 2000, Proterozoic lode gold and (iron)-copper-gold deposits: A comparison of Australian and global examples: Reviews of Society of Economic Geologists, v. 13, p. 69-101.

Pitcher, W.S., 1993, The Nature and Origin of Granite: Blackie Academic and Professional, London, 352 p.

Pollard, P.J., 2000, Evidence of a magmatic fluid source for iron oxide-Cu-Au mineralisation, *in* Porter, T.M., ed., Hydrothermal Iron Oxide Copper-Gold and Related Deposits: A Global Perspective: PGC Publishing, Adelaide, v. 1, p. 27-41.

Porter, T.M. (editor), 2000, Hydrothermal Iron Oxide Copper-Gold and Related Deposits: A Global Perspective, Volume 1: PGC Publishing, Adelaide, 349 p.

——— 2002a, Hydrothermal Iron Oxide Copper-Gold and Related Deposits: A Global Perspective, Volume 2: PGC Publishing, Adelaide, 377 p.

Porter, T.M., 2002b, Iron oxide alteration/mineralising systems and copper-gold and related mineralisation, *in* Porter, T.M., ed., Hydrothermal Iron Oxide Copper-Gold and Related Deposits: A Global Perspective: PGC Publishing, Adelaide, v. 2, p. 3-7.

Puffer, J.H., and Gorring, M.L., 2005, The Edison magnetite deposits in the context of pre-syn-and post-orogenic metallogenesis in the Grenville Highlands of New Jersey: Canadian Journal of Earth Sciences, v. 42, p. 1735-1748.

Ray, G.E., and Lefebure, D.V., 2000, A synopsis of iron oxide ±Cu ±Au ±P ±REE deposits of the Candelaria-Kiruna-Olympic Dam family, *in* Geological Fieldwork 1999: British Columbia Ministry of Energy and Mines, Paper 2000-1, p. 267-272.

Ray, G.E., and Webster, I.C.L., 2000, The Heff prospect at Heffley Lake, south-central B.C. (092INE096): An unusual example of a mafic-ultramafic-related Cu-Au-REE-bearing magnetite skarn, *in* Geological Fieldwork 1999: British Columbia Geological Survey Branch, Paper 2000-1, p. 273-286.

Reardon, N.C., 1990, Altered and mineralized rocks at Echo Bay, N.W.T., and their relationship to the Mystery Island intrusive suite, *in* Current Research, Part C, Canadian Shield: Geological Survey of Canada, Paper 90-1C, p. 143-150.

Reeve, J.S., Cross, K.C., Smith, R.N., and Oreskes, N., 1990, The Olympic Dam copper-uranium-gold-silver deposit, South Australia, *in* Hughes, F., ed., Geology of Mineral Deposits of Australia and Papua New Guinea: Australian Institute of Mining and Metallurgy, Monograph 14, p. 1009-1035.

Requia, K., and Fontboté, L., 2000, The Salobo iron-oxide copper-gold deposit, Carajás, Northern Brazil, *in* Porter, T.M., ed., Hydrothermal Iron Oxide Copper-Gold and Related Deposits: A Global Perspective: PGC Publishing, Adelaide, v. 1, p. 225-236.

Requia, K., Stein, H., Fontboté, L., and Chiaradia, M., 2003, Re-Os and Pb-Pb geochronology of the Archean Salobo iron oxide copper-gold deposit, Carajás mineral province, northern Brazil: Mineralium Deposita, v. 38, p. 727-738.

Reynolds, L.J., 2000, Geology of the Olympic Dam Cu-U-Au-Ag-REE deposit, *in* Porter, T.M., ed., Hydrothermal Iron Oxide Copper-Gold and Related Deposits: A Global Perspective: PGC Publishing, Adelaide, v. 1, p. 93-104.

Roberts, D.E., and Hudson, G.R.T., 1983, The Olympic Dam copper-uranium-gold-silver deposit, Roxby Downs, South Australia: Economic Geology, v. 78, p. 799-822.

Rogers, M.C., Thurston, P.C., Fyon, J.A., Kelly, R.I., and Breaks, F.W., 1995, Descriptive Mineral Deposit Models of Metallic and Industrial Deposit Types and Related Mineral Potential Assessment Criteria: Ontario Geological Survey, Open File Report 5916, 241 p.

Romer, R.L., Martinsson, O., and Perdahl, J.-A., 1994, Geochronology of the Kiruna iron ores and hydrothermal alterations: Economic Geology, v. 89, p. 1249-1261.

Ronze, P.C., Soares, A.D.V., dos Santos, M.G.S., and Barreira, C.F., 2000, Alemão copper-gold (-U-REE) deposit, Carajás, Brazil, *in* Porter, T.M., ed., Hydrothermal Iron Oxide Copper-Gold and Related Deposits: A Global Perspective: PGC Publishing, Adelaide, v. 1, p. 191-202.

Rotherham, J.F., Blake, K.L., Cartwright, I., and Williams, P.J., 1998, Stable isotope evidence for the origin of the Mesoproterozoic Starra Au-Cu deposit, Cloncurry district, northwest Queensland: Economic Geology, v. 93, p. 1435-1449.

Ryan, P.J., Lawrence, A.L., Jenkins, R.A., Matthews, J.P., Zamora, J.C., Marino, E., and Urqueta, I., 1995, The Candelaria copper-gold deposit, Chile, *in* Pierce, F.W., and Bolm, J.G., eds., Porphyry Copper Deposits of the American Cordillera: Arizona Geological Society Digest, v. 20, p. 625-645.

Schandl, E.S., and Davis, D.W., 2000, Geochronology of the Sullivan deposit: U-Pb and Pb-Pb ages of zircons and titanites, *in* Lydon, J., Höy, T., Slack, J.F., and Knapp, M.E., eds., The Geological Environment of the Sullivan Deposit, British Columbia: Geological Association of Canada, Mineral Deposits Division, Special Publication No. 1, p. 127-135.

Schandl, E.S., Gorton, M.P., and Davis, D.W., 1994, Albitization at 1700 ±2 Ma in the Sudbury - Wanapitei Lake area, Ontario: Implications for deep-seated alkalic magmatism in the Southern Province: Canadian Journal of Earth Sciences, v. 31, p. 597-607.

Schärrer, U., Krogh, T.E., Wardle, R.J., Ryan, B., and Gandhi, S., 1988, U-Pb ages of early and middle Proterozoic volcanism and metamorphism in the Makkovik Orogen, Labrador: Canadian Journal of Earth Sciences, v. 25, p. 1098-1107.

Schneiders, B.R., Scott, J.F., Smyk, M.C., and O'Brien, M.S., 2003, Report of Activities 2002, Resident Geologist Program, Thunder Bay South Regional Resident Geologist Report: Thunder Bay South District: Ontario Geological Survey, Open File Report 6112, 55 p.

Selleck, B., McLelland, J., and Hamilton, M.A., 2004, Magmatic-hydrothermal leaching and origin of late- to post-tectonic quartz-rich rocks, Adirondack Highlands, New York, *in* Tollo, R.P., Corriveau, L., McLelland, J., and Bartholomew, M, eds., Proterozoic Tectonic Evolution of the Grenville Orogen in North America: Geological Society of America, Memoir 197, p. 379-390.

Sillitoe, R.H., 2002, Some metallogenic features of gold and copper deposits related to alkaline rocks and consequences for exploration: Mineralium Deposita, v. 37, p. 4-13.

——— 2003, Iron oxide-copper-gold deposits: An Andean view: Mineralium Deposita, v. 38, p. 787-812.

Simard, M., Beaudoin, G., Bernard, J., and Hupé, A., 2006, Metallogeny of the Mont-de-l'Aigle IOCG deposit, Gaspé Peninsula, Québec, Canada: Mineralium Deposita, v. 41, p. 607-636.

Sinclair, G.S., Barr, S.M., Culshaw, N.G., and Ketchum, J.W.F., 2002, Geochemistry and age of the Aillik Group and associated plutonic rocks, Makkovik Bay area, Labrador: Implications for tectonic development of the Makkovik Province: Canadian Journal of Earth Sciences, v. 39, p. 731-748.

Skirrow, R.G., 2000, Gold-copper-bismuth deposits of the Tennant Creek district, Australia: A reappraisal of diverse high-grade systems, *in* Porter, T.M., ed., Hydrothermal Iron Oxide Copper-Gold and Related Deposits: A Global Perspective: PGC Publishing, Adelaide, v. 1, p. 149-160.

——— 2004, Iron oxide Cu-Au deposits: An Australian perspective on their unifying characteristics, *in* McPhie, J., and McGoldrick, P., eds, Dynamic Earth: Past, Present and Future: Abstracts of the 17th Australian Geological Convention, Hobart, Tasmania, February 8 - 13, Geological Society of Australia, Abstract No. 73, p. 121.

Skirrow, R.G., and Walshe, J.L., 2002, Reduced and oxidized Au-Cu-Bi iron oxide deposits of the Tennant Creek Inlier, Australia: An integrated geologic and chemical model: Economic Geology, v. 97, p. 1167-1202.

Skirrow, R.G., Bastrakov, E., Raymond, O.L., Davidson, G., and Heithersay, P., 2002, The geological framework, distribution and controls of Fe-oxide Cu-Au mineralisation in the Gawler Craton, South Australia: Part II - Alteration and mineralisation, *in* Porter, T.M., ed., Hydrothermal Iron Oxide Copper-Gold and Related Deposits: A Global Perspective: PGC Publishing, Adelaide, v. 2, p. 33-47.

Sleigh, D., 2002, The Selwyn Line tabular iron-copper-gold mineralised system, Mount Isa Inlier, NW Queensland, Australia, *in* Porter, T.M., ed., Hydrothermal Iron Oxide Copper-Gold and Related Deposits: A Global Perspective: PGC Publishing, Adelaide, v. 2, p. 77-93.

Smith, R.J., 2002, Geophysics of iron-oxide copper-gold systems, *in* Porter, T.M., ed., Hydrothermal Iron Oxide Copper-Gold and Related Deposits: A Global Perspective: PGC Publishing, Adelaide, v. 2, p. 357-367.

Smith, M., and Chengyu, W., 2000, The geology and genesis of the Bayan Obo Fe-REE-Nb deposit: A review, *in* Porter, T.M., ed., Hydrothermal Iron Oxide Copper-Gold and Related Deposits: A Global Perspective: PGC Publishing, Adelaide, v. 1, p. 271-281.

Smith, M.P., Henderson, P., and Campbell, L.S., 2000, Fractionation of the REE during hydrothermal processes: Constraints from the Bayan Obo Fe-REE-Nb deposit, Inner Mongolia, China: Geochimica et Cosmochimica Acta, v. 64, p. 3141-3160.

Souza, L.H., and Vieira, E.A., 2000, Salobo 3 Alpha deposit: Geology and mineralisation, *in* Porter, T.M., ed., Hydrothermal Iron Oxide Copper-Gold and Related Deposits: A Global Perspective: PGC Publishing, Adelaide, v. 1, p. 213-224.

Stinson, P., and Brown, D.A., 1995, Iron Range deposits, Southeastern British Columbia (82FI I), *in* Fieldwork 1994: British Columbia Geological Survery, Paper 1995-1, p. 127-134.

Swinden, H.S., Wardle, R.J., Davenport, P.H., Gower, C.F., Kerr, A., Meyer, J.R., Miller, R.R., Nolan, L., Ryan, A.B., and Wilton, D.H.C., 1991, Mineral exploration opportunities in Labrador: A perspective for the 1990s: Government of Newfoundland and Labrador, Geological Survey Branch, Current Research Report 91-1, p. 349-390.

Tallarico, F.H.B., McNaughton, N.J., Groves, D.I., Fletcher, I.R., Figueiredo, B.R., Carvolho, J.B., Rego, J.L., and Nunes, A.R., 2004, Geological and SHRIMP II U-Pb constraints on the age and origin of the Breves Cu-Au-(W-Bi-Sn) deposit, Carajás, Brazil: Mineralium Deposita, v. 39, p. 68-86.

Tazava, E., and de Oliveira, C.G., 2000, The Igarapé Bahia Au-Cu-(REE-U) deposit, Carajás mineral province, northern Brazil, *in* Porter, T.M., ed., Hydrothermal Iron Oxide Copper-Gold and Related Deposits: A Global Perspective: PGC Publishing, Adelaide, v. 1, p. 203-212

Teale, G.S., and Fanning, C.M., 2000, The Portia - North Portia Cu-Au(-Mo) prospect, South Australia; timing of mineralisation, albitisation and origin of ore fluid, *in* Porter, T.M., ed., Hydrothermal Iron Oxide Copper-Gold and Related Deposits: A Global Perspective: PGC Publishing, Adelaide, v. 1, p. 137-147.

Thorkelson, D.J., 2000, Geology and Mineral Occurrences of the Slats Creek, Fairchild Lake and "Dolores Creek" Areas, Wernecke Mountains, Yukon Territory (106 D/16, 106 C/13, 106 C/14): Yukon, Indian and Northern Affairs Canada, Exploration and Geological Services Division, Bulletin 10, 73 p.

Thorkelson, D.J., Mortensen, J.K., Davidson, G.J., Creaser, R.A., Perez, W.A., and Abbott, J.G., 2001, Mesoproterozoic intrusive breccias in

Yukon, Canada: The role of hydrothermal systems in reconstructions of North America and Australia: Precambrian Research, v. 111, p. 31-56.

Thorkelson, D.J., Abbott, J.G., and Roots, C.F., 2003, Proterozoic prism arrests suspect terranes: Insights into the ancient Cordilleran margin from seismic reflection data: Comment: GSA Today, p. 18.

Tornos, F., and Casquet, C., 2005, A new scenario for related IOCG and Ni(Cu) mineralization: the relationship with giant mid-crustal mafic sills, Variscan Iberian Massif: Terra Nova, v. 17, p. 236-241.

Tortosa, D., and Moss, R., 2004, Geology and Exploration of the Coppercorp Property, Sault Ste. Marie Mining Division, Ontario: Nikos Explorations Ltd., Unpublished Report, 158 p. http://nikosexplorations.com/_resources/Coppercorp2004Report.PDF.

Venkataraman, G., Babu Madhavan, B., Ratha, D.S., Antony, J.P., Goyal, R.S., Banglani, S., and Sinha Roy, S., 2000, Spatial modeling for base-metal mineral exploration through integration of geological data sets: Natural Resources Research, v. 9, p. 27-42.

Vielreicher, N.M., Groves, D.I., and Vielreicher, R.M., 2000, The Phalaborwa (Palabora) deposit and its potential connection to iron-oxide copper-gold deposits of Olympic Dam type, *in* Porter, T.M., ed., Hydrothermal Iron Oxide Copper-Gold and Related Deposits: A Global Perspective: PGC Publishing, Adelaide, v. 1, p. 321-329.

Wang, D.H., Yang, J., Yan, S., Xu, J., Chen, Y., Pu, G., and Luo, Y., 2001, A special orogenic-type rare earth element deposit in Maoniuping, Sichuan, China: Geology and geochemistry: Resource Geology, v. 51, p. 177-188.

Wanhainen, C., Broman, C., and Martinsson, O., 2003, The Aitik Cu-Au-Ag deposit in northern Sweden: a product of high salinity fluids: Mineralium Deposita, v. 38, p. 715-726.

Webb, D.R., 2001, Report on the Line Cutting, Prospecting and Geological Mapping and Sampling Program on the Cobalt 1 to 5 Claims, Tag Numbers F54375, F54513, F54514, F59519, and F59516, Respectively, Cameron Bay Area, District of Mackenzie, Northwest Territories: Department of Indian and Northern Affairs, Assessment Report No. 08486, 26 p.

Weihed, P., Arndt, N., Billström, K., Duchesne, J.-C., Eilu, P., Martinsson, O., Papunen, H., and Lahtinen, R., 2005, Precambrian geodynamics and ore formation: The Fennoscandian Shield: Ore Geology Reviews, v. 27, p. 273-322.

Western Mining Corporation, 2003, WMC Resources Business Performance Report, Western Mining Corporation, 60 p.

——— 2004, Press release, October 31th.

Whalen, J.B., Currie, K.L., and Chappell, B.W., 1987, A-type granites: Geochemical characteristics, discrimination and petrogenesis: Contributions to Mineralogy and Petrology, v. 95, p. 407-419.

White, D.J., Zwanzig, H.V., and Hajnal, Z., 2000, Crustal suture preserved in the Paleoproterozoic Trans-Hudson orogen, Canada: Geology, v. 28, p. 527-530.

White, D.J., 2005, High-temperature, low-pressure metamorphism in the Kisseynew domain, Trans-Hudson orogen: crustal anatexis due to tectonic thickening?: Canadian Journal of Earth Sciences, v. 42, p. 707-721.

Williams, P.J., and Skirrow, R.G., 2000, Overview of iron oxide-copper-gold deposits in the Curnamona Province and Cloncurry District (Eastern Mount Isa Block), Australia, *in* Porter, T.M., ed., Hydrothermal Iron Oxide Copper-Gold and Related Deposits: A Global Perspective: PGC Publishing, Adelaide, v. 1, p. 105-122.

Williams, P.J., Barton, M.D., Johnson, D.A., Fontboté, Lluis, DeHaller, A., Mark, G., Oliver, N.H.S., and Marschik, R., 2005, Iron oxide copper-gold deposits: Geology, space-time distribution, and possible mode of origin, *in* Hedenquist, J.W., Thompson, J.F.H., Goldfarb, R.J., and Richards, J.P., eds., Economic Geology 100th Anniversary Volume: Society of Economic Geologists, p. 371-405.

Wodicka, N., David, J., Parent, M., Gobeil, A., and Verpaelst, P., 2003, Géochronologie U-Pb et Pb-Pb de la région de Sept-Îles - Natashquan, Province de Grenville, Moyenne-Côte-Nord, *in* Brisebois, D., and Clark, T., eds., Géologie et Ressources minérales de la partie est de la Province de Grenville: Ministère des Ressources naturelles, de la Faune et des Parcs, Québec, DV 2002-03, p. 59-117.

Woodall, R., 1994, Empiricism and concept in successful mineral exploration: Australian Journal of Earth Sciences, v. 41(1), p. 1-10.

Wyborn, L.A.I., 1998, Younger ca. 1500 Ma granites of the Williams and Naraku batholiths, Cloncurry district, eastern Mt Isa Inlier: Geochemistry, origin, metallogenic significance and exploration indicators: Australian Journal of Earth Sciences, v. 45, p. 397-411.

——— 2002, Granites and copper gold metallogenesis in the Australian Proterozoic, *in* Budd, A.R., Wyborn, L.A.I., and Bastrakova, I.V., eds, The Metallogenic Potential of Australian Proterozoic Granites: Geoscience Australia, Record 2001/12, p. 5-33.

Yang, X.-M., Yang, X.-Y., Zheng, X.-Y., and Le Bas, M.J., 2003, A rare earth element-rich carbonatite dyke at Bayan Obo, Inner Mongolia, North China: Mineralogy and Petrology, v. 78, p. 93-110.

Zack, T., Kronz, A., Foley, S.F., and Rivers, T., 2002, Trace element abundances in rutiles from eclogites and associated garnet mica schists: Chemical Geology, v. 184, p. 97-122.

Metallogeny of the Sudbury Mining Camp, Ontario

Doreen E. Ames[1] and Catharine E. G. Farrow[2]
1. Geological Survey of Canada, 601 Booth Street, Ottawa, Ontario K1A 0E8
2. FNX Mining Company Inc., 1300 Kelly Lake Road, Sudbury, Ontario PE3 5P4
Corresponding author's email: dames@nrcan.gc.ca

Abstract

The Paleoproterozoic (1850 Ma) Sudbury Structure, the remnant of a deformed multi-ring impact basin, is host to an enormous amount of Ni-Cu-PGE sulphide mineralization. Sudbury is Canada's principal base metal mining district with 77 producing and past-producing Ni-Cu-PGE deposits and 11 Cu-Ni-PGE advanced exploration projects with a 1.655 billion tonne past production, reserve, and resource. The Sudbury region is also host to significant pre-1850 Ma mineral resources comprising uranium, Ni-Cu-PGE, Au-Cu, and Au deposits and syn-1850 Ma hydrothermal Zn-Pb-Cu deposits.

The1850 Ma world-class Ni-Cu-PGE deposits and minor hydrothermal Zn-Pb-Cu deposits within the basin are related to the Sudbury Igneous Complex, with over half of the Ni-Cu-PGE in Contact deposits, 25% in Offset deposits, approximately 15% in Frood-Stobie breccia belt deposits, and <10% in "footwall-hosted" deposits. Trace element signatures of the ores tend to be related to host rock (As, Sb) or deposit type (Zn, Cd, Pb). The key to effective exploration in the Sudbury camp is a solid understanding of the 3-D geology followed by appropriate EM/RIM geophysical methods that increase the sampling radius of drillholes.

The recognition of the evolution of the Sudbury area from a Ni-based camp to a polymetallic Ni-Cu-PGE producer is illustrated by the fact that over 50% of the exploration expenditures for 2005 focused on the footwall-hosted Cu-Ni-PGE systems due to the high Cu and PGE grades of the sulphide ore. The result was the discovery of new deposits (Levack Footwall Deposit) in a mature mining camp. New exploration models for low-sulphide, high-PGE (PM Deposit) styles of mineralization are being refined for both the North and South ranges. Base metal deposits are regularly discovered in the Sudbury camp with evolving deposit models, exploration techniques, and genetic models.

Résumé

La structure de Sudbury est le vestige d'un bassin d'impact multi-anneaux déformé datant du Paléoprotérozoïque (1 850 Ma), qui recèle d'énormes quantités de minéralisations sulfurées à Ni-Cu-ÉGP. La région de Sudbury constitue le principal district minier pour les métaux communs au Canada. On y compte 77 gisements de Ni-Cu-ÉGP exploités par le passé ou encore aujourd'hui ainsi que 11 gîtes de Ni-Cu-ÉGP rendus à un stade d'exploration avancé; le volume de la production passée, des réserves et des ressources s'élève à 1,655 milliard de tonnes. La région de Sudbury renferme en outre des ressources minérales formées antérieurement à 1,850 Ma, dont des gîtes d'uranium et de Ni-Cu-ÉGP, de Au-Cu et de Au, ainsi que des gîtes hydrothermaux de Zn-Pb-Cu de formation contemporaine de l'impact météoritique à 1,850 Ma.

Les minéralisations contemporaines de l'impact météoritique à 1,850 Ma, formées des gisements de Ni-Cu-ÉGP de classe mondiale et des gîtes mineurs de Zn-Pb-Cu d'origine hydrothermale à l'intérieur du bassin, sont apparentées au Complexe igné de Sudbury. Plus de la moitié des minéralisations de Ni-Cu-ÉGP appartiennent à des gîtes de la zone de contact du complexe, 25 % à des gîtes associées à des apophyses du complexe, environ 15 % à des gîtes de la zone bréchique de Frood-Stobie et moins de 10 % à des gîtes encaissés dans le mur du complexe. Les signatures en éléments traces des minerais ont tendance à refléter la nature de la roche hôte (As, Sb) ou le type de gîte (Zn, Cd, Pb). Dans le camp minier de Sudbury, le succès de l'exploration repose sur une excellente compréhension de la géologie en 3-D et l'application de méthodes géophysiques EM/RIM appropriées, qui permettent d'étendre le rayon d'investigation des sondages.

La reconnaissance du fait que la région de Sudbury soit passée d'un camp fondé sur la mise en valeur du Ni à une région productrice de Ni-Cu-ÉGP est reflétée par les dépenses d'exploration en 2005, dont plus de 50 % ont été consacrées aux systèmes de Cu-Ni-ÉGP dans les roches du mur du complexe, en raison des teneurs élevées du minerai sulfuré en Cu et en ÉGP. Il en résulte que de nouveaux gisements (gisement de Levack Footwall) ont pu être découverts dans un camp minier parvenu à maturité. De nouveaux modèles d'exploration visant les styles de minéralisations à faible concentration de sulfures et à teneur élevée en ÉGP (gisement de PM) font l'objet de raffinements aussi bien sur le flanc nord que sur le flanc sud. L'amélioration des modèles gîtologiques, des méthodes d'exploration et des modèles génétiques rend possible, à intervalles réguliers, la découverte de gîtes de métaux communs dans le camp minier de Sudbury.

Introduction and History

The Sudbury Structure is the remnant of a deformed 1.85 Ga multi-ring impact crater. This geological feature is of extraordinary economic and scientific significance, but remains enigmatic in many important respects. Sudbury, Canada and Noril'sk-Talnakh, Russia are the greatest Ni-producing districts in the world. The Sudbury ore deposits have produced an estimated US$ 280 billion or CDN$ 330 billion (with a US$/CDN$ exchange rate of 0.85) worth of metal in the past century (using 2005 average metal values) from more than 77 Ni-Cu-PGE deposits and two minor, past-producing hydrothermal Zn-Pb-Cu deposits. The continued prosperity of the Sudbury region depends on mining in the Sudbury structure at increasing depths and on the use of new geoscience knowledge created to find valuable hydrothermal, magmatic Cu-PGE vein deposits at shallow levels.

The Sudbury mining camp was discovered in 1856 when Alexander Murray reported sulphides in a Geological

Ames, D.E., and Farrow, C.E.G., 2007, Metallogeny of the Sudbury mining camp, Ontario, *in* Goodfellow, W.D., ed., Mineral Deposits of Canada: A Synthesis of Major Deposit-Types, District Metallogeny, the Evolution of Geological Provinces, and Exploration Methods: Geological Association of Canada, Mineral Deposits Division, Special Publication No. 5, p. 329-350.

FIGURE 1. Discovery of the supergiant Ni-Cu-PGE Sudbury resource in 1883 during construction the Canadian Pacific Railway.

Survey of Canada report (Murray, 1856). However, the significance of the discovery was not realized until 1883 when a blacksmith working on Canada's first transcontinental railway (Fig. 1) identified sulphides at the present day site of the Murray mine, which resulted in a prospecting and staking rush. From 1888 to 1890, Alfred Barlow and Robert Bell completed field work for the first regional bedrock map of the Sudbury area (Bell, 1891).

Geological and Tectonic Setting

The 1.85 Ga Sudbury structure straddles the boundary between Neoarchean granitic and gneissic rocks of the Superior province to the north and east (North Range) and Paleoproterozoic metasedimentary and metavolcanic rocks of the Southern province to the south (South Range; Fig. 2). The northern footwall rocks consist of the high-grade Levack gneiss complex comprising supracrustal and intrusive units that were deformed and metamorphosed at upper amphibolite to granulite facies in the lower crust during the Neoarchean (Card, 1994; Wodicka, 1997). These rocks form a 0.5 to 5 km wide footwall collar to the northern half of the Sudbury Igneous Complex (SIC) (Milkereit and Group, 1992). The Neoarchean rocks were then intruded by the late Archean (2640 Ma) granite of the Cartier batholith and the 2450 Ma Matachewan mafic dyke swarm (Heaman, 1997), and uplifted during the late Archean or early Proterozoic (Card, 1994; Wodicka, 1997) to lower crustal levels where they were strongly affected by the 1850 Ma impact event. The southern footwall rocks consist of the Paleoproterozoic Huronian Supergroup deposited in an intercontinental rift basin, layered gabbro anorthosite intrusives of the 2490 to 2470 Ma East Bull Lake suite (Krogh et al., 1984; Prevec, 1992; Peck et al., 1993) and 2210 to 2217 Ma gabbroic rocks of the Nipissing suite (Corfu and Andrews, 1986; Noble and Lightfoot, 1992).

Since the recognition of the Sudbury structure as an impact crater by Dietz (1964), support for the impact origin, already overwhelming, has increased. Evidence for the impact event at Sudbury includes 1) the iridium anomaly in the upper 800 m of the crater-filling Onaping Formation, 2) pseudotachylite of the Sudbury breccia that decreases outward in abundance, 3) shatter cones distributed in a radial pattern, and 4) shock metamorphosed minerals in both proximal and distal (~800 km) ejecta (French, 1967; Grieve, 1994; Mungall et al., 2004; Addison et al., 2005) (Fig. 3). Collectively, the geological, geochemical, and petrological observations interpreted as impact features since 1962, combined with significant advances in 2004 and 2005 provide convincing support for an impact origin of the Sudbury structure.

The Sudbury structure consists of 1) the prominent elliptical Sudbury Igneous Complex (SIC) and its associated offset dykes; 2) annularly distributed, brecciated, and shock metamorphosed footwall rocks; and 3) the crater-fill breccias

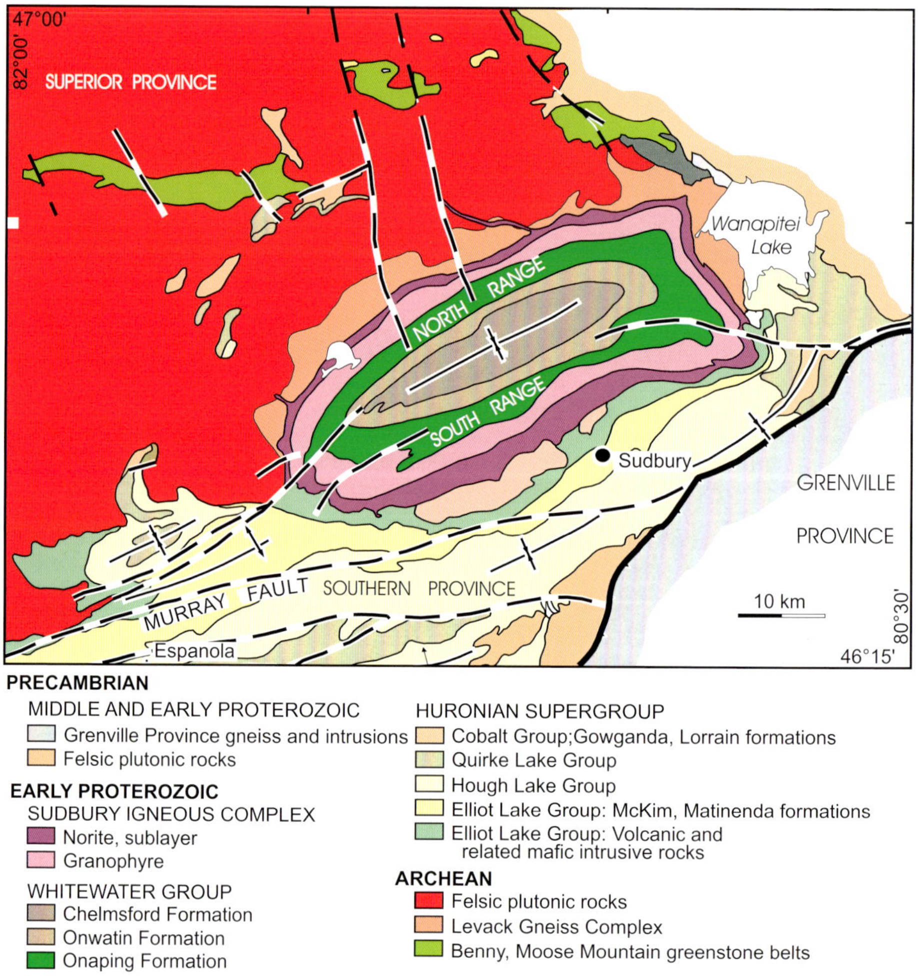

FIGURE 2. Regional geological setting of the Sudbury Structure. The Sudbury Igneous Complex (SIC) straddles the boundary between the Neoarchean Superior Province and the Paleoproterozoic Southern Province.

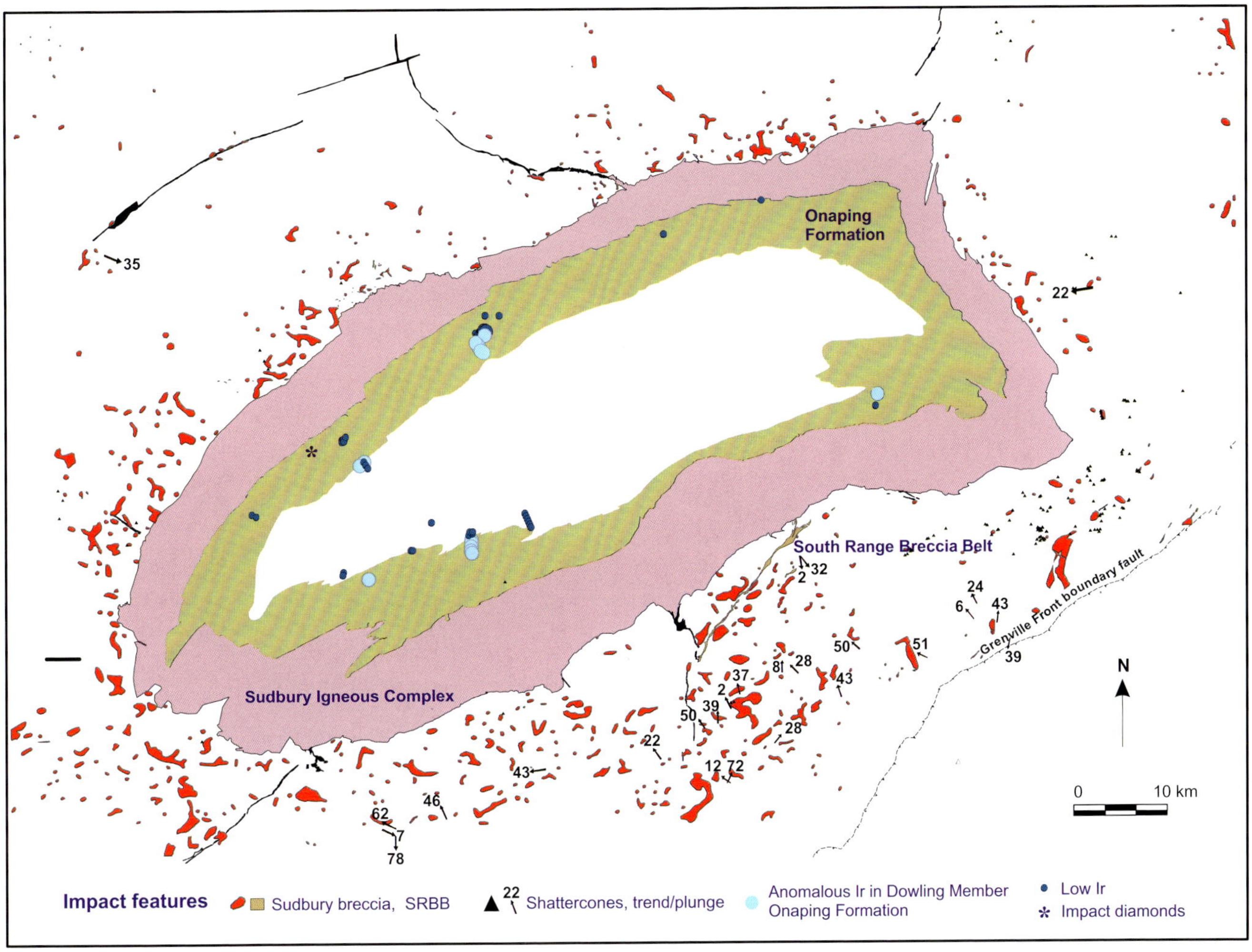

FIGURE 3. Impact-related features associated with the ~250 km diameter, 1850 Ma, Sudbury impact crater. An iridium anomaly is constrained to the three upper units of the Dowling Member within the Onaping Formation and is not present in the overlying Onwatin Formation basinal sediments. The distribution of shattercones and Sudbury breccia pseudotachylite and cataclasite is restricted to the footwall rocks (from Ames et al., 2005).

of the Onaping Formation, which are overlain by sedimentary strata of the Whitewater Group (Figs. 3, 4). The SIC is a Paleoproterozoic (1850±1 Ma), hypabyssal intrusion consisting of a lower ore-bearing sublayer, noritic to gabbroic cumulates, and an upper granophyre produced during impact by melting of the lower crust and carbonaceous sediments (Krogh et al., 1984; Grieve, 1994; Mungall et al., 2004). The initial shock melt is represented by andesite-quartz diorite vitric material preserved in peperitic dykes, bombs, and blocks in the crater-fill that was variably modified through assimilation of country rock fragments to form an evolved shock melt composition for the melt sheet (Ames et al., 2002b).

Offset dykes can extend up to 30 km radially (i.e. maximum represented by the Foy offset) and 14 km concentrically (i.e. Hess offset with a 23 km strike length) from the SIC/footwall contact (Wood and Spray, 1998; Tuchscherer and Spray, 2002) (Fig. 4). The parental composition for the SIC melt sheet, an evolved shock melt, is represented by "quartz diorite" offset dykes (Lightfoot et al., 2001). Though these dykes are traditionally called "quartz diorite", they are compositionally variable between quartz monzodiorite-granodiorite-tonalite (Lightfoot et al., 2001; Ames et al., 2005). Offset structures may or may not contain igneous material as dykes. The Copper Cliff, Worthington, and Foy offsets form a distinct dyke-like feature, whereas others consist of discontinuous pods or segments of igneous material (e.g. proximal Parkin/Whistle, McConnell/Kirkwood, Frood-Stobie breccia belt) with variable amounts of metabreccia, a breccia of uncertain origin, and abundant footwall/Sudbury breccia (Murphy and Spray, 2002).

Pseudotachylite and cataclasite are locally known as Sudbury breccia and extend up to 200 km from the SIC. They consist of brecciated, frictionally comminuted country rocks randomly distributed in a fine-grained, recrystallized matrix. Sudbury breccia is not only critical for understanding crustal processes and estimating the size of the Sudbury impact crater (Spray et al., 2004), but represents an important economic target that is host to Sudbury's largest deposit (Frood-Stobie mine) and Cu-PGE-rich deposits (Fraser-Strathcona Cu zones, McCreedy West Footwall Cu, and McCreedy East 153 zone, Levack Footwall).

Post 1850 Ma igneous events in the Sudbury area include intrusion of the 1745 Ma Eden/Cutler and 1470 Ma Chief Bell granitic suites in the southern footwall rocks and the Sudbury and Grenville diabase dyke swarms at 1235 and 590

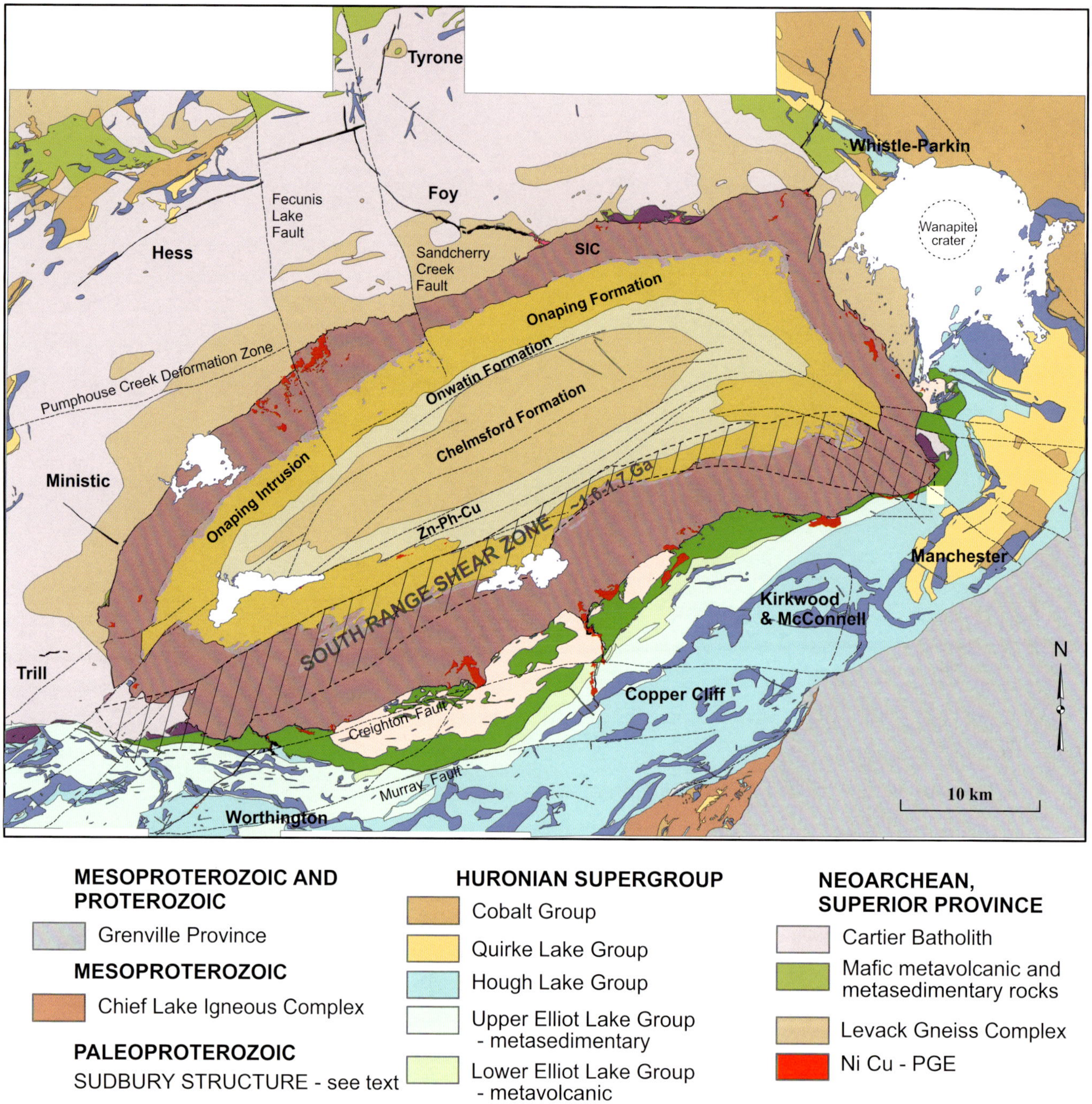

FIGURE 4. Geological map of the Sudbury base metal mining district (Ames et al., 2005) showing the main units of the Sudbury structure (see text on diagram). Offset dykes are labelled in bold.

Ma, respectively (Fig. 5). Both of these diabase dyke swarms intrude the SIC and its orebodies. Deformation that produced the South Range Shear zone strongly affected the South Range of the Sudbury structure (Shanks and Schwerdtner, 1991). It is interpreted as having occurred at 1.6 to 1.7 Ma and is related to the Mazatzal Orogeny (Bailey et al., 2005).

Mineral Deposits

The polymetallic central Ontario metallogenic belt, which is centred in the Sudbury region, hosts pre-1850 Ma uranium, Ni-Cu-PGE, and Au-Cu deposits; syn-1850 Ma magmatic-hydrothermal Ni-Cu-PGE; syn-1850 Ma hydrothermal Zn-Pb-Cu deposits (Errington-Vermilion: Ames et al., 1998, 2005); and several younger, variably sulphide-mineralized, hydrothermal pulses (Figs. 5 to 9). Huronian mineral deposits include the Elliot Lake area paleoplacer pyritic uranium ores, Ni-Cu-PGE sulphides in gabbros of the East Bull Lake suite of intrusions, redbed-hosted Cu in the upper Huronian sequence, and silica in quartz arenites (Bennett et al., 1991). The 2491 to 2475 Ma East Bull Lake suite consists of the East Bull, River Valley, and Agnew intrusions and is interpreted to have crystallized from low-Ti, high-Al, PGE-rich tholeiitic magmas. The ca. 2200 Ma Nipissing gabbros contain sub-economic Ni-Cu-PGE mineralization, yet their intrusion caused remobilization of metals from Archean and Paleoproterozoic host rocks to form Cu, Au,

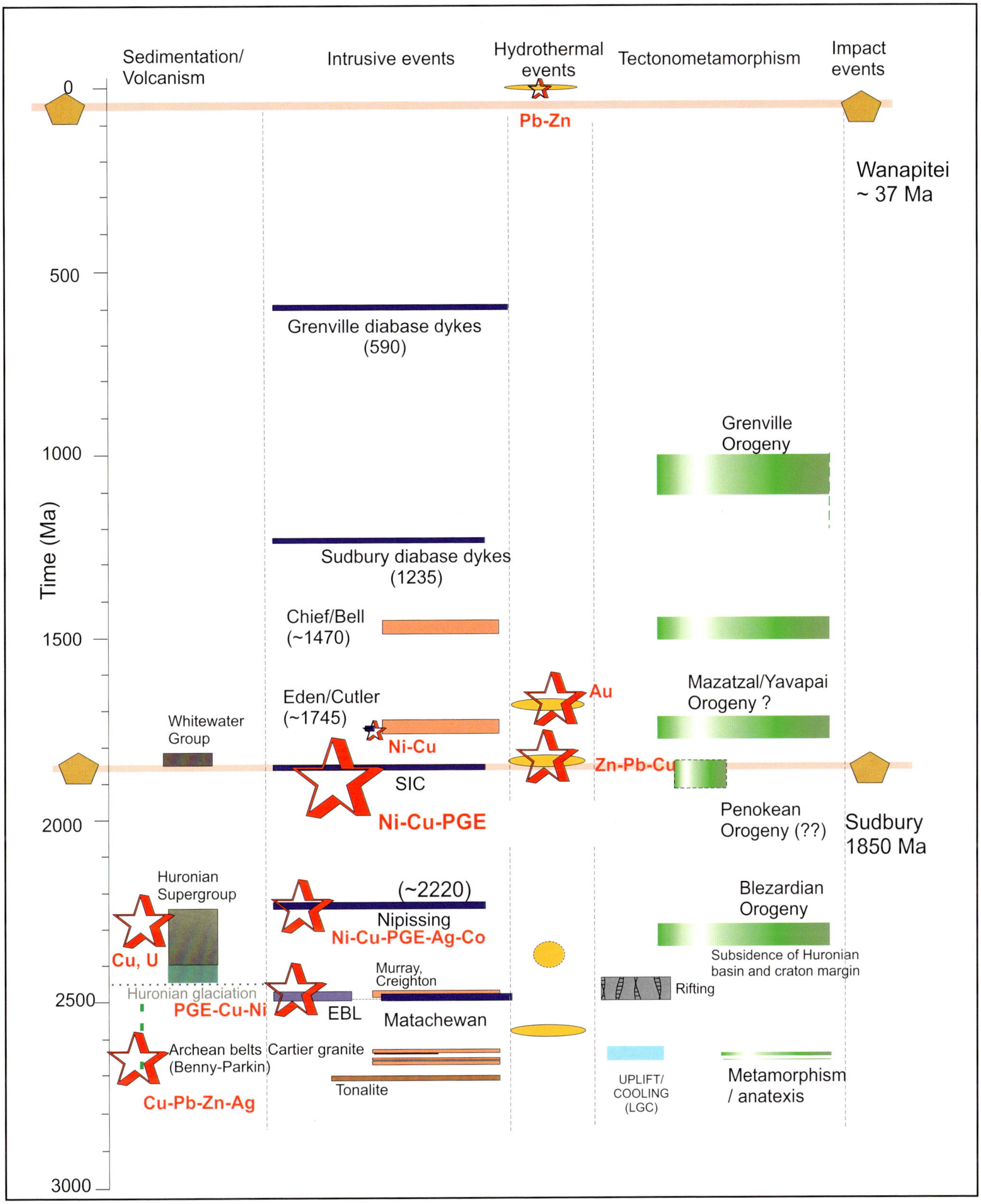

FIGURE 5. Time-stratigraphic chart and geochronological summary of the Sudbury district illustrating the age of major depositional, intrusive, tectonometamorphic, impact, hydrothermal, and metallogenic events (Ames et al., 2005).

Ag, Fe, and W veins. The most famous and economically important of these deposit types are unconformity related, Ag-vein deposits of the Cobalt-Gowganda-Elk Lake area (Andrews et al., 1986; Lightfoot and Naldrett, 1996).

The famous, world-class Ni-Cu-PGE deposits of the 1850 Ma Sudbury structure contain a total of 1500 Mt (past production and present resource) with overall grades of 1% Ni, 1% Cu, and 1 g/t Pd+Pt (Farrow and Lightfoot, 2002). The

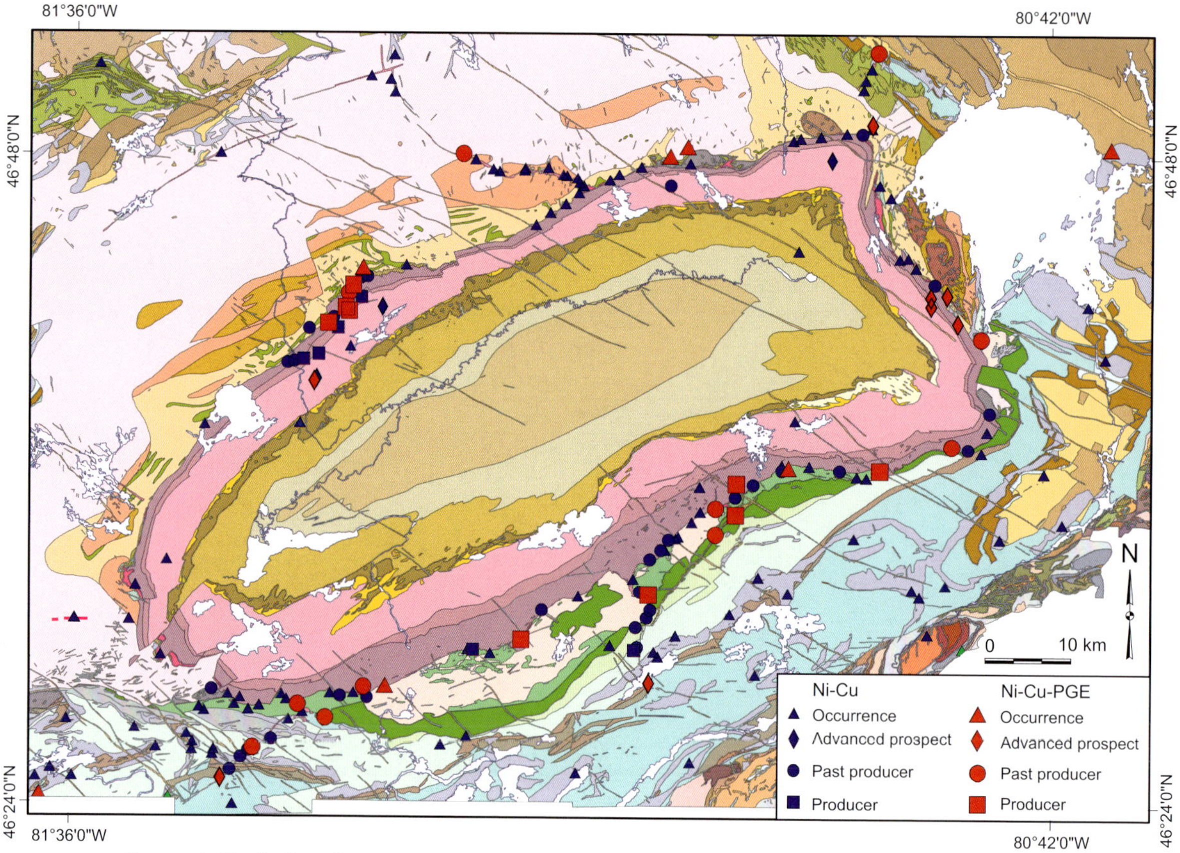

FIGURE 6. Distribution of over 200 Ni-Cu-PGE deposits and occurrences, Sudbury district (Ames et al., 2005).

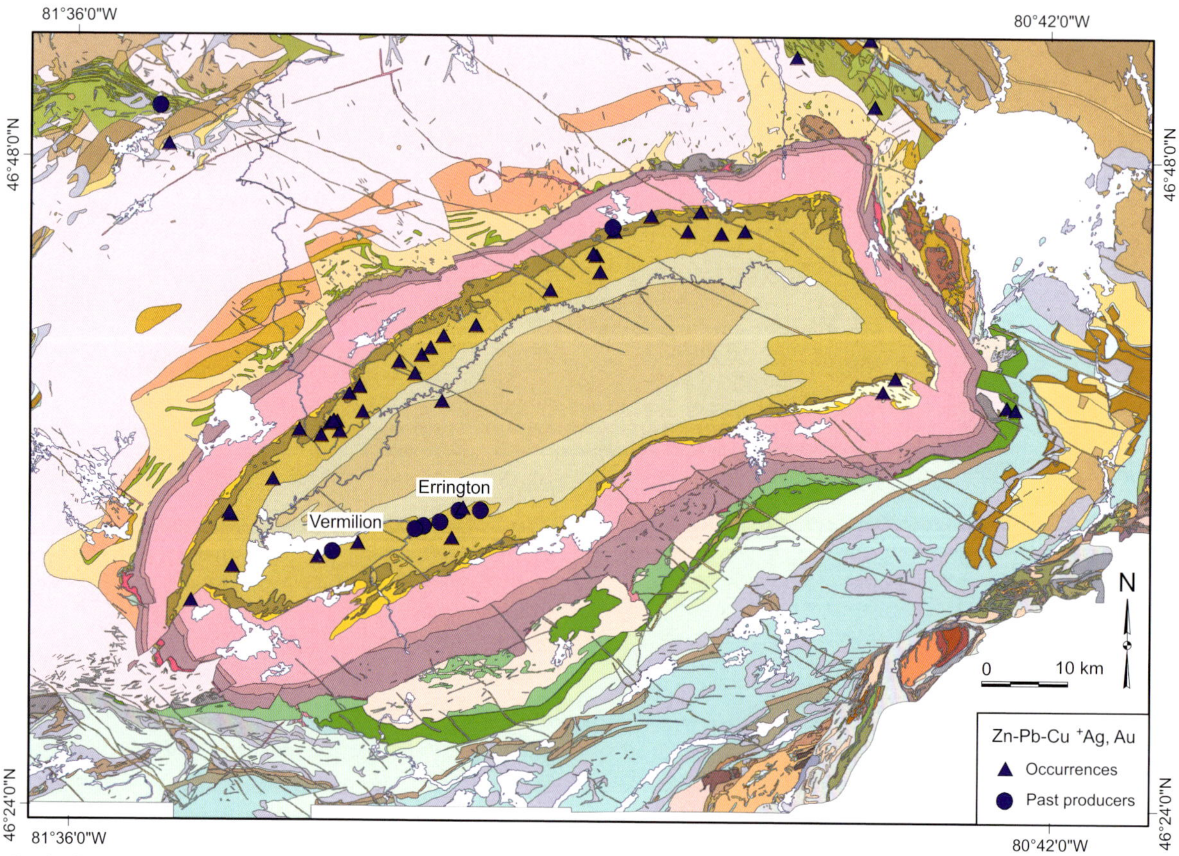

FIGURE 7. Distribution of Zn-Pb-Cu-Ag-Au VMS deposits and occurrences, Sudbury district. Hydrothermal replacement-type massive sulphide deposits (Errington-Vermilion) straddle the Onaping Formation-Vermilion Formation contact, whereas numerous base metal occurrences lie within the hydrothermally altered (ca. 1850 Ma) Onaping Formation (Ames et al., 1998, 2005, 2006a,b).

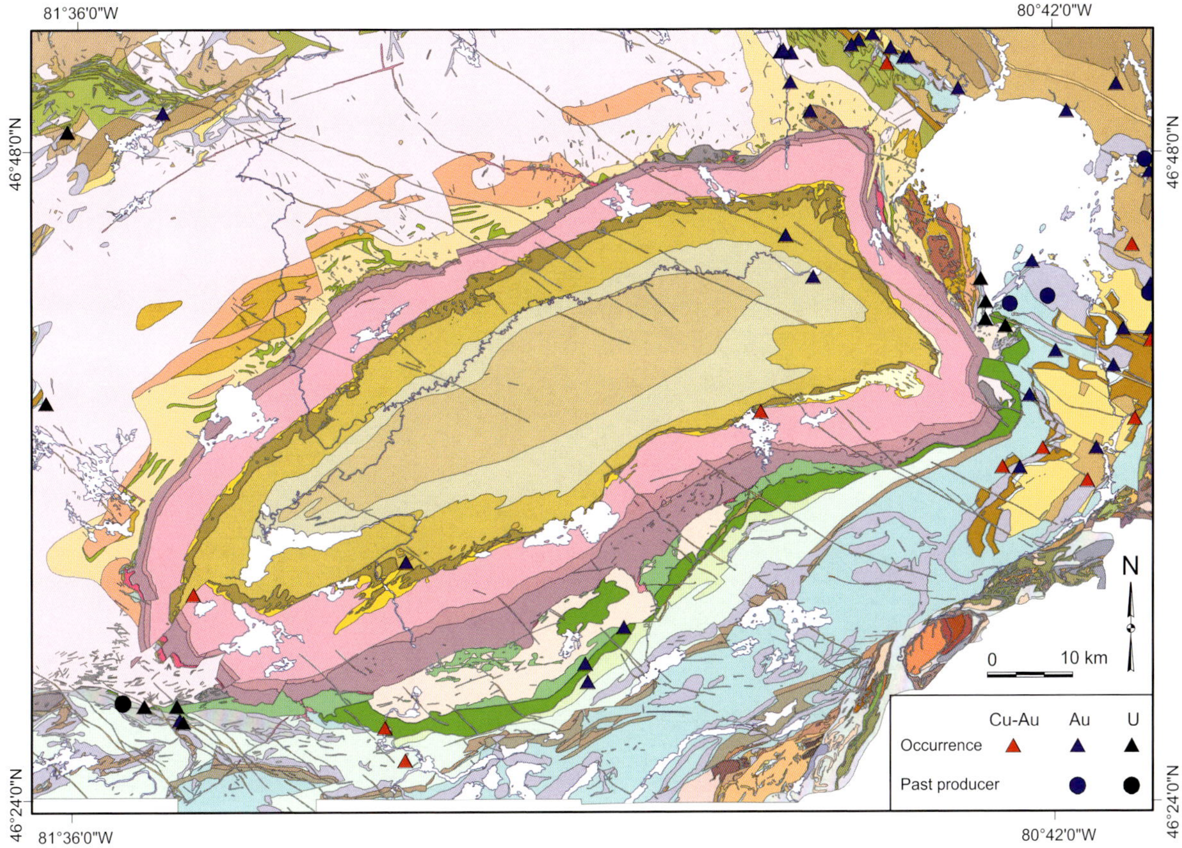

FIGURE 8. Distribution of Au, Au-Cu, and U deposits and occurrences, Sudbury district (Ames et al., 2005).

Sudbury structure presently hosts 77 Ni-Cu-PGE producing and past-producing deposits, 11 Cu-Ni-PGE advanced exploration prospects, and over 140 mineral occurrences (Fig 6; Ames et al., 2005). Regionally important lithologies that host Ni-Cu-PGE deposits are the base of the SIC (sublayer norite, footwall breccia), offset dykes, and variably developed Sudbury breccia within the footwall of the SIC (Fig. 6).

Sudbury ore deposits are divided into three main types (Coats and Snajdr, 1984; Morrison et al., 1994) based on their interrelated environments: contact, offset, and footwall deposits. All contain Ni-Cu-PGE, but the latter have elevated and locally very high PGE contents. Over 50% of the historical Sudbury resource is hosted in contact deposits (Lightfoot et al., 2001), approximately 25% is in offset deposits with approximately 15% from the Frood-Stobie breccia belt, and less than approximately 10% is in footwall-type deposits (Farrow and Lightfoot, 2002). The diverse deposits associated with the 1.85 Ga Sudbury structure span a protracted period of mineralization, and represent a range of ore-forming environments related to cooling of the SIC over one hundred to one million years (Prevec and Cawthorn, 2002). Many deposits in the South Range have been modified by deformation (Owen and Coats, 1984).

Geological Settings of Different Base Metal Deposits

Contact Deposits

Contact deposits are typically pyrrhotite-rich, Ni-Cu ores hosted by depressions or embayments at the base of the SIC at or near the contact with NeoArchean or Paleoproterozoic basement. Contact ores are hosted in a basal igneous "sublayer norite", a discontinuous layer of mafic-ultramafic inclusion-bearing norite to quartz diorite up to 700 m thick (Pattison, 1979; Souch et al., 1969) and/or "footwall breccia", an igneous to metamorphic textured breccia locally developed along the footwall interface between the sublayer and footwall rocks (Coats and Snajdr, 1984; Morrison, 1984).

Kilometre-sized depressions or radial "troughs" contain smaller secondary lateral embayments or "terraces" (Morrison, 1984) that influence the occurrence of the sublayer and the site of some sulphide mineralization (Fig. 10). Thickness variations in topographically controlled debris-flow units in the crater-fill near the base of the Dowling Member, Onaping Formation mimic the depression of the Levack embayment (Fig. 11). The orientation of the Ni-Cu-bearing lateral terraces (Morrison, 1984) is similar to the orientation of syn-crater faults in the Onaping Formation, which are filled with andesitic melt in the crater-fill (Ames, 2002b).

The contact-type deposits are rich in Ni-Ir-Rh, which are hosted by a sulphide and oxide mineral assemblage of pyrrhotite, pentlandite, and chalcopyrite with minor magnetite and pyrite (Appendix 1; Naldrett, 1984). Contact ore Cu/Ni ratios are approximately 0.7, and Pt+Pd+Au values are typically low at <1 g/t (Naldrett and Pessaran, 1992).

Offset Deposits

Offset-type deposits are hosted in radial or concentric "quartz diorite" offset dykes that represent an early phase of quenched impact melt (Lightfoot et al., 2001; Ames, 2002)

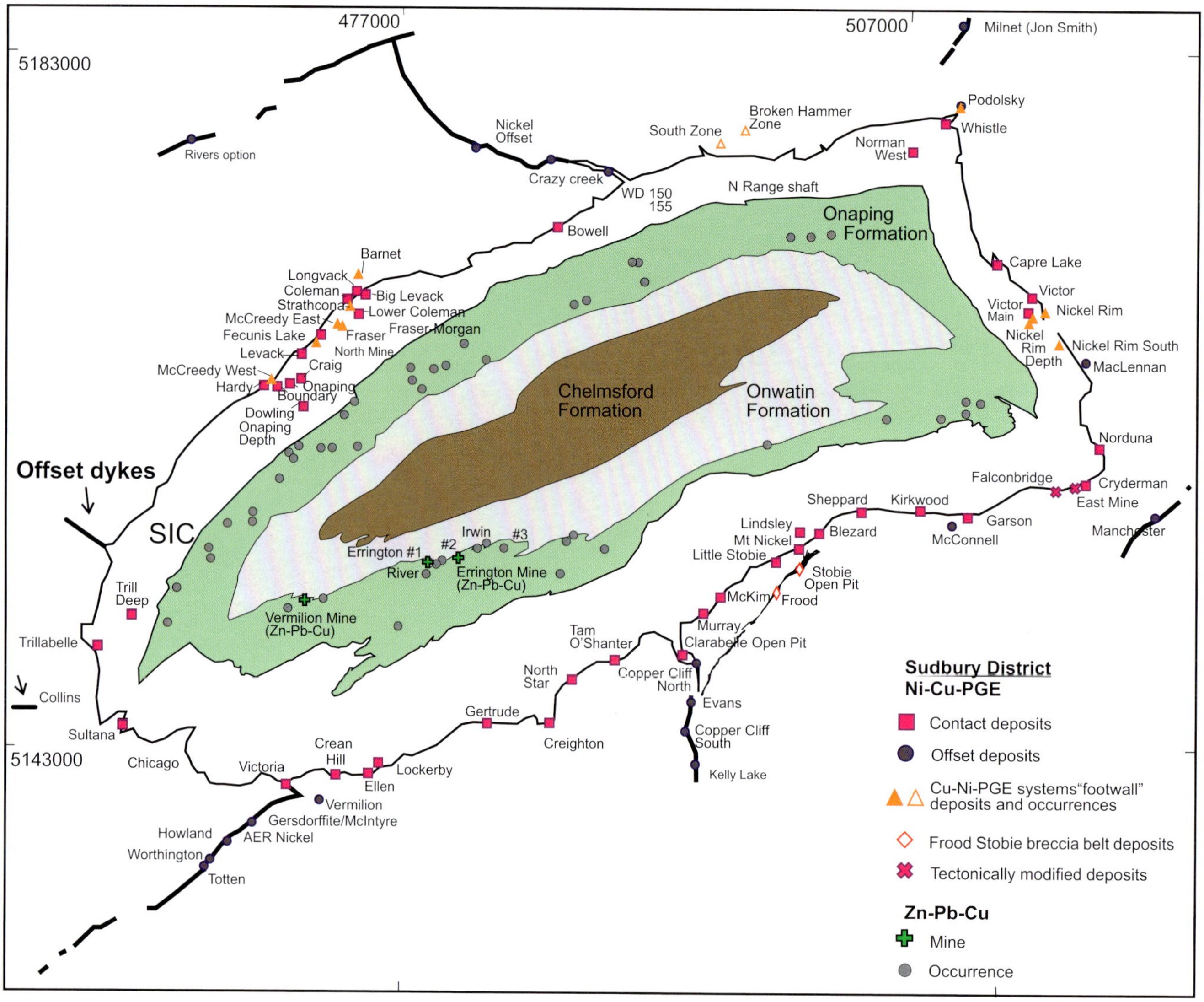

FIGURE 9. Location map of Ni-Cu-PGE deposit types and hydrothermal Zn-Pb-Cu mineralization associated with the 1850 Ma impact event (modified from Ames et al., 2003).

with multiple intrusive pulses (Morris, 1982). Offset dykes locally extend more than 30 km from the SIC (i.e. Foy) with economic mineralization located up to 7 km from the base of the SIC at the Kelly Lake deposits on the Copper Cliff offset dyke; more than 7 km to the Totten deposit on the Worthington offset dyke, and approximately 3.5 km to Ni-PGE mineralization on the newly discovered Trill offset dyke (Fig. 9).

Offset dykes are typically filled by quartz diorite with local inclusion-bearing quartz diorite. The latter phase is interpreted as a secondary pulse of magma and host to offset mineralization. Control on the location of ore deposits along offset dykes may be related to structural jogs; competency contrasts, such as lithology changes along radial offsets (e.g. Copper Cliff, Worthington, Whistle–Parkin); inclusion content (e.g. Worthington); fault displacements (e.g. Copper Cliff); thickening of the offsets (e.g. Totten); and/or Sudbury breccia in the vicinity of the dyke (e.g. Copper Cliff) (Figs. 12, 13) (Cochrane, 1984; Mourre, 2000; Lightfoot and Farrow, 2002; Murphy and Spray, 2002; Stewart, 2002). Offset sulphide ores change in mineral/trace element composition with increasing distance from the SIC along the offset to more PGE-, Cu-, and Au-rich (Farrow and Lightfoot, 2002). Massive to semi-massive, blebby and vein sulphides typically occur in steeply plunging orebodies along the length of the offsets (Figs. 12, 13). Sulphide mineralogy of

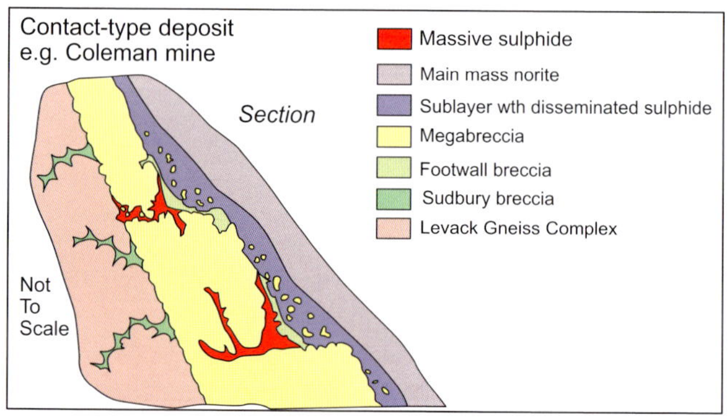

FIGURE 10. Geological cross-section of a contact-type Ni-Cu deposit located at the base of the Sudbury Igneous Complex (SIC) within the sublayer and footwall breccia, Coleman mine (Morrison, 1984).

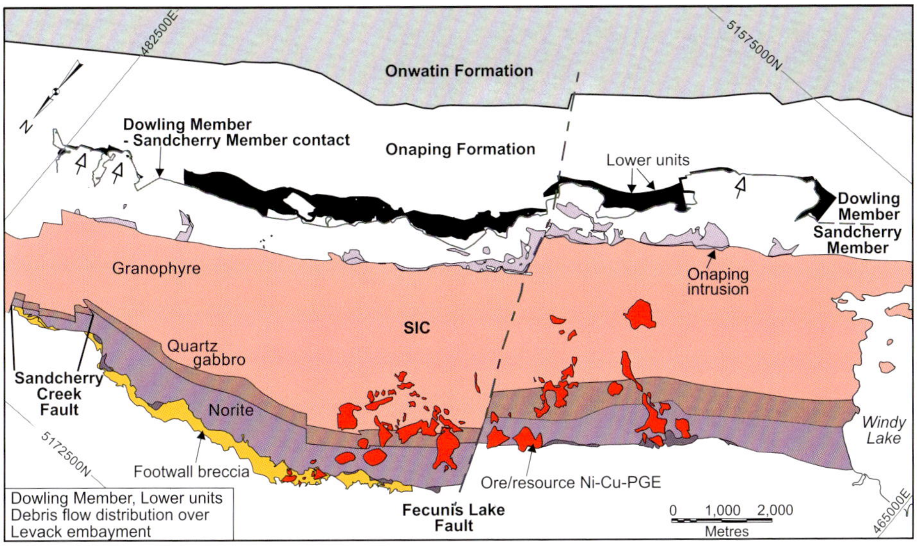

FIGURE 11. Geological section facing south through the Levack embayment and into the overlying crater-filling breccias of the Onaping Formation illustrating the formation of embayments by slumping synchronous with deposition of the Lower units, Dowling Member of the Onaping Formation. Topographically controlled debris flows in the crater-filling Onaping Formation outline the extent of the underlying Ni-Cu-PGE resource-rich Levack embayment. Debris flows of the Lower unit, Dowling Member, in the lower 500 m of the Onaping Formation vary in thickness from <5 m, over paleotopographic highs (positions shown with open black arrows) in the Sandcherry member, to >300 m in the paleotopographic lows mimicking the depression or underlying Levack embayment at the base of the Sudbury Igneous Complex (SIC).

offset-type deposits is dominated by pyrrhotite and less abundant pentlandite and chalcopyrite. Cu/Ni ratios are approximately 1 with Pt+Pd+Au values of more than 2.5 g/t, which is considerably greater than that of the contact deposits.

Cu-Ni-PGE Deposits

Over the past 30 years, Sudbury has evolved into a polymetallic Cu-Ni-PGE mining camp with a steadily increasing production from Cu-Ni-PGE deposits, the greatest sources of PGE in the area. In 2005, almost 19 million grams of Pt, Pd, and Au were produced in Sudbury, placing the area third behind the Bushveld and Noril'sk-Talnahk in world PGE production. This figure is expected to rise sharply through 2010 as the Ni Rim South Footwall, Levack Footwall, and Podolsky 2000 deposits are brought into full production. As a result, over 50% of Sudbury's exploration expenditures were invested in the exploration for this type of deposit in 2005, resulting in a significant new discovery (Levack Footwall Deposit).

These valuable chalcopyrite-rich deposits are distributed around the base of the SIC in a thermally metamorphosed zone proximal to the SIC-basement contact (Morrison et al., 1994). They are hosted by brecciated country rocks termed Sudbury breccia, which are largely devoid of an igneous SIC component (Fig. 14A,B). These historical "footwall-type deposits" (Morrison et al., 1994) have been termed more recently "Cu-Ni-PGE systems" as more diverse deposit-types were discovered in the footwall and offset environments (Farrow et al., 2005). These systems are subdivided based on both qualitative and quantitative criteria into the dominant "sharp-walled" and less common low-sulphide end-members. Cu-Ni-PGE deposits that contain significant tonnages of both end-member styles of mineralization are classified as "hybrid". This new classification is important for the development of new exploration models, geophysical applications, 3-D visualization, resource modeling, mine planning, and grade control (Farrow et al., 2005; King et al., 2006). Deposits in the South Range have a strong tectonic overprint that possibly obscures the definition of these deposit types. As a result, current knowledge is based on North and East Range deposits. Cataclasite and/or pseudotachylite of the Sudbury breccia formed either during initial impact and propagation of the shock wave (Peredery and Morrison, 1984; Thompson and Spray, 1994) or during gravity driven collapse of the impact crater rim along superfaults (Thompson and Spray, 1994; Spray, 1997). The Sudbury breccia development was important ground preparation for the emplacement of the sulphide veins (Farrow et al., 1994; Fedorowich et al., 1999). Footwall Cu-PGE deposits occur as vein,

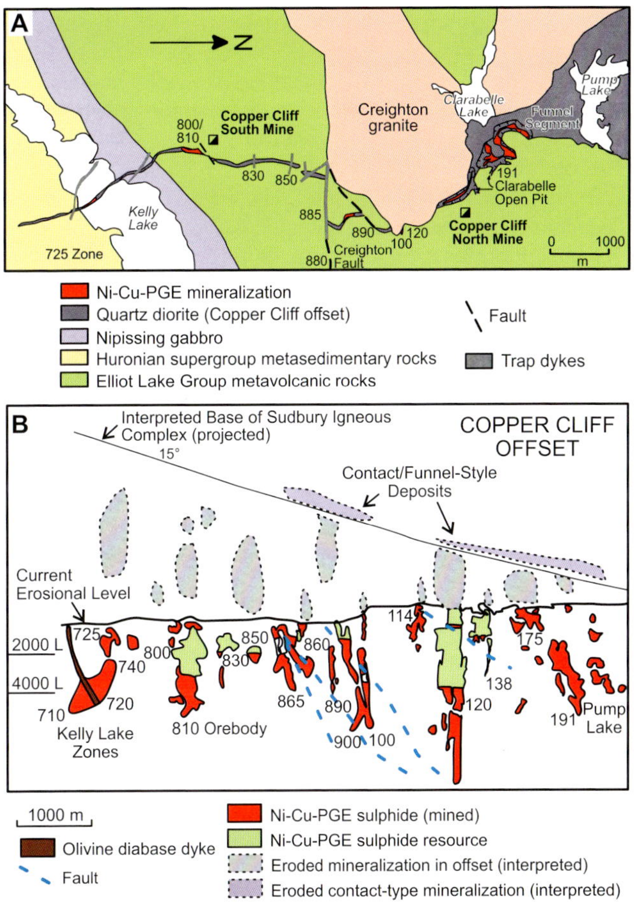

FIGURE 12. **(A)** Offset-type deposits distributed up to 7 km along the Copper Cliff Offset dyke. **(B)** Longitudinal section of the vertically dipping offset-type deposits along the Copper Cliff Offset, extending from the Pump Lake orebodies close to the Sudbury Igneous Complex to the Cu-PGE-rich Kelly Lake orebodies (Farrow and Lightfoot, 2002).

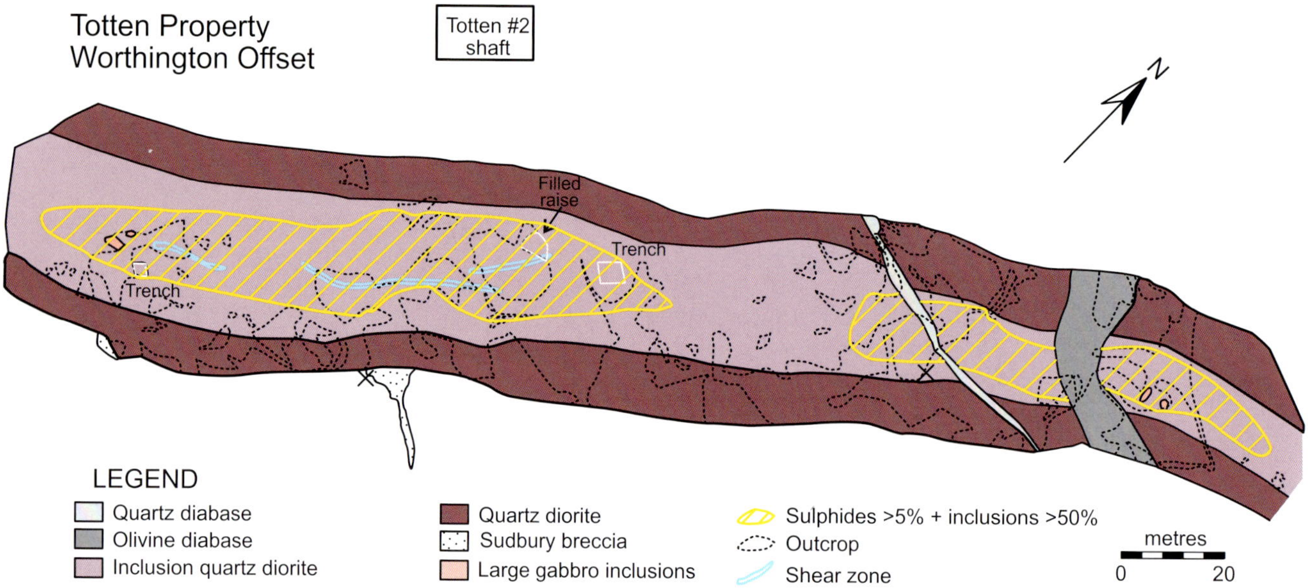

FIGURE 13. Geological map of the Totten deposit, Worthington offset, illustrating the mineralized inclusion quartz diorite in the core of the dyke (Stewart, 2002).

vein stockwork, and disseminated Cu, PGE, Au, Ag, and Ni orebodies (Naldrett and Kullerud, 1967; Li and Naldrett, 1993; Morrison et al., 1994). Textural variations in the widely distributed Sudbury breccia have resulted in the interpretation that thermally metamorphosed Sudbury breccia zones are prospective hosts for Cu-Ni-PGE systems (Morrison et al., 1994; Fedorowich et al., 1999; Hanley and Mungall, 2003). Progressive textural changes in the matrix of the Sudbury breccia from a black, aphanitic matrix and discrete felsic fragments (cold) to a recrystallized, fine-grained leucocratic, equigranular matrix with diffuse felsic fragments and irregular felsic patches (hot) reflect the thermal metamorphism prospective for footwall-hosted Cu-Ni-PGE mineralization. Poikiloblastic biotite and/or amphibole growth, quartz recrystallization, and anatexis with the formation of granophyric patches are characteristic of intense thermal metamorphism of Sudbury breccia that may contain anomalous levels of Cu, Ni, PGE, Au, and Ag in positions up to several hundreds of metres from ore-grade mineralization.

Sharp-walled Cu-Ni-PGE deposits are characterized by chalcopyrite-rich veins with sharp, planar surfaces across which the modal proportion of sulphide drops from greater than 80% to less than 5%. As a result, production grades are directly linked to vein thickness as much as to Ni or PGE tenor of the sulphides. Veins with greater than 30 m strike length are observed to be up to 7 m wide but are typically in the range of centimetres to decimetres. The largest veins are typically oriented subparallel to the SIC contact (e.g. McCreedy East 153 zone), however, some are conjugate systems or are randomly oriented (e.g. McCreedy West Footwall Cu; Fraser-Strathcona Powderhouse zone) (Fig. 14C). The massive chalcopyrite±cubanite vein-style footwall deposits (e.g. McCreedy East 153 zone, McCreedy West Footwall Cu; Fraser-Strathcona Cu zones, Ni Rim Footwall, Victor Deep, and the Levack Footwall deposits) contain minor pyrrhotite, pentlandite, millerite, bornite, and magnetite. Mineral abundances and base and precious metal ratios can vary along the length of individual veins. Metal zonation patterns have been demonstrated to change gradually with increasing Cu/Ni ratios with distance from the SIC contact (i.e. McCreedy West Footwall Cu:Cu/Ni 3.5 to 6.75, Morrison et al., 1994; Fraser-Strathcona Deep Copper zones, Li et al., 1992; Money, 1993; Kormos, 1999). The Cu/Ni

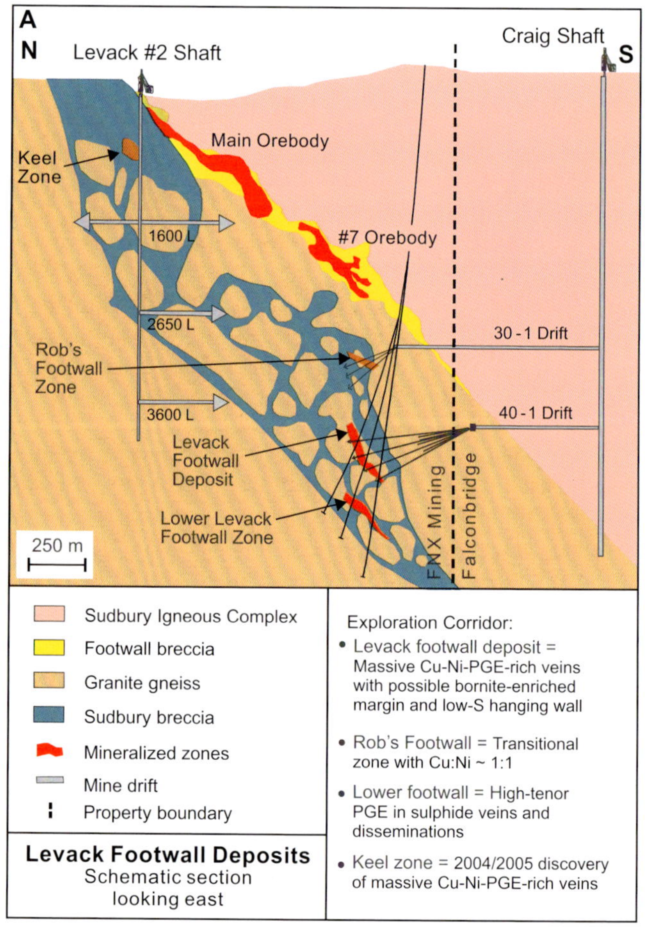

FIGURE 14. Cross-sections of footwall-type mineralization showing the relationship with contact-type mineralization at the base of the Sudbury Igneous Complex. **(A)** Levack footwall deposits.

ratio of the sulphides is typically greater than 6.5 with correspondingly high PGE tenor (Pt+Pd+Au > 7.7 g/t; Farrow and Lightfoot, 2002). Data compiled as part of this study and from Naldrett et al. (1999) show that massive sulphide, sharp-walled vein deposits in production and advanced exploration projects from both the North and East ranges typically grade more than 20% Cu, and vary from 0.70% to more than 4% Ni, and from 10 g/t to more than 30 g/t Pt+Pd+Au. These Cu-Pt-Pd-rich deposits are characteristically Ir-Rh-Os poor (Naldrett et al., 1999).

Low-sulphide PGE footwall mineralization (e.g. McCreedy West PM deposit) occurs as blebs, stringers, and disseminated sulphides and may or may not be economically connected to vein-style footwall mineralization. The low-sulphide McCreedy West PM deposit is the first stand-alone "low-sulphide" system to be put into production in the Sudbury camp. It is hosted in Sudbury breccia and is approximately 350 x 400 x 40 m with an indicated resource of 2.25 million tons at 1.11 wt.% Cu, 0.26 wt.% Ni, and 5.58 g/t Pt+Pd+Au and an inferred resource of 1.06 million tonne at 1.11 wt.% Cu, 0.28 wt.%, Ni, and 6.40 g/t Pt+Pd+Au (Farrow et al., 2005). The low sulphide PGE mineralization of the PM deposit is dominated by chalcopyrite and minor millerite, with a total sulphide content of less than 5 modal percent. The mineralization is defined by a series of stacked sulphide veins (typically <30 cm thick), vein sets, and stockworks, with disseminated sulphide and associated PGE throughout the host Sudbury breccia matrix. To date, only 5% of the resource is composed of massive sulphide veins greater than 30 cm wide (Farrow et al., 2005). A trend of higher Pt/Pd ratios with decreasing S content to below 0.3% has been observed in the extreme southeast extensions of the deposit (Farrow and Jolette, 2004; Farrow et al., 2005).

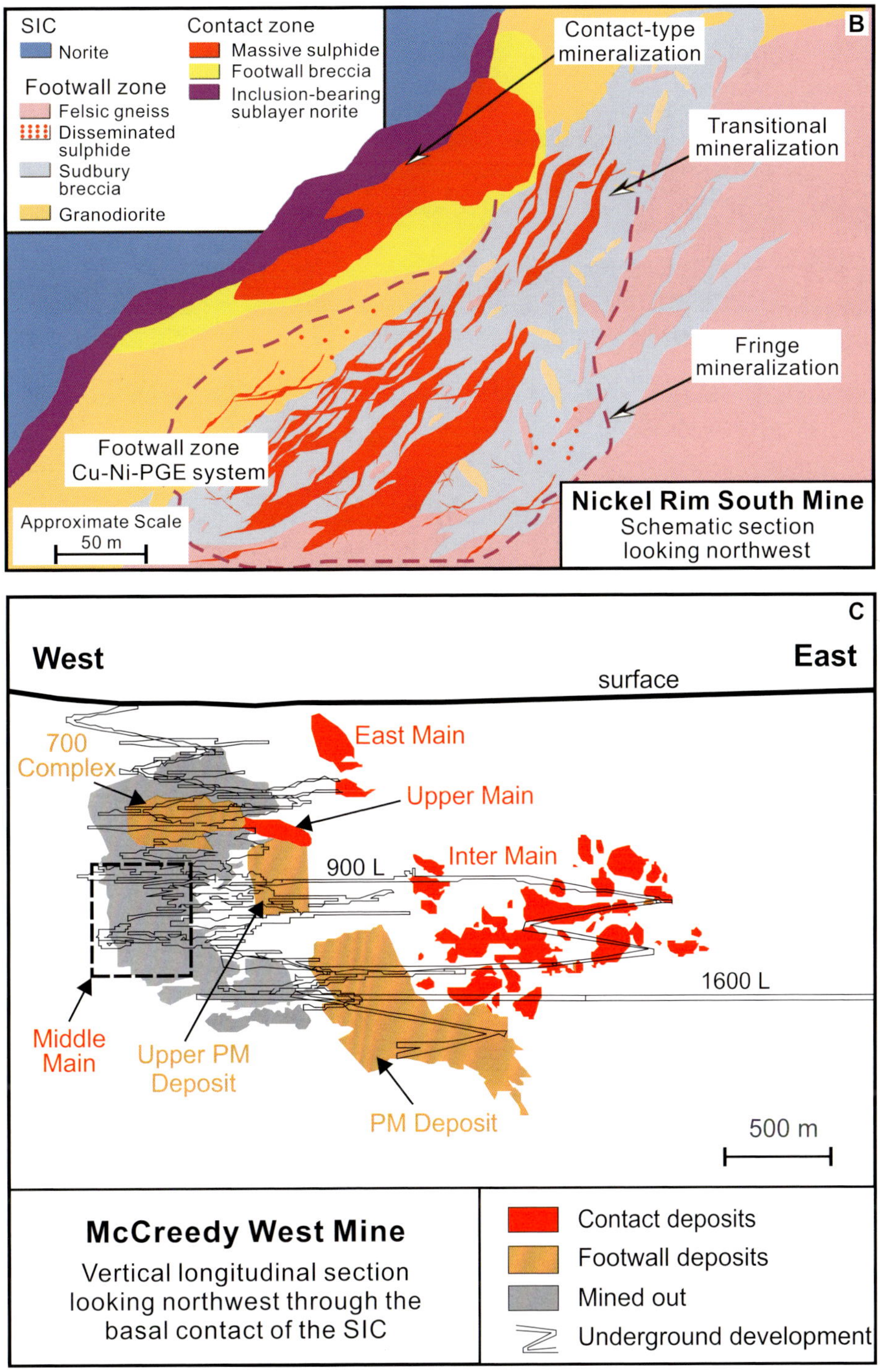

FIGURE 14 CONTINUED. **(B)** Nickel Rim South deposit (after McLean et al., 2005). **(C)** McCreedy West deposit (after Farrow and Lightfoot, 2002). The contact Ni deposits are hosted in the basal units of the Sudbury Igneous Complex (footwall breccia and sublayer norite) and the Cu-(Ni)-PGE deposits are hosted in the footwall Sudbury breccia.

Hybrid Cu-Ni-PGE mineralization is characterized by both sharp-walled and chalcopyrite-rich, irregularly-distributed breccia veins, in addition to domains of significant tonnage of poorly connected stringers and disseminated PGE-bearing sulphides. The best documented of these systems to date are the Podolsky North and 2000 deposits that are hosted by the inclusion-rich quartz diorite and metabreccia-filled Whistle Offset (Farrow et al., 2005). However, it is likely that the mined-out 800-810 Cross-Over zone in the Copper Cliff Offset also represented this hybrid style of Cu-Ni-PGE mineralization. At Podolsky, the sharp-walled and breccia veins are hosted by inclusion-bearing quartz diorite, grey gabbro, and metabreccia, whereas the inclusion-bearing

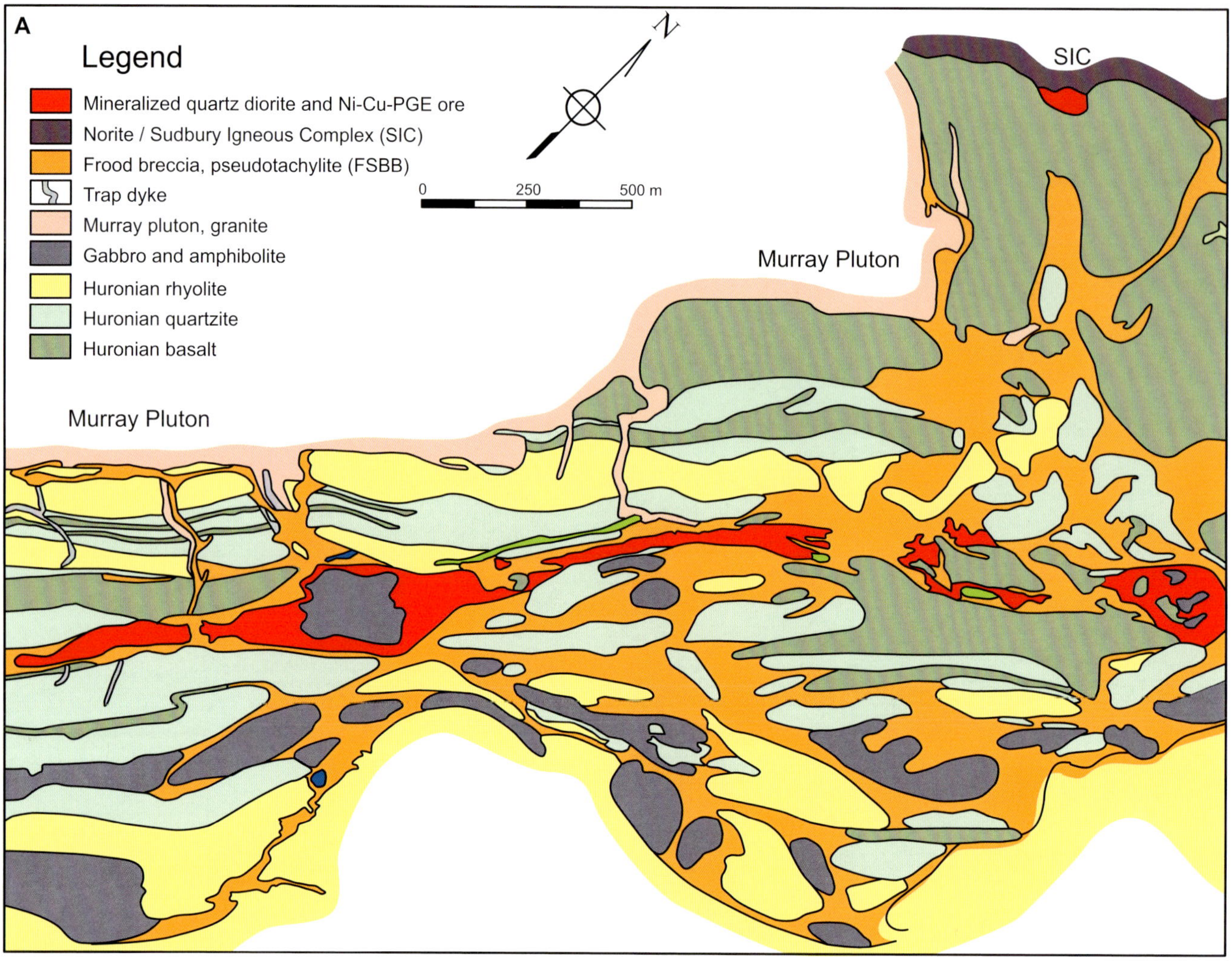

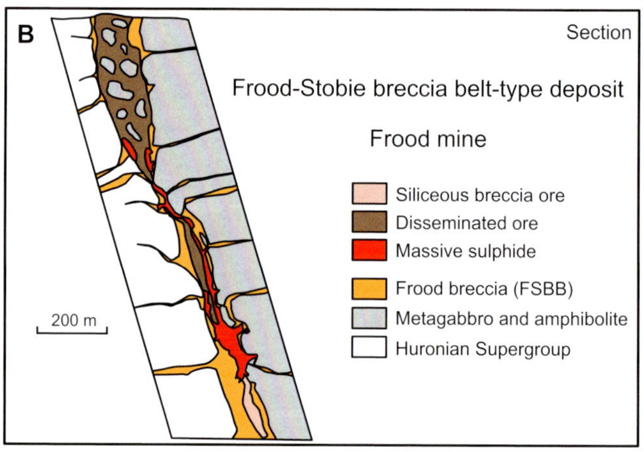

FIGURE 15. Frood-Stobie breccia belt-type deposit. **(A)** Surface geology of the Frood-Stobie mine. **(B)** Coss-section of the Frood-Stobie deposits (after Zurbrigg and staff, 1957).

quartz diorite is the dominant host to the low-sulphide mineralization. Chalcopyrite is typically the dominant sulphide mineral, with minor millerite as the dominant Ni-bearing mineral. Pentlandite is less common than millerite. Assayed intervals of varying sulphide contents indicate that the PGE abundance is not entirely dependent on sulphide content, but that low-sulphide mineralization is significantly higher in PGE tenor than massive chalcopyrite veins (Farrow et al., 2005). The value of these ores is high, as massive chalcopyrite veins with greater than 20% Cu locally contain greater than 1.0% Ni and 10 g/t Pt+Pd+Au. Sharp-walled-style mineralization occurs in a large block of grey gabbro, a rock type most common in South Range mineralized environments, in the hanging wall to the inclusion-bearing quartz-diorite- and metabreccia-hosted part of the 2000 deposit. Current resources at the Podolsky 2000 deposit are reported as indicated resource hybrid = 3,108,200 tons at 3.24% Cu, 0.29% Ni, and 3.62 g/t Pt+Pd+Au; and inferred resource sharp-walled = 4,862,807 tons at 1.16% Cu, 0.15% Ni, and 1.24 g/t Pt+Pd+Au.

Fluid mobility has long been recognized in Sudbury breccia, host rocks to Cu-Ni-PGE ores (Speers, 1957). Alteration haloes enriched in alkali- and halogen-rich minerals and fluid inclusions are recognized around both sharp-walled vein-style footwall mineralization (Farrow and Watkinson, 1992; Farrow et al., 1994; Jago et al., 1994; Molnar et al., 2001; Hanley and Mungall, 2003; Farrow et al., 2005; Watkinson et al., 2005) and low-sulphide mineralization (Killeen, 2006). The hydrothermal alteration is a retrograde overprint on an earlier thermal aureole (Hanley and Mungall, 2003; Hanley et al., 2004). Thin alteration selvages consist of amphibole-epidote-quartz ± chlorite and titanite (Abel et al., 1979; Farrow and Watkinson, 1992; Li et al., 1992; Jago et al., 1994).

South Range Breccia Belt Deposits

The South Range breccia belt is a 100 m to more than 1 km wide zone of pseudotachylite (Sudbury breccia) more than 45 km long in a 1 km wide arc transecting the Paleoproterozoic Huronian Supergroup in the southern part of the Sudbury structure (Spray, 1997; Scott and Spray, 1999, 2000). It is a concave, 70°N dipping zone with <10% quartz-diorite melt pods (Speers, 1957) sub-concentric to the SIC (Fig. 9). Historically, the arcuate zone was known as the Frood-Stobie Offset Dyke (Souch et al., 1969; Grant and Bite, 1984) and more recently termed the South Range breccia belt (Scott and Spray, 1999).

The giant (600 Mt) Frood-Stobie mine has an upper domain of pyrrhotite-rich Ni mineralization and lower domain of Cu-PGE mineralization, termed the "siliceous zone in the South Range Breccia Belt" (Zurbrigg and staff, 1957; Hawley, 1965; Souch et al., 1969; Fleet, 1977; Naldrett, 1984; Farrow and Lightfoot, 2002). Host rocks of the giant sulphide-mineralized system change with depth from inclusion-bearing quartz diorite with disseminated sulphide to South Range footwall breccia with blebby to disseminated sulphide, into Sudbury breccia developed in Huronian rocks with <8 m wide, Cu-PGE veins and disseminations (Fig. 15). Cu/Ni ratios increase with depth with the lower Cu-PGE mineral assemblage of chalcopyrite, cubanite, bornite, and platinum group element minerals. The mineralogy of the Frood-Stobie deposit, parts of which have been closed for decades, are well documented and summarized in Ames et al. (2003). Alteration is dominated by quartz with minor calcite, biotite, amphibole, epidote, and chlorite (Farrow and Lightfoot, 2002).

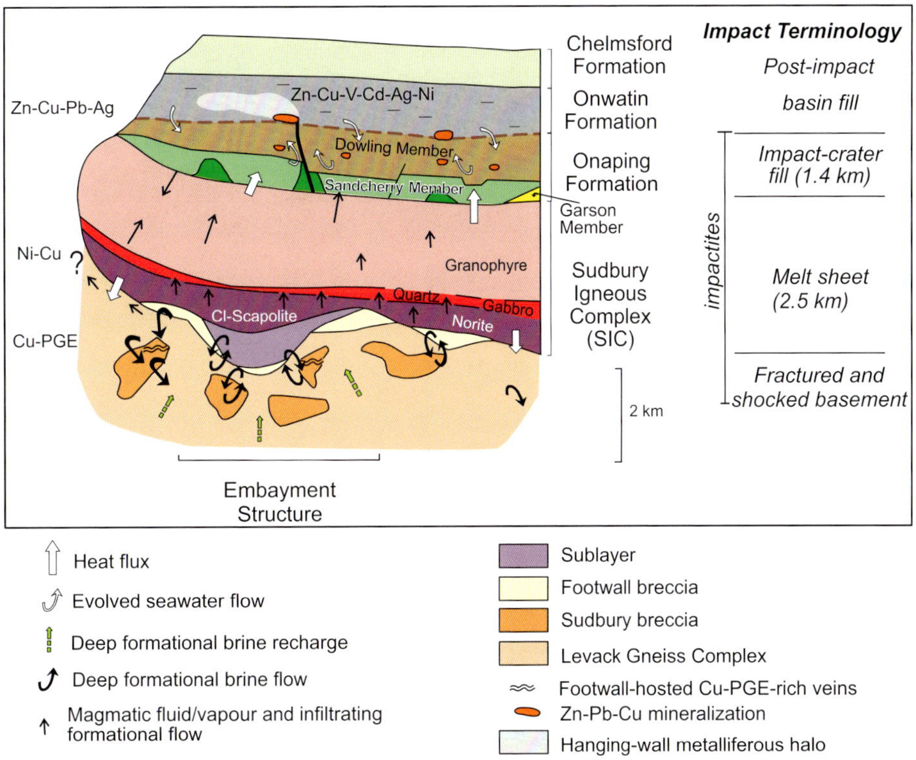

FIGURE 16. Schematic model of the Sudbury impact crater and associated hydrothermal system (after Ames et al., 2006a). Arrows indicate fluid, vapour,and heat-flow paths below, above, and within the melt sheet. The relative positions of the magmatic Ni-Cu, magmatic-hydrothermal Cu-Ni-PGE, and hydrothermal Zn-Pb-Cu ore deposits are shown, as well as a metal-enriched hydrothermal halo in the post-impact argillaceous basin sediments.

Zn-Pb-Cu (Ag, Au) Sudbury Basin / Hydrothermal Massive Sulphide Replacement Deposits

The Sudbury basin hosts two past-producing hydrothermal Zn-Pb-Cu (Ag, Au) deposits hosted by fine tuff at the top of the impactite (Onaping Formation) and the Lower Carbonate member of the Vermilion Formation (Figs. 7, 16). The Errington and Vermilion deposits contain 6.9 and 2.7 Mt, respectively, with grades of approximately 4 to 5 wt.% Zn, 1 to 1.5 wt.% Cu, 1.3 wt.% Pb, 55 g/t Ag, and minor gold. The deposits formed by subseafloor replacement of a fumarolic carbonate mound (Martin, 1957). The Zn-Pb-Cudeposits and over 35 hydrothermal Zn, Pb, and Cu occurrences within the 1.4 km thick Onaping Formation are related to an 1848 Ma hydrothermal system fuelled by the heat from the SIC (Ames et al., 1998, 2005, 2006a) (Figs. 7, 16). The footwall to the deposits is a 1.5 km thick, hydrothermally altered, andesitic breccia sequence of the Onaping Formation. The massive sulphide is a replacement of carbonate mounds of the Vermilion Formation and carbonaceous tuff at the top of the Onwatin Formation. A metalliferous halo has been identified in the post-impact carbonaceous mudstones of the Onwatin Formation above the deposits (Ames et al., 2006a).

The ore lenses occur within tight, inclined, and locally overturned folds with mineralization concentrated in the crest of anticlines, which are associated with deformation from the South Range shear zone. The deposits are marked by an increase in Ba, Si, and Zn near the orebody and Mn depletion within the Vermilion Formation (Gray, 1995). The Errington deposit contains pyrite, chalcopyrite, sphalerite, galena, marcasite, and arsenopyrite, with trace pyrrhotite, freibergite, and argentite. The Vermilion deposit also has trace cuprite, bornite, covellite, pentlandite, hematite, magnetite, gold, and silver.

Although Sudbury is a meteorite impact setting, hydrothermal processes in the basin are reminiscent of those in volcanogenic massive-sulphide-type deposits. In the crater-fill of the Sudbury basin, the Onaping Formation, there are regional, semi-conformable alteration zones (silicification, albitization, and carbonatization), syn-crater faults, paleotopographic highs, and peperite formation immediately below massive sulphide deposits (Ames et al., 2002).

Tectonic Modifications

Most orebodies in the Sudbury area have experienced some degree of tectonic overprint, especially those along the

South Range. This has been interpreted for the most part as relatively local remobilization of sulphides by a combination of mechanical, metamorphic, and hydrothermal/metasomatic processes that have not been well documented. The East Falconbridge and Garson mines represent deformed contact and footwall deposits but have historically been termed "fault-related deposits" (Naldrett, 1984; Owen and Coats, 1984). Tectonic overprinting, due to deformation by the South Range shear zone (1.7 Ga, Bailey et al., 2005), has locally resulted in significant variations in the ore mineralogy by enhancement or depletion of Ni grades. Pyrite-marcasite has locally replaced pyrrhotite and pentlandite due to sulphidation and oxidation, and development of trace mineral assemblages, which include cobaltite-gersdorffite, maucherite, galena, and sphalerite, has been localized. Neotectonic (13 Ma, Marshall et al., 1999) movement along existing long-lived structures in the Sudbury region produced brittle structures, which were subsequently filled by quartz-carbonate-sphalerite and galena veins (Fig. 5). Neotectonic effects on the ore deposits are negligible (Marshall et al., 1999; Farrow and Lightfoot, 2002).

Trace Element and Trace Mineral Signatures

It is widely known that platinum group minerals (PGM) in Ni-Cu-PGE deposits contain the key element(s) S, Se, Te, As, Sb, Bi, Sn, Pb, and Hg (Cabri, 1981, 2002). A number of trace element studies have emphasized the bulk precious metal content of world-class Ni-Cu-PGE ores as at Sudbury and Noril'sk (Keays and Crocket, 1970, 1999; Naldrett et al., 1982, 1994; Distler, 1994). However, little attention has been paid to trace elements that constitute platinum group minerals and accessory minerals to assess their relevance to issues of genesis, metallurgy, or exploitation of resources. Indeed, there are few geochemical data on the trace element signatures of Ni-Cu-PGE sulphide ore deposits (e.g. Jonasson et al., 1987), which is in marked contrast to those known of other base metal deposit types. The regional distribution of trace elements in the Sudbury Cu-Ni-PGE deposits (Ames, 2002) shows the provinciality due to differences in environment and in host rocks from north (Levack Gneiss Complex/Cartier granite) to south (Huronian volcano-sedimentary sequence and Nipissing diabase).

PGE mineralization is associated with all Sudbury ore-types. However, footwall-, offset-, and South Range breccia belt-hosted deposits tend to be PGE-enriched. The footwall and offset environments are the foci of active exploration that is seeking to expand Sudbury's role as a major PGE-producer. Discrete platinum-group minerals, rather than solid solution in base-metal sulphide minerals, account for the majority of the PGE in the Sudbury ores (Cabri, 1984b, 1988; Farrow and Lightfoot, 2002). PGE in Sudbury ores are bound to and therefore spatially associated with Bi, Te, As, Sb, and Sn and their trace minerals.

Arsenic is an important trace element in the Sudbury ores as it is associated with PGE (Hawley et al., 1951). The South Range ore deposits are locally As-bearing but the North and East Range deposits only have low As content (Fig.17A), as indicated by mineral chemical studies (Cabri, 1981; Farrow and Lightfoot, 2002; Ames et al., 2003). Paleoproterozoic mafic intrusions and Huronian volcano-sedimentary rocks that form the South Range footwall stratigraphy are also notably enriched in As (Fig. 17A). Platinum minerals occur in two different styles on the North and South ranges. On the South Range, sperrylite is by far the dominant Pt mineral, where it occurs as individual, subhedral grains or as monomineralic aggregates in silicates or sulphides with little relationship to a specific host mineral assemblage (Cabri, 1988; Farrow and Lightfoot, 2002). Although sperrylite ($PtAs_2$) is present in North Range Cu-PGE footwall deposits and showings (e.g. Fraser-Strathcona Deep Copper, McCreedy West Footwall Cu, McCreedy East 153 zone, and Broken Hammer showing), moncheite, $(Pt,Pd)(Te,Bi)_2$ is the dominant Pt mineral on the North Range (Li and Naldrett, 1993; Farrow and Watkinson, 1997; Ames et al., 2003; Watkinson et al., 2005). Platinum distribution is controlled by arsenides (sperrylite) and associated with sulpharsenide minerals (especially cobaltite-gersdorffite) in the South Range (Cabri, 1988). The Rh, Ir, and Ru minerals (hollingworthite (RhAsS), irarsite (IrAsS), and ruasite (RuAsS)) are typically associated with cobaltite-gersdorffite in South Range deposits (Ames et al., 2003).

A Pd-As trend was observed at Strathcona and Frood with a lower As-Pd-rich siliceous zone and an upper zone with low As-Pd (Keays and Crocket, 1970). The dominant Pd minerals in the South Range are michenerite [(Pd, Pt) BiTe] with less abundant froodite (PdBi2). The Pd minerals occur as complex, polyphase grains spatially associated with the dominant As-bearing mineral in the South Range, a cobaltite-gersdorffite solid solution [(Co,Ni,Fe)AsS]. Cobaltite-gersdorffite has been observed in all deposit types on the South Range of the Sudbury structure (Ames et al., 2003). The As contributed to the South Range ore systems by sperrylite is not significant when compared to that contributed by the cobaltite-gersdorffite solid solution. Only between 1 and 4.5% of the total Pt in the system occurs in solid solution with cobaltite-gersdorffite, whereas 22.3 to 23.3% of the total Pd in the system occurs in the cobaltite-gersdorffite (Cabri, 1981). The association of Pt and Pd with As in the South Range ores suggests that As may be a useful geochemical exploration vector for PGE mineralization in the Sudbury area.

Antimony is an important trace element in the Sudbury ores because it is associated with palladium minerals, including sudburyite [(Pd, Ni)Sb] and mertieite [$Pd_8(Sb,As)_3$] and platinum minerals geversite ($PtSb_2$) at the Copper Cliff and Victor Deep deposits (Appendix 1; Ames et al., 2003). Stibnite (Sb_2S_3) has been identified in the Strathcona footwall ores (Ames et al., 2003). Proportional symbol maps of Sb distribution show that the highest values are restricted to South Range ore deposits (Totten, 430 ppm; Garson, 170-240 ppm; Kelly Lake and Vermilion, 3-17 ppm) although overall the Sb contents are very low (817 ppb, Fig. 17B).

Tin-bearing PGM including niggliite (PtSn) paolovite (Pd_2Sn) and stannopalladinite (Pd$_2$Sn2) have only been reported in the North and East range deposits (Victor Deep, Strathcona, Coleman, and McCreedy East, Ames et al., 2003; Victor Deep, Strathcona, Coleman, McCreedy East, Farrow and Lightfoot, 2002). Cassiterite (SnO_2) was also observed in North Range deposits where it was interpreted to be derived from the local felsic dominated host rocks (Farrow and Lightfoot, 2002). However, the trace element geochemistry indicates that the Sn content of the ores is also

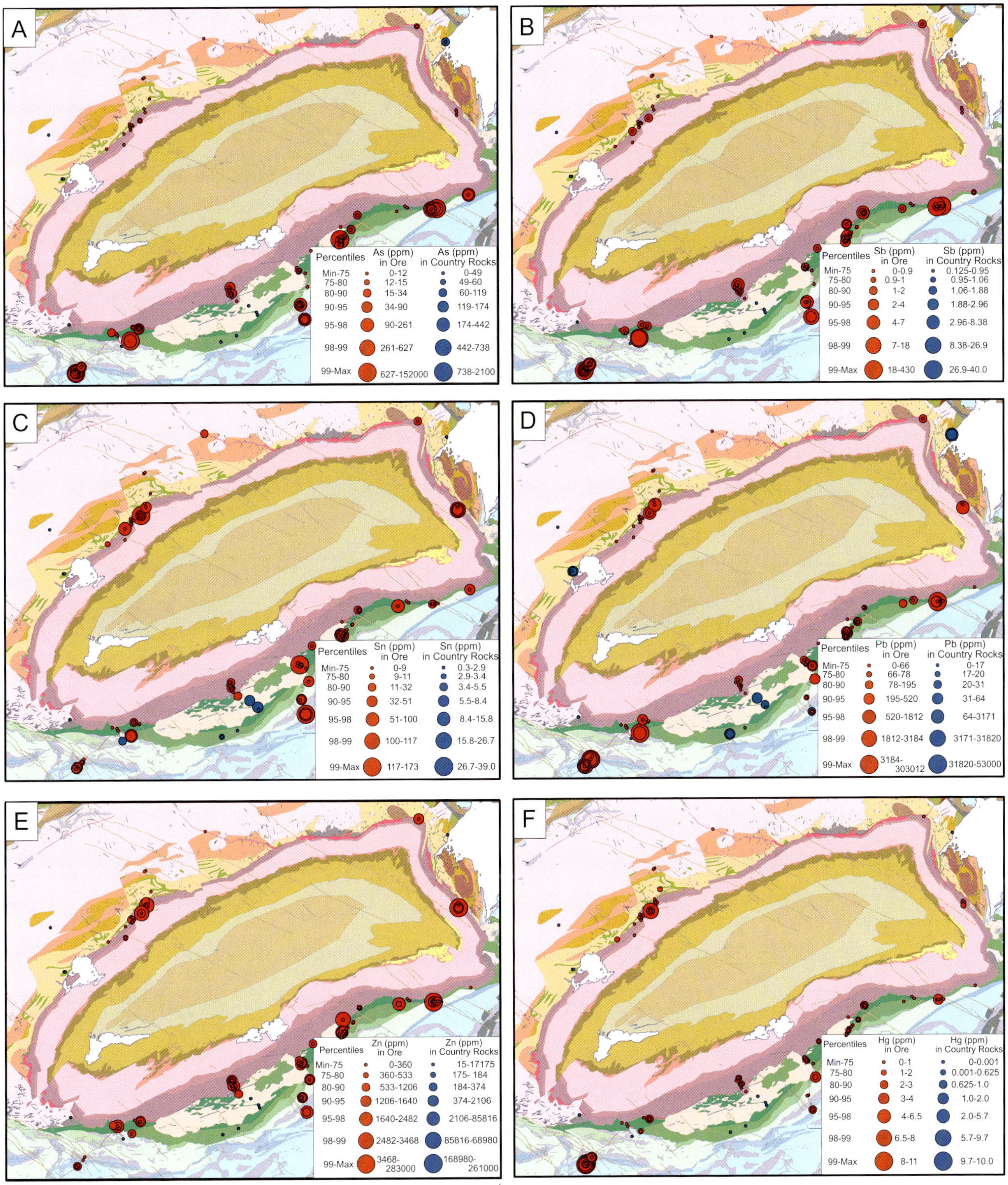

FIGURE 17. Proportional circle element abundance maps of Sudbury Ni-Cu-PGE ore and country rocks. Ore samples n=365 and target (country) rocks n= 143. Geology map (after Ames et al., 2005). **(A)** Arsenic in the Ni-Cu-PGE ores is higher in the South Range deposits reflecting a contribution from Paleoproterozoic mafic intrusions and Huronian volcano-sedimentary country rocks. **(B)** Antimony associated with palladium and platinum minerals shows higher levels in the South Range deposits although the concentrations are generally low. **(C)** Tin, an important element in platinum group minerals in Sudbury ores, is regionally distributed around the entire structure. **(D)** Lead in the Sudbury Ni-Cu-PGE deposits is concentrated in footwall- and offset-type deposits. **(E)** Zinc is enriched in footwall- and offset-type Sudbury deposits in contrast to contact-type deposits. **(F)** Mercury is high (1-10 ppm) in some Sudbury ores independent of ore type or geographic distribution however, some offset-type and footwall-type ores have the highest Hg concentrations.

elevated (50-173 ppm) in South Range deposits (Kelly Lake, Clarabelle pit, Frood-Stobie, Kirkwood footwall, Vermilion and Victoria Cu-Ni-PGE deposits), many of which are not associated with felsic host-rocks at the present day erosional level (Fig. 17C). The Sn-bearing Cu sulphide, stannite, occurs in the Frood and Totten deposits on the South Range (Hawley, 1965; Ames et al., 2003).

Lead is enriched in Sudbury ores, where it occurs as galena and various selenide and telluride minerals associated with chalcopyrite and sphalerite (Appendix 1; Ames et al., 2003). The distribution of Pb is controlled by deposit type and is typically enriched in offset and footwall deposits (Fig. 17D). Contact deposits are characteristically low in Pb (~20 ppm). Deposits highest in Pb (<8600 ppm) include those on the Worthington offset (Worthington, Totten, Victoria), South Range breccia belt deposit (Vermilion), footwall-hosted deposits (Victor Deep, McCreedy East 153 zone, McCreedy West Footwall Cu, and Strathcona) and the strongly deformed Garson deposit with significantly higher Pb (Fig. 17D).

Sphalerite hosts most of the Zn in the Sudbury Cu-Ni-PGE ores and is commonly associated with chalcopyrite (Hawley, 1962; Naldrett et al., 1982; Naldrett, 1984). Trace Zn is heterogeneously distributed in solid solution in chalcopyrite (Appendix 1; Cabri, 1984a). Zinc occurs in all deposit types on both the North and South ranges in concentrations greater than 500 ppm. However, footwall-hosted and offset ore samples tend to be enriched in Zn and Cd (Fig 17E). Contact deposits are generally Zn-Cd-poor. Those deposits with samples notably enriched in Zn (>1000 ppm) include Victor Deep, Strathcona Deep, Little Stobie, McCreedy East153 zone, Kelly Lake, Victoria, Crean Hill, Creighton (footwall deposits, 126 and 403 orebodies), Copper Cliff South, and Vermilion. There is a distinct correlation between Cd and Zn in the Sudbury ores (Fig. 18A), which is controlled by Cd substitution for Zn in sphalerite. A marked feature of the Sudbury ores is the most of the sphalerite is Cd-rich (up to 2.5 wt.%) and contains 5 to 8 wt.% Fe (Fig. 18B). In contrast, Zn in solid solution in chalcopyrite is not associated with cadmium (Cabri, 1984a). Sphalerite in chalcopyrite-millerite- pyrrhotite veins of the McCreedy East 153 zone are Fe-poor with very high Cd (up to 4 wt.%; Fig. 18B). Hawleyite (CdS) was reported in the Fraser-Strathcona deep copper zone (Money, 1993).

Mercury is associated with palladium as the minerals potarite, temagamite, and tischendorfite (Pd-Hg±Te, Se) in PGE deposits (Cabri, 2002) but not in Sudbury ores. Sudbury ores contain Hg in the ppm range (1-11 ppm) in the contact, footwall, and offset deposits (Fig. 17F); the mineralogical site of Hg has not been determined.

Genetic Models

Recent studies have resulted in a significantly improved understanding of the igneous processes involved in the formation of the Sudbury ores from the melt sheet (Lightfoot et al., 2001; Ames et al., 2002), evolution of the ores by monosulphide solid solution (MSS) fractionation (Naldrett et al., 1999; Ballhaus et al., 2001; Mungall et al., 2004) and high-temperature volatiles (Farrow and Watkinson, 1999), and modification by subsequent hydrothermal activity (Li and

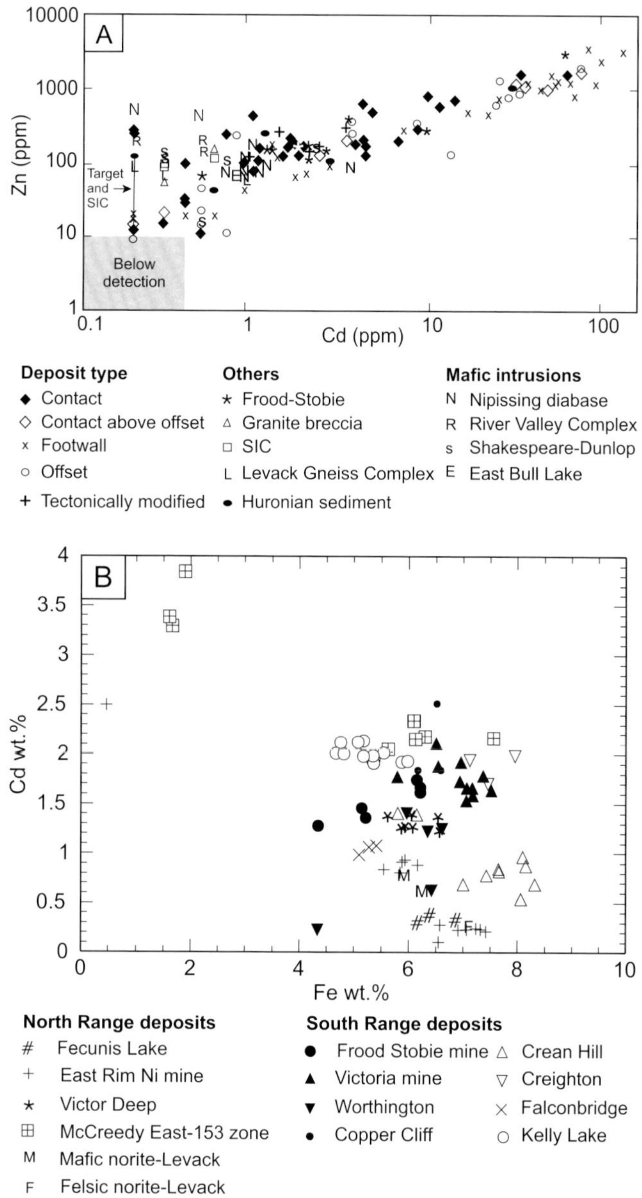

FIGURE 18. Sphalerite mineral chemistry in Sudbury ores. **(A)** Zn-Cd correlation illustrating enrichment in footwall- and offset-type deposits relative to contact-type deposits, the Sudbury Igneous Complex, and country rock sphalerites. **(B)** Cd-Fe in sphalerite shows that most are Cd-rich (up to 2.5 wt.%) and contain 5 to 8 wt.% Fe with a distinct population of Cd-rich Fe-poor sphalerite from Cu-PGE veins from the footwall-type deposit.

Naldrett, 1993; Farrow and Watkinson, 1999; Hanley et al., 2005).

Sudbury Ni-Cu contact-type deposits are interpreted to be of magmatic origin, having formed during differentiation of the SIC followed by sulphide segregation and subsequent collection in topographic lows or "embayments" at the base of the SIC. The origin of the Cu-Ni-PGE systems is controversial with both magmatic and hydrothermal hypotheses having been supported in the literature. More recently, a magmatic-hydrothermal origin was postulated whereby initial magmatic differentiation of the sulphide liquid resulted in the formation of a residual sulphide liquid enriched in Cu, Pt, Pd, and Au. This was then remobilized into structural pathways or permeable zones of brecciated country rock or

Sudbury breccia in the footwall of the SIC. Later hydrothermal mobilization resulted in redistribution of base and precious metals, modification of the ore composition, and the formation of Cl-rich alteration haloes. Halogen-rich haloes have been traced distally to approximately 200 m from the ore (Morrison et al., 1994). The chalcopyrite-rich vein margins have chlorine-rich alteration assemblages with lawrencite ($FeCl_2$), bismoclite (BiOCl), and ferropyrosmallite $[(Fe^{2+},Mn)_8Si_6O_{15}(OH,Cl)_{10}]$ reported at the Fraser mine (Farrow and Watkinson, 1992; Hanley and Mungall, 2003; Hanley et al., 2004). The veins contain low-temperature Pt and Pd minerals (i.e. tellurides, bisthmuthides, and stannides) and metal halide minerals within the PGE-Au-rich, Cu-Ag-bearing, high salinity fluid inclusions (Farrow et al., 1994; Hanley et al., 2005). There is also intense alteration in an "epidote-zone", containing Ni-enriched mineralization along the footwall/SIC contact (i.e. Fraser and McCreedy East mines), further supporting a hydrothermal origin for these ores (Farrow, 1996). Ore metal transport and redistribution involved mixing between regional ground waters and a metal-rich ore fluid with a magmatic component to form a metal-enriched brine that partitioned, causing Au precipitation and phase separation of Cu, Au, Ag, and Bi into a CH^4-bearing fluid (Farrow et al., 1994; Marshall et al., 1999; Hanley, 2006), which further dispersed Cu, Pt, Au, Ag, and Bi (Hanley et al., 2005).

Summary of Effective Exploration Methods in the Sudbury Camp

Geological and 3-D Imaging

Important components of the geological modeling for effective exploration in the Sudbury camp include:

- Three dimensional imaging of the ore-hosting lithologies, including the contact of the SIC - sublayer- offset dykes and the known orebodies (Fig. 19).
- Pseudotachylite (Sudbury breccia) distribution and key textural changes related to its thermal recrystallization in the footwall environment.
- "Footwall or late granite" breccia presence in the basal SIC, contact environment.
- Key textural changes in the Sudbury breccia and footwall lithologies are subtle yet mappable in proximity to mineralization in the footwall environment (Morrison et al., 1994; Hanley and Mungall, 2003; McLean et al., 2005). Distal to mineralization, Sudbury breccia has a black, aphanitic to fine-grained matrix. With increasing proximity to ore the matrix is light greenish grey with 3 to 10% biotite porphyroblasts, fine acicular amphibole, epidote, and chlorite. Closest to the ore, partial melt or anatectic granitic melt zones within the recrystallized matrix comprise 3 to 10% of the Sudbury breccia and other footwall lithologies. Medium-grained granophyric veinlets are common.

Surficial and Bedrock Geochemistry

Over the last hundred years, metal loading from the three Sudbury area smelters at Copper Cliff, Coniston, and Falconbridge has strongly affected the surficial environment (Negusanti, 1995; Friske et al., 1997; Singhroy, 2000; Spiers, 2004). C-horizon till geochemistry has been attempted to distinguish lithologic trends in the North and East ranges in order to avoid this anthropogenic influence (Bajc and Hall, 2000).

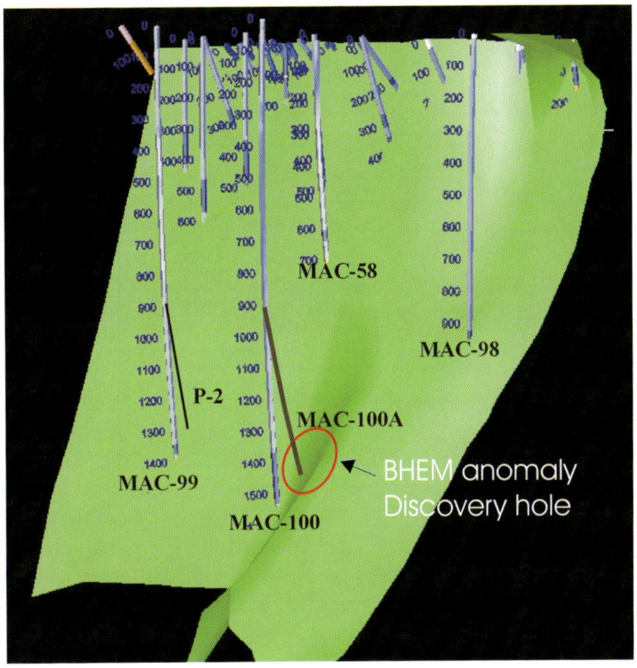

FIGURE 19. Three-dimensional view of the contact of the Sudbury Igneous Complex at the Nickel Rim South deposit, East Range (McLean et al., 2005).

Lithogeochemistry of the norite of the SIC has been used to determine proximity to embayments utilizing the major elements Mg and Ti (Dessureau et al., 2001). Pathfinders, such as halogen, hydrocarbon, or metal enrichment haloes, around the lucrative, but elusive, footwall deposits have only locally been tested (Hanley and Mungall, 2003; Hanley et al., 2005). The Cl content and Cl/(Cl+F) ratios of amphibole and biotite in recrystallized Sudbury breccia has been shown to be elevated within 150 m of mineralization at the Fraser mine (Hanley and Mungall, 2003). Bedrock geochemistry has been used most effectively to improve knowledge of the geological architecture and ore genesis of the Sudbury structure. New exploration techniques for Cu(-Ni)-PGE footwall deposits include trace metal vectoring (i.e. Sn, Zn, Au) along with the halogen content.

Geophysical Techniques

The traditional geophysical exploration tools for Ni deposits, such as magnetic and EM methods, have been successful in locating the historic/classic contact-style, pyrrhotite-rich Ni mineralization for which Sudbury was initially made famous. Indeed, Sudbury has been the test-bed for many non-traditional geophysical techniques in base metal exploration, such as 3-D seismic, audio-magnetotellurics, seismic velocity/cross-hole tomography, and radar imaging of orebodies (Milkereit and Group, 1992; Pflug et al., 1994; King et al., 1996). Cross-hole radio imaging (RIM) was an important definition drilling tool that played an important role in the discovery of two of the most Cu-Ni-PGE-rich deposits discovered in the last decade, the Ni Rim South and Levack Footwall deposits. Although numerous airborne and ground geophysical techniques have been applied in the Sudbury camp, both surficial cultural influences and the increasing depth of new discoveries have rendered modern borehole electromagnetic surveys as most effective in the exploring for highly conductive, pyrrhotite-rich Ni-Cu-PGE orebodies in contact and offset environments (e.g. Victor contact deposit, Chadwick, 1996). The UTEM step response system responds to very conductive targets and increases the target radius of drillholes to greater than 100 m, with optimum coupling of the energized loop, transmitter, and receiver with the conductive body. Sharp-walled and hybrid Cu-Ni-PGE systems where the sulphide veins are connected also respond to borehole EM methods, although the responses may be more difficult to interpret, and have been successfully applied in several recent exploration projects that are now in advanced stages (Victor Depth, Ni Rim South Footwall, Podolsky 2000, Levack Footwall Deposit, see section *Advanced Prospects* of Ames et al., 2005). Low-sulphide Cu-Ni-PGE deposits are difficult to resolve using borehole EM methods and have characteristics more amenable to the application of both surface and borehole induced-polarization (IP) techniques.

Recent Discoveries

Recent exploration success in the Sudbury mining camp is largely due to the linkage of geological knowledge combined with 3-D geological modelling and the integration of advanced geophysical modeling and inversion. The consistent theme in the recent examples of exploration success documented below is that constantly evolving geological knowledge and interpretation has resulted in the identification of target areas that have been refined with appropriate geophysical techniques, typically borehole EM. The recent application of RIM has helped focus definition drilling efforts in deep Cu-Ni-PGE systems.

Stratigraphic drilling of fences of holes in favourable geological environments has been attributed to the discovery of the Victor contact Ni mineralization (1970s) and the McCreedy East 153 zone (1990s), and the Fraser-Morgan contact mineralization (1990s). Additional drilling of the Victor contact mineralization was continued in the 1980s, and a campaign to stratigraphically test the footwall to the SIC contact was initiated. Due to the depth of the footwall holes (greater than 2000 m), borehole EM methods were applied. The subsequent drill testing of the anomalous footwall responses resulted in the first commercial success of the BH-UTEM system developed by Yves Lamontagne. Despite its depth, the Victor contact and footwall-hosted Cu-Ni-PGE environment is once again being evaluated for an advanced exploration program as a result of the nearby Ni Rim South discovery. The McCreedy East 153 zone was discovered in the late 1990s by drill testing a favourable package of Sudbury breccia that hosted the nearby Coleman mine footwall copper zone. The stratigraphic diamond-drill campaign intersected the 153 zone in the fourth hole. Today this zone produces over 400,000 tonnes annually of high-grade Cu, Ni, and PGE ore.

The Kelly Lake Ni-Cu-PGE orebodies in the Copper Cliff Offset were discovered (1995) by the application of new BH-UTEM techniques in holes drilled in previous decades. This “old hole” geophysical program was initiated to test the Copper Cliff Offset distal to the SIC contact since there was known Ni-rich mineralization at surface (740 orebody) and weakly sulphide mineralized intersections in the previously drilled holes. The Kelly Lake orebodies are currently being evaluated for production planning purposes.

The Totten Depth discovery hole was drilled in 1997 to test the down-plunge extension of the previously mined Totten Mine offset Ni-Cu-PGE deposit. Definition drilling was guided by the modeling of BH-UTEM4 responses at depth. Production planning is currently underway at Totten Depth.

The Ni Rim South contact and footwall mineralization was discovered as a result of follow-up drilling in 2001 from a 1995-1996 drill campaign designed to provide deep BH-UTEM platforms in the footwall to the SIC. Borehole MAC-100 intersected minor footwall mineralization and had a weak footwall BH-UTEM response. The resulting wedge hole (MAC-100A) was drilled in late 2001 and intersected both contact Ni (5.30 m of 2.86% Ni, 1.91% Cu, and 1.75 g/t Pt+Pd+Au) and footwall-hosted Cu-Ni-PGE (88.85 m of 2.17% Ni, 6.45% Cu, and 15.40 g/t Pt+Pd+Au, McLean et al., 2005). Definition drilling of the footwall discovery was guided by the application of RIM tomography. Development of a production infrastructure at Ni Rim South was initiated in 2004.

The Podolsky 2000 deposit was initially drilled in the 1980s to test the Whistle Offset inclusion-bearing quartz diorite below the known Whistle Mine Ni deposit. Cu-Ni-PGE mineralization was intersected, but the associated BH-UTEM responses were considered too weak to be of significance. In 2002, follow-up drilling and a new interpretation of borehole geophysics with respect to Cu-Ni-PGE systems resulted in a re-interpretation of the geometry of the sulphide mineralized system. The subsequent drill program from better oriented platforms led to more detailed drilling of the hybrid-type 2000 Cu-Ni-PGE resource and the discovery of significant sharp-walled Cu-Ni-PGE mineralization in the grey-gabbro environment in the hanging wall. High grade mineralization intersected in the hybrid resource was reported as 114.4 feet at 13.8% Cu, 1.0% Ni, and 8.2 g/t Pt+Pd+Au (borehole FNX-4130). Development of a production infrastructure at Podolsky was initiated in late 2004.

The most recent significant discovery in the Sudbury mining district, the Levack Footwall Deposit, was made in 2005. The discovery hole, FNX-6010, was originally drilled in 2003 to a depth of 2000 feet in what was interpreted to be a package of favourable Sudbury Breccia. An intense program of detailed geological modeling of the Levack environment and re-interpretation of weak BH-UTEM responses resulted in the decision to deepen the hole in November of 2004. A decision to further deepen the hole in January 2005 was made as a result of the interpretation of positive Sudbury breccia features and an intensely building BH-UTEM4 anomaly in the bottom of the hole. On February 10 and 11, 2005, two sharp-walled veins were intersected at a depth of approximately 4000 feet with grades of 10.2 feet at 26.2% Cu, 3.0% Ni, and 14.45 g/t Pt+Pd+Au, and 16.4 feet at 26.2% Cu, 3.7% Ni, and 15.45 g/t Pt+Pd+Au. Definition drilling of the sulphide mineralization has been partly guided by the application of RIM tomography and is continuing. At the time of writing, various options to facilitate access to the deposit were being evaluated.

Potential for Known and New Deposit-Types in the Sudbury Camp

The 21st century has witnessed banner years for exploration, active mining claims, and new discoveries in Sudbury with a marked shift in exploration focus to the high-grade Cu-Ni-PGE deposits. The exploration of copper-PGE footwall mineralization is in its infancy in the Sudbury mining camp. The potential for Ni-Cu-PGE discoveries in the Sudbury structure is considered very high, both at surface and at depth, despite its status as a mature Ni-Cu-PGE mining camp. There have been unprecedented exploration expenditures and activities over the past two years by both junior and established mining companies. Recent discoveries of Cu-PGE-rich deposits in the offset and footwall environments, which is difficult to explore, drives the active minerals exploration industry in Sudbury. Recent exploration success by major (INCO, Falconbridge), mid-sized (FNX), and junior (Wallbridge) exploration companies has enhanced the already outstanding resource potential in an area with considerable smelting and refining infrastructure.

The Victor, Ni Rim South, Levack Footwall (>1,200 m), and Onaping depth (2,800 m) discoveries demonstrate the prospectivity at depth. Several recent, relatively shallow, outcropping discoveries include the Podolsky 2000 and the Broken Hammer Cu-(Ni)-PGE (2004) deposits, the new Trill Offset dyke with Cu-Ni-PGE mineralization on or very near surface (2005), and the Segway discovery (2005). The Trill offset dyke was identified in June, 2005 by Wallbridge approximately 4 km from the SIC contact in an area historically mapped as granite (Ames et al., 2006b).

Knowledge Gaps of Thematic and Applied Research

There are key scientific knowledge gaps important in the discovery of new Ni-Cu-PGE deposits in the Sudbury camp that may be addressed by

- geological mapping of footwall lithologies, including Archean "pink areas" and Huronian stratigraphy with excellent stratigraphic markers;
- development of effective mapping methods for determining the regional distribution, types, and thermal recrystallization of pseudotachylite (Sudbury breccia);
- systematic rock geochemistry studies of the matrix of Sudbury breccia to define tracers and determine prospective versus barren pseudotachylite zones;
- determining the regional variation in thermal metamorphism induced by the SIC;
- comprehensive alteration studies around Cu-PGE footwall vein deposits to test for geochemical and/or mineralogical haloes;
- defining the character of the footwall breccia (or "late granite breccia") in the South Range (footwall breccia is the major host to contact Ni ore deposits in the North Range);
- determining the basin- and mine-scale structural and tectonic controls on ore deposition and modification (e.g. establish the amount of Grenvillian deformation north of Grenville front);
- determining impact crater morphology (i.e. determining if there is peak-ring or multi-ring impact, since impact crater architecture influences the structural style and control on mineralization);
- defining the lithogeochemical tracers for Cu-PGE (Ni) magmatic-hydrothermal deposits (e.g. the potential role for the halogens (Cl, F, Br) as pathfinder elements in Sudbury has been successfully tested only at the McCreedy East Footwall Copper, Victor Deep, and Fraser deposits (Jago et al., 1994; Hanley and Mungall, 2003) however, since Cu is a volatile in supercritical brine fluid systems it is also a good pathfinder element at considerably less cost; the application of hydrocarbon mapping around known Cu-PGE vein systems at Sudbury has not yet been tested)
- characterizing and refining the geophysical responses of the various styles of footwall Cu-(Ni)-PGE mineralization;
- establishing a comprehensive, physical rock-properties database for Sudbury for geophysical inversions and 3-D modelling (i.e. density, magnetic susceptibility, remnance, electrical conductivity, and P and S wave velocity).

Acknowledgements

Discussions with Sudbury explorationists, particularly those with Gord Morrison, Ed Pattison, and Mike Sweeny, have been invaluable. We thank James Mungall and Ian Jonasson for their thoughtful reviews.

References

Abel, M.K., Buchan, R., Coats, C.J.A., and Penstone, M.E., 1979, Copper mineralization in the footwall complex, Strathcona mine, Sudbury, Ontario: Canadian Mineralogist, v. 17, p. 275-286.

Addison, W.D., Brumpton, G.R., Vallini, D.A., McNaughton, N.J., Davis. D.W., Kissin, S., Fralick, P.W., and Hammond, A.L., 2005, Discovery of distal ejecta from the 1850 Ma Sudbury impact event: Geology, v. 33, p. 193-196.

Ames, D.E., 2002, Sudbury Targeted Geoscience Initiative (TGI: 2000-2003): Overview and Update, *in* Summary of Field Work and Other Activities 2002: Ontario Geological Survey, Open File Report 6100, p. 17-1 to 17-10.

Ames, D.E., Watkinson, D.H., and Parrish, R.R., 1998, Dating of a regional hydrothermal system induced by the 1850 Ma Sudbury impact event: Geology, v. 26, p. 447-450.

Ames, D.E., Galley, A.G., Legault, D., LaFrance, B., Dubois, A., Parker, J., Benn, K., Kjarsgaard, I., and Zierenberg, R., 2001, Sudbury Targeted Geoscience Initiative (TGI): Role of volatiles, structure and host rocks in the evolution of Sudbury ores, *in* Summary of Field Work 2001: Ontario Geological Survey, Open File Report 6070, p. 25-1 to 25-9.

Ames, D.E., Golightly, J.P., Lightfoot, P.C., and Gibson, H.L., 2002, Vitric compositions in the Onaping Formation and their relationship to the Sudbury Igneous Complex, Sudbury structure: Economic Geology, v. 97, p. 1541-1562.

Ames, D.E., Kjarsgaard, I.M., and Douma, S.L., 2003, Sudbury Ni-Cu-PGE Ore Mineralogy Compilation: Sudbury Targeted Geoscience Initiative (TGI): Geological Survey of Canada, Open File 1787, CD-ROM.

Ames, D.E., Buckle, J., Davidson, A., and Card, K., 2005, Sudbury bedrock compilation; Geology: Geological Survey of Canada, Open File 4570, scale 1:50 000.

Ames, D.E., Jonasson, I.R., Gibson, H.L., and Pope, K.O., 2006a, Impact-generated hydrothermal system- Constraints from the large Paleoproterozoic Sudbury crater, Canada, *in* Cockell, C., Gilmour, I., and Koeberl, C., eds., Biological Processes Associated with Impact Events: Impact Studies: Berlin-Heidelberg, Springer-Verlag, p. 55-100.

Ames, D.E., Singhroy, V., Buckle, J., Davidson, A., and Molch, K., 2006b, Integrated Bedrock Geology-Radarsat-Digital Elevation Data of Sudbury, Ontario: Geological Survey of Canada, Open File 4571, scale 1: 75 000.

Andrews, A.J., Owsiacki, L., Kerrich, R., and Strong, D.F., 1986, The silver deposits at Cobalt and Gowganda, Ontario: I. Geology, petrography and whole-rock geochemistry: Canadian Journal of Earth Sciences, v. 23, p. 1480-1506.

Bailey, J., Lafrance, B., McDonald, A.M., Fedorowich, J.S., Kamo, S., and Archibald, D.A., 2005, Mazatzal-Labradorian-age (1.7-1.6 Ga) ductile deformation of the South Range Sudbury impact structure at the Thayer Lindsley mine, Ontario: Canadian Journal of Earth Sciences, v. 41, p. 1491-1505.

Bajc, A.F., and Hall, G.E.M., 2000, Geochemical Response of Surficial Media, North and East Ranges, Sudbury Basin: Ontario Geological Survey, Open File Report 6033, 265 p.

Ballhaus, C., Tredoux, M., and Späth, A., 2001, Phase relations in the Fe-Ni-Cu-PGE-S system at magmatic temperature and application to massive sulphide ores of the Sudbury Igneous Complex: Journal of Petrology, v. 42, p. 1911-2026.

Bell, R., 1891, Report on the Sudbury Mining District, Districts of Algoma and Nipissing, Province of Ontario, Sudbury Sheet No. 130: Geological Survey of Canada, Separate Report No. 326, 95 p.

Bennett, G., Dressler, B.O., and Robertson, J.A., 1991, The Huronian Supergroup and associated intrusive rocks, Chapter 14 *in* Geology of Ontario: Ontario Geological Survey, Special Volume 4, Part 1, p. 549-591.

Cabri, L.J., 1981, The platinum group minerals, *in* Cabri, L.J., ed., Platinum-Group Elements: Mineralogy, Geology, Recovery: Canadian Institute of Mining and Metallurgy, v. 23, p. 136-138.

——— 1988, Overview on application of platinum mineralogy to mineral exploration and processing, *in* Carson, D.J.T., and Vassiliou, A.H., eds., Process Mineralogy VIII: Applications of Mineralogy to Mineral Beneficiation Technology, Metallurgy and Mineral Exploration and Evaluation, with Emphasis on Precious Metal Ores: Symposium Proceedings, The Minerals, Metals and Materials Society, Phoenix, Arizona, January 25-28, 1988, p. 23-31.

——— 2002, The Platinum group minerals, *in* Cabri, L.J., ed., The Geology, Geochemistry, Mineralogy and Mineral Beneficiation of Platinum Group Element: Canadian Institute of Mining, Metallurgy, and Petroleum, Special Volume 54, p. 13-130.

Cabri, L.J., and LaFlamme, J.H.G., 1976, The mineralogy of the platinum-group elements from some copper-nickel deposits of the Sudbury area, Ontario: Economic Geology, v. 71, p. 1159-1195.

——— 1984, Mineralogy and distribution of platinum-group elements in mill products from Sudbury, *in* Park, W.C., Hausen, D.M., and Hagni, R.D., Proceedings, 2nd International Congress on Applied Mineralogy in the Minerals Industry: American Institute of Mining and Metallurgy, p. 911-922.

Cabri, L.J., Blank, H., El Goresy, A., Nobiling, R., Sizgoric, M.B., and Traxel, K., 1984, Quantitative trace element analyses of sulphides from Sudbury and Stillwater by proton microprobe: Canadian Mineralogist, v. 22, p. 521-542.

Card, K.D., 1994, Geology of the Levack gneiss complex, the northern footwall of the Sudbury structure, Ontario, *in* Canadian Shield: Geological Survey of Canada, Current Research 1994-C, p. 269-278.

Chadwick, J., 1996, Exploring Victor: Mining Magazine, Nov. 1, 1996, p. 270-275.

Coats, C.J.A., and Snajdr, P., 1984, Ore deposits of the North Range, Onaping-Levack area, Sudbury, *in* Pye, E.G., Naldrett, A.J., and Giblin, P.E., eds., The Geology and Ore Deposits of the Sudbury Structure: Ontario Geological Survey, Special Volume 1, p. 328-346.

Cochrane, L., 1984, Ore deposits of the Copper Cliff offset, *in* Pye, E.G. Naldrett, A.J., and Giblin, P.E., eds., The Geology and Ore Deposits of the Sudbury Structure: Ontario Geological Survey, Special Volume 1, p. 347-360.

Corfu, F., and Andrews, A.J., 1986, A U-Pb age for mineralized Nipissing diabase, Gowganda: Canadian Journal of Earth Sciences, v. 23, p. 107-109.

Dessureau, G.R., 2002, Geochemistry of the mafic and felsic norite: Implications for the crystallization history of the Sudbury Igneous Complex, Sudbury Ontario: 9th International Platinum Symposium, Billings, Montana.

Dietz, R.S., 1964, Sudbury structure as an astrobleme: Journal of Geology, v. 72, p. 412-434.

Distler, V.V., 1994, Platinum mineralization of the Noril'sk deposits, in Proceedings of the Sudbury-Noril'sk Symposium: Ontario Geological Survey, Special Volume 5, p. 243-262.

Farrow, C.E.G. 1994, Geology, Alteration, and the Role of Fluids in Cu-Ni-PGE Mineralization of the Footwall Rocks to the Sudbury Igneous Complex, Levack and Morgan Townships, Sudbury District, Ontario: Ph.D. thesis, Carleton University, Ottawa, Ontario, 373 p.

Farrow, C.E.G., and Jolette, C., 2004, Sudbury Cu-PGE environments: Defining 'sharp-walled' and 'low-sulphide' systems: Abstract SS113-06, May 12, 2004, St. Catherines, Ontario, Geological Association of Canada, St-Catherines 2004, p. 248.

Farrow, C.E.G., and Lightfoot, P.C., 2002, Sudbury PGE revisited: Toward an Integrated model, *in* Cabri, L.J., ed., The Geology, Geochemistry, Mineralogy and Mineral Beneficiation of Platinum Group Element: Canadian Institute of Mining, Metallurgy and Petroleum, Special Volume 54, p. 13-130.

Farrow, C.E.G., and Watkinson, D.H., 1992, Alteration and the role of fluids in Ni, Cu and platinum-group element deposition, Sudbury Igneous Complex contact, Onaping-Levack area, Ontario: Mineralogy and Petrology, v. 46, p. 67-83.

——— 1996. Geochemical characteristics of the Epidote Zone: Its development and associated Ni-Cu-PGE remobilization by saline fluids, Fraser Mine, Sudbury, Ontario. Exploration and Mining Geology, v. 5, p. 17-31.

——— 1997, Diversity of precious metal mineralization in footwall Cu-Ni-PGE deposits, Sudbury, Ontario: Implications for hydrothermal models of formation: Canadian Mineralogist, v. 35, p. 817-839.

——— 1999, An evaluation of the role of fluids in Ni-Cu-PGE-bearing, mafic-ultramafic systems, *in* Keays, R.R., ed., Dynamic Processes in Magmatic Ore Deposits and their Application in Mineral Exploration: Geological Association of Canada, Short Course Notes, v. 13, p. 31-67.

Farrow, C.E.G., Watkinson, D.H., and Jones, P.C., 1994, Fluid inclusions in sulfides from North and South range Cu-Ni-PGE deposits, Sudbury Structure, Ontario: Economic Geology, v. 89, p. 647-655.

Farrow, C.E.G. Everest, J.O., King, D.M., Jolette, C., 2005, Sudbury Cu-(Ni)-PGE systems: Refining the classification using McCreedy West mine and Podolsky project case studies, *in* Mungall, J.E., ed., Exploration for Deposits of Platinum-Group Elements: Mineralogical Association of Canada, Short Course Series v. 35, p. 163-180.

Fedorowich, J.S., Rousell, D.H., and Peredery, W.V., 1999, Sudbury breccia distribution and orientation in an embayment environment, *in* Dressler, B.O., and Sharpton, V.L., eds., Large Meteorite Impacts and Planetary Evolution II: Boulder, Colorado: Geological Society of America, Special Paper 33, p. 305-314.

Fleet, M.E., 1977, Origin of disseminated copper-nickel sulphide ore at Frood, Sudbury, Ontario: Economic Geology, v. 71, p. 1449-1456.

French, B.M., 1967, Sudbury structure Ontario: some petrographic evidence for and origin by meteorite impact, *in* French, B.M., and Short, N.M., Shock Metamorphism of Natural Materials: Mono Book Corporation, Baltimore, p. 383-412.

Friske, P.W.B., McCurdy, M.W., and Day, S.J.A., 1997, National Geochemical Reconnaissance, Ontario compilation; Distribution of Nickel in 17 935 Lake Sediment Samples: Geological Survey of Canada, Map 3379a, scale 1:1 500 000.

Grant, R.W., and Bite, A., 1984, Sudbury quartz diorite offset dykes, *in* Pye, E.G., Naldrett, A.J., and Giblin, P.E., eds., The Geology and Ore Deposits of the Sudbury Structure: Ontario Geological Survey, Special Volume 1, p. 275-301.

Gray, M.J., 1995, The Geological Setting of the Vermilion Zn-Cu-Pb-Ag-Au Massive Sulphide Deposit, Sudbury Basin, Canada: M.Sc. thesis, Laurentian University, Sudbury, Ontario, 244 p.

Grieve, R.A.F., 1994, An impact model of the Sudbury Structure, *in* Lightfoot, P.C., and Naldrett, A.J., eds., Proceedings of the Sudbury-Noril'sk Symposium: Ontario Geological Survey, Special Volume 5, p. 245-270.

Hanley, J.J., and Mungall, J.E., 2003, Chlorine enrichment and hydrous alteration of the Sudbury breccia hosting footwall Cu-Ni-PGE mineralization at the Fraser mine, Sudbury, Ontario, Canada: Canadian Mineralogist, v. 41, p. 857-881.

Hanley, J.J., Mungall, J.E., Bray, C.J., and Gorton, P., 2004, The origin of bulk and water-soluble Cl and Br enrichments in ore-hosting Sudbury breccia in the Fraser copper zone, Strathcona embayment, Sudbury, Ontario, Canada: Canadian Mineralogist, v. 42, p. 1297-1318.

Hanley, J.J., Mungall, J.E., Pettke, T., and Spooner, E.T.C., 2005, Ore metal redistribution by hydrocarbon-brine and hydrocarbon-halide melt mixtures, North Range footwall of the Sudbury Igneous Complex, Ontario, Canada: Mineralium Deposita, v. 40, p. 237-256.

Hanley, J., Ames, D., Barnes, J., Sharp, Z., and Pettke, T., 2006: Stable Cl isotope evidence for multiple sources of Cl in hydrothermal and ore fluids at the Sudbury Igneous Complex, Canada. Geological Association of Canada – Mineralogical Association of Canada, Annual Meeting, Montreal, 1996.

Hawley, J.E., 1962, The Sudbury ores: Their mineralogy and origin: Canadian Mineralogist, v. 7, p. 1-207.

——— 1965, Upside-down zoning at Frood, Sudbury, Ontario: Economic Geology, v. 60, p. 529-575.

Hawley, J.E., Lewis, C.L., and Wark, W.J., 1951, Spectrographic study of platinum and palladium in common sulphides and arsenides of the Sudbury District, Ontario: Economic Geology, v. 46, p. 149-162.

Heaman, L. M., 1997, Global mafic magmatism at 2.45 Ga: Remnants of an ancient large igneous province?: Geology, v. 25, p. 299-302.

Jago, B.C., Morrison, G.G., and Little, T.L., 1994, Metal zonation patterns and microtextural and micromineralogical evidence for alkali- and halogen-rich fluids in the genesis of the Victor Deep and McCreedy East footwall copper orebodies, Sudbury Igneous Complex, *in* Lightfoot, P.C., and Naldrett, A.J., eds., Proceedings of the Sudbury-Noril'sk Symposium: Ontario Geological Survey, Special Volume 5, p. 65-75.

Jonasson, I.R., Eckstrand, O.R., and Watkinson, D.H., 1987, Preliminary investigations of the abundance of platinum, palladium and gold in some samples of Canadian copper-nickel ores: Geological Survey of Canada, Paper 87-1A, p. 835-846.

Keays, R.R., and Crocket, J.H., 1970, A study of precious metals in the Sudbury Nickel Irruptive ores: Economic Geology, v. 65, p. 438-450.

Killeen, J.G., 2006, Diversity of PGE mineralization in Cu-Ni-PGE systems within the PM Zone at the McCreedy West Mine, Sudbury, Ontario: B.Sc. (Honours) thesis, Carleton University, Ottawa, Ontario, 75 p.

King, A., Fullagar, P., and LaMontagne, Y., 1996, Borehole geophysics in exploration, development and production; Trends technologies and case histories for the modern explorationist: Prospectors and Developers Association, Short Course, Toronto, March, 1996.

King, D.M., Everest, J.O., and Farrow, C.E.G., 2006, FNX's PM deposit, McCreedy West Mine, Sudbury: The evolution of a 'low-sulphide' Cu-Ni-PGE resource: Canadian Institute of Mining, Metallurgy and Petroleum, Mining Conference and Exhibition, Technical Session,Vancouver, 2006.

Kormos, S.E., 1999. Metal Distribution within Zone 39, a Proterozoic Vein-Type Cu-Ni-Au-Ag-PGE Deposit, Strathcona Mine, Ontario, Canada: M.Sc. thesis, Laurentian University, Sudbury, Ontario, 163 p.

Krogh, T.E., Davis, D.W., and Corfu, F., 1984, Precise U-Pb Zircon and Baddeleyite Ages for the Sudbury Area, Chapter 20 *in* Pye, E.G., Naldrett, A.J., and Giblin, P.E., eds., The Geology and Ore Deposits of the Sudbury Structure: Ontario Geological Survey, Special Volume 1, p. 431-447.

Li, C., and Naldrett, A.J., 1993, High chlorine alteration minerals and calcium-rich brines in fluid inclusions from the Strathcona Deep Copper zone, Sudbury, Ontario: Economic Geology, v. 88, p. 1780-1796.

Li, C., Naldrett, A.J., Coates, C.J.A., and Johannssen, P., 1992, Platinum, palladium, gold and copper-rich stringers at the Strathcona mine, Sudbury: Their enrichment by fractionation of a sulfide liquid: Economic Geology, v. 87, p. 1584-1598.

Lightfoot, P.C., and Farrow, C. E., 2002, Geology, geochemistry and mineralogy of the Worthington Offset Dike: A genetic model for offset dike mineralization in the Sudbury Igneous Complex: Economic Geology, v. 97, p. 1419-1446.

Lightfoot, P.C., and Naldrett, A.J., 1996, Petrology and Geochemistry of the Nipissing Gabbro: Exploration Strategies for Nickel, Copper, and Platinum Group Elements in a Large Igneous Province: Ontario Geological Survey, Study 58, 81 p.

Lightfoot, P.C., Doherty, W., Farrell, K., Keays, R.R., Moore, M., and Pekeski, D., 1997, Geochemistry of the Main Mass, Sublayer, Offsets, and Inclusions from the Sudbury Igneous Complex, Ontario: Ontario Geological Survey, Open File Report 5959, 231 p.

Lightfoot, P.C., Keays, R.R., and Doherty, W., 2001, Chemical evolution and origin of nickel sulfide mineralization in the Sudbury Igneous Complex, Ontario, Canada: Economic Geology, v. 96, p. 1855-1875.

Martin, W.G., 1957, Errington and Vermilion mines: in Structural Geology of Canadian Ore Deposits: 6th Commonwealth Mining and Metallurgical Congress, Special Volume 2, p.363-376.

Marshall, D.D., Watkinson, D.H., Farrow, C.E.G., Molnar, F., and Fouillac, A.-M, 1999, Multiple fluid generations in the Sudbury Igneous Complex: Fluid inclusion, Ar, O, H, Rb and Sr evidence: Chemical Geology, v. 154, p. 1-19.

McLean, S.A., Straub, K.H., and Stevens, K.M., 2005, The discovery and characterization of the Nickel Rim South deposit, Sudbury, Ontario, *in* Mungall, J.E., ed., Exploration for Deposits of Platinum-Group Elements: Mineralogical Association of Canada, Short Course Series, v. 35, p. 358-368.

Milkereit, B., and the Sudbury Working Group, 1992, Deep geometry of the Sudbury Structure from seismic reflection profiling: Geology, v. 20, p. 807-811.

Molnar, F., Watkinson, D.H., and Jones, P.C., 2001, Multiple hydrothermal processes in footwall units of the North Range, Sudbury Igneous Complex, Canada, and implications for the genesis of vein-type Cu-Ni-PGE deposits: Economic Geology, v. 96, p. 1645-1670.

Money, D.P., 1993, Metal zoning in the Deep Copper Zone 3700 Level, Strathcona Mine, Ontario: Exploration Mining Geology, v. 2, p. 307-320.

Morris, W.A., 1982, A paleomagnetic investigation of the Sudbury basin offsets, Ontario, Canada: Tectonophysics, v. 85, p. 291-312.

Morrison, G.G., 1984, Morphological features of the Sudbury Structure in relation to an impact, *in* Pye, E.G., Naldrett, A.J., and Giblin, P.E., eds., The Geology and Ore Deposits of the Sudbury Structure: Ontario Geological Survey, Special Volume 1, p. 513-520.

Morrison, G.G., Jago, B.C., and White, T.L., 1994, Footwall mineralization of the Sudbury Igneous Complex, *in* Lightfoot, P.C., and Naldrett, A.J., eds., Proceedings of the Sudbury-Noril'sk Symposium: Ontario Geological Survey, Special Volume 5, p. 119-132.

Mourre, G., 2000, Geological relationships at discontinuities in the Copper Cliff quartz diorite offset: An investigation into offset dykes and their relationship to the Sudbury Igneous Complex: M.Sc. thesis, Laurentian University, Sudbury, Ontario, 237 p.

Mungall, J.E., Ames, D.E., and Hanley, J.J., 2004, Geochemical evidence from the Sudbury structure for crustal redistribution by large bolide impacts: Nature, v. 429, p. 546-548.

Murphy, A.J., and Spray, J.G., 2002: Geology, mineralization, and emplacement of the Whistle-Parkin offset dike, Sudbury: Economic Geology, v. 97, p. 1399-1418.

Murray, A., 1856, Geological Survey of Canada, 1853-1856, p. 180-181.

Naldrett, A.J., 1984, Mineralogy and composition of the Sudbury ores, *in* Pye, E.G., Naldrett, A.J., and Giblin, P.E., eds., The Geology and Ore Deposits of the Sudbury Structure: Ontario Geological Survey, Special Volume 1, p. 309-326.

Naldrett, A.J., and Kullerud, G., 1967, A study of the Strathcona Mine and its bearing on the origin of the nickel-copper ores of the Sudbury District, Ontario: Journal of Petrology, v. 8, p. 453-531.

Naldrett, A.J., and Pessaran, A., 1992, Compositional variation in the Sudbury ores and prediction of the proximity of footwall copper-PGE ore bodies. Ontario Geological Survey, Miscellaneous Paper, 159, p. 47-62.

Naldrett, A.J., Innes, D.G., Sowa, J., and Gortaon, M.P., 1982, Compositional variations within and between five Sudbury ore deposits: Economic Geology, v. 77, p. 1519-1534.

Naldrett, A.J., Pessaran, R., Asif, M., and Li, C., 1994, Compositional variation in the Sudbury ores and prediction of the proximity of Footwall Copper-PGE orebodies, *in* Lightfoot, P.C., and Naldrett, A.J., eds., Proceedings of the Sudbury-Noril'sk Symposium: Ontario Geological Survey, Special Volume 5, p. 133-143.

Naldrett, A.J., Asif, M., Schandl, E., Searcy, T., Morrison, G.G., Binney, W.P., and Moore, C., 1999, Platinum group elements in the Sudbury ores: Significance with respect to the origin of different ore zones and to the exploration for footwall ores: Economic Geology, v. 94, p. 185-210.

Negusanti, J.J., 1995, Terrestrial metal trends in the Sudbury Area: Mining and the Environment, Conference Proceedings, Sudbury '95, p. 1143-1149.

Noble, S.R., and Lightfoot, P.C., 1992, U-Pb baddeleyite ages for the Kerns and Triangle Mountain intrusions, Nipissing diabase, Ontario: Canadian Journal of Earth Sciences, v. 29, p. 1124-1129.

Owen, D.L., and Coats, C.J.A., 1984, Falconbridge and East mines, *in* Pye, E.G., Naldrett, A.J., and Giblin, P.E, eds., The Geology and Ore Deposits of the Sudbury Structure: Ontario Geological Survey, Special Volume 1, p. 371-378.

Pattison, E.F., 1979, The Sudbury sublayer: Canadian Mineralogist, v. 17, p. 257-274.

Peck, D.C., James, R., and Chubb, P., 1993, Geological environments for PGE-Cu-Ni mineralization in the East Bull Lake Gabbro-Anorthosite Intrusion, Ontario: Exploration and Mining Geology, v. 2, p. 85-104.

Peredery, W.V., and Morrison, G.G., 1984, Discussion of the origin of the Sudbury structure, *in* Pye, E.G., Naldrett, A.J., and Giblin, P.E., eds., The Geology and Ore Deposits of the Sudbury Structure: Ontario Geological Survey, Special Volume 1, p. 491-511.

Pflug, K.A., Killeen, P.G., and Mwenifumbo, C.J., 1994, Acoustic velocity logging at the McConnell nickel deposit, Sudbury area, Ontario: Preliminary in situ measurements, *in* Current Research 1994-C: Geological Survey of Canada, p. 279-286.

Prevec, S., 1992, U-Pb constraints on early Proterozoic mafic magmatism from the southern Superior and western Grenville provinces, Ontario, *in* Radiogenic Age and Isotopic Studies: Report 6: Geological Survey of Canada, Paper 92-2, p. 97-106.

Prevec, S.A., and Cawthorn, R. G., 2002, Thermal evolution and interaction between impact melt sheet and footwall: A genetic model for the contact sublayer of the Sudbury Igneous Complex, Canada: Journal of Geophysical Research, v. 107, p. 14.

Scott, R.G., and Spray, J.G., 1999, Magnetic fabric constraints on friction melt flow regimes and ore emplacement direction within the South Range breccia belt, Sudbury impact structure: Tectonophysics, v. 307, p. 163-189.

——— 2000, The South Range breccia belt of the Sudbury impact structure: A possible terrace collapse feature: Meteoritics and Planetary Science, v. 35, p. 505-520.

Shanks, W.S., and Schwerdtner, W.M., 1991, Structural analysis of the central and southwestern Sudbury Structure, Southern Province, Canadian Shield: Canadian Journal of Earth Sciences, v. 28, p. 411-430.

Singhroy, V., 2000, Remote sensing for monitoring the effects of mining in Sudbury, Canada, *in* Kuehn, I., King, T., Hoerig, B., and Peters, D., eds., Methods in Environmental Geology; Remote Sensing for Site Characterization, Case Studies: Springer Verlag, p. 106-113.

Souch, B.E., Podolsky, T., and Geological staff, 1969, The sulfide ores of Sudbury, *in* Wilson, H.D.G., and Bateman, A.A., eds., Magmatic Ore Deposits: A Symposium: Economic Geology, Monograph 4, p. 252-261.

Speers, E., 1957, The age relation and origin of common Sudbury Breccia: Journal of Geology, v. 65, p. 497-514.

Spiers, G., 2004, Metal Levels in the Soils of the Sudbury Smelter Footprint: Centre for Environmental Monitoring, Laurentian University, 156 p.

Spray, J.G., 1997, Superfaults: Geology, v. 25, p. 275-287.

Spray, J.G., Butler, H.R., and Thompson, L.M., 2004, Tectonic influences on the morphometry of the Sudbury impact structure: implications for terrestrial cratering and modelling: Meteoritics and Planetary Science, v. 37, p. 287-301.

Stewart, M., 2002, Petrology and Mineralogy of Cu-Ni-PGE Ore, Totten Area, Worthington Offset, Sudbury Igneous Complex: M.Sc. thesis, Carleton University, Ottawa, Ontario, 164 p.

Thompson, L.M., and Spray, J.G., 1994, Pseudotachylitic rock distribution and genesis within the Sudbury impact structure, *in* Dressler, B.L., Grieve, R.A.F., and Sharpton, V.L., eds., Large Meteorite Impacts and Planetary Evolution: Geological Society of America, Special Paper 293, p. 275-287.

Tuchscherer, M.G., and Spray, J.G., 2002, Geology, mineralization, and Emplacement of the Foy Offset Dike, Sudbury: Economic Geology, v. 97, p. 1377-1397.

Watkinson, D.H., Jones, P., and Wallbridge staff, 2005, Palladium minerals and hydrous silicate assemblages from the Broken Hammer Zone, Wisner Property of the North Range Footwall, Sudbury Igneous Complex: Geological Association of Canada, Annual Meeting, Halifax.

Wodicka, N., 1997, Sudbury Structure: Northern footwall rocks and Sudbury Igneous complex, *in* Ames., D.E., eds.,Timmins to Sudbury Transect: New Insights into the Regional Geology and Setting of Mineral Deposits: Geological Association of Canada Mineralogical Association of Canada, Joint Annual Meeting, Ottawa '97, Field trip B6 Guidebook, p. 73-93.

Wood, C.R., and Spray, J.G., 1998, Origin and emplacement of offset dykes in the Sudbury Impact Structure: Constraints from Hess: Meteoritics and Planetary Sciences, v. 33, p. 337-347.

Zurbrigg H.F., and Geological staff, 1957, Frood-Stobie mine: In Structural geology of Canadian ore deposits, Congress volume Canadian Institute of Mining and Metallurgy, v. 11, p. 341-350.

Ni-Cu-(PGE) Deposits in the Raglan Area, Cape Smith Belt, New Québec

C.M. Lesher

Mineral Exploration Research Centre, Department of Earth Sciences, Laurentian University, 935 Ramsey Lake Road, Sudbury ON P3E2C6

Corresponding author's e-mail: mlesher@laurentian.ca

Abstract

Nickel-Cu-(PGE) mineralization in the Raglan area occurs within a series of thick (50-200 m) mafic-ultramafic complexes that outcrop discontinuously along the contact between the Proterozoic Chukotat and Povungnituk groups in the east-central part of the Cape Smith Belt of the Ungava Peninsula between Cross Lake and Raglan Lake. All of the ores, host rocks, and country rocks have been regionally metamorphosed to lower greenschist facies, but igneous and volcanic structures and textures are extremely well preserved.

The ultramafic complexes appear to comprise two principal facies assemblages: conduit facies assemblages, which are laterally more restricted and composed primarily of peridotite, and channelized sheet facies assemblages, which comprise a laterally restricted conduit facies composed primarily of peridotite flanked by laterally extensive sheet facies composed of massive gabbro or differentiated peridotite-gabbro. Mineralization occurs exclusively within conduit facies of both assemblages, but the largest deposits occur within the larger conduit facies assemblages. Conduit facies in both assemblage types are relatively massive and undifferentiated, composed primarily of olivine mesocumulate with lesser olivine orthocumulate, thin lower margins of fine-grained pyroxene-porphyritic rock, and thin upper margins of fine-grained pyroxene-porphyritic rock capped by aphyric or microspinifex-textured basalt or basalt breccia. Many units exhibit columnar jointing in their upper and lower parts and polyhedral jointing in their uppermost parts. The lateral margins are interfingered with adjacent sediments and basalts and flanked in some areas by blocky and fluidal peperites. Footwall rocks (sulphidic graphitic semipelites, gabbros, and local basalts) have been eroded thermomechanically, forming larger broader V-shaped first-order embayments in the footwall rocks and superimposed smaller highly irregular (often re-entrant) second-order embayments that localize the Ni-Cu-PGE mineralization. Sediments underlying conduit facies are strongly hornfelsed (recrystallized, bleached, spotted), especially beneath ore-localizing embayments, but sediments and basalts overlying conduit facies, and underlying and overlying sheet facies are only very rarely and very locally contact metamorphosed. The ultramafic complexes have been previously interpreted as feeder sills and lava ponds, but many may represent deeply erosive lava conduits, some may represent invasive (downward burrowing) lava flows, and one may represent a feeder conduit.

The olivine mesocumulate rocks contain up to 40% MgO and rarely preserved relict olivine ranges Fo_{85-88}, but they appear to have formed from magmas originally containing 17 to 19% MgO and Fo_{87-89}. The high MgO and olivine contents suggest that they are petrogenetically related to olivine-phyric basalts in the lower Chukotat Group, but they are variably enriched in highly incompatible lithophile elements (Th-U-LREE) relative to moderately incompatible lithophile elements (Zr-MREE-Ti-Y-HREE) and depleted in Nb-Ta relative to Th compared to Chukotat basalts, consistent with variable degrees of local contamination by Povungnituk Group semipelites.

The Ni-Cu-PGE ores are texturally quite variable, ranging from massive and semimassive through net- and reverse net-textured, and disseminated ores at or near the bases of the complexes (Type I ores) to patchy and uniformly disseminated ores within the ultramafic complexes (Type II ores). Type I ores are localized within second-order embayments within the footwall rocks, which in some cases form linear-trending belts of differing ore tenor. Ore tenors (metals in 100% sulphides) range 4-17% Ni, 1-9% Cu, 3-25 ppm Pd, 1-6 ppm Pd, and 0.1-4 ppm Au, consistent with equilibration with a parental komatiitic basaltic (Chukotat) magma at magma:sulphide ratios (R factors) in the range 300 to 1100, followed by minor fractional crystallization of monosulphide solid solution and local tectonic/metamorphic modification. Sulphur isotope compositions are primarily 4 to 5 per mil, within the range of and consistent with derivation of the majority of the S from sulphides in the footwall semipelites. The ore zones in some areas appear to define multiple trends of differing ore tenor, as observed in other deposits of this type. The ores are interpreted to have formed by thermomechanical erosion of the sulphidic graphitic semipelites at an early stage in the emplacement of the host ultramafic units.

Résumé

La minéralisation en Ni-Cu-(ÉGP) dans la région de Raglan se présente dans une série d'épais (50 à 200 m) complexes mafiques-ultramafiques qui affleurent de manière discontinue le long du contact entre les groupes protérozoïques de Chukotat et de Povungnituk dans la partie centrale est de la zone de Cape Smith de la péninsule d'Ungava entre les lacs Cross et Raglan. Tous les minerais, les roches hôtes et les roches encaissantes ont subi un métamorphisme régional au faciès des schistes verts inférieurs, mais les structures et textures ignées et volcaniques sont extrêmement bien conservées.

Les complexes ultramafiques semblent associer deux principaux assemblages de faciès: des assemblages de faciès de conduits, latéralement moins étendus et principalement composés de péridotite, et des assemblages de faciès de nappes chenalisées, comprenant un faciès de conduit latéralement peu étendu et principalement composé de péridotite qui est flanqué de faciès de nappes étendus composés de gabbro massif ou de péridotite-gabbro différenciés. Seuls les faciès de conduits des deux assemblages sont minéralisés, mais les plus importants gisements se trouvent dans les plus grands assemblages de faciès de conduits. Dans les deux types d'assemblages, les faciès de conduits sont relativement massifs et non différenciés, composés principalement de mésocumulats d'olivine avec des quantités moindres d'orthocumulats d'olivine, de minces marges inférieures de roche porphyrique à grain fin renfermant du pyroxène et de minces marges supérieures de roche porphyrique à grain fin renfermant du pyroxène coiffées de basalte ou de brèche

Lesher, C.M., 2007, Ni-Cu-(PGE) Deposits in the Raglan Area, Cape Smith Belt, New Québec, *in* Goodfellow, W. D., ed., Mineral Deposits of Canada: A Synthesis of Major Deposit-Types, District Metallogeny, the Evolution of Geological Provinces, and Exploration Methods: Special Publication No. 5, Mineral Deposits Division, Geological Association of Canada, p. 351-386.

basaltique de texture aphyrique ou microspinifex. Un grand nombre des unités présentent une structure columnaire à leurs parties supérieure et inférieure et une structure polyédrique à leur partie sommitale. Les marges latérales sont interdigitées des sédiments et des basaltes adjacents et flanquées, dans certaines zones, de pépérites de textures polyédrique et fluidale. Les roches des épontes inférieures (semipélites graphitiques sulfurées, gabbros et basaltes par endroits) ont été thermomécaniquement érodées pour former dans les roches des épontes inférieures les plus grands rentrants de premier ordre en forme de V auxquels sont surimposés de plus petits rentrants très irréguliers de deuxième ordre dans lesquels se situent les minéralisations en Ni-Cu-ÉGP. Les sédiments sous-jacents des faciès de conduits sont intensément transformés en cornéenne (recristallisés, décolorés, tachetés), surtout sous les rentrants renfermant du minerai, mais les sédiments et les basaltes sus-jacents aux faciès de conduits ainsi que sus-jacents et sous-jacents aux faciès de nappes n'ont subi que très localement un métamorphisme de contact. Antérieurement, les complexes ultramafiques ont été interprétés comme étant des filons-couches d'alimentation et des étangs de lave, mais nombre d'entre eux pourraient s'avérer des conduits de lave très érosive, certains des coulées de lave invasives (s'enfouissant) et l'un d'entre eux en particulier pourrait être un conduit d'alimentation.

La roche des mésocumulats d'olivine peut renfermer jusqu'à 40 % de MgO et l'olivine relique rarement conservée se situe dans la plage Fo_{85-88}, mais semble s'être formée à partir de magmas renfermant de 17 à 19% de MgO dans la plage Fo_{87-89}. Les teneur élevées en MgO et en olivine suggèrent qu'elle est pétrogénétiquement reliée aux basaltes à olivine phyriques du Groupe de Chukotat inférieur, mais elle est variablement enrichie en éléments lithophiles très incompatibles (Th-U-éléments de terres rares légers) comparativement aux éléments lithophiles modérément incompatibles (Zr-éléments de terres rares intermédiaires-Ti-Y- éléments de terres rares lourds) et appauvrie en Nb-Ta par rapport au Th comparativement aux basaltes de Chukotat, ce qui est conforme à des degrés variables de contamination par les semipélites du Groupe de Povungnituk.

Les minerais de Ni-Cu-ÉGP sont de textures très variables, allant de massifs à semi-massifs, à des textures réticulées et réticulées inversées, à des minerais disséminés aux bases des complexes ou à leur proximité (minerais de type I) et à des minerais disséminés dans des bancs ou uniformément dans les complexes ultramafiques (minerais de type II). Les minerais de type I se trouvent dans les rentrants de deuxième ordre dans les roches des épontes inférieures qui forment dans certains cas des zones linéaires de minerais de différentes teneurs. Les teneurs des minerais (en métaux dans les minerais à 100 % sulfurés) varient de 4 à 17 % pour le Ni, de 1 à 9 % pour le Cu, de 3 à 25 ppm pour le Pt, de 1 à 6 ppm pour le Pd et de 0,1 à 4 ppm pour l'Au, ce qui est conforme à une équilibration avec un magma basaltique komatiitique d'origine (Chukotat) présentant des rapports magma/sulfure (facteurs R) de l'ordre de 300 à 1100, suivie d'une cristallisation fractionnaire mineure de la solution solide monosulfurée (SSM) puis d'une modification tectonique/métamorphique localisée. Les concentrations en isotopes du S sont principalement de 4 à 5 $^{o}/oo$, c'est-à-dire de l'ordre de celles auxquelles on s'attendrait si la majorité du S des sulfures était dérivé des semipélites de l'éponte inférieure. Dans certains secteurs, les zones de minerai semblent se présenter en multiples bandes de teneurs différentes, tel qu'observé dans d'autres gisements de ce genre. Les minerais se seraient formés par érosion thermomécanique des semipélites graphitiques sulfurées à un stade précoce de la mise en place des unités ultramafiques hôtes.

Introduction

The Ni-Cu-(PGE) deposits in the Raglan area of the 1.9 Ga Cape Smith Belt represent some of the best preserved and best exposed examples of magmatic sulphide mineralization associated with komatiitic rocks. They share many characteristics with other deposits of this type (Lesher, 1989; Lesher and Keays, 2002; Barnes, 2006; Barnes and Lesher, in press): 1) the host units occur at the base of the volcanic sequence and appear to represent the initial expression of komatiitic basaltic volcanism. 2) the host units are the thickest, most magnesian, and most olivine-rich rocks in the sequence, 3) the ores are localized in embayments that transgress footwall rocks, including S-rich sedimentary rocks, and 4) the ores are composed of Type I basal massive/net-textured/disseminated and lesser Type II internal disseminated Fe-Ni-Cu sulphides. However, they are different from other deposits of this type in some respects: 1) they appear to have formed in deeply erosive lava conduits, rather than in extrusive lava conduits (cf. Kambalda: Lesher et al., 1984) or feeder sills (cf. Thompson: Layton-Matthews et al., 2007) the orebodies are more pod-like than the more ribbon-like 'shoots' that characterize many other deposits of this type, 3) the ores have lower Ni/Cu and higher Pd/Ir ratios than deposits associated with high-Mg komatiites (e.g. Kambalda, Perseverance, Thompson: see compilation by Naldrett, 2004), and 4) the ore textural profiles are often much more complex than in other deposits of this type (e.g. Alexo, Kambalda).

The exploration history of the Raglan area has been summarized by Green and Dupras (1999). The first low-grade showings in this region were discovered at the western end of the Cape Smith Belt in 1898 by A.P. Low of the Geological Survey of Canada, and in 1931-1932 were confirmed to extend inland by the Cyril Knight Prospecting Company. However, it was not until 1956 that Harold Kenty and Murray Watts of LeMoyne Explorations discovered the high-grade mineralization at "Deception Creek" (Katinniq #1 showing). Additional exploration was done in the late 1950s and 1960s by a series of companies that would eventually become New Québec Raglan Nickel Mines and which led to the sinking of the Donaldson exploration shaft in 1968 to 1970. Primarily technical work was done in the 1970s, but a major program in 1981 and 1982, led by Colin Coats, mapped the entire length of the mineralized zone between Cross Lake and Donaldson at 1:12,000 scale. Falconbridge Ltd. purchased all of the minority interests in New Québec Raglan Nickel Mines in 1989 and major diamond drilling, geophysical, and mapping programs in 1989 and 1990, overseen by Michel Dufresne, led to an underground exploration program at Katinniq in 1991 and 1992, a feasibility study in 1993, and the construction of the Katinniq concentrator and accommodation complex in 1995 to 1996. Production from the Katinniq underground mine began in December 1997 and continues today under the ownership of Xstrata Ltd. Parts of the Zone 2 and Zone 3 areas have been mined as open pits and are being developed underground. As of

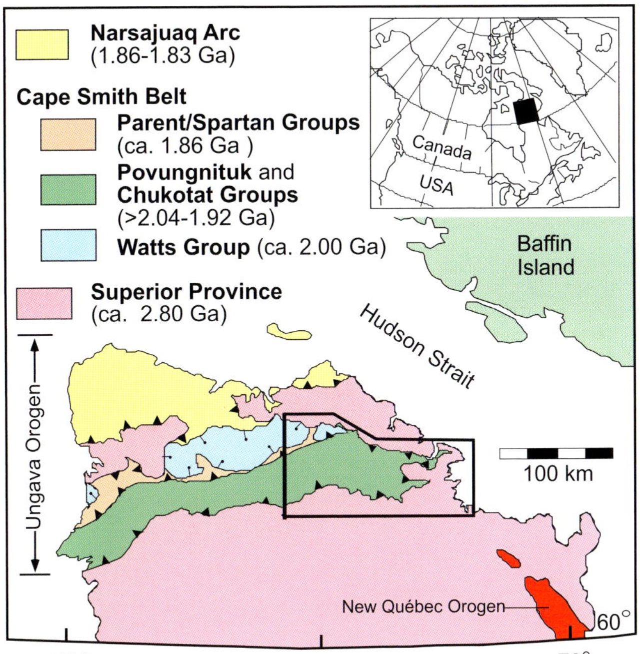

FIGURE 1. Map of the Ungava Peninsula north of 60ºN, showing the major geological elements of the Ungava Orogen (after St-Onge and Lucas, 1993). Area of Figure 2 is outlined.

December 31, 2005, total production was 6.89 Mt at 3.11% Ni and 0.91% Cu, mineral reserves (proven + probable) were 14.85 Mt at 2.80% Ni and 0.77% Cu, with a mineral resource of 3.39 Mt at 2.42% Ni and 0.80% Cu (measured + indicated), and 7.7 Mt at 3.0% Ni and 0.8% Cu (inferred) (Falconbridge Ltd., Annual Report, 2005).

Geologic Setting

The Raglan deposits occur in the early Proterozoic Cape Smith Belt, which extends east-west for 375 km across the Ungava Peninsula of northern Québec, between the Archean Superior Province in the south and a variety of Proterozoic 'suspect' terranes to the north (Fig. 1). The belt is bound to the south, east, and northeast by high-grade gneisses and plutonic rocks of the Superior Province and to the northwest by volcanic rocks of the Parent Group, which are interpreted as a volcanic arc (Picard et al., 1990), fine-grained clastic sedimentary rocks of the Spartan Group which are interpreted as fore-arc basinal deposits (St-Onge and Lucas, 1993), and mafic-ultramafic volcanic and intrusive rocks of the Watts Group, which are interpreted as an ophiolite (Scott et al., 1989). The Cape Smith Belt appears to represent a thin-skinned thrust belt preserved as a stack of klippen (Hoffman, 1985) in a doubly plunging synclinorium, and is interpreted as the preserved part of the foreland thrust belt to the Ungava

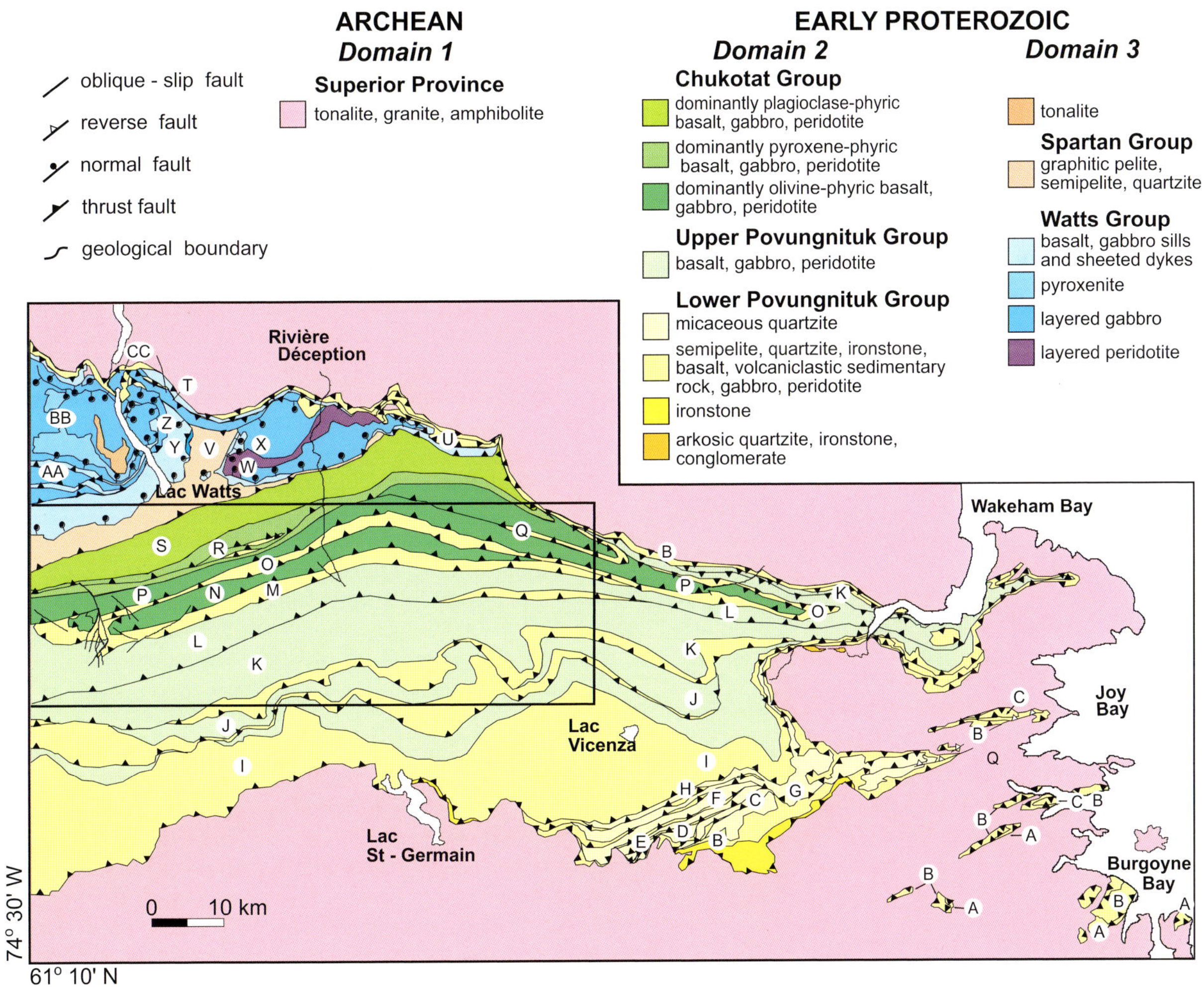

FIGURE 2. Geological compilation map of the eastern part of the Cape Smith Belt (after St-Onge and Lucas, 1993). Area of Figure 3 is outlined.

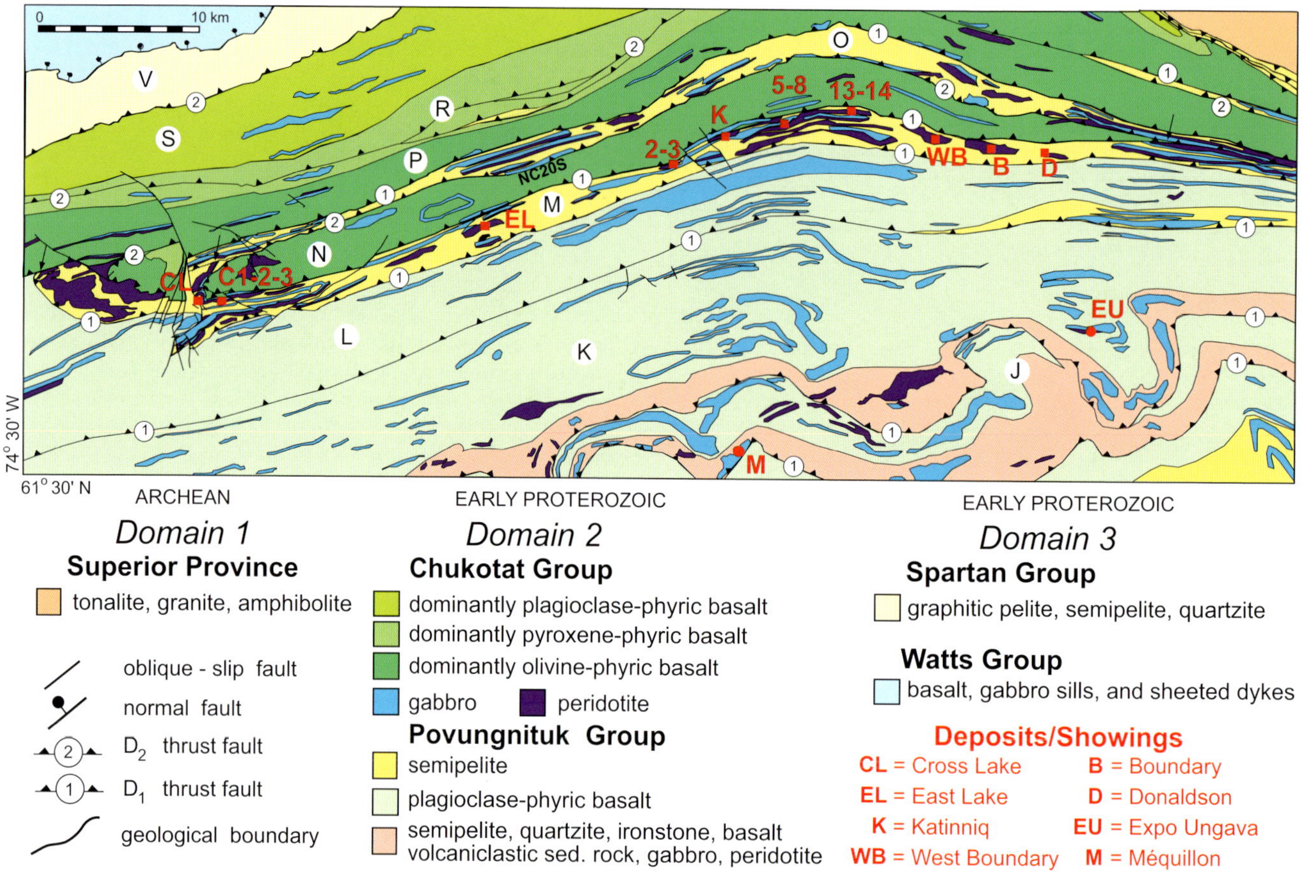

FIGURE 3. Geological compilation map of the Raglan area (modified from St-Onge and Lucas, 1993), showing the locations of major deposits along the Raglan Formation (Cross Lake to Donaldson) and some of those in the Delta horizon (Méquillon and Expo Ungava). NC20S = North Claim sill.

Orogen, an arc-continental collisional zone (Picard et al., 1990; St-Onge and Lucas, 1990).

Tectonostratigraphy

Bergeron (1959) subdivided the rocks in the Cape Smith Belt into two tectonostratigraphic groups: the Povungnituk Group and the Chukotat Group (Fig. 2). The Povungnituk Group comprises a lower sequence of primarily clastic sedimentary rocks and an upper sequence of primarily mafic volcanic and sedimentary rocks intruded by gabbro, pyroxenite, and peridotite sills. The Chukotat Group comprises a thin lower unit of mafic-ultramafic rocks overlain by olivine-phyric, pyroxene-phyric, and plagioclase-phyric basalts. This sequence is interpreted to represent the transition from initial rifting (lower Povungnituk Group) and continental basalt volcanism (upper Povungnituk Group) to opening of an ocean basin (Chukotat Group) (Francis and Hynes, 1979; Hynes and Francis, 1982; Francis et al., 1983; Picard et al., 1990; St-Onge and Lucas, 1993). The major lithological units in the east-central part of the Cape Smith Belt are described below, from structural and stratigraphic base (south) to structural and stratigraphic top (north).

Upper Povungnituk Group

Povungnituk basalt: The mafic volcanic rocks of the upper Povungnituk Group are exposed in the upper part of thrust sheet L (Figs. 2, 3; Table 1). There are excellent outcrops of these rocks along the west shore of Raglan Lake south of the Donaldson Camp and along the Deception River south of Katinniq. They comprise simple and compound massive and pillowed flows of tholeiitic basalt. They are fine- to very fine-grained, light to medium green in colour, and weather to a greenish-grey colour. They are composed primarily of albite-actinolite-chlorite with trace amounts of pyrite, and commonly exhibit a recrystallized intersertal to ophitic texture. The pillows are 0.1 to 1m long and often contain multiple pillow shelves (see Fig. 4G), indicating that they are lava tubes that experienced multiple episodes of lava emplacement and drainage (Sawyer et al., 1983). The contact with the overlying semipelites is poorly exposed and has been mapped as a D_1 regional thrust fault (thrust sheet M: Figs. 2, 3; Table 1) by Hynes and Francis (1979), Coats (1982), and St-Onge and Lucas (1993). However, the fault weaves back and forth across the contact and the sills above and below the contact are very similar, suggesting that the stratigraphic sequence is broadly conformable.

Povungnituk slate: The 1 to 2 km thick sequence of sedimentary rocks at the top of the Povungnituk Group is exposed in the lower part of thrust sheet M, directly underlying the Raglan Formation. These rocks do not outcrop well, but they are exposed discontinuously along the entire length of the Raglan Formation. They are dominated by semipelite, fine-grained graphitic sulphidic slate (see Fig. 4F), and argillite with minor quartzite. The semipelites are composed primarily of fine-grained quartz, white mica, and chlorite, and may contain up to 5% sulphides and significant amounts of graphite. Sedimentary structures, such as graded bedding, are rare. The contact with the overlying Raglan Formation is transgressive and locally sheared, but the uppermost sedimentary rocks are contact metamorphosed discontinuously along the entire length of the Raglan Formation between Cross Lake and Donaldson, and the degree and extent of metamorphism are greater beneath thicker parts of the ultramafic complexes than beneath thinner parts, and greatest beneath transgressive, second-order embayments beneath the ultramafic complexes (e.g. Thacker, 1995; Stilson, 1999). This indicates that the contact is locally unconformable (i.e. thermomechanical erosional), but not tectonic.

Povungnituk sills: Povungnituk basalts and slates have been intruded by mafic-ultramafic sills that range up to several hundred metres in thickness. They occur in thrust sheets I-M and O (Figs. 2, 3). Excellent examples outcrop south of Zone 13-14 (Fig. 3) and east of Cross Lake (Romeo I and II: Fig. 5A). Some are composed of massive gabbro or pyroxenite, or less commonly peridotite, and some are differentiated with thinner lower zones of columnar-jointed peridotite or oikocrystic olivine pyroxenite, and thicker upper zones of layered melanogabbro, mesogabbro, leucogabbro, and ferrogabbro. Some of the undifferentiated ultramafic bodies contain basal accumulations of subeconomic Ni-Cu-(PGE) sulphides (e.g. Expo Ungava, Bravo, Méquillon) and some of the differentiated mafic-ultramafic sills (e.g. Delta, Romeo I) contain narrow PGE-rich zones associated with thin pyroxene-rich pegmatoidal gabbro (e.g. Giovenazzo et al., 1989; Thibert, 1993). Where exposed, the sills have contact metamorphosed both underlying and overlying sedimentary rocks and sometimes contain rafts of overlying sedimentary rocks (St-Onge and Lucas, 1993; Thibert, 1993). These sills have been previously interpreted as feeders to overlying Chukotat volcanic rocks (e.g. Hynes and Francis, 1979; Francis et al., 1981, 1983; Bédard et al., 1983; Giovenazzo et al., 1989). Some of the undifferentiated ultramafic bodies have weighted average MgO contents greater than their chilled margins, indicating an excess olivine component of 30 to 40%, consistent with them being feeder sills (e.g. Bravo: Barnes and Giovennazo, 1990; Méquillon: Tremblay, 1990), but most of the sills have weighted average compositions similar to their chilled margins (Thibert, 1993; C.M. Lesher and R.R. Keays, unpubl. data), suggesting that they did not accumulate significant amounts of olivine and that they represent simple sills rather than subvolcanic feeders. All of the sills analyzed by Burnham et al. (1999) and Lesher et al. (2001) are contaminated, and could not have fed overlying uncontaminated Chukotat basalts (see below).

TABLE 1. Tectonostratigraphic column for the eastern Cape Smith Belt (adapted from St-Onge and Lucas, 1993).

Era	Group/Suite	Lithologies
Late Proterozoic		Diabase dykes
Early Proterozoic	**Narsajuaq**	Tonalite
	Spartan	Semipelite, pelite, quartzite, gabbro
	Watts	Basalt, gabbro sills, sheeted gabbroic dykes
		Pyroxenite
		Layered gabbro
		Layered peridotite
	Chukotat	Dominantly plagioclase-phyric basalt, gabbro
		Dominantly pyroxene-phyric basalt, gabbro
		Dominantly olivine-phyric basalt, gabbro; thick differentiated peridotite-gabbro flows and massive peridotite ± gabbro lava channel complexes; **Ni-Cu-PGE ores**
	Upper Povungnituk	Semipelite, layered gabbro-peridotite sills
		Basalt, volcaniclastic sedimentary rock, rhyolite; minor semipelite and quartzite, gabbro, peridotite, layered gabbro-peridotite sills
	Lower Povungnituk	Micaceous quartzite
		Basalt, volcaniclastic sedimentary rock, rhyolite; minor quartzite, dolomite, calc-silicate rock, gabbro, peridotite, layered peridotite gabbro sills
		Semipelite, pelite, micaceous quartzite, quartzite, conglomerate, ironstone, dolomite, calc-silicate rock; minor basalt and volcaniclastic rocks, gabbro, peridotite, layered peridotite-gabbro sills
		Ironstone, minor quartzite, and semipelite
		Quartzite, ironstone, conglomerate, semipelite
Archean		Tonalite, granite, amphibolite

Raglan Formation

The only economic Ni-Cu-(PGE) sulphide deposits discovered thus far in the Cape Smith Belt occur in the east-central part of the belt in the upper part of thrust sheet M (Figs. 2, 3). The thick mafic-ultramafic complexes that host the ores in this sheet define an apparently discontinuous but stratigraphically distinct, regionally mappable unit defined by Giovenazzo et al. (1989) as the Raglan Horizon and by Lesher et al. (1999) as the Raglan Formation. This unit extends 85 km from Cross Lake in the west, where it is terminated by a D_2 syncline and D_2 thrust fault O, across a D_4 antiform centred between the 5-8 and 13-14 areas, to Wakeham Lake in the east, where it is terminated by the same thrust fault (Figs. 2, 3; see also St-Onge and Lucas, 1994).

There are significant variations in the continuity and quality of outcrops, and in the density of diamond drill core drilling along the Raglan Formation. There are excellent outcrops in the Cross Lake and Katinniq areas, very good outcrops in parts of the Zone 2-3, Zone 5-8, and Boundary areas, and isolated outcrops in other areas, but many areas are covered by frost-heaved outcrop (felsenmeer) and glacial rubble. There is very good diamond drill core information at Katinniq, good drill core information for parts of Zone 2-3, Zone 5-8, and Donaldson, but only moderate drill information for most other areas along the belt. Detailed magnetic and electrical surveys have been done on most of the areas.

FIGURE 4. **(A)** Columnar-jointed peridotite in the lower part of the Katinniq Ultramafic Complex. Columns are ~20-30 cm in diameter. **(B)** Photomicrograph of olivine (lower order interference colours) mesocumulate rock with interstitial clinopyroxene (higher order interference colours), forming a heteradcumulate texture. Width of photo is 7.5 mm. Doubly polarized light. **(C)** Irregular lower contact of between peridotite, pyroxene-phyric basalt, and spotted hornfels along the lower margin of the Katinniq Ultramafic Complex. The contact dips ~45° to the north (right) subparallel to bedding in the upper part of the photo, but bends sharply southward, dips vertical and then ~45° westward, transgressing bedding in lower left part of photo. Underlying/adjacent metasediments are unfolded. **(D)** Lower contact of Katinniq Ultramafic Complex showing contact between lower pyroxene-phyric komatiitic basalt (pinkish grey, above hammer head), 20 cm thick layer of massive, completely-recrystallized semipelite (white, below hammer head), and spotted hornfels (grey, bottom). **(E)** Photomicrograph of pyroxene-phyric komatiitic basalt along lower contact of Katinniq Ultramafic Complex, containing small equant pyroxene phenocrysts (intermediate interference colours) in a matrix of recrystallized sprays of pyroxene (yellow interference colours) and altered interstitial 'glass' (light colours). Width of photo is 8 mm. Doubly polarized light.

Although more details are emerging from detailed exploration diamond drilling by Xstrata Ltd., the 1:2,000 surface mapping that was done prior to mine development (which included outcrops now covered by roads, buildings, and reservoirs) reported in Lesher et al. (1999) indicates that the Raglan Formation contains at least two of the facies assemblages defined by Lesher and Barnes (in press).

Conduit facies assemblage: These units are very thick (up to 200 m in true thickness), laterally restricted, and are composed primarily of relatively massive olivine mesocumulate rocks. Examples include the mineralized peridotite complexes at East Lake (Petch, 1999; Stewart, 2002), Zone 2 and Zone 3 (Mallinson, 1999a,b), Katinniq (Gillies, 1993; Lesher and Charland, 1999), Zone 6 and Zone 8 (Thacker, 1995; Mallinson, 1999c), Zone 13-14 (Vicker and Fedorowich, 1999), West Boundary (Charland, 1999), Boundary (Stilson and Lesher, 1999), and Donaldson (Lesher and Vicker, 1999).

FIGURE 4 CONTINUED. **(F)** Sulphidic graphitic slate ~20 m below lower contact of Katinniq Ultramafic Complex. **(G)** Pillow basalt with 17 shelves (drainage cavities) in upper part of Povungnituk Group, south side of Lac Raglan, Donaldson area.

Channelized sheet facies assemblage: These units are characterized by laterally restricted zones of relatively massive olivine mesocumulates flanked by laterally more extensive zones of differentiated olivine orthocumulates and gabbros. Examples include the Cross Lake – C1-C2-C3 complex (Thibert, 1999; Fig. 5B) and the Zone 2-3 – Katinniq – Zone 5-8 gabbro (Thacker, 1995; Mallinson, 1999a,b,c), which appears to be channelized only in the Zone 5 – Zone 7 area (Figs. 6, 7).

In some areas only a channelized sheet facies assemblage appears to be present (e.g. Cross Lake: Fig. 5), in some areas only a conduit facies assemblage appears to be present (e.g. East Lake: Petch, 1999; West Boundary: Charland, 1999; Donaldson: Lesher and Vicker, 1999), but where both are present, conduit facies assemblages cross-cut underlying channelized sheet facies assemblages (e.g. Zone 2-3: Mallinson, 1999a,b, Katinniq: Lesher and Charland, 1999; Zone 5-8: Thacker, 1995; Mallinson, 1999c), sometimes quite deeply (Figs. 6, 7, 8). The consistent stratigraphic relationships led Lesher et al. (1999) to define the channelized sheet facies assemblage as the Cross Lake Member and the conduit facies assemblage as the Katinniq Member.

Thus far, no mineralization has been found in any of the mafic-ultramafic bodies that occur in thrust sheet O (Figs. 2, 3), which is interpreted to be a duplication of mineralized thrust sheet M (St-Onge and Lucas, 1994). Most appear to represent sheet sills.

D_1 thrust fault N (Figs. 2, 3) sometimes occurs above and sometimes below the discontinuous horizon of sedimentary rock that separates the lowermost Chukotat basalts from the Raglan Formation (see Figs. 5, 6). Consequently, the mafic-ultramafic complexes are conformably overlain by sedimen-

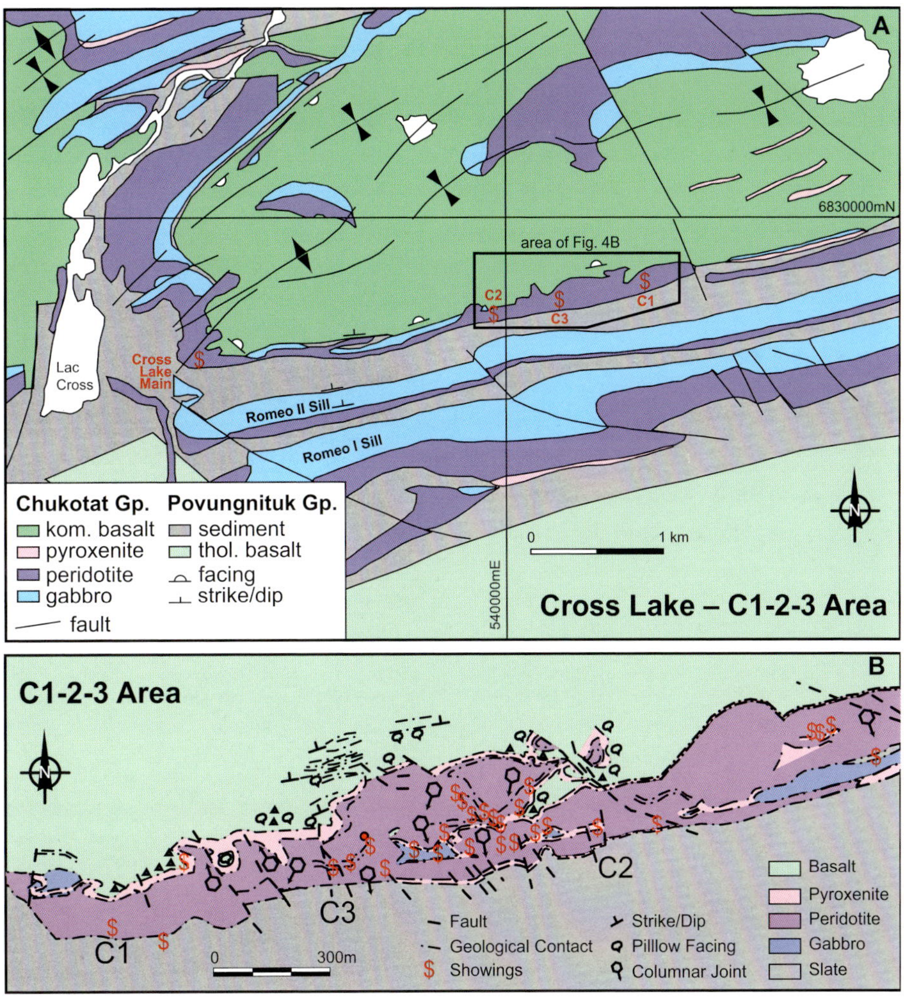

FIGURE 5. **(A)** Simplified geological maps of the Cross Lake area based on 1:12,000 scale mapping by Falconbridge Ltd., showing locations of the Cross Lake Main and C1-C2-C3 areas of the Cross Lake Member, and the Romeo I and Romeo II sills in the upper Povungnituk Group. **(B)** Simplified geological map of the C1-C2-C3 area based on 1:2000 scale mapping by Thibert (1999).

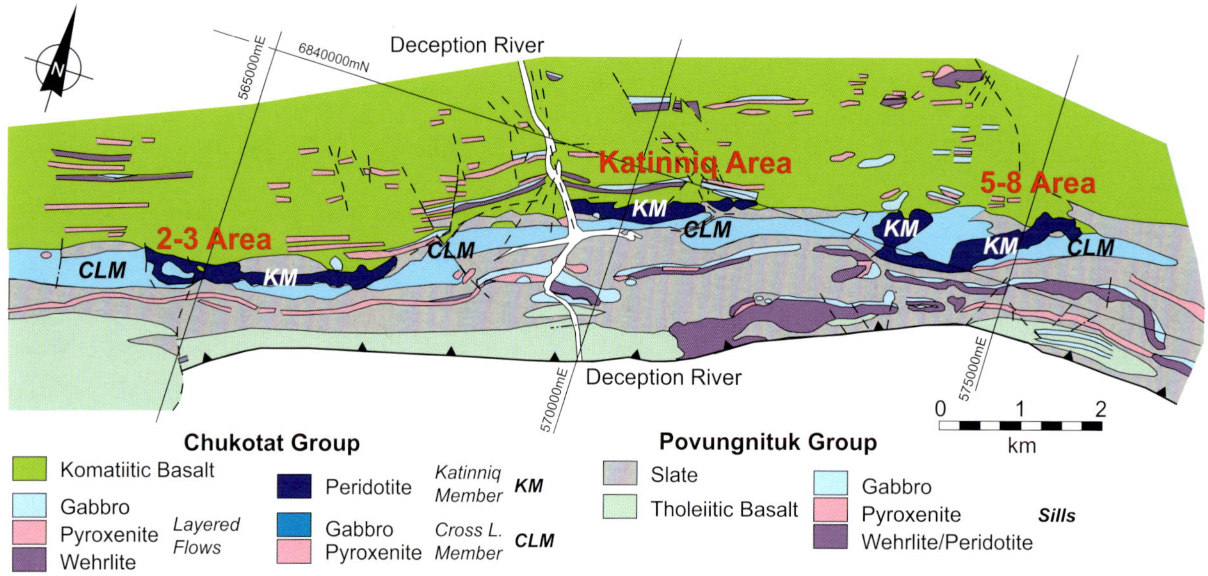

FIGURE 6. Geological compilation map of the Zone 2-3 - Katinniq - Zone 5-8 area showing stratigraphic relationships between the Cross Lake (CLM) and Katinniq (KM) members of the Raglan Formation (modified from 1:12,000 scale mapping by Falconbridge Ltd.).

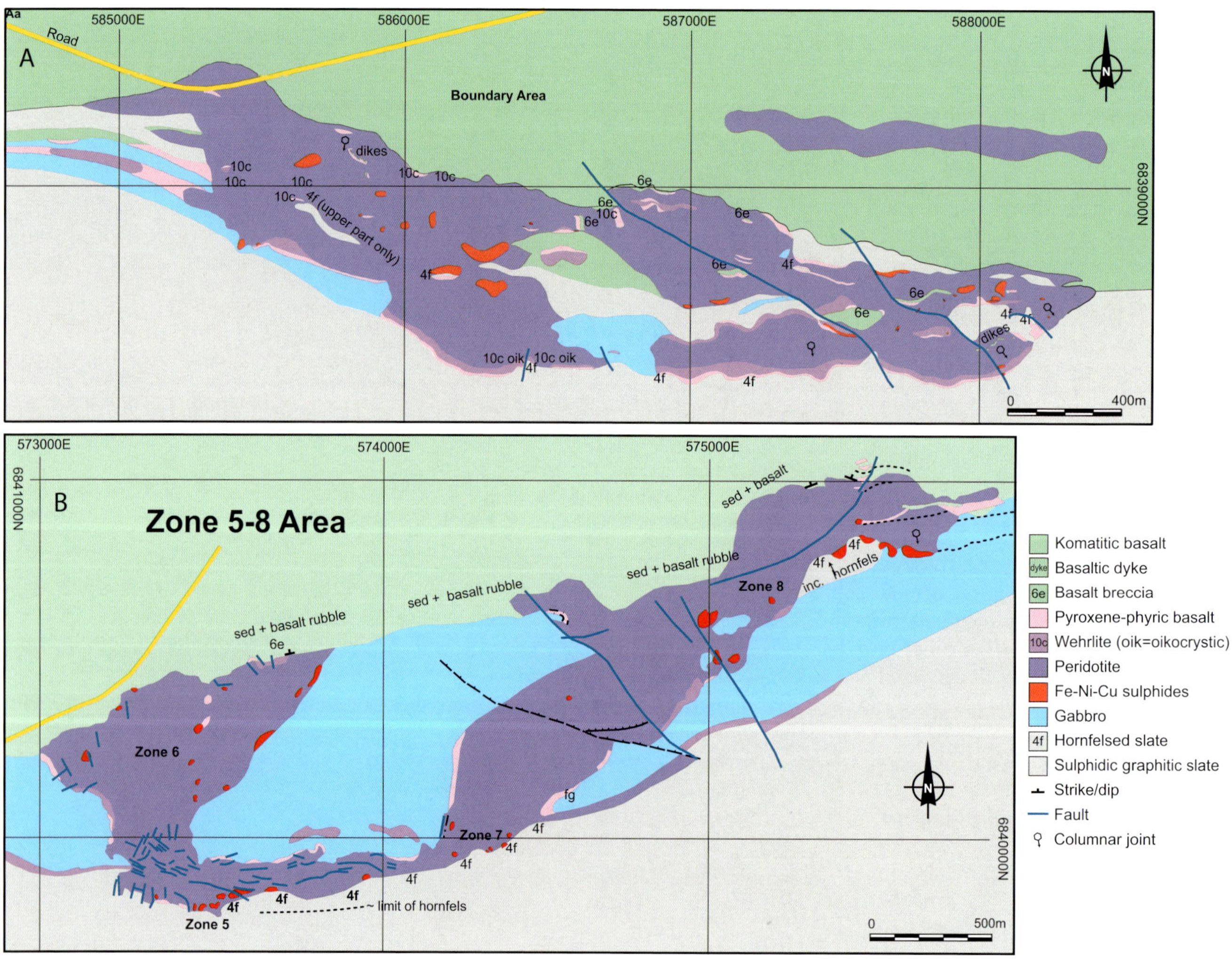

FIGURE 7. **(A)** Simplified geological map of the Boundary Ultramafic Complex based on 1:2,000 scale mapping by C.M. Stilson and C.M. Lesher in 1997. **(B)** Simplified geological map of the Zone 5-8 Ultramafic Complex based on 1:2,000 scale mapping by J.L. Thacker and C.M. Lesher in 1990, and C.M. Lesher and S.L. Gillies in 1991. The apparent discontinuity of some lithologies (e.g. pyroxene-phyric basalts and wehrlites) is an artifact of discontinuous exposure: both areas contain numerous outcrops, but most areas (especially ultramafic rocks) are covered by frost-heaved outcrop and some areas (especially hanging-wall basalts and sediments) are covered by rubble.

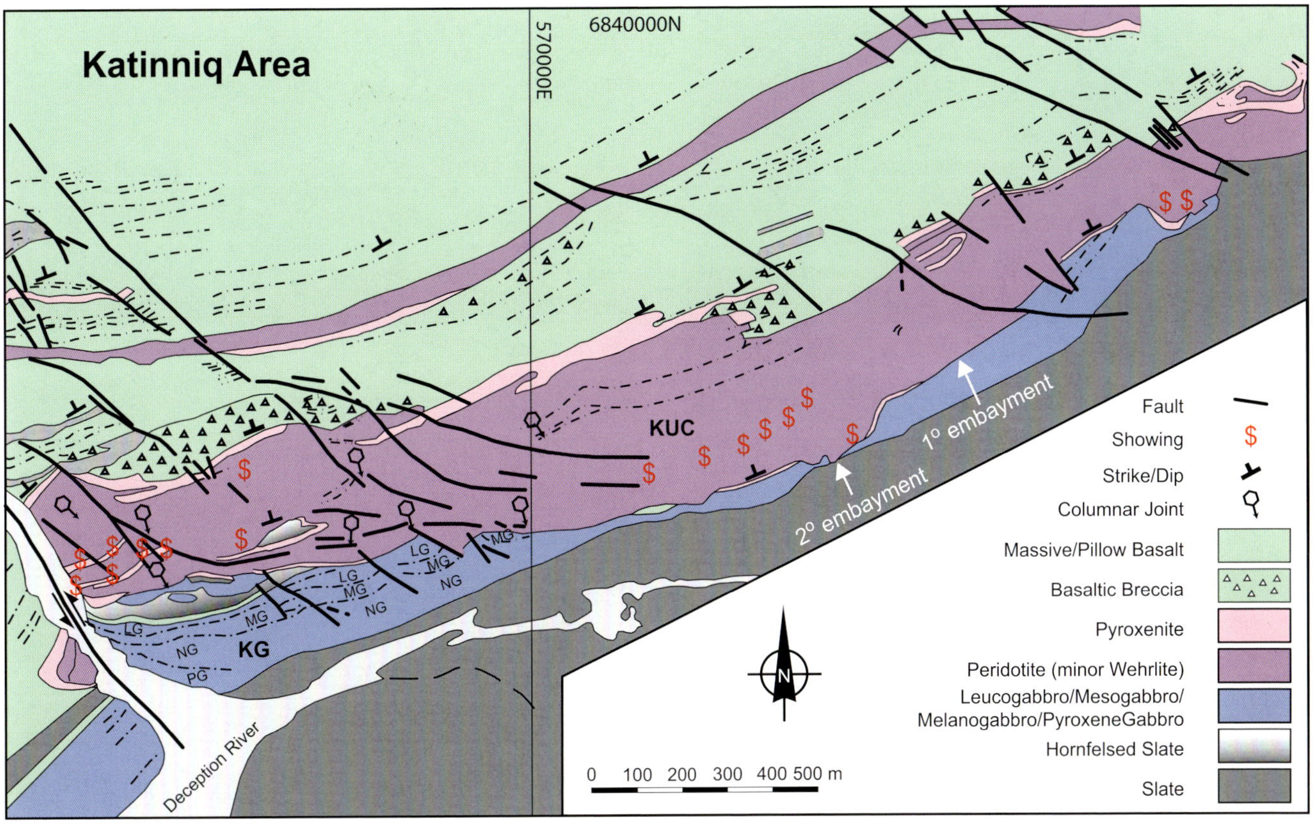

FIGURE 8. Simplified geological map of the Katinniq area based on 1:2,000 scale mapping by C.M. Lesher in 1989-1991 (prior to development of the Katinniq mine site and flooding of the Deception River). KG = Katinniq Gabbro, KUC = Katinniq Ultramafic Complex, LC = leucogabbro, PG = pyroxene gabbro, MG = melanogabbro, NG = mesogabbro.

tary rocks and Chukotat volcanic rocks in some cases and in fault contact with them in other cases. The displacement on this fault has not been determined, but the similarity of the rocks on either side of the fault (olivine-phyric basalts and semipelites) indicates that this part of the stratigraphic sequence is broadly conformable.

Lower Chukotat Group

The lower part of the Chukotat Group is dominated by olivine-phyric basalts that form simple and compound flows with massive, polygonized, pillowed, or brecciated facies. Massive lavas are characterized by ropy surfaces and polyhedral jointing, very fine-grained margins, and fine- to medium-grained central parts. Pillows exhibit indistinct rims and are commonly larger and longer (1-2 m long, 3:1 to 5:1 aspect ratios) in the lower parts of pillowed zones, and smaller and shorter (0.4-0.6m long, 1:1 to 2:1 aspect ratios) in the central and upper parts of pillowed zones (see Fig. 9A). Interpillow spaces are filled with a mixture of basalt breccia, hyaloclastite, pelite, and quartz±calcite. Pillow shelves, representing horizontal drainage cavities (Sawyer et al., 1983) are commonly filled with quartz±calcite and are excellent paleohorizon indicators. Olivine-phyric basalts are composed of actinolite-chlorite-serpentine and contain light green pseudomorphs after olivine phenocrysts and fine disseminated pyrrhotite (see Fig. 9B). The lower part of the Chukotat Group is characterized by the presence of thick sheet flows (up to 100 m), comprising lower olivine cumulate zones of reddish-brown-weathering olivine pyroxenite or pinkish-grey weathering pyroxenite and upper differentiated zones of grey weathering gabbro or basalt (Hynes and Francis, 1982; St-Onge and Lucas, 1993). Olivine-phyric basalts and differentiated flows are exposed in thrust sheet N and in the lower part of thrust sheet P (Figs. 2, 3). There are excellent outcrops in the core of the Cross Lake syncline (Fig. 5) and along the Deception River north of Katinniq (Figs. 6, 8).

Deformation

The rocks in the Cape Smith Belt record four major deformation events (Lucas, 1989; Lucas and St-Onge, 1989; St-Onge and Lucas, 1993): D_1 (<1.87-1.92 Ga) involved southward-directed piggy-back (break backward) thrusting and folding of the Povungnituk and Chukotat groups onto the Superior Craton, increasing the thickness of the Povungnituk Group from a minimum of 7 km (maximum stratigraphic thickness in thrust sheet K: Fig. 2, 3) to a maximum of 20 km (Lucas, 1989). D_1 faults produced a locally pronounced S_1 foliation that pre-dates peak metamorphism. They are most abundant in the southern part of the belt (e.g. thrust sheets A-N and P: Figs. 2, 3). This deformation also produced a penetrative shear fabric adjacent to the basal décollement and retrograde metamorphism of underlying granulite facies rocks of the Superior Province. D_2 (1.83-1.80 Ga) involved southward-directed out-of-sequence (break forward) thrusting and folding of the Povungnituk and Chukotat groups. D_2 structures crosscut D_1 structures and cut down section in the transport direction (e.g. thrust sheets O and Q-V: Figs. 2, 3 and 5). They also cross-cut metamorphic isograds and therefore post-date or are synchronous with peak metamorphism

FIGURE 9. **(A)** Pillow basalt in olivine-phyric member of the Chukotat Group north of the Katinniq mine area. Scale card is 16 cm long. **(B)** Photomicrograph of pyroxene spinifex-textured basalt in the olivine-phyric member of the Chukotat Group north of the Katinniq mine, containing olivine (small light equant hopper crystals in upper left and middle right) phenocrysts, acicular pyroxene spinifex crystals (light, randomly oriented), fine chains of pyroxene (light, between acicular olivine), and altered glass (matrix, tan). Width of photo is 8 mm. Plane-polarized light. **(C)** and **(D)** Flow-top breccias overlying columnar-jointed pyroxene-phyric komatiitic basalt along the upper margin of the lower ultramafic unit in the Zone 2 Ultramafic Complex (C) and the upper margin of the Katinniq Ultramafic Complex (D). **(E)** Pillow breccia along upper margin of the Zone 2 Ultramafic Complex. Pillows in centre of photo are 15 cm long. **(F)** Photomicrograph of pyroxene spinifex-textured basaltic flow-top breccia along the upper margin of the Katinniq Ultramafic Complex, containing olivine (small light equant hopper crystals) phenocrysts, acicular clinopyroxene /orthopyroxene spinifex crystals (light with dark cores, some oriented parallel and some perpendicular to section) in a matrix of altered glass (black). Width of photo is 8 mm. Plane-polarized light. (photos by M. Lévesque).

(St-Onge and Lucas, 1990; Bégin, 1992; St-Onge and Lucas, 1993). This deformation also produced basement imbrications, penetrative shear fabrics, and continued retrograde metamorphism of underlying rocks of the Superior Province. D_3 (ca. 1.76 Ga) involved folding of the allochthonous Cape Smith Belt and the underlying autochthonous Superior Province basement about east-west-trending fold axes (St-Onge and Lucas, 1993). D_4 (<1.74-1.76 Ga) involved east-west compression and folding of the thrust belt and basement about moderately steep northward plunging fold axes, which is most pronounced in the southern part of the belt (Fig. 2).

Although more information is emerging as the deposit is being mined, the structure of the Katinniq Ultramafic Complex is better known than other areas. The complex is transgressed by a series of east-northeast-trending, bedding-parallel faults and north-northwest-trending high-angle faults that break the unit into several lozenge-shaped blocks

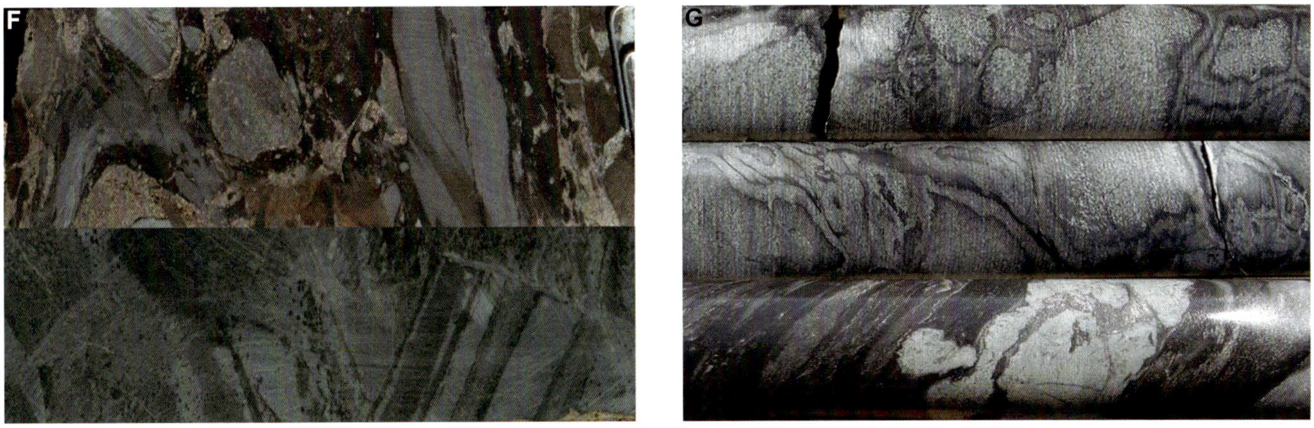

Figure 9 continued. **(G)** Sulphide-rich (top) and sulphide-poor (bottom) sediment breccias along the lateral margin of the Katinniq Ultramafic Complex (Gillies, 1993); the sediment fragments are hornfelsed, but the matrix is not. **(H)** Fluidal peperites along the upper lateral margins of the Zone 2 Peridotite Complex (top two cores) and Katinniq Ultramafic Complex (bottom core) (photos by M. Lévesque).

(Figs. 8, 10), some of which have been rotated slightly (e.g. westernmost block), as indicated by changes in the strike of lithological contacts and plunge azimuths of columnar joints. The Deception River Fault, which offsets Chukotat basalts, differentiated peridotite-gabbro flows, a small tip of the Katinniq Ultramafic Complex, and the Katinniq Gabbro exposed on the west side of the river from the same lithologies exposed of the east side of the river (Fig. 8), is one of the north-northwest-trending faults. Most north-northwest-trending faults offset the hanging wall and footwall contacts of the peridotite slightly, but appear to ramp into and bottom out along east-northeast-trending thrust faults and stratigraphic contacts in overlying sedimentary rocks/basalts and underlying sedimentary rocks/gabbros (Fig. 8). Underground mapping (Chisholm et al., 1999; Chisholm, 2002; Falconbridge Ltd., unpubl. data) indicates that some north-northwest-trending faults crosscut (and therefore postdate) east-northeast-trending faults, but some of the movement on the north-northwest-trending faults may be related to the regional D_4 folding event. The overall style of faulting indicates a brittle regime (Fedorowich, 1999), which is consistent with the relatively low metamorphic grade (see below).

The style of deformation varies with rock type and intensity of deformation, which is very heterogeneous. Pyroxenites typically shear, peridotites typically fracture, and gabbros rarely deform. Most faults in peridotites, for example, comprise numerous closely spaced (1-50 cm) microfaults and shear joints with limited horizontal displacement (<1 m), which fracture the rock but result in little apparent horizontal displacement. Slickenfibres along the north-northwest-trending faults consistently plunge shallowly to moderately northward (i.e. subparallel to dip), consistent with the limited amount of observed horizontal displacement. The minor horizontal component of displacement is normally sinistral, but some fault zones contain narrow blocks exhibiting dextral displacement. The amount of dip-parallel displacement is not yet known. The lithologies are different on either side of some of the faults (e.g. the two westernmost blocks in Fig. 8), but the faults do not appear to displace the shallowly pitching linear ore shoots significantly (see below), suggesting that the amount is minor (10s of metres). The best interpretation at this stage is that the east-northeast- and north-northwest-trending faults represent a compartmentalized D_1-D_2 thrust fault system with east-northeast-trending flats and north-northwest-trending lateral ramps, modified during D_3 and D_4.

The peridotites contain abundant serpentine-rich fractures, which formed during the serpentinization process, very early after emplacement and burial, and that represent the pathways by which fluids were introduced into the rocks and by which certain components (e.g. Cs, Rb, K, Na; Sr, Ba, Ca) were removed (Chisholm, 2002). The serpentine-rich margins were very susceptible to shearing during subsequent deformation and uplift (Chisholm et al., 1999; Chisholm, 2002).

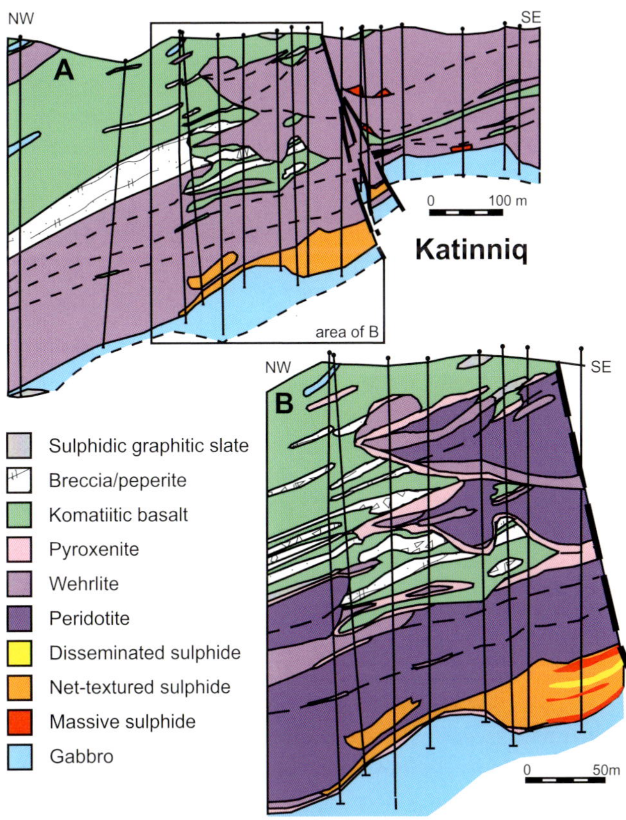

Figure 10. **(A)** Northwest-southeast vertical section through the central part of the Katinniq Ultramafic Complex (Gillies, 1993). Thinner vertical lines are drill cores on which the section is based; thicker dashed lines are faults. **(B)** Details in central part of same section.

TABLE 2. Ultramafic and mafic lithologies in the Raglan area, showing local rock names, igneous textures and mineralogy, metamorphic assemblages, field characteristics, and MgO contents (volatile-free).

Lithology *Local Name*	**Igneous Texture**	**Igneous Mineralogy**	**Metamorphic Assemblage**	**Fresh Colour**	**Fracture**	**Weathered Colour**	**Magnetism**	**MgO** (%)
Ol adcumulate *Dunite*	mg ac Ol	>90% Ol-(Chr) <10% glass	Ant-Mag-Chl±Cpx±Sul	vdk grn blk	chon	lt tan	strong	35-40
Ol mesocumulate *Peridotite*	fg mc Ol cg oik Cpx	75-90% Ol-(Chr) 25-10% glass ± skeletal Cpx	Ant-Mag-Chl±Pyx±Sul	dk grn blk	subch	lt brn	mod	30-35
Ol orthocumulate *Olivine Pyroxenite*	fg oc Ol vcg oik Cpx	10-50% Ol-(Chr) 10-60% Cpx 40-10% glass	Cpx-Ant-Tr-Chl-(Mag)	dk grn gry	uneven	rd brn - gry	sl	25-30
Pyroxenite *Cumulate Pyroxenite*	mg mc	75-90% Cpx 25-20% glass	Cpx-Tr	gry grn	uneven	gry	non	20-25
Pyx - Porphyritic Basalt *Pyroxenite*	fg asx Cpxϕ	15-20% Ol-(Chr) 10-20% Cpx 60-75% glass	Cpx-Ant-Chl-Tr-(Mag)	gry grn	hackly	pnk - gry	non	20-25
Melanogabbro	mg-cg	Cpx-Pl		vdk gry	rough	dk gry, khaki	non	
Mesogabbro	fg-cg	Pl-Cpx		dk gry	rough	med gry	non	
Lecuograbbro	mg-cg	Qtz-Pl-Cpx		med gry	rough	lt gry	non	
Ferrogabbro	vcg-cres	Qtz-Pl-Cpx		lt gry + dk grn	rough	wht + dk grn	non	
Ol-phyric basalt	vfg Olϕ	<10% Ol >90% glass	Ant-Chl-Act ± Cpx ± Sul	lt grn + dk gry	uneven	gry	non	14-18
Pyx-phyric basalt	vfg Pyxϕ	<10% Cpx >90% glass	Act-Chl ± Ant ± Cpx ± Sul	dk grn + gry	uneven	gry	non	10-14
Pl-phyric basalt	vfg Plϕ	<10% Pl >90% glass	Ab-Act-Chl ± Ant ± Cpx ± Sul	wht + lt grn	uneven	gry	non	6-10

Textures: **ac** = adcumulate, **mc** = mesocumulate, **oc** = orthocumulate, **cres** = crescumulate, **oik** = oikocrystic, **ϕ** = phyric, **rsx** = random Ol spinifex, **asx** = acicular pyroxene spinifex, **aph** = aphyric, **dec**= decussate, **fg** = fine-grained, **cg** = coarse-grained. ***Colours***: **vdk** = very dark, **dk** = dark, **lt** = light, **grn** = green/greenish, **gry** = gray/grayish, **brn** = brown, **rd** = reddish, **pnk** = pinkish. ***Fracture***: **chon** = conchoidal, **subch** = subconchoidal. Minerals: **Ol** = olivine, **Cpx** = clinopyroxene, **Chr** = chromite, **Mag** = magnetite, **Chl** = chlorite, **Sul** = Fe-Ni-Cu sulphides, **Tr** = tremolite, **Pl** = plagioclase.

Ductile deformation in the Katinniq area is manifested as a heterogeneous foliation (S_1) subparallel to bedding (S_0), that is well developed locally in semipelites and pyroxenites bordering the ultramafic complex. This is attributed to rheological contrasts at the margins of the peridotite. Drag folding is evident adjacent to most major faults and S_0/S_1 in sediments are locally refolded by D_3 or D_4 deformation, producing secondary crenulations and intersection lineations. Crenulations are well developed in the semipelites, slates, and basalts overlying the complex in the west-central part of the area.

Although there is minor evidence for D_3 folding along the upper and lower contacts of the complex, most of the internal stratigraphic contacts in the Katinniq Ultramafic Complex (Fig. 10) and most of the stratigraphic contacts in the underlying Katinniq gabbro and overlying Chukotat volcanic sequence (Fig. 8) do not exhibit fold geometries and many of the irregularities in the upper contact of the complex are volcanological, representing lateral facies variations (Figs. 8, 10). Thus, there is little evidence that the first-order embayment, which localizes the Katinniq Ultramafic Complex, or the majority of the second-order embayments, which localize the orebodies, were produced by folding, although it is possible that they were modified by folding (Chisholm et al., 1999; Chisholm, 2002).

Metamorphism

The rocks in the Cape Smith Belt have been affected by a complex tectonothermal history involving interaction between i) deformation, ii) uplift and erosion, and iii) thermal equilibration of the thickened crust (St-Onge and Lucas, 1993). The rocks in the southern part of the belt (mainly Povungnituk Group, referred to as an External Domain by St-Onge and Lucas, 1993) are characterized by a higher metamorphic gradient and higher metamorphic grades (hornblende-oligoclase and garnet-oligoclase assemblages), whereas the rocks in the northern part of the belt (mainly Chukotat Group, referred to as an Internal Domain by St-Onge and Lucas, 1993) are characterized by a lower metamorphic gradient and lower metamorphic grades (hornblende-albite and garnet-albite mineral assemblages) (Bégin, 1989, 1992).

The rocks in the east-central part of the Cape Smith Belt have been metamorphosed to lower greenschist facies (actinolite-albite zone: St-Onge and Lucas, 1993) and most mafic-ultramafic rocks now comprise variable proportions of serpentine (predominantly antigorite), clinoamphibole (tremolite in ultramafic lithologies, actinolite in mafic lithologies), chlorite, albite, magnetite, ilmenite, and sulphides. Relict igneous olivine (see below) is rarely preserved, but igneous pyroxene and chromite are commonly preserved. Some of the ultramafic rocks at Donaldson are talc-carbonate altered, but this type of alteration, which is so common in ultramafic rocks in Archean greenstone belts, is rare in the Raglan Formation. Rock structures (pillows, pillow shelves, columnar joints, breccias, sedimentary layering), palimpsest textures (cumulus, porphyritic, skeletal), and relict minerals (clinopyroxene, plagioclase, chromite, and rare olivine) are commonly preserved, so igneous nomenclature is used.

Geochronology

Only sparse geochronological data are available for the eastern Cape Smith Belt (Parrish, 1989). Zircon from a rhyolite in the upper part of the Povungnituk Group ~30 km southwest of Cross Lake, has an age of 1958.6 +3.1/-2.7 Ma. Baddelyite and zircon from a ferrogabbro in the upper part of the Romeo I sill, which is intruded into the uppermost part of the Povungnituk Group east of Cross Lake (Fig. 5A), have a U-Pb age of 1918 +9/-7 Ma. These sills are derived from less magnesian magmas than the mineralized ultramafic complexes and appear to be contaminated by assimilation of Povungnituk sediments, but are petrogenetically most similar to the olivine and pyroxene-phyric basalts in the lower part of the Chukotat Group (see below), so they are probably of broadly similar age. Thus, the mineralized ultramafic complexes in the Raglan Formation are likely between 1959 and 1918 Ma in age.

Host Units

The mineralized ultramafic rocks in the Raglan Formation have been described by Shepherd (1960), Kilburn et al. (1969), Wilson et al. (1969), Miller (1977), Barnes et al. (1982), Coats (1982), Hynes and Francis (1982), Albino (1984), Dillon-Leitch et al. (1986), Giovenazzo et al. (1989), Gillies (1993), St-Onge and Lucas (1993), Thacker (1995), Lesher et al. (1999), Stilson (1999), and Stewart (2002).

Nomenclature and Rock Types

The IUGS terminology for mafic and ultramafic igneous rocks defines all coarse-grained rocks (e.g. gabbro, pyroxenite, olivine pyroxenite, peridotite, and dunite) as plutonic and all fine-grained rocks (e.g. basalt, komatiite) as volcanic; no terminology is provided for fine-grained intrusive rocks or coarse-grained extrusive rocks even though the mineralogy and textures of extrusive, subvolcanic, and plutonic rocks may be otherwise identical. This problem is compounded by the fact that the interpretations of the rocks in the Raglan area vary from being entirely subvolcanic, to being primarily volcanic, to being deeply erosive. Although people working on volcanic and subvolcanic cumulate rocks often avoid using IUGS terminology and simply refer to the ultramafic extrusive/subvolcanic cumulate rocks as olivine orthocumulate (50-75% olivine), mesocumulate (75-95% olivine), and adcumulate (>95% olivine) rocks, this still leaves the problem of what to call the fine-grained pyroxene-rich chilled margins of subvolcanic magma conduits or the medium- to coarse-grained pyroxene-plagioclase-rich rocks in thick massive and differentiated flows. Because the IUGS-type rock names are so firmly entrenched at Raglan, they will be used in this paper to refer to rocks with appropriate mineralogy and textures, but without any genetic implications regarding volcanic or subvolcanic setting (Table 2).

Dunites (olivine adcumulates) are rare in most areas (e.g. occurring only above mineralized mesocumulate rocks at Katinniq: Gillies, 1993), but appear to be more abundant in the East Lake area (Petch, 1999). Unlike the olivine in komatiitic adcumulate rocks in Western Australia (e.g. Perseverance: Barnes et al., 1988), which is normally coarse-grained (>2.5 mm), the olivine in adcumulate rocks in the Raglan area was medium-grained (1.2-2.5 mm).

Peridotites (olivine mesocumulates) are the most common rock type in the Katinniq Member. Unlike the olivine in komatiitic olivine mesocumulate rocks in Western Australia (e.g. Kambalda: Lesher, 1989), which is normally medium-grained, the olivine in mesocumulate rocks in the Raglan area was fine-grained (<1.2 mm). Interstitial areas may contain oikocrystic clinopyroxene, forming heteradcumulate textures (see Fig. 4B), or intercumulus sprays of skeletal pyroxene and altered glass, indicating rapid crystallization and therefore relatively rapid cooling (Gillies, 1993; Thacker, 1995; Stilson, 1999).

Olivine Pyroxenites (clinopyroxene oikocrystic olivine orthocumulates) are common in the Katinniq Member, particularly in the upper parts and in mineralized zones, and in the lower parts of the Cross Lake Member, but they are also common in the lower parts of thick differentiated sills and flows. *Cumulate Pyroxenites* are uncommon, but occur locally as discontinuous layers within the peridotite complexes (Barnes et al., 1982; Gillies, 1993; Thacker, 1995, Stilson, 1999). *Pyroxenites* (pyroxene porphyritic basalts) are common along the margins and less commonly in the internal parts of the Katinniq Member, in the lower parts of the Cross Lake Member, and in the lower parts of non-mineralized sills and layered flows. Those along the upper margins are olivine-phyric and typically exhibit relict random acicular clinopyroxene (pigeonite cores/augite rims) microspinifex textures (see Fig. 9F), whereas those along the lower margins contain fine-grained clinopyroxene phenocrysts (see Fig. 4E). Olivine spinifex textures are rare except in thin apopyses (e.g. in fluidal peperites).

Gabbros range from fine- (transitional with massive basalt) to coarse-grained, and range in composition from melanogabbro through mesogabbro to leucogabbro and Fe-Ti-rich ferrogabbro. They are normally very hard and jointed. In contrast to some of the gabbros in the differentiated sills in the Povungnituk Group, the gabbros in the Raglan Formation are normally massive and are only rarely rhythmically layered. They are most common in the Cross Lake Member and in the upper parts of differentiated sills and flows, but occur locally in the Katinniq Member.

Basalts associated with the Raglan Formation are primarily olivine-phyric basalts with fine-grained random acicular pyroxene microspinifex textures (see Fig. 9B) and rarer fine-grained random olivine microspinifex textures (Barnes and Barnes, 1990; Gillies, 1993; Thacker, 1995; Lévesque et al., 2003). No platy olivine spinifex textures have been reported, consistent with these rocks being derived from a komatiitic basalt magma rather than a komatiite magma (see below). They form 1 to 30 m thick flows and may be massive, pillowed, or brecciated, but breccia facies are most common along the upper margins and in internal parts of the Katinniq Member.

Three types of *breccias* have been recognized: 1) *basalt breccias* composed of angular, cobble-sized fragments of aphanitic and microspinifex-textured basalt, which occur along the upper margins of thick peridotite units, grade downward into polyhedrally-jointed pyroxene-phyric basalt (see Fig. 9C-D), and appear to represent flow-top breccias (Barnes and Barnes, 1990; Lesher et al., 1999), 2) *sedimentary breccias* composed of angular, contact-metamorphosed

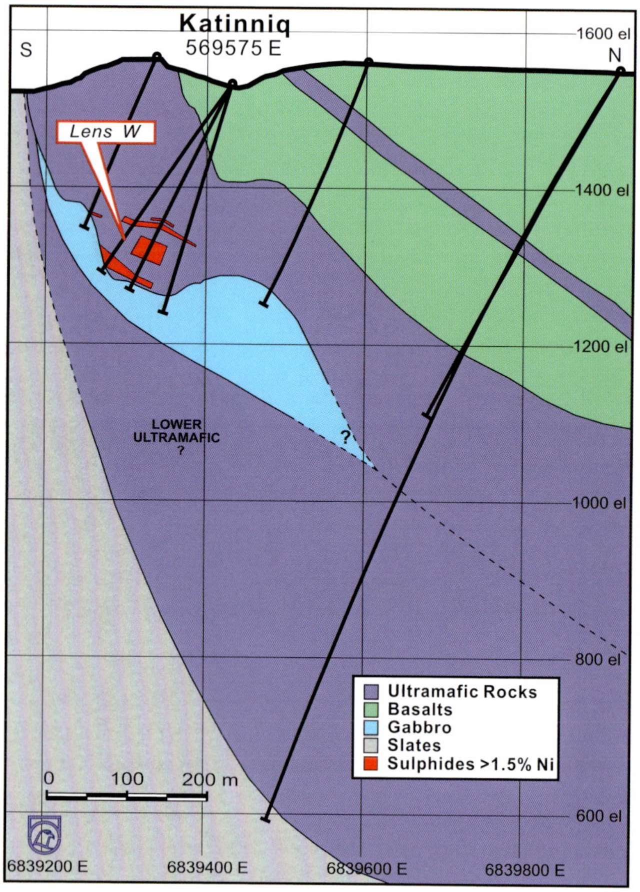

Figure 11. Cross-section (looking west) through lens W in the Katinniq area (from Lesher and Charland, 1999).

sedimentary clasts in a noncontact metamorphosed matrix (see Fig. 9G), interpreted as intraformational collapse breccias (Gillies, 1993), and 3) *basalt-sediment breccias* composed of cobble-sized angular to rounded fragments of aphanitic to microspinifex-textured basalt in a pelitic (sometimes sulphidic) matrix (Fig. 9H), interpreted as peperites (S.W. Beresford, pers. comm., 1999; Lévesque and Lesher, 2002; Lévesque et al., 2003; see also Skilling et al., 2002).

Geology

Cross Lake Member

The mafic-ultramafic units in the Cross Lake Member of the Raglan Formation range up to 100 m in true thickness and up to 10 km or more in strike length (e.g. Cross Lake: Fig. 5, Zone 3 – Zone 2 – Katinniq – Zone 5 – Zone 7 lower gabbro: Fig. 6). They represent a channelized sheet facies assemblage composed of 1) laterally restricted conduit facies containing one or more units of primarily massive peridotite (e.g. Cross Lake Main Zone: Fig. 5A; C1-C2-C3 area: Fig. 5B; Zone 5 and Zone 7 areas: Fig. 7B) and 2) laterally much more extensive flanking/intervening sheet facies composed of differentiated olivine pyroxenite and gabbro (e.g. Cross Lake north and east flanks: Fig. 5A; 5-8 Lower Gabbro: Fig. 7B), differentiated gabbro (e.g. Katinniq Gabbro: Fig. 8), or relatively massive gabbro (e.g. Zone 2-3 Gabbro: Fig. 6; Boundary Lower Gabbro: Fig. 7A). Conduit facies may be capped by flow-top breccias (e.g. Cross Lake and C1-C2-C3: Fig. 5B), but sheet facies are normally capped by massive aphanitic basalt (e.g. Katinniq Gabbro: Fig. 8). Within the structurally disrupted fold nose of the Cross Lake area (Fig. 5A), the peridotites appear to be overlain by highly siliceous sulphide-rich metasedimentary rocks of possible exhalative origin (Thibert, 1999). Only conduit facies of the Cross Lake Member are mineralized (e.g. Cross Lake Main Zone: Fig. 5A; C1-C2-C3: Fig. 5B; Zone 5 and Zone 7: Fig. 7B). Sheet facies of the Cross Lake Member are barren.

Katinniq Member

The mafic and ultramafic rocks in the Katinniq Member of the Raglan Formation range up to 200 m or more in true thickness and up to 2 km in exposed strike length. Many parts are composed almost entirely of peridotite, but they have fine-grained pyroxenitic lower margins, commonly contain internal zones or layers of pyroxenite, gabbro, basalt, and semipelite that correlate along strike and down dip with the stratigraphy in flanking units, sometimes grade laterally into peperites (see below), and upper margins are locally capped by brecciated komatiitic basalts (see Fig. 9C-E) with glassy or microspinifex textures (e.g. East Lake: Petch, 1999; Zone 2-3: Mallinson, 1999a,b; Katinniq: Barnes and Barnes, 1990; Gillies, 1993; Lesher and Charland, 1999; Zone 5-8: Thacker, 1995; Mallinson, 1999c; Boundary: Stilson, 1999; Stilson et al., 1999) (see Figs. 7A,B, 8). No breccias or peperites have been identified along the lower margins. Mass balance calculations indicate that these units contain 70 to 80% excess (cumulate) olivine, consistent with their overall mesocumulate nature.

There are systematic differences in the Katinniq Member along the Raglan Formation. The ultramafic complexes in the East Lake area appear to contain more dunite and some occur deeper in the Povungnituk sequence, leading Petch (1999) and Stewart (2002) to suggest that parts may be a feeder dyke and that the East Lake area may therefore represent an eruptive site. At Zone 2-3, Katinniq, Zone 5-8, Zone 13-14, and West Boundary, the ultramafic complexes are composed primarily of peridotite with only minor amounts of interfingering pyroxenite, gabbro, basalt, and/or semipelite. The Boundary and Donaldson ultramafic complexes contain greater amounts of gabbro, basalt, and semipelite than the other areas, but they are also more penetratively deformed than in the other areas, owing to their location near the eastern end of the Raglan Formation where two D_1 thrust faults converge (Fig. 3).

Several peridotite bodies grade laterally from peridotite through olivine pyroxenite and pyroxenite to gabbro (e.g. Zone 5-8 Central Gabbro: Thacker, 1995; West Boundary: Charland, 1999; Boundary: Stilson, 1999; Stilson et al., 1999), suggesting that they represent different lithofacies of the same unit.

The Katinniq Ultramafic Complex (Fig. 8) was (prior to mining) better exposed than most of the other areas and because of its economic significance has been studied in more detail than the other bodies (Barnes et al., 1982; Lesher and Charland, 1999). It is composed primarily of mesocumulate peridotite and lesser olivine pyroxenite with internal horizons of pyroxenite, basalt, gabbro, and semipelite, and upper and lower margins of olivine pyroxenite and pyroxenite; adcumulate dunites are rare. The stratigraphy of the ultramafic body is complex and comprises at least two major units and several subunits (Figs. 10, 11). The lower unit is not exposed on surface and was only discovered during exploration drilling in 1998; it is composed primarily of massive mesocumulate peridotite capped by gabbro (Fig. 11) and may represent a separate part of the Katinniq Ultramafic Complex or, as at Zone 5 and Zone 7 (see Fig. 7B), a channelized part of the underlying Katinniq Gabbro. The upper unit is well exposed on the surface and comprises a thick lower part of massive and columnar-jointed peridotite (see Fig. 4A), capped locally with pyroxenite± basalt±semipelite, and an upper part composed of massive mesocumulate peridotite and oikocrystic olivine pyroxenite, overlain by columnar- and polyhedrally-jointed pyroxenite (see Fig. 9C-E), capped by aphanitic to microspinifex-textured massive or brecciated basalt (see Fig. 9F). The lower subunits in this unit are thicker (up to 65 m) and laterally more extensive than the middle and upper subunits (up to 30 m). Overlying subunits appear to crosscut underlying subunits in many parts of the complex (Fig. 10). A large lens of hornfelsed argillaceous semipelite occurs within the lower part of the complex (Fig. 8). This semipelite occurs above a narrow zone of sheared peridotite, but is oriented parallel to other stratigraphic contacts, is directly underlain by pyroxenite, and correlates with semipelite horizons intersected in drill core. It may represent i) a fault repetition of footwall rocks, ii) a xenolith (raft), or iii) a partially beheaded horizon of interflow semipelite. Multiple internal horizons of Fe-Ni-Cu sulphide mineralization and interlayering of mesocumulate peridotite, oikocrystic olivine pyroxenite, and pyroxenite (Fig. 10; see also Barnes et al., 1982) indicate variations in magma composition and crystallization rate during emplacement. The uppermost part of the Katinniq Ultramafic Complex is marked locally by a relatively thick (2-10 m, exaggerated on dip slopes) basaltic breccia (Figs. 8, 9D), comprising subrounded to angular, cobble-sized fragments of very fine-grained basalt and fine- to medium-grained pyroxenite in a chloritic matrix (Fig. 9F). The breccia weathers to a tannish-grey colour, disintegrates readily, and forms very rubbly exposures along most of the northern margin of the peridotite (Fig. 8). No basalt breccias have been observed in the footwall of the complex.

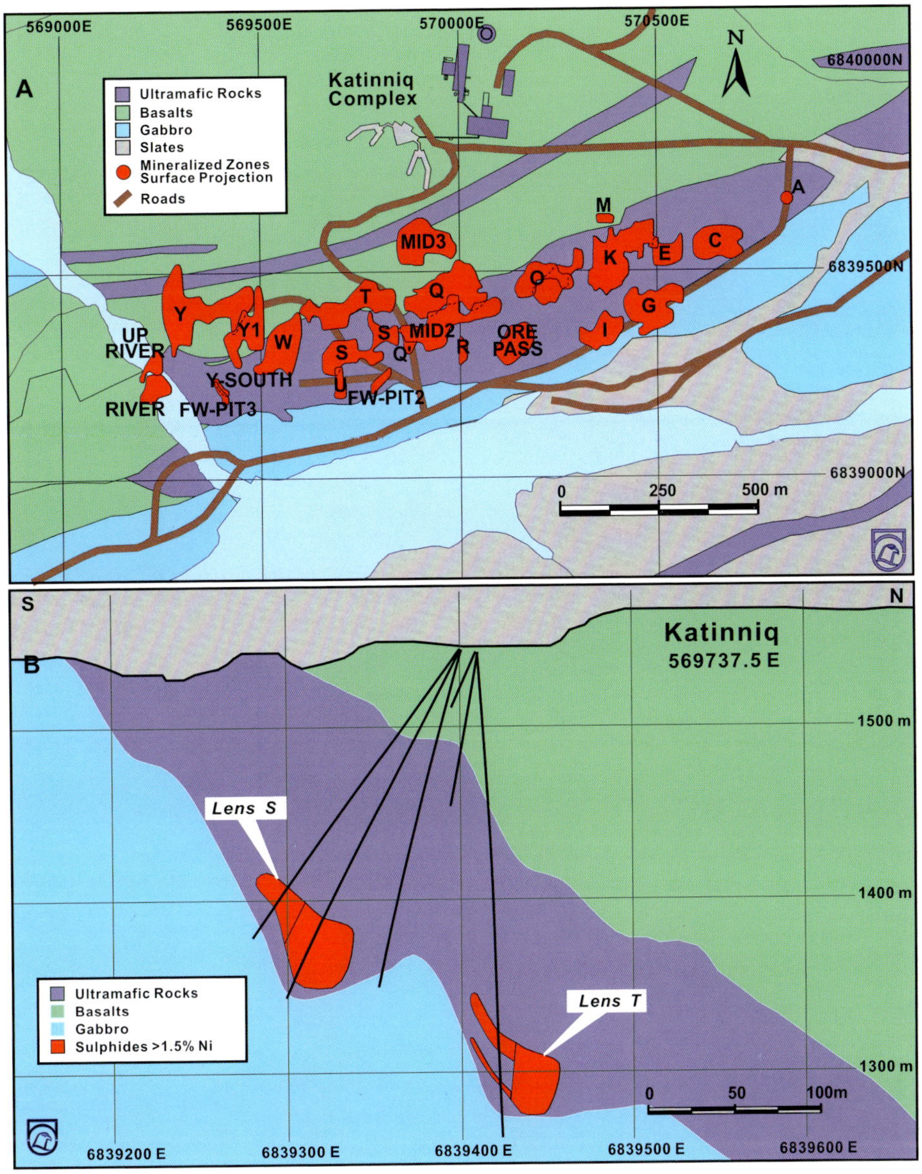

FIGURE 12. (A) Simplified geological map of the Katinniq Ultramafic Complex showing ore lenses in surface projection (from Chisholm et al., 1999). (B) South-north cross-section (looking west) through lenses S and T in the Katinniq Ultramafic Complex at 569737.5E (from Lesher and Charland, 1999).

Contact Relationships

The upper and lower contacts of sheet facies are normally regular and conformable (e.g. Fig. 5A), whereas the upper and lower contacts of conduit facies, irrespective of whether part of a channelized sheet facies assemblage (e.g. Cross Lake Main, C1-C2-C3) or part of a conduit facies assemblage (e.g. Zone 2, Zone 3, Katinniq, Boundary), are quite irregular (e.g. Figs. 5B, 7, 8).

The lower contacts of conduit facies typically define large broad V-shaped first-order embayments in the footwall rocks (e.g. Figs. 6, 7, 8), which almost entirely confine the unit, and smaller more irregular (often re-entrant) second-order embayments (e.g. Figs. 7, 10, 12B), which typically localize Ni-Cu-PGE mineralization. The contacts are markedly transgressive to bedding in sediments or layering in gabbros (Figs. 4C ,8), highly irregular in 3-D, and therefore cannot be attrib-

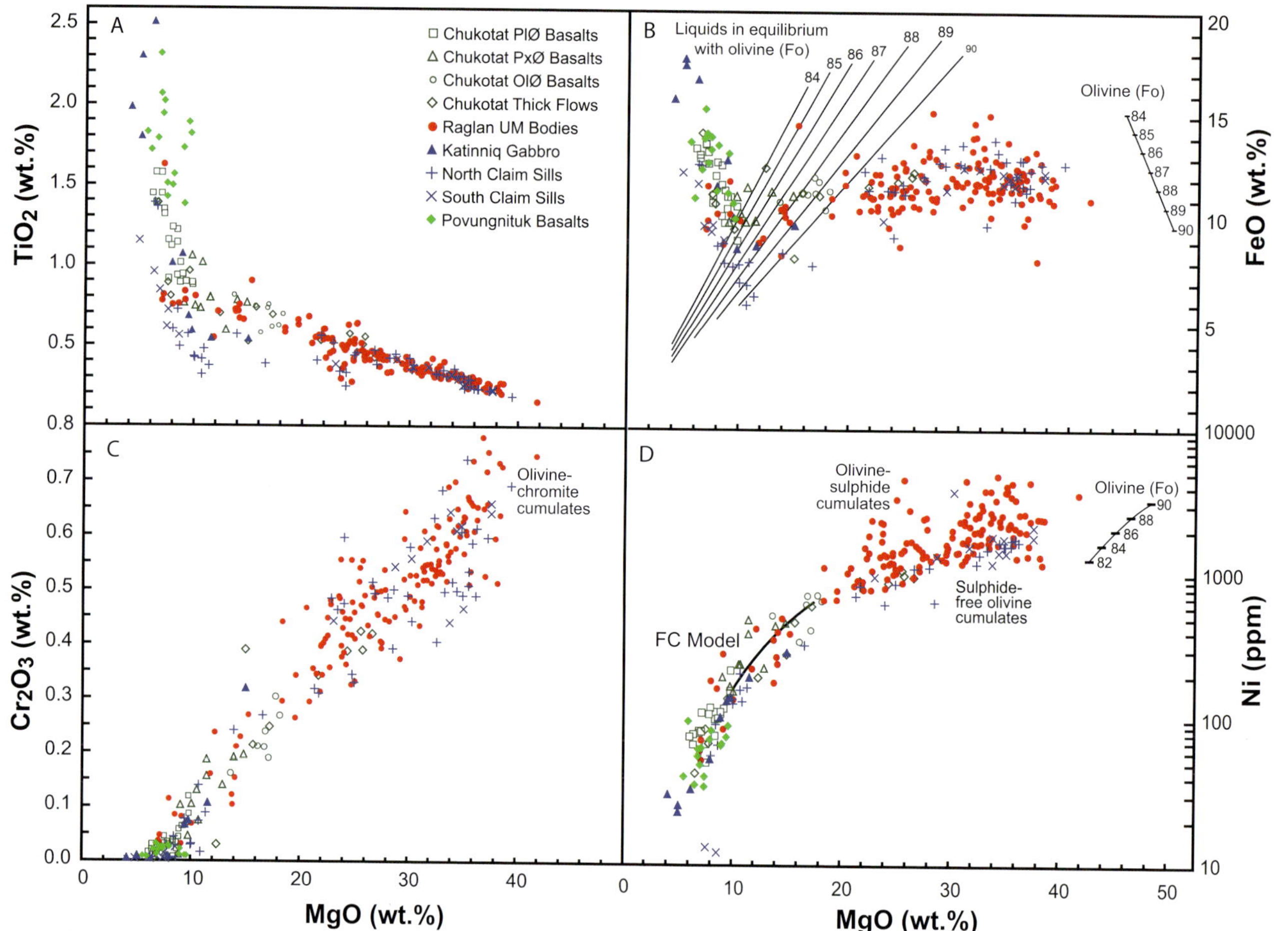

FIGURE 13. (**A**) MgO versus TiO_2 , (**B**) MgO versus FeO_t, (**C**) MgO versus Cr_2O_3, and (**D**) MgO versus Ni variation diagrams for representative volcanic and subvolcanic rocks from the Raglan Block (modified from Burnham et al., 1999). Olivine liquid compositions calculated using K_D = 0.3 with 10% Fe^{3+}.

uted to folding. Some of the contacts are sheared, but many are not, indicating that they are primary igneous features.

The upper contacts of conduit facies are also quite irregular, but most of the irregularities represent mappable stratigraphic units in the peridotite complexes defined by differences in composition (peridotite vs. olivine peridotite vs. pyroxenite), texture (massive vs. oikocrystic), and/or structure (e.g. massive vs. columnar- or polyhedrally jointed vs. breccia) (e.g. Figs. 5B, 8, 7, 10). Some of the upper contacts intersected in drill core along the northern down-dip parts of Katinniq and Zone 2 appear to be intrusive (Lévesque and Lesher, 2002; Lévesque et al., 2003), but most of the contacts exposed in the upper parts of the complexes exposed on surface appear to be conformable (e.g. Cross Lake: Thibert, 1999; Zone 2-3: Mallinson, 1999a,b; Katinniq: Gilles, 1993; Lesher and Charland, 1999; Zone 5-8: Thacker, 1995; Mallinson, 1999c; West Boundary: Charland, 1999; Boundary: Stilson, 1999; Stilson et al., 1999). Conduit facies at Cross Lake, C1-C2-C3, parts of East Lake, Zone 2, Zone 3, Katinniq, Zone 6, Zone 8, and Boundary are capped by aphanitic and microspinifex-textured basalt breccias (e.g. Figs. 5B, 7A,B, 8).

The lateral contacts of conduit facies assemblages are often poorly exposed on surface, but drill-core data indicate that they are very complex, with peridotites and marginal pyroxenites interfingering with adjacent basalts, slates, and peperites (Fig. 10).

Contact Metamorphism

The sedimentary rocks underlying the mineralized peridotite complexes are contact metamorphosed to at least the biotite zone (Williams et al., 1999a). Gabbros are less obviously metamorphosed but basalts can be completely recrystallized (e.g. western end of Katinniq: Fig. 8). The degree and extent of metamorphism are greater beneath thicker parts of the ultramafic complexes than beneath thinner parts, and greatest beneath second-order embayments (e.g. Cross Lake: Thibert, 1999; Katinniq: Lesher and Charland, 1999; Zone 8: Thacker, 1995; Boundary: Stilson, 1999).

Overlying rocks are rarely locally contact metamorphosed (e.g. Katinniq: Lévesque and Lesher, 2002; Lévesque et al., 2003; East Lake Main Body: Stewart, 2002), but even in those areas most are not. For example, most of the hanging-wall slates exposed overlying the peridotite complexes at East Lake, Katinniq, Zone 5-8, and Boundary are dark black and, unlike many footwall slates in the same areas (see Figs. 4A,C, 7, 8), show no evidence of contact metamorphism (Gillies, 1993; Thacker, 1995; Lesher and Charland, 1999; Lesher et al., 1999; Mallinson, 1999a,b,c; Petch, 1999; Stilson, 1999; Stilson et al., 1999).

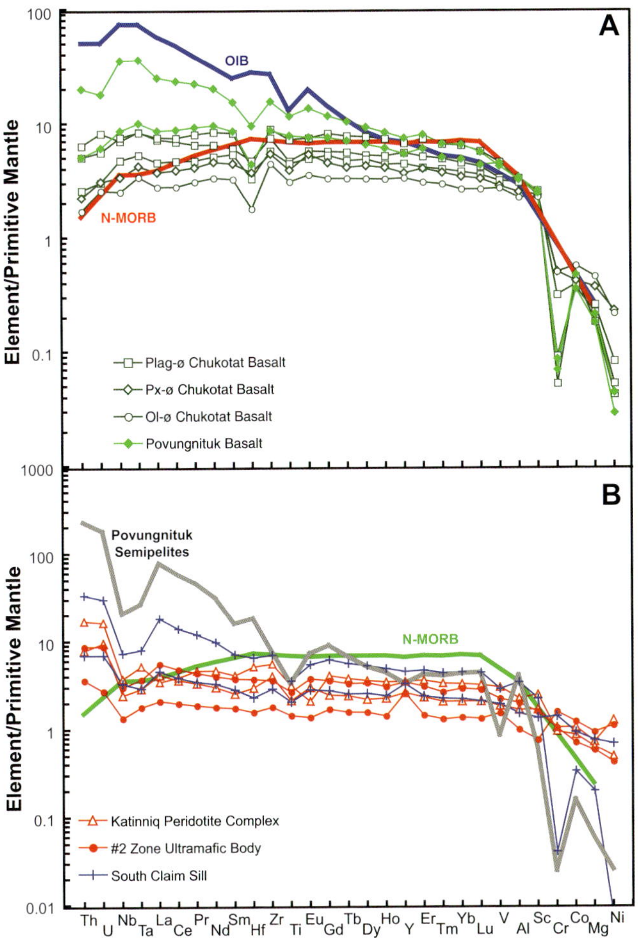

FIGURE 14. (A) Primitive mantle-normalized multi-element patterns of representative Chukotat and Povungnituk basalts from the Raglan Block (Burnham et al., 1999) relative to normal mid-ocean concentrations within olivine-rich peridotites. Primitive mantle values from McDonough and Sun (1995). (B) Primitive mantle-normalized multi-element patterns of representative ultramafic rocks from the Raglan Formation and a differentiated sill within the Povungnituk Group (from Burnham et al., 1999). The data for Zone 2 illustrates the range of compositions typically present within an ultramafic body: the highest concentrations are within pyroxenitic zones and the lowest concentrations within olivine-rich peridotites. Primitive mantle values from McDonough and Sun (1995).

Whole-Rock Geochemistry

The geochemistry of the mineralized peridotite complexes in the Raglan Formation and associated basalts of the Chukotat Group have been studied by Miller (1977), Barnes et al. (1982), Gillies (1993), Thacker (1995), Burnham et al. (1999), Stilson (1999), and Lesher et al. (2001). The geochemistry of the Povungnituk and Chukotat basalts has also been studied by Francis et al. (1981, 1983), Lamothe et al. (1986, 1984), Picard et al. (1990), and Barnes and Picard (1993), and Burnham et al. (1999). The following summary is taken primarily from Burnham et al. (1999) and Lesher et al. (2001).

Chukotat Basalts

Olivine-phyric basalts contain 16-18% MgO, 11-12% FeO_t, ~10% Al_2O_3, 0.6-0.7% TiO_2, 0.05-0.07% P_2O_5, 13-17 ppm Y, 50-55 ppm Zr, and 4-5 Nb. Pyroxene-phyric basalts contain 9-14% MgO, 10-11% FeO_t, 11-13% Al_2O_3, 0.6-0.8% TiO_2, 0.05-0.08% P_2O_5, 14-20 ppm Y, 50-70 ppm Zr, and 4-6 ppm Nb. Plagioclase-phyric basalts contain 6-10% MgO, 9-14% FeO_t, 13-15% Al_2O_3, 0.9-1.6% TiO_2, 0.07-0.15% P_2O_5, 18-32 ppm Y, 60-110 ppm Zr, and 4-11 ppm Nb. Their major element geochemical trends are consistent with fractional crystallization of the observed liquidus phases (Fig. 13). Chukotat basalts and some Povungnituk basalts are depleted in highly incompatible lithophile elements (HILE: U-Th-Nb-Ta-LREE) relative to moderately-incompatible lithophile elements (MILE: Ti-MREE-Y-HREE), broadly similar to modern mid-ocean ridge basalt (Fig. 14A). Negative Cr anomalies confirm that they were saturated in chromite in the source and/or fractionated chromite prior to emplacement.

Raglan Formation Peridotite Complexes

The abundances of most elements, including TiO_2 (Fig. 13A) and other incompatible elements, FeO (Fig. 13B), Cr (Fig. 13C), and Ni (Fig. 13D) in the cumulate rocks from the ultramafic bodies (Table 3) exhibit coherent trends, representing mixtures between a cotectic cumulus assemblage of olivine (Fo_{87}-Fo_{89}), minor chromite, and sulphide, and an intercumulus chromite-saturated komatiitic basaltic magma with a composition similar to that of the most mafic olivine-phyric Chukotat basalt (17-19 wt.% MgO). Barnes et al. (1982) and Barnes and Picard (1993) suggested that a less magnesian parental magma with 15-17% MgO is consistent with both the compositions of fine random pyroxene spinifex-textured rocks immediately above the Katinniq and Zone 2-3 complexes, and with the relatively low Ni/Cu ratios of the ores, and they attributed the higher MgO contents of some of the basalts to the presence (i.e. accumulation) of olivine phenocrysts, but this cannot explain the inferred Fo contents of olivine in the most magnesian cumulate rocks (see Fig. 13B).

Mineralized and nonmineralized cumulate rocks are enriched in HILE relative to MILE and exhibit pronounced negative Nb-Ta-(Ti) anomalies (Fig. 14B). Because U-Th-Nb-Ta-LREE should all be highly incompatible during both mantle partial-melting and any subsequent fractionation of olivine±chromite, the elevated U, Th, and LREE contents and negative Ta and Nb anomalies of the mineralized units cannot be created by variable degrees of melting and/or fractional crystallization of the ultramafic magmas. Barnes et al. (2004) have pointed out that the incompatible lithophile elements in cumulate rocks are potentially more susceptible to modification because they are present in such low abundances, but elements with high charges and small ionic radii, such as U, Th, Nb, Ta, Zr, Hf, REE, Y, and Ti, are rarely very mobile because they are normally housed in alteration-resistant accessory phases and because they are not easily complexed in normal metamorphic fluids. Studies of the mobility of those elements in virtually identical but much more strongly metamorphosed rocks in the Thompson Nickel Belt indicates that all of these elements except La remained relatively immobile during the higher grade upper amphibolite facies metamorphism in this area (Layton-Matthews et al., 2007). As discussed by Lesher et al. (2001), these geochemical characteristics are a signature of contamination by upper continental crustal rocks, and the concentration of the contamination in these units rather than the entire Chukotat basalt sequence and the abrupt changes over relatively short

TABLE 3. Representative geochemical compositions of rocks from the Raglan Block (from Burnham et al., 1999).

Sample	Units	41177 Plag-Phyric Chukotat Basalt	41158 Px-Phyric Chukotat Basalt	41183 Ol-Phyric Chukotat Basalt	41198 Unenriched Povungnituk Basalt	41194 Enriched Povungnituk Basalt	94036 #2 Zone Pyroxenite	41126 Katinniq Orthocumulate Peridotite	94040 #2 Zone Mesocumulate Peridotite	94077 Differentiated Sill Peridotite	94079 Differentiated Sill Mesogabbro
SiO_2	(wt.%)	50.1	49.9	49.1	48.3	47.8	48.0	46.1	44.3	45.8	52.1
TiO_2	(wt.%)	0.894	0.785	0.621	1.57	2.32	0.54	0.432	0.29	0.42	0.72
Al_2O_3	(wt.%)	13.5	11.1	9.87	14.8	14.2	8.9	7.42	4.4	6.8	15.3
Cr_2O_3	(wt.%)	0.120	0.192	0.191	0.026	0.033	0.397	0.404	0.610	0.543	0.016
FeO^t	(wt.%)	10.7	10.6	11.9	13.6	14.0	10.6	10.6	11.5	11.5	9.8
MnO	(wt.%)	0.19	0.18	0.20	0.24	0.21	0.17	0.19	0.16	0.18	0.18
MgO	(wt.%)	9.72	13.9	17.1	8.02	6.78	21.9	27.5	35.3	28.7	7.54
NiO	(wt.%)	0.020	0.057	0.054	0.011	0.007	0.106	0.318	0.277	0.173	0.002
CaO	(wt.%)	12.6	11.9	9.54	10.5	12.1	8.9	6.39	3.1	5.4	10.8
Na_2O	(wt.%)	1.78	1.10	0.99	2.57	1.88	0.14	0.52	0.01	0.28	2.23
K_2O	(wt.%)	0.20	0.12	0.30	0.20	0.27	0.35	0.05	0.02	0.07	1.21
P_2O_5	(wt.%)	0.07	0.07	0.05	0.12	0.29	0.05	0.04	0.04	0.04	0.08
Ba	(ppm)	34	2	93	24	24	28	4	1		208
Sc	(ppm)	42	38	36	41	39	26	27	12	22	36
Cr	(ppm)	823	1312	1307	181	224	2717	2764	4173	3716	107
V	(ppm)	257	232	221	366	343	184	153	126	159	238
Co	(ppm)	41	44	59	49	37	74	111	129	97	35
Ni	(ppm)	161	449	423	85	57	830	2497	2179	1361	13
Cu	(ppm)	88	110	52	95	54	23	266	207	49	28
Zn	(ppm)	81	74	86	110	113	88	64	57	71	85
Y	(ppm)	19	16	14	23	32	15	11	11	14	20
Zr	(ppm)	60	57	47	91	163	38	42	19	31	74
Cs	(ppm)	0.23	0.15	0.27	0.07	0.04	1.57	2.02	0.25	1.60	0.92
Rb	(ppm)	4	7	11	6	8	18	12	8	12	51
Sr	(ppm)	338	189	85	161	562	45	16	18	37	130
Nb	(ppm)	3.1	2.2	1.7	5.7	23.3	2.0	1.6	0.9	2.2	4.9
La	(ppm)	2.96	2.42	1.81	5.58	16.2	3.63	2.74	1.38	3.00	12.0
Ce	(ppm)	7.88	6.53	4.68	14.76	39.2	8.19	6.15	3.35	6.54	23.6
Pr	(ppm)	1.20	1.05	0.79	2.35	5.66	1.13	0.86	0.48	0.89	3.08
Nd	(ppm)	6.42	5.75	4.18	11.96	25.2	5.06	3.81	2.28	4.17	12.4
Sm	(ppm)	2.18	1.80	1.32	3.48	6.23	1.54	1.05	0.72	1.16	2.91
Eu	(ppm)	0.76	0.83	0.55	1.16	2.08	0.58	0.47	0.21	0.44	0.85
Gd	(ppm)	2.86	2.46	1.80	4.09	6.29	1.97	1.35	0.93	1.51	3.39
Tb	(ppm)	0.47	0.41	0.33	0.69	1.02	0.33	0.24	0.16	0.25	0.56
Dy	(ppm)	3.26	2.86	2.21	4.38	6.22	2.30	1.49	1.07	1.74	3.59
Ho	(ppm)	0.65	0.61	0.48	0.90	1.24	0.46	0.34	0.21	0.36	0.73
Er	(ppm)	1.79	1.75	1.33	2.62	3.47	1.36	1.02	0.64	1.05	2.07
Tm	(ppm)	0.27	0.25	0.20	0.34	0.45	0.18	0.14	0.09	0.15	0.30
Yb	(ppm)	1.66	1.50	1.19	2.13	2.87	1.29	0.93	0.61	0.99	1.98
Lu	(ppm)	0.24	0.22	0.18	0.29	0.38	0.19	0.14	0.09	0.14	0.30
Hf	(ppm)	0.93	1.05	0.50	1.24	2.69	1.06	0.85	0.45	0.66	1.86
Ta	(ppm)	0.20	0.16	0.13	0.37	1.33	0.14	0.11	0.07	0.11	0.30
Th	(ppm)	0.21	0.18	0.14	0.40	1.59	0.70	0.63	0.29	0.56	2.71
U	(ppm)	0.06	0.06	0.05	0.12	0.36	0.18	0.20	0.06	0.14	0.62
Ir	(ppb)	0.7	0.8	0.6	0.9	0.15	1.5	3.8	4.2	1.9	0.01
Ru	(ppb)	0.7	2.0	3.5	0.1	0.05	4.3	15.9	17.2	6.1	0.01
Rh	(ppb)	1.2	0.7	1.6		0.02	1.1	7.3	5.8	1.1	
Pt	(ppb)	14	8.8	16.6	0.9	0.13	8.6	42.3	31.1	7.1	0.25
Pd	(ppb)	15	8.2	15.7	2.0	0.84	8.6	115	34.2	11.3	
Au	(ppb)	0.7	0.2		1.8	0.58	0.12	5.4	5.3	0.7	0.54

Major and minor elements analyzed by wavelength-dispersive XRF spectrometry on fused glass disks and pressed powder pellets at the Laurentian University Central Analytical Facility (Dr. J. Huang, analyst). Trace elements analyzed by ICP-MS after mixed acid dissolution at the Ontario Geoscience Laboratories in Sudbury. Noble metals determined by ICP-MS following NiS fire assay preconcentration in the joint LU_OGL low_level PGE laboratory (T. Richardson, analyst). All data normalized to 100% volatile-free, with total Fe as FeO.

distances indicates that the contamination occurred locally, most likely by assimilation of the underlying Povungnituk sediments, which exhibit complementary geochemical characteristics. Thus, the parental magmas must have been more magnesian than the 15 to 17% suggested from the whole-rock and analyzed olivine compositions (see above), but the absence of Cr-poor peridotites and the presence of chromite on the liquidus of almost all rocks places an upper limit of ~20 wt.% MgO on the parental magma composition (Murck and Campbell, 1986; see discussion by Lesher and Stone, 1996).

Burnham et al. (1999) demonstrated that the incompatible element compositions of the majority of the ultramafic complexes may be modelled by the assimilation of sulphidic sediment from the uppermost parts of the Povungnituk Group by a typical olivine-phyric Chukotat Group basalt, and that the amount of material assimilated by mineralized conduit facies (normally <8%) was greater than that assimilated by sheet facies or barren sills (normally <5%) or overlying olivine-phyric basalts (<2%, normally 0%) (Fig. 15). The presence of contamination in mineralized conduit facies, weakly mineralized sheet facies, and nonmineralized at Raglan contrasts with Kambalda, where mineralized conduit facies are less contaminated than flanking barren sheet facies (Lesher and Arndt, 1995; Lesher et al., 2001), but it is similar to that at Perseverance where the mineralized conduit facies appears to be more contaminated than the flanking nonmineralized "overbank" facies (Barnes et al., 1995; Lesher et al., 2001). Such differences may be attributed to differences in the physical volcanology of the units and the nature of the substrate: whereas the principal contaminant at Kambalda appears to have been a thin sulphidic sediment overlying a thick sequence of more refractory basalts, the principal contaminant at Raglan appears to have been the thick sequence of underlying sulphidic sediments and the principal contaminant at Perseverance appears to have been the thick sequence of underlying felsic volcaniclastic rocks. Following thermomechanical erosion of the thin sediment layer at Kambalda, further contamination of the magmas was inhibited by the high melting point of the basalts, and the lava conduits were flushed by relatively uncontaminated magmas that had flowed over only the basaltic substrate (Lesher and Arndt, 1995). In contrast, at Raglan and Perseverance where the substrates were unconsolidated sedimentary and pyroclastic rocks, the magmas could erode deep into the substrate (Williams et al., 1999a,b, 2002), leading to continuous contamination of the magma and the strongest contamination signature within the lava conduits (Barnes et al., 1995; Burnham et al., 1999; Lesher et al., 2001).

The location of the greatest crustal contamination within the thickest, most magnesian units at the base of the Chukotat Group suggests that fluid dynamics was a critical factor in the metallogenesis

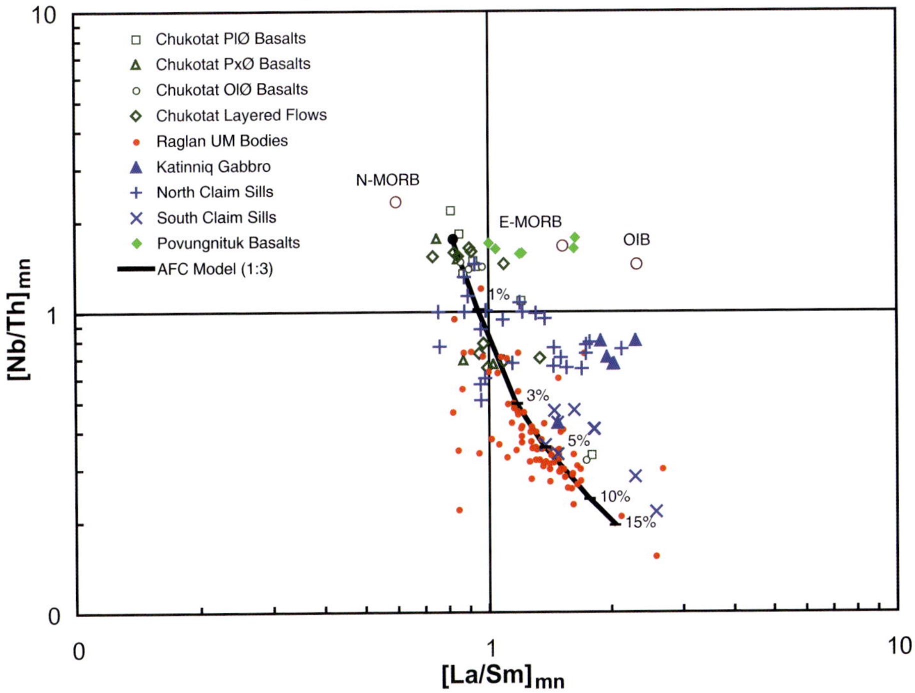

FIGURE 15. Mantle-normalized La/Sm versus Nb/Th ratios for volcanic and subvolcanic rocks in the Raglan area (modified from Burnham et al., 1999). The trace element compositions of some Povungnituk Group basalts resemble those of enriched mid-ocean basalt; the compositions of Chukotat Group basalts and other Povungnituk Group basalts are closer to those of normal mid-ocean ridge basalt (i.e. derived from depleted mantle), and the ultramafic bodies of the Raglan Formation appear to be derived from Chukotat magmas that have assimilated variable amounts of upper Povungnituk Group sedimentary rocks (A:FC = 1:3). N-MORB, E-MORB, and OIB compositions from Sun and McDonough (1989). Mantle-normalization values and primitive mantle composition from McDonough and Sun (1995).

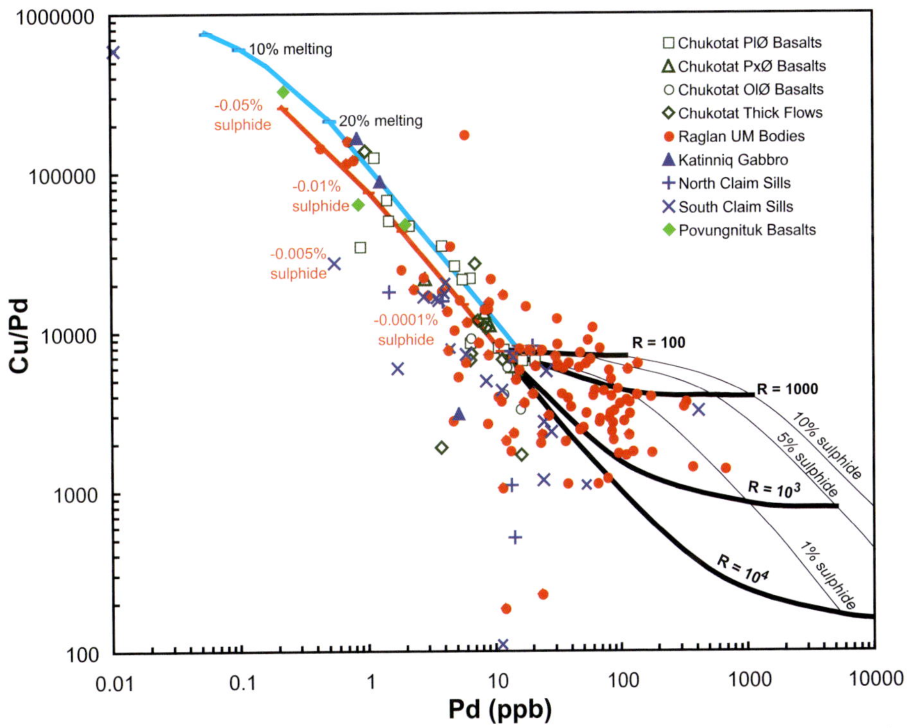

FIGURE 16. Plot of Pd versus Cu/Pd (see Barnes et al., 1993) for volcanic and subvolcanic rocks from the Raglan Block (Burnham et al., 1999) illustrating magma compositions expected for up to 35% partial melt extraction from an undepleted mantle source (solid blue line), the removal of small amounts of sulphide from a sulphide-saturated magma (solid red line) and the compositions of cumulates containing variable amounts of sulphide for R = 100 to 10^4 (thick black lines). Thin dashed lines join samples with equal sulphide contents formed under different R factors. Sulphide compositions calculated for an initial magma composition having lost 0.001% sulphide (Burnham et al., 1999).

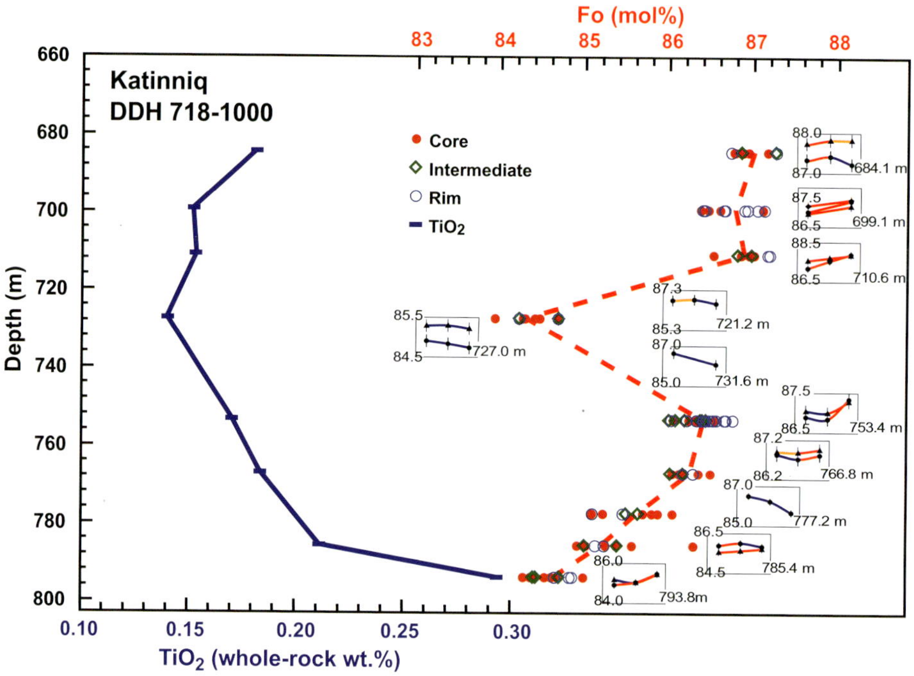

FIGURE 17. Variations in maximum forsterite content (Fo = 100Mg/(Mg+Fe) and zoning profiles of relict igneous olivines in DDH 718-1000 through the lower part of the Katinniq Ultramafic Complex (from Lévesque et al., 2002; Lévesque, in prep.). The rocks in this section are massive peridotites and contain only sparse disseminated sulphides. Red indicates reverse zoning (and therefore increasing Fo content, MgO of magma, and temperature), blue indicates normal zoning (and therefore decreasing Fo content, MgO of magma, and temperature), orange indicates no zoning. Analysts: O.M. Burnham and M. Lévesque.

of the deposits. Owing to their relatively low viscosities, komatiitic magmas may flow turbulently provided the velocity of the magma is sufficient (see discussion by Williams et al., 1998, 1999a,b), facilitating assimilation of wall rocks and significant contamination. Once the flow rate slows, however, a thermal boundary layer may develop between the flow and the substrate, inhibiting thermal erosion and assimilation. Another factor may have been the geometry of the units (Burnham et al., 1999): channelization would have constrained the flows and maintained the high velocities required for turbulent flow and assimilation, however, where the magmas broke out of the conduits to form sheet flows or sheet sills, flow rates would have decreased, resulting in laminar flow and limiting assimilation. A key point is that because the upper Povungnituk black shales are enriched not only in highly incompatible lithophile elements but also sulphide, thermomechanial erosion would have not only contaminated the magma, but would have also melted considerable volumes of barren sulphide with which the magmas could interact to produce Ni-Cu-(PGE) sulphide ores (Lesher et al., 2001).

Povungnituk Sills

Detailed petrographic and geochemical analysis of the Romeo I sill southeast of Cross Lake (Fig. 5A) and the NC20S sill (the lower sill in thrust sheet O north of East Lake: Figs. 2, 3) indicates that these sills formed by fractionation of a magma with a composition similar to the most magnesian Chukotat Group basalts (Thibert, 1993; C.M. Lesher and R.R. Keays, unpubl. data). Although the fractionating phases are similar to those in Chukotat Group basalts (i.e. Ol → Cpx+Ol → Cpx+Plag±Ol: Francis et al., 1983), plagioclase appears earlier in the fractionation history and forms a greater proportion of the phases that fractionated from the most evolved magmas (Thibert et al., 1989; Thibert, 1993).

Most of the differentiated sills are enriched in LREE ($[La/Sm]_{mn}$ = 1.0-2.6) and depleted in Nb-Ta relative to Th ($[Th/Nb]_{mn}$ = 1.0-4.6) (e.g. Figs. 14B, 15), but the depletion in Nb-Ta is normally less than in the mineralized ultramafic units. Although some of the LREE enrichment in the gabbroic zones of the sills might be explained by the removal of significant amounts of LREE-depleted clinopyroxene, similar enrichments are observed in the ultramafic zones, in which olivine and chromite are the only cumulus phases. Because none of these elements are compatible in olivine or chromite, the trace element compositions of the peridotites could not have been produced by fractionation or accumulation of these minerals, and require that the magma was either originally incompatible-element enriched or that fractionation was accompanied by contamination by upper crust and/or adjacent Povungnituk Group sediments (Burnham et al., 1999).

Although the incompatible trace element compositions of the differentiated sills trend broadly toward the local Povungnituk Group sedimentary rocks, the compositions of the uncontaminated magmas appear to be enriched in LREE and Nb relative to the inferred initial composition of the Chukotat basalts, suggesting that some of the sills may have formed from magmas similar to the nearby unenriched Povungnituk Group basalts (Burnham et al., 1999).

Chalcophile Element Geochemistry

Owing to the presence of variable amounts of disseminated sulphides, nearly all of the cumulate rocks from the mineralized ultramafic bodies along the Raglan Horizon are enriched in chalcophile elements relative to sulphide-poor peridotites of similar whole-rock composition from either the nonmineralized differentiated sills within the Povungnituk Group sedimentary rocks or the overlying Chukotat Group basalts (Burnham et al., 1999). Barnes et al. (1992) noted that the mineralized units at Bravo, Delta, and Méquillon are also enriched in chalcophile elements. In Figure 16, the chalcophile element compositions of the ultramafic cumulates are compared to those predicted for cumulates containing different amounts of cumulus sulphide and formed under a range of magma:sulphide mass ratios (R factors), assuming an initial silicate magma composition similar to those of the most magnesian Chukotat Group basalts. Assuming no sulphides were segregated prior to eruption, the compositions of the majority of the ultramafic cumulate

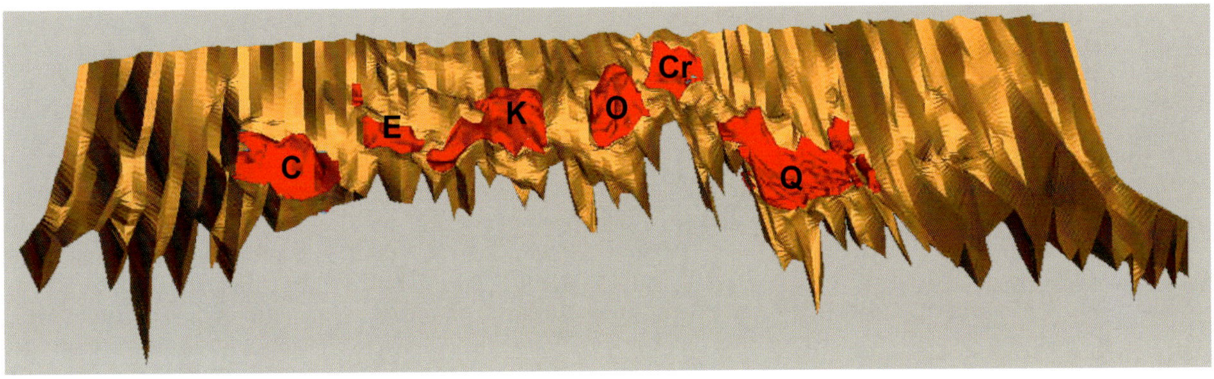

FIGURE 18. Projection (looking south) of the north-dipping footwall ultramafic/gabbro contact in the eastern part of the Katinniq area, showing the localization of major ore zones in irregular, open to re-entrant, second-order embayments in the contact (from Chisholm et al., 1999).

rocks from the Raglan Formation may be modelled by the presence of between 0.1 and 5% sulphides segregated at R = 100 to 10,000 (Burnham et al., 1999). Exceptions to this are the pyroxenitic rocks that occur near flow boundaries in a number of ultramafic bodies along the horizon. These appear to be relatively depleted in Pd and possess higher Cu/Pd ratios than the initial magma, suggesting that the magma may have lost minor amounts of sulphide (<0.001%) prior to eruption. Alternatively, if the initial magma is assumed to be slightly depleted in chalcophile elements prior to emplacement (as proposed by Barnes and Picard, 1993), then the range of chalcophile element compositions observed in the Raglan Horizon ultramafic cumulates would coincide with the compositions expected for olivine cumulates containing smaller amounts of sulphide than those estimated above (i.e. 0.1-2%) and sulphide segregation at higher R factors (1000-10,000). Because the lower sulphide content estimated for the cumulates formed from the depleted magma are close to those observed in the analyzed samples (<2%), a model in which the magmas were marginally depleted in highly chalcophile elements prior to eruption appears most likely for the petrogenesis of the ultramafic bodies of the Raglan Horizon. Although such depletion may be the result of the retention of PGE in the mantle residue, it most likely reflects the segregation of a small amount of sulphide prior to or during eruption (Burnham et al., 1999).

Olivine Chemistry

Relict igneous olivines are rarely preserved in the Raglan Formation, but those in a 108 m-thick section through the lower third of the Katinniq Ultramafic Complex range Fo_{85-87} (Lévesque et al., 2002) and those in grab samples from the Main Sill at East Lake range Fo_{86-88} (Stewart, 2002). Those at Katinniq exhibit an overall upwards increase in maximum Fo content and many are reversely zoned (Fig. 17). Similar increases in the Fo contents of olivine occur at Perseverance (Barnes et al., 1988) and in parts of the thick mesocumulate units at Kambalda (Lesher, 1989), but those olivines are unzoned. There is a broadly negative correlation between Fo content and TiO_2 content in the lower part of the section (M.L. Lévesque, unpubl. data) that is consistent with some of the variation being attributable to re-equilibration with trapped liquid, but the upper section does not show any correlation and the reverse zoning is not consistent with re-equilibration with trapped liquid. The systematic variations in olivine composition can be attributed to 1) in situ crystallization of olivine from progressively more magnesian magmas (Lévesque et al., 2002) or 2) lower Fo primocrysts overgrown by olivine crystallized from a more magnesian magma. It also means that the olivines crystallized and cooled relatively rapidly (Fe-Mg diffusion rates are of the order of 10 μm per day at these temperatures and such thick units require months to crystallize and years to cool) and are therefore unlikely to represent intratelluric phenocrysts. This is also consistent with the relatively high CaO (up to 0.29%) and MnO (up to 0.24%) contents of the olivines (see discussion by Lesher, 1989). The relatively low Fo values indicate that the composition of the magma at Katinniq ranged ~11-14% MgO, depending on FeO content (see Fig. 13C), which is much less than the 17 to 19% MgO composition of the parental magma inferred above and still significantly less than the 15 to 17% MgO in the fine random pyroxene spinifex-textured rocks analyzed by Barnes et al. (1982) and Barnes and Picard (1993). The low Fo contents of the olivines are therefore consistent with the whole-rock geochemical data, which indicate that these rocks have been significantly contaminated by upper crustal rocks.

Most of the olivines at Katinniq and some of the olivines at East Lake have Ni contents of 2000 to 2500 ppm, consistent with crystallization from a sulphide-poor magma, but some of the olivines at Katinniq and some of the olivines at East Lake have Ni contents as low as 900 ppm, consistent with equilibration with a sulphide-bearing magma (Duke and Naldrett, 1978; see discussion by Lesher and Stone, 1996).

Ni-Cu-(PGE) Mineralization

Geology

The geology of the deposits in the Raglan Formation has been described by Thibert (1999: Cross Lake), Petch (1999: East Lake), Mallinson (1999a,b,c: Zone 2, Zone 3, Zone 5-8), Lesher and Charland (1999: Katinniq), Vicker and Fedorowich (1999: Zone 13-14), Charland (1999: West Boundary), Stilson et al. (1999: Boundary), Lesher and Vicker (1999: Donaldson), and Chisholm et al. (1999: Katinniq mine geology), from which much of this section is taken. Those in the Delta Horizon have been described by Giovenazzo et al. (1989), Barnes and Giovennazo (1990), Tremblay (1990), and Giovenazzo (1991).

The deposits in the Raglan Formation typically contain multiple distinct ore zones, which can be subdivided into several types on the basis of location and geometry (Chisholm et al., 1999). The Katinniq deposit, for example, comprises 19 major ore lenses (see Fig. 12A above), 16 of which occur at or near the footwall contact of the basal subunit and three of which occur at or near the base of internal subunits, and the Zone 2 deposit comprises eight major ore lenses, four of which occur at or near the footwall contact of the basal subunit and two of which occur at or near the base of internal subunits.

Contact Ores

Most of the economic Ni-Cu-(PGE) sulphide mineralization in the Raglan Formation is localized in embayments along the footwall contact of the ultramafic complexes (e.g. Figs. 10, 12, 18) and corresponds to the Type I mineralization of Lesher and Keays (2002). Most of the contact ore zones at Katinniq are 15 to 40 m thick, have a strike lengths of 35 to 150 m, dip north-northeast at 30 to 45°, and contain up to 1.2 Mt of ore (Chisholm et al., 1999). Many ores zones exhibit a typical magmatic segregation/capillary infiltration profile: massive sulphides (typically 8-15% Ni, locally up to 20%) overlain by net-textured sulphides (typically 5-7% Ni) grading upwards into disseminated (typically 1.5-4.5% Ni), and then blebby sulphides (typically ~1% Ni) (Gillies, 1993; Chisholm et al., 1999; Chisholm, 2002). However, the C, E, and O ore zones at Katinniq are stratigraphically more complex, comprising two or more 'cycles' of massive, net-textured, and disseminated mineralization (see Fig. 19C). Where the contacts between ore types are gradual, this has been interpreted as evidence for multiple episodes of sulphide emplacement (Gillies, 1993), but where the contacts are sharp, this may also be attributed to fault replication (T. Mallinson, pers. comm., 2006). Contact ores rarely contain subangular to rounded, partly resorbed inclusions of footwall rocks (Chisholm et al., 1999; Chisholm, 2002: see Fig. 19B). Some ores are transgressed and/or offset by faults that have offset magmatic layering and mobilized some of the sulphides, forming ores with strong tectonite fabrics and/or breccia ores (corresponding to the type V mineralization of Lesher and Keays, 2002), but many ores and footwall contacts are otherwise relatively undeformed (Chisholm et al., 1999).

Hanging-Wall Ores

Ni-Cu-(PGE) mineralization also occurs in overlying subunits of the ultramafic complexes, typically forming strata-bound lenses of massive (less abundant than in contact ores), net-textured, and disseminated mineralization. In some cases these occur on or appear to correlate with internal stratigraphic horizons (e.g. Fig. 10) and correspond to the Type II mineralization of Lesher and Keays (2002), but in some cases they represent fault-duplications of Type I contact ores (Chisholm et al., 1999; Chisholm, 2002; T. Mallinson, pers. comm., 2006).

Narrow Vein-Type Ores

Several areas at Katinniq contain thin (1-2 m) tabular or sheet-like zones of mineralization that extend up to 50 m or more along strike, and which have been dubbed "narrow veins" by the miners. Chisholm et al. (1999) and Chisholm (2002) divided them into two types. Type 1 "veins" are deformed magmatic layers or fault slices. They are hosted by ultramafic rocks, occur at or near the N-dipping contact of the gabbro footwall, contain a "normal" magmatic sulphide assemblage (pyrrhotite-pentlandite-chalcopyrite), and comprise a zone of massive sulphides grading upward into disseminated sulphides. The wall rocks bordering the "veins" are altered and the sulphides in the veins are banded. Ore tenors range from 8-15% Ni, 1-2% Cu, and 1-2 g/t Pt+Pd+Au, similar to contact ores. Type 2 veins are true veins. They occur within the footwall gabbro, dip south (i.e. normal to the footwall contact), and are composed entirely of chalcopyrite-rich massive sulphides, consistent with fractional crystallization of overlying contact ores and/or with deformation-induced mobilization of chalcopyrite. In one location (NV1) where cross-cutting relationships were observed, Chisholm et al. (1999) and Chisholm (2002) noted that a Type 2 vein cross cut a Type 1 "vein", consistent with the latter interpretation.

Ore Localization

Type I contact ores are typically localized within second-order footwall embayments in the footwall rocks (Figs. 10, 11, 12A, 18) and many Type II hanging-wall ores within interpreted transgressive units in the complex (Fig. 10). The embayments that host the contact ores have been interpreted as: 1) "paleotopographic" features at the base of a sill (Barnes et al., 1982; see Fig. 20), 2) thermomechanical erosion conduits dislocated by faults (Gillies, 1993; Chisholm et al., 1999; Chisholm, 2002), and 3) folds along the footwall contact (Fedorowich, 1999). The contacts are clearly magmatic (see Fig. 19D) and transgressive (Fig. 8), the massive ores locally contain xenoliths of footwall rocks (Chisholm et al., 1999; Chisholm, 2002: see Fig. 19B), and the geometries of deformed embayments appear to have been modified by all phases of faulting and folding, so the embayments are best interpreted as structurally-modified thermal erosion structures (Chisholm et al., 1999; Chisholm, 2000).

Ore Mineralogy and Textures

The mineral assemblage in most Raglan ores is monoclinic pyrrhotite $Fe_{1-x}S$ – pentlandite $(FeNi)_9S_8$ – chalcopyrite $CuFeS_2$ ± magnetite Fe_3O_4 ± pyrite FeS_2 ± ferrochromite $CrFe_2O_4$ (Coats, 1982b; Gillies, 1993), with trace amounts of sphalerite ZnS, arsenopyrite FeAsS, and platinum-group minerals, including sperrylite $PtAs_2$, merenskyite $(Pd,Pt)(Te,Bi)_2$, sudburyite (Pd,Ni)Sb, moncheite (Pt,Pd)$(Te,Bi)_2$, temagamite Pd_3HgTe_3, kotulskite Pd(Te,Bi), testibiopalladite Pd(Sb,Bi)Te, and possibly hollingworthite (Rh,Pt,Pd)AsS (Dillon-Leitch et al., 1986; Seabrook et al., 2004). Violarite Ni_2FeS_4 and marcasite FeS_2 occur as a supergene phases in some areas, but most ores are remarkably fresh, even at the surface. The relative abundances of pyrrhotite, pentlandite, and chalcopyrite vary from sample to sample, zone to zone, and deposit to deposit, reflecting variations in original Fe, Ni, and Cu contents (see below) and modifications during metamorphism and deformation.

The following primary and secondary ore textures have been identified at Katinniq (modified from Gillies, 1993):

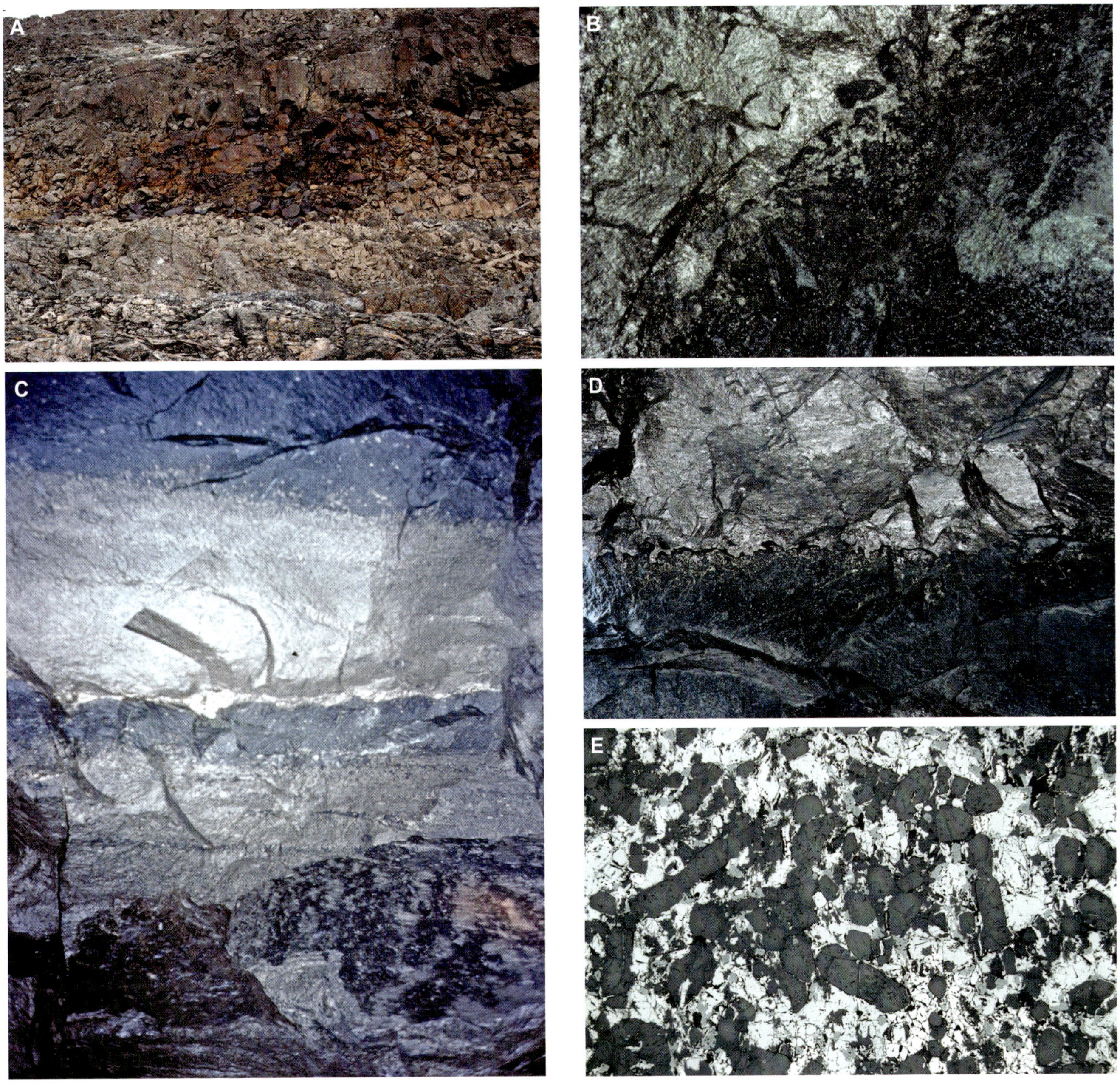

FIGURE 19. (A) Fe-Ni-Cu sulphide showing (rusty red, centre) in the western part of the Katinniq Ultramafic Complex. Overlying rocks are columnar-jointed peridotites (reddish brown), underlying rocks are pyroxenites (pinkish grey) and spotted slates (bottom). Showing is ~2 m thick and is not connected to the main ore zones down dip. (B) Gabbro inclusions (black, centre, largest is 20 cm in length) in massive Fe-Ni-Cu sulphide (yellow metallic, upper left) along lower contact with Katinniq Gabbro (dark green, bottom), C-1400 stope, Katinniq mine. (C) Complex ore profile in C-1400-3 stope, Katinniq mine, comprising net-textured Fe-Ni-Cu sulphides (lowermost part is joint surface), disseminated sulphides (dark green band in centre, ~5 cm thick), massive sulphides (bright metallic, upper centre), and disseminated sulphides (dark green, top). All contacts are undeformed. (D) Irregular primary magmatic contact between massive Fe-Ni-Cu sulphides (metallic, top) and gabbro (dark green, bottom), C-1400-3 stope, Katinniq mine. Width of photo is ~ 1 m. (E) Photomicrograph of net-textured ore containing serpentinized elongate phenocrysts of olivine (dark grey) in a matrix of Fe-Ni-Cu sulphides (bright) and magnetite (grey). Width of photo is 8 mm. Incident light.

Primary

1) interstitial disseminated: <1 to 30% sulphide (pyrrhotite-pentlandite±chalcopyrite±magnetite), fine-grained (1-3 mm), interstitial to serpentinized olivine grains;
2) blebby disseminated: 1 to 5% sulphide (pyrrhotite-pentlandite-magnetite-chalcopyrite), fine to coarse (0.3-2.0 cm), pentlandite partially enclosed by pyrrhotite;
3) leopard-textured: 20 to 40% sulphide (pyrrhotite-pentlandite-magnetite-chalcopyrite), interstitial disseminated sulphides surrounding clinopyroxene oikocrysts
4) patchy net-textured: 20 to 40% sulphide (pyrrhotite-pentlandite-magnetite-chalcopyrite), patches (0.5-2.0 cm), enclosing serpentinized olivine grains; commonly associated with interstitial disseminated sulphides;

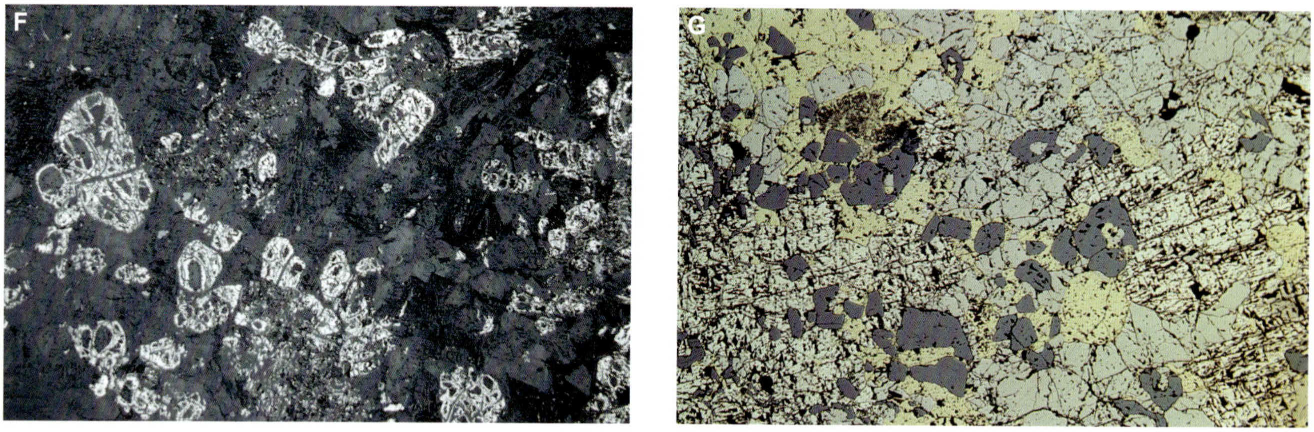

Figure 19 continued. **(F)** Photomicrograph of reverse net-textured ore containing sulphidized (bright), serpentinized phenocrysts of olivine (dark grey). Width of photo is 8 mm. Incident light. **(G)** Photomicrograph of low-tenor massive ore, containing pyrrhotite (light yellowish grey), pentlandite (light yellow, more pronounced partings), chalcopyrite (yellow), and magnetite (grey). Width of photo is 8 mm. Incident light.

5) net-textured: 40 to 70% sulphide (pyrrhotite-pentlandite-magnetite-chalcopyrite), enclosing serpentinized olivine grains (see Fig. 19E);

6) massive: >70% sulphide (pyrrhotite-pentlandite-magnetite-chalcopyrite±pyrite), typically coarse (0.5-2.0 cm) pentlandite in a matrix of pyrrhotite and chalcopyrite (see Fig. 19G).

Secondary

1) cloudy disseminated: <1 to 50% sulphide (pyrrhotite±pentlandite±magnetite±chalcopyrite), microscopic (<0.1 mm), interstitial to and/or within serpentinized olivine grains, typically pyrrhotite-rich;

2) reverse net-textured: 40 to 70% sulphide (pyrrhotite-pentlandite-magnetite-chalcopyrite), enclosing and partially replacing (?) serpentinized olivine grains (see Fig. 19F); less common than net-textured;

3) patchy reverse net-textured: 20 to 40% sulphide (pyrrhotite-pentlandite-magnetite-chalcopyrite), patches (0.5-2.0 cm), enclosing and partially replacing (?) serpentinized olivine grains; commonly associated with patchy net-textured sulphides;

4) vein: deformed sulphide (chalcopyrite-pyrrhotite±pentlandite±magnetite), typically 0.1 to 5.0 cm in width, generally in chlorite, chrysotile, and/or minor calcite veins.

Each of the magmatic textures reflects differences in the timing of sulphide saturation and mechanism of sulphide segregation, and many ore horizons are stratigraphically complex, comprising multiple episodes of mineralization (Fig. 19C). Total sulphide abundances in individual mineralized horizons are relatively constant, but sulphide textures are not, suggesting that the mineralization was emplaced in a series of pulses and that sulphide textures were established locally as a result of local variations in fluid dynamics and crystallization conditions (Gillies, 1993).

Although massive sulphides directly overlie footwall gabbros or hornfelsed sedimentary rocks in some areas (e.g. Fig. 19B,D), many ores overlie a lower pyroxenitic margin (Fig. 19A). This suggests that some of the mineralization segregated after formation of the lower chilled margin.

Reverse net-textured sulphides (Fig. 19F) were first identified at Raglan and appear to have formed by replacement of olivine (Philpotts, 1961; Dillon-Leitch et al., 1986). Although it is possible that this may have occurred by direct sulphidation of olivine, the olivines would have not contained very much Fe (~10% for Fo_{88}), so some Fe would have had to have been added. As such, it is more likely these textures formed by sulphidation of magnetite and precipitation of FeS during infiltration of a S-rich fluid. In any case, the abundance of pyrrhotite and paucity of magnetite in serpentinites containing cloudy disseminated mineralization indicates that substantial amounts of S have been introduced

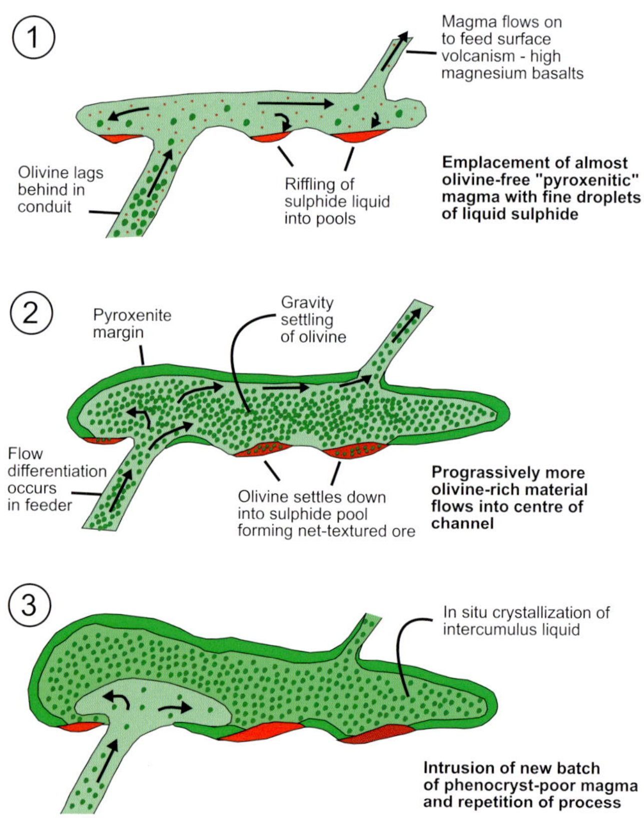

Figure 20. Inflationary sill model for the Katinniq Ultramafic Complex (Barnes et al., 1982).

TABLE 4. Average compositions of Ni-Cu-(PGE) ores in the Cape Smith Belt recalculated to 100% sulphides (from Barnes et al., 1997b; see also Barnes and Lightfoot, 2005).

			%	%	%	ppb	ppb	ppb	ppb	ppb	ppb	ppb	
Locality	**Ore Type**	**n**	**S**	**Ni**	**Cu**	**Au**	**Pd**	**Pt**	**Rh**	**Ru**	**Ir**	**Os**	**Ref**
Raglan Horizon													
Cross Lake	Cu-rich	5	39.5	6.31	3.26	627	8997	2548	320	227	41	45	1
	Cu-poor	3	40.4	4.25	1.31	108	2971	850	603	950	183	139	1
Zone 2-3	disseminated	33	37.7	12.2	2.64	326	7782	3125	505	1133	167	225	2
Katinniq	disseminated	21	38.3	11.4	3.45	525	11878	4032	796	2109	396	467	3
	Cu-rich	16	38.6	10.0	4.57	272	7290	3427	481	988	178	249	3
	Cu-poor	14	38.9	12.5	1.34	135	5122	2190	722	1881	301	445	3
Donaldson	disseminated	6	37.1	17.1	4.16	246	25261	6606	695	1943	362	459	4
	Cu-rich	5	36.0	17.7	9.08	3766	15408	2040	73	77	11	16	4
	Cu-poor	9	40.0	10.1	1.06	43	3153	2791	704	1575	286	337	4
Delta Horizon													
Bravo	disseminated	1	40.0	5.70	1.40	162	2671	1132	218	225	71	44	5
	Cu-rich	2	39.9	2.90	10.2	143	14860	2547	802	177	77	46	5
Delta	disseminated	1	40.5	5.15	0.68	187	3628	2133	326	365	49		1
	Cu-rich	14	39.7	7.63	2.51	49	6042	3085	282	61	29	26	1
	Cu-poor	19	39.6	9.39	1.22	38	1023	1165	528	501	83	72	1

n = number of samples in average; references: 1 = Giovenazzo (1991); 2 = = S.-J. Barnes (unpubl. data); 3 = Barnes et al. (1982), Giovenazzo (1991), and Gillies (1993); 4 = Giovenazzo (1991) and Dillon Leitch (1986); 5 = Barnes and Giovenazzo (1990).

(Gillies, 1993), probably during conversion of sedimentary pyrite to pyrrhotite during diagenesis and metamorphism of the underlying Povungnituk sediments.

Patchy disseminated textures are particularly abundant at Raglan. Although a model involving cotectic precipitation of sulphide and olivine in ~60:1 proportions (Duke, 1986) accounts for the abundances of sulphides in komatiitic dunite-hosted deposits like Dumont (Eckstrand, 1975) and Mt Keith (Groves and Keays, 1979; Grguric and Rosengren, 2004), it does not explain why even in these deposits the sulphide distributions can be locally very heterogeneous and why the sulphides at Raglan are commonly very patchy and heterogeneous. Eckstrand (1975), Groves and Keays (1979), Donaldson (1981), Gillies (1993), and Grguric and Rosengren (2004) have shown that the Ni and S contents of disseminated sulphides can be significantly modified by addition of Ni during serpentinization of olivine and S during metamorphism of adjacent sulphide-bearing sediments, but this should not necessarily affect the distributions of the sulphides. The less magnesian (komatiitic basalt vs. komatiite) composition of the magma at Raglan resulted in clinopyroxene crystallizing as an intercumulus phase, but this produces leopard-textured mineralization, not patchy mineralization. It is also possible that uniformly disseminated ores reflect a more homogeneous sulphide nucleation and precipitation process, whereas patchy ores reflect a more heterogeneous sulphide nucleation and precipitation process, or that patchy sulphides have coalesced from originally more disseminated sulphides. More work will be required to resolve this issue.

Ore Chemistry

The chemistry of Raglan sulphide ores has been studied by Barnes et al. (1992, 1997b), Barnes and Picard (1993), and Gillies (1993). Disseminated sulphides exhibit metal ratios similar to olivine microspinifex-textured komatiitic basalts at Zone 2-3 and Katinniq (Barnes et al., 1992), which indicates that they have been in equilibrium with a liquid of that composition and could have segregated from it. However, the Ni and Cu tenors of the sulphides vary by a factor of ~3 and range over an order of magnitude for the noble metals (Table 4, Figs. 21, 22). The wide range in compositions can be modelled by varying the mass ratio silicate to sulphide liquid (R-factor) between 300 and 1100 (Barnes and Picard, 1993; Gillies 1993, Barnes et al., 1997b). At Katinniq and Donaldson where more data are available, disseminated sulphides are richer in PGE than massive, net-textured, or vein-type sulphides. This suggests that the disseminated ores formed at higher R factors than net-textured and massive sulphides, i.e., that the latter had less opportunity to interact with silicate magma. This may be because the massive and net-textured sulphides were transported as sulphide-rich layers (Lesher and Campbell 1993) and/or because they were trapped at the margins of the flows and froze before they had an opportunity to interact more extensively with the magma (Barnes et al. 1997b). Gillies

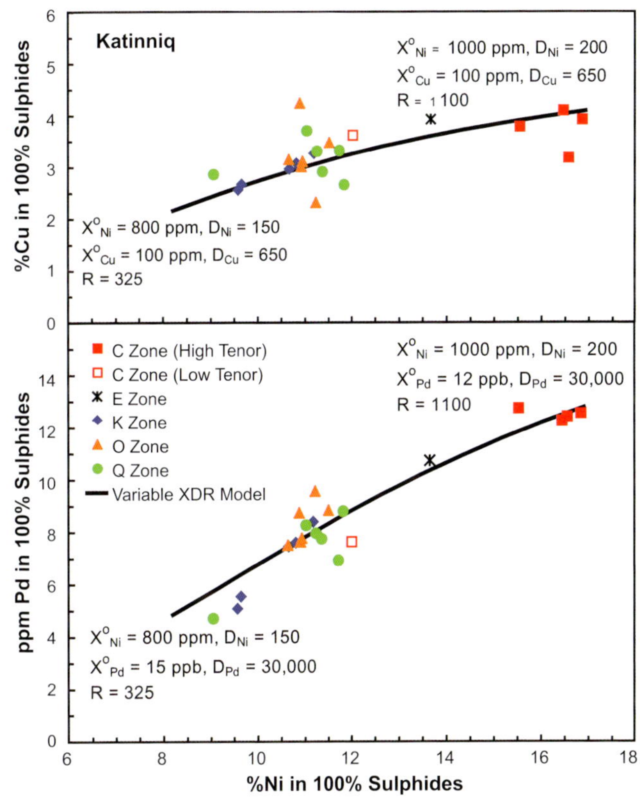

FIGURE 21. Cu-Ni and Pd-Ni models for Katinniq ores (from Gillies, 1993) using the method of Lesher and Campbell (1993).D = sulphide/silicate partition coefficient, Xo = initial magma composition, R = magma:sulphide mass ratio. Calculated R factors vary between 325 for low-tenor ores and 1100 for high-tenor ores, similar to those obtained by Barnes and Picard (1993) for a wider range of ores in the Raglan Formation.

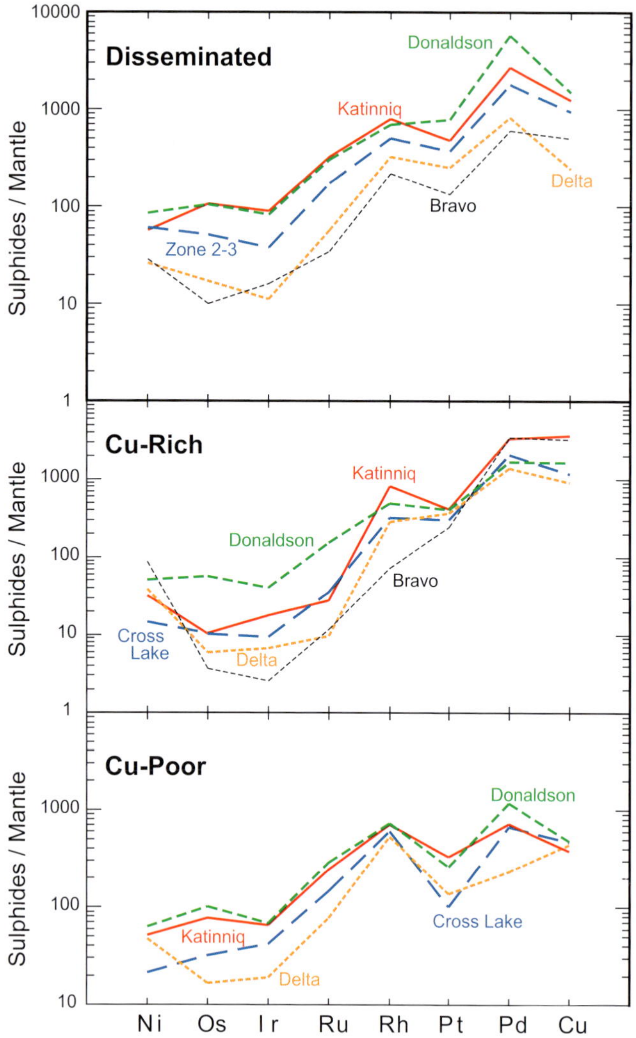

FIGURE 22. Mantle-normalized chalcophile element patters for Raglan ores (from Barnes et al., 1997b; data sources in Table 4).

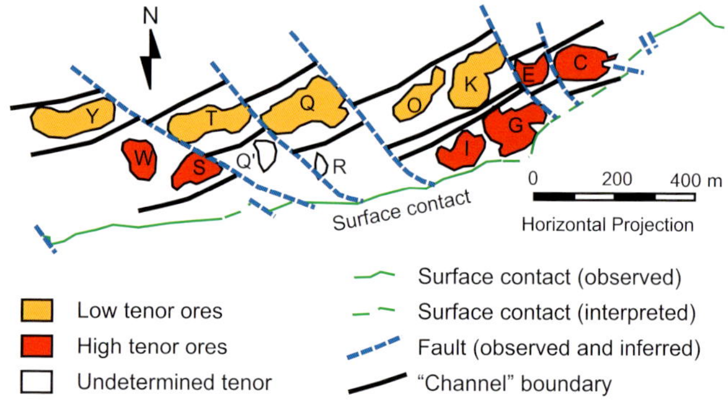

FIGURE 23. Plan projection of major contact ore zones, ore tenor belts, and offsetting faults in the Katinniq Ultramafic Complex (Gillies, 1993). Faults are inferred from those exposed on the surface and interpreted from footwall contour maps. The ore zones define two ore tenor belts, one low tenor (8-12% Ni in 100% sulphides) and one high tenor (13-17% Ni in 100% sulphides).

(1993) showed that the ore zones at Katinniq define two broad belts of differing ore tenor (Fig. 23), a northerly one having a Ni tenor of ~10 and a southerly one having a Ni tenor of ~15. There are also two parallel trends of ore lenses at Zone 3 and Zone 2, a northerlyon having a Ni tenor of ~15 and a southerly one having a Ni tenor of ~10 (Mallinson, 1999a,b). Interestingly, the sulphides hosted by lava conduit facies in the Raglan Formation have higher metal contents than the sulphides hosted by intrusions in the Delta Horizon. This suggests that the intrusions were less dynamic than the flows, which is consistent with the smaller amounts of olivine accumulation and lower degrees of contamination in those bodies (see discussion by Lesher et al., 2001). Barnes et al. (1997b) and Barnes and Lightfoot (2005) have shown that some Raglan ores have higher Cu/Ni and Pd/Ir ratios and that some Raglan ore have lower Cu/Ni and Pd/Ir ratios than disseminated ores, suggesting that they cooled slowly enough to fractionally crystallize monosulphide solid solution (MSS) and to segregate Fe-Ni-IPGE-rich MSS and residual Cu-PPGE-rich sulphide liquid. Most Raglan ores, including most disseminated ores, are depleted in Au relative to Pd and Cu, which may reflect mobilization during metamorphism.

Sulphur Isotope Geochemistry

The $\delta^{34}S$ values of 31 whole rocks and 14 mineral separates (pyrrhotite and chalcopyrite) representing a variety of ore types (disseminated, net-textured, and semi-massive) from eight deposits across the belt were analyzed by Lesher et al. (1999), and are shown in Figure 24. They range from 3 to 6‰ $\delta^{34}S$ and the S isotopic compositions of disseminated and massive sulphide layers in metasedimentary rocks from three localities across the belt are similar to the ores, ranging from 4 to 5‰ $\delta^{34}S$. One analysis of a secondary vein has a duplicated value of +20‰ $\delta^{34}S$. Sulphides within single ore samples and within individual ore horizons are similar (<1‰ differences), suggesting that they are near isotopic equilibrium, whereas samples from different ore horizons within a deposit and from different deposits within the belt are more variable (1-3‰ differences), suggesting that they evolved independently to some degree.

The $\delta^{34}S$ values of the ores (3-6‰ $\delta^{34}S$) are systematically different from mantle values (0±1‰ $\delta^{34}S$) and should not have fractionated significantly at magmatic temperatures. The similarity between the S isotopic compositions of the ores and the metasedimentary rocks supports field relationships suggesting that the S in the ores was derived by melting/assimilation of footwall sediments upstream from

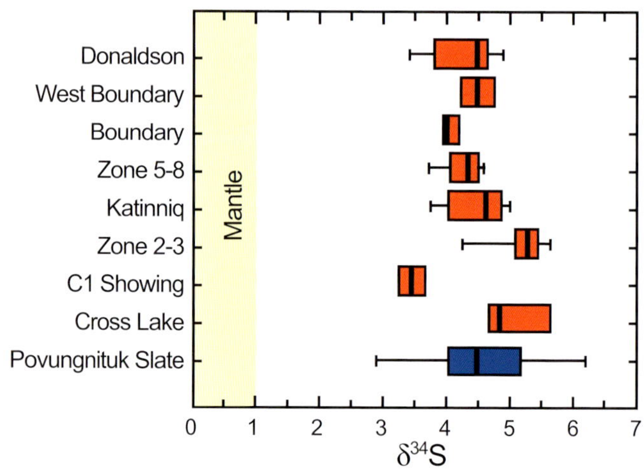

FIGURE 24. Summary of S isotopic data for ores and barren sulphidic semipelites in the Raglan area, Cape Smith Belt (data from Lesher et al., 1999).

TABLE 5. Summary of features consistent with various models for the mineralized ultramafic complexes in the Raglan Formation.

Observed Feature	Lava Pond	Deeply Erosive Lava Conduit	Invasive Flow	High-Level Intrusive
Columnar joints in upper and lower parts	+	+	+	+
Peperites and intrusive upper contacts in *lateral* parts	–	+	+	+
Rapidly-cooled mesostases and high-Ca olivine	+	+	~	~
Conformable upper contacts	+	+	~	~
Flow-top breccias and polyhedral joints	+	+	~	–
Asymmetric erosion and V-shaped embayments	~	+	~	–
Asymmetric contact metamorphism	+	+	–	–
Linear geometry of host units (subregional scale)	–	+	+	–
Linear tends of ore shoots (local scale)	–	+	+	–
Consistent location at the base of Chukotat Group	+	+	+	~

+ consistent, ~ possible, – inconsistent

the site of mineralization (Barnes et al., 1993; Lesher et al., 1999) similar to most other deposits of this type (Lesher, 1989; Lesher and Keays, 2002; Barnes, 2006; Barnes and Lesher, in press).

Re-Os Isotope Geochemistry

The Re-Os isotope systematics of the ores and host rocks have been studied by Luck and Allègre (1984) and Shirey and Barnes (1994, 1995), and are characterized by 0.184-25.6 ppb Re, 0.108-23.9 ppb Os, $^{187}Re/^{188}Os$ ratios of 1.63-9.97, and $^{187}Os/^{188}Os$ ratios of 0.168-0.441. Underlying S-rich Povungnituk slates are characterized by 17 ppb Re, 0.39 ppb Os, $^{187}Re/^{188}Os$ ratios of 216-278, and $^{187}Os/^{188}Os$ ratios of 7.8-9.2. Seventeen samples from the Zone 2-3 and Katinniq deposits in the Raglan Formation and the Bravo and Delta sills in the Povungnituk Group yielded a Re-Os isochron with an age of 1907 ± 138 Ma and a high mean squared weighted deviates (MSWD), that overlaps the 1958 to 1918 Ma U-Pb age suggested from stratigraphic correlations (Parrish, 1989), and an initial $^{187}Os/^{188}Os$ ratio corresponding to a γOs of -1.4 at 1907 Ma.

Shirey and Barnes (1994, 1995) used the near-chondritic Os isotopic compositions of the ores and host rocks and the highly radiogenic compositions of the slates to suggest that the ore deposits could not have formed by incorporation of S from the slates. However, Lesher and Stone (1996) and Lesher and Burnham (2001) have shown that Os isotopic signatures are more rapidly erased than S isotopic signatures in dynamic magmatic systems. For example, using γOs values of 0 and +800 for the parental lava and the sediment, and the mass balance equations of Lesher and Burnham (2001), the γOs value of the contaminated lava is only 0.9 to 2.0 (i.e. only 0.11-0.25% of 800) over the 300 to 1100 range of magma:sulphide ratios determined by Barnes et al. (1993) and Gillies (1993), well within the errors propagated by extrapolating the measured isotopic abundances to 1907 Ma. Using $\delta^{34}S$ values of 0 and +4 for the parental lava and the sediment, and the mass balance equations of Lesher and Burnham (2001), the $\delta^{34}S$ value of the contaminated mixture is between 1.7 and 2.5 (i.e. still 43-63% of 4) over the same range, confirming that the S isotopic system is a more sensitive indicator of contamination than the Os isotopic system in dynamic magmatic systems of this type.

Interpretation

Volcanic Setting

The volcanic settings and modes of emplacement of the mineralized ultramafic complexes in the Raglan Formation are still not completely clear. They have been historically interpreted as sills (e.g. Wilson et al., 1969; Hynes and Francis, 1979; Francis et al., 1981, 1983; Barnes et al., 1982; Bédard et al., 1983; Giovenazzo et al., 1989; St-Onge et al. 1993; 1994), but some have been interpreted as flows that evolved into sills during the evolution of the volcanic pile (Boundary: Albino, 1984), as sills that evolved into lavas (D. Francis, pers. comm., 1990), as lava ponds (Katinniq: Barnes and Barnes, 1990), and as deeply erosive lava conduits (Gillies, 1993; Thacker, 1995; Lesher and Charland, 1999; Lesher et al., 1999; Mallinson, 1999a,b,c; Stilson, 1999; Stilson et al., 1999). Parts have been interpreted as feeder dykes (Zone 2: Miller, 1977; East Lake South: Petch, 1999; Stewart, 2002). It is also possible that some may represent invasive (downward burrowing) lava flows (see e.g., Beresford and Cas, 2001). Some of the differences in interpretation can be attributed to the possibility that different parts of the Raglan Formation formed in different ways and/or in different settings, but most exhibit many of the same features and probably formed in a broadly similar setting.

Although many of the criteria used to distinguish between extrusive, invasive, erosive, and intrusive modes of emplacement can be ambiguous in high-level subvolcanic systems if not extremely well exposed (see discussion by Dann, 2000; Arndt et al., 2005), we have enough information to place some constraints on their setting and mode of emplacement (Table 5).

Flow-top breccias: Hydroclastic breccias often (but not always) form along the rapidly cooled upper margins of submarine lava flows, but appear to rarely (if ever) form along the upper margins of sills or invasive flows. The aphanitic to microspinifex-textured breccias that occur along many parts of the upper margins of conduit facies in most areas of the Raglan Formation contain larger, more uniformly sized fragments than the breccias that characterize most hydroclastic flow-top breccias in basalts and komatiites, but the absence of a sedimentary matrix indicates that they are not peperites. They often grade downwards into polyhedrally jointed pyroxene-phyric basalt and often represent dissagregated equivalents of those rocks rather than true hydroclastic breccias. In any case, their presence along parts of the upper margins of conduit facies suggests that those units are at least partly extrusive, whereas their absence along the upper margins of sheet facies is consistent with them being extrusive, invasive, or intrusive.

Polyhedral and columnar joints: Columnar joints are common both in thick flows and high-level sills, and therefore only provide evidence of high-level emplacement. However, polyhedral joints appear to form only in lava flows (see discussion by Arndt et al., 2004). They are common in

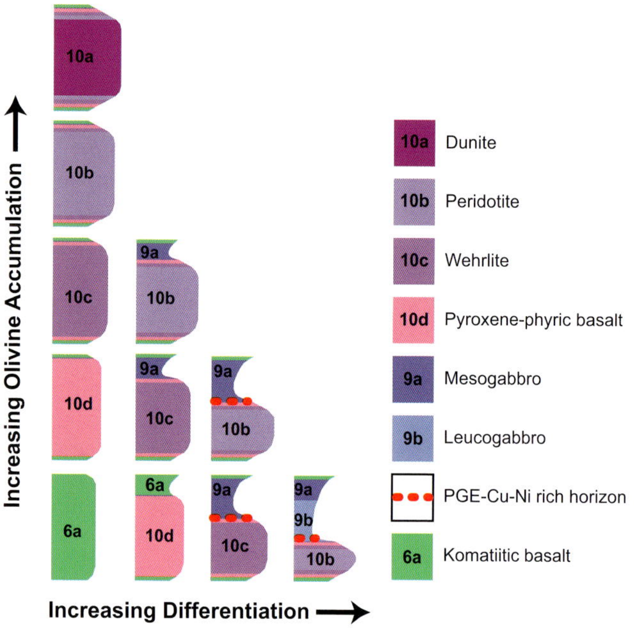

FIGURE 25. Facies variations of volcanic and subvolcanic rocks in the Chukotat Group as a function of degree of differentiation and degree of olivine accumulation (adapted from Lesher et al., 1984). Widths of profiles are proportional to MgO content.

the upper parts of conduit facies and appear to be analogous in many respects to the irregularly oriented jointed entablature zones of thick Columbia River flood basalt flows (Long and Wood, 1986). Polyhedral joints have not been observed in thinner sheet facies in the Raglan Formation, suggesting that those parts might be invasive or intrusive.

Peperites: Peperites often form along the margins of sills or invasive (downward-burrowing) lava flows intruded into wet unconsolidated sediments, but may also form along the lateral margins of deeply erosive lava conduits. The blocky and fluidal peperites along the lateral margins of the Zone 2 and Katinniq ultramafic complexes provide unambiguous evidence that magma from those bodies intruded unconsolidated and most likely wet sediments. This excludes an entirely extrusive setting, but does not distinguish between entirely intrusive, invasive, or deeply erosive settings.

Rapidly cooled mesostases and high Ca in olivine: The microspinifex-textured mesostases in many of the mesocumulate rocks and the high Ca and Mn contents of rare relict olivine indicate rapid cooling. This excludes an intermediate to deep intrusive setting, but does not distinguish between an extrusive, invasive, erosive, or very shallow intrusive setting.

Conformable upper contacts: The upper contacts of sheet facies are regular and conformable, consistent with both sheet flows and sheet sills. The upper contacts of conduit facies are irregular, but also normally conformable. Apophyses of conduit facies rarely intrude overlying rocks, no roof pendants have been observed extending downward from the overlying rocks into the ultramafic complexes, and no dykes have been observed extending upward from the ultramafic complexes into the overlying sediments or basalts. These features are most consistent with an extrusive or deeply erosive origin, but do not preclude an intrusive or invasive origin.

Asymmetric erosion and V-shaped embayments: Conduit facies occupy broad, V-shaped embayments in the footwall rocks. Stratigraphic and structural studies, crosscutting relationships, and rare xenoliths indicate that the embayments formed by thermomechanical erosion and are not structural. There is no reason why overlying sediments and basalts would not also have been eroded unless they were not there, so the V-shaped geometry of most of the embayments favours an extrusive origin. Williams et al. (1998) calculated that a 10 m thick flow could have fluidized and assimilated wet sediment at rates of 5 to 20 m per day at distances of 10 to 100 km from the source. Embayments tens to hundreds of metres deep could have formed over periods of prolonged flow of the order of weeks. It is possible that the embayments represent upward erosion by a sill that evolved into a lava conduit, but the geometry is more consistent with progressive downcutting by a deeply erosive lava conduit (see discussion by Jarvis, 1995).

Asymmetric contact metamorphism: Contact metamorphic aureoles normally form above and below thick sills, above and below thick invasive (downward-burrowing) flows, and beneath thick lava flows, especially if the rocks are unconsolidated sediments, but they may also form above thick, slowly cooling lava ponds or lava conduits if they become covered by sediment or other volcanic rocks. The conduit facies were very thick, very hot (up to 1400ºC based on inferred parental magma compositions), and very dynamic (based on the large amounts of accumulated olivine), and would have cooled very slowly. If they intruded as high-level sills into unconsolidated sediments (required in any case by the marginal peperites), they should have generated very large metasomatic aureoles (see e.g., Polyansky and Reverdatto, 2006). There is little alteration above the Bravo or Méquillon intrusive bodies, but as noted by Barnes and Giovenazzo (1990) and Tremblay (1990), those are orthocumulate bodies containing only 30 to 40% excess cumulus olivine, compared to the 70 to 80% excess cumulus olivine in the mesocumulate bodies in the Raglan Formation. Although the rocks above the East Lake main body are very locally contact metamorphosed (Stewart, 2002), the overall pattern of contact metamorphism associated with the conduit facies of the Raglan Formation is overwhelmingly asymmetric: underlying rocks are strongly modified but overlying rocks are not obviously modified. Such asymmetry is much more consistent with a lava pond or deeply erosive lava channel than with an entirely intrusive body.

Linear conduits and linear ore tenor belts: The linear nature of the conduit facies (indicated by inverse magnetic models and confirmed by diamond drilling: see Exploration Model) and the linear nature of the ore tenor belts identified at Katinniq and Zone 2-3 are more consistent with a lava channel or possibly an invasive flow than with a lava pond or subvolcanic sill.

Stratigraphic location: Although thick olivine pyroxenite-gabbro and thinner pyroxenite sills occur throughout the Povungnituk Group (Fig. 3), all of the thick differentiated units in the lower Chukotat Group appear to be thick lava flows (Francis and Hynes, 1979). The Raglan Formation occupies a very specific stratigraphic horizon at the base of the Chukotat volcanic sequence for 55 km between Cross Lake and Raglan Lake (Fig. 3). The magmas that formed

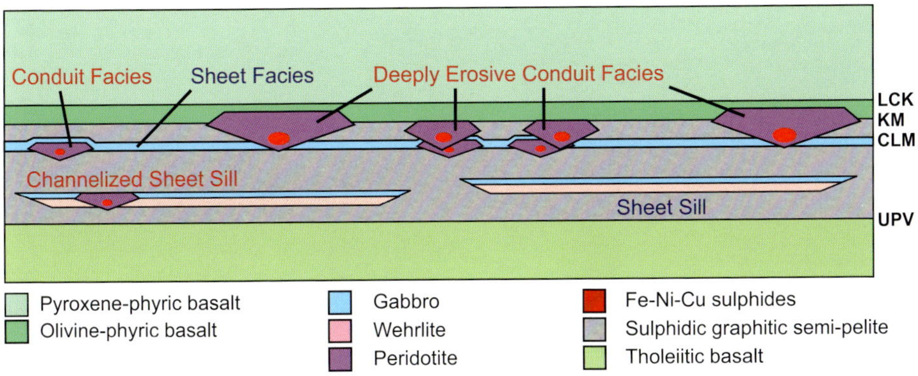

FIGURE 26. Interpretive schematic cross-section through conduit and channelized sheet facies assemblages in the Raglan area, east-central Cape Smith Belt (from Lesher et al., 1999a). Only conduit facies are mineralized. CLM = Cross Lake Member (channelized sheet facies assemblages), KM = Katinniq Member (conduit facies assemblages), LCK = Lower Chukotat Group, UPV = Upper Povungnituk Group.

these complexes are the most magnesian in the sequence, but they are texturally and compositionally gradational with differentiated peridotite-gabbro and pyroxenite-basalt flows in the overlying Chukotat volcanic sequence. The stratigraphic sequence of thick mesocumulate units (Katinniq Member) and differentiated gabbro/orthocumulate units, with thin intercalated sediments overlain by a sequence of lavas containing progressively thinner flow units and progressively lower MgO contents, is similar to many komatiite sequences worldwide (Lesher, 1989; Lesher and Barnes, in press).

The field relationships therefore suggest that the Cross Lake Member, which represents a channelized sheet facies assemblage in the Cross Lake–C1–C2–C3 and Zone 2–Zone 3–Katinniq–Zone5–Zone7 areas, and the Katinniq Member, which represents a conduit facies in the Zone 2–Zone 3, Katinniq, Zone 6, Zone 8, Zone 13-14, West Boundary, and Boundary areas, both formed at very high levels and exhibit features of extrusive, deeply erosive, invasive, and intrusive environments.

Cross Lake Member

The field relationships therefore suggest that the Cross Lake Member originated as 1) a series of very high-level sheet sills that evolved into a channelized sheet facies assemblage (Lesher and Barnes, in press) as flow became focussed within a conduit or 2) a series of erosive lava channels that spread laterally as sills. The presence of greater amounts of contamination within conduit facies compared to sheet facies is more consistent with the former interpretation. The presence of flow-top breccias in the Cross Lake and C1-C2-C3 areas indicates that parts of the conduit facies in that area became extrusive.

Katinniq Member

The field relationships suggest that most parts of the Katinniq Member originated as lava conduits that deeply eroded their substrates, eventually becoming intrusive along their margins. The peperites along the lateral margins of some of the complexes and local intrusive relationships along the margins in lateral areas reflect the latter stages of that process. A schematic model for the volcanic/subvolcanic sequence is given in Figure 26. The presence of fine-grained graphitic, sulphidic clastic metasedimentary rocks underlying the mineralized peridotite units, the presence of pillow basalts directly overlying the mineralized peridotites, and the virtual absence of vesicles in any of the volcanic rocks indicate that the volcanic sequence was erupted into relatively deep water.

Volcanic Evolution

The basalts, gabbros, pyroxenites, olivine pyroxenites, peridotites, and dunites in the Raglan Formation are interpreted to be derived from a parental magma of komatiitic basalt composition, which has produced a wide range of different rock types depending on the degrees of in situ differentiation and olivine accumulation (Fig. 25).

The first phase of Chukotat volcanism (Cross Lake Member) appears to have formed channelized sheet sills and/or invasive sheet sills (Fig. 26) in the Cross Lake-C1-C2-C3 (Fig. 5), Zone 3-Zone 2-Katinniq-Zone 5-Zone 7 (Figs. 6, 7B, 27B), and Boundary (Figs. 7A, 27A) areas.

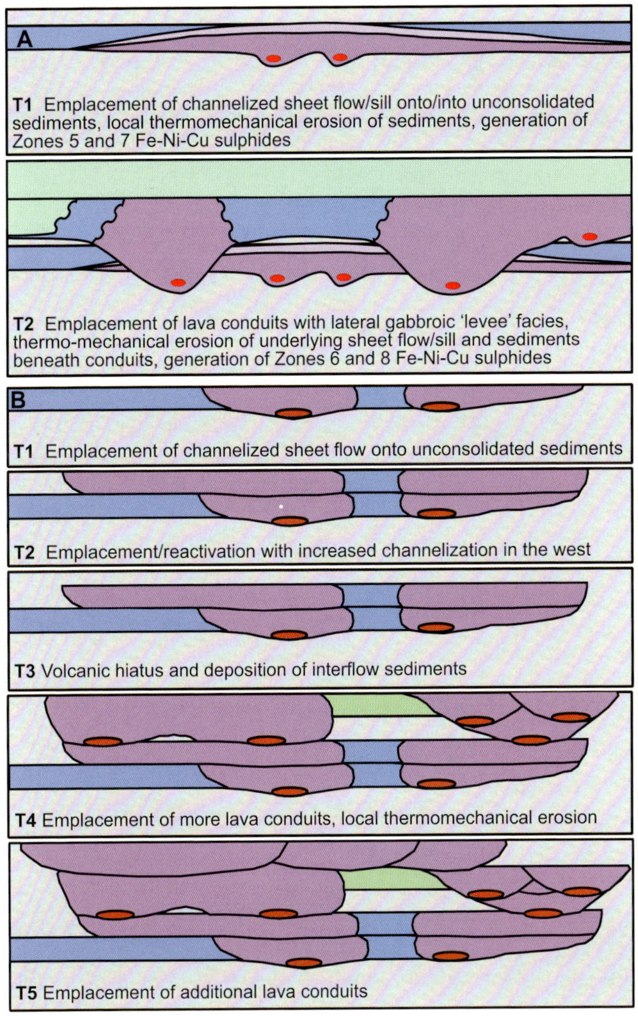

FIGURE 27. Emplacement models for **(A)** the Zone 5-8 Complex (adapted from Thacker, 1995) and **(B)** Boundary Complex (adapted from Stilson, 1999).

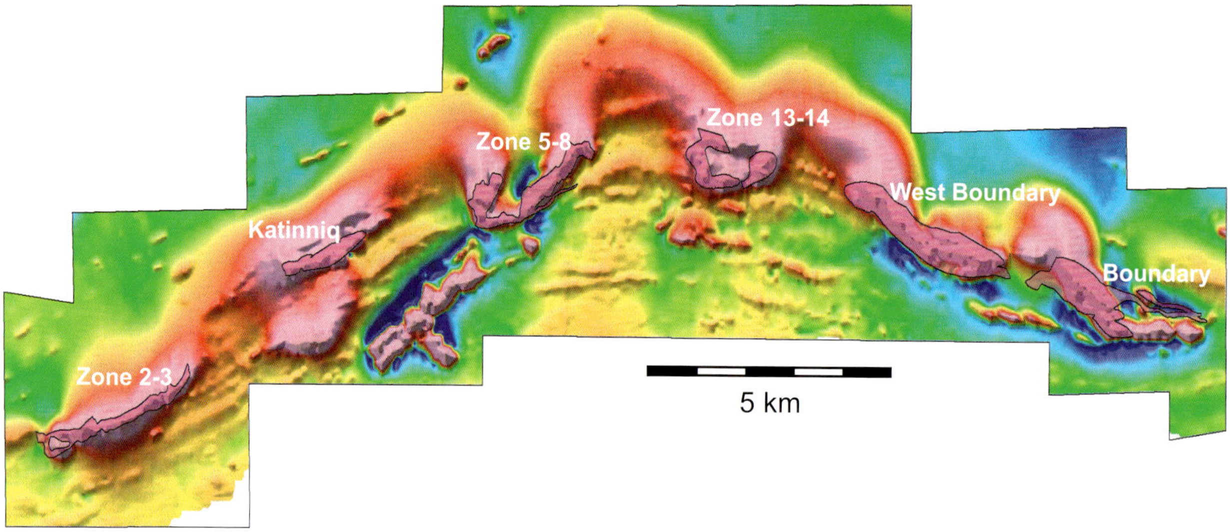

Figure 28. Total field magnetic map for the central part of the Raglan Belt between the Zone 2-3 and Boundary complexes (from Osmond and Watts, 1999), imaging the magnetite-rich serpentinized peridotite complexes. The surface expressions of the complexes are outlined in black and appear to be connected down-dip to the north beneath overlying Chukotat Group basalts. The magnetic bodies in the underlying Povungnituk Group to the south are sills.

Where present, mineralization is restricted to the thickest and most magnesian parts of these units, which are interpreted to represent a conduit facies (e.g. Cross Lake Main Zone, C1-C2-C3 zones, Zone 5, Zone 7). These conduit facies are flanked by thinner, less magnesian zones of differentiated peridotite-gabbro, interpreted to represent a sheet facies (e.g. Cross Lake north and east flanks, C1-2-3 west and east flanks, Zone 5 west flank, Zone 7 east flank, Boundary west flank). A thin layer of sediments is well preserved above the Cross Lake Member in the Katinniq and Boundary areas, but appears to have been extensively eroded by the second phase of volcanism in other areas. Many of the sheet facies are indistinguishable from the sills in the upper part of the underlying Povungnituk Group, providing an important spatial and temporal link between the upper Povungnituk and lower Chukotat Groups in this part of the Cape Smith Belt.

The second phase of Chukotat volcanism (Katinniq Member) appears to formed deeply erosive lava conduits (Fig. 26) in the East Lake, Zone 3, Zone 2, Katinniq, Zone 6, Zone 8, West Boundary, and Boundary areas (Figs. 3, 7A,B, 27A,B). The peridotite zones in these units appear to be composed of multiple, overlapping conduits that are capped by pyroxenites, basaltic breccias, and more rarely pillow breccias. These units appear to have thermomechanically eroded underlying units, leaving in some cases only isolated remnants of the first unit (e.g. Lower Gabbro in Boundary Complex: Fig. 27B). 3-D magnetic inversion models and deep stratigraphic drilling (Figs. 28, 30) indicate that the units are continuous in the subsurface, which has been confirmed by deep stratigraphic drilling, suggesting that they represent long linear lava conduits analogous to the sinuous rilles on Mars, Venus, and the Moon (Williams et al., 1999a).

In the Zone 5-8 area, conduit facies of the Katinniq Member (Zone 6 and Zone 8) are superimposed on conduit facies of the Cross Lake Member (Zone 5 and Zone 7) (Fig. 27A), suggesting that the Katinniq Member may have reactivated the same conduits used by the Cross Lake Member or that the emplacement of the Katinniq Member in that area was controlled by a low related to the conduits in the Cross Lake Member.

Subsequent volcanic episodes become progressively less voluminous, grading from relatively massive peridotite-pyroxenite units through thick olivine pyroxenite-gabbro flows, thin pyroxenite-basalt flows, and thick medium-grained ('gabbroic') komatiitic basalt flows to thin massive and pillowed komatiitic basalt lava lobes (Fig. 29). This sequence of deposition, in which the early phases of volcanism are very voluminous, but discontinuous, and later phases of volcanism are more continuous, but less voluminous, appears to be typical of many areas of komatiitic volcanism (Lesher, 1989; Lesher and Barnes, in press).

Ore Genesis

The Fe-Ni-Cu-(PGE) sulphides in the Raglan Formation have long been interpreted to be of magmatic origin (e.g. Kilburn et al., 1969) and more recent studies have confirmed this interpretation (e.g. Barnes et al., 1982; Dillon-Leitch et al., 1986; Barnes et al., 1992; Gillies, 1993).

Early workers assumed that the sulphides originated in the mantle, that they were transported as immiscible sulphide droplets, and that they settled into depressions at the bases of the ultramafic complexes (Fig. 20). However, magmas with ~18% MgO are derived by moderate degrees of melting (see Barnes and Picard, 1993) and should be moderately undersaturated in sulphide (Keays, 1982; Lesher and Groves, 1986; Arndt et al., 2004). The absence of PGE depletion in most of the host rocks confirms that the magmas were undersaturated in sulphide until immediately prior to or during emplacement. Most peridotites in the Raglan Formation contain ~1% fine disseminated Fe-Ni-Cu sulphides and most basal pyroxenites contain minor blebby sulphides, indicating that the lava was saturated in sulphide on emplacement and remained saturated in sulphide during crystallization. The S isotopic compositions of the ores are consistent with derivation of most of the S from underlying S-rich sediments, not

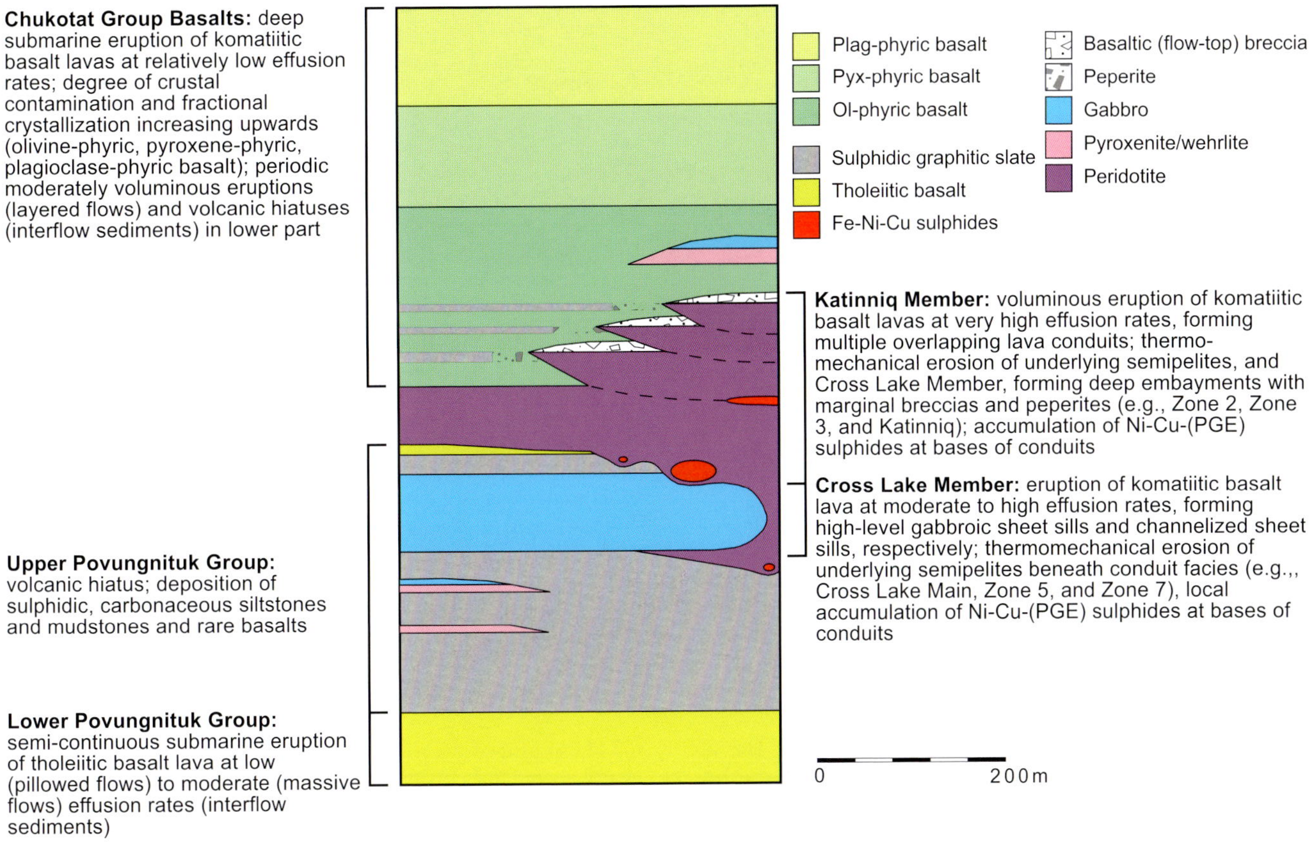

FIGURE 29. Interpretive stratigraphic column for the Raglan Formation.

with derivation from the magma. The mineralized ultramafic complexes appear to have thermomechanically eroded the footwall rocks, so the most logical source of S is the sediments missing in the ore environments and/or upstream from the present ore locations.

Most of the Ni-Cu-(PGE) sulphide ores occur within footwall embayments near the base of the ultramafic complexes or near the bases of internal flow units. Although many contacts are locally sheared, there is no evidence of shearing, faulting, or folding in some places and it is unlikely that faults could produce such complex embayment geometries, so the embayments are best attributed to thermomechanical erosion, modified by faulting and folding (Chisholm, 1999; Lesher et al., 1999).

The amount of sulphides in most zones is much greater than could have dissolved in or exsolved from the overlying host unit, especially given that the ultramafic complexes comprise multiple thin units. This means that the sulphides likely formed by melting of the sulphides in the sediments, not by assimilation and reprecipitation (see discussion by Lesher and Campbell, 1993). A detailed mass balance of the amount of S in the ore environment is complicated by incomplete exposure and the complex geometry of the host units, ore zones, and ore-localizing embayments. However, if we assume that the amount of sediment removed was equal to the 200 m maximum thickness of many host units, that the sediment contained an average of 2% sulphide, and that the magma was already saturated in sulphide at that stage (by melting of sediments upstream), it would produce 8 m of net-textured ore. Most of the ore zones are thicker (10-40 m), but they occupy less than 50% of the contact even in the most mineralized areas, so the amount of S in the ores is greater than, but of the same order of magnitude as the amount of S in the missing sediments. Any shortfall can be attributed to accumulation of sulphide melted 'upsteam' in the system, as interpreted for many other magmatic Ni-Cu-PGE deposits (e.g. Lesher, 1989; Lesher and Keays, 2002; Naldrett, 2004; Barnes and Lightfoot, 2005).

The moderately high R factors indicate that the sulphides equilibrated with and extracted metals from a mass of magma 300 to 1100 times the mass of sulphides. This is consistent with a model in which S is melted from a large volume of sulphidic sediments upstream from the present ore zones, sulphides are transported along the length of the conduits as immiscible layers and/or droplets in the turbulently flowing lava, and sulphides are concentrated in embayments at the bases of the conduits. This model is similar to that proposed for many other deposits of this type (Lesher and Keays, 2002; Barnes and Lesher, in press).

Exploration Model

The development of exploration models for the deposits in the Raglan area has been reviewed by Green and Dupras (1999). The first exploration models were based on accumulation of mantle-derived sulphides that had settled out of mafic/ultramafic magmas into depressions at the bases of ultramafic sills (Barnes et al., 1982: Fig. 20). Many aspects of those models are still valid: an association with thick, poorly differentiated peridotite units and localization within footwall embayments. However, those models and ones

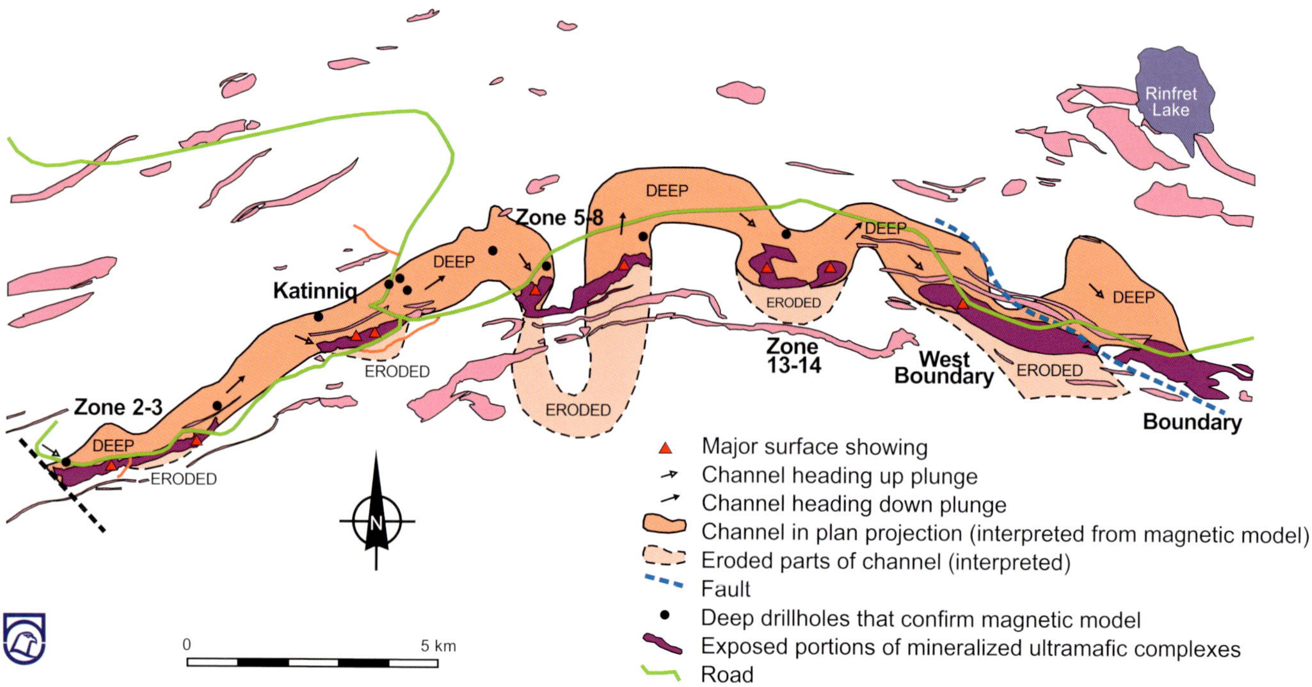

FIGURE 30. Meandering lava channel exploration model for the ores in the Raglan Belt (from Green and Dupras, 1999).

involving ponding of lava near a feeder (Barnes and Barnes, 1990) were not able to predict the geometrics of the host units, nor the locations of the ore-localizing embayments.

Following the first application of a lava conduit model to the komatiite-associated Ni-Cu-PGE deposits at Kambalda (Lesher et al., 1984), a similar model was applied to the Raglan area beginning in 1989. Between 1990 and 1993 the new model successfully predicted continuations and extensions of known mineralization, which led to the decision to develop an underground ramp at Katinniq from which to do underground drilling and ultimately to the decision to mine the deposit (Green and Dupras, 1999).

During this time (1989-1991) outcrop patterns, total field magnetic data, and the asymmetry of the morphologies of the embayments (Thacker, 1995) were used to infer that the lava conduits at Zone 3, Zone 2, and Katinniq plunged broadly northeast, the lava conduit at Zone 6 plunged broadly north, and the lava conduit at Zone 8 plunged broadly northeast. A similar northeast trend has also been inferred for East Lake (Stewart, 2002). It was also recognized that the lava conduits at Katinniq had been dislocated by north-northwest-trending lateral ramps, producing a distribution of orebodies like eggs in an carton, where original northeast-trending lava conduits now trended broadly east-west (Gillies, 1993: Fig. 23).

The exploration model for Raglan was refined with the completion of a regional 3-D magnetic model (Fig. 28: Watts 1997; Watts and Osmond, 1999), which confirmed the plunges inferred in the earlier studies but which also suggested that the conduits were connected in the subsurface and might be part of a single 'meandering' system with an overall east-west trend. Deep stratigraphic drilling between Katinniq and 5-8, between Zone 5-8 and Zone 13-14, and between Zone 13-14 and West Boundary (Fig. 30) in 1997-1998 confirmed the validity of the subsurface interpretations in the model (Green and Dupras, 1999).

The origin of the meandering pattern in the magnetic data is not completely clear. Folding is unlikely, except perhaps in the 5-8 area where a footwall sill appears folded (Fig. 6), because there is no evidence of folding in the footwall or hanging wall rocks in most areas, although intrafolial folding cannot be completely excluded. Green and Dupras (1999) interpreted the pattern as a single meandering lava conduit (Fig. 30). If this interpretation is correct, then parts of the system have been eroded (e.g. between Zones 3 and 2, in the middle of Katinniq, in the middle of Zone 5-8, and in the middle of Zone 13-14: Fig. 30). An alternative interpretation is that one or more conduits have been variably dislocated by north-northwest-trending thrust faults (lateral ramps) like those observed at Katinniq (Fig. 8), between West Boundary and Boundary (Stilson and Lesher, 1999), between Boundary and Donaldson (Stilson and Lesher, 1999), and at Donaldson (Lesher and Vicker, 1999). Both alternatives are consistent with the presence of two parallel trends of ore lenses at Katinniq (Fig. 23) and at Zones 2 and 3 (Mallinson, 1999a,b), and both suggest that additional mineralization might occur in parallel conduits down dip.

As noted by Green and Dupras (1999), an important aspect of this exploration model is that it predicts the amount of prospective host rock that exists in the slice that has been modelled from surface to 1000 metres. They noted that more than 90% of the presently defined resource is shallower than 300 m and even after 50 years of drilling, exploration in the known mineralized areas is far from complete even to that depth. Thus, after the entire slice to 1000 m has been fully explored, there is excellent exploration potential below that depth that is still within reach of mining.

Acknowledgements

This paper is based on the work of many former and current Falconbridge/Xstrata geologists, whom I thank for many stimulating discussions in the office and core shack, particularly Scott Bruce, Colin Coats, Anne Charland, Al Coutts, Michel Dufresne, John Fedorowich, Maria Gabriel, Danielle Giovenazzo, Tony Green, Ron Lemery, Terry Mallinson, George Nemcsok, Richard Osmond, Christine Petch, Mike Sweeny, François Thibert, Christian Tremblay, Phil Vicker, and Tony Watts, and many former researchers whom I thank for helpful discussions and/or for providing copies of maps, reports, and data, particularly George Albino, Sarah-Jane Barnes, Steve Barnes, Don Francis, Andrew Hynes, Marc St-Onge, Christian Picard, and Steve Lucas. I am especially grateful to Mike Knuckey for inviting me to work in the Raglan area, and to J.-P. Cloutier, Michel Dufresene, and Tony Green for supporting the field work and for providing access to field vehicles and helicopters. The logistical assistance of Doug Armstrong, Jacques Harvey, and Daniel Mélançon at the Donaldson Camp was much appreciated. Much of my mapping and sampling was done with former graduate students Joe Thacker (1989 and 1990), Sally Gillies (1991), David Williams (1994), Chuck Stilson (1997), Kevin Chisholm (1998), and Martin Lévesque (2000), with former PDF Marcus Burnham (1999), and with colleague Reid Keays (1993 and 1994). The manuscript benefited greatly from very insightful reviews by Sarah-Jane Barnes, Steve Barnes, and Steve Beresford, and very helpful editorial comments by Wayne Goodfellow. This research has been supported at various stages by grants (1989-1992, 1997-2002) and considerable logistical assistance from Falconbridge Ltd., and by grants from the US National Science Foundation (EAR-9018938), the Canadian Mining Industry Research Organization (94E-04), and the Natural Sciences and Engineering Research Council of Canada (IRC 202522-97, DG 1998-2002 and 2002-2007).

References

Albino, G.V., 1984, Petrology, Geochemistry, and Mineralization of the Boundary Ultramafic Complex, Quebec, Canada: M.Sc. thesis, Colorado State University, 220 p.

Arndt, N.T., Lesher, C.M., and Czamanske, G.K., 2005, Mantle-derived magmas and magmatic Ni-Cu-(PGE) deposits, *in* Hedenquist, J.W., Thompson, J.F.H., Goldfarb, R.J., and Richards, J.P., eds., Economic Geology 100th Anniversary Volume: The Economic Geology Publishing Company, p. 5-23.

Arndt, N.T, Lesher, C.M., Houlé, M.G., and Lacaze, Y., 2004, Volcanic setting of a spinifex-textured komatiite sill in Dundonald Township, Ontario: Journal of Petrology, v. 45, p. 2555-2571.

Barnes, S.-J., and Giovenazzo, D., 1990, Platinum-group elements in the Bravo intrusion, Cape Smith Fold Belt, northern Quebec: Economic Geology, v. 85, p. 1269-1272.

Barnes, S.-J., and Lightfoot, P.C., 2005, Formation of magmatic nickel sulfide ore deposits and processes affecting their copper and platinum group element contents, *in* Hedenquist, J.W., Thompson, J.F.H., Goldfarb, R.J., and Richards, J.P., eds., Economic Geology 100th Anniversary Volume: The Economic Geology Publishing Company, p. 179-214.

Barnes, S.-J., and Picard, C.P., 1993, The behavior of platinum-group elements during partial melting, crystal fractionation, and sulphide segregation: An example from the Cape Smith fold belt, northern Quebec: Geochimica et Cosmochimica Acta, v. 57, p. 79-87.

Barnes, S.-J., Picard, C., Giovenazzo, D., and Tremblay, C., 1992, The composition of nickel-copper sulphide deposits and their host rocks from the Cape Smith Fold Belt, northern Quebec: Australian Journal of Earth Sciences, v. 39, p. 335-347.

Barnes, S.-J., Couture, J.-F., Sawyer, E.W., and Bouchaib, C., 1993, Nickel-copper occurrences in the Belleterre-Angliers Belt of the Pontiac Subprovince and the use of Cu-Pd ratios in interpreting platinum-group element distributions: Economic Geology, v. 88, p. 1402-1418.

Barnes, S.-J., Makovicky, E., Karup-Moller, S., Makovicky, M., and Rose-Hansen, J. 1997a, Partition coefficients for Ni, Cu, Pd, Pt, Rh, and Ir between monosulphide solid solution and sulphide liquid and the implications for the formation of compositionally zoned Ni-Cu sulphide bodies by fractional crystallization of sulphide liquid: Canadian Journal of Earth Sciences, v. 34, p. 366-374.

Barnes, S.-J., Zientek, M., and Sevensen, M., 1997b, Ni, Cu, Au and platinum-group element contents of sulphides associated with intraplate magmatism: A synthesis: Canadian Journal of Earth Sciences, v. 34, p. 337-351.

Barnes, S.J., 2006, Komatiite-hosted nickel sulfide deposits of the Yilgarn Craton: Geology, geochemistry, and genesis, *in* Barnes, S.J., ed., Nickel Deposits of the Yilgarn Craton: Society of Economic Geologists, Special Publication 13, 210 p.

Barnes, S.J., and Barnes, S.-J., 1990, A new interpretation of the Katinniq nickel deposit, Ungava, northern Quebec: Economic Geology, v. 85, p. 1269-1272.

Barnes, S.J., and Lesher, C.M., in press, Chapter 10: Komatiite-associated Ni-Cu-PGE deposits, *in* Arndt, N.T., ed., Komatiites: Cambridge University Press.

Barnes, S.J., Coats, C.J.A., and Naldrett, A.J., 1982, Petrogenesis of a Proterozoic nickel sulfide-komatiite association: the Katinniq sill, Ungava, Quebec: Economic Geology, v. 77, p. 413-429.

Barnes, S.J., Hill, R.E.T., and Gole, M J., 1988, The Perseverance ultramafic complex, Western Australia: The product of a komatiite lava river: Journal of Petrology, v. 29, p. 305-331.

Barnes, S.J., Lesher C.M., Keays R.R., 1995, Geochemistry of mineralized and barren komatiites from the Perseverance nickel deposit, Western Australia: Lithos, v. 34, p. 209-234.

Barnes, S.J., Hill, R.E.T., Perring, C.S., and Dowling, S.E., 2004, Lithogeochemical exploration for komatiite-associated Ni-sulfide deposits: Strategics and limitations: Mineralogy and Petrology, v. 82, p. 259-293.

Bédard, J.H., Francis, D.M., and Hynes, A.J., 1983, Fractionation in the feeder system at a Proterozoic rifted margin: Canadian Journal of Earth Sciences, v. 21, p. 489-499.

Bégin, N.J., 1989. P-T conditions of metamorphism inferred from metabasites of the Cape Smith Belt, northern Quebec: Geoscience Canada, v. 16, p. 151-154.

——— 1992, Contrasting mineral isograd sequences in metabasites of the Cape Smith Belt, northern Quebec, Canada: Three new bathograds for mafic rocks: Journal of Metamorphic Geology, v. 10, p. 685-704.

Beresford, S.W., and Cas, R.A.F., 2001, Komatiitic invasive lava flows, Kambalda, Western Australia: Canadian Mineralogist, v. 39, p. 525-535.

Bergeron, R., 1959. Povungnituk range area, New Quebec: Quebec Department of Natural Resources, Preliminary Report 392, and Preliminary Map 1279, 9 p.

Burnham, O.M., Lesher, C.M., and Keays, R.R., 1999, Geochemistry of mafic-ultramafic complexes and associated basalts in the Raglan Block, *in* Lesher, C.M., ed., Komatiitic Peridotite-Hosted Fe-Ni-Cu-(PGE) Sulphide Deposits in the Raglan Area, Cape Smith Belt, New Québec: Laurentian University, Mineral Exploration Research Centre, Guidebook Series, v. 2, p. 159-173.

Charland, A., 1999, Geology of the West Boundary area, *in* Lesher, C.M., ed., Komatiitic Peridotite-Hosted Fe-Ni-Cu-(PGE) Sulphide Deposits in the Raglan Area, Cape Smith Belt, New Québec: Laurentian University, Mineral Exploration Research Centre, Guidebook Series, v. 2, p. 123-132.

Chisholm, K.M., 2002, Nature and Origin of Ore-Localizing Embayments at the Katinniq Ni-Cu-(PGE) Sulphide Deposit, Cape Smith Belt, New Québec: M.Sc. Thesis, Laurentian University, Sudbury, Ontario, 166 p.

Chisholm, K.M., Welch, M.G., Nemcsok, G., and Lemery, R.G., 1999, Mine geology of the Katinniq and Zone 2 Ni-Cu-(PGE) deposits, *in* Lesher, C.M., ed., Komatiitic Peridotite-Hosted Fe-Ni-Cu-(PGE) Sulphide Deposits in the Raglan Area, Cape Smith Belt, New Québec: Laurentian University, Mineral Exploration Research Centre, Guidebook Series, v. 2, p. 150-158.

Coats, C.J.A., 1982, Geology and nickel sulphide deposits of the Raglan area, Ungava, Quebec: Ministère de l'Energie et des Ressources du Québec, GM 40480.

Dann, J.C., 2000, The 3.5 Ga Komati Formation, Barberton Greenstone Belt, South Africe, Part I: New maps and magmatic architecture: South African Journal of Geology, v. 103, p. 47-68.

Dillon-Leitch, H., Watkinson, D.H., and Coats, C.J.A., 1986, Distribution of platinum-group elements in the Donaldson West deposit, Cape Smith belt, Quebec: Economic Geology, v. 81, p. 1147-1159.

Donaldson, M.J., 1981, Redistribution of ore elements during serpentinization and talc-carbonate alteration of some Archaean dunites, Western Australia: Economic Geology, v. 76, p. 1698-1713.

Duke, J.M., 1986, Petrology and Economic Geology of the Dumont Sill: An Archean Intrusion of Komatiitic Affinity in Northwestern Québec: Geological Survey of Canada, Economic Geology Report 35, 56 p.

Duke, J.M., and Naldrett, A.J., 1978, A numerical model of the fractionation of olivine and molten sulfide from komatiite magma: Earth and Planetary Science Letters, v. 39, p. 255-266.

Dupras, N., and Green, A.H., 1999, Exploration and Development History of the Raglan Ni-Cu-(PGE) Deposits, *in* Lesher, C.M., ed., Komatiitic Peridotite-Hosted Fe-Ni-Cu-(PGE) Sulphide Deposits in the Raglan Area, Cape Smith Belt, New Québec: Laurentian University, Mineral Exploration Research Centre, Guidebook Series, v. 2, p. 1-16.

Eckstrand, O.R., 1975, The Dumont serpentinite: A model for control of nickeliferous opaque assemblages by alternation products in ultramafic rocks: Economic Geology, v. 70, p. 83-201.

Fedorowich, J.S., 1999, Structural geology of the Raglan Belt, *in* Lesher, C.M., ed., Komatiitic Peridotite-Hosted Fe-Ni-Cu-(PGE) Sulphide Deposits in the Raglan Area, Cape Smith Belt, New Québec: Laurentian University, Mineral Exploration Research Centre, Guidebook Series, v. 2, p. 35-42.

Francis, D.M., and Hynes, A.J., 1979, Komatiite-derived tholeiites in the Proterozoic of New Quebec: Earth and Planetary Science Letters, v. 44, p. 473-481.

Francis, D.M., Hynes, A.J., Ludden, J.N., and Bédard, J., 1981, Crystal fractionation and partial melting in the petrogenesis of a Proterozoic high-MgO volcanic suite, Ungava, Quebec: Contributions to Mineralogy and Petrography, v. 78, p. 27-36.

Francis, D.M., Ludden, J.N., and Hynes, A.J., 1983, Magma evolution in a Proterozoic rifting environment: Journal of Petrology, v. 24, p. 556-582.

Gillies, S.L., 1993, Physical Volcanology of the Katinniq Peridotite Complex and Associated Fe-Ni-Cu-(PGE) Mineralization, Cape Smith Belt, Northern Quebec: MSc Thesis, University of Alabama, 146 p.

Giovenazzo, D., 1991, Géologie et caracteristiques géochemique des mineralisations Ni-Cu-EPG de la region de Delta, ceinture de Cape-Smith, Nouveau Québec: Ph.D. Thesis, Université du Québec à Chicoutimi, 238 p.

Giovenazzo, D., Picard, C., and Guha, J., 1989, Tectonic setting of Ni-Cu-PGE deposits in the central part of the Cape Smith Belt: Geoscience Canada, v. 16, p. 134-136.

Green, A.H., and Dupras, N., 1999, Exploration Model for Komatiitic Peridotite-Hosted Ni-Cu-(PGE) Mineralization in the Raglan Belt, *in* Lesher, C.M., ed., Komatiitic Peridotite-Hosted Fe-Ni-Cu-(PGE) Sulphide Deposits in the Raglan Area, Cape Smith Belt, New Québec: Laurentian University, Mineral Exploration Research Centre, Guidebook Series, v. 2, p. 191-199.

Grguric, B.A., and Rosengren, N.M., 2004, MKD5 nickel sulfide deposit, Mount Keith, *in* Neumayr, P., Harris, M., and Beresford, S.W., eds., Gold and Nickel Deposits in the Norseman-Wiluna Greenstone Belt, Yilgarn Block, Western Australia – A Field Guide: Geological Survey of Western Australia, Record 2004/16, p. 93-103.

Groves, D.I., and Keays, R.R., 1979, Mobilization of ore-forming elements during alteration of dunites, Mt. Keith-Betheno, Western Australia: Canadian Mineralogist, v. 17, p. 373-389.

Hoffman, P.F., 1985, Is the Cape Smith Belt (northern Quebec) a klippe?: Canadian Journal of Earth Sciences, v. 22, p. 1361-1369.

Hynes, A.J., and Francis, D.M., 1982, A transect of the Early Proterozoic Cape Smith foldbelt, New Quebec: Tectonophysics, v. 88, p. 23-59.

Jarvis, R.A., 1995, On the cross-sectional geometry of thermal erosion channels formed by turbulent lava flows: Journal of Geophysical Research, v. 100, p. 10127-10140.

Keays, R. R., 1982, Palladium and iridium in komatiites and associated rocks: Application to petrogenetic problems, *in* Arndt N.T., and Nisbet, E.G., eds., Komatiites: George Allen and Unwin, London, p. 435-457.

Kilburn, L.C., Wilson, H.D.B., Graham, A.R., Ogura, Y., Coats, C.J.A., and Scoates, R.F.J., 1969, Nickel sulfide ores related to ultrabasic intrusions in Canada, *in* Wilson, H.D.B., ed., Magmatic Ore Deposits: Economic Geology, Monograph 4, p. 276-293.

Lamothe, D., Gagnon, R., and Clark, T. , eds., 1986, Exploration en Ungava: Données récentes sur la géologie et la gîtologie: Ministère de l'Énergie et des Ressources du Québec, DV 86-16, 81 p.

Lamothe, D., Picard, C., and Moorhead, J., 1984, Région du lac Beauparlant, bande de Cape Smith-Maricourt, Nouveau-Québec. Ministère de l'Énergie et des Ressources du Québec, DP 84-39.

Layton-Matthews, D., Lesher, M., Burnham, O.M., Liwanag, J., Haldend, N.M., Hulberte, L., and Peck, D.C., 2007, Magmatic Ni-Cu-(PGE) deposits of the Thompson Nickel Belt *in* Goodfellow, W.D., ed., Mineral Deposits of Canada: A Synthesis of Major Deposit-types, District Metallogeny, the Evolution of Geological Provinces, and Exploration Methods, Special Publication 5, Mineral Deposits Division, Geological Association of Canada, p. 409-432.

Lesher, C.M., 1989, Komatiite-associated nickel sulfide deposits, *in* Whitney, J.A., and Naldrett, A.J., eds., Ore Deposition Associated with Magmas: Economic Geology Publishing Company, Reviews in Economic Geology, v. 4, p. 45-101.

——— 1999, Regional stratigraphic, structural, and metamorphic setting of the East-Central Cape Smith Belt, *in* Lesher, C.M., ed., Komatiitic Peridotite-Hosted Fe-Ni-Cu-(PGE) Sulphide Deposits in the Raglan Area, Cape Smith Belt, New Québec: Laurentian University, Mineral Exploration Research Centre, Guidebook Series, v. 2, p. 17-34.

Lesher, C.M., and Arndt, N.T., 1995, REE and Nd isotope geochemistry, petrogenesis and volcanic evolution of contaminated komatiites at Kambalda, Western Australia: Lithos, v. 34, p. 127-157.

Lesher, C.M., and Barnes, S.J., in press, Chapter 9: Physical Volcanology, in N.T. Arndt, Komatiites, Cambridge University Press.

Lesher, C.M., and Burnham, O.M., 2001, Multicomponent elemental and isotopic mixing in Ni-Cu-(PGE) ores at Kambalda, Western Australia: Canadian Mineralogist, v. 39, p. 421-446.

Lesher, C.M., and Campbell, I.H., 1993, Geochemical and fluid dynamic modeling of compositional variations in Archean komatiite-hosted nickel sulfide ores in Western Australia: Economic Geology, v. 88, p. 804-816.

Lesher, C.M., and Charland A., 1999, Geology of the Katinniq area, *in* Lesher, C.M., ed., Komatiitic Peridotite-Hosted Fe-Ni-Cu-(PGE) Sulphide Deposits in the Raglan Area, Cape Smith Belt, New Québec: Laurentian University, Mineral Exploration Research Centre, Guidebook Series, v. 2, p. 91-106.

Lesher, C.M., and Groves, D.I., 1986, Controls on the formation of komatiite-associated nickel-copper sulfide deposits, *in* Friedrich, G., Genkin, A.D., Naldrett, A.J., Ridge J.D., Sillitoe, R.H., and Vokes, F.M., eds., Geology and Metallogeny of Copper Deposits: Springer-Verlag, Heidelberg, p. 43-62.

Lesher, C.M., and Keays, R.R., 2002, Komatiite-Associated Ni-Cu-(PGE) Deposits: Mineralogy, Geochemistry, and Genesis, *in* Cabri, L.J., ed., The Geology, Geochemistry, Mineralogy, and Mineral Beneficiation of the Platinum-Group Elements: Canadian Institute of Mining, Metallurgy and Petroleum, Special Volume 54, p. 579-617.

Lesher, C.M., and Stone, W.E., 1996, Exploration geochemistry of komatiites, *in* Wyman, D., ed., Igneous Trace Element Geochemistry: Applications for Massive Sulphide Exploration: Geological Association of Canada, Short Course Notes, v. 12, p. 153-204.

Lesher, C.M., and Vicker, P., 1999, Geology of the Donaldson area, *in* Lesher, C.M., ed., Komatiitic Peridotite-Hosted Fe-Ni-Cu-(PGE) Sulphide Deposits in the Raglan Area, Cape Smith Belt, New Québec: Laurentian University, Mineral Exploration Research Centre, Guidebook Series, v. 2, p. 143-149.

Lesher, C.M., Arndt, N.T., and Groves, D.I., 1984, Genesis of komatiite-associated nickel sulphide deposits at Kambalda, Western Australia: A distal volcanic model, *in* Buchanan, D.L., and Jones, M.J., eds., Sulphide Deposits in Mafic and Ultramafic Rocks: Institution of Mining and Metallurgy, London, p. 70-80.

Lesher, C.M., Thibert F., Gillies, S.L., Stilson, C.M., Thacker, J.L., and Williams, D.A., 1999a, Lithology and physical volcanology of komatiitic peridotite-gabbro complexes in the Raglan Block, *in* Lesher, C.M., ed., Komatiitic Peridotite-Hosted Fe-Ni-Cu-(PGE) Sulphide Deposits in the Raglan Area, Cape Smith Belt, New Québec: Laurentian University, Mineral Exploration Research Centre, Guidebook Series, v. 2, p. 43-60.

Lesher, C.M., Barnes, S.-J., Gillies, S.L., and Ripley, E.M., 1999b, Ni-Cu-(PGE) sulphides in the Raglan Block, *in* Lesher, C.M., ed., Komatiitic Peridotite-Hosted Fe-Ni-Cu-(PGE) Sulphide Deposits in the Raglan

Area, Cape Smith Belt, New Québec: Laurentian University, Mineral Exploration Research Centre, Guidebook Series, v. 2, p. 177-184.

Lesher, C.M., Burnham, O.M., Keays, R.R., Barnes, S.J., and Hulbert, L., 2001, Geochemical discrimination of barren and mineralized komatiites associated with magmatic Ni-Cu-(PGE) sulfide deposits: Canadian Mineralogist, v. 39, p. 673-696.

Lévesque, M., and Lesher, C.M., 2002, Invasive Features of Mafic-Ultramafic Rocks at the Zone 3, Zone 2, and Katinniq Ni-Cu-(PGE) Deposits, Raglan Formation, Cape Smith Belt, Nouveau-Québec: 9th International Platinum Symposium, Billings, Montana, p. 257-260.

Lévesque, M., Lesher, C.M., and Charland, A., 2003, Volcanic or Subvolcanic Emplacement of the Raglan Formation in the Zone 3, Zone 2, and Katinniq Ni-Cu-(PGE) Deposits, Cape Smith Belt, Nouveau-Québec? Canadian Institute of Mining and Metallurgy, Annual Meeting, Montréal.

Lévesque, M., Lesher, C.M., Burnham, O.M., and Vicker, P., 2002, Chemistry of relict igneous olivines in komatiitic peridotites of the Raglan Formation, Cape Smith Belt, Nouveau-Québec: Geological Association of Canada - Mineralogical Association of Canada Annual Meeting, Abstract Volume, Saskatoon, p. 68.

Long, P.E., and Wood., B.J., 1986, Structures, textures, and cooling histories of Columbia River basalt flows: Geological Society of America Bulletin, v. 97, p. 1144-1155.

Lucas, S.B., 1989, Structural evolution of the Cape Smith thrust belt and the role of out-of-sequence faulting in the thickening of mountain belts: Tectonics, v. 8, p. 655-676.

Lucas, S.B., and St-Onge, M.R., 1989, Structural evolution of the Cape Smith Belt from initial thrusting to basement-involved folding: Geoscience Canada, v. 16, p. 122-126.

Luck, J.-M., and Allegre, C.J., 1984, 187Re/187Os investigation in sulfide from Cape Smith komatiites: Earth and Planetary Science Letters, v. 68, p. 205-208.

Mallinson, T., 1999a, Geology of the Zone 2 area, *in* Lesher, C.M., ed., Komatiitic Peridotite-Hosted Fe-Ni-Cu-(PGE) Sulphide Deposits in the Raglan Area, Cape Smith Belt, New Québec: Laurentian University, Mineral Exploration Research Centre, Guidebook Series, v. 2, p. 84-90.

——— 1999b, Geology of the Zone 3 area, *in* Lesher, C.M., ed., Komatiitic Peridotite-Hosted Fe-Ni-Cu-(PGE) Sulphide Deposits in the Raglan Area, Cape Smith Belt, New Québec: Laurentian University, Mineral Exploration Research Centre, Guidebook Series, v. 2, p. 76-83.

——— 1999c, Geology of the Zone 5-8 area, *in* Lesher, C.M., ed., Komatiitic Peridotite-Hosted Fe-Ni-Cu-(PGE) Sulphide Deposits in the Raglan Area, Cape Smith Belt, New Québec: Laurentian University, Mineral Exploration Research Centre, Guidebook Series, v. 2, p. 107-116.

McDonough W. F., and Sun S.-S., 1995, The composition of the Earth: Chemical Geology, v. 120, p. 223-253.

Miller, A.R., 1977. Petrology and geochemistry of the 2-3 ultramafic sill and related rocks, Cape Smith-Wakeham Bay fold belt, Quebec: Ph.D. Thesis, University of Western Ontario, London, Ontario, 219 p.

Murck B. W., and Campbell I. H., 1986, The effects of temperature, oxygen fugacity and melt composition on the behavior of chromium in basic and ultrabasic melts: Geochimica et Cosmochimica Acta, v.50, p.1871-1887.

Naldrett, A.J., 2004, Magmatic Sulfide Deposits: Geology, Geochemistry and Exploration: Springer, 727 p.

Osmond, R., and Watts, A., 1999, 3D geophysical model of the Raglan Belt, *in* Lesher, C.M., ed., Komatiitic Peridotite-Hosted Fe-Ni-Cu-(PGE) Sulphide Deposits in the Raglan Area, Cape Smith Belt, New Québec: Laurentian University, Mineral Exploration Research Centre, Guidebook Series, v. 2, p. 185-190.

Parrish, R.R., 1989, U-Pb geochronology of the Cape Smith Belt and Sugluk block, northern Quebec: Geoscience Canada, v. 16, p. 126-130.

Petch, C., 1999, Geology of the East Lake area, *in* Lesher, C.M., ed., Komatiitic Peridotite-Hosted Fe-Ni-Cu-(PGE) Sulphide Deposits in the Raglan Area, Cape Smith Belt, New Québec: Laurentian University, Mineral Exploration Research Centre, Guidebook Series, v. 2, p. 68-75.

Philpotts, A.R., 1961, Textures of the Ungava nickel ores: Canadian Mineralogist, v. 6, p. 680-689.

Picard, C., Lamothe, D., Piboule, M., and Oliver, R., 1990. Magmatic and tectonic evolution of a Proterozoic basin system: The Cape Smith Thrust-Fold Belt (New-Quebec): Precambrian Research, v. 47, p. 223-249.

Polyansky, O.P., and Reverdatto V.V., 2006, Contact metamorphism and metasomatism near the Talnakh intrusion: Fluid convection and heat transfer modeling on the basis of the finite-difference method: Doklady Earth Sciences, v. 411A, p. 1480–1484.

Sawyer, E.W., Barnes, S.-J., and Buck, M.J., 1983, Pillow shelves: determination of bedding direction and structural facing from shelves in deformed pillow lavas: Canadian Journal of Earth Sciences, v. 20, p. 1483-1487.

Scott, D.J., St-Onge, M.R., Lucas, S.B., and Helmstaedt, H., 1989, The 1998 Ma Purtuniq ophiolite: imbricated and metamorphosed oceanic crust in the Cape Smith thrust belt, northern Quebec: Geoscience Canada, v. 16, p. 144-147.

Seabrook, C.L., Prichard, H.M., and Fisher, P.C., 2004, Platinum-group minerals in the Raglan Ni-Cu-(PGE) sulfide deposit, Cape Smith, Quebec, Canada: Canadian Mineralogist, v. 42, p. 485-497.

Shepherd, N., 1960, The petrography and mineralogy of the Cross Lake area, Ungava, New Quebec: Ph.D. Thesis, University of Toronto, Toronto, Ontario, 176 p.

Shirey, S.B., and Barnes, S.-J., 1994, Re-Os and Sm-Nd isotopic constraints on basaltic volcanism and magmatic sulphide formation in the Cape Smith foldbelt, Quebec: Mineralogical Magazine, v. 58A, p. 835-836.

——— 1995, Os-Nd isotope systematics of ultramafic-mafic magmatism: Cape Smith foldbelt and midcontinent rift system, *in* Proceedings of International Symposium on Petrology and Metallogeny of Volcanic and Intrusive Rocks of the Midcontinent Rift System (IGCP Project 336: Petrology and Metallogeny of Intraplate Mafic and Ultramafic Magmatism), University of Minnesota – Duluth Continuing Education and Extension, Duluth, p. 175-176.

Skilling, I., White, J.D.L., and McPhie, J., 2002, Peperite: Review of magma-sediment mingling: Journal of Volcanology and Geothermal Research, v. 114, p. 1-17.

Stewart, A.J., 2002, The geochemistry and physical volcanology of the East Lake Ultramafic Zone, Cape Smith Belt, northern Quebec: M.Sc. Thesis, McGill University, Montreal, Quebec, 88 p.

Stilson, C.M., 1999, Geology of the Boundary Ultramafic Complex and Genesis of Associated Fe-Ni-Cu-(PGE) Sulfide Mineralization, Cape Smith Belt, New Québec: M.Sc. thesis, University of Alabama, Tuscaloosa, Alabama, 258 p.

Stilson, C.M., Vicker, P., and Lesher, C.M., 1999, Geology of the Boundary area, *in* Lesher, C.M., ed., Komatiitic Peridotite-Hosted Fe-Ni-Cu-(PGE) Sulphide Deposits in the Raglan Area, Cape Smith Belt, New Québec: Laurentian University, Mineral Exploration Research Centre, Guidebook Series, v. 2, p. 133-142.

St-Onge, M.R., and Lucas, S.B., 1990, Evolution of the Cape Smith Belt: Early Proterozoic continental underthrusting, ophiolite subduction, and thick-skinned folding: Geological Association of Canada, Special Paper 37, p. 313-352.

——— 1993, Geology of the Eastern Cape Smith Belt: Parts of the Kangiqsujuaq, Cratère du Nouveau-Quebec, and Lacs Nuvilik Map Areas, Quebec: Geological Survey of Canada, Memoir 438, 110 p.

——— 1994, Controls on the regional distribution of iron-nickel-copper-platinum-group element sulfide mineralization in the eastern Cape Smith Belt, Quebec: Canadian Journal of Earth Science, v. 31, p. 206-218.

St-Onge, M.R., Lucas, S.B., Scott, D.J., Bégin, N. J., Helmstaedt, H., and Carmichael, D.M., 1988, Thin-skinned imbrication and subsequent thick-skinned folding of rift-fill, transitional-crust, and ophiolite suites in the 1.9 Ga Cape Smith Belt, northern Quebec: Geological Survey of Canada, Paper 88-1C, p. 1-18.

St-Onge, M.R., Lucas, S.B., Scott, D.J., and Bégin, N.J., 1989, Evidence for the development of oceanic crust and for continental rifting in the tectonostratigraphy of the Early Proterozoic Cape Smith Belt: Geoscience Canada, v. 16, p. 119-121.

Sun S.-S., and McDonough W. F., 1989, Chemical and isotopic systematics of oceanic basalts: implications for mantle compositions and processes, *in* Saunders, A.D., and Norry. M.J., eds., Magmatism in the Oceanic Basins: Geological Society of London, Special Publication 42, p. 313-345.

Tremblay, C., 1990, Les éléments du groupe du platine dans le dyke de Mequillon, Ceinture de Cape Smith, Nouveau-Québec: MSc Thesis, Université du Québec à Chicoutimi, 76 p.

Thacker, J.L., 1995, Geology of the 5-8 Gabbro-Peridotite Complex, Cape Smith Belt, New Québec: M.Sc. Thesis, University of Alabama, 173 p.

Thibert, F, 1993, Pétrogénese du filon-couche différencié Roméo 1 situé dans la cienture plissée-faillée du Cap Smith, Nouveau-Québec: M.Sc. These, Université de Montréal, 145 p.

——— 1999, Geology of the Cross Lake area, *in* Lesher, C.M., ed., Komatiitic Peridotite-Hosted Fe-Ni-Cu-(PGE) Sulphide Deposits in the Raglan Area, Cape Smith Belt, New Québec: Laurentian University, Mineral Exploration Research Centre, Guidebook Series, v. 2, p. 61-67.

Thibert, F., Picard, C., and Trzcienski, W., 1989, Pétrologie des filons-couches différenciées Roméo 1 et 2 dans la partie centrale de la bande du Cap Smith: Geoscience Canada, v. 16, p. 140-144.

Vicker, P., and Fedorowich, J.S., 1999, Geology of the Zone 13-14 area,*in* Lesher, C.M., ed., Komatiitic Peridotite-Hosted Fe-Ni-Cu-(PGE) Sulphide Deposits in the Raglan Area, Cape Smith Belt, New Québec: Laurentian University, Mineral Exploration Research Centre, Guidebook Series, v. 2, p. 117-122.

Watts, A.H., 1997, Exploring for Nickel in the 90's, or 'Til Depth Us Do Part', *in* Gubbins, A.G., ed., Exploration 97: Geophysics and Geochemistry at the Millennium, Proceedings of the Fourth Decennial International Conference on Mineral Exploration, p. 1003-1014.

Watts, A.H., and Osmond, R., 1999, 3D Geophysical Model of the Raglan Belt, *in* Lesher, C.M., ed., Komatiitic Peridotite-Hosted Fe-Ni-Cu-(PGE) Sulphide Deposits in the Raglan Area, Cape Smith Belt, New Québec: Laurentian University, Mineral Exploration Research Centre, Guidebook Series, v. 2, p. 185-190.

Williams, D.A., Kerr, R.C., and Lesher, C.M., 1998, Emplacement and erosion by Archean komatiite lava flows: Revisited: Journal of Geophysical Research, v. 103, p. 27533-27549.

——— 1999a, Thermal erosion modeling of lava channels in the Raglan Block, *in* Lesher, C.M., ed., Komatiitic Peridotite-Hosted Fe-Ni-Cu-(PGE) Sulphide Deposits in the Raglan Area, Cape Smith Belt, New Québec: Laurentian University, Mineral Exploration Research Centre, Guidebook Series, v. 2, p. 174-176.

——— 1999b, Thermal and fluid dynamics of komatiitic lavas associated with magmatic Ni-Cu-(PGE) sulphide deposits, *in* Keays, R.R., Lesher, C.M., Lightfoot, P.C., and C.E.G. Farrow, C.E.G., eds., Dynamic Processes in Magmatic Ore Deposits and their Application in Mineral Exploration: Geological Association of Canada, Short Course, v. 13, p. 367-412.

Williams, D.A., Kerr, R.C., Lesher, C.M., and Barnes, S.J., 2002, Analytical/numerical modeling of komatiite lava emplacement and thermal erosion at Perseverance, Western Australia: Journal of

THE VOISEY'S BAY DEPOSIT, LABRADOR, CANADA

ANTHONY J. NALDRETT[1] AND CHUSI LI[2]

1. Department of Geology, University of Toronto, 22 Russell Street, Toronto, Ontario M5S 3B1
2. Department of Geological Sciences, Indiana University, Bloomington, Indiana 47405
Corresponding author's email: ajn2306@aol.com

Abstract

The Voisey's Bay Ni-Cu deposit currently comprises 78x10^6 tonnes of proven, probable, indicated, and inferred mineralization grading 2.20 wt.% Ni, 1.70 wt.% Cu, and 0.134.% Co. It was discovered in October 1994, and stimulated an intense program of exploration in 1995, 1996, and 1997 for Ni deposits in neighbouring parts of Labrador. The deposit is associated with two 1.334 Ga troctolite intrusive chambers, the upper Eastern Deeps and lower Reid Brook chambers, connected by a 10 to 100 m wide dyke. The igneous rocks are part of the Nain Plutonic Suite that consists of anorthositic, granitic, troctolitic, and ferrodioritic intrusions emplaced between 2.334 and 1.29 Ga. The mineralization is composed of massive, semimassive, and disseminated pyrrhotite, pentlandite, and chalcopyrite located within the dyke (particularly where it widens) in gneiss adjacent to the dyke and along the line of entry of the dyke to the upper of the two chambers. The sulphides are associated with a magmatic breccia (referred to as feeder breccia or basal breccia) and with troctolite that is commonly close to this breccia. The trace element and Pb and Nd isotope geochemistry of the troctolites indicate that they have interacted with crustal rocks, both at depth and close to the level of the intrusions. Sulphur and oxygen isotopes suggest some interaction with the local gneisses; the radiogenic nature of the Os demands such interaction. The Ni and Fo content of the olivine indicates that early injections of troctolite experienced significant chalcophile depletion, but that later influxes of magma experienced less depletion, and upgraded early sulphides in chalcophile metals. The existence and richness of the Voisey's Bay deposit is attributed to its occurrence within a plumbing system that has served to focus repeated pulses of magma. Other prospects discovered during the exploration rush in Labrador are thought to be of lesser size or grade for one or more of the following reasons: (i) the magma with which they interacted was more fractionated than the Voisey's Bay troctolites, and thus contained less Ni; (ii) there was a lack of adjacent sulphur-rich country rocks with which the magma could interact; and/or (iii) the magma involved was not focussed within a well defined conduit.

Résumé

Le gisement de Ni-Cu de Voisey's Bay renferme actuellement 78 x 10^6 tonnes de concentrations prouvées, probables, indiquées, et impliquées de minerai présentant des teneurs en Ni de 2,20 % en poids, en Cu de 1,70 % en poids, et Co de 0,134% en poids. Sa découverte en octobre 1994 a suscité un program d'exploration intense en 1995, 1996 et 1997 à la recherche de gîtes de Ni dans les parties avoisinantes du Labrador. Le gisement est associé à deux chambres magmatiques de troctolite de 1,334 Ga, la chambre supérieure Eastern Deeps et la chambre inférieure Reid Brook, reliées par un dyke d'une largeur de 10 à 100 m. Les roches ignées font partie de la suite plutonique de Nain qui se compose d'intrusions anorthositiques, granitiques, troctolitiques et ferrodioritiques mises en place entre 2,334 et 1,29 Ga. La minéralisation prend la forme de pyrrhotine, de pentlandite et de chalcopyrite massives, semi-massives et disséminées dans le dyke (en particulier aux endroits où il s'élargit), dans les gneiss adjacents au dyke et le long de la ligne de pénétration du dyke dans la partie supérieure des deux chambres magmatiques. Les sulfures sont associés à une brèche magmatique (appelée brèche nourricière ou brèche basale) et à la troctolite couramment présente à proximité de cette brèche. La géochimie des éléments en traces et des isotopes de Pb et de Nd des troctolites indique qu'elles ont interagi avec les roches crustales aussi bien en profondeur qu'à proximité de leur niveau de mise en place dans la croûte. Les isotopes de S et de O laissent croire qu'il y a eu une certaine interaction avec les gneiss locaux; la nature radiogénique de Os impose une telle interaction. La teneur en Ni et le contenu en Fo de l'olivine révèle qu'il y a eu un important appauvrissement en éléments chalcophiles des injections précoces de troctolite, mais un appauvrissement moindre des apports ultérieurs de magma, qui ont permis d'enrichir les sulfures préexistants en métaux chalcophiles. L'existence et la richesse du gisement de Voisey's Bay sont attribuées à son occurrence dans un réseau de canalisations magmatiques qui ont servi à concentrer des poussées répétées de magma. Les autres gîtes découverts au moment de la ruée vers l'exploration au Labrador seraient de moindres dimensions ou à plus faible teneur pour l'une ou l'autre des raisons suivantes: i) le magma avec lequel il y a eu interaction était davantage fractionné que celui des troctolites de Voisey's Bay et renfermait ainsi moins de Ni; ii) l'absence de roches encaissantes riches en soufre dans les environs immédiats avec lesquelles le magma aurait pu interagir; iii) le magma en cause n'était pas concentré dans un conduit bien défini.

History

In the summer of 1993, two prospectors, Albert Chislett and Chris Verbiski, who had been sent to Labrador by Diamond Fields Resources Ltd. to conduct stream sampling for diamond indicator minerals, discovered a gossan with the blue staining characteristic of copper on a hilltop, 45 km southwest of the village of Nain. An initial 5 hole drill program was conducted to investigate the occurrence in October 1994 and one of the holes intersected a 41.2 m zone of semi-massive to massive sulphide grading 2.96% Ni, 1.89% Cu, and 0.16% Co within a wider 71.0 m zone that graded 2.23% Ni, 1.47% Cu, and 0.12% Co. A major drilling program was started by Diamond Fields Resources Ltd. in January 1995 and by July 1995, 31.7 million tonnes of ore grading 2.83% Ni, 1.68% Cu, and 0.12% Co had been outlined at surface in the area known as the "Ovoid" and "Mini-Ovoid". In October 1995, a second major discovery, known as the "Eastern Deeps", was made during a stratigraphic drilling

Naldrett, A.J., and Li, C., 2007, The Voisey's Bay deposit, Labrador, Canada, *in* Goodfellow, W.D., ed., Mineral Deposits of Canada: A Synthesis of Major Deposit-Types, District Metallogeny, the Evolution of Geological Provinces, and Exploration Methods: Geological Association of Canada, Mineral Deposits Division, Special Publication No. 5, p. 387-407.

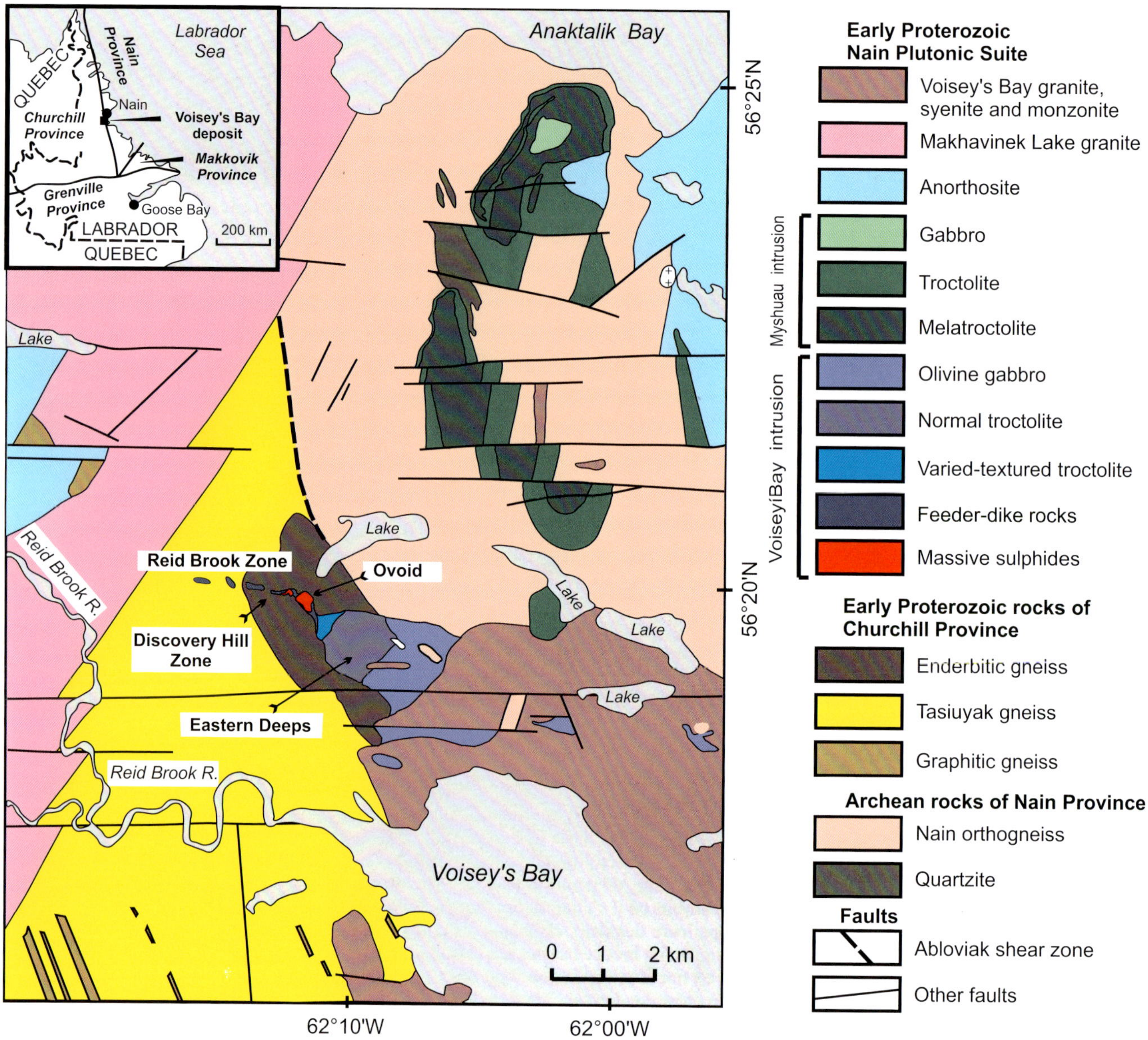

FIGURE 1. Geology of the Voisey's Bay area. Drafted using data from Ryan (1990) and Evans-Lamswood et al. (2000).

program. The deposit was acquired by INCO Ltd. in August 1996. Reserves and resources as of 2005 (Voisey's Bay Nickel Co. website, 27th October 2006) were 32x10^6 tonnes of proven and probable reserve grading 2.75 wt.% Ni, 1.59 wt.% Cu, and 0.14 wt.% Co; 40x10^6 tonnes of indicated resource grading 1.89 wt.% Ni, 1.90 wt.% Cu, and 0.12 wt.% Co; and 6x10^6 tonnes of inferred resource grading 1.9 wt.% Ni, 1.0 wt.% Cu, and 0.2 wt.% Co. The success at Voisey's Bay led to a huge surge in exploration in this part of Labrador in 1995, 1996, and to a lesser extent in 1997. Many new Ni-bearing occurrences were located and explored, but thus far none has proved to be of economic interest.

Geology

Regional Geology

The geology of the Nain area has been described by Ryan (1990; 1996; 2000), Ryan et al. (1995), and Kerr and Ryan (2000). The deposit has been described by Naldrett et al. (1996; 1997), Li and Naldrett (1999), Lightfoot and Naldrett (1999), and Evans-Lamswood et al. (2000); aspects of the geochemistry have been addressed by Lambert et al. (1999; 2000), Ripley et al. (1999; 2000), Amelin et al. (2000). Li and Naldrett (2000), Li et al. (2000), and Naldrett et al. (2000a); Naldrett et al. (2000b) described the sulphide mineralogy.

The deposit is associated with the 1.334 Ga (Amelin et al., 1999) Voisey's Bay intrusion that belongs to the Nain Plutonic Suite and transects the east-dipping 1.85 Ga collisional boundary between the Proterozoic Churchill Province to the west and the Archean Nain Province to the east (Ryan et al., 1995) (Fig.1). The Churchill Province comprises reworked Archean rocks and interbanded sulphide- and graphite-bearing garnet-sillimanite and quartzofeldspathic paragneisses collectively known as "Tasiuyak gneiss", and minor massive to lineated enderbitic gneiss. The Nain Province comprises interbanded granitic, and intermediate and mafic orthogneisses that exhibit retrogressed granulite-

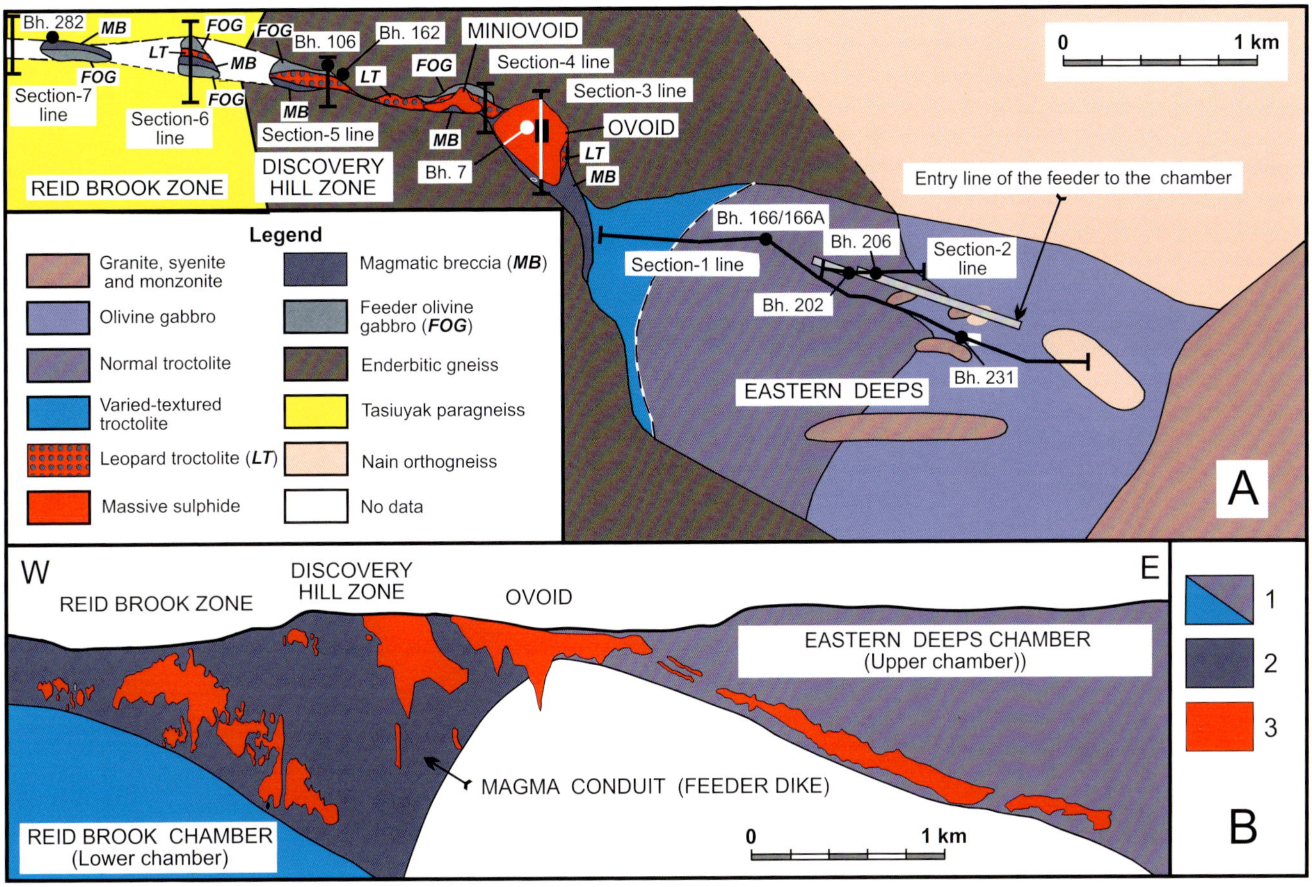

FIGURE 2 (A) Geological map of the Voisey's Bay intrusion (modified after Li and Naldrett, 1999). (B) Generalized longitudinal cross-section - the igneous rocks have been projected onto a vertical plane (modified after Lightfoot and Naldrett, 1999). Legend for cross-section: 1 = gabbro-troctolite bodies (magmatic chambers); 2 = rocks comprising the magma-conduit assemblage; 3 = sulphide mineralization.

and amphibolite-facies mineral assemblages. The latest movement along the collisional zone is marked by the Abloviak shear zone, for which the last movement has been dated at 1.73 to 1.75 Ga (Van Kranendonk, 1996).

Two troctolitic intrusions are present in the Voisey's Bay area (Fig. 1), the Voisey's Bay intrusion and the Mushuau intrusion. Both are members of the Nain Plutonic Suite, which comprises a suite of anorthosite, granite, ferrodiorite (absent in the area shown in Fig. 1) and troctolite bodies that were emplaced in the interval 1.350 to 1.290 Ga (Ryan, 2000). The Voisey's Bay and Mushuau intrusions were originally regarded as a single complex (Ryan et al., 1995) but Amelin et al. (1999) dated the Voisey's Bay intrusion at 1.334 Ga and Li et al. (2000) reported the age of the Mushuau Intrusion as 1.317 to 1.313 Ga. The two intrusions also have distinctly different trace element and isotopic signatures (see below).

Geology of the Voisey's Bay Intrusion

At the present level of understanding (Fig. 2), the Voisey's Bay intrusion comprises five components: (1) an upper chamber (the "Eastern Deeps"), which comprises an olivine gabbro (OG) sequence, a normal troctolite (NT) sequence and a varied-textured troctolite (VT) sequence; (2) this is fed from below by a gabbroic/troctolitic feeder sheet, which is exposed at surface farther west (the Discovery Hill zone); (3) the "Ovoid", which is situated between the Eastern Deeps and the Discovery Hill zone, appears to be the base of the Eastern Deeps exposed at surface; (4) the "Mini-Ovoid" lies directly west of the Ovoid but is separate from it; (5) the feeder sheet has been traced 2.5 km west of the Ovoid, and in its western part, the Reid Brook zone, it dips steeply south and appears to widen out into a lower chamber. These are described in turn below.

The Eastern Deeps

The Eastern Deeps chamber (Fig. 2) is exposed at surface as a troctolite-olivine gabbro body intruded by syenite and monzonite (dated at 1.305 Ga, Amelin et al., 1999). The basal contact of the troctolite dips uniformly at 25° southeast from the vicinity of the Ovoid (Figs. 2B, 3). In this area, the stratigraphic succession comprises a lower sheet of Basal (= Magmatic) Breccia Sequence (BBS) (see description of rock types below), overlain by Varied-textured Troctolite or VT, that is in turn overlain by Normal Troctolite (NT). The uppermost unit in the Eastern Deeps chamber is the Olivine Gabbro (OG), which is likely intruded by the underlying troctolites (see discussion of olivine compositions below). A feeder sheet enters the Eastern Deeps from the north and appears horizontal in the north-south section shown in Figure 4. The entry line of the sheet rakes southeast, following the plunge of the base of the Eastern Deeps, which accounts for the apparent dip of the sheet in Figure 3. Several hundred metres north of the entry line, the sheet assumes a steep dip, similar to that of the feeder to the west. Large blocks of ultramafic rock appear to choke the feeder sheet in a few places. The distribution of the varied-textured troctolite shows a strong relationship to the north wall of the Eastern Deeps chamber (thought to be a growth fault that has controlled emplacement of the chamber) and to the entry line of the feeder sheet (Fig. 4). It thins out and disappears to the south.

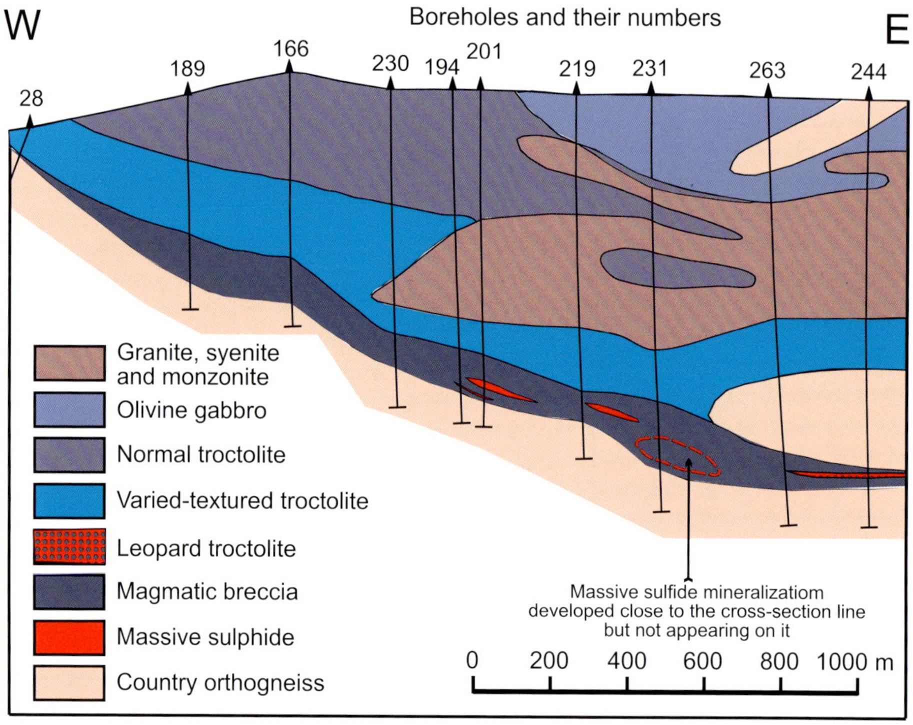

FIGURE 3. Cross-section through Eastern Deeps (modified after Li and Naldrett, 1999). The line of the cross-section is shown in Figure 2A.

Sulphide, which is present in minor amounts throughout most of the varied-textured troctolite, increases to 10 to 15 modal % (in places up to 25%) close to the mouth of the feeder as shown in Figure 4. In this vicinity, the basal breccia sequence contains both disseminated and blotchy sulphide and also lenses of massive sulphide.

Naldrett et al. (1996) have shown that massive sulphide occurs at the base of the Eastern Deeps, both within the feeder sheet (i.e. north of the stippled line representing its line of entry into the chamber in Fig. 2A) and up to 100 m south into the chamber. The basal breccia sequence is absent or thin near the line of entry of the feeder, and thickens away from this in a southwesterly direction, before thinning out again farther southwest. The basal breccia sequence seems to form a "bar" just off the entry line of the sheet, with the best development of massive sulphide lying between this bar and the entry line.

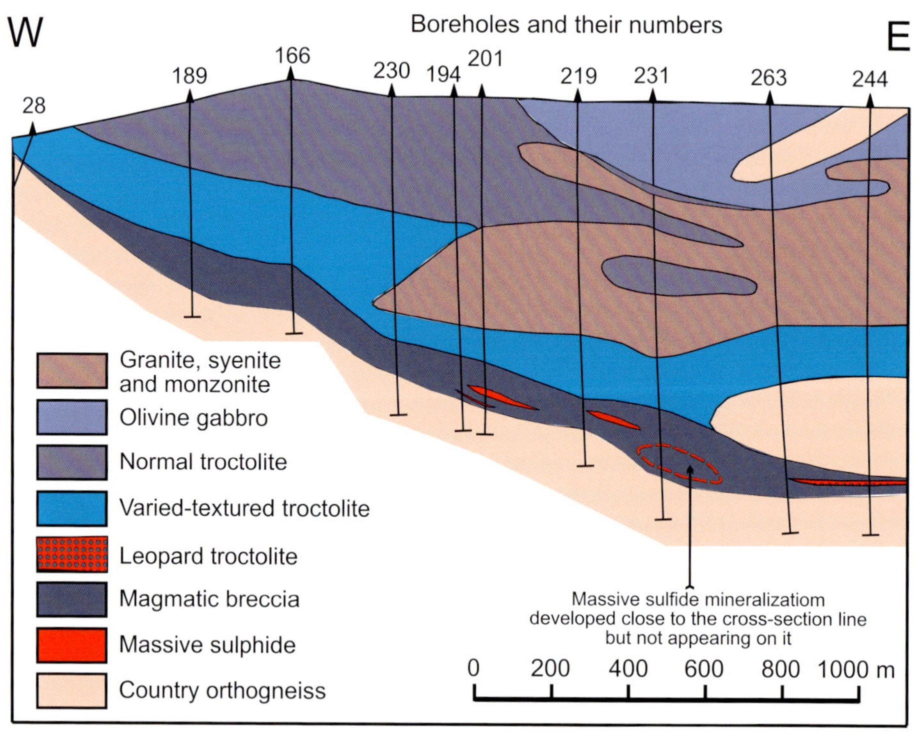

FIGURE 4. North-south cross-section across the northern part of the Eastern Deeps chamber. Note the vertical north wall of the chamber, the horizontal entry of the feeder sheet, the steepening in the dip of the sheet to the north, and the relationship between the entry line of the feeder and the distribution of varied-textured troctolite. This unpublished figure was very kindly provided by P.C. Lightfoot (pers. comm., 2006).

The stratigraphy of the feeder sheet extending north of the chamber is a close match to that observed in the Discovery Hill zone (see below), although it tends to be less thick where it has been studied so far. The upper part of the sheet consists of unmineralized Olivine Gabbro (Feeder Olivine Gabbro or FOG), either in chilled contact with the overlying gneiss, or as a mixed zone of troctolite and gneissic fragments. Sulphides increase downward and much of the disseminated sulphide shows "leopard texture" (Leopard Troctolite or LT) which consists of a troctolite containing interstitial sulphides and oikocrysts of augite (see description below). Sulphide content may increase to about 50%, below which there is an abrupt change in some holes to massive sulphide, which is underlain by mineralized basal breccia sequence. The basal breccia sequence is separated from

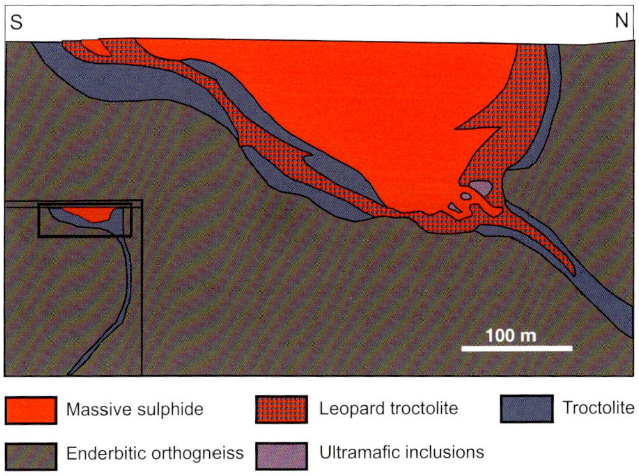

FIGURE 5. Cross-section through the Ovoid (after Lightfoot and Naldrett, 1999).

the gneiss in most places by a thin zone of sulphide-poor, fine-grained ferrogabbro.

The Ovoid and Mini-Ovoid

The Ovoid consists of up to 110 m of massive sulphide overlying a variable thickness of basal breccia sequence and minor leopard troctolite, all located within a bowl-shaped structure and overlain by 10 to 20 m of gravel and clay (Fig. 5). A thin (10-20 m thick) feeder sheet joins it from below. The westward continuation of the Ovoid (referred to as the Mini-Ovoid) is essentially a thickened variant of the Discovery Hill zone (see below). It dips 40°N and displays the stratigraphic succession typical of this zone, plus a thick zone of massive sulphide between the leopard troctolite and basal breccia sequence.

The Discovery Hill Zone

In this zone, the thickness of the feeder sheet varies from 10 to 100 m; the dip changes from 40°N close to surface and close to the Mini-Ovoid to a steep southerly orientation at depth. Where the dip is less than 70°, the stratigraphic succession consists of an upper contact zone of feeder olivine gabbro, which overlies a zone of leopard troctolite that in turn overlies basal breccia sequence (Fig. 6). Mineralization is restricted to the wider parts of the feeder and occurs as a series of elongate lenses, within the plane of the sheet, all raking to the east (Fig. 2B).

The Reid Brook Zone

In the Reid Brook zone, the dip of the feeder sheet is in excess of 70° to the south. The subvertical stratigraphic succession that characterizes the Discovery Hill zone is not well developed here. The breccia sequence that occurs at the basal contacts of both the Eastern Deeps and Discovery Hill zone (i.e. basal breccia sequence) is referred to as feeder breccia in the Reid Brook zone because the unit is not related to the base. The feeder breccia is not restricted to the footwall of the steeply dipping sheet, but can occur anywhere within it. The feeder olivine gabbro and leopard troctolite are concentric with massive sulphides within the feeder in some places.

Evidence of the former presence of the feeder sheet and its mineralization has been traced well west of Reid Brook (see Fig. 1). Large masses (xenoliths) of mineralized troctolite and inclusion-bearing magmatic breccia occur at a prospect known locally as Ashley, within younger Makovenik Lake granite, 7 km west-southwest of the Ovoid.

In February 2005, a figure appeared on the INCO Ltd. website showing a major series of intersections of semimassive to massive sulphide associated with a change in dip of the feeder sheet (Fig. 7). The sulphides occur both within the feeder and as lenses of massive sulphide that apparently follow planes of weakness in the adjacent gneisses (Evans-Lamswood, 2005). No data have been released at the time of writing (October 2005) that tie this discovery to published aspects of the Voisey's Bay structure as a whole, although the available information indicates that a major resource exists in the Reid Brook area at a depth of 200 to 300 m. This

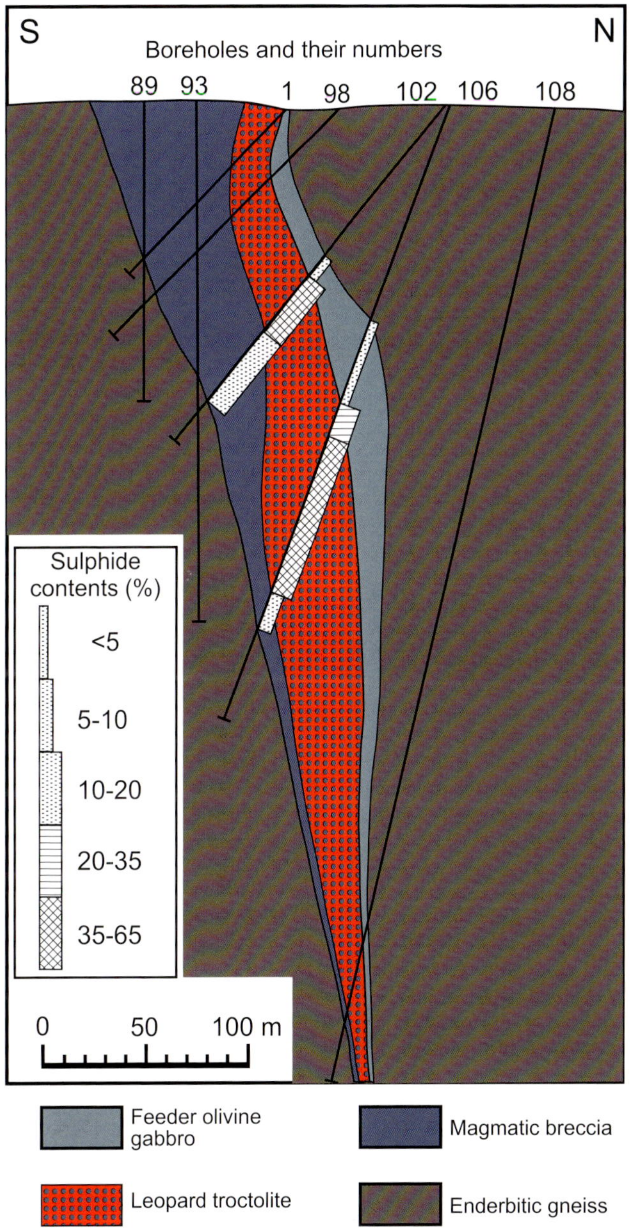

FIGURE 6. North-south cross-section through the Discovery Hill zone along line 5. The location of the section line is shown in Figure 2A. Modified after Li and Naldrett (1999).

is likely to feature prominently in plans for future development at Voisey's Bay.

Principal Rock Types of the Voisey's Bay Intrusion

Olivine Gabbro and Leucotroctolite

Olivine gabbro (Fig. 8C) is a common rock-type that occurs, together with several thin (10-30 m) layers of gabbro, in the olivine gabbro sequence of the Eastern Deeps. It is an olivine-plagioclase cumulate with 50 to 70 modal % plagioclase and 15 to 30 modal % olivine. Augite varies from 10 to 20 modal % and occurs mainly as large oikocrysts enclosing grains of olivine and plagioclase. Reaction rims of orthopyroxene surrounding olivine are not common in the olivine gabbro.

The Leucotroctolite (LUT) in the Reid Brook lower chamber has a ratio of olivine to plagioclase similar to that in the olivine gabbro of the Eastern Deeps. Oikocryst augite however is less abundant in the leucotroctolite (usually less than 5 modal %) than in the olivine gabbro. Reaction rims of orthopyroxene on olivine are more common in the leucotroctolite than in the olivine gabbro.

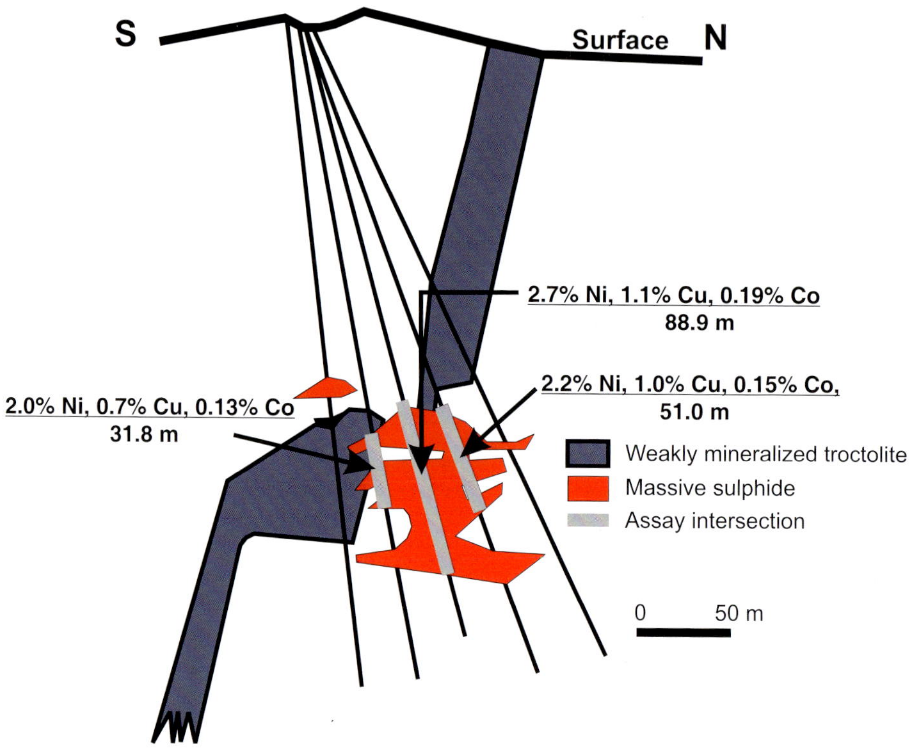

FIGURE 7. Cross-section showing recently discovered mineralization in the Reid Brook area of the Voisey's Bay deposit (INCO Ltd. website, March 2005).

Feeder Olivine Gabbro

Feeder Olivine Gabbro (FOG) is a plagioclase+/-olivine cumulate containing a much higher proportion of intercumulus minerals (between 30-50 modal %) than either the leucotroctolite or olivine gabbro (Fig. 8D). Euhedral to subeuhedral plagioclase crystals (2-10 mm in length) constitute 40 to 50 modal % and are randomly oriented to form a framework for intercumulus minerals. Small amounts of olivine and augite (less than 10 modal % each) often occur as oikocrysts enclosing grains of plagioclase. Intercumulus minerals include orthopyroxene (5-10 modal %), hornblende (8-15 modal %), biotite (up to 15 modal %), ilmenite (around 5 modal %), and small (less than 1 mm) subhedral plagioclase laths (up to 15 modal %).

Normal Troctolite

Normal Troctolite (NT) is a medium-grained troctolite of uniform texture (Fig. 8B). It consists of 20 to 40 modal % cumulus olivine and 40 to 65 modal % cumulus plagioclase 5 to 10 mm in length. The plagioclase laths are randomly oriented and are often partially enclosed by olivine, in some places to the extent that olivine oikocrysts surround plagioclase grains. Oikocryst augite is rare in the normal troctolite, instead small amounts of orthopyroxene (less than 5 modal %) occur as reaction rims on some olivine grains. Rare titanomagnetite occurs as an intercumulus mineral associated with ilmenite. Several bands of ilmenite, 0.5 to 1.5 m thick, occur in the upper part of the normal troctolite. Small amounts of augite are present in a myrmekitic intergrowth with the ilmenite in these bands.

Varied-Textured Troctolite

Varied-textured Troctolite (VT) differs from the normal troctolite in that it contains up to 25 vol.% gneiss inclusions and some blotchy sulphide. Plagioclase varies in size from several millimetres to several centimetres in length. Larger plagioclase crystals occasionally enclose grains of euhedral olivine. Pegmatitic plagioclase laths, more than 15 mm in length, are often observed projecting into patches of sulphide. Mantling of olivine by orthopyroxene is more common and interstitial hornblende, biotite, and apatite are more abundant in the varied-textured troctolite than in the normal troctolite. On the other hand, augite oikocrysts are less abundant and have more often been altered to tremolite, particularly in areas where they are in direct contact with sulphide. Interstitial ilmenite is less abundant in the varied-textured troctolite than in the normal troctolite. Magnetite, which is rare in the normal troctolite, is more common in the varied-textured troctolite where it is often associated with gneissic inclusions and patches of sulphide.

Leopard Troctolite

Leopard Troctolite (LT) contains up to 50 vol.% interstitial sulphide and about 10 modal % oikocrysts of augite. So far leopard troctolite has only been observed within the feeder. Cumulus plagioclase ranges from 40 to 60 and cumulus olivine from 20 to 30 modal %. The augite oikocrysts appear to have grown in situ and, in growing, have pushed interstitial sulphide aside so that they now appear as black

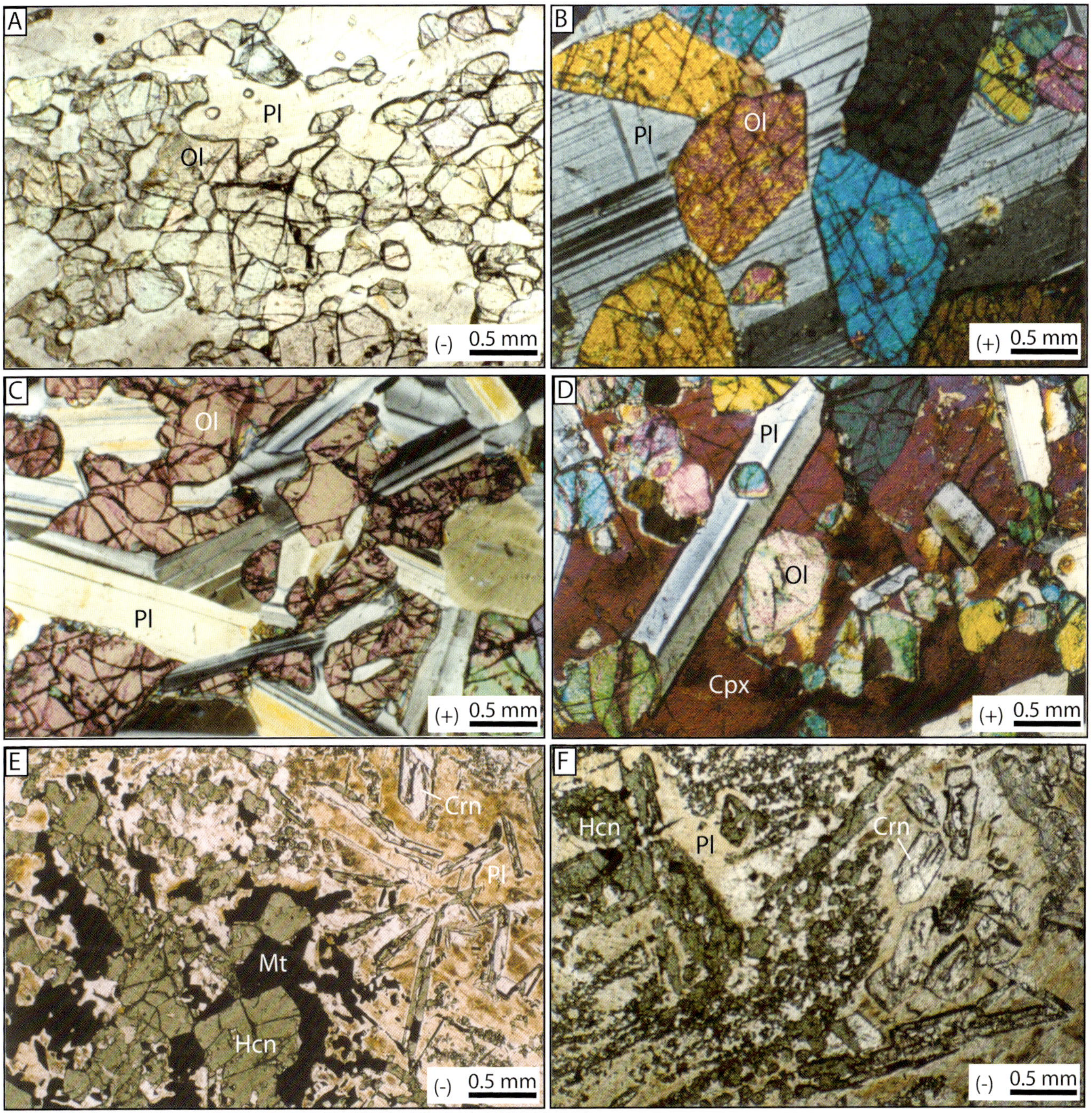

FIGURE 8. Photomicrographs of principal Voisey's Bay rock types. **(A)** Melatroctolite with olivine (Ol) grains enclosed by plagioclase(Pl) oikocrysts. **(B)** Normal troctolite in which plagioclase (Pl) encloses several olivine (Ol) grains poikilitically. **(C)** Olivine gabbro from the Eastern Deeps with olivine in a matrix of plagioclase laths. **(D)** Feeder olivine gabbro with olivine (Ol) grains and plagioclase (Pl) laths poikilitically enclosed in clinopyroxene (Cpx). **(E)** Xenolith of Tasiuyak gneiss from magmatic breccia showing needles of corundum (Crn) in plagioclase (Pl) and hercynite (Hcn) pseudomorphous after garnet with associated magnetite (Mt). **(F)** Xenolith of Tasiuyak gneiss from Magmatic Breccia with corundum (Crn) in plagioclase (Pl) with hercynite (Hcn) pseudomorphing some of the corundum needles.

spots in a matrix with a high sulphide content, thus giving rise to the term 'leopard troctolite'. The oikocrysts themselves contain a high proportion of olivine chadocrysts. With increasing amounts of sulphide, oikocrysts disappear from leopard troctolite and the mineralization grades into interstitial sulphide ore.

Magmatic Breccias

As described above, Basal Breccia Sequence (BBS) and Feeder Breccia (FB) are similar in mineral composition but recognized by different names because of their generally different inclusion content and different distribution within the feeder. Both contain more than 25 vol.% of inclusions and variable amounts of sulphide. Three types of inclusions are present: (i) fragments of gneisses (i.e. Tasiuyak gneiss); (ii) melatroctolite/ultramafic rocks in a few places (Fig. 8A);

and (iii) sulphide-poor troctolite (rare). The troctolite fragments are irregular, several centimetres in diameter, and often have sharp margins; they have experienced only minor alteration including serpentinization of olivine and sericitization of plagioclase. The ultramafic fragments, mainly melatroctolite, wehrlite, and dunite, are irregular in shape with diameters of up to tens of centimetres. Olivine in the ultramafic rock inclusions are often highly serpentinized. The inclusions of gneiss are smaller, ranging from a few millimetres to several centimetres in diameter. They have reacted with the enclosing troctolite; details of this reaction are discussed below.

Sulphides occur in three principal forms in basal breccia sequence/feeder breccia: (i) as zones or stringers of interstitial sulphide enclosing inclusions and crosscutting sulphide-poor breccia ore, (ii) as irregular blotches, and (iii) as stringers of massive sulphide that crosscut other variants of this environment.

Ferrogabbro

This is a fine-grained, noncumulate rock comprising granular pyroxene (less than 10 modal %, mainly clinopyroxene), greenish brown hornblende (25-40 modal %), small plagioclase laths (30-40 modal %), biotite (10-15 modal %), and ilmenite (5-10 modal %) (Fig. 8D). Ferrogabbro is generally massive but occasionally exhibits flow banding particularly in areas where sulphide is present.

Nickel and Forsterite Contents of Olivine in Different Rock Sequences

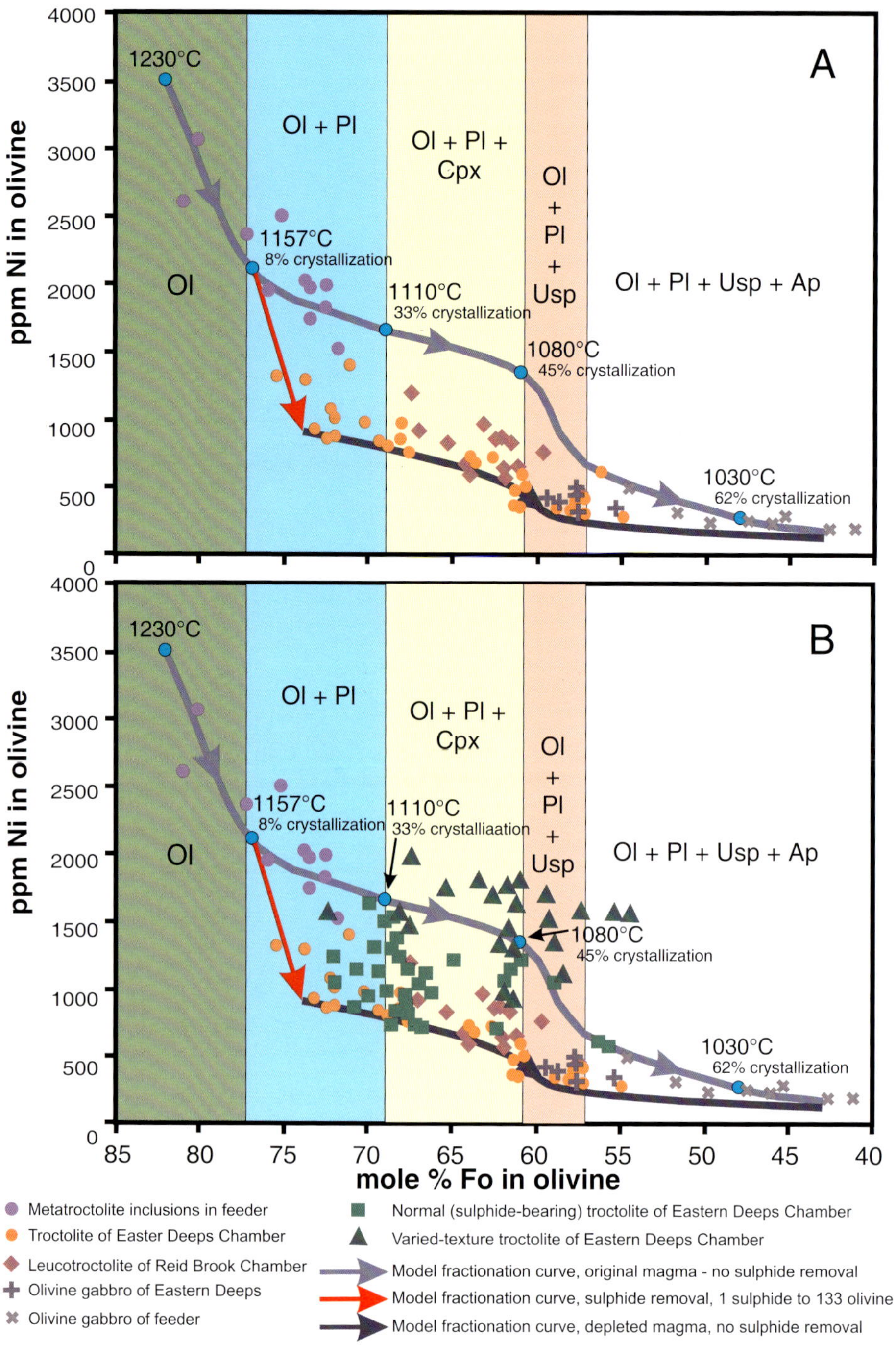

FIGURE 9. Plot of nickel (Ni) content of olivine versus forsterite (Fo) content for **(A)** rocks related to the initial introduction of magma; and for **(B)** rocks related to the later introduction(s). Model curves developed with the use of Ghiorso and Sack's (1995) program MELTS as discussed in the text.

Plots of nickel (Ni) versus forsterite (Fo) content in olivine are shown in Figures 9A and 9B. Olivine in the melatroctolite inclusions observed in the feeder show a steep trend of decreasing Ni content with decreasing Fo, in which Ni decreases from 3100 ppm at about 80 mole % Fo to 1500 ppm at 72 mole % Fo (Fig. 9A). Olivine in the leucotroctolite of the Reid Brook chamber, some troctolites and olivine gabbros of the Eastern Deeps chamber, and Feeder olivine gabbro show a more gentle trend with Ni decreasing from 1500 ppm at 75 mole % Fo to as low as 200 ppm at 41 mole % Fo (Fig. 9A). The model curves shown in the figure have been derived using Ghiorso and Sack's (1995) program MELTS for the fractional crystallization of mafic magmas. A discussion of the method and its application to olivine in the mineralized intrusions at Noril'sk can be found in Li et al. (2003).

The olivine of the melatroctolite inclusions follows the model curve for fractional crystallization of the assumed initial magma. Olivine from sulphide-free troctolites from the Eastern Deeps chamber and leucotroctolites of the Reid Brook chamber, along with olivine gabbros from the Eastern Deeps chamber and the feeder, do not follow the same model fractionation curve but fall below it. Their behaviour is explicable if the magma underwent a period of fractionation involving the removal of sulphide and silicate in a ratio of 1:133 (red line in Fig. 9A), followed by further sulphide-free fractionation of the Ni-depleted magma (purple line in Fig. 9A). Li and Naldrett (1999) proposed that interaction between the troctolite magma and the surrounding, sulphide-bearing Tasiuyak gneiss had caused this sulphide saturation.

Olivine in the normal troctolite and varied-textured troctolite (Fig. 9B), both of which rock types contain minor to significant amounts of sulphide, show no systematic trend but instead form a "cloud" with Fo varying between 73 and 52 mole % and Ni between 2000 and 700 ppm. The cloud extends from the model curve for depleted magma up to and above the curve for undepleted magma. Li and Naldrett (1999) showed that olivine in contact with sulphide react with the sulphide, exchanging Ni and Fe. Figure 10 illustrates the difference in the Ni-Fo covariation in olivine from the Normal and varied-textured troctolites at Voisey's Bay between those in contact with sulphide and those that do not show such contact. Olivines in contact with sulphide are distinctly poorer in Ni than those that occur as inclusions in plagioclase and are insulated from sulphide. This is evidence that at Voisey's Bay reaction between sulphide and olivine has depleted the olivine in Ni, and therefore that the Ni contents shown for the olivine from the Normal and varied-textured troctolites at Voisey's Bay are minimum values and that the original contents were likely higher. Geologic relationships indicate that the Normal and varied-textured troctolites postdate the other intrusive phases, thus providing evidence for the passage of a later, undepleted magma though the system.

Geochemistry of Voisey's Bay Rocks

Major Elements

Li et al. (2000) have shown that major element compositions of the intrusive rocks are controlled essentially by the proportion of cumulus plagioclase, olivine, and trapped silicate liquid. Most samples of normal troctolite and varied-textured troctolite in the Eastern Deeps subchamber and the leucotroctolite of the Reid Brook subchamber fall within or close to the olivine-plagioclase composition array. A few varied-textured troctolite samples show abnormally high Al_2O_3 contents, which is attributed to the presence of high-Al residual gneiss inclusions in the samples. The olivine gabbro in the Eastern Deeps and the feeder olivine gabbro in the conduit are displaced toward samples represented by the chilled margin rocks of the conduit, reflecting their higher content of trapped liquid. The basal breccia sequence samples contain abundant Al_2O_3-rich residual gneiss inclusions and plot, as would be expected, on the Al_2O_3-rich side of olivine-plagioclase array.

Trace Elements

Samples from the conduit and marginal zones of the Eastern Deeps, Ovoid and Mini-Ovoid, Discovery Hill zone, and Reid Brook zone are shown in Figure 11A (Li et al., 2000) as primitive mantle-normalized spidergram plots (Sun and McDonough, 1989). The samples show strong enrichment in the LREE, enrichment in some of the large ion lithophile elements (Rb, Ba, and K), low Th and U contents,

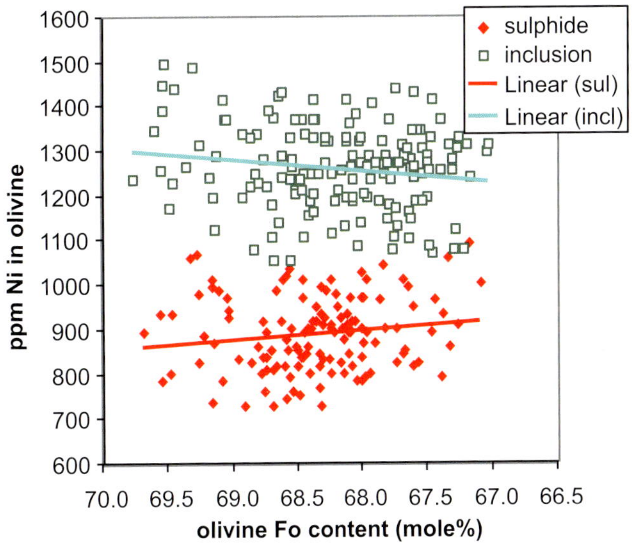

FIGURE 10. The contrasting nickel (Ni) contents of olivine grains in contact with sulphide (in leopard troctolite) and those that occur in inclusions in plagioclase and are therefore out of contact with sulphide.

and a slightly negative Ta+Nb anomaly. The elevated Sr and Eu concentrations reflect the modal plagioclase content; likewise, Ti is present in oxide minerals (ilmenite and titanomagnetite), and P is present in apatite; both the oxide and apatite contents of the conduit and marginal rocks tend to be high but variable. Apart from these elements, the spidergram profiles are very similar, and show no relationship to the nature of the immediate host rocks. The data presented in Figure 11A indicate that the marginal ferrogabbro and olivine gabbro are enriched in trapped liquid relative to the troctolites, and that the compositions of the finer grained (chilled) marginal rocks are not influenced by chemical contributions from the immediate country rocks. The presence throughout the deposit of inclusions that have originated from paragneiss indicates that equilibration of the melt with country rock occurred at depth, and not in situ within the conduit. Indeed, there is every indication that the Voisey's Bay intrusion magma did not react extensively with the wall rocks, although projections of orthogneiss within the Eastern Deeps do show extensive local evidence of assimilation as evidenced by the presence of troctolite-gneiss mélange along the contact. White feldspathic clots within the varied-textured troctolite, 0.5 to 5 cm in diameter, may also be the result of such assimilation.

Average primitive mantle-normalized spidergram patterns of troctolitic rocks and olivine gabbro of the Eastern Deeps and Reid Brook chambers are shown in Figure 11B. With the exception of Pb and Ta, the normalized patterns of the varied-textured troctolite and normal troctolite are essentially indistinguishable. The absolute abundances of trace elements are controlled by the amount of trapped liquid, which is higher in the olivine gabbro than the varied-textured troctolite and normal troctolite. The profiles have a similar shape to the conduit assemblage rocks, with moderate LREE enrichment, and steep negative Th+U and lesser Ta+Nb anomalies.

Figure 11C shows the spidergram patterns of different groups of ultramafic inclusions. The mineralized melatroctolite inclusions have similar spidergram patterns to the varied-

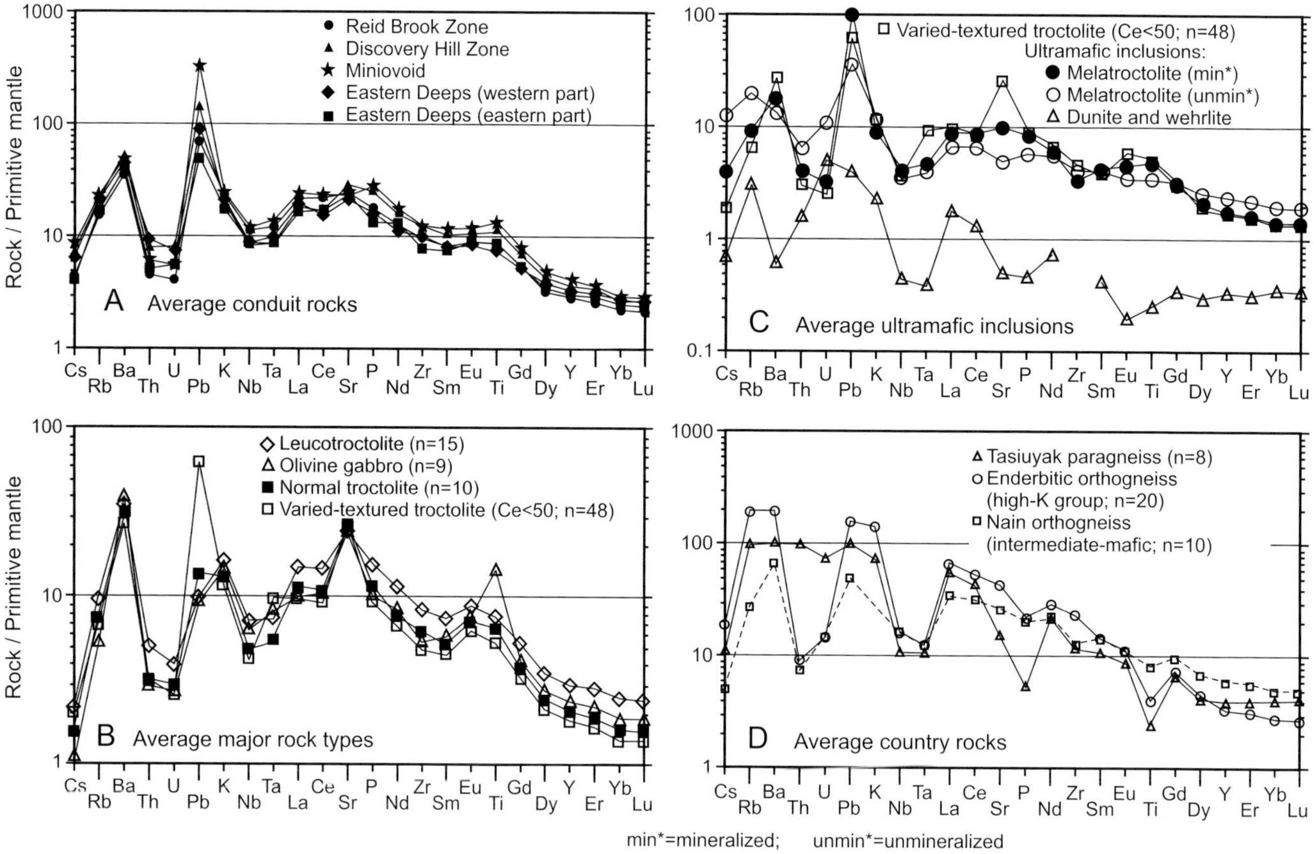

FIGURE 11. Primitive mantle-normalized (Sun and McDonough, 1989) spidergrams for rocks of the Voisey's Bay intrusion (after Li et al., 2000).

textured troctolite, as would be expected if they are ultramafic cumulates that were formed at an early stage with the sulphide saturated magma, and were subsequently broken up and incorporated as cognate xenoliths. A second group of unmineralized melatroctolite inclusions have spidergram patterns slightly flatter than the first group. These rocks may have crystallized from a less fractionated, less contaminated and sulphide-undersaturated magma. The third group of ultramafic inclusions is essentially highly altered rocks that have compositions between dunite and wehrlite, and are characterized by no disseminated sulphide and very different spidergram patterns when compared to the Voisey's Bay Intrusion and other ultramafic inclusions. They are regarded as exotic inclusions picked up by the magma during ascent though the crust. The gabbroic fragments in the breccia sequence are geochemically like the fine-grained ferrogabbroic rocks developed at the walls of the conduit (both have Ce/Yb = 27 to 33 and MgO = 5 to 8 wt.%).

Different phases of the Voisey's Bay Intrusion, the nearby Mushuau Intrusion and the Tasiuyak paragneiss, are shown on a plot of La/Sm versus Th/Nb in Figure 12. Most varied-textured troctolite and conduit samples from the Voisey's Bay Intrusion have La/Sm ratios between 3.0 and 4.0 and Th/Nb ratios of less than 0.1. The normal troctolite samples, which are not shown in the plot, have the ratios of La/Sm and Th/Nb similar to the varied-textured troctolite samples. The Tasiuyak paragneiss is characterized by La/Sm between 4.0 and 7.0 and Th/Nb between 0.4 and 1.0. The basal breccia sequence samples, which contain high proportions of inclusions of what are interpreted to be reacted Tasiuyak paragneiss, extend in array from the field of the varied-textured troctolite of the Voisey's Bay Intrusion and conduit rocks towards that of the Tasiuyak paragneiss, lending support to the interpretation that the inclusions are indeed of Tasiuyak paragneiss. Some conduit and varied-textured troctolite samples also extend toward the field of the Tasiuyak paragneiss, in part due to their content of inclusions. Li et al. (2000) concluded that all rocks of the Voisey's Bay Intrusion are consistent with derivation from mantle-derived picritic magma that underwent an early phase of contamination in which the La/Sm ratio was increased, but the Th/Nb ratio was unaffected, before rising to the present stratigraphic level of the Voisey's Bay Intrusion. Support for their supposition that the initial magma at Voisey's Bay had derived from a primitive mantle comes from the trend of the Voisey's Bay Intrusion unmineralized melatroctolite inclusions, which provide a link between primitive mantle and that of the Voisey's Bay Intrusion conduit rocks.

Cu/Zr Ratios

Both Cu and Zr are incompatible elements during the early stages in the fractional crystallization of sulphide-unsaturated mafic/ultramafic magmas, and thus both increase in concentration as fractionation proceeds, with the Cu/Zr ratio remaining constant. Since in cumulate rocks incompatible elements are contributed by the magma that has become trapped interstitially between cumulus grains, the Cu/Zr ratio of any sample of cumulate will reflect the ratio in the magma that gave rise to the cumulate. Once sulphide immiscibility develops in magma, Cu is no longer incompatible, and the Cu/Zr ratio will decrease as sulphides are removed. In the same way, if the magma or cumulates acquire additional sulphides from elsewhere, the Cu/Zr ratio will be higher than that of the initial, sulphide-unsaturated magma. Lightfoot et al. (1994) pointed out that Noril'sk basalts that were known *not* to have reacted with sulphide (Tuklonsky, Upper Morongovsky, Mokuaelevsky) had Cu/Zr ratios between 1 and 3, while those that were known to have reacted with sulphide had ratios of <1. Li and Naldrett (1999) showed that there is a correlation between Ni depletion shown by olivine and the Cu/Zr ratios of the rocks hosting the olivine. Their data are reproduced in Figure 13 in which it is seen that those rocks with Cu/Zr ratios <1 contain less S and also contain olivine with generally low Ni contents. Thus, at Voisey's Bay, Cu/Zr ratios are a useful indication of rocks that have formed from magma that has interacted with sulphide liquid and has thus become chalcophile depleted.

Reaction between Troctolitic Magma and Inclusions of Tasiuyak Gneiss

Li and Naldrett (2000) have shown that inclusions of gneiss in the basal breccia sequence and feeder breccia have reacted extensively with the enclosing magma (Fig. 15E). In general, the degree of reaction of the inclusions increases from the Reid Brook zone through the Discovery Hill zone to the Eastern Deeps. In the course of this reaction, garnet in the gneiss has been oxidized to form hercynite and magnetite (Fig. 8E) with the release of SiO_2 to the magma; cordierite has been dehydrated to form hercynite with the release of SiO_2; orthopyroxene and K-feldspar have reacted together to produce hercynite with the release of SiO_2 and K_2O; plagioclase has broken down to produce corundum (Fig. 8E, F), losing SiO_2 and Na_2O to the magma and itself becoming enriched in anorthite; and the corundum has subsequently reacted with FeO and MgO from the enclosing magma to form hercynite

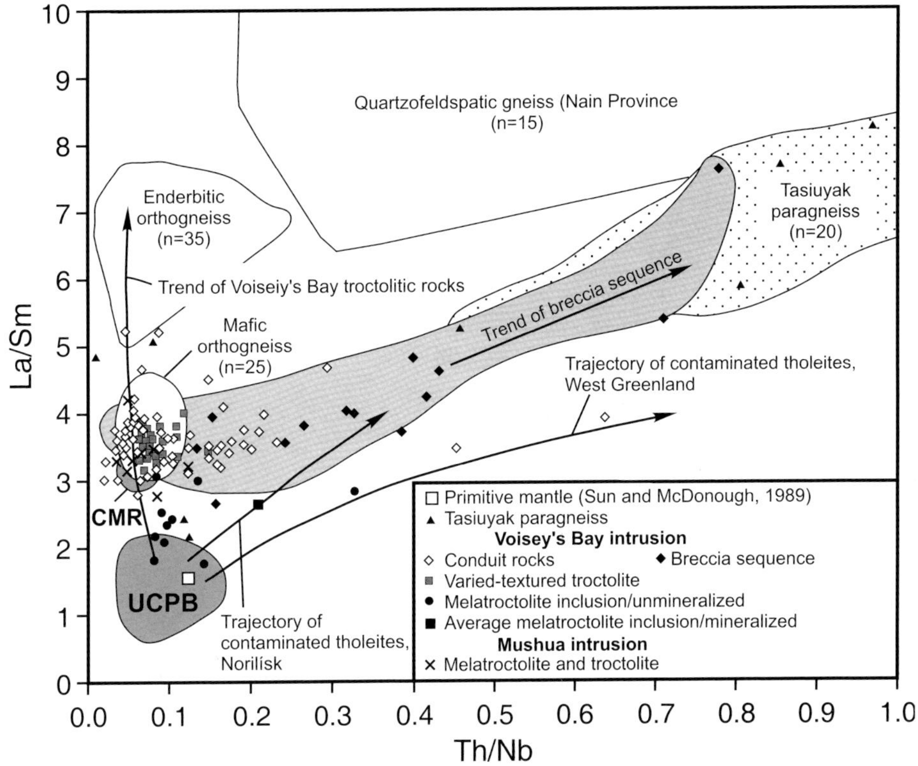

FIGURE 12. Plot of La/Sm versus Th/Nb for rocks of the Voisey's Bay and Mushuau intrusions, and country gneisses (after Li et al., 2000). CMR = chilled margin of the Voisey's Bay intrusion (n=8); UCPB = uncontaminated picritic basalts from Noril'sk and West Greenland (after Lightfoot and Hawkesworth, 1997).

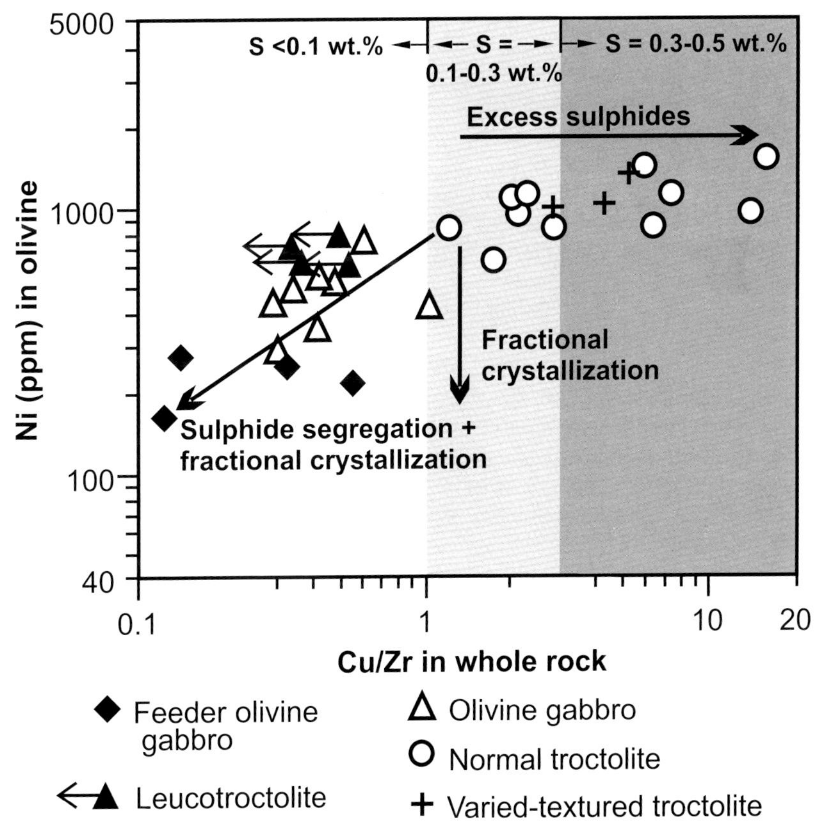

FIGURE 13. Plot of olivine Ni content versus whole rock Cu/Zr ratio (after Li and Naldrett, 1999). Small arrows indicate that Cu contents in these rocks are lower than the detection limit of 20 ppm that is used in calculating the Cu/Zr ratios for these samples.

(Fig. 8F). The colour of the hercynite changes progressively from beige in the Reid Brook zone, through green in the Discovery Hill zone to black in the Eastern Deeps; this colour change reflects an increase in $FeAl_2O_4$ at the expense of the $MgAl_2O_4$ in the hercynite. The inclusions of the Eastern Deeps characteristically show a series of reaction rims ranging from labra-dorite through biotite to orthopyroxene, which is thought to be the consequence of diffusion-controlled activity gradients of the principal oxide components. It is seen that interaction with the gneiss results in the addition of SiO_2, K_2O, and Na_2O to the magma. This felsification of magma would have resulted in a decrease in its ability to dissolve sulphide and therefore would have promoted sulphide immiscibility.

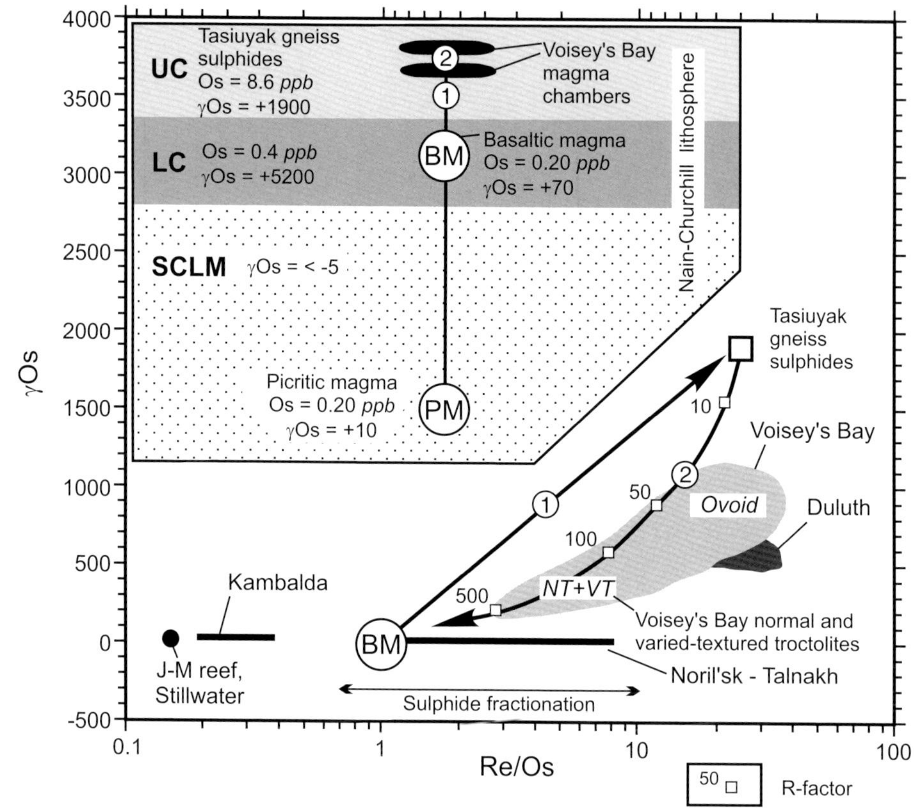

FIGURE 14. Plot of initial γOs vs. Re/Os, showing data for Cu-Ni-Co mineralization from the Voisey's Bay deposit (after Lambert et al., 2000), J-M reef of Stillwater complex (after Lambert et al., 1994), Duluth complex (after Ripley et al., 1999), Noril'sk-Talnakh area (after Walker et al., 1994), and Kambalda camp (after Foster et al., 1996). Modified after Lambert et al. (2000). The inset shows a model in which the picritic magma (PM), derived from a mantle plume, passes through subcontinental lithospheric mantle (SCLM). In the lower crust (LC) it fractionates and forms basaltic magma (BM). Owing to interaction with surrounding rocks, its γOs increases to +70. The basaltic magma (BM) then ascends into the upper crust (UC), where it develops a magmatic chamber in the Tasiuyak gneiss (Reid Brook chamber) and assimilates sulphides from the gneiss (line 1 in inset and main plot). Tasiuyak sulphides contain 8.6 ppb Os with γOs = +1900. The assimilated sulphides saturate the basaltic magma, immiscible Fe-sulphide liquid develops and absorbs Ni, Cu, Co, and PGE causing the magma to become strongly depleted in these elements. Fresh, undepleted magma subsequently enters the magmatic system and forces the previously depleted magma up a vertical dyke to form the Eastern Deeps magmatic chamber. The fresh magma interacts with the early formed sulphides enriching them in chalcophile metals, and decreasing their Re/Os ratio and γOs value (line 2 in the plot). Sulphide compositions of the Voisey's Bay deposit correspond to an R-factor of 50 to 500.

Isotope Geochemistry

U-Pb Geochronology

Amelin et al. (1999) have studied the age of the rocks of the Voisey's Bay area using U-Pb methods on baddeleyite, zircon, and apatite. Baddeleyite from troctolites and gabbros of the Voisey's Bay intrusion gave an age of 1.332 ± 0.001 Ga. Zircons from the normal troctolite and some from the olivine gabbro gave an age coeval with the baddeleyite, but those from varied-textured troctolite and Feeder olivine gabbro gave an age of 1.3050 ± 0.0008 Ga, which is the same as that shown by granite and syenite that cut the Voisey's Bay intrusion, and which Amelin et al. (1999) ascribe to a link between the development of the zircons and metamorphism/metasomatism associated with intrusion of the felsic rocks. Li et al. (2000) obtained an age of 1.3128 ± 0.0013 Ga from baddeleyite and zircon from the Mushuau intrusion, demonstrating that, despite the similarity in rock types, it is 20 Ma younger than the Voisey's Bay intrusion.

Nd-Pb-Sr Isotopic Data

Amelin et al. (2000) have studied the Nb-Pb-Sr isotopic systematics of the Voisey's Bay and nearby Mushuau (1.317-1.313 Ga) intrusions. They found that the Voisey's Bay intrusion has the most mantle-like, least contaminated initial isotopic compositions among the mafic intrusions of the Nain Plutonic Suite: ε_{Nd} = -1 to -2; $^{87}Sr/^{86}Sr_{(1.320Ga)}$ = 0.7034 to 0.7038; $^{206}Pb/^{204}Pb$ = 15.34 to 15.54; $^{207}Pb/^{204}Pb$ = 15.10 to 15.18; and $^{208}Pb/^{204}Pb$ = 35.24 to 35.56. These ratios were found to be uniform in troctolites and gabbros from throughout the intrusion. Isotopic ratios for the Mushuau intrusion are quite distinct from the Voisey's Bay intrusion and are typical for the rest of the Nain Plutonic Suite: ε_{Nd} = -3 to -10, $^{87}Sr/^{86}Sr_{(1.320Ga)}$ = 0.7034 to 0.7052; $^{206}Pb/^{204}Pb$ = 14.21 to 14.55; $^{207}Pb/^{204}Pb$ = 14.63 to 14.77; and $^{208}Pb/^{204}Pb$ = 34.36 to 34.65. Amelin et al. (2000) concluded that the primary magmas of the Voisey's Bay intrusion were either derived from an enriched continental mantle, or contaminated by a small amount of crustal material during ascent through lower-middle crust, and then by a volumetrically minor (probably 8-13%) amount of Tasiuyak gneiss. They attribute the observed variations in the Mushuau intrusion, and the reported variations in other mafic intrusions of the Nain Plutonic Suite to assimilation of 15 to 35% of a U-depleted Archean crust. They suggest that the larger amount of contamination experienced by intrusions other than the Voisey's Bay body is due to the magmas giving rise to these having passed through the crust 15 or more million years later than the Voisey's Bay magma, by which time the crust had become much hotter due to the passage of preceding magma.

TABLE 1. Average composition of mineralization from different environments at Voisey's Bay.

Mineralization type / Deposit Zone	n	S	Ni	Cu	Co	Pd	Pt	Rh	Ru	Ir	Os	Au	Pd/Pt
Massive Mineralization													
Eastern Deeps	5	37.66	3.19	0.85	0.18	145	51	19	34	7.4	16	22	2.82
Ovoid	52	33.76	4.12	2.53	0.18	226	88	7	13	1.8	2	82	2.57
Mini-Ovoid	11	35.65	3.72	2.16	0.18	208	166	8	16	2.3	4	109	1.25
Discovery Hill	1	36.6	2.5	1.02		73	25	9		1.9		36	2.98
Reid Brook	15	37.01	2.84	1.11		97	28	10	11	2.8	4	40	3.42
Leopard Textured Mineralization													
Eastern Deeps	8	15.9	1.45	0.75		89	50	7	13	2.6	6	35	1.79
Mini-Ovoid	7	13.64	1.42	0.9	0.07	78	70	4	5	1	2	73	1.11
Discovery Hill	18	13.09	1.16	1.28	0.06	77	76	3	9	0.7	1	77	1.01
Reid Brook	9	12.32	1.04	0.69		41	45	3	7	0.8	1	55	0.92
Mineralization in Magmatic Breccia													
Eastern Deeps	42	8.86	1	0.47	0.06	69	38	4	10	1.3	3	41	1.81
Ovoid	3	7.43	0.83	0.72	0.04	96	41	5	13	2.2	3	46	2.37
Mini-Ovoid	1	15.3	1.24	1.12	0.06	25	112	4	4	1.1	2	31	0.22
Discovery Hill	34	7.16	0.75	0.39	0.06	63	40	2	6	0.6	1	47	1.56
Reid Brook	23	5.86	0.59	0.3		40	25	2	6	0.5	1	29	1.58
Mineralization in Varied textured troctolite													
Eastern Deeps (>10% sulphide)	21	10.64	1.21	0.75		73	71	6	11	1.9	3	68	1.02

n = number of samples. Contents: S, Ni, Cu, Co – wt.%; Pd, Pt, Rh, Ru, Os, Ir, Au – mg/t (ppb)

Re-Os Isotopic Data

Re-Os data are important in the study of an orebody because the decay product, ^{188}Os, is concentrated in sulphide, and thus provides a direct measurement on the sulphides themselves. Lambert et al. (2000) found that massive sulphide samples from the Voisey's Bay ores have high Re concentrations (148-288 ppb) for their Os concentrations (4.8-28 ppb), yielding high Re/Os ratios (2.9-38) that are similar to those for massive sulphides from Sudbury and the Duluth Complex. Whole rock Re-Os isotopic data exhibit a large spread in $^{187}Re/^{188}Os$ (14-157) but do not define a precise isochron. They attribute this to different "packages" of sulphide having reacted with magma at different proportions of magma to sulphide (i.e. different values of R where R = magma/sulphide ratio). However, large whole rock sulphide samples from the Ovoid yield an imprecise 1320 Ma isochron age that is consistent within error with baddeleyite U-Pb ages from the magmatic system (Amelin et al., 1999).

Lambert et al. (2000) argued that the high initial γOs values (+200 to +1100 = percent deviation in calculated initial $^{187}Os/^{188}Os$ from mantle of the same age) for sulphide-rich samples from the Voisey's Bay intrusion document significant magma interactions with older Nain-Churchill Province crust. They considered the hypothesis that their Re-Os isotopic data is the result of contamination of basaltic magma similar to fine-grained feeder zone olivine gabbros and troctolites with sulphidic/graphitic Proterozoic Tasiuyak paragneiss (γOs = +1900), followed by an R-factor process (300 to >5,000) that improved the tenor (metal concentration in 100% sulphide) of the sulphide liquid during transport in the active Voisey's Bay magma conduit. However, they pointed out that R-factors of this magnitude are inconsistent with Naldrett et al.'s (2000a) Cu, Ni, and PGE data for the mineralization. Their preferred model (see Fig. 14 and associated legend) is that the initial immiscible sulphide liquid interacted with a second, chalcophile-element undepleted (>150 ppt Os) magma. This model gives R-factors (50-500) that are consistent with those of Naldrett et al. (2000a), and conforms to that proposed by Li et al. (2000) on the basis of their geochemical data.

Oxygen Isotope Data

Ripley et al. (2000) have shown that the $\delta^{18}O$ values of Tasiuyak gneiss range from 8.3 to 16.1‰, and the enderbitic and mafic orthogneiss from 5.7 to 8.5 with most values less than 7.5 per mil. Reacted inclusions of Tasiuyak gneiss in basal breccia sequence are depleted in ^{18}O relative to Tasiuyak gneiss outside the intrusion, ranging from 4.7 to 10.6 per mil. Elevated $\delta^{18}O$ values (up to 9.3 per mil) are found in the troctolitic and noritic matrix to the breccias, but are restricted to less than 2 cm from the edges of inclusions. These isotopic exchange profiles have been produced during subsolidus cooling, and do not record the major ^{18}O depletion that has been experienced by the gneiss inclusions. It is thought that major contamination has been diluted by large volumes of later magma that have passed through the system, disrupting and moving breccia, sulphides, and early magma up to a higher level.

Sulphur Isotopes

Ripley et al. (1999; 2002) have shown that $\delta^{34}S$ varies for the different mineralized environments within the Voisey's Bay intrusion: disseminated ore from Eastern Deeps, -2.0 to 2.7‰; massive ore from the Ovoid and Mini-Ovoid, -2.1 to 0‰; and disseminated mineralization from the Reid Brook zone, -4.1 to 1.4‰. Pyrrhotite from the Tasiuyak gneiss ranges from -17 to +18.3‰, while that from the Nain gneiss ranges from 0.2 to 3.3‰. If, as believed, the Tasiuyak gneiss is indeed the source of the sulphur in the Voisey's Bay ores,

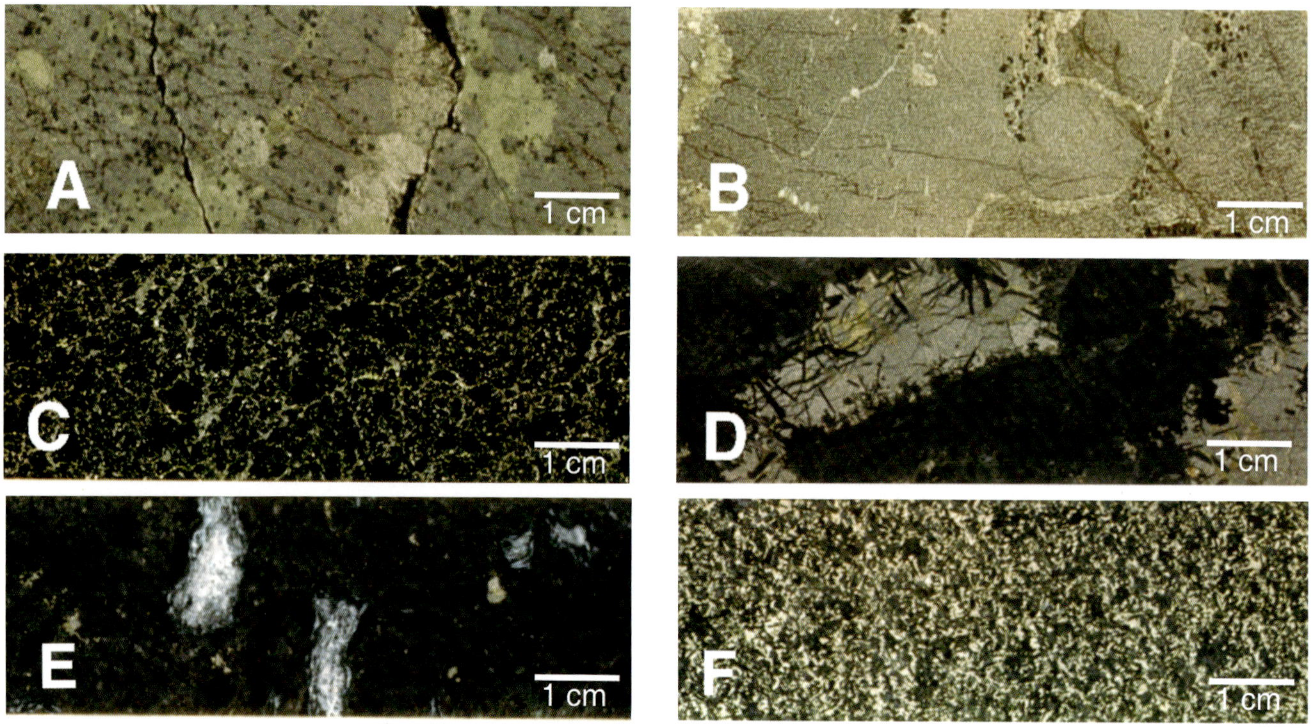

FIGURE 15. **(A)** Typical magmatic sulphide from the Ovoid. Note the large size (fingernail-size) of the pentlandite grains and the lack of pentlandite either encircling pyrrhotite crystals as rims or occurring within pyrrhotite as flames. The small black dots within the sulphides are magnetite. **(B)** Typical magmatic sulphide from the Eastern Deeps. Note that the pentlandite occurs as smaller masses than in photo A and is present both as rims to pyrrhotite grains and within the pyrrhotite as flames oriented parallel to the pyrrhotite 0001 basal parting. **(C)** Typical 'leopard' texture. The sulphides occur interstitial to olivine and plagioclase but have been displaced by oikocrysts of augite, which also occur interstitial to olivine and plagioclase and now form the large (0.5 cm diameter) black areas. The rock is referred to as leopard troctolite. **(D)** Blotchy sulphide within varied-textured troctolite. The crystals growing into the sulphide are plagioclase, which is much coarser when hosted by sulphide than in the silicate-dominant areas. **(E)** Basal (or Magmatic/Feeder) Breccia. The white areas are the remnants of inclusions of Tasiuyak gneiss that have reacted with the enclosing troctolite magma as discussed in the text. Close examination shows that the white areas, composed largely of anorthite, are rimmed by darker areas largely composed of hercynite. **(F)** Typical disseminated ore in which sulphides occur interstitial to olivine and plagioclase. This sample, which contains about 50 vol.% sulphide, is an example of the richest disseminated ore. If the sulphide content exceeds 50%, the sulphides tend to segregate as irregular areas/veins of massive sulphide.

considerable homogenization must have taken place while the sulphur was being incorporated into the magma.

Mineralization within the Voisey's Bay Intrusion

Types of Mineralization

There are four principal types of mineralization at Voisey's Bay; representatives of each are found in most of the mineralized environment: (i) massive sulphides, (ii) leopard textured sulphides, (iii) sulphides in basal breccia sequence, and (iv) disseminated sulphides in varied-textured troctolite. In addition, chalcopyrite- and bornite-bearing Cu-rich veins within the adjacent gneiss constitute what is yet an insignificant mineralization type – time will show whether substantial reserves of this material will be discovered, as has been the case at Sudbury and Noril'sk.

The mineralogy of the different types of mineralization was described in detail by Naldrett et al. (2000b). Brief descriptions of the sulphide morphologies are given below. Average chemical compositions of the ores are presented in Table 1.

Massive Sulphide

This consists of massive (>85% sulphides) accumulations of pyrrhotite (varying proportions of troilite and hexagonal pyrrhotite), pentlandite, chalcopyrite, cubanite, and magnetite. Pyrrhotite occurs as coarse crystals, exceeding 10 cm in diameter in some samples. Naldrett et al. (2000b) pointed out that troilite is missing from the Eastern Deeps and most abundant within the Ovoid. Where present, troilite occurs as fine exsolution lamellae in hexagonal pyrrhotite. Cubanite occurs as discrete grains and exsolution lamellae in chalcopyrite and shows a similar distribution to troilite, being present where troilite is present and absent where it is absent. Massive sulphides of the Ovoid, Mini-Ovoid, and Reid Brook zone are characterized by very coarse grains (1-2 cm) of pentlandite, with little or no pentlandite occurring as rims to pyrrhotite or as lamellae within pyrrhotite (Fig. 15A). Coarse pentlandite grains are rare in the massive sulphide of the Eastern Deeps and most pentlandite forms rims around crystals of pyrrhotite or occurs as lamellae within pyrrhotite (Fig. 15B). Magnetite is present in amounts of about 5% by volume in sulphides at the periphery of the Ovoid, 1% at the centre; it is rare in the massive sulphide of the Eastern Deeps.

Leopard-Textured Sulphide

As discussed above, the term 'leopard texture' is derived from the principal characteristic of this sulphide, which is that of black spots (augite and olivine oikocrysts ~0.5 cm diameter) in a yellow matrix (Fig. 15C). The matrix consists of sulphides (primarily pyrrhotite, pentlandite, and chalcopyrite) that occur interstitial to the principal cumulus silicates (plagioclase and olivine). The proportion of sulphide varies from 20 to 50%. In samples with more than 50% sulphide, the sulphides have separated from the silicate host to form small schlieren of massive material (Fig. 15F). As has been remarked previously, the black spots are due to the development of oikocrysts of augite and olivine, which appear to have grown interstitial to the other silicates, when the sulphides were still liquid, and to have pushed the latter aside. The mineralogy of the leopard textured sulphide is similar to that of adjacent massive sulphide (Naldrett et al., 2000a). The rock showing this texture is referred to as 'leopard troctolite'.

Sulphides in Magmatic Breccias

Mineralization in the basal breccia sequence is much more variable than that of the two preceding mineralization types. In part it consists of small lenses of massive sulphide, and in part of veins of leopard troctolite. Much of the mineralization comprises 'blotches' of sulphide interstitial to the inclusions of gneiss, troctolite, mela-troctolite, and ultramafic rocks, which are common in this rock unit (Li and Naldrett, 1999; Li et al., 2000). Again, the mineralogy tends to reflect that of adjacent massive sulphide.

Disseminated Sulphides in Varied-Textured Troctolite

Sulphides are abundant (approx. 25 vol.%) in the lower part of the varied-textured troctolite, and decrease in amount upward. They occur in two forms; the most common is as irregular "blotches", 10 to 30 cm in diameter, of sulphide intergrown with coarse silicates (Fig. 15D). The silicate minerals in these blotches comprise plagioclase and olivine that are very much coarser than those in the enclosing rock. In addition, coarse rods of plagioclase penetrate the sulphide blotches, growing away from plagioclase in the enclosing varied-textured troctolite. The second form of sulphide comprises irregular patches of disseminated sulphide, generally less than 20% in amount, which are present within varied-textured troctolite that displays a normal, medium-grained, igneous fabric.

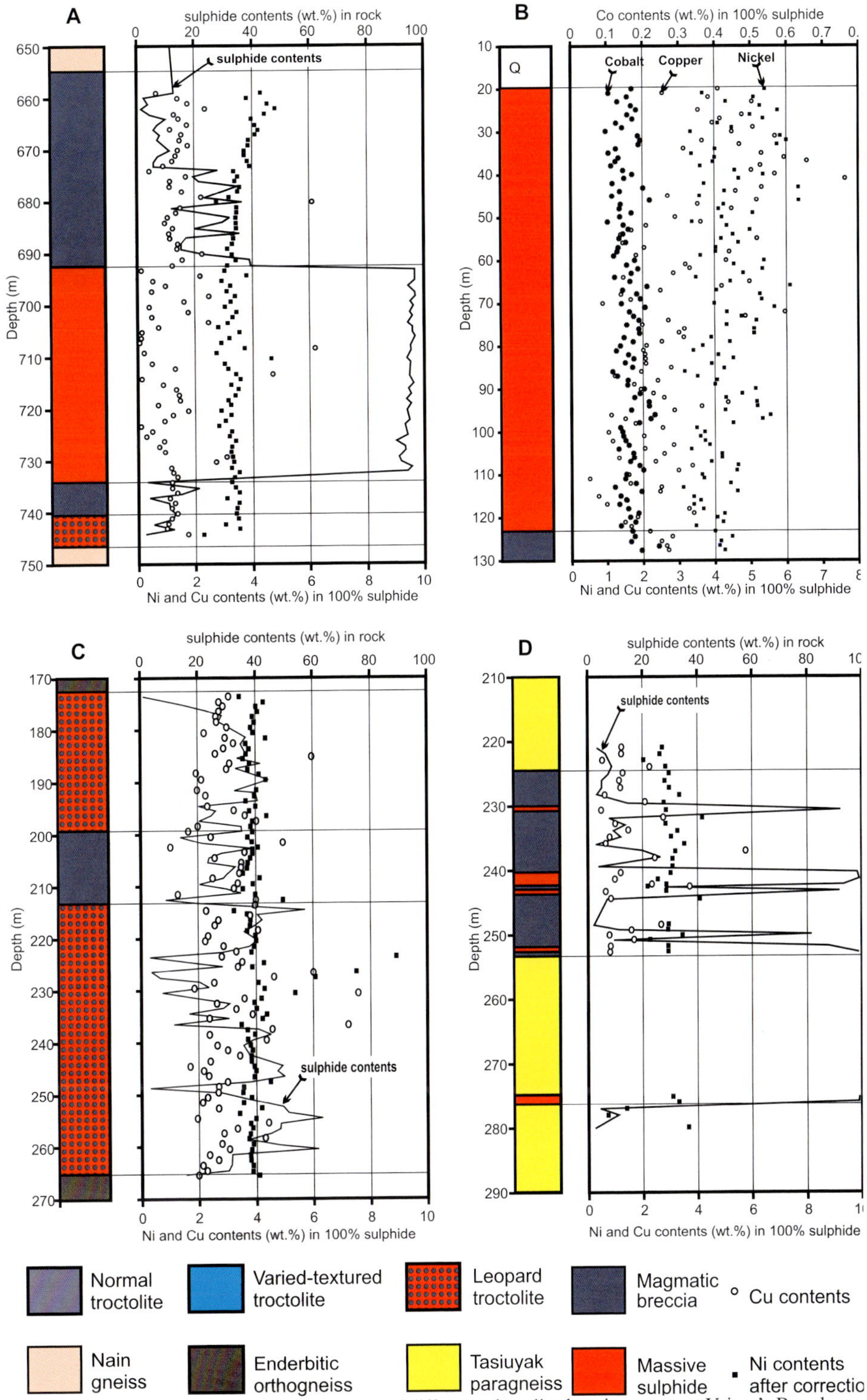

FIGURE 16. Profiles along holes representative of different mineralized environments at Voisey's Bay showing the variation in Ni and Cu content in 100% sulphides, along with the sulphide content of the rock, and, in the case of (B) the Co content of the sulphides. **(A)** Hole 202 through mineralization in the Eastern Deeps. **(B)** Hole 7 through the Ovoid. **(C)** Hole 162 through the Discovery Hill zone. **(D)** Hole 282 through mineralization in the Reid Brook area. Data from Naldrett et al. (2000b).

Nickel, Copper and PGE Tenors in Sulphides from Different Mineralized Environments

Note that all Ni and Cu concentrations quoted in this section refer to metal content in 100% sulphide and have been taken from Naldrett et al. (2000a)[1]. Variation in the Ni and Cu content of sulphides across typical profiles through the Voisey's Bay deposit are illustrated in Figure 16. Figure 16A shows variations along hole 202, which is typical of mineralization in the Eastern Deeps. It was drilled at the mouth of the feeder in a location where a 40 m thick len of massive sulphide occurs in a slight depression in the base of the intrusion adjoining the feeder mouth. Ni tenor in the massive sulphide is relatively constant, averaging 3 to 3.5 wt.%. Cu tenor is scattered and low, averaging about 1 wt.%. Both Ni and Cu rise in sulphides in the basal breccia sequence Sequence that overlies the massive zone, Ni to between 4.5 and 5 wt.%, Cu to between 1 and 2 wt.%. In other holes through the Eastern Deeps, the varied-textured troctolite above the massive and basal breccia-hosted mineralization contains variable amounts of sulphides with tenors that are upward of 5 wt.% Ni and 2 to 3 wt.% Cu.

Hole 7 (Fig. 16B) penetrated the massive sulphide occupying the centre of the Ovoid. Ni tenor increases upward from an average of 3.90 wt.% over the lowermost 20 m of massive sulphide to an average of 4.66 wt.% over the uppermost 20 m. Cu increases even more markedly from an average of 2.06 wt.% over the lowermost 20 m to 4.66 wt.% over the uppermost. In contrast, Co shows a slight decrease from an average of 0.16 wt.% over the lowermost 20 m to 0.15 wt.% over the uppermost. These variations are attributed to fractional crystallization as is discussed below.

Hole 162 (see Fig. 16C) is typical of the mineralization of the Discovery Hill zone. It intersected 55 m (true thickness) of leopard troctolite containing 15 m of basal breccia sequence. The sulphide distribution is variable, generally between 20 and 40 wt.%, rising to 50 to 60 wt.% in many samples near the base of the zone. The Ni tenor is very consistent, averaging 4.04 wt.%. Cu is more variable but shows no consistent variation with either rock type or position and averages 3.15 wt.% over the mineralized intersection.

Hole 282 (see Fig. 16D) is typical of those cutting a mineralized zone in the Reid Brook area that plunges east a short distance above the Reid Brook subchamber. The hole cut a 25 m thick (true thickness) zone of feeder breccia containing a weak dissemination of sulphide interspersed with stringers of massive sulphide, and then a thin stringer of massive sulphide in the Tasiuyak gneiss 20 m below the zone of feeder breccia. The Ni and Cu tenors are more scattered than elsewhere, but average respectively 3.13 wt.% and 1.91 wt.% over the zone of feeder breccia and 3.20 wt.% and 0.86 wt.% for the massive stringer in the gneiss.

In summary, average Ni tenors within different parts of the deposit vary from a low of 3 to 3.5 wt.% in much of the mineralization from the Eastern Deeps and Reid Brook zone, through values of 4 to 4.5 wt.% in the Ovoid and Discovery Hill zone, to 5 to 7 wt.% in sulphides in the varied-textured troctolite. Copper tenors are even more variable. In general, massive sulphide is characterized by low Cu and thus high Ni/Cu ratios in excess of 2.0 (the Ovoid is a marked exception to this generalization), and disseminated sulphides have values less than 2, typically about 1.6.

The results of the analysis of Ni, Cu, and PGE in a total of 315 samples from the Voisey's Bay deposit recalculated to metal content in 100% sulphide (Naldrett et al., 2000a) are summarized in Table 2. The PGE concentrations are distinctly low in comparison with data from many other Ni-Cu magmatic sulphide deposits. Space does not permit a detailed analysis of the variations here, and readers are referred to the original paper.

Naldrett et al. (2000a) demonstrated that there is a marked covariation in Cu, Au, Pd, and Pt within the Ovoid mineralization, which they argue is consistent with that to be expected as a result of the fractional crystallization of monosulphide solid solution from a sulphide melt (see variation in Ni, Cu, and Co with height in hole 7 in Fig. 16B). They concluded that this body of sulphide has cooled upward and inward from the sides, with the crystallization of monosulphide solid solution first occurring adjacent to the margins. The residual liquid became concentrated toward the top and centre of the body, enriched in those elements that are incompatible in monosulphide solid solution, i.e. Cu, Au, Pd, and Pt.

Geological Model

The overall plunge of all components of the Voisey's Bay deposit is about 25° east, with the result that different levels of the system are exposed at different localities (see Fig. 2). The deepest level is exposed to the west, so that the feeder occurs at surface and the exit of the feeder from the lower chamber has been intersected in the drilling. The Discovery Hill zone represents an intermediate level comprising the feeder between the lower and upper chamber, while the upper chamber is preserved as the Eastern Deeps. It is thought (Naldrett et al., 1996) that the Ovoid represents the

1. In calculating the metal content in 100% sulphide, the whole-rock Ni content is taken, on the assumption that it is all in sulphide, and then Ni is calculated as though it were all in pentlandite containing 36 wt.% Ni (a typical value for Sudbury pentlandite). The sulphur required for the pentlandite is then subtracted from the whole rock sulphur content. The same calculation is made for Cu, assuming that it is all present as chalcopyrite. The remaining sulphur is then calculated as pyrrhotite, so that one can obtain the total sulphide content of the rock by summing the amounts of pentlandite, chalcopyrite, and pyrrhotite. Because the amounts of sulphide in many of the samples in hole 166 and some samples from other holes are relatively low, Ni contained in olivine constitutes a significant proportion of the total Ni content of the rock, and has to be corrected for. This has been done by performing the calculation at first with no allowance made for Ni in olivine, and obtaining the composition of the resulting sulphide. The Ni content of the olivine is then calculated using this value and assuming (1) that the $Ni/Fe_{(olivine)} = (Ni/Fe_{(sulphide)})/25$, where 25 is the observed exchange partition coefficient between co-existing sulphide and olivine at Voisey's Bay (see Li and Naldrett, 1999) and (2) that $Ni/Fe_{(sulphide)}$ excludes Fe tied up as chalcopyrite. In making this calculation, it is assumed that the composition of the olivine is 65 mole% Fo (an average value for olivine of the varied-textured troctolite). Again, assuming that the non-sulphide portion of the sample contains 25 wt.% olivine (the observed average content for the varied-textured troctolite), this amount of Ni is then subtracted from the whole rock Ni content, and the calculation repeated for Ni in 100% sulphide. The whole calculation is iterated until the calculated Ni content in sulphides varies by no more than 0.05 wt.% between the last and penultimate iteration. In the case of many samples with less than about 2% sulphide, the calculation never converges, and these samples have been omitted. Several assumptions are required in making this correction (e.g. the Fo content of the olivine and the amount of olivine in the rock) that are not constant in all of the samples. However, one is using thousands of analyses performed in the course of routine exploration analysis, and it is not practical to measure the olivine composition and modal proportion of olivine for every sample. The reader can assure him-self/herself that the error introduced in this way is relatively small, by repeating the correction procedure assuming, for example, that the Fo content is 60 mole% instead of 65 mole%, or that the modal % olivine is 30 instead of 25.

TABLE 2. Average metal contents in 100% sulphide of ores of the Voisey's Bay deposit (standard deviations in parenthesis) (data from Naldrett et al. (2000)).

Mineralization Type, Location, and Number of Samples ()	Ni wt.%	Cu wt.%	Pt ppb	Pd ppb	Rh ppb	Ru ppb	Ir ppb	Os ppb	Au ppb
Varied textural troctolite									
Eastern Deeps >10% sulphide (21)	4.72 (1.03)	2.91 (1.38)	360 (309)	354 (264)	19 (9)	31 (34)	6.4 (3.8)	10 (8)	355 (349)
Eastern Deeps <10% sulphide (25)	6.08 (1.04)	3.22 (0.93)	447 (251)	354 (347)	19 (10)	60 (88)	5.1 (2.7)	4 (7)	546 (387)
Leopard troctolite									
Eastern Deeps (9)	3.41 (0.23)	1.87 (0.50)	107 (75)	154 (126)	16 (3)	23 (20)	6 (1.4)	13 (4)	93 (49)
Miniovoid (7)	3.94 (0.24)	2.77 (0.94)	145 (71)	228 (60)	12 (6)	25 (16)	2.9 (0.91)	7 (2)	224 (80)
Discovery Hill Zone (12)	3.79 (0.27)	2.23 (0.47)	377 (197)	345 (129)	9 (1)	35 (24)	2.0 (0.8)	3 (1)	337 (73)
Reid Brook Zone (20)	3.29 (0.58)	1.98 (1.32)	162 (119)	143 (76)	10 (1)	30 (12)	2.5 (0.5)	4 (2)	189 (119)
Magmatic breccia									
Eastern Deeps (44)	4.40 (0.88)	2.47 (1.09)	242 (220)	264 (180)	13 (6)	37 (26)	4.6 (2.5)	9 (6)	245 (176)
Ovoid (4)	4.14 (1.10)	3.44 (1.31)	430 (159)	521 (136)	23 (5)	70 (16)	9.1 (1.9)	13 (4)	469 (178)
Miniovoid (6)	4.11 (0.85)	3.83 (1.68)	489 (255)	307 (181)	10 (2)	56 (43)	3.6 (2.3)	11 (8)	360 (452)
Discovery Hill Zone (18)	4.06 (0.41)	2.20 (0.78)	266 (154)	346 (163)	11 (4)	31 (17)	2.7 (1.7)	5 (3)	278 (168)
Reid Brook Zone (26)	3.75 (0.72)	1.69 (0.72)	203 (136)	258 (171)	14 (10)	82 (60)	4.1 (4.2)	7 (5)	200 (143)
Massive sulphide									
Eastern Deeps (12)	3.32 (0.99)	1.15 (1.55)	34 (30)	144 (106)	18 (6)	26 (17)	6.6 (2.4)	15 (6)	18 (17)
Ovoid (52)	4.61 (1.08)	2.84 (1.40)	123 (111)	252 (115)	8 (3)	17 (10)	2.0 (1.1)	4 (1.5)	93 (130)
Miniovoid (11)	3.96 (1.23)	2.30 (2.41)	178 (85)	221 (127)	9 (3)	17 (5)	2.5 (0.6)	5 (2)	116 (48)
Reid Brook Zone (16)	2.95 (0.43)	1.17 (0.86)	42 (33)	102 (36)	10 (1)	19 (6)	2.8 (0.8)	4 (3)	60 (67)

base of the upper chamber exposed at surface, although others (cf. Lightfoot, 1997) believe it to be a swelling in the feeder conduit.

As discussed above, the systematic relationship between the Ni and Fo content of olivine from the ultramafic rock inclusions, feeder olivine gabbro at all localities, olivine gabbro of the Eastern Deeps chamber, and leucotroctolite of the Reid Brook chamber that is illustrated in Figure 9 suggests that these rocks are related to each other by fractionation of a single magma. The rapid decline in Ni with decreasing Fo content in the olivine, taken in conjunction with low whole rock Cu/Zr (see Fig. 13) ratios for the same samples, suggests that the magma from which they crystallized was sulphide saturated, and became depleted in Ni due to the separation of sulphide. The appreciably higher Ni content of olivine in the varied-textured troctolite and normal troctolite of the Eastern Deeps, coupled with the high whole-rock Cu/Zr ratios, indicates that their parental magma was less fractionated and less chalcophile-depleted than other sequences. Li and Naldrett (1999) concluded, therefore, that at least two waves of magma were involved at Voisey's Bay. The first crystallized in a lower chamber to form the ultramafic/mafic layers from which the inclusions were derived (Fig. 17A), before fractionating further as it progressed up the feeder to form the feeder olivine gabbro lining the walls of the feeder and ultimately the olivine gabbro of the Eastern Deeps (Fig. 17B). While in the lower chamber, this magma reacted with gneiss fragments derived from the walls and roof. The reaction resulted in sulphide immiscibility developing as fractionation was proceeding, either as a result of felsification of the magma (cf. Li and Naldrett, 1993) and/or due to the addition of sulphur from the gneiss.

The trace element data of Li et al. (2000) indicate that an early magma, most probably of picritic composition, fractionated and interacted with mid-crustal rocks and acquired an elevated, non-mantle, La/Sm ratio. The magma then ascended into rocks with a high Th/Nb ratio (Tasiuyak gneiss) where further reaction and incorporation of gneissic inclusions resulted in this signature also being imposed on the magma. Amelin et al.'s (2000) Nd, Pb, and Sr data support this concept of a two-stage interaction at different crustal levels. The Re-Os data of Lambert et al. (2000) also provide strong support for reaction between an early picritic predecessor of the troctolitic magma and enclosing Tasiuyak gneiss. Ripley et al.'s (1999) S-isotope data are consistent with the derivation of at least part of the sulphur from the gneiss; the oxygen isotope data (Ripley et al., 2000) again support interaction between troctolite and wall rocks.

Li and Naldrett (1999) proposed that the sulphides that segregated from the first wave of magma remained within the lower chamber but were then picked up by the second pulse of magma and carried up along the overlying conduit. Calculations based on recently determined partition coefficients for the distribution of Ni between olivine and sulphide melt (Brenan and Caciagli, 2000) indicate that sulphide liquid in equilibrium with magma giving rise to the olivine found in rocks that formed from the first magma pulse would have contained less than 3.0 wt.% Ni. As Naldrett et al. (2000a) have shown, the Ni content of the sulphide liquid responsible for the ores was between 3.5 and 4.5 wt.%, which argues strongly that this liquid re-equilibrated with a magma of higher Ni content than that first passing through the feeder into the upper chamber. Varying degrees of re-equilibration can account for the variable metal tenors discussed above in the different mineralized environments of the deposit. As the second wave of magma advanced up the conduit into the upper chamber (the Eastern Deeps), sulphides became concentrated in hydrodynamic traps within the conduit (Evans-Lamswood et al., 2000) and at the point where the conduit broadened out (Fig. 17C) as it entered the upper chamber (Naldrett et al., 1996).

Conclusions as to the Genesis of the Voisey's Bay Deposit

At this stage in the studies of the Voisey's Bay deposit, certain key aspects of its genesis have emerged:

1. The deposit is related to two magma chambers, one 1.5 km stratigraphically above the other, connected by a feeder dyke (Fig. 2).
2. Magma, most likely of picritic composition, rose into the crust where it started to crystallize and interact with mid-crustal gneisses, acquiring a crustal trace element and isotopic signature.
3. The magma continued its ascent and swelled out to form the lower (Reid Brook) chamber. Here it reacted with the enclosing Tasiuyak gneiss and eventually became saturated in sulphide. Various degrees of this reaction are preserved in the inclusions of the basal breccia sequence, feeder breccia, and varied-textured troctolite.
4. The initial magma lost much of its Ni as a consequence of olivine fractionation and separation of immiscible sulphide liquid; this liquid also removed other chalcophile elements, including Cu, causing a decrease in the Cu/Zr ratio.
5. The initial magma was forced out of the lower chamber, along with some of the sulphide that had segregated from it, and rose up the feeder sheet to develop an upper chamber about 1 km vertically above the lower chamber. Sulphides became lodged in and near the feeder as the magma rose up it.
6. Fresh, chalcophile element-undepleted magma entered the lower chamber and was probably responsible for the early magma being forced upward. The new magma continued through the system, disrupting early cumulates, picking up sulphides and partially-reacted inclusions, and transporting all of these up the feeder system

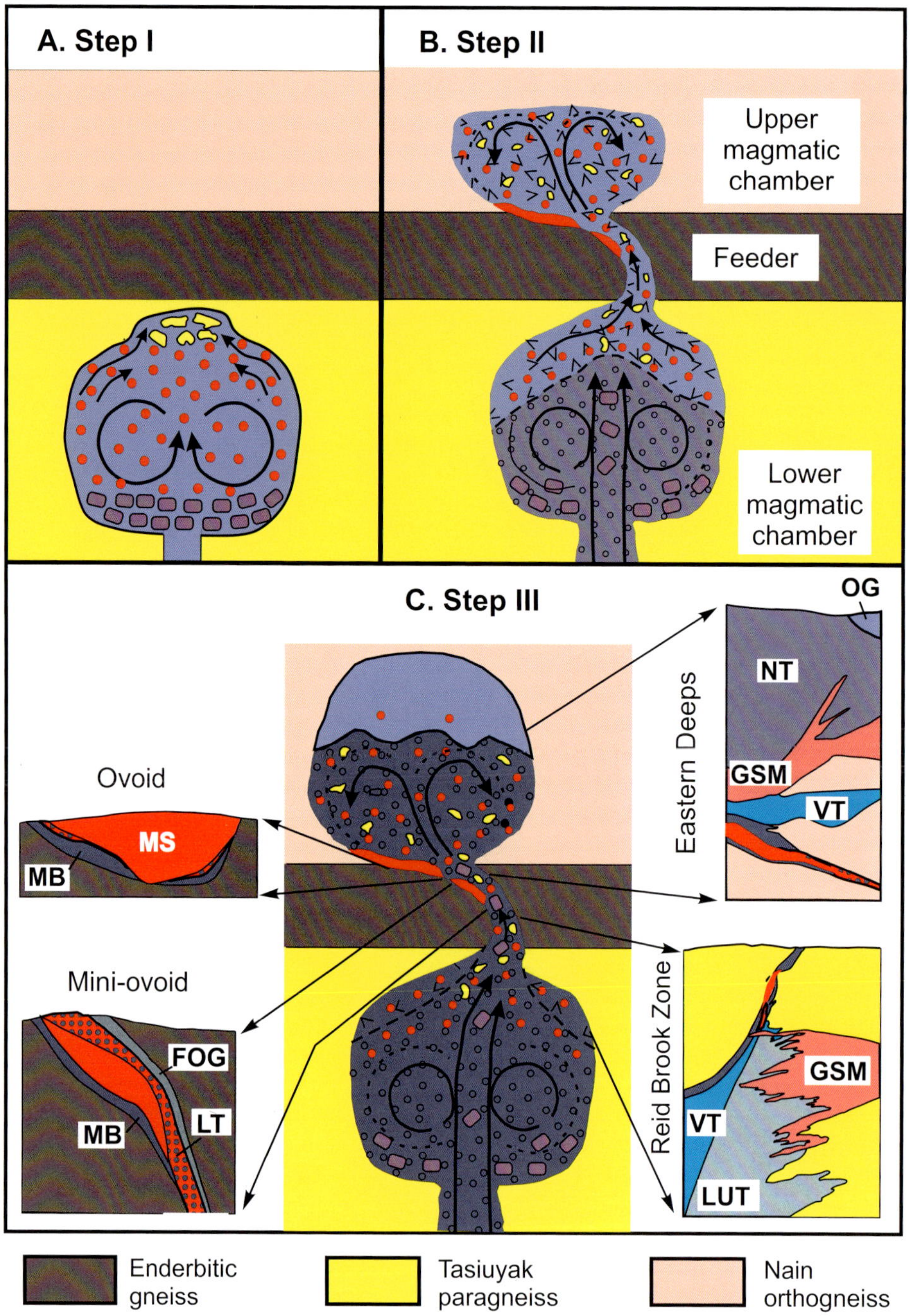

FIGURE 17. Genetic model for the Voisey's Bay deposit. (A) Step I: Magma rose to a lower chamber within Tasiuyak gneiss, olivine crystallization occurred with the formation of ultramafic cumulates. The magma reacted with gneiss fragments from collapsed roof, and became sulphide saturated. (B) Step II: New, Ni-undepleted magma entered the lower chamber, disrupting early cumulates, and pushing up the early magma to an upper chamber in Nain Gneiss. During transportation some sulphide became lodged in and near the feeder as the magma rose. (C) Step III: The later magma continued to flow through the system, picking up sulphides and partially reacted gneiss inclusions and transporting them up the feeder system to spread out over the lower part of the upper chamber. The fresh, undepleted magma enriched the sulphides in chalcophile metals. The insets show the types of sulphide accumulation that developed. Modified after Li and Naldrett (1999). FOG – Feeder olivine gabbro; GSM – Granite, syenite, and monzonite; LT – Leopard troctolite; LUT – Leucotroctolite; MB – Magmatic breccia; MS – Massive sulphide;NT – Normal troctolite; OG – Olivine gabbro; VT – Varied-textured troctolite.

to spread out in the lower part of the upper chamber.

7. As it progressed through the system, the new magma interacted with sulphides that were lodged there, upgrading them in chalcophile elements.

What Voisey's Bay Has to Tell Us about Ni Sulphide Exploration

The Voisey's Bay deposit reflects many of the major factors that have been recognized as important with respect to the formation of Ni sulphide deposits (cf. Naldrett, 1997, 1999): These include

1. Areas that have been intruded by hot mafic/ultramafic magma (e.g. rocks with Fo-rich olivine).
2. Locations where these rocks are in close proximity to, or have cut through, sulphide-bearing country rocks (e.g. the Tasiuyak gneiss at Voisey's Bay).
3. Field evidence of substantial interaction between the country rocks and the intruding magma, such as breccias composed of country rocks in an igneous matrix (e.g. the basal breccia/feeder breccia at Voisey's Bay).
4. Chemical evidence of contamination (e.g. LREE enrichment and high Th/Nb ratios at Voisey's Bay).
5. Chalcophile depletion in the intrusive rocks (e.g. Ni-depleted olivine in the early intrusive phase and low Cu/Zr ratios in the same rocks at Voisey's Bay).
6. Evidence of an input of fresh, less depleted magma that can have enriched the early-forming sulphides (e.g. the normal troctolite and varied-textured troctolite at Voisey's Bay).
7. Structure that has channelized the flow of sulphide-bearing magma and caused the sulphides to concentrate in hydrodynamic traps (e.g. the connecting dyke at Voisey's Bay).

All of these factors have been responsible for the formation of the rich mineralization at Voisey's Bay. Other areas, in which only some of the factors were present, have not proved so far to be of economic interest, as is discussed below.

The intense exploration undertaken on intrusions that form the Nain Plutonic Suite in the aftermath of the Voisey's Bay discovery, coupled with careful documentation of the exploration results by members of the Newfoundland Geological Survey (Ryan et al., 1995; Kerr, 1999, 2003; Kerr and Ryan, 2000), mean that we have an almost unique overview of the different types of deposits, their genesis, and relative importance associated with a single igneous province.

Kerr and Ryan (2000) have subdivided the deposits and prospects into several categories: A) those related to troctolite/olivine gabbro intrusions; B) those related to hybrid pyroxenites at the margins of troctolite/olivine gabbro intrusions; C) those related to anorthosite complexes either a) in norites-gabbros within or close to the anorthosites, or b) within the anorthosites themselves; and D) those related to ferrodiorite.

With respect to (B), it has become clear that sulphide mineralization in hybrid pyroxenites at the margins of large intrusions has not had access to sufficient quantities of magma to concentrate Ni and Cu to any great degree, with the result that the metal tenors of the sulphides are low (Kerr and Ryan, 2000). In addition, the pyroxenitic magmas themselves have not experienced the dynamic flow regimes that are necessary to concentrate the sulphides that they carried.

With respect to (C), sulphides within the anorthosites are the end products of the cooling of small batches of mafic magma that originated during the crystallization of the anorthosites, so that Ni has generally been incorporated into early forming silicates and the Ni tenors of the sulphides are low. Again, these magmas have not experienced the dynamic environment necessary to concentrate sulphides or, indeed, to expose the sulphides to large volumes of magma.

With respect to (D), the ferrodiorites are extreme differentiates within which sulphides developed at a late stage when most of the Ni present in the original magma had been removed in mafic silicates.

With respect to (A), the majority of the troctolite/gabbro bodies lie east of the Nain-Churchill suture and they do not seem to have achieved sulphide saturation at an early stage. If one looks at Figure 1, it is clear that the only troctolites/gabbros that have developed significant amounts of sulphide are those that have intruded the sulphide-bearing Tasiuyak gneiss, namely those at Voisey's Bay and Pants Lake, 80 km south of Voisey's Bay (also know as the Voisey's Bay South occurrence). These magmas have had their sulphur content increased through reaction with the Tasiuyak gneiss. However, at Pants Lake, so much sulphide appears to have developed in the magmas responsible for the intrusions, that these became very chalcophile depleted. This accounts for the fact that any sulphides that have been found associated with them are low in tenor (Kerr and Ryan, 2000; Kerr, 2003). There is no evidence of an input of fresh magma that could have enriched early forming sulphides. Additionally at Pants Lake, the sulphides have also not been in the right dynamic environment to have been concentrated to more than 10 to 20 wt.%.

Acknowledgements

This paper is based to a large extent on the research conducted under the NSERC Collaborative Research and Development grant held by one of us (AJN) from January 1996 to December 1998, the results of which were published in Economic Geology in June/July 2000. I am grateful to Diamond Fields Resources Inc. who supported the initial application and to INCO Ltd. who continued support. A number of distinguished scientists joined our research group, including Ed Ripley (Indiana University), David Lambert (then with Monash University), Yuri Amelin (then with the Royal Ontario Museum), Peter Lightfoot (INCO Ltd.), and Bruce Ryan and Andrew Kerr (Newfoundland Geological Survey). Dr Chusi Li (then at the University of Toronto) undertook the role of leading researcher and project manager, and Mohammed Asif, Sasha Krstic, and Jagmohan Singh assisted greatly while at the University. Our research team met twice a year with geologists working on the deposit, who included Dan Lee, Dawn Evans-Lamswood, Robert Wheeler, and Colin MacKenzie. Ideas were discussed freely and it is as a result of these discussions and the research that they stimulated that we reached the understanding of the deposit that is put forward in Economic Geology Volume 95, Number 4, and that we have attempted to summarize here. We are grateful to all of those named above for one of the most stimulating and enjoyable research collaborations that we have had the good fortune to experience. We would also like to acknowledge, with many thanks, a figure that Peter Lightfoot had used in a presentation in 2001, and that he kindly allowed us to reproduce here as Figure 4.

References

Amelin, Y., Li, C., and Naldrett, A.J., 1999, Geochronology of the Voisey's Bay complex Labrador Canada by precise U-Pb dating of co-existing baddeleyite zircon and apatite: Lithos, v. 47, p. 33-51.

Amelin, Y., Li, C., Valeyev, O., and Naldrett, A.J., 2000, Nd-Pb-Sr isotope systematics of crustal assimilation in the Voisey's Bay and Mushuau intrusions Labrador Canada: Economic Geology, v. 94, p. 815-830.

Brenan, J.M., and Caciagli, N.C., 2000, Fe-Ni exchange between olivine and sulfide liquid: Implications for oxygen barometry in sulfide-saturated magmas: Geochimica et Cosmochimica Acta, v. 64, p. 307-320.

Evans-Lamswood, D.M., Butt, D.P., Jackson, R.S., Lee, D.V., Muggridge, M.G., Wheeler, R.I., and Wilton, D.H.C., 2000, Physical controls associated with the distribution of sulphides in the Voisey's Bay Ni-Cu-Co deposit Labrador: Economic Geology, v. 95, p. 749-770.

Evans-Lamswood, D.M., 2005, A new sulphide domain: Geometric relationships and sulphide distribution within the Reid Brook Zone, part of the Voisey's Bay Ni-Cu-Co deposit: Geological Association of Canada/Mineralogical Association of Canada Joint Annual Meeting, Abstract.

Foster, J.G., Lambert, D.D., Frick, L.R., and Maas, R., 1996, Re-Os isotopic evidence for genesis of Archaean nickel ores from uncontaminated komatiites: Nature, v. 382, p. 703-706.

Ghiorso, M.S., and Sack, R.O., 1995, Chemical mass transfer in magmatic processes. IV. A revised and internally consistent thermodynamic model for the interpolation and extrapolation of liquidus-solidus equilibria in magmatic systems at elevated temperatures and pressures: Contributions to Mineralogy and Petrology, v. 119, p. 197-212.

Kerr, A., 1999, Mafic rocks of the Pants Lake Intrusion and related Ni-Cu-Co mineralisation in north-central Labrador, in Current Research, Newfoundland Department of Mines and Energy, Geological Survey Branch, Report 99-1, p. 215-253.

——— 2003, Nickeliferous gabbroic intrusions of the Pants Lake area Labrador Canada: Implications for the development of magmatic sulfides in mafic systems: American Journal of Science, v. 303, p. 221-258.

Kerr, A., and Ryan, B., 2000, Threading the eye of the needle: Lessons from the search for another Voisey's Bay in Northern Labrador: Economic Geology, v. 95, p. 725-748.

Lambert, D.D., Walker, R.J., Morgan, J.W., Shirey, S.B., Carlson, R.W., Zientek, M.L., Lipin, B.B., Koski, M.S., and Cooper, R.L., 1994, Re-Os and Sm-Nd isotope geochemistry of the Stillwater Complex, Montana: Implications for the petrogenesis of the J-M Reef: Journal of Petrology, v. 35, p. 1717-1753.

Lambert, D.D., Foster, J.G., Frick, L.R., Li, C., and Naldrett, A.J., 1999, Re-Os isotopic systematics of the Voisey's Bay Ni-Cu-Co magmatic ore system Labrador Canada: Lithos, v. 47, p. 69-88.

Lambert, D.D., Frick, L.R., Foster, J.G., Li, C., and Naldrett, A.J., 2000, Re-Os isotopic systematics of the Voisey's Bay Ni-Cu-Co magmatic sulfide system, Labrador, Canada:II. Implications for parental magma chemistry, ore genesis and metal redistribution: Economic Geology, v. 95, p. 867-888.

Li, C., and Naldrett, A.J., 1993, Sulfide capacity of magma: a quantitative model and its application to the formation of sulfide ores at Sudbury Ontario: Economic Geology, v. 88, p. 1253-1260.

——— 1999, Geology and petrology of the Voisey's Bay intrusion: Reaction of olivine with sulfide and silicate liquids: Lithos, v. 47, p. 1-32.

——— 2000, Melting reactions of gneissic inclusions with enclosing magma at Voisey's Bay: Implications with respect to ore genesis: Economic Geology, v. 95, p. 801-814.

Li, C., Lightfoot, P.C., Amelin, Y., and Naldrett, A.J., 2000, Contrasting petrological and geochemical relationships in the Voisey's Bay and Mushuau intrusions, Labrador, Canada: Implications for ore genesis: Economic Geology, v. 95, p. 771-801.

Li, C., Ripley, E.M., and Naldrett, A.J., 2003, Compositional variations of olivine and sulfur isotopes in the Noril'sk and Talnakh intrusions, Siberia: Implications for ore-forming processes in dynamic magma conduits: Economic Geology, v. 98, p. 69–86.

Lightfoot, P.C., 1997, Geological and geochemical relationships in the Reid Brook intrusive complex, Labrador: Implications for the origin of the Voisey's Bay magmatic Ni-Cu-Co ores: EOS, Transactions, American Geophysical Union, v. 78, p. F810.

Lightfoot, P.C., and Hawkesworth, C.J., 1997, Flood basalts and magmatic Ni, Cu, and PGE sulphide mineralization; comparative geochemistry of the Noril'sk (Siberian Traps) and West Greenland sequences: Geophysical Monograph, v. 100, p. 357-380.

Lightfoot, P.C., and Naldrett, A.J., 1999, Geological and geochemical relationships in the Voisey's Bay intrusion Nain Plutonic Suite Labrador Canada, in Keays, R.R., Lesher, C.M., Lightfoot, P.C., and Farrow, C.E.G., eds., Dynamic Processes in Magmatic Ore Deposits and Their Application in Mineral Exploration. Short Course Notes: Geological Association of Canada, v. 13, p. 1-31.

Naldrett, A.J., 1997, Ni-Cu-PGE deposits of the Noril'sk region and other world-class nickel sulfide deposits: Plutonic Ore Deposits Symposium, Canberra, February 1996, Australian Journal of Earth Sciences, v. 44, p. 283-315.

——— 1999, World class Ni-Cu-PGE deposits: Key factors in their genesis: Mineralium Deposita, v. 34, p. 227-240.

Naldrett, A.J., Keats, H., Sparkes, K., and Moore, R., 1996, Geology of the Voisey's Bay Ni-Cu-Co deposit, Labrador, Canada: Exploration and Mining Geology, v. 5, p. 169-179.

Naldrett, A.J., Li, C., and Krstic, S., 1997, The Voisey's Bay Ni-Cu-Co deposit, Labrador, Canada: Implications for exploration elsewhere, in Papunen, H., ed., Mineral Deposits: Research and Exploration Where do They Meet? Balkema, Rotterdam, Proceedings of the Fourth Biennial Society of Geology Applied to Mineral Deposits Meeting, Turku Finland, August 1997, p. 7-8.

Naldrett, A.J., Asif, M., Krstic, S., and Li, C., 2000a, The composition of ore at the Voisey's Bay Ni-Cu sulfide deposit with special reference to platinum-group elements: Economic Geology, v. 95, p. 845-866.

Naldrett, A.J., Singh, J., Krstic, S., and Li, C., 2000b, The mineralogy of the Voisey's Bay Ni-Cu-Co deposit Northern Labrador Canada: Influence of oxidation state on textures and mineral compositions: Economic Geology, v. 95, p. 889-900.

Ripley, E.M., Park, Y.-R., Li, C., and Naldrett, A.J., 1999, Sulfur and oxygen isotopic evidence of country rock contamination in the Voisey's Bay Ni-Cu-Co deposit Labrador Canada: Lithos, v. 47, p. 53-68.

——— 2000, Oxygen isotopic study of the Voisey's Bay Ni-Cu-Co deposit Labrador Canada: Economic Geology, v. 95, p. 831-844.

Ripley, E.M., Li, C., and Shin, D., 2002, Paragneiss assimilation in the genesis of magmatic Ni-Cu-Co sulphide mineralization at Voisey's Bay, Labrador: $\delta^{34}S$, $\delta^{13}C$ and Se/S evidence: Economic Geology, v. 97, p. 1307-1318.

Ryan, B., 1990, Geological Map of the Nain Plutonic Suite and surrounding rocks (Nain-Nutak NTS 14SW): Newfoundland Department of Mines and Energy Geological Survey Branch, Map 90-44, scale 1:500000.

——— 1996, Commentary of the Location of the Nain-Churchill Boundary in the Nain Area: Newfoundland Department of Natural Resources, Geological Survey Report 96-1, 21 p.

——— 2000, The Nain-Churchill boundary and the Nain Plutonic Suite: A regional perspective on the geological setting of the Voisey's Bay Ni-Cu-Co deposit: Economic Geology, v. 95, p. 703-724.

Ryan, B., Wardle, R.J., Gower, C.F., and Nunn, G.A.G., 1995, Nickel-Copper-Sulphide Mineralization in Labrador, the Voisey Bay Discovery and Its Exploration Implications: Newfoundland Department of Natural Resources, Geological Survey Branch, Current Research Report 95-1, 28 p.

Sun, S.S., and McDonough, W.F., 1989, Chemical and isotopic systematics of ocean basalts: Implications of mantle composition and processes, in Saunders, A.D., and Norry, M.J., eds., Magmatism of the Ocean Basins: Geological Society of London, Special Publication No. 42, p. 313-345.

Van Kranendonk, M., 1996, Tectonic evolution of the Paleoproterozoic Torngat orogen: Evidence from pressure-temperature-time deformation paths in the North River map area: Tectonics, v. 15, p. 843-869.

Walker, R.J., Morgan, J.W., Horan, M.F., Czamanske, G.K., Krogstad, E.J., Fedorenko, V.A., and Kunilov, V.E., 1994, Re-Os isotopic evidence for an enriched-mantle plume source for the Noril'sk-type ore-bearing intrusions Siberia: Geochimica et Cosmochimica Acta, v. 58, p. 4179-4197.

Magmatic Ni-Cu-Platinum-Group Element Deposits of the Thompson Nickel Belt

Daniel Layton-Matthews[1,2], C. Michael Lesher[1], O. Marcus Burnham[1,3], Janice Liwanag[4], Norman M. Halden[4], Larry Hulbert[5], and David C. Peck[6]

1. Mineral Exploration Research Centre, Department of Earth Sciences, Laurentian University, Sudbury, Ontario P3E 6B5

2. Current Address: Department of Geological Sciences and Geological Engineering, Queen's University, Kingston, Ontario K7L 3N6

3. Current Address: Ontario Geoscience Laboratories, Willet Green Miller Centre, 933 Ramsey Lake Road, Sudbury, Ontario P3E 6B5

4. Department of Geological Sciences, Wallace Building, University of Manitoba, Winnipeg, Manitoba R3T 2N2

5. Geological Survey of Canada, 601 Booth Street, Ottawa, Ontario K1A 0E8

6. Anglo American Exploration (Canada) Ltd., Suite 800 - 700 West Pender Street, Vancouver, British Columbia V6C 1G8

Corresponding author's email: dlayton@geol.queensu.ca

Abstract

The Circum-Superior Boundary Zone in Manitoba hosts several world-class magmatic Ni-Cu-platinum-group element (PGE) deposits. Operating mines in the Thompson Nickel Belt (TNB) include the Thompson T1, T3, and Birchtree mines. In 2004, the International Nickel Company (Inco) reported proven and probable reserves of 27 Mt grading 2.10% Ni and 0.14% Cu. These deposits are locally concentrated in a narrow, northeast-trending tectonic belt that represents collisional remnants of Archean Superior Province rocks, autochthonous Proterozoic supracrustal sequences (Ospwagan Group, TNB), and allochthonous Proterozoic rocks related to the Trans-Hudson Orogen. Since the discovery of Ni-Cu-(PGE) mineralization by Inco in 1955, it has been recognized that the TNB deposits exhibit many fundamental characteristics of other major magmatic Ni-Cu-(PGE) districts.

Mineralization within the TNB is associated with variably serpentinized ultramafic intrusions that range in composition from dunite to pyroxenite and are hosted by clastic and chemical sediments of the Ospwagan Group. The serpentinized intrusions are generally lensoid to tabular in shape, which reflects the multiple phases of deformation experienced by the host Ospwagan Group during orogenesis. Ore within the TNB is found as 'primary' massive sulphide mineralization within the ultramafic bodies (i.e. Pipe and Birchtree deposits) or, more commonly, as 'secondary' massive sulphide mineralization within metasedimentary rocks of the Ospwagan Formation. Serpentinized intrusions occur at specific stratigraphic levels within the Ospwagan Group. The consistent stratigraphic correlations between ultramafic intrusions, S-rich chemical sedimentary rocks, S isotopic data, and Ni-Cu-(PGE) sulphide mineralization collectively suggest that the mineralization formed by assimilation of S-rich sedimentary rocks by high-temperature ultramafic magmas. Post-ore deformation and metamorphism have significantly modified the primary characteristics of many of the TNB ore deposits.

Résumé

La zone périphérique de transition de la Province du lac Supérieur au Manitoba renferme plusieurs gisements magmatiques de Ni-Cu-éléments du groupe du platine (ÉGP) de classe mondiale. Les mines en exploitation dans la ceinture nickélifère de Thompson (CNT) sont les mines Thompson T1 et T3 et la mine Birchtree. En 2004, l'International Nickel Company (Inco) a signalé des réserves prouvées et probables de 27 millions de tonnes de minerai renfermant 2,10 % de Ni et 0,14 % de Cu. Ces gisements sont concentrés localement dans une étroite zone tectonique d'orientation nord-est (correspondant à la CNT), qui constitue un assemblage vestigial de collision formé de roches archéennes de la Province du lac Supérieur, de séquences de roches supracrustales autochtones du Protérozoïque (Groupe d'Ospwagan, CNT) et de roches allochtones du Protérozoïque associées à l'orogène trans-hudsonien. Depuis la découverte des minéralisations de Ni-Cu-ÉGP par l'Inco en 1955, on a reconnu que les gîtes de la CNT présentent un grand nombre des caractéristiques fondamentales qui définissent les autres principaux districts de minéralisations magmatiques de Ni-Cu-(ÉGP).

Dans la CNT, les minéralisations sont associées à des intrusions ultramafiques serpentinisées à des degrés divers dont la composition varie de la dunite à la pyroxénite et qui sont encaissées dans les roches sédimentaires clastiques et chimiques du Groupe d'Ospwagan. Les intrusions serpentinisées sont généralement de forme lenticulaire à tabulaire, ce qui est un reflet des multiples phases de déformation qu'a subies le Groupe d'Ospwagan pendant l'orogenèse. Dans la CNT, le minerai se présente sous forme de minéralisations « primaires » de sulfures massifs à l'intérieur de corps ultramafiques (p. ex. les gisements de Pipe et de Birchtree) ou, plus couramment, sous forme de minéralisations « secondaires » de sulfures massifs dans les roches métasédimentaires du Groupe d'Ospwagan. Les intrusions serpentinisées se situent à des niveaux stratigraphiques spécifiques dans le Groupe d'Ospwagan. Les corrélations stratigraphiques cohérentes entre les intrusions ultramafiques, les roches sédimentaires chimiques riches en S, les données sur les isotopes du S et les minéralisations sulfurées de Ni-Cu-ÉGP suggèrent collectivement que la minéralisation s'est formée par assimilation de roches sédimentaires riches en S par des magmas ultramafiques de haute température. La déformation et le métamorphisme postérieurs à la formation des minerais ont considérablement modifié les caractéristiques primaires d'un grand nombre des gîtes de la CNT.

Layton-Matthews, D., Lesher, C.M., Burnham, O.M., Liwanag, J., Halden, N.M., Hulbert, L., Peck, D.C., 2007, Magmatic Ni-Cu-platinum-group element deposits of the Thompson Nickel Belt, *in* Goodfellow, W.D., ed., Mineral Deposits of Canada: A Synthesis of Major Deposit-Types, District Metallogeny, the Evolution of Geological Provinces, and Exploration Methods: Geological Association of Canada, Mineral Deposits Division, Special Publication No. 5, p. 409-432.

Introduction

The Thompson Nickel Belt (TNB) represents the remnants of a Neo-Archean - Paleoproterozoic continental margin and lies within the Circum-Superior Boundary Zone (CSBZ) that separates 3.1 to 2.6 Ga autochthonous Archean Superior Province rocks in the east from 1.92 to 1.83 Ga allochthonous Proterozoic domains of the Trans-Hudson Orogen (THO) to the west (Fig. 1 inset). The TNB has long been defined as a narrow, north-east-trending (030°) geological domain that locally hosts magmatic Ni-Cu-(PGE) mineralization (McRitchie, 1995). The TNB constitutes part of the larger Circum-Superior Boundary Zone (CSBZ), where the TNB component of the CSBZ extends from approximately 50 km northeast of Thompson to approximately 125 km southwest of Thompson, from where it is covered by Paleozoic carbonate platform rocks (Fig. 1). Beneath the Paleozoic platform, the geophysical expression of the TNB continues approximately 275 km southward towards the Saskatchewan border, where the Circum-Superior Boundary Zone is interpreted to continue into North Dakota to a depth of 500 to 2000 m beneath Paleozoic and Mesozoic rocks. To the northeast of the exposed TNB, the CSBZ trends east through the Ni-Cu-(PGE)-bearing Fox River Belt and Sutton Inlier, and continues north through the Belcher Islands, the Ottawa Islands, and east through the Ni-Cu-(PGE)-bearing Cape Smith Belt (Lucas et al., 1990).

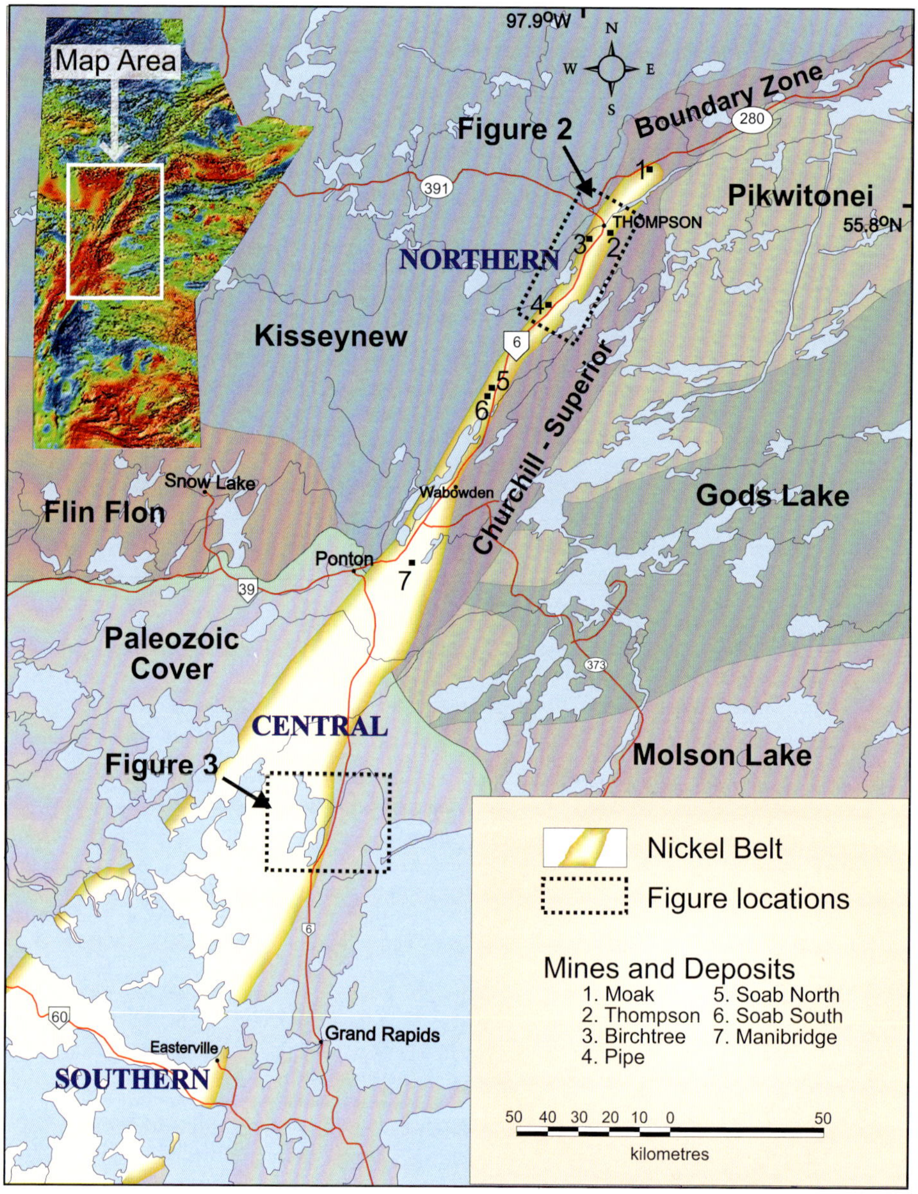

FIGURE 1. General geological map of the Thompson Nickel Belt. Areas of Figures 2 and 3 are outlines in dashed lines.

The TNB hosts several past and current producing world-class magmatic Ni-Cu-(PGE) deposits (Table 1; Fig. 1). Although the deposits of the TNB have been structurally and metamorphically modified, they exhibit many fundamental characteristics of other major magmatic Ni-Cu-(PGE) districts. Over the past fifty years, exploration has concentrated on the exposed section of the northern Thompson Nickel Belt (NTNB) and central Thompson Nickel Belt (CTNB), between Moak and Muhigan Lakes. More recently, exploration has been extended to include the Paleozoic-covered southern Thompson Nickel Belt (STNB).

Inco reports provide some of the earliest descriptions of the Ni-Cu-(PGE) deposits in the Thompson area, but most of these reports remained confidential until the early 1980s (Peredery, 1979, 1982). Inco geologists identified two types of ultramafic bodies: 1) "ultramafites", the ore-hosting metadunitic to metaperidotitic rocks in the TNB, and 2) "ultramafic amphibolites", metagabbroic rocks that occur in close proximity to the ultramafic bodies (terminology after Peredery et al., 1982). Several of the TNB ultramafic bodies have been studied as part of masters and doctoral theses or government directed studies, including the Manibridge ultramafic body (Coats and Brummer, 1971; Coats et al., 1976), Reservation 34 ultramafic bodies (Bliss, 1973), West Manasan and Pipe II ultramafic bodies (De Saboia, 1978), and the Bucko ultramafic body (Good and Naldrett, 1993). Several workers have summarized and classified the TNB ultramafic rocks as a group (Coats, 1966; Scoates, 1971), but it was not until the work of Macek and Bleeker (Macek, 1986; Bleeker, 1990) and the TNB working group (Macek et al., 2002), which expanded on Inco's stratigraphic correlations, that a stratigraphy and tectonostratigraphic model for the genesis of the TNB ultramafic bodies and the Ospwagan Group were developed.

TABLE 1. Mineral resources and deposits of the Thompson Nickel Belt (after Bamburak, 1980; Peredery, 1982; McRitchie, 1995).

Name	Size of UM body (DDH intersection width, m)	Deposit Status	Location UTM (NAD83)	Description
Mel Zone	n/a	Explored Lease	576762.69 E 6203813.21 N	Mineral Resource - n/a
Moak	250-1000	Explored Lease	588635.39 E 6199717.48 N	Mineral Resource - 45 Mt @ 0.7% Ni, widths up to 90 m
Mystery Lake	250-1000	Explored Lease	577874.66 E 6187672.92 N	Mineral Resource - 227 Mt @ 0.6% Ni
Thompson Area				Combined Resource - 150 Mt @ 2.32% Ni, 0.16% Cu, 0.046% Co, 0.83 g/t PGE[1]
Thompson 1C	10-250	Present Producer	571987.65 E 6175512.04 N	Mineral Deposit - 4.5 Mt
Thompson 1D	n/a	Present Producer	n/a	Mineral Deposit - 19 Mt @ 2.5% Ni
Thompson South Pit	n/a	Past Producer	n/a	Mineral Deposit - n/a
Birchtree	10-250	Present Producer	567401.25 E 6173472.14 N	Mineral Deposit - n/a
Pipe I	10-250	Explored Lease	553193.46 E 6150384.44 N	Mineral Deposit - n/a
Pipe II	250-1000	Past Producer	n/a	Mineral Deposit - ~18 Mt[2]
Pipe Deep	n/a	Explored Lease	n/a	Mineral Resource - 3.6 Mt @ 2.32 Ni, 0.1% Cu
Hambone	10-250	Explored Lease	573033.62 E 6127480.23 N	Mineral Resource - 3.27 Mt @ 0.81% Ni and 1.09 Mt @1.10% Ni in north Creek zone
Grass	10-250	Explored Lease	541194.64 E 6121504.84 N	Mineral Resource - low grade
Soab (North & South)	1-10	Past Producer	538042.03 E 6120827.49 N	Mineral Resource - 0.9 Mt grading up to 1.5% Ni
Bowden	n/a	Explored Lease	522859.16 E 6086262.75 N	Mineral Resource - 87.9 Mt @0.627% Ni
Discovery	n/a	Explored Lease	524292.08 E 6084007.59 N	Mineral Resource - 4.5 Mt @ ~1% Ni
Bucko	n/a	Past Producer	522060.43 E 6081418.11 N	Mineral Resource - 18.8 Mt at 1% Ni or 2.5 Mt @ 2.23% Ni and 0.17% Cu. Includes 11.7 m @ 5.1% Ni, 0.40% Cu, and 1.62 g/t PGE[3]
Resting Lake	n/a	Explored Lease	518593.22 E 6079436.51 N	Mineral Resource - 90 Mt @ 0.30% Ni-Cu
Manibridge	n/a	Past Producer	510493.18 E 6061680.43 N	Mineral Deposit - 1.27 Mt @ 2.55% Ni and 0.27% Cu
Minago	n/a	Explored Lease	487790.01 E 5993365.93 N	Mineral Resource - 20.5 Mt @ 1.02% Ni
William Lake	n/a	Explored Lease	474922.46 E 6075387.78 N	Mineral Resource - 0.74% Ni over 32 m

1. Naldrett, 2004
2. Bleeker, 1990
3. Crowflight Minerals Ltd., press release, Sept. 20, 2005
n/a = not available

The regional geologic setting and tectonic evolution of the TNB have been summarized by Kornik and MacLaren (1966), Gibb (1968; 1982), Weber and Scoates (1978), Green (1981), Bleeker (1990), Machado et al., (1990), Weber (1990), Lewry et al., (1994), and White et al., (1999; 2002). Additional geological information are available from the Manitoba Geological Survey bedrock maps and the recently complied series of fifteen maps of the Thompson Nickel Belt (Macek et al., 2001, and references therein). This map-series was a collaborative project involving the Manitoba Geological Survey, Inco Limited, Falconbridge Limited, and Hudson Bay Exploration and Development Company Limited that covered an area from Moak Lake (northeast of Thompson) to Bracken Lake (north of Grand Rapids).

The aim of this manuscript is to provide a current synopsis of the geology and exploration models for the Ni-Cu-(PGE) deposits in the TNB. It is derived largely from the work of the foregoing authors, but includes some of the data and interpretations generated in the recently completed Canadian Mining Industry Research Organization (CAMIRO) Thompson Nickel Belt project (Burnham et al., 2003).

Exploration History

Fraser (1985) provides a complete account of the International Nickel Corporation (Inco) Ltd.'s early explo-

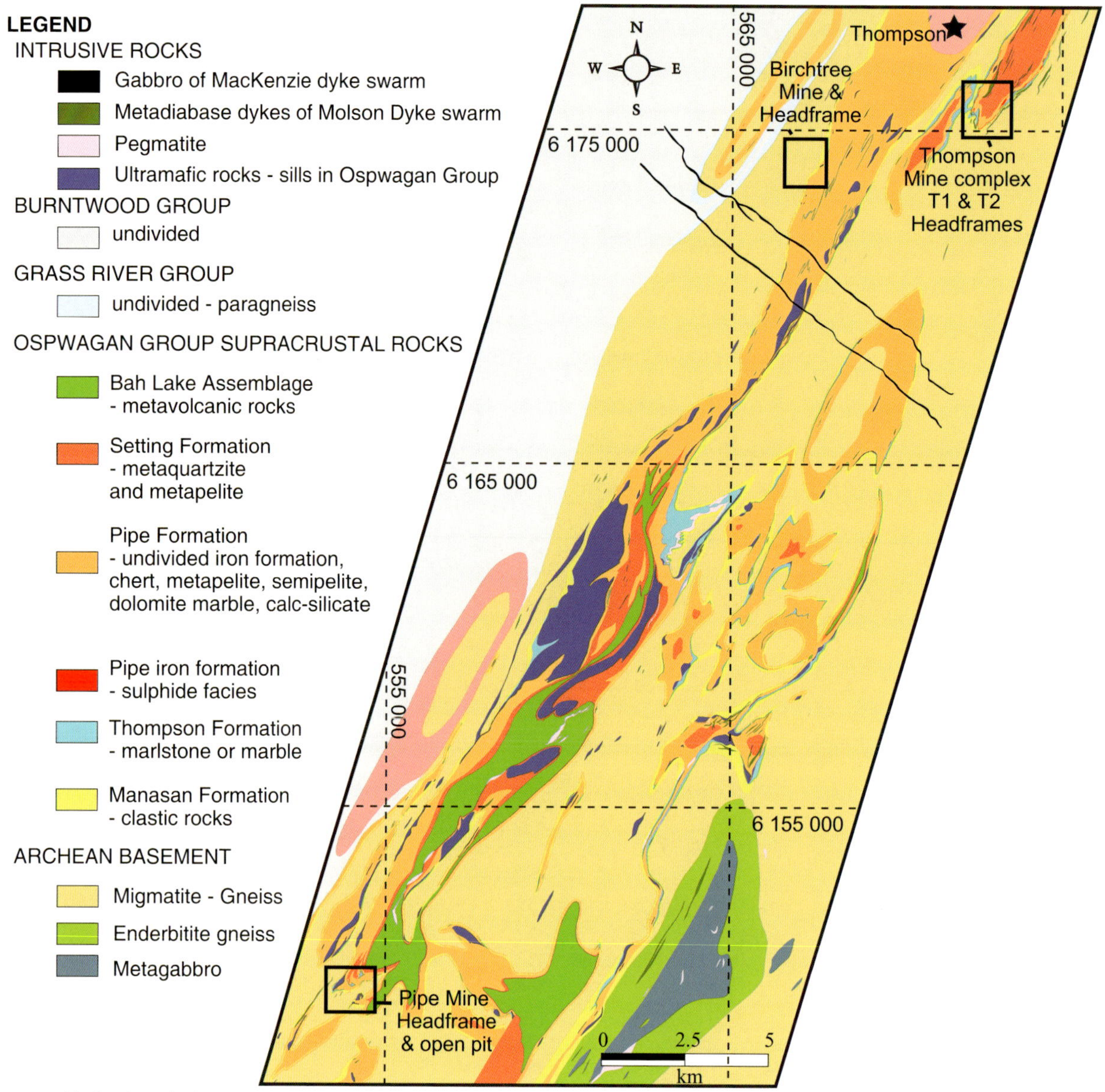

FIGURE 2. Interpreted bedrock geological map from the Thompson to the Pipe deposits in the Northern Thompson Nickel Belt (modified after Macek et al., 2001). Mine locations indicated by black boxes.

ration efforts and their discovery of Thompson Ni deposits in the immediate vicinity of the City of Thompson, Manitoba. Exploration in the TNB area started in 1946, when Inco began a 10 year geophysical and geological exploration program, leading to the discovery of the Moak property in 1955. By the beginning of 1956, over 72 km of exploration diamond drilling had been completed and in February 1956 the main Thompson orebody was discovered. This discovery led to the construction of the current mining complex and the City of Thompson (Fraser, 1985). The Thompson mine officially opened on March 25, 1961, becoming the Thompson Nickel Belt's first active mine. Additional resources were discovered on the Pipe I property south of Upper Ospwagan Lake. By 1961, Inco had 8000 active claims in the TNB, and had conducted an extensive geophysical program that included 128,000 km of aeromagnetic surveys, 112,000 km of airborne electromagnetic surveys, and 17,600 km of ground geophysical surveys, by which time the Pipe II deposit had been discovered. Production from an open pit at Pipe II began in 1969, with production continuing until its closure in 1984. From 1961 to 1971, Inco brought the Thompson, Birchtree, Pipe I, Pipe II, Soab North, and Soab South mines into production (Table 1). In response to falling Ni prices, Inco closed all but the Thompson and Pipe open pit mines in the TNB by 1977. In 1986, the Thompson Open Pit North was brought into production, and mining of the Thompson 1C orebody between the 732 and 975 m levels, reactivation of the Birchtree mine, and construction of the Thompson Open Pit South soon followed. In 1990, Inco announced plans to mine the Thompson 1D orebody to the 1100 m level and to deepen the Birchtree shaft from the 1045 m level to the 1295 m level (Bamburak, 1980; Hulbert et al., 2005). The deepening of the Birchtree mine was completed in 2002, nearly doubling daily production, increasing reserves to 13.6 Mt of 1.79% Ni, and extending Birchtree's mine production life to 2016. Recently, mineralization and host rocks similar to those of TNB (i.e. Ospwagan Group; see below) were discovered north and west of Mystery Lake. The current operating mines in the TNB include the Thompson T1, T3, and Birchtree mines (Figs. 1, 2). In 2004, proven and probable reserves of 27 Mt grading 2.10% Ni and 0.14% Cu were reported (International Nickel Company Limited, Annual Report, 2004) for the active mines in the TNB.

In the Paleozoic-covered southern Thompson Nickel Belt (STNB), Falconbridge conducted a multi-year (1991-2002) (Falconbridge Limited, Annual Report, 1996, 2004) exploration program that identified significant Ni mineralization in the William Lake area (Figs. 1, 3). The principal resource was one of several Ni occurrences that are spread over a distance of approximately 20 km. In 1994, Falconbridge completed extensive airborne electromagnetic (GEOTEM) and magnetic surveys to improve the regional delineation of the more prospective zones on its properties. Several other companies have recently operated, or are actively operating, exploration programs in the southwest extension of the TNB, most notably Hudson Bay Exploration and Development Company Ltd. (Minago River area), Cominco Ltd. (mid-1990s; Winnipegosis Belt), and CrowFlight Minerals Inc. (Wabowden area).

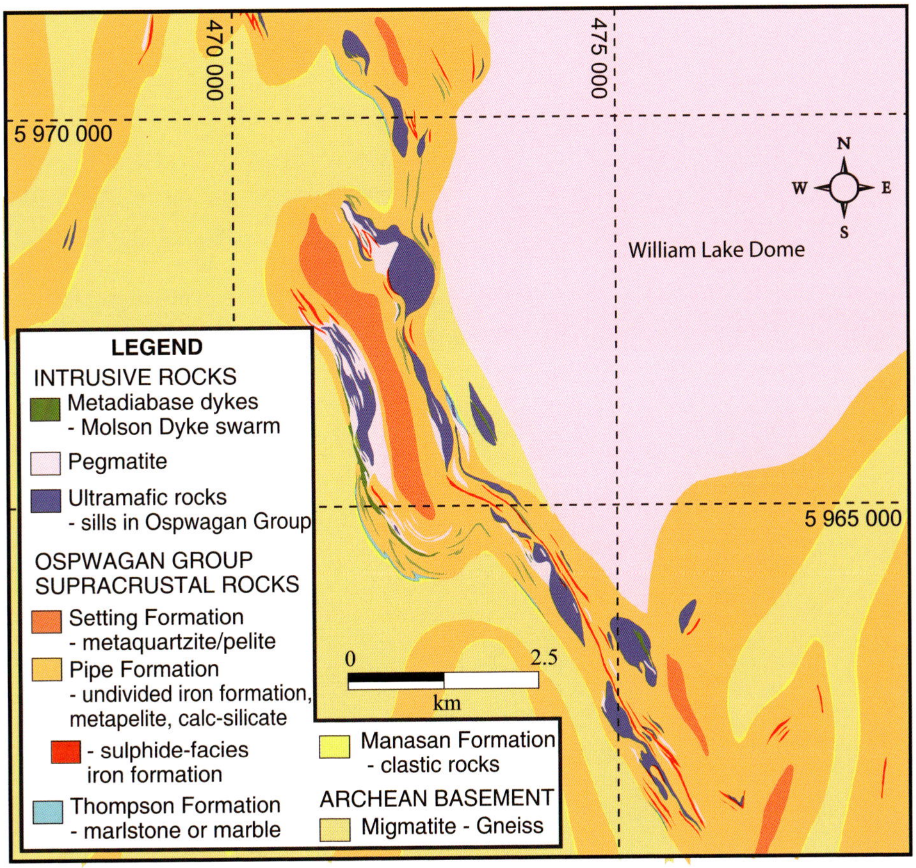

FIGURE 3. Interpreted bedrock geological map of the William Lake area in the southern Thompson Nickel Belt (modified after Macek et al., 2001).

Stratigraphy

The Circum-Superior Boundary Zone in northern Manitoba has been interpreted to represent part of a foreland margin of the Superior Craton that is marked by 100 km scale re-entrants, such as the Winnipegosis Komatiite Belt, the Fox River Belt (Lucas et al., 1990), and the TNB (White et al., 2002). The recognition that the Ni-Cu-(PGE)-bearing ultramafic bodies occur at specific stratigraphic levels within the foreland margin (Pipe Formation of the Ospwagan Group) has been the historical focus of exploration of magmatic ore deposits in the TNB. The stratigraphic framework, formational thickness, and sedimentary characteristics of the Ospwagan Group are presented in Figure 4.

The Ospwagan Group (Scoates et al., 1977) is interpreted to have been deposited near a passively rifted margin on a continental platform that experienced a subsequent period of active rifting and ultramafic to mafic magmatism, as represented by the boudinaged, mineralized, and nonmineralized ultramafic sills. The following sections review the stratigraphy of the Ospwagan Group, as understanding the level of emplacement of the ultramafic intrusions has been critically important in mineral exploration in the TNB (Bleeker, 1989).

Basement Gneisses

The Ospwagan Group unconformably overlies felsic to intermediate migmatitic gneisses that record at least one generation of deformation prior to the deposition of the Ospwagan Group (Figs. 5A,B). The basement gneisses contain relics of granulite facies metamorphism or retrogressed pseudomorphs after ortho- and clinopyroxene (Bleeker, 1990). Gneiss compositions vary in the TNB and likely reflect protolith inhomogeneity. Generally, the gneisses are tonalitic in composition, and exhibit varying degrees of mylonitization and anatexis. Biotite-rich varieties contain garnet and sillimanite porphyroblasts.

Ospwagan Group

The Ospwagan Group consists of a sequence of clastic and chemical sediments separated from underlying migmatitic Archean basement by an angular unconformity. It has been divided into four formations, from oldest to youngest (Bleeker, 1990): basal conglomerates and quartzites of the Manasan Formation; impure calc-silicates of the Thompson Formation; semipelites, sulphidic sediments, and iron formations of the Pipe Formation; and interlayered quartzite and semipelites of the Setting Formation. The Setting Formation is overlain by mafic-ultramafic metavolcanic rocks (Bah Lake Assemblage: Zwanzig, 2004) and/or a younger suite of clastic metasedimentary rocks (Grass River Group: Zwanzig and McRitchie, 1997). Nickel-Cu-(PGE) mineralization in the TNB is associated almost exclusively with, or localized within, ultramafic bodies within the lower part of the Ospwagan Group, in particular the Pipe Formation (Bleeker, 1990).

Manasan Formation

The contact of the Ospwagan Group with the underlying basement is angular to disconformable, and in some locations is marked by a 5 to 10 cm thick sillimanite-rich layer, interpreted by Bleeker (1990) to represent a metaregolith. Two members have been identified in the Manasan Formation: M1 and M2. The M1 basal quartzite and basal conglomerate fine upwards and gradationally increase in phyllosilicate content toward the M1-M2 transition (Fig 5C,D). It is light beige to grey, and subarkosic to arkosic in composition with 85 to 90 wt.% SiO_2. The overlying M2 unit, known as the Manasan semipelite, contains silt-sized microcline and quartz grains in an argillaceous matrix, is light to dark grey in colour, and is interpreted as an argillaceous wacke.

The angular unconformity is evidence that the Archean basement was exposed during the Late Archean to Early Proterozoic. The progression from clastic quartzite (M1) to argillaceous wacke (M2) has been interpreted as a transgressive event in response to a passive margin subsidence (Bleeker, 1990).

Thompson Formation

The Thompson Formation marks the onset of chemical sedimentation in the Ospwagan Group, and has been subdivided into three recognizable members: T1 semipelite with sodic plagioclase, T2 semipelite with calcic plagioclase, and T3 dolomitic marble.

The T1 unit comprises microcline, diopside, hornblende, and biotite that form alternating laminae of microcline-rich and diopside+carbonate+biotite+quartz layers. Leucosome produced during peak metamorphism is pink and comprises K-feldspar with minor quartz (Fig 6A,B). The T2 unit was first defined at the Thompson Open Pit (1C), where it superficially resembles the M2 semipelite; however, diopside and calcic plagioclase are more abundant in T2. Leucosome veins within the T2 member are white and contain abundant plagioclase (Bleeker, 1990). The T3 member consists of dolomitic marble with interlayered 'chert' bands. The T3

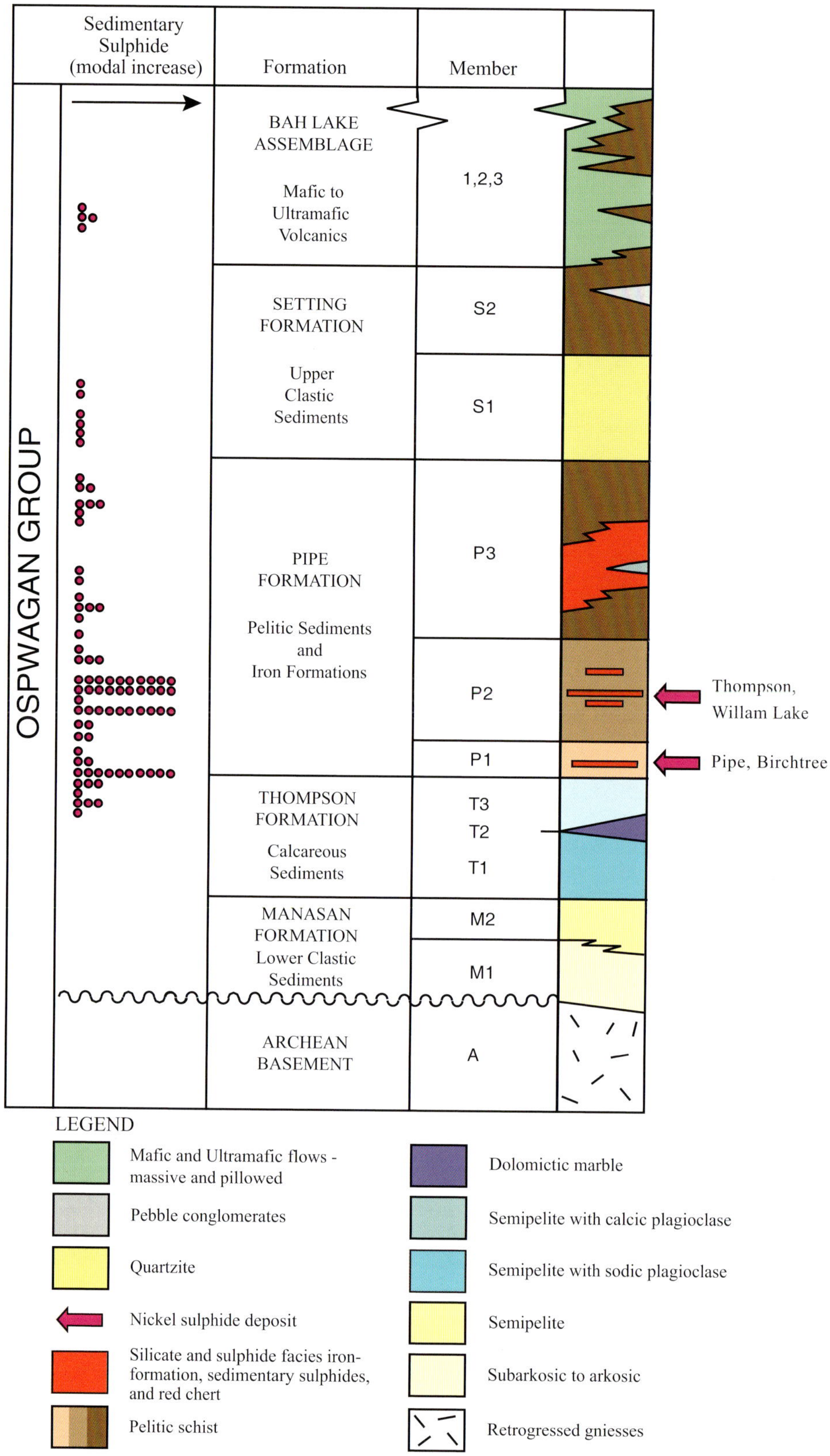

FIGURE 4. Reconstructed Ospwagan Group lithostratigraphy for the Thompson Nickel Belt (modified after Bleeker, 1990). Stratigraphic locations of ore deposits are indicated by pink arrows.

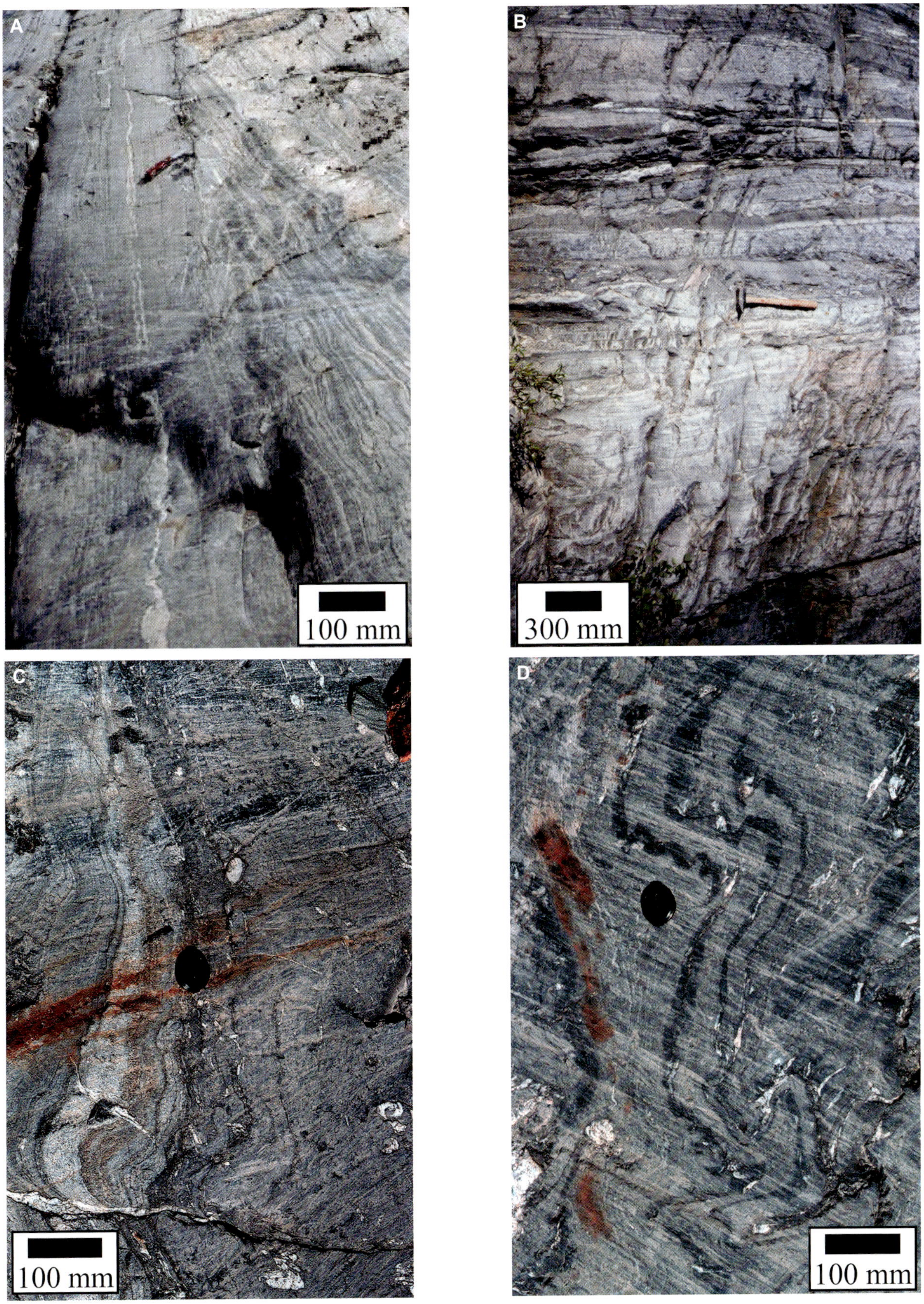

FIGURE 5. Photographs of the Ospwagan Group. **(A)** Angular relationship between basement gneisses (right) and overlying Manasan Formation (left) at the southern margin of the Thompson South Pit. **(B)** Same contact as is shown in photograph A but at Pipe Pit; the contact (hammer parallel to contact, Manasan on top, basement on bottom) has been obscured by deformation. **(C)** The Manasan Formation at Pipe Pit M1/M2 contact. Contact is located at the lens cap, with the M1 unit to the right and the M2 to the left. **(D)** M2 unit at Pipe Pit with pegmatitic leucosome.

FIGURE 6. Photographs of the Ospwagan Group. **(A)** 'Skarns' of the Thompson Formation (T1) from Manasan quarry. **(B)** Thick sequence of Thompson Formation from the southern margin of Thompson south pit.

unit usually has a pale green colour, with varying proportions of fosteritic olivine, clinopyroxene, and calcic amphibole.

The Thompson Formation is interpreted to represent waning clastic sedimentation and waxing chemical sedimentation. This is consistent with continued transgression and the generation of a starved basin (Bleeker, 1990).

Pipe Formation

The Pipe Formation consists of a sequence of chemical and chemical-clastic sedimentary rocks that can be subdivided into three members: P1 silicate- and sulphide-facies iron formation and red chert, P2 pelitic schist, and the P3 pelitic schist with calc-silicate intercalations, silicate-facies iron formation, dolomitic marble, and chert with calc-silicate intercalations.

The Pipe Formation is named after the type locality at Pipe Lake, which has since been drained and mined by open pit (Pipe II Pit). The glacially polished surface exposures in this area are exceptional and the mapping by Bleeker (1990) at Pipe Pit has formed the basis for defining most of the Pipe Formation stratigraphy.

The P1 member comprises pelitic schists that have a gradational lower contact with T3 calcareous sediments. The P1 member contains abundant graphite and moderate amounts of sedimentary sulphides. Sulphide-facies iron formation (known as the lower sulphide iron formation) is characterized by thin pyrrhotite layers intercalated with graphitic chert or schist (Fig 6C). Sulphide-facies iron formations are overlain by hornblende+biotite±grunerite±garnet silicate-facies iron formation (lower silicate-facies iron formation) with minor disseminated sulphides. Some layers within the silicate-facies iron formation contain up to 20% (modal) magnetite (Fig. 6D).

The P2 member is characterized by muscovite-biotite or biotite schists with abundant aluminous porphyroblasts (garnet, staurolite, or sillimanite, depending on metamorphic grade), and represents the main parcel of pelitic schists in the TNB. P2 schists are intercalated with thin layers of chert, calc-silicate, and silicate-facies iron formation (Fig. 7A). The uppermost part of the P2 member is characterized by an increasing abundance of sulphide-rich lamellae and semi-massive to massive pyrrhotite-pyrite layers (known as the upper sulphide-facies iron formation) (Fig. 7B).

The thick succession of pelitic schists of P2 grade into cyclic units of silicate-facies iron formation and layered quartzites of the P3 member. The lower section of the P3 member is garnet- and biotite-rich near the base and these minerals decrease in abundance up section, likely reflecting a reduction in the pelitic component (Fig. 7C). The uppermost part of the Pipe Formation is marked by P3 upper silicate-facies iron formation that comprises quartz, biotite, garnet, calcic amphibole, ortho- and/or clinopyroxene, fayalite, carbonate, and magnetite with minor graphite and sulphide (Fig. 7D).

The Pipe Formation is interpreted to represent periods of prolonged chemical sedimentation and nondeposition of clastic material, alternating with periods of turbiditic sedi-

FIGURE 6 CONTINUED. **(C)** Sulphide-facies iron formation (known as the lower sulphide iron formation) with thin pyrrhotite layers intercalated with graphitic chert or schist from Pipe Pit **(D)** Isoclinally folded lower silicate-facies iron formation from Pipe Pit.

FIGURE 7. Photographs of the Ospwagan Group from Pipe Pit. **(A)** P2 member of pelitic schists with sulphide-rich lamellae and semimassive to massive pyrrhotite-pyrite layers. **(B)** Muscovite-biotite or biotite schists of the P2 member with abundant aluminous porphyroblasts (garnet, staurolite, or sillimanite, depending on metamorphic grade). Commonly intercalated with thin layers of chert, calc-silicate or silicate-facies iron formation.

FIGURE 7 CONTINUED. **(C)** Layered quartzites of the P3 member. **(D)** Upper silicate-facies iron formation of the P3 member.

mentation (Bleeker, 1990).

Setting Formation

The Setting Formation is characterized by the absence of iron formations and is composed of interlayered quartzites and pelitic schists that coarsen upward into chemically immature wackes and, locally, pebbly conglomerates with occasional graded beds. The Setting Formation has been referred to as the "core quartzites" (Fig. 8A), as they represent the core of the downward-facing sequence of the Thompson mine structure (Bleeker, 1990). Minor mafic volcanic rocks (flows and tuff) are intercalated with the clastic rocks near the top of the Setting Formation. The Setting Formation is interpreted to represent a magmatically active foredeep, with clastic material derived from a tectonically active hinterland (Bleeker, 1990).

Bah Lake Assemblage

The mafic metavolcanic rocks overlying the Setting Formation have been referred to as the Ospwagan Formation (Bleeker, 1990), but have recently been renamed the Bah Lake Assemblage (Zwanzig, 2004). Although these rocks were interpreted by Bleeker (1990) to stratigraphically overlie the Setting Formation, many of the diamond-drill core intersections of mafic rocks cannot be stratigraphically correlated with any confidence. Nevertheless, these rocks have been separated into three broad members based on petrographic textures and chemical composition: BL1 metabasalt, BL2 magnesian metabasalt, and BL3 metapicrite (metamorphic olivine and 12-18 wt.% MgO; see discussion by Kerr and Arndt, 2001).

The paucity of continuous exposures of the Bah Lake Assemblage makes inferences about stratigraphic correlation between the members (BL1, BL2, and BL3) difficult. The best exposures of the Bah Lake Assemblage are sporadically located between Setting and Mystery lakes, and consist predominantly of pillowed and massive flows (Fig. 8B,C), but basal units commonly contain heterolithic mafic breccias.

Metabasalts of the BL1 member are pillowed and massive flows with minor fragmental rocks. In outcrop, they are fine grained, grey to black in colour, and are commonly foliated. The BL2 member is often spatially associated with BL1, but is more magnesian in composition. BL2 magnesian metabasalts are commonly pale grey, massive with pillow-like structures, and are generally characterized by abundant porphyroblastic olivine. Metapicrites of the BL3 member are closely associated with the BL1 and BL2 members at Upper and Lower Ospwagan Lakes, forming pillowed to massive flows with a characteristic pale green colour. Pillowed sequences have distinct pale cores with dark green selvages (Fig. 8D). In almost all instances, volcanic rocks of the Bah Lake Assemblage have been metamorphosed to amphibolite facies (Peredery, 1979).

Ultramafic Units

The stratigraphic relationships between the ultramafic bodies of the TNB and the host Ospwagan Group metasedimentary and metavolcanic rocks have been difficult to estab-

FIGURE 8. Photographs of the Ospwagan Group and Thompson Nickel Belt volcanic lithologies. **(A)** 'Core quartzite' of the Setting Formation from Setting Lake. **(B)** Pillowed metabasalts from Ospwagan Lake. **(C)** Pillowed metabasalts from the Mid Lake area. **(D)** Pillowed ultramafic volcanic rocks in the Mystery Lake area.

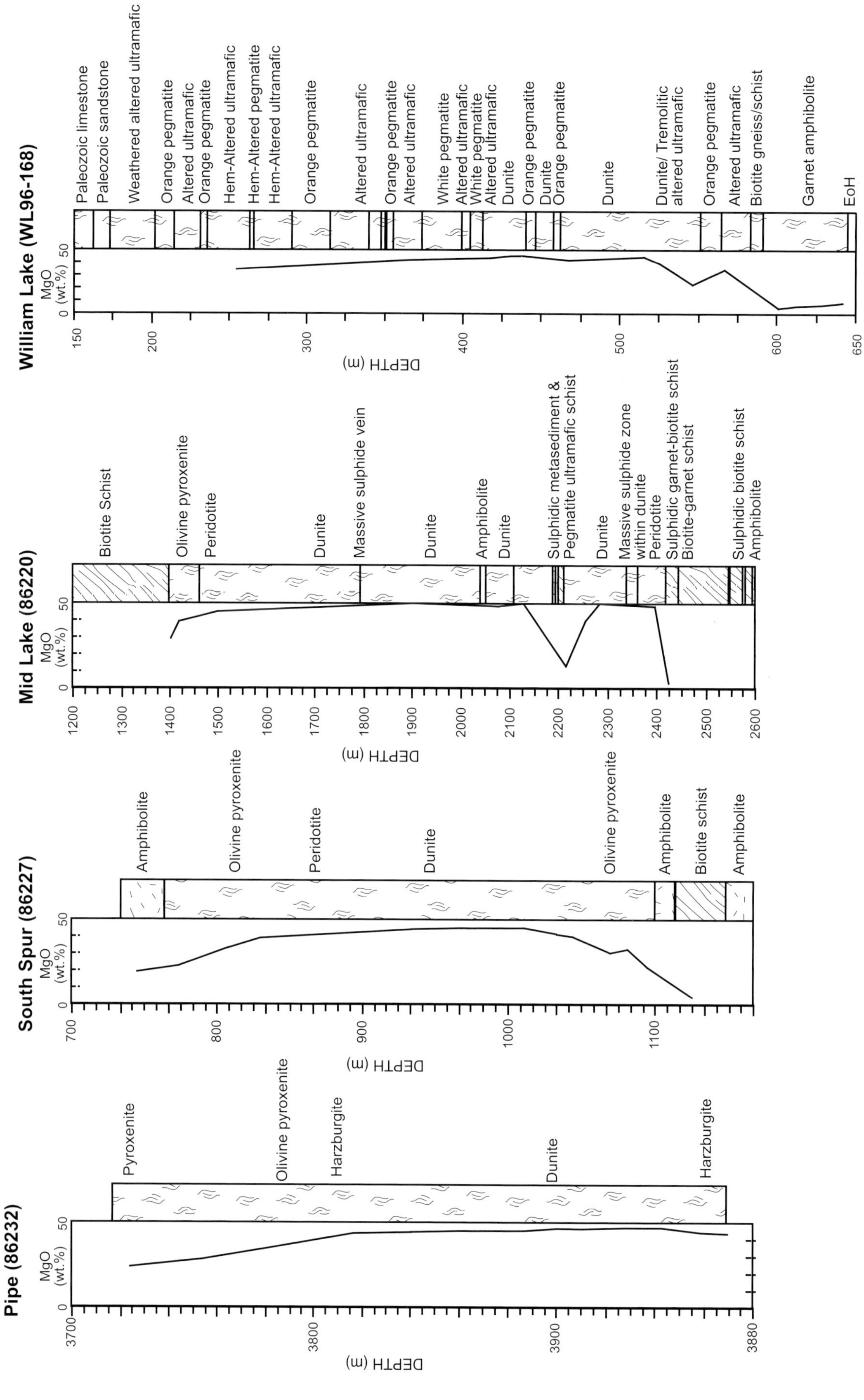

Figure 9. Graphical log and geochemical profile through the William Lake (DDH Wl96-168), Mid Lake (DDH 86220), South Spur (86227), and Pipe (86232) ultramafic bodies from the northern, central, and southern Thompson Nickel Belt. MgO data reflect primary compositional variation within the ultramafic body (modified after Burnham et al., unpub. CAMIRO report, 1999).

lish because of their limited exposure, variable host-rock metamorphism, and discontinuous nature. Owing to their high degree of serpentinization, the ultramafic bodies in the exposed northern and central TNB have experienced intense glacial erosion and form recessive topographic features that are now partially to completely filled by lakes. Extensive drilling based on geophysical surveys by Inco, Falconbridge, and Hudson Bay Mining and Smelting have shown that most ultramafic bodies occur as discrete and disjointed boudin (Figs. 2, 3). The geometry and internal structure of many of the ultramafic bodies is still poorly constrained and, in many cases, only one-dimensional information is available.

Ultramafic bodies are generally lensoid to tabular in shape and have tectonized contacts with the country rocks. Although the morphology of the ultramafic bodies is a product of tectonism, deformation is largely restricted to the margins, with the degree of primary textural preservation increasing away from the margins toward the cores of the ultramafic bodies.

The ultramafic bodies appear to have experienced multiple phases of deformation, similar to the surrounding country rocks, indicating that the ultramafic bodies were emplaced after, or perhaps during, sedimentation but prior to initial phases of deformation (Cumming et al., 1982; Peredery, 1982; Bleeker, 1989, 1990). In most cases, tectonism and pegmatite intrusions along the margins of the ultramafic bodies have obscured original contacts. However, in rare examples, primary intrusive contacts between the ultramafic rocks and the surrounding sediments are preserved.

Larger, less tectonized ultramafic bodies commonly exhibit an asymmetric variation in cumulus olivine content and whole-rock MgO content (Fig. 9) characterized by a thin pyroxenitic basal zone that grades into a thicker lower zone of chromite-bearing olivine peridotites and dunites that become progressively less olivine-rich and grade upward into a thinner pyroxenitic upper zone. Similar asymetric trends have been observed in many intrusive and extrusive ultramafic and mafic bodies in less deformed areas such as Kambalda, Western Australia (Lesher et al., 1981; Donaldson, 1983; Lesher, 1989); Dumont, Quebec (Duke, 1986); Pechenga, Russia (Hanski and Smolkin, 1989); Mouth Keith, Australia (Hopf and Head, 1998); and Forrestania, Western Australia (Porter and Mckay, 1981), where the upward decrease in olivine content correlates with stratigraphic younging direction. In areas of the TNB where the facing direction of the host sedimentary rocks are well established, the petrographic and geochemical variations within the ultramafic bodies are consistent, indicating concordant to semiconcordant emplacement into the sedimentary rocks. Unfortunately, insufficient mineral compositional data are available to establish whether the olivine-rich lower portions reflect flow differentiation of intratelluric olivine or preferential accumulation of in situ crystallized olivine.

Some ultramafic bodies are more strongly differentiated. Within many intrusions from the central TNB, the upper stratigraphic portions grade from chromite-undersaturated olivine cumulate rocks into plagioclase- and garnet-bearing amphibolite horizons that can be traced over several 100s of metres. The plagioclase- and garnet-bearing amphibolite horizons most likely represent more evolved liquids (basaltic/gabbroic) that solidified in the upper parts of the sill and were subsequently metamorphosed to amphibolite.

Mineralogy and Petrology

The ultramafic bodies in the TNB have experienced deformation and metamorphism under upper amphibolite facies and serpentine±carbonate alteration that resulted from the infiltration of post-peak metamorphism volatile-rich fluids from metasedimentary country rocks. Although almost all igneous textures and minerals have been destroyed along the deformed margins of the bodies, primary igneous textures and relict magmatic minerals are commonly preserved in the cores of the ultramafic bodies. However, as even the most deformed rocks in the TNB can be identified on the basis of metamorphic mineralogy and whole-rock geochemical composition, all of the rocks have been classified in terms of the textures and mineralogy of the igneous protoliths.

On the basis of relict igneous textures and/or whole-rock geochemistry, the ultramafic rocks in the TNB can been subdivided into three main lithologies: metadunites, metaperidotites (including harzburgites and rare wehrlites), and metapyroxenites (both orthopyroxenites and clinopyroxenites). Many of the ultramafic bodies appear to have preserved a primary igneous zoning and layering, where olivine modal abundances vary systematically through individual drill-core intersections.

Metadunites

Metadunites in the TNB are generally serpentinized olivine accumulate rocks with >90% fine- to medium-grained serpentine pseudomorphs after olivine in a fine-grained matrix of amphibole±chlorite±serpentine±chromite. Despite pervasive alteration to mesh-textured serpentine, altered olivine grains commonly preserve part of their original morphology, including grain outlines and irregular fracture patterns. Olivine grains are generally equant to elongate, with either curvilinear grain boundaries that meet in triple-point junctions and define a close-packed mosaic texture (Fig. 10A) or with ovoid, occasionally idiomorphic shapes (Fig. 10B). Olivine grain sizes vary between 0.2 and 10 mm, but are generally ~1 mm. Some samples exhibit a weak orientation or flattening of olivine grains. Magnetite defining the rims of altered olivine grains indicates that the dunites initially contained <10% intergranular material. In addition to the fine-grained magnetite generated during serpentinization of olivine, the most altered dunites commonly contain stockworks of magnetite and/or carbonate veins. Most of the metadunites appear to have originally contained minor amounts of fine-grained euhedral intercumulus chromite (Fig. 10C,D). It is unclear if these primary relict chromites experienced Fe-Mg interchange with olivine during cooling. However, due to the small proportion of chromite, it seems unlikely that olivine chemistry would have been substantially altered by Fe-Mg interchange.

Unserpentinized dunite cores occur in the central parts of many of the TNB ultramafic bodies. Such dunites are composed of more than 95% olivine with almost complete boundary contact between the grains and relatively little intergranular material (Fig. 10A,B), suggesting that they might be less altered not only because they are in the cores of the bodies and therefore further from the infiltrating-react-

FIGURE 10. Photomicrographs of textures and mineralogy of olivine cumulate rocks from the Thompson Nickel Belt. **(A)** Olivine cumulate, Mid Lake (DDH 86291) PPL. **(B)** Partially serpentinized ovoid to idiomorphic, cumulate olivines PPL. **(C)** Subhedral to anhedral chromite inclusions within olivine grains, Hambone East (DDH 74289) RL. **(D)** Euhedral intercumulus chromite grains and sulphide bleb, Mid Lake (DDH 86291) RL. **(E)** Serpentinized metaperidotite, South Spur (DDH 86227) PPL. **F)** Partially serpentinized peridotite, North Manasan (DDH 89299) PPL = plane polarized light, TL = reflected light. Ccp = chalcopyrite, Chr = chromite, Pn = pentlandite, Po = pyrrhotite.

ing fluids, but also because they were less permeable than the other rocks. Unaltered dunites commonly contain fine-grained disseminated chromite (1-2%, 60-100 μm), which may occur both as subhedral to anhedral inclusions within the olivine grains and as euhedral intercumulus grains. In addition to coarse chromite inclusions, many of the relict olivines within the dunites contain finely disseminated chromite micro-inclusions or inclusion trails, from which the origin of the relict olivine may be discerned. In many cases, the compositions of these micro-inclusions are similar to those of the cores of the coarse chromite inclusions and intercumulus grains, and coarse-grained intercumulus chromites inclusions from relatively unmetamorphosed komatiite flows (Barnes and Brand, 1999), suggesting that they have not interacted with either serpentinizing fluids or any trapped melt phases. On this basis, the chromite micro-inclusions are interpreted to represent primary igneous inclusions, the unaltered olivines in these samples are inferred to be of magmatic origin, and their compositions are interpreted to reflect the compositions of the parental magmas. In a number of cases, the micro-inclusions are either significantly more aluminous or ferric iron-rich than would be expected for magmatic chromite grains (Fig. 10C). In such cases, the inclusions (and by association the host olivines) are inferred to have formed through dehydration reactions involving alteration minerals formed at lower metamorphic grades, the relict olivines are inferred to be of metamorphic origin, and their compositions can not be used to estimate the compositions of the parental magmas.

Metaperidotites

Metaperidotites in the TNB are composed of 40 to 90% fine- to medium-grained, variably serpentinized subhedral olivine grains in a matrix of fine-grained chlorite±calcic amphibole±serpentine with trace amounts of fine-grained chromite (Fig. 10E). The replacement of the interstitial material by chlorite±amphibole±serpentine suggests that many of the peridotites were originally composed of olivine and pyroxene (O'Hanley, 1996) and the presence of calcic amphibole in these intergrowths indicates that the pyroxene was originally Ca-rich clinopyroxene (i.e. that the rocks were originally wehrlites) rather than Ca-poor orthopyroxene (i.e. the rocks were originally harzburgites). Although komatiitic olivines crystallized in rapidly cooled extrusive environments may contain significant amounts of Ca (see review by Lesher, 1989), the olivines in the ultramafic intrusions in the TNB appear to have crystallized in slowly cooled intrusive environments and have low Ca contents. The high Fo contents of the relict igneous olivines (up to Fo94) suggest that they crystallized from magmas containing up to 22% MgO, indicating that the magmas were truly komatiitic.

Although orthopyroxene-bearing olivine cumulate rocks have been reported in the ultramafic bodies of the TNB by previous workers, predominantly at the Thompson mine (De Saboia, 1978; Peredery, 1979, 1982; Paktunc, 1984; Paktunc et al., 1984), very few harzburgites were identified either petrographically or geochemically in the 2093 samples of ultramafic rocks that were studied in CAMIRO project 97E-02 (Burnham et al., 2003). Those identified were generally poikilitic olivine mesocumulates, in which the orthopyroxene forms oikocrystic grains enclosing cumulus olivine. The paucity of orthopyroxene outside the Thompson mine area may reflect either the higher metamorphic grade in the Thompson area or a difference between the amount of wall-rock assimilation during the formation of the Thompson and other ultramafic bodies. Whereas the ultramafic bodies within the Thompson mine experienced upper amphibolite- to granulite-facies metamorphism, during which the assemblage olivine±talc±anthophyllite or chlorite would have been unstable and reacted to form either orthopyroxene or orthopyroxene±olivine±spinel (Paktunc, 1984), the majority of the ultramafic bodies in other parts of the TNB experienced only middle to upper amphibolite-facies metamorphism, during which olivine±talc, olivine±anthophyllite, or chlorite are stable phases. Alternatively, because most of the TNB ultramafic bodies outside the Thompson mine area (i.e. northern TNB) are either nonmineralized or subeconomically mineralized intrusions, the presence of orthopyroxene in the Thompson ultramafic bodies may reflect greater felsification of the magma via assimilation of wall rocks, potentially inducing sulphide saturation and orthopyroxene crystallization (e.g. Irvine, 1975).

Partially serpentinized peridotites, in which the cores of olivine grains are partially preserved, have been documented in only a few ultramafic bodies (e.g. North Manasan and Spur South: Fig. 10F). These rocks are characterized by <90% fine- to medium-grained, partially serpentinized olivines within a matrix of >10% fine-grained chlorite±amphibole±serpentine. The relict olivines are ovoid, between 0.1 and 3 mm in diameter, and contain opaque micro-inclusions similar to those present in the dunites.

Metapyroxenites

Olivine metapyroxenites occur along the margins of the ultramafic bodies and as layers within the bodies. They are composed primarily of fine- to medium-grained olivine and/or pyroxene altered to either mesh-textured serpentine or porphyroblastic calcic amphibole in a matrix of acicular chlorite±amphibole±serpentine. In most cases, olivine is completely serpentinized. However, relict olivine fragments are preserved in olivine pyroxenites along the margins of some ultramafic bodies.

Although pure orthopyroxenite and clinopyroxenite layers are preserved in a few of the ultramafic bodies, such layers are relatively rare in the TNB. In most cases, the pyroxenites have been altered to an assemblage of either tremolite or hornblende±plagioclase (clinopyroxenites) or tremolite± anthophyllite±orthopyroxene (orthopyroxenites) and can only be easily recognized from their whole-rock chemical compositions.

Structure

Several models have been proposed to explain the structure of the TNB (e.g. Fueten and Robin, 1989; Bleeker, 1990; Fueten, 1992; White et al., 1999; Gapais et al., 2005). The main differences between the models are the relative weights given to the fold geometry of stratigraphic units versus finite strain and kinematic analysis. Interpretations of superimposed fold structures in terms of successive deformation phases has led to the development of a model involving early nappe tectonics followed by northeast-directed transpression (Bleeker, 1990). However, strain and kinematic analysis has led to the development of a southeast-side-up kinematic model (Fueten and Robin, 1989). Both approaches are limited by the paucity of outcrop in the TNB and given that the different models arise from the application of a different range of structural techniques, a consensus on the evolution of the TNB has not been reached to date.

Geochronological data (Machado et al., 1990) are not compatible with a simple nappe model, because the dispersion of ages (1836-1720 Ma) militated against a single major metamorphic melting event of regional extent, as may be expected and has been observed in areas of known thrust tectonics. Nevertheless, a protracted history of polyphase deformation similar to that observed in the adjacent part of the Trans-Hudson Orogen, and characterized by more than one period of metamorphism related to multiple continental collisions, remains a viable tectonic model (Zwanzig, 1999, and references therein). Although an alternative, transpressional

TABLE 2. Summary of dominant Thompson Nickel Belt structure: comparison of Bleeker (1990, shaded) and Burnham et al. (2003, white background).

	Age	Folds	Faults/Shears	Kinematic Indicators	Schistosity	Lineations	Metamorphism	Interpretation
D_1	Predates the mafic dykes in the Pipe Pit	Recumbent, isoclinal, in Ospwagan Group and basement rocks	Major movement zone at top of basement		S_1 schistosity and boudinage are sub-horizontal, layer parallel, axial planar to F_1 folds		Prograde, up to amphibolite facies	Nappe development, SE verging if Moak Lake structure is a syncline
D_1	Predates the mafic dykes in the Pipe Pit	Original style unknown (probably upright), involves mainly Ospwagan Group	Inferred, but not directly observed, east-verging if present		Possible S_1 ENE-trending fracture cleavage in Bah Lake assemblage		Uncertain, probably low grade	Possible recumbent folds may be related to extensional tectonic, and/or upright folds leading to tectonic burial
D_2	Postdates the mafic dykes in the Pipe Pit	Recumbent, tight to isoclinal; folded F_1 folds, mafic dykes, sillimanite lenses and earliest pegmatites	Shear zones mainly in metasedimentary rocks and adjacent basement	Intrafolial fold packets suggest non-coaxial strain	S_2 schistosity present in some fold hinges but coplanar with S_1 in fold limbs		Middle to upper amphibolite	Continued recumbent deformation near peak metamorphism
D_2	Syn- and post-Kisseynew sedimentation S_2 cut by >1818 Ma granite	Recumbent, D_2 has refolded and tightened F_1 folds and may have reoriented them from upright to recumbent	Superior Boundary fault zone, local SW-verging thrust faults along the boundary	Probably dextral, NE-over-SW rotation of early porphyroblasts	S_2 schistosity, gneissosity and boudinage are the regional foliation, subparallel to layering and Archean foliation	L_2 biotite crenulations have variable plunges	Middle to upper amphibolite, possible lower granulite	Development of recumbent structure, SW thrusting of Ospwagan Group over Kisseynew gneisses, vergence changes to SW, possibly W
D_3	1770 Ma pegmatite coeval with F_3 folds	Main set of folds, NE trends, fold the peak metamorphic minerals, axial planes vertical to steeply SE dipping, axes plunge to S and NE	NE-trending mylonites (forming late D_3 and continuing through D_5)	Indications of mainly flattening, shear related to flexural slip	NE-trending fabric, S_3 absent or weakly developed as axial planar to F_3 folds	L_3 crenulation and intersection lineations, parallel to flexural-slip striations	Retrograde, amphibolite to greenschist facies	Strong NW-SE shortening with minor independent dip-slip components; left-stepping, NE-trending, en echelon folds indicate sinistral transpression.
D_3	As above	As above	Mylonites, NNE-trending, steeply SE-dipping. Conjugate faults, N to NNE and ENE	ESE-up shear bands with sinistral component. Sinistral and dextral faults	S_3 NE- trending, steeply SE-dipping, best developed in tight F_3 fold hinges and shear zones	L_{3-4} is a weak down-dip stretching lineation, variable plunges in the southwest	?Granulite to greenschist facies	Mainly NW-SE shortening; regional, SE-side-up sinistral transpression, vertical stretching. Late-D_4 NW-SE compression with NE extension
D_4		As below					Greenschist facies	
D_3-D_4*		Minor NE-trending, sub-vertically plunging folds in F_3 fold limbs and mylonitic foliation (including sheath folds)	NE-trending low temperature mylonites and brittle faults; late SE-trending faults	Flattening and dip-slip, also sinistral shear zones in the west, but late (D_4) E-SE dextral slip	Mylonitic foliation overprinted on F_3 fold limbs	Weak down-dip stretching lineation; late horizontal striae	Retrograde greenschist facies	Modifies older structures, local E-SE dextral faults overprint NE sinistral faults

* F_4 and F_6 of Bleeker (1990) are minor fold phases only.

tectonic model is easier to reconcile with the geochronological data (Gapais et al., 2005), a complete transpressional model for the TNB has not been presented using high-resolution geochronological methods. A comparison of differing tectonic models for the TNB is presented in Table 2.

Mineral Deposits and Occurrences

Sulphides occur in many of the rocks in the Thompson Nickel Belt, but are most abundant in sulphide-facies iron formations and associated metasedimentary rocks of the Pipe Formation and in ultramafic bodies intruded into the Pipe Formation. Because many sulphide phases, particularly chalcopyrite and pyrrhotite, are more ductile at high temperatures compared to many silicate phases, sulphide-rich rocks are more easily deformed than associated silicate-rich rocks (see reviews by Marshall et al., 2000). Although the Ni-Cu-(PGE) sulphides in the TNB should have homogenized to monosulphide solid-solution (MSS) at such high metamorphic grades (upper amphibolite to granulite facies) and recrystallized to low-temperature mineral assemblages during retrograde metamorphism, macroscopic and microscopic sulphide deformation structures, fabrics, and textures are well preserved in the TNB, recording a complex polyphase deformational history (see Bleeker, 1990). Although mineralization in the TNB is more deformed than those in most other magmatic Ni-Cu-(PGE) deposits, and in many cases the mineralization is now hosted within metasedimentary rocks, the mineralogy and geochemistry are similar to other komatiite-associated magmatic Ni-Cu-(PGE) deposits (Lesher, 1989; Lesher and Keays, 2002). Primary (i.e. non-supergene) ores are dominated by pyrrhotite-pentlandite, pyrrhotite-pentlandite-pyrite, and less common pentlandite-pyrite and pentlandite-pyrite-millerite assemblages (Bleeker, 1990; Liwanag, 2000). Chalcopyrite, magnetite, and chromite are ubiquitous minor phases.

Thompson Deposit

The Thompson ore is stratabound and occurs within the pelitic schist unit of the Pipe Formation (Fig. 2). The Thompson structure consists of nearly upright, variably plunging, high-amplitude folds interpreted to have formed during an F3 folding event in the Thompson Nickel Belt (Bleeker, 1990). Reworked basement gneisses occur on the outer limbs of the structure, while the Ospwagan Group supracrustal rocks occur in the core.

Birchtree and Pipe Deposits

The Birchtree and Pipe deposits share similar lithological and structural settings. The ultramafic rocks of the Birchtree deposit are interpreted to be part of a discontinuous array of ultramafic bodies that stretch from the Pipe II open pit to the Birchtree mine (Bleeker, 1990) (Figs. 1, 2). The discontinuous ultramafic bodies from the Birchtree deposit and the Pipe deposits occur in similar stratigraphic positions within the Ospwagan Group supracrustal rocks, i.e., within the graphitic sulphide-facies iron formation of the P1 member of the Pipe Formation. The massive and semimassive sulphide mineralization in the Birchtree and Pipe deposits has undergone a lesser degree of deformation relative to that in the Thompson deposits. Massive sulphide accumulations within the Birchtree and Pipe deposits tend to occur as ultramafic breccias consisting of peridotitic to dunitic clasts in a sulphide matrix. These breccias are typically proximal to ultramafic boudins and are not stretched within the metasedimentary horizon to the same extent as in the Thompson deposits.

William Lake Deposit

The William Lake deposit lies beneath Paleozoic carbonate rocks in the southern portion of the Thompson Nickel Belt (Fig. 1, 3). Intrusion-hosting metasediments include quartzites, pelites, calcareous metasediments, iron formations, and graphitic schists that are similar to the Thompson and Birchtree deposits. As such, the ultramafic bodies are interpreted as being hosted by the Ospwagan Group metasediments (Macek et al., 2002). The ultramafic bodies consist of pyroxenite, peridotite, and dunite. No work has been published to place the exact stratigraphic position of theses ultramafic bodies within the Ospwagan Group lithostratigraphic section of Bleeker (1990), however, maps published by the TNB working group (Macek et al., 2002), indicate that they reside within the Pipe Formation. The ultramafic bodies and the hosting metasediments are concentrated on the southeastern margin of a granitic intrusion, the William Lake Dome (Fig. 3). Numerous centimetre- to metre-scale granitic pegmatite veins, interpreted to be related to the intrusion of the William Lake Dome, cross-cut the older Ospwagan Group rocks and the ultramafic bodies in many drill-core samples (Fig. 9).

The mineralization in the William Lake area typically occurs as disseminated sulphides within the ultramafic bodies, and locally grades to centimetre- to decimetre-scale massive and semimassive sulphide accumulations toward the base of bodies. Banded sulphides, sulphide stringers, and disseminated sulphides within metapelites and iron formations are commonly encountered in diamond-drill cores.

Mineralization Types

Komatiite-associated Ni sulphide deposits, such as those in the TNB, are part of a continuum of lithotectonic associations in the family of magmatic Ni-Cu-(PGE) deposits, which contains a variety of mineralization types traditionally classified by ore morphology and ore host rocks (Lesher and Keays, 2002): Type I stratiform basal massive/net-textured/disseminated mineralization, Type II stratabound internal disseminated mineralization, Type III stratiform 'reef-style' mineralization, Type IVa Ni-enriched metasediment mineralization, Type IVb hydrothermal mineralization, and Type V tectonic 'offset' mineralization.

Deformation within the Thompson Nickel Belt mineralization has led to more detailed classifications based on ore textures (disseminated versus inclusion-bearing versus massive) and host rocks (i.e. ultramafic versus metasediment) (Zurbrigg and Patterson, 1963; Peredery, 1982; Bleeker, 1990; Larocque et al., 1998; Liwanag, 2000). The classification scheme presented in this review reflects the most recent descriptions of macroscopic, mesoscopic, and microscopic textures in sulphide mineralization, and the character of the host rock in which the sulphide mineralization resides (Bleeker, 1990; Liwanag, 2000). The abbreviations for the nomenclature in this classification scheme include a prefix indicating the texture of the sulphides (D = disseminated,

TABLE 3. Sulphide ore types, textures, distributions, and host rocks in the Thompson Nickel Belt (adapted and expanded from Bleeker (1990) and Liwanag (2000).

Sulphide Abundance	Sulphide Texture	Rock Texture	Sulphide Distribution	Host Rock: Meta-Sediment	Host Rock: Ultramafic Breccia	Host Rock: Ultramafic Rock
				Extraparental*		Intraparental*
≤10%	Disseminated	diverse	interstitial	DS-SED D$-SED	D$-UB	D$-U
10-40%	Heavy Disseminated Patchy Disseminated Blebby Disseminated	diverse	interstitial to intercumulus	HDS-SED HD$-SED	HD$-UB PD$-UB	HD$-U PD$-U B$-U
	Layered	laminated to banded	interstitial to intergranular	LS-SED L$-SED	—	—
40-70%	Net-Textured	meso- to adcumulate	intercumulus	—	—	N$-U
	Semi-Massive	schistose, gneissic, mylonitic	intergranular	SMS-SED SM$-SED	SM$-UB	—
>70%	Massive	cataclastic, blastomylonitic	—	MS-SED M$-SED	M$-UB	—

Notes: All textures and abundances are gradational. Abbreviations based on textural type: B = brecciated, D = disseminated, HD = heavily disseminated, L = laminated, M = massive, N = net textured, PD = patchy disseminated, SM = semimassive; mineralization status: $ = mineralized and S = unmineralized; and a suffix indicating the nature of the host rock: U = ultramafic, UB = ultramafic breccia, SED = metasediments.
* nomenclature after Bleeker, 1990.

HD = heavily disseminated, N = net textured, SM = semimassive, M = massive, L = laminated), the uneconomic or economic mineralization (S and $, respectively), and a suffix indicating the nature of the host rock (U = ultramafic, UB = ultramafic breccia, SED = metasediments) (Table 3). Figure 11 shows the general location of the sulphide types in relation to the TNB ore deposit model of Bleeker (1990). In general, the "intraparental" ores of Bleeker (1990) and D$-U, L$-U, and M$-UB ores of Liwanag (2000) correspond to Types I-II ores of Lesher and Keays (2002), whereas the "extraparental" ores of Bleeker (1990) and D$-SED, SM$-SED, and M$-SED of Liwanag (2000) correspond to Type V ores of Lesher and Keays (2002).

Barren disseminated sulphides in metasediments (DS-SED) are composed of primarily fine- to medium-grained, foliated pyrrhotite patches or euhedral grains of pyrite, or both (Fig. 12A). Sulphides adjacent to or near ultramafic rocks may also contain very fine-grained, angular to rounded patches of pentlandite and/or chalcopyrite.

Barren semimassive and massive sulphides in metasediments (SMS-SED and MS-SED) occur as millimetre- to decimetre-scale layers of pyrrhotite (Fig. 12B). Pyrrhotite is typically fine-grained and annealed, and comprises both hexagonal and monoclinic polymorphs. The sulphides do not contain macroscopic pentlandite and are normally graphitic, commonly containing up to 25% graphite. Graphite occurs as foliated single inclusions in sulphides or as aggregates associated with foliated biotite flakes and rounded to subrounded quartz and plagioclase. Chalcopyrite, pyrite, and magnetite are present in minor to trace amounts. Chalcopyrite typically occurs as very fine-grained blebs within pyrrhotite and as 'porous' masses adjacent to silicate inclusions. Pyrite occurs as subhedral aggregates replacing pyrrhotite. Magnetite occurs as randomly distributed, fine-grained, anhedral grains.

Heavily disseminated or mineralized semimassive to massive sulphides in metasediments (HD$-SED, SM$-SED, and M$-SED) occur at the contacts between ultramafic bodies and barren metasediments (interpreted to be largely in place)

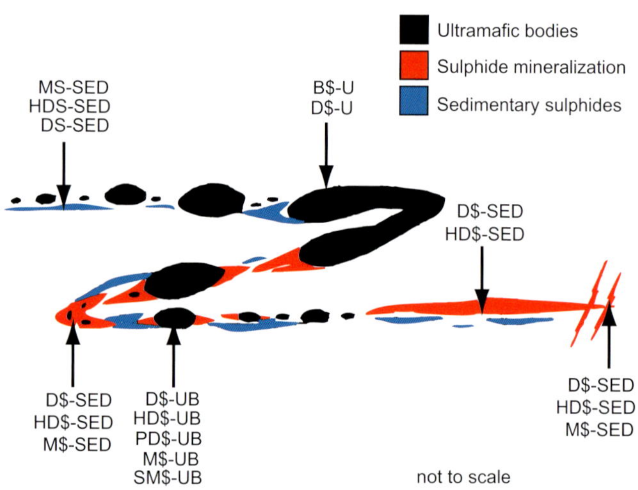

FIGURE 11. Schematic geometry of ultramafic bodies and mineralization in the Thompson Nickel Belt (adapted from Bleeker, 1990). Arrows indicate generalized locations of mineralization types presented in Table 3. Abbreviations based on textural type, host rock, and mineralization status: D = disseminated, HD = heavily disseminated, N = net textured, SM = semimassive, M = massive, L = laminated, PD = patchy disseminated, S = mineralized and $ = nonmineralized, and a suffix indicating the nature of the host rock: U = ultramafic, UB = ultramafic breccia, SED = metasediments.

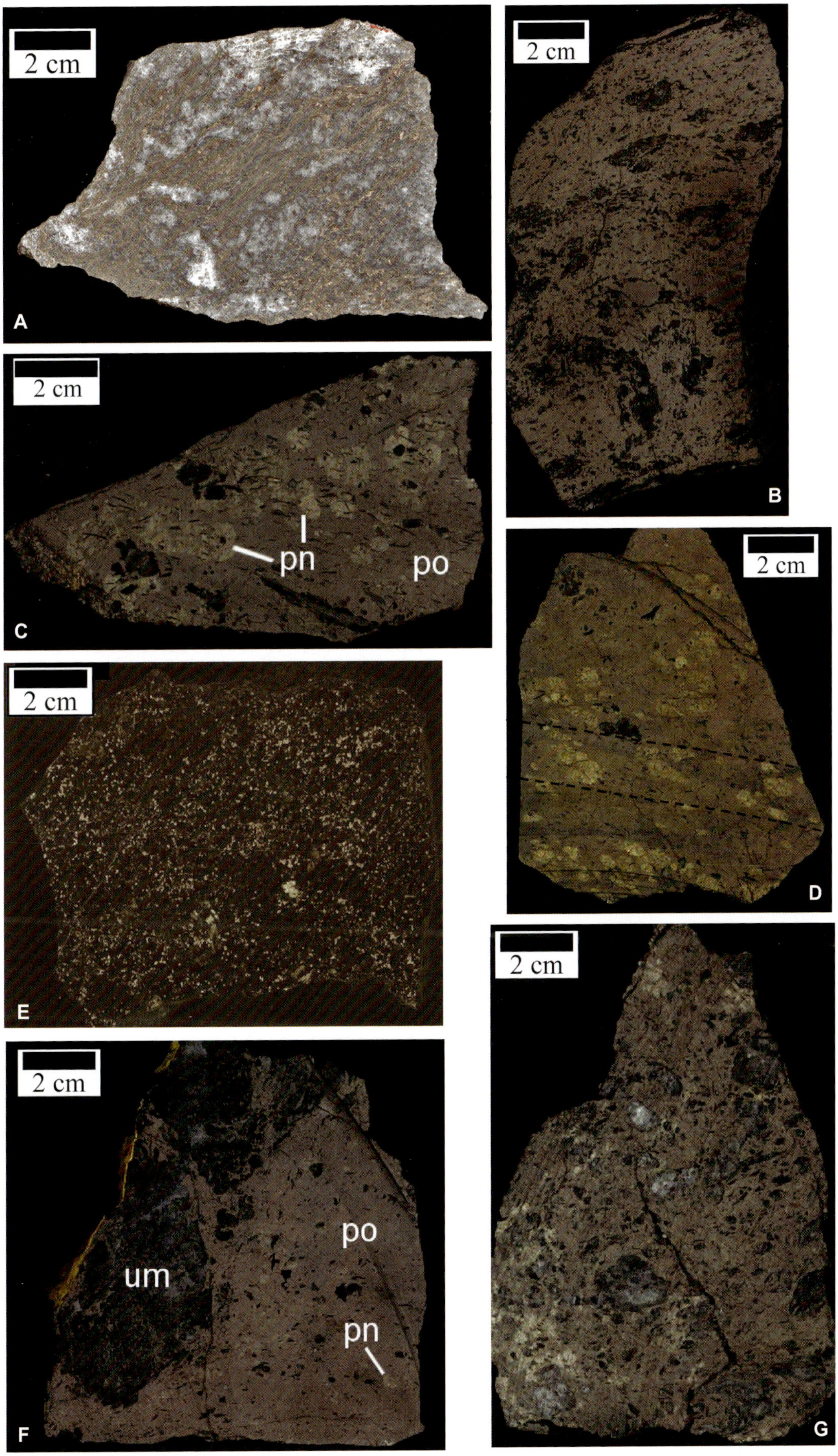

FIGURE 12. Textures representing sulphides in metasedimentary and ultramafic rocks (Photographs taken from Liwanag 2000). **(A)** D$-SED: mineralized biotite schist from the Thompson 1D orebody. **(B)** MS-SED: Thompson T1 mine, showing massive pyrrhotite with foliated single crystal inclusions of graphite and biotite, and ovoid aggregates of plagioclase, quartz, graphite, and biotite. **(C)** M$-SED: Thompson T1 mine, showing coarse-grained annealed pyrrhotite and lighter coloured pentlandite eyes. Inclusions are foliated mafic mica and aggregates of bio -tite, quartz, and plagioclase. **(D)** M$-SED: Thompson 1C orebody showing coarse pyrrhotite and pentlandite eyes oriented parallel to foliated mica inclusions and schistose aggregates (highlighted by dashed lines). **(E)** D$-U: Birchtree mine showing serpentinized peridotite containing interstitial sulphides. **(F)** M$-UB: Birchtree mine showing centimetre-scale inclusions of serpentinite (um) and single crystal inclusions of mafic mica grains, quartz, and feldspar in massive pyrrhotite (po). Pentlandite (pn) occurs as medium-grained light coloured 'eyes'. **(G)** M$-UB: Thompson T1 mine, sample consisting of medium-grained annealed pyrrhotite and pentlandite with inclusions of pelitic schist fragments and serpentinized peridotite.

and isolated within barren metasediments (interpreted to have been mobilized from ultramafic contacts). They are composed of mainly pyrrhotite and pentlandite (Fig. 12C) with minor amounts of chalcopyrite, pyrite, magnetite, chromite, and gersdorffite [NiAsS]. Pyrrhotite occurs as annealed polygonal grains. Pentlandite occurs as i) very fine lamellae; ii) elongated, wispy 'chains'; and iii) coarse-grained, subangular to subrounded 'eyes' that are typically <1 to 5 mm in size, but range up to 2 cm in some horizons. In many massive sulphide horizons in the Thompson mine, pentlandite eyes are often segregated into centimetre-scale bands within a coarse-grained, annealed pyrrhotite matrix (Fig. 12D). Pyrite typically occurs as aggregates of subhedral grains that replace both pentlandite and pyrrhotite. Gersdorffite occurs as very fine, anhedral to subhedral grains. Magnetite and chromite occur as very fine-grained euhedral to subhedral grains randomly oriented in a sulphide matrix. Fine- to coarse-grained chalcopyrite typically occurs adjacent to pentlandite in stringers that project into the metasedimentary rocks along the contact. Chalcopyrite is also concentrated adjacent to silicate inclusions.

Mineralized disseminated and net-textured sulphides in ultramafic rocks (D$-U and N$-U) range from fine-grained interstitial disseminations through lobate intercumulus grains to coarse-grained interstitial networks (Fig. 12E). Disseminated sulphides typically occur in the interior parts of ultramafic bodies, whereas net-textured sulphides typically occur in the exterior parts of the bodies. The sulphides comprise mainly pyrrhotite and pentlandite with minor chalcopyrite and magnetite. Pentlandite and chalcopyrite typically form subangular to subrounded patches along the edges of pyrrhotite grains. Magnetite typically occurs as fine anhedral grains adjacent to sulphides and as lamellae in silicates.

Mineralized semimassive and massive sulphides in ultramafic breccias (SM$-UB and M$-UB) form the matrix component of the breccias (Fig. 12F). Breccia fragments are fine- to very coarse-grained, tectonized inclusions of the host ultramafic rocks. The textures of the sulphides are similar on a microscopic scale to those in metasedimentary rocks (see above). The sulphide mineralogy is dominated by pyrrhotite and pentlandite, with minor amounts of chalcopyrite, pyrite, chromite, and gersdorffite. Pyrrhotite occurs as annealed medium- to coarse-grained polygonal grains. Pentlandite occurs as i) very fine lamellae; ii) elongated, wispy 'chains'; and iii) coarse-grained, subangular to subrounded 'eyes'. Pentlandite eyes are typically <1 to 5 mm in size. Gersdorffite occurs as very fine, anhedral to subhedral grains.

Textural Variations between Ore Deposits

Mineralized semimassive to massive sulphide samples from the Birchtree, T1, 1C, 1D, and William Lake deposits were characterized in terms of host rocks, proximity to ultramafic boudins, inclusion populations, and macroscopic texture of pentlandite (Liwanag, 2000). Ores from the Birchtree mine are semimassive and massive sulphides in ultramafic breccias that occur in ultramafic boudin necks and at the bases of ultramafic bodies. They contain randomly oriented to oriented, medium- to coarse-grained, brecciated ultramafic inclusions and minor pelitic inclusions. Pyrrhotite, the principal sulphide phase, is associated with up to 10% fine- to medium-grained pentlandite (Fig. 12F). Owing to their closer proximity to ultramafic boudins, the Birchtree samples are interpreted to be closer in textural character to 'primary' magmatic massive sulphides compared to the samples from Thompson (Peredery, 1982; Bleeker, 1990; Liwanag, 2000).

Ores from the Thompson mine exhibit a spectrum of textures that vary between the T1 fold nose, the 1C orebody, and the 1D orebody, which reflect the highly tectonized nature of the mineralization. T1 ores are semimassive and massive sulphides in metasediments (SM$-SED and M$-SED), but are texturally similar to the Birchtree mineralization (Liwanag, 2000). Most mineralization is near ultramafic boudins, but some occur as massive sulphide lenses hosted by pelitic schists distant from ultramafic boudins. T1 ores typically contain randomly oriented to oriented, medium- to coarse-grained ultramafic and pelitic inclusions, and contain up to 25% medium- to coarse-grained pentlandite grains (Fig. 12G). Single crystal inclusions also occur in the T1 massive sulphides; the most abundant crystal inclusions are fine- to medium-grained foliated biotite flakes, and rounded quartz and plagioclase grains.

Ores from 1C are dominantly massive sulphides in metasediments (M$-SED), but also include massive sulphides in ultramafic breccias, and net-textured and disseminated sulphides in ultramafic boudins (M$-UB, N$-U, and D$-U). Thompson 1C massive sulphides in metasediments differ from T1 ores in that the pentlandite grains tend to be coarser and distributed into parallel, centimetre-scale bands within the ore horizons (Fig. 12D). Single crystals of biotite, quartz, and plagioclase dominate the inclusion population.

Ores from the 1D orebody are primarily massive sulphides in metasediments (M$-SED), which are hosted by pelitic and quartz-rich metasediments that have been intruded by late pegmatitic granite (Fig. 12G). The 1D orebody has undergone extreme and complex deformation (Bleeker, 1990). The 1D ores are unique compared to most other ores in the TNB; undulating contacts between rock units, numerous metre-scale shear zones, and abundant pegmatitic/tonalitic inclusions indicate more complex deformation. These ores are interpreted to be relatively more distal to ultramafic boudins than T1 and 1C ores; no interpreted ultramafic parent rocks have been observed within 250 m along strike of the 1D orebody (Liwanag, 2000). These ores typically contain up to 40%, well rounded, medium- to coarse-grained pegmatitic/tonalitic inclusions, and up to 10%, fine- to medium-grained pentlandite grains (Fig. 12G). Locally, the pentlandite grains are as coarse grained as in 1C ores, however, they are interpreted to occur in less deformed portions of the orebody.

Mineralization in the William Lake area occurs primarily as disseminated sulphide within ultramafic boudins (D$-U) and has textures similar to those in other stratabound disseminated Ni-Cu-(PGE) deposits, such as Mt. Keith (Groves and Keays, 1979) and Dumont (Eckstrand, 1975). Mineralization occupies interstitial spaces between olivine and occurs as very fine (μm-sized) inclusions in olivine. This suggests that these rocks were saturated in sulphide at an early to intermediate stage of their crystallization history. Minor massive sulphide hosted in ultramafic breccias and metasedimentary rocks (M$-UB, M$-SED) in the William

Lake area have macroscopic textures similar to those at Birchtree, T1, and 1C, but are much less deformed than those at 1D.

Exploration Methods

Historically, Inco Ltd. has been the largest land claim holder within the northern and central TNB. As such, Inco has been continually exploring the Thompson Nickel Belt over the past 50 years in a search for new sources of Ni feed. Early work by Inco geologists indicated that the ultramafic intrusion-hosting Ospwagan Group rocks have been metamorphosed to schists and gneisses and that the complex deformation history has produced complex folding patterns (summarized by Peredery et al., 1982). The realization that the Ni-Cu-(PGE) deposits in the TNB are associated with ultramafic intrusions that occur within a specific horizons in the Ospwagan Group rocks has provided a working exploration model that uses an integrated approach combining geology and geophysics. Preliminary maps using this integrated approach have been recently published for the TNB from Moak Lake in the north to Lamb Lake in the southern Thompson Nickel Belt (Macek et al., 2001).

Using this exploration approach, audio magnetotelluric (AMT) surveys are used to interpret the geology to depths of two kilometres. Promising stratigraphic intervals selected from AMT surveys are then investigated by surface UTEM surveys to a depth of approximately 600 m. Anomalous conductivity occurring in the favourable stratigraphic horizon (i.e. Pipe Formation sulphide-facies iron formation) is tested by diamond-drill holes, followed by borehole electromagnetic surveys (Inco Ltd., Annual report 2004).

Underground exploration within the operating mines in the TNB is influenced by the complex deformation history of the northern TNB. At this scale, exploration includes detailed structural analysis of the host rock and sulphide mineralization. Geological interpretations are tested by exploration diamond drilling supported by in-hole electromagnetic surveys (Inco Ltd., Annual Report, 2004).

Exploration in the southern TNB, using traditional geophysical techniques, is hampered by the southward-thickening Paleozoic cover rocks and the southwestern (south of Grand Rapids, Manitoba) formational brines (Gascoyne et al., 1987; McRitchie, 1995).

Knowledge Gaps

Timing and Level of Emplacement of Ultramafic Sills into the Ospwagan Group

The ultramafic compositions of the intrusions make them difficult to date absolutely, as they do not contain primary minerals suitable for radiometric dating. The only intrusion that has been dated is a differentiated ultramafic-mafic body at Setting Lake, which yielded a U-Pb zircon age of 1880 ± 2 Ma (Hulbert et al., 2005).

The ultramafic bodies occur throughout the lower and central parts of the Ospwagan Group stratigraphy up to the lower member of the Setting Formation. This may indicate that they were emplaced during the deposition of the Setting Formation or that their emplacement was limited to lower levels for physical reasons (e.g. buoyancy, rheology, and/or tectonics). This suggests that they must be younger than the 2.6 Ga age of the youngest detrital zircon in the Setting Formation (Burnham et al., 2003).

Thus, the only constraints on the absolute age for the ultramafic bodies as a group are that they may have been emplaced between 2.6 and 1.80 Ga. Given the magnitude of magmatism indicated by the prevalent 1880 to 1890 Ma mafic-ultramafic 'Molson' dykes in the adjacent Superior Province, it is most likely that the ultramafic sills were emplaced during this event. However, the absence of absolute age dates for the ultramafic intrusive event(s) and a firm understanding of the nature of the intrusive event(s) (i.e. multiple emplacement events and/or multiple levels in the Ospwagan Group), combined with the complex deformational history, hampers the development of more detailed stratigraphic interpretations.

Tectonic Structure and History

Contrasting interpretations of the TNB structure and tectonic history have recently been presented (Bleeker, 1990; Zwanzig and McRitchie, 1997; Zwanzig, 1999; Gapais et al., 2005). Structural mapping using stratigraphic markers (e.g. Bleeker, 1990; Zwanzig and McRitchie, 1997) suggest that the TNB structure is dominated by upright folds superposed on westward-verging isoclinal recumbent folds. However, strain and kinematic analysis of shear zones supported by U-Pb ages of syn-kinematic pegmatite intrusions (e.g. Fueten, 1992; Potrel et al., 2000; Gapais et al., 2005) suggests that the structure is dominated by north-northeast-trending reverse shear zones (southeast-side up) in a transpressional tectonic zone. The progressive, polyphase nature of the deformation in the TNB is the likely cause of inconsistencies between the two models; however, more detailed work is still required before a coherent model will emerge for the tectonic evolution of the TNB.

Host Units

The high MgO and cumulate olivine contents of the ultramafic intrusions could be attributed to emplacement as olivine-magma mushes, accumulation of intratelluric olivine phenocrysts, or accumulation of olivine in situ. Although insufficient mineral chemical data are available to distinguish between the latter two possibilities, the density of a komatiitic magma containing abundant olivine would be much greater than the density of the sedimentary section, even if consolidated, so the first possibility seems unlikely. If so, then the ultramafic intrusions represent dynamic flow-through feeder sills, similar to the lava channels in extrusive komatiitic systems (see discussion by Lesher, 1989).

Ore Genesis and Crustal Contamination Signatures

The consistent stratigraphic correlations between ultramafic bodies, S-rich sedimentary rocks, and Ni-Cu-(PGE) sulphide mineralization (Peredery, 1982), and the large component of crustal sulphur in the ores (Eckstrand et al., 1989) suggest that the majority of the sulphides formed via melting of the S-rich sedimentary rocks. The large degree of crustal contamination (up to 30%: Lesher and Burnham, 2001; Lesher et al., 2001; Burnham et al., 2003) in the host units indicates that large amounts of silicates were assimilated, but the solubility of S in even komatiitic magmas is quite low (Shima and Naldrett, 1975), so it is most likely that the sulphide simply melted, remaining as a 'xenomelt' (see discussion by Lesher and Burnham , 2001) at or near the bases of the sills. This interpretation is supported by the high (normal) PGE contents of some of host rocks (Lesher et al., 2001), which requires that the magmas did not equilibrate with sulphides prior to emplacement. Ironically, geochemical models of the ores suggest that there was too much S in the system and not enough magma, and that the deposits would have had higher metal contents (especially PGE and Cu) if less S had been added or if more magma had interacted with the sulphides (see discussion by Lesher et al., 2001). Nevertheless, much more work is required to establish the nature of contamination (wholesale versus incongruent versus volatilization), the degree of contamination, and the amount of S required to generate an economic deposit.

Deformation during High-Grade Metamorphism

The Ni-Cu-(PGE) mineralization in the TNB is often detached from what is inferred to be the primary ore host (i.e. ultramafic rocks). This tectonic detachment has been described as simple boudinage (e.g. Bleeker, 1990). However, during this mobilization of the sulphide minerals from their ultramafic hosts, fluids generated from the metamorphism of the ore zones and their host rocks, appear to have mobilized many elements away from the primary ore environment, not only Ni (Bleeker, 1990) but particularly Au, Pd, and Cu (Burnham et al., 2003). On the basis of experimental studies, it has been suggested that many elements thought to be chloride-immobile (i.e. PGE) at low temperature (i.e. ~300°C; Wood, 1987, 1996; Pickrell et al., 1996) may be very mobile in chloride solutions at near-magmatic conditions (Hanley et al., 2005). Additionally, at lower temperatures, the PGE may be transported in significant concentrations via bisulphide complexation. Although most other analyzed elements are present in the TNB ores are at predicted abundances (Burnham et al., 2003), it is possible that other, more volatile elements were added to the mineralization during high-grade metamorphism. Future work on the TNB must establish the extent of element mobility, the extent of structural and metamorphic/metasomatic mobilization within regional and mine scales, and the volume of fluids that may have been generated during metamorphic dehydration of the host rocks. Such information is critical in understanding genetic links between the different deposits in the TNB (e.g. Thompson, Pipe, and Birchtree mines).

Acknowledgements

This synopsis reflects our participation CAMIRO Project 97E-02 and includes information summarized by the authors, but also includes insight gained from other project collaborators, including Nuno Machado, Dianne Michalak, Alain Potrel, Kim Toope, Herman Zwanzig, Kevin Ansdell, Chris Böhm, Carl Chandler, Eric Ducharme, Denis Gapais, Reid Keays, Jürgen Kraus, Josef Macek, Peter Theyer, Mark Pacey, Paul Lenton, and John Broome. We have also benefited from discussions with and assistance from Paul Golightly, Patti Tirshman, Larry Larson, Scott Mooney, Dave Owens, John Pearson, Jamie Robertson, Rick Somerville, Gary Sorensen, Rob Stewart, and Kevin Wells.

We are very grateful to the sponsors of the CAMIRO Project 97E-02, including Billiton Ltd., Cominco Ltd., Falconbridge Ltd., Hudson Bay Exploration and Development Ltd., Inco Ltd., and WMC Ltd., for their strong and enthusiastic scientific, logistical, and financial support.

We would also like to thank A. Naldrett and J. Hanley for their constructive reviews of the manuscript.

References

Bamburak, J.D., 1980, Mineral Inventory of Manitoba: Manitoba Energy and Mines, Mineral Resources Division, Economic Geology Report ER79-6, 41 p.

Barnes, S.J., and Brand, N.W., 1999, The distribution Cr, Ni, and chromite in komatiites, and application to exploration for komatiite-hosted nickel sulfide deposits: Economic Geology and the Bulletin of the Society of Economic Geologists, v. 94, p. 129-132.

Bleeker, W., 1989, Structural geology and deformational history of the Thompson nickel orebodies and environs, Thompson, Manitoba, *in* Galley, A.G. (compiler), Investigations by the Geological Survey of Canada in Manitoba and Saskatchewan during the 1984-1989 Mineral Development Agreements: Geological Survey of Canada, Open File 2133, p. 111-115.

——— 1990, Evolution of the Thompson nickel belt and its nickel deposits, Manitoba, Canada: Ph.D. thesis, University of New Brunswick, Fredericton, New Brunswick, 444 p.

Bliss, N.W., 1973, A Comparative Study of Two Ultramafic Bodies at the SW End of the Manitoba Nickel Belt; with Special Reference to the Chromite Mineralogy: Ph.D. thesis, McGill University, Montreal, Quebec, 330 p.

Burnham, O.M., Halden, N., Layton-Matthews, D., Lesher, C.M., Liwanag, J., Heaman, L., Hulbert, L., Machado, N., Michalak, D., Pacey, M., Peck, D., Potrel, A., Theyer, P., Toope, K., and Zwanzig, H. 2003, Geology, Stratigraphy, Petrogenesis, and Metallogenesis of the Thompson Nickel Belt, Manitoba: Final Report for CAMIRO Project 97E-02: Sudbury, Mineral Exploration Research Centre, 410 p.

Coats, C.J.A., 1966, Serpentinized Ultramafic Rocks of the Manitoba Nickel Belt: Ph.D. thesis, University of Manitoba, Winnipeg, Manitoba, 254 p.

Coats, C.J.A., and Brummer, J.J., 1971, Geology of the Manibridge nickel deposit, Wabowden, Manitoba, Special Paper, v. 9: Ottawa, Geological Association of Canada, p. 155-165.

Coats, C.J.A., Green, P.W., and Wilson, H.D.B., 1976, Sulphide mobilization in the Manibridge orebody: Canadian Institute of Mining and Metallurgy Bulletin, v. 69, p. 154-159.

Cumming, G.L., Eckstrand, O.R., and Peredery, W.V., 1982, Geochronologic interpretations of Pb isotope ratios in nickel sulfides of the Thompson Belt, Manitoba: Canadian Journal of Earth Sciences, v. 19, p. 2306-2320.

De Saboia, L.A., 1978, The West Manasan and the Pipe Mine 2 Ultramafic Bodies of the Thompson Nickel Belt: M.Sc. thesis, University of Western Ontario, London, Ontario, 154 p.

Donaldson, J.A., 1983, Progressive Alteration of Barren and Weakly Mineralized Archean Dunites from Western Australia: Ph.D. thesis, University of Western Australia, Perth, Australia, 345 p.

Duke, J.M., 1986, The Dumont nickel deposit; a genetic model for disseminated magmatic sulphide deposits of komatiitic affinity, *in* Gallagher, M.J., Ixer, R.A., Neary, C.R., and Prichard, H.M., eds., Metallogeny of Basic and Ultrabasic Rocks: Institute of Mining and Metallurgy, p. 9-12.

Eckstrand, O.R., 1975, The Dumont serpentinite; a model for control of nickeliferous opaque mineral assemblages by alteration reactions in ultramafic rocks: Economic Geology, v. 70, p. 183-201.

Eckstrand, O.R., Grinenko, L.N., Krouse, H.R., Paktunc, A.D., Schwann, P.L., and Scoates, R.F.J., 1989, Preliminary data on sulphur isotopes and Se/S ratios, and the source of sulphur in magmatic sulphides from the Fox River Sill, Molson Dykes, and Thompson nickel deposits, northern Manitoba, *in* Current Research, Part C: Geological Survey of Canada, Paper 89-1C, p. 235-242.

Fraser, H.S., 1985, A Journey North: The Great Thompson Nickel Discovery: Inco Ltd., 388 p.

Fueten, F., 1992, Tectonic interpretations of systematic variations in quartz c-axis fabrics across the Thompson Belt: Journal of Structural Geology, v. 14, p. 775-789.

Fueten, F., and Robin, P.-Y.F., 1989, Structural petrology along a transect across the Thompson Belt, Manitoba; Dip slip at the western Churchill-Superior boundary: Canadian Journal of Earth Sciences, v. 26, p. 1976-1989.

Gapais, D., Potrel, A., Machado, N., and Hallot, E., 2005, Kinematics of long-lasting Paleoproterozoic transpression within the Thompson nickel belt, Manitoba, Canada: Tectonics, v. 24, p. 16.

Gascoyne, M., Davison, C.C., Ross, J.D., and Pearson, R., 1987, Saline groundwaters and brines in plutons in the Canadian Shield, *in* Fritz, P., Frape, S.K., eds., Saline Water and Gases in Crystalline Rocks: Geological Association of Canada, Special Paper 33, p. 53-68.

Gibb, R.A., 1968, The densities of Precambrian rocks from northern Manitoba: Canadian Journal of Earth Sciences, v. 5, Part 1, p. 433-438.

——— 1982, The gravity signature of the Superior/Churchill structural province boundary: A proposed Proterozoic collision zone: Society of Exploration Geophysicists, Technical Program Expanded Abstracts 1982, p. 283-285.

Good, D.J., and Naldrett, A.J., 1993, Geology and distribution of platinum-group elements, Bucko Lake Intrusion, Thompson Belt, Manitoba: The Canadian Mineralogist, v. 31, p. 45-60.

Green, A.G., 1981, Results of a seismic-reflection survey across the fault zone between the Thompson Nickel Belt and the Churchill Tectonic Province, northern Manitoba: Canadian Journal of Earth Science, v. 18, p. 13-25.

Groves, D.I., and Keays, R.R., 1979, Mobilization of ore-forming elements during alteration of dunites, Mt. Keith-Betheno, Western Australia: The Canadian Mineralogist, v. 17, p. 373-389.

Hanley, J., Mungall, J., Pettke, T., Spooner, E., and Bray, C., 2005, Ore metal transport by hydrocarbon vapour in the footwall of the Sudbury igneous complex, Canada: Geochimica et Cosmochimica Acta, v. 69, p. A738.

Hanski, E.J., and Smolkin, V.F., 1989, Pechenga ferropicrites and other Early Proterozoic picrites in the eastern part of the Baltic Shield: Precambrian Research, v. 45, p. 63-82.

Hopf, S., and Head, D.L., 1998, Mount Keith nickel deposit, *in* Berkman, D.A. and Mackenzie, D.H., eds., Geology of Australian and Papua New Guinean Mineral Deposits: Australasian Institute of Mining and Metallurgy, Monograph 22, p. 307-314.

Hulbert, L.J., Hamilton, M.A., Horan, M.F., and Scoates, R.F.J., 2005, U-Pb zircon and Re-Os isotope geochronology of mineralized ultramafic intrusions and associated nickel ores from the Thompson nickel belt, Manitoba, Canada: Economic Geology, v. 100, p. 29-41.

Irvine, T.N., 1975, Crystallisation sequences of the Muskox intrusion and other layered intrusions. II. Origin of chromitite layers and similar deposits of other magmatic ores: Geochimica et Cosmochimica Acta, v. 39, p. 991-1020.

Kerr, A.C., and Arndt, N.T., 2001, A note on the IUGS reclassification of the high-Mg and picritic volcanic rocks: Journal of Petrology, v. 42, p. 2169-2171.

Kornik, L.J., and MacLaren, A.S., 1966, Aeromagnetic study of the Churchill-Superior boundary in northern Manitoba: Canadian Journal of Earth Sciences, v. 3, p. 547-557.

Larocque, A.C.L., Boulet, M.P., and Hebert, R., 1998, Applications of microbeam analysis of sulfides to the discrimination of ore types at Inco's Birchtree Deposit, Thompson nickel belt, Manitoba: Geological Association of Canada - Mineralogical Association of Canada - Canadian Geophysical Union, Joint Annual Meeting, Program with Abstracts, v. 23, p. 103.

Lesher, C.M., 1989, Komatiite-associated nickel sulfide deposits, *in* Whitney, J.A., and Naldrett, A.J., eds., Ore Deposition Associated with Magmas: Society of Economic Geologists, Reviews in Economic Geology, v. 4, p. 45-101.

Lesher, C.M., and Burnham, O.M., 2001, Multicomponent elemental and isotopic mixing in Ni-Cu-(PGE) ores at Kambalda, Western Australia: Canadian Mineralogist, v. 39, p. 421-446.

Lesher, C.M., and Keays, R.R., 2002, Komatiite-associated Ni-Cu-(PGE) deposits; Mineralogy, geochemistry, and genesis, *in* Cabri, L.J., ed., The Geology, Geochemistry, Mineralogy, and Mineral Beneficiation of the Platinum-Group Elements: Canadian Institute of Mining, Metallurgy and Petroleum, Special Volume 54, p. 579-617.

Lesher, C.M., Lee, R.F., Groves, D.I., Bickle, M.J., and Donaldson, M.J., 1981, Geochemistry of komatiites from Kambalda, Western Australia; I, Chalcophile element depletion - a consequence of sulfide liquid separation from komatiitic magmas: Economic Geology, v. 76, p. 1714-1728.

Lesher, C.M., Burnham, O.M., Keays, R.R., Barnes, S.J., and Hulbert, L., 2001, Geochemical discrimination of barren and mineralized komatiites associated with magmatic Ni-Cu-(PGE) sulfide deposits: Canadian Mineralogist, v. 39, p. 673-696.

Lewry, J.F., Hajnal, Z., Green, A., Lucas, S.B., White, D., Stauffer, M.R., Ashton, K.E., Weber, W., and Clowes, R., 1994, Structure of a Paleoproterozoic continent-continent collision zone: A Lithoprobe seismic reflection profile across the Trans-Hudson Orogen, Canada: Tectonophysics, v. 232, p. 143-160.

Liwanag, J., 2000, Post-Magmatic Modification of the Sulphide Deposits from the Thompson Nickel Belt, Manitoba, Canada: M.Sc. thesis, University of Manitoba, Winnipeg, Manitoba, 207 p.

Lucas, S.B., and St-Onge, M.R., 1989, Structural evolution of the Cape-Smith Belt from initial thrusting to basement-involved folding: Geoscience Canada, v. 16, p. 122-126.

Lucas, S., Picard, C., and St-Onge, M., 1990, Tectonic, magmatic, and metallogenic evolution of the Early Proterozoic Cape Smith Thrust Belt: Geoscience Canada, v. 16, p. 117-163.

Macek, J.J., 1986, Lithological investigations in INCO's Thompson open pit mine, *in* Report of Field Activities: Manitoba Energy and Mines, Mineral Resources Division, p. 167-171.

Macek, J.J., McGregor, C.R., and Zwanzig, H.V., 2001, Thompson Nickel Belt Project; Progress report, *in* Report of Activities 2001: Manitoba Industry, Trade and Mines, Manitoba Geological Survey, p. 47-50.

Macek, J.J., McGregor, C.R., and Coueslan, C.G., 2002, Thompson nickel belt project: Progress report (parts of NTS 63J, 63O and 63P), *in* Report of Activities: Manitoba Industry, Trade and Mines, Manitoba Geological Survey, p. 115-116.

Machado, N., Bleeker, W., and Anonymous, 1990, U-Pb geochronology of the southern Trans-Hudson Orogen in Manitoba, *in* Hajnal, Z., and Lewry, J., eds., Trans-Hudson Orogen Transect Planning Workshop: Lithoprobe Report, v. 12, p. 24-25.

Marshall, B., Vokes, F.M., and Larocque, A.C.L., 2000, Regional metamorphic remobilization; upgrading and formation of ore deposits, *in* Spry, P.G., Marshall, B., and Vokes, F.M., eds., Metamorphosed and Metamorphogenic Ore Deposits: Society of Economic Geologists, Reviews in Economic Geology, v. 11, p. 19-38.

McRitchie, W.D., 1995, Mineral Development Potential in Manitoba - Nickel in the Southwest Extension of the Thompson Nickel Belt: Manitoba Energy and Mines, Economic Geology Report ER95-2, 24 p.

Naldrett, A.J., 2004, Magmatic Sulfide Deposits; Geology, Geochemistry and Exploration: Springer-Verlag, Berlin, 725 p.

O'Hanley, D.S., 1996, Serpentinites: Records of Tectonic and Petrological History, v. 34: Oxford University Press, Oxford Monographs on Geology and Geophysics, No. 34, 277 p.

Paktunc, A.D., 1984, Petrogenesis of ultramafic and mafic rocks of the Thompson Nickel Belt, Manitoba: Contributions to Mineralogy and Petrology, v. 88, p. 348-353.

Peredery, W.V., 1979, Relationship of ultramafic amphibolites to metavolcanic rocks and serpentinites in the Thompson Belt, Manitoba: The Canadian Mineralogist, v. 17, p. 187-200.

——— 1982, Geology and nickel sulphide deposits of the Thompson Belt, Manitoba, *in* Hutchinson, R.W., Spence, C.D., and Franklin, J.M., eds., Precambrian Sulphide Deposits: Geological Association of Canada, Special Paper 25, p. 165-209.

Pickrell, J.M., Wood, S.A., and Anonymous, 1996, Stability constants of palladium-hydroxide complexes as a function of ionic strength (0.1-1m) and temperature (25-85 degrees C): Geological Society of America, Abstracts with Programs, v. 28, p. 148.

Porter, D., J, and Mckay, K., G, 1981, The nickel sulfide mineralization and metamorphic setting of the Forrestania area, Western Australia: Economic Geology, v. 76, p. 1524-1549.

Potrel, A., Peck, D., Machado, N., Wegleitner, C., and Anonymous, 2000, The Thompson nickel belt (Manitoba, Canada): A typical Paleoproterozoic transpressive shear zone: 31st International Geological Congress, Brazil 2000, Abstracts Volume, CD-ROM.

Scoates, R.F.J., 1971, Mineral potential of ultramafic rocks of Manitoba: Canadian Mining Journal, v. 92, p. 67.

Scoates, R.F.J., Macek, J., and Russell, J.K., 1977, Thompson Nickel Belt project, *in* Report of Field Activities: Manitoba Department of Mines, Resources and Environmental Management, p. 47-54.

Shima, H., and Naldrett, A.J., 1975, Solubility of sulfur in an ultramafic melt and the relevance of the system Fe-S-O: Economic Geology, v. 70, p. 960-967.

Weber, W., 1990, Trans-Hudson Orogen: Manitoba perspective in Hajnal, Z., and Lewry, J., eds., Trans-Hudson Orogen Transect Planning Workshop: Lithoprobe Report, v. 12, p. 7-8.

Weber, W., and Scoates, R.F.J., 1978, Archean and Proterozoic metamorphism in the northwestern Superior Province and along the Churchill-Superior boundary, Manitoba, *in* Fraser, J.A., and Heywood, W.W., eds., Symposium on Metamorphism in the Canadian Shield: Geological Survey of Canada, Paper 78-10, p. 5-16.

White, D.J., Jones, A.G., Lucas, S., B, Hajnal, Z., Asudeh, I., White, D., Roberts, R., Forsyth, D., Kay, I., Cartwright, T., Carroll, P., Hajnal, Z., Koperwhats, B., Musacchio, G., and Farrell, D., 1999, Tectonic evolution of the Superior boundary zone from coincident seismic reflection and magnetotelluric profiles: Tectonics, v. 18, p. 430-451.

White, D.J., Lucas, S.B., Bleeker, W., Hajnal, Z., Lewry, J.F., and Zwanzig, H.V., 2002, Suture-zone geometry along an irregular Paleoproterozoic margin: The Superior boundary zone, Manitoba, Canada: Geology, v. 30, p. 735-738.

Wood, S.A., 1987, Thermodynamic calculations of the volatility of the platinum group elements (PGE); the PGE content of fluids at magmatic temperatures: Geochimica et Cosmochimica Acta, v. 51, p. 3041-3050.

——— 1996, The role of humic substances in the transport and fixation of metals of economic interest (Au, Pt, Pd, U, V): Ore Geology Reviews, v. 11, p. 1-31.

Zurbrigg, H.F., and Patterson, J.M., 1963, Thompson mine geology: Canadian Mining and Metallurgical Bulletin, v. 50, p. 11-16.

Zwanzig, H.V., 1999, Structure and stratigraphy of the south flank of the Kisseynew Domain in the Trans-Hudson Orogen, Manitoba: Implications for 1.845-1.77 Ga collision tectonics: Canadian Journal of Earth Sciences, v. 36, p. 1859-1880.

——— 2004, Mafic-ultramafic magmatism of the Bah Lake assemblage, Ospwagan Group, Upper Ospwagan Lake, Manitoba (NTS 63O9), *in* Field Activities: Manitoba Industry, Economic Development and Mines, Manitoba Geological Survey, p. 149-155.

Zwanzig, H.V., and McRitchie, W.D. (prefacer), 1997, Revised stratigraphy of the Setting Lake area (parts of NTS 63O/1 and 2, and 63J/15): Manitoba Energy and Mines, Report of Activities, p. 103-108.

Overview of the Geology and Cr-PGE Potential of the Southern Québec Ophiolite Belt

J.H. Bédard[1], P. Pagé[2], V. Bécu[2], J.-M. Schroetter[2], A. Tremblay[3]

1. Geological Survey of Canada – Québec, 490 de la Couronne, Québec, Québec G1K 9A9
2. Institut National de la Recherche Scientifique-Eau, Terre et Environnement, 490 de la Couronne, Québec, Québec, G1K 9A9
3. Département des Sciences de la Terre, Université du Québec á Montréal, CP 8888, Succursale centre ville, Montréal, Québec, H3C 3P8

Corresponding author's email: jbedard@nrcan-rncan.gc.ca

Abstract

The southern Québec Ophiolite Belt consists of partly dismembered oceanic terranes accreted against the Laurentian margin in the Ordovician, and then reworked by Silurian and Devonian deformation. Mapping, and structural and stratigraphic analysis of the Thetford Mines ophiolitic complex (TMOC), show that many of the complexities in the outcrop patterns can be attributed to a major phase of pre-obduction, synmagmatic extension (seafloor spreading) and tectonic exhumation along paleonormal faults, with erosional degradation of tilted fault blocks. The Thetford Mines ophiolitic complex contains the bulk of the chrome-platinum-palladium mineralization that has been discovered to date in the southern Québec ophiolite belt. Refractory mantle harzburgite hosts podiform chromitite oredobies and similar chromitites occur in the Lac Montjoie Complex to the south. The platinum group element (PGE) contents of harzburgite can be simulated by equilibrium partial melting with olivine + orthopyroxene + chromite residues, implying an absence of residual sulphides. Concentrations of Pt and Pd, as well as La, Ce, and large ion lithophile elements (LILE), may be slightly enriched compared to other trace elements, suggesting metasomatic enrichment by an aqueous fluid derived from the subducting slab. The lowermost crust is dominated by dunite and interbedded chromitite layers, generally 1 to 3 cm thick, but locally exceeding 10 m. The lower crust and mantle chromitites rarely show enrichment in PGEs. The middle crust is dominated by pyroxenites and gabbros. All TMOC cumulate rocks analyzed to date have trace element geochemical signatures that imply a boninitic affinity, and are probably consanguineous with the boninitic dykes and lavas. Whole-rock data indicate systematic stratigraphic enrichment in incompatible elements and depletion in compatible elements, suggesting deposition from a fractionating body of magma. Chromitite veins and breccia fills located in crustal dunite and pyroxenite locally reach grades of 20 g/ton Pt+Pd. Detailed mapping and structural work suggest that the Starchrome Cr-PGE showing, and possibly the Hall Cr-PGE orebody, are associated with pre-obduction normal faults. At Starchrome, there is no apparent link between the abundances of PGEs and those of other incompatible elements, although there are enrichments in light rare earths in the wall rocks to the Cr-PGE ore. The chromite at Starchrome also contains potarite (a low-temperature PdHg amalgam) associated with actinolite in fractured chromite and intergrain sites. Together, these features support a model of metasomatic introduction of PGEs by hydrothermal fluids. The PGEs may have been concentrated in volatile-rich residual melts in structural traps above ponded sub-crustal/intracrustal boninitic intrusions, with subsequent escape of fluid through faults, and deposition of ore during decompression. The existence of potarite suggests that analysis of mercury in soils may represent a useful exploration tool in areas covered by Quaternary deposits. The presence in the TMOC of erosional unconformities that reach down into pyroxenitic and dunitic cumulates suggests the potential for recycling of PGE chromitites and the formation of paleoplacer-type ore within the overlying sedimentary rocks of the St.-Daniel Mélange. Mantle rocks are less voluminous in the coeval Asbestos complex (AOC) to the south, and hence this massif has less potential for podiform chromite ore. The AOC volcanic stratigraphy is polygenetic, with a tholeiitic series at the base and boninites above, which presumably correlate with those of the coeval TMOC, implying that the TMOC and AOC were obducted as a single slab, and hence, have identical PGE-exploration potential. Further south, the Mont Orford complex (MOOC) is dominated by gabbros and tholeiitic lavas, and only subordinate late boninitic dykes are found. The affinity and age of the cumulate rocks of the MOOC are uncertain and its links to the TMOC-AOC cannot yet be evaluated.

Résumé

La ceinture ophiolitique du sud du Québec se compose de terranes océaniques partiellement démembrés, qui se sont accrétés à la marge de la Laurentie à l'Ordovicien, puis ont été remaniés par déformation au Silurien et au Dévonien. La cartographie ainsi que l'analyse structurale et stratigraphique du complexe ophiolitique de Thetford Mines (COTM) montrent que de nombreux éléments de la configuration complexe des affleurements peuvent être attribués à une importante phase de distension synmagmatique (expansion océanique) antérieure à l'obduction, puis à une exhumation tectonique le long de paléofailles normales, laquelle a été accompagnée d'une dégradation des blocs-failles basculés. Le COTM renferme la plupart des minéralisations de chrome-platine-palladium qui ont été découvertes jusqu'ici dans la ceinture ophiolitique du sud du Québec. La harzburgite réfractaire d'origine mantellique contient des corps minéralisés de chromitite podiformes; des chromitites semblables sont aussi présentes dans le complexe du Lac Montjoie, au sud. La teneur en éléments du groupe du platine (ÉGP) de l'harzburgite peut être simulée au moyen d'une réaction de fusion partielle à l'équilibre avec résidus d'olivine, d'orthopyroxène et de chromite, ce qui sous-entend l'absence de sulfures résiduels. Les concentrations de Pt et de Pd, ainsi que celles de La, de Ce et d'éléments lithophiles à grand rayon ionique peuvent être légèrement enrichies comparativement à celles d'autres éléments traces, ce qui suppose un enrichissement métasomatique par un fluide aqueux issu de la plaque plongeante. La partie basale de la croûte se compose principalement de dunite dans laquelle sont interstratifiées des couches de chromitite. L'épaisseur de ces couches est généralement de 1 à 3 cm, mais peut atteindre plus de 10 m, par endroits. Les chromitites du manteau et de la croûte inférieure présentent rarement un enrichissement en ÉGP. La croûte intermédiaire est surtout composée de pyroxénite

Bédard, J.H., Pagé, P., Bécu, V., Schroetter, J.-M., and Tremblay, A., 2007, Overview of the geology and Cr-PGE potential of the southern Québec Ophiolite Belt, *in* Goodfellow, W.D., ed., Mineral Deposits of Canada: A Synthesis of Major Deposit-Types, District Metallogeny, the Evolution of Geological Provinces, and Exploration Methods: Geological Association of Canada, Mineral Deposits Division, Special Publication No. 5, p. 433-448 et de gabbro. Tous les cumulats du COTM analysés jusqu'à maintenant présentent des signatures géochimiques des éléments traces qui témoignent d'affinités boninitiques, ainsi que d'une origine commune probable avec les laves et les dykes de boninite. Suivant la stratigraphie, des données d'analyse sur roche totale indiquent un enrichissement systématique en éléments compatibles, de même qu'un appauvrissement en éléments compatibles, ce qui suppose une mise en place par fractionnement d'un magma. Dans les filons de chromitite et les matériaux de remplissage de brèche logés dans la pyroxénite et la dunite crustales, la teneur combinée en Pt et Pd atteint, par endroits, 20 g/tonne. Des travaux cartographiques et structuraux détaillés laissent croire que l'indice de Cr-ÉGP de Starchrome et, peut-être, le corps minéralisé de Cr–ÉGP de Hall sont associés à des failles normales de formation antérieure à l'obduction. Dans l'indice de Starchrome, il n'existe aucun lien apparent entre les fortes concentrations d'ÉGP et celles d'autres éléments compatibles, malgré un enrichissement en terres rares légères dans les épontes du minerai de Cr-ÉGP. La chromite de l'indice de Starchrome contient aussi de la potarite (alliage de Pd et de Hg de basse température) associée à de l'actinote dans les fractures de la chromite et les matériaux intergranulaires. L'ensemble de ces caractéristiques appuie un modèle d'introduction métasomatique des ÉGP par des fluides hydrothermaux. Les ÉGP se sont peut-être concentrés dans des bains résiduels riches en éléments volatils, au sein de pièges structuraux, au-dessus d'intrusions boninitiques confinées en milieu subcrustal ou intracrustal, après quoi les fluides auraient fui par des failles et le minerai aurait été déposé suite à la décompression. L'existence de potarite laisse supposer que le dosage du mercure dans les sols pourrait constituer un bon outil d'exploration dans les zones recouvertes de dépôts quaternaires. Dans le COTM, la présence de discordances d'érosion jusqu'à la profondeur des cumulats pyroxénitiques et dunitiques laisse croire à un recyclage possible des chromitites à ÉGP et la formation possible d'un minerai de type paléoplacer dans les roches sédimentaires sus-jacentes du mélange de Saint-Daniel. Dans le complexe ophiolitique d'Asbestos (COA), au sud, qui est contemporain du COTM, les roches mantelliques sont moins volumineuses, ce qui se traduit par un potentiel inférieur en minerai de chromite podiforme pour ce complexe. La stratigraphie volcanique du COA est polygénique. Elle est définie à la base par une série tholéiitique, qui est surmontée de boninites, lesquelles peuvent vraisemblablement être mises en corrélation avec celles du COTM, du même âge, ce qui laisse supposer que l'obduction du COTM et du COA s'est produite au sein d'une même plaque et que, par conséquent, leur potentiel est le même, sur le plan de l'exploration ciblant les ÉGP. Plus au sud se trouve le complexe ophiolitique du mont Orford (COMO), qui renferme surtout des gabbros et des laves tholéiitiques et de moindres quantités de dykes boninitiques tardifs. Les affinités et l'âge des cumulats du COMO sont incertains, et les liens de celui-ci avec le COTM et le COA n'ont pas encore été étudiés.

Introduction

The ophiolitic rocks of the southern Québec Ophiolite Belt (Laurent, 1975a, 1975b; Church, 1977; Hébert and Bédard, 2000) represent a historic Cr and asbestos resource. The peak year for use of Canadian asbestos in the United States was 1973, when nearly 700,000 t was imported. By 1996, these imports had dropped to 21,000 t (Ross and Virta, 2001). Chromite deposits were exploited intensively during both world wars, when overseas supplies of this strategic metal were cut off (e.g. Laurent and Kaçira, 1987; Gauthier et al., 1987, 1990). Table 1 shows some production figures and grades, taken from Gauthier et al. (1987). Recently, some of these massifs have been explored for their PGE (platinum group element) potential (Corrivaux and Laflamme, 1990; Morin et al., 1992; Schroetter et al., 2000).

TABLE 1. Chrome produced from the Thetford Mines Ophiolite Complex.

Showing or Mine	Date of Production	Tonage Ore Extracted	Grade $\%Cr_2O_3$	Inferred Reserves
Bennett Martin	1895-1918	2195 t	40.68	900 t
Ward-Chromore	1938-1944	840 t	21.07	
Huard	1894-1918	520 t	54.79	
Mine Montréal	1898-1949	32850 t	16.86	
Lac Breeches	1894-1896	400 t	55	
Brousseau	1897-1916	255 t	31	
Mt.-Caribou	1915-1918	?	41-53	
Parent	1915-1918	?	?	1350 t possible
Parent	1942	50 t	41.8	
Nadeau	189?	100 t	~53	
Lambly-Nadeau	1893-1899	2100 t	~52	
Victoria	1895-1903	>1200 t	~52	
Reed-Bélanger	WW I & II ?	Large	<58	
Mine Caribou	1894-1901	>800 t	?	
Mine Caribou	1916-1920	20814 t	34-40	>40000 t
Vaillancourt	1894-1909	>300 t	38	
Ward-Ross	1897-1901	1400 t	~47	
Ward-Ross	1915-1934	>4500 t	~47	
Dumais	1894-1897	450 t	~53	
American Chrome	1896-1917	1785 t	~48	

The ophiolites in southern Québec occur in the internal part of the orogen (pre-Silurian Appalachians) and were affected by post-obduction deformation, which makes it difficult to correlate facies within massifs and between the different massifs. In southern Québec, the metamorphic and structural evolution of the underlying continental margin is well established (e.g. Tremblay and Pinet, 1994; Castonguay and Tremblay, 2003) and is constrained by $^{40}Ar/^{39}Ar$ isotopic age dating (Whitehead et al., 1995; Castonguay et al., 2001, Tremblay and Castonguay, 2002), providing a framework for the structural analysis of the southern Québec ophiolitic belt and overlying oceanic terranes. This paper summarizes results of a recent field and laboratory re-investigation of these rocks, with the goal of providing insights that may guide exploration for Cr and PGEs.

Geological Context

The southern Québec Appalachians comprise three principal lithotectonic assemblages (Fig. 1): the Cambro-Ordovician Humber and Dunnage zones, and the Silurian-Devonian succession of the Connecticut Valley-Gaspé synclinorium, or trough, located to the southeast of the Guadeloupe Fault (Williams, 1983; Williams and Hatcher, 1983). The Humber Zone represents the vestiges of a passive continental margin sequence, while the Dunnage Zone is an assemblage of oceanic terranes. The contact between the Humber and Dunnage zones is the Baie Verte-Brompton Line (BBL), which is loosely defined as a linear zone of discontinuous serpentinites, dismembered ophiolites and

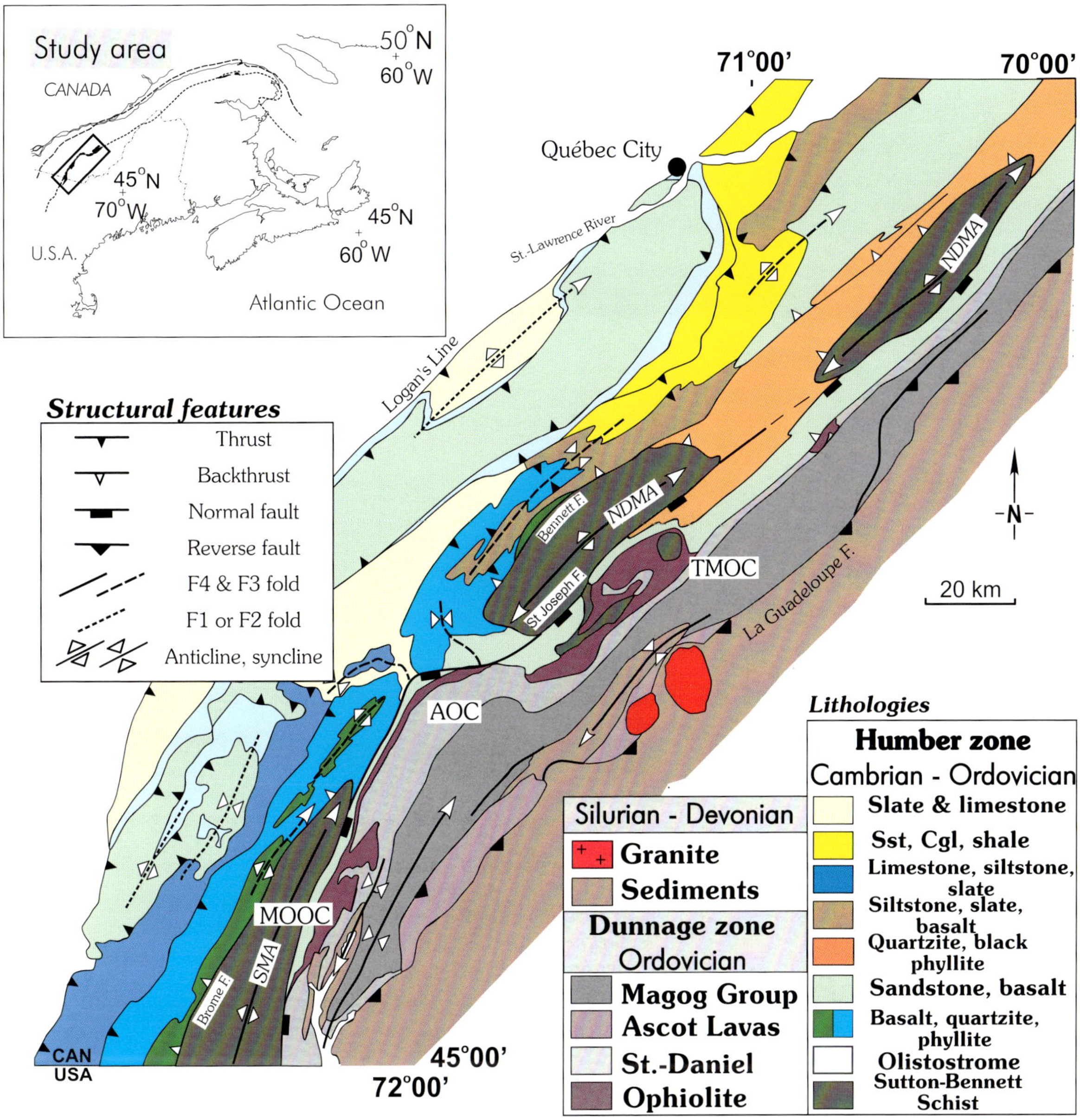

FIGURE 1. Geological map of the southern Québec Appalachians (modified from Schroetter et al., 2005). AOC = Asbestos Ophiolitic Complex; BOG = Bolton Group; Cgl. = conglomerate; F. = Fault; LBOC = Lac Brompton Ophiolitic Complex; MOOC = Mont-Orford Ophiolitic Complex; NDMA = Notre-Dame Mountains Anticlinorium; RPM = Rivière des Plantes Mélange; SMA = Sutton Mountains Anticlinorium; Sst. = sandstone; TMOC = Thetford-Mines Ophiolitic Complex; WV = Ware Volcanics.

mélanges (Williams and St-Julien, 1982). In southern Québec, the Dunnage Zone comprises four major assemblages (Figs. 1, 2): (1) ophiolitic complexes, remnants of oceanic crust formed in pericontinental suprasubduction zone environments (Hébert and Bédard, 2000); (2) the St.-Daniel Mélange, interpreted by Cousineau and St-Julien (1994) as an oceanic accretionary complex, but which we reinterpret as having been deposited upon the ophiolitic rocks (Schroetter et al., 2003, 2006); (3) the Ascot Complex, a composite terrane of volcanic arc sequences (Tremblay et al., 1989); and (4) the Magog Group, a fore-arc sedimentary sequence (Cousineau and St-Julien, 1994; Schroetter et al., 2003, 2006).

The Humber Zone is limited to the southeast by the St-Joseph fault and the Baie Verte-Brompton Line (Pinet et al., 1996, Tremblay and Castonguay, 2002), which together constitute a composite east-dipping normal fault system in southern Québec. In the hanging wall of the normal fault system, continental metamorphic rocks exposed in the cores of Acadian antiforms, formed by the Bécancour and Carineault anticlines, yield Middle Ordovician $^{40}Ar/^{39}Ar$ muscovite ages (469-460 Ma; Whitehead et al., 1995; Castonguay et al., 2001), which are consistent with high-temperature metamorphic ages of the Notre Dame Mountains Anticlinorium, and which are attributed to Ordovician Taconian metamorphism and regional deformation (ca. 490-445 Ma; St-Julien and Hubert, 1975, Stanley

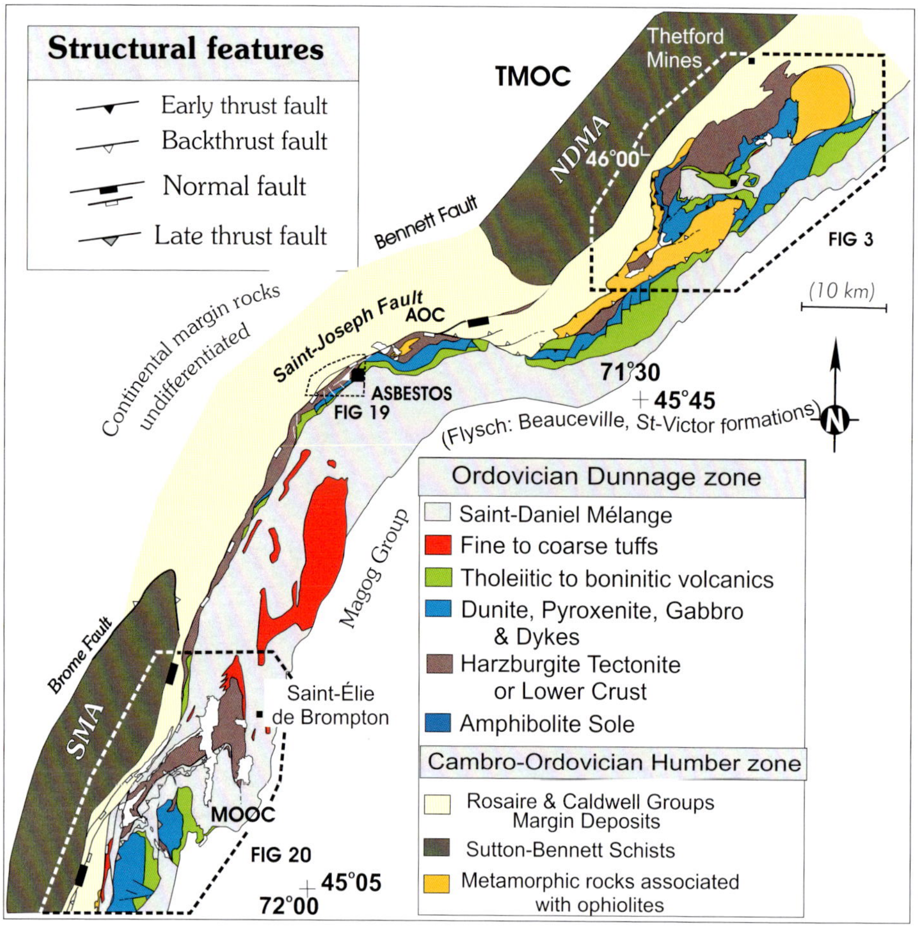

FIGURE 2. Geological map of the southern Québec ophiolitic belt, based on Schroetter et al. (2005) and complemented by data from Beulac (1982), Brassard and Tremblay (1999), Brodeur and Marquis (1995), Cooke (1938, 1950), Hébert (1980), Hébert (1983), Huot (1997), Lamarche (1973), Lavoie (1989), Marquis (1989), Pinet (1995), Riordon (1954), Rodrigue (1979), St-Julien (1963, 1971, 1987), and St-Julien and Slivitzky (1985). See Figure 1 for abbreviations.

and Ratcliffe, 1985, Ratcliffe et al., 1998, Tremblay and Castonguay, 2002). Existing models attribute the Taconian Orogeny to the collision between the Laurentian margin and island-arc terranes (the Taconian arcs) developed over either a southeast-facing (Osberg, 1978; Stanley and Ratcliffe, 1985; Ratcliffe et al., 1998), or a northwest-facing (St-Julien and Hubert, 1975) subduction zone. Pinet and Tremblay (1995a,b) argued that Taconian deformation and metamorphism of the Laurentian margin resulted from oceanic crust obduction.

The presence of Silurian / Early Devonian (431-411 Ma) backthrusts (towards the southeast), and normal faults (southeast-side down) in southern Québec Humber Zone rocks was interpreted in terms of crustal-scale exhumation by extensional collapse of the internal Humber Zone following the Taconian Orogeny (Tremblay and Castonguay, 2002), possibly as a result of slab delamination and isostatic rebound of the subducted plate, and resulted in the development of the Silurian / Early Devonian successor basins of the Connecticut Valley-Gaspé Trough (Tremblay and Pinet, 2005). Middle Devonian (Acadian) deformation heterogeneously overprints both pre- and post-Taconian rocks (Tremblay, 1992; Tremblay et al., 2000), and is characterized by open to isoclinal folds with an axial-planar cleavage and high-angle northwest-verging reverse faults. The Acadian Orogeny is considered to record the final destruction of Iapetus by a continent/continent collision between the Laurentian margin and Gondwana and/or the Composite Avalon Terrane (Osberg, 1978; Williams and Hatcher, 1983; Robinson et al., 1998; van Staal et al., 1998).

The Southern Québec Ophiolite Belt

The southern Québec Ophiolite Belt comprises four major ophiolitic complexes (Figs. 1, 2): the Thetford Mines (TMOC), Asbestos (AOC), Lac Brompton (LBOC), and Mont Chauve/Mont Orford (MOOC) complexes, and numerous smaller slivers (e.g. the Rivière des Plantes ophiolitic mélange). These ophiolite massifs were previously considered to represent kilometre-scale, fault-bounded blocks within the St.-Daniel Mélange, which was interpreted as the remnant of a subduction complex (Cousineau and St-Julien, 1992, 1994). Recent detailed mapping and structural analysis in the Thetford Mines area has challenged this interpretation, suggesting rather that the St.-Daniel is a piggy-back basin deposited on top of the ophiolite as it was being obducted, exhumed, and eroded (Schroetter et al., 2003, 2005, 2006; cf. Dérosier, 1971; Hébert, 1983). The base of the St.-Daniel, the Coleraine Breccia (Hébert, 1981; Schroetter et al., 2003, 2006), contains fragments of ophiolitic (some 10s of metres in size) and continental margin rocks, indicating that both were being exhumed and eroded at this time. The 441 to 460 Ma Ascot Complex (U-Pb zircon ages, Fig. 1; David and Marquis, 1994), dominated by felsic to intermediate volcanic rocks, is interpreted as an arc, built partly on oceanic basement, partly on a detached continental sliver (Tremblay, 1989; Tremblay et al., 1989, 1994).

The Thetford Mines Ophiolitic Complex (TMOC) and Its Orebodies

Mantle and Plutonic Crust

The Thetford Mines ophiolitic complex (TMOC) outcrops as a northeast-trending belt, 40 km in length and 10 to 15 km in width (Fig. 3). The U/Pb zircon ages of trondhjemitic intrusions are 479 ± 3 Ma (Dunning et al., 1986) and 480 ± 2 Ma (Whitehead et al., 2000). The northwestern part of the TMOC has a 5 km thick mantle section (Laurent, 1975b; Laurent et al., 1979), which contains numerous small chromitite pods (Fig. 4; Kaçira, 1971; Laurent and Kaçira, 1987; Schroetter et al., 2000). These podiform chromitites

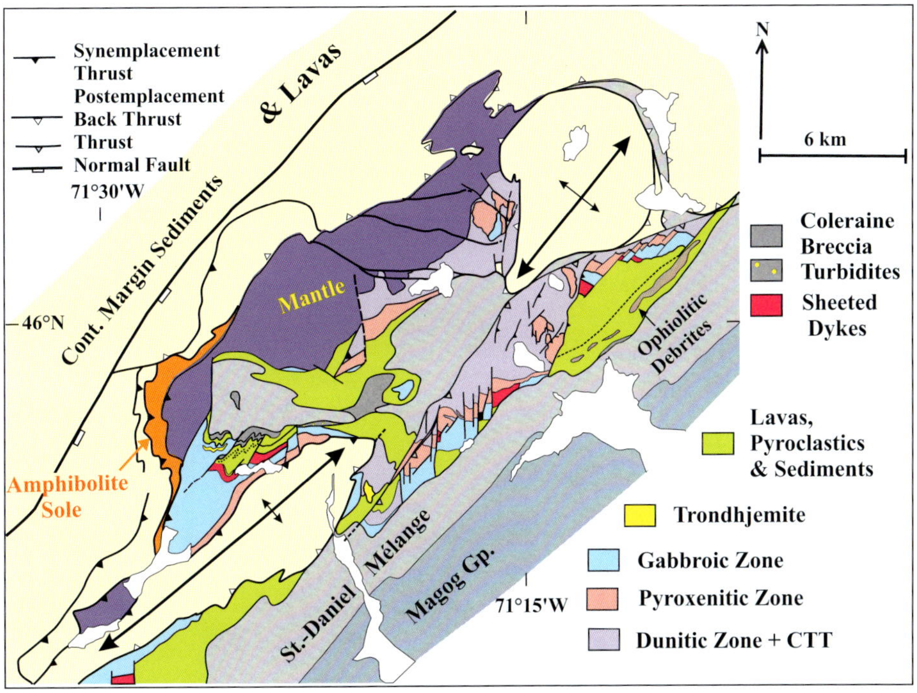

FIGURE 3. Geological map of the Thetford Mines ophiolitic complex. See Figure 2 for location.

are locally orbicular (Fig. 4, cf. Leblanc and Nicolas, 1992), undeformed, and are encased in discordant dunite intrusions. No significant PGE (platinum group element) values have yet been detected in these (Fig. 5), and most of the small orebodies are mined out (Table 1). Major and trace element analyses imply that the TMOC harzburgites are residues from 20 to 25% partial melting (Pagé et al., 2003). Platinum group element (PGE) abundances can be matched by residual assemblages of olivine + orthopyroxene + chromite (Pagé et al., 2003), suggesting that sulphides were not part of the residual assemblage. Model melts are, however, too enriched in Os, Ir, and Ru in comparison to most of the boninitic lavas in the massif. Fractionation of minor amounts of chromite could have caused rapid depletion of these elements and so reconcile models and observations (Pagé et al., 2003, 2004). Highly incompatible elements like La, Ce, LILE (large ion lithophile elements: Rb, Cs, Pb, Sr), Pt, and Pd are commonly slightly enriched in the harzburgites, in comparison to the predicted residues from partial melting, and we have suggested (Pagé et al., 2003) that this records a metasomatic overprint by aqueous fluids derived from the subducting slab.

The plutonic crustal section is 0 to 1.5 km thick (Schroetter et al., 2003), and consists of dunitic, pyroxenitic, and gabbroic zone cumulates and a partly preserved sheeted dyke complex (Fig. 6). Stratiform chromitites (Fig. 7) occur within the Dunitic Zone. Chromitite beds are typically 1 to 3 cm thick, but locally exceed 10 m (Reed-Bélanger Mine, Gauthier et al., 1990). Some of these orebodies still contain significant amounts of chromitite, and could represent an economic resource, although only a few contain high PGE values (Fig. 5). We interpret these chromitites as depositional cumulates, with the abundance of chromite reflecting either sorting (e.g. Wager and Brown, 1968) or pulses of chromite generated by incongruent dissolution of wall rocks and enclaves (e.g. Bédard and Hébert, 1998).

Trace element inversion models (Bédard, 1994, 2001) imply that all cumulate rocks of the TMOC formed from boninitic melts (Fig. 8; Bédard et al., 1992, 2001). Our preliminary results (e.g. Pagé et al., 2004) suggest that rocks of the pyroxenitic and gabbroic zones record differentiation of sizeable melt bodies, of the order of 1 km in thickness, with cumulates showing gradual upward evolution of facies and trace element abundances (Figs. 9, 10). The 'stratigraphy' is punctuated by emplacement of dunitic and lherzolitic sills (1-20 m). We have sampled two profiles in detail. In the Nadeau Hill section (Fig. 9, located immediately to the east of showing #13 of Fig. 5), an example of a sterile section, Pd abundances display a sawtooth profile vs the stratigraphic height; and the uniformly low abundances of Pd (<80 ppb) do not correlate with any other magmatic geochemical signature (Fig. 9).

The Diamond Hills section (Fig. 10) also shows evidence for systematic magmatic differentiation, but in contrast to Nadeau, it hosts a Cr-Pt-Pd showing (Star Chrome, 'St' on Fig. 4, up to 20 g/ton Pt+Pd). The Star Chrome showing is composed of an anastamosing network of massive to antinodular, PGE-enriched chromitite veins (Fig. 11, 1-30 cm; Bécu, 2005), separated from medium-grained foliated websteritic host rocks by a coarse-grained orthopyroxenite reaction rim (2-10 cm). The stratigraphic distribution of PGE abundances in the Diamond Hills section is very spiky (Fig. 6), but abundances overall are markedly enriched relative to

FIGURE 4. Example of an orbicular chromitite in a dunite pipe from within the harzburgitic mantle, Thetford Mines ophiolitic complex, Mine Grand-Père (site 35b on Fig. 5).

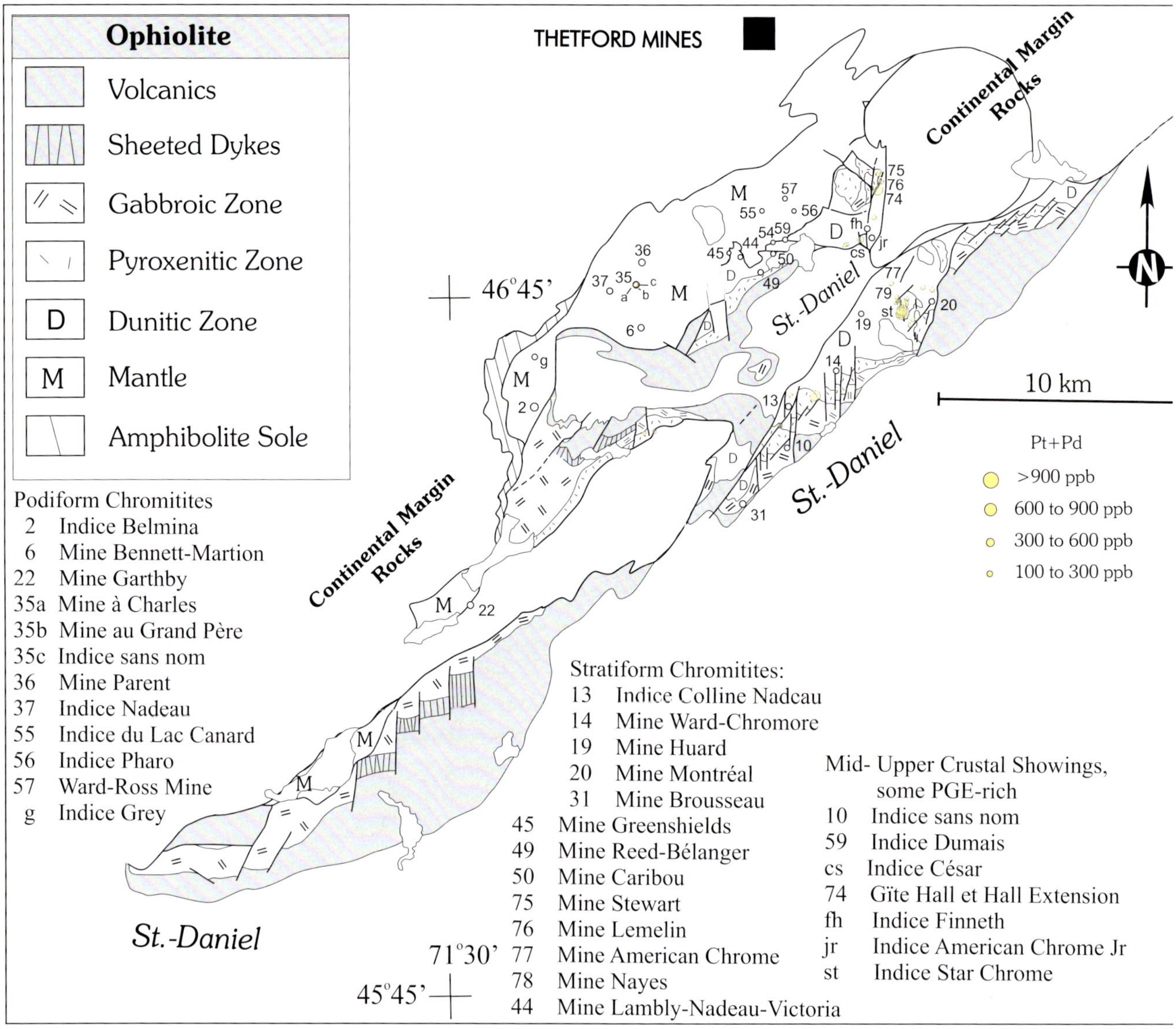

FIGURE 5. Simplified geological map of the Thetford Mines ophiolitic complex showing locations of showings and mines, and whole-rock values for Pd and Pt from Schroetter et al. (2001), including data from the literature (Oshin and Crockett, 1982, 1986b; Corrivaux and Laflamme, 1990; Gauthier et al. 1987, 1990; Tanguay et al., 1990) and our unpublished results.

FIGURE 6. Sheeted dyke complex. All dykes are boninitic and oriented roughly north-south. Near Lac de l'Est.

FIGURE 7. Layered chromitite-dunite from the base of the crustal section, top of a small hill east of Colline Nadeau (site 13 on Fig. 5). Note the small offset, micro-faults filled with websterite.

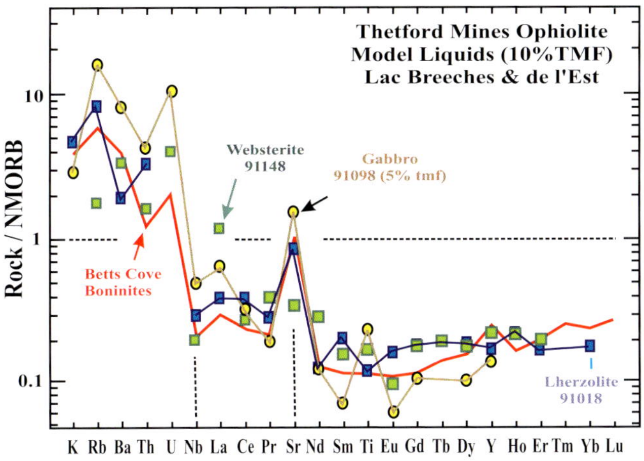

FIGURE 8. Examples of trace element inversion results using the procedure and partition coefficient datasets of Bédard (1994, 2001). Rock modes were estimated from thin sections and norms. Samples are from near Lac Breeches and Lac de l'Est. Assumed trapped melt fractions (tmf) vary from 5 to 10%. The average Betts Cove boninite lava (Low Ti) is from Bédard (1999). Trace element profiles normalized to normal MORB from Sun and McDonough (1989).

Nadeau, with considerable apparent thicknesses of cumulates with Pd abundances >200 ppb, and peaks up to 800 ppb. There seems to be no link between abundance of Pd and indices of magmatic differentiation in any of these rocks. The immediate host rocks to the Star Chrome PGE-Cr ore are enriched in light rare-earths (Fig. 12; Bécu, 2005), suggesting that rare earth data may help in finding orebodies in poorly exposed terrains. Mineralogical analysis (Pagé et al., 2001) has revealed the existence of Potarite (PdHg), a low-temperature mineral that occurs either in interstitial sites adjacent to chromite grains, in association with greenschist-grade minerals, or in cracks within chromite grains (Fig. 13). The textural evidence (potarite), the associated LREE enrichment, and the lack of correlation with magmatic geochemical signatures together suggest that Pt-Pd enrichment in these rocks results from a hydrothermal process. It is also possible that the hydrothermal overprint simply leached the sulphur from a pre-existing magmatic sulphide concentration. Additional research is required to discriminate between these two hypotheses.

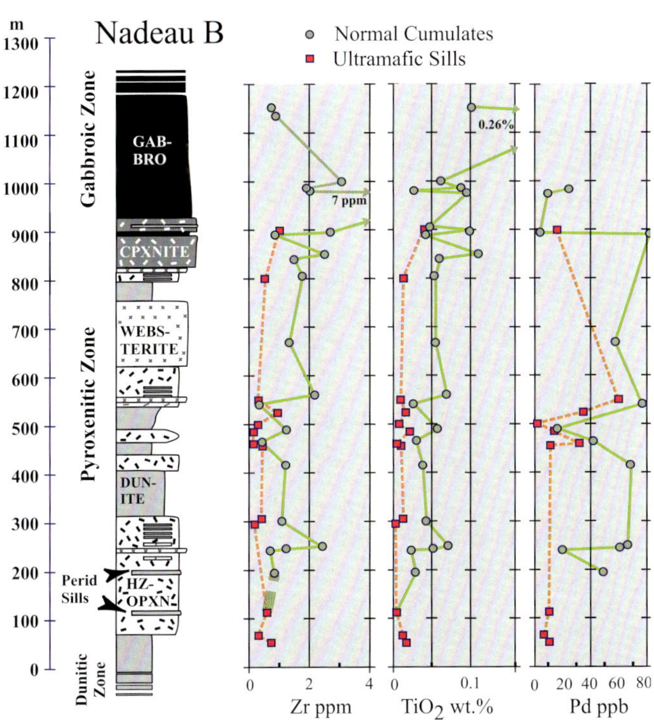

FIGURE 9. Stratigraphic column from the Nadeau Hills section (near showing #13 in Fig. 5), showing abundances of TiO_2, Zr, and Pd (whole rock) for cumulates and sills.

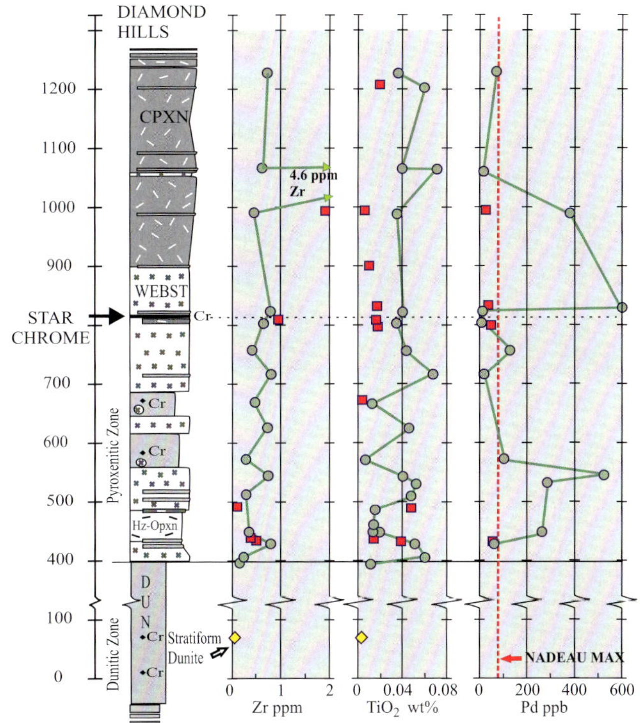

FIGURE 10. Stratigraphic column from the Diamond Hills section (near showing st on Fig. 5), showing abundances of TiO_2, Zr, and Pd (whole rock) for cumulates and sills. Same symbols as Figure 9.

The Hall Mine (site 74 on Fig. 5) occurs as a string of PGE-rich chromitite pods, schlieren, and breccia fills, commonly with antinodular textures (Fig. 14). Chromitite is hosted by an approximately 100 m wide dunite pipe that transects the layered peridotites, pyroxenites, and gabbros of

FIGURE 11. Star Chrome showing (site 'st' on Fig. 5). Note anastomosing network of PGE-rich chromitite veins (Chr) with orthopyroxenite (Opx) haloes. A relict websterite (Wb) is embedded in the orthopyroxenite.

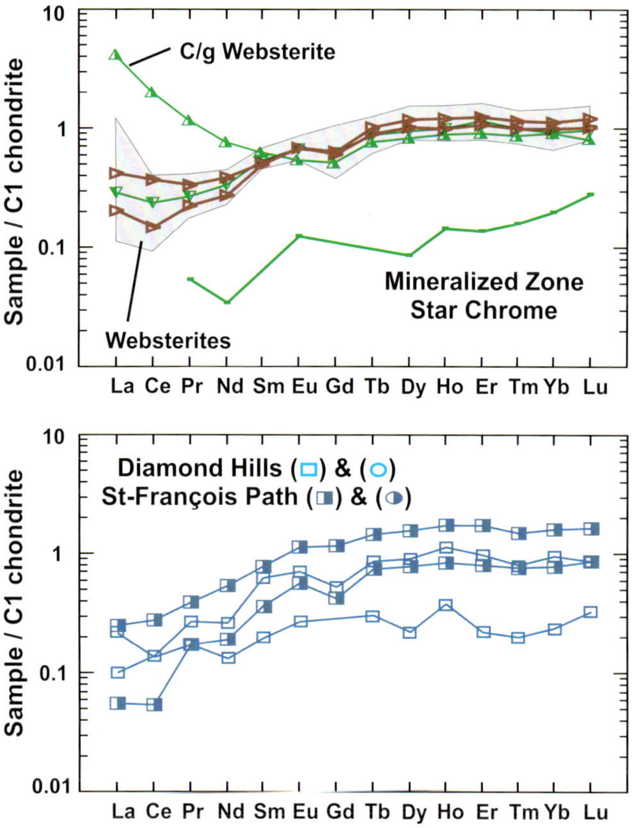

FIGURE 12. Chondrite-normalized (Sun and McDonough, 1989) rare-earth profiles for pyroxenites sampled distally (Diamond Hills and St. François path) and proximally (mineralized zone) to the Star Chrome showing. Note the marked light-rare-earth enrichment shown by proximal samples.

FIGURE 14. Example of antinodular PGE-rich chromitite forming veins, pods, and schlieren in a metasomatic dunite pipe. Hall Mine (site 74 on Fig. 5).

the middle plutonic crust. The chromitite contains a wide variety of PGE minerals (Fig. 15; Corrivaux and Laflamme, 1990), typically as inclusions in chromite. The dunite pipe contains septa of layered peridotite-pyroxenite and is surrounded by an aureole of pyroxene-dunite (Fig. 16). Wispy layering in the pyroxene-dunite aureole is parallel to the foliation seen in the surrounding layered peridotite-pyroxenite, supporting a metasomatic origin for the dunitic aureole. Parasitic intrusions in the vicinity of Hall (Fig. 17) have orientations and mineralogical signatures similar to those of the Hall chromitite-dunite pipe, and probably represent coeval sheets. Some of these parasitic intrusions are synkinematic, and suggest ascent of the main Hall 'magma' along synoceanic normal faults (Schroetter et al., 2003).

Taken together, these data are consistent with a hypothesis (Hébert and Bédard, 2000; cf. Corrivaux and Laflamme, 1990) where underplating, water-bearing boninitic intrusions differentiated at the base of the crust, with PGEs being scavenged and concentrated into Cl-H_2O-rich residual melts that accumulated in a structural trap (Fig. 18). Detailed mapping and structural work suggests that the Star Chrome PGE showing (and possibly the Hall orebody, Corrivaux 1989; Corrivaux and Laflamme, 1990; Gauthier et al., 1990; Morin et al., 1992) are associated with pre-obduction normal faults (see below), which would have channeled ore-carrying melts and fluids derived from the ponded intrusions beneath, and which may have triggered decompressive deposition of Cr and Pt-Pd in and around the faults.

Supracrustal Rocks

The TMOC supracrustal sequence is extremely variable, both in thickness and lithology, but boninitic lava flows and

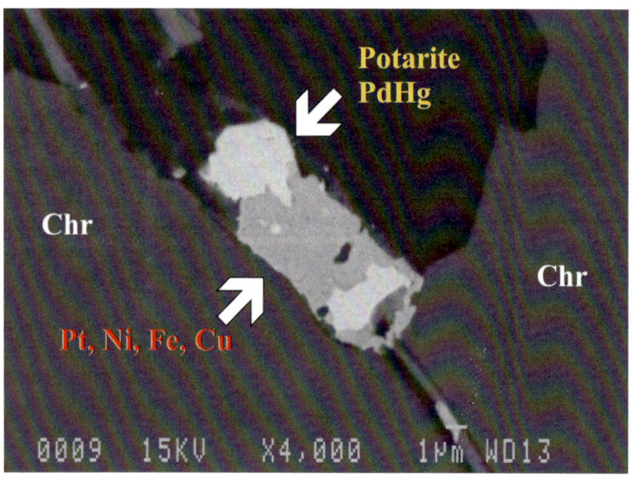

FIGURE 13. Backscattered electron probe image showing a composite grain of potarite (PdHg), and a PtNiFeCu alloy in an intertitial site and filling a fracture in a chromite grain (Chr). Star Chrome showing.

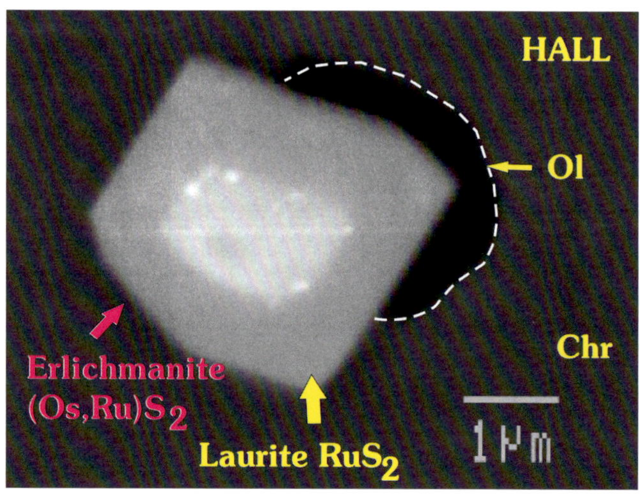

FIGURE 15. Backscattered electron probe image showing a refractory PGE mineral included in chromitite (Chr) from the Hall Mine. Ol = olivine.

FIGURE 16. Wispy pyroxene schlieren in pyroxene-dunite separating the massive metasomatic dunite pipe hosting the Hall orebody, from the surrounding layered websterite-peridotite. Layering orientation in the surrounding peridotite is parallel to the foliation seen in the pyroxene-dunite.

FIGURE 17. Lherzolitic breccia fill in the core of a small fault zone. The matrix peridotite is fresh, while meta-websterite clasts are strongly affected by greenschist-facies hydrothermal metamorphism with development of chlorite + actinolite + prehnite(?) veins. Note that clasts appear to have rotated independently, demonstrating that the peridotite intruded after a complete magmatic hydrothermal cycle and that emplacement may have been guided by faulting. The orientation of this magma-filled fault is subparallel to the orientation of the dunite pipe hostng the Hall orebody; chromite grains in the lherzolitic matrix of the igneous breccia shown here is almost identical to the composition of chromite in the Hall orebody.

felsic pyroclastic rocks dominate (Church, 1977, 1978, 1987; Oshin and Crockett, 1986a). Tholeiitic lavas are locally preserved at the base of the Lac de L'Est section (Oshin and Crockett, 1986a; Laurent and Hébert, 1989) and at Mt. Ham (Hébert, 1980; Olive et al., 1997), while tholeiitic sills cut both series at Mt. Ham (Olive et al., 1997). Boninites are typically extremely magnesium rich and may be high in Cr and Ni, and locally in sulphur and volatiles (e.g. Crawford, 1989; Danyushevsky et al., 1994; Bédard, 1999; Kim and Jacobi, 2002). Their ubiquity in the TMOC suggests a link with Cr and PGE mineralization. The ophiolitic lavas and tuffs are overlain by laterally discontinuous debris flows characterized by centimetre- to metre-scale angular fragments of ultramafic, volcanic, sedimentary, and metasedimentary rocks, known either as the Coleraine Group (Riordon, 1954) or Coleraine breccia (Hébert, 1981). The coarse-grained units wedge out laterally into fine-grained siliciclastic rocks (red argillites and siltstones, green tuffs), and up progressively into turbidites, argillites, siltstones, and pebbly mudstones of the St.-Daniel Mélange (Schroetter et al., 2006). The common and widespread occurrence of felsic pyroclastic rocks at the base of the St.-Daniel Mélange (Figs. 2, 3, 19, 20) and in the potentially correlative Magog Group (Cousineau, 1988; Cousineau and St.-Julien, 1994), suggest that island-arc volcanism was active during and after obduction of the TMOC. Considerable geochronological and geochemical work needs to be done to document how many volcanic events are recorded in this package. The base of the volcano-sedimentary sequence is a major erosional surface in many places, which can penetrate down to lower crustal and mantle levels (Fig. 3). The presence in the TMOC of erosional unconformities that reach pyroxenitic to dunitic cumulates suggests the potential for erosion of crustal PGE-chromites, and its local redeposition as paleoplacers within the overlying sedimentary rocks of the St.-Daniel Mélange.

Structural Evolution

Pre-obduction: We have interpreted crustal thickness variations and juxtaposition of volcanic and deep-crustal facies in the TMOC as the consequences of extension, faulting, and erosion (Schroetter et al., 2003). The evidence for

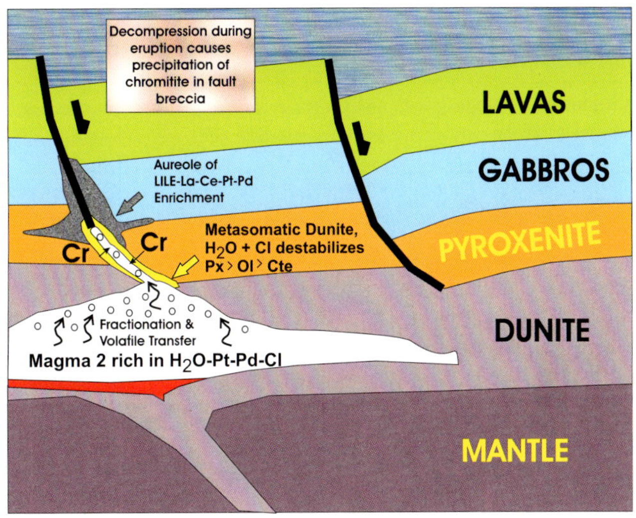

FIGURE 18. Schematic illustration showing a possible model to explain the development of Hall-Star Chrome-type orebodies in the middle-upper crust, with replenishing water-rich (4-8% H_2O) boninitic magma ponds at the base of the crust and fractionates. Fractional crystallization enriches the residual melt in Pt-Pd-H_2O-Cl, and favours volatile-transfer of Cl-Pt-Pd complexes towards the roof zone of the chamber. When a downwardly propagating normal fault penetrates the magma chamber, the PGE-enriched melt escapes through this conduit. Because enrichment of this magma in water destabilizes pyroxene more than olivine, and olivine more than chromite, a metasomatic aureole is created. Dissolution of pyroxene into the melt enriches it in Cr. At some stage, the melt undergoes a sudden decompression and brecciation, possibly linked to movement on the fault, possibly due to propagation of the melt towards the surface. This decompression triggers sudden crystallization of chromite in the open space, with loss of the water and Cl that escapes up along the fault. When chromite crystallizes, enrichments in the siderophile PGEs (Os-Ir-Ru) develop, either as primary metallic inclusions in chromite or as monosulphide blebs later effected by desulphurization reactions. Much of the Pt-Pd budget escapes with the lost volatiles, which ascends along the fault, and could develop cryptic PGE enrichments associated with enrichment in LILEs and LREE. Star Chrome would represent an intermediate facies.

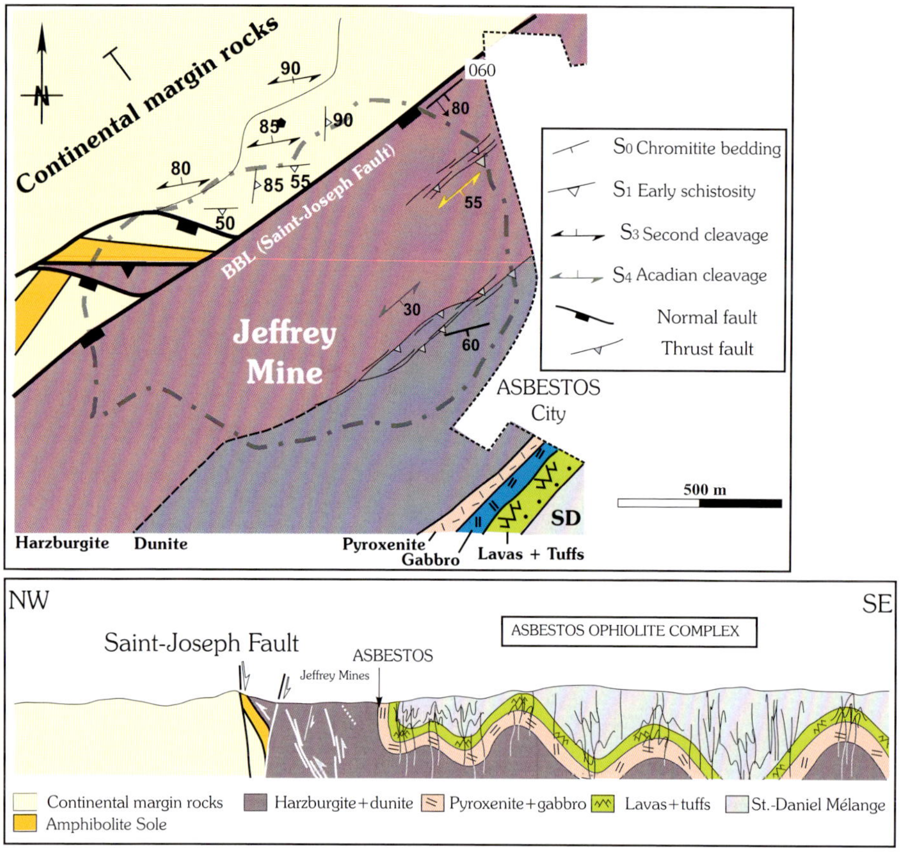

FIGURE 19. Geological map and structural profile of the Jeffrey Mine in the Asbestos area. BBL = Baie Verte-Brompton Line ; SD = Saint-Daniel. See Figure 2 for location. Adapted from Schroetter et al. (2005).

coeval extension and magmatism, and the presence of a sheeted dyke complex (Fig. 6), imply that the TMOC formed by seafloor spreading (Schroetter et al., 2003). The dominance of a boninitic signature in cumulate and volcanic rocks suggests that spreading occurred in a subduction zone environment, possibly in a forearc setting (Bédard and Kim, 2002; cf. Bédard et al., 1998). Our mapping has revealed the presence of numerous subvertically dipping, north-south to 20°-striking synmagmatic (Fig. 17) faults in the TMOC, spaced ~1 km apart on average (Fig. 3; Schroetter et al., 2003). Assuming that rhythmic cumulate layering was originally paleohorizontal, then kinematic analysis implies that these were originally normal faults separating a series of tilted blocks. Swarms of north-south-striking dykes are oriented parallel to the major faults and locally constitute a sheeted complex (Figs. 3, 6).

Obduction: The only structures that can be strictly assigned to obduction processes are those found associated with the metamorphic sole (up to 1 km thick), which occurs at the base of the ophiolite (Fig. 3). Metamorphic temperatures ranged from 850 to 780°C near the contact with the harzburgite tectonite, to 580 to 500°C at the base of the sole; metamorphic pressure ranged from 7 to 5 kbars (Clague et al., 1981; Feininger, 1981). This implies a inverse thermal gradient of 40°C/km (Feininger, 1981). Amphibolitic rocks from the sole yielded $^{39}Ar/^{40}Ar$ ages of 477 ± 5 Ma (Whitehead et al., 1995), whereas the underlying metasediments (i.e. Carineault and Bécancour anticlines) yielded muscovite $^{40}Ar/^{39}Ar$ ages ranging between 469 to 460 Ma (Whitehead et al., 1995; Castonguay et al., 2001). These ages have been interpreted as the result of the intraoceanic detachment (i.e. the amphibolite sole) that culminated with the emplacement of the oceanic crust onto the continental margin (Pinet and Tremblay, 1995a; Tremblay and Castonguay, 2002). The high pressures suggest that Laurentian margin lavas and sediments were subducted down to at least 20 km depth. Two-mica granites with high Sr-isotopic signatures emplaced within mantle rocks of the ophiolite (Clague et al., 1985) yielded U-Pb zircon ages of 469 ± 4 and 470 +5/-4 Ma (Whitehead et al., 2000), suggesting that the thermal anomaly and shear heating associated with juxtaposition of the young, still-hot oceanic mantle against Laurentian margin sediments were sufficient to induce anatexis. The lack of penetrative obduction-related deformation in ophiolitic units situated immediately above the metamorphic sole suggests that it is a localized detachment fault accommodating major displacement.

Post-obduction: Two post-obduction deformational events are recognized in the TMOC. (1) A Silurian southeast-verging folding/backthrusting event, and (2) an Acadian folding/thrust faulting event (Devonian), which forms the dominant regional fabric in the Dunnage Zone of southern Québec (Tremblay and St-Julien, 1990; Tremblay, 1992).

1. Backthrusting structures and fabrics: The northwestern limit of the TMOC is a major backthrust fault, along which greenschist-grade metasediments of the continental margin were thrust over the ophiolite (Fig. 3). The fault trends north-northeast and dips approximately 50° towards the west-northwest. Fabrics indicate a west-over-east sense of shear (Schroetter et al., 2005). Similar southeast-verging (hinterland-directed) thrusts and associated recumbent folds occur elsewhere in the massif (Fig. 3), and are correlated with the D_3 deformation of Tremblay and Pinet (1994), which has been dated at ca. 430 to 415 Ma by Castonguay et al. (2001). Although the TMOC was obducted by 477 to 460 Ma, and was being actively eroded soon afterwards, the common and widespread occurrence of felsic pyroclastic rocks at the base and within the St.-Daniel Mélange (Schroetter et al., 2003, 2006) and in the potentially correlative Magog Group (Cousineau, 1984, 1988, 1994; Cousineau and St.-Julien, 1994), imply that island-arc volcanism was active during and after TMOC obduction. The collision of such an arc (perhaps preserved in the Ascot Complex, Fig. 1) with the Laurentian margin and the accreted ophiolites might account for the transition from regional deformation dominated by obduction-related and thin-

skinned tectonics to a tectonic style dominated by thick-skinned basement tectonics, and would have induced the formation of the southeast-verging backthrusts and backfolds in the metamorphic rocks of the internal Humber Zone and the adjacent ophiolites (Castonguay et al., 1997; Castonguay and Tremblay, 2003; Schroetter et al., 2005).

2. Acadian folding/Reverse faulting: This deformational event is characterized by subvertical folds trending northeast-southwest with an axial-planar cleavage that is developed on a regional scale (St-Julien and Hubert, 1975; Labbé and St-Julien, 1989; Tremblay and St-Julien, 1990; Tremblay and Pinet, 1994). Associated reverse faults trend north-northeast and dip steeply to the southeast. Chromitite beds and peridotitic dykes in dunite can be used as stratigraphic markers and define open folds with axial surfaces that are subparallel to the reverse faults, which suggests to us that the reverse faults and the open folds are cogenetic (Schroetter et al., 2005). In terms of regional structures, these reverse faults and related folds are responsible for: the formation of (a) a dome-and-basin interference pattern (Ramsey Type I) with antiformal culminations corresponding to the Carineault and Bécancour antiforms, (b) a series of northwest-directed reverse faults that juxtaposed the lower crust against St.-Daniel sediments, and (c) folds within the St.-Daniel Mélange and the overlying Magog Group (Fig. 3).

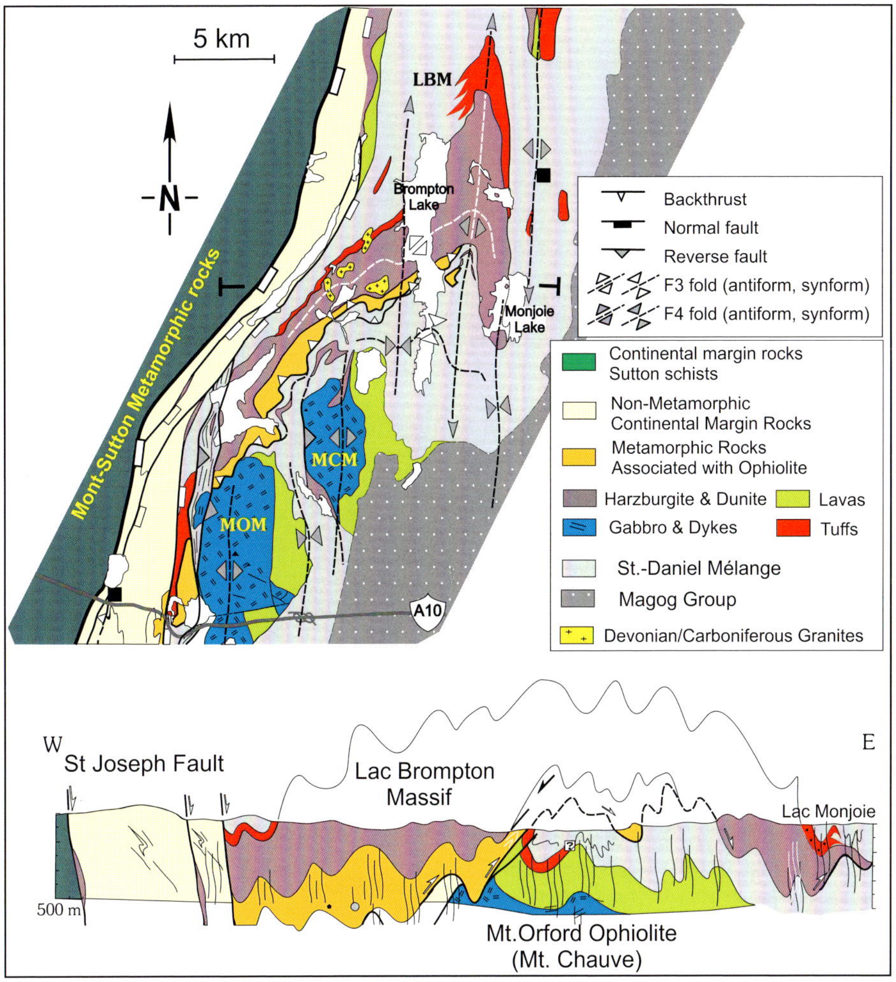

FIGURE 20. Geological map and structural profile of the Lac Brompton (LBOC) and Mont Orford (MOOC) ophiolitic complexes. See Figure 2 for location. Adapted from Schroetter et al. (2005). LBM = Lac Brompton massif; MCM = Mont Chauve massif; MOM = Mont Orford massif.

The Asbestos and Lac Brompton Ophiolitic Complexes

The Asbestos Ophiolite Complex (AOC) is located approximately 20 km to the southwest of the Thetford Mines ophiolitic complex (Figs. 1, 2, 19). The contact between the ophiolite and the internal Humber Zone to the west defines the Baie Verte-Brompton Line (Williams and St-Julien, 1982). Here the synobduction event is restricted to an amphibolitic thrust slice embedded in a major fault at the contact between the Humber Zone and the ophiolite (Fig. 19). The 50 metre-wide fault zone shows well developed SC fabrics indicating down-to-the-east normal faulting (Schroetter et al., 2005). This contact (and hence the Baie Verte-Brompton Line) is therefore interpreted as a major normal fault that has been correlated with the St. Joseph Fault of the Thetford Mines area, and which forms the St. Joseph-Baie Verte-Brompton Line fault system (Tremblay and Castonguay, 2002). It is attributed to the Late Silurian-Lower Devonian southeast-verging deformation and is overprinted by Acadian regional folds and faults.

AOC U/Pb zircon ages of 473 to 480 Ma (Dunning et al., 1986; Whitehead et al. 2000) are similar to those of the TMOC. In comparison to the TMOC, the AOC preserves a thinner (2-2.5 km) but very similar ophiolitic sequence (Fig. 19), consisting of harzburgite tectonite overlain by dunitic, pyroxenitic, and gabbroic cumulates, and capped by diabasic and volcanic rocks (Hébert, 1980). Rhythmic, stratiform chromitite layers occur in the dunitic zone cumulates. The lavas are mostly tholeiitic at the base and dominantly boninitic at the top (Hébert, 1980). The ophiolitic lavas are overlain by fine-grained volcaniclastic rocks and flow breccias, and then by the phyllites of the St.-Daniel Mélange. The AOC crust is chopped up by north-south-trending paleonormal faults that may be equivalent to the pre-obduction faults in the TMOC, which have similar orientations. Because the AOC and the TMOC show the same gross stratigraphy, are both dominated by boninitic magmas, have similar U/Pb zircon formation ages and pre-obduction structures, and occupy similar structural positions in relation to backthrusts (Schroetter et al., 2005), we suggest that they were originally connected and may have been obducted as a single coherent slab.

To the east and south of the AOC (Figs. 2, 20), the dominant structures are north-northeast- to northeast-trending Acadian folds with a subvertical axial-planar cleavage. These folds define a series of antiforms and synforms that dominate the geological map pattern of this region, with ophiolitic rocks occurring locally as antiformal culminations protruding through the St.-Daniel Mélange (Figs. 2, 19). Ophiolitic plutonic rocks occur in the vicinity of the St. François River and consist of dunitic peridotites, gabbros, and volcanic and volcaniclastic rocks that can be correlated with those of the AOC (Figs. 2, 20). This sequence of ophiolitic rocks extends almost continuously southwards toward St.-Élie-de-Brompton where it merges into the Lac Montjoie ophiolitic mélange of the Lac Brompton ophiolitic complex (LBOC; Daoust et al., 2005; DeSouza et al., 2005), previously described as a serpentinite diapir (Lamothe, 1978). Our reconnaissance-scale fieldwork (with additional observations reported by Daoust et al. (2005) and DeSouza et al. (2005)) suggests rather that the Lac Montjoie rocks are part of an ophiolitic sequence (mantle harzburgite, dunite, chromitite and orthopyroxenite dykes, gabbros), that is underlain, perhaps tectonically, by metamorphic rocks of continental affinity to the south. These rocks are stratigraphically overlain by AOC-type boninitic lavas and tuffs and by conglomerates and finer grained sedimentary rocks of the St.-Daniel Mélange, with the whole sequence defining a northward-plunging anticline.

The continuity of lithologies and exposures suggest to us that the AOC crops out discontinuously between Asbestos itself and Brompton. Given the previously discussed resemblance between the TMOC and the AOC, this implies that the ophiolitic rocks of this large area (over 120 km of strike length) may originally have formed a single panel of obducted oceanic crust. The AOC or LBOC have not been systematically investigated in terms of potential PGE mineralization. In view of the similar ages and lithologies, and the interpretted correlation with the TMOC, we infer that the AOC and LBOC constitute a large potential PGE exploration target.

The Mont Orford Ophiolite Complex

South of Saint-Élie-de-Brompton, the ophiolitic sequence of the The Mont Orford Ophiolite Complex (MOOC) forms two principal masses, the Mont Chauve and the Mont Orford-Chagnon massifs (Figs. 1, 2, 20; Rodrigue 1979, Laurent and Hébert, 1989, Huot, 1997; Turcotte, 1999; Huot et al., 2002). A thin basal mélange has a serpentinite matrix and contains a wide variety of lithologies as clasts, including mantle harzburgite tectonite (Huot et al., 2002). Above this mélange, there are 100 m scale tectonic slivers of harzburgitic mantle rocks and dunites (Huot et al., 2002). This deformation was interpreted as being related to obduction by Turcotte (1999). In the Mont Chagnon massif, most of the plutonic crust is composed of inter-penetrating intrusions of gabbro, hornblende-gabbro, diorite and trondhjemite (Huot et al., 2002). Mafic cumulates are more abundant in the Mont Orford massif (Rodrigue, 1979). Volcanic rocks at Mt.Chagnon are mainly basalts and andesites, with localized felsic lava flows and detrital units, overlain by St.-Daniel sediments (Huot et al., 2002). Lava chemistry has been interpreted in terms of arc-backarc (Harnois and Morency, 1989; Laurent and Hébert, 1989; Hébert and Laurent, 1989) or arc-forearc to backarc environments (Huot et al., 2002). Faults and polymictic conglomerates occurring at the base of the supracrustal unit are interpreted as synoceanic faults and associated talus deposits by Huot (1997). The existing age of 504 ± 3 Ma (David and Marquis, 1994) is problematic, since separated zircons are contaminated by grenvillean-age cores. However, a tectonostratigraphic link between the AOC-LBOC and the MOOC can be inferred from structural relationships shown by the ophiolitic rocks in the Saint-Élie-de-Brompton area (Figs. 2, 20) where the southern extremity of the LBOC is separated from the MOOC by an extensive metamorphic unit of micaschists and albite-chlorite laminated greenschist (Fig. 20). As it is the case in the Thetford Mines area, these metamorphic rocks are characterized by three generations of fabrics, documented by St-Julien (1961, 1963, 1970) and Lamothe (1978). The youngest structures are northeast-trending, open to tight folds and are clearly related to the Acadian Orogeny (Tremblay, 1992). The pre-Acadian deformation is more difficult to interpret, but both the MOOC and the overlying St.-Daniel Mélange seem to be structurally overlain by (1) the metamorphic rock unit mentioned above, and (2) ophiolitic rocks of the LBOC. The structural juxtaposition occurs along a northwest-dipping fault zone (Fig. 20) attributed to the southeast-verging back-thrusting event we have documented in the TMOC (Schroetter et al., 2005), and which is refolded by a north-plunging Acadian antiform in the Brompton Lake area.

Implications for Exploration

Mantle and lower crustal rocks in the southern Québec Ophiolite Belt rarely contain high PGE abundances, and so represent poor PGE exploration targets. These facies may, however, still contain significant amounts of chromitite and the largely unmapped parts of the LBOC especially may represent a target for Cr exploration.

All TMOC cumulates analysed to date have yielded boninitic geochemical signatures, and are probably consanguineous with overlying lavas and sheeted dykes. The dominance of boninitic magmas suggests a genetic link with Cr-PGE mineralization, suggesting that the identification of boninite-rich lava or cumulate sections in ophiolites may indicate a favourable environment for Cr-PGE exploration. Alternatively, given the difficulty of identifying boninitic rocks, the presence of abundant dunitic facies may be a useful guide.

Veins and breccia fills of chromitite enriched in PGEs cut and replace dunites (Hall deposit) and pyroxenites (Star Chrome showing) of the middle crust. These ores locally reach grades of over 20 g/ton Pt+Pd. Our mapping and structural work suggest that these PGE ores are associated with pre-obduction paleonormal faults, and may have been generated from Cl-PGE-enriched aqueous fluids that concentrated in structural traps above large boninitic sills that underplated the crust and differentiated (Fig. 18). Movements along some of these faults may have opened up channels, allowing PGE-enriched fluids to escape and migrate up along the faults, possibly depositing ores as a result of decompression, vesiculation, or fluid mixing. Temperatures would have decreased with distance and time, allowing formation of lower temperature minerals such as potarite. As a consequence, it seems possible that these paleonormal faults might be good places to explore for PGE ore.

At Star Chrome, the Pt-Pd enrichment seems to have little to do with any magmatic geochemical signature (S or Cr content). Enrichment in light rare-earth elements has been observed in the cumulate rocks in immediate contact with the ore, and could represent a useful exploration vector. The existence of potarite (PdHg) suggests that analysis of Hg in soils might be a useful exploration tool in areas covered by Quaternary overburden.

The documentation in the TMOC of erosional surfaces that penetrated down to the Dunitic Zone suggests that there exists the potential for erosion and redeposition of Hall-Star Chrome-type PGE ore to produce paleoplacer deposits in the sedimentary rocks that immediately overlie the ophiolite. We suggest that closely spaced sampling of the base of the St.-Daniel might be a useful strategy. A more detailed sedimentological analysis of the lower St.-Daniel is needed to provide insight as to where paleoplacers might best be sought.

Our stratigraphic work (Schroetter et al., 2006) implies that the base of the St.-Daniel is represented by the debris flows previously known as the 'Coleraine Breccias'; and that the St.-Daniel is not an accretionary mélange, but rather the base of a fore-arc flysch, possibly directly correlative to the Magog Group, which developed as a piggy-back basin on top of the ophiolite as it was emplaced onto the continental margin. Felsic tuffs are common at the base of the St-Daniel (Figs. 1, 2, 20; Schroetter et al., 2003, 2006) and in the Magog Group (Cousineau and St.-Julien, 1994). Tuffaceous units are also common to the southeast of the Asbestos complex (Figs. 2, 20), but it is not clear whether these represent (a) younger tuffs unrelated to the basal St.-Daniel rocks, or (b) structural repetitions of the basal St.-Daniel, possibly with associated ophiolitic facies. If hypothesis 'b' is correct, then the sedimentary sequences in and around these tuffs may also host paleoplacer deposits. Much more detailed age dating and geochemical correlation work is needed to clarify the internal stratigraphy of the St.-Daniel and Magog tuffs, and to determine whether tuffs to the southeast of the AOC also represent potential exploration targets. These lithotectonic correlations between the different massifs of the Southern Quebec Ophiolite Belt imply that the high PGE potential of the northernmost massif, the Thetford Mines complex, might also extend to the other massifs in the suite, defining a metallogenetic province over 120 km in length.

Conclusions

The southern Québec Ophiolite Belt consists of a series of partly dismembered oceanic terranes accreted against the Laurentian margin in the Ordovician, and then reworked by Silurian and Devonian events. The TMOC contains the bulk of the known Cr-PGE mineralization. All TMOC cumulate rocks analyzed to date have trace element geochemical signatures that imply a boninitic affinity, implying consanguinity with the overlying boninitic lavas and dykes. The ubiquity of boninites in the TMOC suggests a link with Cr and PGE mineralization. Podiform chromitites occur in the mantle, while stratiform chromitites are common in lower crustal dunites. Chromitite veins and breccia fills located in crustal dunite and pyroxenite locally reach grades of 20 g/ton Pt+Pd, and we strongly suspect that the PGEs and Cr were transported by magmatic fluids or fluid-charged magmas (Fig. 20). The presence in the TMOC of erosional unconformities that reach deep into the crust suggests the potential for paleoplacer-type ore in overlying sedimentary rocks. On the basis of stratigraphic, geochemical, age, and structural arguments, we correlate the TMOC and AOC, which implies that the two were obducted as a single slab, and hence, have identical PGE-exploration potential. This correlation may extend down to the LBOC, which our reconnaissance work suggests is dominated by mantle rocks. The LBOC is severely undermapped, and much detailed structural, stratigraphic, petrological, geochemical, and geochronological work will be required to confirm this hypothesis. Further south, the Mont Chauve and Mont Orford massifs (MOOC) appear to be different from the TMOC-AOC-LBOC, with only rare occurrences of lower crustal or mantle facies, and of boninitic magmas. Additional age dating and other geological work is needed to link the putative TMOC-AOC-LBOC oceanic terrane with the MOOC.

Acknowledgements

Thanks are due to Réjean Hébert, Roger Laurent, and the late Pierre St-Julien for introducing us to the area and sharing their knowledge during earlier field seasons; to Bertrand Brassard and Ressources Allican Inc.; Nicolas Pinet, Yves Hébert, and François Huot for numerous discussions; Pierre Cousineau and the late lamented Giff Kessler for volcanological and sedimentological insights; and to Yildirim Dilek and Peter Robinson for comments on previous efforts. The manuscript benefited from constructive reviews by Ingrid Kjarsgaard, Cees van Staal, Ray Coish, and Jon Kim. This research was supported by the Geological Survey of Canada, by Canadian National Science and Engineering Research Council grant ESS 233685-99, by a Diversification de l'Exploration Minérale au Québec grant provided by Valorisation-Recherche Québec (project 2201-133), and by Ressources Allican Inc. This is Geological Survey of Canada Contribution # 2005300.

References

Bécu, V., 2005: Pétrologie et géochimie de l'indice de chromitites platinifères Star Chrome, complexe ophiolitique de Thetford Mines: M.Sc. thesis, Institut National de la Recherche Scientifique, Québec, PQ, Canada, 177 p.

Bédard J.H., 1994. A procedure for calculating the equilibrium distribution of trace elements in the minerals of cumulate rocks, and the concentration of trace elements in the coexisting liquids: Chemical Geology, v. 118, p. 143-153.

——— 1999, Petrogenesis of boninites from the Betts Cove Ophiolite, Newfoundland, Canada: Identification of subducted source components: Journal of Petrology, v. 40, p. 1853-1889.

——— 2001, Parental magmas of the Nain Plutonic Suite anorthosites and mafic cumulates: a trace element modelling approach: Contributions to Mineralogy and Petrology, v. 141. p. 747-771.

Bédard, J.H., and Hébert, R., 1998, Intra-crustal chromitites, North Arm Mountain Massif, Bay of Islands ophiolite: Formation by incongruent dissolution of feldspar and pyroxene into intrusions of peridotite: Journal of Geophysical Research, v. 103, p. 5165-5184.

Bédard, J.H., and Kim, J., 2002, Boninites in NE Appalachian Notre-Dame Subzone ophiolites: Tectonic models for Iapetus terranes: Geological Society of America NE, Abstracts with Programs, v. 34, p. A60.

Bédard, J.H., Hébert, R., and Constantin, M., 1992, Syn-kinematic boninitic intrusions and hybrid gabbroic breccias from the Thetford Mines ophiolite: Geological Association of Canada/Mineralogical Association of Canada, Wolfville, Program and Abstracts, v. 17, p. A7.

Bédard, J.H., Lauzière, K., Tremblay, A., and Sangster, A., 1998, Evidence from Betts Cove ophiolite boninites for forearc seafloor-spreading: Tectonophysics, v. 284, p. 233-245.

Bédard, J.H., Tremblay, A., Schroetter, J.-M., Bécu, V., and Pagé, P., 2001, Structural and magmatic evolution of the Thetford Mines Ophiolite: Preliminary results from new mapping and analysis: Geological Association of Canada / Mineral Association of Canada, Abstracts with Program, v. 26, p. 11.

Beulac, R.X., 1982, Etude pétrologique du Complexe ophiolitique du lac Nicolet: M.Sc. thesis, Université Laval, Québec, Canada, 79 p.

Brassard, B., and Tremblay, A., 1999, Synthèse géologique et métallogénique de la M.R.C. de l'Amiante. Rapport final préparé pour la société d'aide au développement de la collectivité de l'Amiante (SADC), 1/50 000 scale map, northern and southern sheets of the Amiante MRC.

Brodeur, E., and Marquis, R., 1995, Géologie de la région d'Orford: Ministère des Ressources naturelles du Québec-Ressources naturelles Canada, ET 93-06, 73 p.

Castonguay, S., and Tremblay, A., 2003, Tectonic evolution and significance of Silurian-Early Devonian hinterland-directed deformation in the internal Humber zone of the southern Quebec Appalachians, Canadian Journal of Earth Sciences, v. 40, p. 255-268.

Castonguay, S., Tremblay, A., Ruffet, G., Féraud, G., Pinet, N., and Sosson, M., 1997, Ordovician and Silurian metamorphic cooling ages along the Laurentian margin of the Québec Appalachians: Bridging the gap between New-England and Newfoundland: Geology, v. 22, p. 583-586.

Castonguay, S., Ruffet, G., Tremblay, A., and Féraud, G., 2001, Tectonometamorphic evolution of the southern Quebec Appalachians: $^{40}Ar/^{39}Ar$ evidence for Middle Ordovician crustal thickening and Silurian-Early Devonian exhumation of the internal Humber Zone, Geological Society of America Bulletin, v. 113, p. 144-160.

Church, W.R., 1977, The ophiolites of southern Quebec: oceanic crust of Betts Cove type: Canadian Journal of Earth Sciences, v. 14, p. 1668-1673.

——— 1978, The ophiolites of southern Quebec: oceanic crust of Betts Cove type: Reply: Canadian Journal of Earth Sciences, v. 15, p.1882-1883.

——— 1987, The geochemistry and petrogenesis of ophiolitic volcanic rocks from Lac de l'Est, Thetford Mines Complex, Quebec, Canada: Discussion: Canadian Journal of Earth Sciences, v. 24, p. 1270-1273.

Clague, D., Rubin, J., and Brackett, R., 1981, The age and origin of garnet amphibolite underlying the Thetford Mines ophiolite, Quebec: Canadian Journal of Earth Sciences, v. 18, p. 469-486.

Clague, D., Frankel, C.S., and Eaby, J.S., 1985, The age and origin of felsic intrusions on the Thetford Mines ophiolite, Québec: Canadian Journal of Earth Sciences, v. 22, p. 1257-1261.

Cooke, M.H.-C., 1938, Région de Thetford, de Disraëli et de la moitié orientale de Warwick Québec: Commision Géologique du Canada, Memoir 211, 176 p.

——— 1950, Geology of the Southwestern Part of the Eastern Townships of Québec: Geological Survey of Canada, Memoir 257, 142 p.

Corrivaux, L., 1989, Étude des gîtes de chrome-platine du Canton de Coleraine, Estrie: M,.Sc. thesis, Université du Québec á Montréal, Canada, 190 p.

Corrivaux, L., and Laflamme, J.H.G., 1990, Mineralogy of platinum group elements in chromitite and ophiolite of Thetford Mines, Quebec: Canadian Mineralogist, v. 28, p. 579-595.

Cousineau, P.A., 1984, La Formation de Saint-Daniel et le Groupe de Magog entre Saint-Georges-de-Beauce et le lac Etchemin: Ministère des ressources Naturelles du Québec, MB 84-09, 95 p.

——— 1988, Paléogéographie et évolution tectonique d'une partie de la zone de Dunnage à l'est de la rivière Chaudière, Québec: Unpub. Ph.D. thesis, Université Laval, Québec, Canada, 284 p.

——— 1994, Subaquous pyroclastic deposits in an Ordovician fore-arc basin: An example from the Saint-Victor Formation, Quebec Appalachians, Canada: Journal of Sedimentary Petrology, v. A64, p. 867-880.

Cousineau, P.A., and St-Julien, P., 1992, The Saint-Daniel Mélange: Evolution of an accretionary complex in the Dunnage terrane of the Québec Appalachians: Tectonics, v. 11, p. 898-909.

——— 1994, Stratigraphie et paléogéographie d'un bassin d'avant-arc ordovicien, Estrie-Beauce, Appalaches du Québec: Canadian Journal of Earth Sciences, v. 31, p. 435-446.

Crawford, A.J., ed., 1989, Boninites: Unwyn Hyman, London, 465 p.

Danyushevsky, L.V., Falloon, T.J., Sobolev, A.V., Crawford, A.J., Carroll, M., and Price, R.C., 1994, The petrology and geochemistry of boninite: Journal of Petrology, v. 35, p. 1183-1211.

Daoust, C., DeSouza, S., Tremblay, A., and Gauthier, M., 2005, The Lac-Brompton ophiolitic Complex: Stratigraphy and structure of a newly-recognized ophiolite in the Southern Québec Appalachians, Canada: Geological Society of America NE meeting, Saratoga Springs, NY, Abstracts with Program, v. 37, p. 72.

David, J., and Marquis, R., 1994, Géochronologie U-Pb dans les Appalaches du Québec: Application aux roches de la zone de Dunnage: La Revue géologique du Québec, v. 1, p. 16-20.

De Souza, S., Daoust, C., Gauthier, M., and Tremblay, A., 2005, The Lac-Brompton ophiolitic Complex: Stratigraphy, structure and Cu-Ni mineralization potential, southern Québec Appalachians, Canada: Geological ssociation of Canada Meeting, Halifax, Abstracts, v. 30, p. 42.

Dérosier, C., 1971, Étude géologique des brèches de la région de Thetford Mines (Province de Québec, Canada: Ph.D. thesis, Université de Paris VI, Paris, France, 124 p.

Dunning, G.R., Krogh, T.E., and Pederson, R.B., 1986, U-Pb zircon ages of Appalachian-Caledonian ophiolites: Terra Cognita 6, Abstract L51.

Feininger, T., 1981, Amphibolite associated with the Thetford Mines Ophiolite Complex at Belmina Ridge, Québec: Canadian Journal of Earth Sciences, v. 18, p. 1878-1892.

Gauthier, M., Auclair, M., and Durocher, M., 1987, Synth Étude pétrologique du Complexe Ophiolitiquèse métallogénique de l'Estrie et de la Beauce (Secteur Nord) Annexe 1: Ministère de l'Énergie et des Ressources du Québec, Manuscrit Brut MB 87-39.

Gauthier, M., Corrivaux, L., Trottier, L.J., Cabri J., Laflamme, J.H.G., and Bergeron, M., 1990, Chromitites platinifères des complexes ophiolitiques de l'Estrie-Beauce, Appalaches du Sud du Québec: Mineralium Deposita, v. 25, p. 169-178.

Harnois, L., and Morency, M., 1989, Geochemistry of Mount Orford Ophiolite Complex, Northern Appalachians, Canada: Chemical Geology, v. 77, p. 133-147.

Hébert, R., 1980, Etude pétrologique des roches d'Asbestos et du Mont Ham (Ham Sud), Québec: M.Sc. thesis, Université Laval, Québec, Canada, 182 p.

——— 1981, Conglomérats polygéniques ophiolitiques: anciens éboulis de talus de fond océanique: Canadian Journal of Earth Sciences, v. 1, p. 619-623.

Hébert, R., and Bédard, J.H., 2000, Les ophiolites d'avant-arc et leur potentiel minéral: exemple des complexes ophiolitiques du sud du Québec: Chronique de la Recherche Minière 539, p. 101-117.

Hébert, R. and Laurent, R., 1989, Mineral chemistry of ultramafic and mafic plutonic rocks of the Appalachian ophiolites, Québec, Canada: Chemical Geology, v. 77, p. 265-285.

Hébert, Y., 1983, Étude pétrologique du Complexe Ophiolitique de Thetford Mines, Québec: Ph.D. thesis, Université Laval, Québec, Canada, 426 p.

Huot, F., 1997, Etude pétrologique des roches des processus magmatiques reliés au massif ophiolitique du Mont-Chagnon, Québec, Canada: M.Sc. thesis, Université Laval, Québec, Canada, 105 p.

Huot, F., Hébert, R., and Turcotte, B., 2002, A multistage magmatic history for the genesis of the Orford ophiolite (Quebec, Canada): a study of the Mt Chagnon massif: Canadian Journal of Earth Sciences, v. 39, p. 1201-1217.

Kaçira, N., 1971, Geology of Chromitite Occurences and Ultramafic Rocks of the Thetford Mines-Disraëli Area, Québec: Ph.D. thesis, University of Western Ontario, London, Canada, 246 p.

Kim, J., and Jacobi, R.D., 2002, Boninites: characteristics and tectonic constraints, northeastern Appalachians: Physics and Chemistry of the Earth, v. 27, p. 109-147.

Labbé, J.-Y., and St-Julien, P., 1989, Failles de chevauchement acadiennes dans la region de Weedon, Estrie, Québec: Canadian Journal of Earth Sciences, v. 26, p. 2268-2277.

Lamarche, R.Y., 1973, Complexe ophiolitique d'Asbestos: Ministère de l'Énergie et des Ressources du Québec, DP-144, GM 28558, 9 p.

Lamothe, D., 1978, Analyse structurale du Mélange ophiolitique du Lac Montjoie: M.Sc. thesis, Université Laval, Québec, Canada, 68 p.

Laurent, R., 1975a, Occurences and Origin of the Ophiolites of the Southern Quebec, Northern Appalachians: Canadian Journal of Earth Sciences, v. 12, p. 443-455.

——— 1975b, Petrology of the alpine-type serpentinites of Asbestos and Thetford Mines, Quebec: Schweizerische Mineralogisch und Petrologische Mitteilungen, v. 55 p. 431-455.

Laurent, R., and Hébert, R., 1989, The volcanic and intrusive rocks of the Québec Appalachian ophiolites (Canada) and their island-arc setting: Chemical Geology, v. 77 p. 287-302.

Laurent, R., and Kaçira, N., 1987, Chromite deposits in the Appalachian ophiolites: *in* Stowe, C.W., ed., Evolution of Chromium Ore Fields: Van Nostrand Reinhold Co., N.Y., p. 169-193.

Laurent, R., Hébert, R., and Hébert, Y., 1979, Tectonic setting and petrological features of the Quebec appalachian ophiolites, *in* Malpas, J., and Talkington, R.W., eds., Ophiolites of the Canadian Appalachians and Soviet Urals: International Geological Correlation Project 39, Memorial University of Newfoundland, Report No. 8, p. 53-77.

Lavoie, D., 1989, Géologie de la Formation de Saint-Daniel et du Groupe de Magog, région de Richmond, Rapport final, Estrie: Ministère de l'Énergie et des Ressources du Québec, MB 89-06, 44 p.

Leblanc, M., and Nicolas, A., 1992, Les chromitites ophiolitiques: Chronique de la Recherche Minière 507, p. 3-25.

Marquis, R., 1989, Géologie de la région de Windsor, Estrie: Ministère de l'Énergie et des Ressources du Québec, MB 89-12, 22 p.

Morin, Y., Hébert R., and Bédard, J., 1992, Origin of the chromitites of the Hall orebody in the Thetford Mines ophiolite complex, Québec: Geological Association of Canada/Mineralogical Association of Canada, Abstracts with Program v. 17, p. 79-80.

Olive, V., Hébert, R., and Loubet, M., 1997, Isotopic and trace element constraints on the genesis of boninitic sequence in the Thetford-Mines ophiolitic complex, Québec, Canada: Canadian Journal of Earth Sciences, v. 34, p. 1258-1271.

Osberg, P.H., 1978, Synthesis of the geology of north-eastern Appalachians, USA, *in* Tozer, E.T., and Schenk, P.E., eds., Caledonian-Appalachian Orogen of the North Atlantic region: Geological Survey of Canada, Paper 78-13, p. 137-147.

Oshin, I.O., and Crockett, J.H., 1982, Noble metals in the Thetford Mines ophiolite, Quebec, Canada: Part I., Distribution of gold, iridium, platinum and palladium in the ultramafic and gabbroic rocks: Economic Geology, v. 77, p. 1556-1570.

——— 1986a, The geochemistry and petrogenesis of ophiolitic volcanic rocks from Lac de l'Est, Thetford Mines Complex, Quebec, Canada: Canadian Journal of Earth Sciences, v. 23, p. 202-213.

——— 1986b, Noble metals in the Thetford Mines ophiolite, Quebec, Canada: Part II. Distribution of gold, silver, iridium, platinum and palladium in the Lac de L'Est volcanosedimentary section: Economic Geology, v. 81, p. 931-945.

Pagé, P., Bécu, V., Tremblay, A., Bédard, J.H., and Brassard, B., 2001, Occurrences of Platinum-palladium-enriched chromitites in the Thetford Mines ophiolite, southern Québec: Geological Society of America, Abstracts with Programs v. 33, p. 58-59.

Pagé, P., Bédard, J.H., Tremblay, A., and Schroetter, J.M., 2003, The Thetford Mines Ophiolite Complex: Focus on the petrology, mineralogy and geochemistry (REE, PGE) of a supra-subduction mantle section: EOS Transactions of the American Geophysical Union, v. 84 p. F1538.

Pagé, P., Bédard, J.H., Tremblay, A., and Minarik, W., 2004, Systematics of Platinum-group element distribution in the boninitic Thetford Mines ophiolite complex, Canada: Melting and fractional crystallization effects: EOS Transactions of the American Geophysical Union, v. 85, p. F508.

Pinet, N., 1995, Cartographie du secteur Lac Bisby- Petit Lac Saint-François, Propriété Coleraine. QERDEM B Dossier d'exploration minière: Ressources Minières Coleraine Inc., GM 53715, 33 p.

Pinet, N., and Tremblay, A., 1995a, Tectonic evolution of the Québec-Maine Appalachians : from oceanic spreading to obduction and collision in the north Appalachians: American Journal of Science, v. 295, p. 173-200.

——— 1995b, Is the Taconian orogeny of the southern Québec the result of an Oman-type obduction?: Geology, v. 23, p. 121-124.

Pinet, N., Castonguay, S., and Tremblay, A., 1996, Thrusting and back thrusting in the Taconian internal zone, southern Quebec Appalachians: Canadian Journal of Earth Sciences, v. 33, p. 1283-1293.

Ratcliffe, N.M., Hames, W.E., and Stanley, R.S., 1998, Interpretation of ages of arc magmatism, metamorphism, and collision tectonics in the Taconian orogen of western New England: American Journal of Science, v. 298, p. 791-797.

Riordon, P.H., 1954. Région de Thetford Mines-Black Lake, Province de Québec: Ministère des Mines, Rapport Préliminaire 295, 23 p.

Robinson, P., Tucker, R.D., Bradley, D., Berry, H.N., and Osberg, P.H., 1998, Paleozoic Orogens in New England, USA. Gff (Geologiska Fööreningens i Stockholm Föörhandlingar), v. 120, p. 119-148.

Rodrigue, G., 1979, Étude pétrologique des roches ultramafiques du Mont Orford, Québec: M.Sc. thesis, Université Laval, Québec, Canada, 148 p.

Ross, M., and Virta, R.L., 2001, Occurrence, production and uses of asbestos, *in* Nolan, R.P., Langer, A.M., Ross, M., Wicks, F.J., and Martin, R.F., eds., The Health Effects of Chrysotile Asbestos – Contribution of Science to Risk-Management Decisions: Canadian Mineralogist, Special Publication 5, p. 79-88.

Schroetter, J.-M., Bécu, V., Pagé, P., Tremblay, A., and Bédard, J.H., 2000, Chromitites ophiolitiques de la région de Thetford Mines, Rapport des activités de terrain de la saison Été 2000: Ministère des Ressources Naturelles Québec, GM 58649, 98 p., CD ROM.

Schroetter, J.-M., Pagé, P., Tremblay, A., and Bédard, J.H., 2001, Origine et caractérisation structurale des chromitites ophiolitiques du sud du Québec: Rapport d'activités de la saison de terrain été 2001, pour le compte de la société Allican Inc., 38 p.

Schroetter, J.-M., Pagé, P., Bédard, J.H., Tremblay, A., and Bécu, V., 2003, Forearc extension and seafloor spreading in the Thetford Mines Ophiolite complex, *in* Dilek, Y., and Robinson, P.T., eds., Ophiolites in Earth History: Geological Society of London, Special Publication 218, p. 231-251.

Schroetter, J.-M., Tremblay, A., and Bédard, J.H., 2005, Structural evolution of the Thetford-Mines Ophiolitic Complex, Canada: Implications for the southern Québec Ophiolite Belt: Tectonics, v. 24, TC1001, 20 p., doi: 10.1029/2003TC00160.

Schroetter, J.-M., Tremblay, A., Bédard, J.H., and Villeneuve, M.E., 2006, Syncollisional basin development in the Appalachian Orogen - The St-Daniel Mélange, southern Québec, Canada: Geological Society of America Bulletin, v. 118, p. 109-125.

Stanley, R.S., and Ratcliffe, N.M., 1985, Tectonic synthesis of the Taconian orogeny in western New England: Geological Society of America Bulletin, v. 96, p. 1227-1250.

St-Julien, P., 1961, Rapport préliminaire sur la région du Lac Montjoie: Ministère des Richesses Naturelles, Québec, Canada, RP-464, 14 p.

——— 1963, Rapport préliminaire sur la région de Saint-Élie-d'Orford, Comtés de Sherbrooke et de Richmond: Ministère des Richesses Naturelles, Québec, Canada, RP-492, 13 p.

——— 1970, Région d'Orford-Sherbrooke, compilation géologique: Ministère de l'Énergie et des Ressources du Québec, Carte 1619.

——— 1971, Géologie de la région de Disraeli (moitié est) : Ministère des ressources Naturelles du Québec, RP-587, 23 p.

——— 1987, Géologie des régions de Saint-Victor et de Thetford-Mines (moitié est): Ministère de l'Énergie et des Ressources du Québec, MM 86-01, 66 p.

St-Julien, P., and Hubert, C., 1975, Evolution of the Taconian Orogeny in the Québec Appalachians: American Journal of Science, v. 275, p. 337-362.

St-Julien P., and Slivitsky, A., 1985, Compilation géologique de la region de l'Estrie-Beauce: Ministère de l'Énergie et des Ressources du Québec, carte 2030, scale 1:250 000.

Sun, S.-S., and McDonough, W.F., 1989, Chemical and isotopic systematics of oceanic basalts: implications for mantle compositions and processes: *in* Saunders, A.D., and Norry, M.J. eds., Magmatism in the Ocean Basins: Geological Society of London, Special Publication 42, p. 313-345.

Tanguay, S, Hébert R., and Bergeron, M., 1990, Distribution of PGE in pyroxene-bearing ultramafic cumulates in the Thetford Mines ophiolitic complex, Quebec: Canadian Mineralogist, v. 28, p. 597-605.

Tremblay, A., 1989, Géologie structurale et géochimie des roches volcaniques et sédimentaires du Complexe d'Ascot, Sherbrooke, Québec, Canada: Ph.D. thesis, Université Laval, Québec, Canada, 177 p.

——— 1992, Tectonic and accretionary history of Taconian oceanic rocks of the Québec Appalachians: American Journal of Science, v. 292, p. 229-252.

Tremblay, A., and Castonguay, S., 2002, The structural evolution of the Laurentian margin revisited (southern Quebec): Implications for the

Salinian Orogeny and Appalachian successor basins: Geology, v. 30, p. 79-82.

Tremblay, A., and Pinet, N., 1994, Distribution and characteristics of Taconian and Acadian deformation, southern Québec Appalachians: Geological Society of America Bulletin, v. 106, p. 1172-1181.

——— 2005, Diachronous supracrustal extension in an intraplate setting and the origin of the Connecticut Valley-Gaspé and Merrimack troughs, Northern Appalachians: Geological Magazine, v. 142, p. 7-22.

Tremblay, A., and St-Julien, P., 1990, Structural style and evolution of a segment of the Dunnage Zone from the Québec Appalchians and its tectonic implications: Geological Society of America Bulletin, v. 102, p. 1218-1229.

Tremblay, A., Hébert, R., and Bergeron, M, 1989, Le Complexe d'Ascot des Appalaches du sud du Québec: pétrologie et géochimie: Canadian Journal of Earth Sciences, v. 26, p. 2407-2420.

Tremblay, A., Laflèche, M.R., McNutt, R. H., and Bergeron, M., 1994, Petrogenesis of Cambro-Ordovician subduction-related granitic magmas of the Québec Appalachians, Canada: Chemical Geology, v. 113, p. 205-220.

Tremblay, A., Ruffet, G., and Castonguay, S., 2000, Acadian metamorphism in the Dunnage zone of the southern Québec, northern Appalachians: $^{40}Ar/^{39}Ar$ evidence for collision diachronism: Geological Society of America Bulletin, v. 112, p. 136-146.

Turcotte, B., 1999, Métamorphisme et hydrothermalisme dans le massif ophiolitique du Mont-Chagnon: M.Sc. thesis, Laval University, Québec, Canada, 204 p.

van Staal, C.R., Dewey, J.F., MacNiocaill, C., and McKerrow, W.S., 1998, The Cambrian-Silurian tectonic evolution of the northern Appalachians and British Caledonides: History of a complex, west and southwest Pacific-type segment of Iapetus, *in* Blundell, D.J., and Scott, A.C., eds., Lyell: The Past is the Key to the Present: Geological Society of London, Special Publication 143, p. 199-242.

Wager, L.R., and Brown, G.M., 1968, Layered Igneous Rocks: Oliver and Boyd, Edinburgh, 588 p.

Whitehead, J., Reynolds, P.H., and Spray, J.G., 1995, The sub-ophiolitic metamorphic rocks of the Québec Appalachians: Journal of Geodynamics, v. 19, p. 325-350.

Whitehead, J., Dunning,G.R., and Spray, J.G., 2000, U-Pb geochronology and origin of granitoid rocks in the Thetford Mines ophiolite, Canadian Appalachians: Geological Society of America Bulletin, v. 112, p. 915-928.

Williams, H., 1983, The Appalachian Orogen in Canada: Canadian Journal of Earth Sciences, v. 16, p. 792-807.

Williams, H., and Hatcher, R.D.J., 1983, Appalachian suspect terranes, *in* Hatcher Jr , R.D., Williams, H., and Zietz, I., eds., Geophysics of Mountain Chains, Geological Society of America, Memoir 158, p. 33-53.

Williams, H., and St-Julien, P., 1982, The Baie Verte-Brompton Line: Early Paleozoic continent ocean interface in the Canadian Appalachiens: *in* St-Julien, P., and Béland, J., eds., Major Structural Zones and Faults of the Northern Appalachians: Geological Association of Canada, Special Paper 24, p. 177-207.

Metallogeny of the Bathurst Mining Camp, Northern New Brunswick

Wayne D. Goodfellow

Geological Survey of Canada, 601 Booth Street, Ottawa, Ontario K1A 0E8
Corresponding author's email: wgoodfel@nrcan.gc.ca

Abstract

The Bathurst Mining Camp (BMC) hosts 45 volcanic-sediment hosted massive sulphide deposits and 95 occurrences, including the super-giant Brunswick No. 12 deposit with a geological resource of 230 Mt grading 7.66 wt.% Zn, 3.01 wt.% Pb, 0.46 wt.% Cu, and 91 g/t Ag. Ten of these deposits have been brought into production, with a total production to 1999 of 128 Mt with an average grade of 2.87 wt.% Pb, 6.58 wt.% Zn, 0.93 wt.% Cu, and 82 g/t Ag; only the Brunswick No. 12 deposit remains in production.

BMC deposits formed in a sediment-covered back-arc continental rift, referred to as the Tetagouche-Exploits back-arc basin, during periods when the basin was stratified with a lower anoxic water-column. The basin was subsequently intensely deformed and metamorphosed during multiple collisional events related to east-dipping subduction of the basin.

Four hydrothermal events spanning 12 to 14 million years have been recognized; the Chester (478 Ma), Caribou (472-470 Ma), Brunswick (469-468 Ma), and Stratmat (467-465 Ma) horizons. The Stratmat and Brunswick horizons both occur in the Tetagouche Group, whereas the Caribou and Chester horizons are hosted by the California Lake and Sheephouse Brook groups, respectively.

There are two major types of iron formation in the Bathurst Mining Camp. Type 1 is a carbonate-oxide-silicate iron formation that is spatially associated with most massive sulphide deposits of the Brunswick ore horizon. Type 2 consists of widely distributed Llanvirian Fe-Mn-oxides that are not spatially associated with massive sulphide deposits and formed during a period of oxygenated seawater conditions.

Most deposits are zoned vertically and laterally from a high-temperature, vent-proximal, Cu-Co-Bi-rich veined and brecciated core to vent-distal Zn-Pb-Ag-rich hydrothermal sediments. The vent complex is commonly underlain by a highly deformed sulphide stringer zone that extends hundreds of metres beneath deposits and consists of veins and impregnations of sulphides, silicates, and carbonates that cut chloritized and sericitized volcanic and sedimentary rocks.

The massive sulphide deposits formed from low-salinity and high-temperature (>300°C) buoyant hydrothermal fluids, which explains the vent-proximal nature of most deposits. The reduced sulphur in BMC deposits originated mostly from an ambient reduced water column. Variations among deposits of different age are controlled by the global secular $\delta^{34}S$ curve for sedimentary sulphate and sulphide. The base metals were probably derived from both hydrothermal and magmatic fluids, whereas elements such as Sn, In, Au, As, and Sb probably originated from magmatic fluids.

Résumé

Le camp minier de Bathurst (CMB) compte 95 occurrences et 45 gisements de sulfures massifs logés dans des roches volcano-sédimentaires, y compris le gisement super géant Brunswick no 12, dont les ressources s'élèvent à 230 Mt de minerai titrant 7,66 % massiques de zinc, 3,01 % massiques de plomb, 0,46 % massique de cuivre et 91 g/t d'argent. Dix de ces gisements ont été mis en exploitation, ce qui a permis de produire, jusqu'en 1999, 128 Mt de minerai titrant en moyenne 2,87 % massiques de plomb, 6,58 % massiques de zinc, 0,93 % massique de cuivre et 82 g/t d'argent. Cependant, de ceux-ci, seul le gisement Brunswick no 12 demeure en exploitation.

Les gisements du CMB se sont formés dans un rift continental d'arrière arc recouvert de sédiments, soit le bassin marginal de Tetagouche-Exploits, à l'époque où ce bassin présentait une stratification qui comportait une colonne d'eau anoxique inférieure. Le bassin a ensuite été déformé et métamorphisé au cours de multiples épisodes de collision rattachés à sa subduction à pendage oriental.

Quatre épisodes hydrothermaux qui se sont écoulés sur 12 à 14 Ma ont été identifiés dans les horizons de Chester (478 Ma), de Caribou (de 472 à 470 Ma), de Brunswick (de 469 à 468 Ma) et de Stratmat (de 467 à 465 Ma). Les horizons de Stratmat et de Brunswick se trouvent dans le Groupe de Tetagouche, tandis que ceux de Caribou et de Chester reposent respectivement dans les groupes de California Lake et de Sheephouse Brook.

Le CMB comporte deux principaux types de formations ferrifères. Le premier consiste en des formations ferrières de carbonates-oxydes-silicates associées sur le plan spatial à la plupart des gisements de sulfures massifs de l'horizon minéralisé de Brunswick. Le second consiste en des oxydes de Fe et de Mn llanvirniens couvrant une grande étendue, qui ne sont pas associés sur le plan spatial aux gisements de sulfures massifs et qui se sont formés pendant une période où l'eau de mer était oxygénée.

La plupart des gisements présentent une zonation verticale et latérale qui va d'un noyau bréchifié, filonien, riche en Cu-Co-Bi et voisin d'un évent à haute température jusqu'à des sédiments hydrothermaux riches en Zn-Pb-Ag éloignés de cet évent. Le complexe d'évent se trouve généralement au-dessus d'une zone de filonnets de sulfures fortement déformée qui s'étend sur des centaines de mètres sous les gisements et se compose de filons et d'imprégnations de sulfures, de silicates et de carbonates qui recoupent des roches sédimentaires et volcaniques séricitisées et chloritisées.

Les gisements de sulfures massifs sont issus de fluides hydrothermaux en ébullition à haute température (> 300°C) et dans un milieu à faible salinité, ce qui explique pourquoi la plupart d'entre eux se trouvent à proximité d'un évent. Le soufre réduit dans les gisements du CMB est principalement attribuable à une colonne d'eau réduite ambiante. Les variations parmi les gisements d'un âge différent sont restreintes par la courbe séculaire globale de valeurs de $\delta^{34}S$ rattachée aux sulfates et aux sulfures sédimentaires. Les métaux communs sont probablement issus de fluides hydrothermaux et de fluides magmatiques, tandis que l'étain, l'indium, l'or, l'arsenic et l'antimoine proviennent probablement de fluides magmatiques.

Goodfellow, W.D., 2007, Metallogeny of the Bathurst Mining Camp, northern New Brunswick, *in* Goodfellow, W.D., ed., Mineral Deposits of Canada: A Synthesis of Major Deposit-Types, District Metallogeny, the Evolution of Geological Provinces, and Exploration Methods: Geological Association of Canada, Mineral Deposits Division, Special Publication No. 5, p. 449-469.

Introduction and History

The Bathurst Mining Camp (BMC) is one of Canada's most important base metal mining districts, accounting in 2001 for 30% of Canada's production of Zn, 53% of Pb, and 17% of Ag (Minerals and Metals Sector, Natural Resources Canada). The value of production from the BMC in 2001 exceeded $500 million and accounted for 70% of total mineral production in New Brunswick. Approximately 2000 people are directly employed by the mining industry in the BMC. Without the discovery of new ore reserves, mining will cease in less than ten years at current production rates, eliminating the principal economic activity in northeastern New Brunswick.

Mining has taken place at ten BMC massive sulphide deposits with a total production to 1999 of 128 million tonnes with an average grade of 2.87 wt.% Pb, 6.58 wt.% Zn, 0.93 wt.% Cu, and 82 g/t Ag, but only the Brunswick No. 12 deposit is currently in production (Goodfellow and McCutcheon, 2003). In addition, 0.337 million tonnes of Cu have been mined from the Caribou supergene zone, and 1.254 million tonnes containing 136 g/t Ag and 4 g/t Au from the Caribou, Murray Brook, and Heath Steele gossans (Boyle, 2003; McCutcheon et al., 2003) (Table 1).

TABLE 1. Tonnage and grades of past and present producers, Bathurst Mining Camp.

Deposit	Million tonnes (Mt)	Type	Pb (%)	Zn (%)	Cu (%)	Ag (g/t)	Au (g/t)
Primary Sulfides							
Brunswick No. 12	88.807	present	3.49	8.81	0.34	99.9	
Brunswick No. 6	12.197	past	2.15	5.43	0.40	67.0	
Captain North Ext (CNE)	0.039	past	4.42	9.97		134.7	
Heath Steele ACD Zones	2.472	past	1.73	7.38	0.73	76.7	
Heath Steele B Zone	20.723	past	1.75	4.79	0.98	65.5	
Heath Steele N-5 and Stratmat Boundary	1.137	past	2.98	8.11	0.35	44.0	
Caribou	1.343	past	3.24	6.78	0.32	97.0	
Restigouche	0.231	past	5.49	6.34		132.9	
Wedge	1.504	past	0.65	1.61	2.88	20.6	
Chester	0.003	past			1.46		
Total	128.455	(avg)	2.87	6.58	0.93	82	
Supergene Zones							
Caribou	0.337	past			3.66		
Gossans							
Caribou	0.062	past				171.4	5.35
Heath Steele ACD, B	0.178	past				176	4.77
Murray Brook	1.014	past				61.4	1.79
Total	1.254				(avg)	136.1	3.97

Notes: from McCutcheon et al. (2003)

Approximately 70% of the deposits in the BMC were discovered in the 1950s using geological, geophysical, and geochemical methods (Goodfellow and McCutcheon, 2003). The pace of discovery of major deposits generally waned after the 1960s (McCutcheon et al., 2003). Almost every deposit and occurrence, except for satellite lenses drilled during underground development of major deposits (e.g. Brunswick No. 12 West Zone and Brunswick Northend), were discovered at the surface or subcropping beneath a generally thin and discontinuous cover of glacial sediment. This is not surprising considering that the depth of penetration of airborne electromagnetic systems is limited to a few hundred metres (Keating et al., 2003), and conventional stream and till geochemistry are only effective in detecting deposits that have a near-surface expression (Leybourne et al., 2003; Parkhill and Doiron, 2003).

Geological and Tectonic Setting

The BMC has been subdivided into four approximately coeval groups of volcanic and sedimentary rocks (Tetagouche, California Lake, Sheephouse Brook and Fournier groups) that host most of the deposits and overlie sedimentary rocks of the Miramichi Group (Fig. 1). The different groups comprise blocks and slivers (i.e. the Tetagouche, California Lake, Sheephouse Brook, and Fournier blocks and the blueschist, and Bamford Brook slivers) that represent widely separated, ensialic to ensimatic parts of the Tetagouche-Exploits back-arc basin that were juxtaposed during the Ashgill - Ludlow closure of the Tetagouche-Exploits back-arc basin (van Staal et al., 2003a) (Fig. 2, from van Staal et al., 2003a). The BMC is interpreted to have formed in a Sea of Japan style back-arc basin that opened by rifting of continental crust in the Early Ordovician (Fig. 3).

Most of the deposits occur in the Tetagouche and California Lake groups. The California Lake Group contains approximately equal proportions of felsic and mafic volcanic rocks and probably represents a marginal basin underlain by oceanic and remnant continental crust. The Tetagouche Group, in contrast, consists mostly of felsic volcanic rocks and was probably deposited in a different extensional basin that was underlain by mostly continental crust (van Staal et al., 2003a).

Most BMC massive sulphide deposits are associated with submarine calc-alkalic felsic volcanic rocks that formed from melted continental crust during the initial stages of back-arc rifting. Felsic volcanic rocks consist of flows, submarine pyroclastic, volcaniclastic, and epiclastic deposits that include quartz and quartz-feldspar phyric dacite to rhyolite, aphyric and crystal tuffs, hyaloclastites, autobreccias, and sub-volcanic quartz-feldspar porphyries. Continued back-arc rifting produced alkalic and tholeiitic basalt flows and hyaloclastites, resulting in a bimodal volcanic sequence (Rogers and van Staal, 2003; Rogers et al., 2003). The basalts generally post-date felsic volcanism and associated massive sulphide deposits, and their compositions reflect a progression from low degrees of partial melting to more primitive, MORB-like tholeiites. The eruption of MORB-like basalts of the Tetagouche Group (~470-460 Ma) coincides with the formation of oceanic crust in the adjacent Fournier Group (~465 Ma) and marks the transformation of the Tetagouche-Exploits back-arc rift basin to an oceanic marginal sea (van Staal et al., 2003a).

The ambient seafloor environment in the Tetagouche-Exploits basin is interpreted to have varied from stratified with anoxic bottom waters during the deposition of regionally extensive Arenig black shales of the Knights Brook, Patrick Brook, Nepisiguit Falls, and Spruce Lake formations, to well oxygenated during the deposition of Llanvirnian maroon shales and cherts, and returning to anoxic conditions during deposition of the Caradocian black shales (Goodfellow et al., 2003b). The Llanvirinan maroon shales and cherts post-date the formation of most massive sulphide deposits in the BMC (Fig. 2) and are widely distributed, but are chemically similar to Fe- and Mn-oxyhy-

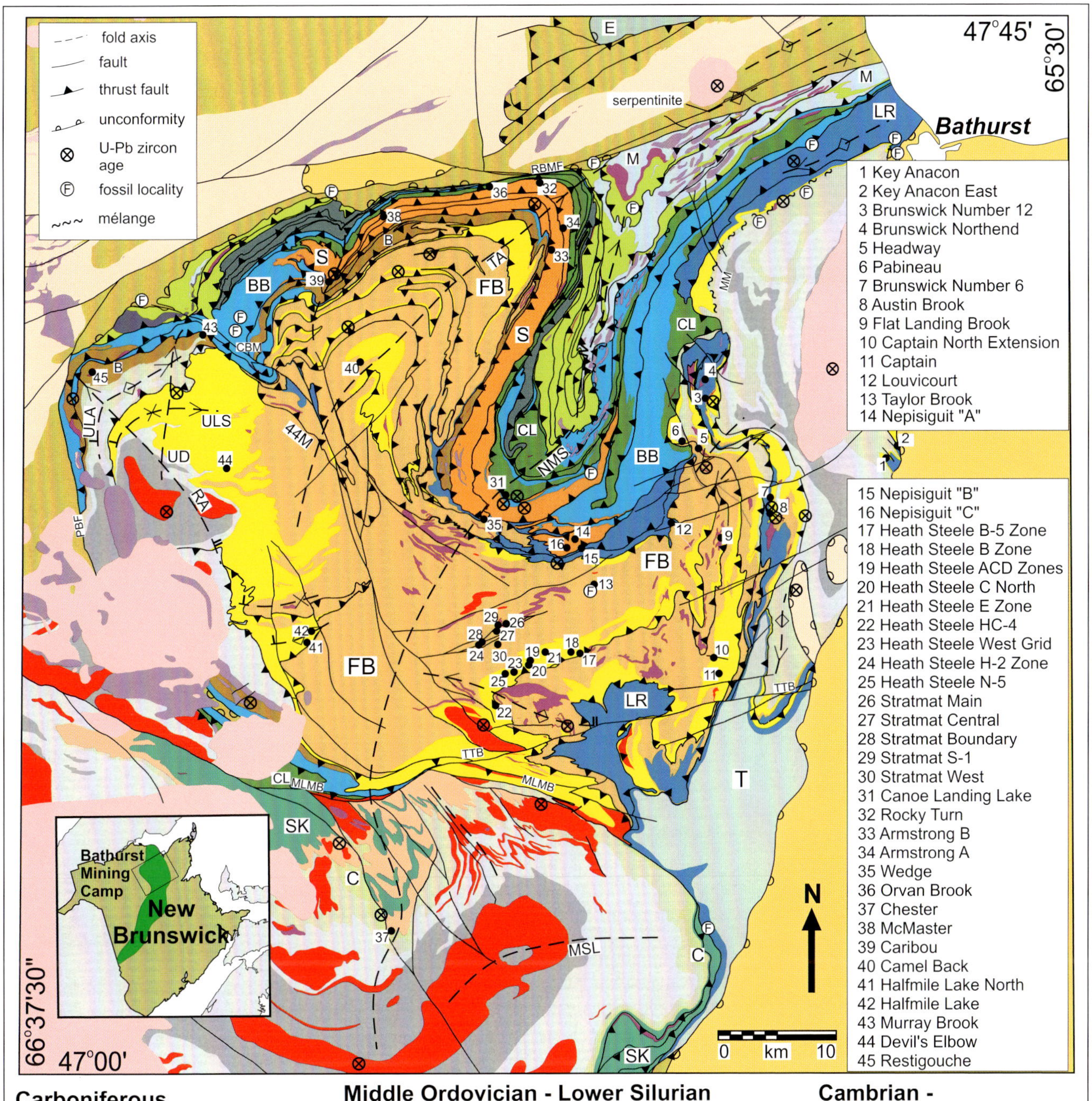

FIGURE 1. General geology of the Bathurst Mining Camp (from van Staal et al., 2003a) and major volcanic-sediment-hosted massive sulphide deposits (from Goodfellow and McCutcheon, 2003). MLMB - Moose Lake-Mountain Brook fault; MSL - Mullin Stream Lake structure; TA - Tetagouche antiform; TTB - Tomogonops - Tozer Brook fault; UD - Upsalquitch dome; ULA - Upsalquitch Lake antiform; ULS - Upsalquitch Lake synform.

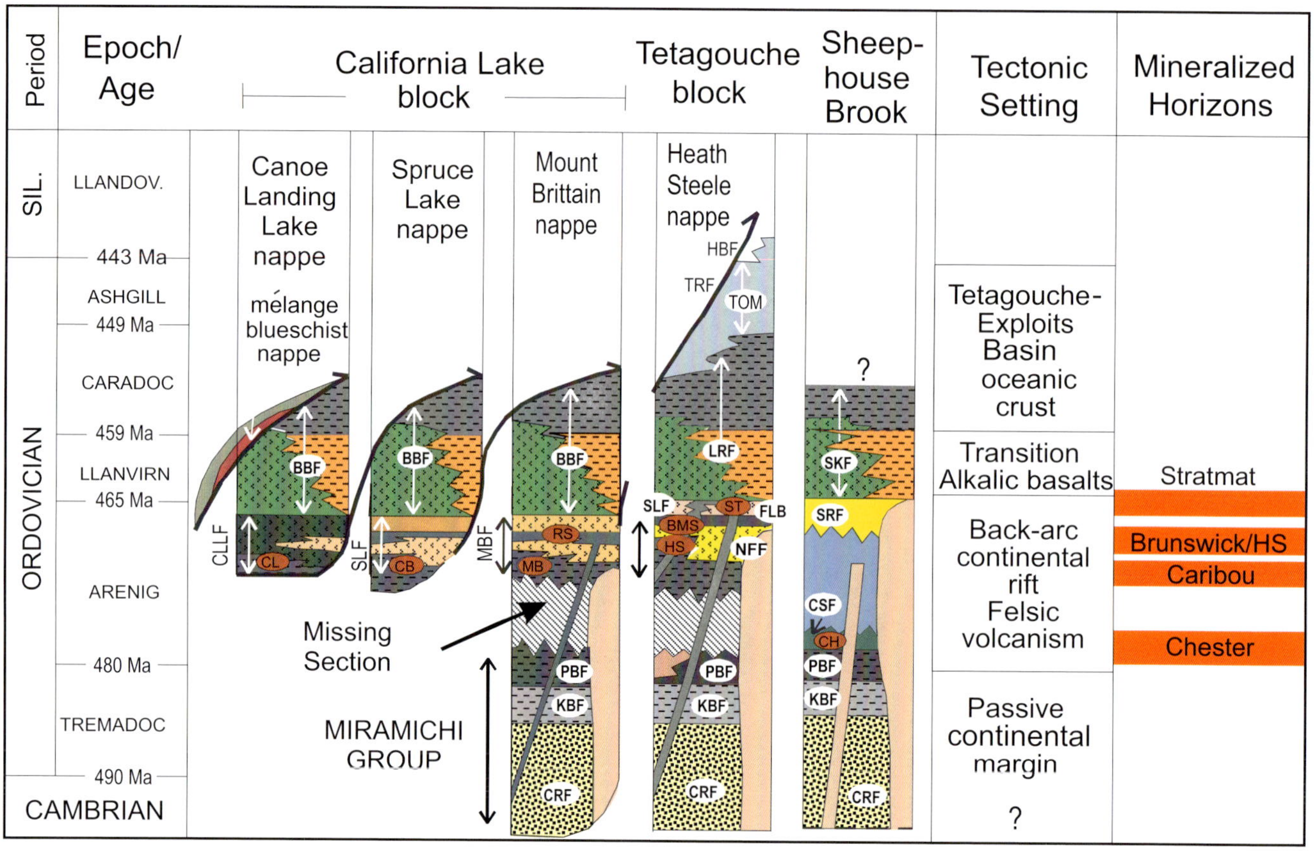

FIGURE 2. Generalized stratigraphic columns showing the age and lithostratigraphy of deposits hosted by the Flat Landing Brook (FLB) and Nepisiguit Falls (NFF) formations, Tetagouche Group, the Spruce Lake (SLF), Mount Brittain (MBF), and formations, California Lake Group, and the Clearwater Stream Formation (CSF), Sheephouse Brook Group (modified from van Staal et al., 2003a). Other formations are Boucher Brook Formation (BBF), Chain of Rocks Formation (CRF), Knights Brook Formation (KBF), Little River Formation (LRF), Patrick Brook Formation (PBF), and Slacks Lake Formation (SKF). Also shown are the tectonic settings and four major ore horizons. Thicknesses of stratigraphic units are not indicated because they have been modified by intense deformation and show rapid lateral changes, particularly for felsic volcanic units. Deposits shown are Brunswick No. 6 and 12 (BMS), Caribou (CB), Chester (CH), Canoe Landing Lake (CL), Heath Steele (HS), Murray Brook (MB), Restigouche (RS), and Stratmat (ST).

droxides associated with modern seafloor hydrothermal vents (Goodfellow et al., 2003b). Their high contents of Fe, Mn, Zn, Co, Ni, and P, and positive Ce/Ce* and Eu/Eu* anomalies provide evidence for major hydrothermal activity during Llanvirnian time. However, most of the metal content of the hydrothermal fluids that vented at the sea floor appears to have been lost to an oxidized ambient water column and dispersed, consistent with the absence of massive sulphide deposits of this age in the BMC.

Mineral Deposits and Occurrences

Deposit Types and Classification

Two major deposit types occur in the BMC: 1) classic volcanic-sediment hosted massive sulphide (VSHMS) deposits, and 2) secondary gold-rich gossans and related supergene zones preserved at some massive sulphide deposits.

The VSHMS deposits are an economically important subgroup of seafloor hydrothermal deposits that are hosted by both sedimentary and bimodal volcanic rocks in sedimented back-arc rifts. VSHMS deposits contain 23.7 Mt of sulphides on average compared to 5.2 Mt for deposits hosted by bimodal volcanic sequences containing minor sediment (Barrie and Hannington, 1999). Furthermore, seafloor hydrothermal deposits formed in sedimented back-arc rifts (i.e. VSHMS and SEDEX deposits) acount for well over 50% of the world's Zn and Pb reserves, and approximately 40% of the world's Zn and Pb production.

Deposit Distribution

Four hydrothermal events spanning 12 to 14 million years have been recognized in the BMC. From oldest to youngest, these are Chester (478 Ma), Caribou (472-470 Ma), Brunswick (469-468 Ma), and Stratmat (467-465 Ma) ore

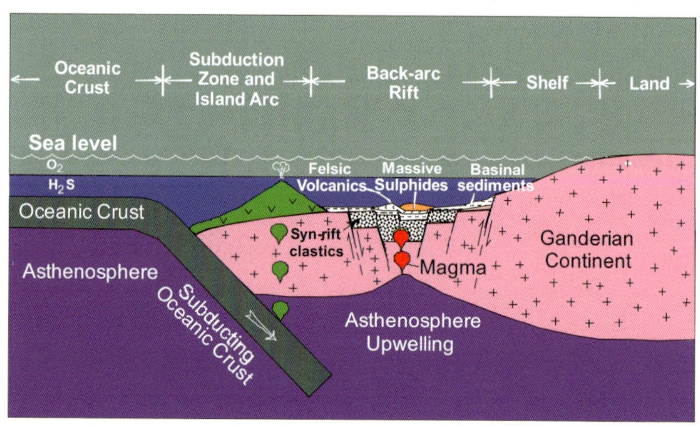

FIGURE 3. Formation of the BMC in a back-arc continental rift partly filled with a clastic sedimentary syn-rift sequence overlain by interbedded black shale and felsic volcanic rocks that host massive sulphide deposits (from Goodfellow and McCutcheon, 2003).

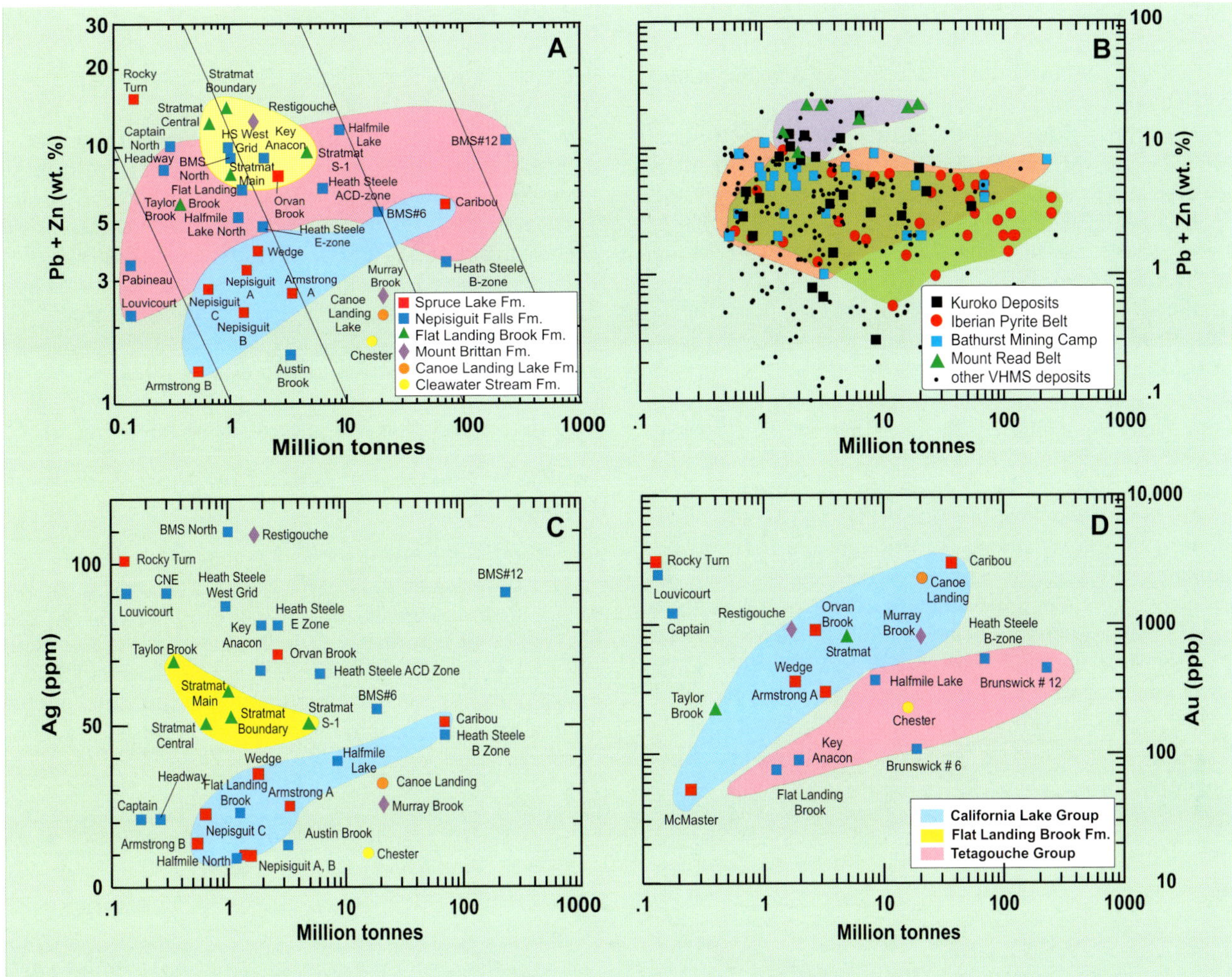

FIGURE 4. Grade versus tonnage plots for massive sulphide deposits. **(A)** Pb+Zn versus millions of tonnes, BMC. **(B)** Pb+Zn versus millions of tonnes for massive sulphide deposits of the BMC, the Kuroko district (Tanimura et al., 1983), the Iberian Pyrite Belt (Sáez et al., 1999), and the Mount Read Belt (Large, 1992). **(C)** Ag versus millions of tonnes, BMC. **(D)** Au versus millions of tonnes, BMC.

horizons (Fig. 2). The term "ore horizon" refers to an episode of hydrothermal activity that led to the formation of massive sulphide deposits hosted at a unique stratigraphic position. The Stratmat and Brunswick horizons both occur in the Tetagouche Group, whereas the Caribou and Chester horizons occur in the California Lake and Sheephouse Brook groups, respectively.

Of the 45 known deposits in the BMC (McCutcheon et al., 2003), 31 are within the Tetagouche Group (Brunswick and Stratmat horizons), 13 are in the California Lake Group (Caribou horizon), and one is in the Sheephouse Brook Group (Chester horizon). Of the 25 deposits with more than 1 Mt of delineated resources, the breakdown by group is 15, 9, and 1, respectively. Clearly, the Tetagouche Group contains the majority of the deposits in the BMC and also contains 63 of the 95 VMS-type occurrences (McCutcheon et al., 2003).

The Brunswick horizon (of the Nepisiguit Falls Formation) is host to most of the deposits in the BMC. This horizon also contains the largest and most economically important deposits. The average tonnage of deposits of the Brunswick, Caribou and Chester horizons, respectively, is between 10 and 20 Mt, although the Chester horizon (Clearwater Stream Formation) contains only one deposit. Deposits hosted by the Stratmat horizon (Flat Landing Brook Formation) are generally smaller with an average size of 1.57 Mt (Goodfellow and McCutcheon, 2003).

Grade and Tonnage Statistics

The massive sulphide deposits in the BMC range in size from <1 Mt to the super-giant Brunswick No. 12 deposit with geological resources of 229 Mt (Fig. 4A and Appendix 1). Massive sulphide deposits average 12.74 Mt with average grades of 4.72 wt.% Zn, 1.78 wt.% Pb, 0.64 wt.% Cu, 51 g/t Ag, and 0.54 g/t Au. Total mineral resources in the BMC for all deposits for which estimates have been made are 496.9 Mt (Goodfellow and McCutcheon, 2003).

Most Bathurst deposits plot near the Pb-Zn axis on a ternary Pb-Zn-Cu diagram, and overlap the field for Canadian Pb-Zn-Cu deposits; only the Louvicourt and Restigouche deposits fall in the Zn-Pb field. The average Pb/(Pb+Zn) and Cu/(Cu+Pb+Zn) ratios for all deposits are 0.27 and 0.20, respectively. Cu/(Cu+Pb+Zn) ratios are highly variable and range from very low values for the Flat Landing Brook deposit (0.004), to moderate values for the Brunswick No. 12 (0.041), and No. 6 (0.074) deposits, and to high values for the Heath Steele B-zone (0.215), Chester (0.309), and Wedge (0.426) deposits. Deposits such as Wedge and Chester with high Cu/(Pb+Zn) ratios are characterized by a vent complex with relatively minor bedded ore facies.

Compared to similar deposits elsewhere, the Pb+Zn contents of deposits of the BMC are generally higher than for deposits of the Iberian Pyrite Belt (Sáez et al., 1999), overlap values for the Kuroko deposits (Tanimura et al., 1983), and are mostly less than those for deposits of the Mount Read District (Large, 1992) (Fig. 4B). Ratios of Pb/(Pb+Zn) for Bathurst deposits are generally higher than those for the Mount Read deposits (Large, 1992) and lower than for deposits of the Iberian Pyrite Belt (Sáez et al., 1999).

Brunswick horizon deposits are also relatively enriched in Pb and Zn compared to deposits in the Caribou horizon (Spruce Lake and Mount Brittain formations) (Fig. 4A); the Stratmat deposits in the Flat Landing Brook Formation have generally the highest Pb+Zn grades (Fig. 4A). These variations in metal grades among the different ore horizons are reflected by a systematic increase of Zn, Pb, and Ag with decreasing age, from the Chester horizon to the Stratmat horizon. Copper generally behaves antithetically to Pb, Zn, and Ag, decreasing with decreasing age, except for the Stramat deposits where Cu is higher than for Brunswick horizon deposits. Pb/(Pb+Zn) ratios of average deposit compositions display a systematic increase from the Caribou to the Stratmat horizon, whereas Cu/(Cu+Pb+Zn) ratios decrease from the Chester to the Brunswick horizon (Goodfellow and McCutcheon, 2003).

Average Ag grades display a wide range, from about 10 up to 110 g/t in the Brunswick North deposit (Fig. 4C); Au is likewise highly variable, ranging from about 50 to about 3000 ppb (Fig. 4D) (Goodfellow and McCutcheon, 2003; McClenaghan et al., 2003). Silver is depleted in deposits of the Spruce Lake Formation compared to those hosted by the Flat Landing Brook and Nepisiguit Falls formations. Gold behaves antithetically to Ag and is enriched in deposits of the Caribou horizon relative to deposits of other ore horizons (Fig. 4D). Unlike Zn and Pb (Fig. 4A), Au generally displays a positive correlation with increasing tonnage (Fig. 4D).

Geology of Volcanic-Sediment Hosted Massive Sulphide Deposits

Morphology and Architecture

The morphology of massive sulphide deposits in the BMC ranges from mounds with low aspect ratios (length: thickness) to tabular bodies with high aspect ratios (Goodfellow and McCutcheon, 2003). Three types of deposits are recognized on the basis of morphology, architecture, textures, mineralogy and metal zoning. Type 1 consists of economi-

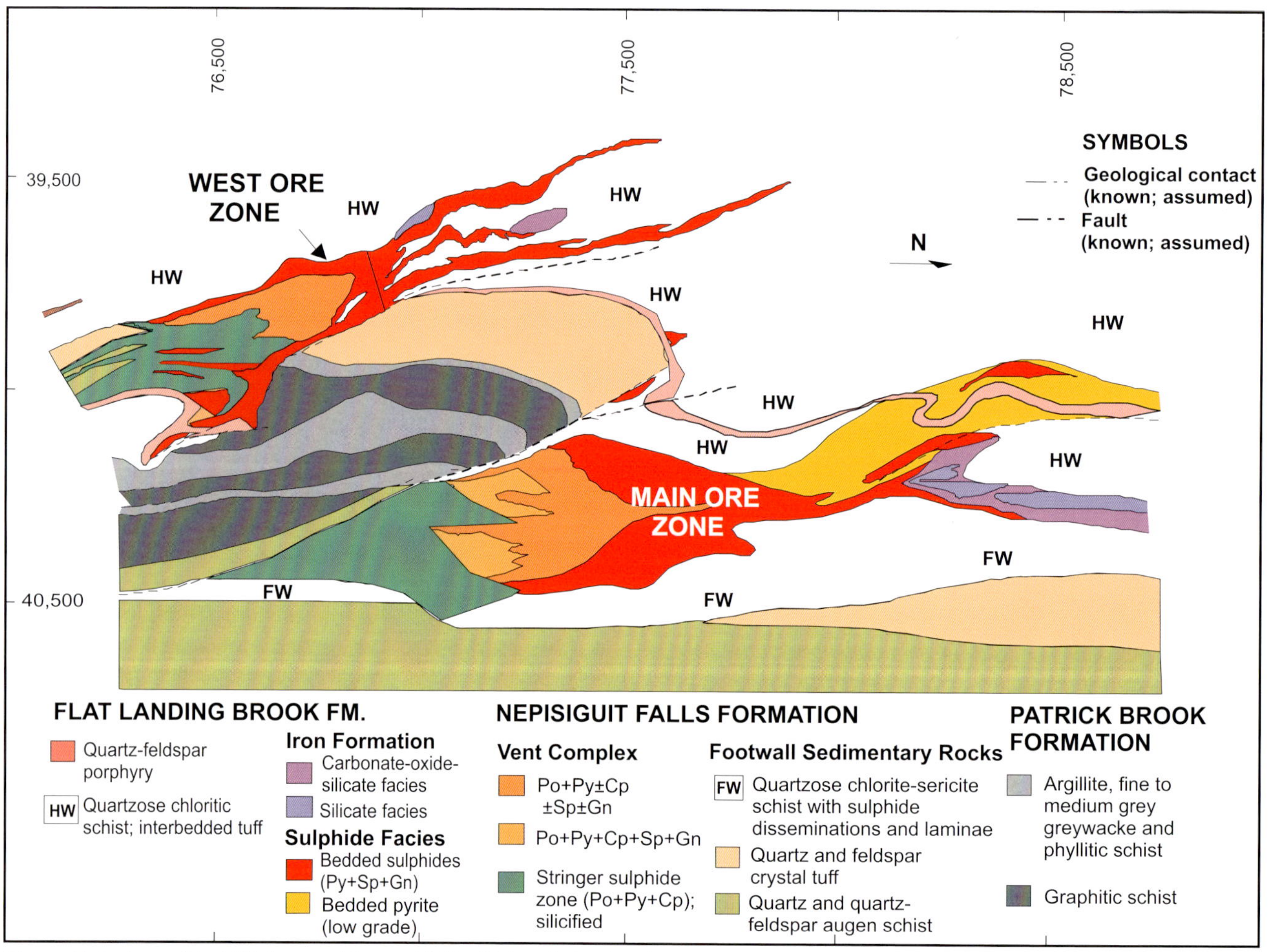

FIGURE 5. Geology of the 1400-foot level, Brunswick No. 12 deposit showing the general stratigraphy and all the different hydrothermal facies (from Luff et al., 1992).

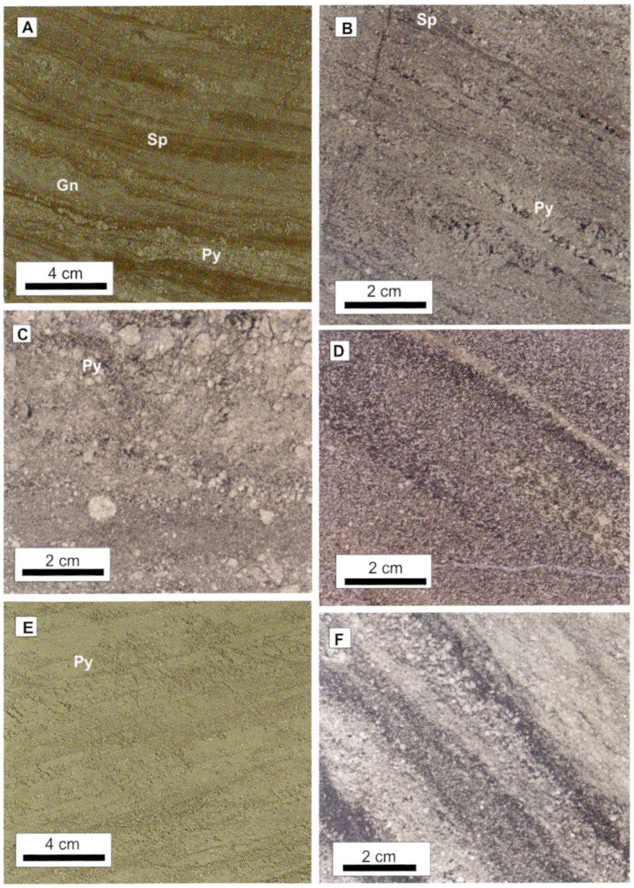

FIGURE 6. Photographs of the bedded ores and bedded pyrite facies in deposits of the BMC. **(A)** Brown sphalerite and grey galena interbanded with medium-grained pyrite, Brunswick No. 12 deposit; **(B)** Fine-grained brown sphalerite interbedded with fine-grained pyrite, Caribou deposit; **(C)** Weakly bedded fragmental sulphides composed of pyrite, sphalerite and galena, Taylor Brook deposit; **(D)** Brown sphalerite and galena interbedded with pyrite, Orvan Brook deposit; **(E)** Laminated Bedded Pyrite facies, Brunswick No. 12 deposit; **(F)** Interbanded dark grey chloritic sediment and pyrite with variable sphalerite and galena, Halfmile Lake and Heath Steele C-zone.

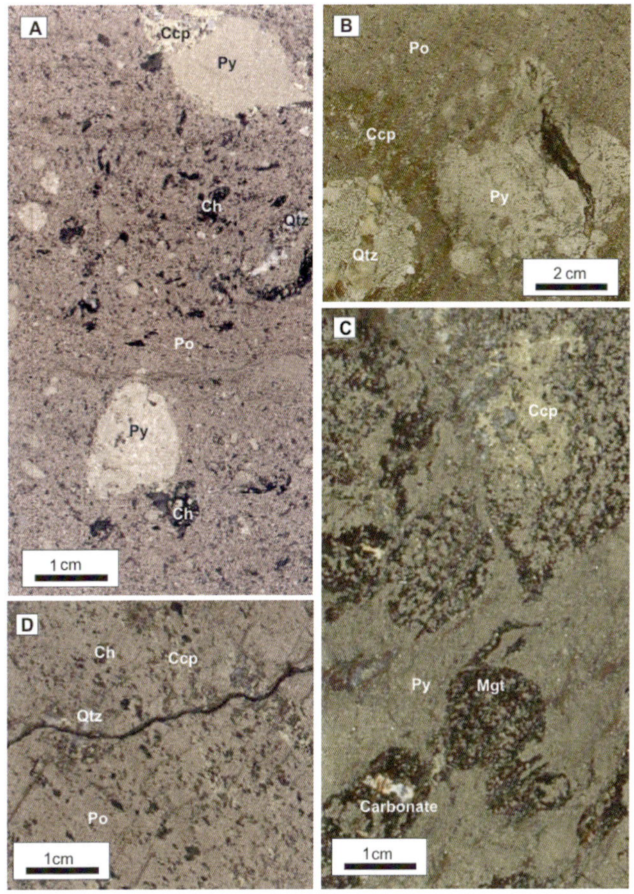

FIGURE 7. Photographs of the vent complex associated with massive sulphide deposits of the BMC. **(A)** Fragments of pyritic sulphides in a matrix of predominantly pyrrhotite (Po), chloritized (Ch) host rocks, quartz (Qtz), and chalcopyrite (Ccp), Heath Steele deposit; **(B)** Fragments of massive pyrite (Py), one of which is cut by a quartz vein, in a matrix of mostly pyrrhotite (Po) containing blebs and disseminations of chalcopyrite (Ccp), Halfmile Lake deposit; **(C)** Clasts comprised of pyrite (Py), magnetite (Mgt), and chalcopyrite (Ccp) in a matrix of pyrite, quartz, and ferroan carbonate, Caribou deposit; **(D)** Clasts of strongly chloritized host rocks (Ch) and quartz (Qtz) in a fine-grained matrix of pyrrhotite (Po) with blebs and disseminations of chalcopyrite (Ccp), Key Anacon deposit.

cally important tabular bodies (e.g. Brunswick No. 12 and No. 6, Heath Steele B-zone, Caribou, Halfmile Lake) that are zoned vertically and laterally from a high-temperature, vent-proximal, veined and brecciated core to vent-distal hydrothermal sediments as follows: 1) vent complex (Po+Mt +Py+Cp+Qtz± Sp±Gn); 2) bedded ores (Py+Sp+Gn±Cp); and 3) bedded pyrite (Py±Sp±Gn). This mineral zonation is reflected by higher values of Cu, Co, Bi, and Cu/(Cu+Pb+Zn) in vent proximal facies and higher values of Zn, Pb, Ag, Au, Cd, Sn, In, As, Sb, Tl, Hg, and Cu/(Cu+Pb+Zn) in vent distal facies. Type 2 deposits are characterized by pod-shaped bodies of massive Cu-rich mineralization interpreted as a vent complex (e.g. Chester, Wedge). It is possible that Type 2 deposits are Type 1 deposits that have lost most of their distal facies by erosion. Type 3 deposits are tabular, unzoned, lack a vent complex, and may represent distal or reworked clastic deposits (e.g. Orvan Brook, Canoe Landing Lake).

The vent complex in Types 1 and 2 deposits is commonly underlain by a highly deformed sulphide feeder zone that extends hundreds of metres beneath the deposit. The feeder zone consists of impregnations and veins of pyrrhotite± pyrite+chalcopyrite+quartz that cut hydrothermally altered volcanic and sedimentary rocks.

Hydrothermal Sulphide Facies

Massive sulphide deposits in the BMC are represented by up to five distinct hydrothermal facies: 1) bedded ores, 2) bedded pyrite, 3) vent complex, 4) sulphide stringer zone, and 5) carbonate-oxide-silicate iron formation. The Brunswick No. 12 deposit is one of the better examples of a deposit where all the facies are well represented in both the main and west ore zones (Fig. 5). The massive sulphides are generally fine- to medium-grained and texturally variable. Pyrite, pyrrhotite, sphalerite, galena, and chalcopyrite comprise over 95% of the sulphides; arsenopyrite, marcasite, stannite, tetrahedrite, freibergite, boulangerite, jamesonite, bournonite, magnetite, cassiterite, siderite and quartz account for most of the remaining hydrothermal minerals (Stanton, 1959; Roy, 1961; Boorman, 1975; Jambor, 1979; Goodfellow, 2003; Goodfellow et al., 2003a).

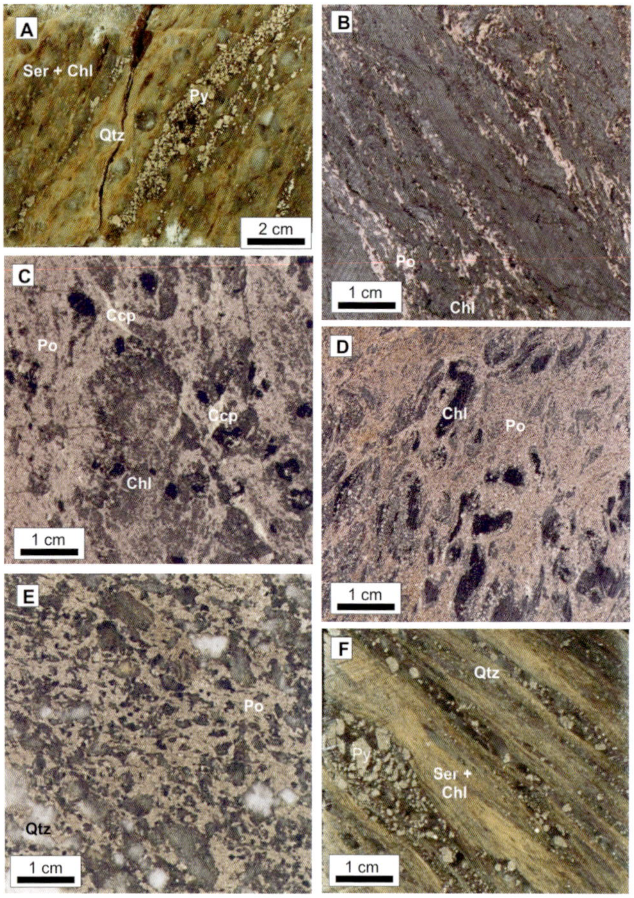

FIGURE 8. Photographs of sulphide stringer zone samples from massive sulphide deposits, BMC. **(A)** Veins of recrystallized pyrite (Py) cutting quartz-feldspar 'porphyry' hydrothermally altered to phengite (Ser) and chlorite (Chl), Nepisiguit Falls Formation, Brunswick No. 12 deposit. **(B)** Pyrrhotite (Po) veins cutting chloritized (Chl) volcaniclastic rocks underlying the Brunswick No. 6 deposit. Veins have been transposed into the penetrative cleavage planes. **(C)** Brecciated host rocks underlying the Key Anacon deposit infilled and veined by pyrrhotite (Po) with disseminated chalcopyrite (Ccp). **(D)** Breccia fragments of altered and deformed host rocks infilled by pyrrhotite, Halfmile Lake deposit. **(E)** Breccia fragments of altered volcanic rocks infilled by pyrrhotite and hydrothermal quartz, Chester deposit. **(F)** Transposed veins of recrystallized pyrite (Py) and quartz (Qtz) cutting intensely sericitized (Ser) and chloritized (Chl) felsic volcanic rocks beneath the Caribou deposit.

Bedded Ores consist generally of fine- to medium-grained, interbanded pyrite, brown sphalerite and galena (Fig. 6A, B, E, F). Minor sulphide minerals include arsenopyrite, marcasite, cassiterite, stannite, tetrahedrite and bournonite. Gangue minerals occur as disseminations throughout, or form bands consisting of siderite, ankerite, dolomite, calcite, barite, magnetite, cassiterite, quartz, chlorite, muscovite, talc, stilpnomelane, minnesotaite and chamosite. In addition to the most common granular aggregates, pyrite displays recrystallized, colloform, framboidal and vuggy textures (Lea and Rancourt, 1958; Stanton, 1959; Boorman, 1975; Chen, 1978; Jambor, 1979; Goodfellow, 2003).

Sphalerite bands are commonly wispy and discontinuous, and are interpreted to represent layers of very fine-grained bedded sulphides that have been recrystallized and intensely deformed. Galena occurs with sphalerite, although it is less abundant than sphalerite and is generally finer grained than either sphalerite or pyrite. Galena typically forms blebs, veins, and disseminations in sphalerite and appears to have replaced sphalerite locally. Chalcopyrite and pyrrhotite are relatively minor in the bedded ore facies compared to the vent complex. Where chalcopyrite does occur in bedded sulphides, it is as veins that cut other sulphides, and as blebs and disseminations that replace sphalerite and pyrite. Chalcopyrite also occurs in the interstices of pyrite grains and as exsolution-like blebs in sphalerite.

The Bedded Pyrite facies (Fig. 6C, D) represents an uneconomic component of many massive sulphide deposits and consists mainly of massive, barren pyrite with minor sphalerite, galena, and chalcopyrite. At the Brunswick No. 12 deposit, it forms a discrete facies in the main ore zone where it both overlies and is lateral to bedded ores (Fig. 5). The pyrite facies displays a range of macrotextures ranging from massive to well bedded and locally collomorphic. At the microscopic scale, the pyrite is euhedral to subhedral, and locally framboidal (Chen, 1978).

The Vent Complex is composed of pyrrhotite and/or pyrite breccia that is infilled, variably replaced and veined by pyrrhotite, pyrite, chalcopyrite, magnetite, chlorite, quartz, and ferroan carbonates, mostly siderite (Fig. 5). Individual clasts are rounded to angular, up to 10 cm across, and consist of pyrrhotite, pyrite and chloritized host rocks (e.g. Heath Steele, Fig. 7A; Brunswick No. 12, Fig. 7B), and pyrite and magnetite (Caribou, Fig. 7C). At the Key Anacon deposit, the pyrrhotite appears massive with interstitial silicates and chalcopyrite (Fig. 7D).

Chalcopyrite veins, blebs and disseminations occur throughout the matrix and locally in clasts. Unlike sphalerite in the bedded ores, sphalerite in the vent complex occurs most commonly as veins and in the matrix of massive pyrrhotite/pyrite breccias. The sphalerite in the Caribou vent complex is generally dark grey to black and is enriched in Fe (mean = 5.6 wt.%), but is depleted in Cd, Mn and Sn compared to sphalerite from the bedded ores. Galena in the Caribou vent complex contains significantly higher Ag contents (mean = 0.11 wt.%) compared to galena from the bedded sulphides (mean = 0.006 wt.%). At Caribou, magnetite commonly forms patches, veins and disseminations in the vent complex and is more abundant than in the bedded ores (Goodfellow, 2003).

The evidence that these breccias formed by vent-related processes includes 1) the absence of bedded textures that are common in the distal part of deposits; 2) the occurrence of breccias above a well developed sulphide stringer zone with a vein mineralogy similar to that in the overlying Vent Complex; and 3) the high chalcopyrite, pyrrhotite, and magnetite content of the massive sulphides, which is typical of vent-proximal sulphides associated with VHMS deposits elsewhere (Large, 1977, 1992; Galley et al., 1995; Sáez et al., 1996; Goodfellow et al., 1999).

The Sulphide Stringer Zone consists of veins and impregnations of sulphides in hydrothermally altered sedimentary and volcanic rocks that underlie the vent complex. Sulphide minerals are predominantly pyrrhotite and/or pyrite with disseminations and veins of chalcopyrite, and traces of sphalerite and galena. Non-sulphide minerals include anhedral to subhedral quartz and ferroan carbonate minerals. The sulphide veins are irregular to anastamosing, continuous to discontinuous, and are characterized by sharp boundaries with alteration selvages composed mostly of chlorite and muscovite (Goodfellow, 2003). Most veins have been transposed sub-parallel to bedding by deformation, giving the sulphide stringer zone a stratabound appearance (de Roo and van Staal, 2003; Luff et al., 1992).

A sulphide stringer zone stratigraphically underlies the Brunswick No. 12 deposit, south of the main and west ore zones (Fig. 5). Here, pyrite veins cut chloritized and sericitized quartz-feldspar porphyry (Fig. 8A) and crystal tuff of the Nepisiguit Falls Formation (Luff et al., 1992). At the Brunswick No. 6 deposit, pyrrhotite±chalcopyrite veins cut chloritized host rocks (Fig. 8B), whereas the Key Anacon (Fig. 8C) and Halfmile Lake (Fig. 8D) stringer zones are composed of clasts of chloritized host rocks suspended in a pyrrhotite chalcopyrite matrix. The sulphide stringer zone at the Chester deposit is represented by a hydrothermal breccia, with interstitial pyrrhotite±chalcopyrite+quartz and chlorite alteration (Fig. 8E), which cuts altered felsic volcanic rocks of the Clearwater Stream Formation (Fyffe, 1995). At the Caribou deposit, this zone is represented by a network of dominantly pyrite veins (Fig. 8F) that cut pervasively chloritized and sericitized sedimentary and felsic volcanic rocks (Goodfellow, 2003).

Hydrothermal Sediments

There are two major types of iron formation in the BMC. Type 1 is the carbonate-oxide-silicate iron formation that is spatially associated with most Arenigian massive sulphide deposits of the Brunswick ore horizon (e.g. Brunswick No. 12 and 6, Austin Brook, Heath Steele, Key Anacon) (Saif, 1980; Peter and Goodfellow, 1996; Peter and Goodfellow, 2003; Peter et al., 2003a,b). None of the other mineralized horizons in the BMC (i.e. Chester, Caribou, and Stratmat) contain Type 1 iron formations. Type 2 consists of widely distributed Llanvirian Fe-Mn oxides that are not spatially or temporally associated with massive sulphide deposits (Goodfellow et al., 2003b). Type 1 iron formations have been divided into four major facies on the basis of mineralogy: carbonate (mostly siderite), magnetite, silicate and hematite.

The carbonate facies is composed of mostly interbedded siderite, chert, and epiclastic sedimentary rocks (Fig. 9A); beds of apatite occur locally. Siderite layers are variable in thickness and commonly display sharp basal contacts and diffuse tops, and in some samples the siderite is delicately laminated (Fig. 9B). The hydrothermal minerals are dominated by siderite, quartz (after chert), calcite, dolomite, and kutnahorite. Minor phases include apatite, magnetite, stilpnomelane, spessartine garnet, pyrrhotite, sphalerite, and galena (Peter and Goodfellow, 1996; Peter et al., 2003b). The apatite locally forms beds whereas the magnetite commonly occurs as euhedral crystals that intersect bedding planes and clearly postdate sedimentation.

The magnetite facies is commonly associated with the carbonate facies and consists of magnetite interbanded with siderite (Fig. 9C, D). Unlike the siderite facies, the magnetite bands do not display internal layering. Major hydrothermal minerals are magnetite, siderite, and quartz (after chert); minor phases include apatite, stilpnomelane, pyrrhotite, chalcopyrite, sphalerite, and galena (Peter and Goodfellow, 1996; Peter et al., 2003b). The lack of distinct layering in magnetite bands and the secondary growth of magnetite in bedded siderite suggest that magnetite formed after sedimentary siderite, either during diagenesis or metamorphism.

FIGURE 9. Photographs of iron formations, BMC. **(A)** Interbedded cream to orange siderite (Sd) and silicate host rocks, proximal facies, Brunswick No. 12 deposit (J. Peter, unpublished). **(B)** Delicately interlaminated cream to grey siderite (Sd) and host sedimentary rocks; **(C)** Interlaminated cream siderite (Sd) and magnetite (Mgt), proximal facies, Brunswick No. 12 deposit (from Peter and Goodfellow, 1996). **(D)** Dark grey interlaminated magnetite and dark grey shale, minor disseminated sulphides, mostly pyrite, intermediate to distal facies (J. Peter, unpublished). **(E)** Pale to medium green chloritic iron formation (Ch) with disseminated carbonate aligned along bedding planes, distal facies (from Peter and Goodfellow, 1996). **(F)** Maroon hematitic (Hem) chert interbedded with alkali basalt, Little River Formation, Brunswick No. 12 area.

The silicate facies is a green to black chloritic schist, locally spotted with siderite (Fig. 9E). It occurs interbedded with other iron formation facies, and is common near the distal margins of the iron formation. It consists primarily of felted chlorite masses and generally minor chert, stilpnomelane, phengite, feldspar, biotite, ilmenite, rutile, epidote, apatite, magnetite, spessartine garnet, sphalerite, galena, chalcopyrite, arsenopyrite, and bornite (Peter et al., 2003b). The hematite facies is restricted to the Austin Brook and Brunswick No. 6 deposits where it occurs near the top of thick iron formation sequences. The hematite is interbedded with carbonate, magnetite, and silicate facies. Major minerals are hematite, chert, and magnetite.

Llanvirian Fe-Mn-oxides in maroon chert (Fig. 9F) and shale, Type 2 iron formation, constitute the most widespread

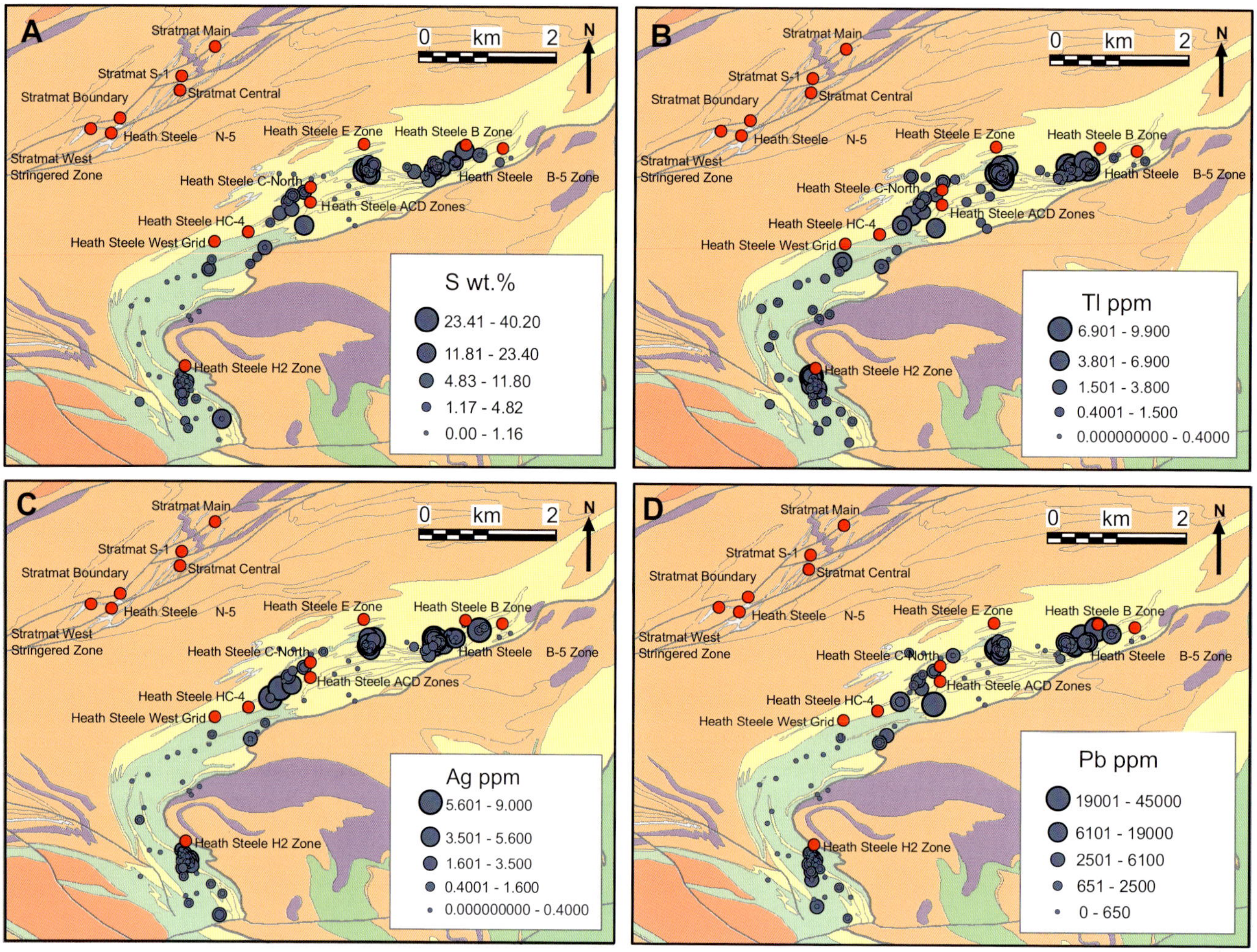

FIGURE 10. Spatial distribution of elements in iron formation associated with massive sulphide deposits, Heath Steele belt, for **(A)** sulphur; **(B)** thallium; **(C)** silver; **(D)** lead (from Peter and Goodfellow, 2003).

chemical sedimentary rocks in the BMC. These rocks have high contents of Fe, Mn, Zn, Co, Ni, and P, and positive Eu/Eu* anomalies when normalized to North American shale composite (Gromet et al., 1984), and are chemically similar to Fe- and Mn-oxyhydroxides associated with modern seafloor hydrothermal vents (Goodfellow et al., 2003b). Any base metals present in the venting hydrothermal fluids have been lost to an oxidized ambient water column and dispersed over large areas of the Tetagouche rift, similar to the fate of metals in modern day black smokers.

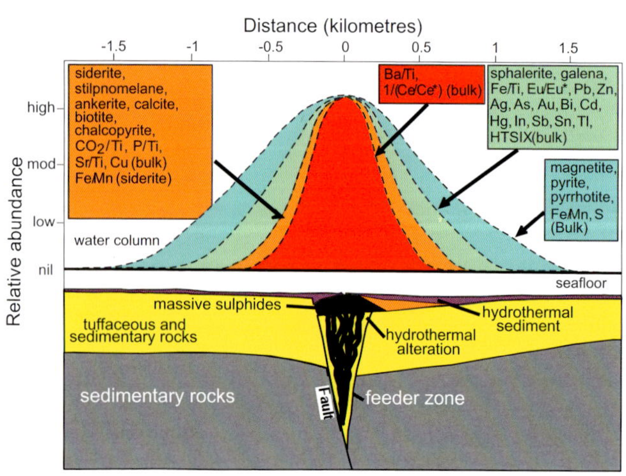

FIGURE 11. Schematic diagram showing the lateral extent of mineral and chemical halos that surround massive sulphide deposits in the BMC (from Peter and Goodfellow, 2003).

Type 1 iron formations provide a valuable stratigraphic marker for locating sequences favourable to exploration and vectors for defining hydrothermal centres and associated deposits along mineralized horizons. In general, Type 1 iron formations are zoned laterally such that the thickness and relative proportion of hydrothermal components increase with proximity to sulphide deposits (Peter and Goodfellow, 1996, 2003). For example, in the Heath Steele belt the contents of siderite, magnetite, and sulphur (Fig. 10A) are highest in the vicinity of the B, E, H2, and Mowat zones (Peter and Goodfellow, 2003). Most chalcophile elements including Tl (Fig. 10B), Ag (Fig. 10C), and Pb (Fig. 10D) are anomalously high near massive sulphide deposits. The most effective vectors include siderite, magnetite, Fe/Mn, Ba/Al, P/Al, Zn, Pb, Cu, Ag, As, Au, Bi, Cd, Co, Mo, Se, Sb, Sn, In, Tl, and Eu/Eu*, all of which increase with proximity to sulphide deposits (Peter and Goodfellow, 1996, 2003). In addition, the composition of minerals also reflects bulk chemical changes of the whole rock; for example, the Fe/Mn ratio in

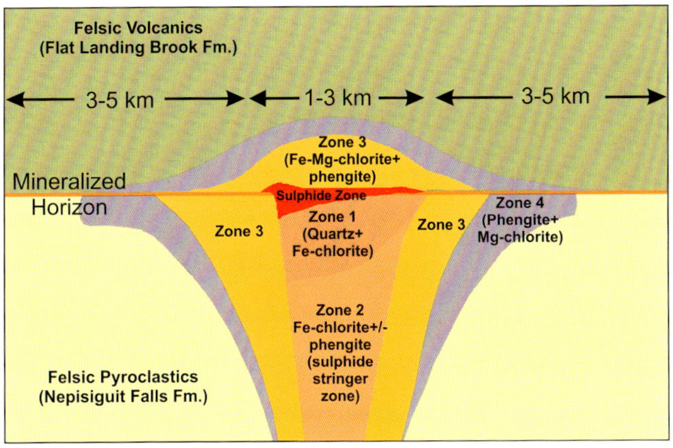

FIGURE 12. Schematic diagram showing the distribution of the different hydrothermal alteration zones at the Brunswick No. 12 deposit, BMC (modified from Lentz and Goodfellow, 1994).

siderite increases toward deposits as the Fe/Mn ratio of the whole rock increases.

The size of mineralogical and chemical halos surrounding massive sulphide deposits is variable and dependent on the particular vector (Peter and Goodfellow, 2003). In general, siderite, stilpnomelane, ankerite, calcite, biotite, chalcopyrite, and Cu are restricted to within about 0.5 km of mineralization, whereas sphalerite, galena, Pb, Zn, Ag, As, Au, Bi, Cd, Hg, In, Sn, Sb, and Tl extend up to 1.5 km from massive sulphide deposits (Fig. 11).

Hydrothermal Alteration

Both pre- and post-ore hydrothermal alteration is associated with most deposits in the BMC, although hydrothermal alteration has been documented in detail for only a few deposits. These include the Brunswick No. 12 (Goodfellow, 1975; Juras, 1981; Luff et al., 1992; Lentz and Goodfellow, 1993, 1996), Brunswick No. 6 (Nelson, 1983; Yang et al., 2003), Heath Steele (Wahl, 1977; Lentz et al., 1997), Halfmile Lake (Adair, 1992; Yang et al., 2003), and Caribou (Goodfellow, 2003) deposits.

The vent complex associated with most deposits is typically underlain by a highly deformed sulphide feeder zone that extends hundreds of metres beneath the deposit. This feeder zone is characterized by impregnations and veins of sulphides, silicates, and carbonates that cut hydrothermally altered volcanic and sedimentary rocks. Hydrothermal alteration is widespread (1-5 km laterally and hundreds of metres vertically) and comprises the following assemblages from the core to the margins of the feeder zone (Fig. 12): Zone 1 - Quartz+Fe-rich chlorite+pyrrhotite+chalcopyrite (Fig. 13A); Zone 2 - Fe-rich chlorite+phengite +/- pyrite (Fig. 13B); Zone 3 - Fe-Mg-chlorite+phengite+albite (Fig. 13C, D); Zone 4 - albite+Mg-rich chlorite (Fig. 13E, F). Volcanic rocks overlying some deposits are also altered to Mg-rich chlorite and albite (Luff et al., 1992; Lentz and Goodfellow, 1993; Goodfellow, 2003). In general, sulphides, chlorite, and phengite increase, and feldspars decrease with proximity to the core of the hydrothermal fluid upflow zone. A schematic diagram showing the spatial distribution of hydrothermal alteration zones typical of the Brunswick No. 12 deposit is presented in Figure 12. Talc is also an important alteration phase in the Stratmat deposits, although the alteration assemblages and their distribution have not been documented in detail (Hamilton, 1992).

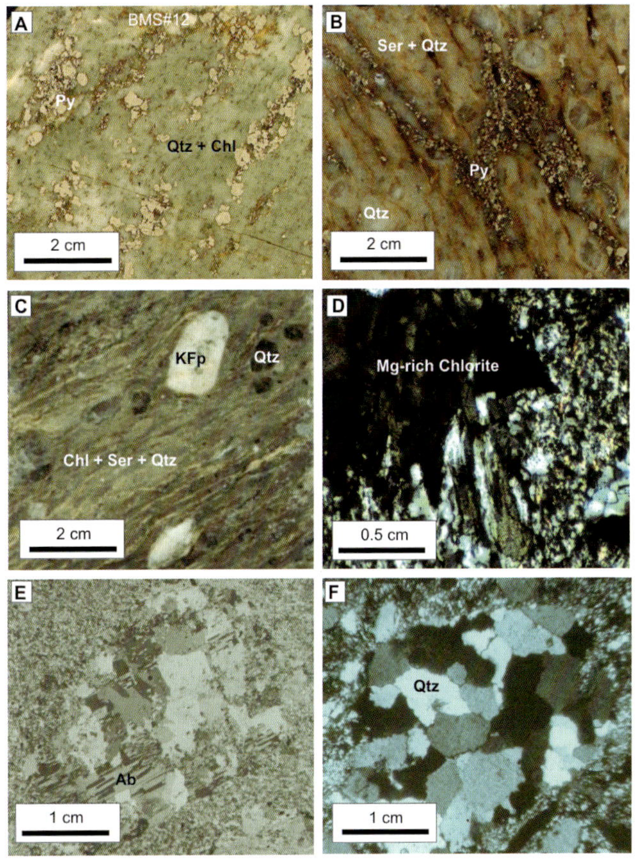

FIGURE 13. Hydrothermal alteration facies, Brunswick No. 12 deposit. **(A)** Photograph of silicified (Qtz) and chloritized (Chl) sedimentary rocks with pyrite (Py) veins and disseminations beneath the vent complex (Zone 1 in Fig. 12). **(B)** Photograph of phengitized (Ser) and silicified quartz (Qtz) phyric pyroclastic volcanic rocks cut by pyrite (Py) stringers (Zone 2). Feldspars have been destroyed by hydrothermal alteration. **(C)** Photograph of quartz (Qtz)-feldspar (Kfp) phyric pyroclastic volcanic rocks that have been altered to Mg-rich chlorite and phengite (Zone 3). **(D)** Photomicrograph of Mg-rich chlorite (Chl) showing a diagnostic olive green colour under polarized light (Zone 4). **(E)** Photomicrograph of chessboard albite (Ab) alteration of what is was probably a microcline phenocryst, Nepisiguit Falls Formation (Zone 4). **(F)** Photomicrograph of a quartz mosaic replacing a K-feldspar phenocryst in the quartz-feldspar "porphyry", Nepisiguit Falls Formation (Zone 3).

Compared to similar unaltered rocks, hydrothermally altered felsic volcanic rocks in Zone 1 are characterized by elevated Si, Fe, CO_2, Cu, Zn, Pb, Ag, Cd, Sn, In, Bi, Tl, As, Sb, and Hg contents, and Eu/Eu* ratios, and are depleted in Na and Ca. The composition of Zone 2 is similar to Zone 1 except for silica, which is not as notably enriched as in Zone 1. Zone 3 is enriched in Mg, Mn, CO_2, S, and base metals, and depleted in Na, Ca, K, Ba, and Rb as a result of widespread hydrothermal alteration of both K-feldspar and plagioclase to chlorite and phengite. Anomalous base metal and CO_2 contents reflect minor sulphide and carbonate disseminations and veinlets in the distal parts of the hydrothermal upflow zone. Zone 4 is enriched in Mg and Na and depleted in Ca, reflecting the chloritization and albitization of K-feldspar, which originally was probably microcline (Luff et al., 1992; Rogers et al., 2003).

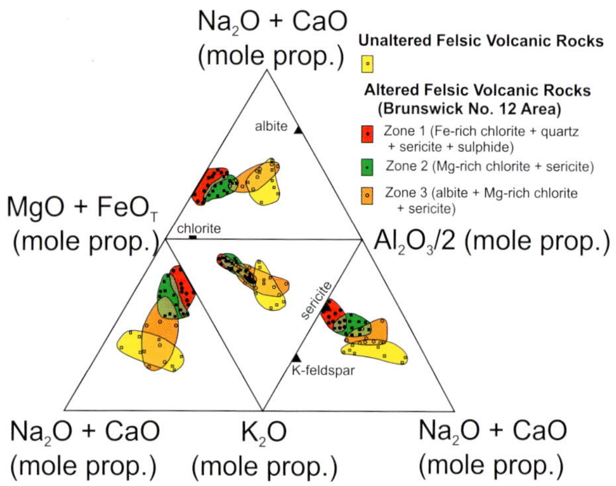

FIGURE 14. Unfolded ($MgO+FeO_T$) - $Al_2O_3/2$ - K_2O - (Na_2O+CaO) tetrahedron showing the chemical changes in major elements in the different hydrothermal zones at the Brunswick No. 12 deposit, BMC (Luff et al., 1992).

The relative changes of major cations in unaltered felsic volcanic rocks and the different alteration zones at the Brunswick No. 12 deposit are illustrated with an unfolded (FeO_T+MgO) - ($CaO+Na_2O$) - Al_2O_3 - K_2O tetrahedron (Fig. 14). Altered felsic volcanic and sedimentary rocks show the same general trend of increasing FeO+MgO at the expense of the alkali elements bound in feldspar.

Sulphide Deformation and Metamorphism

Deposits of the BMC have been intensely deformed and metamorphosed during multiple collisional events related to subduction of the Tetagouche-Exploits back-arc basin. The penetrative polyphase ductile deformation (D_1 to D_4) that affected all rock units in the BMC commenced in the Late Ordovician (Ashgill) and ended in the Late Devonian (van Staal and de Roo, 1995). Peak metamorphic conditions vary from 325 to 400°C and 6 to 7 kb (Currie et al., 2003). The present distribution and morphology of massive sulphide deposits and associated sulphide stringer zones are mainly controlled by D_1 and, to a lesser extent, D_2 structures (de Roo and van Staal, 2003). The D_1 structures formed in response to a progressive, thrust-related deformation associated with incorporation of the various blocks into the southeast-facing Brunswick subduction complex (van Staal, 1994) where most rocks were transformed into high pressure-low temperature metamorphic tectonites (van Staal et al., 1990). D_2 is represented by asymmetrical upright folds and local narrow shear zones that are thought to be related to transpression induced by oblique convergence. In general, D_2 refolded and reoriented the shallow-dipping D_1 structures into "steep belts", producing locally complex fold interference patterns (van Staal et al., 2003a). D_1 and D_2 structures were commonly overprinted by F_3 recumbent folds and kinks, and steeply dipping structures were transposed into D_3 "flat belts" where vertical D_3 shortening strain was high. Flat belts thus represent areas with significant vertical, ductile thinning (de Roo and van Staal, 1994). Where D_3 strain was low, the rocks preserved their overall steep attitude, despite being deformed into open recumbent folds. Most of the Sheephouse Brook block and parts of the Heath Steele nappe experienced relatively high D_3 strain and hence were largely transformed into a flat belt (van Staal et al., 2003a).

Sulphides were recrystallized and underwent major remobilization during D_1 and, to a lesser extent, D_2 events. Massive sulphide deposits folded by D_2 structures, such as the Brunswick No. 6 and No. 12 deposits, commonly were thickened in the hinges of the folds (van Staal and Williams, 1984). The extent of mobilization was dependent on the strain and the ductility of different sulphide minerals. Since the strain is highly heterogeneous, even at the deposit-scale, the degree of mobilization is variable within and between deposits. Sulphide deformation has resulted in the preferential mobilization of galena and sphalerite into the hinges of F_1 and F_2 folds, the injection of galena and chalcopyrite into unmineralized wall rocks, and the transposition of sulphide veins in the sulphide stringer zone into subparallelism with the axial planes of F_1 and F_2 folds.

The flow of ductile sulphides parallel to F_1 axial planes probably contributed to the sulphide banding or tectonic layering that characterizes Pb-Zn-rich zones in virtually all deposits of the BMC. The only primary sulphide textures preserved are finely laminated pyrite in the bedded pyrite zone, pyrrhotite-chalcopyrite breccia in the vent complex (although the latter commonly displays deformational textures), and quartz-sulphide veins in low strain areas within the sulphide stringer zone.

Genetic Models

Although deposits of the BMC are generally associated with felsic volcanic rocks, the relative timing of hydrothermal and felsic magmatic events, and the location of deposits with respect to volcanic centres, are highly variable. The intensity of volcanism may have exerted an important control on the size of individual deposits, since large deposits in the Brunswick (e.g. Brunswick No. 12 and 6) and Heath Steele belts formed during a major hiatus in felsic volcanism that is represented by interbedded volcanic sedimentary and minor tuffaceous rocks. Other major deposits such as the Caribou and Canoe Landing Lake deposits are hosted by mostly shale and clearly formed during a major break in volcanism, although the Caribou deposit was prematurely terminated by volcanism of the Spruce Lake Formation. Smaller deposits such as Restigouche, Chester, Armstrong B, Captain North Extension, Devils Elbow, and Flat Landing Brook are hosted by an almost continuous felsic volcanic sequence with few major breaks in volcanism and probably formed proximal to volcanic centres.

Based on fluid inclusion salinities (3 and 8 wt.% equivalent NaCl, Goodfellow and Peter, 1999) and temperatures >300°C estimated from the pyrrhotite-chalcopyrite-sphalerite vein assemblage (Luff et al., 1992), the hydrothermal fluids were most likely buoyant and behaved as hydrothermal plumes (Fig. 15). With continuing fluid discharge, the neutrally buoyant plume would spread laterally, assuming that there were no major currents in the basin. This dispersion process may account for the systematic lateral zonation of minerals and elements in hydrothermal sediments associated with massive sulphide deposits along the Brunswick (Peter and Goodfellow, 1996) and Heath Steele belts (Peter and Goodfellow, 2003).

The origin of reduced sulfur is a subject of debate (Goodfellow and Peter, 1999). However, the similarity of $\delta^{34}S$ values for the Brunswick No. 12 and Caribou massive sulphide deposits and those of pyrite in host black shales remote from mineralization supports a common and, therefore, ambient seawater origin for most of the reduced sulfur (Goodfellow and Peter, 1996; Goodfellow et al., 2003b). Deposits from the Caribou, Brunswick, and Stratmat horizons fall on the $\delta^{34}S$ evolutionary curve for sedimentary pyrite from the Selwyn Basin, suggesting further that the sulphur in these deposits originated from the global ocean reservoir (Fig. 16). A reduced, stagnant water column may have played several roles that were important to the formation of the Bathurst sulphide deposits. These include 1) providing a ready supply of biogenically reduced sulfur to fix hydrothermal metals, 2) increasing the capture of metals by preventing sulphide oxidation in buoyant hydrothermal plumes (unlike modern black smokers where >90% of the metals is lost to the oxygenated water column), and 3) reducing the dispersion of sulphide from vent sites by strong bottom currents.

The Llanvirian maroon shales and cherts interbedded with alkali basalts of the Little River and Boucher Brook formations, which overlie the felsic volcanic rocks, mark a change from reducing to oxidizing bottom water conditions. A strong hydrothermal signature in these sedimentary rocks indicates that hydrothermal discharge continued well after the formation of most of the massive sulphide deposits in the Camp. The paucity of deposits at this time can be explained by the absence of ambient seawater H_2S to fix hydrothermal metals, and the loss to the water column of most of the metals carried in hydrothermal plumes due to sulphide oxidation and metal dilution (Goodfellow et al., 2003b).

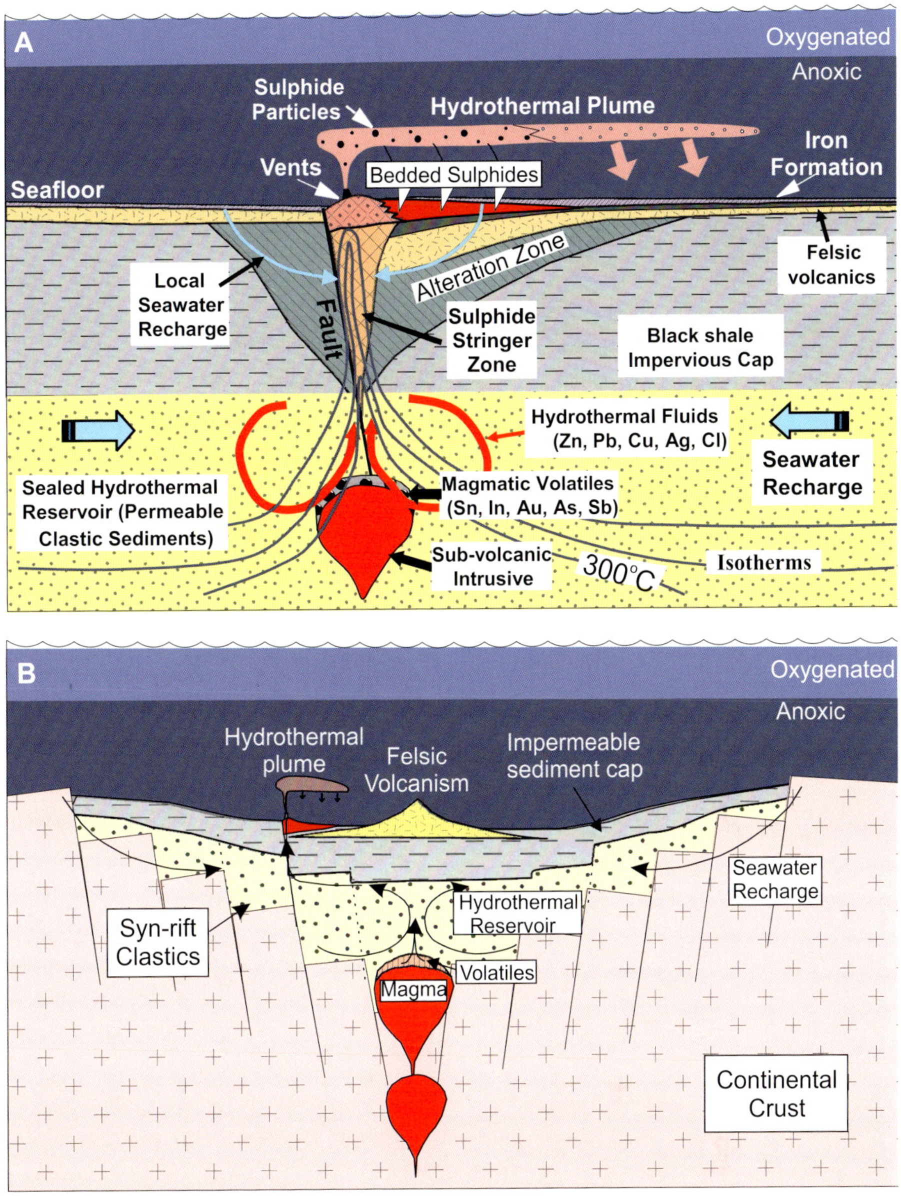

FIGURE 15. Hydrothermal and sulphide depositional models, BMC. **(A)** Detailed view of the hydrothermal system showing magma driven fluid convection and metal leaching within the hydrothermal reservoir, the distribution of isotherms, magmatic input of Sn, In, Au, As, and Sb, hydrothermal/magmatic fluid upflow along reactivated extensional faults, widespread hydrothermal alteration in pre-ore rocks and sediments, venting of hydrothermal fluids at the seafloor forming buoyant plumes, the sedimentation of hydrothermal plume precipitates and the formation of bedded sulphide facies and associated iron formations, and the zone refining of bedded sulphides within the area of fluid venting. **(B)** Back-arc continental rift filled with a syn-rift, highly permeable, coarse-grained clastic sequence sealed by impervious black organic-rich shales of the Knights Brook and Patrick Brook formations. The permeable clastic sequence behaves as a hydrothermal reservoir in areas of high heat flow generated by intrusion of felsic magmas into the underlying attenuated continental crust. Because of the low permeability of wet muds, recharge of the reservoir probably occurred at the margins of the rift where the sediments were thin. Local convection and the heating of reservoir fluids generated isothermal and fluid overpressured conditions. Discharge was likely along reactivated deep extensional faults.

There are two possible sources of metalliferous fluids that formed the BMC massive sulphide deposits - hydrothermal and magmatic fluids. Radiogenic lead isotopes (Fig. 17) combined with high Pb/Pb+Zn ratios indicate a crustal metal source, consistent with a back-arc continental rift setting, although it is unclear whether Pb was leached from continental basement, or was released from magmas that were generated by melting basement rocks. Perhaps the strongest evidence for a magmatic component in deposits is the strong positive correlation of Sn (and In) with total tonnage (Fig. 18 from Goodfellow and McCutcheon, 2003). Gold also correlates positively with deposit size, and displays a high positive correlation with Sb and As (McClenaghan et al., 2003). If Sn enrichment in large deposits was controlled by hydrothermal processes alone, then it follows that the mineralogy and

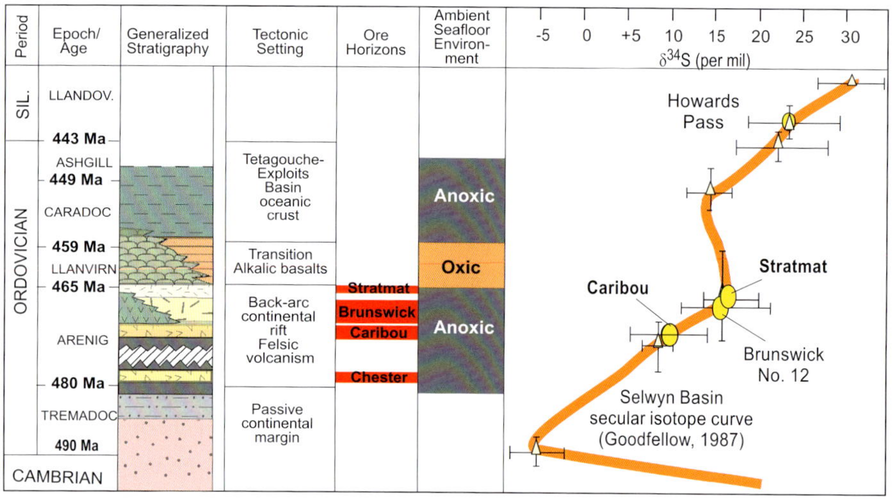

FIGURE 16. Average and range of $\delta^{34}S$ values for massive sulphide deposits from different ore horizons of the Bathurst Mining Camp, plotted on the Selwyn Basin secular curve for sedimentary pyrite (Goodfellow, 1987).

metal ratios of deposits should likewise reflect highly variable fluid compositions among deposits. However, this is not the case, nor is there any correlation of base metal contents and ratios with deposit size. This suggests, therefore, that the large deposits in the BMC are associated with high-level magma bodies that have contributed Sn, In, Au, As, and Sb to the hydrothermal system (e.g. Fig. 15A).

The anatomy of the hydrothermal system that formed the Bathurst deposits has attributes of both a classic convective cell driven by a localized heat source proposed for VHMS deposits, and a permeable hydrothermal reservoir capped by impervious sediments envisaged for SEDEX deposits (Fig. 15B). Because the Bathurst deposits formed in a continental back-arc rift, their proximity to continental sources for sediments ensured that the rift was covered by impermeable and thermally insulating fine-grained clastic sediments. These sediments served to seal the hydrothermal reservoir, reduced conductive heat loss and focused fluid discharge at long-lived vent sites. This type of hybrid hydrothermal system is therefore conducive to the formation of giant deposits and probably explains why VSHMS and SEDEX deposits are generally fewer in number but an order of magnitude larger, on average, than massive sulphide deposits hosted by only volcanic rocks.

A second major factor that contributed to the formation of giant deposits is anoxic ambient seawater that provided both an unlimited supply of reduced sulphur to fix metals, and conditions that prevented the oxidation of sulphide particles in buoyant hydrothermal plumes. This probably explains why most of the giant Phanerozoic VSHMS (e.g. BMC, Iberian Pyrite Belt, Mount Read Belt) and SEDEX deposits (e.g. Red Dog, Howards Pass, Anvil District) formed during periods of global anoxia in the Late Cambrian, Middle Ordovician (Arenig), Late Ordovician-Early Silurian (Caradoc-Llandovery), and Late Devonian-Early Mississippian.

In summary, the large size of many Bathurst deposits, and VSHMS deposits in general, results from a number of factors including a) hydrothermal architecture consisting of a hydrothermal reservoir capped by impervious fine-grained sediments, b) prolonged episodes of hydrothermal activity, c) focussed discharge from long-lived vent sites, d) formation during a major hiatus in volcanism, e) anoxic bottom waters that facilitated the total capture of metals in buoyant hydrothermal fluids, and f) direct magmatic input of metals, particularly in the case of large deposits.

Exploration Methods

Geological

Exploration using geological criteria should focus on the four major horizons that host sulphide deposits in the BMC.

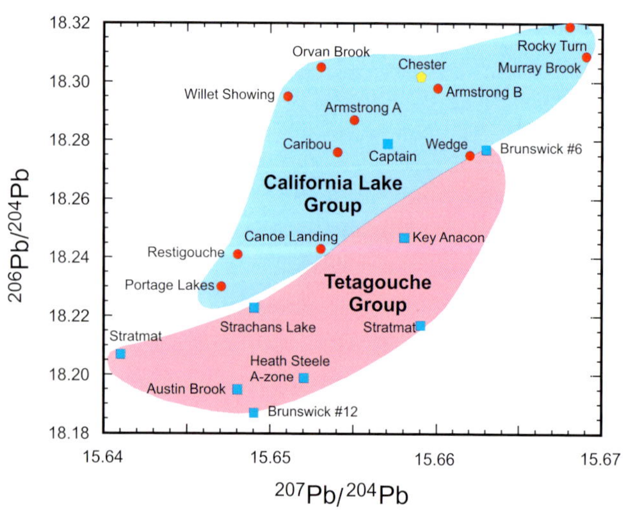

FIGURE 17. $^{207}Pb/^{204}Pb$ versus $^{206}Pb/^{204}Pb$ for massive sulphide deposits of the Bathurst Mining Camp (unpublished data from R.I. Thorpe).

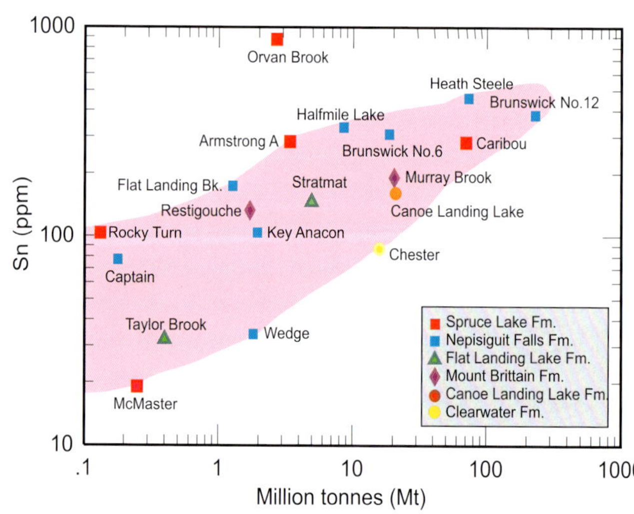

FIGURE 18. Average tin content of massive sulphides versus geological reserves of volcanic-sediment hosted massive sulphide deposits in the Bathurst Mining Camp (Goodfellow and McCutcheon, 2003).

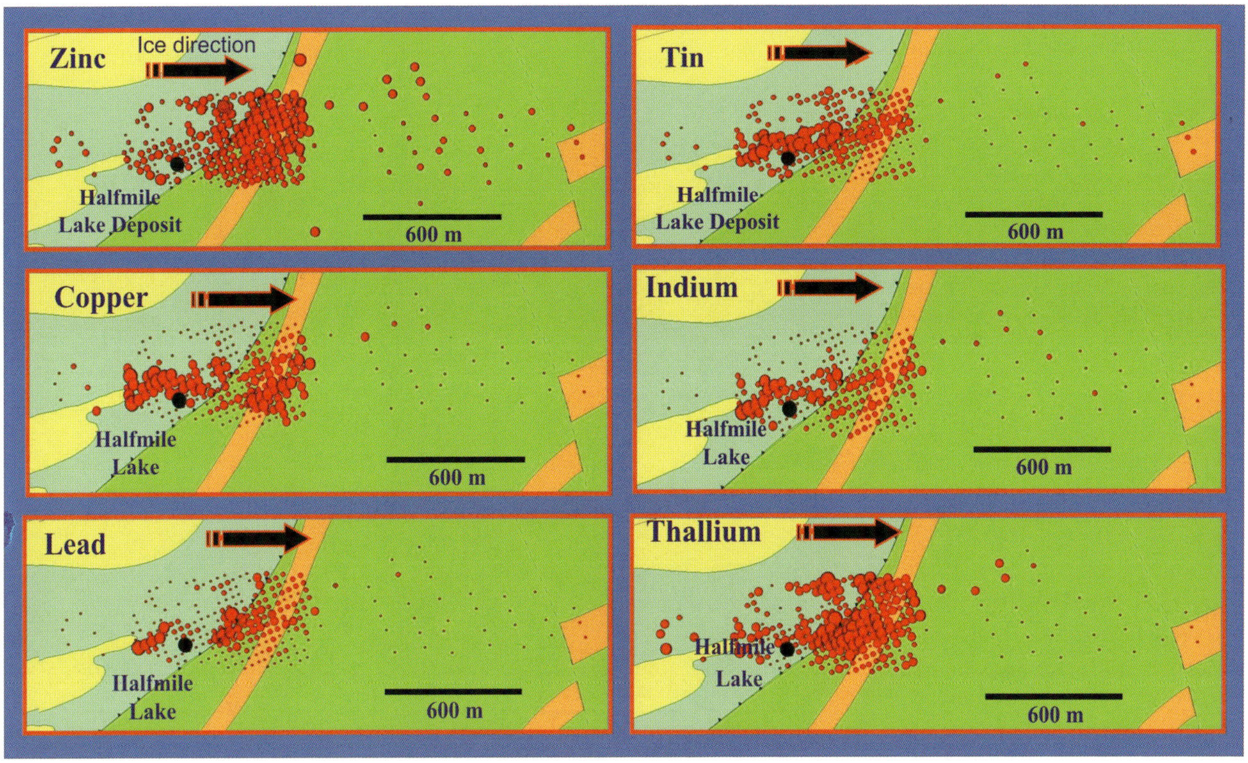

FIGURE 19. Down ice dispersion of Zn, Sn, Cu, In, Pb, and Tl in basal till at the Halfmile Lake deposit, Bathurst Mining Camp (from Parkhill and Doiron, 2003).

These are from oldest to youngest: 1) Chester, 2) Caribou, 3) Brunswick/Heath Steele, and 4) Stramat horizons.

The massive sulphide deposits and occurrences of the Chester horizon occur within the Clearwater Stream, Sevogle River, and Slacks Lake formations. The Clearwater Stream Formation hosts the only known deposit in the group: the Chester deposit with mineral resources of 15.72 Mt. The Sevogle River Formation contains three of the four known occurrences, and the Slacks Lake Formation hosts the fourth occurrence (McCutcheon et al., 2003).

The Caribou horizon occurs in the California Lake Group, which includes the Spruce Lake, Mount Brittain, and Canoe Landing Lake formations (Figs. 1 and 2). The Spruce Lake Formation hosts 10 of the 13 deposits and 15 of the 21 occurrences in the California Lake Group. Caribou is the largest deposit with mineral resources of 69 Mt.

The Brunswick horizon hosts most of the large and productive deposits in the Camp including the Brunswick No. 12 (229 Mt), Heath Steele B-zone (69 Mt), and Brunswick No. 6 (18 Mt) deposits. All deposits occur in altered mudstone and/or volcaniclastic rocks that either postdate or predate the massive quartz-feldspar "porphyry" of the Nepisiguit Falls Formation. Most deposits are also associated with laterally extensive carbonate-oxide-silicate hydrothermal sedimentary rocks that define the ore horizon and display mineral and chemical vectors to mineralization (Peter and Goodfellow, 2003).

The Stratmat horizon, the youngest ore horizon in the BMC, occurs in the Flat Landing Brook Formation (Figs. 1 and 2). The deposits are small; the largest deposit, Stratmat S-1, contains only 4.94 Mt. The Flat Landing Brook Formation consists of aphyric to feldspar (±quartz)-phyric rhyolitic flows, hyaloclastites, and crackle breccias (Rogers et al., 2003), that are interbedded with minor ash tuff, basalt and mudstone.

Deposits of the BMC are commonly associated with laminated black shales and formed when the water column in the Tetagouche-Exploits basin was stratified with anoxic and H_2S-rich bottom waters.

Lithogeochemistry

Iron Formations

As discussed in the section "Hydrothermal sediments", the most effective vectors that increase with proximity to sulphide deposits are contents of siderite, magnetite, Zn, Pb, Cu, Ag, As, Au, Bi, Cd, Co, Mo, Se, Sb, Sn, In, and Tl, and ratios of Fe/Mn, Ba/Al, P/Al, Eu/Eu* (Peter and Goodfellow, 1996, 2003). In general, significant increases in content of siderite, stilpnomelane, ankerite, calcite, biotite, chalcopyrite, and Cu are restricted to within about 0.5 km of mineralization, whereas sphalerite, galena, and notable enrichments in Pb, Zn, Ag, As, Au, Bi, Cd, Hg, In, Sn, Sb, and Tl extend up to 1.5 km from massive sulphides.

Hydrothermal Alteration

Hydrothermal alteration is widespread around individual deposits (1 to 5 km laterally and 100s of m vertically) and is zoned with regard to silicate compositions from the core to the margins of the hydrothermal fluid discharge conduit (Fig. 12). The most visibly obvious manifestations of hydrothermal alteration are green chloritic schists and sericitic rocks, the obliteration of feldspar phenocrysts in porphyritic volcanic rocks, and the occurrence of veinlets and disseminations of sulphides and ferroan carbonate. In the proximal core, the highly silicified facies has a pale green to off-white appearance (Fig. 13).

Within alteration zones, chlorite Fe/(Fe+Mg) ratios and the abundance of most ore-forming and -associated elements generally increase from the outer margins of the alteration zones to the core of hydrothermal fluid upflow zone. The increase of these elements toward the centre of hydrothermal fluid discharge is accompanied by the addition of Si, Mg and K at the expense of Na, Ca and Sr bound in feldspars and what was probably originally glass (e.g. Fig. 14 from Goodfellow and McCutcheon, 2003).

Till, Soil and Stream Sediment Geochemistry

Geochemically anomalous till defines dispersal trains that extend more than 500 m down-ice from the Halfmile Lake, Restigouche, and Stratmat mineral deposits (Parkhill and Doiron, 2003). Indium, Sn, and As, and to a lesser extent Cu, Pb, Ag, and Zn in the <0.063 mm fraction of basal till, are the best indicators of massive sulphide deposits. As an example, the Halfmile Lake deposit is reflected by Zn, Cu, Pb, Sn, In, and Tl dispersion trains that extend approximately 1 km down ice from the mineralization (Fig. 19). The suite and concentration of indicator elements in anomalous tills vary, depending on geological history and the proportions of fresh and preglacially weathered bedrock and mineralized detritus in the glacial sediments (Klassen, 2003). Element contents generally increase with decreasing grain size. Some indicator elements (Sn, Sb) are anomalous only in fractions finer than silt-size (<0.045 mm), likely reflecting the grain size of their host minerals in bedrock.

Lead isotopes have been used for the first time in the BMC for fingerprinting Pb sources in glacial till dispersion trains (Hussein et al., 2003). The Pb isotope ratios ($^{206}Pb/^{204}Pb$) in basal sediments systematically decrease toward the Halfmile Lake deposit, approaching values for massive sulphides ($^{206}Pb/^{204}Pb$ = 18.24, $^{207}Pb/^{204}Pb$ = 15.68, and $^{208}Pb/^{204}Pb$ = 38.20) and define a dispersal train that extends 600 m eastward from the mineralization in a direction broadly consistent with the ice-flow direction (Parkhill and Doiron, 2003). Lead isotope ratios, when plotted against Pb abundance data, define curves that are consistent with the mixing of two end members, one represented by the sulphide deposit and the other by the host rocks (Hussein et al., 2003).

A geochemical study of soils at the Restigouche deposit revealed that conventional leaches (aqua regia) of B-horizon soil for Pb, Cu, Ag and As can be used successfully on a regional scale to locate the Restigouche Zn-Pb deposit (Hall et al., 2003). However, selective extractions, hydroxylamine hydrochloride, and enzyme leaches provide better target delineation and contrast. By far the greatest contrast of anomaly-to-background signal is shown by Pb in B-horizon soil (enzyme>hydroxylamine>total), As in soil by the hydroxylamine leach, and As in humus by aqua regia. The preferred medium to locate deeply buried mineralization is humus digested by a conventional aqua regia attack and analyzed for As, Pb, Ag, and Cu (Hall et al., 2003).

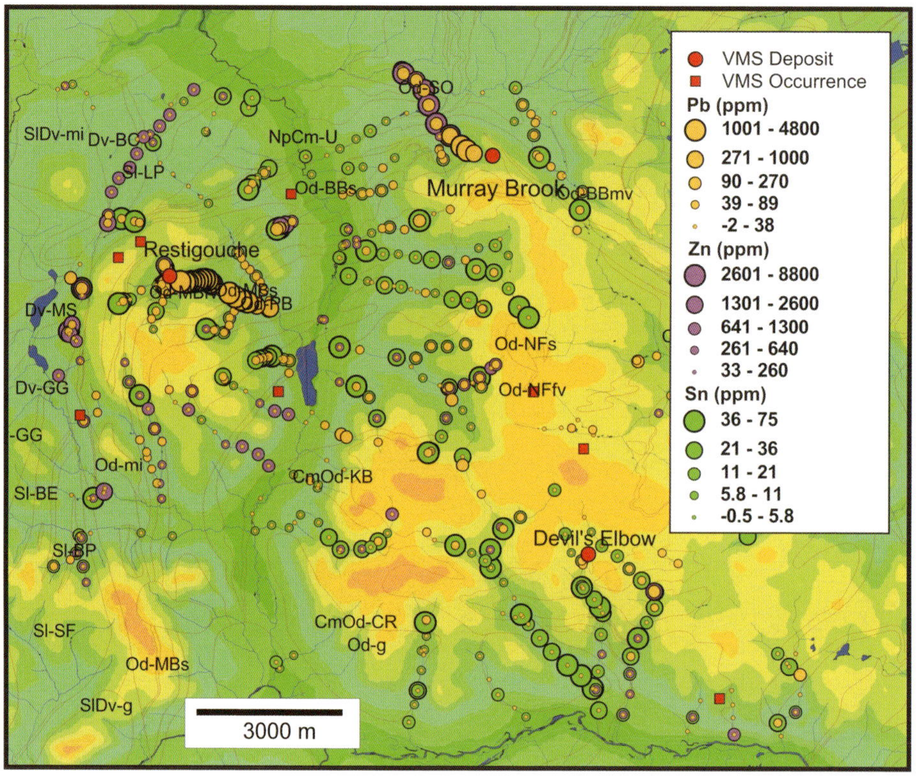

FIGURE 20. Proportional symbol plots of Pb, Zn, and Sn abundances in the minus 80-mesh fraction of sediments from streams draining the Restigouche-Murray Brook area, northwestern Bathurst Mining Camp (from data in Leybourne et al., 2003).

The Restigouche and Murray Brook deposits are both reflected by Pb anomalies in stream sediment; the Murray Brook deposit displays a strong Zn anomaly whereas the Restigouche deposit is characterized by a weak Zn anomaly in sediments (Fig. 20). Tin is elevated in streams draining the Devil's Elbow deposit and in other streams that have no known mineralization. Base metal dispersion from known sulphide deposits is generally restricted (typically less than 1 to 2 km), indicating that sampling should be carried out at a detailed scale (500 m intervals or less along streams).

Aqueous Geochemistry

Stream-water Zn anomalies extend up to 4 km downstream from the Restigouche deposit (< 2 km for Pb) but are more restricted at other massive sulphide deposits in the BMC (Leybourne and Goodfellow, 2003; Leybourne et al., 2003). The type and magnitude of the stream metal anomalies are controlled by 1) the extent of sulphide oxidation, 2) the major ion composition of ground and surface waters, and 3) the local hydrology, which controls the depth of recharge of oxygenated waters and the discharge of ground-

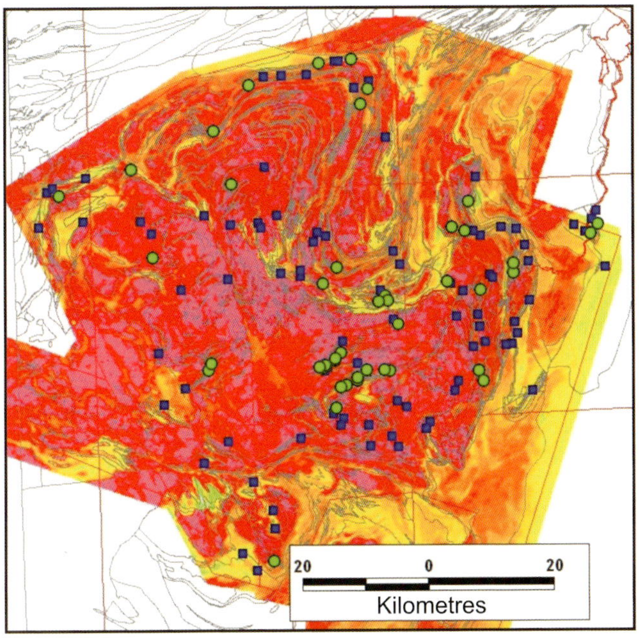

FIGURE 21. Apparent conductivity of the Bathurst Mining Camp generated from the 4433 Hz co-planar electromagnetic data (Keating et al., 2003). Also shown is an outline of the geology (van Staal et al., 2003b), deposits (circles), and occurrences (squares) in the BMC (Goodfellow and McCutcheon, 2003).

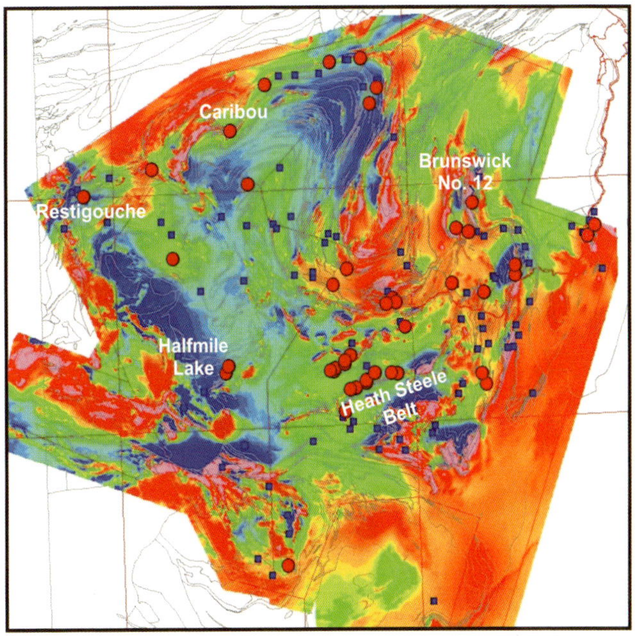

FIGURE 23. Total magnetic field of the Bathurst Mining Camp generated from data collected by a helicopter survey at 200 m line spacing (Keating et al., 2003). Also shown is an outline of the geology (van Staal et al., 2003b), deposits (circles), and occurrences (squares) in the BMC (Goodfellow and McCutcheon, 2003). Scale as in Figure 21.

waters into the stream system. The dispersion distance of metals in stream waters and sediments is controlled by pH, adsorption of elements onto precipitating Mn- and Fe-oxyhydroxides, and dilution by unmineralized stream base-flow. The partial extraction of elements in stream sediments allows for the identification of the most labile phases (adsorbed and bound to amorphous Mn- and Fe-oxyhydroxides) and yields greater anomaly contrast for some metals (e.g. Pb, Zn, Tl), but reveals essentially the same anomalies as identified using total acid extraction techniques.

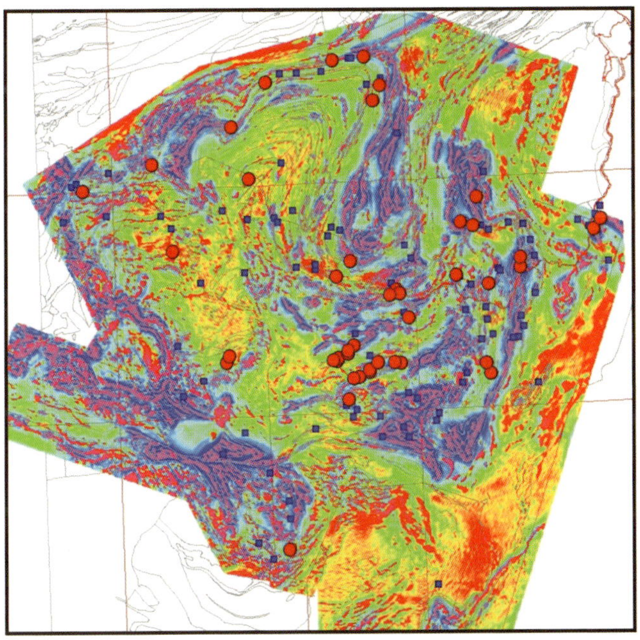

FIGURE 22. Vertical magnetic gradient of the Bathurst Mining Camp generated from data collected by a helicopter survey at 200 m line spacing (Keating et al., 2003). Also shown is an outline of the geology (van Staal et al., 2003b), deposits (circles), and occurrences (squares) in the BMC (Goodfellow and McCutcheon, 2003). Scale as in Figure 21.

Surface waters are an effective exploration tool, even in a mature mining camp such as the BMC, and offer the potential of detecting sulphide deposits concealed at depth. The local hydrology around deposits determines whether a concealed target has a surficial expression (Leybourne and Goodfellow, 2003; Leybourne et al., 2003).

Geophysics

Geophysics has played an important role in the discovery of massive sulphide deposits in the BMC (see McCutcheon et al., 2003). As part of the EXTECH II project, a multi-parameter (magnetic, electromagnetic and radiometric), helicopter-supported and high-resolution (200 m line spacing) survey was flown over the entire camp (Thomas et al., 2000; Chung, 2003; Keating et al., 2003; Shives et al., 2003). The objectives were two-fold: 1) to support the 1:20,000-scale geological mapping of the Camp, and 2) to stimulate exploration by identifying new targets for mineral exploration. Because of the poor rock exposure (1-2%), airborne geophysical data were used extensively in producing a new 1:100,000-scale geological map of the BMC (van Staal et al., 2003b).

The apparent conductivity map generated from the 4433 Hz coplanar electromagnetic data clearly outline bedrock geology (Fig. 21). Conductive units, particularly carbonaceous sedimentary rocks of different age and in different formations, are distinct and very well defined, and have been used to support detailed geological mapping below a thin cover of glacial sediments. The conductivity maps correspond very well with the calculated first vertical derivative of the magnetic field (Fig. 22) and help distinguish between

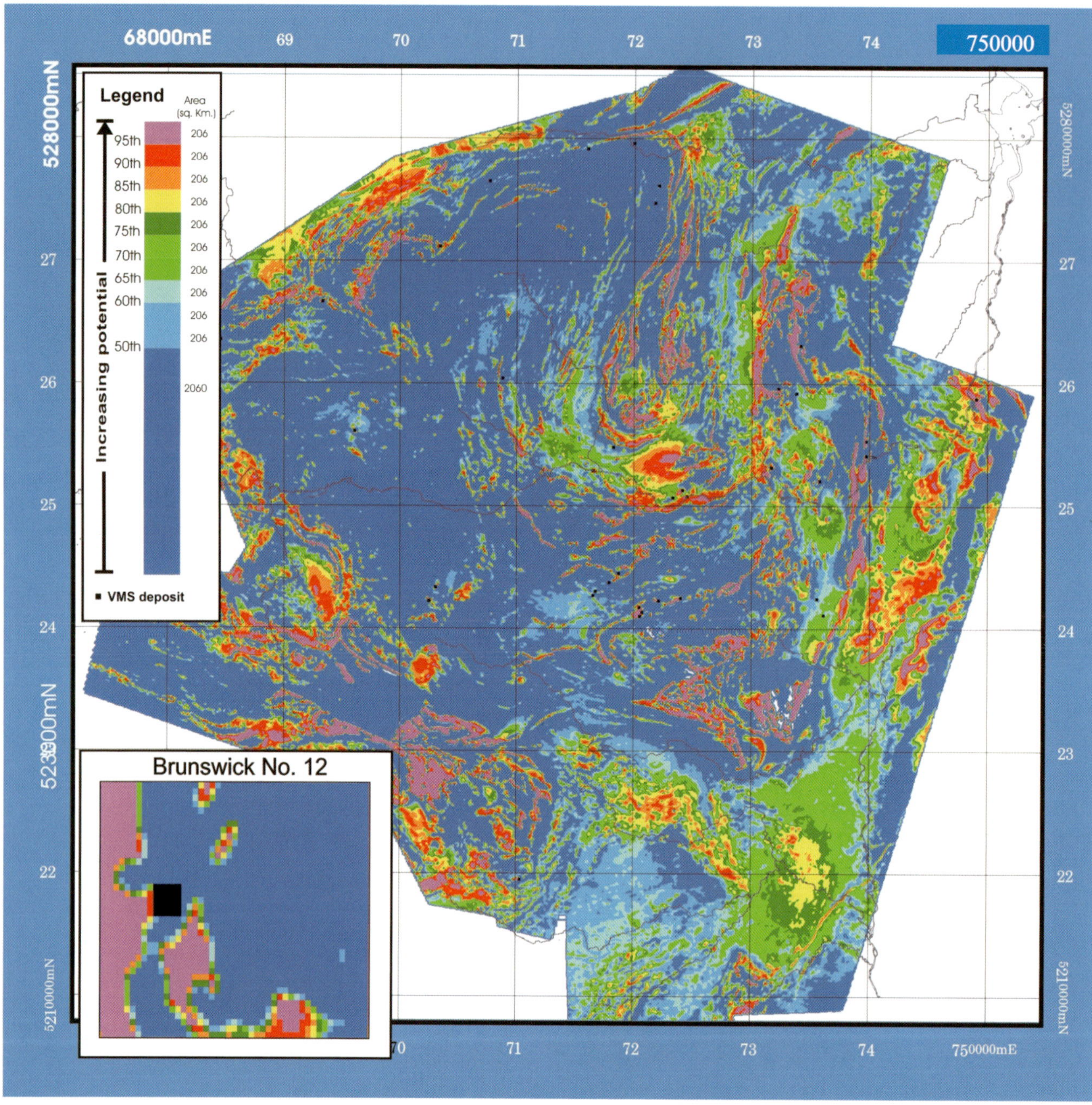

FIGURE 24. Mineral potential prediction map for VMS deposits in the BMC, generated by a discriminant analysis model using five layers: middle and low frequency conductivity, RTP total magnetic field and vertical gradients of RTP magnetic data, and the radiometric Th/K ratio. The 37 discovered deposits are shown as black squares. The inset is a detailed view of the Brunswick No. 12 deposit (from Chung, 2003).

bedrock and overburden conductivity responses (Keating et al., 2003). The vertical gradient is very effective in delineating geological units, fold structures, and fault offsets (van Staal et al., 2003b). The total magnetic field (Fig. 23) is controlled mostly by 1) the distribution of units dominated by basalt, particularly the highly magnetic alkali basalts such as those overlying the Brunswick No. 12 deposit, 2) magnetite-rich iron formation associated with deposits of the Brunswick horizon, and 3) pyrrhotite-magnetite-rich vent complexes and sulphide stringer zones.

The release of the geophysical maps led directly to the discovery of the Camelback massive sulphide deposit and the identification of several new anomalies with high mineral potential that were staked and are currently being evaluated (Fig. 24). Most deposits in the BMC are characterized by coincident, circular to oval positive magnetic and apparent conductivity anomalies of variable intensity. This is particularly true in the northwest part of the BMC where the host rocks are dominated by weakly magnetic and poorly conductive felsic volcanic rocks. The magnetic response reflects magnetite and pyrrhotite, which are concentrated in most massive sulphide deposits in the Camp. The magnetite occurs in vent complexes where it forms a magnetite-pyrite assemblage (Goodfellow and McCutcheon, 2003) or in iron formations that overlie deposits of the Brunswick horizon (Peter and Goodfellow, 1996, 2003). The pyrrhotite is commonly associated with chalcopyrite in vent complexes and sulphide stringer zones of many of the deposits and also occurs in iron formations.

Gravity anomalies related to near-surface deposits range from the exceptionally large 4 mGal amplitude, premining signature of the Brunswick No. 6 deposit to very small anomalies, such as the 0.25 mGal amplitude anomaly associated with the Key Anacon deposit. There is no linear relationship between tonnage and amplitude for the deposits of the BMC, although a linear relationship is apparent between amplitude and deposit thickness (Thomas, 2003). The presence of positive gravity anomalies is primarily a function of the proximity of massive sulphide deposits to the bedrock surface, which explains why most deposits of the BMC have a gravity anomaly. However, the potential of the gravity method for locating deeply buried sulphide deposits in the BMC is demonstrated by discovery of the Maybrun sulphide deposit beneath 180 m of Carboniferous sedimentary cover (Thomas, 2003).

Potential for Known and New Deposit Types

Since most deposits in the BMC either outcrop at the surface or subcrop below thin glacial sediments, there is a high potential for deposits concealed at depth below thick sequences of glacial sediments or within rock formations. There is also a high potential for lode gold deposits in major structures that separate blocks that were juxtaposed in the New Brunswick subduction complex (van Staal et al., 2003a).

Knowledge Gaps

Detailed maps of the Quaternary geology are required to evaluate the massive sulphide potential in areas where the glacial sediments are thick, since virtually all deposits either outcrop or subcrop below <10 m of glacial sediments. Also, detailed basal till geochemistry and mineralogy would be effective in locating deposits concealed beneath thick sequences of till.

- Construction of 3-D geological maps of the BMC that can be used to target mineralized horizons concealed at depth.
- Systematic, detailed and multi-parameter geochemistry of stream and seep sediment and water using MC-ICP-MS technology that allows precise analyses of elements and diagnostic isotopes. This data would not only assist in the detection of concealed deposits but also provide baseline data for the environmental assessment of the impact of mineral development on the surface environment.
- For deposits such as Brunswick No. 12 that are overlain by highly magnetic alkali basalts, the magnetic anomaly associated with massive sulphides is partly masked by anomalies generated by the basalts. Similarly, the highly conductive carbonaceous shales can mask electromagnetic anomalies associated with massive sulphides. This is major problem in exploring for deposits in the BMC because of the close stratigraphic relationship between organic-rich black shales and sulphide mineralization. To address this problem, it will be necessary to better understand the physical properties of organic-rich sedimentary rocks and to develop more effective geophysical methods that can discriminate between different types of conductive rocks.
- Development and testing of deeply penetrating airborne gravity and ground seismic methods for detecting massive sulphide deposits in the BMC.

Acknowledgements

This synthesis of the metallogeny of the BMC is based mostly on new research generated by the EXTECH II project and published recently in Economic Geology Monograph 11, 2003. John Lydon and Steve McCutcheon are thanked for thorough reviews this paper. Jennifer Owen and Ingrid Kjarsgaard are thanked for helpful editorial comments on this and an earlier version of the paper, respectively.

References

Adair, R. N., 1992, Stratigraphy, structure, and geochemistry of the Halfmile Lake massive-sulfide deposit, New Brunswick: Exploration and Mining Geology, v. 1, p. 151-166.

Barrie, C. T., and Hannington, M. D., 1999, Classification of volcanic-associated massive sulfide deposits, *in* Barrie, C. T., and Hannington, M. D., eds., Volcanic-Associated Massive Sulfide Deposits: Processes and Examples in Modern and Ancient Settings: Reviews in Economic Geology, v. 8, Society of Economic Geologists, p. 2-10.

Boorman, R. S., 1975, Mineralogical review of lead-zinc-copper sulphide deposits, Bathurst-Newcastle area, New Brunswick, New Brunswick Research and Productivity Council, Report No. 5128, 249 p.

Boyle, D. R., 2003, Preglacial weathering of massive sulfide deposits in the Bathurst Mining Camp: economic geology, geochemistry and exploration applications, *in* Goodfellow, W. D., McCutcheon, S. R., and Peter, J. M., eds., Massive Sulfide Deposits of the Bathurst Mining Camp, New Brunswick, and Northern Maine: Economic Geology Monograph 11, Society of Economic Geologists, p. 689-721.

Chen, T. T., 1978, Colloform and framboidal pyrite from the Caribou deposit, New Brunswick: Canadian Mineralogist, v. 16, p. 9-16.

Chung, C. F., 2003, Use of airborne geophysical surveys for constructing mineral potential maps, *in* Goodfellow, W. D., McCutcheon, S. R., and Peter, J. M., eds., Massive Sulfide Deposits of the Bathurst Mining Camp, New Brunswick, and Northern Maine: Economic Geology Monograph 11, Society of Economic Geologists, p. 879-891.

Currie, K., van Staal, C. R., Peter, J. M., and Rogers, N., 2003, Conditions of metamorphism of the main massive sulfide deposits and surrounding host rocks in the Bathurst Mining Camp, *in* Goodfellow, W. D., McCutcheon, S. R., and Peter, J. M., eds., Massive Sulfide Deposits of the Bathurst Mining Camp, New Brunswick, and Northern Maine: Economic Geology Monograph 11, Society of Economic Geologists, p. 65-78.

de Roo, J. A., and van Staal, C. R., 1994, Transpression and recumbent folding: steep belts and flat belts in the Appalachian Mobile Belt of northern New Brunswick, Canada: Geological Society of America Bulletin, v. 106, p. 541-552.

——— 2003, Sulfide remobilization and sulfide breccias in the Heath Steele and Brunswick deposits, Bathurst Mining Camp, New Brunswick, *in* Goodfellow, W. D., McCutcheon, S. R., and Peter, J. M., eds., Massive Sulfide Deposits of the Bathurst Mining Camp, New Brunswick and Northern Maine: Economic Geology Monograph 11, Society of Economic Geologists, p. 479-496.

Fyffe, L. R., 1995, Regional geology and lithogeochemistry in the vicinity of the Chester VMS deposit, Big Bald mountain area, New Brunswick, Canada: Exploration and Mining Geology, v. 4, p. 153-173.

Galley, A. G., Watkinson, D. H., Jonasson, I. R., and Riverin, G., 1995, The subsea-floor formation of volcanic-hosted massive sulfide; evidence from the Ansil Deposit, Rouyn-Noranda, Canada: Economic Geology, v. 90, p. 2006-2017.

Goodfellow, W. D., 1975, Major and minor element halos in volcanic rocks at Brunswick No. 12 sulphide deposit, N.B., Canada, *in* Elliot, I. L., and Fletcher, W. K., eds., Geochemical Exploration 1974: Elsevier, Amsterdam, p. 279-295.

——— 1987, Anoxic stratified oceans as a source of sulphur in sediment-hosted stratiform Zn-Pb deposits (Selwyn Basin, Yukon, Canada): Chemical Geology, v. 65, p. 359-382.

——— 2003, Geology and genesis of the Caribou deposit, Bathurst Mining Camp, northern New Brunswick, *in* Goodfellow, W. D., McCutcheon, S. R., and Peter, J. M., eds., Massive Sulfide Deposits of the Bathurst Mining Camp, New Brunswick, and Northern Maine, Economic Geology Monograph 11, Society of Economic Geologists, p. 327-360.

Goodfellow, W. D., and McCutcheon, S. R., 2003, Geological and genetic attributes of volcanic-associated massive sulfide deposits of the Bathurst Mining Camp, northern New Brunswick - a synthesis, *in* Goodfellow, W. D., McCutcheon, S. R., and Peter, J. M., eds., Massive Sulfide Deposits of the Bathurst Mining Camp, New Brunswick, and Northern Maine: Economic Geology Monograph 11, Society of Economic Geologists, p. 245-301.

Goodfellow, W. D., and Peter, J. M., 1996, Sulphur isotope composition of the Brunswick No. 12 massive sulphide deposit, Bathurst Mining Camp, N.B.:Iimplications for ambient environment, sulphur source and ore genesis: Canadian Journal of Earth Sciences, v. 33, p. 231-251.

——— 1999, Reply: Sulphur isotope composition of the Brunswick No. 12 massive sulphide deposit, Bathurst Mining Camp, N.B.: implications for ambient environment, sulphur source and ore genesis: Canadian Journal of Earth Sciences, v. 36, p. 127-134.

Goodfellow, W. D., Zierenberg, R. A., and ODP Leg 169 Shipboard Science Party, 1999, Genesis of massive sulphide deposits at sediment-covered spreading centers, *in* Barrie, C. T., and Hannington, M. D., eds., Volcanic-associated Massive Sulfide Deposits: Processes and Examples in Modern and Ancient Settings, Reviews in Economic Geology, Society of Economic Geologists, v. 8, p. 297-324.

Goodfellow, W. D., McCutcheon, S. R., and Peter, J. M., 2003a, Massive sulfide deposits of the Bathurst Mining Camp and northern Maine; introduction and summary of findings, *in* Goodfellow, W. D., McCutcheon, S. R., and Peter, J. M., eds., Massive Sulfide Deposits of the Bathurst Mining Camp, New Brunswick and Northern Maine: Economic Geology Monograph 11, Society of Economic Geologists, p. 1-16.

Goodfellow, W. D., Peter, J. M., Winchester, J. A., and van Staal, C. R., 2003b, Ambient marine environment during formation of massive sulfide deposits of the Bathurst Mining Camp: importance of reduced bottom waters in metal precipitation and sulfide preservation, *in* Goodfellow, W. D., McCutcheon, S. R., and Peter, J. M., eds., Massive Sulfide Deposits of the Bathurst Mining Camp, New Brunswick, and Northern Maine: Economic Geology Monograph 11, Society of Economic Geologists, p. 129-156.

Gromet, L. P., Dymek, R. F., Haskin, L. A., and Korotev, R. L., 1984, The "North American shale composite": its compilation, major and trace element characteristics: Geochimica et Cosmochimica Acta, v. 48, p. 2469-2482.

Hall, G.E.M., Parkhill, M.A., and Bonham-Carter, G. F., 2003, Conventional and selective leach geochemical exploration methods applied to humus and B horizon soil overlying the Restigouche VMS deposit, Bathurst Mining Camp, New Brunswick, *in* Goodfellow, W. D., McCutcheon, S. R., and Peter, J. M., eds., Massive Sulfide Deposits of the Bathurst Mining Camp, New Brunswick, and Northern Maine: Economic Geology Monograph 11, Society of Economic Geologists, p. 763-782.

Hamilton, A., 1992, Geology of the Stratmat boundary and Heath Steele N-5 zones, Bathurst Camp, northern New Brunswick: Exploration and Mining Geology, v. 1, p. 131-135.

Hussein, A. A., Lochner, C., and Bell, K., 2003, Application of Pb isotopes to mineral exploration in the Halfmile Lake area, Bathurst, New Brunswick, *in* Goodfellow, W. D., McCutcheon, S. R., and Peter, J. M., eds., Massive Sulfide Deposits of the Bathurst Mining Camp, New Brunswick, and Northern Maine: Economic Geology Monograph 11, Society of Economic Geologists, p. 679-688.

Jambor, J. L., 1979, Mineralogical evolution of proximal and distal features in New Brunswick massive sulfide deposits: Canadian Mineralogist, v. 17, p. 79-87.

Juras, S. J., 1981, Alteration and sulfide mineralization in footwall felsic and metasedimentary rocks, Brunswick No. 12 deposit, Bathurst, New Brunswick, Canada: Unpub. M.Sc. thesis, University of New Brunswick, Fredericton, 208 p.

Keating, P., Thomas, M. D., and Kiss, F., 2003, Significance of high resolution magnetic and electromagnetic survey for exploration and geological investigations, Bathurst Mining Camp, *in* Goodfellow, W. D., McCutcheon, S. R., and Peter, J. M., eds., Massive Sulfide Deposits of the Bathurst Mining Camp, New Brunswick, and Northern Maine: Economic Geology Monograph 11, Society of Economic Geologists, p. 783-798.

Klassen, R. A., 2003, The geochemical and physical properties of till, Bathurst Mining Camp, New Brunswick, Canada, *in* Goodfellow, W. D., McCutcheon, S. R., and Peter, J. M., eds., Massive Sulfide Deposits of the Bathurst Mining Camp, New Brunswick, and Northern Maine: Economic Geology Monograph 11, Society of Economic Geologists, p. 661-678.

Large, R. R., 1977, Chemical evolution and zonation of massive sulfide deposits in volcanic terrains: Economic Geology, v. 72, p. 549-572.

——— 1992, Australian volcanic-hosted massive sulfide deposits: features, styles and genetic models: Economic Geology, v. 87, p. 471-510.

Lea, E. R., and Rancourt, C., 1958, Geology of the Brunswick Mining and Smelting orebodies, Gloucester County, New Brunswick: Canadian Institute of Mining and Metallurgy Bulletin, v. 61, p. 95-105.

Lentz, D., and Goodfellow, W. D., 1993, Petrology and mass balance constraints on the origin of quartz augen schist associated with the Brunswick massive sulphide deposits, Bathurst, New Brunswick: Canadian Mineralogist, v. 31, p. 877-903.

——— 1994, Character, distribution, and origin of zoned hydrothermal alteration features at the Brunswick No. 12 massive sulphide deposit, Bathurst Mining Camp, New Brunswick, *in* Abbott, S. A., ed., Current Research: New Brunswick Department of Natural Resources and Energy, Minerals Resources, Information Circular 94-1, p. 94-119.

——— 1996, Intense silicification of footwall sedimentary rocks in the stockwork alteration beneath the Brunswick No. 12 massive sulphide deposit, Bathurst, New Brunswick: Canadian Journal of Earth Sciences, v. 33, p. 284-302.

Lentz, D. R., Hall, D. C., and Hoy, L. D., 1997, Chemostratigraphic, alteration, and oxygen isotopic trends in a profile through the stratigraphic sequence hosting the Heath Steele B Zone massive sulphide deposit, New Brunswick: Canadian Mineralogist, v. 35, p. 841-874.

Leybourne, M. I., and Goodfellow, W. D., 2003, Processes of metal solution and transport in groundwaters interacting with undisturbed massive sulfide deposits, Bathurst Mining Camp, New Brunswick, *in* Goodfellow, W. D., McCutcheon, S. R., and Peter, J. M., eds., Massive Sulfide Deposits of the Bathurst Mining Camp, New Brunswick, and Northern Maine: Economic Geology Monograph 11, Society of Economic Geologists, p. 723-739.

Leybourne, M. I., Boyle, D. R., and Goodfellow, W. D., 2003, Interpretation of stream water and sediment geochemistry in the Bathurst Mining Camp, New Brunswick: application to mineral exploration, *in* Goodfellow, W. D., McCutcheon, S. R., and Peter, J. M., eds., Massive Sulfide Deposits of the Bathurst Mining Camp, New Brunswick, and Northern Maine: Economic Geology Monograph 11, Society of Economic Geologists, p. 741-761.

Luff, W. M., Goodfellow, W. D., and Juras, S. J., 1992, Evidence for a feeder pipe and associated alteration at Brunswick No. 12 massive-sulfide deposit: Exploration and Mining Geology, v. 1, p. 167-185.

McClenaghan, S. H., Goodfellow, W. D., and Lentz, D. R., 2003, Gold in massive sulfide deposits, Bathurst Mining Camp: distribution and genesis, *in* Goodfellow, W. D., McCutcheon, S. R., and Peter, J. M., eds., Massive Sulfide Deposits of the Bathurst Mining Camp, New Brunswick, and Northern Maine: Economic Geology Monograph 11, Society of Economic Geologists, p. 303-326.

McCutcheon, S. R., Luff, W. M., and Boyle, R. W., 2003, The Bathurst Mining Camp, New Brunswick, Canada: history of discovery and evolution of geological models, *in* Goodfellow, W. D., McCutcheon, S. R., and Peter, J. M., eds., Massive Sulfide Deposits of the Bathurst Mining Camp, New Brunswick, and Northern Maine: Economic Geology Monograph 11, Society of Economic Geologists, p. 17-35.

Nelson, G. A., 1983, Alteration of footwall rocks at Brunswick No. 6 and Austin Brook deposits, Bathurst, New Brunswick, Canada: Unpub. M.Sc. thesis, University of New Brunswick, Fredericton, 236 p.

Parkhill, M. A., and Doiron, A., 2003, Quaternary geology of the Bathurst Mining Camp and implications for base metal exploration using drift prospecting, *in* Goodfellow, W. D., McCutcheon, S. R., and Peter, J. M., eds., Massive Sulfide Deposits of the Bathurst Mining Camp, New Brunswick, and Northern Maine: Economic Geology Monograph 11, Society of Economic Geologists, p. 631-660.

Peter, J. M., and Goodfellow, W. D., 1996, Mineralogy, bulk and rare earth element geochemistry of massive sulphide-associated hydrothermal sediments of the Brunswick horizon, Bathurst mining camp, New Brunswick: Canadian Journal of Earth Sciences, v. 33, p. 252-283.

——— 2003, Hydrothermal sedimentary rocks of the Heath Steele Belt, Bathurst Mining Camp, New Brunswick 3. Application of mineralogy and mineral and bulk composition to massive sulfide exploration, *in* Goodfellow, W. D., McCutcheon, S. R., and Peter, J. M., eds., Massive Sulfide Deposits of the Bathurst Mining Camp, New Brunswick, and Northern Maine: Economic Geology Monograph 11, Society of Economic Geologists, p. 417-433.

Peter, J. M., Goodfellow, W. D., and Doherty, W., 2003a, Hydrothermal sedimentary rocks of the Heath Steele Belt, Bathurst Mining Camp, New Brunswick 2. Bulk and rare earth element geochemistry and implications for origin, *in* Goodfellow, W. D., McCutcheon, S. R., and Peter, J. M., eds., Massive Sulfide Deposits of the Bathurst Mining Camp, New Brunswick, and Northern Maine: Economic Geology Monograph 11, Society of Economic Geologists, p. 391-415.

Peter, J. M., Kjarsgaard, I. M., and Goodfellow, W. D., 2003b, Hydrothermal sedimentary rocks of the Heath Steele Belt, Bathurst Mining Camp, New Brunswick 1. Mineralogy and mineral chemistry, *in* Goodfellow, W. D., McCutcheon, S. R., and Peter, J. M., eds., Massive Sulfide Deposits of the Bathurst Mining Camp, New Brunswick, and Northern Maine: Economic Geology Monograph 11, Society of Economic Geologists, p. 361-390.

Rogers, N., and van Staal, C. R., 2003, Volcanology and tectonic setting of the northern Bathurst Mining Camp: Part II - mafic volcanic constraints on back-arc opening, *in* Goodfellow, W. D., McCutcheon, S. R., and Peter, J. M., eds., Massive Sulfide Deposits of the Bathurst Mining Camp, New Brunswick, and Northern Maine: Economic Geology Monograph 11, Society of Economic Geologists, p. 181-201.

Rogers, N., van Staal, C. R., McNicoll, V., and Theriault, R., 2003, Volcanology and tectonic setting of the northern Bathurst Mining Camp: Part I - extension and rifting of the Popelogan Arc, *in* Goodfellow, W. D., McCutcheon, S. R., and Peter, J. M., eds., Massive Sulfide Deposits of the Bathurst Mining Camp, New Brunswick, and Northern Maine: Economic Geology Monograph 11, Society of Economic Geologists, p. 157-179.

Roy, S., 1961, Mineralogy and paragenesis of the lead-zinc-copper ores of the Bathurst-Newcastle district: Geological Survey of Canada Bulletin 72, 31 p.

Sáez, R., Almodovar, G. R., and Pascual, E., 1996, Geological constraints on massive sulphide genesis in the Iberian pyrite belt: Ore Geology Reviews, v. 11, p. 429-451.

Sáez, R., Pascual, E., Toscano, M., and Almodóvar, G. R., 1999, The Iberian type of volcano-sedimentary massive sulphide deposits: Mineralium Deposita, v. 34, p. 549-570.

Saif, S. I., 1980, Petrographic and geochemical investigation of iron formation and other iron-rich rocks in Bathurst District, New Brunswick, Current Research Part A, Paper 80-1A, Geological Survey of Canada, p. 309-317.

Shives, R. B. K., Ford, K. L., and Peter, J. M., 2003, Mapping and exploration applications of gamma ray spectrometry in the Bathurst Mining Camp, northeastern New Brunswick, *in* Goodfellow, W. D., McCutcheon, S. R., and Peter, J. M., eds., Massive Sulfide Deposits of the Bathurst Mining Camp, New Brunswick, and Northern Maine: Economic Geology Monograph 11, Society of Economic Geologists, p. 819-840.

Stanton, R. L., 1959, Mineralogical features and possible mode of emplacement of the Brunswick Mining and Smelting orebodies, Gloucester County, New Brunswick: Canadian Institute of Mining and Metallurgy Bulletin, v. 570, p. 631-642.

Tanimura, S., Date, J., Takahashi, T., and Ohmoto, H., 1983, Geologic setting of the Kuroko deposits, Japan: Part II. Stratigraphy and structure of the Hokuroko District, *in* Ohmoto, H., and Skinner, B. J., eds., The Kuroko and Related Volcanogenic Massive Sulfide Deposits, Economic Geology Monograph 5, Society of Economic Geologists, p. 24-38.

Thomas, M. D., 2003, Gravity signatures of massive sulfide deposits, Bathurst Mining Camp, New Brunswick, Canada, *in* Goodfellow, W. D., McCutcheon, S. R., and Peter, J. M., eds., Massive Sulfide Deposits of the Bathurst Mining Camp, New Brunswick, and Northern Maine: Economic Geology Monograph 11, Society of Economic Geologists, p. 799-817.

Thomas, M. D., Walker, J. A., Keating, P., Shives, R., Kiss, F., and Goodfellow, W. D., 2000, Geophysical atlas of massive sulphide signatures, Bathurst Mining Camp, New Brunswick: Geological Survey of Canada Open File 3887, p. 105 p.

van Staal, C. R., 1994, Brunswick subduction complex in the Canadian Appalachians: Record of Late Ordovician to Late Silurian collision between Laurentia and the Gander margin of Avalon: Tectonics, v. 13, p. 946-962.

van Staal, C. R., and de Roo, J. A., 1995, Mid-Paleozoic tectonic evolution of the Appalachian Central Mobile belt in northern New Brunswick, Canada: collision, extension collapse and dextral transpression, *in* Hibbard, J. P., Staal, C. R. v., and Cawood, P. A., eds., Current perspectives in the Appalachian-Caledonian Orogen: Geological Association of Canada, Special Paper 41, p. 367-389.

van Staal, C. R., and Williams, P. F., 1984, Structure, origin and concentration of the Brunswick 12 and 6 orebodies: Economic Geology, v. 79, p. 1669-1692.

van Staal, C. R., Ravenhurst, C. E., Winchester, J. A., Roddick, J. C., and Langton, J. P., 1990, Evidence for a post-Taconic blueschist suture in northern New Brunswick, Canada: Geology, v. 18, p. 1073-1077.

van Staal, C. R., Wilson, R. A., Fyffe, L. R., Langton, J. P., McCutcheon, S. R., Rogers, N., McNicoll, V., and Ravenhurst, C. E., 2003a, Geology and tectonic setting of the Bathurst Mining Camp and its relationship to coeval rocks in southwestern New Brunswick and adjacent Maine - a synthesis, *in* Goodfellow, W. D., McCutcheon, S. R., and Peter, J. M., eds., Massive Sulfide Deposits of the Bathurst Mining Camp, New Brunswick, and Northern Maine: Economic Geology Monograph 11, Society of Economic Geologists, p. 37-60.

van Staal, C. R., Wilson, R. A., Rogers, N., Fyffe, L. R., Gower, S. J., Langton, J. P., McCutcheon, S. R., and Walker, J. A., 2003b, A New Geological Map of the Bathurst Mining Camp and Surrounding Areas - A Product of Integrated Geological, Geochemical and Geophysical Data, *in* Goodfellow, W. D., McCutcheon, S. R., and Peter, J. M., eds., Massive Sulfide Deposits of the Bathurst Mining Camp, New Brunswick, and Northern Maine: Economic Geology Monograph 11, Society of Economic Geologists, p. 61-64.

Wahl, J. L., 1977, Rock geochemical exploration at the Heath Steele and Key Anacon deposits, New Brunswick: Unpub. Ph.D. thesis, University of New Brunswick, 429 p.

Yang, K., Scott, S. D., and Goodfellow, W. D., 2003, Footwall alteration associated with massive sulfide deposits in the Bathurst Mining Camp, New Brunswick: implication for seafloor hydrothermal mixing processes, *in* Goodfellow, W. D., McCutcheon, S. R., and Peter, J. M., eds., Massive Sulfide Deposits of the Bathurst Mining Camp, New Brunswick, and Northern Maine: Economic Geology Monograph 11, Society of Economic Geologists, p. 435-456.

Volcanic-Hosted Massive Sulphide Deposits of the Finlayson Lake District, Yukon

Jan M. Peter[1], Daniel Layton-Matthews[2,3], Stephen Piercey[4], Geoffrey Bradshaw[5,6], Suzanne Paradis[7], and Amy Bolton[8,9]

1. Geological Survey of Canada, 601 Booth Street, Ottawa, Ontario K1A OE8
2. Department of Geology, University of Toronto, 22 Russell Street, Toronto, Ontario M5S 3B1
3. Present address: Department of Geological Sciences and Geological Engineering, Queen's University, Kingston, Ontario K7L 3N6
4. Department of Earth Sciences and Mineral Exploration Research Centre, Laurentian University, Ramsey Lake Road, Sudbury, Ontario P3E 2C6
5. Department of Earth and Ocean Sciences, University of British Columbia, 6339 Stores Road, Vancouver, British Columbia V6C 1Z4
6. deceased
7. Geological Survey of Canada, 4718 – 9860 Saanich Road, Sidney, British Columbia V8L 4B2
8. School of Earth and Ocean Sciences, P.O. Box 3055, University of Victoria, Victoria, British Columbia V8W 3P6
9. Present address: School of Chemical Engineering and Analytical Science, University Manchester, P.O. Box 88, Sackville Street, Manchester, M60 1QD, UK
Corresponding author's email: jpeter@nrcan.gc.ca

Abstract

The Finlayson Lake District (FLD) is a crescent-shaped 300 km long, and 50 km wide area within southeastern Yukon Territory. It comprises part of the Yukon-Tanana Terrane, one of the innermost terranes of the Canadian Cordillera that underlies much of the central Yukon and parts of Alaska and British Columbia, and part of the Slide Mountain Terrane. The FLD is consists of several fault- and unconformity-bound groups and formations of early Mississippian to early Permian metamorphosed plutonic, volcanic, and sedimentary rocks. The rocks were formed in a variety of tectonic settings, including rifted frontal arc, continental back-arc, and oceanic back-arc that range in age from 365 to 275 Ma.The FLD hosts numerous base metal sulphide deposits that collectively contain in excess of 30 Mt. The Fyre Lake Cu-Co is a Besshi-type deposit containing 15.4 Mt grading 1.2% Cu, 0.8% Co, and 0.46 g/t Au that occurs at the transition from mafic volcanic rocks to overlying turbiditic sedimentary rocks in a fore-arc setting. The Kudz Ze Kayah Zn-Pb-Cu and GP4F Zn-Pb Kuroko-type deposits are hosted by felsic volcanic and volcaniclastic rocks in the immediate footwall whereas the Wolverine Zn-Pb-Cu deposit is hosted by graphitic shales and felsic volcanic and volcaniclastic rocks and is best classified as a volcanic-sediment-hosted massive sulphide (VSHMS) deposit. These latter three deposits are thought to have formed in ensialic back-arc rift or back-arc basin environments. The deposits have strongly hydrothermally altered rocks in the stratigraphic footwalls, with alteration styles varying from chlorite (Fyre Lake, Kudz Ze Kayah, Wolverine, Ice), sericite (Kudz Ze Kayah, GP4F, Wolverine), silica/quartz (Kudz Ze Kayah, Wolverine), carbonate (Wolverine) and albite (Kudz Ze Kayah). Feeder zones/stringer veins are present at the Fyre Lake, Kudz Ze Kayah, Wolverine and Ice deposits. Hanging-wall alteration is present at Fyre Lake (weak chlorite) and Wolverine (weak sericite). Wolverine has a laterally extensive (>10 km strike length) chemical exhalative sedimentary iron formation horizon in the hanging wall that is not present at Kudz Ze Kayah or GP4F. Much of the Kudz Ze Kayah deposit likely formed as a mound on the seafloor, as did at least some of the Wolverine deposit. However, subsurface replacement was important at Wolverine, and likely was entirely responsible for the formation of GP4F. The Kudz Ze Kayah deposit contains an inferred geological resource of 13 Mt at 5.55% Zn, 1.00% Cu, 1.30% Pb, 125 g/t Ag, and 1.2 g/t Au. GP4F contains an inferred resource of 1.5 Mt grading 6.4% Zn, 3.10% Pb, 0.10% Cu, 89.7 g/t Ag, and 2.0 g/t Au. Wolverine has measured and indicated reserves of 4.51 Mt grading 12.04% Zn, 1.15% Cu, 1.57% Pb, 351.5 g/t Ag, and 1.68 g/t Au; with an additional inferred resource of 1.69 Mt grading 12.16% Zn, 1.23% Cu, 1.74% Pb, 385.1 g/t Ag, and 1.71 g/t Au. The Ice Cyprus-type deposit is hosted by mafic volcanic rocks at the transition from underlying brecciated pillow basalt to overlying massive basalt and is thought to have formed in a back-arc basin/ocean basin setting. The geological resource at Ice is 4.56 Mt grading 1.48% Cu; some intervals contain about 1% Zn, but a Zn resource has not been calculated.

There are numerous other showings and prospects that require additional exploration work. The relatively nascent level of exploration in the FLD suggests that prospecting, coupled with traditionally successful exploration methods, such as surficial geochemistry, and airborne and ground geophysics (magnetics, EM) focused on the prospective rock units, are likely to result in new discoveries in the coming years.

Résumé

Le district de Finlayson Lake (DFL) est une étendue falciforme longue de 300 km et large de 50 km dans la partie sud-est du Yukon. Elle se compose d'un segment du terrane de Yukon-Tanana, l'un des plus intérieurs des terranes de la Cordillère canadienne qui s'étend à une bonne partie du centre du Yukon et à des parties de l'Alaska et de la Colombie-Britannique, et d'un segment du terrane de Slide Mountain. Le DFL comprend plusieurs groupes et formations limités par des failles ou des discordances qui se composent de roches plutoniques, volcaniques et sédimentaires métamorphisées datant du Mississippien précoce au Permien précoce. Ces roches se sont formées entre 365 et 275 Ma dans toute une gamme de cadres tectoniques, dont des arcs frontaux en distension, des arrière-arcs continentaux et des arrière-arcs océaniques. Le DFL recèle de nombreux gisements de sulphures de métaux communs qui renferment

Peter, J.M., Layton-Matthews, D., Piercey, S., Bradshaw, G., Paradis, S., and Boulton, A., 2007, Volcanic-hosted massive sulphide deposits of the Finlayson Lake District, Yukon, *in* Goodfellow, W.D., ed., Mineral Deposits of Canada: A Synthesis of Major Deposit-Types, District Metallogeny, the Evolution of Geological Provinces, and Exploration Methods: Geological Association of Canada, Mineral Deposits Division, Special Publication No. 5, p. 471-508.

ensemble plus de 30 Mt de minerai. Le gisement de Cu-Co de Fyre Lake est un gîte de type Besshi renfermant 15,4 Mt de minerai titrant 1,2 % de Cu, 0,8 % de Co et 0,46 g/t de Au. Ce gisement se présente à la transition entre un empilement de roches volcaniques mafiques et une succession sus-jacente de roches sédimentaires turbiditiques dans un cadre d'avant-arc. Les gisements de Zn-Pb-Cu de Kudz Ze Kayah et de Zn-Pb de GP4F, des gîtes de type Kuroko, sont encaissés dans des roches volcaniques et volcanoclastiques felsiques tout au sommet de l'empilement volcanique, alors que le gisement de Zn-Pb-Cu de Wolverine est encaissé dans des shales et des roches volcaniques et volcanoclastiques felsiques et s'insère mieux dans la catégorie des gîtes de sulfures massifs dans des roches volcano-sédimentaires. On pense que ces trois derniers gisements se sont formés dans des milieux ensialiques de rift d'arrière-arc ou de bassin d'arrière-arc. Ils présentent aux épontes stratigraphiques inférieures des roches ayant subi une intense altération hydrothermale, dont les styles variés sont caractérisés par la présence de chlorite (Fyre Lake, Kudz Ze Kayah, Wolverine, Ice), de séricite (Kudz Ze Kayah, GP4F, Wolverine), de silice/quartz (Kudz Ze Kayah, Wolverine), de carbonates (Wolverine) et d'albite (Kudz Ze Kayah). Des zones nourricières/filoniennes sont présentes aux gisements de Fyre Lake, de Kudz Ze Kayah, de Wolverine et d'Ice. Il y a altération dans l'éponte supérieure aux gisements de Fyre Lake (faible chloritisation) et de Wolverine (faible séricitisation). Le gisement de Wolverine présente dans son éponte supérieure un horizon latéralement étendu (>10 km d'étendue longitudinale) de formation de fer chimique d'origine exhalative qui est absent des gisements de Kudz Ze Kayah et de GP4F. Le gisement de Kudz Ze Kayah s'est en grande partie vraisemblablement formé sous la forme d'un monticule sur le fond marin, comme, à tout le moins en partie, le gisement de Wolverine. Cependant, le remplacement en subsurface a été important au gisement de Wolverine et il est vraisemblablement entièrement responsable de la formation du gisement de GP4F. Le gisement de Kudz Ze Kayah renfermerait des ressources présumées de 13 Mt titrant 5,55 % de Zn, 1,00 % de Cu, 1,30 % de Pb, 125 g/t de Ag et 1,2 g/t de Au. Au gisement de GP4F, les ressources présumées s'élèveraient à 1,5 Mt renfermant 6,4 % de Zn, 3,10 % de Pb, 0,10 % de Cu, 89,7 g/t de Ag et 2,0 g/t de Au. Au gisement de Wolverine, les ressources mesurées et indiquées s'établissent à 4,51 Mt titrant 12,04 % de Zn, 1,15 % de Cu, 1,57 % de Pb, 351,5 g/t de Ag et 1,68 g/t de Au, avec des ressources présumées additionnelles s'élevant à 1,69 Mt renfermant 12,16 % de Zn, 1,23 % de Cu, 1,74 % de Pb, 385,1 g/t de Ag et 1,71 g/t de Au. Le gisement d'Ice, de type Chypre, est encaissé dans des roches volcaniques mafiques à la transition entre une succession inférieure de basalte en coussins bréchifié et une succession supérieure de basalte massif, et il se serait formé dans un cadre de bassin d'arrière-arc/bassin océanique. Au gisement d'Ice, les ressources minérales sont de 4,56 Mt renfermant 1,48% de Cu; certains intervalles renferment environ 1% de Zn, mais les ressources en Zn n'ont pas été calculées.

De nombreux autres indices minéralisés et prospects doivent faire l'objet de travaux d'exploration additionnels. Le niveau d'exploration relativement embryonnaire du DFL suggère que la prospection associée aux méthodes classiques éprouvées d'exploration comme la géochimie des matériaux superficiels et les levés géophysiques aériens et au sol (magnétiques, EM) focalisés sur les unités lithologiques prometteuses permettront vraisemblablement de nouvelles découvertes au cours des années à venir.

Introduction

The discovery of a number of volcanic-hosted massive sulphide (VHMS) deposits in the Finlayson Lake District (FLD) within a relatively short timespan since 1995 facilitated the need to understand the stratigraphy and tectonic history of the region. Until recently, the age, stratigraphy and tectonic settings of the known VHMS deposits were poorly constrained, and this hindered exploration efforts in this part of the North American Cordillera. Recent work has provided much new information on the geology (Murphy and Timmerman, 1997; Murphy, 1998, 2001; Murphy and Piercey, 1998, 1999a,b,c, 2000a,b, 2002; Piercey and Murphy, 2000, 2001; Murphy et al., 2002, 2006), paleotectonic settings and metallogeny (Piercey, 2001; Piercey et al., 1999, 2001a,b,c, 2002a,b, 2003, 2004, 2006, in press), and the geology and origin of VHMS deposits (Bradshaw et al., 2001, in press; Layton-Matthews et al., 2001, in press; Boulton, 2002; Bradshaw, 2003; Layton-Matthews, 2005). This paper summarizes current knowledge of the massive sulphide deposits since the seminal in-depth report of Hunt (2002). The paper focuses discussion of the five largest and best known deposits in the area: 1) Fyre Lake, 2) Kudz Ze Kayah, 3) GP4F, 4) Wolverine, and 5) Ice.

Tectonic Setting

Pericratonic terranes contain elements of the continental margin that are of uncertain paleogeography (Wheeler and McFeely, 1991). Until a few years ago, the belt of pericratonic terranes, which lies between autochthonous miogeoclinal strata of ancestral North America and oceanic and island-arc terranes that were accreted during the Mesozoic, was one of the most poorly understood elements of the North American Cordillera. These pericratonic terranes span the entire length of the Cordillera, from Mexico to Alaska, and record contrasting depositional and tectonic evolutionary histories along different segments of the orogen (Colpron et al., 2006a; Nelson et al., 2006). In places, the pericratonic terranes are depositionally tied to ancestral North America (Colpron and Price, 1995).

One of these pericratonic terranes, the Yukon-Tanana Terrane is one of the innermost terranes of the Canadian Cordillera and underlies much of central Yukon, and parts of Alaska and British Columbia (Colpron et al., 2006a). The Yukon-Tanana Terrane is bounded to the northeast by rocks of the Slide Mountain Terrane, with the latter separating the Yukon-Tanana Terrane from North American continental margin rocks (Nelson et al., 2006) (Fig. 1). The Yukon-Tanana Terrane consists of polydeformed metamorphosed sedimentary, volcanic, and plutonic rocks derived from a variety of igneous and sedimentary protoliths of Late Devonian to Early Permian age (Mortensen and Jilson, 1985; Mortensen, 1992b; Colpron et al., 2006a). The Yukon-Tanana Terrane contains pre-Devonian continental basement, the Snowcap assemblage (Colpron et al., 2006a), and is the product of periodic continental arc and rift magmatism. The Yukon-Tanana Terrane has isotopic and provenance ties to Proterozoic cratonic source regions, suggesting that it evolved, at least in part, in proximity with the ances-

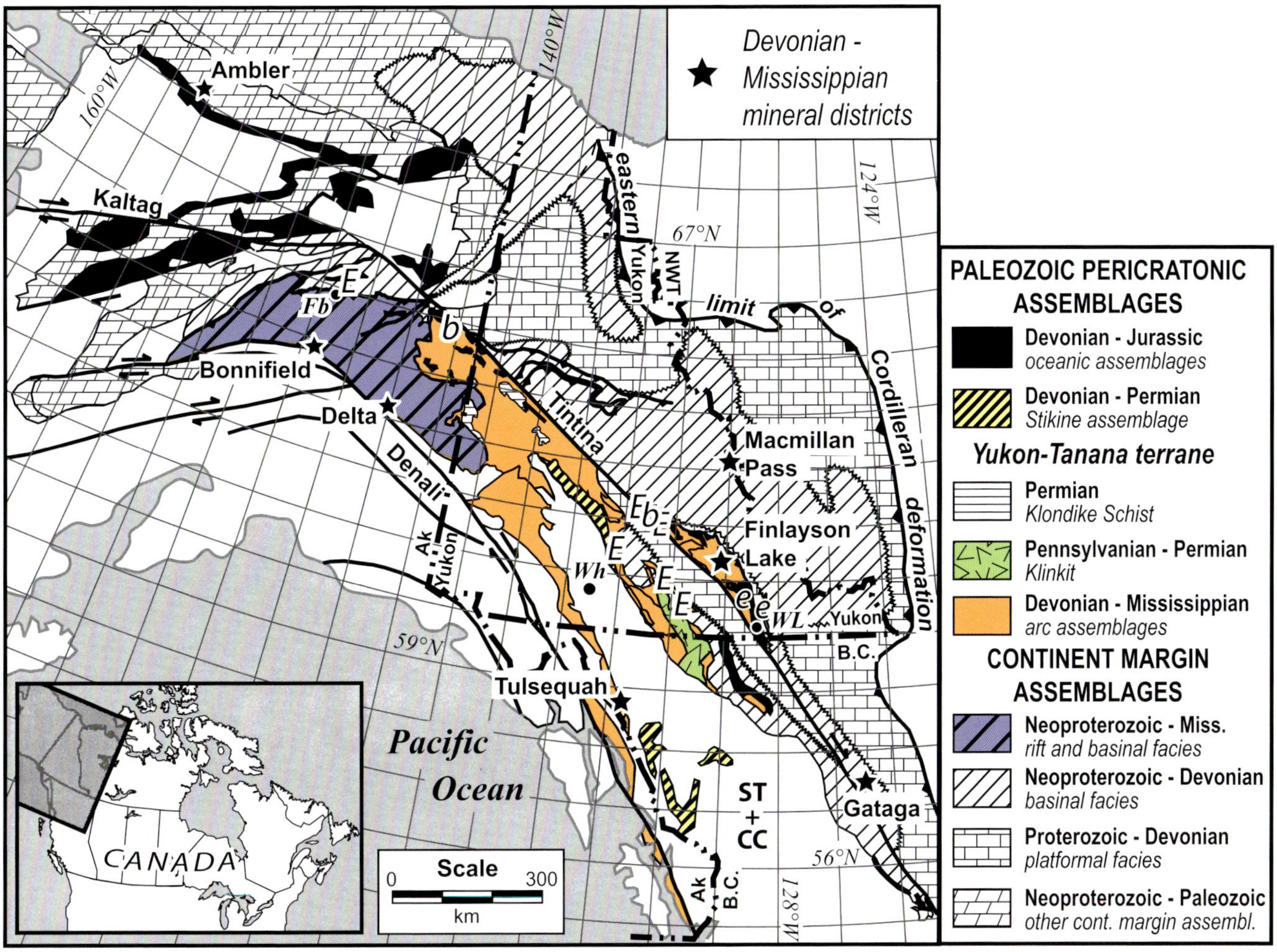

FIGURE 1. Late Paleozoic to early Mesozoic tectonic assemblages of the northwestern North American Cordillera. The Finlayson Lake District and other mineral areas are indicated with black stars. b=blueschist, age unknown; CC=Cache Creek Terrane; D=Dawson; E=Permian ecologite; e=Mississippian ecologite; Fb=Fairbanks, Alaska; ST=Stikinia Terrane, Wh=Whitehorse; WL=Watson Lake (modified from Murphy et al., 2006).

tral margin of North America (Paradis et al., 1998; Nelson et al., 2002, 2006; Colpron et al., 2006a; Mortensen et al., 2006; Piercey et al., 2006). It has been suggested that the Yukon-Tanana Terrane was perhaps at one time linked to other similar terranes found in the Cordillera, such as the Kootenay Terrane in south central British Columbia, as part of an extensive single arc and arc/basement assemblage, which has now been broken apart and displaced by transcurrent faults such as the Tintina Fault (Mortensen, 1992b; Colpron et al., 2006a; Nelson et al., 2006).

The Slide Mountain Terrane in the FLD comprises Mississippian to Lower Permian basinal clastic rocks, and chert and mafic and felsic volcanic rocks of the Carboniferous Fortin Creek Group that have been deposited on oceanic crust in the Yukon-Tanana Terrane back-arc region. These rocks are cut by ca. 273 to 274 Ma intrusions that are likely comagmatic with Campbell Range Formation volcanic rocks. Basalt and chert of the Lower Permian Campbell Range Formation, assigned to the Slide Mountain Terrane, unconformably overlap rocks of the Yukon-Tanana Terrane, the Fortin Creek Group, and the terrane boundary between Yukon-Tanana Terrane and Slide Mountain Terrane. The FLD is juxtaposed against Proterozoic and Paleozoic strata of the ancient North American continental margin along the Tintina Fault to the southwest, and to the northeast the Inconnu Thrust separates the combined Yukon-Tanana and Slide Mountain terranes from North America (Mortensen and Jilson, 1985; Plint and Gordon, 1996, 1997; Tempelman-Kluit, 1979; Fig. 2).

Geology of the Finlayson Lake District

Stratigraphy

The FLD forms a crescent-shaped area approximately 300 km long and 50 km wide that extends from Ross River in the north to Watson Lake in the south (Fig. 1), and comprises Devonian-Mississippian volcanic, intrusive, and sedimentary rocks. The FLD is juxtaposed against Proterozoic and Paleozoic strata of the ancient North American continental margin along the Tintina Fault to the southwest, and to the northeast the Inconnu Thrust separates the combined Yukon-Tanana and Slide Mountain terranes from North America (Mortensen and Jilson, 1985; Plint and Gordon, 1996, 1997; Tempelman-Kluit, 1979; Fig. 2). The Jules Creek Fault separates the Yukon-Tanana Terrane from the Slide Mountain Terrane in the FLD. The FLD is contiguous with the main part of the Yukon-Tanana Terrane, which underlies most of west central Yukon, after restoration of 425

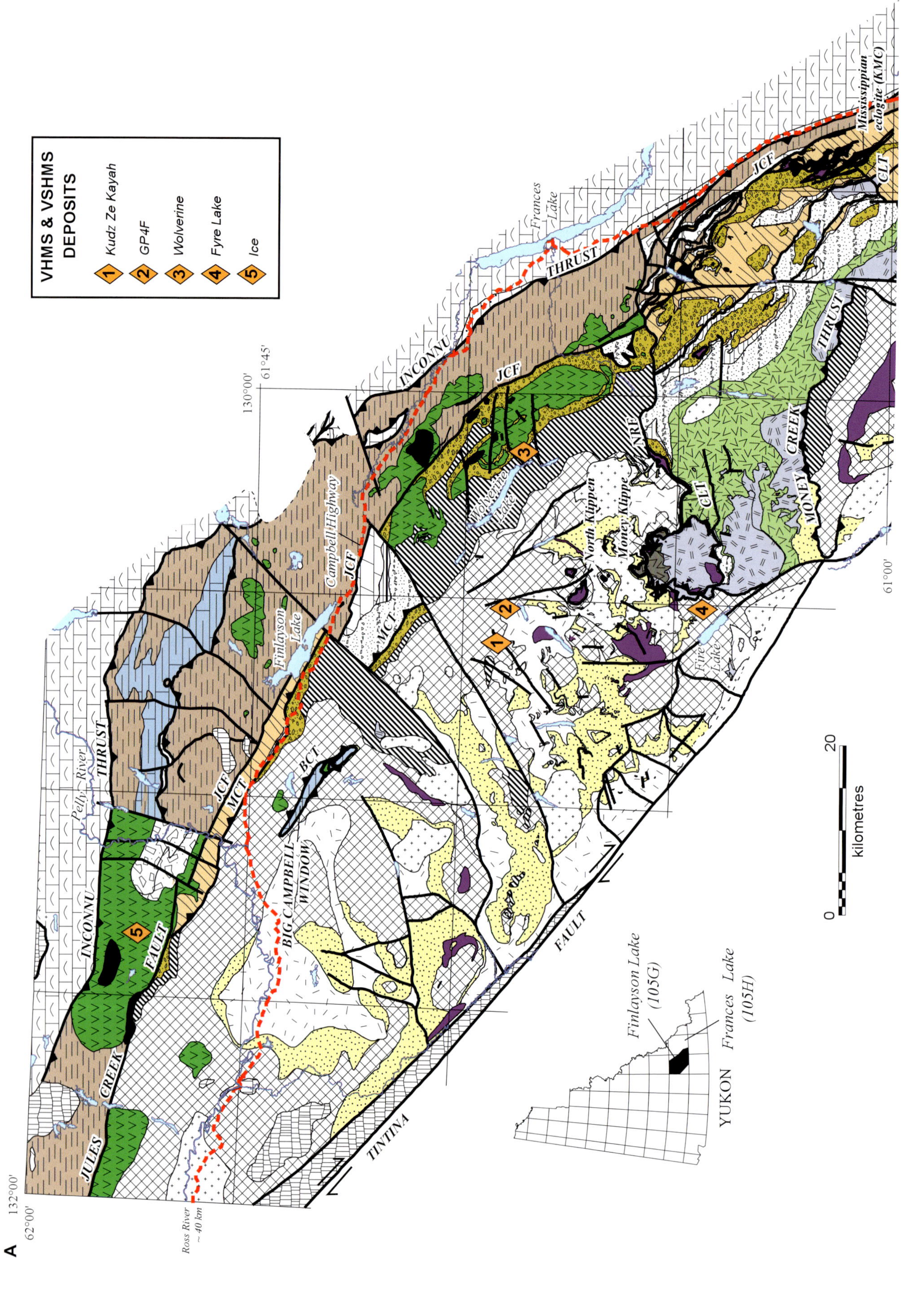

FIGURE 2. A) Geology map of the Finlayson Lake District and environs.

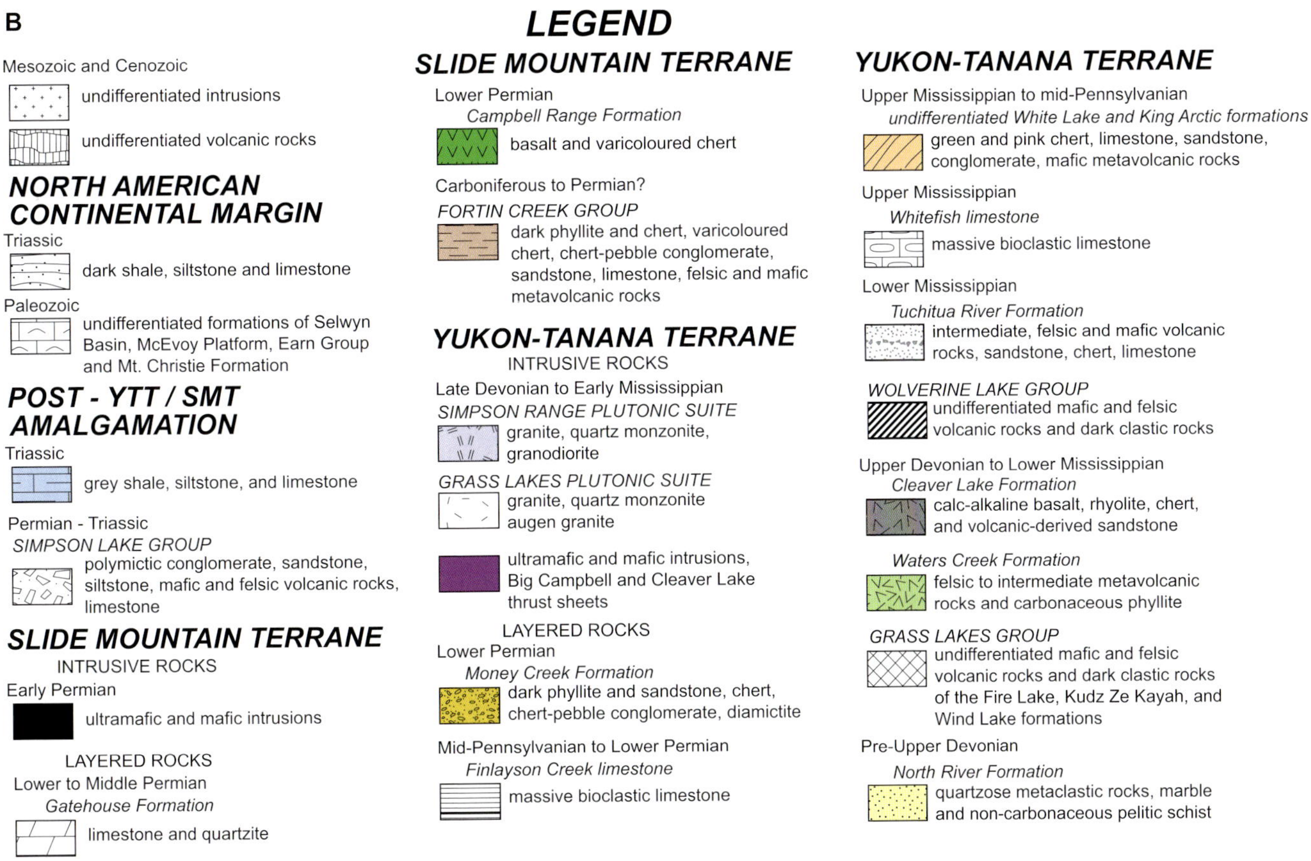

FIGURE 2. B) Legend for map (from Murphy et al., 2006). Abbreviations are as follows: JCF=Jules Creek Fault; MCT=Money Creek Thrust; SMT=Slide Mountain Terrane; YTT=Yukon-Tanana Terrane.

km of Late Cretaceous right-lateral, strike-slip movement on the Tintina Fault (e.g. Mortensen, 1992b).

Rocks of the FLD comprise several fault- and unconformity-bound groups and formations of early Mississippian to Early Permian metamorphosed plutonic, volcanic, and sedimentary rocks (i.e. Cleaver Lake, Money Creek, and Big Campbell thrust sheets; Murphy et al., 2006; Fig. 3). Massive sulphide deposits occur solely within the Big Campbell thrust sheet (Fig. 2) (with one exception, the Ice Cu-Au-Co deposit, which is hosted by metabasalts of the Campbell Range Formation of the Slide Mountain Terrane). Rocks of the Big Campbell thrust sheet include Pre-Late Devonian quartz-rich metasedimentary rocks of the North River Formation, which are the structurally deepest rocks of the Yukon-Tanana Terrane; mafic and felsic volcanic, and carbonaceous metaclastic rocks of the Upper Devonian Grass Lakes Group; Late Devonian to Early Mississippian granitic metaplutonic rocks of the Grass Lakes plutonic suite; carbonaceous metaclastic and mafic and felsic volcanic rocks of the Lower Mississippian Wolverine Lake Group; and carbonaceous metaclastic rocks and chert of the Lower Permian Money Creek Formation (Murphy et al., 2006) (Fig. 3).

The Grass Lakes Group consists of strongly foliated and lineated layered metasedimentary and volcanic rocks that crop out in a roof setting above and between bodies of Early Mississippian granitic orthogneiss and weakly foliated mid-Cretaceous granite (Murphy, 1998). The Grass Lakes Group has been subdivided into three formations, from oldest to youngest, the Fire Lake, Kudz Ze Kayah, and Wind Lake formations. The Upper Devonian (ca. 365 Ma) Fire Lake Formation is a mafic volcanic sequence composed mainly of chloritic phyllite with some carbonaceous phyllite and rare muscovite-quartz phyllite of probable felsic volcanic protolith. Intrusions and sills of mafic and serpentinized ultramafic metaplutonic rocks occur within the Fire Lake Formation. Stratigraphically overlying the Fire Lake Formation is the Kudz Ze Kayah Formation, which is a Late Devonian (ca. 360-356 Ma) sequence dominated by felsic volcanic and volcaniclastic and sedimentary rocks. It consists predominantly of feldspar-muscovite-quartz phyllite and augen phyllite of probable felsic volcanic and volcaniclastic origin, and lesser fine-grained carbonaceous and siliciclastic sedimentary rocks. The Wind Lake Formation forms the uppermost unit of the Grass Lakes Group and consists of carbonaceous phyllite, quartzite, and chloritic phyllite of probable alkalic mafic volcanic and intrusive protolith.

Coeval with the Kudz Ze Kayah and Wind Lake formations are the peraluminous meta-plutonic granitoids of the Grass Lakes Suite. These granitoids are the subvolcanic intrusive equivalents to the felsic volcanic host rocks to the Kudz Ze Kayah deposit, and are as old as 363 ± 3.3 Ma (Mortensen, 1992b). These plutonic rocks are deformed and were themselves re-intruded by younger, late-kinematic plutonic rocks before deposition of the Wolverine Lake Group.

The Grass Lakes Group is unconformably overlain by rocks of the Wolverine Lake Group and consists of a basal unit of conglomerate, grit, sandstone, and carbonaceous argillite, a middle unit of quartz-feldspar phyric felsic volcanic rocks, rare chert and sandstone, and an upper unit of aphyric rhyolite, argillite, magnetite iron formation, and mafic volcanic and intrusive rocks (Murphy et al., 2006). A second unconformity separates the Wolverine Lake Group from the overlying carbonaceous metaclastic rocks (carbonaceous phyllite, chert-pebble conglomerate, quartzofeldspathic sandstone to pebble conglomerate, and locally, matrix-supported diamictite) and dark grey to black chert of the Lower Permian Money Creek Formation. Both the Grass Lakes Group and Wolverine Lake Group occur in the footwall of the Money Creek thrust and record two cycles in the evolution of a Late Devonian to early Mississippian ensialic back-arc (Murphy and Piercey, 2000a; Piercey et al., 2001a, 2006). These groups are separated by an unconformity marking a period of deformation, uplift, and erosion.

Uranium-Pb geochronology places an upper age limit of 356.9 ± 0.5 Ma for the host rocks to the Wolverine Deposit and other zones (see below) (Mortensen, 1992b; Piercey et al., in press), and the immediate stratigraphic hanging wall is dated at 346 ± 2.2 Ma (Piercey, 2001), indicating that Wolverine is younger than Kudz Ze Kayah.

The Campbell Range Formation, which unconformably overlies rocks of the Wolverine Lake Group (Fig. 3), is a mafic-dominated sequence consisting of basalt, chert, and argillite (Fig. 2). The age of this assemblage is Pennsylvanian to Permian, based on radiolarians and ca. 273 to 274 Ma U-Pb ages on gabbros and plagiogranites (Murphy et al., 2006). These rocks are the youngest strata hosting VHMS deposits in the FLD.

Petrochemistry and Chemostratigraphy

The mafic metavolcanic rocks that host the Fyre Lake deposit are strongly depleted in the light rare earth elements (LREE) and high field strength elements (HFSE), have elevated Mg, Ni, and Cr contents, and chemically are similar to boninites present in some suprasubduction zone ophiolites (Sebert and Hunt, 1999; Sebert et al., 2004). Boninites are typically thought to originate as melts from a depleted mantle in rift, forearc, and arc-rift settings. The presence of continentally derived clastic sedimentary rocks intercalated with these boninites, as well as a LREE- and HFSE-enriched transitional subalkaline basalt immediately beneath the deposit, and the presence of mafic volcanic rocks with distinctive arc signatures (Ta, Nb, and Ti depletions) in the area, indicate that the deposit formed in an evolved island arc setting constructed on transitional crust or continental arc that evolved to a back-arc setting (Piercey et al., 2001c, 2004). At Fyre Lake, boninites are overlain by N-MORB to transitional subalkaline basalts (Sebert et al., 2004).

The felsic volcanic rocks that host the Kudz Ze Kayah deposit have elevated HFSE and REE (Zr>500 ppm; Ti/Sc=313-345; Zr/TiO_2=630-2,185; Zr/Sc=15.3-190.3; Zr/Y=3.3-17.7; Nb/Ta=11.6-15.8; Nb/Y=0.4-1.1; Piercey et al., 2001a). They plot in the fields for within-plate, crustally derived, A-type felsic rocks and are interpreted to represent evolving products of an ensialic back-arc rift or back-arc basin (Piercey et al., 2001a, 2006, and references therein). The mafic volcanic rocks of the Wind Lake Formation are weakly alkalic with chemical signatures similar to ocean island basalts (Piercey et al., 2002a).

The host rocks to the GP4F deposit are of similar geochemical composition to the nearby Kudz Ze Kayah deposit. The felsic flows and tuffs that host the GP4F deposit are of rhyolitic composition and fall within the transitional alkaline and subalkaline fields, they have high HFSE and REE contents, high HFSE ratios (e.g. avg Ti/Sc=325; Zr/TiO_2=1157; Zr/Sc=62), and within-plate (A-type) signatures. Their Nb/Ta ratios (8.5-16.4; avg. 13.5) are similar to rocks derived from the continental crust (~11-12), which suggests that the GP4F felsic rocks are derived by melting of continental crust (Boulton, 2002). Mafic dykes are of intermediate to basaltic composition and alkaline in character (Boulton, 2002) and resemble ocean island basalts (OIB) contaminated by continental crust; their composition is similar to mafic flows of the Wind Lake Formation (Piercey et al., 2002a). On the basis of the geochemical similarities of the felsic and mafic volcanic host rocks to those of Kudz Ze Kayah, the GP4F deposit is thought to have formed in a similar paleotectonic environment.

The felsic rocks from the stratigraphic footwall to the Wolverine deposit also have elevated HFSE and REE contents (Ti/Sc=248-417; Zr/TiO_2=559-1,220; Zr/Sc=40.4-84.7, Zr/Y=3.4-14.7; Nb/Ta=10.4-17.1; Nb/Y=0.5-1.1) and their composition is consistent with formation in an ensialic back-arc setting (Piercey et al., 2001a). Feldspar porphyritic rhyolites in the immediate footwall have even higher HFSE and REE contents (avg. values, n=24: Ti/Sc =302; Zr/Ti=0.21; Zr/Sc=61.6; Zr/Y=9.9; Nb/Ta=16.5; Nb/Y=0.67; Piercey et al., in press). Aphyric rhyolites in the hanging wall of the deposit, however, have low HFSE contents. The Wolverine Lake Group rocks are capped by MORB-type basalt indicative of transition to true back-arc spreading and ocean crust formation (Piercey et al., 2002b). The presence of an unconformity between carbonaceous phyllites of the Wind Lake Formation and overlying Wolverine Lake Group rocks indicates uplift, then resumption of ensialic back-arc magmatism and evolution to a spreading ridge, as evidenced by the presence of basaltic mafic volcanic rocks in the uppermost part of the Wolverine Lake Group (e.g. Murphy et al., 2006).

Basalts that host the Ice deposit display homogeneous MORB signatures with a possible influence from subcontinental lithospheric mantle in the lowermost basalts (Piercey et al., 1999). A permissible interpretation is that the deposit was formed in a back-arc basin/ocean basin setting (Piercey et al., 1999). The footwall basalts to the Money (Julia) prospect, 5 km east of Wolverine (see *Money (Julia) Prospect* below), have an E-MORB signature (Piercey et al., 1999).

Deformation and Metamorphism

In the Fire Lake area, host to the Fyre Lake deposit, the presence of biotite and garnet in the metasedimentary and metavolcanic rocks indicate upper greenschist metamorphic conditions (Sebert et al., 2004). Rocks in the vicinity of the deposit have been gently folded along a northerly trending anticlinal to monoclinal structure. Rocks have a strong F_1 schistosity parallel to compositional layering, and there is a

FIGURE 3. Structural and stratigraphic relationships in the Finlayson Lake District (modified from Murphy et al., 2006). Abbreviations are as follows: FC=Finlayson Creek limestone; KA=King Arctic Formation; KMC=Klatsa metamorphic complex; NR=North River Formation; WF=Whitefish limestone; WL=White Lake Formation. Timescale is from Okulitch (2002).

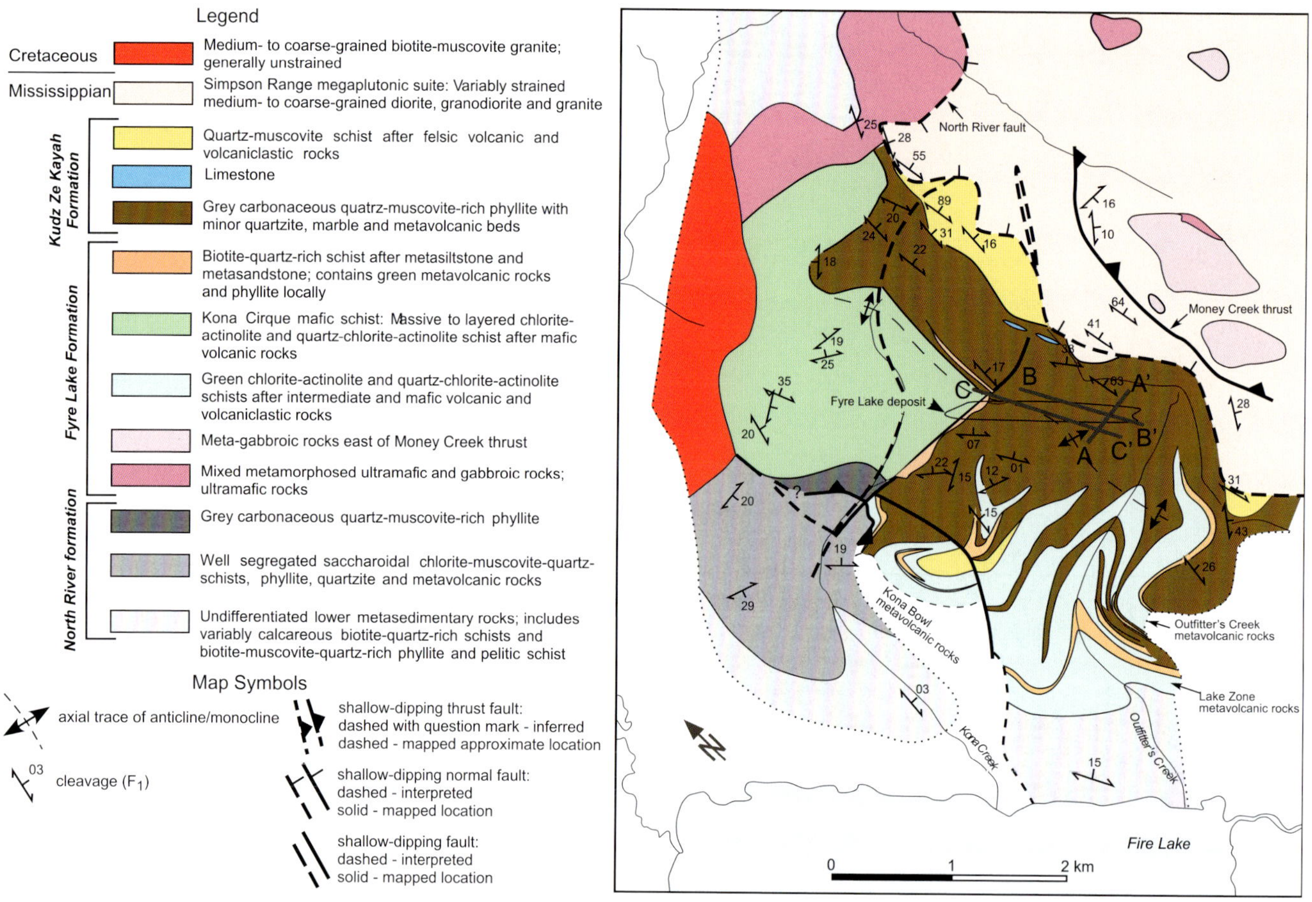

FIGURE 4. Geology map of the Fyre Lake deposit area. The surface projection of the Fyre Lake massive sulphide deposit is shown in the area outlined by the solid black line (modified after Sebert et al., 2004). Also shown are section lines for cross section through East Kona and West Kona zones (A-A'), longitudinal section through East Kona zone (B-B'), and longitudinal section through West Kona zone (C-C') depicted in Figure 7A-C.

strong L_1 mineral lineation trending 120 to 140° and plunging shallowly to the southeast in the plane of the F_1 foliation. The host rocks and the Fyre Lake deposit have been deformed in Permian, Cretaceous, and Mississippian times.

At Kudz Ze Kayah, the rocks and mineralization have undergone upper greenschist to lower amphibolite facies metamorphism and have been intensely deformed. At least three phases of deformation are recognized (Murphy and Timmerman, 1997). The deposit lies within an F_1 fold, has been thickened by F_2 folds, and imbricated by F_3 deformation. D_1 deformation was intense and penetrative, resulting in an S_0-S_1 composite fabric (Schultze, 1996). F_1 folds are isoclinal and recumbent, and accompanied by attendant limb thrust faulting.

At GP4F, the mineralization and all host rocks, except the mafic dykes, have been metamorphosed to upper greenschist facies, as evidenced by the presence of garnet, biotite, chlorite, albite, and titanite. However, the sulphides at GP4F are significantly less recrystallized than at Kudz Ze Kayah due to a lower metamorphic overprint. The GP4F deposit contains trace to minor gahnite, and this mineral is believed to have formed by desulphidation of sphalerite during metamorphism (Spry and Scott, 1986) A major deformation event in the region has obliterated most primary volcanic and sedimentary features in the rocks. The sequence is strongly foliated with one dominant S_2 foliation (MacRobbie and Holroyd, 2000).

The immediate host rocks to the Wolverine deposit are less strongly metamorphosed and deformed than those at Kudz Ze Kayah, and metamorphic grades at Wolverine are middle greenschist based on the mineral assemblage of actinolite, albite, chlorite, and biotite. A major deformational event is recorded as a prominent S_1 foliation that trends northwest and dips gently to the northeast. F_1 folds in the vicinity of the deposit verge to the southwest, indicating that the Wolverine deposit is on the eastern limb of an open, upright antiform. Deformation has resulted in extensive gouge zones in ductile rocks such as argillite and intense fracturing in brittle rocks including rhyolite.

Volcanic-Hosted Massive Sulphide Deposits

Grade and Tonnage

To date, forty-one VHMS deposits and occurrences have been discovered at different stratigraphic levels within the FLD, and about 34 Mt of massive sulphides have been delineated since the district was recognized in the mid-1990s. The combined resource of Kudz Ze Kayah, Wolverine, and GP4F, the polymetallic massive sulphide deposits with the greatest economic potential, is 21.5 Mt grading 8.2 % Zn, 0.97 % Cu, 1.7 % Pb, 203 g/t Ag, and 1.6 g/t Au.

The Fyre Lake deposit has resources of 15.4 Mt at 1.2% Cu, 0.8% Co, and 0.46 g/t Au using the kriging method and a cut off of 0.5 % Cu, or 8.2 Mt at 2.1 % Cu, 0.11 % Co, and 0.73 g/t Au using a sectional block method and cut off of

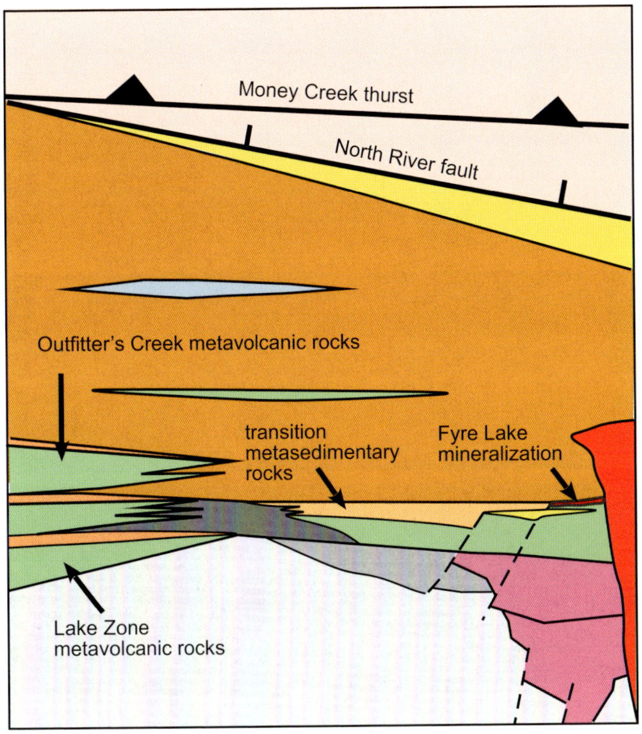

FIGURE 5. Generalized stratigraphic section of the Fyre Lake deposit area depicting the spatial relationships of the major rock units (from Sebert et al., 2004); legend as in Figure 4.

1.0 % Cu that have been delineated within the East and West Kona zones (Columbia Gold Mines Ltd., 1999).

The Kudz Ze Kayah deposit contains an inferred resource of 13 Mt grading 5.55% Zn, 1.00% Cu, 1.30% Pb, 125 g/t Ag, and 1.2 g/t Au within the ABM zone (Schultze, 1996), including an open pit mineable resource of 11.1 Mt at 5.61% Zn, 0.85% Cu, 1.56% Pb, 136.9 g/t Ag, and 1.33 g/t Au (Expatriate Resources Ltd., 2000). Another zone, the Fault Creek zone, intersected in 8 drillholes, contains a further inferred resource of 50,000 tonnes at 7.1% Zn, 1.0% Pb, 4.7% Cu, 130 g/t Ag, and 2.0 g/t Au (Expatriate Resources Ltd., 2000; Hunt, 2002).

The GP4F deposit contains an inferred resource of 1.5 Mt grading 6.4% Zn, 3.10% Pb, 0.10% Cu, 89.7 g/t Ag, and 2.0 g/t Au (MacRobbie and Holroyd, 2000), as outlined in nine drillholes.

At the end of the 1997, 71 drillholes had intersected mineralization at the Wolverine deposit and a geological resource of 6.237 Mt grading 12.66% Zn, 1.33% Cu, 1.55% Pb, 370.9 g/t Ag, and 1.76 g/t Au was reported (Tucker et al., 1997). Measured and indicated reserves have subsequently been revised to incorporate new drilling to give 4.51 Mt grading 12.04% Zn, 1.15% Cu, 1.57% Pb, 351.5 g/t Ag, and1.68 g/t Au, with an additional inferred resource of 1.69 Mt grading 12.16% Zn, 1.23% Cu, 1.74% Pb, 385.1 g/t Ag, and 1.71 g/t Au (Yukon Zinc Corporation, 2006a).

The discovery hole at the Sable Zone (see below) intersected 0.6 m of exceptionally high-grade massive sulphides grading 13.3% Zn, 0.8% Pb, 0.8% Cu, 416 g/t Ag, and 1.9 g/t Au, similar to that at the periphery of the Wolverine deposit. However, there is currently no calculated resource for this occurrence.

In late 2004, Yukon Zinc Corporation drilled two holes into the Fisher zone (see below), and drillhole WV95-06 intersected 2.2 m of massive sulphide with an average grade of 2.8% Zn, 1.4% Pb, 0.12% Cu, 62 g/t Ag, and 0.14 g/t Au. However, a resource has not been calculated for this zone.

The Ice deposit has an indicated resource of 4.56 Mt grading 1.48 % Cu, including 3.4 Mt open pittable mineralization at 1.48% Cu (Expatriate Resources Ltd., 1999). The massive sulphide body is overlain by Cu-oxide mineralization for which there is insufficient data to estimate a geological resource; however, 2.7 Mt of this fall within the open pittable resources (Expatriate Resources Ltd., 1999; Hunt, 2002, Yukon Zinc Corporation, 2006b). There are also some Zn-rich (~1%) massive sulphide zones, but the Zn resource has not been quantified.

The Skyblaze zone on the Goal Net North property contains 16 m of greater than 1% Zn and 0.73 m of semi-massive sulphide mineralization grading 3.0 % Zn, 1.85 % Pb, 0.14% Cu, 63 g/t Ag and 0.2 g/t Au (Yukon Zinc Corporation, 2005). There are currently too few drillholes to calculate a resource. Although a resource for the Thunderstruck zone has not been calculated, drill intercepts over a thickness of 0.30 m reportedly grade 13.4% Zn, 5.07% Pb, 0.34% Cu, 40.7 g/t Ag, and 0.04 g/t Au with low Se contents (Yukon Zinc Corporation Ltd., 2004b, 2005).

Geological Settings and Deposit Descriptions

In the FLD, VHMS deposits are hosted in Late Devonian to Early Mississippian arc and back-arc successions of the Yukon-Tanana Terrane and the Slide Mountain Terrane, northeast of the Tintina Fault and southwest of the Inconnu Thrust (Hunt, 2002; Murphy et al, 2006). Prior to the discovery of Kudz Ze Kayah, GP4F, Wolverine, and Ice in the FLD in the mid-1990s, VHMS deposits were not a widely recognized and significant deposit type in Yukon.

The major (significant) massive sulphide deposits of the district occur at different stratigraphic levels within the FLD stratigraphy (Fig. 3; Murphy and Piercey, 1999a,b,c, 2000b; Hunt, 2002). Below are descriptions of each of these VHMS deposits.

Fyre Lake Deposit

The Fyre Lake deposit (also referred to as Kona) is located about 160 km northwest of Watson Lake and 140 km southeast of Ross River. The Fyre Lake deposit occurs within the ca. 365 Ma Fire Lake Formation of the Grass Lakes Group (Fig. 4). The Fire Lake Formation consists of a metamorphosed volcanic sequence composed mainly of chloritic phyllite with some carbonaceous phyllite, and rare muscovite-quartz phyllite of probable felsic volcanic origin (Figs. 4, 5). The Fire Lake Formation lies on clastic rocks of the North River Formation (Fig. 5).

The immediate host rocks to the deposit are chlorite-quartz and chlorite-actinolite-quartz schists (mafic volcanic and volcaniclastic rocks; Fig. 6A), and turbiditic sedimentary rocks form the immediate stratigraphic hanging wall (Fig. 6B). The metamorphosed sedimentary rocks are predominantly finely laminated carbonaceous phyllite and these overlie the volcanic rocks (Figs. 4, 5). Intercalated quartz-biotite schists (mafic volcaniclastic) and chlorite-mica-

FIGURE 6. Photographs of Fyre Lake deposit host rocks and mineralization. **(A)** Footwall mafic volcaniclastic rocks. **(B)** Hanging-wall intercalated mafic volcaniclastics and sedimentary rocks. **(C)** Chalcopyrite-rich massive sulphide from East Kona zone (Hunt, 2002). **(D)** Fine- to medium-grained blebby to disseminated pyrite with interstitial chalcopyrite in a matrix of grey, quartz-rich (silicified wall rock) matrix, West Kona zone. **(E)** Foliation-parallel quartz-sulphide-magnetite bands interpreted to be stringer veins from feeder zone that have been tectonically transposed parallel to F_1 foliation, immediately below West Kona zone (DDH 97-115; 746.8 m). **(F)** Fine-grained magnetite fragments in matrix of pyrite-pyrrhotite-chalcopyrite-quartz from West Kona zone (DDH 97-95; 331 m). **(G)** Patchy to discontinuously banded pyrite-quartz crosscutting fragmental, fine-grained magnetite, West Kona zone (DDH 97-95; 341.8 m). **(H)** Wispy, banded magnetite (dark grey), chlorite (dark green), and semimassive sulphide (pyrite-chalcopyrite) with notable absence of chlorite, from immediately below the upper horizon of the East Kona zone (DDH 96-35; 67.8 m). **(I)** Finely bedded to laminated magnetite iron formation horizon from top of massive sulphide, lower horizon of East Kona zone (DDH 96-34; 72.3 m). Unless stated, the locations of individual samples are not specified. Images **D, E, F, G, H,** and **I** are from Sebert et al. (2004). The diameter of the core in photos (C) to (H) is 4.5 cm.

quartz schists (wackes) occur at the base of the sedimentary sequence and grade upwards into the carbonaceous sedimentary rocks.

The Fyre Lake deposit consists of three parallel, northwest-trending tabular, sheet-like massive sulphide horizons that are grouped into two zones, termed the East Kona (Upper and Lower horizons) and West Kona zones (Fig. 7A-C). The zones are separated by an inferred, steeply dipping, reverse fault that down-drops the west side by about 100 m (Fig. 7A; Foreman, 1998).

East Kona consists of an upper and a lower horizon, each with an average thickness of 8 to 12 m and an average width of 100 to 125 m (Fig. 7B). The upper horizon occurs at the contact between the underlying volcanics and overlying sedimentary rocks. The lower horizon occurs 40 to 70 m stratigraphically beneath the upper horizon, within mafic volcanic rocks (Fig. 7B). East Kona massive sulphides are composed primarily of massive pyrite, with lesser pyrrhotite, chalcopyrite, sphalerite, and local lenses of massive magnetite (Fig. 6C).

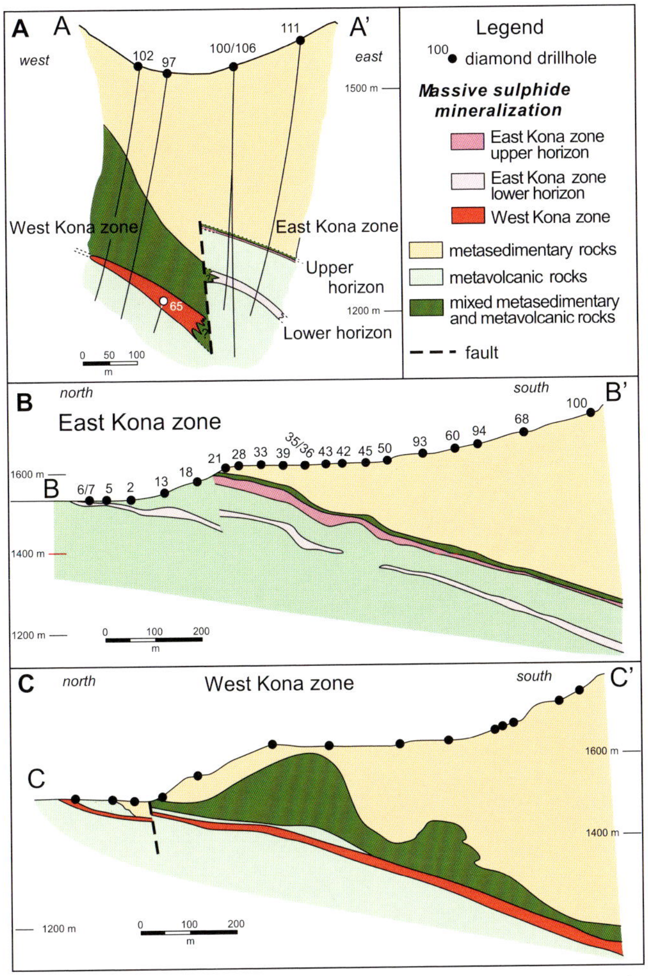

FIGURE 7. Schematic stratigraphic sections through the Fyre Lake deposit. **(A)** Cross-section along line A-A' in Figure 4. **(B)** Longitudinal section of the East Kona zone along line B-B' in Figure 4. **(C)** Longitudinal section of West Kona zone along line C-C' in Figure 4. Figures are modified from Hunt (2002), which are based on Foreman (1998).

The West Kona zone has a strike length of 1,500 m, dips gently to the northeast, and remains open at both ends along strike (Fig. 7C). Although the massive sulphide lenses are elongated parallel to the local lineation, an inferred step fault with a facies change across it separates the East and West Kona zones; this fault is suggested to be synvolcanic in origin and may have served as the conduit for the mineralizing fluids (Sebert et al., 2004).

In general, the massive sulphides are zoned, with pyrite±sphalerite commonly occurring at the top part of the lenses, and pyrrhotite and chalcopyrite being more common in the lower parts of the lenses as well as comprising a significant portion of semimassive and disseminated mineralization that lies stratigraphically beneath the lenses. West Kona massive sulphides are dominated by magnetite-pyrite-chalcopyrite in a grey siliceous matrix in the east, and transitioning to pyrite-chalcopyrite (Fig. 6D) going westward, and massive pyrrhotite in the western-most part.

The semimassive and disseminated mineralization contains chlorite and quartz. Quartz-rich bands within mineralization (e.g. Fig. 6E) are interpreted to be stringer veins of a footwall feeder zone that have been tectonically transposed parallel to the prominent foliation, with the chlorite representing altered wallrock (Sebert et al., 2004).

A unit of fine-grained chert with a few percent sulphides (chalcopyrite, pyrite) and magnetite locally marks the top of massive sulphide lenses of the West Kona zone and the lower horizon of the East Kona zone (Fig. 6I). This unit, typically 15 to 40 cm thick, but ranging to several metres, is thought to be an exhalative sediment (Columbia Gold Mines Ltd., 1999; Sebert et al., 2004).

Although there have been several microscopic investigations of the Fyre Lake mineralization, the origin of the magnetite remains unclear and enigmatic. The high degree of deformation of the deposit indicates that the present textural association of magnetite with other sulphides and alteration minerals may largely be due to deformational processes. Textural relationships in Figure 6G and H indicate that magnetite predates sulphide, and sulphides have brecciated and partly replaced magnetite, with subsequent deformation imparting a crudely banded texture of alternating magnetite- and sulphide-rich layers.

Kudz Ze Kayah Deposit

The Kudz Ze Kayah deposit occurs within a sequence of predominantly felsic volcanic and volcaniclastic rocks, and minor metasedimentary rocks of the Kudz Ze Kayah Formation (Fig. 8). The felsic volcanic and volcaniclastic rocks are generally fine-grained, though centimetre-sized clasts occur in some intervals. Schultze (1996), in a brief early report of the deposit indicated that the rocks were emplaced as flows, and lapilli and ash tuffs. Limited petrographic studies of these rocks by company personnel note the presence of embayed feldspar crystal fragments in some porphyritic samples (e.g. Fig. 9A,D), quartz phenocrysts or phenoclasts in others, and quartz and feldspar in others, indicative of a tuffaceous origin. However, an epiclastic origin cannot presently be ruled out, and further work is needed to unambiguously ascertain the volcanic setting and mode of emplacement of these rocks. The host-rock sequence is stratigraphically overlain by carbonaceous sedimentary rocks (argillite) and alkalic mafic volcanic rocks of the Wind Lake Formation (Fig. 8). A biotite-chlorite-calcite schist (Fig. 9E) is interpreted to be a strongly deformed mafic intrusive body that was emplaced as a sill or dyke after formation of the deposit.

The main part of the Kudz Ze Kayah deposit, termed the ABM zone (Fig. 8), is a 500 m long, isoclinally folded lens of massive to semimassive sulphide (Figs. 10, 11) that occurs within a syncline that is overturned to the south and is half of a south-vergent anticline-syncline pair. The mineralization extends down dip about 400 m, averages 18 m thick, and reaches a maximum thickness of 34 m at the nose of the fold. The smaller southern part of the mineralization, referred to as the Fault Creek zone, has been downdropped by faulting (Fig. 8).

The stratigraphic sequence in the upper limb of the syncline is overturned, and hydrothermally altered rocks in the stratigraphic footwall now form the structural hanging wall to the deposit (Fig. 10). Both stratigraphic footwall and hanging-wall contacts comprise volcaniclastic rocks (Fig. 9A-D) and what are interpreted to be rhyolite flows. The most

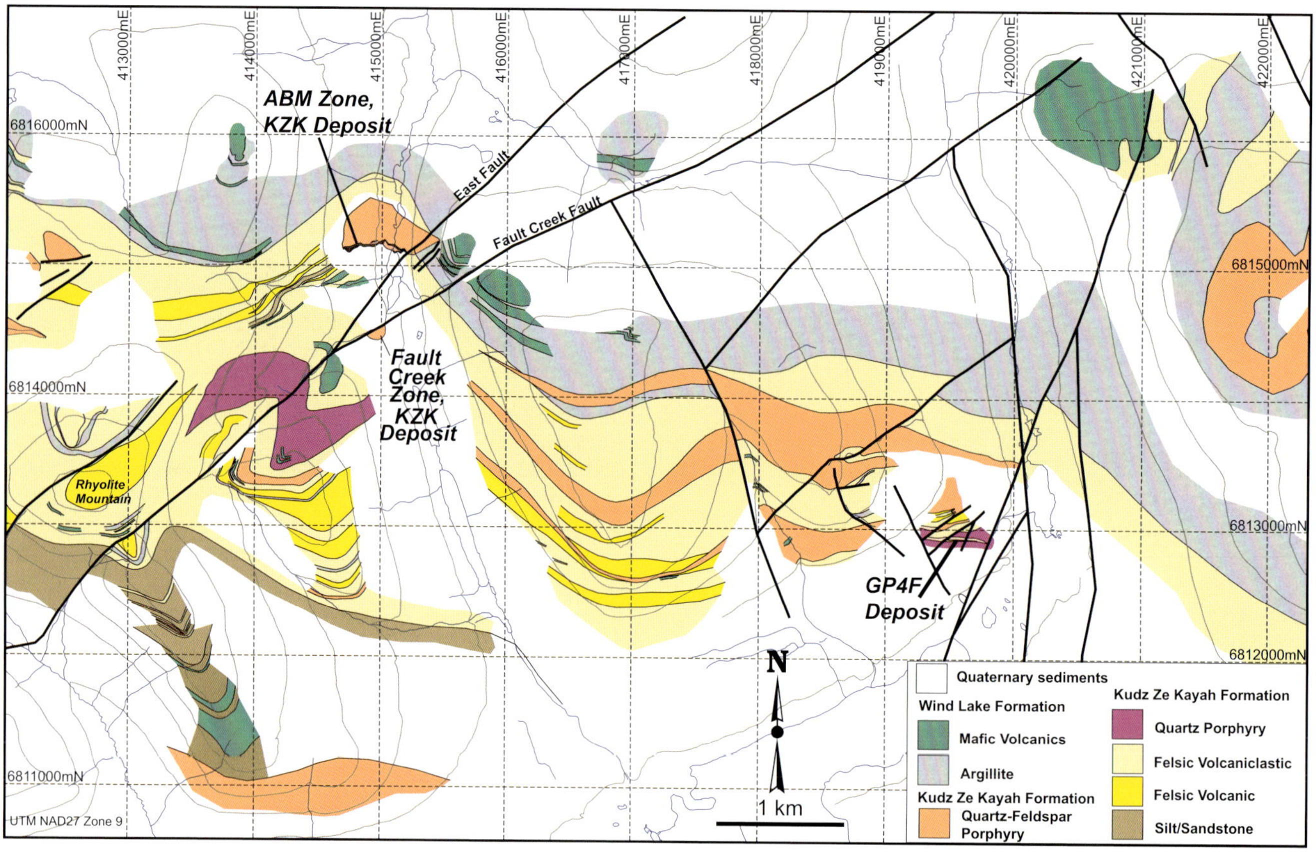

FIGURE 8. Surface geological map of the Kudz Ze Kayah and GP4F deposit area (after Cominco Ltd., unpub. data, 1997).

strongly hydrothermally altered rocks are situated near the fold axis. Nearly symmetric base-metal zoning within the two limbs of the folded sulphide body supports the interpretation that the deposit is overturned. Figure 11A is a representation of the present, post-deformational configuration of the Kudz Ze Kayah deposit, including the location of the feeder zone and the adjacent mafic intrusive body, and Figure 11B presents the interpreted predeformational configuration.

The mineralogy of the sulphide bodies is dominated by pyrite (e.g. Fig. 9F,G), followed by sphalerite, pyrrhotite, galena, chalcopyrite, and trace arsenopyrite, tetrahedrite-tennantite, boulangerite, and electrum. Gangue minerals include chlorite, magnetite (e.g. Fig. 9F,G), quartz (e.g. Fig. 9H), Fe-carbonate, sericite, albite, and barite (Fig. 9I). Where present, barite is generally concentrated in the stratigraphic upper parts of the sulphide lens. In places, the sulphides have been strongly deformed and recrystallized, and many primary textures obliterated due to a strong foliation (e.g. Fig. 9F,G).

GP4F Deposit

The GP4F deposit is located over 5 km southeast of the Kudz Ze Kayah deposit (Fig. 8). The deposit is within the same package of Late Devonian-Early Mississippian rocks as Kudz Ze Kayah and is hosted by a sequence of alternating aphyric and porphyritic felsic volcanic rocks and felsic volcaniclastic rocks (collectively 90 vol.%) that is crosscut by mafic dykes (10 vol.%). Mineralization at GP4F occurs in an overturned homoclinal sequence as a single semimassive to massive sulphide len up to 3.2 m thick that dips to the north at 30 to 35° N (Fig. 12), has a strike length of 200 m, extends 350 m down dip, and remains open to the northeast and east (MacRobbie and Holroyd, 2000; Boulton, 2002).

The aphanitic, aphyric felsic volcanic rocks are white to cream coloured and composed of subangular aggregates of quartz and interstitial muscovite in a cryptocrystalline albite-quartz matrix (Fig. 13A). Based on the presence of flow contacts in places, these rocks were likely emplaced predominantly as coherent rhyolite flows.

Felsic volcaniclastic rocks are fine- to medium-grained, grey, equigranular, and composed of quartz, muscovite, and chlorite with minor disseminated tourmaline (Fig. 13B). Some intervals contain abundant sub-centimetre- to centimetre-sized clasts of fine-grained felsic rock fragments (e.g. Fig. 13C). The volcanic setting and emplacement mechanism of the volcaniclastic rocks has not been determined.

Immediately below the massive sulphide lens is a massive, homogenous, foliated quartz-feldspar porphyry (Fig. 12) with 5 to 10% slightly bluish 2 to 5 mm quartz phenocrysts, 15 to 20% 3-12 mm feldspars in a fine-grained matrix of quartz-feldspar-sericite and muscovite (Fig. 13D). Other porphyritic intervals contain only feldspar phenocrysts (2-3%, 1-3 mm) in a quartz-feldspar-muscovite-biotite matrix or only bluish grey quartz phenocrysts (10-15% 1-4 mm) in a matrix of feldspar-biotite-sericite with trace quantities of disseminated tourmaline. The mode of emplacement of the porphyritic felsic rocks has not been determined; however, these relatively little-deformed, blue quartz-phyric porphyries yielded ages of 347 to 345 Ma (Murphy et al., 2006), suggesting that they are intrusive and part of the Wolverine suite.

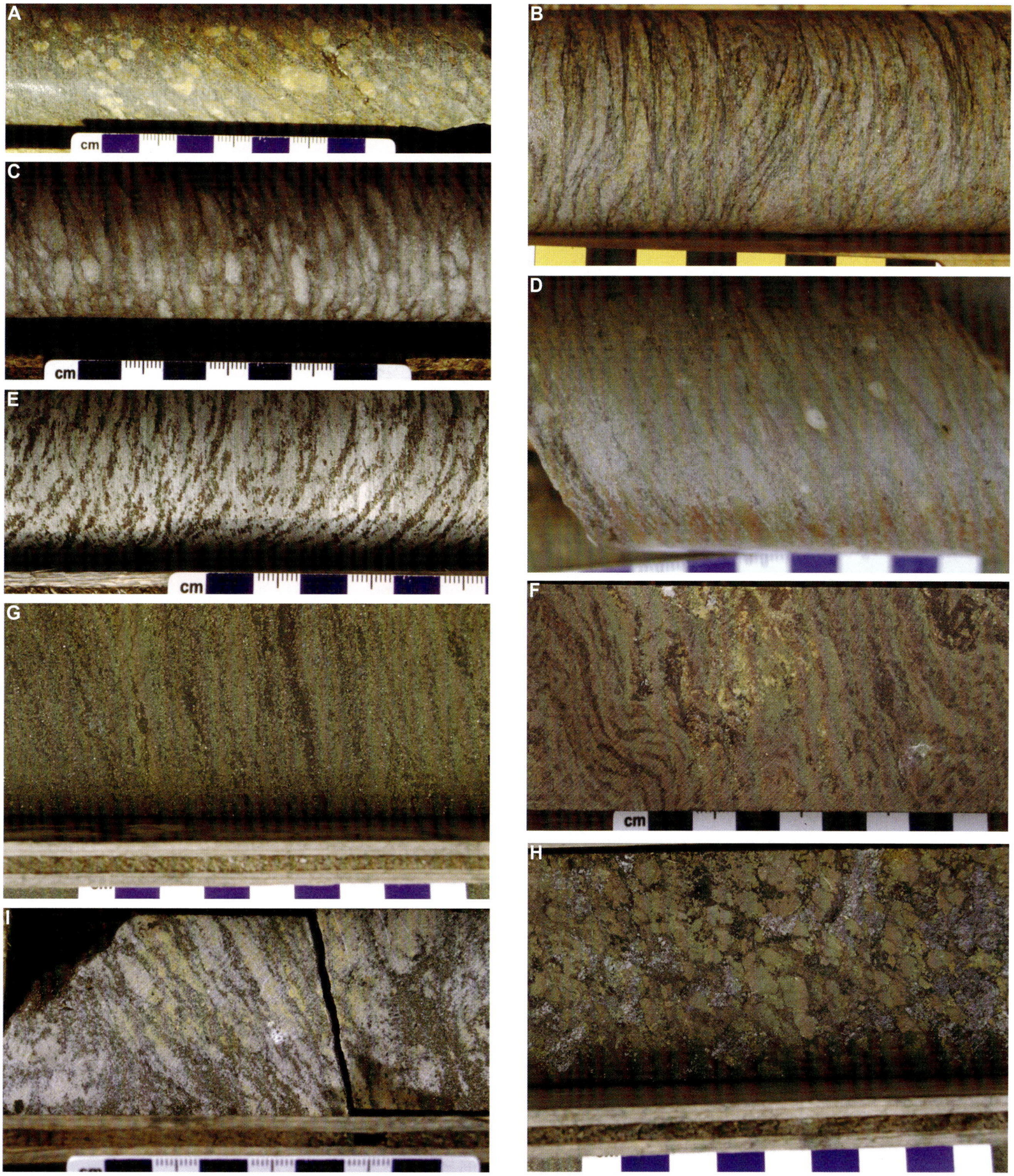

FIGURE 9. Photographs of Kudz Ze Kayah deposit host rocks and mineralization. **(A)** K-feldspar porphyry from stratigraphic hanging wall. **(B)** Quartz-sericite schist from stratigraphic footwall. **(C)** Felsic fragmental volcaniclastic rock from stratigraphic footwall. **(D)** K-feldspar-phyric quartz-sericite schist from stratigraphic footwall. **(E)** Biotite-chlorite-calcite mafic schist of probable intrusive origin (post-mineralization). **(F)** Banded pyrite-magnetite from massive sulphide lens. **(G)** Recrystallized pyrite-magnetite from massive sulphide lens. **(H)** Chalcopyrite-pyrrhotite network sulphide with quartz gangue from stratigraphic base of sulphide lens. **(I)** Barite-pyrite from stratigraphic upper part of massive sulphide lens.

Mafic dykes are dark grey/brown, fine- to medium-grained, and consist of quartz, biotite, sericite, calcite (in veinlets), and minor titanite; they are the most unaltered rock type present and display no obvious foliation.

The GP4F mineralization is relatively Cu-poor and Pb-rich in comparison to Kudz Ze Kayah (see Grade and Tonnage section above). Mineralization is crudely banded and consists of medium- to coarse-grained buckshot pyrite

FIGURE 9 CONTINUED. Photographs of Kudz Ze Kayah deposit host rocks and mineralization. **(J)** Strongly chloritized felsic volcaniclastic rock, stratigraphic footwall (DDH K97-182; 42.5-48.2 m). The diameter of the core is 4.5 cm. **(K)** Sericite-altered felsic volcaniclastic rock, stratigraphic footwall. **(L)** Chlorite-albite altered felsic volcaniclastic rock from immediate stratigraphic footwall to the massive sulphides. **(M)** Fe-carbonate-altered felsic volcaniclastic rock, stratigraphic hanging wall. Unless stated, the locations of individual samples are not specified.

(50-60 vol.%) in a matrix of sphalerite, pyrrhotite, galena, magnetite (locally) and trace chalcopyrite (Fig. 13E-H). Gangue minerals are quartz, feldspar, chlorite, carbonate, sericite, and/or biotite.

A 'marker tuff' unit occurs almost universally in the structural hanging wall and is a light to medium grey/beige colour, homogenous, finely banded, fine-grained rock with a strong sericite – and lesser biotite – dark green chlorite-quartz-rich (± tourmaline and garnet) matrix and 10 to 30% disseminated to blebby pyrite, pyrrhotite, sphalerite, chalcopyrite, and gahnite (Fig. 13G,H). Mineralization in the marker tuff formed by partial replacement of the original host rocks.

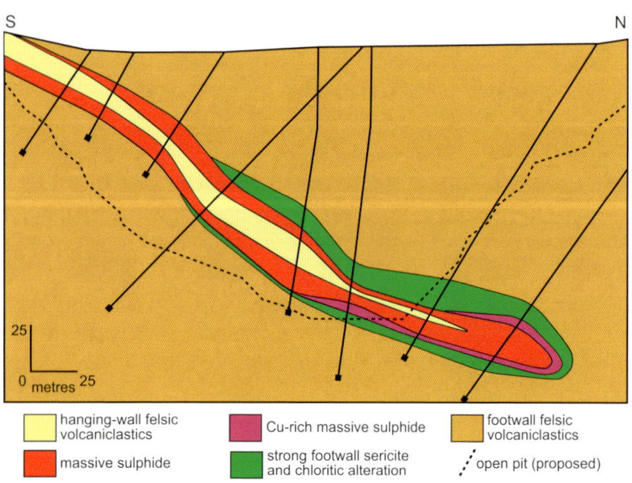

FIGURE 10. Geological cross-section 414850N through the Kudz Ze Kayah deposit (modified from Cominco Ltd., unpublished data, 1997, with additional information from J. Peter, unpublished data).

<u>Wolverine Deposit</u>

The Wolverine deposit is hosted by rocks of the Wolverine Lake Group, dominated by felsic volcanic and volcaniclastic and carbonaceous sedimentary rocks and capped by mafic volcanic rocks (Fig. 14). The Wolverine Lake Group has a lower conglomerate with fragments of the Kudz Ze Kayah Formation within it and unconformably overlies the Kudz Ze Kayah Formation (Murphy and Piercey, 1998, 1999a).

The host-rock sequence (Fig. 15; Bradshaw et al., 2000) consists of black to dark grey, fine-grained carbonaceous argillite, fine- to coarse-grained, ±quartz-feldspar-phyric volcaniclastic rocks (Fig. 16A); K-feldspar-phyric porphyry (Fig. 16B); and massive flow-banded aphanitic, aphyric rhyolite (Fig. 16C); fine-grained felsic volcaniclastic rock with abundant centimetre-sized flattened clasts of aphyric rhyolite (Fig. 16D); and felsic volcaniclastic rock with abundant centimetre-sized flattened clasts of rhyolite and argillite (Fig. 16E). The volcanic setting and mode of emplacement of the felsic volcanic and volcaniclastic rocks has not been studied in detail. Also, within the host sequence are two intervals of magnetite iron formation (Fig. 16F) and a narrow discontinuous interval of carbonate exhalite (Fig. 16G); these are discussed further below.

Mineralization at Wolverine occurs as two steeply dipping coalesced, elongate to tabular, massive and semimassive sulphide lenses named the Lynx and Wolverine zones (Figs. 17, 18). The deposit has a strike length of approximately 750 m, dips steeply to the north, and remains open down-dip to the northeast where it crosses onto claims owned by Teck Cominco Ltd. The massive sulphides are up to 16 m thick (Figs. 17, 18). Stringer veins in the stratigraphic footwall beneath the massive sulphides (Fig. 18) demarcate fluid

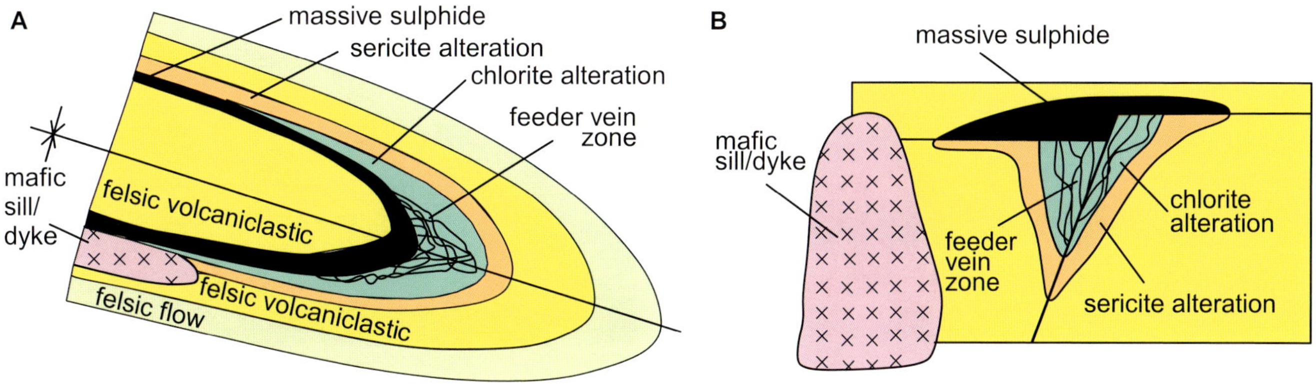

FIGURE 11. (A) Generalized schematic cross-section of present configuration of Kudz Ze Kayah deposit and host rocks. **(B)** Generalized schematic cross-section of original Kudz Ze Kayah deposit morphology and host rock configuration.

upflow zones and define the feeder system that channeled the mineralizing fluids. Deformation has modified the original morphology of the Wolverine deposit, but the overall metal zoning and relationships between the lenses and their feeder zones (Fig. 18) suggest that much of the original deposit architecture remains intact. Thicker sulphide intersections are underlain by felsic volcaniclastic rocks (Fig. 16A,B,D,E). These rocks commonly contain extensive footwall replacement mineralization (Fig. 16K-M) and alteration (Fig. 16P-S) and/or stringer mineralization of pyrite, sphalerite, chalcopyrite, and galena (Fig. 16N,O).

According to Bradshaw et al. (in press) there are three types of mineralization: 1) banded/layered massive sulphides (Fig. 16I); 2) semimassive replacement-style sulphides (Fig. 16K-M); and 3) stringer sulphide veins (Fig. 16N,O). The massive sulphides are predominantly pyrite and sphalerite with minor pyrrhotite, chalcopyrite, galena, trace tetrahedrite and arsenopyrite, and rare marcasite, meneghinite ($Pb_{13}CuSb_7S_{13}$), boulangerite ($Pb_5Sb_4S_{11}$), bournonite ($PbCuSbS_3$), miargyrite ($Ag_2Sb_2S_3$), native Au, and electrum. Sulphides are generally fine grained and recrystallized in response to deformation and metamorphism. However, in a few places, colloform and framboidal pyrite textures are preserved, particularly in the uppermost parts of the sulphide lenses near the contact with hanging-wall argillite. A zone of

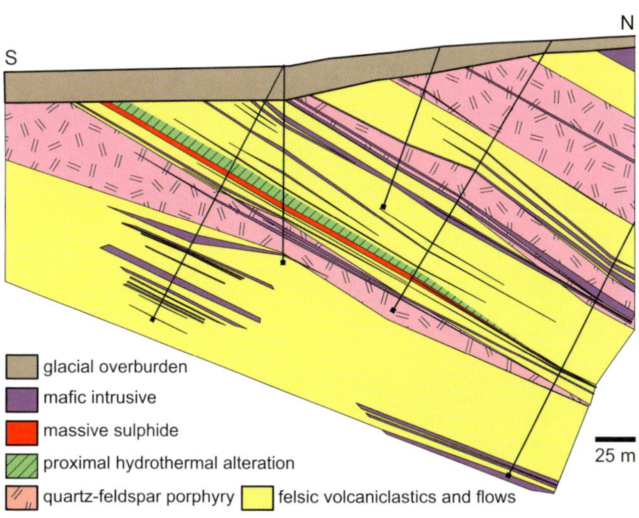

FIGURE 12. Cross section 419450E through the GP4F deposit (modified from Teck Cominco Ltd., unpub. data, 2000 and Boulton, 2002).

replacement-style semimassive mineralization less than 3 m thick consists of sulphide minerals (chalcopyrite, pyrrhotite, sphalerite) that have partly to completely replaced permeable felsic volcaniclastic host rocks in patches that range from several centimetres to 1 m in diameter. This style of mineralization occurs immediately adjacent to stringer vein zones and extends up to about 100 m laterally outward from the periphery of the massive sulphide lenses (Fig. 18). Stringer vein zones, up to 25 m wide, consist of millimetre to 25 cm wide veins of pyrite, sphalerite, and lesser chalcopyrite, pyrrhotite, and arsenopyrite, together with common gangue minerals quartz, calcite, dolomite, ankerite, siderite, chlorite, biotite, and muscovite (Fig. 16N,O).

Within the Wolverine deposit there is no well preserved vertical base metal zoning from footwall to hanging-wall contacts, likely due to the deformational effects. However, high Cu areas dominated by chalcopyrite are confined predominantly to the base of the massive sulphide lenses within massive and replacement-style mineralization (e.g. in the area between the Lynx and Wolverine zones; Fig. 19A). High Zn and Pb areas are in the massive sulphides at the outer fringes of the deposit (Fig. 19B,C). Silver resides mostly within tetrahedrite.

Ice Deposit

The Ice deposit is located in the northern part of the FLD (Fig. 2) about 60 km east of Ross River. Host rocks to the deposit are basalts, cherts, and mudstones (Figs. 20, 21, 22) of the Campbell Range Formation. The immediate host rocks to the deposit are massive basalts, porphyritic-pillowed basalts (Fig. 23A) and variably autobrecciated pillowed basalts that are plagioclase porphyritic immediately below the deposit (Fig. 23B,C) (Becker, 1997, 1998; Eaton and Pigage, 1997; Pigage, 1997). These are interbedded with black, grey, green, and red ribbon cherts (e.g. Fig. 23D), and minor greywackes and carbonaceous mudstones.

The massive sulphide lens at Ice is up to 28 m thick (Fig. 20), and is underlain by a zone of stringer sulphide veins (Figs. 22, 23E). Primary features and textures are much better preserved at Ice than in the Kudz Ze Kayah, GP4F, and Wolverine deposits. The lens comprise a Cu-rich basal core of chalcopyrite, bornite, and Cu-rich sphalerite (e.g. Fig. 23F) that is situated stratigraphically directly above the stringer vein zone. The basal Cu-rich sulphides are surrounded by thinner, lower grade Cu mineralization. Stringer sulphide

FIGURE 13. Photographs of GP4F deposit host rocks and mineralization. **(A)** Rhyolite flow (DDH GP4F98-193; 81.3 m). **(B)** Fragmental rhyolite volcaniclastic rock. **(C)** Fragmental rhyolite volcaniclastic rock. **(D)** Quartz-feldspar porphyry (DDH GP4F98-188; 92.3 m). **(E)** Sphalerite-rich 'buckshot' pyrite massive sulphide (DDH K98-188; 77.6 m). **(F)** Pyrite-rich 'buckshot' massive sulphide (DDH K98-188; 78.5 m). **(G)** Sphalerite-rich replacement mineralization with fine-grained chlorite and sericite (DDH GP4F98-193; 174.8 m). **(H)** Sphalerite-rich replacement mineralization with fine-grained chlorite and sericite (DDH K98-189; 84.4 m). Unless stated, the locations of individual samples are not specified. Scales are in centimetres.

vein mineralization is locally present within the footwall-brecciated, porphyritic basalt. Stringer veins are predominantly pyrite-quartz-chalcopyrite and specular hematite. The massive sulphides are largely overlain by a thin (~5 cm - 0.5 m) siliceous hematitic chert unit that passes upwards into a hanging wall of massive basalt flows and interlayered chert. The massive sulphide mineralization consists of low Pb, and very low As, Sb, Hg, and Se contents, consistent with a setting dominated by mafic volcanic host rocks.

Secondary mineralization is confined to the zone of near surface weathering that typically ranges between 5 and 50 m below surface, and extends to almost 80 m depth along fractures. Secondary mineralogy includes minerals that have wholly or partially replaced primary sulphide minerals and others that were precipitated from groundwater.

Fisher Zone

The Fisher Zone, discovered in 1995 by Westmin Resources Ltd., is located some 8 km northwest of the Wolverine deposit, and on the same apparent stratigraphic horizon, referred to as the Wolverine Belt (Fig. 14). The Fisher zone consists of numerous narrow bands or lenses of sphalerite-pyrite-galena within sericite-quartz±chlorite altered felsic volcanic and volcaniclastic rocks (Tucker, 1999). In the footwall, there is a K-feldspar ± quartz porphyry that is similar in appearance and composition to that in the Wolverine and Lynx zones (Piercey et al., 2001b). This porphyry has been dated at 346.0 ± 2.2 Ma, similar to the 347.8 ± 1.3 Ma age for Wolverine (Piercey et al., in press). The area contains a magnetite iron formation that has been thickened due to repetition by folding.

Sable Zone

The Sable Zone is located 1.6 km southeast of and along strike from the Wolverine deposit, along the Wolverine Belt (Fig. 14). Mineralization consists of thin intersections of high-grade massive sulphide and strongly silica-pyrite and Fe-carbonate-altered rhyolite. A K-feldspar-quartz porphyry in the footwall has been dated at 352.4 ± 1.5 Ma (Piercey et al., in press).

Puck Zone

Puck is located 2 km southeast of the Wolverine deposit along the Wolverine Belt (Fig. 14) and consists of sphalerite±galena±chalcopyrite stringer veinlets in variably sericite-altered quartz-phyric felsic volcanic rocks. In places, there is a well developed calcite-pyrite exhalite, magnetite iron formation, and barite. There is a narrow (6.1 m) interval of K-feldspar±quartz porphyritic rocks in the footwall, similar to that at the Wolverine deposit and the Fisher zone (Piercey et al., 2001b), and this has been dated at 356.9 ± 0.5 Ma (Piercey et al., in press).

Goal Net Property

The Goal Net property of Yukon Zinc Corporation is a 10 by 6 km block of claims within rocks of the Grass Lakes Group situated equidistantly between the Fyre Lake deposit to the south and the GP4F deposit to the north (Murphy and Piercey, 1999b,c) (Fig. 24). The Goal Net property contains at least twelve exploration targets, only several of which have been tested. In the Goal Net North area, Zn-Pb-Cu semimassive sulphides

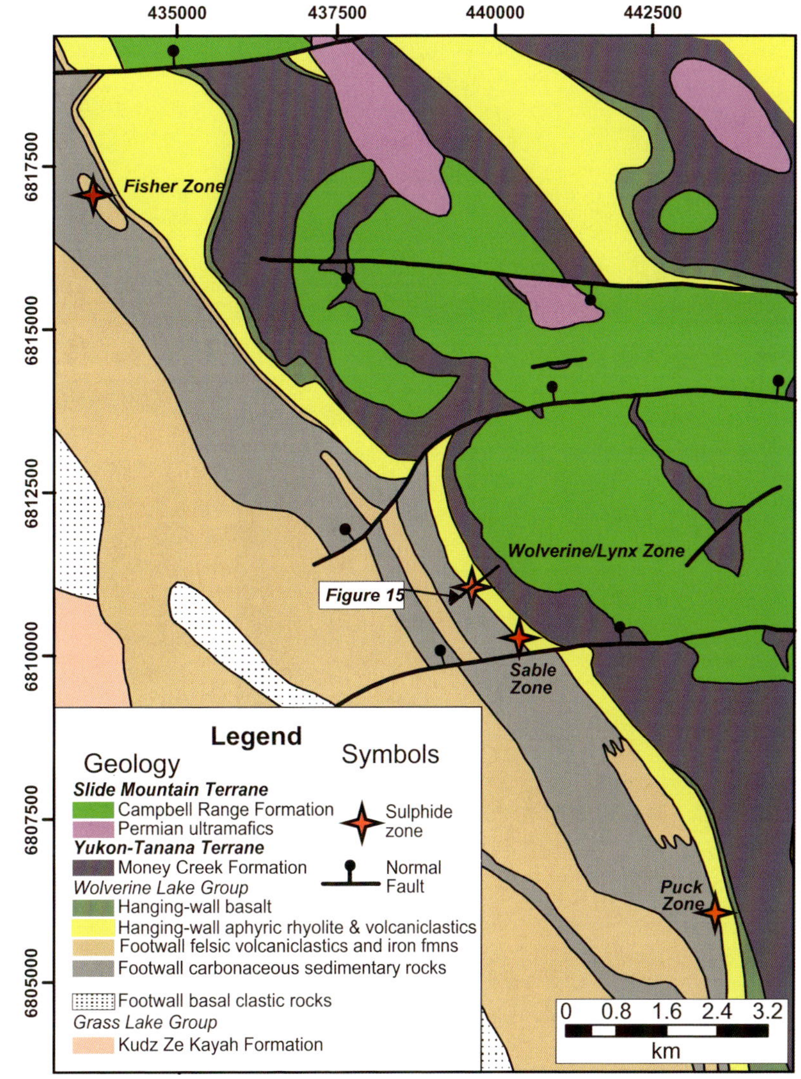

FIGURE 14. Surface geology map of the Wolverine Belt and Wolverine deposit area (modified from Murphy et al., 2006; Bradshaw et al., in press; Expatriate Resources Ltd., unpub. data). Also shown is the cross section line for Figure 15.

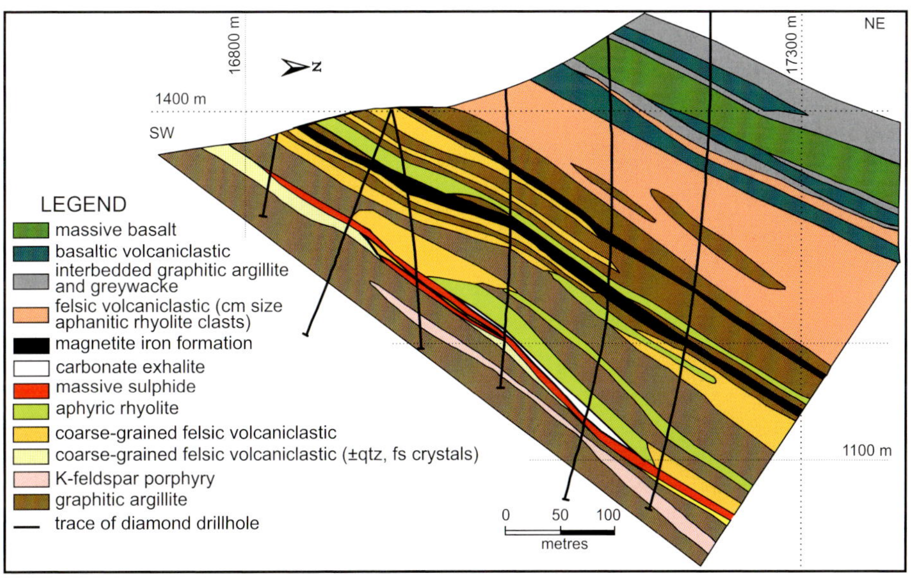

FIGURE 15. Cross-section 16250E through the Wolverine deposit (modified from Expatriate Resources Ltd., unpub., 2001; Bradshaw et al., 2001).

FIGURE 16. Photographs of Wolverine deposit host rocks, mineralization, and hydrothermal alteration. **(A)** Footwall quartz-phyric felsic volcaniclastic (DDH WV97-102). **(B)** Footwall feldspar-porphyritic rhyolite. **(C)** aphyric rhyolite forming the immediate hanging wall. **(D)** Hanging-wall felsic volcaniclastic with rhyolite breccia clasts (DDH WV97-112). **(E)** Hanging-wall rhyolite and argillite clast breccia. **(F)** Magnetite iron formation, from stratigraphic hanging wall to Wolverine Zone. **(G)** Calcite-pyrite exhalite that forms the immediate hanging wall to the massive sulphides in places. **(H)** Chalcopyrite-pyrrhotite-rich sulphides from base of massive sulphide lens (DDH WV00-120). **(I)** Banded, sphalerite-rich massive sulphides from upper part of massive sulphide lens.

FIGURE 16 CONTINUED. Photographs of Wolverine deposit host rocks, mineralization, and hydrothermal alteration. **(J)** Possible primary massive sulphide breccia, with pyrite-rich clasts in a sphalerite-rich matrix (DDH WV95-12). **(K)** Chalcopyrite-rich replacement-style mineralization from immediately below the massive sulphide lens (DDH WV95-21). **(L)** Chalcopyrite-rich replacement-style mineralization from immediately below the massive sulphide lens, with abundant silver-grey chlorite (DDH WV96-56; 333 m). **(M)** Replacement-style mineralization from immediately below the massive sulphide lens, with dark green porphyroblasts in a matrix of light grey felted chlorite (DDH WV96-71). **(N)** Sphalerite-chalcopyrite-pyrite-quartz stringer veins crosscutting felsic volcaniclastic, from feeder zone; note silicified vein margins (DDH WV95-25). **(O)** Sphalerite-chalcopyrite-pyrite-quartz stringer veins in graphitic argillite, from feeder zone; veins have been transposed parallel to prominent foliation (DDH WV96-25). **(P)** Intensely silicified felsic volcaniclastic with deformed pyrite-quartz veinlets. **(Q)** Strongly chloritized felsic volcaniclastic (DDH WV96-27). **(R)** Fe-carbonatized felsic volcaniclastic. **(S)** Strongly sericitized felsic volcaniclastic (DDH WV96-39). Unless stated, the locations of individual samples are not specified.

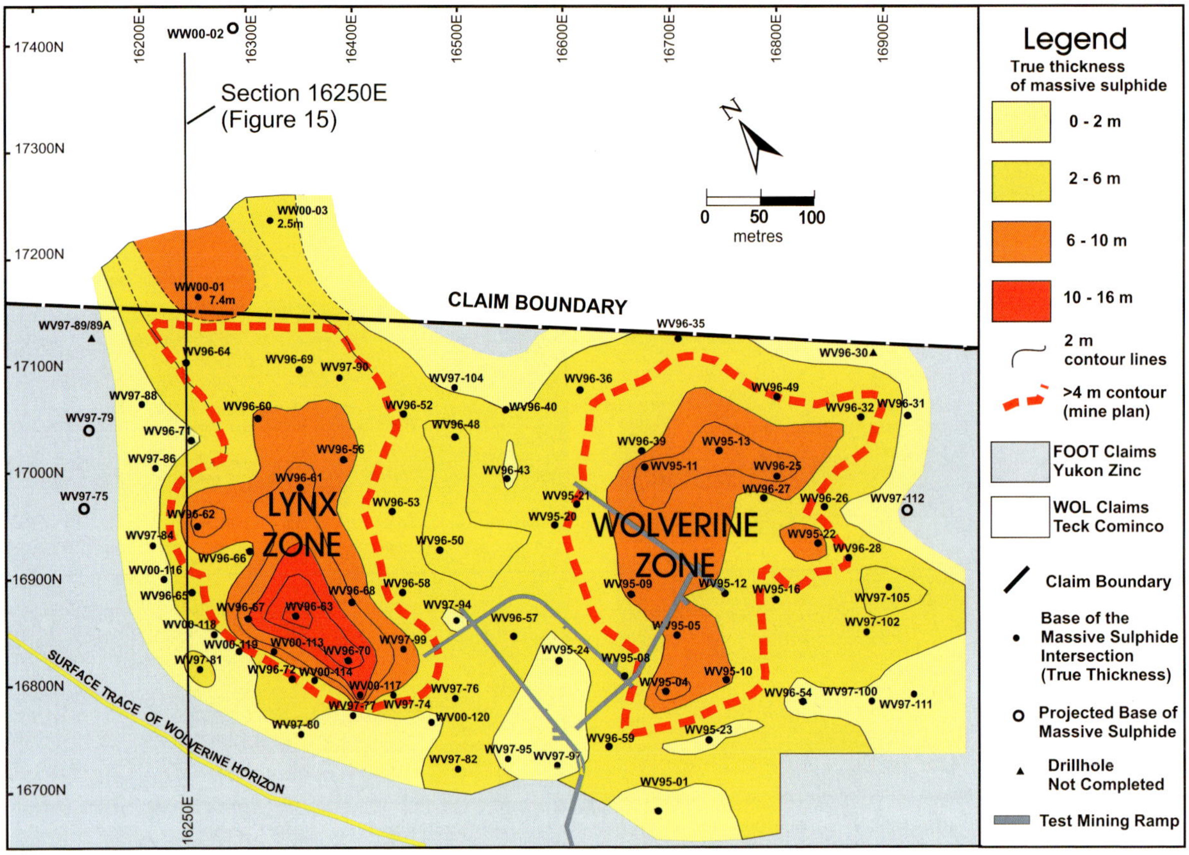

FIGURE 17. Geological plan of the Wolverine deposit depicting the two thickened zones of massive sulphide (Lynx and Wolverine zones), diamond drillhole collars, and contours of massive sulphide thickness (Yukon Zinc Corporation, 2004a).

and disseminated sulphides are associated with pyritic sericite-altered felsic volcanic and volcaniclastic rocks (interpreted to be rhyolite tuff and quartz porphyry rhyolite, respectively), as well as disseminated pyrite-chalcopyrite with biotite-chlorite alteration. Low-grade stratiform Zn mineralization also occurs over 16 m in flat-lying argillaceous felsic volcaniclastic rocks within the Skyblaze zone. In the Goal Net South area, the Thunderstruck zone, 4.3 km south of Skyblaze (Fig. 24), is a gently southerly dipping continuous sheet of thin, high-grade pyrite-pyrrhotite-sphalerite-bearing semi-massive to massive sulphides in chlorite-sericite-altered felsic schist that extends from outcrop down-dip for over 550 m (Yukon Zinc Corporation, 2004b, 2005). The zone is thought to be located at the same time stratigraphic position as both the Kudz Ze Kayah and GP4F deposits, within the basal portion of the Kudz Ze Kayah Formation.

FIGURE 18. Geological plan of the Wolverine deposit depicting the distribution of feeder zone (as defined by the presence of stringer veins) and replacement-style mineralization/alteration; also shown are 6 to 10 m and 10 to 16 m massive sulphide thickness contours (for comparison with Fig. 17). Inset shows an idealized, schematic section. Figure constructed from J. Peter (unpub. data, 2001).

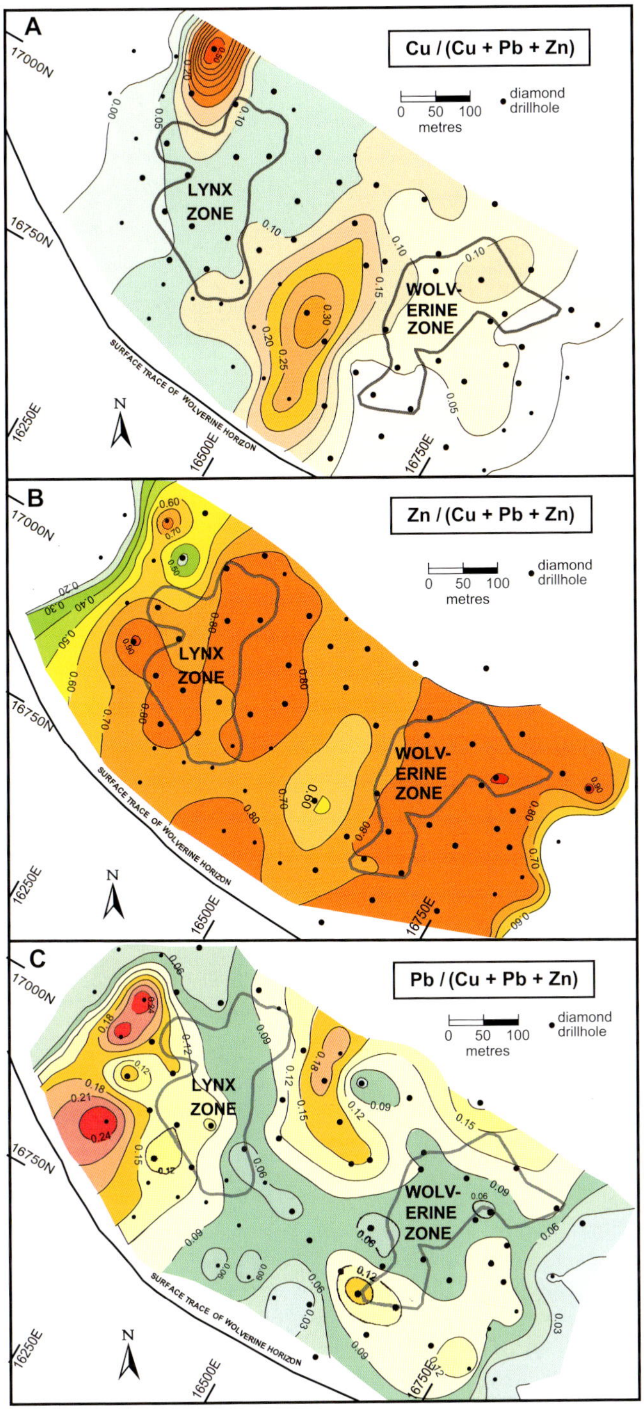

FIGURE 19. Plan contour maps of the Wolverine deposit illustrating the lateral distribution of the metals Cu, Pb, and Zn in comparison to zones of greatest sulphide thickness: **(A)** Cu/ (Cu+Pb+Zn). **(B)** Zn/(Cu+Pb+Zn); and **(C)** Pb/(Cu+Pb+Zn) (data from Expatriate Resources Limited, 2000). The positions of the individual drillholes are shown as in Figures 17, 18. Outlined zones indicate a massive sulphide thickness >6 m (from Fig. 17).

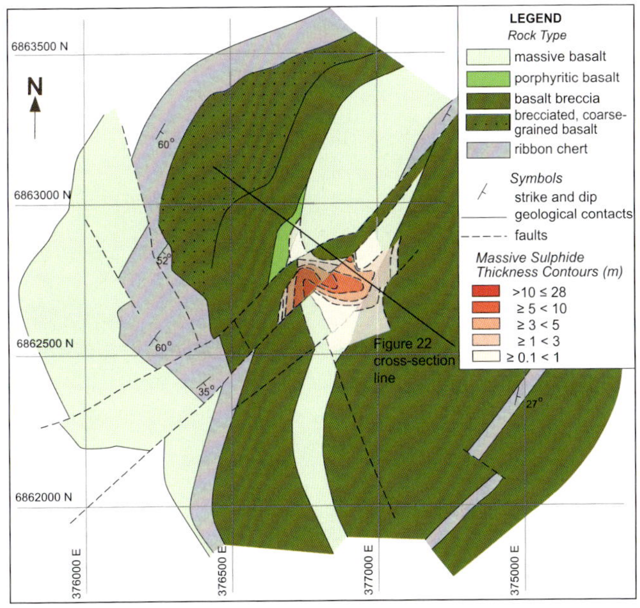

FIGURE 20. Surface geological map of the Ice deposit area (modified from Pigage, 1997; Becker, 1998; Expatriate Resources Ltd., 1999; and Hunt, 2002). Also shown is the cross-section line for Figure 22.

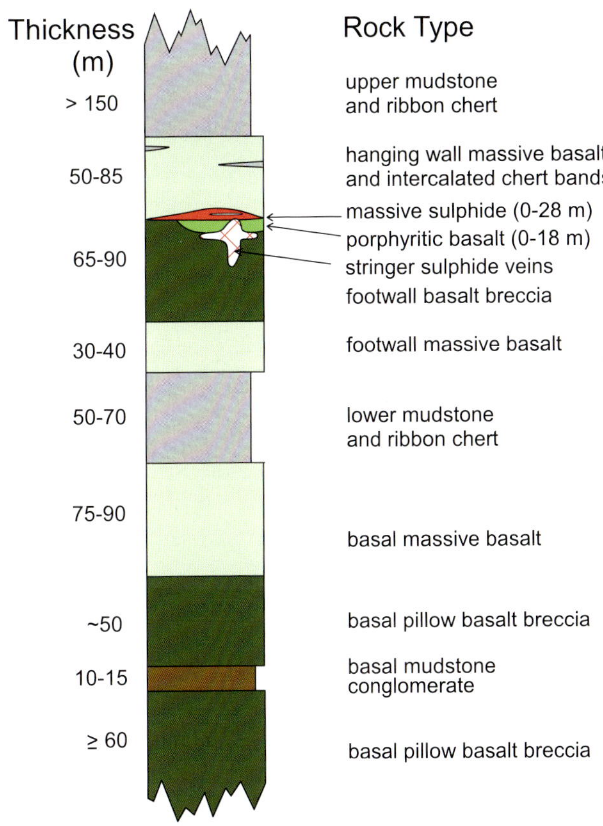

FIGURE 21. Generalized stratigraphic column for the Ice deposit and host rocks (modified from Pigage, 1997 and Hunt, 2002).

Money (Julia) Prospect

The Money (also referred to as Julia) prospect is located about 5 km east of the Wolverine Deposit in a thick sequence of pillowed and pillow breccia basalts intercalated with mudstone and chert of the Campbell Range Formation (Hunt, 1998). The deposit consists of a tabular lens of massive banded and siliceous pyritic sulphides that has a strike length of at least 53 m, extends down dip for at least 130 m, and averages 1 m in thickness. The footwall host rocks include pillow basalt, pillow basalt breccia with maroon mudstone matrix, chert, mudstone, and maroon shale (Baknes, 1997). The basalt and basaltic tuff have been hydrothermally altered to quartz-sericite-pyrite.

FIGURE 23. Photographs of Ice deposit host rocks and mineralization. **(A)** Porphyritic basalt (immediate footwall). **(B)** Basalt flow hyaloclastite. **(C)** Volcanic conglomerate. **(D)** Hematitic chert interpillow sediment. **(E)** pyrite veinlets in strongly chlorite altered volcanic rock from feeder zone; **(F)** Chalcopyrite-bornite-hematite massive sulphide. Unless stated, the locations of individual samples are not specified.

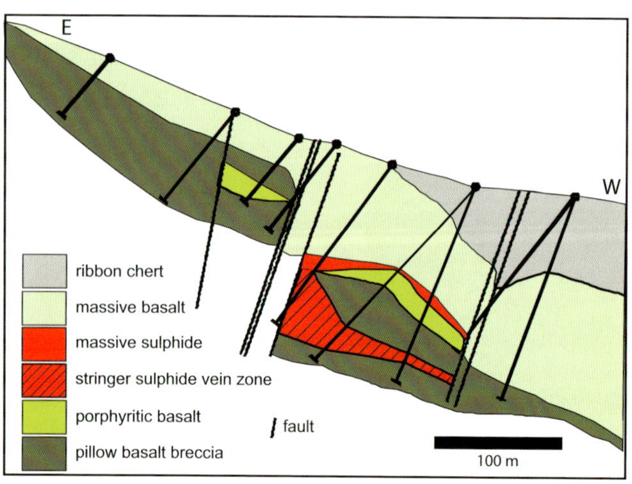

FIGURE 22. Geological cross-section 11400E through the Ice deposit (modified from Becker, 1998).

Other Prospects

In addition to the deposits and occurrences described above, there are a number of others that have been recognized in the FLD. Details on these properties that include the Pack (or Pak), Blueline, Pneumonia (Mony), Akhurst (Expo), Ant (Hat Trick), Red Line, Cobb, League, and Nad properties within the Grass Lakes Group, the Hoo (Argus), El (Tin), and Eldorado properties in North River Formation rocks, are given by Hunt (2002).

Deposit Classification

Host rocks, and metal association and tenor (Cu-Co-Au), indicate that the Fyre Lake deposit is a Besshi-type VHMS deposit (e.g. Fox, 1984; Slack, 1993). A hallmark of Besshi-type deposits is that sedimentary rocks are a significant component of the host stratigraphy in addition to mafic volcanic rocks. Such deposits are also typically enriched in Co and Au.

The Kudz Ze Kayah deposit can be classified as a Kuroko-type deposit that formed as a mound on the paleoseafloor. At least part of the deposit likely formed by subsurface replacement of host strata (e.g. albite-altered rocks and mineralization).

The GP4F deposit, only 5 km away from Kudz Ze Kayah, is also best classified as a Kuroko-type VHMS deposit and is genetically related to rift-related felsic volcanism within a back-arc basin as suggested by Piercey et al. (2001a) for the nearby Kudz Ze Kayah deposit. GP4F has no feeder zone characterized by stringer quartz-sulphide veins and associated intense hydrothermal alteration, and this indicates that the preserved portions of the deposit are the distal part of the seafloor hydrothermal system (perhaps with subsurface replacement), with the proximal Cu-rich mound and feeder zone eroded away.

The Wolverine deposit has many characteristics of a polymetallic Kuroko-type deposit, but with some characteristics common to SEDEX deposits. In this respect, Wolverine also has many geological, geochemical, and metallogenic similarities with the world-class Brunswick No. 12, No. 6, and Heath Steele deposits in the Bathurst Mining Camp of northern New Brunswick (see Goodfellow and McCutcheon, 2003). Among these similarities are 1) felsic volcaniclastic and black shale host rocks; 2) relatively Zn-rich (up to 30 wt.% in some intersections); and 3) presence of spatially associated and laterally extensive iron formations. Although the term "Bathurst-type" deposit has been coined previously, Goodfellow (2001) argued that volcanic-sediment-hosted massive sulphide (VSHMS) is a more meaningful classification, and we also favour such a moniker.

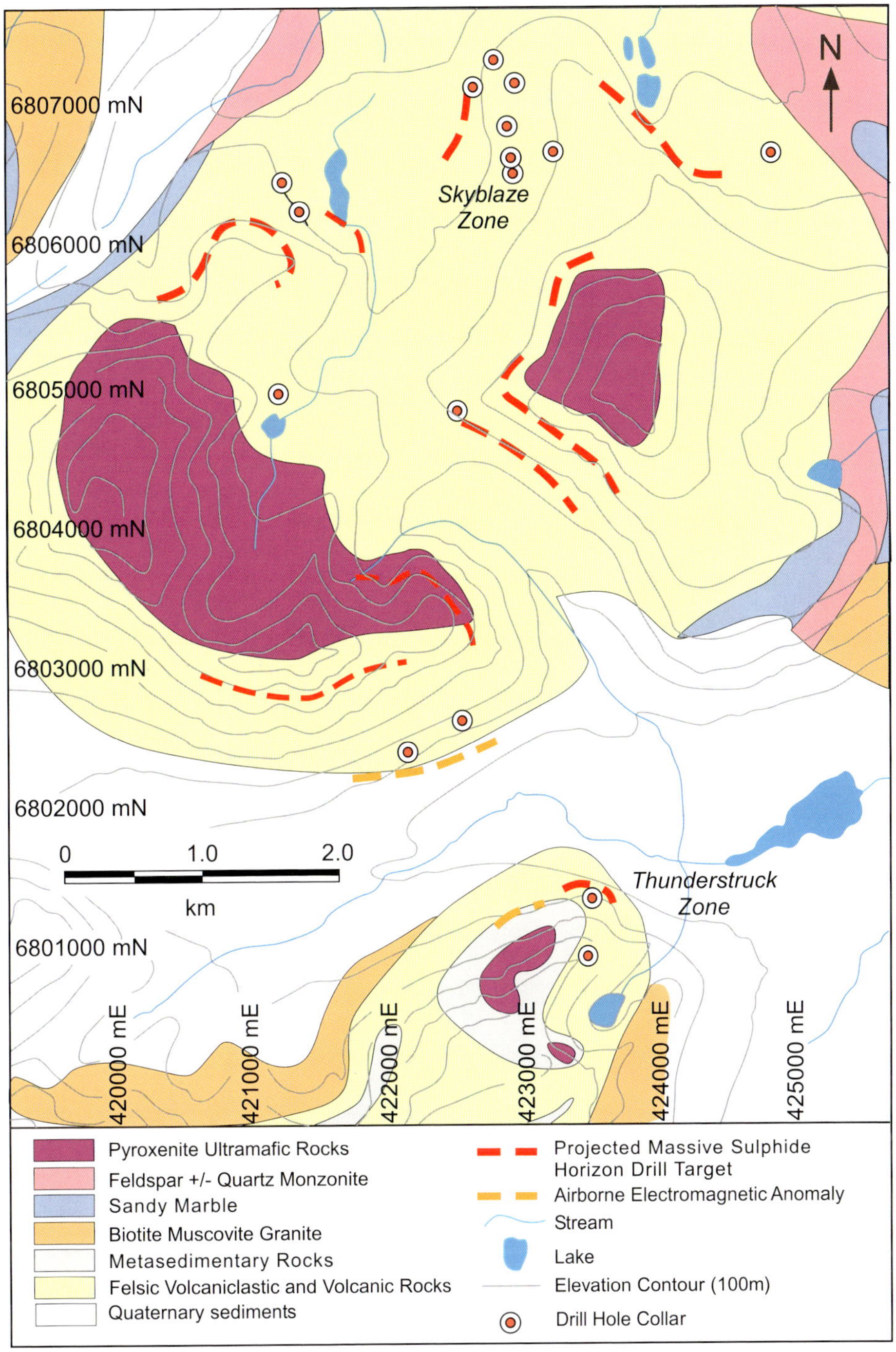

FIGURE 24. Surface geology map of the Goal Net property Shown are elevation contours, geology, drillhole collar locations, locations of massive sulphide mineralization (Thunderstruck and Skyblaze zones), and airborne electromagnetic anomalies (modified from Yukon Zinc Corporation, 2005).

The Ice deposit, together with the Money occurrence, are the only known Cyprus-type (Cu-Zn) VHMS deposits in the FLD; distinguishing criteria for this deposit-type are the pillowed to mafic volcanic host rocks, and the Cu-rich and Pb-Zn-poor nature of the mineralization.

Alteration Mineralogy and Lithogeochemistry

The VHMS deposits of the FLD are associated with hydrothermally altered host rocks. In the stratigraphic footwalls, alteration styles varying from chlorite (Fyre Lake, Kudz Ze Kayah, Wolverine, Ice), sericite (Kudz Ze Kayah, GP4F, Wolverine), silica/quartz (Kudz Ze Kayah, Wolverine), carbonate (Wolverine) and albite (Kudz Ze Kayah). Feeder zones/stringer veins are present at the Fyre Lake, Kudz Ze Kayah, Wolverine and Ice deposits. Hanging-wall alteration is present at Fyre Lake (weak chlorite) and Wolverine (weak sericite).

At Fyre Lake, the mafic volcanic rocks stratigraphically beneath and peripheral to the massive sulphides have been hydrothermally altered to chlorite and quartz (Sebert et al.,

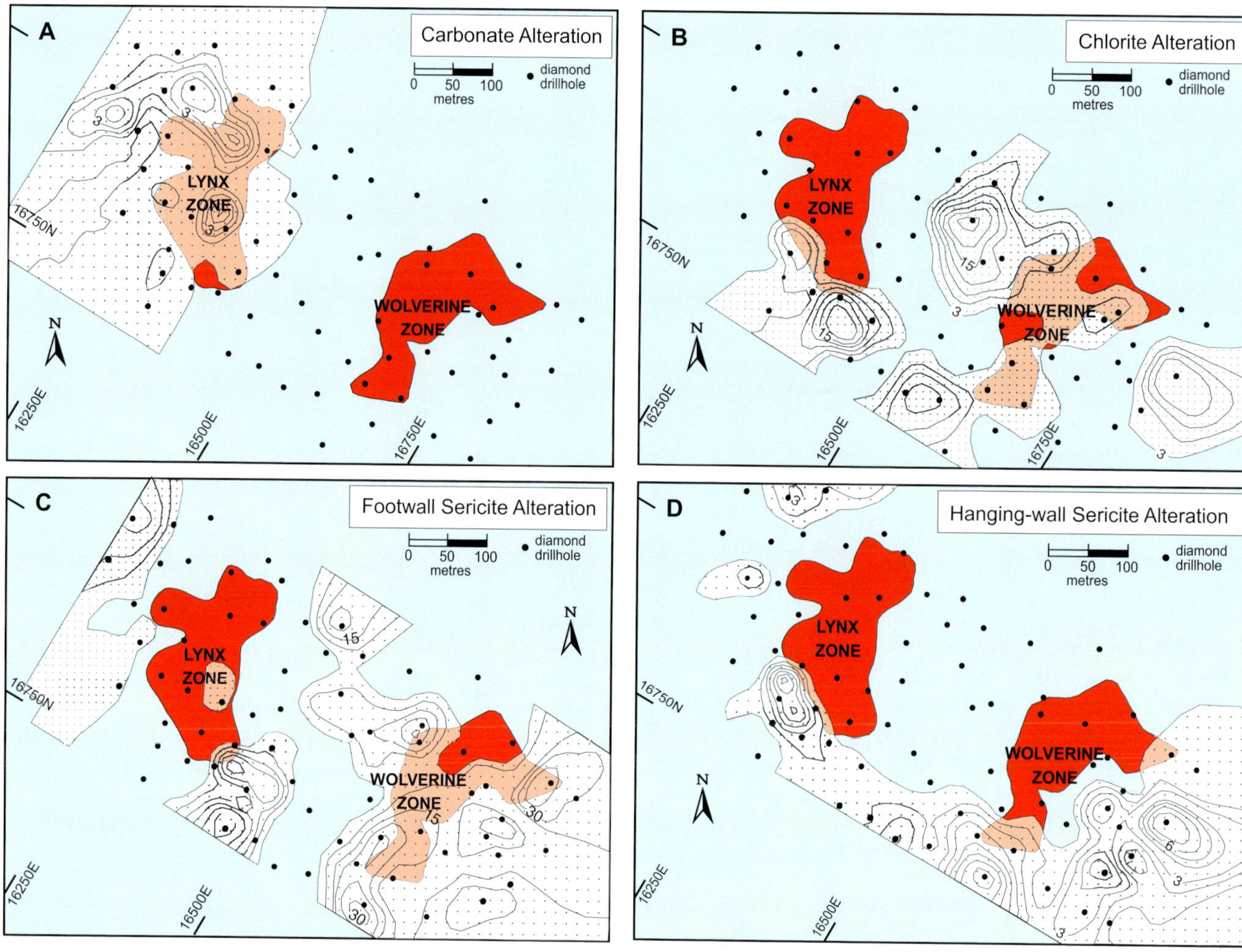

FIGURE 25. Plan maps of the distribution of hydrothermally altered rock in the Wolverine deposit. Contours represent the true thickness of altered rocks in drill core that contain greater than 5 vol.% alteration minerals, as determined by visual estimation. **(A)** Footwall carbonate alteration; contoured areas contain calcite or ankerite in the footwall to the massive sulphide lenses (contour interval = 1 m). Carbonate alteration minerals occur only in the Lynx zone and are associated with relatively thin zones of sericite altered rock. **(B)** Footwall chlorite alteration; contoured areas contain a thickness of at least 3 m of chlorite-altered rock (contour interval = 3 m). The thickest zones of chlorite altered rock are generally associated with zones of 'replacement-style' semimassive sulphide. **(C)** Footwall sericite alteration; contoured areas include a thickness of at least 5 m of sericite-altered rock (contour interval = 5 m). The thickest zones of sericite-altered rock occur on the edges of the Wolverine zone, lateral to the thickest zones of chlorite-altered rock. **(D)** Hanging-wall sericite alteration; contoured areas include a thickness of at least 1 m of sericite-altered rock (contour interval = 1 m). Sericite-altered rock in the hanging wall occurs on the edges of the deposit and generally correlates with sericite altered rock in the footwall. Red coloured areas (the Wolverine and Lynx zones) indicate a true thickness of massive sulphide greater than 6 m (from Fig. 17).

2004). The hanging-wall mafic volcanic rocks have also been weakly chloritized (Sebert et al., 2004). In general, Sebert et al. (2004) indicate that there is a greater Mg content in altered rocks relative to unaltered rocks.

At the Kudz Ze Kayah deposit, structural hanging-wall (stratigraphic footwall) rocks have been hydrothermally altered to sericite, chlorite, and albite, and distributed in discrete zones; hydrothermal alteration intensity increases with proximity to sulphides. The most proximal alteration at the immediate base of the massive sulphides is dominated by albite (Fig. 9L), presumably because of intense replacement of detrital sedimentary and volcaniclastic rocks. This gives way to Fe chlorite-dominated alteration (Fig. 9J) that is commonly associated with chalcopyrite and pyrrhotite blebs and disseminations, and pyrrhotite±chalcopyrite±pyrite±sphalerite±galena veins in the feeder zone. This, in turn gives way to a more distal sericite-dominated (±disseminated pyrite and/or pyrrhotite) alteration in the outermost zones (Fig. 9K). Silicification is developed only locally in a few places in the chlorite zone associated with sulphide replacement and feeder veins. Ankerite and siderite are ubiquitously developed in almost all of the host rocks in and around the deposit and occurs as Fe carbonate disseminations that have preferentially replaced felsic clasts in volcaniclastic rocks (Fig. 9M). Structural footwall (stratigraphic hanging-wall) rocks, however, are not visibly hydrothermally altered.

At GP4F, there is a narrow zone of sericite-chlorite hydrothermal alteration within felsic rocks immediately adjacent to (structurally above) the massive sulphides in the stratigraphic footwall (so-called marker tuff) (Boulton, 2002). The alteration intensity is weaker than is observed at Kudz Ze Kayah and is characterized by the destruction of feldspars. Quartz-feldspar porphyry occurs (with one exception) structurally below or stratigraphically above the massive sulphide lens. Stratigraphically below the lens, feldspars phenocrysts in porphyritic rocks have been completely altered to sericite or secondary albite, resulting in a rock containing only quartz phenocrysts.

At Wolverine, hydrothermal alteration effects are largely confined to the permeable felsic volcaniclastic rocks that are the immediate footwall of the massive sulphide lenses, though they extend (in the footwall) to the periphery of the massive sulphide lenses and to a lesser extent the hanging wall. Four hydrothermal alteration styles are recognized (Fig. 25A-D; Peter, 2001; Bradshaw et al., in press): silicic alteration (Fig. 16P; distribution is not shown but restricted to close proximity to the feeder zone-see Fig. 18), carbonate alteration (Figs. 16R, 25A), chlorite alteration (Figs. 16Q, 25B), and sericite alteration (Figs. 16S, 25C,D).

Stringer sulphide veins typically have a 5 to 10 cm wide quartz alteration selvage (Fig. 16N,P). Zones of silicified rocks also extend up to 20 m below the base of sulphide lenses. Chalcopyrite- and pyrrhotite-rich semimassive replacement-style mineralization has attendant chlorite and Fe-carbonate, and more sphalerite-rich, semimassive replacement-style mineralization has associated sericite and ankerite alteration of the host rocks.

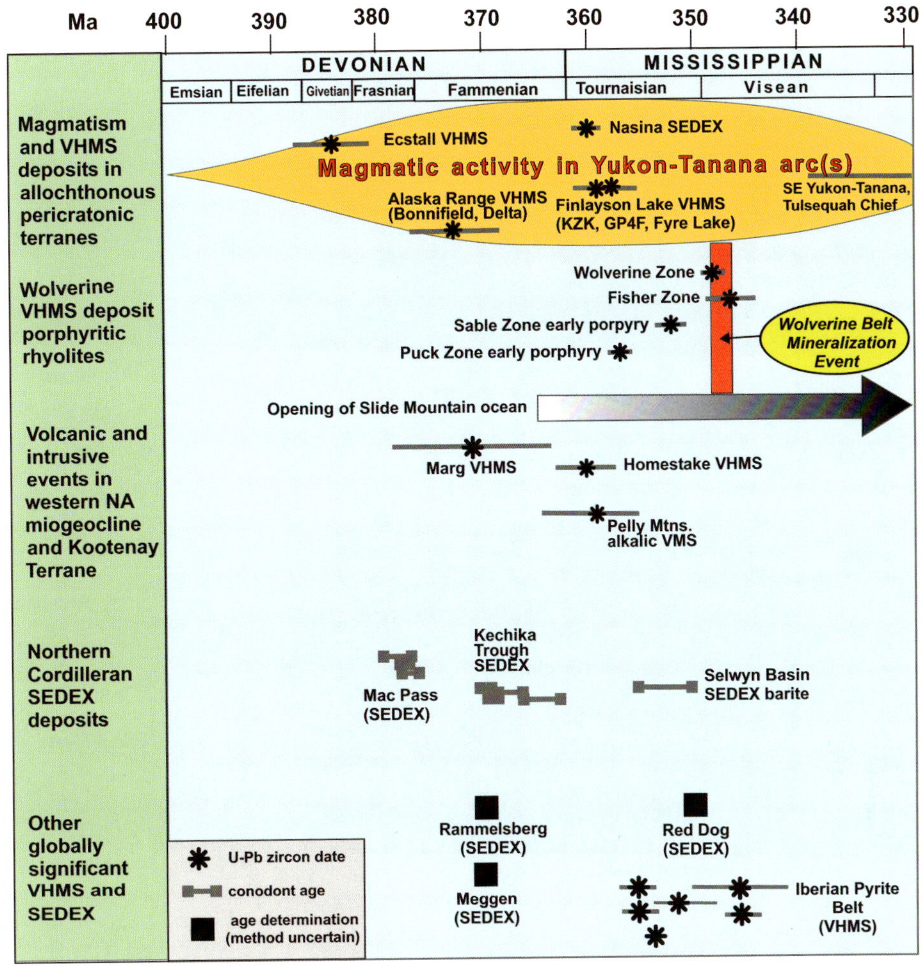

FIGURE 26. Age distribution for VHMS and VSHMS deposits in the FLD. Also shown are ages for other deposits in the Yukon-Tanana Terrane, and VHMS and SEDEX deposits from the Kootenay Terrane and the North American miogeocline, and globally significant (>50 Mt) VHMS and SEDEX deposits/camps that formed during the Devonian-Mississippian (modified from Piercey et al., in press and Nelson et al., 2002).

Carbonate alteration consists of various carbonates, ranging from creamy white calcite to orange-brown ankerite to brown siderite. The carbonate, comprising up to 30 vol.% of the rock, consists of porphyroblasts up to 2 cm in diameter. Carbonate alteration associated with replacement-style mineralization consist of 1 to 2 m wide zones, is most prevalent beneath the Lynx zone lens (Fig. 25A). This carbonate is generally ankeritic with up to 38 mole% Mg, whereas carbonate gangue within massive sulphide is generally end-member siderite although it can contain up to 26 mole% Mg (Bradshaw, 2003). Carbonate in sericite-altered rock is generally calcite.

Chlorite alteration zones consist of abundant fine-grained chlorite (up to 50 vol.%) in coarse-grained volcaniclastic rocks. They generally flank silicic alteration zones and can extend up to 30 m below the base of the sulphide lenses. There is commonly a transitional change from chlorite-dominant to sericite-dominant alteration mineralogy. Electron microprobe analyses of chlorite show two compositional varieties: 1) dark green magnesian clinochlore, which forms fine-grained, massive aggregates; and 2) less common ferroan clinochlore, which forms local clots in carbonate altered rocks (Bradshaw et al., in press).

Sericite altered rocks occur within stratabound zones in the footwall up to 40 m wide and are generally peripheral to and below chlorite altered rocks. Sericite altered footwall rocks typically contain 40 to 60 vol.% sericite, with the abundance of the mineral decreasing with increasing distance from the sulphide lenses. Sericite alteration is developed best in the Wolverine zone and in the area between the Wolverine and Lynx zones (Fig. 25C). Minor sericite alteration occurs in hanging-wall rocks at the western, eastern, and up-dip edges of the sulphide lenses (Fig. 25D). Electron microprobe analyses indicate that the sericite is Ba-rich phengite (Bradshaw et al., in press).

Lithogeochemical indicators of footwall hydrothermal alteration at Wolverine are elevated Al_2O_3/Na_2O ratios, and low Na_2O on both a regional and local scale (G. Bradshaw, 2003, unpub. data; S. Piercey, unpub. data). Locally, the rocks at Wolverine and also Kudz Ze Kayah and GP4F have elevated K_2O due to the presence of abundant sericite, MgO-FeO due to chlorite and Fe carbonates, and CO_2-LOI (loss on ignition) due to the presence of Fe-carbonate, sericite, and chlorite.

At the Ice deposit, hydrothermal alteration is restricted to the stringer zone, where the rocks are dark green and intensely chlorite altered. Only a few metres from the stringer zone, the rocks are relatively unaltered (Becker, 1997, 1998; Eaton and Pigage, 1997; Pigage, 1997). To date there has been no systematic study of the hydrothermal alteration associated with the Ice deposit, but a reconnaissance study of basalt samples from the stringer zone indicates that they exhibit typical Na_2O and CaO depletions with enrichments in Fe_2O_3 and K_2O (S. Piercey, unpub. data).

Exhalites

The narrow (15-40 cm wide), fine-grained chert with a few percent sulphides (chalcopyrite, pyrite) and magnetite that locally marks the top of massive sulphide lenses of the Fyre Lake deposit (Fig. 6I) is thought to be an exhalative sediment (Columbia Gold Mines Ltd., 1999; Sebert et al., 2004). However, the horizon does not appear to extend significantly beyond the mineralization. At the Ice deposit, red hematitic cherts occur at the mineralized horizon as narrow horizons and as interpillow sediment (Fig. 23F). As in other Cyprus-type deposits, these exhalites are generally of restricted areal extent (e.g. Peter, 2003).

In the FLD, exhalites of widespread areal distribution are only known to be associated with deposits in the area extending from the Fisher zone to the Sable zone, the Wolverine deposit, and the Puck Zone, referred to informally as the Wolverine Belt. There are three exhalite horizons in the vicinity of the Wolverine deposit: a narrow (less than 1 m wide) grey to white carbonate-predominant rock consisting of up to 90 vol.% patchy to massive carbonate (calcite> ankerite>siderite) with lesser magnetite, chlorite, and pyrite (Figs. 14, 16G) that locally overlies the massive sulphides, and two horizons up to several metres wide of a grey to black, magnetite-predominant iron formation contain 5 to 60 vol.% of disseminated to massively layered magnetite with lesser carbonate, chlorite, quartz, and white mica (Figs. 14, 16F). The layering is interpreted to be primary bedding. Monomineralic magnetite layers are interbedded with other minerals on a millimetre-scale. This magnetite-predominant iron formation occurs 80 to 100 m stratigraphically above the Wolverine massive sulphides and is laterally extensive with a strike length in excess of 12 km. It displays mineralogical and geochemical similarities with iron formations of the Heath Steele Belt in the Bathurst District, New Brunswick (Peter, 2003; Peter et al., 2003a,b).

It is important to note that the Kudz Ze Kayah and GP4F deposits, with many characteristics similar to Wolverine, have no associated exhalites.

Metallogeny

Within the pericratonic terranes along the western North American margin from Alaska to southern British Columbia, VHMS mineralization is associated with Devonian-Mississippian arc magmatism and associated back-arc basins The episode(s) of extension within the arcs, ensialic back-arc basin formation, and felsic magmatism coincided with VHMS deposits in the FLD and elsewhere in the Yukon-Tanana Terrane (Fig. 26; Mortensen 1992a; Piercey et al. 2001a, 2002a, 2006; Dusel-Bacon et al. 2006; Nelson et al., 2006) and the initial opening of the Slide Mountain Ocean between the pericratonic terranes and the North American continent at about the Devonian-Mississippian boundary (Nelson, 1993; Nelson and Bradford 1993; Creaser et al. 1999). This period of arc and back-arc rifting and felsic to intermediate volcanism in the pericratonic terrane is also

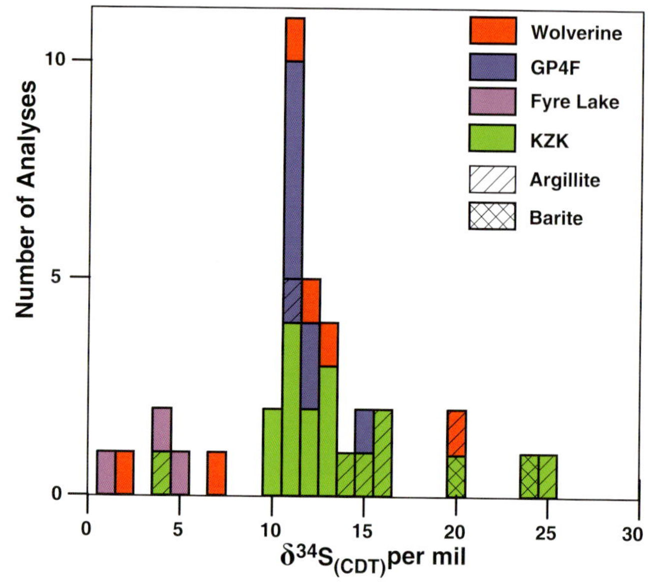

FIGURE 27. Histogram of bulk and laser sulphur isotope data for sulphides, sulphates and host rocks from the Fyre Lake, Kudz Ze Kayah, GP4F, and Wolverine VHMS deposits. Bulk data from J. Peter (unpub. data), Layton-Matthews (2005), and laser data from Bradshaw (2003) and Bradshaw et al. (in press).

coeval with extension, alkalic magmatism, and VHMS and SEDEX deposit formation in rocks of the North American miogeocline during the mid-Paleozoic (e.g. Paradis and Nelson 2000; Nelson et al. 2002; Nelson and Colpron, 2007). The metallogenic evolution of the FLD is outside the scope of this paper, and the reader is referred to Piercey et al. (2001a, 2002a, 2006) and Nelson et al. (2002, 2006) for more detailed discussions.

Genetic Models

Hydrothermal Fluid Characteristics

Fluid Temperatures

The presence of hydrothermal sericite at Kudz Ze Kayah, GP4F, and Wolverine indicates that the mineralizing fluids were moderately acidic, and the Cu-sulphide mineralogy for Kudz Ze Kayah and Wolverine is consistent with precipitation from high-temperature (≈350ºC) fluids. Homogenization temperatures for primary two-phase fluid inclusions in quartz from quartz-sulphide stringer veins at Wolverine give temperatures of 235 to 353ºC (Bradshaw et al., in press). This range is similar to geothermometric estimates for hydrothermal alteration chlorite that yields formation temperatures of 273 to 288°C, and arsenopyrite geothermometric estimates of 215 to 336°C (Bradshaw et al., in press). The presence of sphalerite-galena-chalcopyrite stringer veinlets in variably sericite-altered quartz-phyric felsic volcanic rocks at the Puck zone 2 km southeast of Wolverine suggests that this area was the locus of high-temperature upflow.

The paucity of Cu at GP4F indicates that this deposit likely formed at lower temperatures than Kudz Ze Kayah, based on the solubility considerations of Cu, Pb, and Zn as chloride complexes (Hannington et al., 1999a). Further, no feeder zone characterized by stringer quartz-sulphide veins and associated intensely altered rocks have been identified, and this suggests that the deposit is the distal part of a seafloor hydrothermal system.

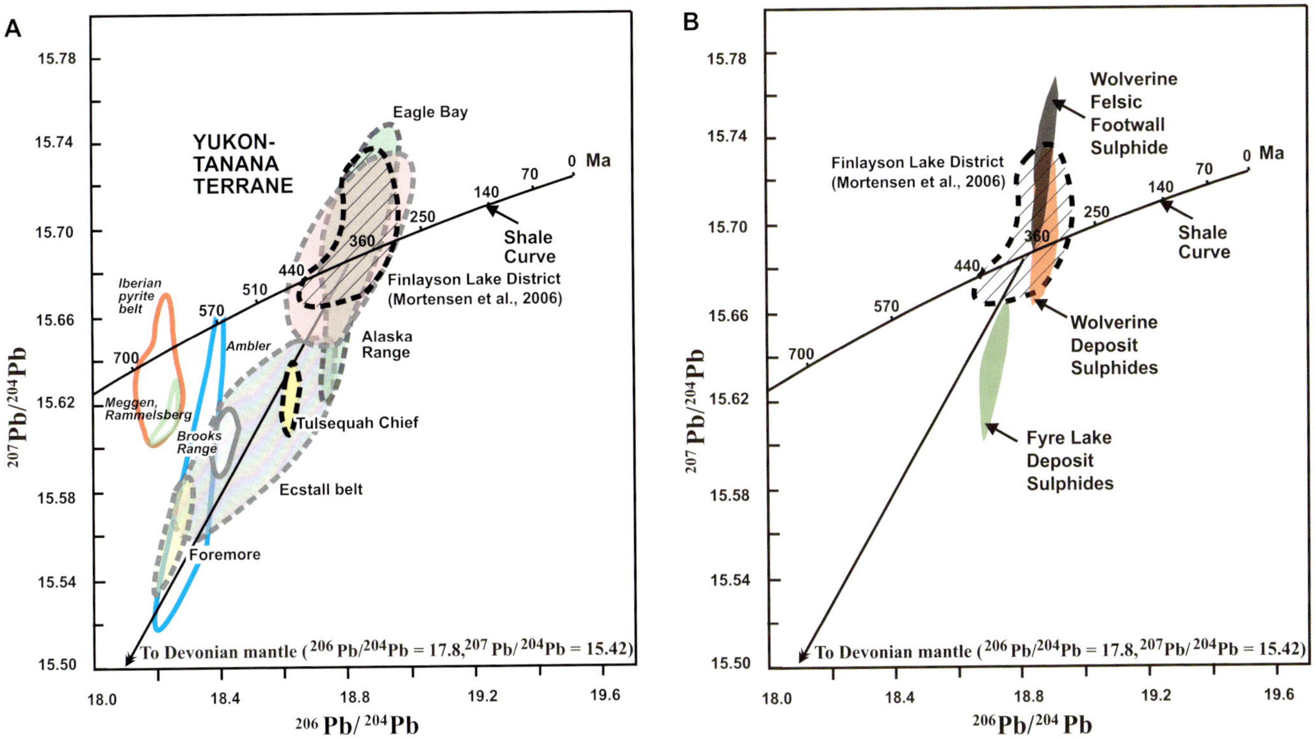

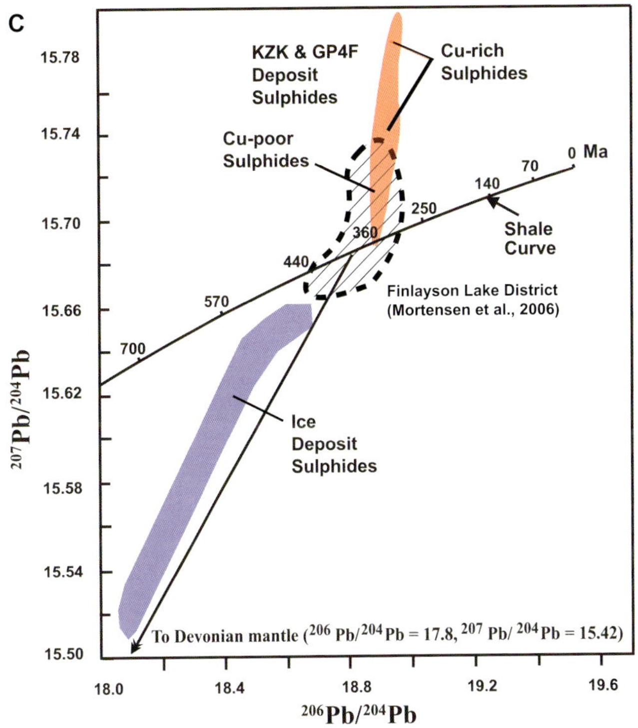

FIGURE 28. $^{206/204}$Pb versus $^{207/204}$Pb isotope plot showing: **(A)** data fields for Devonian-Mississippian VHMS deposits of the FLD from Mortensen et al. (2006); the Ecstall Belt (Alldrick et al., 2003; D. Alldrick, unpub. data, 2005); para-autochthonous deposits of the Alaska Range (Church et al., 1987), Kootenay Terrane (Mortensen et al., 2006), and Stikinia (Tulsequah Chief: Childe, 1997; Foremore: Logan et al., 2000); Devonian-Mississippian VHMS deposits of the Ambler district, Alaska, and the Red Dog Devonian-Mississippian SEDEX deposit (Church et al., 1987); Devonian-Mississippian Meggen and Rammelsberg (Germany) SEDEX deposits (Wedepohl et al., 1978); and Devonian-Mississippian VHMS deposits of the Iberian Pyrite Belt (Marcoux, 1998). Also shown are the Shale Curve of Godwin and Sinclair (1982) and the Devonian mantle value (Doe and Zartman, 1979). Figure is modified from Nelson et al. (2006). **(B)** Wolverine and Fyre Lake deposit sulphides; data are from J. Peter (unpub. data, 2005) and D. Layton-Matthews (2005). Also shown are data field for Devonian-Mississippian VHMS deposits of the FLD from Mortensen et al. (2006), Shale Curve, and Devonian mantle. **(C)** Kudz Ze Kayah, GP4F, and Ice deposit massive sulphides; data are from J. Peter (unpub. data, 2005) and D. Layton-Matthews (2005). Also shown are data field for Devonian-Mississippian VHMS deposits of the FLD from Mortensen et al. (2006), Shale Curve, and Devonian mantle.

Based on Cu solubility considerations, the Cu-rich nature of portions of the Ice and Fyre Lake deposits indicate that the Cu-rich portions of these deposits formed at temperatures of about 300 to 350°C (Crerar and Barnes, 1976). However, fluid inclusion homogenization studies have not been done on either of these deposits.

Fluid Salinities and Behaviour

Primary fluid inclusions in quartz from stringer veins at Wolverine contain low-salinity (2.1-8.5 wt.% NaCl equivalent; mean 6.0) fluids, indicating that the mineralizing fluids are predominantly hydrothermally modified seawater (Bradshaw et al., in press). The fluid inclusions show no evidence of fluid boiling and indicate that the deposit formed at a water depth of at least 1000 m. Based on temperature-density considerations of fluids of various salinities (Turner and Campbell, 1987), the venting mineralizing fluids behaved buoyantly at the seafloor. There are no other direct measurements of fluid salinities from other deposits in the FLD, but fluids are not expected to have been highly saline and have formed a brine pool as has been suggested for some VHMS deposits.

Fluid Redox and Evolution

The mineral zonation of the massive sulphides at Fyre Lake (basal pyrrhotite, upward to pyrite-chalcopyrite, and uppermost magnetite-pyrite) is likely a primary feature related to redox variations attendant with the evolution of the primary sulphide mound in response to zone refining, with the magnetite deposited late in the paragenetic history of the deposit as the (originally reduced) fluids became more oxidizing. This interpretation, however, contradicts that of Sebert et al. (2004) who suggest that the intercalations of magnetite and sulphides are the result of multiple cycles of fluid evolution from oxidizing to reducing.

The coexistence of pyrite, pyrrhotite, and chalcopyrite and the preservation of Fe-rich sphalerite near the bases of the massive sulphide lenses in the Kudz Ze Kayah, GP4F and Wolverine deposits indicates a predominantly low and narrow range of log fO_2 (-35 to -43) and log fS_2 (-8 to -13) conditions within the sulphide mounds and replacement zones (Toulmin and Barton, 1964; Barton and Toulmin, 1966; Helgeson, 1969; Barton, 1978; Barton et al., 1979; Bowers et al., 1985; Hannington et al., 1999a). The absence of pyrrhotite and chalcopyrite, and the presence of Fe-poor sphalerite near the tops of the polymetallic sulphide deposit mounds in the FLD suggest a somewhat higher log fO_2 (-35 to -30) and lower depositional temperatures (<250°C).

The presence of magnetite-bearing exhalite (Fig. 16F) along the Wolverine Belt, extending from the Fisher to the Puck zones (a distance of over 14 km; Fig. 14) indicates that these rocks are largely the product of hydrothermal plume fallout and widespread settling on the seafloor. However, geochemical data indicate that these rocks were formed from lower temperature fluids than the massive sulphides (Peter, 2003), following massive sulphide deposition.

At the Ice deposit, field identification of paragenetic sequences (L. Pigage and S.J. Piercey, unpub. data) and preliminary reflected light microscopy (S.J. Piercey, unpub. data) show complex paragenetic relationships between sulphide minerals and a changing fluid history through the evolution of the deposit. The deposit was likely formed as a pyrite-rich mound on the seafloor, but zone refining during fluid circulation through the mound and venting of high-temperature hydrothermal fluids resulted in the replacement of the core of the mound pyrite by chalcopyrite (and quartz), and finally bornite-sphalerite (±magnetite). The presence of specular hematite within the stringer veins also indicates that the hydrothermal fluids from which the massive sulphides were precipitated evolved from reducing and high temperature, to oxidizing and lower temperature (Barton, 1984).

Duration of Hydrothermal Activity

The age and composition of igneous rocks suggest that high heat flow existed within the greater FLD from 365 to 346 Ma, a 19 million year time span (e.g. Piercey et al., 2006). Zircon saturation temperatures indicate that the footwall porphyritic rhyolites at Wolverine were the products of high temperature (>900°C) partial melting of continental crust (Piercey et al., in press) and this high heat flow was responsible for initiating and maintaining the Wolverine deposit hydrothermal circulation system. Magmatic heat was present beneath the basin during formation of the Wolverine deposit from at least 352 to 346 Ma (Piercey et al., in press).

Because of the low sedimentation rate for graphitic shales (e.g. 0.4-13 cm per 1,000 years; Goodfellow and Jonasson, 1987; Goodfellow and Turner, 1989) that form the immediate hanging wall to massive sulphides in parts of Wolverine (e.g. see Fig. 15), the cessation of hydrothermal activity is estimated to be on the order of 400,000 to several million years after Wolverine was formed and prior to resumption of hydrothermal activity marked by the iron formations.

Sources of Sulphur, Lead, Selenium, and Other Constituents

Sulphur Isotopes

Sulphur isotope values ($\delta^{34}S$) have been used extensively to discern the source(s) of S in sulphides and sulphates in VHMS deposits (e.g. Ohmoto, 1996, and references therein). Possible sources include 1) abiogenic S reduced from seawater, whose S isotope composition varies through time, but is strongly positive, e.g., ≈20 per mil at present (Claypool et al., 1980); 2) igneous rocks - close to 0‰ for MORB to -8‰ for rhyolite (Ohmoto and Rye, 1979); 3) biogenically reduced from seawater - typically strongly negative (Saelen et al., 1993). In most modern seafloor sulphide deposits, seawater is entrained, downdrawn and heated, and seawater sulphur is largely removed by precipitation of anhydrite in the subsurface (Seyfried and Bischoff, 1981), such that vent fluids largely contain minimal seawater sulphate (Shanks and Seyfried, 1987). In sedimented seafloor hydrothermal environments, biogenic reduction of seawater sulphate can be a major source of S, and this process can cause strong excursions in the isotopic composition of the ambient seawater sulphate, particularly in restricted anoxic basins (Saelen et al., 1993).

To date, there are 37 bulk S isotope analyses of bulk sulphide, barite, and host lithologies from the Fyre Lake, Kudz Ze Kayah, GP4F, and Wolverine deposits. The data for sulphides range from $\delta^{34}S_{CDT}$ 0.5 to 25‰ (Fig. 27; Layton-Matthews, 2005; J. Peter, unpublished data). Additionally, there is a dataset of 86 laser spot analyses of sulphides from 15 samples from the Wolverine deposit (Bradshaw, 2003; Bradshaw et al., in press) obtained using the technique of Beaudoin and Taylor (1994); however, these are not shown in Figure 27.

Fyre Lake deposit sulphides range from $\delta^{34}S_{CDT}$ 0.5 to 5.1‰ (n=3). These values reflect derivation of S largely from igneous sources, likely by leaching of magmatic sulphides within the host mafic volcanic rocks, together with a small amount of reduced seawater sulphate.

The GP4F deposit exhibits a relatively narrow range for massive sulphides and host rocks of $\delta^{34}S_{CDT}$ 11.1 to 14.6‰ (n=8). For the Kudz Ze Kayah deposit, there is a much wider overall range of 9.8 to 25.0‰, with the sulphides ranging from 9.8 to 13.0‰ (n=11), barite within the massive sulphide ranging from 23.7 to 20.0‰ (n=2); barite from the stratigraphic upper part of Kudz Ze Kayah has a $\delta^{34}S_{CDT}$ value of 21.9‰ (n=1), and argillite ranges from 13.6 to 25.0‰ (n=5), with one outlier of 3.8‰. At the time of formation of Kudz Ze Kayah in the lower Mississippian (Tournaisian), the S isotopic composition of seawater is estimated to have been ≈22‰ (Claypool et al., 1980), and the analyses of Kudz Ze Kayah barite are in close agreement with this value. The ranges for GP4F and Kudz Ze Kayah sulphides are relatively narrow and suggest derivation of S from partial reduction of seawater sulphate and S leached from volcanic rocks (e.g. Ohmoto and Goldhaber, 1997).

Sulphur isotope compositions of sulphides (bulk and laser) from Wolverine give $\delta^{34}S_{CDT}$ values of -4.9 to 18.4‰. Black shale from the Sable prospect located along strike and to the west of Wolverine has a bulk $\delta^{34}S_{CDT}$ value of 20.4‰, similar to estimates for seawater at that time (Claypool et al., 1980). However, Bradshaw (2003) and Bradshaw et al. (in press), found that pyrite within hanging-wall argillite analyzed by laser has a negative value (-5.7‰). This indicates that the heavy bulk value may reflect hydrothermal modification/introduction and may not be of sedimentary (likely negative) origin through biogenic sulphate reduction (e.g. Ohmoto and Goldhaber, 1997). However, in situ microanalyses of isolated pyrite grains can give unrepresentative analyses, and more work needs to be done to resolve these questions. Nevertheless, the wide range of sulphur isotope compositions for Wolverine mineralization, with lighter values from the tops of the lenses, and heavier values in the lower parts and in the stringer veins indicates multiple sources. Sulphur in sulphides from the tops of the lenses was likely derived mainly from bacterial reduction of seawater sulphate, whereas sulphur in the stringer veins and at the base of the lenses was likely derived from the inorganic reduction of seawater sulphate in the subsurface.

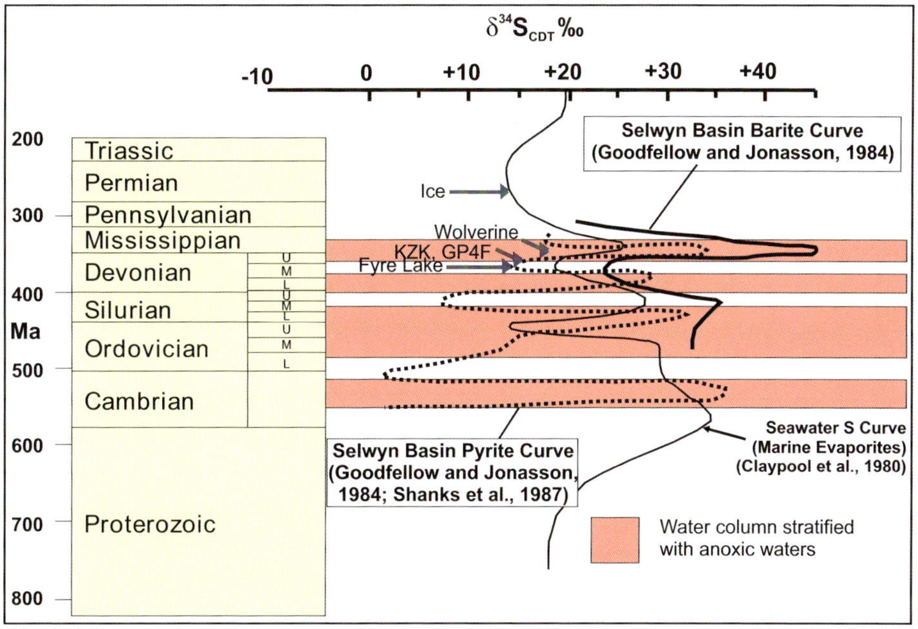

FIGURE 29. Secular variation in $\delta^{34}S$ in seawater sulphur in the northern Cordillera, as indicated by the Selwyn Basin pyrite and barite curves (Goodfellow and Jonasson, 1984; Goodfellow, 1987; Shanks et al., 1987), and for global seawater sulphate (Claypool et al., 1980). The Wolverine deposit, and perhaps also Kudz Ze Kayah and GP4F, and some other VHMS and SEDEX deposits in the northern Cordillera and elsewhere in the World occur at one of the largest $\delta^{34}S$ excursions in Earth's history suggesting the importance of anoxic ocean conditions in their genesis. Figure is modified from Goodfellow et al. (1993).

Lead Isotopes

Lead isotopes have long been used as a tracer of Pb and, by inference other metals, in VHMS deposits (e.g. Doe and Zartman, 1979). The small size of VHMS circulation cells and total fluid fluxes (in comparison to SEDEX deposits) dictates that the Pb isotopic ratios of the sulphides are sensitive to the isotopic compositions of the various lithologies in the subsurface reaction zone. There are three possible sources of Pb in the FLD VHMS deposits: 1) mantle and mantle-derived mafic rocks (basalts, gabbros); 2) sedimentary rocks; and 3) crustally-derived volcanic and igneous rocks.

Mortensen et al. (2006) report four Pb isotope analyses for each of the Fyre Lake and Ice mafic-hosted VHMS deposits, and one and two analyses for each of the felsic-hosted Wolverine and Kudz Ze Kayah VHMS deposits, respectively. Layton-Matthews (2005), Layton-Matthews et al. (in press), and J. Peter (unpub. data) provide data for GP4F and have conducted additional Pb isotope analyses for the Fyre Lake, Kudz Ze Kayah, and Wolverine deposits to examine within-deposit Pb isotope variability. Figure 28A-C presents the lead isotope analyses for the Fyre Lake, Wolverine, Kudz Ze Kayah, and GP4F deposits, and the data of Mortensen et al. (2006) for the Ice deposit. For comparison the shale growth curve of Godwin and Sinclair (1982) is given. Figure 28A plots the lead isotope compositions in $^{207}Pb/^{204}Pb$ versus $^{206}Pb/^{204}Pb$ space for FLD sulphides determined by Mortensen et al. (2006) and various other SEDEX and VHMS deposits from the northern Cordillera of similar age (see figure caption for details).

The Fyre Lake deposit gives the least radiogenic ratios in comparison to the felsic-hosted VHMS deposits (Fig. 28B). The mafic volcanic rocks that host the Fyre Lake deposit are interpreted to be derived by partial melting of the mantle (Piercey et al., 2001c), and the low Pb isotopic ratios for the sulphides reflect a mixed source of mantle-derived magmatic and crustal Pb (either from sedimentary rocks in the subsurface, or magmatic rocks that have been contaminated by continental crust) (Layton-Matthews, 2005). The Ice analyses range from very non-radiogenic (n=2), reflecting mantle-derived lead, to more radiogenic values (n=2) (Fig. 28C; Mortensen et al., 2006). The deposit and its host volcanic rocks are interpreted to have formed in a back-arc basin that developed between the Yukon-Tanana Terrane and the North American Miogeocline (Murphy et al., 2006), but this basin is thought to have been geographically close to sources of continent-derived detritus from the Yukon-Tanana Terrane on the west (Patchett and Gehrels, 1998; Murphy et al., 2006; Nelson et al., 2006; Piercey et al., 2006). Therefore, Mortensen et al. (2006) concluded that at least a minor amount of Pb in the sulphides was derived from relatively Pb-rich continental sediments, and the Pb isotopic compositions of Ice mineralization reflect a mixing line between primitive mantle and a much more radiogenic Yukon-Tanana Terrane.

Lead isotope ratios of Wolverine sulphides lie on the shale growth curve at ca. 350 Ma and trend toward much more radiogenic values than the field for felsic-hosted VHMS deposits of the FLD (Mortensen et al., 2006; hatched field shown in Fig. 28A,B). This likely reflects Pb derived from volcanic and intrusive rocks that originated from the melting of old radiogenic crustal rocks at Wolverine. For the Kudz Ze Kayah and GP4F deposits, the Pb isotopic data do not intersect the shale curve but form a cluster over a wide range of $^{207}Pb/^{204}Pb$ and narrow $^{206}Pb/^{204}Pb$ that falls above the shale curve. As pointed out by Layton-Matthews (2005) and Layton-Matthews et al. (in press), the differences between the highest $^{207}Pb/^{204}Pb$ ratios for Wolverine and Kudz Ze Kayah are statistically significant and outside analytical error. This is likely due to sourcing of greater amounts of radiogenic lead from sedimentary rocks that are present in abundance in the vicinity of Kudz Ze Kayah and GP4F.

Modeling of the Kudz Ze Kayah and GP4F data using a two-stage growth curve of Stacey and Kramers (1975) indicates a μ value of 11 and extraction ages of 3.14 Ga. This implies that the Pb in these deposits is very old and was likely sourced from intrusive and extrusive felsic rocks that formed by partial melting of 3.1 Ga crust. Detrital zircons of this or similar age (3.34 Ga) have been noted in the Coast Mountains of southeastern Alaska, rocks that are thought to be the basement for Yukon-Tanana Terrane (Gehrels and Kapp, 1998; Gehrels, 2002). Also, Devine et al. (2006) have found 2.82 Ga inherited zircon ages in the Klatsa metamorphic complex (Yukon Tanana Terrane), about 50 km south of Kudz Ze Kayah. Further, Archean Nd-Hf depleted mantle model ages for the Snowcap assemblage that underlies and forms the basement to the Yukon-Tanana Terrane (Piercey and Colpron, unpublished data) indicate Archean age crust that is likely of detrital origin and derived from the Slave craton.

Similar modeling for Wolverine suggests a different and less radiogenic reservoir (μ=12, extraction age of 1.7 Ga; Layton-Matthews, 2005; Layton-Matthews et al., in press). Thus, hydrothermal fluids predominantly sourced Pb from younger, ca. 1.7 Ga rocks. These results are in agreement with detrital zircon ages for rocks thought to represent the Yukon-Tanana Terrane basement (e.g. Gehrels and Kapp, 1998; Gehrels, 2002) and with the detrital zircon attributes of Yukon-Tanana Terrane (Colpron et al., 2006b; Murphy et al., 2006), with a main peak at ca. 1.8 to 1.7 Ga Wopmay age, and lesser peaks at older ages. Additionally, Nd model ages of granites and felsic rocks are largely Paleoproterozoic (ca. 1.8-1.6 Ga; Piercey et al., 2003) also indicating melting of a mixed 1.8 Ga crust and Archean sedimentary basement.

Selenium

The three polymetallic deposits (Kudz Ze Kayah, GP4F and Wolverine) are variably enriched in Se, with Wolverine being the most enriched (1,100 ppm average), Kudz Ze Kayah having intermediate enrichments (210 ppm weighted average), and GP4F (8 ppm average) having the lowest contents (Layton-Matthews, 2005; Layton-Matthews et al., in press). This attribute is relatively uncommon for VHMS deposits (e.g. Huston et al., 1995; Hannington et al., 1999b; Layton-Matthews et al., in press). Selenium in these deposits is resident within clausthalite (PbSe), galena, chalcopyrite, sphalerite, pyrite, and pyrrhotite.

Selenium isotope ratios ($\delta^{82/76}Se_{MERCK}$) for Kudz Ze Kayah, GP4F, and Wolverine sulphides indicate that Se originated by magmatic degassing of SeO_4, rapid reduction to H_2Se, entrainment in the mineralizing fluid, mixing with Se leached from felsic volcanic rocks (for Kudz Ze Kayah) and graphitic shales (for Wolverine), and deposition at high temperatures ($\approx$350°C) (Layton-Matthews, 2005). Selenium in the deposits is mobilized and transported under high-temperature conditions as H_2Se and concentrated in the Cu-rich portions of deposits; mixing of high-temperature fluids with ambient seawater in the brecciated margins of the mounds causes ore fluid oxidation and can also result in a clausthalite-pyrite-barite and Se-rich matrix to the breccia clasts at Kudz Ze Kayah. Recrystallization of pre-existing mineralization during zone refining by a high Se/S ore fluid results in increased selenium tenor of all sulphides. Generally, with decreasing ore fluid temperature, there is a decrease in $\Sigma Se/\Sigma S$ and Se concentration in the ore fluid. Mass balance considerations indicate that the Kudz Ze Kayah and Wolverine deposits require a large Se reservoir in addition to that from magmatic degassing and the felsic volcanic rocks, and this Se was probably provided by leaching from contemporaneous volcanic rocks and graphitic shales hosting the Kudz Ze Kayah and Wolverine deposits, respectively.

Ambient Seafloor Environment

The FLD VHMS deposits show a wide variety of characteristics reflective of deposition under widely different redox conditions. The relatively low $\delta^{34}S$ values for Fyre Lake sulphides, together with the presence of abundant disseminated magnetite of likely primary origin in the mineralization indicate that this deposit was formed under oxic ambient conditions. Similarly, the presence of a narrow hematite-quartz rock immediately overlying the mineralization at Ice indicates that this deposit was formed under similarly oxic ambient conditions.

The presence of black carbon-rich graphitic shales suggest that the ambient bottom water in the basin in which the Wolverine deposit formed may have been anoxic. Heavy sulphur isotopic values of the Wolverine mineralization and of graphitic shales (argillite) far away from the influence of the mineralizing processes (Fig. 27) provide supporting evidence that anoxic bottom waters at the site of massive sulphide deposition may have controlled the sulphur isotopic composition of the sulphides (Eastoe and Gustin, 1996). Seawater sulphate was organically reduced to sulphide in the water column and this was then fixed with venting metals.

Throughout Paleozoic time, there are several periods of time where the World's oceans were anoxic. H_2S in a reduced water column has been shown to be a critical requirement for the formation of SEDEX deposits (Fig. 29; e.g., Goodfellow and Jonasson, 1984; Goodfellow, 1987; Shanks et al., 1987; Goodfellow et al., 1993) and some VSHMS (e.g. some deposits in the Bathurst Mining Camp, northern New Brunswick: Goodfellow and Peter, 1996). Of note is the close relationship of other so-called VSHMS to periods of global anoxia (e.g. Iberian Pyrite Belt, Spain and Portugal). Wolverine, and perhaps Kudz Ze Kayah and GP4F, was also formed within a period of global anoxia, and the relatively heavy sulphur isotope compositions of the sulphides (Fig. 27) might be explained by such a source.

Further evidence for anoxia comes from the iron formations at the Wolverine deposit. The absence of bedded barite and the occurrence of Ba in carbonates, stilpnomelane, sericite, and biotite in laminated iron formations that overlie and are lateral to massive sulphides (J. Peter, unpub. data) indicate that Ba was introduced into the basin by hydrothermal fluids but was not precipitated as barite. Since Ba forms an insoluble sulphate (Blount, 1977), the lack of barite probably reflects very low sulphate contents in the ambient seawater at the site of venting. Such a low-sulphate environment may represent ambient marine conditions where bacterial sulphate reduction went to completion, or a low-sulphate environment.

Graphitic shales are not present in the immediate footwall to the Kudz Ze Kayah and GP4F deposits, but they are present in the hanging wall at the former where they are part of the Wind Lake Formation. The strongly positive sulphur isotope values for mineralization at Kudz Ze Kayah and GP4F would also suggest that anoxia may have played an important role in the mineralizing processes there also.

Deposit Morphologies and Architecture

The sulphide lens morphologies together with mineral textures at Kudz Ze Kayah, GP4F, Ice, and Fyre Lake indicate that these deposits were principally formed as discrete mounds on the paleoseafloor. Some mineralization occurs as replacements of volcaniclastics rocks in the shallow subsurface in and around the fluid upflow zones (e.g. Kudz Ze Kayah, GP4F, and Wolverine). The Fyre Lake, Wolverine, Kudz Ze Kayah, and Ice deposits have well preserved though deformed feeder vein zones (e.g. Figs. 6E, 16N,O, 23E). At Wolverine, the distribution of these veins, as well as replacement style mineralization, indicate that the Lynx and Wolverine zones likely represent two coalesced seafloor mounds (Fig. 18, inset). At Wolverine, in thinner mineralized intersections on the fringes of the Lynx and Wolverine zones, carbonaceous argillite forms the immediate footwall (Fig. 15), implying either contemporaneous onlapping by sulphide mass wasting, hydrothermal plume fallout over a protracted period of sedimentation, or subseafloor replacement (Bradshaw et al., in press). The semimassive replacement style mineralization is thought to have formed very near the paleoseafloor where the fluids were unfocused and where the rocks were not indurated.

Exploration Methods

Historical

The Fyre Lake deposit was discovered by Cassiar Asbestos Corporation in 1960 during follow-up work to locate the source of massive sulphide float boulders. Exploration efforts have included prospecting, geologic mapping, horizontal loop (Max-Min) and ground magnetics and electromagnetic (EM) surveys, and soil and silt geochemistry. Twenty-three shallow packsack (224 m) and 20 AX (1423 m) drillholes were completed between 1960 and 1991 by various companies. In 1996 and 1997, Pacific Ridge Explorations Ltd. (formerly Columbia Gold Mines Ltd.) drilled 115 holes (23,200 m), upon which the current resource is based.

The Fetish claims were staked in the Wolverine deposit area in 1973 by the Finlayson Joint Venture (Chevron Canada Ltd., Union Oil Company of Canada Ltd., and Marietta Resources International Ltd.). The exploration syndicate discovered numerous multi-element geochemical anomalies in soils over a gossanous area during a grid soil sampling survey, and conducted mapping and trenching that year. The company drilled 2 short holes and intersected low-grade Cu and Zn-sulphide mineralization. However, these claims lapsed in 1982, whereupon they were restaked by Archer, Cathro, and Associates (1981) Ltd., and further ground magnetic and Max-Min EM surveys were conducted that year.

In 1993, Equity Engineering Ltd. conducted exploration in the Wolverine deposit area (Fig. 14) after concluding that the region held promise for hosting VHMS-style mineralization. On behalf of Atna Resources Ltd., Equity staked claims over the Fetish sulphide mineralization and carried out an exploration program that consisted of geological mapping, prospecting, and rock/soil geochemistry. In 1995, the property was optioned from Atna by Westmin Resources Ltd. and by the end of that year Westmin had earned a 60% interest in the project and entered into a joint venture with Westmin as operator. The latter undertook an exploration program of detailed geological mapping, systematic grid soil geochemistry, and diamond drilling. This fieldwork led to the identification of several additional multi-element geochemical anomalies, and the intersection of massive sulphide mineralization in the Wolverine Zone. The first follow-up drillhole intersected 8.4 m of massive sulphide grading 7.63 g/t Au, 1,358.3 g/t Ag, 0.56% Cu, 3.45% Pb, and 14.22% Zn. This was followed by additional drilling in the period 1995 through 1997.

In 1996, Westmin carried out metallurgical testing on the Wolverine deposit. Results confirmed the presence of unusually high levels of Se (average of 1035 ppm Se, Expatriate Resources Ltd., 2000, unpub. data), a deleterious contaminant that could significantly impact the value of mineral concentrates. In 1997, recognition of hydrothermal alteration in drill core led to the discovery of the Sable Zone 1.6 km to the southeast of Wolverine. In early 1998, Boliden Ltd. acquired the assets of Westmin Resources Ltd. and, following the takeover, Expatriate concluded a letter of agreement to purchase Boliden's interests in the Wolverine Project; by June 1999 a new Wolverine Joint Venture (Expatriate Resources Ltd. and Atna Resources Ltd.) conducted metallurgical and marketing investigations of the ore. Drilling programs by Expatriate in 2000 expanded the existing resource and further defined the known mineralization on the Wolverine property. In 2005, Expatriate Resources (now Yukon Zinc Corporation) constructed a decline into the massive sulphides and undertook a test mining program. Further, they completed an infill drilling program on the Wolverine deposit. Currently, the company intends to bring the deposit into production in the near future.

Cominco Ltd. originally carried out a surficial geochemical survey in the vicinity of the Kudz Ze Kayah deposit in 1977 and anomalous base metal anomalies in a subsequent Geological Survey of Canada regional geochemistry silt survey (Hornbrook and Friske, 1988) attracted their attention. Prospecting by Cominco led to the discovery of a cobble of layered massive sulphide. A subsequent UTEM geophysical survey in 1993 identified an EM feature in an area with soil geochemical anomalies, and the presence of polymetallic sulphide float and felsic volcanic rocks. In 1994, the geophysical anomaly was drilled and the first hole intersected 22.5 m of massive sulphides in two zones. This outlined mineralization at Kudz Ze Kayah was named the ABM zone in recognition of A.B. Mawer, the discoverer of the mineralized cobble. That year, Cominco completed 52 diamond drillholes (8500 m) and flew 15,000 km of airborne geophysical surveys. In 1995, intensive exploration was done, which included 15,000 m of diamond drilling in 120 holes, engineering and environmental studies, and the construction of an approximately 20 km long access road to the Robert Campbell Highway. During 1996 and 1997, Cominco drill-tested airborne geophysical targets in the Kudz Ze Kayah area. To the end of 1997, 168 exploration drillholes, totaling 24,663 m, were drilled. In 1998, Cominco carried out geological work and diamond drilling in and around Kudz Ze Kayah. In 1994 a number of geophysical anomalies were identified during a regional airborne geophysical survey. Follow-up drilling in 1998 of geophysical anomaly '4F' led to the discovery of the GP4F deposit.

The Ice deposit was the first to be discovered in the Campbell Range Formation. High-grade oxide Cu mineralization was discovered in 1996 and claims were staked by Archer, Cathro, and Associates Ltd. Geological mapping, grid and reconnaissance soil geochemistry, and airborne and ground magnetic and electromagnetic surveys were conducted in 1996 and 1997. Thirty-four diamond drillholes (2,704 m) were completed in 1996 and 87 holes (7,880 m) were completed in 1997. No exploration work has been carried out since 1997. The deposit is now owned by Yukon Zinc Corporation.

In the Goal Net North area, soil geochemical surveys by Expatriate Resources Ltd. in 1998 and 1999 identified localized strong multi-element anomalies that coincided with induced polarization chargeability anomalies. Semimassive sulphides were discovered in 1999 in surface pits and this discovery was followed up by diamond drilling in 2000 and 2001 that helped to delineate the extent of mineralization. Further drilling in 2004 identified massive sulphide mineralization of the Skyblaze zone. The Thunderstruck zone in the Goal Net South area was discovered by Yukon Zinc Corporation, the successor to Expatriate Resources Ltd., during property mapping.

The Money prospect was discovered by the recognition of ferricrete and Cu-Zn-Au-Ag-bearing massive sulphide boulders in creeks.

Prospecting

The geological and regional stratigraphic framework has been vastly improved over the last few years through the work of Murphy and coworkers (e.g. Murphy et al., 2006, and references therein), and this will enable effective prospecting and exploration on a regional scale outside of the known areas of mineralization. Much exploration for VHMS deposit in the FLD remains to be done by prospecting for outcropping mineralization or associated hydrothermally altered rock. This is because the area has received relatively little exploration attention to date in comparison to the mature base metal mining camps of Canada. The initial staking rush in the mid 1990s generated by the discovery of the Kudz Ze Kayah deposit resulted in relatively few companies staking large blocks of ground. The limited resources available to these companies caused them to focus their efforts on property scale exploration, and grassroots work was curtailed. Grassroots exploration should utilize lithogeochemical methods to recognize geochemically favourable volcanic rocks (petrochemistry and chemostratigraphy), hydrothermally altered volcanic, volcaniclastic and sedimentary rocks, and exhalites.

Petrochemistry and Chemostratigraphy

VHMS deposits of then FLD are known to occur at lithological contacts between rocks of markedly different character (e.g. volcanic and sedimentary facies) or composition (e.g. rhyolite basalt). Within the FLD, rocks favourable for hosting VHMS deposits include those formed within an arc rift, back-arc rift, or at the transition from arc to back-arc magmatism.

Exploration for Cyprus- and Besshi-type deposits should be focussed on areas of high-temperature mafic magmatism as defined by lithogeochemistry. In the FLD, the presence of boninitic rocks is an indication of a depleted mantle source typical of nascent to back-arc regimes (e.g. Piercey et al., 2001c). Such rocks require high-temperature melting that is capable of initiating a stable, long-lived hydrothermal system. Boninites of the Fire Lake Formation are host to the Fyre Lake deposit in the FLD. Depleted arc tholeiites, mid-ocean ridge basalts, (MORB) and back-arc basin basalts (BABB) are also favourable high-temperature mantle melt targets (Swinden 1991, 1997; Syme et al., 1999; Wyman et al., 1999; Piercey et al., 2004). MORB can form not only at mid-ocean ridges but also in back-arc basins, and these rocks have flat REE patterns with low Nb and Th contents (Ewart et al., 1994; Hawkins and Allan, 1994; Hawkins, 1995). BABB are chemically similar to MORB but have minor negative Nb anomalies (Ewart et al., 1994; Hawkins and Allan, 1994; Hawkins, 1995) and are formed within back-arc basins.

For the polymetallic Kuroko-type VHMS and VSHMS deposits, the felsic magmas emplacement at high levels in the crust provided the magmatic heat source that initiated hydrothermal convective circulation, making high-temperature felsic volcanic rocks the best targets. Such rocks typically fall in the within-plate, A-type fields on discrimination plots and are characterized by very high HFSE (Nb, Zr > 300 ppm, Y), REE, high Zr/Sc, and negative chondrite-normalized Eu anomalies: $(La/Yb)_n < 7$, $(Gd/Yb)_n < 2$, and Y/Zr < 7. They also have higher Zr saturation temperatures than rhyolites that are not associated with the VHMS deposits (e.g. Barrie, 1995) (FLD range 750-1000ºC; mean 899ºC; n=55 (Piercey et al., 2001a, 2006; S. Piercey, unpub. data, 2002).

Felsic or felsic-sedimentary rock sequences underlain by continental crust typically have alkalic and MORB mafic rocks associated with felsic volcanic rocks. Although the VHMS deposits are commonly hosted by the high-temperature felsic and sedimentary rocks, they are usually overlain by alkalic basalts and MORB. Such an evolution from felsic to mafic magmatism is attributed to the onset of decompression melting of the mantle following thinning, rifting, and partial melting of the overlying crust (e.g. Piercey et al., 2002a). Alkalic basalts have high Nb, Th, and LREE, and are the product of lithospheric mantle melting induced by rifting, whereas MORB formed from upwelling asthenosphere due to basin opening. Within the FLD, alkalic rocks of the Wind Lake Formation crosscut and overlie the Kudz Ze Kayah and GP4F deposits (Piercey et al., 2002a). Similarly, at the Wolverine deposit, rocks of the uppermost part of the Wolverine Lake Group are basaltic flows and sills with E-MORB to BABB affinities (Piercey et al., 2002b).

Hydrothermal Alteration

The recognition of proximal focused hydrothermal upflow-related siliceous, chloritic, and sericitic alteration can further focus exploration. Both regional and local hydrothermal alteration effects can be recognized using mineral assemblage identification by X-ray diffraction, short-wavelength infrared spectrometry, conventional whole rock lithogeochemistry, and oxygen isotope mapping.

Hydrothermally altered rocks generally have high LOI, H_2O, and CO_2 contents, due to the presence of hydrated alteration minerals such as chlorite and sericite, and/or carbonates. The formation of alteration minerals typically involves reactions in which Al_2O_3 is conserved, whereas other elements such as Na_2O are lost to the hydrothermal fluids (e.g. Date et al., 1983). Unaltered rocks typically have Na_2O contents of 2 to 5 wt.%, and those with less than 2 wt.% are altered by feldspar destruction; those with more than 5 wt.% Na_2O are characterized by albite or paragonite alteration and Na_2O addition (e.g. Giorgetti et al., 2003). For these reasons, the Al_2O_3/Na_2O index of Spitz and Darling (1978) is an effective indicator of hydrothermal alteration in a sample. Typically, rocks that have been strongly hydrothermally altered rocks have very high Al_2O_3/Na_2O ratios (>50-100) and indicate a loss of alkali elements and a residual gain in Al_2O_3; unaltered rocks have ratios that are typically less than 10. Anomalously high MgO, Fe_2O_3, and FeO contents in felsic rocks commonly reflect the addition of these elements via chlorite and pyrite alteration (e.g. Hajash and Chandler, 1981). A useful graphical tool to recognize hydrothermal alteration in rocks is the alteration box plot of Large et al. (2001), because this plot indicates where least altered rocks should plot and also provides fields and vectors for altered variants. The presence of high contents of metals (Cu, Pb, Zn, Au, Ag, Co, Se, etc.) in rocks relative to unaltered equivalents suggest that the rocks are hydrothermally altered.

Exhalites

Distal hydrothermal sediments (exhalites) composed of finely layered quartz, magnetite, chlorite, carbonate, and trace sulphides are direct products of seafloor hydrothermal activity (e.g. Spry et al., 2000; Peter, 2003). These rocks are typically formed during the nascent and/or waning stages of hydrothermal venting in ancient hydrothermal systems and can be used as vectors toward mineralization (e.g. Peter and Goodfellow, 1996, 2003). Some exhalites display mineralogical and geochemical zonation around known associated VHMS deposits (see Peter, 2003, and references therein). Minerals indicative of close proximity to mineralization include magnetite, siderite, ankerite, calcite, dolomite, stilpnomelane, apatite, and pyrite. Elements that are indicative of close proximity to mineralization include Fe, Mn, Cu, Pb, Zn, Ag, Au, As, Bi, Cd, Se, Hg, Sb, Sn, In, and Tl, and others. Certain elements (e.g. Ti, Al) are indicative of detrital input, and the ratio of hydrothermal to detrital element (e.g. Fe/Ti) can provide a simple indicator of proximity to a hydrothermal vent. High hydrothermal:detrital ratios signal a close proximity to the vent. Combinations of these elements, perhaps given different weightings of importance, can be arranged to give an index that may be more effective than individual elements or ratios (e.g. Peter, 2003; Peter and Goodfellow, 2003).

Geophysics

The high magnetic susceptibility of some massive sulphide deposits (e.g. Fyre Lake - abundant pyrrhotite; Ice - abundant magnetite), imparted by the presence of pyrrhotite and magnetite indicates that ground and airborne electromagnetics would be highly effective exploration methods in the FLD. The Kudz Ze Kayah deposit is indicated by both a strong magnetic and electromagnetic anomaly (Holroyd and Klein, 1998) although the Wolverine deposit does not have a strong magnetic response. However, the iron formations that extend from the Fisher zone southeast to Wolverine and beyond have a strong magnetic signature that is readily identified by airborne surveys (Yukon Zinc Corp., unpub. data, 2000; R. Shives, pers. comm., 2007). Further, altered hanging-wall rocks at Wolverine are identified by radiometric techniques. The disseminated to massive nature of the sulphides suggests that induced polarization methods should be effective.

Surficial Geochemistry

The original Fetish claims (now the Wolverine deposit) and the Fisher Zone were found using soil geochemistry, and the Kudz Ze Kayah deposit area was outlined by a stream sediment Pb-Zn anomaly in a geochemical survey undertaken by the GSC (Hornbrook and Friske, 1988). The GP4F deposit has been metamorphosed to higher metamorphic grade and is known to contain gahnite that probably have formed by desulphidation of sphalerite during metamorphism (Spry and Scott, 1986). Drift prospecting for gahnite and other diagnostic minerals may therefore be effective in focusing exploration efforts in the FLD. A regional till geochemistry survey and Quaternary geology investigations by Bond and Plouffe (2002) showed ice flow patterns during glacial maxima in the FLD trend towards the west-northwest. Basal till is widespread across plateau surfaces and glaciofluvial sand and gravel dominate in low-lying areas. Multi-element anomalies are present in till near known mineralized zones, and till down-ice from the Kudz Ze Kayah is enriched in Pb and Zn. In addition, they found base metal anomalies northwest of Wolverine Lake and southwest of Finlayson Lake where no mineral occurrences are known.

Knowledge Gaps

Although there has been a plethora of new knowledge gained on the tectonic setting, geological setting, and genesis of the VHMS deposits of the FLD, there are a number of unresolved questions. 1) What are the volcanic and sedimentary settings of the deposits? Detailed reconstructions of the volcanic stratigraphy are needed. 2) What is the exact nature of the long time span between Kudz Ze Kayah aged mineralization (ca. 362 Ma) and Wolverine aged mineralization (ca. 346 Ma)? Recent age dating in the Wolverine area indicates that hydrothermal circulation was active during this time span within a condensed stratigraphic section (Piercey et al., in press). 3) What role did anoxia play in the formation of the polymetallic VHMS and VSHMS deposits of the FLD? 4) What role did magmatic fluids and volatiles play in the development of these deposits? Did they contribute metals to the ore-forming system? 5) Additional information is needed on the nature and extent of regional scale hydrothermal alteration in the FLD, particularly detailed studies of alteration associated with the hydrothermal upflow zones associated with different VHMS deposits. An understanding of the three-dimensional distribution and timing of alteration is needed along with the isotopic and geochemical attributes of this alteration. 6) What is the exact nature and distribution of Se enrichment in polymetallic VHMS deposits in FLD, and how does this relate to such events responsible for the enrichment of Se in base metal deposits in the Selwyn Basin (e.g. Wolf deposit)? 7) Finally, the applicability and effectiveness of unconventional surficial methods (e.g. heavy minerals in tills, mobile metal ions, enzyme leaches, and partial extractions) in exploration need to be evaluated.

Acknowledgements

Funding came from Geological Survey of Canada (GSC) Proposal Approval System Project 1017; GSC project X15 of the CCGKN Program and project TG6007 of the Targeted Geoscience Initiative III Program are also gratefully acknowledged for time and resources to write this manuscript. Discussions with Wayne Goodfellow and Don Murphy are gratefully acknowledged. Ingrid Kjarsgaard conducted some of the mineralogical and microprobe analyses summarized herein. Constructive reviews were provided by Fernando Tornos, Thomas Monecke, and the Editor, Wayne Goodfellow. This is GSC Contribution Number 20060409. DLM gratefully acknowledges financial support from NSERC and the Mineralogical Association of Canada through postgraduate scholarships. Steve Piercey's research in the FLD has been funded by the Yukon Geological Survey, the Geological Survey of Canada, Yukon Zinc Corporation, Atna Resources Ltd., and a Natural Sciences and Engineering Research Council of Canada Discovery Grant. Bradshaw, Peter, and Layton-Matthews also thank Expatriate Resources Ltd., Yukon Zinc Corporation, and Teck Cominco Ltd. for their generous logistical and/or financial support.

References

Abbott, J.G., and Turner, R.J.W., 1991, Character and paleotectonic setting of Devonian stratiform sediment-hosted Zn, Pb, Ba deposits, Macmillan fold belt, Yukon, *in* Abbott, J.G., and Turner, R.J.W., eds., Mineral Deposits of the Northern Canadian Cordillera, Yukon, Northeastern British Columbia (Field Trip 14): Geological Survey of Canada, Open File 2169, p. 99-136.

Alldrick, D., Nelson, J.L., Gabites, J.E., and Childe, F., 2003, Lead isotopes from the Ecstall greenstone belt, Coast Mountains, northwest British Columbia - significance for global VMS exploration and crustal evolution of the Yukon-Tanana Terrane: Geological Association of Canada/Mineralogical Association of Canada, Program with Abstracts, v. 28, abstract 620.

Baknes, M., 1997, Money, *in* Yukon Exploration and Geology 1996: Whitehorse, Yukon, Yukon Geology Program, p. 38-40.

Barrie, C.T., 1995, Zircon thermometry of high-temperature rhyolites near volcanic-associated massive sulfide deposits, Abitibi subprovince, Canada: Geology, v. 23, p. 169-172.

Barton, P.B., Jr., 1978, Ore textures involving sphalerite from the Furutobe mine, Akita Prefecture, Japan: Mining Geology, v. 28, p. 293-300.

Barton, P.B., Jr., 1984, Redox reactions in hydrothermal fluids, *in* Henley, R.W., Truesdell, A.H., and Barton, P.B., Jr., eds., Fluid-Mineral Equilibria in Hydrothermal Systems: Economic Geology, Reviews in Economic Geology, v. 1, p. 99-113.

Barton, P.B., Jr., and Toulmin, P., III, 1966, Phase relations involving sphalerite in the Fe-Zn-S system: Economic Geology, v. 61, p. 815-849.

Barton, P.B., Jr., Skinner, B.J., and Barnes, H.L., 1979, Sulfide mineral stabilities, *in* Barnes, H.L., ed., Geochemistry of Hydrothermal Ore Deposits, 2nd edition: John Wiley and Sons, p. 278-403.

Becker, T.C., 1997, Geological mapping, prospecting, soil geochemistry, and airborne geophysical surveys on the Ice property: Assessment Report 093839 written for Expatriate Resources. Ltd.

——— 1998, Assessment report describing geological mapping, prospecting, soil geochemistry, and diamond drilling on the Ice property in the Watson Lake Mining District, Yukon Territory: Assessment Report 093869 written for Expatriate Resources. Ltd.

Beaudoin, G., and Taylor, B.E., 1994, High precision and spatial resolution sulfur isotope analysis using miles laser microprobe: Geochimica et Cosmochimica Acta, v. 58, p. 5055-5063.

Blount, C.N., 1977, Barite solubilities and thermodynamic properties up to 300°C and 1400 bars: American Mineralogist, v. 62, p. 942-957.

Bond, J.D., and Plouffe, A., 2002, Finlayson Lake Targeted Geoscience Initiative (southeastern Yukon). Part 2: Quaternary geology and till geochemistry, *in* Emond, D.S., Weston, L. H., and Lewis, L.L., eds., Yukon Exploration and Geology 2001: Yukon Region, Indian and Northern Affairs Canada, p. 209 228 plus 10 fold-out maps.

Boulton, A., 2002, GP4F polymetallic volcanic-hosted massive sulphide (VHMS) deposit, Finlayson Lake District, Yukon Territory: B.Sc. Thesis, University of Victoria, Victoria, British Columbia, 56 p.

Bowers, T.S., Von Damm, K.L., and Edmond, J.M., 1985, Chemical evolution of mid-ocean ridge hot springs: Geochimica et Cosmochimica Acta, v. 49, p. 2239-2252.

Bradshaw, G.D., 2003, Geology and Genesis of the Wolverine Polymetallic Volcanic Rock-Hosted Massive Sulphide (VHMS) Deposit, Finlayson Lake District, Yukon, Canada: M.Sc. Thesis, University of British Columbia, Vancouver, British Columbia, 181 p.

Bradshaw, G.D., Tucker, T., Peter, J.M., Paradis, S., and Rowins, S.M., 2001, Geology of the Wolverine polymetallic volcanic-hosted massive sulphide deposit, Finlayson Lake district, Yukon Territory, Canada, *in* Emond, D.S., and Weston, L.H., eds., Yukon Exploration and Geology 2000: Indian and Northern Affairs Canada, p. 269-287.

Bradshaw, G.D., Rowins, S.M., Peter, J., and Taylor, B., in press, Genesis of the Wolverine deposit, Finlayson Lake district, Yukon Territory: Mineralogical, mineral chemical, fluid inclusion, and sulfur isotopic evidence: Economic Geology.

Childe, F., 1997, Timing and Tectonic Setting of Volcanogenic Massive Sulphide Deposits in British Columbia: Constraints from U/Pb Geochronology, Radiogenic Isotopes and Geochemistry: Ph.D. thesis, University of British Columbia, Vancouver, British Columbia, 298 p.

Church, S.E., Gray, J.E., Delevaux, M.H., and LeHuray, A.P., 1987, Pb-isotope signatures of Devonian-Mississippian massive sulphide deposits in Alaska and their significance to mineral exploration, *in* Elliott, I.L., and Smee, B.W., eds., Geoexpo/86-Exploration in the North American Cordillera: Rexdale, Ontario, Association of Exploration Geochemists, p. 132-141.

Claypool, G.E., Holser, W.T., Kaplan, I.R., Sakai, H., and Zak, I., 1980, The age curves of sulfur and oxygen isotopes in marine sulfate and their mutual interpretation: Chemical Geology, v. 28, p. 199-260.

Colpron, M., and Price, R.A., 1995, Tectonic significance of the Kootenay terrane, southeastern Canadian cordillera: an alternative model: Geology, v. 23, p. 25-28.

Colpron, M., Nelson, J.L., and Murphy, D.C., 2006a, A tectonostratigraphic framework for the pericratonic terranes of the Northern Cordillera, *in* Colpron, M., and Nelson, J., eds., Paleozoic Evolution and Metallogeny of Pericratonic Terranes at the Ancient Pacific Margin of North America: Geological Association of Canada, Special Publication 45, p. 1-24.

Colpron, M., Mortensen, J.K., Gehrels, G.E., and Villeneuve, M., 2006b, Basement complex, Carboniferous magmatism and Paleozoic deformation in Yukon-Tanana Terrane of central Yukon: Field, geochemical and geochronological constraints from Glenlyon map area, *in* Colpron, M., and Nelson, J., eds., Paleozoic Evolution and Metallogeny of Pericratonic terranes at the Ancient Pacific Margin of North America: Geological Association of Canada, Special Publication 45, p. 131-152.

Columbia Gold Mines Ltd., 1999, Fyre Lake Project: Joint venture participation sought: unpublished internal report and news release No. 99-02, March 1, 1999.

Crerar, D.A., and Barnes, H.L., 1976, Ore solution chemistry V. Solubilities of chalcopyrite and chalcocite assemblages in hydrothermal solution at 200°C to 350°C: Economic Geology, v. 71, p. 772-794.

Creaser, R.A., Goodwin, B.J.A.S., and Erdmer, P., 1999, Geochemical and Nd isotopic constraints for the origin of eclogite protoliths, northern Cordillera; implications for the Paleozoic tectonic evolution of the Yukon-Tanana Terrane: Canadian Journal of Earth Sciences, v. 36, p. 1697-1709.

Date, J., Watanabe, Y., and Saeki, Y., 1983, Zonal alteration around the Fukazawa Kuroko deposits, Akita Prefecture, northern Japan: Economic Geology, Monograph 5, p. 365-386.

Devine, F., Carr, S.D., Murphy, D.C., Davis, W.J., Smith, S., and Villeneuve, M., 2006, Geochronological and geochemical constraints on the origin of the Klatsa metamorphic complex: Implications for Early Mississippian high-pressure metamorphism within Yukon-Tanana Terrane, *in* Colpron, M., and Nelson, J., eds., Paleozoic Evolution and Metallogeny of Pericratonic terranes at the Ancient Pacific Margin of North America: Geological Association of Canada, Special Publication 45, p. 107-131.

Doe, B.R., and Zartman, R.E., 1979, Plumbotectonics I, the Phanerozoic, *in* Barnes, H. L., ed., Geochemistry of Hydrothermal Ore Deposits, 2nd edition: John Wiley and Sons, p. 22-69.

Dusel-Bacon, C., Hopkins, M.J., Mortensen, J.K., Dashevsky, S.S., Bressler, J.R., and Day, W.C., 2006, Paleozoic tectonic and metallogenic evolution of the pericratonic rocks of east-central Alaska and adjacent Yukon, *in* Colpron, M., and Nelson, J., eds., Paleozoic Evolution and Metallogeny of Pericratonic Terranes at the Ancient Pacific Margin of North America: Geological Association of Canada, Special Publication 45, p. 25-74.

Eastoe, C.J., and Gustin, M.M., 1996, Volcanogenic massive sulfide deposits and anoxia in the Phanerozoic oceans: Ore Geology Reviews, v. 10, p. 179-197.

Eaton, D., and Pigage, L.C., 1997, Geological mapping, prospecting, soil geochemistry, geophysical surveys and diamond drilling on the Ice property: unpublished internal report for Archer, Cathro and Associates (1981) Ltd.

Ewart, A., Bryan, W.B., Chappell, B.W., and Rudnick, R.L., 1994, Regional geochemistry of the Lau-Tonga arc and backarc systems, *in* Hawkins, J.W., Parson, L.M., Allan, J.F., Abrahamsen, N., Bednarz, U., Blanc, G., Bloomer, S.H., Boe, R., Bruns, T.R., Bryan, W.B., Chaproniere, G. C. H., Clift, P.D., Ewart, A., Fowler, M.G., Hergt, J.M., Hodkinson, R.A., Lavoie, D.L., Ledbetter, J.K., MacLeod, C.J., Nilsson, K., Nishi, H., Pratt, C.E., Quinterno, P.J., Reynolds, R.R., Rothwell, R.G., Sager, W. W., Schoeps, D., Soakai, S., and Styzen, M.J., eds., Proceedings of the Ocean Drilling Program: Scientific Results, v. 135, p. 385-425.

Expatriate Resources Limited, 1999, Ice deposit resource estimate, news release, November 30, 1998, 2 p.

Expatriate Resources Limited, 2000, Expatriate acquires Kudz Ze Kayah and creates major new zinc-silver exploration and development project. News release, March 21, 2000, 5 p., available at: http://www.yukonzinc.com/Pa037848.pdf.

Foreman, I., 1998, The Fyre Lake project 1997: Geology and mineralization of the Kona massive sulphide deposit, Yukon Exploration and Geology, 1997: Indian and Northern Affairs Canada, p. 105-113.

Fox, J.S., 1984, Besshi-type volcanogenic sulphide deposits - a review: Canadian Institute of Mining and Metallurgy Bulletin, v. 77, p. 57-67.

Gehrels, G.E., 2002, Detrital zircon geochronology of the Taku Terrane, Southeast Alaska: Canadian Journal of Earth Sciences, v. 39, p. 921-931.

Gehrels, G.E., and Kapp, P.A., 1998, Detrital zircon geochronology and regional correlation of metasedimentary rocks in the coast mountains, southeastern Alaska: Canadian Journal of Earth Sciences, v. 35, p. 269-279.

Giorgetti, G., Monecke, T., Kleeberg, R., and Herzig, P.M., 2003, Intermediate sodium-potassium mica in hydrothermally altered rocks of the Waterloo Deposit, Australia; a combined SEM-EMP-XRD-TEM study: Contributions to Mineralogy and Petrology, v. 146, p. 159-173.

Godwin, C.I., and Sinclair, A.J., 1982, Average lead isotope growth curves for shale-hosted zinc-lead deposits, Canadian Cordillera: Economic Geology, v. 77, p. 675-690.

Goodfellow, W.D., 1987, Anoxic stratified oceans as a source of sulphur in sediment-hosted stratiform Zn-Pb deposits (Selwyn Basin, Yukon, Canada): Chemical Geology, v. 65, p. 359-382.

——— 2001, Genesis of massive sulphide deposits in the Bathurst Mining Camp, northern New Brunswick, Canada: North Atlantic Minerals Symposium, Extended Abstracts Volume, p. 51-57.

Goodfellow, W.D., and Jonasson, I.R., 1984, Ocean stagnation and ventilation defined by $\delta^{34}S$ secular trends in pyrite and barite, Selwyn Basin, Yukon: Geology, v. 12, p. 583-586.

——— 1986, Environment of formation of the Howards Pass (XY) Zn-Pb deposit, Selwyn Basin, Yukon, *in* Morin, J. A., ed., Mineral Deposits of Northern Cordillera: Canadian Institute of Mining and Metallurgy, Special Volume 37, p. 19-50.

Goodfellow, W.D., and McCutcheon, S.R., 2003, Geologic and genetic attributes of volcanic sediment-hosted massive sulfide deposits of the Bathurst Mining Camp, northern New Brunswick - a Synthesis, *in* Goodfellow, W.D., McCutcheon, S.R., and Peter, J.M., eds., Massive Sulfide Deposits of the Bathurst Mining Camp, New Brunswick, and northern Maine: Economic Geology, Monograph 11, p. 245-310.

Goodfellow, W.D., and Peter, J.M., 1996, Sulphur isotope composition of the Brunswick No. 12 massive sulphide deposit, Bathurst Mining Camp, New Brunswick: implications for ambient environment, sulphur source, and ore genesis: Canadian Journal of Earth Sciences Canadian Journal of Earth Sciences, v. 33, p. 231-251.

Goodfellow, W.D., and Turner, R.J.W., 1989, Sulfur isotope variability in sediment-hosted massive sulphide deposits as determined using the ion microprobe SHRIMP: 1. An example from the Rammelsberg orebody - a discussion: Economic Geology, v. 84, p. 451-452.

Goodfellow, W.D., Lydon, J.W., and Turner, R.J.W., 1993, Geology and genesis of stratiform sediment-hosted (SEDEX) zinc-lead-silver sulphide deposits, *in* Kirkham, R.V., Sinclair, W.D., Thorpe, R.I., and Duke, J.M., eds., Mineral Deposit Modeling: Geological Association of Canada, Special Paper 40, p. 201-251.

Hajash, A, and Chandler, G.W. 1981, An experimental investigation of high-temperature interactions between seawater and rhyolite, andesite, basalt and peridotite. Contributions to Mineralogy and Petrology, v. 78, p. 240-254.

Hannington, M.D., Bleeker, W., and Kjarsgaard, I., 1999a, Sulfide mineralogy, geochemistry, and ore genesis of the Kidd Creek Deposit; Part I, North, Central and South orebodies: Economic Geology, Monograph 10, p. 163-224.

Hannington, M.D., Bleeker, W., and Kjarsgaard, I., 1999b, Sulfide mineralogy, geochemistry, and ore genesis of the Kidd Creek Deposit; Part II, The bornite zone: Economic Geology, Monograph 10, p. 225-266.

Hawkins, J.W., 1995, Evolution of the Lau Basin - Insights from ODP Leg 135, *in* Taylor, B., and Natland, J., eds., Active Margins and Marginal Basins of the Western Pacific: American Geophysical Union, Geophysical Monograph, v. 88, p. 125-173.

Hawkins, J.W., 1994, Petrologic evolution of Lau Basin sites 834 through 839, *in* Allan, J.F., Hawkins, J.W., Parson, L.M., Allan, J.F., Abrahamsen, N., Bednarz, U., Blanc, G., Bloomer, S.H., Boe, R., Bruns, T.R., Bryan, W.B., Chaproniere, G.C.H., Clift, P.D., Ewart, A., Fowler, M.G., Hergt, J.M., Hodkinson, R.A., Lavoie, D.L., Ledbetter, J.K., MacLeod, C.J., Nilsson, K., Nishi, H., Pratt, C.E., Quinterno, P.J., Reynolds, R.R., Rothwell, R.G., Sager, W.W., Schoeps, D., Soakai, S., and Styzen, M.J., eds., Proceedings of the Ocean Drilling Program, Scientific Results: v. 135, p. 427-470.

Helgeson, H.C., 1969, Thermodynamics of hydrothermal systems at elevated temperatures and pressures: American Journal of Science, v. 267, p. 729-804.

Holroyd, R.W., and Klein, J., 1998, Geophysical aspects of the Kudz Ze Kayah massive sulfide discovery, southeastern Yukon, *in* Walton, G., and Jambor, J., eds., Pathways 98: Yukon Chamber of Mines and Society of Economic Geologists Inc., Extended Abstracts Volume, p. 181-184.

Hornbrook, E.H.W., and Friske, P.W.B., 1988, Regional Stream and Water Geochemical Reconnaissance Data, southeastern Yukon: Geological Survey of Canada, Open File 1648, 70 p.

Hunt, J.A., 1998, The setting of volcanogenic massive sulphide deposits in the Finlayson Lake district, *in* Yukon Exploration and Geology 1997, p. 99-104.

Hunt, J.A., 2002, Volcanic-associated massive sulphide (VMS) mineralization in the Yukon-Tanana Terrane and coeval strata of the North American miogeocline, *in* The Yukon and Adjacent Areas: Indian and Northern Affairs Canada, Bulletin 12, 107 p.

Huston, D.L., Sie, S.H., Suter, G.F., Cooke, D.R., and Both, R.A., 1995, Trace elements in sulfide minerals from eastern Australian volcanic-hosted massive sulfide deposits: 1. Proton microprobe analyses of pyrite, chalcopyrite, and sphalerite, and 2. Selenium levels in pyrite: Comparison with delta S-34 values and implications for the source of sulfur in volcanogenic hydrothermal systems: Economic Geology, v. 90, p. 1167-1196.

Large, R.R., Gemmell, J.B., Paulick, H., and Huston, D.L., 2001, The alteration box plot: A simple approach to understanding the relationship between alteration mineralogy and lithogeochemistry associated with volcanic-hosted massive sulfide deposits: Economic Geology, v. 96, p. 957-971.

Layton-Matthews, D., 2005, Genesis of Se in Volcanic-Hosted Massive Sulfide Systems: Wolverine, Kudz Ze Kayah, and GP4F VMS Deposits, Finlayson Area, Yukon, Canada: Ph.D. Thesis, University of Toronto, Toronto, Ontario, 247 p.

Layton-Matthews, D., Peter, J.M., Scott, S.D., Paradis, S., Tucker, T.L., and Bradshaw, G.D., 2001, Distribution of Se and other trace elements in massive sulphides of the Wolverine Zn-Pb-Cu-Ag-Au volcanic-hosted massive sulphide deposit, Finlayson Lake district, Yukon, Canada: Geological Association of Canada/Mineralogical Association of Canada Joint Annual Meeting, St John's, Newfoundland, May 2001, Abstracts with Program, v. 26, p. 83.

Layton-Matthews, D., Peter, J.M., Scott, S.D., Leybourne, M.I., and Paradis, S., in press, Selenium in sulfide minerals from the Kudz Ze Kayah, Wolverine, and GP4F volcanogenic massive sulfide deposits: Distribution and implications for deposit genesis: Economic Geology.

Logan, J.M., Drobe, J.R., and McClelland, W.C., 2000, Geology of the Forrest Kerr-Mess Creek Area, northwestern British Columbia: British Columbia Ministry of Energy and Mines, Bulletin 104, 164 p.

MacRobbie, P., and Holroyd, R.W., 2000, 1998 and 1999 Year End Report, TAG Property (KZK Project). Cominco Ltd. Exploration Canada: unpublished Internal Report, 41 p.

Marcoux, E., 1998, Lead isotope systematics of the giant massive sulphide deposits in the Iberian Pyrite Belt: Mineralium Deposita, v. 33, p. 45-58.

Mortensen, J.K., 1992a, New U-Pb ages for the Slide Mountain Terrane in southeastern Yukon Territory, *in* Radiogenic and Isotopic Studies, Report 5 (91-2: Geological Survey of Canada p. 167-173

——— 1992b, Pre-mid-Mesozoic tectonic evolution of the Yukon-Tanana Terrane, Yukon and Alaska: Tectonics, v. 11, p. 836-853.

Mortensen, J.K., and Jilson, G.A., 1985, Evolution of the Yukon-Tanana Terrane; evidence from southeastern Yukon Territory: Geology, v. 13, p. 806-810.

Mortensen, J.K., Dusel-Bacon, C., Hunt, J., and Gabites, J., 2006, Lead isotopic constraints on the metallogeny of middle and late Paleozoic syngenetic base-metal occurrences in the Yukon-Tanana and Slide Mountain/Seventymile terranes and adjacent portions of the North American miogeocline, *in* Colpron, M., and Nelson, J., eds., Paleozoic Evolution and Metallogeny of Pericratonic Terranes at the Ancient Pacific Margin of North America: Geological Association of Canada, Special Publication 45, p. 261-279.

Murphy, D.C., 1997, Preliminary geology map of Grass Lakes area, Pelly Mountains, southeastern Yukon (105G/7): Yukon Geology Program Open File, v. 1997-3.

Murphy, D.C., 1998, Stratigraphic framework for syngenetic mineral occurrences, Yukon-Tanana Terrane south of Finlayson Lake: a progress report, Yukon Exploration and Geology 1997: Indian and Northern Affairs Canada, p. 51-58.

——— 2001, Yukon-Tanana Terrane in southwestern Frances Lake area, southeastern Yukon, *in* Yukon Exploration and Geology 2000: Indian and Northern Affairs Canada, p. 217-233.

Murphy, D.C., and Piercey, S.J., 1998, Preliminary geological map of northern Wolverine Lake area, Pelly Mountains, southeastern Yukon (NTS 105G/8, north half): Indian and Northern Affairs Canada, Open File 1998-4, 1:50,000 scale.

——— 1999a, Finlayson Project: Geological evolution of Yukon-Tanana Terrane and its relationship to Campbell Range belt, northern Wolverine Lake map area, southeastern Yukon, *in* Roots, C. F., and Emond, D. S., eds., Yukon Exploration and Geology 1998: Indian and Northern Affairs Canada, p. 47-62.

——— 1999b, Geological map of Wolverine Lake area (105G/8), Pelly Mountains, southeastern Yukon: Indian and Northern Affairs Canada, Open File 1999-3 (1:50 000 scale).

——— 1999c, Geological map of Finlayson Lake area, southeast quarter (105G/7, 8 and parts of 1,2 and 9), southeastern Yukon: Indian and Northern Affairs Canada, Open File 1999-4 (1:100 000 scale).

——— 2000a, The Money Creek Thrust, Yukon-Tanana Terrane, southeastern Yukon; intra-terrane shortening by thrust re-activation of a syn-depositional fault, *in* Abstracts with Programs: Geological Society of America, Cordilleran Section and associated societies, 96th annual meeting; Geological Society of America, v. 32, p. 57.

——— 2000b, Syn-mineralization faults and their re-activation, Finlayson Lake massive sulphide belt, Yukon-Tanana Terrane, southeastern Yukon, *in* Emond, D.S., and L.H. Weston, eds., Yukon Exploration and Geology 1999: p. 55-66.

Murphy, D.C., and Timmerman, J.R.M., 1997, Preliminary geology of the northeast third of Grass Lakes map area (105G/7), Pelly Mountains, southeastern Yukon, *in* Yukon Exploration and Geology 1996: p. 62-73.

Murphy, D.C., Colpron, M., Roots, C.F., Gordey, S.P., and Abbott, J.G., 2002, Finlayson Lake Targeted Geoscience Initiative (southeastern Yukon), Part 1: Bedrock geology, *in* Emond, D.S., Weston, L.H., and Lewis, L.L., eds., Yukon Exploration and Geology 2001: Indian and Northern Affairs Canada, p. 189-207.

Murphy, D.C., Mortensen, J. K., Piercey, S. J., Orchard, M. J., and Gehrels, G. E., 2006, Mid-Paleozoic to early Mesozoic tectonostratigraphic evolution of Yukon-Tanana and Slide Mountain terranes and affiliated overlap assemblages, Finlayson Lake massive sulphide district, southeastern Yukon, *in* Colpron, M., and Nelson, J., eds., Paleozoic Evolution and Metallogeny of Pericratonic Terranes at the Ancient Pacific Margin of North America: Geological Association of Canada, Special Publication 45, p. 75-106.

Nelson, J.L., 1993, The Sylvester Allochthon: upper Paleozoic marginal-basin and island-arc terranes in northern British Columbia: Canadian Journal of Earth Sciences, v. 30, p. 631-643.

Nelson, J., and Colpron, M., 2007, Tectonics and metallogeny of the British Columbia, Yukon and Alaskan Cordillera, 1.8 Ga to present, *in* Goodfellow, W. D., ed., Mineral Deposits of Canada: A Synthesis of Major Deposit-types, District Metallogeny, the Evolution of Geological Provinces, and Exploration Methods: Special Publication 5, Mineral Deposits Division, Geological Association of Canada, p. 755-791.

Nelson, J.L., and Bradford, J., 1993, Geology of the Midway-Cassiar area, northern British Columbia (104O, 104P): Ministry of Energy, Mines and Petroleum Resources Bulletin 83, 94 p.

Nelson, J.L., Paradis, S., Christensen, J., and Gabites, J., 2002, Canadian Cordilleran Mississippi valley-type deposits; a case for Devonian-Mississippi back-arc hydrothermal origin: Economic Geology, v. 97, p. 1013-1036.

Nelson, J.L., Colpron, M., Piercey, S.J., Dusel-Bacon, C., Murphy, D.C., and Roots, C.F., 2006, Paleozoic tectonic and metallogenetic evolution of pericratonic terranes in Yukon, northern British Columbia and eastern Alaska, *in* Colpron, M., and Nelson, J., eds., Paleozoic Evolution and Metallogeny of Pericratonic Terranes at the Ancient Pacific Margin of North America: Geological Association of Canada, Special Publication 45, p. 323-360.

Ohmoto, H., 1996, Formation of volcanogenic massive sulfide deposits: the Kuroko perspective: Ore Geology Reviews, v. 10, p. 135-177.

Ohmoto, H., and Rye, R.O., 1979, Isotopes of Sulfur and Carbon, *in* Barnes, H.L, ed., Geochemistry of Hydrothermal Ore Deposits, 2nd edition: John Wiley and Sons, Inc. p. 509-567.

Ohmoto, H., and Goldhaber, M.B., 1997, Sulfur and carbon isotopes, *in* Barnes, H.L., ed., Geochemistry of Hydrothermal Ore Deposits, 3rd edition: John Wiley and Sons, Inc., p. 517-611.

Okulitch, A.V., 2002, Geological time chart, 2002: Geological Survey of Canada, Open File 3040.

Paradis, S., and Nelson, J.L., 2000, The Devonian-Mississippian metallogenetic history of Western Canada, from leading plate margin to continent interior, *in* Gemmell, J.B., and Pongratz, J. eds., Volcanic Environments and Massive Sulfide Deposits; Program and Abstracts: International

Conference and Field Meeting (3), University of Tasmania, Hobart, Australia, p. 139.

Paradis, S., Nelson, J.L., and Irwin, S.E.B., 1998, Age constraints on the Devonian shale-hosted Zn-Pb-Ba deposits, Gataga district, northeastern British Columbia, Canada: Economic Geology, v. 93, p. 184-200.

Patchett, P.J., and Gehrels, G.E., 1998, Continental influence on Canadian Cordilleran terranes from Nd isotopic study, and significance for crustal growth processes: Journal of Geology, v. 106, p. 269-280.

Peter, J.M., 2001, Brief notes on alteration mapping of the Wolverine deposit: Unpublished report to Expatriate Resources Ltd., 3 p.

Peter, J.M., 2003, Ancient iron-rich metalliferous sediments (iron formations): their genesis and use in the exploration for stratiform base metal sulphide deposits, with examples from the Bathurst Mining Camp, *in* Lentz, D.R., ed., Geochemistry of Sediments and Sedimentary Rocks: Secular Evolutionary Considerations to Mineral Deposit-Forming Environments: Geological Association of Canada, GEOtext v. 4, p. 145-173.

Peter, J.M., and Goodfellow, W.D., 1996, Mineralogy, bulk and rare earth element geochemistry of massive sulphide-associated hydrothermal sediments of the Brunswick Horizon, Bathurst Mining Camp, New Brunswick: Canadian Journal of Earth Sciences, v. 33, p. 252-283.

——— 2003, Hydrothermal sedimentary rocks of the Heath Steele Belt, Bathurst Mining Camp, New Brunswick. 3. Application of hydrothermal sediment mineralogy and mineral and bulk composition to the exploration for concealed massive sulfide mineralization, *in* Goodfellow, W.D., McCutcheon, S.R., and Peter, J.M., eds., Massive Sulphide Deposits of the Bathurst Mining Camp, New Brunswick, and Northern Maine: Economic Geology, Monograph 11, p. 417-433.

Peter, J.M., Kjarsgaard, I.M., and Goodfellow, W.D., 2003a, Hydrothermal sedimentary rocks of the Heath Steele Belt, Bathurst Mining Camp, New Brunswick: Part 1. Mineralogy and mineral chemistry, *in* Goodfellow, W.D., McCutcheon, S.R., and Peter, J.M., eds., Massive Sulphide Deposits of the Bathurst Mining Camp, New Brunswick, and Northern Maine: Economic Geology, Monograph 11, p. 361-390.

Peter, J.M., Goodfellow, W.D., and Doherty, W., 2003b, Hydrothermal sedimentary rocks of the Heath Steele Belt, Bathurst Mining Camp, New Brunswick: Part 2. Bulk and rare earth element geochemistry and implications for origin, *in* Goodfellow, W. D., McCutcheon, S. R., and Peter, J. M., eds., Massive Sulphide Deposits of the Bathurst Mining Camp, New Brunswick, and Northern Maine: Economic Geology, Monograph 11, p. 391-415.

Piercey, S.J., 2001, Petrology and Tectonic Setting of Felsic and Mafic Volcanic and Intrusive Rocks in the Finlayson Lake Volcanic-Hosted Massive Sulphide (VHMS) District, Yukon, Canada; A record of Mid-Paleozoic Arc and Back-Arc Magmatism and Metallogeny: Ph.D. Thesis, University of British Columbia, Vancouver, British Columbia, 305 p.

Piercey, S.J., and Murphy, D.C., 2000, Stratigraphy and regional implications of unstrained Devono-Mississippian volcanic rocks in the Money Creek thrust sheet, Yukon-Tanana Terrane, Southeastern Yukon, *in* Emond, D. S., and Weston, L. H., eds., Yukon Exploration and Geology 1999: Indian and Northern Affairs Canada, p. 67-78.

Piercey, S.J., and Murphy, D.C., 2001, Contrasting arc and non-arc mafic volcanic rocks in the Fyre Lake unit, Yukon-Tanana Terrane, Finlayson Lake region, Yukon: Slave-Northern Cordilleran Lithospheric Experiment (SNORCLE) Newsletter, v. 79, p. 261-266.

Piercey, S.J., Murphy, D.C., Mortensen, J.K., and Hunt, J.A., 1999, Geochemistry of metavolcanic rocks from the Yukon-Tanana Terrane and Campbell Range Belt, Finlayson Lake Region, southeastern Yukon Territory: preliminary results: Slave-Northern Cordillera Lithospheric Experiment (SNORCLE) Newsletter, v. 69, p. 312.

Piercey, S.J., Paradis, S., Murphy, D.C., and Mortensen, J.K., 2001a, Geochemistry and paleotectonic setting of felsic volcanic rocks in the Finlayson Lake volcanic-hosted massive sulfide district, Yukon, Canada: Economic Geology, v. 96, p. 1877-1905.

Piercey, S.J., Peter, J.M., Bradshaw, G.D., Tucker, T., and Paradis, S., 2001b, Geological characteristics of high-level subvolcanic porphyritic intrusions associated with the Wolverine Zn-Pb-Cu volcanic-hosted massive sulphide deposit, Finlayson Lake District, Yukon, Canada, *in* Emond, D.S., and Weston, L.H., eds., Yukon Exploration and Geology 2000: Indian and Northern Affairs Canada, p. 335-346.

Piercey, S.J., Murphy, D.C., Mortensen, J.K., and Paradis, S., 2001c, Boninitic magmatism in a continental margin setting: Geology, 29, p. 731-734.

Piercey, S.J., Mortensen, J.K., Murphy, D.C., Paradis, S., and Creaser, R.A., 2002a, Geochemistry and tectonic significance of alkalic mafic magmatism in the Yukon-Tanana Terrane, Finlayson Lake region, Yukon: Canadian Journal of Earth Sciences, v. 39, p. 1729-1744.

Piercey, S.J., Paradis, S.J., Peter, J.M., and Tucker, T.L., 2002b, Geochemistry of basalt from the Wolverine volcanic-hosted massive-sulphide deposit, Finlayson Lake District, Yukon Territory: Geological Survey of Canada, Current Research, 2002-A3, 11 p.

Piercey, S.J., Mortensen, J.K., and Creaser, R.A., 2003, Neodymium isotope geochemistry of felsic volcanic and intrusive rocks from the Yukon-Tanana Terrane in the Finlayson Lake region, Yukon, Canada: Canadian Journal of Earth Sciences, v. 40, p. 77-97.

Piercey, S.J., Murphy, D.C., Mortensen, J.K., and Creaser, R.A., 2004, Mid-Paleozoic initiation of the northern Cordilleran marginal backarc basin; geologic, geochemical, and neodymium isotope evidence from the oldest mafic magmatic rocks in the Yukon-Tanana Terrane, Finlayson Lake District, southeast Yukon, Canada: Geological Society of America Bulletin, v. 116, p. 1087-1106.

Piercey, S.J., Nelson, J.L., Colpron, M., Dusel-Bacon, C., Simard, R.-L., and Roots, C. F., 2006, Paleozoic magmatism and crustal recycling along the ancient Pacific margin of North America, northern Cordillera, *in* Colpron, M., and Nelson, J., eds., Paleozoic Evolution and Metallogeny of Pericratonic Terranes at the Ancient Pacific Margin of North America: Geological Association of Canada, Special Publication 45, p. 261-280.

Piercey, S.J., Peter, J.M., Mortensen, J.K., Paradis, S., Murphy, D.C., and Tucker, T.L., in press, Geological, geochemical and U-Pb age constraints on the origin of footwall porphyritic rhyolites from the Wolverine volcanic-hosted massive sulfide (VHMS) deposit, Finlayson Lake District, Yukon, Canada: Economic Geology.

Pigage, L., 1997, Mapping and stratigraphy at Ice: Unpublished internal report, Expatriate Resources Ltd., 5p.

Plint, H.E., and Gordon, T.M., 1996, Structural evolution and rock types of the Slide Mountain and Yukon-Tanana terranes in the Campbell Range, southeastern Yukon Territory: Geological Survey of Canada, Current Research, 1996-A, p. 19-28.

——— 1997, The Slide Mountain Terrane and the structural evolution of the Finlayson Lake Fault Zone, southeastern Yukon: Canadian Journal of Earth Sciences, v. 34, p. 105-126.

Saelen, G., Raiswell, R., Talbot, M.R., Skei, J.M., and Bottrell, S.H., 1993, Heavy sedimentary sulfur isotopes as indicators of super-anoxic bottom-water conditions: Geology, v. 21, p. 1091-1094.

Schultze, H.C., 1996, Summary of the Kudz Ze Kayah project, volcanic hosted massive sulphide deposit, Yukon Territory, *in* Yukon Exploration and Geology, 1995: Indian and Northern Affairs Canada, p. 29-31.

Sebert, C., and Hunt, J.A., 1999, A note on preliminary lithogeochemistry of the Fire Lake area, Yukon, *in* Exploration and Geology 1998: Whitehorse, Yukon Geology Program, p. 139-142.

Sebert, C., Hunt, J.A., and Foreman, I.J., 2004, Geology and Lithogeochemistry of the Fyre Lake Copper-Cobalt-Gold Sulphide-Magnetite Deposit, southeastern Yukon: Yukon Geological Survey, Open File 2004-17, 46 p.

Seyfried, W.E., Jr., and Bischoff, J.L., 1981, Experimental seawater-basalt interaction at 300°C and 500 bars: chemical exchange, secondary mineral formation, and implications for the transport of heavy metals: Geochimica et Cosmochimica Acta, v. 45, p. 135-147.

Shanks, W.C., III and Seyfried, W.E., Jr., 1987, Stable isotope studies of vent fluids and chimney minerals, southern Juan de Fuca Ridge; sodium metasomatism and seawater sulfate reduction: Journal of Geophysical Research, v. 92, p. 11387-11399.

Shanks, W.C., III, Woodruff, L.G., Jilson, G.A., Jennings, D.S., Modene, J. S., and Ryan, B.D., 1987, Sulfur and lead isotope studies of stratiform Zn-Pb-Ag deposits, Anvil Range, Yukon: Basinal brine exhalation and anoxic bottom-water mixing: Economic Geology, v. 82, p. 600-634.

Slack, J. F., 1993, Descriptive and grade-tonnage models for Besshi-type massive sulphide deposits, *in* R.V. Kirkham, W.D. Sinclair, R.I. Thorpe and J.M. Duke, eds., Mineral Deposit Modeling: Geological Association of Canada, Special Paper 40, p. 343-371.

Spitz, G., and Darling, R., 1978, Major and minor element lithogeochemical anomalies surrounding the Louvem copper deposit, Val d'Or, Quebec: Canadian Journal of Earth Sciences, v. 15, p. 1161-1169.

Spry, P.G., and Scott, S.D., 1986, The stability of zincian spinels in sulphide systems and their potential as exploration guides for metamorphosed massive sulphide deposits. Economic Geology, v. 81, p. 1146-1163.

Spry, P.G., Peter, J.M., and Slack, J.F., 2000, Meta-exhalites as exploration guides to ore, *in* Spry, P.G., Marshall, B., and Vokes, F.M. eds., Metamorphosed and Metamorphogenic Ore Deposits: Economic Geology, Reviews in Economic Geology, v. 11, p. 163-201.

Stacey, J.S., and Kramers, J.D., 1975, Approximation of terrestrial lead isotope evolution by a two-stage model: Earth and Planetary Science Letters, v. 26, p. 207-221.

Swinden, H.S., 1991, Paleotectonic settings of volcanogenic massive sulphide deposits in the Dunnage Zone, Newfoundland Appalachians:

——— 1997, The application of volcanic geochemistry to the metallogeny of volcanic-hosted sulfide deposits in central Newfoundland, *in* Wyman, D.A., ed., Trace Element Geochemistry of Volcanic Rocks: Applications for Massive Sulfide Exploration: Geological Association of Canada, Short Course Notes, v. 12, p. 329-358.

Syme, E.C., Lucas, S.B., Bailes, A.H., Stern, R.A., and Lucas, S.B., 1999, Contrasting arc and MORB-like assemblages in the Paleoproterozoic Flin Flon Belt, Manitoba, and the role of intra-arc extension in localizing volcanic-hosted massive sulphide deposits: Canadian Journal of Earth Sciences, v. 36, p. 1767-1788.

Tempelman-Kluit, D.J., 1979, Transported cataclasite, ophiolite and granodiorite in Yukon: evidence for arc-continent collision: Geological Survey of Canada, Paper 79-14, p. 27.

Toulmin, P., III and Barton, P.B., Jr., 1964, A thermodynamic study of pyrite and pyrrhotite: Geochimica et Cosmochimica Acta, v. 28, p. 641-671.

Tucker, T.L., 1999, Summary report on the geological, geochemical, geophysical, geotechnical, metallurgical, environmental and diamond drilling surveys on the Wolverine property, Yukon Territory, Canada: Expatriate Resources Ltd., 67 p.

Tucker, T.L., Turner, A.J., Terry, D.A., and Bradshaw, G.D., 1997, Wolverine massive sulfide project, Yukon, *in* Yukon Exploration and Geology 1996: p. 53-55.

Turner, J.S., and Campbell, I.H., 1987, Temperature, density and buoyancy fluxes in "black smoker" plumes, and the criterion for buoyancy reversal: Earth and Planetary Science Letters, v. 86, p. 85-92.

Wedepohl, K.H., Delevaux, M.H., and Doe, B.R., 1978, The potential source of lead in the Permian Kupferschiefer Bed of Europe and some selected Paleozoic mineral deposits in the Federal Republic of Germany: Contributions to Mineralogy and Petrology, v. 65, p. 273-281.

Wheeler, J.O., and McFeely, P., 1991, Tectonic assemblage map of the Canadian Cordillera: Geological Survey of Canada Map 1712A.

Wyman, D.A., 1999, Paleoproterozoic boninites in an ophiolite-like setting, Trans-Hudson Orogen, Canada: Geology, v. 27, p. 455-458.

Wyman, D.A., Bleeker, W., and Kerrich, R., 1999, A 2.7 Ga komatiite, low Ti tholeiite, arc tholeiite transition, and inferred proto-arc geodynamic setting of the Kidd Creek Deposit; evidence from precise trace element data, *in* Hannington, M. D., and Barrie, C. T., eds., The Giant Kidd Creek Volcanogenic Massive Sulfide Deposit, Western Abitibi Subprovince, Canada: Society of Economic Geologists, Monograph 10, p. 511-528.

Yukon Zinc Corporation Limited, 2004a, Annual Report-2004, 16 p.. Available at: http://www.yukonzinc.com/documents/2004YukonZinc AnnualReport-s.pdf.

——— 2004b, news release, September 8, 2004. Available at: http://www.yukonzinc.com/ news_archive.cfm

——— 2005, news release, January 7, 2005. Available at: http://www.yukonzinc.com/documents/YZC%20NR% 202005-01-07.pdf.

——— 2006a, Yukon Zinc Corporation confirms Wolverine deposit resources, news release, 4 p. Available at:http://www.yukonzinc.com/ documents/YZC-NR-2006-01-10.pdf

——— 2006b, Ice deposit (website). Available at: http://www.yukonzinc.com/ ice.htm

Metallogeny of the Paleoproterozoic Flin Flon Belt, Manitoba and Saskatchewan

Alan G. Galley[1], Ric Syme[2], and Alan H. Bailes[2]

[1]*Geological Survey of Canada, 601 Booth Street, Ottawa, Ontario K1A 0E8*
[2]*Manitoba Geological Survey, 360-1395 Ellice Avenue, Winnipeg, Manitoba R3G 3P2*
Corresponding author's email:agalley@nrcan.gc.ca

Abstract

For the past 75 years, the Paleoproterozoic Flin Flon Belt has been a significant contributor of base metals and Au to the Canadian economy. It is host to 27 significant volcanogenic massive sulphide deposits that originally contained over 154 Mt of ore, including the 62 Mt Flin Flon deposit. It is also host to several past-producing orogenic Au deposits, including the second >1 million ounce producer (New Britannia) within a Canadian Paleoproteroic terrane, the first being the Flin Flon VMS deposit. The Flin Flon Belt is part of the dominantly juvenile core of the Trans-Hudson Orogen, and consists of a series of 1.92 to 1.87 Ga accreted oceanic arcs and intervening back-arc basins, all intruded by successor arc plutons. The resulting collage has undergone several phases of deformation associated with subduction-related collision, oblique compression, and uplift during the closing of the original ocean basin. During the formation of its various domains and succeeding tectono-metamorphic history, the Flin Flon Belt was host to three main metallogenic events. The first is characterized by the formation of several types of VMS deposits within the various oceanic rifted arc terranes, with mafic-bimodal being the dominant VMS type. A period of successor arc and magmatism included development of mafic gabbroic intrusions with Ni-Cu-PGE and PGE potential, plus subeconomic Cu-Mo porphyry-type mineralization. Crustal thickening during continued subduction-related collision was accompanied by formation of orogenic, shear-hosted Au mineralization.

The extension of the Flin Flon arc and ocean basin collage under the Phanerozoic cover rocks leaves much of this Paleoproteroic juvenile terrane open to further discovery of VMS and orogenic Au deposits. This will require a better understanding of the sub-Phanerozoic geology through high-resolution geophysics, a better understanding of deformation history and resulting structural architecture, geochemical and mineralogical zoning of large-scale VMS hydrothermal systems, and geochemical remote sensing.

Résumé

Au cours des 75 dernières années, la ceinture de Flin Flon du Paléoprotérozoïque a fourni une contribution importante en métaux communs et en or à l'économie canadienne. Elle recèle 27 importants gisements de sulfures massifs volcanogènes qui renfermaient à l'origine plus de 154 Mt de minerai, dont le gisement de Flin Flon de 62 Mt. On y trouve également plusieurs gisements de Au orogéniques dont l'exploitation a cessé, incluant le deuxième plus important producteur de plus d'un million d'onces (New Britannia) dans un terrane paléoprotérozoïque au Canada, le premier ayant été le gisement de SMV de Flin Flon. La ceinture de Flin Flon fait partie du cœur principalement juvénile de l'orogène trans-hudsonien et se compose d'une série d'arcs océaniques et de bassins d'arrière arc intercalaires accrétés datant de 1,92 à 1,87 Ga et tous pénétrés par des plutons d'arcs successeurs. Le collage qui en résulte a subi plusieurs phases de déformation associées à la subduction liée à la collision, à une compression oblique et à un soulèvement pendant la fermeture du bassin océanique d'origine. Pendant la formation de ses divers domaines et l'histoire tectonométamorphique qui a suivi, la ceinture de Flin Flon a été le siège de trois principaux événements métallogéniques. Le premier est caractérisé par la formation de plusieurs types de gîtes de SMV à l'intérieur des divers terranes océaniques d'arcs en distension, les gîtes mis en place dans des assemblages bimodaux à dominante mafique étant la variété dominante des gîtes de type SMV. Une période de formation d'arcs successeurs et de magmatisme a permis la mise en place d'intrusions mafiques (gabbroïques) présentant un potentiel pour les minéralisations de Ni-Cu-ÉGP et de ÉGP, ainsi que de minéralisations subéconomiques de Cu-Mo de type porphyrique. Un épaississement crustal pendant la subduction liée à la collision qui s'est poursuivie s'est accompagné de la formation de minéralisations orogéniques de Au dans des cisaillements.

Le prolongement du collage d'arcs et de bassins océaniques de Flin Flon sous les roches de couverture du Phanérozoïque fait en sorte qu'une grande partie de ce terrane juvénile du Paléoprotérozoïque pourrait receler d'autres gîtes de SMV et de Au orogéniques. L'exploration à la recherche de ces gîtes exigera une meilleure compréhension de la géologie sub-phanérozoïque par l'application de méthodes géophysiques haute résolution, une meilleure compréhension de l'histoire de la déformation et de l'architecture tectonique résultante, la zonalité géochimique et minéralogique des grands réseaux hydrothermaux à l'origine des gîtes de SMV ainsi que la télédétection géochimique.

Introduction and Mining History

The Paleoproterozoic Flin Flon Belt of central Manitoba and Saskatchewan (Fig. 1) contains both syngenetic, polymetallic base metal and orogenic Au deposits, but it is best known for its two volcanogenic massive sulphide (VMS) camps: Flin Flon and Snow Lake (Fig. 2). Tom Creighton and the Mosher and Dion brothers first discovered sulphide mineralization along the Manitoba-Saskatchewan border in the Flin Flon area in 1914. This mineralization was the surface expression of the 62 Mt Flin Flon VMS deposit. The deposit lay dormant until 1930 due to the lack of economic methods

Galley, A.G., Syme, E.C., and Bailes, A.H., 2007, Metallogeny of the Paleoproterozoic Flin Flon Belt, Manitoba and Saskatchewan, *in* Goodfellow, W.D., ed., Mineral Deposits of Canada: A Synthesis of Major Deposit Types, District Metallogeny, the Evolution of Geological Provinces, and Exploration Methods: Geological Association of Canada, Mineral Deposits Division, Special Publication No. 5, p. 509-531.

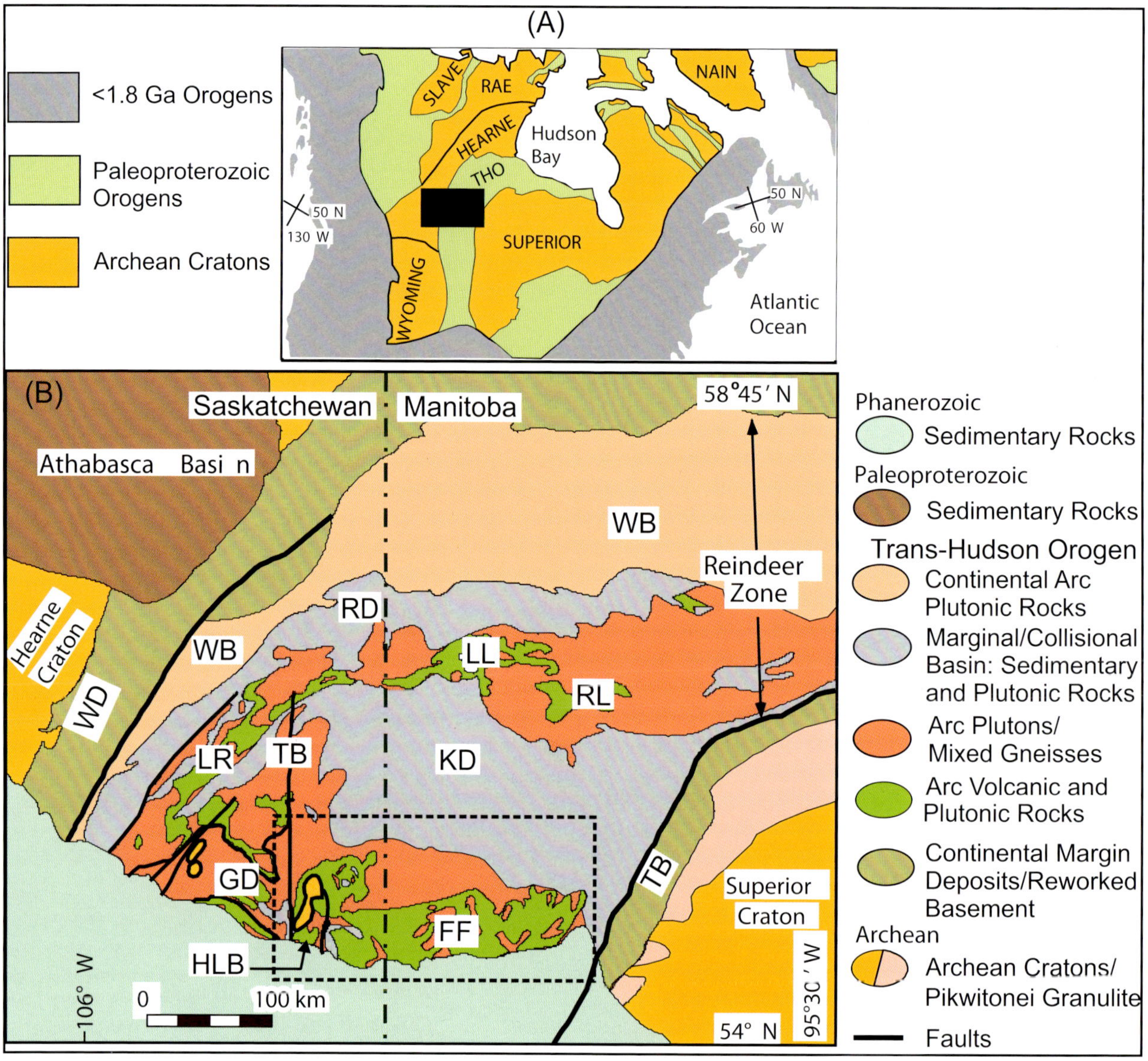

FIGURE 1. Location map after Hoffman (1989) illustrating the position of the Flin Flon Belt in relation to the Precambrian geology of North America **(A)** and central Canada **(B)**. The latter illustrates the components of the dominantly juvenile core to the Trans-Hudson Orogen (THO) in relation to the bordering Archean terranes, middle Proterozoic Athabasca Basin, and overlying Phanerozoic strata of the western Canadian sedimentary basin. FF=Flin Flon Belt; GD=Glennie Domain; HLB=Hanson Lake Block; LL-LR=Lynn Lake-LaRonge Belt; KD=Kisseynew Domain; RD=Rottenstone Domain; RL=Rusty Lake Belt; TH=Thompson Belt; TB=Tabbernor fault zone; WB=Wathaman-Chipewayan Batholith; WD=Wollaston Domain. Modified from Lucas et al. (1996), from the original by Hoffman (1989). Dashed box represents area shown in more detail in Figure 2.

for separating Zn from sphalerite. The Mandy VMS deposit on the shore of nearby Schist Lake was the first to produce ore in the region (1916) from a high-grade chalcopyrite lens. The development of an electrolytic method to produce Zn, coupled with the ability to produce inexpensive hydroelectric power from the nearby Saskatchewan River, resulted in the construction of a smelter complex and opening of the Flin Flon deposit. The town that grew up on the Manitoba side of the border was named "Flin Flon" by prospectors in 1915, after the character Josiah Flintabbatey Flonatin in the dime novel "The Sunless City." The twin town on the Saskatchewan side of the border was named Creighton after Tom Creighton, co-discoverer of the Flin Flon deposit.

Prospecting from the 1920s through to the 1950s utilized the extensive system of interconnected lakes along the Grassy River system to push east and west along the length of the exposed Flin Flon greenstone belt. This resulted in the discovery of a number of VMS deposits, including the Sherritt-Gordon (1928), Don Jon (1930), and North Star (1949). With the introduction of ground and airborne electromagnetic surveys in the early 1950s, Hudson Bay Exploration and Development Ltd. (HBED) discovered the majority of the VMS deposits found from 1952 until the present. This includes the seven VMS deposits developed in the Snow Lake area, most of which were noted on a 1955 geology map by Russell (1957) as sulphide occurrences prior to their discovery.

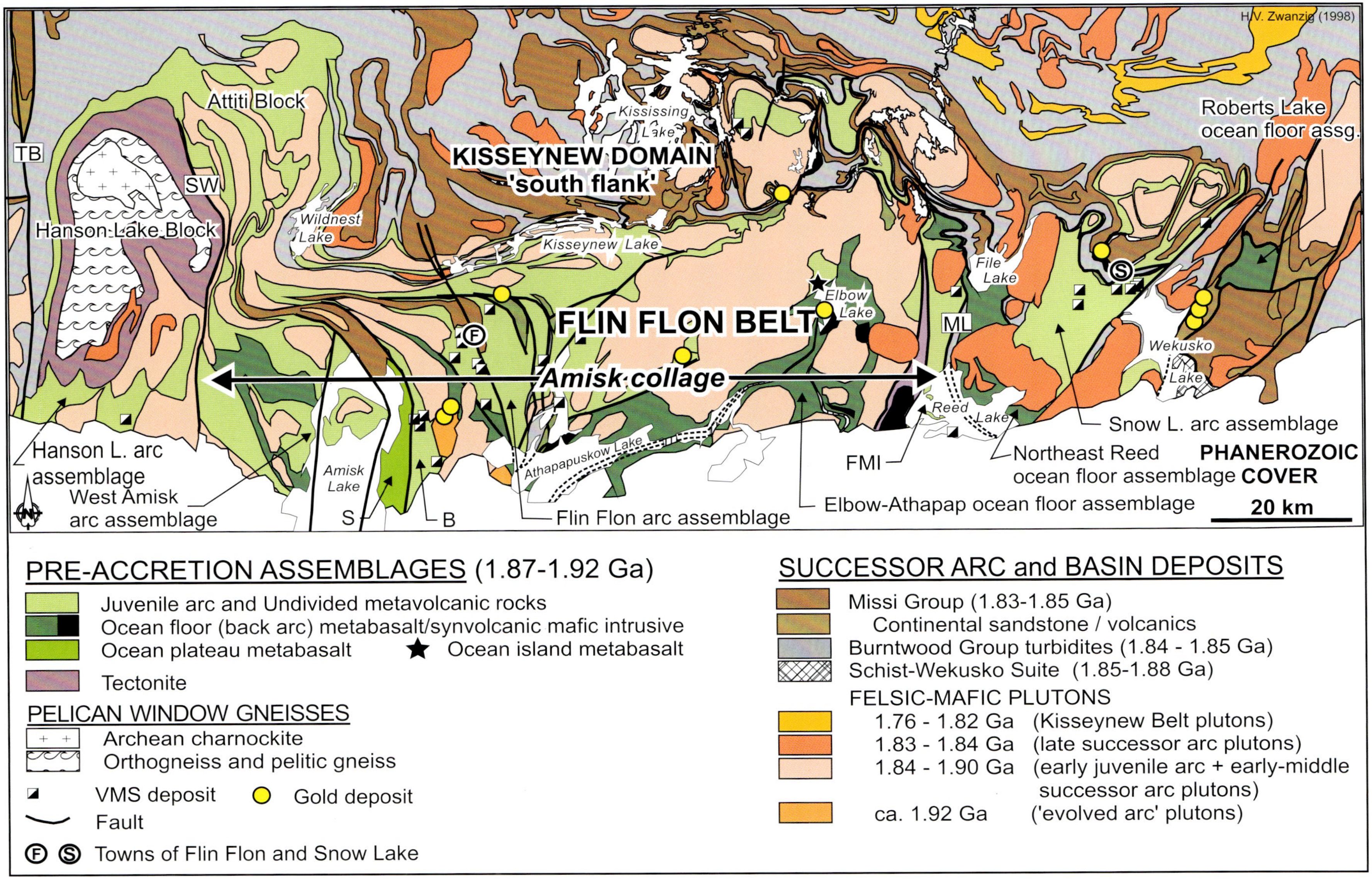

FIGURE 2. Map of the Flin Flon Belt, illustrating the tectonicstratigraphic assemblages, the location of the various accretionary assemblages, and major mineral deposits. B=Birch Lake assemblage; FMI=Fourmile Island assemblage; ML=Morton Lake fault zone; S=Sandy Bay assemblage; TB=Tabernor fault zone; SW=Sturgeon-Weir fault zone. Modified from Zwanzig (1999) and Lucas et al., (1996).

To date, 27 deposits have been mined in the Flin Flon Belt (Table 1). Many of these deposits are polymetallic and contain significant contents of precious metals. The Flin Flon VMS deposit alone produced over 120 tonnes Au. The most recent discoveries include Callinan (1984), Photo Lake (1994), Konuto (1994), and Triple 7 (1993; Table 1). As of 2007, there are five operating base metal mines in the Flin Flon Belt, with HudBay Minerals Ltd. pursuing an aggressive exploration program investigating some of the numerous untested SPECTRUM® airborne electromagnetic anomalies.

At approximately the same time that prospectors were discovering polymetallic sulphide occurrences in the western part of the Flin Flon Belt, others were discovering Au showings 200 km to the east at Wekusko Lake (Fig. 2). The first Au discoveries were made in 1914 by M.J. Hackett and R. Woosey on the east shore of Wekusko Lake (formerly Herb Lake; Stockwell, 1937). By 1918, the small mining town of Herb Lake had sprung up around several small arsenopyrite-bearing shear-hosted lode Au vein deposits. Less than 10,000 ounces of Au was produced from this camp, the largest being the Rex-Laguna, which produced 7,122 ounces between 1918 and 1925 (Table 2). In 1926, Chris Parres discovered auriferous quartz-carbonate-arsenopyrite veins outcropping along the north shore of Snow Lake at what was to become the Nor Acme Au deposit. Mining began at Nor Acme in 1949 and continued until its closure in 1956. In 1999 it re-opened as the New Britannia mine, and became the first and only million-ounce primary Au producer in the Flin Flon Belt. The 1980s saw production from several other vein-hosted Au deposits, none of which remain open today (Table 2). These include the Puffy Lake, Tartan Lake, and Rio deposits.

Other minor deposit types represented in the Flin Flon Belt include the Dion Lake porphyry Cu-Mo mine southwest of Flin Flon. There is also potential for Ni-Cu-PGE mineralization in mafic layered intrusion and gabbroic intrusions, as

TABLE 1. Grade and tonnages, and discovery and production dates, Flin Flon deposits*.

Deposit	Discovery	Opened	Closed	Tonnes**	Au (g/t)	Ag (g/t)	Cu (%)	Zn (%)
Birch Lake Back-arc Assemblage								
Konuto	1994	1998	2005	1,948,500	2.07	8.64	4.13	1.41
Centennial	1969	1975	1986	1,473,767	1.58	20.23	1.41	2.48
Flexar	1952	1969	1972	305,939	1.30	6.51	3.76	0.50
Birch Lake	1950	1957	1960	272,898	0.10	4.11	6.21	Tr
Flin Flon Arc Assemblage								
Flin Flom	1915	1930	1992	62,520,625	2.64	41.49	2.17	4.13
Trout Lake	1976	1982		21,269,476	0.99	16.05	1.60	4.93
Triple 7	1993	1998		14,495,288	2.12	31.08	2.64	4.98
Callinan	1984	1989		8,395,667	2.05	25.55	1.29	4.02
Schist lake	1947	1950	1976	1,846,656	1.30	37.03	4.30	7.27
West Arm	1973	1977	1997	1,394,149	1.58	17.49	3.21	1.48
Coronation	1953	1960	1965	1,281,726	2.06	5.14	4.25	0.20
White Lake	1963	1972	1982	849,784	0.72	27.09	1.98	4.64
Cuprus	1941	1948	1954	462,096	1.30	28.8	3.25	6.40
Birch lake	1950	1957	1960	272,898	0.10	4.11	6.21	Tr
North Star	1950	1955	1958	241,692	0.34	8.57	6.11	Tr
Mandy	1915	1917	1944	102,252	3.05	55.54	5.63	13.69
Don Jon	1950	1955	1957	79,329	0.96	15.09	3.09	Tr
Four Mile Arc Assemblage								
Spruce Point	1973	1981	1992	1,865,095	1.68	19.54	2.06	2.40
Dickstone	1935	1971	1975	775,504	0.41	12.00	2.46	3.12
Snow Lake Arc Assemblage								
Chisel Lake	1956	1960	1994	7,153,536	1.77	44.77	0.54	10.60
Stall lake	1956	1964	1994	6,381,129	1.41	12.34	4.41	0.50
Chisel North	1985	2000		2,847,194	0.40	21.54	0.15	9.36
Osborne	1953	1968	1983	2,807,471	0.27	4.11	3.14	1.50
Anderson	1963	1970	1988	2,513,290	0.62	7.54	3.40	0.10
Rod	1970	1984	1991	735,219	1.71	16.11	6.63	2.90
Photo Lake	1994	1995	1998	689,885	4.87	29.49	4.58	6.35
Lost & Ghost	1956	1972	1988	581,437	1.20	39.09	1.34	8.60

* Data from HudBay Minerals unpublished data.
** Tonnages and grades are not National Instrument 43-101 compliant.

TABLE 2. Historical Statistics for Flin Flon Belt Gold Mines.

Deposit	Discovery	Opened	Closed	Tonnes Milled	Oz Au Produced	Production & Reserves*	Grade (g/t)	Source
New Britannia	1926	1949	2004		1,323,000	4,994,871	3.9	Northern Miner, 2004
Laguna (Rex)	1914	1918	1940		58,962	unknown	unknown	Richardson & Ostry, 1996
Gurney	1919	1937	1939	94,650	28,045	unknown	unknown	Richardson & Ostry, 1996
Tartan Lake	1931	1987	1988	592	18,510	375,000	6.27	Richardson & Ostry, 1996
Ferro-Rainbow	1923	1932	1933	1,006	730	66,000	13	Richardson & Ostry, 1996
Moose Horn (Ballast)	1914	1917	1918	42	194	unknown	unknown	Richardson & Ostry, 1996
Bingo	1915	1926	-	-	128	unknown	unknown	Richardson & Ostry, 1996
Gold Hill	1934	1935	1963	103	147	unknown	unknown	Richardson & Ostry, 1996
Century	1919	1941	1942	-	59.6	272,000	12	Richardson & Ostry, 1996
Apex	1918	-	-	-	-	758,000	3.63	Richardson & Ostry, 1996
Newcor	-	-	1947	-	-	43,000	10.56	Coombe Geoconsultants Ltd., 1984
Rio	-	1984	1985	-	-	124,000	12.54	Coombe Geoconsultants Ltd., 1984
Henney-Maloney	-	-	1941	-	-	15,000	14.85	Coombe Geoconsultants Ltd., 1984

* Not instrument 43-101 compliant.

well as Ta mineralization in pegmatites.

Geological Setting of the Flin Flon Belt

The Paleoproterozoic Flin Flon Belt is approximately 200 km in strike length and has an exposed width of up to 70 km. The Flin Flon Belt is overlain to the south by Paleozoic sandstone, limestone, and dolomite of the Western Canada Sedimentary Basin, and is bordered to the north by high-grade paragneiss and granitoid rocks of the Kisseynew Domain (Fig. 1).

The Flin Flon Belt is interpreted to be an accreted assemblage of oceanic to continental margin arc terranes, interspersed with oceanic basins representing back-arc, fore-arc, and oceanic settings (Lucas et al., 1996; Syme et al., 1996). It is part of the Reindeer zone, a largely juvenile portion of the Trans-Hudson Orogen separating the Archean Superior and Hearne provinces (Fig. 1). Recent tracer isotope studies have confirmed the presence of >3.0 Ga Archean crust (the Sask Craton) below parts of the Trans-Hudson (Lucas et al., 1996). The Shield Margin National Mapping Program (NATMAP; Lucas et al., 1996) traced the Flin Flon Belt assemblages below the Phanerozoic to the south and recognized highly metamorphosed and deformed Flin Flon volcanic and sedimentary formations (Zwanzig, 1990, 1999) within the Kisseynew Domain to the north. To the east, the Flin Flon Belt is separated from the Paleoproterozoic Thompson Nickel Belt by Kisseynew Domain rocks. To the west, the Flin Flon Belt is terminated against the Tabernor fault zone (Fig. 1).

The recent Geological Survey of Canada (GSC)-Manitoba-Saskatchewan NATMAP Shield Margin Project and LITHOPROBE Trans-Hudson Orogen transect built on an extensive existing geological database that led to a much improved understanding of the components and evolution of the southeastern Reindeer zone, including the Flin Flon Belt (e.g., Lucas et al., 1996). These investigations have shown that, at a crustal scale, the Flin Flon "greenstone" belt is only one of three components in a northeast-dipping stack, juxtaposed during 1.84 to 1.80 Ga collisional deformation (Lucas et al., 1999):

1. At the lowest structural level (exposed in the Pelican Window, Fig. 2, within the Hanson Lake Block): metaplutonic rocks and paragneisses (3.20–2.40 Ga) of the "Sask craton" (Corrigan et al., 2007).

2. At intermediate structural levels: Flin Flon Belt (now defined to include the Attitti Block and Paleoproterozoic rocks in the Hanson Lake Block) and Glennie Domain, shown in Figures 1 and 2, (together comprising the "Flin Flon-Glennie Complex" [FFGC]; Lucas et al., 1996).

3. At the highest structural levels: marine turbidites (Burntwood Group; 1.85–1.84 Ga) and distal facies of alluvial-fluvial sandstones (Missi Group) in the Kisseynew Domain (Fig. 1).

Geological Evolution of the Flin Flon Belt

The stratigraphy of the Flin Flon Belt has been previously subdivided into two major groups, the Amisk Group metavolcanic rocks and Missi Group continental metasedimentary rocks (Bruce, 1918; Harrison, 1951). The Flin Flon Belt is now recognized as consisting of several 1.9 to 1.88 Ga terranes comprised of four main tectono-stratigraphic assemblages that represent both juvenile and continentally underlying oceanic segments of a Paleoproterozoic ocean basin that were accreted during formation of the Trans-Hudson orogen (Syme, 1990; Syme and Bailes, 1993; Stern et al., 1995a,b; Lucas et al., 1999; Figs. 1, 2). The orogen was formed by oblique collision between the Superior and Hearne Archean terranes, and the resulting collage is separated into a number of assemblages distinguished by unique tectono-stratigraphy, and dismembered by fault systems that were originally thrust surfaces (Syme, 1995, Lucas et al., 1996). Each tectonostratigraphic assemblage is a distinct package of rocks in terms of its stratigraphy, geochemistry, isotopic signature, age, and inferred plate tectonic setting (see below; Syme and Bailes, 1993; Lucas et al., 1996; Corrigan et al., 2007).

The tectonostratigraphic assemblages were juxtaposed in

an accretionary complex at ca. 1.88 to 1.87 Ga, probably as a result of arc–arc collision (D_1; Lucas et al., 1996; Stern et al., 1999; Fig. 3). Accretionary collage-bounding structures were largely obliterated by subsequent deformation and metamorphic events (D_2–D_5), but are inferred where juxtaposed terranes are "stitched" together by calc-alkaline plutons related to a 1.866 to 1.838 Ga successor arc formation (Whalen et al., 1999). Coeval subaerial volcanism is recorded in ca. 1.87 to 1.85 Ga calc-alkaline to shoshonitic volcaniclastic sequences (Syme, 1988; Bailes and Syme, 1989; Lucas et al., 1996; Stern et al., 1996). Unroofing of the accretionary collage, development of a paleosol, and deposition of alluvial-fluvial sedimentary rocks (Missi suite; Bailes and Syme, 1989; Holland et al., 1989) occurred ca. 1.85 to 1.84 Ga (Ansdell, 1993). These events were coeval with the waning stages of post-accretion arc magmatism (Stern and Lucas, 1994; Whalen and Hunt, 1994; Lucas et al., 1996). Development of the Kisseynew turbidite basin (now part of the Kisseynew Domain, Figs. 1, 2) was synchronous with continental sedimentation in the Flin Flon Belt (Ansdell, 1993; David et al., 1993, 1996; Machado and Zwanzig, 1995; Connors, 1996; Connors et al., 1999).

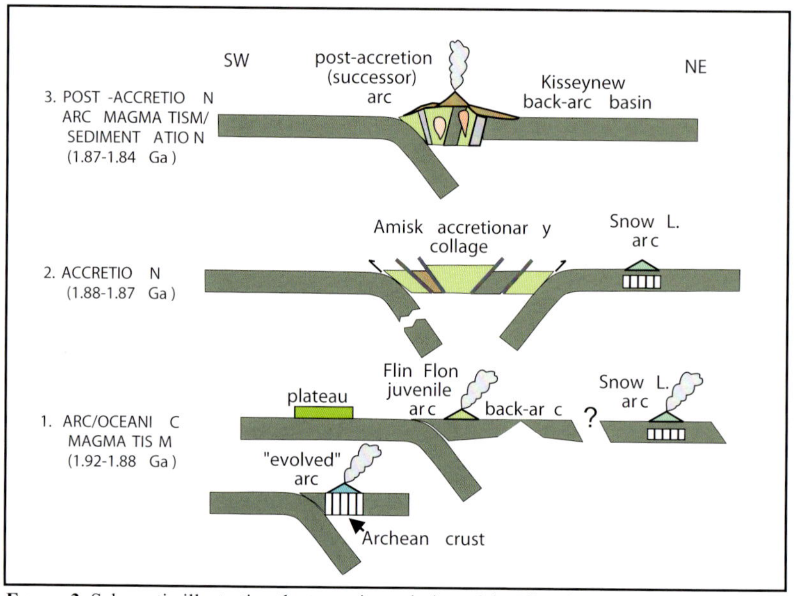

FIGURE 3. Schematic illustrating the tectonic evolution of the Flin Flon district (modified from Lucas et al., 1996).

The transition from Kisseynew basin extension to collisional collapse occurred rapidly at about 1.840 Ga, although sedimentation and magmatism continued through to ~1.830 Ga (Ansdell and Norman, 1995; Machado and Zwanzig, 1995; David et al., 1996; Connors et al., 1999). The Kisseynew Domain was thrust over the Amisk collage along the southern flank of the Kisseynew Domain (Harrison, 1951; Zwanzig, 1990; Lucas et al., 1994; Connors, 1996; Connors et al., 1999; Zwanzig, 1999). Following collisional thickening and peak metamorphism at 1.83 to 1.80 Ga, the Flin Flon Belt experienced protracted intracontinental deformation to ca. 1.69 Ga (Lucas et al., 1996; Stern et al., 1999).

The Flin Flon Accreted Terranes

The Flin Flon Belt consists of two principal segments (Amisk collage and Snow Lake arc assemblage) that were juxtaposed during southwest-verging continent–continent collision between 1.84 and 1.82 Ga. To the west of the Amisk collage is a volcano-sedimentary domain of similar age (Hanson Lake arc assemblage; Fig. 2). Although not traditionally designated as part of the Flin Flon Belt, the Hanson Lake arc assemblage will be included here due to the presence of volcano-sedimentary rock units and several VMS occurrences and deposits of similar age (Maxeiner et al., 1993, 1999).

Although each of these three tectonic segments has a distinct character, their metallogenic tenor is a product of the same three tectonically controlled evolutionary stages of the region. The first is represented by syngenetic polymetallic base metal and precious metal deposits and occurrences that formed during a pre-accretionary stage within distinct oceanic, suprasubduction environments. The second consists of post-accretion intrusion-related mineralization associated with successor arc formation and extensional magmatism. The third is comprised of shear zone-related orogenic Au deposits that formed during periods of collision, oblique compression, and crustal thickening.

Hanson Lake Block

The Hanson Lake arc assemblage structurally overlies Archean crust and contains coeval volcanic and sedimentary assemblages that are found within the neighbouring Flin Flon Belt. It is also host to numerous VMS deposits and showings. The fault-bounded area containing the Hanson Lake arc assemblage and underlying Archean crust is known as the Hanson Lake Block (HLB) (Figs. 1, 2).

The HLB is composed of a highly deformed and metamorphosed assemblage of 1.91 to 1.85 Ga volcanic and sedimentary rock, and 1.86 to 1.81 Ga syntectonic intrusions and migmatitic gneisses that have been thrust over the ca. 2.5 Ga Neoarchean charnokitic and enderbitic intrusive rocks known as the Peleican Window (surface expression of the Sask Craton; Ashton et al., 1987; Maxeiner et al., 1993, 1999; Ashton and Lewry, 1994; Fig. 2). The HLB is terminated to the west by the Tabernor fault zone and to the east by the Sturgeon-Weir fault zone, which separates the HLB from those assemblages traditionally included within the Flin Flon Belt, and extends southward below the Phanerozoic cover. The principal reason for the original exclusion of the arc assemblages of the HLB from those of the Flin Flon Belt was that they were originally believed to have formed upon Archean crust and were, therefore, not considered part of the Flin Flon oceanic suprasubduction suite.

Supracrustal rocks of the HLB are dominated by metavolcanic and metasedimentary rocks. Volcanism and sedimentation are coeval from 1910 to 1880 Ma, with sedimentation continuing to at least 1850 Ma (Maxeiner et al., 1993, 1999).

Volcanic strata are dominantly tholeiites (Maxeiner et al., 1993, 1999) and include pillowed basalt overlain by intermediate to felsic flows and volcaniclastic rocks. Also present is a large felsic hypabyssal intrusive/extrusive complex (Maxeiner et al., 1993). The volcanic assemblage is in contact with calc-silicate-carbonate-rich strata, silicate-facies iron formation, and polymictic conglomerate, and overlain by psammitic greywacke and mafic wacke.

The supracrustal assemblages of the HLB are intruded by numerous synvolcanic intrusions, ranging in composition from ultramafic through gabbro and quartz diorite to rhyolitic. Large antiformal domes of migmatitic gneiss are accompanied by lit-par-lit injection into the supracrustal formations. Metamorphic grade generally increases from south to north, from upper greenschist to upper amphibolite facies, with regional metamorphism peaking between 1810 and 1806 Ma. A major folding event took place between 1860 and 1850 Ma, and was followed by 1810 to 1800 Ma continental collision that caused the thrusting of this terrane over Archean basement (Ashton and Lewry, 1994). Deformation that accompanied crustal thickening and post-peak metamorphism continued until 1770 Ma.

The HLB contains numerous metamorphosed synvolcanic hydrothermal alteration zones and spatially and genetically associated sulphide mineralization. These include felsic aluminous schists and zones of cordierite-anthophyillite and garnet-amphibole alteration in more mafic rocks. Two significant VMS deposits are associated with the felsic volcanic strata. The Western Nuclear Pb-Zn deposit (1967–69) produced 147 Mt grading 9.99 percent Zn, 0.51 percent Cu, and 5.83 percent Pb, with minor Au and Ag. The McIlvenna Bay deposit has estimated reserves (pre-National Instrument 43-101) of 13.08 Mt grading 4.95 percent Zn, 1.26 percent Cu, 0.53 g/t Au, and 24.3 g/t Ag (CAMECO, 1992).

Amisk Collage

The Amisk collage is comprised of a series of fault-bounded tectonostratigraphic assemblages (Syme, 1995; Lucas et al., 1996; Figs. 2, 4). These are intruded by post-accretionary plutons and are overlain by fluvial-alluvial sedimentary rocks of the Missi Group. The collage is bounded to the west by the Sturgeon-Weir fault system, to the east by the Morton Lake fault zone, and to the north by the southern flank of the Kisseynew domain. It extends to the south below the Phanerozoic cover. The Amisk collage contains the West Amisk, Birch Lake, Flin Flon, and Fourmile Island oceanic arc assemblages, and the Sandy Bay and Elbow-Athapapuskow back-arc basin basalt assemblages (Stern et al., 1999; Syme et al., 1999; Figs. 2, 4).

The West Amisk arc assemblage consists of island arc tholeiitic basalt overlain by the West Amisk differentiated calc-alkalic suite (high-Al basalt to rhyolite; Figs. 2, 4). The felsic component of the calc-alkaline suite forms a series of rhyolite domes, associated volcaniclastic facies, and hypabyssal intrusions that are overlain by a Hawaiian-type emerging shield volcano of basalt and andesite flows, and scoriaceous volcaniclastic facies (Ayres et al., 1991). The arc tholeiite and calc-alkaline suites are overlain to the north by mafic greywacke-argillite and minor pebble conglomerate. The West Amisk arc assemblage contains the hybrid VMS-epithermal Laurel Lake deposit (Fig. 4) and numerous small auriferous occurrences of arsenopyrite-rich quartz-carbonate veins and vein systems (Byers and Dahlstrom, 1954; Walker and McDougall, 1987).

The West Amisk assemblage is in fault contact to the east with the Sandy Bay assemblage (Fig. 4), which consists of pillowed to massive basalt interpreted to represent an oceanic plateau succession that formed either in an oceanic or back-arc basin (Reilly et al., 1994; Watters et al., 1994; Ryan and Williams, 1999; Syme et al., 1999).

The Birch Lake assemblage is flanked to the west by the Mosher Lake shear zone (Fig. 4). It consists of a 2 to 8 km-thick succession of tholeiitic to boninitic eruptive flows and associated subvolcanic sills (Watters et al., 1994; Stern et al., 1995a,b; Wyman, 2003). The volcanic succession contains little to no volcaniclastic material and no known felsic extrusive strata. There are four known Cu-Zn VMS deposits, Coronation, Birch, Flexar, and Konuto (Table 1), located at the contacts between these massive to pillowed mafic flow units.

The Flin Flon assemblage consists of a number of fault-bounded structural blocks that have little or no stratigraphic correlation (Syme and Bailes, 1993; Stern et al., 1995a; Syme, 1995; Lucas et al., 1996; Fig. 4).

- The Flin Flon block (Fig. 4, 5a, b) is bounded to the west by the West Arm fault and to the east by the Ross Lake fault, and is host to five rhyolite-associated Zn-Cu-Au VMS deposits (Flin Flon, Triple 7, Callinan, Schist, and Mandy). The block is a primitive, bimodal oceanic arc succession dominated by a number of tholeiitic basalt and basaltic andesite units, including mafic flows and volcaniclastic rocks, and minor rhyolite flows and domes, the latter including the Flin Flon mine rhyolite (1903 +7/–5 Ma; Stern et al., 1999; Syme et al., 1999).

- The Hook Lake suite is bounded to the west by the Ross lake fault and to the east by the Inlet Arm fault (Bailes and Syme, 1989; Fig. 4). The block comprises basalt, mafic volcaniclastic rocks, and minor rhyolite and volcanic conglomerate. It contains numerous base metal showings but no significant deposits to date.

- The Bear Lake suite consists of a four km-thick arc succession bounded to the west by the Inlet Arm fault and to the east by the Northeast Arm fault (Figs. 4, 6A), and is interpreted to be part of a subaqueous shield volcano and associated cauldron that is overlain by an intra-arc rift succession and associated intrusions (Bailes and Syme, 1989, Syme and Bailes, 1993; Fig. 6A). The shield volcano is composed of a 1 km-thick succession of massive to pillowed flows, minor flow breccia, and several synvolcanic gabbro/diorite sills (Fig. 6B). A cauldron fill succession of reworked dacitic tuff, interlayered andesite lapilli tuff, and finely laminated turbidite is host to the Cuprus and White Lake Zn-Cu-Pb VMS deposits (Fig. 6A). Minor rhyolite dome construction occurred prior to the emplacement of cauldron sediment fill. This arc-rift succession is overlain by 1886 Ma ferrobasalt

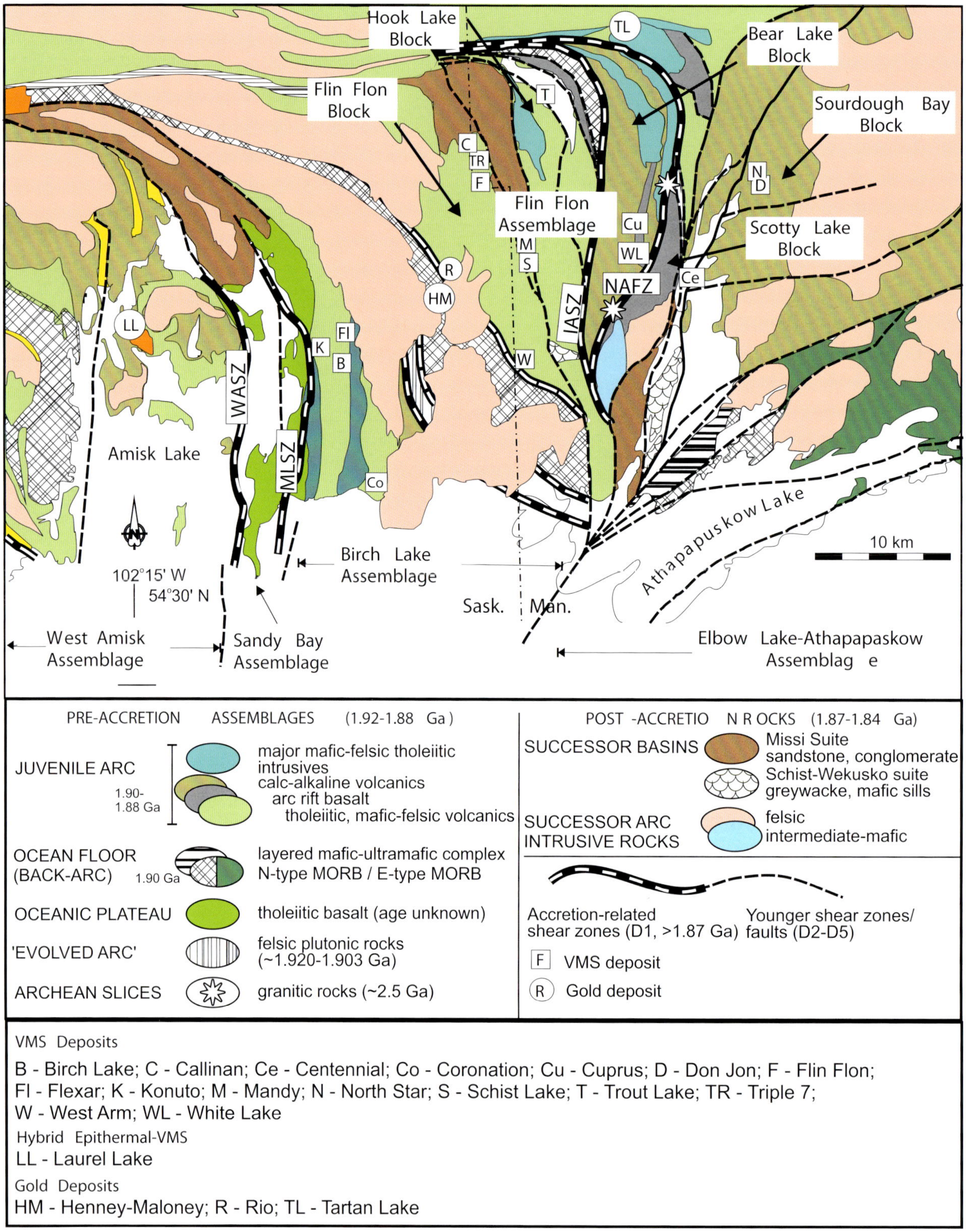

FIGURE 4. Tectonic assemblage map of the central part of the Flin Flon Belt illustrating the pre-accretion tectonostratigraphic assemblages, post-accretion plutons, volcano-sedimentary basins and faults. Also shown are the present- and past-producing VMS and Au deposits. Accretion-realted shear zones include: WASZ: West Amisk shear zone; MLSZ: Mosher Lake shear zone; IASZ: Inlet Arm shear zone; NASZ: Northeast Arm shear zone. Modified from Syme et al., 1999.

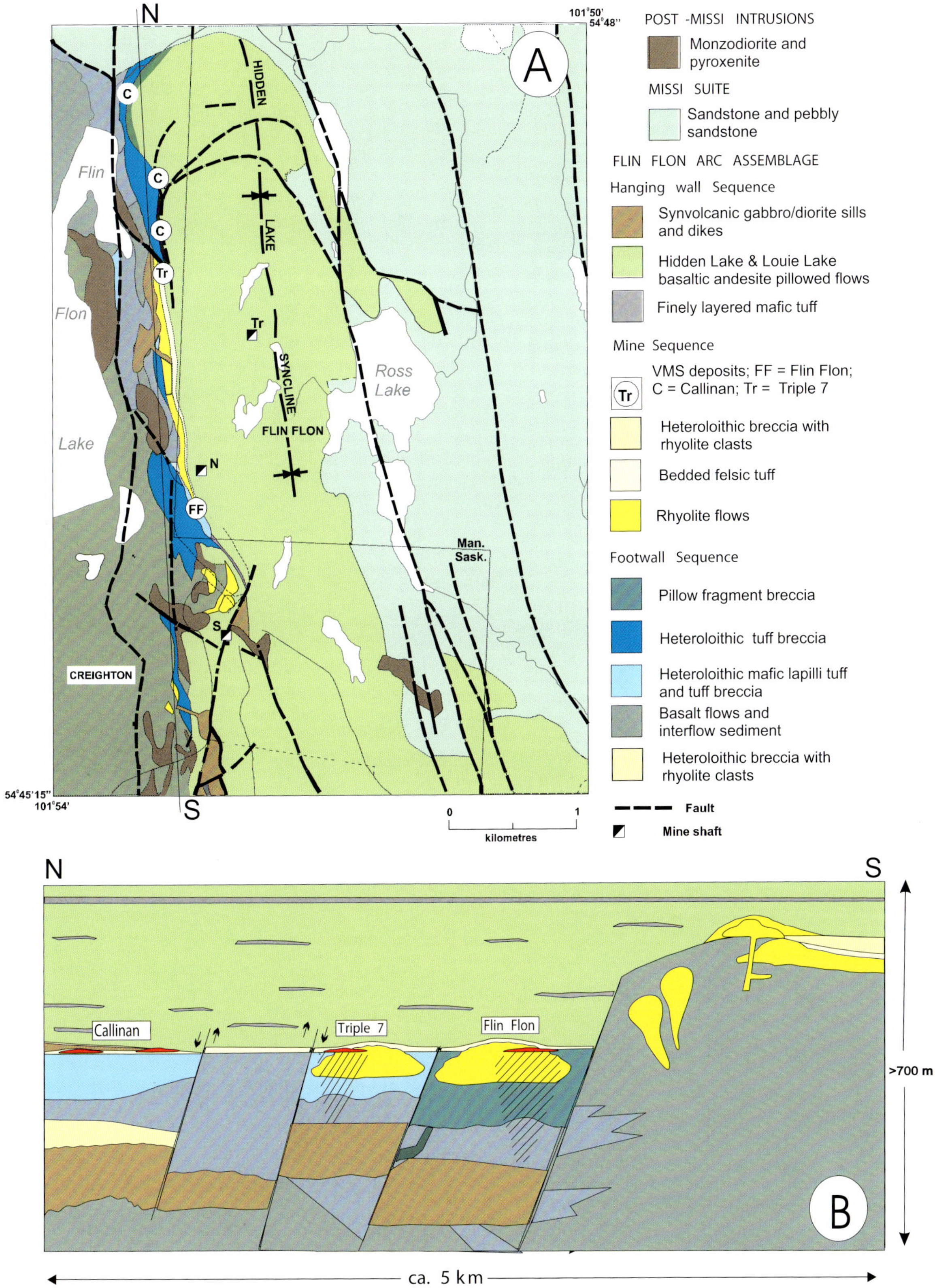

FIGURE 5. (A) Geology in the immediate host geology for the Flin Flon-Triple 7-Callinan VMS deposits showing their location along a stratigraphic interval containing rhyolite flows, felsic heterolithic breccia, and finely layered felsic tuff (Stockwell, 1960; Bailes and Syme, 1989; Devine et al., 2002). **(B)** Schematic cross-sectional reconstruction of the host stratigraphy to the deposits illustrating the presence of a cauldron structure. Modified from Syme and Bailes (1993); Ames et al., 2002; and Devine et al., (2002).

and 1885 Ma shoshonitic scoriaceous tuff, mudstone, and chert. Although the Bear Lake basaltic andesite and overlying the cauldron infill sequence represents rifting of a 1.9 Ga oceanic arc, the overlying ferrobasalt and shoshonite indicate back-arc basin development (Bailes and Syme, 1989; Syme et al., 1999).

- The Sourdough Bay volcanic suite is the most easterly VMS-hosting arc suite in the Flin Flon arc assemblage, and contains the Centennial, Pine Bay, North Star, and Don Jon Cu-Zn VMS deposits (Bailes and Syme, 1989; Mitchinson, 2004; Fig. 4). It is bounded to the west by the Centennial fault and to the east by the North Arm fault (Bailes and Syme, 1989). This volcanic suite is composed of transitional tholeiitic to calc-alkalic basalt and basaltic andesite, along with several large rhyolite dome complexes.

- The Elbow-Athapapuskow assemblage is a semi-continuous belt of juvenile ocean floor basalt and co-magmatic gabbro sills and mafic-ultramafic intrusions (Syme and Bailes, 1993; Stern et al., 1995b; Fig. 2). This assemblage is up to 25 km wide and over 100 km in length, and is in fault contact with flanking arc assemblages. The assemblage consists principally of N-MORB and subordinate E-MORB (Syme and Bailes, 1993; Stern et al., 1995b; Syme et al., 1999). The absence of VMS deposits in this back-arc assemblage led Syme and Bailes (1993) to postulate that the hydrothermal conditions necessary for VMS formation were not present in the Amisk collage.

- The Fourmile Island assemblage is a suite of tholeiitic intermediate to felsic flows and heterolithologic volcaniclastic rocks that host the Dickstone and Spruce Point Zn-Cu VMS deposits (Table 1). This 5.5 km-thick ocean arc volcanic succession represents the eastern limit of the Amisk collage. Its eastern limit is defined by the Morton Lake shear zone (Fig. 4), an early syn-accretion structure (Syme, 1990; Stern et al., 1995b; Lucas et al., 1996).

Snow Lake Area

The arc and ocean floor assemblages in the eastern part of the Flin Flon Belt are collectively sufficiently distinct from the Amisk collage arc assemblages to suggest that they represent remnants of unrelated arc terranes (Lucas et al., 1996; Syme et al., 1996). This eastern part of the belt is characterized by a number of allochthons in a thrust stack that is bordered to the west by the Morton Lake fault zone and to the east and north by the overthrust Kisseynew Domain (Bailes et al., 1994; Syme, 1995). These allochthons are comprised of the Snow Lake arc and the Northeast Reed and Roberts Lake ocean floor assemblages (Fig. 2) that are separated by major bounding fault systems. The 1.89 Ga Snow Lake arc assemblage (David et al., 1996; Bailes and Galley, 1999) is the only one that contains significant VMS mineralization. It is exposed in a thrust stack that includes several structurally imbricated slivers of 1.84 to 1.83 Ga post-accretion sedimentary strata of the Burntwood suite (Stern et al., 1995a; Connors et al., 1999). The >6 km-thick dominantly juvenile oceanic to crustally contaminated arc succession of the Snow Lake arc assemblage consists of three conformable volcanic successions that record the evolution from nascent or primitive arc (Anderson sequence), through mature arc (Chisel sequence), to rifting and opening of a back-arc basin (Snow Creek sequence; Fig. 7).

Anderson Sequence

The lowest recognized volcanic cycle is a compositionally bimodal primitive tholeiitic arc sequence dominated by pillowed to massive basaltic to basaltic andesite and subordinate Ca-boninite flows. This mafic succession contains at least three rhyolite flow complexes consisting of massive to lobe-hyaloclastite flows that are intruded by the 22 km-long subvolcanic Sneath Lake tonalite-trondhjemite intrusive complex (Bailes and Galley, 1999; Fig. 7). The last rock type within the primitive arc succession is a 1 to 5 m-thick, finely layered, sulphidic, mixed argillite-epiclastic chert Foot-Mud horizon (Fig. 7).

Chisel Sequence

The primitive nascent arc Anderson succession is conformably overlain by a mature arc sequence of intercalated flows, volcaniclastic rocks, and several discrete rhyolite flow complexes with associated monolithic debris flows of the Chisel sequence. The mafic flows in this sequence are typically more fractionated than those in the primitive arc (Stern et al, 1995a; Bailes and Galley, 1999). The volcaniclastic rocks consist of turbiditic and coarser-grained heterolithologic mafic, and mixed mafic and felsic debris flow deposits. The lower part of the sequence contains numerous synvolcanic intrusions, is extensively hydrothermally altered, and is capped by a series of discrete rhyolite flow complexes that host the Chisel Lake, Lost, Ghost, Chisel North, and Photo Lake Zn-Pb-Cu-Ag-Au-rich VMS deposits (Fig. 7). Overlying the VMS deposits are fine- to coarse-grained heterolithologic debris flows and associated mafic effusive flows.

Snow Creek Sequence

The debris flows of the upper part of the mature Chisel sequence arc succession are abruptly overlain by a thick succession of massive to pillowed basalt flows and associated basalt sills of the Snow Creek sequence (Fig. 7). This monotonous flow succession contains minor interflow hyaloclastite or flow breccia, indicating very rapid deposition. The N-MORB composition of these flows suggests they formed by rifting of the Snow Lake arc assemblage and development of back-arc ocean floor (Stern et al., 1995a; Bailes and Galley, 1999). The rift succession is in thrust contact with a structurally overlying succession that includes an imbricated 1.84 Ga Burntwood metagreywacke and a structural panel of arc volcanic rocks that resemble the upper part of the Chisel sequence. No VMS mineralization is known in this back-arc sequence.

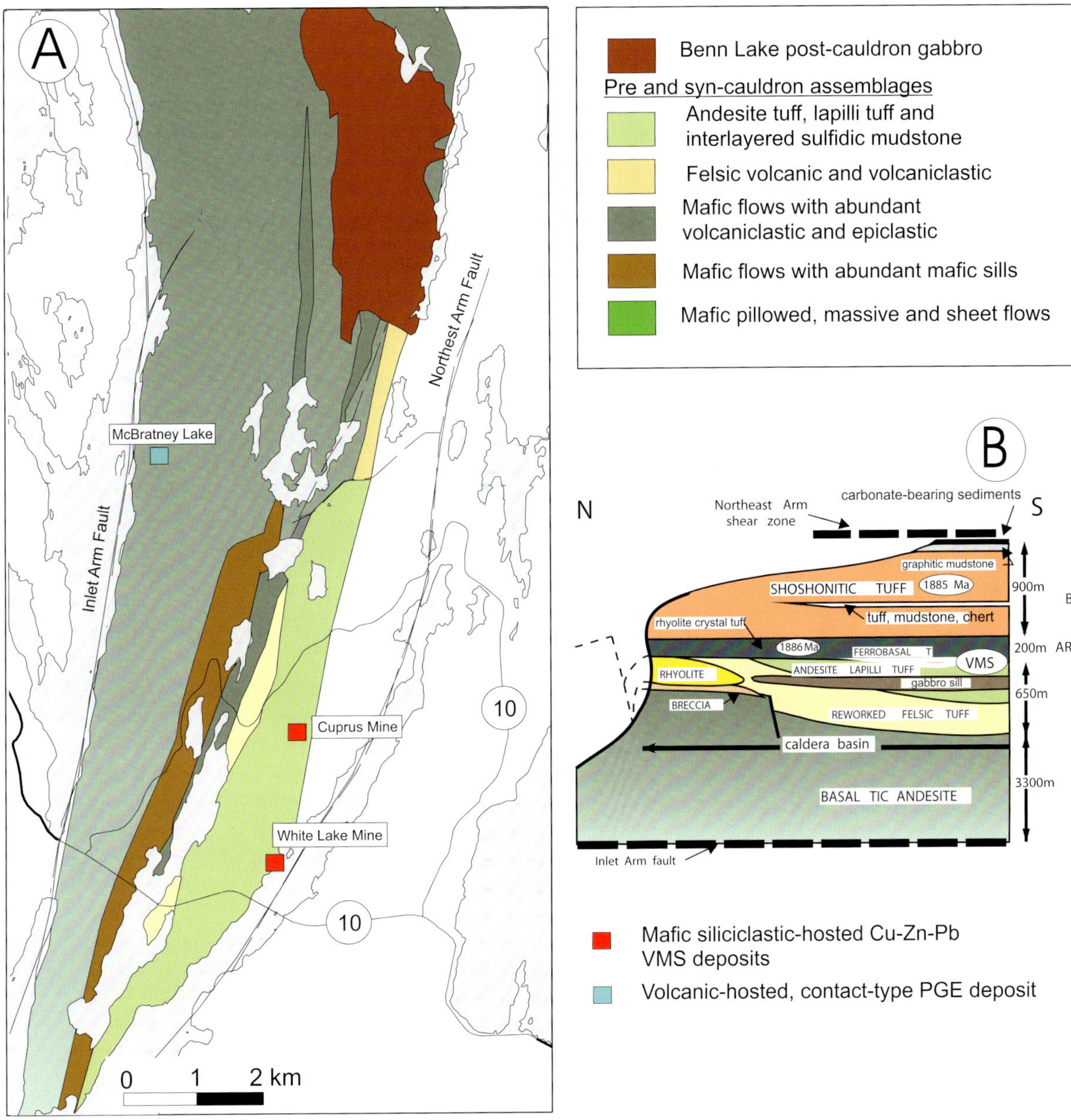

FIGURE 6. **(A)** Simplified geology of the Bear Lake suite in the vicinity of the White Lake and Cuprus Cu-Zn-Pb VMS deposits. Modifed from Bailes and Syme (1989). **(B)** Schematic cross section of the Bear Lake cauldron. From Syme and Bailes (1993).

Metallogeny of the Flin Flon Belt

The major mineralizing events recognized in the Flin Flon belt took place during the three main stages of crustal development: pre-accretion, post-accretion, and continent–continent collision. The pre-accretionary stage is represented by syngenetic base metal and Au deposits. The syn- to post-accretionary stage is characterized by several examples of intrusion-hosted base and precious metal deposits, and the continental collision stage by the development of orogenic Au deposits and REE-enriched pegmatites.

Pre-Accretion Mineralization

All the pre-accretion base and precious metal deposits and occurrences occur within the oceanic arc assemblages (Syme and Bailes, 1993). These include VMS deposits and at least one hybrid VMS-epithermal deposit. The VMS mineralization is classified by lithologic setting, which in most cases is unique to particular stages of arc development (Barrie and Hannington, 1999; Franklin et al., 2005), and includes VMS deposits hosted within mafic-back-arc, bimodal-mafic, bimodal-felsic, pelitic-mafic, and felsic-siliciclastic. The host lithologies comprise tectonostratigraphic assemblages that are commonly bound unconformable footwall and hanging-wall contacts and formed during specific volcano-magmatic or sedimentary events (Franklin et al., 2005).

The Flin Flon Belt VMS deposits display a wide range of

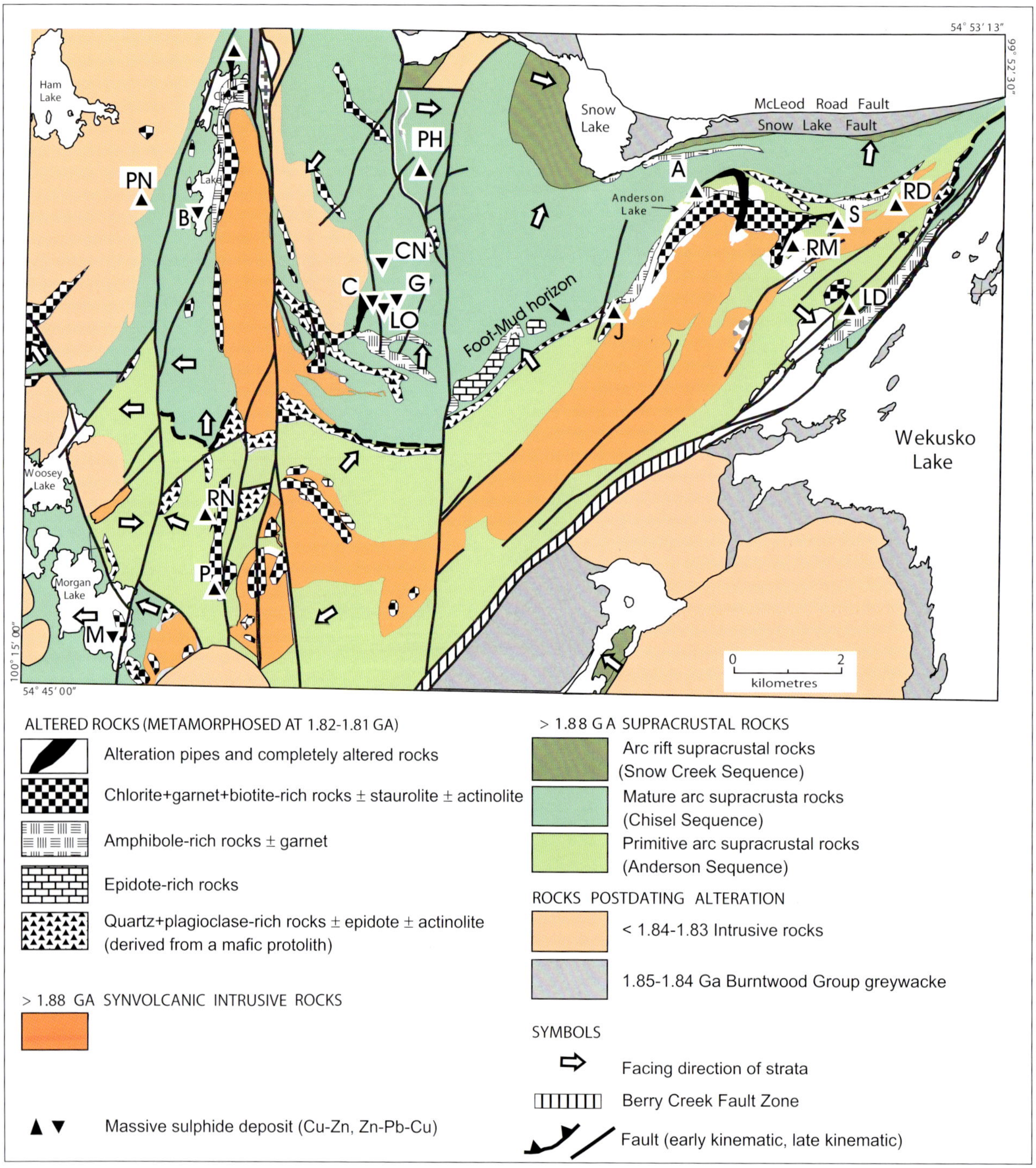

FIGURE 7. Generalized geology map of the Snow Lake arc assemblage, including the major synvolcanic and post-accretionary intrusions, large-scale metamorphosed hydrothermal alteration zones, and VMS deposits (underlined) and major occurrences. A=Anderson; B=Bomber zone; C=Chisel Lake; CN=Chisel North; G=Ghost; J=Joannie zone; LO=Lost; LD=Linda zone; M=Morgan Lake zone; P=Pot Lake zone; PH=Photo Lake; PN=Pen zone; RD=Rod; RM=Ram Zone; RN=Raindrop zone; S=Stall Lake. Modified from Bailes and Galley (1999).

metal contents and associations (Cu, Cu-Zn, Zn-Cu, and Zn-Cu-Pb VMS types; Fig. 8A), typical of the range of arc environments contained within this accretionary collage. Most of the Flin Flon VMS deposits contain between 1 and 2 g/t Au, which is in the range for most Canadian VMS deposits (Fig. 8B). Photo Lake, Coronation, Triple 7, and Flin Flon could be classified as Au-rich VMS deposits according to the classifications of Hannington et al. (1999) and Dubé et al. (2007). The Flin Flon Zn-Cu deposit produced over 120 tonnes (almost 4 million ounces) of Au, making it the largest Au producer in the belt and the third largest VMS producer of Au in Canada after the Horne and LaRonde deposits (Galley et al., 2007 and references therein)

Due to the extensive deformation history of the Flin Flon Belt, these pre-accretion syngenetic seafloor VMS deposits have undergone varying degrees of deformation, so that pri-

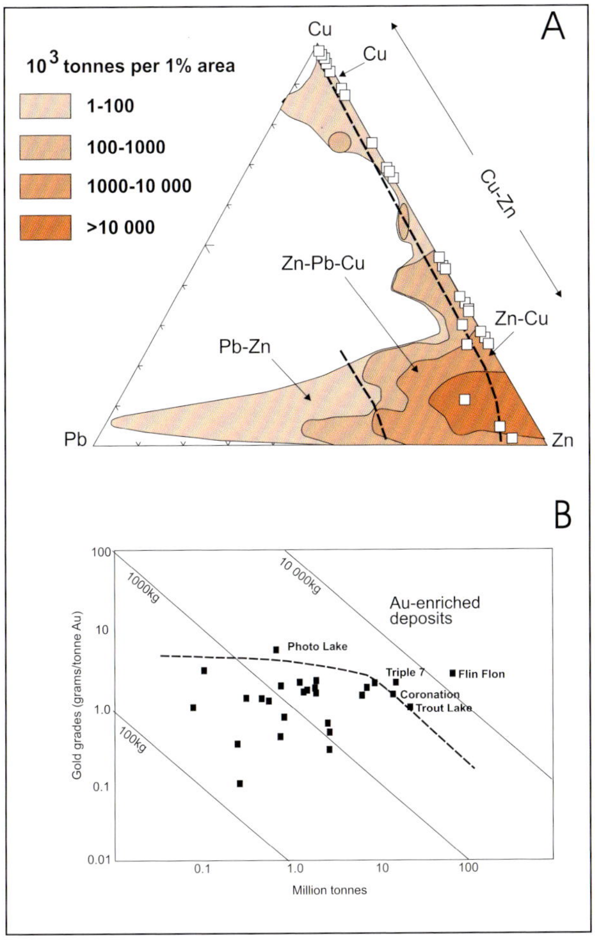

Figure 8. (A) Tonnage and grade of Flin Flon VMS deposits plotted on a Cu-Zn-Pb ternary plot for all Canadian VMS deposits as calculated by Galley et al. (2007). Data from Table 1. **(B)** Binary plot of Au content vs. tonnage for the Flin Flon VMS deposits. Although four of the deposits could be defined as Au-rich, almost all have Au contents that fall within the normal range of 1–2 g/t Au (Franklin et al., 2005).

mary features such as bedding are only rarely preserved (Fig. 9A,B). In most cases, the stratigraphic horizons along which these deposits formed acted as the foci for strain due to the presence of extensive phyllosilicate-rich alteration and the ductile nature of the sulphides, and because deposits commonly lie along stratigraphic contacts between competent volcanic formations. In intensely deformed cases, such as at the Osborne Lake deposit in the Snow Lake arc, originally discordant footwall alteration zones are transposed semi-parallel to stratigraphic contacts. Individual massive sulphide lenses are elongate and plunge parallel to the strong regional lineation. Other characteristics include the development of foliation (locally strong) in both host rocks and deposits, with tectonic compositional banding (Fig. 9C), sulphide blastesis (Fig. 9D), durchbewegung (tectonic breccias; Fig. 9E), sulphide cataclasis, and mylonitization (Fig. 9C), and secondary sulphide vein formation within and around the massive sulphide lenses.

In the case of deposits now hosted within lower amphibolite grade rocks, metamorphic and structural remobilization of the massive sulphides in lenses can produce coarse-grained sphalerite (Fig. 9F) and finer-grained pyrite surrounded by finer-grained massive to banded sphalerite-pyrite (Fig. 9D). Deposit wall rocks contain coarse-grained veins of chalcopyrite-arsenopyrite or galena-sphalerite, both with high Au (up to 10s of ppm) and Ag (up to 100s of ppm) contents (Galley et al., 1993). Where massive sulphide bodies are hosted in upper amphibolite grade rocks, there is also textural evidence for sulphide melt development. The mobility of base and precious metals during metamorphism and deformation has radically changed the original metal zonation in these VMS deposits.

Back-arc Mafic VMS Deposits

Mafic-dominated tectonostratigraphic assemblages are defined as those that consist primarily of effusive mafic flows with minimal interflow volcaniclastic units and <5 volume percent felsic volcanic rocks (Franklin et al., 2005). Synvolcanic hypabyssal mafic intrusions, in the form of sills and dikes, are common. These mafic volcanic successions are most common at ocean spreading centres where they are characterized by N-MORB basalts, or in nascent, fore-, or back-arc environments where the volcanic rocks can include N-MORB, low-Ti basalt, boninite, and arc tholeiitic basalts. Eruptive events may be separated by a mixture of iron-rich chemical sediment and fine-grained water-lain tuff commonly referred to as exhalite (Galley and Koski, 1999, and references therein).

In the Flin Flon Belt, the Sandy Bay and Birch Lake assemblages of the Amisk collage and certain parts of the Elbow-Athapapuskow back-arc and ocean floor assemblage are dominated by basalt flows (Figs. 2, 4). Of these, only the Birch Lake assemblage contains significant VMS mineralization (the Birch, Flexar, Coronation, and Konuto deposits; Table 1; Fig. 4). The deposits are localized at the contacts between either mafic effusive flows or between mafic flows and overlying flow breccias and mafic tuff, and consist of multiple massive sulphide lenses and footwall vein stockworks enclosed in strongly chloritized and silicified mafic rock. The sulphide mineralogy is dominated by chalcopyrite, pyrrhotite, and pyrite with subordinate magnetite and sphalerite (Byers et al., 1965; Tourigny et al., 2002). The Konuto deposit is composed principally of sulphide remobilized into syntectonic quartz and carbonate veins containing high concentrations of both Cu and Au (Tourigny et al., 2002).

Bimodal-Mafic VMS Deposits

Mafic-bimodal volcanic successions are dominated by mafic flows and associated autoclastic units and subordinate discrete felsic flow centres (Franklin et al., 2005). The rock assemblages hosting VMS deposits may contain significant volumes of mafic volcaniclastic strata and subordinate felsic autoclastic debris associated with dome complexes. Exhalite horizons are associated with the VMS mineralization and can be laterally extensive. Synvolcanic intrusions are common as dike swarms that are spatially associated with centres of seafloor hydrothermal activity and as larger quartz diorite-tonalite-trondhjemite sill-like intrusive complexes (Galley, 2003; Franklin et al., 2005). High abundances of footwall volcaniclastics, isolated felsic flow complexes, and discrete dike swarms are all characteristic of extensional graben or cauldron features that commonly host the bimodal-mafic-type VMS deposits.

Bimodal-mafic VMS deposits, the principal type of VMS

FIGURE 9 Photograph of representative hydrothermal facies. **(A)** Sulphide layers within the argillite unit hosting the West Arm VMS deposit; **(B)** Argillite-chert layers hosting the White Lake deposit are replaced by pyrite, pyrrhotite, and sphalerite; **(C)** Strong mylonitization and cataclasis of pyrite-sphalerite ore from the Mandy deposit; **(D)** Well-developed sucrose texture due to the development of metamorphic blastesis in sphalerite-pyrite ore from the Chisel VMS deposit. **(E)** Well-developed durchbewegung texture in pyrite-chalcopyrite-pyrrhotite ore from the Callinan deposit. Clasts include bedded chert, altered rhyolite, and quartz vein material; **(F)** Coarse-grained development of sphalerite and pyrite blastesis in the Ghost deposit. Note large, twinned sphalerite crystal (arrow).

deposit in the Flin Flon Belt, tend to form in clusters, a feature typical of mafic-bimodal VMS deposits. The pyrite-sphalerite-pyrrhotite-chalcopyrite massive sulphide lenses that are underlain by chlorite-sericite-quartz-rich alteration zones containing pyrrhotite-chalcopyrite-pyrite vein stockworks (Franklin et al., 2005; Galley et al., 2007). The Flin Flon (62.5 Mt), Callinan (8.4 Mt), and newly developed Triple 7 (14.2 Mt) deposits occur in a single felsic volcanic-dominated interval (Table 1; Fig. 5A) within a 4 km-thick tholeiitic basalt-dominated sequence (Bailes and Syme, 1989;

Thomas, 1992; Syme and Bailes, 1993). Mafic breccias infill a cauldron structure that is at least 4 km in diameter (Syme and Bailes, 1993; Devine et al., 2002) and is truncated to the north by a fault contact with overlying fluvial sedimentary rocks of the Missi Group. The three orebodies lie along a 3 km strike length containing several felsic flows, domes, and associated heterolithologic debris flows. The felsic volcanic centres are underlain by a thick sequence of mafic debris flows that are interlayered with mafic effusive flows. The VMS deposits and associated felsic volcanic strata are overlain by a sequence of shallowly emplaced basaltic andesite peperitic sills in finely layered mafic tuff. These are in turn overlain by a thick sequence of pillowed mafic flows.

The Mandy and Schist Lake deposits (Fig. 4) are located to the southeast of and have a footwall stratigraphy similar to the Flin Flon-Callinan-Triple 7 deposit cluster, but their stratigraphic hanging wall is faulted away (Simard, 2006). Other examples of bimodal-mafic-hosted VMS deposits include the Dickstone and Spruce Lake deposits in the Fourmile Island arc assemblage (Fig. 2), and the Anderson-Stall-Rod-Linda cluster of VMS deposits in the Snow Lake arc assemblage (Fig. 7). The latter are hosted within a tholeiitic felsic rhyolite complex near the top of a >3 km succession of subaqueous low-Ti tholeiitic basalt to basaltic-andesite flows that includes a unit of high-Ca boninite (Stern et al., 1995a). High-Ca boninites are relatively rare in modern settings, having only been identified within the northern end of the Tongan forearc (Falloon et al., 1992) and in the upper Pillow Lavas of the Cretaceous Troodos ophiolite (Duncan and Green, 1987). On the basis of the high-Ca boninite and low-Ti basalt association, Stern et al. (1995a) suggested that the primitive bimodal volcanic phase occurred in a nascent arc tectonic setting. This primitive arc succession is intruded by a 15 km-long quartz diorite-tonalite-trondhjemite intrusive complex (Fig. 7). The deposits are spatially associated with a sulphidic exhalite known as the Foot-Mud horizon that extends for 15 km (Bailes and Galley, 1999). The Anderson, Stall, Rod, and Linda deposits are all localized along different internal flow contacts with an 8 km-long felsic dome complex that is intruded by the latest phase of the Sneath Lake subvolcanic intrusive complex.

Bimodal-Felsic VMS Deposits

VMS-hosting bimodal felsic assemblages are defined as those that include equal proportions of mafic and felsic volcanic strata, with a large component of both mafic and felsic volcaniclastic rocks typical of more mature island-arc assemblages (Franklin et al., 2005). The VMS deposits are spatially associated with felsic flow complexes underlain by differentiated tholeiitic to transitional tholeiite-calc alkaline basalt-andesite-dacite assemblages. The deposits are Zn-Pb-Cu-Ag-Au-rich and characterized by sphalerite-pyrite-galena-rich massive sulphides and a pyrrhotite-chalcopyrite-sphalerite-rich footwall vein and stockwork complex. The formation of seafloor carbonate appears to pre-date the generation of the massive sulphides, which partially replaces the carbonate, forming skarnoid alteration assemblages (Galley et al., 1993). The presence of carbonate is a characteristic of this deposit type in the Paleoproterozoic in several terranes worldwide (Galley, 1996). Footwall alteration generally includes extensive zones comprising dominant sericite-quartz-pyrite and more restricted chlorite-quartz-pyrite. This bimodal-felsic-type of VMS deposits occurs within the Chisel mature arc sequence of the Snow Lake arc assemblage (Fig. 7), and is represented by the Chisel Lake, Chisel North, Lost, and Ghost deposits (Table 1).

In the Hanson Lake Block (Fig. 2), a fractionated arc assemblage of tholeiitic oceanic arc rocks contains a large felsic complex consisting of dacite and rhyolite flows, pyroclastic, and autoclastic flows. The eruptive complex is intruded by quartz-feldspar porphyry and overlain by mixed volcaniclastic units followed by well-bedded greywacke. Mixed oxide-silicate-sulphide facies iron formation is also present in places near the base of the greywacke succession. Massive sulphide mineralization in located within and on top of the felsic sequence and includes both Pb-Zn-Cu (Western Nuclear) and Cu-Zn (McIlvenna Bay) massive sulphide deposits (Maxeiner et al., 1993).

Hybrid Epithermal VMS Deposits

There are several examples in the literature of a sub-set of bimodal-felsic VMS deposit types that share some of the characteristics of epithermal deposits (Hannington et al., 1999; Galley et al., 2007). These deposits are commonly associated with rhyolite dome complexes, have high precious metal contents, and are dominated by aluminous proximal alteration assemblages, such as quartz-sericite-pyrite. They are postulated to have formed in shallow-water environments in which the hydrothermal fluids boiled in the subsurface (Sillitoe et al., 1996; Hannington et al., 1999). This resulted in the oxidation and acidification of the rocks and formation of aluminous alteration mineral assemblages. The deposits are commonly pyrite-rich, with abundant tennantite-tetrahedrite, and galena, and have high As, Sb, and Hg, along with Au and Ag. There may also be a magmatic volatile component involved in the formation of the deposit as a result of degassing of an underlying, shallow felsic intrusion (Sillitoe et al., 1996).

The Laurel Lake Au-base metal deposit hosted by the West Amisk arc assemblage is perhaps the only known example of a hybrid epithermal-VMS hydrothermal system in the Flin Flon Belt (Fig. 4; Walker and McDougall, 1987; Ansdell and Kyser, 1991). The Laurel Lake deposit occurs within a mature oceanic arc, fractionated basalt-dacite-rhyolite sequence of rocks. The Au-Ag mineralization is hosted within a cluster of hydrothermally altered porphyritic rhyolite and rhyodacite extrusive complexes, and associated hypabyssal stocks and dikes. Although the host volcanic suite is believed to be approximately 1.88 Ga, the nearby hydrothermally altered syn- to post-accretionary Missi Island quartz diorite, trondhjemite, and quartz monzonite intrusive suite is considerably younger (1860–1837 Ma; Ansdell and Kyser 1991). The deposit consists of a number of anastomosing quartz-sulphide veins containing quartz, muscovite, pyrite, tennanite, chalcopyrite, sphalerite, galena, electrum, and carbonate within an alteration halo of predominantly sericite-pyrite. Gold and Ag are contained within the electrum and tennanite. The veins are deformed and the alteration assemblage is metamorphosed and crosscut by D_2 deformation features. The deposit forms the core of a broader sericite-pyrite-rich alteration zone that contains several other auriferous quartz-sulphide stockwork systems (Byers and Dahlstrom, 1954). The mineralizing fluids were saline (<10.3 wt. % NaCl equivalent) and CO_2-bearing, and are believed to represent dominantly seawater that was modified at temperatures of around 300°C (Ansdell and Kyser, 1991).

Pelitic-Mafic VMS Deposits

Pelite-mafic VMS deposits are a category that includes seafloor deposits that formed in volcanic basins or rifts partly infilled by sediment. This can include discrete cauldrons or larger back-arc rifts (Franklin et al., 2005). The infill sediment are commonly a mixture of pelite and turbidite, which are usually intruded by mafic sills. There is no spatially associated effusive mafic or felsic volcanic strata. The deposits consist typically of laterally extensive interlayered sulphide and clastic sediment, and include the Cuprus and White Lake Zn-Cu-Pb-Ag deposits. These deposits are hosted within the shaley laminated turbidites in a cauldron infill sequence floored by reworked dacitic tuff and are overlain by andesitic lapilli tuff (Bailes and Syme, 1989; Fig. 6). The deposits are composed of both massive sulphide and finely bedded to interbedded sulphide-shale. The sulphides are composed of pyrite, sphalerite, pyrrhotite, chalcopyrite, and galena and are underlain by chlorite-quartz-rich hydrothermal alteration pipes and vein stockworks.

The West Arm Cu-Zn deposit (Fig. 4) is a stratiform, massive to finely bedded ore lens hosted by a sequence of graphitic argillite, pyritic siltstone, and banded chert-hematite iron formation (Syme, 1987). This deposit is underlain by variably silicified and carbonatized pillowed basalt, and related pillow fragment breccia.

VMS-related Hydrothermal Alteration

One of the distinctive features of VMS deposits is the scale and composition of accompanying hydrothermal alteration (Galley, 1993; Franklin et al., 2005; Galley et al., 2007). VMS deposits form on or near the seafloor from the focused discharge of hot, saline, low-pH hydrothermal fluids (Galley, 1993 and references therein). These ore-forming fluids are generated by the interaction of circulating seawater and sub-seafloor strata at various temperatures, with the heat commonly supplied by shallowly emplaced subvolcanic intrusions (Galley, 2003). The result of this seawater-rock interaction at temperatures that range from approximately 140°C to >400°C is the formation of alteration zones of varying composition and size (Galley, 1993 and references therein; Alt, 1995 and references therein). These zones may range in size from several tens of kilometres to several tens of metres, depending on whether they represent widespread diffuse recharge or discharge of hydrothermal fluid, or focused discharge. Each alteration facies has a particular mineralogy and chemistry depending on rock composition, fluid composition, and temperature of rock-fluid reaction. Because of their close genetic and spatial relationship to VMS deposits, alteration zones provide an effective vector in locating massive sulphide mineralization.

In the Flin Flon Belt, seafloor hydrothermal alteration zones is commonly accentuated through mineralogical changes that accompany regional metamorphism. Where the original alteration is characterized by hydrated mineral assemblages, the secondary metamorphic mineral assemblages can be quite distinctive. Large-scale distal alteration can take the form of epidotization, silicification, or Mg-Fe metasomatism. At Snow Lake, the top of the primitive arc Anderson sequence is affected by a zone of silicification and bleaching over 15 km long (Figs. 10; 11B). Extensive zones of Fe-Mg metasomatism focused about the rhyolite flow complexes are characterized by a chlorite-aluminosilicate-rich assemblage. In the Bear Lake Block (Fig. 4), there is a zone of epidote alteration >10 km in strike length and up to 4 km thick (Bailes and Syme, 1989; Fig. 11A). Zones of silicification (Fig. 11C), Fe-Mg metasomatism, and garnet-amphibole alteration (Fig. 11D) underlie the Chisel-Lost-Ghost VMS deposits within the Chisel sequence (Skirrow and Franklin, 1994; Bailes and Galley, 1999; Figs. 6, 10). Subvolcanic intrusion in both of these arc sequences are also affected by hydrothermal alteration (Fig. 11E,F).

Post-accretion Mineralization

The pre-accretion 1.92 to 1.88 Ga oceanic volcanic arcs and intervening oceanic basin successions were intruded by 1.87 to 1.84 Ga post-accretion granitoid intrusions associated with successor arc magmatism (Lucas et al., 1996). Of these, several contain occurrences of Cu-Mo mineralization (Baldwin, 1980; Galley and Franklin, 1987). The 1.84 Ga Phantom-Boot Lake intrusive suite is spatially associated with numerous occurrences of Au, Cu, Mo, and W (Galley and Franklin, 1987). The intrusive complex contains a series of sills, stocks, and dike swarms of diorite, hornblende porphyritic quartz monzodiorite, monzodiorite, quartz monzonite, and microcline porphyritic granite. Mineralization is present as: 1. chalcopyrite, scheelite molybdenum and Au in quartz breccias associated with hematite-quartz-ferrodolomite-epidote alteration, and 2. Au-scheelite-chalcopyrite mineralization in fracture-controlled, sericite-quartz-carbonate-pyrite alteration zones. Mineralized vein and breccia-vein systems are hosted either by the various phases of the intrusive complex or by the surrounding volcanic strata where they are intruded by feldspar porphyritic granite dikes.

Also formed during this tectonic stage in the evolution of the Flin Flon Belt are Ni-Cu-PGE-rich ultramafic to differentiated gabbroic intrusions that are pre-Missi Group sedimentation but post-date arc accretion and compression. Although poorly constrained, they appear to have been emplaced about 45 million years after arc formation between 1.84 and 1.845 Ga (Bailes and Theyer, 2006). These intrusions are commonly small (<1000 m long and 500 m wide) and occur within several tectonic assemblages of the Flin Flon Belt. These include the Namew Lake intrusion with associated a Ni-Cu-rich pyrrhotite-pentlandite-violerite massive sulphide lens (Menard et al., 1996) and the Wonderland gabbro suite (Oliva et al., 2002; Bailes and Theyer, 2006). The Namew Lake intrusion is an 1847±6 Ma komatiitic amphibolite that intruded into an upper amphibolite gneiss terrane, believed to be equivalent to Flin Flon Belt strata, underlying Phanerozoic cover rocks. The McBratney Lake sulphide showing is spatially associated with a small gabbro body that is part of the Wonderland in-

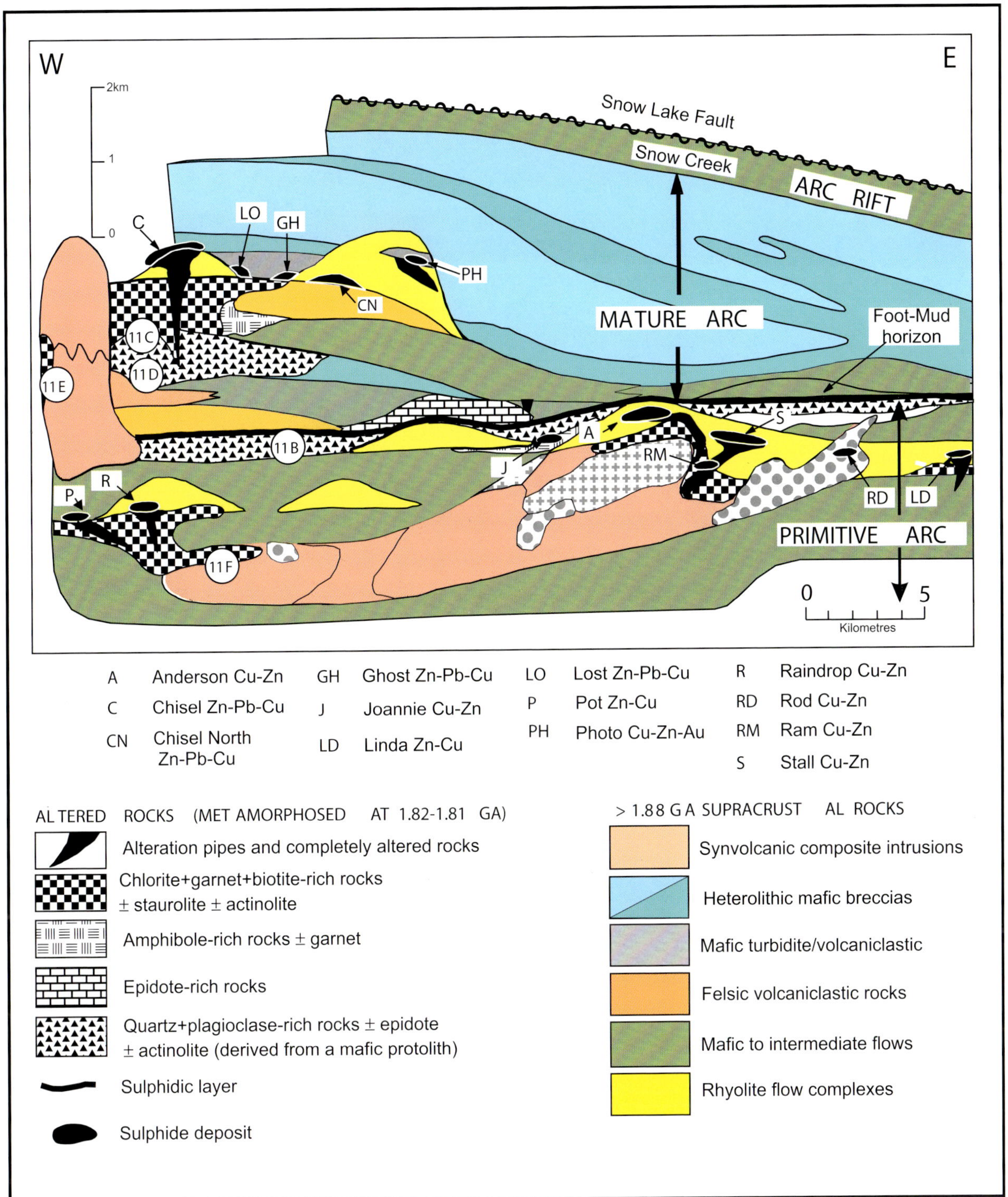

FIGURE 10. Schematic cross section through the Snow Lake arc assemblage illustrating the distribution and types of metamorphosed hydrothermal alteration zones and their relationship to known VMS deposits. Photos in Figure 11 are indicated. Modified from Bailes and Galley (1999).

trusive suite dated at 1839±3.9 Ma (Bailes and Theyer, 2006). This PGE-Au-rich showing consists of veins containing chalcopyrite, Ni-pyrrhotite, pentlandite, millerite, and magnetite, plus several PGE-bearing Te-As-Sb minerals (Oliva et al., 2002).

Collisional Tectonic Mineralization

The Flin Flon Belt is host to numerous shear zone-hosted orogenic Au deposits and occurrences that have been exploited at various times over the last 100 years. Only sev-

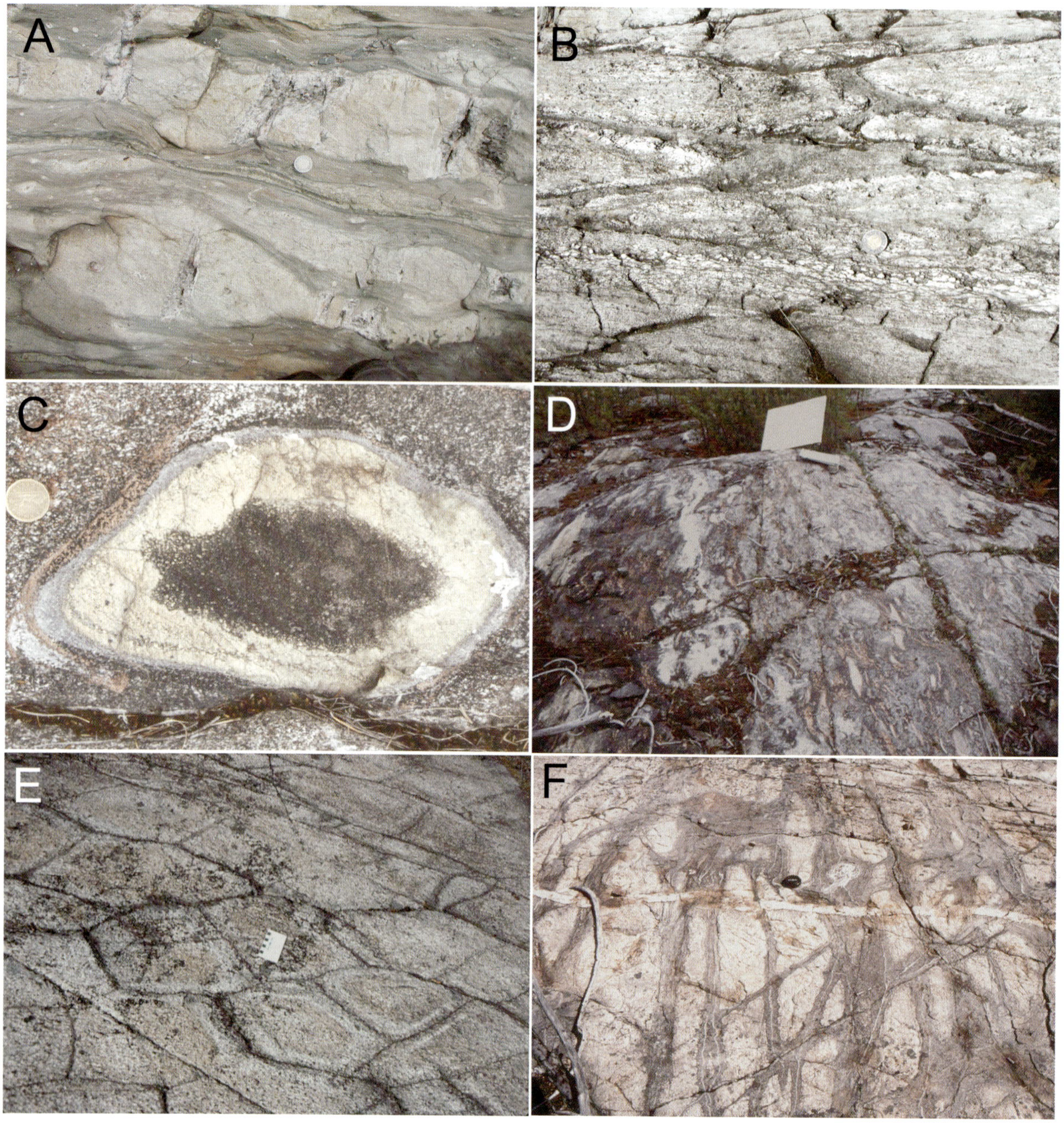

FIGURE 11. Photographs of representative alteration types. **(A)** Pillowed basalt flows strongly altered to an epidote-quartz-plagioclase assemblage within a zone of extensive hydrothermal alteration below the Cuprus and White Lake VMS deposits in the Bear Lake Block; **(B)** Silicified margins on pillows from the basal pillowed basalt formation of the Bear Lake Block. **(C)** Close-up of the complex alteration of a clast of feldspar-phyric basalt, now represented by the metamorphic assemblage of plagiocalse-actinolite-epidote-garnet. The mafic breccia is part of the silicified Edwards Formation, Chisel sequence, in the deep footwall to the Chisel, Lost, and Ghost Zn-Pb-Cu VMS deposits, Snow Lake; **(D)** Intense quartz-feldspar-amphibole-garnet alteration at the contact between finely layered mafic volcaniclastic and overlying mafic volcanic breccia of the Edwards Lake Formation. These altered strata lie approximately 1,200 m below the Chisel-Lost-Ghost VMS deposit horizon and are interpreted to represent the metamorphosed equivalent of a sub-seafloor high-temperature reaction zone (Bailes and Galley, 1999); **(E)** Columnar-jointed facies in a trondjemite phase of the Sneath Lake subvolcanic intrusion, Anderson sequence (Fig. 6). Fracture surfaces are defined by a quartz-feldspar-staurolite metamorphic assemblage typical of felsic igneous rocks affected by extensive sub-seafloor Si-Al-Fe-enriched alteration, and then metamorphosed to biotite-almandine facies; **(F)** Fracture pattern within the margin of the Sneath Lake syn-volcanic intrusion. This trondhjemite phase of the Richard Lake subvolcanic intrusion, Chisel sequence (Fig. 6), was hydrofractured by circulating hydrothermal fluids. Subsequent regional metamorphism formed a (Mg-chlorite)-(Mg-biotite)-staurolite-quartz-rich mineral assemblage. Modified from Bailes and Galley (1999).

en deposits have recorded production, of which only the New Britannia mine has produced over 30 tonnes (1 million ounces) of Au (Table 2). This compares to the Flin Flon VMS deposit, which alone has produced over 120 tonnes of Au (Table 1; Fig. 8B).

Orogenic Au mineralization in the Flin Flon Belt is localized within ductile oblique-slip, high-angle shear zones and southwest-verging thrusts that developed during D_3 peak

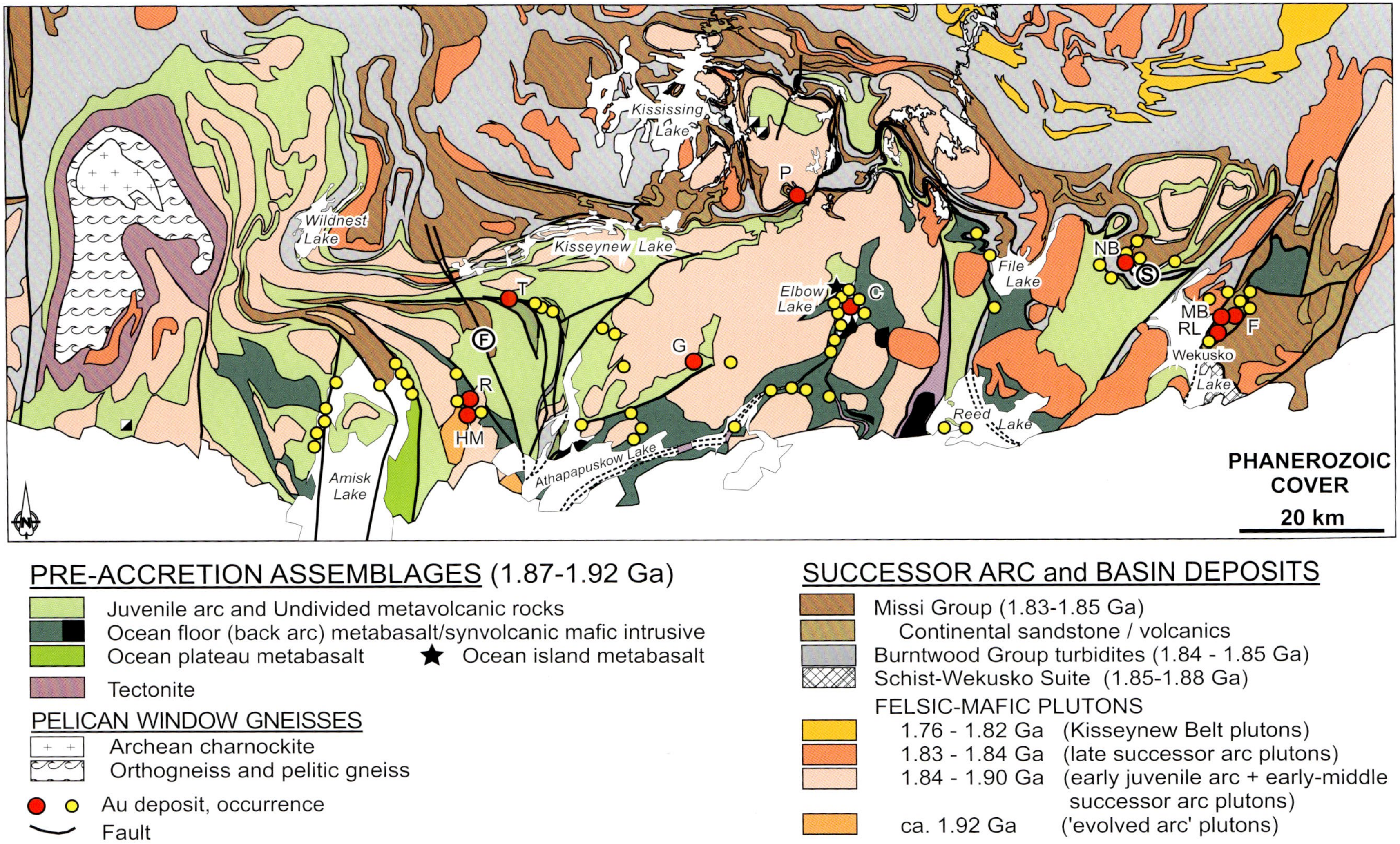

FIGURE 12. Distribution of Au occurrences (yellow dots) and past producers (red dots) in the Flin Flon Belt. The Au mineralization tends to form as clusters of occurrences along major fault systems. Gold mines: C=Century; F=Ferro; G=Gurney; HM=Henney-Maloney; MB=Moosehorn-Ballast; NB=New Britannia; P=Puffy Lake; R=Rio; RL=Rex-Laguna; T=Tartan Lake. From Richardson and Ostry (1996).

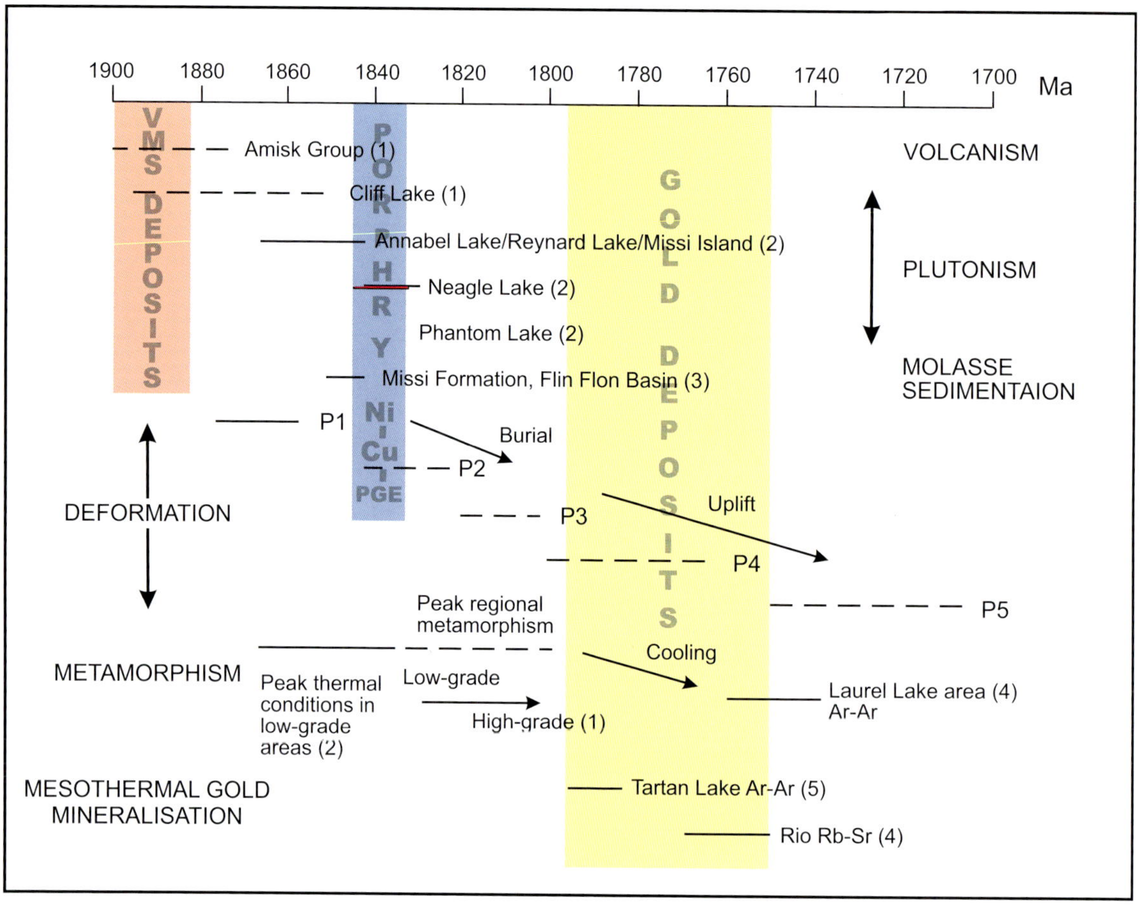

FIGURE 13. Summary of tectonic evolution and major deposit-generating events, Flin Flon Belt. Modified from Ansdell and Kyser (1992).

metamorphic conditions between 1840 and 1805 Ma (Ansdell and Kyser, 1991; Fedorowich et al., 1995). These D_3 shears formed as a result of regional collisional shortening, crustal thickening, and uplift (Ansdell and Kyser, 1991; Fedorowich et al., 1995; Lucas et al., 1996). Many of these shear zones were then reactivated under post-peak metamorphic brittle-ductile conditions during the D_4, dominantly strike-slip faulting. D_4 brittle-ductile shearing and faulting was accompanied by intense refolding of the D_3 shear foliations, with steep plunges on the fold axes (Fedorowich et al., 1995). Most of the shear zone-hosted Au occurrences and deposits form in clusters about these reactivated fault systems. These include the Wekusko Lake, Snow Lake, Northstar, Elbow Lake, Phantom Lake, East, and West Amisk Lake clusters (Fig. 12).

Major Au deposits and occurrences are localized at lithologic contacts that are characterized by a large competency contrast. In the case of the Tartan Lake and Rio deposits (Figs. 4, 12), Au is concentrated along volcanic-intrusive contacts where there is a dilation or jog in the host shear zones, or an apophysis of volcanic or sedimentary rock in the intrusion (Pearson et al., 1986; Fedorowich et al., 1995). Several of the Wekusko Lake deposits lie along the contact between Missi Group sedimentary strata and shallowly emplaced quartz porphyry sills (Stockwell, 1937). At the New Britannia deposit the ore lenses lie along a folded basalt-rhyolite contact near the apex of a fold (Galley et al., 1988). In the case of the Ferro Mine of the Wekusko Lake camp, the deposit is hosted by a shear zone occupying the axis of a F_2 isoclinal fold (Stockwell, 1937).

The principal auriferous vein systems are oriented parallel to the steeply dipping C-fabric within the shear zones, and have a crack-and-seal texture with thin discontinuous bands commonly represented by oriented and sheared fragments of mineralized wall rock (Ansdell and Kyser, 1991). Secondary mineralized veins formed as steeply dipping extensional features crosscutting the shear-parallel veins. Auriferous zones within the veins commonly occur near the intersection of anastomosing shears with moderate to steep plunges parallel to the intersection lineation (Fedorowich et al., 1995). The auriferous vein events are commonly pre-dated by the formation of quartz-tourmaline-carbonate veins and overprinted by shallowly dipping quartz veins believed to be associated with development of D_5 brittle fault systems.

Gold mineralization formed over an extended period of post-peak metamorphic cooling between 1790 Ma and 1740 Ma, as indicated by the Ar-Ar and Rb-Sr dating of muscovite and tourmaline from auriferous Au veins (Ansdell and

Kyser, 1991; Fedorowich et al., 1995). The O and H isotope systematics of hydrothermal minerals associated with Au mineralization are compatible with the formation of auriferous fluids by devolatilization reactions during prograde metamorphism (Ansdell and Kyser, 1992). This is supported by H_2O-CO_2-NaCl fluid inclusion compositions and pressure-temperature calculations indicating temperatures and pressures of vein quartz formation from 250° to 420°C and 1.2 to 2.4 kbars (4.3–8.6 km), respectively. Orogenic Au mineralization began to form in the Flin Flon Belt at approximately 1805 Ma and continued for several tens of millions of years during regional metamorphic cooling with accompanying transition from ductile to brittle deformation.

Summary

There are three major metallogenetic periods in the Flin Flon Belt. The period of pre-accretion mineralization included formation of VMS deposits within a variety of submarine oceanic arc and extensional arc settings formed between 1.91 and 1.88 Ga (Fig. 13). The differing character and distribution of the Flin Flon Belt VMS deposits can be related to different extensional arc settings.

The arc and intra-arc basins were compressed together to form a series of accreted arc assemblages between 1.88 and 1.87 Ga. This post-accretionary period (1.87–1.84 Ga) was characterized by the emplacement of various types of Tonalite-Trondjhemite-Granodiorite (TTG) granitoids associated with successor arc magmatism and tholeiitic to alkalic intermediate to mafic igneous rocks related to faulting and extension. Non-economic Cu-Mo mineralization is recognized in some of the former, whereas the latter includes intrusions and related Ni-Cu and PGE mineralization.

The following period of terrane collision and crustal thickening between 1.84 and 1.80 Ga was accompanied by the formation of fault- and shear-related orogenic Au deposits. The deformation and metamorphic history of these dominantly vein-hosted Au systems indicate formation during faulting related to peak regional metamorphic conditions, with continued vein deformation and Au deposition during post-peak cooling conditions.

Exploration Guidelines

Paleoproterozoic volcano-sedimentary terranes are most commonly explored for VMS deposits. Next to the Devonian, the Paleoproterozoic is the most prolific era for the generation of VMS deposits (Franklin et al., 2005). As with most widely prospected VMS terranes, there is little known of the geology of the Flin Flon Belt below the surface. Furthermore, much of the southern part of the Flin Flon Belt is overlain by Phanerozoic sedimentary rocks of varying thickness, which further complicates evaluating the VMS potential of the belt. The most successful method to date for discovering VMS deposits in the Belt has been airborne electromagnetic surveys. An increased knowledge of the tectonostratigraphy has allowed companies to use lithogeochemistry to focus on bimodal primitive rifted arc sequences, which host most of the VMS deposits. These tectonostratigraphic domains are now being traced northwards into high-grade metamorphic terranes originally believed to be part of the sediment-rich Kisseynew Domain. The presence of amphibolite regional metamorphism has resulted in the recognition of large-scale hydrothermal alteration zones because of their distinctive mineralogy. Acquiring an understanding of the relationship between producing these metamorphic mineral assemblages and alteration geochemistry will result in a better understanding of how to vector towards VMS mineralization, particularly within the sub-Phanerozoic sections of the arc assemblages. Better assessment of VMS potential below the Phanerozoic cover should include high-resolution aeromagnetic surveys and wildcat drilling to characterize the magnetic signatures, thereby allowing known tectonostratigraphic domains to be traced in the subsurface. As metamorphism has remobilized many of the more volatile metals commonly associated with Paleoproterozoic VMS deposits (Hg, As, Sb), geochemical methods should be pursued to detect possible hanging-wall geochemical anomalies within the overlying Phanerozoic strata.

Orogenic Au deposits are not uncommon, but are typically smaller than their Archean counterparts, usually producing <300 tonnes of Au. The structural controls necessary to develop these deposits is the same for most terranes, regardless of age. These include subsidiary faults and shears parallel to major regional lineaments, volcanic-sedimentary reverse fault contacts, competency contrast linked with jogs and offsets in fault/shear zones, and chemical traps. There is some potential for hybrid epithermal-VMS Au deposits, but the volumes of the required shallow-water to subaerial terranes are very limited.

Acknowledgements

This summary of the present knowledge of the tectonic and metallogenic evolution of the Flin Flon Belt is the product of the work of many different geologists, whose contributions we hope are appropriately referenced in the text. This summary also benefited from conversations with D. Ames, K. Ansdell, C. Beaumont-Smith, C. Devine, M. Fedikow, H. Gibson, K. Gilmore, D. Zeihlke, H. Zwanzig and many others. We very much appreciated the thorough and thoughtful reviews of the draft manuscript by J. Peter, S. Piercey, and W. Goodfellow. Their comments did much to improve the clarity of the manuscript.

References

Alt, J.C., 1995, Subseafloor processes in midocean ridge hydrothermal systems: AGU Geophysical Monograph 91, p. 85–114.

Ames, D.E., Tardif, N., MacLachlan, K., and Gibson, H.L., 2002, Geology and hydrothermal alteration of the hanging-wall stratigraphy to the Flin Flon-777-Callinan volcanogenic massive sulphide horizon (NTS 63K12NW and 13SW), Flin Flon area, Manitoba: Manitoba Industry, Trade and Mines, Manitoba Geological Survey, Report of Activities 2002, p. 20–34.

Ansdell, K.M., 1993, U-Pb zircon constraints on the timing and provenance of fluvial sedimentary rocks in the Flin Flon and Athapapuskow basins, Flin Flon Domain, Trans-Hudson Orogen, Manitoba and Saskatchewan: Geological Survey of Canada, Paper 93-2, p. 49–57.

Ansdell, K.M., and Kyser, T.K., 1991, The geochemistry and fluid history of the Proterozoic Laurel Lake Au-Ag deposit, Flin Flon greenstone belt: Canadian Journal of Earth Science, v. 28, p. 155–177.

——1992, Geochemistry of granitoids in the western Flin Flon domain: Saskatchewan Geological Survey, Saskatchewan Energy and Mines, Miscellaneous Report 92-4, p. 149–157.

Ansdell, K.M., and Norman, A.R., 1995, U-Pb geochronology and tectonic development of the southern flank of the Kisseynew Domain, Trans-

Hudson Orogen, Canada: Precambrian Research, v. 72, p. 147–167.

Ashton, K.E., and Lewry, J.F., 1994, Vergence of the "Pelican Slide" and Sturgeon-weir Shear zone: LITHOPROBE, Report 38, p. 12–17.

Ashton, K.E., Wilcox, K.H., Wheatley, K.J., Paul, D., and de Tombe, J., 1987, The boundary zone between the Flin Flon Domain, Kisseynew gneisses, and Hanson Lake Block, northern Saskatchewan: Saskatchewan Geological Survey, Saskatchewan Energy and Mines, Miscellaneous Report 87-4, p. 131–134.

Ayres, L.D., Van Wagoner, N.A., and Ferreira, W.S., 1991, Voluminous shallow water to emergent phreatomagmatic basaltic volcaniclastic rocks, Proterozoic (1886 Ma) Amisk Lake composite volcano, Flin Flon greenstone belt, Canada: SEPM, Special Publication No. 47, p. 175–187.

Bailes, A.H., and Galley, A.G., 1999, Evolution of the Paleoproterozoic Snow Lake arc assemblage and geodynamic setting for associated volcanic-hosted massive sulphide deposits, Flin Flon Belt, Manitoba, Canada: Canadian Journal of Earth Science, v. 36, p. 1789–1805.

Bailes, A.H., and Syme, E.C., 1989, Geology of the Flin Flon—White Lake area: Manitoba Energy and Mines, Geological Report GR87-1, 313 p.

Bailes, A.H., and Theyer, P., 2006, Wonderland gabbro: U-Pb age and geological implications (NTS63K/12NE): Manitoba Science, Technology, Energy and Mines, Manitoba Geological Survey, Report of Activities 2006, p. 15–19.

Baldwin, D.A., 1980, Porphyritic intrusions and related mineralization in the Flin Flon volcanic belt: Manitoba Department of Energy and Mines, Mineral Resources Division, Economic Geology Report, ER79-4, p. 23.

Barrie, C.T., and Hannington, M.D., 1999, Introduction: Classification of VMS deposits based on host rock composition: Economic Geology Series, v. 10, p. 2–10.

Bruce, E.L., 1918, Amisk-Athapapuskow Lake district: Geological Survey of Canada, Memoir 105, 91 p.

Byers, A.R., and Dahlstrom, C.D.A., 1954, Geology and mineral deposits of the Amisk-Wildnest Lakes area, Saskatchewan: Saskatchewan Department of Mineral Resources, Report No. 14, 177 p.

Byers, A.R., Kirkland, S.J.T., and Pearson, W.J., 1965, Geology and mineral deposits of the Flin Flon area, Saskatchewan: Saskatchewan Department of Mineral Resources, Report No. 62, 95 p.

CAMECO, 2002, Annual Shareholders Report, 2002: Cameco Corporation, Saskatoon, Saskatchewan.

Connors, K.A., 1996, Unravelling the boundary between turbidites of the Kisseynew Domain and volcano-plutonic rocks of the Flin Flon domain in the eastern Trans-Hudson Orogen, Canada: Canadian Journal of Earth Sciences, v. 33, p. 811–829

Connors, K.A., Ansdell, K.M., and Lucas, S.B, 1999, Coeval sedimentation, magmatism, and fold-thrust development in the Trans-Hudson Orogen: Propagation of deformation into an active continental arc setting, Wekusko Lake area, Manitoba: Canadian Journal of Earth Sciences, v. 36, p. 275–291.

Coombe, W., 1984, Gold in Saskatchewan: Saskatchewan Geological Survey, Open File Report 84-1, 134 p.

Corrigan, D., Galley, A., and Pehrsson, S., 2007, Tectonic evolution and metallogeny of the southwestern Trans-Hudson Orogen: Geological Association of Canada, Mineral Deposits Division, Special Publication 5, p. 881-902.

David, J., Machado, N., Bailes, A., and Syme, E., 1993, U-Pb geochronology of the Proterozoic Flin Flon-Snow Lake Belt: New results: LITHOPROBE, Report 34, p. 84–87.

David, J., Bailes, A.H., and Machado, N., 1996, Evolution of the Snow Lake portion of the Paleoproterozoic Flin Flon and Kisseynew belts, Trans-Hudson Orogen, Manitoba, Canada: Precambrian Research, v. 79, p. 107–124.

Devine, C.A., Gibson, H.L., Bailes, A.H., MacLachlan, K., Gilmore, K., and Galley, A.G., 2002, Stratigraphy of volcanogenic massive sulphide-hosting volcanic and volcaniclastic rocks of the Flin Flon Formation, Flin Flon (NTS 63K12 and 13), Manitoba and Saskatchewan: Manitoba Industry, Trade, and Mines, Manitoba Geological Survey, Report of Ativities 2002, p. 9–19.

Dubé, B., Gosselin, P., Mercier-Langevin, P., Hannington, M., and Galley, A., 2007, Gold-rich volcanogenic massive sulphide deposits, *in* Goodfellow, W.D., ed., Mineral deposits of Canada: A synthesis of major deposit types, district metallogeny, the evolution of geological provinces, and exploration methods: Geological Association of Canada, Mineral Deposits Division, Special Publication 5, p. 75-94.

Duncan, R.A., and Green, D.H., 1987, The genesis of refractory melts in the formation of oceanic crust: Contributions to Mineralogy and Petrology, v. 96, p. 326–342.

Falloon, T.J., Malahoff, A., Zonenshain, L.P., and Bogdanov, Y. 1992, Petrology and geochemistry of back-arc basin basalts from Lau Basin spreading ridges at 15°, 18°, and 19°: Contributions to Mineralogy and Petrology, v. 47, p. 1–35.

Fedorowich, J.S., Kerrich, R., and Stauffer, M.R., 1995, Geodynamic evolution and thermal history of the central Flin Flon Domain, Trans-Hudson Orogen: Constraints from structural development, $^{40}Ar/^{39}Ar$, and stable isotope geothermometry: Tectonics, v. 14, p. 472–503.

Franklin, J.M, Gibson, H.L., Jonasson, I.R., and Galley, A.G., 2005, Volcanogenic massive sulphide deposits: Economic Geology, 100th Anniversary Volume, p. 523–560.

Galley, A.G., 1993, Semi-conformable alteration zones in volcanogenic massive sulphide districts: Journal of Geochemical Exploration, v. 48, p. 175–200.

——1996, Paleoproterozoic volcanic-related massive sulphide deposits: Tectonic and depositional environments, The Gangue, no. 54, p. 10–13.

——2003, Composite synvolcanic intrusions associated with Precambrian VMS-related hydrothermal systems: Mineralium Deposita, v. 38, p. 443–473.

Galley, A.G., and Franklin, J.M., 1987, Gold-tungsten-copper-molybdenum mineralization in the Phantom Lake area: Saskatchewan Geological Survey, Saskatchewan Energy and Mines, Miscellaneous Report 86-4, p. 141–143.

Galley, A.G., and Koski, R.A., 1999, Setting and characteristics of ophiolite-hosted volcanogenic massive sulfide deposits: Reviews in Economic Geology, v. 8, p. 215–236.

Galley, A.G., Ames, D.E., and Franklin, J.M., 1988, Geological setting of gold mineralization, Snow Lake, Manitoba: Geological Survey of Canada, Open File 1700, map with accompanying notes, scale: 1:5000.

Galley, A.G., Bailes, A.H., and Kitzler, G., 1993, Geological setting and hydrothermal evolution of the Chisel Lake and North Chisel Zn-Pb-Ag-Au massive sulphide deposit, Snow Lake, Manitoba: Exploration and Mining Geology, v. 2, p. 271–295.

Galley, A.G., Hannington, M., and Jonasson, I., 2007, Volcanogenic massive sulphide deposits, *in* Goodfellow, W.D., ed., Mineral deposits of Canada: A synthesis of major deposit types, district metallogeny, the evolution of geological provinces, and exploration methods: Geological Association of Canada, Mineral Deposits Division, Special Publication 5, p. 141-161.

Hannington, M.D., Poulsen, K.H., Thompson, J.F.H., Sillitoe, R.H., 1999, Volcanogenic gold in the massive sulfide environment: Economic Geology Series, v. 10, p. 325–356.

Harrison, J.M., 1951, Precambrian correlation and nomenclature, and problems of the Kisseynew gneisses in Manitoba: Geological Survey of Canada, Bulletin 20, 53 p.

Hoffman, P.F., 1989, Precambrian geology and tectonic history of North America, *in* A.W. Bailey, and Palmer, A.R., eds., The Geology of North America—An overview: Geological Society of America, Boulder, Co., p. 447–512.

Holland, H.D., Feakes, C.R., and Zbinden, E.A., 1989, The Flin Flon paleosol and the composition of the atmosphere 1.8 BYBP: American Journal of Science, v. 289, p. 362–389.

Lucas, S.B., White, D., Hajnal, Z., Lewry, J., Green, A., Clowes, R., Zwanzig, H., Ashton, K., Schledewitz, D., Stauffer, M., Norman, A., Williams, P.F., and Spence, G., 1994, Three-dimensional collisional structure of the Trans-Hudson orogen, Canada: Tectonophysics, v. 232, p. 161–178.

Lucas, S.B., Stern, R.A., Syme, E.C., Reilly, B.A., and Thomas, D.J., 1996, Intraoceanic tectonics and the development of continental crust; 1.92–1.84 Ga evolution of the Flin Flon Belt, Canada: Bulletin of the Geological Association of America, v. 108, no. 5, p. 602–629.

Lucas, S.B., Syme, E.C., and Ashton, K.E., 1999, New perspectives on the Flin Flon Belt, Trans-Hudson Orogen, Manitoba and Saskatchewan: An introduction to the special issue on the NATMAP Shield Margin Project,

Part 1: Canadian Journal of Earth Sciences, v. 36, p. 135–140.

Machado, N., and Zwanzig, H., 1995, U-Pb geochronology of the Kisseynew Domain in Manitoba: Provenance ages for the metasediments and timing of magmatism: LITHOPROBE, Report 38, p. 133–142.

Maxeiner, R.O., Sibbald, T.I.I., and Watters, B.R., 1993, Geology of the Hanson Lake area (part of NTS 63L/10): Saskatchewan Geological Survey, Saskatchewan Energy and Mines, Miscellaneous Report 93-4, p. 40–49.

Maxeiner, R.O, Sibbald, T.I, Slimmon, W.L., Heaman, L.M., and Watters, B.R., 1999, Lithogeochemistry of volcano-plutonic assemblages of the southern Hanson Lake Block and southeastern Glennie Domain, Trans-Hudson Orogen: Evidence for a single island arc complex: Canadian Journal of Earth Science, v. 36, p. 209–225.

Menard, T., Lesher, C.M., Stowell, H.H., Price, D.P., Pickell, J.R., Onstott, T.C., and Hulbert, L., 1996, Geology, genesis, and metamorphic history of the Namew Lake Ni-Cu deposit, Manitoba: Economic Geology, v. 91, p. 1394–1413.

Mitchinson, D.E., 2004, The effectiveness of lithogeochemistry versus X-ray diffeaction-defined mineralogy in outlining areas of volcanogenic massive sulphide-related alteration: A comparative study within the Paleoproterozoic Baker Patton felsic complex, Flin Flon, Manitoba, Canada: Unpublished M.Sc. thesis, Sudbury, Laurentian University, 94 p.

Northern Miner, 2004, New Britannia drags down High River gold: Northern Miner, v. 90, no. 14, p. 1.

Oliva, G., Theyer, P., and Burszytn, N., 2002, Platinum Group Element investigations in the Flin Flon greenstone belt: Petrology and mineralogy of the McBratney Lake PGE-Au occurrence (63K/13), Manitoba: Manitoba industry, Trade, and Mines, Manitoba Geological Survey, Report of Activities 2002, p. 94–99.

Pearson, J.G., McDougall, F.H., and Galley, A.G., 1986, Geology and evolution of gold occurrences in the Flin Flon-Amisk Lake area, Saskatchewan: Canadian Institute of Mining and Metallurgy, Special volume 38, p. 399.

Reilly, B, Slimmon, W.L., Harper, C.T., Ashton, K.E., and Watters, B.R., 1994, Contrasting lithotectonic assemblages from the western Flin Flon Domain: LITHOPROBE, Report 38, p. 105–111.

Richardson, D.J., and Ostry, G., 1996, Gold deposits of Manitoba, 2nd edition: Manitoba Energy and Mines, Economic Geology Report ER86-1, p. 55–90.

Russell, G.A., 1957, Structural studies of the Snow Lake-Herb Lake area, Herb Lake Mining Division, Manitoba: Manitoba Mines Branch Publication 55-3, 120 p.

Ryan, J.J., and Williams, P.F., 1999, Structural evolution of the eastern Amisk collage, Trans-Hudson Orogen, Manitoba: Canadian Journal of Earth Sciences, v. 36, p. 251–273.

Sillitoe, R.H., Hannington, M.D., and Thompson, J.F.H., 1996, High sulfidation deposits in the volcanogenic massive sulfide environment: Economic Geology, v. 91, p. 204–212.

Simard, R-L., 2006, Geology of the Schist Lake–Mandy mines area, Flin Flon, Manitoba (part of NTS 63K12): Manitoba Trade and Industry, Manitoba Geological Survey, Report of Activities 2006, p. 9–21.

Skirrow, R.G., and Franklin, J.M., 1994, Silicification and metal leaching in subconcordant alteration zones beneath the Chisel Lake massive sulphide deposit, Snow Lake, Manitoba: Economic Geology, v. 89, no. 1, p. 31–50.

Stern, R.A., and Lucas, S.B., 1994, U-Pb zircon constraints on the early tectonic history of the Flin Flon accretionary collage, Saskatchewan: Geological Survey of Canada, Current Research 1994-F, Report 8, p. 75–86.

Stern, R.A., Syme, E.C., Bailes, A.H., and Lucas, S.B., 1995a, Paleoproterozoic (1.86-1.90 Ga) arc volcanism in the Flin Flon Belt, Trans-Hudson Orogen, Canada: Contributions to Mineralogy and Petrology, v. 119, p. 117–141.

Stern, R.A., Syme, E.C., and Lucas, S.B., 1995b, Geochemistry of 1.9 Ga MORB- and OIB-like basalts from the Amisk collage, Flin Flon Belt, Canada: Evidence for an intra-oceanic origin: Geochimica et Cosmochimica Acta, v. 59, p. 3131–3154.

Stern R.A., David, J., Lucas, S.B., and Syme, E.C., 1996, Chronology of crustal growth: 1.9–1.8 Ga magmatism and accretion tectonics in the Flin Flon Belt, Trans-Hudson Orogen (Canada): Precambrian '95 International Conference on Tectonics and Metallogeny of Early/Mid Precambrian Orogenic Belts, August 28 to September 1, 1995, Montreal, Quebec, Special Issue, p. 21–27.

Stern, R.A., Machado, N., Syme, E.C., Lucas, S.B., and David, J., 1999, Chronology of crustal growth and recycling in the Paleoproterozoic Amisk collage (Flin Flon Belt), Trans-Hudson Orogen, Canada: Canadian Journal of Earth Science, p. 1807–1827.

Stockwell, C.H., 1937, Gold deposits of Herb Lake area, northern Manitoba: Geological Survey of Canada, Memoir 208, p. 46.

——1960, Flin Flon-Mandy Lake Area, Manitoba and Saskatchewan: Geological Survey of Canada, Map 17078A, with descriptive notes, scale: 1:12,000.

Syme, E.C., 1987, Athapapuskow Lake Project: Manitoba Energy and Mines, Minerals Division, Report of Field Activities, 1987, p. 30–39.

——1988, Athapapuskow Lake Project: Manitoba Energy and Mines, Minerals Division, Report of Field Activities, 1988, p. 20–34.

——1990, Stratigraphy and geochemistry of the Lynn Lake and Flin Flon metavolcanic belts, Manitoba: Geological Association of Canada, Special Paper 37, p. 143–161.

——1995, 1.9 Ga arc and ocean floor assemblages and their bounding structures in the central Flin Flon Belt: LITHOPROBE, Report No. 48, p. 261–272.

Syme, E.C., and Bailes, A.H., 1993, Stratigraphic and tectonic setting of volcanogenic massive sulphide deposits, Flin Flon, Manitoba: Economic Geology, v. 88, p. 566–589.

Syme, E.C., Bailes, A.H., and Lucas, S.B., 1996, Tectonic assembly of the Paleoproterozoic Flin Flon Belt and setting of VMS deposits: Geological Association of Canada/Mineralogical Association of Canada Joint Annual Meeting, Winnipeg, Manitoba, May 27-29, 1996, Field Trip Guidebook B1, 131 p.

Syme, E.C., Lucas, S.B., Bailes, A.H., and Stern, R.A., 1999, Contrasting arc and MORB-like assemblages in the Paleoproterozoic Flin Flon Belt, Manitoba, and the role of intra-arc extension in localizing volcanic-hosted massive sulphide deposits: Canadian Journal of Earth Sciences, v. 36, p. 1767–1788.

Thomas, D.J., 1992, Highlights of investigations around Flin Flon Mine: Reassessment of the structural history: Saskatchewan Geological Survey, Saskatchewan Energy and Mines, Miscellaneous Report 92-4, p. 3–15.

Tourigny, G., Gilmore, K., Howland, L., and Levers, J., 2002, Relationships between ore and overprinting structures at the Konuto Lake VMS deposit, Flin Flon domain: Saskatchewan Geological Survey, Saskatchewan Industry and Resources, Miscellaneous Report 2002-4.2, v. 2, http://www.ir.gov.sk.ca/adx/asp/adxGetMedia.asp.

Walker, T., and McDougall, F., 1987, Geology of the Laurel Lake gold-silver deposit: Saskatchewan Geological Society, Special Publication 8, p. 44–53.

Watters, B.R., Dostal, J., Slimmon, W.L., and Thomas, D.J., 1994, Geochemistry, petrogenesis and tectonic setting of Early Proterozoic volcanic rocks of the Flin Flon Domain, Saskatchewan (Canada): Oceanic back-arc volcanism: Neues Jahrbuch fur Mineralogie Monatshefte, v. 9, p. 416–432.

Whalen, J.B., and Hunt, P.A., 1994, Geochronological study of granitoid rocks in the Elbow Lake map area, Manitoba: A portion of the Flin Flon Domain of the Trans-Hudson Orogen: Geological Survey of Canada, Current Research 1994-F, Report 8, p. 1–10.

Whalen, J.B., Syme, E.C., and Stern, R.A., 1999, Geochemical and Nd isotopic evolution of Paleoproterozoic arc-type granitoid magmatism in the Flin Flon Belt, Trans-Hudson orogen, Canada: Canadian Journal of Earth Science, v. 36, p. 227–250.

Wyman, D.A., 2003, Paleoproterozoic boninites in an ophiolite-like setting, Trans-Hudson Orogen, Canada: Geology, v. 27, p. 455–458.

Zwanzig, H.V., 1990, Kisseynew gneiss belt in Manitoba: stratigraphy, structure, and tectonic evolution: Geological Association of Canada, Special Paper 37, p. 95–120.

——1999, Structure and stratigraphy of the south flank of the Kisseynew Domain in the Trans-Hudson Orogen, Manitoba: Implications for 1.845–1.77 Ga collision tectonics: Canadain Journal of Earth Science, v. 36, p. 1859–1880.

Volcanogenic Massive Sulphide Deposits of the Archean, Noranda District, Quebec

Harold L. Gibson[1] and Alan G. Galley[2,1]

1. *Mineral Exploration Research Centre, Department of Earth Sciences, Laurentian University, Sudbury, Ontario, Canada P3E 2C6*

2. *Geological Survey of Canada*

601 Booth Street, Ottawa, Canada, K1S 5B6

Corresponding author's email: hgibson@laurentian.ca

Abstract

The Noranda mining district is one of Canada's nation builders, having been the key for opening up northern Quebec for economic expansion and regional development. Over an 85-year period, 20 economic volcanogenic massive sulphide (VMS) deposits have been discovered by prospecting, geological, geophysical and lithogeochemical techniques. Besides VMS deposits, the Noranda District is also host to 19 orogenic Au deposits and several intrusion-hosted Cu-Mo deposits and occurrences. VMS deposits of the Noranda District occur in the Noranda formation of the Archean Blake River Group of the Abitibi Subprovince. The majority of deposits are hosted within the Noranda Cauldron, an asymmetric volcanic depression filled with effusive basalt and basaltic andesite flows and subordinate rhyolite flow-dome complexes. The Noranda Cauldron is floored by a large, multiphase subvolcanic intrusive complex that is hypothesized to be the heat engine responsible for development of much of the VMS and porphyry mineralization in the district. The VMS deposits are hosted by two distinct lithofacies, flow and volcaniclastic, where volcaniclastic includes primary pyroclastic deposits, and those that are re-deposited and syneruptive. Of the 20 past producing VMS deposits, 17 are hosted by flows (8 with mafic and 9 with felsic flows) whereas volcaniclastic rocks host the Horne, Bouchard-Hebert and Corbet deposits. The largest massive sulphide deposits occur within the felsic volcaniclastic strata at the margins of or directly overlying the Noranda Cauldron.

The Noranda VMS deposits typify the bimodal-mafic type, where the majority of the host stratigraphy is composed of a mafic volcanic flow succession in which discrete felsic dome complexes are situated over major synvolcanic faults. These fault systems are also defined within the Noranda Cauldron by discrete dyke swarms originating from the underlying subvolcanic intrusive complex. The inferred primitive rifted arc environment typical of this deposit-type is best characterized by a large volcanic edifice within the present-day Kermadec and Tonga-Fiji suprasubduction tectonic zone of the western Pacific.

Résumé

Le district minier de Rouyn-Noranda est l'une des régions édificatrices du Canada; elle a constitué une des clés de l'expansion économique et du développement régional du Québec septentrional. Pendant un intervalle de 85 ans, 20 gisements rentables de SMV ont été découverts par des méthodes de prospection géologiques, géophysiques et lithogéochimiques. En plus des gisements de SMV, le district de Rouyn-Noranda recèle également 19 gisements d'Au orogéniques et plusieurs gisements de Cu-Mo porphyres. Les gisements de SMV du district de Rouyn-Noranda se situent dans les roches du sous-groupe archéen de Blake River de la sous-province d'Abitibi. La majorité des gisements ont été découverts dans la subsidence en chaudron de Noranda, une dépression volcanique asymétrique principalement comblée de coulées d'andésite basaltique et de complexes de rhyolite secondaires sur un grand complexe intrusif subvolcanique soupçonné d'être le moteur thermique ayant engendré la formation d'une bonne partie des minéralisations en SMV et porphyres dans le district. Dans le district de Rouyn-Noranda, on trouve les gisements de SMV dans deux types de lithofaciès distincts, des faciès de coulées et volcanoclastiques dont les roches volcanoclastiques sont principalement constituées de dépôts pyroclastiques et des faciès dont les roches ont été redéposées et sont synéruptives. Parmi les 20 gisements de SMV ayant été exploités par le passé, 17 se trouvaient dans des coulées (8 dans des coulées mafiques et 9 dans des coulées felsiques), alors que les roches volcanoclastiques n'en renfermaient que 3 – les gisements de Horne, de Bouchard-Hebert et de Corbet. Les plus importants gisements de sulfures massifs se présentent dans les strates volcanoclastiques aux marges du chaudron de Noranda ou directement sur celui-ci.

La minéralisation en sulfures massifs de Noranda caractérise les gisements de SMV de type bimodal, dans lesquels la majorité de la stratigraphie hôte se compose d'une succession de coulées volcaniques mafiques où des complexes en forme de dômes felsiques discrets sont présents au-dessus des failles synvolcaniques majeures. Ces réseaux de failles sont également définis à l'intérieur du chaudron de Noranda par la présence d'essaims de filons intrusifs émanant du complexe subvolcanique intrusif sous-jacent. L'environnement primitif type d'arc de divergence des gisements de ce genre est au mieux caractérisé par l'actuel grand édifice volcanique présent dans la région tectonique de la zone suprasubduction de Kermadec et Tonga-Fidji dans le Pacifique occidental.

Introduction

The Noranda District contains the most thoroughly studied and documented volcanogenic massive sulphide (VMS) deposits of any Archean volcanic complex. Over an 85-year period, 20 economic VMS deposits have been discovered by prospecting, geological, geophysical and lithogeochemcial techniques. Research conducted on these well preserved, ancient deposits has, and will continue to make, a significant contribution to our understanding of VMS deposits and the

Gibson, H., and Galley, A., 2007, Volcanogenic massive sulphide deposits of the Archean, Noranda District, Quebec, *in* Goodfellow, W. D., ed., Mineral Deposits of Canada: A Synthesis of Major Deposit-types, District Metallogeny, the Evolution of Geological Provinces, and Exploration Methods, Special Publication No. 5, Mineral Deposits Division, Geological Association of Canada, p. 533-552.

development of genetic and exploration models for this deposit-type (Gibson et al., 2000). The Noranda District is also host to 19 small, orogenic Au deposits whose total combined production is eclipsed by the almost 10 million ounces (Moz) of Au produced by the Horne VMS deposit (Gibson et al., 2000). Other types of mineralization in the Noranda District include the intrusion-hosted Don-Rouyn Cu-Mo deposit and St. Jude Cu-Mo occurrence. In this paper, we describe the geological setting and characteristics of VMS deposits in the Noranda District as well as the exploration history of this historic and influential base and precious metal district.

Regional and District Geology

The Archean Abitibi Subprovince is host to 83 VMS deposits with an aggregate geological tonnage of approximately 730 Mt (million tonnes) that represent an in-ground metal content of approximately 9.8 Mt Cu, 22.7 Mt Zn and 37 Moz of Au (Gibson et al., 2005; Franklin et al., 2005). Six of the deposits contain >20 Mt and two of these, the 53.7 Mt Horne and the 69.2 Mt LaRonde deposits, account for almost 20 Moz of Au. Abitibi VMS deposits range in age from 2696-2701 Ma (e.g., Noranda and Doyon-Bousquet-LaRonde) to 2730Ma (Matagami Lake, Normetal), and are clustered within 13 districts comprised of one (e.g., Quevillon and Normetal) to 20 deposits (e.g., Noranda).

VMS deposits of the Noranda District occur in the Archean Blake River Group of the Abitibi Subprovince, Superior province (Percival, 2007). Stratigraphic subdivision and nomenclature for subunits within the Blake River Group are undergoing revision as summarized in Table 1 and illustrated in Fig. 1. In this paper we use the most recent stratigraphic nomenclature for the Blake River Group in Quebec as outlined in Table 1. Geochronological data (Table 1) indicate that volcanism, which constructed the Blake River Group, spanned approximately 8 Ma (2696 to 2704 Ma). Errors associated with the individual ages indicate that there is considerable overlap in the age of the different formations that comprise the Blake River Group. This overlap in age plus the absence of unconformities or thick epiclastic sedimentary successions between or within formations suggest that volcanism was located distal from subaerial or above storm-wave base landmasses, and that volcanism was continuous and not punctuated by a pronounced hiatus.

The structural style of the Blake River Group in the Noranda District is not uniform as illustrated in Fig. 1. North-south structural shortening within the more central Duprat-Montbray and Noranda formations is accommodated by broad, generally northeast-trending, open folds and north-south to north-northwest trending, east dipping, reverse faults that result in the generally east-facing and shallow east-dipping strata that characterize the Noranda for-

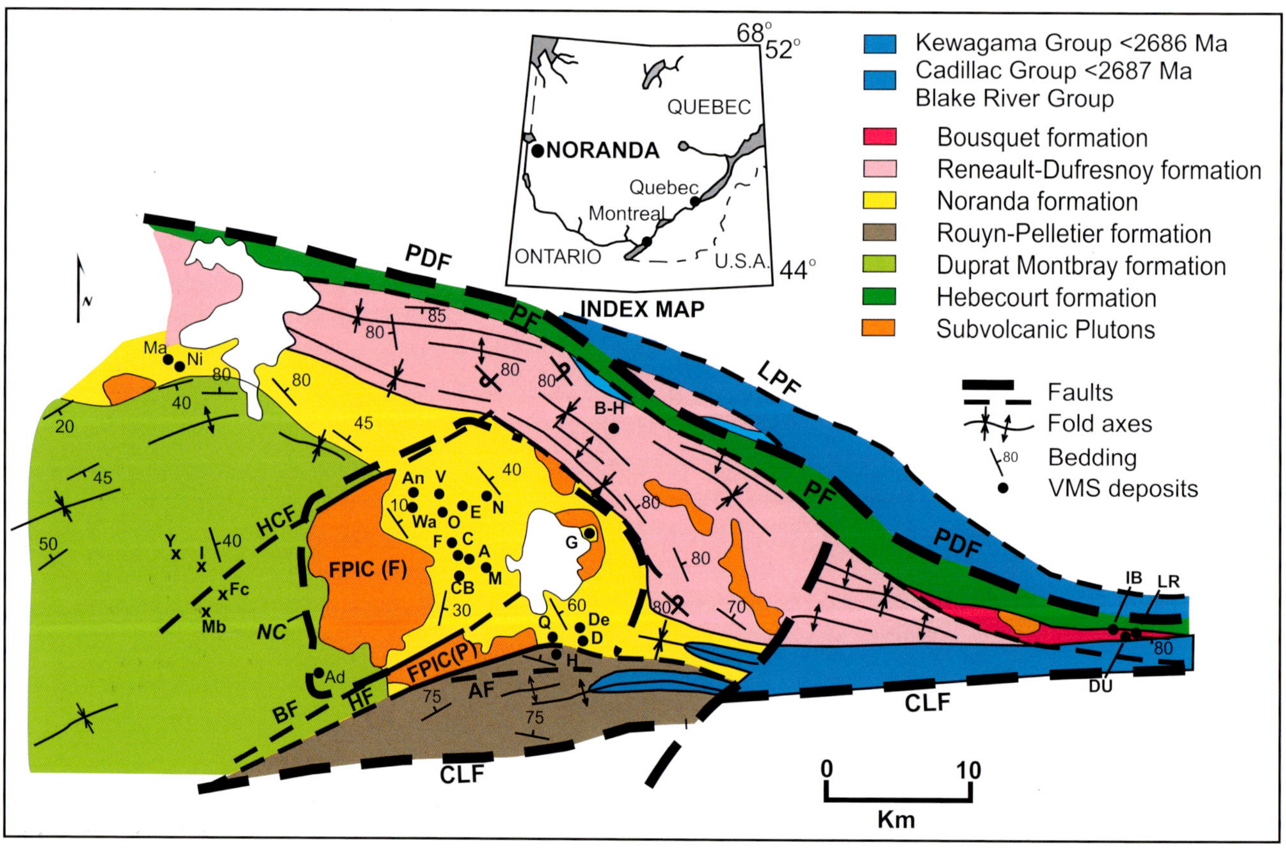

FIGURE 1. Stratigraphic subdivision and structure of the Blake River Group showing the location of the Noranda Cauldron and VMS deposits (modified after Goutier et al., 2006; Peloquin, 2005). (PDF = Porcupine Destor fault, PF = Parfouru fault, LPF = La Pause fault, CLF = Cadillac Larder Lake fault, HCF = Hunter Creek fault, BF = Beauchastel fault, HF = Horne fault, AF = Andesite fault, NC = inferred structural margin of the Noranda Cauldron. FPIC(F) signifies the Flavrian segment of the Flavrian-Powell Intrusive Complex, whereas FPIC(P) signifies the faulted-off Powell segment of the subvolcanic intrusive complex. Refer to Table 2 for abbreviations used to denote VMS deposits and significant occurrences.

TABLE 1. A comparison of the stratigraphic subdivision and geochronology of the Blake River Group in Quebec using informal formations with the subgroup stratigraphic terminology of Ontario (modified after Mortensen 1987; 1993; Tremblay et al., 1995; Mercier-Langevin et al., 2004; Lafrance et al., 2005; Ayer et al., 2005; Peloquin, 2005; Gauthier et al., 2006; David et al., 2006).

Quebec Formations	Age Ma	Ontario Subgroups	Age Ma	
Bousquet	2696.9+/-1 to 2698.6+/-1.5			Upper Blake River Group 2696-2701
Reneault-Dufresnoy	2696+/-1.1 to 2698+1.3/-0.7			
Noranda	2698.5+/-2			
Rouyn-Pelletier	2701			
Duprat-Montbray	2697-2701	Misema	2696.6+/-1.3 to 2701+/-2	Lower Blake River Group 2701-2704
Hebecourt	2701-2704	Garrison	2701-2704	

mation. In the more peripheral Hebecourt, Rouyn-Pelletier, and Reneault-Dufresnoy formations, structural shortening is accommodated by steep faults and steep, isoclinal, west-verging folds whose fold axes strike parallel to the adjacent Destor-Porcupine and Cadillac-Larder Lake faults that bound the Blake River Group (Fig. 1). This deformation produced steeply dipping strata that parallel the major bounding faults within these formations. In general, deformation within the Hebecourt, Rouyn-Pelletier, and Reneault-Dufresnoy formations decreases with increasing distance from the bounding faults. However, the change to shallow dipping strata within the Noranda and Duprat-Montbray formations is abrupt and occurs across steeply dipping faults.

It is uncertain if the different formations illustrated in Fig. 1 represent individual, overlapping volcanoes that evolved separately during discrete episodes of volcanism, and collectively define the Blake River Group in Quebec or if the formations collectively define a single volcano? Given the uncertainties regarding the extent and distribution of the formations due to their overlap in age, and the absence of bounding unconformities, both interpretations are reasonable. In this regard, Spence and deRosen-Spence (1975), deRosen-Spence (1976), Dimroth et al., (1982), Gibson (1990) and Peloquin et al., (1990) interpreted VMS deposits of the Noranda District to have formed a large, 7 to 9 km-thick, shield-like, bimodal, volcanic edifice comprised of one or more volcanoes dominated by basaltic and rhyolitic flows with subordinate, localized deposits of volcaniclastic rocks referred to as the Noranda Volcanic Complex (NVC) (Fig. 2A). The NVC consists of distinct fault blocks or sectors that now include the Noranda formation and portions of the Reneault-Dufresnoy, Rouyn-Pelletier and Duprat-Montbray formations (Fig. 1; Table 1; Gibson and Watkinson, 1990; Peloquin et al., 1990). Spence (1967), deRosen-Spence (1976), Gibson (1990) and Peloquin et al. (1990) subdivided what are now the Noranda and Reneault-Dufresnoy formations of the NVC into five conformable cycles. Cycles I through V young to the east (Fig. 2A), with each cycle consisting of an andesitic basalt to basaltic lower unit (Fig. 3A) and a rhyolitic (Fig. 3B) or a rhyolitic with lesser basaltic andesite upper unit. VMS deposits, subeconomic deposits, and occurrences are associated with the upper, rhyolitic part of the cycles. Ages for the different formations suggest that the entire NVC may have formed in as little as 3 Ma (Table 1).

A 15 x 20 km area of synvolcanic subsidence, referred to as the Noranda Cauldron, is interpreted to occur within the centre of the Noranda formation of the NVC (deRosen-Spence, 1976; Dimroth et al. 1982; Gibson, 1990) (Fig. 2B). The interpreted structural margins of the cauldron are defined by the Hunter Creek and Horne Creek faults to the north and south, respectively, and by the western margin of the Flavrian Powell Intrusive Complex (FPIC), and the d'Alembert shear and the synvolcanic d'Alembert Pluton to the west and east, respectively (Fig. 2A; Lichtblau and Dimroth, 1980; Dimroth et al., 1982; Gibson and Watkinson, 1990). The Noranda Cauldron is centred on a northeast trending rift, the Old Waite Paleofissure (Fig. 2B), which is defined by a 1 km-wide, sheeted intrusive complex that contains basaltic, rhyolitic, and composite dykes and sills (Gibson, 1990; Gibson and Watkinson, 1990). The Noranda Cauldron also contains several younger and smaller nested subsidence structures referred to as the Despina and Delbridge cauldrons (Fig. 2B; Gibson and Watkinson, 1990). Recently, Pearson (2005) and Pearson and Daigneault (2006) hypothesized that the Noranda Cauldron may be part of two older and larger calderas. The first is the New Senator caldera, which encompasses most of the Noranda formation, and the second and oldest the Misema caldera, which encloses the entire Blake River Group (Pearson, 2005).

Within the Noranda Cauldron subsidence was passive, episodic, and is interpreted to have occurred in response to the emplacement and partial evacuation of underlying magma chambers, now represented by comagmatic phases of the 2697 to 2701 Ma synvolcanic FPIC (Goldie, 1979; Mortensen, 1987; 1993; Paradis et al., 1988; Galley and van Breemen, 2002; Galley, 2003) (Figs. 2A-2B). The FPIC occupies the centre of this subsidence structure, where it was emplaced within its own volcanic equivalents through sequential emplacement of a series of sill complexes (Richard, 1999; Galley, 2003) and generated the convective hydrothermal system responsible for cauldron-hosted VMS deposits and associated hydrothermal alteration (Kennedy, 1985; Gibson, 1990).

Mineral Deposits, Occurrences and Grade/Tonnage Statistics

The Blake River Group in the Noranda District is host to 20 past-producing VMS deposits and 10 subeconomic VMS deposits and/or significant base metal occurrences (Table 2). The distribution of past-producers and subeconomic deposits within the Noranda District is not uniform. Eighteen past-producing deposits and five subeconomic deposits occur within the Noranda formation (Cycles III and IV), one past producer and one subeconomic deposit occurs within the Rouyn-Pelletier formation, one past producer occurs in the Reneault-Dufresnoy formation (Cycle V), and four subeconomic deposits occur within the lowermost, Duprat-Montbray formation (Cycles I and II) (Fig. 2A; Table 2). Clearly, the Noranda formation has been the most favourable

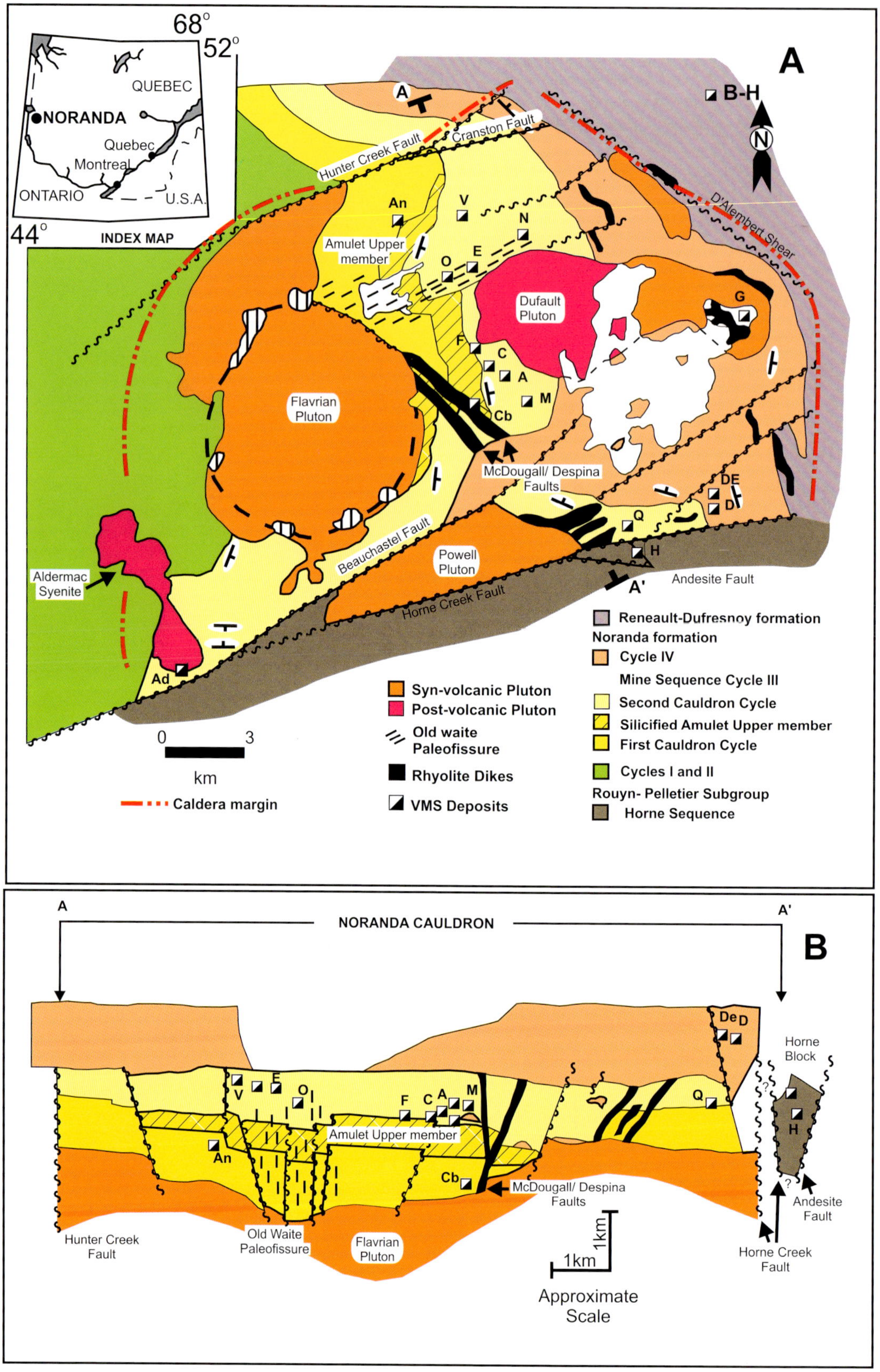

FIGURE 2. **(A)** Geologic map of the central portion of the Noranda Volcanic Complex (NVC; Noranda formation) showing the location and limits of the central area of subsidence, the Noranda Cauldron, between the Horne Creek and Horne (Andesite) faults, and the subvolcanic Flavrian-Powell Intrusive Complex and Old Waite Paleofissure which occur in the core of the subsidence structure (modified after Gibson, 2005). Refer to Table 2 for abbreviations used to denote VMS deposits and significant occurrences; **(B)** Reconstructed north-south geologic cross-section through the Noranda Cauldron showing the stratigraphic and structural control on VMS deposits (modified after Gibson, 2005).

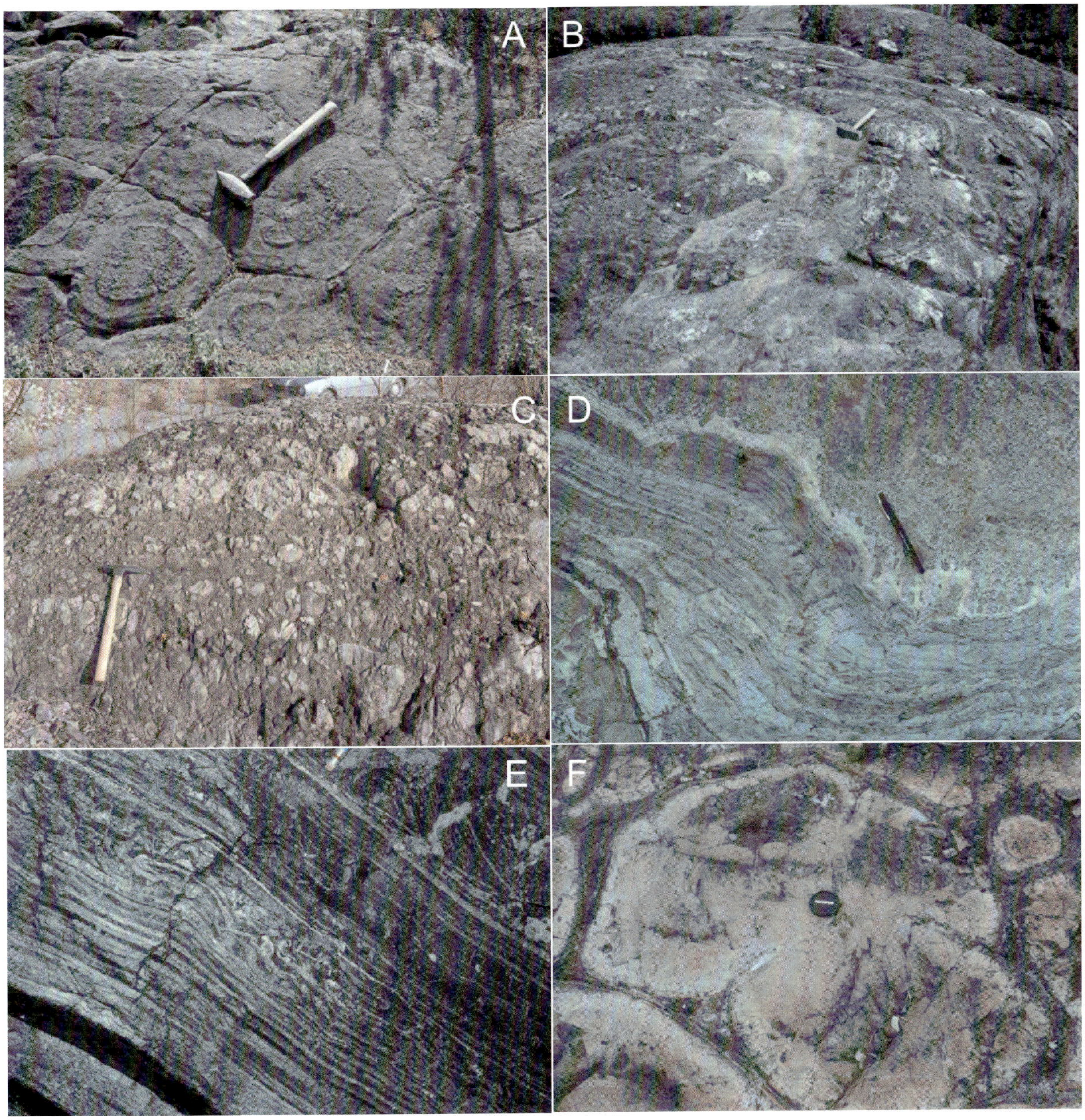

FIGURE 3. Photographs of rocks in the Noranda District, Quebec. **(A)** Columnar-jointed basalt from the Rusty Ridge formation typical of the flow lithofacies flows that define the base of each eruption cycle within the Noranda Cauldron; **(B)** Typical lobe-hyaloclastite rhyolite flow facies of the Amulet L member that formed from fissure-fed felsic eruptions that define the upper parts of the Noranda cauldron cycles; **(C)** Coarse-bedded and graded monolithologic rhyolite block breccia formed from dome collapse of the Cycle IV Delbridge rhyolite; **(D)** Main Contact Tuff from the Corbet Mine area with pillowed mafic flows in the footwall and overlying massive mafic flows. Typical example of the thin exhalite tuff units that characterize several of the VMS-hosting hori,zons within the Noranda Cauldron; **(E)** Mineralized Main Contact Tuff with pyrite and sphalerite in proximity with and along strike from the D68 ore bodies (Fig. 2A). Note fine interlayering of chert-sulphide (white) and mafic aquagene tuff (dark green); **(F)** Silicified pillowed flow from Upper Amulet andesite that directly underlies the C Contact Tuff horizon as example of low temperature silicification in conjunction with silica precipitation on the paleo-seafloor surface.

time-stratigraphic interval for the formation of VMS deposits as it contains 90 % of the past producing deposits and 50 % of the subeconomic deposits, whereas the older Duprat-Montbray formation was less favourable. As will be discussed later, the Noranda formation is unique in that it was erupted and emplaced during a period of cauldron subsidence and high heat flow.

The number of economic deposits is not the only measure of stratigraphic favourability. The Rouyn-Pelletier formation, because of the Horne Upper and Lower H orebodies, accounts for 60 % of the total tonnage produced from the Noranda District as well as nearly 60 % of the Cu and almost 80 % of the Au; this does not include the metal contained within the much larger, more Zn-rich, and subeconomic Horne No. 5 zone (Kerr and Mason, 1990; Table 2). Although there is some variation in the Zn and Cu content between the Noranda VMS deposits, the most significant variation in grade is reflected by Au content (Table 2). The

TABLE 2. Grade and tonnage characteristics, lithofacies and discovery methods for VMS deposits of the Noranda District. Deposits in italics are subeconomic or are significant occurrences. The Bouchard-Hebert total mineral inventory is an estimate of the total resources, including ore reserves (modified after Boldy, 1979; Gibson and Watkinson, 1990; Kerr and Gibson, 1993; Franklin et al., 2005).

Formations and VMS deposits	Dominant Lithofacies	Discovery Date	Depth (m)	Discovery Methods	Mt Ore (Geol.)	CU%	ZN%	AU (g/t)	AG (g/t)	Cu/ (Cu + Zn)
Reneault-Dufresnoy										
Bouchard-Hebert - Main Lenses (B-H) (formerly Mobrun)	Volcaniclastic-Felsic	1956	9	G-Geophysics	1.35	0.7	2.51	2.27	27.67	22
Bouchard-Hebert 1100 lens (B-H)	Volcaniclastic-Felsic	1988		In-mine	11.50	0.77	5.24	1.48	36.90	13
Bouchard-Hebert - Total Mineral Inventory	Volcaniclastic-Felsic				20.33	0.76	3.42	1.28	31.65	18
Noranda										
Cycle 3 - Mine Sequence										
Aldermac (Ad)	Flow-Felsic	1925	9	G-Geophysic, Prospecting	2.86	1.54	4.12	0.48	31.20	27
Amulet Lower A (A)	Flow-Mafic	1938	213	Geology	4.69	5.14	5.28	1.43	44.10	49
Amulet Upper A (A)	Flow-Mafic	1925	Surface	Prospecting	0.19	2.37	6.12	2.00	46.00	28
Amulet C (C)	Flow-Mafic	1925	Surface	Prospecting	0.57	2.2	8.5	0.60	86.70	21
Amulet F Shaft (F)	Flow-Mafic	1929	38	Geology	0.27	3.4	8.6	0.30	46.30	28
Amulet No. 11 (A)	Flow-Mafic		600	Mine Exploration	0.45	3.6	2.4	0.70	22.00	60
Ansil (An)	Flow-Felsic	1980	1280	Geology	1.58	7.22	0.94	1.60	26.50	88
Corbet (Cb)	Volcaniclastic-Mafic	1974	700	Geology, L-Geochemistry	2.78	2.92	1.62	1.00	21.00	64
East Waite (E)	Flow-Felsic	1949	396	Geology	1.50	4.1	3.25	1.80	31.00	56
Millenbach (M)	Flow-Felsic	1966	700	Geology	3.56	3.46	4.33	1.00	56.20	44
Norbec (including D orebody) (N)	Flow-Felsic	1961	335	Geology, Prospecting	4.47	2.75	4.75	0.91	44.30	37
Old Waite (O)	Flow-Mafic	1925	Surface	Prospecting	1.12	4.7	2.98	1.10	22.00	61
Quemont (Q)	Flow-Felsic	1945	61	G-Geophysics	16.65	1.2	1.8	5.50	18.00	40
Vauze (V)	Flow-Felsic	1957	7	Geology	0.35	2.9	0.94	0.70	24.00	76
West Ansil (Wa)	Flow-Mafic	2003	250	Geology, Computer Modelling						
Bedford	Flow-Felsic	1945	Surface	Prospecting	0.90	0.89				
Ribago	Flow-Mafic	1983	700m	Geology, L-Geochemistry						
Moosehead	Flow-Mafic	1925	Surface	Prospecting						
Cycle 4										
Delbridge (D)	Flow-Felsic	1965	91	Geology, L-Geochemistry	0.36	0.55	8.6	2.40	68.60	6
Deldona (De)	Flow-Felsic	1947	152	Geology	0.09	0.3	5	4.10	26.00	6
Gallen (G) formerlyWest MacDonald	Flow-Felsic	1944	7	G-Geophysics	8.10	0.08	3.36	0.06	2.40	2
Pinkos 3 (P)	Flow-Felsic	2007	?	Geology? -New Discovery						
Cycle3-4?										
Magusi River (Ma)	Flow-Felsic	1972	15	A-Geophysics	3.73	1.2	3.55	1.10	31.20	25
New Insco (Ni)	Flow-Mafic	1973	15	A-Geophysics	0.89	2.59		0.90	20.57	
Rouyn-Pelletier										
Horne - No. 5 Zone (H)	Volcaniclastic-Felsic			In-Mine	150.00	0.1	0.9			10
Horne-H Orebodies (H)	Volcaniclastic-Felsic	1923	Surface	Pospecting	54.30	2.22		6.10	13.00	
Duprat-Montbray										
Four-corners (Fc)	Flow-Felsic		Surface	Pospecting						
Inmont (I)	Flow-Felsic		Surface	Pospecting						
Montbray (Mb)	Flow-Felsic	2004	110	A,B-Geophysics						
Yvannex (Y)	Flow-Felsic		Surface	Pospecting						

average Au content of 16 of the past-producing deposits is <2 g/t, whereas the Horne Upper and Lower H orebodies, and the Quemont ore bodies have average Au content of 6.1 and 5.5 g/t Au and can be classified as Au-rich VMS deposits (Poulsen et al., 2000). The smaller Deldona and Delbridge deposits and Bouchard-Hebert Main lens have average Au grades >2 g/t and are enriched in Au relative to most VMS deposits of the Norand District (Table 2). Except for the Bouchard-Hebert main lens, all of the Au-rich VMS deposits are localized along the Horne Creek Fault, a splay of the Cadillac-Larder Lake Fault located 6 km to the south. Some of the most visibly spectacular concentrations of Au and telluride minerals at the Horne deposit are hosted by quartz veins associated with quartz syenite dykes. These auriferous quartz veins, interpreted as a local remobilization phenomenon (Kerr and Mason, 1990), together with the location of the Horne mine between second-order splays off the Cadillac-Larder Lake Fault (the Horne Creek and Andesite Faults), suggest that Au at the Horne deposit, and possibly at the Quemont, Delbridge and Deldona deposits located along the same second-order fault, was introduced by a post-volcanic, epigenetic or orogenic event. However, the syngenetic Au model is now generally accepted and will be discussed below (Kerr and Mason, 1990; MacLean and Hoy, 1991; Barrett et al., 1991; Cattalani et al., 1993; Gibson et al., 2000; Dubé et al., 2007a, Mercier-Langevin et al., 2007).

Although not discussed further within this paper, the Noranda District is also host to 19 small, orogenic Au deposits. All Au deposits, except the Chadborne deposit, are hosted by quartz veins in shear zones and can be subdivided into three types based on their spatial relationship to the Cadillac-Larder Lake Fault and morphology. The first type (14 deposits) includes structurally controlled, disseminated and quartz-vein lode deposits associated with parallel to subparallel splays of the Cadillac-Larder Lake fault (Robert, 1990; Gauthier et al., 1990). The ore zones occur within volcanic rocks of the Blake River Group and metasedimentary rocks of the Timiskaming and Pontiac groups where they are spatially related to either megascopic z-folds that refold earlier east-west shears or to areas of intersection between east-west shears and later northeast striking faults (Gauthier et al., 1990). The second type (5 deposits) is hosted by northeast-trending, shear zone-hosted quartz veins within various phases of the subvolcanic FPIC (Robert, 1990; Riverin et al., 1990; Carriere et al., 2000). The third type (Chadbourne deposit) is hosted by a vertical, diatreme breccia pipe where the Au occurs, along with quartz, pyrite and siderite, within the breccia matrix (Walker, 1980; Gibson et al., 1984).

Two types of magmatic-related mineralization also occur within the Noranda District. The Don-Rouyn Cu-Mo deposit is hosted within an early trondhjemite phase of the FPIC, where it is crosscut by a late phase tonalitic dyke swarm. The deposit, interpreted to be an Archean porphyry deposit, consists of a bornite-pyrite-rich strongly silicified core changing outwards to a chalcopyrite-pyrite vein stockwork, and collectively constitutes a mineral resource of approximately 36 Mt with an average grade of 0.15 % Cu (Kotila, 1975; Goldie et al., 1979, Jebrak et al., 1997, Pelletier and Jébrak, 1994). Along the western margin of the FPIC a quartz diorite-rhyolite dyke swarm hosts numerous pyrite-galena-Au and pyrite-magnetite vein occurrences (Kennedy, 1985; Richard, 1999). This vein-dyke system terminates at its northern end at the St. Jude Cu-Mo-enriched, magmatic-hydrothermal breccia pipe (Kennedy, 1985, Carriere et al., 2000). An intramineral aplite dyke associated with this mineralizing phase has a U-Pb zircon age of 2697 ± 2 Ma (Galley and van Breemen, 2002). This age is a similar to that of the youngest trondhjemite phase in the core of the FPIC, and to the Cycle V Clericy rhyolite (Galley and van Breemen, 2002).

Deposit Classification and Description

VMS deposits have been classified on the basis of their composition, tectonic setting, and host rock composition (e.g., Franklin et al., 1981; Hannington et al., 1999; Sawkins, 1976; Morton and Franklin, 1987; Barrie and Hannington, 1999). In this paper we use the lithostratigraphic scheme proposed by Franklin et al. (2005) that classifies VMS deposits into five types based on the principal volcanic and sedimentary lithological units that formed concurrently with the deposits within the entire district; bimodal-mafic, mafic, pelitic-mafic, bimodal-felsic, and siliciclastic-felsic. As this classification is based on a larger and more representative stratigraphic interval it has the potential to more confidently relate VMS types to their geodynamic setting, a comparison that cannot be made on deposit-scale geological or geochemical characteristics alone.

VMS deposits of the Noranda District formed within a bimodal succession comprised principally of tholeiitic to transitional basaltic to basaltic andesite flows, with lesser volumes of high-temperature, high–silica FIII and FII rhyolitic flow domes (Ludden et al., 1982; Lesher et al., 1986; Barrie et al., 1993; Kerr and Gibson, 1993), and lesser and more localized FII (FIII?) rhyolitic to dacitic volcaniclastic rocks. Based on this lithological association the Noranda VMS deposits fall within the bimodal-mafic type of Franklin et al. (2005). The bimodal-mafic type also includes most other Archean VMS districts (e.g., Kamiskotia, Kidd Creek, Matagami), Paleoproterozoic VMS districts of eastern Fennoscandia and the Canadian Trans-Hudson (e.g., Flin Flon), Paleozoic VMS districts of the mid and south Urals (e.g., Tarnyer; Sibai), and Mesozoic VMS deposits of Peru and Ecuador (e.g.,Tambo Grande, La Plata) (Franklin et al. 2005 and references therein).

Lithofacies Association

Within each of the five lithostratigraphic types, the lithofacies that directly host the VMS deposits may vary extensively, and the bimodal-mafic VMS deposits of the Noranda District are not an exception. VMS deposits of the Noranda district are hosted by two distinct lithofacies, flow (Figs. 3A-3B) and volcaniclastic (Fig. 3C), where volcaniclastic includes primary pyroclastic deposits and those that are re-deposited and syneruptive (Table 2) (Gibson et al., 1999). Of the 20 economic VMS deposits, 17 are hosted by flows (8 with mafic and 9 with felsic flows) whereas volcaniclastic rocks host only the Horne, Bouchard-Hebert and Corbet deposits. Although volcaniclastic strata host the Corbet and part of the Ansil deposit, the volcaniclastic units are volumetrically minor and are interpreted to be a product of mild fire fountaining eruptions (<1 km^3; Gibson et al., 1993). Part of the Ansil deposit, considered as flow lithofacies deposit, is hosted by the channelized Cranston rhyolite block and ash flow, which originated as a pyroclastic facies of the Northwest rhyolite composite flow (Galley et al., 1995). On the other hand, the felsic volcaniclastic lithofacies that hosts the Horne and Bouchard-Hebert deposits are voluminous and consist of primary and syneruptive volcaniclastic deposits of probable pyroclastic origin with minor felsic flows, domes, and cryptodomes (Kerr and Gibson, 1993).

Deposit Morphology

Noranda VMS deposits consist of a concordant lens of massive sulphide that is either cone- or mound-shaped where hosted by a mafic or felsic flow lithofacies (Fig. 4), or more tabular and sheet-like where hosted by a felsic volcaniclastic lithofacies (Figs. 5-6). The massive sulphide lens, where hosted by flows, overlies a vertically extensive, discordant stockwork of stringer mineralization that, in some cases (e.g., Ansil, Corbet, and Amulet Lower A; Kerr and Gibson, 1993; Galley et al.,1995) extends up through the massive sulphide lens and into the hanging wall rocks for several hundred meters (Figs. 4 and 7). In volcaniclastic hosted deposits (e.g., Horne, Bouchard-Hebert) the massive sulphide lens is underlain by a diffuse and vertically restricted or semiconformable stockwork of stringer and disseminated sulphides. The VMS deposits are polymetallic, and the minerals present in all deposits include pyrrhotite (the dominant sulphide in flow lithofacies deposits), pyrite (the dominant sulphide in volcaniclastic lithofacies deposits), with variable proportions of chalcopyrite, sphalerite, galena, and magnetite. Accessory elements and minerals commonly include native Au and Ag, electrum, tetrahedrite, tennantite, telluride minerals and, in some deposits, cassiterite (e.g., Corbet and Millenbach; Knuckey et al., 1982; Kunckey and Watkins, 1982).

The massive sulphide lens and underlying stringer zone display a distinct mineral and metal zonation characterized by a more Cu-rich, chalcopyrite-pyrrhotite (-magnetite) core that grades outward and upward to a more Zn-rich, pyritic or pyrite (pyrrhotite)-sphalerite fringe (Fig. 8). The proportion of more Cu-rich to Zn-rich ore, expressed by the (Cu/(Cu +Zn)) x 100 ratio in Table 2, is variable between deposits but in general felsic volcaniclastic hosted deposits such as Bouchard-Hebert (and Horne if the No.5 zone is included) are more Zn-rich, whereas in mafic- or felsic flow - hosted deposits the proportion of Zn- to Cu-rich ores is more variable (e.g., Ansil versus Amulet Upper A), but commonly they are more Cu-rich. Gold and Ag within the deposits can be associated with Cu-rich ore (e.g., Horne; Kerr and Mason, 1990), Zn-rich ore (e.g. Corbet; Knuckey and Watkins, 1982), or with both (e.g. Ansil; Galley et al., 1995).

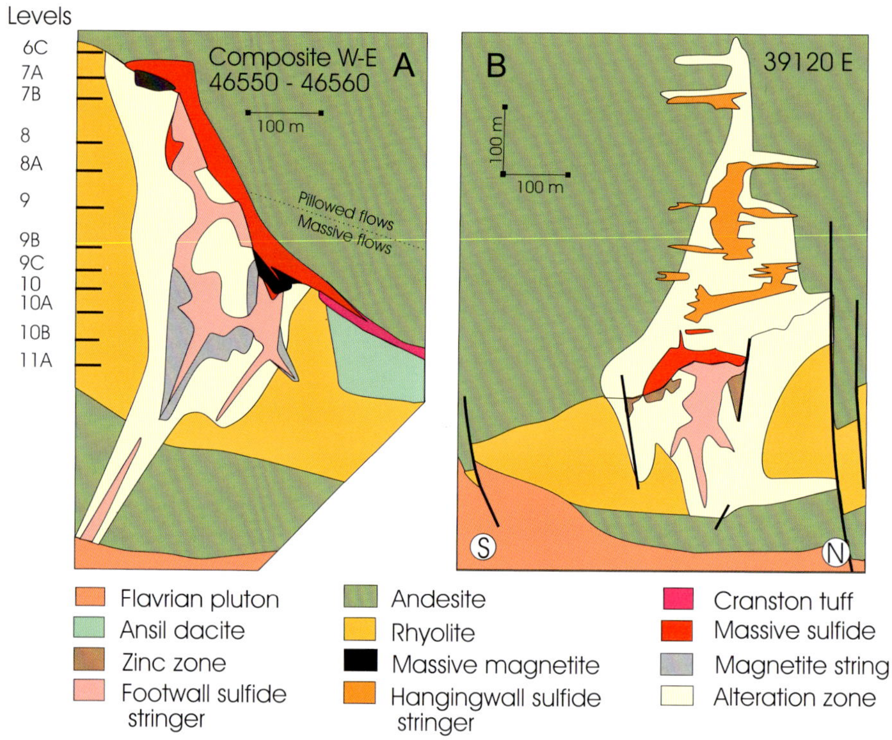

FIGURE 4. Cross-section and long-section through the Ansil deposit as an example of a typical cone- or mound-shaped massive sulphide lens hosted by a felsic flow lithofacies footwall and overlain by mafic flow lithofacies hanging wall. From Galley et al., 1995. Note the extensive nature of the discordant footwall and hanging wall alteration zones typical of many of Noranda's lithofacies-hosted VMS deposits.

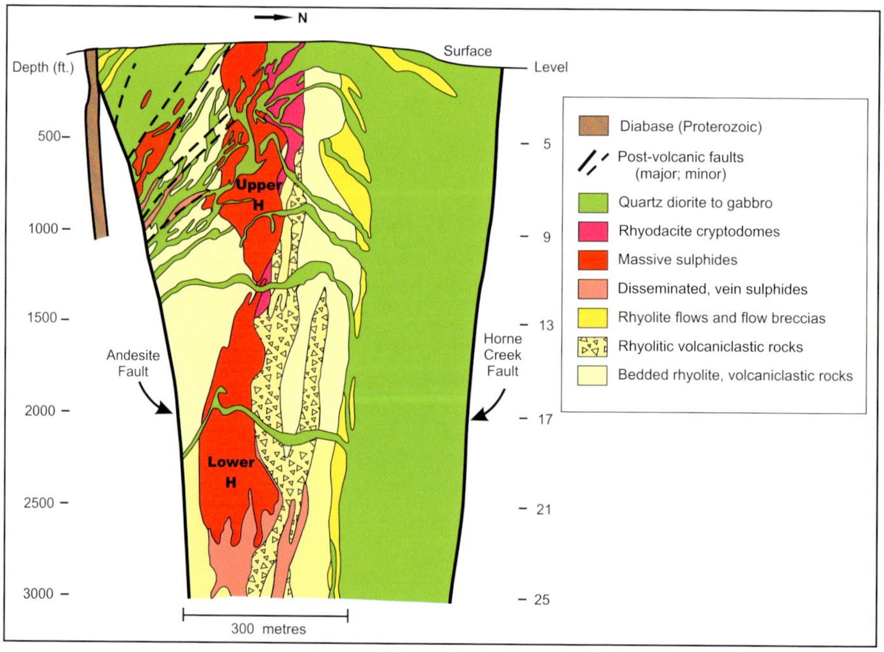

FIGURE 5. Vertical, north-south cross-section through the Horne deposit showing the Upper H (UH) and Lower H (LH) orebodies (modified from Kerr and Gibson, 1993; see Figure 2 for location of the Horne deposit). This is an example of a volcaniclastic lithofacies-hosted VMS deposit from the Noranda mining camp.

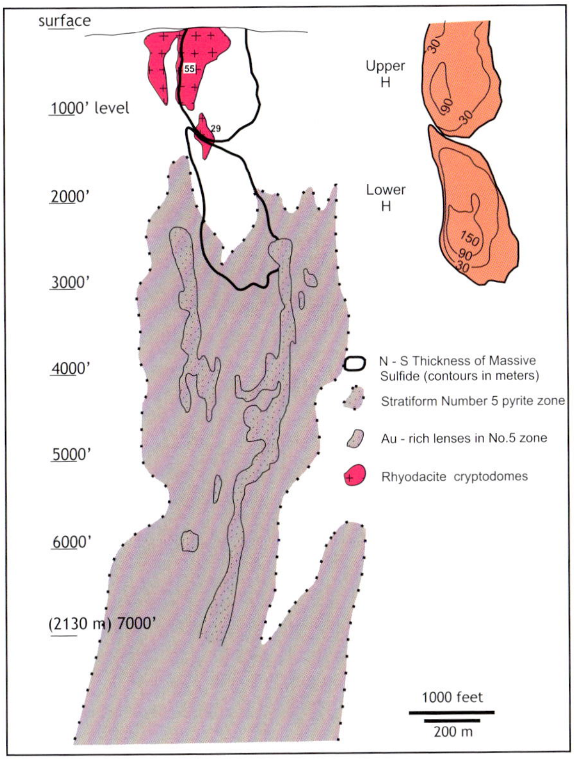

Figure 6. Vertical cross section looking north through the Upper and Lower H orebodies and the No. 5 zone of the Horne Deposit. Isopach maps of the Upper and Lower H are presented on the right. Note the tabular and sheet like form of the No. 5 zone that is hosted by a predominately volcaniclastic lithofacies (modified after Kerr and Mason, 1990).

Metal Zonation

The zonation of Cu, and Zn within the massive sulphide deposits was attributed by Gibson and Watkinson (1990), Kerr and Gibson (1993) and Galley et al. (1995) to zone refining processes whereby a lower temperature pyrite-sphalerite assemblage is replaced by a higher temperature chalcopyrite-pyrrhotite (-magnetite) assemblage, as described by Eldridge et al., (1983) for the Kuroko VMS deposits, by Large (1992) for VMS deposits of the Mount Read Volcanics, and by Hannington and Scott (1989) and Herzig and Hannington (1995), for modern seafloor sulphide deposits. The replacement process is interpreted to reflect a progressively higher temperature for the hydrothermal fluid ascending through each stringer zone and sulphide lens (Figs. 9A through 9C). In Archean VMS deposits the increase in fluid temperature is interpreted to result from self-sealing processes, such as sulphide (silica) precipitation and hydrothermal alteration within the hydrothermal vent, which progressively insulated and inhibited the ascending higher temperature hydrothermal fluid from interacting with down-drawn, cooler ambient seawater (Lydon, 1988; Gibson, 1990; Alt, 1995; Hannington et al.,1999; Gibson, et al., 1999). The occurrence of stringer mineralization and hydrothermal alteration extending for significant distances into the hanging wall strata above some deposits (e.g., for 300 m at the Ansil and Amulet Lower A deposits; Figs. 3 and 7), and the more Cu-rich character of deeper deposits stacked

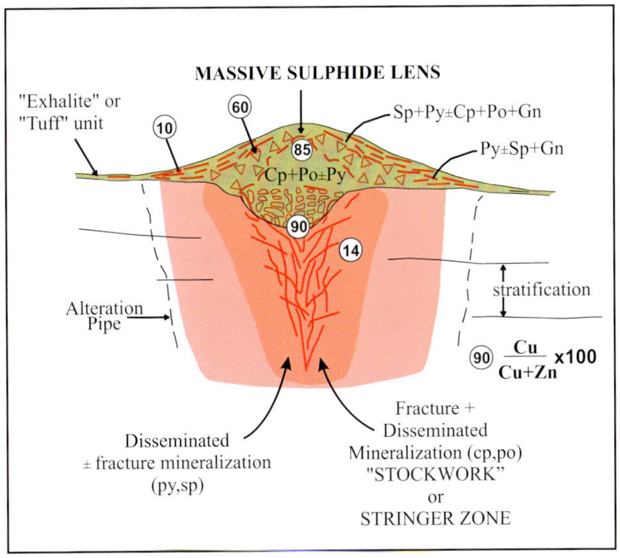

Figure 8. Typical sulphide textures, structures and metal zoning within an idealized Noranda flow hosted VMS deposit (modified from Gibson and Kerr, 1993).

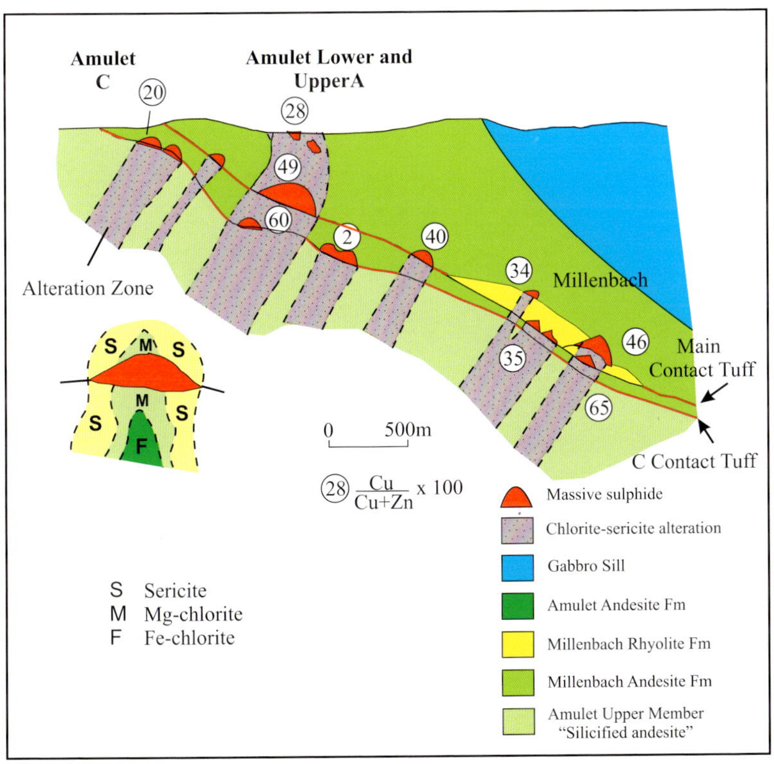

Figure 7. Cross-section through the Amulet and Millenbach VMS deposits illustrating the stratigraphic stacking of deposits within basaltic and rhyolitic vent areas, and metal zoning within and between individual VMS deposits as illustrated by the Cu/(Cu+Zn) x 100 ratio. The progressively more Cu-rich character of VMS deposits at depth within the Amulet alteration pipe from 28 at the Upper A, 49 at Lower A, and 60 at the #11 deposit is interpreted to reflect sub-seafloor zone refining due to the continued passage of ascending hydrothermal fluid that dissolved Zn from the lower deposits (replaced by chalcopyrite) to form a more Zn-rich deposit at the seafloor (Upper A). Refer to Figure 2 for location of deposits (modified from Kerr and Gibson, 1993).

within a single alteration pipe (e.g., Amulet Lower and Upper A; Fig. 7), has been interpreted to indicate that zone refining occurred below the seafloor as the deposits were buried by a significant thickness of flows (Knuckey et al., 1982; Kerr and Gibson, 1993; Galley et al.,1995).

The Cu-Au or Zn-Au (Ag) association within the Noranda VMS deposits has also been attributed to zone refining processes described by Hannington and Scott (1989), Hannington et al. (1999), Huston and Large, (1989); Large et al. (1989), Kerr and Gibson (1993), Gibson and Kerr (1993); Gibson et al. (2000) and Dubé et al. (2007a). However, the processes responsible for forming Au-rich VMS deposits such as the Horne (6.1 g/t Au) and Quemont (5.5 g/t Au) deposits, and for Au-enrichment at Delbridge (2.4 g/t Au), Deldona (4.1 g/t Au) and at the Bouchard Hebert main lens (2.27 g/t Au) are not well understood. At Horne, discordant zones of Fe-chlorite alteration and Cu-Au mineralization crosscut earlier sericite-quartz alteration and were interpreted by Kerr and Mason (1990), MacLean and Hoy (1991), and Barrett et al. (1991) to be the product of a later, higher temperature stage of zone refining. These authors also attribute this later, higher temperature stage to be responsible for the replacement of lower temperature Fe and Zn sulphides within the Upper and Lower H ore bodies by Au- and Cu-bearing assemblages dominated by pyrrhotite, pyrite, chalcopyrite and magnetite (Barrett et al., 1991; Kerr and Gibson, 1993). In this model, the large, stratiform ("clastic") No. 5 Zone may represent a pyrite-dominated, "unrefined",

FIGURE 9. Photographs of mineralized rocks. **(A)** Sphalerite-rich remnant of Cranston Tuff surrounded by massive pyrrhotite-chalcopyrite on Level 9A of the Ansil Mine; **(B)** Cranston Tuff completely replaced by pyrrhotite-chalcopyrite due to a mature stage of zone refining; **(C)** Mature stage of zone refining with massive pyrrhotite-chalcopyrite replaced with massive magnetite; **(D)** Groundmass within a coarse volcaniclastic unit is replaced by massive pyrrhotite-chalcopyrite at the Corbet deposit; **(E)** A typical footwall sulphide stringer zone in which sulphide stockwork veining infills fractures in massive, chlorite-altered basalt in the core of the Corbet deposit footwall alteration pipe; **(F)** Sericite-altered rhyolitic volcaniclastic flow breccia from footwall to the Horne deposit.

low grade Zn deposit (Fig. 5; Sinclair, 1971; Gibson et al., 2000). At the Ansil deposit, Au appears to have been more highly concentrated in the Cu-rich orebodies due to remobilization of both Cu and Au during the partial replacement of the lower half of the orebody by massive magnetite and associated hydrothermal skarn (Galley et al., 1995; Galley et al., 2000; Fig. 9C). On the other hand, the 2698 Ma Au-rich VMS deposits of the Bousquet formation of the Doyon-Bousquet-LaRonde camp located 50 km east of the Noranda District are interpreted to have acquired their high Au grades from a magmatic fluid contribution to a seawater derived hydrothermal fluid (Dubé et al., 2007b; Mercier-Langevin et al., 2007).

Thus, the anomalously high Au grade at Horne and Quemont and the elevated Au grade at Delbridge, Deldona, and the Bouchard-Hebert Main lens may have resulted from a highly efficient zone refining process. Alternatively, other processes that can result in a significantly higher Au concentration in the hydrothermal fluid and deposits include a direct magmatic contribution of Au (Cu) and/or boiling of the hydrothermal fluid (Gibson and Watkinson, 1990; Gibson et al., 1999, Hannington et al., 1999; Dubé et al., 2007a and references therein). Boiling can occur at various depths depending on the salinity and temperature of the hydrothermal fluid (e.g. <1500 m at 350°C; Hannington et al., 1999). Although Lichtblau and Dimroth (1980) have interpreted above storm wave base conditions during emplacement of the Powell Tuff unit north of the Horne Creek Fault, there is no evidence reported from strata of the Horne block to infer a shallow, above storm wave base environment during formation of the Horne deposit.

Mechanisms of Formation

Based on sulphide textures, structures, and metal zoning patterns that are similar to those of modern and actively forming seafloor sulphide deposits, the Noranda mafic and felsic flow-hosted deposits are interpreted to have formed on the seafloor through processes of chimney collapse, hydraulic fracturing, and precipitation of sulphides within the lens, like their modern counterparts.

In contrast, VMS deposits hosted by volcaniclastic lithofacies such as Horne, Bouchard-Hebert, and parts of the Corbet and Ansil deposits, are interpreted to have formed in whole or in part below the seafloor. Features such as remnants of volcaniclastic material within massive sulphide (Fig. 9A), partial sulphide replacement of volcanic lapilli and clasts (Fig. 9B), sulphide cement of coarser volcaniclastic deposits (Fig. 9D), and transgressive contacts between sulphides and their host volcaniclastic rocks are interpreted to indicate sub seafloor mineralizing processes (Galley et al., 1995; Doyle and Allen, 2003) such as: 1) sulphide precipitation in void spaces within unconsolidated and permeable volcaniclastic deposits, 2) hydraulic jacking or displacement of beds with precipitation of sulphide within the lens, and 3) replacement (Kerr and Mason, 1990; Gibson and Kerr, 1993; Kerr and Gibson, 1993; Galley et al., 1995; Hannington et al., 1999; Gibson et al., 2000). These deposits are not typically associated with exhalative tuffs and include the two largest VMS deposits in the district, Horne and Bouchard-Hebert (Figs. 4 and 5).

Most VMS deposits within the Noranda Cauldron are associated with a number of thin (typically <0.5m), extensive, continuous, and well bedded water lain tuff deposits that formed at several stratigraphic intervals throughout the Noranda Cauldron (e.g., Main Contact and C Contact tuffs; Knuckey et al., 1982; Figs. 3D, E and 7); these are referred to as exhalites or exhalative tuffs (Kalogeropoulos and Scott, 1983; 1989) These exhalites have a fine-grained clastic component, derived through suspension sedimentation of ash, and a chemical component composed of chert and sulphides, derived from mineral precipitates from vent plumes that were contemporaneous with formation of the sulphide deposits (Comba, 1975; Kalogeropoulos and Scott, 1983;1989).

Within the Noranda Cauldron, the exhalite units represent a significant hiatus in volcanism and are commonly of large extent (i.e., up to 10 km). The C Contact tuff, in particular (Figs. 3D, E), is underlain by up to several hundreds of metres of weakly to strongly silicified basaltic to andesitic flows (Fig. 3F). Here, the silicification is interpreted to be a product of relatively low temperature convection of compositionally modified seawater that was enriched in Si and other elements (Fe, Mg,) through interaction with glass-rich, basaltic-rhyolitic flows (Gibson et al., 1983). Venting of heated, Si-enriched seawater released Si, Fe and other elements to form the hydrothermal component of the overlying C Contact tuff (Gibson et al., 1983; Gibson, 1990; Paquette-Mihalasky, 1999). At Noranda and elsewhere, the presence of these exhalative units are used to define prospective contacts, and the distribution and tenor of their contained metals have been used to define exploration targets along these contacts (Peter and Goodfellow, 1996; 2003).

Alteration Assemblages

Typical alteration assemblages associated with the Noranda deposits are chlorite and sericite. Calcite and Fe-carbonate occur in proximity to some deposits (e.g. Delbridge, Deldona), but the relationship of carbonate alteration to the VMS deposits is uncertain. The proportion of sericite versus chlorite alteration and the morphology of the alteration zone differ between VMS deposits hosted by a flow lithofacies versus those hosted by a volcaniclastic lithofacies (Figs. 10-11). The discordant stockwork of stringer sulphide that occurs below and locally above mafic or felsic, flow-hosted, massive sulphide deposits are enveloped within vertically extensive, well-defined, pipe-like zones of chloritized and sericitized rock (Riverin and Hodgson, 1980; Fig. 9E). In contrast to the marked vertical extent of the alteration pipes their diameter is rarely larger than the massive sulphide lenses, except where the immediate footwall unit is a volcaniclastic unit (e.g., Corbet; Gibson et al., 1993). The alteration pipes are characterized by a mineralogical and compositional zoning from an inner core dominated by quartz and Fe-chlorite, with subordinate minnasotite and Fe-talc. This changes progressively outward (and in some cases upward) to an assemblage dominated by Mg-chlorite which, in turn, changes to an outer zone dominated by sericite (Fig. 10). Although the proportion of chlorite versus sericite alteration is variable between deposits, chlorite-rich alteration is typically dominant.

Broad, diffuse, semiconformable alteration zones dominated by sericite and quartz characterize the felsic volcaniclastic-hosted Horne and Bouchard-Hebert deposits (Caumartin and Caille, 1990; Kerr and Mason, 1990; Kerr and Gibson, 1993) (Figs. 9F, 11). Within these alteration zones chlorite is less common but where present, crosscuts sericite-quartz alteration as narrow zones rich in Fe-chlorite (Fig. 11). In both flow and volcaniclastic hosted deposits, the sericite (+/-quartz) alteration is characterized by a depletion in Na_2O and CaO, and enrichment in K_2O (+/- SiO_2) whereas the chlorite alteration is typically depleted in Na2O, CaO, K_2O and SiO_2, and enriched in MgO (Mg-chlorite) and/or FeO (Fe-chlorite); in all cases chlorite is later and cross-cuts sericite (Gibson and Kerr, 1993).

In addition to chlorite and sericite alteration that is spatially and temporally related to the VMS deposits, the Noranda District has undergone synvolcanic, regional semiconformable alteration that has resulted in the development of several key alteration types (Gibson et al., 1983; Gibson, 1990; Santaguida, 1999; Hannington et al., 2003). The first is spilitization, a texturally non-destructive alteration that has affected all volcanic rocks within the Blake River Group. Spilitization is characterized by higher Na_2O values, lower CaO values, and a greenschist assemblage of chlorite, quartz, epidote and albite in mafic rocks and sericite (+/-chlorite), quartz, and albite in felsic volcanic rocks (Gibson, 1990). In subgreenschist facies volcanic rocks, spilitization is characterized by a prehnite and pumpelleite assemblage that typically occurs within amgydules of both mafic and felsic volcanic rocks. The second is epidote-quartz alteration, which manifests itself as irregular patches and amoeboid forms decimeters to metres in size within mafic volcanic rocks. Epidote-quartz alteration is characterized by an increase in CaO and a depletion in mobile elements such as FeO, MgO, K^2O, Zn and Cu; SiO_2 is conserved (Gibson, 1990; Santaguida, 1999). The alteration is texturally destructive and results in a granular mosaic of clinozoisite and quartz +/-amphibole. The alteration occurs within all mafic volcanic rocks of the Blake River Group, but is most intense within mafic volcanic rocks, dykes, and sills proximal to synvolcanic structures located within and peripheral to the Noranda Cauldron. On a regional scale there is a change from epidote-dominant to clinozoisite-dominant in proximity to the hanging wall contact of the subvolcanic FPIC (Hannington et al., 2003). The last type of regional-scale alteration is silicification, a texturally non-destructive alteration characterized by a quartz-albite assemblage and elemental changes marked by the addition of SiO_2 and Na_2O, and a depletion of mobile elements such as FeO, MgO, K_2O, and Zn (Gibson et al., 1983; Paquette-Mihalasky, 1999). Pervasive and widespread silicification is restricted to permeable areas (amygdule zones, flow tops) within andesitic flows of the Upper member of the Amulet formation, a distinct marker unit that underlies most of the VMS deposits and associated exhalite units within the Noranda Cauldron (Fig. 3F) (Gibson, 1990).

Gibson et al., (1983) and Gibson (1990) suggested that the morphology, elemental changes, and distribution of the semiconformable alteration types indicated that the alteration was caused by fluid-rock interaction within a large geothermal-hydrothermal system centred on the Noranda Cauldron. This was supported by a regional oxygen isotope study by Cathles (1993), which delineated the size and distribution of lower temperature (<300°C) seawater-rock interaction zones. Furthermore, this study defined the presence of discrete, higher temperature (>300°C) hydrothermal alteration zones that are spatially associated with the upper contact of the subvolcanic FPIC, and with more linear zones containing clusters of VMS deposits and occurrences within the Noranda Cauldron. Santaguida et al., (1998), Santaguida (1999) and Hannington et al. (2003) documented grain-scale compositional heterogeneity of chlorite, sericite, and epidote within these assemblages, and a systematic change from a Mg-rich to more Fe-rich chlorite, and from a more Fe-rich to Al-rich clinozoisite within the assemblages in volcanic rocks peripheral to and within the Noranda Cauldron and its underlying subvolcanic intrusive complex. They interpreted that the change in mineral chemistry indicated a hydrothermal as opposed to a metamorphic origin for the semi conformable alteration assemblages.

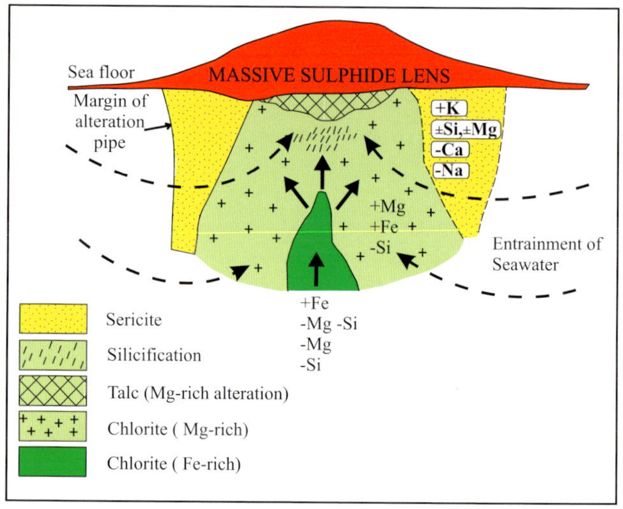

FIGURE 10. Idealized distribution of alteration assemblages and compositional gains and losses associated with Noranda VMS deposits hosted by flow lithofacies (modified from Gibson and Kerr, 1993).

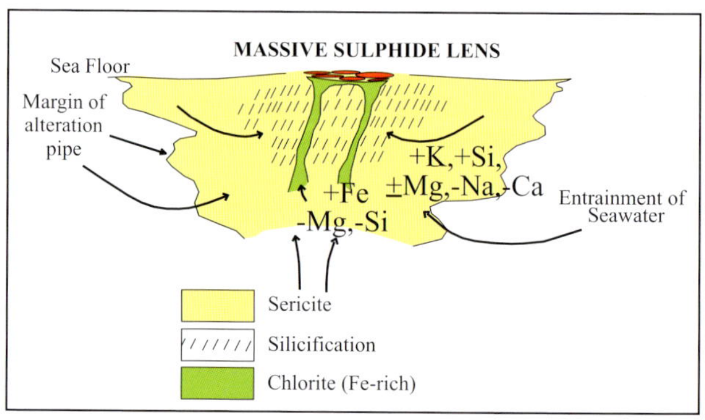

FIGURE 11. Idealized distribution of alteration assemblages and compositional gains and losses associated with Noranda VMS deposits hosted by volcaniclastic lithofacies (e.g., Horne and Bouchard-Hebert; modified from Gibson and Kerr, 1993).

Distribution of VMS Deposits

At the deposit-scale, Knuckey et al. (1982), Gibson, (1990), Gibson et al. (1993), and Kerr and Gibson (1993) have demonstrated a pronounced spatial association between mafic and felsic volcanic centres, synvolcanic faults and massive sulphide deposits in the Noranda District (e.g., Amulet-Millenbach deposits; Fig. 12). Swarms of rhyolite, basalt and composite dykes and lesser sills of the same define the synvolcanic faults; these structures not only controlled the location of volcanic centres, vents, and VMS deposits but also, in some cases, accommodated subsidence (e.g. Old Waite Paleofissure, Fig. 2). At the district-scale, Gibson (1990), Gibson and Watkinson, (1990). and Kerr and Gibson (1993) have shown that the VMS deposits are preferentially located in time and space to the development of the Noranda Cauldron (Fig. 2A and Table 2).

Cycle I (2701 Ma), located west of the FPIC (Fig. 2A) contains the subeconomic Inmont, Four-corners, Ivanex and Montbray VMS deposits (Fig.1, Table 1). The Four-corners and Inmont deposits are chalcopyrite-pyrrhotite stringer zones that are interpreted to be the erosional remnants of former VMS deposits. The Ivanex deposit is a bedded, pyritic, Zn-rich transported sulphide deposit, and the Montbray deposit consists of a small massive sulphide lens and stringer sulphide mineralization (Martin and Masson, 2005). All Cycle I VMS deposits are hosted by rhyolitic flow domes and Cycles I and II are interpreted to have been erupted before subsidence of the Noranda Cauldron (2697 Ma to 2701 Ma, Table 1; Gibson, 1990). However, the 2696.9 +/-3.4 Ma age for a Cycle II rhyolite (David et al., 2006) is comparable in age to Cycle III (2698.5 +/-2 Ma; David et al., 2006), the Mine Sequence that, along with Cycle IV, hosts 18 of the economic massive sulphide deposits in the district.

Noranda formation Cycles III and IV are interpreted to have been erupted during incremental, piece-meal subsidence to produce the "trap door-like" Noranda Cauldron that is centred on a northeast-trending rift referred to as the Old Waite Paleofissure (Gibson and Watkinson, 1990; Fig. 2A, B). U-Pb zircon ages of 2698.5 (+/- 2 Ma) for the Millenbach Rhyolte and 2696.5 +/-2.4Ma for the rhyolitic Fishroe synvolcanic intrusion are the only constraints on the timing and duration of Cycle III and IV volcanism and cauldron development (David et al., 2006; Table 1). Sixteen of the VMS deposits within and along the margins of the Noranda Cauldron are spatially associated with coincident rhyolitic and basaltic volcanic centres marked by rhyolitic flow dome complexes, thick ponded basaltic flows successions and dyke swarms (Gibson, 1990; Gibson and Watkinson, 1990; Kerr and Gibson, 1993). The remaining two deposits (New Insco and Magusi River; Table 2), although located outside of the northern structural margin of the Noranda Cauldron (Fig. 1), occur within rhyolitic and basaltic strata interpreted to be part of the Noranda formation (Cycle III time equivalent?). The Noranda formation VMS deposits, with the exception of the Quemont and Gallen deposits, are typically small and contain <5 Mt of ore (Table 2).

The giant, Au-rich Horne deposit is located immediately south of the subvertical Horne Creek Fault and within the Rouyn-Pelletier formation (Figs. 2A,B). The Horne deposit consists of three main orebodies, the Upper H, Lower H, and the No. 5 (Figs. 5,6); only the Upper and Lower H orebodies were mined (Table 2). The Upper H and Lower H orebodies consist of massive to semi-massive sulphides with a mineralogy dominated by pyrite and pyrrhotite with lesser amounts of chalcopyrite, magnetite and, locally, sphalerite, plus trace amounts of native Au (electrum) and Au-Ag telluride minerals (Kerr and Mason, 1990; Fig. 5). The larger No.5 orebody consists of massive to semi-massive pyrite (subordinate pyrrhotite) and lesser sphalerite within volcaniclastic units containing clasts of massive sulphide (Fig. 6). Sericite, as a broad semi-conformable zone, forms the dominant footwall alteration and, below the Upper and Lower H orebodies, is crosscut by local narrow zones of Fe-chlorite alteration. The Horne orebodies are hosted by a southeast

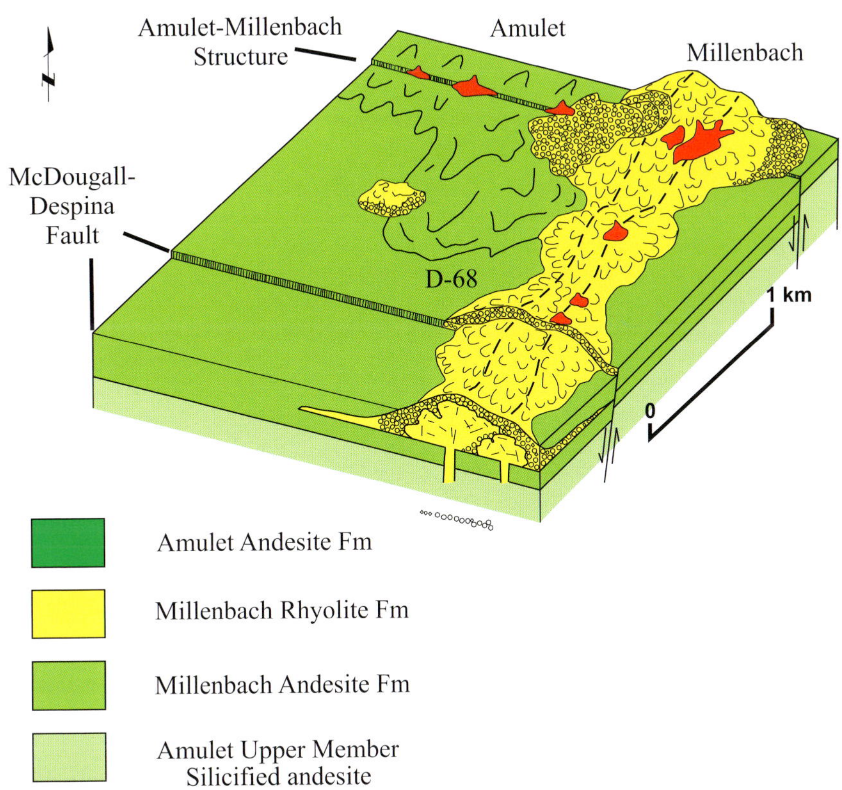

FIGURE 12. Volcanic reconstruction of the Millenbach dome complex showing the pronounced structural control on volcanism and hydrothermal discharge that results in a coincidence between volcanic vents and VMS deposits. The Amulet Lower A deposit is localized along a fissure for basalt volcanism whereas the Millenbach VMS deposits occurs along the top of the rhyolite ridge and directly above and along the feeding fissure (modified from Gibson et al., 1999). Refer to Figure 1 for deposit location and to Figure 6 for a cross-section.

striking, sub-vertical, north-facing, 900m-thick rhyolitic volcaniclastic lithofacies with subordinate, coherent, rhyolite lithofacies occurring as flows, domes and cryptodomes. The rhyolitic succession and the ore bodies are crosscut by a swarm of basaltic dykes and, within the upper part of the succession, basaltic sills. The timing of the basalt dyke/sill swarm has not yet been established, but if synvolcanic it could define a major rift along the inferred south margin of the cauldron that is similar to but narrower than the dyke swarm associated with the Old Waite Paleofissure located within the core of the cauldron.

The Horne deposit lies within the Horne Block, an east-west-trending "structural wedge" that is bound to the north and south by the Horne Creek and Andesite Faults (Figs. 2A, B). These faults, plus the lack of geochronological data, have hampered attempts to confidently correlate the Horne volcanic succession with Cycle III and IV volcanic units of the Noranda formation to the north or with volcanic units of the Rouyn-Pelletier formation to the south (2701 +/- 1.6 Ma, David et al., 2006; Table 1; Fig. 2A). Spence and deRosen-Spence (1976) interpreted the Horne and Quemont (Noranda formation) deposits to be time-stratigraphic equivalent based on a tentative correlation of volcaniclastic units across the Horne Creek fault. Gibson (1990), Gibson and Watkinson (1990) and Kerr and Gibson (1993) suggested that the Horne deposit may have formed upon older volcanic strata of the Rouyn-Pelletier formation within a fault block located on the southern structural margin of the Noranda Cauldron but that the deposit itself may be the time-stratigraphic equivalent of the entire Mine Sequence or Cycle III of the Noranda formation. Recent interpretations by Pearson (2005) would tentatively place the Horne deposit and the entire Noranda Cauldron within the larger New Senator caldera.

The Bouchard-Hebert deposit is the only economic VMS deposit discovered thus-far within the Reneault-Dufresnoy formation (Cycle V of the NVC); a U-Pb zircon age of 2698 +/- 1Ma has been established for rhyolitic units hosting the deposit (Mortensen, 1993). The deposit consists of several Zn-rich, pyritic massive sulphide orebodies, the largest being the 10 Mt 1100 lens (Table 2; Caumartin and Caille, 1990; Barret et al., 1992; Larocque et al., 1993). The Bouchard Hebert orebodies are similar to those at the Horne deposit in that they are: 1) hosted within a volcanic succession dominated by a rhyolitic volcaniclastic lithofacies with subordinate coherent lithofacies as flows or domes; 2) associated with a broad, semiconformable zone of sericite alteration that is locally cross cut by narrower and aerially restricted zones of chlorite alteration; and 3) pyrite-rich.

Genetic Models and Geodynamic Setting

VMS deposits are syngenetic, stratabound accumulations of sulphide minerals that formed in spatial, temporal and genetic association with contemporaneous volcanism and/or plutonism. VMS deposits are interpreted to have formed on and immediately below the seafloor by the discharge of a high temperature, evolved, seawater-dominated hydrothermal fluid (Franklin et al., 1981; Lydon, 1984, Franklin et al., 2005, and references therein). The metals are interpreted to have been leached from hydrothermal reactions zones by heated, compositionally modified seawater convecting under a steep geothermal gradient through the volcanic pile with or without a magmatic contribution of metals (Galley, 1993; Franklin et al., 2005). However, to understand the fundamental processes responsible for the formation of VMS deposits at Noranda it is important to step-back from the details of individual deposits and examine the larger-scale controls that could potentially affect the VMS hydrothermal system.

VMS deposits of the Noranda District formed within a bimodal, basalt-dominated, central volcanic complex, the NVC, of primitive arc-like geochemical affinity that may have consisted of one or more volcanic edifices (Ludden et al. 1982; Gelinas et al., 1984; Peloquin, 2000; Peloquin, 2005). DeRosen-Spence (1976), Dimroth et al. (1982), Gibson, 1990, Gibson and Watkinson (1990), and Kerr and Gibson (1993) have shown that 19 of the 20 past producing VMS deposits formed within or in proximity to the Noranda Cauldron, and that the 16 VMS deposits that occur in the cauldron formed during hiatuses that separate at least 3 cauldron-subsidence cycles (Gibson, 1990; Gibson and Watkinson, 1990; Fig. 2A). The Noranda Cauldron is interpreted to be a seafloor manifestation of a rift that is now represented by the Old Waite Paleofissure, an extensive, sheeted dyke swarm upon which the cauldron is centred. Other evidence for rifting include: 1) the emplacement of the sill-like, multiphase, FPIC within the Old Waite Paleofissure and core of the cauldron, and 2) trondhjemite phases of the Flavrian-Powel plutons which, like the high temperature FIII rhyolites they intrude, are interpreted to be a product of partial melting of hydrated oceanic crust within a shallow crustal environment that is typical of rifts (Lesher et al., 1986; Galley, 2003; Hart et al., 2004).

As discussed, a large number of factors such as volcanic environment (flows versus volcaniclastic lithofacies), depositional processes (zone refining), and synvolcanic structures influenced the location, formation, and preservation of VMS deposits at Noranda. However, rifting provides the two essential components required to form VMS hydrothermal systems. First, it provides an aerially restricted but vertically extensive heat source, i.e., a thermal corridor that is large enough, hot enough, and at an appropriate level within the crust to not only produce crustally-derived magmas (e.g., FIII rhyolite and tronhjemites at Noranda; Lesher et al., 1986) but also to generate and sustain the long-lived, high-temperature seawater convective hydrothermal system required to form the Noranda VMS deposits. The restriction of high-temperature, semiconformable alteration assemblages to the Noranda Cauldron attests to the high heat- and fluid-flow that was focused within this rift (Gibson, 1990; Cathles, 1993; Santaquida et al. 1999; Hannington et al., 2003). Second, rifting is associated with long-lived, re-activated and deeply penetrating faults that provide pathways for magma ascent into upper crustal magma chambers or to feed surface eruptions (i.e., enhanced heat flow), and the cross-stratal permeability required to permit and facilitate hydrothermal circulation and to maintain and focus hydrothermal discharge (Cathles et al., 1997; Barrie and Hannington, 1999). Rifting may also allow for a direct contribution of magmatic metals from fluids exsolved from magmas emplaced at various levels within the rift that caused or resulted from partial melting (e.g., Tomkins and Mavrogenes, 2003; Hart et al., 2004).

Thus, rifting of the NVC resulted in high heat flow, cross-stratal permeability, and the development of a high temperature VMS hydrothermal system that was focused in time and space to a vertical, long-lived thermal corridor defined by the rift. Hydrothermal discharge and the formation of VMS deposits at or near the seafloor were restricted to areas within and adjacent to Noranda Cauldron, the seafloor manifestation of the rift. The stratigraphic relationship of the Bouchard-Hebert deposit to VMS deposits within the Noranda Cauldron is uncertain. However, the overlap in age between the Mine Sequence (Millenbach Rhyolite 2698.5 +/2 Ma; David et al., 2006) and the Bouchard-Hebert rhyolite (22698 +/-1 Ma; Mortensen, 1993) suggests that VMS deposits within the Noranda (except for the Magusi and New Insco deposits whose absolute age is unknown) and Reneault-Dufresnoy formations formed within a narrow time interval (<1Ma) and perhaps within the same thermal corridor that developed during rifting and cauldron subsidence. On an even larger scale, the similarity in age between the Bousquet formation (2696+/-1.1 to 2698+1.3/-0.7 Ma; Mercier-Langevin et al., 2004; Lafrance et al., 2005) that is host the LaRonde, Bousquet and Dumagami VMS deposits and VMS deposit in the Noranda and Reneault-Dufresnoy formations suggests that VMS deposits within both districts formed within a <1 Ma period at 2698 Ma. The age of the Au-rich Horne deposit and its stratigraphic relationship and timming to the Noranda and Reneault-Dufrenoy formations is uncertain.

Viewed in this context, the Noranda VMS deposits may have formed within a volcanic complex that consists of several separate volcanic edifices and in a geodynamic environment that was, perhaps, similar to a modern primitive arc or back arc environment that was undergoing extension and rifting. Modern examples of these environments include the Tonga Ridge-Lau basin system (Gill, 1987; Kerrich and Wyman, 1996; Hannington et al., 2006) and the Kermadec arc (Stoffers et al., 1999a, 1999b). For example, the Monowai arc volcano (Monowai Cone) and caldera located along the Kermadec arc are shown at the same scale as the Noranda Cauldron and Central Volcanic Complex in Figure 13. Besides the similarity in scale between the Monowai caldera and Noranda Cauldron there are other features of this modern rifted arc that have implications for volcanic reconstruction and the timing of VMS deposit formation at Noranda that include: 1) the Monowai caldera formed during a period of arc extension (rifting) that was superimposed upon a cluster of several arc volcanoes, the Monowai cone being just one of those early volcanoes. The Noranda Cauldron is interpreted to have formed during a rifting event that represented the final evolutionary stage of the Noranda Volcanic Complex (Gibson, 1990; Gibson et al., 2005); 2) The Monowai cone (Fig. 13) and other arc volcanoes represent contemporaneous, separate but overlapping individual volcanoes that may have had a similar or slightly different evolution. However if tilted on end, deformed and viewed in cross section, the individual Monowai arc volcanoes could not be distinguished but would collectively define a volcanic complex composed of different formations, perhaps like those that comprise the Noranda Volcanic complex; and 3) the Monowai caldera, like the Noranda Cauldron, is the focus of high-temperature hydrothermal activity and contains several active hydrothermal vents. In both cases, processes responsible for the development of the high temperature hydrothermal systems responsible for the VMS deposits at Noranda and the active vents at Monowai are linked to high heat flow and structural permeability that are restricted in time and space to caldera development.

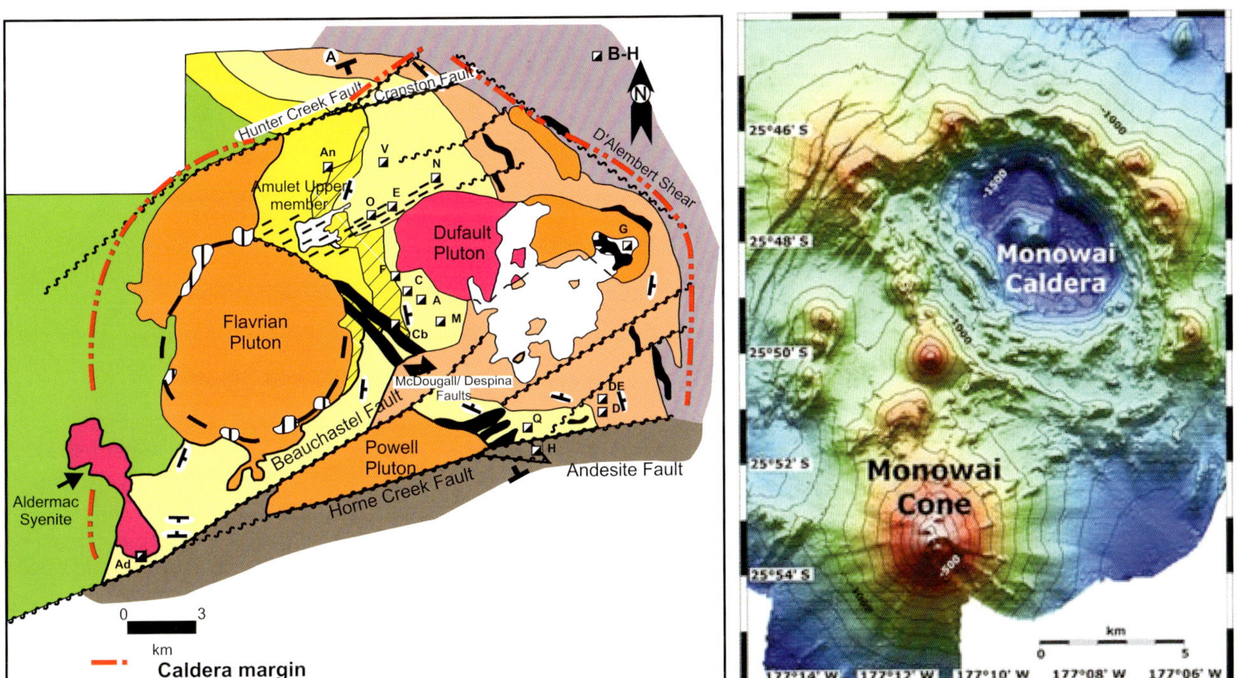

FIGURE 13. Comparison of the Archean, Noranda Volcanic Complex (NCV) and cauldron with the Monowai arc volcano and caldera, Kermadec Arc (see Fig. 2 for Noranda Volcanic Complex legend).

Summary of Effective Exploration Methods

The Noranda District has a long and colourful exploration history that continues today. Prospector Edmund Horne discovered the Horne deposit, the largest VMS deposit in the district, in 1923 and Alexis Minerals and Noranda discovered the West Ansil deposit in 2003 (Table 2). Exploration in the early years was difficult with no existing infrastructure or access except canoe routes, and in most respects was more challenging and costly than exploration in the remote areas of today. The Horne discovery not only opened up the exploration and settlement of northwestern Quebec but also resulted in the creation of Noranda, formerly one of Canada's largest mining companies (Lulin, 1990). It can be considered a nation builder. In contrast, the West Ansil deposit was discovered within a now mature Noranda camp and resulted from a comprehensive compilation and reinterpretation of existing geological, geochemical and geophysical data using sophisticated earth modeling software (Table 2; Martin, 2005). In this respect, exploration in the Noranda District records an evolution in discovery methods that Boldy (1979) referred to as the "Exploration Life Cycle of a mining district". Boldy (1979) divided the Exploration Life Cycle of the Noranda District into 3 stages: 1) the Early Years, 1920-1935; 2) the Middle Years, 1936-1955; and 3) the Later Years, 1955-1977, which now can be extended to 2006.

The Early Years were dominated by four prospecting discoveries (Horne, Amulet Upper A and C, and Old Waite), one geophysical discovery (Aldermac, dip needle), one geological discovery (Amulet F), and one in-mine exploration discovery (Horne Lower H) (Table 2). The Middle Years were characterized by three discoveries using empirical geological models (recognized association of VMS deposits with favourable lithologies, alteration, and structure; Amulet Lower A, Deldona, and East Waite), two ground EM and magnetic discoveries (Gallen, and Quemont), and one in-mine discovery (Horne No.5 zone) (Table 2). The Later Years are subdivided into a pre- and post-1981 periods; the latter is referred to the "current or mature" period. The pre-1981 discoveries resulted from the use of a progressively more sophisticated volcanogenic and syngenetic geological model for VMS deposits (4 discoveries; Vauze, Norbec, Millenbach and Ansil), the introduction of the use of geology coupled with lithogeochemistry (2 discoveries; Delbridge and Corbet), and more sophisticated ground EM (1 discovery; Mobrun) and, for the first time, airborne EM geophysical discoveries (Magusi River and New Insco) (Table 2). Exploration from 1920 to 1981 yielded 19 economic and 2 subeconomic VMS deposits over a 60-year period, with 9 discoveries made during the pre-1981 Later Years, versus 6 and 7 discoveries during the Middle and Early Years. However, because of the enormous size of the Horne deposit, more than 60% of the total tonnage of the Noranda District was discovered during the Early Years.

Except for the in-mine discovery of the 1100 lens at Bouchard-Hebert and the small subeconomic Ribago deposit, exploration during the post-1981, mature period was characterized by a 22-year discovery gap until discovery of the West Ansil deposit in 2003, the Montbray deposit in 2004 and the Pinkos 3 discovery in 2007 (Alexis Minerals, Press Release, January 09th, 2007). The West Ansil discovery resulted from the use of a computer aided geological, geochemical and geophysical compilation of the Noranda District, sophisticated 3-D computer visualization tools, and a more sophisticated and refined genetic model for VMS deposits (Martin and Masson, 2005). The Montbray discovery resulted from the use of a new, more deeply penetrating airborne EM (Megatem) and borehole EM geophysical surveys (Martin and Masson, 2005). Although not economic, these three latest discoveries illustrate that during the mature stage of a mining district new discoveries will most likely result from a re-evaluation of existing data aided by sophisticated visualization software and by new or improved geophysical or geochemical technologies. The three recent discoveries in the Noranda District (West Ansil, Montbray, Pinkos 3) also occur at shallow depths and reverse a general trend of progressively deeper discoveries with time (Table 2).

Exploration Potential

The Noranda District is now in the mature stage of its Exploration Life cycle, following a long and successful exploration and development history. However, there are opportunities for new discoveries and, as the most recent discoveries indicate, these will likely result from a re-evaluation and interpretation of existing geological, geochemical, and geophysical data using modern 3-D computer techniques, the development of new deeper penetrating airborne, ground and borehole geophysical techniques, and new geochemical techniques to explore beneath glacial deposits.

If we assume that the present exposure of the Noranda District represents a somewhat random section or sections through the volcanic edifice that is host to the VMS deposits, then the distribution of known VMS deposits and occurrences can be taken as an indication of the prospectivity for different parts of the NVC. In this regard areas to be considered for future exploration may include:

1. The area west of the Flavrian-Powell intrusive complex (Cycles I and II) is less explored, but has only yielded 4 subeconomic deposits and occurrences, and may therefore have a lower mineral potential (Fig. 2).

2. With most of the VMS deposits and significant occurrences and subeconomic deposits located within the Noranda Cauldron, or in proximity to it, this area warrants further exploration. The area north of the Noranda Cauldron needs to be re-examined as it is less explored and it is poorly exposed (glacial cover); however, it may be less favorable as the deposits discovered to date (New Insco and Magusi) are small and low grade. There is still room for smaller (1-5 Mt), undiscovered deposits within the cauldron (e.g., along the northern margin and within or adjacent to the old Waite Dyke swarm); however they will likely be deep (>700m). Cycle IV volcanics that host the Delbridge and Deldona deposits have not been thoroughly explored and excellent potential exists especially to north and northwest of these deposits as evidenced by the recent discovery of the Pinkos 3 zone. The Horne block (south of the Horne Creek Fault) contains the largest and most Au-rich deposit in the district, yet it is poorly understood and this block has not been as

thoroughly explored. The Horne block and the area immediately south of the Andesite fault, which should contain the southern continuation of the Horne stratigraphy, seem the most favorable for the discovery of a new deposit or additional lenses peripheral to the main Horne deposit (Fig. 2).

3. Although Cycle V felsic volcaniclastics of the Reneault-Dufresnoy formations have only yielded one economic deposit, Bouchard-Hebert, this is the second largest VMS deposit in the district and the Main lens was enriched in Au (Fig. 1; Table 2). This portion of the NVC is arguably the least explored, the most poorly exposed due to thicker, more extensive glacial cover and, based on new U-Pb zircon ages (2698 +/1 Ma and 2698.3 +/1 0.8/2697.8 +/-1 Ma respectively, Table 1), is the time equivalent and perhaps the continuation of the Bousquet formation that contains extensive felsic volcaniclastic rocks that host the Bousquet 1, Bousquet 2-Dumagami and LaRonde Penna, Au-rich VMS deposits (see Mercier-Langevin et al., 2007).

Knowledge Gaps

Despite an exploration and research history that spans more than 85 years, there remain significant knowledge gaps in our understanding of the Noranda Volcanic Complex and associated VMS deposits. What we believe to be the most significant knowledge gaps are listed below.

1) What is the time-stratigraphic relationship of the Horne deposit to VMS deposits immediately north of the Horne Creek fault and to strata south of the Andesite fault?

2) What are the processes responsible for Au-enrichment at the Horne, Quemont, Delbridge and Deldona VMS deposits, which are located along the Horne Creek fault?

3) What is the time-stratigraphic relationship between the Reneault-Dufresnoy and Bousquet formations?

4) Is the Noranda Volcanic complex a single, large volcanic edifice or does it represent several separate edifices? If the latter interpretation is correct, is the Noranda Cauldron confined to a single edifice or was it superimposed on several edifices.

5) Is the Noranda Cauldron part of nested caldera that include larger structures such as the proposed New Senator and Misema caldera? And if so what are the metallogenic and exploration implications?

6) What is the large-scale geodynamic setting of the Blake River Group and the Noranda Volcanic complex?

Acknowledgements

This paper is dedicated to all who have explored or have conducted research at Noranda, as our knowledge of this district stems from the results of their work. Many of our colleagues have influenced our thoughts on the Noranda District and on VMS deposits in general and we would to thank Bankie Bancroft, Dave Comba, Benoit Dubé, Jim Franklin, Jean Goutier, Mark Hannington, David Kerr, Mike Knuckey, Thomas Monecke, Gerald Riverin, and David Watkinson. We would also like to thank a host of dedicated students that we have been fortunate to have worked with, including, Suzanne Paradis, Shirley Peloquin, Frank Santaguida, Francine Paquette-Mihilasky, Renee Turmel, and Steve Zubowski. Thorough reviews by Patrick Mercier-Langevin, Benoit Lafrance, and Wayne Goodfellow improved the manuscript and their efforts are gratefully acknowledged. HLG acknowledges an NSERC Discovery Grant, which has supported his VMS research at Noranda and elsewhere.

References

Alt, J. C., 1995, Subseafloor processes in mid-ocean ridge hydrothermal systems: Seafloor Hydrothermal Systems: Physical, Chemical, Biological, and Geological Interactions, American Geophysical Union, Geophysical Monograph 91, p. 85-114.

Ayer, J.A., Thurston, P.C., Bateman, R., Dubé, B., Gibson, H.L., Hamilton, M.A., Hathway, B., Hocker, S.M ., Houlé, M.G., Hudak, G., Ispolatov, V.O., Lafrance, B., Lesher, C.M., MacDonald, P.J., Péloquin, A.S., Piercey, S.J., Reed, L.E., and Thompson, P.H., 2005, Overview results from the Greenstone Architecture Project; Discover Abitibi Initiative: Ontario Geological Survey, Open File Report 6154, 146 p.

Barrett, T.J., MacLean, W.H. and Cattalani, S., 1991, Massive sulphide deposits of the Noranda area Quebec. I. The Horne mine: Canadian Journal of Earth Sciences, v. 28, p. 465-488.

Barrett, T.J., Maclean, W.H., and Cattalani, S., 1992. Massive sulphide deposits of the Noranda area, Quebec, II. The Mobrun mine: Canadian Journal of Earth Sciences, v.29, p. 1349-1374.

Barrie, C. T., and Hannington, M.D., 1999, Introduction; Classification of VMS deposits based on host rock composition, *in* C. T. Barrie, and Hannington, M.D., ed., Volcanic-Associated Massive Sulphide Deposits: Processes and Examples in Modern and Ancient Settings, Reviews in Economic Geology, v. 8, p. 2-10.

Barrie, C.T., Ludden, J.N. and Green, T.H., 1993, Geochemistry of volcanic rocks associated with Cu-Zn and Ni-Cu deposits in the Abitibi subprovince: Economic Geology, v. 88, p. 1341-1358.

Barrie, C.T., Cathles, L.M., Erendi, A., Schwaiger, H., Murray, C., Barrie, C.T., and Hannington, M.D., 1999, Heat and fluid flow in volcanic-associated massive sulfide-forming hydrothermal systems, in Volcanic-associated massive sulfide deposits; processes and examples in modern and ancient settings: Reviews in Economic Geology, v. 8, p. 201-219.

Boldy, J., 1979, Exploration discoveries, Noranda District, Quebec, *in* Hood,.P.J., ed., Geophysics and geochemistry in the search for metallic ores: Geological Survey of Canada, Report 31, p. 593-604.

Carriere, C., Jébrak, M., Angelier, J., and Holyland, P. 2000, The Silidor Deposit, Rouyn-Noranda District, Abitibi Belt: Geology, Structural Evolution, and Paleostress Modeling of an Au Quartz Vein-Type Deposit in an Archean Trondhjemite: Economic Geology, v. 95, p. 1049 - 1065.

Cathles, L. M., 1993, Oxygen Isotope Alteration in the Noranda Mining District, Abitibi Greenstone Belt, Quebec: Economic Geology, v. 88, p. 1483-1511.

Cathles, L.M., Erendi, A.H., and Barrie, T., 1997, How long can a hydrothermal system be sustained by a single intrusive event: Economic Geology, v. 92, p.766-771.

Cattalani, S., Barrett. T.J., MacLean, W.H., Hoy, L., Hubert, C. and Fox, J.F., 1993, Métallogènese des gisements Horne et Quemont: Ministère de l'Énergie et des Ressources du Québec, Report ET 90-07, 121 p.

Caumartin C., and Caille, M.-F., 1990. Volcanic stratigraphy and structure of the Mobrun mine, *in* Rive, M., Verpaelst, P., Gagnon, Y., Lulin, L.M., Riverin, G., and Simard, A., eds., The Northwestern Quebec Polymetallic Belt: A summary of 60 years of mining exploration: Canadian Institute of Mining Metallurgy, Special Volume 43, p.133-142.

Comba, C.D.A., 1975, Copper-Zinc zonation in tuffaceous exhalites, Millenbach Mine, Noranda, Quebec: Msc. thesis, Queen's University, Kingston, Ontario, 135p.

David, J., Dion, C., Goutier, J., Roy, P., Bandyayera, D., Legault, M., and Rheaume, P., 2006, Datations U-Pb effectuées dans la Sous-province de l'Abitibi à la suite des travaux de 2004-2005: Ministère des Ressources naturelles et de la Faune, RP 2006-04, 22 p.

de Rosen-Spence, A.F., 1976, Stratigraphy, development and petrogenesis of the central Noranda volcanic pile, Noranda, Quebec: Ph.D. thesis, University of Toronto, Toronto, Ontario, 166 p.

Dimroth, E., Imreh. L., Rocheleau, M. and Goulet, N., 1982, Evolution of the south central part of the Archean Abitibi Belt, Quebec. Part 1: Stratigraphy and paleogeographic model: Canadian Journal of Earth Sciences, v. 19, p. 1729-1758.

Doyle, M.G., and Allen, R.L., 2003. Sub-sea floor replacement in volcanic-hosted massive sulphide deposits: Ore Geology reviews, v.23, p.183-222.

Dubé, B., Gosselin, P., Mercier-Langevin, P., Hannington, M., and Galley, A., 2007a, Gold-rich Volcanogenic Massive Sulphide Deposits, *in* Goodfellow, W. D., ed., Mineral Deposits of Canada: A Synthesis of Major Deposit-types, District Metallogeny, the Evolution of Geological Provinces, and Exploration Methods, Special Publication 5, Mineral Deposits Division, Geological Association of Canada, p. 75-94.

Dubé, B., Mercier-Langevin, P., Hannington, M.D., Lafrance, B., Gosselin, P., And Gosselin, G., 2007b, The LaRonde Penna Au-rich volcanogenic massive sulfide deposit, Abitibi, Quebec: Mineralogy and geochemistry of alteration and implications for genesis and exploration, Economic Geology, in press.

Eldridge, C.S., Barton, P.B. and Ohmoto, H., 1983, Mineral textures and their bearing on formation of the Kuroko orebodies: Economic Geology, Monograph 5, p. 241-281.

Franklin, J.M., Sangster, D.F. and Lydon, J.W., 1981, Volcanogenic massive sulphide deposits: Economic Geology 75th Anniversary volume, p. 485-627.

Franklin, J.M, Gibson, H.L., Jonasson, I.R. and Galley, A.G., 2005, Volcanogenic Massive Sulphide Deposits, *in* Hedenquist, J.W., Thompson, J.F.H., Goldfarb, R.J., and Richards, J.P., eds., Economic Geology, 100th Anniversary Volume, The Economic Geology Publishing Company, p. 523-560.

Galley, A.G., 1993, Semi-conformable alteration zones in volcanogenic massive sulphide districts: Journal of Geochemical Exploration, v. 48, p. 175-200.

Galley, A.G., 2003, Composite synvolcanic intrusions associated with Precambrian VMS- related hydrothermal systems: Mineralium Deposita, v. 38, p. 443-473.

Galley, A. G., Watkinson, D. H., Jonasson, I. R., and Riverin, G., 1995, The subseafloor formation of volcanic-hosted massive sulphide; evidence from the Ansil deposit, Rouyn-Noranda, Canada: Economic Geology, v. 90, p. 2006-2017.

Galley, A.G., and van Breemen, O., 2002, Timing of synvolcanic magmatism in relation to base-metal mineralization, Rouyn-Noranda, Abitibi volcanic belt, Quebec: Radiogenic Age and Isotopic Studies, Report 15, Geological Survey of Canada, Current Research 2002-F8, 9 p.

Galley, A.G., Jonasson, I.R., and Watkinson, D.H., 2000, Magnetite-rich calc-silicate alteration in relation to a synvolcanic intrusion at the Ansil volcanogenic massive sulphide deposit, Rouyn-Noranda, Quebec, Canada: Mineralium Deposita, v. 35, p. 619-637.

Gauthier, N., Rocheleau, M., Kelly, D., and Gagnon, Y. 1990, Controls on the location of gold mineralization within the Cadillac tectonic Zone, Rouyn-Beauchastel segment, Abitibi Belt, Quebec, *in* Rive, M., Verpaelst, P., Gagnon, Y., Lulin, L.M., Riverin, G., and Simard, A., eds., The Northwestern Quebec Polymetallic Belt: A summary of 60 years of mining exploration., Canadian Institute of Mining and Metallurgy, Special Volume 43, p. 185-198.

Gelinas, L., Trudel, P. and Hubert, C., 1984, Chemostratigraphic subdivision of the Blake River Group, Rouyn-Noranda Area, Abitibi, Quebec: Canadian Journal of Earth Sciences, v. 21, p. 220-231.

Gibson, H.L., 1990, The Mine Sequence of the Central Noranda Volcanic Complex: Geology, alteration, massive sulphide deposits and volcanological reconstruction: Ph.D. thesis, Carleton University, Ottawa, Ontario, 715 p.

Gibson, H.L., 2005, Volcanic-hosted ore deposits, in Marti, J., and Ernst, G. J., eds., Volcanoes and the Environment. Cambridge University Press, p. 333-386.

Gibson, H.L. and Kerr, D.J., 1993, Giant volcanic-associated massive sulphide deposits: with emphasis on Archean examples: Economic Geology, Special Publication 2, p. 319-348.

Gibson, H.L. and Watkinson, D.H., 1990, Volcanogenic massive sulphide deposits of the Noranda Cauldron and Shield Volcano, Quebec, *in* Rive, M., Verpaelst P., Gagnon, Y., Lulin, L.M., Riverin, G., and Simard, A., eds., The Northwestern Quebec Polymetallic Belt: A summary of 60 years of mining exploration, Canadian Institute of Mining and Metallurgy, Special Volume 43, p. 119-132.

Gibson, H.L., Watkinson, D.H. and Comba, C.D.A., 1983, Silicification: Hydrothermal alteration in an Archean geothermal system within the Amulet Rhyolite formation, Noranda, Quebec: Economic Geology, v. 78: p. 954-971.

Gibson, H.L., Walker, S.D. and Coad, P.R., 1984, Surface geology and volcanogenic base metal massive sulphide and gold deposits of Noranda and Timmins: Guide Book for Field Trip #14, Geological Association of Canada-Mineralogical Association of Canada, Annual General Meeting, London, Ontario, 46p.

Gibson, H.L., Watkinson, D.H., Watkins, J. J., Labrie, M. and Doiron, G., 1993, Volcanological reconstruction of the Corbet breccia pile and Cu-Zn massive sulphide deposit, Noranda, Quebec: Exploration and Mining Geology, v. 2, p. 1-16.

Gibson, H.L., Morton, R.L. and Hudak, G.J., 1999, Submarine volcanic processes, deposits and environments favorable for the location of volcanic-associated massive sulphide deposits, *in* Barrie, C.T., and Hannington, M.D., eds., Volcanic-Associated Massive Sulphide Deposits: Processes and Examples in Modern and Ancient Settings. Reviews in Economic Geology, v. 8, p. 13-51.

Gibson, H.L., Kerr, D.J., and Cattalani, S., 2000, The Horne mine: history, influence on genetic models, and a comparison to the Kidd Creek Mine: Exploration and Mining Geology, v. 9, pp. 91-111.

Gibson, H.L., Franklin, J.M., Hannington, M.D., and Piercey, S. 2005, Abitibi volcanogenic massive sulphide deposits: the spectrum on environments, modern analogs and implications for exploration, *in* Program with Abstracts, Quebec Exploration 2005, Quebec City, p. 6.

Gill, J.B., 1987, Early geochemical evolution of an oceanic island arc and backarc: Fiji and South Fiji Basin: Journal of Geology, v. 95, p. 589-615.

Goldie, R., 1979, Consanguineous Archean intrusive and extrusive rocks, Noranda, Quebec. Chemical similarities and differences: Precambrian Research, v. 9, p. 275-287.

Goldie, R., Kotila, B., and Seward, D., 1979, The Don Rouyn mine: an Archean porphyry copper deposit near Noranda, Quebec: Economic Geology, v. 74, p. 1680-1684.

Goutier, J., Ayer, J., and Thurston, P., 2006, Synthèse lithologique et stratigraphique de la Sous-province de l'Abitibi: Ministère des Ressources naturelles et de la Faune – Association de l'exploration minière du Québec, Québec, DV 2006-03, p. 24.

Hannington, M.D., and Scott, S.D., 1989, Gold mineralization in volcanogenic massive sulphides: implications of data from active hydrothermal vents on the modern sea floor, *in* The Geology of Gold Deposits: The Perspective in 1988, Economic Geology Monograph 6, pp. 491-507.

Hannington, M. D., Poulsen, K.H., Thompson, J.F.H. Sillitoe, R.H. 1999, Volcanogenic gold in the massive sulphide environment, *in* C. T. Barrie, and Hannington, M.D., ed., Volcanic-Associated Massive Sulphide Deposits: Processes and Examples in Modern and Ancient Settings, Reviews in Economic Geology, v. 8, p. 325-356.

Hannington, M.D., Santaguida, F., Kjarsgaard, I.M., and Cathles, L.M., 2003, Regional greenschist facies hydrothermal alteration in the central Blake River Group, western Abitibi subprovince, Canada: Mineralium Deposita, v. 38, p. 393-422.

Hannington, M.D., Gibson. H.L., and Dubé, B., 2006, The metallogeny of modern submarine volcanic arcs and implications for ancient greenstone belts, *in* Program with Abstracts, annual meeting of the Geological Association of Canada and Mineralogical Association of Canada, vol. 31, p. A-13.

Hart, T., Gibson, H. L. and Lesher, C. M., 2004, Trace element geochemistry and petrogenesis of felsic volcanic rocks associated with volcanogenic Cu-Zn-Pb massive sulphide deposits: Economic Geology, v. 99, p. 1003-1013.

Herzig, P.M., and Hannington, M.D., 1995, Polymetallic massive sulphides at the modern seafloor: A review: Ore Geology Reviews, v. 10, p. 95-115.

Huston, D. L., and Large, R.R., 1989, A chemical model for the concentration of gold in volcanogenic massive sulphide deposits: Ore Geology Reviews, v. 4, p. 171-200.

Jébrak, M., Harnois, L., Carriere, A., and Lafrance, J., 1997, The Don Rouyn Cu-Au porphyry system, in Robert, J.-F. C. a. F., ed., Atypical gold deposits in the Abitibi Belt, B5: Ottawa, GAC/MAC, p. 72-80.

Kalogeropoulos, S. I., and Scott, S. D., 1983, Mineralogy and geochemistry of tuffaceous exhalites (tetsusekiei) of the Fukazawa Mine, Hokuroku District, Japan, *in* Ohmoto, H., and Skinner, B. J., eds., The Kuroko and

Related Volcanic Massive Sulphide Deposits, Economic Geology Monograph 5, p. 412-432.

Kalogeropoulos, S.I. and Scott, S.D., 1989, Mineralogy and geochemistry of an Archean tuffaceous exhalite: The Main Contact Tuff, Millenbach Mine area,, Noranda, Quebec: Canadian Journal of Earth Sciences, v. 26, p. 88-105.

Kennedy, L.P., 1985, The geology and geochemistry of the Archean Flavrian Pluton, Noranda, Quebec: Ph.D. thesis. University of Western Ontario, 469p.

Kerr, D.J. and Gibson, H.L., 1993, A comparison of the Horne volcanogenic massive sulphide deposit and intracauldron deposits of the Mine Sequence, Noranda, Quebec: Economic Geology, v. 88, p. 1419-1442.

Kerr, D.J. and Mason, R., 1990, A re-appraisal of the geology and ore deposits of the Horne Mine Complex at Rouyn-Noranda, Quebec, *in* Rive, M., Verpaelst P., Gagnon, Y., Lulin, L.M., Riverin, G., and Simard, A., eds., The Northwestern Quebec Polymetallic Belt: A summary of 60 Years of Mining Exploration, Canadian Institute of Mining and Metallurgy Special Volume 43 , p. 153-165.

Kerrich, R., and Wyman, D., 1996, The trace element systematics of igneous rocks in mineral exploration: an overview, *in* D.A. Wyman, ed., Trace Element Geochemistry of Igneous Rocks: Application to Massive Sulphide Exploration, Geological Association of Canada, Short Course Notes, v. 12, p. 1-50.

Knuckey, M.J. and Watkins, J. J., 1982, The geology of the Corbet massive sulphide deposit, Noranda District, Quebec, Canada: Geological Association of Canada, Special Paper 25, p. 255-296.

Knuckey, M. S., Comba, C. D. A., and Riverin, G., 1982, The Millenbach deposit, Noranda District, Quebec - an update on structure, metal zoning and wall rock alteration: Geological Association of Canada, Special Paper 25, p. 297-318.

Kotila, B., 1975. An Archean porphyry copper deposit: the Don Rouyn mine, Noranda, Quebec: B.Sc. thesis, Queen University, 29 p.

Lafrance, B., Davis, D.W., Goutier, J., Moorhead, J., Pilote, P., Mercier-Langevin, P., Dubé, B., Galley, A.G., and Mueller, W.U., 2005, Nouvelles datations isotopiques dans la portion Québécoise du Groupe de Blake River et des unités adjacentes: Ministères des Ressources naturelles et de la Faune, Québec; RP 2005-01, 15 p.

Large, R.R., 1992, Australian volcanic-hosted massive sulphide deposits: features, styles, and genetic models: Economic Geology, v. 87, p. 469-478.

Large, R.R., Huston, D.L., McGoldrick, P.J., Ruxton, P.A., and McArthur, G., 1989, Gold distribution and genesis in Australian volcanogenic massive sulphide deposits and their significance for gold transport models, in The Geology of Gold Deposits: The perspective in 1988, Economic Geology Monograph 6, pp. 520-536.

Larocque, A.C.L., Hodgson, C.J. and Lafleur, P.-J., 1993, Gold distribution in the Bouchard-Hebert volcanic-associated massive sulphide deposit, Noranda, Quebec: A preliminary evaluation of the role of metamorphic remobilization: Economic Geology, v. 88, p. 1443-1459.

Lesher, C.M., Goodwin, A.M., Campbell, I. H., and Gorton, M.P., 1986, Trace-element geochemistry of ore-associated and barren, felsic metavolcanic rocks in the Superior Province, Canada: Canadian Journal of Earth Sciences, v. 23, p. 222-237.

Lichtblau, A.P., and Dimroth, E., 1980, Stratigraphy and facies at the south margin on the Archean Noranda caldera, Noranda, Quebec, in Current Research, Part A, Geological Survey of Canada, Paper 80-1A, p. 69-76.

Ludden, J.N., Gelinas, L., and Trudel, P., 1982. Archean metavolcanics from the Rouyn-Noranda District, Abitibi Greenstone belt, Quebec: Mobility of trace elements and petrogenetic constraints: Canadian Journal of Earth Sciences, v. 19, p. 2276-2287.

Lulin, J.-M., 1990, Une analyse du développement minier du Nord-Ouest Québécois, *in* Rive, M., Verpaelst P., Gagnon, Y., Lulin, L.M., Riverin, G., and Simard, A., eds., The Northwestern Quebec Polymetallic Belt: A summary of 60 years of mining exploration, Canadian Institute of Mining and Metallurgy, Special Volume 43, p. 17-34.

Lydon, J.W., 1988, Volcanogenic massive sulphide deposits. Part 2: genetic models: Geoscience Canada, v. 15, p. 43-65.

MacLean, W.H. and Hoy, L.H., 1991, Geochemistry of hydrothermally altered rocks at the Horne mine, Noranda, Quebec: Economic Geology, v. 86, p. 506-528.

Martin, L., and Masson, M., 2005, Towards new discoveries thanks to the application of 3D modeling in the Noranda Camp, in Program with Abstracts, Quebec exploration 2005, Quebec City, p. 14.

Mercier-Langevin, P., Dubé, B., Hannington, M. D., Davis, D. W., and Lafrance, B., 2004, Contexte Geologique et Structural des Sulfures Massifs Volcanogenes Auriferes du Gisement LaRonde, Abitibi: Ministère des ressources naturelles et de la Faune, ET 2003-03, 49 p.

Mercier-Langevin, P., Dubé, B., Lafrance, B., Galley, A., Moorhead, J., and Gosselin, P., 2007, Metallogeny of the Doyon-Bousquet-LaRonde Mining Camp, Abitibi Greenstone Belt, Quebec, *in* Goodfellow, W.D., ed., Mineral Deposits of Canada: A synthesis of Major Deposit-types, District Metallogeny, the evolution of Geological Provinces and Exploration Methods, Special Publication 5, Mineral Deposits Division, Geological Association of Canada, p. 673-701.

Morton, J. L., and Franklin, J.M., 1987, Two-fold classification of Archean volcanic-associated massive sulphide deposits: Economic Geology, v. 82, p. 1057-1063.

Mortenson, J.K., 1987, Preliminary U-Pb zircon ages for volcanic and plutonic rocks of the Noranda-Lac Abitibi area, Abitibi Subprovince, Quebec, in Current Research, Part A, Geological Survey of Canada, Paper 87-1A, p. 581-590.

Mortensen, J.K., 1993, U-Pb geochronology of the eastern Abitibi Subprovince. Part 2: Noranda-Kirkland Lake area: Canadian Journal of Earth Sciences, v.30, p. 29-41.

Paquette-Mihalasky, F.I., 1999, Silicification: Oxygen isotope mapping of an Archean geothermal system at Noranda, Quebec: M.Sc thesis, Laurentian University, Sudbury, Ontario, 171p.

Paradis, S., Ludden, J. and Gelinas, L., 1988, Evidence for contrasting compositional spectra in comagmatic intrusive and extrusive rocks of the late Archean Blake River Group, Abitibi, Quebec: Canadian Journal of Earth Sciences, v. 25, p. 134-144.

Pearson, V., 2005, The Blake River Group: an imbricated caldera complex, in Program with Abstracts, Quebec Exploration 2005, Quebec City, p. 15.

Pearson, V., and Daigneault, R., 2006, A giant Archean caldera complex, the Blake River Group, Abitibi Subprovince, in Program with Abstracts, Annual meeting of the Geological Association of Canada and Mineralogical Association of Canada, v. 31, p. A-23.

Pelletier, C., and Jébrak, M., 1994, Au-(Cu-Mo) porphyries in the Noranda area, Blake River Group, southern Abitibi greenstone belt, Quebec, in Program with Abstracts, Geological Association of Canada; Mineralogical Association of Canada; Canadian Geophysical Union, Joint Annual Meeting (1994), p. A-19.

Peloquin, A.S., 2000, Reappraisal of the Blake River Group stratigraphy and its place in the Archean volcanic record: PhD. Thesis, University of Montréal, Quebec, 189p.

Peloquin, A.S., 2005, Geology and base metal mineralization in Ben Nevis, Katrine and Clifford townships: Discover Abitibi initiative: Ontario Geological Survey, Open File Report 6161, 86p.

Peloquin, A. S., Potvin, R., Paradis, S., Lafleche, M.R., Verplaelst, Gibson, H.L., 1990, The Blake River Group, Rouyn-Noranda area, Quebec: A stratigraphic synthesis, in Rive, M., Verpaelst P., Gagnon, Y., Lulin, L.M., Riverin, G., and Simard, A., eds., The Northwestern Quebec Polymetallic Belt: A Summary of 60 Years of Mining Exploration, Canadian Institute of Mining and Metallurgy, Special Volume, p. 107-118.

Percival, J.A., 2007, Geology and metallogeny of the Superior Province, Canada, *in* Goodfellow, W.D., ed., Mineral Deposits of Canada: A Synthesis of Major Deposit-types, District Metallogeny, the Evolution of Geological Provinces and Exploration Methods, Special Publication 5, Mineral Deposits Division, Geological Association of Canada, p. 903-928.

Peter, J. M., and Goodfellow, W. D., 1996, Mineralogy, bulk and rare earth element geochemistry of massive sulphide-associated hydrothermal sediments of the Brunswick horizon, Bathurst mining camp, New Brunswick: Canadian Journal of Earth Sciences, v. 33, p. 252-283.

Peter, J. M., and Goodfellow, W. D., 2003, Hydrothermal Sedimentary Rocks of the Heath Steele Belt, Bathurst Mining Camp, New Brunswick: Part 3: Application of mineralogy and mineral bulk compositions to massive sulphide exploration, *in* Goodfellow, W. D., McCutcheon, S. R., and Peter, J. M., eds., Massive sulphide deposits of the Bathurst Mining Camp, New Brunswick and Maine, Society of Economic Geology, Monograph 11, p. 417-433.

Poulsen, K.H., Robert, F., and Dubé, B., 2000, Geological Classification of Canadian Gold Deposits: Geological Survey of Canada, Bulletin 540, 106 p.

Richard, M.G., 1999, Evolution of the Flavrian Pluton and its association with VHMS deposits and granitoid-hosted gold deposits of the Noranda Cauldron: Rouyn-Noranda, Quebec, Canada. Ph.D. thesis, Université de Montreal, Montreal, Québec, 318p.

Riverin, G. and Hodgson, C.J., 1980, Wall-rock alteration at the Millenbach Cu-Zn mine, Noranda, Quebec: Economic Geology, v. 75, p. 424-444.

Riverin, G., Bernard, D. and Boily, B., 1990, The Donalda gold deposit, Rouyn-Noranda, Québec, *in* Rive, M., Verpaelst P., Gagnon, Y., Lulin, L.M., Riverin, G., and Simard, A., eds., The Northwestern Quebec Polymetallic Belt: A Summary of 60 Years of Mining Exploration, Canadian Institute of Mining and Metallurgy, Special Volume 43, p. 199-209.

Robert, F., 1990. An overview of gold deposits in the Eastern Abitibi belt, *in* Rive, M., Verpaelst P., Gagnon, Y., Lulin, L.M., Riverin, G., and Simard, A., eds., The Northwestern Quebec Polymetallic Belt: A Summary of 60 Years of Mining Exploration, Canadian Institute of Mining and Metallurgy, Special Volume 43, p.93-105.

Santaguida, F., 1999, The paragenetic relationships of epidote-quartz hydrothermal alteration within the Noranda Volcanic Complex, Quebec: PhD thesis, Carleton University, Ontario, 302 p.

Santaguida, F., Gibson, H.L., Watkinson, D.H., and Hannington, M.D., 1998, Semi-conformable epidote-quartz hydrothermal alteration in the Central Noranda Volcanic Complex: Relationship to volcanic activity and VMS mineralization: CAMIRO Project 94E07, The Use of Regional-Scale Alteration and Subvolcanic Intrusions in the Exploration for Volcanic-Associated Massive Sulphide Deposits, unpublished report, p. 139-180.

Sawkins, F. J., 1976, Massive sulphide deposits in relation to geotectonics, in Strong, D. F., ed., Metallogeny and plate tectonics, Geological Association of Canada, Special Paper 14, p. 222-240

Sinclair, W.D., 1971, A volcanic origin for the No.5 zone of the Horne mine, Noranda, Quebec: Economic Geology, v. 66, p. 1225-1231.

Spence, C.D., 1967, The Noranda Area: Canadian Institute of. Mining and Metallurgy, Centennial Field Excursion Guidebook, p. 36-39.

Spence, C.D., and de Rosen-Spence, A.F., 1975, The place of sulphide mineralization in the volcanic sequence at Noranda, Quebec: Economic Geology, v. 70, p. 90-101.

Stoffers, P., Wright, I., de Ronde, C., Hannington, M., Villinger, H., Herzig, P., and Shipboard Scientific Party, 1999a, Little-studied arc-backarc system in the spotlight: EOS (Transactions American Geophysical Union), v. 80, no. 32, p. 353-359.

Stoffers, P., Wright, I., and Shipboard Scientific Party, 1999b, Havre Trough-Taupo Volcanic Zone: Tectonic, magmatic and hydrothermal processes: Cruise Report SO-135: Institut fur Geowissenschaften, Christian-Albrechts Universitat, Kiel, Germany, 77 p.

Tomkins, A.G. and Mavrogenes, J.A., 2003, Generation of metal-rich felsic magmas during crustal anatexis: Geology, v. 31, p. 765-768.

Tremblay, A., Tourigny, G. and Machado, N., 1995, Zircon U/Pb age constraints on deformation and gold mineralization of Mooshla pluton, southern Abitibi belt, Canada: Geological Society of America annual meeting, New Orleans, Abstract with Programs, p. A-163.

Walker, S.D., 1980, The Chadbourne Mine, Noranda, Quebec: A gold bearing breccia: Canadian Institute of Mining and Metallurgy, Special Volume 24, p.58-68.

Base Metal Metallogeny of the Selwyn Basin, Canada

Wayne D. Goodfellow

Geological Survey Of Canada, 601 Booth Street, Ottawa, Ontario K1A 0E8
Corresponding author's email:wgoodfel@nrcan.gc.ca

Abstract

The sedimentary exhalative (SEDEX) deposits of the Selwyn Basin occur in four major districts: MacMillan Pass and Gataga (Late Devonian), Howards Pass (Early Silurian), and Anvil (Late Cambrian). In addition, there are the Clear Lake (Late Devonian) and Mel (Cambrian) deposits that do not fall inside the districts above. There are twelve major SEDEX deposits with proven reserves that include past producers Faro, Vangorda, and Grum. The deposits range in size from the giant Howards Pass deposit at 115 Mt to the relatively small Driftpile deposit at 2.4 Mt. Metal grades and ratios are highly variable between deposits, with values ranging from 4.7 to 16.2% Zn, 2.0 to 5.5% Pb, and up to 110 g/t Ag. The average Cu and Au contents are generally very low except for the Vangorda and DY deposits in the Anvil District where values range up to 0.3% and 0.8 g/t, respectively. Stratiform barite deposits also occur in the basin.

The Selwyn Basin is an elongate rift-controlled, epicratonic sedimentary basin, extending from Alaska through the Yukon, western Northwest Territories, and British Columbia, which is flanked to the east and north by carbonate platforms that host co-genetic Mississippi Valley-type deposits. The sedimentary architecture of this continental rift consists of a Late Proterozoic thick synrift, mostly oxidized, clastic sequence (Windermere Supergroup), which is overlain by Paleozoic sag or postrift basinal, reduced facies shales and cherts that host the SEDEX deposits.

SEDEX deposits were formed by the reaction of metals in basinal saline fluids with biogenic H_2S in the ambient anoxic water column. Based on available fluid inclusion data, the morphology of stratiform sulphide bodies, and vent textures, hydrothermal fluids that vented into seawater behaved either as buoyant plumes or bottom-hugging brines. SEDEX deposits formed from buoyant plumes are mound-shaped and characterized by rapid thickness changes away from the vent. Because deposits formed from dense brines commonly occupy depressions remote from vent sites, they tend to be more uniform in thickness, laterally widespread, delicately bedded, and not strongly zoned.

Résumé

Les gîtes de sulfures exhalatifs dans des roches sédimentaires (gîtes sedex) du bassin de Selwyn reposent dans quatre principaux districts, soit ceux de MacMillan Pass et de Gataga (Dévonien tardif), celui de Howards Pass (Silurien précoce) et celui d'Anvil (Cambrien tardif). En outre, il faut mentionner les gisements de Clear Lake (Dévonien tardif) et de Mel (Cambrien), qui ne font pas partie des districts susmentionnés. On compte 12 importants gisements sedex qui comportent des réserves prouvées, dont les gisements de Faro, de Vangorda et de Grum, qui ont autrefois été exploités. La taille de ces gisements va de celle du gisement géant de Howards Pass (115 Mt) à celle de l'assez petit gisement de Driftpile (2,4 Mt). Les concentrations de métaux et leurs proportions relatives varient considérablement d'un gisement à l'autre, les teneurs allant de 4,7 à 16,2 % pour le zinc, de 2,0 à 5,5 % pour le plomb et atteignent jusqu'à 110 g/t pour l'argent. Les teneurs moyennes en cuivre et en or sont généralement très faibles, sauf dans les gisements de Vangorda et de DY du district d'Anvil, où elles atteignent jusqu'à 0,3 % pour le cuivre et jusqu'à 0,8 g/t pour l'or. En outre, le bassin de Selwyn compte aussi des gîtes stratiformes de barytine.

Le bassin de Selwyn est un bassin sédimentaire épicratonique de forme allongée dont l'existence est liée à celle d'un rift. Il s'étend de l'Alaska à la Colombie-Britannique, en passsant par le Yukon et l'ouest des Territoires du Nord Ouest. Il est bordé à l'est et au nord par des plates formes carbonatées, qui renferment des gîtes cogénétiques de type Mississippi-Valley. La succession sédimentaire de ce rift continental est constituée à la base d'une épaisse séquence de roches détritiques généralement oxydées (Supergroupe de Windermere), dont le dépôt, au Protérozoïque tardif, est contemporain de la formation du rift. Cette séquence est recouverte d'une succession à faciès réduit de chert et de shale de bassin, dans laquelle on trouve les gîtes sedex, qui s'est déposée, au Paléozoïque, lors de l'affaissement du rift ou postérieurement au rifting.

Les gîtes sedex sont issus de la réaction des métaux présents dans les fluides salins de bassin avec le H_2S biogène de la colonne d'eau anoxique ambiante. Selon des données sur les inclusions fluides, la morphologie des corps sulfurés stratiformes et la texture des évents, les fluides hydrothermaux libérés dans l'eau de mer par les évents ont réagi à la manière de saumures longeant le fond océanique ou de panaches en suspension. Les gîtes sedex issus des panaches en suspension adoptent la forme de monticules et se caractérisent par une forte diminution d'épaisseur en s'éloignant des évents. Les gîtes formés par des saumures denses occupent généralement des dépressions éloignées des évents, si bien qu'ils présentent habituellement une épaisseur plus uniforme, une grande étendue latérale, une stratification délicate et une faible zonalité.

Introduction and History

The Selwyn Basin (Fig. 1) is one of the most productive basins for Zn-Pb-Ag sedimentary exhalative (SEDEX) and stratiform barite deposits in the world (Goodfellow and Lydon, 2007). This basin hosts twelve major deposits with proven reserves, including past producers Faro, Vangorda, and Grum (Table 1). Mississippi Valley-type (MVT) Zn-Pb-Ag deposits occur in the adjacent Mackenzie and Macdonald carbonate platforms (e.g. Paradis and Nelson, 2007; Paradis et al., 2007). Sedimentary barite is common in rocks of different ages in the Selwyn Basin where it is an important constituent of Late Devonian and Cambrian SEDEX deposits.

The Selwyn Basin is a large, lenticular sedimentary basin that occupies the western margin of ancestral North America and extends for 1000s of kilometres from Alaska through the Yukon, western Northwest Territories and British Columbia,

Goodfellow, W.D., 2007, Base metal metallogeny of the Selwyn Basin, Canada, *in* Goodfellow, W.D., ed., Mineral Deposits of Canada: A Synthesis of Major Deposit-Types, District Metallogeny, the Evolution of Geological Provinces, and Exploration Methods: Geological Association of Canada, Mineral Deposits Division, Special Publication No. 5, p. 553-579.

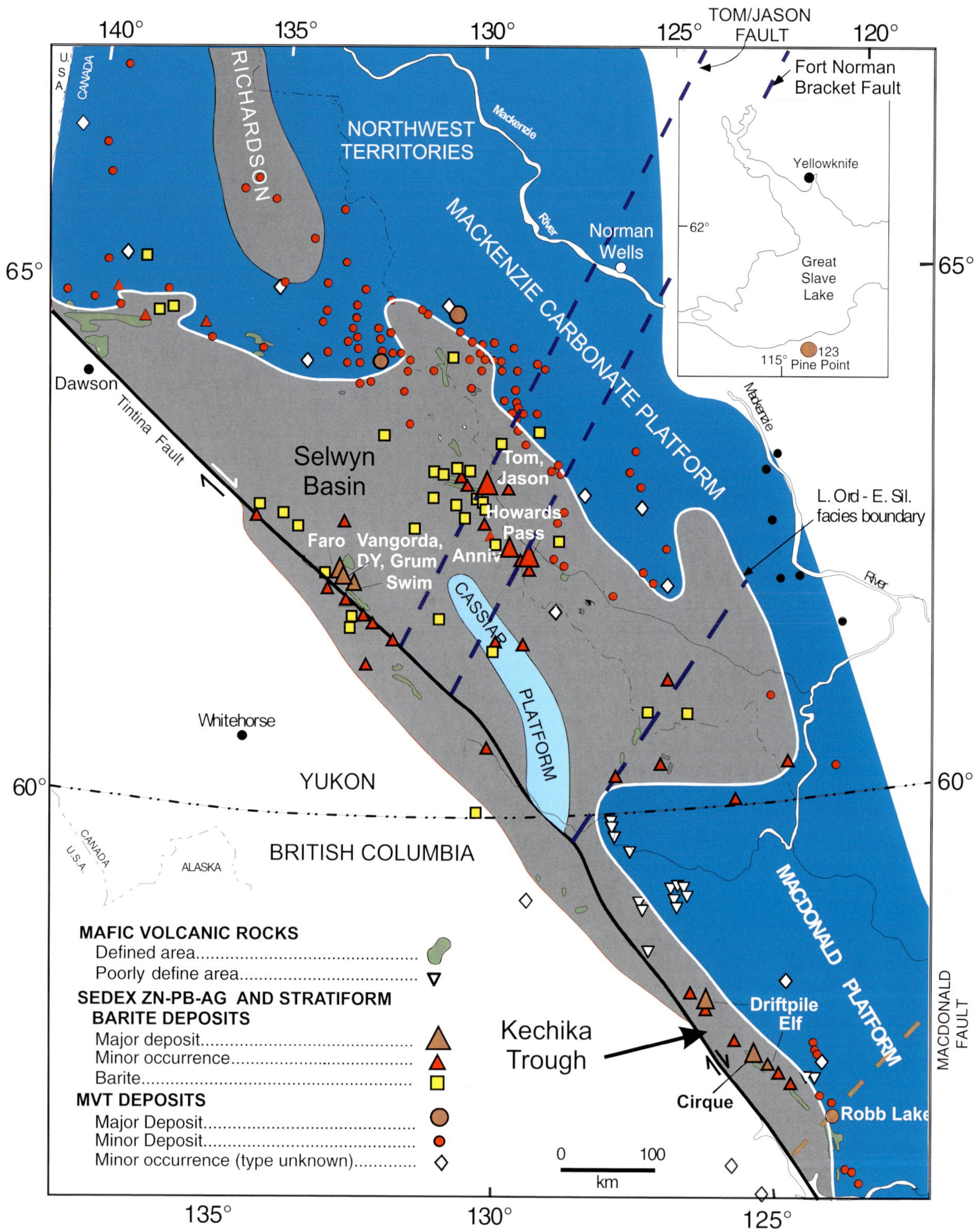

FIGURE 1. Distribution of lower Paleozoic sedimentary facies in the northern Canadian Cordillera (from Cecile, 1982). Sedimentary exhalative, Mississippi Valley-type, and stratiform barite deposits are also plotted. Northeast-oriented dashed lines are inferred transform faults that influenced sedimentation during the Paleozoic (from Goodfellow et al., 1993).

and into the United States (Nelson and Colpron, 2007). Rocks of the Selwyn Basin are generally well preserved except for fold and thrust deformation, and subgreenschist metamorphism. The complete sedimentary architecture of an intracratonic rift is represented, including a Late Proterozoic thick synrift clastic sequence of mostly oxidized and locally derived coarse- and fine-grained clastic rocks, and a Paleozoic sag or postrift basinal sedimentary sequence consisting mostly of black and carbonaceous pelagic and hemipelagic shale and chert. Endogenic fault-scarp sedimentary breccias occur in local second- and third-order basins, particularly in the Late Devonian.

TABLE 1. Grade and tonnage of important sedimentary exhalative deposits, Selwyn Basin.

Deposit	Location	Geological Resources Size (Mt)	Zn %	Pb %	Zn+Pb %	Cu %	Ag (g/t)	Au (g/t)	$BaSO_4$	Age	Discovery Date	Discovery Method	References
Late Devonian													
Tom	MacMillan Pass, Selwyn Basin, Yukon	15.7	7.0	4.6	11.6		49.1		yes	L. Devonian	1951	Stream gossan	McClay and Bidwell, 1986; Goodfellow and Rhodes, 1990; Yukon MinFile 2004
Jason	MacMillan Pass, Selwyn Basin, Yukon	10.1	7.4	6.5	11.9		79.9		yes	L. Devonian	1974	Soil geochemistry	Turner, 1990; *Northern Miner*, May 28th, 1990, p. 21; Yukon MinFile 2004
Cirque	Gataga District, N.E. British Columbia	52.2	8.0	2.0	10.0				yes	L. Devonian	1977	Stream geochemistry	Jefferson et al., 1983; Canadian Mines Handbook 1991-2, p. 127
Driftpile	Gataga District, N.E. British Columbia	2.4	11.9	3.1	15.0				yes	L. Devonian	1974	Stream geochemistry	MacIntyre, 1983
Clear Lake	Clear Lake District, Yukon	5.6	11.4	2.0	13.4		38.1		yes	Devonian	1965	Geophysics, EM	Grapes, 1987
Early Silurian													
Howards Pass (XY, Anniv)	Howards Pass District, Selwyn Basin, Yukon	115.4	5.38	2.08	7.46				no	E. Silurian	1968	Stream geochemistry	Placer Development Annual Report, January 1983
XY	Howards Pass District, Selwyn Basin, Yukon	60.0	5.46	2.36	7.82				no	E. Silurian	1968	Stream geochemistry	Placer Development Annual Report, January 1983
Anniv	Howards Pass District, Selwyn Basin, Yukon	55.4	5.29	1.77	7.06				no	E. Silurian	1968	Stream geochemistry	Placer Development Annual Report, January 1983
Early Cambrian													
Faro	Anvil District, Yukon	57.6	5.7	3.4	9.1		36.0		yes	Cambrian	1965	Prospecting	Jennings and Jilson, 1986; Yukon MinFile 2004
Dy	Anvil District, Yukon	21.1	6.7	5.5	12.2	0.1	84.0	1.0	yes	Cambrian	1976	Drilling	Jennings and Jilson, 1986; Yukon MinFile 2004
Grum	Anvil District, Yukon	31.0	4.9	3.1	8.0		49.0		yes	Cambrian	1954	Prospecting	Jennings and Jilson, 1986; Yukon MinFile 2004
Swim	Anvil District, Yukon	4.3	4.7	3.8	8.5		42.0		yes	Cambrian	1965	Mineralized float	Jennings and Jilson, 1986; Yukon MinFile 2004
Vangorda	Anvil District, Yukon	7.5	4.9	3.8	8.7	0.3	54.0	0.8	yes	Cambrian	1953	Prospecting	Jennings and Jilson, 1986; Yukon MinFile 2004
Mel	Mel District	6.8	7.1	2.1	9.2				yes	Cambrian	1967	Prospecting	Miller and Wright, 1986;Yukon MinFile 2004

The initial discovery of SEDEX deposits in the Selwyn Basin was made at MacMillan Pass where a red-stained creek draining mineralization led to the discovery of the Devonian Tom deposit by Hudson Bay Exploration and Development Company Limited in 1951 (Table 1). The adjacent Jason deposit was not discovered until 1974 when the Ogilvie joint venture staked the area based on stream and soil geochemical anomalies. About two years after the discovery of the Tom deposit, Al Kulan uncovered sulphide float in a stream bed that ultimately led to the discovery of the Vangorda deposit in the Anvil District. This set off one of the largest staking rushes in the Yukon that resulted in the discovery of the Grum deposit in 1954 followed by the Faro and Swim deposits in 1965 (Deklerk, 2003). The Dy deposit was not discovered until 1976. Elsewhere in the Selwyn Basin, the follow-up of coincident Zn and Pb stream geochemical anomalies led to the discovery of the Howards Pass deposits (XY, Anniv, and OP) by Placer Development Limited in 1968, and the Driftpile and Cirque deposits in 1974 and 1977, respectively (Deklerk, 2003). Several additional mineralized zones have been discovered more recently by Pacifica Resources Limited including the Brodel, HC, Don, and Pelly North (http://www.pacifica-resources.com/exp_na_selwyn.cfm).

There are other deposit types that fall within the geographic area represented by the Selwyn Basin but most of these postdate the Selwyn Basin and are typically associated with Mesozoic and Cenozoic tectonic and related magmatic events. Since these deposits are unrelated to the continental rifting that led to the formation of the Selwyn Basin, they are not considered in this synthesis of Selwyn Basin metallogeny.

Geologic and Tectonic Setting

The Selwyn Basin is an elongate Paleozoic marine basin represented by mostly black shales and cherts that hosts SEDEX and stratiform barite deposits (Fig. 1). Mississippi Valley-type (MVT) deposits are common in flanking shallow water carbonates of the Mackenzie and MacDonald carbonate platforms. The Selwyn Basin has a complex history of extensional tectonism from Late Proterozoic to Mississippian, which gave rise to a Paleozoic epicontinental margin characterized by a basin and arch morphology (Goodfellow et al., 1993).

The oldest rocks exposed in the Selwyn Basin consist of a thick (4-6 km) sequence of Hadrynian-Cambrian clastic sedimentary rocks comprising the Windermere Supergroup (Eisbacher, 1981) (Fig. 2). These clastic rocks are interpreted as a synrift sequence that formed by the erosion of Hudsonian crystalline basement during continental breakup, initiated about 760 Ma ago (Eisbacher, 1981). Rapid subsidence of the continental margin during the Cambrian indicates that the continent rifted apart and formed a marginal ocean basin to the west at this time (Bond et al., 1988).

The Windermere rocks are overlain by deep-water Cambro-Ordovician carbonate rocks of the Rabbitkettle Formation, which are in turn overlain by basinal facies chert

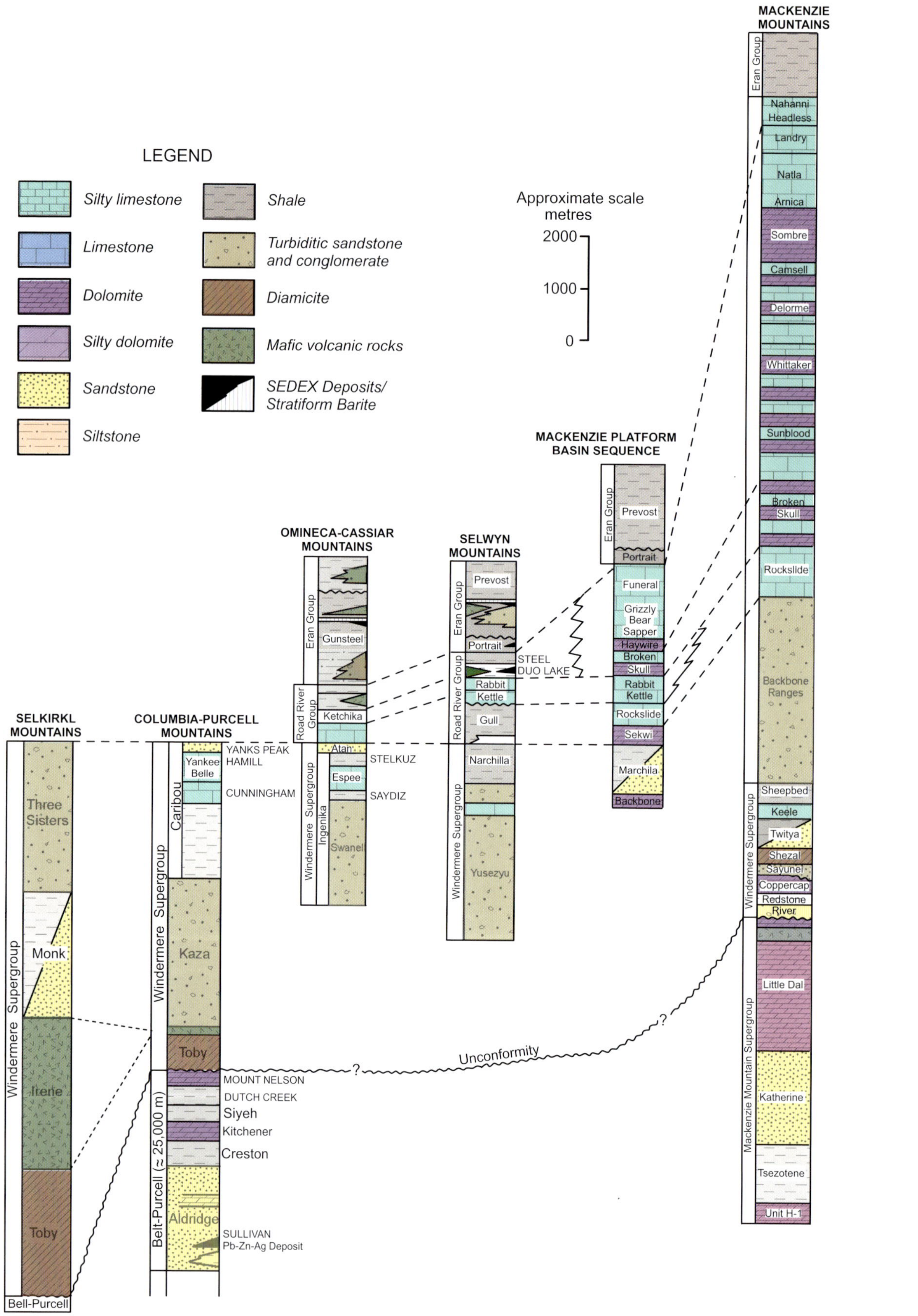

FIGURE 2. Stratigraphic cross-section of the Northern Canadian cordillera showing the facies and thickness relationships between the Mackenzie carbonate platform, the Selwyn Basin (Selwyn Mountains), the Gataga Trough in northeastern British Columbia (Omineca-Cassiar Mountains), and the Belt-Purcell Basin in southeastern British Columbia (from Goodfellow, 2004).

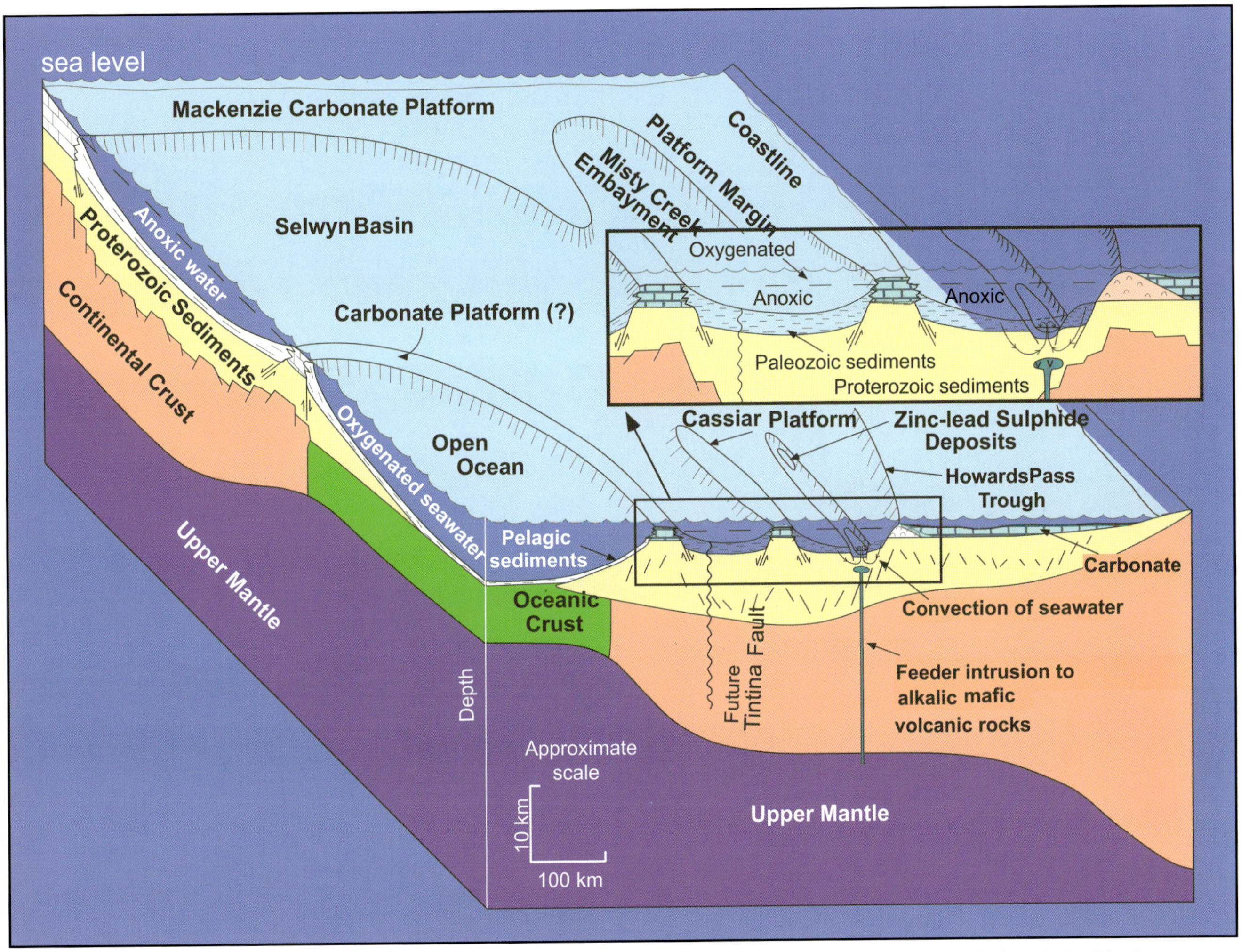

FIGURE 3. Block diagram of the rifted Paleozoic continental margin of northwestern Canada showing the stratigraphic and facies relationships between synrift clastic and postrift or sag basinal and platformal sequences, magmatism, hydrothermal activity, and ambient anoxic seawater conditions (Goodfellow et al., 1993).

and shale of the Road River Group (Ordovician-Devonian) and chert and autochthonous black clastic rocks of the Earn Group (Devono-Mississippian) (Fig. 2) (Gordey et al., 1988). The basinal sedimentary facies is bounded by the Mackenzie and Macdonald carbonate platforms to the east and north, and is dissected in its south-central part by the intrabasinal shelf facies of the Cassiar Platform (Fig. 1). Although the western margin is obscured by a later collisional event and right lateral displacement on the Tintina Fault, it has been interpreted by Goodfellow (1985) to be the outer margin of the Proterozoic continental shelf (Fig. 3). This is in general agreement with more recent tectonic reconstructions (see Nelson and Colpron, 2007, and references therein).

Three major episodes of mafic volcanism have been identified in the Selwyn Basin (i.e. Early Cambrian, Middle Ordovician, and Middle to Late Devonian). The volcanic rocks form discontinuous, lenticular belts that parallel rift-bounding faults or occur as isolated volcanic piles, and consist dominantly of mafic volcanic flows, dykes, and tuffs. Their incompatible element chemistry is similar to that of alkaline basalts formed in continental rifts and is interpreted to represent low-percentage partial melts derived from a previously metasomatized mantle (Goodfellow et al., 1995).

In most cases, there appears to be a close temporal and/or spatial relationship between volcanic centres and SEDEX deposits of the Selwyn Basin (Goodfellow et al., 1993) (Fig. 4). The Late Cambrian SEDEX deposits of the Anvil District are overlain by the Menzie Creek volcanic rocks although the thickest portion of the Menzie Creek mafic volcanic sequence lies 10 to 20 km northwest of the stratiform deposits (Jennings and Jilson, 1986). At McMillan Pass, Yukon, the Tom and Jason deposits (early Late Devonian) occur at the same stratigraphic horizon as a belt of Middle to early Late Devonian alkalic mafic flows and volcaniclastic rocks (Abbott and Turner, 1991). The thickest accumulations of volcanic rocks lie 10 to 40 km west of the SEDEX deposits, although mafic volcanic rocks have been drilled near the Tom deposit (D. Rhodes, pers. comm., 2006). The Early Silurian Howards Pass deposits are not spatially associated with volcanic rocks, although mafic tuffs of this age occur about 70 to 80 km to the north-northwest along the trend of the Selwyn Basin. In the Gataga District, northeastern British Columbia, Devonian volcanic rocks are both temporally and spatially associated with Late Devonian SEDEX deposits (MacIntyre, 1992; Goodfellow et al., 1995).

The tectonic setting of the Selwyn Basin has been a subject of debate. Differential subsidence curves of Bond et al.

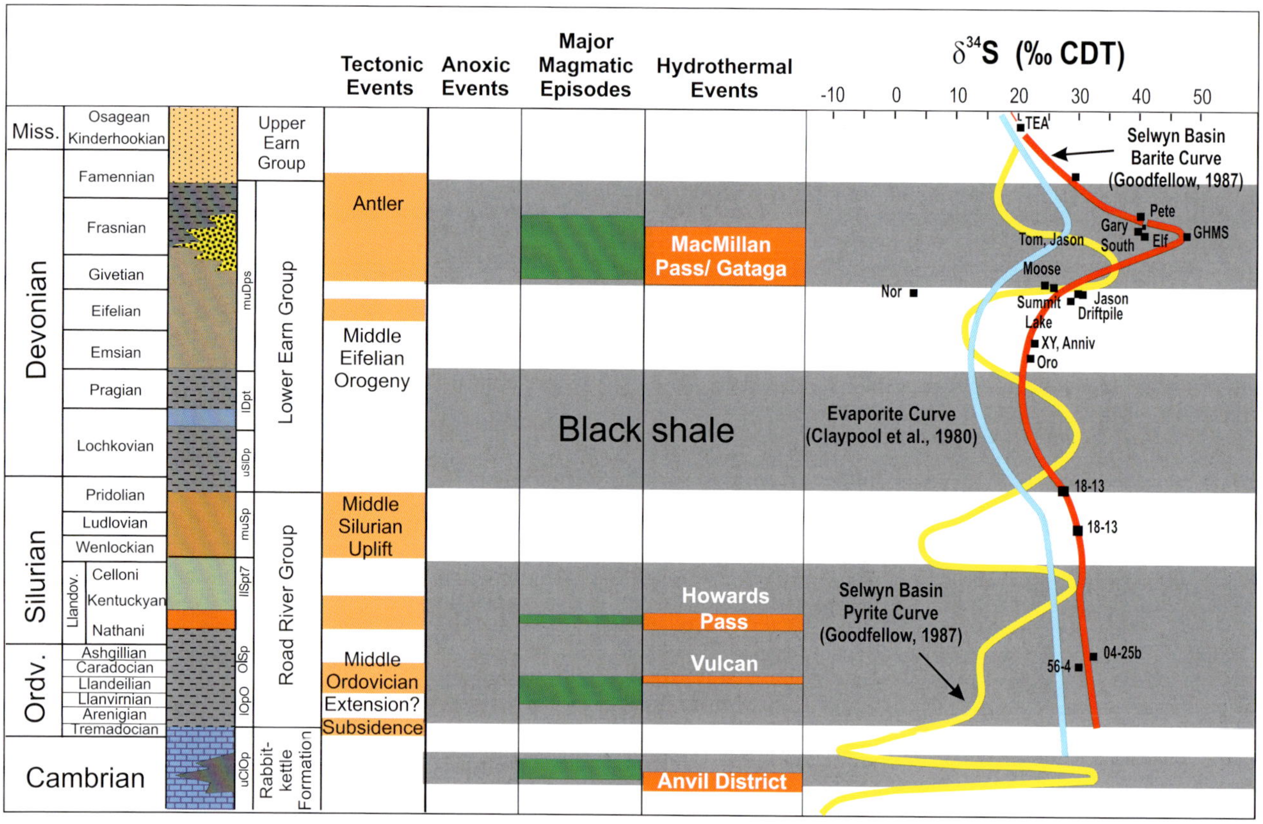

FIGURE 4. Tectonic (Goodfellow et al., 1986), anoxic (Goodfellow, 1987), magmatic (Goodfellow et al., 1995), and hydrothermal events are shown along with secular sulphur isotope curves for pyrite and barite (Goodfellow, 1987; Shanks et al., 1987) in the Selwyn Basin, and the global sulphur isotope curve for evaporites (Claypool et al., 1980).

(1988) indicate rapid subsidence in the Cambrian, consistent with major rifting and the formation of a Paleozoic passive margin on the eastern side of a newly formed ocean. However, the Late Devonian SEDEX deposits in the Selwyn Basin are interpreted to have formed in a continental back-arc rift that was generated by the roll back of the oceanic slab beneath the western margin of the North American craton (Fig. 5, from Nelson et al., 2002). Therefore, the Selwyn Basin is a continental rift that formed by intra-continental rifting and was subsequently reactivated in a far field, back-arc extensional basin during the Devonian (Nelson et al., 2002), similar to that proposed for the McArthur River Basin in northern Australia (Betts et al., 2003). The Devonian extensional event that affected the western North American margin is commonly referred to as the Antler Orogeny (Trexler et al., 2004).

Seafloor Environments

Most SEDEX and stratiform barite deposits in the Selwyn Basin are hosted by laminated carbonaceous shale and chert (Goodfellow et al., 1986; Jennings and Jilson, 1986; Turner and Einaudi, 1986; MacIntyre, 1992). Atomic S/C ratios greater than 0.13 for basinal sedimentary rocks of the Selwyn Basin indicate that the water column

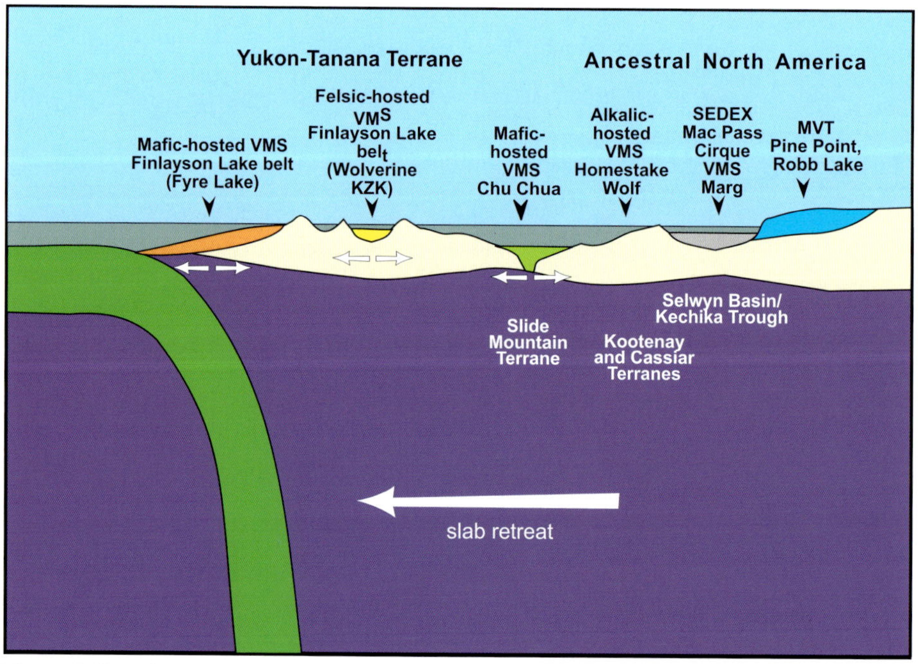

FIGURE 5. Tectonic model for the Devono-Mississippian margin of western North America, showing a complex history of ocean plate subduction, back-arc rifting, and the formation of sedimentary exhalative and Mississippi Valley-type deposits in the northern Canadian Cordillera (from Nelson et al., 2002).

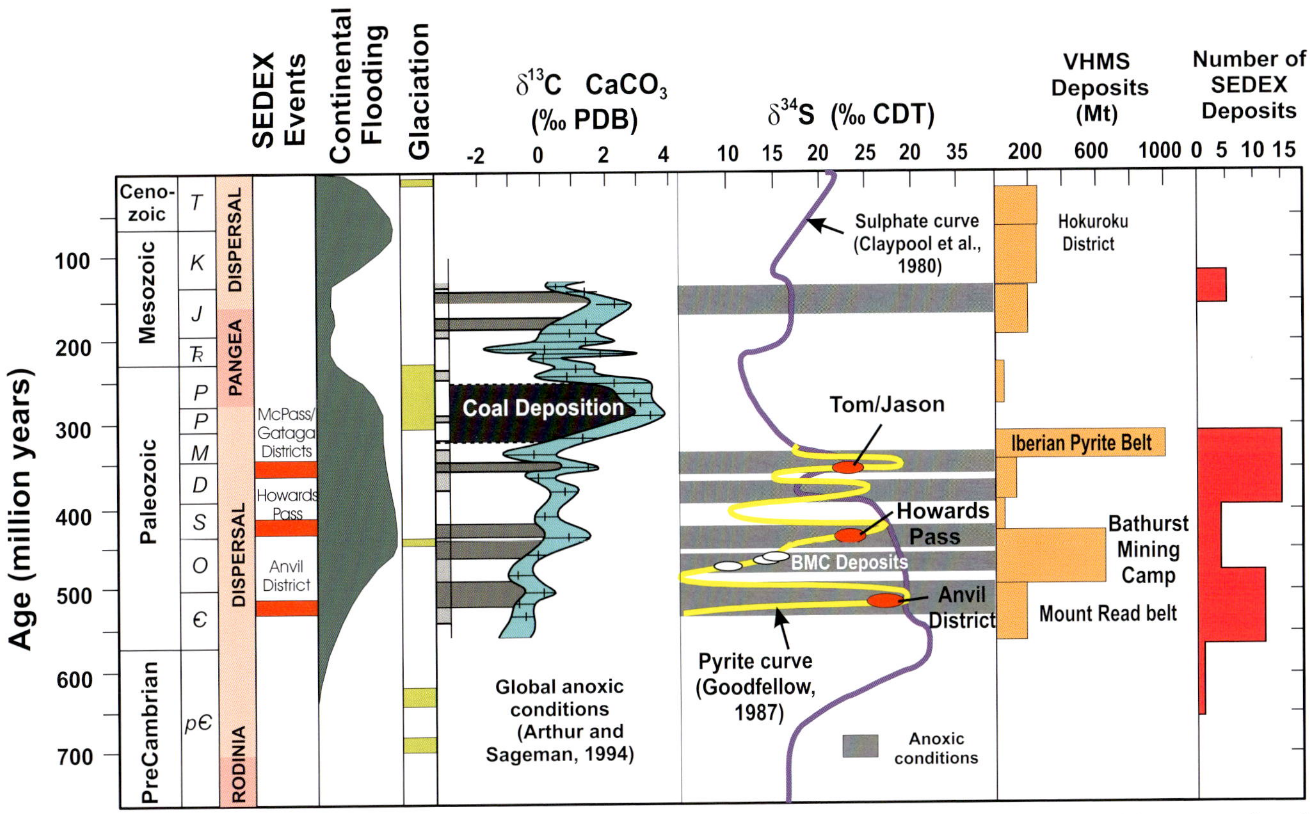

FIGURE 6. Late Proterozoic and Phanerozoic tectonic (continental accretion and dispersion) (Fischer, 1984) and glacial (Baker, 1997) events, continental flooding (Vail et al., 1977), and sulphur isotope secular curves for evaporites (Claypool et al., 1980), carbonates (Arthur and Sageman, 1994), and pyrite (Goodfellow, 1987). Also shown are global (Arthur and Sageman, 1994) and Selwyn Basin (Goodfellow, 1987) anoxic periods, and the temporal distribution of volcanogenic massive sulphide (Mosier et al., 1983) and sedimentary exhalative (Goodfellow et al., 1993) deposits.

was probably stratified with anoxic bottom water during formation of SEDEX deposits at MacMillan Pass and Howards Pass (Goodfellow, 1984, 1987). This is consistent with the delicately laminated, nonbioturbated character of sedimentary rocks, which indicates bottom water conditions hostile to bioturbating organisms.

Faunal assemblages and the presence of fossiliferous limestone suggest that most Phanerozoic SEDEX deposits formed within 30º of the equator (Goodfellow et al., 1993). For example, most Middle Devonian to Mississippian SEDEX deposits plotted on a continental reconstruction by Scotese (1984) for the Late Devonian straddle the equator.

The $\delta^{34}S$ curves for pyrite and barite (Figs. 4, 6) established for the Selwyn Basin (Goodfellow and Jonasson, 1984; Shanks et al., 1987) record four major anoxic episodes: Early Cambrian, Late Ordovician to Early Silurian, Early Devonian, and Late Devonian. Most SEDEX deposits in the Selwyn Basin formed during these anoxic episodes. Because the secular curve for barite from the Selwyn Basin is similar in shape to Claypool et al.'s (1980) global evaporite curve, and considering that the anoxic cycles can be correlated across the Selwyn Basin and recognized in other Paleozoic basins of the Cordillera and Arctic islands (Goodfellow, 2004), it is suggested that these S isotope cycles reflect global anoxia as opposed to local basin stagnation and closed-system S isotope fractionation. The fact that the Selwyn Basin $\delta^{34}S$ curve for barite is consistently 10 to 15‰ more positive than the global curve for marine evaporites may indicate partly restricted seawater circulation between the Selwyn Basin and the open ocean, or between anoxic bottom waters and an overlying oxygenated water column in a stratified sea (Goodfellow, 1987).

Mineral Deposits

Deposit Types and Classification

The Selwyn Basin hosts three major deposit types that are products of basin-forming processes: 1) SEDEX Zn-Pb-Ag; 2) stratiform barite and Ba-carbonate; and 3) Mississippi Valley-type (MVT) Pb-Zn deposits. The SEDEX and barite deposits are mostly synsedimentary, whereas the MVT deposits formed subseafloor in mostly carbonates of the adjacent Mackenzie and MacDonald platforms (Fig. 1). MVT deposits are included here because of their spatial association with the hinge line separating platformal and basinal facies rocks, their temporal association with SEDEX deposits of the Selwyn Basin, their formation from similar basinal metalliferous fluids, and because they have been interpreted as co-genetic with SEDEX deposits of the Selwyn Basin (Fig. 7) (Goodfellow, 1987; Nelson et al., 2002; Paradis et al., 2007).

There are three major episodes during which SEDEX deposits formed in the Selwyn Basin: 1) Late Cambrian Anvil District, 2) Early Silurian Howards Pass District, and 3) Late Devonian MacMillan Pass and Gataga Districts. In addition, there are the Clear Lake (Late Devonian) and Mel (Cambrian) deposits that do not fall inside the above districts.

SEDEX deposits in each of these districts are hosted by reduced, carbonaceous, fine-grained basinal facies sediments that are now represented by mostly black carbonaceous mudstone, shale, chert, and, in the case of the highly metamorphosed Anvil District, graphitic schist. Higher energy clastic sediments, including within basin black clastics of the Earn Group, are most abundant in the MacMillan Pass District (e.g. Gordey et al., 1987; Abbott and Turner, 1991).

Stratiform barite deposits in the Selwyn Basin (Fig. 1) span the Cambrian to Mississippian but are most common in the Devonian. In the Early Silurian Howards Pass district, barite is absent from the sulphide zone where the Ba content is typically less than 500 ppm (Goodfellow et al., 1986). Stratiform barite deposits are barren of base metals, typically well bedded, and are considered to have formed from low-temperature and shallowly derived fluids (Lydon, 1986). Stratiform Ba-carbonate deposits are less common than barite deposits and are best represented by the Cathy deposits at MacMillan Pass, Yukon (Turner and Goodfellow, 1990). The barium carbonate deposits are interpreted to have formed during diagenesis by the dissolution of barite by bacterial sulphate reduction that was followed during barium carbonate precipitation within sulphate-depleted organic-rich sediment.

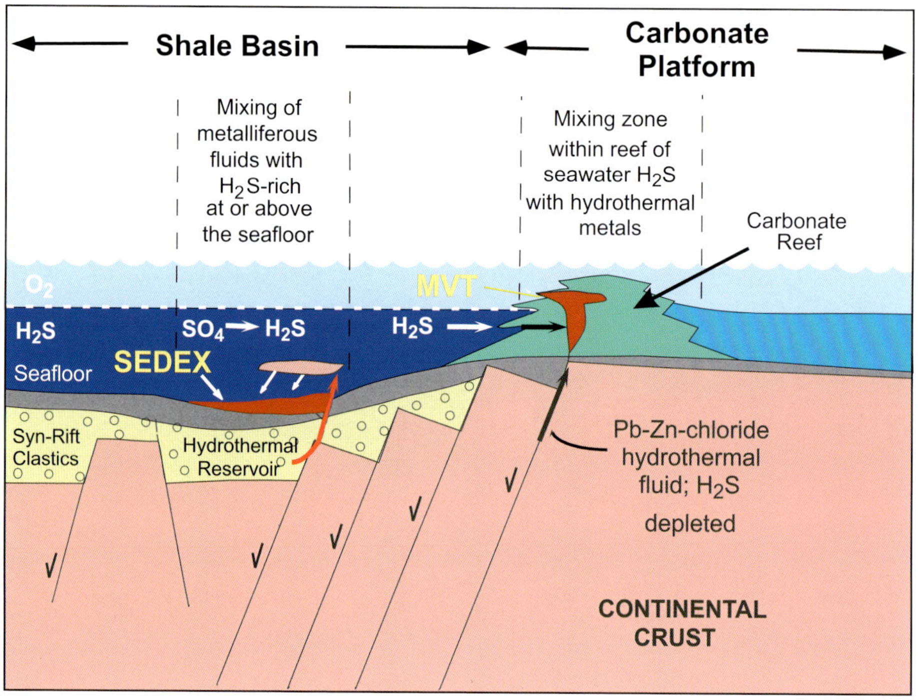

FIGURE 7. Model for the cogenetic formation of Late Devonian sedimentary exhalative (SEDEX) and Mississippi Valley-type (MVT) deposits in the Selwyn Basin and Mackenzie Platform. This model indicates that both SEDEX and MVT deposits formed by the mixing of metalliferous fluids with ambient anoxic waters at the seafloor (i.e. SEDEX deposits) and within permeable carbonates (i.e. MVT deposits). The model is supported by the close temporal association between Late Devonian SEDEX and MVT deposits, the spatial association of both SEDEX and MVT deposits with reactivated extensional faults, and similar metal ratios and $\delta^{34}S$ values for sulphides (Goodfellow et al., 1993).

Mississippi Valley-type (MVT) deposits occur typically within platformal carbonate sequences and formed by the infilling and replacement of carbonates (Paradis et al., 2007). The deposits are therefore epigenetic, morphologically irregular, and characterized by sulphide precipitation in open space. MVT deposits that have been dated by Rb/Sr isotopic methods include Pine Point (Nakai et al., 1993), Polaris (Christensen et al., 1995), Robb Lake (Nelson et al., 2002), and Nanisivik (Sherlock et al., 2004). Paleomagnetic dates are not discussed here because of uncertainty regarding the source of magnetism and the high susceptibility of this method to resetting during later metamorphic and hydrothermal events. Using isotopic ages, the Pine Point, Robb Lake, and Polaris deposits formed in the Late Devonian, at about the same time SEDEX formed in the adjacent Selwyn Basin. These MVT ages have important genetic implications since they appear to contradict gravity-driven fluid models related to much later orogenies proposed for the Pine Point deposit in particular (Garven, 1984) and MVT deposits in general (Bradley and Leach, 2003).

Grade and Tonnage

The grade, tonnage, and ages of major SEDEX deposits in North America are given in Table 1.

Past producers include the Faro, Vangorda, and Grum deposits. The deposits range in size from the giant Howards Pass deposit at 115 million tonnes to the relatively small Driftpile deposit at 2.4 million tonnes (Table 1; Fig. 8). The Howards Pass deposit, however, is generally low grade, containing 5.38% Zn, and 2.08% Pb, and is not economic. Metal contents and ratios are highly variable between deposits, with values ranging from 4.7 to 16.2% Zn, 2.0 to 5.5% Pb, and up to 110 g/t Ag. The average Cu and Au contents are generally very low except for the Vangorda and DY deposits in the Anvil District where values range up to 0.3% and 1.0 g/t, respectively (Table 1).

Compared to SEDEX deposits worldwide, those in the Selwyn Basin span the size range of global deposits and, except for the Howards Pass deposits, contain less than 100 million tonnes of sulphide ores (Fig. 8). Most of the giant current or past-producing SEDEX deposits are located in the McArthur River/Mount Isa, Broken Hill, and Belt-Purcell Proterozoic basins of Australia and Canada, and the Devonian-Carboniferous Brooks Range northern Alaska.

Deposit Distribution

MacMillan Pass

The Pb-Zn deposits of the MacMillan Pass district fall within the Central Block, a 30 km wide by 60 km long belt of Late Proterozoic to Triassic strata that trends west-northwest (Fig. 9), which has been referred to as the Macmillan Pass graben (Abbott and Turner, 1991). All three SEDEX deposits at MacMillan Pass are hosted within Devonian lower Earn Group rocks that consist of interbedded chert-pebble conglomerates, diamictites, black laminated shale and chert, and alkalic mafic volcanic rocks. The Tom and Jason deposits occur along an extensional synsedimentary fault that is defined by sedimentary breccias, paleofacies distributions, and hydrothermal fluid upflow zones (Goodfellow and Rhodes, 1990; Abbott and Turner, 1991). These relationships are illustrated in a geological cross-section of the Jason deposit (Fig. 10). The Boundary Creek or Nidd deposit is hosted in fault structures and chert-pebble conglomerate, and formed epigenetically in a subseafloor environment (Turner and Rhodes, 1990).

All known mineralization at MacMillan Pass occurs in areas of major sedimentary facies and thickness changes, demonstrating the relationship of mineralization to areas of maximum syndepositional tectonism (Winn and Bailes, 1987). It is common in these areas for multiple chert pebble conglomerate horizons to occur with locally derived diamictites composed of grit to boulder size clasts floating in a mudstone matrix. Alkalic basaltic flows and volcaniclastics overlap in time with the formation of the Tom, Jason, and Boundary Creek deposits, and are spatially associated with these deposits (Goodfellow et al., 1995).

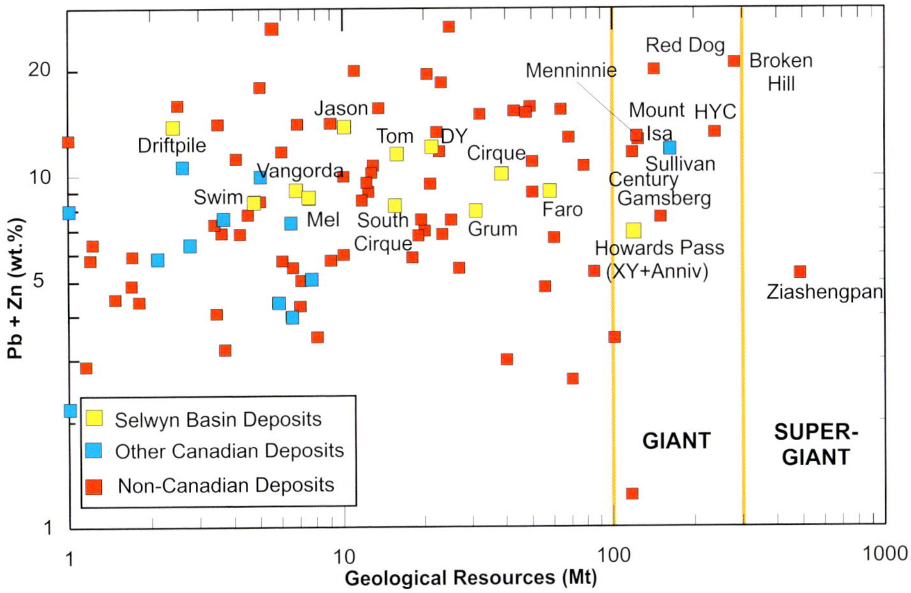

FIGURE 8. Grade (Pb+Zn wt.%) versus geological resources (million tonnes) for global, Canadian, and Selwyn Basin sedimentary exhalative deposits.

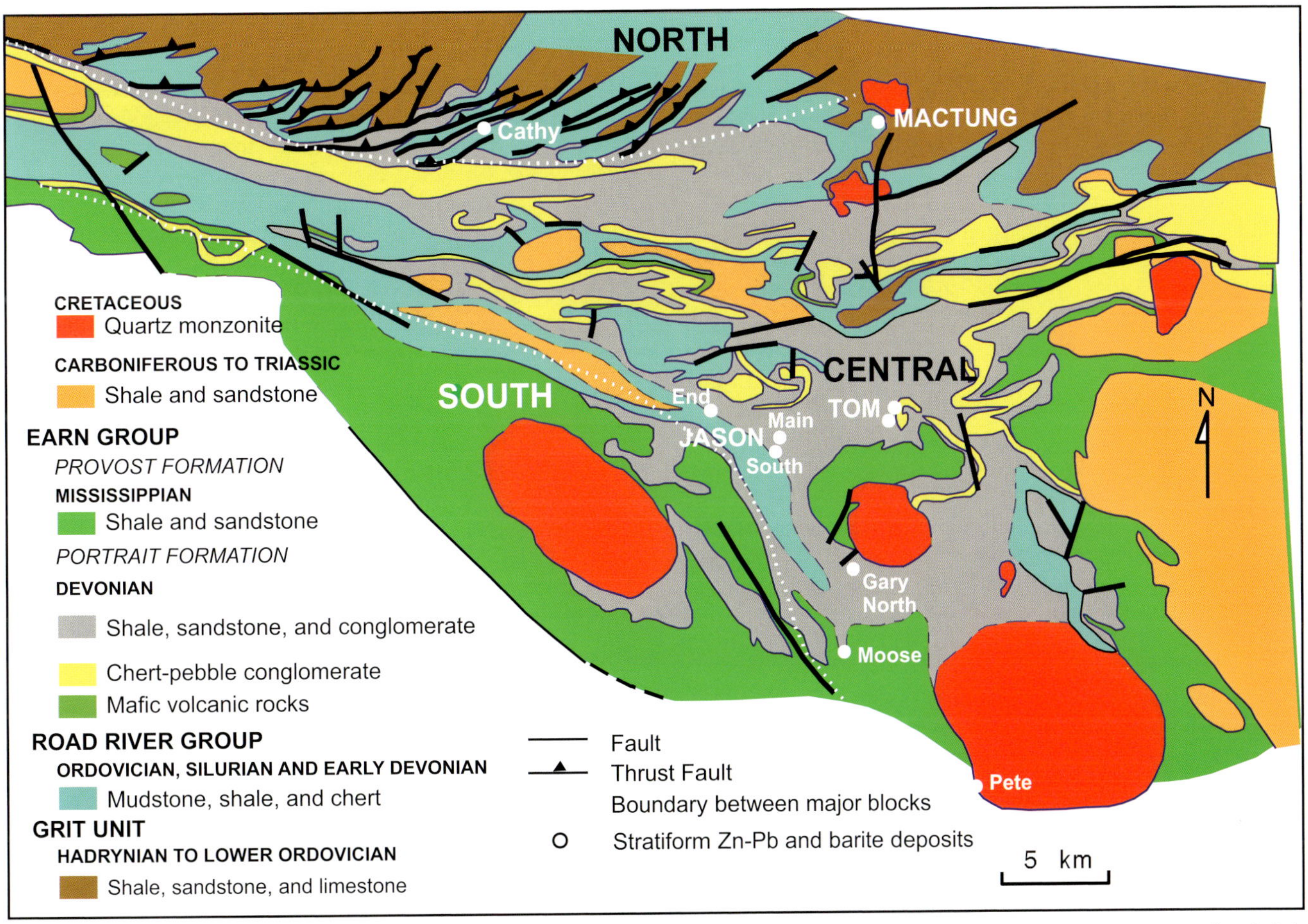

FIGURE 9. Geology of MacMillan Pass District showing the location of Zn-Pb sedimentary exhalative and stratiform barite, and Ba-carbonate (i.e. Cathy) deposits, and their relationship to sedimentary facies and mafic volcanic rock (Abbott and Turner, 1991).

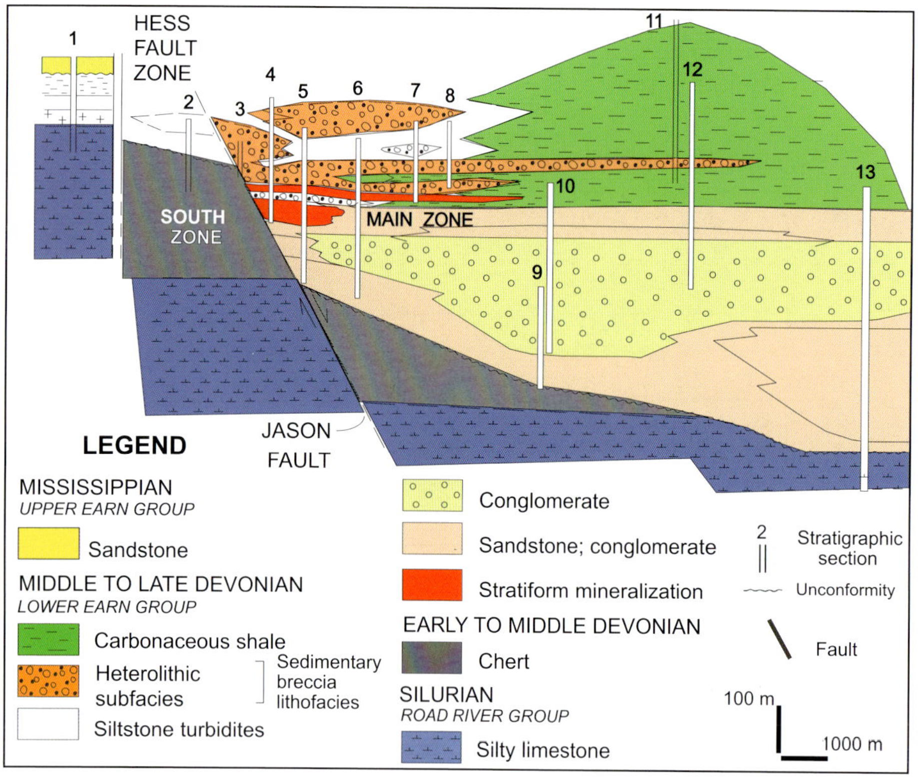

Figure 10. Geological cross-section of the Jason deposit showing the relationship between the synsedimentary Jason Fault and fault scarp breccias, the general stratigraphy, and the sulphide zones (Turner, 1990).

calcareous dolomitic shale, which is in turn overlain by highly carbonaceous, black cherty shale. This organic carbon-rich unit forms the stratigraphic footwall to the Howards Pass deposits, ranges up to 50 m in thickness, and is characterized by delicately laminated framboidal pyrite and abundant graptolites (Goodfellow et al., 1986).

The bedded sulphides are overlain by a very distinctive black phosphatic chert that is up to 50 m thick near the centre of the Howards Pass Basin (Goodfellow et al., 1986). The apatite is thinly bedded or delicately interlaminated with black carbonaceous chert. This unit extends for over 10 km from the Howards Pass deposit, and the thickness and phosphate content decreases away from the Zn-Pb mineralization. Near the Howards Pass deposit, the phosphatic chert contains both concentrically and radially zoned carbonate concretions up to 1 m in diameter. The concretions are commonly rimmed by pyrite, which weathers a rusty brown colour in surface exposures.

The phosphatic chert is overlain by an orange-weathering, bioturbated, wispy laminated mudstone of Middle to Late Silurian age. This mudstone can be readily recognized in the field and represents a valuable marker unit for separating black shales of the Road River Group from those of the Earn Group.

In the area of the Howards Pass deposit, the bioturbated mudstones are overlain by Middle to Upper Devonian rusty weathering, baritic mudstone and siltstone, and coarse-grained clastic rocks of the Earn Group.

Howards Pass

This district consists of the XY, Anniv, and OP deposits, which lie in what has been referred to as the 'zinc corridor', a local second-order basin that extends for approximately 35 km. Recent exploration and drilling by Pacifica Resources Ltd. has located several new zones in this corridor, including Brodel, HC, Don, and Pelly North (http://www.pacifica-resources.com/exp_na_selwyn.cfm). This corridor is oriented northwest-southeast and runs parallel to the hinge line that defines the edge of the Mackenzie Carbonate Platform to the west (Fig. 11). The XY, Anniv, and OP deposits are interpreted as occurring in local third-order basins within the zinc corridor (Morganti, 1979).

The local geology consists of a starved sequence of Middle Ordovician to Early Silurian dark grey- to light grey-weathering black and highly carbonaceous shale and chert, which is overlain by Middle to Late Silurian orange-weathering bioturbated mudstone of the Road River Group. The Road River Group is underlain by impure basinal carbonates of the Rabbitkettle Formation and overlain by Devono-Mississippian black shale and chert, and clastic units of the Earn Group that hosts the Tom and Jason deposits at MacMillan Pass (Gordey, 1981).

The Road River Group in the Howards Pass area consists of five distinct units referred to informally as "members" by Morganti (1979). The lowermost shale unit sits on a transitional unit separating pyritic shales of the Road River Group from limestone of the Rabbitkettle Formation. The pyritic shales range up to 30 m in thickness and are dark grey and typically finely laminated. This unit is overlain by up to 60 m of massive to weakly laminated dark grey, weakly pyritic,

Anvil District

The Anvil District is located on the western margin of the Selwyn Basin near the Tintina Fault (Fig. 12). The district consists of five deposits with defined reserves, the Faro, Grum, Vangorda, DY, and Swim deposits (Table 1).

The Anvil District is underlain by polydeformed, Late Proterozoic to Upper Paleozoic sedimentary and volcanic rocks intruded by Cretaceous granitic plutons (Jennings and Jilson, 1986). The lowest exposed unit, the Late Proterozoic 'Grit Unit' of the Windermere Supergroup, consists of a thick synrift sequence composed of interbedded coarse- and fine-grained clastic rocks that were deposited during the earliest stages of continental rifting.

Shales at the top of the Grit Unit pass upward into Early and Middle Cambrian brown and black shales of the Mt. Mye Formation (Jennings and Jilson, 1986). The Mt. Mye Formation consists of noncalcareous phyllite and schist, marble and calc-silicate lenses, carbonaceous schist, and

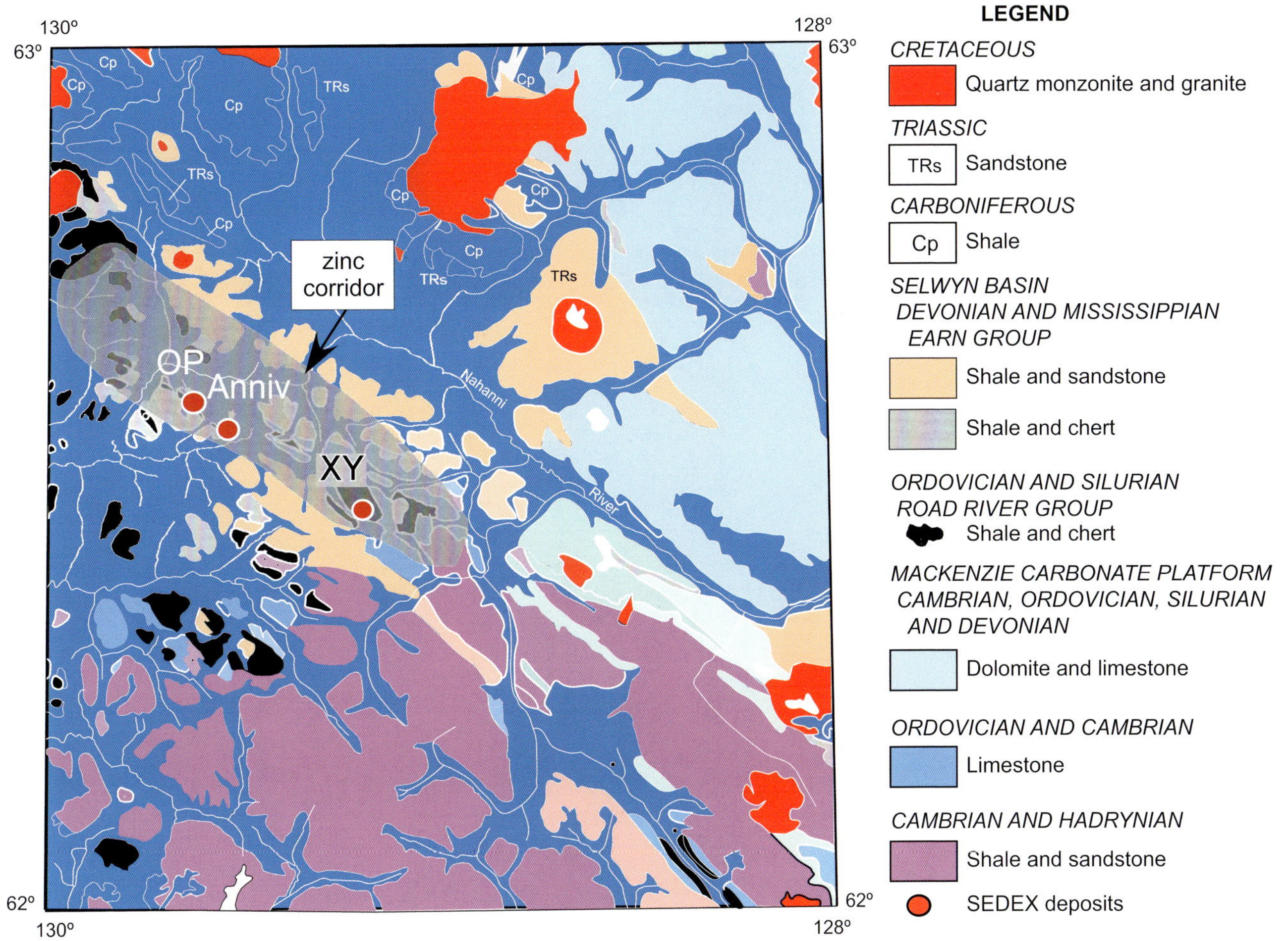

FIGURE 11. Geology of the Nahanni map-area (N.T.S. 105I) showing the distribution of shales and cherts of the Road River Group that host sedimentary exhalative deposits of the Howards Pass District (geology from Gordey et al., 2000).

minor psammitic schist and basalt. This formation is overlain by over 1000 m of calcareous phyllite and schist, calc-silicate, basalt, carbonaceous phyllite, and minor marble of the Vangorda Formation. The Vangorda Formation has been correlated by Jennings and Jilson (1986) with the Rabbitkettle Formation, which underlies the Road River Group further to the east in the Selwyn Basin.

The Menzie Creek Formation overlies the Vangorda Formation and is composed of over 1000 m of pillowed and massive flows, volcanic breccia tuff, and interbedded black shale (Jennings and Jilson, 1986). This formation is considered correlative with the Road River Group that hosts the Howards Pass deposits. Massive sulphide deposits of the Anvil District are hosted by carbonaceous schist/phyllite near the contact between the Mount Mye and Vangorda formations.

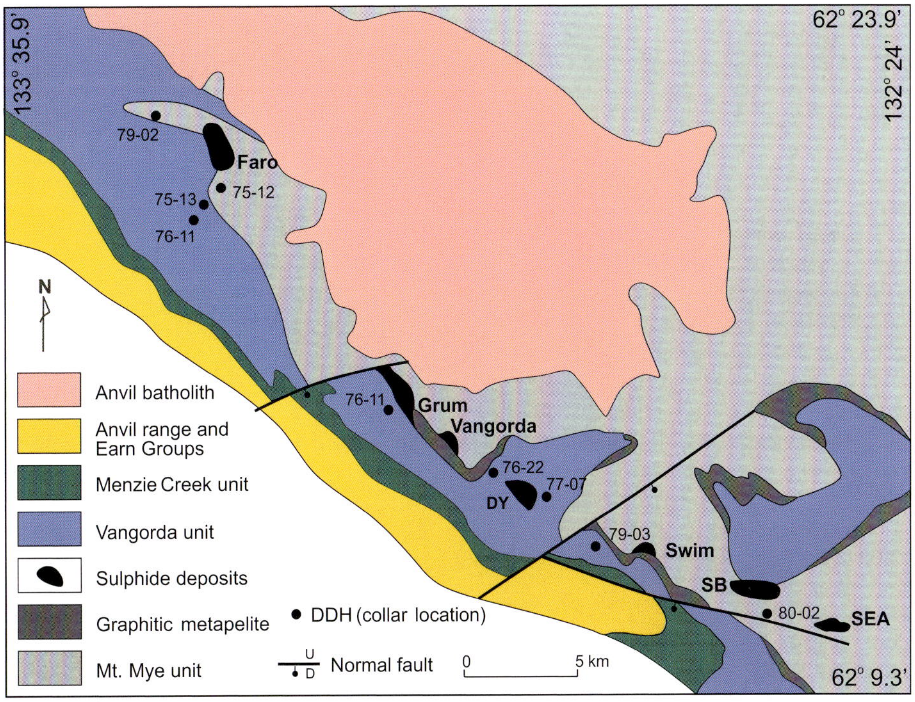

FIGURE 12. Generalized geology of the Anvil District, Selwyn Basin, Yukon, showing the location of the Faro, Grum, Vangorda, DY, Swim, SB, and SEA deposits (from Jennings and Jilson, 1986).

The Anvil District is structurally complex and variably metamorphosed from greenschist to amphibolite grade. The rocks exhibit evidence of five periods of deformation of which the first two are penetrative, regionally developed, and accompanied by regional metamorphism (Jennings and Jilson, 1986).

Gataga District

The Gataga District is located in northeastern British Columbia and forms part of the Kechika trough that hosts several SEDEX and stratiform barite deposits, including the Driftpile, Bear, Cirque, Fluke, Akie, and Elf deposits (Fig. 13, from MacIntyre, 1992). The largest deposit is the Cirque with reserves in excess of 38 Mt of 8% Zn, 2.2% Pb, and 47.2 g/t Ag for the main Cirque, and 15.5 Mt of 6.9% Zn, 1.4% Pb, and 32 g/t Ag for the south Cirque (MacIntyre, 1992). Deposits of the Gataga District are hosted by Upper Devonian (Famennian) carbonaceous shale and chert of the Gunsteel Formation, lower Earn Group (Jefferson et al., 1983; Paradis et al., 1998).

The Kechika Trough forms part of the Selwyn Basin and represents a long-lived, rift-controlled, intracratonic basin bordered by carbonate platforms. To the east, the basinal facies is flanked by carbonates of the MacDonald Platform. The stratigraphy within the basin is represented by mostly black, carbonaceous shale and chert that were deposited in a quiet, deep-water, anoxic marine environment.

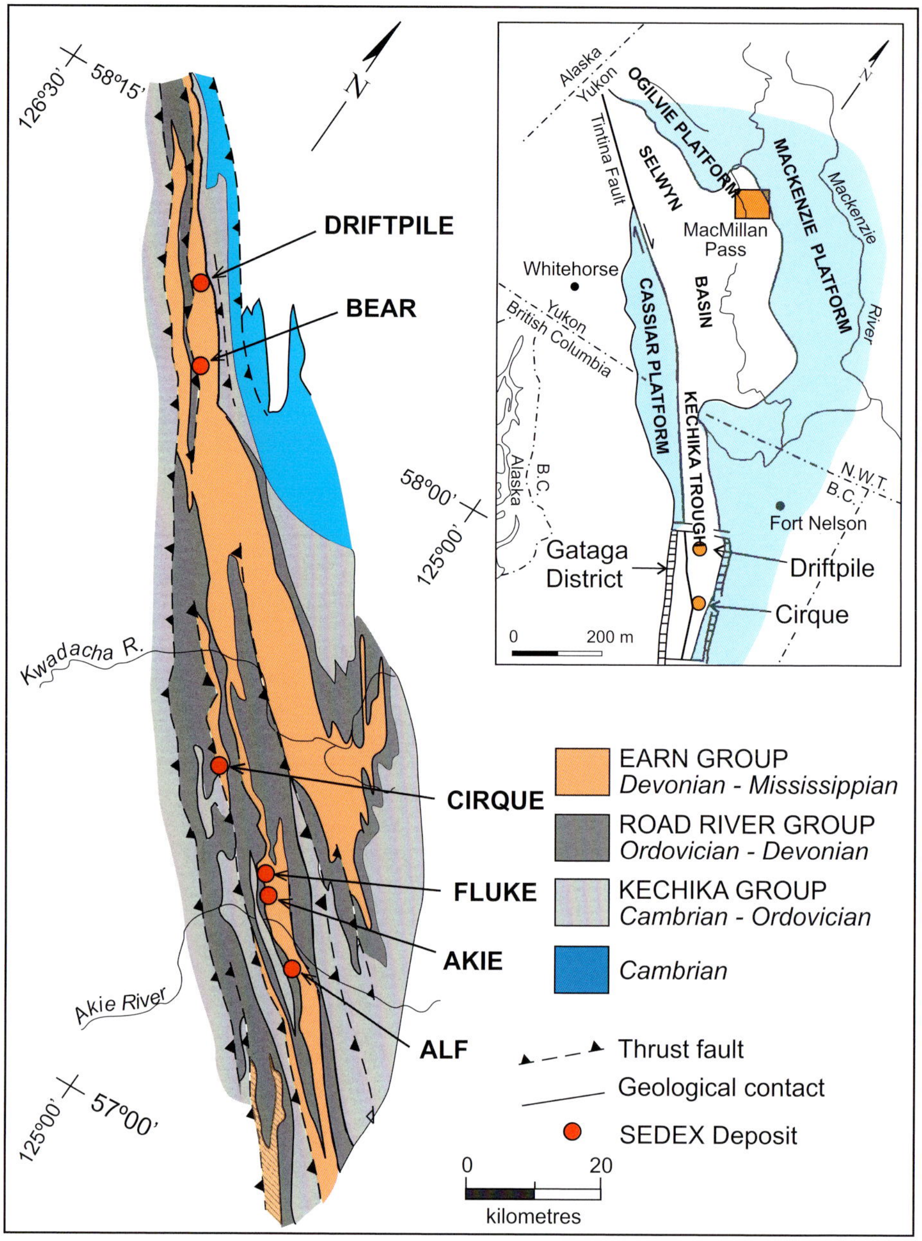

FIGURE. 13. Regional geology of the Kechika Trough showing the location of Late Devonian sedimentary exhalative deposits (from McClay et al., 1989; MacIntyre, 1992).

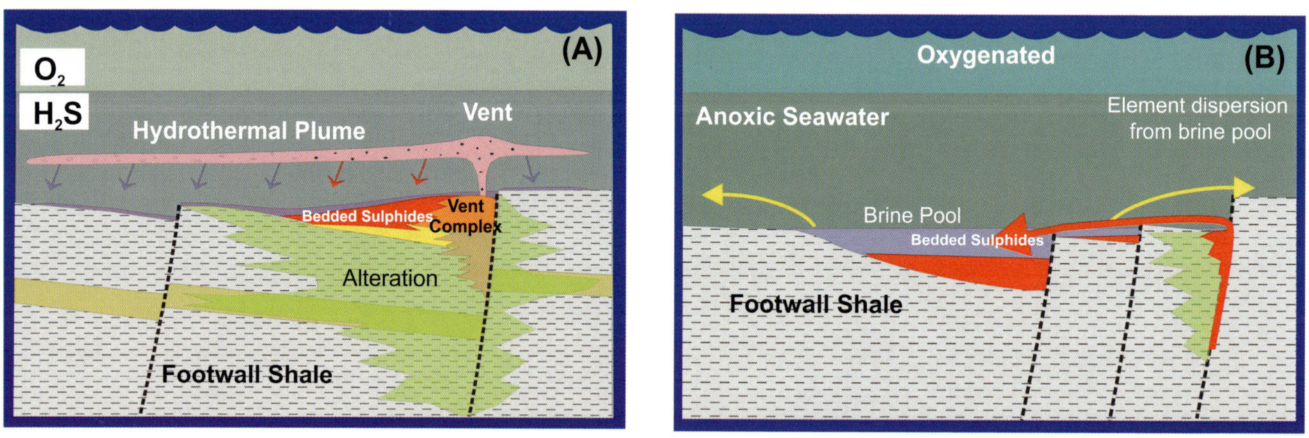

FIGURE 14. Genetic models for sedimentary exhalative deposits. **(A)** Vent-proximal deposits formed from buoyant hydrothermal plumes (e.g. Tom and Jason, Yukon). **(B)** Vent-distal deposit formed from bottom-hugging brines (e.g. Howards Pass and Anniv, Yukon).

The organic-rich shale and chert are typically pyritic and most of the pyrite occurs as disseminations and/or laminae. At the microscopic level, the pyrite is commonly framboidal, which is consistent with a biogenic origin. Stratiform Zn-Pb (SEDEX) and barite deposits formed during three major time periods in the history of the Kechika Trough: Middle Ordovician, Early Silurian and Late Devonian (MacIntyre, 1992). However, the youngest mineralizing episode is the most important in terms of the number and size of SEDEX deposits. Both the Middle Ordovician and Late Devonian deposits are temporally and, in many cases, spatially associated with alkalic basalts (Goodfellow et al., 1995).

The Gataga area was deformed into a northeast-verging fold and thrust belt during Jurassic-Cretaceous shortening of the passive margin. McClay et al. (1988) described five major thrust panels, each consisting of rocks of different stratigraphic levels and/or lithofacies. The strata within each thrust panel are folded and stacked in thrust duplexes.

Morphology and Architecture of Sedimentary Exhalative Deposits

There are two major types of SEDEX deposits in the Selwyn Basin: 1) vent-proximal with laterally extensive bedded ores (e.g. Tom, Jason, and Faro), and 2) distal stratiform (e.g. Howards Pass and Driftpile deposits) with no evidence of a vent system (Fig. 14). The bulk of the ore is contained in a tabular stratiform sulphide body that typically has a high aspect ratio, i.e., the ratio of the lateral extent of the body to its maximum stratigraphic thickness. However, some of the highest grades in the case of the Tom deposit are near the transition from the vent complex to the bedded ore facies (Goodfellow and Rhodes, 1990). This suggests that zone refining in proximal deposits is an important process for upgrading the metal content of SEDEX deposits (Goodfellow et al., 1993).

Vent-proximal deposits overlie vents, display a mound-like morphology, are typically zone refined, and formed in most cases from buoyant hydrothermal fluids (Fig. 14A). Although it is possible for deposits that formed from bottom-hugging fluids to be proximal in cases where a seafloor depression coincided with a vent, this setting is considered to be rare because of the absence of a vent complex in most distal deposits. Distal SEDEX deposits, by contrast, are remote from vents, display a sheet-like morphology, are delicately laminated, and typically contain no barite. Vent-distal deposits are interpreted to have formed from bottom-hugging metalliferous brines (Fig. 14B).

The Devonian and Cambrian SEDEX deposits of the MacMillan Pass (Tom and Jason), Gataga (Cirque and Driftpile), and Anvil (Faro, Vangorda, Swim, Grum, and DY) districts are mostly vent proximal and formed therefore from fluids that probably behaved bouyantly upon discharge into seawater. The Howards Pass (XY and Anniv) deposits, however, are interpreted to have formed from bottom-hugging metalliferous brines (Goodfellow et al., 1986).

Hydrothermal Facies in Sedimentary Exhalative Deposits

Bedded Ores

The bedded ore facies are composed typically of regularly interlayered hydrothermal minerals interspersed with beds of the host lithologies (e.g. Figs. 15A-D, 16A-C). Hydrothermal laminae are commonly monomineralic in Devonian deposits, whereas sphalerite and galena are closely intermixed at Howards Pass (Jonasson and Goodfellow, 1986). Bedding contacts are typically very sharp, particularly between beds of hydrothermal minerals and host lithologies. The most common hydrothermal minerals are pyrite, sphalerite, galena, chert, and carbonates. Barite is a major constituent of most Selwyn Basin deposits, with the notable exception of those in the Howards Pass district (see Table 1 and Goodfellow, 2004).

Distal Hydrothermal Sediments

The outer margins of the stratiform body consist typically of bedded hydrothermal products that have no direct economic value but provide widespread targets for mineral exploration (Goodfellow et al., 1993). These distal hydrothermal sediments are composed of the more mobile or more stable and/or more oxidized, non-ore components of the bedded ore facies. For example, the distal sedimentary facies at Tom and Jason consists of barite and minor sulphides (Fig. 16D) that contain anomalously high Zn, Cd, and Hg, and extends several kilometres from the deposits (Bailes et al., 1986; Goodfellow and Rhodes, 1990). At the XY deposit (Howards Pass), the distal and overlying fringes of the ore zone are composed of a pyritic phosphatic chert (Fig. 15F) with carbonate concretions (Fig. 15E), and are characterized by Zn, Cd, Hg, and As anomalies that extend several kilometres from the sulphide zone. Footwall shales are enriched in Cu, Co, Ni, Mo, As, Sb, Zn, Cd, Hg, and Pb relative to correlative rocks remote from mineralization. Halos of pyrite enriched in Cu, Ni, and Co also surround the mineralized zone at the XY deposit (Goodfellow et al., 1983, 1986).

Vent Complex

The distinctive and economically important zone of interaction between upflowing hydrothermal fluids and hydrothermal sediments is termed the "vent complex" (Goodfellow et al., 1993). In contrast to the regularly layered appearance of the bedded ore facies, the vent complex is heterogeneous in nature and is generally composed of massive zones, replacement patches, irregular veins, and/or disseminations of sulphides, carbonates, and silicates (mostly quartz). The mineral assemblage is dominated by pyrite, pyrrhotite, galena, sphalerite, calcite, siderite, ankerite, dolomite, Ba-carbonates, and quartz and lesser muscovite, chlorite, Ba-silicates, chalcopyrite, arsenopyrite, and sulphosalt minerals (e.g. Tom, Fig. 16E). In deposits in which the vent complex has been well documented, the contact between the stratabound vent complex and adjacent sedimentary-hydrothermal facies is a discordant replacement contact. Both the Tom (Goodfellow and Rhodes, 1990) and Jason deposits (Turner, 1990) in the MacMillan Pass district have well documented vent complexes. The dominant process within vent complexes is the reaction of upflowing hydrothermal fluids with hydrothermal and host sediment, which results in the filling of open space and the replacement of the lower temperature sedimentary minerals by higher temperature hydrothermal assemblages.

FIGURE 15. Photographs of rocks from the Howards Pass (XY) deposit. **(A)** Active Member, lower unit: sphalerite, galena, and pyrite delicately interlaminated with pale grey cherty limestone and dark grey carbonaceous chert. The sample shows intense folding with a pressure dissolution axial planar cleavage filled with sphalerite and galena. **(B)** Active Member, upper unit: sphalerite and galena with minor pyrite delicately interlaminated with black carbonaceous chert and cut by axial planar pressure dissolution cleavages filled with mostly sphalerite and galena. **(C)** Active Member, upper unit: high-grade sphalerite and galena (pale grey) interbedded with a darker grey unit consisting of interbedded sphalerite, galena, pyrite, and chert cut of a pressure dissolution cleavage. **(D)** Active Member, upper unit: thin interval consisting of delicately interlaminated framboidal pyrite and black carbonaceous chert. Laminated pyrite flanked by bedded sphalerite and galena cut by a pressure dissolution cleavage. **(E)** Phosphatic chert unit: Carbonate concretion with a pyrite rim hosted by black, carbonaceous phosphatic chert. **(F)** Phosphatic chert unit: carbonate fluorapatite (pale grey) interbedded with black, carbonaceous chert.

FIGURE 16. Photographs of the different hydrothermal facies, Tom deposit. **(A)** Pink facies: interbanded white barite, pink, pale green and black sphalerite, galena, and grey chert. **(B)** Grey facies: interbedded white barite, cream sphalerite, and dark grey mudstone. Galena occurs finely disseminated in sphalerite layers. **(C)** Black facies: black carbonaceous chert finely laminated with barite and sphalerite. Framboidal pyrite and galena also occur along bedding planes. **D)** Distal facies: black carbonaceous mudstone with blebs of barite and Ba-feldspar; framboidal pyrite and sphalerite also disseminated throughout. **(E)** Vent complex: mixture of medium-grained galena and ferroan carbonate with circular patches of crystalline pyrite. **(F)** Stringer zone: black silicified argillite cut by veins of ferroan carbonate containing pyrite, pyrrhotite, galena, and arsenopyrite. Later tectonic quartz veins also present.

Feeder Pipe and Associated Alteration

The nature and extent of the hydrothermal alteration and mineralization in the 'feeder pipe' depend to a large degree on the mineralogy and physical properties of footwall lithologies, the temperature and chemical composition of the hydrothermal fluids, and hydrostatic pressure or water depth. Host siliciclastic or carbonate sediments within the feeder pipe are commonly altered, veined, and/or brecciated (Fig. 16F). Alteration minerals that have been reported for the Selwyn Basin SEDEX deposits include quartz, muscovite, chlorite, ankerite, siderite, and sulphides (Goodfellow and Rhodes, 1990; Turner, 1990). The abundance of sulphides in alteration zones is typically low, but pyrite, pyrrhotite, galena, sphalerite, chalcopyrite, tetrahedrite, and arsenopyrite are commonly present.

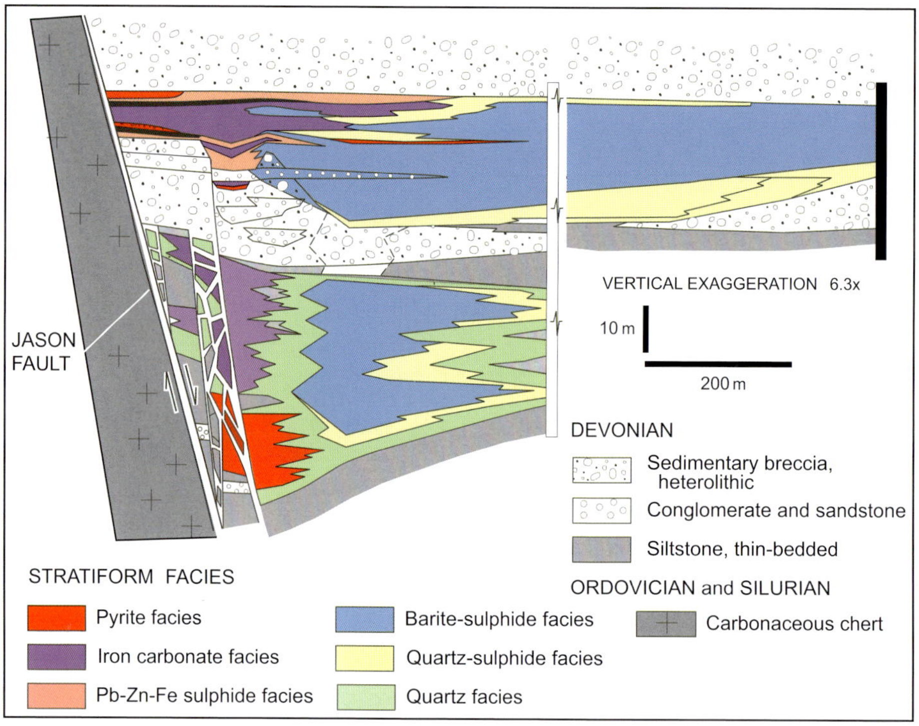

FIGURE 17. Geological cross-section of the Jason deposit showing the different hydrothermal facies and the relationship of these to the Jason fault (Turner, 1990).

The feeder zones at the Tom and Jason deposits appear to be rooted in a synsedimentary extensional fault (McClay and Bidwell, 1986; Rhodes and Matthew, 1993). Fault-scarp breccias, debris flows, and abrupt facies changes associated with the Jason and Tom deposits provide compelling evidence that the fault was active before, during, and after sulphide formation (Winn and Bailes, 1987; Rhodes and Matthew, 1993). At the Jason deposit, the presence of soft-sediment deformation textures and sulphide-bearing ferroan carbonate veins within the fault indicate that hydrothermal fluids migrated along this fault during formation of the deposit (Turner, 1990).

The feeder pipes underlying the Tom and Jason deposits are both represented by silicified mudstone, siltstone, and conglomerate that have been fractured, brecciated, infilled, veined, and partly replaced by silica, ferroan carbonate, micaceous, and sulphide minerals (Goodfellow and Rhodes, 1990; Turner, 1990). The sediments have been intensely brecciated near the core of the feeder pipe, either by hydrofracturing or by tectonic movement along adjacent syndepositional faults. Pyrrhotite, pyrite, chalcopyrite, and sphalerite occur in the feeder zone, but constitute less than 1 wt.% of the rock. Because of their high permeability, coarse-grained clastic units are more hydrothermally altered to carbonate and clay minerals remote from the centre of the fluid discharge than are less permeable strata, and this has resulted in the formation of local stratabound zones of hydrothermal alteration.

Hanging-Wall Mineralization and Alteration

Hydrothermal sediments and hydrothermal alteration that extend 100s of metres into the hanging wall of many SEDEX deposits provide large exploration targets for concealed deposits. At the Tom and Jason deposits, metal-rich beds of barite-pyrite extend up to 500 m into the overlying sedimentary sequence and provide evidence of waning hydrothermal activity over a significant time span (Bailes et al., 1986; Goodfellow and Rhodes, 1990). The Howards Pass deposit is also overlain by up to 50 m of laminated phosphatic chert that contains elevated pyrite contents and the anomalous enrichment of most ore-forming and ore-associated chalcophile elements (Goodfellow et al., 1983, 1986).

Mineral and Element Zonation

One of the most characteristic features of the stratiform facies is the lateral zonation of hydrothermal textures, minerals, and elements around centres of hydrothermal fluid discharge (e.g. Fig. 17, Jason deposit). This zonation reflects a change from open space filling and replacement textures in the vent complex to sedimentary textures outward from the vent, and is accompanied by a reduction in the thickness of the stratiform body and a decrease in the number and/or thickness of individual beds of hydrothermal minerals.

A decrease of Pb/(Zn+Pb) ratios away from the vent complex is the most pronounced and consistent feature of SEDEX deposits in general, and has been well documented for the Tom (Goodfellow and Rhodes, 1990), Jason (Turner, 1990), Cirque (Jefferson et al., 1983), and Faro (Jennings and Jilson, 1986) deposits. Other chemical zonation patterns include increases in Pb:Ag, (Zn+Pb):Cu, Fe:Zn, Ba:Zn, and SiO_2:Zn ratios away from the vent complex. These ratios of hydrothermal elements also increase upwards through the bedded ore facies in some deposits. At the Tom deposit, subsurface replacement processes have produced a marked textural, mineralogical, and chemical zonation from the core of the vent complex to the distal fringes (Goodfellow and Rhodes, 1990). This has resulted in the systematic zonation of Zn, Pb, Cu, As, Sb, Ba, Fe, Mn, Ca, and CO_2 within and about the vent complex, and an increase in Pb/(Pb+Zn) ratios towards the centre of hydrothermal fluid discharge.

Genetic Models

Although the tectonic setting and favourable stratigraphic intervals are useful criteria for selecting sedimentary basins that represent the most promising exploration targets, another set of criteria is needed to decide where to focus an exploration program in a sedimentary basin. These criteria require an understanding of processes controlling the positioning of ore-forming hydrothermal systems within a sedimentary basin, the generation of ore-forming hydrothermal fluids, the focusing of hydrothermal fluids at the seafloor, the precipitation of metals, and the preservation of a SEDEX deposit at the seafloor.

Composition of Hydrothermal Fluids

Measurable primary fluid inclusions are rare in SEDEX deposits due to the inherent fine grain size of primary hydrothermal sediments and the recrystallization of hydrothermal products during diagenesis and burial metamorphism, which entrap formational but not hydrothermal fluids in the recrystallized ore minerals. Constraints on the composition of ore fluids that formed SEDEX deposits are therefore provided by a very limited fluid inclusion database, experimental and thermochemical modeling of mineral-solution equilibria in probable reaction zones for SEDEX deposits, isotopic ratios in hydrothermal products, and by analogy to modern metalliferous hydrothermal systems.

There are only about three or four published studies of fluid inclusions in SEDEX deposits that are sufficiently detailed to allow reasonable conclusions as to the composition and temperature of the ore-forming fluids. At the Tom (Ansdell et al., 1989) and Jason (Gardner and Hutcheon, 1985) deposits, homogenization temperatures average about 260°C and salinities average about 9 wt.% equivalent NaCl, or about 2.5 times the salinity of modern seawater. Fluids measured in inclusions from the Silvermines deposit, Ireland, record salinities between 8 and 28 wt.% equivalent NaCl and homogenization temperatures between 50 and 260°C (Samson and Russell, 1983). The negative correlation between salinity and homogenization temperature for inclusions in quartz was interpreted by Samson and Russell (1983) to indicate the mixing of 230°C, 8 to 12 wt.% equivalent NaCl salinity fluids from a deep source, with lower temperature, high-salinity fluids of seafloor brine pools and/or shallow pore fluids of the carbonate sediments formed by contemporaneous evaporitic processes. This limited data suggest that the salinities of some SEDEX ore-forming fluids are several times greater than that of seawater, but less saline than compositions reported for most MVT deposits (Leach et al., 2005).

The Pb, Sr, H, and O isotopic compositions of hydrothermal minerals in SEDEX deposits support a sialic crustal origin for the hydrothermal fluids. In a Pb isotopic study of SEDEX deposits located within the Selwyn Basin, Godwin and Sinclair (1982) concluded that the most likely source of Pb was the Hudsonian basement or clastic sediments derived directly from this basement. Initial $^{87}Sr/^{86}Sr$ ratios for hydrothermal barite and calcite in the Selwyn Basin SEDEX deposits are highly radiogenic (Goodfellow et al., 1986; Goodfellow and Rhodes, 1990; Turner, 1990). The Sr isotopes in barite and calcite associated with Pb-Zn SEDEX deposits are more radiogenic than Sr isotopes in barren sedimentary barite deposits, which are typically similar to seawater. A mixing model for Sr in hydrothermal fluids and seawater shows that Sr in barite from SEDEX deposits formed from hydrothermal fluids (>200°C), whereas Sr in barite deposits probably formed by the mixing of ambient seawater with cooler Ba-rich, reduced fluids with low metal contents (Goodfellow, 2004). This interpretation is consistent with a low-temperature origin for barren barite of the Selwyn Basin proposed by Lydon et al. (1985).

Thermochemical modeling indicates that the solubility of sphalerite and galena in an aqueous fluid depends primarily on the temperature, pH, fO_2, and chloride:sulphide ratio of the fluid (e.g. Anderson, 1973; Cooke et al., 2000). The solubility products of both sphalerite and galena, and the stabilities of Zn and Pb aqueous chloride complexes, increase with increasing temperature so that the total solubility of sphalerite and galena in chloride solutions increases exponentially with increasing temperature. Increased concentrations of sulphide in saturated solutions decrease Zn and Pb concentrations by precipitating sphalerite and galena, so that the highest concentrations of metals are achieved in solutions that have high metal:sulphide ratios, i.e., are reduced sulphur deficient compared to stoichiometric ratios of Zn:S and Pb:S for sphalerite and galena, respectively. Above the SO_4/H_2S boundary, there is a marked increase in the solubility of sphalerite and galena due to the low activity of reduced sulfur (e.g. Lydon, 1983). The lack of significant amounts of Cu in most SEDEX deposits suggests that the temperature of the ore-forming fluids was generally below 300°C. This is because the solubility of chalcopyrite is much more temperature dependent than are the solubilities of sphalerite and galena in a chloride solution. At seawater chloride concentrations and a fO_2 within the stability fields of sulphides, the solubility of Cu is less than 1 ppm below 300°C (Lydon, 1983).

These theoretical and experimental conclusions are corroborated by the compositions of natural hydrothermal fluids that contain at least several ppm Zn+Pb but little Cu, are typically below 300°C, have metal:sulphide ratios in excess of 10:1, and contain more than 100,000 ppm chloride (i.e. >2.7 molal Cl or more than five times seawater salinity). For example, the deeper high-salinity brines of the Salton Sea, which contain 506 ppm Zn, 95 ppm Pb, and 6 ppm Cu, have a chloride concentration of 154,000 ppm and a temperature of 305°C (McKibben et al., 1988). A brine from the Raleigh oil field, central Mississippi, contains 320 ppm Zn and 106 ppm Pb at a chloride concentration of 193,800 ppm and an estimated temperature of 135°C (Carpenter et al., 1974; Hanor, 1996).

The high Ba contents of many SEDEX deposits also reflect high chloride contents of the ore-forming fluids. In pure water, the stoichiometric solubility curve for barite goes through a maximum of 1 to 2 ppm at just above 100°C. However, at chloride concentrations greater than 3 molal, the solubility of barite continuously increases with temperature due to the formation of Ba chloride complexes, so that the stoichiometric solubility of barite can reach tens of ppm at temperatures greater than 150°C (Blount, 1977). The solubility of barite is also enhanced by low sulphate activity, which in turn is a reflection of low total S content, low fO_2 (where the dominant aqueous S species present in the fluid is H_2S or HS^- rather than SO_4^{2-}), and/or, to some extent, high salinity (when sulphate activity is reduced by sulphate complexing with Na or Ca cations). The paucity of barite in the feeder pipe of Ba-rich SEDEX deposits (e.g. Tom deposit: Goodfellow and Rhodes, 1990; Jason deposit: Turner, 1990) suggests that the hydrothermal fluids that formed these deposits contained minor sulphate.

The origin of sulphide S in SEDEX deposits is a subject of considerable interest, because the amount of Zn and Pb that a hydrothermal fluid can carry is inversely proportional to its sulphide S content, and therefore the more hydrothermal sulphide S delivered to the depositional site, the less metal that can be delivered by the hydrothermal fluids. Goodfellow (1987), Shanks et al. (1987), and Goodfellow and Rhodes (1990) have presented evidence that indicates most of the reduced S in stratiform deposits of the Selwyn Basin was derived from biogenic sulphide in the ambient water column at the time of sulphide deposition. At Howards Pass, the S isotope composition of bedded pyrite from the mineralized zone falls on the S isotope evolutionary curve for pyrite in the Selwyn Basin (Fig. 4). This suggests a common and therefore ambient seawater origin of reduced S in the Howards Pass deposit. Sulphur isotope values for sulphides from the Tom deposit likewise fall on a pyrite evolutionary curve for the Late Devonian that extends 100s of metres above the mineralized zone (Figs. 4, 6).

One of the key questions that has plagued geologists is the origin of reduced S in MVT deposits. Beales and Jackson (1966) were perhaps the first to propose a two-fluid model to explain the origin of this type of deposit. In their model, they envisaged the mixing of a metal-rich, S-poor basinal fluid with a S-rich fluid at the site of sulphide formation within carbonate rocks. More recently, Anderson and McQueen (1982) have argued on the basis of ore textures and thermodynamic constraints that the hydrothermal fluids responsible for the formation of some MVT deposits are unlikely to carry substantial reduced S contents in the presence of base metals. Goodfellow (1987), in a study of SEDEX deposits of the Selwyn Basin, suggested that reduced S in MVT deposits may also have been derived from an ambient reduced water column. This would account for a second S-rich fluid postulated by Anderson and McQueen (1982), and explain the heavy S isotopic values in many MVT deposits, the spatial association of some MVT deposits with the rifted margins of anoxic basins such as the Selwyn Basin (Fig. 1), and the temporal overlap of Phanerozoic SEDEX and MVT deposits (Goodfellow, 1987; Sangster, 1990). Furthermore, a reduced lower water column in adjacent rift-controlled sedimentary basins would be expected to reflux through relatively porous and permeable reef complexes, and thereby provide an inexhaustible supply of reduced sulphur to precipitate metals in hydrothermal fluids migrating upwards along reactivated extensional structures (Fig. 7). According to this model, SEDEX and MVT deposits are both coeval and cogenetic with the former precipitated from vent fluids at or just below the seafloor and the latter formed below the paleo-seafloor within carbonate platformal sequences from metalliferous fluids originating in a common hydrothermal reservoir (Goodfellow et al., 1993).

Fluid Flow

Several models have been proposed to explain the upflow of hydrothermal fluids to the seafloor of sedimentary basins hosting SEDEX deposits. They may be broadly divided into two end-member categories: 1) static reservoir models, and 2) models that involve convection cells.

The stratal aquifer model formulated by Lydon (1983) and applied to specific SEDEX deposits (e.g. Lydon, 1986) identifies the optimal conditions for generating metalliferous hydrothermal fluids. A synrift clastic sequence capped by impervious basinal shales is an ideal setting for a geothermal reservoir regardless of the heat source. The low thermal conductivity and low permeability of the capping argillaceous sediments would reduce both conductive and convective heat loss, thereby establishing elevated temperatures within the permeable reservoir and sustaining the hydrothermal system for a prolonged time period. The low permeability of the argillaceous cap retards dewatering during compaction of the lower arenaceous sequence, thereby preserving high porosities and promoting geopressuring of the hydrothermal reservoir. The discharge of ore-forming solutions to the seafloor takes place when and where the cap rock is ruptured by extensional faulting.

The ore fluids in this model have been termed formational waters, basinal brines, or connate waters. Their essential characteristic is that they are an integral part of the sedimentary sequence, and have a long residence time in the reservoir zone. As mathematically modeled by Cathles and Smith (1983), discharge from such a geopressured hydrothermal system would tend to be episodic over a long period of time, due to the cyclic pore fluid pressure increase and release during compaction, and fluid discharge would coincide with tectonic movement on faults.

The main proponents for a convection cell model are Russell (1983) for the Irish deposits and more recently Yang et al. (2004) for the McArthur River deposit. Russell's model assumes a downward propagating convective cell along extensional faults that mine the regionally elevated heat in the rifted continental crust. The hydrothermal fluids consist of heated seawater that has reacted with sialic rocks along the flow paths of the convection cell. Yang et al. (2004) envisage the sinking of cold saline and dense seawater along basin margin extensional faults and the upflow along faults with discharge of heated metalliferous and saline fluids into a local fault-controlled basin.

Siting of the Hydrothermal Fluids

The spatial association of SEDEX deposits with syndepositional faults has been well documented for many SEDEX deposits. In the Selwyn Basin, late Devonian basinal strata are characterized by syndepositional tectonism. In the MacMillan Pass area, the Tom and Jason SEDEX deposits occur within an extensional basin or rhombochasm, 20 km long by 7 km wide, possibly related to strike-slip movement on west-trending faults (Fig. 9). Evidence for syndepositional faulting consists of mud-clast breccia and diamictite, rapid change of facies and thickness of stratigraphic units, mineralized faults with a soft sediment deformation, changes in the level of erosion below a Devonian unconformity, and marked changes in the style and orientation of Mesozoic structures.

Both the Tom and Jason deposits are spatially and temporally associated with extensional faults that played an important role in the siting of hydrothermal vents. The Jason SEDEX deposit is interbedded with and overlain by a sequence of diamictite, siltstone, and shale that is 200 m thick (Turner, 1990). The Jason fault that bounds the orebodies is interpreted as a syndepositional structure because diamictite lobes thicken and coarsen toward it, a breccia pipe and the vent complex of the deposit occur along it, and the fault displays syndepositional textures. The nearby Tom deposit also occurs adjacent to the north-trending Tom fault and overlies diamictite, resedimented conglomerate derived from the Tom fault, and slumped interbedded argillite and sandstone (Goodfellow and Rhodes, 1990).

Another possible control on the siting of hydrothermal discharge is coeval magmatism. As noted earlier, there is a temporal and spatial association of mafic magmatism with some SEDEX deposits in the Selwyn Basin (e.g. MacMillan Pass District, Fig. 9). Conductive heat loss from magmas accumulated at the base of attenuated crust is probably the main source of anomalously high heat flow in intracontinental rift environments and, although its role is much less direct than for volcanogenic massive sulphide deposits, deep-seated magmatism may be important to the formation of major SEDEX deposits.

Longevity of Hydrothermal Systems

Hydrothermal systems responsible for the formation of many SEDEX deposits appear to be long-lived (Goodfellow and Lydon, 2007). The basal contact between the stratiform sulphides and underlying sedimentary rocks is commonly sharp, whereas the top of the sulphide zone grades into an overlying sequence consisting of lower temperature hydrothermal sediments such as barite, phosphate, silica, carbonate, and pyrite. This change in the composition of hydrothermal sediments with time suggests that the hydrothermal system decayed slowly over a protracted time period.

At the Howards Pass deposit, the onset of hydrothermal activity is marked by an abrupt change from graptolitic shales to hydrothermal carbonate containing laminated sulphides. The mineralized zone spans two conodont zones, indicating that it formed over an interval of several million years (Goodfellow et al., 1986). The sulphides are overlain by 50 metres of Zn-rich phosphatic chert, which continued to form for millions of years after the sulphides were deposited. The phosphatic chert is in turn overlain by Ba-rich Middle to Late Silurian bioturbated mudstone, and Middle to Late Devonian stratiform barite. The long-lived hydrothermal system at Howards Pass has led to the formation of stacked deposits of base metals, barite, and apatite from Early Silurian to Late Devonian, indicating several stages of reactivation.

The sedimentary rocks overlying the Jason and Tom deposits are also enriched in barite, sphalerite, and pyrite that extend for up to 500 m into the stratigraphic hanging wall.

Timing of Sulphide Formation

Laminated or bedded sulphides associated with most SEDEX deposits are interpreted as seafloor hydrothermal sediments. Evidence for a sedimentary origin includes the following:

1. Sedimentary textures, such as compositional layering that parallels bedding in the enclosing sediments, and graded and crosslaminated sulphide minerals.
2. Hydrothermal products that have clearly been reworked by sedimentation. Fault scarp sulphide breccias composed of clasts of stratiform and vent facies at the Jason deposit clearly support a synsedimentary origin (Turner, 1990).
3. Feeder pipes that are restricted to the stratigraphic interval containing the stratiform sulphide zone and/or to sedimentary units directly underlying the deposits. This stratigraphic relationship indicates that the time of hydrothermal activity corresponds to the time of formation of the sulphide deposit, and that the sulphides must be synsedimentary in origin, otherwise the feeder zone would have continued into the overlying sedimentary sequence.
4. The asymmetric stratigraphic zonation of ore-forming and ore-associated hydrothermal elements, and sulphur isotopes, in vent-proximal deposits that is not consistent with a replacement origin.
5. Faunal populations in interbedded sedimentary rocks that are different from those in the enclosing rocks and indicate an unusual local environment at the time of sulphide formation (e.g. stunted fauna associated with the Howards Pass, Goodfellow et al., 1986).
6. Sulphur isotope values in layered sulphides that are the same as those for pyrite from coeval sedimentary rocks remote from mineralization thus supporting an ambient water column source of sulphur and, therefore, a sedimentary origin for metal sulphides (Goodfellow, 1987).

On a microscopic scale, the form and distribution of sulphide minerals reflect the overprinting effects of deformation and metamorphism (McClay, 1983). Even in weakly deformed and metamorphosed (subgreenschist) deposits, such as the Howards Pass (XY), the sulphides have recrystallized, and some are locally mobilized by pressure dissolution and transposed parallel to the principal cleavage direction (Jonasson and Goodfellow, 1986). Pyrite framboids have been variably infilled, replaced, and overgrown by pyrite and sphalerite to produce atoll replacement textures. Although the compositional layering in SEDEX deposits reflects depositional processes, the detailed form and distribution of most sulphide minerals in ancient deposits is probably controlled by secondary processes.

Fluid Behaviour at the Seafloor

Based on the density-temperature curves of Sato (1972), and the filling temperatures and salinities of fluid inclusions at the Tom and Jason deposits (Ansdell, 1985; Gardner and Hutcheon, 1985), the hydrothermal fluids that formed these deposits probably behaved as buoyant plumes, at least initially (Fig. 18). The same fluid inclusion data indicate that the minimum water depths necessary to prevent boiling

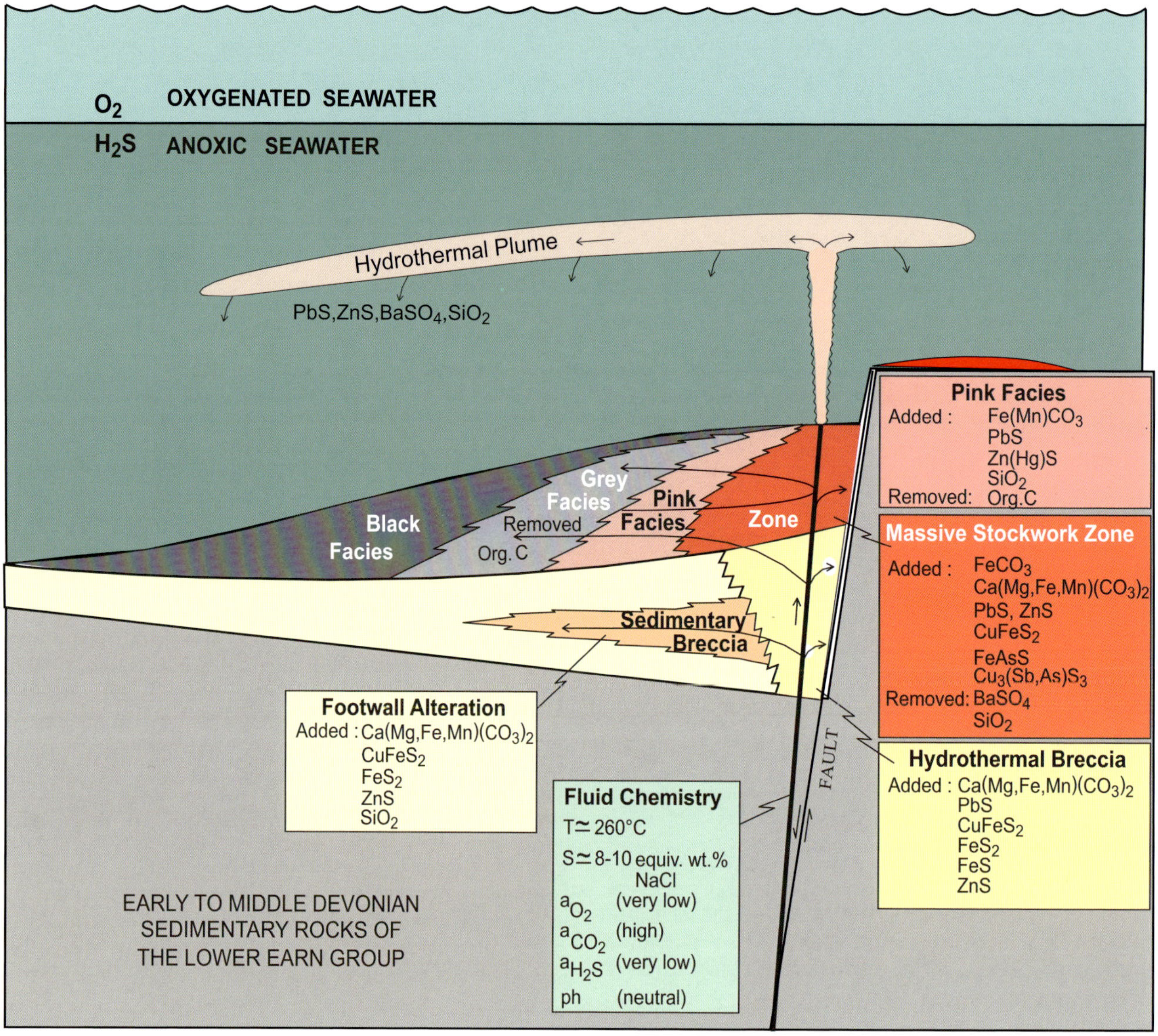

FIGURE 18. Genetic model for the Tom deposit, MacMillan Pass, Yukon (from Goodfellow and Rhodes, 1990).

range between 250 and 800 m for the Tom and Jason deposits (Haas, 1971).

In the case of the Howards Pass deposits, there are several lines of evidence that suggest that these deposits formed from brines that accumulated in local third-order basins (Fig. 19). This evidence includes the sheet-like morphology that thickens in the centre of the deposit and thins gradually toward the margins, the delicately laminated textures, and the lack of a pronounced zoning that characterize vent-proximal deposits such as Tom and Jason. Also, the occurrence of several deposits near Howards Pass that are stratigraphically, texturally, mineralogically, and geochemically indistinguishable indicates one major hydrothermal event with many pulses. Furthermore, these bottom-hugging fluids flowed into several third-order subbasins within an elongated northeast-southwest second-order basin more than 35 km in length controlled by extensional faults oriented parallel to the hinge line separating the Selwyn Basin from the Mackenzie carbonate platform to the east.

Exploration Methods

Geological Methods

There are several geological attributes that are clearly important for the formation of SEDEX deposits in the Selwyn Basin; these are summarized below.

Timing: SEDEX deposits formed during three major events in the Selwyn Basin – Late Cambrian, Early Silurian, and Late Devonian. The Middle Ordovician is a fourth relatively minor event during which the Vulcan deposit formed.

Seafloor environment: All SEDEX deposits are hosted by black to dark grey basinal facies sedimentary rocks consisting of mostly shale and chert.

Synsedimentary faulting: All deposits formed in what are interpreted as local second- and third-order basins, and many are associated with penecontemporaneous extensional faults and associated fault-scarp breccias.

Volcanism: All Selwyn Basin SEDEX deposits are temporally related to mafic magmatism, and most Cambrian and Late Devonian deposits are both temporally and spatially associated with mafic volcanic rocks.

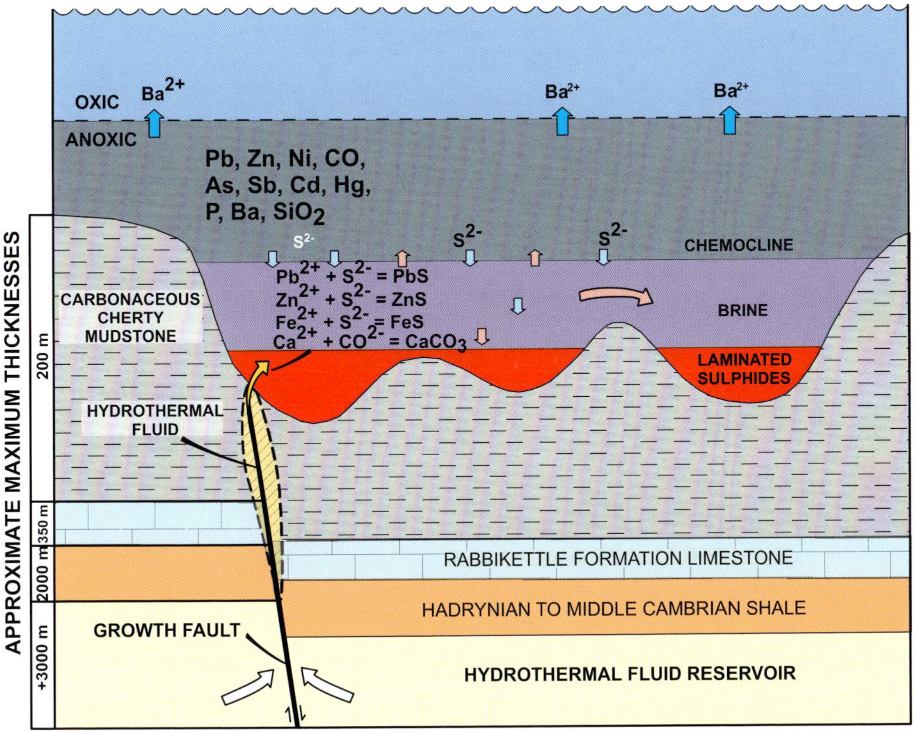

FIGURE 19. Genetic model for the Howards Pass (XY) deposit, MacMillan Pass, Yukon (from Goodfellow, 2004).

increase of Zn, Pb, Cu, Ni, Co, As, Sb, Se, Cd, and Hg with proximity to the XY deposit. Analyses of sulphide separates show that Ni, Cu, Co, As, and Sb reside mostly in pyrite, whereas Zn, Cd, and Hg are bound by sphalerite. Copper and Ni in pyrite separates range up to 2000 ppm in units hosting the XY deposit, compared to less than 500 ppm in coeval rocks remote from mineralization. Therefore, the rocks hosting the XY deposits are enriched not only in pyrite but a pyrite that is anomalously enriched in Ni, Cu, Co, As, and Sb (Goodfellow et al., 1983).

Middle to Late Silurian bioturbated mudstone is unusual in its Ba content, which is rarely less than 5000 ppm in the Selwyn Basin. Barium is also laterally zoned ranging up to an average value of 2.8 wt.% in bioturbated mudstone in DDH 99 that intersects the XY deposit near the centre of the subbasin, where the Ba occurs as barite and hyalophane. Barium is also enriched in Earn Group black clastics that host Zn-Pb-barite and barren barite deposits at MacMillan Pass and Gataga districts. Barite in this unit in the XY deposit area is reflected by elevated Ba contents in section XY0179 (Goodfellow, 1984).

The carbonaceous mudstone underlying the Howards Pass (XY) deposit is highly enriched in K compared to coeval rocks remote from mineralization. K-feldspar/illite ratios in this unit at the XY deposit is up to five times greater than in

Related deposit types: Many SEDEX deposits are temporally and spatially associated with stratiform barite and sedimentary phosphatic deposits, and cogenetic Mississippi Valley-type deposits that formed in the adjacent carbonate platform.

Lithogeochemical Methods

Rocks hosting the Howards Pass deposits are characterized by widespread geochemical halos that extend several kilometres laterally from the mineralized zone and up to 100s of metres into the stratigraphic hanging wall (Fig. 20). One of the characteristics of the Howards Pass deposits is that, although the Pb-Zn mineralization is of Early Silurian age, there is clear evidence in the sedimentary rocks overlying the XY deposit that hydrothermal activity continued until the Late Devonian, producing extensive postore geochemical halos.

The pyrite content of shale and chert hosting Howards Pass (XY) deposit shows a systematic increase towards Zn-Pb mineralization. This is reflected by total S that increases with proximity to mineralization in most units hosting the deposit, particularly the footwall carbonaceous shale and overlying phosphatic chert (Fig. 20). The increase in S and pyrite is accompanied by an

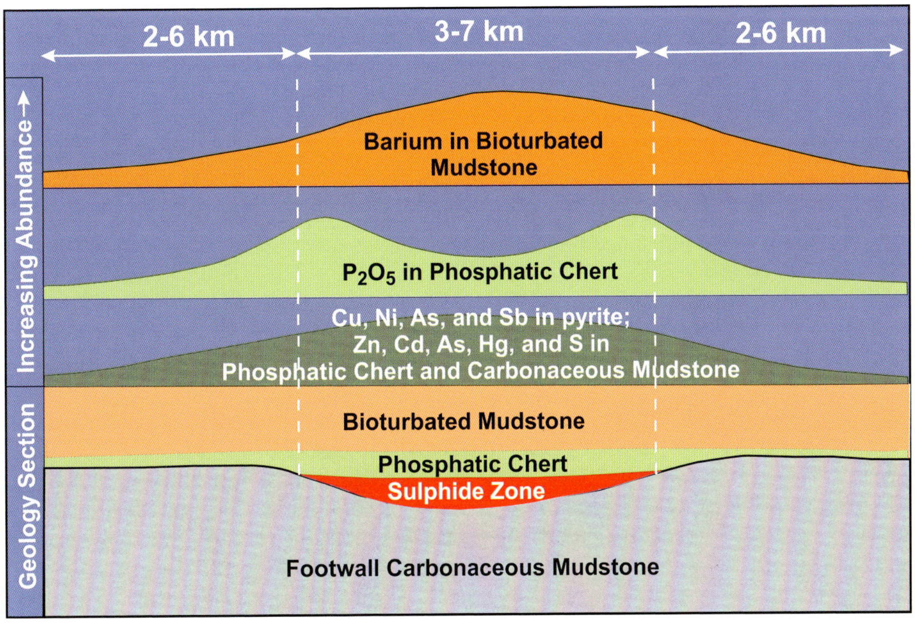

FIGURE 20. Lateral and vertical zonation of elements and minerals about the Howards Pass (XY) deposit, Selwyn Basin, Yukon (from Goodfellow, 2004).

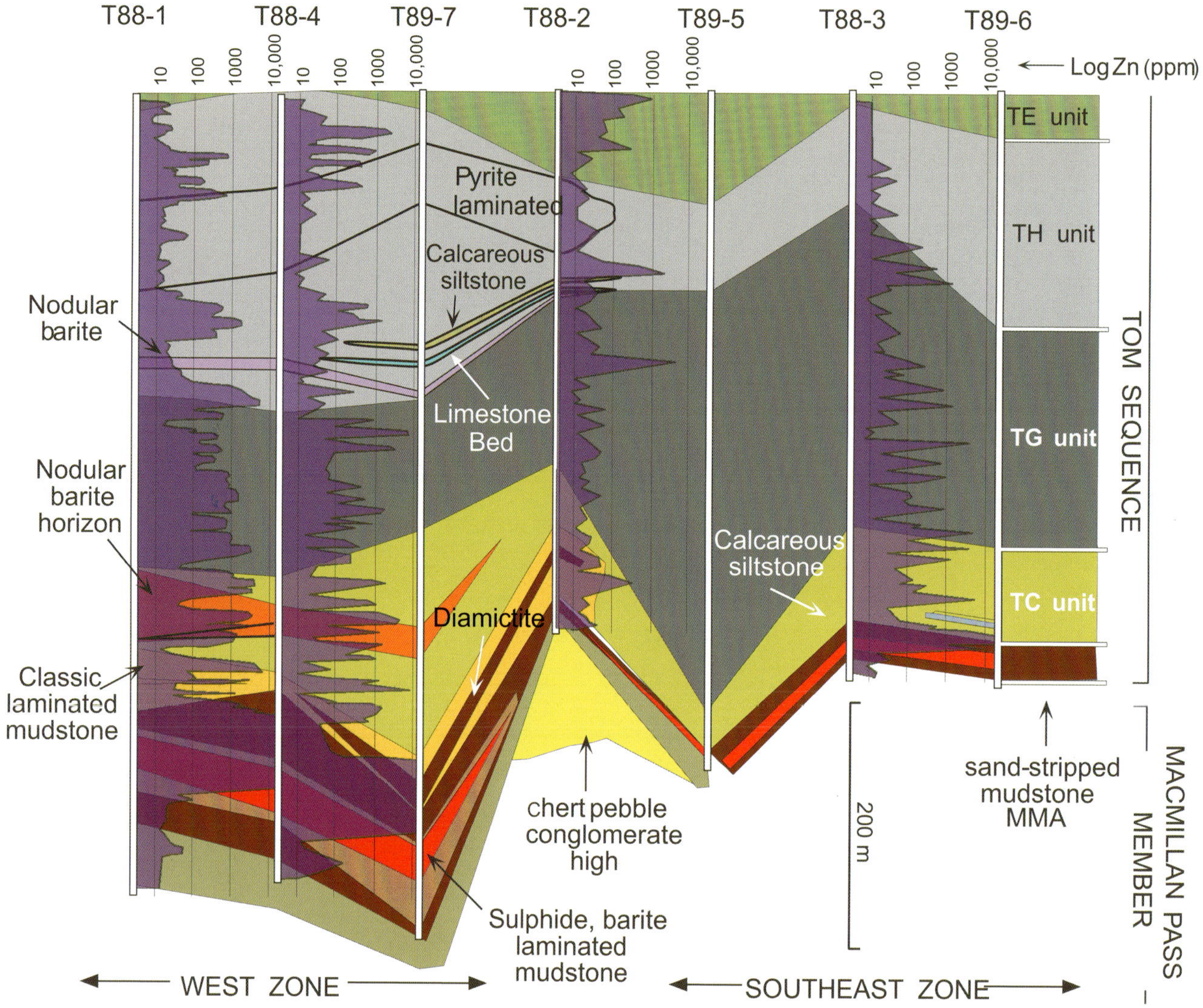

FIGURE 21. Zinc content of core from drillholes spanning the Tom Sequence plotted on a longitudinal section of the West and Southeast Zones (based on analyses of samples taken every 3 or 4.5 m) (from Goodfellow and Rhodes, 1990).

correlative rocks from section XY0279 situated remote from mineralization. Most of the K in this unit is bound in illite, microcline, and hyalophane. Microprobe analyses of feldspar that commonly total less than 100% suggest that NH_4^+ or some other volatile compound is present. The NH_4^+ content of the footwall carbonaceous mudstone ranges from 0.17 to 0.26 wt.%. This footwall mudstone is also enriched in V and TOC (total organic carbon), which display a strong positive correlation. Unlike K, however, V and TOC are enriched in this unit throughout the Selwyn Basin, and do not show any systematic lateral variation with proximity to the Howards Pass deposits.

Postore Geochemical Halos

The Howards Pass deposits are overlain by a black, organic carbon-rich (average TOC = 8 wt.%) chert containing large carbonate concretions (Fig. 15E) and fluorapatite beds (Fig. 15F). The carbonate concretions range up to 1 m in diameter, are both concentrically and radially zoned, and are most abundant in the area of mineralization. The carbonate-fluorapatite is thinly bedded and typically pyritic.

The phosphatic chert unit reaches thicknesses of 50 m near the centre of the XY subbasin and thins to less than 5 m in the Summit Lake area remote from the deposit. This change in thickness is accompanied by a decrease in the content of P_2O_5 from average values approaching 6 wt.% in DDH 42 to values of 1.2 wt.% in section SL7980 remote from mineralization. The carbonate-fluorapatite is interpreted as a hydrothermal product because of its close spatial and temporal association with the Howards Pass deposits (Goodfellow, 1984). Pyrite from the phosphatic chert contains elevated contents of Ni, As, Sb, and Se relative to pyrite remote from mineralization. Most of the Ni resides in Py with values up to 5449 ppm although millerite has also been identified (Goodfellow et al., 1986).

The Tom deposit is overlain by a thick sequence of mostly mudstone and chert of the Tom Sequence. These sediments are enriched in nodular barite, laminated pyrite, and sphalerite relative to coeval Devonian rocks remote from mineralization. The high sphalerite content is reflected by Zn values that exceed 1000 ppm up to 300 m above the Tom West and Southwest sulphide zones (Fig. 21), whereas rocks over-

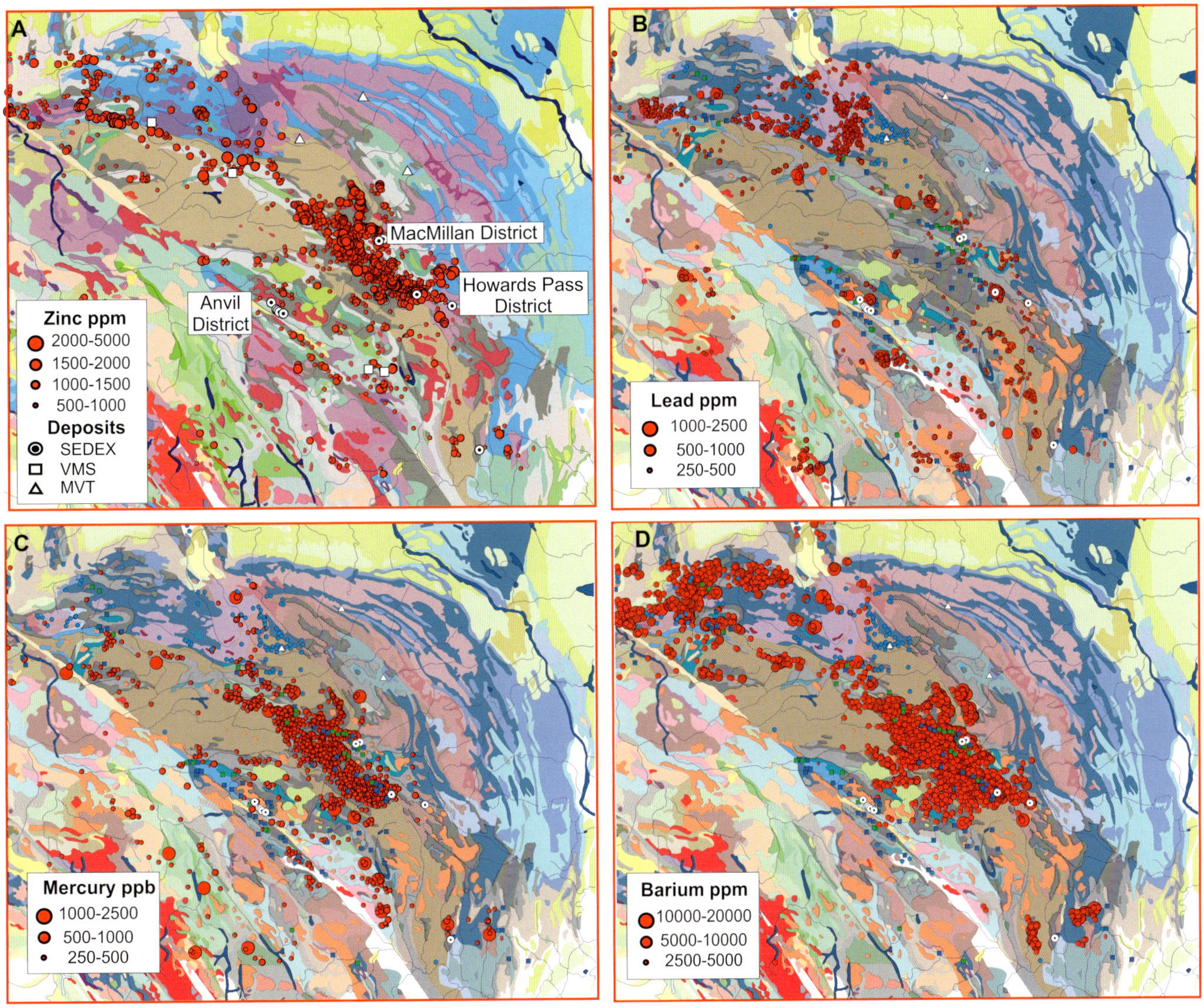

FIGURE 22. Proportional circle maps showing the distribution of **(A)** Zn, **(B)** Pb, **(C)** Hg, and **(D)** Ba in sediments from streams draining the Selwyn Basin, Yukon.

lying the conglomerate high separating these zones display background values of less than 100 ppm.

The hydrothermal systems at the Tom and Howards Pass deposits were long-lived, reaching a maximum during the formation of the bedded sulphide ores and decaying gradually over the millions of years represented by the overlying cherty mudstones and cherts. As a result, the hanging-wall sequence is characterized by intervals of laminated pyrite, phosphate, and barite. Zinc contents, which range up to thousands of ppm, are significantly elevated compared to background shales from the Selwyn Basin, and probably formed by the discharge of low-temperature hydrothermal fluids into the basin. From an exploration viewpoint, the mineralogically and geochemically anomalous sedimentary rocks overlying the SEDEX deposits of the Selwyn Basin represent a large exploration target.

Stream and Seep Geochemistry

Of the ore-forming elements, Zn is the most widespread, whereas Pb is more spatially restricted to the immediate area of the sulphide deposits. This is shown by the distribution of Zn (Fig. 22A) and Pb (Fig. 22B) in sediment from streams draining SEDEX deposits in the Selwyn Basin. Mercury (Fig. 22C) is distributed similar to Zn, owing to its residence in sphalerite, whereas Ba (Fig. 22D) is concentrated in areas underlain by Devonian shales and cherts of the Earn Group and Early to Middle Silurian orange-weathering mudstone of the Road River Group. Both the Howards Pass deposits (XY, Anniv, and OP) and MacMillan Pass deposits (Tom and Jason) are reflected by stream sediments that are highly enriched in Zn, Pb, Hg, and Ba.

The widespread distribution of Zn is controlled by two major factors: 1) laterally extensive Zn enrichment in Road River shales that host the Howards Pass deposits (Goodfellow et al., 1983); and 2) the high solubility of Zn in alkaline streams draining calcareous shales of the Road River Formation. Lead is more spatially restricted to the immediate area of the sulphide deposits because Pb has a very low solubility in alkaline stream waters and is therefore dispersed from mineralized zones as clastic grains. This is consistent with the low Pb but high Zn content of groundwater seeps at the Howards Pass and Anniv deposits (Goodfellow, 1989).

FIGURE 23. Lime-green zinc-rich moss overlying the Howards Pass (XY) deposit, Selwyn Basin, Yukon.

A secondary Zn zone consisting mostly of hemimorphite, smithsonite, and wurtzite (Jonasson et al., 1987) with associated lime-green Zn-rich moss overlies the mineralization at the XY deposit (Fig. 23). Secondary Zn mineralization was also discovered in a stream west of Anniv where both the sediments and water are highly enriched in Zn (Goodfellow, 1989).

Geophysics

Airborne geophysics has not been extensively used in the exploration for SEDEX deposits of the Selwyn Basin. The paucity of magnetic minerals, such as pyrrhotite and magnetite, in most deposits renders magnetic methods ineffective, and electromagnetic methods are hampered by the high conductivity of black shales. Ground gravity has proven effective in the Anvil District in locating buried deposits along strike from the Faro deposit (Brock, 1973).

Potential for Known and New Deposit Types

The potential for SEDEX and other deposit-types in the Selwyn Basin is considered high. There are several areas of high potential:

- Buried SEDEX deposits in existing mining districts, e.g., MacMillan Pass graben, Anvil District, and Howards Pass Zn corridor.
- SEDEX deposits outside known districts, particularly in areas drained by streams that are highly anomalous in Zn and Pb, and SEDEX-associated elements such as Hg and Ba (Fig. 22).
- Nickel-PGE deposits in Devonian black shales that are correlative with the NICK deposit.

Summary and Conclusions

- There are three major ages of SEDEX deposits in the Selwyn Basin: 1) Late Cambrian, 2) Early Silurian, and 3) Late Devonian.
- SEDEX deposits of the Selwyn Basin occur within fault-bounded grabens that formed during episodes of rifting of continental margins.
- There is a close temporal relationship, and in some cases spatial relationship, between SEDEX deposits and mafic alkali volcanic rocks. These volcanic rocks formed from low volume magma discharge from the mantle along deep-seated structures.
- Sedimentary rocks hosting most SEDEX deposits consist of autochthonous carbonaceous chert, shale, siltstone, carbonates, and fault-scarp sedimentary breccias and related clastic rocks. Highly carbonaceous nonbioturbated laminated sediments, the absence of benthic fauna, anomalously high S/C ratios, and upwardly increasing positive $\delta^{34}S$ trends in pyrite in ambient sedimentary rocks all suggest that SEDEX deposits of the Selwyn Basin formed during periods when the oceans were stratified with reduced H_2S-rich bottom waters.
- Most SEDEX deposits formed within half-grabens near the down-faulted margins (if the hydrothermal fluid was buoyant), or in nearby bathymetric depressions (if the fluid behaved as a bottom-hugging brine). Hydrothermal vents are typically located along extensional faults.
- There are two types of SEDEX deposits in the Selwyn Basin – vent-proximal and vent-distal. Vent-proximal deposits are mound-shaped, intensely vertically and laterally zoned, and characterized by rapid thickness changes away from the vent. Vent-distal deposits tend to be more uniform in thickness, laterally widespread, delicately bedded, and not strongly zoned.
- Most SEDEX deposits are composed of two major hydrothermal facies, a sedimentary-hydrothermal facies and a vent facies. The absence of a vent facies in some deposits, e.g., Howards Pass, is probably due to either poor preservation of the vent complex or the accumulation of metalliferous brines and sulphide products in depressions remote from the vent sites.
- Sedimentary textures (e.g. graded and cross-laminated beds, sulphide sedimentary breccias, and slump structures), the asymmetrical zonation in stratiform sulphides, the occurrence of vent complexes below the sulphide zone, the growth of limited and stunted faunal populations during sulphide formation, and the similarity of $\delta^{34}S$ values in coeval pyrite within and remote from SEDEX deposits all indicate that most of the sulphides were deposited as sediments on the seafloor.
- The hydrothermal systems that generate SEDEX deposits in the Selwyn Basin are typically long-lived, build to a temperature maximum quickly, and decay slowly over intervals that may span millions of years.
- Mineral assemblages, fluid inclusions, and stable isotope data indicate that most SEDEX deposits formed from Zn-Pb (± Cu, Fe, Ag, Au, Ba)-rich, low to moderate temperature, weakly to highly saline basinal fluids that were generated by reaction with Windermere clastic sediments eroded from continental crust. The reduced S contents of these fluids were typically depleted.

Knowledge Gaps

The following is a summary of the key knowledge gaps that are considered important in the discovery of new SEDEX deposits in the Selwyn Basin and the steps that need to be taken to close them.

Geology of the Selwyn Basin: Targeted geological mapping, stratigraphy, and geochronology is needed in areas that are poorly understood to define host sequences with mineral potential, local second- and third-order basins, and synsedimentary extensional faults.

Geochemistry: Targeted stream and seep geochemistry should be undertaken using highly sensitive methods of analyzing water for SEDEX-associated elements and isotopes.

Systematic and comprehensive stratigraphic geochemistry of sedimentary sequences that have a high potential for buried SEDEX deposits need to be completed.

Geophysics: Deeply penetrating ground gravity survey should be completed in SEDEX districts with a high potential for concealed deposits.

Ore Genetic Studies: Better understanding of the hydrothermal architecture and hydrology, sources of metals and reduced sulphur, focusing mechanism, and seafloor depositional processes is needed. Also, new vectors to ore should be established by mapping the mineralogical and chemical halos lateral to, below, and above known SEDEX deposits.

Acknowledgements

I wish to thank Ian Jonasson and Suzanne Paradis for thorough reviews of this paper. Ingrid Kjarsgaard is thanked for checking and editing an earlier version.

References

Abbott, J.G., and Turner, R.J.W., 1991, Character and paleotectonic setting of Devonian stratiform sediment-hosted Zn, Pb, Ba deposits, Macmillan fold belt, Yukon, *in* Abbott, J.G., and Turner, R.J.W., eds., Mineral Deposits of the Northern Canadian Cordillera, Yukon, Northeastern British Columbia (Field Trip 14): Geological Survey of Canada, Open File 2169, p. 99-136.

Anderson, G.M., 1973, The hydrothermal transport and deposition of galena and sphalerite near 100ºC: Economic Geology, v. 68, p. 480-492.

Anderson, G.M., and McQueen, R.W., 1982, Ore deposit Models 6, Mississippi Valley-type lead-zinc deposits: Geoscience Canada, v. 9, p. 108-117.

Ansdell, K.M., 1985, Fluid Inclusion and Stable Isotope Study of the Tom Ba-Pb-Zn Deposit, Yukon Territory, Canada: M.Sc. thesis, University of Alberta, Edmonton, Alberta, 123 p.

Ansdell, K.M., Nesbitt, B.E., and Longstaffe, F.J., 1989, A fluid inclusion and stable-isotope study of the Tom Ba-Pb-Zn deposit, Yukon Territory, Canada: Economic Geology, v. 84, p. 841-856.

Arthur, M., and Sageman, B.B., 1994, Marine shales: Depositional mechanisms and environments of ancient deposits: Annual Review of Earth and Planetary Sciences, v. 22, p. 499-551.

Bailes, R.J., Smee, B.W., Blackadar, D.W., and Gardner, H.D., 1986, Geology of the Jason lead-zinc-silver deposits, Macmillan Pass, eastern Yukon, *in* Morin, J.A., ed., Mineral Deposits of Northern Cordillera: Canadian Institute of Mining and Metallurgy,. Special Volume 37, p. 87-99.

Baker, V.R., 1997, Megafloods and glaciation, *in* Martini, I. P., ed., Late Glacial and Postglacial Environmental Changes; Quaternary, Carboniferous-Permian, and Proterozoic: Oxford University Press, New York, p. 98-108.

Beales, F.W., and Jackson, S.A., 1966, Precipitation of lead-zinc ores in carbonate reservoirs as illustrated by Pine Point ore field, Canada: Transactions of the Institution of Mining and Metallurgy, v. B75, p. 278-285.

Betts, P.G., Giles, D., and Lister, G.S., 2003, Tectonic environment of shale-hosted massive sulfide Pb-Zn-Ag deposits of Proterozoic northeastern Australia: Economic Geology, v. 98, p. 557-576.

Blount, C.N., 1977, Barite solubilities and thermodynamic properties up to 300ºC and 1400 bars: American Mineralogist, v. 62, p. 942-957.

Bond, C.B., Kominz, M.A., and Grotzinger, J.P., 1988, Cambro-Ordovician eustacy: Evidence from geophysical modelling of subsidence in Cordilleran and Appalachian passive margins, *in* Kleinspehn, K.L., and Paola, C., eds., New Perspectives in Basin Analysis, Frontiers in Sedimentary Geology: Springer-Verlag, New York, p. 129-160.

Bradley, D.C., and Leach, D.L., 2003, Tectonic controls of Mississippi Valley-type lead-zinc mineralization in orogenic forelands: Mineralium Deposita, v. 38, p. 652-667.

Brock, J.S., 1973, Geophysical exploration leading to the discovery of the Faro deposit: Canadian Mining and Metallurgical Bulletin, v. 66, p. 97-116.

Carpenter, A.B., Trout, M.L., and Pickett, E.E., 1974, Preliminary report on the origin and evolution of lead- and zinc-rich oil field brines in central Mississippi: Economic Geology, v. 69, p. 1191-1206.

Cathles, L.M., and Smith, A.T., 1983, Thermal constraints on the formation of Mississippi Valley-type lead-zinc deposits and their implications for episodic basin dewatering and deposit genesis: Economic Geology, v. 78, p. 983-1002.

Cecile, M.P., 1982, The Lower Paleozoic Misty Creek Embayment, Selwyn Basin, Yukon and Northwest Territories: Geological Survey of Canada, Bulletin 335, 78 p.

Christensen, J.N., Halliday, A.N., Leigh, K.E., Randell, R.N., and Kesler, S.E., 1995, Direct dating of sulfides by Rb-Sr; a critical test using the Polaris mississippi valley-type Zn-Pb deposit: Geochimica et Cosmochimica Acta, v. 59, p. 5191-5197.

Claypool, G.E., Holser, W.T., Kaplan, I.R., Sakai, H., and Zak, I., 1980, The age curves of sulfur and oxygen isotopes in marine sulfate and their mutual interpretation: Chemical Geology, v. 28, p. 3-4.

Cooke, D.R., Bull, S.W., Large, R.R., and McGoldrick, P.J., 2000, The importance of oxidized brines for the formation of Australian Proterozoic stratiform sediment-hosted Pb-Zn (Sedex) deposits: Economic Geology, v. 95, p. 1-18.

Deklerk, R.C., 2003, Yukon MINFILE: Whitehorse, Canada, Yukon Geological Survey.

Eisbacher, G.H., 1981, Sedimentary Tectonic and Glacial Record in the Windermere Supergroup, Mackenzie Mountains, Northwestern Canada, Geological Survey of Canada, Paper 80-27, 40 p.

Fischer, A.G., 1984, Biological innovations and the sedimentary record, *in* Holland, H.D., and Trendall, A.F., eds., Patterns of Change in Earth Evolution: Springer-Verlag, Berlin, p. 145-157.

Gardner, H.D., and Hutcheon, I., 1985, Geochemistry, mineralogy and geology of the Jason Pb-Zn deposits, Macmillan Pass, Yukon, Canada: Economic Geology, v. 80, p. 1257-1276.

Garven, G., 1984, The role of regional flow in the genesis of the Pine Point deposit, Western canada Sedimentary Basin: Economic Geology, v. 80, p. 307-324.

Godwin, C.I., and Sinclair, A.J., 1982, Average lead isotope growth curves for shale-hosted zinc-lead deposts: Economic Geology, v. 77, p. 675-690.

Goodfellow, W.D., 1984, Geochemistry of rocks hosting the Howards Pass (XY) strata-bound Zn-Pb deposit, Selwyn Basin, Yukon Territory, Canada: Proceedings of the Sixth Quadrennial IAGOD Symposium: Stuttgart, Germany, E. Schweizerbart'sche Verlagsbuchhandlung (Nägele u. Obermiller), p. 91-112.

——— 1985, Geochemistry unveils "blind" deposits: GEOS, v. 14, p. 17-20.

——— 1987, Anoxic stratified oceans as a source of sulphur in sediment-hosted stratiform Zn-Pb deposits (Selwyn Basin, Yukon, Canada): Chemical Geology, v. 65, p. 359-382.

——— 1989, Interpretation of stream geochemistry leading to the discovery of a secondary zinc deposit, Pelly River, Nahanni map area, Yukon, *in* Current Research 98-1E: Geological Survey of Canada, p. 31-50.

——— 2004, Geology, genesis and exploration of SEDEX deposits, with emphasis on the Selwyn Basin, Canada, *in* Deb, M., and Goodfellow, W.D., eds., Sediment-Hosted Lead-Zinc Sulphide Deposits: Attributes and Models of Some Major Deposits of India, Australia and Canada: Narosa Publishing House, Delhi, India, p. 24-99.

Goodfellow, W.D., and Jonasson, I.R., 1984, Ocean stagnation and ventilation defined by $\delta^{34}S$ secular trends in pyrite and barite, Selwyn Basin, Yukon: Geology, v. 12, p. 583-586.

Goodfellow, W.D., and Lydon, J.W., 2007, Sedimentary-exhalative (SEDEX) deposits, *in* Goodfellow, W. D., ed., Mineral Deposits of Canada: A Synthesis of Major Deposit-types, District Metallogeny, the Evolution of Geological Provinces, and Exploration Methods: Geological Association of Canada, Mineral Deposits Division, Special Publication 5, p. 163-183.

Goodfellow, W.D., and Rhodes, D., 1990, Geological setting, geochemistry and origin of the Tom stratiform Zn-Pb-Ag-barite deposits, *in* Abbott, J.G., and Turner, R.J.W., eds., Mineral Deposits of the Northern Canadian Cordillera, Yukon, Northeastern British Columbia (Field Trip 14): Geological Survey of Canada, Open File 2169, p. 177-244.

Goodfellow, W.D., Jonasson, I.R., Morganti, J.M., and Parslow, G.R., 1983, Zonation of chalcophile elements about the Howard's Pass (XY) Zn-Pb deposit, Selwyn Basin, Yukon. in Geochemical Exploration 1982: Journal of Geochemical Exploration, v. 19, p. 503-542.

Goodfellow, W.D., Jonasson, I.R., and Morin, J.A., 1986, Environment of formation of the Howards Pass (XY) Zn-Pb deposit, Selwyn Basin, Yukon, *in* Morin, J.A., ed., Mineral Deposits of Northern Cordillera: Canadian Institute of Mining and Metallurgy, Special Volume 37, p. 19-50.

Goodfellow, W.D., Lydon, J.W., and Turner, R.W., 1993, Geology and genesis of stratiform sediment-hosted (SEDEX) Zn-Pb-Ag sulphide deposits, *in* Kirkham, R.V., Sinclair, W.D., Thorpe, R.I., and Duke, J.M., eds., Mineral Deposit Modeling: Geological Association of Canada, Special Paper 40, p. 201-251.

Goodfellow, W.D., Cecile, M.P., and Leybourne, M.I., 1995, Geochemistry, petrogenesis and tectonic setting of lower Paleozoic alkalic and potassic volcanic rocks, northern Canadian Cordilleran Miogeocline: Canadian Journal of Earth Sciences, v. 32, p. 1236-1254.

Gordey, S.P., 1981, Geology of the Nahanni map-area, Yukon Territory and District of Mackenzie: Geological Survey of Canada, Open File 780, 7 maps, scale 1:50 000.

Gordey, S.P., Abbott, J.G., Tempelman, K.D.J., and Gabrielse, H., 1987, "Antler" clastics in the Canadian Cordillera: Geology, v. 15, p. 103-107.

Gordey, S.P., McMillan, N.J., Embry, A.F., and Glass, D.J., 1988, Devono-Mississippian clastic sedimentation and tectonism in the Canadian Cordilleran Miogeocline, *in* Devonian of the World; Proceedings of the Second International Symposium on the Devonian System; Volume II, Sedimentation: Memoir: Canadian Society of Petroleum Geologists, v. 14, p. 1-14.

Gordey, S.P., Makepeace, A.J., and (compilers), 2000, Yukon Digital Geology: Geological Survey of Canada, v. 2 Open File D3826, CD-ROM.

Grapes, K.L., 1987, Lithologic and Textural Study of the Clear Lake Fe-Zn-Pb,Ag Massive Sulphide Deposit: M.Sc. thesis, Carleton University, Ottawa, Ontario, 329 p.

Haas, L., Jr., 1971, The effect of salinity on the maximum thermal gradient of a hydrothermal system at hydrostatic pressure: Economic Geology, v. 66, p. 940-946.

Hanor, J.S., 1996, Controls on the solublization of lead and zinc in basinal brines, *in* Sangster, D.F., ed., Carbonate-Hosted Lead-Zinc Deposits: Economic Geology, Special Publication No. 4, p. 483-500.

Jefferson, C.W., Kilby, D.B., Pigage, L.C., and Roberts, W.J., 1983, The Cirque barite-zinc-lead deposits, northeastern British Columbia, *in* Sangster, D.F., ed., Short Course in Sediment-Hosted Stratiform Lead-Zinc Deposits: Mineralogical Association of Canada, Short Course Handbook No. 8, p. 121-139.

Jennings, D.S., and Jilson, G.A., 1986, Geology and sulphide deposits of the Anvil Range, Yukon, *in* Morin, J.A., ed., Mineral Deposits of Northern Cordillera: Canadian Institute Mining and Metallurgy, Special Volume 37, p. 339-361.

Jonasson, I.R., and Goodfellow, W.D., 1986, Sedimentary and diagenetic textures, and deformation structures within sulphide zone of Howards Pass (XY) Zn-Pb deposit, Yukon and Northwest Territories, *in* Morin, J.A., ed., Mineral Deposits of Northern Cordillera: Canadian Institute of Mining and Metallurgy, Special Volume 37, p. 51-70.

Jonasson, I.R., Goodfellow, W.D., Walker, D.A., Jackson, L.E., Jr., and Anonymous, 1987, Secondary zinc mineralization in post-glacial sediments and talus at Howards Pass, Yukon, Canada; mineralogy, geochemistry and genesis, *in* Augustithis, S. S., ed., Weathering; Its Products and Deposits:Theophrastus Publications, Athens, 30 p.

Leach, D.L., Sangster, D.L., Kelly, K.D., Large, R.R., Garven, G., Allen, C.R., Gutzmer, J., and Walter, S., 2005, Sediment-hosted lead-zinc deposits: a global perspective, *in* Hedenquist, J.W., Thompson, J.F.H., Goldfarb, R.J., and Richards, J.P., eds., One Hundredth Anniversary Volume 1905-2005: Society of Economic Geologist, p. 561-6007.

Lydon, J.W., 1983, Chemical parameters controlling the origin and deposition of sediment-hosted stratiform lead-zinc deposits, *in* Sangster, D.F., ed., Sediment-Hosted Stratiform Lead-Zinc Deposits: Mineralogical Association of Canada, Short Course Handbook v. 8, p. 175-250.

——— 1986, Models for the generation of metalliferous hydrothermal systems within sedimentary rocks and their applicability to the Irish Carboniferous Zn-Pb deposits, *in* Andrew, C.J., Crowe, R.W.A., Finlay, S., Pennell, W.M., and Pyne, J.F., eds., Geology and Genesis of Mineral Deposits in Ireland: Irish Association for Economic Geology, p. 555-577.

Lydon, J.W., Goodfellow, W.D., and Jonasson, I.R., 1985, A general genetic model for stratiform baritic deposits of the Selwyn Basin, Yukon Territory and District of Mackenzie, in Current Research Part A: Geological Survey of Canada, Paper 85-1A, p. 651-660.

MacIntyre, D.G., 1982, Geologic setting of recently discovered stratiform barite-sulphide deposits in Northeast British Columbia: Canadian Institute of Mining and Metallurgy Bulletin, v. 75, p. 99-113.

——— 1992, Geological setting and genesis of sedimentary exhalative barite and barite-sulfide deposits, Gataga District, northeastern British Columbia: Exploration and Mining Geology, v. 1, p. 1-20.

MacIntyre, D.G., and Sangster, D.F., 1983, Geology and stratiform barite-sulphide deposits of Gataga District, northeast British Columbia, *in* Sangster, D.F., ed., Short Course in Sediment-Hosted Stratiform Lead-Zinc Deposits: Mineralogical Association of Canada, Toronto, v. 9, p. 85-119.

McClay, K.R., 1983, Deformation of stratiform lead-zinc deposits, in Sangster, D.F., ed., Sediment-Hosted Stratiform Lead-Zinc Deposits: Mineralogical Association of Canada, Short Course Handbook v. 9, p. 283-309.

McClay, K.R., and Bidwell, G.E., 1986, Geology of the Tom deposit, MacMillan Pass, Yukon, *in* Morin, J.A., ed., Mineral Deposits of the Northern Cordillera: Canadian Institute of Mining and Metallurgy, Special Volume 37, p. 100-114.

McClay, K.R., Insley, M.W., Way, N.A., Anderton, R., and Anonymous, 1988, Tectonics and mineralization of the Kechika Trough, Gataga area, northeastern British Columbia, in Current Research, Part E, Cordillera and Pacific Margin: Geological Survey of Canada, Paper 88-1E, p. 1-12.

McClay, K.R., Insley, M.W., Anderton, R., Cooper, M.A., and Williams, G.D., 1989, Inversion of the Kechika Trough, northeastern British Columbia, Canada: Geological Society, Special Publications, v. 44, p. 235-257.

McKibben, M.A., Andes, J.P.J., and Williams, A.E., 1988, Active ore formation at a brine interface in metamorphosed deltaic lacustrine sediments: The Salton Sea geothermal system, California: Economic Geology, v. 83, p. 511-523.

Miller, D., and Wright, J.V., 1986, Mel barite -zinc-lead deposit - an exploration case history, *in* Morin, J.A., ed., Mineral Deposits of Northern Cordillera: Canadian Institute of Mining and Metallurgy, Proceedings of the Mineral Deposits of Northern Codillera Symposium, December 1983, Special Volume 37, p. 19-50.

Morganti, J.M., 1979, The Geology and Ore Deposits of the Howards Pass Area, Yukon and Northwest Territories: The Origin of Basinal Sedimentary Stratiform Sulphide Deposits: Ph.D. thesis, University of British Columbia, Vancouver, British Columbia, 317 p.

Mosier, D.L., Singer, D.A., and Salem, B.B., 1983, Geologic and Grade-Tonnage Information on Volcanic-Hosted Copper-Zinc-Lead Massive Sulfide Deposits: United States Geological Survey, Open File Report 83-89, p. 77p.

Nakai, S.I., Halliday, A.N., Kesler, S.E., Jones, H.D., Kyle, J.R., and Lane, T.E., 1993, Rb-Sr dating of sphalerites from Mississippi Valley-type (MVT) ore deposits: Geochimica et Cosmochimica Acta, v. 57, p. 417-427.

Nelson, J., and Colpron, M., 2007, Tectonics and metallogeny of the British Columbia, Yukon and Alaskan Cordillera, 1.8 Ga to the present, *in* Goodfellow, W.D., ed., Mineral Deposits of Canada: A Synthesis of Major Deposit-types, District Metallogeny, the Evolution of Geological Provinces, and Exploration Methods: Geological Association of Canada, Mineral Deposits Division, Special Publication 5, p. 755-791.

Nelson, J., Paradis, S., Christensen, J., and Gabites, J., 2002, Canadian Cordilleran Mississippi Valley-type deposits; a case for Devonian-Mississippi back-arc hydrothermal origin: Economic Geology, v. 97, p. 1013-1036.

Paradis, S., and Nelson, J., 2007, Metallogeny of the Robb Lake carbonate-hosted zinc-lead district, northeastern British Columbia, *in* Goodfellow, W.D., ed., Mineral Deposits of Canada: A Synthesis of Major Deposit-types, District Metallogeny, the Evolution of Geological Provinces, and Exploration Methods: Geological Association of Canada, Mineral Deposits Division, Special Publication 5, p. 633-654.

Paradis, S., Nelson, J.L., and Irwin, S.E.B., 1998, Age constraints on the Devonian shale-hosted Zn-Pb-Ba deposits, Gataga District, northeastern British Columbia, Canada: Economic Geology, v. 93, p. 184-200.

Paradis, S., Hannigan, P., and Dewing, K., 2007, Mississippi Valley-type (MVT) lead-zinc deposits, *in* Goodfellow, W. D., ed., Mineral Deposits of Canada: A Synthesis of Major Deposit-types, District Metallogeny, the Evolution of Geological Provinces, and Exploration Methods: Geological Association of Canada, Mineral Deposits Division, Special Publication 5, p. 185-203.

Rhodes, D., and Matthew, I.G., 1993, Geology and genesis of the Tom Deposit, Yukon, Canada, *in* World Zinc '93; Proceedings of the International Symposium on Zinc: Australasian Institute of Mining and Metallurgy, v. 7/93, p. 95-110.

Russell, M.J., 1983, Major sediment-hosted exhalative zinc + lead deposits: Formation from hydrothermal convection cells that deepen during crustal extension, *in* Sangster, D.F., ed., Sediment-Hosted Stratiform Lead-Zinc Deposits: Mineralogical Association of Canada, Short Course Handbook v. 9, p. 251-282.

Samson, I.M., and Russell, M.J., 1983, Fluid inclusion data from Silvermines base-metal-baryte deposits, Ireland: Transactions of the Institution of Mining and Metallurgy, v. B92, p. 67-71.

Sangster, D.F., 1990, Mississippi Valley-type and sedex lead-zinc deposits: A comparative examination: Transactions of the Institution of Mining and Metallurgy, v. B99, p. 21-42.

Sato, T., 1972, Behaviors of ore-forming solutions in seawater: Mining Geology, v. 22, p. 31-42.

Scotese, C.R., 1984, An introduction to this volume: Paleozoic paleomagnetism and the assembly of Pangea, *in* Van der voo, R., Scotese, C.R., and Bonhommet, N., eds., Plate Reconstruction from the Paleozoic Paleomagnetism: American Geophysical Union, Geodynamic Series, v. 12, p. 1-10.

Shanks, W.C., III, Woodruff, L.G., Jilson, G.A., Jennings, D.S., Modene, J.S., and Ryan, B.D., 1987, Sulfur and lead isotope studies of stratiform Zn-Pb-Ag deposits, Anvil Range, Yukon: Basinal brine exhalation and anoxic bottom-water mixing: Economic Geology, v. 82, p. 600-634.

Sherlock, R.L., Lee, J.K.E., and Cousens, B.L., 2004, Geological and geochronological constraints on the timing of mineralization at the Nanisivik zinc-lead Mississippi Valley-type deposit, Northern Baffin Island, Nunavut, Canada: Economic Geology, v. 99, p. 279-293.

Trexler, J.H., Jr., Cashman, P.H., Snyder, W.S., and Davydov, V.I., 2004, Late Paleozoic tectonism in Nevada; timing, kinematics, and tectonic significance: Geological Society of America Bulletin, v. 116, p. 525-538.

Turner, R.J.W., 1990, Jason stratiform Zn-Pb-barite deposit, Selwyn Basin, Canada (NTS 105-O-1): Geological setting, hydrothermal facies and genesis, *in* Abbott, J.G., and Turner, R.J.W., eds., Mineral Deposits of the Northern Canadian Cordillera, Yukon, Northeastern British Columbia (Field Trip 14): Geological Survey of Canada, Open File 2169, p. 137-175.

Turner, R.J.W., and Einaudi, M.T., 1986, The geological setting and genesis of the south zone stratiform Pb-Zn-barite deposits, MacMillan Pass, Yukon, *in* Turner, R.J.W. and Einaudi, M.T., eds., The Genesis of Stratiform Sediment-Hosted Lead and Zinc Deposits: Stanford University Publications in the Earth Sciences, v. 20, p. 5-12.

Turner, R.J.W., and Goodfellow, W.D., 1990, Barium carbonate bodies associated with the Walt [Cathy] stratiform barium deposit, Selwyn Basin, Yukon; a possible vent complex associated with a Middle Devonian sedimentary exhalative barite deposit, *in* Current Research, Part E: Cordillera and Pacific Margin: Geological Survey of Canada, Paper 90-1E, p. 309-319.

Turner, R.J.W., and Rhodes, D., 1990, Boundary Creek zinc deposit [Nidd Property], Macmillan Pass, Yukon; Sub-seafloor sediment-hosted mineralization associated with volcanism along a Late Devonian syndepositional fault, *in* Current Research, Part E: Cordillera and Pacific Margin: Geological Survey of Canada, Paper 90-1E, p. 321-335.

Vail, P.R., Mitchum, R.M., Jr., and Thompson, S., III, 1977, Seismic stratigraphy and global changes of sea level; Part 4, Global cycles of relative changes of sea level, *in* Payton, C.E., ed., Seismic Stratigraphy: Applications to Hydrocarbon Exploration: American Association of Petroleum Geologists, Memoir 26, p. 83-97.

Winn, R.D., Jr., and Bailes, R.J., 1987, Stratiform lead-zinc sulfides, mudflows, turbidites: Devonian sedimentation along a submarine fault scarp of extensional origin, Jason deposit, Yukon Territory, Canada: Geological Society of America Bulletin, v. 98, p. 528-539.

Yang, J., Large, R.R., and Bull, W., 2004, Factors controlling free thermal convection in faults in sedimentary basins: Implications for the formation of zinc-lead mineral deposits: Geofluids, v. 4, p. 237-247.

Geology and Metallogeny of the Belt-Purcell Basin

John W. Lydon
Geological Survey of Canada, 601 Booth Street, Ottawa, Ontario K1A 0E8
Corresponding author's email: jlydon@nrcan.gc.ca

Abstract

The Mesoproterozoic Belt-Purcell Basin outcrops over an area of about 200,000 km^2 in Idaho, Montana, and southeastern British Columbia, and preserves an 18 to 20 km thickness of sediments that were deposited in an intracontinental rift. The Prichard (U.S.A.) and Aldridge (Canada) Formations consist of 10 to12 km of marine turbidites with intercalated tholeiitic sills and reflect a combination of high accumulation rates (>500 m/my) and magmatic activity along the axial part of the rift. Relatively small strata bound Cu-Co massive sulphide deposits in central Idaho arguably represent Besshi-type mineralization in the axial zone of the rift, but the Proterozoic metallogenic events with the greatest metal endowment are associated with basinal dewatering during the rift fill stage of basin evolution. These events produced the world-class Sullivan Zn-Pb-Ag SEDEX deposit of British Columbia, hosted by the Aldridge Formation, and the world-class Coeur d'Alene Ag-Pb vein district of Idaho, hosted by Prichard- to Revette-aged rocks. The Revette Formation, corresponding to the earliest part of the rift sag stage, also hosts the Spar Lake, Rock Creek, and Montanore Red Bed Cu deposits of Montana, which together have a proven resource of more than 330 million tonnes of approximately 0.7% Cu and 50 g/t Ag. Reports of other Proterozoic mineral deposit types, including IOCG, Cu-Ni-PGE, and rift-sag stage SEDEX deposits, suggest that the mineral potential of the Belt-Purcell Basin has not yet been fully realized.

Résumé

Le bassin de Belt-Purcell du Mésoprotérozoïque affleure sur quelque 200 000 km^2 en Idaho, au Montana et dans le sud de la Colombie-Britannique. Il contient une succession de 18 à 20 km d'épaisseur de roches sédimentaires déposées dans un rift intracontinental. Les formations de Prichard (É.-U.) et d'Aldridge (Canada) consistent en une succession de 10 à 12 km de turbidites marines au sein desquelles s'intercalent des filons-couches tholéiitiques. Cet ensemble témoigne d'un taux d'accumulation sédimentaire élevé (> 500 m/Ma) et de processus magmatiques actifs le long de la partie axiale du rift. Dans le centre de l'Idaho, de relativement petits gîtes stratoïdes de sulfures massifs à Cu Co représentent vraisemblablement des minéralisations de type Besshi formées dans la partie axiale du rift. Cependant, les épisodes métallogéniques du Protérozoïque qui ont donné lieu aux plus importantes ressources en métaux sont associés, dans l'évolution du bassin sédimentaire, à la déshydratation du bassin pendant la phase de remplissage du rift. Ces épisodes ont produit le gisement de classe mondiale de Sullivan, en Colombie Britannique, un gîte de sulfures exhalatifs dans des roches sédimentaires (gîte sedex) à minéralisation de Zn-Pb-Ag au sein de la Formation d'Aldridge, de même que le district à filons de Ag-Pb de calibre mondial Coeur d'Alène, en Idaho, où la minéralisation est contenue dans des roches dont l'âge va de celui de la Formation de Prichard à celui de la Formation de Revette. La Formation de Revette, qui correspond à la partie initiale de la phase d'affaissement du rift, comprend aussi les gisements cuprifères de type redbeds de Spar Lake, de Rock Creek et de Montanore, au Montana, dont les ressources prouvées totalisent plus de 330 Mt de minerai titrant environ 0,7 % de Cu et 50 g/t de Ag. Des rapports sur d'autres types de gîtes minéraux du Protérozoïque, y compris des gîtes d'oxydes de fer-Cu-Au, des gîtes de Cu-Ni-ÉGP et des gîtes de sulfures exhalatifs dans des roches sédimentaires (gîtes sedex) associés à la phase d'affaissement du rift, laissent supposer que le potentiel minéral du bassin de Belt Purcell n'a pas encore été totalement déterminé.

Introduction and History

The Belt-Purcell Basin is a Mesoproterozoic intracontinental rift filled by marine and fluviatile sediments. Rocks of the basin comprise the Belt-Purcell Supergroup, which is up to 20 km thick (Winston and Link, 1993) and outcrops over an area of approximately 200,000 km^2. Only about 10% of this area is in Canada, the greater part being in the states of Montana and Idaho of the U.S.A. (Fig. 1). The portion of the basin and its sedimentary succession that occurs in the U.S.A is called the Belt Basin and Belt Supergroup, respectively, whereas their northward extensions in Canada are known as the Purcell Basin and Purcell Supergroup, respectively. The term Belt-Purcell refers to the combined American and Canadian segments and is used here when referring to the basin as a single entity.

Mining interest in the Belt-Purcell Basin started in the 1860s with the development of placer gold deposits, especially in the Bannack and Virginia City areas of Montana and the Boise basin area of Idaho. During the 1880s, prospecting away from the diminishing resources of the placer gold mining camps led to the discovery of the world-class Coeur d'Alene silver-lead district and the world-class copper-silver-gold-molybdenum porphyry-epithermal system at Butte (e.g. Klein, 2004). A similar history of mining development took place in Canada. Mining interest in southeastern British Columbia started with the Kootenay gold rush of 1864 to the placer deposits of Wild Horse Creek at the eastern edge of the Purcell Basin. Prospecting west of the placer gold fields led to the discovery in 1892 of the outcropping lead-silver ores of the North Star and the nearby Sullivan deposits. Although mining and smelting of lead-silver ores of the Sullivan deposit had started in 1900, its immense size was not fully realized until 1912, following its purchase in 1910 by the Consolidated Mining and Smelting Company of Canada (subsequently renamed Cominco) (Hamilton et al., 2000).

Up to 2001, the Coeur d'Alene district had produced about 7,400,000 tonnes of Pb, 2,900,000 tonnes of Zn, and 35,600 tonnes of Ag (data from Long, 1998; Larsen et al.,

Lydon, J.W., 2007, Geology and metallogeny of the Belt-Purcell Basin, *in* Goodfellow, W.D., ed., Mineral Deposits of Canada: A Synthesis of Major Deposit-Types, District Metallogeny, the Evolution of Geological Provinces, and Exploration Methods: Geological Association of Canada, Mineral Deposits Division, Special Publication No. 5, p. 581-607.

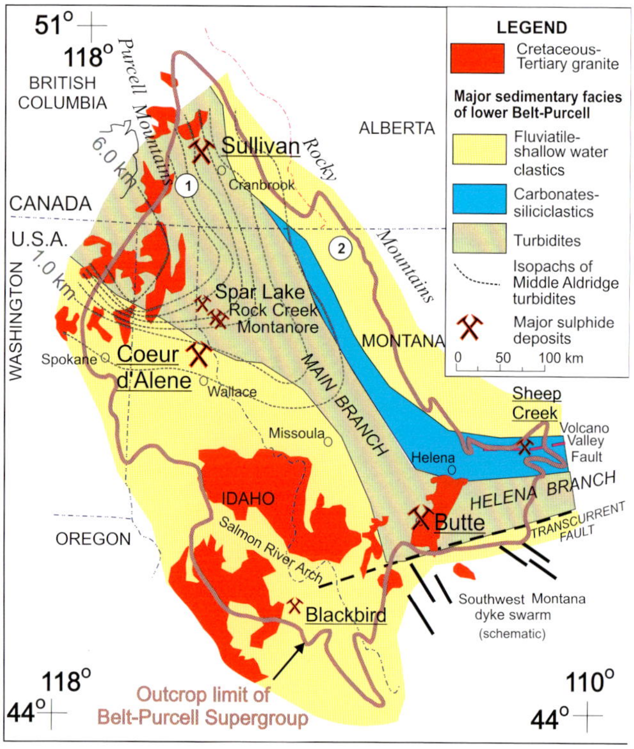

FIGURE 1. Map showing the outcrop extent of the Belt-Purcell Basin, the locations of major mineral deposits, and the simplified distribution of sedimentary facies of the lower part of the Belt-Purcell Supergroup (i.e. Aldridge-Pritchard formations and stratigraphic equivalents). Numbers refer to profiles in Figure 3. The locations of minor mineral deposits occurring in Canada are shown in Figure 4.

2004) and the Butte camp about 9,700,000 tonnes of Cu, 2,500,000 tonnes of Zn, 1,700,000 tonnes Mn, and 22,600 tonnes of Ag, as well as significant amounts of Mo, Pb, and Au (Robin McCulloch, Montana Bureau of Mines, pers. comm., 2005). Mining from the Sullivan deposit continued until the end of 2001, producing a total of 8,412,077 tonnes of Pb, 7,944,446 tonnes of Zn, and 9,264 tonnes of Ag plus significant quantities of Sn, Cd, Sb, Cu, Bi, and Au from 150 million tonnes of ore (British Columbia Geological Survey, Minfile 082FNE052).

The early establishment of this phenomenal metal endowment of the area has supported an almost permanent mineral exploration interest, and subsequent years have seen the discovery of a variety of mineral deposit types, including seafloor copper-cobalt deposits of the Blackbird mining district (Nash and Hahn, 1989; Bending and Scales, 2001) and the Sheep Creek deposit near White Sulphur Springs, Montana (Himes and Petersen, 1990), Red-Bed Cu-Ag deposits of the Montana Copper Belt, including the Spar Lake (Troy) mine (Harrison, 1972; Garlick, 1987; Hayes et al., 1989), and lode-Au deposits (e.g. Perry and Childs, 2003). More recently, the potential of the area for iron oxide copper-gold (IOCG) deposits (Stinson and Brown, 1994; Gillerman et al., 2003) and PGE mineralization in mafic intrusions (Buckley et al., 1994) has become apparent.

This range of mineral deposit types hosted by rocks of the Belt-Purcell Supergroup is the result of two main metallogenic intervals. The first was during the Mesoproterozoic accumulation of the Belt-Purcell Supergroup and consisted mainly of base metal seafloor deposits (SEDEX and Besshi types) and diagenetic veins and replacements. The ore-forming processes were driven by interplay of extensional tectonics with sediment compaction, and the emplacement of rift-related mafic magmas, its related hydrothermal convection. The second was during the Cretaceous-Tertiary Laramide Orogeny and consisted of porphyry and epithermal styles of mineralization related to the emplacement of felsic magmas, and mesothermal veins associated with compressional and transtensional tectonics. This synopsis focuses on the first metallogenic interval and on the Canadian portion of the Belt-Purcell Basin, highlighting the one world-class deposit so far discovered in this area – the Sullivan SEDEX-type deposit.

Geological and Tectonic Setting

Global Setting

Continental reconstruction puts the Belt-Purcell Basin near the continental margin of Laurentia, one of the continents formed by the Neoproterozoic break-up of the Rodinia supercontinent. The best constrained reconstruction (Sears and Price, 1978, 2000, 2003a,b) fits Laurentia adjacent to Siberia, and is supported by the matching of structural trends, zircon ages, and the alignment of the Mesoproterozoic Taimyr and Uzdha troughs with the Belt-Purcell Basin (Fig. 2). Other reconstructions fit Laurentia adjacent to Australia and Antarctica (Dalziel, 1991;

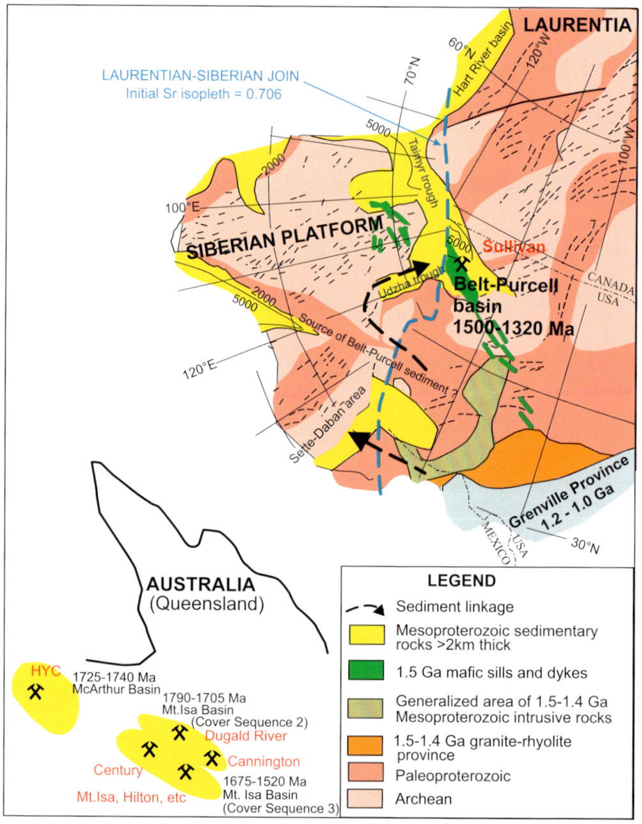

FIGURE 2. Continental reconstruction for the Mesoproterozoic after Sears and Price (2000, 2003a,b). Note that a different reconstruction (Hoffman, 1991) based on the correlation of Grenville-age tectonic belts places Antarctica contiguous to the western edge of Laurentia, with Australia contiguous to the Hart River Basin.

Hoffman, 1991; Moores, 1991). It is interesting to note that in both reconstructed continental configurations, the world's most productive Mesoproterozoic sedimentary basins for SEDEX deposits (the Australian Broken Hill, Mt. Isa, and McArthur basins and the Laurentian Belt-Purcell), were all at low latitudes and over the same segment of the earth's mantle during sedimentation.

Tectonic Setting

The Belt-Purcell Supergroup fills part of an interconnected system of Mesoproterozoic rifts that extended from western North America to Labrador and were linked to the opening of the Grenville ocean (Gower and Tucker, 1994). The Belt-Purcell rift is the passive type caused by lithospheric tension, as opposed to the active type caused by the ascent and lateral spreading of a mantle plume (Chandler, 2000). The evidence for this is the preservation of basal quartzites (Fort Steele and Neihart formations) and the absence of pre-rift volcanism, which indicates a lack of thermal doming associated with active-type rifts. The REE chemistry of rift-related tholeiitic sills suggests magma derivation by partial melting of the mantle beneath a highly attenuated crust 35 to 40 km thick (Anderson and Goodfellow, 2000), consistent with a passive-type intracontinental rift.

Synsedimentary Faulting

Extensional faulting and sporadic tholeiitic to alkaline magmatism continued over most of the approximately 180 million years of sediment accumulation in the Belt-Purcell Basin, from about 1500 Ma to about 1320 Ma. Synsedimentary faults associated with rifting had their most profound effect in the lower part of the Belt-Purcell stratigraphic succession. This faulting is responsible for the contrast in thickness between the turbidites and intercalated sills of the Aldridge/Prichard rift-fill sequence and the stratigraphically equivalent shelf facies of the Greyson, Newland, Chamberlain, and Neihart formations (Fig. 3). The rift-fill sequence is about 12 km thick in the Purcell Mountains (Höy et al., 2000), in contrast to the cumulative thickness of 2.5 km for the shelf sequence at the eastern end of the Helena embayment (Chandler, 2000). Two directions of synsedimentary faults have been recognized: approximately north- to northwest-trending rift-parallel faults, and east to north-east-trending transfer faults (Höy et al., 2000; Chandler, 2000; Turner et al., 2000a).

A prime example of a rift-parallel synsedimentary fault is the structure that separates the shallow-water facies of the Fort Steele Formation and the overlying condensed lower Middle Aldridge in the Hughes Range, from the deep-water thick turbidite sequence of the of the Lower and Middle Aldridge in the Purcell Mountains (see Figs. 3, 4). However, this structure has little or no affect on the thickness of the upper part of the Middle Aldridge and younger Mesoproterozoic rocks. Other examples include faults that control the north-trending Sullivan Corridor, various faults northeast of Moyie Lake, the Iron Range Fault northeast of Creston, and the north-trending elongation of many mud-volcano vents ("discordant fragmentals") (Höy et al., 2000; Turner et al., 2000b). Examples of east- to northeast-trending synsedimentary faults include antecedents to the Moyie-

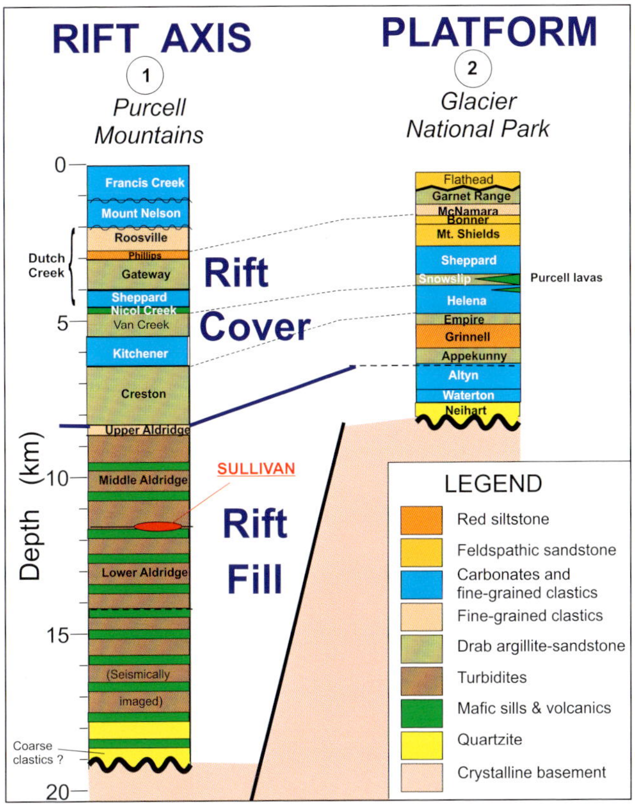

FIGURE 3. Stratigraphic correlation and comparison of average thicknesses between formations of the rift-fill, rift-cover, and platform sedimentary sequences of the Belt-Purcell Supergroup. Locations of profiles are shown in Figure 1.

Dibble Creek and St. Mary-Boulder Creek fault systems (see Fig. 4), across which facies and thickness of Aldridge rocks change, and in proximity to which there is an anomalous thickness of Moyie sills (Höy et al., 2000).

Stratigraphy and Sedimentology of the Belt-Purcell Supergroup

The Belt-Purcell Rift consists of two branches (Fig. 1). The main or Purcell branch, which contains the Sullivan deposit, trends northwest through the Purcell Mountains of southeastern British Columbia and is characterized by a basal, 12 km thick, turbidite-sill complex – the Aldridge Formation in Canada and the Prichard Formation in the U.S.A. To the northwest, rocks of the Purcell branch are covered by Neoproterozoic and Phanerozoic strata and to the southeast they are truncated against an east-northeast-trending transfer fault. The east-northeast-trending or Helena branch extends along the northern side of this transfer fault to form the Helena embayment (Fig. 1).

Stratigraphic relationships of the Belt-Purcell are shown in Figure 5. The lower part of the Supergroup consists of marine turbidites, which infilled the rift grabens, and stratigraphically equivalent shallow marine to fluvial sandstones, mudstones, and carbonates that were deposited on the surrounding rift platform. These synrift sequences are overlain by shallow marine to lacustrine and fluviatile mudstones, carbonates, and sandstones that extend over the rifted and platformal areas alike, forming a rift-cover or rift-sag

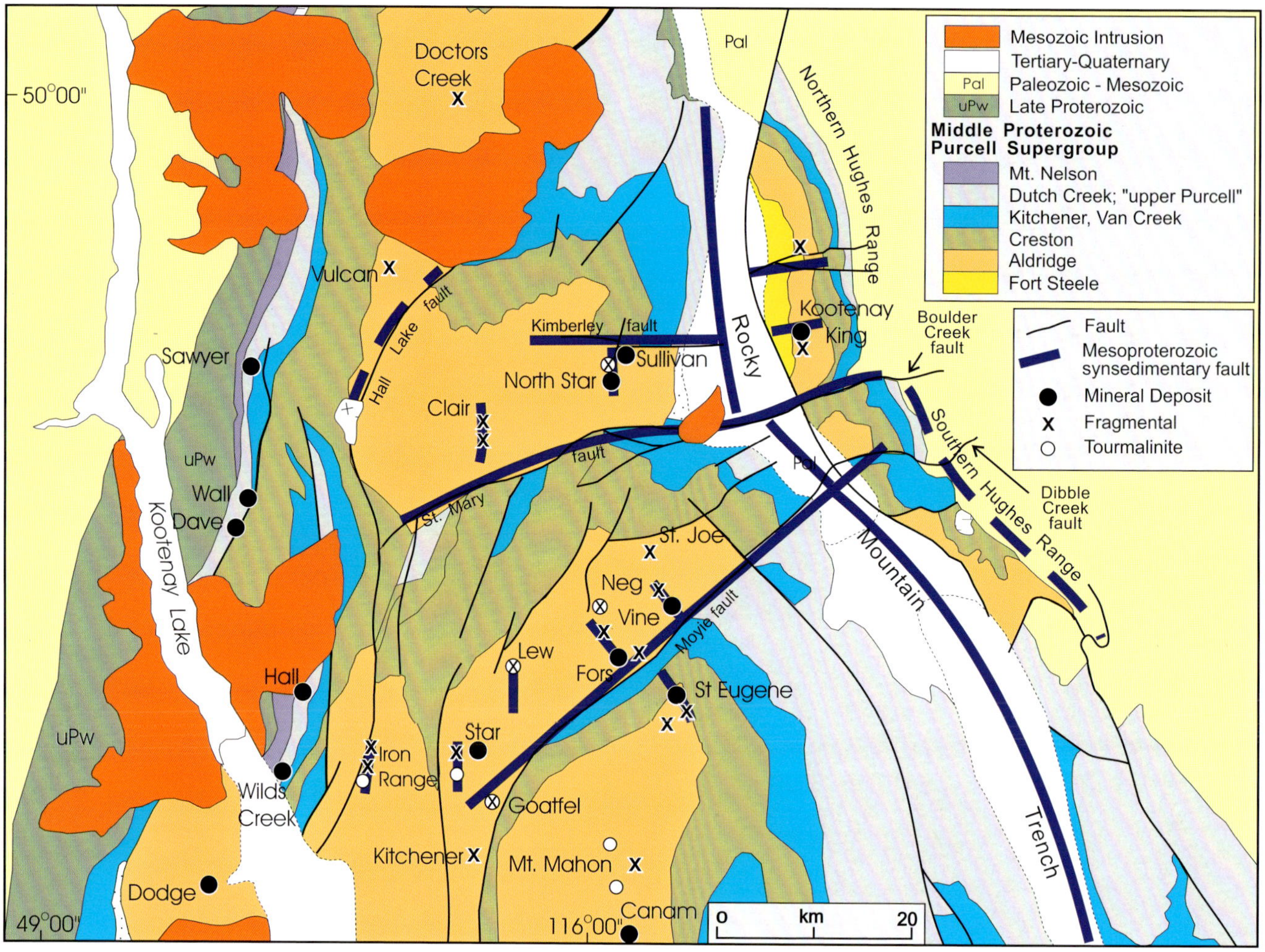

FIGURE 4. Geological map of the southern part of the Purcell anticlinorium showing the locations of Mesoproterozoic sulphide deposits, tourmalinite occurrences, sedimentary fragmental bodies, and major synsedimentary faults. Thick blue lines indicate the approximate position of faults active during the Mesoproterozoic. Lithologies of geological units are shown on Figure 3. After Höy et al. (2000).

sequence. Strata of the Belt-Purcell Basin can thus be divided into three main facies groups (Fig. 3):

1. *Basinal facies of rift-fill sequence* consisting mainly of deep-water turbidites in the Purcell Branch (Aldridge and Prichard formations) and deep-water calcareous argillite and turbidites that shoal upwards to midshelf carbonates and siliciclastics (Newland Formation) in the Helena Branch.
2. *Shallow-water platformal and fan-delta facies* deposited at the margins of the rift and surrounding shelf, and approximately synchronous with turbidite deposition within the rift. Lithologies include fluviatile and deltaic quartz-rich arenites (Fort Steele and Neihart formations), fan-delta complexes containing coarse-grained debris flows shed from fault scarps (Lahood Formation), distal deep-water argillite-siltite debris flows (Greyson Formation), and shallow-water platformal carbonates (Waterton and Altyn formations).
3. S*hallow-water, mud flat, fluvial, lagoonal, alluvial, and playa facies of rift-cover or rift-sag sequence* that covers both the rift and its adjacent platforms, and forms the upper part of the Belt-Purcell Supergroup. Rock types include red, purple, and green argillites and siltites of the Ravalli Group (Creston Formation in Canada), a transgressive carbonate-rich sequence of the Middle Belt Carbonate (Kitchener Formation in Canada), and northward and eastward deepening fine-grained clastics of the Missoula Group (Sheppard, Gateway, Phillips, Roosville, and Mount Nelson formations in Canada) that range from a large sandy alluvial apron in the southwest through marginal marine sand and mud flats, to shallow-marine sediments composed of siliceous and carbonate mud in the north and east.

Depositional Environment of the Belt-Purcell Supergroup

At the time of deposition of the lower rift-fill sequence, the Belt-Purcell rift was a deep, elongate but relatively narrow water-filled trough into which one or more very large rivers drained at its southern head forming a fluviatile-deltaic complex. Turbidites derived from the delta complex fill moved northwestward during most of the rift-fill stage (Chandler, 2000; Höy et al., 2000). This transport direction, deduced from current direction indicators and sedimentary facies distributions, is consistent with the age profiles of detrital zircons that match the ages of basement terranes of the North American craton to the south and to the east of the

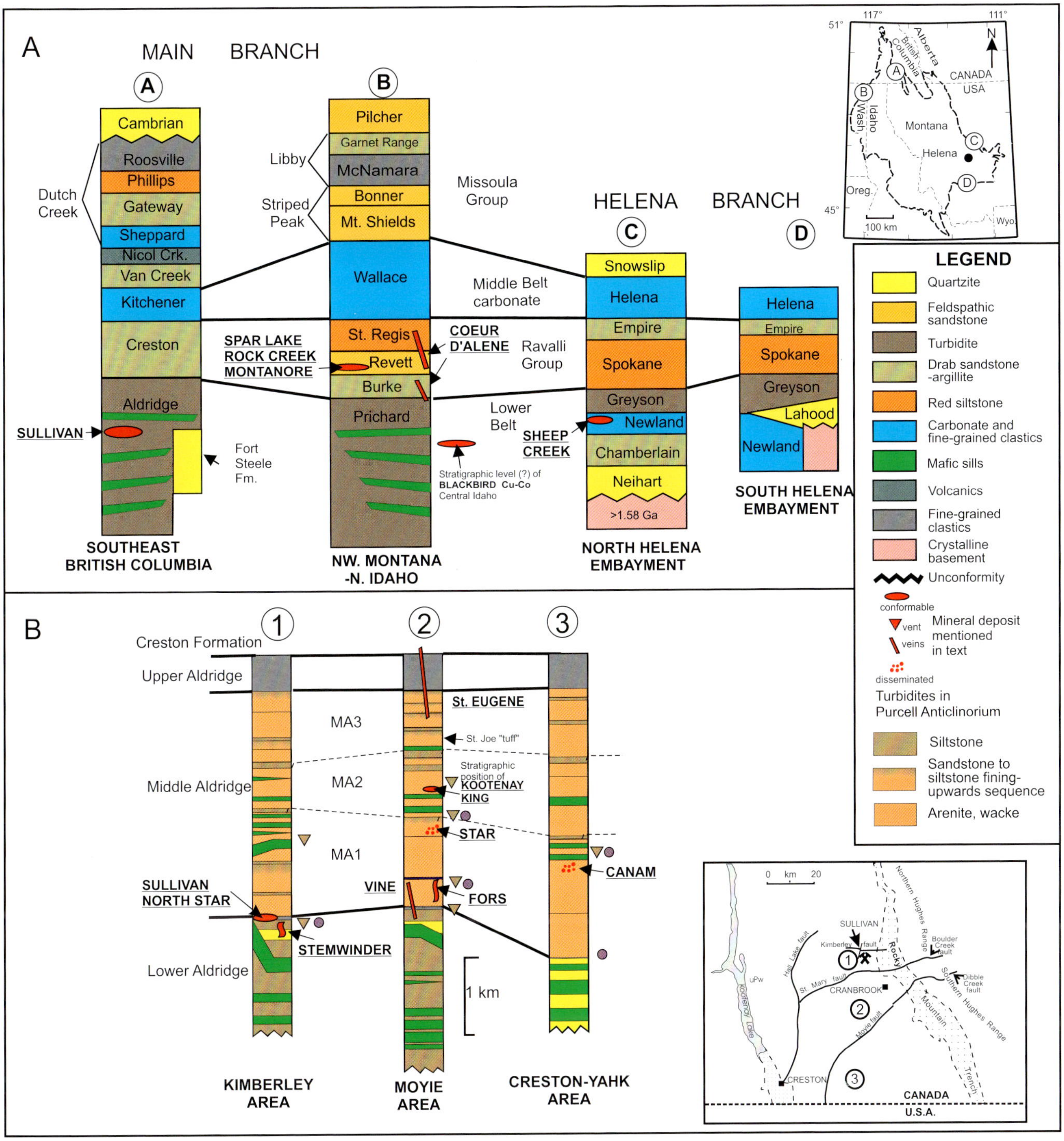

FIGURE 5. Stratigraphic correlation of the formations of the Belt-Purcell Supergroup showing the stratigraphic positions of the major mineral deposits. Note that there is no vertical scale. Modified after Chandler (2000).

Belt-Purcell Basin (Ross et al., 1992; Ross and Villeneuve, 1999). During the latter part of the rift-fill stage, and continuing throughout the rift-sag phase, sediment supply to the Belt Basin was mainly from the southwest (Chandler, 2000). Zircons in this sediment have age characteristics consistent with an Australian (Ross and Villeneuve, 1999) or Siberian (Sears and Price, 2000) source. Although the rift-cover sequence was deposited in a dominantly shallow-water environment, it contains deep-water facies deposited in local areas of greater differential subsidence. This local differential subsidence reflects a continuation of rift activity, albeit on a diminished scale, to at least the stratigraphic level of the Sheppard and Gateway formations (Höy, 1993).

The combination of various lines of evidence suggests that the Belt-Purcell Basin was marine rather than lacustrine:

1. Salt casts, evaporitic nodules, oolites, and syneresis cracks in the shallow-water facies of the lower Belt-Purcell and in strata of the rift-cover sequence indicate that the permanent water column was saline (Chandler, 2000).

2. Chamosite pellets, most common in shallow marine waters (Johnson, 1978), occur in the Fort Steele and Newland formations (Schieber, 1993; Chandler, 2000), and pelletal phosphorite, unknown as a Precambrian lacustrine mineral, occurs in the Spokane Formation (Gulbrandsen, 1966).
3. Strong tidal action, which is generally indicative of a marine environment, is suggested by herringbone crossbeds and ripples in the Fort Steele, Prichard, Altyn, Yellowjacket, Newland, and Creston formations, and by extremely elongated stromatolites in the Altyn Formation (Chandler, 2000).
4. Strontium concentrations of the Newland Formation and boron content of the Middle Belt Carbonate (Schieber, 1993) are typical for marine sediments.

From the time of the Creston Formation onwards, there is evidence for both marine and lacustrine environments (see discussion in Lydon, 2000a). During this period of time, the rift had been filled by fluviatile, deltaic, turbidite, and hemipelagic sediments, and its surrounding platform had become covered with fluviatile to shallow-water sandstones, fine-grained-clastics, and carbonates.

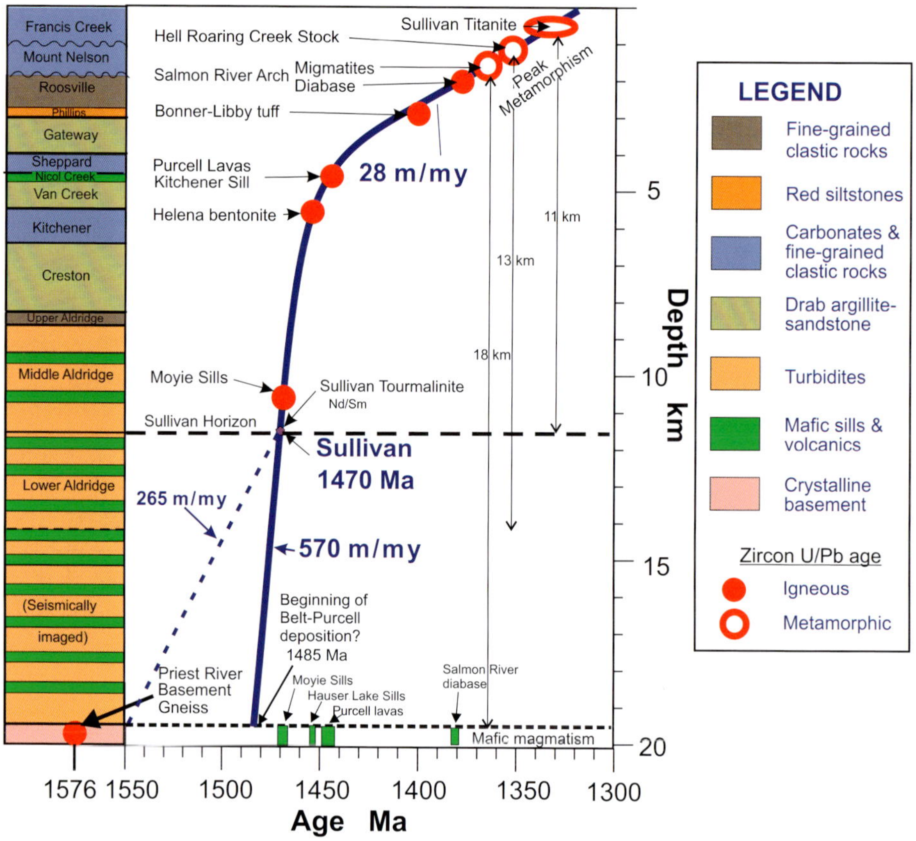

FIGURE 6. Average thickness of stratigraphic divisions of the Purcell Supergroup (Höy, 1993) and stratigraphic or depth-controlled events dated by U-Pb zircons or Pb-Pb titanite methods (see Lydon, 2000a, for sources of data). The thick dark blue line is a smoothed joining of the geochronological data points, the slope of which gives averaged accumulation rates for the Belt-Purcell Basin. The data suggest an average accumulation rate of 570 m/my for the Lower and Middle Aldridge Formation. The dotted blue line indicates the minimal accumulation rates for the Lower Aldridge, assuming a minimal erosion period of 25 my for exposure of the Priest River basement gneiss. Modified after Lydon (2000a).

During deposition of the Aldridge-Pritchard turbidites, the deepest part of the water column may have been anoxic. Perhaps the strongest evidence for this is that the hemipelagic component of the sedimentary succession, which is preserved as inter-turbidite silty argillite, contains up to 3 wt.% organic carbon (Goodfellow, 2000). However, sulphur/carbon ratios of Aldridge sedimentary rocks with >0.3% organic carbon span the range for normal marine sediments, with only 30% plotting in the field for sediments deposited in an anoxic marine water column (see Fig. 12-19, Goodfellow, 2000). Intermittent stratigraphic excursions of sulphur isotope ratios of disseminated iron sulphides to heavy values in the Aldridge Formation (Goodfellow, 2000) and the Newland Formation (Lyons, 1993; Lyons and Luepke, 1995; Lyons et al., 2000) are interpreted by the authors to indicate periodic restriction of marine circulation and the development of a stratified water column. The great majority of Aldridge sedimentary rocks have MnO contents of <0.1%, similar to anoxic black sediments (Goodfellow, 2000), but this evidence is somewhat ambiguous because fine-grained Aldridge sedimentary rocks (argillites) have low concentrations of all elements (except Si, Al, and K) compared to average shales, due to dilution by an unusually high quartzo-feldspathic content (Lydon et al., 2000b).

Accumulation Rates of the Belt-Purcell

Stratigraphically controlled U-Pb ages of zircons from volcanic rocks (Evans et al., 2000) include 1454 Ma for a bentonite near the top of the Helena Formation, 1443 Ma for felsic tuffs at the stratigraphic level of the Purcell lavas and equivalent Nicol Creek volcanics, and 1401 Ma for airfall tuff at the contact between the Bonner and Libby formations. These are plotted at their Purcell stratigraphic equivalents to the formations of the Belt Supergroup in Figure 6. Moyie sills in the Aldridge (Anderson and Davis, 1995; Schandl and Davis, 2000) and the Pritchard (Sears et al., 1998) formations have the same age of 1468+3/-2 Ma, and are plotted in Figure 6 midway between the base of the Middle Aldridge and the top of the MA2 sedimentary cycle (Fig. 5) of the Middle Aldridge Formation (Höy et al., 2000), which is the stratigraphic interval over which the 1468 Ma suite of sills were emplaced (see discussion in the section on "Stratigraphic interval corresponding to the 1468 Ma Moyie sill event" below). Metamorphic migmatites of the Salmon River Arch (Fig. 1), dated at 1370 Ma, are plotted at their calculated thermobarometric depth (Doughty and Chamberlain, 1996), and metamorphic titanites from the

Sullivan deposit dated at 1325 to 1330 Ma (Schandl and Davis, 2000) are plotted at the 11 km depth of its peak Mesoproterozoic metamorphic pressure (De Paoli and Pattison, 2000). A smoothed curve through these seven points extrapolates to the seismically-imaged base of the Aldridge Formation at 1485 Ma, and gives average accumulation rates of 570 m/my from the beginning of Purcell sedimentation up to the top of the Kitchener Formation (Lydon, 2000a). These accumulation rates are lower than those calculated by Evans et al. (2000). However, Evans et al. (2000) appear to have assumed that only about 500 m of Prichard-Aldridge Formation had accumulated by 1468 Ma, the time at which the Moyie sills were being emplaced. A Sm/Nd date of 1470 Ma for the footwall tourmalinite (Jiang et al., 2000), which predates the Sullivan deposit, is only slightly older than the 1468 Ma U/Pb zircon age for the mine sill that post-dates the Sullivan deposit, and supports the 1470 Ma age for the Sullivan deposit deduced from the accumulation rates.

Metamorphism and Tectonic Deformation

Rocks of the Purcell Supergroup have been affected by at least three distinct episodes of regional metamorphism and deformation, which have been termed the East Kootenay Orogeny (1350-1300 Ma), the Goat River Orogeny (900-800 Ma), and Jurassic-Cretaceous (160-60 Ma) metamorphism, deformation, and plutonism (McMechan and Price, 1982). In addition, the rocks have been affected by Paleozoic faulting.

East Kootenay Orogeny

The regional metamorphic grade of rocks in the Purcell anticlinorium (Reesor, 1973; McMechan and Price, 1982) and stratigraphically equivalent rocks of the Belt Supergroup in the U.S.A. (Maxwell and Hower, 1967; Eslinger and Sellars, 1981) generally increases with stratigraphic depth. This progression has been interpreted to reflect burial diagenesis and metamorphism (Maxwell and Hower, 1967; Eslinger and Sellars, 1981; Doughty and Chamberlain; 1996; Lydon et al., 2000b). Peak metamorphic conditions of 1370 Ma migmatites of the Salmon arch yields estimates of 680°C at 450 MPa (14 km burial) and 690°C at 650 MPa (20 km burial) (Doughty and Chamberlain, 1996). These thermobarometric estimates are consistent with the maximum thickness of the Belt Supergroup of 15 to 20 km (Harrison, 1972) or at least 18 km (Winston and Link, 1993) estimated from field mapping. The peak metamorphic conditions at Sullivan calculated to be 450°C and 380 MPa (De Paoli and Pattison, 1995, 2000), equivalent to 11 km of burial, are consistent with a geothermal gradient of 38 to 45°C/km during accumulation of the Belt-Purcell Supergroup.

A Mesoproterozoic age for this regional metamorphism has been well established by radiometric dating of metamorphic titanite (1330-1360 Ma) from a Moyie sill in the U.S.A. (Ross et al., 1992), hydrothermally altered rocks at Sullivan, which give Pb-Pb ages of 1325 to 1329 Ma (Schandl et al., 1993; Schandl and Davis, 2000), and zircon U-Pb ages of 1370 Ma from migmatitic leucosome at the base of the Belt Supergroup (Doughty and Chamberlain, 1996). A foliation accompanying burial metamorphism, commonly axial planar to large folds, is associated with most Belt-Purcell rocks and tends to increase in intensity with depth (Höy, 1993). Mesoproterozoic tectonics was dominated by extensional (rift-parallel) and transfer (transverse to the rift trend) faults associated with the development of the Belt-Purcell rift. The East Kootenay Orogeny reflects the burial metamorphism of a thick sedimentary pile in the high geothermal gradient of an actively rifting environment.

The Goat River Orogeny

The Goat River Orogeny was proposed by McMechan and Price (1982) to describe an 800 to 900 Ma episode of uplift, block faulting, and metamorphism that marked the onset of deposition of the Windermere Supergroup. Evidence for a regional metamorphic event at this time is restricted to resetting of K-Ar and palaeomagnetic signatures, indicating that associated regional metamorphism was a low-grade event (McMechan and Price, 1982). On the basis of U-Pb dating of titanites and zircons, Anderson and Davis (1995) suggested a thermal disturbance at 1030 to 1120 Ma that Anderson and Parrish (2000) correlate with Grenville-age metamorphism.

Paleozoic Tectonics

The major impact of Paleozoic tectonics on the area of the Purcell anticlinorium was a 5 to 10 km downthrow to the northwest along the antecedent to the Moyie-Dibble Creek Fault between Early Cambrian and Late Devonian time. The vertical fault movement that uplifted the area to the south ("Montania") resulted in a 240 km right-hand deflection of the boundary between the Cordilleran miogeocline and the North American cratonic platform (Price and Sears, 2000).

Jurassic-Cretaceous and Cretaceous-Paleogene Orogenies

Most of the major structural features of the Purcell anticlinorium are of Jurassic-Tertiary age and are associated with the development of the Rocky Mountain fold and thrust belt. However, many of the major faults and other deformational features can be related to antecedent structures that date back to Proterozoic and Archean structures (Höy et al., 2000; Price and Sears, 2000). Mesozoic-Cenozoic deformation and metamorphism of Purcell rocks are related to a Late Jurassic-Early Cretaceous episode of left-lateral transpression and a Late Cretaceous-Paleocene episode of right-lateral transpression, both being due to the oblique collision of the tectonic collage of Cordilleran accreted terranes ("Quesnellia") with the North American craton (Price and Sears, 2000). These two periods of transpression were separated by the intrusion of mid-Cretaceous (90-115 Ma) granite plutons. The movement of the different Jurassic-Paleocene thrust panels has been modeled as a rotation of approximately 30° about an Euler pole situated near Helena, Montana by Price and Sears (2000). Much of the relative lateral movement between the different thrust panels was along reactivated transfer faults of the Mesoproterozoic rift, notably the Moyie-Dibble Creek, St. Mary, and Hall Lake faults, and divides the Purcell anticlinorium into structural blocks. Palinspastic reconstruction, based on compensating for the different relative movements of the thrust panels, restores rocks of the Belt-Purcell Basin to their configuration prior to the Jurassic collisional event. Although this reconstruction changes the orientation of the Purcell Branch of the

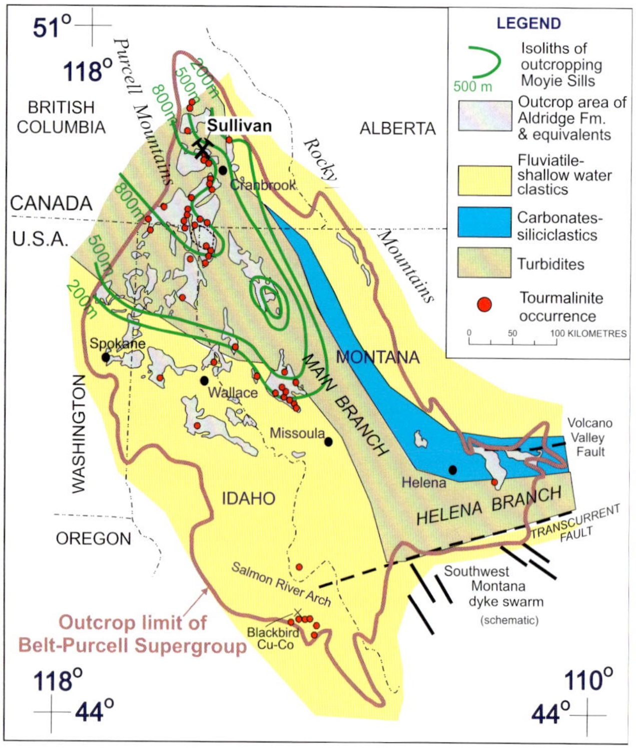

FIGURE 7. Map of the outcrop area of the Belt-Purcell Supergroup showing that the occurrence of Aldridge-Pritchard-age tourmalinite occurrences are largely coextensive with the tholeiitic Moyie sills. Note that the isoliths are for outcropping Moyies sills of the Middle Aldridge and the uppermost part of the Lower Aldridge Formation, and do not reflect the volume of sills seismically imaged in the lower part of the rift-fill sequence (see Fig. 6). Data from Höy et al. (2000) and sources cited therein.

rift to a more westerly trend, at less of an angle to the Helena Branch, geological relationships within each structural block are not affected.

Stratigraphy and Lithologies of the Aldridge Formation

Most mineral exploration of the Belt-Purcell Supergroup in Canada has focused on the Aldridge Formation, the target being SEDEX deposits similar to the Sullivan deposit. The Aldridge Formation of southeastern British Columbia and the Prichard Formation in the contiguous U.S.A. (Cressman, 1989) form the basal part of the Belt-Purcell Supergroup and, as shown in Figure 6, are essentially a sediment-sill complex that infilled an actively spreading intracontinental rift (Höy et al., 2000). Aldridge sedimentary rocks are dominantly turbidites, which most commonly consist of an arenite base and a siltite or argillite top, representing the A and E parts, respectively, of the Bouma turbidite model, with the B, C, and D parts commonly being indistinct (Hamilton et al., 1982). The Aldridge Formation has been subdivided by Cominco into three informal, but widely used and accepted members (Reesor, 1958, 1973).

The Lower Aldridge Formation is a sequence of thin- to medium-bedded, pyrrhotite-rich, distal argillaceous turbidites, whereas to the south, near Creston, more proximal, thick-bedded quartzites of the Ramparts Facies are dominant (Höy et al., 2000). Only about 2 km of Lower Aldridge turbidites are exposed at the surface, but seismic imaging suggests that the sediment-sill complex extends into the subsurface for an additional 6 km (Cook and van der Velden, 1995). East of the Rocky Mountain Trench, the deltaic to fluviatile Fort Steele Formation is considered to be the shallow-water equivalent of the upper part of the Lower Aldridge Formation (Höy et al., 2000).

The Middle Aldridge Formation consists of up to 2400 m of medium-bedded quartzitic turbidites with inter-turbidite intervals of siltstone or argillite, which extend both over the central or axial part of the rift basin in the Purcell Mountains and the more marginal areas of the basin in the Northern Hughes Range east of the Rocky Mountain Trench. Within the Middle Aldridge Formation are intervals of laminated carbonaceous siltite consisting of alternating light and dark layers, 1 to 10 mm thick, whose relative thicknesses remain constant over distances in excess of 300 km (Huebschman, 1973). The barcode-like patterns formed by the alternating light and dark layers (see Lydon, 2000c for details) are diagnostic of a specific stratigraphical horizon, and have been used by Cominco to establish stratigraphical correlations within the otherwise monotonous sequence of turbidites. More than a dozen of these "Marker" intervals have been recognized (Hamilton et al., 2000). The Middle Aldridge Formation has been divided into three upward-fining cycles, the bottom of each representing a period of tectonic extension and basinal subsidence (Höy et al., 2000). These cycles are termed, from stratigraphically lowest to stratigraphically highest, MA1, MA2, and MA3, respectively (Fig. 5). The top of MA1, in particular, has a high incidence of sedimentary fragmental rocks, tourmalinites, and sulphide deposits, including the Kootenay King SEDEX deposit and the disseminated sulphide deposits of Canam and Star (Fig. 4).

The Upper Aldridge Formation is approximately 300 m thick and consists of thin-bedded to laminated, pyrrhotite-rich argillite and siltite, and represents a shallowing basin (Höy et al., 2000). The laterally most extensive stratiform tourmalinization in the Purcell Basin occurs in the Upper Aldridge Formation in the Doctor Creek-Findlay Creek area (Fig. 4) (Greig et al., 2001).

Moyie Sills

The gabbroic Moyie sills occupy about 40 vol.% of the Aldridge stratigraphic sequence in the area immediately to the west of the Sullivan deposit, and between 5 and 30 vol.% elsewhere in the Aldridge Formation. Isoliths of sills within the exposed rocks indicate over 1 km of aggregate sill thickness along the axis of the Belt-Purcell rift (Fig. 7) (Höy et al., 2000), implying a minimum total volume of more than 50,000 km^3 of magma emplacement (Anderson and Goodfellow, 2000). Inclusion of the deeper part of the Prichard-Aldridge sediment-sill complex, which has only been seismically imaged, would greatly increase this estimate.

U-Pb dates on zircons from mafic sills associated with the rifting have given ages of 1469 Ma (Sears et al., 1998), 1468 Ma (Anderson and Davis, 1995; Schandl and Davis, 2000), 1457 Ma (Sears et al., 1998), 1445 Ma (Höy, 1989), 1449 Ma (D.W. Davis cited in Brown and Woodfill, 1998), 1433 Ma (Zartman et al., 1982), and 1379 Ma (Doughty and Chamberlain, 1996), suggesting recurrent mafic magmatic upflow over the period 1470 to 1430 Ma, with perhaps major episodes at about 1470 and 1445 Ma, and a final upflow at about 1380 Ma (see Fig. 6). Additional U-Pb ages of felsic magmatic products associated with the mafic magmatism in the Purcell lavas (stratigraphic equivalent to the Nicol Creek volcanics) of 1443 Ma (Aleinikoff et al., 1996; Evans et al., 2000), of porphyritic granite in the Salmon River Arch of 1370 Ma (Evans and Zartman, 1990), and the Hell Roaring Creek stock of 1340/1360 Ma, indicates either association of felsic differentiates with mafic magmatic upflow (Doughty and Chamberlain, 1996) or anatectic melts generated by mafic magma emplacement at high metamorphic temperatures.

Hydrous autometamorphism of most sills, the formation of peperite and "granofels" along sill margins, and the local disruption, fluidization, and expulsion upwards of adjacent sediments as mud-volcano eruptions, indicates that the sills were emplaced at shallow depth into wet, semiconsolidated sediment (Höy, 1989; Buckley and Sears, 1998; Anderson and Höy, 2000; Höy et al., 2000) and were synchronous with Middle Aldridge Formation sedimentation. Not uncommonly, and especially near the top of the Lower Aldridge (Doug Anderson, pers. comm., 1995), the sills have partially melted or incorporated wet sediment forming "granophyre" (Cominco terminology), "granofels" (Turner et al., 2000b), or "granosediment" (Poage et al., 2000) at their margins.

Mud Volcanoes and Tourmalinites

The Aldridge and Prichard formations record a high incidence of mud-volcano activity that is manifest as bodies of sedimentary fragmental rocks (Figs. 4, 7). In the majority of cases, these discordant and concordant bodies respectively represent the vent breccias of the conduits and the eruption breccias (or wastage aprons of the extrusive mounds) of mud volcanoes. On a local scale, discordant sedimentary fragmental bodies are rooted in disrupted and fragmented sedimentary strata that have been fluidized along the upper contact of Moyie sills (Anderson and Höy, 2000; Höy et al., 2000). Individual bodies of sedimentary fragmental rocks vary in maximum dimension from a few tens to hundreds of metres and occur over a stratigraphic interval of at least 3000 m, with the greatest volume occurring at or just below the Lower to Middle Aldridge contact, and a second incidence near the top of the MA1 sedimentary cycle (Anderson and Höy, 2000). By far the largest mud-volcano field is that of the Sullivan-North Star corridor, which is at least 7 km long and up to 3 km wide (Ransom and Lydon, 2000; Turner et al., 2000a). Discordant fragmental, or vent-breccia bodies, are commonly elongated along rift-parallel synsedimentary faults. Some of the larger concordant sedimentary fragmental bodies may be debris flows related to the slumping of unstable submarine slopes.

Tourmalinite in the Belt-Purcell Basin is a rock consisting of >30 vol.% tourmaline that was formed by the replacement of the aluminosilicate fraction of a sedimentary rock by tourmaline (Slack et al., 2000). Most tourmalinite occurrences are in the Aldridge or Prichard formations, and most are within or immediately adjacent to sedimentary fragmental rocks, particularly discordant bodies (Anderson and Höy, 2000). About 30% of all bodies of sedimentary fragmental rocks in the Aldridge Formation are at least partially tourmalinized (Anderson and Höy, 2000). The genetic association of mud volcanoes, tourmalinization, and Moyie sill emplacement is reflected by the three being co-extensive on a regional scale (Fig. 7). The great majority of tourmalinite occurrences in the Prichard-Aldridge formations and the sedimentary fragmental bodies, with which they are associated, are restricted to those areas where there is a significant cumulative thickness of Moyie sills, especially along the presumed locus of gabbroic magma upflow at the rift axis (Lydon, 2000b).

Stratigraphic Interval of the 1468 Ma Moyie sill Event

As discussed by Anderson and Parrish (2000), the predominance of sills over dykes and lava flows, together with sediment fluidization and the formation of "granofels" at sill margins, indicate that the magmas of the Moyie sill event were emplaced into water-saturated, unconsolidated sediments at the level of neutral magma buoyancy. Under these conditions, sill emplacement tends to migrate to successively higher stratigraphic levels, because the compaction and thermal induration of unconsolidated sediments surrounding a newly emplaced sill reduces the capacity of the sediments to accommodate subsequent magma emplacement by lateral flow. As noted above, mud volcanos in the Aldridge are spatially and genetically associated with sill emplacement, and reflect the release of pressure created by compaction of sediments by sill emplacement. The stratigraphically lowest expressions of mud-volcano activity are the sedimentary fragmental rocks just below the Sullivan Horizon in the Sullivan-North Star corridor and near the top of the Lower Aldridge at other localities (Anderson and Höy, 2000). The stratigraphically highest expression is the St. Joe "tuff" (Fig. 5B) near the base of the MA3 fining-upward cycle of the Middle Aldridge (Anderson and Parrish, 2000; Höy et al., 2000).

The stratigraphic level of the lowest mud-volcano eruption is not conclusive evidence for the timing of initiation of the Moyie sill magmatic event, because sill emplacement into wet, unconsolidated sediment does not necessarily produce mud-volcano activity at the seafloor. The stratigraphic separation between the St.Joe "tuff" and the base of the Middle Aldridge is about 2.2 km in the Moyie area, which, assuming the accumulation rate of 570 m/my of Figure 5, represents a time interval of 3.85 my. The maximum duration of Moyie sill emplacements is the 5 my encompassed by the analytical uncertainty (1468 +3/-2 Ma) for the U/Pb dating of zircons from both Lower Aldridge and Middle Aldridge Moyie sills (Anderson and Davis, 1995). Therefore, Moyie sill emplacement must have commenced no more than 1.15 my prior to initiation of Middle Aldridge sedimentation. Because accumulation rates of the Lower Aldridge cannot have been higher than 570 m/my (Fig. 5), the lowest stratigraphical level corresponding to the time at which emplacement of Moyie sills commenced is therefore about 650 m below Lower-Middle Aldridge contact. However, because the Sullivan Horizon, just below the Lower-Middle Aldridge contact, represents a period of sedimentation starvation with very much lower average accumulation rates, it is likely that the 1.15 my time period is represented by a much smaller thickness than 650 m. Initiation of sill emplacement was therefore likely very close in time to the tectonic event that caused a deepening of the basin in the Purcell anticlinorium, which in turn caused the sediment starvation at the top of the Lower Aldridge (Höy et al.,

Table 1. A selection of grade and tonnage data for Mesoproterozoic deposits of the Belt-Purcell Basin. See text for classification criteria.

Deposit type	Deposit name	Cu %	Co %	Zn %	Pb %	Ag g/t	Au g/t	Tonne	Comments	Reference
Seafloor Sulphide deposits										
SEDEX										
	Kootenay King	0.00		15.60	5.35	67	0.00	13,000	Mineral resource	BC Minfile 082FGNW009
	North Star	0.00		6.12	35.50	673	0.00	61,000	Production	BC Minfile 082FNE053
	Sullivan	0.00		5.86	6.08	67	0.00	161,970,000	Original mineral resource	Hamilton et al., 1983
	Sheep Creek Lower Zone	4.00						4,000,000	Mineral resource	Zieg, 1993
	Sheep Creek Upper Zone	2.50	0.10					4,500,000	Mineral resource	Zieg, 1993
Besshi										
	Blackbird	1.50	0.60					960,099	Production 1951-1959	Nash and Hahn, 1989
	Ram & Sunshine	0.54	0.68				0.56	1,922,568	Proven, probable and inferred resources	Bending and Scales, 2001
Vent Complex deposits										
	Stemwinder	0.00		15.60	3.70	76	0.00	25,000	Production	BC Minfile 082FNE116
Stratabound disseminated sulphide deposits										
Redbed Copper deposits										
	Spar Lake	0.70				43		80,000,000	Pre-mining mineral resource	Calculated from data in Reipas and Couture, 2005
	Rock Creek	0.72				45		124,200,000	Mineral resource	Calculated from data in Couture Couture and Tanaka, 2005
	Montanore	0.78				60		136,000,000	Mineral resource	Long et al., 1998
Veins										
Coeur d'Alene district										
	Bunker Hill	0.01	0.00	3.72	0.28	45	0.01	1,279,729	Production 1981-90	Bennett and Singer, 1991
	Sunshine	0.36	0.00	0.00	0.01	706	0.01	1,499,832	Production 1981-90	Bennett and Singer, 1991
	Coeur	0.75	0.00	0.00	0.00	559	0.08	1,357,406	Production 1981-90	Bennett and Singer, 1991
	Galena	0.58	0.00	0.00	0.00	621	0.08	1,797,867	Production 1981-90	Bennett and Singer, 1991
	Lucky Friday	0.22	0.00	1.42	11.29	582	0.30	1,520,362	Production 1981-90	Bennett and Singer, 1991
	Star-Morning	0.00	0.00	5.90	4.24	95	0.00	497,068	Production 1981-90	Bennett and Singer, 1991
	Coeur d'Alene district 1884-1997	0.00	0.00	2.62	6.66	313		109,496,626	Production 1884-1997 Grades estimated	Long, 1998
Purcell anticlinorium										
	St. Eugene			0.98	7.66	124	0.05	1,475,266	Production	BC Minfile 082GSW025
	Vine			2.40	4.60	52	1.80	547,000	Mineral resource	Höy and Pighin, 1995

2000). The metallogenetic significance of this timing is that it means that the Sullivan ore deposit formed during the early stages of sill emplacement into Lower Aldridge turbidites.

Mineral Deposits and Occurrences

Overview of Deposit Types and Classification

The Belt-Purcell Basin contains a variety of base metal mineral deposits and occurrences (Figs. 1, 4). Höy et al. (2000) have classified Mesoproterozoic deposits occurring in the Purcell anticlinorium into four main types, which, with the addition of Besshi-type deposits of Idaho and Redbed Copper-type deposits of Montana, make up the variety of Proterozoic mineral deposit types that occur in Belt-Purcell rocks. These six deposit types can be classified into the three groups that are briefly described below. More detailed descriptions of some of these deposits are described in later sections.

Seafloor Sulphide Deposits

SEDEX deposits represent the accumulation of iron, zinc, and lead sulphides on the seafloor, or just below the sediment surface, around hydrothermal vents of sedimentary basins. The hydrothermal fluids, which in general cannot be directly linked to magmatism, are thought to generally have a temperature <250°C and represent the discharge of formational brines from a compacting sedimentary pile (Lydon, 1996). Sullivan is the prime example in the Belt-Purcell Basin and is described in detail below; other examples include the Sheep Creek (Fig. 1) and Kootenay King (Fig. 4) deposits.

Besshi-type deposits are stratiform Cu-bearing, massive sulphide deposits that typically have a very high length:thickness ratio and occur in sill-sediment complexes of oceanic extensional environments. They formed from high temperature (>300°C) seafloor hydrothermal systems of seawater convection cells driven by the heat of subsurface mafic magmatic intrusions. They are a variant of the volcanogenic massive sulphide (VMS) class of deposits. Deposits of the Idaho Cu-Co belt, including Blackbird (Fig. 1), Iron Creek, and Black Pine have been interpreted to be of this type (e.g. Nold, 1990).

Vent complex / feeder pipe deposits are interpreted to represent the upflow zone of seafloor hydrothermal vents that did not develop a substantial overlying concordant sulphide deposit, or alternatively the root zone of a SEDEX deposit from which the concordant stratiform part has been eroded. The Stemwinder and Fors (Fig. 4) deposits are examples.

Stratabound Disseminated Sulphide Deposits

Redbed Copper deposits form by deposition of copper sulphides from oxic groundwaters at a reducing interface, such as black shales, usually during the early burial and compaction history of a sedimentary succession. Redbed Copper deposits of the Revette Formation in Montana contain three major deposits of this type (Fig. 1): Troy (Spar Lake), which has already been mined, Rock Creek, which is in its final stages of permitting, and Montanore.

Disseminated sphalerite and galena occupy large volumes in arenaceous beds over stratigraphic intervals of tens of metres in the Middle Aldridge. The Star and Canam deposits (Fig. 4) are examples.

Veins

Mesozoic veins. Vein deposits in Aldridge rocks have been divided into Cu types, Pb-Zn-Ag types, and Au types (Höy, 1993). Many of these deposits are associated with Mesozoic intrusions and are not considered further here.

Mesoproterozoic veins. A Mesoproterozoic age for the galena-sphalerite and galena-tetrahedrite veins of the Coeur d'Alene district of Idaho (Fig. 1), and the St. Eugene and Vine (Cominco data) sphalerite-galena veins of the Purcell anticlinorium (Fig. 4) is indicated by the Pb isotope compositions of galena (Beaudoin, 1977).

Deposit Distribution

The geographical distribution of major Mesoproterozoic deposits is shown in Figure 1. Smaller deposits of the Canadian portion of the Belt-Purcell Basin are shown in Figure 4. All the major Mesoproterozoic mineral deposits occur in the lower part of Belt-Purcell stratigraphy (Fig. 5). Seafloor sulphide deposits appear to be restricted to Aldridge (Pritchard) host rocks and their stratigraphic equivalents, and fall into the time interval of rift fill sedimentation. Epigenetic deposits span the range from Aldridge to Creston Formation time (see age-stratigraphic relationships in Fig. 6 and Canada-U.S.A stratigraphic correlations in Fig. 5). It is interesting to note that vein deposits hosted by Aldridge-age rocks dominantly have a Pb-Zn-Ag metal association, whereas epigenetic deposits in Creston age (Ravalli Group) rocks, whether they be veins (e.g. tetrahedrite-bearing veins of the Coeur d'Alene district) or stratabound disseminated sulphides (e.g. Spar Lake, Rock Creek) dominantly have a Cu-Ag metal association.

Grade and Tonnage Statistics

A selection of grade and tonnage data for Mesoproterozoic mineral deposits of the Belt-Purcell Basin are shown in Table 1.

Production

The greatest production of metals has come from the Sullivan deposit, amounting to over 16,000,000 tonnes of Pb+Zn and 9,000 tonnes of Ag. The aggregate base metal production from the Coeur d'Alene district is a little over half that from Sullivan, but its aggregate silver production is about three times that of Sullivan. As reflected in the statistics of Table 1, the bulk of silver production from the Coeur d'Alene district is from the tetrahedrite-bearing set of veins rather than the sphalerite-galena set of veins. During 2004, about 5.5 million ounces (approximately 171 tonnes) of Ag was produced from the Coeur d'Alene district, mainly from the Galena, Coeur, and Lucky Friday mines. Production of Pb, Zn, and associated Ag from other deposits, such as the St. Eugene vein, North Star SEDEX, and Stemwinder Vent Complex deposits has been very small (Table 1) in comparison to Sullivan or the Coeur d'Alene district. Production from the Spar Lake Cu-Ag Red Bed Copper type of deposit from 1981 until suspension of mining in 1993 was 33,742,281 tons (approximately. 30.6 million tonnes) of ore at an average grade of 0.66% Cu and 1.50 opt Ag (approximately. 42 g/t) (Reipas and Couture, 2005). Production from the Blackbird Cu-Co mine for the period 1951 to 1959 is reported as 6350 tons of Co from an average ore grade of 0.6% Co and 1.5% Cu by Nash and Hahn (1989) (i.e. 960,099 tonnes ore treated if 100% Co recovery), but Bending and Scales (2001) state that the mine produced >5,000,000 tons of ore during the period 1949 to 1965.

Resources

Mining at Sullivan ceased in 2001 with depletion of its reserves. Exploration for the continuation of the Sullivan deposit on the north side of the Kimberley fault (Fig. 8) has so far met with some, though limited, success. At the end of 2004, proven and probable reserves at the Galena, Coeur, and Lucky Friday mines of the Coeur d'Alene district amounted to about 1,337,000 tonnes with an average grade of approximately 472 g/t Ag (data from 2005 annual reports of the Coeur d'Alene Mines Corporation and the Hecla Mining Company). Proven and probable reserves remaining at the Troy deposit at resumption of production in 2004 were estimated as 8.7 million tons (approximately. 7.9 million tonnes) at 0.65% Cu and 1.57 opt (approximately 44 g/t) Ag (Reipas and Couture, 2005). The Rock Creek deposit, which is at an advanced stage of production feasibility study, is estimated to contain a total inferred mineral resource of about 124,000,000 tonnes at 0.72% Cu and 45 g/t Ag. A marginally greater resource is inferred for the nearby Montanore deposit, which is currently undergoing a permitting procedure in anticipation of a future mining operation. Between 1978 and 1982, exploration by Noranda Ltd. defined 5 million tons of "mining reserves" and 7.3 million tons of "indicated reserves" of Cu-Co ores in the Blackbird mining district (Nash and Hahn, 1989), but it is uncertain whether these resources include the approximately 2 million tonnes listed in Table 1 for the Ram and Sunshine deposits. Other Cu-Co resources have been delineated at the Sheep Creek deposit (Table 1).

Geology of the Sullivan Deposit

Within the Canadian portion of the Belt-Purcell Basin, the Sullivan deposit is by far the most important mineral deposit. At the time of its closure at the end of 2001 it was Canada's longest-lived continuous mining operation and had produced metals worth over $20 billion in terms of 2005 metal prices. Only the Sullivan deposit will be described in detail here. Detailed descriptions of the Sullivan deposit have previously been given by Freeze (1966), Ethier et al. (1976), Ransom (1977), Campbell et al. (1980), Hamilton et al. (1982, 1983), Campbell and Ethier (1983), McClay (1983), Conly et al. (2000), Lydon et al. (2000a), Owens, (2000), Ransom and Lydon (2000), and Lydon (2004).

Geological Setting

The Sullivan deposit occurs just below the contact between the Lower and Middle Aldridge Formation. On a regional scale, the top of the Lower Aldridge is marked by up to 20 m of a black laminated mudstone that is termed "CWL" (carbonaceous wacke laminite) by Cominco (Ransom and Lydon, 2000). This interval, called the Sullivan Horizon, records a period of sediment starvation following a sudden tectonic downdrop of the basin (Höy et al., 2000), which in effect caused a regression of the delta complex to the south and a removal of this part of the basin beyond the

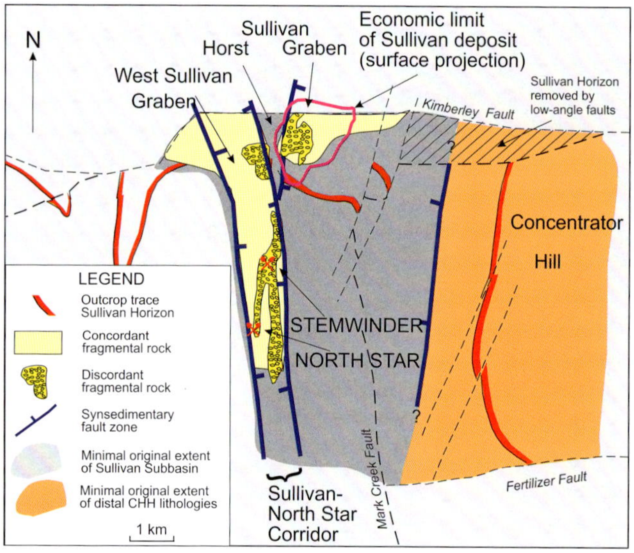

FIGURE 8. Map of the major geological features of the Sullivan Subbasin and the Sullivan-North Star Corridor. CHH refers to the laminated muscovite and pyrrhotite-rich laminated facies of the Concentrator Hill Horizon. After Ransom an d Lydon (2000).

northern limit of its turbidite fan. The subsequent northward progradation of the turbidite fan marks the beginning of the Middle Aldridge. In the Kimberley area, the stratigraphic interval occupied by the CWL is up to 200 m thick due to the intercalation of products from contemporaneous mud-volcano and hydrothermal activity (Ransom and Lydon, 2000). The area of stratigraphic thickening and mud-volcano activity is called the Sullivan Subbasin and measures at least 13 km in a north-south direction and 3 to 5 km in an east-west direction (Fig. 8). In cross-section, the Subbasin is asymmetric and is thickest in the west where it is marked by a 1 km wide, north-south trending zone of mud volcanoes, synsedimentary faults, disrupted sediments, and hydrothermal alteration called the Sullivan-North Star Corridor (Turner et al., 2000a, b). The Subbasin gradually thins to the east and terminates against an interpreted, west-facing synsedimentary fault that separates the Subbasin proper from the stratigraphically equivalent plume deposits of the Concentrator Hill Horizon (see Fig. 12J in Ransom and Lydon, 2000).

Rocks of the Sullivan-North Star corridor have been affected by a variety of hydrothermal alteration types that are similar to those occurring in the footwall and hanging wall of the Sullivan deposit itself (Leitch et al., 2000; Turner et al., 2000c). The most extensive is sericite alteration, which is strongly controlled by the distribution of coarse sedimentary fragmental rocks stratigraphically lower than the Sullivan Horizon. Sericite alteration is only marginally enriched in K_2O (as measured by K_2O/Al_2O_3 ratios) compared to unaltered Aldridge sedimentary rocks, and its typical pale yellowish grey-green colour is due to the absence of biotite that gives the brown colour to unaltered Aldridge sedimentary rocks. Other alteration types include tourmalinization, chloritization, and albitization (Turner et al., 2000c).

The Sullivan Subbasin is close to the axis of maximum Aldridge turbidite thickness (Fig. 1) and maximum aggregate thickness of Moyie sills (Fig. 7) (Price and Sears, 2000), and is interpreted to be a local structurally controlled half-graben or graben related to the spreading of the Belt-Purcell rift. The Sullivan deposit is situated at the northern end of the Sullivan-North Star Corridor (Fig. 8). It has been suggested (Höy et al., 2000; Turner et al., 2000b) that the Sullivan deposit was localized by the intersection of the Sullivan Corridor and an ancestral Kimberley Fault, which is parallel to synsedimentary rift transfer faults identified elsewhere in the Aldridge Basin (Fig. 4) (Höy et al., 2000; Price and Sears, 2000). However, if this ancestral Kimberley Fault existed as a synsedimentary fault, it seems to have had little effect on subbasin lithofacies, because they can be traced in drill core across the present Kimberley Fault without any major change in thicknesses or lithologies (Ransom and Lydon, 2000). On the other hand, at the time of Moyie sill emplacement, there was an east-west structural grain to subbasin rocks, because dykes rising from the Mine sill follow this orientation. Furthermore, pyrite-chlorite-carbonate alteration, which appears to be genetically related to emplacement of the mine sill (Turner and Leitch, 1992), extensively overprints the Sullivan orebody along the present Kimberley Fault (see Fig. 9).

Morphology and Architecture of the Sullivan Deposit

The Sullivan deposit is a semiconcordant sulphide body whose economic portion is approximately 2,000 m by 1,600 m in area (Figs. 9, 10), and up to 100 m thick (Fig. 10) (Hamilton et al., 1982) that accumulated in the crater or on the flanks of a mud volcano (Ransom and Lydon, 2000; Turner et al., 2000b). The orebody can be divided into a thicker, more massive, western "Vent Complex" and a thinner, bedded eastern part, the two being separated by a structurally complex Transition Zone (Fig. 9, 10). The western

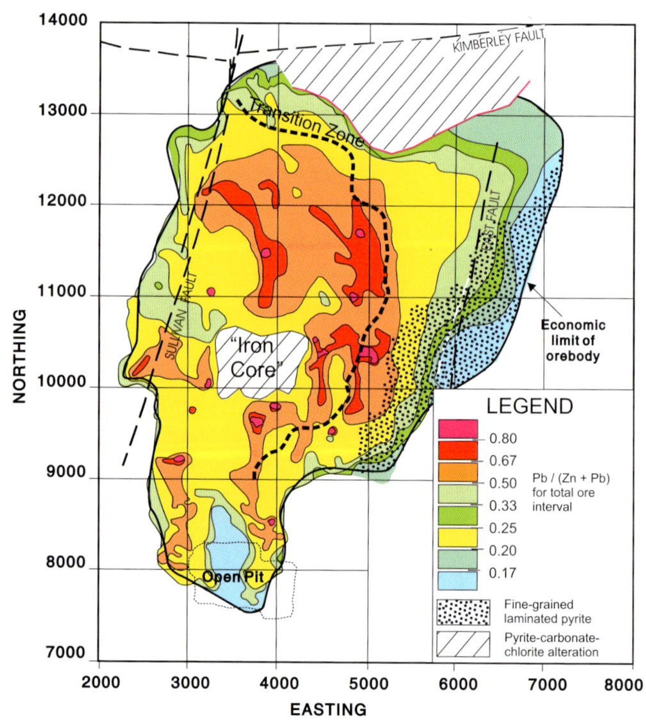

FIGURE 9. Map of the Sullivan deposit showing the average Pb/(Zn+Pb) ratio for the entire ore thickness (after Hamilton et al., 1982) and the approximate distribution of pyrite-rich sulphides. Note that the grid coordinates are in feet.

part overlies a subvertical, hydrothermal upflow zone and alteration pipe, and is termed the Vent Complex. The main sulphide minerals are pyrrhotite, pyrite, sphalerite, galena, and minor boulangerite, with associated magnetite, cassiterite, manganiferous carbonate, and garnet. The ore horizon has been traced about 7 km east and 12 km south of Sullivan (Fig. 8) as the Concentrator Hill Horizon, a 3 to 7 m thick interval that consists of thin pyrrhotite layers and disseminations in laminated quartz-muscovite siltites and argillites containing geochemically elevated values of Pb and Zn (Fig. 13E) (Ransom and Lydon, 2000).

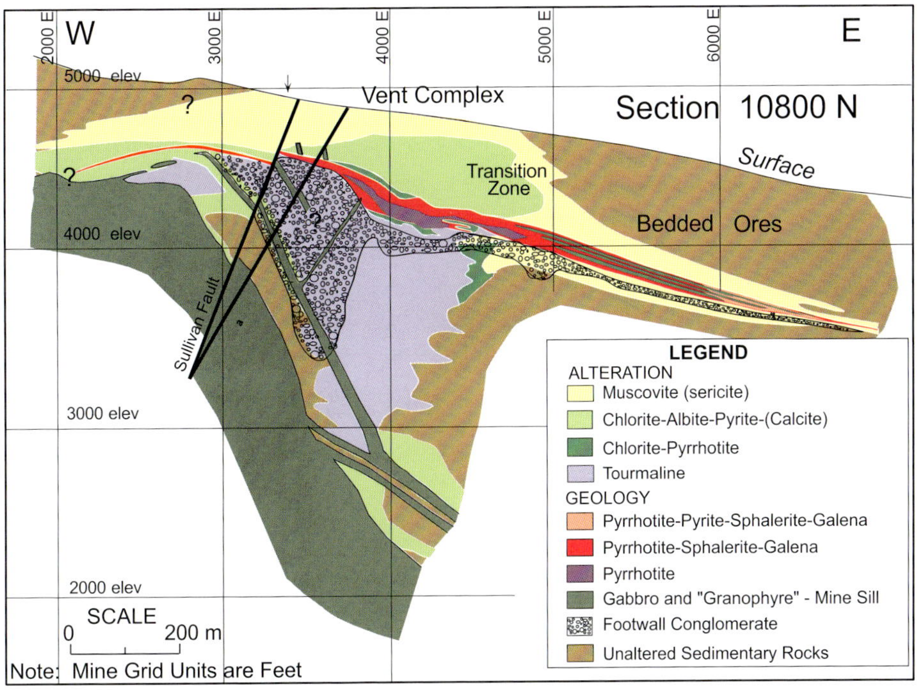

FIGURE 10. Semi-schematic west-east cross-section through the Sullivan deposit showing the distribution of major ore types and major alteration types (based on Hamilton et al., 1982).

Footwall

The immediate footwall to the Sullivan deposit consists of vent breccias, extrusion breccias and mudflows of a mud volcano ("Footwall Conglomerate"), which is more than 80 m thick under the Vent Complex but tapers eastwards and northwards to a few metres of bedded sandstones and siltstones. Under the Vent Complex, the Footwall Conglomerate and footwall turbidites have been tourmalinized (Fig. 11A) to form an impressive alteration pipe about 1 km in diameter that can traced to a depth of 450 m below the orebody where it is cut by the gabbro and granofels of the Mine Sill (Fig. 10). Over an area that is more extensive than the tourmalinite pipe, footwall rocks are cut by a variety of sulphide-bearing veins and are impregnated with sulphide disseminations that locally almost entirely replace the host rock (Fig. 11A). Veins containing pyrrhotite, with minor sphalerite, galena, arsenopyrite, chalcopyrite, and quartz and/or calcite as gangue minerals are locally abundant beneath the massive sulphides of the Vent Complex (Hamilton et al., 1982).

Sullivan Ore Zone

Vent Complex. Massive sulphides of the Vent Complex comprise about 70% of Sullivan ore tonnage. The main sulphide body is up to 100 m thick and is coarsely layered, consisting of a lower uneconomic pyrrhotite-rich sulphide zone (Fig. 11B), a middle zone of tectonically foliated pyrrhotite-galena-sphalerite (Fig. 11C), and an upper zone of galena-sphalerite-pyrrhotite massive sulphides that are intercalated with lithic units.

Bedded ores. The Bedded Ores consist of five concordant layers of interlaminated sulphides and argillite (Fig. 11D, E, F), termed the "Ore Bands" by Cominco, that are separated by beds of arenaceous to argillaceous mud flows, termed the "Waste Bands" (Fig. 12). From base to top, the principal sulphide layers have been termed the "Main", "A", "B", "C", and "D" Ore Bands, and the intervening siliciclastic layers, the "A", "B", "C", and "D" Waste Bands (Fig. 12). The Bedded Ores are more than 30 m thick near the Transition Zone and gradually taper to <10 m at the economic limit of the orebody. The bottom part of the Main Band is fragmental in texture, consisting of subangular to rounded clasts of wall rock, pyrite, coarsely crystalline calcite, sphalerite, and quartz in a matrix dominated by pyrrhotite (Fig. 13A). This fragmental sulphide layer is a zone of durchbewegt sulphides, formed along a tectonic décollement at the base of the sulphide body during Cretaceous deformation.

Stratigraphic correlation between the Bedded Ores and the Vent Complex is lost in the Transition Zone, which is characterized by spectacular, convolute folding of the Bedded Ores (Fig. 13B). This deformation developed during the eastward movement of the Vent Complex over the Bedded Ores along reverse faults, which depressed the nose of the Bedded Ore sequence downward to the base of the Vent Complex massive sulphides. Galena-rich sulphides have been injected upward into the stratigraphic level of the "HU" hanging-wall ores (Paakki et al., 1995).

Hanging-wall ores. The hanging wall sequence to the main Sullivan orebody consists of three graded beds, the "I", "H", and "HU", each up to 10 m thick and marked by a basal, quartz-rich arenite and an upper interval of laminated, pyrrhotitic and carbonaceous argillite up to about 3 m thick. Above the Transition Zone and arcing around the northern part of the hanging wall albitite zone, the laminated, pyrrhotitic, carbonaceous argillite intervals of the "I", "H", and "HU" graded beds contain bodies of laminated sulphides, very similar in appearance to those of the "A" through "D" Ore Bands (Fig. 11E). Parts of these sulphides have been mined. Conly et al. (2000) interpret these ores to reflect synsedimentary deposition but Paakki et al. (1995) point out that at least some of the Hanging Wall Ores represent galena-rich tectonic mobilization from the main orebody into the hanging wall.

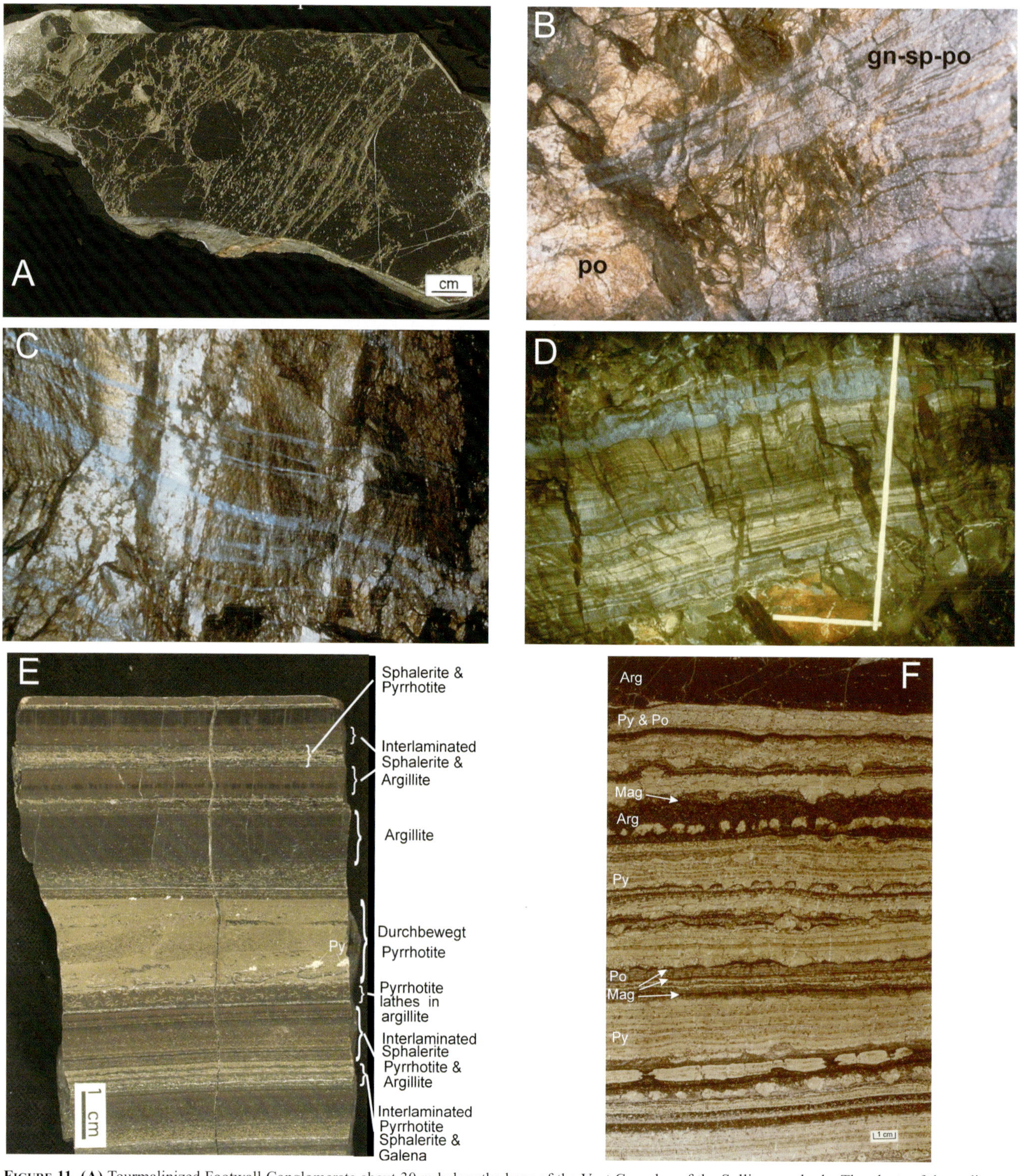

FIGURE 11. (A) Tourmalinized Footwall Conglomerate about 30 m below the base of the Vent Complex of the Sullivan orebody. The clasts of the sedimentary fragmental rock are clearly preserved. The tourmalinite is cut by a network of pyrhhotite veinlets and disseminated grains, which are concentrated in the matrix and highlight the sedimentary fabric of the rock. **(B)** Contact relationships between a remnant block of galena-sphalerite-pyrrhotite banded ore in massive "barren" pyrrhotite that replaces the basal part of the orebody in the Vent Complex. 39-R-14 area. Width of photo about one metre. Photo by Cominco, from Leitch et al. (2000). **(C)** Middle part of the Vent Complex massive sulphide body showing the strong tectonic foliation marked by galena-rich schlieren of high-strain zones. Vertical field of view about 1.5 metres. 34-R stope. Photo by Cominco. From Lydon et al. (2000a). **(D)** Interlayered laminated sphalerite-galena-pyrrhotite sulphides and argillite (dark layers) in the Bedded Ores "A" Band. Note limits of Ore Band are marked by white chalk, below and above which are the mud flows of the "A" and "B" Waste Beds, respectively. Width of photo about two metres. Photo by Cominco. **(E)** Polished slab from the "B" Band, 3200 SE Crosscut, Sullivan mine, showing the typically laminated nature of the Bedded Ores. Note the layer of durchbewegt pyrrhotite-rich sulphides, which reflects bedding-parallel tectonic movement – a common feature of the Bedded Ores. From Lydon and Reardon (2000). **(F)** Laminated pyrite interlayered with argillite, and thin layers of coarsely crystalline magnetite, in the Ramp Extension, Southeastern Fringe of the Sullivan orebody. Some of the pyrite layers have been partially replaced by pyrrhotite. The small "spots" in the laminated pyrite layers are porphyroblasts of manganiferous garnet. From Lydon et al. (2000a).

Hydrothermal Alteration

There are at least five "stages" and more than seven types of hydrothermal alteration associated with the Sullivan deposit (Leitch et al., 2000):

1. *Tourmaline alteration* of the large footwall "tourmalinite pipe" (Fig. 10; 11A) and also small discontinuous bodies that are preserved around the periphery of albite-chlorite in the hanging wall.
2. *Chlorite-pyrrhotite alteration* occurs principally as a semiconformable body at the contact of the massive sulphides and footwall tourmalinite (Fig. 10) and also in discontinuous bodies at the hanging wall contact of the massive sulphides in the western part of the deposit (Shaw et al., 1993a). Argillite laminae within the Bedded Ores are almost invariably chloritized (Lydon et al., 2000a).
3. *Chlorite-albite-pyrite alteration* is most abundant in the immediate hanging wall of the Sullivan deposit (Fig. 10, 13C), vertically above the tourmalinite pipe (Shaw et al., 1993b), where there is a general zonation from an albite-rich core to a chlorite-pyrite-rich periphery. Chlorite-albite-pyrite alteration also occurs along the western side of the Sullivan-North Star corridor and along contacts of gabbro sills (Turner et al., 2000c).
4. *Pyrite-carbonate alteration.* Immediately below the body of hanging wall albite-chlorite-pyrite alteration the main sulphide body has been replaced by a pyrite-chlorite-carbonate assemblage (Fig. 13D) (Shaw et al., 1993a), which is most intense in the "iron-core" (Fig. 9) (Hamilton et al., 1982). Along the Kimberley Fault, sulphides of the Ore Bands have been replaced by a pyrite-chlorite-calcite assemblage (Fig. 9) (Hamilton et al., 1982; Owens, 2000).
5. *Carbonate alteration.* Especially near the southern part of the Transition Zone, Waste Bands and to a lesser extent the Ore Bands, are heavily impregnated with calcitic carbonate. In some cases the waste bands are so heavily carbonated that they may be classified as an impure limestone.
6. *Muscovite (sericitic) alteration* forms an envelope around the entire deposit (Fig. 10) (Shaw and Hodgson, 1980; Leitch et al., 2000; Turner et al., 2000c), and also occurs along the extent of the Sullivan Corridor (Turner et al., 2000a). The envelope of sericite alteration around the deposit but the lack of significant potassium enrichment of sericitised rocks, suggests that it may be a metamorphic effect, in which biotite is sulphidized to iron sulphide and muscovite.
7. *Other alteration types* are of only local or sporadic extent, including chlorite-biotite-garnet and carbonate-muscovite near the eastern margin of the tourmalinite zone (Shaw et al., 1993a) and silicification near zones of muscovite-altered tourmalinite (Leitch et al., 2000).

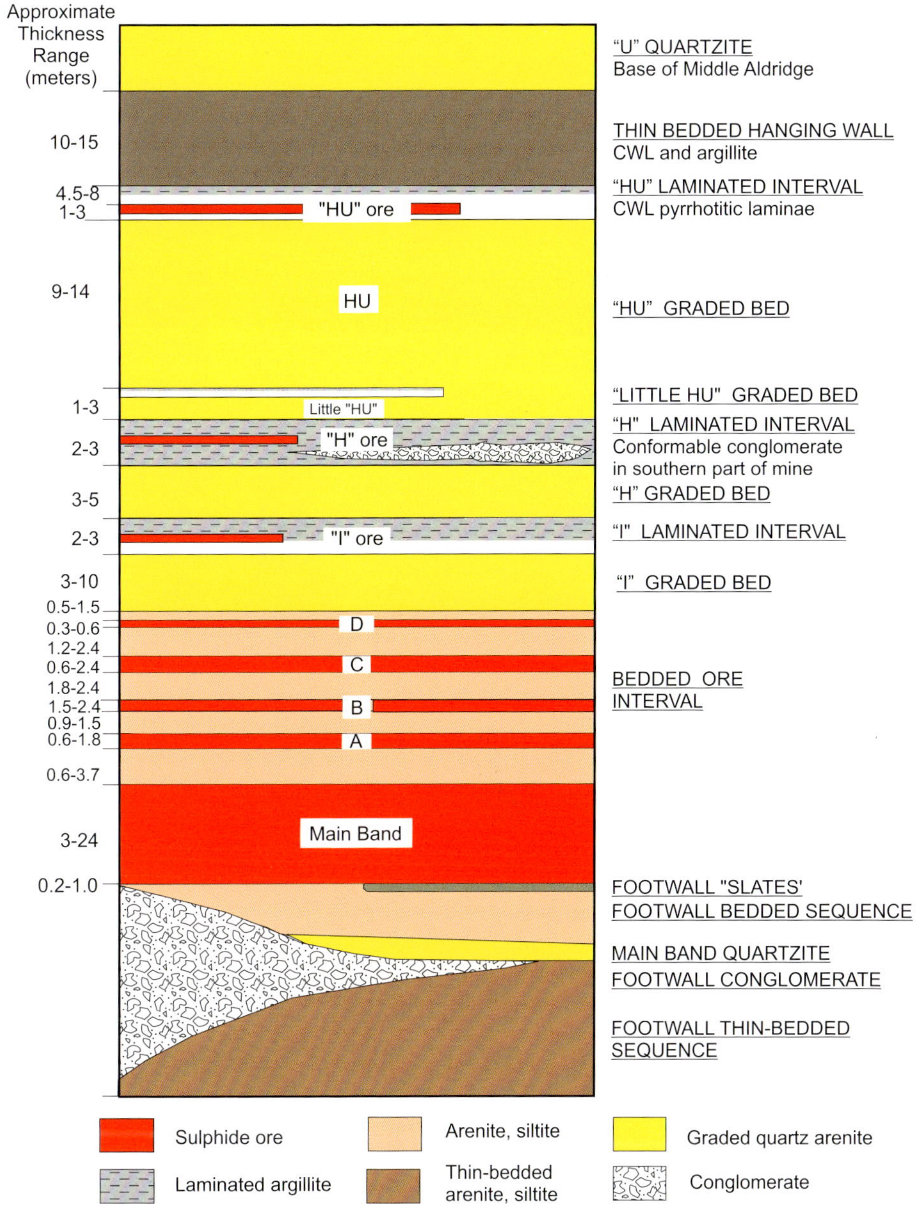

FIGURE 12. Idealized mine stratigraphy in the Bedded Ores. Modified after Hamilton et al. (1982).

These various types of alteration span the history of hydrothermal events at Sullivan, but most are not directly related to the main ore-forming event. The tourmalinite alteration pre-dates ore formation and may represent the venting of "vapour" from the phase separation of formational waters heated to high temperatures by sill emplacement at depth in

FIGURE 13. (A) Durchbewegt sulphide, basal part of Main Band, DDH 7246, 136 ft. Rounded clasts of pyrite, argillite, and coarse-grained calcite in a pyrrhotite matrix. Durchbewegung structure is produced by autogenous milling of the more mechanically competent fragments during tectonic flow of the mechanically incompetent matrix. (B) Convolute folding of galena-rich bedded ore in the Transition Zone. Dark layers are argillite. Note pyrrhotite-rich sulphides at bottom of photograph. Width of photo about two metres. 4250-P Scram 3. Photo by Cominco. From Lydon et al. (2000a). (C) Interlayer albitised (white) areanceous part and chlorite-pyrite-albite-altered (dark coloured) siltite-argillite part of thin-bedded turbidites from 45 Shaft in the hanging wall of the Sullivan deposit. Field of view about two metres. Photo by Cominco. (D) Pyrite-carbonate alteration of the pyrrhotite-bearing massive sulphides of the main orebody in the "Iron Core" of the Vent Complex. Vertical extent of photograph is about one metre. Photo by Cominco. (E) Polished slab of drill core from the distal hydrothermal sediments of the Concentrator Hill Horizon (CHH). The laminated rock consists mainly of a muscovite-rich argillite and siltite with thin pyrhhotite layers and is geochemically anomalously high in zinc and lead contents. (F) Drill-core intersection of the Upper Ore Zone, Sheep Creek, showing typically interbedded argillite layers (light grey), calcareous microturbidites (dark grey), and fine-grained pyrite (brown). Evidence that the pyrite layers represent replacement of calcareous microturbidites is the absence of calcareous microturbidites in pyretic intervals and the incomplete replacement of calcareous microturbidite layers by pyrite from the top down at locations indicated by the white arrows.

the Lower Aldridge (Lydon, 2000b). The main role of the tourmalinite pipe in the genesis of the deposit may therefore have been to prepare a rigid hydrothermal conduit that channeled the upflow of later hydrothermal systems. Only the chlorite-pyrrhotite alteration has been linked to the main ore-forming event (Shaw and Hodgson, 1986). The chlorite-albite-pyrite, pyrite-carbonate, and carbonate alteration are all probably linked to the emplacement of the Mine Sill (Turner and Leitch, 1992).

Deformation

Although the Sullivan deposit appears to have preserved its overall original architecture in remarkably pristine fashion, there is intense deformation within the sulphide body. During Jurassic-Cretaceous deformation, horizontal compressive stresses produced horizontal movement of the hanging wall with respect to the footwall along décollements in the sulphide body (Campbell et al., 1980; McClay, 1983). The major décollement was along the base of the Main Band, which is marked by a zone of durchbewegung structure up to several metres thick (Fig. 13A). Movement along low-angle thrusts and bedding-parallel surfaces are expressed in the Vent Complex as gneissic banding and schlieren textures (Fig. 11C) and in the Bedded Ores as chloritic, slickensided, sharp contacts of sulphide layers with Waste Bands and bedding-parallel zones of durchbewegt sulphides (e.g. Fig. 11E). The Bedded Ores commonly contain zones of bedding-parallel compression (e.g. tightly folded layers) juxtaposed against zones of bedding-parallel extension (e.g. boudinaged or dismembered layers).

Zonation

Thickness

The thickness of the Sullivan deposit generally decreases in all directions away from the central part of the Vent Complex. The main interruption to this trend is due to tectonic thinning along the Transition Zone where sulphides of the main body have been squeezed out laterally by the downward movement of the Bedded Ore sequence along a reverse fault system towards the base of the main sulphide body. This squeezing produced the convolute folding typical of the eastern margin of the Transition Zone (Fig. 13B). Generally, the eastward decrease in thickness of the "A" through "D" Ore Bands is very gradual, although tectonic effects locally influence the isopachs of individual layers of the Bedded Ore sequence (Ransom and Merber, 2000). Thickness of the Waste Bands tends to mimic the thickness of the overlying Ore Band.

Grade and Composition

The Sullivan Deposit as a whole is strongly concentrically zoned with respect to Pb/(Zn+Pb) ratios, which gradually decrease outwards (Fig. 9) (Hamilton et al., 1982). Contours for Pb/(Zn+Pb) crosscut stratigraphy of the Bedded Ores and are steepest near the Transition Zone. In three dimensions, these contours form a concentric shell that is convex upwards about the Vent Complex. Grades of both Pb and Zn decrease outwards, the grade of Pb dropping off at a greater rate than that of zinc, so that the Pb/(Zn+Pb) ratios of the ore decrease both outwards and stratigraphically upwards. Tin, occurring mainly as cassiterite (Hamilton et al., 1982), and Cu are concentrated along zones marked by the highest Pb/(Zn+Pb) ratios. Copper minerals include tetrahedrite and chalcopyrite (Hamilton et al., 1982). Arsenic, which occurs in arsenopyrite and tetrahedrite, is highest in a zone surrounding the Iron Core (Freeze, 1966), whereas Sb is concentrated in a zone outside the high tin zone. The zone of high Sb content corresponds to high boulangerite content in the Bedded Ores. Manganese has more than twice the concentration in the Bedded Ores than it does in the Vent Complex ores but does not show a simple concentric zonation like Zn or Pb (Lydon et al., 2000a).

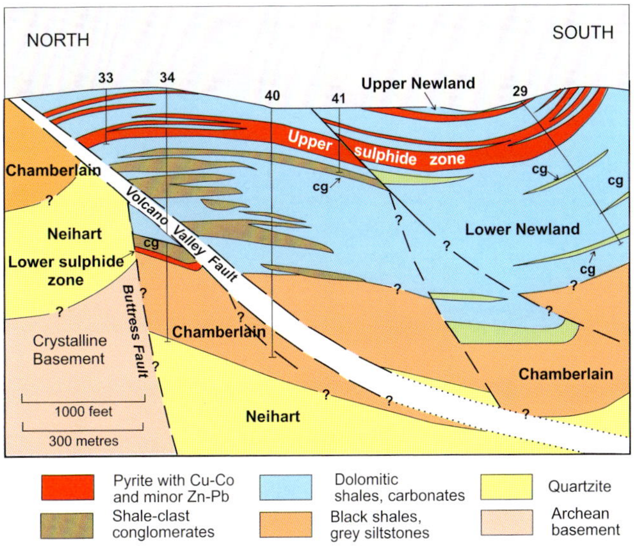

FIGURE 14. Cross-section through the Sheep Creek deposit (modified after Himes and Peterson, 1990).

Mineralogy

There is a pronounced zonation of the main ore-forming sulphide minerals outwards from the base of the Vent Complex. In general, this zonation is primarily a decrease in the proportion of pyrrhotite relative to other sulphides of the ores away from the "barren" basal part of the Vent Complex where it is by far the dominant sulphide. The proportions of sphalerite and galena in the ores are greatest above and lateral to the barren pyrrhotite zone in the Vent Complex. Laterally away from the Vent Complex, as the grade of sphalerite and galena decrease, the ratio of fine-grained pyrite to pyrrhotite in the ores increases. Magnetite is intimately associated with this fine-grained pyrite both as layers and as porphyroblasts (Fig. 11E), the latter suggesting that some of the magnetite is of a retrograde metamorphic origin (Lydon and Reardon, 2000). Superimposed on this concentric zonation is a replacement of pyrrhotite-bearing assemblages by a late coarse-grained pyrite-carbonate-chlorite assemblage in both the core of the Vent Complex and along the Kimberley Fault (Fig. 9).

Geology of Other Mineral Deposit-Types of the Belt-Purcell Basin

SEDEX Deposits

Sheep Creek

The Volcano Valley Fault is close to the northern limit of Belt-Purcell rocks of the Helena embayment and marks the locus of a Mesoproterozoic synsedimentary fault system. Along the southern side of the fault, laminated calcareous shales of the Newland Formation contain sedimentary debris flows and massive to laminated stratiform pyritic layers and lenses. The most extensive zone of pyritic beds occur near the upper contact of the lower Newland Formation , and is up to 100 m thick, has a strike length of about 15 miles (24 km), and extends up to 5 miles (8 km) south of the Volcano Valley Fault (Zieg, 1993). Gossans formed by weathering of these pyritic bodies were mined for iron in the 19th century. Locally, sphalerite, galena, and pyrrhotite replace the pyritic layers to form ore textures similar to the bedded ores of the Sullivan deposit. Although the pyritic Newland Formation at the Sheep Creek property (Fig. 13F) was originally explored for its Zn-Pb potential, exploration attention became focused on its Cu-Co content. The Sheep Creek deposit consists of a stratigraphically lower pyrite zone associated with debris flows containing 4.0 million tonnes at 4% Cu and a stratigraphically upper baritic pyrite zone containing 4.5 million tonnes at 2.5% Cu and 0.1% Co (Fig. 14) (Himes and Petersen, 1990; Rankin and Zieg, 1990; Zieg, 1993; Zieg and Leitch, 1993). The Cu occurs as chalcopyrite in thin layers along microfractures and interstitial to pyrite; the Co occurs as cobaltian pyrite, which also has an anomalously high nickel content.

Other SEDEX Deposits

Within the Canadian portion of the Belt-Purcell Basin, the North Star deposit occurs at the same stratigraphic level and about 4 km south of Sullivan. It is an erosional remnant from which about 60,000 tons of Pb ore was mined from 1895 to 1929. Kootenay King is a small stratiform massive sulphide lens occurring in the Middle Aldridge Formation on the eastern side of the Rocky Mountain Trench, and from which about 13,000 tons of ore was mined from 1952 to 1953. Barite-base metal sulphide mineralization at the Wilds Creek (Leg) and Dave deposits along the western edge of the Purcell anticlinorium have been mapped as occurring in the Dutch Creek or Kitchener Formation (Fig. 5), and may also be of the SEDEX type (Brown and Klewchuck, 1995). The Mineral King deposit, from which about 2.1 million tonnes of baritic Zn-Pb ores were mined from 1954 to 1967, was also suggested to be a SEDEX deposit in the rift-sag sequence of the Purcell Basin (Lydon, 1996), but it has since been shown to occur in a thrust slice of Devonian rocks (Lydon and Graf, 2000).

Within the American portion of the Belt-Purcell Basin, the Soap Gulch deposit near Melrose, Montana, may, like Sheep Creek, represent Mesoproterozoic SEDEX mineralization.

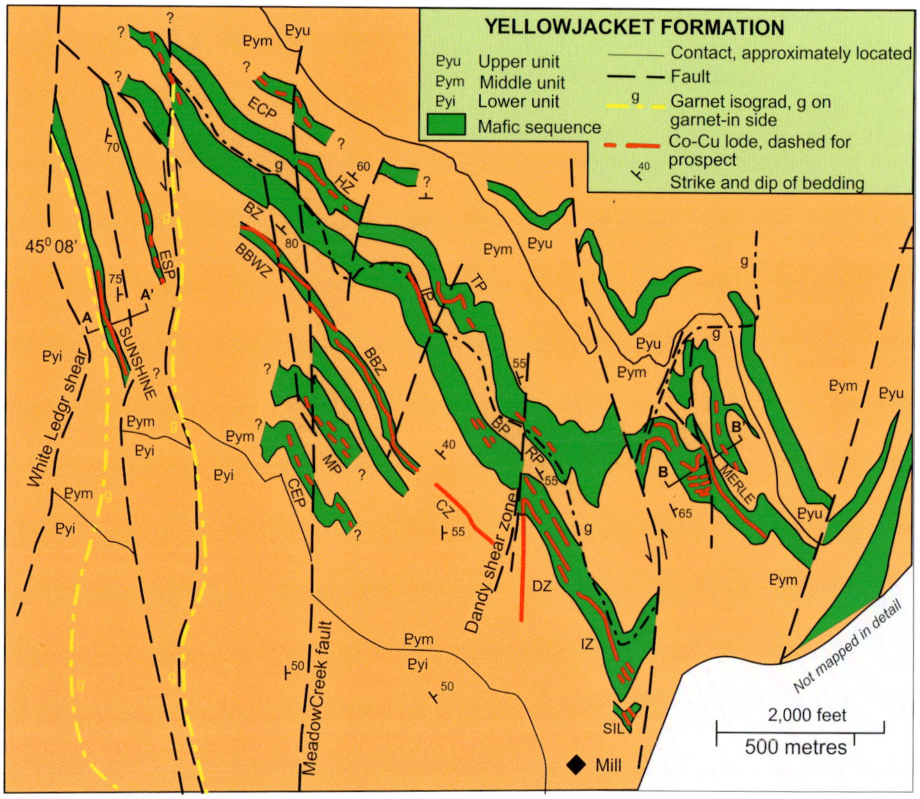

FIGURE 15. Geological map of the Blackbird mine area, Idaho, showing locations of Cu-Co bodies. Note middle unit of Yellowjacket Formation is about 1000 m thick. Cu-Co sulphides occur over its entire thickness. From Nash and Hahn (1989).

Besshi-Type Deposits

Blackbird District

The Cu-Co deposits of the Blackbird District (Fig. 15) occur in the middle unit of the Yellowjacket Formation, as a 5,200 m thick coarsening upward sequence of argillites, siltites, and quartzites (Nash and Hahn, 1989). Mafic dykes and sills are also common in the area (Nash and Hahn, 1989). The Cu-Co mineralization occurs as stratiform layers of cobaltite, chalcopyrite, and pyrite associated with biotite-garnet ±chloritoid layers (interpreted to be mafic tuffs or volcaniclastics) and tourmaline-bearing chemical sediments in the middle part of the sequence (Nash and Hahn, 1989; Bending and Scales, 2001). Some of the biotite-rich layers have a high Cl content (Nash and Connor, 1993). Cobaltite-bearing, tourmaline-cemented breccias are common in the lower part of the sequence, but are generally of low Co grade (Nash and Hahn, 1989). Based on the association of the sulphide deposits with rocks interpreted to be mafic volcanics, the occurrence of both stratiform and discordant (vent or feeder zones) sulphide mineralization and hydrothermal alteration, and metal associations similar to Besshi-type deposits, the sulphide deposits are interpreted to be of

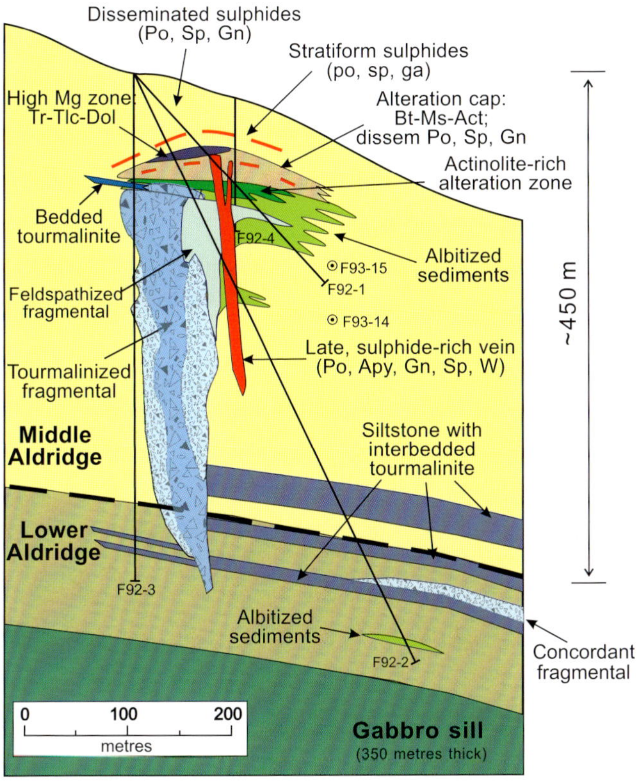

FIGURE 16. Schematic cross-section of the Fors deposit. Abbreviations: Apy – arsenopyrite; Gn – galena; Po – pyrrhotite; Sp – sphalerite; W – scheelite. From Höy et al. (2000), after Höy and Pighin (1995).

seafloor origin (Nash and Hahn, 1989; Nold, 1990; Bending and Scales, 2001). At the Iron Creek deposit, 28 km southeast of the Blackbird deposit, the mineralization consists mainly of cobaltiferous pyrite associated with bedded magnetite.

Most authors correlate the Yellowjacket Formation with the Prichard Formation (Chandler, 2000). However, based on a comparison of detrital zircon ages, Link and Fanning (2003) suggested the Yellowjacket Formation is comparable to the Missoula Group (see stratigraphic correlations in Fig. 5) of Montana. This might imply that the volcanism may be associated with the episode of magmatism that also produced the Purcell lavas of the Snowslip Formation (see Figs. 3, 5) and Nichol Creek volcanics (see Fig. 5) of the Purcell anticlinorium, and not the magmatic event that produced the Moyie sills.

Vent Complex / Feeder Pipe Deposits

Fors

The Fors prospect (Britton and Pighin, 1995) is a blind deposit discovered by drill-testing low-grade surface mineralization. Fors is a pipe-shaped body of discordant sedimentary fragmentals that extends from below the Lower-Middle Aldridge contact to approximately 380 m above this horizon with a small overlying concordant cap (Fig. 16). The discordant sedimentary fragmental is variably altered to tourmalinite, and the concordant cap consists of actinolite-, talc-, and muscovite-alteration assemblages with disseminated to bedded sulphides. A 2 m thick semi-massive sulphide vein of pyrrhotite with variable amounts of sphalerite and galena, with accessory scheelite, chalcopyrite, and bismuthinite, cuts most of the actinolite-rich alteration zone and underlying albitized sediments. Although similar in many respects to the feeder pipe at Sullivan, Fors is at a higher stratigraphic level, contains abundant magnesium-rich calcsilicate and biotite alteration, and has elevated values of Ag, Au, W, As, and Bi. No resource has been calculated for this small deposit. The highest grade drill intercept was 1 m of massive sulphides with 9.35% Pb, 16.4% Zn, 0.09% Cd, and 98 g/t Ag, and 42 m of disseminated Pb-Zn mineralization, including 7 m of pyrrhotite that resembles the Concentrator Hill Horizon near Sullivan.

Other Vent Complex / Feeder Pipe Deposits

The Stemwinder deposit, about 2 km south of Sullivan, consists of discordant pyrrhotite, sphalerite, and galena associated with tourmalinized fragmental rocks. Production and reserves are about 150,000 tonnes at 16.6% Zn, 3.1% Pb, and 82 g/t Ag.

Stratabound Disseminated Sulphide Deposits

Redbed Copper Deposits

The Spar Lake, Rock Creek, and Montanore deposits of northwest Montana are hosted by the Revette Formation, a sequence of sandstones in the middle part of the Ravalli Group, which are the stratigraphic equivalent of the Creston

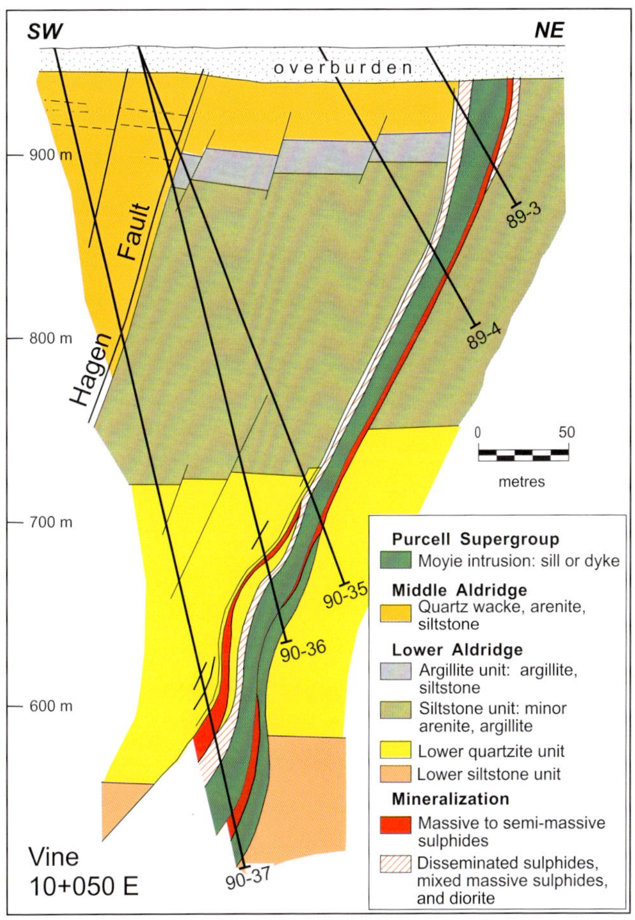

FIGURE 17. Schematic cross-section of the Vine vein deposit. From a diagram in Höy et al., 2000, after Britton and Pighin (1995).

Formation in Canada (Fig. 5). The Spar Lake deposit is laterally zoned, with the dominantly bornite-chalcocite-calcite assemblage of ore-grade mineralization fringed by lower grade zones of chalcopyrite, galena, and pyrite (Hayes et al., 1989). The ores are thought to have formed by the migration of warm (120°C) fluids along bedding or along vertical fluid-escape structures with deposition of metal sulphides in quartzite pore spaces at a pre-existing pyrite-leucoxene interface before any major compaction of the rocks (Hayes and Einaudi, 1986; Hayes et al., 1989).

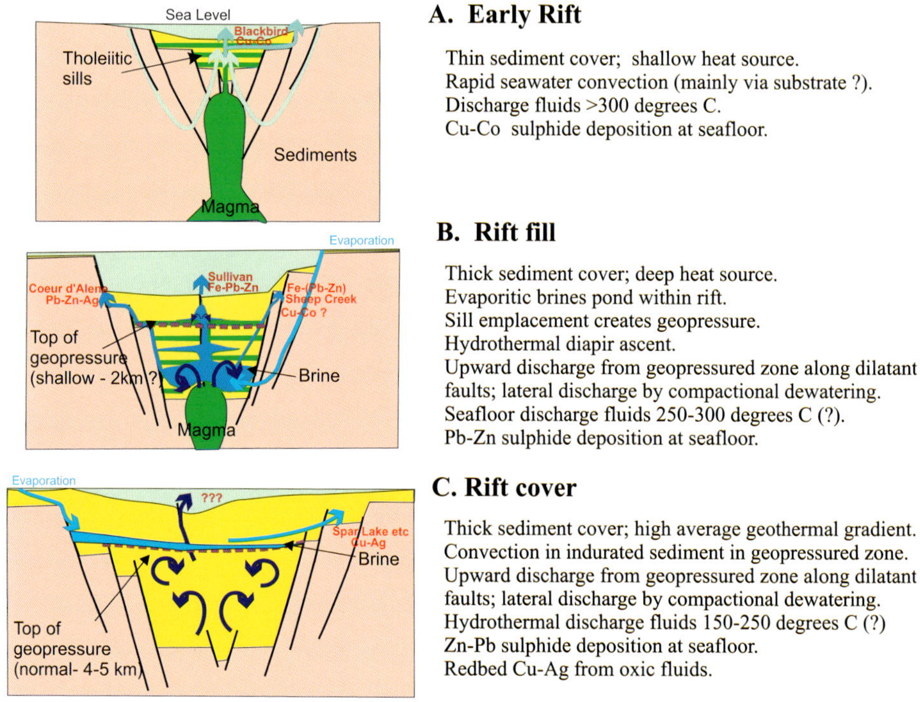

FIGURE 18. Schematic representation of the evolution of the Belt-Purcell Basin and associated metallogenetic events.

Disseminated Sphalerite and Galena

The Canam and Star mineral claims in the southern part of the Purcell anticlinorium are of this type. Höy et al. (2000) state that published Cominco lead isotope data of Canam galena indicates that it is of similar lead isotopic composition as Aldridge SEDEX and vein deposits (Beaudoin, 2000). Höy et al., (2000) suggest that the mineralization represents diagenetic sulphides deposited from hydrothermal fluids rising along active growth faults and spread laterally along permeable arenaceous layers

Veins

Coeur d'Alene District

The quartz-Fe carbonate-galena-sphalerite-tetrahedrite veins of the Coeur d'Alene district form one of the world's most productive silver camps, having produced over 34,000 metric tonnes of silver up until the late 1990s (Long, 1998). The veins of the district can be divided into of two types based on their ore mineralogy:a low-silver galena-sphalerite-pyrrhotite-pyrite type and a high-silver galena-tetrahedrite type (Leach et al., 1998). Veins hosted by the Prichard and Burke formations are typically of the low-silver type, whereas veins hosted by the Revette and St. Regis formations are typically of the high-silver type (Leach et al., 1988). Most veins occur within two kilometres of a 40 km long segment of the Osburn fault, a major west-northwest-trending structure with a 26 km right-lateral displacement within the Lewis and Clark structural zone (Fryklund, 1964; Bennett and Venkatakrishnan, 1982; Leach et al., 1998). The Lewis and Clark structural zone marks a major crustal structure that has been intermittently active since the Proterozoic (Harrison, 1972; Winston, 1986; White, 2000). The age of the mineralization is controversial: Pb/Pb isotope model ages for galena give ages of 1400 to 1500 Ma, (Zartman and Stacey, 1971) suggesting genesis during burial diagenesis, but Ar/Ar and Rb/Sr dating of gangue minerals give ages as young as Cretaceous suggesting a second metallogenetic event associated with emplacement of the Idaho and Kaniksu batholiths (Leach et al., 1998; Fleck et al., 2002). Leach et al. (1998) suggest that the galena-sphalerite-pyrrhotite-pyrite type of veins formed during the Proterozoic, and the galena-tetrahedrite type formed during the Late Cretaceous to early Tertiary.

St. Eugene

The St. Eugene deposit (Fig. 4), the largest vein deposit in the Purcell Supergroup, consists of galena, sphalerite, pyrite, pyrrhotite, and minor chalcopyrite and tetrahedrite with a gangue of quartz and calcite that infilled a fracture system over a vertical extent of about 1300 m and over a strike length of about 3,000 m. About 1.5 million tonnes of ore were mined from the St. Eugene vein prior to 1916 from which about 113,000 tonnes of Pb, 15,000 tonnes of Zn, 180,000 kg of Ag, and 80 kg of Au were recovered (Höy et al., 2000). The St. Eugene deposit shares many similarities with veins of the Coeur d'Alene district, including Pb isotope ratios, (Beaudoin, 1997, 2000), mineralogy and oxygen isotope values of quartz (Joncas and Beaudoin, 2002). Joncas and Beaudoin (2002) suggest that the St. Eugene vein formed from metamorphic fluids with a heavy oxygen isotope signature prior to the Mesoproterozoic regional metamorphism.

Vine

The Vine deposit (Höy and Pighin, 1995) consists of massive pyrrhotite, sphalerite, and galena, with minor chalcopyrite and arsenopyrite and an associated quartz-calcite-chlorite gangue that follows a Moyie dyke over a vertical extent of about 800 m (Fig. 17). Lead isotope compositions of galena are similar to other Mesoproterozoic vein deposits of the Belt-Purcell Basin (Beaudoin, 2000), and Höy et. al. (2000) suggest that the fluid flow that precipitated the mineralization is associated with Aldridge extensional faulting and magmatism.

Genetic Models

The different types of Proterozoic mineral deposits of the Belt-Purcell Basin can all be related to different stages in the sediment filling of an intracontinental rift (Fig. 18).

Early Rift – Besshi Deposits

During initial rifting, when the sediment cover is thin (Fig. 18A), high-temperature (>300°C) seawater convection cells are able to develop within the competent rock of the crystalline basement, driven by the heat of mafic magma chambers. If discharged at the seafloor, they may form Besshi-type volcanogenic massive sulphide deposits. The stratiform Cu-Co deposits of the Blackbird mining district are in this category (Nash and Hahn, 1989; Bending and Scales, 2001). Their occurrence in the Yellowjacket Formation, which might correlate with the mid-Belt Missoula Group (Link and Fanning, 2003) and hence the 1443 Ma mafic magmatism, may only reflect the southward propagation of the Belt Purcell rift (Burtis et al., 2003) so that the onlapping Belt-Purcell sediments were close to basement at this southern extremity of the preserved parts of the basin.

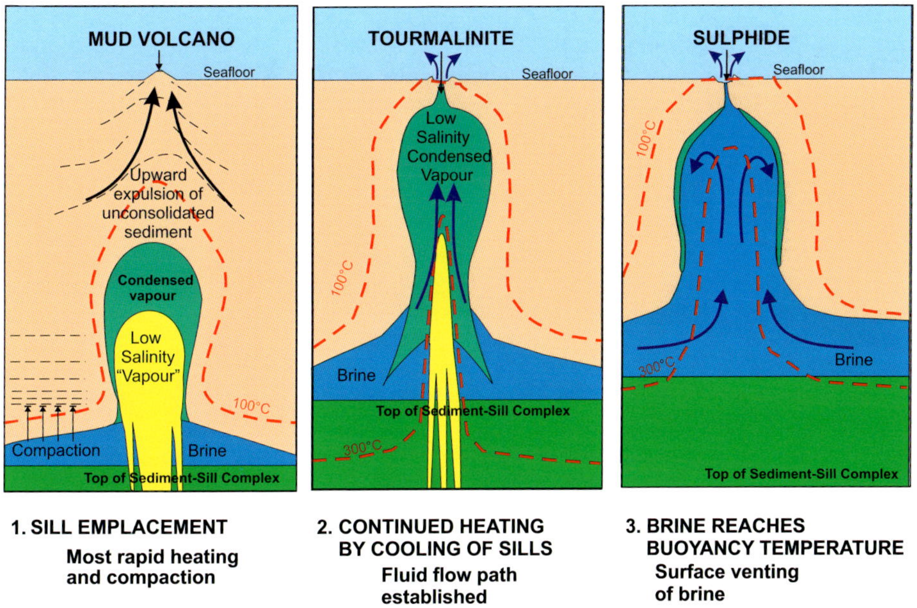

FIGURE 19. Model involving the segregation of a hydrothermal diapir driven by the emplacement of thick tholeiitic sills into a thick succession of unconsolidated marine turbidites to explain the three main sequential seafloor hydrothermal events at Sullivan: 1) mud-volcano activity with minor deposition of manganese carbonates/oxides; 2) the intense tourmalinization of the mud-volcano vent and surrounding sediments; and 3) the deposition of sulphides. From Lydon (2004).

Rift Fill - SEDEX Deposits

The most favourable environment for the formation of SEDEX deposits has long been recognized to be intracratonic rifts that have been filled by marine sediments, with the deposits being spatially associated with synsedimentary faults (Large, 1983; Lydon, 1996). Climatic conditions that produce evaporitic environments, and consequently highly saline formational waters, are an integral part of the classic genetic model for SEDEX deposits (Lydon, 1983), and therefore intracontinental rifts of low latitudes are the optimal geological setting. The ore fluids for SEDEX deposits are thought to be these saline formational waters that leached metals from argillaceous sediments during the smectite-illite transformation during burial diagenesis (Lydon, 1983) and were expelled to the seafloor as a consequence of basinal dewatering, either by the breaching of geopressured reservoirs by synsedimentary faulting (Lydon, 1983; 1986), by compaction-driven stratal migration (Cathles and Smith, 1983; Cathles, 1986) or by thermally-driven convection (Russell, 1983; Yang et al., 2004).

Typically, SEDEX deposits occur in the rift cover sedimentary sequence deposited during the rift-sag stage (Lydon, 1996), such as the SEDEX deposits of the Mt. Isa and McArthur basins of Australia, the Paleozoic deposits of the Selwyn Basin in Canada, and those of the Central Irish Basin. Sullivan, like Broken Hill of Australia, however, occurs in the rift-fill sequence, and is associated with the tectonically and magmatically most active stage of rift development (Fig. 18B). However, in all other aspects, Sullivan conforms to the classic SEDEX model: the Belt-Purcell rift developed in a low-latitude environment (Chandler, 2000), there is direct fluid inclusion evidence for highly saline formational brines (Leitch and Lydon, 2000); and it is associated with the synsedimentary faulting and the sedimentary fragmental rocks of the Sullivan-North Star Corridor (Turner et al., 2000a,b).

To explain 1) the spatial association of one of the world's largest Zn-Pb deposits and one of the world's largest tourmaline deposits (the alteration pipe below Sullivan); 2) the spatial association between tourmalinization and mud volcanos at Sullivan and elsewhere in the Belt-Purcell Basin; 3) the co-extent of mud volcanos and Moyie sills in the Belt-Purcell Basin; and 4) the formation of Sullivan during the early stages of the Moyie sill magmatic event, Lydon (2004) proposed a segregated hydrothermal diapir genetic model (Fig. 19). The essence of this model is that initiation of sill emplacement prematurely both compacted and superheated the Lower Aldridge turbidite sequence. The superheating caused sill emplacement into a low-density low-salinity fluid and a high-density hypersaline fluid, which, because of their large density contrasts, physically segregated into different hydrothermal reservoirs. Ascent of hydrothermal fluids from this reservoir system is envisaged as a slowly rising diapir. The initial release of formational fluid overpressure created by sill-induced compaction resulted in mud-volcano activity (Fig. 19, Stage 1). The overpressure created by phase separation of saline formational fluids was relieved by the ascent and venting of the low-density fluid, which utilized the conduits afforded by the mud-volcano vents (Fig. 19, Stage 2). This low-density low-salinity fluid tourmalinized the substrate along its ascent path, indurating the unconsolidated sediments and forming a mechanically competent cross-stratal fluid conduit that served to channel the subsequent upflow of the high-salinity fluid. This high-salinity fluid, or brine (Fig. 19), was presumably a mixture of connate saline formational waters and the hypersaline fluid formed near sill contacts by high-temperature phase separation of super-heated formational fluids. Ascent of the brine diapir took place only after the hydrothermal reservoirs had been evacuated of their low-density fluid, and only after the brine had been heated to buoyancy with respect to shallow formational fluids (Fig. 19, Stage 3). Venting of these brines at the seafloor produced the sulphide deposit. The hanging wall albitite is related to the local heating of ambient saline groundwaters surrounding the deposit by the Mine Sill, after the Sullivan deposit had been buried to at least 200 m, and the level of neutral buoyancy for sill emplacement had progressed upwards into the upper kilometre of Lower Aldridge sediments. The sulphide of the ores was likely both of hydrothermal origin and locally produced by the bacteriogenic reduction of the water column and/or hydrothermal sulphate (Campbell et al., 1978; Goodfellow, 2000; Taylor and Beaudoin, 2000).

The Sheep Creek deposit has all of the geological characteristics of a SEDEX deposit, except for its high Cu and Co content. Based on textural and fluid inclusion evidence, Himes and Petersen (1990) suggested that the deposit originated as a seafloor to early diagenetic pyritic body, that was invaded by hot (>250°C), moderately saline fluids before lithification of adjacent sediments. The expulsion of hot fluids was likely triggered by synsedimentary fault movement (Rankin and Zieg, 1990). The position of the Sheep Creek deposit at the edge of the basin suggests a stratal migration of fluids from the thicker and higher temperature part of the axial zone of the rift.

Rift Cover – Redbed Copper, Veins, SEDEX (?)

The exact ages of emplacement of the Redbed Copper deposits of the Revette Formation in Montana or the veins of the Coeur d'Alene district and the Purcell anticlinorium are not known. Both deposit types are epigenetic and both occur in rocks of the Ravalli Group (Fig. 5). Sulphides of the Spar Lake Cu-Ag deposit were deposited prior to any significant compaction (Hayes and Einaudi, 1986; Hayes et al., 1989), which suggests that they formed during or before deposition of the overlying Wallace Formation carbonates (Fig. 5). Major galena-bearing veins with Proterozoic Pb isotope signatures do not occur in rocks younger than Ravalli age (Creston age in Canada), which suggests that veins of this type also formed no later than Wallace-Kitchener time. If so, both Redbed Copper deposits and Pb-Zn-Ag vein deposits are related to migrations of saline basinal fluids during the earliest stages of the rift-cover phase (Fig. 3; Fig. 18C) following the very high accumulation rates of Creston and older sedimentary rocks (Fig. 6).

If direct comparison is made to the most productive part of the stratigraphy of other highly productive intracratonic rifts, the rift-cover sequence of the Sheppard Formation and younger strata should be the most prospective part of the Belt-Purcell Supergroup for SEDEX deposits. As noted above, the Mineral King SEDEX deposit, which was previously thought to be an example of this SEDEX potential in the upper part of Belt-Purcell stratigraphy (Lydon, 1996) actually occurs in a thrust slice of Devonian age rocks (Lydon and Graf, 2000). Other showings of stratiform barite-base metal mineralization on the western limb of the Purcell anticlinorium that have been mapped as occurring in Dutch Creek or Kitchener Formation (Brown and Klewchuck, 1995), including the Wild Creek (Leg) and Dave deposits, may be indicative of a SEDEX metallogenetic event in the upper part of Belt-Purcell stratigraphy. However, barite from the Dave deposit has sulphur isotope ratios similar to the Mineral King deposit, and likewise may also be in a Devonian thrust slice (Lydon and Graf, 2000).

Potential for SEDEX and New Deposit-Types in the Belt-Purcell Basin

SEDEX Deposits

There has been no research on what is the maximum amount of base metal that can be mobilized in an intracratonic sedimented rift and concentrated as SEDEX deposits. The fact that no second major deposit has been discovered despite decades of exploration in both the Broken Hill, Australia, and Sullivan areas, both of which are very large Pb-rich deposits of a rift-fill sequence, does not bode well for the discovery of another Sullivan in the axial part of the rift within the Aldridge-Pritchard formations. However, the occurrence of the Sheep Creek deposit at the margin of the Belt-Purcell Basin shows that the basin margins, as for example in the case of the Central Irish Basin, have potential that is deserving of investigation. Pb-Zn-Ag veins and Cu-Ag Redbed deposits formed in Ravalli Group (Creston equivalent) rocks shows that metalliferous fluids circulated in at least some parts of the basin during the lower part of the rift cover sequence. Höy (1993) documented synsedimentary faulting in the Sheppard and Gateway formations, subsequent to the magmatic event that formed the Nichol Creek volcanics and Purcell lavas. This combination of evidence for metalliferous basinal fluids, tectonic activity, and high heat flow during accumulation of the rift-cover sequence are positive indicators for SEDEX potential in the upper part of the Belt-Purcell stratigraphy.

Redbed Cu Deposits

With two recently discovered large deposits waiting to come into production, the potential for the discovery of more Redbed Copper deposits in the Revett Formation of Montana is obvious. The quartzite facies, which has trapped the Cu-Ag mineralization, does not appear to continue northwards as part of the Creston Formation in Canada. However, the green and purple colours of the Creston Formation do indicate intraformational redox fronts, so there is potential for at least small deposits of this type.

Iron Oxide Copper-Gold (Olympic Dam type) Deposits

If the fundamental conditions for the genesis the Olympic Dam and Cloncurry subtypes of iron oxide-copper-gold (IOCG) deposits (e.g. Hitzman et al., 1992; Kerrich et al., 2000; Barton and Johnson, 2002) is met by the emplacement of a felsic pluton into a sedimentary basin containing saline formational waters, then there is potential for this deposit type in the Belt-Purcell Basin. Attention to this possibility has already been drawn with respect to the magnetite-hematite deposits of the Iron Range Mountains in southeastern British Columbia (see Fig. 4) (Stinson and Brown, 1994) and the iron oxide-copper-thorium-REE veins of the Lemhi Pass area in Idaho, which have given a 1055 Ma age (Gillerman et al., 2003). The apparent occurrence of large 1370 Ma granites in the Salmon River area (Evans and Zartman, 1990) support potential for granite emplacement into Belt-Purcell rocks while they still possessed saline formational waters.

Ni-Cu-PGE Deposits

The Mesoproterozoic tholeiitic and alkaline sills represent an upflow of a large volume of mafic magma that has been partly contaminated by assimilation of crustal rocks (Anderson and Goodfellow, 2000). The assimilation of sulphide in crustal rocks is a key component of genetic models for such Cu-Ni deposits as Noril'sk (Eckstrand, 1996). The upflow zones of Belt-Purcell magmas, and perhaps the thickest sills (up to 500 m), are therefore good candidates for magmatic sulphide segregations containing copper, nickel, and PGE. The hydrous autometamorphism of the sills by interaction with wet sediments adds the possibility of Pd-rich hydrothermal Cu-PGE deposits. That the process has taken place, at least on a small scale, is illustrated by the occurrence near Dixon, Montana of a 3 metres thick deposit in a gabbro that post-dates the Ravalli group, which grades about 3 ppm PGE over a strike length of 8 km (Buckley et al., 1994; Lauer 1994), and by samples geochemically enriched in PGE from the Yahk area of British Columbia (Höy, 1989).

Knowledge Gaps of Thematic and Applied Research at the District and Deposit Scale

In order to more completely understand the timing and location of ore-forming processes and hence realize the full mineral potential of the Belt-Purcell Basin, particularly the Canadian portion, it is necessary to address the following knowledge gaps:

1. The need for direct radiometric dating of ore-forming events in order to relate the causes of ore formation to other geological events. Very few deposits have been radiometrically dated.

This can be carried out by direct dating of the ores by Nd/Sm of carbonate and sulphides; Re/Os on pyrite and other sulphides; Rb/Sr of carbonates, sulphides, and fluid inclusions; U/Pb on xenotime; etc. or by bracketing the age of the ore horizon by U/Pb dating of host rocks, e.g. intercalated volcanic ash horizons. Important among these are the assumed Mesoproterozoic vein deposits and the assumed SEDEX deposits on the western limb of the Purcell anticlinorium.

2. The need to develop prognostic criteria as to the favourability of a sedimentary basin to contain SEDEX deposits.

This is a problem that is applicable to SEDEX deposits in general. The investigation of the Belt-Purcell Basin, and its successor rift-controlled Widermere Basin (note the recent discovery of Windermere sills and dykes in Montana (Burtis et al., 2003)), would involve comparative studies of basinal evolution on productive and non-productive basins. Essential information would include

i) Sediment accumulation rates (needs radiometric dating control).

ii) Evolutionary paths of mineralogical and lithogeochemical changes during diagenesis and burial metamorphism to understand the timing and controls of metal-leaching windows. This would require stratigraphically controlled lithogeochemical and mineralogical profiles supported by radiometric dating of diagenetic and burial metamorphic minerals.

iii) Evolutionary paths of pore fluids in the basin, with emphasis on determining the stage or circumstances under which pore fluids become metalliferous. This would have to rely on measuring the chemical composition of fluid inclusions in cements and diagenetic minerals.

iv) History of tectonic and magmatic events within the basin. The recent discovery of rift-related magmatic events high in the stratigraphy of the Belt-Purcell Basin, and the continuation of these events in the succeeding Windermere Basin (e.g. Burtis et al., 2003; Chamberlain et al., 2003), shows that the history of the Belt-Purcell Basin is still incompletely understood. These knowledge gaps hinder the recognition of the full potential of the basin for the occurrence of mineral deposits.

v) Paleoclimatic and water column history of the basin. This would require various geochemical and isotopic (e.g. sulphur, carbon, and strontium) profiles through the basin and systematic observations on lithological and sedimentological environmental indicators.

3. Need to develop prognostic criteria as to the most favourable stratigraphic interval(s) and the most favourable spatial position(s) for SEDEX deposits in a sedimentary basin.

This requires coming to a fundamental understanding as to the circumstances that lead to a focused expulsion of formational waters in a sedimentary basin. This would require developing expertise on the hydrology of modern sedimentary basins and the factors that control the migration of large volumes of basinal fluids. Much of this could be accomplished by compilation (especially of data generated by the oil exploration industry) and mathematical modeling of fluid flow, using rock property parameters for the Belt-Purcell Basin.

Acknowledgements

Most of the author's awareness of the Belt-Purcell Basin was developed in conjunction with the 1991-96 Sullivan Project carried out by the Geological Survey of Canada in collaboration with Cominco Ltd. (now Teck Cominco), the British Columbia Geological Survey Department, the United States Geological Survey, and various universities. In particular Trygve Höy, Doug Anderson, Paul Ransom, Ray Price, and Jerry Zieg are thanked for sharing their knowledge of the Belt-Purcell Basin, both during the project and subsequently. Cominco personnel who had worked at the Sullivan mine, especially Owen Owens, John Hamilton, Paul Ransom, and Norris Del Bel Belluz, are thanked for sharing their knowledge and experience, and for supplying some of the photographs used in this paper. Karen Kelly and Wayne Goodfellow are thanked for their careful reviews of an earlier version of this paper and for making suggestions to improve its clarity and quality.

References

Aleinikoff, J.N., Evans, K.V., Fanning, M.C., Obradovich, J.D., Ruppel, E.T., Zieg, J.A., and Steinmetz, J.C., 1996, SHRIMP U-Pb ages of felsic igneous rocks, Belt Supergroup, western Montana: Geological Society of America, 28th Annual Meeting, Abstracts with Programs, v. 28; 7, p. 376.

Anderson, H.E., and Davis, D.W., 1995, U-Pb geochronology of the Moyie sills, Purcell Supergroup, southeastern British Columbia: Implications for the Mesoproterozoic geological history of the Purcell (Belt) Basin: Canadian Journal of Earth Sciences, v. 32, p. 1180-1193.

Anderson, H.E., and Goodfellow, W.D., 2000, Geochemistry and isotope chemistry of the Moyie sills: Implications for the early tectonic setting of the Mesoproterozoic Purcell Basin, Chapter 17 *in* Lydon, J.W., Höy, T., Slack, J.F., and Knapp, M., eds., The Geological Environment of the Sullivan Deposit, British Columbia: Geological Association of Canada, Mineral Deposits Division, Special Publication No. 1, p. 302-321.

Anderson, D., and Höy, T, 2000, Fragmental sedimentary rocks of the Aldridge Formation, Purcell Supergroup, British Columbia, Chapter 14 *in* Lydon, J.W., Höy, T., Slack, J.F., and Knapp, M., eds., The Geological Environment of the Sullivan Deposit, British Columbia: Geological Association of Canada, Mineral Deposits Division, Special Publication No. 1, p. 259-271.

Anderson, H.E., and Parrish, R.R., 2000, U-Pb geochronological evidence for the geological history of the Belt-Purcell Supergroup, southeastern British Columbia, Chapter 7 *in* Lydon, J.W., Höy, T., Slack, J.F., and Knapp, M., eds., The Geological Environment of the Sullivan Deposit, British Columbia: Geological Association of Canada, Mineral Deposits Division, Special Publication No. 1, p. 113-126.

Barton, M.D., and Johnson, D.A., 2002, Alternative brine sources for Fe-oxide (-Cu-Au) systems: Implications for hydrothermal alteration and metals, *in* Porter, T.M., ed., Hydrothermal Iron Oxide Copper-Gold and Related Deposits: A Global Perspective: PGC Publishing, Adelaide, v. 1, p. 43-60.

Beaudoin, G., 1997, Proterozoic Pb isotope evolution in the Belt-Purcell Basin: Constraint from syngenetic and epigenetic sulfide deposits: Economic Geology 92, p. 343-350.

——— 2000, The lead isotope composition of the Sullivan deposit, Chapter 31 *in* Lydon, J.W., Höy, T., Slack, J.F., and Knapp, M., eds., The Geological Environment of the Sullivan Deposit, British Columbia: Geological Association of Canada, Mineral Deposits Division, Special Publication No. 1, p. 574-581.

Bending, J.S., and Scales, W.G., 2001, New production in the Idaho Cobalt Belt: A unique metallogenetic province: Transactions of the Institute of Mining and Metallurgy, Section B, Applied Earth Sciences 110, p. 81-87.

Bennett, E.H., and Springer, D., 1991, Mine Production in the Coeur d'Alene District, 1981-1990: Idaho Geological Survey GeoNote 20, 1 p.

Bennett, E.H., and Venkatakrishnan, R., 1982, A palinspastic reconstruction of the Coeur d'Alene mining district based on ore deposits and structural data: Economic Geology, v. 77, p. 1851-1866.

Britton, J.M., and Pighin, D.L., 1995, Fors - A Proterozoic sedimentary exhalative base metal deposit, Purcell Supergroup, southeastern British Columbia: British Columbia Ministry of Energy, Mines, and Petroleum Resources, Geological Fieldwork 1994, Paper 1995-1, p. 99-109.

Brown, D.A., and Klewchuck, P., 1995, The Wilds Creek (Leg) zinc-lead-barite deposit, southeastern British Columbia: Preliminary ideas: British Columbia Ministry of Energy, Mines, and Petroleum Resources, Geological Fieldwork 1994, Paper 1995-1, p. 157-164.

Brown, D.A., and Woodfill, R.D., 1998, The Moyie industrial partnership project: Geology and mineralization of the Yahk-Moyie Lake area, southeastern British Columbia, *in* Geological Fieldwork 1997: British Columbia Ministry of Employment and Investment, Paper 1998-1, p. 9-1 - 9-22.

Buckley, S.N., and Sears, J.W., 1998, Shallow emplacement of sills into Wet Belt Supergroup sediments at Perma, western Montana, *in* Berg, R.B., ed., Proceedings of Belt Symposium III: Montana Bureau of Mines and Geology, Special Publication 112, p. 32-43.

Buckley, S.N., Sears, J.W., Ryan, P.C., and Lauer, D., 1994, Base-metal and PGE mineralization, sedimentation, and mafic magmatism related to rifting of the Middle Proterozoic Belt Supergroup, Western Montana: Northwest Geology, v. 23, p. 13-18.

Burtis, E.W., Sears, J.W., and Chamberlain, K.R., 2003, Windermere sill and dike complex in northwestern Montana: Implications for the Belt-Purcell Basin: Northwestern Geology, v. 32, p. 180-185.

Campbell, F.A., and Ethier, V.G., 1983, Environment of deposition of the Sullivan orebody: Mineralium Deposita, v. 18, p. 39-55.

Campbell, F.A., Ethier, V.G., Krouse, H.R., and Both, R.A., 1978, Isotopic composition of sulfur in the Sullivan orebody, British Columbia: Economic Geology, v. 73, p. 246-268.

Campbell, F.A., Ethier, V.G., and Krouse, H.R., 1980, The massive sulfide zone, Sullivan orebody: Economic Geology, v. 75, p. 916-926.

Cathles, L.M., 1986, A tectonic/ hydrodynamic view of basin-related mineral deposits, *in* Turner, R.J.W., and Einaudi, M., eds., The Genesis of Stratiform Sediment-Hosted Lead and Zinc Deposits; Conference Proceedings: Stanford University Publications, Geological Sciences, v. 20, p. 171-176.

Cathles, L.M., and Smith, A.T., 1983, Thermal constraints on the formation of Mississippi Valley-type lead-zinc deposits and their implications for episodic dewatering and deposit genesis: Economic Geology, v. 78, p. 983-1002.

Chamberlain, K.R., Sears, J.W., and Doughty, P.T., 2003, The case for episodic rifting, subsidence and deposition in the Belt-Purcell Basin, 1470-1370 Ma: Northwest Geology, v. 32, p. 190-191.

Chandler, F.W., 2000, The Belt-Purcell Basin as a low-latitude passive rift: Implications for the geological environment of Sullivan type deposits, Chapter 6, *in* Lydon, J.W., Höy, T., Slack, J.F., and Knapp, M., eds., The Geological Environment of the Sullivan Deposit, British Columbia: Geological Association of Canada, Mineral Deposits Division, Special Publication No. 1, p. 82-112.

Conly, A.G., Goodfellow, W.D., Taylor, R.P., and Lydon, J.W., 2000, Geology, geochemistry, and sulphur isotope geochemistry of the hanging wall sulphide zones and their related hydrothermal alteration, Sullivan Zn-Pb-Ag deposit, Chapter 30 *in* Lydon, J.W., Höy, T., Slack, J.F., and Knapp, M., eds., The Geological Environment of the Sullivan Pb-Zn-Ag Deposit, British Columbia: Mineral Deposits Division of the Geological Association of Canada, Special Publication 1, p. 541-573.

Cook, F.A., and van der Velden, A.J., 1995, Three-dimensional crustal structure of the Purcell anticlinorium in the Cordillera of southwestern Canada: Geological Society of America, Bulletin 107, p. 642-664.

Couture, J.-F., and Tanaka, W., 2005, Independent Technical Report on the Rock Creek Cu-Ag Project, Montana, USA, amended January 27, 2005, www.SEDAR.com filing Feb.4 2005, PDF document, 81 p.

Cressman, E.R., 1989, Reconnaissance stratigraphy of the Prichard Formation (Middle Proterozoic) and the early development of the Belt Basin, Washington, Idaho, and Montana: United States Geological Survey, Professional Paper 1490, 80 p.

Dalziel, I.W.D., 1991, Pacific margins of Laurentia and east Antarctica-Australia as a conjugate rift pair: Evidence and implications for an Eocambrian supercontinent: Geology, v. 19, p. 598-601.

De Paoli, G.R., and Pattison, R.M., 1995, Constraints on temperature-pressure conditions and fluid composition during metamorphism of the Sullivan orebody, Kimberley, British Columbia from silicate-carbonate equilibria: Canadian Journal of Earth Sciences, v. 32, p. 1937-1949.

——— 2000, Thermobarometric calculation of peak metamorphic conditions of the Sullivan Deposit, Chapter 15 *in* Lydon, J.W., Höy, T., Slack, J.F., and Knapp, M., eds., The Geological Environment of the Sullivan Pb-Zn-Ag Deposit, British Columbia: Geological Association of Canada, Mineral Deposits Division, Special Publication 1, p. 272-280.

Doughty, P.T., and Chamberlain, K.R., 1996, Salmon River Arch revisited, new evidence for 1370 Ma rifting near the end of deposition in the Middle Proterozoic Belt Basin: Canadian Journal of Earth Sciences, v. 33, p. 1037-1052.

Eckstrand, O.R., 1996, Magmatic nickel-copper-platinum group elements, *in* Eckstrand, O.R., Sinclair, W.D., and Thorpe, R.I., eds., Geology of Canadian Mineral Deposit Types, Geology of Canada, No. 8: Geological Survey of Canada (also Geological Society of America, The Geology of North America P-1), p. 583-614.

Eslinger. E.V., and Sellars, B., 1981, Evidence for the formation of illite from smectite during burial metamorphism of the Belt Supergroup, Clark Fork, Idaho: Journal of Sedimentary Petrology 51, p.203-216.

Ethier, V.G., Campbell, F.A., Both, R.A., and Krouse, H.R., 1976, Geological setting of the Sullivan orebody and estimates of temperatures and pressures of metamorphism: Economic Geology, v. 71, p. 1570-1588.

Evans, K.V., and Zartman, R.E., 1990, U-Th-Pb and Rb-Sr geochronology of middle Proterozoic granite and augen gneiss, Salmon River Mountains, east-central Idaho: Geological Society of America, Bulletin 102, p. 63-73.

Evans, K.V., Aleinikoff, J.N., Obradovich, J.D., and Fanning, M.C., 2000, SHRIMP U-Pb geochronology of volcanic rocks, Belt Supergroup, western Montana: Evidence for rapid deposition of sedimentary strata: Canadian Journal of Earth Science, v. 37, p. 1287-1300.

Fleck, R.J., Criss, R.E., Eaton, G.F., Cleland, R.W., Wavra, C.S., and Bond, W.D., 2002, Age and origin of base and precious metal veins of the Coeur d'Alene mining district, Idaho: Economic Geology, v. 97, p. 23-42.

Freeze, A.C., 1966, On the origin of the Sullivan orebody, Kimberley, British Columbia, *in* A Symposium on the Tectonic History and Mineral Deposits of the Western Cordillera: Canadian Institute of Mining and Metallurgy, Special Volume 8, p. 263-294.

Fryklund, V.C. Jr., 1964, Ore Deposits of the Coeur d'Alene district, Shoshone County, Idaho: United States Geological Survey, Professional Paper 445, 103 p.

Garlick, W.G., 1989, Genesis of the Spar Lake strata-bound copper-silver deposit, Montana; Part 1, Controls inherited from sedimentation and preore diagenesis; a discussion: Economic Geology, v. 82, p. 1967-1968.

Gillerman, V.S., Lund, K., and Evans, K.V., 2003, Stratigraphy, structure and mineral deposits of the Lemhi Pass area, Central Beaverhead Mountains, Eastern Idaho: Northwest Geology, v. 32, p. 134-146.

Goodfellow, W.D., 2000, Anoxic Conditions in the Aldridge Basin during Formation of the Sullivan Zn-Pb Deposit: Implications for the Genesis of Massive Sulphides and Distal Hydrothermal Sediments; Chapter 12 *in* Lydon, J.W., Höy, T., Slack, J.F., and Knapp, M., eds., The Geological Environment of the Sullivan Pb-Zn-Ag Deposit, British Columbia: Geological Association of Canada, Mineral Deposits Division, Special Publication No. 1, p. 218-250.

Gower, C.F., and Tucker, R.D., 1994, Distribution of pre-1400 Ma crust in the Grenville Province: Implications for rifting in Laurentia-Baltica during Geon 14: Geology, v. 22, p. 827-830.

Greig, C.J., Termuende, T.J., and Downie, C.C., 2001, Exploration in the Greenland Creek-Doctor Creek-Findlay Creek area: potential for Sedex-type deposits in the Aldridge Formation other than at "Sullivan Time", *in* Sullivan Geological Meeting, Program with Abstracts, Kimberley, British Columbia, November 2001, p. 17-18.

Gulbrandsen, R.A., 1966, Precambrian phosphorite in the Belt Series in Montana: United States Geological Survey, Professional Paper 550-D, p. D199-D202.

Hamilton, J.M., Bishop, D.T., Morris, H.C., and Owens, O.E., 1982, Geology of the Sullivan orebody, Kimberley, B.C., Canada, *in* Hutchinson, R.W., Spence, C.D., and Franklin, J.M., eds., Precambrian Sulphide Deposits: Geological Association of Canada, Special Paper 25, p. 597-665.

Hamilton, J.M., Delaney, G.D., Hauser, R.L., and Ransom, P.W., 1983, Geology of the Sullivan Deposit, Kimberley, B.C., Canada, *in* Sangster, D.F., ed., Sediment-Hosted Stratiform Lead-Zinc Deposits: Mineralogical Association of Canada, Short Course Handbook 8, p. 31-83.

Hamilton, J.M., McEachern, R.G., and Owens, O.E., 2000, A history of geological investigations at the Sullivan deposit, Chapter 2 *in* Lydon, J.W., Höy, T., Slack, J.F., and Knapp, M., eds., The Geological Environment of the Sullivan Pb-Zn-Ag Deposit, British Columbia: Geological Association of Canada, Mineral Deposits Division, Special Publication No. 1, p. 4-12.

Harrison, J.E., 1972, Precambrian Belt Basin of northwestern United States: Its geometry, sedimentation and copper occurrences: Geological .Society of America, Bulletin 83, p. 1215-1240

Hayes, T.S., and Einaudi, M.T., 1986, Genesis of the Spar Lake strata-bound copper-silver deposit, Montana; Part 1, Controls inherited from sedimentation and preore diagenesis: Economic Geology, v. 81, p. 1899-1931.

Hayes, T.S., Rye, R.O., Whelan, R.O., and Landis, G.P., 1989, Geology and sulphur-isotope geothermometry of the Spar Lake stratabound Cu-Ag deposit in the Belt Supergroup, Montana, *in* Boyle, R.W., Brown, A.C., Jefferson, C.W., Jowett, E.C., and Kirkham, R.V., eds., Sediment-Hosted Stratiform Copper Deposits: Geological Association of Canada, Special Paper 36, p. 319-338.

Himes, M.D., and Petersen, E.U., 1990, Geological and mineralogical characteristics of the Sheep Creek Cu-Co sediment-hosted stratabound sulfide deposit, Meagher County, Montana, *in* Proceedings of the Gold '90 Symposium, Salt Lake City, Utah: p. 533-546.

Hitzman, M.W., Oreskes, N., and Einaudi, M.T., 1992, Geological characteristics and tectonic setting of Proterozoic iron oxide (Cu-U-Au-LREE) deposits: Precambrian Research, v. 58, p. 241–287.

Hoffman, P.F., 1991, Did the breakout of Laurentia turn Gondwanaland inside-out? Science, v. 252, p. 1409-1412.

Höy, T., 1989, The age, chemistry, and tectonic setting of the Middle Proterozoic Moyie sills, Purcell Supergroup, southeastern British Columbia: Canadian Journal of Earth Sciences, v. 26, p. 2305-2317.

——— 1993, Geology of the Purcell Supergroup in the Fernie West-Half Map Area, Southeastern British Columbia: British Columbia Ministry of Energy, Mines, and Petroleum Resources, Geological Surveys Branch, Bulletin 84, 157 p.

Höy, T and Pighin, D., 1995, Vine – A Middle Proterozoic massive sulphide vein, Purcell Supergroup, southeastern British Columbia: British Columbia Ministry of Energy, Mines and Petroleum Resources, Geological Fieldwork 1994, Paper 1995-1, p. 85-98.

Höy, T., Anderson, D., Turner, R.J.W., and Leitch, C.H.B., 2000, Tectonic, magmatic, and metallogenic history of the early synrift phase of the Purcell Basin, southeastern British Columbia, Chapter 4 *in* Lydon, J.W., Höy, T., Slack, J.F., and Knapp, M., eds., The Geological Environment of the Sullivan Pb-Zn-Ag Deposit, British Columbia, Geological Association of Canada, Mineral Deposits Division, Special Publication No. 1, p. 32-60.

Huebschmann, R.P., 1973, Correlation of fine carbonaceous bands across a Precambrian stagnant basin: Journal of Sedimentary Petrology, v. 43, p. 688-699.

Jiang, S-Y., Slack, J.F., and Palmer, M.R., 2000, Sm-Nd dating of the giant Sullivan Pb-Zn-Ag deposit, British Columbia: Geology, v. 28, p. 751-754.

Johnson, H.D., 1978, Shallow siliciclastic seas, Chapter 9 *in* Reading, H., ed., Sedimentary Environments and Facies: Elsevier, New York, p. 207-258.

Joncas, I., and Beaudoin, G., 2002, The St. Eugene deposit, British Columbia: A metamorphosed Ag-Pb-Zn vein Proterozoic Belt-Purcell Rocks: Economic Geology, v. 97, p. 11-22.

Kerrich, R., Goldfarb, R., Groves, D., and Garwin, S., 2000, The geodynamics of world-class gold deposits: Characteristics, space-time distribution and origins: Reviews of the Society of Economic Geologists, v. 13, p. 501–551.

Klein, T.L., 2004, Mineral Deposit Data for Epigenetic Base- and Precious-Metal and Uranium-Thorium Deposits in South-Central and Southwestern Montana and Southern and Central Idaho: United States Geological Survey, Open File Report 2004-1005, 16 p. plus database.

Large, D.E., 1983, Sediment-hosted massive sulphide lead-zinc deposits: an empirical model, *in* Sangster, D.F., ed., Sediment-Hosted Stratiform Lead-Zinc Deposits: Mineralogical Association of Canada, Short Course Handbook 8, p. 1-29.

Larsen, J.C., Long, K.R., Assmus, K.C. and Zientek, M.L., 2004, Mineral Production and Mining Trends for Selected Non-Fuel Commodities in Idaho and Montana, 1905-2001: United States Geological Survey, Open-File Report 2004-1414, 85 p.

Lauer, D.D., 1994, Copper and platinum group mineralization in a mafic dike-sill body of the Revais Creek mining district, Flathead Indian Reservation, Montana; Northwest Geology, v. 23, p. 41-42.

Leach, D.L., Landis, G.P., and Hofstra, A.H., 1988, Metamorphic origin of the Coeur d'Alene base- and precious metal veins in the Belt Basin, Idaho and Montana: Geology, v. 16, p. 122-125.

Leach, D.L., Hofstra, A.H., Church, S.E., Snee, L.W., Vaughn, R.B., and Zartman, R.E., 1998, Evidence for Proterozoic and Late Cretaceous-Early Tertiary ore-forming events in the Coeur d'Alene district, Idaho and Montana: Economic Geology, v. 93, p. 347-359.

Leitch, C.H.B., and Lydon, J.W., 2000, Fluid inclusion petrography and microthermometry of the Sullivan Deposit and surrounding area, Chapter 33, *in* Lydon, J.W., Höy, T., Slack, J.F., and Knapp, M., eds., The Geological Environment of the Sullivan Pb-Zn-Ag Deposit, British Columbia: Geological Association of Canada, Mineral Deposits Division, Special Publication No. 1, p. 617-632.

Leitch, C.H.B., Turner, R.J.W., Ross, K.V., and Shaw, D.R., 2000, Wallrock alteration at the Sullivan deposit, British Columbia, Canada, Chapter 34, *in* Lydon, J.W., Höy, T., Slack, J.F., and Knapp, M., eds., The Geological Environment of the Sullivan Pb-Zn-Ag Deposit, British

Columbia: Geological Association of Canada, Mineral Deposits Division, Special Publication No. 1, p. 633-651.

Link, P.K., and Fanning, C.M., 2003, Detrital zircon ages from the Yellowjacket, Apple Creek and Gunsight Formations, Blackbird Mining Districts, Salmon River Mountains, Central Idaho: Northwest Geology v. 32, p. 206-207.

Long, K.R., 1998, Production and Disposal of Mill Tailings in the Coeur d'Alene Mining Region, Shoshone County, Idaho; Preliminary Estimates: United States Geological Survey, Open-File Report 98-595, 14 p.

Long, K.R., DeYoung, J., and Ludington, S.D., 1998, Database of Significant Deposits of Gold, Silver, Copper, Lead, and Zinc in the United States: United States Geological Survey, Open-File Report 98-206, SD4 Data Base, Online version 1, Excel spreadsheet.

Lydon, J.W., 1983, Chemical parameters controlling the origin and deposition of sediment-hosted stratiform lead-zinc deposits, *in* Sangster, D.F., ed., Sediment-Hosted Stratiform Lead-Zinc Deposits: Mineralogical Association of Canada, Short Course Handbook 8, p. 175-250.

——— 1986, Models for the generation of metalliferous hydrothermal systems within sedimentary rocks, and their applicability to the Irish Carboniferous Zn-Pb deposits, *in* Andrew, C.J., Crowe, R.W.A., Finlay, S., Pennell, W.A., and Pyne, J.F., eds., Geology and Genesis of Mineral Deposits in Ireland: Irish Association for Economic Geology, p. 555-578.

——— 1996, Sedimentary exhalative sulphides (SEDEX), *in* Eckstrand, O.R., Sinclair, W.D., and Thorpe, R.I., eds., Geology of Canadian Mineral Deposit Types, Geology of Canada, No. 8: Geological Survey of Canada (also Geological Society of America, The Geology of North America P-1), p. 130-152.

——— 2000a, A synopsis of the current understanding of the geological environment of the Sullivan deposit, Chapter 3 *in* Lydon, J.W., Höy, T., Slack, J.F., and Knapp, M., eds., The Geological Environment of the Sullivan Pb-Zn-Ag Deposit, British Columbia: Geological Association of Canada, Mineral Deposits Division, Special Publication No. 1, p. 12-31.

——— 2000b, Controls of albitite and tourmalinite deposition, Purcell Supergroup, *in* Höy, T., Brown, D., Lydon, J., Cominco Mine Staff, and others, eds., Sullivan and other Zn-Pb Deposits, Southeastern British Columbia and Northern Washington: Geological Association of Canada, Field Trip Guidebook No. 17, p. 133-139.

——— 2000c, Meadowbrook Marker, *in* Höy, T., Brown, D., Lydon, J., Cominco Mine Staff, and others, eds., Sullivan and other Zn-Pb Deposits, Southeastern British Columbia and Northern Washington: Geological Association of Canada, Field Trip Guidebook No. 17, p. 55-59.

——— 2004, Genetic models for Sullivan and other Sedex deposits, *in* Deb, M., and Goodfellow, W.D., eds., Sediment-Hosted Lead-Zinc-Sulphide Deposits: Narosa Publishing House, New Delhi, p. 149-190.

Lydon, J.W., and Graf, C., 2000, Mineral King - The age of the deposit; *in* Höy, T., Brown, D., Lydon, J., Cominco Mine Staff, and others, eds., Sullivan and other Zn-Pb Deposits, Southeastern British Columbia and Northern Washington: Geological Association of Canada, Field Trip Guidebook No. 17, p. 133-139.

Lydon, J.W., and Reardon, N.E., 2000, Sphalerite compositions of the Sullivan Deposit and their implications for its metamorphic history, Chapter 16 *in* Lydon, J.W., Höy, T., Slack, J.F., and Knapp, M., eds., The Geological Environment of the Sullivan Pb-Zn-Ag Deposit, British Columbia: Geological Association of Canada, Mineral Deposits Division, Special Publication No. 1, p. 281-301.

Lydon, J.W., Paakki, J., Anderson, H.E., and Reardon, N.C., 2000a, An overview of the geology and geochemistry of the Sullivan Deposit, Chapter 27, *in* Lydon, J.W., Höy, T., Slack, J.F., and Knapp, M., eds., The Geological Environment of the Sullivan Pb-Zn-Ag Deposit, British Columbia: Geological Association of Canada, Mineral Deposits Division, Special Publication No. 1, p. 505-522.

Lydon, J.W., Walker, R., and Anderson, H.E., 2000b, Lithogeochemistry of the Aldridge Formation and the chemical effects of burial diagenesis, Chapter 9 *in* Lydon, J.W., Höy, T., Slack, J.F., and Knapp, M., eds., The Geological Environment of the Sullivan Pb-Zn-Ag Deposit, British Columbia: Geological Association of Canada, Mineral Deposits Division, Special Publication No. 1, p. 136-179.

Lyons, T.W., 1993, Geochemical constraints on paleoenvironments within the Belt Supergroup (Middle Proterozoic), Montana: Preliminary results: Belt Symposium III, Informal Program and Abstracts, Whitefish, Montana, 3 p.

Lyons, T.W., and Luepke, J.J., 1995, Geochemical constraints on paleoenvironments within the lower Belt Supergroup (middle Proterozoic), Montana: Geological Society of America, 1995 Annual Meeting, v. 27, p. 463.

Lyons, T.W., Luepke, J.J., Schreiber, M.E., and Zieg, G.A., 2000, Sulfur geochemical constraints on Mesoproterozoic restricted marine conditions: lower Belt Supergroup, northwestern United States; Geochimica et Cosmochimica Acta, v. 64, p. 427-437.

Maxwell, D.T. and Hower, J., 1967, High-grade diagenesis and low-grade metamorphism of illite in the Precambrian Belt Series: American Mineralogist, v. 52, p. 843-857.

McClay, K.R., 1983, Structural evolution of the Sullivan Fe-Pb-Zn-Ag orebody, Kimberley, British Columbia, Canada: Economic Geology, v. 78, p. 1398-1424.

McMechan, M.E., and Price, R.A., 1982, Superimposed low-grade metamorphism in the Mount Fisher area, southeastern British Columbia - Implications for the East Kootenay Orogeny: Canadian Journal of Earth Sciences, v. 19, p. 476-489.

Moores, E.M., 1991, Southwest U.S.-Antarctic (SWEAT) connection, a hypothesis: Geology, v. 19, p. 425-428.

Nash, J.T., and Connor, J.J., 1993, Iron and chlorine as guides to stratiform Cu-Co-Au deposits, Idaho Cobalt Belt, U.S.A., Mineralium Deposita, v. 28, p. 99-106.

Nash, J.T., and Hahn, G.A., 1989, Stratabound Co-Cu deposits and mafic volcaniclastic rocks in the Blackbird mining district, Lemhi County, Idaho, *in* Boyle, R.W., Brown, A.C., Jefferson, C.W., Jowett, E.C., and Kirkham, R.V., eds., Sediment-Hosted Stratiform Copper Deposits: Geological Association of Canada, Special Paper 36, p. 339-356.

Nold, J.L., 1990, The Idaho cobalt belt, northwestern United States; A metamorphosed Proterozoic exhalative ore district: Mineralium Deposita, v. 25, p. 163-168.

Owens, O.E., 2000, Sullivan ore mineralization, Chapter 28 *in* Lydon, J.W., Höy, T., Slack, J.F., and Knapp, M., eds., The Geological Environment of the Sullivan Pb-Zn-Ag Deposit, British Columbia: Geological Association of Canada, Mineral Deposits Division, Special Publication No. 1, p. 523-533.

Paakki, J.J., Lydon, J.W., and Del Bel Belluz, N., 1995, Durchbewegt sulphides, piercement structures, and gabbro dyke displacement in the vent complex of the Sullivan Pb-Zn deposit, British Columbia, *in* Cordillera and Pacific Margin: Geological Survey of Canada, Current Research 1995-A, p. 81-90.

Perry, R.V., and Childs, J.F., 2003, On the way to Beartrack (Parts I to IV): Northwest Geology, v. 32, p. 86-106.

Poage, M.A; Hyndman, D.W., and Sears, J.W., 2000, Petrology, geochemistry, and diabase-granophyre relations of a thick basaltic sill emplaced into wet sediments, western Montana: Canadian Journal of Earth Sciences, v. 37, p. 1109-1119.

Price, R.A., and Sears, J.W., 2000, A preliminary palinspastic map of the Mesoproterozoic Belt-Purcell Supergroup, Canada and USA; Implications for the tectonic setting and structural evolution of the Purcell Anticlinorium and the Sullivan deposit, Chapter 5 *in* Lydon, J.W., Höy, T., Slack, J.F., and Knapp, M.E., eds., The Geological Environment of the Sullivan Deposit, British Columbia: Geological Association of Canada, Mineral Deposits Division, Special Publication No. 1, p. 61-81.

Rankin, P.W. and Zieg, G.A., 1990, Sheep Creek copper deposit, Montana, U.S.A.: Pacific Rim Congress 90, Australian Institute of Mining and Metallurgy, Publication 90/3, p. 491-492.

Ransom, P.W., 1977, Geology of the Sullivan Orebody, *in* Höy, T., ed., Lead-Zinc Deposits of Southeastern British Columbia: Geological Association of Canada, Field Trip No.1 Guidebook, p. 7-21.

Ransom, P.W., and Lydon, J.W., 2000, Geology, sedimentology, and evolution of the Sullivan sub-basin; Chapter 23 *in* Lydon, J.W., Höy, T., Slack, J.F., and Knapp, M., eds., The Geological Environment of the Sullivan Pb-Zn-Ag Deposit, British Columbia: Geological Association of Canada, Mineral Deposits Division, Special Publication No. 1, p. 440-469.

Ransom, P.W., and Merber, D., 2000, Low-angle thrusts in the southeast fringe of the Sullivan mine, Chapter 29 *in* Lydon, J.W., Höy, T., Slack, J.F., and Knapp, M., eds., The Geological Environment of the Sullivan Pb-Zn-Ag Deposit, British Columbia: Geological Association of Canada, Mineral Deposits Division, Special Publication No. 1, p. 534-540.

Reesor, J.E., 1958, Dewar Creek Map-Area with Special Emphasis on the White Creek Batholith, British Columbia: Geological Survey of Canada, Memoir 292, 78 p.

——— 1973, Geology of the Lardeau Map-Area, East-Half, British Columbia: Geological Survey of Canada, Memoir 369, 129 p.

Reipas, K., and Couture, J.-F., 2005, Independent Technical Report on the Troy Cu-Ag Project, Montana, USA, amended January 27, 2005, www.SEDAR.com filing Feb.4 2005, PDF document, 223 p.

Ross, G.M., and Villeneuve, M.E., 1999, The Belt Basin: Provenance insights offer tectonic hindsight: Geological Society of America, Abstracts with Programs, v. 31, no. 7, p. A-300.

Ross, G.M., Parrish, R.R., and Winston, D., 1992, Provenance and U-Pb geochronology of the Mesoproterozoic Belt Supergroup (northwestern United States): Implications of age of deposition and pre-Panthalassa plate reconstructions: Earth and Planetary Science Letters, v. 113, p. 57-76.

Russell, M.J., 1983, Major sediment-hosted exhalative zinc + lead deposits; formation from hydrothermal convection cells that deepen during crustal extension, *in* Sangster, D.F., ed., Sediment-Hosted Stratiform Lead-Zinc Deposits: Mineralogical Association of Canada, Short Course Handbook 9, p. 251-282.

Schandl, E.S., and Davis, D.W., 2000, Geochronology of the Sullivan deposit: U-Pb and Pb-Pb ages of zircons and titanites, Chapter 8 *in* Lydon, J.W., Höy, T., Slack, J.F., and Knapp, M., eds., The Geological Environment of the Sullivan Pb-Zn-Ag Deposit, British Columbia: Geological Association of Canada, Mineral Deposits Division, Special Publication No. 1, p. 127-135.

Schandl, E.S., Davis, D.W., and Gorton, M.P., 1993, The Pb-Pb age of metamorphic titanite in the chlorite-pyrite altered footwall of the Sullivan Zn-Pb SEDEX deposit, B.C., and its relationship to ore: Geological Association of Canada-Mineralogical Association of Canada, Program with Abstracts, no. 18, p. A93.

Schieber, J., 1993, Survey of sedimentologic, geochemical and mineralogical features of the Belt Series and their bearing on the lacustrine vs. marine debate, *in* Belt Symposium III, Informal Program and Abstracts: Whitefish, Montana, 3 p.

Sears, J.W., and Price, R.A., 1978, The Siberian connection: A case for the Precambrian separation of the North American and Siberian cratons: Geology, v. 6, p. 267-270.

——— 2000, New look at the Siberian connection; No SWEAT: Geology, v. 28, p. 423-426.

——— 2003a Tightening the Siberian connection to western Laurentia: Bulletin of the Geological Society of America, v. 115, p. 943-953.

——— 2003b, The Udzha hypothesis: Siberian spillway to the Belt-Purcell Basin: Northwest Geology, v. 32, p. 167-171.

Sears, J.W., Chamberlain, K.R., and Buckley, S.N., 1998, Structural and U-Pb geochronological evidence for 1.47 Ga rifting in the Belt Basin, western Montana: Canadian Journal of Earth Sciences, v. 35, p. 467-475.

Shaw, D.R., and Hodgson, C.J., 1980, Wallrock alteration at the Sullivan Mine, Kimberley, British Columbia (abstract): Canadian Institute of Mining and Metallurgy Bulletin, v. 73, p. 75.

——— 1986, Wallrock alteration at the Sullivan Mine, Kimberley, British Columbia, *in* Turner, R.J.W., and Einaudi, M.T., eds., The Genesis of Stratiform Sediment-Hosted Lead and Zinc deposits: Conference Proceedings, Stanford University Publications, School of Earth Sciences, v. 20, p. 13-21.

Shaw, D.R., Hodgson, C.J., Leitch, C.H.B., and Turner, R.W., 1993a, Geochemistry of tourmalinite, muscovite and chlorite-garnet-biotite alteration, Sullivan Zn-Pb deposit, British Columbia, *in* Current Research, Part A: Geological Survey of Canada, Paper 93-1A, p. 97-107.

——— 1993b, Geochemistry of albite-chlorite-pyrite and chlorite-pyrhhotite alteration, Sullivan Zn-Pb deposit, British Columbia. *in* Current Research, Part A: Geological Survey of Canada, Paper 93-1A, p. 109-118.

Slack, J.F., Shaw, D.R., Leitch, C.H.B., and Turner, R.J.W., 2000, Tourmalinites and coticules from the Sullivan Pb-Zn-Ag Deposit and vicinity, British Columbia: Geology, geochemistry, and genesis, Chapter 39 *in* Lydon, J.W., Höy, T., Slack, J.F., and Knapp, M., eds., The Geological Environment of the Sullivan Pb-Zn-Ag Deposit, British Columbia, Geological Association of Canada, Mineral Deposits Division, Special Publication No. 1, p. 736-767.

Stinson, P., and Brown, D.A., 1994: Iron mineralization and ultramafic dikes on Iron Range Mountain, southeastern British Columbia: Northwest Geology, v. 23, Addendum, p. 99-101.

Swanson, C.O., and Gunning, H.C., 1945, Geology of the Sullivan Mine: Canadian Institute of Mining and Metallurgy Transactions, v. 48, p. 645-667.

Taylor, B.E., and Beaudoin, G., 2000, Sulphur isotope stratigraphy of the Sullivan Pb-Zn-Ag deposit, B. C.: Evidence for hydrothermal sulphur, and bacterial and thermochemical sulphate reduction, Chapter 37 *in* Lydon, J.W., Höy, T., Slack, J.F., and Knapp, M., eds., The Geological Environment of the Sullivan Pb-Zn-Ag Deposit, British Columbia: Geological Association of Canada, Mineral Deposits Division, Special Publication No. 1, p. 696-719.

Turner, R.J.W., and Leitch, C.H.B., 1992, Relationship of albitic and chloritic alteration to gabbro dykes and sills in the Sullivan mine and nearby area, southeastern British Columbia, *in* Current Research, Part E; Geological Survey of Canada, Paper 92-1E, p. 95-105

Turner, R.J.W., Leitch, C.H.B., and Delaney, G.D., 2000a, Syn-rift structural controls on the paleoenvironmental setting and evolution of the Sullivan orebody, Chapter 32 *in* Lydon, J.W., Höy, T., Slack, J.F., and Knapp, M., eds., The Geological Environment of the Sullivan Pb-Zn-Ag Deposit, British Columbia: Geological Association of Canada, Mineral Deposits Division, Special Publication No. 1, p. 582-616.

Turner, R.J.W., Leitch, C.H.B., Höy, T., Ransom, P.W., Hagen, A., and Delaney, G.D., 2000b, Sullivan graben system: District-scale setting of the Sullivan Deposit, Chapter 21 *in* Lydon, J.W., Höy, T., Slack, J.F., and Knapp, M., eds., The Geological Environment of the Sullivan Pb-Zn-Ag Deposit, British Columbia: Geological Association of Canada, Mineral Deposits Division, Special Publication No. 1, p. 370-207.

Turner, R.J.W., Leitch, C.H.B., Ross, K.V., and Höy, T., 2000c, District-scale alteration associated with the Sullivan Deposit, British Columbia, Canada, Chapter 22 *in* Lydon, J.W., Höy, T., Slack, J.F., and Knapp, M., eds., The Geological Environment of the Sullivan Pb-Zn-Ag Deposit, British Columbia: Geological Association of Canada, Mineral Deposits Division, Special Publication No. 1, p. 408-439.

White, B.G., 2000, Coeur d'Alene mining district: Product of preconcentrated source deposits and tectonism within the Lewis and Clark line, *in* Roberts, S., and Winston, D., eds., Geologic Field Trips, Western Montana and Adjacent Areas: Geological Society of America, p. 95-101.

Winston, D.K., 1986, Tectonics and sedimentation of the middle Proterozoic Belt Basin, and their influence on Cretaceous compression and Cenozoic extension in western Montana and northern Idaho, *in* Peterson, J.A., ed., Paleotectonics and Sedimentation in the Rocky Mountain Region, United States: American Association of Petroleum Geologists, Memoir 41, p. 87-118.

Winston, D., and Link, P.K., 1993, Middle Proterozoic rocks of Montana, Idaho and eastern Washington, the Belt Supergroup, *in* Reed, J.C., Jr., Bickford, M.E., Houston, R.S., Link, P.K., Rankin, D.W., Sims, P.K., and Van Schmus, W.R., eds., Precambrian, Conterminous U.S., The Geology of North America C-2: Geological Society of America, p. 487-521.

Yang, J., Large, R.R., and Bull, S.W., 2004, Factors controlling free thermal convection in faults in sedimentary basins: implications for the formation of zinc–lead mineral deposits; Geofluids, v. 4, p. 237-247.

Zartman, R.E., and Stacey, J.S., 1971, Lead isotopes and mineralization ages in Belt Supergroup rocks, northwestern Montana and northern Idaho: Economic Geology, v. 66, p. 849-860.

Zartman, R.E., Peterman, Z.E., Obradovich, J.D., Gallego, M.D., and Bishop, D.T., 1982, Age of the Crossport C Sill near Eastport, Idaho, *in* Reid, R.R., and Williams, G.A., eds., Society of Economic Geologists, Coeur d'Alene Field Conference Proceedings, Idaho Bureau of Mines and Geology, Bulletin 24, p. 61-69.

Zieg, G.A , 1993, The geology of the Sheep Creek copper deposit, Meagher County, Montana, *in* Belt Symposium III, Informal Program and Abstracts: Whitefish, Montana, p. 2-4.

Zieg, G.A., and Leitch, C.H.B., 1993, The geology of the Sheep Creek copper deposits, Meagher County, Montana, *in* Berg, R.B., comp., Belt Symposium III Abstracts, 1993: Montana Bureau of Mines and Geology, Open-File Report 1999, p. 67-69.

Metallogeny of the Pine Point Mississippi Valley-Type Zinc-Lead District, Southern Northwest Territories

PETER HANNIGAN

Geological Survey of Canada, 3303 - 33 Street N. W., Calgary, Alberta T2L 2A7
Corresponding author's email: phanniga@nrcan.gc.ca

Abstract

The Pine Point Mississippi Valley-type lead-zinc deposit constitutes one of Canada's principal base-metal mining districts with initial geological resources representing near 9% of the Pb and 6% of the Zn endowment in Canada. The deposit is located on the eastern flank of the Western Canada Sedimentary Basin in southern Northwest Territories near the southern shore of Great Slave Lake. It occurs in a mid-Devonian carbonate barrier complex called the Presqu'ile barrier.

Subtle tectonic adjustments along three northeast-trending hinge lines contributed to the evolution of numerous paleo-environmental facies within and adjacent to the barrier complex. The hinge lines are approximate projections of the underlying Precambrian McDonald-Great Slave Lake fault system. They also are lines of weakness in the bedrock where preferential karstification, coarse-crystalline dolomitization (replacement and cement), and sulphide mineralization are concentrated.

At Pine Point, all Mississippi Valley-type deposits are hosted in middle Devonian carbonate strata within or adjacent to a dolomitized barrier complex. The mineralization is pervasive in interconnected paleokarst networks that are directly related to distinct lithofacies. The ore deposits are strongly controlled by individual stratal horizons, although some orebodies are discordant.

Epigenetic sulphide mineralization occurs as open-space cavity fills and local replacement of internal sediments in carbonate strata. The main ore mineralization stage at Pine Point encompassed a complex overlap and repetition of diagenetic phases or processes including deposition of sulphides, hydrothermal dissolution, fracturing and collapse of host rocks, internal clastic sedimentation, saddle dolomite cementation, and thermal alteration of host rock and associated bitumen.

Mineral potential in the Pine Point district was evaluated by applying critical genetic attributes controlling the location of Mississippi Valley-type deposits to empirical evidence of sphalerite and/or galena occurrences.

Résumé

Le gisement de Pine Point, une minéralisation de type Mississippi-Valley, forme l'un des principaux districts miniers de métaux communs du Canada, ses ressources géologiques initiales comptant pour près de 9% des richesses en plomb et 6% des richesses en zinc du Canada. Il se situe dans le flanc oriental du Bassin sédimentaire de l'Ouest du Canada, dans le sud des Territoires du Nord-Ouest, près de la rive sud du Grand lac des Esclaves. Il est encaissé dans un complexe de roches carbonatées du Dévonien moyen, le complexe barrière de Presqu'île.

Des compensations tectoniques subtiles survenues le long de trois charnières de direction nord-est ont contribué à l'évolution de nombreux faciès paléoenvironnementaux au sein et au voisinage du complexe barrière. Les charnières constituent le prolongement approximatif d'un réseau de failles précambrien sous-jacent, soit le système de McDonald-Grand lac des Esclaves. Elles représentent aussi les zones de faiblesse du substratum rocheux, là où se concentrent une karstification préférentielle, une dolomitisation à cristaux grossiers (remplacement et cimentation) et une minéralisation sulfurée.

Dans le district de Pine Point, tous les gîtes de type Mississippi-Valley sont encaissés dans des strates carbonatées du Dévonien moyen, au sein ou au voisinage d'un complexe barrière dolomitisé. La minéralisation présente un caractère envahissant au sein de réseaux de paléokarsts interconnectés qui montrent des liens directs avec des lithofaciès distincts. Les gîtes sont délimités de façon nette par des horizons distincts, mais certains corps minéralisés sont discordants.

Des minéralisations sulfurées épigénétiques remplissent des cavités et remplacent, par endroits, des sédiments internes au sein des strates carbonatées. Dans le district de Pine Point, la principale période de minéralisation a englobé une série complexe de processus ou de phases diagénétiques, qui se sont répétés et chevauchés dans le temps, parmi lesquels on peut mentionner le dépôt de sulfures, la dissolution hydrothermale, la fracturation et l'effondrement des roches encaissantes, la sédimentation détritique interne, la cimentation dolomitique à cristaux en selle de cheval et l'altération thermique des roches encaissantes et du bitume associé.

Le potentiel minéral du district de Pine Point a été déterminé en soumettant des indices empiriques formés d'occurrences minéralisées de sphalérite ou de galène à un examen fondé sur les attributs génétiques clés qui régissent l'existence des gîtes de type Mississippi-Valley.

Introduction and History

The Pine Point Zn-Pb ore deposit is located in platformal carbonate rocks on the eastern margin of the Western Canada Sedimentary Basin (WCSB) in southern Northwest Territories (Fig. 1). Several orebodies are hosted in middle Devonian carbonate strata within or adjacent to the dolomitized Presqu'ile barrier complex. The orebodies are listed in Appendix 1 and their locations are illustrated in Figure 2.

Historically, the Pine Point Mississippi Valley-type (MVT) Zn-Pb district has been one of Canada's most important base-metal mining districts, accounting for a large proportion of the country's Pb and Zn production through much of its mine life. During production at Pine Point (1964-

Hannigan, P., 2007, Metallogeny of the Pine Point Mississippi Valley-type zinc-lead district, southern Northwest Territories, *in* Goodfellow, W.D., ed., Mineral Deposits of Canada: A Synthesis of Major Deposit-Types, District Metallogeny, the Evolution of Geological Provinces, and Exploration Methods: Geological Association of Canada, Mineral Deposits Division, Special Publication No. 5, p. 609-632.

1988), approximately 26% of the nation's Pb and 17% of its Zn were extracted (Department of Energy Mines Resources Canada, 1967-1973; Energy Mines Resources Canada, 1974-1990; Rhodes et al., 1984; Brophy, 1987; Ellis and Hearn, 1990; Ellis, 1995). During peak production periods (1966-1970; 1986-1987), Pb and Zn extracted at Pine Point accounted for 36 and 22% of the nation's production, respectively. The total geological resource in the Pine Point district (Pine Point and Great Slave Reef properties combined) constitutes near 9% of the lead and 6% of the zinc in Canada (Gurmendi, 2002; Natural-Resources-Canada, 2004). The value of Pine Point ore milled during the final year of significant production (1987) was nearly $500 million in 1987 dollars and represented 50% of the total value of mineral production in the Northwest Territories (Ellis and Hearn, 1990). Direct employment at Pine Point during peak production was nearly 700 personnel. Due to economics, production ceased at Pine Point in 1988. Reclamation has taken place and the town of Pine Point, as well as the railway, have been removed.

Fifty of the 100 drill-defined orebodies were brought into production in the Pine Point district (Appendix 1). Total production from 1964 to 1988 was nearly 64.3 million tonnes (Mt) at an average grade of 3.0% Pb and 7.0% Zn. All production took place on the Pine Point property (Fig. 2). Open-pit mining methods were employed on all orebodies, except for two underground operations. No mining has taken place to date on the Great Slave Reef property, which is located west of Pine Point property (Fig. 2).

Surface showings of lead and zinc mineralization near Pine Point on the southern shore of Great Slave Lake were first brought to the attention of Klondike-bound prospectors by local Indians in 1898. Bell (1902) visited the showings in 1899 and reported the mineralization in Devonian limestones adjacent to numerous sinkholes. Several assays from this showing did not reveal anomalous silver content, and the lack of precious metals in this very remote area precluded significant development during the ensuing decades. In the early part of the twentieth century, various field parties of the Geological Survey of Canada visited the Pb-Zn exposures (Camsell, 1915; Cameron, 1917, 1918). Cameron (1917) identified the mineralized host rock as coarse-crystalline vuggy dolomite with cavities occupied by curved rhombohedral dolomite crystals ('saddle dolomite'). This host rock was called 'Presqu'ile' dolomite. Cameron also found similar coarse-crystalline vuggy dolomite at Windy Point on the southwest shore of Great Slave Lake and correlated these rocks with host rocks at Pine Point. Scattered crystals of galena in a cross-cutting calcite vein and active tar springs and bitumen were noted at Windy Point.

Initial concerted exploration at Pine Point began in 1929 when the Consolidated Mining and Smelting Company merged with the Atlas Exploration Company. Work, which included digging test pits, drilling, and sinking an exploratory shaft, commenced on the mining claims encompassing the Pb-Zn showings (Bell, 1929). At this time, five high-grade orebodies containing fine-crystalline cubical galena intergrown with botryoidal sphalerite were discovered, and two mineralized trends were identified. Drilling indicated the Zn-Pb ore was nearly 500,000 tons (0.45 Mt). Numerous similarities between the Pine Point occurrence and MVT deposits in the Tri-State region of the Mississippi Valley were observed at this time (Bell, 1929).

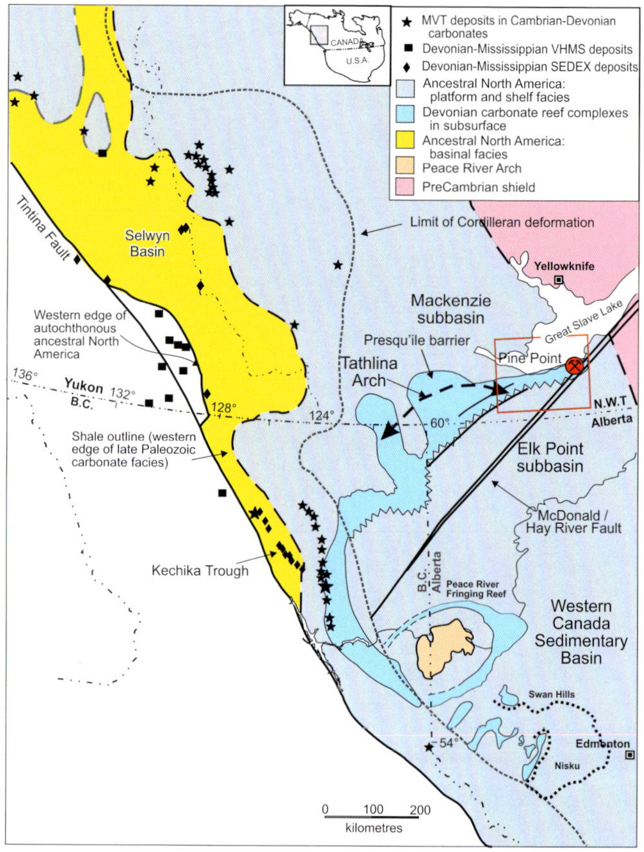

Figure 1. Devonian-Mississippian tectonic framework of the western Canadian continental margin, and location of the Pine Point mineral deposit on the eastern flank of the Western Canada Sedimentary Basin (adapted from Nelson et al., 2002). The Devonian Presqu'ile barrier complex separating two depositional subbasins is shown. The Mackenzie subbasin north of the barrier exhibits normal marine sedimentation and the Elk Point subbasin to the south contains a restricted carbonate-evaporite shelf and evaporite sedimentary succession. Tathlina Arch is depicted. The location of the detailed geological map shown in Figure 3 is outlined by the red box.

In 1947, Consolidated Mining and Smelting Company (Cominco) staked a large concession in the Pine Point area encompassing all known mineralization. The concession was acquired in order to test new geological concepts with respect to orebody location, such as whether significant Pb and Zn occurrences have a specific stratabound control similar to classic MVT deposits, and if they are structurally controlled by major Precambrian basement faults that may extend beneath the property from the cratonic Shield to the southwest. Between 1948 and 1963, extensive exploration work was undertaken by Cominco within the concession. This work delineated 8.8 million tonnes of ore averaging 2.6% Pb and 5.9% Zn in several orebodies.

During mining and production phases at Pine Point, which commenced in 1964, additional exploration took place on adjacent and nearby properties. New orebodies were discovered on the Pyramid claims (X-15 and W-17 in 1965, Thorpe, 1966); Buffalo River Exploration property (A-55 in 1966, Thorpe, 1972); Coronet claims (R-61 and S-65 in 1966, Thorpe, 1972) and the Yellowknife Base Metals property (YBM orebody) in 1966 (Thorpe, 1972). Pine Point Mines Ltd., a subsidiary of Cominco, formed to finance Pine

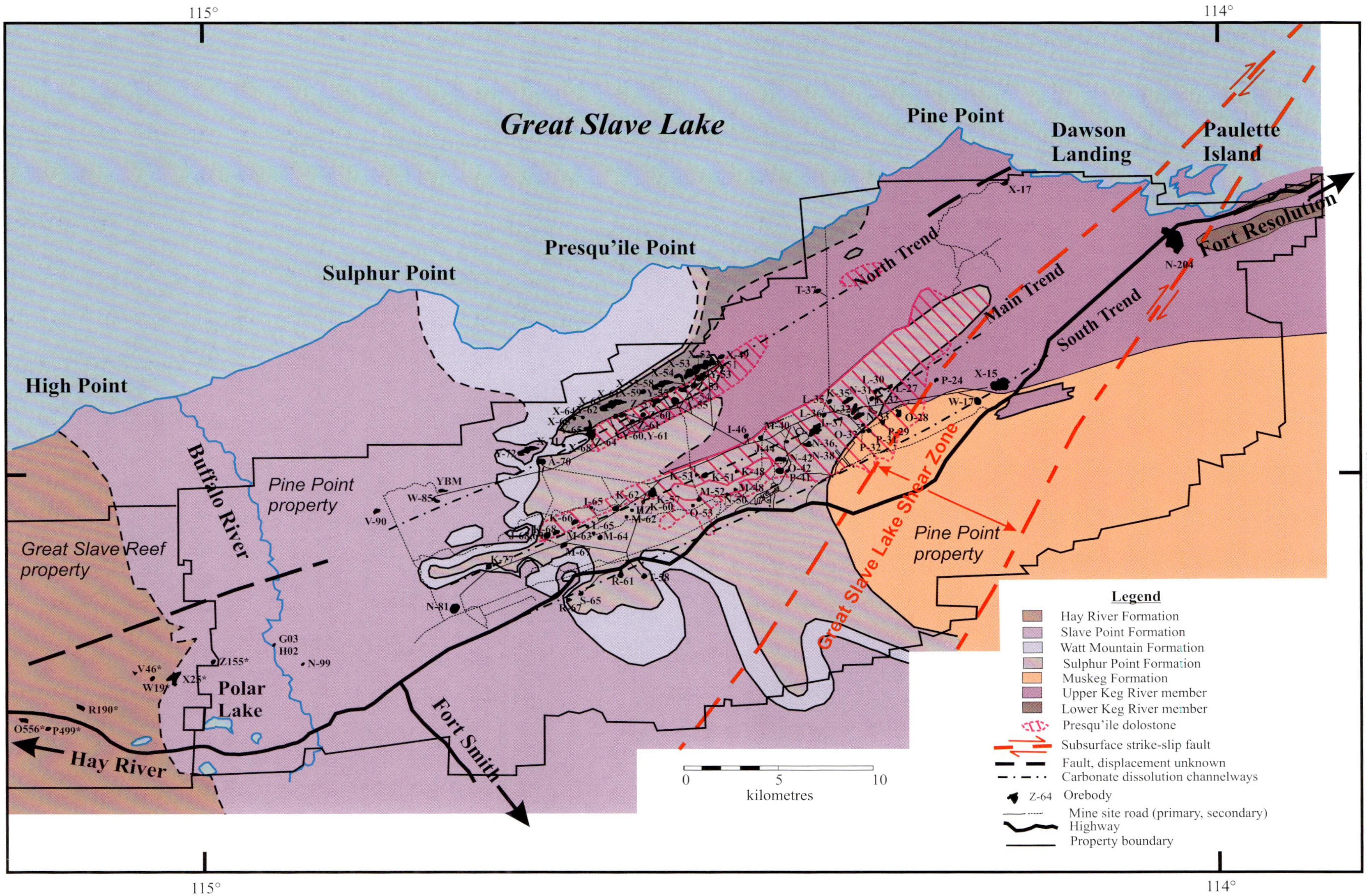

FIGURE 2. General geology of the Pine Point mining camp with orebody locations in the Pine Point and Great Slave Reef properties (starred orebodies in Great Slave Reef property) (adapted from Rhodes et al., 1984). Major mineralized trends (i.e. North, Main, and South trends) on carbonate dissolution channelways are shown.

Point mine production, purchased these properties and added them to their reserve and production inventory (Gibbins et al., 1977).

Lead and Zn exploration also took place westward, along the Presqu'ile barrier outside of the Pine Point property. The Great Slave Reef (GSR) project was located immediately to the west of the Pine Point property on a group of claims owned by Western Mines (Westmin) (see Fig. 2 for property location). Although the area was previously staked in 1965, prior to the start of production at Pine Point, little work was done on the property because favourable stratigraphic units were too deep for effective conventional induced-polarization geophysical exploration. Western Mines acquired the property in 1975 and commenced an extensive drilling program. The drill program was based on the premise that the 'Main Trend', along which many of the ore deposits at Pine Point are located, extends westward into the GSR property (Fig. 2). Drilling completed between 1975 and 1981 outlined seven additional Pb-Zn orebodies on the GSR property (Fig. 2). Estimated resource potential of the seven orebodies on the property is 7.25 Mt at a grade of 10.3 percent combined Pb and Zn (2% Pb+Zn cut-off) (Beales et al., 2002). No production has taken place to date on these orebodies, principally due to greater depths and extensive groundwater discharge into the area.

Presqu'ile-type dolomite also outcrops on the northwest shore of Great Slave Lake at Windy Point and northward along the eastern shore of Prairie Lake (Fig. 3; Norris, 1965), and consequently, extensive Pb-Zn mineral exploration has taken place in this area. Williams (1977) described the reef-like carbonate at Windy Point as lithologically identical to the ore-bearing reef complex at Pine Point and correlation between the two has been proposed. In 1955, more than two thousand mining claims were staked in the area by Windy Point Mining Company (McGlynn, 1971). These claims were geochemically prospected, mapped, and drilled in 1956. Lead-zinc mineralization found in the drillholes proved to be uneconomic and no zones of significant size were found. Between 1965 and 1980, geophysical work and diamond drilling continued in areas of interest throughout the Windy Point property, but only minor mineralization was encountered. There is no evidence of major faulting in the region, although several minor faults with small displacements cutting the Paleozoic dolostones have been observed (Gibbins, 1984). Most sulphides occur in coarse-crystalline dolomites (Presqu'ile-type dolomite) in the reefal complex. Galena and sphalerite appear as isolated crystals, sparse irregular patches, or veinlets in brecciated fine-crystalline dolomite (Gibbins, 1984). The mineralization is often associated with oil staining and coarse vuggy dolospar.

North of Windy Point, another property called Qito was explored by Cominco Ltd. Zinc and Pb soil anomalies were tested by geophysical surveys and drilling between 1976 and 1980. Trace to minor amounts of galena and sphalerite clots and disseminations were found within karsted dolomite in these drillholes.

Adjacent to the southern shore of Great Slave Lake, the continuation of the favourable diagenetic Presqu'ile dolomite facies was confirmed by extensive and widely spaced drilling on the Hay West property owned and operated by Cominco Limited. This property bounds the Great Slave Reef project area on its western boundary and straddles the Hay River. The property was staked in 1978 and worked until 1981. Forty-two drillholes, totalling appoximately 19,500 m, were completed in the region. Outcrop on the property is very sparse, mostly confined to the Hay River Escarpment and Gorge. Therefore, drilling, down-hole geophysical logging, and seismic surveys acquired much of the geological information known in the area. The Hay West property straddles the westward continuation of the Presqu'ile barrier complex. Recrystallization of Sulphur Point and upper Pine Point carbonate strata to coarse-crystalline dolomite was confirmed in these holes and appears to follow northeasterly trends subparallel to the Great Slave Lake Shear Zone (GSLSZ) (Gibbins, 1984). The reconnaissance diamond-drilling program was successful in defining the approximate boundaries of the Muskeg evaporite facies, Presqu'ile barrier complex, and the boundary of the carbonate front toward the western edge of the property (Gibbins, 1984). Several holes encountered hydrogen sulphide gas, and a blowout of crude oil, methane, and hydrogen sulphide occurred in one of these holes. Three holes encountered thick sequences of collapse breccia with minor quantities of galena and sphalerite. Seismic surveys were successful in identifying collapse structures on certain horizons of interest. One of these seismically identified, collapsed brecciated structures was confirmed by drilling (Gibbins, 1984).

Another mineral property (Tathlina) was acquired by Gulf Minerals Canada in 1978 directly west of the Hay West property. An abandoned oil well on this property (NWT Desmarais Lake No. 1 C-19) was reported to have encountered minor Pb and Zn mineralization. The drilling program completed on this property confirmed the presence of Presqu'ile dolomite and Pb-Zn mineralization. Electromagnetic geophysical work, drilling, and down-hole geophysical logging outlined an embayment of the carbonate-shale facies front (Gibbins, 1984). These claims occur on the northern flank of the Tathlina Uplift and south of the northeasterly trending Tathlina Fault. In this area, the entire Pine Point Group grades laterally to the west into shales of the Horn River Formation. Macqueen and Ghent (1975) discovered significant Pb and Zn in shale samples from the Horn River Formation and the bituminous member of the Pine Point Group. They speculated that these shales might be a source of mineralization in the region. Visible mineralization was encountered in three of the eleven drillholes completed by Gulf Minerals. Rock geochemical anomalies were encountered in most drillholes but these were not related to any visible sulphides (Gibbins, 1984).

Geological and Tectonic Setting

Geological Setting

Phanerozoic strata in the Interior Plains of the WCSB consist of a northeasterly tapering wedge of supracrustal rocks overlapping the westward extension of the Precambrian craton (Fig. 1) (Porter et al., 1982; Price, 1994). The wedge of sediments thickens southwestwards from zero at the craton/Phanerozoic boundary to nearly six kilometres adjacent to the deformed belt of the Cordilleran Orogen (Porter et al., 1982; Wright et al., 1994). Filling the WCSB are two major

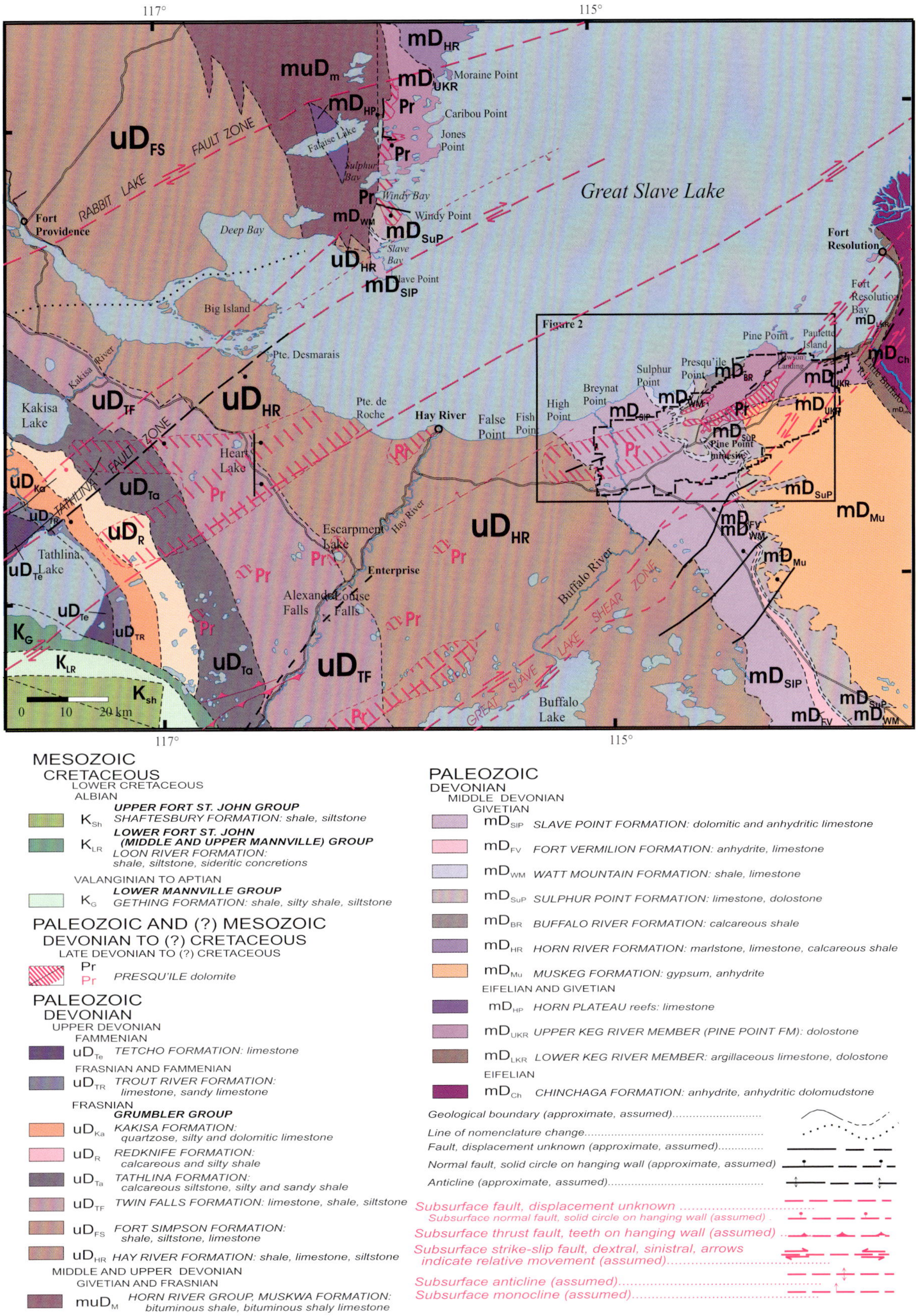

FIGURE 3. Geological map of the Pine Point district (adapted from Okulitch, 2006). Surface (black) and interpreted subsurface (magenta) geological contacts and structures are shown. The location of the Presqu'ile dolostone in Pine Point mine-site area (solid hachures) is from Rhodes et al. (1984) and the subsurface distribution of Presqu'ile dolostone (broken hachures) is from Janicki (2006b). The area confined by the box indicates the limits of the Pine Point mine property map depicted in Figure 2.

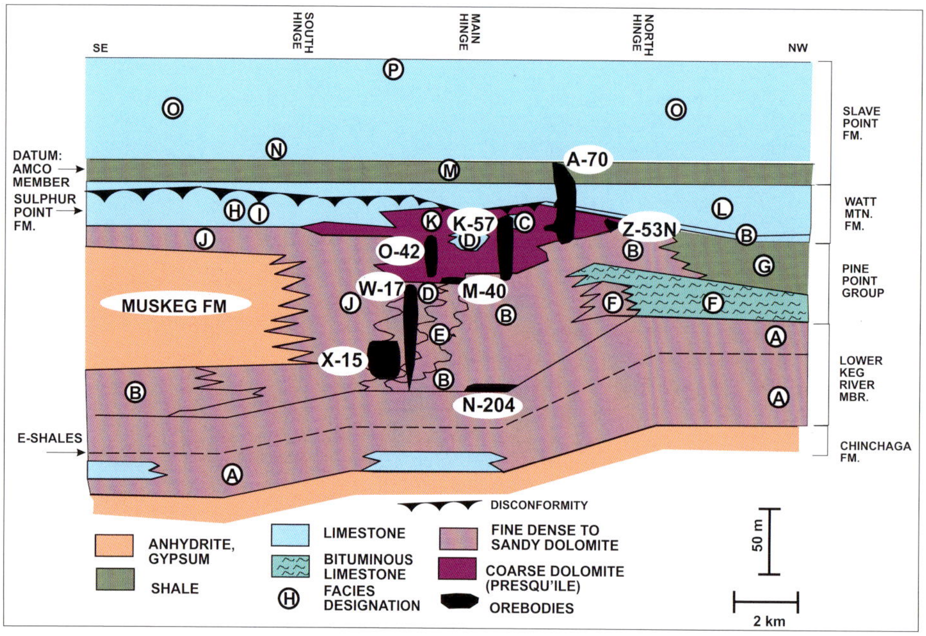

FIGURE 4. A schematic diagram of Middle Devonian stratigraphic relationships in the Presqu'ile barrier and their relation with selected orebodies at Pine Point (adapted from Skall, 1975). Complex diachronous relationships of middle Givetian stratigraphy required the establishment of various lithofacies units. These lithofacies units originally described by Skall (1975) and refined by Pine Point Mines were established on the basis of distribution with respect to the barrier complex and depositional environment. The stratigraphic units have been subdivided into the following lithofacies (Skall, 1975; Rhodes et al., 1984): the Tentaculites (F) and Buffalo River shale (G) in the basin north of the barrier complex; the off-reef (B), shallow fore-reef (C), and clean arenite (E) in the fore-reef; the organic barrier (D) in the barrier complex itself; and the gastropod (H), Amphipora (I) and south flank back-reef (J) to the south of the barrier. Skall (1975) subdivided the Slave Point Formation into four facies, the M (Amco member), N (Fort Vermilion-equivalent), O, and P facies. The location of orebodies A-70, K-57, M-40, N-204, O-42, X-15, W-17, and Z-53N are also shown.

sedimentary successions of approximately equal volume derived from distinct tectonic settings. The older succession consists of Late Proterozoic to mid-Jurassic dominantly marine sediments. These easterly derived rocks were deposited on a stable passive continental margin and the adjacent craton by oceanward progradation (Podruski et al., 1988) and consist of episodic transgressive and regressive miogeoclinal deposits consisting of open marine platformal carbonates, basinal shales, and restricted evaporitic sediments. A second major sedimentary succession of westerly derived shallow marine and nonmarine clastic sediments of Late Jurassic to Tertiary age buried the 'cratonic platform' succession. Several cratonic arches evolving prior to or during sedimentation profoundly affected environments of deposition, resulting in complex sedimentary facies patterns adjacent to these basement highs (Peace River and Tathlina arches, Fig. 1). The host rocks for MVT mineralization at Pine Point are Middle Devonian reef complex and platformal carbonates. Within the Givetian carbonate succession, a constricted linear buildup of carbonates up to 200 m thick formed the Presqu'ile barrier complex (Kyle, 1981; Rhodes et al., 1984). The barrier separates two depositional subbasins; the Mackenzie subbasin to the north, where shales and argillaceous limestones were deposited, and the Elk Point subbasin to the south, where evaporites and lesser carbonates were laid down (Fig. 1). The barrier complex exposed on the surface within the Pine Point mine area dips gently to the west, and is traceable in the subsurface through the southern portion of Great Slave Plain in Northwest Territories into northwestern Alberta and northeastern British Columbia. Skall (1975) demonstrated that subtle tectonic adjustments along three N65°E hinge lines parallel to the barrier complex were responsible for the development of paleokarst and a number of paleo-environmental facies within and adjacent to the barrier (Figs. 2, 4 (marked as North, Main, and South trends)). These hinge lines seem to be approximate projections of the underlying Precambrian McDonald-Great Slave Lake fault system, although Skall (1975) pointed out that the trend of the hinges and the fault system are not parallel. The hinges trace lines of weakness where preferential karstification, coarse-crystalline dolomitization (Presqu'ile dolomite), and mineralization are concentrated.

Stratigraphy

Precambrian basement in the Pine Point district consists of granite, granite gneiss, and quartzite of Aphebian age (Fig. 5, Norris, 1965). Early Proterozoic collisional tectonics caused major discontinuities to form in the basement, such as the Great Slave Lake Shear Zone (GSLSZ) (Fig. 3). Potential field geophysical signatures can trace many of these major discontinuities beneath the basinal sediments. Other major faults and discontinuities identified in the district include the Tathlina Fault Zone and Rabbit Lake Fault Zone (Fig. 3).

A major episode of tectonic uplift and wide-ranging erosion effectively removed almost all pre-Devonian strata in northern Alberta and southern Northwest Territories. Gentle folding of the pre-Devonian surface produced a series of basins and arches that profoundly effected subsequent Devonian sedimentation. A major positive feature effecting sedimentation in southern Northwest Territories is the Tathlina Arch (Figs. 1, 5), cored by Precambrian granitic rock. The intracratonic arch prevented marine encroachments onto the craton from the west (Reinson et al., 1993), producing a restricted marine environment during Emsian and Eifelian time. Rocks of the Lower Elk Point Group (Basal Red Beds, Mirage Point, Cold Lake, and Chinchaga formations) consist of interbedded red beds, evaporites, shallow marine clastics, and minor carbonates (100-350 m thick) (Fig. 5). The Eifelian Upper Chinchaga Member contains evaporites and lesser dolostones and limestones, and unconformably overlies the older succession (Fig. 5).

Subsequent to initially restricted Devonian deposition, an open-marine carbonate-depositing sea transgressed the platform (Reinson et al., 1993; Kent, 1994). Basin-wide deposition of normal marine carbonates in a ramp-to-platform setting forms the Lower Keg River Member ('Keg River

Platform'; Facies A from Skall (1975) in Fig. 5). At Pine Point, Lower Keg River strata underlie back-barrier, fore-barrier, and barrier strata. The succession maintains a constant thickness of nearly 65 m in the Pine Point district (Rhodes et al., 1984). This homogeneous unit consists of dense to sucrosic dolostone with carbonaceous to argillaceous wisps.

Continued subsidence and marine transgression led to the development of an extensive upper Keg River barrier complex on the southern flank of the Tathlina Arch. The complex, referred to as the Presqu'ile barrier, extends across the basin from the Pine Point minesite near its eastern margin to northeastern British Columbia at the Cordilleran deformation front (Fig. 1). Southeast of the barrier, abundant isolated reefs, reef mounds, and banks of the Upper Keg River Member were deposited on Lower Keg River platform deposits. North and west of the barrier, Horn Plateau isolated reefs grew on the Lonely Bay carbonate platform (Fig. 5; Meijer Drees, 1993). The Presqu'ile barrier effectively restricted the sea south of the barrier and normal marine sedimentation ceased. In this back-barrier region, the regressive phase of the Upper Elk Point cycle is represented by the development of a carbonate-evaporite shelf succession and evaporite basinal deposits constituting the Elk Point subbasin (Fig. 1). This evaporitic succession, defined as the Muskeg Formation in southern Northwest Territories, contains halite, anhydrite, and evaporitic dolostone (Figs. 3, 4, 5). Normal marine sedimentation continued, however, north of the barrier in the Mackenzie subbasin, where forereef deep-water 'bituminous shale and limestone beds' (Meijer Drees, 1993) of the Horn River Formation (Figs. 3, 5) were deposited.

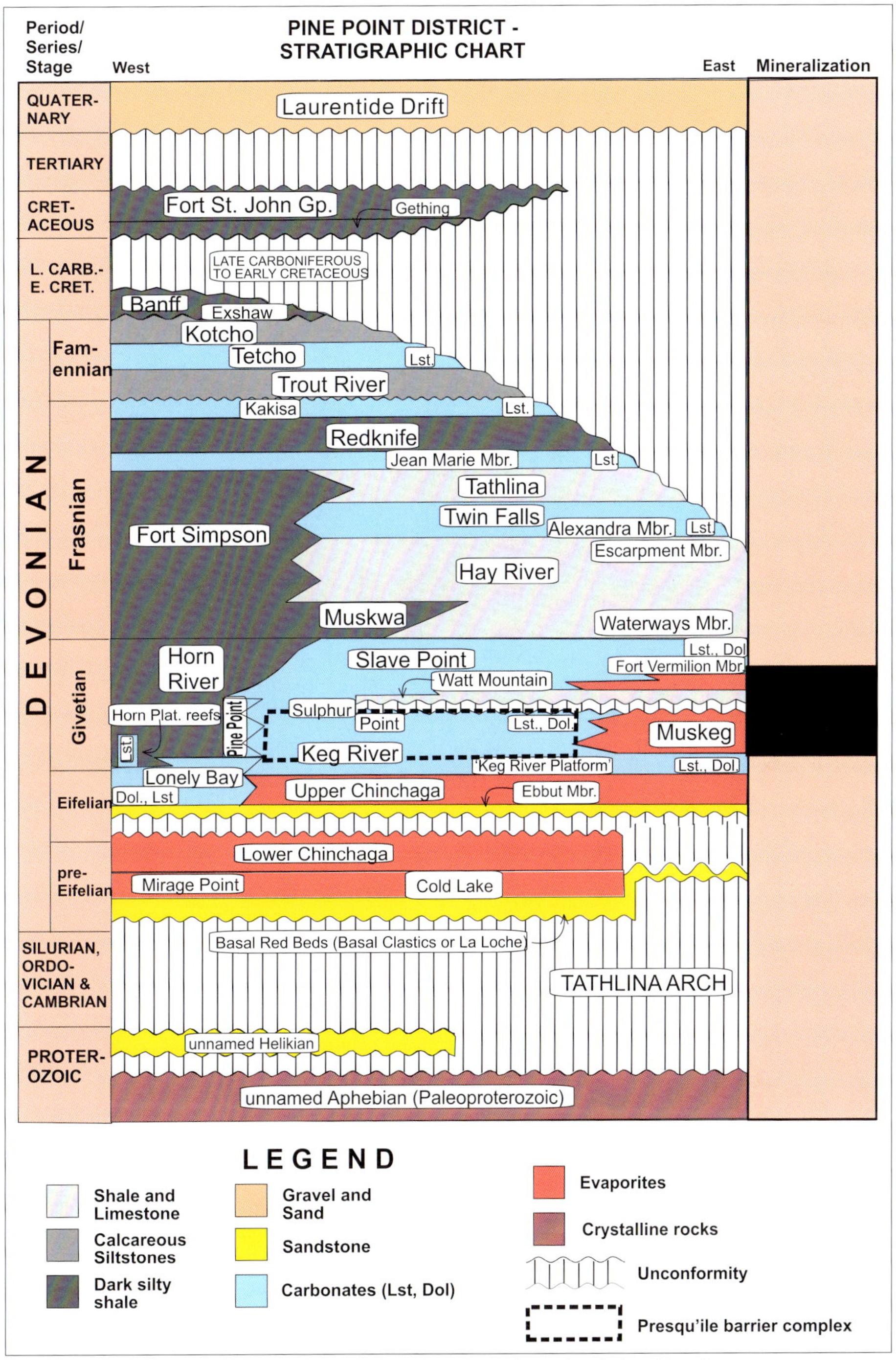

FIGURE 5. Stratigraphic chart illustrating nomenclature and schematic relationships along an east west transect across the Slave Plain in southern Northwest Territories (adapted from Morrow et al., 2002). Also shown is the schematic representation of the position of the Presqu'ile barrier complex and the temporal distribution of MVT mineralization within the stratigraphic framework.

The Presqu'ile barrier at Pine Point consists of two superimposed buildups, the lower or Upper Keg River Member buildup (lower Pine Point Group in Skall, 1975, Pine Point Formation) and the succeeding Sulphur Point buildup (upper Pine Point Group in Skall, 1975; Figs. 4, 5). The lower buildup constitutes approximately two-thirds of the Presqu'ile barrier and is a flat lensoid accumulation of dominantly carbonate sands and muds. As the barrier complex grew, the various facies in the lower buildup prograded northwards. The superimposed Sulphur Point buildup is a tabular carbonate accumulation containing local bioherms with related bioclastic limestones and carbonate sands (Rhodes et al., 1984). Near the end of the cycle, a marine incursion into northwestern Alberta and southern Northwest Territories resulted in the deposition of shelf carbonates of the Sulphur Point Formation, south and east of the barrier as well as on the barrier itself.

During Middle Devonian time, the North, Main, and South hinge lines or trends delineated blocks of carbonate sediment that underwent dissimilar subsidence leading to abrupt transitions in lithofacies as well as rapid changes in thickness. The stratigraphic units representing the unique depositional environment at Pine Point have been subdivided into the following lithofacies (Skall, 1975; Rhodes et al., 1984): the Tentaculites (F) and Buffalo River shale (G) in the basin north of the barrier complex; the off-reef (B), shallow fore-reef (C), and clean arenite (E) in the fore-reef; the organic barrier (D) in the barrier complex itself; and the gastropod (H), Amphipora (I) and south flank back-reef (J) to the south of the barrier (Fig. 4).

The Upper Keg River Member (lower Pine Point Group) lies conformably on the carbonate platform (Fig. 5). The formation is composed of a diverse assemblage of lithofacies that has been extensively dolomitized (Skall, 1975). The maximum thickness of the assemblage is 175 m at the axis of the reef complex and thins to the north and south forming a lens-like accumulation (Rhodes et al., 1984).

Rhodes et al. (1984) emphasized in their discussion on barrier lithofacies that the Presqu'ile barrier is not an organic reef because only about 10 percent of the sediments are organic floatstones and no boundstones or framestones are present. The lower carbonate buildup consists of carbonate sands bounded by carbonate muds to the north. According to Rhodes et al. (1984), it is more appropriate to describe the barrier as an organic bank or sediment pile. Even though there are substantial north-south and vertical variations of lithofacies in the carbonate bank, there is, however, a consistency of facies along the strike of the barrier.

The schematic diagram of the barrier shown in Figure 4 is drawn with great vertical exaggeration. The slope of the barrier and the angle of progradation in real space are nearly horizontal (about one or two degrees). The nearly flat geometry indicates that the height of the barrier was never great and that the lithofacies successions were deposited in water depths ranging from zero to 30 m. These depths are ideal for the evolution of contrasting depositional environments that are expressed by their diverse lithofacies (Rhodes et al., 1984).

The Sulphur Point Formation (facies C, D, H, and I, Fig. 4) conformably succeeds the Upper Keg River Member on the barrier, whereas to the south it transitionally overlies the Muskeg Formation (facies J, Fig. 4). Thicknesses of the Sulphur Point Formation range considerably; from 65 m in the barrier complex to 30 m on the south side of the barrier, and 9 m far to the south in Wood Buffalo National Park. The formation thins to the north of the barrier and abruptly terminates. Similar to Upper Keg River Member, the Sulphur Point Formation exhibits lateral and vertical facies variations at Pine Point.

South of the barrier complex, Muskeg Formation dolostones and evaporites are present. The Muskeg Formation is equivalent to the Upper Keg River Member and the basal Sulphur Point Formation (Figs. 4, 5). The evaporites and carbonates of the Muskeg Formation in the Elk Point subbasin are 350 to 400 m thick. This succession is thicker than time-equivalent strata within the barrier due to greater subsidence and rapid sedimentation rates in the subbasin. At Pine Point, equivalent strata consisting of fine-crystalline dolostones predominate (facies J, Fig. 4). The thickness of the dolostones ranges from 18 to 170 m. This facies was deposited in a back-barrier restricted environment. The South hinge coincides with the northern limit of evaporites of the Muskeg Formation.

In the Mackenzie subbasin north of the barrier, marine fissile calcareous shales with disseminated iron sulphides were deposited. These rocks are partly equivalent to the upper part of the carbonate barrier complex. This formation, called Buffalo River (facies G; Fig. 4) (Skall, 1975), thickens northward to a maximum thickness of 60 m. The North hinge coincides with the southern limit of Buffalo River Formation deposition. Deposition of the Buffalo River shales commenced during Upper Keg River sedimentation and continued beyond the Sulphur Point depositional interval (Rhodes et al., 1984).

The Presqu'ile dolomite was originally interpreted as a reefal development and consequently assigned formational status (e.g. Norris, 1965; Campbell, 1966). Skall (1975) recognized that the Presqu'ile facies (facies K; Fig. 4) is not a dolomitized reef core, but a spectacular later stage diagenetic dolomitization event that affected a broad array of primary carbonate lithofacies (Figs. 3, 4). The Presqu'ile dolostone is now interpreted as a coarse-crystalline vuggy dolomite facies not confined by environmental conditions or timing of deposition of the original carbonates. The coarse texture of this dolomite destroyed much of the original carbonate fabric, making environmental interpretation difficult. The vugs are filled or lined with late-stage white coarse-crystalline dolomite and lesser calcite, sulphides, altered bitumen, and native sulphur (Krebs and Macqueen, 1984). Rhodes et al. (1984) estimated that 60 to 70% of the Sulphur Point Formation on the Pine Point property has been altered to coarse-crystalline dolomite. Generally, Presqu'ile dolomitization extends upward from the Upper Keg River/Sulphur Point contact to the top of the Sulphur Point Formation (Fig. 4). Occasional remnants of Sulphur Point limestone are preserved in the barrier complex (Fig. 4). Presqu'ile dolomite is occasionally found in the underlying Upper Keg River Member. Infrequently, it extends upward across the sub-Watt Mountain disconformity into the overlying Slave Point Formation. The main areal distribution of the Presqu'ile dolomite follows the North and Main hinges (Figs. 3, 4). It is thickest along the Main hinge, where it varies between 60 and 70 m. Many of the Pine Point orebodies occur in karsted carbonates that are altered to Presqu'ile dolostone (e.g. M-40, O-42, and K-57, part of A-70 in Fig. 4), but deposits are not necessarily restricted to this diagenetic facies (e.g. X-15, W-17, and N-204; Fig. 4).

Upper Elk Point deposition was terminated by a significant base-level drop defined by a major unconformity upon which coastal marine and continental shales, sandstones, and limestones of the Watt Mountain Formation were deposited (Figs. 3, 4, 5) (Meijer Drees, 1993; Reinson et al., 1993). Skall (1975) described a regression occurring near the Middle-Upper Givetian boundary resulting in a partial disconformity and subaerial karst development that affected the topographically higher parts of the barrier. Skall (1975) speculated that the main part of the barrier was exposed up to 30 m above sea level. To the north of the barrier, however, marine sedimentation continued unabated as shown by the termination of the disconformity and the stratigraphic conformity between the underlying B-facies and overlying Watt Mountain Formation (Fig. 4). The conformable relationship is confirmed by thicker accumulations of Watt Mountain carbonates north of the barrier.

Kyle (1981) interpreted extensive karstic development on the barrier, as indicated by the development of abundant collapse breccias filled with detritus and resulting subaerial sinkholes formed by percolation of vadose waters. Karstification is considered to be an important ground-preparation process in the host rock for subsequent MVT sulphide mineralization (Leach and Sangster, 1993).

Following emergence of the barrier at the beginning of Late Givetian time, Watt Mountain micritic limestones with occasional waxy shales or clay layers were deposited on the disconformity above and south of the barrier complex (facies L, Fig. 4; Skall, 1975). This formation averages less than 10 m in thickness where it disconformably overlies the barrier complex, but increases to near 33 m north of the barrier. The Watt Mountain Formation was deposited in tidal flat, lacustrine, and less commonly lagoonal and restricted environments.

Subsequent sedimentation continued with deposition of peritidal anhydrites and carbonates of the Fort Vermilion Member. The Fort Vermilion Member is partly laterally equivalent and also overlain by open-marine platform carbonates of the Slave Point Formation (Figs. 3, 5), which conformably overlies the Watt Mountain Formation at Pine Point. Skall (1975) subdivided the Slave Point Formation into four facies, the M (Amco member), N (Fort Vermilion-equivalent), O, and P facies (Fig. 4). Generally, the Slave Point Formation in the Pine Point area maintains a uniform and layered character. The Slave Point platform carbonates in southern Northwest Territories are overlain by shale and argillaceous limestone of the Hay River Formation (Waterways Member) (Figs. 3, 5).

Upper Devonian sedimentation continued west of the Pine Point property. Numerous transgressive-regressive sequences of prograding platform or shelf carbonates interlayered with westwardly prograding siliciclastic wedges were deposited (Figs. 3, 5).

The broad shallow-water ramps that developed throughout the basin during Late Devonian time persisted to the Early Carboniferous. In southern Northwest Territories, Carboniferous sediments are preserved in the western subsurface (Okulitch, 2006). A major Late Carboniferous to Early Cretaceous unconformity overlying the Banff Formation occurs across much of the district.

Structure and Epeirogenic Tectonism

Epeirogenic tectonic adjustments along three parallel hinge lines (i.e., North, Main, and South trends, Fig. 2) following the strike of the barrier profoundly impacted facies distribution at Pine Point (Skall, 1975) and the distribution of MVT orebodies. These subtle tectonic movements presumably occurred during mid-Givetian time in the cratonic platform sedimentary succession. The hinge lines are parallel to interpreted Middle Devonian faults and fractures (Douglas and Norris, 1974) but are divergent from the trend of the major Precambrian fault system east and south of the property (i.e., N65°E for the hinge lines; N45°E for the McDonald-GSLSZ fault system).

Rhodes et al. (1984) argued against profound structural control for the distribution of orebodies at Pine Point due to this observed divergence of structural trends. They also observed that Middle Proterozoic dyke swarms crossing the McDonald fault in the East Arm of Great Slave Lake to the northeast of Pine Point are not dislocated, and the Lower Keg River and Chinchaga stratigraphic units underlying the barrier complex at Pine Point form uniform blankets of sediment with no sign of vertical displacement. These observations suggest that little or no reactivation occurred within the McDonald-GSLSZ fault system since Proterozoic time, and this lack of movement likely indicates the system was effectively sealed, preventing the movement of metal-rich fluids upward along the faults into the Phanerozoic carbonate cover. However, one can argue that the subparallelism of the hinge lines with facies fronts, pinchouts, and depositional limits indicate a certain degree of syndepositional structural control (Skall, 1975). Also, the divergent trends do not exclude a kinematic link where parallel en echelon normal faults trending N65°E are subsidiary extensional splays of the main vertical brittle fault (Pana, 2006).

Faulting and fracturing are also present in the Middle Devonian succession and are concentrated along the hinge zones but not restricted to them (Skall, 1975). These hinge zones remained as lines of weakness where post-Devonian reactivation may have occurred. Movement on various post-Amco faults also affected the mid-Givetian structures. Skall (1975) pointed out that contemporaneous faulting and fracturing could also be caused by compaction or volume shrinkage due to diagenesis. Rhodes et al. (1984) identified from structural contour maps, subtle undulations post-dating Amco deposition trending both north-south and east-west. They speculated that these undulations were caused by differential compaction or loading and warping of underlying lower Elk Point evaporite successions. In some ore zones, high-angle faults with displacements of 2 to 10 m are present. These young structures displace all the strata (Rhodes et al., 1984).

Similar northeast and east-northeast linear trends recognized by aeromagnetic contours, structural contours, isopach maps, facies changes, and coarse-crystalline dolomite belts in bordering areas to the west of Pine Point, such as the Tathlina Lake and Rabbit Lake faults, have been mapped (Morrow et al., 2002; Janicki, 2006b). This pattern of linear structures could represent a series of horsts and grabens occurring at the west end of Great Slave Lake (Gibbins, 1988).

Mineral Deposits and Occurrences

Deposit Types and Classification

A single major deposit type has been recognized in the Pine Point district, the classic Mississippi Valley-type (MVT) Zn-Pb sulphide deposit.

MVT deposits account for 25 percent of the world's lead and zinc resource (Tikkanen, 1986). In Canada, MVT mineral production occurred between 1964 and 2002. During this time, about 30 percent of Canada's annual lead and zinc

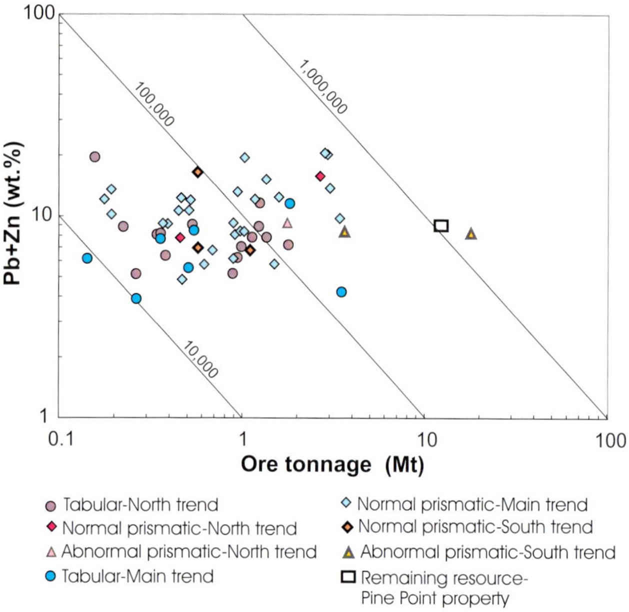

FIGURE 6. Grade versus tonnage plot of total production for individual MVT orebodies as well as for nonproduced orebodies with reported current proven reserves at Pine Point according to major mineralized trend and orebody type (production data supplied by D. Rhodes of Teck-Cominco). Estimated grade and tonnage of the remaining geological resource in the Pine Point district is also depicted.

production was derived from MVT deposits (Sangster, 1996). At the end of 2002, all MVT mines in Canada were closed, thus reducing the contribution of Pb and Zn from MVT deposits to nil with respect to Canadian mineral production. In the Pine Point district, 100 orebodies are distributed over about 1600 square km (Fig. 2).

Orebody Distribution

All MVT deposits or orebodies in the Pine Point district are hosted in middle Devonian Givetian carbonate strata within and adjacent to a dolomitized carbonate barrier complex. Most of the 100 known Pb-Zn deposits or orebodies in the Pine Point district are hosted by more than one carbonate stratal unit (Appendix 1). The Sulphur Point Formation hosts in whole or in part the majority of orebodies at the Pine Point mine site. In the Pine Point district (outside of Pine Point and Great Slave Reef properties), there are also 92 occurrences of visible Pb and Zn mineralization in Middle and Upper Devonian carbonate strata (Hannigan, 2005). Most individual orebodies are associated with the major diagenetic dolomite facies (Presqu'ile dolomite).

Orebodies in the Pine Point district are preferentially distributed along zones of weakness or hinge lines where extensive interconnected paleokarst networks developed at the lower limit of Presqu'ile-type coarse dolomitization of upper barrier porous limestone (Figs. 2, 4). The paleokarst networks are most extensively developed along two major linear features called the North and Main trends, where Presqu'ile dolomitization was pervasive (Fig. 2). The South Trend is less well defined, consisting of only a limited number of scattered orebodies accompanied by very little Presqu'ile dolomitization. In the Pine Point district, 57 orebodies are distributed along the Main trend, 36 occur along the North trend and seven are located along the South trend (Fig. 2).

Grade and Tonnage

Individual orebodies in the Pine Point mining camp range in size from less than 100,000 tonnes to nearly 17.5 Mt, with an average grade of 2.0% Pb and 6.2% Zn in the largest prismatic orebody with the mine designation of X-15 (Appendix 1). Figure 6 illustrates the grade-tonnage distribution for orebodies with reported proven reserves as well as past producing orebodies according to orebody type and trend. Not all Pine Point orebodies are shown due to unavailability of reserve or production information. However, an estimation of remaining geological resource on the Pine Point property was determined and is depicted. Tonnage and grade distributions are similar along each trend. Deposits averaged 1.32 Mt of ore grading near 7.0% Zn and 3.0% Pb (district average in Fig. 7). Metal contents of individual orebodies ranged from 0.4 to 8.8% Pb and 2.8 to 14.1% Zn (Fig. 7). The iron content of individual orebodies varied from less than 0.5 to nearly 10.5%; the district average is near 3.5% (Kyle, 1981). Total geological resource (produced resource and remaining proven reserves) for the Pine Point district is estimated by the author to be nearly 94.5 Mt. The Pb/(Pb+Zn) ratios for Pine Point district orebodies range from 0.06 to 0.56, giving a district average of 0.26. Orebodies of the normal prismatic variety are generally Pb-rich, averaging about 0.35. Normal prismatic orebodies in the Main Trend are generally more Pb-rich, averaging nearly 0.37, whereas orebodies in the North and South trends average about 0.25. Tabular and abnormal prismatic orebodies in all trends are relatively Zn-rich, averaging 0.25 (Fig. 7).

Prismatic orebodies tend to have coarser crystalline galena occurring peripherally to the main ore zone. Some of the prismatic bodies are zoned with a Pb-rich core, passing outward to a Zn-rich zone, and enveloped by a Fe-rich aure-

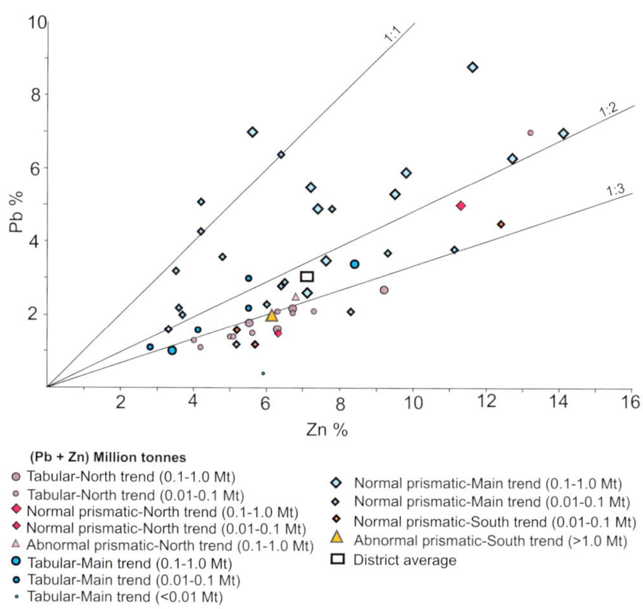

FIGURE 7. Weight percent of Pb and Zn in major orebodies at Pine Point according to mineralized trend, orebody type, and tonnage (tonnages supplied by D. Rhodes of Teck-Cominco). The district average is also shown.

ole (Kyle, 1981). The areal distribution of iron-sulphides is wider than galena and sphalerite, and the Fe/(Fe + Zn + Pb) ratio increases outward from prismatic orebodies. There are relatively small barren sulphide bodies on the property consisting mostly of iron sulphides with minor lead and zinc. Tabular orebodies that do not have associated prismatic ore shoots seem to have a more regular grade.

Weighted averages of Pb and Zn in various orebodies from the upper (Sulphur Point buildup) and lower (Keg River buildup) portions of the Presqu'ile barrier indicate weak vertical metal zoning. In the upper barrier, a weighted average of Pb/(Pb + Zn) is 0.20, indicating that these orebodies are slightly richer in zinc compared to orebodies in the lower barrier (0.26). Kyle (1981) reported a district-wide pattern of metal zoning as determined from ore-reserve data, with Pb/(Pb+Zn) increasing from about 0.2 in the southeast to about 0.5 in the northwest in zones parallel to the Main and North trends. New information collected by the author on ore located to the southwest of Kyle's (1981) study reveals the ratio decreasing southwesterly to near 0.2. A weighted average calculation by the author of Pb/(Pb+Zn) on orebodies along the entire length of the Main Trend shows a consistent ratio (0.36), indicating there is no district-wide lateral zonation.

Orebody Morphology

Nature of Sulphide Bodies

Rhodes et al. (1984) cited karstification of the barrier as a major control for ore deposition at Pine Point. They pointed out that the trend and geometry of these paleokarst bodies are similar to the orebodies and thus, are strongly related. At Pine Point, karst refers to both surface and subsurface solution features and pore spaces. The mineralization is pervasive in interconnected paleokarst networks. The networks are mainly developed in the porous barrier complex along the North and Main hinge zones and are directly related to distinct lithofacies. At Pine Point, karstification not only occurs along these distinct trends, but is embodied in four distinct karstic orebody-types: tabular, prismatic, abnormal prismatic, and B-spongy type (Rhodes et al., 1984). These orebody types are illustrated in Figure 8. Ore deposition is strongly controlled by discrete stratal horizons, although individual prismatic orebodies are discordant.

Tabular karst is the most common and widespread solution network in the Pine Point district. The karstic network occurs along a crude stratabound horizon that coincides with the base of the Presqu'ile dolomite near the Sulphur Point/Upper Keg River contact (Fig. 8). The tabular ore deposits are elongate flat-lying bodies associated with an interconnected network of karst channels (see Fig. 9A, where an example of stratabound karstic or breccia dolostone is shown representing a potential host for tabular-type ore deposits). These karst channels can attain thicknesses up to 12 m and widths up to 240 m (Rhodes et al., 1984).

Most past production at the Pine Point mine exploited normal prismatic deposits, which are chimney-like vertically elongate orebodies associated with major cavity dissolution and prismatic openings extending upward from the tabular horizon (Fig. 8; Rhodes et al., 1984). These local zones of enhanced dissolution appear to coincide with areas of greater

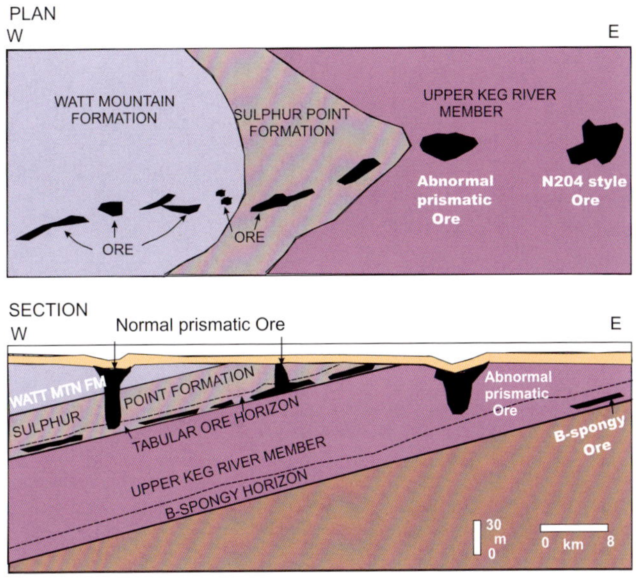

FIGURE 8. Schematic plan and vertical section of orebody types and their relation to the regional stratigraphy (adapted from Rhodes et al., 1984).

lithologic weakness where more intense jointing and fracturing increased the porosity and permeability. Dissolution caused sagging and collapse of the overlying strata, sometimes affecting up to 100 m of sediment. These collapse structures seem to be related to structurally elevated areas (ridges or domes). The collapse process allowed the resedimentation of the karstic openings by clastic carbonates ranging from well laminated internal sediments to breccias incorporating surrounding and overlying material at and above the tabular karst horizon (Rhodes et al., 1984). The breccias show fine fragmentation at the base of the cavity structures and grade upward into coarser and fewer distorted fragments. The finer crystalline internal sediments and overlying breccias contain the sulphide mineralization.

There is a type of karstic structure of similar geometry to normal prismatic bodies that is not rooted in the tabular ore horizon. These 'abnormal' prismatic orebodies are karstic features separated from the dominant paleokarst network hosting tabular and normal prismatic deposits. It is believed that these abnormal structures or unconsolidated collapses formed as a result of tabular or prismatic bodies originally hosted in overlying strata collapsing into underlying strata. The abnormal orebodies are affected by continuous dissolution and are thought to represent modern karst or sinkholes (Fig. 8; Rhodes et al., 1984). Some of these modern sinkhole structures are mineralized (e.g. X-15 and W-17 orebodies, Fig. 4).

B-spongy mineralization (i.e. disseminated mineralization restricted to the porous horizon called the B-spongy member (Rhodes et al., 1984)) has been discovered in a single orebody at Pine Point, the N-204 orebody located near the northeastern corner of the property near surface (Figs. 2, 4, 8). The B-spongy member consists of fine-crystalline dolomite containing fossil moldic porosity. On the eastern part of the property, this open porosity is accompanied by fine crackle brecciation (Rhodes et al., 1984). Carbonate dissolution associated with karsting in this horizon created substantial vuggy, fracture and intergranular porosity.

FIGURE 9. **(A)** Dolomite boulder from the rubble pile at the Pine Point mine site. The very porous partly cemented rubble packbreccia layer is underlain by laminated dolostone strata. The breccia layer, a horizon of enhanced porosity and permeability, may form tabular karst structures that commonly host tabular sulphide concentrations at Pine Point. Photo courtesy of D. W. Morrow. **(B)** Ore from the Pine Point mine. Collapse breccia containing open-space filling of white sparry dolomite (SDol) surrounding broken angular fragments of dolomitic internal sediment impregnated by fine-crystalline dark brown sphalerite. Late white and coarse calcite (Cal) with associated native sulphur fill the core of the former cavity. Fine-crystalline dark brown sphalerite replaces laminated sandy internal sediment (IS) deposited on the floor (?) of the cavity. Nickel for scale (2 cm). **(C)** Aggregates of sphalerite crystals and white sparry dolomite filling small fractures in brecciated carbonate strata at Pine Point (courtesy of S. Paradis). **(D)** Dolostone boulder with abundant coarse-crystalline saddle or hydrothermal dolomite filling cavities and fractures, Pine Point mine site.

Mineralization is disseminated in the porous and brecciated strata.

Dimensions of Orebodies

Dimensions of the 100 known orebodies in the Pine Point district vary from 40 to 2000 m in length, 15 to 1000 m in width, and 0.5 to 100 m thick (Appendix 1). Tabular orebodies range in length from 100 to 2000 m, width from 15 to 1000 m, and are 1 to 30 m thick. Normal prismatic orebodies vary in length from 40 to 560 m, width from 15 to 185 m, and thickness from 0.5 to 100 m. Abnormal prismatic orebodies show dimensions of 130 to 800 m in length, 130 to 400 m in width, and 14 to 75 m in thickness.

Mineralogy and Textures

Characteristic Mississippi Valley-type deposit features, such as karst development, dolomitization, and genetic association with evaporites and hydrocarbons, are all present in the Pine Point district (Figs. 3, 5). Mineralization is epigenetic.

Sphalerite and galena are the only economic minerals at Pine Point. Ore-associated hydrothermal minerals include marcasite, pyrite, native sulphur, and bitumen and occasionally pyrrhotite, celestite, barite, gypsum, anhydrite, and fluorite. Gangue minerals are dolomite and calcite.

Typical sulphide textures at Pine Point include open-space filling of breccias, fractures, and vugs (Fig. 9B). Massive replacement of host rocks, internal sediments, and sulphide disseminations are also observed in the Pine Point district. Much of the ore occurs as very fine-crystalline laminated aggregates of botryoidal or colloform sphalerite, often intergrown with dendritic or skeletal galena (Leach and Sangster, 1993). Commonly, botryoidal banded sphalerite was followed by precipitation of euhedral to anhedral sphalerite crystals forming irregular seams (Krebs and Macqueen, 1984). Sphalerite also occurs as aggregates of coarse-grained crystals associated with white sparry dolomite precipitated in small fractures (Fig. 9C). A mosaic breccia texture is also common at Pine Point where sphalerite and hydrothermal dolomite cement breccia clasts (Fig. 9B,C). On the roofs of cavities and along breccia fragment bases, stalactitic sphalerite containing hollow axial canals, sometimes filled with galena, have formed. This stalactitic ore habit is indicative of sulphide precipitation in a vadose environment (Krebs and Macqueen, 1984).

Mineral Chemistry

The only common metals in the mineralizing solution are Pb, Zn, and Fe and they are present at a ratio of 2:5:3 (Kyle, 1981). Copper, Mn, and Cd are present locally as trace elements in the common sulphides (Kyle, 1977; Gleeson and Gromek, 2006); separate minerals have not been recognized containing these metals.

In the Pine Point district, homogenization temperatures (Th) of primary aqueous fluid inclusion assemblages in coarse sphalerite crystals from various orebodies give a range of 51 to 108°C (Roedder, 1968; Kyle, 1977; Turner, 2003). These fluid inclusion filling temperatures likely represent minimum temperatures of the ore-forming solution, because the most suitable inclusions for homogenization temperatures determination occur within later slow-forming coarse sphalerite crystals. Homogenization temperatures, therefore, would be biased toward lower temperatures assuming that the temperature decreases during ore deposition (Krebs and Macqueen, 1984). Higher temperature banded and botryoidal fine-grained sphalerite crystals generally do not contain suitable fluid inclusions for the accurate determination of filling temperatures (Anderson, 1975; Kyle, 1977). Fluid inclusion assemblage measurements in coarse-crystalline matrix and saddle dolomites at Pine Point display higher homogenization temperatures (86-126°C) than sphalerite (Kyle, 1981; Turner, 2003; Coniglio et al., 2006; Paradis et al., 2006). An increase of homogenization temperature in fluid inclusions in sphalerite and saddle dolomite was observed from the eastern Pine Point mineral property westward to mineral properties near Hay River and at Windy Point on the south-western shore of Great Slave Lake (Pine Point property, 60-90°C; Hay River and Windy Point areas, 97-113°C) (Turner, 2003). These paleo-temperatures exceed values attributed solely to their maximum burial depth in the barrier complex under Paleozoic cover. During Late Paleozoic time, the barrier complex in the Pine Point district was thought to have been buried under approximately 1500 m of Paleozoic cover (Krebs and Macqueen, 1984). At this depth and assuming a geothermal gradient of 30°C/km, the resulting temperature should have been approximately 60°C, much lower than measured homogenization temperatures. Burial alone cannot produce the observed fluid-inclusion filling temperatures and therefore other thermal sources must be present in the Pine Point region (Krebs and Macqueen, 1984). An investigation of thermal maturity of organic matter at Pine Point (Macqueen and Powell, 1983) indicates a substantial contrast between thermally immature indigenous organic material in country rock carbonates and mature to overmature bitumen proximal to and within orebodies, demonstrating that thermal anomalies were present in the district and were directly associated with Pb-Zn mineralization.

Early work at Pine Point revealed that freezing temperatures derived from fluid inclusions in sphalerite range from -10 to -35°C, indicating that the ore-bearing solution was highly saline (15 to more than 23%; up to 38% total salts, NaCl-equivalent) (Roedder, 1968). Turner (2003) recorded ice-melting temperatures in sphalerite inclusions between -22 and -33°C and in saddle dolomite inclusions between -13 and -32°C. These temperature ranges correspond to average salinities of 27 wt.% NaCl-equivalent for sphalerite and 25.5 wt.% NaCl-equivalent for dolomite (Turner, 2003). The presence of hydrohalite ($NaCl.H_2O$) with eutectic melting temperatures ranging between -40 and -56°C at Pine Point indicates that the salt hydrates are dominantly composed of $CaCl_2$ (Turner, 2003). West of Pine Point district on the Presqu'ile barrier, homogenization temperatures in saddle dolomite are generally much higher (mean = 137°C), fluids are less saline (10-28 wt.% NaCl eq.) and are dominantly NaCl- and $CaCl_2$-rich with subordinate KCl and $MgCl_2$ (Qing, 1991; Qing and Mountjoy, 1994b) compared to saddle dolomites at Pine Point.

The relatively low temperatures (compared to many other mineral deposit types) and high salinities associated with fluid inclusions in Pine Point dolomites and sphalerites are characteristic of MVT deposits and indicate derivation from deeply buried basinal fluids affected by low-temperature diagenetic or metasomatic processes rather than from magmatic or volcanogenic sources. Chemical analyses of fluid inclusions in carbonate-hosted deposits in the Pine Point district indicate that mineralizing solutions were dense brines dominated by chlorine, sodium, calcium, potassium, and magnesium in decreasing order of abundance (Kyle, 1981; Leach and Sangster, 1993). The bulk of the fluids have Cl/Br values that are less than that of seawater, indicating the fluids are Br-enriched (Gleeson and Gromek, 2006). These Br-rich fluid inclusions suggest that the parent fluids evolved from highly evaporated seawater rather than common seawater. Pine Point brines are similar in chemistry to oil-field brines (Kyle, 1981; Haynes and Kesler, 1987; Leach and Sangster, 1993). K/Na values of fluid inclusions in the MVT deposits are greater than the ratios found in oil-field brines, however, which suggests either that MVT ore fluids are more evolved from seawater or are at least partly composed of interstitial fluids from evaporite beds enriched in residual K (Anderson and Macqueen, 1988).

The parent fluids of coarse-crystalline dolomites, saddle dolomites, and ore minerals in the Pine Point district have been characterized using stable and radiogenic isotopes. A spatial trend was identified by Qing (1991) where $\delta^{18}O_{PDB}$ values of saddle dolomite decrease westward along the Presqu'ile barrier from an average of approximately -9.0‰ at Pine Point to about -14.0‰ in northeastern British Columbia. This decrease in $\delta^{18}O$ value was attributed to a westward increase of the temperature of the dolomitizing fluid. Similar trends were also observed for the homogenization temperatures of fluid inclusions in saddle dolomite (92 to 178°C downdip and westward) (Qing, 1991) and for $^{87}Sr/^{86}Sr$ ratios that gradually increase westward from 0.7083 for saddle dolomite at Pine Point to 0.7095 in northeastern British Columbia (Qing and Mountjoy, 1994b). Recent work challenges these identified isotopic trends and shows rather variable and inconsistent spatial distribution of stable and radiogenic isotopes (Coniglio et al., 2006; Paradis et al., 2006). The nonmineralized sulphide-free dolomites do not show a westward decrease in $\delta^{18}O$ values and mineralized dolomites actually show a westward increase. A significant increase of $^{87}Sr/^{86}Sr$ in close proximity to the Great Slave Lake Shear Zone suggests a key role ascribed to the shear zone by providing a site for upward circulation of hydrothermal fluids from radiogenic basement rocks and precipitation of hydrothermal dolomite and sulphides in overlying porous carbonates (Paradis et al., 2006).

Sulphur isotopes of galena, sphalerite, marcasite, and pyrite from numerous Pine Point orebodies display a low variability (+12.6‰ to +24.1‰ $\delta^{34}S_{CD}$; average of +20.1‰ with a standard deviation of 2.6) (Sasaki and Krouse, 1969). Sverjensky (1986) reported sulphur isotopes for ore fluids depositing galena and sphalerite as +22 and +23‰, respectively. Middle Devonian anhydrites present in the adjacent Elk Point subbasin to the south of the barrier complex have mean $\delta^{34}S_{CD}$ values of +19 to +20‰, very similar to values determined for sulphides at Pine Point. These narrow sulphur isotope ranges are consistent with derivation from a single source or several sources that have been well homogenized. Evans et al. (1968) indicated that $\delta^{34}S_{CD}$ values at Pine Point are very similar to oils in the WCSB. The most reasonable source for sulphur in Pine Point sulphides is seawater sulphate (Sasaki and Krouse, 1969; Dunsmore, 1973). Powell and Macqueen (1984) proposed that thermochemical sulphate reduction occurred at Pine Point because bitumens became more altered as indicated by increasing S/C contents in proximity to orebodies. Also, there was an increase in $\delta^{34}S_{CD}$ with increasing sulphur content in the bitumens. These arguments along with the common association of saddle dolomites, pyrobitumen, sulphides, and native sulphur in vuggy ore were presented to support thermochemical sulphate reduction (Powell and Macqueen, 1984). The uniform sulphur isotopic composition in the sulphides suggests that biogenic sulphate reduction is not likely a major contributor, because this process is generally characterized by large isotopic fractionation (Krebs and Macqueen, 1984), except under closed or semi-closed conditions in a reduced water column (Goodfellow et al., 1993). Fowler (2000) and Grasby et al. (2002) speculated that thermochemical sulphate reduction may have formed H_2S elsewhere in the basin that subsequently migrated to the site of ore deposition, because thermal maturation measurements of organic-rich facies proximal to orebodies indicate that these rocks were not buried deeply enough to experience temperatures required for thermochemical sulphate reduction. Large accumulations of sour gas are present downdip in the Presqu'ile barrier, and the H_2S may have migrated to the site of ore deposition and mixed with hydrothermal metal-rich fluids upwelling from basement faults (GSLSZ-McDonald fault system) (Fowler, 2000; Grasby et al., 2002).

Alteration Mineralogy/Chemistry

The MVT deposits in the Pine Point district feature such diagenetic processes as dolomitization, recrystallization, dissolution, and hydrothermal brecciation.

Dolomitization and recrystallization

At Pine Point, dolostones are differentiated into three main dolomite-types: fine- to medium-crystalline matrix or replacement dolomite, coarse-crystalline matrix dolomite (Presqu'ile), and coarse-crystalline white dolospar (saddle dolomite) (Skall, 1975; Kyle, 1981; Krebs and Macqueen, 1984; Rhodes et al., 1984).

Much of the lower carbonate succession in the Presqu'ile barrier has been recrystallized to a fine- to medium-crystalline matrix dolomite on a property-wide basis (Krebs and Macqueen, 1984). Sedimentary structures and fossils are recognizable, although fossils have been leached forming moldic porosity. The dolostone consists of a very uniform mosaic of fine-grained dolomite crystals. This early-stage dolomitization event is associated with the development of the South hinge and the resulting deposition of Muskeg evaporites and carbonates in the restricted shelf to the south of the barrier, and was thought to have been caused by a reflux mechanism (Adams and Rhodes, 1960; Skall, 1975; Krebs and Macqueen, 1984). The reflux model is consistent with widespread distribution of the dolomite that formed during Middle Givetian time. Refluxing Mg-brines derived from Muskeg evaporite pans not only dolomitized the adjacent back-reef limestones, but limestones in the barrier and the fore-reef as well. Kyle (1981) objected to the reflux model due to hydrological considerations, and proposed a mechanism involving evaporative pumping of seawater through carbonate sediments and subsequent mixing in the subsurface with meteoric water at the time of cessation of South Hinge tectonic adjustments ('Dorag' dolomitization mode) (Badiozamani, 1973) as a more appropriate model.

As discussed above, much of the coarse-crystalline Presqu'ile dolomite affects a large proportion of the Sulphur Point Formation and occasionally part of the Upper Keg River Member, and rarely occurs above the disconformity into overlying Watt Mountain and Slave Point formations (Krebs and Macqueen, 1984). In contrast to the finely crystalline dolomite, the original fabrics of the precursor carbonate are destroyed or occur as relicts (Krebs and Macqueen, 1984). The vuggy habit is characteristic of this dolomite and vugs range in size from several millimetres to a few centimetres, and locally up to metre-sized cavities. The vugs are filled or partly lined with white dolomite, calcite, sulphides, bitumen, and/or native sulphur; these vug-fills are clearly post-Presqu'ile dolomite accumulations. Isolated remnants of limestone are preserved in the upper barrier complex of the Sulphur Point Formation within coarse-crystalline dolomite contrasting from the underlying pervasive fine-crystalline dolostone succession. The contacts between coarse-crystalline dolomite and limestone remnants are sharp, irregular, and transect bedding (Kyle, 1981). The main areal distribution of Presqu'ile dolomite on the Pine Point property follows the North and Main hinges, where they also reach their maximum thickness.

Fritz and Jackson (1972) and Kyle (1981) concluded that subsurface mixing of fresh and marine water ('Dorag' dolomitization) was the mechanism that formed this coarse-crystalline dolomite facies. Krebs and Macqueen (1984) argued against this model because the distribution of the dolomite should, therefore, be restricted to below the Watt Mountain disconformity. Warm ascending solutions likely played at least a subsidiary role in the dolomitization process. Krebs and Macqueen (1984) suggested that the Presqu'ile dolomite and its associated white void-filling dolomites, or saddle dolomites, were formed by various, superimposed complex processes or stages including subsurface mixing of fresh and marine waters and hot ascending hydrothermal solutions. Repeated invasion of these dolomitizing waters produced the large aureoles of Presqu'ile dolomite along the porous hinge lines.

The third dolomite type present at Pine Point is the white 'vein' dolosparite or saddle dolomite cements that occurs in association with infilled solution cavities and sulphides. This dolomite also includes white specks developed as a mottled pattern in the Presqu'ile dolomite. The dolomite is widespread, coarsely crystalline and possesses curved crystal faces and cleavage (Fig. 9D). The saddle dolomite has complex age relationships with ore, varying from pre-ore, syn-ore, to post-ore (Krebs and Macqueen, 1984). Regardless of the age relationships of the dolomite, the fluid inclusion filling temperatures of the dolomite suggest a hydrothermal origin. Hot ascending solutions, possibly derived from deeply buried Precambrian faults, heated the groundwater in open fractures and voids and precipitated these dolomite cements (Krebs and Macqueen, 1984).

Dissolution and brecciation

Internal karstification and collapse occurred during the pre-ore stage and the sulphide mineralization stage. It is impossible to differentiate the magnitude of dissolution that occurred in each stage. During pre-ore dissolution, the undolomitized remnants of limestone were dissolved between the dolomitized fractures and fragment margins (Krebs and Macqueen, 1984). This dissolution process resulted in the development of cellular carbonates containing voids of different shapes and sizes separated by thin walls of fine-crystalline dolomite. The resultant breccia-moldic porosity formed both boxwork dolomites and possibly zebra dolomites at Pine Point (Figs. 9A, 10A; Dunsmore, 1973). Internal sediments were deposited at the bottom of voids during this pre-ore dissolution stage.

The enigmatic zebra dolomites that are often intimately associated with MVT mineralization are not necessarily a dissolution phenomenon. Wallace et al. (1994) proposed that zebra textures were formed as a consequence of fracturing and sheet cavity development of unaltered carbonates with, in some cases, solution of cavity margins, followed by replacement by diagenetic dolomite and penecontemporaneous cementation. Nielsen et al. (1998) observed clear inclusion-poor saddle dolomite cements interlayered with white inclusion-rich and dark to black replacement dolomites and suggested the formation of zebra dolomites occurred within an overpressured fluid system.

During the main ore mineralization stage, further solution processes enlarged older dissolution macropores and created new collapse structures, thus preparing the brecciated host rock for subsequent sulphide mineralization and internal sediment filling (Krebs and Macqueen, 1984). Boundaries between massive orebodies and barren dolomites are generally sharp and distinct (Skall, 1975; Kyle, 1977, 1981). In the peripheral zones of larger collapse centres, well developed fractures and veins occur parallel and oblique to bedding planes, indicating an increase of brecciation intensity as the mineralized zone is approached (Krebs and Macqueen, 1984). These fractures and veins are lined with white coarse-crystalline saddle dolomite. The veins and fractures commonly form a halo around central mineralized collapse zones. Similarly, calcite flooding produced halos around many orebodies at Pine Point, giving the carbonate host rocks a coarsely granular appearance. Interior to these calcite flooding zones, very fine calcite and coarse-crystalline white sparry dolomite veins and vug-fillers are present.

Other alteration features

Soft sticky bitumen masses (Fig. 10B) often with associated heavy oil, and altered solid splintery pyrobitumen with conchoidal fracture commonly occur in the Pine Point district. The heavy oil fills remnant voids and vugs in Presqu'ile dolomite. Many levels of the barrier complex show an intense impregnation by bitumen where small pores have been filled by droplets (Krebs and Macqueen, 1984). Near some orebodies, hydrocarbons have been altered to splintery pyrobitumen that forms irregular concentrations in large vugs, glossy globules, and thin coatings on carbonate crystals (Krebs and Macqueen, 1984). Fluid inclusion studies of dolomites and sphalerite in the district reveal a population of inclusions filled with petroleum (N. Wilson, pers. comm., 2004).

Lithogeochemical patterns have been mapped in the Pine Point district (Turner et al., 2002). One such pattern depicts the combined densities of the major metallic elements generally associated with MVT deposits, that is, Fe, Pb, and Zn. At Pine Point, Fe-Pb-Zn anomalous zones display pronounced concentric distributions around orebodies in the Upper Keg River Member and Sulphur Point Formation. The anomalous zones decrease gradationally from maximum density coincident with high-grade prismatic cores to barren country rocks. The anomalies are widespread in the Upper Keg River and Sulphur Point units, negligible in the Watt Mountain Formation, and are confined to major collapse solution features in the Slave Point Formation. Also, along the Main trend, high Fe anomalies tend to be displaced north of the deposits. At the deposit scale, Fe is most widely distributed, Zn is intermediate and Pb occurs near the centre of the orebodies. Strontium distribution is very irregular but spatially overlaps the deposits (Turner et al., 2002).

Paragenesis

A detailed paragenetic sequence in the Pine Point district and along the Presqu'ile barrier is difficult to surmise due to the highly variable areal distribution and scale of diagenetic events. Despite these difficulties, Krebs and Macqueen (1984) established a comprehensive paragenetic sequence for the Pine Point mineral property and Qing (1991) refined and presented a generalized paragenesis for the entire Presqu'ile barrier (Fig. 11).

The diagenetic features that occur in carbonate rocks along the Presqu'ile barrier developed in three depositional environments: submarine, subaerial, and subsurface or burial. Submarine features include micrite envelopes, micrite, syntaxial cement, fibrous cement, and fine- to medium-crystalline dolomite (Qing, 1991, 1998). Subaerial diagenesis includes the precipitation of minor local pendant cements as well as dissolution and brecciation. Submarine and subaerial diagenetic events predate the main sulphide mineralization or ore stage. The most significant diagenetic alteration within the barrier complex was brought about by large-scale dissolution, due to the invasion of hydrothermal fluids into the subsurface subsequent to burial (Krebs and Macqueen, 1984; Qing, 1991; Qing and Mountjoy, 1992, 1994a,b). Subsurface diagenesis occurring during the ore and post-ore

FIGURE 10. **(A)** Wall of pit N-42, Pine Point mine site. Selective replacement of bedding-parallel bands by white dolomite forms zebra texture (photo courtesy of W.A. Turner). **(B)** Soft sticky bitumen in carbonate boulders, Pine Point mine site. Bitumen starts to flow during hot summer days.

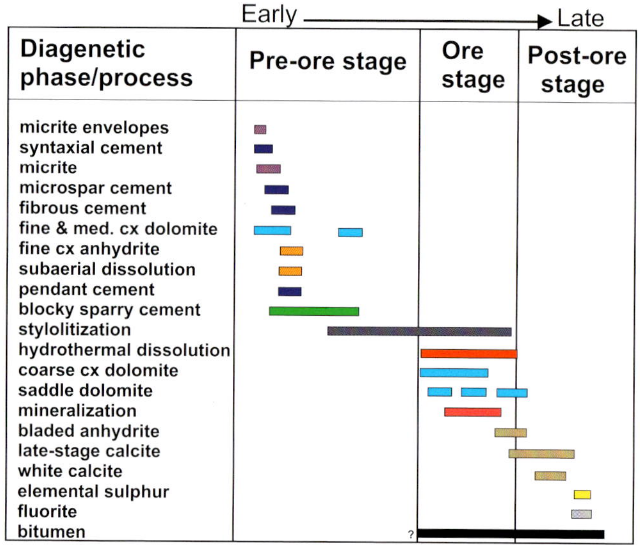

FIGURE 11. Paragenetic sequence at Pine Point. Pre-ore, ore, and post-ore stage diagenetic phases and processes are depicted. Adapted from Krebs and Macqueen (1984) and Qing (1991). (cx = crystalline)

stage intervals produced blocky sparry calcite cement, compaction and stylolitization features, coarse-crystalline and saddle dolomites, sulphide mineralization, late-stage calcite, fluorite, anhydrite, native sulphur, and bitumen (Fig. 11).

Genetic Models

Several models of fluid flow have been proposed for delivery of metal-bearing fluids to various sites where MVT mineral deposits were precipitated.

Jackson and Beales (1967) proposed the sedimentary-diagenetic origin model using the Pine Point deposit as their primary example. The model is based on the theory that basin-derived fluids or brines acquire heat, metals, and other solutes during the migration process and deposit sulphides in the host carbonates when conditions are favourable for precipitation. Fluid flow was driven by sediment compaction. The movement of warm brine from well compacted sediments into strata of higher porosity is very similar to the process of migration of petroleum from their source beds along permeable conduits or strata to various trap configurations. Jackson and Beales' (1967) model speculates that fluids leached metals from host rocks and transported them by means of chloride or organic complexes. The metals in the solution precipitate when hydrogen sulphide is encountered. Nearby evaporites (e.g. Muskeg Formation at Pine Point) provided sulphate, and biogenic or thermogenic sulphate reduction occurs at the site of mineral precipitation in the presence of petroleum.

Cathles and Smith (1983) and Bethke (1985) demonstrated that the steady-state compaction of basinal sediments invoked in Jackson and Beales' (1967) burial compaction model does not provide sufficient rates of discharge to produce the heat necessary for the elevated temperatures recorded in fluid inclusions of the host rocks or ore minerals in the ore districts. One possible mechanism to overcome this problem is the process of releasing fluids episodically in overpressured zones of compacting sediments (Cathles and Smith, 1983). This 'episodic-compaction' model requires rapid sedimentation and swift transport of fluids through the aquifers, thereby producing the thermal anomalies prevalent in the ore districts. A significant limitation of the compaction-driven fluid flow system is the requirement that mineral deposits only form during early stages in the geological history of the basin (Anderson and Macqueen, 1988), not likely the situation interpreted at Pine Point.

Another mechanism of formation fluid flow is the topographically driven fluid-flow model conceptualized by Garven and Freeze (1984). This model involves recharged groundwater in the uplifted orogen on the flank of the basin migrating downward through shale units into the underlying carbonate unit by cross-formational fluid flow (Fig. 12). As the fluid migrates updip through the deep portion of the basin along the carbonate aquifer, it acquires heat and dissolved components. The fluids or brines are eventually discharged by means of this immense hydrothermal circulation system on the basin's cratonic flank. Metals leached from shale sources spread throughout the basin, but concentrate at

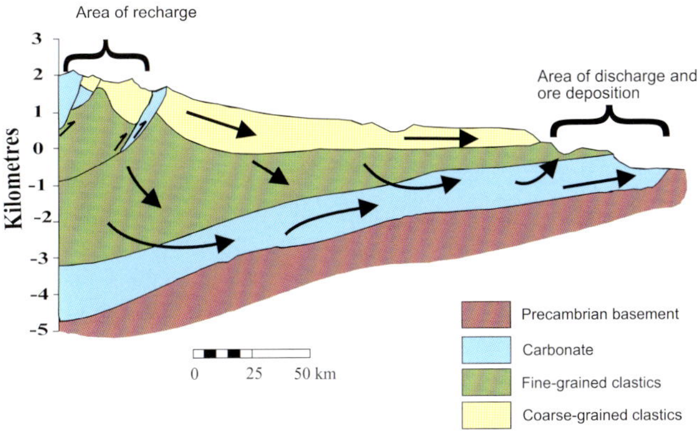

FIGURE 12. Topographically induced fluid-flow by Garven and Freeze (1984).

the discharge end where conditions are most favourable for precipitation of ore. This model suggests fluid circulation was initiated as a result of the Cretaceous-Tertiary Laramide Orogen and mineralization took place during or in the immediate aftermath of that tectonic episode. Later work by Nesbitt and Muehlenbachs (1994, 1995), Root (2001), and Nelson et al. (2002) refuted this Laramide MVT mineralization event, and speculated that mineralization and dolomitization in the WCSB were likely associated with the Late Devonian-Early Mississippian Antler Orogeny.

Another means of moving large volumes of dolomitizing and/or mineralizing solutions through large rock masses is the hydrothermal convection model (Morrow et al., 1990; Morrow, 1998). Buoyancy forces related to temperature and salinity variations drive convection. The continual vertical recycling of connate formational fluids derived from halite-saturated residual brines in the Elk Point subbasin provides greater opportunity for the growth of coarse open-space-filling dolospar crystals, such as saddle dolomite (Morrow, 1998). There is more than enough Mg present in the Elk Point evaporites to form all Devonian dolomite in the WCSB (Shields and Brady, 1995). The Presqu'ile dolomite at Pine Point displays vertically oriented chimney-like breccia bodies cemented by white sparry dolomite and containing laminated internal sediments. Upward and downward fluid flow may be focussed preferentially along fault zones (Morrow et al., 1990). Nelson et al. (2002) discussed the applicability of this thermal convection model to the distribution of volcanogenic, sedimentary exhalative and MVT deposits such as Robb Lake and Pine Point in the WCSB.

The fluid-flow models cited above give appropriate mechanisms for the delivery of metal-bearing fluids to the site of deposition, but there remains a need to define the chemical processes required to localize the deposition of sulphides. Possible models are the reduced sulphur model, the sulphate reduction model, and various fluid mixing models (Leach and Sangster, 1993).

The reduced sulphur model assumes that metals and reduced sulphur are transported together in the same fluid. Sulphide deposition is achieved by means of cooling, dilution of the brine by meteoric waters or less saline groundwater, or by changes in pH (Anderson, 1975; Leach and Sangster, 1993). Sverjensky (1984) indicated that pH values of 4.5 or lower are required in order for reduced sulphur and metals to be transported in the same fluid. These acidic conditions are incompatible if the fluid is transported through a carbonate aquifer as proposed in the topographic recharge fluid-flow model. Anderson (1975) showed that metal solubilities are too low in solutions containing reduced sulphur for metal transport to occur. Evidence consistent with ore deposition by means of the reduced sulphur model is the presence, in some instances, of district-wide uniform colour-banding in sphalerites (McLimans et al., 1980). This feature suggests that local sources of sulphur external to the fluid likely do not occur since individual sources would not produce such a consistent feature (McLimans et al., 1980). This so-called 'sphalerite stratigraphy' was not observed at Pine Point, however. Similarly, the uniform sulphur isotope composition of ore sometimes observed in MVT districts is consistent with a single metal/sulphur-bearing fluid (Leach and Sangster, 1993). However, the uniform colour banding in sphalerite and similar heavy sulphur isotope compositions may also signal the presence of a well homogenized sulphur source that is most likely external due to mass balance considerations (Anderson and Macqueen, 1988). The external sulphur-rich source requires a second fluid in the system for metal precipitation, due to the limiting effect of reduced sulphur on the solubility of metals in low-temperature fluids buffered at high pH (Goodfellow et al., 1993).

The sulphate reduction model is based on the premise that the reduced sulphur concentration is increased at the deposition site due to sulphate reduction processes (Leach and Sangster, 1993). The model requires delivering both metal and sulphate to the precipitation site, where methane or other organic matter are present. The organic matter reduces the sulphate to precipitate the sulphides. The spatial association of evaporites and bitumen in some MVT deposits (such as Pine Point, Powell and Macqueen, 1984) indicates that biogenic or thermogenic sulphate reduction are feasible mechanisms (Machel, 1989).

Mixing models imply that sulphide is precipitated where metal-rich brine mixes with a H_2S-rich fluid (Anderson and Macqueen, 1988; Leach and Sangster, 1993). The mixing of fluids will produce high degrees of supersaturation, large chemical and/or concentration gradients, and relatively rapid rates of precipitation of sulphides. These chemical processes result in precipitation of fine-crystalline, dendritic, and colloform ores, all of which are found at Pine Point. Fluid mixing can also yield large crystals if the chemical gradients are small and mixing is very slow, as for example, by means of diffusion of H_2S through the wall rock. Adams et al. (2000) demonstrated that mixing of two fluids took place at Pine Point due to distinctive shifts in the chemistry of the parent fluid at the ore deposit site. Major differences in fluid inclusion homogenization temperatures in matrix dolomites, saddle dolomites, and sphalerite, and significant changes in salinities, brine composition, and ^{18}O enrichment in parent hydrothermal brines all favour the introduction of a second fluid to the hydrologic system at Pine Point.

Age of Pine Point Mineralization

The timing of mineralization in MVT deposits is particularly difficult to determine because of a number of factors. These include 1) the occurrence of MVT deposits in stable tectonic areas, precluding the use of geological field observations of post-ore events in order to determine relative timing of mineralization; 2) the general lack of magnetic minerals in the strata making paleomagnetic methods untenable; and 3) the lack of igneous material or feldspars in the immediate region from which one can apply radiometric dating methods. Despite these restrictions, numerous dating studies have been completed at Pine Point. The paleomagnetic method of dating MVT stratabound ore deposits was attempted by Beales et al. (1974). In this paper, it was concluded that magnetic intensities were too low subsequent to demagnetization to lead to any specific conclusions with respect to the age of mineralization. Another attempt using the paleomagnetic technique was described by Symons et al. (1993), who speculated that the observed remnant magnetism provided a cluster of magnetic measurements with respect to a unique pole position. This pole position indicates that ores at Pine Point are Late Cretaceous to Eocene in age, suggesting that metal-rich brines were possibly forced into the Presqu'ile barrier during the Laramide Orogeny. Enkin et al. (2000) observed that all carbonate rocks in the Rocky Mountain thrust belt and foreland area have been remagnetized in Late Cretaceous-Early Tertiary time and any paleomagnetic remnant magnetism is not preserved. This Late Cretaceous-Early Tertiary remagnetization event (Gillen et al., 1999) has been recognized elsewhere in the WCSB extending eastward into the Elk Point subbasin of Saskatchewan (Koehler et al., 1997). Identical poles were found in unaltered limestones and secondary dolomites, so the remagnetization event is not directly related to the dolomitization episode or the sulphide-mineralizing event. Thus, there must be some doubt with respect to the accuracy of paleomagnetic age determinations in strata affected by this regional Cretaceous-Tertiary remagnetization event.

Cumming et al. (1990) studied lead isotopes in the Pine Point district and observed a remarkable lead isotope homogeneity occurring throughout the area. The lead isotopes at Pine Point conform to the normal or N-type leads giving model ages near that of the host rocks (Baadsgaard et al., 1966; Cumming and Robertson, 1969; Kyle, 1977, 1981). The distinct homogeneous character allows one to utilize the single-stage lead-evolution model procedure for age determination of lead precipitation. This method gives a mineralization age of 290 Ma. This Late Pennsylvanian age is younger than the host rocks but much older than the Laramide tectonic event. Cumming et al. (1990) also noted that Pine Point leads are much less radiogenic than those reported in Devonian carbonate-hosted Pb-Zn deposits in the Cordillera to the west. This difference in lead isotopic signature likely specifies unique paleogeographic positions for the two lead types, thus indicating separate and distinct mineralization events, fluid sources, and fluid pathways in the two regions.

Apatite fission-track analysis in the Pine Point district was undertaken by Arne (1991) in order to determine the regional thermal history of the region. The analysis and frequency distribution of annealed fission-track lengths in apatite revealed that a heating event occurred subsequent to the deposition of Devonian rocks. The heating event, attributed to deep burial, reached its maximum during Cretaceous time before a subsequent cooling episode took place when the deeply buried Devonian strata were uplifted and eroded. The maximum attained paleo-temperatures fall in the same range of estimated temperatures of ore formation, suggesting that these deposits formed during deep burial in Cretaceous time. This interpretation indicates that convective transport of heat is not necessary as heat is generated solely by deep burial.

Rb-Sr dating of sphalerites in the Pine Point district defines an age of 361 ±13 Ma, indicating that mineralization occurred shortly after deposition of the Middle Devonian host carbonate (Nakai et al., 1993). This age is not consistent with the tectonic-driven fluid-flow model of Garven (1985), which ascribes dolomitization and accompanying mineralization to effects generated by Laramide Cordilleran deformation, but is consistent with fluid flow and mineralization brought about by tectonic compression during the Late Devonian-Mississippian Antler Orogeny on the western margin of the continent (Nelson et al., 2002).

Exploration Methods

Favourable Geological Features

Strata of exploration interest are carbonate successions developed on the flanks of large and deep sedimentary basins. Unconformities and paleokarsts are important diagnostic features to examine in these favourable strata. Arching or flexing of the underlying basement produces topographic highs on the basement surface. These highs may define zones of increased fluid movement or heat flow, and basement faults can provide escape routes for metal-bearing fluids and conduits for increased heat flow. MVT deposits are commonly found in dolostone sequences associated with evaporitic strata. Carbonate fronts dividing basinal shales and platform carbonates are important features to map. Other associated features indicating MVT potential are highly permeable zones, such as reefs, or abundant open-fractures or faults, white sparry dolomite associated with the vuggy dolostone, and collapse breccias. All of these favourable indicators are present at Pine Point.

Additional local geological exploration criteria observed in the Pine Point district include 1) dolostone is not common in the Slave Point Formation, but where it does occur, it is almost always near mineralization; 2) tabular orebodies tend to follow bioclastic beds such as floatstones in the Sulphur Point Formation; 3) 'Presqu'ilized' dolomite and white sparry dolomite vugs and veins are only present near ore in the Watt Mountain Formation; and 4) blue-grey dolomite is closely associated with both tabular and prismatic ore. Recent work by Turner et al. (2003) indicates that blue-grey dolomite, along with other carbonate alteration types, exhibits unique spectral signatures in the Pine Point mining camp. Turner et al.'s (2003) study also determined that blue-grey dolomites contain greater sulphide concentrations compared to their white counterparts, and commonly envelop orebodies. It was also observed that disseminated lead and zinc sulphides generally are uncommon in the Slave Point Formation but their rare occurrences almost always indicate proximity to prismatic ore deposits. Another observation is that crackle brecciation is fairly extensive in the Slave Point Formation directly over prismatic orebodies. Tabular orebodies, however, do not have extensive associated crackle zones. At Pine Point, the highest grade of base metal mineralization occurs in karst structures hosting the greatest amount of infilling internal sediment (Rhodes et al., 1984). In prismatic orebodies, internal sediments are thicker near the base of the karst network where tabular karst begins.

Surficial Geochemistry

Most MVT deposits have minor geochemical signatures because of the limited primary dispersion of elements bounded in sphalerite and galena into surrounding carbonate rocks (Lavery et al., 1994). When weathering of the sulphides occurs and minerals such as limonite, cerussite, anglesite, smithsonite, hemimorphite, and pyromorphite are formed, the soil and stream sediments surrounding these deposits may show anomalous concentrations of Pb, Zn, or Fe. At Pine Point, Pb and Zn geochemical dispersion surveys in lake sediments, soils, and tills were undertaken (Brabec, 1983). Zinc gives larger and more contrasting anomalies in lake sediments and soils; however, not all Zn anomalies are associated with an orebody. Major shale units, especially organic-rich ones, display Zn background values similar in magnitude to anomalies associated with orebodies (Macqueen and Ghent, 1975; Hannigan, 2005). These same shales usually have low Pb contents. Therefore, areas of coincident Pb-Zn anomalies may indicate Pb-Zn sulphide mineralization. Lead, however, generally gives poorly defined anomaly patterns due to its low mobility.

Geophysics

At Pine Point, induced polarization (IP) proved to be a powerful exploration tool for finding orebodies, but only when sufficient conductive minerals occur. If an orebody at Pine Point dominantly consists of sphalerite, with low abundance of galena, marcasite, or pyrite, the IP response is generally too weak for positive identification of an orebody. Most of the Pine Point MVT deposits contain sufficiently high concentrations of conducting minerals to be good IP targets. There may be a number of undetectable sphalerite-rich orebodies (with respect to IP geophysics) that have not been discovered in the Pine Point region. Chargeability anomalies at Pine Point are generally on the order of 10 milliseconds above background (Lajoie and Klein, 1979). Normal background variations of chargeability in volcanic terranes, for example, are of this magnitude, so 10 milliseconds above background is not detectable in these areas. The success of the IP method at Pine Point, however, depends greatly on the uniform low chargeability background of the host carbonates. Lajoie and Klein (1979) also demonstrated that weak variations of chargeability in IP surveys at Pine Point could sometimes be correlated with changes in lithology. Electromagnetic (EM) and self-potential (SP) electrical methods were tested over a known orebody at Pine Point (N-42, Lajoie and Klein, 1979). Poor electrical responses were observed in these tests due to the open-space filling nature of the ore where gangue minerals and sphalerite interrupted the electric paths between conductive minerals (galena, pyrite, and marcasite).

Other geophysical methods tested and/or utilized in the Pine Point region include magnetic, gravity, reflection seismic, and borehole geophysics. In some Pine Point deposits, there are small amounts of the weakly magnetic iron sulphide mineral, pyrrhotite, associated with pyrite and marcasite mineralization. Because Pine Point is located in the auroral zone where severe magnetic storms are commonplace, the weak magnetic signature associated with orebodies containing pyrrhotite may not be discernable. Gravity has been proven to be useful as a complementary tool with IP in locating sphalerite-rich orebodies at Pine Point, because of the significant contrast in density between metallic and gangue minerals. The degree of porosity, often directly associated with mineralization, can vary significantly in the host rock at Pine Point. Lajoie and Klein (1979) indicated that this high contrast in porosity might be detectable by gravity surveys. Borehole geophysical logs can also detect differentiation of porous and nonporous horizons, specifically, sonic, and density survey logs. Seismic and gravity methods are also useful for delineating subsurface topography and faults.

Mineral Potential Mapping

The mineral potential in the Pine Point district was evaluated conceptually by combining previously discussed critical genetic attributes and features controlling the location of MVT deposits with empirical evidence of mineralization, such as orebodies, visible lead and/or zinc mineralization in outcrop and core, and geochemical anomalies in soils and rocks. Various diagnostic and permissive criteria related to fluid-flow pathways, depositional traps, and fluid outflow zones were analyzed in the district in order to derive a mineral potential or prospectivity map (Fig. 13). All of these criteria were derived either directly from geological and structural data or indirectly by means of geophysical and geochemical data. Five categories of mineral potential were defined dependent on the number of critical diagnostic features present in a particular region. High potential represents areas containing orebodies or visible mineral occurrences with an additional four to seven diagnostic factors. High to moderate potential indicates areas of geochemical anomalies and/or four to seven diagnostic factors. Moderately prospective areas have three diagnostic or permissive factors, low to moderate two factors, and low potential areas have one or no factors. Single diagnostic features that are repeated in an area (for example, intersection of faults) also increased the potential on the map. It was not possible, however, to assign a weighting of critical features based on their relative importance, with respect to location of potential mineral deposits.

Critical components of a regional-scale MVT system are all related to fluid flow during dynamic basin evolution. The five components that describe a regional hydrothermal system include energy, ligand source, metal source, transport pathways, and trap zones (D'Ercole et al., 2000). The first three components cannot be predicted due to the unresolved controversy of the nature of these sources in the genetic models. Transport pathways and trap zones, however, can be predicted and these particular components are related to specific structural and stratigraphic features associated with MVT ore deposition. Specific geological features associated with these mappable components include major basement lineaments and faults that may have acted as structural conduits for migration of ore fluids. Garven and Freeze (1984) indicated that in some cases certain geological units, such as carbonate, act as aquifers. Basement highs are zones along which fluids are restricted from further lateral flow and thus represent possible depositional sites. Basement highs are also responsible for localization of reef growth, which are

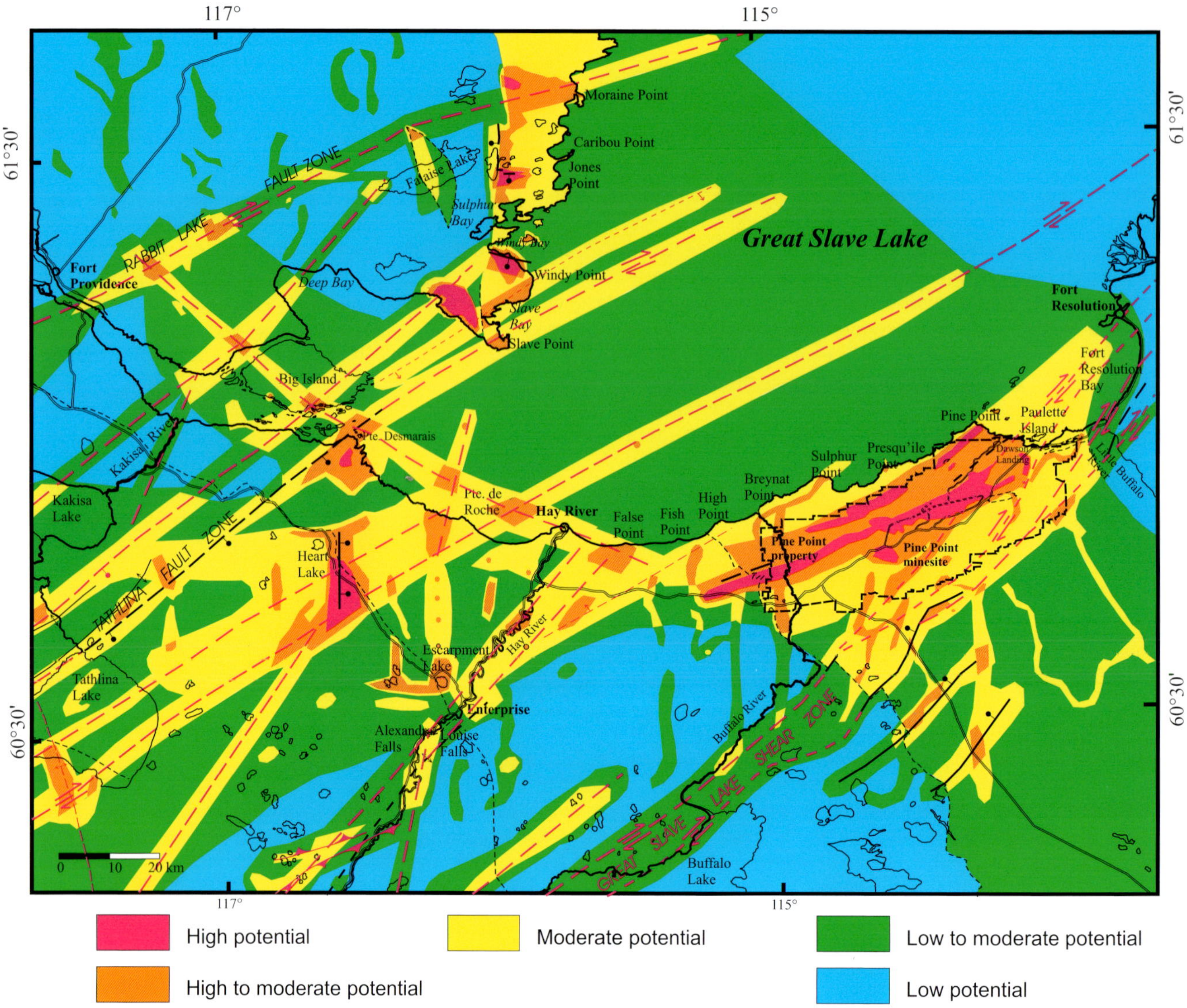

FIGURE 13. A mineral potential map of the Pine Point district (identical map area to Fig. 3). Potential rating based on a combination of known mineral occurrences and number and intensity of various diagnostic and permissive factors recognized in the area that are directly or indirectly related to MVT mineralization. See text for further discussion.

often potential host rocks for MVT mineralization. Fluid outflow or discharge is measured by mineralogical zoning of ore or alteration minerals around the deposit. Alteration zones at Pine Point are rather narrow and quite abrupt and dispersion is quite limited. At the regional district scale, the zones are insignificant and were not incorporated into the metallogenic map.

Specific diagnostic criteria for MVT mineralization at Pine Point include 1) proximity to faults and fractures (Morrow et al., 2002; MacLean, 2006; Okulitch, 2006), 2) proximity to major basement lineaments (Morrow et al., 2002), 3) spatial relationship to subsurface basement highs, 4) anomalously high amounts of base metals (Hannigan, 2005), 5) occurrences of reefal carbonate barrier complexes on carbonate platforms (Meijer Drees, 1993), 6) spatial locations of paleokarst networks (Rhodes et al., 1984), and 7) Presqu'ile dolomitization (Rhodes et al., 1984; Janicki, 2006b; Okulitch, 2006). Permissive criteria for mineralization at Pine Point include 1) their relation to carbonate aquifers (Janicki, 2006a), 2) proximity to evaporites (Norris, 1965; Rhodes et al., 1984; Meijer Drees, 1993), 3) relationships with carbonate or shale aquicludes or cap-rocks (Norris, 1965; Janicki, 2006a), 4) associations with hydrocarbons (Hannigan, 2005), specifically pyrobitumen, and 5) marcasite and pyrite haloes (Turner et al., 2002). Outcrop of potential host carbonate versus subcrop (6) (Norris, 1965) was also considered to be an important permissive criterion in mapping mineral potential due to their widely varying potential mining costs.

Figure 13 suggests significant potential in the immediate Pine Point mine-site along the Main, North, and South hinge zones continuing further west under Upper Devonian carbonate cover along the trend of the Main Hinge. Areas of high potential are also indicated in the Heart Lake area and near Pte. Desmarais on the southern shore of Great Slave Lake. Significant MVT potential is also expected in the Slave Point/Windy Point/Sulphur Bay area on the southwestern shore of Great Slave Lake. High to moderate potential is observed in the Escarpment Lake area, along the Hay River, and in the Tathlina Lake area. A significant area of high to moderate potential is also interpreted in the Moraine Point area of southwestern Great Slave Lake. Significant regions of moderate potential surround the Pine Point mining camp, encompass the lower Hay River, and occur in proximal areas to major fault zones in the western portion of the district underlain by the Presqu'ile barrier complex.

Potential for Known and New Deposit Types

The single known deposit type discovered in the Pine Point mining district is the Mississippi Valley-type lead-zinc sulphide accumulation. MVT deposits belong to a spectrum of sediment-hosted base metal deposits, so there may be potential for these types of deposits in the district. This spectrum includes sedimentary exhalative (SEDEX), carbonate-hosted F-Ba deposits, sandstone-hosted Pb deposits, "Irish-type" Zn-Pb deposits, carbonate-hosted Cu-Pb-Zn or Kipushi-type deposits, carbonate-hosted manto Ag-Pb-Zn deposits, and Broken Hill Pb-Zn deposits. There is also potential for the discovery of diamond-bearing kimberlite pipes in the district.

Knowledge Gaps

Although much new knowledge on the origin of MVT deposits has been achieved in the last 60 years, specifically the recognition that MVT deposits and districts are products of immense fluid-flow systems associated with extensional tectonic processes, there are many key knowledge gaps remaining with respect to ore deposition. The following is a list of specific knowledge gaps at Pine Point:

1) What is the nature and extent of structural control of mineralization at Pine Point?
2) What is the regional extent of dolomitizing and mineralizing fluids and what pathways did they take?
3) What are the most critical ground-preparation processes in the host rocks for initiation and localization of sulphide deposition?
4) Why do these deposits contain only lead and zinc in significant quantities? What is the chemical mechanism that preferentially deposited these metals?
5) Are evaporites critical to the origin of these deposits?
6) What is the contribution of organic matter to sulphide deposition? Is it critical for mineralization?
7) What causes metal zoning in some deposits and not in others?
8) What are the sources of metals and sulphur?
9) What complex compounds transported the metals and sulphur and what were these ligand sources?
10) What are the temporal and genetic similarities of Pine Point to MVT deposits on the western flank of the WCSB (Robb Lake, Bear-Twit, Gayna River, and Monarch-Kicking Horse)? Are they all part of the same paleo-hydrological system?

Acknowledgements

The work of many geologists in industry, universities, and government agencies contributed to the data and interpretations presented in this paper. In particular, I would like to acknowledge Dereck Rhodes of Teck-Cominco, Mario Coniglio of University of Waterloo, Sarah Gleeson of University of Alberta, Allan Turner and Ed Janicki of C.S. Lord Northern Geoscience Centre, and David Morrow, Suzanne Paradis, Warner Miles, Bernie MacLean, and Nicholas Wilson for assisting with studies in the Pine Point district.

This paper benefited from thorough reviews by Suzanne Paradis, Wayne Goodfellow, and David Morrow. I am also grateful to Ingrid Kjarsgaard for editorial assistance.

References

Adams, J.E., and Rhodes, M.L., 1960, Dolomitization by seepage refluxion: American Association of Petroleum Geologists, Bulletin, v. 44, p. 1912-1920.

Adams, J.J., Rostron, B.J., and Mendoza, C.A., 2000, Evidence for two-fluid mixing at Pine Point, NWT: Journal of Geochemical Exploration, v. 69-70, p. 103-108.

Alldrick, D.W., Batchelor, E.W., Collins, J.A., Lantos, E.A., Rhodes, D., Webb, R.J., and West, J.M., 1981, Pine Point lead-zinc deposits - District of Mackenzie, N.W.T., *in* Thompson, R.I., and Cook, D.G., eds., Field Guides to Geology and Mineral Deposits: Geological Association of Canada, Mineralogical Association of Canada, and Canadian Geophysical Union, Calgary '81 Annual Meeting, p. 155-157.

Anderson, G.M., 1975, Precipitation of Mississippi Valley-type ores: Economic Geology, v. 70, p. 937-942.

Anderson, G.M., and Macqueen, R.W., 1988, Mississippi Valley-Type lead-zinc deposits, *in* Roberts, R. G., and Sheahan, P. A., eds., Ore Deposit Models: Geological Association of Canada, Geoscience Canada Reprint Series 3, p. 79-90.

Arne, D.C., 1991, Regional thermal history of the Pine Point area, Northwest Territories, Canada, from apatite fission track analysis: Geology, v. 86, p. 428-435.

Baadsgaard, H., Campbell, F.A., Cumming, G.L., Evans, T., Kanasewich, E., Krouse, R.H., and Robertson, D.K., 1966, Isotopic data from the Cordillera and Liard Basin in relation to the genesis of the Pine Point lead-zinc deposits: Transactions of Royal Society of Canada, v. Sections I, II and III, Fourth Series-Volume III, p. 16.

Badiozamani, K., 1973, The Dorag dolomitization model - application to the Middle Ordovician of Wisconsin: Journal of Sedimentary Petrology, v. 43, p. 965-984.

Beales, F.W., Carracedo, J.C., and Strangway, D.W., 1974, Paleomagnetism and the origin of Mississippi Valley-type ore deposits: Canadian Journal of Earth Sciences, v. 11, p. 211-223.

Beales, P., Woodward, R., and Campbell, C., 2002, Great Slave Reef, *in* A Guide to Mineral Deposits of the Northwest Territories: Minerals, Oil and Gas Division, Resources, Wildlife and Economic Development, Government of Northwest Territories, p. 82.

Bell, J.M., 1929, The lead-zinc deposits near Pine Point, Great Slave Lake: Transactions of the Canadian Institute of Mining and Metallurgy, v. 32, p. 122-139.

Bell, R., 1902, Mackenzie District: Geological Survey of Canada, Annual Report 12, Part A, 8 p.

Bethke, C.M., 1985, A numerical model of compaction-driven groundwater flow and heat transfer and its application to the paleohydrology of intracratonic sedimentary basins: Journal of Geophysical Research, v. 80, p. 6817-6828.

Brabec, D., 1983, Evaluation of soil anomalies by discriminant analysis in geochemical exploration for carbonate-hosted lead-zinc deposits: Economic Geology, v. 78, p. 333-339.

Brophy, J.A., 1984, Pine Point Mine, Chapter 2 *in* Brophy, J.A., and Ellis, C.E., eds., Operating Mines, Mineral Industry Report 1980-81, Northwest Territories: Indian and Northern Affairs Canada, p. 55-60.

——— 1987, Pine Point Mine, *in* Ellis, C. E., ed., Operating Mines, 1984-85, Mineral Industry Report 1984-85, Northwest Territories: Indian and Northern Affairs Canada, p. 15-31.

Cameron, A.E., 1917, Reconnaissance on Great Slave Lake, North West Territories: Geological Survey of Canada, Summary Report 1916, 11 p.

——— 1918, Explorations in the Vicinity of Great Slave Lake: Geological Survey of Canada, Summary Report 1917, Part C, 8 p.

Campbell, N., 1966, The lead-zinc deposits of Pine Point: Canadian Mining and Metallurgical Bulletin, Transactions, v. 59, p. 288-295.

Camsell, C., 1915, An Exploration of the Region between Athabaska and Great Slave Lakes, Alberta and Northwest Territories: Geological Survey of Canada, Summary Report 1914, 6 p.

Carter, K., 1983, Geology of the N-81 orebody, Pine Point, N.W.T.: CIM Bulletin, v. 76, p. 38.

Cathles, L.M., and Smith, A.T., 1983, Thermal constraints on the formation of Mississippi Valley-type lead-zinc deposits and their implications for episodic basin dewatering and deposit genesis: Economic Geology, v. 78, p. 983-1002.

Coniglio, M., Morrow, D.W., and Wilson, N., 2006, Reassessment of Middle Devonian dolomites, Presqu'ile barrier, Northwest Territories, *in* Hannigan, P.K., ed., Potential for Carbonate-Hosted Lead-Zinc Mississippi Valley-Type Mineralization in Northern Alberta and Southern Northwest Territories. Geoscience Contributions, Targeted Geoscience Initiative: Geological Survey of Canada, Bulletin 591, in press.

Cumming, G.L., and Robertson, D.K., 1969, Isotopic composition of lead from the Pine Point deposit: Economic Geology, v. 64, p. 731-732.

Cumming, G.L., Kyle, J.R., and Sangster, D.F., 1990, Pine Point: A case history of lead isotopic homogeneity in a Mississippi Valley-type district: Economic Geology, v. 85, p. 133-144.

Department of Energy Mines and Resources Canada, Canadian Minerals Yearbook, Mineral Report: issued by Mineral Resources Division, 1967-1968, Mineral Resources Branch, 1969-1972, Department of Energy, Mines and Resources, 1973: Annual Report.

D'Ercole, C., Groves, D.I., and Knox-Robinson, C.M., 2000, Using fuzzy logic in a Geographic Information System environment to enhance conceptually based prospectivity analysis of Mississippi Valley-type mineralisation: Australian Journal of Earth Sciences, v. 47, p. 913-927.

Douglas, R.J.W., and Norris, A.W., 1974, Geology, Great Slave, District of Mackenzie: Geological Survey of Canada, Map 1370A, scale 1:500 000.

Dunsmore, H.E., 1973, Diagenetic processes of lead-zinc emplacement in carbonates: Transactions of the Institution of Mining and Metallurgy, v. 82, p. B168-B173.

Ellis, C.E., 1995, Operating Mines, Chapter 2, *in* Beales, P., ed., Mineral Industry Report 1988-89: Northwest Territories, Indian and Northern Affairs Canada, p. 13-24.

Ellis, C.E., and Hearn, K., 1990, Operating Mines, Chapter 2, *in* Ellis, C.E., ed., Mineral Industry Report 1986-87: Northwest Territories, Indian and Northern Affairs Canada, p. 11-32.

Energy Mines and Resources Canada, 1974-1990, Canadian Minerals Yearbook, Mineral Report: Annual Report.

Enkin, R.J., Osadetz, K.G., Baker, J., and Kisilevsky, D., 2000, Orogenic remagnetizations in the Front Ranges and inner Foothills of the southern Cordillera: Chemical harbinger and thermal handmaiden of Cordilleran deformation: Geological Society of America Bulletin, v. 112, p. 929-942.

Evans, T.L., Campbell, F.A., and Krouse, H.R., 1968, A reconnaissance study of some western Canadian lead-zinc deposits: Economic Geology, v. 63, p. 349-359.

Fowler, M.G., 2000, Long distance migration of hydrogen sulfide and the origin of the Pine Point Pb-Zn deposit: GeoCanada 2000-The Millenium Geoscience Summit, Abstract 344, 3 p.

Fritz, P., and Jackson, S.A., 1972, Geochemical and isotopic characteristics of Middle Devonian dolomites from Pine Point, northern Canada: 24th International Geological Congress, Section 6, Stratigraphy and Sedimentology, p. 230-243.

Garven, G., 1985, The role of regional fluid flow in the genesis of the Pine Point deposit, Western Canada Sedimentary Basin: Economic Geology, v. 80, p. 307-324.

Garven, G., and Freeze, R.A., 1984, Theoretical analysis of the role of groundwater flow in the genesis of stratabound ore deposits. 2. Quantitative results: American Journal of Science, v. 284, p. 1125-1174.

Gibbins, W., 1978, The Great Slave Plain, Western Churchill and Great Slave Plain. North of 60, Mineral Industry Report 1975: Northwest Territories: Department of Indian Affairs and Northern Development, EGS 1978-5, p. 35-39.

——— 1984, The Great Slave Plain, Chapter 5 *in* Brophy, J.A., and Ellis, C.E., eds., Southeast Mackenzie District, Mineral Industry Report 1980/81, Northwest Territories: Indian and Northern Affairs Canada, EGS 1984-5, p. 253-265.

——— 1987, The Great Slave Plain, Chapter 6 *in* Ellis, C.E., ed., Southeast Mackenzie District, Mineral Industry Report 1984-85, Northwest Territories: Indian and Northern Affairs Canada, p. 147-148.

——— 1988, Metallic mineral potential of the Western Interior Platform of the Northwest Territories: Geoscience Canada, v. 15, p. 117-119.

——— 1990, The Great Slave Plain, Chapter 6 *in* Ellis, C.E., ed., Southeast Mackenzie District, Mineral Industry Report 1986-87, Northwest Territories: Indian and Northern Affairs Canada, p. 116.

Gibbins, W.A., Seaton, J.B., Laporte, P.J., Murphy, J.D., Hurdle, E.J., and Padgham, W.A., 1977, Mineral industry report 1974, Northwest Territories: Indian and Northern Affairs Canada, EGS 1977-5, 267 p.

Gillen, K.P., van der Voo, R., and Thiessen, J.H., 1999, Late Cretaceous-Early Tertiary remagnetization of the Devonian Swan Hills Formation recorded in carbonate cores from the Caroline gas field: American Association of Petroleum Geologists, Bulletin, v. 83, p. 1223-1235.

Gleeson, S. A., and Gromek, P., 2006, The origin of hydrothermal sulphide and dolomite mineralizing fluids in southern Northwest Territories and northern Alberta, *in* Hannigan, P.K., ed., Potential for Carbonate-Hosted Lead-Zinc Mississippi Valley-Type Mineralization in Northern Alberta and Southern Northwest Territories. Geoscience Contributions, Targeted Geoscience Initiative: Geological Survey of Canada, Bulletin 591, in press.

Goodfellow, W.D., Lydon, J.W., and Turner, R.J.W., 1993, Geology and genesis of stratiform sediment-hosted (SEDEX) zinc-lead-silver sulphide deposits, *in* Kirkham, R.V., Sinclair, W.D., Thorpe, R.I., and Duke, J.M., eds., Mineral Deposit Modeling: Geological Association of Canada, Special Paper 40, p. 201-251.

Goodwin, J.A., 1983, Pine Point Mines, Operating Mines, Chapter II. Mineral Industry Report 1978, Northwest Territories: Indian Affairs and Northern Development, EGS 1983-2, p. 14-16.

Grasby, S. E., Fowler, M., and Stasiuk, V., 2002, Sulphur compounds in bitumens and pyrobitumens in relation to microstructure and their role in forming the Pine Point Pb-Zn deposit, NWT: Geological Association of Canada-Mineralogical Association of Canada, Joint Annual Meeting, Saskatoon, Abstract Volume, p. 41.

Gurmendi, A.C., 2002, The mineral industry of Canada: U.S. Geological Survey Minerals Yearbook-2002, p. 5.1-5.18.

Hannigan, P.K., 2005, A GIS dataset compilation of lead and zinc occurrences for the Targeted Geoscience Initiative project: Potential for carbonate-hosted Pb-Zn (MVT) deposits in northern Alberta and southern Northwest Territories (NTS 74, 84, 85): Geological Survey of Canada, Open File 5002, 1 CD-ROM.

Haynes, F.M., and Kesler, S.E., 1987, Chemical evolution of brines during Mississippi Valley-type mineralization: Evidence from East Tennessee and Pine Point: Economic Geology, v. 82, p. 53-71.

Hurdle, E.J., and Gibbins, W.A., 1978, Pine Point Mines, Operating Mines. North of 60, Mineral Industry Report 1975, Northwest Territories: Department of Indian Affairs and Northern Development, EGS 1978-5, p. 111-114.

Jackson, S.A., and Beales, F.W., 1967, An aspect of sedimentary basin evolution: The concentration of Mississippi Valley-type ores during the late stages of diagenesis: Bulletin of Canadian Petroleum Geology, v. 15, p. 393-433.

Janicki, E.P., 2006a, Hydrodynamic study of the Middle Devonian of the southeastern Great Slave Plain, Northwest Territories, *in* Hannigan, P.K., ed., Potential for Carbonate-Hosted Lead-Zinc Mississippi Valley-Type Mineralization in Northern Alberta and Southern Northwest Territories. Geoscience Contributions, Targeted Geoscience Initiative: Geological Survey of Canada, Bulletin 591, in press.

——— 2006b, Distribution of Presqu'ile dolomite in the Great Slave Plain, *in* Hannigan, P.K., ed., Potential for Carbonate-Hosted Lead-Zinc Mississippi Valley-Type Mineralization in Northern Alberta and Southern Northwest Territories. Geoscience Contributions, Targeted Geoscience Initiative: Geological Survey of Canada, Bulletin 591, in press.

Kent, D. M., 1994, Paleogeographic evolution of the cratonic platform - Cambrian to Triassic, *in* Mossop, G.D., and Shetsen, I., eds., Geological Atlas of the Western Canada Sedimentary Basin: Canadian Society of Petroleum Geologists/Alberta Research Council, p. 69-86.

Koehler, G., Kyser, T.K., Enkin, R., and Irving, E., 1997, Paleomagnetic and isotopic evidence for the diagenesis and alteration of evaporites in the Paleozoic Elk Point Basin, Saskatchewan, Canada: Canadian Journal of Earth Sciences, v. 34, p. 1619-1629.

Krebs, W., and Macqueen, R., 1984, Sequence of diagenetic and mineralization events, Pine Point lead-zinc property, Northwest Territories, Canada: Bulletin of Canadian Petroleum Geology, v. 32, p. 434-464.

Kyle, J.R., 1977, Development of Sulfide-Hosting Structures and Mineralization, Pine Point, Northwest Territories: Unpub. Ph.D. thesis, University of Western Ontario, London, Ontario, 226 p.

——— 1981, Geology of the Pine Point lead-zinc district, Chapter 11 *in* Wolf, K.H., ed., Handbook of Strata-Bound and Stratiform Ore Deposits, Regional Studies and Specific deposits: Elsevier Publishing Company, New York, v. 9, p. 643-741.

Lajoie, J.J., and Klein, J., 1979, Geophysical exploration at the Pine Point Mines Ltd. zinc-lead property, Northwest Territories, Canada, *in* Hood, P. J., ed., Geophysics and Geochemistry in the Search for Metallic Ores: Geological Survey of Canada, Report 31, p. 653-664.

Lavery, N.C., Leach, D.L., and Saunders, J.A., 1994, Lithogeochemical investigations applied to exploration for sediment-hosted Zn-Pb ore deposits, *in* Fontebotè, L., and Boni, M., eds., Sediment Hosted Zn-Pb Ores: Springer-Verlag, p. 393-428.

Leach, D.L., and Sangster, D.F., 1993, Mississippi Valley-type lead-zinc deposits, *in* Kirkham, R.V., Sinclair, W.D., Thorpe, R.I., and Duke, J.M., eds., Mineral Deposit Modeling: Geological Association of Canada, Special Paper 40, p. 289-314.

Machel, H. G., 1989, Relationships between sulphate reduction and oxidation of organic compounds to carbonate diagenesis, hydrocarbon accumulations, salt domes, and metal sulphide deposits: Carbonates and Evaporites, v. 4, p. 137-151.

MacLean, B.C., 2006, The sub-Phanerozoic basement surface under the Great Slave Plain of the Northwest Territories and its influence on overlying strata, *in* Hannigan, P.K., ed., Potential for Carbonate-hosted Lead-zinc Mississippi Valley-type Mineralization in Northern Alberta and Southern Northwest Territories. Geoscience Contributions, Targeted Geoscience Initiative: Geological Survey of Canada, Bulletin 591, in press.

Macqueen, R.W., and Ghent, E.D., 1975, Occurrence of Zinc in Devonian Metalliferous Shales, Pine Point region, District of Mackenzie: Geological Survey of Canada, Report of Activities Paper 75-1, Part B, 5 p.

Macqueen, R.W., and Powell, T.G., 1983, Organic geochemistry of the Pine Point lead-zinc ore field and region, Northwest Territories, Canada: Economic Geology, v. 78, p. 1-25.

McGlynn, J.C., 1971, Metallic Mineral Industry, District of Mackenzie, Northwest Territories: Geological Survey of Canada, Paper 70-17, 194 p.

McLimans, R.K., Barnes, H.L., and Ohmoto, H., 1980, Sphalerite stratigraphy of the Upper Mississippi Valley Zinc-Lead District, southwest Wisconsin: Economic Geology, v. 75, p. 351-361.

Meijer Drees, N.C., 1993, The Devonian Succession in the Subsurface of the Great Slave and Great Bear Plains, Northwest Territories: Geological Survey of Canada, Bulletin 393, 222 p.

Morrow, D.W., 1998, Regional subsurface dolomitization; Models and constraints: Geoscience Canada, v. 25, p. 57-70.

Morrow, D.W., Cumming, G.L., and Aulstead, K.L., 1990, The Gas-Bearing Devonian Manetoe Facies, Yukon and Northwest Territories: Geological Survey of Canada, Bulletin 400, 54 p.

Morrow, D.W., MacLean, B.C., Tzeng, P., and Pana, D., 2002, Subsurface Paleozoic Structure and Isopach Maps and Selected Seismic Lines in Southern Northwest Territories and Northern Alberta: Implications for Mineral and Petroleum Potential: Geological Survey of Canada, Open File 4366, 22 p.

Nakai, S., Halliday, A.N., Kesler, S.E., Jones, H.D., Kyle, J.R., and Lane, T.E., 1993, Rb-Sr dating of sphalerites from Mississippi Valley-type (MVT) ore deposits: Geochimica et Cosmochimica Acta, v. 57, p. 417-427.

Natural Resources Canada, 2004, The MVT deposits database, Beta release 3.6; World Minerals Geoscience Database Project, Geological Survey of Canada, Mineral Resources Division: http://www.nrcan.gc.ca/gsc/mrd/wmgdb/priv/globaldb/mvtbr36/mvtbr36.htm.

Nelson, J., Paradis, S., Christensen, J., and Gabites, J., 2002, Canadian Cordilleran Mississippi Valley-type deposits: A case for Devonian-Mississippian back-arc hydrothermal origin: Economic Geology, v. 97, p. 1013-1036.

Nesbitt, B.E., and Muehlenbachs, K., 1994, Paleohydrology of the Canadian Rockies and origins of brines, Pb-Zn deposits and dolomitization in the Western Canada Sedimentary Basin: Geology, v. 22, p. 243-246.

——— 1995, Importance of paleo-fluid systems in the southern Canadian Rockies in the genesis of mineral deposits in the Rockies and Western Canada Sedimentary Basin, *in* Ross, G. M., ed., 1995 Alberta Basement Transects Workshop, LITHOPROBE Report #47: LITHOPROBE Secretariat, University of British Columbia, p. 250-253.

Nielsen, P., Swennen, R., Muchez, P., and Keppens, E., 1998, Origin of Dinantian zebra dolomites south of the Brabant-Wales Massif, Belgium: Sedimentology, v. 45, p. 727-743.

Norris, A., 1965, Stratigraphy of Middle Devonian and Older Paleozoic Rocks of the Great Slave Lake Region, Northwest Territories: Geological Survey of Canada, Memoir 322, 180 p.

Okulitch, A.V., 2006, Phanerozoic bedrock geology, Slave River, District of Mackenzie, Northwest Territories: Geological Survey of Canada, National Earth Science Series, Geological Atlas, Open File 5281, map scale 1:1 000 000.

Padgham, T., Caine, T.W., Hughes, D.R., Jefferson, C.W., Kennedy, M.W., and Murphy, J.D., 1974, North of 60, Mineral Industry Report 1969 and 1970: Northwest Territories: Indian and Northern Affairs, Volume 3 of 3, 168 p.

Pana, D., 2006, Unravelling the structural control of "Mississippi Valley-type" deposits and prospects in carbonate sequences of the Western Canada Sedimentary Basin, *in* Hannigan, P.K., ed., Potential for Carbonate-Hosted Lead-Zinc Mississippi Valley-Type Mineralization in Northern Alberta and Southern Northwest Territories, Geoscience Contributions, Targeted Geoscience Initiative: Geological Survey of Canada, Bulletin 591, in press.

Paradis, S., Turner, W.A., Coniglio, M., Wilson, N., and Nelson, J.L., 2006, Stable and radiogenic isotopic signatures of mineralized Devonian carbonates of the northern Rocky Mountains and the Western Canada Sedimentary Basin, *in* Hannigan, P.K., ed., Potential for Carbonate-Hosted Lead-Zinc Mississippi Valley-Type Mineralization in Northern Alberta and Southern Northwest Territories. Geoscience Contributions, Targeted Geoscience Initiative: Geological Survey of Canada, Bulletin 591, in press.

Podruski, J.A., Barclay, J.E., Hamblin, A.P., Lee, P.J., Osadetz, K.G., Procter, R.M., and Taylor, G.C., 1988, Conventional Oil Resources of Western Canada (Light and Medium); Part I: Resource Endowment: Geological Survey of Canada, Paper 87-26, 126 p.

Porter, J.W., Price, R.A., and McCrossan, R.G., 1982, The Western Canada Sedimentary Basin: Philosophical Transactions of the Royal Society of London, v. A 305, p. 169-192.

Powell, T.G., and Macqueen, R.W., 1984, Precipitation of sulfide ores and organic matter: Sulfate reactions at Pine Point, Canada: Science, v. 224, p. 63-66.

Price, R. A., 1994, Cordilleran tectonics and evolution of the Western Canada Sedimentary Basin, *in* Mossop, G.D., and Shetsen, I., eds., Geological Atlas of the Western Canada Sedimentary Basin: Canadian Society of Petroleum Geologists/Alberta Research Council, p. 13-24.

Qing, H., 1991, Diagenesis of Middle Devonian Presqu'ile dolomite, Pine Point NWT and Adjacent Subsurface: Unpub. Ph.D. thesis, McGill University, 292 p.

——— 1998, Petrography and geochemistry of early-stage, fine- and medium-crystalline dolomites in the Middle Devonian Presqu'ile Barrier at Pine Point, Canada: Sedimentology, v. 45, p. 433-446.

Qing, H., and Mountjoy, E., 1992, Large-scale fluid flow in the Middle Devonian Presqu'ile barrier, Western Canada Sedimentary Basin: Geology, v. 20, p. 903-906.

——— 1994a, Origin of dissolution vugs, caverns, and breccias in the Middle Devonian Presqu'ile barrier, host of Pine Point Mississippi Valley-type deposits: Economic Geology, v. 89, p. 858-876.

——— 1994b, Formation of coarsely crystalline, hydrothermal dolomite reservoirs in the Presqu'ile barrier, Western Canada Sedimentary Basin: American Association of Petroleum Geologists, Bulletin, v. 78, p. 55-77.

Randall, A.W., Barr, D.A., and Giroux, G.M., 1985, Geological aspects of ore reserve estimates, Pine Point district zinc-lead deposits, N.W.T., *in* Ranta, D.E., ed., Applied Mining Geology; Ore Reserve Estimation: Society of Mining Engineers of AIME, p. 175-197.

Reinson, G.E., Lee, P.J., Warters, W., Osadetz, K.G., Bell, L.L., Price, P.R., Trollope, F., Campbell, R.I., and Barclay, J.E., 1993, Devonian Gas Resources of the Western Canada Sedimentary Basin, Part I: Geological Play Analysis and Resource Assessment: Geological Survey of Canada, Bulletin 452, 128 p.

Rhodes, D., Lantos, E.A., Lantos, J.A., Webb, R.J., and Owens, D.C., 1984, Pine Point orebodies and their relationship to the stratigraphy, structure, dolomitization, and karstification of the Middle Devonian barrier complex: Economic Geology, v. 79, p. 991-1055.

Roedder, E., 1968, Temperature, salinity, and origin of the ore-forming fluids at Pine Point, Northwest Territories, Canada, from fluid inclusion studies: Economic Geology, v. 63, p. 439- 450.

Root, K.G., 2001, Devonian Antler fold and thrust belt and foreland basin development in the southern Canadian Cordillera: Implications for the Western Canada Sedimentary Basin: Bulletin of Canadian Petroleum Geology, v. 49, p. 7-36.

Sangster, D. F., 1996, Mississippi Valley-type lead-zinc, *in* Eckstrand, O.R., Sinclair, W.D., and Thorpe, R.I., eds., Geology of Canadian Mineral Deposit Types: Geological Survey of Canada, Geology of Canada, no. 8, p. 253-261.

Sasaki, A., and Krouse, H.R., 1969, Sulfur isotopes and the Pine Point lead-zinc mineralization: Economic Geology, v. 64, p. 718-730.

Shields, M.J., and Brady, P.V., 1995, Mass balance and fluid flow constraints on regional-scale dolomitization, Late Devonian, Western Canada sedimentary basin: Bulletin of Canadian Petroleum Geology, v. 43, p. 371-392.

Skall, H., 1975, The paleoenvironment of the Pine Point lead-zinc district: Economic Geology, v. 70, p. 22-47.

Sverjensky, D.A., 1984, Oil field brines as ore-forming solutions: Economic Geology, v. 79, p. 23-27.

——— 1986, Genesis of Mississippi Valley-type lead-zinc deposits: Annual Review of Earth and Planetary Sciences, v. 14, p. 177-199.

Symons, D.T.A., Pan, H., Sangster, D.F., and Jowett, E.C., 1993, Paleomagnetism of the Pine Point Zn-Pb deposits: Canadian Journal of Earth Sciences, v. 30, p. 1028-1036.

Thorpe, R.I., 1966, Mineral Industry of the Northwest Territories 1965: Geological Survey of Canada, Paper 66-52, 66 p.

——— 1972, Mineral Exploration and Mining Activities, Mainland Northwest Territories, 1966 to 1968 (excluding the Coppermine River Area): Geological Survey of Canada, Paper 70-70, 204 p.

Tikkanen, G.D., 1986, World resources and supply of lead and zinc, in Bush, W. R., ed., Economics of Internationally Traded Minerals: Society of Mining Engineers, Inc., p. 242-250.

Turner, W.A., 2003, Microthermometric study of fluids associated with Pb-Zn mineralization in the vicinity of the Pine Point mining camp: C.S. Lord Northern Geoscience Centre, NWT Open file 2003-06, 30 p.

Turner, W.A., Pierce, K.L., and Cairns, K.A., 2002, Great Slave Reef (GSR) project drillhole database. A Compilation of the Drillhole Locations, drill Logs, and Associated Geochemical Data for the Great Slave Reef Joint Venture Project; Interior Platform, Northwest Territories, Canada (85B11 to 14): C.S. Lord Northern Geoscience Centre, Yellowknife, NWT, Open Report 2002-001, CD-ROM.

Turner, W.A., Laamrani, A., and Rivard, B., 2003, Laboratory reflectance spectra of hydrothermally altered carbonate facies, Pine Point mining camp, NWT, Canada: Geochemistry: Exploration, Environment, Analysis, v. 3, p. 369-379.

Wallace, M.W., Both, R.A., Morales Ruano, S., Hach-Ali, P.F., and Lees, T., 1994, Zebra textures from carbonate-hosted sulfide deposits: Sheet cavity networks produced by fracture and solution enlargement: Economic Geology, v. 89, p. 1183-1191.

Williams, G.K., 1977, The Hay River Formation and Its Relationship to Adjacent Formations, Slave River Map Area, N.W.T.: Geological Survey of Canada, Paper 75-12, 17 p.

Wright, G.N., McMechan, M.E., and Potter, D.E.G., 1994, Structure and architecture of the Western Canada Sedimentary Basin, *in* Mossop, G. D., and Shetsen, I., eds., Geological Atlas of the Western Canada Sedimentary Basin: Canadian Society of Petroleum Geologists/Alberta Research Council, p. 25-40.

Metallogeny of the Robb Lake Carbonate-Hosted Zinc-Lead District, Northeastern British Columbia

Suzanne Paradis[1] and JoAnne L. Nelson[2]

1. Geological Survey of Canada, 9860 West Saanich Road, Sidney, BC V8L 4B2

2. British Columbia Ministry of Energy and Mines, Geological Survey Branch, P.O. Box 9320 Station Provincial Government, Victoria, BC V8W 9N3

Corresponding author's email: suparadi@NRCan.gc.ca

Abstract

The Robb Lake Zn-Pb district consists of numerous carbonate-hosted sulphide deposits of the classic Mississippi Valley-type (MVT). Most deposits are hosted by deformed and thrust-faulted Silurian to Middle Devonian platform dolostones of the Muncho-McConnell, Stone, and Dunedin formations, which are found adjacent to the shelf front of the northern Canadian Rocky Mountains. The Robb Lake deposit, the most significant deposit in the district, consists of 19 known stratabound Zn-Pb showings that are hosted in interconnecting bedding-parallel and crosscutting breccia bodies in as much as 200 m of the stratigraphic sequence. The main sulphide minerals are sphalerite, galena, and pyrite, occurring as open-space fillings in breccias, fractures, and vugs, as well as disseminated clusters of crystals that occupy intergranular pore spaces.

The timing of the Robb Lake sulphide mineralization is constrained by paleomagnetism, which gives an Eocene age, and by Rb-Sr isotope data on sphalerite, which suggest a Paleozoic age, although there is a considerable scatter in the results due to heterogeneous $^{87}Sr/^{86}Sr$ values in primary fluids. Fluid inclusion and isotopic data suggest that sulphide mineralization formed from saline (16 to >23 wt.% NaCl equiv.) NaCl-$CaCl_2$ brines ranging in temperature from 87 to 154°C and that these fluids interacted with adjacent or underlying siliciclastic rocks. Such fluids are commonly linked to deep burial sedimentary and tectonic compaction or to topographic or gravity-driven fluid flow during the Cretaceous-Tertiary Laramide Orogeny. Alternately, the fluids may have originated as hydrothermal solutions channeled along reactivated back-arc intrabasinal faults and permeable stratified units. In such a scenario, MVT mineralization would be the far-field consequence of subduction and slab rollback that took place on the western margin of North America during the Devonian-Mississippian and generated back-arc and intra-arc spreading and related exhalative activity.

Résumé

Le district à gîtes de Zn-Pb de Robb Lake renferme de nombreux gisements du type Mississippi-Valley classique, qui se composent de minéralisations sulfurées encaissées dans des roches carbonatées. La plupart de ces gisements sont encaissés dans les dolomies de plate-forme des formations de Muncho-McConnell, de Stone et de Dunedin, qui sont déformées et recoupées par des failles de chevauchement. Ces dolomies, dont l'âge varie du Silurien au Dévonien moyen, reposent au voisinage du front du milieu de plate-forme continentale du nord des Rocheuses canadiennes. Le gisement de Robb Lake, qui est le plus important du district, englobe 19 occurrences stratoïdes de Zn Pb. La minéralisation se présente dans des corps bréchiques interconnectés, qui présentent une attitude stratiforme ou discordante et s'étendent sur une épaisseur stratigraphique pouvant atteindre 200 m. Les principaux minéraux sulfurés sont la sphalérite, la galène et la pyrite, qui se présentent sous différentes formes allant des matériaux de remplissage de brèches, de fractures et de cavités aux amas de cristaux disséminés dans des pores intergranulaires.

Les limites chronologiques de la formation des minéralisations sulfurées du district de Robb Lake se situent à l'Éocène, d'après le paléomagnétisme, et au Paléozoïque, selon des données sur les isotopes de Rb et de Sr de la sphalérite. Cependant, les résultats d'analyse sont très dispersés en raison de l'hétérogénéité des valeurs du rapport $^{87}Sr/^{86}Sr$ dans les fluides primaires. Les données sur les inclusions fluides et les données isotopiques laissent supposer que les minéralisations sulfurées se sont formées à partir de saumures de NaCl-$CaCl_2$ (teneur en équivalent poids de NaCl allant de 16 à plus de 23%) dont les températures variaient de 87 à 154°C, et que ces fluides ont intéragi avec des roches silicoclastiques adjacentes ou sous-jacentes. De tels fluides sont généralement rattachés à une compaction sédimentaire et tectonique en profondeur ou à l'écoulement gravitaire de fluides durant l'orogenèse du Laramide au Crétacé-Tertiaire. Par contre, ils pourraient avoir apparu sous forme de solutions hydrothermales qui se sont écoulées le long d'unités stratifiées perméables et de failles réactivées à l'intérieur d'un bassin d'arrière-arc, ce qui impliquerait que les minéralisations de type Mississippi-Valley sont le résultat lointain de la subduction et du retour en arrière d'une dalle à la marge ouest de l'Amérique du Nord, au Dévonien-Mississippien, ce qui aurait entraîné une expansion océanique dans les milieux arrière-arc et intra-arc et une activité exhalative concomitante.

Introduction and History

The Robb Lake Zn-Pb district is located in platform carbonate rocks of northeastern British Columbia (Fig. 1). It includes several deposits of lead and zinc mineralization within the northern Rocky Mountains Mississippi Valley-type (MVT) belt. The known MVT deposits are listed in Table 1 and Appendix 1 and their locations are shown on Figure 2. The Robb Lake MVT Zn-Pb deposit (Macqueen and Thompson, 1978; Paradis et al., 1999; Nelson et al., 2002) is the most extensively exposed and most thoroughly explored deposit in the northern Rocky Mountains MVT belt (56°56'N, 123°43'W; 94B/13; Fig. 2), and because of that, reference to this deposit is common in this paper. It contains a geological resource of 6.5 million tonnes at 7.11% combined lead and zinc, (2.4 m mining width, 5% cutoff grade; Consolidated Barrier Reef Resources, Rights Offering Circular, November 29, 1984). Exposed in more than 19

Paradis, S., and Nelson, J.L., 2007, Metallogeny of the Robb Lake carbonate-hosted zinc-lead district, northeastern British Columbia, *in* Goodfellow, W.D., ed., Mineral Deposits of Canada: A Synthesis of Major Deposit-Types, District Metallogeny, the Evolution of Geological Provinces, and Exploration Methods: Geological Association of Canada, Mineral Deposits Division, Special Publication No. 5, p. 633-654.

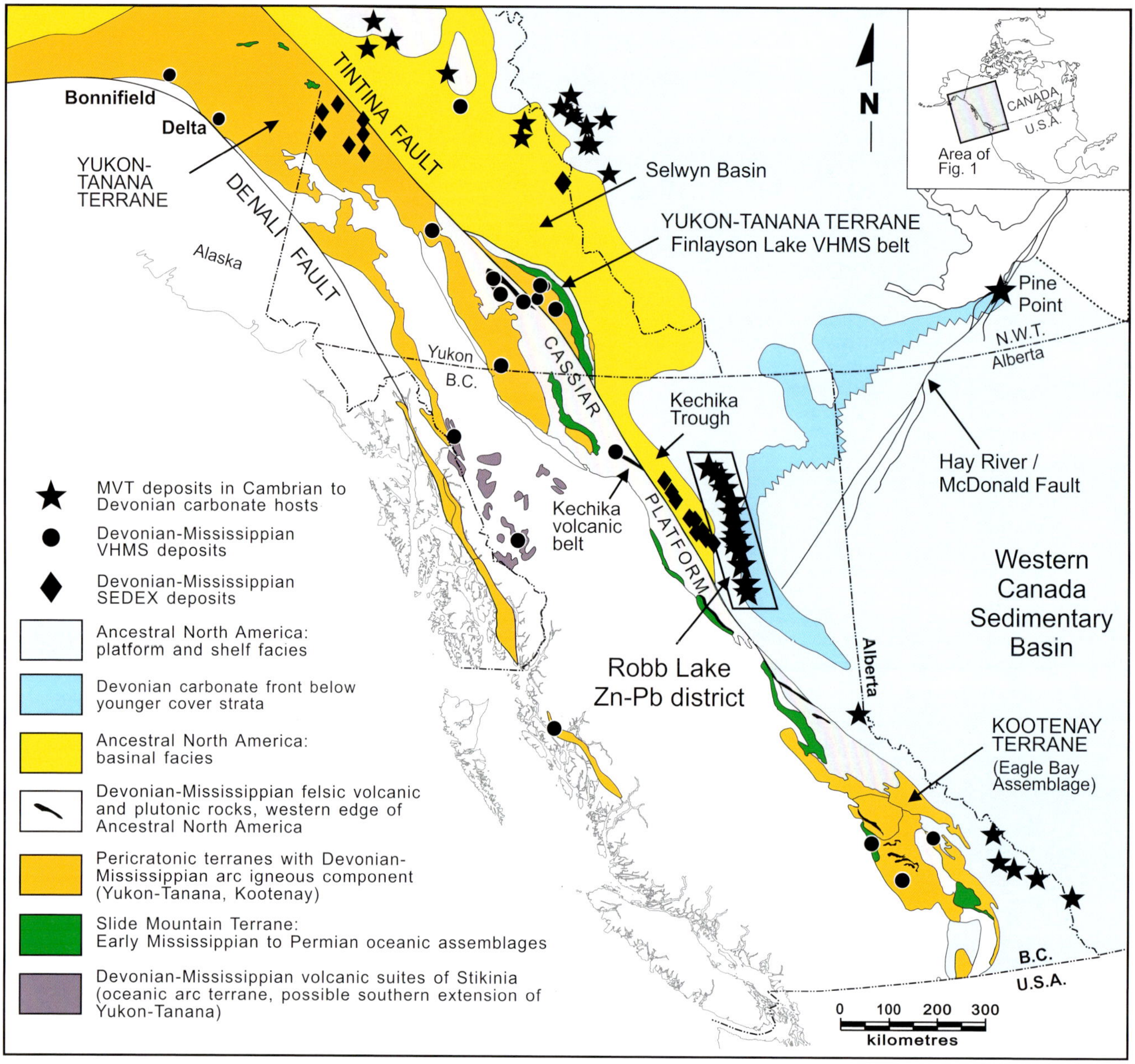

FIGURE 1. Devonian-Mississippian tectonic and metallogenetic framework of the ancestral western Canadian continental margin, and location of the Robb Lake Zn-Pb district. Note the general trend from the volcanic-hosted massive sulphide (VHMS) deposits dominant in outermost pericratonic terranes, through sedimentary exhalative (SEDEX) deposits dominant in continent-margin basins, to Mississippi Valley-type (MVT) deposits in carbonate platforms of the continental interior (from Nelson et al., 2002). The Rocky Mountains Mississippi Valley-type belt is defined by a linear trend of stratabound and stratiform Zn-Pb deposits hosted in Paleozoic carbonate rocks. MVT deposits from the Robb Lake district are tabulated in Appendix 1.

showings in a 40 km^2 area, the deposit consists of a series of interconnected bedding-parallel and crosscutting breccia bodies and peripheral veins and stockworks with white sparry dolomite cement, sphalerite, galena, pyrite, quartz, calcite, and pyrobitumen. Most of the other occurrences are small showings.

Lead-zinc mineralization was discovered near the Robb Lake deposit in 1971 by Arrow-Interamerican Corporation, Barrier Reef Resources Limited, and Ecstall Mining Ltd. This discovery sparked an exploration staking rush in the northern Rocky Mountains of British Columbia that led to the discovery of numerous Zn-Pb deposits and recognition of a potential new Zn-Pb belt. Other mineral properties in the district include Richards Creek, Cosburn, Southface, CYN, Perkings, Reff, Poco, Coral, Coral #2, Nabe, Wind, Dev, Ron, Marini 1, ZRR, Nick, Peace, Smith, BV, Ospika, and WL (Table 1 and Appendix 1).

Drilling at the Robb Lake deposit between 1972 and 1975 led to the incomplete delineation of three prospective orebodies, the "Lower zone", "East Webb Ridge zone", and "West Webb Ridge zone" (Fig. 3). Drilling was continued by Texasgulf in the summers of 1980 and 1981. Since then, no exploration work has been done. Core is stored at an airstrip at the confluence of Mississippi Creek and the Halfway River. The property presently belongs to Doublestar Resources Ltd.

In 1959, Irish (1970) began reconnaissance mapping of the Halfway River map area (94B, latitude 57°00'N to

TABLE 1. Zn-Pb deposits of the Robb Lake district.

Deposit Name[1]	Relevant Showings	BC MINFILE	Commodity	Status Mining[2]	Geological Resources				Comments
					Zn (%)	Pb (%)	$BaSO_4$	Tonnes	
Robb Lake	Canyon, Waterfall, North Face, Lower, Upper, Downstream, Talis, Cascade, Webb Ridge East and West Webb), Monocline, Camp, North Texas, Tennessee Mountain 1, T.G.S.	094B 005	Zn, Pb	Developed prospect	7.11			6,449,481 (resource estimated - 1984) at 5% cutoff grade	Geological resources are measured over a minimum mining width of 2.4 m combining 3 separate showings (Lower, East Webb, West Webb).
Robb Lake	Tennessee Mountain 2, Bear, Sheep Creep	094B 005	Zn, Pb	Developed prospect					
Richards Creek		094G 022	?	Occurrence					
Cosburn (Redfern Lake)		094G 036	Ba, Zn, Pb	Occurrence					
Southface (Redfern Lake)		094G 018	Zn, Pb	Occurrence					A grab sample from the southern end graded 7.8% Zn and 0.14% Pb; 700 metres to the west a second sample returned 1.8% Zn and 0.04% Pb (Assessment Report 5551). A grab sample from a 3 by 1-metre lens of massive limonite, sphalerite, smithsonite, galena and minor pyrite graded 44.0% Zn and 14.6% Pb (Assessment Report 5551).
RB (June, Mount Bertha)		094G 004	Ba, Zn, Pb	Occurrence					Composition of the pods and veins range from 100% calcite to 50% barite.
CYN (Nabesche)		094B 019	Zn, Pb	Occurrence					A chip sample assayed 1.037% Zn and 2.17% Pb (Assessment Report 5470).
Perkins (Laurier)	Perkins Main, Northeast Perkins	094B 015	Zn, Pb	Occurrence					A grab sample from the Perkins Main contains 0.05% Pb and 26.1% Zn (Assessment Report 5468). The Northeast Perkins showing consists of talus boulders rich in sphalerite.
Reef (Marini 2)		094B 003	Zn, Pb, Ag	Occurrence					A grab sample contained 0.0912% Pb, 0.16% Zn and 3.5 g/t Ag (Assessment Report 4772).
Poco	Poco Main, Poco North, West Rusty, East Rusty	094B 007	Pb, Zn	Occurrence					A continuous chip sample from the West Rusty zone averaged 8.5% Zn and 3.2% Pb over 12.5 metres; a three-metre chip from a trench on the Poco Main showing assayed 13.8% Zn and 5.5% Pb (Assessment Report 5745). A five hole drill program in 1975 failed to intersect significant grades.
Coral		094B 008	Zn, Pb	Occurrence					A sample of weathered rubble from a 1-metre trench excavated in 1984 assayed 1.45% Pb and 3.47% Zn (Assessment Report 13724).
Coral #2		094B 021	Zn	Occurrence					A rock sample, collected in 1985, assayed 2.36% Zn, 0.01% Pb and 0.343 g/t Ag (Assessment Report 15040). Follow-up drilling intersected 0.34% Zn and 0.01% Pb in a 20-centimetre zone of dolomite breccia (Assessment Report 15040).
Nabe		094B 009	Zn	Occurrence					
Wind (Blow)		094B 010	Zn, Pb	Occurrence					
Dev	2 unnamed showings	094B 011	Zn, Pb, Ag, Cu, Cd	Occurrence					A grab sample assayed 1.5 g/t Ag, 0.0272% Cu, 0.0030% Pb, 1% Zn and 0.0760% Cd (Assessment Report 4791).
Ron		094B 012	Zn, Pb, Fe	Occurrence					
Marini I		094B 014	Zn	Occurrence					
ZRR		094B 017	Zn	Occurrence					3.1% Zn was obtained from a 4-metre chip sample of brecciated Dunedin dolomite containing finely disseminated sphalerite (Assessment Report 5246). A piece of float graded greater than 50% Zn.
Nick		094B 031	Pb, Zn	Occurrence					
Peace		094B 032	Pb, Zn	Occurrence					Two holes drilled on the occurrence failed to intersect significant grades.
Smith (Brin)		094B 033	Zn, Pb	Occurrence					A grab sample of dolomite breccia contained 0.09% Zn and less than 0.01% Pb (Assessment Report 5370).
BV (Muncho Lake barite, Sulphur Creek)		094N 002	Ba	Developed prospect			65%	100,000,000 tonnes	One zone assayed 80% $BaSO_4$ over 3 metres, and another 18-metre wide zone averaged 65% $BaSO_4$ (Assessment Report 1682). Based on these results, and assumptions regarding the shape of the deposit and minimum mining width, an inferred "tonnage" or resource of 100 Mt was derived, with an indicated grade of 65% $BaSO_4$ (Assessment Report 1682, pages 22, 23).
Ospika (Witch)		094B 018	Zn	Occurrence					
WL		094B 006	Zn, Pb, Cu	Occurrence					A grab sample from the gossan contained 0.1% Cu, 0.114% Pb and 0.0013% Zn. A three-metre chip sample from a trench contained 0.0024% Cu, 0.1470% Pb, and 0.0216% Zn (Assessment Report 5246).

[1] ( ): other name(s) for the deposit.

[2] Deposit = Any natural, but locally restricted, concentration of minerals in the Earth's crust. Developed prospect = Occurrence on which exploration and progressed to a stage that allows a reasonable estimate of the amount(s) of one or more of the potentially mineable commodities. Occurrence = Anomalous concentration of minerals that are common to ore minerals elsewhere, but are uneconomic.

56°00'N and longitude 124°00'W to 122°00'W), which includes the Robb Lake deposit and most of the other Zn-Pb deposits (except Richards Creek, Cosburn, and Southface). This area is located between the Peace (latitude 56°00'N) and the Halfway (latitude 57°00'N) rivers. Four field seasons led to the first published geological map of the Halfway River area. This map was immediately used by mineral explorationists, who in the late summer of 1971 discovered lead-zinc mineralization at Robb Lake. Intense exploration continued through 1974. By that time, a host of new geological problems, requiring investigation at a more refined scale, needed attention, and Thompson (1989) began mapping in 1975. M.Sc. and Ph.D. theses were done in the course of the following years on specific geological problems. Taylor (1977) studied the carbonate stratigraphy and petrology, and suggested a peritidal environment for the sedimentary succession, which was also indicated by Macqueen and Thompson (1978). Taylor concluded that the formation of the mineralized breccias was due to intrastratal karst and that the mineralization occurred subsequent to brecciation from formation waters that were derived from compacting shales. Manns (1981) studied the stratigraphic aspects of the Silurian-Devonian sequence hosting the Robb Lake Zn-Pb mineralization. He concluded that the host breccias were collapse structures due to solution of evaporitic dolostone platform at an unknown time prior to the Laramide Orogeny. Sangster (1973, 1979) and Sangster and Lancaster (1976) favoured a paleokarst solution-collapse origin, linked to an unconformity within and above the Silurian-Devonian carbonate sequence. Macqueen and Thompson (1978) rejected this hypothesis on the basis of the form and distribution of the breccias at Robb Lake as then known, as well as the absence of any evidence of the former presence of evaporites, which would be precursors to evaporite solution brecciation. They suggest a hydraulic fracturing mechanism that resulted from compaction and dewatering of the shale basin to the west.

Fluid inclusion microthermometric analyses were performed by Sangster and Carrière (1991) on sphalerite samples from the Robb Lake deposit. Paleomagnetic dating of the Zn-Pb mineralization at Robb Lake has been done by Smethurst et al. (1999). An attempt to date the mineralization at Robb Lake by Rb-Sr isotopic analysis on sphalerite samples was undertaken by Nelson et al. (2002).

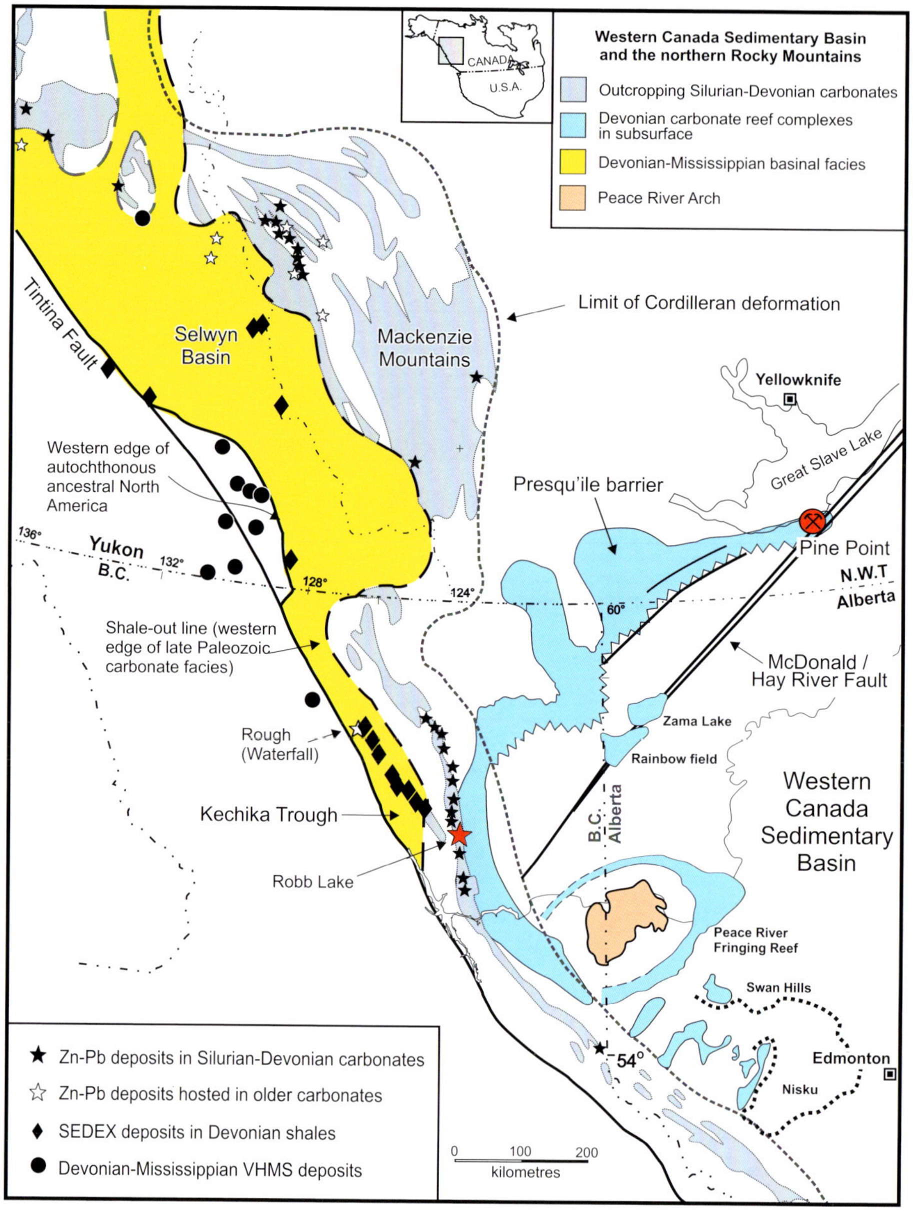

FIGURE 2. Location of the Zn-Pb carbonate-hosted deposits within the northern Canadian Rocky Mountains and the Western Canada Sedimentary Basin, the sedimentary exhalative (SEDEX) deposits within the Kechika Trough and the Selwyn Basin, and the volcanic-hosted massive sulphide (VHMS) deposits in the outermost continental margin. Diagram is compiled from Morrow et al. (1990), Olson et al. (1994), and Mossop and Shetsen (1994), and modified from Nelson et al. (2002). MVT deposits from the Robb Lake district are tabulated in Appendix 1.

Geological and Tectonic Setting

Cambrian to Middle Devonian shelf strata are exposed in the northern Rocky Mountains of British Columbia and in the Mackenzie Mountains of eastern Yukon and western Northwest Territories. In the Rocky Mountains, Paleozoic carbonate rocks host a linear trend of stratabound and stratiform Zn-Pb deposits called the "Rocky Mountains MVT belt" (Fig. 1). This trend of deposits is part of a series of belts including the Devonian-Mississippian (and older) volcanic-

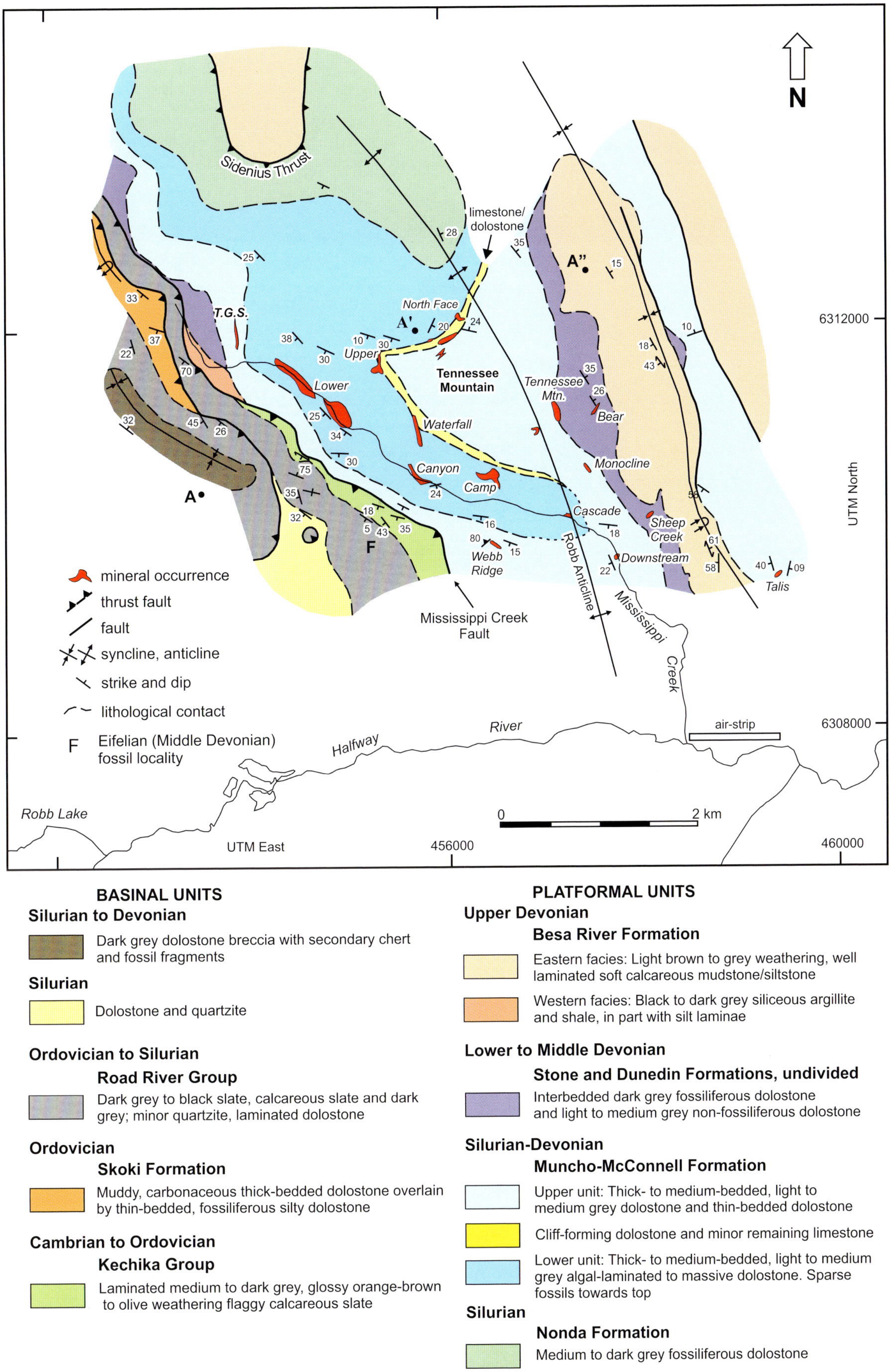

FIGURE 3. Geological map of the Robb Lake deposit area (Nelson et al., 1999; Paradis et al., 1999). Cross-section A-A" is illustrated in Figure 4.

TABLE 2. Stratigraphic setting and characteristics of the Zn-Pb deposits within the Robb Lake district.

Formation/Group	Deposit name	Showings	Minerals	Mineralization
Muncho-McConnell Formation	Robb Lake	Canyon, Waterfall, North Face, Lower, Upper, Downstream, Talis, Cascade, Webb Ridge (East and West Webb), Monocline, Camp, North Texas, Tennessee Mountain 1, T.G.S.	Sphalerite, galena, pyrite, marcasite, bitumen	Open-space filling of breccias (crackle, mosaic and rubble, and rock-matrix), fractures, and vugs. The breccias are bedding-parallel and crosscutting.
Muncho-McConnell Formation	Ron		Sphalerite, galena	Fine sphalerite and coarse galena occur in a brecciated horizon of medium to thick-bedded dolomite.
Dunedin, Besa River, Slave Point formations	CYN (Nabesche)		Sphalerite	Minor disseminated sphalerite in fractures and brecciated zones of the Dunedin, Besa River and Slave Point formations.
Stone Formation	Nick		Sphalerite, galena	Less than 2% disseminated galena and sphalerite are hosted in Stone Formation dolomite which is locally recrystallized to a white sparry dolomite.
Stone and Dunedin formations	Robb Lake	Tennessee Mountain 2, Bear, Sheep Creep	Sphalerite, galena, pyrite, marcasite, bitumen	Open-space filling of breccias (crackle, mosaic and rubble, and rock-matrix), fractures, and vugs. The breccias are bedding-parallel and crosscutting.
Stone and Dunedin formations	Richards Creek		Pyrite, marcasite, quartz, bitumen (?)	Open-space filling of fractures and veins, small massive pods partially replaced the host dolomite.
Stone and Dunedin formations	Coral		Sphalerite, galena	Red and yellow sphalerite with minor galena disseminated in an irregular zone of dolomite breccia.
Stone and Dunedin	Peace		Galena, sphalerite	Disseminated galena and sphalerite in Stone or Dunedin Formation dolomite
Dunedin Formation	Cosburn (Redfern Lake)		Barite, fluorite, minor galena	Body of massive barite (±fluorite, ±galena) hosted in limestone breccia of the Dunedin Formation. Approximately 250 metres to the south, a second body of barite crosscuts bedding in the limestone.
Dunedin Formation	Southface (Redfern Lake)		Sphalerite, galena, limonite, smithsonite, minor pyrite	Two stratiform zones of disseminated sphalerite, galena, quartz, and calcite in open-space fillings, fractures, and replacements of the fossiliferous-granular dolostone. A 3 m by 1 m lens of massive limonite, sphalerite, smithsonite, galena and minor pyrite occurs within the upper mineralized zone.
Dunedin Formation	RB (June, Mount Bertha)		Barite, calcite, galena, minor sphalerite	Several occurrences of stratabound pods up to 75 m by 30 m in size, sheet-like veins, and pipe-like masses of calcite-barite mineralization. Occurrences of galena and minor sphalerite are associated with the calcite-barite pods.
Dunedin Formation	Perkins (Laurier)	Perkins Main, Northeast Perkins	Sphalerite, minor galena, limonite,	Disseminated in coarse-grained brecciated dolomite.
Dunedin Formation	Poco	Poco Main, Poco North, West Rusty, East Rusty	Galena, sphalerite, hydrozincite	Galena and sphalerite occur as veins and disseminations in Devonian reefal dolomite breccias. Hydrozincite is common in the dolomite.
Dunedin Formation	Reef (Marini 2)		Hematite	Finely disseminated hematite and a moderate amount of limonite forming conspicuous red stains in Dunedin Formation dolomite. No primary sulphides, but anomalous values in Pb, Zn, Ag.
Dunedin Formation	Nabe		Sphalerite	Disseminated sphalerite in dolostone.
Dunedin Formation	Wind (Blow)		Sphalerite, galena	Sphalerite and minor galena disseminated in limestone, dolomite, and dolomite breccias.
Dunedin Formation	Dev	2 unnamed showings	Sphalerite, galena	The first showing consists of pale yellow sphalerite disseminated in calcite/white dolomite veins and matrix of a Dunedin Formation limestone bed. The second showing at the base of the Dunedin possibly in the Stone Formation (upper unit) dolomite/mudstone consists of galena associated with white dolomite and minor quartz as the matrix of a solution breccia, occasionally replacing the host.
Dunedin Formation	ZRR		Smithsonite/hemimorphite, sphalerite, pyrite and/or marcasite	Smithsonite/hemimorphite coats fractures and occurs as breccia cement; sphalerite is disseminated in Dunedin Formation limestone; pyrite and/or marcasite occurs as pods, veins, and disseminations in brecciated dolostone near fault.
Dunedin Formation	Coral #2		Smithsonite, sphalerite.	Dolomite breccia cemented with smithsonite and minor disseminated sphalerite.
Dunedin Formation	Smith (Brin)		Sphalerite, galena	Minor disseminated sphalerite and galena in Dunedin Formation dolomite breccia.
Skoki Formation	Ospika (Witch)		Hydrozincite, pyrite	Minor hydrozincite in fractures of dolomite breccias; disseminated pyrite in silicified dolomite. Thompson (1989) mentioned sphalerite and smithsonite in fractured dolostone of the Ordovician Skoki Formation.
Skoki Formation	WL		Limonite, rare chalcopyrite	Three limonitic pyritic gossans overlying massive pyrite/marcasite pods, veins, and disseminations.
Road River Group	Marini 1		Minor sphalerite	Fractures containing minor sphalerite crosscut Upper Silurian or Lower Devonian calcareous quartzite and shale within a thrust slice of the Road River Group.

( ): other name(s) for the deposit.

hosted massive sulphide (VHMS) deposits of the pericratonic terranes, the sedimentary exhalative (SEDEX) deposits of the Kechika Trough and Selwyn Basin, and the MVT deposits in Silurian-Devonian platform dolomites of the outcropping carbonate platform (Thompson, 1989). Below the late Paleozoic-Mesozoic cover, the carbonate platform extends eastward and becomes the Presqu'ile Barrier, which hosts the Pine Point MVT Zn-Pb deposits in the Northwest Territories (Rhodes et al., 1984; Hannigan, 2007).

Figures 1 and 2 illustrate the Devonian-Mississippian tectonic and metallogenic framework of the northern Canadian Rocky Mountains and Western Canada Sedimentary Basin (WCSB). The Paleozoic stratigraphy of the northern Canadian Cordillera is discussed in Thompson (1989), McMechan and Thompson (1993), and Ferri et al. (1999). Most of the known MVT deposits in the northern Rocky Mountains are located between the Peace and the Muskwa rivers. They occur within a belt of Silurian to Middle Devonian carbonate rocks that are situated close to areas in which the thrust-faulted shelf-basin boundary, i.e. the carbonate front, occurs. The mineralized units are part of an extensive lower and middle Paleozoic platform assemblage

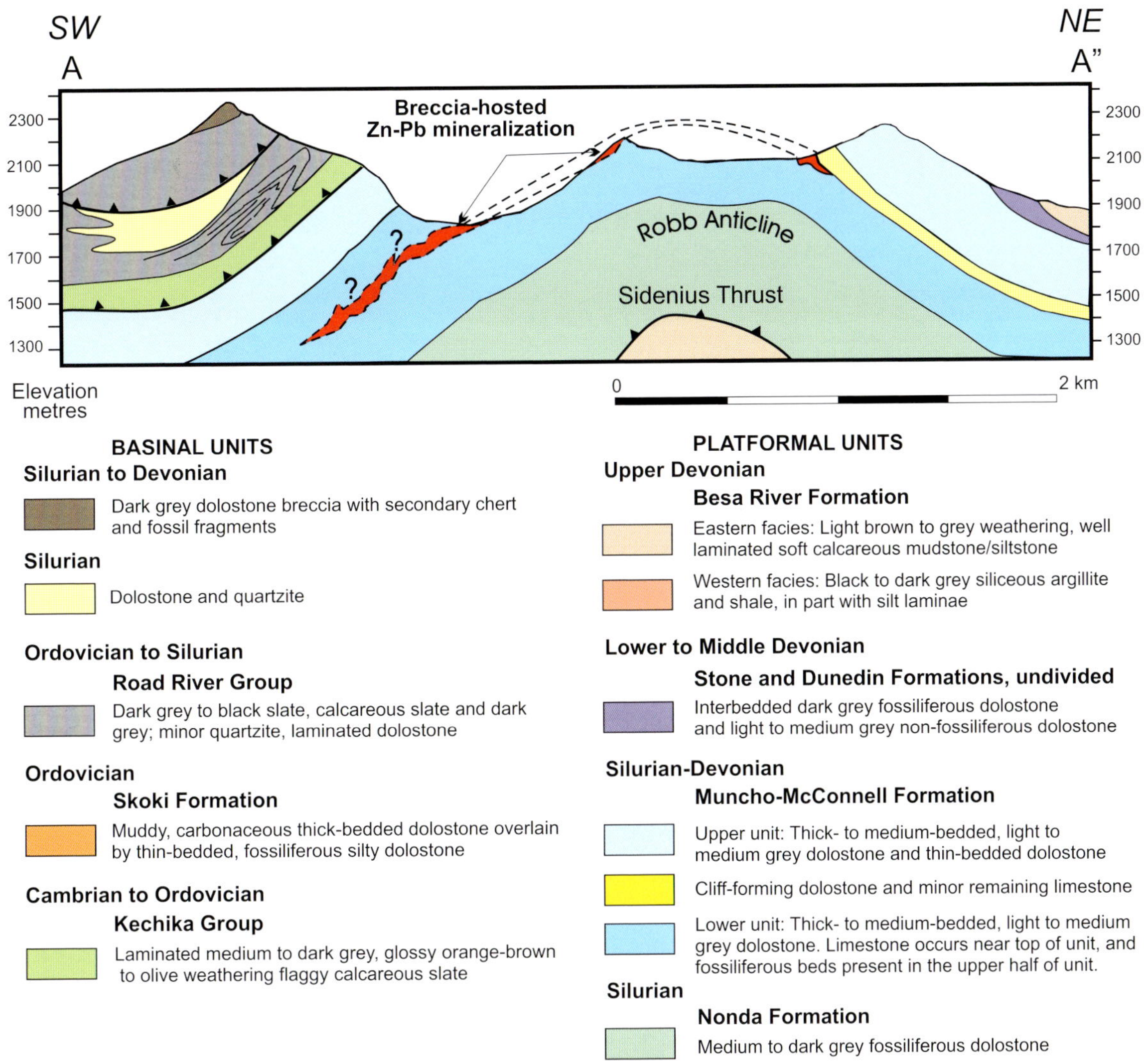

FIGURE 4. Cross-section of Mississippi Creek, looking northwest. The section A-A" is located on Figure 3 (Nelson et al., 1999; Paradis et al., 1999).

of shallow-water carbonate sediments that accumulated above a miogeoclinal wedge of upper Proterozoic and Lower Cambrian terrigenous rocks (Wheeler and Gabrielse, 1972).

The Kechika Trough is an intracontinental basin located immediately west of the exposed carbonate MVT belt (Fig. 2). It is characterized by Lower Cambrian to Mississippian deep-water sediments (McClay et al., 1989) including Late Devonian Earn Group strata that host SEDEX Zn-Pb-barite deposits (Paradis et al., 1998). The Selwyn Basin in the Yukon is analogous to the Kechika Trough, also containing SEDEX deposits of Cambrian through to Early Mississippian age. West of the Kechika Trough and Selwyn Basin lies the Cassiar platform (Fig. 1), which is composed of shelf and shelf-slope sedimentary facies that were deposited oceanward of ancestral North America. The westernmost tectonic elements of the Canadian Cordillera are the allochthonous and para-autochthonous pericratonic terranes, which comprised the active margin of North America in mid to late Paleozoic time. The pericratonic terranes (Fig. 1) contain widespread Devonian-Mississippian arc and arc-rift volcanic and plutonic suites with accompanying volcanic-hosted massive sulphide (VHMS) deposits (Nelson et al., 2002).

Mineral Deposits

Deposit Types and Classification

A single major deposit type has been recognized in the Robb Lake district, i.e. the classic MVT Zn-Pb sulphide deposits. This economically important deposit type is hosted in Silurian to Middle Devonian carbonate rocks deposited in shallow-water environments, which have been subsequently dolomitized, brecciated, and have undergone some dissolution.

Globally, MVT deposits are generally small; most are less than two million tonnes and are zinc-dominant. Grades seldom exceed 10% Pb+Zn (Leach and Sangster, 1993). MVT deposits account for 35% of the world's lead and zinc resource (Leach, 2005). In Canada, MVT mineral production occurred between 1964 and 2002. During this time, about 30% of Canada's annual lead and zinc production was derived from this deposit type (Sangster, 1996). At the end of 2002, all MVT mines in Canada were closed. In the Robb Lake district, approximately 22 deposits are distributed over approximately 2000 km^2, but none of these were ever put into production.

MVT deposits characteristically occur in clusters called districts, and each deposit may contain several showings. For

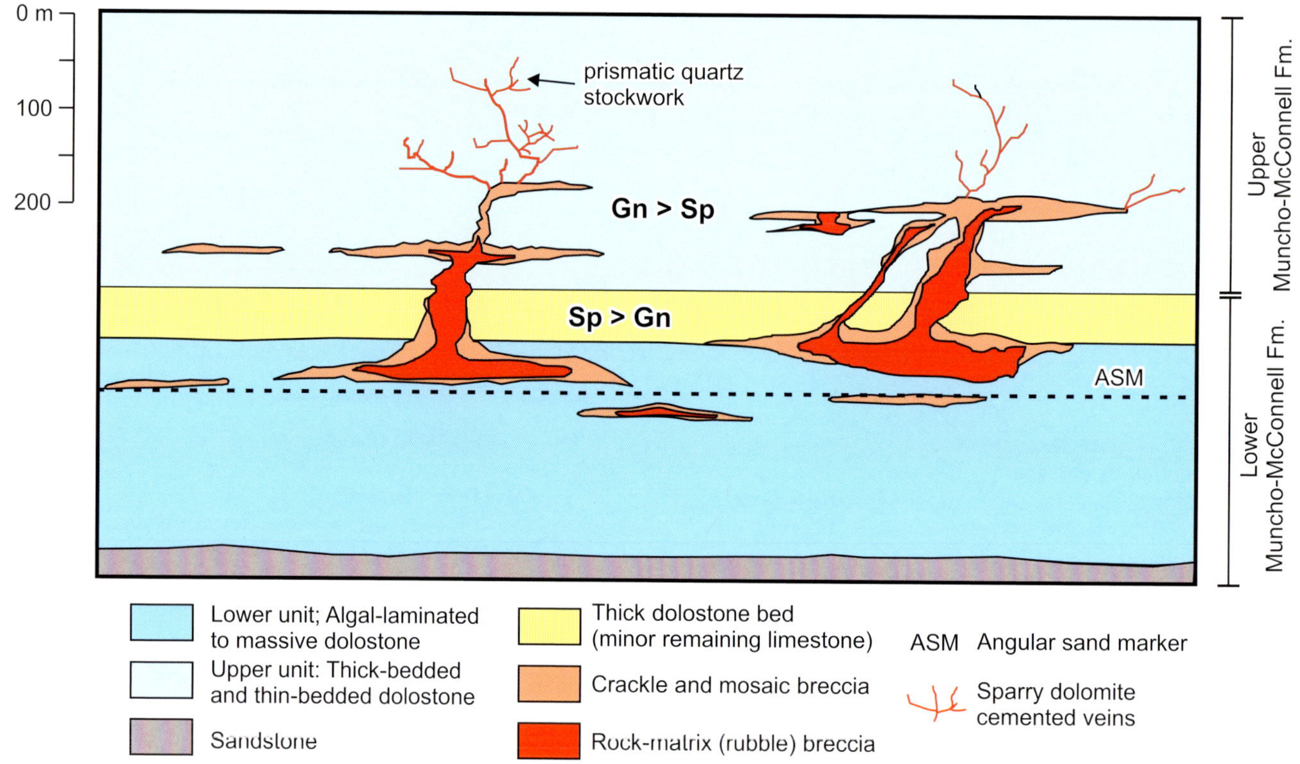

FIGURE 5. Schematic diagram illustrating the main breccia types, their relationships, and stratigraphic location within the Robb Lake deposit (Nelson et al., 1999; Paradis et al., 1999).

example, the Robb Lake deposit contains approximately 19 showings, and the Pine Point deposit in the Northwest Territories contains 100 drill-defined orebodies. Each of these orebodies can contain up to several hundred million tonnes of sulphides distributed over tens to hundreds to thousands of square kilometres.

Deposit Distribution

In the Robb Lake district, most of the deposits are hosted in Silurian to Middle Devonian platform dolostones of the Muncho-McConnell and Dunedin formations (Table 2), and are in close proximity to the carbonate platform edge. Only three deposits, Opiska, WL, and Marini 1, are hosted in Ordovician carbonaceous dolostone of the Skoki Formation and the Ordovician to Silurian calcareous slate of the Road River Group.

At the Robb Lake deposit, approximately 19 known stratabound Zn-Pb showings are hosted by breccias of the Silurian-Devonian platform carbonate succession of the Muncho-McConnell Formation. A few mineralized bodies lie within the overlying sequence attributed to the Stone-Dunedin formations. Most of the showings occur along the valley of Mississippi Creek and the adjacent mountain slopes (Figs. 3, 4).

Grade and Tonnage

The Robb Lake deposit has been the most thoroughly explored MVT Zn-Pb deposit in the northern Rockies MVT belt, and it is the only one with calculated geological resources, which are 6.5 million tonnes at 7.11% combined lead and zinc (2.4 m mining width, 5% cutoff grade; Consolidated Barrier Reef Resources Ltd., Rights Offering Circular, November 29, 1984).

Deposit Morphology

Nature of Sulphide Orebodies

Mineralization in the district ranges from open-space fillings of breccias, fractures, and vugs to disseminated clusters of crystals that occupy intergranular pore spaces. Sulphide-hosting structures are most commonly zones of highly brecciated dolomite; in some instances (e.g. Robb Lake deposit) these zones are arranged in linear patterns suggesting a tectonic control. These breccia zones may range from more or less concordant tabular structures, controlled by individual strata, to discordant cylindrical structures within tens of metres of sedimentary sequences (Fig. 5).

At the Robb Lake deposit, the breccias are interconnected, bedding-parallel and/or crosscutting bodies in as much as 200 m of stratigraphic section (Fig. 5). Not all the breccias are mineralized. The favourable mineralized sections form a broad stratabound zone occupying the upper 200 m of the lower unit of the Muncho-McConnell Formation and the lower 130 m of the upper unit; a 70 m-thick barren section separates the mineralized zones (Boronowski and James, 1982). The mineralized zones have a northwesterly elongation and alignment, parallel to the strike direction and to the structural and paleofacies features. The breccia bodies form relatively thin and narrow horizons and pods that are both parallel to and crosscut bedding. Several extend for more than 300 m along bedding and crosscut more than 50 m of section. The shape and extent of several breccia bodies are well exposed on the south and north slopes of Tennessee Mountain (Fig. 6A) and along Mississippi Creek (Fig. 6B,

C). These slope exposures demonstrate that the showings along the creek, such as the Lower, Canyon, and Cascade zones, are stratigraphically equivalent to the showings on the south side of Tennessee Mountain (the Upper, Waterfall, and Camp zones, and potentially the North Face zone), i.e. the south slope of Tennessee Mountain is more or less a dip slope. This is well illustrated on the cross-section of the valley of Mississippi Creek (Fig. 4).

The mineralized breccias at Robb Lake described below are of several textural types: sparry dolomite-cemented crackle, mosaic and rubble breccias, and rock-matrix (trash) breccias (Fig. 5). In general, the rock-matrix (trash) breccias occur at the base of the breccia bodies and are overlain and fringed by rubble, mosaic, and crackle breccias that are cemented by white sparry dolomite. Also, in the overall system, rock-matrix (trash) breccias are preferentially located within the lower unit of the Muncho-McConnell Formation. The dolomite-cemented rubble, mosaic, and crackle breccias tend to overlie the trash breccias and are more abundant in the upper unit of the Muncho-McConnell Formation.

Crackle and mosaic breccias. The crackle breccias show little displacement of fragments in coarse-grained sparry white dolomite cement (Fig. 7A, B). The fragments consist exclusively of highly angular clasts of the variably altered crosscut dolostone host rock. The mosaic breccias have fragments that are largely but not wholly displaced (Fig. 7C, D). The fragments also consist of highly angular clasts of the crosscut dolostone host rock in white, coarse-crystalline dolomite cement. Generally, mosaic breccias grade out through crackle breccias into unfractured dolostone. It is notable that in these breccias, individual clasts tend to be of somewhat uniform size, typically 5 to 20 cm across; although the fragments can range in size from 1 mm to several metres in longest dimension. Megabreccias of this type are rare. A substantial portion of the mineralization at Robb Lake occurs within the dolomite cement of the mosaic and crackle breccias.

Rubble breccias. In contrast to the crackle and mosaic breccias, the rubble breccias are polymictic, and the fragments are completely displaced and show no match with each other (Fig. 7E, F). By textural definition, this category includes the rock-matrix (trash) breccias (Fig. 8), which are rubble breccias with fragmental matrix instead of dolomite cement. The rubble breccias are by far the most volumetrically important type of breccias and the most varied. They contain a variety of clasts that are included in a fine-grained dark grey fragmental matrix, in white sparry dolomite cement, or in a mix of both. The breccias are generally grain-supported and locally matrix-supported. The fragments consist of a variety of altered and unaltered dolostone lithology (90 vol.%), white sparry dolomite (i.e. subangular fragments and thin curviplanar segments of vein selvages, 5 vol.%), shale/mudstone (2 vol.%), finely laminated shaly carbonate (1 vol.%), chert (1 vol.%), pyrobitumen-rich material (<1 vol.%), sulphides (<1 vol.%), and fossils (brachiopods and gastropods, <1 vol.%). The fragments are highly angular to subangular, and can vary in size from less than 1 mm to several metres. However, most breccias are dominated by fragments less than 10 cm in diameter. The matrix of the trash breccias is a dark grey to black carbonaceous dolostone. On close inspection, this matrix is composed of fine fragments,

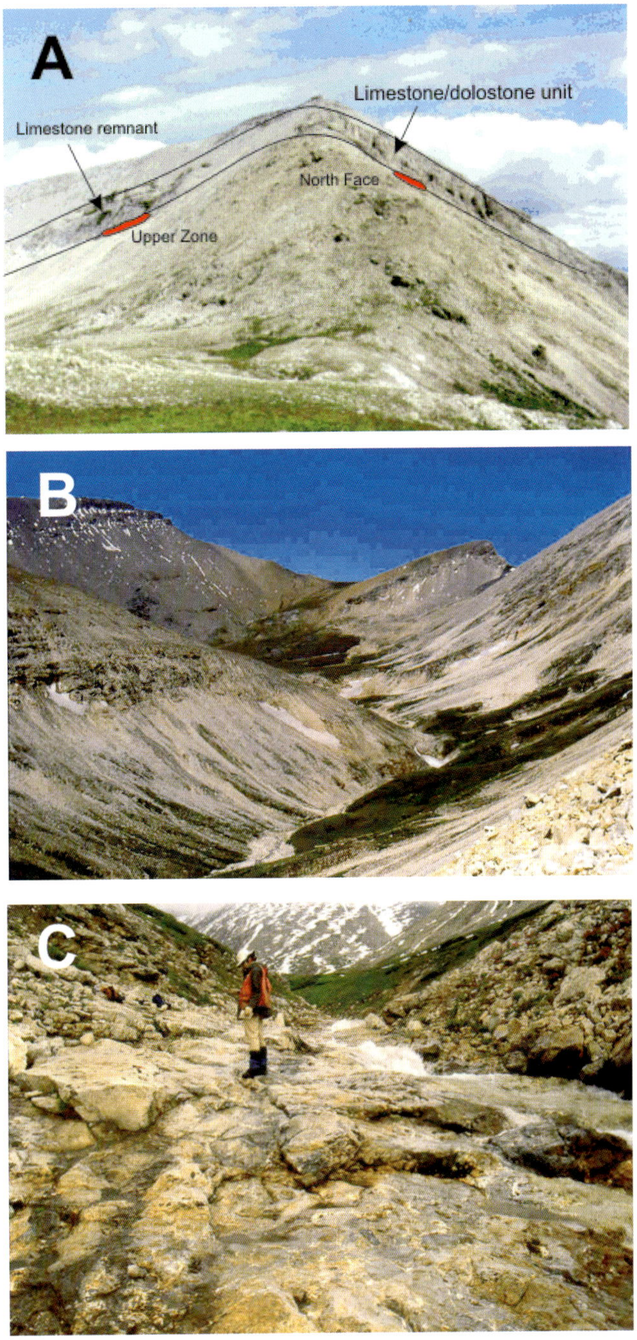

Figure 6. (A) View looking northwest at the North Face and Upper mineralized zones on Tennessee Mountain, Robb Lake deposit. This view is the same as shown in Figure 4. The mineralized zones lie immediately below the cliff-forming limestone/dolostone unit that forms the uppermost unit of the lower Muncho-McConnell Formation. A remnant of limestone lies above the North Face zone. **(B)** View of the Mississippi Creek valley looking northeast. **(C)** View of the Lower showing at the Robb Lake deposit looking northwest. The Lower showing is exposed along Mississippi Creek.

cemented by overgrowths and interstitial dolospar and quartz. Some of the fine fragments are possibly pyrobituminous shale, however much of the dark colour of the matrix can be accounted for by the dark to medium grey dolostone fragments, occasional shale fragments, and the finely disseminated pyrite. The sulphides, i.e. sphalerite, galena and

FIGURE 7. **(A)** Crackle breccia with sphalerite crystals in white dolomite cement. **(B)** White sparry dolomite, sphalerite, and galena filling fractures within crackle breccia. **(C)** Mosaic breccia with sphalerite crystals (red) in white sparry dolomite cement. **(D)** Mosaic breccia with fine-grained pyritic mud supporting the dolostone fragments. **(E)** Rubble breccia consisting of variably altered dolostone fragments and shale fragments in white dolomite cement. **(F)** Rubble breccia with a dolomite cement passing into a dark grey fragmental matrix in the upper part of the photo. Abbreviations: Dol=dolomite, Sp=sphalerite, Gn=galena, and Py=pyritic mud.

pyrite, are concentrated in the matrix as individual grains and clusters, and as fracture filling material.

The spatial relationships between these breccias are illustrated on Figure 5. The rock-matrix (trash) breccias crosscut the dolomite-cemented mosaic and crackle breccias, and locally the dolomite cement of the crackle and mosaic breccias enclose fragments of the rock-matrix (trash) breccia. We interpret this to mean that these breccias originated as an overlapping, multi-episodic sequence of events prior to and during mineralization. The incorporation of sulphide clasts in the rock-matrix (trash) breccias is evidence that at least some brecciation and mineralization were contemporaneous.

Dimensions

Dimensions of mineralized bodies within most deposits of the Robb Lake district are not known or have not been recorded. At Robb Lake, several bodies extend for more than 300 m along bedding and crosscut more than 50 m of stratigraphic section; others are thin and narrow bodies and pods parallel to bedding (Paradis et al., 1999).

FIGURE 8. Various photos of the rock-matrix (trash) breccias. **(A)** Dolostone (D), shale (S), and white sparry dolomite (Dol) fragments in dark grey fragmental matrix. **(B)** Subrounded white sparry dolomite fragments and angular thin and elongated white sparry dolomite (Dol) fragments in a dark grey fragmental matrix. Note the pyrite (Py) aggregates replacing the sparry dolomite fragments and the matrix. **(C)** Various fragment types and shapes in a fine-grained fragmental matrix. **(D)** Dolostone fragments with zebra texture (Z) in the rock-matrix breccia. **(E)** Rock-matrix (trash) breccias with intact and broken fossil fragments (F) in the matrix. Note the intact gastropod (G) in the middle of the photo. **(F)** Dolostone, shale, and white sparry dolomite fragments in dark grey fragmental matrix. Note the dolostone fragments with zebra texture and the fine elongated white dolomite fragments.

Host Rocks

The deposits are hosted in carbonate rocks, usually dolostone and less commonly limestone. The dolostone consists of medium- to coarse-grained white sparry dolomite that has replaced a fine-grained dolostone host, which itself has replaced a limestone host.

Most of the Zn-Pb deposits are found in Silurian to Middle Devonian platform dolostones of the Muncho-McConnell and the Dunedin formations that form part of the outcropping Paleozoic carbonate platform (Thompson, 1989). Only three deposits, Opiska, WL, and Marini 1, are found in the basinal facies of the Ordovician Skoki Formation and the Ordovician to Silurian Road River Group (Table 2, Appendix 1).

The main stratigraphic units encountered in the Zn-Pb deposit areas are described below. They are divided into two disparate stratigraphic sequences (basinal and platformal units) separated by major thrust faults. Basinal units include

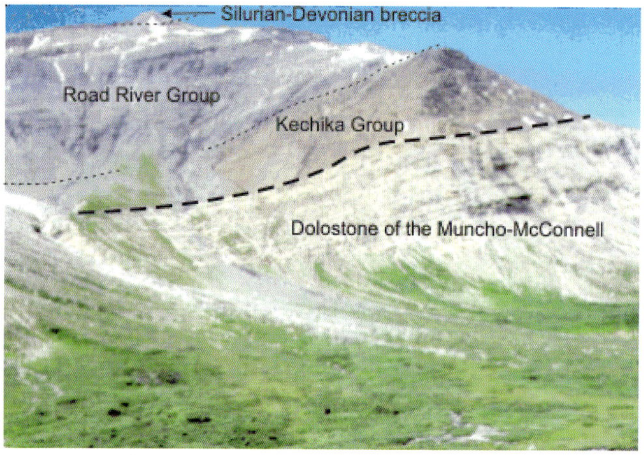

FIGURE 9. View looking west of the mountain slope immediately west of Mississippi Creek, showing structural succession. A major thrust fault, the Mississippi Creek Fault (heavy dashed line), exists between the dolostone of the Muncho-McConnell Formation and rocks of the Kechika Group.

the Cambrian-Ordovician Kechika Group, the Ordovician Skoki Formation, the Ordovician-Devonian Road River Group, and unnamed units of Silurian-Devonian quartzite-dolostone and Silurian dolostone breccia. The platformal succession includes the Silurian Nonda Formation, Silurian-Devonian Muncho-McConnell Formation, undivided Lower to Middle Devonian Stone and Dunedin formations, and Upper Devonian Besa River Formation.

Kechika Group. Cambrian to Ordovician rocks of the Kechika Group consist of grey, orange-brown to olive weathered calcareous shale and slate characterized by distinct colour laminations in the millimetre to centimetre range. Rocks are well cleaved, forming a papery to slabby cleavage. In the Mississippi Creek area of the Robb Lake deposit, they are overlain by rocks of the Road River Group (Fig. 9).

Skoki Formation. The Ordovician Skoki Formation is a thick dolostone succession that can be traced more or less continuously from the western Main Ranges of the southern Rocky Mountains (Norford, 1969) to the north into the Ware (94F) and the Trutch (94G) map areas (Thompson, 1989). In the Robb Lake deposit area, the Skoki Formation, which outcrops on the southwest wall of the cirque that heads Mississippi Creek, is subdivided into two distinct units. The lower part of the Skoki consists of 70 m of thickly bedded, cliff-forming dolostone. It is light to medium grey, muddy and carbonaceous. The upper Skoki is a recessive unit of light brown to chalky orange weathering, light to medium grey, thinly bedded silty dolostone. Beds are up to 50 cm thick. This unit contains bivalves, abundant burrow structures, and rare trilobites.

Road River Group. Road River strata consist of a thick succession of off-shelf clastic rocks of Ordovician to Silurian age (Gordey and Anderson, 1993; Ferri et al., 1999). The group makes up the Laurier Anticlinorium between Halfway River (latitude 56°N) and Lady Laurier Lake (latitude 56°43'N, longitude 123°45'W), and forms parts of the more westerly thrust sheets between Balden Creek (approximate latitude 56°35'N) and the outlet of Wicked River (approximate latitude 56°05'N). In the Robb Lake deposit area, the Road River Group is dominated by non-fossiliferous flaggy slate and calcareous slate and well cleaved, dark grey to black carbonaceous limestone. Less abundant are white quartzite and thickly bedded, highly fossiliferous carbonaceous limestone densely packed with fragments of corals and brachiopods. This highly heterogeneous section is typical of the Road River Group near shelf margins. It is overlain by an unnamed Silurian-Devonian dolostone and quartzite unit in the lower of the two thrust sheets, and by the Silurian breccia in the upper one.

Silurian breccia. The Silurian breccia is a thin, cliff-forming dolostone breccia that extends for 30 km from Lady Laurier Lake (south of Robb Lake) to Mt. Kenny immediately west of Robb Lake (Thompson, 1989). The breccia overlies the Road River Group on the ridge tops at the headwaters of Mississippi Creek (Fig. 9). Breccia beds consist of medium to dark grey, angular to subangular dolostone and black chert fragments in grey dolomite cement. They also contain abundant and well preserved fossil fragments of halysitid, favositid, and colonial corals. The dolostone fragments exhibit grey-black colour laminations on the millimetre to centimetre scale. The breccia itself is bedded on the centimetre to 10 metre scale; sets of thin dolostone calcilutite and calcarenite, and small-clast breccia alternate with thick units of unsorted, matrix-supported megabreccia. According to Thompson (1989), the fragments are all derived from the Nonda Formation, and the unit is interpreted as a debris flow (or succession of debris flows) deposited on the foreslope of a Nonda reef.

Silurian-Devonian dolostone and quartzite. This unnamed unit lies in exposed stratigraphic contact above the Road River Group on the ridge southwest of Mississippi Creek in the vicinity of the Robb Lake deposit. It consists of a basal thick-bedded grey quartzite, overlain by light grey thick-bedded dolostone and quartzite. No fossils were seen. It is correlated with the Silurian-Devonian shelf dolostone-sandstone units farther northeast.

Nonda Formation. The Nonda Formation is a distinctive medium grey to dark grey weathering, fossiliferous dolostone of Early Silurian age. The formation is 200 to 660 m thick and can be traced from northern Pine Pass map area (93O) to the British Columbia-Yukon border (Thompson, 1989). Fossils include crinoids, and halysitid and favositid corals. The alternating medium to dark grey coloured beds make the Nonda Formation unmistakable. Bedding thickness ranges from 0.15 to 2 m. Locally, fine calcareous laminations are observed. Chert nodules and silicified fossils are common throughout the unit. The Nonda Formation exposed at the Robb Lake deposit forms the core of the Robb Anticline.

Muncho-McConnell Formation. The Silurian-Devonian Muncho-McConnell Formation forms most of the high peaks adjacent to the mountain front (Thompson, 1989). It consists almost entirely of light grey, weathering-resistant dolostone and sandy dolostone, except for a few beds of dark grey micritic limestone. The Muncho-McConnell is over 500 m thick in the Robb Lake deposit area, and consists of fine-grained crystalline dolomite that formed largely in a low energy, intertidal to supratidal environment. The Muncho-McConnell Formation can be subdivided into upper and lower units based on bedding thickness, and outcrop and bedding characters. The lower unit is a light to medium grey, thick- to medium-bedded dolostone approximately 250 m thick. On closer inspection, the lower unit consists of a cyclic repetition of thick dolostone beds, and thinner, often laminated and more recessive dolostone beds. A sandstone unit <50 m thick is present at the base of the lower unit. A marker bed, known as the "angular sand marker" (ASM), is present within the dolostone sequence at about 110 m below the top of the unit (Manns, 1981). It corresponds to a local disconformity. The uppermost 30 to 50 m of the lower unit are particularly thick-bedded and often form major cliffs. On Tennessee Mountain (i.e. near the North Face and Camp showings of the Robb Lake deposit), parts of this upper section consist of thick-bedded to very thick-bedded limestone pods that pass laterally and vertically into the typical thick dolostone beds. Sedimentary textures within the lower unit include abundant fenestrae, less common cryptalgal laminites, burrows, and rare intervals that show current laminations and rip-up clast breccias. Large brachiopods (average 5 to 10 cm long) are common within the thick dolostone beds of the upper half of the unit. Gastropods, amphiporids, stromatolites, fragments of bivalves, and other unidentified fossils were also observed in the lower unit. Taylor (1977) linked the presence of fossils within the Muncho-McConnell near Robb Lake to the proximity of the Silurian-Devonian facies front. The upper unit is 250 to 300 m thick, and consists of thick-to medium-bedded light to medium grey dolostone that alternates with thin-bedded medium grey dolostone in a characteristic banded or stepped outcrop pattern. The rhythmic sequence of the upper unit is better developed than in the lower unit, and the thin bedded dolostone intervals are more extensive. The basal sequence (0-25 m) of the upper unit is composed of laminated dark grey silty dolomite mudstone with local sand concentrations. A marker bed, known as the "pale sand marker" (PSM; 1-2 m thick), was identified in drillholes in Webb Ridge (Manns, 1981). Fragments of limestone beds are observed at the base of the mineralized breccia at the T.G.S. showing of the Robb Lake deposit. Taylor (1977) and Manns (1981) put a disconformity at the top of the lower unit and at the top of the upper unit. These disconformities indicate the presence of significant erosion surfaces, as marked by the presence of sand markers (ASM and PSM), scour surfaces, and soft sediment deformation features.

Stone and Dunedin formations. The Lower to Middle Devonian Stone Formation consists of a lower member of striped light grey and brownish grey dolostone and sandy dolostone, and an upper member of massive, light grey to whitish dolostone and sandy dolostone (Thompson, 1989). The overlying strata of the Dunedin Formation exhibit rapid lateral changes in thickness and facies and are exposed discontinuously across the Halfway River map area. Where it can be distinguished, the Dunedin Formation consists of dark grey, medium- to thick-bedded, fine-crystalline dolostone, wackestone, and grainstone in which stromatoporoids and crinoid ossicle fragments are common (Thompson, 1989). The Stone and Dunedin formations could not be distinguished in the Robb Lake deposit area, so they are described together in this section. They consist of interbedded medium to dark grey fossiliferous, partly calcareous dolostone and light to medium grey-buff weathering, fossiliferous and non-fossiliferous dolostone. Some fossiliferous limestone is black (reef-front?) breccia. Fossils include brachiopods, gastropods, crinoids, and rugose and colonial corals.

Besa River Formation. The Upper Devonian Besa River Formation consists of black weathering, calcareous and non-calcareous shale, silty carbonaceous limestone, and brown weathering, argillaceous siltstone and mudstone. It outcrops in three regions of the Robb Lake deposit area: in the core of a syncline east of the main showing area at Robb Lake, in the footwall of the Sidenius thrust fault in the core of the Robb Anticline, and in a narrow band in the immediate footwall of the major thrust fault in the headwaters of Mississippi Creek (Fig. 3). In its eastern exposures, the Besa River Formation consists of soft brown-grey weathering, finely laminated dark grey to black calcareous to noncalcareous argillite/shale and siltstone. In its western exposure, it consists of dark grey to black, siliceous argillite and shale with silt laminae.

Mineralogy and Textures

MVT deposits in the district have simple sulphide assemblages that consist of sphalerite, galena, pyrite, and marcasite. Gangue minerals are dolomite, calcite, and quartz; barite, bitumen, and fluorite are less common. There are a variety of accessory minerals present in some deposits, but absent in most. For example, smithsonite, hemimorphite, hydrozincite, hematite, and limonite have been reported in a few deposits. Organic material, such as pyrobitumen, is common, though volumetrically small, in most deposits. It occurs within various breccia types where it fills small cavities and fractures. It potentially also forms a component of the rock-matrix (trash) breccia, giving the dark grey colour to the matrix.

Sulphide textures. Taylor et al. (1975) provided initial published observations and interpretations of the Robb Lake mineralized breccias; recent observations and interpretations were provided by Paradis et al. (1999), Nelson et al. (1999), and Nelson et al. (2002).

Sulphide textures are mostly related to open-space filling of breccias, fractures, and vugs (Fig. 10). Replacement of carbonate host rocks and internal sediments, and sulphide disseminations are also observed (e.g. Robb Lake deposit). Sulphides and white coarse-grained and saddle dolomites constitute the cement of the breccias, fractures, and vugs. The sulphides are massive, disseminated, and banded. Massive sulphides are found in replacement zones of the carbonate host rocks where aggregates of sphalerite (and lesser amounts of pyrite and galena) replaced the host carbonates. Massive replacement mineralization is generally not common in deposits of the Robb Lake district. Disseminated sulphides are more common and occur as fine to coarse crystals of yellow, dark orange, red, or brown sphalerite and galena overlain by, or intergrown with, white sparry dolomite cement.

At the Robb Lake deposit, sphalerite occurs as fine to coarse single crystals or aggregates of anhedral crystals (1 mm to 1 cm) in white sparry dolomite cement of the crackle, mosaic, and rubble breccia, and in the fine-grained carbonaceous dolostone matrix of the rock-matrix (trash) breccia. Anhedral crystals or aggregates of crystals of spha-

FIGURE 10. (A) Reddish sphalerite crystals and white sparry dolomite cement fill fractures of the crackle breccia. (B) Sphalerite partially surrounding dolostone fragments. (C) Disseminated sphalerite crystals in white sparry dolomite cement of a mosaic breccia. (D) Aggregates of massive yellowish sphalerite crystals and white sparry dolomite cement replacing the dolostone; scale = 5 cm. Abbreviations: Dol=dolomite, Sp=sphalerite, and Gn=galena.

lerite occur as rims on one or several sides of the angular dolostone fragments, or as scattered crystals within the white sparry dolomite cement. According to Manns (1981), geopetal sphalerite (i.e. snow-on-the-roof texture) is abundant at Robb Lake; we observed sphalerite lining the bottoms of cavities and/or coating indiscriminately any or all sides of fragments. Sphalerite crystals and aggregates range from less than 1 mm to 1.5 cm. Very fine-grained sphalerite (i.e. <1 mm) is only present in the matrix of the rock-matrix (trash) breccia. Sphalerite crystals are usually anhedral and are commonly fractured. Colloform sphalerite has been reported at the Waterfall showing (Macqueen and Thompson, 1978; Manns, 1981). Overall, galena is less abundant than sphalerite and pyrite. It commonly occurs as euhedral crystals (<1cm to 2 cm) and less commonly as anhedral aggregates in the white sparry dolomite cement of the rubble, mosaic, and crackle breccias, and in the carbonaceous dolomite matrix of the rock-matrix (trash) breccia. Galena also occurs as fracture fillings with or without sphalerite and quartz. Sangster and Carrière (1991) observed galena veinlets crosscutting sphalerite crystals and grains of sphalerite completely enclosed by galena. In showings located in the lower unit of the Muncho-McConnell, galena is typically associated with sphalerite, and may have co-precipitated or precipitated slightly later. In the upper unit, galena frequently occurs by itself, as euhedral and anhedral crystals disseminated in the white sparry dolomite, and is often associated with quartz. When quartz is present and associated with galena, it occurs as fractures cutting galena or as inclusions in galena (Sangster and Carrière, 1991). Pyrite occurs as disseminated fine-grained euhedral to subhedral crystals and aggregates in the rock-matrix (trash) breccia, and as massive aggregates of fine-grained pyrite along stylolites, bedding planes, and fracture fillings.

Mineral zonation. A broad mineral zonation is recognized in the distribution of the sulphides and other minerals at the Robb Lake deposit, but has not been recognized elsewhere within the district. Sphalerite and galena are preferentially concentrated in the lower unit with approximate ratios of zinc to lead of 7:1 at the Lower showing, 1.5:1 at the West Webb showing, and 4.5:1 at the East Webb showing (Manns, 1981). Galena becomes more abundant in breccia zones located stratigraphically up section. It occurs without sphalerite in dolomite-drusy quartz veins high on the southeast ridge of Tennessee Mountain, and in the dolomite matrix of a breccia peripheral to a sphalerite-rich trash breccia in the North Face showing. In general, the galena-sphalerite ratio appears to increase in higher and more distal parts of the system. Drusy, crystalline quartz (Fig. 11A) is more distal yet. It occurs as extensions to dolomite-galena veins on the southeast ridge of Tennessee Mountain, and is abundant elsewhere within open vugs and fractures in the upper unit. The best grade mineralization (hole 113-82 intersected 15.02% Pb+Zn over 3 m, Boronowski and James, 1982), which could reach a Zn:Pb ratio of 2:1, is predominantly associated with the pyrobitumen- and carbonaceous-rich matrix of the rock-matrix (trash) breccias.

Host Rock and Mineral Deposit Chemistry

Limited analytical chemistry has been done on the deposits within the Robb Lake district. Macqueen and Thompson (1978) reported analyses of sulphide isotopes on various deposits and metal content of the Nonda, Stone, and Muncho-McConnell formations. Nelson et al. (2002) and

TABLE 3. Carbon, oxygen, and strontium isotope values of various carbonate types from the Robb Lake deposit.

Sample #	Orebody/Zone	Description of sample	Carbonate Type	$\delta^{13}C_{PDB}$	$\delta^{18}O_{PDB}$	$\delta^{18}O_{SMOW}$	$^{87}Sr/^{86}Sr$
4-5	North face Zone	Limestone	Lim	1.28	-10.97	19.61	0.70917
6-1	Camp Zone	Fine-grained dolostone	FCD	0.76	-10.12	20.48	0.70974
72-4(1)	Downstream Zone	Fine-grained dolostone	FCD	1.58	-10.36	20.23	
11-2	Canyon Zone	Coarse-grained white sparry dolomite in concordant mineralized breccia	CCD	-0.86	-13.81	16.68	0.71444
11-4	Canyon Zone	Coarse-grained white sparry dolomite	CCD	-1.37	-14.83	15.62	0.71267
1B-5	Lower Zone	Coarse-grained white sparry dolomite in rubble mineralized breccia	CCD	-1.12	-14.32	16.15	0.71342
2-2(2)	Upper Zone	Coarse-grained sparry dolomite in mosaic mineralized breccia	CCD	-1.49	-15.39	15.04	0.71779
3-5	Waterfall Zone	Coarse-grained white sparry dolomite in mosaic mineralized breccia	CCD	-1.66	-14.93	15.52	0.71199
72-4(2)	Downstream Zone	Coarse-grained white sparry dolomite in vein	CCD	-1.64	-15.58	14.85	
73RGV-38	Webb Ridge Zone	Coarse-grained white sparry dolomite	CCD	-1.47	-15.08	15.37	0.71181
7-5	TGS Zone	Coarse-grained white sparry dolomite in rubble mineralized breccia	CCD	-1.43	-14.98	15.47	0.71550
7-6	TGS Zone	Coarse-grained white sparry dolomite in crackle mineralized breccia	CCD	-0.80	-14.22	16.25	
84-80	Webb Ridge Zone	Coarse-grainedwhite sparry dolomite in crackle mineralized breccia	CCD	-1.48	-15.20	15.24	
98JN3-1(1)	Canyon Zone	White sparry dolomite in rubble mineralized breccia	CCD	-1.19	-15.18	15.27	

[1] Data are from Nelson et al. (2002).

All Sr isotopic data are presented relative to a value of 0.710245 for the NIST Sr isotopic standard SRM987; data are corrected for variable mass fractionation to a value of 0.1194 for $^{86}Sr/^{88}Sr$. Uncertainty in $^{87}Sr/^{86}Sr$ is expressed as ±2 sigma of the mean (2sm), equivalent to 2 standard errors (2SE).

All results for carbon and oxygen isotopes are reported in the usual permil notation relative to the PDB and SMOW standards; precision and accuracy is ±0.2 permil for $\delta^{13}C$ and $\delta^{18}O$, respectively

CCD = coarse-crystalline dolomite, FCD = Fine-crystalline dolomite, Lim = limestone.

Paradis et al. (2006) reported stable and radiogenic isotope analyses of sulphides and carbonate gangue minerals from the Robb Lake deposit (Table 3) and other sulphide occurrences within the WCSB. Location and description of the Robb Lake samples and the analytical methods can be found in Nelson et al. (2002) and Paradis et al. (2006). Results on the Robb Lake samples are discussed below.

Carbon, oxygen, and strontium isotopes. Carbon, oxygen, and strontium isotopes have been measured on pre-mineralization carbonate lithofacies and main-mineralization sparry dolomite cement of the Robb Lake deposit. The pre-mineralization host dolostone and limestone and the main-mineralization sparry dolomite cement at Robb Lake form two separate populations. Pre-mineralization dolostone and limestone have $\delta^{18}O_{PDB}$ values of -10.1‰, -10.4‰, and -11.0‰, and $\delta^{13}C_{PDB}$ values of 0.8‰, 1.6‰, and 1.3‰. The main-mineralization sparry dolomite cement have $\delta^{18}O_{PDB}$ values ranging from -15.6 to -13.8‰ and $\delta^{13}C_{PDB}$ values ranging from -1.7 to -0.8‰. These values are lower than the host dolostone and limestone, and the Middle Devonian marine carbonate values (Hurley and Lohmann, 1989). $\delta^{13}C_{PDB}$ values reflect slight ^{12}C-enriched CO_2 produced by organic matter oxidation related to hydrocarbon generation. Low $\delta^{18}O_{PDB}$ values point to the involvement of hydrothermal fluids.

The $^{87}Sr/^{86}Sr$ ratio of the host limestone is 0.7092, and that of the host dolostone is slightly higher at 0.7097. Both are higher than Devonian seawater ratios, which range from 0.7078 to 0.7083 (Burke et al., 1982). This indicates that the host carbonates at Robb Lake have undergone alteration during diagenesis, probably by modified seawater containing slightly more radiogenic strontium derived from compaction of adjacent or underlying siliciclastic rocks. Significantly radiogenic strontium is found in the main-mineralization sparry dolomite cement (0.7118 to 0.7178). This implies that "radiogenic" strontium (i.e. ^{87}Sr) was preferentially introduced into the carbonates at Robb Lake during dolomitization and mineralization. Solutions with strontium isotope ratios in excess of 0.7100 can develop through the interaction of deeply circulating fluids with fine-grained siliciclastic rocks (Morrow et al., 1990), which contain alkali feldspars rich in radiogenic strontium. Low $\delta^{18}O$ values

FIGURE 11. (A) Open space crystalline quartz from vein occurrences in the Muncho-McConnell Formation near the summit plateau of Tennessee Mountain. **(B)** Selective replacement of bedding-parallel bands of white sparry dolomite forms a zebra texture.

along with high $^{87}Sr/^{86}Sr$ ratios in dolomite cement associated with sulphides at Robb Lake, and the location of Robb Lake within the Rocky Mountains can be ascribed, respectively, to high temperature fluids and the influence of siliciclastic sources.

Fluid inclusions. Fluid inclusion studies have been done on the white sparry dolomite and sphalerite from the Robb Lake deposit (Sangster and Carrière, 1991). Fluid inclusions in sphalerite and dolomite are typically 5 to 20 μm in size and are aqueous two-phase inclusions at room temperature. Sphalerite contains inclusions having moderately elevated ThL (homogenization temperature to liquid phase) values of 87 to 154°C, averaging 119°C. Ice melting temperatures (Tm_{ice}) indicate high salinity (16 to >23 wt.% NaCl equiv.) $NaCl\text{-}CaCl_2$ fluids.

Lead isotopes. Lead isotope analyses of sulphide samples from the Robb Lake district are highly radiogenic and fall along the linear Cordilleran carbonate lead trend (Morrow and Cumming, 1982; Nelson et al., 2002). The Cordilleran carbonate lead trend is characteristic of the northern Rocky Mountains MVT belt. Highly radiogenic leads of this character are found within the western Presqu'ile Barrier and the Peace River Arch area of the WCSB (Paradis et al., 2006). This linear trend intersects the shale curve of Godwin and Sinclair (1982) in the Devonian-Mississippian range, and overlaps the field for Devonian SEDEX deposits of the Selwyn Basin and Kechika Trough.

The Cordilleran carbonate lead trend is consistent with northeasterly fluid flow paths from the miogeocline into the Devonian carbonate aquifers of the WCSB. It is similar to the lead isotopic trends shown by classic MVT deposits of the central United States, which are interpreted as mixing lines between ordinary upper crustal lead and an anomalous lead that is far more radiogenic than any known average crustal source (Heyl et al., 1974). Many authors have advocated regional fluid flow in sandstone aquifers as the source of lead in carbonate-hosted deposits (cf. Goldhaber et al., 1995). Others have suggested the underlying Precambrian basement as the source for lead. With decreasing fluid temperatures eastward, away from the Rocky Mountain front, the fluids may have become much more enriched in radiogenic lead due to preferential extraction of radiogenic lead from metamict zircons rather than equilibrium extraction of combined radiogenic and common leads from feldspars (cf. Nelson et al., 2002). The Cordilleran carbonate lead trend therefore appears to define a mixing line between the "least" radiogenic values represented by the Devonian SEDEX deposits and an extremely radiogenic end-member of unknown origin (Nelson et al., 2002; Paradis et al., 2006).

Alteration

Megascopically, three distinct phases of dolomitization affected the carbonate rocks in the Robb Lake district. An early phase consists of a light grey weathering, medium and dark grey replacive dolostone composed of fine-grained crystalline dolomite. This dolomite dominates the sequence regionally and is considered to be the result of very early diagenetic alteration of primary calcite. A later, coarse sparry and prismatic white to grey dolomite coats some dolostone fragments and is associated with so-called zebra texture (Fig. 11B). Zebra texture is a common feature of carbonate rocks associated with MVT deposits. It consists of thin dolostone layers (mm to cm scale) identical to, and connecting with, the surrounding homogeneous dolostone, but separated from each other by coarse sparry and prismatic white dolomite layers (mm to cm scale) or, in some cases, by voids of the same shape as the white layers, by pyrobitumen, or by a combination of the three. The layering is either oriented parallel to bedding or in herringbone patterns that resemble cross bedding. It never extends continuously either vertically or laterally for more than 30 cm or so. Zebra texture affects all the dolostone units of the Robb Lake area, and fragments of the zebra texture are incorporated in the mineralized breccias (see Fig. 8D,F). The third dolomite in the Robb Lake area is a white sparry dolomite that forms the cement of the breccias, and fills the joints, fractures, vugs, and primary porosity such as fenestrae and fossil interiors (Figs. 7A, C, E, F, and 10A, B, C, D).

With the help of cathodoluminescence petrography, Manns (1981) was able to observe several generations of dolomite: (1) fine-grained inert dolomite forming the regional host-rock dolostone, (2) inert coarse white sparry and prismatic dolomite lining cavities, coating dolostone fragments in breccias, and associated with the zebra texture, (3) luminescent dolomite forming the cement of the mineralized breccias and associated with sphalerite, galena, pyrite, and local pyrobitumen, and (4) inert dolomite filling the central portions of large cavities.

Structures

The primary control on the location of mineralization within the Robb Lake district is related to open-space filling of breccias, fractures, and vugs. Brecciation and open-space filling are therefore important processes in the formation of the Robb Lake deposits, as in most MVT deposits. However, the temporal and genetic relationships between brecciation and mineralization at the Robb Lake deposit are controversial (Macqueen and Thompson, 1978; Nelson et al., 1999; Paradis et al., 1999). Are the breccias that host mineralization the result of wall-rock dissolution by MVT mineralizing brines or were they formed by other, earlier fluids, possibly related to regional weathering and karstification, which created an extensive paleoaquifer system for later MVT brines (Sangster, 1988)? Did alteration and dissolution of carbonate wall rock in the breccias cause sulphide deposition? These questions remain unanswered at Robb Lake, as indeed at other MVT deposits around the world.

In some MVT districts, it is documented that solution of carbonate by ore-related brines has allowed collapse and brecciation of the overlying beds, and that gravity was an important force in the shattering. However, the fine median size and high degree of angularity of the fragments within the trash breccias and of the breccia bodies at Robb Lake suggest that other forces, such as hydraulic fracturing (Macqueen and Thompson, 1978), may have supplemented simple collapse.

Age of Mineralization

MVT Zn-Pb deposits in the northern Canadian Rocky Mountains and the WCSB have been linked alternatively to the Late Devonian SEDEX deposits of the Selwyn Basin and the Kechika Trough (Sangster, 1990; MacIntyre, 1991; Goodfellow et al., 1993) or to Cretaceous-Tertiary regional fluid flow related to the Laramide Orogeny (Garven, 1985; Symons et al., 1993; Symons et al., 1998; Smethurst et al., 1999). Knowledge of absolute age of MVT mineralization is important for the understanding of genetic relationships between Zn-Pb mineralization and regional- to global-scale tectonic events.

Dating by paleomagnetism of the Robb Lake and Pine Point deposits indicates that sulphide mineralization is Eocene in age at the 47 ± 10 Ma Robb Lake deposit (Smethurst et al., 1999) and formed between mid Late Cretaceous and Paleocene at the 71 ± 13 Ma Pine Point deposit (Symons et al., 1993). In contrast, radiometric dating by Rb-Sr sphalerite isochron at Pine Point gives Middle Devonian ages of 361 ± 13 Ma (Nakai et al., 1993) and 374 ± 21 Ma (Brannon et al., 1995). Radiometric dating by Rb-Sr sphalerite isochron at Robb Lake has been attempted by Nelson et al. (2002). Eight sphalerite samples were collected from the Lower, Upper, Webb Ridge, and Canyon showings. Analytical procedures and information on the samples can be found in Nelson et al. (2002). The combined sphalerite and fluid inclusion analyses from the Robb Lake samples do not form a single isochron with a well defined age on a Rb-Sr diagram (see Fig. 6 of Nelson et al., 2002). The data show considerable scatter due to inhomogeneity in the initial $^{87}Sr/^{86}Sr$ ratios of the metal-bearing fluids. However, three of the Robb Lake sphalerite residues fall on the Pine Point isochron and give a calculated apparent age of 350 ± 100 Ma. And the Rb-Sr isotope data for Robb Lake sphalerite and fluid inclusions form a steep array that points down toward the fluid inclusion data for Pine Point, which form a well constrained linear isochron at 362 ± 9 Ma (Nakai et al., 1993). This relationship suggests that the Robb Lake fluid inclusions represent a mixture of two populations of metal-precipitating fluids, one end-member fluid with values of $^{87}Sr/^{86}Sr$ similar to the Pine Point ore fluid and the other containing highly radiogenic strontium. The results of the radiometric dating by Rb-Sr sphalerite isochron at Robb Lake, although not conclusive, suggest that the sulphide mineralization is of Paleozoic age. Significantly, the data set does not lend itself to interpretations of a Cretaceous-Tertiary age for mineralization.

Genetic Models

The origin of the enigmatic MVT deposits have been the subject of a number of reviews, including Jackson and Beales (1967), Anderson and Macqueen (1988), Leach and Sangster (1993), Paradis et al. (2006), and others. Recent advances in understanding large-scale fluid flow in the crust, coupled with new radiogenic, geochemical, and geological studies of MVT districts, have established that most MVT mineral districts are the products of regional- or sub-continental-scale hydrological processes. It is believed that deposits formed from hot to warm, saline, aqueous solutions (similar to oil-field brines) that migrated out of sedimentary basins, through aquifers, to the basin periphery and into the platform carbonate sequences. To effect this movement of ore-bearing brines, at least three different fluid-flow processes have been proposed:

- The topographic or gravity-driven fluid-flow model (Garven and Freeze, 1984; Garven, 1985; Bethke and Marshak, 1990; Garven and Raffensperger, 1997; Yao and Demicco, 1997).
- The sedimentary and tectonic compaction model (Jackson and Beales, 1967; Sharp, 1978; Cathles and Smith, 1983; Bethke, 1985; Oliver, 1986).
- The Hydrothermal Convection Model (Morrow, 1998).

The first model involves flushing of subsurface brines out of a sedimentary basin by groundwater flow from recharge areas in elevated regions of a foreland basin to discharge areas in lower elevated regions during the Cretaceous-Tertiary Laramide Orogeny (Garven and Freeze, 1984; Garven, 1985; Bethke and Marshak, 1990; Garven and Raffensperger, 1997). In this model, groundwater migrates through shale units into the underlying carbonate units by cross-formational fluid flow. Considerable geologic evidence listed by Leach and Sangster (1993) support this model. The model has been proposed for several MVT districts in the world, particularly those of the US mid-continent and the Pine Point district in Canada.

The second model considers that sedimentary and/or tectonic compaction of sediments in a subsiding basin drives a continuous outward flow of pore fluids laterally along aquifers (Jackson and Beales, 1967). This model has also been applied to the Pine Point district. However, maintaining sufficient rates of discharge in order to maintain high initial fluid temperatures during transport of up to hundreds of kilometres from basin source to platform depositional site is difficult. A variation of this model, known as episodic outward flow, was therefore proposed by Cathles and Smith (1983). This revised model involves overpressuring of subsurface aquifers by rapid sedimentation, followed by rapid and episodic release of basinal fluids. Another variation of the second model involves tectonic loading and compression of sediments during the development of orogenic thrust belts, which may cause the rapid expulsion of formational fluids outward into the foreland basins with the thrust belts behaving like giant squeegees (Oliver, 1986).

The third model involves deep convection circulation of hydrothermal brines due to buoyancy forces related to temperatures and salinity variations (see Fig. 6 of Anderson and Macqueen, 1988; Morrow, 1998). The model supports long-lived flow systems that are capable of recycling subsurface solutions many times through the rock mass. This model has been invoked to explain the regional hydrothermal dolomitization in the WCSB (Morrow, 1998; Coniglio et al., 2006), Manetoe facies of the Northwest Territories (Aulstead et al., 1988; Morrow et al., 1990), Ordovician gas-producing carbonates of the Michigan Basin (Coniglio et al., 1994), and MVT deposits of the northern Canadian Rocky Mountains (Paradis and Nelson, 2000; Nelson et al., 2002).

In the latter case, Devonian-Mississippian Rb-Sr ages for the sphalerite from the Robb Lake and Pine Point carbonate-hosted Zn-Pb mineralization (Nakai et al., 1993; Nelson et al., 2002) suggest that the deposits formed during Devonian-Mississippian continental tectonic activity on the western margin of North America (Paradis and Nelson, 2000; Nelson et al., 2002). High heat flow and the development of deep hydrothermal convection systems were generated in an actively rifting back-arc setting eastward of an easterly dipping subduction zone that evolved along the western continental margin in early Devonian time, creating the arc and back-arc volcanic-plutonic-sedimentary complexes of the Canadian Cordillera. These arc and back-arc systems were affected by episodic rifting, which generated hydrothermal activity and formation of VHMS deposits in the outermost continental margin (i.e. the pericratonic terranes), SEDEX deposits in a more inland belt of continent-margin basins (i.e. Kechika Trough and Selwyn Basin), and MVT deposits on the continental platform. The carbonate-hosted Zn-Pb mineralization and hydrothermal dolomites of the Rocky Mountains and WCSB (north of the Peace River Arch) are interpreted as the products of this far-reaching subduction effects. The genesis of hydrothermal activity resulted in fluids being driven along reactivated back-arc structures, faults, and permeable strata (Paradis and Nelson, 2000; Nelson et al., 2002; Paradis et al., 2006).

Goodfellow et al. (1993) have also argued for a connection between northern Cordilleran SEDEX and MVT deposits, based on the spatial association of MVT deposits with the rifted margins of an anoxic basin, the similarity of metal ratios, and the similar heavy and variable sulphur isotope values. They envisaged a mixing model in which reduced, sulphur-bearing fluids circulating in porous and permeable carbonates encountered metal-bearing fluids that rose along basement structures to the site of ore deposition. Major fault systems, such as the McDonald-Hay River Fault, Rabbit Lake Fault Zone, and the Hay River Fault Zone, are likely candidates for this role at Pine Point. The Robb Lake MVT district lies approximately on the extrapolated southwest extension of the Middle Devonian Slave Point Formation (which hosts Zn-Pb mineralization at Pine Point), and near the subsurface projection of the McDonald-Hay River Fault (Fig. 2). Also, as noted by Macqueen and Thompson (1978), Thompson (1989), and by us, the deposits within the Robb Lake district are located next to the tectonically compressed shelf-slope facies boundary, and the carbonate platformal strata that host the Robb Lake deposit form the immediate footwall of a major thrust fault, the Mississippi Creek Fault, which has deep water, early Paleozoic basinal strata in its hanging wall. This outlines the close relationships of the Robb Lake MVT district to large-scale tectonic features.

At the deposit scale, Zn-Pb mineralization occurs as breccia-, vein-, and fracture-fillings, with breccias being the most important. The origin and timing of the various types of breccias versus mineralization are controversial. Taylor (1977) suggested that the formation of the mineralized breccias was due to intrastratal karst and that the mineralization occurred subsequent to brecciation from formation waters that were derived from compacting shales. Manns (1981) concluded that the host breccias were collapse structures due to solution of evaporitic dolostone platform at an unknown time prior to the Laramide Orogeny. Sangster (1973, 1979) and Sangster and Lancaster (1976) favoured a paleokarst solution-collapse origin, linked to an unconformity within and above the Silurian-Devonian carbonate sequence. This interpretation does not explain the presence of Zn-Pb deposits within the Stone and Dunedin formations, and other features/textures related to karsting and solution-collapse breccias, which are absent at Robb Lake (Taylor et al., 1975). These include intrakarst sediments, sagging and collapse of the overlying strata, and open-space sulphide textures such as stalactites and sulphide-carbonate laminites (Nelson et al., 1999). Macqueen and Thompson (1978) rejected all these hypotheses, and ascribed the breccias to hydraulic fracturing that resulted from compaction and dewatering of the shale basin to the west during Laramide deformation.

Exploration Methods

Mineral exploration in the Robb Lake district started with prospecting and the discovery of the Lower Zn-Pb showing at the Robb Lake deposit. A staking rush ensued, which resulted in the discovery of numerous Zn-Pb deposits in the entire Silurian-Devonian outcrop belt of northeastern British Columbia during the early 1970s. Several drilling programs ensued at the Southface, Poco, Coral #2, Peace, BV, Perkins, and Robb Lake deposits. Following these activities, an induced polarization survey was done in 1973, and geological mapping of the Halfway River area (94 B) was undertaken by R. I. Thompson in 1975. By about that time, the interest of mining and exploration companies in the district had waned, and exploration work had stopped. More recently a geological map of the Robb Lake property (scale 1:20,000) was done by Nelson et al. (2000). This property-scale map focused on the distribution of the geological units and the Zn-Pb mineral deposits.

As a result of these field studies, the following features are outlined that are key exploration criteria for the Robb Lake district:

Tectonic setting: The deposits are carbonate-hosted and located close to the shelf-slope facies boundary; therefore, carbonate fronts dividing basinal shales and platform carbonates are promising areas.

Stratigraphic setting: At the Robb Lake deposit, most of the mineralized bodies occur within the Muncho-McConnell Formation, with a few deposits located within a thin overlying sequence of the Stone and Dunedin formations. Within the Muncho-McConnell Formation, the mineralized breccias form a broad stratabound zone occupying the upper 200 m of the lower unit and the lower 130 m of the upper unit.

Structures: The deposits are localized in the immediate footwall strata of a major thrust fault, the Mississippi Creek Fault, which has deep water, Early Paleozoic basinal strata in its hanging wall. Identification of conduits (e.g. faults, breccias, barrier reef complexes), traps, and prospective stratigraphy can provide a means of predicting the potential locations of undiscovered massive sulphide bodies in the district.

Textures: Sulphides accompanied by white sparry dolomite cement occur as open-space fillings of breccias, fractures, and vugs; rarely as replacement of the host carbonate. Disseminated sulphides in the carbonate rocks may indicate the proximity to orebodies.

Hydrothermal alteration: Hydrothermal dolomitization of precursor carbonate rocks is common and mostly appears as medium- to coarse-crystalline dolomite and saddle dolomite cement in sulphide-bearing breccias, veins, fractures, and vugs. Silicification is also locally present and appears as open-space crystalline quartz in vugs and veins of the upper Muncho-McConnell Formation.

Geochemical anomalies: The presence of bitumen and native sulphur occurs in mineralized and non-mineralized dolostone; stream sediment and water are locally anomalous in deposit-forming and ore-associated elements (i.e. Zn, Fe, Ag, ±Pb).

Potential for Known and New Deposit Types

Although the Robb Lake district is known for its Zn-Pb MVT mineralization associated with platform carbonate rocks of the Silurian-Devonian Muncho-McConnell Formation, there is potential for MVT Zn-Pb mineralization in any of the Silurian to Middle Devonian carbonate formations in the district. At the Robb Lake deposit, all the carbonate strata of the Muncho-McConnell Formation and the Stone and Dunedin formations are favourable for MVT Zn-Pb mineralization.

Potential for other types of deposits, such as "Irish-type", carbonate-hosted Cu-Pb-Zn (or Kipushi-type), skarns, and carbonate-hosted manto-type Ag-Pb-Zn also exist. Producing gas fields are known in strata of the Alberta Plateau along the eastern boundary of the Halfway River map area (94 B); therefore the potential for natural gas fields also exist in the deformed strata at deeper levels beneath the Foothills and eastern margin of the Rocky Mountains.

Knowledge Gaps

Mineral exploration within the Robb Lake district remains overall at the grass-roots level (except at the Robb Lake deposit), and additional prospecting and exploration work needs to be done to find deposits of economic status. The current knowledge gaps in our understanding of MVT deposits within the Robb Lake district are: 1) the role of tectonic processes, 2) the chemical processes involved, and 3) the age of MVT ore-forming events.

Tectonic Processes

It is becoming clear that there are genetic links between MVT mineralization and regional- and global-scale tectonic processes. MVT deposits are products of enormous hydrothermal systems that left trace mineralization over wide areas (2000 km^2 in the case of the Robb Lake district). The nearly ubiquitous occurrence of MVT deposits at the shelf-basin facies boundary (i.e. Canadian Cordillera) and anywhere within the interior of the carbonate platform (i.e. central United States) reflects focused migration of deep-basin brines along basement structures into shelf-carbonate sequences. Thus, understanding the regional hydrogeologic framework of the Robb Lake district is of paramount importance in the evaluation of the district to contain large MVT deposits.

MVT deposits have been attributed to large-scale migration of fluids during compressional and extensional processes. The most popular models, the topographically driven fluid-flow (or gravity-driven fluid-flow) model (Garven and Freeze, 1984; Garven, 1985; Bethke and Marshak, 1990; Garven and Raffensperger, 1997; Yao and Demicco, 1997) and the sedimentary and tectonic compaction model (Jackson and Beales, 1967; Sharp, 1978; Cathles and Smith, 1983; Oliver, 1986), are associated with ore fluid migration in compressive tectonic regimes and have often been cited to explain the MVT mineralization in the US mid-continent and western Canada. A more recent model involves deep convection circulation of hydrothermal brines due to buoyancy forces related to temperatures and salinity variations (Morrow, 1998) and rifting in the back-arc and arc regions of the western margin of North America during the Devonian-Mississippian (Nelson et al., 2002). This model would take place in an extensional tectonic regime, and may require fluid-driving mechanisms other than compression in a convergent-margin setting; it is poorly constrained at the moment.

Other questions, such as the ones listed below, become important in understanding formation of MVT deposits in the northern Rocky Mountains MVT belt, and indirectly guide the exploration for undiscovered MVT deposits in the Robb Lake district.

- What is the role of continental extension in the genesis of these MVT deposits?
- What are the fluid-driving mechanisms for MVT sulphide formation in extensional regimes?
- What types of ground-preparation processes are required for sulphide deposition and also what governs the location of sulphide bodies in the district?
- Are the deposits of the Robb Lake belt, Pb-Zn mineralization in the subsurface Presqu'ile Barrier and Pine Point deposit cogenetic, i.e. the result of a single eastward flush of MVT brines, or are they polygenetic? Are the MVT deposits of the Robb Lake belt genetically related to the SEDEX deposits of the Kechika Trough?
- How does the local and regional hydrology and paleohydrology of the adjacent shale basin (i.e. Kechika Trough) relate to dolomitization as well as mineralization? What flow paths are involved and what is the duration of their operation?
- What controls the hydrology of the carbonate platform and the adjacent shale basin? Is it related to the distribution of fractures and faults in the crystalline basement rocks and overlying sediments? Do some faults serve to recharge and others to discharge fluids?
- Why do the mineralized and unmineralized breccias preferentially favour specific stratigraphic intervals?
- What are the origins of the breccias within the district?
- How did the rock-matrix (trash) breccias of the Robb Lake deposit form? How does solution collapse promote the fine clast size and highly angular clast shapes seen in these; and by what collapse mechanism are entire vein selvages detached? On the other hand, what other mechanism of brecciation could be operative in the context of relatively cool brines (87 to 154°C with a mean of 119°C, Sangster and Carrière, 1991)?

Chemical Processes

Chemical processes that localized deposition of sulphides are critical to the development of models for MVT deposits, yet the specific chemical reaction(s) that led to sulphide deposition remains one of the most controversial aspects of MVT deposits. Several questions remain:

- What are the sources of lead, zinc, and sulphur?

- What is the mechanism that selects the lead and zinc?
- What causes metal zoning at the Robb Lake deposit?
- What is the contribution of organic matter to sulphide deposition? Are hydrocarbons critical for mineralization to occur within the district?
- What controls the chemical composition of sulphide-forming fluids?
- Were evaporites necessary for the generation of metalliferous brines?
- What are the chemical attributes of hydrothermal reaction zones that have generated ore-forming fluids?
- What alteration vectors are most effective in exploring for MVT deposits in the Robb Lake district?

Age of MVT Ore-Forming Events

Rb-Sr data on sphalerite from the most significant deposit in the northern Rockies MVT belt, the Robb Lake Zn-Pb deposit, although showing considerable scatter due to heterogeneous $^{87}Sr/^{86}Sr$ values in primary fluids, suggest that the mineralization is of Paleozoic age. These data are consistent with the published Pine Point Rb-Sr isochron, which indicates a Late Devonian age (362 ± 9 Ma, Nakai et al., 1993; 374 ± 21 Ma, Brannon et al., 1995) of Zn-Pb mineralization, and favour Devonian-Mississippian over Cretaceous orogenic models.

Even if advances in age dating have been achieved at Robb Lake, much still remains to be done, as no other deposits in the northern Rocky Mountains have been dated and there are still many questions to be answered:

- What controls the temporal distribution of MVT deposits within the Robb Lake district?
- What are the temporal and genetic links between MVT deposits of the Robb Lake belt and SEDEX deposits of the Kechika Trough? Is there a genetic/temporal link? If yes, could they have formed from the same hydrothermal system?
- What are the temporal and genetic links between MVT deposits of the northern and southern Rocky Mountains?

Acknowledgements

Field and analytical work on the Robb Lake deposit was conducted as part of the Central Foreland National Mapping Project, in collaboration with the Geological Survey of Canada (Calgary). Gilles St-Jean of the G.G. Hatch Isotope Laboratories of the University of Ottawa performed the oxygen and carbon isotope analyses on carbonates and Brian L. Cousens of the Herzberg Laboratories of Carleton University performed the strontium isotope analyses on carbonates. We are grateful to R.I. Thompson and R.W. Macqueen who reviewed the preliminary version of the manuscript; their suggestions greatly improved the manuscript. Drafting was done by Richard Franklin.

References

Anderson, G.M., and Macqueen, R.W., 1988, Mississippi Valley-Type lead-zinc deposits, in Roberts, R.G., and Sheahan, P.A., eds., Ore deposit models: Geological Association of Canada, v. Geoscience Canada Reprint Series 3, p. 79-90.

Aulstead, K.L., Spencer, R.J., and Krouse, H.R., 1988, Fluid inclusion and isotopic evidence on dolomitization, Devonian of Western Canada: Geochimica et Cosmochimica Acta, v. 52, p. 1027-1035.

Bethke, C.M., 1985, A numerical model of compaction-driven groundwater flow and heat transfer and its application to the paleohydrology of intracratonic sedimentary basins: Journal of Geophysical Research, v. 80, p. 6817-6828.

Bethke, C.M., and Marshak, S., 1990, Brine migration across North America - the plate tectonics of groundwater: Earth and Planetary Science Letters, Annual Review, v. 18, p. 287-315.

Boronowski, A.J., and James, S.C., 1982, Review of the Robb Lake Programs 1971-1982 and Review and Interpretation of Webb Ridge Geology with emphasis on the East Webb Zone Drilling 1982, Unpublished Report, Kidd Creek Mines Ltd, 41 p.

Brannon, J.C., Cole, S.C., Podosek, F.A., and Misra, K.C., 1995, Radiometric dating of ancient calcite, Th-Pb and U-Pb isochrons for ore-stage and late-stage calcite from central Tennessee zinc district, an Appalachian-Ouachita age MVT deposit: Geological Society of America, 1995, Abstracts with Programs, v. 27, p. 118.

Burke, W.H., Denison, R.E., Hetherington, E.A., Koepnick, R.B., Nelson, H.F., and Otto, J.B., 1982, Variations of seawater 87Sr/86Sr throughout Phanerozoic time: Geology, v. 10, p. 516-519.

Cathles, L.M., and Smith, A.T., 1983, Thermal constraints on the formation of Mississippi Valley-type lead-zinc deposits and their implications for episodic basin dewatering and deposit genesis: Economic Geology, v. 78, p. 983-1002.

Coniglio, M., Sherlock, R., Williams-Jones, A.E., Middleton, D., and Frape, S.K., 1994, Burial and hydrothermal diagenesis of Ordovician carbonates from the Michigan Basin, Ontario, Canada, in Purser, B., Tucker, M., and Zenger, D., eds., A volume in honor of Dolomieu: International Association of Sedimentologists, v. Special Publication 21, p. 231-254.

Coniglio, M., Morrow, D.W., and Wilson, N., 2006, Reassessment of Middle Devonian dolomites, Presqu'ile barrier, Northwest Territories, *in* Hannigan, P.K., ed., Potential for Carbonate-hosted lead-zinc Mississippi Valley-type Mineralization in Northern Alberta and Southern Northwest Territories: Geological Survey of Canada, Geoscience Contributions, Targeted Geoscience Initiative, Bulletin 591, in press.

Ferri, F., Rees, C., Nelson, J., and Legun, A., 1999, Geology and mineral deposits of the northern Kechika Trough between Gataga River and the 60th parallel: British Columbia Ministry of Energy and Mines, Bulletin 107, 122 p.

Garven, G., 1985, The role of regional fluid flow in the genesis of the Pine Point deposit, Western Canada Sedimentary Basin: Economic Geology, v. 80, p. 307-324.

Garven, G., and Freeze, R.A., 1984, Theoretical analysis of the role of groundwater flow in the genesis of stratabound ore deposits. 2. Quantitative results: American Journal of Science, v. 284, p. 1125-1174.

Garven, G., and Raffensperger, J.P., 1997, Hydrogeology and geochemistry of ore genesis in sedimentary basins, *in* Barnes, H.L., ed., Geochemistry of hydrothermal ore deposits: Wiley, New York, p. 125-189.

Godwin, C.J., and Sinclair, A.J., 1982, Average lead isotope growth curves for shale-hosted zinc-lead deposits, Canadian Cordillera: Economic Geology, v. 77, p. 675-690.

Goldhaber, M.B., Church, S.E., Doe, B.R., Aleinikoff, J.N., Brannon, J.C., Podosek, F.A., Mosier, E.L., Taylor, C.D., and Gent, C.A., 1995, Lead and sulfur isotope investigation of Paleozoic sedimentary rocks from the southern midcontinent of the United States: implications for paleohydrology and ore genesis of the southeast Missouri lead belts: Economic Geology, v. 90, p. 1875-1910.

Goodfellow, W.D., Lydon, J.W., and Turner, R.J.W., 1993, Geology and genesis of stratiform sediment-hosted (SEDEX) zinc-lead-silver sulphide deposits, *in* Kirkham, R.V., Sinclair, W.D., Thorpe, R.I., and Duke, J.M., eds., Mineral Deposit Modeling: Geological Association of Canada, Special Paper 40, p. 201-251.

Gordey, S.P., and Anderson, R.G., 1993, Evolution of the Northern Cordilleran Miogeocline, Nahanni map area (105I), Yukon and Northwest Territories: Geological Survey of Canada, Memoir 428, 214 p.

Hannigan, P.K., 2007, Metallogeny of the Pine Point Mississippi Valley-Type Zinc-Lead District, Southern Northwest Territories, *in* Goodfellow, W.D., ed., Mineral Resources of Canada: A Synthesis of Major Deposit Types, District Metallogeny, the Evolution of Geological Provinces, and Exploration Methods: Mineral Deposits

Synopsis of the Polaris Zn-Pb District, Canadian Arctic Islands, Nunavut

Keith Dewing[1], Robert J. Sharp[2], and Elizabeth Turner[3]

1. Geological Survey of Canada, 3303-33rd Street NW, Calgary, Alberta T2L 2A7
2. Trans Polar Geological, 60 Hawkmount Heights NW, Calgar,y Alberta T3G 3S5
3. Department of Earth Science, Laurentian University, Sudbury, Ontario, P3E 2C6
Corresponding author's email: kdewing@NRCan.gc.ca

Abstract

The Polaris Mine was a Mississippi Valley-type (MVT) deposit hosted in dolomitized Middle Ordovician limestone. Total production was 20.1 Mt at 13.4% Zn and 3.6% Pb. There are about 80 showings in the district, which stretches from Somerset Island to the Grinnell Peninsula. There are two deposit types in the Polaris District: 1) structurally controlled, carbonate-hosted Zn-Pb-Fe deposits typical of MVT deposits, and 2) structurally and stratigraphically controlled, carbonate-hosted Cu deposits enriched by later supergene removal of Fe and S.

Mineralization is paragenetically simple, with sphalerite and galena as the ore minerals, and with dolomite and marcasite as the main gangue minerals. The deposits formed from brines at about 90 to 100°C. The age of the mineralization is constrained to post-Late Devonian folding and may be associated with the last stages of the Ellesmerian Orogeny or the opening of the Sverdrup Basin. Copper-rich mineralization is known from four showings, is associated with zinc-lead mineralization and is confined to a single interval in the Silurian.

The metallogenic model for Polaris invokes a source of metal ions within the stratigraphic column since strontium shows no indication of basement involvement. Metals are then carried in sulphate-rich brines through deep aquifers. Driven by an orogenic process, circulating fluids rose along faults until they encountered organic-rich, permeable limestone overlain by impermeable shale. Bacterial sulphate reduction reduces sulphate to H_2S using organic matter as a reductant. The resulting H_2S has a shift in $\delta^{34}S$ values of -15‰. The H_2S reacts inorganically with zinc, lead, and iron ions to produce sulphides with a $\delta^{34}S$ value of about 10‰. Released hydrogen ions dissolve carbonate resulting in dissolution breccias and other dissolution fabrics. Excess sulphate was expelled from the system and produced a halo of barite up to 10 km from the deposit with $\delta^{34}S$ of 40 to 60‰.

Résumé

La mine Polaris consistait en un gisement de type Mississippi-Valley encaissé dans un calcaire dolomitisé de l'Ordovicien moyen. Au total, elle a produit 20,1 Mt de minerai titrant 13,4 % de Zn et 3,6 % de Pb. Le district de Polaris compte quelque 80 indices minéralisés et s'étend de l'île Somerset jusqu'à la péninsule Grinnell. On y trouve deux types de gîtes, à savoir des gîtes de Zn-Pb-Fe encaissés dans des roches carbonatées, qui affichent un contrôle structural et sont représentatifs des gîtes de type Mississippi-Valley, et des gîtes de Cu logés dans des roches carbonatées, qui montrent des contrôles structuraux et stratigraphiques et un enrichissement supergène ultérieur par l'élimination du Fe et du S.

Du point de vue paragénétique, les minéralisations sont simples. Elles présentent la sphalérite et la galène comme minéraux métallifères, ainsi que la dolomite et la marcasite comme principaux minéraux de gangue. Les gîtes du district se sont formés à partir de saumures affichant des températures de 90 à 100°C environ. L'âge maximal des minéralisations est défini par un plissement du Dévonien tardif, qui pourrait être rattaché aux dernières phases de l'orogenèse Ellesmérienne ou à l'ouverture du bassin de Sverdrup. Quatre indices minéralisées ont révélé des minéralisations riches en cuivre, qui sont associées aux minéralisations de zinc-plomb et confinées à un seul intervalle du Silurien.

Le modèle métallogénique proposé pour le district de Polaris établit une source d'ions métalliques dans la colonne sédimentaire, car la composition isotopique du strontium ne témoigne d'aucune contribution du socle. Les métaux ont donc été transportés dans des saumures riches en sulfates circulant dans de profonds aquifères. Un phénomène orogénique a fait en sorte que les fluides en circulation se sont élevés le long de failles jusqu'à ce qu'ils atteignent des horizons perméables de calcaire riche en matière organique, qui sont recouverts de shale imperméable. Des bactéries ont réduit les sulfates en H_2S en se servant de la matière organique comme agent réducteur. Le H_2S résultant présente un décalage des valeurs de $\delta^{34}S$ de -15 ‰. Le H_2S réagit de manière inorganique avec les ions de zinc, de plomb et de fer pour produire des sulfures dont la valeur de $\delta^{34}S$ est d'environ 10 ‰. Les ions d'hydrogène émis dissolvent les carbonates pour produire des brèches de dissolution et d'autres fabriques de dissolution. Les sulfates excédentaires ont été expulsés du système et ont produit une auréole de barytine qui s'étend jusqu'à 10 km du gisement et les valeurs de $\delta^{34}S$ de ce minéral varient de 40 à 60 ‰.

Introduction and History

The Polaris Zn-Pb District lies in the central Arctic Islands, Nunavut, and spans an area that is roughly 450 km north-south by 130 km east-west (Fig. 1). The southern limit is the southern edge of Somerset Island; the northern edge is the northern shore of the Grinnell Peninsula. The eastern limit runs along the longitude of Wellington Channel between Cornwallis and Devon islands and the western limit is on eastern Bathurst Island.

There are about 80 individual Zn-Pb showings occur within the study area (Fig. 1, Table 1). The largest of these is the Polaris deposit, a carbonate-hosted Zn-Pb deposit of about 20 million tonnes grading about 17% Zn+Pb. The deposit was mined from 1982 to 2002, when it shut down due to depletion of reserves. During its latter days, Polaris employed a workforce of 235 employees on an 8 week onsite and 4 week offsite rotation. Sphalerite and galena were separated from the gangue in a complex known as "the barge"

Dewing, K., Sharp, R.J., Turner, E., 2007, Synopsis of the Polaris Zn-Pb District, Canadian Arctice Islands, Nunavut, *in* Goodfellow, W.D., ed., Mineral Deposits of Canada: A Synthesis of Major Deposit-Types, District Metallogeny, the Evolution of Geological Provinces, and Exploration Methods: Geological Association of Canada, Mineral Deposits Division, Special Publication No. 5, p. 655-672.

which in addition to the concentrator, housed offices, warehouse, powerhouse, and heavy equipment and other repair shops. The concentrator complex was built on a barge in Trois Rivières, Québec, and towed to Little Cornwallis Island where it was floated into an excavated lagoon and surrounded by rockfill. The annual mining rate of 1,040,000 tonnes per year produced 200,000 tonnes of zinc and 30,000 tonnes of lead concentrates that were stored in a large concentrate shed. Ocean-going ships docked near the concentrate shed and were loaded directly with a conveyor system. Freighters arrived in early July and departed up until late October.

Exploration in the Polaris District took place in five phases (Fig. 2): 1) a reconnaissance exploration period (1960-1970) during which most surface showings on Cornwallis and Little Cornwallis islands were found; 2) a discovery period (1971-1979) during which the Polaris orebody was drilled based on a gravity anomaly and feasibility studies were completed. Also, new showings were discovered, and many showings received limited drill testing; 3) a production period (1980-1988), dominated by drilling and mining at the Polaris Mine; 4) an ore-replacement exploration period (1989-2001) during which showings close to Polaris were extensively drilled, showings on Cornwallis Island drill tested, and new showings found and drilled on Somerset and Bathurst islands and on the Grinnell Peninsula; and 5) a reclamation period (2002-2004) during which the mine site was restored and the infrastructure removed.

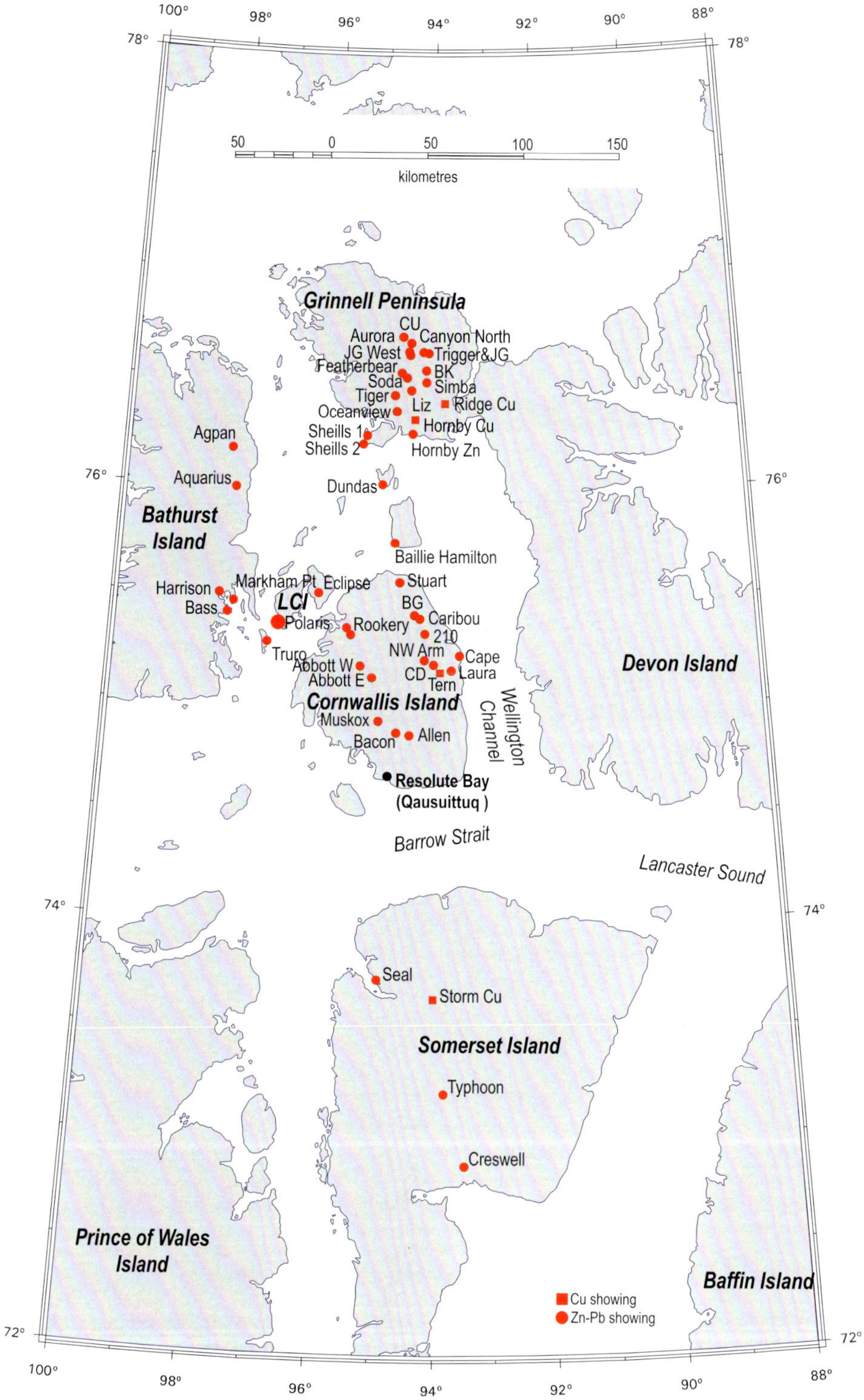

FIGURE 1. Lead-zinc-copper showings in the Polaris District, central Arctic Islands, Nunavut. LCI is Little Cornwallis Island. Some dots represent clusters of showings. Locations and sources are listed in Appendix 1.

Early studies on the Polaris deposit included a brief description of the geology and mineralization by Muraro (1974), fluid inclusion work by Jowett (1975), and lead isotope work on the Truro Island showing by Heal (1976). Kerr (1977) proposed a metallogenic model that required formation of karst during the Early Devonian Boothia Uplift, then migration of metal-bearing and sulphur-bearing fluids to the site of deposition during the Late Devonian. The karst model guided exploration during much of the 1970s and 1980s. Randell introduced a hydrothermal karst model (Randell, 1994; Randell and Anderson, 1997; Randell et al., 1997), and controversial ideas related to the timing of folding. Sharp et al. (1995a,b) produced a field-trip guide for the Polaris Mine. Savard et al. (2000) reported on fluid inclusion and isotopic data from the deposit and Dewing and Turner (2003) looked at the relative ages of faults near Polaris. Mineralization at Polaris was dated by paleomagnetics (Symons and Sangster, 1992), and Rb-Sr dating of sphalerite (Christensen et al., 1995). Randell (1994), Disnar and

TABLE 1. Distribution of showings by formation.

Formation	# showings	% of showings	Com.
Bird Fiord (Dbi)	3	3.75	Zn
Blue Fiord (Dbl)	15	18.75	Zn, Pb
Disappointment Bay (Ddb)	6	7.5	Zn, Pb
Prince Alfred (Dpa)	2	2.5	Zn, Pb, Cu
Barlow Inlet (Sbi)	2	2.5	Zn, Pb
Cape Storm (Scs)	5	6.25	Cu
Allen Bay (Osa)	8	10	Zn, Pb
Thumb Mountain (Oct)	34	42.5	Zn, Pb
Bay Fiord (Ocb)	2	2.5	Zn, Pb
Eleanor River (Oe)	2	2.5	Zn
Ship Point (Osp)	1	1.25	Zn
Total	80	100%	

Héroux (1995), and Héroux et al. (1996, 1999) examined organic matter alteration from Polaris.

Studies of individual showings apart from Polaris are restricted to showings in Devonian strata on Baillie Hamilton Island (Thorsteinsson, 1984), showings in Devonian strata on Bathurst Island (Harrison and de Freitas, 1996; Rose, 1999), the Rookery Creek showing on Cornwallis Island (Dewing and Turner, 2003), and the 'Bermuda' showing on Grinnell Peninsula (Mitchell, 2000).

Geological and Tectonic Setting

Cambrian to Devonian strata of the Arctic Islands are a mixed sedimentary-igneous succession, collectively referred to as the *Franklinian succession*, which were deposited on a passive-to-convergent cratonic margin (Fig. 3; Trettin, 1991). The upper boundary of this succession is constrained as latest Devonian or earliest Carboniferous, but the oldest strata are not exposed. Trettin (1991) suggested that the margin initiated following a widespread Proterozoic igneous event at 723 Ma (see Heaman et al., 1992). The oldest exposed strata are late Neoproterozoic in age (Dewing et al., 2004).

Following Precambrian continental breakup, thick Lower Cambrian clastics and carbonate rocks accumulated along the rifted margin. The shelf-to-basin transition was established by Early Cambrian time (Trettin, 1994; Harrison, 1995). During the Cambrian and Ordovician, the Franklinian shelf accumulated carbonates, evaporites, and continent-derived siliciclastic rocks. During this time, marked differences in subsidence rates between shelf and basin produced an escarpment at the shelf margin (Trettin, 1994). The Cambrian-Ordovician carbonate platform was bordered to the north by the shale-dominated Hazen Basin.

Cambro-Ordovician platformal strata pass southeast and south into the thin, undisturbed cratonic sedimentary cover. The boundary between these depositional realms occurs approximately at the southern and eastern boundaries of the Parry Island – Central Ellesmere fold belts. A hinge line exists between Somerset and Cornwallis islands; the Cambro-Ordovician sedimentary succession thickens rapidly northward across this feature (de Freitas and Mayr, 1993; de Freitas et al., 1999).

During latest Ordovician, the shelf margin retreated substantially towards the south and southeast, forming the Cape Phillips Embayment. The embayment received mostly graptolitic shale and carbonates, but isolated microbial and coral-microbial banks within the embayment kept up with sea level rise at least until Early Silurian time (Wenlock). The shelf-to-basin transition was a ramp immediately following shelf retreat, but developed a steeper rimmed profile by late Llandovery time (Early Silurian). The facies front is well exposed halfway up the eastern coast of Cornwallis Island.

Vertical movement along a narrow, north-south intracratonic uplift running from Boothia Peninsula to Grinnell Peninsula during Late Silurian to Middle Devonian (Givetian) time produced unconformities and syntectonic clastic sediments (the Boothia Uplift). This deformation probably was caused by stress transmitted cratonward from the Caledonian Mobile Belt on East Greenland (Miall, 1986; Okulitch et al., 1986). Structures within the Boothia Uplift are characterized by broad synclines and narrow, evaporite-cored anticlines (Thorsteinsson, 1986; Mayr et al., 1998). The margins of the anticlines contain abundant thrust faults (Fig. 4; Henrichsen, 2003; Jober, 2005). Dense, anastamosing arrays of subvertical normal faults with significant vertical offset characterize the western margin of the uplift on eastern Bathurst Island.

By early Middle Devonian time, the effects of plate convergence were widespread over most of the Arctic Islands. Shallow-marine and nonmarine syntectonic clastic rocks were deposited in a foreland basin adjacent to a southeastward- and southward-advancing deformation front. The maximum preserved thickness of this clastic wedge is about 4000 m. Thermal maturity indicators in the area of the Boothia Uplift are generally low (conodont alteration index of <2 and graptolite-based vitrinite equivalent reflectances of about 1% (Gentzis et al., 1996; Héroux et al., 1999)) indicating that the clastic wedge could not have attained a great thickness over the Boothia Uplift.

In the Arctic Islands, the youngest preserved strata are of Famennian age. The Middle Devonian to earliest Carboniferous phase of deformation is known as the Ellesmerian Orogeny (Thorsteinsson and Tozer, 1970). East-west-trending folds that resulted from the Ellesmerian Orogeny characterize Melville, Bathurst, and southern Ellesmere Islands. In the area of the pre-existing Boothia

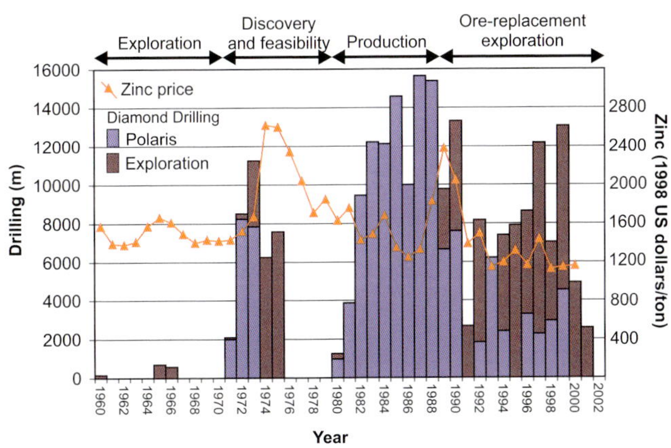

FIGURE 2. Exploration and ore-definition drilling in the Polaris District between 1960 and 2002. High zinc prices in 1974 and 1990 coincide with new rounds of regional exploration. Historical zinc prices from Planchy (1998).

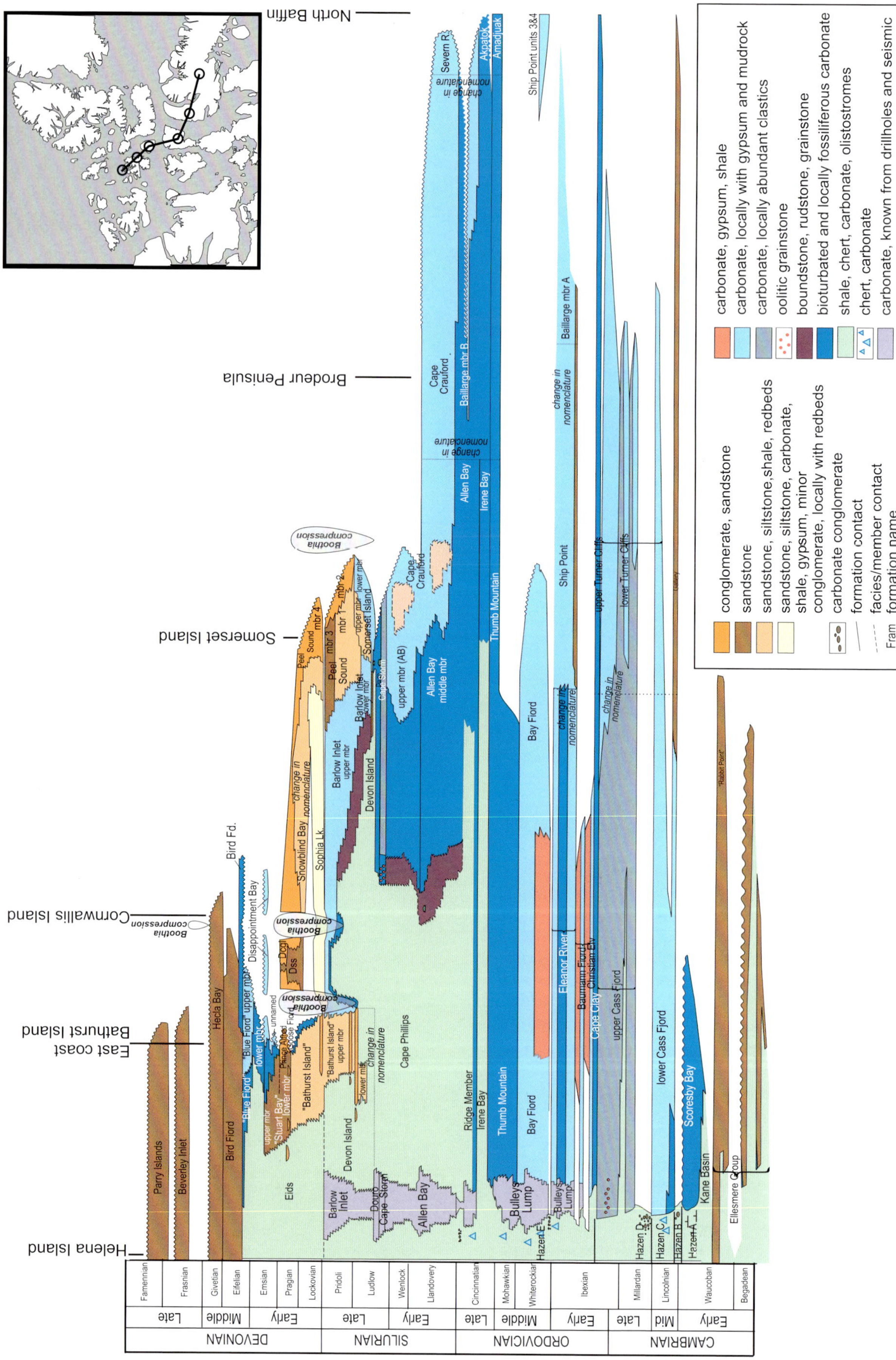

FIGURE 3. Time-stratigraphic diagram showing a south to north transect from northwestern Baffin Island to northern Bathurst Island. The age range of the formations is based on data in Trettin (1969), Mayr (1978), Nentwich (1987), Stewart (1987), Thorsteinsson and Mayr (1987), Nentwich and Jones (1989), Mayr et al. (1994, 1998), de Freitas et al. (1999), and Sanford and Grant (2000). Formations below the upper Cass Fjord Formation are not exposed on the line of section and the presence is based on occurrences elsewhere in the Arctic Islands.

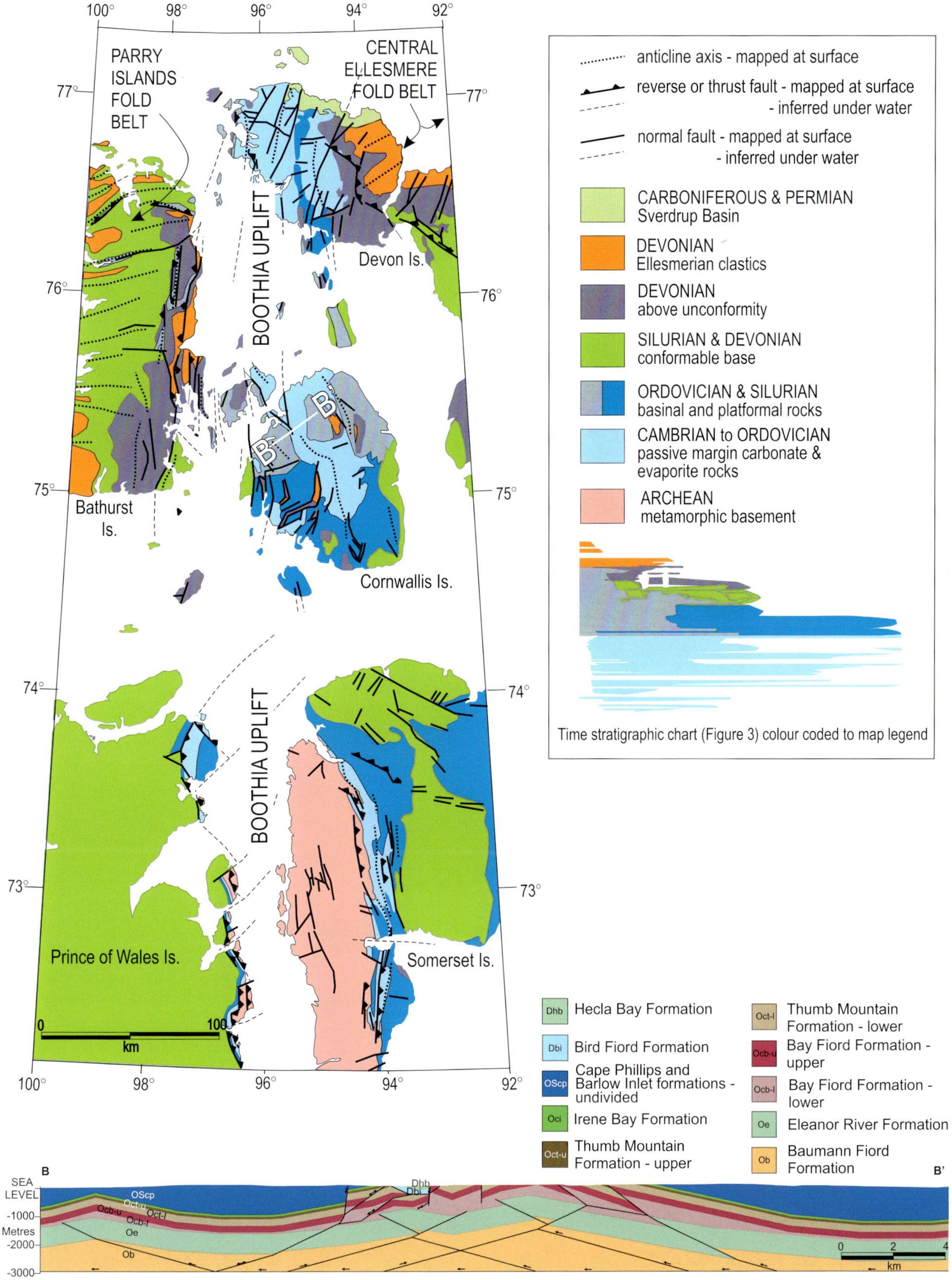

FIGURE 4. Generalized geological map of the central Arctic Islands, Nunavut, (after Okulitch, 1991). The base of the legend shows the time-stratigraphic diagram in Figure 3 colour coded to match this map legend. At the base of the diagram is a hypothetical cross-section B-B' (shown by white line in the centre of the map). This presents a conceptual model (Henrichsen, 2003) that shows a Boothia structure with a thick, evaporite-filled core and thrusted carbonates on the flanks. No attempt was made to balance sub-Baumann Fiord units. Later extension reactivated some of the thrusts as normal faults and created small half grabens.

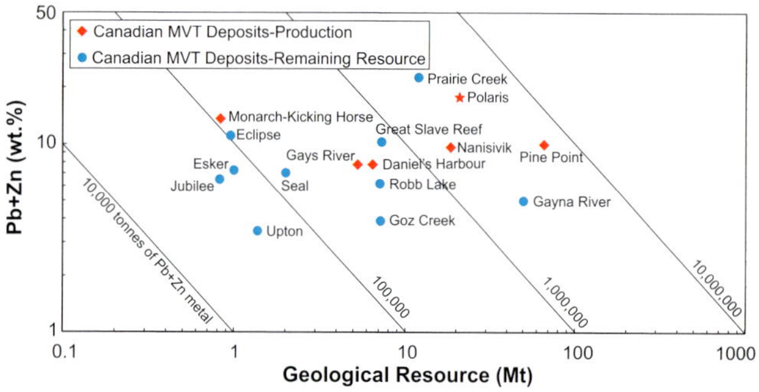

FIGURE 5. Resources versus grade diagram for Canadian Mississippi Valley-type deposits and resources. Production at the Polaris Mine is marked by red star. From Paradis et al. (this volume).

Uplift, the Ellesmerian compression produced low-amplitude east-west folds locally forming an interference fold pattern with the pre-existing north-south Boothia folds.

Rifting during the Early Carboniferous (Viséan) produced the Sverdrup Basin to the north of the lower Paleozoic shelf margin. Regional extension related to the formation of the Sverdrup Basin may have reactivated older thrust faults within the Boothia Uplift as extensional faults creating half grabens locally on the leading edge of the older thrusts.

Mineral Deposits and Occurrences

Deposit Types and Classification

There are two deposit types in the Polaris District: 1) structurally controlled, carbonate-hosted Zn-Pb-Fe deposits typical of the Mississippi Valley-type (MVT) deposit, and 2) structurally and stratigraphically controlled, carbonate-hosted Cu deposits enriched by later supergene removal of Fe and S. This deposit type is not discussed in detail herein.

Deposit Distribution

The Ordovician Thumb Mountain Formation hosts most of the Zn-Pb mineralization and includes the largest deposits at Polaris, Eclipse, and Truro. Consequently, the Thumb Mountain Formation has been extensively prospected. Significant mineralization was found in the Devonian Blue Fiord Formation only in 1995 (Harrison and de Freitas, 1996) and 14 of the 15 showings in that unit were found between 1996 and 2001. Copper mineralization is restricted to the Cape Storm, Prince Alfred, and Douro formations.

Grade and Tonnage

Before mining, the Polaris orebody contained a diluted in situ ore reserve of 26 million tonnes grading 3.7% Pb and 13.9% Zn based on an 8% combined Pb+Zn (cutoff over five metres). Total production at the end of mine life was 20.1 Mt of ore at 13.4% Zn and 3.6% Pb. This places Polaris in the upper echelon of individual carbonate-hosted Zn-Pb mines in the world (Fig. 5).

No other showing in the district was found to be economic so their grade and tonnage are speculative. Teck Cominco estimated the Eclipse showing to contain 765,000 (undiluted) tonnes grading 0.89% Pb and 11.84% Zn, with a possible additional 165,000 tonnes grading 0.07% Pb and 8.83% Zn. Eclipse contains both sulphide and oxide ore making the recovery more difficult. Estimates of the size of the Seal showing indicate up to 2 Mt (diluted) at 7% Zn. No other showings had ore-grade intersections that could be correlated to nearby drillholes

Age

The age of sphalerite crystallization at Polaris has been established as 366 ± 15 Ma, (Late Devonian to Early Carboniferous) based upon Rb/Sr isotopic ratios in sphalerite (Christensen et al., 1995). This age is supported by a Late Devonian magnetic paleopole for the ore (Symons and Sangster, 1992). Recent, as yet unpublished, work by Christensen indicates some showings (e.g. Dundas Island) have a Viséan (Early Carboniferous, 337 Ma) age, synchronous with the opening of the Sverdrup Basin.

Deposit Morphology

Host Rocks

Four stratigraphic units are exposed around Polaris (Figs. 6, 7, 8). The oldest is the Middle Ordovician Bay Fiord Formation, a 440 m thick succession of anhydrite and dolostone. This is overlain by the Middle to Upper Ordovician Thumb Mountain Formation, which, where unaltered, consists of 340 m of carbonate. These shallow-water carbonates and evaporites are overlain by deeper water shale of the Irene Bay (60 m) and Cape Phillips (600 m) formations. The Thumb Mountain Formation hosts the mineralization at the Polaris Mine.

The lower member of the Thumb Mountain Formation is approximately 250 m thick and consists of dolomitic limestone or dolostone. The basal 35 to 45 metres are composed of thick-bedded, burrow-mottled, lime- or dolomudstone and wackestone. Overlying strata of the lower member (approximately 180 m) consist of medium- to thick-bedded, pale grey to brown, lime- or dolomudstone, peloidal grainstone, with lesser wackestone. Metre-scale cycles of mottled dolomudstone or wackestone overlain by plane-laminated or fenestral lime- or dolomudstone occur. The uppermost 10 to 30 m of the lower member of the Thumb Mountain Formation, known as the *Tetradium* Interval, contain rare to abundant *Tetradium* (characterized by calcite-filled tubes with a 0.5 to 2.0 mm square cross-section).

The upper Thumb Mountain Formation is approximately 80 m thick. A basal chert marker (4-10 m thick) consists of thickly bedded, medium to dark brown, burrow-mottled skeletal wackestone with white-weathering silicified burrows several centimetres in diameter. Above the Chert Marker, the upper Thumb Mountain Formation consists of 30 to 40 m of burrow-mottled, medium-bedded skeletal wackestone with lesser interbeds of nodular, argillaceous lime mudstone and wackestone. This is followed by 15 to 20 m of massive to thick-bedded, fossiliferous wackestone containing abundant macrofossils. The uppermost part of the Thumb Mountain Formation consists of 20 to 25 m of argillaceous, nodular, skeletal wackestone with interbedded green terrigenous mud.

Within the upper Thumb Mountain at Polaris, and on western Cornwallis Island, is the organic-rich "Vr" Marker, a 1 m thick wackestone bed with total organic carbon of between 0.5 and 1.5%, and containing abundant, dark orange to dark brown algal colonies up to 1.5 cm long and 0.5 cm high that are internally massive or faintly laminated.

The Irene Bay Formation is a 55 m thick Upper Ordovician sequence gradationally overlying the Thumb Mountain Formation. It consists of interbedded green mudstone, argillaceous grey-green limestone, and massive to nodular grey limestone.

The Upper Ordovician to Silurian Cape Phillips Formation has a lower Ridge Member (a thickly bedded dolomitic or calcareous wackestone that is 7-10 m), overlain by 600 to 2200 m of shale, argillaceous limestone, and chert. A facies front on central Cornwallis Island marks the transition from basinal Cape Phillips Formation to the north to platformal strata of the Allen Bay, Cape Storm, Douro, and Barlow Inlet formations to the south.

Unconformably overlying Ordovician or older Silurian strata are several Upper Silurian - Lower Devonian alluvial conglomerate units that record the Boothia Uplift (Fig. 3; Miall and Gibling, 1978; Thorsteinsson and Uyeno, 1980; Muir and Rust 1982; de Freitas and Mayr, 1993; Mayr et al. 1998). Conglomerate units are restricted to regions in front of each Boothia-aged thrust fault. The Early Devonian Disappointment Bay Formation overlies the conglomerate, or forms the lowest Devonian unit where the conglomerate is absent. This is, in turn, overlain by the dolomitic Blue Fiord and Bird Fiord formations. Devonian units sit with an angular unconformity on underlying Ordovician and Silurian units on Cornwall Island (Thorsteinsson, 1986), Little Cornwallis Island (Turner and Dewing, 2004), and Grinnell Peninsula (Mayr et al., 1998).

Dimensions

Dimensions of the Polaris deposit are 800 m strike length, 300 m at its widest point, and a thickness ranging from 20 m in the west to 100 m in the central portion. The deposit is blind, with the topmost ore lying 50 m below surface on the west and up to 160 m

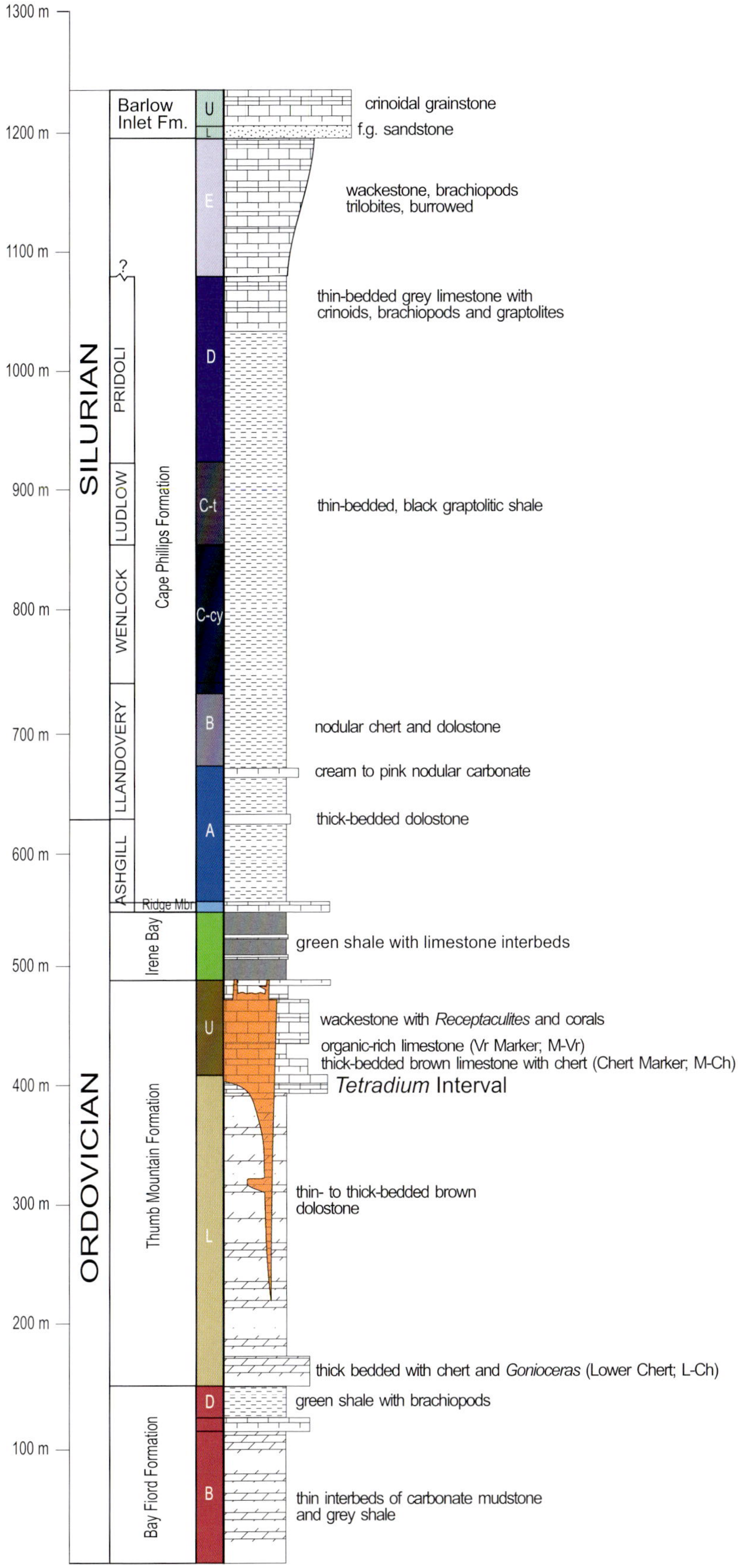

Figure 6. Stratigraphic column for Little Cornwallis Island. Ore at the Polaris Mine is hosted in the Thumb Mountain Formation (red overlay). Cape Phillips Formation member C is divided on the basis of graptolites into a cyrtograptid-bearing "C-cy" unit and a *Bohemograptus tenuis* bearing C-t unit.

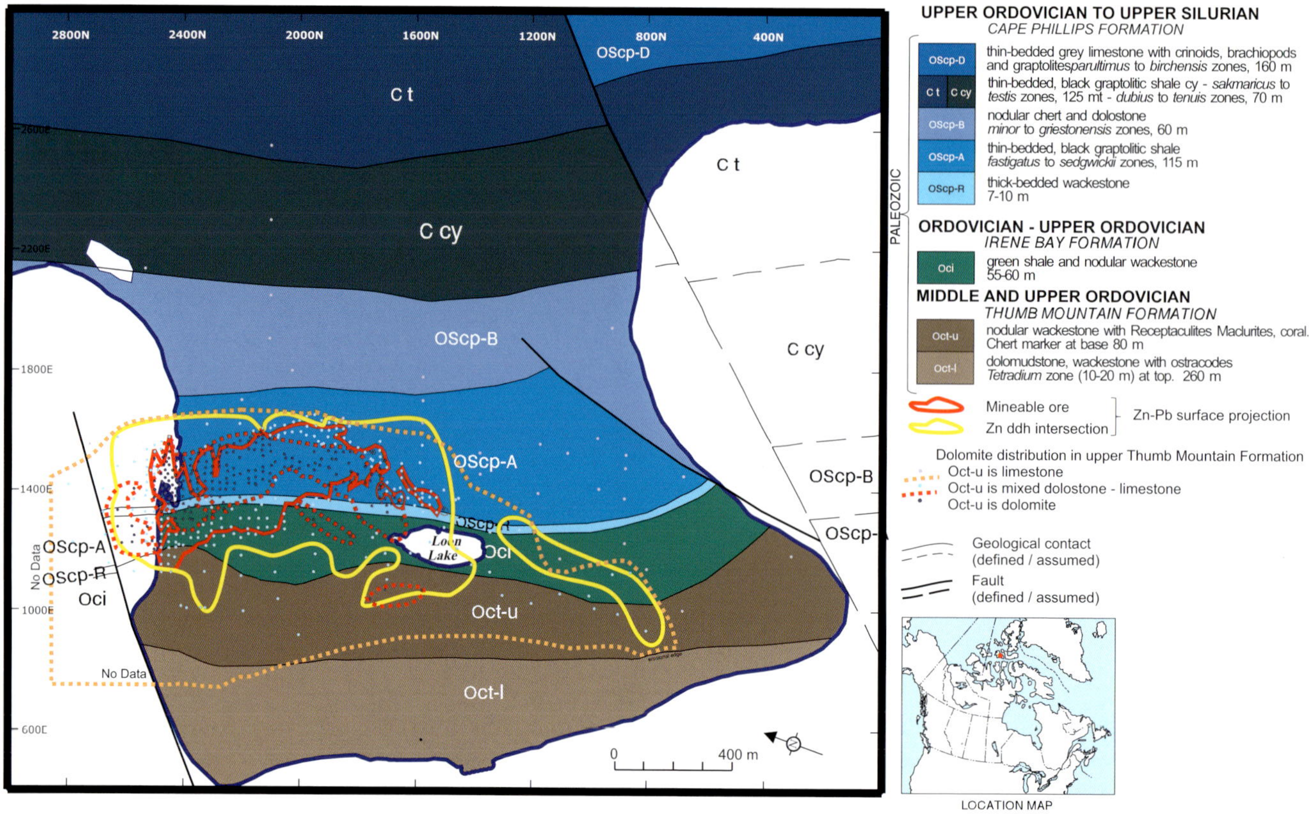

FIGURE 7. Geological map of Polaris Peninsula, Little Cornwallis Island. Plan of the mined ore shown in solid red. The limit of the drillholes containing an interval of >1% Zn are shown by a yellow line. The limits of completely and incompletely dolomitized upper Thumb Mountain Formation are shown by dashed lines and coloured dots. The mine grid is 23° east of true north. After Turner and Dewing (2004).

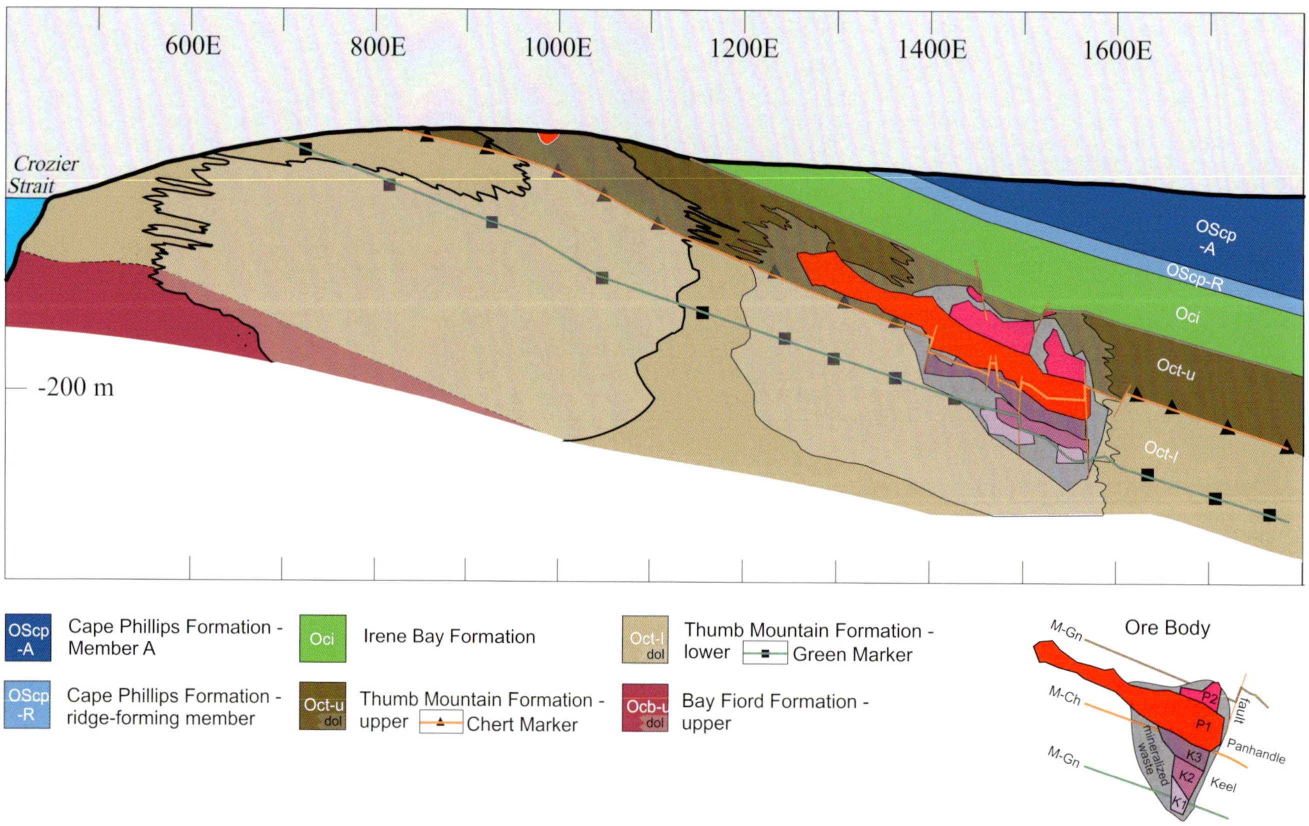

FIGURE 8. Section through the Polaris Mine on mine grid 2100 N showing the outline of the mined ore, mineralized waste and the collapse of the marker beds in the orebody. The dolomitization outlines are after Randell (1994). No vertical exaggeration. Dominant lithotypes for each unit are on the legend for Figure 7.

TABLE 2. Ore grade by division within Polaris Mine.

Zone	Pb+Zn	Fe	Zn/Pb
P2	16.5	4.6	3.1
P1	22.4	4.5	3.5
K3	19.1	3.1	5.6
K2	17.3	1.7	4.3
K1	15.5	0.8	8.1

below surface on the east side. It forms a continuous sulphide body.

The Eclipse showing is 700 m long, with a 300 metre long western portion that is about 70 to 100 m wide and an eastern portion about 250 m long and 25 to 50 m wide, separated by a 150 m long central portion that appears to be poorly mineralized. The Seal Showing has a strike length of about 400 m, daylights to the southwest, and is 50 to 100 m wide and 12 to 20 m thick. Other showings range in size from 100 by 80 m (Dundas Island) to many that are <10 m across. Showings often occur in clusters of 5 to 10 spread parallel to strike over roughly 10 km.

Nature of Sulphide Orebodies

The Polaris orebody consists of two parts (Fig. 8). The Panhandle Zone forms the upper part of the deposit and is an elongate tabular Zn>Fe>Pb body, roughly concordant with the upper Thumb Mountain Formation, in which carbonate is almost completely replaced by sulphide. The Keel Zone occupies the lower and eastern part of the orebody and consists of a Zn>Pb>Fe vein stockwork. Five ore types were recognized: two ore types occur in the upper Thumb Mountain and three in the lower Thumb Mountain (Sharp et al., 1995b).

P1 ore was the economically most important and formed a 10 to 30 m thick, high-grade Zn-Fe-Pb, tabular unit hosted in the basal part of the upper Thumb Mountain Formation. P1 ore consisted of massive, carbonate replacement, breccia-fill and vein sulphide averaging 4 to 5% Fe with up to 30% Fe (see Table 2). Overlying the P1 ore, above the major north-south structures defining the boundaries of the Keel graben, lie blocks designated as P2 ore. This ore is composed of numerous thin to 1m thick veins of sphalerite, marcasite, and galena oriented vertically, sub-perpendicular to bedding. Marcasite halos surround most P2 veins. Directly underlying the P1 tabular body is the high-grade K3 ore composed of complex, cross-cutting veins along with massive to disseminated sulphide replacing Tetradium beds with 3 to 5% Fe. Differentiation between K3 and P1 ore was based mainly on lower iron contents of K3 versus the P1 ore. K2 ore underlies the K3 unit and is composed of fracture-filling and vein sulphides with lesser replacement of carbonate, and local pods of mineralized brecciation carrying grades of 8 to 20% Pb+Zn and 1 to 2% Fe. The lowermost ore unit is K1 ore composed of fracture-filling and vein sulphides, is discontinuous and of lower grade (10 to 15% Pb+Zn, 1 % Fe). This ore unit follows either the north-south (mine grid) axis of mineralization or a northeast-southwest trend and is centred on lower stratigraphic horizons around a green marker bed in the lower Thumb Mountain Formation. To the east and west of the orebody, sulfide mineralization dies out rapidly within 15 m of the ore boundaries. Sulfide feathers out into small, scattered lenses and sulfide vein systems to the north and south of the orebody.

Only the Eclipse and Seal showings have the pervasive replacement of carbonate by sulphides, which is typical of the high-grade ore at Polaris, but the solution breccias that are so prominent at Polaris are poorly developed at Eclipse, Seal, and at other showings. Similarly, the abundant marcasite that characterized Panhandle mineralization at Polaris is not present at Eclipse, Seal, Truro, or most other showings. Only the JG showing on Grinnell Peninsula contained a long drill intersection of marcasite with minor sphalerite, but adjacent drillholes contained little or no sulphides.

Mineralogy, Textures, and Ore Chemistry

The Polaris orebody exhibits multiple generations of dolomite and sulphide precipitation and of carbonate and sulphide dissolution. Details of the alteration are after

FIGURE 9. High-grade ore samples from the Polaris Mine. Designators for mine location in the keel mining areas are based on elevation and mine grid coordinates. They are level, in metres, with 1000 being sealevel and decreasing with depth, followed by, in metres, the latitude coordinate divided by 10. Hence 850-176 refers to 850 level at 1760 north latitude. Keel stopes, all oriented east-west, mined the keel ore zone from 1720N or 172 stope in the south to 237 stope in the north at 2370 north latitude. **(A)** Marcasite replaced carbonate and banded sphalerite. 880 level by waterline cutout. 7 cm wide. **(B)** Sphalerite-galena vein showing banded ore on either side and crystalline galena and sphalerite in the centre. 820 Bypass, 8 cm wide.. **(C)** Colloform and banded ore, 820-229, 8 cm wide. **(D)** Mineralized collapse breccia, 820-218, 8 cm wide. **(E)** Marcasite-rich ore showing collapsed fragments of banded sphalerite 880-185, 11 cm high. **(F)** Colloform sphalerite with dendritic galena and minor dolospar, 730-205, 9 cm high.

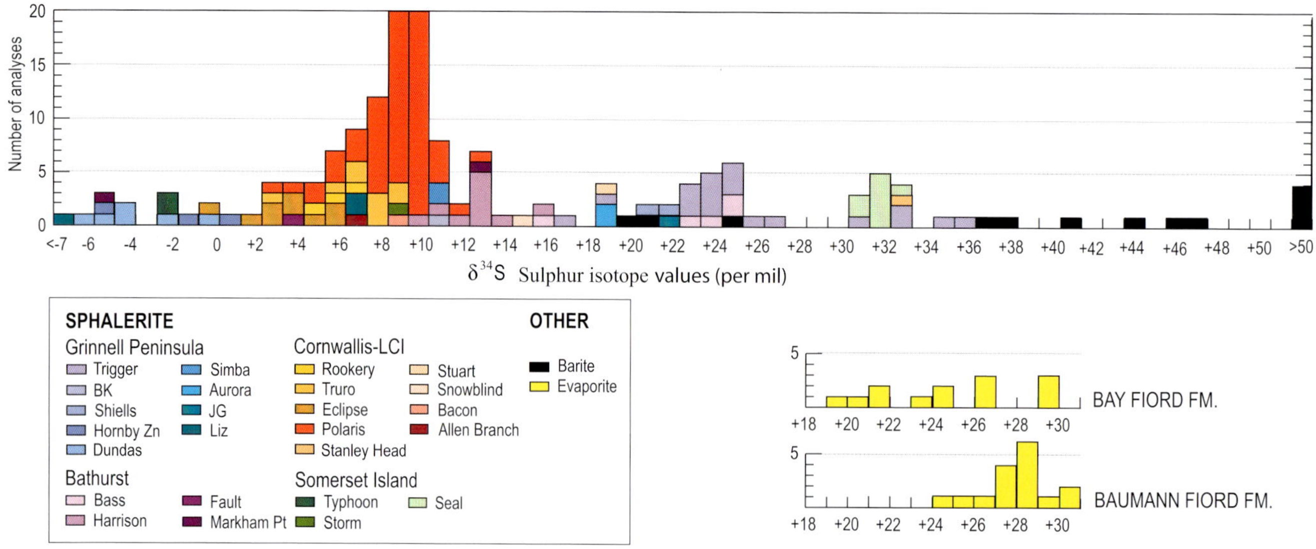

Figure 10. Sulphur isotopes from sphalerite in the Polaris District colour coded by location. Data from Davies and Krouse (1975), Randell (1994), and Mitchell (2000), and as well as unpublished GSC data.

Savard et al. (2000). The alteration that extends farthest from the orebody is brown, Fe-poor dolomite that replaces the original limestone of the host Thumb Mountain Formation. Crystal size of the replacement dolomite increases towards the core of the deposit. The replacement dolomite always contains some sphalerite or galena, thus is considered part of the mineralizing event. Two types of sparry dolomite are recognized: 1) white, saddle dolomite, which fills vugs and fractures or that can grow replacively from finer grained replacement dolomite. It is Fe poor and inclusion rich and the bulk of the fluid inclusion homogenization temperatures range from 85 to 105°C; 2) white to pink, Fe-rich, inclusion-poor saddle dolomite that fills vugs and appears to post-date the white saddle dolospar. Most homogenization temperatures fall in the range of 80 to 105°C. The $\delta^{18}O$ (SMOW) and $\delta^{13}C$ (PDB) values for the replacement dolomite and the two types of saddle dolomite cluster tightly with $\delta^{18}O$ = -10.0 to –8.5‰ and $\delta^{13}C$ = -1.5 to +1.0‰. $^{87}Sr/^{86}Sr$ ratios for replacement dolomite and the two sparry dolomite phases range from 0.70822 to 0.70948. This is above the range for Middle Ordovician seawater (0.7078 to 0.7083; Savard et al., 2000).

Sphalerite occurs as small subhedral crystals disseminated in the host dolostone; within brecciated dolostone as fine aggregates, yellow, grey, brown, and black crystals, and light to dark brown colloform masses and banded encrustations (Fig. 9). Sphalerite $\delta^{34}S$ values range from +2 to +13‰ with a median of +9‰ (Fig.10; Randell, 1994). Laser ablation studies on $\delta^{34}S$ in colloform sphalerite from Polaris (Jaroslaw Nowak, unpublished data, see Fig. 9B) indicate that each band shows a systematic variation from 8‰ in the colloform centre of the mass to 12‰ at the outer crystalline rim. Homogenization temperatures for sphalerite range from 65 to 145°C with most of the data falling between 90 and 105°C. Ice melting temperatures indicate salinity in the range of 24.6 to 31.3 wt.% (NaCl + $CaCl_2$). Hydrocarbon inclusions are rare (<0.07 mol %; Savard et al., 2000).

Galena occurs as small dendritic and skeletal intergrowths within colloform sphalerite, and as cubes and truncated (octahedral) cubes, or as thin, polycrystalline veins. Galena $\delta^{34}S$ ranges from 1.0 to 9.4‰ with a median of 4.5‰. Low α lead occurs in galena within the Panhandle ore. This material was separated and sold at a premium to silicon chip manufacturers who used it to make solder in which the electrons in the outer orbitals would not jump to the next-closest solder. Marcasite typically replaces carbonate, is massive, and green-brass, but also occurs as small needles on dolomite, or as submillimetre blades and aggregates. Limited sampling of marcasite for $\delta^{34}S$ gives a range of 4.9 to 8.9‰.

Pyrobitumen is a common, though volumetrically small, component of the deposit occurring in vugs as crusts or blobs up to 1 cm in diameter. It occurs on top of sulphides and the two sparry dolomite phases, but predates calcite phases. Native sulphur ranges up to 4 wt.% (S. Grasby, GSC-Calgary, unpublished data). Dating of pyrobitumen by Re-Os gives 368 ± 15 Ma, indicating that it is part of the mineralizing event (Selby et al., 2005). Other evidence for hydrocarbon generation at Polaris is found in abundant bitumen (68% of 125 organic matter separates contain bitumen), hydrocarbon fluid inclusions, thermal maturity values indicating passage into the oil window, and petrographic evidence for localized, direct transformation of alginite into bitumen (L.V. Stasiuk, pers. comm., 1992; Randell, 1994).

Barite is rare within the Polaris Mine, occasionally occurring as 1 to 3 mm crystals sparsely coating sphalerite in vugs, but commonly occurs in veins, along faults, or as pore-fill in crinoidal grainstone for up to 10 km from the deposit. Barite $\delta^{34}S$ values range from +36 to +58‰. It commonly occurs with quartz.

There are at three phases of calcite that post-date the main mineralizing event. The first is a fine-grained, grey calcite that appears to have grown at a fluid interface and is associated with small needles of pyrite. The second phase of calcite consists of small, clear crystals less than 5 mm in length. The third phase consists of 2 to 5 cm, often doubly terminated, brown crystals. Small crystals of brown sphalerite rarely occur between the second and third phases of calcite. Homogenization temperatures for calcite range from 75 to

175°C with most of the data above 120°C. Ice melting temperatures indicate salinity in the range of 0.0 to 0.9 wt.% NaCl equiv. Calcite has $\delta^{18}O$ and $\delta^{13}C$ values (PDB) of 5.0 to 25.4‰ and -25.1 to -5.1‰, respectively. $^{87}Sr/^{86}Sr$ ratios for replacement calcite range from 0.70796 to 0.70883 (Savard et al., 2000).

Other Zn-Pb showings have simple paragenetic sequences and a subset of the ore textures seen at Polaris (Fig. 11). The smallest showings only contain fracture-filling crystalline sphalerite (± galena, marcasite); larger showings exhibit some replacement of carbonate by sphalerite in addition to the fracture-filling phases; the largest showings (Eclipse, Truro) exhibit barite, marcasite replacement of carbonate, minor collapse breccias, banded aggregate and colloform sphalerite in addition to fracture filling and replacement crystalline sphalerite.

Figure 11. Alteration and sulphide textures from the Polaris District. *Somerset Island:* **(A)** Crackle and mosaic breccia with chalcocite and calcite. Storm showing ddh ST60 76.4 m. **(B)** Massive chalcocite cementing brecciated Allen Bay Formation. Storm showing ddh ST60, 75.65 m **(C)** Solution breccia, Storm showing ddh ST60, 138.4 m **(D)** Storm Showing. Chalcopyrite and bornite in dolomite-lined vug being replaced by digenite. Storm showing ddh ST60, 45 m x10. Photo by N.S.F. Wilson, GSC Calgary. View is 5 mm wide. **(E)** Pseudobrecciated Turner Cliffs Formation. Seal showing ddh AB95-2, 161 to 215 m. *Dundas Island:* **(F)** Sphalerite and marcasite on collapse breccia clast. *Grinnell Peninsula:* **(G)** Marcasite and trace sphalerite replacing carbonate. JG showing ddh 99-10, 34.0 m **(H)** Marcasite replacing brecciated carbonate and cemented by dolospar. Liz showing ddh LZ01, 18.4 m

Alteration Mineralogy/Chemistry

Geology of Dolomitized Body

The dolomite halo shows sharp margins on its downdip (east) and southern limits, but extends 550 m up dip where it is exposed on surface on the Polaris Peninsula (Figs. 7, 12; Sharp and Dewing, 2004). Carbonate beds in the overlying Irene Bay Formation are dolomitized directly over the orebody, but the upper surface of the dolomite halo drops to lower stratigraphic levels with increasing distance from the orebody. Thinning of the upper Thumb Mountain Formation and collapse of the overlying Irene Bay Formation starts at the edge of the dolomitized halo indicating the removal of some calcite material during the dolomitization. Collapse continues towards the centre of the deposit indicating that upper Thumb Mountain Formation continued to be thinned even after it was dolomitized. The Thumb Mountain Formation is thinned by 50% in the orebody (Figs. 8, 12). Few beds within the dolomite halo escape dolomitization; even apparently impermeable lime mudstone beds are dolomitized. All beds outside of the dolomitized halo are limestone in the vicinity of Polaris.

Breccias occur only within the dolomitized halo around the orebody and always contain at least some mineralization. Brecciation is thus considered to be part of the mineralizing event. Breccias are concentrated in certain stratigraphic intervals away from the orebody, but become thicker and coalesce towards the orebody. There is a general (though not infallible) progression from pseudobreccia and crackle breccia to collapse or solution breccia towards the core of the orebody (Fig. 12). Collapse breccias are common within the orebody but pervasive overprinting and replacement by zinc-lead-iron sulphide mineralization makes it difficult to recognize carbonate textures.

The gradation from pseudobreccia to crackle breccia to collapse or solution breccia towards the orebody is presum-

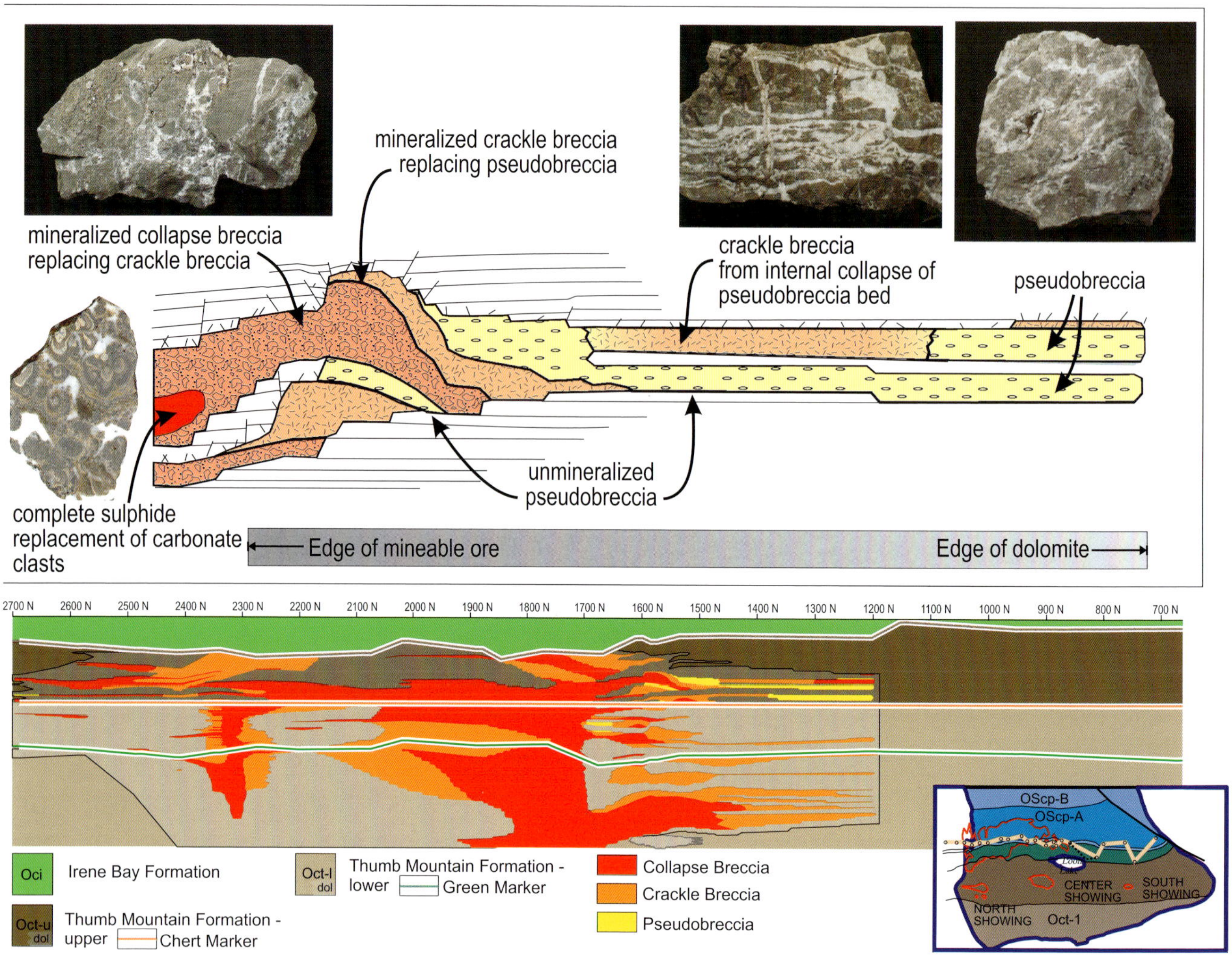

FIGURE 12. *Lower part:* Long section through Polaris Mine hung on the base of the Chert Marker (M-Ch) showing distribution of dolomitization and breccia types. Note the thinning of the upper Thumb Mountain Formation starting at the edge of the dolomitization and continuing towards the centre of the orebody. Section length 2000 m, location shown on inset map. No vertical exaggeration. *Upper part:* Schematic figure showing the distribution of breccia fabrics within the upper Thumb Mountain Formation at Polaris Mine (roughly between 1400 and 1650 N). Pseudobreccia forms in porous beds far from the orebody and grades into crackle and collapse breccia closer to the ore. x2 vertical exaggeration.

ably a result of increasing dissolution. Pseudobreccia is formed in porous and permeable beds where acidic fluid migrating away from the site of ore deposition began to remove carbonate. Removal of carbonate reduced the strength of the pseudobrecciated beds and these began to collapse. Crackle breccia formed as blocks of pseudobreccia crumbled and settled into the newly created void space and as undissolved beds adjusted due to the overall thinning of the Thumb Mountain Formation. The increasingly porous and permeable crackle breccia zones were then further dissolved creating solution and collapse breccias (Fig. 12). Pseudobreccia is only recognized on the fringes of the dolomitized halo as it has been overprinted by more intense brecciation and mineralization closer to the orebody.

Clay Mineral Assemblages

There is a zonation of clay mineral assemblages within the host rocks away from the ore (Fig. 13; Héroux et al., 1996, 1999). Six clay mineral assemblages are present: 1) the background sedimentary clay assemblage in the Cape Phillips Formation consists of illite>interstratified illite/smectite>chlorite; 2) the background clay mineral assemblage for the Thumb Mountain Formation is illite>chlorite>interstratified illite/chlorite; 3) the Irene Bay Formation and less commonly the uppermost part of the Thumb Mountain Formation contains a background chlorite-corrensite assemblage (chlorite+corrensite+chlorite/smectite mixed layer)>illite, interstratified with illite/smectite; 4) directly over the main orebody in the upper part of the Thumb Mountain and in the Irene Bay formations is a pure illite facies (with traces of chlorite) containing well crystallized illite with few swelling layers; 5) within 250 m of the ore in the Thumb Mountain Formation is a coarse kaolinite assemblage. It contains kaolinite and illite, with some traces of chlorite. The kaolinite to illite ratio is always much higher in the 2 to 16 micron fraction than in the <2 micron fraction. Scanning electronic microscopy shows that this kaolinite is present as large vermicules filling the porosity in dolostone; 6) within 1000 m of the orebody in the Thumb Mountain Formation is a fine kaolinite assemblage containing illite and kaolinite with trace chlorite. The illite to kaolinite ratio is

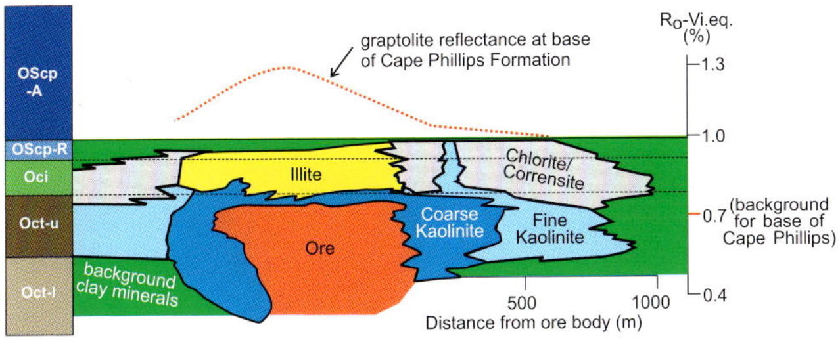

FIGURE 13. Schematic long section through the Polaris Mine showing the distribution of clay mineral assemblages and graptolite reflectance in the overlying Cape Phillips Formation. Stratigraphic column shown on the left (north) side. Details in Randell (1994) and Héroux et al. (1996).

about the same in the less than 2 micron fraction as in the 2 to 16 micron fraction.

Organic Matter Alteration

Reflectance of organic matter (R_O) in the impermeable Cape Phillips and Irene Bay formations near the ore shows higher values than away from the ore (Fig. 13; Randell, 1994; Héroux et al., 1996, 1999). Above the orebody, graptolites in the Cape Phillips Formation exhibit a R_O of 1.35%, twice the expected background value (R_O=0.7%) for this lithostratigraphic unit as determined from remote diamond drillholes and outcrops. R_O values from the Irene Bay Formation are also 50% higher than background above the ore zone. These values decrease away from the ore, attaining background values at 1.2 km radius from the mine. In contrast, Ro values from the upper Thumb Mountain Formation within the ore deposit are 20% lower than the expected R_O background (R_O=0.5% rather than 0.75%). This "suppressive" effect is also shown by decreasing Ro values with depth. The difference between the lowest and the highest R_O values, occurring respectively near the bottom and the top of at the same diamond drill, is twice the lower value.

Cu-Rich Mineralization and Associated Alteration

Alteration at the copper-rich showings at Storm and Hornby Zn includes fracturing to crackle breccia; small zones of mosaic packbreccias with calcite, pyrite, and subordinate chalcocite cement; pervasive hematite staining; and rare malachite/chalcocite on fractures. Mineralization is usually accompanied by medium to coarsely crystalline dolomitization and local dedolomitization of the host lithology; minor sparry dolomite; pockets of silicification; limonite staining and hematization; crackle and mosaic breccias with solution-rounding clasts. Chalcocite and bornite are the most common copper sulphides, with lesser chalcopyrite (Fig. 11). Accessory copper minerals include cuprite, covellite, and native copper. Malachite and azurite occur as surface stain and are associated with iron oxides in drill core. Pyrite is the main non-copper sulphide but small veins of sphalerite and galena occur. Fluid inclusion homogenization temperatures on sparry dolomite yield 100 to 120°C and ice melting temperatures indicate salinities >35 wt.% NaCl eq. Sphalerite has a $\delta^{34}S$ value of 8.5 ‰, whereas two samples of chalcocite yielded $\delta^{34}S$ values of –9.5 and +10.7 ‰.

Two diagenetic events are documented at the Storm showing. The first was dolomitization accompanied by precipitation of pyrite, chalcopyrite, and lesser bornite. Silica, in the form of cryptocrystalline quartz, minor sphalerite, and galena, were deposited at the margins of the alteration system. The second event was minor dissolution and dedolomitization of dolomite, dissolution of galena, and precipitation of zinc oxides, covellite, chalcocite, bornite, native copper, iron oxide, and malachite. Bornite, covellite and chalcocite replace pyrite and chalcopyrite and fill fissures within the rock. These copper minerals were subsequently altered to malachite and azurite. Calcite precipitation accompanied the second event.

The copper mineralization in the Polaris district is thought to be directly associated with the main zinc mineralizing event because 1) Zn-Pb mineralization occurs with or overlaps copper mineralization and 2) both were formed by saline brines of similar composition and temperature. Copper mineralization occurs at the contact between the Allen Bay and Cape Storm formations, possibly reflecting a local source of metals.

Structure and Deformation

The primary control on the location of showings within the Polaris District is structural. Mineralization occurs in four distinct structural settings: 1) along strike-slip and tear faults that separate the east margin (Harrison, Agpan, Aquarius showings) and west margin (Liz-Tiger-Simba, BK, Trigger) of the Boothia Uplift from the adjacent Parry Island and Central Ellesmere fold belts; 2) in grabens and half grabens along extensionally reactivated, Boothia-aged thrust faults (Eclipse, Rookery, Hornby, Dundas, Abbott, Bacon); 3) in the axial region of east-west trending anticlines (Stuart River, Caribou, Baillie Hamilton, high-grade zones within Polaris Mine); and 4) on an east-west-trending fault on Somerset Island marked by a line of small grabens and facies change (Seal, Storm). This structure directly lines up with the southern bounding fault of the Proterozoic Borden Basin on Baffin Island and thus is presumably the reactivation of basement faults reflected in the overlying Paleozoic strata.

There is a second-order, stratigraphic control on the location of mineralization within the sedimentary pile and on the mineralogy of the showing. Limestone units that are overlain by less permeable shale, evaporite, or argillaceous dolostone units are preferentially mineralized. These include the Thumb Mountain, Blue Fiord, and Allen Bay formations. Cu-Fe±Zn-Pb mineralization (Storm Cu, Hornby Cu, Eclipse, Tern Lake) occurs only within Silurian strata, either at the contact between the Allen Bay and overlying Cape Storm formations (Storm, Tern Lake), or where Early Devonian synorogenic clastic strata overlie Silurian strata (Hornby Cu, Eclipse).

Two showings on eastern Cornwallis Island located along the shelf margin contact between the basinal Cape Phillips Formation and the platformal Barlow Inlet Formation (Laura Lakes and Cape) have no obvious associated structures and appear to be stratigraphically controlled showings.

Genetic Models

The model for the Polaris deposit (Fig. 14) invokes a source of metal ions within the stratigraphic column, since strontium or lead isotopes show no indication of basement or basal clastics being involved. Metals are then carried in sul-

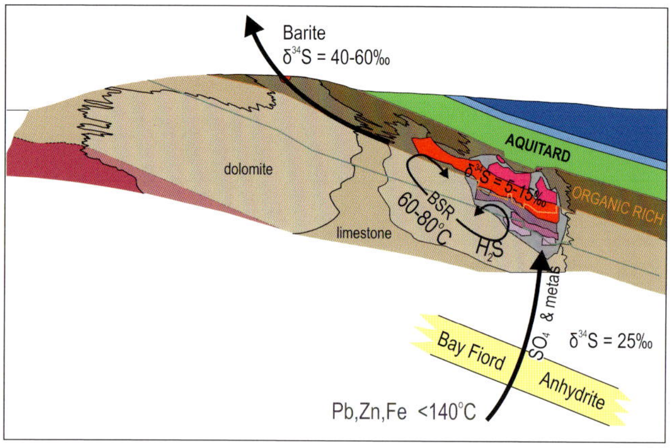

FIGURE 14. Genetic model for the Polaris Mine showing the origin of metal and sulphate-rich fluids in the sedimentary pile, transport as a single fluid to site of ore precipitation, reduction and fractionation by bacterial sulphate reduction (BSR), combination of isotopically light H_2S and metal ions to form ore and create acidic fluids, and expulsion of isotopically heavy sulphate to produce a barite halo. Legend as for Figure 8.

phate-rich brines through aquifers deep in the stratigraphic column. Driven by an orogenic process, either the Late Devonian Ellesmerian Orogeny or the mid-Carboniferous opening of the Sverdrup Basin, circulating fluids rose along faults until they encountered the organic-rich, permeable Thumb Mountain Formation limestone overlain by the impermeable Irene Bay Formation shale.

Bacterial sulphate reduction reduces sulphate to H_2S using the abundant algal organic matter in the upper Thumb Mountain Formation as a reductant. The resulting H_2S has a shift in $\delta^{34}S$ values of -15‰ (Fig. 10). The H_2S reacts inorganically with zinc, lead, and iron ions to produce sulphides with a $\delta^{34}S$ value of about 10‰. Released hydrogen ions dissolve carbonate resulting in dissolution breccias and other dissolution fabrics. It is proposed that excess sulphate was expelled from the system and produced a halo of barite up to 10 km from the deposit with $\delta^{34}S$ values of 40 to 60‰. Evidence of a temporal link between barite and ore is based on one sample from the 930-124 stope at Polaris that has sphalerite that pre- and post-dates barite, proximity to Polaris (barite is common only on southern Little Cornwallis Island), heavy sulphur isotope values, and occurrence with marcasite.

Temperatures measured from fluid inclusions and inferred from the thermal maturity data indicate a range of 80 to 105°C. This is the appropriate range for bacterial sulphate reduction (Machel et al., 1995). Thermochemical sulphate reduction is unlikely to have been active since 1) the temperatures are too low (≥140°C is needed); 2) there is no blocky calcite with strongly negative $\delta^{13}C$ values, which is typically produced during thermochemical sulphate reduction; 3) there is fractionation of the sulphur isotopes, a process that does not occur during thermochemical sulphate reduction (Machel et al., 1995); and 4) the $\delta^{34}S$ in individual sphalerite masses is inhomogeneous, typically showing a variation from 8‰ in the colloform centre of the mass to 12‰ at the outer crystalline rim. Presumably this reflects pulses of fluid activity, with bacterial sulphate reduction producing a gas cap with strongly fractionated H_2S, followed by the introduction of metal-rich brine, rapid precipitation of amorphous colloform sphalerite, followed by slower precipitation of crystalline sphalerite as the gas cap is depleted and gas slowly produced by ongoing bacterial sulphate reduction.

This model fails to explain many of the showings on Grinnell Peninsula, at the Seal deposit on Somerset Island, or at Bass Point on Bathurst Island, where the $\delta^{34}S$ of sphalerite is about the same as marine sulphate (25 to 30‰). These showings also lack barite and some have sphalerite in calcite veins. Grinnell and Somerset are beyond the northern or southern limits of the Baumann Fiord and Bay Fiord evaporites and may reflect the long distance transport of thermochemical sulphate reduction generated gas. If so, the $\delta^{34}S$ of individual crystals should be homogeneous.

Exploration Methods

Historical

Exploration cycles in the district typically started with prospecting, stream sediment heavy mineral sampling, and more recently, airborne hyperspectral surveys. These activities have been followed by property-scale mapping, soil sampling, clay and thermal maturity analysis, gravity, IP, and magnetic surveys. Targets identified were then drilled.

Geological techniques have focused on property-scale mapping of alteration indicators, such as the distribution of dolomite, carbonate dissolution fabrics, barite, and sulphide occurrences and textures. Stream sediment heavy mineral sampling has proven effective in locating showings, but soil sampling often gives results that are hard to interpret, presumably due to the lack of soil development and chemical weathering in the permafrost zone. Clay mineral and organic matter alteration techniques (described above; Héroux et al., 1999) are effective in locating showings and providing vectors to the area of maximum alteration. Airborne hyperspectral surveys employed by Noranda on Grinnell Peninsula located numerous new showings, but other airborne techniques (e.g. EM) failed to locate any new showings. The Polaris Mine responded well to IP and gravity surveys, with a whopping 2.2 milligal residual gravity anomaly over the deposit, although the gravity anomaly is influenced more by the large body of dolomite within limestone rather than sulphide within dolomite. IP located the Polaris deposit, but

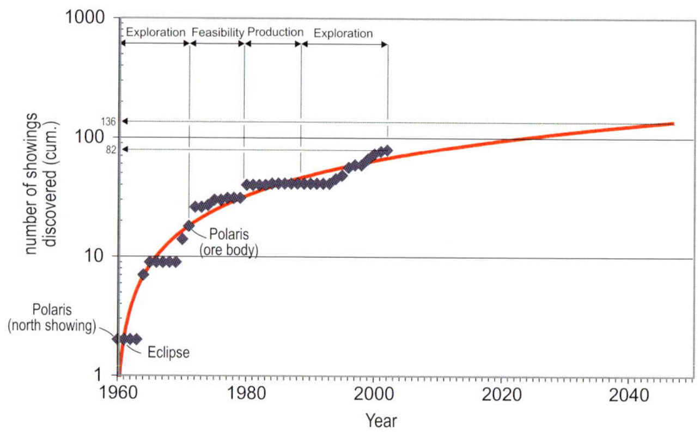

FIGURE 15. Graph of number of known showings versus year of discovery. Eighty-two showings are known and the best-fit curve indicates that as many as fifty showings remain to be found.

elsewhere in the district has provided numerous false positives due to salt-water lenses (taliks) within the permafrost and remains of uncertain value.

Potential for New Targets

Exploration potential exists for additional deposits within the Polaris District. The curve showing number of known showings versus year indicates that many showings remain to be found (Fig. 15). The difficult logistics and short summer season often governed exploration instead of geology, so undiscovered showings could be significant.

The district may respond well to a second exploration cycle built on more sophisticated exploration techniques. Exploration starting with hyperspectral surveys to identify areas of iron oxides, barite, and clay mineral alteration could be followed by thermal maturity sampling spatially associated with potential structurally controlled fluid conduits, pathfinder geochemical surveys (Zn, Pb, Cd, Cu, Ba, Fe), and then the more traditional methods described above.

Highest priorities for additional exploration (Fig. 16) should be 1) poorly explored, Devonian-filled half grabens at Rookery Creek, Taylor River, and Eleanor River on Cornwallis Island; on Baillie Hamilton Island; and along the Icefield fault on Grinnell Peninsula; 2) Liz-Tiger-Simba trend on Grinnell Peninsula; 3) carbonate-hosted, copper-rich showings associated with the Allen Bay-Cape Storm formational contact in the vicinity of faults northeast of Resolute Bay, and on structures parallel to the Aston-Batty Line on Somerset Island and on southern Devon Island; and 4) unexplained thermal anomalies at Victor Creek northeastern Cornwallis Island described by Héroux et al. (1999), and 5) buried Thumb Mountain Formation on eastern Bathurst Island

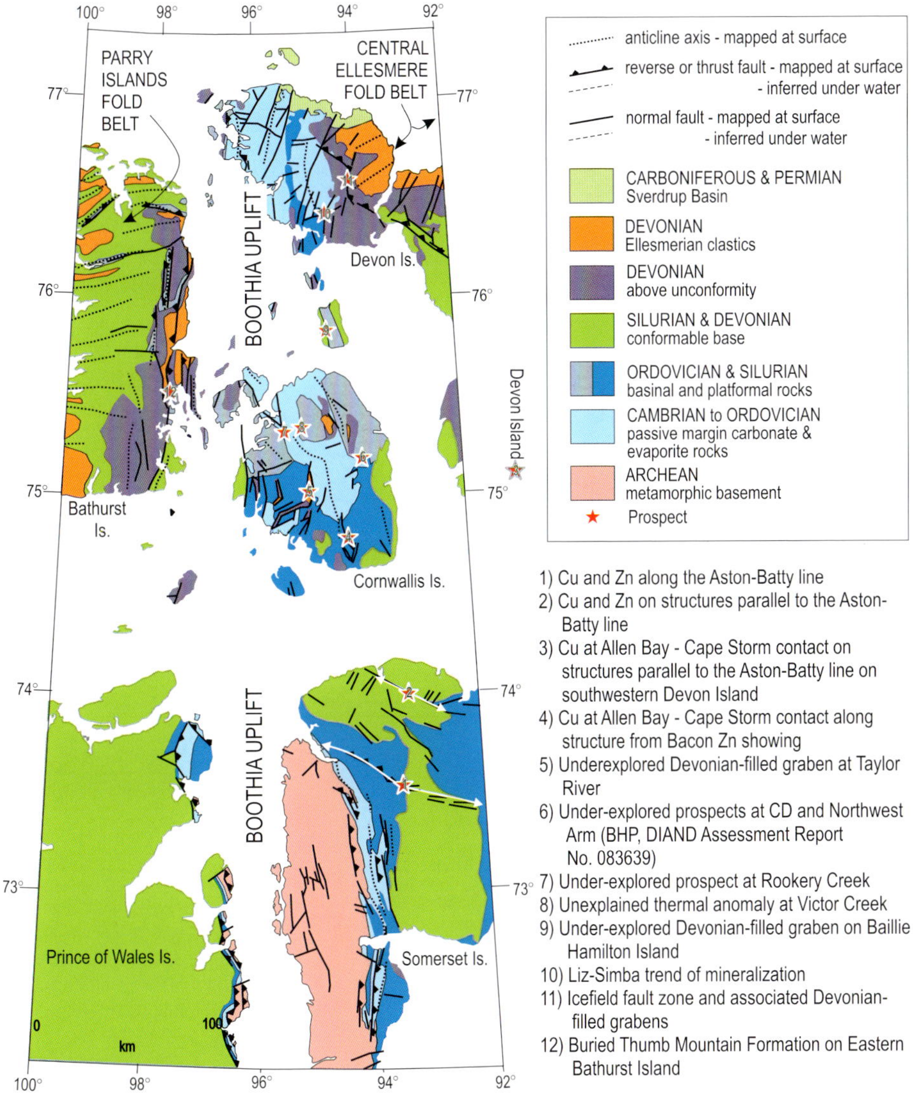

FIGURE 16. Prospects for Zn-Pb or Cu potential within the Polaris District that these authors consider to have high potential.

There is another, as yet untested, deposit model for a separate type of lead-zinc mineralization in the district. Absolute dating of the Nanisivik deposit on northern Baffin Island (Sherlock et al., 2003) has yielded a Middle Ordovician age (461 Ma or Chayzan) for the mineralization and associated alteration of volcanic dykes. This result, while controversial, indicates that some potential exists for iron-rich, carbonate-hosted deposits in areas that overlie the ancestral Borden Basin (Fig. 17). The age of mineralization as determined by Sherlock et al. (2003) is synchronous with the deposition of the Bay Fiord Formation. The intraplatformal basin that developed and accommodated evaporites of the Bay Fiord Formation may have resulted from the onset of subduction under the North American continent and the associated drag from the down-going slab. The "Melville Arch" (Figs. 3, 17) might have been the resulting peripheral bulge.

Knowledge Gaps

The following knowledge gaps need to be addressed for effective future exploration:

Mapping

- 1:50 000 mapping is required on Cornwallis Island to locate small, Ellesmerian-aged structures that are beyond the resolution of the current 1:250 000 map. These structures controlled fluid flow at Polaris and on northeast Cornwallis Island (Jober, 2005). Locating these potential fluid conduits would help target exploration.
- 1:50 000 mapping is required on Somerset Island in the

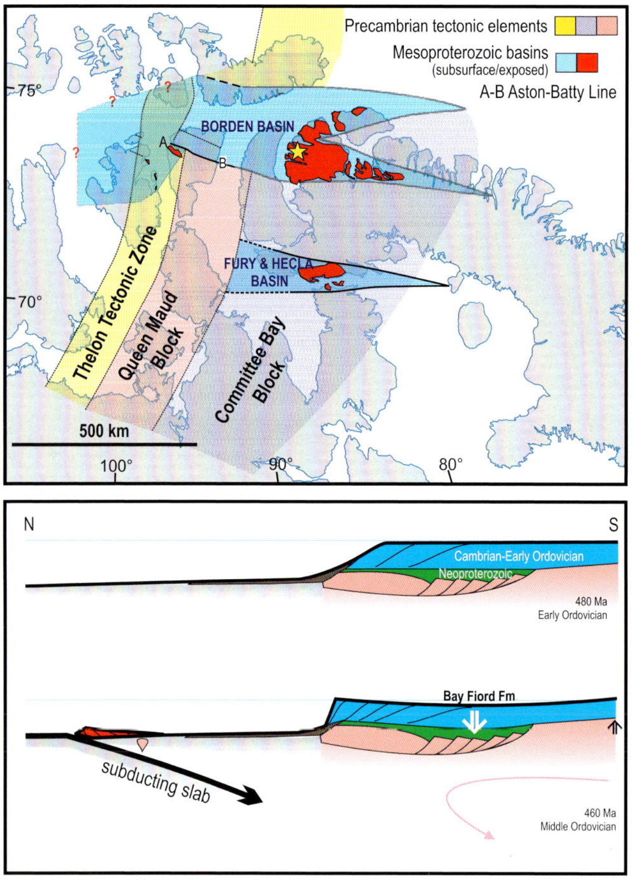

FIGURE 17. Top. Distribution of Precambrian tectono-stratigraphic elements and their inferred distribution below the Paleozoic cover (modified from Jackson, 2000). Outcrops of Proterozoic sedimentary basins in red. Location of Proterozoic sedimentary basins in blue. Nanisivik Mine shown by yellow star. Line A-B shows location of Aston Batty Line on northern Somerset Island. *Bottom.* Possible configuration of the northern margin of Laurentia in the Early and Middle Ordovician showing the postulated onset of subduction below Laurentia and the formation of an intraplatformal basin (Bay Fiord Formation) and a peripheral bulge.

area of the Aston-Batty Line to better define its location and look for additional alteration indicators.

- Mapping and prospecting in stratigraphy younger than the Middle Ordovician Thumb Mountain Formation. The Allen Bay-Cape Storm contact in the vicinity of faults on Somerset, Cornwallis, Grinnell, and Devon islands may be analogous with the Storm copper showing and should be revisited.
- The Devonian stratigraphy is complex and poorly understood. New data on the age of the base of each subbasin are desirable. Many of the Devonian-filled half grabens on Cornwallis, Grinnell, and Bathurst islands have never been mapped in detail or prospected.

Geochemistry

- Geochemical characterization of the alteration halo should be improved as a guide to exploration: 1) chemistry of dolomite associated with alteration versus regional burial dolomite; 2) barite distribution; 3) hyperspectral, PIMA and ASTER response of alteration products (barite, dolomite, clay); and 4) thermal alteration of organic matter across the district
- Organic matter studies to 1) characterize hydrocarbon generation around the Polaris deposit and perform mass balance calculations; and 2) continue mapping thermal maturity indicators to look for buried orebodies in areas of exploration interest.

Geophysics

- Higher resolution regional gravity and aeromagnetic surveys would better locate basement structures that partially control structures in overlying Paleozoic strata.
- Shallow seismic surveys of Pb-Zn prospects using a low-energy source may offer an alternate way to pinpoint drill targets.

Acknowledgements

The authors thank Michael Gunning, Victoria Yehl, Chris Harrison, and Wayne Goodfellow for the many suggestions they made to improve this paper. TeckCominco Ltd. is warmly thanked for providing data on the Polaris, Storm, Truro, and Eclipse deposits, as well as for logistical support in 2001. Polar Continental Shelf Project and Technical Field and Support Services, both of Natural Resources Canada, helped with logistical support.

References

Christensen, J.N., Halliday, A.N., Leigh, K.E., Randell, R.N., and Kesler, S.E., 1995, Direct dating of sulphides by Rb-Sr: A critical test using the Polaris Mississippi-Valley-type Zn-Pb deposit: Geochimica et Cosmochimica Acta, v. 59, p. 5191-5197.

Davies, G.R., and Krouse, H.R., 1975, Sulphur isotope distribution in Paleozoic sulphate evaporites, Canadian Arctic Archipelago: Geological Survey of Canada, Report of Activities 75-1B, p. 221-225.

de Freitas, T.A., and Mayr, U., 1993, Middle Paleozoic tear faulting, basin development, and basin uplift, central Canadian arctic: Canadian Journal of Earth Science, v. 30, p. 603-620.

de Freitas, T.A., Trettin, H.P., Dixon, O.A., and Mallamo, M., 1999, Silurian System of the Canadian Arctic Archipelago: Bulletin of Canadian Petroleum Geology, v. 47, p. 136-193.

Dewing, K., and Turner, E., 2003, Structural Setting of the Cornwallis Lead-Zinc District, Arctic Islands, Nunavut: Geological Survey of Canada, Current Research. B-4, 9 p.

Dewing, K., Harrison, J.C., Pratt, B.R., and Mayr, U., 2004, A probable Late Neoproterozoic age for the Kennedy Channel and Ella Bay formations, northeastern Ellesmere Island and its implications for passive margin history of the Canadian Arctic: Canadian Journal of Earth Sciences, v. 41, p. 1013-1025.

Disnar, J.R., and Héroux, Y., 1995, Dégradation et lessivage des hydrocarbures de la formation ordovicienne de Thumb Mountain encaissant le gîte Zn-Pb de Polaris (Territoires du Nord-Ouest, Canada): Canadian Journal of Earth Sciences, v. 32, p. 1017-1034.

Gentzis, T., de Freitas, T., Goodarzi, F., Melchin, M., and Lenz, A.C., 1996, Thermal maturity of lower Paleozoic sedimentary successions in Arctic Canada: American Association of Petroleum Geologists Bulletin, v. 80, p. 1065-1084.

Harrison, J.C., 1995, Melville Island's Salt-Based Fold Belt, Arctic Canada: Geological Survey of Canada, Bulletin 472, 331 p.

Harrison, J.C., and de Freitas, T., 1996, New showings and new geological settings for mineral exploration in the Arctic Islands: Geological Survey of Canada, Current Research 1996-B, p. 81-91.

Heal, G.E.N., 1976, The Wrigley-Lou and Polaris-Truro lead zinc deposits, Northwest Territories: Unpub. M.Sc. thesis, University of Alberta, Edmonton, Alberta, 172 p.

Heaman, L.M., LeCheminant, A.N., and Rainbird, R.H., 1992, Nature and timing of Franklin igneous events, Canada; Implications for a late Proterozoic mantle plume and the break-up of Laurentia: Earth and Planetary Science Letters, v. 109, p. 117-131.

Henrichsen, M., 2003, Structural geometry and kinematics of deformation in the Stanley Head Region, western Cornwallis Island, Nunavut: Unpub. M.Sc. thesis, University of British Columbia, Vancouver, British Columbia, 147 p.

Héroux, Y., Chagnon, A., and Savard, M., 1996, Organic matter and clay anomalies associated with base-metal sulfide deposits: Ore Geology Reviews, v. 11, p. 157-173.

Héroux, Y., Chagnon, A., Dewing, K., and Rose, H.R., 1999, The carbonate-hosted base-metal sulphide Polaris deposit in the Canadian Arctic: Organic matter alteration and clay diagenesis, *in* Glikson, M., and Mastalerz, M., eds., Organic Matter and Mineralisation: Thermal Alteration, Hydrocarbon Generation and Role in Metallogenesis: Kluwer Academic Publisher, p. 260-295.

Jackson, G.D., 2000, Geology of the Clyde-Cockburn Land map area, north-central Baffin Island, Nunavut: Geological Survey of Canada, Memoir 440. 303 p.

Jober, S.-A. 2005, Structural geology of the Stuart River – Caribou Lake region, eastern Cornwallis Island, Nunavut: Unpub. M.Sc. thesis, University of New Brunswick, Fredricton, New Brunswick, 104 p.

Jowett, C., 1975, Nature of the ore-forming fluids of the Polaris lead-zinc deposit, Little Cornwallis Island, N.W.T., from fluid inclusion studies: Canadian Institute of Mining and Metallurgy Bulletin, v. 68, p. 124-129.

Kerr, J.W., 1977, Cornwallis lead-zinc district; Mississippi Valley-type deposits controlled by stratigraphy and tectonics: Canadian Journal of Earth Sciences, v. 14, p. 1402-1426.

Machel, H.G., Krouse, H.R., and Sassen R., 1995, Products and distinguishing criteria of bacterial and thermochemical sulfate reduction: Applied Geochemistry, v. 10, p. 373-389.

Mayr, U., 1978, Stratigraphy and Correlation of Lower Paleozoic Formations, Subsurface of Cornwallis, Devon, Somerset, and Russell Islands, Canadian Arctic Archipelago: Geological Survey of Canada, Bulletin 276, 55 p.

Mayr, U., Packard. J.J., Goodbody, Q.H., Okulitch, A.V., Rice, R.J., Goodarzi, F., and Stewart, K.R., 1994, The Phanerozoic Geology of Southern Ellesmere and Northern Kent Islands, Canadian Arctic Archipelago: Geological Survey of Canada, Bulletin 470, 298 p.

Mayr, U., de Freitas, T., Beauchamp, B., and Eisbacher, G., 1998, The Geology of Devon Island North of 76°, Canadian Arctic Archipelago: Geological Survey of Canada, Bulletin 526, 500 p.

Miall, A.D., 1986, Effects of Caledonian tectonism in Arctic Canada: Geology, v. 14, p. 904-907.

Miall, A.D., and Gibling, M., 1978, The Siluro-Devonian clastic wedge of Somerset Island, Arctic Canada, and some regional paleogeographic implications: Sedimentary Geology, v. 21, p. 85-127.

Mitchell, I., 2000, The Mineralising Fluids, Stable Sulphur Isotopes, and Radiogenic Age of the Bermuda Pb-Zn showing, Nunavut, Canada: Unpub. B.Sc. thesis, University of Waterloo, Waterloo, Ontario, 65 p.

Muir, I.D, and Rust, B.R., 1982, Sedimentology of a Lower Devonian coastal alluvial fan complex; the Snowblind Bay Formation of Cornwallis Island, Northwest Territories, Canada: Bulletin of Canadian Petroleum Geology. v. 30, p. 245-263.

Muraro, T.W., 1974, Lead-zinc mining on Little Cornwallis Island, *in* Wren, A.E., and Cruz, R.B., eds., Proceedings of the 1973 National Convention, Canadian Society of Exploration Geophysists, p. 230-234.

Nentwich, F.W., 1987, Stratigraphy and Sedimentology of the Ordovician and Silurian Brodeur Group, Northern Brodeur Peninsula, Baffin Island, Unpub. Ph.D. thesis, University of Alberta, Edmonton, Alberta, 457 p.

Nentwich, F.W., and Jones, B., 1989, Stratigraphy and sedimentology of Ordovician and Silurian strata, northern Brodeur Peninsula, Baffin Island: Bulletin of Canadian Petroleum Geology, v. 37, 428-442.

Okulitch, A.V., comp., 1991, Geology of the Canadian Arctic Archipelago, Northwest Territories and North Greenland, Figure 2 in Trettin, H.P., ed., Geology of the Innuitian Orogen and Arctic Platform of Canada and Greenland: Geological Survey of Canada, Geology of Canada No. 3 (also Geological Society of America, Geology of North America, v. E). 1:2,000,000 scale, 1 sheet.

Okulitch, A.V., Packard, J.J., and Zolnai, A.I., 1986, Evolution of the Boothia Uplift, Arctic Canada: Canadian Journal of Earth Sciences, v. 23, p. 350-358.

Planchy, J., 1998, Zinc, in Sachs, J., ed., Metal prices in the United States through 1998: minerals.usgs.gov/ minerals/pubs/commodity/zinc/720798.pdf

Randell, R.N., 1994, Geology of the Polaris Zn-Pb Mississippi Valley-Type Deposit, Canadian Arctic Archipelago: Unpub. Ph.D. thesis, University of Toronto, Toronto, Ontario, 736 p.

Randell, R.N., and Anderson, G.M., 1997, Geology of the Polaris Zn-Pb deposit and surrounding area, Canadian Arctic Archipelago, *in* Sangster, D.F., ed., Carbonate-Hosted Lead-Zinc Deposits: Society of Economic Geologists, Special Publication No. 4, p. 307-319.

Randell, R.N., Héroux , Y., Chagnon, A., and Anderson, G.M., 1997, Organic matter and clay minerals at the Polaris Zn-Pb deposit, Canadian Arctic Archipelago, in Sangster, D.F., ed., Carbonate-Hosted Lead-Zinc Deposits, Society of Economic Geologists, Special Publication No. 4, p. 320-329.

Rose, S.R.A., 1999, Sedimentology and diagenesis of the lower Blue Fiord Formation carbonates in a prospective Mississippi-Valley-type (Pb-Zn) setting, Bathurst Island, N.W.T.: Unpub. M.Sc. thesis, University of Calgary, Calgary, Alberta, 172 p.

Sanford, B.V., and Grant, A.C., 2000, Geological framework of the Ordovician System in the southeast Arctic Platform, Nunavut, *in* McCracken, A.D., and Bolton, T.E., eds., Geology and Paleontology of the Southeast Arctic Platform and Southern Baffin Island, Nunavut: Geological Survey of Canada, Bulletin 557, p. 13-38.

Savard, M.M., Chi, G., Sami, T., Williams-Jones, A.E., and Leigh, K., 2000, Fluid inclusion and carbon, oxygen, and strontium isotope study of the Polaris Mississippi Valley-type Zn-Pb deposit, Canadian Arctic Archipelago: Implications for ore genesis: Mineralium Deposita, v. 35, p. 495-510.

Selby, D., Creaser, R.A., Dewing, K., and Fowler, M., 2005. Evaluation of bitumen as a 187Re-187Os geochronometer for hydrocarbon maturation and migration: A case study from the Polaris MVT deposit, Canada: Earth and Planetary Science Letters, v. 235, p. 1-15.

Sharp, R.J., and Dewing, K., 2004, Dolomitization and brecciation at the Mississippi-Valley type Zn-Pb Polaris Mine, central Arctic Islands, Nunavut, *in* Packard, J.J., and Davies, G.R., eds., Dolomites: The Spectrum - Mechanisms, Models, Reservoir Development: Canadian Society of Petroleum Geology Core Conference, p. 27.

Sharp, R.J., Ste-Marie, C.P., and Lorenzini, C., 1995a, Field study of Polaris Mine area, NWT, Canada, *in* Misra, K.C., ed., Carbonate-Hosted Lead-Zinc-Fluorite-Barite Deposits of North America: Guidebook Series, v. 22, p. 38-41.

Sharp, R.J., Ste-Marie, C.P., Lorenzini, C., Leigh, K.E., Dewing, K., Héroux , Y., and Chagnon, A., 1995b, A field guide to the geology of the Polaris Mine, Little Cornwallis Island, Northwest Territories, Canada, *in* Misra, K.C., ed., Carbonate-Hosted Lead-Zinc-Fluorite-Barite Deposits of North America: Guidebook Series, v. 22, p. 19-37.

Sherlock, R.L., Lee, J.K.E., and Cousens, B., 2003, Geological and geochronological constraints on the timing of mineralization at the Nanisivik Zinc-Lead Mississippi Valley-type deposit, Northern Baffin Island, Nunavut, Canada: Geological Association of Canada, Annual Meeting, Vancouver, Abstract no. 124, 1 p.

Stewart, W.D., 1987, Late Proterozoic to Early Tertiary Stratigraphy of Somerset Island and Northern Boothia Peninsula, District of Franklin, N.W.T.: Geological Survey of Canada, Paper 83-26, 78 p.

Symons, D.T.A., and Sangster, D.F., 1992, Late Devonian paleomagnetic age for the Polaris Mississippi Valley-type Zn-Pb deposit, Canadian Arctic Archipelago: Canadian Journal of Earth Sciences, v. 29, p. 15-25.

Thorsteinsson, R., 1984, A sulphide deposit containing galena, in the Lower Devonian Disappointment Bay Formation on Baillie Hamilton Island, Canadian Arctic Archipelago: Geological Survey of Canada, Paper 84-1B, p. 269-274.

——— 1986, Geology of Cornwallis Island and neighbouring smaller islands, Canadian Arctic Archipelago: Geological Survey of Canada, Map 1626A, scale 1:250,000.

Thorsteinsson, R., and Mayr, U., 1987, The Sedimentary Rocks of Devon Island, Canadian Arctic Archipelago: Geological Survey of Canada, Memoir 411, 182 p.

Thorsteinsson, R., and Tozer, E.T., 1970, Geology of the Arctic Archipelago, in Douglas, R.J.W., ed., Geology and Economic Minerals of Canada: Geological Survey of Canada, Economic Geology Report no. 1, p. 547-590.

Thorsteinsson, R., and Uyeno, T.T., 1980, Stratigraphy and Conodonts of Upper Silurian and Lower Devonian Rocks in the Environs of the Boothia Uplift, Canadian Arctic Archipelago: Geological Survey of Canada, Bulletin 292, 75p.

Trettin, H.P., 1969, Lower Paleozoic sediments of northwestern Baffin Island, District of Franklin. Geological Survey of Canada, Bulletin 157, 70 p.

Trettin, H.P. (ed.) 1991: Geology of the Innuitian Orogen and Arctic Platform of Canada and Greenland: Geological Survey of Canada, Geology of Canada no. 3 (also Geological Society of America, Geology of North America v. E), 569 p.

Trettin, H.P., 1994, Pre-Carboniferous Geology of the Northern Part of the Arctic Islands: Hazen Fold Belt and Adjacent Parts of central Ellesmere Fold Belt, Ellesmere Island: Geological Survey of Canada, Bulletin 430, 248 p.

Turner, E.C., and Dewing, K. 2004, Geology, Little Cornwallis Island and Rookery Creek (Cornwallis Island), High Arctic, Nunavut: Geological Survey of Canada, Open File 1780, 1:50 000 scale.

METALLOGENY OF THE DOYON-BOUSQUET-LARONDE MINING CAMP, ABITIBI GREENSTONE BELT, QUEBEC

PATRICK MERCIER-LANGEVIN[1], BENOÎT DUBÉ[1], BENOÎT LAFRANCE[2], MARK HANNINGTON[3], ALAN GALLEY[4], JAMES MOORHEAD[5], AND PATRICE GOSSELIN[1]

1. Geological Survey of Canada, 490 rue de la Couronne, Québec, Quebec G1K 9A9
2. Cogitore Resources, 1300 Saguenay, Rouyn-Noranda, Quebec J9X 7C3
3. Department of Earth Sciences, University of Ottawa, Marion Hall, 140 Louis Pasteur,Ottawa, Ontario K1N 6N5
4. Geological Survey of Canada, 601 Booth Street, Ottawa, Ontario K1A 0E8
5. Ministère des Ressources naturelles et de la Faune, 400 Lamaque, Val-d'Or, Quebec J9P 3L4
Corresponding author's email: pmercier@nrcan.gc.ca

Abstract

The Doyon-Bousquet-LaRonde (DBL) mining camp deposits are hosted by the volcanic and intrusive rocks of the 2701-2696 Ma Bousquet Formation located in the upper part of the 2703-2694 Ma Blake River Group in the Abitibi greenstone belt. Since the 1930s, mining has taken place at eight Au ± base metal deposits in the DBL camp for a cumulative production, at the end of 2005, of 67 million tonnes (Mt) of ore at 5.67 g/t Au (381 tonnes Au, or 12.3 Moz). The total production, reserves, and estimated resources amount to 133 Mt of ore at 5.23 g/t Au (694 t Au, or 22.3 Moz).

Three main Au±Cu-Zn-Ag deposit types are recognized in the DBL mining camp: 1) Au-rich volcanogenic massive sulphide deposits (LaRonde Penna, Bousquet 2-Dumagami, Bousquet 1, Westwood, Ellison, Warrenmac); 2) epizonal 'intrusion-related' Au-Cu sulphide-rich vein systems (Doyon); and 3) shear zone-hosted ('orogenic') Au-Cu sulphide-rich veins (Mouska, Mic Mac, Mooshla), which share some characteristics with the first two ore types.

The origin of the Au in the DBL camp has been extensively debated in the past and three models have been proposed: synvolcanic or syngenetic, multi-stage, and syndeformation. In these models, Au in the volcanogenic massive sulphide (VMS) and sulphide-rich veins of the camp is considered to be 1) a primary component of the hydrothermal fluids at the origin of the VMS (synvolcanic model) and of the epizonal intrusion-related veins, or 2) in part structurally introduced within and in part remobilized from massive sulphides into high-strain zones (multi-stage model), or 3) of metamorphic origin (syndeformation model). Recent studies of the LaRonde Penna deposit and the Mooshla synvolcanic pluton-hosted vein systems, combined with the geological synthesis and reconstruction of the DBL mining camp geology, have provided further insights into the synvolcanic or syngenetic model for the introduction of the Au, highlighting the fact that deformation is not a prerequisite for the formation of Au±Cu-Zn-Ag, Doyon-Bousquet-LaRonde-type deposits, as previously thought.

Résumé

Les gisements du camp minier Doyon-Bousquet-LaRonde (DBL) sont encaissés par les roches volcaniques et intrusives de la Formation de Bousquet (2701-2696 Ma) se retrouvant au sommet du Groupe de Blake River (2703-2694 Ma) dans la ceinture de roches vertes de l'Abitibi. Depuis les années 30, huit gisements d'Au ± métaux usuels ont été exploités pour une production totale, à la fin de 2005, de 67 Mt de minerai à 5,67 g/t Au (381 t Au, ou 12,3 M onces). Le total de la production, des réserves et des ressources estimées est évalué à 133 Mt de minerai à 5,23 g/t Au (694 t Au, ou 22,3 M onces).

Trois principaux types de dépots à Au±Cu-Zn-Ag sont reconnus dans le camp minier DBL: 1) SMV riches en Au (LaRonde Penna, Bousquet 2-Dumagami, Bousquet 1, Westwood, Ellison, Warrenmac), 2) systèmes de veines épizonales à Au-Cu riches en sulfures et reliées à une intrusion (Doyon), et 3) veines à Au-Cu riches en sulfures associées à des cisaillements ("orogéniques": Mouska, Mic Mac, Mooshla), lesquelles montrent des similitudes avec les deux premiers types de dépôts.

L'origine de l'or dans le camp DBL a été amplement débatu par le passé et trois modèles ont été proposés : synvolcanique ou syngénétique, multiphasé et syn-déformation. Dans ces modèles, l'or dans les SMV et les veines riches en sulfures du camp est considéré soit comme une composante primaire des fluides hydrothermaux à l'origine des SMV (model synvolcanique), soit en partie comme une composante tardive introduite structuralement et possiblement en partie remobilisé de sulfures massifs pré-existants dans des couloirs de déformation (modèle multiphasé), ou encore comme une composante associée aux fluides métamorphiques reliés aux épisodes de déformation régionale (modèle syn-déformation). Les études récentes du gisement LaRonde Penna et des systèmes de veines reliés au pluton synvolcanique de Mooshla, ainsi que la synthèse géologique du camp DBL ont mis en évidence plusieurs éléments clés supportant le modèle synvolcanique ou syngénétique pour l'introduction de l'or, soulignant le fait que la déformation n'est pas un élément déterminant dans la formation de dépôts à Au ± Cu-Zn-Ag de type Doyon-Bousquet-LaRonde tel que suggéré par le passé.

Introduction and History

The Doyon-Bousquet-LaRonde (DBL) mining camp is located in the eastern part of the Blake River Group in the Southern Volcanic Zone of the Archean Abitibi greenstone belt, between the Cadillac-Larder Lake fault zone to the south and the Lac Parfouru fault to the north (Fig. 1). A first phase of prospecting in this area in the 1920s and 1930s led to the discovery of the Mic Mac and Mooshla Au-Cu deposits, which were mined between 1939 and 1947. Several other Au and base metal occurrences were also found at or

Mercier-Langevin, P., Dubé, B., Lafrance, B., Hannington, M., Galley, A., Moorhead, J., and Gosselin, P., 2007, Metallogeny of the Doyon-Bousquet-LaRonde mining camp, Abitibi greenstone belt, Quebec, *in* Goodfellow, W.D., ed., Mineral Deposits of Canada: A Synthesis of Major Deposit-Types, District Metallogeny, the Evolution of Geological Provinces, and Exploration Methods: Geological Association of Canada, Mineral Deposits Division, Special Publication No. 5, p. 673-701.

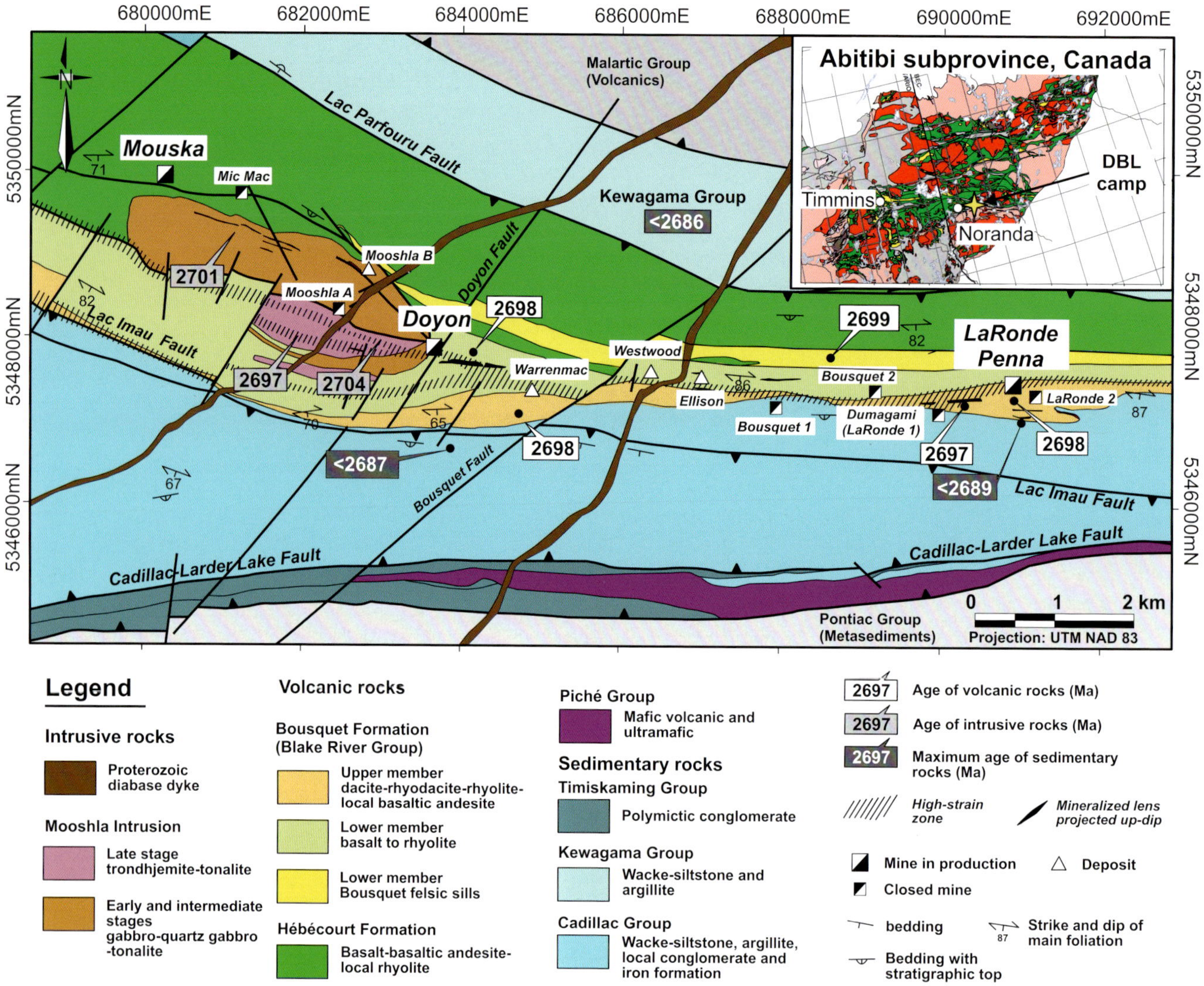

FIGURE 1. Location of the Doyon-Bousquet-LaRonde mining camp in the Abitibi greenstone belt (inset map) and simplified geological map of the host sequence showing the location of the deposits relative to the major faults in the area. Modified from Lafrance et al. (2003a,b) and Dubé et al. (2004). See text for U-Pb dating references.

near surface during this period (e.g. Thompson-Bousquet, Westwood). A period of limited exploration activity followed from 1940 to about 1970. A second phase of significant exploration, comprising drilling and geophysics, led to the discovery and development of the Doyon (Silverstack), Bousquet 1, and Dumagami Au (-Cu) deposits in the late 1970s and 80s (Lulin, 1990). Subsequent surface exploration on the mining properties of this area, or underground from the existing mines, helped define several satellite lenses and extensions to known ore zones (e.g. Bousquet 2 and Mouska). A third major period of underground and surface exploration (stripping, drilling, and geophysics) started about 1990 and led to the discovery in 1992-1993 of the LaRonde Penna Au-Zn-Cu-Ag volcanogenic massive sulphide deposit. Since then, aggressive underground exploration continued around the LaRonde Penna, Doyon, and Mouska mines, as well as on surface elsewhere in the camp. Except for the LaRonde Penna deposit, which extends from 800 to more than 3000 m below surface, all the mined ore lenses have been discovered from surface or near surface exploration around outcrop showings and follow-up drilling, and by ground geophysics in areas covered by overburden. The LaRonde Penna deposit was found through drilling at depth from the Dumagami mine, guided by inferred geological potential.

Since the 1930s, mining has taken place at eight Au ± base metal deposits in the DBL camp for a cumulative production, at the end of 2005, of 67 million tonnes (Mt) of ore at an average grade of 5.67 g/t Au (381 tonnes (t) Au, equivalent to 12.3 Moz Au; Table 1). Significant amounts of Ag and base metal also have been produced in the camp since 1999 from the LaRonde Penna deposit (Table 1). Overall, production from the DBL camp accounted for 9.9% of the Canadian Au production in 2005 (Table 2), 14.1% of Ag, 1.3% of Cu, and 12.3% of Zn, representing one of the largest mining camps and Au producing districts in the country. It is the most economically important mining camp in Quebec (Table 2) with three mines currently in operation in the camp; LaRonde Penna, Doyon and Mouska (Fig. 1).

The total past production, current reserves, and estimated resources for the DBL camp amount to 694 t Au, ranking fifth among the Canadian Au districts (Table 3). The camp

TABLE 1. Total production, reserves, and geological resources for the Doyon-Bousquet-LaRonde mining camp.

		Production						Reserves and geological resources						Total for each deposit					
Deposit	Year	Mt	Au (g/t)	Au (tonnes)	Ag (g/t)	Cu (%)	Zn (%)	Mt	Au (g/t)	Au (tonnes)	Ag (g/t)	Cu (%)	Zn (%)	Mt	Au (g/t)	Au (tonnes)	Ag (g/t)	Cu (%)	Zn (%)
Au-rich volcanogenic massive sulphide deposits																			
Dumagami	1988-1999	7.33	6.27	46.0	11.7	0.47	0.35							7.33	6.27	46.01	11.7	0.47	0.07
LaRonde Penna*(1)	2000-(2005)	12.3	3.53	43.44	53.7	0.28	2.66	46.46	4.51	209.5	42.7	0.34	2.04	58.76	4.31	252.97	45.0	0.33	2.17
Bousquet 1	1978-1996	7.45	5.3	39.5				1.82	5.97	10.9				9.27	5.447	50.36			
Bousquet 2	1990-2002	8.14	8.14	66.25		0.56								8.14	8.14	66.25		0.56	
Ellison								0.25	5.67	1.4				0.25	5.67	1.4			
Westwood								8.61	5.19	44.7				8.61	5.19	44.72			
Warrenmac								0.34	5.1					0.34	5.1				
	TOTAL	35.22	5.54	195.19	38.0	0.41	1.69	57.48	4.66	266.5				92.7	5.00	461.71			
Epizonal "intrusion-related" Au-Cu vein systems																			
Doyon*	1980-(2005)	29.78	5.55	165.14				7.92	5.17	41.0	2.4			37.7	5.47	206.09			
	TOTAL	29.78	5.55	165.14				7.92	5.17	41.0				37.7	5.47	206.09			
Shear zone-hosted Au-Cu vein systems																			
Mouska*	1989-(2005)	1.51	11.55	17.45				0.38	14.4	5.5	3.1			1.89	12.13	22.95			
Mic Mac	1942-1947	0.72	4.62	3.34										0.72	4.62	3.34			
Mooshla	1939-1940	0.05	29.66	0.13										0.05	29.66	0.13			
	TOTAL	2.24	9.35	20.93				0.38	14.4	5.5				2.62	10.09	26.42			
TOTAL DBL mining camp		67.24	5.67	381.26				65.78	4.78	313.0				133.02	5.23	694.23			

* Active mines
Mt=million tonnes
g/t=gram per tonne
(1) Comprises the LaRonde II mine project reserves and resources

also hosts two of the fifteen largest Canadian Au deposits (Table 4), and LaRonde Penna is the second largest Au deposit of the VMS class after the Horne deposit (Fig. 2) (see also Dubé et al., 2007a, Galley et al., 2007; Gibson and Galley, 2007).

Although LaRonde Penna is a VMS deposit, its Au content, in grams per tonne, exceeds that of the combined base metal content in weight percent, and therefore corresponds to the classification proposed by Poulsen et al. (2000) for Au-rich VMS deposits. The main characteristics of Au-rich VMS deposits are described in detail in Poulsen and Hannington (1996), Hannington et al. (1999), and Dubé et al. (2007a). Numerous other deposits of the DBL camp are also considered Au-rich VMS deposits: Bousquet 1, Bousquet 2-Dumagami, Westwood, Ellison, and Warrenmac (Table 1 and Fig. 2). The cumulative production, reserves, and resources of these VMS deposits is 92 Mt of ore at an average grade of 5 g/t Au for a total of 462 t of Au (or 14.8 Moz Au). In terms of tonnage, the DBL camp ranks amongst the largest Canadian VMS districts (c.f. Galley et al., 2007) and first among the Au-rich VMS districts subtype (Gosselin and Dubé, 2005a; Dubé et al., 2007a). Four of the largest Au-rich VMS mines presently known in the world are located in the DBL camp (Bousquet 1, Bousquet 2, Dumagami, and LaRonde Penna; Fig. 2), and six of these deposits (DBL camp deposits, Horne, and Quemont) are found in the Blake River Group in the Abitibi (Fig. 2).

Despite the fact that most of the reserves and estimated resources in the camp are hosted in VMS deposits, about half of the Au produced in the camp to date comes from vein-type deposits (e.g. Doyon, Mouska, Mic Mac, and Mooshla A; Table 1). Vein-type deposits consist of multi-ounce, centimetre- to metre-wide, sulphide-rich vein networks, largely transposed within the regional foliation. As discussed below, most of the vein systems comprising these deposits are thought to be related to volcanism and magmatism but were clearly strongly affected by the regional deformation events.

Regardless of the type of mineralization, the origin of the Au in the camp has been a source of much controversy in the past. Many authors proposed a synvolcanic or syngenetic origin for the Au (e.g. Filion et al., 1977; Valliant and Barnett, 1982; Valliant and Hutchinson, 1982; Valliant et al., 1982, 1983; Bateman, 1985; Stone, 1990, 1991; Tourigny et al., 1993; Teasdale et al., 1996; Gosselin, 1998). However, the intense although heterogeneous deformation (Tourigny et al., 1988; Tourigny and Tremblay, 1997; Belkabir et al., 1998) that has affected all of the deposits (see below) led some authors to conclude that at least a portion of the Au present in the deposits of the district had been structurally introduced and/or remobilized into high-strain zones (multi-

TABLE 2. Contribution of the Doyon-Bousquet-LaRonde mining camp mines to the Quebec and Canada production in 2005.

Production in 2005	Au (t)	Qc(1)	Can(2)	Ag (t)	Qc(1)	Can(2)	Cu (t)	Qc(1)	Can(2)	Zn (t)	Qc(1)	Can(2)
LaRonde Penna	6.8	28.6%	5.7%	150	76.2%	14.1%	7378	18.9%	1.3%	76545	72.8%	12.3%
Doyon	3.3	13.9%	2.8%									
Mouska	1.7	7.1%	1.4%									
Total DBL (3)	11.8	49.6%	9.9%	150	76.2%	14.1%	7378	18.9%	1.3%	76545	72.8%	12.3%

(1) Total production, in 2005, for Quebec (23.8 t Au, 197 t Ag, 39090 t Cu, 105136 t Zn) from National Resources Canada, Minerals and Mining statistics

(2) Total Canadian production, in 2005 (119.7 t Au, 1068 t Ag, 570619 t Cu, 623101 t Zn) from National Resources Canada, Minerals and Mining statistics

(3) Total DBL = Total production in 2005 for the Doyon-Bousquet-LaRonde mining camp active mines

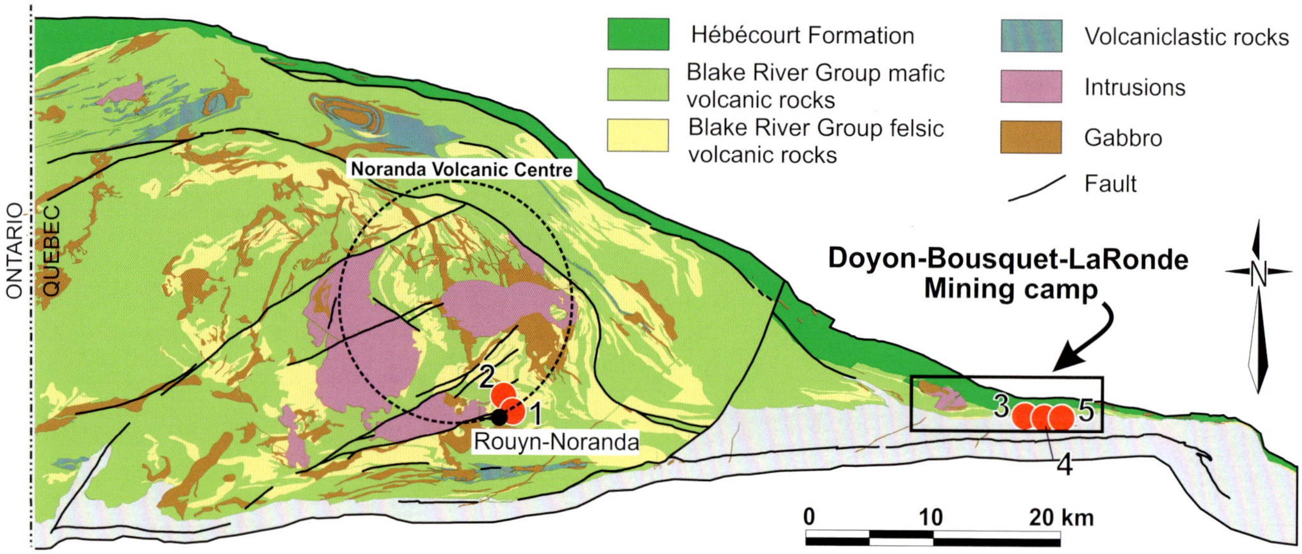

Data for selected Au-rich VMS deposits

No.	Deposit	Tonnage (t)	Gold (t)
1	Horne	54 300 000	331.1
2	Quémont	13 924 000	66.0
3	Bousquet 1	9 264 940	50.4
4	Bousquet 2-Dumagami	15 475 827	112.3
5	LaRonde Penna	58 757 662	253.0
6	Eskay Creek	3 121 697	118.6
7	Big Missouri	2 453 924	7.1
8	Agnico-Eagle	6 928 978	35.9
9	Montauban	632 774	2.8
10	Boliden	8 300 000	125.3
11	Hassai	7 176 035	69.9
12	Mt. Morgan	116 040 700	321.1

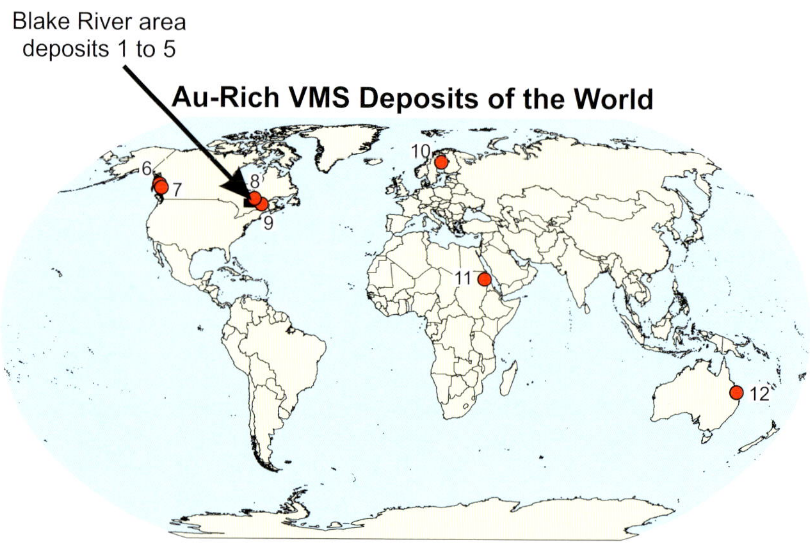

FIGURE 2. Location and statistics of a selection of Au-rich VMS deposits of the world with emphasis on the unique concentration of this type of deposit in the Blake River Group of the Abitibi greenstone belt that comprises five of these deposits. Three of these five deposits are part of the Doyon-Bousquet-LaRonde mining camp. Statistics from Gosselin and Dubé (2005b). Map of the Blake River Group modified from Lamothe et al. (2005). World map modified from Dubé et al. (2007a).

stage model: e.g., Guha et al., 1982; Tourigny et al., 1989a,b; Belkabir and Hubert, 1995; Belkabir et al., 2004), whereas others proposed an epigenetic (syndeformation) model in which the Au was deposited during regional deformation and metamorphism, and superimposed onto previously formed alteration and sulphide-rich zones (e.g. Marquis et al., 1990a,b,c; Hoy et al., 1990; Savoie et al., 1990; Marquis et al., 1992; Trudel et al., 1992). More recently, ideas about the genesis of Au-rich VMS deposits have evolved due to an improved understanding of massive sulphide deposits forming on the present-day seafloor (e.g. Hannington et al., 2005, and references therein) and within fossil hydrothermal systems (e.g. Sillitoe et al., 1996; Hannington et al., 1999; Huston, 2000). Recent studies of the LaRonde Penna deposit (Dubé et al., 2004, 2007b; Mercier-Langevin, 2005; Mercier-Langevin et al., 2004, 2007a,b; Hannington et al., 2007) and Mooshla-hosted vein systems (Galley and Pilote, 2002; Galley et al., 2003; Galley and Lafrance, 2007), and the geological synthesis and reconstruction of the DBL mining camp geology (Lafrance et al., 2003a), have provided further insights into the synvolcanic or syngenetic model for the introduction of the Au.

Here, we review the main geological characteristics of the DBL mining camp and present the key attributes for each type of ore deposit found in this location. The deposit classifications proposed are simplified for the purpose of compilation, as most of the deposits in the camp comprise various ore types. The descriptions that are presented are meant to illustrate the wide spectrum of ore and alteration styles recognized in the DBL mining camp, in order to draw some general guidelines for exploration applicable in the Abitibi subprovince and perhaps elsewhere in Archean greenstone belts and possibly younger volcanic sequences. The geological setting and the Au-rich volcanogenic massive sulphide deposits sections presented here are taken, in a large part, from Mercier-Langevin et al. (2007a,b,c), Dubé et al. (2007b), and Hannington et al. (2007).

Geological Setting

The DBL mining camp is an approximately 10 km long, east-west-trending succession of volcanic rocks located 45 km east of Rouyn-Noranda (Fig. 1, insert map). The mining camp is situated in the eastern part of the Blake River Group in the Late Archean Abitibi greenstone belt of the Superior Province. The Blake River Group is restricted in this area to a narrow east-west-striking, south-facing homoclinal volcanic sequence between sedimentary rocks of the <2686 Ma (Davis, 2002) Kewagama and the <2687 Ma (Lafrance et al., 2005) Cadillac Groups (Lafrance et al., 2003a) (Fig. 1). Rocks of the DBL camp (Bousquet Formation) represent the upper part of the 2703 to 2694 Ma Blake River Group (Péloquin et al., 1990; Barrie et al., 1993; Mortensen, 1993; Ayer et al., 2002; Lafrance et al., 2005). This sequence had previously been divided into several lithotectonic domains from north to south by Tourigny et al. (1988, 1989a,b). Lafrance et al. (2003a) later formally divided the sequence into the Hébécourt Formation in the north and the Bousquet Formation in the south.

TABLE 3. Major Canadian gold districts.

District	Location	Number of deposits	Total gold (production, reserves, and estimated resources) Tonnage (Mt)	Au grade (g/t)	Au (t) (tonnes)	rank
Timmins	ON	20	436.7	4.93	2151.4	1
Red Lake	ON	9	59.8	16.52	987.9	2
Kirkland Lake	ON	8	67.7	12.81	867.4	3
Val d'Or	QC	18	180.2	4.50	810.5	4
DBL camp	QC	7	132.7	5.23	694.2	5
Rouyn-Noranda	QC	17	101.5	5.78	586.1	6
Yellowknife	NT	2	28.2	15.94	449.4	7
Larder Lake	ON	3	42.9	9.17	393.3	8
Kemess-Toodoggone	BC	2	822.7	0.47	375.7	9
Malartic	QC	5	62.0	4.87	301.9	10

BC=British Columbia. NT=Northwerstern Territories. ON=Ontario. QC=Quebec
Source: Gosselin and Dubé (2005a)

TABLE 4. Major Canadian gold deposits.

Deposit	Total gold (production, reserves and estimated resources) Tonnage (t) (tonnes)	Au grade (g/t)	Au (tonnes)	rank
McIntyre-Hollinger-Coniaurum	104,215,929	9.47	987.0	1
Campbell Red Lake	37,549,765	21.28	799.2	2
Kirkland Lake	53,407,239	14.94	797.7	3
Hemlo	118,911,867	6.07	721.4	4
Dome	111,400,224	4.578	508.5	5
Sigma - Lamaque	90,350,488	4.92	444.4	6
Kemess	821,990,788	0.45	369.4	7
Horne *(VMS)*	54,300,000	6.10	331.1	8
Kerr Addison	35,976,261	9.10	327.4	9
LaRonde Penna	58,757,662	4.31	253.0	10
Pamour	86,482,555	2.89	249.6	11
Giant-Lolor-Supercrest	15,854,675	15.65	248.2	12
Tundra, FAT	113,057,000	2.07	233.7	13
Doyon	37,703,649	5.47	206.1	14
Negus - Nerco Con	12,345,202	16.30	201.3	15
Eskay Creek *(VMS)*	3,121,664	37.99	118.6	21
Dumagami-Bousquet 2	15,475,827	7.25	112.3	22
Quémont *(VMS)*	13,924,000	4.74	66.0	39
Bousquet	9,264,940	5.44	50.4	52
Westwood	8,613,000	5.19	44.7	58
Mouska	1,892,771	12.13	23.0	90

Source: Gosselin and Dubé (2005a)

Hébécourt Formation

The Hébécourt Formation comprises laterally extensive tholeiitic basalt and basaltic andesite flows, cogenetic gabbroic sills, and small isolated tholeiitic rhyolite flows (Fig. 1). The mafic to intermediate flows are massive to pillowed (Fig. 3A) and some are characteristically magnetic (Lafrance et al., 2003a; Belkabir et al., 2004). The pillows, up to 2.5 m in diameter, are locally variolitic and glomeroporphyritic. The flows are intercalated with tholeiitic microgabbro sills of up to a few tens of metres in thickness. The tholeiitic rhyolites are concentrated near the Mouska mine and are generally very fine grained and schistose. The zones 07, 08, and 22 at Mouska and the Mic Mac ore zones are hosted in the Hébécourt Formation (Fig. 1).

Bousquet Formation

The Hébécourt Formation is overlain by the Bousquet Formation, which hosts most of the known ore zones of the camp. The Bousquet Formation is further subdivided into two members: the lower member (2699-2698 Ma) and the upper member (2698-2697 Ma, Lafrance et al., 2005; Mercier-Langevin et al., 2007b). The lower member is composed of tholeiitic to transitional (between tholeiitic and calc-alkaline) mafic to felsic strata, whereas the upper member is dominated by transitional to calc-alkaline, intermediate (basaltic andesite and andesite) to felsic (dacite, rhyodacite, and rhyolite) rocks (Lafrance et al., 2003a; Mercier-Langevin et al., 2004, 2007a,b).

Lower Member of the Bousquet Formation

The laterally extensive (Fig. 1) mafic to intermediate volcaniclastic units (scoriaceous tuffs, Fig. 3B) of the lower member are interpreted to be of possible subaqueous pyroclastic origin (Stone, 1990; Lafrance et al., 2003a). These block, lapilli, and ash tuffs are stratified and cut by coeval and cogenetic dykes and sills, and also by felsic feeder dykes related to overlying felsic volcanic rocks of the upper member (Mercier-Langevin, 2005). In the Doyon area (Fig. 1), this tuffaceous sequence is overlain by a glomeroporphyritic lobe-hyaloclastite dacitic flow dated at 2698.3 ± 0.9 Ma (Lafrance et al., 2005) and by a dacitic to rhyolitic flow breccia (Savoie et al., 1991). Elsewhere in the camp, the tuffaceous sequence is overlain by a heterogeneous unit (Fig. 1) composed of pillowed basalt and/or andesite that is intercalated or grades laterally into brecciated flows. The heterogeneous unit is overlain in the Warrenmac area (Fig. 1) by a strongly altered and deformed rhyodacite (Lafrance et al., 2003a). Tholeiitic to transitional, rhyolitic, blue quartz-bearing sills (Fig. 3C) intruded the interface between the

A
Pillow margins
25 cm
B
20 cm
C
Blue quartz
2 cm
D
Felsic lobe
Andesitic hyaloclastite
2 m
E
25 cm
F
Blue quartz
2 cm
G
2 cm
H
10 cm

Hébécourt Formation and the base of the lower member of the Bousquet Formation (Lafrance et al., 2003a). The thickest portion of these sills, dated at 2698.6 ± 1.5 Ma (Lafrance et al., 2003a), is located north of the Bousquet 1, Bousquet 2, and Dumagami mines (Fig. 1; Dubé et al., 2004).

Upper Member of the Bousquet Formation

The units of the upper member of the Bousquet Formation are characterized by an irregular distribution throughout the DBL mining camp (Lafrance et al., 2003a; Fig. 1), but are thickest in the LaRonde Penna mine area. The lowermost unit of the upper member consists of dacitic to rhyodacitic lobes and flow-breccia deposits (Fig. 3D). This dacitic to rhyodacitic flow breccia interval is intercalated with andesitic flows and hyaloclastites, and it is cut by mafic sills and dykes, as demonstrated in the Bousquet 2-LaRonde Penna area (Mercier-Langevin et al., 2007a). The andesite flows are locally overlain by fine-grained hyaloclastite of the same composition and thin beds of fine-grained volcanogenic sedimentary rocks and graphitic argillite. The intercalation suggests the emplacement of intermittent felsic autoclastic flows and andesitic flows within restricted basins or depressions where local or intermittent sedimentation occurred prior to burial by felsic volcanic rocks (Mercier-Langevin et al., 2007a). This andesite-dacite interval is overlain by rhyodacitic to rhyolitic flow breccia (Fig. 3E) and coeval rhyolitic domes or cryptodomes mainly developed in the LaRonde Penna area and in the Ellison-Warrenmac segment (Fig. 1). The rhyodacitic to rhyolitic flow breccia, dated at 2698 ± 1 Ma in the Doyon mine area (Lafrance et al., 2003a), now completely covers consanguineous domes and/or cryptodomes that are dated at 2698.3 ± 0.8 Ma in the LaRonde Penna mine area (Mercier-Langevin et al., 2007a). A subsequent break in volcanic activity is indicated by the deposition of thin graphitic argillite beds between the LaRonde Penna and Bousquet 2-Dumagami deposits and by the formation of concomitant Zn-rich massive sulphides (LaRonde Penna 20 North Zn zone) along the same horizon as the argillite (Mercier-Langevin, 2005), above the rhyodacitic to rhyolitic flow breccia. The latter unit is in contact with the Cadillac Group sedimentary rocks in the central part of the camp (Fig. 1) where the upper member is thinner, but elsewhere it is overlain either by a feldspar- and blue quartz-phyric rhyolite (Fig. 3F), a basaltic andesite, or an autoclastic rhyolite (Fig. 1). The latter is the uppermost unit of the upper member of the Bousquet Formation and hosts the feldspar- and blue quartz-phyric rhyolite and the basaltic andesite. The feldspar- and blue quartz-phyric rhyolite was emplaced, at least in part, as a cryptodome within the uppermost autoclastic rhyolite at 2697.8 ± 1 Ma (Mercier-Langevin et al., 2007a). The basaltic andesite (Fig. 3G) forms a sill and dyke complex that was also emplaced within the autoclastic rhyolite, locally cutting through the feldspar- and blue quartz-phyric rhyolite (Fig. 4). The sill and dyke complex was probably transported through the same structures as the feldspar- and quartz-phyric rhyolite, as suggested by the close spatial association and the cross-cutting relationships (Fig. 4). They were emplaced more or less synchronously with the surrounding autoclastic rhyolite in the upper part of the upper member of the Bousquet Formation. The autoclastic rhyolite is present only in the LaRonde Penna mine area and is mostly composed of block- and lapilli-sized flow breccia (Fig. 3H) cut by lobes a few metres in length. Thin layers of crystal tuff are locally intercalated with the flow breccia.

Mooshla Pluton

The polyphase Mooshla synvolcanic intrusion was sequentially emplaced into the upper part of the Hébécourt Formation and into the overlying lower member of the Bousquet Formation (Valliant and Hutchinson, 1982; Gaudreau, 1986; Langshur, 1990; Fig. 1). It hosts the Mooshla A and B Au deposits and parts of the Doyon and Mouska Au (-Cu) deposits. This intrusive complex was emplaced as three differentiated intrusive phases: 1) early tholeiitic to transitional dioritic sills; 2) intermediate tholeiitic to transitional gabbro-tonalite that forms the northern half of the intrusion (stages 1 and 2 combined on Fig. 1); and 3) late transitional to calc-alkaline trondhjemite and porphyry, representing the upper part (south) of the intrusion (Galley and Lafrance, 2007). The intermediate phase of gabbro-tonalite is coeval with the lower member of the Bousquet Formation and the later trondhjemitic phase is contemporaneous and possibly comagmatic with the upper member (Langshur, 1990; Galley and Lafrance, 2007). The late intrusive phase gave a reliable U-Pb zircon age of 2696.9 ± 1 Ma (Lafrance et al., 2005). The intermediate intrusive phase gave a more or less reliable age of 2701 ± 1 Ma (Zhang et al., 1993), and an aplitic dyke cutting this intermediate intrusive phase has been imprecisely dated at 2704 ± 4 Ma (Tremblay et al., 1995).

The Mooshla pluton is associated with a series of hydrothermal events responsible for the formation of the numerous ore zones that comprise the Mouska, Mooshla A and B, Mic Mac, and Doyon orebodies. These various styles of mineralization are representative of synvolcanic and pre-intrusion VMS-style mineralization (Mic Mac), late 'early intrusive'-stage VMS and sulphide-rich vein-type mineralization (Mouska), syn 'late-intrusive'- stage quartz and sulphide vein-type mineralization (Doyon and Mooshla A), and post-intrusion or syndeformation quartz ± tourmaline-carbonate-sulphide vein-type mineralization (Mooshla B and Mouska; Belkabir et al., 2004; Galley and Lafrance, 2007).

Cadillac Group

The Bousquet Formation of the Blake River Group is overlain by the Cadillac Group turbidite units that have been

FIGURE 3. **(A)** Pillowed basalts of the Hébécourt Formation, northwest of the LaRonde Penna mine. **(B)** Scoriaceous tuff of the lower member of the Bousquet Formation, northeast of the LaRonde Penna mine. **(C)** Tholeiitic quartz and feldspar porphyritic Bousquet felsic sills of the lower member of the Bousquet Formation, emplaced in the upper part of the Hébécourt Formation, north of the LaRonde Penna mine. **(D)** Upper member of the Bousquet Formation; dacite-rhyodacite lobes emplaced within fine-grained andesitic hyaloclastite, LaRonde Penna mine. **(E)** Upper member of the Bousquet Formation; rhyodacite-rhyolite dome/cryptodome with in situ breccia, south of the LaRonde Penna mine. **(F)** Upper member of the Bousquet Formation; feldspar and blue quartz-phyric rhyolite, LaRonde Penna mine. **(G)** Upper member of the Bousquet Formation; lower part of the basaltic andesite sills characterized by a glomeroporphyritic texture, east of the LaRonde Penna mine. **(H)** Rhyodacitic to rhyolitic block and lapilli tuff of the upper felsic unit in the LaRonde Penna mine host sequence, upper member of the Bousquet Formation.

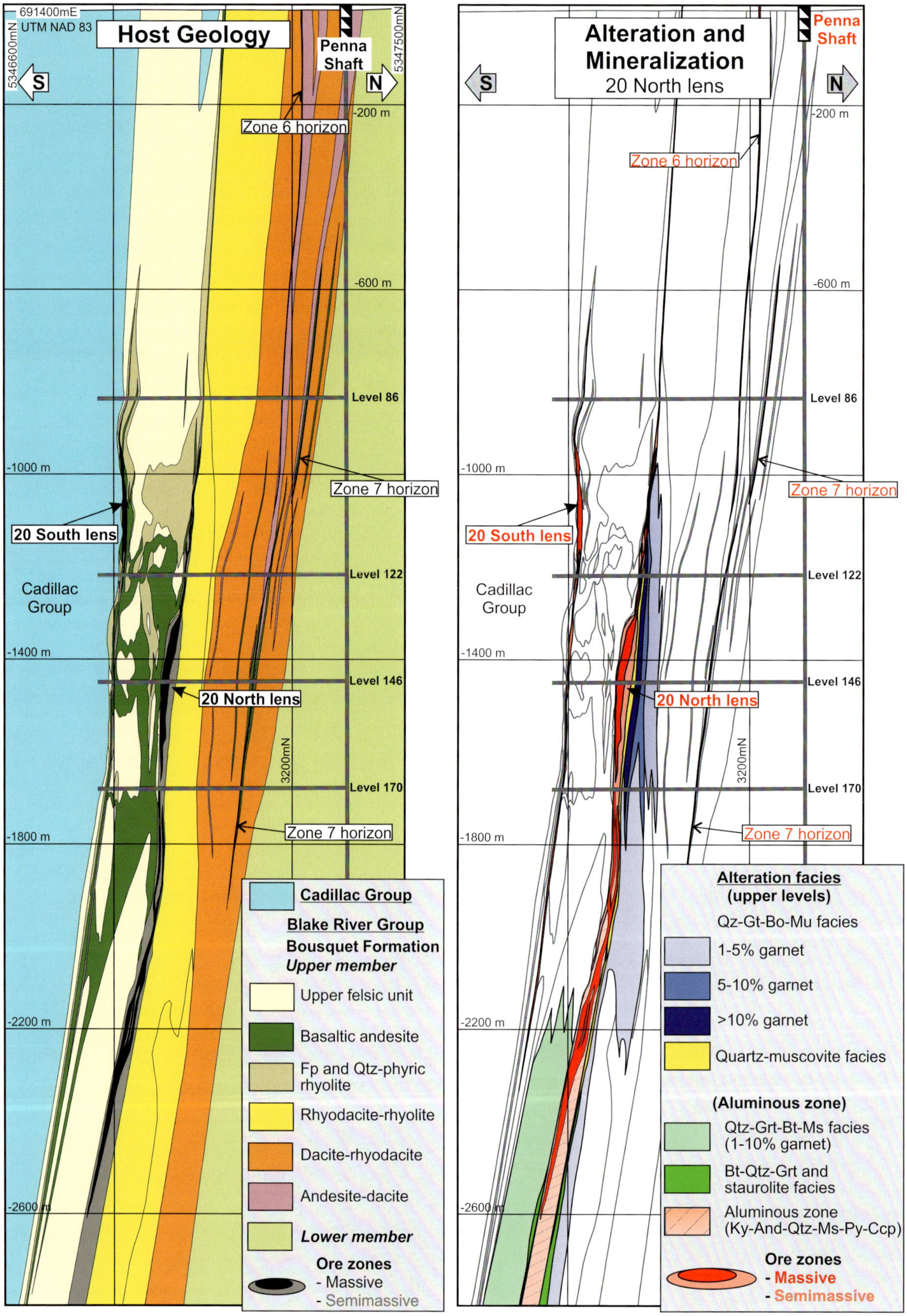

FIGURE 4. Geological section through the LaRonde Penna deposit showing the distribution of the upper member units of the Bousquet Formation, the zones 7 and 6, the 20 North and 20 South lenses, and the distribution-zonation of the principal alteration assemblages developed along the 20 North lens from surface to a depth of 2700 m. Modified from Mercier-Langevin et al. (2007a). And = andalusite, Bt = biotite, Ccp = chalcopyrite, Fp = feldspar, Grt = garnet, Ky = kyanite, Ms = muscovite, Py = pyrite. Qtz = quartz.

described by Dimroth et al. (1982), Lajoie and Ludden (1984), Stone (1990), and Davis (2002). This turbiditic sequence is younger than 2687 Ma (Davis, 2002). The contact between the volcanic and sedimentary rocks is not exposed on surface in the LaRonde Penna area but it has been intersected in many drillholes. This contact has been interpreted variously as structural (Dumagami fault of Tourigny et al., 1988, 1989b, 1993; Marquis et al., 1990a) and as conformable (Valliant and Hutchinson, 1982; Stone, 1990). In the LaRonde Penna mine area, the contact is slightly discordant (i.e. interpreted to be erosional) to subconcordant as illustrated in section (Fig. 4). A thin horizon (≤20 cm) of semimassive to massive pyrrhotite and pyrite is commonly present at or near the contact between the volcanic and sedimentary rocks of the Blake River Group volcanic sequence and the Cadillac Group sedimentary sequence. It is locally slightly anomalous in Zn (Dubé et al., 2004). This district-wide sulphide-rich interval may result from the interaction between spent or late hydrothermal fluids and seawater during a period of nondeposition along the unconformity (Dubé et al., 2007b; Mercier-Langevin et al., 2007a) as the sedimentary rocks were emplaced at least 5 to 10 m.y. after cessation of volcanism, as indicated by the age of the youngest Cadillac Group detrital zircon at LaRonde Penna (2689 Ma: Mercier-Langevin et al., 2007a) and at Doyon (2687.4 Ma: Lafrance et al., 2005).

Deformation and Metamorphism

The Blake River Group is characterized by three main stages of deformation related to north-south convergence, followed by oblique dextral shearing (Dimroth et al., 1983; Hubert et al., 1984). The D1 event caused regional folding (Hubert et al., 1984) and was overprinted by D2, which is the main deformation event in the DBL mining camp (Tourigny et al., 1988; Marquis et al., 1990a, Mercier-Langevin et al., 2004, 2007a). An east-west-trending, steeply south-dipping penetrative schistosity (regional S2) is present everywhere in the camp and is responsible for strong flattening, stretching, folding, and shearing of the primary features in most deposits (e.g. Bousquet 1: Tourigny et al., 1989b; Bousquet 2: Tourigny et al., 1993; Dumagami: Marquis et al., 1990b, c; Doyon: Savoie et al., 1990; 1991). A stretching lineation steeply plunging west is locally developed on the S2 schistosity. A north-northeast-trending cleavage is locally superimposed on the main schistosity and late sinistral north-northeast-trending faults are locally developed (e.g. Doyon fault: Savoie et al., 1991; Gosselin, 1998). Two episodes of metamorphism have been recognized in the region: a prograde upper greenschist- to lower amphibolite-facies episode associated with the main deformation event D2, and a subsequent retrograde greenschist-facies event (Powell et al., 1995). The same metamorphic episodes have been recognized throughout the DBL mining camp (Tourigny et al., 1989a; Marquis et al., 1990a; Lafrance et al., 2003a; Mercier-Langevin, 2005; Dubé et al., 2007b).

Mineral Deposits

Deposit Types

The classification of the Au ± Cu-Zn-Ag deposits of the DBL mining camp has evolved with the different interpretations and the continual advancement of knowledge about lode Au deposits, VMS deposits, and epithermal deposits. The Bousquet 1 deposit and surrounding ore zones were classified by Valliant and Hutchinson (1982) as stratiform volcanic-sedimentary exhalative pyritic bodies and subconformable sulphidic veins, whereas the Doyon ore zones were interpreted as epigenetic (syndeformation). Epigenetic veins (e.g. Doyon) and epigenetic sulphide-rich zones (e.g. Bousquet 1) were considered by Savoie et al. (1990) and by Trudel et al. (1992) to characterize the mineralization of the Bousquet district. The Doyon deposit was later described as porphyry-like by Poulsen (1996), as an intrusion-hosted Au system by Poulsen and Hannington (1996) and Hannington et al. (1999), and as an intrusion-related Au-Cu sulphide-rich vein-type deposit by Poulsen et al. (2000), whereas the Bousquet deposit, as well as the Bousquet 2-Dumagami and LaRonde Penna orebodies, were considered to be synvolcanic pyritic Au deposits by Poulsen and Hannington (1996) and Hannington et al. (1999), and submarine Au-rich VMS deposits by Poulsen et al. (2000). The reader is referred to these contributions for a discussion of the characteristics of each deposit and its proposed classification.

The classification presented here is a simplified scheme based on the descriptions presented in the publications cited above and on more recent detailed work conducted in the camp (e.g. Galley and Lafrance, 2007; Dubé et al., 2007b; Hannington et al., 2007; Mercier-Langevin et al., 2007a,b,c). Three main deposit types are recognized in the DBL mining camp: 1) the Au-rich VMS deposits that comprise the LaRonde Penna lenses, the Bousquet 2-Dumagami orebody, the Bousquet 1 orebody, and the Westwood, Ellison, and Warrenmac ore zones; 2) the epizonal 'intrusion-related' Au-Cu vein systems that comprise the Doyon deposit; 3) the shear zone-hosted ('orogenic') Au-Cu veins that include the Mouska, Mic Mac, and Mooshla A and B deposits. The latter are classified as shear zone-hosted deposits as they occur almost entirely in major structures related to the main regional deformation episode (D2). However, they share some characteristics with the first two ore types and therefore could in part consist of transposed and/or remobilized epizonal intrusion-related Au-Cu vein systems or Au-rich volcanogenic sulphides.

Au-Rich Volcanogenic Massive Sulphide Deposits

The Au-rich VMS deposits of the DBL mining camp (i.e. LaRonde Penna, Bousquet 2-Dumagami, Bousquet 1, Ellison, Westwood, and Warrenmac) have accounted for more than half of the total tonnage and Au produced in the camp to the end of 2005, with 195 t of Au extracted from 35 Mt of ore at an average grade of 5.54 g/t Au (Table 1).

The Au-rich VMS deposits consist of stacked ore lenses (or mine zones) at different stratigraphic intervals in the Bousquet Formation and represent different episodes of sulphide precipitation in a long-lived hydrothermal system (Dubé et al., 2007b; Mercier-Langevin et al., 2007a). The relative position of the ore lenses in each deposit is illustrated in the schematic stratigraphic columns presented in Figure 5. Overall, the various lenses are located higher in the stratigraphy in the eastern part of the camp (e.g. LaRonde Penna mine) than in the western part of the camp (e.g. Warrenmac and Westwood lenses: Fig. 5). All the VMS ore zones show the same attitude, with a steeply dipping plunge

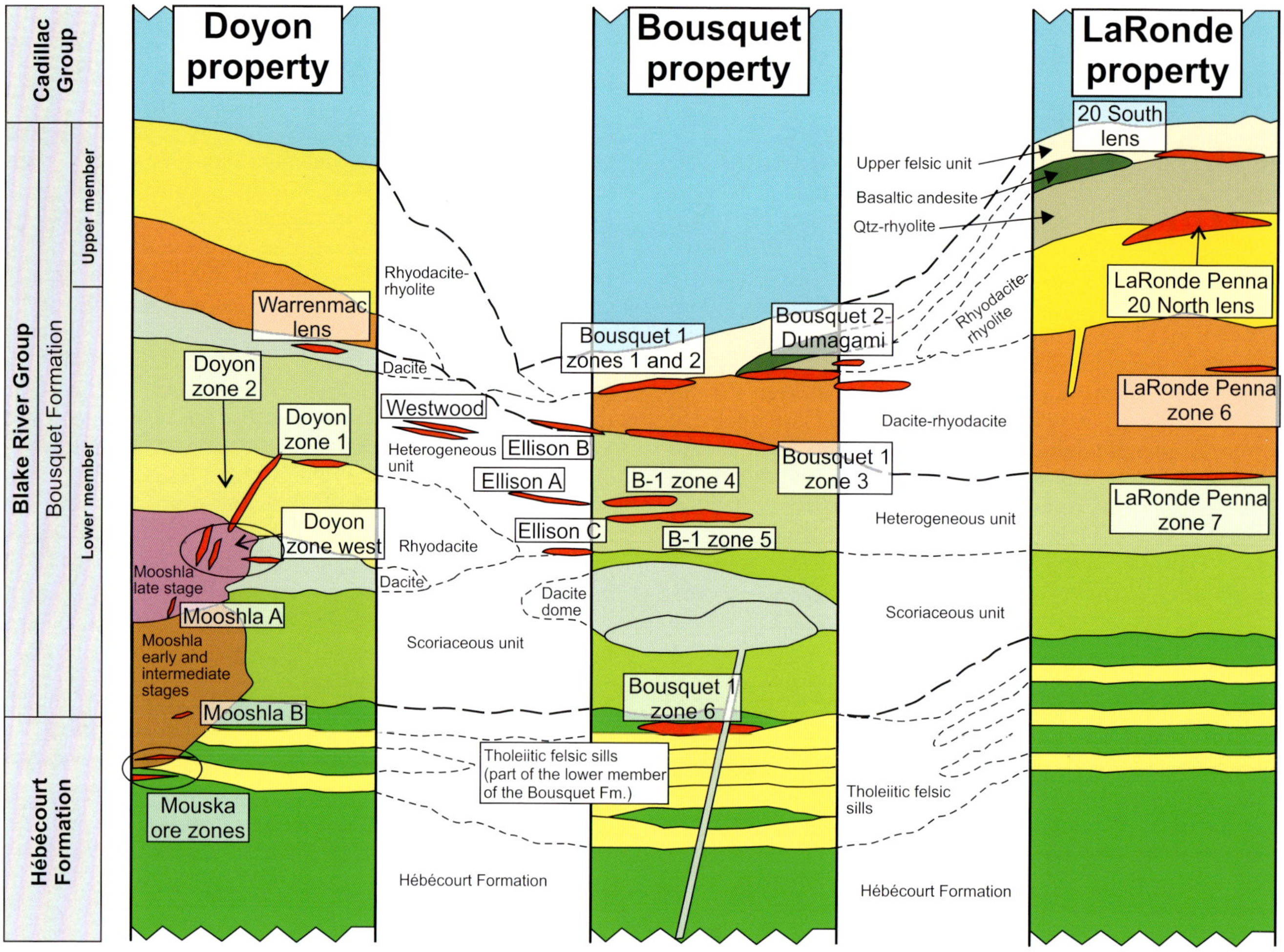

FIGURE 5. Simplified stratigraphy of the Doyon-Bousquet-LaRonde mining camp illustrating the stratigraphic setting of the main ore lenses in the Doyon, Bousquet 1 and LaRonde Penna areas. This figure shows the thickening of the upper member of the Bousquet Formation towards the east (LaRonde Penna area), the stacking of ore lenses, and the general upward distribution of the VMS lenses towards the east. Modified from Lafrance et al. (2003a), Mercier-Langevin et al. (2007a), and Mercier-Langevin et al. (2007c).

to the west (Fig. 6) that represents the primary elongation axis of the mineralization along inferred, regularly spaced, synvolcanic faults (Mercier-Langevin, 2005, Mercier-Langevin et al., 2007a). This general attitude is mimicked at a smaller scale by the distribution of volcanic units at depth, hydrothermal alteration zonation, and metal zonation within the lenses (Mercier-Langevin, 2005).

Mineralization

The VMS mineralization of the DBL mining camp is characterized by semimassive to massive sulphide lenses and thin zones of transposed sulphide veins and veinlets, commonly associated with variable amounts of disseminated sulphides. Individual ore zones or lenses are characterized by different ore types that are gradational from disseminated and vein sulphides to semimassive to massive sulphides. The ore zones discussed here are presented in an order that is based on the stratigraphic level, beginning with the lowermost lenses.

At Bousquet 1, the ore is concentrated in five different zones (1 to 5, from south to north) with about 80 percent of the ore in zone 3, which is located at the interface between the lower and upper members of the Bousquet Formation (Fig. 7). Tourigny et al. (1989a) divided the ore from these zones in disseminated-type and vein-type mineralization with foliation-parallel veins and foliation-oblique veins. The foliation-parallel veins are largely dominant in zone 3 and form centimetre- to metre-thick pinch-and-swell structures along the main east-west regional foliation. These veins consist of sulphides with lesser amounts of quartz and muscovite. Pyrite is the main constituent of the sulphides with chalcopyrite and minor amounts of sphalerite, pyrrhotite, galena,bornite, rutile, tellurides (altaite and calaverite), and native Au (Tourigny et al., 1989a). The foliation-oblique veins are mainly developed in zones 1, 2, 4, and 5, and consist of folded, transposed, and discontinuous stringers that form zones hundreds of metres long (Tourigny et al., 1989a). They are composed predominantly of pyrite and quartz in felsic host rocks (zones 1 and 2) and of pyrite and carbonate in mafic host rocks (zones 4 and 5), with local concentrations of chalcopyrite. The foliation-oblique veins also contain trace minerals such as sphalerite, galena, arsenopyrite, pyrrhotite, bornite, magnetite, rutile, gudmundite, stannite, and native Au. The auriferous disseminations are present in zones 3 and 5 within disseminated pyrite (up to 10 vol.%) and trace amounts of chalcopyrite, sphalerite, pyrrhotite, and

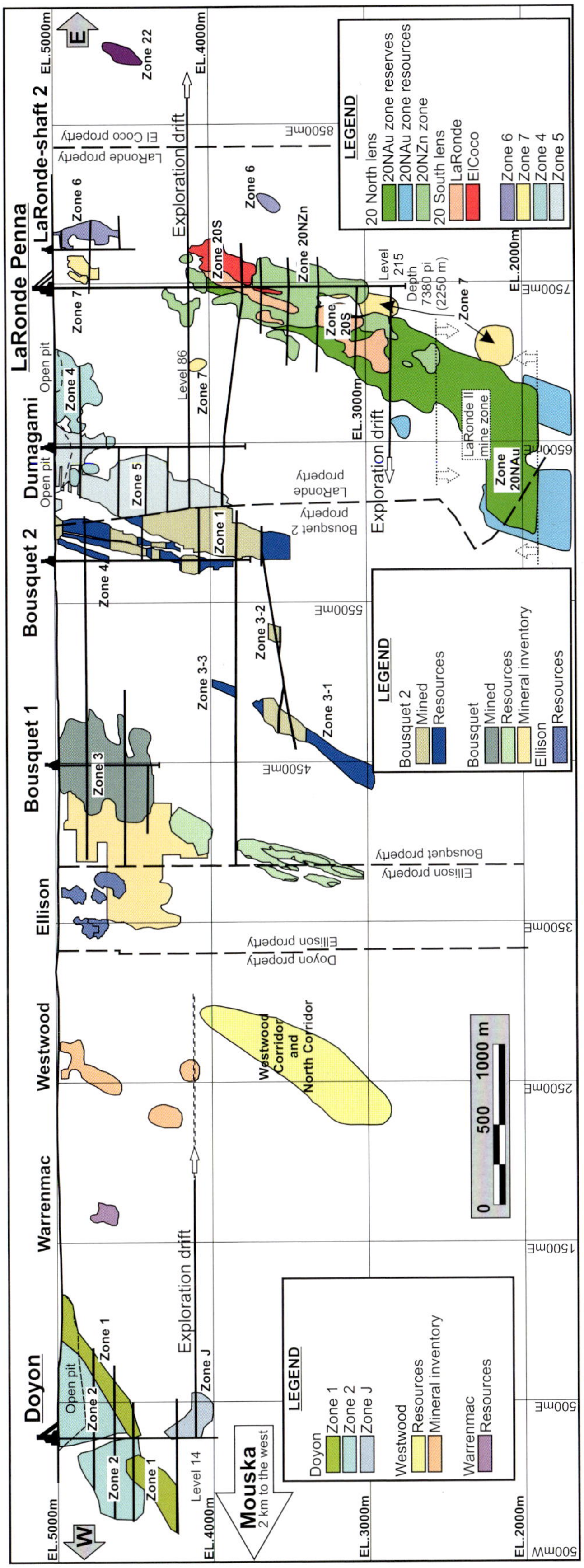

rare tellurides, and native Au is present around the vein-type ore (Tourigny et al., 1989a). The mineralization at Ellison and in zones 3-1 and 3-3 (Fig. 6) can be compared to the vein-type ore of the Bousquet 1 mine, as it forms narrow, transposed stringers of semimassive to massive sulphide veins (pyrite with chalcopyrite±sphalerite and trace amounts of pyrrhotite, galena, electrum, and native Au) as shown on Figure 8A.

At Westwood, west of Bousquet 1 (Fig. 6), two mineralized corridors have been defined: the North Corridor and the Westwood Corridor, which are in the same stratigraphic interval as zones 3, 4, and 5 of the Bousquet 1 mine, and zones A, B, and C at Ellison (Fig. 5) in the upper part of the lower member of the Bousquet Formation (Lei, 2004) or at the base of the upper member. Both the North and Westwood corridors consist of metre-wide auriferous disseminated sulphide zones cut by centimetre- to decimetre-thick semimassive to massive sulphide veins forming transposed stringers in schistose rocks. The disseminated and vein sulphides are largely dominated by granoblastic pyrite with traces of pyrrhotite and interstitial chalcopyrite (Fig. 8B). Trace amounts of sphalerite, galena, tellurides, and native Au can be found in the sulphide veins. This area is characterized by a Au-Cu association, whereas the Warrenmac lens (Figs. 5, 6) is characterized by a Au-Zn association (Lei, 2004). The Warrenmac lens is located near the interface between the lower and upper members of the Bousquet Formation (Fig. 5), and consists of semimassive to massive sulphides dominated by pyrite with variable amounts of sphalerite, chalcopyrite, and pyrrhotite, and trace amounts of galena, electrum, and native Au. Other VMS Zn-Cu anomalies occur at this stratigraphic interval south-southwest of the Mooshla intrusion but none contain ore-grade material.

In the eastern part of the camp (Bousquet 2-Dumagami, and LaRonde Penna, Fig. 1), there are many stacked lenses that mostly sit higher in the stratigraphic sequence than the ore zones at Warrenmac, Westwood, Ellison, and Bousquet 1 (Fig. 5). The two lowermost ore zones in this area (zones 6 and 7 in Figs. 4, 5), located near the base of the upper member of the Bousquet Formation, consist of semimassive to massive sulphides that form numerous isolated economic lenses as shown on Figure 6. Some lenses of zones 6 and 7 were mined to surface (LaRonde shaft 2) and can be traced to a depth of more than 2300 m, with zone 7 currently being mined at depth at LaRonde Penna (Fig. 6). These zones are composed of pyrite, chalcopyrite, and sphalerite with minor amounts of pyrrhotite, magnetite, tellurides, and native Au (Hannington et al., 2007). A talus breccia containing Au-rich massive sulphide clasts at the margin of zone 6 was mined close to surface from shaft 2. This mineralized clast-bearing talus breccia, developed laterally from the zone 6 massive sulphide mound, probably formed during a volcanism hiatus

FIGURE 6. Composite longitudinal map (looking north) of the Doyon-Bousquet-LaRonde mining camp ore zones from surface to a depth of 3.2 km. The map excludes the Mouska mine ore zones that are located about 2 km west of the Doyon mine. Some ore zones of the Bousquet 1 mine and Ellison deposit are not outlined for clarity. The relatively regular spacing of the ore lenses is attributed to a major synvolcanic faults system that controlled the distribution of the metals and of some volcanic and intrusive units through the camp. Modified from Mercier-Langevin et al. (2007c).

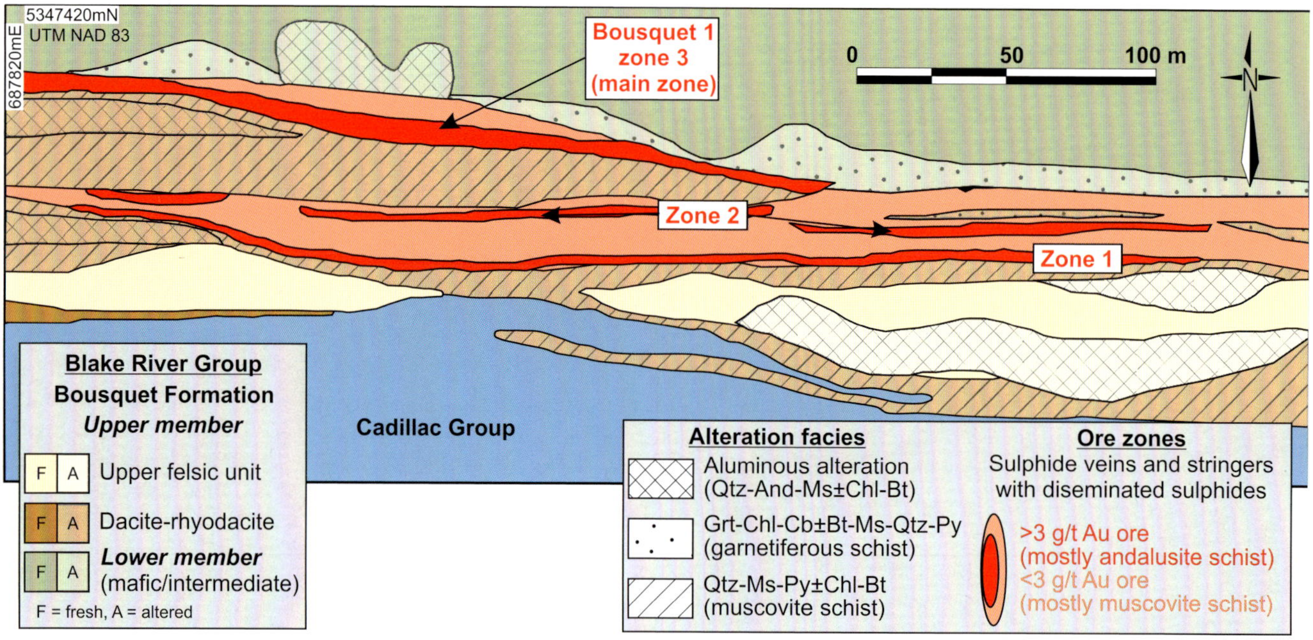

FIGURE 7. Schematic geology of the Bousquet 1 mine area illustrating the distribution of the zones 1, 2, and 3 at the base of the upper member of the Bousquet Formation. It also shows the overall distribution of the main alteration assemblages relative to the ore lenses. Modified from Valliant and Barnet (1982), Valliant et al. (1983), and Tourigny et al. (1992). And = andalusite, Bt = biotite, Cb = carbonate, Chl = chlorite, Grt = garnet, Ms = muscovite, Py = pyrite, Qtz = quartz.

in or near local subbasins (Dubé et al., 2007; Mercier-Langevin et al., 2007a). The LaRonde Penna 20 North lens occurs above zones 6 and 7 in the upper half of the upper member of the Bousquet Formation (Figs. 4, 5, 6). This lens extends from a depth of 800 m to more than 3200 m (Fig. 6). The lens is currently mined to a depth of 2450 m and it is going to be mined to a depth of 3110 m from an internal shaft (LaRonde II mining project: Fig. 6). The 20 North lens locally reaches more than 40 m in thickness (Fig. 4) and it has been subdivided into two zones: the 20 North Au zone at the base (north) and the 20 North Zn zone on top (south) (Fig. 4). The 20 North Au zone is composed of auriferous pyrite and chalcopyrite veins and veinlets forming a dense stockwork, with local semimassive centimetre- to metre-thick lenses, which is now strongly flattened and partially transposed in the main foliation. This Au-rich stockwork formed by subseafloor replacement in the upper part of a rhyodacitic to rhyolitic flow breccia (Fig. 8C) and represents the feeder zone to the 20 North Zn zone massive sulphide lens (Dubé et al., 2007b, Hannington et al., 2007, Mercier-Langevin et al., 2007a). The 20 North Zn zone consists of massive pyrite, sphalerite, and chalcopyrite with minor galena. The sphalerite is more abundant in the upper part (south) of the 20 North Zn zone and forms centimetre- to decimetre-scale massive bands parallel to the main regional foliation alternating with massive pyrite (Fig. 8D). Galena occurs in localized, narrow veins and veinlets with sphalerite near the top of the lens. A low-grade to barren zone of up to 2 m in thickness is locally developed between the 20 North Au and the 20 North Zn zones. It is composed of coarse pyrite and quartz, and represents the interface between high-temperature mineralization in the footwall (Au-Cu) and low-temperature mineralization (Zn-Pb) above (Mercier-Langevin et al., 2004). The 20 North Zn zone also contains dismembered graphitic argillite beds that contain nodular pyrite. The presence of sedimentary strata within the Zn-rich part of the lens (upper part) indicates episodes of seafloor exhalative activity at this interface.

At depth in the mine, the 20 North Au zone changes gradually into a zone of disseminated to semimassive to locally massive sulphides (Fig. 8E) composed of auriferous pyrite and chalcopyrite hosted in a quartz, kyanite, andalusite, staurolite, and muscovite schist, referred to as the 'aluminous zone' (Dubé et al., 2007b; Mercier-Langevin et al., 2007a). The 20 North Zn zone gradually disappears at depth as the aluminous zone develops (Fig. 4). A massive sulphide lens composed of auriferous pyrite, chalcopyrite, and sphalerite has been intersected in the deepest part of the 20 North lens

FIGURE 8. (A) Narrow, transposed stringers of semimassive to massive sulphide veins of zone 3-1 at Bousquet 1. The stringers are composed of pyrite and chalcopyrite with minor amounts of sphalerite, pyrrhotite, galena, electrum, and native gold in schistose felsic rocks of the upper member of the Bousquet Formation typical of Bousquet 1 and Ellison ore zones, East wall. **(B)** Semimassive sulphide band or vein, which is part of a transposed stringer in the Westwood North corridor. The sulphides consist mainly of granoblastic pyrite with minor amounts of chalcopyrite and pyrrhotite in a quartz-muscovite matrix. **(C)** Transposed auriferous pyrite and chalcopyrite stringer of the LaRonde Penna 20 North lens (20 North Au zone) emplaced within silicified and sericitized rhyodacitic to rhyolitic flow breccia. **(D)** Massive sulphides of the LaRonde Penna 20 North lens (20 North Zn zone) composed of pyrite bands and sphalerite bands. The banding represents the main foliation in the massive sulphides. **(E)** Disseminated to semimassive sulphides in association with the aluminous alteration developed at depth in the LaRonde Penna mine along the 20 North lens. Ore in the aluminous zone is composed mostly of pyrite and chalcopyrite in a quartz, kyanite, andalusite, and muscovite matrix. **(F)** Sulphide ore breccia of the Bousquet 2-Dumagami orebody. This ore breccia is composed of semimassive to massive pyrite and chalcopyrite supporting silicified wall-rock fragments, West wall. **(G)** Sulphide ore breccia of the Bousquet 2-Dumagami orebody with interstitial massive bornite±tennantite and tellurides associated with elevated Au and Cu grades. **(H)** Late north-south fractures coated with native Au within the Bousquet 2-Dumagami auriferous sulphide zones. And = andalusite, Bn = bornite, Ccp = chalcopyrite, Ky = kyanite, Py = pyrite, Qtz = quartz, Sp = sphalerite.

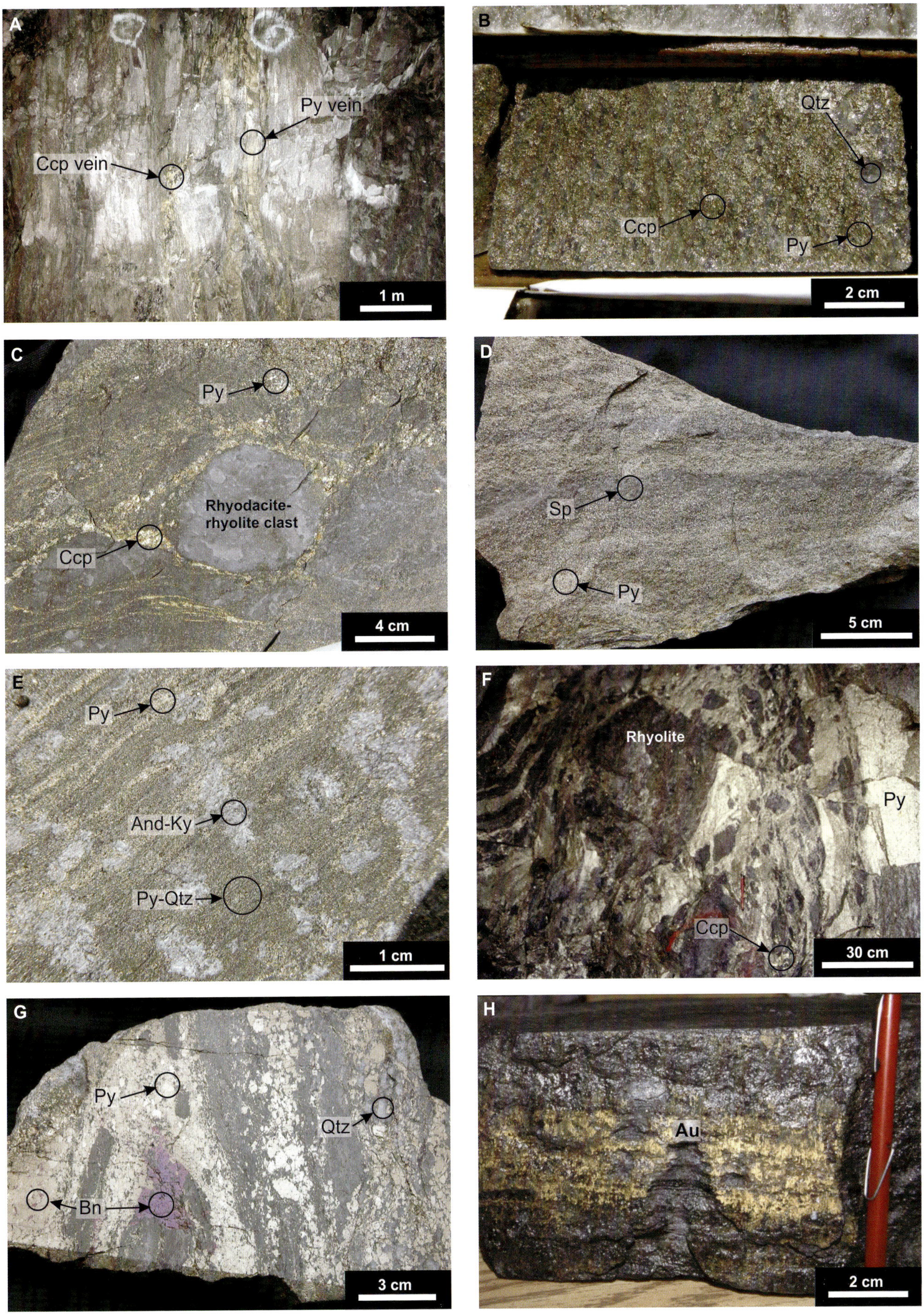
A
Py vein
Ccp vein
1 m
B
Qtz
Ccp
Py
2 cm
C
Py
Rhyodacite-
rhyolite clast
Ccp
4 cm
D
Sp
Py
5 cm
E
Py
And-Ky
Py-Qtz
1 cm
F
Rhyolite
Py
Ccp
30 cm
G
Py
Qtz
Bn
3 cm
H
Au
2 cm

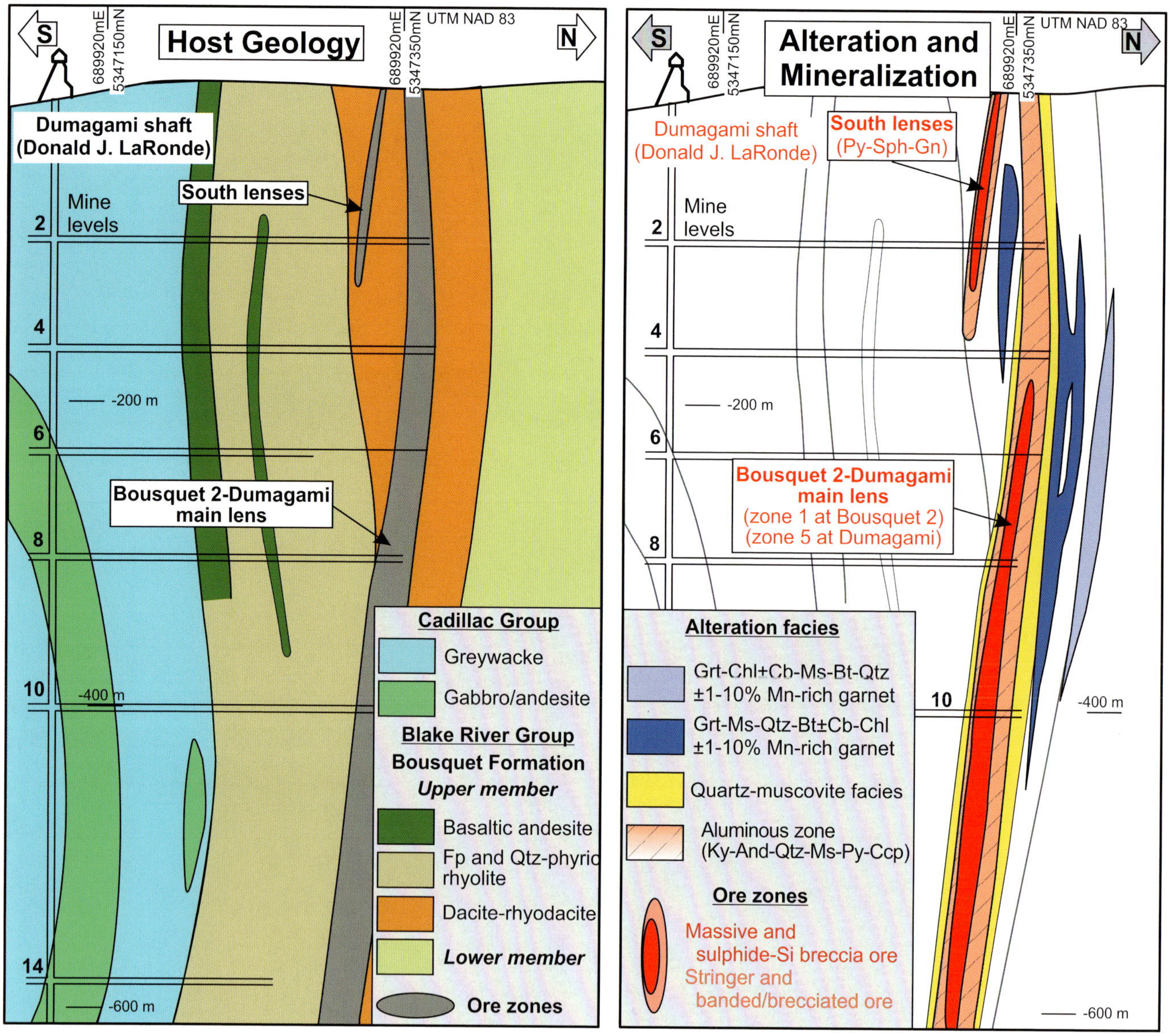

FIGURE 9. Simplified geological section through the Bousquet 2-Dumagami orebody showing the distribution of the upper member units of the Bousquet Formation, the main lens that was defined as zone 1 at Bousquet 2 and zone 5 at Dumagami, the Au-Zn south lenses, and the distribution-zonation of the principal alteration assemblages developed along the main lens. Modified from Marquis et al. (1992). And = andalusite, Bt = biotite, Cb = carbonate, Chl = chlorite, Ccp = chalcopyrite, Fp = feldspar, Gn = galena, Grt = garnet, Ky = kyanite, Ms = muscovite, Py = pyrite, Qtz = quartz, Si = silica, Sp = sphalerite.

and could suggest the development of a second Au-Cu-Zn-Ag zone associated with aluminous alteration at the lowest levels in the LaRonde Penna mine. A complex assemblage of accessory and trace minerals such as arsenopyrite, tetrahedrite, stannite, magnetite, Bi and Au tellurides, Pb, Sb and Ag sulphosalts have been identified in the LaRonde Penna mine ores (Dubé et al., 2004; Hannington et al., 2007). Gold is mostly present as electrum and auriferous tellurides.

The Bousquet 2-Dumagami orebody occurs on the same stratigraphic horizon as the LaRonde Penna 20 North lens. However, the rhyodacitic to rhyolitic flow breccia unit in the footwall of the 20 North lens thins towards the west and is not present in the Bousquet 2-Dumagami footwall (Fig. 5). This single orebody was developed as two separate mines (Dumagami, later renamed Donald J. LaRonde, and Bousquet 2) as it was bisected by a property boundary (Fig. 6). The Au- and Cu-rich orebody or lens is characterized by enrichment of Zn and Pb towards the stratigraphic top (Marquis et al., 1990a; Tourigny et al., 1993; Teasdale et al., 1996). Three principal sulphide facies were recognized in Bousquet 2 by Tourigny et al. (1993) from the base to the top of the lens (north to south: Fig. 9). These facies are 1) stringer ore; 2) massive pyrite and sulphide ore breccia; and 3) banded and brecciated sulphide. The stringer ore is composed of 15 to 35 vol.% of partially transposed anastomosing veins and veinlets of massive pyrite with minor amounts of chalcopyrite, bornite, and tennantite, and traces of digenite and sphalerite (Tourigny et al., 1993). Massive pyrite and sulphide ore breccia facies occur in the central part of the orebody (Fig. 9) and consist of discontinuous massive pyrite lenses of up to 5 m thick interfingered with pyrite stringers (30 to 60 vol.%) supporting wall-rock fragments (Fig. 8F). In the ore, variable amounts of bornite, chalcopyrite, sphalerite, pyrrhotite, galena, tennantite, and chalcocite are associated with the pyrite; trace amounts of magnetite, digenite, mawsonite, renierite, tellurides, and colusite are also found (Marquis et al., 1990a; Tourigny et al., 1993). In this facies, elevated Au and Cu grades are associated with the presence of bornite interstitial to the granoblastic pyrite (Fig. 8G). The banded and brecciated sulphide facies, mostly developed in the eastern upper part of the lens (Marquis et al., 1990a, Fig. 9), consist of narrow, strongly transposed, and discontinuous semimassive to massive pyrite and/or sphalerite veins and lenses containing variable amounts of chalcopyrite, bornite, pyrrhotite, and galena (Marquis et al., 1990a, Mercier-Langevin, 2005). The Bousquet 2-Dumagami orebody is characterized by late north-south fractures locally filled with sulphides and coated by native Au (Fig. 8H). This feature, restricted to the Au-rich sulphide zones, was originally attributed to a late introduction of Au in the deposits (syn-brittle deformation) but it has since been shown that most of the Au associated with base metals in the camp was part of the synvolcanic hydrothermal system responsible for the formation of the VMS lenses (see Dubé et al., 2007b; Hannington et al., 2007; Mercier-Langevin et al., 2007a,b). The formation of Au-coated fractures in massive sulphides is apparently the result of local remobilization of the Au along fracture planes, as first suggested by Tourigny et al. (1993).

The uppermost lens in the DBL mining camp is the 20 South lens at LaRonde Penna (Figs. 1, 5, 6). This lens is located above the 20 North lens (Fig. 4) and generally about 10 to 15 m below (north) the Cadillac Group sedimentary strata. The lens is up to 200 m wide and has a known vertical extent of at least 1300 m (Fig. 4), from 900 to 2180 m depth. The economic mineralization is concentrated mainly in two zones: 1) at 900 to 1580 m depth; and 2) at 1700 to 2180 m depth. The lens is thicker (up to 10 m) in the upper levels of the mine and gets thinner with increasing depth. The massive sulphide portion of the 20 South lens in the upper part of the mine hosts the highest Au grades (as high as 30 g/t Au with 8% Zn, 0.5% Cu, and more than 250 g/t Ag, making it one of the richest zones in the camp: Mercier-Langevin et al., 2004; Dubé et al., 2007b). Elsewhere, the mineralized zone consists of centimetre- to metre-wide Au- and Zn-rich massive sulphides. Pyrite, pyrrhotite, sphalerite, chalcopyrite, and galena are the most abundant phases in the 20 South lens and are accompanied by minor amounts of tellurides and electrum. Sphalerite and galena are concentrated towards the upper part (south portion) of the lens. The 20 South lens is slightly discordant to the ore-hosting lithological units, suggesting that it was formed, at least in part, by replacement. Towards the centre of the deposit, the lens is hosted by basaltic andesite, whereas to the east the ore is hosted by the rhyodacitic to rhyolitic flow-breccia units that are present in both the footwall and hanging wall (Fig. 4).

Hydrothermal Alteration

The Au-rich VMS deposits of the DBL mining camp are associated with upper greenschist-facies metamorphic assemblages that are derived from synvolcanic hydrothermal alteration of the host volcanic rocks (Dubé et al., 2007b). The hydrothermal alteration is widespread and, owing to the stacking of lenses, hanging-wall alteration commonly merges with the footwall alteration of the overlying lenses. However, a distinct zonation associated with individual lenses is still preserved.

The mafic rocks of the lower member of the Bousquet Formation are chloritized and locally epidote-quartz-magnetite altered away from the ore zones and around the Mooshla pluton. Biotite-rich bands and amphibole-rich bands are developed in the footwall of some of the ore lenses that are located above or hosted in mafic rocks (e.g. Westwood, Bousquet 1, Bousquet 2, and zones 6 and 7 at LaRonde Penna). The biotite-rich bands also contain feldspar and muscovite, whereas the amphibole-rich bands contain chlorite, feldspar, and tourmaline (Fig. 10A). An assemblage of plagioclase, chlorite (clinochlore), biotite, carbonate (ankerite), muscovite, quartz, and manganese-rich garnet is present in the mafic rocks underlying the Westwood and Bousquet 1 ore zones (Fig. 10B; Valliant and Barnett, 1982; Elioupoulos, 1983; Tourigny et al., 1989a). The garnet porphyroblasts are more abundant near the centre of the alteration zone, immediately beneath the ore lens, rather than at the periphery, and the garnets contain up to 25% MnO (e.g. Bousquet 1 zone 3: Valliant and Barnett, 1982). Millimetre-sized Mn-rich garnet is also present in the mafic footwall rocks of the Bousquet 2-Dumagami orebody (Marquis et al., 1990a) and locally in the footwall of zone 7 at LaRonde Penna (Dubé et al., 2007b). The relative abundance of garnet in the mafic rocks is variable from one lens to another but tends to increase with proximity to the ore, together with the abundance of disseminated pyrrhotite and pyrite. The carbonate alteration (ankerite, calcite, and dolomite) in the mafic rocks is more intense towards the Bousquet 1 deposit and is well developed in the footwall and immediate hanging wall of zone A at Ellison in association with light green chlorite (Fig. 10C).

Manganese-rich garnet alteration also occurs within the footwall felsic rocks at Dumagami (Marquis et al., 1990a,b), and especially at LaRonde Penna in the footwall rhyodacite-rhyolite flow breccia of the 20 North lens (Dubé et al., 2007b; Mercier-Langevin et al., 2007a) where it forms a large, discordant to distal, semiconformable quartz-biotite-garnet assemblage that grades into a proximal quartz-garnet-biotite-muscovite zone mainly developed in the upper and intermediate levels of the mine (~750-2000 m below surface, Fig. 4) below the Zn-rich part of the lens (Dubé et al., 2007b). The abundance (5-30 vol.%) and size (1-15 mm) of the Mn-rich garnet porphyroblasts increase towards the 20 North lens, where they occur locally in millimetre- to centimetre-wide bands of garnet, epidote, clinozoisite, muscovite, pyrite, and pyrrhotite (Fig. 10D). The garnet is a Mn-rich Fe-Ca spessartine or almandine with up to 16.3 wt.% MnO (Dubé et al., 2007b). In addition to biotite, quartz and muscovite, and variable amounts of chlorite, amphibole, pyroxene, clinozoïsite, epidote, and chloritoid occur in the garnet alteration zone in the footwall of the 20 North lens (Dubé et al., 2007b). This alteration assemblage gradually disappears with depth in the LaRonde Penna mine as the aluminous alteration develops (see below), but a similar, although less intense, Mn-rich garnet-bearing alteration assemblage is present in the felsic rocks in the hanging wall of the 20 North lens at depth in the mine (Fig. 4; Dubé et al., 2007b). Otherwise, the hanging-wall alteration of the 20 North lens at LaRonde Penna is dominated by muscovite, quartz, and variable amounts of fine-grained biotite in felsic rocks. A distinctive pinkish alteration assemblage also is

A
Hbl-Chl-Pl
Pl-Bt-Ms-Qtz
Bt-Ms-Pl
3 cm
B
Grt
1.5 cm
C
Chl-Cb
3 cm
D
Grt
Qtz
1 cm
E
Py
Po
5 cm
F
And
CENTIMETRES
3 cm
G
Rhyolite
(>95% SiO2)
Py
2 cm
H
Py
Qtz
2 cm

developed in a mafic sill and dyke complex located in the hanging wall of LaRonde Penna 20 North lens and in the footwall of the 20 South lens (Dubé et al. 2007b; Mercier-Langevin et al., 2007a,b). This alteration consists of plagioclase (albite), hornblende, and chlorite replaced by a fine-grained assemblage of quartz, biotite, rutile-titanite, pyrrhotite, and pyrite with minor amounts of tourmaline, calcite, epidote, and muscovite (Fig. 10E; Dubé et al., 2007b).

Parts of Bousquet 1 (Fig. 7), Bousquet 2-Dumagami (Fig. 9), and LaRonde Penna 20 North lens (Fig. 4) are associated at depth with a distinctive proximal aluminous alteration assemblage composed of variable amounts of muscovite, aluminosilicates (kyanite, andalusite, and pyrophyllite), and quartz. At LaRonde Penna, the gradual transition from quartz-garnet-biotite-muscovite footwall alteration to the aluminous alteration at depth is marked by the replacement of garnet and biotite by staurolite, kyanite, and andalusite, and also by increasing amounts of quartz and pyrite towards the core of the aluminous alteration zone (Dubé et al., 2007b). Muscovite (±paragonite) is abundant at the periphery of the aluminous zone in association with minor amounts of aluminosilicates (Fig. 10F) but gradually decreases in abundance towards the ore zone where the mineralization is hosted in silica-rich (>95 wt.% SiO_2) breccias as at Bousquet 2 (Fig. 10G). Silica-rich alteration is also developed at Westwood but is of lesser extent and not associated with significant amounts of aluminosilicates (Fig. 10H). The aluminous alteration developed at Bousquet 1, Bousquet 2-Dumagami, and LaRonde Penna is thought to represent the metamorphosed equivalent of an advanced argillic alteration assemblage caused by leaching of the host rock by acidic hydrothermal solutions (Tourigny et al., 1993; Teasdale et al., 1996; Dubé et al., 2007b).

The DBL camp Au-rich VMS lenses are associated with different alteration assemblages that can be of regional extent or restricted to a single deposit (i.e. focussed versus unfocused fluid flow). The 20 North lens at LaRonde Penna also illustrates how a single lens can be associated with a number of different alteration types and metal associations. These important variations in alteration style and distribution should be considered in any exploration models targeting Au-rich VMS deposits.

Epizonal 'Intrusion-Related' Au-Cu Vein Systems

Epizonal 'intrusion-related' Au-Cu veins in the DBL mining camp occur mainly in the Doyon mine but possibly also in parts of the Mouska deposit as discussed below. The epizonal intrusion-related nomenclature proposed here refers to the inferred link between the hydrothermal activity responsible for the formation of the sulphide-rich Au-Cu veins and the emplacement, at shallow depth, of late intrusive phases of the Mooshla synvolcanic pluton (see Galley and Lafrance, 2007).

The epizonal intrusion-related Au-Cu vein systems of the Doyon mine (Table 1), and some parts of the Mouska mine, include sulphide-rich (75-80 vol.%) veins and veinlets, pyrrhotite-pyrite-chalcopyrite to pyrite-chalcopyrite disseminations, and quartz-sulphide veins (Savoie et al., 1991; Trudel et al.,1992; Belkabir and Hubert, 1995; Gosselin, 1998; Galley and Lafrance, 2007). They have been interpreted as epigenetic (syndeformation) or mesothermal by Hoy et al. (1990), Marquis et al. (1990b,c), Savoie et al. (1986, 1990, 1991), as syngenetic or synmagmatic by Filion et al. (1977), Valliant and Hutchinson (1982), Arseneau (1995), and Gosselin (1998), and as remobilized synvolcanic mineralization by Guha et al. (1982), Belkabir and Hubert (1995), Galley et al. (2003), and Galley and Lafrance (2007).

At Doyon, the mineralization is concentrated in three different zones (Figs. 11, 12): zones 1 and 2 at the southeastern margin of the Mooshla pluton, and the West zone within the southeastern part of the intrusion. Zones 1 and 2 are hosted by highly altered, schistose felsic ± mafic volcanic rocks in the upper part of the lower member of the Bousquet Formation (Fig. 13), and form slightly discordant orebodies (zone 2, Figs. 12, 13) and intensely transposed vein stockworks in highly schistose rocks (zone 1). The West zone consists of steeply west-dipping northwest-southeast, north-south, and north-northeast – south-southwest oriented veins (Fig. 14) that cut a tonalitic part of the intermediate intrusive phase (Fig. 11) of the Mooshla pluton (Gosselin, 1998; Galley and Lafrance, 2007). The West zone is also part of a slightly Au-enriched east-west to west-northwest – east-southeast corridor that traverses the entire intrusion (Fig. 11).

Mineralization

Doyon zone 1 consists of numerous pyrite and quartz veinlets or laminations and contains about 20 vol.% sulphides. The veinlets vary from a few millimetres to a few centimetres in width and can contain more than 70 vol.% pyrite (Savoie et al., 1991). They represent sulphide-rich vein networks that have been transposed into discrete east-west deformation zones along the southern margin of the Mooshla intrusion (Fig. 15A). Minor amounts of chalcopyrite, sphalerite, galena, arsenopyrite, and chalcocite are associated with the pyrite (Guha et al., 1982). Native Au is finely disseminated and interstitial to pyrite grains. Trace minerals, including auriferous tellurides, are also recognized (calaverite, petzite, and altaite: Guha et al., 1982). The gangue minerals (<30 vol.%) are mainly quartz with muscovite, carbonates, and chlorite (Guha et al., 1982; Savoie et al., 1990, 1991).

Figure 10. Photographs of representative alteration assemblages related to Au-rich VMS lenses of the DBL mining camp. **(A)** Biotite, feldspar, and muscovite-rich bands and amphibole, chlorite, feldspar ± carbonate bands developed in the footwall of the Westwood ore lenses and typical of the other ore lens located in the lower member of the Bousquet Formation or just above it such as the Bousquet 1 ore zones and the LaRonde Penna zone 7. **(B)** Proximal alteration assemblage composed of plagioclase, chlorite, biotite, carbonate, muscovite, quartz, and manganese-rich garnet developed in the upper part of the lower member of the Bousquet Formation in the Westwood-Bousquet 1 area. **(C)** Carbonate, light-green chlorite, muscovite, and quartz alteration assemblage developed in the Ellison zone A horizon. **(D)** LaRonde Penna 20 North lens alteration assemblage composed of Mn-rich garnet, biotite, chlorite, epidote, muscovite, pyrite, and pyrrhotite developed in the upper levels of the mine in association with the Zn-rich part of the deposit (20 North Zn zone). **(E)** LaRonde Penna 20 North lens hanging-wall alteration and 20 South lens footwall alteration. This assemblage, composed of quartz, biotite, rutile-titanite, pyrrhotite, and pyrite with minor amounts of tourmaline, calcite, epidote, and muscovite, is developed in the basaltic andesite located between the 20 North and South lenses. **(F)** Quartz-muscovite±aluminosilicate assemblage developed at the margin of the aluminous alteration zones of Bousquet 2-Dumagami (photo) and LaRonde Penna. **(G)** Silica-rich breccia of the Bousquet 2-Dumagami orebody. This alteration is host of the mineralization and is related to an intense acid leaching of the wallrocks, leaving only silica. **(H)** Proximal quartz-muscovite alteration at Westwood. And = andalusite, Bt = biotite, Cb = carbonate, Chl = chlorite, Grt = garnet, Hbl = hornblende, Ms = muscovite, Pl = plagioclase, Po = pyrrhotite, Py = pyrite. Qtz = quartz.

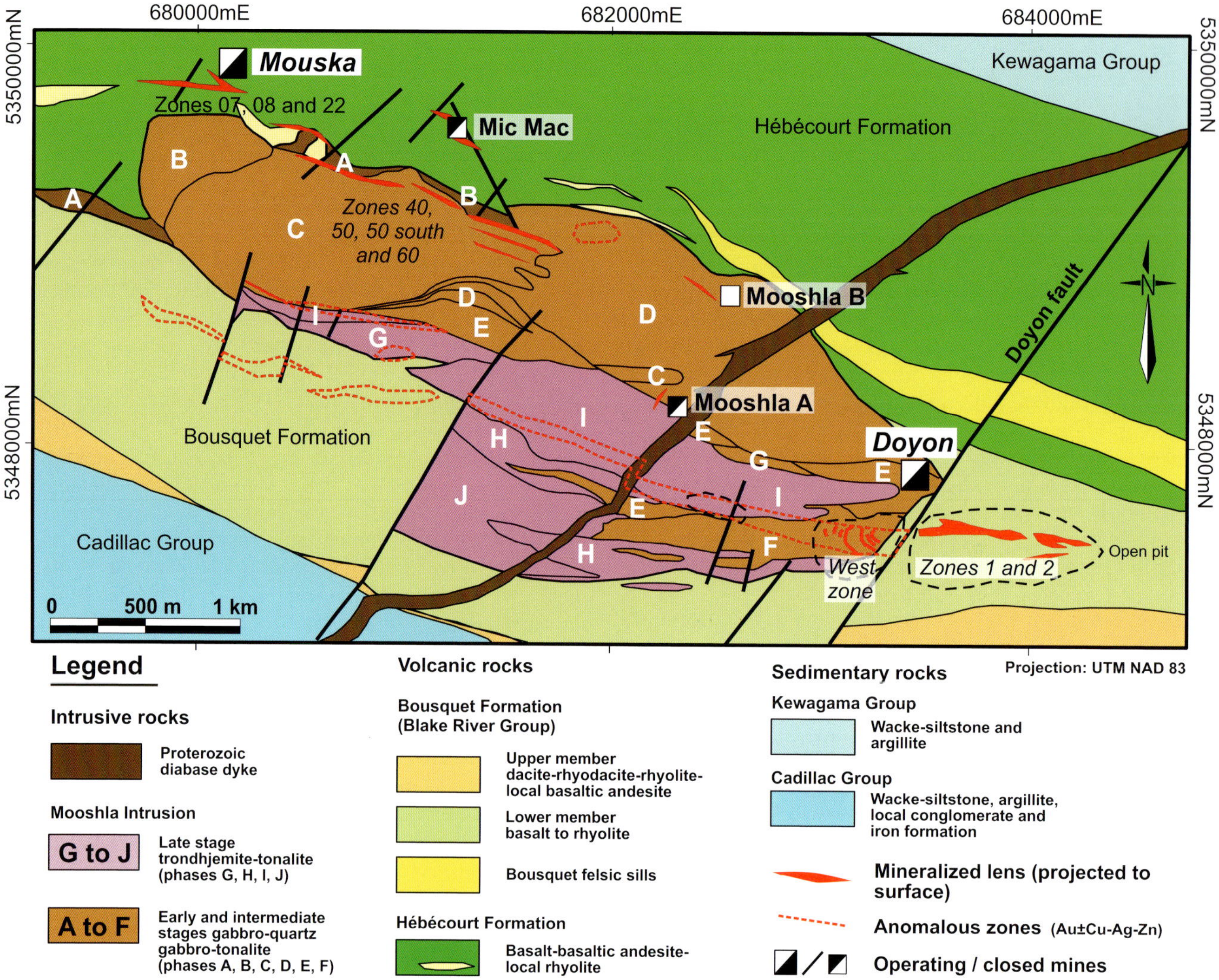

FIGURE 11. Simplified geological map of the Mooshla synvolcanic intrusion. The Mooshla pluton hosts parts of the mineralizations of the Doyon-Bousquet-LaRonde mining camp: zones 40, 50, 50 south, and 60 at Mouska, Mooshla A and B deposits, and a large part of the West zone at Doyon. From Galley and Lafrance (2007).

Doyon zone 2 and the West zone are characterized by similar vein types that are discordant to the stratigraphy and the regional foliation (Figs. 12, 14, 15B). These veins, characterized by a brecciated texture, vary in thickness from a few centimetres to a few decimetres, and contain between 10 and 75 vol.% sulphides (Fig. 15C) with quartz, dolomite, and calcite as the main gangue minerals and chlorite, tourmaline, and rutile as trace constituents (Savoie et al., 1990, 1991). Overall, pyrite is the main sulphide in the veins but chalcopyrite can locally be as abundant as pyrite (Fig. 15D). The pyrite forms coarse-grained, semimassive to massive aggregates within the veins. These aggregates are commonly cut by irregular chalcopyrite veins and veinlets that are restricted to the quartz and sulphide veins. Chalcopyrite is also present as micro-inclusions within pyrite crystals (Savoie et al., 1991; Gosselin, 1998). Sulphides other than pyrite and chalcopyrite are relatively rare but include sphalerite, pyrrhotite, galena, arsenopyrite, bornite, and chalcocite, which are closely associated with the chalcopyrite (Savoie et al., 1991). Native Au is locally abundant in the veins of zone 2 and the West zone, and is commonly associated with tellurides (calaverite, tetradymite, tellurobismuthite, petzite, and altaite: Savoie et al., 1990, 1991; Fig. 15E).

Hydrothermal Alteration

At Doyon, zones 1 and 2, which are hosted mainly in volcanic rocks, are associated with large discordant to subconcordant alteration zones (schists) in which the alteration gradually intensifies towards the veins (Fig. 13). The distal alteration consists of a propylitic assemblage composed of chlorite, muscovite, carbonate, oligoclase, rutile, and pyrite with minor amounts of garnet, chloritoid, and biotite in mafic rocks, and by an assemblage of quartz, plagioclase, muscovite, rutile, and pyrite, and trace amounts of chloritoid and clinozoisite in felsic rocks (Savoie et al., 1991). These assemblages are gradually replaced, towards the veins, by quartz, chlorite, muscovite, rutile, and pyrite schists in mafic rocks, and by quartz, muscovite, rutile, and pyrite schists in felsic rocks (Fig. 15F). The proximal alteration that hosts the mineralization of zones 1 and 2 is composed of quartz, muscovite, aluminosilicates, rutile, pyrite, and chlorite, with the chlorite being more abundant in mafic rocks (Fig. 13).

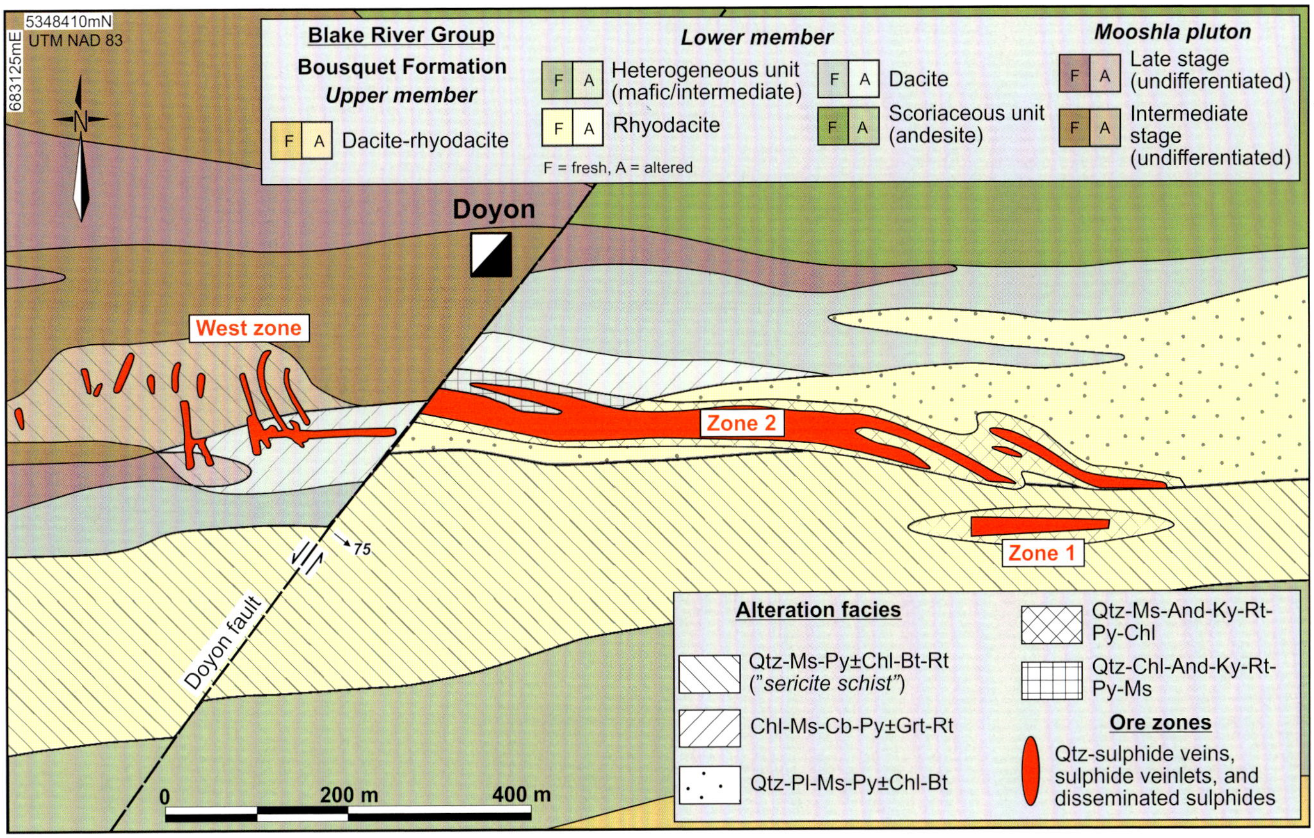

FIGURE 12. Simplified geological map of the Doyon deposit showing the distribution of the volcanic and intrusive units of the lower and upper members of the Bousquet Formation and of the Mooshla pluton. The map also shows the approximate position of the main ore zones projected to surface and a schematic representation of the main alteration assemblages developed around the ore zones. Modified from Savoie et al. (1991). And = andalusite, Bt = biotite, Cb = carbonate, Chl = chlorite, Grt = garnet, Ky = kyanite, Ms = muscovite, Pl = plagioclase, Py = pyrite, Qtz = quartz, Rt = rutile.

Locally, an assemblage of quartz, aluminosilicates, rutile, and pyrite without micas is developed (aluminous alteration of Savoie et al., 1991). The aluminosilicates consist of andalusite and kyanite porphyroblasts in part retrograded to kaolinite and pyrophyllite.

In contrast, the veins of the West zone at Doyon are hosted by intrusive rocks that do not exhibit the extensive alteration developed in zones 1 and 2. The alteration is mostly concentrated in the vein selvages, from a few centimetre to a few decimetres from the veins (Savoie et al., 1991), especially in the more 'mafic' phases of the intrusion (Gosselin, 1998). The distal alteration is dominated by chlorite, biotite, and epidote (Fig. 15G), which replace primary ferromagnesian minerals, and by muscovite±carbonate that partially replace plagioclase. Anhydrite-quartz-pyrite-chalcopyrite veins are present at the margin of the Doyon mine mineralization (Savoie et al., 1991; Galley and Lafrance, 2007). The proximal alteration is dominated by muscovite, pyrite, quartz, and carbonate with minor amounts of chlorite and biotite (Fig. 14) and is characterized by the complete destruction of the primary textures (Fig. 15H). Pyrite can constitute up to 15 vol.% of the proximal alteration assemblage (Gosselin, 1998), where it is commonly hosted within miarolitic cavities (Galley and Lafrance, 2007).

Shear Zone-Hosted Au-Cu Vein Systems

The shear zone-hosted (and/or remobilized) Au-Cu vein systems of the DBL mining camp comprise the Mic Mac, Mooshla A and B, and parts of the Mouska deposit (Table 1). The Mouska deposit consists of a series of subparallel, subvertical east-west to northwest-southeast-trending sulphides and quartz-sulphides vein systems and associated narrow, high-angle reverse ductile and brittle ductile shear zones superimposed on the regional foliation (Belkabir and Hubert, 1995; Tourigny and Tremblay, 1997; Belkabir et al., 1998). The deformation is partitioned into high-strain zones that host the veins, and this deformation may have been responsible for remobilization of ore that formed prior to deformation and that resemble the epizonal intrusion-related sulphide-rich veins (e.g. Belkabir, 1995; Belkabir and Hubert, 1995; Belkabir et al., 2004; Galley and Lafrance, 2007). Zones 07, 08, and 22 at Mouska and the Mic Mac deposit are hosted in mafic to intermediate volcanic rocks of the Hébécourt Formation north of the Mooshla pluton (Figs. 11, 16), and zones 40, 50, 50 south, and 60 at Mouska and Mooshla A and B ore zones are hosted in the northern part of the intrusion (Fig. 11). A number of subsidiary zones and mineralized structures exist around these principal ore zones in the intrusion, especially in its western half, south of Mouska and west of Doyon (Fig. 11).

Mineralization

Three main types of mineralization are recognized: 1) stringer zones composed of sulphide veins, veinlets, and disseminations; 2) quartz-sulphide veins; and 3) quartz-sulphides ± tourmaline-carbonate-chlorite veins (Belkabir and

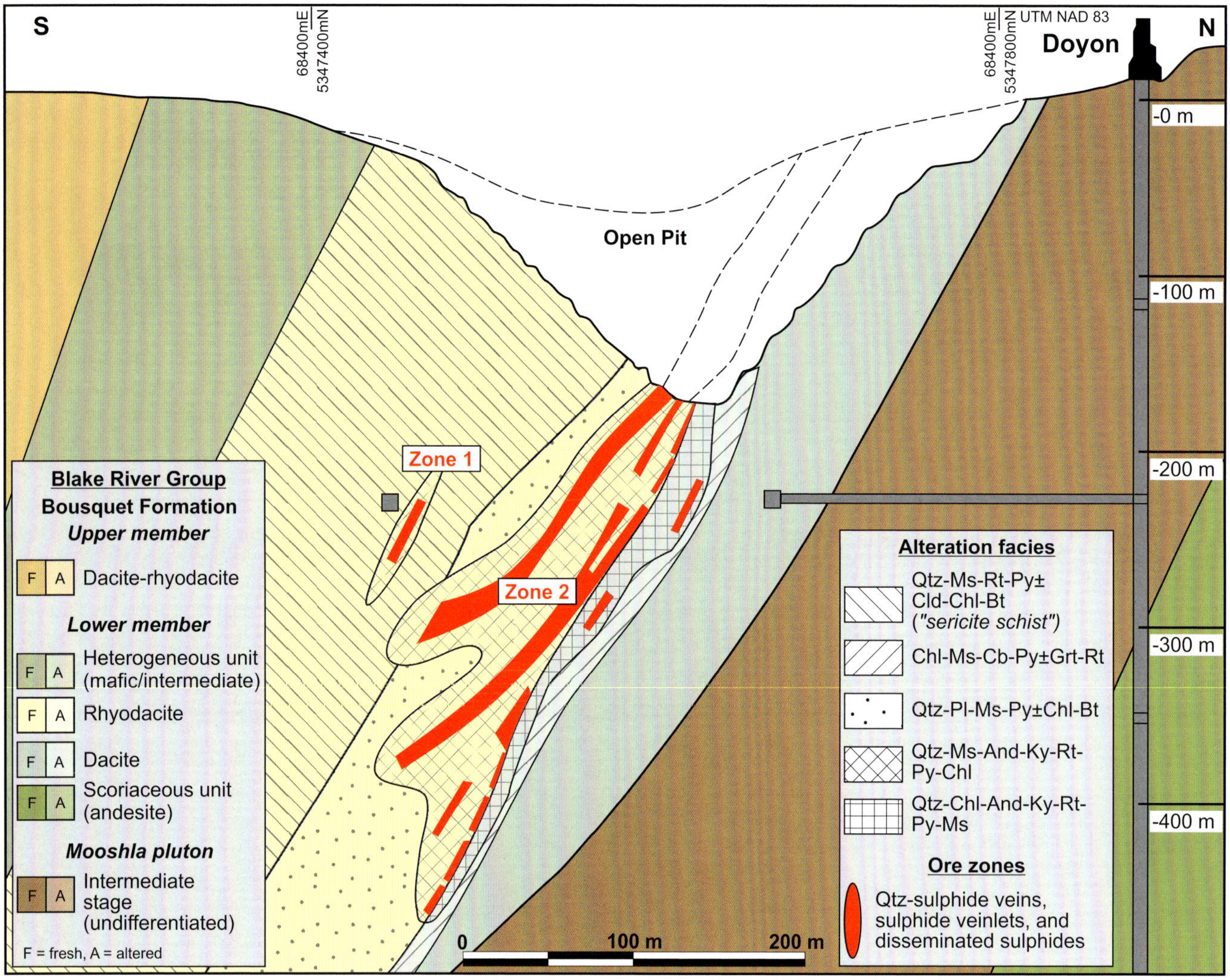

FIGURE 13. Simplified composite geological section through the zones 1 and 2 of the Doyon mine showing the distribution of the volcanic and intrusive units of the lower and upper members of the Bousquet Formation and of the Mooshla pluton. The map also shows a schematic representation of the main alteration assemblages developed around the ore zones. Modified from Savoie et al. (1991). And = andalusite, Bt = biotite, Cb = carbonate, Chl = chlorite, Cld = chloritoid, Grt = garnet, Ky = kyanite, Ms = muscovite, Pl = plagioclase, Py = pyrite, Qtz = quartz, Rt = rutile.

Hubert, 1995; Galley and Lafrance, 2006). These veins typically vary in thickness between a few centimetres to a few decimetres (Fig. 17A) and rarely are more than a metre in width. The sulphide veins and veinlets (type 1) contain variable amounts sulphides in a quartz-dominated matrix. The veins are also associated with disseminated sulphides and abundant chlorite in the wall rocks. The vein sulphides consist mainly of chalcopyrite, pyrrhotite, and pyrite, which are accompanied by trace amounts of magnetite, ilmenite, hematite, sphalerite, electrum, and tellurides of Au, Ag, Pb, and Bi (Belkabir and Hubert, 1995; Belkabir et al., 2004). The disseminated sulphides and stringers (type 1), common in zone 22 at Mouska (Fig. 16), are enclosed within pervasively chlorite-altered and sheared volcanic±intrusive rocks and consist mostly of pyrite (up to 50 vol.%: Belkabir and Hubert, 1995; Belkabir et al., 2004) with quartz. The second vein-type (quartz-sulphide: type 2), common in zones 07, 08, 40, and 50 at Mouska (Figs. 11, 16), typically contains less than 75 vol.% of sulphides. Quartz in these veins is greyish (Fig. 17A), in contrast to other late, barren to weakly mineralized white quartz-bearing veins. Pyrrhotite and chalcopyrite are dominant (Fig. 17B) and are associated with minor amounts of pyrite and trace amounts of magnetite, ilmenite, electrum, and tellurides of Ag and Pb (Belkabir and Hubert, 1995). The type 2 veins resemble the epizonal intrusion-related Au-Cu veins such as those found at Doyon. The third vein type (quartz-sulphides ± tourmaline-carbonate-chlorite: type 3) is developed mostly in zones 40 and 50 at Mouska (Belkabir and Hubert, 1995) and at Mooshla B (Fig. 17C; Arseneau, 1995; Gosselin, 1998), and is characterized locally by a laminated, crack-and-seal, or banded textures (Galley and Lafrance, 2007). These veins contain much less sulphide (<20 vol.%, mainly chalcopyrite, pyrrhotite, and pyrite). They form as straight-walled veins both parallel to discrete east-west shears and as associated subhorizontal veins. The quartz-sulphides ± tourmaline-carbonate-chlorite veins locally follow the margins of dioritic dykes that transect the Mooshla intrusion and its volcanic host rocks (e.g. Mooshla A and B deposits). Gold in these three types of mineralization occurs as micro-inclusions in sulphides, as free grains ,or filling late microfractures in quartz and/or sulphides (Belkabhir et al., 2004). This type of vein shares

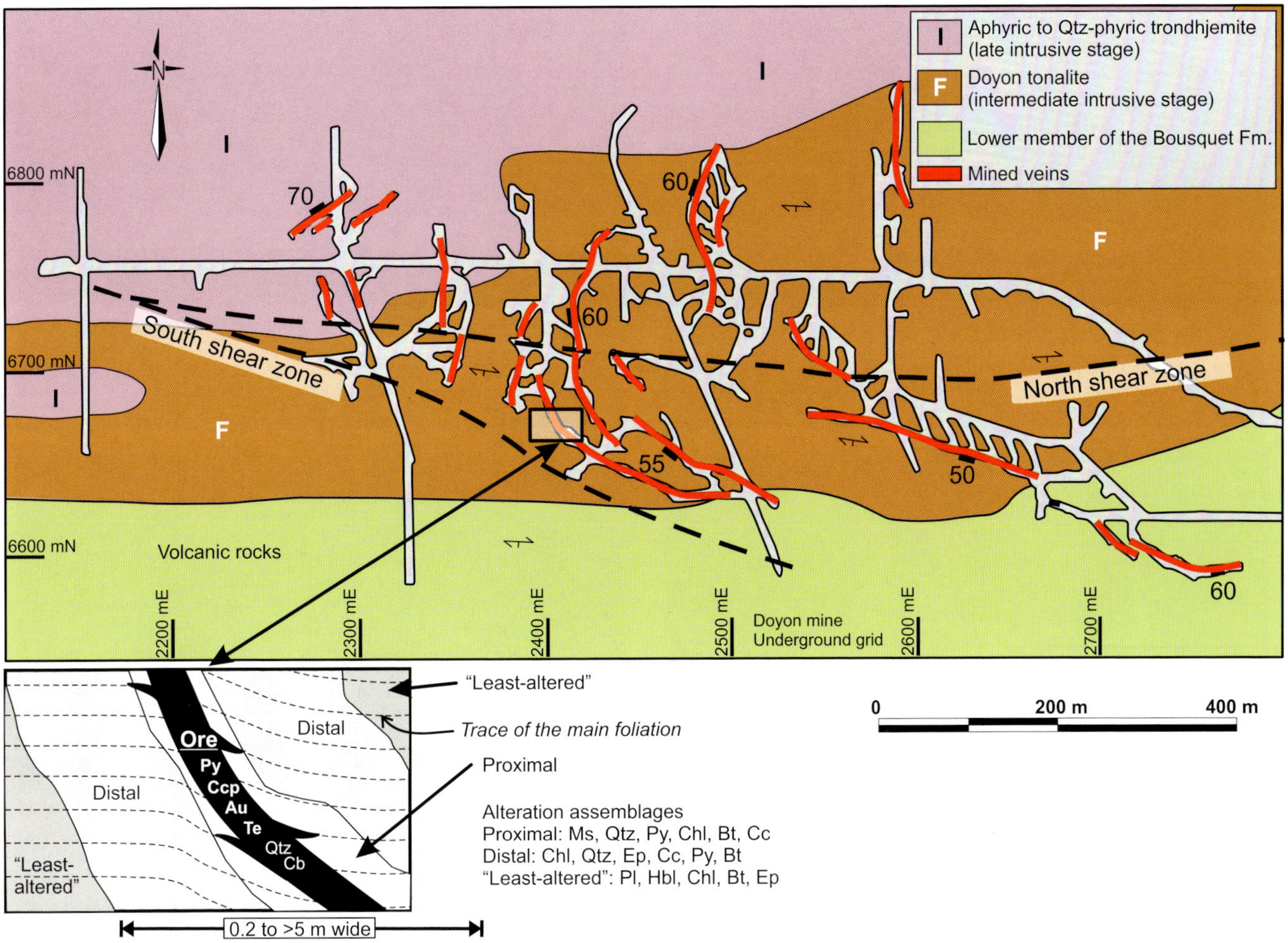

FIGURE 14. Simplified geological map of the West zone veins on level 4-0 (230 m depth) at Doyon that illustrates the complex distribution of the veins that are locally at high angle with the east-west regional foliation. These veins are mostly hosted by the Mooshla synvolcanic intrusion. A schematic illustration of the alteration assemblages developed in the vein selvages is given in the inset. Modified from Gosselin (1998). Bt = biotite, Cb = carbonate, Cal=calcite, Chl=chlorite, Ccp = chalcopyrite, Ep = epidote, Hbl=hornblende, Ms=muscovite, Pl=plagioclase, Py=pyrite, Qtz=quartz, Te=tellurides.

many analogies with orogenic-style or mesothermal-style auriferous quartz ± tourmaline-carbonate veins.

The disseminated sulphides and the semimassive to massive sulphides in veins show textures typical of remobilization and recrystallization, such as annealing and durchbewegung breccia textures (Belkabir and Hubert, 1995). The sulphides locally occur in ribbons thought to represent local transposition in the regional foliation, and chalcopyrite is locally concentrated in small-scale fold hinges associated with late reverse movements. The veins themselves are boudinaged (Fig. 17D) and folded along the main schistosity, which is concentrated in the vein selvages and very weakly developed in the host rocks away from the veins (Fig. 17E).

Hydrothermal Alteration

The Au-Cu vein systems of the Mouska, Mic Mac, and Mooshla A and B orebodies (Fig. 11) are associated with metre-wide alteration haloes around the veins, with more intense hydrothermal alteration adjacent to the veins. The shearing spatially associated with the ore zones is generally confined to corridors defined by the proximal alteration along the veins (Fig. 17E, F). Away from the veins, the alteration is mainly characterized by chlorite with subsidiary amounts of carbonate, epidote, muscovite, and quartz with minor amounts of disseminated sulphides (Belkabir et al., 2004). The carbonate alteration is stronger in mafic volcanic rocks than in intrusive rocks. The transition from distal to proximal alteration envelopes is marked by a gradual decrease in the abundance of plagioclase, hornblende, epidote, actinolite, and chlorite abundances, which are replaced by an assemblage of biotite, quartz, muscovite, disseminated sulphides, and iron oxides (Fig. 17F).

Genetic and Exploration Models

The Bousquet Formation is interpreted to represent the remnants of a composite volcano, about 10 km in diameter, distal to the larger Noranda Volcanic Complex (Fig. 1), north of Rouyn-Noranda (Lafrance et al., 2003a). The volcanic and intrusive rocks comprise a continuous, differentiated succession generated by partial melting of depleted upper mantle and/or juvenile material (mafic crust). The magmatic activity is thought to have been associated with the progression from depleted upper mantle diapirism and mafic-ultramafic magma underplating and assimilation, to magmatic differentiation (assimilation–fractional crystallization) at mid-crustal levels in subsidiary magma chambers within relatively thick,

A
20 cm
B
Ccp
Py
Qtz
Main foliation
15 cm
C
Ccp
Py
Qtz
5 cm
D
Ccp
Qtz
15 cm
E
Te
Au
2 cm
F
Qtz-Ms-Py
Rt-Ms
5 cm
G
2 cm
H
Ms-Qtz-Py
5 cm

FIGURE 15. Photographs of representative alteration and ore assemblages related to epizonal "intrusion-related" Au-Cu veins. **(A)** Highly transposed stockwork of sulphide-quartz veins and veinlets of zone 1 at Doyon. The sulphides consist mostly of pyrite with minor amounts of chalcopyrite, sphalerite, galena, arsenopyrite, chalcocite, and tellurides, East wall. **(B)** Quartz, pyrite, and chalcopyrite veins of zone 2 at Doyon. The zone 2 and West zone sulphide-rich veins are locally discordant to the stratigraphy, at a high angle with the main foliation and partially boudinaged, plane view. **(C)** Foliation-parallel, centimetre-wide quartz and sulphide vein of the West zone at Doyon. The veins contain up to 75 vol.% sulphides, mainly pyrite and chalcopyrite, East wall. **(D)** Intrusion-hosted chalcopyrite-rich vein of the West zone at Doyon. Sulphide-rich veins typical of zone 2 and West zone at Doyon commonly show brecciated textures with pyrite and/or chalcopyrite isolating quartz ± carbonate-chlorite fragments, plane view. **(E)** Native Au and greyish tellurides in a quartz-sulphide vein of West zone at Doyon. **(F)** Proximal alteration assemblage composed of quartz, aluminosilicates, muscovite, rutile, and pyrite that is developed in felsic rocks hosting the zones 1 and 2 mineralization at Doyon, East wall. **(G)** Example of the distal alteration developed in the Mooshla intrusion near the West zone veins. This alteration is characterized by a gradual replacement of primary ferromagnesian minerals by chlorite, biotite, and epidote and by the partial destruction of plagioclase by muscovite±carbonate, plane view. **(H)** Example of the proximal alteration developed in the Mooshla intrusion along the West zone vein selvages. This alteration is characterized by an assemblage of muscovite, pyrite, quartz, calcite ± chlorite and biotite. Ccp = chalcopyrite, Py = pyrite, Qtz = quartz. Ms = muscovite, Rt = rutile, Te = tellurides.

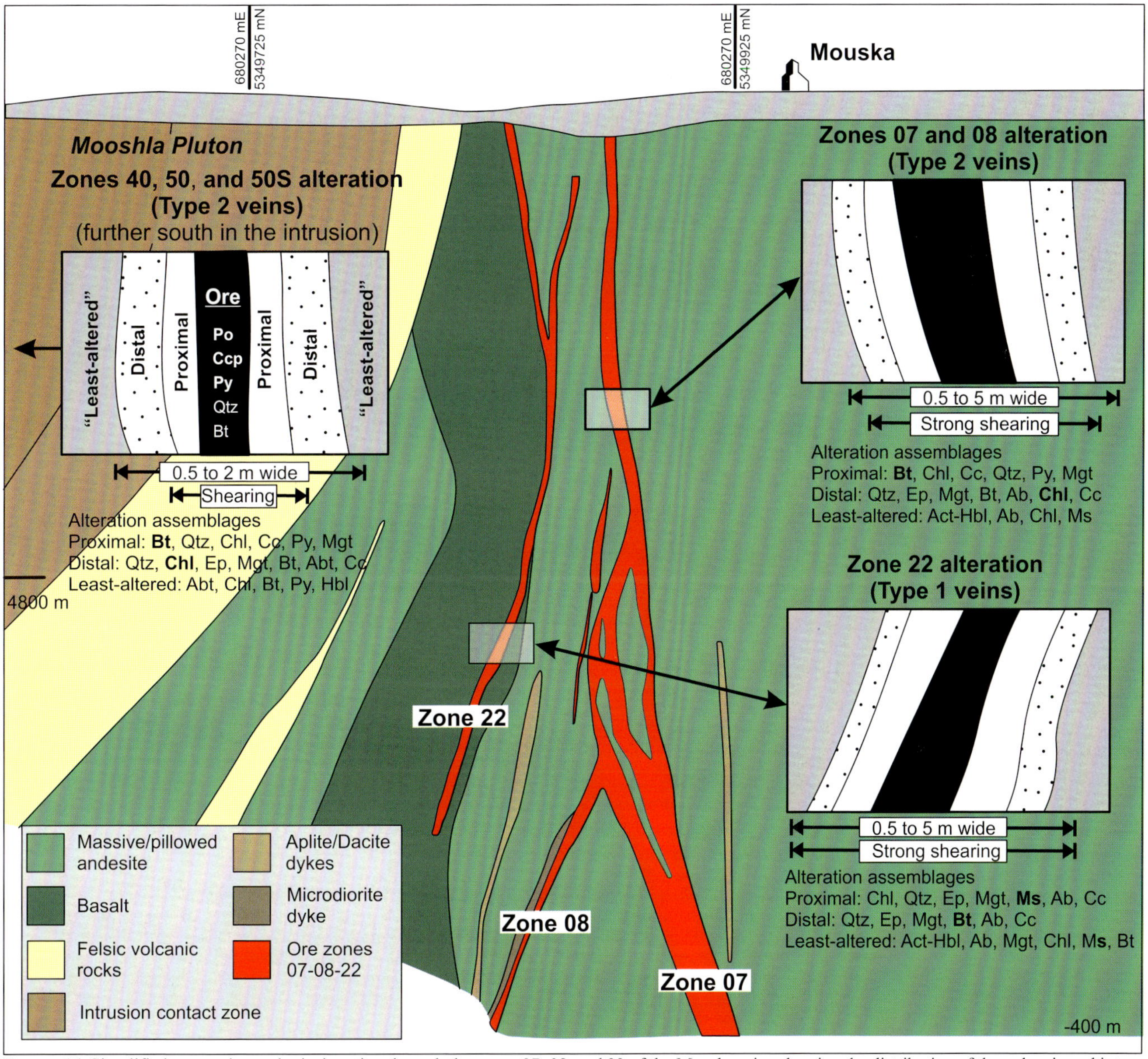

FIGURE 16. Simplified composite geological section through the zones 07, 08, and 22 of the Mouska mine showing the distribution of the volcanic and intrusive units of the Hébécourt Formation and of the Mooshla pluton. Schematic illustrations of the alteration assemblages developed in the vein selvages is given in insets for zones 07, 08 (type 2), and 22 (type 1) that are hosted by andesitic to basaltic volcanic rocks of the Hébécourt Formation and for zones 40, 50, and 50 south (type 2) that are hosted by the Mooshla synvolcanic intrusion. Modified from Belkabir and Hubert (1995) and Belkabir et al. (2004). Ab = albite, Act = actinolite, Bt = biotite, Cc = calcite, Chl = chlorite, Ccp = chalcopyrite, Ep = epidote, Hbl = hornblende, Mgt = magnetite, Ms = muscovite, Po = pyrrhotite, Py = pyrite, Qtz= quartz.

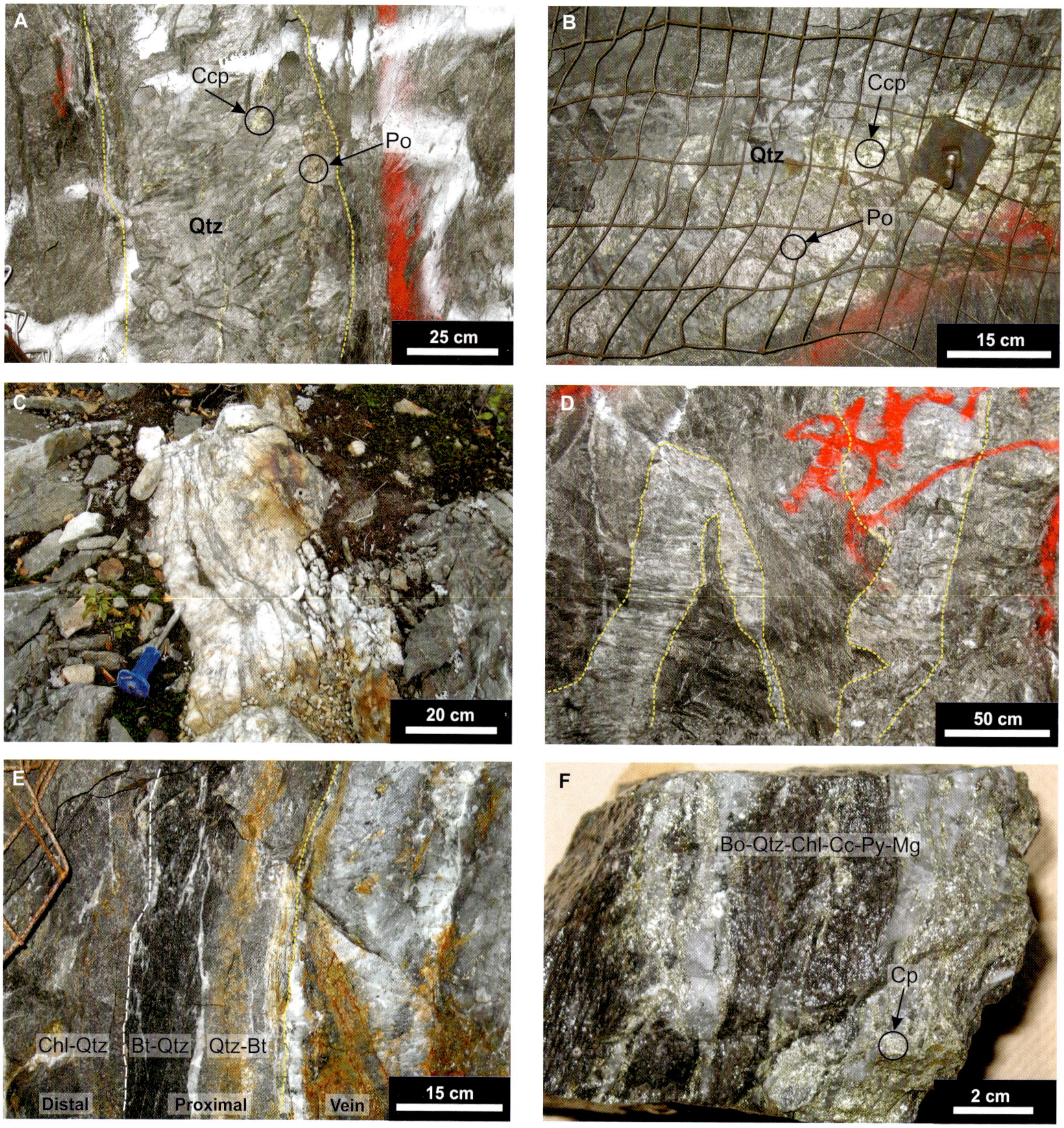

FIGURE 17. Photographs of representative alteration and ore assemblages related to shear zone-hosted Au-Cu veins. **(A)** Centimetre- to decimetre-wide, subvertical sulphide-rich grey quartz veins hosting the mineralization of zone 50 at the Mouska mine, West wall. **(B)** Sulphide-rich quartz vein of zone 50 at the Mouska mine. The mineralized veins contain up to 75 vol.% sulphides with mainly chalcopyrite and pyrrhotite with minor amounts of pyrite, magnetite, ilmenite, electrum, and tellurides, plane view. **(C)** Quartz-sulphide ± tourmaline-carbonate-chlorite vein, Mooshla B deposit. **(D)** Folded and boudinaged sulphide-rich quartz vein in a narrow shear zone following apparent reverse movement on the shear plane, zone 50, Mouska mine, East wall. **(E)** A proximal (centimetre- to decimetre-wide) alteration halo is developed along the veins of zone 50 at the Mouska mine. The proximal alteration is composed of a mineral assemblage dominated by quartz and biotite, with minor amounts of chlorite, calcite, pyrite, and magnetite. The biotite becomes more abundant a few centimetres away from the vein selvages (biotite- and quartz-dominated assemblage) and is gradually replaced away from the vein by chlorite (chlorite- and quartz-dominated assemblage). Strain is concentrated in the proximal alteration zone, West wall. **(F)** Narrow sulphide-rich grey quartz veins of zone 50 at the Mouska mine associated with a characteristic proximal alteration assemblage that is strongly foliated and composed predominantly of biotite and quartz with variable amounts of chlorite, calcite, pyrite, and magnetite. The proximal alteration assemblage is restricted to the vein selvages and grades outward a few centimetres to a decimetre to a weakly foliated chlorite-dominated alteration assemblage. Bt = biotite, Cc = calcite, Chl = chlorite, Ccp = chalcopyrite, Mgt = magnetite, Py = pyrite, Po = pyrrhotite, Qtz = quartz.

juvenile or immature mafic±felsic crust (see Mercier-Langevin et al., 2007b). This geodynamic context is different from the extensional back-arc setting proposed for the Noranda Volcanic Complex in the central Blake River Group west of the DBL mining camp (Fig. 2) as older and thicker crust is involved in the geodynamic setting and petrogenesis of the DBL volcanic terrain. The inferred setting (i.e. high-pressure felsic volcanic rocks in a rifted volcanic terrain over a relatively thick juvenile lithosphere) is considered to have been favourable for the generation of long-lived, deep-seated hydrothermal fluid circulation, with possible variable contributions to the hydrothermal solutions from the crust, from degassing magma and from metasomatized upper mantle as discussed in Mercier-Langevin et al. (2007b). The felsic rocks generated in this context (dacite, rhyodacite, rhyolite, and intrusive equivalents in the polyphased Mooshla pluton) are transitional to calc-alkaline and correspond to Archean FII-type rhyolites, which have been considered to be of limited prospectivity in the literature (Lesher et al., 1986; Hart et al., 2004). The unique concentration of Au-rich VMS deposits and epizonal intrusion-related Au-Cu vein systems in association with transitional to calc-alkaline felsic rocks in the DBL camp demonstrates the elevated prospectivity of such felsic volcanic sequences and also suggests that Archean FII-type felsic rocks and the inferred geodynamic setting in which they are produced could be responsible, at least in part, for the elevated Au content of the associated deposits and therefore represent favourable exploration targets for primary Au deposits (Mercier-Langevin et al., 2007b).

The upper member of the Bousquet Formation is notably thickened in the LaRonde Penna mine area, which indicates the development of a local effusive centre. Such effusive centres in the camp are probably important sites for the focus of intense hydrothermal activity (Mercier-Langevin et al., 2007a). Despite late deformation and metamorphic events, the locations of synvolcanic faults can be inferred from the distinct elongation of felsic dome complexes, associated mafic sill and dyke swarms, and metal zonation within the lenses, especially in the LaRonde Penna and Bousquet 2-Dumagami areas (Mercier-Langevin 2005; Mercier-Langevin et al., 2007a). The distribution of the volcanic units and the relatively regular spacing (±500 to 800 m) between the ore lenses at camp scale strongly suggests the systematic development of synvolcanic faults (e.g. Fig. 6). The sulphides were deposited locally on the seafloor above these faults (e.g. LaRonde Penna 20 North Zn zone) but were also formed by concomitant extensive subseafloor replacement (e.g. 20 North Au zone at LaRonde Penna), which explains the disseminated, vein type and semimassive nature and distribution of the ore at Bousquet 1, Ellison, and Westwood, for example. This emphasizes the fact that cross-strata permeability was a key factor in controlling fluid discharge in this area during protracted volcanism-magmatism and hydrothermal activity. The epizonal intrusion-related Au-Cu veins developed in zone 2 and West zone at Doyon are thought to be related to syn- to late-magmatic faulting of the late intrusive phases of the Mooshla pluton upon cooling, caused by emplacement at shallow depth (Galley and Lafrance, 2007). These volatile-rich late intrusive phases may have been responsible for the formation of the auriferous quartz-sulphide veins and the aluminous alteration associated with it (Galley and Lafrance, 2007).

The different alteration and mineralization styles developed in the DBL mining camp are best explained by variations along strike in the style of volcanism and magmatic activity, as exemplified by LaRonde Penna (Fig. 18) (cf. Dubé et al., 2007b; Mercier-Langevin et al., 2007a). Metamorphosed propylitic alteration caused by near-neutral, seawater-dominated hydrothermal fluids in the volcaniclastic units in the upper levels of the LaRonde Penna mine and advanced argillic alteration caused by acidic fluids proximal to shallow felsic subvolcanic intrusions or cryptodomes at greater depth in the mine coexist along the same mineralized lens (20 North lens, Fig. 4). A similar situation was proposed for so-called high-sulphidation deposits formed in the VMS environment (Sillitoe et al., 1996). This situation seems to contrast with subaerial volcanic centres in which near-neutral and intensely acidic conditions occur in separate volcanic-hydrothermal systems unless telescoped (e.g. Sillitoe, 1994). These different conditions appear to be possible in different parts of the same hydrothermal system in submarine environments, in large part due to variable degrees of mixing between magmatic volatile-rich fluids and hydrothermally convected seawater as shown schematically in Figure 18. The model proposed is supported by the gradual transition between the two ore styles at LaRonde and by similar variations developed at the camp scale (e.g. Au and Zn mineralizations at Warrenmac versus Au and Cu zones at Westwood-Ellison-Bousquet 1, with and without intense acid leaching). Propylitic alteration (i.e. Mn-rich garnet ± quartz, biotite, chlorite, carbonate) commonly represents the distal footprint of the hydrothermal system at Bousquet 1, Dumagami, and Westwood, but is proximal to the mineralization at LaRonde Penna (20 North lens) and Warrenmac. The aluminous alteration (advanced argillic), which is thought to be produced by fluids rich in magmatic volatiles, is a key vector towards Cu-Au ore at depth at LaRonde-Penna or at Bousquet 2-Dumagami and Bousquet 1 (Dubé et al., 2007b). These variations in hydrothermal assemblages reflect significant variations in the hydrothermal fluids within the volcanic complex and correspond to different types of mineralization.

The eruption of the Bousquet Formation upper member, the emplacement of the Mooshla pluton, and the formation of associated precious and base metal deposits at ca. 2698 Ma correspond to a particularly fertile episode of Au-rich VMS formation in the Blake River Group. This could suggest a possible correlation between the petrogenetic evolution of the Abitibi greenstone belt and the Au endowment of some deposits (e.g. Mercier-Langevin et al., 2007a).

Although the regional north-south convergence and oblique dextral shearing strongly affected the orebodies, deformation is not a prerequisite for the formation of Au-rich, Doyon-Bousquet-LaRonde-type deposits, as previously thought. However, the localization of high-strain zones outside known tectonic corridors, such as major faults, could indicate the presence of syngenetic alteration zones where subsequent deformation was focussed. Deformation can also be responsible for the redistribution and possible concentration of Au in narrow structural corridors. As proposed by Belkabir and Hubert (1995), Belkabir et al. (2004), and

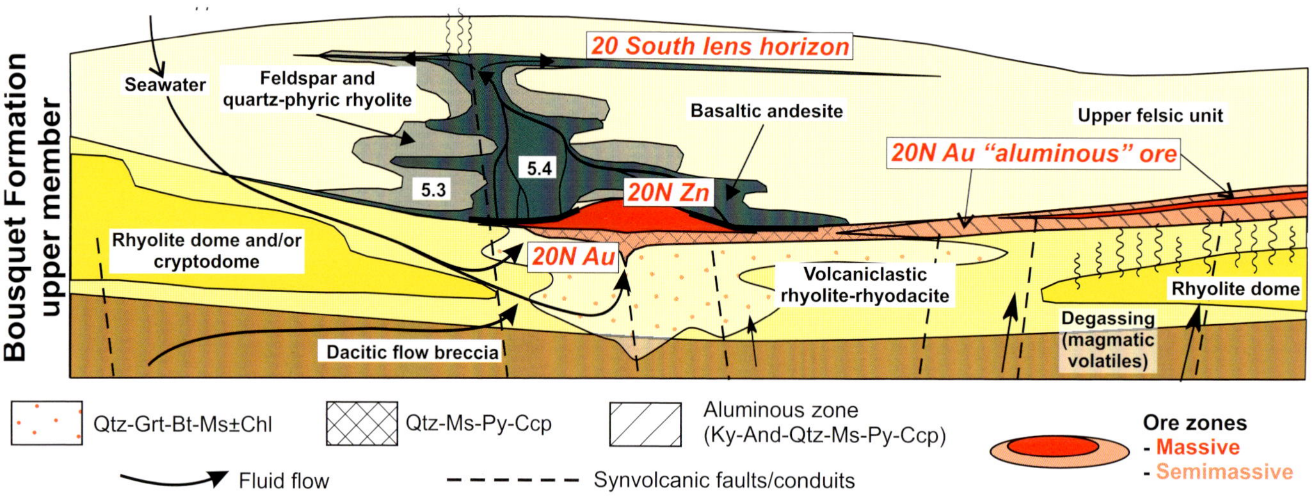

FIGURE 18. Schematic model of the Bousquet 2-LaRonde Penna hydrothermal system. The variations in ore and alteration style observed along the 20 North lens are illustrated as an example to show the existing alteration styles developed in the Doyon-Bousquet-LaRonde mining camp. These alteration styles can vary from one deposit to another, from one lens to another in a single deposit, or along a single lens as is the case along the 20 North lens at LaRonde Penna. The initiation of the hydrothermal activity responsible for the formation of most of the stacked Au-rich VMS lenses and of the epizonal "intrusion-related" Au-Cu vein systems of the Doyon-Bousquet-LaRonde mining camp is thought to be coincident with a shift in magmatic activity as suggested by the spatial association between the mineralizations and the felsic-dominated calc-alkaline volcanic-intrusive rocks of the upper member of the Bousquet Formation and the late Mooshla intrusive stage. The calc-alkaline felsic volcanism-plutonism of the upper member of the Bousquet Formation marks the gradual magmatic evolution from regional tholeiitic to transitional, mafic-dominated volcanism-plutonism of the Hébécourt Formation and of the lower member of the Bousquet Formation. See discussion in the text. Modified from Mercier-Langevin et al. (2007a). And = andalusite, Bt = biotite, Chl = chlorite, Ccp = chalcopyrite, Grt = garnet, Ky = kyanite, Ms = muscovite, Py = pyrite, Qtz = quartz.

Galley and Lafrance (2007), low-grade zones could have been 'upgraded' around and within the Mooshla pluton.

Specific Objectives of Current and Future Research

A number of challenges remain in developing a practical metallogenic model for the Doyon-Bousquet-LaRonde mining camp. Specific objectives of current and future research include the following.

1. Better constraints on the geological setting and relationships between the various ore zones at the camp scale to define the main controls on the location and deposition of ore (e.g. in relation to the position of the main synvolcanic faults). New mapping and 3-D compilations will improve exploration drilling at depth, as the potential is still very high to find new ore lenses or vein systems.
2. Establish the link between the Mooshla synvolcanic intrusion, its epizonal 'intrusion-related' Au-Cu veins and the surrounding Au-rich VMS deposits and occurrences in an integrated volcanic, magmatic, and hydrothermal model.
3. Further constrain the timing relationships between volcanism, plutonism, hydrothermal activity, deformation, and metamorphism using high-precision U-Pb and Re-Os geochronology on volcanic, intrusive, and sedimentary rocks as well as on sulphides and graphitic argillite beds.
4. Trace the hydrothermal upflow zones in the stratigraphic footwall and hanging-wall sequences using alteration minerals, geochemistry, volcanology, and stable isotope mapping. The stable isotope signature of the various alteration zones will also provide insights into the hydrothermal conditions that prevailed during ore formation and possibly contribute to answering the question about the source(s) of the mineralizing fluids.
5. Characterize the isotopic signature of the Au-rich VMS mineralization and the epizonal 'intrusion-related' Au-Cu vein systems versus the shear zone-hosted or remobilized Au-Cu vein systems to better constrain the source(s) of the metals and sulphur.
6. Better understand the origin of the barren sulphides located at the contact between the Bousquet Formation volcanic rocks and the younger Cadillac Group sedimentary rocks, and their relationship to productive hydrothermal systems.
7. Document the relationship between the DBL mining camp and the rest of the Blake River Group, including other Au-rich VMS (e.g. Horne and Quemont) of the Blake River Group as well as older Au-rich VMS deposits of the Abitibi greenstone belt.

Acknowledgements

This paper synthesizes more than 30 years of observations and careful descriptions made by numerous exploration and mining geologists, research scientists, and dedicated students who had to deal with major tectonic disturbances of primary textures and key geological relationships. Those geoscientists have to be acknowledged for their efforts and their input into the understanding of the DBL mining camp. This mining camp is now one of the major mining areas in Canada, thanks to the mining and exploration companies that continue to believe in the high exploration potential of the Abitibi greenstone belt.

The authors also wish to express their sincere appreciation to Agnico-Eagle Mines Limited and to IAMGOLD Corporation (formerly Cambior Inc.) for their essential scientific contribution, the authorization to publish, and for their keen interests in geoscience. M. Boutin and K. Lauzière contributed to the drawing of some figures. This paper was substantially improved by helpful comments and careful constructive reviews by J. H. Bédard, I. R. Jonasson, M. Legault, and the Editor, W. Goodfellow.

Geological Survey of Canada Contribution No. 20060546.

References

Arseneau, G.J., 1995, Metallogeny and Associated Alteration of Gold Occurrences, Bousquet Township, northwestern Quebec: Ph.D. thesis, Colorado School of Mines, Golden, Colorado, 168 p.

Ayer, J., Amelin, Y., Corfu, F., Kamo, S., Ketchum, J., Kwok, K., and Trowell, N., 2002, Evolution of the southern Abitibi greenstone belt based on U-Pb geochronology: autochthonous volcanic construction followed by plutonism, regional deformation and sedimentation: Precambrian Research, v. 115, p. 63-95.

Barrie, C.T., Ludden, J.N., and Green, T.H., 1993, Geochemistry of volcanic rocks associated with Cu-Zn and Ni-Cu deposits in the Abitibi subprovince: Economic Geology, v. 88, p. 1341-1358.

Bateman, P.W., 1985, Rock Alteration at the Bousquet Gold Mine, Quebec: M.Sc. thesis, University of Western Ontario, London, Ontario, 159 p.

Belkabir, A., 1995, Structure et Métallogénie du Secteur Ouest du District Aurifère Archéen de Bousquet, Abitibi, Canada: Ph.D. thesis, Université de Montréal, Montreal, Quebec, 220 p.

Belkabir, A., and Hubert, C., 1995, Geology and structure of a sulfide-rich gold deposit: An example from the Mouska gold mine, Bousquet district, Canada: Economic Geology, v. 90, p. 1064-1079.

Belkabir, A., Hubert, C., and Hoy, L., 1998, Fluid-rock reactions and resulting change in rheological behaviour of a composite granitoid: the Archean Mooshla stock, Canada: Canadian Journal of Earth Sciences, v. 35, p. 131-146.

——— 2004, Gold emplacement and hydrothermal alteration in metabasic rocks at the Mouska mine, Bousquet district, Abitibi, Quebec, Canada: The Canadian Mineralogist, v. 42, p. 1079-1096.

Davis, D.W., 2002, U-Pb geochronology of Archean metasedimentary rocks in the Pontiac and Abitibi subprovinces, Quebec, constraints on timing, provenance and regional tectonics: Precambrian Research, v. 115, p. 97-117.

Dimroth, E., Imreh, L., Rocheleau, M., and Goulet, N., 1982, Evolution of the south-central part of the Archean Abitibi belt, Quebec. Part I: Stratigraphy and paleogeographic model: Canadian Journal of Earth Sciences, v. 19, p. 1729-1758.

Dimroth, E., Imreh, L., Goulet, N., and Rocheleau, M., 1983, Evolution of the south-central part of the Archean Abitibi belt, Quebec. Part III: Plutonic and metamorphic evolution and geotectonic model: Canadian Journal of Earth Sciences, v. 20, p. 1374-1388.

Dubé, B., Mercier-Langevin, P., Hannington, M., Davis, D., and Lafrance, B., 2004, Le Gisement de Sulfures Massifs Aurifères Volcanogènes LaRonde, Abitibi, Québec: Altérations, Minéralisations et Implications pour l'ENploration: Ministère des Ressources naturelles, de la Faune et des Parcs du Québec, MB 2004-03, 112 p.

Dubé, B., Gosselin, P., Mercier-Langevin, P., Hannington, M., and Galley, A., 2007a, Gold-rich volcanogenic massive sulphide deposits, *in* Goodfellow, W.D., ed., Mineral Deposits of Canada: A synthesis of Major Deposit-Types, District Metallogeny, the Evolution of Geological Provinces, and Exploration Methods: Geological Association of Canada, Mineral Deposits Division, Special Publication 5, p. 75-94.

Dubé, B., Mercier-Langevin, P., Hannington, M.D., Lafrance, B., Gosselin, P., and Gosselin, G., 2007b, The LaRonde Penna world-class Au-rich volcanogenic massive sulfide deposit, Abitibi, Quebec: Mineralogy and geochemistry of alteration and implications for genesis and exploration: Economic Geology, in press.

Elioupoulos, E.G., 1983, Geochemistry and Origin of the Dumagami Pyritic Gold Deposit, Bousquet Township, Quebec: M.Sc. thesis, University of Western Ontario, London, Ontario, 263 p.

Filion, M., Vallée, M., and Lavoie, C., 1977, Les gisements d'or de la SOQUEM-Silverstack, Canton Bousquet, Québec: Canadian Institute of Mining and Metallurgy Bulletin, v. 70, p.159-172.

Galley, A.G., and Lafrance, B., 2007, Évolution et métallogénie du pluton de Mooshla: Ministère des Ressources naturelles et de la Faune, in press.

Galley, A.G., and Pilote, P., 2002, Géologie et métallogénie de l'intrusion de Mooshla, district minier de Bousquet, Cadillac, Québec: Ministère des Ressources naturelles, DV 2002-10, p. 37.

Galley, A.G., Pilote, P., and Davis, D., 2003, Gold-related subvolcanic Mooshla intrusive complex, Bousquet Mining District, P.Q. in Ore Deposits at Depth: Challenges and Opportunities: The Canadian Institute of Mining and Metallurgy, 2003 Field Conference, September 23-26, 2003. Technical Sessions Abstract Volume, p. 16.

Galley, A., Hannington, M., and Jonasson, I., 2007, Volcanogenic massive sulphide deposits, *in* Goodfellow, W.D., ed., Mineral Deposits of Canada: A synthesis of Major Deposit-types, District Metallogeny, the Evolution of Geological Provinces, and Exploration Methods: Geological Association of Canada, Mineral Deposits Division, Special Publication 5, p. 141-161.

Gaudreau, R., 1986, Intrusion Synvolcanique et Minéralisation Aurifère: Exemple du Pluton de Mooshla, Canton de Bousquet, Abitibi: M.Sc. thesis, Université Laval, Quebec, 42 p.

Gibson, H.L., and Galley, A.G., 2007, Volcanogenic massive sulphide deposits of the Archean Noranda District, Quebec, *in* Goodfellow, W.D., ed., Mineral Deposits of Canada: A synthesis of Major Deposit-types, District Metallogeny, the Evolution of Geological Provinces, and Exploration Methods: Geological Association of Canada, Mineral Deposits Division, Special Publication 5, p. 533-552.

Gosselin, G., 1998, Veines de Quartz Aurifères Précoces à la Zone Ouest de la Mine Doyon, Canton de Bousquet, Preissac, Abitibi: M.Sc. thesis, Université du Québec à Chicoutimi, 125 p.

Gosselin, P., and Dubé, B., 2005a, Gold deposits of Canada: Distribution, Geological Parameters and Gold Content: Geological Survey of Canada, Open-File Report 4896, 1 CD-ROM.

——— 2005b, Gold Deposits of the World: Distribution, Geological Parameters and Gold Content: Geological Survey of Canada, Open-File Report 4895, 1 CD-ROM.

Guha, J., Gauthier, A., Vallée, M., Descarreaux, J., and Lange-Brard, F., 1982, Gold mineralization at the Doyon mine (Silverstack), Bousquet, Quebec, *in* Hodder, R.W., and Petruk, W., eds., Geology of Canadian Gold Deposits: Canadian Institute of Mining and Metallurgy, Special Volume 24, p. 50-57.

Hannington, M.D., Poulsen, K.H., Thompson, J.F.H., and Sillitoe, R.H., 1999, Volcanogenic gold in the massive sulfide environment: Reviews in Economic Geology, v. 8, p. 325-356.

Hannington, M.D., de Ronde, C.E.J., and Petersen, S., 2005, Sea-floor tectonics and submarine hydrothermal systems: Economic Geology 100th Anniversary Volume, p. 111-141.

Hannington, M.D., Mercier-Langevin, P., Dubé, B., and Gosselin, G., 2007, Mineralogy, geochemistry and isotopic studies of the gold-rich ores of the LaRonde Penna deposit: Economic Geology, in prep.

Hart, T.R., Gibson, H.L., and Lesher, C.M., 2004, Trace element geochemistry and petrogenesis of felsic volcanic rocks associated with volcanogenic massive Cu-Zn-Pb sulfide deposits: Economic Geology, v. 99, p. 1003-1013.

Hoy, L.D., Trudel, P., Tourigny, G., Kheang, L., Savoir, A., and Crépeau, R., 1990, Isotopic and fluid inclusion constraints on the origin of vein Au mineralization at Bousquet township, Quebec, *in* Rive, M., Verpaelst, P., Gagnon, Y., Lulin, J.-M., Riverin, G., and Simard, A., eds., The Northwestern Quebec Polymetallic Belt: A Summary of 60 Years of Mining Exploration: Canadian Institute of Mining and Metallurgy, Special Volume 43, p. 413-423.

Hubert, C., Trudel, P., and Gélinas, L., 1984, Archean wrench-fault tectonics and structural evolution of the Blake River Group, Abitibi belt, Quebec: Canadian Journal of Earth Sciences, v. 21, p. 1024-1032.

Huston, D.L., 2000, Gold in volcanic-hosted massive sulfide deposits: distribution, genesis, and exploration: Reviews in Economic Geology, v. 13, p. 400-426.

Lafrance, B., Moorhead, J., and Davis, D., 2003a, Cadre Géologique du Camp Minier de Doyon-Bousquet-LaRonde: Ministère des Ressources Naturelles, de la Faune et des Parcs du Québec, ET 2002-07, 43 p.

Lafrance, B., Mercier-Langevin, P., Dubé, B., Galley, A.G., Hannington, M.D., Davis, D.W., Moorhead, J., Pilote, P., and Mueller, W.U., 2003b, Carte Synthèse de la Formation de Bousquet: Partie ouest: Ministère des Ressources Naturelles, de la Faune et des Parcs; DV 2003-08, échelle 1: 20 000.

Lafrance, B., Davis, D.W., Goutier, J., Moorhead, J., Pilote, P., Mercier-Langevin, P., Dubé, B., Galley, A., and Mueller, W.U., 2005, Nouvelles Datations Isotopiques dans la Portion Québécoise du Groupe de Blake River et des Unités Adjacentes: Ministère des Ressources Naturelles, de la Faune et des Parcs du Québec, RP 2005-01, 15 p.

Lajoie, J., and Ludden, J., 1984, Petrology of the Archean Pontiac and Kewagama sediments and implications for the stratigraphy of the southern Abitibi belt: Canadian Journal of Earth Sciences, v. 21, p. 1305-1314.

Lamothe, D., Harris, J.R., Labbé, J.-Y., Doucet, P., Houle, P., Moorhead, J., Dion, C., Savard, R., and Melançon, M., 2005, Évaluation du Potentiel en Minéralisations de Type Sulfures Massifs Volcanogènes (SMV) pour l'Abitibi: Ministère des Ressources Naturelles, de la Faune et des Parcs, EP 2005-01.

Langshur, A., 1990, The Geology, Geochemistry and Structure of the Mooshla Intrusion, Bousquet Mining Centre, Quebec: M.Sc. thesis, Université d'Ottawa, 172 p.

Lei, Y., 2004, Secteur Doyon-Mouska: Progression des travaux d'exploration et nouvelles découvertes: Ministère des Ressources Naturelles, Faune et Parcs, Québec Exploration 2004, Résumé des Conférences et des Photoprésentations, p. 12.

Lesher, C.M., Goodwin, A.M., Campbell, I.H., and Gorton, M.P., 1986, Trace-element geochemistry of ore-associated and barren, felsic metavolcanic rocks in the Superior Province, Canada: Canadian Journal of Earth Sciences, v. 23, p. 222-237.

Lulin, J.-M., 1990, Une analyse du développement minier du nord-ouest québécois, *in* Rive, M., Verpaelst, P., Gagnon, Y., Lulin, J.-M., Riverin, G., and Simard, A., eds., The Northwestern Quebec Polymetallic Belt: A summary of 60 Years of Mining Exploration: Canadian Institute of Mining and Metallurgy, Special Volume 43, p. 17-34.

Marquis, P., Brown, A.C., Hubert, C., and Rigg, D.M., 1990a, Progressive alteration associated with auriferous massive sulfide bodies at the Dumagami mine, Abitibi greenstone belt, Quebec: Economic Geology, v. 85, p. 746-764.

Marquis, P., Hubert, C., Brown, A.C., 1990b, Gold deposits in deformed massive sulphide and pyretic alterites in the Bousquet district, southern Abitibi Subprovince, *in* Ho, S.E., Robert, F., and Groves, D.I. (eds)., Gold and Base-Metal Mineralization in the Abitibi Subprovince, Canada, with Emphasis on the Quebec Segment: University of Western Australia, Short Course Notes, Publication No. 24, p. 243-273.

Marquis, P., Hubert, C., Brown, A.C., and Rigg, D.M., 1990c, An evaluation of genetic models for gold deposits of the Bousquet district, Quebec, based on their mineralogic, geochemical, and structural characteristics, *in* Rive, M., Verpaelst, P., Gagnon, Y., Lulin, J.-M., Riverin, G., and Simard, A., eds., The Northwestern Quebec Polymetallic Belt: A Summary of 60 Years of Mining Exploration: Canadian Institute of Mining and Metallurgy, Special Volume 43, p. 383-399.

Marquis, P., Hubert, C., Brown, A.C., Scherkus, E., Trudel, P., and Hoy, L.D., 1992, Géologie de la Mine Donald J. LaRonde: Ministère de l'Énergie et des Ressources du Québec, ET 89-06, 106 p.

Mercier-Langevin, P., 2005, Géologie du Gisement de Sulfures Massifs Volcanogènes Aurifères LaRonde, Abitibi, Québec: Ph.D. thesis, Institut National de la Recherche Scientifique, Eau, Terre et Environnement, Université du Québec, Québec, 694 p.

Mercier-Langevin, P., Dubé, B., Hannington, M.D., Davis, D.W., and Lafrance, B., 2004, Contexte Géologique et Structural des Sulfures Massifs Volcanogènes Aurifères du Gisement LaRonde, Abitibi: Ministère des Ressources Naturelles, de la Faune et des Parcs du Québec, ET 2003-03, 47 p.

Mercier-Langevin, P., Dubé, B., Hannington, M.D., Davis, D.W., Lafrance, B., and Gosselin, G., 2007a, The LaRonde Penna Au-rich volcanogenic massive sulfide deposit, Abitibi greenstone belt, Quebec: Part I. Geology and geochronology: Economic Geology, submitted.

Mercier-Langevin, P., Dubé, B., Hannington, M.D., Richer-Laflèche, M., and Gosselin, G., 2007b, The LaRonde Penna Au-rich volcanogenic massive sulfide deposit, Abitibi greenstone belt, Quebec: Part II. Lithogeochemistry and paleotectonic setting: Economic Geology, submitted.

Mercier-Langevin, P., Dubé, B., Lafrance, B., Hannington, M.D., Galley, A., and Moorhead, J., 2007c, A group of papers devoted to the LaRonde Penna Au-rich volcanogenic massive sulfide deposit, eastern Blake River Group, Abitibi greenstone belt, Quebec – Preface: Economic Geology, submitted.

Mortensen, J.K., 1993, U-Pb geochronology of the eastern Abitibi Subprovince. Part 2: Noranda – Kirkland Lake area: Canadian Journal of Earth Sciences, v. 30, p. 29-41.

Péloquin, A.S., Potvin, R., Laflèche, M.R., Verpaelst, P., and Gibson, H.L., 1990, The Blake River Group, Rouyn-Noranda area, Quebec: A stratigraphic synthesis, *in* Rive, M., Verpaelst, P., Gagnon, Y., Lulin, J.-M., Riverin, G., and Simard, A., eds., The Northwestern Quebec Polymetallic Belt: A Summary of 60 Years of Mining Exploration: Canadian Institute of Mining and Metallurgy, Special Volume 43, p. 107-118.

Poulsen, K.H., 1996, Lode gold, *in* Eckstrand, R.O., Sinclair, W.D., and Thorpe, R.I., eds., Geology of Canadian Mineral Deposit Types: Geological Survey of Canada, Geology of Canada no. 8, p.323-328.

Poulsen, K.H., and Hannington, M.H., 1996, Volcanic-associated massive sulphide gold, *in* Eckstrand, R.O., Sinclair, W.D., and Thorpe, R.I., eds., Geology of Canadian Mineral Deposit Types, Geological Survey of Canada, Geology of Canada no. 8, p.183-196.

Poulsen, K.H., Robert, F., and Dubé, B., 2000, Geological Classification of Canadian Gold Deposits: Geological Survey of Canada, Bulletin 540, 106 p.

Powell, W.G., Carmichael, D.M., and Hodgson, C.J., 1995, Conditions and timing of metamorphism in the Southern Abitibi greenstone belt: Canadian Journal of Earth Sciences, v. 32, p. 768-786.

Savoie, A., Perrault, G., and Fillion, G., 1986, Geological setting of the Doyon gold deposit, Bousquet Township, Abitibi, Québec, Canada, *in* MacDonald, A.J., ed., Proceedings on Gold'86, An International Symposium on the Geology of Gold: Toronto, Ontario, p. 97-107.

Savoie, A., Trudel, P., Sauvé, P., and Perrault, G., 1990, Géologie de la mine Doyon, Cadillac, Québec, *in* Rive, M., Verpaelst, P., Gagnon, Y., Lulin, J.-M., Riverin, G., and Simard, A., eds., The Northwestern Quebec Polymetallic Belt: A Summary of 60 Years of Mining Exploration: Canadian Institute of Mining and Metallurgy, Special Volume 43, p. 401-411.

Savoie, A., Trudel, P., Sauvé., P., Hoy, L., and Lao, K., 1991, Géologie de la Mine Doyon (Région de Cadillac): Ministère des Ressources Naturelles du Québec, ET 90-05, 80 p.

Sillitoe, R.H., 1994, Erosion and collapse of volcanoes: causes of telescoping in intrusion-centered ore deposits: Geology, v. 22, p. 945-948.

Sillitoe, R.H., Hannington, M.D., and Thompson, J.F.H., 1996, High sulfidation deposits in the volcanogenic massive sulfide environment: Economic Geology, v. 91, p. 204-212.

Stone, W.E., 1990, Archean volcanism and sedimentation in the Bousquet gold district, Abitibi greenstone belt, Quebec: Implications for stratigraphy and gold concentration: Geological Society of America Bulletin, v. 102, p. 147-158.

——— 1991, Archean volcanism and sedimentation in the Bousquet gold district, Abitibi greenstone belt, Quebec: implications for stratigraphy and gold concentration: Alternative interpretation and reply, Reply: Geological Society of America Bulletin, v. 103, p. 1256-1257.

Teasdale, N., Brown, A., and Tourigny, G., 1996, Gîtologie de la mine Bousquet 2: Ministère des Ressources Naturelles, MB 96-37, 43 p.

Tourigny, G., and Tremblay, A., 1997, Origin and incremental evolution of brittle/ductile shear zones in granitic rocks: natural examples from the southern Abitibi belt, Canada: Journal of Structural Geology, v. 19, p. 15-27.

Tourigny, G., Hubert, C., Brown, A.C., and Crépeau, R., 1988, Structural geology of the Blake River at the Bousquet mine, Abitibi, Quebec: Canadian Journal of Earth Sciences, v. 25, p. 581-592.

Tourigny, G., Brown, A.C., Hubert, C., and Crépeau, R., 1989a, Synvolcanic and syntectonic gold mineralization at the Bousquet mine, Abitibi greenstone belt, Quebec: Economic Geology, v. 84, p. 1875-1890.

Tourigny, G., Hubert, C., Brown, A.C., and Crépeau, R., 1989b, Structural control on gold mineralization at the Bousquet mine, Abitibi, Quebec: Canadian Journal of Earth Sciences, v. 26, p. 157-175.

Tourigny, G., Hubert, C., Brown, A.C., Crépeau, R., Trudel, P., Hoy, L., and Kheang, L., 1992, Géologie de la mine Bousquet: Ministère de l'Énergie et des Ressources du Québec, ET 89-09, 99 p.

Tourigny, G., Doucet, D., and Bourget, A., 1993, Geology of the Bousquet 2 mine: An example of a deformed, gold-bearing polymetallic sulfide deposit: Economic Geology, v. 88, p. 1578-1597.

Tremblay, A., Tourigny, G., and Machado, N., 1995, Zircon U/Pb age constraints on deformation and gold mineralization of the Mooshla pluton, southern Abitibi belt, Canada: Geological Society of America, Annual Meeting, New Orleans, Abstract with Programs, p. A-163.

Trudel, P., Sauvé, P., Tourigny, G., Hubert, C., and Hoy, L., 1992, Synthèse des Caractéristiques Géologiques des Gisements d'Or de la Région de Cadillac (Abitibi): Ministère de l'Énergie et des Ressources du Québec: MM 91-01, 106 p.

Valliant, R.I., and Barnett, R.L., 1982, Manganiferous garnet underlying the Bousquet gold orebody, Quebec: Metamorphosed manganese sediment as a guide to gold ore: Canadian Journal of Earth Sciences. v. 19, p. 993-1010.

Valliant, R.I., and Hutchinson, R.W., 1982, Stratigraphic distribution and genesis of gold deposits, Bousquet region, Northwestern Quebec, *in* Hodder, R.W., and Petruk, W., eds., Geology of Canadian Gold Deposits: Canadian Institute of Mining and Metallurgy, Special Volume 24, p. 27-40.

Valliant, R.I., Mongeau, C., and Doucet, R., 1982, The Bousquet pyritic gold deposit, Bousquet region, Quebec: Descriptive geology and preliminary interpretations on genesis, *in* Hodder, R.W., and Petruk, W., eds., Geology of Canadian Gold Deposits: Canadian Institute of Mining and Metallurgy, Special Volume 24, p. 41-49.

Valliant, R.I., Barnett, R.L., and Hodder, R.W., 1983, Aluminium silicate-bearing rock and its relation to gold mineralization; Bousquet mine, Bousquet Township, Quebec: Canadian Institute of Mining and Metallurgy Bulletin, v. 76, p. 81-90.

Zhang, Q., Machado, N., Ludden, J., and Moore, D., 1993, Geotectonic constraints from U-Pb ages for the Blake River Group, the Kinojévis Group and the Normétal mine area, Québec: Geological Association of Canada – Mineralogical Association of Canada, Joint Annual Meeting, Program and Abstracts, v. 18, p. A-114.

Base Metal and Gold Deposits of the Betts Cove Complex, Baie Verte Peninsula, Newfoundland

Alan L. Sangster[1], Stephanie L. Douma[2], and Jamie Lavigne[3]

1. Geological Survey of Canada (retired), 986 Parkhurst Blvd., Ottawa, ON K2A 3M9

2. 446 Hartleigh Avenue, Ottawa, Ontario K2B 5J4

3. 203 - 210 Cedar Street, Sudbury, Ontario P3B 1M6

Corresponding author's email: alsangster@rogers.com

Abstract

The Betts Cove complex outcrops on the northeastern tip of the Baie Verte Peninsula, Newfoundland. It consists of a lower ophiolite sequence composed of a basal cumulate unit (now mostly talc-carbonate schist), pillowed basalt, and pillow breccia, which are intruded by small gabbro bodies. The ophiolitic rocks are overlain by tholeiitic pillow basalt and pillow breccia, and are capped by red-green argillite. The upper part of the sequence consists of mafic to intermediate volcanic and derived epiclastic rocks.

The Tilt Cove copper deposits have been mined out. The east deposits appear to have consisted of massive to disseminated chalcopyrite with pyrite and pyrrhotite, which replaces pillow breccia throughout the tholeiitic basalt immediately above its contact with boninitic volcanic rocks. The west deposit is more complex and may replace shear zones in chlorite-altered pillowed basalts. Contents of other base metals and gold are low. The deposit is regarded as 'ophiolite-hosted'-type, even though it occurs stratigraphically above ophiolite *sensu strictu*.

Other smaller copper occurrences are primarily in shear zones that bring pillowed basalt and basalt breccia into contact with transitional (upper) zone sheeted dyke rocks. These occurrences are dominantly pyrite with variable amounts of chalcopyrite (e.g. Mount Misery). The Nudulama occurrence contains isolated spectacular gold values, whereas the Betts Cove occurrence contains high-grade gold values that vary directly with ore-grade zinc values.

The Nugget Pond gold deposit occurs in a red/green distal turbidite horizon stratigraphically above the tholeiitic pillowed basalt unit. The ore occurs at three stratigraphic locations, 1) a magnetite-rich horizon at the base of the sedimentary section, 2) a magnetite-rich horizon near the top of the massive red sandstone, and 3) in a sulphide-, magnetite-rich chloritic unit in the base of a green laminated siltstone. The gold mineralization is co-extensive with the intrusion of en echelon quartz-feldspar-carbonate veining. The ore formed by sulphidation of magnetite during a fluid flow event, probably during the Devonian.

Other gold occurrences in the complex are small but related to the same strata as the Nugget Pond horizon. Occurrences at Castle Rock and Long Pond east are both associated with talc-carbonate schist in the local sections.

Résumé

Le complexe de Betts Cove affleure sur l'extrémité nord de la péninsule de la Baie Verte, Terre-Neuve. Il est formé d'une séquence ophiolite inférieure composée d'un cumulat basal (à présent surtout du schiste à talc carbonaté), de basalte en coussins et d'une brèche en coussins qui sont pénétrés par de petits massifs gabbroïques. Les roches ophiolitiques sont recouvertes du basalte en coussins et de la brèche en coussins tholéiitiques et coiffées d'une unité d'argilite rouge/verte extensive en superficie. La partie supérieure de la séquence se compose de roches volcaniques mafiques à intermédiaires et de roches épiclastiques dérivées.

Les gisements de cuivre de Tilt Cove sont épuisés. Les gisements de l'est semblent être composés de chalcopyrite massive à disséminée avec de la pyrite et de la pyrrhotite qui remplacent la brèche en coussins dans le basalte tholéiitique, immédiatement au-dessus de son contact avec les roches volcaniques boninitiques. Le gisement de l'ouest est plus complexe et peut remplacer les zones de cisaillement dans les basaltes en coussins altérés en chlorite. L'altération se compose surtout de chlorite et les teneurs en autres métaux communs et en or sont faibles. Le gisement est considéré comme encaissé dans l'ophiolite, même s'il est présent stratigraphiquement, *sensu strictu*, au-dessus de l'ophiolite.

D'autres petites occurrences de cuivre se trouvent essentiellement dans les zones de cisaillement qui mettent en contact les basaltes en coussins et les brèches basaltiques avec la zone transitionnelle (supérieure) de roches filoniennes feuilletées. Ces occurrences se composent surtout de pyrite, avec des quantités variables de chalcopyrite (par ex., Mount Misery). L'occurrence de Nudulama contient des quantités isolées spectaculaires d'or, tandis que l'occurrence de Bette Cove contient une forte teneur en or qui varie directement avec la teneur en minerai de zinc.

Le gisement de Nugget Pond est présent dans un horizon de turbidite distal rouge/vert stratigraphiquement au-dessus de l'unité basalte tholéiitique. Le minerai est présent dans trois emplacements stratigraphiques : 1) un horizon riche en magnétite à la base de la section sédimentaire, 2) un horizon riche en magnétite près du dessus du grès rouge massif et 3) une unité chloritique riche en sulfures et en magnétite à la base d'une siltite stratifiée verte. La minéralisation aurifère est co-extensive avec l'intrusion de filons quartz-feldspath-carbonate en échelon. Le minerai s'est formé par sulfuration de la magnétite durant l'écoulement des fluides, probablement durant le Dévonien.

D'autres occurrences aurifères sont petites mais reliées à la même couche que l'horizon de Nugget Pond. Les occurrences à Castle Rock et à l'est de Long Pond sont toutes les deux associées au schiste intrusif à talc carbonaté dans les sections locales.

Background

Base metal and gold production from the Ordovician volcanic rocks of the Betts Cove complex was an important contributor to the Newfoundland economy. Copper was first discovered at Tilt Cove in 1857 and production first began at the West Mine in 1864 (Hibbard, 1983). The East Mine or

Sangster, A.L., Douma, S.M., and Lavigne, J., 2007, Base metal and gold deposits of the Betts Cove complex, Baie Verte Peninsula, Newfoundland, *in* Goodfellow, W.D., ed., Mineral Deposits of Canada: A Synthesis of Major Deposit-Types, District Metallogeny, the Evolution of Geological Provinces, and Exploration Methods: Geological Association of Canada, Mineral Deposits Division, Special Publication No. 5, p. 703-721.

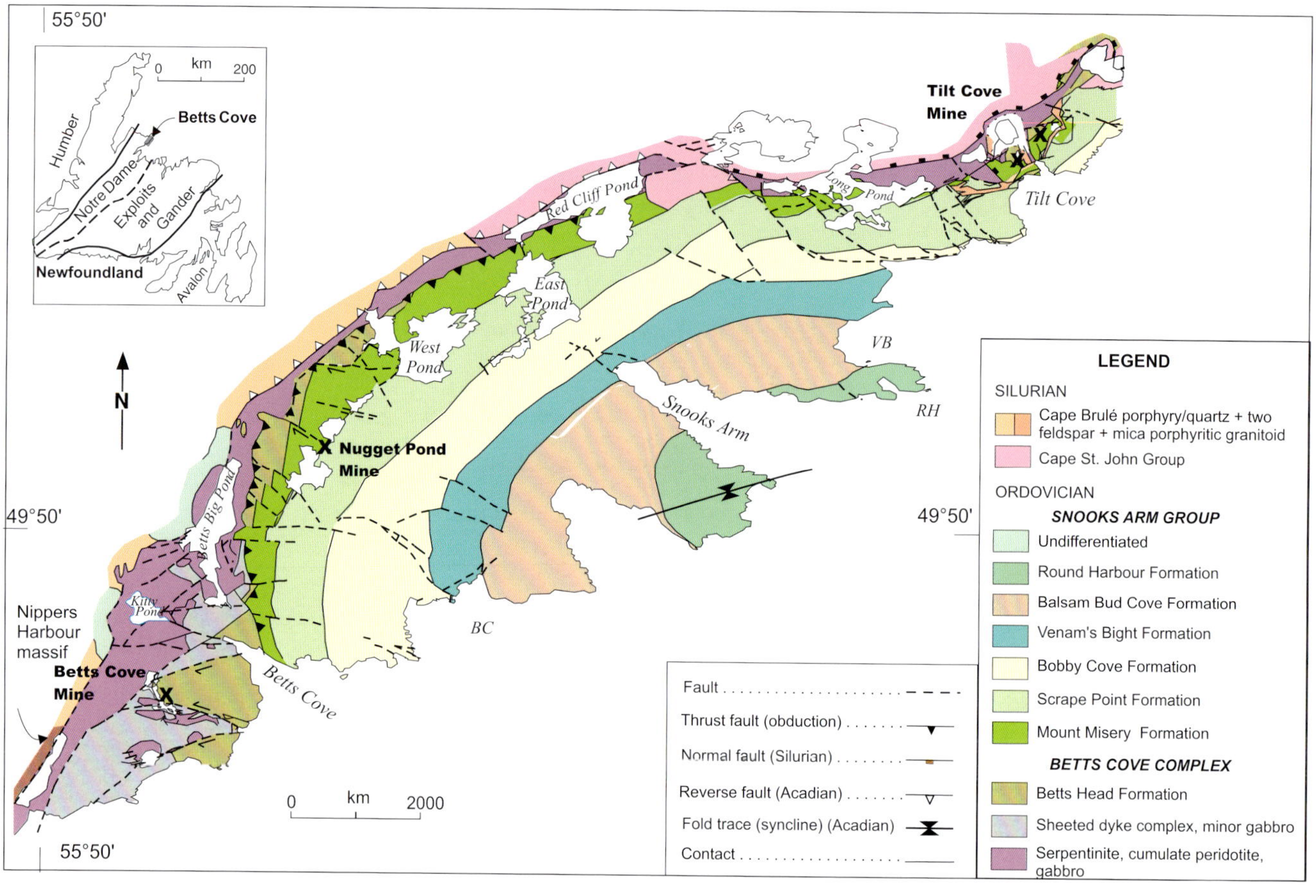

FIGURE 1. Geology of the Betts Cove complex (after Bédard et. al., 2000).

Main Mine was discovered in 1886, and produced more or less continuously until 1917. Renewed exploration in 1954 increased ore reserves at both the West and Main mines and production resumed in 1957. The mines finally closed in 1967. In total, about 8.2 million tons were mined at Tilt Cove mostly between 1957 and 1967, when approximately 7.4 million tons of ore grading 1.24% Cu was mined. About 42,500 oz of gold was extracted during this last period of mining (Strong, 1984).

The Betts Cove Mine was discovered about 1864 and production began in 1875. The mine was prematurely closed in 1886 when the hanging-wall volcanic rocks caved into the mine workings. Production tonnage figures are not known precisely, but 130,682 tons of hand-cobbed ore, grading up to 10% Cu, were shipped (Strong, 1984).

Exploration for gold mineralization associated with the mafic and ultramafic rocks in Newfoundland increased in the mid to late 1980s, and resulted in the discovery of the small, high-grade, Nugget Pond gold deposit in 1988/89. Richmont Mines began production in April 1997 with a reserve of 488,000 short tons of ore grading 0.357 oz of gold per ton. The mine closed in September 2001 after producing 167,028 oz of gold (Wardle, 2005).

Overview of the Geology and Mineral Occurrences of the Betts Cove Complex

The Betts Cove complex (Bédard et al., 2000) is one of several accreted volcanic terranes that form part of the Notre Dame Subzone of the Dunnage Zone of northern and central Newfoundland. The Betts Cove complex is well preserved because it has undergone much less deformation and metamorphism than many others, and it may provide a more accurate picture of the protoliths of other metamorphosed ophiolitic complexes. Regional mapping of the Baie Verte Peninsula and a summary of previous work is provided by Hibbard (1983). Detailed mapping of the Betts Cove complex was completed by Upadhyay (1973, 1980) accompanied by detailed local studies by Coish (1977a,b, 1989) and Jenner (1977). Recent remapping of the ophiolitic complex (Bédard and Lauziere, 1998; Bédard et al., 1998a,b, 1999, 2000) provides the basis for maps in this article.

The Betts Cove complex consists of the Betts Cove ophiolite and the overlying Snooks Arm Group, both of Middle Ordovician age (Fig. 1). The Betts Cove complex is in fault contact with Silurian subaerial ignimbrites, related metasedimentary rocks, and mafic flows of the Cape St. John Group to the north. The basal unit of the Betts Cove ophiolite is a cyclic and rhythmic cumulate sequence, ranging in composition from peridotite through pyroxenite to norite (Bédard et al., 2000). These primary rock types can only be seen in the portion of the ophiolite north and west of Betts Cove. Throughout most of the remainder of the ophiolite, the cumulate unit has been strongly altered to a talc-carbonate-magnetite schist with local remnants of serpentinite. The basal cumulate unit is overlain by the sheeted dyke unit, which is present only in the western part of the area. Sheeted dykes are in turn overlain by about 1 km of pillow basalt and basalt breccia. The lower part of the basalt unit (Betts Head

Formation) is boninitic (high Mg) in composition and can be identified in the field by a distinctive red weathering. The upper part of the basalt unit (Mount Misery Formation) is composed of similar lithologies but is of tholeiitic composition. The contact between the ophiolite sequence and the overlying Snooks Arm Group has been placed at the transition from boninitic to tholeiitic basalt (Bédard et al., 2000), corresponding to the Betts Head-Mount Misery formational boundary. A variable suite of medium- to coarse-grained gabbronorite intrudes these rocks. Much of the gabbronorite occupies the stratigraphic position of the Mount Misery pillowed volcanic units and xenoliths of volcanic rocks are common.

The overlying Scrape Point Formation consists of a lower metasedimentary unit, commonly referred to as the Nugget Pond Horizon (NPH), and an upper basalt unit (Bédard et al., 2000). The metasedimentary unit is a complex mixture of green distal turbiditic rocks derived from volcanic detritus. In part, particularly in the Nugget Pond Mine area, these rocks are oxidized to a brick red colour and locally enriched in detrital and/or chemically precipitated magnetite. True magnetite-hematite-chert iron formation is locally present near the base of the unit. East of Nugget Pond, the Nugget Pond Horizon is intruded by ferrogabbro sills that increase in quantity to the east and are most prominent at Tilt Cove. The overlying volcanic rocks consist of massive tholeiitic glomerophorphyritic and vesicular basalt that have a much lighter colour index than the underlying Mount Misery basalt. The Scrape Point Formation is overlain by the Bobby Cove Formation consisting of tuffaceous conglomerate and sandstones, volcanogenic turbidite, and purple mudstones. These rocks are overlain by the Venam's Bight Formation consisting of tholeiitic pillow basalt. The Venam's Bight Formation is overlain by the Balsam Bud Formation, consisting of pelagites, basalt, rhyolite tuff tuffaceous sandstone, and volcanic debris flows. The highest unit in the Snooks Arm Group is the Round Harbour Formation consisting of massive epidotized pillow basalt.

Mineralization in the Betts Cove complex is recognized only in the volcanic rocks of the Betts Head and Mount Misery formations and the sedimentary rocks of the Scrape Point Formation.

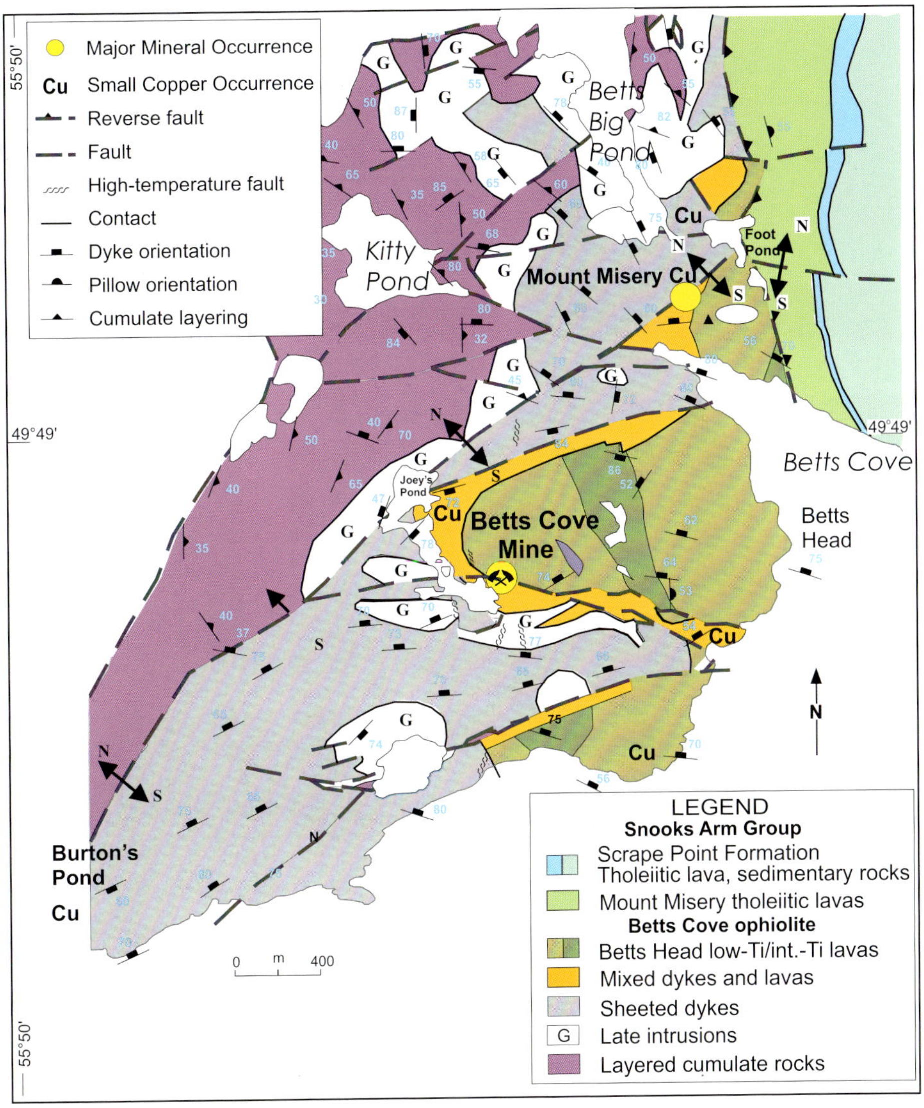

FIGURE 2. Location and geological setting of the Betts Cove area (modified after Bédard et. al., 2000)

There is extensive literature on the mineral occurrences of the Betts Cove complex, mainly based on thesis studies. The soil and rock geochemistry of gold and a number of small gold occurrences hosted by the talc-carbonate schist were studied by Beischer (1988). Al (1990) and Lavigne (1993) concluded that they are listwaenitic-type gold occurrences in which gold is thought to be mobilized during alteration by ultramafic rocks

The base metal occurrences are well described and have been firmly placed in the classification of Cyprus-type copper occurrences (Donoghue et al., 1959; Papezik, 1964; Squires, 1981; Strong, 1984; Saunders, 1985, 1990; Saunders and Strong, 1986, 1988; Hudson, 1988; Strong and Saunders, 1988). The association of anomalous gold with some of these occurrences has been documented by Hurley (1982), Hurley and Crocket (1985), Santaguida and Hannington (1996), Bédard et al. (2000), and Sangster and Pollard (2001).

FIGURE 4. Photographs from the Betts Cove area. **(A)** Two to four metre diameter orange-brown-weathering pillows in the hanging wall of the Betts Cove Mine. The colour is typical of the weathering of boninitic volcanic rocks. **(B)** Sheeted dyke unit stratigraphically below the Betts Cove Mine. The dykes average 50 cm to 1 m in thickness. Basalt dykes form more than 90% of this unit with interstitial basalt breccia. **(C)** Ore dump sample composed mainly of massive pyrite with an "eye" or chalcopyrite. Darker mineral in lower right is sphalerite. **(D)** Ore dump sample containing well banded pyrite containing layers rich in disseminated sphalerite and some chalcopyrite. In thin section, the pyrite texture is entirely cataclastic and cemented by the other sulphide minerals. **(E)** Ore dump sample containing cataclastic pyrite and sphalerite. The texture is interpreted as mechanical segregation during shearing. **(F)** Waste dump sample containing pyrite and pyrrhotite with traces of chalcopyrite in a small fragment basalt breccia. This type of breccia is typical of the interdyke breccias in the sheeted dyke unit. (All photographs by the author).

and pillow basalts and towards the western end it is entirely within the sheeted dyke unit. The occurrence contains variable but commonly low levels of disseminated pyrite and chalcopyrite. The best mineralization is on the height of land between Betts Cove and Foot Pond (Fig. 5B). Here, the sheared basalts host a zone up to 50 m wide of disseminated pyrite and chalcopyrite in sheared and altered sheeted dykes. Massive pyrite veins up to 10 cm thick are exposed in a pit at the west end of the occurrence. On the cliff above Betts Cove, the zone is only 2 m thick and is composed of silicified and pyritized sheared basalt. The dump nearest to Betts Cove contains basalt breccia with a quartz matrix and minor chalcopyrite and pyrrhotite.

Burton's Pond Cu Occurrence

The Burton's Pond occurrence (Fig. 2) has been studied by Snelgrove (1929), Douglas et al (1940), and Hudson (1988). The occurrence is small, consisting of a 100 m long irregular mineralized zone containing vein quartz and locally heavy sulphide mineralization consisting of pyrrhotite,

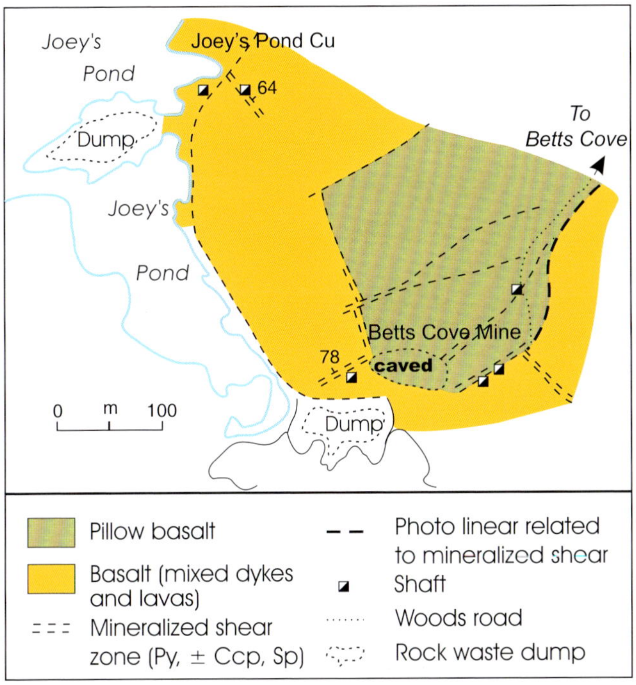

FIGURE 3. Geology of the Betts Cove-Joeys Pond mine area (modified after Bédard et al., 2000).

The Betts Cove Area

The Betts Cove Cu-Zn-Au Occurrence

Betts Cove geology and the Betts Cove deposit have been extensively discussed by Snelgrove (1929), Douglas et al. (1940), Baird (1951), Snelgrove and Baird, (1953), Upadhyay et al. (1971), Upadhyay (1973), Upadhyay and Strong (1973), Coish (1977b), Saunders (1985, 1990), Saunders and Strong (1986, 1988), Sangster et al. (1995), and Bédard et al. (2000).

The Betts Cove Mine and the nearby Joey's Pond occurrence outcrop in the upper part of the sheeted dyke unit about 1.25 km southwest of the head of Betts Cove (Fig. 2). At the Betts Cove Mine, several outcropping chloritic shear zones, ranging from 1 to 3 m in thickness, contain mainly pyrite with low concentrations of copper and gold (Fig. 3; Appendix 1). The ore zone comprised a concordant high-grade massive pyrite (Fig. 4C,D,E,F) lens containing chalcopyrite, sphalerite, and gold, which is no longer visible in outcrop as it is buried below the caved area. However, ore samples are preserved in the dumps (Fig. 4). Two sets of mineralized shear zones can be seen on surface; one striking about 315 to 345° and a second striking 050 to 060°, which offset the northwest-southeast shears.

The mineralized shear zones form the faulted contact between the sheeted dyke (Fig. 4B) and pillowed basalt units (Fig. 4A). The shear zones are composed mainly of chlorite with associated highly strained quartz and locally abundant anthophyllite, tremolite, calcite, and stilpnomelane. The Betts Cove occurrence is the most polymetallic occurrence in the area containing abundant zinc, minor lead, and high levels of gold (Appendix 1). The opaque minerals consist mainly of pyrite, which is highly brecciated, chalcopyrite, and sphalerite. Au is present as free gold with up to 20% Ag, but the amount of gold present varies directly with the zinc grade of the ore. Traces of galena, PbTe, and AgTe are also present. The pronounced banding observed in hand specimens of ore is of tectonic origin and is not a sedimentary exhalative or detrital feature, although Upadhyay and Strong (1973) interpreted it as primary.

At Joey's Pond, 300 m north of the Betts Cove Mine (Figs. 2, 3), a shear zone with a similar orientation to the northwest-southeast shear zone at Betts Cove contains stringer pyrite-chalcopyrite mineralization and may be the strike extension of the ores mined at Betts Cove. The occurrence is much lower in the sheeted dyke unit than at Betts Cove, suggesting a cross-cutting relationship.

Upadhyay and Strong (1973) interpreted the ore mined at Betts Cove as a northwest-southeast-striking exhalative massive sulphide lens that was modified by northwest-southeast shearing and subsequently affected by southwest -northeast shear zones into which some of the earlier sulphides were remobilized. However, southeast of the Betts Cove Mine, the northwest-southeast shear has recently been traced by drilling (Noveder Inc.), across the sheeted dyke unit for several hundred metres, again indicating that the mineralized shear is discordant rather than concordant. Also, the interpretation of sulphide banding as used by Upadhyay and Strong (1973) is precarious in light of recent work on seafloor sulphides and studies of layered textures in older massive sulphides. Modern seafloor analogues of massive sulphides commonly have a core of unbanded sulphide, mainly composed of intergrown and mechanically eroded sulphide chimneys, with some occurrences exhibiting an apron of bedded sulphides that have formed lateral to the mound (Herzig and Hannington, 1995). However, in these examples the layered sulphides form a relatively small part of a sulphide lens. Large (1992) has described textures observed in Australian massive sulphides and described various cases of sulphide banding that formed as aprons as a result of turbidite flows. However, Large (1992) indicates that in many examples of deformed sulphides the sulphide banding is commonly of metamorphic origin, as the textures of the sulphides in the Betts Cove occurrence indicate.

Thesis work by Saunders (1985), and reported in Saunders and Strong (1986), interpreted the shear zones as synoceanic faults and used the alteration zonation halo around the deposit to conclude that the ore was deposited from a subseafloor hydrothermal convection cell with flow focussed within the porous fault zones. Hydrothermal flow was most probably driven by heat from associated gabbroic sheets. Tremblay et al. (1997) found evidence of reactivation of these faults during obduction of the ophiolite.

The Mount Misery Cu-Au Occurrence

The Mount Misery occurrence (Fig. 2, 5; Appendix 1) occupies an east-west shear zone (Fig. 5A) that forms a strong airphoto linear feature between the Joeys Pond occurrence at Betts Cove and Foot Pond north of Betts Cove (Bédard et al., 2000). Earlier work has been done by Snelgrove (1929), Douglas et al. (1940), Upadhyay and Strong (1973), and reported in Hibbard (1983). The mineralized part of the shear zone extends from the head of Betts Cove to the northeast for about 500 m. At its eastern end, the shear zone forms the contact between the sheeted dyke unit

FIGURE 5. Photographs from the Mount Misery area. **(A)** The photograph shows the north side of Betts Cove on the flank of Mount Misery. The Mount Misery occurrence occupies a sheared zone that can be traced in outcrop between the two arrows. The shear zone extends over the brow of the hill where the best mineralization occurs about 250 metres north of (behind) the upper arrow. The entire outcrop on the barren hills is sheeted dyke with the exception of the two rounded hillcrests right of the upper arrow, which are pillowed basalts. The height of the cliff is 200 metres. **(B)** Highly sheared sheeted dyke unit in the core of the mineralized area. Mineralization consists mainly of disseminated pyrite with minor chalcopyrite. Stringers of massive pyrite to 10 cm in thickness are present. (All photographs by the author.)

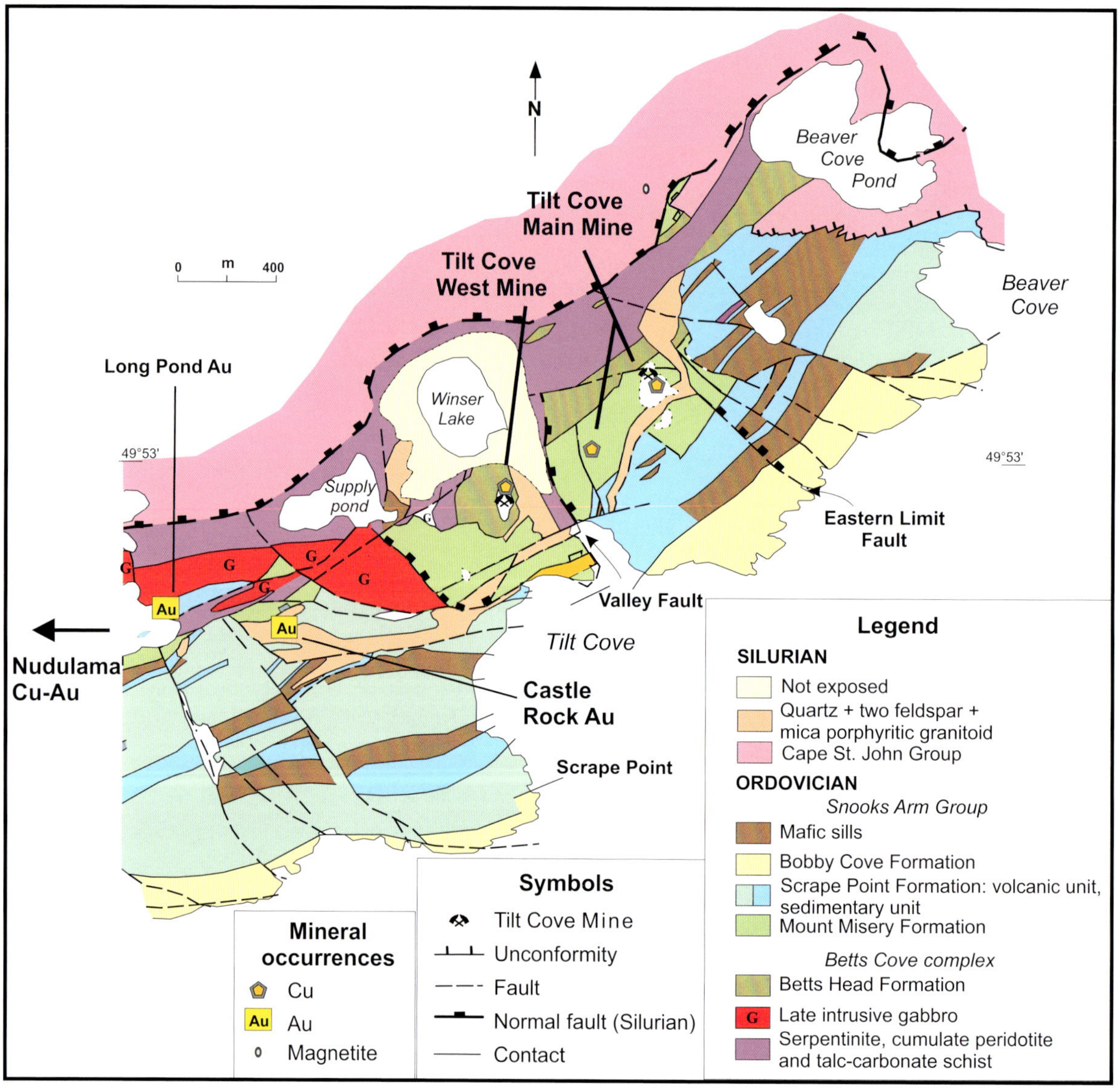

FIGURE 6. Location and geological setting of the Tilt Cove Mine (modified after Bédard et. al., 2000).

pyrite, and chalcopyrite. The occurrence is commonly known as a Cu-Ni occurrence but nickel is noticeably absent in any quantity in the analyses in Appendix 1. Some gold is present.

Tilt Cove Area

The Tilt Cove Deposit

The Tilt Cove deposit is the largest and most productive deposit within the Betts Cove complex. There was very little research carried out on the deposit until after production ceased in 1967. Early commentary and description was by Murray and Howley (1881), Snelgrove (1931), Douglas et al. (1940), and Bichan (1958). The complexity of structure in the immediate mine area was noted by Neale (1967) and Papezik (1964) described a small occurrence of unusual nickel mineralization in the ores. Regional mapping was reported by Upadhyay (1973) and Upadhyay and Strong (1973) and the immediate mine area and deposit was mapped and described by Squires (1981). Strong (1984) and Strong and Saunders (1988) reviewed the stratigraphy and alteration in detail, and modelled the deposit as an ophiolite-hosted copper deposit. Bédard et al. (2000) remapped the Betts Cove complex and modelled its formation.

FIGURE 7. Photograph of the east side of the valley at Tilt Cove showing a section through the Betts Cove complex. The cumulate unit, now totally altered to talc-carbonate rock and serpentinite, occupies the valley on the north (left) side of the photograph. It is in fault contact to the north with the Silurian Cape St. John Group and to the south with the volcanic section of the complex. The first 50 metres or so of cliff face is boninite, composed mainly of basalt breccia with extensive basaltic dykes. It has been called a sheeted dyke unit but there is some question if weather of the dyking is sufficiently dense to meet the definition of sheeted dyke *sensu strictu*. The unit contains uneconomic stringer pyrite-chalcopyrite mineralization. The tholeiitic basalt unit here is composed entirely of vesicular, variolitic pillow basalt talus breccia. The lower part of the unit is rusty weathering and the Main Zone orebodies lie 50 to 300 metres east of the cliff face. On the right, red and green cherty argillite with late ferro-gabbro sills form the base of the overlying Snooks Arm Group. This is the lateral equivalent of the Nugget Pond Horizon. (Photo by the author.)

The sulphide deposits at Tilt Cove occur within a 400 m thick basalt unit consisting of massive and pillowed basalt, and basalt breccia (Fig. 6, 7, 8A) of boninitic to tholeiitic affinity belonging to the Betts Head and Mount Misery formations (Figs. 6, 7). In most of the area, the basalt is in structural contact with the underlying ultramafic unit, which consists mainly of serpentinite and talc-carbonate schist. The sheeted dyke unit is absent, except for a thin wedge north of the open pit on the west ore zone. The ore-bearing unit is overlain by a sedimentary sequence belonging to the Scrape Point Formation. This includes a basal, locally pyrite-bearing and/or magnetite-bearing, red argillite and chert, and overlain by grey chert and greywacke that are extensively intruded by fine-grained ferrogabbro sills. The sedimentary and volcanic units are cut by numerous dacitic feldspar-quartz porphyry sills and dykes of Silurian age.

Mineral production at Tilt Cove came from several zones distributed both east and west of the village site. The west zone (Fig. 6) outcropped on the west side of the valley at Tilt Cove. The deposit consisted mainly of disseminated and stringer chalcopyrite-pyrite mineralization forming a steeply dipping pipe-like body in massive and pillowed basalt (Squires, 1981). The deposit directly overlies rocks of sheeted dyke affinity, which form the ridge immediately north of the pit. A small lens of massive sulphide occurred just south of the west zone at depth and was truncated by the valley fault. Examples of various ore types are shown in Figure 8B, C, D, E, and F.

The East or Main Zone is separated from the west zone by the valley fault. This fault zone is 50 to 75 m wide and is occupied by extensive serpentine and talc-carbonate schist and also by Silurian quartz-feldspar porphyry dykes that intrude west zone mineralization. Movement on the fault has been interpreted as normal with east side down (Squires, 1981).

The Main Mine area lies east of the valley fault. It consists of low-grade stockwork mineralization in the "A" Zone and the adjacent massive Cliff Zone, "Bogen's Lode", and farther east the Main Zone, which itself is composed of a number of massive sulphide lenses (Hibbard, 1983) (Fig. 9). Surface basalt exposure in the Main Mine is entirely basalt breccia, however, sections of the workings, described by Donoghue et al. (1959) indicate that the breccia in the Main Zone gives way to "andesite" (mine terminology) at a shallow depth. The east end of the Main Zone is terminated by a major fault that puts the host Betts Head and Mount Misery formation volcanic rocks in contact with the overlying sedimentary rocks of the Scrape Point Formation. The fault zone is occupied by sheared Silurian quartz-feldspar porphyry.

Appendix 1 indicates compositional variation of various ore types. It is notable that both gold and zinc contents are low. The chemistry of the deposits is rather simple. Copper is the main element present with very low quantities of both lead and zinc. Gold production was significant but the gold

FIGURE 8. Tilt Cove ore-types – all samples are from dump material. **(A)** Talus breccia in the cliff face on the east side of the valley is the typical host for much of the Main Zone ore. The breccia is composed of 100% fragments of vesicular pillowed basalt cemented and veined by white carbonate. **(B)** Disseminated streaky chalcopyrite and pyrite in a chlorite matrix. This ore type is most commonly found around the open cut of the West Mine. **(C)** Pillow fragment with the vesicles filled with chalcopyrite. In the lower part of the photograph is pyrite that was probably either fracture filling or interstitial between pillow fragments. **(D)** Magnetite was locally abundant in the deposit. This unlocated sample is massive magnetite with minor pyrite and chalcopyrite. **(E)** Mainly pyrite with minor chalcopyrite from dump material at the Main Mine. The arcuate sulphide band is interpreted to be a product of replacement of the rim of a pillow fragment. **(F)** Chalcopyrite stringers in basalt. (All photographs by the author.)

concentrations of the samples analysed are generally low with occasional values in the order of 1 to 2 ppm. Nickel and cobalt concentrations are also low and do not reflect the presence of a small amount of nickel ore that was produced from the west zone in the period between 1869 and 1876. The nickel mineralization appears to be a secondary product, as it was associated with the contact between talc-carbonate schist (altered peridotite) and basalt. The nickel mineralization consisted of nickel sulpharsenides, principally niccolite, maucherite, chloanthite, gersdorffite, arsenopyrite, and millerite (Papezik, 1964). Appendix 1 indicates elevated concentrations of As in many of the samples analysed.

Alteration minerals associated with the deposits are dominantly chlorite with associated carbonate and some stilpnomelane. Intense alteration is restricted to a relatively narrow zone around the sulphide deposits (Fig. 5). On the east side of the Main Zone, alteration associated with the East Limit Fault has produced an assemblage of specular hematite-chlorite-sulphide.

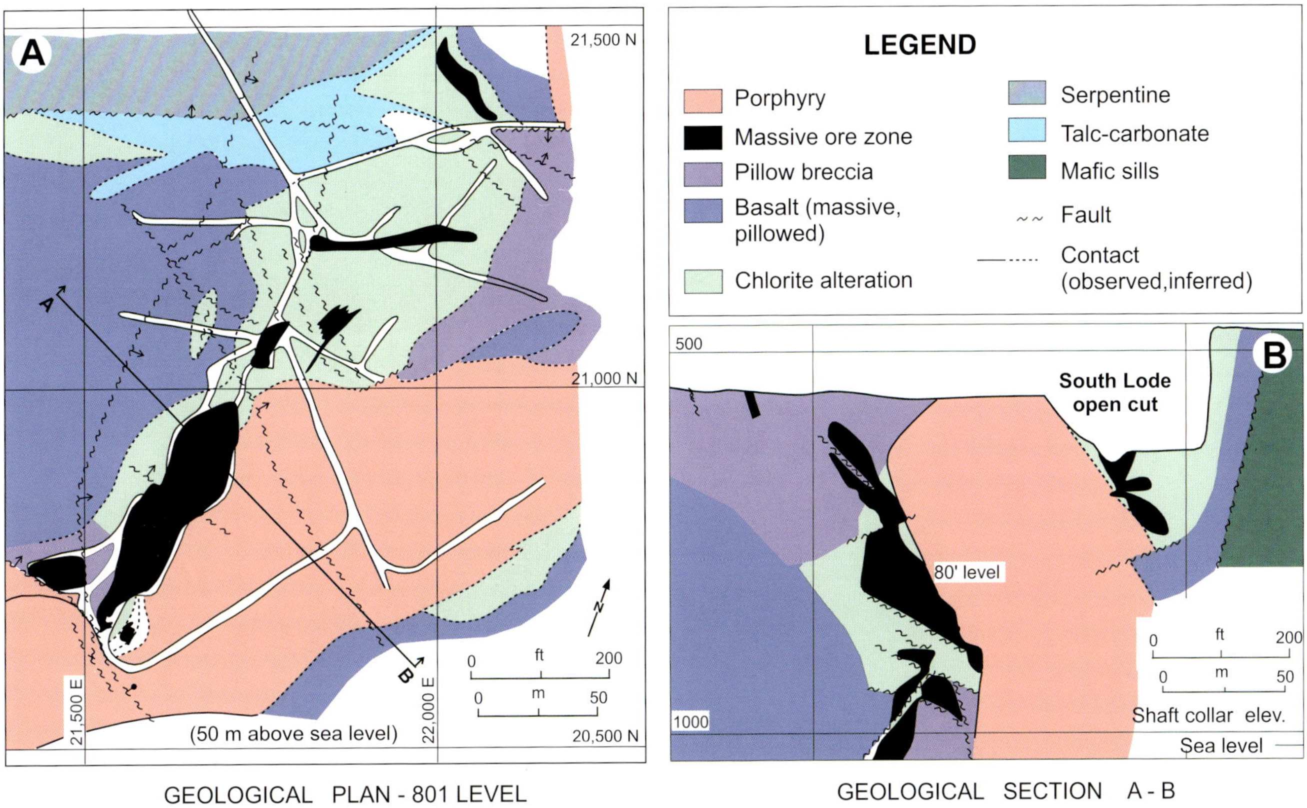

FIGURE 9. Geological plan (A) and section (B) of the Tilt Cove Mine (modified after Squires, 1981).

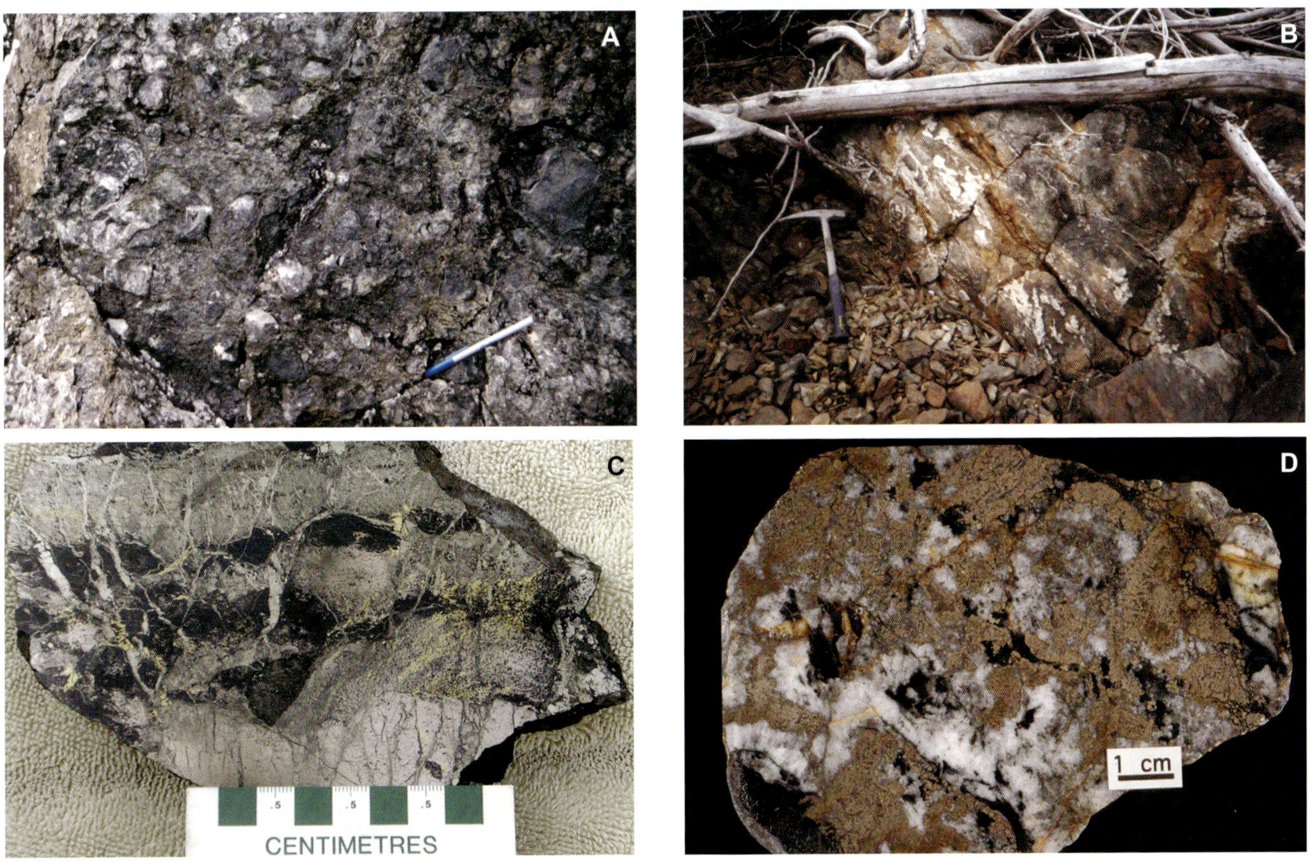

FIGURE 10. Nudulama occurrence plates. (A) Basalt pebble conglomerate near the Nudulama occurrence. (B) A small exposure of sheared basalt with traces of sulphide mineralization. (C) Dump sample containing sheared and brecciated basalt containing clasts and stringers of pyrrhotite, pyrite, and chalcopyrite. (D) Dump specimen containing pyrrhotite-chalcopyrite with vein quartz.

Without access to underground workings and with little mineralization available on surface for study, it is difficult to model the Tilt Cove deposit. Squires (1981) interpreted the various lenses as part of a single ophiolite-type copper-rich volcanic massive sulphide system. The West and "A" zones were considered to be the deepest and intermediate portions of the feeder zone respectively, and the Main Zone massive sulphides were interpreted to represent the overlying massive lenses, separated from the "A" Zone by a system of flat faults.

Nudulama Occurrence

The Nudulama Occurrence (Appendix 1) lies on a small peninsula (Sugarloaf) in Long Pond, about two kilometres west of the Tilt Cove Mine (Fig. 6). The local rock types include a sheeted dyke unit east of the occurrence and basalt breccia and basalt conglomerate (reworked breccia) (Fig. 10A) west of the occurrence. Most of what can be seen now is on a dump beside an old shaft. The dump contains mainly chalcopyrite as filling in brecciated, sheared (Fig. 10B), chloritic basalt (Fig. 10C,D). The mineralization appears to be from a sheared zone that separates basalt pillow breccia from the underlying sheeted dyke complex. The occurrence is notable for some high gold values.

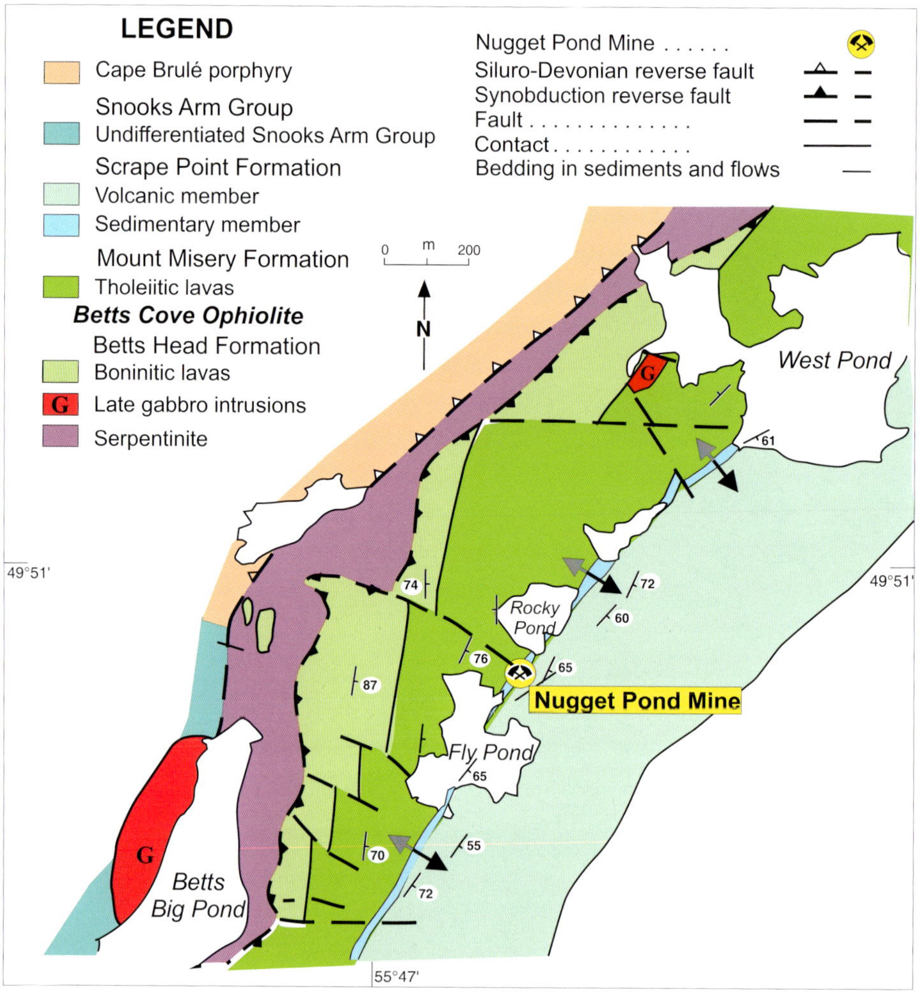

FIGURE 11. Location and geological setting of the Nugget Pond Mine (modified after Bédard et al., 2000).

The Nugget Pond Gold Deposit

The Nugget Pond deposit is a small, high-grade gold deposit that exhibits many unusual features. It is the only mine that has seen recent production in the Betts Cove complex. The deposit was found by Bitech Corporation in 1988 and achieved production by Richmont Mines in early 1997 with a defined reserve of 488 000 tonnes at 0.357 ounces of gold per ton. The gold deposit consists of manto-style disseminated pyrite-gold mineralization within a thin metaturbidite in the basal unit of the Scrape Point Formation that overlies pillowed basalt of the Mount Misery Formation (Swinden et al., 1990; Sangster et al., 1994a,b, 1995) (Figs. 11, 12). The sedimentary host rocks are termed the Nugget Pond Horizon (NPH).

Stratigraphic Setting

The Nugget Pond deposit is hosted by a red and green turbidite sequence that lies between pillowed basalt of the Mount Misery Formation and overlying massive porphyritic basalt of the Scrape Point Formation (Figs. 13, 14). The sedimentary sequence is typically 40 to 60 m thick and dips 40 to 60° to the southeast in the area of the mine.

The underlying Mount Misery basalt is commonly variolitic and vesicular, and varies from fresh to moderately altered. Relatively fresh basalt consists primarily of plagioclase, clinopyroxene, chlorite, actinolite, and minor to trace amounts of leucoxene, calcite, epidote, and pyrite. Distinctive dark basalt is associated with inter- and intra-pillow metasomatic alteration. This basalt is composed of plagioclase, chlorite, and magnetite with the latter forming up to 15 modal percent of the rock. Three types of inter- and intra-pillow alterations have been identified. The first alteration type consists of epidocites and related alteration of quartz-chlorite-calcite-epidote-pyrite replacements or veining. The second alteration type consists of diffuse replacement of the basalts by aggregates of quartz-calcite-muscovite-albite-chalcopyrite-pyrrhotite. The third alteration type consists of quartz-calcite with chlorite, muscovite, magnetite, and hematite and is referred to as interpillow jasper zones. They typically lie within 5 m of the upper contact of the basalt.

The Nugget Pond Horizon is divided into a lower red clastic unit and an upper green clastic unit (Fig. 13). The red clastic unit is subdivided into a lower magnetic unit, a middle brick red clastic unit, and an upper interlaminated red-green clastic unit that is also magnetic. The upper green clastic unit consists of a lower sheared chloritic unit containing

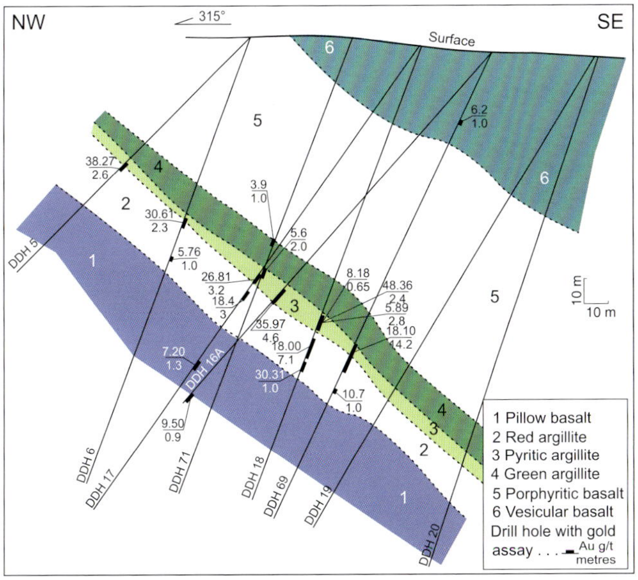

FIGURE 12. Drill section of the Nugget Pond Au deposit (modified after Bédard et. al., 2000)

pyrrhotite, pyrite, and magnetite, and an upper unit of medium to light green finely laminated and graded turbidite.

The basal red unit is very complex and is here referred to as an iron formation for simplicity. It is typically 3 to 5 m although locally up to 18 m thick. The unit comprises four subunits that are not universally present or consistent in character. The lowest is a chaotic horizon containing clasts of Mount Misery pillow basalt and chert that are cemented by cherty quartz and magnetite. The second unit is also a breccia but contains clasts of bedded sedimentary rocks, chert, and quartz cemented by chert and quartz and locally by magnetite and quartz (Fig. 14A). Some of these textures probably represent hydrothermal breccias related to ore deposition while others are clearly of sedimentary origin. The third subunit is a very fine-grained, very dark red siltstone that contains abundant, very fine-grained, bedded magnetite. The fourth subunit is a brick red to brown-black very fine-grained siltstone that also contains abundant magnetite.

The largest part of the red unit (Fig. 13) is a coarsening upward sequence of mudstone, siltstone, sandstone, and pebbly sandstone. The lowest part of the unit consists of an aphanitic, red, massive to very thinly laminated mudstone (Fig. 14B) composed of quartz and chlorite with feldspar and sericite (XRD analysis) and an evenly distributed hematite matrix. Higher in the succession, the proportion of silt-sized particles increases resulting in a fine- to medium-laminated siltstone and mudstone Near the top of the red succession, a coarser grained, mainly green, sand-sized sediment component results in a unit consisting of banded red siltstone-mudstone and green sandstone (Fig. 14C). The sandstone may contain millimetre- to centimetre-scale basalt clasts. The red-green unit commonly contains laminae of dark red magnetite-rich mudstone (Fig. 14D).

The upper green metasedimentary unit consists of two rock types (Fig. 13). The lower unit directly overlying the red unit is dark green, and is composed of chlorite, muscovite, quartz, and carbonate. It is moderately to well foliated and contains small-scale deformation structures. It is distinctive in that it has abundant foliation-parallel bands of sieve-textured pyrite and pyrrhotite that varies from about 15% to locally submassive sulphide (Fig. 14E). The lower part of the unit is strongly enriched in magnetite. The non-auriferous sulphides in this unit are biogenic with $\delta^{34}S$ values less than -15‰ (Sangster et al., 2001).

The uppermost and thickest portion of the green unit is a sequence of thinly laminated and graded, mainly green sandstone, siltstone, and mudstone forming Bouma A or B to E(t) sequences (Fig. 14F,G). The colour of individual metasedimentary rocks varies with the grain size. Sandstone is typically darker green, whereas siltstone is lighter green to tan or red in colour. The coarser clasts in the sandy facies indicate derivation from a basaltic protolith. The pyrite content is always <<1%. Near the contact with the overlying basalt, there is an increase in shearing accompanied by alteration and an increase in pyrite, epidote, and magnetite.

The overlying basalt is massive and is characterized by 10 to 20 percent plagioclase phenocrysts, which occur as glomerophorphyritic clusters in a fine-grained basaltic groundmass (Fig. 14H).

Structural Setting

The Nugget Pond gold deposit is hosted by the basal sedimentary unit of the Scrape Point Formation at its intersection with a strong northwest-southeast airphoto linear feature in the footwall rocks that does not appear to extend into the hanging-wall rocks. The interpreted fault has not been found

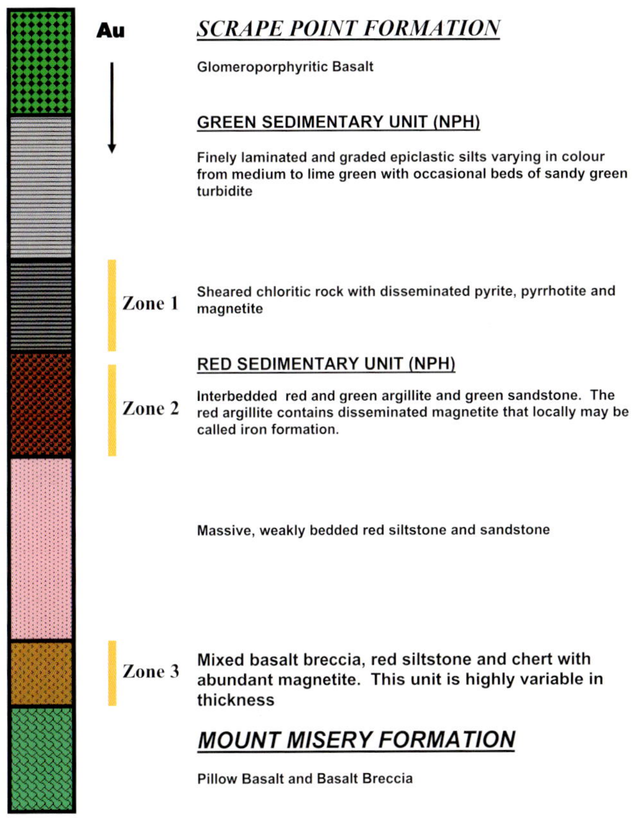

FIGURE 13. Generalized stratigraphic column of the host rocks at the Nugget Pond gold deposit showing the stratigraphic location of the mineralized zones.

FIGURE 14. Regional rock types at the Nugget Pond gold deposit. **(A)** This sample from the basal unit of the red sedimentary sequence consists of red chert fragments in a matrix of grey, magnetite-bearing quartz. This unit is host to Zone 3 mineralization. **(B)** The main part of the red sedimentary unit is a bright red, moderately well bedded to massive siltstone and mudstone. Magnetite and pyrite are not common in this unit. **(C)** In the upper part of the red unit (Fig. 13B), the colour changes to a dark red, bedding becomes finer and more pronounced and centimetre-scale beds of greenish coarse sandstone are common. The darker red colour is caused by the presence of very fine-grained detrital magnetite beds. This unit hosts Zone 2 mineralization. **(D)** Thin section showing the distribution of magnetite in a dark red sandstone. Length of the photograph is 4 cm. **(E)** Sheared chloritic rock from the basal unit of the green sedimentary sequence. At Nugget Pond, this unit may be highly sulphidic (upper core) but nonauriferous outside the deposit. It contains abundant magnetite disseminated with the pyrite-pyrrhotite and rarely may contain greater than 50% magnetite without sulphides (lower core). Some quartz veining is present (middle core). This unit hosts Zone 1 ore. **(F)** Core from the main portion of the green sedimentary unit. Magnetite was not identified in this unit and only minor pyrite was noticed, as seen on the left end of the upper core.

in underground mine workings, but there is a modest thickening in Scrape Point sedimentary rocks at the appropriate location, suggesting the presence of a synvolcanic fault in the Mount Misery Formation and other footwall units. Detailed structural studies have not been conducted in the deposit area, but examination of drill core and examination underground at an early stage of the development suggests that the mineralization and alteration is genetically related to an extensional array of hydrothermal quartz-albite-carbonate-pyrite veins within the metasedimentary rocks (Figs. 15, 16).

The Nugget Pond deposit is not exposed on surface. A mine development roadcut, 50 m north of the subcrop of the west end of the deposit, revealed stratified red argillite, green argillite of the ore section, and glomeroporphyritic basalt. These exposures are totally unaltered demonstrating that the

Figure 14 continued. **(G)** The green turbidite unit may be thickly bedded or very thinly bedded, as seen in the upper and lower cores. **(H)** Massive glomeroporphyritic basalt that overlies the Nugget Pond Horizon. This unit is not mineralized but its competence may have acted as a butress promoting fracturing in the rocks below those that host the deposit.

Figure 15. Typical ore samples from the Nugget Pond Mine. **(A)** Zone 1 Au ore consisting of well banded pyrite cut by quartz-albite-carbonate veining. **(B)** Low-grade ore consisting of large (1 to 10 cm), zoned pyrite metacrysts in finely banded, altered red-green sediment. This ore type was common around the fringes of the deposit. **(C)** Banded ore from Zone 2 resulting from the replacement of magnetite by pyrite. The red-brown colour has been digitally enhanced for clarity. **(D)** Banded ore from Zone 2 showing the co-existence of replacement ore with late (?) pyrite metacrysts.

mineralization and alteration effects on the host rocks are local.

Hydrothermal Quartz-Albite-Carbonate Veins, Mineralization, and Alteration

The gold mineralization and alteration at Nugget Pond are co-extensive with a network of quartz-albite-carbonate-pyrite veins that cut rocks of the Nugget Pond Horizon. The veins are texturally variable and can be divided into three principle modes of occurrence: 1) veins with sharp to locally diffuse contacts (Fig. 15A,E,F), 2) volumes of rock with irregular impregnations of vein material (Fig. 15G), and 3) irregular patches of massive vein material dominated by carbonate (Fig. 15G).

The extensional vein array is composed of quartz-albite-carbonate ± pyrite veins that strike parallel to the deposit with shallow to moderate dips to the south. They range from 1 to about 5m in dip dimension and vary in thickness from 2 to 25 cm. In addition to the veins, the host rock commonly contains similar material irregularly distributed throughout

Figure 15 continued. (E) Ore from the core of Zone 2 showing quartz-albite-carbonate veins forming the matrix to stilpnomelane altered red argillite (dark at top of specimen). **(F)** Tension gash veins filled with quartz-albite-carbonate cutting massive black stilpnomelane. Note the irregular distribution of pyrite in the altered rock. Length of the field is 3 metres. **(G)** Irregular blebs of quartz-feldspar-carbonate in chloritic argillite. An area of coalesced vein material dominated by carbonate can be seen in the lower right corner of the photograph. Length of the field is 3 metres. **(H)** Specimen of spectacular gold ore hosted by coarse-grained carbonate. (Photos by the author with the exeption of Plate H, which is courtesy of Andy Kerr, Newfoundland Geological Survey.)

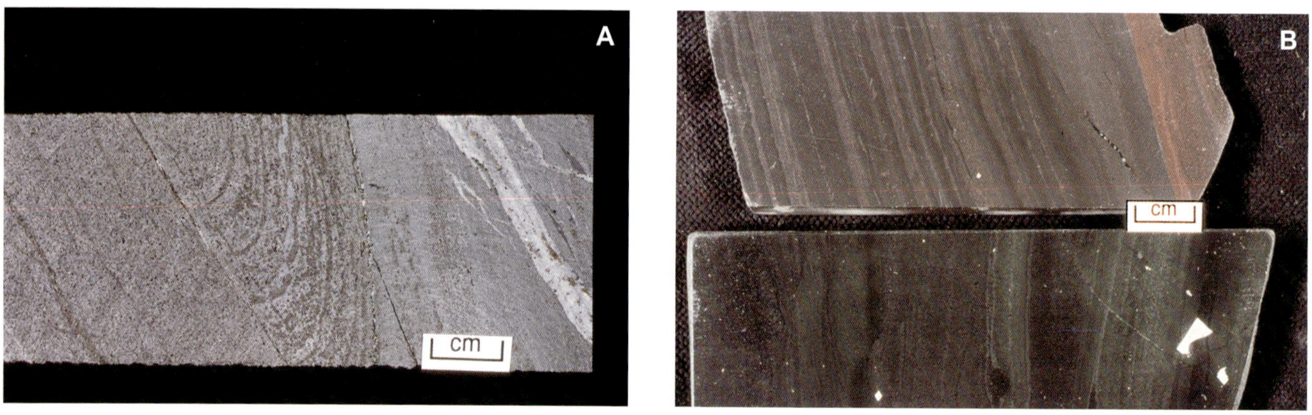

Figure 16. Alteration of host rocks at Nugget Pond. Plate. **(A)** Green argillite with very fine-grained stilpnomelane rosettes distributed throughout. **(B)** Approaching the Nugget Pond ore zone, the red-green banded argillite may appear bleached due to the presence of very fine-grained stilpnomelane.

the altered rock as 1 to 10 cm diameter clots of coarse-grained quartz-albite-carbonate, locally disseminated in altered host rock, commonly forming up to 25% of the rock by volume. Where this dispersed material is abundant, it may coalesce into irregular pods of massive carbonate-rich vein material. Spectacular gold "shows" have been found in this coarse carbonate (Fig. 15H). A single age determination on a single grain of xenotime from the quartz-albite-carbonate veins (R. Parish and A. Sangster, unpublished data) gave a Devonian age of 374 ± 8 Ma.

Mineralization in the Nugget Pond gold deposit is dominantly pyrite with very subordinate chalcopyrite, galena, Ag-telluride, and native silver. Pyrite is commonly coarse grained and cubic with an average grain size between 0.5 and 2.0 cm, but locally may be up to 10 cm. The pyrite may overgrow pre-existing synsedimentary pyrite within ores hosted by the lower green sedimentary unit, and it commonly overgrows magnetite in both the lower green unit and the red unit. The amount of magnetite in the pyritic rocks of the ore zones is less than that seen in the adjacent unmineralized

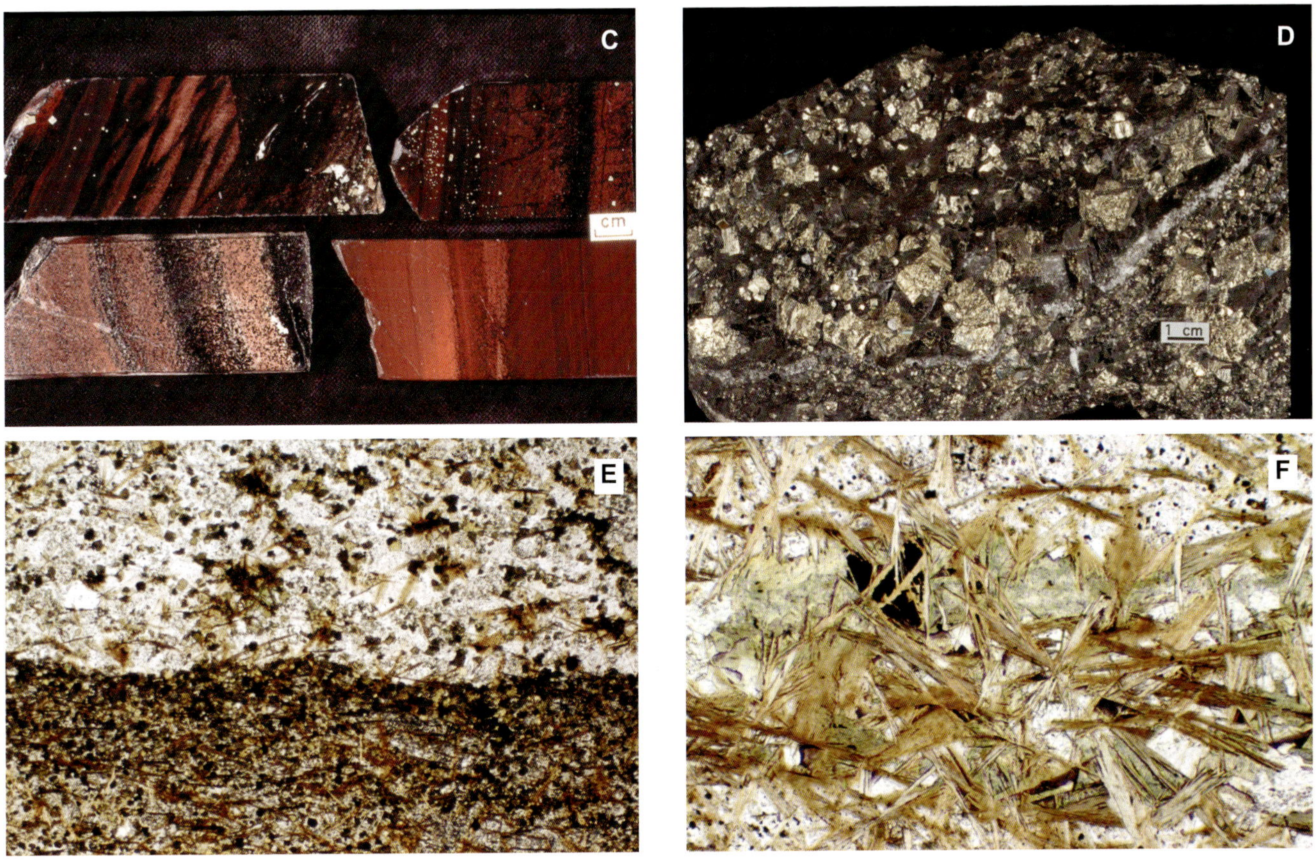

FIGURE 16 CONTINUED. **(C)** Varying degrees of development of black stilpnomelane alteration in red argillite from very weak (lower left), to moderate (lower right), to intense (upper). The initial stages of alteration are strongly controlled by bedding (lower) and cleavage (upper right). **(D)** Massive black stilpnomelane with metacrystic pyrite from the fringes of the Zone 2. **(E)** Thin section showing weak bladed stilpnomelane (upper half) and heavier bladed stilpnomelane on opposite sides of a bedding contact. The bedding control on stilpnomelane formation suggests that the composition of the sedimentary rock, probably the iron content, is a controlling factor (length of slide is 5 mm). **(F)** Coarse rosettes of light brown stilpnomelane overgrowing green chlorite in red argillite. Length of the field is 1 cm.

rocks, indicating that much of the pyrite has formed by sulphidation of magnetite. Pyrite is also abundant in the quartz-albite-carbonate veins associated with the deposit and also occurs in similar veins stratigraphically below and above the deposit, in the lower basalt, and overlying green turbidite and basalts respectively.

Gold in the ores occurs as free grains of native gold filling fractures in pyrite, coating pyrite grains, or as small irregular-shaped grains within stilpnomelane-altered rock. Some spherical gold inclusions are present within pyrite. The spectacular gold-bearing samples are restricted to the coarse carbonate veining. Accessory minerals include ubiquitous but minor (<0.1%) chalcopyrite and rare galena, which occur mainly as small inclusions in pyrite and as aggregates of coarser grains in the carbonate phase of the hydrothermal quartz-albite-carbonate veins. The silver content of the ores is contained both within the gold (5 to 15% Ag) and as <5 micron sized grains of an Ag-telluride that occurs as inclusions in pyrite. One grain of native silver was observed during SEM examinations of samples. Table 1 lists typical assay values for Nugget Pond ore samples.

The principal alteration mineral at Nugget Pond is stilpnomelane (Fig. 16), the iron-rich brittle mica (Swinden et al., 1990; Sangster et al. 1994a). This developed in preference to biotite because of the iron-rich character of the host rocks. The bulk of the alteration is in the sedimentary rocks. In weakly altered rocks, most sedimentary features are preserved (Fig. 16A,B,C,F). Strongly altered rocks consist almost exclusively of stilpnomelane and pyrite, and all vestiges of sedimentary layering and colour have been destroyed (Fig. 16D). Initial indications of alteration on the fringes of the orebody consist of bleaching of the red and green argillite (Fig. 16D) and the appearance of a "spotted" texture due to the presence of millimetre-scale rosettes of stilpnomelane (Fig. 16A,C,F). These are often restricted to particular centimetre-scale sedimentary layers (Fig. 16C). In the ore zone, the assemblage stilpnomelane calcite, ilmenite, sphene, pyrite ± rutile predominates. Biotite is present in minor amounts in the fringes of the alteration zone and is overgrown by stilpnomelane. In areas with the most intense alteration, massive stilpnomelane is intimately associated with the hydrothermal quartz-albite-carbonate veins (Fig. 15A,E) and it forms veins that cut the sedimentary rock-types In these areas the stilpnomelane is accompanied by albite ± quartz.

In the footwall basalt, alteration is locally present but is much less extensive than in the sedimentary rocks. The principal alteration mineral is biotite, which overgrows the regional chlorite alteration. In more altered areas, an assemblage of stilpnomelane-albite-calcite-quartz-pyrite is present with the stilpnomelane overgrowing biotite.

TABLE 1. Principle mineral occurrences of the Betts Cove complex.

Name	Host Rock	Age of Host Rocks	Age of Mineralization	Minerals	Type	Size	Grade	Status
Copper Occurrences								
Tilt Cove	Basalt, talus breccia	Middle Ordovician	Middle Ordovician	Ccp, Py, Po	Sub-seafloor Replacement	8,200,000 tons	~1% Cu See Appendix 1	Past producer
Nudulama	Basalt breccia and sheared zone	Middle Ordovician	Middle Ordovician	Ccp, Py, Po, +/- Au	Sub-seafloor Replacement	Occurrence	See Appendix 1	Prospect
Mount Misery	Basalt sheared zone	Middle Ordovician	Middle Ordovician	Ccp, Py, Po	Sub-seafloor Replacement	Occurrence	See Appendix 1	Prospect
Betts Cove	Basalt sheared zone	Middle Ordovician	Middle Ordovician	Py, Ccp, Sp, Au	Sub-seafloor Replacement	Small	See Appendix 1	Past producer
Burtons Pond	Gabbro	Middle Ordovician	Middle Ordovician	Po, Py, Ccp, +/- Au	Sub-seafloor Replacement	Occurrence	See Appendix 1	Prospect
Gold Occurrences								
Nugget Pond	Red-green argillite	Middle Ordovician	Devonian	Py, Au	Replacement Manto	130,000 tons	10% hand cobbed	Past producer
Castle Rock	Argillite and talc carbonate	Middle Ordovician	Devonian	Py, Au	Replacement shear zone	Occurrence	See Appendix 1	Prospect
Long Pond	Argillite and talc carbonate	Middle Ordovician	Devonian	Py, Au	Replacement	Occurrence	See Appendix 1	Prospect

Note: In addition to the occurrences listed here there are a large number of "indications" of mineralization throughout the area listed in Hibbard (1983) and in mineral assessment files.

Ore Distribution

There are three individual ore zones in the mine (Fig. 13): the hanging-wall zone, or "Zone 1" that occurs within the dark green chloritic sediment at the base of the green unit; "Zone 2" located in the magnetite-rich portion of the upper part of the red unit; and the footwall zone or "Zone 3" that occurs in the magnetite-rich basal portion of the red unit. Details of variation in individual zones are provided below.

Zone 1 is located within the dark green sediment at the base of the green, turbidite unit. It is black in colour due to the presence of stilpnomelane and contains 10 to 15 percent pyrite. The fine- to medium- grained pyrite occurs as semi-laminated to laminated bands with a barren grey-green sandstone layer often located within the zone. The immediate footwall contains 4 to 6 percent disseminated pyrite, of less than 8 mm grain size, over a width of 0.5 m. Zone 1 is 3 to 5 m thick and is commonly crosscut by calcite-albite veins less than 15 cm wide (Fig. 15A). Locally, the sedimentary rocks exhibit evidence of ductile deformation. Immediately above Zone 1 is a hanging-wall fault that consists of highly fractured grey-green sedimentary rocks or fault breccia/gouge up to 1.5 m wide.

Zone 2 is located within red and green interbedded sandstone and siltstone that contains up to 10 percent disseminated, coarse-grained euhedral pyrite (<24 mm) with abundant discontinuous irregular pink albite and calcite veins (Fig. 15B,C,D). Stilpnomelane alteration within this zone is commonly less intense than in Zone 1.

Zone 3 is located at the base of the sediments above the footwall contact with the pillow lavas. In the west, the sediments are tuffaceous and contain light green chlorite with moderate calcite impregnations and 8 to 10 percent pyrite. Locally, the zone may host remnant blocks of a cherty fragmental rock within a magnetite matrix. In the east (known as Zone 3 upper), the sediments are darker in colour with a finer grained sulphide occurring at the contact between the metasedimentary rocks and the pillow lavas. Immediately above the pyrite is a 10 to 30 cm band of fine-grained magnetite. Representative analyses of Nugget Pond ore are presented in Table 4.

Origin of the Deposit

The bulk of the geological evidence indicates that the deposit is a product of sulphidation of magnetite-rich zones in the sedimentary horizons by the fluids that generated the quartz-albite-carbonate veins 374 ± 8 Ma ago. The deposit is spatially coincident with the extent of veining, and occurs where the veining intersects the magnetite-rich horizons. Quartz-albite-carbonate veins commonly have halos of pyrite and microscopically the pyrite can be seen to replace magnetite or overgrow earlier syngenetic pyrite in Zone 'A' (Sangster et al., 1994a,b; Sangster and Pollard, 2001). It is interpreted that the hydrothermal fluids accessed the sedimentary horizon along a reactivated synvolcanic fault that appears on the surface as a strong north-northwest-striking linear feature. This feature has not yet been clearly identified in the mine nor has the source of the hydrothermal fluids been determined. However, it is felt that the fluids are most probably the same ones that altered the underlying cumulate unit to talc-carbonate schist and mobilized trace gold from the ultramafic rocks to the site of ore deposition. In support of a gold source from the alteration of the cumulate unit, indications of gold are widespread in the talc carbonate as centimetre-scale quartz veins with a few flecks of pyrite (Al, 1990; Beischer, 1988; Lavigne, 1993).

Nugget Pond Horizon Mineralization

Distal to the Nugget Pond Mine, the Nugget Pond Horizon commonly contains disseminated pyrite in the green laminated argillite and magnetite in the red argillite. The units are not continuous and the presence of magnetite is quite patchy. Two geologically interesting gold occurrences

FIGURE 17. Sample of high-grade mineralization from the Castle Rock showing. The football-sized boulder was recovered from a backhoe trench in a talc-carbonate replacement along a sheared zone cutting basalt.

are related to the Nugget Pond Horizon in an area of very complex structure immediately west of Tilt Cove. The Long Pond occurrence (Fig. 6) is at the east end of Long Pond and consists of geochemically elevated gold values (Appendix 1) in erratically pyritic green argillite associated with cherty iron formation bordering shear-zone-hosted talc carbonate. Both the argillite and the talc carbonate are auriferous.

The Castle Rock occurrence (Fig. 6) consists of heavily disseminated pyrite in a magnetic red argillite band in green argillite (Fig. 17). The occurrence is also associated with a shear zone containing talc carbonate occurring within mafic volcanic breccia. Gold assays are very high (see Appendix 1) but the occurrence is of limited size.

Mineral Exploration Techniques

For the purposes of mineral exploration, the Betts Cove complex presents three mineralized settings for consideration when developing exploration programs.

Gold deposits are exemplified by the Nugget Pond deposit, which is a small, disseminated pyrite deposit in a sedimentary sequence that contains units that are magnetite-rich. The most important controls are the presence of a favourable sedimentary horizon that is cut by a significant fault, as occurs at Nugget Pond. The deposit was discovered using soil geochemistry for gold and this is perhaps the most direct screen of a geological target area. Magnetic and induced polarization surveys were of limited use at Nugget Pond because of the small size of the deposit. However both types of surveys should be useful in exploring for larger deposits. Much of the Nugget Pond Horizon subcrop is covered by lakes, which provides an additional challenge for explorationists.

The base metal deposit types occur as massive replacements of basalt breccia and as disseminations in shear zones that juxtapose the sheeted dyke unit and the overlying boninitic and tholeiitic basalt. The degree to which mineralized shear zones as seen at Betts Cove and Mount Misery may be feeders to the massive and disseminated sulphides as found at Tilt Cove is not known, but a link should be considered in any exploration model. Geochemical surveys with Cu, Zn, and Au as essential elements would be part of any exploration program. The standard electromagnetic and induced polarisation techniques would also be useful in both cases but with the realization that very complex structure in some areas might result in difficulties in interpretation.

Summary

The Betts Cove complex consists of a basal ophiolite suite of rocks consisting of peridotite, sheeted dykes, boninitic pillowed basalt and basalt breccia, and younger gabbros that intrude the basal unit. The rocks are fresh and unaltered except for the peridotite, which is extensively altered to serpentinite and talc-carbonate schist. The ophiolitic rocks are overlain by the Snooks Arm Group that is composed of basal tholeiitic basalt and basalt breccia that is in turn overlain by volcanically derived turbiditic metasedimentary rocks and a series of mafic and intermediate volcanic rocks and related epiclastic volcanic and related epiclastic sedimentary rocks.

All of the known significant mineral occurrences in the Betts Cove ophiolite and Snooks Arm Group are hosted within the ophiolitic rocks, and the basal volcanic rocks and lowermost sedimentary rocks (Nugget Pond Horizon) of the Snooks Arm Group. Two broad groups of mineral occurrences of different ages are recognized, 1) copper dominated base metal occurrences that are located within the boninitic and basal tholeiitic basalts and 2) gold occurrences that are mainly within the sedimentary rocks of the Nugget Pond Horizon (Scrape Point Formation metasedimentary rocks).

1) The base metal deposits have long been classified as ophiolite-type occurrences (Upadhyay and Strong, 1973; Strong, 1984; Saunders, 1985, 1990; Saunders and Strong, 1986, 1988; Hudson, 1988; Strong and Saunders, 1988). Most of the smaller base metal occurrences lie near the contact between the sheeted dyke unit and the overlying boninitic basalts. Some are concordant (e.g. Betts Cove) but others are clearly discordant and form mineralized shear zones cutting the volcanic rocks (e.g. Mount Misery, Nudulama, and Betts Cove). The Tilt Cove deposit, on the other hand, though roughly concordant, contains elements that reflect a substantial element of structural control (stringer zones, breccia replacement) as well as zones of possible exhalative mineralization. However, this zone is contained within the lowest tholeiitic lavas of the Mount Misery basalt and is underlain by the boninitic lavas of the Betts Head Formation. If the mineralized shear zones within the ophiolite *sensu strictu* are interpreted as feeders to Tilt Cove-type mineralization, then much of the mineralization in the Betts Cove complex may be younger than the Betts Head Formation that defines the upper limit of the ophiolitic rocks of Bédard et al. (1998a,b, 1999, 2000).
2) The Nugget Pond gold deposit is the only substantial gold occurrence with others several orders of magnitude smaller. The one common denominator among the smaller gold occurrences is the presence of associated talc-carbonate schist (e.g. Castle Rock, Long Pond). The relationship of the Nugget Pond deposit with the talc-carbonate is tenuous at best and is restricted to a strong

air photo linear connecting the deposit area to the talc-carbonate-altered peridotite at the base of the ophiolite. As well, Beischer (1988), Al (1990) and Lavigne (1993) have documented geochemically anomalous gold and a number of small gold occurrences in and near the talc-carbonate horizon. In the absence of known Devonian intrusions in the area, the Nugget Pond and related gold occurrences are interpreted to have an origin related to the fluid-flow event that altered the peridotite to talc carbonate, probably during the Devonian.

Acknowledgements

This project was originally funded under a Geological Survey of Canada Industrial Partners Program project with Bitech Corporation and successor companies on the property Noveder Inc., Sulliden Exploration Inc., and Dianor Resources Inc. The authors wish to acknowledge the collaboration of Bill Hamilton and Jim Wade of Bitech, Alain Morissette of Noveder and Jacques Trottier of Sulliden and their staffs for contributions to the success of the project. The authors also wish to thank the people of the Tilt Cove-Betts Cove area who provided assistance to the project in many ways. Terry Rideout, Ray Wimbleton and Paul Winsor served as boatsmen along the rugged coastline. Don Collins provided access to a large collection of early photographs of operations at Tilt Cove as well as stories from the early days to add colour to the black and white images. A.L.S is particularly grateful to a gentleman of La Scie, whose name I do not know, who made a special trip to the gas station, while I was refueling, to make sure I had a plate of freshly baked stuffed squid for lunch. Such is Newfoundland hospitality.

Reviews by Andrew Kerr and Benoit Dubé greatly improved and clarified the manuscript.

References

Al, T.A., 1990, The Character and Setting of Gold Mineralization Associated with the Betts Cove Ophiolite; Unpub. M.Sc. thesis, Memorial University of Newfoundland, St. John's, Newfoundland, 151 p.

Baird, D.M., 1951, The Geology of the Burlington Peninsula, Newfoundland: Geological Survey of Canada, Paper 51-21, 70 p.

Bédard, J.H., and Lauziere, K., 1998, The Betts Cove Ophiolite and Its Cover Rocks: Geological Association of Canada – Mineralogical Association of Canada Joint Meeting, Geological Association of Canada, Guidebook 1998 B5, 37 p.

Bédard, J.H., Lauziere, K., Boisvert, E., Sangster, A.L., Tellier, M., Tremblay, A., and Dec, T., 1998a, Geological Map of the Betts Cove Ophiolite and Its Cover Rocks: Geological Survey of Canada, Open File Map 3271, scale 1:20 000.

Bédard, J.H., Lauziere, K., Tremblay, A., Sangster, A.L., Tellier, M., 1998b, Evidence from Betts Cove ophiolite boninites for forearc seafloor spreading: Tectonophysics, v. 284, p. 233-245.

Bédard, J.H., Lauziere, K., Boisvert, E., Sangster, A.L., Tellier, M., Tremblay, A., and Dec, T., 1999, Geological Map of the Betts Cove Ophiolite and Its Cover Rocks: Geological Survey of Canada, Map 1969A, scale 1:20 000.

Bédard, J.H., Lauziere, K., Tremblay, A., Sangster, A.L., Douma, S., and Dec, T., 2000, The Betts Cove Ophiolite and Its Cover Rocks: Geological Survey of Canada, Bulletin 550, 76 p.

Beischer, G., 1988, Report on geological, geochemical, and geophysical surveys, Betts Cove project, May to November, 1987: Unpub. company report, Newfoundland Assessment File # 2E/13 (584 and maps 2E/13 (584), p. 1-54.

Bichen, W.J., 1958, Summary Report on the Ore Zones, Structural Controls for Mineralization, the Post-Ore Dislocations and the Overall Economic Potential of the Tilt Cove Mine: Unpub. Report, Maritimes Mining Corporation Limited, 16 p.

Coish, R.A., 1977a, Petrology of the Mafic Units of West Newfoundland Ophiolites: Unpub. Ph.D. thesis, University of Western Ontario, London, Ontario, 227 p.

——— 1977b, Ocean floor metamorphism in the Betts Cove ophiolite, Newfoundland: Contributions to Mineralogy and Petrology, v. 60, p. 255-270.

——— 1989, Boninitic lavas in Appalachian ophiolites: A review, *in* Crawford, A.J., ed., Boninites and Related Rocks: Unwyn Hyman, London, p. 264-287.

Donoghue, H.G., Adams, W.S., and Harpur, C.E., 1959, Tilt Cove operations of the Maritimes Mining Corporation Limited; Canadian Institute of Mining and Metallurgy Bulletin, Proceedings, v. 52, p. 150-169.

Douglas, V.G., Williams D., Rove, O.N. and others., 1940, Copper Deposits of Newfoundland: Geological Survey of Newfoundland, Bulletin 20, 176 p.

Herzig, P.M., and Hannington, M.D., 1995, Polymetallic massive sulphides at the modern seafloor; A review: Ore Geology Reviews, v.10, p. 95-115.

Hibbard, J., 1983, Geology of the Baie Verte Peninsula, Newfoundland: Newfoundland Department of Mines and Energy, Mineral Development Division, Memoir 2, 277 p.

Hudson, K.A., 1988, Gold and Base Metal Mineralization in the Nippers Harbour Ophiolite, Newfoundland: Unpub. M.Sc. thesis, Memorial University of Newfoundland, St. John's, Newfoundland, 305 p.

Hurley, T.D., 1982, Metallogeny of a Volcanogenic Gold Deposit, Cape St.-John Group, Tilt Cove, Newfoundland: Unpub. M.Sc. thesis, McMaster University, Hamilton, Ontario, 69 p.

Hurley, T.D., and Crocket, J.H., 1985, A gold-sphalerite association in a volcanogenic base-metal-sulphide deposit near Tilt Cove, Newfoundland: Canadian Mineralogist, v. 23, p. 423-430.

Jenner, G.A., 1977, Geochemistry of the Upper Snooks Arm Group, Newfoundland: Unpub. M.Sc. thesis, University of Western Ontario, London, Ontario, 134 p.

Jenner, G.A., and Fryer, B.J., 1980, Geochemistry of the upper Snooks Arm Group basalts, Burlington Peninsula, Newfoundland: Evidence against formation in an island arc: Canadian Journal of Earth Sciences, v.17, p. 888-900.

Large, R.R., 1992, Australian volcanic-hosted massive sulphide deposits; Features, styles, and genetic models: Economic Geology, v. 87, p. 471-510.

Lavigne, J., 1993, The Geology and Geochemistry of Gold Mineralization in the Betts Big Pond Area, Newfoundland: Unpub. M.Sc. thesis, University of Ottawa, Ottawa, Ontario, 216 p.

Murray, A., and Howley, J.P., 1881, Report of the Geological Survey of Newfoundland from 1864 to 1880: Geological Survey of Newfoundland, 536 p.

Neale, E.R.W., 1967, Burlington (Baie Verte) Peninsula, Newfoundland, *in* Jenness, S.E., ed., Report of Activities, Part A: Geological Survey of Canada, Paper 67-1A, p. 183-186.

Papezik, V.S., 1964, Nickel minerals at Tilt Cove, Notre Dame bay, Newfoundland: Geological Association of Canada, Proceedings, v. 15, pt. 2, p. 27-32.

Sangster, A.L., and Pollard, D., 2001, Field guide to the mineral occurrences in the Betts Cove and Tilt Cove areas, Newfoundland, *in* Evans, D.T.W. and Kerr, A., eds., Geology and Mineral Deposits of the Northern Dunnage Zone, Newfoundland Appalachians: Geological Association of Canada-Mineralogical Association of Canada-Canadian Society of Petroleum Geologists, Annual Meeting, St. John's, Newfoundland, Field Trip Guide A2, p. 37-51.

Sangster, A.L., Lavigne, J., and Douma, S., 1994a, Geological studies of the Nugget Pond gold deposit, Baie Verte Peninsula, Newfoundland: Unpublished Industrial Partners Program Report, Geological Survey of Canada, 88 p.

Sangster, A.L., Lavigne, J.G., Douma, S., and Hamilton, W., 1994b, Geology, alteration and stable-isotope geochemistry of the Nugget Pond gold deposit: Newfoundland Department of Mines, Report of Activities, p.34-37.

Sangster, A.L., Lauzière, K., and Bédard, J.H., 1995, Mineral deposit studies in the Betts Cove ophiolite: Newfoundland Department of Mines, Report of Activities, p.20-21.

Sangster, A.L., Evans, D., Pollard, D., and Bradley, P., 2001, Sulphur isotopic relationships in the Nugget Pond gold deposit and regional sediment-hosted gold occurrences, northern Newfoundland, *in* North Atlantic Minerals Symposium, Extended Abstracts Volume, St. John's, Newfoundland, p. 144-148.

Santaguida, F., and Hannington, M.D., 1996, Characteristics of gold mineralization in volcanogenic massive sulphide deposits of the Notre Dame Bay area, central Newfoundland: Canadian Journal of Earth Sciences, v. 33, p. 316-334.

Saunders, C.M., 1985, Controls of Mineralization in the Betts Cove Ophiolite: Unpub. M.Sc. thesis, Memorial University of Newfoundland, St. John's, Newfoundland, 200 p.

——— 1990, A field guide to mineralization in the Betts Cove area, *in* Swinden, H.S., Evans, D.T.W., and Keen B.F., eds., Metallogenic Framework of Base and Precious Metal Deposits, Central and Western Newfoundland - Field Trip 1: Geological Survey of Canada, Open File 2156, p. 194-200.

Saunders, C.M., and Strong, D.F., 1986, Alteration-zonation related to variations in water/rock ratio at the Betts Cove ophiolitic massive sulphide deposit, Newfoundland, Canada; *in* Gallagher, M.J., Ixer, R.A., Neary, G.R., and Prichard, H.M., eds., Metallogeny of Basic and Ultrabasic Rocks: London Institute of Mining Metallurgy, Conference Proceedings, p.161-175.

——— 1988, Geological setting of mineralization at Tilt Cove, Betts Cove Ophiolite, Newfoundland: *in* Swinden, H.S., and Keen, B.F., eds., The Volcanogenic Sulphide Districts of Central Newfoundland, A Guidebook and Reference Manual for Volcanogenic Sulphide Deposits in the Early Palezoic Oceanic Volcanic Terranes of Central Newfoundland: Geological Association of Canada, Mineral Deposits Division, p. 54-61.

Snelgrove, A.K. 1929, Geological report on the Betts Cove Copper mine: Unpublished report.

——— 1931, Geology and ore deposits of the Betts Cove-Tilt Cove area, Notre Dame Bay, Newfoundland: Canadian Mining and Metallurgical Bulletin, v. 24, no. 4, 43 p.

Snelgrove, A,K, and Baird, D.M., 1953, Mines and Mineral Resources of Newfoundland: Newfoundland Department of Mines and Resources, 149 p.

Squires, G.C., 1981, The Distribution and Genesis of the Tilt Cove Sulphide Deposits: Unpub. Ph.D. thesis, Memorial University of Newfoundland, St. John's, Newfoundland, 115 p.

Strong, D.F., 1984, Geological relationships of alteration and mineralization at Tilt Cove, Newfoundland, *in* Swinden, H.S., ed., Mineral Deposits of Newfoundland: A 1984 Perspective: Newfoundland Department of Mines and Energy, Mineral Development Division, p. 81-90.

Strong, D.F. and Saunders, C.M., 1988, Ophiolite sulphide mineralization at Tilt Cove, Newfoundland; Controls by upper mantle and crustal processes: Economic Geology, v. 83, p. 239-255.

Swinden, H.S., Jenner, G.A., Kean, B.F., and Evans, D.T.W., 1989, Volcanic rock geochemistry as a guide for massive sulphide exploration in central Newfoundland, *in* Current Research: Newfoundland Department of Mines and Energy, v. 89-1, p. 201-219.

Swinden, S., McBride, D.M., and Dubé, B., 1990, Preliminary geological and mineralogical notes on the Nugget Pond Gold Deposit, Baie Verte Peninsula, Newfoundland, *in* Current Research: Newfoundland Department of Mines and Energy, Geological Survey Branch, Report 90-1, p. 201-215.

Tremblay, A., Bédard, J.H. and Lauziere, K., 1997, Structural history of the Betts Cove complex and cover rocks, Canadian Appalachians: From obduction to exhumation and folding of an Ordovician oceanic crust: Journal of Geology, v. 105, p. 701-716.

Upadhyay, H.D., 1973, The Betts Cove Ophiolite and Related Rocks of the Snooks Arm Group, Newfoundland: Unpub. Ph.D. Thesis, Memorial University of Newfoundland, St.John's, Newfoundland, 224 p.

——— 1980, Ordovician komatiites and associated boninite-type magnesian lavas from Betts Cove, Newfoundland, *in* Arndt, N.T., and Nisbet, E.G., eds., Komatiites: George Allen and Unwin, London, p.187-198.

Upadhyay, H.D., and Strong, D.F., 1973, Geological setting of the Betts Cove copper deposits, Newfoundland: An example of ophiolite sulphide mineralization: Economic Geology, v. 68, p. 161-167.

Upadhyay, H.D., Dewey, J.F., and Neale, E.R.W., 1971, The Betts Cove ophiolite complex, Newfoundland Appalachian oceanic crust and mantle: Geological Association of Canada, Proceedings, v. 24, p. 27-34.

Wardle, R., comp., 2005, Mineral Commodities Brochure #004, Gold: Newfoundland Geological Surveys Branch, Brochure.

Metallogenic Summary of the Meguma Gold Deposits, Nova Scotia

Alan. L. Sangster[1] and Paul K. Smith[2]

1. Geological Survey of Canada (retired), 986 Parkhurst Blvd., Ottawa, ON K2A 3M9
2. Nova Scotia Department of Natural Resources, 1701 Hollis Street Founders Square, 3rd. Floor, Halifax, Nova Scotia B3J 3M8
Corresponding author's email: alsangster@rogers.com

Abstract

The lower Ordovician Meguma Group of Nova Scotia consists of a lower thick unit of Goldenville Formation greywacke and an upper unit of black sulphidic Halifax Formation slate. The transition between the two units contains an Mn-rich horizon and Tremadocian fossils. The auriferous Goldenville formation comprises thick beds of light to dark grey sandy metagreywacke that contain thin beds of green chloritic silt and clay on the upper contact of each Bouma cycle. Occasionally at this stratigraphic horizon, the green slaty beds are replaced by a sequence of black sulphidic slate, which occurs abundantly throughout local stratigraphic sections.

Within these carbonaceous sections, auriferous ribbon (laminated, crack-seal) veins occur, commonly at the upper contacts of slate beds with the overlying metagreywacke. They may extend for several kilometres in length and 10s to 100s of bedding-parallel veins may occur in a single district. Also occurring are en echelon veins, various types of angular veins (smaller), pegmatitic veins, and cross cutting veins (larger). Veins that have undergone extensive deformation are sometimes juxtaposed with undeformed veins, indicating that the vein-forming process occurred over a substantial period of time during deformation. The veins and wall rocks both contain arsenopyrite with accessory pyrite, pyrrhotite, galena, and chalcopyrite; carbonate is ubiquitous. Isotopic studies suggest a biogenic origin for the sulphide and an origin for the carbonate as oxidized organic carbon. $^{40}Ar/^{39}Ar$ analyses of muscovite and other minerals related to the veins suggested an age circa 360 to 380 Ma and an origin related to Devonian granite emplacement has been suggested. More recently however, Re/Os dating of arsenopyrite suggests that the vein formation began with the onset of metamorphism circa 408 Ma. Other ages at about 375 Ma suggest a second vein-forming episode and perhaps continuous or episodic mineralization throughout deformation and intrusion of granites.

Résumé

Le Groupe de Meguma de l'Ordovicien inférieur en Nouvelle-Écosse se compose d'une épaisse unité inférieure de grauwacke attribuée à la Formation de Goldenville et d'une unité supérieure d'ardoise sulfurée de couleur noire rapportée à la Formation de Halifax. La zone de transition entre les deux unités renferme un horizon riche en manganèse et des fossiles trémadociens. La formation aurifère de Goldenville comprend d'épaisses couches de métagrauwacke gréseux de couleur gris pâle à gris foncé qui contiennent de fines couches de silt et d'argile chloriteuses de couleur verte formant le contact supérieur de cycles de Bouma individuels. Occasionnellement le long de cet horizon stratigraphique, les couches d'ardoiseuse verte sont remplacées par une séquence d'ardoise sulfurée noire qui est abondamment présente dans les coupes stratigraphiques locales.

À l'intérieur de ces intervalles carbonés, des filons rubanés (laminés, de scellement de fissures) à minéralisation aurifère sont couramment présents aux contacts supérieurs des lits d'ardoise avec les couches de métagrauwacke sus-jacentes. Ils peuvent se prolonger sur plusieurs kilomètres et des dizaines voire des centaines de ces filons parallèles à la stratification peuvent être présents dans un même district. Des filons en échelon, divers types de filons disposés à angle (plus petits), des filons pegmatitiques et des filons transversaux (plus gros) peuvent également être présents. Des filons qui ont subi une intense déformation sont parfois juxtaposés à des filons non déformés indiquant que le processus de formation des filons s'est échelonné sur une période substantielle au cours de la déformation. De l'arsénopyrite et, en quantités accessoires, de la pyrite, de la pyrrhotine, de la galène et de la chalcopyrite sont observées à la fois dans les filons et dans les épontes, et la présence de carbonates est généralisée. Des études isotopiques laissent croire que les sulfures sont biogènes et que les carbonates proviennent de l'oxydation de carbone organique. Les analyses $^{40}Ar/^{39}Ar$ de la muscovite et d'autres minéraux associés aux filons révèlent un âge d'environ 380 à 360 Ma et une origine liée à la mise en place des granites du Dévonien. Plus récemment, cependant, la datation Re/Os de l'arsénopyrite mène à croire que la formation des filons a commencé au début du métamorphisme aux environs de 408 Ma.. D'autres datations ayant livré un âge d'environ 375 Ma laissent croire qu'il y aurait eu un deuxième épisode de formation de filons et peut-être que la minéralisation a été épisodique ou continue tout au long de la déformation et de la mise en place des granites.

Introduction and History

The first confirmed bedrock discovery of gold in Nova Scotia occurred in September 1858 when Captain C. L'Estrange noticed the yellow metal in quartz outcrops at Mooseland on the Tangier River while hunting moose in the area. Two years later in May 1860, John Pulsiver of Musquodoboit was shown the Mooseland occurrence by Joe Paul, one of L'Estranges Indian guides. He traveled to Halifax to register his find and pointed out other favourable sites to Peter Mason who found gold at the head of Tangier Harbour in October 1860. Mooseland and Tangier were proclaimed "Gold Districts" and were surveyed under provincial legislation in April 1861. This sparked Canada's first gold rush and during the summer of 1861, prospecting based on similarity of rock lithologies to the Tangier and Mooseland areas resulted in the discovery of many other districts including Sherbrooke (Goldenville), Wine Harbour, Lawrencetown, Oldham, and Waverly (Malcolm 1929, 1976) (Fig. 1).

Sangster, A.L., and Smith, P.K., 2007, Metallogenic summary of the Meguma gold deposits, Nova Scotia, *in* Goodfellow, W.D., ed., Mineral Deposits of Canada: A Synthesis of Major Deposit-Types, District Metallogeny, the Evolution of Geological Provinces, and Exploration Methods: Geological Association of Canada, Mineral Deposits Division, Special Publication No. 5, p. 723-732.

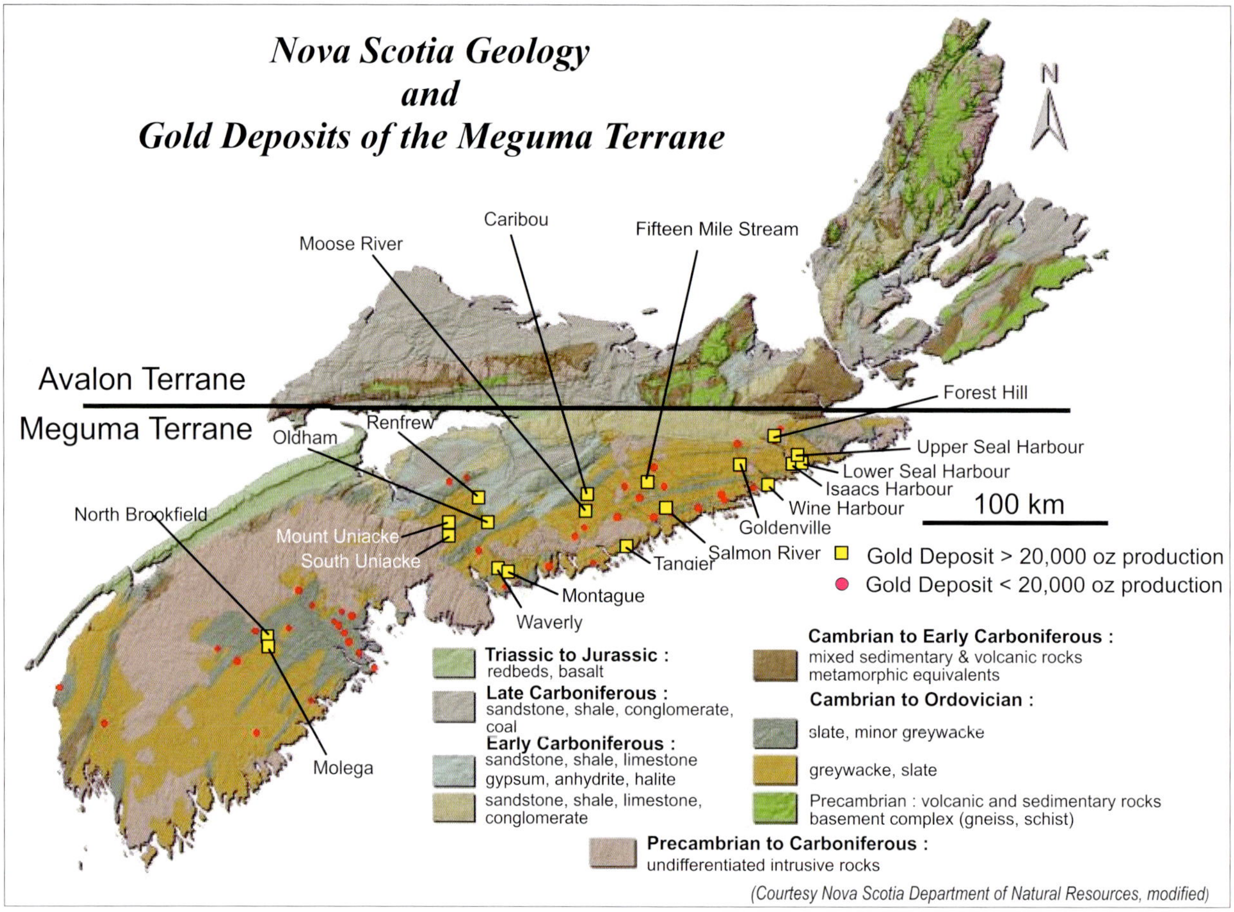

FIGURE 1. Regional geology of Nova Scotia and location of the 20 largest past-producing gold deposits. For smaller gold occurrences, the reader is referred to Malcolm and Faribault (1912) and Malcolm (1976) (courtesy Nova Scotia Department of Natural Resources).

Gold production from Nova Scotia gold veins has never reached the levels of districts discovered much later in the Abitibi, Kirkland Lake, or Timmins. Production in Nova Scotia remained between 15,000 and 30,000 troy ounces per year between 1862 and 1905 and dropped to less than 10,000 ounces per year after 1910 (Fig. 2). There have been brief periods of production at a few locations particularly late in the depression and during periods of higher gold prices between 1980 and the present (2006), but figures on total production for this period are not available.

Regional Geology and Tectonic Setting

Nova Scotia is divided geologically into two distinct parts, the Avalon Terrane to the north and the Meguma Terrane to the south (Fig. 1). The two terranes are separated by the east-west-trending Minas Geofracture (commonly referred to as the Cobequid-Chedabucto Fault System). Docking of the two terranes was accompanied by major sinistral, transcurrent motion along this fault, followed by minor dextral movement. Overlying Devono-Carboniferous sediments, which are common on both sides of the Minas Geofracture, stitch these two terranes together (Schenk, 1995).

The exposed Meguma Terrane is a 480 km long by 120 km wide (maximum) wedge of Lower Paleozoic metasedimentary rocks (Meguma Group) that were folded into long east-

TABLE 1. Production of the largest 20 gold deposits, Meguma Group, Nova Scotia.

Gold District	Years Active*	ounces	grams
Goldenville	1862-1941	209383	6511821
Caribou	1869-1968	91336	2840543
Oldham	1862-1946	85178	2649020
Waverly	1862-1940	72567	2256821
Montague	1863-1940	65197	2027624
Upper Seal Harbour	1893-1958	57846	1799001
Renfrew	1862-1958	51596	1604620
North Brookfield	1887-1936	43148	1341887
Wine Harbour	1862-1939	42347	1316976
Salmon River (Dufferin)	1881-1939	41805	1300148
Isaacs Harbour	1862-1958	39694	1234493
Lower Seal Harbour	1894-1949	34188	1063253
Molega	1888-1950	33460	1040612
Mount Uniacke	1867-1941	27737	862621
Tangier	1862-1919	26287	817510
Moose River*	1888-1939	25917	806025
Forest Hill	1895-1957	25102	780685
Fifteen Mile Stream	1879-1941	21220	659930
South Uniacke	1888-1948	20762	645701
Lake Catcha	1887-1961	17962	558603
Other		165889	5159154
Total		1198619	37277048

* Moose River production prior to 1888 included with Caribou

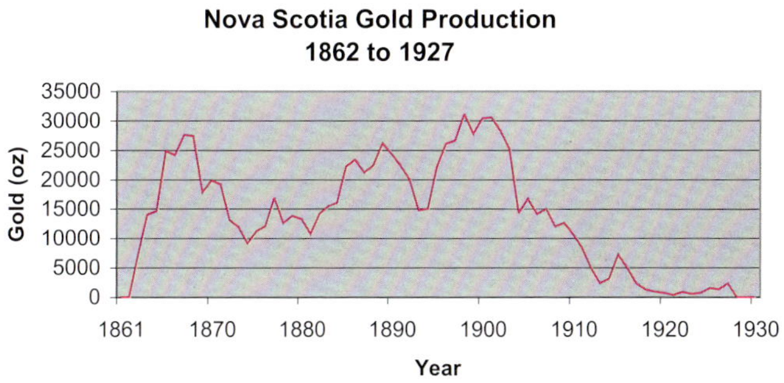

FIGURE 2. Nova Scotia Gold production, 1862 to 1927.

west trending, doubly plunging folds and regionally metamorphosed to greenschist and locally to amphibolite facies during the Devonian Acadian Orogeny (ca 400Ma). These metasediments were then intruded by voluminous Devonian per-aluminous granitoids (ca. 375Ma) and were subsequently overlain by carbonate and clastic sedimentary rocks and evaporates of the Horton and Windsor groups respectively. For the purpose of this summary, only the Meguma Group is discussed herein. The Meguma Group consists of the Cambrian Goldenville Formation metagreywacke, which is apporoximately 6.7 km thick with an unknown base. It is conformably overlain by black slate of the Halifax Formation, which, in one section near Halifax, is 11.8 km thick. Silurian volcanic and clastic rocks overlie these strata disconformably. The Goldenville Formation consists of massive, thick-bedded metagreywacke that is dark grey (carbonaceous) to light grey (calcareous) in colour. The thick coarser beds are commonly separated by thin slatey horizons that may either be chloritic or very carbonaceous. The Goldenville Formation grades upwards through manganese-rich strata into a basal unit of very carbonaceous sulphidic black slate. This in turn is overlain by typical Halifax Formation slate that consists of about 75% black carbonaceous sulphidic slate and 25% thinly bedded (~10 cm) to cross-laminated metasiltstone. The upper Halifax Formation most often consists of grey-green slates and siltstones. The proportions of the individual units is variable and most of the Halifax Formation seen in outcrop is carbonaceous and sulphidic.

Geology of the Gold Districts

Concordant and discordant gold-bearing veins occur throughout the Meguma Group, although most of the recorded 37 metric tons of gold has been recovered from the eastern part of the Meguma Terrane. Gold deposits are present throughout much of the exposed stratigraphy of the Goldenville Formation (Fig. 3), but are most common in the upper 1.5 km and between 3.5 and 4.5 km below the upper contact (Ryan and Smith, 1998). Disseminated gold mineralization is locally present (Moose River, North Brookfield). The auriferous veins occur either on or near the crests of regional anticlinal domes (Fig. 4A,B). Most of the districts occur within greenschist-facies rocks (Taylor and Schiller, 1966) and east of the South Mountain Batholith. However, several significant districts (e.g. Cochrane Hill, Forest Hill) are within amphibolite-grade metamorphic rocks that are spatially associated with numerous Devonian-Carboniferous granitic intrusions in the easternmost part of the province.

Most of the gold deposits and occurrences are associated with thicker than normal, interstratified slate and metasiltstone within the Goldenville Formation. Within these gold districts, the fine-grained lithologies are variably argillaceous, silty, carbon-rich and sulphidic with abundant pyrrhotite and arsenopyrite (Fig. 5A,B,C; Brunton, 1928; Malcolm, 1929; Smith, 1984) with pervasive carbonate alteration.

Concordant, auriferous quartz veins, which include bedding-parallel, stratiform, and stratabound geometry, are located within or immediately below the upper margins of incompetent, impermeable argillite horizons in the Goldenville Formation. Many of the districts are located on the steeper, sometimes overturned limb of anticlinal folds (e.g. Cochrane Hill, Beaver Dam, Fig. 6; Goldenville, Fig. 7; Tangier) (Smith and Kontak, 1986) or in parasitic second-order structures on the limbs of larger folds (e.g. Mt. Uniacke, Fig. 9: Dufferin, Fig. 8). Henderson and Henderson

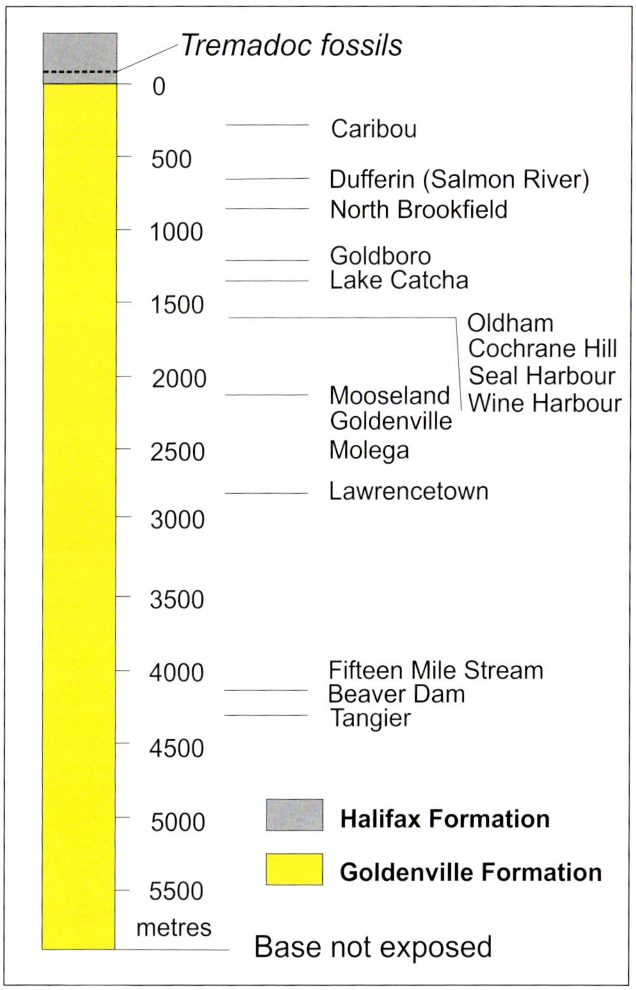

FIGURE 3. Stratigraphic location of selected gold deposits, Meguma Terrane, Nova Scotia. There is a concentration of occurrences at greater than 1500 metres and greater than 2000 metres below the upper Goldenville Formation contact, but these occurrences are widely separated and can not be demonstrated to be at the same stratigraphic horizon. Depths data was acquired by measuring the stratigraphic depth from sections on regional geological maps of Faribault (1898, 1901).

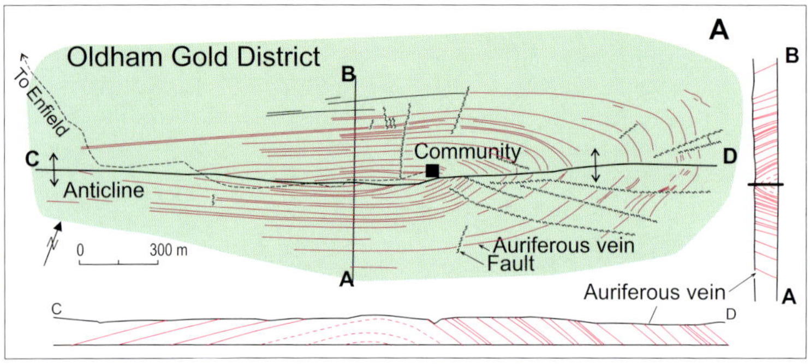

FIGURE 4. Geology of typical Nova Scotia gold districts. (A) Gold vein distribution in the Oldham gold district are more or less symmetrically distributed about the dome of a doubly plunging anticline. Black shale beds underlying the veins are too thin to show at this scale (modified after Faribault, 1898). (B) Gold veins only on the south side of the anticline at the Mount Uniacke gold district. It is probable that the veins formed on the crest of the fold and were left stranded on the south flank after migration of the fold hinge to the north (modified after Faribault, 1901).

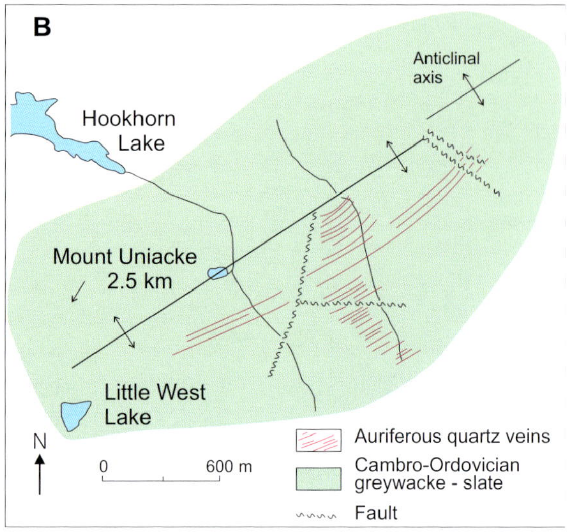

(1986) pointed out that many workers mistakenly refer to the Meguma deposits as saddle reefs. True saddle reefs (Herman, 1914 in Boyle, 1979; Ramsay, 1980) are restricted to fold hinges and are considerably thickened in the axial zones with little down-limb extension. Meguma veins normally maintain constant thickness around fold hinges and have significant down-dip extent (e.g. Goldenville). However, the Dufferin deposit represents a significant exception in that large saddle-like, auriferous veins have been mined successfully on a small scale (Fig. 8).

Ore-grade gold mineralization commonly occurs in several settings including (1) along vein-wall rock contacts, as medium- to coarse-grained aggregates, (2) in massive white quartz, (3) less commonly as fine-grained films and wires between crack-seal laminae of concordant veins, and (4) at the intersections of discordant and concordant veins (Fig. 9). Where gold occurs with arsenopyrite, pyrrhotite, galena, and chalcopyrite, it is most commonly present as fracture fillings and coatings or as small blebs within or attached to individual crystals.

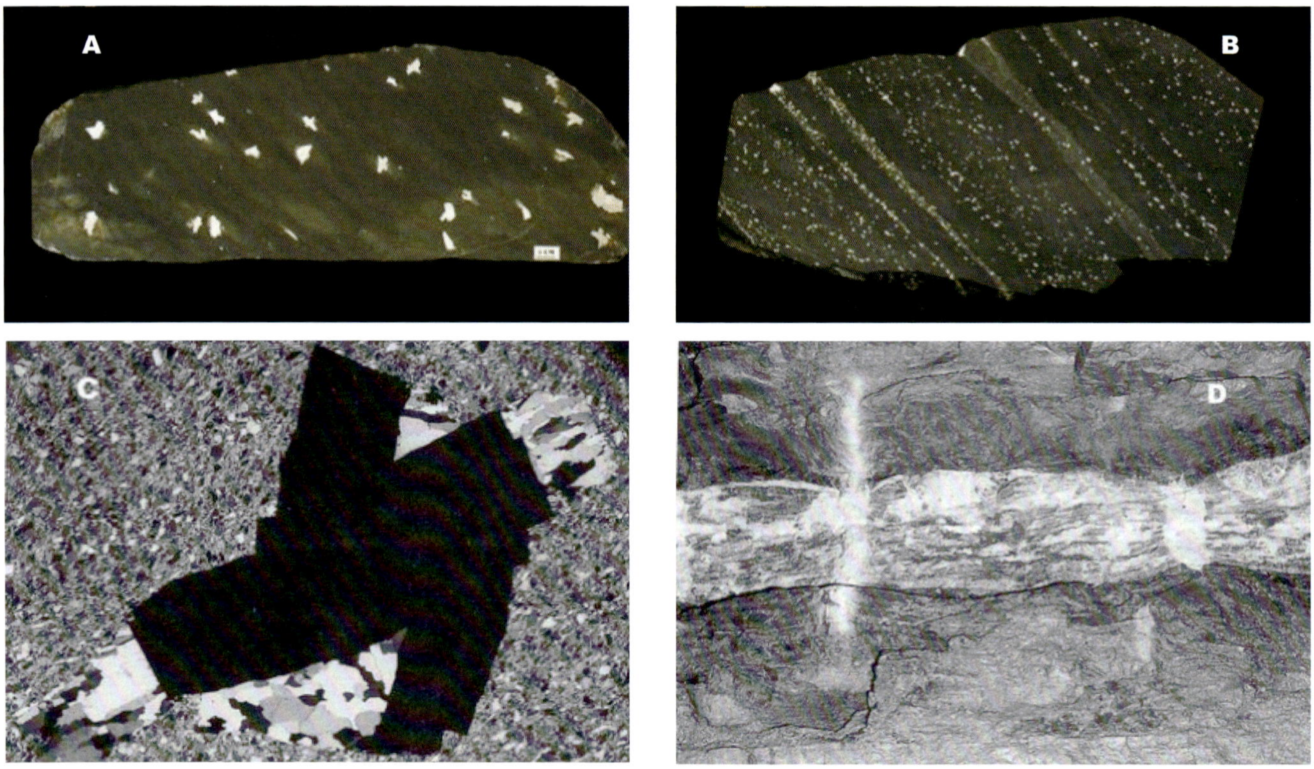

FIGURE 5. Geological features of Meguma gold Deposits. (A) Cruciform arsenopyrite crystals in metagreywacke from the Caribou deposit alteration aureole. Note lighter coloured silicification marbling the sample, particularly around arsenopyrites. (B) Bedding-parallel bands of disseminated arsenopyrite in black slate associated with veins at the Lawrencetown occurrence. Both lithologies in "A" and "B" may be auriferous. (C) Quartz pressure shadows on arsenopyrite crystals indicate crystallization prior to the last deformation. (D) Laminated "crack-seal" quartz (25 cm thick) from the Marker vein, Tangier mine. This is the most common vein-type and usually consists of laminated quartz overprinted by white bull quartz. (from Sangster, 1990).

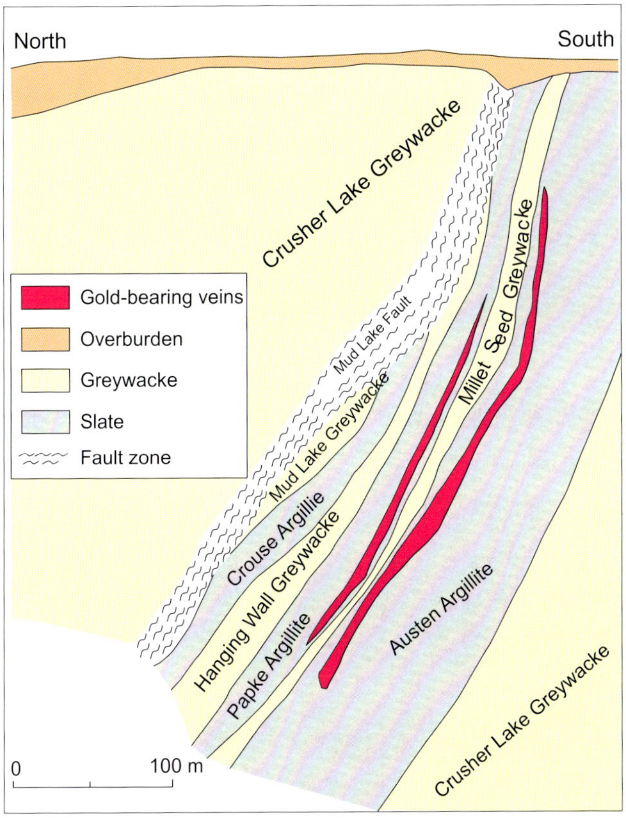

FIGURE 6. Section through en echelon gold veins at the Beaver Dam deposit, Nova Scotia (courtesy Seabright Resources Inc.).

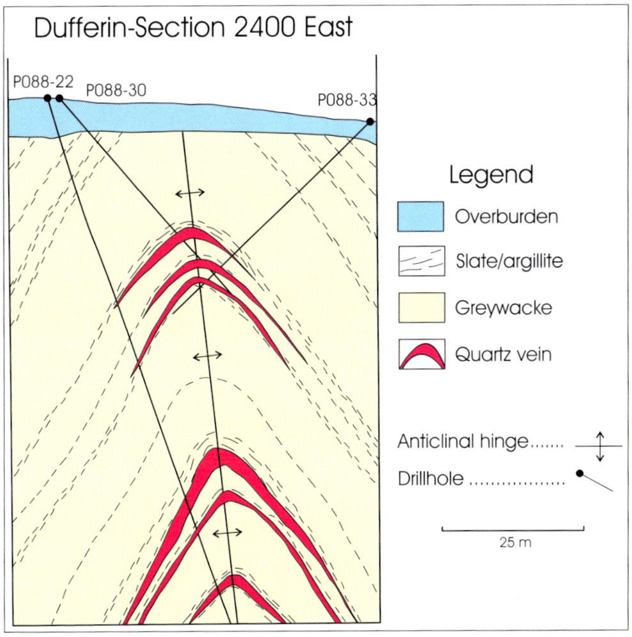

FIGURE 8. True stacked saddle reefs at the Dufferin extension deposit, Salmon River gold district. This deposit is the most recent discovery in the Meguma Group and is in production. It is the extension of the Old Dufferin Mine that has been faulted about 2 km north of the historic district, on a northwest fault (modified after Ryan and Smith, 1998).

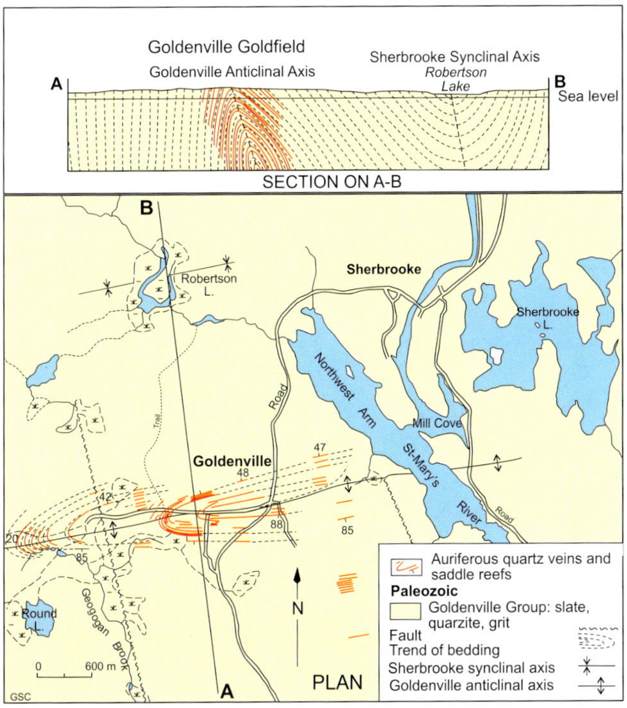

FIGURE 7. Geological plan of the Goldenville gold district. The best vein density occurred only on the western plunge of the culmination of the anticline (modified after Malcolm and Faribault, 1912).

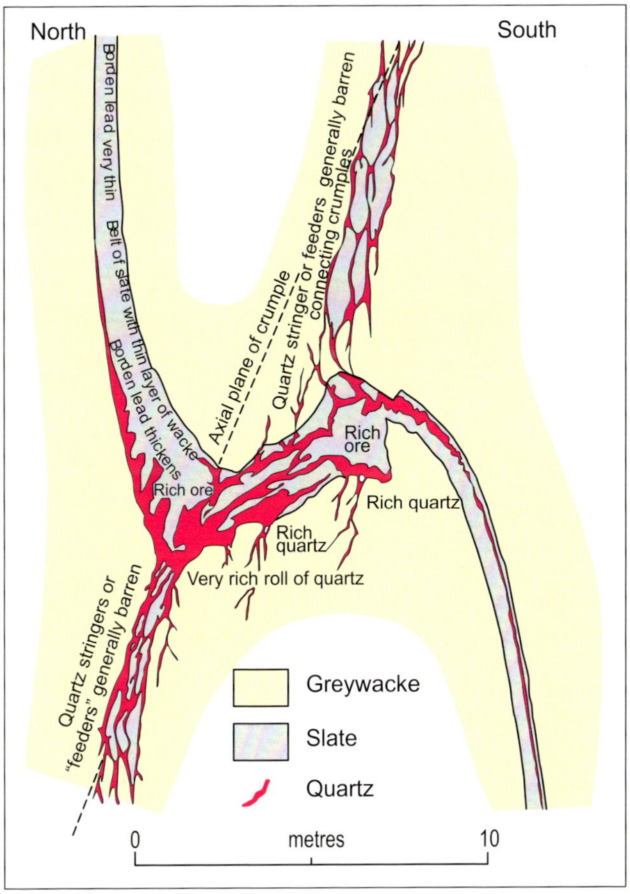

FIGURE 9. Geological section of a folded auriferous bedding-parallel vein intersected by a late barren vein breccia from the Mount Uniacke gold district. Note the gold enrichment in the bedding-parallel vein near the intersection with the late vein (after Malcolm and Faribault, 1912).

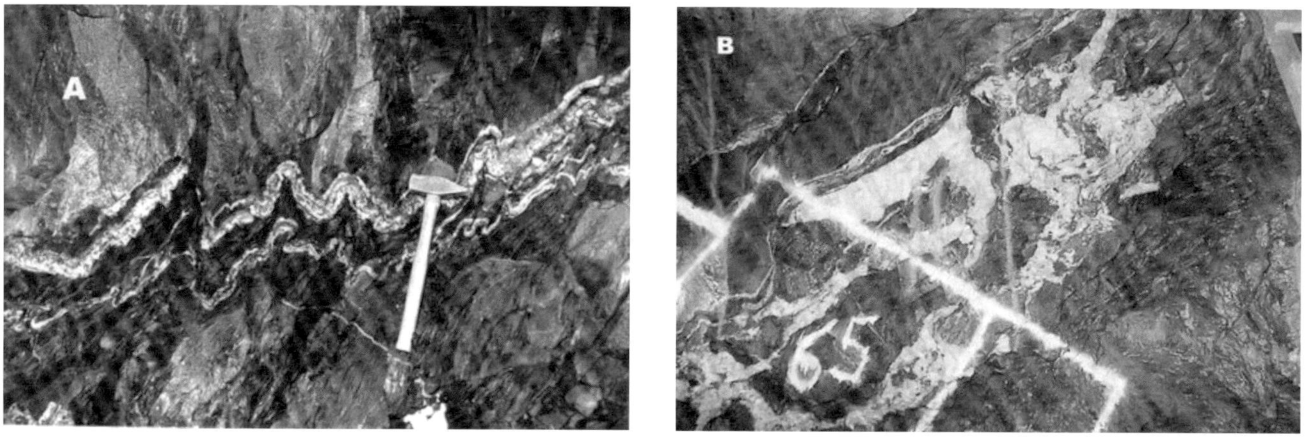

FIGURE 10. Geological features of Meguma gold deposits. (A) Twin vein, tangier gold district showing a pair of buckled, highly auriferous, quartz veins near the crest of the Tangier anticline. Also visible is the relationship of the veins to the overlying metagreywacke and underlying black slate. (B) Nugget vein, Tangier gold mine, showing chaotic quartz vein textures resulting from the development of superimposed, variably rotated en echelon veins in black, organic-rich, arsenopyrite-rich slate. Painted numbers are 25 cm high (from Sangster, 1990).

Throughout the Meguma Terrane quartz, ankerite (with variable calcite and ferroan dolomite), plagioclase, chlorite, biotite, and muscovite form abundant vein minerals (Kontak and Smith, 1988a). Arsenopyrite is variably present in all gold districts ranging up to ~15% locally in some wall rock lithologies. It is less abundant within the veins and constitutes an average of ~2% for the individual deposits. The gold distribution varies directly with the sulphide (pyrrhotite, chalcopyrite, pyrite, and galena) content. Scheelite is common at the Moose River deposit and locally present at the Beaver Dam, Mt. Uniacke, and Pleasant River Barrens deposits and is a minor accessory phase in several other gold occurrences. At the West Gore deposit, gold is genetically associated with antimony mineralization (Smith and Kontak, 1986).

Deposit Description and Classification

Several types of veins have been described at individual districts (Henderson, 1983; Smith, 1983b; Haynes 1986, 1987; Henderson et al., 1986a; Ryan and Smith, 1998).

Stratiform Veins ("Leads")

In general, the earliest vein structures are concordant, laminated veins with crack-seal texture (Ramsay, 1980), which are interpreted to have formed during episodic hydraulic overpressure (Fig. 5D) (Faribault, 1899; Graves, 1976; Graves and Zentilli, 1982; Henderson, 1983; Mawer, 1985, 1986, 1987; O'Brien, 1985; Henderson et al., 1986b; Henderson and Henderson, 1990). These so-called "laminated", BP, or "interbedded" veins are regionally concordant but crosscut bedding in detail and often extend from one stratigraphic horizon to another. On anticlinal hinges in the presence of an axial planar pressure solution cleavage, the veins are "buckled" (Fig. 10A) and gold grades are substantially higher than elsewhere. The ribbon veins commonly maintain their thickness over fold hinges and are not true saddle-reef veins.

In a few deposits, Dufferin (Fig. 8) being the best example, true saddle-reef veins are present with thick, massive, concordant accumulations of vein-quartz in fold noses that pinch out down-dip on fold limbs.

Also included in this category are "stratabound veins that consist of irregular accumulations of vein quartz that are restricted to one slate horizon (Fig. 10B). Gold veins in these veins are highly irregular

En echelon Veins

En echelon veins commonly form in slate or metasiltstone beds between massive metagreywacke units on the limbs of major folds. The veins fill extensional fractures that formed as a result of bedding-parallel shear during folding (Anderson et al., 1986a).

Angular Veins

These veins have discordant geometry and may be laterally extensive. They may attain a thickness of several metres and in gross scale resemble en echelon veins, except that they may crosscut several stratigraphic horizons at any one location and are typically continuous both above and below the concordant veins.

Pegmatitic Veins

Pegmatitic veins are restricted to areas around granitic intrusions and areas of higher grade metamorphism. At Cochrane Hill they occur both within and outside the mineralized horizons. These veins contain quartz, orthoclase, garnet, plagioclase, muscovite, biotite, andalusite, and rare traces of gold (Smith, 1983a,b).

AC Veins

Late extension parallel to regional fold axes has produced quartz-filled ac joints in the metagreywacke (Henderson, 1983; Williams and Hy, 1990). These veins are regionally present and occur irrespective of the presence of gold-bearing veins. However, these veins may contain trace amounts of gold where they occur in gold districts.

Northwest Faults and Veins

Subvertical quartz-calcite veins, vein breccias, and sinistral kink bands are parallel to regional, northwest-trending sinistral faults that cut all fold structures in the Meguma Terrane as well as the Devono-Carboniferous granitoids. They are thought to have a significant effect on gold distribution in some districts (e.g. Beaver Dam, Moose River). Gold grades in some concordant veins at the Tangier deposit increase near intersections with northwest faults (D. Forgeron, pers. comm.). In the Leipsigate, Brookfield, Caribou, and Rawdon districts (among others) the northwest faults host younger, goldbearing veins.

Miscellaneous Structures

Second-order structures located along regional folds are spatially and temporally associated with high-grade gold mineralization. At the Mount Uniacke gold district, gold ore was found at intersections of concordant and barren veins developed along a normal fault associated with a second-order parasitic fold (Fig. 9). At the Caribou mine, significant gold was mined from concordant veins and a younger, northwest-trending stockwork zone on the flank of the major fold (Fig. 11). At both the Leipsigate and Ecum Secum gold districts, gold mineralization occurs in faults lying subparallel to the regional fold axes.

Disseminated Non-Vein Gold I

In addition to vein-hosted gold, the Moose River district contains a substantial reserve of low-grade gold ore that is not directly associated with veining. At the Touquoy zone in the Moose River district (Bierlein and Smith, 2003), an indicated plus inferred resource of 6 million tonnes grading about 2 g/t gold has been identified by exploration in slate and metagreywacke in the vicinity of the auriferous veins. (Bierlein and Smith, 2003).

Individual deposits varied somewhat in the relative proportions of various mineralized features. For complete descriptions of most deposits the reader is referred to Malcolm and Faribault (1912), reprinted without photographs as Malcolm (1976).

Grade and Tonnage

Reliable grade and tonnage figures are not available for the Meguma Group gold deposits. Firstly, most of the mining occurred in the 19th and early 20th century and was effectively exploration by mining. “Winzes” were sunk on veins every 30 metres or so and ore-grade material encountered was mined laterally from these entries. The mining process itself was unorthodox by modern standards in that the easily mined slate in the footwall to the veins was removed and then the vein quartz was peeled off with a hammer and chisel. Grade figures available generally referred to tons of quartz crushed and not tones mined. In addition, research on old records by the Nova Scotia Department of Natural Resources indicates a possible average quartz vein grade of about 0.48 oz/ton. However, this figure does not account for the fact that at many mines, miners were paid in gold that did not form part of the recorded production and, as is the case of many locations that mine coarse-grained gold, much gold may have left the mine in the pockets of the miners.

The Goldenville district (Fig. 7) was the largest single producer in the Meguma Terrane with historical production of 6,535,730 g of gold (Sangster, 1990). The district included 105 veins, principally as concordant veins along an asymmetrical, east trending anticline. The veins are exposed over

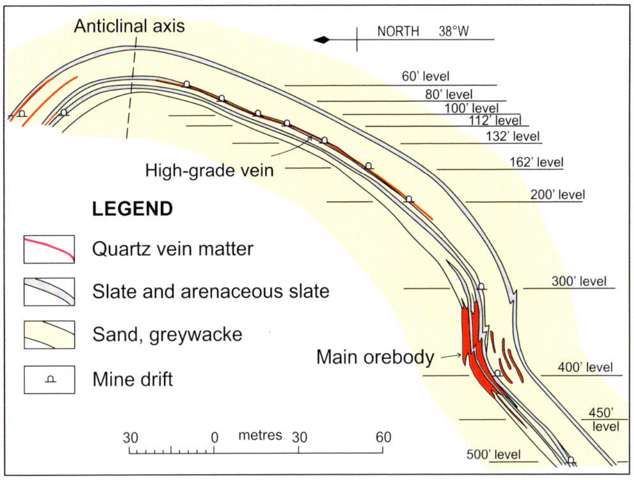

Figure 11. Geological section of the Caribou Mine, Nova Scotia. Note the two ore types: 1) a bedding-parallel, crack-seal vein in the upper part of the fold, and 2) quartz-healed breccia ore associated with a “kink” in the south flank of the fold (modified after Bell, 1948).

a 518 m section across the anticline (Malcolm, 1976) and occur primarily along the boundaries of slate beds and locally within the adjacent metagreywacke. Both bedding-parallel ("leads") and crosscutting ("angulars") veins are present. In addition to gold, the veins contain minor galena, sphalerite, pyrrhotite, pyrite, and arsenopyrite.

The Caribou Mine produced 2,841,950 g of gold between 1869 and 1945. Gold occurs in three settings. Like the Goldenville district, there are high grade concordant leads within slaty beds or at bedding contacts within quartzites as well as crosscutting lenses (angulars) branching off from the concordant veins. In addition, there is a large zone of network quartz veins in the metagreywacke confined to a flexure in the southern limb of the northeast-trending Caribou anticline (Douglas, 1948) (Fig. 11), which produced substantial amounts of gold.

Genetic Models

Three general mechanisms have been proposed for the origin of the gold veins: 1) syngenetic, hydrothermal deposition on the seafloor, 2) early syntectonic deposition from hydrothermal fluids of diverse origins, and 3) late syntectonic deposition from magmatic or deep crustal hydrothermal fluids. The genesis of the Meguma gold veins remains controversial. The syngenetic theories seem to have lost supporters but the early and late syntectonic theories still have substantial support.

Hunt (1868) was the first to propose a syngenetic origin for the quartz veins within the enclosing metasediments. Hind (1869, 1870, 1872) supported Hunt’s proposal for the laminated veins, but attributed other concordant veins to later processes. McBride (1978) reported a pyrite+carbonate-rich stratum 1300 m below the Goldenville-Halifax transition zone (GHT) that contains several small gold occurrences and proposed a syngenetic origin for the gold. Haynes (1983, 1986, 1987) proposed a complex, hydrothermal, hot spring model. Haynes (1987) interpreted the laminated, concordant veins as fumarole related geyserite, and the crosscutting and semiconcordant mineralized veins as subsurface feeders and parasites of the hydrothermal system.

Documentation of structural features (Henderson and Henderson, 1986; Henderson et al., 1986a; Mawer, 1986, 1987; Williams and Hy, 1990) does not support a synsedimentary origin for these veins, though it has been argued that synsedimentary components have been mobilized to form the veins.

Early syntectonic theories have been commonly used to explain the origin of the veins. Early workers, including Campbell and Poole (1862), Campbell (1863), Selwyn (1872), Gilpin (1888), Faribault (1899), Malcolm and Faribault (1912), and Malcolm (1929) stressed the importance of structure as a controlling factor and proposed the derivation of vein materials from the enclosing sedimentary rocks, and "ascending" or "descending" hydrothermal fluids as transporting agents. Boyle (1966, 1979) supported a lateral secretion model. Modern variants of these theories support an early syntectonic, pre-granite, and pre- to syn-folding origin for the veins, with the enclosing host rocks providing the vein source material mobilized by prograde metamorphic fluids (Graves, 1976; Graves and Zentilli, 1982; Henderson, 1983; Henderson and Henderson, 1986; Henderson et al., 1986b; Sangster, 1990, 1992).

Late syntectonic theories of vein origin include magmatic hydrothermal processes related to Acadian (Devono-Carboniferous) granites and vein formation from deep, crustally derived metamorphic fluids synchronous with Acadian plutonism, albeit, not directly from a magmatic source. A magmatic source for both quartz veins and gold was proposed by Rickard (1927), Newhouse (1936), Emmons (1937), Bell (1948), and Douglas (1948). Recent proposals allow for a source of vein materials either directly from a granitic source, from the lower crust or mantle, or from the enclosing sedimentary rocks, mobilized by magmatic or deep crustal metamorphic fluids driven by magmatic heat (Smith and Kontak, 1986; Kontak and Smith, 1987, 1988b, 1989a,b, 1993; Mawer 1985, 1986, 1987; Hy and Williams, 1986; Kontak et al., 1990, 1993; Kontak and Archibald, 2002). In support of this model, $^{40}Ar/^{39}Ar$ dates between 380 and 362 Ma have been measured for vein-fill amphibole and mica in several deposits (Kontak et al., 1990; 1993, 1998), which is consistent with models invoking protracted mineralization from metamorphogenic fluids broadly related to Acadian orogenesis and magmatism.

Many researchers contend that several generations of veins are present (Graves, 1976; Haynes and Smith, 1983; Henderson et al., 1986a,b; Mawer, 1986) representing semi-continuous vein development throughout an extended period of time. They point out the coexistence of undeformed crack-seal (ribbon) veins with highly deformed veins that have undergone extension on fold limbs. Sangster (1992) has argued, based on the presence of heavy sulphur biased vein sulphides, and host rocks, together with the fact that the vein carbonate is oxidized organic carbon, that gold was preconcentrated in Goldenville Group carbonaceous sediments in an anoxic oceanic environment and later remobilized and reconcentrated to form the gold deposits. Sangster (1992) suggested that the first gold veins to form would have been concordant veins during initial deformation and metamorphism. Vein formation and mineralization may have then proceeded episodically (perhaps including the period of 380 to 360 Ma suggested by Kontak et al. (1990)) in response to continuing deformation and intrusion of granite plutons, resulting in later generations of veins.

Recently, Morelli et al. (2005) published Re-Os ages on arsenopyrite for two deposits. The Ovens deposit on the coast 100 km west of Halifax gave an age of 407 ± 4 Ma similar to the age of onset of the initial age of metamorphism. The saddle reef veins at the Dufferin deposit gave an age of 380 ± 3 Ma, similar to former $^{40}Ar/^{39}Ar$ ages and the age of intrusion of the Devonian monzogranites of the Meguma Terrane. These data indicate at least two ages of mineralization and potentially episodic mineralization over a period of some 15 to 30 Ma.

Exploration Methods

Exploration for and discovery of these deposits has been mainly a boot and hammer process both for the original discoveries and for exploration in recent years. Most of the historical deposits were discovered by prospecting in outcrop or in float followed by trenching and, if promising, sinking regular declines along the vein.

Recent exploration has been based on historical records, prospecting, and undertaking of till geochemistry surveys and heavy mineral analyses. Some use has been made of induced polarization geophysical surveys with mixed results. The host rocks have a low magnetic relief, however gradient magnetic surveys have shown that the natural magnetism of the Goldenville Formation is often destroyed in the vicinity of gold districts but that there is a weak magnetic anomaly over the deposit due to the association of pyrrhotite with the gold-bearing veins.

Knowledge Gaps

1. The body of opinion on the origin of the Meguma gold vein districts prior to about 1990 was that the veins formed from hydrothermal fluids from a deep crustal source or fluids derived from within the Meguma Group prior to the intrusion of the Devonian granites. During the 1990s, $^{40}Ar/^{39}Ar$ dating of muscovite and other minerals gave consistent ages of about 370 Ma, which was interpreted by some to indicate that the deposits were related to the emplacement of Devonian granites, and by others to indicate that this age could not explain field observations indicating that the deposits are, at least in part, related to the onset of metamorphism circa 400 Ma.

   Recent Re-Os dating at several deposits shows that there are at least two distinct ages of mineralization (408 Ma and 376Ma). This is the first tangible isotopic evidence supporting field observation of distinct ages of vein formation. It is proposed that the careful, systematic Re-Os dating of key gold districts should aid greatly in establishing the paragenetic sequence of gold mineralizing events between these ages and potentially identify the better mineralizing events within this time range.
2. There is a growing realization that dark grey meta-greywacke is the unaltered variety of Goldenville Formation and that the regional lighter grey meta-greywacke may represent weak distal carbonate alteration that is prevalent at all gold districts in the Meguma

Terrane. Mapping the distribution of this lighter grey facies would assist greatly in understanding the genesis of these enigmatic deposits and would provide new criteria for exploration and evaluation of size and distribution potential of these gold districts. The mapping should be accompanied by lithochemical studies to provide a basis for comparison of the metagreywacke facies and alteration facies proximal to the gold districts.

Acknowledgements

The authors wish to acknowledge reviews by Bob Ryan of the Nova Scotia Departmant of Natural Resources and Wayne Goodfellow of the Geological Survey of Canada, which improved this manuscript.

References

Bell, L.V., 1948, Caribou mine, in Structural Geology of Canadian Ore Deposits: Canadian Institute of Mining and Metallurgy, Jubilee Volume, p. 927-936.

Bierlein, F.P., and Smith, P.K., 2003, The Touquoy Zone Deposit: An example of "unusual" orogenic gold mineralization in the Meguma Terrane, Nova Scotia, Canada: Canadian Journal of Earth Sciences, v. 40, p. 447-466.

Boyle, R.W., 1966, Origin of gold and silver deposits of the Meguma Series, Nova Scotia: Geological Association of Canada/Mineralogical Association of Canada, Program, Annual Meeting, p. 7-8.

——— 1979, The Geochemistry of Gold and Its Deposits: Geological Survey of Canada, Bulletin 280, 584 p.

Brunton, S. Sir, 1928, Some observations on a so-called interbedded vein at Tangier, N.S.: Mining Society of Nova Scotia, Annual Meeting, p. 784-803.

Campbell, J.S., 1863, Nova Scotia Gold Fields: A report to the Honourable Joseph Howe, Provincial Secretary, Halifax, 12 p.

Campbell, J.S., and Poole, H., 1862, Report on the goldfields: A report to the commissioner of the Crown Lands: Halifax, 36 p.

Douglas, G.V., 1948, Structure of the gold veins of Nova Scotia, in Structural Geology of Canadian Ore Deposits: Canadian Institute of Mining and Metallurgy, Jubilee Volume, p. 919-926.

Emmons, E.R., 1937, Gold Deposits of the World: McGraw-Hill, New York, 562 p.

Faribault, E.R., 1898, Map of the Oldham Gold District: Geological Survey of Canada, Map 642.

——— 1899, The gold measures of Nova Scotia and deep mining: Journal of the Canadian Mining Institute, v. 2, p. 119-129.

——— 1901, Map of the Mount Uniacke Gold District: Geological Survey of Canada, Map 709.

Gilpin, E., 1888, Notes on the Nova Scotia gold veins: Transactions of the Royal Society of Canada, v. VI, sec. IV, p. 63-70.

Graves, M.C., 1976, The formation of gold-bearing quartz veins in Nova Scotia: Unpub. M.Sc. Thesis, Dalhousie University, Halifax, Nova Scotia, 166 p.

Graves, M.C., and Zentilli, M., 1982, A review of the geology of gold in Nova Scotia, *in* Hodder, R.W., and Petruk, W., Geology of Canadian Gold Deposits: Canadian Institute of Mining and Metallurgy, Special Volume 24, p. 233-242.

Haynes, S.J., 1983, Typomorphism of turbidite-hosted auriferous quartz veins, southern Guysborough County, *in* Report of Activities, 1982: Nova Scotia Department of Mines and Energy, Mines and Minerals Branch, Report 83-1, p. 183-224.

——— 1986, Geology and geochemistry of turbidite-hosted gold deposits, greenschist facies, eastern Nova Scotia, *in* Duncan Keppie, J., Boyle, R.W., and Haynes, S.J., eds., Turbidite-Hosted Gold Deposits: Geological Association of Canada, Special Paper 32, p. 161-178.

——— 1987, Classification of quartz veins in turbidite-hosted gold deposits, greenschist facies, eastern Nova Scotia: Canadian Institute of Mining and Metallurgy Bulletin, v. 80, p. 37-51.

Haynes, S.J., and Smith, P.K., 1983, Gold potential of the Meguma Group: New concepts: Maritime Sediments and Atlantic Geology, v. 19, p. 96.

Henderson, J.R., 1983, Analysis of structure as a factor controlling gold mineralization in Nova Scotia, in Report of Activities 1982, Nova Scotia Department of Mines and Energy, Mines and Minerals Branch, Report 83-1, p. 265-282.

Henderson, M.N., and Henderson, J.R., 1986, Constraints on the origin of gold in the Meguma zone, Ecum Secum area, Nova Scotia, *in* Contributions to the Geology of the Meguma Terrane, Nova Scotia: Maritime Sediments and Atlantic Geology, v. 22, p. 1-14.

——— 1990, Crack-seal texture in bedding-parallel, gold-bearing, columnar-quartz veins: evidence of fossil water-sills, *in* Sangster, A.L., ed., Mineral Deposit Studies in Nova Scotia, Volume 1: Geological Survey of Canada, Paper 90-8, p. 163-168.

Henderson, J.R., Henderson, M.N., Mawer, C.K., Crocket, J.H., Clifford, P.M., and Fueten, F., 1986a, Timing and origin of auriferous quartz veins, Meguma Terrane, Nova Scotia: Poster Volume, Gold '86, Geological Association of Canada, Mineral Deposits Division, p. 70-71.

Henderson, J.R., Wright, T.O., and Henderson, M.N., 1986b, A history of cleavage and folding: an example from the Goldenville Formation, Nova Scotia: Geological Society of America Bulletin, v. 97, p. 1354-1366.

Hind, H.Y., 1869, Notes on the structure of the Nova Scotia gold district: Transactions of the Nova Scotia Institute of National Science, v. II, pt. 3, p. 102-109.

——— 1870, On two gneissoid series in Nova Scotia and New Brunswick, supposed to be the equivalents of the Huronian (Cambrian) and Laurentian: Geological Society of London, v. XXVI, p. 468-479.

——— 1872, Report on the Mt. Uniacke, Oldham, and Renfrew gold mining districts: Halifax, p. 136.

Hy, C., and Williams, P.F., 1986, Meguma quartz veins, Tangier area: 10th Annual Open House and Review of Activities, Nova Scotia Department of Mines and Energy, Information Series Number 10, p. 202-205.

Kontak, D.L, and Archibald, D.A., 2002, $^{40}Ar/^{39}Ar$ dating of hydrothermal biotite from high-grade gold ore, Tangier gold deposit, Nova Scotia: further evidence for 370 Ma gold metallogeny in the Meguma Terrane: Economic Geology, v. 97, p. 619-628.

Kontak, D.K. and Smith, P.K., 1987, Alteration haloes and their implications for gold mineralization in the Meguma zone, Nova Scotia, *in* Report of Activities, 1987: Nova Scotia Department of Mines and Energy, Report 87-1, p. 65-74.

——— 1988a, Meguma gold studies IV: Chemistry of vein mineralogy, *in* Report of Activities 1987 Part B: Nova Scotia Department of Mines and Energy, Mines and Minerals Branch, Report 88-1, p. 85-101.

——— 1988b, Meguma gold studies VI: Integrated model for the genesis of Meguma-hosted lode gold deposits, *in* Report of Activities 1987, Part B: Nova Scotia Department of Mines and Energy, Mines and Minerals Branch, Report 88-1, p. 111-119.

——— 1989a, Fluid inclusion study of quartz vein polytypes in the Beaver Dam gold deposit, Meguma terrane, Nova Scotia, Canada: Pan-American Conference on Research on Fluid Inclusions (Second Biennial: PACROFI 2), v. 2, p. 38.

——— 1989b, Sulphur isotopic composition of sulphides from the Beaver Dam and other Meguma Group-hosted gold deposits, Nova Scotia: Canadian Journal of Earth Sciences: v. 26, p. 1617-1629.

——— 1993, An overview of Meguma gold deposits in the Meguma Terrane of southern Nova Scotia: Exploration and Mining Geology, v. 2, p. 418-421.

Kontak, D.J., Smith, P.K., Reynolds, P., and Taylor, K., 1990, Geological and $^{40}Ar/^{39}Ar$ geochronological constraints on the timing of quartz vein formation in Meguma Group lode-gold deposits, Nova Scotia: Atlantic Geology. v. 26, p. 201-227.

Kontak, D.J., Smith, P.K., Reynolds, P.H., 1993, Geology and $^{40}Ar/^{39}Ar$ geochronology of the Beaver Dam gold deposit, Meguma Terrane, Nova Scotia, Canada: Evidence for mineralization at 370 Ma: Economic Geology, v. 88, p. 139-170.

Kontak, D.J., Horne, R.J., Sandeman, J., Douglas, A., and Lee, J.K.W., 1998, $^{40}Ar/^{39}Ar$ dating of ribbon-textured veins and wallrock material from Meguma lode gold deposits, Nova Scotia: implications for timing and duration of vein formation in slate-belt-hosted vein gold deposits: Canadian Journal of Earth Sciences, v. 35, p. 746-761.

Malcolm, W., 1929, Gold Fields of Nova Scotia: Geological Survey of Canada, Memoir 156, 253 p.

——— 1976, Gold Fields of Nova Scotia: Geological Survey of Canada, Memoir 385 (reprint of Memoir 129 without plates or maps), 253 p.

Malcolm, W., and Faribault, E.R., 1912, Gold Fields of Nova Scotia: Geological Survey of Canada, Memoir 20-E, 331 p.

Mawer, C.K., 1985, Origin of bedding-concordant auriferous quartz veins, Meguma Terrane, Nova Scotia: Geological Survey of Canada, Unpub. Internal Report, 19 p.

——— 1986, The bedding-concordant gold-quartz veins of the Meguma Group, Nova Scotia, *in* Keppie, J.D., Boyle, R.W. and Haynes, S.J., eds., Turbidite-Hosted Gold Deposits: Geological Association of Canada, Special Paper 32, p. 135-148.

——— 1987, Mechanics of formation of gold-bearing quartz veins, Nova Scotia, Canada: Tectonophysics, v. 135, p. 99-119.

McBride, D.E., 1978, Geology of the Ecum Secum Area, Halifax and Guysborough Counties, Nova Scotia: Nova Scotia Department of Mines and Energy, Report 78-1, 12 p.

Morelli, R.M., Creaser, R.A., Selby, D., Kontak, D.J., and Horne, R,J., 2005, Rhenium-osmium geochronology of arsenopyrite in Meguma Group gold deposits, Meguma Terrane, Nova Scotia, Canada: Evidence for multiple gold-mineralizing events: Economic Geology, v. 100, p. 1229-1242.

Newhouse, W.H., 1936, A zonal gold distribution in Nova Scotia: Economic Geology, v. 31, p. 805-831.

O'Brien, B.H., 1985, The Formation of Veins in Greenschist Facies Rocks and the Early Deformation of the Meguma Group, Eastern Nova Scotia: Nova Scotia Department of Mines and Energy, Paper 85-2, 35 p.

Ramsay, J.G., 1980, The crack-seal mechanism of rock deformation: Nature, v. 284, p. 135-139.

Rickard, T.A., 1927, Gold mining in Nova Scotia, *in* Report on the Mines, 1926: Nova Scotia Department of Public Works and Mines, p. 163-177.

Ryan, R.J., and Smith, P.K., 1998, A review of the mesothermal gold deposits of the Meguma Group, Nova Scotia, Canada, *in* Ramsay, W.R.H., Bierlin, F.P., and Arne, D.C., eds., Mesothermal Gold Mineralization in Space and Time: Ore Geology Reviews, v. 13, p. 153-183.

Sangster, A.L., 1990, Metallogeny of the Meguma Terrane, Nova Scotia, *in* Sangster, A.L., ed., Mineral Deposit Studies in Nova Scotia, Volume 1: Geological Survey of Canada, Paper 90-8, p. 115-162.

——— 1992, Light stable isotope evidence for a metamorphogenic origin for bedding-parallel, gold-bearing veins in Cambrian flysch, Meguma Group, Nova Scotia, *in* Jonasson, I.R., ed., International Association on the Genesis of Ore Deposits, Eighth Quadrennial Symposium: Exploration and Mining Geology, v. 1, p. 69-79.

Schenk, P.E., 1995, Meguma Zone, Chapter 3, *in* Williams, E.H., ed., Geology of the Appalachian-Caledonian Orogen in Canada and Greenland: Geological Survey of Canada, Geology of Canada, no. 6, p. 261-277.

Selwyn, A.R.C., 1872, Notes and observations on the goldfields of Quebec and Nova Scotia, *in* Report of Progress, 1870-71: Geological Survey of Canada, p. 252-282.

Smith, P.K., 1983a, Stop 3-5: Cochrane Hill gold mine, *in* Gold Deposits of the Meguma Terrane of Nova Scotia: Canadian Institute of Mining and Metallurgy, CIM Geology Division Excursion Guidebook, p. 72-104.

——— 1983b, Geology of the Cochrane Hill gold deposit, Guysborough County, Nova Scotia, *in* Report of Activities 1982: Nova Scotia Department of Mines and Energy, Mines and Minerals Branch, Report 83-1, p. 225-256.

——— 1984, Geology and lithogeochemistry of the Cochrane Hill gold deposit - an indication of metalliferous source beds: Nova Scotia Department of Mines and Energy, Report 84-1, p. 203-214.

Smith, P.K., and Kontak, D.J., 1986, Meguma gold studies: Advances in geological insight as an aid to gold exploration: Tenth Annual Open House and Review of Activities, Program and Summaries, Nova Scotia Department of Mines and Energy, Information Series No. 12, p. 105-114.

Taylor, F.C., and Schiller, E.A., 1966, Metamorphism of the Meguma Group of Nova Scotia: Canadian Journal of Earth Sciences, v. 3, p. 959-974.

Williams, P.F., and Hy, C., 1990, Origin, deformational and metamorphic history of gold-bearing quartz veins on the Eastern Shore of Nova Scotia, *in* Sangster, A.L., ed., Mineral Deposit Studies in Nova Scotia, Volume 1: Geological Survey of Canada, Paper 90-8, p. 165-194.

Geological History, Mineral Occurrences and Mineral Potential of the Sedimentary Rocks of the Canadian Arctic Archipelago

Keith Dewing[1], Elizabeth Turner[2], and J.C. Harrison[1]

1. Geological Survey of Canada, 3303-33rd Street NW, Calgary, Alberta T2L 2A7

2. Department of Earth Sciences, Laurentian University, Sudbury, Ontario P3E 2C6

Corresponding author's email: kdewing@NRCan.gc.ca

Abstract

The Canadian Arctic Islands contain many areas that have a high mineral potential. Five geological assemblages are recognized: crystalline Archean to Paleoproterozoic basement; Mesoproterozoic sedimentary basins and volcanic rocks; Lower Paleozoic passive to convergent margin sedimentary rocks, Upper Paleozoic to Mesozoic rift basin sediments and volcanic rocks; and Tertiary passive margin sediments. To date, only two mines have been put into production in the islands: Paleozoic Mississippi Valley-type (MVT) carbonate-hosted Zn-Pb Polaris Mine on Little Cornwallis Island and Proterozoic MVT carbonate-hosted Zn-Pb Nanisivik Mine on northern Baffin Island. There is high mineral exploration potential around these past producers, as well as around known showings of Proterozoic-hosted redbed copper on Victoria Island and Paleozoic-hosted polymetallic (Zn-Pb-Ag-Au) veins on northern Ellesmere Island. There is also potential for as-yet-unfound i) Zn-Pb SEDEX deposits in the Proterozoic on northern Baffin Island and in the Paleozoic on Melville, Bathurst, and Ellesmere islands; ii) sediment-hosted uranium in the Proterozoic of northern Baffin Island, in the Paleozoic of Melville Island, and in Mesozoic and Tertiary of Banks and Prince Patrick islands; and iii) Noril'sk-style Ni-Cu-PGE deposits in the Proterozoicon Victoria Island and in the Mesozoic on Axel Heiberg Island.

The Canadian Arctic is logistically challenging, making it expensive to explore and develop mines. Mineral exploration is also hampered by a lack of local prospectors, large areas that are mapped in insufficient detail, and the lack of digitally available topography, bedrock and surficial mapping, airborne geophysics, and hyperspectral surveys.

Résumé

Dans les îles de l'Arctique canadien, un grand nombre de régions présentent un fort potentiel minéral. On reconnaît cinq assemblages géologiques : un socle cristallin de l'Archéen-Paléoprotérozoïque, des bassins sédimentaires et des roches volcaniques du Mésoprotérozoïque, des roches sédimentaires de cadre de marge passive à marge de convergence du Paléozoïque inférieur, des roches sédimentaires et des roches volcaniques de bassin de rift du Paléozoïque supérieur-Mésozoïque et des roches sédimentaires de marge passive du Tertiaire. Jusqu'à maintenant, seulement deux mines ont été mises an exploitation dans l'archipel : la mine Polaris, un gisement de Zn-Pb de type Mississippi-Valley dans des roches carbonatées du Paléozoïque, dans la Petite île Cornwallis et la mine Nanisivik, un autre gisement de Zn-Pb de type Mississippi-Valley dans des roches carbonatées du Protérozoïque, dans le nord de l'île de Baffin. Il existe un potentiel élevé pour l'exploration minière aux environs de ces anciennes exploitations ainsi qu'aux environs d'indices connus de cuivre dans des roches sédimentaires de type redbeds du Protérozoïque dans l'île Victoria et de filons à minéralisation polymétallique (Zn-Pb-Ag-Au) dans des roches du Paléozoïque dans le nord de l'île d'Ellesmere. Il existe des possibilités de gîtes non encore découverts des trois types ci-après : i) des gîtes sedex de Pb-Zn dans les roches protérozoïques du nord de l'île de Baffin et dans les roches paléozoïques des îles Melville, Bathurst et Ellesmere, ii) des gîtes d'uranium dans des roches sédimentaires du Protérozoïque du nord de l'île de Baffin, dans les roches paléozoïques de l'île Melville, et dans les roches du Mésozoïque et du Tertiaire des îles Banks et Prince Patrick et iii) des gîtes de Ni-Cu-ÉGP du genre Noril'sk dans les roches protérozoïques de l'île Victoria et dans les roches mésozoïques de l'île Axel-Heiberg.

L'Arctique canadien présente un défi logistique qui rend coûteuses l'exploration à la recherche de gisements et leur mise en exploitation. L'absence de prospecteurs locaux nuit en outre à l'exploration minière dans de grandes régions qui ne sont pas cartographiées de manière suffisamment détaillée, exploration que ne facilite pas non plus la non disponibilité sous forme numérique de cartes topographiques, des cartes géologiques du substratum rocheux ou des matériaux superficiels ainsi que l'inexistence de levés géophysiques et hyperspectraux aériens.

Introduction

The Arctic Archipelago includes a land area of 780,000 km^2 covering much of the Northwest Territories and Nunavut (Fig. 1). It contains three of the ten largest islands in the world (Baffin, Victoria, and Ellesmere) plus twenty-two others of appreciable size. Exploration for minerals in this vast area has been hampered by difficult logistics, short summer, high costs, and lack of local prospectors and infrastructure. To date, only two mines have been put into production in the islands: the Paleozoic carbonate-hosted Mississippi Valley-type (MVT) Zn-Pb Polaris Mine on Little Cornwallis Island and Proterozoic carbonate-hosted MVT Zn-Pb Nanisivik Mine on northern Baffin Island (Fig. 1). The Polaris District is covered in detail in another contribution in this volume (Dewing et al., 2006), and a description of the Nanisivik deposit is given herein.

There is high mineral exploration potential around both these past producers, as well as around known showings of Proterozoic-hosted redbed copper and banded iron formation iron ore, and Paleozoic-hosted polymetallic (Zn-Pb-Ag-Au) veins. There is also potential for undiscovered, Zn-Pb Sedimentary Exhalative (SEDEX), sediment-hosted uranium, and Noril'sk-style Ni-Cu-PGE deposits.

Dewing, K., Turner, E., and Harrison, J.C., 2007, Geological history, mineral occurrences and mineral potential of the sedimentary rocks of the Canadian Arctic Archipelago, *in* Goodfellow, W.D., ed., Mineral Deposits of Canada: A Synthesis of Major Deposit-Types, District Metallogeny, the Evolution of Geological Provinces, and Exploration Methods: Geological Association of Canada, Mineral Deposits Division, Special Publication No. 5, p. 733-753.

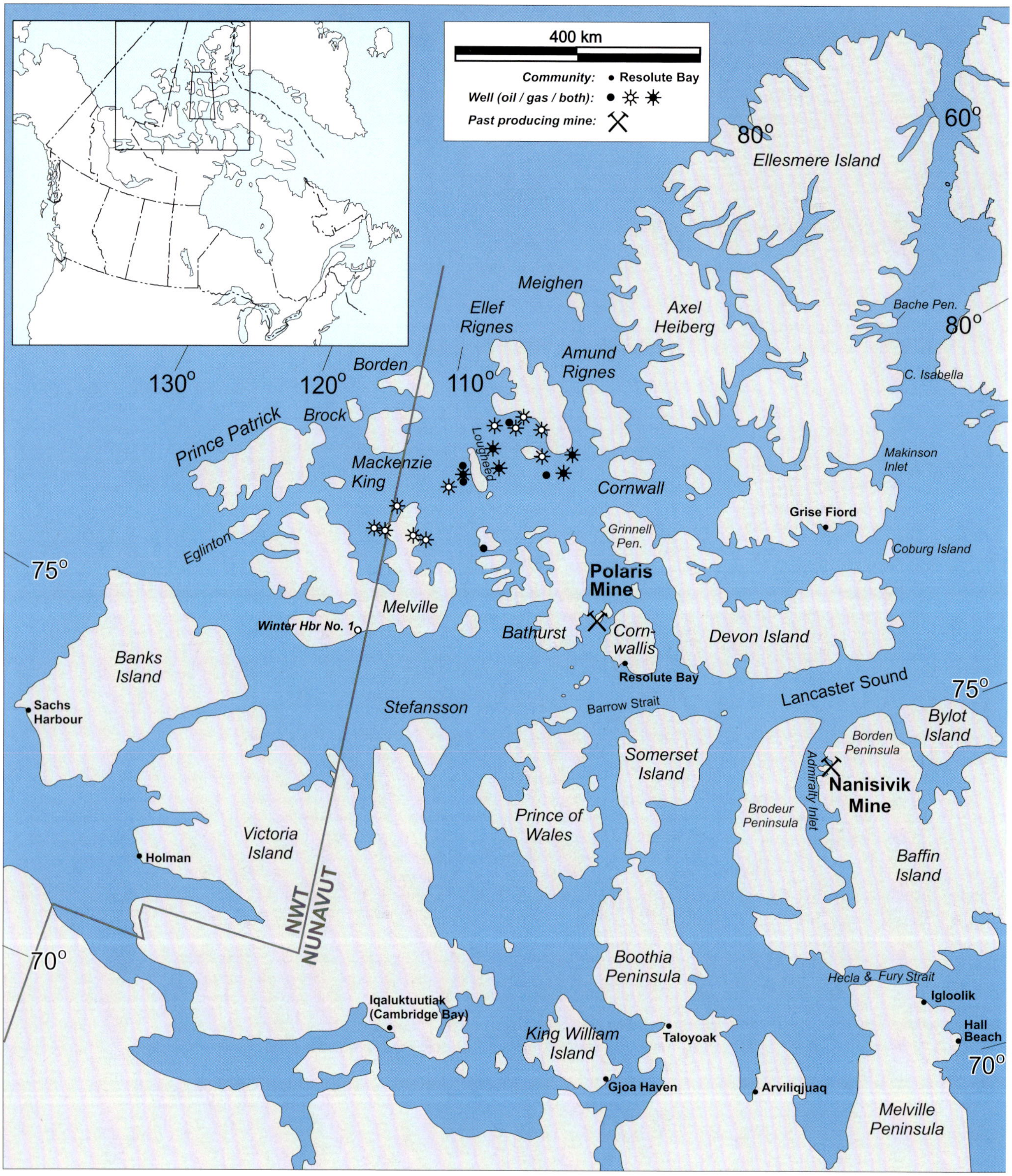

FIGURE 1. The Canadian Arctic Islands showing the location of past-producing Zn-Pb mines, oil and gas wells, communities, and the locations referred to in the text.

This contribution is arranged chronologically, starting with the oldest sedimentary strata exposed in the Arctic Islands and proceeding to the youngest. The geological setting for each basin is summarized, and known and likely mineral deposit-types are identified. The geological provinces are shown on Figure 2 and a time-stratigraphic correlation diagram (Fig. 3) ties the mineralizing events to the tectono-stratigraphic assemblages.

Crystalline Basement

The metamorphic grade of Paleoproterozoic and Archean rocks ranges from upper amphibolite to granulite facies

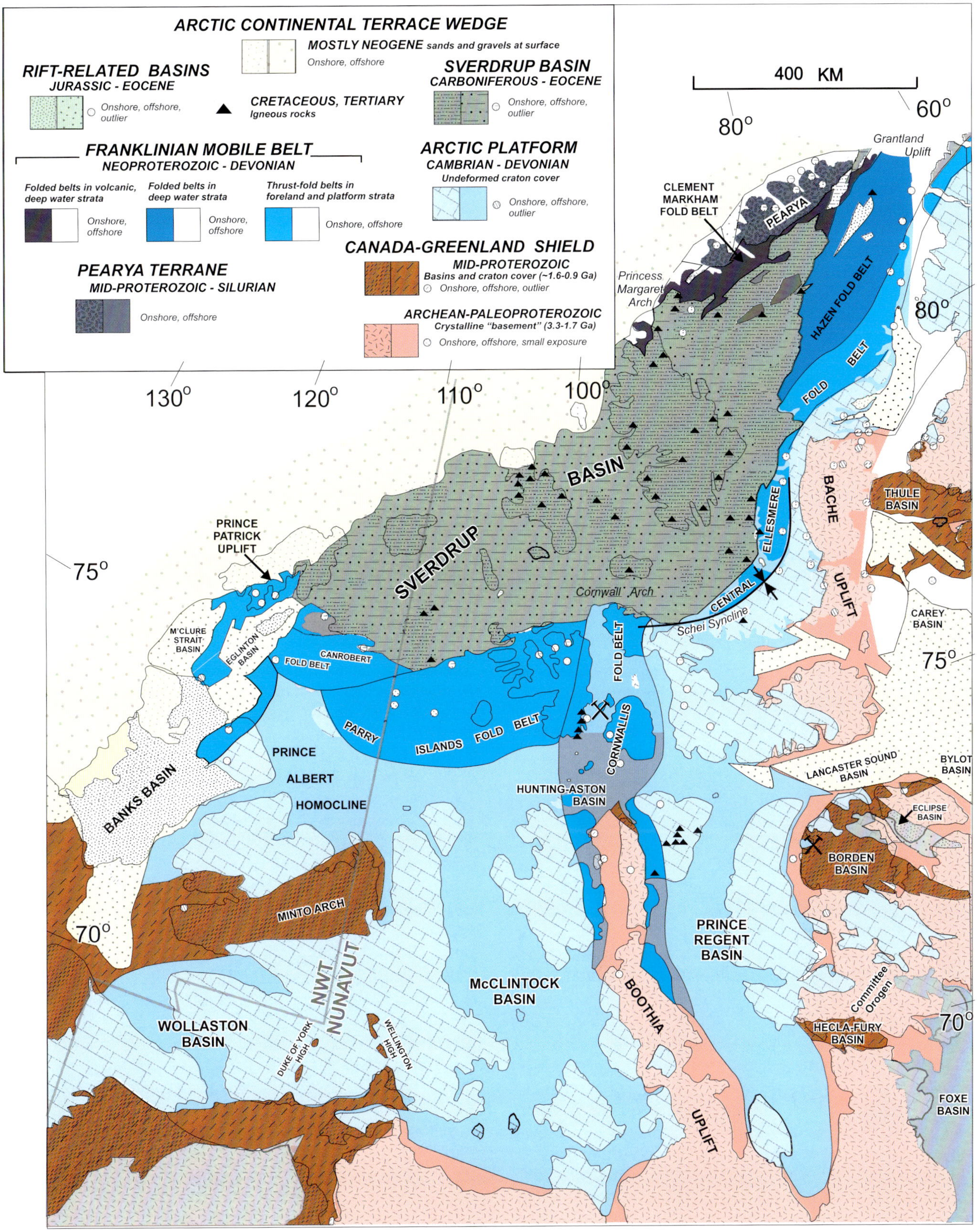

FIGURE 2. The major tectono-stratigraphic divisions of the Canadian Arctic Islands. Simplified from Okulitch (1991).

throughout the Arctic Islands. Paleoproterozoic regional metamorphism and tectonic events have affected all rocks to varying extent, and the nature and location of contained terrane boundaries is uncertain. Rocks as old as 3.3 Ga occur in the Victoria Fjord inlier in central North Greenland (Hansen et al., 1987), but there are very few reliable radiometric ages.

Mesoarchean and Neoarchean rocks are most widely represented in the "Committee Orogen", a northeasterly trending terrane that extends from northern Baffin Island and Bylot Island to the Thule and Inglefield Bay area on northwest Greenland (Figs. 2, 3, Jackson, 2000). Supracrustal units of the Committee Orogen include the Mary River Group of northern Baffin Island. Typical rocks in the Mary River Group include clastic metasedimentary rocks and mafic metavolcanic rocks, metamorphosed iron formation and meta-anorthositic rocks. More widespread are orthogneisses, felsic to intermediate metaplutonic rocks, and other gneisses of uncertain origin. Banded iron formation deposits in the Mary River Group are currently being assessed for their iron ore potential by Baffinland Iron Ore Mine Ltd. Radiometric ages are 2.72 to 2.76 Ga for the supracrustal rocks and 2.7 to 2.95 Ga for the associated gneisses (Jackson, 2000). Younger Archean ages have been found on southern Devon Island (Frisch, 1988) and, considered together with detrital zircon ages from Prudhoe Land (Dawes, 2004), indicate a belt of 2.2 to 2.6 Ga Neoarchean basement located northwest of the Committee Orogen.

The Committee Orogen is bordered to the south by intensely deformed metasedimentary rocks of the Paleoproterozoic Piling Group (Jackson, 2000). These units hold the potential for metamorphosed carbonate-hosted Zn deposits in the Flint Lake Formation and SEDEX Zn-Pb deposits in the graphitic pelites of the Astarte River Formation (Scott et al., 2003). The Black Angel deposit of west Greenland (Pederson, 1980) is hosted in evaporite-bearing marbles of similar age to the Piling Group.

Jackson (2000) followed the lead of Hoffman (1989) in assigning the crystalline rocks of much of eastern Devon Island, southeastern Ellesmere Island, and Inglefield Land to the Thelon Tectonic Zone. The Greenland nomenclature refers to these rocks as "Inglefield mobile belt" (Dawes et al., 2000; Dawes, 2004; Dawes and Garde, 2004). Four distinct lithofacies belts are recognized containing varying proportions of supracrustal rocks (Etah Group of Dawes, 2004) and granitoid intrusives (Etah meta-igneous complex of Dawes et al., 2000). Subsequent retrogressive regional metamorphism is documented, for example, along the western edge of the shield on southeastern Ellesmere Island (Frisch, 1988).

From northwest to southeast, there are four recognizable facies belts of the Inglefield mobile belt: 1) a granitoid magmatic arc with granulite-grade aluminous metasedimentary rocks; 2) a belt of aluminous metasedimentary rocks, calcitic marble, sulphidic paragneiss, and granitoid orthogneiss; 3) aluminous metasedimentary rocks, calcitic and dolomitic marble, quartzite, and granitoid orthogneiss; and 4) aluminous metasedimentary rocks, granitoid orthogneiss, and minor quartzite. The supracrustal rocks of the Inglefield mobile belt were assigned to the Etah Group by Dawes (2004).

The magmatic arc is exposed from the south coast of Ellesmere Island to northern Inglefield Land via western Makinson Inlet and western Bache Peninsula. Igneous compositions range from quartz diorite and monzodiorite to granite and syenite. There are also metamorphosed mafic dykes. Cordierite, sillimanite, and garnet are common in the granitoid rocks, and sillimanite gneiss of aluminous sedimentary origin is extensive and gradationally interleaved with the plutonic rocks (Harrison, 1984; Frisch, 1988; Dawes, 2004). Granites that are interleaved with Etah Group supracrustals have SHRIMP U-Pb ages of 1.95 to 1.91 Ga in Inglefield Land (Nutman et al., reported in Dawes, 2004), and radiometric ages of 1.91 to 1.97 Ga on southeastern Ellesmere Island (Frisch, 1988). A younger phase of granites and migmatite in Inglefield Land has SHRIMP U-Pb ages of 1.75 to 1.74 Ga (Nutman et al., reported in Dawes, 2004).

Smaller granitoid bodies occur throughout the other three supracrustal belts. The Etah Group belt of calcitic marble with aluminous, sulphidic, and graphitic paragneiss extends from the mouth of Makinson Inlet to northern Inglefield Land via eastern Bache Peninsula. The sulphidic metasedimentary rocks form locally spectacular gossans that contain podiform, pyrrhotite-rich massive sulphide with less common copper and nickel sulphide (Harrison, 1984; Frisch, 1988; Dawes, 2004). Detrital zircons from paragneiss lenses in coastal northern Inglefield Land yield SHRIMP U-Pb ages of 2.0 to 1.98 Ga (Nutman et al., reported in Dawes, 2004).

Southward, the Tah Group belt with sulphidic metasedimentary rocks gives way to Etah Group quartzite, calcitic ± dolomitic marble, calcsilicate, and aluminous paragneiss. The line of "facies change" coincides with the mouth of the Makinson Inlet (Harrison, 1984; Frisch, 1988). Quartzite is common with marble southward throughout southeasternmost Ellesmere Island, Coburg Island (Harrison, 1984; Frisch, 1988), and on northern Devon Island (Harrison, 1984). Marble with minor quartzite (but no sulphidic metasediment) also occurs in a distinct belt extending from Cape Isabella on coastal eastern Ellesmere Island to Sunrise Pynt in northwest Greenland; thence northeastward through southern and central Inglefield Land (Frisch, 1988; Dawes, 2004). Detrital zircons from an Etah Group quartzite in southern Inglefield Land yield SHRIMP U-Pb ages of 3.25 to 2.35 Ga with a dominant population in the 2.6 to 2.45 Ga range (Nutman et al., reported in Dawes, 2004).

The fourth and last of the facies belts of the Inglefield mobile belt, featuring aluminous metasedimentary rocks with minor quartzite, lies throughout much of southern Devon Island and in southern Inglefield Land and Prudhoe Land. It has been suggested that these quartzitic strata represent the lowest Paleoproterozoic metasedimentary rocks preserved above a basement complex (Harrison, 1984; Dawes et al., 2000), distinguished as "Prudhoe Land granulite complex" in Prudhoe Land (Dawes, 2004). A SHRIMP U-Pb dating of Prudhoe Land complex igneous protolith yielded an age of 1.98 Ga, and associated quartzite in the Thule area contains detrital zircons in the 3.2 to 2.2 Ga age range (Nutman et al., reported in Dawes, 2004).

Assuming that the Prudhoe Land granulite complex is the basement below Etah Group, then deposition of these supracrustals would appear to be from 1.98 to 1.92 Ga, simi-

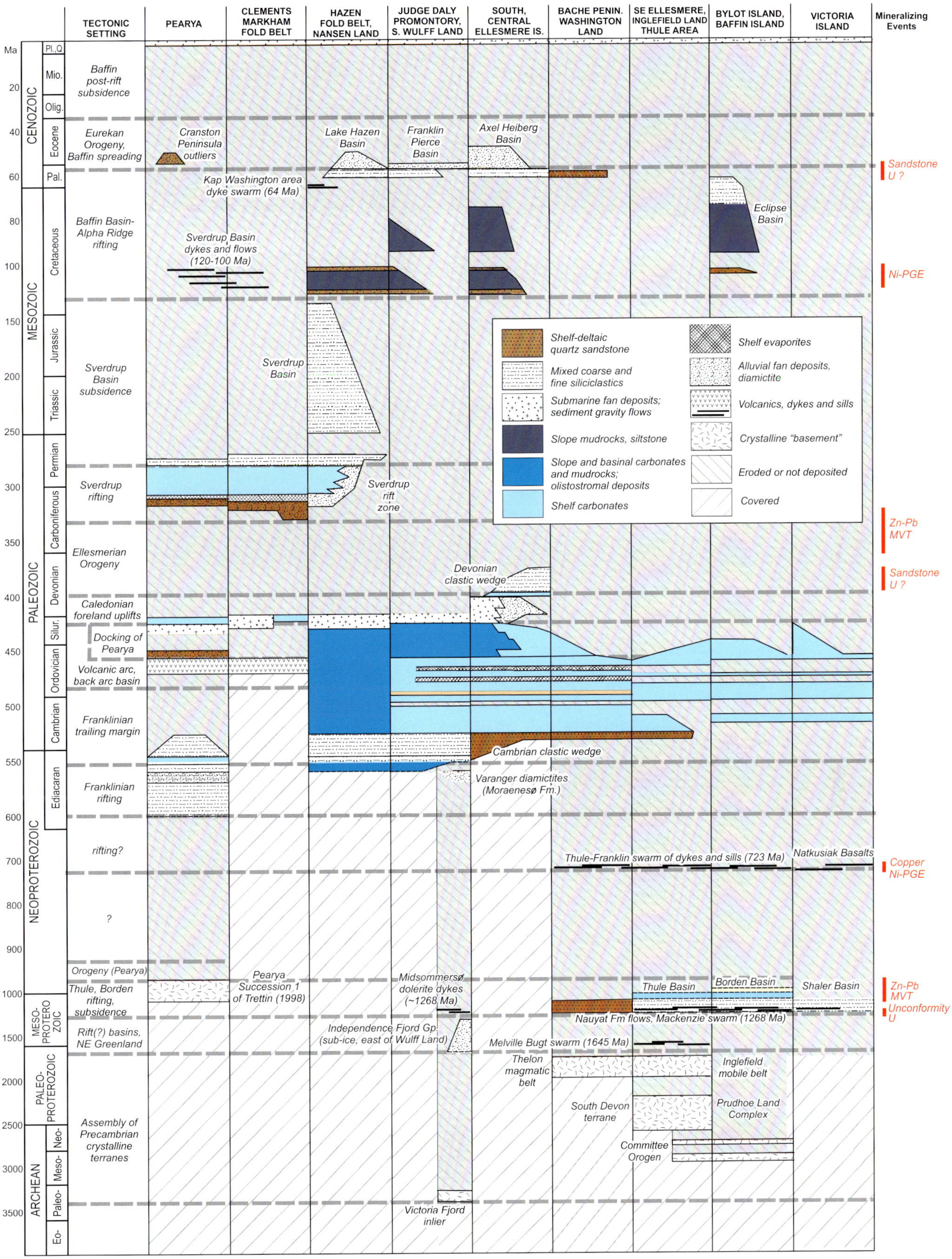

FIGURE 3. The temporal relationship of the major tectono-stratigraphic successions in the Arctic Islands and the timing of mineralizing events.

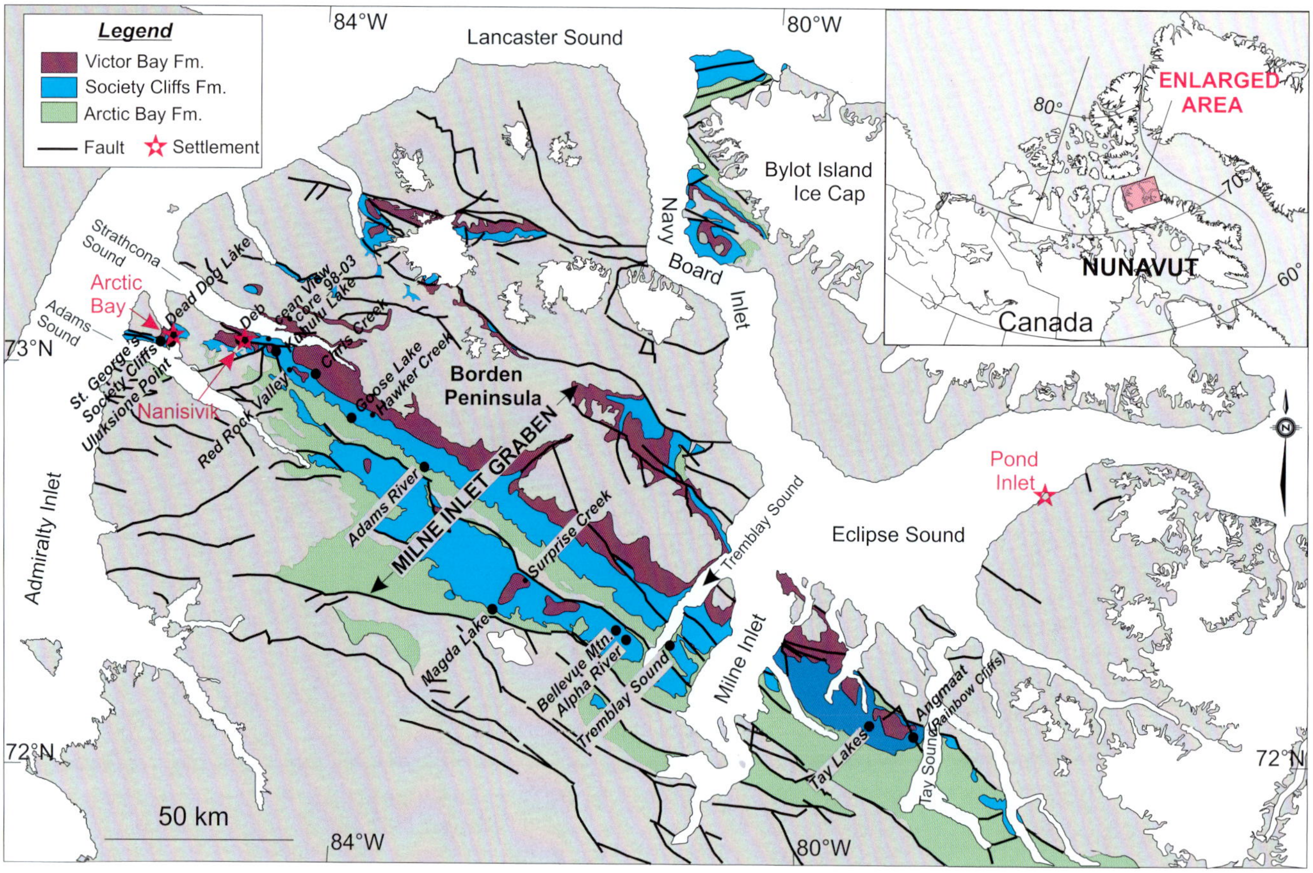

FIGURE 4. Location map of Baffin Island, the Mesoproterozoic Bylot Basins of Nunavut, and the Milne Inlet Graben and its base metal showings.

lar to the Piling Group of central Baffin Island, which Jackson (2000) considers to have accumulated at about 1.9 Ga.

The lower age limit for deformation associated with the Inglefield mobile belt within the map area is provided by dolerite dykes of the Melville Bugt swarm with a U-Pb baddeleyite age of 1.645 Ga (reported by Mike A. Hamilton in Dawes, 2004).

Proterozoic Basins

Unmetamorphosed and virtually undeformed Mesoproterozoic strata are preserved in several disparate areas of the central and eastern Canadian Arctic islands (Fig. 1). Basins on northern and northwestern Baffin Island, Somerset Island, and eastern Ellesmere Island (and nearby parts of Greenland) have been broadly correlated, and likely represent remnants of a formerly contiguous continental margin that was undergoing rifting during sediment accumulation (Jackson and Iannelli, 1981). It is possible that other Mesoproterozoic basin remnants, or continuations of known basins, are present beneath younger rock cover elsewhere in the islands (e.g. Mayr et al., 2004). All of these basins are considered to be approximately 1.2 Ga, although published radiometric dates are sparse. Faults that accommodated rifting in some of these basins have been episodically reactivated since the Mesoproterozoic, and some may be implicated in the development of structures in overlying Paleozoic rocks.

Borden Basin – Baffin Island

The Borden Basin is an aulacogen comprising three distinct, syndepositional grabens that cover an area approximately 200 km x 300 km on and near the Borden Peninsula (Baffin Island) and Bylot Island. The Milne Inlet Graben is the largest of these grabens, and is exposed as a 300 km long, southeast-tapering wedge (Fig. 4). Small inliers of Borden Basin rocks are also present beneath basal Paleozoic strata on the east side of Brodeur Peninsula (Scott and de Kemp, 1998) and on southern Devon Island (Thorsteinsson and Mayr, 1987). The Borden Basin contains widespread carbonate-hosted Pb+Zn±Cu±Ag mineralization, including the Nanisivik Mine orebody.

The 6 km thick Bylot Supergroup of the Borden Basin accumulated over Archean metamorphic basement of the Rae craton in grabens that subsided along a series of northwest-trending normal faults (Jackson and Iannelli, 1981). It is divided into three groups (Fig. 5): 1) the Eqalulik Group, recording rifting (Nauyat Formation basalts and minor associated sedimentary rocks) and subsequent deposition of fluvial to marginal marine, mature quartz sand (Adams Sound Formation), and shallow marine to basinal shale (Arctic Bay Formation); 2) the Uluksan Group, consisting of the Society Cliffs and Victor Bay formations. Diverse peritidal to basinal dolostones of the Society Cliffs Formation, host to most of the base-metal mineralization in the district, were subaerially eroded in most locations prior to deposition of Victor Bay Formation shale and carbonate rocks. Overlying the Victor Bay Formation is 3) a succession of alluvial to marine

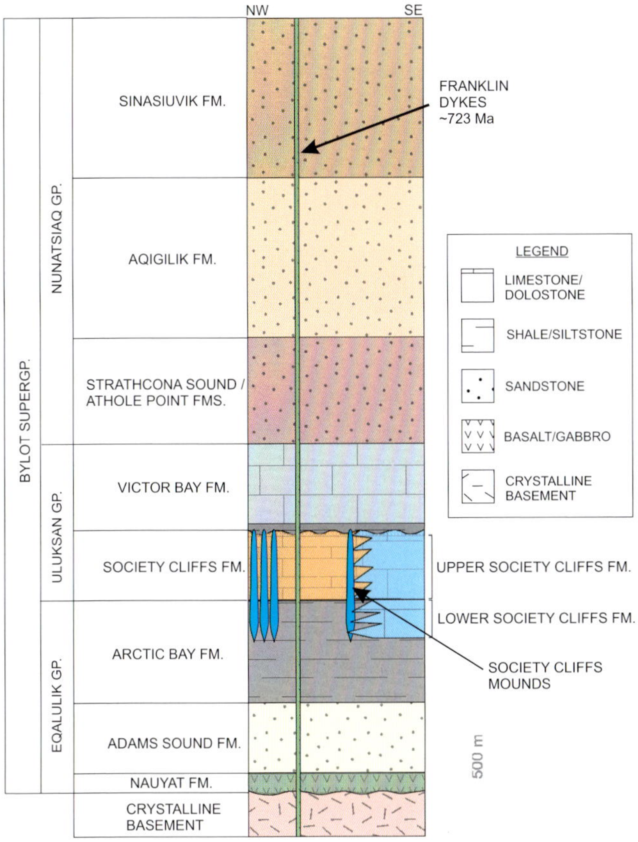

FIGURE 5. Stratigraphy of the Bylot Supergroup, northern Baffin Island.

terrigenous sandstones and carbonate lithoclast conglomerates (Athole Point, Strathcona Sound, Aqigilik, and Sinasiuvik formations belonging to the Nunatsiau Group). These poorly understood formations have been inferred to record renewed tectonic uplift of basement blocks within and adjacent to the basin. Lithofacies distributions and paleocurrents in most formations indicate overall basin deepening (as opposed to intragraben deepening) to the west or northwest, consistent with the inferred presence of an ocean to the northwest (Jackson and Iannelli, 1981) and with the known distribution of cratonic basement. The present-day orientation of this inferred Mesoproterozoic paleocontinental margin is, therefore, roughly the same as that of the Phanerozoic continental margin; both were constrained by a northeast-trending (present-day coordinates) limit to the cratonic basement.

Thus the early rifting (Nauyat and Adams Sound formations) is now understood to have been followed by a protracted interval of episodic extensional tectonism (Arctic Bay, Society Cliffs, and Victor Bay formations, Turner, 2003a, 2004a) and a final stage of basin inversion associated with distal compressive stress generated in a foreland fold-thrust belt affecting the Coppermine River Group, some 1000 km to the (present-day) west of the Borden Basin, and bracketed between 1.27 Ga and deposition of the Shaler Group at about 1.0 Ga (Hildebrand and Baragar, 1991; Sherman et al., 2002). The Bylot Supergroup is crosscut by Franklin gabbro dykes (ca. 723 Ma, Pehrsson and Buchan, 1999), and unconformably overlain by flat-lying lower Paleozoic strata.

The Nanisivik Mine and Other Mineral Showings

Numerous Pb-Zn showings are located on the Borden Peninsula and nearby areas on northern Baffin Island (Fig. 1). The Nanisivik Mine (operation 1976-2002, located at approximately 73°03'N, 84°30'W) produced approximately 17.9 million metric tonnes of ore grading approximately 9.0% Zn, 0.7% Pb, and 35 g/t Ag. The orebody and associated showings are loosely classified as MVT deposits, although anomalously high temperatures and very high Zn:Pb ratios, along with an enormous amount of pyrite (100 Mt), indicate an unusual mode of formation (Sutherland and Dumka, 1995; St. Marie et al., 2001).

Stratigraphic relations among the units relevant to mineralization throughout the graben are considerably more complex than indicated by the stratigraphic nomenclature and early correlations (Turner, 2003a). This is meaningful for prospecting for base metals because there is a significant amount of stratigraphic and lithofacies control on the disposition of mineralization throughout the graben. Similarly, contacts between and lateral correlations among units have been revised, changes that may be extremely important to mineralization models, particularly if mineralization is synsedimentary or in any way karst-related.

The Nanisivik orebody and almost all associated showings are hosted by the Society Cliffs Formation (Figs. 3, 4). The Society Cliffs Formation is a dolostone with great lithostratigraphic and geographic facies variation, and spans the entire strike length of the exposed Milne Inlet Graben. In the westernmost two-thirds of the Milne Inlet Graben, the host dolostone is a deep-water laminite locally punctuated by giant deep-water carbonate mounds (Turner, 2004a,b,c). Stromatolites, microbial laminites, and shallow-water to peritidal features are completely absent. This observation contrasts with the content of previous papers in which the laminated Nanisivik host rock was inferred to be of peritidal or sabkha origin (e.g. Geldsetzer, 1973a,b; Jackson and Iannelli, 1981; Ghazban et al., 1992). Although freshly broken laminite rock surfaces have a bituminous odour, rock-evaluation analysis shows that the rock contains little organic carbon (M. Fowler, pers comm., 2004).

In the eastern one-third of the graben, the formation consists of nonbituminous peritidal cycles, each on the order of ~10 to 50 m thick (Kah, 1997; Kah et al., 1999). Between the two is a platform-margin to outer-ramp area. The formation is in excess of 1000 m thick in the east and approximately 250 m thick in the west (Turner, 2004c), where an unknown but possibly significant amount of post-depositional erosion, together with a possibly low primary accumulation rate of the deep-water laminite resulted in a comparatively thin succession. A depositional age of ca. 1199 Ma for the Society Cliffs Formation is provided by Pb-Pb dating on dolostone (Kah, as cited in Samuelsson et al., 1999).

Normal faults, along which the grabens accommodated subsidence, have been repeatedly reactivated since the Mesoproterozoic, but significant tectonic events in the interval since deposition of the Society Cliffs Formation are limited to a poorly known east-directed compressional event that coincided with the latter part of basin-filling at the end of deposition of the Victor Bay Formation and throughout deposition of Nunatsiaq Group (Sherman et al., 2002). The

Mesoproterozoic strata in the Milne Inlet Graben dip gently northeast (~10-20°) and are generally undeformed except where they have been drag-folded near reactivated normal faults (cf. Scott and de Kemp, 1998), compactionally domed over deep-water mounds (Turner, 2004c), or possibly subtly folded by Mesoproterozoic tectonism (Sherman et al., 2002). Normal faults are common and are particularly pervasive in the Nanisivik area (cf. Scott and de Kemp, 1998), where they deviate slightly from the dominant northwest fault trend.

The main ore zone at Nanisivik is entirely enclosed by laminated dolostone and deep-water mound facies of the middle to upper Society Cliffs Formation (Turner, 2004c). This zone, along with other iron sulphide bodies on the mine property and areas lying to the east, are located in the apices of subtle compaction-related structural highs (Fig. 6; Clayton and Thorpe, 1982; Patterson et al., 2003). The orebody is in a narrow horst within a broader zone of predominantly east-trending high-angle normal faults. The orebody is elongate, with an east-west length of approximately 3 km and width of approximately 100 m (Fig. 6, Clayton and Thorpe, 1982; Sutherland and Dumka, 1995). The lower part of the orebody forms a narrow keel some 65 m deep and 5 to 30 m wide that follows a stockwork of faults and fractures (Clayton and Thorpe, 1982, 1982; Sutherland and Dumka, 1995). The uppermost part of the orebody expands into a tabular mass 100-200 m wide and 10 to 30 m thick. The upper surface of the orebody is strikingly horizontal, varying in present-day elevation by no more than several metres over its length, and crosscuts the gently north-dipping layering of the host rock (Fig. 6). Ore and host are crosscut by a diabase dyke (the "mine dyke") presumed to be of Franklin age (723 Ma), as are all other gabbro dykes in the Borden Basin (Heaman et al., 1992; Pehrsson and Buchan, 1999).

The main-zone sulphide body is largely massive, barren pyrite with minor dolospar gangue, but with

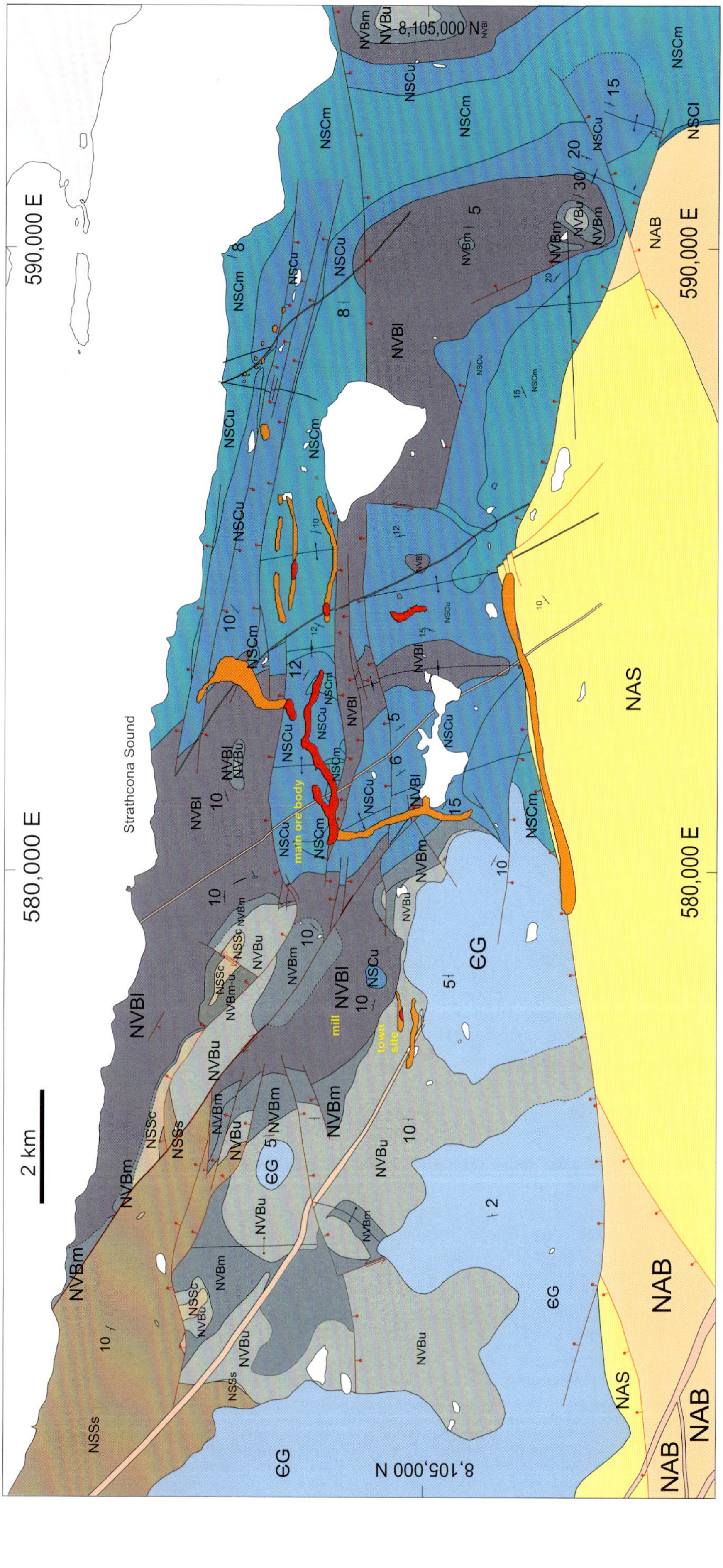

FIGURE 6 Map of Nanisivik Mine area, northern Baffin Island (see Fig. 4 for location). Geology after Patterson et al. (2003).

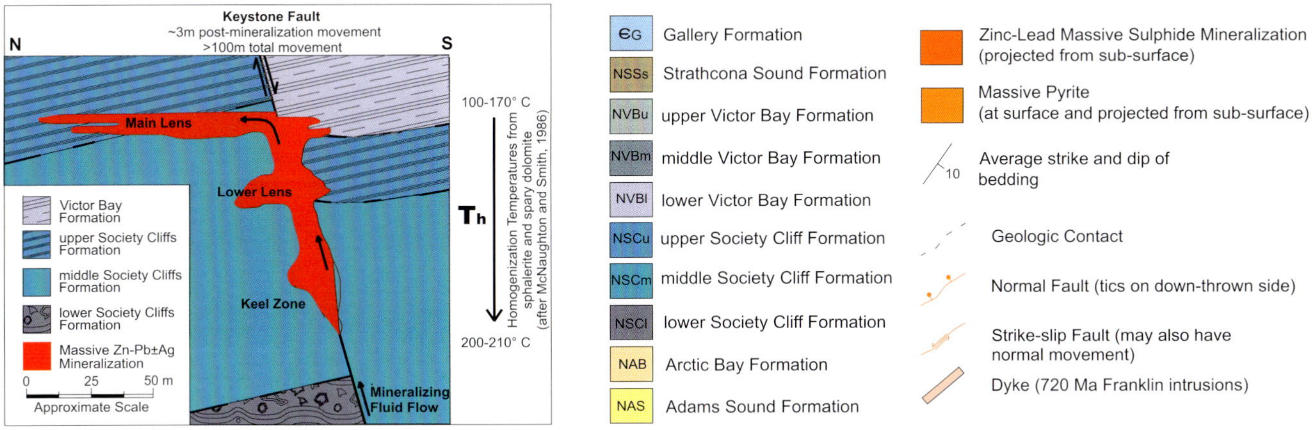

FIGURE 6 CONTINUED. Schematic cross-section through the Nanisivik Main Lens and Keel Zone. Note association of Keystone Fault and Keel Zone mineralization. Homogenization temperatures are from fluid inclusions trapped within sphalerite and sparry dolomite (data from McNaughton and Smith, 1986).

sphalerite-rich zones containing galena, pyrite, and dolospar in the lower part. A succession of mineralization pulses has been identified using crosscutting relationships (Arne et al., 1991). Ore textures range from massive to centimetrically banded (Fig. 7) and are generally coarsely crystalline.

In addition to the main ore zone, other subeconomic sulphide bodies are present on the mine property. All are elongate and spatially associated with predominantly east-trending normal faults.

Nanisivik: Mode of Formation

Nanisivik has been classified as an MVT deposit. Field-based models of sulphide bodies in the Nanisivik area focused on early void-filling mineralization in porosity formed by meteoric karstification (Clayton and Thorpe, 1982; Olson, 1984). This interpretation was based on the inferred cave-like external shape of the orebody, a known depositional hiatus and erosion interval post-dating deposition of the Society Cliffs Formation, and the absence of obvious replacement textures in the ore. Later work, however, identified ore textures ("diagenetic crystallization rhythmites", Arne and Kissin, 1989) characteristic of progressive replacement of host rock under pulsating fluid influx that only initially exploited small karstic fluid conduits (Arne and Kissin, 1989; Arne et al., 1991). Ore banding and orebody geometry was then reinterpreted as the result of mineral precipitation at the interface between a gas cap (H_2S and CH_4) trapped in an antiform that was capped by Victor Bay Formation shale, and oxidized metalliferous brine, within progressively dissolving host dolostone (Arne et al., 1991; Sutherland and Dumka, 1995). Anomalous metal ratios and unusually high ore-precipitation temperatures (>200°, McNaughton and Smith, 1986; Arne et al., 1991) suggest that the Nanisivik orebody might be more closely related to SEDEX-type rather than MVT deposits. Fluid-inclusion data indicate that ore emplacement took place at ≥1600 m burial depth (McNaughton and Smith, 1986), yielding an oldest possible depositional time coeval with deposition of the middle and upper Nunatsiaq Group.

Nanisivik: Age of Mineralization

Sphalerite Rb-Sr dating of Nanisivik ore yielded an age of ca. 1.1 Ga (Christensen et al., 1993). Paleomagnetic data from hydrothermally altered dolostone around the Nanisivik orebody suggest a mineralization age of ca. 1095 Ma (Symons et al., 2000), although there is some dispute as to the meaning of the paleomagnetic data in light of the gas-cap mineralization model of Arne et al. (1991) and Sutherland and Dumka (1995). A Mesoproterozoic mineralization age is consistent with the timing of subtle compressive tectonism that began after deposition of the Victor Bay Formation and continued through deposition of the >1500 m thick Nunatsiaq Group This event reversed the basin's formerly west-deepening polarity (Sherman et al., 2002) and filled it with southwesterly derived, texturally immature sediment (Knight and Jackson, 1994).

In contrast, feldspathic (adularia) alteration along the margins of a gabbro dyke that exhibits complex crosscutting relations with the orebody, host strata, and a normal fault, has been dated at ca. 461 Ma (Middle Ordovician), which is presumed by Sherlock et al. (2004) to be the age of mineralization. The validity of this conclusion is challenged by fluid-inclusion data (McNaughton and Smith, 1986) showing strikingly elevated fluid-inclusion homogenization temperatures in the immediate vicinity of the mine dyke. These observations are consistent with dyke emplacement into a structure that crosscut the orebody and concomitant ore recrystallization in the vicinity of the dyke, rather than 500 m.y. later ore emplacement around a pre-existing dyke.

There is only limited evidence for any tectonic or fluid activity in the Arctic during the Middle Ordovician. The thick, monotonous, shallow-water Ordovician carbonate succession was interrupted by evaporite-filled intraplatformal basins in the Early Ordovician, a widespread mid-Ordovician disconformity over the southern part of the Arctic Islands, and evidence of folding and faulting of the succession below the mid-Ordovician unconformity on Prince of Wales Island (section A-A', Figure 9 of Mayr et al., 2004). This means that there may have been driving forces or heat for Ordovician mineralizing fluids as proposed by Sherlock et al. (2004). Although poorly known, Ordovician, subduction-related volcanism occurs in the Clements Markham fold belt on northern Ellesmere Island (Trettin, 1987; Trettin et al., 1987). The onset of a subduction beneath North America may have had some tectonic effects such as the creation of a subtle back-arc setting and peripheral bulge.

FIGURE 7. Zn-Pb showings, northeast Ellesmere Island. **(A)** Sphalerite in dolostone from Great Plains Ltd. Zn-Pb prospect on Judge Daly Promontory, northeast Ellesmere Island (sample GSC C-412161b). **(B)** Dolomitized Scoresby Bay Formation near Carl Ritter Bay, northeast Ellesmere Island (sample GSC C-412151). **(C)** Galena and sphalerite from Great Plains Ltd. Zn-Pb prospect on Judge Daly Promontory, northeast Ellesmere Island (sample GSC C-412161a). **(D)** Crackle breccia with sphalerite from Great Plains Ltd. Zn-Pb prospect on Judge Daly Promontory, northeast Ellesmere Island.

Base metal showings have not yet been found in Paleozoic rocks on Baffin Island.

Milne Inlet Graben – Base Metal Mineralization on Northern Baffin Island

Regional studies of mineralization throughout the Milne Inlet Graben indicate that although mineralization is strongly fault-related, stratigraphic and lithofacies controls commonly constrain the location and geometry of ore bodies (Turner, 2003b, 2004c). Although there are mineralogical differences among the showings (specifically, the amount of sphalerite, presence/absence of both Cu minerals and fluorite, and amount of Fe minerals), work on regional trends in metal ratios, trace elements, temperature, and fluid composition has hitherto been minimal, and thus regional patterns in fluid flow and sulphur/metal source remain unrecognized. Economic mineralization potential in the area has been summarized by Sangster (1998; 1999), who catalogued the mineral occurrences as well as discussed the potential for SEDEX, MVT, and sandstone-hosted Cu-Pb-(Zn) on northern Baffin Island.

Little is publicly known about the controls on mineralization throughout the Milne Inlet Graben outside of assessment files submitted by PetroCanada in the 1980s and Cominco in the late 1990s (e.g. Harrison, 1984). New exploration strategies could take advantage of 1) the recognition that stratigraphy and lithofacies are important to geographic location, stratigraphic position, and geometry of sulphide bodies in the Milne Inlet Graben; 2) enormous advances in the understanding of Precambrian carbonate rocks, such that the host rocks to mineralization in the Milne Inlet Graben and their contribution to mineralization are now much better understood; and 3) the possibility that more detailed exploration could locate SEDEX-type mineralization in the Milne Inlet Graben that was missed by industry reconnaissance exploration.

Hunting-Aston Basin – Somerset and Prince of Wales Islands

The Hunting-Aston Basin is 300 km northwest of the westernmost exposures of the Borden Basin (Fig. 2). The Hunting and Aston formations are exposed as a small wedge (12 km x 45 km) on northwestern Somerset Island and in outliers on eastern Prince of Wales Island and nearby small islands. Strata are northeast-dipping, and are overlain to the northeast by Paleozoic rocks. The ~1400 m thick Aston Formation lies unconformably on basement rocks and consists of conglomerate, quartz arenite to arkose, with minor siltstone and dolostone (Tuke et al., 1966; Dixon et al., 1971; Reinson et al., 1976; Mayr et al., 2004). The Aston Formation is intruded by diabase sills on Prince of Wales Island that yield a U-Pb date of 1268 Ma (Mayr et al., 2004). The preserved portion of the overlying Hunting Formation is probably about 1100 m thick, although a complete section has never been measured owing to complications presented by normal faults. Estimates of thickness from seismic sections on Prince of Wales indicate up to 6200 m (Mayr et al., 2004). The Hunting Formation consists of three members (Dixon, 1974): 1) a lower <100 m thick member of mixed terrigenous mudstone and dolomudstone; 2) a middle member at least 300 m thick consisting of peritidal dolostone, including silicified seafloor dendrites and other abiogenic precipitates; and 3) an unnamed upper member, approximately 700 m thick and dominated by pink-weathering dolostones in part consisting of supratidal tepee cycles, which, in the upper 200 m, become quartz-sandy and exhibit thick, isopachous, fibrous cements that fill fracture systems that are both bedding-parallel and at high angles to bedding (Turner, unpublished data). The formation is overlain by lower Paleozoic strata.

Hunting Formation member 2 lithofacies are strikingly similar to peritidal dolostones of the upper member of the Society Cliffs Formation in the southeast Milne Inlet Graben. This, together with the approximate colinearity and northwest structural trend of the Borden Basin grabens and Aston-Hunting basin, and the known distribution of cratonic basement, lends support to the idea that they are part of a formerly contiguous, approximately northeast-trending, locally rifted continental margin (Jackson and Iannelli, 1981). Paleocurrent directions in the Aston Formation, extending north and west from the southern part of the Boothia Uplift, suggest that the uplift was an active high in the Mesoproterozoic.

If the Hunting-Aston Basin is contemporaneous and colinear with the Borden Basin, then potential for Nanisivik-style deposits exists in the Hunting Formation as well as below Paleozoic cover on Somerset and Prince of Wales islands and the Brodeur Peninsula of northern Baffin Island. The lack of carbonate mounds, the small areal extent, and the lack of mineral showings, however, reduces the attractiveness of the Huntington Formation for Nanisivik-style deposits.

Fury and Hecla Basin – Baffin Island

The Fury and Hecla basin is exposed in an area 85 km x 15 km on northwestern Baffin Island, on nearby small islands to the south in Fury and Hecla Strait, and as small outliers on the northern margin of Melville Peninsula (Chandler, 1988). Approximately 6000 m of gently south-dipping strata lie unconformably on granitic basement, and comprise sandstones and basalts of the Nyeboe, Sikosak, and Hansen formations (approximately 650 m total), sandstone, shale, and minor dolostone of the Agu Bay Formation (approximately. 600 m), Whyte Inlet Formation quartz arenite (to ~3000 m), and sandstone, siltstone, and shale of the Autridge Formation (~2000 m). The succession is loosely correlated lithostratigraphically with the lower part of the Bylot Supergroup, 200 km to the north.

Uraniferous granites occur in the basement that lies unconformably below the Fury and Hecla Basin (Ciesielski and Maley, 1980; Maurice, 1982). Five occurrences of uranium mineralization are adjacent to faults that separate granites and pegmatites from redbeds of the Nyeboe Formation (Ciesielski and Maley, 1980). Maurice (1982) demonstrated that uranium mineralization post-dates deposition of the Proterozoic strata and may be unconformity related.

Strata in the Fury and Hecla Basin, which are inferred to be correlative to the mineralized Society Cliffs Formation in the Borden Basin, are mainly clastic with only a small carbonate component (Chandler, 1988), limiting the potential for Nanisivik-style Zn-Pb mineralization.

Thule Basin – Northeast Ellesmere Island

The Thule Basin is exposed on northwest Greenland, and as small remnants (10-20 km maximum diameter) on southeastern Ellesmere Island (Fig. 1) (Frisch and Christie, 1982; Jackson, 1986; Dawes, 1997). There, basement rocks are overlain by volcanic and terrigenous rocks of the Wolstenholme Formation (approximately 1100 m). On the Greenland side of Nares strait, terrigenous strata of the Dundas Formation and overlying dolostones of the Narssârssuk Formation are also present. The succession has been lithostratigraphically correlated with the Nauyat to Victor Bay formations in the Borden Basin, 350 km to the south (Jackson, 1986). Only small parts of the Thule Basin are exposed on the Canadian side of Nares Strait and Davis Strait and no mineral occurrences have been noted.

Shaler Supergroup – Victoria Island

The Neoproterozoic Shaler Group on Victoria Island is exposed in a structurally simple uplift called the Minto Arch (Fig. 2). The Shaler Supergroup consists of up to 4 km of shallow water clastic, carbonate, shale, and evaporitic rocks (Rainbird et al., 1994). These are divided, in ascending order, into the Rae Group, Reynold's Point Group, Minto

Inlet, Wynniatt, and Killian formations, and were deposited on an epicontinental shallow-marine basin in an arid environment.

The units with the greatest exploration potential are the Rae and Reynold's Point groups. The Rae Group (previously the Glenelg Formation) has been divided into four members, but these were given formational status by Rainbird et al. (1994). In ascending order, these are (i) a lower, deep-water clastic unit (Escape Rapids Formation, 600+ m), (ii) a shallow-water, cherty carbonate unit (Mikkelsen Island Formation, 500+ m), (iii) an upper fluvial clastic unit (Nelson Head Formation, 150-350 m), and (iv) an orange-weathering stromatolitic dolostone unit (Aok Formation, 35-50 m).

The Shaler Supergroup is unconformably overlain by the Natkusiak Formation, a thick sequence of subaerial, plateau-type basalts. These lavas are assumed to be synchronous with gabbroic sills and dykes that intrude the underlying units. These sills are up to 50 m thick and extend for tens of kilometres. The Natkusiak Formation is part of the 725 Ma Franklin Dyke Swarm (Dupuy et al., 1995).

Three styles of copper mineralization occur in the Rae Group (Darch et al., 1991): 1) widespread stratabound disseminated sulphides (chalcopyrite and pyrite with associated malachite) within the quartzite-dominated Nelson Head Formation; 2) contact metamorphic mineralization (massive chalcopyrite) that occurs along the base of diabase sills near the Nelson Head – Aok formational boundary; and 3) zones of massive sulphide replacement associated with karsting of the Aok Formation. There is also potential for disseminated copper sulphides at the base of the Shaler succession as are found in the Coppermine Homocline on the mainland (Roscoe, 1984).

Mineralization appears related to a regional-scale lineament that runs from northeast Banks Island, across northern Victoria Island, and southeast towards King William Island (Darch et al., 1991). This structure appears to have affected paleocurrent directions at the time of deposition of the Rae Group, as well as the orientation of diabase dykes and even the glacial history and shoreline trends. This 310° trending structure is also parallel to the trend of kimberlite bodies on central Victoria Island (Galaxy and King Eider trends, Diamonds North Ltd web page.).

Large sheets of native copper occur in the Natkusiak basalts (Jefferson et al., 1985), especially near the boundaries of red-weathering sediments and pyrite-rich tuffs and flows, and disseminated within amygdaloidal flows. There are 138 documented mineral occurrences and anomalies of this type in the NORMIN database. These targets were explored in the 1980s by Panarctic Oils (Nelson, 1984) and by Noranda in the 1990s (Darch et al., 1991).

Jefferson et al. (1988) also suggested potential for unconformity and sandstone-hosted uranium at or near the base of the Rae Group. These deposit types have yet to be found within the Shaler Supergroup.

Potential for Noril'sk-style Ni-PGE targets related to the Natkusiak volcanics has been suggested by Jefferson et al. (1994) and Hulbert (2005). Exploration in the late 1990s by Aber Resources, however, failed to locate any nickel-PGE mineralization on Victoria Island (Hopkins et al., 1997).

Potential for mineral deposits in the basement below the thin Paleozoic cover must be similar to that described by Roscoe (1984) for the Bathurst Inlet region to the south, and include volcanic massive sulphide (VMS) base metals and vein- and iron formation-hosted Au.

Mid-Neoproterozoic to Late Silurian Passive to Convergent Margin

Arctic Islands

Up to 6000 m of Neoproterozoic to Silurian, clastic, carbonate and evaporite sediments were deposited on the Arctic continental margin following mid-Neoproterozoic rifting at about 723 Ma (Heaman et al., 1992). Five episodes of deposition are known.

The oldest stratal package that developed on the rifted margin is only known from reflection seismic profiles on Melville Island (Harrison, 1995). It consists of up to 630 m of probable clastic rocks that sit with an angular unconformity on deformed, presumably Proterozoic, strata.

Neoproterozoic strata of the second depositional cycle are known from northern Ellesmere and Melville islands (Harrison, 1995; Dewing et al., 2004). These consist of over 1000 m of clastic shelf and slope sediments overlain by carbonate platform strata. These strata extend farther south than Melville Island, but no Neoproterozoic sediments exist on southern or central Ellesmere Island, Devon, Baffin, Prince of Wales, Somerset, or Victoria islands.

Neoproterozoic strata are unconformably overlain by Lower Cambrian clastic strata of the third depositional cycle (Trettin, 1969; Thorsteinsson and Mayr, 1987; Packard and Mayr, 1994; Trettin, 1994; Harrison, 1995; Dewing et al., 2004). About 2500 m of sandstone and shale occur near the shelf edge on northeast Ellesmere Island, but Lower Cambrian clastic rocks thin rapidly towards the continent and the shale portions of the sequence are lost due to facies change in most areas of the southern and central Arctic Islands. These strata were deposited in a fluvial-deltaic-shoreface setting. Local changes in thickness, as described on north Baffin (Trettin, 1969) and the subsurface of Prince of Wales Island (Mayr et al., 2004) indicate that there was some localized tectonic activity.

The fourth depositional cycle was initiated with a transgression of late Early Cambrian age. A distinct, carbonate ramp became differentiated from the condensed, deep-water basinal shales to the (present day) north. The platform maintained a ramp configuration during most of the Cambrian.

The fifth depositional cycle started after the hiatus that marks the Cambrian-Ordovician boundary. Up to 2300 m of carbonate and gypsum were deposited on the platform during the Ordovician. During the same time period, only 350 m of basinal strata accumulated. The shelf-margin escarpment is marked by an abrupt lateral change from shallow-water sediments to deep-water shales; there are no slope deposits that may have indicated the existence of a carbonate ramp (Trettin, 1994). A distinctive, shallow-water to peritidal carbonate shoal developed on the edge of the escarpment for most of Ordovician time (de Freitas and Mayr, 1995). Behind the shelf-margin shoal was an intraplatformal basin that was the locus of deposition for the Ordovician strata. These strata are dominantly shallow subtidal to peritidal in origin and comprise the Cape Clay, Christian Elv, Baumann Fiord, Eleanor River, Bay Fiord, Thumb Mountain, and Irene Bay formations. The change from ramp to rimmed platform is thought to be related to the change from a passive margin setting in the Cambrian to a convergent margin in the Ordovician and the onset of upwelling currents onto the platform.

During latest Ordovician, the shelf margin retreated substantially towards the south and southeast, forming an embayment that received mostly graptolitic shale and carbonates. Isolated coral-microbial banks within the embayment kept up with sea level rise at least until Early Silurian time. The shelf-to-basin transition was a ramp immediately following shelf retreat, but developed a steeper rimmed profile by Early Silurian time.

Neoproterozoic to Late Silurian Mineral Potential

Although the Late Devonian Zn-Pb mineralization at the Polaris deposit is hosted in Middle Ordovician strata, no mineral deposits are known to have formed in the Arctic Islands during the Middle Ordovician. SEDEX Zn-Pb deposits occur in latest Ordovician to Early Silurian shales in northeastern Greenland, and in Early Silurian shales in the Selwyn Mountains, Yukon, and Northwest Territories. The deposit at Citronen Fiord on northern Greenland (van der Stijl and Mosher, 1998) occurs as a stratiform massive sulphide deposit at least 3 km long and 500 m wide with a resource of about 20 million tonnes of 7% Zn in the Amundsen Land Group (correlative with the Cape Phillips Formation in the Arctic Islands). The Howard's Pass deposit in the Selwyn Mountains contains a drill-indicated resource of 59 million tonnes grading 2.1% Pb and 5.4% Zn (see Goodfellow and Jonasson, 1983).

Basinal strata of Silurian age on northeast Ellesmere Island have small pyrite-rich zones (Harrison et al., 1999a) and on Melville Island have elevated zinc and phosphate values (showing 4 of Harrison, 1995). Limited exploration on Melville Island (Noakes, 1998), Bathurst Island (Anglin and Harrison, 1999), and central Ellesmere (Gibbins, 1985) failed to find significant showings.

Ordovician-Silurian Intracratonic Basins

Large intracratonic basins in Hudson Bay and Foxe Basin formed at the same time as the passive to convergent margin in the Arctic Islands. The stratigraphic section of the Hudson Bay Basin consists of about 2500 m of Ordovician, Silurian, and Devonian rocks, with some probable Cretaceous rocks in the centre of Hudson Bay. The basin is roughly circular, with a diameter of 500 km. Strata are exposed on surface in northeastern Manitoba, northern Ontario, and on Southampton Island at the north end of Hudson Bay (Sanford et al., 1993). The potential for MVT deposits is very low because the stratigraphic section is very thin, it has very low thermal maturity, and there are no reported hydrothermal dolomite occurrences or Zn showings.

The Foxe Basin, located between Melville Peninsula and Baffin Island, extends onshore in southeastern Baffin Island. The Bell Arch separates it from the Hudson Bay Basin to the south. Paleozoic strata are also preserved in two parallel northwest-oriented rift systems. The northern rift is less developed and extends across Baffin Island (Amajuak Lake) into Cumberland Sound. The southern rift system, north of Southampton Island and beneath Hudson Strait and Foxe Channel, also contains Mesozoic clastic strata. This basin has limited potential for mineral deposits as there are no identified mechanisms for moving and focusing large amounts of fluid.

Pearya

Pearya is a composite exotic terrane that was emplaced on the northwest margin of northern Ellesmere Island and the northern tip of neighbouring Axel Heiberg Island, probably during the Late Silurian (Trettin, 1987). Pearya is divisible into five successions (see Fig. 3). The first is a late Mesoproterozoic - early Neoproterozoic crystalline basement consisting of granitoid gneiss with minor amounts of schist, amphibolite and metasedimentary rocks. The second succession is metamorphosed sedimentary and volcanic rocks. The third succession consists of rift-related and passive margin sediments and minor mafic and siliceous volcanics ranging in age from late Proterozoic to latest Cambrian or early Ordovician. These units are followed by carbonate and shale, mafic volcanics, and ultramafic-mafic complexes, of probable Early to Middle Ordovician age, which appear to have been thrust over parts of the third succession during the mid-Ordovician M'Clintock Orogeny. The fifth succession comprises upper Middle Ordovician to Upper Silurian sedimentary and volcanic strata that unconformably overlap the fourth succession and parts of the third succession.

Pearya has the potential for a variety of mineral deposits, but the area has been of little interest to industry because it is accessible only by aircraft or icebreaker, and because a National Park was established over most of its extent in 1988. An assessment of metallic mineral potential by Sangster (1981) identified potential for Zn, Cu, and Au deposits within the area of the park.

Late Silurian to Middle Devonian Boothia Uplift

Vertical movement along a 450 km long and 130 km wide, north-south oriented intracratonic uplift running from Boothia Peninsula to Grinnell Peninsula is termed the Boothia Uplift. The Boothia Uplift was active from the Late Silurian to Middle Devonian (Givetian) and affected both the Precambrian basement and the lower Paleozoic sequence. The southern segment (Boothia Peninsula, eastern Prince of Wales, and Somerset Island) contains the lower structural level, which is characterized by faulted basement rocks. The northern segment of the uplift (Cornwallis Island, Grinnell Peninsula, eastern Bathurst Island) contains the upper structural level, which is characterized by broad synclines and narrow, evaporite-cored anticlines (Kerr, 1977; Thorsteinsson, 1986; Okulitch et al., 1991; de Freitas and Mayr, 1993; Mayr et al., 1998, 2004). Although no Precambrian rocks are exposed north of Barrow Strait, airborne magnetic and gravity geophysics surveys show that the north-south basement trend continues north of Barrow Strait as far as Cornwallis Island (Miles et al., 2000a,b).

Mapping and gravity surveys indicate westerly directed thrust faults in both Precambrian and Paleozoic strata (Berkhout, 1973; Mortensen and Jones, 1986). Miall (1986) and Okulich et al. (1986; 1991) suggested a link to Late Silurian to Middle Devonian Caledonian contraction on Greenland, with which the Boothia Uplift is roughly synchronous. Several Upper Silurian – Lower Devonian conglomerate units record the beginning of the Boothia Uplift (Miall and Gibling, 1978; Thorsteinsson and Uyeno, 1980; Muir and Rust, 1982; de Freitas and Mayr, 1993; Mayr et al., 1998).

No mineral showings are known from strata related to the Boothia Uplift, although structures created during this event appear to have acted as channels for fluids during the Late Devonian mineralizing event in the Polaris District (Dewing et al., 2006).

Late Devonian – Early Carboniferous Ellesmerian Orogeny

By early Middle Devonian time, the effects of plate convergence were widespread over most of the Arctic Islands. Shallow-marine and nonmarine syntectonic clastic rocks were deposited in a foreland basin adjacent to a southeastward- and southward-advancing deformation front. The foreland basin was subsequently folded and faulted by the advancing orogenic front. In the Arctic Islands, the youngest preserved strata are of Famennian age. This Middle Devonian to earliest Carboniferous continent-continent collision is known as the Ellesmerian Orogeny (Thorsteinsson and Tozer, 1970).

East-west-trending folds that resulted from the Ellesmerian Orogeny characterize Melville and Bathurst islands (Parry Islands Fold Belt) and southern Ellesmere Island (central Ellesmere Fold Belt). Harrison (1995) demonstrated, using mapping and stratigraphic studies as well as drilling and reflection seismic surveys, that thick units of evaporites and shale profoundly influenced the style of contractional tectonics within the thin-skinned Parry Islands Fold Belt on Melville Island. He documented two main phases of the Ellesmerian Orogeny on Melville Island; a first phase of long-wavelength folding above the evaporite décollement layer, and a second phase of sinistral transpressive compression that created deep-seated folds oblique to the trend of the surface folding.

The easternmost portion of the Parry Islands Fold Belt, located on eastern Bathurst Island, intersects and overlaps the north-trending Boothia Uplift. Boothia structures were reactivated during Ellesmerian deformation and this area is characterized by dramatic interference structures (Harrison et al., 1993; Harrison and de Freitas, 1999), and by a pronounced, structurally disharmonic relationship between folds above and below the detachments of the Bay Fiord Formation evaporite (Fox, 1985). Structures in the higher levels belong to the east-striking Parry Islands Fold Belt, whereas structures present below the detachment belong to the north-striking Boothia Uplift (Fox, 1985). The main part of the Boothia Uplift was interpreted to act as a buttress during south-directed Ellesmerian deformation, although Jober (2005), and Henrichsen (2003) demonstrated localized interference folding within lower Paleozoic rocks on Cornwallis Island. Ellesmerian structures do not occur south of Barrow Strait (Okulitch et al., 1991).

Late Devonian Sandstone Uranium Potential

Gregory et al. (1961) reported on aerial radiometric profiles on Bathurst and Melville islands, including a Devonian sandstone sample yielding 30 ppm uranium. Embry and Klovan (1976) discuss the potential for sandstone-hosted uranium within the clastic wedge that formed in front of the Ellesmerian Orogen and noted anomalously high gamma ray values in the Weatherall Formation around the 550 m depth (1775-1826 ft) in the Winter Harbour No. 1 well. Uranium (as well as gold) paleo-placer potential was noted by Anglin and Harrison (1999) on western Bathurst Island and adjacent smaller islands, particularly in areas underlain by the upper part of the Hecla Bay Formation.

Late Devonian – Early Carboniferous Zn-Pb Mississippi Valley-Type Potential

The Polaris deposit and other showings in the Polaris district formed during the waning stages of the Ellesmerian Orogeny or first stages of the opening of the Sverdrup Basin (Dewing et al., 2006). Because Zn-Pb showings are common in the Polaris District, there should be potential for similar carbonate-hosted Zn-Pb elsewhere in the Arctic Islands. Zn-Pb showings occur in Silurian carbonates on central Ellesmere Island (Harrison, 1984; Gibbins, 1985; de Freitas et al., 1995) and in Ordovician carbonates on Melville Island (Harrison, 1995), but neither area has been subject to intensive industry exploration. The young age for mineralization (post-Ellesmerian folding and related to transpression and/or orogenic collapse at the terminal stages of the Ellesmerian Orogeny) in the Polaris District (Dewing et al., 2006) indicates that the bounding faults of grabens and half grabens with Carboniferous fill on Melville and Bathurst islands, as well as on the Grinnell Peninsula are prospective, but have received limited prospecting. Other areas of interest are 1) the large lineament that underlies northern and central Victoria Island (see section on Cu mineralization in the Shaler Supergroup, above); 2) a lineament that follows the trend of the Bathurst Fault Zone has never been prospected where it intersects lower Paleozoic strata, although fluorite (Gibbins et al., 1986) and anomalous till chemistry (Nixon, 1988; Sharpe, 1992) have been reported; 3) the earliest Ordovician Cape Clay Formation on northeast Ellesmere Island has Zn-Pb showings (Harrison et al., 1999a) but the unit has not been widely prospected on Ellesmere Island or elsewhere (e.g. Devon Island or the correlative Turner Cliffs Formation on northern Baffin Island). Copper mineralization on Somerset Island and the Grinnell Peninsula within the Polaris District are described by Dewing et al. (2006).

Late Devonian – Early Carboniferous Zn-Pb-Ag-Au Vein Potential

There are three types of Zn-Pb showings on NE Ellesmere Island (Harrison et al., 1999a). From basin to platform these are 1) carbonate-hosted Pb-Zn-Ag-Au-Cu (Sb-As) quartz veins; 2) Zn-Pb-Fe showings close to the shelf margin; and 3) carbonate-hosted Zn-Pb showings in platformal strata. The three types are mineralogically, stratigraphically, and spatially distinct but they show enough geochemical similarities that a common event can be postulated for their origin.

The first showing type occurs in breccias in Lower Cambrian basinal shales. The showings occur as quartz-dolomite veins that occupy fold hinge axial planar fractures and fold limb transverse fractures. Metallic minerals include sphalerite, galena, boulangerite, arsenopyrite, wurtzite, tetrahedrite, and pyrite with silver over 1000 g/t, gold to 20 g/t, and 1 to 6% each of Pb and Zn in selected hand specimens. Copper sulphosalt mineralization occurs in the Lower Cambrian Grantland Formation on northwest Ellesmere Island (Mosher, 1981; Trettin, 1994), and Harrison (1995) reported similar vein-style mineralization in the basinal Ibbett Bay Formation on Melville Island.

The second type of showing occurs in Ordovician shales of the basinal Hazen Formation. These showings occur as veins, in brecciated carbonate lenses or as replacement of carbonate. The dominant mineral is pyrite, but crystalline sphalerite and galena occurs with or crosscuts the pyrite and can form greater than 5% of the rock.

The third type of showing is hosted in Cambrian and earliest Ordovician platformal carbonates. Alteration consists of extensive dolomitization and replacement by dolospar, locally with coarsely-crystalline sphalerite, galena, and minor Cu sulphide minerals. Similar occurrences of Zn-Pb mineralization are reported from Greenland (Dawes et al., 2000) where Pb-Zn-Ag occurs along faults in the Cape Clay and Cass Fjord formations in Washington Land.

All three showing types have elevated arsenic and silver compared to the trace element distribution in the more typical MVT Polaris deposit. Antimony, Mo, and P levels in the platformal carbonate showings are also exceptionally high when compared to MVT deposits. Lead isotopes from the Great Plains showing in the Cape Clay Formation and from the Greenland showings are, unlike the Polaris deposit, highly radiogenic again indicating different modes of origin for prospects of northeast Ellesmere and the central Cornwallis region.

A model that genetically links the three types of showings would require the generation of heat and source of metal from small igneous intrusions of Devonian, Carboniferous, or Cretaceous-early Tertiary age under the basinal Hazen Formation. Mineralizing fluids would have then migrated outward and upward and cooled, thereby depositing their dissolved ions. The hottest fluids and most complex mineralogy would have formed deposits in dilational space in rigid units of the Hazen Formation closest to the heat source. Fluids could migrate either along faults (the second type of mineralization) or up dip along aquifers into platformal carbonate strata (the third type of showing).

Carboniferous to Cretaceous Sverdrup Basin

The Ellesmerian orogenic belt collapsed to form fault-bounded lakes and a narrow, saline inland sea during the Early Carboniferous (Viséan, Beauchamp et al., 2001). This later opened to form a seaway (the Sverdrup Basin) to the north of the lower Paleozoic shelf margin. Rifting in the Carboniferous and Early Permian was followed by thermal subsidence in the Late Permian and Mesozoic. The Sverdrup Basin accumulated sediments throughout the Viséan and Mesozoic.

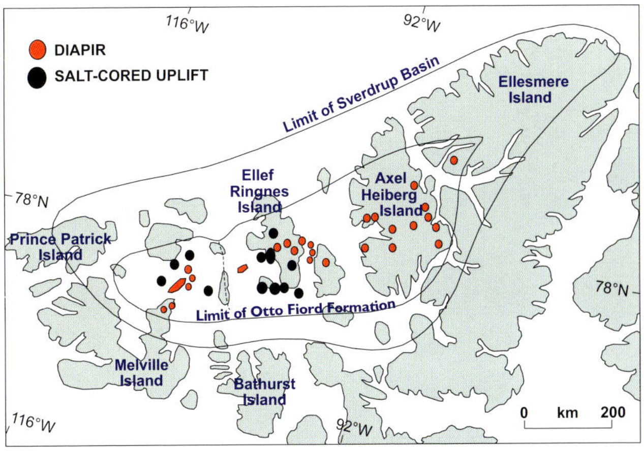

FIGURE 8. Sverdrup Basin showing the distribution of evaporites in the Carboniferous Otto Fiord Formation and evaporitic diapirs.

The oldest strata in the Sverdrup Basin are organic-rich lacustrine shales, sandstones, and conglomerates deposited in grabens. As the Sverdrup Basin subsided and expanded during the Late Carboniferous, thick evaporites (Otto Fiord Formation) were deposited in two subbasins east and west of Lougheed Island (Fig. 8). Terrestrial and nearshore clastic strata were deposited around the margins of the basin.

Marine transgression in the Late Carboniferous and Early Permian created favourable conditions for carbonates and reefs to form around the edge of the basin. Argillaceous limestone and shale were deposited in the basin centre. Changing climate conditions in the Late Permian drastically curtailed carbonate deposition. Instead, shale (van Hauen Formation) was deposited across almost the entire basin.

The Permo-Triassic boundary is marked by an unconformity at the basin margins, but was conformable across the centre of the basin. Fluctuating sea level and basin tectonics caused the repeated advance and retreat of deltaic strata during the Early Triassic to Lower Jurassic. The largest deltaic system deposited over 1500 m of stacked delta-front, and delta-plain sandstones that form the principle hydrocarbon reservoir in the Arctic Islands (Lower Jurassic Heiberg Group). Transgression drowned the Heiberg delta system and deposited widespread Middle and Upper Jurassic shale.

By the mid-Jurassic, the basin was restricted to between the emergent Sverdrup Rim on the northwest side of the Sverdrup Basin and the southern margin located on Ellesmere-Bathurst-Melville islands. Nearshore sandstones were deposited on either side of this northeast-trending seaway. A Late Jurassic to Early Cretaceous transgression deposited shales (Deer Bay Formation) across the basin. This was followed by uplift and erosion associated with the onset of rifting in the Canada Basin in the Early Cretaceous, followed by deposition of fluvial and deltaic sandstones of the Isachsen Formation across the basin and onto the newly forming continental margin.

Starting in the Early Cretaceous and lasting about 60 Ma, flood basalts, gabbroic sheets and dykes were emplaced in the eastern part of the Sverdrup Basin (Fig. 9). The peak igneous activity was synchronous with the accumulation of volcanic rocks on the Alpha Ridge and spreading of the Canada Basin in the Hauterivian. This activity occurred where the north-northeast-striking late Paleozoic rifts of the

Sverdrup Basin intersect the northeasterly trending rifted margin of the Canada Basin.

Subsequent to the Aptian (Early Cretaceous), deposition on the newly formed continental margin outstripped that in the Sverdrup Basin.

Sverdrup Mineral Potential

Redbeds of the Carboniferous Canyon Fiord Formation on Melville Island have stratabound pyrite and copper sulphide occurrences reminiscent of continental redbed copper occurrences elsewhere (Harrison, 1995). Industry exploration failed to find any commercially significant mineralization (Rees, 1999).

Low-grade, sediment-hosted uranium has been reported by Jonasson and Dunsmore (1979) in Carboniferous limestone breccias adjacent to evaporite diapirs on Ellef Rignes Island. Anecdotal reports, by Geological Survey of Canada field officers, of iron sulphide occurrences adjacent to evaporite-cored domes on Ellef Rignes and Axel Heiberg islands (Fig. 8) have also been recorded.

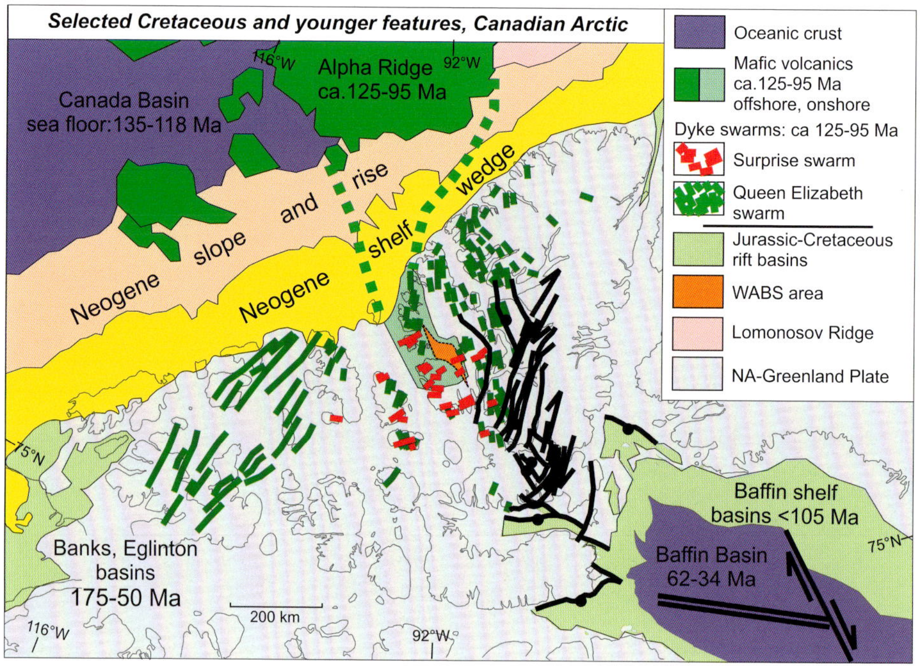

FIGURE 9. Sverdrup Basin showing the distribution of volcanic rocks. Dyke swarms after Buchan and Ernst (2004); Baffin Bay: Chalmers and Pulvertaft (2001).

Flood basalts of Early Cretaceous age likely had a mantle plume origin (Fig. 9, Harrison et al., 1999b). This, along with the presence of evaporites and black shale in the Sverdrup Basin has led Williamson (2000) to speculate on the high potential for Noril'sk-style Ni-Cu-PGE in the eastern Sverdrup Basin. No exploration for this target type has been undertaken.

Jurassic to Tertiary Arctic Continental Margin

Four cratonic basins developed southwest of the Sverdrup Basin starting in the mid-Jurassic in response to rifting in the Arctic Ocean. These basins are roughly elliptical in map view and have gently inwardly dipping strata. These basins continued to develop until the Early Cretaceous when they were covered by the sedimentary prism of the continental margin (Harrison and Brent, 2005). The largest of these basins forms a north-south trending belt of rocks through the central portions of Banks Island and small extensions to both the north and south of the island.

The Arctic Continental Margin is represented onshore by a narrow strip of sediments that extends from Banks Island to Meighen Island. Upper Cretaceous and Paleogene sediments unconformably overly strata of the Sverdrup Basin. The wedge continues for about 175 km northwest across the continental shelf and seismic refraction studies north of Axel Heiberg Island reveal a 10 km thick sedimentary succession on the outer shelf.

Deposition of Kanguk Formation shale in the Late Cretaceous reflects the flooding of continental margins during a worldwide sea-level rise. The Eureka Sound Group, consisting of poorly consolidated, fine- to coarse-grained sandstones having abundant coal, was deposited in alluvial, deltaic, and estuarine settings. These range in age from Campanian or Maastrichtian to mid-Eocene (Harrison et al., 1999b).

Ruzicka (1977) and Jefferson et al. (1988) identified the potential for roll-front, sandstone-hosted uranium in the Isachesen, Christopher, and bentonite-bearing Kanguk formations. No systematic exploration has yet been undertaken for this deposit type.

Tertiary Eurekan Orogeny

The final stage of the development of the geology of the Arctic Islands resulted from Early Paleogene widening of the north Atlantic and the rotation and translation of Greenland away from North America. Rotation of Greenland initially caused widespread extension across the Arctic Islands including the development of a failed rift arm underlying Lancaster Sound and grabens containing Tertiary strata on Somerset, Cornwallis, Devon, and Bylot islands (Stewart, 1987; Thorsteinsson and Mayr, 1987; Okulitch and Trettin, 1991).

Extension was followed in the Eocene by translation and compression as Greenland moved north and west relative to North America. This compressive event is termed the Eurekan Orogeny (Okulitch and Trettin, 1991; Harrison et al., 1999b). Deformation generated long wavelength folds and reactivated salt diapirs. Folding extended as far west as Prince Patrick Island (Harrison and Brent, 2005). Folds and thrusts related to the Eurekan Orogeny also occur on Bathurst and Melville islands as well as the Grinnell Peninsula of Devon Island.

Northeast-trending, Cretaceous tholeiitic dykes and sills (Osadetz and Moore, 1988) and Lower Cretaceous tholeiitic volcanic flows (Embry and Osadetz, 1988) occur on Axel Heiberg Island and western and northwestern Ellesmere Island. Igneous rocks of Eocene age crop out on southeast-

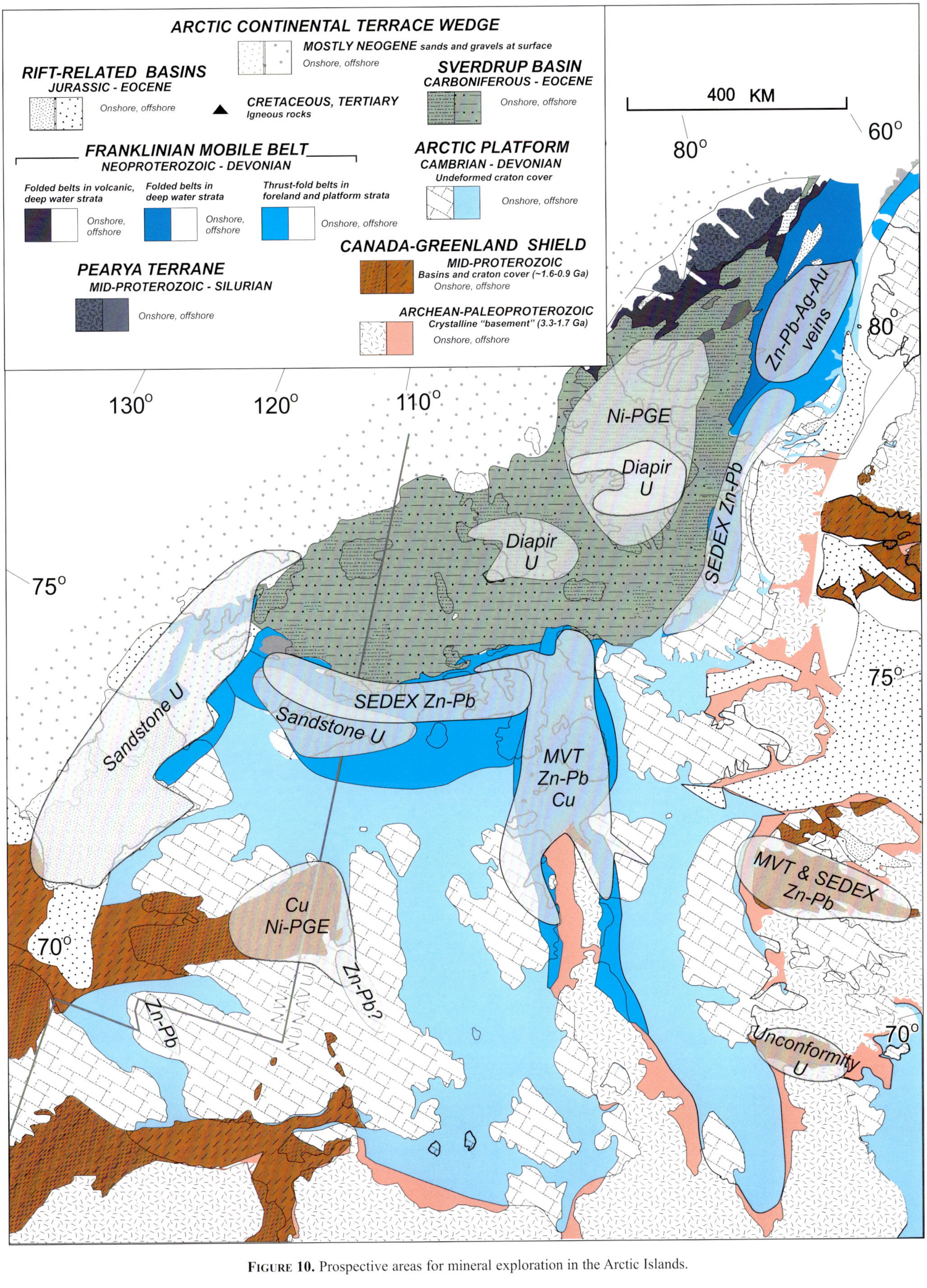

FIGURE 10. Prospective areas for mineral exploration in the Arctic Islands.

ern Bathurst Island and Grinnell Peninsula, indicating that structures related to Eurekan Orogeny rifting locally extend to the mantle (Mitchell and Platt, 1984). The small size and areal extent of these volcanics limits the potential for Noril'sk-type Ni-PGE.

Knowledge Gaps

The Arctic Islands have been under represented by mineral exploration due in large part to the remoteness, high cost of transportation and drilling, short exploration season, and lack of infrastructure. The feasibility of developing a deposit is also lessened by the high costs of fuel, drilling, aircraft, and labour, along with the short shipping season and environmental sensitivities. These factors likely outweigh the gaps in geoscience knowledge in continuing to limit mineral exploration. The Arctic Islands do, however, present some opportunities: 1) the Arctic Islands are poorly explored, so the chance of a new, near-surface discovery is probably higher than in southern regions; 2) the lack of vegetation allows rapid assessment of areas using hyperspectral surveys; 3) base and precious metal exploration can likely piggy-back on exploration for high-value commodities, such as diamonds, oil, and gas. There are several key geoscience gaps that could encourage future exploration:

- Airborne potential field geophysics.
- Geological mapping of poorly known areas such as Victoria Island and southern-central Ellesmere.
- Prospecting, geochemical and alteration studies and detailed mapping of specific targets such as the Sverdrup basin volcanics, Somerset-Brodeur-north Baffin copper trend, Victoria Island copper, and sandstone-hosted uranium.
- Continuing digital compilation of information to enable topography, bedrock, and surficial mapping, as well as comprehensive assessment files.

Conclusion

The Arctic Islands present a challenging exploration area, but one with a potentially large pay-off. Sediment-hosted Zn-Pb and sediment-hosted Cu have proven potential and lower geological risk, but these targets have received at least one round of exploration. Ni-PGE targets have higher geological risk, but are generally poorly explored, especially within the Sverdrup Basin. Sandstone-hosted and diapir-related uranium have received next to no exploration in the Arctic Islands and thus have unknown potential (see Fig. 10 for summary).

Acknowledgements

The authors thank Robert Sharp, Michael Gunning, and Victoria Yehl, for the many discussions about Arctic geology. Comments and suggestions by Wayne Goodfellow and reviewers Cameron Allen and Hamish Sandeman greatly improved the paper.

References

Anglin, C.D., and Harrison, J.C., 1999, Bathurst Island Mineral and Energy Resource Assessment Project; Mineral and Energy Resource Assessment of Bathurst Island Area, Nunavut (Parts of NTS 68G, 68H, 69A, 69B and 79A): Geological Survey of Canada, Open File 3714, 242 p.

Arne, D.C., and Kissin, S.A., 1989, The significance of 'diagenetic crystallization rhythmites' at the Nanisivik Pb-Zn-Ag deposit, Baffin Island, Canada: Mineralium Deposita, v. 24, p. 230-232.

Arne, D.C., Curtis, L.W., and Kissin, S.A., 1991, Internal zonation in a carbonate-hosted Zn-Pb-Ag deposit, Nanisivik, Baffin Island, Canada: Economic Geology, v. 86, p. 699-717.

Beauchamp, B., Harrison, J.C., Utting, J., Brent, T.A., and Pinard, S., 2001, Carboniferous and Permian Subsurface Stratigraphy, Prince Patrick Island, Northwest Territories, Canadian Arctic: Geological Survey of Canada, Bulletin 565, 93 p.

Berkhout, A.W.J., 1973, Gravity in the Prince of Wales, Somerset, and northern Baffin islands region: Symposium on the Geology of the Canadian Arctic, Proceedings, p. 63-79.

Buchan, K.L., and Ernst, R.E., 2004, Diabase dyke swarms and related units in Canada and adjacent regions [with accompanying notes]: Geological Survey of Canada, Map 2022A, 1:5,000,000.

Chalmers, J.A., and Pulvertaft, T.C.R., 2001, Development of the continental margins of the Labrador Sea: A review, *in* Wilson, R.C.L., Whitmarsh, R.B., Taylor, B., and Froitzheim, N., eds., Continental Margins: A Comparison of Evidence from Land and Sea: Geological Society of London, Special Publication 187, p. 77-105.

Chandler, F.W., 1988, Geology of the Late Precambrian Fury and Hecla Group, northwest Baffin Island, District of Franklin: Geological Survey of Canada, Bulletin 370, 30 p.

Christensen, J.N., Halliday, A.N., Kesler, S.E., and Sangster, D.F., 1993, Further evaluation of the Rb-Sr dating of sphalerite: The Nanisivik Precambrian MVT deposit, Baffin Island, Canada: Geological Society of America, Abstracts with Programs, p. 471.

Ciesielski, A., and Maley, J., 1980, Basement geology, Agu Bay, Gifford River, Baffin Island, District of Franklin: Geological Survey of Canada, Open File 687, 2 sheets, scale 1:125,000.

Clayton, R.H., and Thorpe, L., 1982, Geology of the Nanisivik zinc-lead deposit, *in* Hutchinson, R.W., Spence, C.D., and Franklin, J.M., eds., Precambrian Sulphide Deposits: Geological Association of Canada, Special Paper 25, p. 739-758.

Darch, W., Maré, M., and Walmsley, T., 1991, Summary Report on the Victoria Island Copper Project, District of Franklin, Northwest Territories: Department of Indian and Northern Development, Assessment Report 83088, 271 p.

Dawes, P.R., 1997, The Proterozoic Thule Supergroup, Greenland and Canada: History, Lithostratigraphy, and Development: Geology of Greenland Survey, Bulletin 174, 150 p.

——— 2004, Explanatory Notes to the Geological Map of Greenland, 1:500,000, Humboldt Gletscher, sheet 6: Geological Survey of Denmark and Greenland, Map Series 1, 48 p.

Dawes, P.R., and Garde, A.A., 2004, Geological Map of Greenland, Humboldt Gletscher, Sheet 6: Geological Survey of Denmark and Greenland, scale 1:500,000.

Dawes, P.R., Frisch, T., Garde, A.A., Iannelli, T.R., Inseon, J.R., Jensen, S.M., Pirajno, F., Sonderholm, M., Stemmerik, L., Stouge, S., Thomassen, B., and van Gool, J.A.M., 2000, Kane Basin: Geology of Greenland Bulletin, v. 186, p. 11-28.

de Freitas, T.A., and Mayr, U., 1993, Middle Paleozoic tear faulting, basin development, and basin uplift, central Canadian arctic: Canadian Journal of Earth Sciences, v. 30, p. 603-620.

——— 1995, Kilometre-scale microbial buildups in a rimmed carbonate platform succession, Arctic Canada: New insight on Lower Ordovican reef facies: Bulletin of Canadian Society of Petroleum Geologists, v. 43, p. 407-432.

de Freitas, T., Harrison, J.C., Mayr, U., and Thorsteinsson, R., 1995, Stratigraphic Controls on Potential Hydrocarbon and Mineral Accumulations of central Ellesmere Island: Geological Survey of Canada, Open File 3058, 6 p.

Dewing, K., Harrison, J.C., Pratt, B.R., and Mayr, U., 2004, A probable late Neoproterozoic age for the Kennedy Channel and Ella Bay formations, northeastern Ellesmere Island and its implications for passive margin history of the Canadian Arctic: Canadian Journal of Earth Sciences, v. 41, p. 1013-1025.

Dewing, K., Sharp, R.J., and Turner, E., 2006, Synopsis of the Polaris Zn-Pb District, Canadian Arctic Islands, Nunavut, *in* Goodfellow, W.D., ed., Mineral Resources of Canada: A Synthesis of Major Deposit-Types, District Metallogeny, the Evolution of Geological Provinces, and Exploration Methods: Special Publication xx, Mineral Deposits Division, Geological Association of Canada, p. 655-672.

Dixon, J., 1974, Revised stratigraphy of the Hunting Formation (Proterozoic), Somerset Island, Northwest Territories: Canadian Journal of Earth Sciences, v. 11, p. 635-642.

Dixon, O.A., Williams, S.R., and Dixon, J., 1971, The Aston Formation (Proterozoic) on Prince of Wales Island, Arctic Canada: Canadian Journal of Earth Sciences, v. 8, p. 732-742.

Dupuy, C., Michard, A., Dostal, J., Dautel, D., and Baragar, W.R.A., 1995, Isotope and trace-element geochemistry of Proterozoic Natkusiak flood basalts from the northwestern Canadian Shield: Chemical Geology, v. 120, p. 15-25.

Embry, A.F., and Klovan, J.E., 1976, The Middle-Upper Devonian clastic wedge of the Franklinian geosyncline: Bulletin of the Canadian Society of Petroleum Geologists, v. 24, p. 485-639.

Embry, A.F., and Osadetz, K.G., 1988, Stratigraphy and tectonic significance of Cretaceous volcanism in the Queen Elizabeth Islands, Canadian Arctic Archipelago: Canadian Journal of Earth Sciences, v. 25, p. 1209-1219.

Fox, F.G., 1985, Structural geology of the Parry Islands foldbelt: Bulletin of the Canadian Society of Petroleum Geologists, v. 33, p. 306-340.

Frisch, T., 1988, Reconnaissance Geology of the Precambrian Shield of Ellesmere, Devon and Cobourg Islands, Canadian Arctic Archipelago: Geological Survey of Canada, Memoir 409, 102 p.

Frisch, T., and Christie, R.L., 1982, Stratigraphy of the Proterozoic Thule Group, southeastern Ellesmere Island, Arctic Archipelago: Geological Survey of Canada, Paper 81-19, 13 p.

Geldsetzer, H., 1973a, The Tectono-Sedimentary Development of an Algal-Dominated Helikian Succession on northern Baffin Island, N.W.T.: Geological Association of Canada, Memoir 19, 28 p.

——— 1973b, Syngenetic dolomitization and sulfide mineralization, *in* Amstutz, G.G., and Bernard, A.J., eds., Ores and Sediments: Springer, Berlin, p. 115-127.

Ghazban, F., Schwarcz, H.P., and Ford, D.C., 1992, Multistage dolomitization in the Society Cliffs Formation, northern Baffin Island, Northwest Territories, Canada: Canadian Journal of Earth Sciences, v. 29, p. 1459-1473.

Gibbins, W.A., 1985, Arctic Islands Region, *in* Brophy, J.A., ed., Mineral Industry Report: Northwest Territories, Indian and Northern Affairs, EGS1985-4, p. 95-113.

Gibbins, W.A., Boucher, D., and Skublak, W., 1986, Fluorite from Princess Royal Islands; historic and possible economic significance: Arctic, v. 39, p. 89-91.

Goodfellow, W.D., and Jonasson, I.R., 1986, Environment of formation of the Howard's Pass (XY) Zn-Pb deposit, Selwyn Basin, Yukon, *in* Morin, J.A., ed., Mineral Deposits of Northern Cordillera: Canadian Institute of Mining and Metallurgy, Special Volume 37, p. 19-50.

Gregory, A.F., Bower, M.E., and Morley, L.W., 1961, Geologic Interpretation of Aerial Magnetic and Radiometric Profiles, Arctic Archipelago, Northwest Territories: Geological Survey of Canada, Bulletin 73, 148 p.

Hansen, B.T., Kalsbeek, F., and Holm, P.M., 1987, Archean age and Proterozoic metamorphic overprinting of the crystalline basement at Victoria Fjord, North Greenland: Geologiske Undersogelse Rappport, v. 133, p. 159-168.

Harrison, J.C., 1984, Eastern Ellesmere, Coburg and eastern Devon Island Mineral Inventory, NTS 29G, 38F and G, 39B,C,D,E,F,G and H, District of Franklin, N.W.T.: Department of Indian and Northern Development, Assessment Report 81743, 125 p.

Harrison, J.C., 1995, Melville Island's salt-based fold belt, Arctic Canada: Geological Survey of Canada, Bulletin 472, 331 p.

Harrison, J.C., and Brent, T.A., 2005, Basins and Fold Belts of Prince Patrick Island and Adjacent Areas, Canadian Arctic Islands: Geological Survey of Canada, Bulletin 560, 197 p.

Harrison, J.C., and de Freitas, T., 1999, Overview of bedrock geology, *in* Anglin, C.D., and Harrison, J.C., eds., Mineral and Energy Resource Assessment of Bathurst Island Area, Nunavut: Geological Survey of Canada, Open File 3714, p. B1-B40.

Harrison, J.C., de Freitas, T., and Thorsteinsson, R., 1993, New Field Observations on the Geology of Bathurst Island, Arctic Canada: Part B, Structure and Tectonic History: Geological Survey of Canada, Paper 93-1B, 11 p.

Harrison, J.C., Dewing, K., Lee, C.C., and Stasiuk, L.D., 1999a, New Mineral Occurrences on Northeastern Ellesmere Island and New Opportunities for Mineral Exploration in Northern Nunavut: Geological Survey of Canada, Open File 3822, 42 p.

Harrison, J.C., Mayr, U., McNeil, D.H., Sweet, A.R., McIntyre, D.J., Eberle, J.J., Harington, C.R., Chalmers, J.A., Dam, G., and Nohr-Hansen, H., 1999b, Correlation of Cenozoic sequences of the Canadian Arctic region and Greenland; Implications for the tectonic history of northern North America: Bulletin of Canadian Society of Petroleum Geology, v. 47, p. 223-254.

Heaman, L.M., LeCheminant, A.N., and Rainbird, R.H., 1992, Nature and timing of Franklin igneous events, Canada: Implications for a Late Proterozoic mantle plume and the break-up of Laurentia: Earth and Planetary Science Letters, v. 109, p. 117-131.

Henrichsen, M., 2003, Structural Geometry and Kinematics of Deformation in the Stanley Head Region, western Cornwallis Island, Nunavut: Unpub. M.Sc. thesis, University of British Columbia, 147 p.

Hildebrand, R.S., and Baragar, W.R.A., 1991, On folds and thrusts affecting the Coppermine River Group, northwestern Canadian Shield: Canadian Journal of Earth Sciences, v. 28, p. 523-531.

Hoffman, P.F., 1989, Precambrian geology and tectonic history of North America, *in* Bally, A.W., and Palmer, A.R., eds., The Geology of North America - An Overview: Geological Society of America, Boulder, Colorado, A, p. 447-512.

Hopkins, R., Vivian, G., Liebenberg, L., Armstrong, K., Gall, Q., Sharp, B., and Mason, I., 1997, Report on Exploration Activities - Including Geological Mapping, Sampling, Airborne Magnetic and Electromagnetic Surveys and Ground HLEM Surveys, Northeastern Victoria Island, NWT: DIAND, Assessment file 83981, 41 p.

Hulbert, L.J., 2005, Map of Mafic and Ultramafic Bodies Related to the Franklin Magmatic Event, Minto Inlier, Victoria Island, District of Franklin, Northwest Territories and Nunavut: Geological Survey of Canada, Open File 4928, 1 sheet, scale 1:500,000.

Jackson, G.D., 1986, Notes on the Proterozoic Thule Group, northern Baffin Bay: Geological Survey of Canada, Current Research, Part A, Paper 86-1A, 12 p.

——— 2000, Geology of the Clyde-Cockburn Land map area, north-central Baffin island, Nunavut: Geological Survey of Canada, Memoir 440, 303 p.

Jackson, G.D., and Iannelli, T.R., 1981, Rift-Related Cyclic Sedimentation in the Neohelikian Borden Basin, northern Baffin Island: Geological Survey of Canada, Paper 81-10, p. 269-302.

Jefferson, C.W., Nelson, W.E., Kirkham, R.V., Reedman, J.H., and Scoates, R., 1985, Geology and Copper Occurrences of the Natkusiak Basalts, Victoria Island, District of Franklin: Geological Survey of Canada, Paper 85-1A, 12 p.

Jefferson, C.W., Scoates, R.F.J., and Smith, D.R., 1988, Evaluation of the Regional Non-Renewable Resource Potential of Banks and northwestern Victoria Islands, Arctic Canada: Geological Survey of Canada, Open File 1695, 63 p.

Jefferson, C.W., Hulbert, L.J., Rainbird, R.H., Hall, G.E.M., Gregoire, D.C., and Grinenko, L., 1994, Mineral resource assessment of the 723 Ma Franklin igneous event, Arctic Canada; Comparison with the Palaeozoic Noril'sk Talnakh Ni-Cu-PGE deposits of Russia: Minerals Colloquium, p. 22.

Jober, S.-A., 2005, Structural Geology of the Stuart River - Caribou Lake Region, eastern Cornwallis Island, Nunavut: Unpub. M.Sc. thesis, University of New Brunswick, Fredricton, New Brunswick, 104 p.

Kah, L.C., 1997, Sedimentological, Geochemical, and Paleobiological Interactions on a Mesoproterozoic Carbonate Platform: Society Cliffs Formation, northern Baffin Island, Arctic Canada: Unpub. Ph.D. thesis, Harvard University, 190 p.

Kah, L.C., Sherman, A.G., Narbonne, G.M., Knoll, A.H., and Kaufman, A.J., 1999, ä13C stratography of the Proterozoic Bylot Supergroup, Baffin Island, Canada: Implications for regional lithostratigraphic correlations: Canadian Journal of Earth Sciences, v. 36, p. 313-332.

Kerr, J.W., 1977, Cornwallis Fold Belt and the mechanism of basement uplift: Canadian Journal of Earth Sciences, v. 14, p. 1374-1401.

Knight, R.D., and Jackson, G.D., 1994, Sedimentology and Stratigraphy of the Mesoproterozoic Elwin Subgroup (Aqigilik and Sinasiuvik Fms), uppermost Bylot Supergroup, Borden Rift Basin, northern Baffin Island: Geological Survey of Canada, Bulletin 455, 48 p.

Maurice, Y.T., 1982, Uraniferous Granites and Associated Mineralization in the Fury and Hecla Strait Area, Baffin Island, N.W.T.: Geological Survey of Canada, Paper 81-23, 13 p.

Mayr, U., de Freitas, T., Beauchamp, B., and Eisbacher, G., 1998, The Geology of Devon Island North of 76 Degrees, Canadian Arctic Archipelago: Geological Survey of Canada, Bulletin 526, 500 p.

Mayr, U., Brent, T.A., de Freitas, T., Frisch, T., Nowlan, G.S., and Okulitch, A.V., 2004, Geology of Eastern Prince of Wales Island and Adjacent Smaller Islands, Nunavut: Geological Survey of Canada, Bulletin 574, 88 p.

McNaughton, K., and Smith, T.E., 1986, A fluid inclusion study of sphalerite and dolomite from the Nanisivik lead-zinc deposit, Baffin Island, Northwest Territories, Canada: Economic Geology, v. 81, p. 713-720.

Miall, A.D., 1986, Effects of Caledonian tectonism in Arctic Canada: Geology, v. 14, p. 904-907.

Miall, A.D., and Gibling, M.R., 1978, The Siluro-Devonian clastic wedge of Somerset Island, Arctic Canada, and some regional paleogeographic implications: Sedimentary Geology, v. 21, p. 85-127.

Miles, W.F., Roest, W.R., and Vo, M.P., 2000a, Gravity Anomaly Map, Canada: Geological Survey of Canada, Open File 3830a, scale 1:7 500 000.

——— 2000b, Magnetic Anomaly Map, Canada: Geological Survey of Canada, Open File 3829a, scale 1: 7 500 000.

Mitchell, R.H., and Platt, R.G., 1984, The Freemans Cove volcanic suite; field relations, petrochemistry, and tectonic setting of nephelinite-basanite volcanism associated with rifting in the Canadian Arctic Archipelago: Canadian Journal of Earth Sciences, v. 21, p. 428-436.

Mortensen, P.S., and Jones, B., 1986, The role of contemporaneous faulting on Late Silurian sedimentation in the eastern M'Clintock Basin, Prince of Wales Island, Arctic Canada: Canadian Journal of Earth Sciences, v. 23, p. 1401-1411.

Mosher, G.Z., 1981, Geological report on prospecting permits 796, 797 Ellesmere Island, District of Franklin, Northwest Territories: Department of Indian and Northern Development, Assessment Report 81480, 30 p.

Muir, I.D., and Rust, B.R., 1982, Sedimentology of a Lower Devonian coastal alluvial fan complex; the Snowblind Bay Formation of Cornwallis Island, Northwest Territories, Canada: Bulletin of Canadian Petroleum Geology, v. 30, p. 245-263.

Nelson, W.E., 1984, Geological Report of the Field Programs on Mineral Prospecting Permit Numbers 974-997: Department of Indian and Norther Development, Assessment File 81858, 87 p.

Nixon, F.M., 1988, Till Sampling Program and Presentation of Physical and Geochemical Data from Western Victoria Island, Northwest Territories: Geological Survey of Canada, Paper 88-15, 36 p.

Noakes, S.B., 1998, Assessment Report: Prospecting and Sediment Sampling, Permit Numbers 2035, 2036, 2039-2045, west Melville Island Area, North Mining district, N.W.T.: Department of Indian and Northern Development, Report 84021, 24 p.

Okulitch, A.V., Packard, J.J., and Zolnai, A.I., 1986, Evolution of the Boothia Uplift, Arctic Canada: Canadian Journal of Earth Sciences, v. 23, p. 350-358.

Okulitch, A.V., 1991, Geology of the Canadian Arctic Archipelago and North Greenland, *in* Trettin, H.P., ed., Innuitian Orogen and Arctic Platform, Canada, and Greenland: Geological Survey of Canada, Geology of Canada, v. 3, scale 1:2 000 000.

Okulitch, A.V., and Trettin, H.P., 1991, Late Cretaceous - Early Tertiary deformation, Arctic Islands, Chapter 17 *in* Stott, D.F., and Aitkin, J.D., eds., Sedimentary Cover of the Craton in Canada: Geological Survey of Canada, Geology of Canada, no. 3, p. 469-489.

Okulitch, A.V., Packard, J.J., and Zolnai, A.I., 1991, Late Silurian-Early Devonian deformation of the Boothia Uplift, *in* Trettin, H.P., ed., Innuitian Orogen and Arctic Platform, Canada, and Greenland: Geological Survey of Canada, Geology of Canada, v. 3, p. 302-307.

Olson, R.A., 1984, Genesis of paleokarst and strata-bound zinc-lead sulfide deposits in a Proterozoic dolostone, northern Baffin Island, Canada: Economic Geology, v.79, p. 1056-1103.

Osadetz, K.G., and Moore, P.R., 1988, Basic Volcanics in the Hassel Formation (Mid-Cretaceous) and Associated Intrusives, Ellesmere Island, District of Franklin, Northwest Territories: Geological Survey of Canada, Paper 87-21, 19 p.

Packard, J.J., and Mayr, U., 1994, Middle Cambrian to Upper Ordovician carbonate platform, *in* Mayr, U., ed., The Phanerozoic Geology of Southern Ellesmere and North Kent Islands, Canadian Arctic Archipelago: Geological Survey of Canada, Bulletin 470, p. 30-67.

Patterson, K.M., Powis, K., Sutherland, R.A., and Turner, E.C., 2003, Stratigraphy and Structural Geology of the Nanisivik Area, Northern Baffin Island: Geological Survey of Canada, Open File 1552, 1 sheet, scale 1:100 000.

Pederson, F.D., 1980, Remobilization of the massive sulfide ore of the Black Angel Mine, central West Greenland: Economic Geology, v. 75, p. 1022-1041.

Pehrsson, S.J., and Buchan, K.L., 1999, Borden dykes of Baffin Island, Northwest Territories: A Franklin U-Pb Baddeleyite Age and a Paleomagnetic Reinterpretation: Canadian Journal of Earth Sciences, v. 36, p. 65-73.

Rainbird, R.H., Jefferson, C.W., Hildebrand, R.S., and Worth, J.K., 1994, The Shaler Supergroup and Revision of Neoproterozoic Stratigraphy in Amundsen Basin, Northwest Territories: Geological Survey of Canada, Paper 1994-C, 10 p.

Rees, M., 1999, Report of 1997 Geology and Geochemistry at Kitson River and Sabine Peninsula, Melville Island, N.W.T.: Department of Indian and Northern Development, Assessment Report 84176, 54 p.

Reinson, G.E., Kerr, J.W., and Stewart, W.D., 1976, Stratigraphic Field Studies, Somerset Island, District of Franklin (58B to F): Geological Survey of Canada, Paper 76-1A, 3 p.

Roscoe, S.M., 1984, Assessment of Mineral Resource Potential in the Bathurst Inlet Area, Including the Proposed Bathurst Inlet National Park: Geological Survey of Canada, Open File 788, 75 p.

Ruzicka, V., 1977, Assessment of Selected Uranium Occurrences and Areas Favourable for Uranium Mineralization in Canada: Geological Survey of Canada, Paper 1977-1A, 6 p.

Samuelsson, J., Dawes, P.R., and Vidal, G., 1999, Organic-walled microfossils from the Proterozoic Thule Supergroup, Northwest Greenland: Precambrian Research, v. 96, p. 1-23.

Sangster, D.F., 1981, Part 3.1 Metallic minerals, mineral and hydrocarbon resource potential of the proposed northern Ellesmere Island national park, District of Franklin, N.W.T.: Geological Survey of Canada, Open File 786, p. 9-12.

——— 1998, Mineral Deposits Compilation and Metallogenic Domains, Northern Baffin Island and Northern Melville Peninsula, Northwest Territories: Geological Survey of Canada, Open File 3635, scale 1:1,000,000.

——— 1999, Metallogenic estimate of undiscovered mineral resource potential, northern Baffin Island and northern Melville Peninsula, Northwest Territories (Nunavut): Geological Survey of Canada, Open File 3637, p. 43-58.

Scott, D.J., and de Kemp, E.A., 1998, Bedrock Geology Compilation, Northern Baffin Island and Northern Melville Peninsula, Northwest Territories: Geological Survey of Canada, Open File 3633, 3 sheets, scale 1:500,000.

Scott, D.J., St-Onge, M.R., and Corrigan, D., 2003, Geology of the Archean Rae Craton and Mary River Group and the Paleoproterozoic Piling Group, Central Baffin Island, Nunavut: Geological Survey of Canada, Current Research 2003-C26, 12 p.

Sharpe, D.R., 1992, Quaternary Geology of Wollaston Peninsula, Victoria Island, Northwest Territories: Geological Survey of Canada, Memoir 434, 91 p.

Sherlock, R.L., Lee, J.K.W., and Cousens, B.L., 2004, Geologic and geochronologic constraints on the timing of mineralization at the Nanisivik Zn-Pb Mississippi Valley-Type deposit, Northern Baffin Island, Nunavut, Canada: Economic Geology, v. 99, p. 279-293.

Sherman, A.G., James, N.P., and Narbonne, G.M., 2002, Evidence for reversal of basin polarity during carbonate ramp development in the Mesoproterozioc Borden Basin, Baffin Island: Canadian Journal of Earth Sciences, v. 39, p. 519-538.

St.Marie, J., Kesler, S.E., and Allen, C.R., 2001, Origin of iron-rich Mississippi Valley-type deposits: Geology, v. 29, p. 59-62.

Stewart, W.D., 1987, Late Proterozoic to Early Tertiary Stratigraphy of Somerset Island and Northern Boothia Peninsula, District of Franklin, N.W.T.: Geological Survey of Canada, Paper 83-26, 78 p.

Sutherland, R.A., and Dumka, D., 1995, Geology of Nanisivik mine, N.W.T., Canada, *in* Misra-Luca, C., ed., Carbonate-Hosted Lead-Zinc-Fluorite-Barite Deposits of North America: Society of Economic Geologists, Guidebook Series 22, p. 4-12.

Symons, D.T.A., Symons, T.B., and Sangster, D.F., 2000, Paleomagnetism of the Society Cliffs dolostone and the age of the Nanisivik zinc deposits, Baffin Island, Canada: Mineralium Deposita, v. 35, p. 672-682.

Thorsteinsson, R., 1986, Geology of Cornwallis Island and Neighbouring Smaller Islands, Canadian Arctic Archipelago: Geological Survey of Canada, Map 1626A, 1:250,000.

Thorsteinsson, R., and Mayr, U., 1987, The Sedimentary Rocks of Devon Island, Canadian Arctic Archipelago: Geological Survey of Canada, Memoir 411, 182 p.

Thorsteinsson, R., and Tozer, E.T., 1970, Geology of the Arctic Archipelago, *in* Douglas, R.J.W., ed., Geology and Economic Minerals of Canada: Geological Survey of Canada, Economic Geology Report no. 1, p. 547-590.

Thorsteinsson, R., and Uyeno, T.T., 1980, Stratigraphy and Conodonts of Upper Silurian and Lower Devonian Rocks in the Environs of the Boothia Uplift, Canadian Arctic Archipelago: Geological Survey of Canada, Bulletin 292, 75 p.

Trettin, H.P., 1969, Lower Paleozoic Sediments of Northwestern Baffin Island, District of Franklin: Geological Survey of Canada, Bulletin 157, 70 p.

——— 1987, Pearya: A composite terrane with Caledonian affinities in northern Ellesmere Island: Canadian Journal of Earth Sciences, v. 24, p. 224-245.

——— 1994, Pre-Carboniferous Geology of the Northern Part of the Arctic Islands: Hazen Fold Belt and Adjacent Parts of Central Ellesmere Fold Belt, Ellesmere Island: Geological Survey of Canada, Bulletin 430, 248 p.

Trettin, H.P., Parrish, R., and Loveridge, W.D., 1987, U-Pb age determinations on Proterozoic to Devonian rocks from northern Ellesmere Island, Arctic Canada: Canadian Journal of Earth Sciences, v. 24, p. 246-256.

Tuke, M.F., Dineley, D.L., and Rust, B.R., 1966, The basal sedimentary rocks in Somerset Island, N.W.T.: Canadian Journal of Earth Sciences, v. 3, p. 697-711.

Turner, E.C., 2003a, New Contributions to the Stratigraphy of the Mesoproterozoic Society Cliff Formation, Borden Basin, Northern Baffin Island, Nunavut: Geological Survey of Canada, Open File 2003-B3, 13 p.

——— 2003b, Lead-Zinc Showings Associated with Debrite Layer Shed from Synsedimentary Faults, Mesoproterozoic Society Cliff Formation, Northern Baffin Island, Nunavut: Geological Survey of Canada, Open File 2003-B2, 7 p.

——— 2004a, Stratigraphy pf the Mesoproterozoic Society Cliff Formation (Borden Basin, Nunavut): Correlation between Northwestern and Southeastern Areas of the Milne Inlet Graben: Geological Survey of Canada, Open File 2004-B3, 11 p.

———2004b, Origin of Basinal Carbonate Laminites of the Mesoproterozoic Society Cliff Formation (Borden Basin, Nunavut) and Implications for Base-Metal Mineralization: Geological Survey of Canada, Open File 2004-B2, 12 p.

——— 2004c, Kilometre-Scale Carbonate Mounds in Basinal Strata: Implications for Base-Metal Mineralization in the Mesoproterozoic Arctic Bay and Society Cliffs Formations, Borden Basin, Nunavut: Geological Survey of Canada, Open File 2004-B4, 12 p.

van der Stijl, F.W., and Mosher, G.Z., 1998, The Citronen Fjord massive sulphide deposit, Peary Land, north Greenland: Discovery, mineralization, and structural setting: Geology of Greenland Survey Bulletin, v. 179, p. 40.

Williamson, M.-C., 2000, The potential of flood basalts for hosting magmatic sulphide deposits: an application of exploration criteria to the Sverdrup Basin, Nunavut, Arctic Canada: Atlantic Geology, v. 36, p. 175-176.

Tectonics and Metallogeny of the British Columbia, Yukon and Alaskan Cordillera, 1.8 Ga to the Present

JoAnne Nelson[1] and Maurice Colpron[2]

1. British Columbia Geological Survey, Box 9333 Stn Prov Govt, 5th Floor 1810 Blanshard St., Victoria, British Columbia V8W 9N3

2. Yukon Geological Survey, P.O. Box 2703 (K-10), Whitehorse, Yukon Y1A 2C6

Corresponding author's email: JoAnne.Nelson@gov.bc.ca

Abstract

The northern Cordilleran Orogen of western Canada and Alaska comprises rocks that attest to over 1.8 billion years of tectonic history, from cratonization of the Laurentian continental core to current subduction and transform motion off the west coast today. Evolving tectonic styles, ranging from Proterozoic intracratonic basin formation through Paleozoic rifting through the construction of Mesozoic and younger intraoceanic and continent margin arcs, has led to the wide variety of metallogenetic styles that define the mineral wealth of the northern Cordillera.

The northern Cordillera is made up of five large-scale tectonic provinces: the Laurentian craton and its deformed margins; allochthonous terranes of the peri-Laurentian realm, that represent offshore rifted continental fragments, arcs, and ocean basins formed in a setting similar to the modern western Pacific province; the Arctic and Insular terranes, a group of crustal fragments that originated in the Arctic realm between Laurentia and Siberia and were transported separately southwards to impinge on the outer peri-Laurentian margin in Mesozoic time; and the late-accreted Pacific terranes, Mesozoic to Cenozoic accretionary prisms that developed along an active Pacific plate margin with a configuration much like the present one. Each tectonic province carries its own metallogenetic signature. Superimposed post-accretionary magmatic arcs and compressional and extensional tectonic regimes have also given rise to important mineral deposit suites. Seafloor hotspring deposits forming presently along the Juan de Fuca Ridge off the southwest coast of British Columbia show the continuation of Cordilleran metallogeny into the foreseeable future.

Résumé

La partie nord de l'orogène de la Cordillère, qui s'étire le long de la côte Ouest du Canada et en Alaska, comprend des roches témoignant d'une évolution tectonique longue de plus de 1,8 milliard d'années. Amorcée par la cratonisation du noyau continental laurentien, cette évolution se poursuit de nos jours par le déplacement de la côte Ouest par subduction et coulissage. Des styles tectoniques en évolution, depuis la formation de bassins intracratoniques au Protérozoïque jusqu'à la distension continentale au Paléozoïque puis à l'édification d'arcs intra-océaniques et d'arcs de marge continentale au Mésozoïque et plus récemment, ont engendré toute une gamme de styles métallogéniques déterminant la richesse minérale du nord de la Cordillère.

La Cordillère septentrionale comprend cinq grandes provinces tectoniques : le craton laurentien et ses marges déformées; les terranes allochtones du domaine péri-laurentien, constitués de fragments continentaux, qui se sont détachés par rifting et dispersés au large, d'arcs et de bassins océaniques formés dans un cadre similaire à celui de l'actuelle province du Pacifique occidental; les terranes arctiques et insulaires, un groupe de fragments crustaux émanant du domaine arctique entre la Laurentie et la Sibérie, transportés séparément vers le sud et poussés contre la marge péri-laurentienne au Mésozoïque; et les terranes du Pacifique d'accrétion tardive, constitués de prismes d'accrétion datant du Mésozoïque au Cénozoïque, qui se sont formés le long d'une marge active de la plaque du Pacifique présentant une configuration très semblable à la marge actuelle. Chacune de ces provinces présente une signature métallogénique qui lui est propre. Les arcs magmatiques post-accrétionnaires et les régimes tectoniques de compression et d'extension ont en outre engendré d'importantes successions de gîtes minéraux. Les dépôts de sources chaudes qui s'accumulent actuellement sur le fond marin le long de la dorsale Juan de Fuca, au large de la côte sud-ouest de la Colombie-Britannique, attestent de la continuation de la métallogénie cordillérienne dans l'avenir prévisible.

Introduction

This synthesis spans the northern Cordilleran Orogen of British Columbia, Yukon, and Alaska (Fig. 1). The modern Cordillera was founded upon the western margin of the Laurentian craton (later North America), a continental margin that has been tectonically active from its inception in Late Precambrian time up to the present. Because the northern Cordillera has evolved as an active belt over such a protracted period of time, through diverse tectonic styles and paleogeographic configurations, it provides an excellent case study of how metallogenesis is governed by tectonics. The purpose of this article is to weave the stories of metallic mineral deposits into the overall tectonic history of the northern Cordillera, and also to serve as a portal for more detailed sources of information, for instance Nokleberg et al. (2000, 2005).

Metallic mineral deposits preserved within the northern Cordillera range in age from 1.6 billion to less than 20 million years. The wide variety of deposit types reflects tectonic processes such as intra-continental and intra-arc extension, and the multistage development of pericratonic fringing magmatic arcs. Of most economic significance at present are SEDEX (sedimentary exhalative), VMS (volcanogenic massive sulphide), porphyry, and mesothermal (orogenic), epithermal and intrusion-related gold deposits. Many other deposit types, including Irish-type syngenetic sulphide, MVT (Mississippian Valley-type Pb-Zn), skarn, and iron-oxide copper-gold, are also well represented.

The northern Cordillera is a highly fertile metallogenetic environment that hosts world-class orebodies. The most notable mined examples include the 160 Mt Sullivan, the 142 Mt Red Dog, and the 120 Mt Anvil Range SEDEX

Nelson, J., and Colpron, M., 2007, Tectonics and metallogeny of the British Columbia, Yukon and Alaskan Cordillera, 1.8 Ga to the present, *in* Goodfellow, W.D., ed., Mineral Deposits of Canada: A Synthesis of Major Deposit-Types, District Metallogeny, the Evolution of Geological Provinces, and Exploration Methods: Geological Association of Canada, Mineral Deposits Division, Special Publication No. 5, p. 755-791.

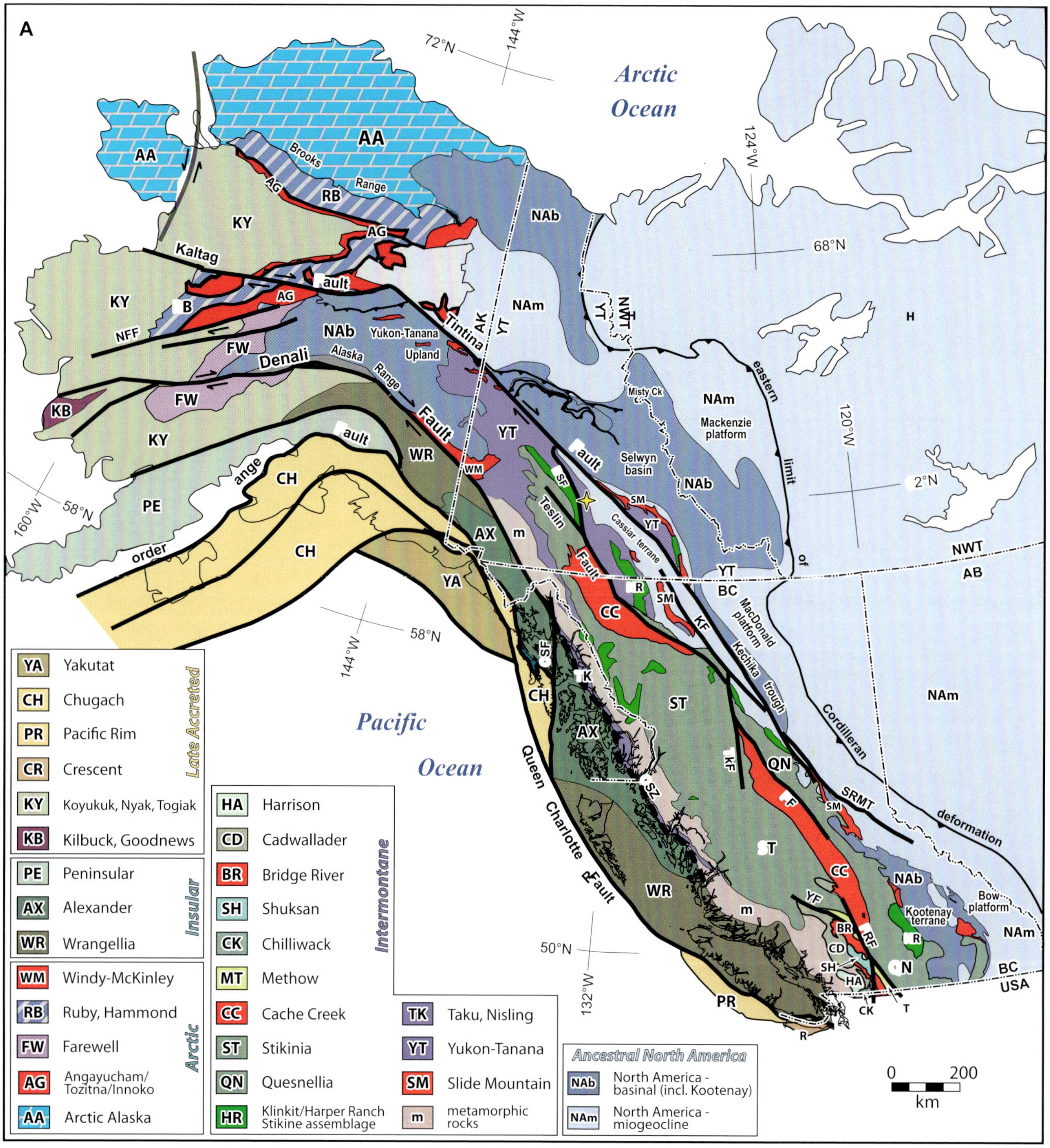

FIGURE 1. Terranes (**A**) and tectonic realms (**B** – facing page) of the Canadian-Alaskan Cordillera. Geological features/regions and major strike-slip fault systems discussed in the text and restored in Figure 2 are shown. Yellow star marks the location of Solitary Mountain in south-central Yukon. Fault abbreviations: BSF – Big Salmon fault; CSF – Chatham Strait fault; CSZ – Coast shear zone; FRF – Fraser River fault; KF – Kechika fault; NFF – Nixon Fork-Iditarod fault; PF – Pinchi fault; SMRT – southern Rocky Mountain trench; TkF – Takla-Finlay-Ingenika fault system; YK – Yalakom fault. Other abbreviations: AB – Alberta; AK – Alaska; BC – British Columbia; NWT – Northwest Territories; YT – Yukon Territory. Sources: Wheeler et al. (1991); Silberling et al. (1994); Colpron (2006).

deposits; the long-lived, billion-tonne Highland Valley Cu-Mo porphyry deposits; as well as a host of other producing or past-producing Cu±Au±Mo porphyries that are a vital part of the economy of interior British Columbia (location, grade, and tonnage data on all cited deposits are listed in Appendix 1). Other world-class prospects are moving towards mine stage at present. These include British Columbia's Galore Creek and Alaska's Pebble porphyry deposits, which contain resources of 385 million and 1 billion tonnes, respectively.

The set of deposits included in this discussion is, of necessity, a small subset, slightly over 130 mines and major undeveloped mineral resources, of the 13 000 occurrences listed

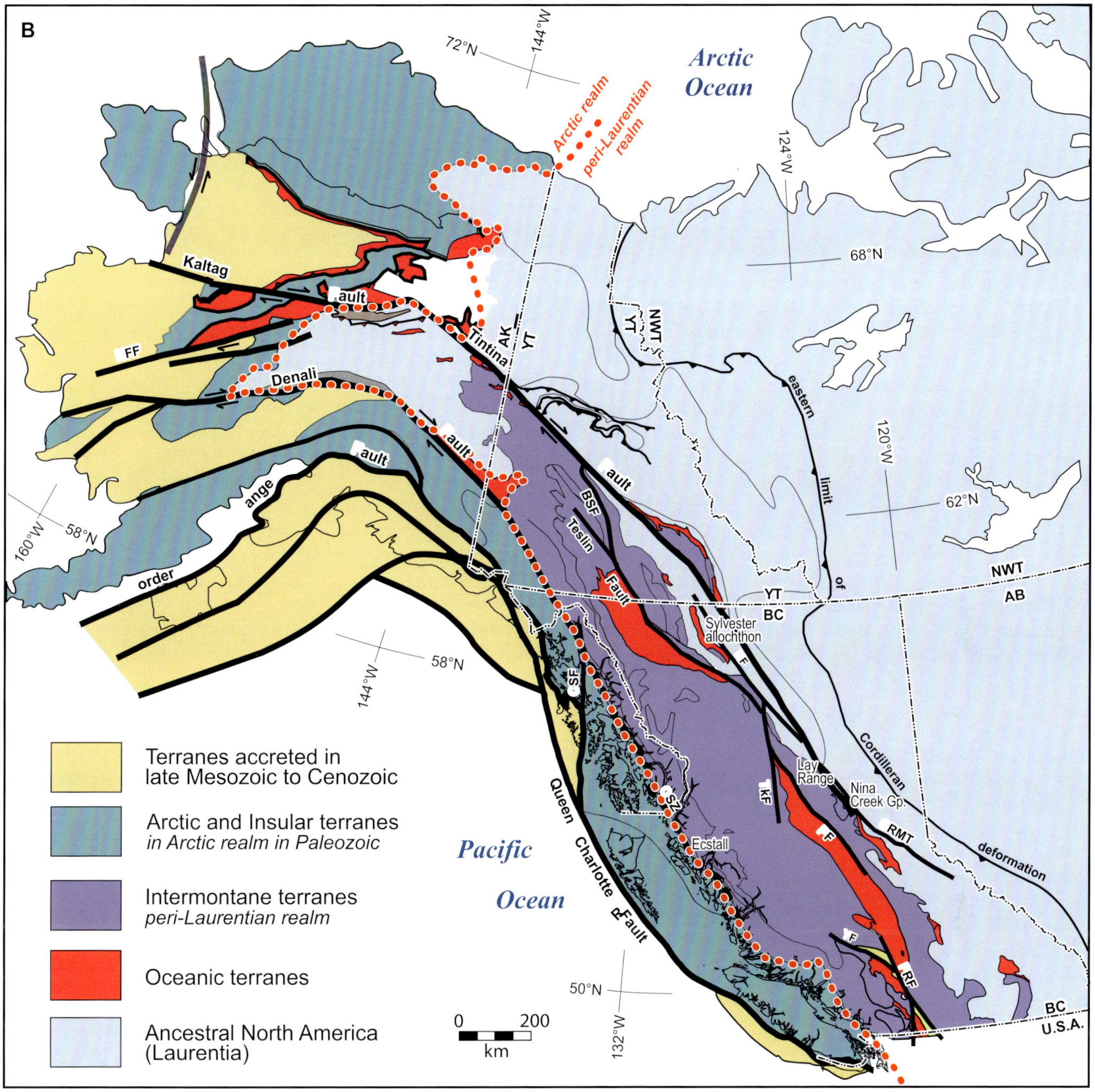

FIGURE 1 CONTINUED.

in British Columbia's Minfile, the over 2600 in Yukon's Minfile, the more than 7100 occurrences listed in the Alaska Resource Data File, and the 329 occurrences in the NORMIN database for the Mackenzie Mountains in Northwest Territories. The list was generated partly based on size and grade of individual deposits and of cumulative economic potential of belts and camps, but also on less easily quantifiable criteria, such as perceived economic potential and linkage to large-scale tectonic events. In the end, we have made a subjective selection; we thank our respective organizations for their input, but the ultimate choices were our own.

Overview of Cordilleran Tectonics

Within the northern Cordillera, Proterozoic to Triassic miogeoclinal, mainly sedimentary, platformal to basinal strata of the Laurentian continental margin (NAm and NAb on Fig. 1A) extend into eastern British Columbia and Yukon. These strata are incorporated into the fold and thrust belt of the eastern Cordillera. Farther west, most of British Columbia, Yukon, and Alaska are made up of Paleozoic to Mesozoic volcanic, plutonic, sedimentary, and metamorphic assemblages that represent magmatic arcs, microcontinents and ocean basins, accreted to western Laurentia in Mesozoic and younger time. These, along with the parautochthonous deformed belt and the undisturbed platform of Alberta, are overlain by syn- and post-accretionary clastic deposits. The western and inner parts of the orogen are pierced by post-accretionary plutons and in places overlain by thick accumulations of relatively young volcanic strata.

The variety of pre-accretionary assemblages, many of them juxtaposed across faults, led to the conceptualization of the Cordillera as a collage of tectonostratigraphic terranes – independent, fault-bounded entities with "suspect" or unspecified relationships to each other as well as to the continent (Coney et al., 1980; Jones et al., 1983). After a quarter of a century of subsequent research, the terranes on Figure 1A are remarkably similar in outline to those originally proposed – a testament to the insight and inspiration of that work. On the other hand, considerable progress has been made in refining the origins, relationships, and individual and collective histories of these entities. The result is an emerging image of Cordilleran evolution that is simpler, clearer, and more explicit in its historical implications.

Although the northern Cordillera has been subdivided into a multiplicity of terranes (Fig. 1A), it can be viewed more globally as consisting of five first-order tectonic entities: 1) Ancestral North America (Laurentia); 2) the allochthonous marginal pericratonic terranes (Intermontane terranes); 3) the exotic Insular and Farewell terranes; 4) Arctic Alaska; and 5) Mesozoic and younger arc and accretionary terranes that form a western and southern fringe to the older elements. These are shown on Figure 1B, which is a simplified and, in places, updated version of the terrane maps of Wheeler et al. (1991), Silberling et al. (1994), and Nokleberg et al. (2000).

Ancestral North America (or Laurentia) includes the western craton margin, the miogeocline with its platforms (Mackenzie, McDonald, and Bow) and basins (Selwyn, Kechika, and Misty Creek) and its fringing, parautochthonous pericratonic terranes (Cassiar and Kootenay) (Fig. 1A). The western, outboard boundary of this autochthonous to parautochthonous belt is marked by discontinuous slivers and slices of the Slide Mountain oceanic terrane, which were derived from a marginal rift basin of Late Devonian to Permian age that once lay between the continent and a mixed belt of rifted pericratonic and Devonian through Jurassic arc terranes (Yukon-Tanana, Quesnellia, and Stikinia). The belt of pericratonic terranes was originally bounded on its outer, oceanward margin by an accretionary zone comprised by what is now referred to as the Cache Creek terrane, a relict fore-arc assemblage that includes slivers of high-pressure metamorphic assemblages, as well as blocks containing Permian fusulinid and coral faunas of very exotic, Tethyan (Asian) affinity (Ross and Ross, 1983). The terranes of the Intermontane belt have been referred to as the Intermontane superterrane by Monger et al. (1982), who regarded them as a product of Triassic-Jurassic tectonic amalgamation. In this discussion, we describe initial and ongoing relationships that span the entire period of the existence of these terranes. In keeping with Appalachian terminology (Zagorevsky et al., 2006), these terranes could be said to constitute the peri-Laurentian realm (Fig. 1B). The position of the exotic Cache Creek terrane, enclosed within the pericratonic belt, is a constructional anomaly that may be best explained by oroclinal enclosure that developed as the Intermontane terranes amalgamated and accreted to the continent (Mihalynuk et al., 1994a).

In contrast to the Intermontane terranes, the Insular terranes (Wrangellia and Alexander) and the Farewell terrane of central Alaska, although long-lived (Precambrian to Triassic) and in part of pericratonic origin, show no evidence of early relationships to the western margin of North America. Instead, their early faunal and isotopic affinities are with Siberia and Barentia (Bazard et al., 1995; Nokleberg et al., 2000; Bradley et al., 2003). The Arctic Alaska composite terrane, although continental to pericratonic, is anomalous with respect to adjacent western North America (Patrick and McClelland, 1995; Amato, 2004), and it is commonly considered displaced and/or rotated; it too was probably mobile within the Arctic realm in pre-Cretaceous time. It bears stratigraphic similarities to the Chukotka peninsula of the Russian Far East and it has been proposed that they were contiguous throughout most of their history (Natal'in, 2004; Amato et al., 2006). Together, these terranes constitute an original set of detached crustal fragments, along with subsequent Paleozoic and Mesozoic arcs and basins, which developed mainly within the Arctic realm: they are referred to here collectively as the Arctic/Insular terranes (Fig. 1B).

Early in the history of Wrangellia, a Pennsylvanian pluton linked it to the Alexander terrane (Gardner et al., 1988). By Late Triassic to Early Jurassic time, at least Wrangellia seems to have been transported to a more southerly paleolatitude west of the North American margin, prior to mid-Jurassic accretion with the Intermontane terranes (Aberhan, 1999; Smith et al., 2001). After accretion, but prior to 430 km of Eocene dextral displacement on the Tintina fault (Gabrielse et al., 2006), this set of terranes bounded the combined North American margin and Intermontane terranes to the west (Fig. 2).

The outermost belt of terranes contains relatively young, Mesozoic to Paleogene assemblages, including the accreted Yukon-Koyukuk and Talkeetna arcs, the accreted Paleocene-Eocene seamounts of the Crescent terrane, and the Chugach, Pacific Rim, and Yakutat terranes, which are accretionary complexes dominated by trench sediments. Older crustal fragments, such as the Precambrian Kilbuck terrane (Box et al., 1990), are also involved in this mass of late-accreted crustal material. These arc and accretionary assemblages developed within the eastern Pacific realm near or on the developing Cordilleran margin.

Realm Boundaries

If the major realms of the Cordillera – the autochthonous North American, the peri-Laurentian, and Arctic realms – are first-order tectonic features, then the nature of the boundaries between them becomes of particular interest in understanding Cordilleran evolution. The boundary between the edge of ancestral North America and the Intermontane (peri-Laurentian) terranes lies within the Slide Mountain terrane (Fig. 1A,B), a Late Devonian to Middle Permian oceanic assemblage of subtly expressed pericratonic affinity (Nelson, 1993). The Slide Mountain terrane exhibits Paleozoic stratigraphic linkages both to North America (Klepacki and Wheeler, 1985) and to the Yukon-Tanana terrane (Murphy et al., 2006); it contains interbedded MORB (mid-ocean ridge) basalts and Mississippian chert-quartz sandstones and conglomerates that contain Precambrian detrital zircons (Nelson, 1993, and unpublished data).

The boundary between the Intermontane and Arctic/Insular terranes (peri-Laurentian and Arctic realms) is more variable in nature and is marked by very few oceanic

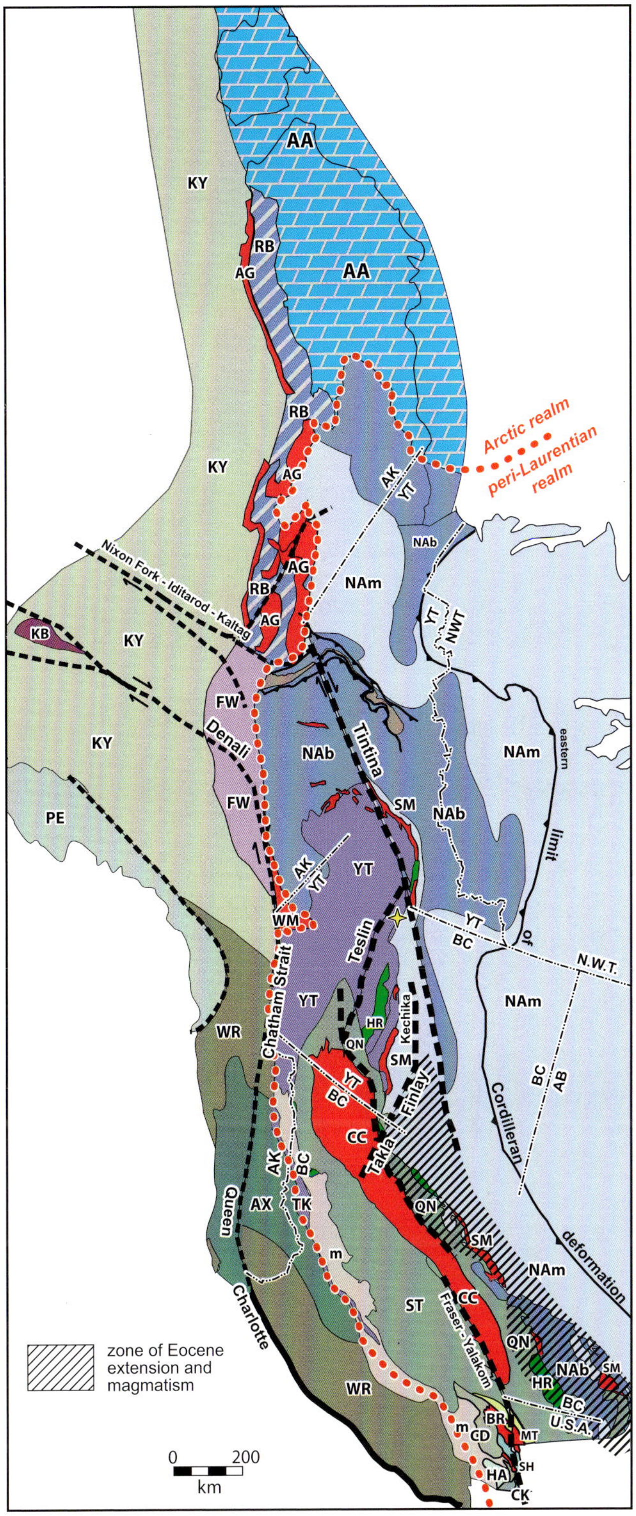

FIGURE 2. An approximate pre-Late Cretaceous restoration of the Canadian-Alaskan Cordillera. Constraints on the restoration of the dextral strike-slip faults are discussed in the text (e.g. 430 km on Tintina [Gabrielse et al., 2006]; 370 km on Denali [Lowey, 1998]; 150 to 200 km on Chatham Strait [Hudson et al., 1982]; 130 km on Kaltag/Tozitna [Patton and Tailleur, 1977]). Southward projections of the Tintina fault are assumed to lie between Stikinia and southern Quesnellia. Additional motion on the Teslin and related faults is required to restore the two "halves" of the Cache Creek terrane (Gabrielse, 1985). Southern projection of the Denali fault is located on the Chatham Strait fault. Restoration of this bend results in considerable clockwise rotation. To compensate, the Insular terrane block has been rotated counterclockwise to lie adjacent to the Coast Mountains. The Brooks Range is rotated clockwise 30°, and the Seward Peninsula is aligned with it. In general, only rigid-body displacements have been restored. For instance, Eocene extension related to dextral transtension in central and southern British Columbia has not been restored in this diagram. Terranes of the Pacific realm are not shown, as their locations at this time are poorly constrained. Terrane legend as in Figure 1A. Yellow star marks the approximate location of Solitary Mountain.

remnants. In the Coast Mountains of British Columbia and southeast Alaska, it is a collisional suture between the Alexander/Wrangellia terranes and the Yukon-Tanana terrane, onlapped by the Gravina belt, a mid-Jurassic to Cretaceous arc and clastic sequence (Gehrels, 2001). In western Yukon, the intervening oceanic Windy-McKinley terrane shows intriguing similarities with the Alexander terrane (Mortensen and Israel, 2006): it may be a fragment of intervening oceanic crust. In central Alaska, the Livengood belt of the Farewell terrane is in thrust contact with western North American strata with intervening Jurassic-Cretaceous flysch and minor ultramafic rocks (Beaver Creek belt of Dover, 1994). This structural setting is complicated by displacement on the Tintina fault and its splays, and the Z-shaped Alaska orocline (Patton and Tailleur, 1977; Churkin et al., 1980; Moore et al., 1987; Coe et al., 1989; Stone, 1989; Dover, 1994, among many others); a feature related both to northerly motion of the block west of the Tintina fault and, in many models, to counterclockwise rotation of the Brooks Range (compare Figs. 1 and 2).

The Ruby terrane is presently an enigma. It is pericratonic, and intruded by Devonian plutons. It may be a North American salient, or it could be affiliated with the Hammond terrane in the southern Brooks Range, a part of the Arctic Alaska terrane that also contains Late Devonian plutons. The Arctic Alaska composite terrane in the Brooks Range is separated from western North America to the east by a thrust belt that involves Cretaceous overlap strata (Dover, 1994; Gordey and Makepeace, 2001). Its outer (southern) margin is in the footwall of a thrust fault that carries the Devonian to Jurassic Angayucham oceanic terrane in its hanging wall. Cretaceous blueschists within the footwall of this fault zone are evidence for subduction of the Arctic Alaska terrane beneath the Angayucham Ocean and the Jurassic-Cretaceous Yukon-Koyukuk arc (Moore et al., 2004). The Ruby terrane exhibits similar structural relationships and timing; however, it faces northwest. The pericratonic Hammond and Ruby terranes, coupled with the Angayucham oceanic terrane, presently enclose the Yukon-Koyukuk terrane in a west-facing V-shaped frame (Fig. 1A,B).

Structure of this Paper

As a prologue to the tectonic/metallogenetic histories presented here, we lay out the existing constraints for historical reconstruction of the evolving Cordillera, starting with the restoration of estimated Late Cretaceous to Eocene strike-slip displacements and their consequences (Fig. 2). This reconstruction emphasizes the integrity of the fundamental components of the Cordillera: the amalgamated Intermontane terranes, the Insular and Farewell terranes as micro-continental blocks; the continuity of the Ruby and Arctic Alaska terranes; the not-yet-accreted Pacific terranes. Figure 2 incorporates the best available estimates and interpretations of displacements and is offered as a reasonable post-accretion, pre-strike-slip dismemberment interpretation of the Cordillera and the large-scale entities within it.

The main part of the paper is a chronological treatment of tectonics and metallogeny of western North America (Laurentia) and the peri-Laurentian realm; the Arctic/Insular realm; and then syn- and post-accretionary events and deposits. We first describe the development of Paleoproterozoic to Neoproterozoic intracratonic basins of Laurentia and their contained deposits, which are mostly older than and unrelated to the opening of the western margin of the continent during the breakup of Rodinia. Second, we describe, starting with that rift event, the tectonic development and metallogeny of the combined western North American (Laurentian) margin and Intermontane terranes, as an evolving southwest Pacific-style active margin. Its timeline spans Late Proterozoic time through the last Early Jurassic manifestation of pre-accretionary arc development in offshore Quesnellia and Stikinia. Third, we separately describe the tectonic development and metallogeny of the Arctic/Insular terranes. As their origins and "terrane trajectories" are less well understood, and each may well have developed independent of the others, this treatment is of necessity more piecemeal.

Finally, we cover Late Jurassic and younger tectonic patterns and metallogeny, which developed during and after accretion of the Arctic/Insular terranes to the Intermontane terranes and continent fringes, when a simple west-facing subduction system was imposed on the new, amalgamated, and much broader western continent margin. The Mesozoic and younger arcs of the outer terranes are considered as intra-oceanic extensions of those superimposed on the growing Cordillera. This approach was first applied successfully by Nokleberg et al. (2000), who emphasized the role of paired belts (arcs and their accretionary complexes) in interpreting the evolution of the circum-Pacific region.

Reconstructing the Pre-Late Cretaceous Cordillera: Faults, Cut-offs, and Transverse Features versus Stipulations of Paleomagnetic Data

Throughout this paper, the present-day distribution of terranes on Figures 1A and 1B is used as a framework for locating mineral deposits and districts. However, it is useful first to visualize the different sets of terranes as they were just after accretion but prior to their dismemberment by Cretaceous and younger transcurrent faults. Figure 2 represents a graphical re-organization of the present distribution of terranes on Figure 1A. It follows the same lines as the reconstruction of Pavlis (1989), with added constraints from recently published geological evidence and revised interpretations. Its configuration in mid-Cretaceous time is largely constrained by:

1. <u>Restoration of 430 to 490 km of Eocene dextral motion on the Tintina fault</u>, most of which occurred in Eocene time (Gabrielse et al., 2006). The amount and timing of this displacement is constrained by the following: offset of the northeastern section of Yukon-Tanana terrane from its main exposures southwest of the fault; offset of a set of Cretaceous thrust imbricates north of Fairbanks, which are considered equivalent to the Robert Service and Tombstone thrusts in Yukon (Fig. 1A); offset of the ca. 92 to 90 Ma Tombstone plutonic suite from its Alaskan equivalent, the Livengood suite; and offset of Late Cretaceous (70-65 Ma) Sn-bearing peraluminous intrusions (McQuesten plutonic suite in Yukon). North of Fairbanks, the Tintina fault abruptly bends and splays into the Kaltag, Tozitna, and other faults that have a cumulative dextral displacement of approximately 130 km (Patton and Tailleur, 1977). The remaining displacement was probably taken up in compression and rotation.
2. <u>Restoration of approximately 100 km of Eocene dextral motion on the Fraser and Yalakom faults, to restore offset imbricate fault panels</u> (Umhoeffer and Schiarizza, 1996). As Price and Carmichael (1986) pointed out, the amount of Eocene dextral fault offset is far less in the southern Cordillera than it is on the Tintina fault and much of the relative motion was probably expressed as extension.
3. <u>Restoration of approximately 370 km of Eocene and younger dextral motion on the Denali fault</u> (Lowey, 1998). In this reconstruction, we use the Chatham Strait fault as a southern continuation of the Denali fault (Hudson et al., 1982). This results in an overall shortening and widening of the pre-displacement Insular terrane.
4. <u>Rotation of Arctic Alaska 30°+ clockwise</u> and restoration of the Seward Peninsula to align with the Brooks Range (Patton and Tailleur, 1977). The motion history of Arctic Alaska during Cretaceous opening of the Canada Basin is controversial, with models ranging from extreme counterclockwise rotations up 66 degrees from an original position against the Canadian Arctic Islands (Grantz et al., 1990; Toro et al., 2004), to a complex opening history of the Arctic Ocean that involved spreading in several senses, as well as strike-slip displacement (Lane, 1992, 1997). Our restoration represents a compromise. In its favour, we point out that this restoration, in concert with back-rotation of the Ruby terrane as displacement along Tintina fault is restored, results in a fairly straight continental margin, and eliminates the anomalous V-shaped enclosure of the Yukon-Koyukuk arc.

Figure 2, although a cartoon and not unique, is a more coherent image of the Cordillera than the current one (Fig. 1A,B), which has been shaped by post-accretionary strike-slip tectonics, development of the arcuate southern Alaska orocline (Coe et al., 1989), the contortion of Alaskan tectonic trends into the Z-shaped present day configuration (Patton and Tailleur, 1977; Pavlis, 1989), and the extrusion of terranes into the Bering Sea region (Scholl, 1991; Redfield et al., 2005). In Figure 2, the two offset parts of the Yukon-Tanana terrane are joined, as are the two separate belts of the Cache Creek terrane. The northwestern tip of the ancestral North American continent is gently curvilinear and in part bounded by a Cretaceous thrust complex. The Farewell and Insular terranes are depicted as separate accreted micro-continents.

The key message embodied in Figure 2 is that the Cordillera can be viewed as a set of very few, large-scale tectonic entities – the realms – of which individual, smaller terranes form a part. The history of each of these entities involves interaction between its components over a long period of time, except in the case of the very young, outermost Pacific terranes. In each of these histories, metallogeny closely follows on tectonic developments, particularly rift zone development and the growth of superimposed magmatic arcs.

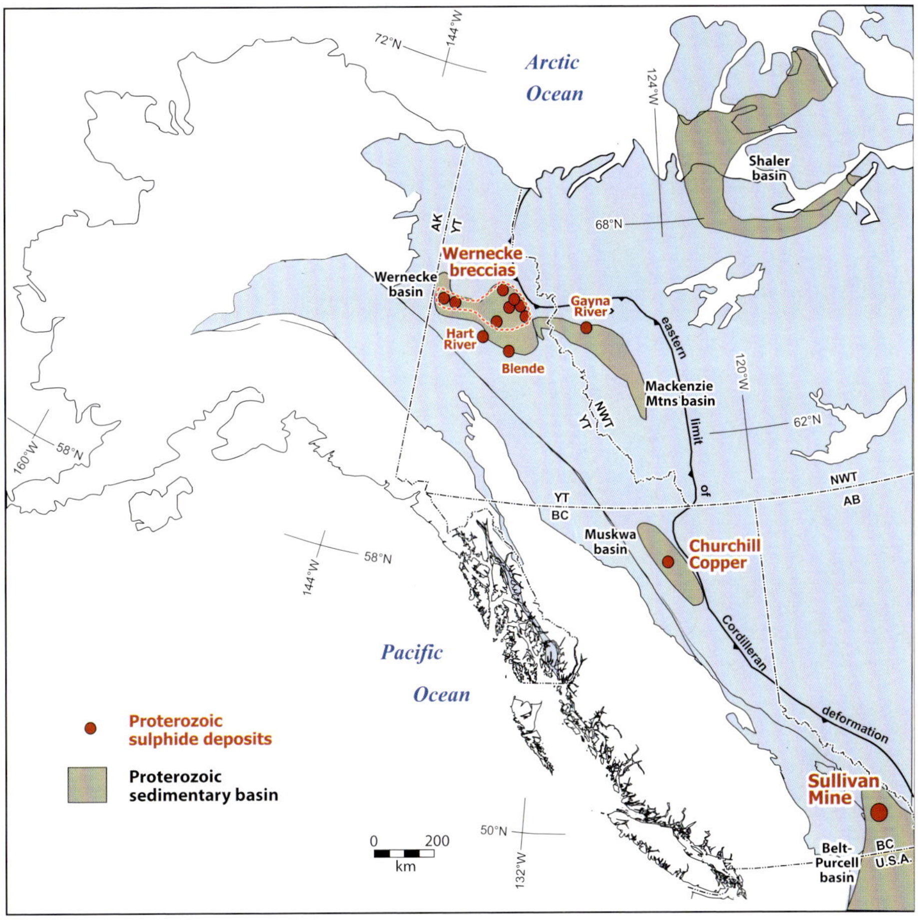

FIGURE 3. Proterozoic basins and mineral deposits of northwestern Laurentia. The basins are generalized from Wheeler and McFeely (1991); Wernecke deposits are from Lefebure et al. (2005).

Assessment of Cretaceous Paleomagnetic Data and Its Relevance to Tectonic Reconstructions

The geologically based reconstructions that underlie the configuration shown in Figure 2 are at odds with a literal paleogeographic interpretation of the Cretaceous paleomagnetic data set for the Canadian and Alaskan Cordillera (Enkin, 2006, and references therein). In this latest, well considered synthesis, ten high-quality poles from bedded rocks between 95 and 65 Ma, and one from a batholith (Mt. Stuart) in which post-crystallization tilting is demonstrably minimal, are overall about 15° far-sided (shallower inclination) compared to the North American reference pole. The difference is ascribed to 2000 km of post-Cretaceous northward translation of the entire Cordillera, as far inboard as the Rocky Mountain fold and thrust belt (Enkin, 2006). The involvement of the entire width of the Cordillera in megatranslation contrasts with earlier models, in which the western Cordilleran terranes were seen as more paleomagnetically discordant than those of the central Cordillera (cf. Irving and Wynne, 1990).

This new, more radical translational model is dictated by the discordant paleomagnetic results from a recent study on comparatively young volcanic rocks, the 70 Ma Carmacks Group, which overlie the easternmost Yukon-Tanana terrane at Solitary Mountain in south-central Yukon, next to the Tummel fault – a steeply dipping fault zone juxtaposing Yukon-Tanana and Cassiar terranes (Figs. 1A, 2; Enkin et al., 2006). Detailed geological mapping in this area has shown that the contact aureole of the mid-Cretaceous (ca. 105 Ma) Glenlyon batholith, which intrudes Cassiar terrane east of the fault, overprints the Tummel fault and eastern Yukon-Tanana terrane, thereby limiting the amount of possible Late Cretaceous dextral displacement along this fault (Colpron et al., 2005). This constraint effectively disqualifies the Tummel fault as a candidate for continent-scale displacement younger than 70 Ma. The nearby Tintina fault has a well supported Eocene offset that falls some 1500 km short of that apparently required by the paleomagnetic data. One is then forced to seek a cumulative dextral offset three times that on the Tintina, within the parautochthonous Selwyn Basin and Mackenzie/Rocky Mountains – regions that are dominated by well defined, regionally discontinuous thrust faults and related folds and where cross-structures such as the east-west-trending southern margin of the Selwyn Basin and associated northeast-trending lineaments (Liard and Fort Norman lines of Cecile et al., 1997), the west-northwest trend of the mid-Cretaceous Tombstone-Tungsten plutonic suites (Hart et al., 2004), and the inferred projection of northeast-trending Proterozoic and Archean domains from the Alberta basement into the Cordilleran Orogen of southern British Columbia and northern Idaho (Malton, Monashee, and Priest River complexes; Crowley, 1999) apparently defeat any possible location for continent-scale transcurrent faults (e.g. Gabrielse et al., 2006). A definitive presentation and evaluation of the geological constraints on large-scale transcurrent motions is urgently needed.

In the end, it may be more productive to re-examine the assumptions behind a strict paleogeographic interpretation of the anomalously shallow Cretaceous paleomagnetic inclinations, as did Butler et al. (2001). Their careful analysis suggests that a combination of moderate translation with minor tilting and sedimentary and/or tectonic compaction could account for the inclination anomalies, which average about 15 degrees. Since the original magnetic inclinations are very steep, almost any tectonic effect would result in flattening. Because paleolatitude is related to inclination

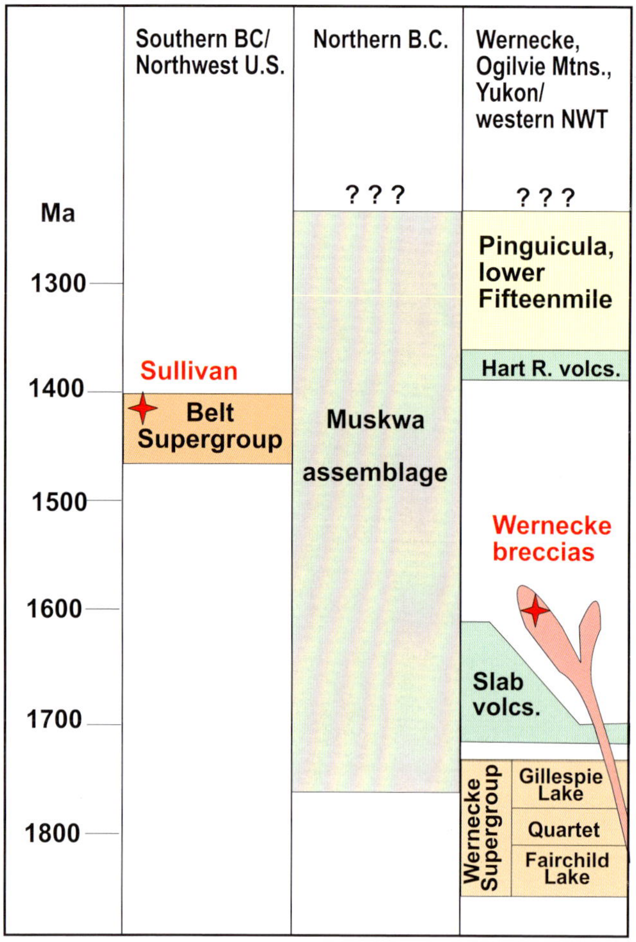

FIGURE 4. Stratigraphy of the Wernecke, Muskwa, and Belt-Purcell basins. From Thorkelson et al. (2001a).

through a tangent function (tan I = 2 tan L), the proximity of the northern Cordillera to the Cretaceous paleopole means that each degree of inclination difference results in more than a degree of apparent latitudinal translation. This kind of compromise probably points towards the direction in which the long-standing "Baja BC" controversy will eventually be resolved; however, a full treatment is outside the scope of this paper.

Ancestral North America (Laurentia) and Its Fringing Terranes

Proterozoic Basins of Western Laurentia (1.84 to 0.78 Ga)

Within the deformed belt of the parautochthonous continental margin are a series of separate sedimentary basins that developed after regional cratonization at ca. 1850 Ma, but well prior to formation of the open western margin of Laurentia at ca. 750 to 570 Ma. They occur in three main depocentres which are, from north to south, the Wernecke-Mackenzie Mountains, and the Muskwa and Belt-Purcell basins (Fig. 3; see also Fig. 9 in Thorkelson et al., 2001a). All three areas contain significant mineral deposits, each of distinct types (Nokleberg et al., 2005). The Belt-Purcell Basin contains the giant Sullivan SEDEX deposit; rocks of the Muskwa Basin are host to the Churchill Copper vein systems; and the Cu-Au-U-Co-enriched Wernecke breccias crosscut older strata of the Wernecke Supergoup. Mississippi Valley-type mineralization occurs in younger strata of the Mackenzie Mountains Supergroup (Gayna River deposits).

The northernmost of these Proterozoic depocentres, in the Wernecke and Mackenzie mountains, records multiple episodes of basinal sedimentation punctuated by orogenesis, magmatism, and prolonged subaerial exposure (Thorkelson et al., 2005). During deposition of the 13 km thick Wernecke Supergroup (post-1.84, pre-1.71 Ga; Thorkelson et al., 2001a; Fig. 4), the Wernecke basin was a broadly subsiding marine basin built on extending crystalline basement. The hydrothermal Wernecke breccias were emplaced ca. 1.6 Ga into the sequence after it was deformed by the southeast-verging Racklan Orogeny (Thorkelson et al., 2001b). The breccias (Fig. 3, Appendix 1) form an extensive, curvilinear, east-west-trending province that corresponds to Wernecke Supergroup exposures in the Wernecke and Ogilvie mountains of north-central Yukon (Fig. 3); they are considered IOCG-type (iron-oxide copper-gold; Lefebure et al., 2005; Corriveau, 2007). Their tectonic setting was certainly post-orogenic and probably anorogenic. The precursor Racklan Orogeny is seen as a farfield effect of a remote orogeny, perhaps localized and focused in the relatively thin crust below the basin (Thorkelson et al., 2001a), and their proximity to continental margins is not known. Significant parallels can be drawn between the Wernecke breccias and the giant Olympic Dam deposit in Australia, and Thorkelson et al. (2001b) offer an intriguing pre-Rodinia continental reconstruction that places them less than 1000 km from each other in a single, contiguous Mesoproterozoic province. However, pre-Rodinian and Rodinian reconstructions are contentious and poorly constrained (see below); and in contrast to the Wernecke breccias, the Olympic Dam deposit formed in an active tectonic setting associated with continental arc magmatism (Ferris and Schwartz, 2003).

Following an approximately 220 m.y. hiatus, marine sedimentation resumed in northern Yukon with deposition of the 3.5 km thick Pinguicula Group and associated intrusion of the ca. 1.38 Ga Hart River sills and eruption of volcanic equivalents (Abbott, 1997; Thorkelson et al., 2005). The Hart River massive sulphide deposit (Fig. 3, Appendix 1) is the only known significant occurrence associated with this phase of deposition. The youngest of the precursor Proterozoic basins of the northern Cordillera, the Mackenzie Mountains Supergroup (1.0-0.78 Ga; up to 4 km-thick), forms part of a widespread Neoproterozoic depocentre in northwestern Laurentia (e.g. Rainbird et al., 1996). The lower part of the Mackenzie Mountains Supergroup consists of shallow marine to fluvial mudstone, siltstone, and quartz arenite, which contains abundant Mesoproterozoic detrital zircons (1.25-1.0 Ga); these strata are interpreted to form part of a pan-continental braided-river system with sources in the Grenville Orogen of southeastern Laurentia (Rainbird et al., 1997). Overlying carbonates and minor evaporites of the Little Dal Group record deposition in an epicratonic basin. Reef mounds in the basinal facies of the Little Dal contain the only known significant mineralization in the Mackenzie Mountains Supergroup, the Gayna River deposits, a cluster of MVT deposits totalling some 50 Mt averaging 4.7% Zn (Fig. 3, Appendix 1). The age of the Gayna River deposits is not well constrained. Kesler (2002) relates their formation to penecontemporaneous basinal fluids, in which case mineralization would have been early Neoproterozoic.

The age of the westward-thickening sedimentary prism in the Muskwa basin is loosely constrained between 1766 Ma, the age of its youngest detrital zircons, and cross-cutting 780 Ma diabase dykes (Ross et al., 2001). Its detrital zircon signature allies it with the Wernecke Supergroup, although their stratigraphies do not directly correspond. Like the Wernecke basin, it has been interpreted as an intracontinental rift basin (Long et al., 1999). The Churchill Cu-bearing quartz-ankerite veins (Appendix 1) are associated with abundant diabase dykes in a generally northeast-striking array that postdates a phase of northeast-vergent folding. It is suggested that both veins and dykes relate to the onset of Neoproterozoic rifting that fragmented Rodinia.

The Belt-Purcell Basin is host to one of Canada's stellar deposits, the 160 Mt Sullivan Pb-Zn-Ag orebody (Fig. 3, Appendix 1), which was mined continuously from 1914 until closure in 2001. At 1.47 to 1.40 Ga (Ross and Villeneuve, 2003), or more broadly 1.50 to 1.32 Ga (Lydon, 2007), the Belt Supergroup is significantly younger than the Wernecke Supergroup (Fig. 4). It also shows a much more active style of extension than the Wernecke and Muskwa basins, with rapid rates of sedimentation (18-20 km thick; Lydon, 2007), mafic volcanic rocks and sill complexes, and well defined intrabasinal syndepositional faults (Ross and Villeneuve, 2003). Facies and paleocurrent indicators, as well as detrital zircon populations of 1.61 to 1.5 Ga that are not represented in the nearby Laurentian basement, are interpreted by Ross and Villeneuve (2003) to indicate the presence of non-North American western source terranes that were subsequently removed in Late Proterozoic or earlier continental rifting. The presence of detrital grains identical in age to sedimentation in the Belt Basin suggests that it may have abutted an active magmatic belt on its western side (Ross et al., 1992; Ross and Villeneuve, 2003). This intra-continental rift basin provided the tectonic setting for large-scale accumulation of syngenetic, sediment-hosted massive sulphide ore. Sullivan itself, which has been dated at 1.47 to 1.45 Ga by Sm-Nd methods (Jiang et al., 2000), is associated with a northeast-striking synsedimentary fault array, including the St. Mary, Kimberley, and Moyie faults (Höy et al., 2000), and with voluminous synsedimentary mafic sills, the Moyie sills. The northeast-striking faults parallel basement structures, at a high angle to the north-northwesterly-trending basin margin, as well as to trends of rift grabens. Lead-zinc-silver mineralization, with its characteristic associated tourmaline alteration, is localized where the cross-faults offset rift basins (Höy et al., 2000).

Worldwide, there are no other large SEDEX deposits comparable in age to Sullivan. Broken Hill, Mt. Isa, and HYC in Australia are ca. 1.69 to 1.67 Ga, 200 Ma older than the Belt-Purcell Basin. Sullivan, it seems, stands alone, not as part of a later fragmented global metallotect. The challenge of finding a western continuation of the Belt-Purcell Basin on some other continent, with its promise of prospective ground for the discovery of a second Sullivan, is part of a broader tectonic question: which continent or continents represent North America's missing twin(s), following the breakup of Rodinia? Four candidates have been proposed: Antarctica/Australia ("SWEAT" connection; Hoffman, 1991; Moores, 1991; Ross et al., 1992); Australia ("AUSWUS" connection; Burrett and Berry, 2000), Siberia (Sears and Price 2000, 2003a,b), and the Yangtze block (Li et al., 1995). The reconstruction of Sears and Price (2003b) is particularly appealing in that it aligns a ca. 1.5 Ga mafic sill/dyke swarm from the southern Wyoming craton, through the Moyie sills, and into the northern Siberian platform and shows the Taimyr trough as a western continuation of the Belt-Purcell Basin (Fig. 5A). If this is correct, then the continental rift zone that hosted the Sullivan orebody did not follow the later outline of the western North American margin, but instead cut across it at a fairly high angle. According to Sears and Price (2000, 2003b), the 1610 to 1500 Ma detrital zircons in the Belt-Purcell Basin were derived from southwestern Laurentia, but transported eastwards along the Udzha trough (Fig. 5A). In this reconstruction, the Wernecke basin in northern Laurentia is not assigned a counterpart (Sears and Price, 2003a,b, Fig. 1).

Tectonics and Metallogeny of the Rifted North American Continental Margin and Its Fringing Arcs (780 to 180 Ma)

Phase 1: Initial Rifting and After (Cambrian to Silurian)

Following rifting in the Belt-Purcell Basin and formation of the Sullivan deposit, a period of tectonic quiescence ensued throughout much of western Laurentia that would last for nearly 700 million years. There are hints of Grenvillian-age tectonism buried beneath the western Cordillera (cf. Anderson and Parrish, 2000; Doughty, 2004; Lund et al., 2004; Ross et al., 2005), but nothing to compare with the construction of the Grenville Orogen of eastern and southern Laurentia; there are no major episodes of contraction or extension. Then, in Neoproterozoic to Early Cambrian time, the cratonal nugget of Laurentia broke out of the Rodinia supercontinent, its western side opening first to initiate the formation of the ocean that would, in time, become Panthalassa – the world ocean (Scotese, 2002). The western Laurentian margin was born.

The creation of the margin involved at least two separate episodes of rifting (Fig. 5B; see Colpron et al., 2002). An earlier phase (younger than ca. 723 Ma) is shown in coarse, immature clastic facies of the Windermere Supergroup, and a later, final phase in the late Neoproterozoic to Early Cambrian Hamill/Gog Group of southeastern British Columbia. Rifting occurred concurrently with at least two global Neoproterozoic glaciations (Rapitan [Sturtian] and Ice Brook [Marinoan]; 'Snowball Earth' events of Hoffman et al., 1998; Hoffman and Schrag, 2002). In the south, thermal subsidence associated with the later rifting event began about 575 Ma (Bond and Kominz, 1984). The lower part of the Hamill/Gog Group lies unconformably on the Windermere, and strong facies/thickness variations along with significant local accumulations of basalt suggest that it was deposited in a series of north-trending rift grabens (Warren, 1997). A dyke related to this phase of magmatism is ca. 570 Ma, which dates the second pulse of rifting in the southern Cordillera (Colpron et al., 2002). During this time, the northern Cordillera was apparently the locus of passive margin sedimentation. Intermittent extension between these two major episodes of rifting is indicated by sporadic alka-

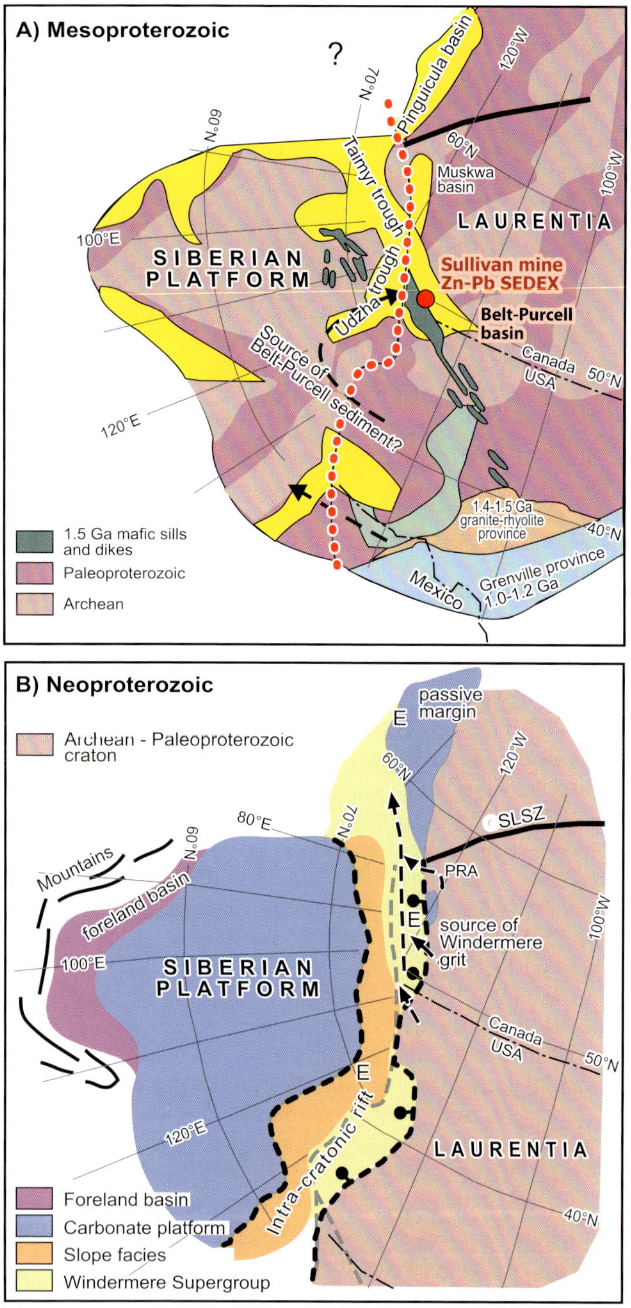

FIGURE 5. (A) Possible Mesoproterozoic reconstruction of western Laurentia and Siberia. From Lydon (2007); adapted from Sears and Price (2000, 2003a,b). (B) Neoproterozoic breakup of Rodinia according to the Sears and Price (2000, 2003) reconstruction. In this interpretation, coarse clastic sedimentation of the Windermere Supergroup occurred in an intracratonic rift between Laurentia and Siberia (see also Colpron et al., 2002). Final separation of Laurentia and Siberia took place in latest Neoproterozoic to Early Cambrian time. "E" marks ediacaran fauna localities. GSLSZ – Great Slave Lake shear zone; PRA – Peace River arch.

line magmatism along the length of the Cordilleran margin (Fig. 6; Lund et al., 2003; Pigage and Mortensen, 2004). To account for the differences in rift history between the northern and southern Canadian Cordillera, Colpron et al. (2002) offer two alternative scenarios that each depict the detachment of two separate continents from the North American margin. Figure 5B depicts the rifting scenario most compatible with the Rodinia reconstruction of Sears and Price (2000, 2003a,b) (Fig. 5A).

Although it marks a major continental rift event, the Windermere Supergroup hosts only limited syngenetic mineralization, all of which is located in the far northern Cordillera of Yukon and Northwest Territories. Mineralization in the Redstone copper belt, which includes the Coates Lake deposit (Fig. 6; Appendix 1) in the Mackenzie Mountains, is hosted by carbonate of the Coates Lake Group – a succession of evaporite, redbed (sandstone, conglomerate), and carbonate deposited in fault-bounded grabens that developed during the early stages of Windermere rifting (Jefferson and Ruelle, 1986). The copper mineralization is diagenetic in origin and concentrated at the transition zone between redbed and carbonate. The source of metal is inferred to be the volcanic-rich sedimentary rocks of the basal Coates Lake Group (Jefferson and Ruelle, 1986). Overlying glaciogenic strata of the Rapitan Group (diamictite, mudstone) are host to one of the largest and most unusual iron deposits in North America: the more than 5.5 billion tonne Crest deposit (Appendix 1; Yeo, 1986). The Rapitan banded iron formations consist of hematite-jasper rhythmites, locally with dropstones, that are interpreted as chemical precipitates during a major marine transgression in the aftermath of the Sturtian snowball event (Hoffman and Schrag, 2002). Iron was dissolved in the anoxic seawater that existed beneath the near-global sea-ice, and the iron formations were deposited at the end of glaciation by the oxidation of ferrous Fe, when ocean and atmosphere once again interacted.

In contrast with the Neoproterozoic, there are numerous syngenetic deposits of Cambrian age in both the northern and southern Canadian Cordillera. The Faro deposit, which was mined for nearly 30 years (1970-1998), is part of the Anvil district, a northwesterly belt of SEDEX Zn-Pb-Ag deposits hosted in the western Selwyn Basin (Fig. 6; Appendix 1). The Selwyn Basin was a persistent, rift-controlled deep-marine embayment in the northern continental margin throughout Paleozoic time, bounded to the west by the relatively high-standing McEvoy and Cassiar platforms that were probably contiguous prior to Eocene displacement on the Tintina fault. The Anvil deposits are of Cambrian age, hosted in siliceous graphitic units in the uppermost part of the Mt. Mye Formation and lowermost Vangorda Formation (Jennings and Jilson, 1986). Deposits are localized near the northeastern margin of the graphitic facies. Jennings and Jilson (1986) infer that the alignment of the deposits with the edge of the reduced sedimentary facies is due to control by a second-order extensional basin. Redox changes, due to the mixing of anoxic bottom waters, were probably important within it (Shanks et al., 1987).

In southeastern British Columbia, thick Lower Cambrian limestone of the Badshot Formation forms part of the outer continental margin succession in the Kootenay arc – a curvilinear belt that comprises the most westerly exposure of parautochthonous early Paleozoic strata. This limestone is host to a number of stratiform, stratabound, laminated to massive sulphide bodies (Fig. 6; Nelson, 1991). They have many features in common with Irish-type deposits and, like them, probably arose through fluid venting along seafloor growth faults. The presence of carbonate leads to much more extensive subsurface replacement mineralization than in typ-

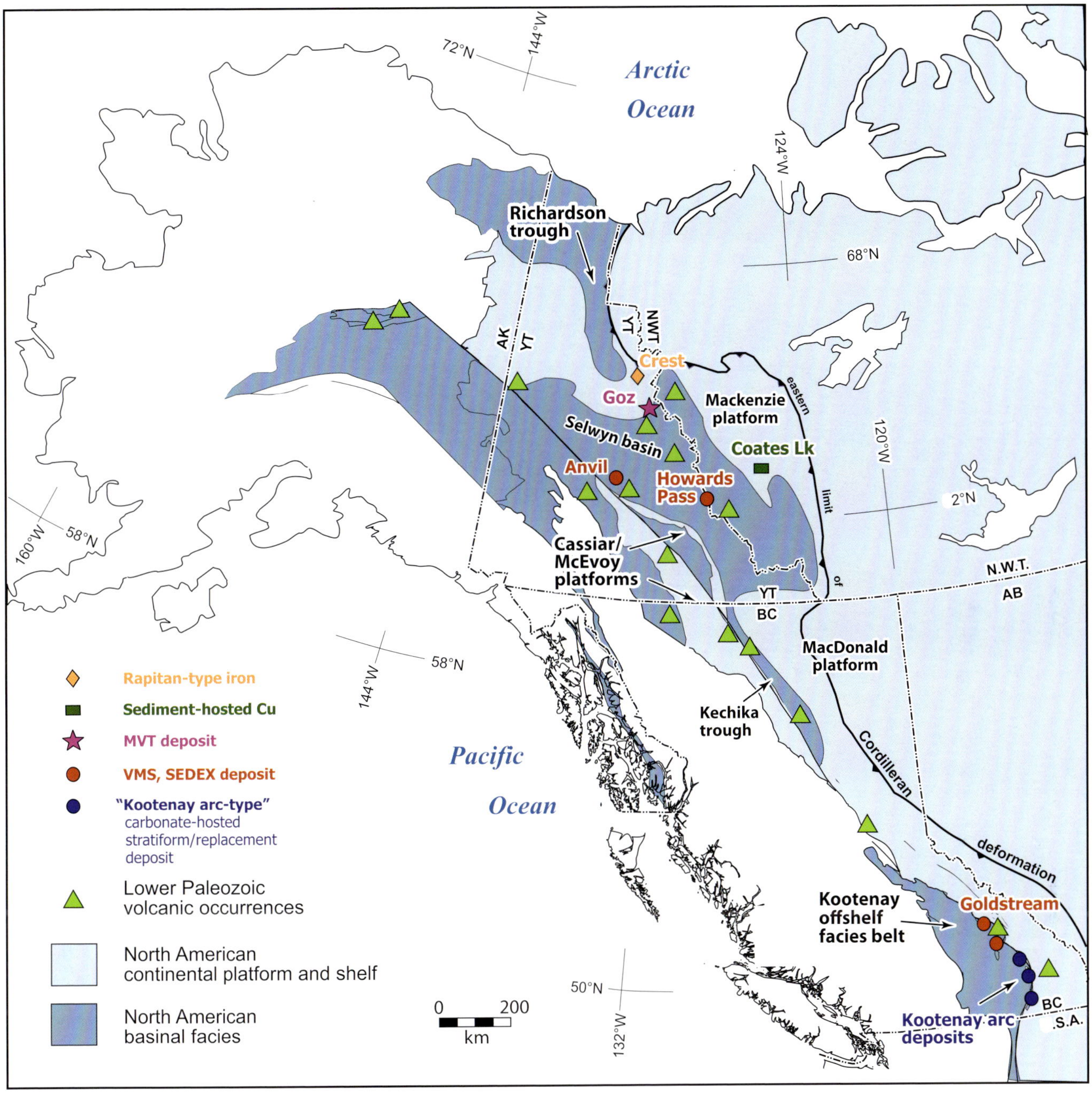

FIGURE 6. Distribution of early Paleozoic deposits and paleogeographic elements of the Laurentian margin. Volcanic occurrences from Goodfellow et al. (1995). Deposit locations from Abbott et al. (1986), Nelson (1991), Massey (2000a,b), and Logan and Colpron (2006).

ical SEDEX deposits. The Badshot limestone is overlain by the Lardeau Group, a >3.5 km-thick lower Paleozoic siliciclastic succession that does not correspond to coeval units of the continental shelf to the east and is thought to represent deposition in a deep trough along the outer continent margin (Logan and Colpron, 2006). In it, a belt of Cambro-Ordovician Besshi-type, Cu-Zn-rich VMS deposits are associated with fine-grained sedimentary and MORB-type mafic volcanic rocks, and laterally extensive manganese- and boron-enriched exhalative horizons (Fig. 6; Logan and Colpron, 2006). The most notable of these is the Goldstream deposit (Fig. 6; Appendix 1), which was mined in the early 1990s. Both the Lardeau Group itself and its contained deposits are thought to have formed within an intracontinental rift zone characterized by episodic rapid subsidence, OIB ("ocean island basalt" = enriched intraplate basalt) to MORB magmatism, and local hydrothermal activity (Logan and Colpron, 2006). The persistence of this strongly extensional tectonic style along the western Laurentian margin, long after its separation from adjoining continents, remains an outstanding mystery of Cordilleran development.

A further significant episode of extension of the so-called passive margin occurred in Ordovician time, as documented by the strongly alkalic Jowett Formation in the Lardeau Group (Logan and Colpron, 2006), as well as elsewhere in the Cordillera. Alkalic to ultrapotassic volcanic rocks occur throughout the Selwyn Basin and its southern extension, the

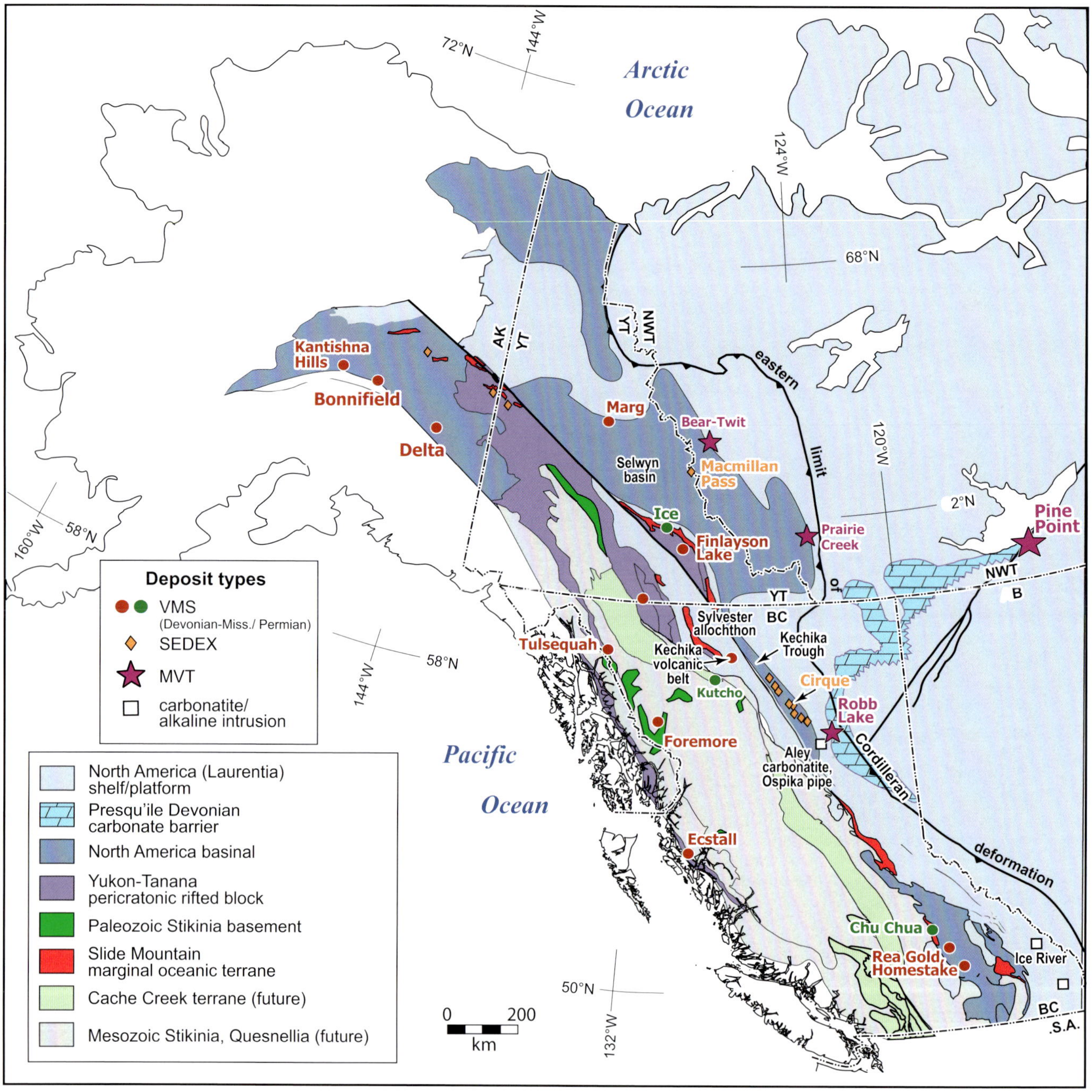

FIGURE 7. Devonian-Mississippian tectonics and metallogeny of the western Laurentian margin. From Nelson et al. (2002, 2006).

Kechika trough, and within basinal successions of northern Yukon (Fig. 6; Goodfellow et al., 1995). From their extensive compilation, these authors infer three major volcanic pulses, Cambrian, Early to Middle Ordovician, and Middle to Late Devonian. Silurian volcanic rocks and diatremes are also present. Ordovician to Early Devonian SEDEX-type deposits occur in the Selwyn Basin and Kechika trough, hosted by black shales of the Road River Group and other carbonaceous strata (Abbott et al., 1986; MacIntyre, 1991). Most are small; however, the very large Howards Pass district (Fig. 6; Appendix 1) is hosted by Early Silurian strata in the Selwyn Basin. In contrast with most other Cordilleran SEDEX deposits, those at Howards Pass did not apparently form in an active extensional regime. The deposits consist of laterally extensive, finely laminated sphalerite, galena, and pyrite associated with laminated limestone, cherty mudstone, and carbonaceous chert (Goodfellow and Jonasson, 1986). They infer a stable, starved, anoxic basinal setting, based on high $\delta^{34}S$ values in pyrite, low sedimentation rates, and the absence of local scarp-related facies or volcanic activity.

Phase 2: Establishment, Rifting, and Separation of the Continent-Margin Arc (Devonian to Mississippian)

There is very limited evidence for offshore or continent-margin arcs on the western side of Laurentia prior to Devonian time. The apparently long interval after initiation of the margin contrasts sharply with Appalachian history, where Laurentia was interacting with offshore arcs by

Middle Ordovician time (van Staal, 2007). Hints of Silurian arc activity are seen in the Trinity ophiolite/Yreka terrane in the western Klamath Mountains and the Shoofly Complex in the northern Sierras of California (Saleeby, 1990); there are a few documented Silurian plutons in the Yukon-Tanana terrane in the Coast Mountains of southeastern Alaska (Saleeby, 2000). How widespread and significant these early arcs were, and even where they were located with respect to the western margin of Laurentia, are unclear. Could the Silurian-Devonian deformation seen in the Shoofly Complex have been a factor in the initiation of margin-long subduction under the continent by Late Devonian time?

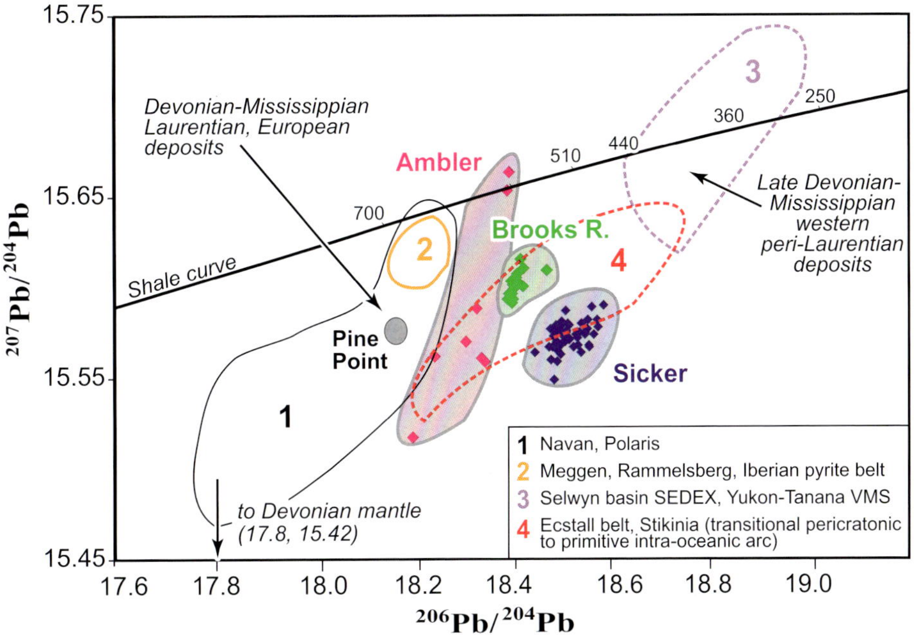

FIGURE 8. Pb isotopic signature of Devonian-Mississippian syngenetic deposits of the Cordillera compared with Arctic and European deposits. Data sources: Sicker from Robinson et al. (1996); Pine Point and European deposits from compilation of Nelson et al. (2002) and references therein; all other data from Mortensen et al. (2006).

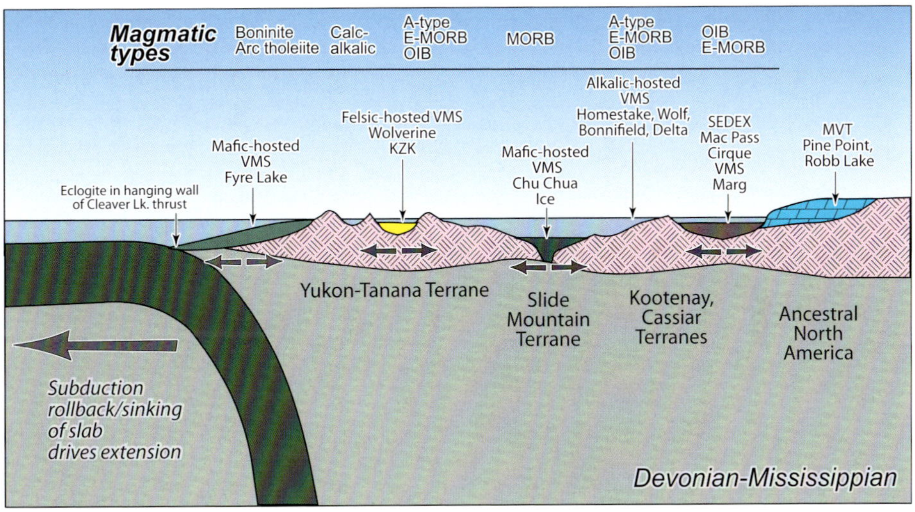

FIGURE 9. Schematic cross section of the western Laurentian margin in Devonian-Mississippian time with variation in magmatic styles and associated deposits. From Nelson et al. (2006). Although of Permian age, the mafic-volcanic-hosted VMS deposits of the Slide Mountain terrane (Chu Chua, Ice) are shown here in relation to other late Paleozoic geodynamic elements of the peri-Laurentian realm. Note that these deposits formed during the mature stage of the Slide Mountain Ocean; the nascent stage of this marginal basin is depicted here.

Whatever its precursors, the Middle to Late Devonian saw a flowering of arc-related magmatism along the length of the North American Cordillera that spanned both the autochthonous margin and the allochthonous peri-Laurentian realm (Rubin et al., 1990; Colpron et al., 2006). This event is well expressed in the western parautochthonous regions of the Canadian and Alaskan Cordillera, notably the Kootenay terrane and Yukon-Tanana upland, as well as the allochthonous Yukon-Tanana and Stikine terranes (Fig. 7; for redefinition of parts of the former Yukon-Tanana terrane as parautochthonous, see Dusel-Bacon et al., 2006; Nelson et al., 2006). High-quality U-Pb ages on volcanic and intrusive bodies range from ca. 390 to ca. 320 Ma, with the major magmatic peak at 360 to 350 Ma (Nelson et al., 2006). Throughout this interval, magmatism was strongly bimodal, with felsic components showing strong continental crustal influence and mafic components showing a broad compositional spectrum including arc tholeiite, calc-alkalic, MORB, E-MORB (enriched MORB), OIB, boninitic, and BABB (back-arc basin basalt) variants (Piercey et al., 2006). The pericratonic arcs were developed on attenuating Laurentian continental and adjacent crust, as shown by common detrital zircon and isotopic signatures (Fig. 8; Nelson et al., 2006). For instance, detrital zircon populations in the Yukon-Tanana terrane show dominant 1.8 to 2.0 Ga peaks that are comparable to the basement ages of northwestern Laurentia as well as to detrital zircon populations in autochthonous strata of the northwestern continental margin (Gehrels et al., 1995; Gehrels and Ross, 1998). During Early Mississippian time, the Slide Mountain Ocean opened to separate a large portion of the frontal arc from the continental margin: this rifted Laurentian sliver became the Yukon-Tanana terrane (Figs. 9, 10).

The Devonian-Mississippian development, attenuation, and detachment of pericontinental arcs gave rise to a wide variety of mineral deposits. Most important from this period are syngenetic sulphide deposits that range from volcanogenic in immediate back-arc and intra-arc settings, to

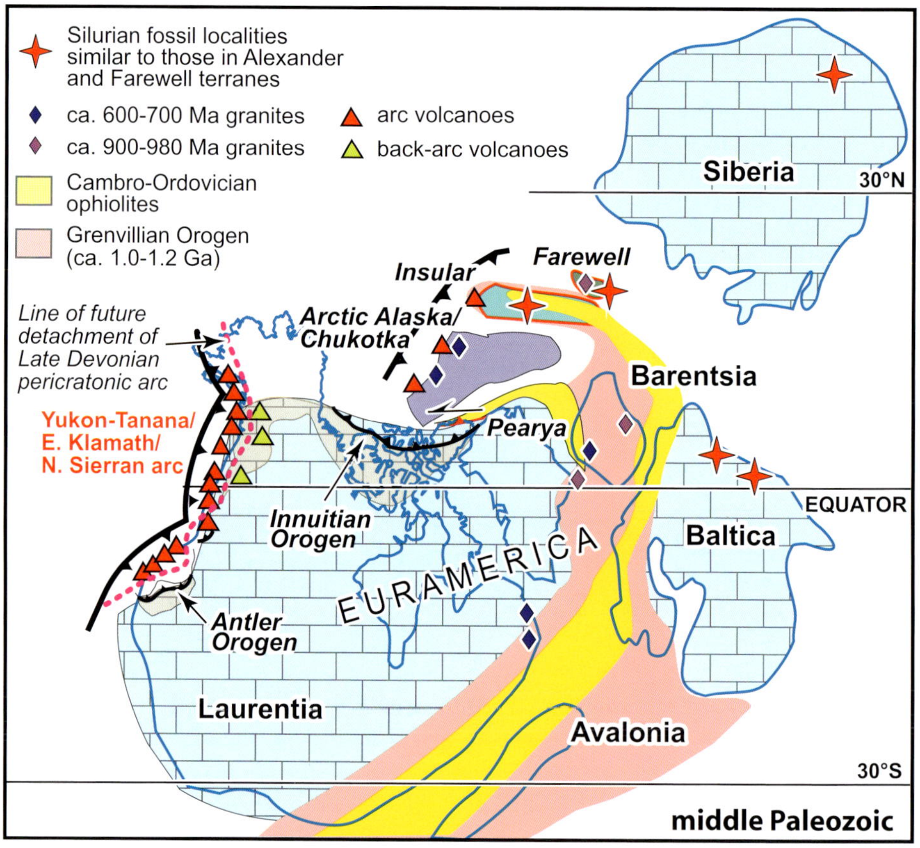

FIGURE 10. Schematic reconstruction of the peri-Laurentian and Arctic realm terranes in mid-Paleozoic time. Continental reconstruction is from Scotese (2002). Caledonides from Trettin (1987) and Gee (2005). Grenvillian basement from Trettin (1987). The locations of Precambrian granitoids are from Patrick and McClelland (1995) and Gee (2005). Silurian fossil localities are from Soja and Antoshkina (1997). The Farewell terrane is placed most proximal to the future Early Permian Uralian orogenic belt, which will affect it. The Alexander terrane is placed on trend with the Caledonian Orogeny and in the zone of ca. 1.0 Ga basement. Arctic Alaska/Chukotka is shown adjacent to the Arctic Islands. Red dashed line along the western Laurentia indicates the future site of the opening of the Slide Mountain Ocean.

SEDEX styles in the more distant back-arc regions of the Selwyn Basin and Kechika trough (Fig. 7; Nelson et al., 2002). Notable VMS districts include the Finlayson Lake belt in the Yukon-Tanana terrane, Tulsequah Chief in far northwestern Stikinia, the Delta and Bonnifield districts in the parautochthonous Alaska Range, and the Eagle Bay deposits in Kootenay terrane of southeastern British Columbia (Dusel-Bacon et al., 2006; Murphy et al., 2006; Paradis et al., 2006a; Piercey et al., 2006). Of these, Tulsequah Chief, and Samatosum, Homestake, and Rea Gold in the Eagle Bay district have been producers; Wolverine in the Finlayson Lake district is currently in mine development stage (Fig. 7; Appendix 1). The largest and best known Devonian SEDEX deposits are those at Macmillan Pass in the eastern Selwyn Basin and Cirque in the Kechika trough. In 2005, drilling on the less explored Akie prospect, south of the Cirque, has returned some notable intersections, including 37 m at 11.3% Zn, 2.65% Pb, and 21 g/t Ag (Schroeter et al., 2006). All of these are hosted by black, carbonaceous, siliceous strata of the Earn Group. They are associated with facies transitions and synsedimentary faults related to an extensional tectonic regime that affected the entire continental margin in Middle to Late Devonian time. Yet farther inboard, Devonian MVT deposits, such as Robb Lake and even Pine Point, formed when relatively cool hydrothermal fluids interacted with carbonate strata (Fig. 7; Nelson et al., 2002). Prairie Creek is an Irish-type polymetallic deposit, located in a faulted basin-to-platform transitional setting on the eastern margin of the Selwyn Basin, partly carbonate-hosted and partly structurally controlled (Paradis et al., 2006b).

Phase 3: Offshore Arcs, Quiet Margin (Pennsylvanian-Permian)

Arc-related sequences continued to build upon the rifted continental fragment of the Yukon-Tanana terrane throughout late Paleozoic time (Colpron et al., 2006; Nelson et al., 2006; Piercey et al., 2006). Extensions of these arcs may have been founded on oceanic substrates as well. The character of volcanism and sedimentation changed. Instead of widespread bimodal rhyolite-basalt volcanism associated with carbonaceous, basinal strata that characterized Devonian to Mississippian arcs, Pennsylvanian-Permian sequences were dominated by intermediate volcanic and volcaniclastic deposits and lesser enriched basalts, epiclastic beds, limestone, siliciclastic sedimentary strata, and chert (Simard et al., 2003; Gunning et al., 2006). These late Paleozoic units are depositionally linked both to older parts of the Yukon-Tanana terrane, and to Mesozoic strata of Stikine and Quesnel terranes (Read and Okulitch, 1977; Jackson et al., 1991; Mihalynuk et al., 1994b; Mihalynuk, 1999; Simard et al., 2003; Nelson and Friedman, 2004). Triassic/Jurassic intrusive suites are continuous across all three (Wheeler and McFeely, 1991). Together, the pericratonic Yukon-Tanana terrane and its overlying late Paleozoic cover form the underpinnings of the main Intermontane arc terranes.

Faunal affinities suggest that in Early to mid-Permian time, Stikine, Quesnel, and Eastern Klamath terranes probably lay 2000 to 3000 km west of North America (Belasky et al., 2002). Voluminous basalt – evidence of strong late Pennsylvanian-Early Permian spreading – and paucity of siliciclastic sedimentation in the Slide Mountain terrane are consistent with a broad ocean between the peri-Laurentian arc terranes and the continent at this time (Fig. 11; Nelson et al., 2006). Continental margin tectonics were more subdued during the late Paleozoic than at any time before or since: Permian strata are thin and dominated by limestone and chert, rather than siliciclastic sediments.

The late Paleozoic also saw a lull in formation of mineral deposits, which separated the previous era of widespread syngenetic sulphide mineralization from the subsequent flowering of porphyry and epigenetic mineralization that would dominate the Mesozoic. Tectonically, continental rifting in the peri-Laurentian realm was finished, and the crustal thickening that characterized Mesozoic arc history had not yet begun. The only deposits of significance that formed during this interval are modest Cyprus-type Cu-rich massive sulphide deposits such as Ice in the Finlayson belt and Chu Chua in southeastern British Columbia, both within the Slide Mountain ocean basin (shown on Fig. 7).

Phase 4: Marginal Basin Closure, Overlapping Arcs, and Accretionary Prisms (Middle Permian to Early Jurassic)

From its Devonian inception through the Early Permian, easterly dipping subduction took place under the western Laurentian margin, and later under its fringing pericratonic arc terranes. The Middle to Late Permian record in the Yukon-Tanana terrane documents reversal of this pattern. The Klondike assemblage, dominated by felsic, continentally influenced volcanic and intrusive rocks, is restricted to the eastern and northern side of the terrane; east of it lies a belt of coeval high-pressure metamorphic rocks and synorogenic clastic deposits (Fig. 12; Colpron et al., 2006; Nelson et al., 2006). Evidently, the Slide Mountain marginal ocean closed by westerly subduction beneath Yukon-Tanana terrane and remnants of the Slide Mountain Ocean that remained attached to Yukon-Tanana. Up to this point, there has been considerable controversy about which part of the continental margin the Yukon-Tanana and its affiliated terranes accreted to in Permo-Triassic time. A recent study by Beranek and Mortensen (2006) shows characteristic Yukon-Tanana detrital zircons in autochthonous Triassic strata in Yukon, including those derived from the spatially limited Klondike assemblage. These observations are strong evidence that the Yukon-

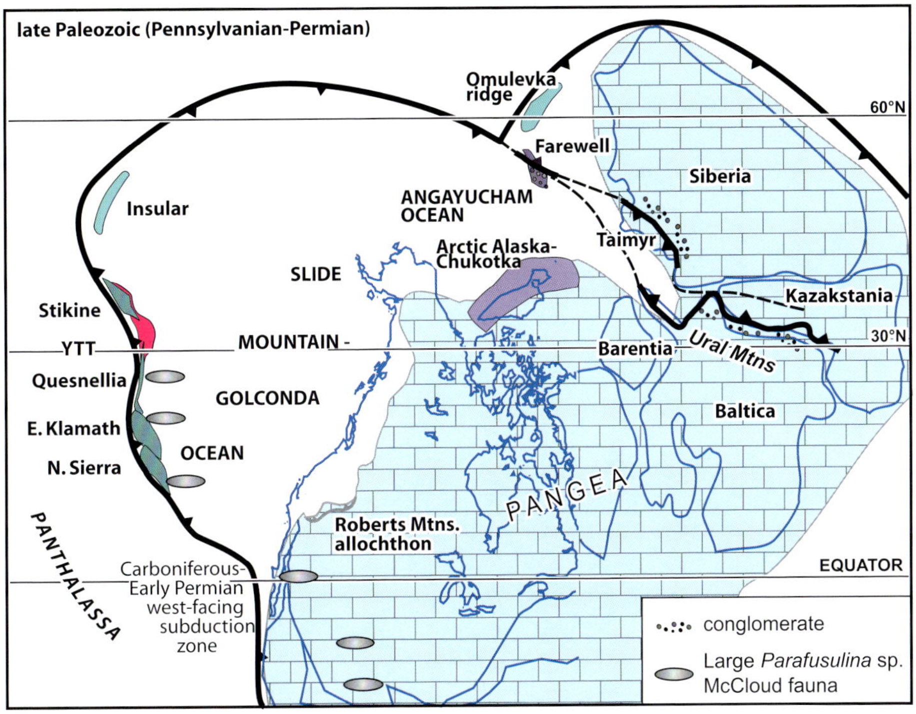

FIGURE 11. Late Paleozoic (Pennsylvanian-Early Permian) schematic reconstruction. Continents from Scotese (2002); northern continents and Uralian-Taimyr orogen from Bradley et al. (2003). Location of Insular and Intermontane terranes based largely on Belasky et al. (2002). Location of Farewell terrane based conceptually on Bradley et al. (2003). Omulevka Ridge from Nokleberg et al. (2005). YTT – Yukon-Tanana terrane.

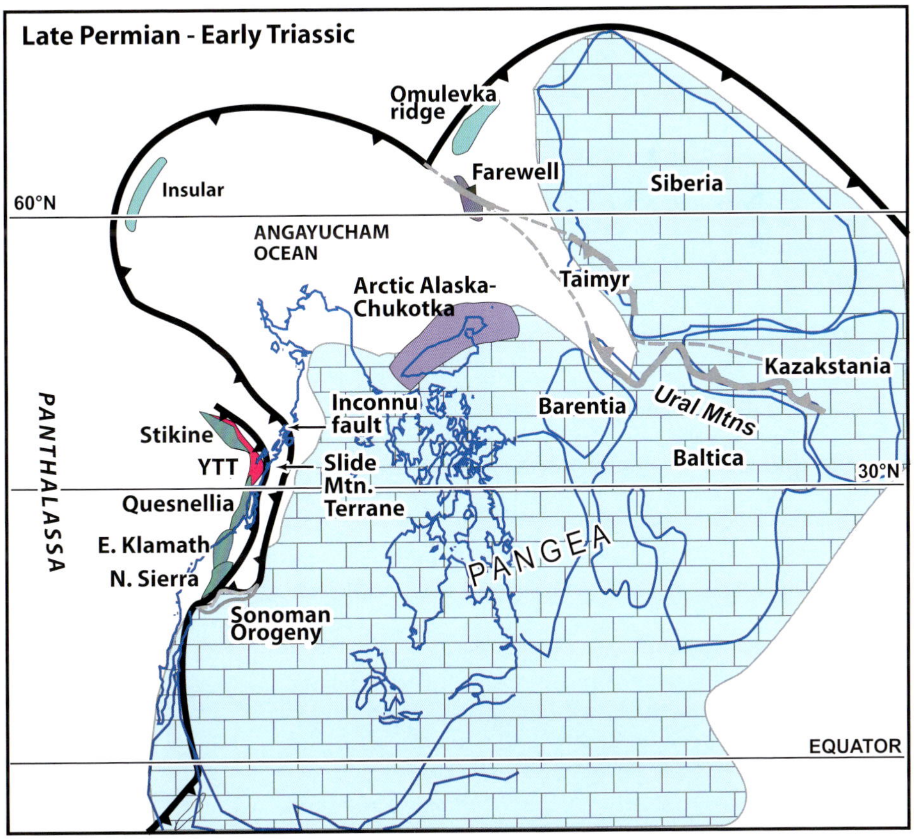

FIGURE 12. End Paleozoic (Late Permian – Early Triassic) schematic reconstruction. Sources as in Figure 11. YTT – Yukon-Tanana terrane.

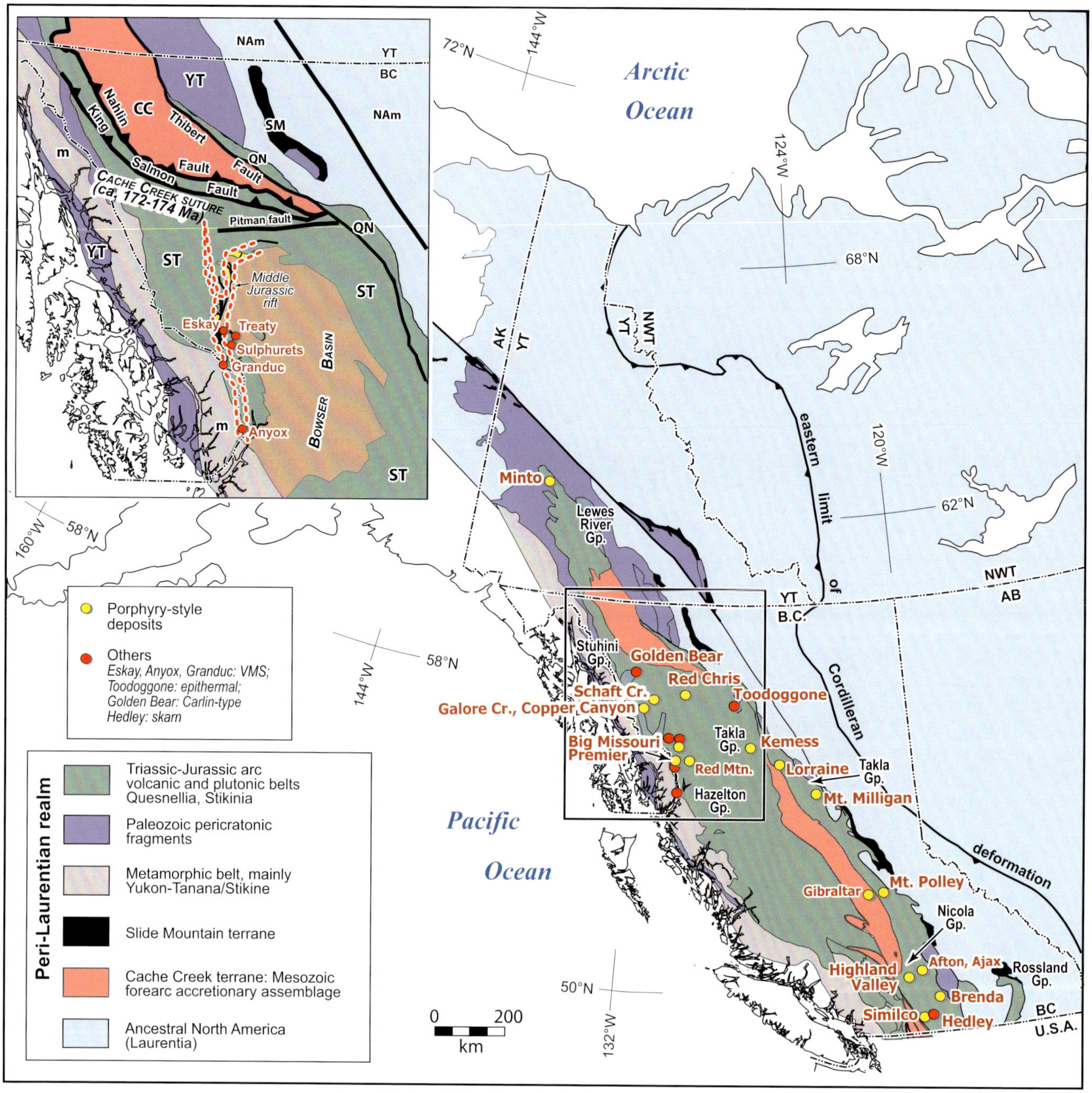

FIGURE 13. Triassic to Middle Jurassic magmatic belts and associated deposits of the Intermontane terranes. Deposit locations are from McMillan et al. (1995). Inset shows details of the mid-Jurassic Eskay rift and associated VMS deposits (from Alldrick et al., 2005; note that Granduc is Late Triassic but lies within the later rift zone). Bowser basin is the Middle Jurassic to Early Cretaceous foreland basin to the Cache Creek accretionary complex. Terrane labels as in Figure 1A.

Tanana terrane must have arrived near its present position in the Triassic (Fig. 12).

Permo-Triassic time marked a major tectonic, magmatic, and depositional break in the Intermontane, peri-Laurentian arc terranes. Except for the Klondike arc, Paleozoic arc magmatic activity was extinguished in the Intermontane terranes by Early (Gunning et al., 2006) to Middle Permian time.

The oldest Mesozoic volcanic and intrusive rocks in Quesnel and Stikine terranes are Middle Triassic, but voluminous arc-related buildups with coeval, cogenetic plutonism began in the Late Triassic (Fig. 13; Anderson, 1991). They include the Takla and Nicola groups northeast of the Cache Creek terrane and the Takla and Stuhini groups to the southwest (the oceanic, accretionary Cache Creek assemblage is used to demarcate between similar Mesozoic strata assigned to Quesnel and Stikine terranes; see Wheeler et al., 1991). Late Triassic, dominantly augite-(plagioclase-) phyric volcanogenic units on both sides of the Cache Creek assemblage are strongly similar in field characteristics, arc geochemistry, and primitive isotopic signatures (Dostal et al., 1999). In Quesnel terrane, the main Triassic volcanic accumulations lie west of the belt of Paleozoic pericratonic exposures, which are onlapped by thinner, volcanic-poor siliciclastic units with continentally influenced isotopic signatures (Unterschutz et al., 2002) that also overlie Yukon-Tanana terrane and the western continental margin. Following Permo-Triassic accretion of the innermost pericratonic terranes (Slide Mountain, Quesnel, and Yukon-Tanana), it is likely that the axis of the new west-facing arc migrated outboard from the collision zone. In the Stikine terrane, there is no evidence for a shift in the arc axis: the Stuhini and Takla groups are developed in a broad region on top of variably deformed Paleozoic arc units.

On both sides of the Cache Creek assemblage, newly configured Early Jurassic arcs were superimposed on Triassic architecture; contact relationships range regionally from disconformable to deeply unconformable on folded and thrust-faulted older strata. In Quesnel and Yukon-Tanana terranes, the Jurassic magmatic belt migrated eastwards towards the continent, as shown by abundant 200 to 185 Ma plutons in the Yukon-Tanana terrane, and volcanic strata of the Lower Jurassic Rossland Group, which lie well east of the Triassic Nicola Group in southern British Columbia (Fig. 13; Wheeler and McFeely, 1991). In the Stikine terrane, Lower Jurassic volcanogenic strata of the Hazelton Group are widespread and voluminous. The favoured tectonic model for this terrane in Early Jurassic time is a microplate with subduction under both east and west sides (present coordinates), which generated two arcs separated by a marine trough (Marsden and Thorkelson, 1992).

Late Triassic to Early Jurassic Cu-Au and Cu-Mo porphyry deposits of Stikine and Quesnel terranes (Fig. 13) are collectively the most important group of deposits in British Columbia; this belt extends into the Yukon-Tanana terrane of Yukon (Minto and Williams Creek; Tafti and Mortensen, 2004; Fig. 13). It includes long-time producers such as Highland Valley, Gibraltar, Copper Mountain, Brenda, and Afton; newer mines such as Mt. Polley and Kemess, probable future mines such as Galore, and unexploited large resources such as Mt. Milligan, Red Chris, and Schaft Creek (Appendix 1). Volumes, literally, have been written about these deposits – Canadian Institute of Mining and Metallurgy Special Volumes 15 (1976, edited by A. Sutherland Brown) and 46 (1995, edited by T.G. Schroeter) – a sign of their enduring significance. Host intrusions range from ca. 210 Ma (Galore, Highland Valley) to ca. 183 Ma (Mt. Milligan); most common ages are ca. 205 to 202 Ma (Mortensen et al., 1995), which corresponds to the end stage of Triassic arc activity and the pause before the onset of renewed Jurassic volcanism. The host intrusions tend to be high level, even subvolcanic (Nelson and Bellefontaine, 1996), except for those in the Yukon-Tanana terrane, which are deeply exhumed (Tafti and Mortensen, 2004). They range from calc-alkalic to mildly alkalic (shoshonitic) in affinity: this tendency is probably due to melting of upper mantle that was previously metasomatised during repeated Devonian to Triassic subduction events (Nelson and Bellefontaine, 1996). At the time that they were emplaced, the overriding plate(s) – the Quesnel and Stikine arcs – were in a state of compression (or sinistral transpression; P. Schiarizza, pers. comm., 2005). Evidence for crustal thickening includes Triassic-Jurassic folding and reverse-faulting, Triassic-Jurassic unconformities, a transition from predominantly marine conditions in the Triassic to widespread subaerial and shallow submarine volcanism in the Jurassic, and the continentward migration of the Quesnel arc axis. The porphyry deposits were accompanied by other epigenetic deposit types, including gold skarn at Hedley, gold-silver in structurally controlled carbonate replacements at Golden Bear, and transitional vein to stratabound ores that were mined at Big Missouri (Fig. 13; Appendix 1). The abundance of porphyry and other deposits marks Stikinia and Quesnellia as remarkably rich metallotects, comparable to modern arc settings like Papua New Guinea.

In contrast to the porphyry deposits that occur throughout Stikinia and Quesnellia, the now-closed Granduc mine in northwestern British Columbia exploited a series of stratiform Late Triassic, copper-rich VMS lenses. The sulphide lenses are associated with magnetite iron formation. They overlie a section of primitive tholeiitic basalts that have been dated as ca. 223 Ma by U-Pb methods (Childe, 1997). A primitive arc or back-arc setting is inferred for this unusual deposit. It is located along the Unuk River shear zone, which later became the setting of the Eskay and related mid-Jurassic VMS deposits (see below).

The Cache Creek terrane intervenes structurally between the nearly identical Stikine and Quesnel arc terranes. It is an accretionary assemblage that contains late Paleozoic to Lower Jurassic oceanic strata; the remnants of late Paleozoic oceanic plateaus; limestones with Permian Tethyan fusulinid faunas that most closely resemble those of China and Japan; melange belts; a peculiar, primitive Late Permian arc, the Kutcho arc; and Late Triassic and mid-Jurassic blueschists. It marked a major destructive plate boundary. Correlative assemblages that occur west of pericratonic arc terranes in Oregon and California are considered to mark the paleo-Pacific outer margin of North America and its marginal terranes in late Paleozoic to early Mesozoic time (Miller, 1988).

The role of the Cache Creek terrane in the Canadian Cordillera is more complex. Triassic fore-arc facies in central Quesnellia, including ultramafic-gabbro clast conglomerates, occur adjacent to the eastern Triassic blueschist belt of the Cache Creek terrane (Paterson, 1977; Struik et al., 2001): they are clear evidence that Cache Creek formed the fore-arc accretionary wedge for a west-facing Quesnel arc, analogous to arc terranes farther south (Miller, 1988). Conversely, Lower Jurassic clastic strata of probable fore-arc affinity occur in northeastern Stikinia and adjacent Cache Creek terrane (Mihalynuk, 1999), and the Late Triassic Granduc orebody may have formed in a back-arc environment in western Stikinia, evidence of an easterly facing arc. Apparently, opposing arcs in Quesnel and Stikine terranes faced the same, open, still-subducting ocean in early Mesozoic time (Mihalynuk et al., 1994a). In Figures 14 and 15, the Stikinia block is rotated clockwise to restore this scenario. Of the two alternate options shown in Nokleberg et al. (2000, p. 60), the enclosure model is preferred. The other option of restoring a hypothetical mid-Jurassic sinistral fault with more than 1000 km offset does not account for observed continuity in Jurassic and older units in the Yukon-Tanana terrane to the north or for the east-facing Early Jurassic Stikine arc, and it would juxtapose fundamentally disparate tectonic facies such as thick Triassic volcanic sequences with thin sedimentary intervals, and the Early

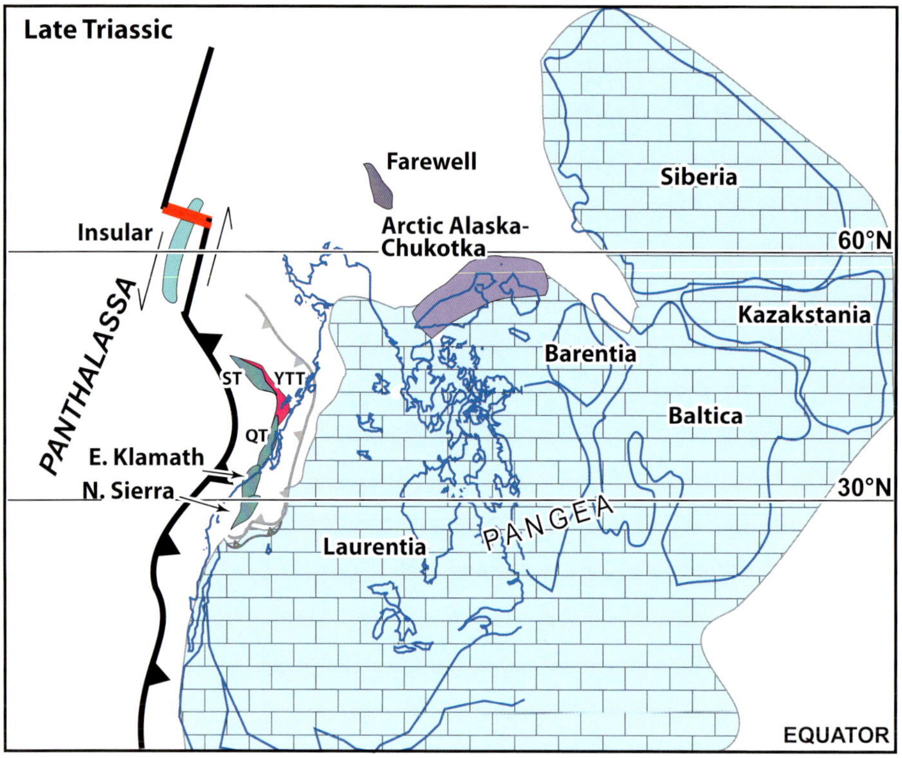

FIGURE 14. Late Triassic schematic reconstruction. Sources as in Figure 11. QT – Quesnel terrane; ST – Stikine terrane; YTT – Yukon-Tanana terrane.

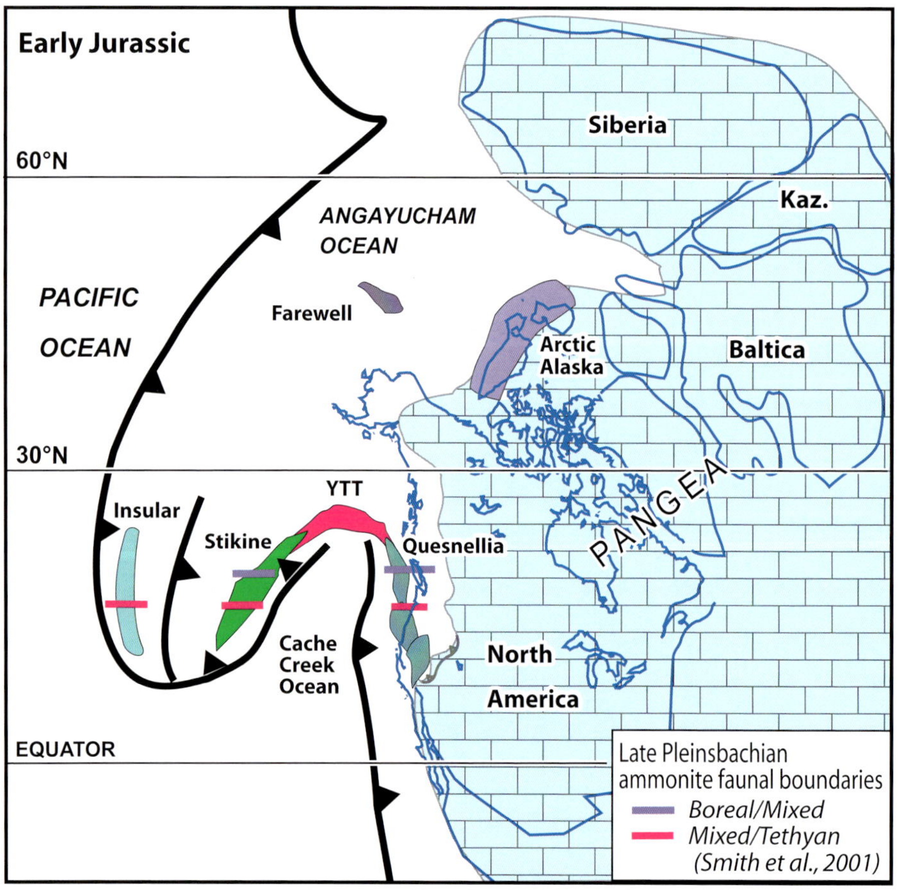

FIGURE 15. Early Jurassic schematic reconstruction. Fossil constraints from Smith et al. (2001). Continental reconstruction from Scotese (2002). YTT – Yukon-Tanana terrane.

Jurassic Hazelton arc or arcs with coeval metamorphic belts and coarse clastic accumulations.

Phase 5: Final Accretion of the Intermontane Terranes and the Eskay Event (180 to 170 Ma)

In mid-Jurassic time, the Stikine crustal block collided with the already accreted Quesnel terrane and western North America, killing both the Quesnel and eastern Stikine (Hazelton) arcs and trapping the Cache Creek accretionary prism and remaining Cache Creek ocean floor between (Fig. 16). The timing for this event is well constrained by:

- Youngest oceanic strata - radiolarian cherts in the Cache Creek terrane: Pleinsbachian-Toarcian (Cordey et al., 1987)
- Youngest arc magmatism in Quesnellia: late Pleinsbachian-early Toarcian (186-183 Ma; Nelson and Bellefontaine, 1996)
- Pinning plutons across Quesnel/North America boundary: Toarcian (ca. 185-181 Ma; Murphy et al., 1995)
- Youngest arc magmatism in eastern Stikinia (186 Ma; Larry Diakow, pers. comm., 2006)
- Oldest detritus from collision zone in Bowser basin: Bajocian, with interpreted Aalenian foreland depression (Ricketts et al., 1992)
- Oldest post-kinematic intrusions: Fourth of July suite near Atlin, ca. 172 Ma (Mihalynuk et al., 1992); Nelson plutonic suite, ca. 175 Ma, and younger (Murphy et al., 1995).

As a result of the final stages of collision, the Cache Creek terrane was obducted onto northeastern Stikinia. One panel, the French Range, contains coherent blueschist sheets that have been precisely dated as 172 Ma both for peak metamorphism (U-Pb zircon) and cooling ($^{40}Ar/^{39}Ar$ plateau ages; Mihalynuk et al., 2004). They model the final collision as a Halmahera-like arc-arc collision (Fig. 16B,C). Evenchick et al. (2006) suggest that North America – specifically the orogen in southern British Columbia – formed a tectonic hinterland during west-southwesterly thrusting and foreland basin sedimentation in the Bowser basin (Fig. 13 inset).

FIGURE 16. **A)** Middle Jurassic schematic reconstruction (after Mihalynuk et al., 1994a). Continental reconstruction from Scotese (2002). CC – Cache Creek terrane; Ins. – Insular terranes; ST – Stikinia. Nutesyn arc from Nokleberg et al. (2005). **B)** Middle Jurassic schematic cross-section across the Intermontane realm (after Mihalynuk et al., 2004). **C)** Tectonic map and cross-section of the modern Molucca Sea region.

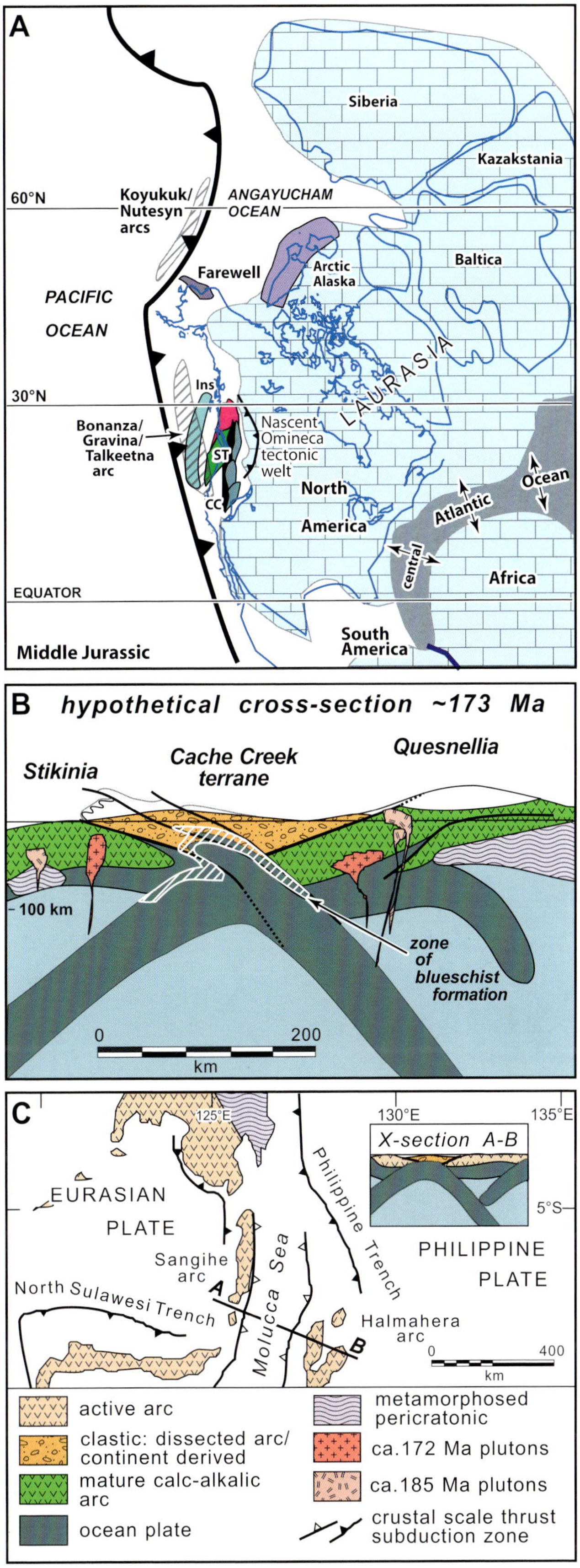

Just prior to final tightening, the narrow, laterally continuous, north-trending Eskay rift zone cut across the Stikine crustal block, at a high angle to the incipient zone of collision as it is defined by the Nahlin and King Salmon faults (Fig. 13 inset). The rift filled with clastic sedimentary strata ranging from fanglomerate to conglomerate to sandstone/argillite (Alldrick et al., 2005), and a bimodal volcanic suite of locally thick pillow basalts and small rhyolite bodies that give ages of ca. 175 Ma (MacDonald et al., 1996). The Eskay Creek precious metal-rich massive sulphide deposit (Fig. 13; Appendix 1), which has been the highest grade VMS mine in the world since its opening in 1998, occurs in carbonaceous sedimentary strata above one of the rhyolites and below pillow basalts. Anyox, a Cu-rich, basalt-hosted VMS-style mined deposit, lies to the south along strike of the rift zone. It is considered to be coeval with Eskay, but of a more primitive character due to underlying crust, more advanced rifting, or both (Evenchick and McNicoll, 2002).

The Eskay rift zone is a back-arc rift in the sense that it lies north of a zone of subdued arc magmatic activity that succeeded the western mainstage Hazelton arc. However, it represents a complete departure from earlier tectonic/magmatic styles and trends. A sinistral shear sense is recorded on its bounding faults and in adjacent rocks (Alldrick et al., 2005). We speculate that it developed as a cross-fracture related to a shear couple across the Stikine block, in response to collisional stresses particularly on its northeastern margin (Fig. 13 inset). This would explain its anomalous northerly trend, its short-lived but intense development as a rift zone, and the lateral displacement across it.

One of the thrust sheets in the Cache Creek terrane comprises a Late Permian, primitive island arc assemblage, the Kutcho arc, along with its Triassic and Jurassic clastic cover (Childe et al., 1998). This arc is dissimilar to Stikinia and Quesnellia in its chemical composition, extremely primitive isotopic character, and age, and is regarded as a fragment of a nascent volcanic arc that formed within the paleo-Pacific ocean and was accreted along with the remainder of the Cache Creek terrane. The Kutcho Creek Cu-rich massive sulphide deposit (shown on Fig. 7; Appendix 1) is associated with felsic volcanic rocks in the northern part of the Kutcho arc. As this fragment is recognized along the length of the Cache Creek terrane (Childe et al., 1998), there is considerable strike potential for Kutcho equivalents.

The Arctic and Insular Terranes

We have followed the development of the Intermontane terranes, from the first separation of ensialic arc fragments from the continent in the Devonian, through their history in the peri-Laurentian realm, to the verge of re-accretion with North America. We now take up the story of the terranes of the Arctic realm, beginning with their conjectural time(s) and place(s) of origin, up to the arrival of the Insular terranes outboard of the peri-Laurentian realm, also in Jurassic time.

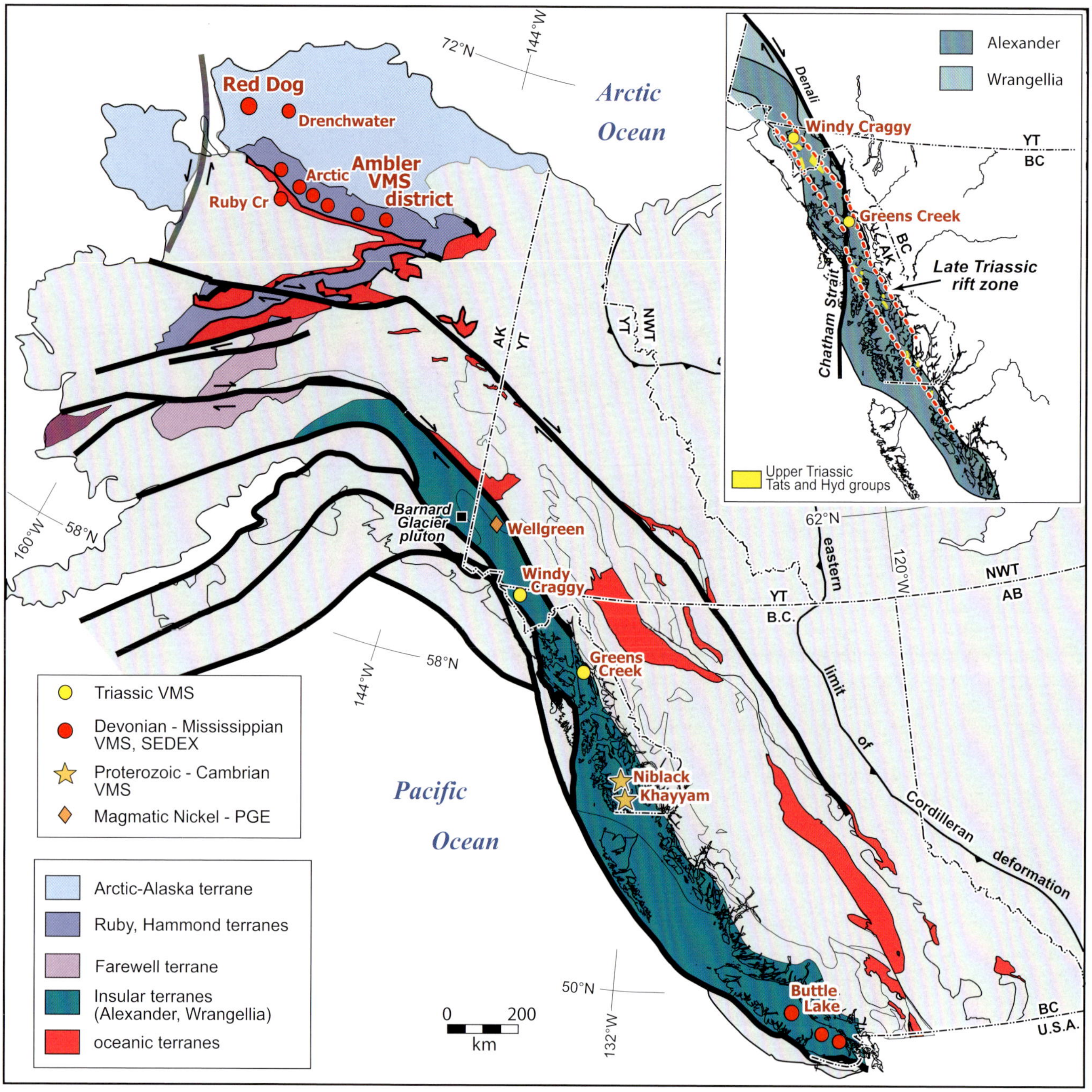

FIGURE 17. Mineral deposits and key tectonic features of the Arctic realm terranes. Deposit locations from Massey (2000a) and Nokleberg et al. (1994). Location of the Pennsylvanian Barnard Glacier pluton, that stitches Wrangellia and Alexander terrane, is from Gardner et al. (1988). Inset shows details of the Late Triassic rift zone in Alexander terrane and associated VMS deposits. Late Triassic outcrops compiled from Mihalynuk et al. (1993), Gehrels et al. (1987) and Silberling et al. (1994).

The Arctic realm comprises three major terrane affiliations (Figs. 1, 17):

- The Arctic Alaska terrane includes both Late Precambrian-Paleozoic continental shelf-slope strata of the northern Brooks Range, and Late Proterozoic and younger pericratonic rocks of the southern Brooks Range ("schist belt") and Seward Peninsula. The poorly studied Ruby terrane of central Alaska shows some affinities with the Brooks Range, but also some similarities with western Laurentia; it may be a structural composite of elements of both (Roeske et al., 2006).
- The Farewell terrane, formerly subdivided into the Nixon Fork, Dillinger, and parts of the Mystic terrane, comprises Proterozoic basement overlain by Paleozoic shelf and slope strata and Permian clastic wedge deposits (Bradley et al., 2003).
- The Insular terranes include the Alexander terrane and Wrangellia. The Alexander terrane is a crustal fragment with a regionally variable character. In southeastern Alaska, it consists of Late Precambrian to early Paleozoic volcanic arc rocks without obvious continental underpinnings, but with some continentally derived sedimentary strata that contain Precambrian detrital zircon populations (Gehrels et al., 1996). In far northwestern

British Columbia, southern Yukon, and far eastern Alaska, it comprises lower Paleozoic continental shelf-platform strata (Mihalynuk et al., 1993), overlain by late Paleozoic arc sequences that it shares with Wrangellia (Gardner et al., 1988). Wrangellia is a long-lived, multi-episodic, Devonian and younger arc terrane. Both Alexander and Wrangellia experienced non-arc mafic volcanism in Late Triassic time. In Wrangellia, this was expressed as the widespread Carnian-age Nikolai-Karmutsen basaltic event, which has been interpreted as the product of oceanic plateau volcanism (Lassiter et al., 1995). In the Alexander terrane, slightly younger, Norian basalts are confined to a single rift zone that hosts the giant Windy Craggy Besshi-type VMS deposit and the Greens Creek VMS mine (Fig. 17; Appendix 1).

Origins and Early Histories (Late Precambrian - Early Paleozoic)

Arctic Alaska terrane: In the northern Brooks Range, siliciclastic and carbonate strata are as old as late Neoproterozoic (Moore et al., 1994). Early Paleozoic deep-water, volcanic-rich and arc-related facies are also present, indicating the presence of offshore basinal and arc assemblages that were deformed and then truncated by pre-Middle Devonian and pre-Mississippian unconformities (Moore et al., 1994). Some Cambrian to Mississippian faunas are of Laurentian and some of Siberian affinities (Blodgett et al., 2002; Dumoulin et al., 2002; Dumoulin and Harris, 2006). In the southern Brooks Range, a single granitic orthogneiss body has been dated at ca. 970 Ma (McClelland et al., 2006). Orthogneiss and volcanic ages range from 750 to 540 Ma, both there and in the Seward Peninsula (Amato et al., 2006). On the Seward Peninsula, deformed igneous and related rocks are overlain by sandstone containing detrital zircons as young as ca. 540 Ma, which are overlain in turn by unmetamorphosed fossiliferous Early Ordovician carbonate. This Cambro-Ordovician deformational event may have been of significance throughout Arctic Alaska (Amato et al., 2006). Igneous ages of 750 to 540 Ma are rare in northern Laurentia, but compare to arcs and collisional orogens of northern Siberia, Barentia, and Baltica (Patrick and McClelland, 1995; Amato et al., 2006). Thus, Arctic Alaska and its western extension into Chukotka may be a crustal fragment that rifted from Siberia during the Neoproterozoic breakup of Rodinia (Dumoulin and Harris, 2006), but continued to interact with continental margins of the Arctic realm well into Paleozoic time. Detrital zircons from Triassic strata in Chukotka, Wrangel Island and westernmost Arctic Alaska (Lisburne Hills) share important Devonian-Carboniferous (380-320 Ma) and Permo-Triassic (ca. 250 Ma) components with the Verkhoyansk fold belt of eastern Siberia (Miller et al., 2004); Uralian granites and metamorphic rocks and silicic magmatism related to the Siberian traps are likely sources. Detrital grains with ages between 2030 to 890 Ma are reported from Seward Peninsula (Amato et al., 2004), a limited data set that nevertheless suggests parallels with the Farewell terrane (see below).

Farewell Terrane

The oldest crystalline rocks of the Farewell terrane include ca. 1200 Ma and 980 to 920 Ma rhyolites and ca. 980 to 850 Ma orthogneisses (Bradley et al., 2003, 2006; McClelland et al., 2006). The overlying carbonate and siliciclastic strata were deposited along a late Neoproterozoic continental margin; minor Silurian tuffs suggest arc activity (Bradley et al., 2006). Early Paleozoic faunas show the influence of both Siberian and Laurentian provinces (Blodgett et al., 2002; Bradley et al., 2006). Detrital zircon peaks of ca. 2050, 1375, and 950 Ma are unlike Laurentia (Bradley et al., 2006; compare with western Laurentian data of Gehrels et al., 1995 and Gehrels and Ross, 1998). The Early Permian (ca. 285 Ma) Browns Fork Orogeny is a distinctive event of deformation, metamorphism, and deposition of a clastic wedge; it has been linked to Uralian tectonism associated with collision between Baltica, Siberia, and Taimyr (Bradley et al., 2003). Like Arctic Alaska/Chukotka, the Farewell terrane seems to have been a player in the Arctic region during much of Paleozoic time.

The Kilbuck-Idono terranes are small pericratonic blocks located in southwestern Alaska. Kilbuck basement has been dated as ca. 2070 to 2040 Ma, an age range that may fit Siberia or other cratons more than western Laurentia (Box et al., 1990). Bradley et al. (2006) point out the similarity of ca. 2050 Ma detrital zircons in the Farewell terrane to the age of the Kilbuck terrane.

Alexander Terrane

The oldest known rocks in the southern Alexander terrane belong to the latest Proterozoic to Early Cambrian, arc-related Wales Group (U-Pb ages of ca. 595 and ca. 554 Ma; Gehrels et al., 1996). They host a number of VMS deposits, notably the Khayyam and Niblack on Prince of Wales Island in southeastern Alaska (Fig. 17; Appendix 1). These strata were deformed in the pre-Early Ordovician Wales Orogeny, coeval with deformation recorded in the Arctic Alaska terrane on the Seward Peninsula (Amato et al., 2006). In southern Alexander terrane, arc magmatism resumed in Early Ordovician-Early Silurian time, followed by clastic and carbonate sedimentation and pluton emplacement in the later Silurian. The Klakas Orogeny was marked by thrust imbrication, metamorphism, ductile deformation, and deposition of the Early Devonian Karheen Formation clastic wedge (Gehrels et al., 1996). Timing of Ordovician-Silurian arc building and Silurian-Devonian orogeny are similar to the Appalachian and Caledonian orogens (Bazard et al., 1995). By contrast, Paleozoic strata in the northern Alexander terrane are generally platformal, dominated by thick shallow-water carbonate sequences with lesser clastic strata and, particularly in the Cambrian section, basalt flows and sills (Mihalynuk et al., 1993).

Detrital zircon ages from the Karheen Formation show Precambrian peaks at 1.0 to 1.2 (dominant), 1.35 to 1.39, 1.48 to 1.53, 1.62 to 1.68, 1.73 to 1.77, 1.8 to 2.0, and 2.5 to 3.0 Ga, a pattern very unlike northwestern Laurentia, but showing strong Grenvillian influence and a possible connection with Baltica (Bazard et al., 1995; Gehrels et al., 1996; compare with dominant 1.8-2.0 Ga northwestern Laurentian peaks in Gehrels et al., 1995 and Gehrels and Ross, 1998). Silurian stromatolite faunas are very similar to those in the Ural Mountains, as well as those of the Farewell terrane (Soja and Antoshkina, 1997). This evidence, combined with paleomagnetic and detrital zircon constraints and northern

Alexander platformal stratigraphy, corroborates the Arctic, continent-proximal setting of the terrane in mid-Paleozoic time (Fig. 10).

For all their uniqueness, these three terranes share a number of themes that bear on their origins and early history. Detrital zircon signatures and, in the case of the Farewell terrane, ca. 1200 to 850 Ma basement ages, require a non-western Laurentian cratonal connection, with Baltica and Siberia as favoured alternates. Metarhyolites of 980 to 920 Ma are found in both Arctic Alaska and the Farewell terrane. Arctic Alaska shares a ca. 750 Ma plutonic event with Siberia and Barentia, and 565 to 544 Ma igneous activity with Baltica as well as the southern Alexander terrane. The sub-Early Ordovician unconformity is common to parts of Arctic Alaska, Chukotka, and the southern Alexander terrane. The stable platforms and basins of the northern Alexander terrane and the Farewell terrane were not affected by this event. Paleozoic faunas in all three terranes resemble each other (Soja and Antoshkina, 1997; Blodgett et al., 2002; Dumoulin et al., 2002; Dumoulin and Harris, 2006) and show mixed North American and Siberian affinities, with particularly strong Siberian/Baltic connections in Silurian time for Alexander and Nixon Fork, and Cambrian through Permian for the Farewell terrane. Figure 10 is a cartoon showing the proposed positions for the Farewell and Alexander terranes in mid-Paleozoic time.

Devonian-Mississippian Syngenetic Deposits

The Arctic Alaska terrane is host to two different types and ages of syngenetic sulphide deposits: the gigantic Red Dog SEDEX deposit, the largest Zn producer in the world (Fig. 17; Appendix 1) and other similar deposits in the western Brooks Range; and the Ambler district VMS deposits in the southern Brooks Range. Red Dog and its smaller equivalents are located within black shales of the Mississippian Kuna basin, and are related to extension and isolation of the basin by a seaward landmass (Young, 2004). The immense stratabound orebody has been dated at ca. 338 Ma by Re-Os methods on pyrite (Morelli et al., 2004), in accord with Osagean-Chesterian (mid- to Late Mississippian) radiolarian ages from the host Kuna Formation (Dumoulin et al., 2001).

The Ambler district lies within pericratonic rocks along the southern fringe of the Arctic Alaska terrane. The deposits are associated with metarhyolite bodies that have been dated at ca. 380 Ma by U-Pb methods (McClelland et al., 2006). Mineralization is hosted within a Middle Devonian ensialic arc sequence that reconstructs to the south (present coordinates) of the main Brooks Range. Unlike the Selwyn Basin-Yukon Tanana connection, there is no direct connection between the Ambler and Red Dog districts: they are separated in time by 40 m.y., in addition to their occurrence in separate allochthonous panels. However, their Pb-isotopic character is similar (Fig. 8), consistent with the involvement of similar crustal sources.

The oldest rocks in Wrangellia are exposed far to the south of its juncture with the Alexander terrane. They belong to the Devonian Sicker Group on Vancouver Island, which comprises island arc and related strata. The Myra Falls deposit (Fig. 17; Appendix 1), dated as ca. 370 Ma by U-Pb methods on associated rhyolites (Juras, 1987), consists of a series of lens-shaped massive sulphide bodies hosted by basaltic to rhyolitic flows and volcaniclastic rocks, lesser epiclastic rocks, argillites, and cherts of the Upper Devonian Myra Formation. The Myra Formation overlies feldspar-pyroxene porphyritic andesite flows, flow breccias, and minor pyroclastic deposits of the Price Formation. The Myra Falls (Buttle Lake) deposits are interpreted to lie within an arc-rift zone (Juras, 1987). Facies and thickness changes across later faults could be interpreted as a structurally inverted basin (Nelson, 1997; also see cross-sections of the Price zone in Juras, 1987). Lead isotopic data from Myra Falls form a distinct cluster that is significantly more radiogenic than the Ambler district and overlaps the least radiogenic end of the range (Stikinia) for western Cordilleran pericratonic Devonian-Mississippian deposits (Robinson et al., 1996; Fig. 8).

Pennsylvanian-Permian Insular Terrane Linkage and other Tectonic Events

Relationships between the Devonian Sicker Group and the Alexander terrane are undefined. In northern Wrangellia, the oldest exposed strata are of an arc-related succession of Pennsylvanian age, the Station Creek Formation. The Middle Pennsylvanian Barnard Glacier pluton (ca. 309 Ma) intrudes both the Station Creek Formation and nearby Paleozoic schists (the Kaskawulsh Group) that locally form the basement of the Alexander terrane (Gardner et al., 1988). It also seals the contact between them. Pennsylvanian-Permian arc deposits are widespread in Alaskan Wrangellia (Skolai arc; see Nokleberg et al., 1994), whereas in southern Wrangellia and in the Alexander terrane, Upper Mississippian to Permian strata are thin and non-volcanogenic (Mihalynuk et al., 1993; Massey, 1995; Gehrels et al., 1996). Given these relationships, the Barnard Glacier locality could be a point at which the Pennsylvanian-Early Permian Skolai arc was anchored on an older Insular crustal fragment, as the modern Aleutian arc is anchored to the Alaska Peninsula.

The Browns Fork Orogeny affected the Farewell terrane in Pennsylvanian-Early Permian time. This ca. 285 Ma metamorphic event is documented by $^{40}Ar/^{39}Ar$ muscovite plateau ages in the northern part of the terrane; thick Pennsylvanian-Permian coarse clastic deposits, interpreted as a foreland wedge, are widespread (Bradley et al., 2003). They speculate that the collision zone in the Urals and Taimyr could have been linked to a system of convergent plate boundaries such as that documented by the Skolai arc. A possibly correlative (or younger?) deformational event affects strata as young as Lower Permian but predates the deposition of Triassic rocks in Wrangellia of southwestern Yukon (Israel et al., 2006).

Statistical analysis of marine faunas – brachiopods, corals, and fusulinids – suggests that in Early Permian time, the Insular terranes lay north of the peri-Laurentian terranes (Quesnellia, Stikinia, and the Eastern Klamaths), and perhaps some few thousands of kilometres west of the northern Laurentian continental margin (Fig. 11; Belasky et al., 2002). This suggests that by late Paleozoic time, the Insular terranes had been transported out of the Arctic realm, and were already enroute to their more southerly Mesozoic location.

Late Triassic Rifting, Flood Basalts, and Associated Deposits

Through the end of Paleozoic time, the story of the Arctic and Insular terranes involved pericratonic crustal fragments and arcs developing near them, or even stepping onto them. In the Late Triassic, a new tectonic style affected the Insular terranes. During a very short time interval in the Carnian (ca. 230 Ma), flood basalts of non-arc, enriched, tholeiitic affinity inundated the older arc basement of Wrangellia (Lassiter et al., 1995). The short duration (approximately 5 m.y.) and high volume of this event, along with the chemical and isotopic signatures of the Karmutsen and Nikolai basalts, has led to its interpretation as a plume-related event, intermediate between true oceanic plateaus and continental flood basalts such as the ca. 250 Ma Siberian traps (Lassiter et al., 1995; see also Reichow and Saunders, 2005). Mafic-ultramafic sills and dykes intruding Paleozoic strata of Wrangellia in southwestern Yukon and adjacent Alaska are thought to be feeders to the Nikolai basalts (Hulbert, 1997); they host a number of Ni-Cu-PGE occurrences, including the Wellgreen deposit (Fig. 17; Appendix 1).

In Norian time, a rift zone developed in the northern and eastern part of the Alexander terrane. It is defined by a narrow, discontinuous belt of rift-fill clastic and volcanic strata of the Tats (Mihalynuk et al., 1993) and Hyd (Gehrels et al., 1987) groups. Volcanic rocks in the northern end of the rift zone are predominantly calc-alkaline basalts (Mihalynuk et al., 1993); farther south, intermediate and felsic units become important (Taylor et al., in press). The very large Windy Craggy Besshi-type Cu-Co-rich deposit (Peter et al., 1999; Fig. 17; Appendix 1) is hosted in interbedded argillite and basalt in the northern part of the rift. The Greens Creek mine (Fig. 17; Appendix 1), a long-producing VMS deposit with enhanced precious metal contents, is associated with rhyolite in the southern part of the rift. Taylor et al. (in press) infer a south to north transition from incipient rifting of thick, older crust, associated with low-volume mixed volcanic rocks and polymetallic VMS mineralization, to wider, more complete rifting, accompanied by basalt and Cu-enriched deposits.

These Triassic volcanic, tectonic, and metallogenetic events were unique to the Insular terranes. During that time, the Farewell and Arctic Alaska terranes were the sites of stable, clastic and carbonate sedimentation, and although there is evidence of rifting as old as 185 to 190 Ma in the northern Arctic Alaska terrane, an Early Cretaceous breakup unconformity is taken as the initial opening of the Canada Basin and beginning of significant rotation there (Moore et al., 1994). Ammonite species distribution suggests that, by ca. 186 Ma, Wrangellia had reached a position significantly south of its present location (Smith et al., 2001). It is tempting to relate the Triassic Wrangellian flood basalts and rifting within the Alexander terrane to their southward translation from the Arctic realm, perhaps crossing a plume track. It is also significant that during the Triassic there is no evidence of arc development in the Insular terranes, and thus no deep subduction zone root to stabilize their position. They could have been completely coupled to a fast-moving oceanic plate (Fig. 14).

Bringing the Realms Together: The Middle Jurassic

There is good evidence for the evolving relationship of the Insular terranes to the western North American pericratonic realm in Jurassic time. Variations in latitude of the present boundaries between late Pleinsbachian boreal amaltheid, mixed, and southern Tethyan ammonite taxa show that Quesnellia has been displaced about 500 km northward since that time (ca. 185 Ma), Stikinia and southern Wrangellia about twice that, and the Peninsular terrane in south-central Alaska has travelled north several thousand kilometres (Fig. 15; Smith et al., 2001). Distribution of high- vs. low-latitude Sinemurian and Pleinsbachian pectinoid bivalves shows similar patterns (Aberhan, 1999). By this time, the Insular terranes – once of high-Arctic provenance – were outboard of, and in part co-latitudinal with, Stikinia.

Their initial accretion to the pericratonic belt has been documented at several localities in the northern and central Coast Mountains. In southeastern Alaska, the Alexander terrane lies structurally beneath the Taku terrane (=Yukon-Tanana/Stikine terrane; see Gehrels, 2001, 2002) along a large-magnitude, low-angle ductile fault system that is crosscut by Late Jurassic (162-139 Ma) dykes (Saleeby, 2000). In this region, Middle Jurassic (ca. 177-168 Ma) volcanic rocks and Upper Jurassic-Lower Cretaceous strata of the Gravina belt overlie both terranes (Gehrels, 2001). Farther north, the inboard margin of the Alexander terrane is involved in a broad dextral shear zone of mid-Jurassic age (McClelland and Gehrels, 1990). In the central Coast Mountains, intrusion, ductile deformation, and metamorphism occurred ca. 160 to 155 Ma; this is interpreted to reflect collision between the Insular and Intermontane terranes (van der Heyden, 1992).

Similar timing for collision between the Insular terranes and the outer margin of the Intermontane terranes is seen in the 205 to 156 Ma Talkeetna arc in the Peninsular terrane of southwestern Alaska (Clift et al., 2005). It is modelled as south-facing and continuous with the Bonanza arc of southern Wrangellia. Uplift, cessation of magmatism, and deposition of conglomerates at about 160 Ma are taken to record its collision with the Intermontane terrane by north-dipping subduction. The absence of continental signatures in Talkeetna arc igneous rocks argues against the alternative of south-dipping polarity, which would have involved subduction of pericratonic material (Clift et al., 2005). The subduction zone under the western edge of Stikinia would correspond to the western Hazelton arc, which underwent profound changes in configuration at ca. 175 Ma (see above).

As outlined above, Stikinia, Quesnellia, and the North American margin were involved in west-vergent collision by ca. 175 Ma, probably centred on the latitude of southern British Columbia. The Insular terranes were beginning to impinge on the outer fringe of the Intermontane terranes at roughly the same time. Most likely, the two collisional events were related to each other. They also coincide with the initial opening of the Atlantic Ocean and the beginning of strong westward motion of the North American plate (May and Butler, 1987). Thus, accretion of the Insular and Intermontane terranes was not due to the serendipitous arrival of exotic crustal blocks, but rather to impingement of the western continent margin on its fringing elements.

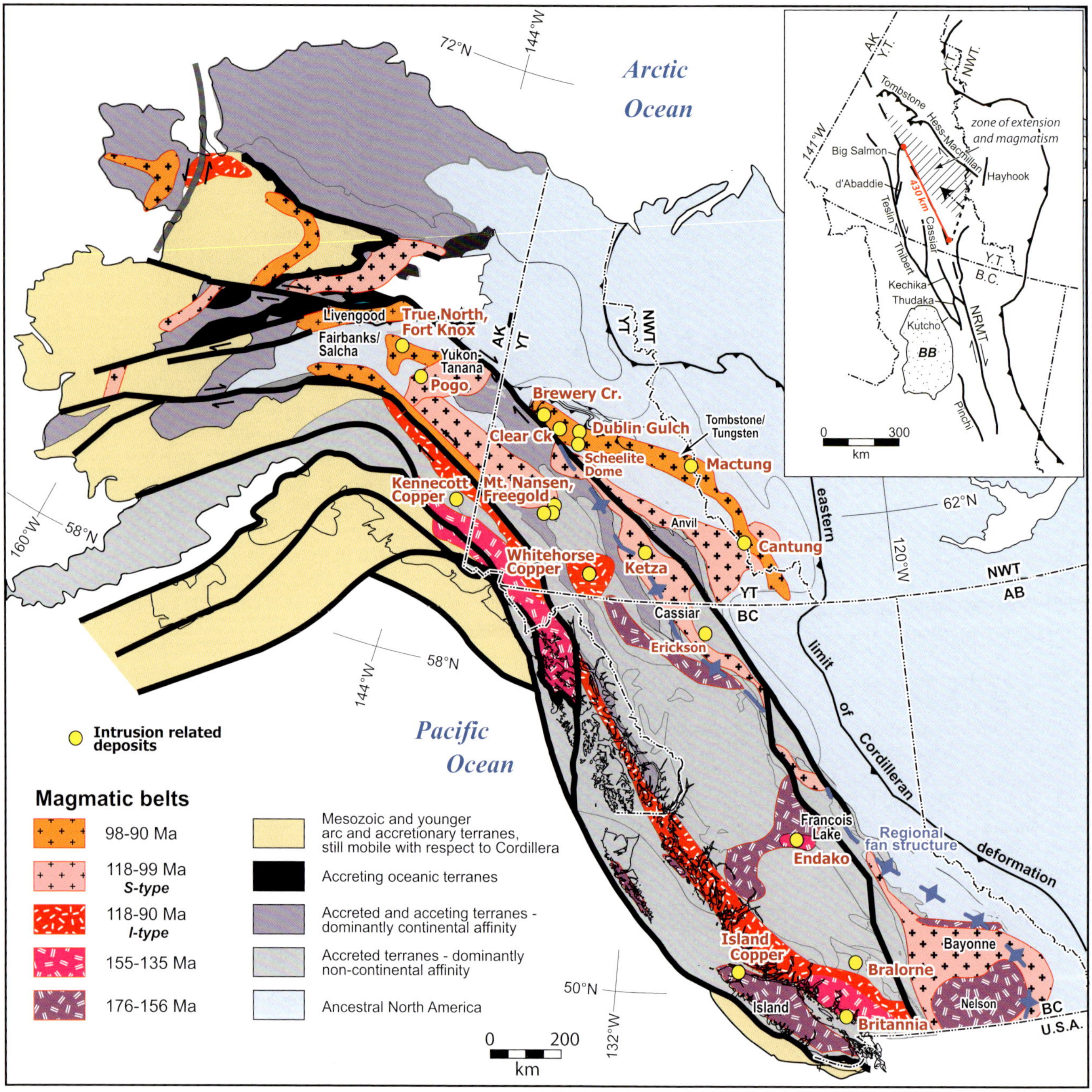

FIGURE 18. Middle Jurassic to mid-Cretaceous magmatism, tectonics and associated deposits. Deposit locations from Hart et al. (2002), Panteleyev (1991), Nokleberg et al. (1994), and BC MINFILE. Magmatic belts from Hart et al. (2004) and Massey et al. (2005). Inset shows relationship of Early Cretaceous transcurrent faults and inferred zone of extension and magmatism in northern B.C. and southern Yukon (after Gabrielse et al., 2006). Dextral strike-slip displacement along the Northern Rocky Mountain Trench (NRMT) and related faults is inferred to transfer into northwest-directed extension in southeast Yukon and compression along Tombstone thrust in western Yukon, when 430 km of displacement is restored along the Eocene Tintina fault (long dash). BB – Bowser basin.

Collision was accompanied by a strong incursion of magmatic activity across the newly accreted terranes and the continental margin. It is represented by the voluminous ca. 175 to 150 Ma Nelson plutonic suite in southern British Columbia. Intrusions of this suite were emplaced during and after development of west-vergent structures in the west flank of a regional fan structure (Murphy et al., 1995; Colpron et al., 1996; Gibson et al., 2005; Fig. 18), and show pronounced crustal influence in their isotopic character (Armstrong and Ghosh, 1990; Murphy et al., 1995). Mid-Jurassic plutonism is also prevalent in a belt that includes insular southeastern Alaska, Vancouver Island (Island plutonic suite, ca. 177-160 Ma; G. Nixon, pers. comm., 2006) and the Queen Charlotte Islands (Breitspreicher and Mortensen, 2004) of southern Wrangellia, and crosses the Insular/Intermontane suture into the central and southern Coast Plutonic Complex and the Skeena Arch of central Stikinia. This arc hosts the now-closed Island Copper mine near Port Hardy on Vancouver Island (Fig. 18; Appendix 1), a Cu-Mo porphyry deposit associated with a ca. 168 Ma body, one of the Island intrusions.

In a coeval but probably unrelated event, the Arctic Alaska and at least parts of the Ruby terrane subducted abortively under the Yukon-Koyukuk arc during Middle Jurassic to Early Cretaceous time. Earliest evidence is $^{40}Ar/^{39}Ar$ cooling ages of 169 to 163 Ma for granulites and peridotites structurally high within the oceanic Angayucham and Tozitna terranes (Wirth et al., 1993; Ghent et al., 2001), and north-vergent deformation in the Brooks Range accompanied by exhumation of blueschists in the hinterland at about 145 to 112 Ma (Moore et al., 2004). Rifting of roughly this age is shown by Jurassic failed-rift deposits overlain by an interpreted Early Cretaceous rift-drift unconformity in northern Arctic Alaska (Grantz et al., 1994). Detachment of Arctic Alaska/Chukotka from the Canadian Arctic Islands by initial opening of the Canada Basin has been linked to southward collision (Lane, 1992, 1997). It is proposed that these events were driven by slab-pull towards the Yukon-Koyukuk subduction zone and its continuation as the South Anyui suture in the Russian Far East (Lane, 1992). Toro et al. (2006) show that there was probably a change in subduction polarity, from southward along the Yukon-Koyukuk zone to northward, with a component of dextral transpression at the Chukotka/South Anyui boundary.

Post-Amalgamation, Cordilleran-Wide Tectonics, Arc Development, and Mineralization

The mid-Jurassic accretionary event fundamentally changed the nature of the emerging Canadian Cordillera from a set of loosely connected arc and pericratonic entities to a thickening transpressional/transtensional orogenic belt on which successor arcs would be built. The developing orogen would prove fertile ground for a variety of epigenetic deposits, with increasing influence of subjacent continental sources on contained metals (e.g. Mo, W), and increasing propensity for the concentration of precious metals. From Late Jurassic through Tertiary time, the combined allochthons impinged eastward as a tectonic wedge over the cratonal margin. Offscraped shelf and foreland sedimentary strata were incorporated into the wedge, while the denuded continental basement drove deep under the advancing allochthons, entering a zone in which widespread crustal melting could occur. Arc-axial igneous activity, as well as broader zones of crustal melting, contributed heat and fluids to the generation of ore deposits. Major types include Cu-Mo(-Au) and Mo porphyries, a variety of intrusion-related gold, silver-lead-zinc and tungsten deposits, and epithermal and mesothermal Au veins. There were four main metallogenetic events during this period: Late Jurassic-Early Cretaceous, mid-Cretaceous, Late Cretaceous, and Paleocene-Eocene (see Nokleberg et al., 2005). The unique character of each is probably related to changing subduction styles, rates, and geometries, as well as to temporal variations in stress regimes and heat transfer into the overlying plate.

Late Jurassic-Early Cretaceous (150-135 Ma)

This earliest post-accretionary magmatic episode represents a continuation of synaccretion (175-150 Ma) intrusion and volcanism, with which it is partly cospatial, although of more limited extent (Fig. 18). The notable deposit associated with this suite is the Endako molybdenite mine (Fig. 18, Appendix 1), for which ca. 144 Ma Re-Os dates on molybdenite coincide with U-Pb dates on the host Endako pluton, part of the Francois Lake suite (Selby and Creaser, 2001; see also Whalen et al., 2001). The orebody is a series of major east-striking veins oriented en echelon to form a zone of approximately 3360 m by 370 m, elongated in a northwesterly direction (British Columbia MINFILE). The veins contain molybdenite, pyrite, and magnetite, with minor chalcopyrite and traces of sphalerite, bornite, specularite, and scheelite. The mine has been in periodic production since 1965, producing over 43 million pounds of Mo concentrate from 309 million tons of ore milled.

Mid-Cretaceous (120 to 90 Ma)

The mid-Cretaceous saw very extensive plutonic activity in the Canadian Cordillera, with the beginning of development of the Coast Plutonic Complex as the prominent linear belt that it appears today, as well as more inboard intrusive belts and clusters, including the metallogenetically important Tombstone/Tungsten and Fairbanks suites (Fig. 18). This episode followed on a Cordilleran-wide magmatic hiatus (Armstong, 1988) that marked, among other changes, the extinction of magmatic activity in most of the Insular belt. With renewed igneous activity in the Coast Plutonic Complex, the region between it and the western plate margin became an essentially amagmatic arc-trench gap.

The mid-Cretaceous plutonic and volcanic rocks of the Coast Plutonic Complex and also the Kluane arc in Alaska (Nokleberg et al., 2005) are clearly the products of a magmatic arc, as shown by their unevolved isotopic character (Friedman and Armstrong, 1995) and linear distribution parallel to, and 100 to 150 km inboard of, the current plate margin. The more inboard plutonic belts were probably not directly related to slab melting. Many of the larger bodies are peraluminous to weakly metaluminous, ilmenite-series granitoids with evolved isotopic signatures; they are interpreted as melts that have interacted substantially with continental crust (Hart et al., 2004). They include the Cassiar suite, plutons in the Yukon-Tanana upland, the Ruby terrane, and an extensive set of plutons in southeastern British Columbia.

Clusters and belts of smaller and comparatively younger plutons tend to be of enhanced metallogenetic significance. The 94 to 90 Ma Fairbanks-Salcha plutonic suite is associated with a variety of intrusion-related Au deposits, including the 7 million ounce Fort Knox mine and the True North mine (Fig. 18; Appendix 1; Hart et al., 2004). Its eastern continuation across the Tintina fault, the 97 to 92 Ma Mayo-Tungsten suites, host Au mineralization at Dublin Gulch, Scheelite Dome, and Clear Creek, and world-class tungsten deposits, including Mactung (57 Mt of 0.95% WO_3) and Cantung (9 Mt of 1.4% WO_3). The ca. 92 Ma Livengood-Tombstone belt hosts the Brewery Creek gold deposit, as well as several Au skarns. These suites are of both ilmenite and magnetite series, and variably reduced.

Sumitomo and Teck Cominco's newly opened Pogo mine, which has reserves of 7.7 Mt at slightly less than 0.5 oz/ton Au (Szumigala and Hughes, 2005), is a 104 Ma mesothermal (orogenic-type) vein deposit that may or may not relate to the mid-Cretaceous plutonic event (Fig. 18; Appendix 1;

Hart et al., 2004). It is located near plutons of the Yukon-Tanana uplands suite, but a direct connection cannot be demonstrated. The mesothermal veins of the Erickson (Table Mountain) mine near Cassiar, British Columbia, have been related to fluid movement in the aureole of a buried mid-Cretaceous intrusive body (Nelson, 1990). Manto deposits in southern Yukon, such as Ketza River and Sa Dena Hes, are also distally related to subjacent Cretaceous intrusions.

The tectonic setting of the Canadian-Alaskan Cordillera during the mid-Cretaceous magmatic climax is complex, and shows a great degree of regional variability. In the southern Coast belt, a compressional, possibly sinistral transpressive event (Monger et al., 1994) closed the Bridge River Ocean and led to southwesterly vergent thrusting of the slightly older (ca. 114 Ma) Gambier arc and its contained Britannia VMS deposit (Lynch, 1991). The Bralorne mesothermal (orogenic-type) vein system developed at a late stage in this event, within a sinistral-reverse shear zone in the southeastern Coast Plutonic Complex (Leitch, 1990). In the Wrangell Mountains of southern Alaska, northeasterly vergent, compressional structures formed during a protracted event during and after accretion of the Peninsular terrane (150-100 Ma). The Kennecott copper deposits, the highest grade copper deposits ever mined, formed as stratabound bodies in the Triassic Chitistone limestone controlled by synorogenic structures (Price et al., 2006).

In the central Coast Plutonic Complex, thick-skinned, west-vergent deformation at ca. 90 to 85 Ma thickened the crust in a zone centred on the Insular/Intermontane tectonic contact (Stowell and Crawford, 2000). In the interior of northern British Columbia and southern Yukon, the ca. 110 Ma Cassiar batholith was emplaced along a synplutonic, northwesterly striking dextral fault. Across the Tintina fault, this motion transfers via a north-striking zone of extension and northwestwardly striking dextral strike-slip faults into the Tombstone strain zone and northwesterly vergent Tombstone and Robert Service thrusts (Murphy, 1997; Gabrielse et al., 2006; Fig. 18). In the Yukon-Tanana upland, in the presumed hinterland of these thrusts, the predominant mid-Cretaceous regime was of northwest-southeasterly extension in the previously-thickened orogen (Hansen and Dusel-Bacon, 1998). The somewhat younger Tombstone, Mayo, and Tungsten belt intrusions may have been lithospheric melts emplaced during a final stage of extension localized along basement structures (Mair et al., 2006).

Late Cretaceous (90 to 65 Ma)

Following the mid-Cretaceous, magmatic activity declined generally across the Cordillera, but with some important exceptions. Along the axis of the Coast Mountains, plutonism shifted northeasterly into the central gneiss belt (Fig. 19; Stowell and Crawford, 2000). In Alaska, Late Cretaceous volcanic-plutonic zones are located generally to the west of the mid-Cretaceous Fairbanks district. The ca. 75 to 67 Ma Kuskokwim district of southwestern Alaska is host to the Donlin Creek intrusion-related gold deposit, with its approximately 12 million ounce resource (Hart et al., 2002; Nokleberg et al., 2005). The gigantic (1 billion tonne plus) Pebble porphyry deposit is located farther southwest, on the Alaska Peninsula near Lake Iliamna (Fig. 19; Appendix 1). Mineralization there, which is hosted by a Late Cretaceous granodiorite body, has been dated at ca. 87 Ma (Schrader et al., 2001).

Elsewhere in the Cordillera, Late Cretaceous intrusions are sparser; although there are some significant deposits of this age. In southern Yukon, the Red Mountain (Boswell River) porphyry Mo deposit (Appendix 1) is contained within a Late Cretaceous quartz-monzonite stock associated with nearby 80 Ma volcanic rocks. The Casino porphyry Cu-Mo-Au deposit in the central Yukon is focused on a ca. 73 Ma complex of subvolcanic intrusions and intrusive breccias (Bower et al., 1995). The hypogene zone of this large but undeveloped deposit contains 445 Mt of 0.23% Cu, 0.024% Mo, and 0.27 g/t Au, and the supergene zone contains 86 Mt at 0.43% Cu, 0.031% Mo, and 0.41 g/t Au (Bower et al., 1995). The Adanac (Ruby Creek) near Atlin and Glacier Gulch near Smithers are both porphyry Mo deposits that focus on small ca. 70 Ma quartz-monzonite stocks, probably late-stage crustal melts. Current high Mo prices have created a new interest in these deposits, which are currently in mine permitting stage. In central British Columbia, the currently producing Huckleberry mine is associated with two small ca. 82 Ma porphyritic granodiorite stocks, part of the Bulkley intrusive suite (Fig. 19).

Paleocene-Eocene (65 to 37 Ma)

The Eocene was the last great metallogenetic epoch in the northern Cordillera. It was characterized by a sharp break from previous magmatic and tectonic patterns, although the influence of previous events can be identified. The most stable element remained the Coast Plutonic Complex. There, the axis of plutonism once again shifted east, but developed adjacent to the older Cretaceous belt of intrusions (Fig. 19). Vigorous volcanic and intrusive activity flared over a short time interval throughout southern British Columbia, from the Skeena arch south into eastern Washington and Idaho. In southern Stikinia, these rocks are termed the Ootsa Lake Group and in the southern Okanagan, the Kamloops and Princeton groups. Compositions are variable, with some tendency to mildly alkalic (shoshonitic); felsic units are abundant. Volcanism accompanied crustal extension and the exhumation of core complexes in the southern and central Omineca belt and southern Stikinia (Price and Carmichael, 1986; Struik, 1993). Sometime during this interval, the Tintina fault accommodated some 430 km of dextral motion, along with ca. 100 km of mostly Late Eocene motion on the Fraser fault. Dextral motion on the Denali fault also began. Overall, the tectonic framework for this interval is viewed as one of rapidly emerging, dextral transtension accompanied by extension and collapse of previously thickened crust, particularly in the southern Omineca belt. This tectonic and magmatic setting created a fertile environment for epithermal deposits, which concentrate within the volcanic fields and near transcurrent faults, and for porphyry Cu-Au-Mo and Mo deposits, which concentrate in the Skeena arch of central Stikinia.

The Skeena arch porphyry deposits are associated with small plutons of the Babine Igneous Suite that are very abundant in the area, in many cases localized along northwest- and northeast-striking faults (Carter et al., 1995; Wojdak and Stock, 1995). The belt contains two past producers, Bell and Granisle, with a total of 130 Mt milled (Carter et al., 1995),

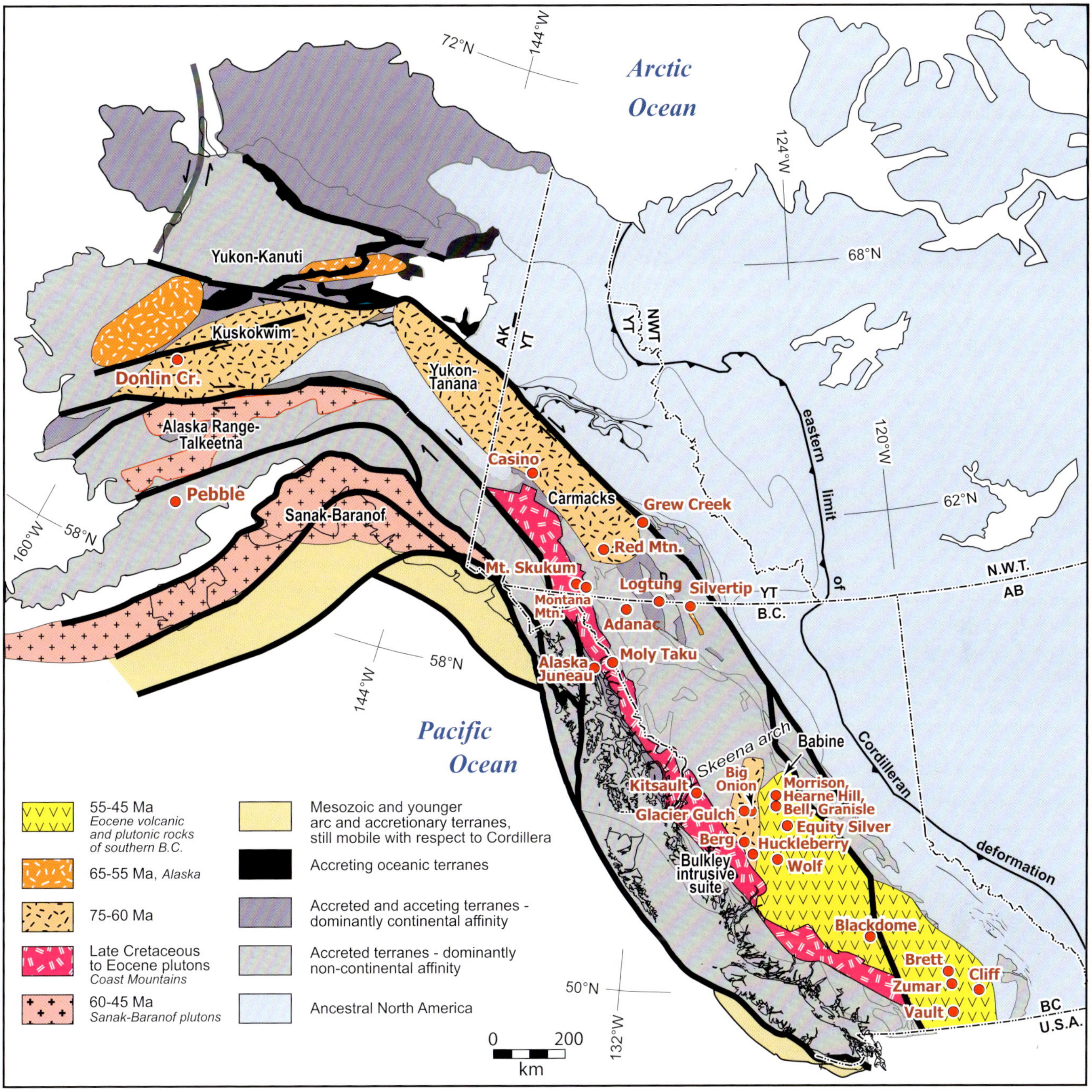

FIGURE 19. Late Cretaceous to Eocene magmatism, tectonics and associated deposits. Volcanic fields of Alaska from Moll-Stalcup (1994) and Hudson (1994). Volcanic fields in British Columbia from Massey et al. (2005). Deposit locations from Hart et al. (2002), Panteleyev (1991), Nokleberg et al. (1994), and BC MINFILE.

and significant resources at the Berg, Morrison/Hearne Hill, and Big Onion deposits (Fig. 19; Appendix 1). Late-stage basin-and-range-style block faulting is apparent in the Skeena arch, as shown by local, fault-controlled differences in erosion levels (Wojdak and Stock, 1995). These faults also controlled the emplacement of the porphyries, suggesting a genetic link. The Skeena arch marks an abrupt northern edge of the voluminous Eocene magmatic belt: it must have been a fundamental structural or thermal boundary in the upper mantle/lower crust.

The Kitsault porphyry molybdenite deposit, which was mined briefly in 1960 to 1972 and 1981 to 1983, is located farther north, on the eastern margin of the Coast Plutonic Complex. There are a few other molybdenite occurrences known in this eastern Eocene belt, such as Moly Taku in far northwestern British Columbia and showings in the Carpenter Creek pluton near Terrace. Given the large concentration of Eocene plutons in the eastern Coast Plutonic Complex, it is surprising that more substantial molybdenum resources have not been found. The Logtung porphyry/skarn W-Mo deposit, located on the British Columbia-Yukon border, has been dated at ca. 58 Ma by U-Pb methods on zircon considered to be of hydrothermal origin (Mihalynuk and Heaman, 2002).

Eocene Au-Ag epithermal deposits are distributed widely but sparsely in the volcanic-rich areas of the Okanagan and southern Stikinia. A few have been mined, such as Blackdome, which is associated with Eocene volcanic rocks near the Fraser fault, and the deposits of the Republic graben in Washington State south of the British Columbia border. Grew Creek, which lies adjacent to the Tintina fault in central Yukon, is a low-sulphidation vein system hosted by Eocene volcanic rocks. Mt. Skukum, a past producer, is a set of epithermal quartz veins in an isolated volcanic centre near the British Columbia-Yukon border. There are a number of developed prospects in the Okanagan, notably the Brett, Zumar, Cliff, and Vault. Exploration in the Nechako Plateau of southern Stikinia has borne less fruit so far. The Wolf property is a swarm of low-sulphidation veins with reported values in one trench of 8.49 g/t Au and 42.21 g/t Ag over 7.5 metres. The Trout comprises patches and veins of chalcedonic quartz with erratic Au and Ag values. In this area, extensive Miocene basalt flows and Quaternary deposits thwart prospecting and the quartz-vein targets do not have pronounced geophysical signatures. Airborne remote sensing and multi-element surveys may be of aid.

In southeastern Alaska, auriferous mesothermal veins of the Juneau gold belt were emplaced along and near reactivated older contractional structures, the Sumdum and Fanshaw faults, at ca. 57 to 53 Ma (Miller et al., 2000). This corresponds to a time of rapid exhumation in the Coast Plutonic Complex, and it is proposed that metamorphic fluids were channelled upwards along the faults (Miller et al., 2000). The Juneau gold belt extends over 300 km in strike length, and has produced over 200 million grams of gold (Nokleberg et al., 1995).

Late Eocene to Present (<37 Ma)

In the Canadian Cordillera, the broad, intense magmatic belts of the Eocene dwindled by 45 Ma and were extinct by 40 Ma. In particular, the Coast Plutonic Complex, which had been an axis of arc activity for over 60 m.y., abruptly shut off, probably because subduction ceased off the west coast of Canada. This part of the plate margin is now the locus of the dextral transcurrent Queen Charlotte fault (Fig. 20). Within the North American plate, the main locus of dextral motion shifted outboard from the Tintina to the Denali fault. The main volcanic arcs shifted to the Aleutian arc of western Alaska and adjoining islands, and south into the Cascades of Washington and Oregon (Fig. 20; see also Nokleberg et al., 2005). This overall pattern would endure throughout the Cenozoic, sustained by easterly subduction of the Juan de Fuca plate and northerly subduction of the Pacific plate. Catface on western Vancouver Island (Fig. 20; Appendix 1) is an anomalously young late Eocene Cu-Mo porphyry deposit, associated with one of a belt of small, ca. 37 Ma plutons (Madsen et al., 2003). Although the setting of Catface has been interpreted as a fore-arc, by the time of its emplacement the ca. 50 Ma arc to the east was extinguished. Alternatively, it could represent the northern tip of the newly established Cascade arc.

Quartz Hill, near Ketchikan in southeastern Alaska (Fig. 20; Appendix 1), is a low-fluorine-type porphyry Mo deposit with one of the world's largest reserves of molybdenum, 1.7 billion tonnes at 0.136% MoS_2 (Nokleberg et al., 1994, 2005). It is associated with a small, high-level, granite-latite-quartz-feldspar porphyry stock, ca. 27 Ma, emplaced into upper-amphibolite-grade metamorphic rocks. This late-stage, post-orogenic body seemingly reaped the rewards of a long process of prior crustal thickening and melting in the Coast Plutonic Complex. By contrast, the earlier voluminous Cretaceous intrusions of the Coast Plutonic Complex tend to be relatively barren, perhaps because they are too deeply eroded.

The Cinola (Spegogna) vein system on eastern Graham Island (Haida G'wai; Fig. 20; Appendix 1) is a large hot-spring-type, low-sulphidation epithermal deposit related to active Miocene displacement on the Sandspit fault. Mineable reserves estimated to June, 1997 are 33.5 Mt grading 2.11 g/t Au at a cutoff grade of 1.20 g/t Au (British Columbia MINFILE). It is the youngest known deposit with significant reserves in the Canadian Cordillera and, significantly, it formed in conjunction with the transcurrent, post-subduction fault regime that now prevails there. Yet more recent seafloor massive sulphide deposits can be found at hydrothermal vents located along the axis of the Juan de Fuca ridge offshore.

By contrast with the now metallogenetically sleepy south, the Cenozoic to Recent Aleutian arc is host to a variety of epithermal and polymetallic vein and porphyry Cu and Mo deposits over an 800 km long belt (Nokleberg et al., 1994, 2005), associated with Tertiary and Quaternary volcanic centres and their hypabyssal roots. A suite of large, low-grade porphyry Cu-Mo deposits occupies the transitional zone, where the young Aleutian arc onlaps from oceanic substrate to the older rocks of the Peninsular terrane (Nokleberg et al., 1994, 2005). The most significant of these is the Pyramid deposit, with 113 Mt of 0.4% Cu and 0.05% Mo (Fig. 20; Appendix 1). Their occurrence points to the importance of previously constructed and thickened arc crust to the genesis of deposits of this type, as was the case with the Triassic-Jurassic porphyries of the Canadian Cordillera.

Discussion: Mechanisms and Models

The history of mineralization in the Canadian-Alaskan Cordillera is very long, spanning over 1.6 billion years, from Mesoproterozoic to the present time. The earliest events – the deep-water, rift-related hydrothermal system that gave rise to Sullivan, and the enigmatic Wernecke breccias – took place in intracontinental settings at unspecified distances from their margins, at a time when plate tectonics in the current sense may not have operated (Stern, 2004). The main tectonic and metallogenetic development of the Cordillera dates from the breakup of the Rodinia supercontinent in late Neoproterozoic to Early Cambrian time. This is particularly true of the western continental margin and the peri-Laurentian terranes, but arc and rift development in the Arctic and Insular terranes also dates to this time. The Paleozoic might be called "the age of syngenetic sulphides", in the sense that VMS deposits associated with rifting arcs and SEDEX deposits associated with rifting continental margins, characterize this phase of tectonic history. In particular, both peri-Laurentian and exotic pericratonic terranes of the Cordillera contain deposits that are part of a worldwide peak in syngenetic sulphide formation and preservation during Devonian-Mississippian time, probably because of dual factors of rift-related tectonics and ocean anoxia (Goodfellow, 2007). Following a weak period (Pennsylvanian to mid-

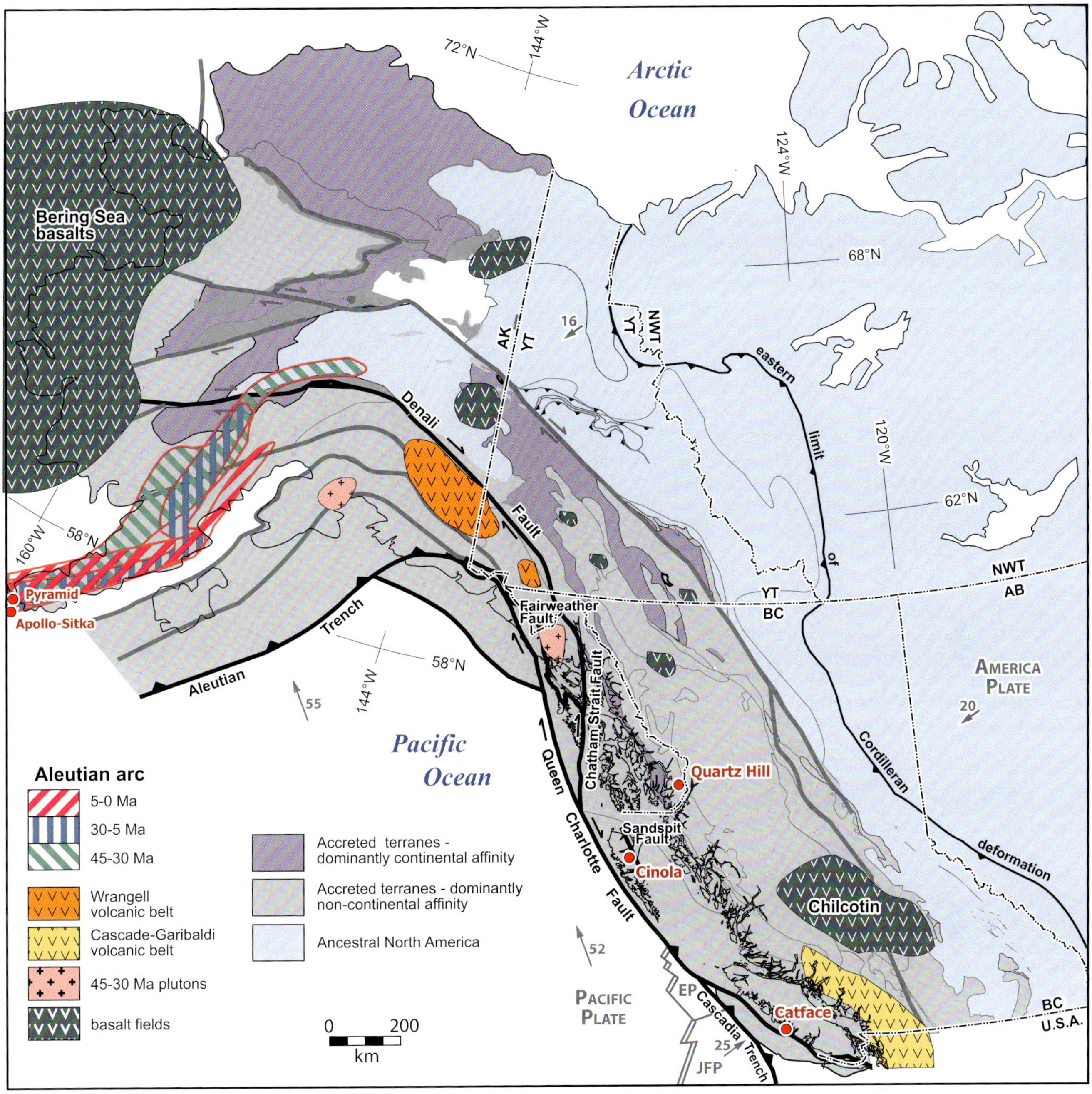

Figure 20. Oligocene to present tectonics and deposits. Volcanic fields of Alaska from Moll-Stalcup (1994). Modern plate configuration and relative motions (vs. fixed hotspot reference; in mm/a) from Riddihough and Hyndman (1991). EP – Explorer plate; JFP – Juan de Fuca plate.

Triassic) in the generation of syngenetic deposits, a new compressional and accretionary tectonic regime led to what could be called "the age of porphyries", which became the premier style of large Cordilleran deposit beginning in Late Triassic time in the pericratonic Intermontane arcs – a trend that would continue through the formation of the giant Pebble deposit within a successor arc during Late Cretaceous time, and only diminished as subduction faltered at the end of Eocene.

Overall, the history of the northern Cordillera depicts the dominance of extension-related tectonics throughout the Paleozoic, followed by increasingly compression-dominated tectonics from the Late Triassic through the Paleocene, with a brief episode of crustal transtension in the southern Canadian Cordillera in the Late Paleocene-Eocene. Monger and Price (2002) have related this fundamental evolution in tectonic style to a change from an offshore arc system dominated by slab rollback, to one in which the overriding North American plate advanced towards its subduction zone as the Atlantic Ocean opened. The initial propagation of magmatic arcs along the western margin of Laurentia (ca. 400-360 Ma) coincided with the final closing of Iapetus. Their rapid subsequent detachment (ca. 360-345 Ma) was probably due to slab rollback, in a process comparable to the Miocene to present opening of the Japan Sea, with concurrent formation of the Kuroko and younger VMS deposits in extending arc and back-arc settings. The Middle Jurassic and younger crustal thickening event that incorporated pericratonic blocks, arcs and accretionary prisms as well as the continental margin itself, was coeval with the opening of the Atlantic Ocean. Thus, a megacycle driven by major changes in continental plate motion can account for the broadest aspects of Cordilleran tectonic history.

Some phases are less clearly related to global-scale plate motions. The initial Permo-Triassic impingement of offshore arcs against North America – the margin-long Sonoman Orogeny – occurred some 70 m.y. before the initial opening of the Atlantic. Did Pangaea as a whole begin to track westwards prior to its breakup? Following this collisional event, new Late Triassic Intermontane arcs seem to have stepped seaward, perhaps due to renewal of slab rollback. At the same time, extension and widespread intraplate basaltic activity prevailed in the Insular terranes. Perhaps these crustal blocks then lay in a back-arc region, along with the Arctic terranes and the rest of the post-Sonoman continental margin. By Early Jurassic time, arcs were reestablished in the Insular terranes (Bonanza, Talkeetna), signifying that probably by then they were part of the circum-Pacific arc system, along with the Hazelton and Takla arcs.

Conclusions

There never was a truly "passive" western margin of Laurentia/North America. It was actively rifting and metallogenetically alive throughout early Paleozoic time. From Late Devonian through Early Jurassic, the seemingly passive miogeocline was protected from the effects of continent-ocean interactions by a moving wall of partly pericratonic island arcs. The large-scale cycle that we observe in the Cordillera, from Paleozoic extensional tectonics to Mesozoic-Paleogene accretion and compression, was largely caused by changes in relative motion of the North American plate with respect to its western subduction zone. The gross aspects of this cycle are clearly reflected in metallogenetic patterns, from the dominance of Paleozoic rift-related syngenetic sulphide deposits, through Cu-Au porphyries in early Mesozoic time, to porphyries and precious metal deposits with increasing continental influence as amalgamation and crustal thickening progressed in Cretaceous and later time. Overall, the duration and complexity of northern Cordilleran tectonics have led to a rich and varied endowment of metallic mineral deposits. Spectacular recent exploration successes such as Pogo (the Cordillera's newest mine at time of writing), Eskay Creek, and Pebble show that the mineral potential of this region is far from a closed book, and that the deposits we know now are also an intimation of future discoveries.

For More Information

Cordilleran study in Canada is vastly aided by excellent modern geological and tectonic assemblage compilations: Canadian Cordillera (Wheeler and McFeely, 1991); Yukon (Gordey and Makepeace, 2001); British Columbia (Massey et al., 2005); and a recent update for the pericratonic terranes of Yukon and northern British Columbia (Colpron, 2006). Large-scale geological and tectonic assemblage maps for Alaska include Silbering et al. (1994) and Nokleberg et al. (1997); a GIS compilation is available for northern Alaska and surrounding areas (Klemperer et al., 2002).

This short synthesis can serve as an introduction to more extensive and detailed metallogenetic/ tectonic syntheses such as that of the Canadian Cordillera by Dawson et al. (1991), and the Canadian-Alaskan Cordillera and Russian Far East of Nokleberg et al. (2005). The interested explorationist is also pointed to online databases in British Columbia and Yukon MINFILE, and the British Columbia mineral deposit profiles (Lefebure, 2005). MAPPLACE (www.em.gov.bc.ca/mining/Geolsurv/MapPlace/) and MapGallery (www.geology.gov.yk.ca/gallery/index.html) are integrated, online GIS-based map data systems for British Columbia and Yukon, respectively, that offer easy integration of spatial data sets, such as geological maps, geophysical and geochemical surveys, and mineral deposit information, in the MINFILE databases.

Acknowledgements

This paper could not have come into being without the support of our managers, Dave Lefebure, Brian Grant, and Grant Abbott, who over the years encouraged us to embark on wide-ranging studies of the Canadian Cordillera. Our ideas have been influenced by many colleagues and co-workers, including Cynthia Dusel-Bacon, Hu Gabrielse, Mitch Mihalynuk, Fil Ferri, Jim Monger, Don Murphy, Suzanne Paradis, Steve Piercey, George Gehrels, Ray Price, Charlie Roots, and John Bradford. Don Murphy contributed the diagram in inset on Figure 18. We thank Brian Grant, Dave Lefebure, and Don Murphy for useful presubmission reviews of the manuscript. Formal reviews by Stephen Piercey, Warren Nokleberg, and volume editor Wayne Goodfellow are also gratefully acknowledged.

References

Abbott, G., 1997, Geology of the upper Hart River area, eastern Ogilvie Mountains, Yukon Territory (116A/10, 116A/11): Yukon Geological Survey, Bulletin 9, 92 p.

Abbott, J.G., Gordey, S.P. and Tempelman-Kluit, D.J., 1986, Setting of stratiform, sediment-hosted lead-zinc deposits in Yukon and northeastern British Columbia, *in* Morin, J.A., ed., Mineral Deposits of Northern Cordillera: Canadian Institute of Mining and Metallurgy, Special Volume 37, p. 1-18.

Aberhan, M., 1999, Terrane history of the Canadian Cordillera: Estimating amounts of latitudinal displacement and rotation of Wrangellia and Stikinia: Geological Magazine, v. 136, p. 481-492.

Alldrick, D.J., Nelson, J.L. and Barresi, T., 2005, Geology and mineral occurrences of the upper Iskut River area: Tracking the Eskay rift through northern British Columbia, *in* Geological Fieldwork 2004, British Columbia Ministry of Energy and Mines, Paper 2005-1, p. 1-30.

Amato, J., 2004, Crystalline basement ages, detrital zircon ages, and metamorphic ages from Seward Peninsula: Implications for Proterozoic and Cambrian-Ordovician paleogeographic reconstructions of the Arctic-Alaska terrane: Geological Society of America Abstracts with Programs, Annual Meeting, Denver, Colorado, v. 36, no. 5, p. 22.

Amato, J., Toro, J., Miller, E. and Gehrels, G.E., 2006, Late Proterozoic magmatism in Alaska and its implications for paleogeographic reconstructions of the Arctic Alaska-Chukotka plate: Geological Society of America, Abstracts with Programs, Cordilleran Section, Anchorage, Alaska, May, 2006, v. 38, no. 5, p. 13.

Anderson, H.E. and Parrish, R.R., 2000, U-Pb geochronological evidence for the geological history of the Belt-Purcell Supergroup, southeastern British Columbia, *in* Lydon, J.W., Höy, T., Slack, J.F., and Knapp, M.E., eds., The Geological Environment of the Sullivan Deposit, British Columbia, Geological Association of Canada, Mineral Deposits Subdivision, Special Publication No. 1, p. 113-126.

Anderson, R.G., 1991, A Mesozoic stratigraphic framework for northwestern Stikinia (Iskut River area), northwestern British Columbia, Canada, *in* Dunne, G. and McDugall, K., eds., Mesozoic Paleogeography of the

Western United States II: Pacific Section, Society of Economic Paleontologists and Mineralogists, v. 71, p. 477-494.

Armstrong, R.L., 1988, Mesozoic and Cenozoic magmatic evolution of the Canadian Cordillera, *in* Clark, S.P., Burchfiel, B.C. and Suppe, J., eds., Processes in Continental Lithospheric Deformation: Geological Society of America Special Paper 218, p. 55-91.

Armstrong, R.L. and Ghosh, D.K., 1990, Westward movement of the 87Sr/86Sr line in southern B.C. from Triassic to Eocene time: Monitoring the tectonic overlap of accreted terranes on North America: Geological Association of Canada Program with Abstract, v. 15, p. A4.

Bazard, D.R., Butler, R.F., Gehrels, G.E. and Soja, C.M., 1995, Early Devonian paleomagnetic data from the Lower Devonian Karheen Formation suggest Laurentia-Baltica connection for the Alexander terrane: Geology, v. 23, p. 707-710.

Belasky, P., Stevens, C.H. and Hanger, R.A., 2002, Early Permian location of western North American terranes based on brachiopod, fusulinid and coral biogeography: Palaeogeography, Palaeoclimatology, Palaeoecology v. 179, p. 245-266.

Beranek, L.P. and Mortensen, J.K., 2006, A Triassic link between Yukon-Tanana terrane and North America: New detrital zircon age, geochemical and biostratigraphic data: Geological Society of America, Abstracts with Programs, Cordilleran Section, v. 38, p. 5.

Blodgett, RB., Rohr, D.M. and Boucot, A.J., 2002, Paleozoic links among some Alaskan accreted terranes and Siberia based on megafossils, *in* Miller, E.L. Grantz, A. and Klemperer, S.L., eds., Tectonic evolution of the Bering Shelf - Chukchi Sea - Arctic margin and adjacent landmasses: Geological Society of America Special Paper 360, p. 273-290.

Bond, G.C. and Kominz, M.A., 1984, Construction of tectonic subsidence curves for the early Paleozoic miogeocline, Southern Canadian Rocky Mountains: Implications for subsidence mechanisms, age of breakup and crustal thinning: Geological Society of America Bulletin, v. 95, p. 155-173.

Bower, B., Payne, J., DeLong, C. and Rebagliati, C.M., 1995, The oxide-gold, supergene and hypogene zones at the Casino gold-copper-molybdenum deposit, west-central Yukon, *in* Schroeter, T.G., ed., Porphyry Deposits of the Northwestern Cordillera of North America: Canadian Institute of Mining, Metallurgy and Petroleum, Special Volume 46, p. 352-366.

Box, S.E., Moll-Stalcup, E.J., Wooden, J.L. and Bradshaw, J.Y., 1990, Kilbuck terrane: Oldest rocks in Alaska: Geology, v. 18, p. 1219-1222.

Bradley, D.C., Dumoulin, J., Layer, P., Sunderlin, D., Roeske, S., McClelland, W., Harris, A.G., Abbott, G., Bundtzen, T. and Kusky, T., 2003, Late Paleozoic orogeny in Alaska's Farewell terrane: Tectonophysics, v. 372, p. 23-40.

Bradley, D.C., Dumoulin, J.A., Blodgett, R.B., Harris, A.G., Roeske, S.M., McClelland, W.C. and Layer, P.W., 2006, Geology and affinity of Alaska's Farewell terrane: Geological Society of America, Abstracts with Program, Cordilleran Section Meeting, v. 38, p. 12.

Breitsprecher, K. and Mortensen, J.K., 2004, BCAge 2004 - a database of isotopic age determinations for rock units from British Columbia: British Columbia Ministry of Energy and Mines, Geological Survey Branch, Open File 2004-3, 7099 records.

Burrett, C. and Berry, R., 2000, Proterozoic Australia - Western United States (AUSWUS) fit between Laurentia and Australia: Geology, v. 28, p. 103-106.

Butler, R.F., Gehrels, G.E. and Kodama, K.P., 2001, A moderate translation alternative to the Baja British Columbia hypothesis: GSA Today, v. 11, p. 4-10.

Carter, N.C., Dirom, G.E. and Ogryzlo, P.L., 1995, Porphyry copper-gold deposits, Babine Lake area, west-central British Columbia, *in* Schroeter, T.G., ed., Porphyry Deposits of the Northwestern Cordillera of North America, Canadian Institute of Mining, Metallurgy and Petroleum, Special Volume 46, p. 247-289.

Cecile, M.P., Morrow, D.W. and Williams, G.K., 1997, Early Paleozoic (Cambrian to Early Devonian) tectonic framework, Canadian Cordillera: Bulletin of Canadian Petroleum Geology, v. 45, p. 54-74.

Childe, F.C., 1997, Timing and tectonic setting of volcanogenic massive sulphide deposits in British Columbia: Constraints from U-Pb geochronology, radiogenic isotopes and geochemistry: Ph.D. thesis, The University of British Columbia, 298 p.

Childe, F.C., Thompson, J.F.H., Mortensen, J.K., Friedman, R.M., Schiarizza, P., Bellefontaine, K. and Marr, J.M., 1998, Primitive Permo-Triassic volcanism in the Canadian Cordillera: Tectonic and metallogenetic implications: Economic Geology, v. 93, p. 224-231.

Churkin, M., Carter, C. and Trexler, J.H., 1980, Collision-deformed Paleozoic continental margin of Alaska - foundation for microplate accretion: Geological Society of America Bulletin, v. 91, p. 648-654.

Clift, P.D., Pavlis, T., Debari, S.M., Draut, A.E., Rioux, M. and Kalemen, P.B., 2005, Subduction erosion of the Jurassic Talkeetna-Bonanza arc and the Mesozoic accretionary tectonics of western North America: Geology, v. 33, p. 881-884.

Coe, R.S., Globerman, B.R. and Thrupp, 1989, Rotation of central and southern Alaska in the early Tertiary: Oroclinal bending by megakinking? in Kissel, C. and Laj, C., eds., Paleomagnetic rotations and continental deformation, Kluwer Academic Publishers, p. 327-342.

Colpron, M. (compiler), 2006, Tectonic assemblage map of Yukon-Tanana and related terranes in Yukon and northern British Columbia (1:1,000,000 scale): Yukon Geological Survey, Open File 2006-1.

Colpron, M., Price, R.A., Archibald, D.A. and Carmichael, D.M., 1996, Middle Jurassic exhumation along the western flank of the Selkirk fan structure: Thermobarometric and thermochronometric constraints from the Illecillewaet synclinorium, southeastern British Columbia: Geological Society of America Bulletin, v. 108, p. 1372-1392.

Colpron, M., Logan, J. and Mortensen, J.K., 2002, U-Pb age constraint for late Neoproterozoic rifting and initiation of the lower Paleozoic passive margin of western Laurentia: Canadian Journal of Earth Sciences, v. 39, p. 133-143.

Colpron, M., Gladwin, K., Johnston, S.T., Mortensen, J.K. and Gehrels, G.E., 2005, Geology and juxtaposition history of Yukon-Tanana, Slide Mountain and Cassiar terranes in the Glenlyon area of central Yukon: Canadian Journal of Earth Sciences, v. 42, p. 1431-1448.

Colpron, M., Nelson, J.L. and Murphy, D.C., 2006, A tectonostratigraphic framework for the pericratonic terranes of the northern Canadian Cordillera, *in* Colpron, M. and Nelson, J.L., eds., Paleozoic Evolution and Metallogeny of Pericratonic Terranes at the Ancient Pacific Margin of North America, Canadian and Alaskan Cordillera: Geological Association of Canada Special Paper 45, p. 1-23.

Coney, P.J., Jones, D.L. and Monger, J.W.H., 1980, Cordilleran suspect terranes: Nature, v. 288, p. 329-333.

Cordey, F., Mortimer, N., DeWever, P. and Monger, J.W.H., 1987, Significance of Jurassic radiolarians from the Cache Creek terrane, British Columbia: Geology, v. 15, p. 1151-1154.

Corriveau, L., 2007, Iron oxide-copper-gold: A Canadian Perspective, *in* Goodfellow, W.D., ed., Mineral Deposits of Canada: A Synthesis of Major Deposit-Types, District Metallogeny, the Evolution of Geological Provinces, and Exploration Methods: Geological Association of Canada, Mineral Deposits Division, Special Publication No. 5, p. 307-328.

Crowley, J.L., 1999, U-Pb geochronologic constraints on Paleoproterozoic tectonism in the Monashee complex, Canadian Cordillera: Elucidating an overprinted geologic history: Geological Society of America Bulletin, v. 111, p. 560-577.

Dawson, K.M., Panteleyev, A., Sutherland Brown, A. and Woodsworth, G.J., 1991, Regional metallogeny, Chapter 19 *in* Gabrielse, H. and Yorath, C.J., eds., Geology of the Cordilleran Orogen in Canada: Geological Survey of Canada, Geology of Canada, no. 4; also Geological Society of America, The Geology of North America, v. G-2, p. 707-768.

Dostal, J., Gale, V. and Church, B.N., 1999, Upper Triassic Takla Group volcanic rocks, Stikine Terrane, north-central British Columbia: geochemistry, petrogenesis, and tectonic implications: Canadian Journal of Earth Sciences, v. 36, p. 1483-1494.

Doughty, P.T., 2004, New SHRIMP evidence for multiple metamorphic events in the northern U.S. Cordillera: Eocene, Cretaceous and Late Precambrian (Grenville) events: Geological Society of America, Abstracts with Programs, Annual Meeting, v. 36, p. 271.

Dover, J.H., 1994, Geology of part of east-central Alaska, Chapter 5 *in* Plafker, G. and Berg, H.C., eds., The Geology of Alaska: Geological Society of America, The Geology of North America, v. G-1, p. 153-204.

Dumoulin, J.A. and Harris, A.G., 2006, Paleogeographic affinities of Carboniferous and older strata in northern Alaska: Geological Society of America, Abstracts with Program, Cordilleran Section Meeting, v. 38, p. 13.

Dumoulin, J., Harris, A.G., Young, L.E. and Blome, C., 2001, Sedimentologic and paleontologic constraints on the setting and age of the Red Dog Zn-Pb-Ag massive sulfide deposit, northern Brooks Range, Alaska: Geological Society of America, Abstracts with Programs, v. 33, p. 271.

Dumoulin, J.A., Harris, A.G., Gagiev, M., Bradley, D.C. and Repetski, J.E., 2002, Lithostratigraphic, conodont, and other faunal links between northeastern Russia and the North American Cordillera; in Miller, E.L. Grantz, A. and Klemperer, S.L., eds., Tectonic Evolution of the Bering Shelf - Chukchi Sea - Arctic Margin and Adjacent Landmasses: Geological Society of America Special Paper 360, p. 291-312.

Dusel-Bacon, C., Hopkins, M.J., Mortensen, J.K., Williams, I., Heslop, K., Dashevsky, S.S., Bressler, J.R. and Day, W.C., 2006, Paleozoic tectonic and metallogenetic evolution of the Yukon-Tanana terrane, east-central Alaska, *in* Colpron, M., and Nelson, J.L., eds., Paleozoic Evolution and Metallogeny of Pericratonic Terranes at the Ancient Pacific Margin of North America, Canadian and Alaskan Cordillera: Geological Association of Canada Special Paper 45, p. 25-74.

Enkin, R.J., 2006, Paleomagnetism and the case for Baja British Columbia, *in* Haggart, J.W., Monger, J.W.H., and Enkin, R.J., eds., Paleogeography of the North American Cordillera: Evidence For and Against Large-Scale Displacements: Geological Association of Canada, Special Paper 46, p. 233-253.

Enkin, R.J., Johnston, S.T., Larson, K.P., and Baker, J., 2006, Paleomagnetism of the 70 Ma Carmacks Group at Solitary Mountain, Yukon, confirms and extends controversial results: Further evidence for the Baja-British Columbia model: *in* Haggart, J.W., Monger, J.W.H., and Enkin, R.J., eds., Paleogeography of the North American Cordillera: Evidence For and Against Large-Scale Displacements: Geological Association of Canada, Special Paper 46, p. 221-232.

Evenchick, C.A. and McNicoll, V.J., 2002, Stratigraphy, structure, and geochronology of the Anyox Pendant, Northwest British Columbia, and implications for mineral exploration: Canadian Journal of Earth Sciences, v. 39, p. 1313-1332.

Evenchick, C.A., McMechan, M.E., McNicoll, V.J. and Carr, S.D., 2006, A synthesis of the Jurassic-Cretaceous tectonic evolution of the central and southeastern Canadian Cordillera: Exploring links across the orogen: Cordilleran Tectonics Workshop, Simon Fraser University, February 17-19, Program with Abstracts, p. 30-31.

Ferris, G.M. and Schwartz, M.P., 2003, Proterozoic gold province of the central Gawler craton: MESA Journal , v. 30, p. 4-12.

Friedman, R.M. and Armstrong, R. L., 1995, Jurassic and Cretaceous U-Pb geochronometry of the southern Coast Belt, British Columbia, 49o-51o N, *in* Miller, D.M. and Busby, C., eds., Jurassic Magmatism and Tectonics of the North American Cordillera: Geological Society of America Special Paper 299, p. 95-139.

Gabrielse, H., 1985, Major dextral transcurrent displacements along the Northern Rocky Mountain Trench and related lineaments in north-central British Columbia: Geological Society of America Bulletin, v. 96, p. 1-14.

Gabrielse, H., Murphy, D.C. and Mortensen, J.K., 2006, Cretaceous and Cenozoic dextral orogen-parallel displacements, magmatism and paleogeography, north-central Canadian Cordillera, *in* Haggart, J.W., Monger, J.W.H., and Enkin, R.J., eds., Paleogeography of the North American Cordillera: Evidence For and Against Large-Scale Displacements: Geological Association of Canada Special Paper 46, p. 255-276.

Gardner, M.C., Bergman, S.C., Cushing, G.W., MacKevett, E.M. Jr., Plafker, G., Campbell, R.B., Dodds, C.J., McClelland, W.C. and Mueller, P.A., 1988, Pennsylvanian pluton stitching of Wrangellia and the Alexander terrane, Wrangell Mountains, Alaska: Geology, v. 16, p. 967-971.

Gee, D.G., 2005, Scandivanian Caledonides (with Greenland), *in* Selley, R.C., Cocks, L.R.M. and Plimer, I.R., eds., Encyclopedia of Geology: Elsevier, online at www.encyclopediaofgeology.com, p. 64-74.

Gehrels, G.E., 2001, Geology of the Chatham Sound region, southeast Alaska and coastal British Columbia: Canadian Journal of Earth Sciences, v. 38, p. 1579-1599.

——— 2002, Detrital zircon geochronology of the Taku terrane, southeast Alaska: Canadian Journal of Earth Sciences, v. 39, p. 921-931.

Gehrels, G.E. and Ross, G. M. 1998, Detrital zircon geochronology of Neoproterozoic to Permian miogeoclinal strata in British Columbia and Alberta. Canadian Journal of Earth Sciences, v. 35, p. 1380-1401.

Gehrels, G.E., Saleeby, J.B. and Berg, H.C., 1987, Geology of Annette, Gravina and Duke islands, southeastern Alaska: Canadian Journal of Earth Sciences, v. 24, p. 866-881.

Gehrels, G.E., Dickinson, W.R., Ross, G.M., Stewart, J.H., and Howell, D.G. 1995, Detrital zircon reference for Cambrian to Triassic miogeoclinal strata of western North America. Geology, v. 23, p. 831-834.

Gehrels, G.E., Butler, R.F. and Bazard, D.R., 1996, Detrital zircon geochronology of the Alexander terrane, southeastern Alaska: Geological Society of America Bulletin, v. 108, p. 722-734.

Ghent, E.D., Roeske, S., Stout, M.Z., Bradshaw, J.Y. and Snee, L., 2001, Mesozoic granulite facies metamorphism of the Pitka mafic-ultramafic complex, northern Alaska: Geological Society of America, Abstracts with Programs, Annual Meeting, v. 33, paper no. 105-0.

Gibson, H. D., Brown, R. L. and Carr, S. D., 2005, U–Th–Pb geochronologic constraints on the structural evolution of the Selkirk fan, northern Selkirk Mountains, southern Canadian Cordillera: Journal of Structural Geology, v. 27, p. 1899-1924.

Goodfellow, W., 2007, Base Metal Metallogeny of the Selwyn Basin, Canada, *in* Goodfellow, W.D., ed., Mineral Deposits of Canada: A Synthesis of Major Deposit-Types, District Metallogeny, the Evolution of Geological Provinces, and Exploration Methods: Geological Association of Canada, Mineral Deposits Division, Special Publication No. 5, p. 553-579.

Goodfellow, W.D. and Jonasson, I.R., 1986, Environment of formation of the Howards Pass (XY) Zn-Pb deposit, Selwyn basin, Yukon, *in* Morin, J.A., ed., Mineral Deposits of Northern Cordillera: Canadian Institute of Mining and Metallurgy, Special Volume 37, p. 19-50.

Goodfellow, W.D., Cecile, M.P. and Leybourne, M.I., 1995, Geochemistry, petrogenesis and tectonic setting of lower Paleozoic alkalic and potassic volcanic rocks, northern Canadian Cordilleran miogeocline: Canadian Journal of Earth Sciences, v. 32, p. 1236-1254.

Gordey, S.P. and Makepeace, A.J. (compilers), 2001, Bedrock geology, Yukon Territory: Yukon Geological Survey, Open File 2001-1 and Geological Survey of Canada, Open File 3754, 1:1 000 000 scale.

Grantz, A., May, S.D., Taylor, P.T. and Lawver, L.A., 1990, Canada Basin, *in* Grantz., A., Johnson, L. and Sweeney, J.F., eds., The Arctic Ocean Region: Geological Society of America, The Geology of North America, v. L, p. 379-402.

Grantz., A., May, S.D. and Hart, P.E., 1994, Geology of the Arctic continental margin of Alaska, *in* Plafker, G., and Berg, H.C., eds., The Geology of Alaska: Geological Society of America, The Geology of North America, v. G-1, p. 17-48.

Gunning, M.H., Hodder, R.W.H. and Nelson, J.L., 2006, Contrasting volcanic styles within the Paleozoic Stikine assemblage, western Stikine terrane, northwestern British Columbia, *in* Colpron, M. and Nelson, J.L., eds., Paleozoic Evolution and Metallogeny of Pericratonic Terranes at the Ancient Pacific Margin of North America, Canadian and Alaskan Cordillera: Geological Association of Canada Special Paper 45, p. 201-227.

Hansen, V.I and Dusel-Bacon, C., 1998, Structural and kinematic evolution of Yukon-Tanana upland tectonites, east-central Alaska: A record of late Paleozoic to early Mesozoic crustal assembly: Geological Society of America Bulletin, v. 110, p. 211-230.

Hart, C.J.R., McCoy, D.T., Goldfarb, R.J., Smith, M., Roberts, P., Hulstein, R., Bakke, A.A. and Bundtzen, 2002, Geology, exploration and discovery in the Tintina gold province, Alaska and Yukon: Society of Economic Geologists, Special Publication 9, p. 241-274.

Hart, C.J.R., Goldfarb, R.J., Lewis, L.L. and Mair, J., 2004, The northern Cordilleran mid-Cretaceous plutonic province: ilmenite-magnetite series granitoids and intrusion-related mineralization: Resource Geology, The Ishihara Volume, v. 54, p. 253-280.

Hoffman, P.J., 1991, Did the breakout of Laurentia turn Gondwanaland inside out? Science, v. 252, p. 1409-1412.

Hoffman, P.F. and Schrag, D.P., 2002, The Snowball Earth hypothesis: testing the limits of global change: Terra Nova, v. 14, p. 129-155.

Hoffman, P.F., Kaufman, A.J., Halverson, G.P. and Schrag, D.P., 1998, A Neoproterozoic Snowball Earth: Science, v. 281, p. 1342-1346.

Höy, T., 1991, Volcanogenic massive sulphide deposits in British Columbia, *in* Ore Deposits, Tectonics and Metallogeny in the Canadian Cordillera: British Columbia Ministry of Energy, Mines and Petroleum Resources, Paper 1991-4, p. 89-123.

Höy, T., Anderson, D., Turner, R.J.W. and Leitch, C.H.B., 2000, Tectonic, magmatic and metallogenetic history of the early synrift phase of the Purcell basin, southeastern British Columbia, *in* Lydon, J.W., Slack, J.F. and Knapp, M.E., eds., The Geological Environment of the Sullivan

Deposit: Geological Association of Canada, Mineral Deposits Division, Special Publication No. 1, p. 32-60.

Hudson, T.L., 1994, Crustal melting events in Alaska, *in* Plafker, G. and Berg, H.C., eds., The Geology of Alaska: Geological Society of America, The Geology of North America, v. G-1, p. 657-670.

Hudson, T.L., Plafker, G. and Dixon, K., 1982, Horizontal offset history of the Chatham Strait fault, *in* Conrad, W.L., ed., The United States Geological Survey in Alaska: Accomplishments during 1980: US Geological Survey Circular 844, p. 128-132.

Hulbert, L., 1997, Geology and metallogeny of the Kluane mafic-ultramafic belt, Yukon Territory, Canada: Eastern Wrangellia - A new Ni-Cu-PGE metallogenic terrane: Geological Survey of Canada, Bulletin 506, 265 p.

Hunt, J.A., Baker, T. and Thorkelson, D.J., 2005, Regional-scale Proterozoic IOCG-mineralized breccia systems: examples from the Wernecke Mountains, Yukon, Canada: Mineralium Deposita, v. 40, p. 492-514.

Irving, E. and Wynne, P.J., 1990, Palaeomagnetic evidence bearing on the evolution of the Canadian Cordillera: Philosophical Transactions of the Royal Society, v. A331, p. 487-509.

Israel, S., Tizzard, A.M. and Major, J., 2006, Bedrock geology of the Duke River area, parts of NTS 115G/2, 3, 4, 6 and 7, southwestern Yukon, *in* Emond, D.S., Bradshaw, G.D., Lewis, L.L. and Weston, L.H., eds., Yukon Exploration and Geology 2005: Yukon Geological Survey, p. 139-154.

Jackson, J.L., Gehrels, G.E., Patchett, P.J. and Mihalynuk, M.G., 1991, Stratigraphic and isotopic link between the northern Stikine terrane and an ancient continental margin assemblage, Canadian Cordillera: Geology, v. 19, p. 1177-1180.

Jefferson, C.W. and Ruelle, J.C.L., 1986, The Late Proterozoic Redstone copper belt, Mackenzie Mountains, Northwest Territories, *in* Morin, J.A., ed., Mineral Deposits of Northern Cordillera: Canadian Institute of Mining and Metallurgy, Special Volume 37, p. 154-168.

Jennings, D.S. and Jilson, G.A., 1986, Geology and sulphide deposits of Anvil Range, Yukon: in Morin, J.A., ed., Mineral Deposits of Northern Cordillera: Canadian Institute of Mining and Metallurgy, Special Volume 37, p. 319-371.

Jiang, S-Y, Slack, J.F. and Palmer, M.R., 2000, Sm-Nd dating of the giant Sullivan Pb-Zn-Ag deposit, British Columbia: Geology, v. 28, p. 751-754.

Jones, D.L., Howell, D.G., Coney, P.J. and Monger, J.W.H., 1983, Recognition, character and analysis of tectono-stratigraphic terranes in western North America, *in* Hashimoto, M. and Uyeda, S., eds., Accretion Tectonics in the Circum-Pacific Region: Terra, Tokyo, p. 21-35.

Juras, S.S., 1987, Geology of the polymetallic volcanogenic Buttle Lake camp, with emphasis on the Price Hillside, Central Vancouver Island, British Columbia, Canada: Ph.D. Thesis, The University of British Columbia, 279 p.

Kesler, S., 2002, Basin-related MVT deposits: Possible relation to early hydrospheric and atmospheric conditions: Geological Society of America, Abstracts with Program, Annual Meeting, v. 34, paper 226-9.

Klemperer, S.L., Greninger, M.L. and Nokleberg, W.J., 2002, Geographic information systems compilation of geophysical, geologic and tectonic data for the Bering Shelf, Chukchi Sea, Arctic margin and adjacent landmasses, *in* Miller, E.L., Grantz, A. and Klemperer, S.L., eds., Tectonic Evolution of the Bering Shelf - Chukchi Sea - Arctic Margin and Adjacent Landmasses: Geological Society of America Special Paper 360, p. 359-374.

Klepacki, D.W. and Wheeler, J.O., 1985, Stratigraphic and structural relations of the Milford, Kaslo and Slocan Groups, Goat Range, Lardeau and Nelson map areas, British Columbia: Geological Survey of Canada, Paper 85-1A, p. 277-286.

Lane, L.S., 1992, A new plate kinematic model of Canada Basin evolution, *in* Thurston, D.K. and Fujita, K., eds., Proceedings of the International Conference on Arctic Margins, Anchorage, Alaska, U.S. Department of the Interior, Alaska Outer Continental Shelf Region, p. 283-288.

——— 1997, Canada Basin, Arctic Ocean: Evidence against a rotational origin: Tectonics, v. 16, p. 363-387.

Lassiter, J.C., DePaolo, D.J. and Mahoney, J.J., 1995, Geochemistry of the Wrangellia flood basalt province: Implications for the role of continental and oceanic lithosphere in flood basalt genesis: Journal of Petrology, v. 36, p. 983-1009.

Lefebure, D.V., 2005, British Columbia Mineral Deposit Profiles, version 2.1, http://www.em.gov.bc.ca/mining/GeolSurv/MetallicMinerals/MineralDepositProfiles/default.htm

Lefebure, D.V., Fonseca, A. and Hunt, J.A., 2005, Wernecke breccia D07: B.C. Mineral Deposit Profiles, modified for Yukon, *in* Fonseca, A. and Bradshaw, G., compilers, Yukon Mineral Deposit Profiles, Yukon Geological Survey Open File 2005-5.

Leitch, C.H.B. 1990, Bralorne; a mesothermal, shield-type vein gold deposit of Cretaceous age in southwestern British Columbia: Canadian Institute of Mining and Metallurgy Bulletin, v. 83, no. 941, p. 53-80.

Li, Z-X, Zhang, L. and Powell, C.M., 1995, South China in Rodinia: Part of the missing link between Australia-East Antarctica and Laurentia?: Geology, v. 23, p. 407-410.

Logan, J.M. and Colpron, M., 2006, Stratigraphy, geochemistry, syngenetic sulphide occurrences and tectonic setting of the lower Paleozoic Lardeau Group, northern Selkirk Mountains, British Columbia, *in* Colpron, M. and Nelson, J.L., eds., Paleozoic Evolution and Metallogeny of Pericratonic terranes at the Ancient Pacific Margin of North America, Canadian and Alaskan Cordillera: Geological Association of Canada Special Paper 45, p. 361-382.

Long, D.G.F., Devaney, J.R. and Pratt, B.R., 1999, Tectonostratigraphic framework of the Mesoproterozoic Muskwa assemblage, northern British Columbia: Slave-Northern Cordillera Lithospheric Evolution (SNORCLE) and Cordilleran Tectonics Workshop, Lithoprobe Report 69, p. 112-121.

Lowey, G.W., 1998, A new estimate of the amount of displacement on the Denali fault system based on the occurrence of carbonate megaboulders in the Dezadeash Formation (Jura-Cretaceous), Yukon, and the Nutzotin Mountains sequence (Jura-Cretaceous), Alaska: Bulletin of Canadian Petroleum Geology, v. 46, p. 379-386.

Lund, K., Aleinikoff, J.N., Evans, K.V. and Fanning, C.M., 2003, SHRIMP U-Pb geochronology of Neoproterozoic Windermere Supergroup, central Idaho: Implications for rifting of western Laurentia and synchroneity of Sturtian glacial deposits: Geological Society of America Bulletin, v. 115, p. 349-372.

Lund, K., Aleinikoff, J.N., Evans, K.V. and Kunk, M.J., 2004, Proterozoic basins and orogenic belts of central Idaho: Geological Society of America, Abstracts with Programs, Annual Meeting, v. 36, no. 5, p. 271.

Lydon, J.W., 2007, Geology and metallogeny of the Belt-Purcell Basin, *in* Goodfellow, W.D., ed., Mineral Deposits of Canada: A Synthesis of Major Deposit-Types, District Metallogeny, the Evolution of Geological Provinces, and Exploration Methods: Geological Association of Canada, Mineral Deposits Division, Special Publication No. 5, p. 581-607.

Lynch, J.V.G., 1991, Georgia Basin Project: Stratigraphy and structure of Gambier Group rocks in the Howe Sound - Mamaquam River area, southwest Coast Belt, British Columbia: Geological Survey of Canada, Report of Activities, Paper 91-1A, p. 49-58.

MacDonald, J.A., Lewis, P.D., Thompson, J.F.H., Nadaraju, G., Bartsch, R.D., Bridge, D.J., Rhys, D.A., Roth, T., Kaip, A., Godwin, C.A. and Sinclair, A.J., 1996, Metallogeny of an Early to Middle Jurassic arc, Iskut River area, northwestern British Columbia: Economic Geology, v. 91, p. 1098-1114.

MacIntyre, D.G., 1991, SEDEX-Sedimentary exhalative deposits, *in* Ore Deposits, Tectonics and Metallogeny in the Canadian Cordillera: British Columbia Ministry of Energy, Mines and Petroleum Resources, Paper 1991-4, p. 25-70.

Madsen, J., Thorkelsen, D., Friedman, R., Marshall, D. and Anderson, R., 2003, Slab windows drive Eocene forearc magmatism on Vancouver Island: Geological Society of America, Abstracts with Programs, v. 35, p. 428.

Mair, J.L., Hart, C.J.R. and Stephens, J.R., 2006, Deformation history of the northwestern Selwyn basin, Yukon, Canada: Implications for orogen evolution and mid-Cretaceous magmatism: Geological Society of America Bulletin, v. 118, p. 304-323.

Marsden, H. and Thorkelson, D.J., 1992, Geology of the Hazelton volcanic belt in British Columbia: Implications for the Early the Middle Jurassic evolution of Stikinia: Tectonics, v. 11, p. 1266-1287.

Massey, N.W.D., 1995, Geology and mineral resources of the Duncan area (092B/13): British Columbia Ministry of Energy, Mines and Petroleum Resources Paper 1992–4, 112 p.

——— 2000a, Volcanogenic massive sulphide deposits of British Columbia: British Columbia Ministry of Energy, Mines Open File 1999-2 (revised 2000), 1: 2 000 000.

——— 2000b, Sedimentary hosted exhalative deposits of British Columbia: British Columbia Ministry of Energy, Mines Open File 2000-22, 1: 2 000 000.

Massey., N.W.D., MacIntyre, D.G., Desjardins, P.J., and Cooney, R.T., compilers, Geology of British Columbia: British Columbia Ministry of Energy, Mines and Petroleum Resources, Geoscience Map 2005-3, scale 1:1 000 000.

May, S.R. and Butler, R.F., 1987, North American Jurassic apparent polar wander: Implications for plate motion, paleogeography and Cordilleran tectonics: Journal of Geophysical Research, v. 91, p. 11,519-11,544.

McClelland, W.C. and Gehrels, G.E., 1990, The Duncan Canal shear zone: a right-lateral shear zone of Jurassic age along the inboard margin of the Alexander terrane: Geological Society of America Bulletin, v. 102, p. 1378-1392.

McClelland, W.C., Schmidt, J.M. and Till, A.B., 2006, New SHRIMP ages from Devonian felsic volcanic and Proterozoic plutonic rocks of the southern Brooks Range, Alaska: Geological Society of America, Abstracts with Program, Cordilleran Section Meeting, v. 38, no. 5, p. 12.

McMillan, W.J., Thompson, J.F.H., Hart, C.J.R. and Johnston, S.T., 1995, Regional geological and tectonic setting of porphyry deposits in British Columbia and Yukon Territory, *in* Schroeter, T., ed., Porphyry Deposits of the Northwestern Cordillera of North America: Canadian Institute of Mining, Metallurgy and Petroleum, Special Volume 46, p. 40-57.

Mihalynuk, M.G., 1999, Geology and mineral resources of the Tagish Lake (NTS 104M/8,9,10E and 104N/12W) area, northwestern British Columbia: British Columbia Ministry of Energy, Mines and Petroleum Resources Bulletin 105, 217 p.

Mihalynuk, M.G. and Heaman, L.M., 2002, Age of mineralized porphyry at the Logtung deposit W-Mo-Bi-Be (beryl, aquamarine), northwest B.C., *in* Geological Fieldwork 2001, British Columbia Ministry of Energy and Mines, Paper 2002-1, p. 35-39.

Mihalynuk, M.G., Smith, M.T., Gabites, J.E., Runkle, D. and Lefebure, D., 1992, Age of emplacement and basement character of Cache Creek terrane as constrained by new isotopic and geochemical data: Canadian Journal of Earth Sciences, v. 29, p. 2463-2477.

Mihalynuk, M.G., Smith, M.T., MacIntyre, D.G. and Deschênes, M., 1993, Tashenshini Project Part B: Stratigraphic and magmatic setting of mineral occurrences, *in* Geological Fieldwork 1992, British Columbai Ministry of Energy, Mines and Petroleum Resources Paper 1993-1, p. 189-228.

Mihalynuk, M.G., Nelson, J.L. and Diakow, L.J., 1994a, Cache Creek terrane: oroclinal paradox within the Canadian Cordillera: Tectonics, v. 13, p. 575-595.

Mihalynuk, M.G., Smith, M.T., Hancock, K.D. and Dudka, S., 1994b, Regional and economic geology of the Tulsequah River and Glacier areas, *in* Geological Fieldwork 1993, British Columbia Ministry of Energy and Mines, Paper 1994-1, p. 171-197.

Mihalynuk, M.M., Erdmer, P., Ghent, E.D., Cordey, F., Archibald, D.A., Friedman, R.M. and Johannson, G.G., 2004, Coherent French Range blueschist: Subduction to exhumation in <2.5 m.y.?: Geological Society of America Bulletin, v. 116, p. 910-922.

Miller, E., Gehrels, G., Amato, J., Toro, J., Prokopiev. A., Moore, T.E., Grantz, A. and Embry, A., 2004, Paleogeography and basin development in the circum-Arctic region: Single-grain U-Pb dating of detrital zircon populations from Triassic sandstones in NE Russia and comparison with Alaska and Canada: Geological Society of America, Abstracts with Programs, Annual Meeting, v. 36, p. 22.

Miller, L.D., Stowell, H.H. and Gehrels, G.E., 2000, Progressive deformation associated with mid-Cretaceous to Tertiary contractional tectonism in the Juneau gold belt, Coast Mountains, southeastern Alaska, *in* Stowell, H.H. and McClelland, W.C., eds., Tectonics of the Coast Mountains, southeastern Alaska and British Columbia: Geological Society of America, Special Paper 343, p. 193-212.

Miller, M.M., 1988, Displaced remnants of a northeast Pacific fringing arc: Upper Paleozoic terranes of Permian McCloud faunal affinity, western U.S.: Tectonics, v. 6, p. 807-830.

Moll-Stalcup, E.J., 1994, Latest Cretaceous and Cenozoic magmatism in mainland Alaska, *in* Plafker, G. and Berg, H.C., eds., The Geology of Alaska: Geological Society of America, The Geology of North America, v. G-1, p. 589-620 and Plate 5.

Monger, J. and Price, R., 2002, The Canadian Cordillera: Geology and tectonic evolution: Canadian Society of Exploration Geophysicists Recorder, February 2002, p. 17-36.

Monger, J.W.H., Price, R.A. and Tempelman-Kluit, D.J., 1982, Tectonic accretion and the origin of the two major metamorphic and tectonic welts in the Canadian Cordillera: Geology, v. 10, p. 70-75.

Monger, J.W.H., van der Heyden, P., Journeay, J.M., Evenchick, C.A. and Mahoney, J.B., 1994, Jurassic-Cretaceous basins along the Canadian Coast belt: Their bearing on pre-mid-Cretaceous sinistral displacements: Geology, v. 22, p. 175-178.

Moore, T.E., Roeske, S.M. and Grantz, A., 1987, Crustal structure of central Alaska along geodynamic continent/ocean transect (A-3): Geological Society of America, Abstracts with Programs, v. 19, p. 434.

Moore, T.E., Wallace, W.K., Bird, K.J., Karl, S.M., Mull, C.G. and Dillon, J.T., 1994, Geology of northern Alaska, *in* Plafker, G. and Berg, H.C., eds., The Geology of Alaska: Geological Society of America, The Geology of North America, v. G-1, p. 49-140.

Moore, T.E., Potter, C.J. and O'Sullivan, P.J., 2004, The Brooks Range, Alaska: The consequence of four distinct tectonic events: Geological Society of America, Abstracts with Programs, v. 36, no. 5, p. 270.

Moores, E.M., 1991, Southwest U.S. - East Antarctic (SWEAT) connection: A hypothesis: Geology, v. 19, p. 425-428.

Morelli, R.M., Creaser, R.A., Selby, D., Kelley, K.D, Leach, D.L. and King, A. R., 2004, Re-Os sulfide geochronology of the Red Dog sediment-hosted Zn-Pb-Ag deposit, Alaska, *in* Kelley, K.D. and Jennings, S.A., eds. Barite and Zinc-Lead-Silver Deposits in the Red Dog District, Western Brooks Range, Northern Alaska: Economic Geology, v. 99, p. 1569-1576.

Mortensen, J.K. and Israel, S., 2006, Is the Windy-McKinley terrane a displaced fragment of Wrangellia? Evidence from new geological, geochemical and geochronological studies in western Yukon: Geological Society of America, Abstracts with Programs, Cordilleran Section, v. 38, p. 93.

Mortensen, J.K., Ghosh, D.K. and Ferri, F., 1995, U-Pb geochronology of intrusive rocks associated with copper-gold porphyry deposits in the Canadian Cordillera, *in* Schroeter, T.G., ed., Porphyry Deposits of the Northwestern Cordillera of North America: Canadian Institute of Mining, Metallurgy and Petroleum, Special Volume 46, p. 142-158.

Mortensen, J.K., Dusel-Bacon, C., Hunt, J. and Gabites, J. 2006, Lead isotopic constraints on the metallogeny of middle and late Paleozoic syngenetic base metal occurrences in the Yukon-Tanana and Slide Mountain/Seventymile terranes and adjacent portions of the North American miogeocline, *in* Colpron, M. and Nelson, J.L., eds., Paleozoic Evolution and Metallogeny of Pericratonic Terranes at the Ancient Pacific Margin of North America, Canadian and Alaskan Cordillera: Geological Association of Canada Special Paper 45, p. 261-279.

Murphy, D.C., 1997, Geology of the McQuesten River region, northern McQuesten and Mayo map areas, Yukon Territory: Yukon Geological Survey, Bulletin 6, 95 p.

Murphy, D.C., van der Heyden, P., Parrish, R.R., Klepacki, D.W., Mcmillan, W., Struik, L.C. and Gabites, J., 1995, New geochronological constraints on Jurassic deformation of the western edge of North America, southeastern Canadian Cordillera, *in* Miller, D.M. and Busby, C., eds., Jurassic Magmatism and Tectonics of the North American Cordillera: Geological Society of America Special Paper 299, p. 159-171.

Murphy, D.C., Mortensen, J.K., Piercey, S.J., Orchard, M.J. and Gehrels, G.E., 2006, Mid-Paleozoic to early Mesozoic tectonostratigraphic evolution of Yukon-Tanana and Slide Mountain terranes and affiliated overlap assemblages, Finlayson Lake massive sulphide district, southeastern Yukon, *in* Colpron, M. and Nelson, J.L., eds., Paleozoic Evolution and Metallogeny of Pericratonic Terranes at the Ancient Pacific Margin of North America, Canadian and Alaskan Cordillera: Geological Association of Canada Special Paper 45, p. 75-105.

Natal'in, B.A., 2004, Phanerozoic tectonic evolution of the Chukotka-Arctic Alaska block: Problems of the rotational model: American Geophysical Union, Fall Meeting, Paper GP43C-04 (online abstract).

Nelson, J.L., 1990, Evidence for a cryptic intrusion Beneath the Erickson-Taurus gold-quartz vein system, near Cassiar, B.C. (104P/4, 5), *in* Geological Fieldwork 1989, British Columbia Ministry of Energy, Mines and Petroleum Resources Paper 1990-1, p. 229-236.

——— 1991, Carbonate-hosted lead-zinc(±silver, gold) deposits of British Columbia, *in* Ore Deposits, Tectonics and Metallogeny in the Canadian Cordillera: British Columbia Ministry of Energy, Mines and Petroleum Resources, Paper 1991-4, p. 71-88.

——— 1993, The Sylvester allochthon: upper Paleozoic marginal basin and island arc terranes in northern British Columbia: Canadian Journal of Earth Sciences, v. 30, p. 631-643.

——— 1997, The quiet counter-revolution: Structural control of syngenetic deposits: Geoscience Canada, v. 24, no. 2, p. 91-98.

Nelson, J.L. and Bellefontaine, K.A., 1996, The geology and mineral deposits of north-central Quesnellia; Tezzeron Lake to Discovery Creek, central British Columbia: British Columbia Ministry of Energy, Mines and Petroleum Resources, Bulletin 99, 112 p.

Nelson, J.L. and Friedman, R., 2004, Superimposed Quesnel (late Paleozoic-Jurassic) and Yukon-Tanana (Devonian-Mississippian) arc assemblages, Cassiar Mountains, northern British Columbia: Field, U-Pb and igneous petrochemical evidence: Canadian Journal of Earth Sciences, v. 41, p. 1201-1235.

Nelson, J.L., Paradis, S., Christensen, J. and Gabites, J., 2002, Canadian Cordillera Mississippi Valley-type deposits: A case for Devonian-Mississippian back-arc hydrothermal origin: Economic Geology, v. 97, p. 1013-1036.

Nelson, J.L., Colpron, M., Piercey, S.J., Dusel-Bacon, C., Murphy, D.C. and Roots, C., 2006, Paleozoic tectonic and metallogenetic evolution of pericratonic terranes in Yukon, northern British Columbia and eastern Alaska; *in* Colpron, M. and Nelson, J.L., eds, Paleozoic Evolution and Metallogeny of Pericratonic Terranes at the Ancient Pacific Margin of North America, Canadian and Alaskan Cordillera: Geological Association of Canada Special Paper 45, p. 323-360..

Nokleberg, W.J., Brew, D.A., Grybeck, D., Yeend, W., Bundtzen, T., Robinson, M.S. and Smith, T.E., 1994, Metallogeny and major mineral deposits of Alaska, Chapter 29 *in* Plafker, G. and Berg, H.C., eds., The Geology of Alaska: Geological Society of America, The Geology of North America, v. G-1, p. 855-903.

Nokleberg, W.J., Bundtzen, T.K., Brew, D.A. and Plafker, G., 1995, Metallogenesis and tectonics of porphyry copper and molybdenum (gold, silver) and granitoid-hosted gold deposits of Alaska, *in* Schroeter, T., ed., Porphyry Deposits of the Northwestern Cordillera: Canadian Institute of Mining, Metallurgy and Petroleum, Special Volume 46, p. 103-141.

Nokleberg, W.J., Parfenov, L.M., Monger, J.W.H., Baranov, B.V., Byalobzhesky, S.G., Bundzten, T.K., Feeney, T.D., Fujita, K., Gordey, S.P., Grantz, A., Khanchuk, A.I., Natal'in, B.A., Natapov, L.M., Norton, I.O., Patton, Jr., W.W., Plafker, G., Scholl, D.W., Skolov, F.D., Sosunov, G.M., Stone, D.B., Tabor, R.W., Tsukanov, N.V. and Vallier, T.L., 1997, Summary Circum-North Pacific tectonostratigraphic terrane map: U.S. Geological Survey Open-File Report 96-727; also Geological Survey of Canada Open File 3428, scale 1:10,000,000.

Nokleberg, W.J., Parfenov, L.M., Monger, J.W.H., Norton, I.O., Khanchuk, A.I., Stone, D.B., Scotese, C.R., Scholl, D.W. and Fujita, K., 2000, Phanerozoic tectonic evolution of the circum-North Pacific: U.S. Geological Survey Professional Paper 1626, 122 p.

Nokleberg, W.J., Bundtzen, T., Eremin, R.A., Ratkin, V.V., Dawson, K.M., Shpikerman, V.V., Goryachev, N.A., Byalobzhesky, S.G., Frolov, Y.F., Khanchuck, A.I., Koch, R.D., Monger, J.W.H., Pozdeev, A.A., Rozenblum, I.S., Rodionov, S.M., Parfenov, L.M., Scotese, C.R. and Sidorov, A.A., 2005, Metallogenesis and tectonics of the Russian Far East, Alaska and the Canadian Cordillera: U.S. Geological Survey Professional Paper 1697, 397 p.

Panteleyev, A., 1991, Gold in the Canadian Cordillera: A focus on epithermal and deeper environments, *in* Ore Deposits, Tectonics and Metallogeny in the Canadian Cordillera: British Columbia Ministry of Energy, Mines and Petroleum Resources Paper 1991-4, p. 163-212.

Paradis, S., Bailey, S.L., Creaser, R.A., Piercey, S.J. and Schiarizza, P., 2006a, Paleozoic magmatism and syngenetic massive sulphide deposits of the Eagle Bay assemblage, Kootenay terrane, southern British Columbia, *in* Colpron, M. and Nelson, J.L., eds, Paleozoic Evolution and Metallogeny of Pericratonic Terranes at the Ancient Pacific Margin of North America, Canadian and Alaskan Cordillera: Geological Association of Canada Special Paper 45, p. 383-414.

Paradis, S., Taylor, B.E.H., Falck, H. and Taylor, A., 2006b, Prairie Creek: An Irish-type carbonate-hosted Zn-Pb-Ag deposit in sedimentary basin of the northern Canadian Cordillera: Geological Association of Canada, Abstracts with Programs, v. 31, p. 114.

Paterson, I.A., 1977, The geology and evolution of the Pinchi fault zone at Pinchi Lake, central British Columbia: Canadian Journal of Earth Sciences, v. 14, p. 1324-1342.

Patrick, B.E. and McClelland, W.C., 1995, Late Proterozoic granitic magmatism on Seward Peninsula and a Barentian origin for Arctic Alaska-Chukotka: Geology, v. 23, p. 81-84.

Patton, W.R. and Tailleur, I.L., 1977, Evidence in the Bering Strait region for differential movement between North America and Eurasia: Geological Society of America Bulletin, v. 88, p. 1298-1304.

Pavlis, T., 1989, Middle Cretaceous orogenesis in the northern Cordillera: A Mediterranean analogue of collision-related extensional tectonics: Geology, v. 17, p. 947-950.

Peter, J.M., Scott, S.D., Barrie, C.T. and Hannington, M.D., 1999, Windy Craggy, northwestern British Columbia: The world's largest Besshi-type deposit, *in* Volcanic-Associated Massive Sulphide Deposits: Processes and Examples in Modern and Ancient Settings: Reviews in Economic Geology, v. 8, p. 261-295.

Piercey, S.J., Nelson, J.L., Colpron, M., Dusel-Bacon, C., Murphy, D.C., Simard, R.-L. and Roots, C.F., 2006, Paleozoic magmatism and crustal recycling along the ancient Pacific margin of North America, northern Cordillera, *in* Colpron, M. and Nelson, J.L., eds, Paleozoic Evolution and Metallogeny of Pericratonic Terranes at the Ancient Pacific Margin of North America, Canadian and Alaskan Cordillera: Geological Association of Canada Special Paper 45, p. 281-322.

Pigage, L.C. and Mortensen, J.K., 2004, Superimposed Neoproterozoic and Early Tertiary alkaline magmatism in the LaBiche River area, southeast Yukon: Bulletin of Canadian Petroleum Geology, v. 52, p. 325-342.

Price, J.B., Hitzman, M., Nelson, E. and Humphrey, J., 2006, An exploration model for Wrangellia's high grade copper deposits at Kennecott, Alaska: Geological Society of America, Abstracts with Programs, Cordilleran Section, v. 38, no. 5, p. 26.

Price, R.A. and Carmichael, D.M., 1986, Geometric test for Late Cretaceous-Paleogene intracontinental transform faulting in the Canadian Cordillera: Geology, v. 14, p. 468-471.

Rainbird, R.H., Jefferson, C.W. and Young, G.M., 1996, The early Neoproterozoic sedimentary Succession B of northwestern Laurentia: Correlations and paleogeographic significance: Geological Society of America Bulletin, v. 108, p. 454-470.

Rainbird, R.H., McNicoll, V.J., Thériault, R.J., Heaman, L.M., Abbott, J.G., Long, D.G.F. and Thorkelson, D.J., 1997, Pan-continental river system draining the Grenville orogen recorded by U-Pb and Sm-Nd geochronology of Neoproterozoic quartz arenites and mudrocks, northwestern Canada: Journal of Geology, v. 105, p. 1-17.

Read, P.B. and Okulitch, A.V., 1977, The Triassic unconformity of south-central British Columbia: Canadian Journal of Earth Sciences, v. 14, p. 606-638.

Redfield, T.F., Scholl, D.W., Beck, M.E. and Fitzgerald, P.G., 2005, Escape tectonics and plate kinematics in the south central Alaskan orocline: Geological Society of America Abstracts with Programs, v. 37, p. 79.

Reichow, M. and Saunders, A., 2005, The Siberian traps - Homepage: University of Leicester, website http://www.le.ac.uk/gl/ads/ SiberianTraps/Index.html

Ricketts, B.D., Evenchick, C.A., Anderson, R.G. and Murphy, D.C., 1992, Bowser basin, northern British Columbia: Constraints on the timing of initital subsidence and Stikinia-North America terrane interactions: Geology, v. 20, p. 1119-1122.

Riddihough, R.P. and Hyndman, R.D., 1991, Modern plate tectonic regime of the continental margin of western Canada, Chapter 13 *in* Gabrielse, H., and Yorath, C.J., eds., Geology of the Cordilleran Orogen in Canada: Geological Survey of Canada, Geology of Canada, no. 4, p. 435-455.

Robinson, M., Godwin, C.I. and Stanley, C.R., 1996, Geology, lithogeochemistry and alteration of the Battle Zone, Buttle Lake volcanogenic mining camp, Vancouver Island, British Columbia: Economic Geology, v. 91, p. 527-548.

Roeske, S.M., McClelland, W.C. and Bradley, D.C., 2006, Ruby terrane, north-central Alaska: Composite of continental affinity rocks: Geological Society of America, Abstracts with Program, Cordilleran Section Meeting, v. 38, no. 5, p. 12.

Ross, C.A. and Ross, J.R.P., 1983, Late Paleozoic accreted terranes of western North America, *in* Stevens, C.H., ed., Pre-Jurassic Rocks in Western North American Suspect Terranes: Pacific Section, Society of Economic Paleontologists and Mineralogists, Los Angeles, California, p. 7-22.

Ross, G.M. and Villeneuve, M., 2003, Provenance of the Mesoproterozoic (1.45 Ga) Belt basin (western North America): Another piece in the pre-Rodinia paleogeographic puzzle: Geological Society of America Bulletin, v. 155, p. 1191-1217.

Ross, G.M., Parrish, R.R. and Winston, D., 1992, Provenance and U-Pb geochronology of the Mesoproterozoic Belt Supergroup (northwestern United States): Implications for age of deposition and pre-Panthalassa plate reconstructions: Earth and Planetary Science Letters, v. 113, p. 57-76.

Ross, G.M, Friedman, R. and Mortensen, J.K., 2005, Detrital zircon and monazite from the Hyland Group (northern Canadian Cordillera and Alaska): Evidence for intra-Cordilleran "Grenville" basement: Geological Society of America, Abstracts with Programs, Cordilleran Section, v. 37, no. 4, p. 56.

Ross, G.M., Villeneuve, M.E. and Theriault, R.J., 2001, Isotopic provenance of the lower Muskwa assemblage (Mesoproterozoic, Rocky Mountains, B.C.): New clues to correlation and source area, *in* Bartley, J. and Kah, L.C., eds., The Mesoproterozoic Earth-Ocean System of Rodinia: Precambrian Research Special Volume 111, p. 57-77.

Rubin, C.M., Miller, M.M. and Smith, G.M., 1990, Tectonic development of mid-Paleozoic volcanic-plutonic complexes: Evidence for convergent margin tectonism; in Harwood, D.S. and Miller, M.M., eds., Paleozoic and Early Mesozoic Paleogeographic Relations; Sierra Nevada, Klamath Mountains, and Related Terranes: Geological Society of America Special Paper 255, p. 1-16.

Saleeby, J.B., 1990, Geochronological and tectonostratigraphic framework of Sierran-Klamath ophiolitic assemblages, *in* Harwood, D.S. and Miller, M.M., eds., Paleozoic and Early Mesozoic Paleogeographic Relations; Sierra Nevada, Klamath Mountains, and Related Terranes: Geological Society of America Special Paper 255, p. 43-56.

——— 2000, Geochronologic investigations along the Alexander-Taku terrane boundary, southern Revillagigedo Island to Cape Fox areas, southeast Alaska, *in* Stowell, H.H. and McClelland, W.C., eds., Tectonics of the Coast Mountains, Southeastern Alaska and British Columbia: Geological Society of America, Special Paper 343, p. 107-144.

Scholl, D.W., 1991, Exploring the idea that Early Tertiary evolution of the Alaska orocline and the Aleutian-Bering Sea region is a manifestation of Kula-plate-driven Cordilleran tectonism and escape tectonics: Geological Society of America Abstracts with Programs, v. 23, p. 435.

Schrader, C.M., Crowe, D., Turner, K. and Stein, H., 2001, $^{40}Ar/^{39}Ar$ and Re-Os geochronology of the Pebble Copper Cu-Au-Mo porphyry deposit, southwest Alaska: Geological Society of America, Abstracts with Programs, v. 33, paper 173-0.

Schroeter, T., (ed.), 1995, Porphyry Deposits of the Northwestern Cordillera: Canadian Institute of Mining, Metallurgy and Petroleum, Special Volume 46, 888 p.

Schroeter, T., Cathro, M., Grieve, D., Lane, R., Pardy, J. and Wojdak, P., 2006, British Columbia mining and mineral exploration overview 2005: British Columbia Ministry of Energy and Mines, 20 p. (available digitally at http://www.em.gov.bc.ca/Mining/Geolsurv/Publications/expl_bc/2005/toc.htm)

Scotese, C.R., 2002, Paleomap project: http://www.scotese.com/earth.htm

Sears, J.W. and Price, R.A., 2000, A new look at the Siberian connection: No SWEAT: Geology, v. 28, p. 423-426.

——— 2003a, Tightening the Siberian connection to western Laurentia: Geological Society of America Bulletin, v. 115, p. 943-953.

——— 2003b, The Udzha hypothesis: Siberian spillway to the Belt-Purcell Basin: Northwest Geology, v. 32, p. 167-171.

Selby, D. and Creaser, R.A., 2001, Re-Os geochronology and systematics in molybdenite from the Endako porphyry molybdenum deposit, British Columbia, Canada: Economic Geology, v. 96, p. 197-204.

Shanks, W.C. III, Woodruff, L.J., Jilson, G.A., Jennings, D.S., Modene, J.S. and Ryan, B.D., 1987, Sulphur and lead isotope studies of stratiform Zn-Pb-Ag deposits, Anvil Range, Yukon: Basinal brine exhalation and anoxic bottom-water mixing: Economic Geology, v. 82, p. 600-634.

Silberling, N.J., Jones, D.L., Monger, J.W.H., Coney, P.J., Berg, H.C. and Plafker, G., 1994, Lithotectonic terrane map of Alaska and adjacent parts of Canada, *in* Plafker, G. and Berg, H.C., eds., The Geology of Alaska: Geological Society of America, The Geology of North America, v. G-1, Plate 3, scale 1:2,500,000.

Simard, R-L., Dostal, J. and Roots, C.F., 2003, Development of Late Paleozoic volcanic arcs in the Canadian Cordillera: an example from the Klinkit Group, northern British Columbia and southern Yukon: Canadian Journal of Earth Sciences, v. 40, p. 907-924.

Sinclair, W.D., 2007, Porphyry deposits, *in* Goodfellow, W.D., ed., Mineral Deposits of Canada: A Synthesis of Major Deposit-Types, District Metallogeny, the Evolution of Geological Provinces, and Exploration Methods: Geological Association of Canada, Mineral Deposits Division, Special Publication No. 5, p. 223-243.

Smith, P.L., Tipper, H.W. and Ham, D.M., 2001, Lower Jurassic Amaltheidae (Ammotina) in North America: paleobiogeography and tectonic implications: Canadian Journal of Earth Sciences, v. 38, p. 1439-1449.

Soja, C.M. and Antoshkina, A.I., 1997, Coeval development of Silurian stromatolite reefs in Alaska and the Ural Mountains: Implications for paleogeography of the Alexander terrane: Geology, v. 25, p. 539-542.

Stern, R.J., 2004, Subduction initiation: Spontaneous and induced: Earth and Planetary Science Letters, v. 226, p. 275-292.

Stone, D.B., 1989, Paleogeography and rotations of Arctic Alaska - An unresolved problem; in Kissel, C. and Laj, C., eds., Paleomagnetic Rotations and Continental Deformation, Kluwer Academic Publishers, p. 343-364.

Stowell, H.H. and Crawford, M.L., 2000, Metamorphic history of the Coast Mountains orogen, western British Columbia and southeastern Alaska, *in* Stowell, H.H. and McClelland, W.C., eds., Tectonics of the Coast Mountains, Southeastern Alaska and British Columbia: Geological Society of America, Special Paper 343, p. 257-283.

Struik, L.C., 1993, Intersecting intracontinental Tertiary transform fault systems in the North American Cordillera: Canadian Journal of Earth Sciences, v. 30, p. 1262-1274.

Struik, L.C., Schiarizza, P., Orchard, M.J., Cordey, F., Sano, H., MacIntyre, D.G., Lapierre, H. and Tardy, M., 2001, Imbricate architecture of the upper Paleozoic to Jurassic oceanic Cache Creek terrane, central British Columbia: Canadian Journal of Earth Sciences, v. 38, p. 495-514.

Sutherland Brown, A., (ed.), 1976, Porphyry Deposits of the Canadian Dordillera: Canadian Institute of Mining, Metallurgy and Petroleum, Special Volume 15, 510 p.

Szumigala, D.J. and Hughes, R.A., 2005, Alaska's mineral industry 2004: A summary: Alaska Division of Geological and Geophysical Survey, Information Circular 51, 18 p.

Tafti, R. and Mortensen, J.K., 2004, Early Jurassic porphyry(?) copper(-gold) deposits at Minto and Williams Creek, Carmacks copper belt, western Yukon, *in* Emond, D.S. and Lewis, L.L., eds., Yukon Exploration and Geology 2003: Yukon Geological Survey, p. 289-303.

Taylor, C.D., Philpotts, J., Premo, W.R., Meier, A.L. and Taggart, J.E., Jr., in press, Triassic metallogenic setting of the Greens Creek massive sulfide deposit in southeastern Alaska, *in* Piercey, S.J., Peter, J.M. and Mortensen, J.K., eds., Continent Margin Massive Sulphide Deposits and their Geodynamic Environments: Special Issue of Economic Geology.

Thorkelson, D.J., Mortensen, J.K., Creaser, R.A., Davidson, G.J. and Abbott, J.G., 2001a, Early Proterozoic magmatism in Yukon, Canada: constraints on the evolution of northwestern Laurentia: Canadian Journal of Earth Sciences, v. 38, p. 1479-1494.

Thorkelson, D.J., Mortensen, J.K., Davidson, G.J., Creaser, R.A., Perez, W.A. and Abbott, J.G., 2001b, Mesoproterozoic intrusive breccias in Yukon, Canada: the role of hydrothermal systems in reconstructions of North America and Australia, *in* Bartley, J. and Kah, L.C., eds., The Mesoproterozoic Earth-Ocean System of Rodinia: Precambrian Research, Special Volume 111, p. 31-56.

Thorkelson, D.J., Abbott, J.G., Mortensen, J.K., Creaser, R.A., Villeneuve, M.E., McNicoll, V.J. and Layer, P.W., 2005, Early and Middle Proterozoic evolution of Yukon, Canada: Canadian Journal of Earth Sciences, v. 42, p. 1045-1071.

Toro, J., Toro, F.C., Bird, K.J. and Harrison, C., 2004, The Arctic Alaska - Canada connection revisited: Geological Society of America, Abstracts with Programs, Annual Meeting, v. 36, no. 5, p. 22.

Toro, J., Miller, E. and Katkov, S., 2006, Does the Brooks Range orogen extend into Chukotka? : Geological Society of America, Abstracts with Programs, Cordilleran Section, v. 38, no. 5, p. 34.

Trettin, H.P., 1987, Pearya: A composite terrane with Caledonian affinities in northern Ellesmere Island: Canadian Journal of Earth Sciences, v. 24, p. 224-245.

Umhoefer, P.J. and Schiarizza, P., 1996, Latest Cretaceous to early Tertiary dextral strike-slip faulting on the southeastern Yalakom fault system, southeastern Coast belt, British Columbia: Geological Society of America Bulletin, v. 108, p. 768-785.

Unterschutz, J.L.E., Creaser, R.A., Erdmer, P., Thompson, R.I. and Daughtry, K.L., 2002, North American margin origin of Quesnel terrane strata in the southern Canadian Cordillera: Inferences from geochemical and Nd isotopic characteristics of Triassic metasedimentary rocks: Geological Society of America Bulletin, v. 114, p. 462-475.

van der Heyden, P., 1992, A Middle Jurassic to Early Tertiary Andean-Sierran arc model for the Coast Belt of British Columbia: Tectonics, v. 11, p. 82-97.

van Staal, C., 2007, Pre-Carboniferous metallogeny of the Canadian Appalachians, *in* Goodfellow, W.D., ed., Mineral Deposits of Canada: A Synthesis of Major Deposit-Types, District Metallogeny, the Evolution of Geological Provinces, and Exploration Methods: Geological Association of Canada, Mineral Deposits Division, Special Publication No. 5, p. 793-818.

Warren, M.J., 1997, Crustal extension and subsequent crustal thickening along the Cordilleran rifted margin of ancestral North America, western Purcell Mountains, southeastern British Columbia: Ph.D. thesis, Queen's University, Kingston, Ontario, 361 p.

Whalen, J.B., Anderson, R.G., Struik, L.C. and Villeneuve, M.E., 2001, Geochemistry and Nd isotopes of the François Lake plutonic suite, Endako batholith: host and progenitor to the Endako molybdenum camp, central British Columbia: Canadian Journal of Earth Sciences, v. 38, p. 603-618.

Wheeler, J.O. and McFeely, P., 1991, Tectonic assemblage map of the Canadian Cordillera and adjacent parts of the United States of America: Geological Survey of Canada, Map 1712A, scale 1:2,000,000.

Wheeler, J.O., Brookfield, A.J., Gabrielse, H., Monger, J.W.H., Tipper, H.W. and Woodsworth, G.J., 1991, Terrane map of the Canadian Cordillera: Geological Survey of Canada Map 1713A, scale 1: 2 000 000.

Wirth, K.R., Bird, J.M., Blythe, A.E. and Harding, D.J., 1993, Age and evolution of western Brooks Range ophiolites, Alaska: Results from $^{40}Ar/^{39}Ar$ thermochronometry: Tectonics, v. 12, p. 410-432.

Wojdak, P. and Stock, G.C., 1995, Big Onion: A supergene-altered porphyry copper-gold deposit, west-central British Columbia, *in* Schroeter, T.G., ed., Porphyry Deposits of the Northwestern Cordillera of North America: Canadian Institute of Mining, Metallurgy and Petroleum, Special Volume 46, p. 410-421.

Yeo, G.M., 1986, Iron-formation in the late Proterozoic Rapitan Group, Yukon and Northwest Territories, *in* Morin, J.A., ed., Mineral Deposits of Northern Cordillera: Canadian Institute of Mining and Metallurgy, Special Volume 37, p. 142-153.

Young, L.E., 2004, A geologic framework for mineralization in the western Brooks Range, Alaska: Economic Geology, v. 99, p. 1281-1306.

Zagorevski, A., Roger, N., van Staal, C.R., McNicoll, V., Lissenberg, C.J. and Valverde-Vaquero, P., 2006, Lower to Middle Ordovician evolution of peri-Laurentian arc and backarc complexes in Iapetus: Constraints from the Annieopsquotch accretionary tract, central Newfoundland: Geological Society of America Bulletin, v. 118, p. 324-342.

Pre-Carboniferous Tectonic Evolution and Metallogeny of the Canadian Appalachians

Cees R. van Staal

Geological Survey of Canada, 605 Robson Street, Suite 101, Vancouver, British Columbia V6B 5J3
Corresponding author's email: Cees.Van_Staal@nrcan-rncan.gc.ca

Abstract

The Canadian Appalachians comprise a complex collage of Early Paleozoic peri-Laurentian and peri-Gondwanan oceanic suprasubduction zone and ribbon-shaped microcontinental terranes. The suprasubduction-zone terranes represent infant arc, extensional arc and back-arc settings, whereas the microcontinents were rifted off from the Laurentian (Dashwoods) and Gondwanan (Ganderia, Avalonia, and Meguma) margins.

Sequential accretion of the oceanic and continental terranes to one another and eventually to Laurentia during closure of the Iapetus and Rheic oceans caused several collisional events between the Late Cambrian and Permian. Three distinct Late Cambrian to Late Ordovician accretionary events of peri-Laurentian infant arc and continental arc terranes to Laurentia caused the Taconic Orogeny. The latter was terminated with Late Ordovician arrival of the peri-Gondwanan Popelogan-Victoria arc.

Early Ordovician collision between the Penobscot arc and Ganderia as a result of closing the intervening back-arc basin in the periphery of Gondwana caused the Penobscot Orogeny. accretion of Ganderia to Laurentia during the Silurian by closing the Tetagouche-Exploits back-arc basin caused the Salinic Orogeny. The Early Devonian Acadian Orogeny was caused by closure of the seaway between composite Laurentia and Avalonia. Accretion of Meguma to Laurentia caused the Middle Devonian to Early Carboniferous Neoacadian Orogeny. Orogenesis was terminated by the Alleghenian Laurentia-Gondwana collision, forming the Pangea supercontinent.

The Appalachian Orogen is rich in volcanic hosted massive sulphide deposits (VHMS) and, to a much less extent, sediment-hosted massive sulphide deposits. Most of the host rocks to the sulphide deposits formed in extensional suprasubduction-zone settings, which are generally sites of enhanced heatflow and hydrothermal circulation. VHMS deposits mainly formed between ca. 515 and 460 Ma, during intraoceanic subduction related to the closure of the main Iapetan oceanic tract that separated the peri-Laurentian and peri-Gondwanan arc terranes. The latter terranes sutured during the Late Ordovician terminal Taconic Orogeny along the Red Indian Line. After closure of the main Iapetan tract, convergence (Salinic) continued through closure of the Tetagouche-Exploits back-arc basin situated behind the Popelogan-Victoria arc and the oceanic seaway that separated Avalonia from Ganderia (Acadian). Both the Salinic and Acadian orogenies were accompanied by important gold mineralization in the central portion of the Appalachian Orogen (central mobile belt). At least some of the gold may be associated with magmatism generated during slab breakoff.

The Rheic Ocean opened during the Early Ordovician due to Avalonia's progressive separation from Gondwana. The subsequent departure and drift of the Meguma microcontinent across the Rheic Ocean caused its Neoacadian accretion to Laurentia at ca. 395 Ma. Meguma's accretion was accomplished by tectonic wedging and shortly, followed by extensive Middle Devonian granite plutonism that formed during continuous flat slab subduction. This tectonic setting probably was responsible for Meguma's extensive gold mineralization.

Résumé

Les Appalaches canadiennes renferment un assemblage complexe de terranes océaniques péri-laurentiens et péri gondwaniens formés au-dessus de zones de subduction, ainsi que de microterranes continentaux en forme de rubans du Paléozoïque précoce. Les terranes qui surmontaient des zones de subduction représentent des milieux d'arc naissant, d'arc de distension et d'arrière-arc, alors que les microterranes continentaux proviennent du rifting des marges de la Laurentie (Dashwoods) et du Gondwana (Gandérie, Avalonie et Meguma).

L'accrétion séquentielle des terranes continentaux et des terranes océaniques entre eux et, ultimement, à la Laurentie, durant la fermeture des océans Iapetus et Rhéique, a provoqué plusieurs épisodes de collision entre le Cambrien tardif et le Permien. Entre le Cambrien tardif et l'Ordovicien tardif, trois épisodes distincts d'accrétion des terranes péri-laurentiens d'arc continental et d'arc naissant à la Laurentie ont causé l'orogenèse taconique, qui s'est terminée par l'arrivée de l'arc péri-gondwanien de Popelogan-Victoria, à l'Ordovicien tardif.

Au cours de l'Ordovicien précoce, à la périphérie du Gondwana, l'arc de Penobscot et la Gandérie sont entrés en collision en raison de la fermeture du bassin d'arrière-arc intermédiaire, ce qui a entraîné l'orogenèse de Penobscot. Au Silurien, l'accrétion de la Gandérie à la Laurentie, causée par la fermeture du bassin d'arrière-arc de Tetagouche-Exploits, a provoqué l'orogenèse salinique. L'orogenèse acadienne du Dévonien précoce résulte de la fermeture du passage océanique entre le continent composite laurentien et l'Avalonie. L'accrétion du terrane de Meguma à la Laurentie a produit l'orogenèse néoacadienne qui s'est déroulée du Dévonien moyen au Carbonifère précoce. L'orogenèse s'est conclue par la collision alléghanienne de la Laurentie et du Gondwana, créant le supercontinent de la Pangée.

L'orogène des Appalaches est riche en gîtes de sulfures massifs encaissés dans des roches volcaniques et, dans une bien moindre mesure, en gîtes de sulfures massifs logés dans des roches sédimentaires. La majeure partie des gîtes de sulfures massifs se sont formés dans des milieux de distension surmontant une zone de subduction, où, en général, les flux thermiques sont plus élevés et la circulation hydrothermale plus intense. Les gîtes de sulfures massifs encaissés dans des roches volcaniques se sont surtout formés entre 515 et 460 Ma environ, au cours de la subduction intra-océanique liée à la fermeture de la principale partie de l'Océan Iapetus qui séparait les terranes d'arc péri-laurentiens et péri-gondwaniens. Les terranes péri-gondwaniens se sont soudés le long de la ligne de Red Indian pendant les derniers stades de l'orogenèse taconique à l'Ordovicien tardif. Après la fermeture de la principale partie de l'Océan

van Staal, C.R., 2007, Pre-Caroniferous tectonic evolution and metallogeny of the Canadian Appalachians, *in* Goodfellow, W.D., ed., Mineral Deposits of Canada: A Synthesis of Major Deposit-Types, District Metallogeny, the Evolution of Geological Provinces, and Exploration Methods: Geological Association of Canada, Mineral Deposits Division, Special Publication No. 5, p. 793-818.

Iapetus, la convergence (orogenèse salinique) s'est poursuivie entraînant la fermeture du bassin d'arrière-arc de Tetagouche-Exploits, derrière l'arc de Popelogan-Victoria, et de l'étendue océanique qui séparait l'Avalonie de la Gandérie (orogenèse acadienne). L'orogenèse salinique et l'orogenèse acadienne ont toute deux été accompagnées d'une importante minéralisation en or, dans la partie centrale de l'orogène des Appalaches (ceinture mobile centrale). Une partie, au moins, de l'or pourrait être associée au magmatisme produit durant la fragmentation de la plaque plongeante.

L'Océan Rhéique s'est ouvert durant l'Ordovicien précoce en raison de la séparation progressive de l'Avalonie du Gondwana. Plus tard, la séparation et la dérive du microcontinent de Meguma dans l'Océan Rhéique ont provoqué l'accrétion de celui-ci à la Laurentie, à 395 Ma environ (déformation néoacadienne). L'accrétion du microcontinent par un écaillage tectonique a été suivie, peu après, au Dévonien moyen, par un important plutonisme granitique, qui résulte d'une subduction continue de plaque plate. La forte minéralisation en or du terrane de Meguma est probablement attribuable à ce cadre tectonique.

Introduction

The Canadian Appalachians stretch from the northernmost tip of Newfoundland through Nova Scotia, Prince Edward Island, and New Brunswick into southern Quebec (Fig. 1).

The metallogeny and its relationships to tectonic evolution of the Canadian Appalachians is the topic of this paper. New data acquired during the last fifteen years or so of tectonic research in the Canadian Appalachians indicate that its tectonic evolution was much more complex than previously thought (e.g. van Staal et al., 1998). Hence the tectonic architecture and evolution of the Canadian Appalachians will be discussed first in some detail, because it has important implications for understanding the tectonic setting of the mineral deposits. This is followed by a description of the metallogeny organized according to the tectono-stratigraphic subdivisions (architecture) and orogenic history. The section concerning the tectonic architecture and evolution is summarized from van Staal (2006) and van Staal et al. (2006).

The Iapetus and Rheic Oceans and Orogenesis in the Appalachian Orogen

The Iapetus Ocean started to open during the late Neoproterozoic and had achieved a significant width during the Early Cambrian (Johnson et al., 1991; Trench et al., 1992; MacNiocaill et al., 1994; Cawood et al., 2001; Hodych et al., 2004).

The closure of Iapetus, including related seaways and marginal basins, was largely responsible for the formation of the Canadian Appalachians and terminated with the docking of the Avalonian microcontinent (see below). The subsequent closure of the Rheic Ocean caused accretion of Meguma and terminated with assembly of Gondwana and Laurentia into the Pangea supercontinent (van Staal, 2005). The latter was accompanied by large-scale Alleghenian-Variscan tectonic events during the Carboniferous-Permian. With the exceptions of localized deformation and plutonism associated with major strike-slip fault zones, the Canadian Appalachians largely escaped the penetrative effects of the Alleghenian terminal orogenic events and hence, these will not be discussed herein.

Early Paleozoic Tectono-Stratigraphic Zones

The Canadian Appalachians, based on the Early Paleozoic and older geology, have been subdivided into tectono-stratigraphic zones and subzones (Williams, 1979, 1995; Williams et al., 1988; Williams and Grant, 1998). From west to east, these are the Humber, Dunnage, Gander, Avalon, and Meguma zones (Fig. 1). The Humber Zone represents the leading edge of Laurentia's margin. The Gander, Avalon, and Meguma zones, represent peri-Gondwanan microcontinents (Ganderia, Avalonia, and Megum,a respectively) that were sequentially accreted to Laurentia during the Middle Paleozoic (450-380 Ma). The Dunnage Zone mainly contains accreted arc terranes that formed within the Iapetus Ocean. The Dunnage Zone has been subdivided into the peri-Laurentian Notre Dame and peri-Gondwanan Exploits subzones (Williams et al., 1988).

Humber Zone

The Canadian Humber Zone (Fig. 1) represents the most western part of the Canadian Appalachians and is mainly underlain by Late Neoproterozoic to Ordovician rocks deposited on Grenvillian crystalline basement as a result of rifting and passive margin development. Grenvillian basement is mainly exposed in western Newfoundland in a series of structural inliers (Fig. 1). Small bodies of Grenvillian basement occur also in Quebec and northwestern Cape Breton Island (Miller and Barr, 2000).

The western boundary of the Humber Zone is the Appalachian structural front. The eastern limit of the Humber Zone is represented by the Baie Verte-Brompton Line (BBL, Williams and St. Julien, 1982).

Rift-related activity in the Canadian segment of east Laurentia occurred between 615 and 540 Ma (e.g. Kamo et al., 1989). It resulted mainly in mafic magmatism and siliciclastic sediments being deposited in fault-bounded grabens. Paleomagnetic data indicate that Iapetus must have opened by at least 570 Ma at the end of the first phase of rifting (Cawood et al., 2001). Another phase of late Neoproterozoic rift-related magmatism (565-550 Ma) took place after this event and was followed shortly by deposition of a Lower Cambrian transgressive sequence (Figs. 2, 3), generally interpreted to represent a rift-drift transition (Williams and Hiscott, 1987; Lavoie et al., 2003). Waldron and van Staal (2001) related this event to the separation of a ribbon-shaped microcontinent from Laurentia, which they named Dashwoods. Dashwoods became the foundation of the continental, Early to Middle Ordovician phases of the Notre Dame arc (Whalen et al., 1997a; van Staal et al., 1998; 2006). Sparse faunal and paleomagnetic data suggests that Dashwoods remained close to Laurentia (Johnson et al., 1991; Nowlan and Neuman, 1995) from which it was separated by the Humber seaway (Waldron and van Staal, 2001). Similar lines of evidence suggest that Dashwoods, or an equivalent ribbon continent along strike, is also present in the subsurface of the Dunnage Zone of southern Quebec (Tremblay et al., 1994; Gerbi et al., 2006).

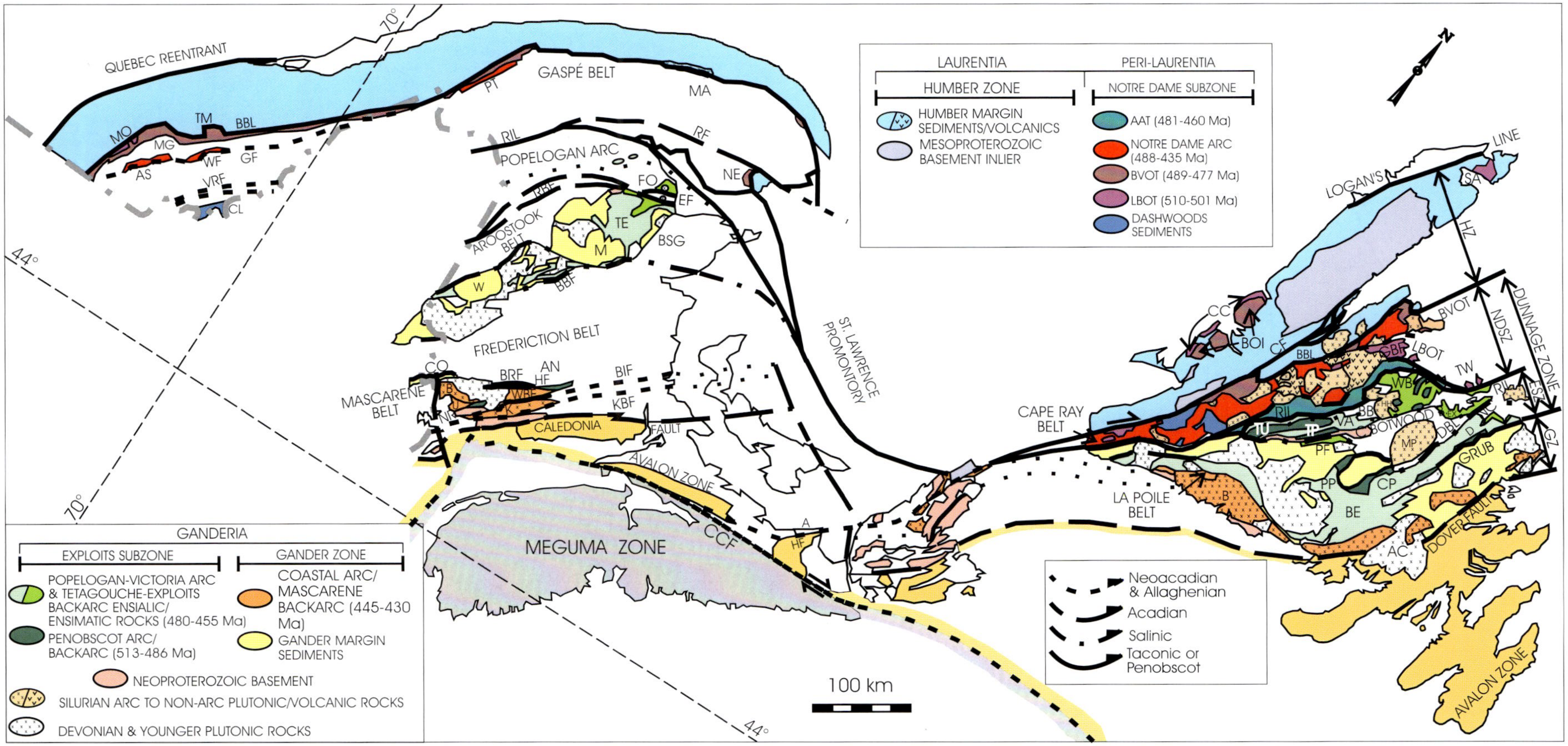

FIGURE 1. Tectonic map of the Canadian Appalachians with the distribution of the Early Paleozoic tectono-stratigraphic zones, subzones and other major tectonic elements (coloured) discussed in text. Middle Paleozoic belts are also indicated but not coloured. Adapted from van Staal (2006). AAT: Annieopsquotch accretionary tract; AC: Ackley granite; AN: Annidale belt; AS: Ascott Complex; B: Burgeo batholith; BB: Badger belt; BBF: Bamford Brook fault; BBL: Baie Verte Brompton Line; BE: Baie d'Espoir Group; BIF: Belleisle fault; BOI: Bay of Island Complex; BVOT: Baie Verte oceanic tract; BRF: Basswood Ridge fault; BSG: Bathurst Supergroup; CB: Cripple Back-Valentine Lake plutons; CC: Coastal Complex; CCF: Cobequid-Chedabucto fault; CF: Cabot fault; CL: Chain Lakes Massif; CO: Cookson Group; CP: Coy Pond Complex; D: Davidsville Group; DBL: Dog Bay Line; EF: Elmtree fault; ESZ: Exploits Subzone; EX: Exploits Group; FO: Fournier Group; GBF: Green Bay fault; GF: Guadeloupe fault; GRUB: Gander River ultrabasic belt; GZ: Gander Zone; HF: Hollow fault; HH: Hodges Hill Pluton; HZ: Humber Zone; K: Kingston belt; KBF: Kennebacasis fault; LBOT: Lushs Bight oceanic tract; M: Miramichi Group; MA: Mont Albert ophiolite; MG: Magog Group; MO: Mount Orford ophiolite; MP: Mount Peyton pluton; NC: Noggin Cove Formation; NE: Neckwick Formation; NDSZ: Notre Dame Subzone; NR: New River belt; PF: Pine Falls Formation; PP: Pipestone Pond Complex; PT: Pointe aux Trembles Formation; RBF: Rocky Brook-Millstream fault system; RF: Restigouche fault; RIL: Red Indian Line; SA: St Anthony Complex; TE: Tetagouche Group; TM: Thetford Mines ophiolite; TP: Tally Pond Group; TU: Tulks Group; TW: Twillingate trondhjemite; VA: Victoria arc; VRF: Victoria River fault; W: Woodstock Group; WB: Wild Bight Group; WBF: Wheaton Brook fault; WF: Weedon Fromation.

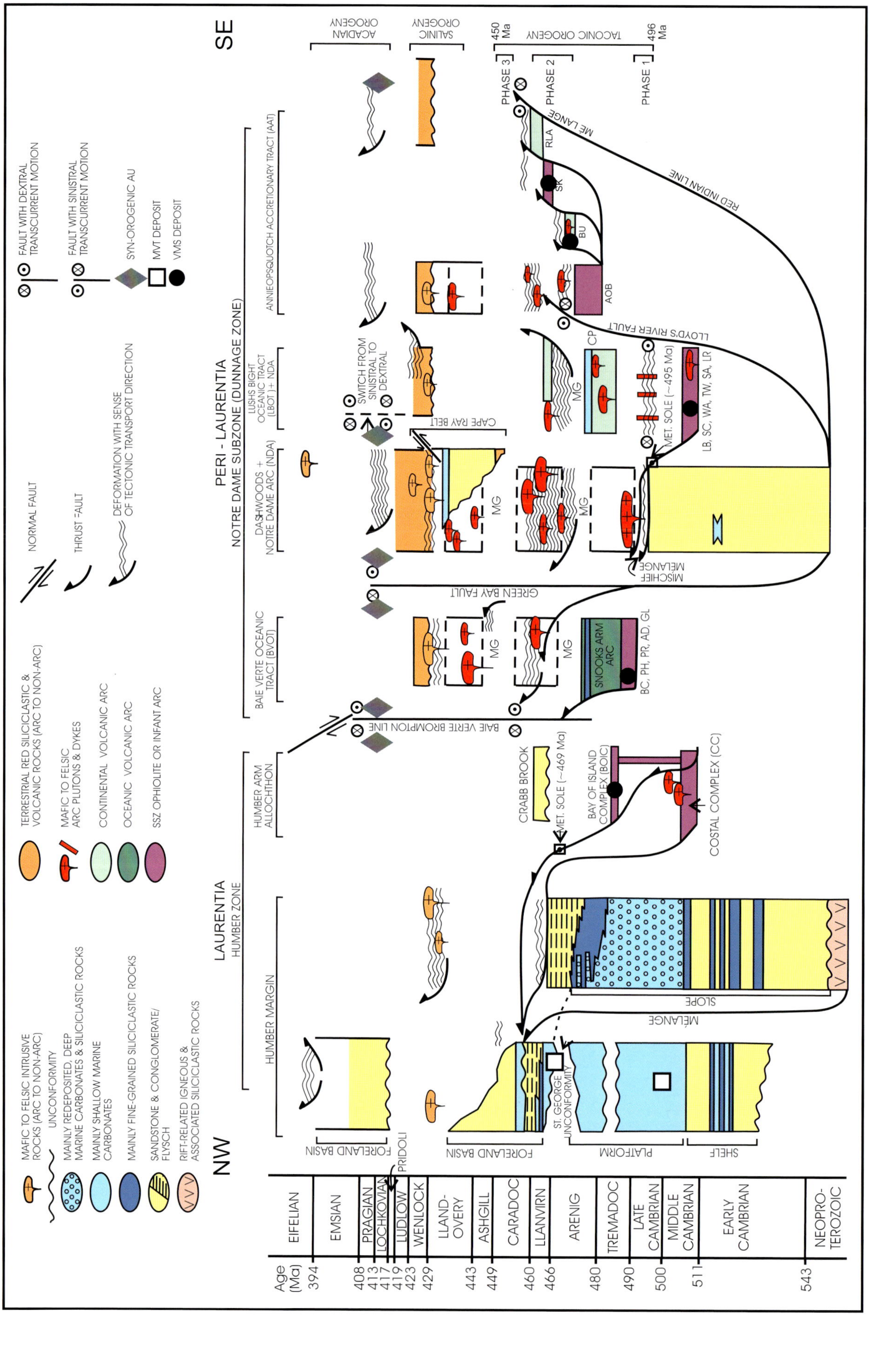

FIGURE 2. Summary of the stratigrapic and tectonic evolution of the Humber Zone and Notre Dame Subzone in Newfoundland and relevant mineral deposits. Modified from van Staal et al. (2006). AD: Advocate complex; AOB: Annieopsquotch ophiolite belt; BC: Betts Cove Complex; BU: Buchans and Robert Arm groups; GL: Grand Lake Complex; LB: Lushs Bight Group; LR: Long Range mafic-ultramafic complex; Met. Sole: metamorphic sole; MG: magmatic gap; PH: Pacquet Harbour Group; PR: Point Rousse Complex; RLA: Red Indian Lake arc; SA: St. Anthony Complex; SC: Sleepy Cove Group; SK: Skidder basalts; WA: Western Arm Group.

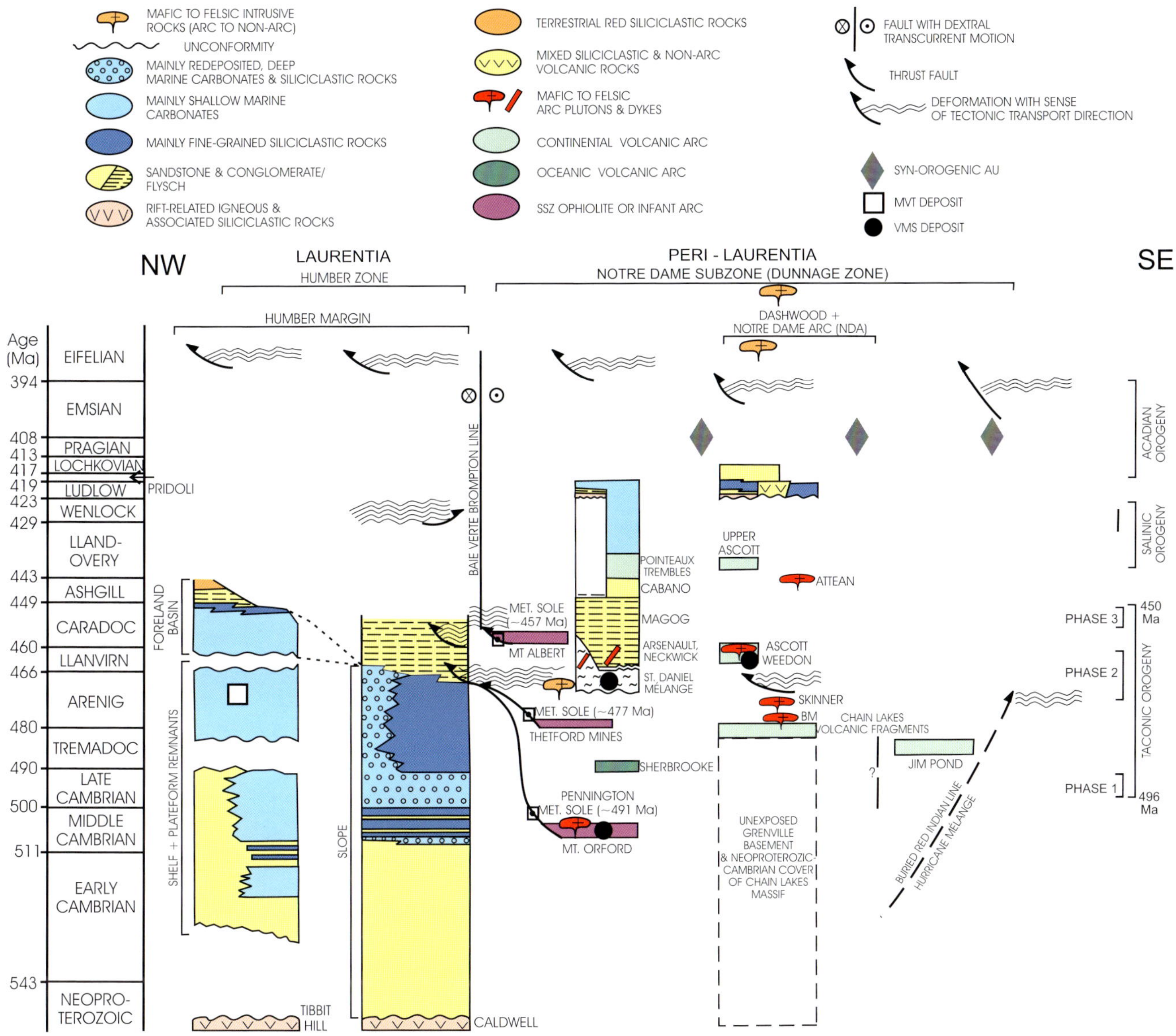

FIGURE 3. Summary of the stratigrapic and tectonic evolution of the Humber Zone and Notre Dame Subzone in Quebec and relevant mineral deposits. Modified from van Staal (2006).

During the Arenig, the passive Humber margin in Newfoundland and Quebec was converted into a convergent margin as a result of progressive loading by an overriding composite oceanic terrane and a trailing arc terrane. In Newfoundland, the oceanic terrane comprises the ophiolitic Coastal and Bay of Islands complexes (Fig. 2); the trailing arc terrane comprises the Notre Dame and Snooks Arm arcs (van Staal et al., 2006). The closure of the Humber Seaway heralds the main phase of the Taconic Orogeny. Tectonic loading of the margin as a result of obduction created a marine foreland basin. Taconic loading of the outboard part of the passive margin appears to have started slightly later in the Quebec reentrant (Fig. 3) (Malo et al., 2001; Lavoie et al., 2003).

Dunnage Zone

The Dunnage Zone mainly contains the remnants of Cambro-Ordovician infant arc (Stern and Bloomer, 1992) and more mature arc terranes that existed within the Iapetus Ocean. The Dunnage Zone is best preserved in Newfoundland, but important segments are also exposed in New Brunswick and southern Quebec (Fig. 1). Paleomagnetic, fossil, and other geological evidence indicate that these terranes either have a peri-Laurentian provenance (Notre Dame Subzone) or a peri-Gondwanan provenance (Exploits Subzone). This suggests that there were several subduction zones active within Iapetus at the same time (van Staal et al., 1998).

Notre Dame Subzone

The Notre Dame Subzone (Fig. 1) is exposed in Newfoundland and Southern Quebec. It lies immediately to the east of the Baie Verte-Brompton Line, which is its tectonic boundary with the adjacent Humber Zone. Its eastern boundary with the Exploits Subzone is represented by the Red Indian Line, a major suture that juxtaposes rocks formed on opposite sides of the Iapetus Ocean (Figs. 2, 4). The Red Indian Line is well exposed in Newfoundland (Williams et

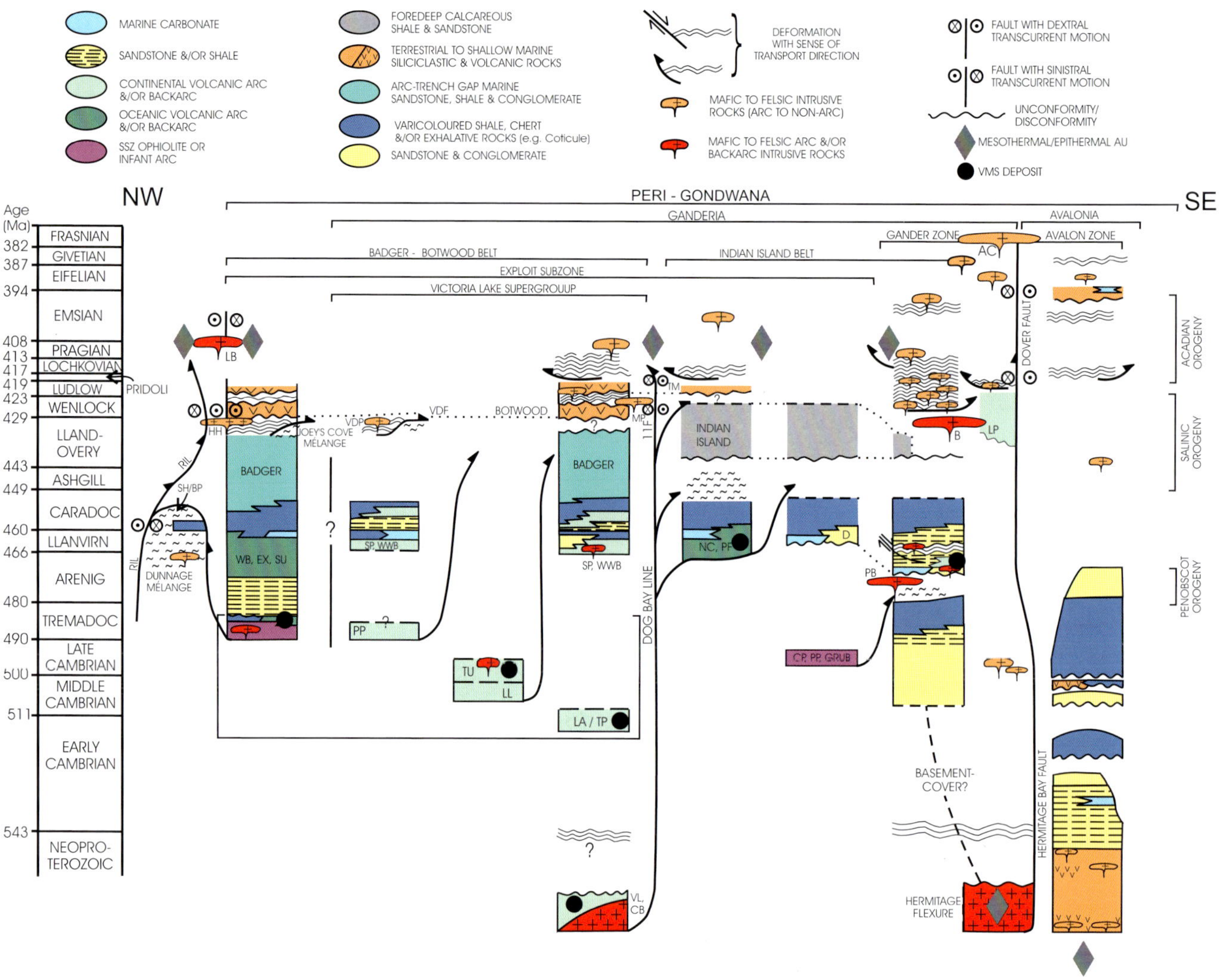

FIGURE 4. Summary of the tectono-stratigraphic evolution of the Exploits Subzone, and Gander and Avalon zones in Newfoundland and relevant mineral deposits. Modified from van Staal (2006). AC: Ackley granite; B: Burgeo granite; CB: Crippleback pluton; CP: Coy Pond Complex; D: Davidsville Group; EX: Exploits Group; GRUB: Gander River ultrabasic belt; HH: Hodges Hill pluton; LA/TP: Lake Ambrose Formation/Tally Pond Group; LB: Loon Bay pluton; LL:Long Lake Group; LP: La Poile Group; MP: Mount Peyton pluton; NC: Noggin Cove Formation; PA: Pats Pond Group; PB: Partridgeberry granite; PF: Pine Falls Formation; PP: Pipestone Pond complex; RIL: Red Indian Line; SH/BP: Sops Head/Boones Point Complex; SP: Sutherlands Point Group; SU: Summerford Group; TU: Tulks Group; WB: Wild Bight Group; WWB: Wigwam Brook Group; VDP: Victoria Delta porphyry; VDF: Victoria Delta fault; VL: Valentine Lake pluton.

al., 1988), but is hidden beneath Middle Paleozoic cover sequences of the Gaspé belt in northernmost New Brunswick near the border with Quebec (Fig 1, see below).

The Notre Dame Subzone in Newfoundland comprises three distinct Cambrian to Middle Ordovician (507-462 Ma) oceanic terranes and a continental magmatic arc (the Notre Dame arc) built on Dashwoods (Waldron et al., 2001). The oldest oceanic rocks occur in the Middle to Upper Cambrian (510-501 Ma) Lushs Bight oceanic tract (LBOT) (Elliott et al., 1991; Szybinski, 1995; Swinden et al., 1997), which primarily includes the Lushs Bight, Western Arm, Cutwell, Moreton's Harbour, and Sleepy Cove groups and several gabbroic to trondhjemitic intrusions (Fig. 2). The Cambrian Coastal Complex (CC) of Karson and Dewey (1978) (Fig. 1; Cawood and Suhr, 1992) and the ophiolitic St. Anthony Complex (Jamieson, 1988; G. Dunning, pers. comm., 2004), which occur as obducted sheets in the Humber Zone of central and northern Newfoundland, respectively, are coeval and probably correlatives of the LBOT (Fig. 2). The ca. 504 Ma suprasubduction-zone Mt. Orford ophiolite in southern Quebec (David and Marquis, 1994; Huot et al., 2002) may also be equivalent (Figs. 1, 3).

The LBOT is characterized by an ophiolitic association of pillow basalts, sheeted dykes, gabbro, and rare ultramafic rocks (Kean et al., 1995). An abundance of boninite and primitive island arc tholeiite (Swinden, 1996; Swinden et al., 1997) combined with the presence of consanguineous juvenile granitoids (Williams and Payne, 1975; Fryer et al., 1992) suggest that this tract represents an infant arc terrane formed during subduction initiation at ca. 510 Ma (van Staal et al., 1998). Paleomagnetic data (Johnson et al., 1991) and low ε_{Nd} values suggest the LBOT formed near the Laurentian margin (Swinden et al., 1997). Field relationships and isotopic evidence suggest that the LBOT and its correlatives in the southern and central parts of the Notre Dame Subzone (e.g. Long Range mafic-ultramafic complex (Hall and van Staal, 1999)) were deformed and emplaced onto Dashwoods (Fig. 5) between 500 and 490 Ma (Szybinski,

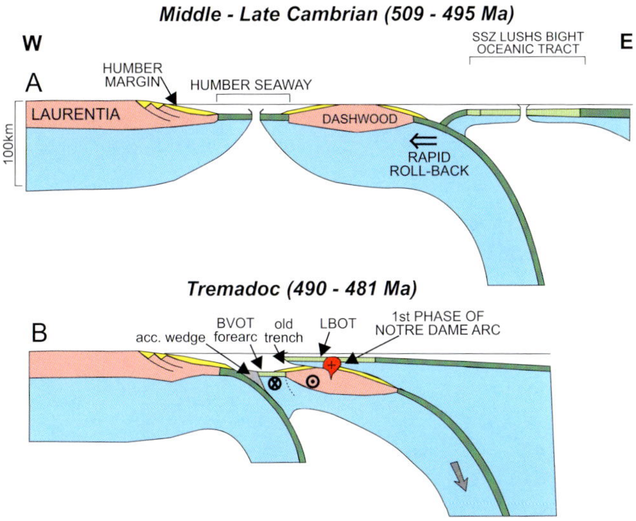

FIGURE 5. (A) Cambrian-Early Ordovician tectonic evolution of the Humber margin and outboard peri-Laurentian terranes. Modified from van Staal et al. (2006). Rapid hinge retreat of the east-dipping (present coordinates) Dashwoods plate is responsible for formation of the infant arc terrane represented by the Lushs Bight oceanic tract. **(B)** Stepping-back of the subduction zone in the Humber Seaway produced the Baie Verte oceanic tract, and the Snooks Arm and Notre Dame arcs.

1995; Swinden et al., 1997; van Staal et al., 1998, 2006; Waldron and van Staal, 2001) shortly after their formation (Fig. 2). After obduction of most of the LBOT onto Dashwoods, a new suprasubduction zone oceanic tract was formed during stepping back of the east-dipping subduction zone into the Humber Seaway (Fig. 5). The remnants of this oceanic tract have been preserved mainly as a narrow fault-bounded wedge along the Baie Verte Brompton Line (Fig. 1) and are referred to as the Baie Verte oceanic tract (BVOT).

The ophiolitic component of the BVOT is significantly younger than its counterpart in the LBOT and has yielded two identical ages of ca. 490 Ma (Dunning and Krogh, 1985; Cawood et al., 1996). The BVOT ophiolites became basement to the oceanic Snooks Arm arc-back-arc complex (Bédard et al., 2000) during the Tremadoc and early Arenig at 487 to 476 Ma (Williams, 1992; Ramezani et al., 2002), coeval with the first phase of the continental Notre Dame arc (Fig. 2). The Notre Dame and Snooks Arm arcs probably formed part of a once continuous arc system, possibly like the present day Sunda (continental) and Banda (oceanic) arcs in Indonesia. The bulk of the southern Quebec ophiolite belt and oceanic elements of the second oceanic arc of Tremblay (1992) in southern Quebec (Sherbrooke arc, Fig. 3) probably correlate with the ophiolitic foundation of the BVOT and its Snooks Arm arc suprastructure. The BVOT was emplaced diachronously onto the Humber margin during the Early Ordovician with collision beginning along first and second order promontories in the margin. Syn-collisional spreading in reentrants, due to rollback and/or transtension, during obduction of the BVOT (Cawood and Suhr, 1992; Schroetter et al., 2003) may have been responsible for the generation of anomalously young suprasubduction zone ophiolites such as the Bay of Islands, Thetford Mines, and Mt. Albert complexes (Figs. 2, 3, also see below).

The continental Notre Dame magmatic arc in Newfoundland existed intermittently from ca. 488 to 435 Ma and is represented by three major magmatic pulses separated by two significant gaps, which coincide with collisional events (Fig. 2; van Staal et al., 2006). Arc magmatism was followed by a 433 to 429 Ma mixed arc/non-arc-like bimodal suite related to slab-breakoff (Whalen et al., 1996, 2006; van Staal et al., 2003a, 2004a). Equivalents of the Notre Dame arc are poorly preserved in Quebec, probably because a large part is buried beneath the Siluro-Devonian cover sequences of the Gaspé belt (Fig. 1). The Middle Ordovician-Early Silurian volcanic and plutonic rocks of the Ascot-Weedon continental arc complex (Tremblay, 1992; David and Marquis, 1994) and Pointe aux Trembles Formation (David and Gariepy, 1990) are correlatives of the second and third phase of the Notre Dame arc. The undated gabbro-diorite sills and tuffaceous rocks in the Middle-Upper Ordovician Magog Group (Tremblay et al., 1995) probably also form part of the Notre Dame arc (Fig. 3).

The Arenig-Llanvirn (480-462 Ma) Annieopsquotch accretionary tract (AAT) (van Staal et al., 1998; Lissenberg et al., 2005b; Zagorevski et al., 2006a) is the youngest oceanic terrane in Newfoundland. The AAT is sandwiched between the Lloyd's River-Hungry Mountain-Lobster Cove fault system and the Red Indian Line (Fig. 2; Colman-Sadd et al., 1992a; van Staal et al., 1998; Lissenberg and van Staal, 2002). The AAT comprises a tectonic collage of 480 to 473 Ma infant arc ophiolite (e.g. Annieopsquotch ophiolite belt, Lissenberg et al., 2005a), arc and back-arc terranes (e.g. Buchan-Robert Arm belt, Swinden et al., 1997; Zagorevski et al., 2006a) that formed as a result of west-directed subduction of Iapetus outboard of Dashwoods. The AAT is not exposed in southern Quebec.

Exploits Subzone

The Exploits Subzone is exposed in central Newfoundland, between the Red Indian Line and the GRUB Line-Day Cove fault system (Figs. 1, 4). The oldest Paleozoic volcanic rocks are Lower Cambrian to Tremadoc (514-486 Ma) and represent the remnants of a peri-Gondwanan arc/back-arc complex (Colman-Sadd et al., 1992b; Jenner and Swinden, 1993), named the Penobscot complex by van Staal et al. (1998). The Cambrian arc elements of the Penobscot complex occur mainly in the ensialic Victoria Lake Supergroup (Dunning et al., 1991; Rogers et al., 2006), which also includes a fault-bounded belt of Late Neoproterozoic (565 Ma) arc plutonic and volcanic rocks (Evans et al., 1990). Zircon inheritance of the latter in the Lower Cambrian volcanic rocks suggests a basement-cover relationship. Ensialic arc activity continued intermittently until at least 486 Ma (O'Brien et al., 1997; Zagorevski et al., 2004). The arc elements of the Penobscot complex are solely preserved in the western half of the Exploits Subzone in Newfoundland (Fig. 1), comprising both ensialic and ensimatic segments (Figs. 1, 4). Related back-arc ophiolitic rocks (GRUB, Coy Pond, and Pipestone Pond complexes) are Upper Cambrian (ca. 494 Ma) (Dunning and Krogh, 1985; Jenner Swinden, 1993) and restricted to the eastern half of the Exploits Subzone. They are localized along the faulted boundary with the Middle Cambrian-Tremadoc arenites and shales of the Gander Zone (e.g. Currie, 1992), which were deposited on the leading edge of the Gondwanan Gander margin (van Staal, 1994).

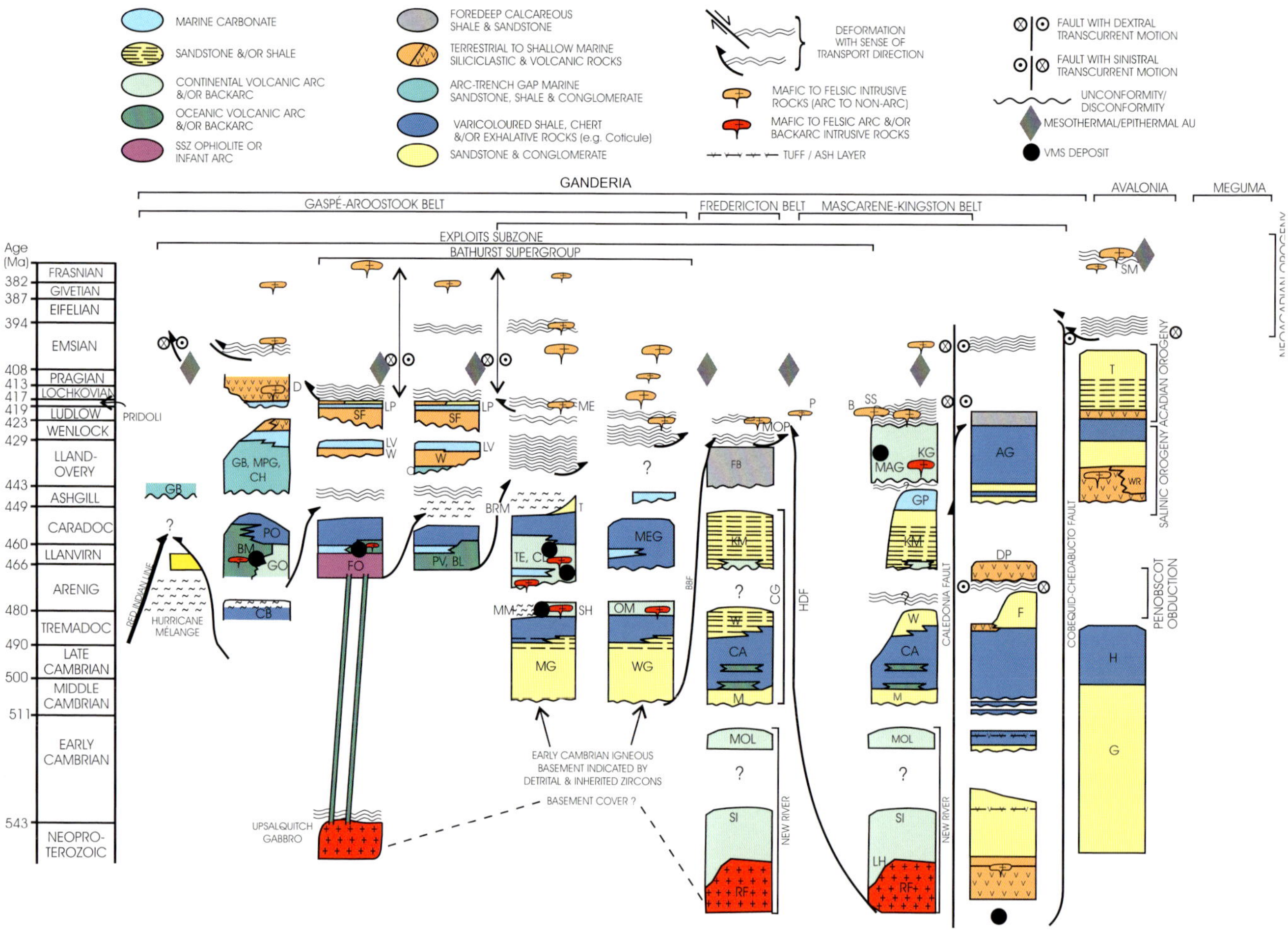

FIGURE 6. Summary of the tectono-stratigraphic evolution of the Exploits Subzone, and Gander and Avalon zones in Maritime Canada and Maine. Modified from van Staal (2006). AG: Arisaig Group; BBF: Bamford Brook fault; BL: Blueschist nappe; BM: Bald Mountain volcanic sequence of Winterville Formation (Maine); BRM: Belledune River mélange; C: Clemville Formation; CA: Calais Formation; CB: Chase Brook Formation (Maine); CG: Cookson Group; CH: Chaleurs Group; CL: California Lake Group; D: Dalhousie Group; DP: Dunn Point volcanics; F: Ferrona Formation; FB: Fredericton Belt sequence; FO: Fournier Group; G: Goldenville Group; GB: Grog Brook Group; GO: Goulette Brook Formation; GP: Goss Point Formation; H: Halifax Group; HDF: Honeydale fault; KBF: Kennebecasis fault; KG: Kingston Group; KM: Kendall Mountain Formation; LB: Lawson Brook schist; LP: La Plante Formation; LV: La Vieille Formation; M: Megunticook Formation; MAG: Mascarene Group; ME: Mount Elisabeth pluton; MEG: Meductic Group; MG: Miramichi Group; MM: Miramichi mélange; MOL: Mosquito Lake Formation; MOP: Mohannes pluton; MPG: Matapedia Group; OM: Oak Mountain Formation; P: Pocomoonshine gabbro; PO: Popelogan Formation; PV: Pointe Verte Formation; RF: Ragged Falls pluton; SF: Simpsons Field Formation; SH: Sheephouse Brook Group; SI: Simpsons Island Formation; SM: South Mountain batholith; SS: St. Stephen gabbro; T: Tomogonops Formation; TE: Tetagouche Group; TO: Torbrook Group; W: Woodland Formation; WE: Weir Formation; WG: Woodstock Group; WR: White Rock Formation.

Correlative volcanic rocks of the Penobscot arc/back-arc complex also occur in New Brunswick in the 497 to 493 Ma Annidale belt (Fig. 1; McLeod et al., 1992) and the adjacent New River Belt (Johnson et al., 1996). Neoproterozoic basement also occurs in the Exploits Subzone of northern New Brunswick (Figs. 1, 6; van Staal et al., 1996).

During the Arenig-Caradoc, arc volcanism recommences in the western half of the Exploits Subzone, whereas the eastern half is dominated by deposition of volcanogenic sandstones and shales with minor felsic and rare mafic volcanic rocks (Fig. 4, Valverde-Vaquero et al., 2006). A lithological contrast between these two parts of the Exploits Subzone persists into the Late Silurian. The boundary between the western and eastern halfs of the Exploits Subzone is the Dog Bay Line (Williams et al., 1993), which is an Early to Late Silurian suture formed after closing a wide back-arc basin, named the Exploits basin in Newfoundland and the Tetagouche basin in New Brunswick (van Staal, 1994). The western half of this back-arc basin represents the remnants of its active side (Popelogan-Victoria arc), the eastern half the passive side (Davidsville margin). The Popelogan-Victoria arc is represented by the post-Tremadoc parts of the Victoria Lake Supergroup, and Wild Bight and Exploits groups in Newfoundland, and the Balmoral Group and Bathurst Supergroup in New Brunswick (van Staal et al., 1998, 2003b). The Davidsville margin comprises the Davidsville, Baie d'Espoir, Bay du Nord, and Harbour le Cou groups in Newfoundland (Valverde-Vaquero et al., 2006), and the Meductic and post-Tremadoc part of the Cookson Group in New Brunswick (van Staal et al., 2003b).

Gander Zone

The Gander Zone (Figs. 1, 4, 6) comprises a distinct sequence of Lower Cambrian to Tremadoc (~520-480 Ma) arenites, siltstones, and/or and shales, generally considered to represent the outboard part of a passive margin (Gander margin, van Staal, 1994). The Gander margin rocks can be traced from northeast Newfoundland into New Brunswick. Its rocks are known by various names: Gander Group and Spruce Brook Formation in eastern and central Newfoundland, and Miramichi, Woodstock, and Cookson groups in northern, eastern, and southern New Brunswick, respectively. The Gander Zone in Newfoundland and New Brunswick is separated from the adjacent Avalon Zone by the Dover- Hermitage Bay- Caledonia Fault system (Figs. 1, 6). Detrital zircons suggest that all three lithological units in New Brunswick's Gander Zone have an identical provenance and are approximately coeval (van Staal et al., 2004b). The detrital zircon data, combined with field relationships and sparce fossils, indicate a dominantly Middle Cambrian to Tremadoc age; the Tremadoc part of the sequence mainly represented by dark grey or black shales (van Staal and Fyffe, 1995; van Staal et al., 2003b). Field relationships and detrital zircon and titanite data indicate a similar age range in Newfoundland (O'Neill, 1991; Colman-Sadd et al., 1992b).

The Neoproterozoic rocks of the Hermitage flexure in southern Newfoundland (O'Brien et al., 1996) are generally inferred to represent basement to the Paleozoic sedimentary rocks of the Gander Zone (Dunning and O'Brien, 1989; O'Brien et al., 1993), although a basement-cover relationship has not been recognized. Such a relationship may be preserved in southern New Brunswick. Here quartz-rich arenite and conglomerate of the Matthews Lake Formation with a detrital zircon content similar to that of the Miramichi, Woodstock, and Cookson groups, disconformably overlie Cambrian and Neoproterozoic igneous rocks of the New River Belt (Johnson and McLeod, 1996). This line of evidence supports isotopic and geological arguments that the Gander and Avalon zones represent two separate peri-Gondwanan basement blocks (van Staal et al., 1996; Samson et al., 2000; Barr et al., 2002), respectively, named Ganderia and Avalonia (van Staal et al., 1998). Because their Neoproterozoic evolution is rather similar, Ganderia and Avalonia are mainly distinguished on basis of their distinctly different Early Paleozoic tectonic histories (Figs. 4, 6).

Avalon Zone

The Avalon Zone (Fig. 1) mainly comprises a distinctive belt of Neoproterozoic, largely juvenile, arc-related, volcano-sedimentary sequences and associated plutonic rocks, which underwent a long-lived tectonic history, including orogenesis before deposition of a Cambrian-Ordovician shale-rich platformal sedimentary succession (Figs 4, 6) (Kerr et al., 1995; Landing, 1996; O'Brien et al., 1996). The Avalon Zone in Canada comprises eastern Newfoundland, and the Mira and Caledonia terranes of Nova Scotia and New Brunswick, respectively (Barr and Kerr, 1997; Barr et al., 1998).

Middle Cambrian to Middle Ordovician rift-related volcanic rocks are presumably mainly related to Avalonia's rifting and departure from Gondwana.

Meguma Zone

The Meguma Zone is the most outboard terrane in the Canadian Appalachians and is exposed on land in southern Nova Scotia (Fig. 1). However, its regional extent is much larger and continues offshore (Hutchinson et al., 1988; Keen et al., 1991; Pe-Piperand Jansa, 1999).

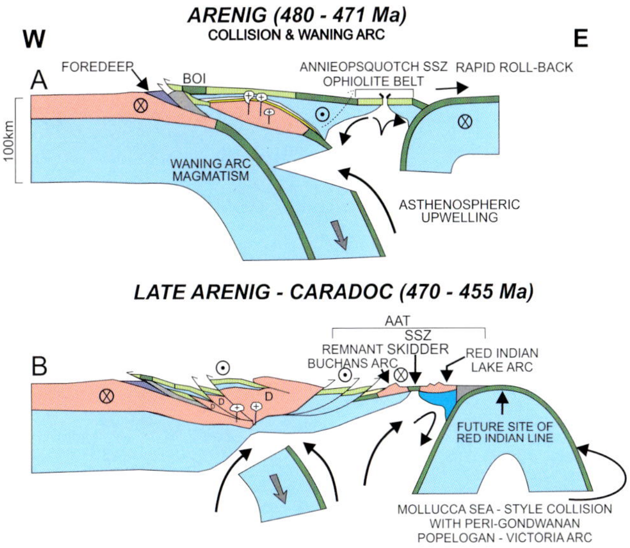

FIGURE 7. **(A)** Taconic 2 collision of Dashwoods with the Humber margin and subduction initiation west of Dashwoods was responsible for formation of the Annieopsquotch ophiolite belt; **(B)** Collisional thickening of the Notre Dame arc, breakoff of the Humber margin slab, and start of accretion of the Annieopsquotch ophiolite belt. Modified from van Staal et al. (2006).

The oldest exposed part of the Meguma Zone comprises a thick (<10 km) Cambrian to Early Ordovician turbiditic sandstone-shale sequence of the Goldenville Group (Fig. 6), which was largely deposited on the continental rise and/or slope to outer shelf of a Gondwanan passive margin. The Goldenville Group is overlain by the Lower Ordovician Halifax Group, which represents a shoaling succession (Schenk, 1997). Together, the Goldenville and Halifax groups constitute the Meguma Supergroup. The Meguma Supergroup is disconformably overlain by the Upper Ordovician to Early Devonian, dominantly shallow marine shelf siliciclastic sedimentary rocks of the Annapolis Supergroup. Rift-related bimodal volcanic rocks of the Late Ordovician-Lower Silurian (~442-438 Ma) White Rock Group occur at the base of the Annapolis Supergroup (Schenk, 1997; Keppie and Krogh, 2000; MacDonald et al., 2002). These rift-related volcanic rocks may be related to the onset of rifting and departure of Meguma from Gondwana. The top of the Annapolis Supergroup comprises the Lower Devonian (Lochkovian to Emsian) Torbrook Group, which in turn is unconformably overlain by Triassic red beds.

Middle Paleozoic Sedimentary and Volcanic Belts

Middle Paleozoic sedimentary and volcanic rocks have been grouped into belts following the nomenclature of Williams (1995). They were deposited in synorogenic basins and hence, impose important constraints on the tectonic setting of the Canadian Appalachians during the Middle Paleozoic.

Gaspé, Cape Ray, and Badger Belts

The Late Ordovician-Early Devonian Gaspé belt (GAB) underlies most of northern and central New Brunswick and adjacent Quebec (Chaleurs Bay, Matapedia, and Gaspé). The

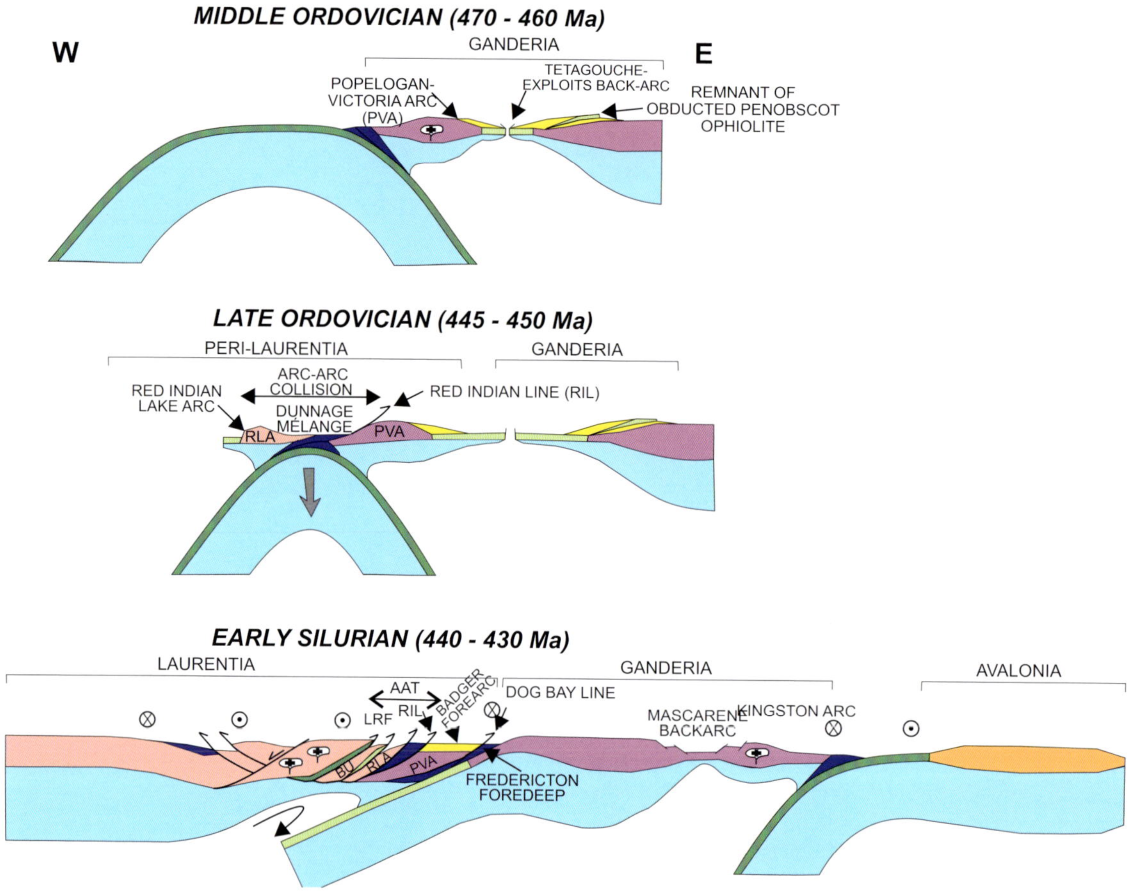

FIGURE. 8. Tectonic evolution of the Popelogan-Victoria arc –Tetagouche-Exploits back-arc system. Modified from van Staal (2006). Arc magmatism shuts off in the Caradoc due to a collision with the peri-Laurentian Red Indian Lake arc. Convergence continues due to stepping back of the subduction zone into the Tetagouche-Exploits back-arc basin, the closure of which results in the Salinic Orogeny.

Ordovician structures associated with the Red Indian Line in Maritime Canada are buried beneath it (Fig. 1).

The Upper Ordovician-Upper Silurian rocks of the GAB are mainly represented by fore-arc siliciclastic turbidites grading upwards into more calcareous rocks to be finally replaced by Wenlock-Ludlow red beds, basalt, and rhyolite (Fig. 6; van Staal and de Roo, 1995; Walker and McCutcheon, 1995; Wilson et al., 2004), immediately prior to Late Silurian Salinic tectonism. Correlative rocks in Newfoundland occur in the Cape Ray and Badger belts (Chorlton et al., 1995; Williams et al., 1995; Dubé et al., 1996). All three belts contain marine sedimentary rocks and Lower to Upper Silurian terrestrial rocks, which have been interpreted as a late Early Silurian overstep sequence (Chandler et al., 1987) that links all tectono-stratigraphic zones and subzones west of the Dog Bay Line in Newfoundland and its equivalent in New Brunswick (Bamford Brook Fault, Fig. 1).

Fredericton and Indian Island Belts

The Fredericton belt is exposed in central New Brunswick (Figs. 1, 6). It comprises a thick sequence of Llandovery to Ludlow marine, locally slightly calcareous turbidites that nowhere are interlayered with volcanic rocks. They are in part coeval with the terrestrial rocks in the Gaspé, Cape Ray, and Badger belts.

Deposition of the Fredericton belt is also coeval with formation of the east-facing Brunswick subduction complex preserved in the Miramichi Highlands (van Staal, 1994; van Staal et al., 2003b). Detritus of the Brunswick subduction complex occurs in the turbidites of the Fredericton belt, providing a tectonic linkage. The Fredericton belt turbidites became strongly deformed, locally with marked easterly overturned structures (Park and Whitehead, 2003), before the end of the Silurian (West et al., 1992, 2003; Tucker et al., 2001). The Fredericton belt has been interpreted as a foredeep basin formed during Early to Late Silurian loading of the passive, Davidsville margin by the overriding Brunswick subduction complex (van Staal and de Roo, 1995), from which the latter is separated by the Bamford Brook Fault in New Brunswick (Figs. 1). The Fredericton belt has been correlated (van Staal, 2005) with lithologically similar rocks in the Indian Islands belt in northeastern and central Newfoundland (Fig. 4), immediately east of the Dog Bay Line (Williams et al., 1993).

Kingston, Mascarene, and La Poile Belts

The Kingston and adjacent Mascarene belts in southern New Brunswick (Figs. 1, 6) comprise tectonically-related sequences of Ashgill-Pridolian volcanic and sedimentary rocks (e.g. Fyffe et al., 1999), which are geologically and geophysically linked to Ganderia (Samson et al., 2000; Johnson, 2001; McLeod et al., 2001; King and Barr, 2004;

van Staal et al., 2004b). The volcanic rocks are predominantly Llandovery in age (442-435 Ma) (Barr et al., 2002; Miller and fyffe, 2002). The Lower Silurian volcanic rocks have compositions indicative of arc and back-arc settings (Johnson and McLeod, 1996; Barr et al., 2002).

Elements of the Kingston, and Mascarene belts continue into Cape Breton Island (e.g. Barr and Jamieson, 1991; Price et al., 1999; Barr et al., 2002) and can also be traced into Newfoundland. The Silurian volcanic and sedimentary rocks of the La Poile belt (O'Brien et al., 1991) and associated arc intrusive rocks (e.g. ca. 430 Ma Burgeo granite, Kerr et al., 1995) along Newfoundland's south coast occupy a similar tectonic position to the Kingston and Mascarene belts (Figs. 1, 4).

The Kingston, Mascarene, and La Poile belts were strongly shortened during the latest Silurian-Early Devonian Acadian Orogeny; the start of Acadian Orogeny being defined by inversion of the Mascarene and La Poile basins, which started during the latest Silurian around 422 Ma (O'Brien et al., 1991; Fyffe et al., 1999; van Staal, 2005).

Arisaig Belt

The Arisaig belt in western Nova Scotia (Figs. 1, 6) is one of the few occurrences of pre-Acadian Siluro-Devonian rocks deposited on Avalonia and exposed on land (Murphy and Keppie, 1995). Here Silurian-Early Devonian (Lochkovian, Boucot et al., 1974), dominantly platformal siliciclastic sediments disconformably overlie Middle Ordovician volcanic rocks of the Dunn Point Formation (Hamilton and Murphy, 2004), which themselves unconformably overlie Cambrian-Lower Ordovician sedimentary rocks. Subsidence analysis suggests a passive margin setting during the Llandovery and Ludlow. The Pridolian-Lochkovian rocks at the top of the Arisaig belt, on the other hand, record a significant increase in subsidence rate, indicative of tectonic loading and foreland basin formation (Waldron et al., 1996). This tectonic loading coincides with initiation of the Acadian Orogeny and the time of Avalonia's accretion to Laurentia. Hence, Avalonia was the lower plate during its collision with Laurentia, which is in accord with the presence of Silurian arc/back-arc volcanic rocks on the southern margin of Ganderia (Kingston, Mascarene, and La Poile belts), which at this time represented Laurentia's leading edge. This also agrees with seismic interpretations (van der Velden et al., 2004).

Tectonic Evolution

The orogenic history of the Appalachian Orogen (500-250 Ma) was complex. Orogenesis started in the Cambrian and continued into the Permian. The orogenic classification used herein is based on grouping of tectonic events that are temporally, spatially and kinematically related.

Taconic Orogeny

The Taconic Orogeny is due to accretionary events in the peri-Laurentian realm between the Late Cambrian and Late Ordovician (500-450 Ma). It comprises three events called Taconic 1, 2 and 3 (Figs. 2, 3; van Staal et al., 2006).

Taconic 1 represents obduction of the suprasubduction zone oceanic lithosphere of the LBOT on the Dashwoods microcontinent in Newfoundland (see above) between 500 and 493 Ma (Fig. 5). Following obduction of the LBOT onto Dashwoods in Newfoundland, subduction stepped back into the Humber Seaway at ca. 490 Ma, behind the now composite Dashwoods-LBOT lithosphere. This new subduction zone is responsible for the upper Cambrian-Tremadoc volcanic and plutonic rocks of the ensialic Notre Dame arc, the suprasubduction-zone oceanic lithosphere of the BVOT, and the Snooks Arm arc (Fig. 5). The ophiolitic component of the BVOT is compositionally similar to the LBOT (Swinden et al., 1997; Bédard et al., 1998) and is for identical reasons interpreted to have formed during subduction initiation. Obduction of Humber seaway oceanic lithosphere onto the Humber margin and accretion of the trailing Dashwoods microcontinent caused Taconic 2. Taconic 2-related orogenesis produced high-grade metamorphism and polyphase deformation in large parts of the Notre Dame Subzone during the Middle Ordovician (Fig. 7; van Staal et al., 2006).

Deceleration of convergence between the Humber margin and Dashwoods as a result of the onset of collision at promontories transfered convergence to the Iapetus Ocean immediately outboard (east) of Dashwoods at ca. 481 Ma by forming a new, west-dipping subduction zone in Iapetus (Fig. 7; Lissenberg et al., 2005a; van Staal et al., 2006). The new subduction zone generated the Arenig (480-473 Ma) Annieopsquotch infant arc ophiolite belt (AOB) and subsequently the Llanvirn (465-460 Ma) Red Indian Lake arc in Newfoundland (Zagorevski et al., 2006a). These arc terranes were accreted to Dashwoods between 470 and 455 Ma forming the Annieopsquotch Accretionary tract (AAT, Lissenberg et al., 2005b). The main Iapetan tract that was situated between the AAT and the peri-Gondwanan Popelogan-Victoria arc (PVA, Figs. 2, 4, 6) was now being rapidly closed by two outwardly-dipping subduction zones (Figs. 7, 8). This convergence terminated in a Moluccan Sea-style arc-arc collision in the late Caradoc (455-450 Ma), sutured along the Red Indian Line (van Staal et al., 1998). This collision terminated the Taconic Orogeny and only left subductable Iapetan oceanic lithosphere in the Tetagouche-Exploits back-arc basin and the narrow seaway between Ganderia and Avalonia (Figs. 6, 8; van Staal, 1994, 2005).

Penobscot Orogeny

The Early Ordovician Penobscot Orogeny is, in part, coeval with the early phases of Taconic orogenesis in the peri-Laurentian realm (Figs. 2, 4). However, it is restricted to Ganderian rocks (Neuman, 1967; Colman-Sadd et al., 1992b), which at that time were situated in the periphery of Gondwana at high southerly latitudes (Liss et al., 1994; van Staal et al., 1996, 1998) on the opposite side of the Iapetus Ocean.

The Penobscot Orogeny mainly involves Tremadoc and/or earliest Arenig obduction of Upper Cambrian (ca. 494 Ma) back-arc ophiolites (Figs. 1, 4; Coy Pond and Pipestone Pond complexes (Jenner and Swinden, 1993)) onto the Gander margin (Colman-Sadd et al., 1992b; van Staal, 1994). The ophiolites formed in a back-arc setting with respect to the Cambrian-Tremadoc volcanic rocks of the Penobscot arc (lower parts of the Victoria Lake Supergroup, Wild Bight and Exploits groups (Fig. 4)) that lie immediately east of the Red Indian Line (Fig. 1). The Penobscot arc/back-arc complex formed above an east-dipping subduction zone (Rogers et al., 2006; Zagorevski et al., 2006b).

Salinic Orogeny

The Salinic Orogeny is a predominantly Silurian orogenic event (Dunning et al., 1990) that is distinct from the Early Devonian Acadian Orogeny in Newfoundland and New England (van Staal, 2005). The prelude to Salinic orogenesis starts in the Ashgill after the Caradoc arc-arc collision that terminated the Taconic Orogeny (see above). Sinistral-oblique convergence between Ganderia and composite Laurentia continued and was accommodated by stepping back of the west-dipping subduction zone into the Tetagouche-Exploits back-arc basin (Figs. 1, 4, 8; van Staal et al., 1998). The Tetagouche-Exploits back-arc comprised a complex mosaic of small oceanic basins separated by extended continental ridges, similar to the modern Sea of Japan (van Staal et al., 2003b; Valverde-Vaquero et al., 2006). Magmatic rocks related to this phase of subduction are represented by the third, ca. 445 to 435 Ma phase of the Notre Dame arc (Figs. 2, 4, 8).

Lower Silurian sediments on both sides of the subduction zone remained distinct (Figs. 4, 8) and had different source areas (Williams et al., 1993; Pollock et al., 2003) until the Wenlock (ca. 425 Ma) in Newfoundland, when the basin closed along the Dog Bay Line (Fig. 4; Williams et al., 1993). Cross-cutting plutons indicate that the marine fore-deep sequences of the Fredericton belt immediately south-east of the Salinic suture in New Brunswick and adjacent Maine were deformed between 425 and 422 Ma (West et al., 1992; Tucker et al., 2001). An inferred angular unconformity, between the strongly folded, steeply dipping Lower Silurian Botwood Group and the overlying, gently-dipping ca. 423 Ma Stony Lake volcanics (Anderson and Williams, 1970; Dunning et al., 1990) west of the Dog Bay Line, indicates that the terminal phase of the Salinic Orogeny was approximately coeval in Newfoundland. The deep crustal architecture of the Salinic Orogen and the remnants of its west-dipping subduction channel have been imaged on reprocessed seismic lines in Newfoundland (van der Velden et al., 2004). Deformation and metamorphism were particularly intense in the Gander Zone; however, orogenesis also affected Laurentia's hinterland (Cawood et al., 1994; Castonguay and Tremblay, 2003).

Acadian Orogeny

The Acadian Orogeny is due to the collision between Laurentia and Avalonia (e.g. Bird et al., 1970; Bradley, 1983; van Staal, 2005). The Gander-Avalon terrane boundary lies along the Hermitage Bay-Dover Fault in Newfoundland and Caledonia Fault in southern New Brunswick (Figs. 1, 4, 8) (Barr et al., 2003; van Staal et al., 2004b).

Syn- to late tectonic, collision-related Late Silurian-Early Devonian (420-400 Ma) granitoids and associated high-grade metamorphism, pervasive in the leading edge of eastern Laurentia (Ganderia, Figs. 4, 5; e.g. Dunning et al., 1990; Burgess et al., 1995; Schofield and D'Lemos, 2000; Valverde et al., 2000; Valverde-Vaquero et al., 2003), represent a minimum age for the Acadian Orogeny because they are coeval with or postdate cleavage development and dextral shear in the Avalonia-Ganderia boundary zone (Dallmeyer and Nance, 1994; Holdsworth, 1994). The best time constraints are given by the Late Silurian (ca. 422 Ma) inversion of the Mascarene and La Poile basins (O'Brien et al., 1991; McLaughlin et al., 2003). The Mascarene belt is interpreted to represent a back-arc basin to the Lower Silurian arc rocks in the adjacent Kingston belt (Fyffe et al., 1999; Barr et al., 2002). Hence, subduction was to the north-west beneath Laurentia's leading edge (Ganderia) and Avalonia was situated on the lower plate (Fig. 8).

Neoacadian Orogeny

The Neoacadian Orogeny comprises Middle Devonian to Early Carboniferous deformation and metamorphism (Robinson et al., 1998). These tectonic events are related to the docking and orogenesis of the Meguma terrane in Canada (Culshaw and Reynolds, 1997; Hicks et al., 1999; Keppie et al., 2002; van Staal, 2005). Meguma's accretion was dextral oblique and accommodated by the Cobequid-Chedabucto fault system on land in Nova Scotia (Fig. 1). Van Staal (2006) proposed that Meguma was situated on a shallowly dipping downgoing plate (Murphy et al., 1999) and transferred to the overriding Laurentian plate as a result of wedging. This model also explains the absence of Devonian arc magmatism in the Laurentian upper plate.

Metallogeny

Rocks of the Canadian Appalachians are host to a large variety of mineral deposits, ranging from syngenetic volcanic-hosted massive and stockwork sulphide (VMS) deposits to chromite, PGE, gold, pluton-associated mineralization and industrial minerals (mainly asbestos, talc, and building stone). Herein mainly Mississippi Valley-type, VMS, and the various types of mesothermal and epithermal gold mineralization specific to each of the various tectono-stratigraphic zones and synorogenic belts will be described. Their tectonic setting will be discussed in light of new ideas concerning Appalachian tectonic evolution. For example, mesothermal and epithermal gold mineralization mainly formed during post-Ordovician orogenic events in the tectonically active core of the orogen (central mobile belt) as a result of the Silurian-Devonian accretion of the peri-Gondwanan microcontinents to Laurentia. Mineralization is commonly associated with major, accretion-related faults and/or reactivated older faults west of the Red Indian Line (Fig. 1).

The other types of mineral occurrences in the Canadian Appalachians are discussed in other contributions within this project. Extensive, relevant descriptions can also be found in Swinden and Dunsworth (1995).

Humber Zone Placer and Mississipi Valley-Type Mineralization

Mineral deposits in the Humber Zone (excluding pre-Appalachian mineral deposits occurring in exposed Grenvillian basement) can be divided into those that occur in Neoproterozoic-Cambrian rift-related sedimentary and volcanic rocks, and those that occur in the overlying dominantly calcareous rocks of the passive margin phase.

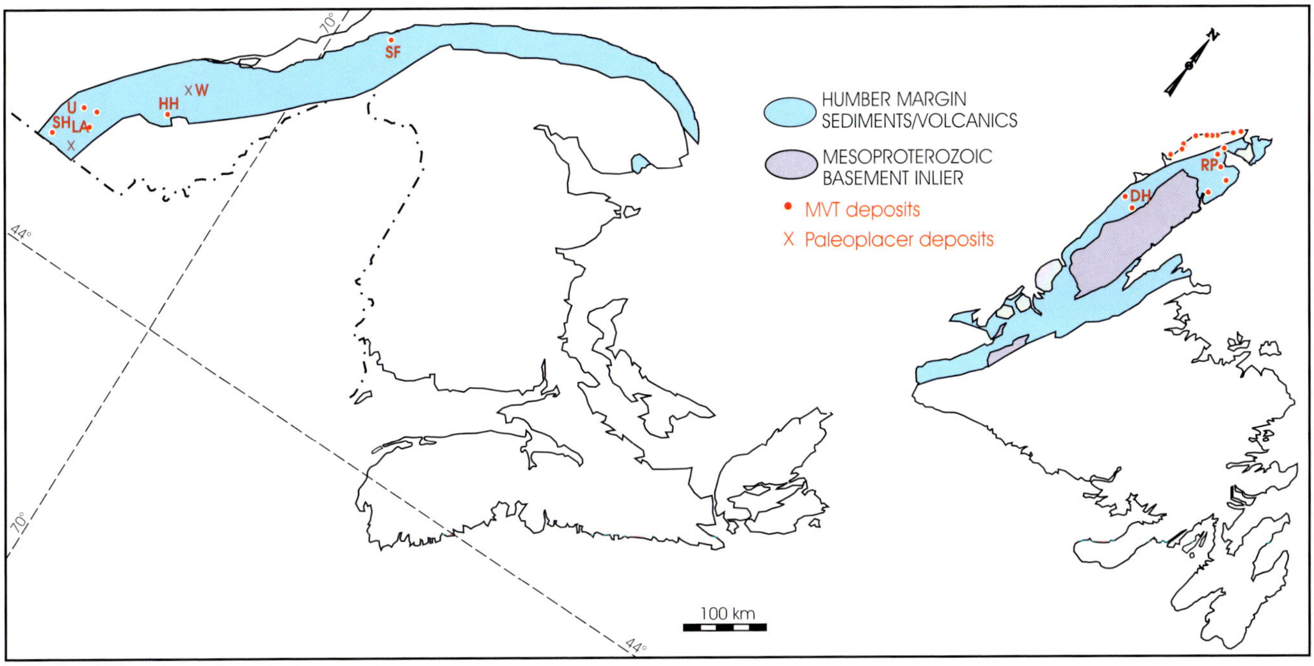

FIGURE 9. The Humber Zone with the distribution of the various mineral deposits discussed in the text. DH: Daniel's Harbour; HH: Harvey Hill; LA: Lord Aylmer; RP: Round Pond; SF: St. Fabien; SH: St. Armand-Highwater; U: Upton; W: Wares.

Rift-Related Rocks

Mineral deposits in the rift-related sandstones include marine Fe-Ti-Zr paleoplacer deposits that accumulated along the beaches and shoals of the Humber margin. They occur principally in Quebec (e.g. Ware deposit, Fig. 9) and are commonly associated with late Neoproterozoic rift-related volcanic rocks, which may have been an important source of the heavy minerals. Titaniferous magnetite, ilmenite, titanite, rutile, zircon, monazite, and tourmaline are the principal heavy minerals found in the deposits.

The rift-related sedimentary and volcanic rocks in southern Quebec are locally also host to important Cu mineralization (Fig. 10). Chalcopyrite and bornite are the principal ore minerals. The Harvey Hill deposit in southern Quebec has been mined and has yielded 400,000 tonnes of ore grading 1.2% Cu and 3.6 g/t Ag. Various syngenetic to epigenetic or hybrid models have been proposed to explain this mineralization.

Passive Margin-Related Rocks

The Cambrian-Ordovician shelf carbonate-dominated successions of the Humber margin are host to important Zn±Pb±Ba mineralization, both in Newfoundland and southern Quebec (Figs. 2, 3). The mineral deposits are generally considered to have formed by Mississippi Valley-type (MVT) metal-bearing brines generated and expelled during the Early to Late Ordovician Taconic Orogeny as a result of the emplacement of large thrust sheets onto the Humber margin (Figs. 2, 3, 7). In Newfoundland, Zn±Pb mineral deposits are commonly hosted in strata that experienced extensive dolomitization, karstification, and brittle fracturing. They principally occur in fractures, stylolites, or veinlets in Cambrian dolostone and as disseminated to semi-massive bodies in pseudobreccias and open spaces in dolomitized Ordovician limestone. The largest and best known deposit is the Daniel's Harbour mine (Fig. 9), which was mined between 1975 and 1990 and contained 6.6 million tonnes of ore grading 7.9% Zn (Swinden and Dunsworth, 1995). Another important occurrence is the Round Pond deposit west of Hare Bay (Fig. 9) with 400,000 tonnes grading 2% Zn.

In Quebec, the stratabound Upton Ba-Zn-Pb deposit (Fig. 9) with 960,000 tonnes of ore grading 46.5% Ba, 1.94% Zn, 0.59% Pb, 0.15% Cu, and 13.5 g/t Ag is the best known Mississippi Valley-type occurrence (Paradis et al., 2004). The deposit is hosted by Lower Ordovician crinoidal limestone and the sulphides occur in disseminations and aggregates filling fractures and open spaces, along stylolites and barite and calcite grain boundaries. Basinal metal-bearing brines migrated through the margin strata during Taconic loading and sulphides were deposited when the fluids encountered H_2S-rich traps in the carbonate host rock (Paradis et al., 2004). The Saint-Fabien deposit is an important MVT occurrence (Beaudoin et al., 1989). Here the mineralization mainly occurs as stratabound barite-sphalerite-galena veins and as disseminations in cavities in partially dolomitized limestone conglomerates and sandstones of the Upper Cambrian Saint-Damase Formation (Lavoie et al., 2003).

Notre Dame Subzone VMS Deposits

The Lower to Middle Cambrian (510-501 Ma) ensimatic rocks of the Lushs Bight oceanic tract (LBOT) are the oldest known rocks in the Notre Dame Subzone (Figs.1, 2, 10). They mainly represent the upper crustal remnants (basalts and sheeted dykes) of an obducted ophiolitic infant arc terrane that was emplaced on the Dashwoods microcontinent during the Late Cambrian, and where the latter was absent, directly onto the Humber margin during the Early Ordovician. The LBOT is host to numerous economically

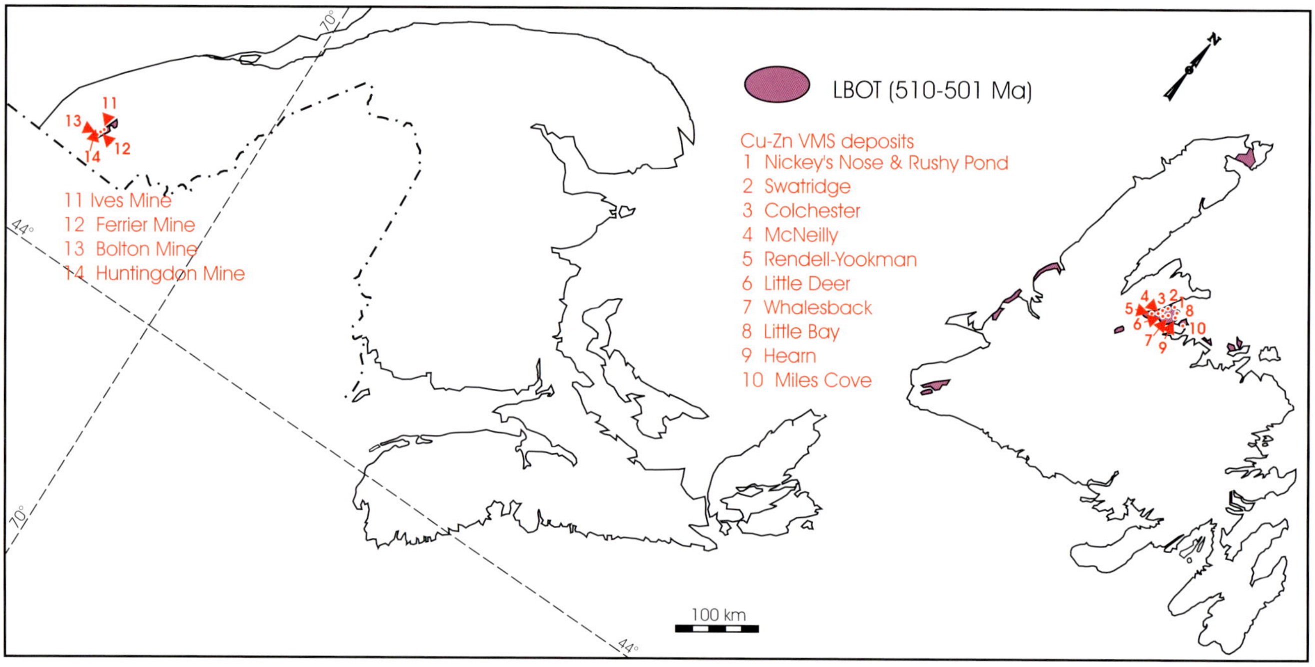

FIGURE 10. VMS deposits in the Lushs Bight oceanic tract (LBOT).

significant syngenetic Cu±Zn±Au VMS deposits (Fig. 10), which are commonly spatially associated with occurrences of boninitic lavas and/or dykes. In the Lushs Bight type area of northern Newfoundland, the Little Bay (>3 M tonnes of 0.8-2.5% Cu), Whalesback (3.8 M tonnes of ~1% Cu), Little Deer (~75,000 tonnes of 1.3% Cu), Swatridge, Colchester, McNeilly and Miles Cove deposits have been mined in the past (Swinden et al., 1995).

The only known potential equivalent of the LBOT in southern Quebec is the ca. 504 Ma Mt. Orford ophiolite complex (Figs. 1, 3, 10) (David and Marquis, 1994; Huot et al., 2002), which is host to four mined Cu±Zn±Au VMS deposits: the Huntington (11 M tonnes of 0.9% Cu, 0.062 g/t Au, and 0.62 g/t Ag with unmined reserves of 800,000 tonnes of 0.88% Cu), Bolton, Ferrier, and Ives mines (Swinden and Dunsworth, 1995).

The Baie Verte oceanic tract (BVOT) in Newfoundland is a significantly younger Early Ordovician (ca. 490-476 Ma) belt of ophiolitic infant arc rocks with a cover consisting of arc/back-arc volcanic rocks (Snooks Arm arc). The BVOT formed during Early Ordovician closure of the Humber Seaway (Figs. 5, 7) and is mainly preserved as a fault bounded wedge along the Baie Verte-Brompton Line (BBL) between Grand Lake and the Baie Verte peninsula (Fig. 11). The tectonically related Bay of Islands Complex was emplaced as part of a large coherent allochthon (Humber Arm allochthon) onto the Humber margin during the Arenig and occurs to the west of the BBL (Figs. 1, 11). The ophiolitic members of BVOT host numerous economically important Cu±Zn±Au±Ag VMS deposits in Baie Verte Peninsula (Fig. 11), of which Tilt Cove (> 8 M tonnes), Betts Cove (118,000 tonnes of 10% Cu), and Rambler-Ming (4.5 M tonnes) have been mined (Hibbard, 1983). The VMS deposits are commonly associated with boninitic volcanic rocks. The Bay of Island Complex is also host to numerous cupriferous VMS deposits of which the York Harbour mine has yielded 91,000 tonnes of ore with reserves of 200,000 tonnes grading 2.68% Cu, 8.25% Zn, 35 to 70 g/t Ag, and < 1 g/t Au. Gregory River, Steep Rock, and Crabb Brook are related deposits (Swinden et al., 1995).

Several small cupriferous massive sulphide deposits hosted by the boninite and island arc tholeiite lavas of the Sherbrooke domain of the Ascott Complex (Tremblay et al., 1989; Tremblay, 1992) represent examples of BVOT–related VMS mineralization in southern Quebec (Swinden and Dunsworth, 1995).

There is no known significant mineralization in the infrastructure of the Notre Dame arc in Newfoundland. The calc-alkaline ensialic volcanic rocks of the Middle Ordovician Ascott Complex and Weedon Formation in southern Quebec (Figs. 1, 3, 12; Tremblay et al., 1989), however, are host to several important Zn±Pb±Cu VMS deposits. Deposits mined in the past include Eustis-Albert (ca. 1.6 M tonnes of ore grading 2.7% Cu), Suffield (508,000 tonnes grading 7% Zn, 0.52% Pb, 0.91% Cu), the Aldermac-Moulton Hill (300,000 tonnes grading 5.32% Zn, 1.89% Pb, 1.17% Cu), and Cupra-d'Estrie (2,180,000 tonnes of ore grading 2.4% Zn, 0.2% Pb, 1.9% Cu, 37.7 g/t Ag, and 0.5 g/t Au) (Swinden and Dunsworth, 1995).

The Middle to Upper Ordovician Magog Group, generally thought to represent the upper part of a syncollisional fore-arc sequence (Cousineau and St. Julien, 1992; Schroetter et al., 2003), contains several VMS deposits hosted by black shales and felsic tuffs. The largest of these is the Champagne deposit with reserves of 172,000 tonnes grading 2.68% Zn, 0.45% Pb, 0.4% Cu, 19.7 g/t Ag, and 2.1 g/t Au. The Bolton volcanic and sedimentary rocks in the BVOT, which occur further south near Lake Memphremagog, supposedly represent the lower part of this fore-arc sequence, because they occur within the St. Daniel mélange. They host the polymetallic (Zn-Cu-Pb-Sn-Sb-Ag) Memphremagog massive sulphide deposit (Fig. 11; Trottier et al., 1991). Fore-arc

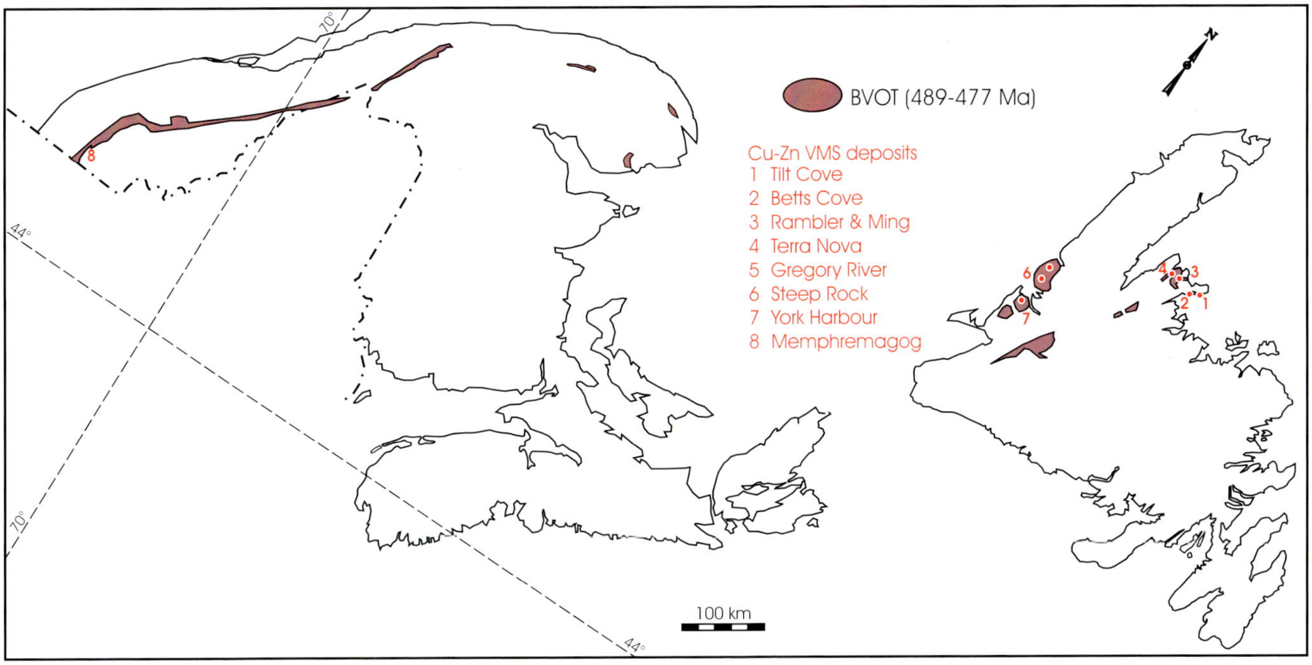

FIGURE 11. VMS deposits in the Baie Verte oceanic tract.

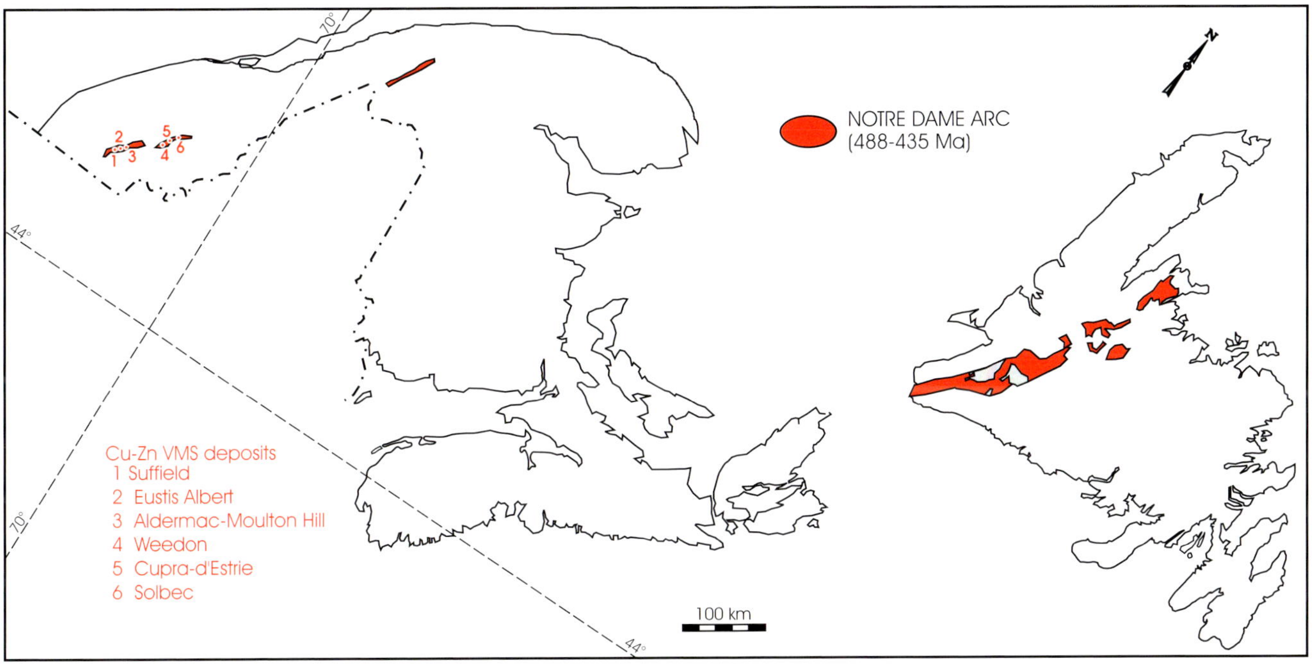

FIGURE 12. VMS deposits in the Notre Dame arc.

basins, particularly those formed during obduction and collision, are not typically volcanically active. The igneous rocks and associated VMS deposits of the St. Daniel mélange and Magog Group are therefore better interpreted as a rift sequence formed in either a back-arc basin (Kim et al., 2003) or a transtensional basin (Trottier et al., 1991); the latter possibly associated with slab break-off.

No VMS deposits are known in the ca. 480 Ma Annieopsquotch ophiolite belt of the Annieopsquotch accretionary tract (AAT). The ca. 473 Ma ensialic Buchans Arc, formed during west-directed subduction beneath a tectonic sliver of Dashwoods lithosphere incorporated in the AAT (Figs. 2, 5; Zagorevski et al., 2003, 2006a), is host to the important Buchans VMS deposit (Fig. 13), which yielded 16,196,876 tonnes of high grade ore (14.51% Zn, 7.56% Pb, 1.33% Cu, 126 g/t Ag, and 1.37 g/t Au) during mining between 1928 and 1984 (Thurlow, 1990). Pilley's Island, Gullbridge (3 M tonnes of ore with 1.1% Cu), Lake Bond, and Shamrock (Fig. 13) are other VMS deposits in this arc sequence (Swinden and Dunsworth, 1995). The Buchans arc was rifted during the Middle Ordovician, which led to the opening of a small back-arc basin (Skidder back-arc, Fig. 7)

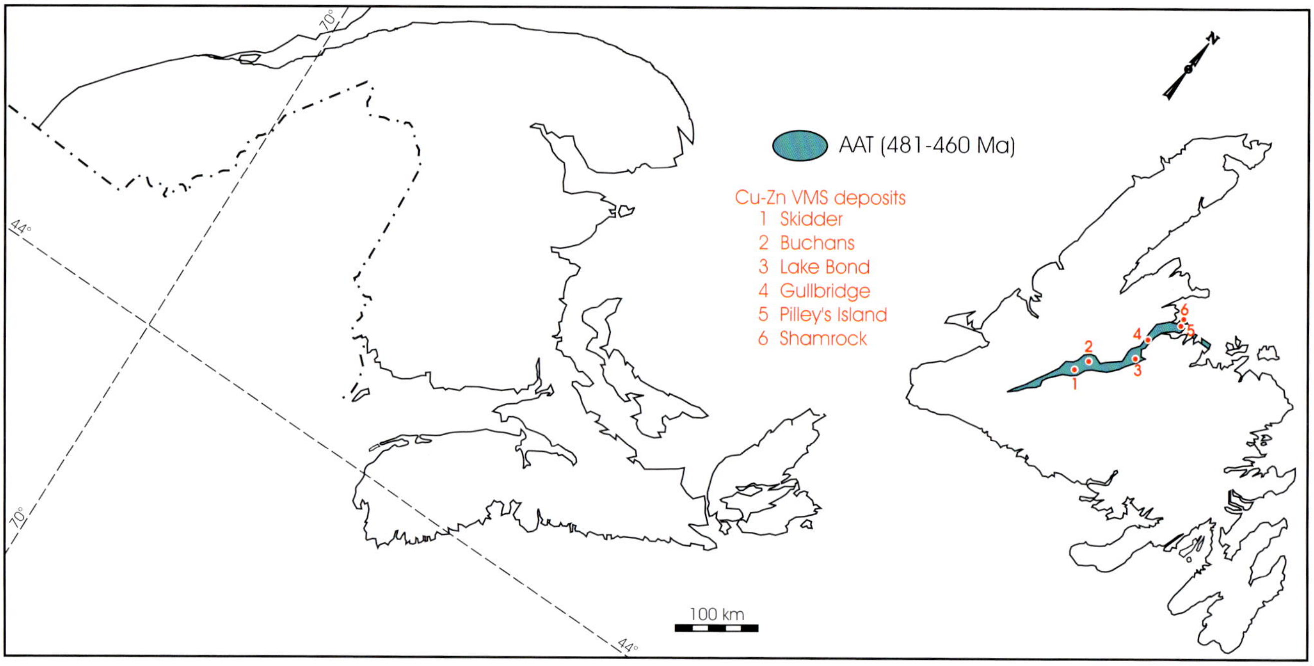

Figure 13. VMS deposits in the Annieopsquotch accretionary tract.

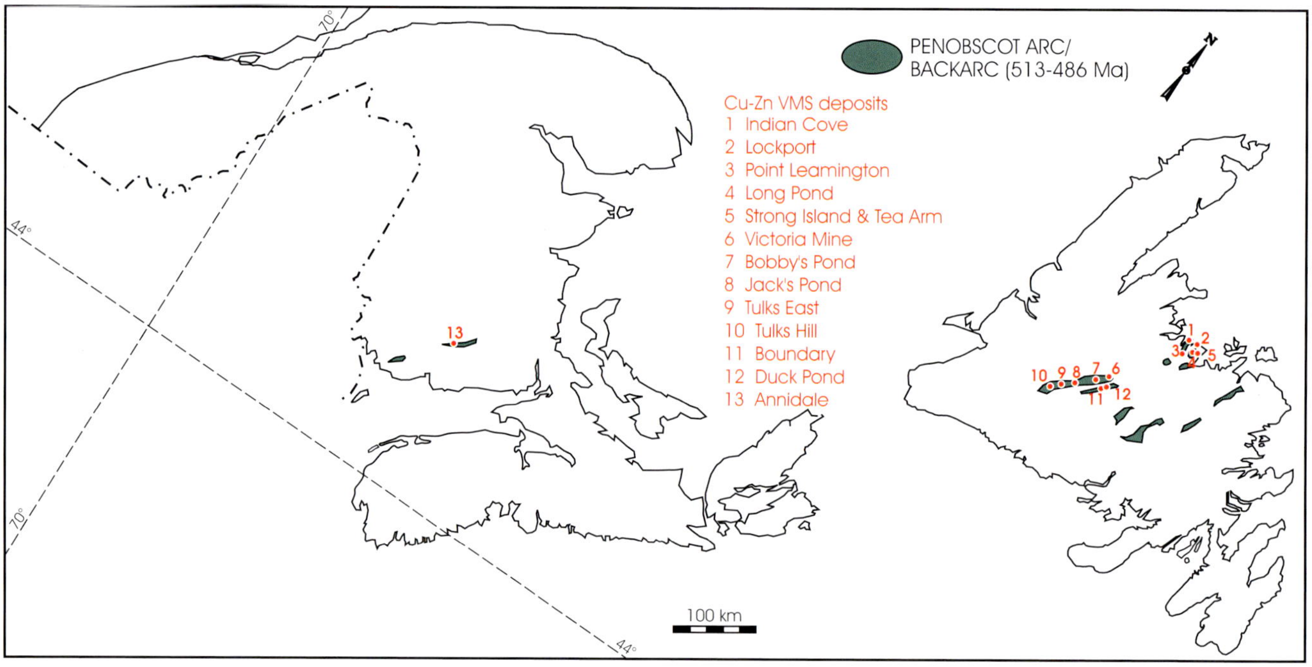

Figure 14. VMS deposits in the Penobscot arc.

at ca. 465 Ma and migration of the active part of the arc trenchwards. The ensimatic Skidder basalt is host to a cupriferous VMS deposit with reserves of approximately 200,000 tonnes grading approximately 2% Cu and 2% Zn.

Exploits Subzone VMS Deposits

The oldest known VMS deposits of the Exploits Subzone occur in the Lower Cambrian (ca. 513 Ma) Tally Pond volcanic rocks (Penobscot arc) of the Victoria Lake Supergroup in central Newfoundland (Figs. 1, 4). Felsic volcanic rocks host the Duck Pond and Boundary deposits (Fig. 14), of which the Duck Pond has reserves of approximately 4.3 M tonnes grading 3.58% Cu, 6.63% Zn, 1.05% Pb, 68.31 g/t Ag, and 1 g/t Au. A slightly younger phase of the Penobscot arc is represented by the Upper Cambrian predominantly felsic volcanic rocks of the Tulks Group (ca. 498 Ma), which host the Tulks Hill, Tulks East, Bobby's Pond and Victoria Mine VMS deposits. The economically most important, the Tulks Hill deposit, contains 750,000 tonnes grading 5 to 6% Zn, 2% Pb, 1.3% Cu, 41 g/t Ag, and 0.4 g/t Au. Equivalent rocks occur in the Annidale belt in southern New Brunswick, which hosts several cupriferous massive sulphide deposits.

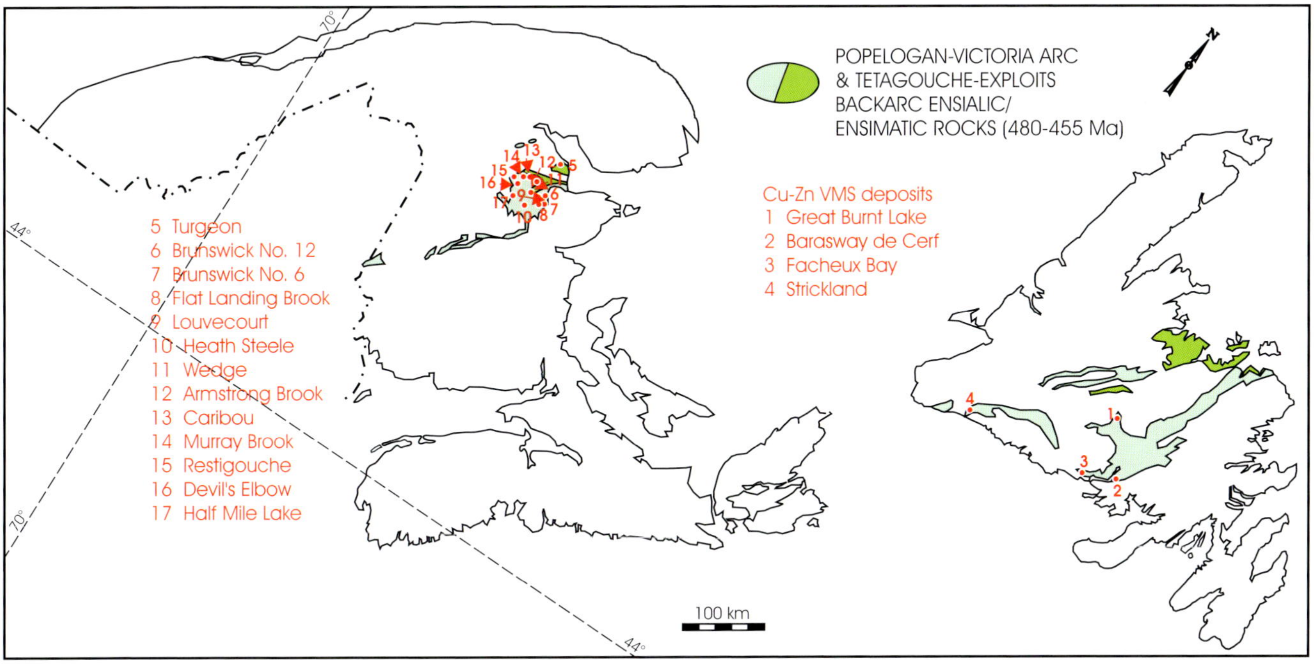

FIGURE 15. VMS deposits in the elements of the Popelogan/Victoria arc – Tetagouche/Exploits back-arc system.

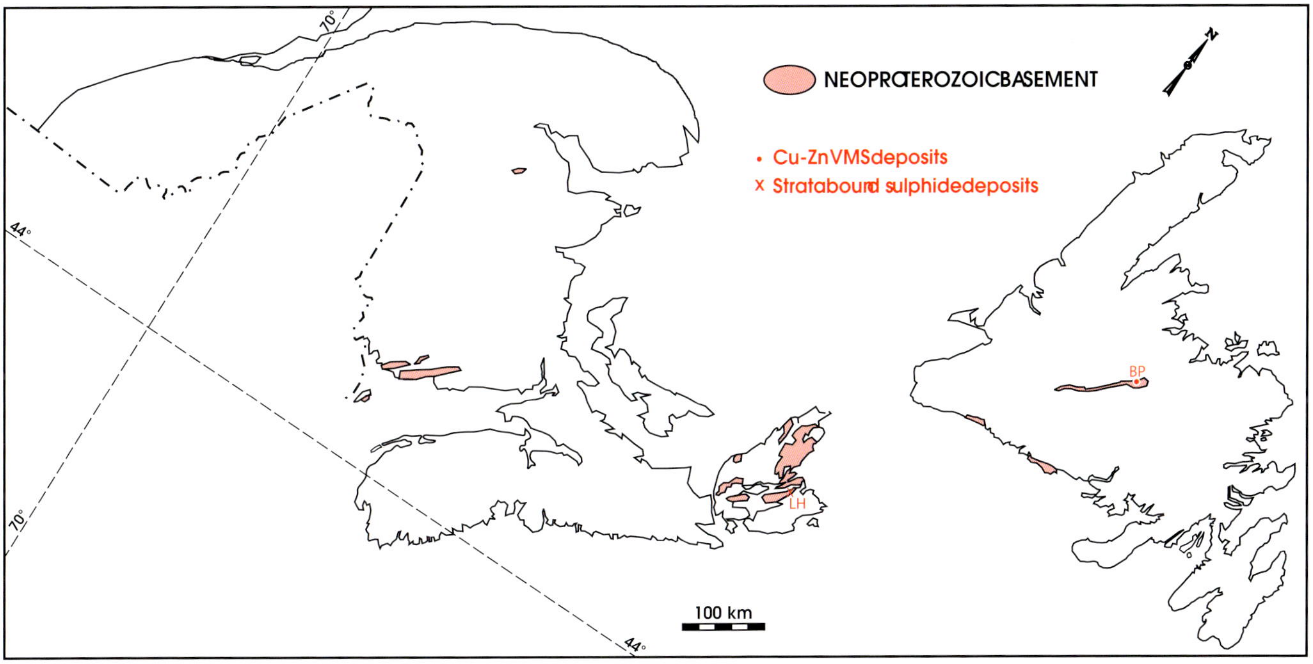

FIGURE 16. Mineral deposits in Ganderian Neoproterozoic basement.

The Annidale belt probably formed in the Penobscot back-arc basin (McLeod et al., 1992).

Early Tremadoc (488-485 Ma), predominantly mafic volcanic rocks of the Pats Pond (Victoria Lake Supergroup), the Wild Bight and Exploits groups represent the youngest phase of the Penobscot arc. The Wild Bight Group contains the Point Leamington, Lockport, Indian Cove, and Long Pond deposits; the Exploits Group contains the Tea Arm and Strong Island deposits (Fig. 14). Point Leamington, the largest of these deposits, is hosted by rhyolite and contains 13.8 M tonnes of massive sulphide grading 1.92% Zn, 0.48% Cu, 18.1 g/t Ag, and 0.9 g/t Au. The VMS deposits are generally associated with refractory volcanic rocks (Swinden and Dunsworth, 1995).

After closure of the Penobscot back-arc basin, arc activity resumed during the Arenig both in Newfoundland and New Brunswick. The composition of the volcanic rocks suggests that the arc was highly extensional, and this phase is referred to as the Popelogan arc in New Brunswick and the Victoria arc in Newfoundland. Its products have been preserved in the stratigraphically upper parts of the Victoria Lake Supergroup, Wild Bight and Exploits groups in

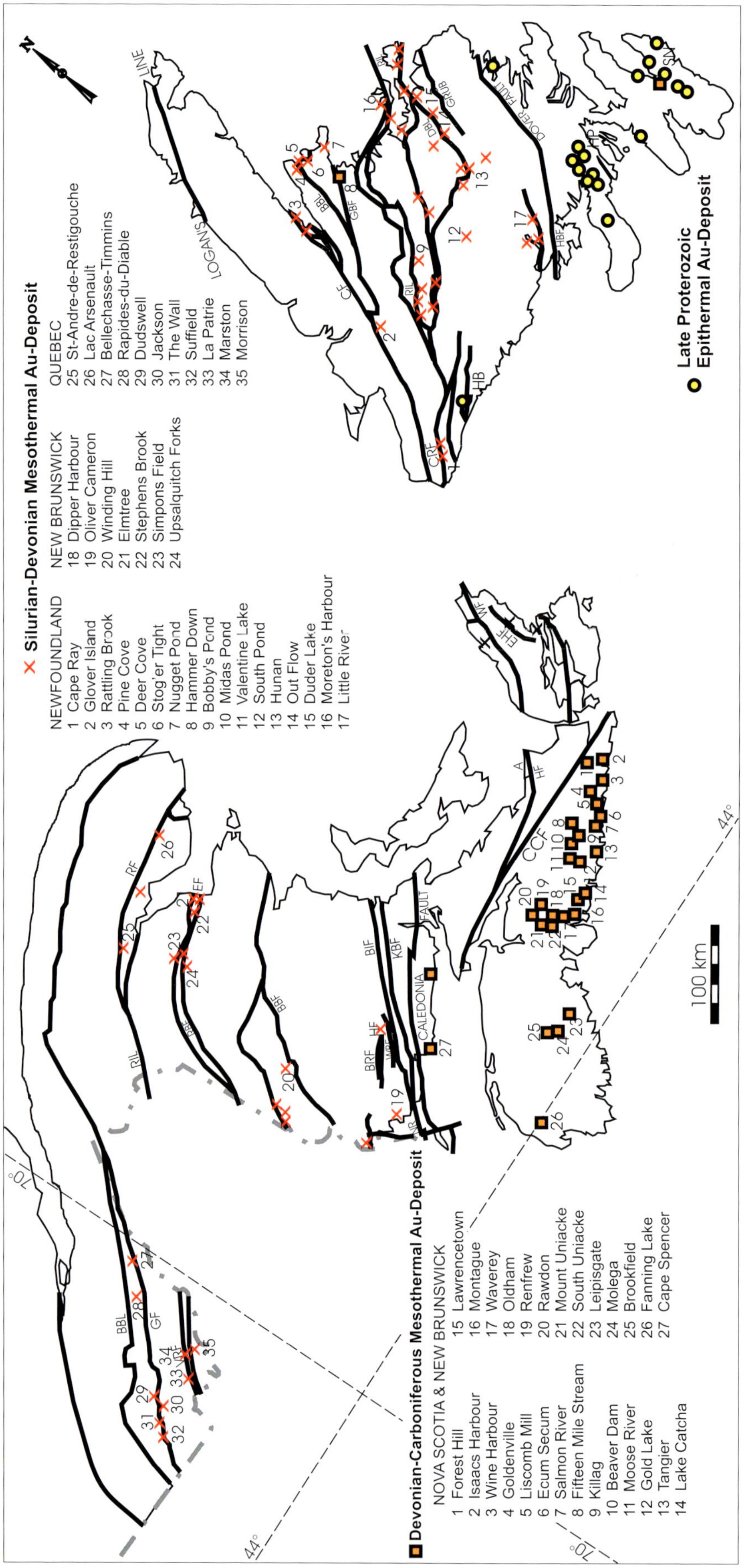

FIGURE 17. Epithermal and mesothermal gold deposits in the Canadian Appalachians. HB: Hope Brook mine; HP: Hickey's Pond prospect; SN: Steep Nap prospect.

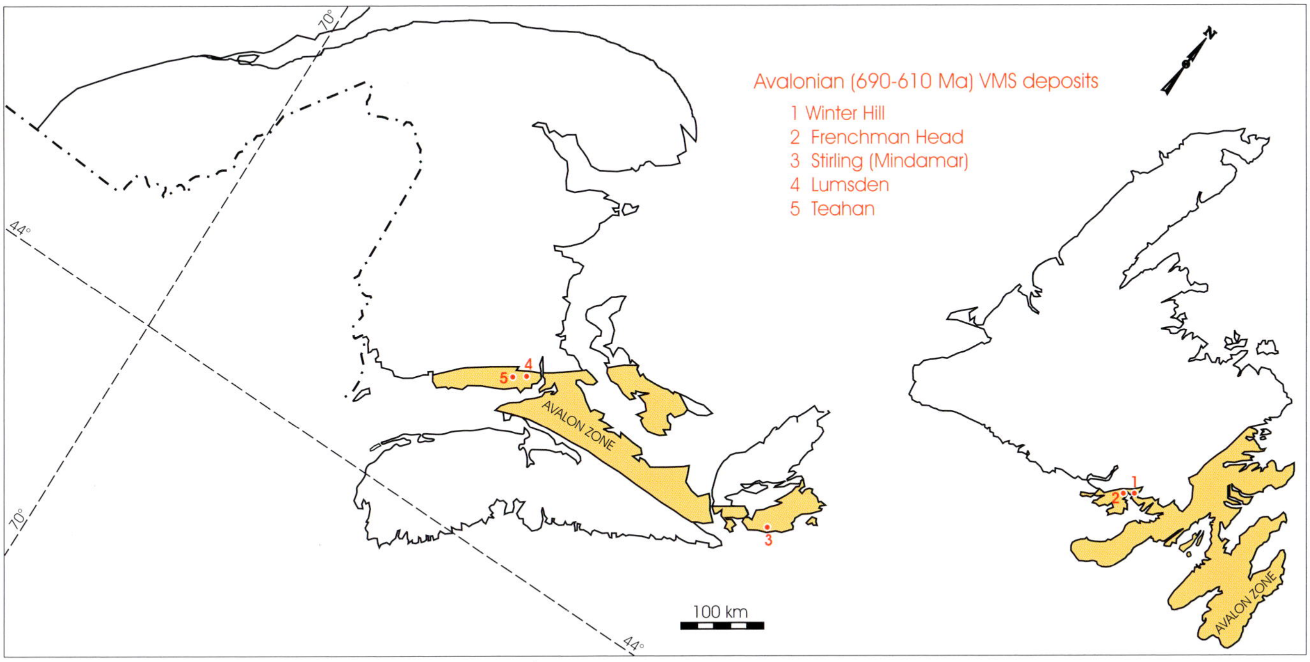

Figure 18. Mineral deposits in the Avalon Zone (gold deposits excluded).

Newfoundland. The Oak Mountain Formation (Fig. 6; Meductic Group) and Goulette Brook Formation (Balmoral Group) represent the Popelogan arc in New Brunswick. Neither of these units contains significant VMS mineralization. Consanguineous calc-alkaline plutons in New Brunswick, however, contain Cu-Mo-Au porphyry mineralization. Economically important VMS deposits are abundant in the associated Tetagouche-Exploits back-arc basin, both in Newfoundland and New Brunswick. In Newfoundland, these include the Strickland (260,000 tonnes grading 5.25% combined Pb and Zn), Facheux Bay, and the Barasway de Cerf deposits, hosted by the Bay du Nord and Baie d'Espoir groups (Figs. 1, 4, 15). In New Brunswick, they include the economically very important VMS deposits of the Bathurst Mining Camp (van Staal et al., 2003b). The giant Brunswick No. 12 Mine contained 229 M tonnes of ore grading 7.66% Zn, 3.01% Pb, 0.46% Cu, 91 ppm Ag, and 0.46% Au. The VMS deposits hosted by ensialic back-arc volcanic rocks are relatively rich in Pb. Those hosted by enismatic mafic volcanic rocks, representing remnants of back-arc oceanic crust, are cupriferous. The Great Burnt Lake deposit (680,000 tonnes grading 2-3% Cu) hosted by oceanic basalts of the Cold Spring Pond Formation of central Newfoundland and the Turgeon deposits (1-2 M tonnes grading 4% Zn and 1.5% Cu in Powerline Zone) hosted by the ophiolitic Fournier Group in northern New Brunswick belong to this group (Figs. 1, 4, 8, 15).

Pre-Silurian Mineral Deposits of the Gander Zone

The Cambrian-Ordovician sedimentary and volcanic rocks typical of Ganderia show little or no significant mineralization other than post-accretion, mesothermal gold deposits and pluton-associated mineralization that occur throughout the central mobile belt (see below). However, Ganderian Proterozoic basement rocks locally contain important mineralization that presumably formed while still part of Gondwana. The volcanic rocks enveloping the ca. 563 Ma Crippleback monzodiorite host the Burnt Pond VMS occurrence (BP in Fig. 16). In Cape Breton, Proterozoic marble in the Bras d'Or terrane hosts the Lime Hill stratabound sulphide deposit (LH in Fig. 16), mainly consisting of bands of massive and disseminated spahalerite, pyrrhotite, and pyrite. It has reserves of 136,000 tonnes grading 9% Zn or 1.8 M tonnes grading 2% Zn. The deposit is either a skarn or a syngenetic, carbonate-hosted massive sulphide deposit (Swinden et al., 1995). Polymetallic skarn deposits, however, are locally present in the Bras d'Or terrane where they are associated with latest Neoproteozoic (ca. 560 Ma) calc-alkaline plutons. Late Neoproterozoic arc plutons and volcanic rocks of the Hermitage flexure host the Hope Brook gold mine (Fig. 17), which contained 11,200,000 tonnes of ore grading 4.54 g/t Au. The gold is probably Precambrian in age.

Pre-Silurian Mineral Deposits of the Avalon Zone

Avalonia contains several types of mineral deposits that formed before its late Silurian accretion to Laurentia. The oldest are several VMS deposits (Fig. 18) that are hosted by Neoproterozoic volcanic rocks, which formed during subduction processes in the periphery of Gondwana, before opening of the Iapetus Ocean.

Avalonian VMS occurrences in Newfoundland comprise the Winter Hill and Frenchman Head deposits hosted by the ca. 680 Ma Tickle Point Formation. In Cape Breton, the coeval Stirling volcanic rocks host the Mindamar or Stirling deposit of which 1.1 M tonnes of ore, averaging 6.4% Zn, 1.5% Pb, 0.74% Cu, 75.2 g/t Ag, and 1.03 g/t Au were mined. In New Brunswick's Avalon Zone, the younger, ca. 630 to 600 Ma Broad River Group host the cupriferous Teahan and Lumsden deposits (Fig. 18).

Late Neoproterozoic (ca. 620 Ma), calc-alkaline granitoids host disseminated Cu and Mo porphyry mineralization in Newfoundland (e.g. Holyrood Granite) and Cape Breton

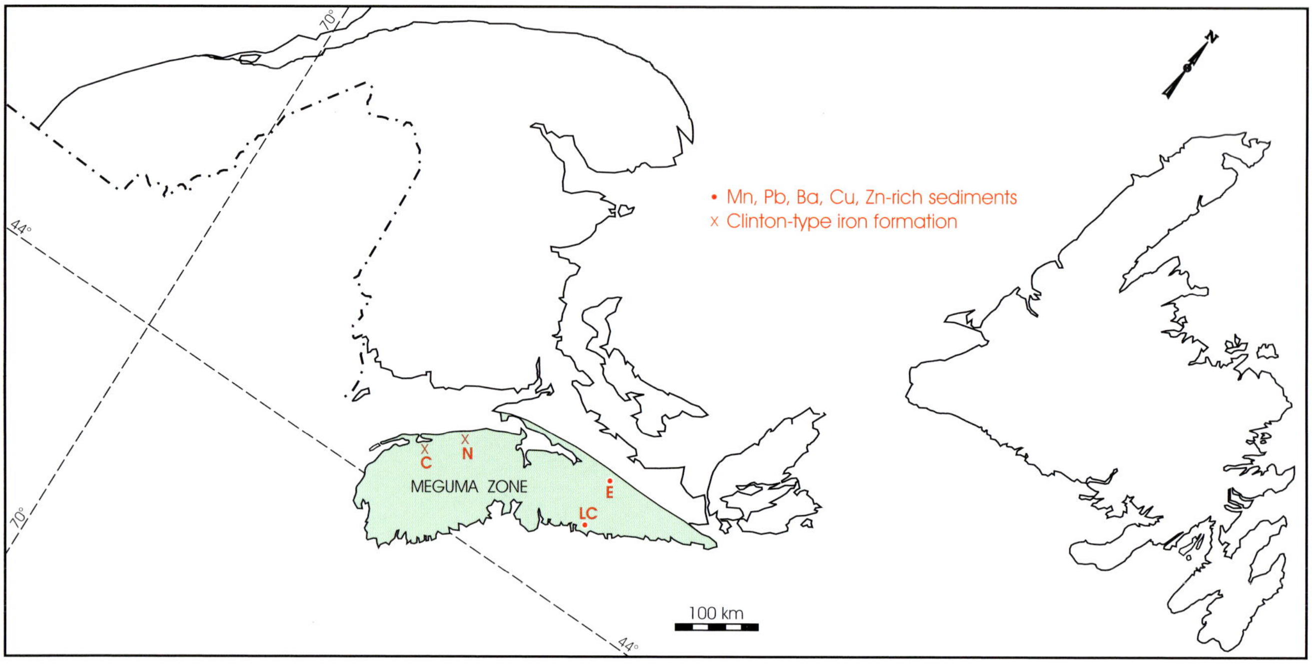

FIGURE 19. Mineral deposits in the Meguma Zone (gold deposits excluded). C: Clementsport; E: Eastville; LC: Lake Charlotte; N: Nictaux-Torbrook.

Island (e.g. Coxheath Hills pluton). Circa 620 Ma pyrophyllite alteration zones developed in predominantly felsic volcanic rocks throughout Newfoundland's Avalon Zone (Fig. 17) are associated with epithermal gold-silver mineralization (e.g. O'Brien et al., 1999).

The Lower to Middle Cambrian sedimentary strata of the Avalon Zone in Newfoundland contain Mn-rich horizons, locally reaching grades of 33.35% Mn. The manganese is probably derived from the weathering of the underlying Neoproterozoic rocks (Swinden and Dunsworth, 1995). The Lower Ordovician siliciclastic rocks of the Bell Island and Wabana groups host the Clinton-type, oolitic hematite iron deposits, which were mined between 1895 and 1966. Potential reserves beneath Conception Bay probably exceed 2 billion tones (Miller, 1983).

Pre-Middle Devonian Mineral Deposits of the Meguma Zone

Meguma contains several types of mineral deposits that formed before its late Early Devonian accretion to Laurentia. The Cambrian-Ordovician Meguma Supergroup underlies most of the exposed Meguma Zone. The Meguma Supergroup is generally regarded as the remnant of a Gondwanan passive margin sequence, probably deposited off the coast of Late Proterozoic-Early Paleozoic North Africa (Schenk, 1997). The transition from the sandy Goldenville to the shaly Halifax groups contains a manganese-rich unit (Fig. 19) that is also enriched in Ba, Cu, Pb, and Zn (Swinden and Dunsworth, 1995). Clinton-type, sedimentary iron deposits occur in the Lower Devonian Torbrook Group (Fig. 19).

Salinic-Acadian Gold Mineralization in the Central Mobile Belt

Mesothermal, and to a much less extent epithermal, gold deposits are common in the Dunnage and Gander zones, and also the eastern part of the Humber Zone (Fig. 17). Gold commonly occurs in sulphide-rich quartz and quartz +carbonate veins or as disseminations in pervasively altered wall rock (Dubé, 1990). Gold in the Humber Zone commonly occurs in veins associated with shear zones developed in the sedimentary and volcanic rocks of the Silurian Sops Arm Group of the Springdale belt, which represents part of an overstep sequence deposited above the Humber Zone and Notre Dame Subzone (Williams, 1995). The shear zones are spatially associated with a splay of the Cabot fault system. The Cabot fault accommodated strike-slip movements as early as the Late Ordovician (Brem et al., 2003) and remained active into the Carboniferous. Mineralization is Late Silurian or younger.

The most significant vein deposits in the Notre Dame Subzone are the Cape Ray, Deer Cove, Pine Cove, and Hammer Down (Fig. 17). The mineralized veins are localized in brittle-ductile faults associated with regional Silurian to Devonian transpression (e.g. Dubé et al., 1996). The Stog'er Tight deposit in the Baie Verte Peninsula, hosted by a pervasively altered Tremadoc (ca. 483 Ma) gabbro sill (Ramezani et al., 2002) related to the Snooks Arm arc is the best example of the altered wall-rock-type (Fig. 17).

The Lac Arsenault and Saint-André-de-Restigouche are the most important mesothermal vein deposits in the Notre Dame Subzone in the Quebec Gaspésie (Fig. 17). Mineralization is associated with the Grand Pabos transcurrent (dextral) fault system, which cuts Upper Ordovician to Lower Silurian siliciclastic and calcareous rocks of the Honorat and Matapedia groups of the Gaspé belt. This fault system was predominantly active during the Devonian. Skarn mineralization is also associated with this fault system. The largest vein of the Lac Arsenault deposit contains approximately 40,000 tonnes grading 15.42 g/t Au, 197 g/t Ag, 6.6% Pb, and 3.5% Zn (Swinden and Dunsworth, 1995). Mesothermal deposits in southeastern Quebec are also associated with major fault zones, particularly the BBL,

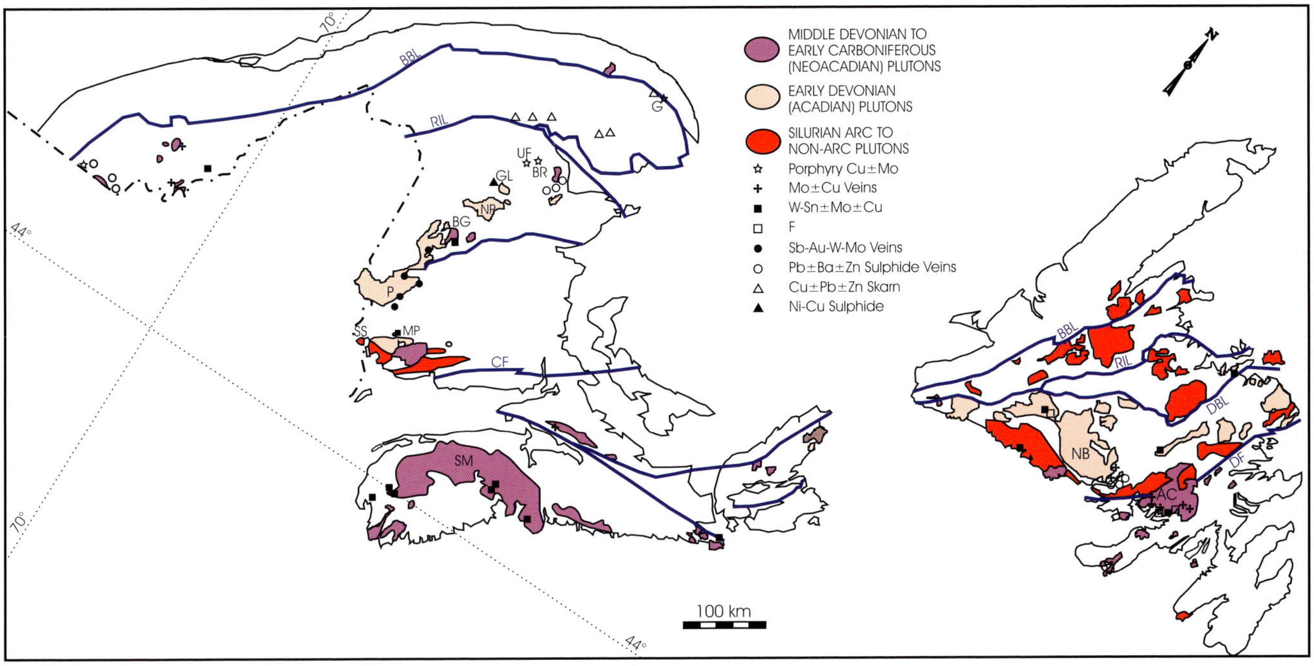

Figure 20. Mineral deposits associated with the Acadian and Neoacadian plutons. AC: Ackley granite; BBL: Baie Verte-Brompton Line; BG: Burnthill granite; BR: Benjamin River porphyry; CF: Caledonia fault; DBL: Dog Bay Line; DF: Dover fault; G: Gaspé Mine; GL: Goodwin Lake gabbro; MP: Mt. Pleasant; NB: North Bay batholith; NP: North Pole granite, P: Pokiok batholith; RIL: Red Indian Line; SS: St. Stephen gabbro; UF: Upsalquitch Forks porphyry.

Guadeloupe, and Victoria River fault zones (Fig. 17). The noteworthy Bellechasse-Timmins deposit, with estimated reserves of 1.2 M tonnes of 2 g/t Au, occurs in quartz-carbonate veins hosted by diorite sills that have intruded the Magog Group.

Mesothermal gold deposits are common throughout the Exploits Subzone and Gander Zone of the Canadian Appalachians. In Newfoundland, there are numerous shear zone-related auriferous veins in the Victoria Lake Supergroup. In particular, the unconformable contact between the late Neoproterozoic plutons at the base of the Victoria Lake Supergroup and the Lower to Upper Silurian Botwood Group was a favourable site for strain localization and formation of auriferous veins. In general, strain localization along the contacts of volcanic units in the Victoria Lake Supergroup with the overlying Badger Group led to shear zones, intense alteration, and formation of auriferous veins. The Midas Pond gold deposit is an example of this (Fig. 17). Shearing and vein formation are associated with Silurian and/or Early Devonian transpression, which was accommodated both by reverse and strike-slip faults. Many of these faults are spatially associated with, and/or intersected by, Siluro-Devonian felsic plutonic rocks. Hence, the source of the ore-bearing fluids may be partly magmatic; whereby the fluids were channeled along the faults.

An extensive belt of auriferous veins, locally also containing antimony (e.g. Hunan deposit, Fig. 17), is associated with the Dog Bay Line and also occurs east of it in rocks of the Exploits Subzone and Gander Zone. The veins formed mainly during latest Silurian or Early Devonian dextral transpression associated with Acadian docking of Avalonia (Holdsworth, 1994) and widespread felsic plutonism. Many of the gold occurrences have characteristics of Carlin-type deposits.

Mesothermal gold deposits in the Exploits Subzone and Gander Zone of New Brunswick are mainly associated with segments or splays of the Rocky Brook-Millstream, Woodstock, Bamford Brook, Honeydale, Basswood Ridge, and Wheaton Brook fault systems (Fig. 1). An example is the relatively large Elmtree deposit in northern New Brunswick, which comprises quartz-carbonate auriferous veins hosted by sheared and altered Ordovician gabbro affected by the Elmtree fault, a splay of the Rocky Brook-Millstream fault. The deposit contains 350,000 tonnes grading 4.46 g/t Au (Tremblay and Dubé, 1991). Most of the above mentioned faults accommodated significant transcurrent and reverse movements during Acadian dextral transpression (van Staal and de Roo, 1995). Auriferous quartz veins occur also in the Gander Zone of Cape Breton Island (Aspy Zone), where they are associated with Acadian transpression localized in fault zones such as the Eastern Highlands shear zone and the Wilkie Brook fault.

Neoacadian Gold Mineralization in the Meguma Zone

Mesothermal, gold-bearing veins in the Meguma Zone generally occur in the Goldenville Group (Fig. 17), where relatively thick beds of fine-grained, sulphide-rich shale are interlayered with sandstone. The veins have been mined since 1862. The gold-bearing veins occur in several orientations, but were preferentially emplaced in the hinges of anticlines that were modified during a late-stage amplification of the regional upright folding. This was synchronous with ca. 370 Ma granite plutonism and low-pressure regional metamorphism (Keppie et al., 2002).

Siluro-Devonian (Acadian) Pluton-Related Mineralization

The central mobile belt of the Canadian Appalachians, particularly the Exploits Subzone and the Gander Zone, contains a large volume of Late Silurian-Early Devonian (422-

395 Ma) felsic to intermediate and minor mafic plutons. Plutons of this age range are rare or absent in the Notre Dame Subzone. They have a broad spectrum of compositions and, not surprisingly, also produced a wide variety of mineral deposits (Fig. 20).

In Newfoundland, peraluminous granites in the Gander Zone and adjacent Exploits Subzone, such as the large North Bay batholith (Fig. 20), which comprises several distinct intrusive phases, are associated with molybdenite and scheelite-bearing quartz veins and/or pegmatites, and veins containing barite, galena, and sphalerite. In New Brunswick, large batholiths like the Late Silurian - Early Devonian Pokiok are associated with quartz veins containing various combinations of stibnite, native antimony, gold, and scheelite (e.g. Lake George deposit). Others, like the North Pole granite, contain base metal sulphide veins. Some high-level, subvolcanic granite stocks (e.g. Benjamin River and Upsalquitch forks) contain porphyry Cu-Mo mineralization. Early Devonian layered mafic-ultramafic intrusions in New Brunswick, such as the St. Stephen and Goodwin Lake gabbroic bodies, locally contain appreciable Ni-Cu sulphide deposits.

Middle Devonian to Early Carboniferous (Neoacadian) Pluton-Related Mineralization

Neoacadian plutons, predominantly granitic, occur in all tectono-stratigraphic zones of the Canadian Appalachians but are most voluminous in the Meguma, Avalon, and Gander zones. Their petrogenesis is linked to the tectonic processes responsible for Meguma's accretion and subsequent convergence that led to the final closure of the Rheic Ocean (e.g. Murphy et al., 1999).

The large, Upper Devonian Ackley granite in Newfoundland, which stitches the Gander-Avalon boundary (Figs. 1, 4, 20), contains several economically interesting molybdenite occurrences hosted by aplite-pegmatite phases. Fluorite (e.g. St. Lawrence Mine), scheelite, and tin mineralization are generally associated with the most evolved, silica-rich granite phases.

In Nova Scotia, the most evolved phases of the voluminous, Upper Devonian South Mountain batholith (Meguma Zone) and other smaller coeval plutons host several deposits rich in Sn, W, Mo, Cu, U, F, As, Ta, and Nb (Fig. 20).

The Upper Devonian-Lower Carboniferous phases of the Saint George batholith in southern New Brunswick host several significant Sn, W, Mo, and Bi deposits (e.g. Mt. Pleasant). The Burnthill, Trouth Hill, and Dungarvon granites in central New Brunswick are host to W, Sn, Mo, Be, Mo, and F deposits, which mainly occur in quartz veins or as disseminations near the margins of the contact aureole of the plutons.

Cu-rich orebodies in the form of skarns (e.g. Needle Mountain at Gaspé Mine), mesothermal veins (e.g. Madeleine), and porphyry-type (e.g Copper Mountain at Gaspé Mine) mineralization are related to the Middle Devonian McGerrigle Pluton and related sattelite bodies in the Gaspésie of the Quebec Appalachians. The Gaspé Mine produced ca. 124 M tonnes of ore with 0.64% Cu and recoverable Mo, Ag, Bi, Se, and Au between 1955 and 1991.

Mineralization related to Neoacadian granites is also present in southeastern Quebec mainly involving Mo-Cu-, Mo-Bi-Ag-Pb-Zn-, Ag-Pb-Zn-Cu-, and W-Ag-Au-Bi-Pb-Zn-bearing veins that mainly occur in the contact aureole.

Acknowledgements

This paper benefited from critical reviews by Hank Williams, Jim Hibbard, and Wayne Goodfellow.

References

Anderson, F.D., and Williams, H., 1970, Gander Lake (West Half), Newfoundland: Geological Survey of Canada, Map 1195A (final map with descriptive notes), scale: 1:250 000.

Barr, S.M., and Jamieson, R.A., 1991, Tectonic setting and regional correlation of Ordovician-Silurian rocks of the Aspy terrane, Cape Breton Island, Nova Scotia: Canadian Journal of Earth Sciences, v. 28, p. 1769-1779.

Barr, S.M., and Kerr, A., 1997, Late Precambrian plutons in the Avalon terrane of New Brunswick, Nova Scotia and Newfoundland, *in* Sinha, A.K., Whalen, J.B., and Hogan, J.P., eds., The Nature of Magmatism in the Appalachian Orogen: Geological Society of America, Memoir 191, p. 45-74.

Barr, S.M., Raeside, R.P., and White, C.E., 1998, Geological correlations between Cape Breton Island and Newfoundland, northern Appalachian orogen: Canadian Journal of Earth Science, v. 35, p. 1252-1270.

Barr, S.M., White, C.E., and Miller, B.V., 2002, The Kingston Terrane, southern New Brunswick, Canada: Evidence for an Early Silurian volcanic arc: Geological Society of America Bulletin, v. 114, p. 964-982.

——— 2003, Age and geochemistry of Late Neoproterozoic and Early Cambrian igneous rocks in southern New Brunswick: similarities and contrasts: Atlantic Geology, v. 39, p. 55-73.

Beaudoin, G., Schrijver, K., Marcoux, F., and Calvez, J.Y., 1989, A vein and disseminated Ba-Zn-pb deposit in the Appalachian thrust belt, St. Fabien, Quebec: Economic Geology, v. 84, p. 799-816.

Bédard, J., Lauziere, K., Tremblay, A., and Sangster, A., 1998, Evidence for forearc seafloor spreading from the Betts Cove ophiolite, Newfoundland: Oceanic crust of boninitic affinity: Tectonophysics, v. 248, p. 233-245.

——— 2000, Betts Cove ophiolite and its cover rocks, Newfoundland: Geological Survey of Canada, Bulletin 550, 76 p.

Bird, J.M., and Dewey, J.F., 1970, Lithosphere plate-continental margin tectonics and the evolution of the Appalachian Orogen: Geological Society of America, Bulletin 81, p. 1031-1060.

Boucot, A.J., Dewey, J.F., Dineley, D.L., Fletcher, R., Fyson, W.K., Griffin, J.G., Hickox, C.J., McKerrow, W.S., and Zeigler, A.M., 1974, Geology of the Arisaig area, Antigonish County, Nova Scotia: Geological Society of America, Special Paper 139, 191 p.

Bradley, D.C., 1983, Tectonics of the Acadian Orogeny in New England and adjacent Canada: Journal of Geology, v. 91, p. 381-400.

Brem, A.G., Lin, S., and van Staal, C.R., 2003, Structural relationships south of Grand Lake Newfoundland: Newfoundland Department of Mines and Energy, Current Research 2002, Geological Survey Report 2002-1, 14 p.

Burgess, J.L., Brown, M., Dallmeyer, R.D., and van Staal, C.R., 1995, Microstructure, metamorphism, thermochronology and P-T-t deformation history of the Port aux Basques gneisses, south-west Newfoundland, Canada: Journal of Metamorphic Geology, v. 13, p. 751-776.

Castonguay, S., and Tremblay, A., 2003, Tectonic evolution and significance of Silurian-Early Devonian hinterland-directed deformation in the internal Humber Zone of the southern Quebec Appalachians: Canadian Journal of Earth Sciences, v. 40, p. 255-268.

Cawood, P.A., and Suhr, G., 1992, Generation and obduction of ophiolites: constraints from the Bay of Islands Complex, Western Newfoundland: Tectonics, v. 11, p. 884-897.

Cawood, P.A., Dunning, G.A., Lux, D., and van Gool, J.A.M., 1994, Timing of peak metamorphism and deformation along the Appalachian margin of Laurentia in Newfoundland: Silurian, not Ordovician: Geology, v. 22, p. 399-402.

Cawood, P.A., van Gool, J.A.M., and Dunning, G.R., 1996, Geological development of eastern Humber and western Dunnage zones: Corner

Brook-Glover Island region, Newfoundland: Canadian Journal of Earth Sciences, v. 33, p. 182-198.

Cawood, P.A., McCausland, P.J.A., and Dunning, G.A., 2001, Opening Iapetus: Constraints from the Laurentian margin in Newfoundland: Geological Society of America Bulletin, v. 113, p. 443-453.

Chandler, F.W., Sullivan, R.W., and Currie, K.L., 1987, The age of the Springdale Group, western Newfoundland, and correlative rocks-evidence for a Llandovery overlap assemblage in the Canadian Appalachians: Transactions of the Royal Society of Edinburgh, Earth Sciences, v. 78, p. 41-49.

Chorlton, L.B., Williams, H., and Wilton, D.H.C., 1995, Cape Ray Belt, *in* Williams, H., ed., The Appalachian/Caledonian Orogen: Canada and Greenland: Geological Survey of Canada, Geology of Canada, No. 6, p. 397-403.

Colman-Sadd, S.D., Stone, P., Swinden, H.S., and Barnes, R.P., 1992a, Parallel geological development in the Dunnage Zone of Newfoundland and the Lower Paleozoic terranes of southern Scotland: an assesment: Transactions of the Royal Society of Edinburgh, Earth Sciences, v. 83, p. 571-594.

Colman-Sadd, S.P., Dunning, G.R., and Dec, T., 1992b, Dunnage-Gander relationships and Ordovician orogeny in central Newfoundland: A sediment provenance and U/Pb study: American Journal of Science, v. 292, p. 317-355.

Cousineau, P.A., and St. Julien, P., 1992, The Saint-Daniel Melange: Evolution of an accretionary complex in the Dunnage Terrane of the Quebec Appalachians: Tectonics, v. 11, p. 898-909.

Culshaw, N., and Reynolds, P., 1997, $^{40}Ar/^{39}Ar$ age of shear zones in the southwest Meguma Zone between Yarmouth and Meteghan, Nova Scotia: Canadian Journal of Earth Sciences, v. 34, p. 848-853.

Currie, K.L., 1992, A new look at Gander-Dunnage relations in Carmanville map area, Newfoundland: Geological Survey of Canada, Paper 92-1D, p. 27-33.

Dallmeyer, R.D., and Nance, R.D., 1994, $^{40}Ar/^{39}Ar$ whole rock phyllite ages from Late Precambrian rocks of the Avalon composite terrane, New Brunswick: Evidence of Silurian-Devonian thermal rejuvenation: Canadian Journal of Earth Sciences, v. 31, p. 818-824.

Dallmeyer, R.D., and Williams, H., 1975, $^{40}Ar/^{39}Ar$ ages from the Bay of Islands metamorphic aureole: their bearing on the timing of Ordovician ophiolite obduction: Canadian Journal of Earth Sciences, v. 12, p. 1685-1690.

David, J., and Gariepy, C., 1990, Early Silurian orogenic andesites from the central Quebec Appalachians: Canadian Journal of Earth Sciences, v. 27, p. 632-643.

David, J., and Marquis, R., 1994, Géochronologie U-Pb dans les Appalaches du Quebec: Application aux roches de la zone de Dunnage: la Revue Géologique, v. 1, p. 16-20.

Dubé, B., 1990, Contrasting styles of gold: only deposits in western Newfoundland (a preliminary report): Geological Survey of Canada, Paper 90-1B, p. 77-90.

Dubé, B., Dunning, G.R., Lauziere, K., and Roddick, J.C., 1996, New insights into the Appalachian Orogen from geology and geochronology along the Cape Ray fault zone, southwest Newfoundland: Geological Society of America Bulletin, v. 108, p. 101-116.

Dunning, G.R., and Cousineau, P.A., 1990, U/Pb ages of single zircons from Chain Lakes Massif and a correlative unit in ophiolitic mélange in Quebec: Geological Society of America, NE-section, 1990, Abstracts with Programs, p. 1.

Dunning, G.R., and Krogh, T.E., 1985, Geochronology of ophiolites of the Newfoundland Appalachians: Canadian Journal of Earth Sciences, v. 22, p. 1659-1670.

Dunning, G.R., and O'Brien, S.J., 1989, Late Proterozoic-Early Paleozoic crust in the Hermitage flexure, Newfoundland Appalachians: U/Pb ages and tectonic significance: Geology, v. 17, p. 548-551.

Dunning, G., O'Brien, S.J., Colman-Sadd, S.P., Blackwood, R., Dickson, W.L., O'Neill, P.P., and Krogh, T.E., 1990, Silurian orogeny in the Newfoundland Appalachians: Journal of Geology, v. 98, p. 895-913.

Dunning, G.R., Swinden, H.S., Kean, B.F., Evans, D.T.W., and Jenner, G.A., 1991, A Cambrian island arc in Iapetus: geochronology and geochemistry of the Lake Ambrose volcanic belt, Newfoundland Appalachians: Geological Magazine, v. 128, p. 1-17.

Elliott, C.G., Dunning, G.R., and Williams, P.F., 1991, New constraints on the timing of deformation in eastern Notre Dame Bay, Newfoundland, from U/Pb zircon ages of felsic intrusions: Geological Society of America Bulletin, v. 103, p. 125-135.

Evans, D.T.W., Kean, B.F., and Dunning, G.R., 1990, Geological Studies, Victoria Lake Group, Central Newfoundland: Newfoundland Department of Mines and Energy, Geological Survey Branch, Current Research, Report 90-1, 15 p.

Fryer, B.J., Kerr, A., Jenner, G.A., and Longstaffe, F.J., 1992, Probing the crust with plutons: Regional isotopic geochemistry of granitoid intrusions across insular Newfoundland: Newfoundland Department of Mines and Energy, Geological Surveys Branch, Current Research, Report 92-1, 21 p.

Fyffe, L.R., Pickerill, R.K., and Stringer, P., 1999, Stratigraphy, sedimentology and structure of the Oak Bay and Waweig formations, Mascarene basins: implications for the Paleotectonic evolution of southwestern New Brunswick: Atlantic Geology, v. 35, p. 59-84.

Gerbi, C., Johnson, S., Aleinikoff, J., Bedard, J., Dunning, G., and Fanning, 2006, Early paleozoic orogenesis in the Maine-Quebec Appalachians 11: Development of the Boundary Mountain region: Canadian Journal of Earth Sciences, submitted.

Hall, L., and van Staal, C.R., 1999, Geology of Southern End of Long Range Mountains (Dashwoods Subzone), Newfoundland: Geological Survey of Canada, Geology of Southern End of Long Range Mountains (Dashwoods Subzone), Newfoundland, scale: 1: 50 000.

Hamilton, M.A., and Murphy, J.B., 2004, Tectonic significance of a Llanvirn age for the Dunn Point volcanic rocks, Avalon terrane, Nova Scotia, Canada: Implications for the evolution of the Iapetus and Rheic oceans: Tectonophysics, v. 379, p. 199-209.

Hibbard, J., 1983, Geology of the Baie Verte peninsula, Newfoundland: Mineral Development division Department of Mines and Energy, Government of Newfoundland and Labrador, Memoir 2, 279 p.

Hicks, R.J., Jamieson, R.A., and Reynolds, P., 1999, Detrital and metamorphic $^{40}Ar/^{39}Ar$ ages from muscovite and whole-rock samples, meguma Supergroup, southern Nova Scotia: Canadian Journal of Earth Sciences, v. 36, p. 23-32.

Hodych, J.P., Cox, R.A., and Kosler, J., 2004, An equatorial Laurentia at 550 Ma confirmed by Grenvillian inherited zircons dated by LAM ICP-MS in the Skinner Cove volcanics of western Newfoundland: implications for inertial interchange true polar wander: Precambrian Research, v. 129, p. 93-113.

Holdsworth, R.E., 1994, Structural evolution of the Gander-Avalon terrane boundary: a reactivated transpression zone in the NE Newfoundland Appalachians: Journal of the Geological Society of London, v. 15, p. 629-646.

Huot, F., Hébert, R., and Turcotte, B., 2002, A multistage magmatic history for the genesis of the Orford ophiolite (Quebec, Canada): A study of the Mont Chagnon Massif: Canadian Journal of Earth Sciences, v. 39, p. 1201-1217.

Hutchinson, D.R., Klitgord, K.D., Lee, M.W., and Trehu, A.M., 1988, U. S. Geological Survey deep seismic reflection profile across the Gulf of Maine: Geological Society of America Bulletin, v. 100, p. 172-184.

Jamieson, R.A., 1988, Metamorphic P-T-t data from western Newfoundland and Cape Breton-implications for Taconian and Acadian tectonics: Geological Society of Canada, Program with Abstracts, v. 13, p. A60.

Jenner, G.A., and Swinden, H.S., 1993, The Pipestone Pond Complex, central Newfoundland: Complex magmatism in an eastern Dunnage Zone ophiolite: Canadian Journal of Earth Sciences, v. 30, p. 434-448.

Johnson, R.J.E., van der Pluijm, B.A., and van der Voo, R., 1991, Paleomagnetism of the Moreton's Harbour Group, northeastern Newfoundland Appalachians: Evidence for an Early Ordovician island arc near the Laurentian margin of Iapetus: Journal of Geophysical Research, v. 96, p. 11689-11701.

Johnson, S.C., 2001, Contrasting geology in the Pocologan River and Long Reach areas: Implications for the New River belt and correlations in southern New Brunswick and Maine: Atlantic Geology, v. 37, p. 61-79.

Johnson, S.C., and McLeod, M.J., 1996, The New River Belt: A unique segment along the western margin of the Avalon composite terrane, southern New Brunswick, Canada, *in* Nance, R.D., and Thompson, M.D., eds., Avalonian and Related Peri-Gondwanan Terranes of the Circum-north Atlantic: Geological Society of America, Boulder, Colorado, Special Paper 304, p. 149-164.

Kamo, S.L., Gower, C.F., and Krogh, T.E., 1989, Birthdate for the Iapetus Ocean? A precise U-Pb zircon and baddeleyite age for the Long Range dikes, southeast Labrador: Geology, v. 17, p. 602-605.

Karson, J., and Dewey, J.F., 1978, Coastal complex, western Newfoundland: an Early Ordovician fracture zone: Geological Society of America Bulletin, v. 89, p. 1037-1049.

Kean, B.A., Evans, D.T.W., and Jenner, G.A., 1995, Geology and Mineralization of the Lushs Bight Group: Newfoundland Department of Natural Resources, Report 95-2, 204 p.

Keen, C.E., Kay, W.A., Keppie, D., Marillier, F., Pe-Piper, G., and Waldron, J.W.F., 1991, Deep seismic reflection data from the Bay of Fundy and Gulf of Maine: Tectonic implications for the northern Appalachians: Canadian Journal of Earth Sciences, v. 28, p. 1096-1111.

Keppie, J.D., and Krogh, T.E., 2000, 440 Ma igneous activity in the Meguma terrane, nova Scotia, Canada: part of the Appalachian overstep sequence? American Journal of Science, v. 300, p. 528-538.

Keppie, D.F., Keppie, J.D., and Murphy, J.B., 2002, Saddle reef auriferous veins in a conical fold termination (Oldham anticline, meguma terrane, nova scotia, Canada): reconciliation of structural and age data: Canadian Journal of Earth Sciences, v. 39, p. 53-63.

Kerr, A., Jenner, G.J., and Fryer, B.J., 1995, Sm-Nd isotopic geochemistry of Precambrian to Paleozoic granitoid suites and the deep-crustal structure of the southeast margin of the Newfoundland Appalachians: Canadian Journal of Earth Sciences, v. 32, p. 224-245.

Kim, J., Coish, R., Evans, M., and Dick, G., 2003, Supra-subduction zone extensional magmatism in Vermont and adjacent Quebec: Implications for Early Paleozoic Appalachian tectonics: Geological Society of America Bulletin, v. 115, p. 1552-1569.

King, M.S., and Barr, S.M., 2004, Magnetic and gravity models across terrane boundaries in southern New Brunswick, Canada: Canadian Journal of Earth Sciences, v. 41, p. 1027-1047.

Landing, E., 1996, Avalon: Insular continent by the latest Precambrian, *in* Nance, R.D., and Thompson, M.D., eds., Avalonian and Related Peri-Gondwanan Terranes of the Circum-North Atlantic: Geological Society of America, Special Paper 304, p. 29-63.

Lavoie, D., Burden, E., and Lebel, D., 2003, Stratigraphic framework for the Cambrian-Ordovician rift and passive margin successions from southern Quebec to western Newfoundland: Canadian Journal of Earth Sciences, v. 40, p. 177-205.

Liss, M.J., van der Pluijm, B.A., and van der Voo, R., 1994, Avalonian proximity of the Ordovician Miramichi terrane, northern New Brunswick, northern Appalachians: Paleomagnetic evidence for rifting and back-arc basin formation at the southern margin of Iapetus: Tectonophysics, v. 227, p. 17-30.

Lissenberg, C.J., and van Staal, C.R., 2002, The Relationships Between the Annieopsquotch Ophiolite Belt, the Dashwoods Block and the Notre Dame arc in Southwestern Newfoundland: Newfoundland Department of Mines and Energy, Current Research 2001-1, 10 p.

Lissenberg, C.J., Zagorevski, A., McNicoll, V.J., van Staal, C.R., and Whalen, J.B., 2005a, Assembly of the Annieopsquotch Accretionary Tract, southwest Newfoundland: Age- and geodynamic constraints from syn-kinematic intrusions: Journal of Geology, v. 113, p. 553-570.

Lissenberg, C.J., van Staal, C.R., Bédard, J.H., and Zagorevski, A., 2005b, Geochemical constraints on the origin of the Annieopsquotch ophiolite belt, southwest Newfoundland: Geological Society of America Bulletin, v. 117, p. 1413-1426.

MacDonald, L.A., Barr, S.M., White, C.E., and Ketchum, J.W.F., 2002, Petrology, age, and tectonic setting of the White Rock Formation, Meguma terrane, Nova Scotia: Evidence for Silurian continental rifting: Canadian Journal of Earth Sciences, v. 39, p. 259-277.

MacNiocaill, C., and Smethurst, M., 1994, Palaeozoic palaeogeography of Laurentia and its margins: A reassessment of Palaeomagnetic data: Geophysical Journal International, v. 116, p. 715-725.

Malo, M., Cousineau, P.A., Sacks, P.E., Riva, J.F.V., Asselin, E., and Gosselin, P., 2001, Age and composition of the Ruisseau Isabelle Mélange along the Shickshock sud fault zone: constraints on the timing of mélanges formation in the Gaspé peninsula: Canadian Journal of Earth Sciences, v. 38, p. 21-42.

McLaughlin, R.J., Barr, S.M., Hill, M.D., Thompson, M.D., Ramezani, J., and Reynolds, P.H., 2003, The Moosehorn plutonic suite, southeastern Maine and southwestern New Brunswick: Age, petrochemistry, and tectonic setting: Atlantic Geology, v. 39, p. 123-146.

McLeod, M.J., Ruitenberg, A.A., and Krogh, T.E., 1992, Geology and U-Pb geochronology of the Annidale Group, southern New Brunswick: Lower Ordovician volcanic and sedimentary rocks formed near the southeastern margin of Iapetus Ocean: Atlantic Geology, v. 28, p. 181-192.

McLeod, M.J., Pickerill, R.K., and Lux, R.D., 2001, Mafic intrusions on Campobello Island: implications for New Brunswick-Maine correlations: Atlantic Geology, v. 37, p. 17-40.

Miller, B.V., and Barr, S.M., 2000, Petrology and isotopic composition of a Grenvillian basement fragment in the Northern Appalachian Orogen: Blair River Inlier, Nova Scotia, Canada: Journal of Petrology, v. 41, p. 1777-1804.

Miller, B.V., and Fyffe, L.R., 2002, Geochronology of the Letete and Waweig formations, mascarene Group, southwestern New Brunswick: Atlantic Geology, v. 38, p. 29-36.

Miller, H.G., 1983, Geophysical constraints on the size and extent of the Wabana hematite deposit: Economic Geology, v. 78, p. 1017-1021.

Murphy, J.B., and Keppie, J.D., 1995, Arisaig Belt, *in* Williams, H., ed., The Appalachian/Caledonian Orogen: Canada and Greenland: Geological Survey of Canada, Geology of Canada, No. 6, p. 360-365.

Murphy, J.B., van Staal, C.R., and Keppie, J.D., 1999, Middle to Late Paleozoic Acadian Orogeny in the northern Appalachians: A Laramide-style plume-modified orogeny: Geology, v. 27, p. 653-656.

Neuman, R.B., 1967, Bedrock Geology of the Shin Pond and Stacyville Quadrangles, Maine: U.S. Geological Survey, Professional Paper 524-I, 47 p.

Nowlan, G.S., and Neuman, R.B., 1995, Palaeontological contributions to Paleozoic paleogeographic and tectonic reconstructions, *in* Williams, H., ed., Geology of the Appalachian-Caledonian Orogen in Canada and Greenland: Geological Survey of Canada, Geology of Canada, No. 6, p. 817-842.

O'Brien, B., O'Brien, S., and Dunning, G.R., 1991, Silurian cover, Late Precambrian-Early Ordovician basement, and the chronology of Silurian orogenesis in the Hermitage Flexure (Newfoundland Appalachians): American Journal of Science, v. 291, p. 760-799.

O'Brien, B., O'Brien, S., Dunning, G.R., and Tucker, R.D., 1993, Episodic reactivation of a Late Precambrian mylonite zone of the Gondwanan margin of the Appalachians, southern Newfoundland: Tectonics, v. 12, p. 1043-1055.

O'Brien, B., Swinden, H.S., Dunning, G.R., Williams, S.H., and O'Brien, F., 1997, A peri-Gondwanan arc-back arc complex in Iapetus: Early-Mid Ordovician evolution of the Exploits Group, Newfoundland: American Journal of Science, v. 297, p. 220-272.

O'Brien, S.J., O'Brien, B.H., Dunning, G.R., and Tucker, R.D., 1996, Late Neoproterozoic Avalonian and related peri-Gondwanan rocks of the Newfoundland Appalachians, *in* Nance, R.D., and Thompson, M.D., eds., Avalonian and Related Peri-Gondwanan Terranes of the Circum-North Atlantic: Geological Society of America, Special Paper 304, p. 9-28.

O'Brien, S.J., Dubé, B., and O'Driscoll, C.F., 1999, High-sulphidation, epithermal-style hydrothermal systems in Late Neoproterozoic Avalonian rocks on the Burin Peninsula, Newfoundland: Implications for gold exploration: Newfoundland Department of Mines and Energy, Geological Survey, Current Research 99-1, 22 p.

O'Neill, P.P., 1991, Geology of the Weir's Pond area, Newfoundland (NTS 2E/1): Newfoundland Department of Mines and Energy, Geological Survey Branch, Report 91-3, 144 p.

Paradis, S., Chi, G., and Lavoie, D., 2004, Fluid inclusions and isotopic evidence for the origin of the Upton Ba-Zn-Pb deposit, Quebec Appalachians, Canada: Economic Geology, v. 99, p. 807-817.

Park, A.F., and Whitehead, J., 2003, Structural transect through Silurian turbidites of the Fredericton Belt southwest of Fredericton, New Brunswick: the role of the Fredericton Fault in late Iapetus convergence: Atlantic Geology, v. 39, p. 227-237.

Pe-Piper, G., and Jansa, L.F., 1999, Pre-mesozoic basement rocks offshore Nova Scotia, Canada: New constraints on the accretion history of the Meguma terrane: Geological Society America Bulletin, v. 111, p. 1773-1791.

Pollock, J.C., Morrissey, K.D., Wilton, D.H.C., van Staal, C.R., and Tubrett, M.N., 2003, Laser ablation U-Th-Pb geochronology of detrital zircons from the Botwood and Indian Islands Belts: Accretionary tectonic evolution of the northeast Exploits subzone, Dunnage Zone, Newfoundland Appalachians: Geological Society of America, 38th Annual Meeting, NE-section, 2003, Abstracts with Programs, p. 79.

Price, J.R., Barr, S.M., Raeside, R.P., and Reynolds, P.H., 1999, Petrology, tectonic setting, and $^{40}Ar/^{39}Ar$ (hornblende) dating of the Late Ordovician-Early Silurian Belle Cote Road orthogneiss, western Cape Breton Highlands, Nova Scotia: Atlantic Geology, v. 35, p. 1-17.

Ramezani, J., Dunning, G.R., and Wilson, M.R., 2002, Geologic setting, geochemistry of alteration, and U-Pb age of hydrothermal zircon from the Silurian Stog'er Tight gold prospect, Newfoundland Appalachians, Canada: Exploration and Mining Geology, v. 9, p. 171-188.

Robinson, P., Tucker, R.D., Bradley, D., Berry, H.N., and Osberg, P.H., 1998, Paleozoic orogens in New England, USA: Geologiska Foreningens Stockholm Forhandlingar, v. 120, p. 119-148.

Rogers, N., van Staal, C.R., McNicoll, V., Pollock, J., Zagorevski, A., Whalen, J., and Kean, B., 2006, Neoproterozoic and Cambrian arc magmatism along the eastern margin of the Victoria Lake Supergroup: a remnant of Ganderian basement in central Newfoundland? Precambrian Research, in press.

Samson, S.D., Barr, S.M., and White, C.E., 2000, Nd isotopic characteristics of terranes within the Avalon Zone, southern New Brunswick: Canadian Journal of Earth Sciences, v. 37, p. 1039-1052.

Schenk, P.E., 1997, Sequence stratigraphy and provenance on Gondwana's margin: The Meguma zone (Cambrian to Devonian) of Nova Scotia, Canada: Geological Society America Bulletin, v. 109, p. 395-409.

Schofield, D.I., and D'Lemos, R.S., 2000, Granite petrogenesis in the Gander Zone, NE Newfoundland: Mixing of melts from multiple sources and the role of lithospheric delamination: Canadian Journal of Earth Sciences, v. 37, p. 535-547.

Schroetter, J.M., Pagé, P., Bédard, J.H., Tremblay, A., and Bécu, V., 2003, Forearc extension and seafloor spreading in the Thetford Mines ophiolite complex, *in* Dilek, Y., and Robinson, P.T., eds., Ophiolites in Earth History: Geological Society of London, Special Publication 218, p. 231-251.

Stern, R.J., and Bloomer, B.S.H., 1992, Subduction zone infancy: examples from the Eocene Izu-Bonin-Mariana and Jurassic California arcs: Geological Society America Bulletin, v. 104, p. 1621-1636.

Swinden, H.S., 1996, Geochemistry of Volcanic Rocks in the Moreton's Harbour-Twillingate Area, Notre Dame Bay: Newfoundland Department of Mines and Energy, Geological Surveys Branch, Current Research 96-1, 20 p.

Swinden, H.S., and Dunsworth, S.M., 1995, Metallogeny, *in* Williams, H., ed., The Appalachian/Caledonian Orogen: Canada and Greenland: Geological Survey of Canada, Geology of Canada, No. 6, p. 681-814.

Swinden, H.S., Jenner, G.A., and Szybinski, Z.A., 1997, Magmatic and tectonic evolution of the Cambrian-Ordovician Laurentian margin of Iapetus: Geochemical and isotopic constraints from the Notre Dame Subzone, Newfoundland, *in* Sinha, K., Whalen, J.B., and Hogan, J.P., eds., The Nature of Magmatism in the Appalachian Orogen: Geological Society of America, Memoir 191, p. 367-395.

Szybinski, Z.A., 1995, Paleotectonic and Structural Setting of the Western Notre Dame Bay Area, Newfoundland Appalachians: Unpub. Ph.D. thesis, Memorial University of Newfoundland, St. John's, Newfoundland, 690 p.

Thurlow, J.G., 1990, Geology the Buchans orebodies- A 1990 summary *in* Swinden, H.S., Evans, D.T.W., and Kean, B.F, Metallogenic Framework of Base and Precious Metal Deposits, Central and Western Newfoundland: 8th IAGOD symposium, Field Trip No. 1 Guidebook. Geological Survey of Canada, Open File 2156, p. 84-91.

Tremblay, A., and Dubé, B., 1991. Structural relationships between some gold occurrences and fault zones in the Bathurst area, northern New Brunswick, *in* Current Research, Part D: Geological Survey of Canada, Paper 90-1D, p. 89-100.

Tremblay, A, Hebert, R., and Bergeron, M., 1989. Le Complexe d'Ascot des Appalaches du sud du Quebec – petrologie et geochimie: Canadian Journal of Earth Sciences, v. 26, p. 2407-2420.

Tremblay, A., Lafleche, M..R., McNutt, R.H., and Bergeron, M., 1994, Petrogenesis of Cambro-Ordovician subduction-related granitic magmas of the Quebec Appalachians, Canada: Chemical Geology, v. 113, p. 205-220.

Tremblay, A., Malo, M., and St. Julien, P., 1995, Dunnage Zone-Quebec, Chapter 3 *in* Williams, H., ed., Geology of the Appalachian-Caledonian Orogen in Canada and Greenland: Geological Survey of Canada, Geology of Canada, No. 6, p. 179-197 (also Geological Society of America, The Geology of North America, v. F1)

Trench, A., Torsvik, T.H., and McKerrow, W.S., 1992, The palaeogeographic evolution of southern Britain during early Palaeozoic times: A reconciliation of palaeomagneitc and biogeographic evidence: Tectonophysics, v. 201, p. 75-82.

Trottier, J., Brown, A.C., and Gauthier, M., 1991, An Ordovician rift environment for the Memphremagog polymetallic massive sulphide deposit, Appalachian Ophiolite Belt, Quebec: Canadian Journal of Earth Sciences, v. 28, p. 1887-1904.

Tucker, R.D., Osberg, P.H., and Berry, V.H.N., 2001, The geology of a part of Acadia and the nature of the Acadian orogeny across central and eastern Maine: American Journal of Science, v. 301, p. 205-260.

Valverde, P., Dunning, G., and van Staal, C.R., 2000. The Margaree orthogneiss (Port aux basques Complex, SW Newfoundland) evolution of a peri-Gondwanan, Mid Arenig-Early Llanvirn, mafic-felsic igneous complex: Canadian Journal of Earth Sciences, v. 37, p. 1691-1710.

Valverde-Vaquero, P, van Staal, C.R., and van der Velden, A., 2003. Acadian orogenesis and high grade metamorphism in the Central Mobile Belt of central Newfoundland: Geological Society of America, 38th Annual Meeting, NE-section, Abstracts with Programs, p. 23.

Valverde-Vaquero, P., van Staal, C.R., McNicoll, V., and Dunning, G., 2006. Middle Ordovician magmatism and metamorphism along the Gander margin in Central Newfoundland. Journal of the Geological Society London 163, p. 347-362.

van der Velden, A.J., van Staal, C.R., and Cook, F.A., 2004, Crustal structure, fossil subduction and the tectonic evolution of the Newfoundland Appalachians: Evidence from a reprocessed seismic reflection survey: Geological Society America Bulletin, v. 116, p. 1485-1498.

van Staal, C.R., 1994, The Brunswick subduction complex in the Canadian Appalachians: record of the Late Ordovician to Late Silurian collision between Laurentia and the Gander margin of Avalon: Tectonics, v. 13, p. 946-962.

——— 2005, The Northern Appalachians, in Selley, R.C., Robin, L., Cocks, M., and Plimer, I.R., eds., Encyclopedia of Geology: Elsevier, Oxford, v. 4, p. 81-91.

——— 2006, Tectonic evolution of the Canadian Appalachians, *in* Percival, J., and Cook, F., eds., Variations in Tectonic Styles in Canada Revisited following Lithoprobe: Geological Association of Canada, Special Paper, in press.

van Staal, C.R., and de Roo, J.A., 1995, Mid-Paleozoic tectonic evolution of the Appalachian Central Mobile Belt in northern New Brunswick, Canada: Collision, extensional collapse, and dextral transpression, *in* Hibbard, J., van Staal, C.R., and Cawood, P., eds., Current Perspectives in the Appalachian-Caledonian Orogen: Geological Association of Canada, Special Paper 41, p. 367-389.

van Staal, C.R., and Fyffe, L.R., 1995, The New Brunswick Dunnage Zone, *in* Williams, H., ed., The Appalachian/Caledonian Orogen: Canada and Greenland: Geological Survey of Canada, Geology of Canada, No. 6, p. 166-167.

van Staal, C.R., Sullivan, R.W., and Whalen, J.B., 1996, Provenance and tectonic history of the Gander Margin in the Caledonian/Appalachian Orogen: implications for the origin and assembly of Avalonia, *in* Nance, R.D., and Thompson, M.D., eds., Avalonian and Related Peri-Gondwanan Terranes of the Circum-North Atlantic: Geological Society of America, Special Paper 304, p. 347-367.

van Staal, C.R., Dewey, J.F., MacNiocaill, C., and McKerrow, S., 1998, The Cambrian-Silurian tectonic evolution of the northern Appalachians: history of a complex, southwest Pacific-type segment of Iapetus, *in* Blundell, D.J., and Scott, A.C., eds., Lyell: the Past is the Key to the Present: Geological Society, Special Publication 143, p. 199-242.

van Staal, C.R., Wilson, R.A., Rogers, N., Fyffe, L.R., Langton, J.P., McCutcheon, S.R., McNicoll, V., and Ravenhurst, C.E., 2003a, Geology and tectonic history of the Bathurst Supergroup and its relationships to coeval rocks in southwestern New Brunswick and adjacent Maine - A synthesis, *in* Goodfellow, W.D., McCutcheon, S.R., and Peter, J.M., eds., Massive Sulfide Deposits of the Bathurst Mining Camp, New Brunswick, and Northern Maine, Economic Geology Monograph 11, p. 37-60.

van Staal, C., Whalen, J.B., Pehrson, S., and McNicoll, V., 2003b, Tectonic evolution of the Notre Dame magmatic arc, Newfoundland Appalachians: AGU Fall Meeting, 2003, p. F1579.

van Staal, C.R., McNicoll, V., Valverde-Vaquero, P., Barr, S., Fyffe, L.R., and Reusch, D.N., 2004a, Ganderia, Avalonia, and the Salinic and Acadian orogenies: Geological Society of America, 39th Annual Meeting, NE-section, 2004, Abstracts with Programs, p. 128-129.

van Staal, C.R., Lissenberg, C.J., Zagorevski, A., McNicoll, V., Whalen, J., Bedard, J., and Pehrsson, S., 2004b, A new look at the tectonic processes involved in the Ordovician Taconic Orogeny in the Northern Appalachians: evidence for multiple accretion of (infant) arc terranes: Geological Society of America, 39th Annual Meeting, Denver, 2004, Abstracts with Programs.

van Staal, C.R., Whalen, J.B., McNicoll, V.J., Pehrsson, S.J., Lissenberg, C.J., Zagorevski, A., van Breemen, O., and Jenner, G.A., 2006, The Notre Dame arc and the Taconic Orogeny in Newfoundland, *in* Hatcher, J., Carlson, M.P., McBride, J.H., and Martínez Catalán, J.R., eds., The 4D Framework of Continental Crust: Geological Society of America, Special Paper, in press.

Waldron, J.W.F., and van Staal, C.R., 2001, Taconic Orogeny and the accretion of the Dashwoods block: a peri-Laurentian microcontinent in the Iapetus Ocean: Geology, v. 29, p. 811-814.

Waldron, J.W.F., Murphy, J.B., Melchin, M.J., and Davis, G., 1996, Silurian tectonics of Western Avalonia: strain-corrected subsidence history of the Arisaig Group, Nova Scotia: Journal of Geology, v. 104, p. 677-694.

Walker, J.A., and McCutcheon, S.R., 1995, Siluro-Devonian Stratigraphy of the Chaleur Bay Synclinorium, Northern New Brunswick: New Brunswick Department of Natural Resources and Energy, Minerals and Energy Division, Current Research 1994, Miscellaneous Report 18, 20 p.

West, D.P., Ludman, A., and Lux, D.R., 1992, Silurian age for the Pocomoonshine Gabbro-Diorite, southeastern Maine and its regional tectonic implications: American Journal of Earth Sciences, v. 292, p. 253-273.

West, D.P., Beal, H.M., and Grover, T., 2003, Silurian deformation and metamorphism of Ordovician arc rocks of the Casco Bay Group, south-central Maine: Canadian Journal of Earth Sciences, v. 40, p. 887-905.

Whalen, J.B., Jenner, G.A., Longstaffe, F.J., Robert, F., and Gariépy, C., 1996, Geochemical and Isotopic (O, Nd, Pb and Sr) constraints on A-type granite petrogenesis based on the Topsails igneous suite, Newfoundland Appalachians: Journal of Petrology, v. 37, p. 1463-1489.

Whalen, J.B., Jenner, G.A., Longstaffe, F.J., Gariépy, C., and Fryer, B.J., 1997a, Implications of granitoid geochemical and isotopic (Nd, O, Pb) data from the Cambrian-Ordovician Notre Dame arc for the evolution of the Central Mobile belt, Newfoundland Appalachians, *in* Sinha, K., Whalen, J.B., and Hogan, J.P., eds., The Nature of Magmatism in the Appalachian Orogen: Geological Society of America, Memoir 191, p. 367-395.

Whalen, J.B., McNicoll, V.J., van Staal, C.R., Lissenberg, C.J., Longstaffe, F.J., Jenner, G.A., and van Breemen, O., 2006, Spatial, temporal and geochemical characteristics of Silurian collision-zone magmatism: An example of a rapidly evolving magmatic system related to slab break-off: Lithos, in press.

Williams, H., 1979, Appalachian Orogen in Canada: Canadian Journal of Earth Sciences, v. 16, p. 792-807.

——— 1995, Geology of the Appalachian-Caledonian Orogen in Canada and Greenland: Geological Survey of Canada, Geology of Canada No. 6, 944 p.

Williams, H., and Grant, A.C., 1998, Tectonic Assemblages, Atlantic Region, Canada: Geological Survey of Canada, Tectonic Assemblages, Atlantic Region, Canada, Open File 3657, scale: 1:3,000,000.

Williams, H., and Hiscott, R.N., 1987, Definition of the Iapetus rift-drift transition in western Newfoundland: Geology, v. 15, p. 1044-1047.

Williams, H., and Payne, J.G., 1975, The Twillingate granite and nearby volcanic rocks: an island arc complex in northeast Newfoundland: Canadian Journal of Earth Sciences, v. 12, p. 982-995.

Williams, H., and St. Julien, P., 1982, The Baie Verte Brompton Line: Early Paleozoic continent ocean interface in the Canadian Appalachians, *in* St. Julien, P., and Beland, J., eds., Major Structural Zones and Faults of the northern Appalachians: Geological Association of Canada, Special Paper 24, p. 177-208.

Williams, H., Colman-Sadd, S.P., and Swinden, H.S., 1988, Tectonostratigraphic subdivisions of central Newfoundland: Geological Survey of Canada, Current Research, Part B, Paper 88-1B, 9 p.

Williams, H., Currie, K.L., and Piasecki, M.A.J., 1993, The Dog Bay Line: A major Silurian tectonic boundary in northeast Newfoundland: Canadian Journal of Earth Sciences, v. 30, p. 2481-2494.

Williams, H., Lafrance, B., Dean, P.l., Williams, P.F., Pickering, K.T., and van der Pluijm, B.A., 1995, Badger Belt, *in* Williams, H., ed., The Appalachian/Caledonian Orogen: Canada and Greenland: Geological Survey of Canada, Geology of Canada No. 6, p. 403-413.

Williams, S.H., 1992, Lower Ordovician (Arenig-llanvirn) graptolites from the Notre Dame subzone, central Newfoundland: Canadian Journal of Earth Sciences, v. 29, p. 1717-1733.

Wilson, R., Burden, E.T., Bertrand, R., Asselin, E., and McCracken, A.D., 2004, Stratigraphy and tectono-sedimentary evolution of the Late Ordovician to Middle Devonian Gaspe belt in northern New Brunswick: Evidence from the Restigouche area: Canadian Journal of Earth Sciences, v. 41, p. 527-551.

Zagorevski, A., Rogers, N., and van Staal, C.R., 2003, Tectono-Stratigraphic Relationships in the Lloyd's River and Tulks Brook Valleys, Central Newfoundland: A Geological Link between Red Indian and King George IV Lakes: Newfoundland Department of Mines and Energy, Geological Survey, Current Research 2002, Report 2002-1, 10 p.

Zagorevski, A., van Staal, C.R., McNicoll, V., and Rogers, N., 2004, Pats Pond and Wigwam brook groups: Record of episodic Peri-Gondwanan Upper Cambrian-Caradoc island arc activity in the Exploits Subzone of Central Newfoundland: Geological Society of America, 39th Annual Meeting, NE-section, 2004, Abstracts with Programs, p. 129.

Zagorevski, A., Rogers, N., van Staal, C.R., McNicoll, V., Lissenberg, C.J., and Valverde-Vaquero, P., 2006a, Lower to Middle Ordovician evolution of peri-Laurentian arc and back-arc complexes in the Iapetus: constraints from the Annieopsquotch Accretionary Tract, Central Newfoundland: Geological Society of America Bulletin, in press.

Zagorevski, A., van Staal, C.R., McNicoll, V.C., and Rrogers, N., 2006b, Upper Cambrian to Upper Ordovician peri-Gondwanan island arc activity in the Victoria Lake Supergroup, Central Newfoundland: tectonic development of the Ganderian margin: American Journal of Science, submitted.

Prospective Metallogenic Settings of the Grenville Province

Louise Corriveau[1], Serge Perreault[2], and Anthony Davidson[3]

1. Geological Survey of Canada, 490 de la Couronne, Quebec G1K 9A9

2. Ministère des Ressources naturelles et de la Faune du Québec, 545 boul. Crémazie est, bureau 1110, Montreal, Quebec H2M 2V1

3. Geological Survey of Canada (retired), 601 Booth St., Ottawa, Ontario K1A 0E8

Corresponding author's email: lcorrive@nrcan.gc.ca

Abstract

The Grenville Province exposes the deep crustal root of a Himalayan-type collisional orogen. However, most of its components predate collisional orogeny, and the main episodes of crustal build-up are Andean in type. Many of the ensuing volcano-plutonic arc belts are now parts of the gneiss complexes that typify large segments of the Grenville Province. Ages of inferred magmatic arcs, intra-arc rifts, aborted back-arcs and island-arcs components range from Archean to Mesoproterozoic, most being Labradorian (1.71-1.60 Ga), Pinwarian (1.52-1.46 Ga), or Elzevirian (ca. 1.35-1.19 Ga). Sedimentary basins and intrusive suites, including those with anorthosite massifs, host base metal (Zn), Fe- and hemo-ilmenite (Fe-Ti), and Fe oxide Cu-Au-REE-U deposits, and could collectively host Carlin, sedimentary exhalative, magmatic Ni-Cu-Co sulphides, PGE, Fe-Ti-V-P, and rare metal deposits. Recognized volcano-plutonic belts host some volcanogenic massive sulphide deposits or share hallmarks of volcanogenic massive sulphide, Fe oxide Cu-Au-REE-U, epithermal Au, and porphyry Cu deposits. The main aim of this synthesis is to inform and, through that information, disparage the wasteland paradigm that has afflicted the Grenville Province for so long. Compared to the Superior Province, the Grenville Province is under-explored, under-prospected, and under-mapped but as it is, to a large extent, a reworked equivalent of the Andes, there is no reason why it should not be as prospective. Like the known deposits, undiscovered deposits are likely to be metamorphosed and perhaps to have undergone ore mineral beneficiation providing a means to further increase the attractiveness of the Grenville Province.

Résumé

La Province de Grenville met à jour les racines profondes d'un ancien orogène de collision de type himalayen. Toutefois, la plupart de ses composantes se sont formées bien avant l'orogenèse de collision dans des milieux géotectoniques de type andin. Les ceintures volcano-plutoniques d'arc qui se sont formées dans de tels milieux font maintenant partie des complexes gneissiques qui caractérisent de grands segments de la Province de Grenville. Les âges des arcs magmatiques, des rifts intra-arc, des bassins d'arrière-arc avortés et des arcs insulaires déduits s'étalent de l'Archéen au Mésoprotérozoïque, la plupart étant du Labradorien (1,71-1,60 Ga), du Pinwarien (1,52 -1,46 Ga) ou de l'Elzévirien (env. 1,35-1,19 Ga). Les bassins sédimentaires et les suites intrusives, y compris celles englobant des massifs d'anorthosite, renferment des gîtes de métaux usuels (Zn), de fer, d'hémo-ilménite (Fe-Ti) et d'oxydes de fer-Cu-Au-ÉTR-U, et pourraient, dans leur ensemble, être les hôtes de gîtes de type Carlin, de gîtes sedex, de gîtes de sulfures magmatiques à Ni-Cu-Co, de gîtes d'ÉGP et de Fe-Ti-V-P ou de gîtes de métaux rares. Les ceintures volcano-plutoniques connues renferment quelques gisements de sulfures massifs volcanogènes ou présentent des caractéristiques métallogéniques propres à ces gîtes ou à celles des gîtes d'oxydes de fer-Cu-Au-ÉTR-U, des gîtes d'or épithermaux ou des gîtes porphyriques de cuivre. Cette synthèse vise à fournir de l'information qui devrait contribuer à changer le paradigme d'une Province de Grenville stérile. Comparée à la Province du lac Supérieur, la Province de Grenville a été insuffisamment explorée et cartographiée. Toutefois, dans une perspective de contextes géologiques de type andin, il n'y a aucune raison qui justifierait qu'elle ne soit pas métallifère. De la même manière que le sont les gîtes découverts, les gîtes non encore décelés ont de fortes chances d'être métamorphisés et leurs minerais auront peut-être subi un enrichissement métamorphique, ce qui augmente d'autant l'attrait pour l'exploration de la Province de Grenville.

Introduction

The Grenville Province (Fig. 1), with its predominantly high-grade metamorphic terranes and spectacular deep-level thrust stacks along ductile shear zones, has epitomized for years the roots of a Himalayan-style collisional orogen (Dewey and Burke, 1973; Rivers, 1983; Davidson, 1984). Recognition of the highly deformed and metamorphosed nature of the rocks and failure to discover mineral deposits among promising targets (e.g. the eastern extension of the Abitibi greenstone belt) brought intense scepticism about the mineral potential of the province and its gneissic "pinkstone" belts. The notion arose that metals must have been leached from original ore settings during high-grade metamorphism, leaving the province largely sterile even though it hosts Fe, Ti, Nb, U, Mo, and Zn deposits, some of them world-class (Fig. 2), and is renowned for its industrial mineral potential.

A new paradigm is emerging from the recognition that the main crustal build-up of the Grenville Province is Andean in type (Rivers, 1997), that metamorphism is mostly isochemical beyond devolatilization (H_2O, CO_2, and halides), and that undifferentiated gneiss complexes conceal volcano-plutonic belts fertile in Cu, Au, U, REE, etc. Although tectonically buried deep in the crust at some stage of their history, many components of the Grenville Province were formed initially at or near surface, or in the mid-crust (see Fig. 1 in Bonnet and Corriveau, 2007). Hence, even among gneisses at granulite facies, such as those of the La Romaine Supracrustal Belt (Fig. 1), pyroclastic units with polygenic lappilistones and volcanic breccia can be preserved providing clues to the presence of volcanosedimentary belts in otherwise fairly monotonous gneissic domains (Fig. 3). In this review, such geological terranes are considered globally as prospective as any others originally formed at low- to mid-

Corriveau, L., Perreault, S., and Davidson, A., 2007, Prospecitive metallogenic settings of the Grenville Province, *in* Goodfellow, W.D., ed., Mineral Deposits of Canada: A Synthesis of Major Deposit-Types, District Metallogeny, the Evolution of Geological Provinces, and Exploration Methods: Geological Association of Canada, Mineral Deposits Division, Special Publication No. 5, p. 819-847.

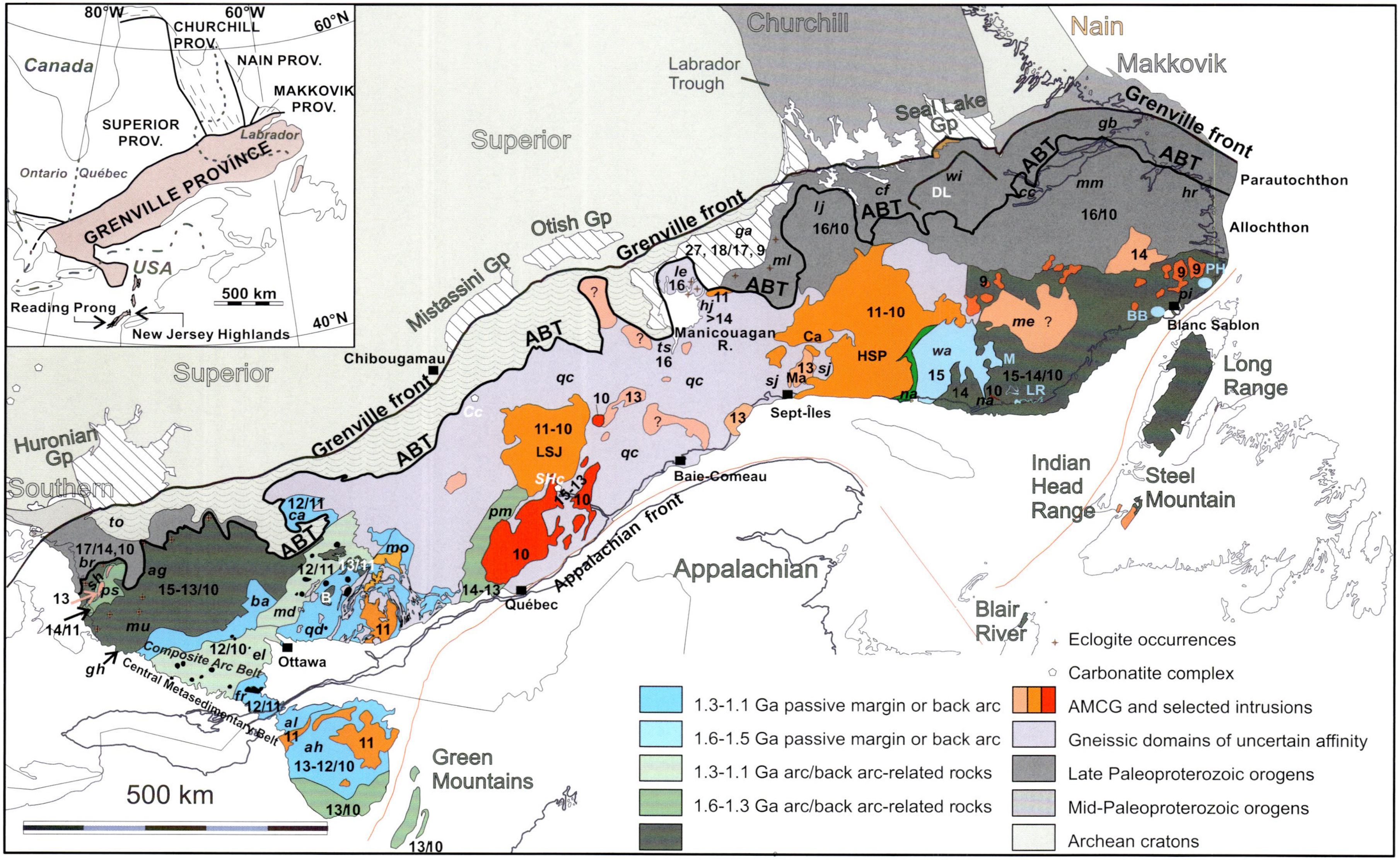

FIGURE 1. Geological subdivisions of the Grenville Province with deposition and intrusion age followed by metamorphic age in geon time scale. Terranes (t.) and domains (d.): ag, Algonquin d.; ah, Adirondack Highlands; al, Adirondack Lowlands; ba, Bancroft t.; br, Britt d.; ca, Cabonga t.; cc, Cape Caribou River allochthon; cf, Churchill Falls t.; el, Elzevir t.; fr, Frontenac t.; ga, Gagnon t.; gb, Groswater Bay t.; gh, Go Home d.; hj, Hart Jaune t.; hr, Hawke River t.; le, Lelukuau t.; lj, Lake Joseph t.; M, Musquaro Lake extension of wa; md, marble-rich d.; me, Mécatina d.; ml, Molson Lake t.; mm, Mealy Mountains t.; mo, Morin t.; mu, Muskoka d.; na, Natashquan Domain; PH, Pitts Harbour Group; pi, Pinware t.; pm, Portneuf-Mauricie d. including the Montauban Group; ps, Parry Sound d.; qc, Quebecia; qd, quartzite-rich d.; sh, Shawanaga d.; to, Tomiko t.; ts, Tshenukutish t.; wa, Wakeham Group; wl, Wilson Lake t. Supracrustal units with hydrothermal alteration zones discussed in text: B, Bondy Gneiss Complex; BB, Baie de Brador assemblage; DL, Disappointment Lake paragneiss; LR, La Romaine Supracrustal Belt. Potassic alkaline plutons of the Kensington-Skootamatta suite are the plutons in black in the Central Metasedimentary Belt. Cc, Crevier carbonatite; Shc, Saint-Honoré carbonatite.

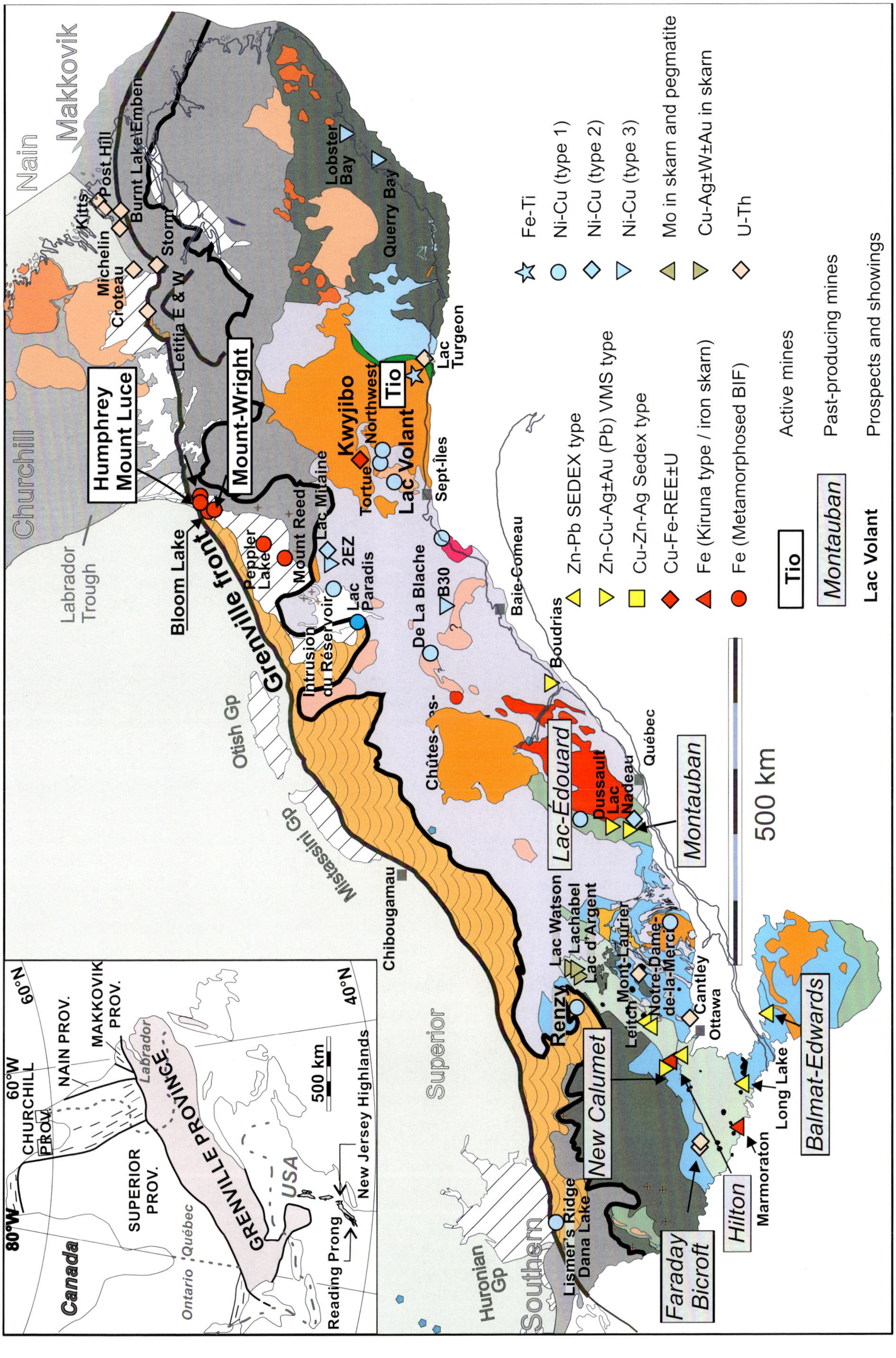

FIGURE 2. Location of mines and key prospects in the Grenville Province. Abbreviations as in Figure 1.

FIGURE 3. Granulite-facies lapillistone of the La Romaine Supracrustal Belt. **(A)** Polygenic lapillistone. **(B)** Monogenic rhyolitic lapillistone. **(C)** Shattered lapilli. **(D)** Metarhyodacitic lapillistone. Note that the smaller size lapillis are flattened and defined a fabric whereas the larger one is not flattened and slightly oblique to the fabric. This points to primary fabric development during deposition of hot pyroclastic material likely in a subaerial setting. **(E)** Accretionary lapilli.

crustal depths. That they are now tectonically juxtaposed and metamorphosed may hamper their recognition but should not affect significantly their potential in terms of resources. Variability in mapping coverage and quality of information, including map designations such as "undifferentiated gneiss", remains, however, a major impediment to exploration (see historical perspective in Davidson, 1998a and Parsons et al., 2005). New knowledge of ore deposits, including metamorphosed and metamorphogenic ones, recent protolith studies in high-grade metamorphic terranes, and new research on continent-scale processes of metallogenic significance provide means of identifying geotectonic environments prospective for mineral deposits in the Grenville Province. They illustrate the fertility of upper amphibolite- to granulite-facies terranes, even for gold deposits, and contribute new vectors to ore settings that are bound to improve the effectiveness of grass-roots exploration across the orogen (discussed in Bonnet and Corriveau, 2007).

In this overview of the Canadian Grenville Province, the detailed geological picture provided by Davidson (1998a) for the *Decade of North American Geology, Volume 7* is not duplicated. Instead, the focus is placed on geological contexts that, we think, have particular significance for assessing the exploration potential of the orogen. Wherever possible, a page link is provided to more exhaustive descriptions and data in review papers that target the tectonic evolution and the geology of the orogen (e.g. Easton, 1992; Rankin et al., 1993; Davidson, 1995, 1998a, b; Rivers, 1997; Carr et al., 2000; Rivers and Corrigan, 2000; Gower and Krogh, 2002, 2003; Tollo et al., 2004) and its current mineral resources (Easton and Fyon, 1992; Sangster et al., 1992; Clark, 2000, 2001a; Gauthier and Chartrand, 2005). The reader can also consult landmark volumes *The Grenville Problem* (Thompson, 1956; Osborne, 1956, p. 13), *The Grenville Province* (Wynne-Edwards, 1972), *The Grenville Event in the Appalachians and Related Topics* (Bartholomew, 1984), *The Grenville Province* (Moore, 1986; Moore et al., 1986), *Mid-Proterozoic Laurentia-Baltica* (Gower et al., 1990), and *Proterozoic Tectonic Evolution of the Grenville Orogen in North America* (Tollo et al., 2004), as well as recent special issues stemming from the Lithoprobe program and government and academia initiatives (Ludden and Hynes, 2000a; Wardle and Hall, 2002; Corriveau and Clark, 2005) and the International Geological Correlation Program (IGCP) Project No. 440 on the *Assembly and Break-up of*

Rodinia (Meert and Powell, 2001). Metamorphism across the Grenville Province and its orogenic significance are covered in Berman et al. (2000), Easton (2000), and Rivers et al. (2002) and are not discussed in detail here. However, the positive aspects of metamorphism on ore deposits (i.e. to upgrade the value of a deposit and facilitate its discovery), the major advances in understanding appropriate pathfinders for metamorphosed ore deposits, and the largely isochemical nature of metamorphism are taken into account in discussing mineral potential (see Bonnet and Corriveau, 2007). Finally, the Grenville Province hosts many industrial mineral deposits (apatite, brucite, calcite, dolomite, feldspar, ferrian ilmenite, graphite, kyanite, magnesite, mica, nepheline, rutile, sillimanite, talc, wollastonite; Hébert, 1993; Jacob and Bélanger, 2001; Bellemare and Jacob, 2004). While addressing herein the metallogenesis of the Grenville Province, we also recognize the importance of exploration for industrial minerals in the orogen, but this subject is not covered per se in this synthesis.

Tectonic Subdivisions

Several lithotectonic and timing nomenclatures, reviewed in Davidson (1998a, p. 207-212) and Tollo et al. (2004), are currently in use for the Grenville Orogen. Among them, two pan-Grenville Province schemes are principally used: the tectonically based subdivisions of Rivers et al. (1989; adapted to take into account current understanding of the orogen) and the tectono-geographic subdivision of Wynne-Edwards (1972). In the former, the major lithotectonic elements of the Grenville Province are divided into two main, semicontinuous, orogen-parallel, stacked belts known as the parautochthonous belt and the structurally overlying allochthonous polycyclic belt, as well as into a series of supracrustal-dominated belts formerly grouped as the allochthonous monocyclic belt (Fig. 1).

The parautochthonous belt (parautochthon) consists of supracrustal and plutonic rocks of the proto-Laurentian craton that were reworked to a major extent during the Grenvillian Orogeny, i.e., during the interval 1080 to 980 Ma (timing scheme of Gower and Krogh, 2002). These reworked Grenville foreland components include 1) the Archean greenstone and gneiss belts of the Superior and Nain cratons, 2) the intervening Paleoproterozoic southeastern Churchill Province with reworked Archean core zones (2780-1740 Ma; Wardle and Hall, 2002), and 3) the Paleoproterozoic Southern, Central Plains and Makkovik provinces and the early Mesoproterozoic Eastern Granite-Rhyolite Province (details in Davidson 1998a, p. 215-225). Individual supracrustal units and intrusive suites of the foreland can, in many cases, be correlated directly with those of the parautochthon, though metamorphic and structural reworking may be severe (Daigneault and Allard, 1994). In some cases, unit-to-unit correlations point to complexities that include potential sedimentary facies changes and severe disruption of foreland units (Davidson, 2001; LaFlèche et al., 2005). In other cases, foreland units are to a large extent obscured by voluminous intrusion of Late Paleoproterozoic and Mesoproterozoic plutons and batholiths (Gower and Krogh, 2003).

Grenville Province rocks have been thrust over and/or reversely faulted against those of the foreland along a series of ductile to brittle faults and shear zones that truncate the structural trends of the proto-Laurentian craton (see review in Ludden and Hynes, 2000b). These faults and shear zones define the Grenville Front and mark the northwest boundary of the Grenville Province (Davidson, 1998a, p. 218; Davidson, 2001). They are in some cases reactivated from pre-existing Archean to Proterozoic structures that have acted as persistent crustal weaknesses (Corrigan et al., 2000; Krogh et al., 2002; Bandyayera et al., 2004). A well developed foreland fold-and-thrust belt occurs in western Labrador (Rivers et al., 1993) but evidence is mounting that early ductile thrusting was also significant in western Quebec and preceded brittle reverse faulting (Bandyayera et al., 2004; Cadéron, written comm., 2005). Gravity and aeromagnetic signatures along the front are anomalous (e.g. major gravity low) and their trends truncate foreland gravity and aeromagnetic lineaments. Even though the front is not a suture, it is a major structure with crust-scale reverse faults, thrusts, and ramps (imaged in seismic transects) locally associated with a jump in Moho depth (imaged in a teleseismic transect) (Green et al., 1988; Davidson, 1998a, p. 251-253; Ludden and Hynes, 2000b; Martignole et al., 2000; Rondenay et al., 2000). One of the marked changes in crustal thickness occurs where the Grenville front zone extends to the Moho and into the mantle southward along the Abitibi-Grenville seismic profile (Martignole et al., 2000). Above this zone, a large region of very low reflectivity is observed in the lower crust. This combination of features is also present directly under the world-class Olympic Dam iron-oxide copper-gold (IOCG) deposit in the Gawler craton of Australia (Lyons et al., 2004) and may indicate a particularly fertile crust at the southern end of the Cabonga terrane in the southwest Grenville Province (Fig. 1). Archean crust is currently interpreted to underlie much of the orogen, gradually tapering to the southeast except in Ontario where it is interpreted to form a thin substrate that extends across much of the orogen (Ludden and Hynes, 2000b).

The allochthonous polycyclic belt includes Paleo- and Mesoproterozoic rocks that have been subjected to more than one orogeny and thrust onto the parautochthon along the Allochthon boundary thrust (Fig. 1). These rocks, displaced with respect to their formation sites, are not for the most part exotic with respect to Laurentia. In fact, the allochthonous belt was largely built through late Paleoproterozoic and Mesoproterozoic magmatic events along the Laurentian margin. In Quebec and Labrador, the Allochthon boundary thrust is clearly delineated by an abrupt change in aeromagnetic signature (Davidson, 1998a, p. 225). In Ontario, the location of this boundary fault is difficult to assess based on aeromagnetic signature, but is now well established through mapping. In particular, difference in chemistry between 1.16 Ga coronitic metagabbro bodies in the allochthonous belt and remants of the 1.24 Ga Sudbury dyke swarm in the parautochthon served to refine its position (Ketchum and Davidson, 2000). A high-pressure belt, with eclogite-facies and high-pressure granulite-facies assemblages, forms semicontinuously the structurally lowest segments of the allochthonous polycyclic belt (Ludden and Hynes, 2000b; Rivers et al., 2002). Recognition of this high-pressure belt is of metallogenic significance for Ti deposits (Cox and Indares, 1999; Gauthier and Chartrand, 2005).

The monocyclic belt was defined as consisting of the Central Metasedimentary Belt, Morin terrane, Adirondack Highlands, and Wakeham Group on the basis that they were thought to have formed coevally during the 1.35 to 0.95 Ga "Grenvillian orogenic cycle" as defined by Moore and Thompson (1980) (Rivers et al., 1989). This subdivision has been superseded following age refinements for its proposed constituents. In this contribution, we follow Tollo et al. (2004) after Rivers et al. (1989) and Carr et al. (2000) and refer to the Wakeham terrane, the Composite Arc Belt and the Frontenac-Adirondack-Morin Belt. The Composite Arc Belt, defined in Ontario, consists of 1.30 to 1.25 Ga carbonate and clastic sediments deposited coevally with mafic to felsic volcanic rocks and associated intrusion of tonalitic plutons (Carr et al., 2000). It encompasses Elzevir and Bancroft terranes in the Central Metasedimentary Belt , and perhaps the Parry Sound Domain in the Central Gneiss Belt (tectonic subdivision of Wynne-Edwards, 1972). In Bancroft terrane, hydrothermally altered volcanic rocks are now recognized as having precursors with geochemical and isotopic signatures diagnostic of Composite Arc Belt volcanic rocks (Peck and Smith, 2005), strengthening this interpretation. In contrast, the inclusion of the Parry Sound Domain is more conjectural, as its supracrustal component may be much older based on a poorly constrained 1350 Ma age for the Whitestone anorthosite and if so may be coeval with the Montauban Group as portrayed in Figure 1 (van Breemen et al., 1986). The northern extension of the Composite Arc Belt into Quebec following Tollo et al. (2004) encompasses the entire Quebec segment of the Central Metasedimentary Belt, whereas in Carr et al. (2000) it encompasses solely its western margin. This discrepancy is attributable to some major knowledge gaps in the age and paleotectonic setting(s) of the supracrustal rocks in western Quebec. Currently, a 1.45 to 1.3 Ga volcano-plutonic continental arc and island arc have been documented and dated within structural windows of the marble-rich and quartzite-rich domains of the Central Metasedimentary Belt in Quebec and within the La Bostonnais Plutonic Complex and Montauban Group of the Portneuf-Mauricie Domain (Fig. 1; Nadeau and van Breemen, 1994; Nadeau et al., 1999; Nantel and Pintson, 2002; Blein et al., 2003; Wodicka et al., 2004). Dated arc components within the 1.30 to 1.25 Ga age range are restricted to a 1.28 Ga tonalite sheet at the southeastern end of the quartzite-rich domain (Cieselski and Sharma, 1995). Williams (1992) also reported arc tholeiite and boninitic basalt in the region of the Calumet deposit and interpreted them as an extension of the Elzevir terrane in Ontario. That they also pointed out strong similarities with the Montauban Group rocks, now known to be 1.45 Ga in age, brings into question the age of the Calumet deposit and its host supracrustal rocks. Previously considered coeval with the Composite Arc Belt, the age of the Wakeham terrane is now bracketed between 1.6 and 1.5 Ga with upper stratigraphic levels being bracketed between 1.52 and 1.50 Ga (Wodicka et al., 2003; van Breemen and Corriveau, 2005).

The parautochthonous and allochthonous belts and the Composite Arc Belt are subdivided into a number of terranes (defined as tectonically bounded segments of orogenic crust distinguished from adjacent terranes on the basis of metamorphic, structural, and age characteristics; e.g. Davidson, 1995) or domains. Details of the various terrane and domain subdivisions and their settings can be found in Easton (1992), Wardle et al. (1997), Carr et al. (2000), and Gower and Krogh (2002). The various terranes are separated by major Grenvillian shear zones that were active during thrust assembly and subsequently reactivated in some cases as extensional shear zones (Davidson, 1984; Hanmer, 1988; Culshaw et al., 1997). In some allochthons, such as those of Ontario, high-strain gneiss is not restricted to domain boundaries but extends within the domains themselves, recording deep crustal ductile flow subsequent to crustal thickening (Culshaw, 2005). Key structural elements associated with buckling and shearing during ductile flow and lineated and foliated (L-S) tectonites formed during shearing help decipher the contractional and extensional phases of the Grenville Orogeny (Culshaw, 2005; Schwerdtner et al., 2005).

Paleo- to Mesoproterozoic Magmatic Arcs and Related Events

The main crustal build-up of the Grenville Province occurred through prolonged, 1.8 to 1.24 Ga, Andean-type continental arc and intracontinental back-arc magmatism with some lateral accretion of magmatic arcs (Rivers, 1997; Hanmer et al., 2000; Gower and Krogh, 2002). The main magmatic arc events currently documented are associated with the late Paleoproterozoic Labradorian (1.71-1.60 Ga) and the Mesoproterozoic Pinwarian (1.52-1.46 Ga; Gower and Krogh, 2002) and Elzevirian (ca. 1.35-1.19 Ga; McLelland et al., 1996) orogenies. Pre-Labradorian 1.8 to 1.7 Ga rocks occur within and south of the parautochthon in the eastern Grenville Province (Gower and Krogh, 2003) but appear to be restricted to the parautochthon in the western Grenville Province (Ketchum and Davidson, 2000). Such rocks are coeval with plutonism in the northern Central Plains Orogen of the Midcontinent, which itself is the eastward extension of the Yavapai Orogen of the southwest United States (Karlstrom et al., 2001). As such, this time period was one of intense crustal build-up along the pre-Grenvillian Laurentian margin and its counterpart within the Grenville Province may be obscured by intense reworking (e.g. units of uncertain affinity in Fig. 1 of Ludden and Hynes, 2000b). Among the crustal elements preserved and of interest are the Killarney volcanoplutonic complex and coeval granites that intruded the Huronian Supergroup and rocks of the southwestern Grenville Province in Ontario (e.g. Davidson and van Breemen, 1994; Rogers et al., 1995). Leucogranite, such as the Killarney granite and allied hypabyssal and volcanic rocks, are only found at the southwest extremity of the Grenville Front near Killarney, whereas granodioritic plutons are distributed over a larger area across the Southern and Grenville provinces. The plutonic rocks of the Killarney complex contain alteration zones of interest for mineral exploration, in particular for Fe oxide Cu-Au-U deposits, and are coeval with A-type granitoids in the Makkovik Province that host several U prospects (Fig. 2; Davidson, 1986; Corriveau, 2007).

In Labrador, 1.68-1.65 Ga juvenile arc terranes are now represented by strongly foliated or gneissic calc-alkaline intermediate to felsic rocks that have a distinct northward polarity of emplacement and are interpreted as having formed outboard of Laurentia above a south-dipping subduction zone (e.g. Lake Melville, Hawke River, and Groswater Bay terranes; Fig. 1). These rocks were accreted to pre-Grenvillian Laurentia, metamorphosed at high grade, migmatized and deformed, all before 1.65 Ga. The high pressure-temperature conditions reached led to local crystallization of sapphirine in pre-Labradorian supracrustal belts (Gower and Krogh, 2003, p. 156). The inferred suture then became the site of 1.65 Ga transitional calc-alkaline to alkali-calcic felsic plutonism, volcanism and minor sedimentation. These rocks constitute the Trans-Labrador batholith (Groswater Bay terrane) and associated northern supracrustal sequences. Minor, coeval, and related felsic volcanic and plutonic rocks are documented to the south in the Pinware terrane. Coeval with the Trans-Labrador batholith is the onset of large-scale, 1.65 to 1.62 Ga mafic-anorthositic-monzogranitic magmatism (Mealy Mountains Intrusive Suite and coeval mafic layered intrusions in the Mealy Mountains terrane; Fig. 1). Contemporaneous emplacement of these distinct magmatic suites led Gower and Krogh (2003, p. 167) to reassess the origin of the Trans-Labrador magmatism as having formed through crustal thickening after collision and arrested subduction instead of being the product of continental arc magmatism above a north-dipping subduction zone. Duchesne et al. (1999) proposed a crustal tongue-melting model for the formation of these anorthosite massifs, layered intrusions, and associated intrusive rocks. Arrested subduction and development of a passive margin after the 1.7 to 1.6 Ga Labradorian Orogeny is compatible with the currently observed lapse in magmatism during geon 15 and the deposition of the Wakeham Group, assuming it is a passive margin basin (Gower and Krogh, 2002, 2003). However, that interpretation conflicts with the presence of 1.61 to 1.55 Ga inherited magmatic-like zircons in volcanic rocks stratigraphically overlying metasediments of the Wakeham Group (deposited at ca. 1.51 Ga) and contemporaneous inheritance in nearby plutons (van Breemen and Corriveau, 2005). These geon 16 to 15 xenocrysts and the presence of the Pinwarian magmatic arc volcanic rocks at the upper stratigraphic level of the Wakeham Group suggest a link between Wakeham Group sedimentation and arc tectonics and on-going subduction for geon 15 (van Breemen and Corriveau, 2005).

Both Pinwarian and Labradorian plutons intruded the eastern Grenville Province south of the Trans-Labrador batholith and their lithologic similarity commonly makes them indistinguishable without systematic dating (Gower and Krogh, 2002; Heaman et al., 2004). Labradorian mafic magmatism is also documented in the Lelukuau terrane and the Tshenukutish Domain (Fig. 1; Cox et al., 1998). Along the St. Lawrence coast, no Labradorian plutons have been found to date but inheritance in Pinwarian igneous rocks and Nd model ages suggest that concealed Labradorian crust extends as far southwest as Sept-Îles, west of the Wakeham Group, and as far southeast as the Long Range inlier in the Appalachian Orogen of Newfoundland (Dickin, 2000, 2004; Heaman et al., 2002; van Breemen and Corriveau, 2005; Fig. 1). Labradorian rocks are also documented in the western Grenville Province of Ontario (e.g. Slagstad et al., 2004).

In contrast to Labradorian plutonism, major Pinwarian 1.52 to 1.46 Ga plutonism is now known to extend from the Pinware terrane of Labrador west to the Saguenay area in the central Grenville Province, north to the Lake Melville and Mealy Mountain terranes of Labrador, and east to the Long Range inlier (Fig. 1; Tucker and Gower, 1994; Wasteneys et al., 1997; Corrigan et al., 2000; James et al., 2001; Nadeau and James, 2001; Gower and Krogh, 2002; Heaman et al., 2002, 2004; Dickin, 2004; Hébert and van Breemen, 2004). This plutonic activity is interpreted as a manifestation of an Andean-type continental margin magmatic arc above a northwest-dipping subduction zone. The approximately 100 000 km^2 of Quebecia 'terrane' stretching from the arc-related Portneuf-Mauricie Domain to the St-Jean Domain of Gobeil et al. (2003) has homogeneous 1.55 Ga T_{DM} model ages and may represent an oceanic arc segment accreted to Laurentia during the Pinwarian Orogeny (Martin and Dickin, 2005). Orogen-parallel rifts associated with the Pinwarian arc led to extrusion of 1.5 Ga intra-arc mafic to felsic volcanic and pyroclastic rocks and deposition of sediments (La Romaine Supracrustal Belt and southeastern extension of the Wakeham Group at Musquaro Lake; Corriveau and Bonnet, 2005). Subsequently these volcanic belts were subjected to intense, fault- and lithologically controlled hydrothermal alterations including Fe-oxide alteration and Cu mineralization (Bonnet et al., 2005). Farther north, Pinwarian mafic intrusions and mafic dyke swarms, such as the Michael and Shabogamo dyke swarms and the Rigolet diorite, are interpreted as delineating an aborted intracontinental back-arc basin, whereas xenocrystic zircons suggest that basement may have an Archean component (Corrigan et al., 2000).

Pinwarian, ca. 1.48 to 1.45 Ga arc and back-arc magmatism is also documented in the allochthonous belt of Ontario and was associated with the emplacement of A-type granitoids, suggesting intra-arc extension (e.g. Britt, Algonquin and Muskoka domains; van Breemen et al., 1986; Corrigan et al., 1994; Nadeau and van Breemen, 1998; Ketchum and Davidson, 2000; Slagstad et al., 2004). As in the Pinware terrane, Pinwarian arc rocks were metamorphosed at ca. 1.45 Ga in Ontario (Ketchum et al., 1994; Gower and Krogh, 2002). Pinwarian crust also extends southward at least as far as the Blair River inlier in the Appalachian Orogen of Nova Scotia (Dickin, 2004; Fig. 1). The documented Pinwarian components attest to a major period of juvenile crustal addition in a magmatic arc along the Laurentian margin in geons 15 to 14 with a potential genetic link to the Eastern Granite-Rhyolite Province of the mid-continental United States (cf. van Breemen and Davidson, 1988; Nadeau and van Breemen, 1998; Bickford et al., 2000; Rivers and Corrigan, 2000; Culshaw and Dostal, 2002; Martin and Dickin, 2005). Slagstad et al. (2004) pointed out that the back-arc related A-type magmatism in the Muskoka Domain could be analogous to the 1.48 Ga A-type silicic and minor mafic magmatism in the Granite-Rhyolite Province that host former Fe-oxide mines with IOCG affinities in the St. François Mountains of Missouri (Lowell and Young, 1999; Lowell and Noll, 2001; Menuge et al., 2002). Renewed magmatism took place in extensional arc settings at 1.36 and 1.38 Ga in the parautochthonous and allochthonous belts in Ontario and Quebec (Sand Bay gneiss and Red Pine Chute orthogneiss; van Breemen et al., 1986; Slagstad et al., 2004; van Breemen and Currie, 2004) furthering the linkage with the mid-continental United States and the renewed 1.38 Ga magmatism in the St. François Mountains (Lowell and Noll, 2001). The 1.36 Ga Sand Bay gneiss association in the Shawanaga Domain (Fig. 1) consists of felsic to intermediate gneiss, amphibolite with minor marble, calc-silicate and quartzite layers interpreted as a bimodal volcanic suite with epiclastic, and Laurentia-derived clastic sediments (Culshaw and Dostal, 1997; Rivers and Corrigan, 2000). Rhyolitic rocks from this belt and the Red Pine Chute orthogneiss display unusual birdwing-shape rare earth element (REE) patterns diagnostic of hydrothermal alterations. In the case of the Red Pine Chute orthogneiss, this hydrothermal event has been dated at 1.03 Ga and is associated with the Kipawa syenite (van Breemen and Currie, 2004).

In the Central Metasedimentary Belt of Quebec, a series of granitic to tonalitic gneiss complexes form tectonic windows of 1.4 to 1.35 Ga, arc-related volcano-plutonic suites among the marble and quartzite domains (Corriveau and Morin, 2000). These gneiss complexes in the southern half of the belt include high-silica rhyolite and metabasalt with a back-arc signature and gneissic rocks with island-arc affinity (Bondy Gneiss Complex, Fig. 1). They are interpreted as part of a former island arc built on a thin continental substrate (Blein et al., 2003; Wodicka et al., 2004). Coeval gneisses farther north have a continental arc signature (Nantel and Pintson, 2002). Such a juxtaposition of island arc and continental arc material occurs farther east but at much lower metamorphic grade. In the 1.45 Ga Montauban Group, island arc volcanic rocks are particularly well preserved, whereas the 1.41 to 1.38 Ga, calc-alkaline La Bostonnais Plutonic Complex provides evidence for continental arc magmatism along Laurentia (Corrigan and van Breemen, 1997; Nadeau et al., 1999; Hanmer et al., 2000). The latter magmatic event is interpreted to extend farther north into the Saguenay area (Hébert and van Breemen, 2004). Taken together, these data record a long-lived arc along the southeastern margin of Laurentia. Information derived from surface geology, seismic reflection lines, and felsic to ultramafic xenoliths suggests that the intermediate and lower crust of the Central Metasedimentary Belt in Quebec corresponds to a stack of arc-related gneiss complexes interleaved with non-exposed quartzite-bearing supracrustal assemblages, mylonites, and mafic to ultramafic intrusive bodies (Corriveau and Morin, 2000; Morin et al., 2005). The marble and quartzite domains of the Central Metasedimentary Belt of Quebec may extend to the east into the Morin terrane and, if this conjecture was correct, they should be grouped with the Frontenac-Adirondack-Morin Belt. In that case, there may be little of the Central Metasedimentary Belt of Quebec that could be included in the Composite Arc Belt of Carr et al. (2000) assuming that these components are exotic, an issue that remains open to debate (cf. Hanmer et al., 2000).

In the Composite Arc Belt, Elzevirian arc volcanic, plutonic, and mineralizing events have been well documented and summarized in Easton and Fyon (1992) and Carr et al. (2000). The belt is interpreted as consisting of 1.29 to 1.28 Ga primitive volcanic arc, 1.28 to 1.27 Ga calc-alkaline volcano-sedimentary island arc/back-arc and related tonalitic plutons, and 1.26 to 1.25 Ga bimodal rifted volcanic arc (back-arc) amalgamated between 1.25 and 1.24 Ga (poorly constrained deformation and metamorphism) and intruded by 1.24 to 1.22 Ga bimodal plutons. A series of mineral deposits are related to that last magmatic event. Evidence of coeval emplacement of tonalitic plutons in adjacent metasedimentary terranes is rare and consists of a 1.28 Ga tonalite among Grenville Supergroup metasedimentary rocks of the Central Metasedimentary Belt of Quebec and 1.32 to 1.30 Ga tonalites in the Adirondack Highlands.

Andean-type crust normally consists of widespread rather than isolated volcanic, volcano-plutonic, and volcano-sedimentary belts in association with plutonic activities in arcs, successor arcs, intracontinental back arcs, and accreted arcs. As illustrated above, widespread arc-related volcanic activity is well documented in the low-grade metamorphic terranes of the Grenville Province (Composite Arc Belt and Montauban Group; Nadeau et al., 1999; Carr et al., 2000). In contrast, extensive volcanic belts related to the successive arc build-up of the Grenville Province are currently rare in the high-grade metamorphic terranes but increasingly there are hints that the Paleo- to Mesoproterozoic arc components are not solely plutonic or emplaced at mid- to deep crustal levels. Conclusive evidence of volcanism is found in association with the 1.74 Ga Killarney granite, the Trans-Labrador batholith, the southeastern margin of the Wakeham Group, the La Romaine Supracrustal Belt (as illustrated in Fig. 3), and the Bondy Gneiss Complex, and geochemically inferred rhyolitic rocks occur in the Shawanaga Domain (Fig. 1; Davidson, 1986; Gower, 1996; Blein et al., 2003; Slagstad et al., 2004; Corriveau and Bonnet, 2005). Of these, three belts host fertile hydrothermal alteration zones and will be described in further detail in subsequent sections. Late- to post-tectonic magmatic successor events to these arc events are also proving fertile.

Sedimentary Basins

Paleo- to Mesoproterozoic sedimentary basins preserved in the Grenville Province include (1) the extensions of older basins from the Grenville foreland; (2) the Paleoproterozoic Disappointment Lake gneisses and correlative units in the Lac Joseph, Churchill Falls, and Wilson Lake terranes (Gower et al., 2002; Gower and Krogh, 2003, p. 152-153; Korhonen and Stout, 2004); (3) the supracrustal rocks in the Tomiko terrane (≤1.68 to ≥1.25 Ga; Easton, 2003a); (4) the 1.6 to 1.5 Ga Wakeham Group (Gobeil et al., 2003); and (5) the ca. 1.3 to 1.2 Ga Grenville Supergroup in the Composite Arc Belt, the Frontenac-Adirondack-Morin Belt, and the New Jersey Highlands and Reading Prong inliers (Fig. 1; Wynne-Edwards, 1972). Metasedimentary rocks also occur as irregular belts among Labradorian, Pinwarian, and younger gneissic rocks. These include the Montauban Group, its migmatitic counterparts and the St-Boniface metasedimentary rocks in the Portneuf-Mauricie Domain (Nadeau et al., 1992; Corrigan and van Breemen, 1997), the 1.5 Ga Saguenay Gneiss Complex near Saguenay (Hébert and van Breemen, 2004), and tracts of paragneiss in the Manicouagan, Baie-Comeau, Sept-Îles, Manitou, and Blanc Sablon area in Quebec, Parry Sound, Britt, Algonquin, and Go Home domains in Ontario, and Pinware terrane and Mecatina Domain in Labrador (Davidson, 1998a; Gobeil, 1997; Gower et al., 2002; Gower and Krogh, 2003, p. 158; Fig. 1E-1 in Perreault and Moukhsil, 2003). The Flinton Group in the Composite Arc Belt (Moore and Thompson,

1980) is much younger (ca. 1.1 Ga; Sager-Kinsman and Parrish, 1993).

Among the Grenville foreland extensions, those of the Paleoproterozoic Huronian and Kaniapiskau supergroups and the Archean Pontiac Group are the most extensive. They were deeply infolded into the crust long before the Grenville Orogeny and this may have fostered their preservation. In contrast, the Otish Group only extends into the Grenville Province as small faulted remnants whereas the adjacent Mistassini Group stops at the Grenville Front (Davidson, 1998a, p. 215-225).

In Labrador, the Paleoproterozoic Disappointment Lake gneiss, and its correlatives, form a 400 km long, tectonically dismembered, arc-related supracrustal package found within the Lac Joseph, Churchill Falls, and Wilson Lake terranes (Thomas et al., 2000; Gower et al., 2002). This supracrustal belt comprises pelitic to semipelitic rocks with minor quartzite and calc-silicate rocks in granulite facies, and rare amphibolite with textural evidence for an origin as pillow basalt. Deposition age is currently bracketed between 1.8 and 1.7 Ga.

Metasedimentary rocks in the Tomiko terrane of Ontario comprise quartz arenite, calc-silicate rock, quartz-muscovite gneiss, feldspathic gneiss, and locally marble and Mn-rich magnetite-chert Fe formation associated with amphibole gneiss, aluminous gneiss, and B- and Mn-rich gneiss with garnet and tourmaline. Complex relationships among mineral parageneses suggest polyphase hydrothermal activity to form the Mn-, B-, Al-, and Fe-rich units. Amphibolites occur but are of uncertain origin (mafic sills or volcanic flows). The area is considered prospective for Broken Hill-type mineralization (Easton, 2003a). This metasedimentary sequence extends northeastward into Quebec. The deposition age is currently bracketed between 1.68 and 1.25 Ga. However, a Paleoproterozoic age is favoured based on the restricted zircon population of a sample dated at 1687 ± 20 Ma (which may point to a volcaniclastic origin) and on Nd/Sm model ages, which are 400 to 500 million years older than those of the Grenville Supergroup in Frontenac terrane (Krogh, 1989; Easton, 2003a, p. 17).

The Wakeham Group in the eastern Grenville Province forms a large sedimentary basin (Fig. 1) interpreted as originally extending farther north with current septa within the Mecatina Domain, farther west as the Manitou and Baie Comeau sediments and farther east at least as far as the La Romaine Supracrustal Belt and likely beyond into the Saint-Augustin complex of the Pinware terrane (James et al., 2002; Perreault and Heaman, 2003; Corriveau and Bonnet, 2005). East of St. Augustin, in Pinware terrane, dated metasedimentary units have a Labradorian age (Gower and Krogh, 2002) but without systematic U-Pb geochronology it is unclear to what age groups the various metasedimentary rocks of this terrane belong.

Formerly ranked as a supergroup and subdivided into two distinct volcano-sedimentary groups, 1.24 Ga in age (Martignole et al., 1994), the Wakeham Group is now known to be essentially arenaceous, largely devoid of volcanic rocks, and either in tectonic or intrusive contact with 1.5 Ga gneiss complexes and their intrusive suites (Gobeil et al., 2003). Isotopic signatures and 1.6 to 2.6 Ga detrital zircon age distributions are similar for the two former groups and, with field relationships, make a two-fold subdivision unnecessary (Gobeil et al., 2003; Wodicka et al., 2003; Y. Larbi, pers. comm., 2004). Based on the above age constraints, the main sedimentary sequence must be older than 1.5 Ga and younger than 1.6 Ga (Larbi et al., 2003; Wodicka et al., 2003). In the Musquaro Lake sector, at the eastern margin of the Wakeham Group, the presence of Pinwarian 1.52 Ga detritus in arenaceous sediments overlain by 1.51 and 1.50 Ga felsic volcanic and pyroclastic rocks, permits the age of sedimentation to be bracketted between 1.52 and 1.50 Ga (van Breemen and Corriveau, 2005). The nature of the sandstones and the age of detrital zircons indicate a largely cratonic source comprising Archean and Proterozoic rocks with local volcanic or plutonic detritus of 1.52 Ga (Wodicka et al., 2003; van Breemen and Corriveau, 2005). The sedimentary package of the Wakeham Group is not diagnostic of a unique tectonic setting. Sedimentation in a passive margin setting is commonly invoked but deposition in an aborted back-arc is an attractive alternative as volcanic rocks overlying sediments bear a continental arc magmatic signature (Rivers and Corrigan, 2000; Y. Larbi, pers. comm., 2004; Corriveau and Bonnet, 2005). The group has Cu-Au-Ag epigenetic mineralization at the interface between quartzite and mafic rocks in association with aluminous gneiss and tourmalinite (Clark, 2003).

The Grenville Supergroup encompasses several metasedimentary successions for which basin correlation remain uncertain (Moore, 1986; Davidson, 1998a, p. 235) and age of deposition is either unconstrained, bracketed between 1.30 (detrital zircon) and 1.18 Ga (oldest pluton from the Frontenac plutonic suite) as in Ontario, or to pre-1.2 Ga in Quebec and pre-1.23 Ga in Ontario (Sager-Kinsman and Parrish, 1993; Davidson, 1998a, p. 238; Corriveau and van Breemen, 2000). In Quebec, a 1.2 Ga zircon age, interpreted as a maximum age of deposition (Friedman and Martignole, 1995), has been reinterpreted as metamorphic in origin based on other regional data (Corriveau and Morin, 2000, p. 249). Hence, the age of the metasedimentary rocks in western Quebec still needs to be established. In Ontario, the Flinton Group overlies the Grenville Supergroup and 1.25 Ga plutonic rocks and carries detrital zircon as young as ca. 1.15 Ga (Sager-Kinsman and Parrish, 1993), providing evidence for distinctly younger sedimentation in this part of the orogen.

The former La Romaine Domain of the eastern Grenville Province (Perreault and Heaman, 2003) was interpreted as a roughly 600 km^2 metasedimentary belt with "meta-arkose" and rare pelite, quartzite, conglomerate, and marble. Many of the markers, likely used to suggest a sedimentary origin for the domain gneisses, can be demonstrated to consist of granitic dykes with striking magmatic layering (Fig. 4), dioritic intrusive breccia and cogenetic enclaves, orthogneiss with spectacular deformation features, a series of metamorphosed lapillistone, felsic tuff with aluminous nodules and veins, as well as aluminous gneiss, nodular gneiss, ironstone, calc-silicate, and carbonaceous rocks of hydrothermal origin and subsequently metamorphosed (Corriveau et al., 2002; Corriveau and Bonnet, 2005). Reassessment of this domain as a series of continental arc-related plutons with coeval narrow intra-arc volcano-sedimentary basins illustrates the current state of knowledge in many of the gneiss complexes of

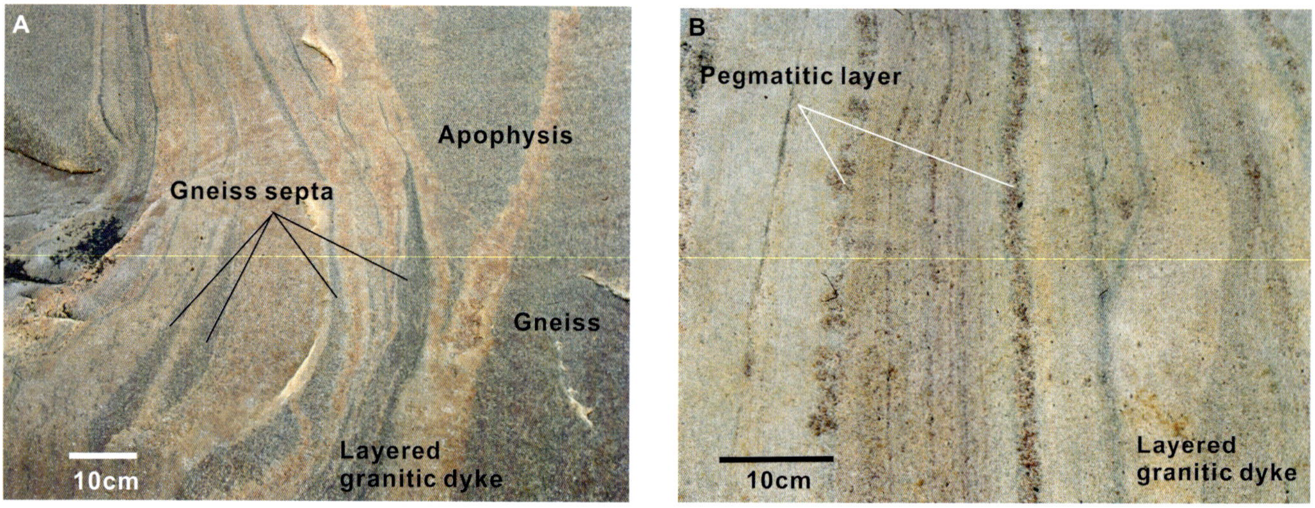

FIGURE 4. Granitic dykes with magmatic layering presenting an appearance of meta-arkoses. (A) Apophyses are found in a few dykes. (B) Some of the layering is coarse grained but most are fine grained.

the Grenville Province. This volcano-plutonic environment is discussed below in the section on prospective arc settings.

Large-Scale Magmatic Events

Anorthosite-Mangerite-Charnockite-Granite, and Mafic Magmatism

Emplacement of large anorthosite massifs and coeval batholiths of mangerite-charnockite-granite (AMCG suites) marks periods of post-orogenic activity or reactivation following the Labradorian, Pinwarian, Elzevirian and Grenvillian orogenies (e.g. McLelland et al., 1996; Gower and Krogh, 2003). In most cases, coeval layered mafic intrusions also occur (Fig. 5). For example, the oldest AMCG suite, the post-orogenic, 1.65 to 1.62 Ga Mealy Mountains Intrusive Suite, is coeval with the Grady Island layered mafic intrusion, the White Bear Arm complex, the Ossok Mountain intrusive suite, and the Alexis River layered anorthosite (Gower and Krogh, 2003, p. 160-163). Although coeval, distribution of the layered intrusions is commonly spatially distinct from the AMCG intrusions, and their cogenetic character is uncertain. Potentially coeval with Mealy Mountain magmatism and intruded by Pinwarian-age mafic to ultramafic, layered intrusions are the granulite-facies mafic sills of the Hart Jaune terrane (e.g. 1.51 Ga Réservoir intrusion; Clark, 2000; Indares and Dunning, 2004). A mafic rim is observed around some anorthosite massifs, such as in the Havre St.-Pierre, Pentecôte, De La Blache, and Tortue anorthosites and components of the Lac St Jean anorthosite (e.g. Gobeil et al., 1999, 2002). The 1.37 to 1.35 Ga Pentecôte anorthosite massif, its marginal mafic to ultramafic intrusion, and associated felsic plutons are coeval with the 1.37 Ga Matamec mafic-silicic layered intrusion and mineralized 1.35 Ga gabbro dykes (Martignole et al., 1993; Saint-Germain and Corriveau, 2003; Nabil et al., 2004), the ca. 1.35 Ga Whitestone anorthosite in the Parry Sound Domain, and the 1.33 Ga De La Blache anorthosite massif (Gobeil et al., 2002). Though lesser magmatic activity is currently reported for geon 13 than geons 16, 15, and 12, more and more mafic and anorthosite intrusions of that age range are being dated as coeval with the calc-alkaline intrusions of the Dysart Suite, La Bostonais Complex and other gneiss complexes in the southwestern and central Grenville Province (Rivers and Corrigan, 2000, p. 364).

Emplacement of the very large, 1.17 to 1.13 Ga Marcy, Morin, Lac St. Jean, Havre St. Pierre, and Fournier AMCG suites implies significant addition to the crust in geon 11 in the central part of the Grenville Province (Martignole et al., 2000; Wodicka et al., 2003; Hamilton et al., 2004). Host rocks were strongly deformed and metamorphosed prior to emplacement for the Morin and Lac St. Jean AMCG suites (Martignole et al., 2000; Hébert and van Breemen, 2004). In contrast, host and members of the Havre St. Pierre and Marcy AMCG suites were subjected to high-grade metamorphism and deformation related to the Grenville Orogeny after anorthosite massif emplacement (Wodicka et al., 2003; Hamilton et al., 2004). However, in the Havre St. Pierre anorthosite, field evidence indicates that anorthosite recrystallization is in part related to emplacement. A few mafic dykes have delaminated their anorthositic hosts and the septa display ptygmatitic and sheath folds solely related to their entrainment by the mafic magmas. Such field observations, among others, illustrate the extremely ductile behaviour of anorthosite at high temperature, and by inference, the ease with which they recrystallize during emplacement.

The anorthosite massifs discussed above belong to the labradorite type but a younger suite of andesine anorthosite was also emplaced in the ca. 1.06 to 1.01 Ga interval (Owens and Dymek, 2005). In areas where waning of the Grenville Orogeny occurred by ca. 1.05 Ga, late to post-orogenic anorthosite suites are documented (e.g. 974 Ma Vieux-Fort Anorthosite, Heaman et al., 2004). Some of them have magmatic Fe-Ti mineralization (e.g. Rivière au Tonnerre anorthosite, Gobeil et al., 2003; 1.02 Ga Labrieville anorthosite suite, Hébert et al., 2005).

Three models are currently invoked for the formation of AMCG suites in the Grenville Province: (1) lithospheric delamination associated with orogenic collapse (e.g. McLelland et al., 1996; Corrigan and Hanmer, 1997), (2) delamination-related melting of underthrust mafic crust (Gauthier et al., 2004c following the model of Duchesne et al., 1999), and (3) mantle plume (Gobeil et al., 2003). The two late- to post-orogenic models challenge long-invoked

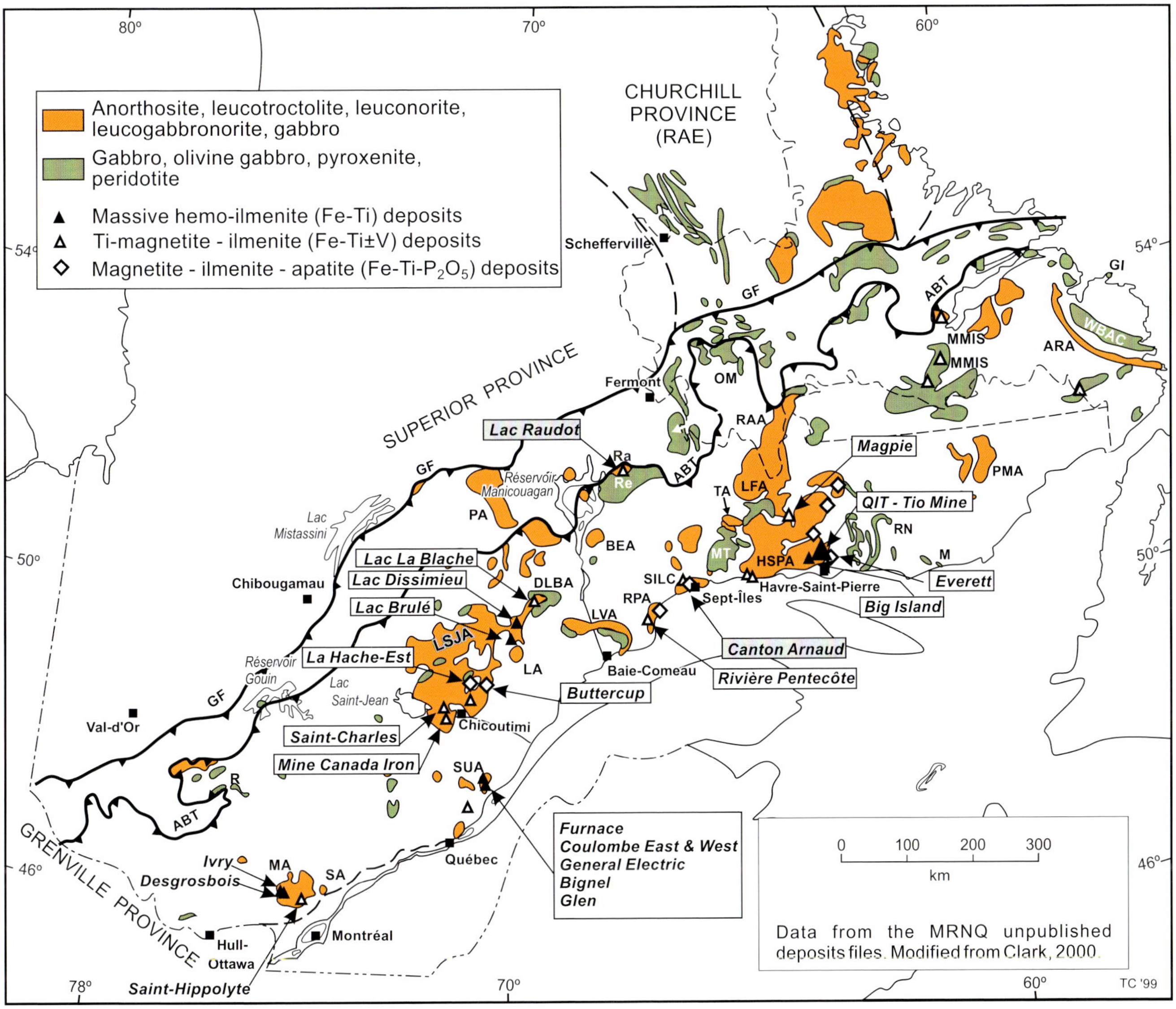

FIGURE 5. Distribution of anorthosite massifs and major mafic intrusions and their Fe-Ti deposits in the Quebec and Labrador segments of the Grenville Province. Anorthosite massifs: ARA, Alexis River; BEA, Berté; LA, Labrie; DLBA, De La Blache; HSPA, Havre-Saint-Pierre; LFA, Lac Fournier; LSJA, Lac Saint-Jean; LVA, Lac Vaillant; MA, Morin; PA, Pambrun; PMA, Petit-Mécatina; RAA, Atikonak River; RPA, Rivière Pentecôte; SA, Shawinigan; SUA, Saint-Urbain; TA, Tortue; . Mafic intrusions and intrusive complexes: GI, Grady Island; M, Musquaro; MMIS, Mealy Mountains; MT, Matamec; OM, Ossok Mountain; R, Raudot; SILC, Sept-Îles; WBA, White Bear Arm. Sills: R, Renzy; RN, Robe Noire. GF, Grenville Front.

anorogenic models stemming from Emslie (1978), and convergent rather than purely extensional settings have been proposed for the Lac St Jean and Morin anorthosites (Higgins and van Breemen, 1996; Corriveau and Morin, 2000).

Other mafic intrusions (Fig. 5) form (1) swarms of mafic dykes such as the 1.47 Ga Shabogamo, 1.43 Ga Michael gabbro, 1.25 Ga Mealy, and 1.24 Ga Sudbury dyke swarms (Connelly et al., 1995; Bethune, 1997; Emslie et al., 1997; Rivers and Corrigan, 2000; Davidson, 2001); (2) mafic to ultramafic sills such as the Robe noire and Lac Renzy sills (Clark, 2000; Scherrer, 2003); (3) metamorphosed, deformed, folded, or tilted layered intrusions such as the 2.47 Ga River Valley anorthosite-gabbro intrusion, 1.63 Ga White Bear Arm complex, the 1.29 Ga Blanc-Sablon gabbroic complex, and the 1.07 Ga Musquaro intrusion (Corriveau et al., 2002; Easton, 2003b; Gower and Krogh, 2003; Heaman et al., 2004); (4) coronitic metagabbro such as those of the allochthonous belt in Ontario and Quebec (Ketchum and Davidson, 2000); (5) syntectonic mafic-felsic sheet intrusions emplaced in steep deformation zones such as those of the Chevreuil suite (Corriveau and van Breemen, 2000); (6) mafic intrusions with primary vertical layering such as the Montjoie and Diable intrusions (ibid) or coeval gabbro and anorthosite bodies, some with layering, in the Frontenac plutonic suite (Davidson and van Breemen, 2000); and (7) collapsed mafic-silicic layered intrusions (MASLI) such as the Matamec complex (Saint-Germain and Corriveau, 2003). Many of these different intrusion types have Ni-Cu magmatic sulphide potential (Wilson, 1993; Clark, 2000). A few of the mafic dyke swarms extend into the Grenville from the foreland where they are not deformed: Shabogamo, Sudbury and probably Michael gabbro dyke swarms.

Other Magmatic Events

The late-stage magmatic suites in the Grenville Province are numerous and document key elements in the evolution of the orogen. Many are prospective for IOCG, Mo, U, and other base, precious, strategic, and rare metals (Haynes, 1986; Gower, 1992; Eckstrand et al., 1996). For example, late magmatic events with A-type granitic affinity gave rise to a spectrum of 1.05 Ga U-Th-Nb-REE and Mo-enriched pegmatites and skarns in the Central Metasedimentary Belt of Ontario and Quebec, to IOCG-type 1.04 Ga mineralization in the Adirondack Highlands, and to the ca. 1 Ga IOCG Kwyjibo deposit in the Canatiche Complex of eastern Quebec (Gauthier et al., 2004a; Selleck et al., 2004; Clark et al., 2005; Lentz and Creaser, 2005; D. Lentz, per. comm., 2004). Alkaline rocks, including nepheline syenite, form (1) series of intrusions such as those of the 1.09 to 1.07 Ga Kensington-Skootamatta potassic alkaline plutonic suite (Corriveau and Gorton, 1993), (2) metasomatic rock complexes such as the 1.03 Ga Kipawa syenite (van Breemen and Currie, 2004), and (3) lenses among gneisses such as the one in the Manitou gneiss complex (Chevé et al., 1999) and the 1.17 Ga Cabonga Nepheline Syenite Complex in the Cabonga terrane (Hudon et al., 2006). Details of some late intrusion-related metallogenic settings are described in the metallogenic section. Finally, post-Grenvillian carbonatite intrusions (Crevier, Saint-Honoré, Oka and Manitou Islands) and the 564 ± 4 Ma Sept-Îles mafic layered intrusion are host to subeconomic to currently mined strategic metals deposits (including Nb at the St-Honoré, formerly Niobec, mine, and ilmenite and apatite at Sept-Îles) but are not discussed further in this synthesis (see Richardson and Birkett, 1996; Cimon and McCann, 2000; Higgins, 2005).

Grenvillian Tectonics

The Grenvillian (1.08-0.98 Ga) Orogeny, as defined by Gower and Krogh (2002), is the only one that affected nearly the entire orogen, from the eastern Grenville Province to the Adirondack Highlands and the Appalachian inliers, south to Texas and Mexico (Tollo et al., 2004 and references therein). This event encompasses the Ottawan and Rigolet events of Rivers and Corrigan (2000). It gave rise to a mountain belt with a strike length the scale of the Himalaya and to the current-day collisional architecture. In contrast, the Elzevirian and Shawinigan (1.18-1.13 Ga) events mostly affected the Grenville Province in Ontario and western Quebec (Davidson, 1995; Carr et al., 2000; Rivers and Corrigan, 2000).

During the Grenvillian Himalayan-type event, extensive crustal thickening and tectonic extrusion led to widespread high-grade metamorphism (Ludden and Hynes, 2000b). Areas not pervasively reworked in the process commonly preserve earlier metamorphic imprints (e.g. Archean metamorphism in parts of the Parautochthon of Quebec; Gariépy et al., 1990; Berclaz et al., 1995; Bandyayera et al., 2004). In the Central Metasedimentary Belt of Quebec, only minor shearing and a weak resetting of metamorphosed marble and post-peak metamorphism skarns are observed after 1.1 Ga (Corriveau et al., 1998; Martignole et al., 2000; Peck et al., 2005). In Labrador, Grenvillian tectonics involved thick-skinned imbrication of Labradorian terranes. As a result, large segments of the parautochthonous and allochthonous belts have not been pervasively reworked and do not have a strong tectonothermal record of the Grenville Orogeny except along major shear zones (e.g. Mealy Mountain terrane, Gower, 1996; Cape Caribou River allochthon, Krauss and Rivers, 2004). These areas and many others across the Grenville Province provide evidence that strain partitioning has strongly controlled the extent of preservation of pre-Grenvillian rocks and metamorphic imprints (e.g. Ketchum et al., 1994; Corriveau et al., 1998; Boggs and Corriveau, 2004).

Geological Environments of Metallogenic Deposits

In the Grenville Province, SEDEX and volcanogenic massive sulphide deposits associated with metasupracrustal belts, Fe-Ti oxides and Ni-Cu sulphides associated with anorthositic rocks (AMCG suites) and gabbro, and various types of mineralization (Fe, U, Mo, REE, Nb, etc.) associated with skarn, pegmatite, and carbonatite intrusions anchor our collective view of the prospective metallogenic settings. Sixty-two deposits are listed in Eckstrand et al. (1996, Fig. 3A) for the Grenville Province; twenty-three are in Ontario and none are in Labrador, although Baltic shield metallogeny of 1.7 to 1.5 Ga terranes in Norway, Sweden, or Finland is strongly relevant to Labrador and supports mineral potential of the area (Gower, 1992). In this section, we review settings that the authors think warrant investigation or reinvestigation of their mineral potential. These are outlined in two groups: environments that extend from the foreland into the Grenville Province, and those specific to the province itself.

Extension of Metallogenic Districts from the Grenville Foreland

Metallogenic settings within the various geologic provinces northwest of the Grenville Front extend from the foreland into the parautochthonous belt of the Grenville Province. In contrast, many of their known mining districts abut against the Grenville Front. A key exception is the world-class Mount Wright Fe mine (Eckstrand et al., 1996; Stanaway, 1996). We would advocate that exploration potential of these extensions in terms of pre-existing ore deposits should largely mimic that of the foreland itself and that the extensions remain in many cases under-explored (e.g. Allard, 1978, 1979).

Archean Greenstone Belts

Archean greenstone belts that extend from the Superior into the Grenville Province have been deformed and intruded by mafic magmas during the Mesoproterozoic and are now considered prospective for Archean to Mesoproterozoic Cu-Ni-Co-PGE deposits, Archean volcanogenic massive sulphide deposits and Archean or Mesoproterozoic mesothermal gold deposits (Allard, 1978, 1979; Cadéron et al., 2004). If present, Archean orebodies will be metamorphosed to amphibolite or granulite facies as those in northeast Superior Province (Archer et al., 2004) and exploration strategies in such non-classical exploration territories will need to be adapted accordingly (Bonnet and Corriveau, 2007).

The southeast limit of Archean crust at depth as outlined in Rivers (1997) and following interpretations of Ludden and Hynes (2000b) are interpreted by Gauthier et al. (2004a, c)

FIGURE 6. The open pit of the Mount Wright mine in Quebec.

FIGURE 7. Metamorphosed iron formation of the Wabush Formation (pencil for scale).

to have metallogenic significance for magmatic Ni-Cu-Co sulphide, platinum-group elements (PGE), Fe-Ti oxides, and Fe oxide Cu-Au (IOCG) deposits.

Early Paleoproterozoic Platinum Group Element Mineralization

At the boundary between the Superior and Southern provinces, rift-related 2.49 to 2.44 Ga mafic intrusions (East Bull Lake intrusive suite) are mineralized with platinum group elements (PGE; Vogel et al., 1998; James et al., 2002; Vaillancourt et al., 2003). This plutonic suite extends eastward into the Grenville Province of Ontario, where the 2.45 Ga River Valley anorthosite-gabbro intrusion is now undergoing active exploration for PGE (Dana Lake and Lismer's Ridge prospects; Easton, 2003b). This magmatism is coeval with volcanism at the base of the Huronian Supergroup and is related to the plume event that gave rise to the major Matachewan dyke swarm. By analogy, Clark (2001b, slide 36) points out that the area south of the 2.47 Ga Mistassini dyke swarm in the foreland may also host such intrusions prospective for PGE mineralization. However, to date that area remains largely uncharted.

Metamorphosed Iron Formation

The Paleoproterozoic Kaniapiskau Supergroup extends from the New Quebec Orogen (Labrador Trough) into Gagnon terrane of the Grenville parautochthon. The Wabush-Lac Jeannine Fe ore belt, notably the Mount Wright Fe mine, is hosted within this southward extension (Figs. 2, 6; Clarke, 1977). In the New Quebec Orogen, most of the iron formations, referred to as taconite (Gross, 1996), are found in the Sokoman Formation. The Wabush Formation is its metamorphosed equivalent south of the Grenville Front (Figs. 1, 7; Neal, 2000). What was originally a continuous belt of Fe formations with quartzite, calcareous and dolomitic carbonate, and shale becomes across the front a dismembered belt of folded and oval-shaped Fe orebodies associated with quartzite, calcitic and dolomitic marble, calc-silicate rock, and aluminous and graphitic paragneiss distributed, in a northeast-southwest trend, over their Archean tonalitic basement in the parautochthonous Gagnon terrane. A northeast to southwest increase in metamorphic grade from greenschist facies north of the Grenville Front to upper amphibolite - lower granulite facies to the south led to significant ore benefitiation with coarsening of hematite and magnetite grain size and production of magnetite through the breakdown of Fe carbonates at the contact with silicates (Clarke, 1977; Klein, 1978; Gross, 1996 p. 62-67). High temperature-pressure metamorphism took place at around 985 Ma, but an early metamorphic event of unknown conditions also occurred at 1740 to 1720 Ma (Jordan et al., 2006). Refolded Fe formations reach several hundreds of metres in structural thickness (Roach and Duffel, 1974). Fe oxides tend to concentrate in fold hinges and the deposits reach hundreds of million tons with 25 to 45% Fe. The Fe ore is easy to separate from its friable gangue minerals and produces, by gravity or magnetic separation, a high-grade Fe concentrate of 66 to 67% Fe (Gross, 1996; Neal, 2000).

Late Paleoproterozoic and Early Mesoproterozoic Mineralization

The Makkovikian and Labradorian arc rocks of the Makkovik Province host several U deposits (Kitts-Michelin, Labrador; Gandhi, 1986) and, like the extension of rocks of similar age in the Baltic Shield (Weihed, 2001; Eilu et al., 2003), have been explored extensively (Marshall et al., 2003). Rocks of the Makkovik Province extend in the parautochthon and may also be of interest for exploration, including for IOCG and U deposits. Likewise, at the southwest end of the Grenville Province in Ontario, late Paleoproterozoic granitoid and other rocks in the parautochthon may deserve exploration for similar deposits.

Fe oxide-REE deposits of the St. François Mountains, Missouri (Pea Ridge, Pilot Knob, Bourbon, Camels Hump, Katz Spring, and Iron Mountain; Seeger et al., 2001) are associated with felsic volcanic rocks and granite (Eastern Granite-Rhyolite Province) of early Mesoproterozoic age (e.g. 1473 ± 3 Ma; Van Schmus et al., 1996). Magnetite and hematite are associated with REE minerals (apatite, monazite, xenotime, bastnaesite, and britholite) in breccia pipes (Seeger et al., 2001). Renewed volcanic, subvolcanic plutonism and hydrothermal activity took place at 1.38 Ga with extrusion of rhyolite ash-flow tuffs and caldera collapse (Menuge et al., 2002). The volcano-plutonic complex that hosts the deposits is coeval with the Pinwarian magmatic arc of the Grenville Province, and deformed and metamorphosed volcanic and granitoid rocks are well documented in both the parautochthon and allochthon (Central Gneiss Belt) in Ontario, as previously described.

Metallogenic Environments and Districts Specific to the Grenville Province

Metallogenic settings specific to the Grenville Province host past-producing mines such as the world-class Balmat-Edwards Zn deposit, the New Calumet and Montauban Zn-Pb-Ag-Au (±Cu) mines, the Long Lake zinc mine, the Renzy Lake and Lac Edouard Ni-Cu deposits, the Hilton and Marmoraton Fe mines, the Faraday, Bicroft and other U mines near Bancroft, as well as many, small Fe, Au, Mo, Zn, and U deposits, some of them also formerly mined (de Lorraine and Dill, 1982; Jourdain et al., 1990; Eckstrand et al., 1996; Lentz, 1996; Clark, 2000). Most of these occur within the supracrustal belts of the orogen. In contrast, the gneiss terranes that dominate the Grenville Province and that may comprise the volcano-plutonic arc settings are largely untapped in terms of mineral resources (Tollo et al., 2004). In this section, we highlight the known settings in supracrustal belts and then emphasize the importance of the volcano-plutonic arc components as potential hosts to metamorphosed, non-traditional, arc-related mineral deposits such as metamorphosed IOCG, VMS Cu-Au, epithermal Au and porphyry Mo, Cu, or Cu-Au. Subsequent sections explore how fertile crust, produced through arc magmatism and crustal-scale fault zones, can also be critical elements in the development of post-collisional magmatic Ni-Cu-Co or Fe-Ti ores, hydrothermal IOCG, skarns, U, and lode Au deposits (cf. Tornos and Casquet, 2005).

Volcano-Plutonic Belts Prospective for Arc-Related Mineral Deposits

At least four metamorphosed volcanogenic massive sulphide (VMS) deposits are known in accreted arc terranes of the Grenville Province (Fig. 2). The New Calumet Zn-Pb-Ag-Au-Cu mine at the northern end of the Bancroft terrane (3.8 Mt at 5.8% Zn, 1.6% Pb, 65 g/t Ag, and 0.4 g/t Au) and the Au-Ag-Zn-Pb Tétreault deposit in Montauban (2.49 Mt at 6.68% Zn, 2.27% Pb, 1.13 g/t Au, and 131.3 g/t Ag) consist of massive sulphide lenses and disseminated sulphides associated with cordierite-anthophyllite-bearing gneisses that are metamorphosed equivalent of hydrothermally altered rocks. In addition, amphibolite and felsic rocks in the Montauban area preserve volcanic textures (Bernier et al., 1987 and references therein; Villeneuve, 1988; Williams, 1992; Bernier and MacLean, 1993; Nadeau et al., 1999; Gauthier et al., 2004b). At the Dussault showing, semimassive Zn-Pb-Cu-Ag-Au mineralization is hosted by hornblende- and diopside-bearing calc-silicate rock associated with orthopyroxene-bearing amphibolite, garnet-bearing quartzofeldspathic gneiss, biotite quartzofeldspathic gneiss, aluminous gneiss (with garnet, sillimanite, and cordierite), pyrite- and cordierite-bearing quartzite, gedrite-orthopyroxene gneiss, and phlogopite-sapphirine-corundum and gahnite-bearing schists. According to Bernier (1993), these lithologic assemblages are the upper amphibolite- to lower granulite-facies equivalents of volcanogenic sulphides and alteration pipes, and alteration zones associated with mafic and felsic volcanism. The fourth occurrence is the Boudrias showing among supracrustal rocks of the Grande Bergeronne area, southeast of Baie Comeau. The locations of smaller showings of this type in the Ontario Central Metasedimentary Belt are given by Easton and Fyon (1992, Fig. 24.9a).

Upper amphibolite- and granulite-facies arc settings presenting alteration assemblages typical of IOCG, Cu-Au VMS, or epithermal Au deposits have also been identified in the 1.4 Ga Bondy Gneiss Complex, the 1.5 Ga La Romaine Supracrustal Belt, and the Bancroft terrane (Figs. 1, 3, 8; Blein et al., 2004; Corriveau and Bonnet, 2005; Peck and Smith, 2005). These and the 1.64 Ga felsic volcaniclastic (?) rocks of the Pitts Harbour Group with multi-element Cu-Au, Ag-As-Mo lake sediment anomalies (Fig. 2; Gower et al., 1995) are considered prospective exploration targets.

The 1.4 to 1.35 Ga arc-related Bondy Gneiss Complex hosts a volcanic-hosted hydrothermal system at granulite facies (Blein et al., 2003, 2004; Fu et al., 2003; Boggs and Corriveau, 2004; Wodicka et al., 2004). High-Mg cordierite-orthopyroxene white gneiss, garnet-biotite-sillimanite gneiss locally with preserved lapilli textures, and orthopyroxene-magnetite-rich gneiss anomalous in Au occur next to well laminated K-enriched felsic gneiss and layered amphibolite. Together these units signal paleo-chlorite, argillic and potassic alteration zones within a felsic volcanic dome with interspersed mafic volcanic (lavas/sills?) rocks. These alteration facies grade into others interpreted as metamorphosed Ca-leached argillic, advanced argillic, potassic to sericitic alteration zones and finally calcic alteration zones with Cu mineralization. Tourmalinite, a few kilometres north of the main alteration zones, indicates the presence of meta-exhalite distal to the main system. Dissemination, network and layers (veins?) of magnetite concordant with the gneissosity occur over an area of 9 by 6 km irrespective of the rock types present and are interpreted to record a pervasive Fe oxide alteration superimposed on the carbonate, argillic and chloritization alteration zones prior to metamorphism. The magnetite-rich units together with garnetite and calc-silicate rocks form the main hosts to Cu mineralization, whereas anomalous Au values were found in sillimanite-rich gneiss typical of metamorphosed high-sulphidation zones. The alteration zones share many hallmarks with those associated with Australian Proterozoic Cu-Au volcanogenic massive sulphide deposits (Large et al., 2001). Cu-Au mineralization in the Bondy Gneiss Complex is associated with HREE, Zr, and Hf enrichments and pervasive birdwing-shaped REE profiles whose distribution is decoupled from major element variations and signatures. The trace element behaviour implies rock interaction with reducing hydrothermal solutions enriched in ionic complexes, and large amounts of fluorine-rich fluids and suggests, with the influx of Fe oxide, polyphase hydrothermal activity not unlike what is observed in many IOCG deposits (Corriveau, 2007).

The amphibolite- to granulite-facies La Romaine Supracrustal Belt and a southeastern volcano-sedimentary extension of the Wakeham Group that reaches Musquaro Lake in eastern Quebec represent two fertile, 1.5 Ga volcano-sedimentary intra-arc basins of the continental magmatic arc of the Pinware terrane with hydrothermally altered and Cu mineralized felsic volcanic centres and subvolcanic batholiths (Fig. 8; Bonnet et al., 2005; Corriveau and

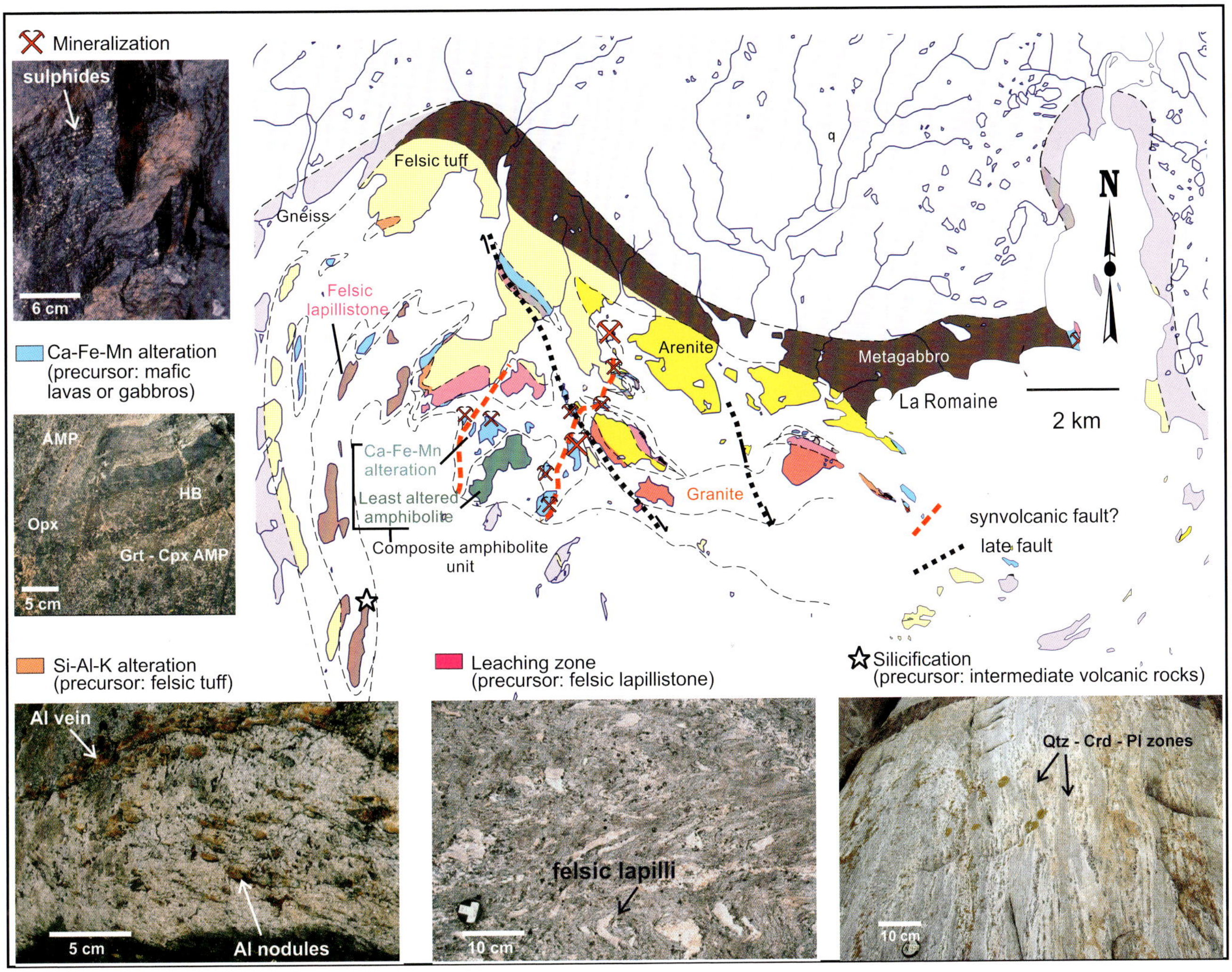

FIGURE 8. Metamorphosed hydrothermal alteration zones within and adjacent to a felsic volcanic centre in the La Romaine Supracrustal Belt of Quebec (Bonnet et al., 2005). All units are steeply dipping. Photos: mineralization - disseminated sulphide, including chalcopyrite, in amphibolite from the composite mafic unit; Ca-Fe-Mn alteration (precursor: mafic lavas or gabbros) - enclave of garnet- and clinopyroxene-bearing (Grt, Cpx) mafic layered gneiss in an amphibolite of the composite amphibolite units; Si-Al-K alteration (precursor: felsic tuff) - series of boudinaged aluminous veins in a pink quartzofeldspathic gneiss; leaching zone (precursor: felsic lapillistone) - hydrothermally-altered lapillistone metamorphosed to a garnet - biotite- sillimanite gneiss with fine-grained homogeneous quartzofeldspathic fragments of similar size and shape to the deformed equivalent lapillistone of Figure 3A; silicification (precursor: intermediate volcanic rocks) - quartz-rich gneiss with cordierite (see Bonnet et al., 2005 for detailed description).

Bonnet, 2005). The volcanic centres host coarse 1.50 Ga rhyolitic to dacitic pyroclastic rocks (Fig. 3) locally intruded by 1.49 Ga quartz-feldspar porphyry or associated with a composite amphibolite unit (lava and/or sill), mineralized in Cu, as well as with nodular quartzofeldspathic gneiss, aluminous gneiss, garnetite, ironstone, carbonate-rich rock, and calc-silicate rock. The latter units display distributions, parageneses, and modes diagnostic of metamorphosed exhalites and hydrothermal alteration zones (e.g. sericitic, argillic, advanced argillic, carbonate, and Fe oxide alterations). The preservation of lapilli demonstrates that precursors were at least in part volcanic in origin whereas abrupt changes in unit thickness and distribution of hydrothermal alteration and mineralization signal the presence of potential synvolcanic faults (Fig. 8). Sporadic mafic lapilli and magma mingling structure point to coeval mafic and felsic magmatism, whereas the volcanic textures and the presence of acid alteration point to vesicular volcanism and hydrothermal activity in a subaerial to shallow marine environment.

The recognition of volcanic textures and metamorphosed hydrothermal alteration is a vindication of mapping strategies developed to explore for volcanic-hosted oxide or sulphide deposits and epithermal deposits in frontier gneiss terranes (Bonnet and Corriveau, 2007). The results illustrate again that units resembling metasedimentary rocks, in particular metapelite, may be first-order targets for the search of metamorphosed hydrothermal deposits in gneiss terranes (Bonnet and Corriveau, 2007). This premise is not new: Gauthier et al. (1985) noted that sillimanite-rich units were a key mineral indicator for gold exploration, and Allard and Carpenter (1988) provided a list of minerals whose unusual abundance and paragenesis formed a promising exploration tool in metamorphosed terranes. Nevertheless, from project implementation to field interpretation, these tools are rarely applied. In this light, four other occurrences of unusual aluminous gneiss in Newfoundland and Quebec are of interest.

Sodic sapphirine-bearing gneiss in the Indian Head Range inlier of Newfoundland (Fig. 1) signals widespread premeta-

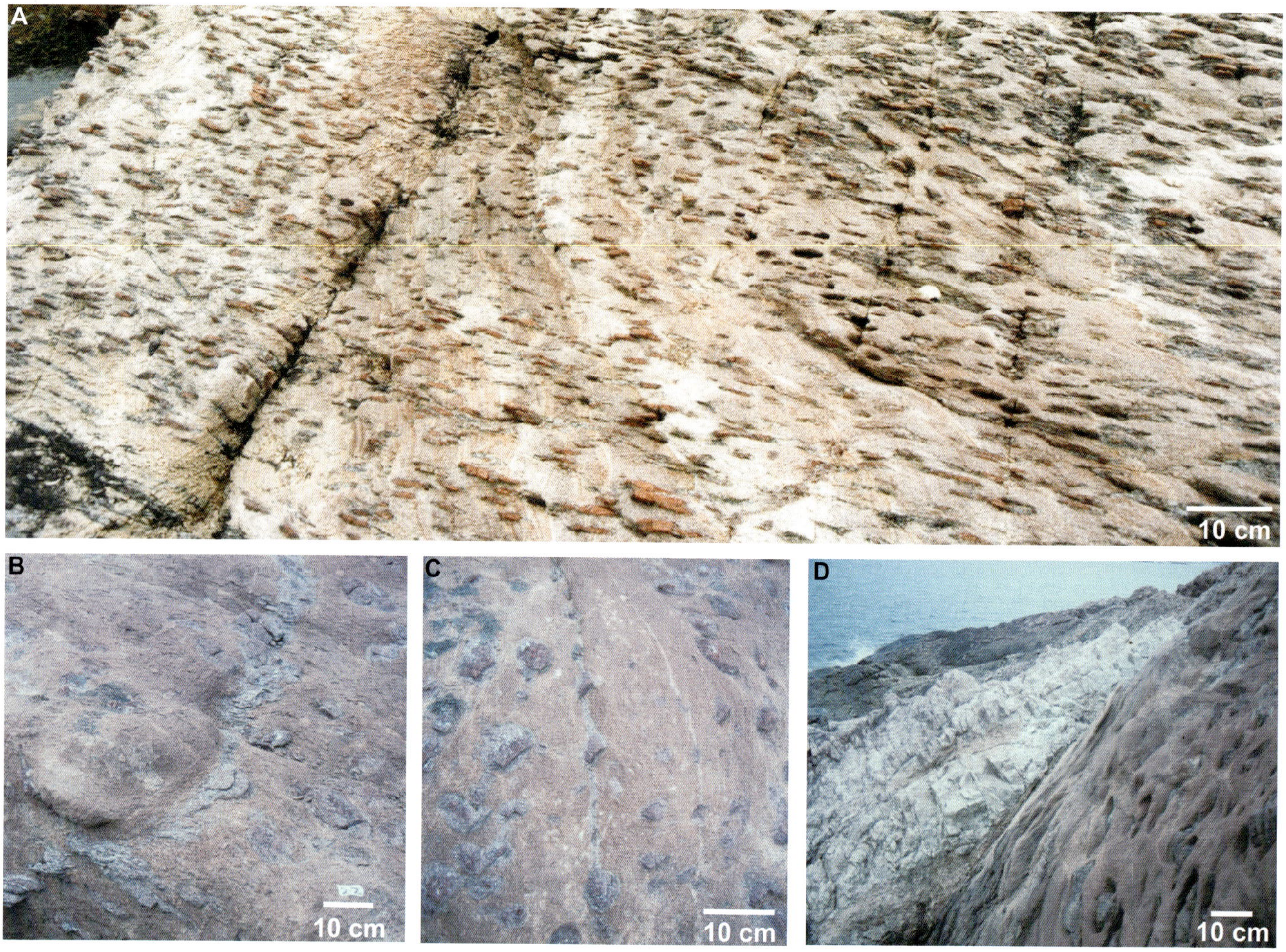

FIGURE 9. Sillimanite rich nodules in quartzofeldspathic gneiss next to a quartz-rich gneiss. **(A)** Veins and disseminated nodules interpreted as zones of argilic or sericitic alteration; **(B, C)** Trains of nodules interpreted as boudinaged veins; **(D)** Quartz-rich gneiss interpretred as a metamorphosed silicification zone.

morphic albitization of "sediments" and occurs with garnetiferous gneiss and dioritic gneiss locally rich in magnetite (Owen et al., 2003). This regional albitization is of unknown origin but offers a target for (IOCG?) exploration. To the north, cordierite-, gedrite-, and sapphirine-bearing gneisses in the Long Range inlier display anomalous enrichment in Mg and Al and are impoverished in Ca. These features are typical of metamorphosed chlorite alteration zones, although other precursors are possible (Owen and Greenough, 1995; Owen et al., 2003). Nodular sillimanite gneiss and siliceous gneiss typical of metamorphosed sericitic to advanced argillic alteration zones also outcrop within quartzofeldpathic gneiss of uncertain origin and supracrustal rocks in the Baie de Brador area near Blanc Sablon (Figs. 1, 9). The nodules form disseminations or trains that mimic boudinaged aluminous veins (Fig. 9A-C) and are found adjacent to a siliceous gneiss (Fig. 9D). These rocks can be traced for over 10 km. North of the Brador Fault, the nodular and siliceous gneisses are bounded by graphitic aluminous (garnet+silimanite±cordierite) gneiss, containing discrete, deformed pyritiferous quartz veins, and a layer of white quartzite, with biotite-garnet lenses (coticule ?), locally with garnet-bearing amphibolite. These features share many characteristics of the sericitic and argillic alteration zones of the La Romaine Supracrustal Belt and Musquaro Lake area of the Wakeham Group as well as those of the IOCG Lyon Mountain setting in the Adirondack Highlands (discussed below) (Figs. 1, 8). This interpretation differs significantly from the original diatexitic model in Perreault and Heaman (2003) and builds on the new knowledge acquired by identifying metamorphosed hydrothermal settings as a means to suggest new territories for mineral exploration (Bonnet and Corriveau, 2007. Finally, arkose units with sillimanite-quartz nodules have also been reported west of Musquaro Lake in the Wakeham Group, some of which crop out in the vicinity of magnetite and molybdenite occurrences (Sharma, 1973; map unit 1). These units have not been revisited since the late 1970s, but descriptions bear striking similarities to those of pyroclastic units at Musquaro Lake and their sericitic alteration.

Supracrustal Belts: Host to Sedimentary Exhaltive and Iron Oxide Copper-Gold Base Metal Deposits

Mesoproterozoic metasedimentary belts of the Grenville Orogen contain the marble-hosted Zn Balmat-Edwards-Pierrepont mining district of the Adirondack Lowlands (Fig. 2) and the Franklin Furnace-Sterling Hill district in New Jersey. Both have inspired mineral exploration for decades (Sangster et al., 1992). Massive Zn sulphide orebodies were formed about 1.3 Ga ago in a sequence of evaporate-bearing siliceous and dolomitic marble of the Grenville Supergroup and subsequently metamorphosed to upper amphibolite facies and polydeformed. Both districts have been recently reclassified as potential McArthur-type SEDEX deposits known to be emplaced within a carbonate evaporitic platform that also contains hematitic clastic sediments (see Gauthier and Chartrand, 2005). Such settings may be also prospective for Broken Hill-type, IOCG and Carlin-type lode gold deposits.

The Balmat orebodies occur within an overturned isocline into which anhydrite of evaporitic provenance flowed. The large ductility contrasts between anhydrite and calc-silicate units led to the development of axial planar macrofractures that evolved into tectonic slides (de Lorraine and Dill, 1982; de Lorraine, 2001, pers. comm., 2005). Conformable massive sphalerite-pyrite-quartz lenses and layers contained within three stratigraphic units probably served as the loci for development of some of the minor or parasitic folds and the tectonic slides associated with the main syncline. Sphalerite was extremely mobile during peak metamorphism (650-700ºC and 6.5 kb) and locally flowed from the massive sulphide "parent" bodies into thin but extensive, crosscutting "macrofractures" at points where macrofractures happened to transect the parent bodies. Spectacular excursions or remobilizations of "daughter" ore into macrofractures occurred locally with sulphide apparently flowing radially outward as far as 2 km away from source bed or parent ore masses. The bulk of remobilized sulphides remained within 0.5 to 1 km from the parent orebodies. Such spectacular ore remobilization may have framed current perception that metamorphosed orebodies are remobilized significantly, however, this is not the norm as exemplified by well preserved stratigraphy even in marble successions (Gauthier et al., 2004b).

In New Jersey, Grenville Supergroup quartzofeldspathic gneiss, marble, and calc-silicate rock host IOCG and Zn oxide deposits, and overlie (structurally?) a calc-alkaline volcanoplutonic belt inferred to be ca. 1.3 to 1.25 Ga in age and to belong to a continental magmatic arc (Puffer and Volkert, 1991; Volkert et al., 2000b; Volkert, 2004a). The Edison magnetite deposits comprise magnetite-quartz layers with variable concentrations of K-feldspar, biotite, garnet, sillimanite, and apatite and some zones of disseminated sulphides (magnetite, bornite, ferrian ilmenite, chalcopyrite, covellite, pyrite, and molybdenite). They are associated with potassic alterations and gneisses recently reinterpreted as hydrothermally altered felsic volcanic and/or pyroclastic rocks (Puffer and Gorring, 2005). The Zn oxide Franklin-Sterling Hill deposits formed through precipitation of zincian carbonates, oxides, and silicates from oxidized and H_2S-poor brines in a shallow marine basin with evaporites (Volkert et al., 2000b; Johnson and Skinner, 2003). A ca. 1.12 Ga rift environment, with volcanic and subvolcanic intrusions, is invoked as the paleo-environment for the variety of ore deposits in the area. The volcano-sedimentary package would have been deposited shortly before or coevally with the emplacement of 1.12 to 1.09 Ga A-type granitoids along crustal-scale faults (Volkert et al., 2000b; Puffer and Gorring, 2005). If this age is correct, then the metasedimentary rocks in New Jersey are younger than and cannot belong to the same basin as the >1.27 Ga Grenville Supergroup in the Central Metasedimentary Belt. Nevertheless, the paleo-environments may be similar. It is of interest to note, however, that they may be coeval with the <1.15 Ga Flinton Group in Ontario, in which small Zn showings occur.

In the Central Metasedimentary Belt of Quebec, several Zn sulphide and Zn oxide deposits are associated with metamorphosed Zn-Mg siderite horizons and stratiform magnetite-bearing Zn-Mn marble units, respectively (Gauthier and Brown, 1986; Gauthier et al., 2004b; Gauthier and Chartrand, 2005). Paleo-environments share many hallmarks of the paleo-basins hosting the Balmat-Edward and Franklin-Sterling Hill deposits and are considered prospective for marble-hosted McArthur-type SEDEX deposits. This mineralization type reflects the presence of hot, oxidized saline fluids of evaporitic origin in a paleo-basin (Cooke et al., 2000), a type of fluid and setting that are also important in IOCG metallogenesis (Marshall and Oliver, 2006). The presence of the Hilton and Marmoraton Fe mines in the same general area is of interest in this respect (Fig. 2; Gilbert, 1959; Gross, 1967, 1996). The former Hilton Fe mine comprises magnetite, hematite, pyrite, and pyrrhotite ore that, with their associated K-feldspar-, amphibole-, and/or magnetite-bearing skarns, veins, pods, or shoots of massive magnetite and biotite-rich schists, are closely related to biotite-granite (Wilson, 1926; Deland, 1959; Gross, 1967). These and other features signal a large intrusion-related hydrothermal system with Fe ore and sodi-calcic and potassic alterations of country rock characteristic of IOCG deposits. The Marmoraton Fe mine, of contact metasomatic skarn type, formed in marble at the contact of the 1.24 Ga Deloro granite stock; it produced over 25 million tonnes of low-P Fe ore averaging 42.8% Fe between 1955 and 1978 (Easton and Fyon, 1992).

Another example of a marble-hosted Zn deposit is the small, defunct Long Lake mine in Ontario (Fig. 2; Wolff, 1979). Here a marble septum within the 1.15 Ga Mountain Grove gabbro (Davidson, 2000) hosted sphalerite-rich ore with minor pyrite, pyrrhotite, chalcopyrite, galena, and magnetite. In its 22 months of operation, this mine produced approximately 90 000 tonnes of ore averaging 11.6% Zn (Carter, 1984). The mineralization may have been derived via remobilization from low-grade SEDEX deposits at depth along the projected interface between mafic-felsic volcanic rocks and marble; pyrite-rich deposits at this interface are exposed in the same region and commonly carry tourmaline. The nature and location of other, smaller marble-hosted Zn deposits in the Central Metasedimentary Belt are outlined in Easton and Fyon (1992, Fig. 24-14.)

The arc-related Paleoproterozoic granulite-facies Disappointment Lake unit in the Wilson Lake terrane of Labrador bears a very high positive aeromagnetic anomaly associated with titanhematite- and magnetite-rich felsic and aluminous gneisses and metamorphosed Fe oxide lenses up to ten metres thick, as well as Fe oxide-dominant aluminous veins (Kletetschka and Stout, 1998; Korhonen and Stout, 2004). The Fe oxide lenses locally crosscut gneissosity. The most distinctive unit is described as sapphirine+quartz and orthopyroxene+sillimanite+quartz restitic layers in the

FIGURE 10. Non-descript biotite quartzofeldspathic gneiss host to the Cu-Ag and Zn-Ag mineralization of the Manitou showing. The quartzofeldspathic gneiss is cut by late granitic pegmatite.

paragneiss. Abundant coarse-grained magnetite coexisting with titanhematite, exsolved titanhematite in orthopyroxene, exsolved hematite in sillimanite, and high Fe^{3+} in sillimanite and corundum, provide a record of high fO_2 (Korhonen and Stout, 2004). This association of felsic, aluminous, and ferruginous gneisses has, like many others, been interpreted as consisting of metasediments and restites (Arima et al., 1986; Thomas et al., 2000), however, a lateritic origin has been proposed for the aluminous gneiss (Leong and Moore, 1972), and a metasomatic origin for crosscutting Fe oxide lenses (Currie and Gittins, 1988). These units bear striking similarities with those resulting from premetamorphic hydrothermal sericitic and Fe oxide alterations of volcanic rocks and an alternative is that the area was subjected to severe Fe oxide and sericitic alterations and that this strong oxidation zone constitutes an exploration target for IOCG deposits. Documentation that abundant Fe oxides also occur in mafic dykes crosscutting the Fe-rich gneisses point to an oxidizing hydrothermal event either coeval or following the emplacement of mafic dykes and prior to metamorphism at 1626 ± 10 Ma (Korhonen and Stout, 2006).

A metasedimentary package, likely Pinwarian in age, the Manitou Metamorphic Complex of the St-Jean Domain hosts the Manitou and Bottine showings where sulphide mineralization was remobilized during Grenvillian metamorphism and deformation. Two generations of mineralization are recognized: 1) Cu-Ag and Zn-Cu mineralization associated with biotite-bearing quartzofeldspathic gneiss (Fig. 10) and 2) Cu-Ag mineralization associated with calc-silicate rocks and late deformation (Perry and Raymond, 1996; Clark, 2003). Results of 1.98% Cu over 1.7 m were returned for the Cu mineralized zones (Clark, 2003). The second generation of Cu mineralization is associated with calc-silicate layers, interpreted as metamorphosed alteration zones, and also forms thin veins of chalcopyrite-bornite-pyrite discordant to the gneissosity, suggesting a post-ductile deformation. The origin of the mineralization is interpreted to be of a SEDEX type with later remobilization (Clark, 2003).

Finally, a striking similarity between the Central Metasedimentary Belt and the Carlin-type gold deposits setting of Nevada, including transition from marble-rich to quartzite-rich rock assemblages, tectonic reworking including thrusting and subsequent extension, and emplacement of lamprophyre dykes, may be worth considering as a new venue for exploration (Corriveau and van Breemen, 2000; cf. Yigit and Hofstra, 2003).

Granitoid-Related Iron Oxide Copper-Gold Deposits and Alterations and Prospective Rare Earth Element Settings

Three IOCG districts are currently known in the Grenville: the Manitou district with the Kwyjibo deposit and Marmont prospect in the eastern Grenville; the Lyon Mountain area in the Adirondack Highlands; and the Mid-Atlantic Fe belt (Friehauf et al., 2002; Clark et al., 2005; Gauthier and Chartrand, 2005; Corriveau, 2007). These districts encompass a large spectrum of low-Ti magnetite and/or hematite-rich breccias, veins, disseminations, and massive bodies with polymetallic enrichments, especially Cu, Au, Ag, U, and REE. Their deposits and prospects are genetically associated with large-scale, A-type granite plutons, as well as with sodi-calcic (e.g. Kwyjibo; Clark et al., 2005), potassic (Edison mine area, New-Jersey: Puffer and Gorring, 2005) and argillic (Lyon Mountain: Selleck et al., 2004) alteration zones. At Lyon Mountain, leaching of country rock and early leucogranite by circulating Na- and Fe-rich magmatic and surface-derived hydrothermal fluids to form the alteration zones was coeval with granite emplacement between approximately 1045 and 1030 Ma (McLelland et al., 2002; Selleck et al., 2004). Post-orogenic, 1.00 to 0.96 Ga, A-type granites and related pegmatites of the Hudson and New Jersey Highlands are proximal to major fault zones and associated with magnetite-U-REE mineralization (e.g. Vassiliou and Puffer, 1985; Volkert et al., 2000a; Volkert, 2004b). This age interval also corresponds to emplacement of the post-orogenic granitoid hosting the Kwyjibo deposit, a replacive, Cloncurry-type IOCG deposit in the eastern Grenville Province. In this deposit, the main polymetallic stage comprises chalcopyrite, pyrite, fluorite, molybdenite, and REE-bearing minerals in hydrothermal veins and stockworks. A titanite age of 975 Ma has been obtained from a mineralized sample. The main ore-forming stage is superimposed on pre-existing (pre-1030 Ma), locally folded, massive, layered or brecciated tabular bodies or stockworks of magnetite ironstone replacing an 1175 ± 4 Ma leucogranite along its fabric. The geodynamic setting is inferred to be an early ca. 1.5 Ga magmatic arc with renewed granitic magmatism at 1.1 and 1.0 Ga, in a successor back-arc extensional setting with major fault zones (Gobeil et al., 2003; Wodicka et al., 2003).

Rare earth element-bearing Fe skarn is also associated with the peralkaline granite Deloro pluton in the Composite Arc Belt in Ontario (Easton, 1989), and calcic metasomatic fronts and veins in the syenite carapace of the potassic alkaline plutons of the Kensington-Skootamatta suite (Fig. 1; Corriveau, 1989) display similar habit and mineral assemblages to those of the REE prospects in the syenite carapace of the Eden Lake carbonatite complex in Manitoba (Mumin and Corriveau, 2004). Also of interest with regards to the potential of the Grenville Province are the 1.03 Ga Kipawa nepheline syenite for its REE anomalies (van Breemen and Currie, 2004); two other occurrences of similar nepheline syenite are known in the Ontario parautochthon.

TABLE 1. Fe-Ti deposits in the Grenville Province of Quebec.

Deposit name	Type*	Main ore assemblage	Size in km length/width/depth	Resources (tons)	Tonnage mined	Grade Fe	Grade TiO_2	Grade P_2O_5	Grade V_2O_5	Grade Cr	Period of mining
Morin Anorthosite											
Ivry	2	Ferrian Ilmenite	0.228 / 0.018 / 056	0.16 Mt	52 000	42.50%	33.23%				1912–18, 1958–59
Degrosbois	3	Ti-Magnetite, Ilmenite		5 Mt	10 800	40.80%	11%				1912–49
Saint-Hyppolyte	3	Ti-Magnetite, Ilmenite		10 Mt		27%	20%				
Saint-Urbain Anorthosite											
Coulombe East and West	2	Ferrian Ilmenite (Rt)	0.24 / 0.03	6.71 Mt	110 000	36%	39%				1910–11, 1939–48
Furnace	2	Ferrian Ilmenite	0.120 / 0.060	2 Mt		26%	27%				1871–75
General Electric	2	Ferrian Ilmenite (Rt)	0.12 / 0.02/ 0.04	6.5 Mt	13 000	27.30%	29.20%				1918
Bignel	2 + 4	Ferrian Ilmenite, Apatite	0.16 / 0.06	3.44 Mt	397 000	34.10%	34.70%				1940–46, 1957–66, 1977
Lac Saint-Jean Anorthosite											
Saint-Charles	3	Ti-Magnetite, Ilmenite, Apatite	1.8 / 0.336	5.4 Mt		29.50%	10.10%	9.25%	0.10%		
La Hache-Est	2	Ti-Magnetite, Apatite, Ilmenite		20.3 Mt		24.75%	5.12%	5.20%			
Buttercup	3	Ti-Magnetite + Ilmenite		3.5 Mt		49%	19%		0.67%		
Labrieville Anorthosite											
Lac Brûlé	2	Ferrian Ilmenite	0.120 / 0.003–0.012**	5.8 Mt		42%	35%				
Pambrum Anorthosite											
Lac Pambrum	3	Ti-Magnetite, Ilmenite		49 Mt		47%	11%				
Lac La Blache Anorthosite											
Hervieux East,			1.13 / 0.04–0.105 / 0.05								
Hervieux West,	2	Ti-Magnetite, Ilmenite	2.4 / 0.08 / 0.05	71.7 Mt		48%	20.10%				
Schmoo			0.365 / 0.09								
Lac Dissimieu	1	Ilmenite (10–15%), Apatite	4 / 0.25	unknown			45%	4.50%			
Havre-Saint-Pierre Anorthosite											
Lac Tio	2	Ferrian Ilmenite	1.09 / 1.04 / 0.11	>125 Mt (1,2)	>90 Mt	38.80%	33.60%				1950–...
Everett	3	Ti-Magnetite, Ilmenite, Apatite	3 / 0.3	294 Mt		16.20%	9.75%	4%			
Lac Fournier Anorthosite											
Magpie (1, 2, 3 et 4)	3	Ti-Magnetite, Ilmenite	3.66 / 0.425 / 0.2	810 Mt		43.10%	10.60%		0.2% V	1.55%	
Sept-Îles layered complex											
Canton Arnaud	4	Apatite, Ilmenite, Ti-Magnetite	2 / 0.175	107 Mt (1)			8.41%	6.20%			

(1) Estimated reserves are based on inferred resources except for the Tio Mine and Canton Arnaud, based on proven and inferred reserves.
(2) Proven and inferred reserves from 1950. Current resources and reserves are kept confidential by QIT. * See text for mineralization type subdivision; ** more than one lens.
(3) Rt, rutile.

Magmatic and Remobilized Ni-Cu Sulphide and Platinum Group Element Deposits

Currently known Cu-Ni deposits (Renzy and Edouard mines, McNickel showing) are largely of low grade (<1% Ni and <1% Cu), richer ones, such as the Lac Volant prospect, are limited in size, and the most significative PGE mineralization known is associated with the 2.49 to 2.44 Ga mafic intrusions that extends from the Southern to the Grenville provinces in Ontario (Wilson, 1993; Clark, 2000, 2001a; Thériault et al., 2003; Vaillancourt et al., 2003).

Cu-Ni (±Co±PGE) sulphides associated with mafic to ultramafic and anorthosite intrusions occur as dissemination to massive lenses with magmatic textures and net veins within mafic-ultramafic rocks at, or close to, contact with supracrustal country rocks (type 1 in Fig. 2). Examples include the Renzy and Lac Edouard mines, the Lac Paradis prospect, and mineralization associated with the Intrusion du Réservoir (Table 1, Fig. 2). Cu-Ni showings in the Mountain Grove gabbro intrusion in Ontario (Wolff, 1979) are also of this type. Magmatic Cu-Ni mineralization is also associated with mafic sills or feeder dykes such as the Lac Volant prospect in the Matamec complex (Fig. 2). This type is considered the most promising. Finally, examples of magmatic Cu-Ni sulphides associated with pyroxenite or melanogabbro emplaced near the margin of anorthosite massifs include Notre-Dame-de-la Merci in the Morin anorthosite, McNickel and Chûtes-des-Passes in the Lac-Saint-Jean anorthosite, a variety of showings in the Lac-de-la-Blache, Lac Tortue, and Northwest Massif of the Havre-Saint-Pierre anorthosites, and the B-20 showing in the Rivière-Pentecôte anorthosite (e.g. Chevé et al., 1999; Gobeil et al., 1999).

Magmatic hydrothermal PGE-dominant (±Cu±Ni) deposits are associated with mafic to ultramafic intrusions (type 2 in Fig. 2; Wilson, 1993; Jobin-Bevans, 2002; Thériault et al., 2003). PGE minerals and sulphides are usually disseminated in the ultramafic facies of these intrusions although lenses of semimassive sulphides and breccia zones may also be present. Examples include the Lac-Nadeau and Lac Mitaine showings (Fig. 2).

A third type consists of remobilized magmatic Cu-Ni sulphides and epigenetic Cu-Ni mineralization (type 3 in Fig. 2). Magmatic Cu-Ni sulphides are remobilized during high-grade metamorphism and concentrated in sulphide-rich veins or lenses along the gneissosity. Epigenitic Cu-Ni sulphides occur in veins along ductile or brittle shear zones and are found in metamorphosed mafic-ultramafic intrusions and adjacent supracrustal rocks. Examples include the B30 showing north of Baie-Comeau, the 2EZ showing in the Hart-Jaune terrane, the Lobster Bay showing in the Pinware terrane, and a gossan zone in the Saint-Augustin Complex. The latter is composed of Fe hydroxide, hematite with pyrite, pyrrhotite, minor chalcopyrite, and pentlandite. Fresh mineralized samples are composed of massive pyrrhotite and pyrite with minor chalcopyrite and pentlandite. This showing is an example of magmatic sulphides emplaced at a contact and remobilized by tectono-metamorphic processes.

Some of the magmatic Ni-Cu mineralization is associated with mafic intrusions characterized by steeply dipping layering and foliation. In some cases, layering was transposed and foliation formed tectonically (e.g. Musquaro intrusion; Corriveau et al., 2002), in others vertical layering and mineral foliation is primary (e.g. the vertically layered Montjoie intrusion; Corriveau and van Breemen, 2000). Steep igneous layering in mafic layered intrusions is most common along marginal border zones of classical layered intrusions (Irvine et al., 1998) and is attributed to sidewall crystallization or to sinking due to tectonism (e.g. Loney and Himmelberg, 1983). It also occurs in collapsed mafic-silicic layered intrusions (MASLI; Wiebe and Collins, 1998) and in syntectonic sheet-like intrusions (Corriveau and van Breemen, 2000). As vertically layered or foliated mafic intrusions may bear traps for magmatic sulphides (e.g. mineralized ultramafic breccia in the Grand-Remous syntectonic mafic sheet intrusion) that are distinct from those associated with classical, horizontally layered intrusions (Clark, 2000), discussing potential style of emplacement provides a mean to better anchor exploration.

Two emplacement models are currently put forward for the Lac Volant massive sulphide prospect: emplacement as a gently inclined feeder dyke subsequently transposed tectonically or emplacement of a vertical feeder dyke after collapse of the early stage mafic and silicic layers of a host MASLI intrusion. The subvertical dyke that hosts the Lac Volant Ni-Cu prospect intrudes the Matamec mafic-felsic complex along a subvertical internal contact of the intrusion (Perry and Roy, 1997). Intrusive sheets, magmatic and solid-state foliation, and intraplutonic shear zones of the complex define a steeply dipping synform dissected by some major thrust faults (Chevé et al., 1999; Gobeil et al., 1999). These authors interpret the synform as a result of folding after emplacement of the Lac Volant dyke. Unfolding the intrusion implies that the Lac Volant dyke was originally mildly inclined (Clark, 2003; Nabil et al., 2004). An alternative interpretation is that the dyke is a late-stage mafic dyke emplaced vertically along the western flank of the Matamec intrusion. The internal verticality of the complex would have resulted from infolding during the collapse of the mafic-silicic layered intrusion (Saint-Germain and Corriveau, 2003) following the model of Wiebe and Collins (1998). This interpretation is based on the field evidence that the mafic and felsic intrusive sheets were formerly horizontal based on sedimentary-like magmatic structures observed between the heavier mafic and lighter felsic magmatic sheets. Inferred polarities are consistent with infolding of the Matamec Complex. The dyke could have served as a feeder to late-stage, overlying magma chambers and may reflect processes that may have taken place during the build-up of the intrusion itself.

Ni-Cu mineralization, such as the one of the Grand Remous ultramafic breccia, can be associated with syntectonic mafic intrusions. Such intrusions can take the aspect of gneissic, hence metamorphic, and tectonically transposed intrusions as emplacement is commonly accompanied by the mingling of mafic and felsic magmas, crystal sorting, repeated injections of magmas, formation of striking magmatic foliation and layering, or recrystallization of the early formed crystals whereas residual magmas were percolating to form veins along adjacent shear zones (Corriveau and van Breemen, 2000). Vertically layered intrusions displaying systematic subvertical modal layering, igneous foliation, erosional troughs and surfaces, contact-parallel, crude, concentric, subvertical layering are distinct from tilted or folded classically layered intrusions in which layering was originally mostly horizontal. Tilting cannot be responsible for systematic subvertical layering of circular stocks, especially not for cylindrical igneous foliation pattern. In such cases, verticality of layering is primary, and likely a result of sidewall crystallization in a subvertical magma conduit. Magmatic sulphides, if present, are likely to gravitationally sink to the bottom of the intrusion. Whether magmatic sulphides would collect and remain preserved in mafic-silicic-layered intrusions, syntectonic sheet intrusions, and vertically layered mafic intrusions is uncertain but knowledge that a targeted intrusion is not a classical horizontally layered intrusion may bear implications on the exploration strategies to be adopted.

Magmatic Fe-Ti-(P-V) Deposits Associated with Anorthositic Suites and Mafic Intrusions

Ilmenite ($FeTiO_3$) and rutile (TiO_2) ores are mined worldwide from mafic and anorthositic plutonic suites, placer deposits, and high temperature and pressure metamorphic rocks (Stanaway, 2003). In the Grenville Province, titaniferous magnetite, magnetite-ilmenite-apatite, and massive ferrian ilmenite deposits have been known since the mid-1850s but only the massive ilmenite Tio deposit in the Havre-Saint-Pierre anorthosite massif is currently mined (Fig. 2; Rose, 1969; Gross, 1996; Perreault and Hébert, 2003). The ore is composed of a dense coarse-grained aggregate of ferrian ilmenite. In general, the deposits occur within tabular intrusions, stocks, sills, or dykes in anorthosite massifs. Locally, they consist of stratiform accumulations in layered segments of anorthosite massifs or in layered mafic intrusions. Five distinct styles of mineralization are recognized:

1. Magmatic disseminated Fe-Ti oxides (titaniferous magnetite+ilmenite) in the host anorthosite or mafic intrusions (e.g. Raudot Complex, Fig. 5).
2. Massive ferrian ilmenite or massive titaniferous magnetite layers, dykes, tabular or irregular bodies locally associated with oxide-rich norite, jotunite or oxide-apatite gabbronorite and in sharp or diffuse contacts with the host anorthosite. In two cases (St-Urbain and Havre-Saint-Pierre anorthositic suites), rutile is present in significant amounts in the ferrian ilmenite ore. Massive titaniferous ores are usually associated with labradorite-bearing anorthosite and massive ferrian ilmenite ores with andesine-bearing anorthosite. According to available data, ferrian ilmenite mineralization occurs only in anorthositic plutons younger than 1160 Ma in the Grenville Province (Hébert et al., 2005).
3. Layers and lenses of nelsonite (magnetite-ilmenite-apatite rich rock), interlayered with magnetitite in layered mafic complexes (e.g. Eocambrian Sept-Îles Layered Complex (Fig. 5; Cimon and McCann, 2000) or with oxide-apatite gabbronorite, oxide-rich norite (jotunite) or ilmenitite in anorthositic complexes.
4. Late injections of magnetitite, ilmenitite, or oxide-rich norite in an already cooled and solidified anorthosite (see Gross, 1996).
5. Magnetite concentration in late monzonite or mangerite dykes cutting anorthosite. Due to their small size, this type has no economic potential.

An overview of the main deposits with tonnage and grade value is presented in Table 1. Among the other known deposits in the area, the most promising ones are the Grader (which was mined in 1950 for a short period), Springer, and Lac-au-Vent. All these deposits and smaller showings are composed of massive ilmenite dykes emplaced in the andesine-bearing anorthosite. One exception is the Big Island deposit composed of a rutile and sapphirine-bearing massive ferrian ilmenite dyke with fragments of oxide-rich norite and anorthosite. Rutile can be as high as 15 modal per cent.

Skarns- and Pegmatite-Related Mineralization

Molybdenite deposits are common in the Central Metasedimentary Belt and are hosted in pegmatite-related skarns (Halo, Hunt, Spain, and Zenith deposits), skarnoid rocks (Liedke, Kirkham, and Bain deposits), apatite-calcite vein and dykes (Kirkham, Que.), and 1.05 Ga pegmatite-aplite bodies (Moss deposit) (D. Lentz, per comm., 2004). Pegmatites intruded the Grenville Supergroup after peak metamorphism or, for the Moss deposit, an A-type syenite intrusion that may mark the late stages of the Kensington-Skootamatta potassic alkaline suite (Lentz and Creaser, 2005). Mineralization occurs as dissemination in the quartz-rich part of the pegmatites.

Skarn-type Cu-Ag±W±Au mineralization occurs among calc-silicate rocks associated with calcitic/dolomitic marble and paragneiss, and felsic intrusions (e.g. Nantel, 2003). Cu sulphides (bornite, chalcocite, chalcopyrite, digenite) and associated Fe sulphides (pyrrhotite and pyrite) are disseminated along veinlets or in centimetre-size pockets (Ortega, 2003; Nantel et al., 2004). Sulphide contents rarely exceed 10% with Cu grade rarely exceeding 1% and Ag values up to 5 ppm (Nantel et al., 2004). Mineralization initially formed through intrusion-related fluid circulation in the supracrustal hosts is, in most cases, remobilized by tectonometamorphic processes. In some cases, mineralization is metamorphogenic with no clear relation to felsic plutonism.

Uranium Mineralization

The Grenville Province is a past producer of U with the Faraday, Bicroft, Canadian Dyno, and Greyhawk mines in the Bancroft terrane of the Central Metasedimentary Belt (Fig. 2). U mineralization occurs in zoned and unzoned pegmatite veins, sulphide-rich siliceous marble, calc-silicate skarn, and calcite and pyroxene veins (Sangster et al., 2006). U and Th mineralization (uraninite, uranothorite, and Th-REE silicates) is also common in metasediment-hosted late-stage pegmatoid veins among migmatites in the Central Metasedimentary Belt of Quebec and in pegmatite associated with late-stage granite that intrudes the Wakeham Group and adjacent gneissic terranes such as those of the Lac Turgeon granite (Baldwin, 1970; Hauseux, 1977; Gauthier et al., 2004b). Other types include 1) local concentrations of U oxides (that may contain up to 15 wt.% ThO_2) associated with magnetite in confined layers of quartzofeldspathic gneiss (meta-arkose?) (Baldwin, 1970; Hébert, 1995); 2) pyrochlore and other REE-bearing minerals in carbonatite and alkaline complexes such as the Quinville and Cantley carbonatites (Hébert, 1995); U-thorite in pyroxenite in the exocontact of the Kipawa peralkaline intrusion; 3) Fe oxide Cu-REE-U mineralization such as the Kwyjibo deposit

(Clark, 2003; Gauthier et al., 2004a, Magrina et al., 2006); and 4) late secondary remobilization along fractures. Though the potential for conglomerate-type (e.g. Elliot Lake) or sandstone-hosted hydrothermal U deposits (Athabasca Basin) exists in parts of the Grenville Province that have only seen lower greenschist-facies metamorphism (e.g. the Wakeham Group), most of the exploration is oriented at the Rössing-type of U mineralization associated with felsic intrusives in supracrustal rocks. The recent recognition of U-bearing IOCG mineralization and metallogenic environments in the Grenville Province opens new prospects for U exploration in this orogen and current research is targeting metamorphosed settings as new frontiers (Cuney and Kish, 2004).

Knowledge Gaps

Synthesis of mineral deposits and districts across Canada in this volume underscore the barren nature of the high-grade metamorphic terranes of the Grenville Province in terms of large mining camps. A few active mines in several million square kilometres raises serious questions such as (1) have we applied the appropriate metallogenic models during past exploration, (2) were the targeted environments appropriately known prior to exploration, and (3) was enough money spent to search efficiently for mineral deposits and over a long enough period to find them?

Effective mineral exploration is anchored on modern geoscience knowledge and guided with the most appropriate metallogenic models and exploration strategies. But in frontier terranes, it all starts with fieldwork; if one does not properly assess the nature of the gneiss it is targeting, exploration has a low chance of success. Worldwide, volcano-plutonic belts are key targets for the search for base, precious, and strategic metals. In gneiss terranes, their recognition remains a challenge as primary volcanic textures become obliterated or difficult to identify where metamorphism was accompanied by significant deformation. Traditional geological markers, such as alteration zones associated with sulphide or oxide mineralizing hydrothermal (and hydrothermal-magmatic) systems, change aspects once metamorphosed at high grade. Being able to recognize hydrothermal alterations zones at an early stage of mapping is essential to future discoveries; so is the understanding of the style of magma emplacement in potentially fertile, nontraditional mafic and mafic-felsic intrusions.

With modern strategies, fieldwork conducted by highly trained field geologists, and adequate funding, the nature of gneiss domains currently construed as sterile can be reassessed. A prime example is the discovery of the Cu-Ni Lac Volant prospect and other Cu-Ni and Cu-Zn showings during a regional mapping project in the St.-Jean Domain (Perreault et al., 1997). What was initially a four-unit map area became a series of maps with intricate settings that range from those hosting SEDEX mineralization to those hosting magmatic Ni-Cu-Co and IOCG mineralization, prospects, and deposits, opening a huge territory to exploration (Gobeil et al., 1999). Another example is the reassessment of the nature of the former La Romaine Domain during a Targeted Geoscience Initiative in the eastern Grenville Province. During this project, a 600 km^2 area of gneisses interpreted as a metasedimentary basin was reapparaised as a metamorphosed continental arc plutonic belt with kilometre-scale wide intra-arc, rift-related volcano-sedimentary successions hosting Cu sulphides and Fe oxides mineralizing systems.

Many geologists involved in mineral exploration see high-grade metamorphism and ductile deformation as destructive events. But without key ore beneficiation of Paleoproterozoic taconite through grain coarsening, recrystallization into purer end-members of higher metallurgical values, and concentration of sulphides in structural traps, the Fe ores in Labrador City and Mount Wright might not be the mines they are today.

Conclusions

Disparaged as a sterile root of a collisional orogen, the high-grade metamorphic terranes of the Grenville Province are increasingly recognized as a complex juxtaposition of Andean-type volcanic and plutonic arc environments, a concept that has direct applications for revisiting the exploration prospectivity of the orogen. Archean, Paleoproterozoic, and Mesoproterozoic surface and near-surface volcano-plutonic magmatic arc, intra-arc rift and island-arc settings share characteristics of VMS, IOCG, epithermal Au, porphyry Cu, and orogenic Au environments. The back-arc settings, the platformal sedimentary basins, and the within-plate orogenic to late- or post-orogenic mafic and felsic intrusions provide targets for Carlin-type and SEDEX deposits, magmatic Ni-Cu-Co sulphides, PGE, Fe-Ti, Fe-Ti-V-P, and rare metals. However, many 1.7 to 1.6 Ga Labradorian, 1.5 Ga Pinwarian, and 1.3 to 1.2 Ga Elzevirian arc components may still remain unidentified among the large tracts of undifferentiated gneiss complexes. Magmatic Ni-Cu-PGE targets are also expected in some of the uncharted terranes and significant traps for magmatic sulphides may still be discovered in orogenic and late-orogenic mafic and anorthositic intrusions.

Exploration targets are already available in settings similar to those hosting Broken Hill Pb-Zn, VMS-Cu-Au, Olympic Dam IOCG, and Voisey Bay Ni-Cu deposits but, to result in ore discoveries, the specificity of gneissic "pinkstone" belts and ore settings in syntectonic intrusions need to be accounted for. This will, in some cases, require the development of alternative models to the now classic ones derived from decades of successful exploration in "greenstone" belts and anorogenic plume- or rift-related magmatic settings. We argue that deposits may be overlooked in the gneiss terranes; first, significantly less exploration is being conducted and less knowledge acquired than those associated with decades of extensive exploration in the large mining camps of the Archean greenstone belts; second, when exploration is being conducted, greenstone belt tools, models, and views are used, these may not be adequate for mineral exploration of Proterozoic gneissic orogens. Will the pinkstone belts of the Grenville Province do in the 21st century what the greenstone belts did in the 20th: significantly sustain rejuvenation and replenishment of Canada's mineral resources? If so, do we currently have the knowledge required to make proper assessments for the sustainable development of such terranes?

Acknowledgements

The authors wish to thank colleagues and Friends of the Grenville, in particular T. Clark, S. Cadéron, W. de Lorraine,

R.M. Easton, C.F. Gower, and J. Stout for sharing public-domain information and personal knowledge of the Grenville Province, as well as Wayne Goodfellow and John Lydon, Geological Survey of Canada, Mineral Synthesis project leader and co-leader. W. de Lorraine kindly provided text for the Balmat-Edwards deposit and M. Cuney revised the uranium section. The manuscript greatly benefited from the constructive and detailed review by R.M. Easton and W. Goodfellow (editor). LC and SP also want to acknowledge A. Davidson for the thoroughness and extent of his original review and for his subsequent input as a co-author.

References

Allard, G.O., 1978, Pétrologie et potentiel économique du prolongement du sillon de roches vertes de Chibougamau dans la Province de Grenville: Ministère des Richesses naturelles, Québec, DPV-604, 46 p.

——— 1979, Prolongement du Complexe du Lac Doré dans la Province de Grenville, à l'est de Chibougamau: Ministère de l'Énergie et des Ressources, Québec, DPV-685, 21 p.

Allard, G.O., and Carpenter, R.H., 1988, Mineralogical anomalies in metamorphosed terrains, a neglected but promising exploration tool: International Conference on the Geochemical Evolution of the Continental Crust, Pocos de Cladas, Brazil, p. 229-236.

Archer, P., Chapdelaine, M., and Huot, F., 2004, Discovery of high-grade metamorphosed volcanogenic massive sulphides in the Coulon belt, James Bay: Québec Exploration 2004, Abstracts of Oral Presentations and Posters, Québec, p. 73.

Arima, M., Kerrich, R., and Thomas, A., 1986, Sapphirine-bearing paragneiss from the northern Grenville Province in Labrador, Canada: Protolith composition and metamorphic P-T conditions: Geology, v. 14, p. 844-847.

Baldwin, A.B., 1970, Uranium and thorium occurences on the North Shore of the Gulf of St.-Lawrence: Canadian Institute of Mining Bulletin, v. 63, p. 699-707.

Bandyayera, D., Rhéaume, P., Cadéron, S., Giguère, E., and Sharma, K.N.M., 2004, Géologie de la région du lac Lagacé (32B/14): Ministère des Ressources naturelles et de la Faune, Québec, RG-2004-02, 31 p.

Bartholomew, M.J., ed., 1984, The Grenville Event in the Appalachians and Related Topics: Geological Society of America, Special Paper 194, 287 p.

Bellemare, Y., and Jacob, H.-L., 2004, Chapitre 2: Pierre architecturale et minéraux industriels: Ministère des Ressources naturelles, de la Faune et des Parcs, Québec, DV 2004-01, p. 59-70.

Berclaz, A., Hébert, R., and Rocheleau, M., 1995, La zone tectonique du front du Grenville à l'est de Louvicourt, Québec: exhumation de la croûte archéenne pendant l'orogénie grenvillienne: Canadian Journal of Earth Sciences, v. 32, p. 1899-1920.

Berman, R., Easton, R.M., and Nadeau, L., 2000, A new metamorphic map of the Canadian Shield; introduction: The Canadian Mineralogist, v. 38, p. 277-285.

Bernier, L., 1993, Étude métallogénique du gîte polymétallique Dussault, canton de Lapeyrère, Province de Grenville: Ministère de l'Énergie et des Ressources, Québec, MB 93-53, 161 p.

Bernier, L.R., and MacLean, W.H., 1993, Lithogeochemistry of a metamorphosed VMS alteration zone at Montauban, Grenville Province: Exploration and Mining Geology, v. 2, p. 367-386.

Bernier, L., Pouliot, G., and MacLean, W.H., 1987, Geology and metamorphism of the Montauban North Gold Zone: A metamorphosed polymetallic exhalative deposit, Grenville Province, Quebec: Economic Geology, v. 82, p. 2076-2090.

Bethune, K.M., 1997, The Sudbury dyke swarm and its bearing on the tectonic development of the Grenville Front, Ontario, Canada: Precambrian Research, v. 85, p. 117-146.

Bickford, M.E., Soegaard, K., Nielsen, K.C., and McLelland, J.M., 2000, Geology and geochronology of Grenville-age rocks in the Van Horn and Franklin Mountains area, west Texas: Implications for the tectonic evolution of Laurentia during the Grenville: Geological Society of America Bulletin, v. 112, p. 1134-1148.

Blein, O., LaFlèche, M.R., and Corriveau, L., 2003, Geochemistry of the granulitic Bondy Gneiss Complex: A 1.4 Ga arc in the Central Metasedimentary Belt, Grenville Province, Canada: Precambrian Research, v. 120, p. 193-218.

Blein, O., Corriveau, L., and LaFlèche, M.R., 2004, Cordierite-orthopyroxene white gneiss: A key to unveiling pre-metamorphic hydrothermal activity in the Bondy gneiss complex, Grenville Province, Québec, *in* Tollo, R., Corriveau, L., McLelland, J., and Bartholomew, M., eds., Proterozoic Tectonic Evolution of the Grenville Orogen in eastern North America: Geological Society of America, Memoir 197, p. 19-33.

Boggs, K.J.E., and Corriveau, L., 2004, Granulite P-T-t paths and retrograde cation diffusion from the Mont-Laurier area, southwestern Grenville Province, Quebec, *in* Tollo, R., Corriveau, L., McLelland, J., and Bartholomew, M., eds., Proterozoic Tectonic Evolution of the Grenville Orogen in eastern North America: Geological Society of America, Memoir 197, p. 35-64.

Bonnet, A.-L. and Corriveau, L., 2007, Alteration vectors to metamorphosed hydrothermal systems in gneissic terranes, *in* Goodfellow, W.D., ed., Mineral Deposits of Canada: A Synthesis of Major Deposit-Types, District Metallogeny, the Evolution of Geological Provinces, and Exploration Methods: Geological Association of Canada, Mineral Deposits Division, Special Publication No. 5, p. 1035-1049.

Bonnet, A.-L., Corriveau, L., and LaFlèche, M.R., 2005, Chemical imprint of highly metamorphosed volcanic-hosted hydrothermal alterations in the La Romaine Supracrustal Belt, eastern Grenville Province, Quebec: Canadian Journal of Earth Sciences, v. 42, p. 1783-1814.

Cadéron, S., Roy, P., and Bandyayera, D., 2004, New metallogenic, geological and metamorphic data from the Grenville Front: Québec Exploration 2004, Abstracts of Oral Presentations and Posters, p. 111.

Carr, S.D., Easton, R.M., Jamieson, R.A., and Culshaw, N.G., 2000, Geologic transect across the Grenville orogen of Ontario and New York: Canadian Journal of Earth Sciences, v. 37, p. 193-216.

Carter, T.R., 1984, Metallogeny of the Grenville Province, southeastern Ontario: Ontario Geological Survey, Open File Report 5515, 422 p.

Chevé, S., Gobeil, A., Clark, T., Corriveau, L., and Perreault, S., 1999, Géologie de la région du lac Manitou (SNRC 22I/14): Ministère des Ressources naturelles, Québec, RG 99-02, 69 p.

Ciesielski, A., and Sharma, K.N.M., 1995, Geochemistry and petrogenesis of 1280 Ma tonalitic orthogneisses of the Central Metasedimentary Belt of the Grenville Province, Quebec: International Conference on Tectonics and Metallogeny of Early/Mid Precambrian orogenic Belt, Precambrian 1995, Program with Abstracts, p. 307.

Cimon, J., and McCann, J., 2000, Le gîte d'apatite-ilménite du complexe mafique lité d'âge Cambrien de Sept-Îles, Québec: Chronique de la Recherche minière, v. 539, p. 63-83.

Clark, T., 2000, Le potentiel en Cu-Ni±Co±EGP du Grenville québécois: exemples de minéralisations magmatiques et remobilisées: Chronique de la Recherche minière, v. 539, p. 85-100.

——— 2001a, Distribution and exploration potential of platinum-group elements in Québec: Ministère des Ressources naturelles, Québec, PRO 2001-06, 13 p.

——— 2001b, Distribution et potentiel des EGP au Québec: Ministère des Ressources naturelles, Québec, http://www.mrn.gouv.qc.ca/publications/mines/potentiel/Cf200106.pdf.

——— 2003, Métallogénie des métaux usuels et précieux, des éléments radioactifs et des éléments des terres rares, région de la moyenne Côte-Nord, *in* Brisebois, D., and Clark, T., eds., Géologie et Ressources minérales de la Partie Est de la Province de Grenville: Ministère des Ressources naturelles de la Faune et des Parcs, Québec, DV 2002-03, p. 269-326.

Clark, T., Gobeil, A., and David, J.,2005, Iron oxide-Cu-Au-type and related deposits in the Manitou Lake area, eastern Grenville Province, Quebec: variations in setting, composition, and style: Canadian Journal of Earth Sciences, v. 42, p. 1829-1847.

Clarke, P.-J., 1977, Gagnon Area: Ministère des Richesses naturelles, Québec, Geological Report 178, 79 p.

Connelly, J.N., Rivers, T., and James, D.T., 1995, Thermotectonic evolution of the Grenville Province of western Labrador: Tectonics, v. 14, p. 202-217.

Cooke, D.R., Bull, S.W., Large, R.R., and McGoldrick, P.J., 2000, The importance of oxidized brines for the formation of Australian Proterozoic stratiform sediment-hosted Pb-Zn (Sedex) deposits: Economic Geology, v. 95, p. 1-17.

Corrigan, D., and Hanmer, S., 1997, Anorthosites and related granitoids in the Grenville orogen: a product of convective thinning of the lithosphere?: Geology, v. 25, no. 1, p. 61-64.

Corrigan, D., and van Breemen, O., 1997, U-Pb age constraints for the lithotectonic evolution of the Grenville Province along the Mauricie transect, Quebec: Canadian Journal of Earth Sciences, v. 34, p. 299-316.

Corrigan, D., Culshaw, N.G., and Mortensen, J.K., 1994, Pre-Grenvillian evolution and Grenvillian overprinting of the Parautochthonous Belt in Key Harbour, Ontario: U-Pb and field constraints: Canadian Journal of Earth Sciences, v. 31, p. 583-596.

Corrigan, D., Rivers, T., and Dunning, G., 2000, U-Pb constraints for the plutonic and tectonometamorphic evolution of Lake Melville terrane, Labrador and implications for basement reworking in the northeastern Grenville Province: Precambrian Research, v. 99, p. 65-90.

Corriveau, L., 1989, Potassic alkaline plutonism in the southwestern Grenville Province: PhD thesis, McGill University, Montréal, 263 p.

——— 2007, Iron oxide copper-gold deposits: A Canadian perspective, *in* Goodfellow, W.D., ed., Mineral Deposits of Canada: A Synthesis of Major Deposit-Types, District Metallogeny, the Evolution of Geological Provinces, and Exploration Methods: Geological Association of Canada, Mineral Deposits Division, Special Publication No. 5, p. 307-328.

Corriveau, L., and Bonnet, A.-L., 2005, Pinwarian (1.5 Ga) volcanism and hydrothermal activity at the eastern margin of the Wakeham Group, Grenville Province, Quebec: Canadian Journal of Earth Sciences, v. 42, p. 1749-1782.

Corriveau, L., and Clark, T., 2005, The Grenville Province: A geological and mineral resources perspective derived from government and academic research initiatives: Canadian Journal of Earth Sciences, v. 42, p. 1637-1642.

Corriveau, L., and Gorton, M.P., 1993, Coexisting K-rich alkaline and shoshonitic magmatism of arc affinities in the Proterozoic: A reassessment of syenitic stocks in the southwestern Grenville Province: Contributions to Mineralogy and Petrology, v. 113, p. 262-279.

Corriveau, L., and Morin, D., 2000, Modelling 3D architecture of western Grenville from xenoliths, styles of magma emplacement and Lithoprobe reflectors: Canadian Journal of Earth Sciences, v. 37, p. 235-251.

Corriveau, L., and van Breemen, O., 2000, Docking of the Central Metasedimentary Belt to Laurentia in geon 12: evidence from the 1.17-1.16 Ga Chevreuil intrusive suite and host gneisses, Quebec: Canadian Journal of Earth Sciences, v. 37, p. 253-269.

Corriveau, L., Rivard, B., and van Breemen, O., 1998, Rheological controls on Grenvillian intrusive suites: implications for tectonic analysis: Journal of Structural Geology, v. 20, p. 1191-1204.

Corriveau, L., Brouillette, P., Scherrer, G., and Bonnet, A.-L., 2002, Extension orientale des roches volcaniques du Groupe de Wakeham et l'intrusion litée troctolitique de Musquaro, Province de Grenville, Basse-Côte-Nord, Québec: Commission géologique du Canada, Étude 02-C29, 11 p.

Cox, R.A., and Indares, A., 1999, Transformation of Fe-Ti gabbro to coronite, eclogite and amphibolite in the Baie du Nord segment, Manicouagan imbricate zone, eastern Grenville Province: Journal of Metamorphic Geology, v. 17, p. 537-555.

Cox, R.A., Dunning, G.R., and Indares, A., 1998, Petrology and U-Pb geochronology of mafic, high-pressure, metamorphic coronites from the Tshenukutish Domain, eastern Grenville Province: Precambrian Research, v. 90, p. 59-83.

Culshaw, N.G., and Dostal, J., 1997, Sand Bay gneiss association, Grenville Province, Ontario: a Grenvillian rift- (and -drift) assemblage stranded in the Central Gneiss Belt?: Precambrian Research, v. 85, p. 97-113.

——— 2002, Amphibolites of the Shawanaga Domain, Central Gneiss Belt, Grenville Province, Ontario: Tectonic setting and implications for relations between the Central Gneiss Belt and Midcontinental USA: Precambrian Research, v. 113, p. 65-85.

Culshaw, N.G., 2005, Intersection of the Shawanaga and Parry Sound shear zones, Central Gneiss Belt, Grenville Province, Ontario: buckling and shearing related to flow in the deep crust: Canadian Journal of Earth Sciences, v. 42, p. 1907-1925.

Culshaw, N.G., Jamieson, R.A., Ketchum, J.W.F., Wodicka, N., Corrigan, D., and Reynolds, P.H., 1997, Transect across the northwestern Grenville Orogen, Georgian Bay, Ontario: Polystage convergence and extension in the lower orogenic crust: Tectonics, v. 16, p. 966-982.

Cuney, M., and Kish, L., 2004, Minéralisations uranifères et fusion crustale: l'exemple de Mont Laurier (Québec, Canada). 72ième congrès de l'ACFAS, C-202 Premières journées De Launay, Programme et résumés, http://www.acfas.ca/congres/congres72/C3598.htm.

Currie, K.L., and Gittins, J., 1988, Contrasting sapphirine parageneses from Wilson Lake, Labrador and their tectonic significance: Journal of Metamorphic Geology, v. 6, p. 603-622.

Daigneault, R., and Allard, G.O., 1994, Transformation of Archean structural inheritance at the Grenvillian foreland parautochthon transition zone, Chibougamau, Quebec: Canadian Journal of Earth Sciences, v. 31, p. 470-488.

Davidson, A., 1984, Identification of ductile shear zones in the southwestern Grenville Province of the Canadian Shield, *in* Kröner, A. and Greiling, R., eds., Precambrian tectonics illustrated: Schweitzerbart'sche Verlagsbuchhandlung, Stuttgart, p. 263-279.

——— 1986, Grenville Front relationships near Killarney, Ontario, *in* Moore, J.M., Davidson, A., and Baer, A.J., eds., The Grenville Province: Geological Association of Canada, Special Paper 31, p. 107-117.

——— 1995, A review of the Grenville orogen in its North American type area: AGSO Journal of Australian Geology and Geophysics, v. 16, p. 3-24.

——— 1998a, An overview of Grenville Province geology, Canadian Shield; Chapter 3 *in* Lucas, S.B., and St-Onge, M.R., coordinators, Geology of the Precambrian Superior and Grenville Provinces and Precambrian Fossils in North America: Geological Survey of Canada, Geology of Canada, no. 7, p. 205-270 (also Geological Society of America, The Geology of North America, v. C-1).

——— 1998b, Geological map of the Grenville Province, Canada and adjacent parts of the United States of America: Geological Survey of Canada, Map 1947A, scale 1:2,000,000.

——— 2000, Geology of the Mountain Grove pluton, Central Metasedimentary Belt, Grenville Province, Ontario: Geological Survey of Canada, Current Research 2000-C23, 7 p.

——— 2001, The Chief Lake complex revisited, and the problem of correlation across the Grenville Front south of Sudbury, Ontario: Precambrian Research, v. 107, p. 5-29.

Davidson, A., and van Breemen, O., 1994, U-Pb ages of granites near the Grenville Front: Geological Survey of Canada, Current Research 1994-F, p. 107-114.

——— 2000, Age and Extent of the Frontenac Plutonic Suite in the Central Metasedimentary Belt, Grenville Province, southeastern Ontario; Geological Survey of Canada, Paper 2000-F, 7 p.

de Lorraine, W.F., 2001, Metamorphism, polydeformation, and extensive remobilization of the Balmat zinc orebodies, northwest Adirondacks, New York: Society of Economic Geologists, Guidebook Series, v. 5, p. 25-54.

de Lorraine, W.F., and Dill, D.B., 1982, Structure, stratigraphic controls and genesis of the Balmat zinc deposits, NW Adirondacks, New York, *in* Hutchinson, R.W., Spence, C.D., and Franklin, J.M., eds., Precambrian Sulphide Deposits: Geological Association of Canada, Special Paper 25, p. 571-596.

Deland, A.N., 1959, Report on development work: Ministère des Ressources naturelles, Québec, GM 08334, 2 p.

Dewey, J.F., and Burke, K.C., 1973, Tibetan, Variscan and Precambrian basement reactivation: Products of continental collision. Journal of Geology, v. 81, p. 683-692.

Dickin, A.P., 2000, Crustal formation in the Grenville Province; Nd-isotope evidence: Canadian Journal of Earth Sciences, v. 37, p. 165-181.

——— 2004, Mesoproterozoic and Paleoproterozoic crustal growth in the eastern Grenville Province: Nd isotope evidence from the Long Range inlier of the Appalachian orogen, *in* Tollo, R.P., Corriveau, L., McLelland, J., and Bartholomew, M., eds., Proterozoic Tectonic Evolution of the Grenville Orogen in North America: Geological Society of America, Memoir 197, p. 495-504.

Duchesne, J.-C., Liégeois, J.P., Vander Auwera, J., and Longhi, J., 1999, The crustal tongue melting model and the origin of massive anorthosites: Terra Nova, v. 11, p. 100-105.

Easton, R.M., 1989, Regional alteration patterns and mineralization associated with the Deloro Granite, Grenville Province, Madoc area, *in* Summary of Field Work and Other Activities, 1989: Ontario Geological Survey, Miscellaneous Paper 146, p. 158-168.

——— 1992, The Grenville Province and the Proterozoic history of central and southern Ontario, *in* Thurston, P.C., Williams, H.R., Sutcliffe, R.H., and Stott, G.M., eds., Geology of Ontario: Ontario Geological Survey, Special Volume 4(2), p. 715-904.

——— 2000, Metamorphism of the Canadian shield, Ontario, Canada. II. Proterozoic metamorphic history: The Canadian Mineralogist, v. 38, p. 319-344.

——— 2003a, Project unit 03-009. Reconnaissance study of the geology and mineral potential of the eastern Tomiko terrane, Grenville Province, *in* Summary of Field Work and Other Activities 2003: Ontario Geological Survey, Open File Report 6120, p. 16-1-16-25.

——— 2003b, Geology and Mineral Potential of the Paleoproterozoic River Valley Intrusion and Related Rocks, Grenville Province: Ontario Geological Survey, Open File Report 6123, 172 p.

Easton, R.M., and Fyon, J.A., 1992, Metallogeny of the Grenville Province, *in* Thurston, P.C., Williams, H.R., Sutcliffe, R.H., and Stott, G.M., eds., Geology of Ontario: Ontario Geological Survey, Special Volume 4(2), p. 1217-1254.

Eckstrand, O.R., Sinclair, W.D., and Thorpe, R.I., 1996, Introduction, *in* Eckstrand, O.R., Sinclair, W.D., and Thorpe, R.I., eds., Geology of Canadian Mineral Deposit Types: Geological Survey of Canada, Geology of Canada, no. 8, p. 1-8.

Eilu, P., Sorjonen-Ward, P., Nurmi, P., and Niiranen, T., 2003, A review of gold mineralization styles in Finland: Economic Geology, v. 98, p. 1329-1353.

Emslie, R.F., 1978, Anorthosite massifs, rapakivi granites and late Proterozoic rifting of North America: Precambrian Research, v. 7, p. 61-98.

Emslie, R.F., Hamilton, M.A., and Gower, C.F., 1997, The Michael gabbro and other Mesoproterozoic lithospheric probes in southern and central Labrador: Canadian Journal of Earth Sciences, v. 34, p. 1566-1580.

Friedman, R., and Martignole, J., 1995, Mesoproterozoic sedimentation, magmatism, and metamorphism in the southern part of the Grenville Province (western Quebec): U-Pb geochronological constraints: Canadian Journal of Earth Sciences, v. 32, p. 2103-2114.

Friehauf, K.C., Smith II, R.C., and Volkert, R.A., 2002, Comparison of the geology of Proterozoic iron oxide deposits in the Adirondack and Mid-Atlantic belt of Pennsylvania, New Jersey and New York, *in* Porter, T.M., ed., Hydrothermal Iron Oxide Copper-Gold and Related Deposits: A Global Perspective: PGC Publishing, Adelaide, v. 2, p. 247-252.

Fu, W., Corriveau, L., LaFlèche, M.R., and Blein, O., 2003, Birdwing-shaped REE profiles and /or isovalent Nb/Ta, Zr/Hf ratios in the Bondy Gneiss Complex, Grenville Province, Québec: Sensitive geochemical markers of a fossil hydrothermal system in mineral exploration: Canadian Institute of Mining Industry Conference and Exhibition, Montreal 2003, Canadian Institute of Mining, Technical Paper (CD-ROM), 17 p.

Gandhi, S.S., 1986, Uranium in Early Proterozoic Aillik Group, Labrador, *in* Evans, E.L., ed., Uranium Deposits of Canada, Canadian Institute of Mining and Metallurgy, Special Volume 33, p. 70-82.

Gariépy, C., Verner, D., and Doig, R., 1990, Dating Archean metamorphic minerals southeast of the Grenville front, western Quebec, using Pb isotopes: Geology, v. 18, p. 1078-1081.

Gauthier, M., and Brown, A.C., 1986, Zinc and iron metallogeny in the Maniwaki-Gracefield district, southwestern Quebec: Economic Geology, v. 81, p. 89-112.

Gauthier, M., and Chartrand, F., 2005, Metallogeny of the Grenville Province revisited: Canadian Journal of Earth Sciences, v. 42, p. 1719-1734.

Gauthier, M., Morin, G., and Marcoux, P., 1985, Minéralisations aurifères de la partie centrale de la Province de Grenville, Bouclier canadien: Canadian Institute of Mining Bulletin, v. 78, p. 60-69.

Gauthier, M., Chartrand, F., Cayer, A., and David, J., 2004a, The Kwyjibo Cu-REE-U-Au-Mo-F property, Quebec: A Mesoproterozoic polymetallic iron oxide deposit in the Northeastern Grenville Province: Economic Geology, v. 99, p. 1177-1196.

Gauthier, M., Corriveau, L., and Chouteau, M., 2004b, Metamorphosed and Metamorphogenic Ore Deposits of the Central Metasedimentary Belt, southwestern Québec and southeastern Ontario Grenville Province: Premières Journées De Launay, Post-symposium Field Trip Guidebook, 20 p.

Gauthier, M., Gardoll, S., and Duchesne, J.-C., 2004c, Quest for palladium: Application of the Groves hypothesis to Precambrian terranes of north-east North America: Geology, v. 32, p. 593-596.

Gilbert, J.E., 1959, Bristol Township-Hilton Mines Ltd, *in* Description of Mining Properties Examined in 1956 and 1957 (exclusive of producing mines) and Outline of Geology and Exploration Work: Department of Mines, Québec, Preliminary Report 390, p. 14-15.

Gobeil, A., 1997, Géologie de la Région de la Rivière Sainte-Marguerite (Phase II): Ministère des Ressources Naturelles, Québec, MB 96-43, 12 p.

Gobeil, A., Chevé, S., Clark, T., Corriveau, L., Perreault, S., and Dion, D.-J., 1999, Géologie de la Région du Lac Nipisso (SNRC 22I/13): Ministère des Ressources Naturelles, Québec, RG 98-19, 60 p.

Gobeil, A., Hébert, C., Clark, T., Beaumier, M., and Perreault, S., 2002, Géologie de la Région du lac De La Blache (22K03/ et 22K/04): Ministère des Ressources naturelles, Québec, RG 2002-01, 49 p.

Gobeil, A., Brisebois, D., Clark, T., Verpaelst, P., Madore, L., Wodicka, N., and Chevé, S., 2003, Synthèse géologique de la région de Manitou-Wakeham (moyenne-côte-nord), *in* Brisebois, D., and Clark, T., eds., Géologie et Ressources minérales de la Partie est de la Province de Grenville: Ministère des Ressources naturelles de la Faune et des Parcs, Québec, DV 2002-03, p. 9-58.

Gower, C.F., 1992, The relevance of Baltic Shield metallogeny to mineral exploration in Labrador, *in* Current Research: Newfoundland and Labrador Geological Survey, Report No. 92-1, p. 331-366.

——— 1996, The evolution of the Grenville Province in eastern Labrador, Canada, *in* Brewer, T.S., ed., Precambrian Crustal Evolution in the North Atlantic Region: Geological Society Special Publication, v. 112, p. 197-218.

Gower, C.F., and Krogh, T.E., 2002, A U-Pb geochronological review of the Proterozoic history of the eastern Grenville Province: Canadian Journal of Earth Sciences, v. 39, p. 795-829.

——— 2003, A U-Pb geochronological review of the Pre-Labradorian and Labradorian geological history of the eastern Grenville Province, *in* Brisebois, D., and Clark, T., eds., Géologie et Ressources minérales de la Partie est de la Province de Grenville: Ministère des Ressources Naturelles de la Faune et des Parcs, Québec, DV 2002-03, p. 147-172.

Gower, C.F., Ryan, A.B., and Rivers, T., 1990, Mid-Proterozoic Laurentia-Baltica: an overview of its geological evolution and a summary of the contributions made by this volume, *in* Gower, C.F., Rivers, T., and Ryan, A.B., eds., Mid-Proterozoic Laurentia-Baltica: Geological Association of Canada, Special Paper 38, p. 1-20.

Gower, C.F., McConnell, J.W., and van Nostrand, T., 1995, New mineral-exploration targets in the Pinware Terrane, Grenville Province, south-west Labrador, *in* Current Research: Newfoundland and Labrador Geological Survey, Report No. 95-1, p. 15-24.

Gower, C.F., Krogh, T.E., and James, D.T., 2002, Correlation chart of the eastern Grenville Province and its northern foreland: Canadian Journal of Earth Sciences, v. 39, p. 897.

Green, A.G., Milkereit, B., Davidson, A., Spencer, C., Hutchinson, D.R., Cannon, W.F., Lee, M.W., Agena, W.F., Behrendt, J.C., and Hinze, W.J., 1988, Crustal structure of the Grenville Front and adjacent terranes: Geology, v. 16, p. 788-792.

Gross, G.A., 1967, Iron deposits in the Appalachians and Grenville regions of Canada, *in* Volume II, Geology of Iron Deposits in Canada: Geological Survey of Canada, Economic Geology Report 22, p. 76-80.

——— 1996, Stratiform iron, *in* Eckstrand, O.R., Sinclair, W.D., and Thorpe, R.I., eds., Geology of Canadian Mineral Deposit Types: Geological Survey of Canada, Geology of Canada, no. 8, p. 47-90.

Hamilton, M.A., McLelland, J., and Selleck, B., 2004, SHRIMP U-Pb zircon geochronology of the anorthosite-mangerite-charnockite-granite suite, Adirondack Mountains, New York: Ages of emplacement and metamorphism, *in* Tollo, R., Corriveau, L., McLelland, J., and Bartholomew, M., eds., Proterozoic Tectonic Evolution of the Grenville Orogen in eastern North America: Geological Society of America, Memoir 197, p. 337-355.

Hanmer, S., 1988, Ductile thrusting at mid-crustal level, southwestern Grenville Province: Canadian Journal of Earth Sciences, v. 25, p. 1049-1059.

Hanmer, S., Corrigan, D., Pehrsson, S., and Nadeau, L., 2000, SW Grenville Province, Canada: The case against post-1.4 Ga accretionary tectonics: Tectonophysics, v. 319, p. 33-51.

Hauseux, M.A., 1977, Mode of uranium occurence in a migmatitic granite terrain, Baie-Johan-Beetz, Quebec: Canadian Institute of Mining Bulletin, v. 70, p. 110-116.

Haynes, S.J., 1986, Metallogenesis of U-Th, Grenville Supergroup, Peterborough County, Ontario, *in* Moore, J.M., Davidson, A., and Baer,

A.J., eds., The Grenville Province: Geological Association of Canada, Special Paper, v. 31, p. 271-280.

Heaman, L.M., Erdmer, P., and Owen, J.V., 2002, U-Pb geochronologic constraints on the crustal evolution of the Long Range Inlier, Newfoundland: Canadian Journal of Earth Sciences, v. 39, p. 845-865.

Heaman, L.M., Gower, C.F., and Perreault, S., 2004, The timing of Proterozoic magmatism in the Pinware terrane of southeast Labrador, easternmost Quebec and northwest Newfoundland: Canadian Journal of Earth Sciences, v. 41, p. 127-150.

Hébert, C., and van Breemen, O., 2004, Mesoproterozoic basement, the Lac-St. Jean anorthosite suite and younger Grenvillian intrusions in the Saguenay region (Quebec): Structural relationships and U-Pb geochronology, *in* Tollo, R., Corriveau, L., McLelland, J., and Bartholomew, M., eds., Proterozoic Tectonic Evolution of the Grenville Orogen in eastern North America: Geological Society of America, Memoir 197, p. 65-79.

Hébert, C., Cadieux, A.-M., and van Breemen, O., 2005, The Labrieville, Lac St. Jean, De La Blache, and St. Urbain anorthosites, Central Grenville, Quebec, Canada: Comparison of U-Pb ages and Ti-Fe-P occurrences: Canadian Journal of Earth Sciences, v. 42, p. 1865-1880.

Hébert, Y., 1993, Principales Sources de Silicates Alumineux au Québec: Ministère des Ressources Naturelles, Québec, MB 93-59, 56 p.

——— 1995, Les Gîtes de Terres rares et Éléments associés dans les Districts miniers de Montréal-Laurentides, Estrie-Laurentides et Côte-Nord-Nouveau-Québec: Ministère des Ressources naturelles, Québec, MB 94-17, 140 p.

Higgins, M.D., 2005, A new interpretation of the structure of the Sept Iles Intrusive suite, Canada: Lithos, v. 83, p. 199- 213.

Higgins, M.D., and van Breemen, O., 1996, Three generations of AMCG magmatism, contact metamorphism and tectonism in the Saguenay-Lac-St. Jean region, Grenville Province, Canada: Precambrian Research, v. 7, p. 327-346.

Hudon, P., Friedman, R.M., Gauthier, G., and Martignole, J., 2006, Age of the Cabonga nepheline syenite, Grenville Province Western Quebec: Canadian Journal of Earth Sciences, v. 43, p. 1237-1249.

Indares, A., and Dunning, G., 2004, Crustal architecture above the high-pressure belt of the Grenville Province in the Manicouagan area: New structural, petrologic and U-Pb age constraints: Precambrian Research, v. 130, p. 199-228.

Irvine, T.N., Andersen, J.C.O., and Brooks, C.K., 1998, Included blocks (and blocks within blocks) in the Skaergaard intrusion: Geologic relations and the origin of rhythmic modally graded layers: Geological Society of America Bulletin, v. 110, p. 1398-1447.

Jacob, H.-L., and Bélanger, M., 2001, Les minéraux industriels au Québec, *in* Dunlop, S., and Simandl, G., eds., Industrial Minerals in Canada: Canadian Institute of Mining, Special Volume 53, p. 187-198.

James, D.T., Kamo, S., Krogh, T., and Nadeau, L., 2001, Preliminary U-Pb geochronological data from Mesoproterozoic rocks Grenville Province, southern Labrador, *in* Current Research: Newfoundland and Labrador Department of Mines and Energy, Geological Survey, Report 2001-1, p. 45-54.

James, R.S., Easton, R.M., Peck, D.C., and Hrominchuck, J.L., 2002, The East Bull Lake intrusive suite: Remnants of an approximately 2.48 Ga large igneous and metallogenic province in the Sudbury area of the Canadian Shield: Economic Geology, v. 97, p. 1577-1606.

Jobin-Bevans, L.S., 2002, Geology and PGE mineralisation along the northern contact of the River Valley intrusion, Sudbury region, Ontario: 9th International Platinum Conference, Billings, USA, 2 p.

Johnson, C.A., and Skinner, B.J., 2003, Geochemistry of the Furnace magnetite bed, Franklin, New Jersey, and the relationship between stratiform iron oxide and stratiform zinc oxide silicate ores in the New Jersey Highlands: Economic Geology, v. 98, p. 837-854.

Jordan, S.L., Indares, A., and Dunning, G., 2006, Partial melting of metapelites in the Gagnon terrane below the high pressure belt in the Manicouagan area (Grenville Province): Pressure-temperature (P-T) and U-Pb age constraints and implications: Canadian Journal of Earth Sciences, v. 43, p. 1309-1329.

Jourdain, V., Gauthier, M., and Guha, J., 1990, Métallogénie de l'Or dans le sud-ouest de la Province de Grenville: Geological Survey of Canada, Open File Report 2287, 50 p.

Karlstrom, K.E., Åhäll, K.-I., Harlan, S.S., Williams, M.L., McLelland, J., and Geissman, J.W., 2001, Long-lived (1.8-1.0 Ga) convergent orogen in southern Laurentia, its extensions to Australia and Baltica, and implications for refining Rodinia: Precambrian Research, v. 111, p. 5-30.

Ketchum, J.W.F., and Davidson, A., 2000, Crustal architecture and tectonic assembly of the Central Gneiss Belt, southwestern Grenville Province, Canada: A new interpretation: Canadian Journal of Earth Sciences, v. 37, p. 217-234.

Ketchum, J.W.F., Jamieson, R.A., Heaman, L.M., Culshaw, N.G., and Krogh, T.E., 1994, 1.45 Ga granulites in the southwestern Grenville Province: Geologic setting, P-T conditions, and U-Pb geochronology: Geology, v. 22, p. 215-218.

Klein, C., 1978, Regional metamorphism of Proterozoic iron formation, Labrador Trough, Canada: American Mineralogist, v. 63, p. 898-912.

Kletetschka, G., and Stout, J.H., 1998, The origin of magnetic anomalies in lower crustal rocks, Labrador: Geophysical Research Letters, v. 25, p. 199-202.

Korhonen, F.J., and Stout, J.H., 2004, Low variance sapphirine-bearing assemblages from Wilson Lake, Grenville Province of Labrador, *in* Tollo, R.P., Corriveau, L., McLelland, J., and Bartholomew, M., eds, Proterozoic Tectonic Evolution of the Grenville Orogen in North America: Geological Society of America, Memoir 197, p. 81-104.

——— 2006, Silicate-oxide equilibria in the Wilson Lake terrane, Labrador - evidence for a pre-metamorphic oxidizing event. American Geophysical Union, 2006 Joint Assembly Meeting, Program with Abstract, http://www.agu.org/meetings/sm06/sm06-sessions /sm06_U21A.html

Krauss, J.B., and Rivers, T., 2004, High-pressure granulites in the Grenvillian Grand Lake thrust system, Labrador: P-T conditions and tectonic evolution, *in* Tollo, R.P., Corriveau, L., McLelland, J., and Bartholomew, M., eds, Proterozoic Tectonic Evolution of the Grenville Orogen in North America: Geological Society of America, Memoir 197, p. 105-134.

Krogh, T.E., 1989, Provenance and metamorphic ages in the Grenville (NW): Lithoprobe Abitibi-Grenville project workshop, March 1989, p. 5-7.

Krogh, T.E., Kamo, S., Gower, C.F., and Owen, J.V., 2002, Augmented and reassessed U-Pb geochronological data from the Labradorian-Grenvillian front in the Smokey archipelago, eastern Labrador: Canadian Journal of Earth Sciences, v. 39, p. 831-843.

LaFlèche, M.R., Birkett, T., and Corriveau, L., 2005, Crustal development at the pre-Grenvillian Laurentian margin: A record from contrasting geochemistry of mafic and ultramafic orthogneisses in the Chochocouane River area, Quebec: Canadian Journal of Earth Sciences, v. 42, p. 1653-1675.

Larbi, Y., Stevenson, R., Verpaelst, P., Brisebois, D., and Madore, L., 2003, Caractérisation isotopique (Sm-Nd) et géochimie du groupe de Wakeham: Un bassin sédimentaire protérozoïque dans la Province de Grenville, *in* Brisebois, D., and Clark, T., eds., Géologie et Ressources Minérales de la Partie est de la Province de Grenville: Ministère des Ressources naturelles de la Faune et des Parcs, Québec, DV 2002-03, p. 247-268.

Large, R.R., McPhie, J., Gemmel, J.B., Herrmann, W., and Davidson, J., 2001, The spectrum of ore deposit types, volcanic environments, alteration halos, and related exploration vectors in submarine volcanic successions: Some examples from Australia: Economic Geology, v. 96, p. 913-938.

Lentz, D.R., 1996, U, Mo, and REE mineralization in the late-tectonic granitic pegmatites, southwestern Grenville Province, Canada: Ore Geology Reviews, v. 11, p. 197-227.

Lentz, D.R., and Creaser, R., 2005, Re-Os model age constraints on the genesis of the Moss molybdenite pegmatite-aplite deposits, southwestern Grenville Province, Quyon, Quebec, Canada: Exploration and Mining Geology, v. 14, p. 95-103.

Leong, K.M., and Moore, J.M., 1972, Sapphirine-bearing rocks from Wilson Lake, Labrador: The Canadian Mineralogist, v. 11, p. 777-790.

Loney, R.A., and Himmelberg, G.R., 1983, Structure and petrology of the La Perouse gabbro intrusion, Fairweather range, southeastern Alaska: Journal of Petrology, v. 24, p. 377-423.

Lowell, G.R., and Noll, P.D., Jr, 2001. Fe-Cu-Au-bearing scapolite skarn in moat sediments of the Taum Sauk Caldera, southeastern Missouri, USA: Mineralogical Magazine, v. 65, p. 373-396.

Lowell, G.R., and Young, G.J., 1999, Interaction between coeval mafic and felsic melts in the St. Francois Terrane of Missouri, USA: Precambrian Research, v. 95, p. 69-88.

Ludden, J., and Hynes, A., 2000a, The Abitibi-Grenville Lithoprobe transect Part III: Introduction: Canadian Journal of Earth Sciences, v. 37, p. 115-116.

——— 2000b, The Lithoprobe Abitibi-Grenville transect: Two billion years of crust formation and recycling in the Precambrian Shield of Canada: Canadian Journal of Earth Sciences, v. 37, p. 459-476.

Lyons, P., Goleby, B., Jones, L., Drummond, B., and Korsch, R., 2004, Imaging the tectonic environment of the giant Olympic Dam iron oxide-copper-gold deposit, South Australia: 11th Deep Seismix Conference, Program with Abstracts, p. 72.

Magrina, B., Jébrak, M., and Cuney, M., 2005, Le magmatisme de la région de Kwyjibo, Province du Grenville (Canada) : intérêt pour les minéralisations de type fer-oxydes associées: Canadian journal of Earth Sciences, v. 42, p. 1849-1864.

Marshall, L.J., and Oliver, N.H.S., 2006, Monitoring fluid chemistry in iron oxide-copper-gold-related metasomatic processes, eastern Mt Isa Block, Australia: Geofluids, v. 6, p. 45-66.

Marshall, L.J., Smith, R., Wilton, D., and Butler, R., 2003, Geology and preliminary exploration results, Central Mineral Belt, Labrador, Canada: Altius Minerals Corporation and Fronteer Development Group, Internal Report, 52 p.

Martignole, J., Machado, N., and Nantel, S., 1993, Timing of intrusion and deformation of the Rivière-Pentecôte anorthosite (Grenville Province): The Journal of Geology, v. 101, p. 652-658.

Martignole, J., Machado, N., and Indares, A., 1994, The Wakeham Terrane: A Mesoproterozoic terrestrial rift in the eastern part of the Grenville Province: Precambrian Research, v. 68, p. 291-306.

Martignole, J., Calvert, A.J., Friedman, R., and Reynolds, P., 2000, Crustal evolution along a seismic section across the Grenville Province (western Quebec): Canadian Journal of Earth Sciences, v. 37, p. 291-306.

Martin, C., and Dickin, A.P., 2005, Styles of Proterozoic crustal growth on the southeast margin of Laurentia: Evidence from the central Grenville Province northwest of Lac St.-Jean, Quebec: Canadian Journal of Earth Sciences, v. 42, p. 1643-1652.

McLelland, J., Daly, S., and McLelland J.M., 1996, The Grenville Orogenic Cycle (ca 1350-1000 Ma): An Adirondack perspective: Tectonophysics, v. 265, p. 1-28.

McLelland, J., Morrison, J., Selleck, B., Cunningham, B., Olson, C., and Schmidt, K., 2002, Hydrothermal alteration of late- to post-tectonic Lyon Mt. Granitic Gneiss, Adirondack Highlands, New York: Origin of quartz-sillimanite segregations, quartz-albite lithologies, and associated Kiruna-type low-Ti Fe-oxide deposits: Journal of Metamorphic Geology, v. 20, p. 175-190.

Meert, J.G., and Powell, C.M., 2001, Assembly and break-up of Rodinia: Introduction to the special volume: Precambrian Research, v. 110, p. 1-8.

Menuge, J.F., Brewer, T.S., and Seeger, C.M., 2002, Petrogenesis of metaluminous A-type rhyolites from the St Francois Mountains, Missouri and the Mesoproterozoic evolution of the southern Laurentian margin: Precambrian Research, v. 113, p. 269-291.

Moore, J.M., 1986, Introduction: The 'Grenville Problem' then and now, *in* Moore, J.M., Davidson, A., and Baer, A.J., eds., The Grenville Province: Geological Association of Canada, Special Paper 31, p. 9-11.

Moore, J.M., and Thompson, P., 1980, The Flinton group: A late Precambrian metasedimentary sequence in the Grenville Province of eastern Ontario: Canadian Journal of Earth Sciences, v. 17, p. 1685-1707.

Moore, J.M., Davidson, A., and Baer, A.J., eds., 1986, The Grenville Province: Geological Association of Canada, Special Paper 31, 358 p.

Morin, D., Hébert, R., and Corriveau, L., 2005, Mesoproterozoic deep K-magmatism recorded in a megacryst- and xenolith-bearing minette dyke, western Grenville Province: Canadian Journal of Earth Sciences, v. 42, p. 1881-1906.

Mumin, A.H., and Corriveau, L., 2004, The Eden deformation corridor and polymetallic mineral belt: Trans Hudson Orogen, Leaf Rapids District, Manitoba: Manitoba Geological Survey, Report of Activities 2004, p. 69-91.

Nabil, H., Clark, T., and Barnes, S.-J., 2004, A Ni-Cu-Co-PGE massive sulphide prospect in a gabbronorite dike at Lac Volant, eastern Grenville Province, Quebec, *in* Tollo, R.P., Corriveau, L., McLelland, J., and Bartholomew, M., eds., Proterozoic Tectonic Evolution of the Grenville Orogen in North America: Geological Society of America, Memoir 197, p. 145-162.

Nadeau, L., and James, D.T., 2001, Preliminary note on the lithogeochemistry and petrogenesis of intrusive rock suites from the Minipi Lake region (NTS map area 13C/south) Grenville Province, Labrador, *in* Current Research: Newfoundland and Labrador Department of Mines and Energy, Geological Survey, Report 2001-1, p. 75-87.

Nadeau, L., and van Breemen, O., 1994, Do the 1.45-1.39 Ga Montauban Group and the La Bostonnais complex constitute a Grenvillian accreted terrane? Geological Association of Canada-Mineralogical Association of Canada, Program with Abstracts, v. 19, p. A81.

——— 1998, Plutonic ages and tectonic settings of the Algonquin and Muskoka allochthons, Central Gneiss Belt, Grenville Province, Ontario: Canadian Journal of Earth Sciences, v. 35, p. 1423-1438.

Nadeau, L., Brouillette, P., and Hébert, C., 1992, Geology and structural relationships along the east margin of the St. Maurice tectonic zone, north of Montauban, Grenville orogen, Quebec, *in* Current Research, Part C, Canadian Shield: Geological Survey of Canada, Paper 92-1C, p. 139-146.

——— 1999, New observations on relict volcanic features in medium grade gneisses of the Montauban Group, Grenville Province, Quebec: Geological Survey of Canada, Paper 99-1E, p. 149-160.

Nantel, S., 2003, Carte de la partie nord de la Ceinture centrale des métasédiments, Province de Grenville, *in* Cartes préliminaires en couleurs des travaux de cartographie et des études 2002-2003: Ministère des Ressources naturelles du Québec; DV 2002-11, 1:100 000 map.

Nantel, S., and Pintson, H., 2002, Géologie de la Région du Lac Dieppe (31O/3): Ministère des Ressources Naturelles, Québec, RG 2001-16, 36 p.

Nantel, S., Giguère, E., and Clark, T., 2004, Géologie de la Région du Lac Duplessis (31 O/06): Ministère des Ressources Naturelles, de la Faune et des Parcs du Québec, RG 2003-01, 51 p.

Neal, H.E., 2000, Iron deposits of the Labrador Trough: Exploration and Mining Geology, v. 9, p. 113-121.

Ortega, J., 2003, Rapport sur les Travaux d'Exploration effectués en 2002 - Projet Gatineau JV (1510): Ministère des Ressources Naturelles, Québec, GM 59949, 58 p.

Osborne, F.F., 1956, The Grenville region of Quebec, *in* Thompson, J.E., ed., The Grenville Problem: The Royal Society of Canada, Special Publications, v. 1, p. 3-13.

Owen, J.V., and Greenough, J.D., 1995, Petrology of cordierite + gedrite granulite from the Grenvillian Long Range Inlier, Newfoundland: Canadian Journal of Earth Sciences, v. 32, p. 1035-1045.

Owen, J.V., Longstaffe, F.J., and Greenough, J.D., 2003, Petrology of sapphirine granulite and associated sodic gneisses from the Indian Head Range, Newfoundland: Lithos, v. 68, p. 91-114.

Owens, B.E., and Dymek, R.F., 2005, Rediscovery of the Mattawa Anorthosite Massif, Grenville Province, Quebec: Canadian Journal of Earth Sciences, v. 42, p. 1699-1718.

Parsons, S., Nadeau, L., Keating, P., and Chung, C.-J., 2006, Optimizing the use of aeromagnetic data for predictive geological interpretation: An example from the Grenville Province, Quebec: Computers and Geosciences, v. 32, p. 565-576.

Peck, W.H., and Smith, M.S., 2005, Cordierite-gedrite rocks from the Central Metasedimentary Belt boundary thrust zone (Grenville Province, Ontario): Mesoproterozoic metavolcanic rocks with affinities to the Composite Arc Belt: Canadian Journal of Earth Sciences, v. 42, p. 1815-1828.

Peck, W.H., De Angelis, M.T., Meredith, M.T., and Morin, E., 2005, Polymetamorphism of marbles in the Morin terrane, Grenville Province, Quebec: Canadian Journal of Earth Sciences, v. 42, p. 1949-1965.

Perreault, S., and Heaman, L.M., 2003, Géologie et géochronologie de la Basse-Côte-Nord (entre Chevery et Blanc Sablon) dans la province géologique de Grenville, *in* Brisebois, D., and Clark, T., eds., Géologie et Ressources Minérales de la Partie est de la Province de Grenville: Ministère des Ressources Naturelles de la Faune et des Parcs, Québec, DV 2002-03, p. 119-146.

Perreault, S., and Hébert, C., 2003, Review of Fe-Ti±V and Fe-TiP_2O_5±V deposits associated with anorthositic suites in the Grenville Province, Québec, *in* Duchesne, J.-C., and Korneliussen, A., eds., Ilmenite Deposits and their Geological Environment: Norges geologiske undersokelse, Special Publication 9, p. 83-84.

Perreault, S., and Moukhsil, A., 2003, Territoire de la Province de Grenville: Ministère des Ressources naturelles, de la Faune et des Parcs, Québec, DV 2004-01, p. 45-51.

Perreault, S., Clark, T., Gobeil, A., Chevé, S., Dion, D.-J., Corriveau, L., Nabil, H., and Lortie, P., 1997, The Cu-Ni-Co Potential of the Sept-Îles

Region: The Lac Volant showing: Ministère des Ressources Naturelles, Québec, PRO 97-03, 10 p.

Perry, C., and Raymond, D., 1996, Le Projet Nipissis de SOQUEM - IOC: un Nouveau Type de Minéralisations Cuprifère sur la Côte-Nord: Ministère des Ressources Naturelles, Québec, DV 96-02, p. 16.

Perry, C., and Roy, I., 1997, Propriété Manitou 1088-1. Rapport sur les Travaux Menés en 1995: Ministère des Ressources Naturelles, Québec, GM-55005, 431 p.

Puffer, J.H., and Gorring, M.L., 2005, The Edison magnetite deposits in the context of pre-syn-and post-orogenic metallogenesis in the Grenville Highlands of New Jersey: Canadian Journal of Earth Sciences, v. 42, p. 1735-1748.

Puffer, J.H., and Volkert, R.A., 1991, Generation of trondhjemite from partial melting of dacite under granulite facies conditions; an example from the New Jersey Highlands, USA: Precambrian Research, v. 51, p. 115-125.

Rankin, D.W., Chiarenzelli, J.R., Drake, A.A., Jr., Goldsmith, R., Hall, L.M., Hinze, W.J., Isachen, W.J., Lidiak, E.G., McLelland, J., Mosher, S., Ratcliffe, N.M., Secor, D.T., Jr., and Whitney, P.R., 1993, Proterozoic rocks east and southeast of the Grenville front, Chapter 5 *in* Reed, J.C., Bickford, M.E., Houston, R.S., Link, P.K., Rankin, D.W., Sims, P.K., and Van Schmus, W.R., eds., Precambrian: Conterminous U.S.: Geological Society of America, The Geology of North America, v. C-2, p. 335-461.

Richardson, D.G., and Birkett, T.C., 1996, Carbonate-associated deposits, *in* Eckstrand, O.R., Sinclair, W.D., and Thorpe, R.I., eds., Geology of Canadian Mineral Deposit Types: Geological Survey of Canada, Geology of Canada, no. 8, p. 541-558.

Rivers, T., 1983, The northern margin of the Grenville Province in western Labrador - anatomy of an ancient orogenic front: Precambrian Research, v. 22, p. 41-73.

——— 1997, Lithotectonic elements of the Grenville Province: review and tectonic implications: Precambrian Research, v. 86, p. 117-154.

Rivers, T., and Corrigan, D., 2000, Convergent margin on southeastern Laurentia during the Mesoproterozoic: Tectonic implications: Canadian Journal of Earth Sciences, v. 37, p. 359-383.

Rivers, T., Martignole, J., Gower, C.F., and Davidson, A., 1989, New tectonic divisions of the Grenville Province, southeast Canadian Shield: Tectonics, v. 8, p. 63-84.

Rivers, T., van Gool, J.A.M., and Connelly, J.N., 1993, Contrasting tectonic styles in the northern Grenville Province: implications for the dynamics of orogenic fronts: Geology, v. 21, p. 1127-1130.

Rivers, T., Ketchum, J.W.F., Indares, A., and Hynes, A., 2002, The High Pressure belt in the Grenville Province: Architecture, timing, and exhumation: Canadian Journal of Earth Sciences, v. 39, p. 867-893.

Roach, R.A., and Duffel, S., 1974, Structural analysis of the Mount Wright map-area, southernmost Labrador-Trough, Quebec, Canada: Geological Society of America Bulletin, v. 85, p. 947-962.

Rogers, M.C., Thurston, P.C., Fyon, J.A., Kelly, R.I., and Breaks, F.W., 1995, Descriptive Mineral Deposit Models of Metallic and Industrial Deposit Types and Related Mineral Potential Assessment Criteria: Ontario Geological Survey, Open File Report 5916, 241 p.

Rondenay, S., Bostock, M.G., Hearn, T.M., White, D.J., and Ellis, R.M., 2000, Lithospheric assembly and modification of the SE Canadian Shield: Abitibi-Grenville teleseismic experiment: Journal of Geophysical Research, v. 105, p. 13735-13754.

Rose, E.R., 1969, Geology of Titanium and Titaniferous Deposits of Canada: Geological Survey of Canada, Economic Geology Report 25, 177 p.

Sager-Kinsman, E.A., and Parrish, R.R., 1993, Geochronology of detrital zircons from the Elzevir and Frontenac terranes, Central Metasedimentary Belt, Grenville Province, Ontario: Canadian Journal of Earth Sciences, v. 30, p. 465-473.

Saint-Germain, P., and Corriveau, L., 2003, Évolution magmatique et géochimique du Complexe de gabbronorite et de monzonite de Matamec, *in* Brisebois, D., and Clark, T., eds., Géologie et Ressources Minérales de la Partie est de la Province de Grenville: Ministère des Ressources Naturelles de la Faune et des Parcs, Québec, DV 2002-03, p. 179-212.

Sangster, A., Gauthier, M., and Gower, C.F., 1992, Metallogeny of structural zones, Grenville Province, northeastern North America: Precambrian Research, v. 58, p. 410-426.

Sangster, P.J., Laidlaw, D.A., Papertzian, V.C., Steele, K.G., Lee, C.R., Barua, M., and Carter, T.R., 2006, Report of Activities 2005, Resident Geologist Program, Southern Ontario Regional Resident Geologist Report: Southeastern and Southwestern Ontario Districts, Mines and Minerals Information Centre, and Petroleum Resources Centre; Ontario Geological Survey, Open File Report 6186, 77 p.

Scherrer, G., 2003, Géochimie et pétrogenèse des roches métagabbroïques du domaine de Natashquan, secteur oriental de la Province de Grenville, Québec: M.Sc. thesis, INRS-ETE, Québec, 150 p.

Schwerdtner, W.M., Riller, U.P., and Borowik, A., 2005, Structural testing of tectonic hypotheses by field-based analysis of distributed tangential shear: Examples from major high strain zones in the Grenville Province and other parts of the southern Canadian Shield: Canadian Journal of Earth Sciences, v. 42, p. 1927-1947.

Seeger, C.M., Nuelle, L.M., Day, W.C., Sidder, G.B., Marikos, M.A., and Smith D.C., 2001, Geologic maps and cross sections of mine levels at the Pea Ridge Iron Mine, Washington County, Missouri: U.S. Geological Survey Miscellaneous Field Studies Map MF-2353, Version 1.0.

Selleck, B., McLelland, J., and Hamilton, M.A., 2004, Magmatic-hydrothermal leaching and origin of late- to post-tectonic quartz-rich rocks, Adirondack Highlands, New York, *in* Tollo, R.P., Corriveau, L., McLelland, J., and Bartholomew, M., eds., Proterozoic Tectonic Evolution of the Grenville Orogen in North America: Geological Society of America, Memoir 197, p. 379-390.

Sharma, K.N.M., 1973, Géologie de la Région du Lac Victor, Comté de Duplessis: Ministère des Richesses Naturelles, Québec, RP 607, 11 p.

Slagstad, T., Culshaw, N.G., Jamieson, R.A., and Ketchum, J.W.F., 2004, Early Mesoproterozoic tectonic history of the southwestern Grenville Province, Ontario: Constraints from geochemistry and geochronology of high-grade gneisses, *in* Tollo, R.P., Corriveau, L., McLelland, J., and Bartholomew, M., eds., Proterozoic Tectonic Evolution of the Grenville Orogen in North America: Geological Society of America, Memoir 197, p. 209-242.

Stanaway, K., 1996, The eastern North American titanium province—A review: Lithology and Mineral Resources, v. 31, p. 509-517.

——— 2003, Titanium mining provinces of the world: Canadian Institute of Mining Industry Conference and Exhibition, Montreal, Technical Paper, CD-ROM.

Thériault, R., Clark, T., Beaumier, M., Dion, D.J., Garneau, C., and Leduc, M., 2003, Ni-Cu-PGE Mineralizations in Quebec: Ministère des Ressources Naturelles, de la Faune et des Parcs, Québec, DV 2003-06, 1 map,scale 1:250 000.

Thomas, A., Blomberg, P., Jackson, V., and Finn, G., 2000, Bedrock geology map of the Wilson Lake area (13E/SE), Labrador: Newfoundland and Labrador Department of Mines and Energy, Geological Survey, Map 2000-01, scale 1:100000, Open File 013E/0061.

Thompson, J.E., 1956, Preface, *in* Thompson, J.E., ed., The Grenville Problem: Royal Society of Canada, Special Publications, v. 1, p. 3-13.

Tollo, R.P., Corriveau, L., McLelland, J., and Bartholomew, M.J., 2004, Proterozoic tectonic evolution of the Grenville orogen in North America: An introduction, *in* Tollo, R.P., Corriveau, L., McLelland, J., and Bartholomew, M., eds., Proterozoic Tectonic Evolution of the Grenville Orogen in North America: Geological Society of America, Memoir 197, p. 1-18.

Tornos, F., and Casquet, C., 2005, A new scenario for related IOCG and Ni(Cu) mineralization: the relationship with giant mid-crustal mafic sills, Variscan Iberian Massif: Terra Nova, v. 17, p. 236-241.

Tucker, R.D., and Gower, C.F., 1994, A U-Pb geochronological framework for the Pinware terrane, Grenville Province, southeast Labrador: Journal of Geology, v. 102, p. 67-78.

Vaillancourt, C., Sproule, R.A., MacDonald, C.A., and Lesher, C.M., 2003, Investigation of Mafic-Ultramafic Intrusions in Ontario and Implications for Platinum Group Element Mineralization: Operation Treasure Hunt: Ontario Geological Survey, Open File Report 6102, 335 p.

van Breemen, O., and Corriveau, L., 2005, U-Pb age constraints on arenaceous and volcanic rocks of the Wakeham Group, eastern Grenville Province: Canadian Journal of Earth Sciences, v. 42, p. 1677-1697.

van Breemen, O., and Currie, K.L., 2004, Geology and U-Pb geochronology of the Kipawa Syenite Complex - a thrust related alkaline pluton - and adjacent rocks in the Grenville Province of western Quebec: Canadian Journal of Earth Sciences, v. 41, p. 431-455.

van Breemen, O., and Davidson, A., 1988, Northeast extension of Proterozoic terranes of mid-continental North America: Geological Society of America Bulletin, v. 100, p. 630-638.

van Breemen, O., Davidson, A., Loveridge, W.D., and Sullivan, R.W., 1986, U-Pb zircon geochronology of Grenville tectonites, granulites, and igneous precursors, Parry Sound, Ontario, *in* Moore, J.M., Davidson, A., and Baer A.J., eds., The Grenville Province: Geological Association of Canada, Special Paper 31, p. 191-207.

Van Schmus, W.R., Bickford, M.E., and Turek, A., 1996, Proterozoic geology of the east-central Midcontinent basement, *in* van der Pluijm, B.A., and Catacosinos, P., eds., Basement and Basins of eastern North America: Geological Society of America, Special Paper 308, p. 7-32.

Vassiliou, A.H., and Puffer, J.H., 1985, Uranium, rare earth and iron mineralization in pegmatite at the Bemco Mine prospect, Cranberry Lake, New Jersey, *in* Hausen, D.M., and Kopp, O.C., eds., Mineralogy - Applications to the Minerals Industry: New York: American Institute of Mining, Metallurgical, and Petroleum Engineers, Paul F. Kerr Memorial Symposium Volume, p. 71-87.

Villeneuve, D., 1988, The Calumet Project, *in* Vallières A., ed., Gîtes métallifères dans le sud du Grenville - Livret-guide d'excursion: Ministère de l'Énergie et des Ressources du Québec, MB 88-10, p. 35-41.

Vogel, D.C., James, R.S., and Keays, R.R., 1998, The early tectono-magmatic evolution of the Southern Province: Implications from the Agnew Intrusion, central Ontario, Canada: Canadian Journal of Earth Sciences, v. 35, p. 854-870.

Volkert, R.A., 2004a, Mesoproterozoic rocks of the New Jersey Highlands, north-central Appalachians: Petrogenesis and tectonic history, *in* Tollo, R.P., Corriveau, L., McLelland, J., and Bartholomew, M., eds., Proterozoic Tectonic Evolution of the Grenville Orogen in North America: Geological Society of America, Memoir 197, p. 1-18.

——— 2004b, Characteristics, age, geochemistry, and mineralization of a postorogenic felsic magmatic suite, New Jersey Highlands: Geological Society of America, Northeastern-Southeastern Section, Tysons Corner, VA, Abstracts with Programs, v. 36, p. 51.

Volkert, R.A., Feigenson, M.D., Patino, L.C., Delaney J.S., and Drake Jr., A.A., 2000a, Sr and Nd isotopic compositions, age and petrogenesis of A-type granitoids of the Vernon Supersuite, New Jersey Highlands, USA: Lithos, v. 50, p. 325-347.

Volkert, R.A., Johnson, C.A., and Tamashausky, A.V., 2000b, Mesoproterozoic graphite deposits, New Jersey Highlands: Geologic and stable isotope evidence for possible algal origins: Canadian Journal of Earth Sciences, v. 37, p. 1665-1675.

Wardle, R.J., and Hall, J., 2002, Proterozoic evolution of the northeastern Canadian Shield: Lithoprobe eastern Canadian Shield onshore-offshore transect (ECSOOT), introduction and summary, *in* Wardle, R.J., and Hall, J., eds., Proterozoic Evolution of the northeastern Canadian Shield: Lithoprobe Eastern Canadian Shield Onshore-Offshore Transect: Canadian Journal of Earth Sciences, v. 39, p. 563-567.

Wardle, R.J., Gower, C.F., Ryan, B., Nunn, G.A.G., James D.T., and Kerr, A., 1997, Geological map of Labrador: Geological Survey, Department of Mines and Energy, Government of Newfoundland and Labrador, Map 97-07, scale 1 : 1 000 000.

Wasteneys, H., Kamo, S., Moser, D., Krogh, T.E., Gower, C.F., and Owen, J.V., 1997, U-Pb geochronological constraints on the geological evolution of the Pinware terrane and adjacent areas, Grenville Province, southeast Labrador, Canada: Precambrian Research, v. 81, p. 101-128.

Weihed, P., 2001, A review of Paleoproterozoic intrusive hosted Cu-Au-Fe-oxide deposits in northern Sweden, *in* Weihed, P., ed., Economic Geology Research, Volume 1 1999-2000: Sveriges Geologiska Undersökning, v. C-833, p. 4-32.

Wiebe, R.A., and Collins, W.J., 1998, Depositional features and stratigraphic sections in granitic plutons: Implications for the emplacement and crystallization of granitic magma: Journal of Structural Geology, v. 20, p. 1273-1289.

Williams, P.J., 1992, Metamorphosed boninitic basalts, arc tholeiites, and cryptic volcanic stratigraphy from the Elzevir Terrane of the Grenville Province, Calumet mine, Quebec: Canadian Journal of Earth Sciences, v. 29, p. 26-34.

Wilson, M.E., 1926, Arnprior-Quyon and Maniwaki Areas, Ontario and Quebec: Geological Survey of Canada, Memoir 136, 152 p.

Wilson, G.C., 1993, Mafic-Ultramafic Intrusions, Base-Metal Sulphides, and Platinum Group Element Potential of the Grenville Province in southeastern Ontario: Ontario Geological Survey, Open File Report, 195 p.

Wodicka, N., David, J., Parent, M., Gobeil, A., and Verpaelst, P., 2003, Géochronologie U-Pb et Pb-Pb de la région de Sept-Îles - Natashquan, Province de Grenville, Moyenne-Côte-Nord, *in* Brisebois, D., and Clark, T., eds., Géologie et Ressources Minérales de la Partie est de la Province de Grenville: Ministère des Ressources Naturelles, de la Faune et des Parcs, Québec, DV 2002-03, p. 59-118.

Wodicka, N., Corriveau, L., and Stern, R., 2004, SHRIMP U-Pb zircon geochronology of the Bondy gneiss complex: Evidence for ca. 1.40 Ga arc magmatism and polyphase Grenvillian metamorphism in the Central Metasedimentary Belt, Grenville Province, Quebec, *in* Tollo, R.P., Corriveau, L., McLelland, J., and Bartholomew, M., eds., Proterozoic Tectonic Evolution of the Grenville Orogen in North America: Geological Society of America, Memoir 197, p. 243-266.

Wolff, J.M., 1979, Geology of the Long Lake Area, Frontenac County: Ontario Geological Survey, Open File Report 5271, 174 p.

Wynne-Edwards, H.R., 1972, The Grenville Province, *in* Price, R.A., and Douglas, R.J.W., eds., Variations in Tectonic Styles in Canada: Geological Association of Canada, Special Paper 11, p. 263-334.

Yigit, O., and Hofstra, A.H., 2003, Lithogeochemistry of Carlin-type gold mineralization in the Gold Bar district, Battle Mountain-Eureka trend, Nevada: Ore Geology Reviews, v. 22, p. 201-224.

The Slave Craton: Geological and Metallogenic Evolution

Wouter Bleeker[1] and Brian Hall[2]

1. *Geological Survey of Canada, 601 Booth Street, Ottawa, Ontario K1A 0E8*

2. *Department of Earth Sciences, Laurentian University, 935 Ramsey Lake Road, Sudbury, Ontario P3E 2C6*

Corresponding author's email: wbleeker@nrcan.gc.ca

Abstract

The Slave craton of the northwestern Canadian Shield is one of the oldest and most distinct building blocks of North American cratonic lithosphere. It hosts Earth's oldest intact rocks, the Acasta gneisses. These ancient gneisses are embedded in a large Mesoarchean to Hadean basement complex that underlies the west-central parts of the craton. Although itself poorly mineralized, the basement complex is overlain by Neoarchean supracrustal sequences, and is heavily intruded and cannibalized by plutonic suites that range in age from 2720-2670 Ma synvolcanic plutons to 2595-2585 Ma late-orogenic batholithic granites. Supracrustal sequences, collectively known as the Yellowknife Supergroup, are represented by an early cover sequence comprising quartzite and banded iron formation (ca. 2800 Ma), a thick dominantly tholciitic greenstone sequence (ca. 2700 Ma), younger arc-like sequences (ca. 2690-2660 Ma), extensive turbidite blankets (ca. 2680-2620 Ma), and finally synorogenic conglomerates that were deposited at ca. 2600 Ma or shortly thereafter. The early cover sequence and the overlying tholeiites represent subaerial exposure and then volcanic-dominated rifting of the basement. Arc-like sequences formed in part on top of the attenuated basement and in progressively widening, juvenile, back-arc-like basins and contain some of Canada's largest undeveloped volcanogenic massive sulphide deposits. After 2680 Ma, much of the Slave craton became overlain by the Burwash Basin, one of the largest and best preserved Archean turbidite basins in the world and comparable in size and setting to the Japan Sea. During orogenesis, supracrustal sequences were telescoped, thickened, and multiply folded between ca. 2650 and 2580 Ma, with a peak in crustal anatexis between 2595 and 2585 Ma (the "granite bloom"). Numerous orogenic gold deposits formed throughout the Slave craton, either as shear- or vein-hosted deposits in deformed greenstones or within the chemical traps provided by banded iron formations in the turbidites. Proterozoic rift-related magmatic suites and arcs around the margins of the craton host a variety of mineral deposits. Finally, the craton was intruded by several hundred Phanerozoic kimberlite pipes, some of which support Canada's first diamond mines.

Résumé

Le craton des Esclaves, qui occupe la partie nord-ouest du Bouclier canadien, est l'un des plus vieux et des plus distinctifs des éléments constitutifs de la lithosphère du craton nord-américain. Il renferme les plus anciennes roches intactes de la Terre, les gneiss d'Acasta. Ces gneiss anciens sont inclus dans un grand complexe de socle datant du Mésoarchéen à l'Hadéen, qui s'étend aux parties centrale et occidentale du craton. Peu minéralisé en soi, le complexe de socle est recouvert de séquences supracrustales du Néoarchéen et est intensément traversé et cannibalisé par des suites plutoniques variant de plutons synvolcaniques âgés de 2720 à 2660 Ma à des batholites de granite tardiorogéniques datant de 2595 à 2585 Ma. Les séquences supracrustales, collectivement attribuées au Supergroupe de Yellowknife, consistent en une séquence de couverture initiale composée de quartzite et de formation de fer rubanée (env. 2800 Ma), en une épaisse séquence de roches vertes à dominante tholéiitique (env. 2700 Ma), en séquences plus récentes apparentées à celles des arcs (env. 2690 à 2660 Ma), en couvertures étendues de turbidites (env. 2680 à 2620 Ma) et, enfin, en conglomérats synorogéniques déposés à 2600 Ma environ, ou peu après. La séquence de couverture initiale et les tholéiites sus-jacentes témoignent d'une exposition subaérienne puis d'un rifting du socle, qui s'est accompagné de la mise en place de matériaux à prédominance volcanique. Les séquences apparentées à celles des arcs se sont formées en partie sur un socle aminci et en partie dans des bassins juvéniles apparentés à des bassins d'arrière-arc s'élargissant progressivement. Ces séquences renferment certains des plus vastes gîtes de sulfures massifs volcanogéniques inexploités du Canada. Après 2680 Ma, une bonne partie du craton des Esclaves a été enfouie sous les dépôts du bassin de Burwash, un des plus grands bassins de turbidites archéennes au monde, aux dépôts parmi les mieux conservés, qui est comparable par sa taille et son cadre à l'actuelle mer du Japon. Pendant l'orogenèse, les séquences supracrustales se sont télescopées et épaissies, et ont été plissées à plusieurs reprises entre environ 2650 Ma et 2580 Ma, avec une période d'intensité maximale d'anatexie crustale de 2595 Ma à 2585 Ma (l'« efflorescence granitique »). De nombreux gîtes d'or orogéniques se sont formés un peu partout dans le craton des Esclaves, soit sous forme de gîtes encaissés dans des zones de cisaillement ou de gîtes filoniens dans des roches vertes déformées, soit sous forme de minéralisations contenues dans les pièges chimiques constitués par les formations de fer rubanées au sein des turbidites. Les suites et les arcs magmatiques liés au rifting, qui sont apparus au Protérozoïque le long des marges du craton renferment toute une gamme de gîtes minéraux. Enfin, le craton a été percé au Phanérozoïque par plusieurs centaines de cheminées de kimberlite, dont certaines ont permis la mise en exploitation des premières mines de diamants au Canada.

Introduction

The Archean Slave craton (Figs. 1, 2, 3) (Henderson, 1981; Padgham and Fyson, 1992; King and Helmstaedt, 1997; Bleeker and Davis, 1999a) is a major building block of the Canadian Shield. It is one of approximately 35 Archean cratons preserved around the world (Bleeker, 2003). Amalgamation of the Slave craton with the Rae craton started at ca. 2 Ga, initiating the climactic growth of Laurentia from 2.0 to 1.8 Ga (Hoffman, 1988, 1989), probably within the broader context of the formation of Earth's first modern supercontinent, Nuna. Much of the Slave craton is old, and within the context of the Laurentian collage it can be considered, for all practical purposes, as a far-travelled, if not exotic, fragment of crust relative to other well known

Bleeker, W., and Hall, B., 2007, The Slave Craton: Geology and metallogenic evolution, *in* Goodfellow, W.D., ed., Mineral Deposits of Canada: A Synthesis of Major Deposit-Types, District Metallogeny, the Evolution of Geological Provinces, and Exploration Methods: Geological Association of Canada, Mineral Deposits Division, Special Publication No. 5, p. 849-879.

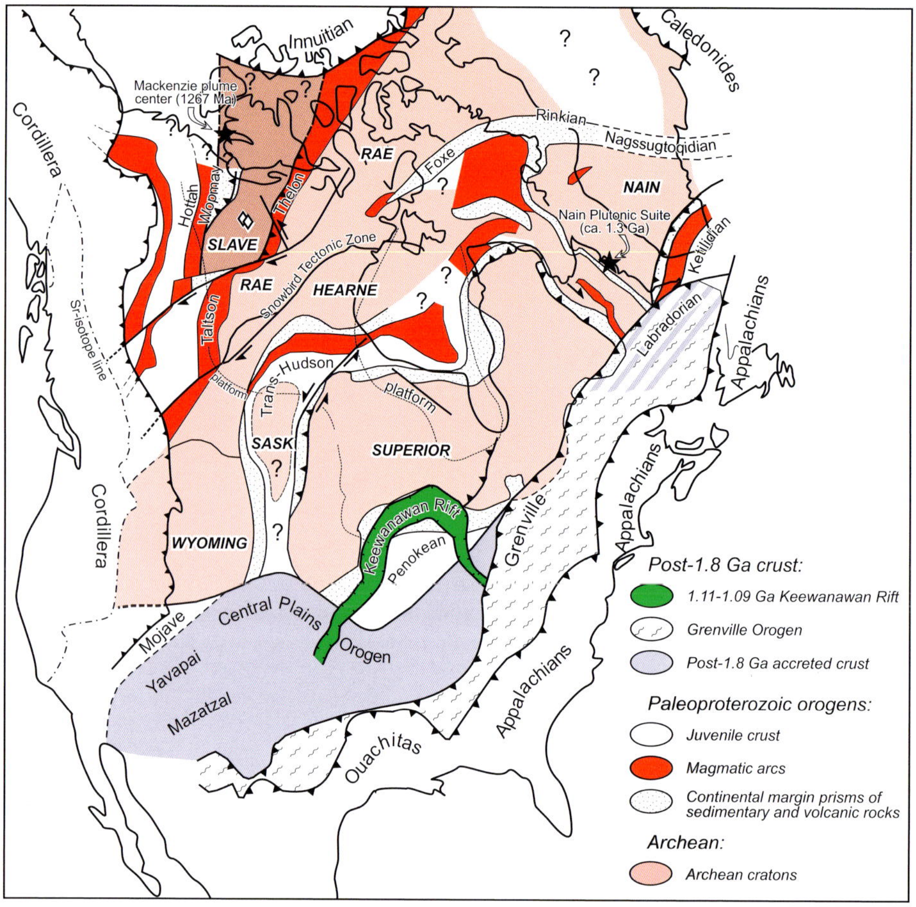

FIGURE 1. Tectonic map of the Precambrian basement of North America, showing the location of the Archean Slave craton relative to other first-order crustal elements. Greenland is shown in a pre-drift position (modified after Hoffman, 1988; Ross et al., 1991).

ments of Sclavia from among the 35 extant cratons. One such Slave craton-wide event was a voluminous "granite bloom" between ca. 2595 and 2585 Ma (Davis and Bleeker, 1999). This singular event in the craton's history transferred, irreversibly, a significant fraction of heat-producing elements and lower crustal fluids to the upper crust, thus allowing slow cooling and stiffening of the lower crust and setting the stage for cratonization and long-term preservation (Bleeker, 2002).

Predating this cratonization event, the Slave crust preserves a complex and spatially heterogeneous record of crustal growth spanning more than 1.5 billion years (e.g. Kusky, 1989; Isachsen and Bowring, 1994; Bleeker and Davis, 1999a; 1999b and references therein; Sircombe et al., 2001; Ketchum et al., 2004; see Iizuka et al. (2006) for a discussion on the recent discovery of a 4.2 Ga zircon xenocryst in Acasta gneiss).

The present paper briefly summarizes this crustal growth history and the overall geological evolution of the Slave craton, from the formation of early sialic basement, to the development of the dominant

cratons in Laurentia (such as the Superior, Hearne, Rae, and Nain, see Bleeker, 2003, 2004).

As a mere fragment of ancient crust (e.g. Hoffman, 1988; Isachsen and Bowring, 1994; Bleeker, 2003), surrounded by Paleoproterozoic rifted margins, the Slave craton originated from the break-up of a much larger late Archean landmass, possibly the speculative late Archean supercontinent Kenorland (Williams et al., 1991) or, perhaps more likely, a smaller landmass referred to as the supercraton Sclavia (Bleeker, 2003). The late Archean and earliest Proterozoic development and evolution of Slave crust should thus be viewed within the context of the growth and subsequent break-up of this larger Sclavia supercraton, even though the shape and size of this supercraton remain currently unknown. The salient point is that ancient cratons, like the Slave, preserve only parts of the much larger tectonic systems in which they were generated.

In agreement with this conceptual view, latest Archean events[1] are remarkably homogeneous across the Slave craton and may be used, together with pre-2.0 Ga Proterozoic mafic dyke swarms[2], to help identify neighbouring frag-

ca. 2.7 Ga greenstone sequences, major orogenesis at ca. 2.6 Ga, and final cratonization at ca. 2.55 Ga, thus providing a framework for discussion of important metallogenic events within the craton. Pertinent information concerning the location, mineralogy, structure, size, grade, age, and stratigraphic and tectonic settings for all significant mineral deposits within or adjacent to the craton is summarized in the Appendix, along with short summaries for some additional metallogenic events. Figure 4 shows the locations of these mineral deposits and occurrences on a geological map of the craton.

Ancient Basement Complex

An ancient and largely crystalline basement complex underlies much of the central and western parts of the craton (Fig. 5, see also Figs. 2, 3) (Baragar and McGlynn, 1976; Kusky, 1989; Bleeker et al., 1999a,b; Ketchum and Bleeker, 2001; Ketchum et al., 2004). It is referred to as the Central Slave Basement Complex (Bleeker et al., 2000).

Along the Acasta River, this basement complex consists of polymetamorphic gneisses of tonalitic to gabbroic composition (Fig. 6A) that yield protolith ages up to 4.03 Ga[3]

1. In the present context, events younger than ca. 2680 Ma.
2. See Bleeker (2004) and Bleeker and Ernst (2006) for general methodology.
3. All ages quoted in this paper are U-Pb zircon ages, unless stated otherwise. These were determined either by isotope-dilution thermal ionization mass spectrometry (ID-TIMS) or sensitive high-resolution ion microprobe (SHRIMP) methods. Generalized ages, or those with poor precision, are given at the billion year (Ga) level, whereas more precise ages and their errors are given at the million year (Ma) level.

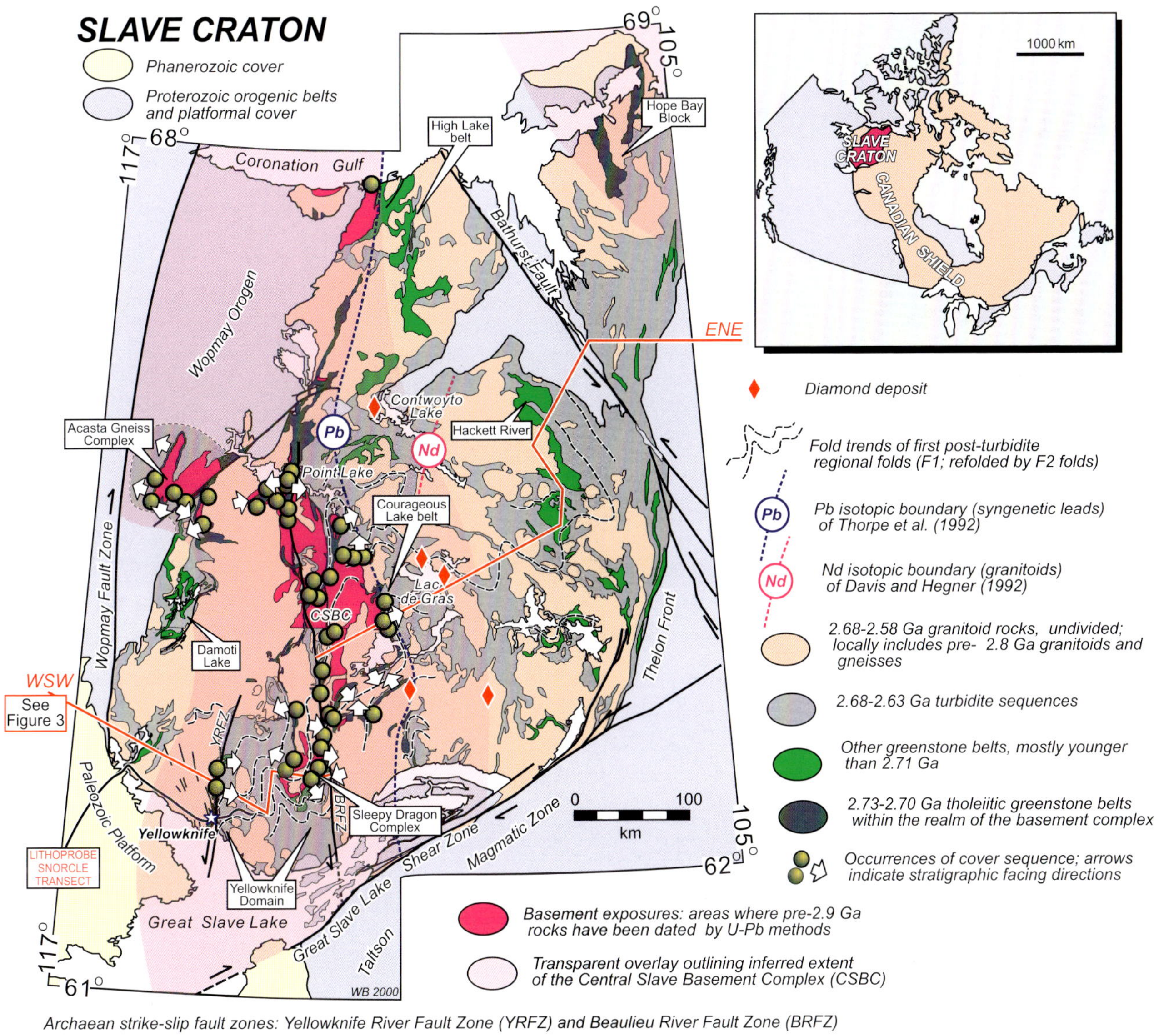

FIGURE 2. Simplified geological map of the Slave craton. Localities mentioned in the text are highlighted. Cross-section line (ENE-WSW) refers to the craton-wide structural section shown in Figure 3.

(Bowring et al., 1989; Stern and Bleeker, 1998; Bowring and Williams, 1999). The record of inheritance extends back to 4.2 Ga (Iizuka et al., 2006). The Acasta gneisses were essentially a chance discovery (Bowring et al., 1989; M. St-Onge, pers. comm., 2000) and no other rocks of this age have yet been found. Apart from a central core of the craton, with sporadic ages >3.5 Ga (Acasta to Point Lake), the Central Slave Basement Complex is mostly younger with important age modes[4], from detrital and protolith U-Pb zircon ages, around 3400 Ma, 3150 Ma, 2950 Ma, and 2826 Ma (Fig. 5B) (e.g. Sircombe et al., 2001; see also Bleeker and Davis (1999b) for a compilation of basement ages).

Interestingly, complementary data from the mantle suggest that at least part of the subcontinental lithospheric mantle below the central part of the craton may be of similar antiquity (e.g. Aulbach et al., 2004). A crude age zonation can be recognized in the basement complex (Ketchum and Bleeker, 2001), although no easily interpretable tectonic pattern has yet emerged. Pre-2.9 Ga supracrustal rocks have been found at the base of some greenstone belts (Bleeker and Davis, 1999b and references therein; Ketchum et al., 2004), but form only a very small component of the craton's geological inventory. An important example of such an occurrence is a felsic metavolcanic rock, dated at 3118 +11/-8 Ma, at the base of the supracrustal succession on the eastern flank of the Winter Lake belt (Hrabi et al., 1995; Hrabi and Grant, 1999). Elsewhere, enclaves of supracrustal origin are known to occur within pre-2.8 Ga basement but have yet to be systematically dated.

Mineralization

There are few, if any, known mineral occurrences of note within the Central Slave Basement Complex, a statistic that is generally mirrored by other Mesoarchean and older gneis-

4. More or less well defined peaks on age histograms, e.g., see Sircombe et al. (2001).

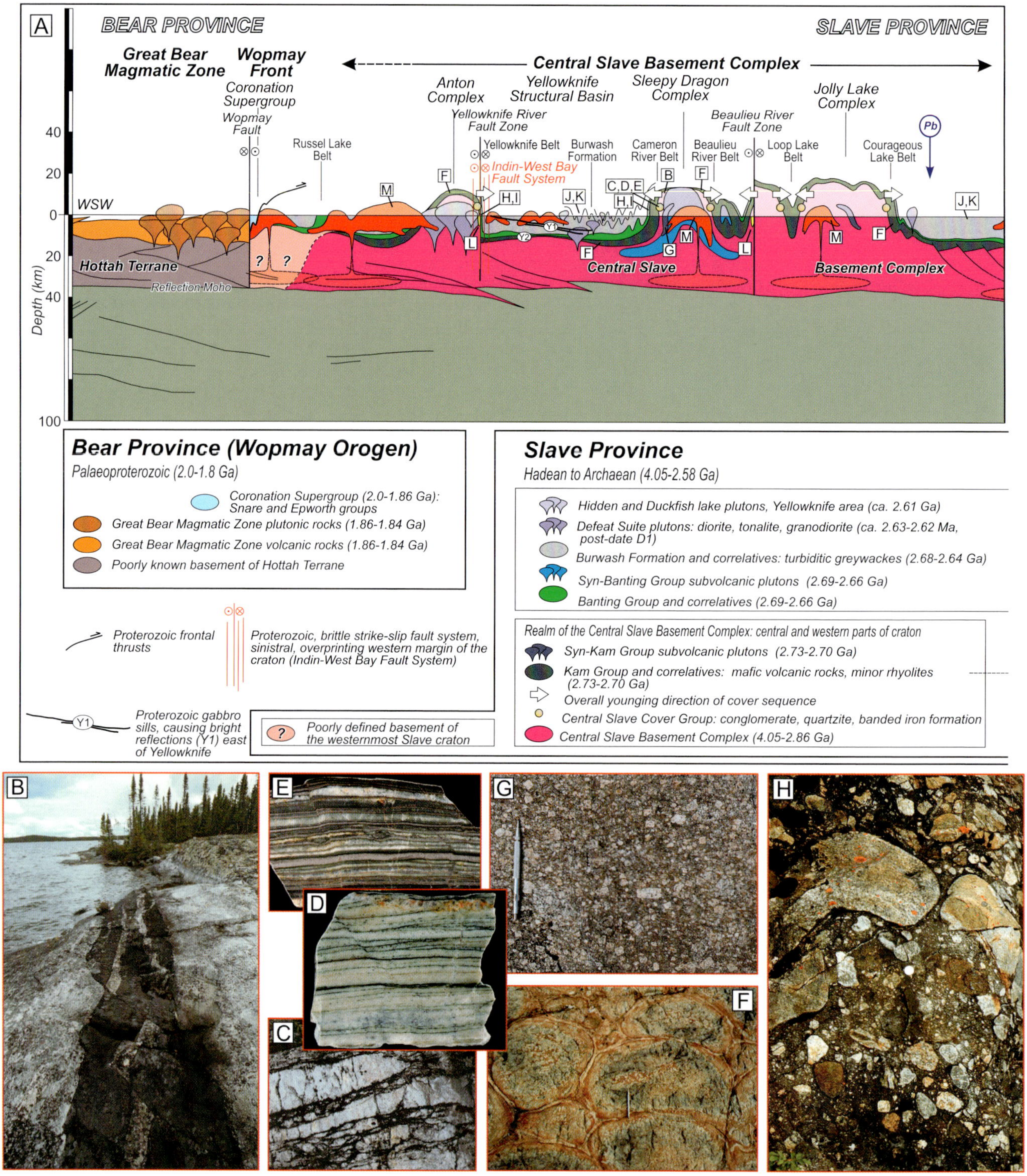

Figure 3. **(A)** West-southwest to east-northeast, structural-stratigraphic section across the Slave craton (after Bleeker, 2002, see Fig. 2 for location of profile). Letters (B,C…M) refer to illustrations below section. No vertical exaggeration with present erosion level at ca. 0 km depth; some units are shown above this level in lighter tones for ease of interpretation. Deep structure in the western part of the profile interpreted from LITHOPROBE's SNORCLE seismic reflection profile (e.g. Cook et al., 1999). Volcanic units <2690 Ma and overlying turbidites overlap the boundary between isotopically different basement domains. Field photos: **(B)** Typical 2.95 Ga foliated tonalites of the Central Slave Basement Complex with transposed 2734 Ma mafic dykes. **(C, D, E)** Basal quartz pebble conglomerate, fuchsitic quartzite, and banded iron formation of the Central Slave Cover Group that overlies the basement complex (~20 cm tall polished slabs). **(F)** Variolitic pillow basalts of the Kam Group, Yellowknife (pen for scale). **(G)** Syn-Kam Group K-feldspar porphyritic granodiorite pluton in basement below greenstone belts (pen for scale). **(H)** Polymict conglomerate, including 10 to 30 cm granitoid cobbles, which occurs locally at the base of the younger, 2690 to 2660 Ma, volcanic cycle.

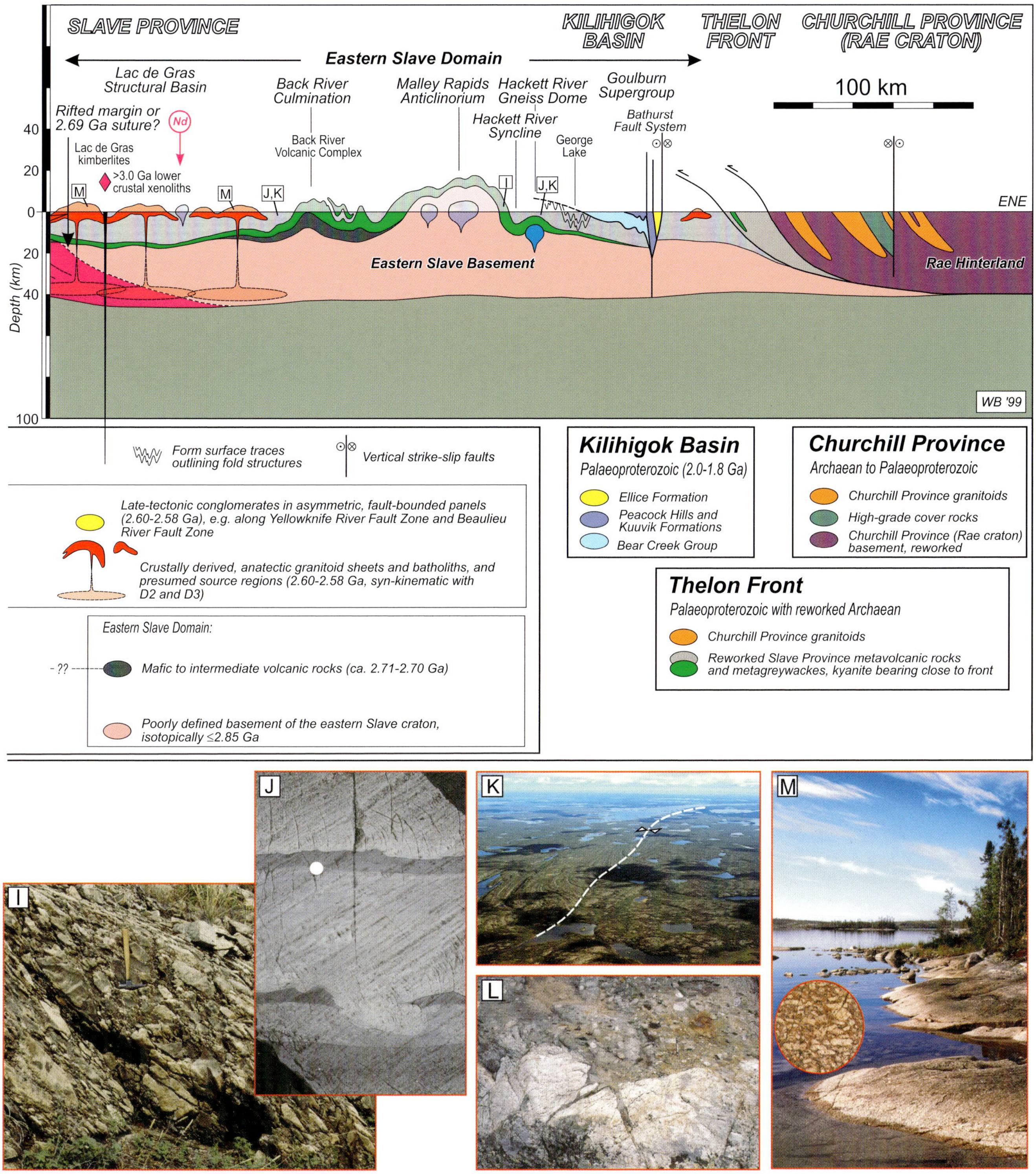

Figure 3 continued. **(I)** Intermediate to felsic volcaniclastic deposits typical of the younger volcanic cycle (hammer for scale). **(J)** Well preserved sub-biotite-grade turbidites in the core of the Yellowknife structural basin, showing graded bedding and load casts. **(K)** Aerial photo of large-scale, upright F_1 fold structures in turbidites of the Yellowknife structural basin. **(L)** Synorogenic conglomerates, <2600 Ma, unconformably overlying unroofed granitoid rocks, Point Lake (pebbles ~5-10 cm in size). **(M)** Late-orogenic, ca. 2585 Ma K-feldspar megacrystic granites of the Morose Suite in the core of the domal Sleepy Dragon Complex; inset shows 1 to 2 cm large K-feldspar megacrysts.

sic complexes in cratons around the world. To a first degree, this poor endowment is thought to correlate with the generally mid- to lower crustal erosion depths within such crystalline basement complexes and, consequently, the virtual lack of supracrustal rocks.

Indirectly, however, the presence of the ancient basement complex may have influenced or exerted important controls on several classes of younger mineral deposits:

- Seafloor volcanogenic massive sulphide (VMS) mineralization within the Neoarchean bimodal rift volcanic

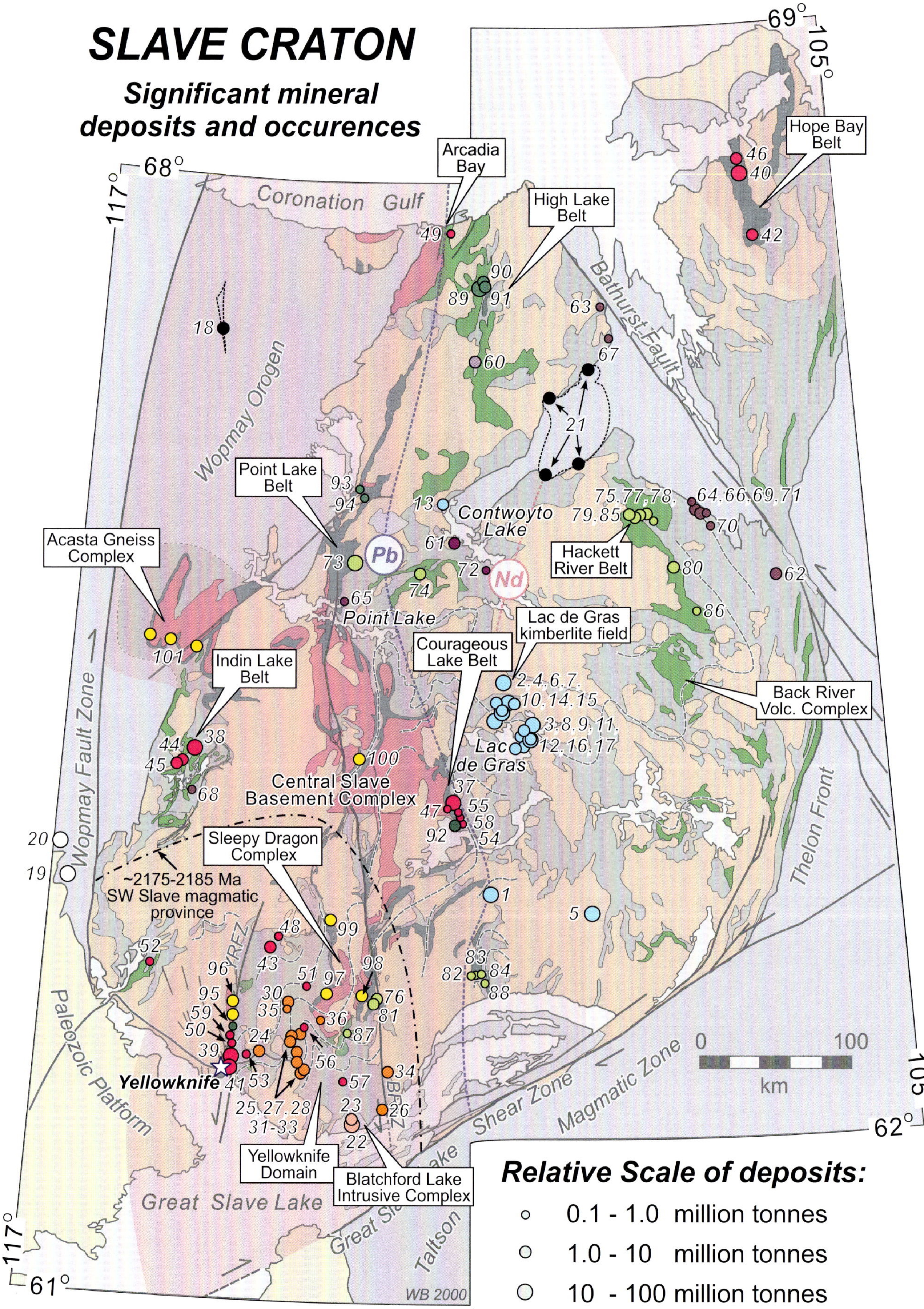
SLAVE CRATON
Significant mineral deposits and occurences
Arcadia Bay
High Lake Belt
Hope Bay Belt
Coronation Gulf
Bathurst Fault
Wopmay Orogen
Point Lake Belt
Contwoyto Lake
Hackett River Belt
Acasta Gneiss Complex
Point Lake
Lac de Gras kimberlite field
Courageous Lake Belt
Indin Lake Belt
Back River Volc. Complex
Lac de Gras
Central Slave Basement Complex
Wopmay Fault Zone
Sleepy Dragon Complex
~2175-2185 Ma SW Slave magmatic province
Thelon Front
Paleozoic Platform
Yellowknife
Yellowknife Domain
Blatchford Lake Intrusive Complex
Great Slave Lake
Great Slave Lake Shear Zone
Taltson Magmatic Zone
WB 2000
Relative Scale of deposits:
0.1 - 1.0 million tonnes
1.0 - 10 million tonnes
10 - 100 million tonnes

Eocambrian to Eocene diamond-bearing and allied rocks

1 Snap Lake
2 Ekati (Koala)
3 Ekati (Jay)
4 Ekati (Fox)
5 Gahcho Kue (Kennady Lake)
6 Ekati (Panda)
7 Ekati (Sable)
8 Diavik (A154-South)
9 Diavik (A-418)
10 Ekati (Pigeon)
11 Diavik (A-21)
12 Ekati (Misery)
13 Jericho
14 Ekati (Beartooth)
15 Ekati (Koala North)
16 Ekati (Lynx)
17 Diavik (A154-North)

Proterozoic mafic intrusions with Ni, Cu, Cr and reef-style PGE mineralization

18 Muskox Intrusion, Mackenzie event

Paleoproterozoic hydrothermal Cu-Au (IOCG) deposits

19 NICO
20 Sue-Diane

Paleoproterozoic mafic intrusion-related Ni, Cu, PGEs, V

(21) Booth River Complex, Lac de Gras event

Approximate extent of Booth River Complex underneath cover of Kilihigok Basin

Paleoproterozoic alkaline intrusion-related rare element deposits

22 Thor (Lake Zone)
23 Thor (T Zone)

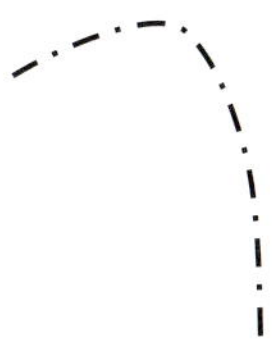

Minimum extent of ca. 2175-2185 Ma magmatic province in the southwestern Slave craton, comprising:

-Dogrib mafic dyke swarm (ca. 2185-2189 Ma)
-Duck Lake sill (2180+/-2 Ma)
-Blatchford Lake Intrusive Complex (2176+/-2 Ma)
-Big Spruce alkaline complex NW of Yellowknife
-Squalus Lake alkaline complex (2180+/-2 Ma)

FIGURE 4. Map of the Slave craton (as Fig. 2) annotated with locations of significant mineral deposits and occurrences. Deposits are grouped by genetic class and age, and listed on the facing page. See Appendix 1 for more complete information on these deposits and occurrences.

Late Archean rare element-enriched pegmatites

24 Murphy
25 F1 (Main Dyke)
26 Moose No. 2
27 F1 (Southwest Dyke)
28 Shorty
29 Ann
30 VO (Cota)
31 Paint
32 Jake
33 K1 (Dyke)
34 Echo Tor
35 Bill
(36) Peg Tantalum

Archean Au deposits - Vein or shear zone-hosted

37 Tundra (Fat)
38 Colomac
39 Giant Yellowknife
40 Madrid
41 Con
42 Boston
43 Discovery
44 Kim
45 Cass
46 Doris
47 Tundra (Carbonate)
48 Nicolas Lake
49 Arcadia
50 Supercrest
51 MAHE
52 Mosher Lake
53 Ptarmigan &Tom
54 Saucer Lake
55 Salmita
56 Thompson-Lundmark
57 Bullmoose
58 Tundra Mine
59 Crestaurum

Archean Au deposits - Intrusion related

60 Ulu (Flood)

Archean Au deposits - BIF or turbidites-hosted

61 Lupin
62 Goose Lake (South)
63 Turner Lake
64 George Lake (Locale 1)
65 REN
66 George Lake (Locale 2)
67 Pistol Lake
68 Damoti Lake (Horseshoe Zone)
69 George Lake (Lone Cow Pond)
70 George Lake (GH)
71 George Lake (Slave)
72 Butterfly Lake

Archean VMS in arc-like enviornments

73 Izok Lake
74 Gondar
75 Hackett (A Zone)
76 Sunrise
77 Hackett (East Cleaver)
78 Hackett (Boot Lake)
79 Hackett (Cleaver Low Grade)
80 Yava
81 Bear
82 B B
83 Kennedy Lake
84 Kennedy Lake (Copper Zone)
85 Hackett (Jo)
86 Musk
87 Turback Lake (XL)
88 Susa Lake

Archean VMS in bimodal rift environments

89 High Lake (West Zone)
90 High Lake (A / B Zone)
91 High Lake (D Zone)
92 Deb
93 Hood River (#10)
94 Hood River (#41)
(95) Homer Lake showings, Yellowknife greenstone belt

Archean BIFs at top of Central Slave Cover Group

(96) Dwyer and Bell Lake
(97) Patterson Lake
(98) Amacher Lake
(99) Brown Lake
(100) Winter Lake belt
(101) Acasta

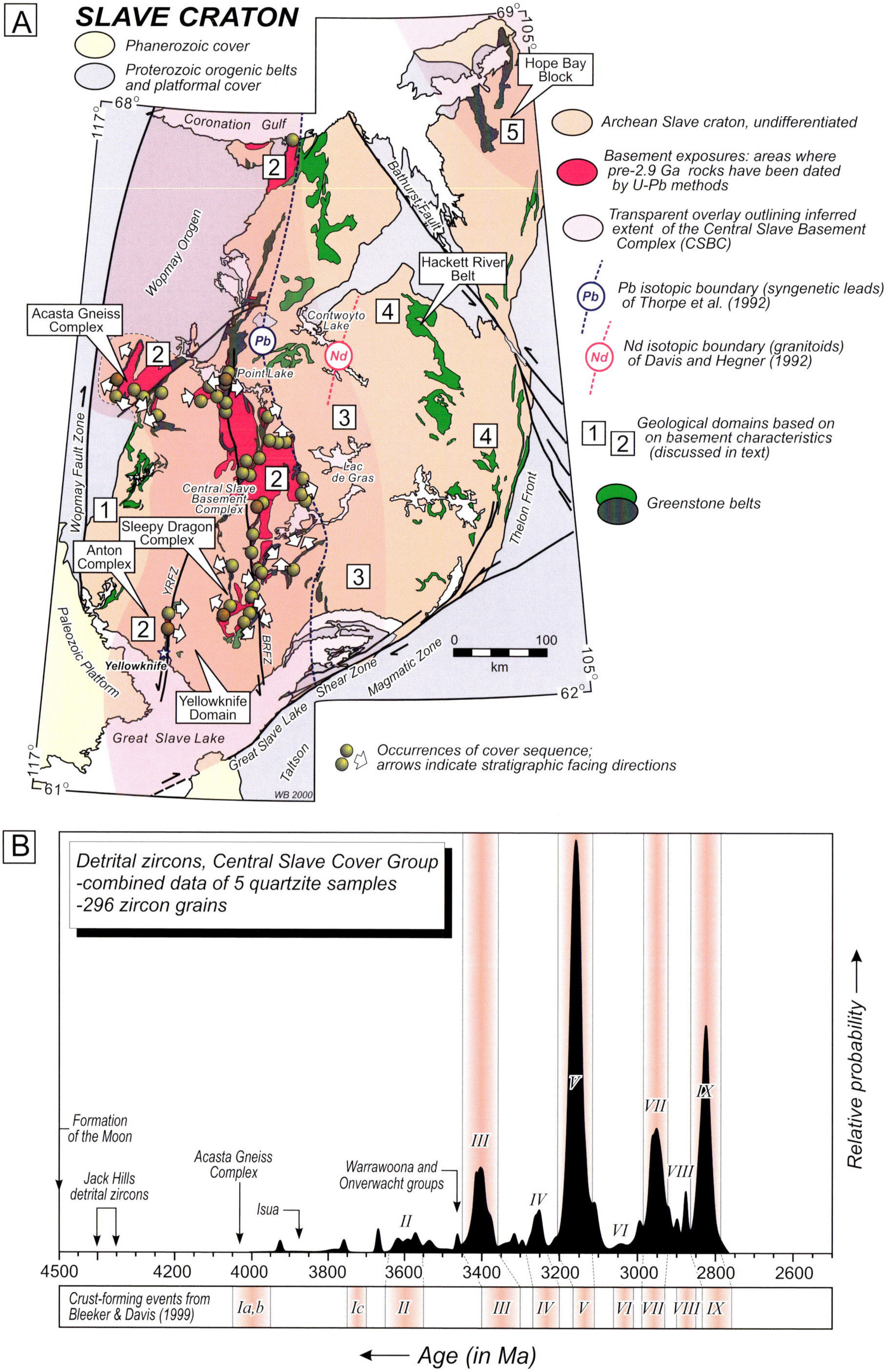
A
SLAVE CRATON
Phanerozoic cover
Proterozoic orogenic belts and platformal cover
Archean Slave craton, undifferentiated
Basement exposures: areas where pre-2.9 Ga rocks have been dated by U-Pb methods
Transparent overlay outlining inferred extent of the Central Slave Basement Complex (CSBC)
Pb isotopic boundary (syngenetic leads) of Thorpe et al. (1992)
Nd isotopic boundary (granitoids) of Davis and Hegner (1992)
Geological domains based on on basement characteristics (discussed in text)
Greenstone belts
Occurrences of cover sequence; arrows indicate stratigraphic facing directions
Coronation Gulf
Hope Bay Block
Bathurst Fault
Wopmay Orogen
Hackett River Belt
Acasta Gneiss Complex
Contwoyto Lake
Point Lake
Lac de Gras
Central Slave Basement Complex
Thelon Front
Wopmay Fault Zone
Sleepy Dragon Complex
Anton Complex
YRFZ
BRFZ
Yellowknife
Yellowknife Domain
Paleozoic Platform
Great Slave Lake
Great Slave Lake Shear Zone
Magmatic Zone
Taltson
km
0
100
68°
69°
117°
105°
62°
61°
WB 2000
B
Detrital zircons, Central Slave Cover Group
-combined data of 5 quartzite samples
-296 zircon grains
Formation of the Moon
Jack Hills detrital zircons
Acasta Gneiss Complex
Isua
Warrawoona and Onverwacht groups
II
III
IV
V
VI
VII
VIII
IX
Relative probability
4500
4200
4000
3800
3600
3400
3200
3000
2800
2600
Crust-forming events from Bleeker & Davis (1999)
Ia,b
Ic
II
III
IV
V
VI
VII
VIII
IX
Age (in Ma)

FIGURE 5. (A) Simplified map showing minimum extent of the Hadean to Mesoarchean basement of the Central Slave Basement Complex (CSBC). Small shaded spheres (yellow or orange) highlight locations where the diagnostic basement to cover stratigraphy has been observed. Based on basement characteristics, five tectono-stratigraphic domains can be recognized across the craton: 1) a westernmost domain, with basement of unknown age; no diagnostic evidence for ancient basement; 2) the main extent of the Central Slave Basement Complex; 3) the buried eastern edge of the ancient basement complex; 4) the eastern Slave "Hackett River domain", which lacks stratigraphic and isotopic evidence for ancient basement; and 5) the "Hope Bay block", which has subtle characteristics that are reminiscent of the Yellowknife area and may well be underlain by ancient basement; currently, data from this area are insufficient to resolve this question. **(B)** Age distribution of the Central Slave Basement Complex, as sampled by 296 concordant to near-concordant detrital zircon grains from five quartzite samples across the basement complex (orange spheres in Fig.5A: Yellowknife, Cameron River belt, northern Beaulieu belt, Point Lake, and quartzite overlying Acasta basement, see Sircombe et al., 2001). Although biased by detrital sampling and variable preservation of basement zircons, this age spectrum allows a rapid assessment of major age components of the basement complex. Note significant age peaks starting at ca. 3400 Ma. The depositional age of these quartzites is ca. 2850 to 2800 Ma. Roman numerals refer to major crust-forming events in the basement recognized from U-Pb protolith ages (see Bleeker and Davis, 1999b).

rocks that overlie faulted basement, such as the basal greenstones of the Yellowknife and Courageous Lake greenstone belts.

- Gold deposits hosted by quartz veins and/or shear zones that developed along the flanks of major basement uplifts.
- Late Archean evolved granites and their associated pegmatite swarms, some of which are enriched in rare elements (e.g. Li, Be, Sn, Nb, Ta). Such enrichment typically correlates with a presence of ancient and multiply recycled sialic crust.
- Diamondiferous kimberlite pipes in the Lac de Gras region (Fig. 7), which appear to have ascended through the edge of the Central Slave Basement Complex (e.g. see cross-section of Fig. 3).

The Cover Sequence

The contiguous nature of the basement complex, by at least 2.9 Ga, is indicated by a thin (typically 2-200 m) but widespread ca. 2.9 to 2.8 Ga cover sequence of quartzite, rare rhyolite, and banded iron formation (Fig. 8, see also Fig. 6B) collectively known as the Central Slave Cover Group (see also Covello et al., 1988; Roscoe et al., 1989; Padgham, 1992; Bleeker et al., 1999b, 2000). This sequence, which is locally intruded by ultramafic sills (Fig. 8), marks the onset of the Neoarchean cycle of supracrustal development (Bleeker et al., 1999b), known collectively as the Yellowknife Supergroup (Henderson, 1970).

The supermature and commonly fuchsitic quartzites that are characteristic of this sequence overlie a regional unconformity that marks the emergence and erosional unroofing of the basement complex in what was probably an aggressive, CO_2-rich, Archean atmosphere (e.g. Kasting, 1993). The quartzites, which locally preserve cross-bedding, mark the progressive drowning of this unconformity in a tide-influenced coastal setting. Overlying banded iron formations (BIFs) are indicative of deeper water, more outboard sedimentation in response to continued subsidence.

Abundant detrital chromite in the quartzites (Figs. 8C,E,F) may suggest contemporaneous komatiitic volcanism. Similar fuchsitic quartzite sequences occur in many other cratons worldwide, particularly between ca. 3.1 Ga and 2.8 Ga. After 2.4 Ga, mature quartzites are rarely fuchsitic, reflecting a lesser role for detrital chromite (and komatiites) in the post-Archean world.

Mineralization

The Central Slave Cover Group hosts some of the more prominent BIFs of the Slave craton (Fig. 4), although most are thin (1-10 m) and variable in composition along strike, changing from oxide-iron formation into silicate-rich varieties, or merely ferruginous cherts. Locally, however, folding has thickened highly magnetic Fe-oxide BIFs into substantial thicknesses (e.g. at Amacher Lake, on the eastern flank of the Sleepy Dragon Complex), resulting in some of the highest amplitude total field magnetic anomalies in the Slave craton. Overall, the BIFs appear of low economic value, although some may possibly host epigenetic gold mineralization and may be under-explored for this commodity. However, across the craton, most of the known iron formation-hosted gold mineralization is associated with younger BIFs that occur intercalated within low- to medium-grade turbidite packages (Fig. 4).

Fuchsitic quartzites that occur stratigraphically below the iron formations are enriched in detrital heavy minerals, including highly stable species such as chromite, zircon, and rutile. Individual, detrital, small black chromite grains are a characteristic feature of these otherwise white to grey quartzites (Bleeker et al., 1999b) and, where recrystallization is strong, allow these clastic rocks to be distinguished from metacherts. During metamorphism and deformation the detrital chromite grains reacted to varying degrees with surrounding minerals to produce bright green fuchsitic mica[5] (Figs. 8E,F). In a few localities, chromite was concentrated enough to form seams of "black sand". These occurrences, although of scientific interest, are too small to be of economic value. If road access were available, some of the green-white quartzite would make attractive building or decorative stone. In Greenland, India, and Australia, similar quartzites are commonly quarried for this purpose. Elsewhere in the world, quartzites similar to those described here contain paleoplacer deposits of gold and/or uranium (e.g. the Witwatersrand quartzites of South Africa; basal Huronian quartzites near Elliot Lake, Canada) (e.g. Roscoe and Minter, 1993). An initial survey of such potential in the Slave craton was carried out by Roscoe (1990, 1992), indeed resulting in anomalous values of gold and uranium.

Ultramafic sills (or flows?) have intruded the cover sequence in several places, and locally contain seams or veins of magmatic chromite (Covello et al., 1988). Economic concentrations have not been found. In one remote locality, on the south shore of Desteffany Lake, one

5. Fuchsite is a Cr-bearing white mica in which Cr^{3+} replaces Al^{3+}. In the quartzites, it formed as a product of the following generalized reaction: detrital chromite ± K-feldspar + fluid → fuchsitic mica + fluid (e.g. Bleeker et al., 2000). Metamorphic breakdown and hydration of detrital K-feldspar was aided by shearing.

A
B
C
D
E
F
G
H
I

FIGURE 6 . Field photographs illustrating some key elements of the geology of the Slave craton. **(A)** Cleaned exposures of the Acasta gneisses at their discovery site. Ancient tonalites (4.03 Ga) occur on left side of the picture, and are intruded by highly deformed younger granite sheets and mafic dykes. **(B)** Basal quartzites of the Central Slave Cover Group overlying basement of the Central Slave Basement Complex. Low terrain to the right is low-weathering basement gneiss; dark ridge in background consists of ca. 2.7 Ga basalts overlying the quartzites. **(C)** Syn-Kam Group quartz-porphyritic tonalite (ca. 2713 Ma), intruding into the northern part of the Yellowknife greenstone belt (geologist Val Jackson for scale). The large sill-like body is cut by somewhat younger mafic dykes that likely fed the upper part of the greenstone belt stratigraphy. Inset shows close-up of altered quartz-porphyritic tonalite. **(D)** Quartz-porphyritic rhyolite breccia with carbonate matrix, typical for the uppermost part of 2690 to 2660 Ma felsic volcanic edifices. Highly angular (proximal) fragments range is size from 1 to 10 cm. **(E)** Massive sulphide mineralization of the Sunrise deposit, associated with ca. 2670 Ma felsic volcanic rocks just below the interface with the Burwash Formation turbidites. **(F)** Thick-bedded sandy turbidites typical of the Burwash Formation in its type area east of Yellowknife. The oblique areal photo (approximately 500 m across, looking towards northeast) shows a typical mushroom interference pattern of an F_1 syncline refolded by north-northwest trending F_2 folds. **(G)** Silicate-facies iron formation interlayered with turbiditic greywackes, George Lake, northeastern Slave. This banded iron formation hosts significant epigenetic gold mineralization (pickets ca. 50 cm high). **(H)** Passive margin strata of the Coronation Supergroup (Epworth Group) overlying the western margin of the rifted Slave craton, structurally at the base of Wopmay Orogen externides (east facing cliff is approximately 100 m high). **(I)** Dense Proterozoic mafic dyke swarms cutting extended Slave crust and its cover (individual dykes of approximately 20-40 m across).

of the authors found sulphide concentrations adjacent to ultramafic rocks within the cover sequence. Overall, however, the volume of komatiitic rocks and the potential for substantial massive Ni-Cu mineralization appears limited, not only at this stratigraphic interval but throughout the Slave craton.

Ca. 2.73-2.70 Ga Tholeiitic Volcanism

Wherever the thin cover sequence is recognized, it is overlain by a thick and extensive sequence of tholeiitic basalts, with minor komatiite and rhyolite tuff intercalations (Fig. 8). In the Yellowknife greenstone belt, this basalt-dominated volcanic sequence (Fig. 9) is known as the Kam Group (Helmstaedt and Padgham, 1986; Bleeker et al., 1999b). Possible correlative basalt successions are known across the basement domain (Fig. 10), as far east as the Courageous Lake belt, and at least as far north as around the Exmouth antiform in the Acasta area (Bleeker et al., 2000). This basalt sequence typically consists of several hundred metres to several kilometres of pillowed and massive flows, intercalated with thin felsic volcaniclastic horizons, and intruded by numerous dykes and sills of multiple generations (e.g. Henderson and Brown, 1966).

Well dated components of this basalt-dominated sequence yield ages from 2722 to 2697 Ma (Davis and Bleeker, unpublished data, Isachsen and Bowring, 1997). A mafic dyke cutting across the lower part of the sequence north of Yellowknife, dated at 2738 Ma (J. Ketchum, pers. comm., 2004), demonstrates that parts of this basalt sequence are even older. In Yellowknife, the top of the sequence is represented by voluminous basaltic flows and intercalated felsic volcanic rocks of the Yellowknife Bay Formation, dated at ca. 2700 Ma (Fig. 9). Similar 2700 Ma ages have been obtained from the Courageous Lake and Acasta areas and strongly support the overall regional correlation (Bleeker et al., 1999b). Stratigraphy, dense dyke swarms, and isotopic data link the basalt sequence to the basement (Henderson, 1985; Bleeker et al., 1999a,b; Northrup et al., 1999; Cousens, 2000; Bleeker, 2002 and references therein).

If the broad regional correlation of these basalts is valid, the magnitude of volcanism (areal distribution >100,000 km^2, typical thickness 1-6 km) approaches large igneous province (LIP) proportions (Coffin and Eldholm, 1994, 2001;

FIGURE 7. North America's first diamond producer, the Ekati Mine of the central Slave craton. Main picture shows the flat barren lands of the central Slave craton, with several pipe-like kimberlite bodies being excavated in circular open pits. Inset (upper left) shows a close-up of one of the partially excavated pipes. Several millimetre-size gem quality diamond octahedra are shown on upper right (photos courtesy of D. Snyder and G. Lockhart).

Ernst et al., 2005). The widespread basaltic volcanism probably accompanied protracted rifting of the basement complex, possibly assisted by mantle plume activity. The stratigraphic succession in the Yellowknife area is compatible with such a rifting interpretation. At the top of the Kam Group, bimodal volcanic rocks of the Yellowknife Bay Formation become progressively more intercalated with volcaniclastic sediments, before final intrusion by thick tholeiitic sills. One of these, the Kam Point gabbro sill, has a preliminary baddeleyite age of ca. 2697 Ma (Davis and Bleeker, unpublished data; see Fig. 9).

From a mantle perspective, it seems likely that events associated with the voluminous basaltic volcanism across the ancient basement terrain must have involved thinning or at least modification of the lithospheric mantle below the Central Slave Basement Complex. Large-scale melting was probably triggered by adiabatic rise of asthenospheric mantle. Perhaps, then, the ca. 2.7 Ga basaltic volcanism may have contributed to the highly depleted mantle compositions underlying the core of the craton (e.g. Griffin et al., 1999; Grütter et al., 1999; Kopylova and Russell, 2000; Carbno and Canil, 2002).

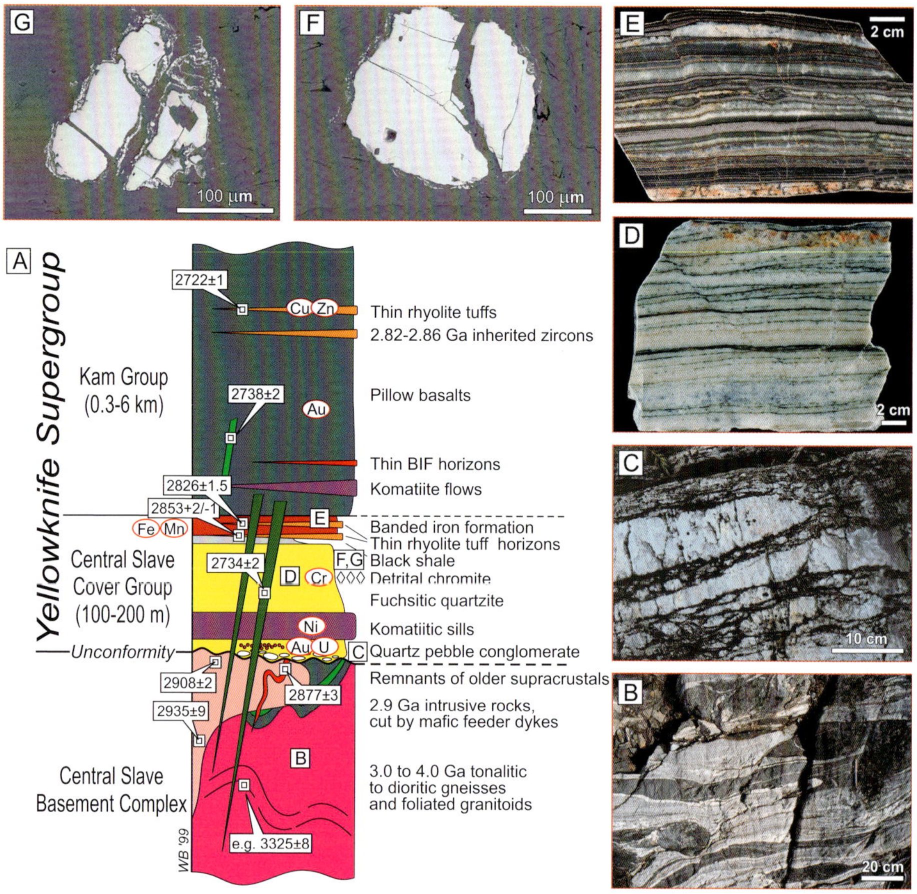

FIGURE 8. **(A)** Generalized stratigraphic column of the Central Slave Cover Group, the autochthonous cover of the Central Slave Basement Complex (Bleeker et al., 1999a,b). Relevant age data are from Isachsen and Bowring (1997), Bleeker et al. (1999a,b), and Ketchum and Bleeker (2000). Photos **(B)** to **(E)** illustrate characteristic lithologies and are keyed to the column. **(F)** Electron back-scatter image of rounded detrital chromite grain in fuchsitic quartzite of photo **(D)**. **(G)** As in photo **(E)**, but showing more advanced metamorphic transformation to fuchsite.

Mineralization

A volcanically dominated rift environment, characterized by bimodal volcanism and minor aprons of volcaniclastic sedimentary rocks, is a highly favourable environment for seafloor hydrothermal activity and the formation of volcanogenic massive sulphide (VMS) deposits. Indeed, numerous showings of sulphidic horizons occur throughout the basalt-dominated greenstone belts of the central and western Slave craton (Fig. 4).

Of particular interest are intercalated felsic volcanic flows and/or sills, which are direct indicators of proximity to a differentiated magmatic centre, and thus a long-lived subvolcanic heat source. The Bell Lake quartz-porphyritic tonalite sill (Fig. 6C) and the rhyolitic Townsite Formation, dated at 2713 ± 2 Ma and ca. 2709 Ma, respectively (Davis et al., 2004), are examples of such proximal felsic volcanic rocks in the Yellowknife greenstone belt. Hydrothermal alteration and minor sulphide mineralization are known from the Yellowknife Belt (e.g. the Homer Lake showings, see Fig. 4), but to date no deposits of potential economic interest have been found. Similarly, despite at least a first wave of exploration across the other basaltic greenstone belts of the west-central Slave craton, the authors are not aware of any major discoveries.

Although possibly unrelated, it is worth mentioning here the VMS deposits of the High Lake greenstone belt of the north-central Slave craton. These deposits, the High Lake A/B, D, and West Zone (Figs. 4, 11), along with several other sulphide-rich horizons (Wolfden Resources Inc., 2005), are associated with bimodal volcanic rocks dated at 2705 Ma (Henderson et al., 2000). Thus the setting and age are compatible with a mafic volcanic-dominated rift environment similar to that of the Kam Group. However, basement rocks have not been identified in the immediate area. Metal ratio plots (Fig. 12) discriminate the High Lake deposits from most other VMS deposits in the Slave, suggesting a similarity to Cyprus-type deposits (the "mafic class" of Barrie and Hannington, 1999), which form in mafic volcanism-dominated extensional settings.

Post-2.70 Ga Volcanism

Following ca. 2.7 Ga basaltic volcanism and rifting, most areas in the Slave craton show a transition to calc-alkaline volcanism characterized by abundant felsic and intermediate volcanic rocks, tholeiitic to calc-alkaline basaltic rocks, and intercalated volcaniclastic sedimentary rocks (Fig. 10). In nearly all areas, these rocks are stratigraphically overlain by turbiditic greywacke-mudstones (Fig. 10). Ages for the arc-like volcanic rocks and their plutonic counterparts typically

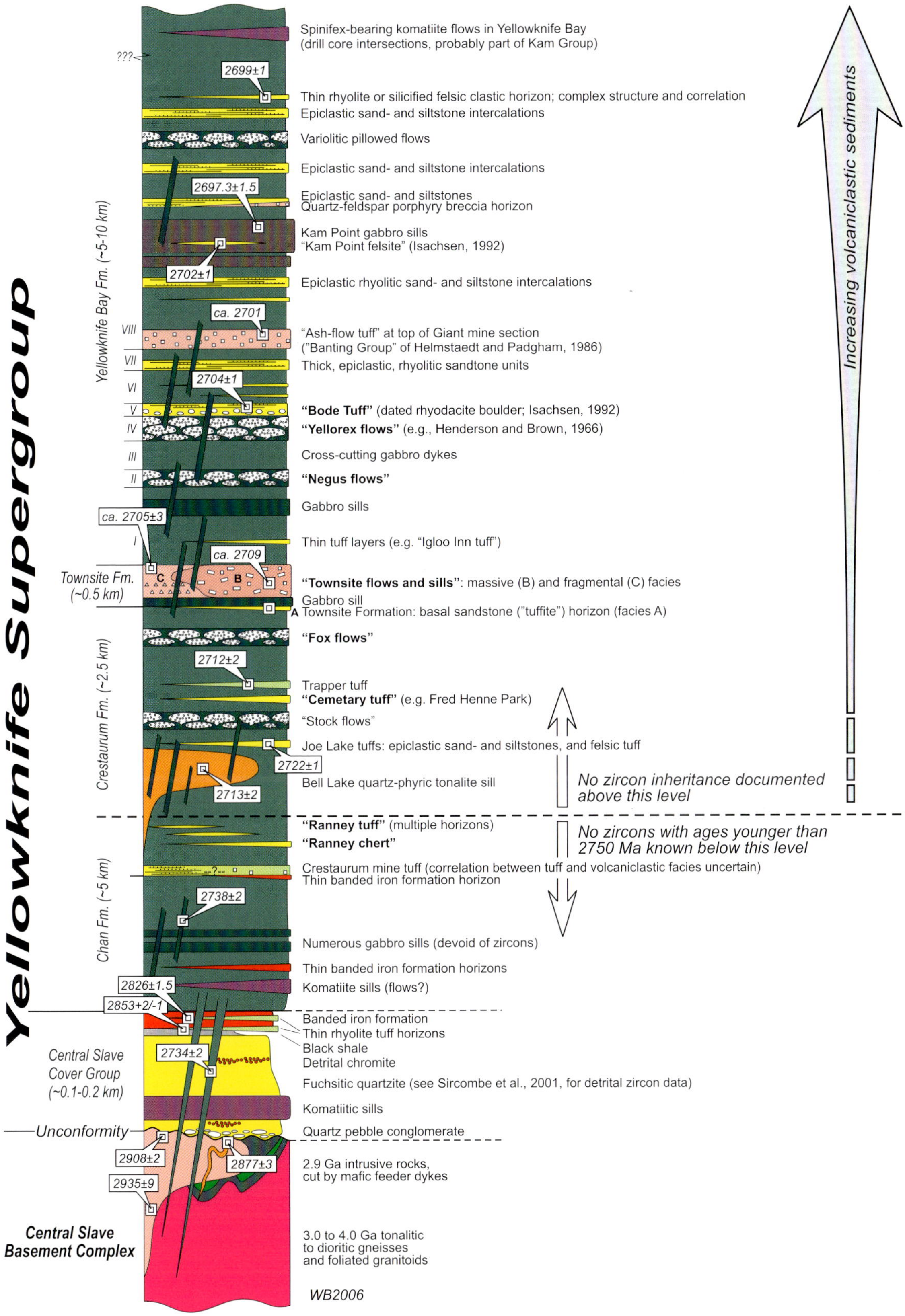

FIGURE 9. Stratigraphy and geochronology of the Kam Group, Yellowknife greenstone belt. All major units except for the basal Chan Formation have now been dated, showing a monotonic, younging upwards age progression through the volcanic pile. Sources of age data: Isachsen et al. (1991), Isachsen (1992), Isachsen and Bowring (1997), Bleeker et al. (1999b), Sircombe et al. (2001), Ketchum and Bleeker, unpublished data, Davis et al. (2004). A crosscutting gabbro dyke in the Chan Formation, dated recently at 2738 Ma (J. Ketchum, pers. comm., 2004), shows that much of the Chan Formation must be >2738 Ma. The youngest event recognized to date is the large intrusive gabbro sills at Kam Point, with a preliminary baddeleyite age of 2697 Ma. A large quartz-porphyritic tonalite sill (see Fig. 6C) has been dated at 2713 Ma and acted as a heat engine for extensive seafloor hydrothermal activity and associated alteration.

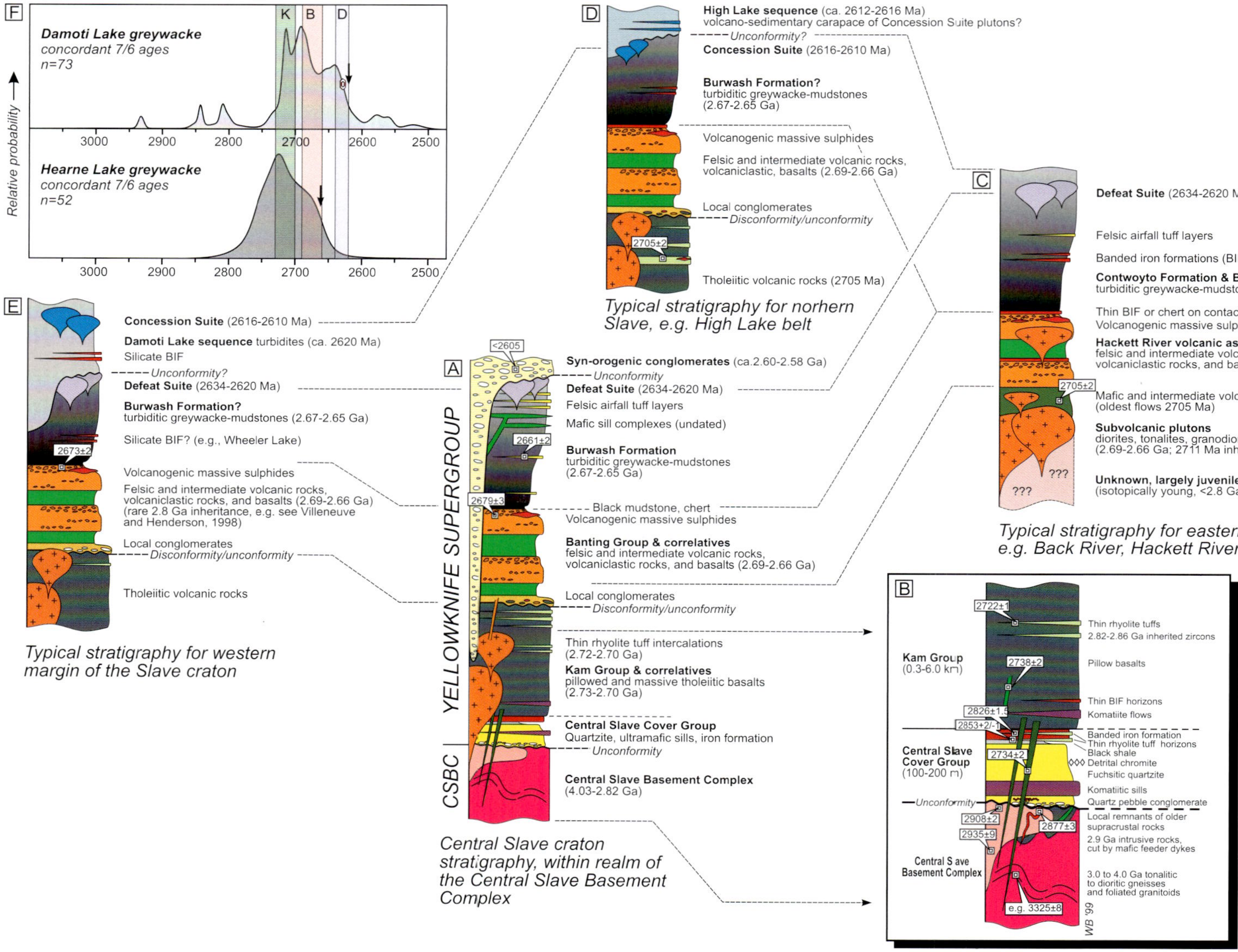

Figure 10. Generalized stratigraphic columns for different parts of the Slave craton. **(A)** Typical stratigraphy for much of the central part of the craton, with Yellowknife Supergroup rocks overlying the Central Slave Basement Complex. **(B)** Enlargement of the basement and its autochthonous cover. **(C)** Typical stratigraphy for much of the eastern Slave (e.g. Hacket River, Back River), lacking the association with ancient basement. **(D)** Typical stratigraphy for the north-central Slave, e.g.,the High Lake belt. An interesting feature is the occurrence of a ca. 2612 to 2616 Ma volcano-sedimentary package distinct from nearby Burwash-type metagreywackes. This "High Lake assemblage" is tentatively interpreted as the volcano-sedimentary carapace of Concession Suite magmatism. **(E)** Generalized column for the westernmost Slave craton. Much of the stratigraphy is similar to that overlying the CSBC (as in Fig. 10A), but is capped by a younger turbidite sequence with abundant silicate-facies iron formations. **(F)** Detrital zircon data for typical Burwash Formation turbidites (lower panel), and the younger greywacke sequence of Damoti Lake (upper panel). Note that the latter sample contains a robust younger age peak, indicating a significant contribution from erosional unroofing of Defeat Suite-age plutons. Data obtained by SHRIMP (Sircombe and Bleeker, unpublished). Shaded intervals highlight typical age ranges for Kam Group (K), Banting Group (B), and Defeat Suite (D) magmatism. Arrows indicate depositional ages. Zircon symbol highlights a single, concordant, and precise zircon age determination obtained by TIMS (Pehrsson and Villeneuve, 1999).

range from 2690 to 2660 Ma (e.g. Mortensen et al., 1988; Isachsen, 1992; Mortensen et al., 1992a,b; Villeneuve and Henderson, 1998; Northrup et al., 1999; Pehrsson and Villeneuve, 1999; Ketchum et al., 2004).

The arc-like volcanic rocks tend to be geochemically juvenile (e.g. Davis and Hegner, 1992; Cousens et al., 2002). They dominate the eastern part of the craton, where they lack any apparent association with older basement, its cover, and/or the basalt-dominated rift sequence discussed above. These observations have led to models in which the eastern Slave represents an exotic juvenile arc (the "Hackett River arc") that collided with the basement domain in the west (e.g. Kusky, 1989, 1990). However, similar arc-like rocks, with identical ages, stratigraphically overlie the basement domain and its cover in the west-central parts of the craton (Fig. 10A), where they can be tied to the basement and bimodal rift volcanic rocks by means of unconformities (Bleeker, 2001), crosscutting feeder dykes, and subvolcanic intrusions (see Fig. 13) (Bleeker et al., 1999a).

It thus appears that, if these rocks were generated in an arc-like setting, this arc was constructed marginal to, and on top of, the highly extended continental crust of the Central Slave Basement Complex. This suggests a marginal to continental arc setting. The arc must have been actively extending, evolving into a back-arc basin that was ultimately filled with turbiditic sediments (Figs. 10, 13). The geochemistry of the volcanic rocks and the associated subvolcanic plutons, although isotopically juvenile, typically shows arc-like signatures (negative Nb and Ta anomalies, light rare earth and large-ion lithophile element enrichment, and relative depletions of heavy rare earths) compatible with enriched sources in a supra-subduction zone setting and final equilibration with garnet-bearing residues. Alternative models invoke partial melting of a mafic underplate, perhaps accompanying delamination (e.g. Cousens et al., 2002).

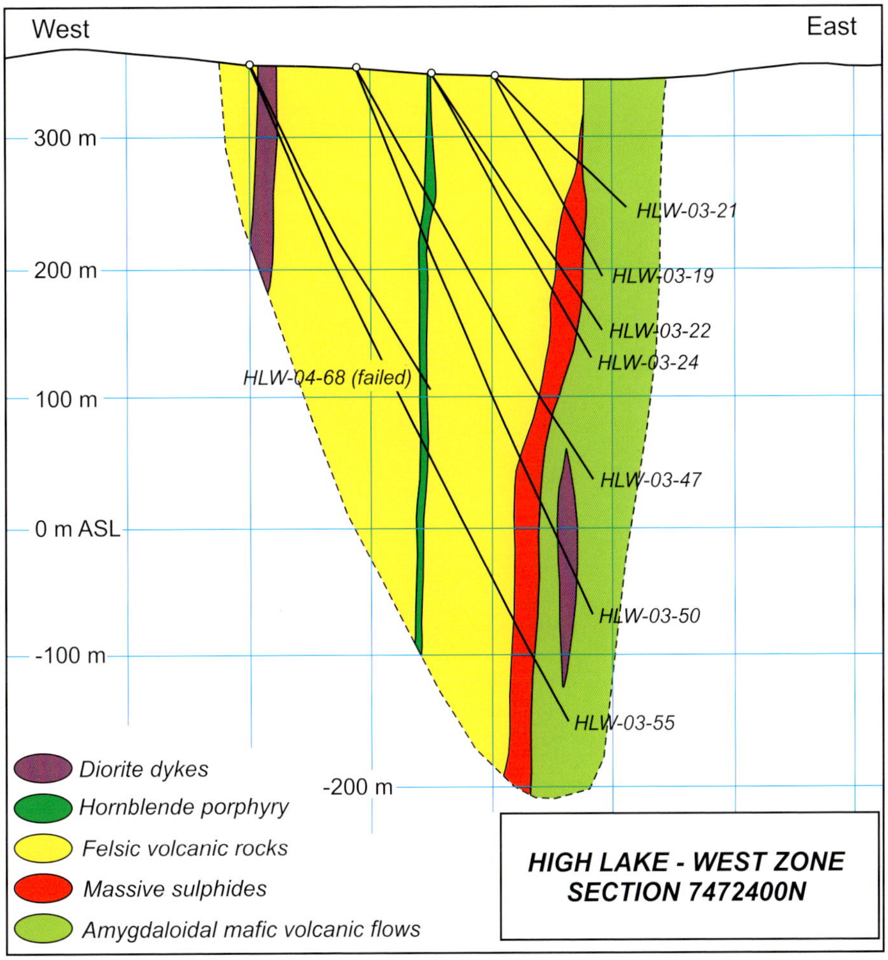

FIGURE 11. Cross-section through one of the High Lake massive sulphide deposits (after Wolfden Resources Inc., 2005).

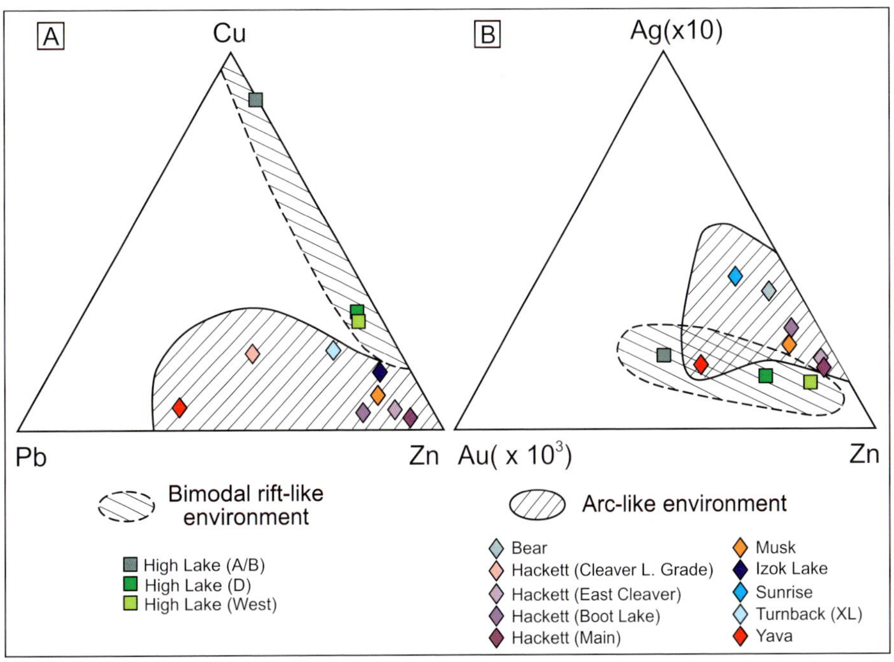

FIGURE 12. Ternary plots of metal ratios for VMS deposits in the Slave craton. The metal ratios define two fields: one for Cu-rich, Pb-poor deposits associated with the older ca. 2.7 Ga basalt-dominated bimodal volcanism, and another characterized by higher Pb (Ag) values for deposits hosted by the younger ca. 2.69 to 2.66 Ga arc-like sequences.

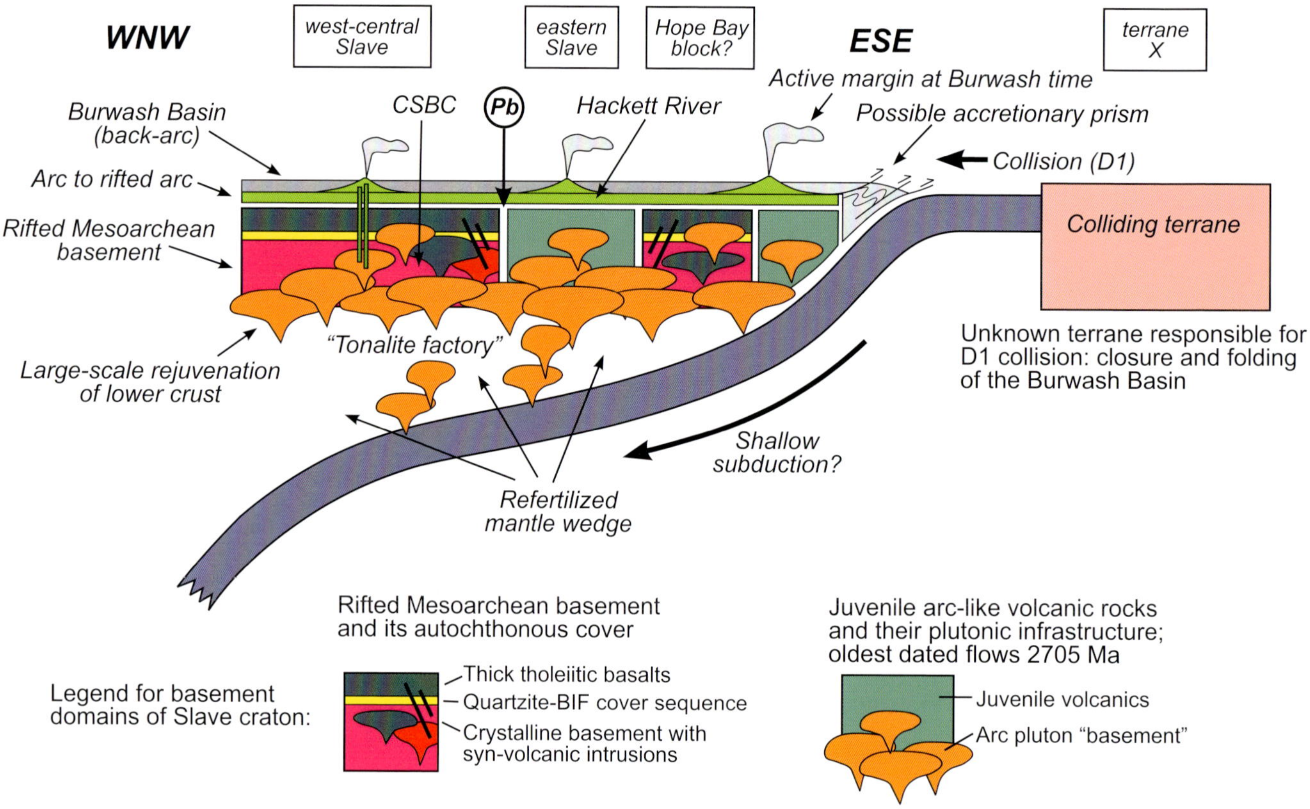

FIGURE 13. Tectonic model for the general setting of the Slave craton between ca. 2690 and 2660 Ma. The entire Slave appears to have been situated in a supra-subduction zone setting, with abundant and widespread calc-alkaline volcanism and plutonism. Arc-like assemblages (e.g. Hackett River, HR) were built across rifted basement and evolved into a large back-arc basin filled with turbidites: the Burwash Basin. Stratigraphically lowest, pre-2687 Ma components of the eastern arc-like domain could be exotic; alternatively, these rocks could represent juvenile volcanics in narrow back-arc rifts. The Hope Bay block may represent a rifted fragment of the Central Slave Basement Complex (CSBC). Voluminous tonalite intrusions rejuvenated much of the lower and mid-crust. Collision of an unknown terrane (X), between 2650 and 2630 Ma, led to closure and F_1 folding of the Burwash Basin.

Mineralization

The arc-like volcanic sequences spanning the interval 2687 to 2660 Ma host the Hackett River, Izok Lake, Sunrise, Gondor and Kennedy Lake deposits (Fig. 4). An exception is the High Lake deposits, which are associated with older, ca. 2705 Ma, bimodal volcanic rocks (discussed above).

Nearly all the VMS deposits of this group are associated with proximal felsic volcanic rocks at or near the transition to overlying turbiditic metasedimentary rocks. This transition is characterized by rhyodacite to rhyolite complexes, volcaniclastic sediment aprons, thin sulphidic chert horizons, and, in some localities, banded iron formations. Carbonate rocks (calc-arenites) are associated with some of the felsic complexes (Fig. 6D). This typical stratigraphic evolution, from shallow water or emergent felsic volcanic complexes to deep-water turbidite sedimentation, suggests active extension, tectonic subsidence of the arc environment, and creation of significant accommodation space that was then filled by deep-water sediments (Fig. 13). Such an environment of active faulting, attenuated lithosphere, active volcanism, and high heat flow, has long been recognized as a favourable setting for development of large VMS deposits.

Other characteristic features of this high heat flow regime are the presence of high-level synvolcanic intrusions, for instance in the Hackett River belt (see Fig. 14; the Sandy Hill Pluton, Hanimor Gneiss Dome, and smaller intrusive bodies that lie within the Malley Rapids Anticlinorium). The larger VMS occurrences of the Hackett River belt (A, East Cleaver, Boot Lake, Cleaver Low Grade, Yava, and Musk) exhibit a strong spatial relationship, and hence inferred genetic relationship to these intrusions. It is possible that these synvolcanic intrusions contributed magmatic fluids to the ascending hydrothermal solutions, resulting in the unusually high silver and lead contents for these deposits. Stratiform carbonate rocks are a common feature associated with the Hackett River deposits as well as the BB, Bear, and Turnback Lake (XL) deposits. A likely origin for these carbonate deposits is the venting of Ca-saturated hydrothermal fluids onto the seafloor and rapid mixing with cold, higher pH seawater (i.e. analogues to "white smokers" in modern seafloor settings). If correct, these carbonate deposits represent excellent marker horizons and vectors in the search for new volcanogenic massive sulphide deposits.

Ca. 2.68-2.66 Ga Sedimentation

Starting at ca. 2680 Ma, a broad turbidite basin — the Burwash Basin — developed across much of the craton and progressively buried the volcanic substrate (e.g. Henderson, 1985; Ferguson et al., 2005). The transition from volcanic and volcaniclastic rocks to deep-water sediments is commonly conformable or disconformable[6]. In a number of localities, however, the transition is marked by a well developed unconformity or nonconformity[7] that cuts down into underlying crystalline rocks and is overlain by shallow-water clastic rocks, including conglomerates with granitoid cobbles (Henderson, 1985; Bleeker et al., 1997; Bleeker, 2001). One such locality occurs along the southwestern flank of the Sleepy Dragon Complex, another at Point Lake, and they demonstrate an autochthonous nature of the sedimentary sequence (Bleeker, 2001). Detrital zircons in the basal clastic rocks indicate that deposition began sometime after 2683 Ma (Bleeker et al., 1997).

A persistence of volcanic intercalations up-section and late mafic sill complexes suggest a volcanically active extensional setting, perhaps best compared with modern back-arcs. The minimum size of this basin was ca. 400 x 800 km (Fig. 15A), making it the largest and possibly best preserved Archean turbidite basin in the world, comparable in size to the Japan Sea. Like the modern Japan Sea, the Burwash Basin was largely ensialic, in agreement with inferences by early workers (e.g. Henderson, 1985).

The Burwash Basin fill consists largely of immature greywackes and mudstones, deposited below wave base, and locally may be up to 10 km thick (Bleeker and Beaumont-Smith, 1995). Intercalated tuff layers have been dated at 2661 Ma (e.g. Bleeker and Villeneuve, 1995). Across the Slave craton, the greywacke turbidites have been given different formational names: the classical Burwash Formation in the Yellowknife Domain (Henderson, 1972); the Contwoyto Formation in central and northern Slave, identical in essentially all aspects to the Burwash Formation further south, except for the presence of intercalated iron formations; the Itchen Formation, a more mud-rich facies in the north-central Slave; and the Beechey Lake Group in the northeastern Slave, which also contains iron formations. Many of the turbidite beds, particularly those of the Burwash and Contwoyto Formation, are sand dominated with only thin silt to mud intervals at the top of the graded beds. In the Yellowknife Domain, thick amalgamated sand beds (2-10 m) are not uncommon (Fig. 6F). Petrography, detrital zircons, and geochemical analysis indicate that the greywacke detritus consists of a mixture of mafic and felsic volcanic rocks and uplifted plutonic infrastructure, with minor input from ancient basement rocks

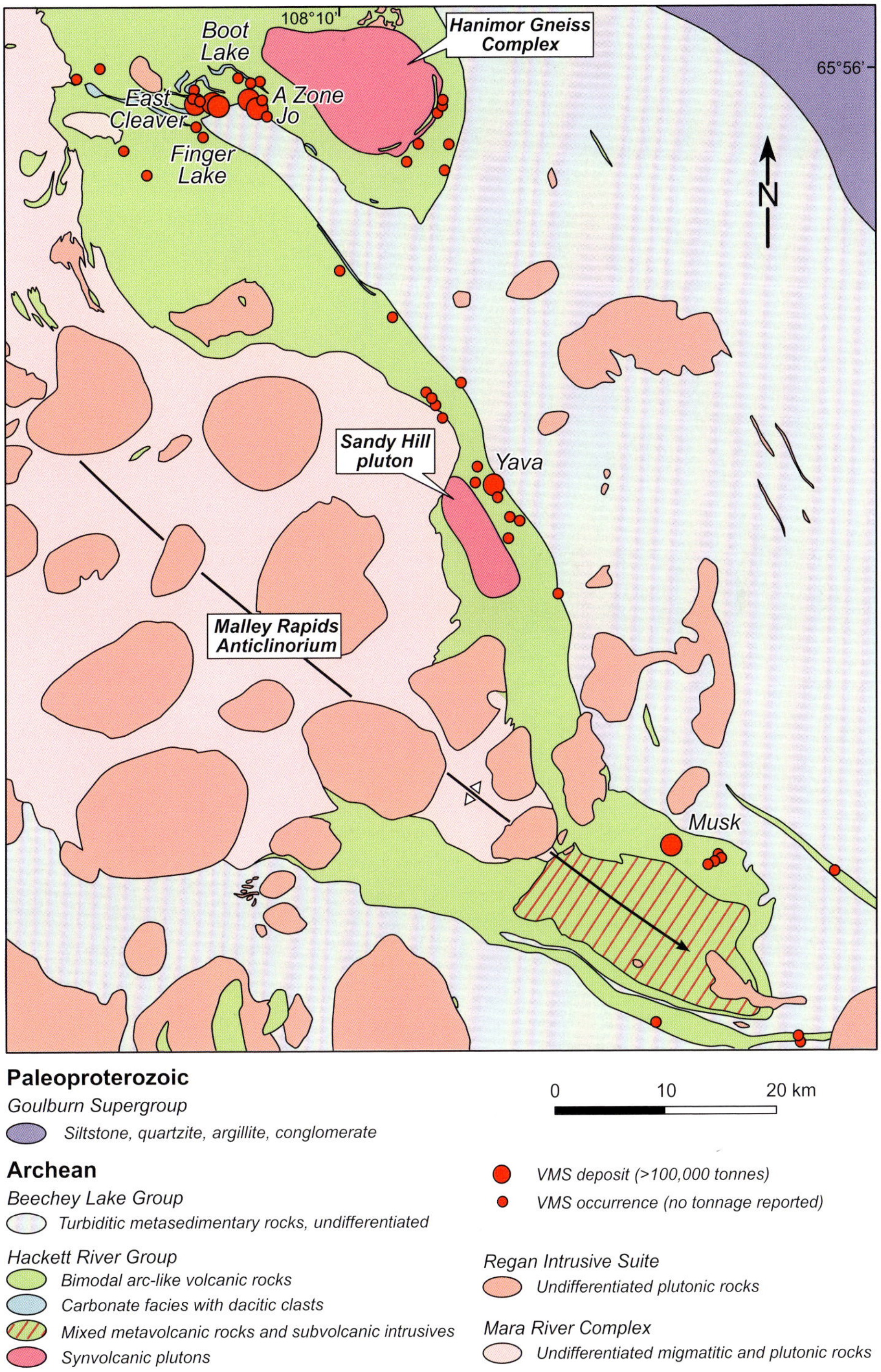

FIGURE 14. Geological map of the Hackett River area (modified from Frith and Fryer, 1985), showing the locations of major VMS deposits as well as smaller occurrences in relation to synvolcanic plutons. Also shown is the carbonate unit that bears a close spatial relationship to the VMS deposits.

6. A disconformable contact is conformable but with a considerable interval of nondeposition, i.e.,a significant time hiatus or diastem.

7. A nonconformity: an unconformity in which a younger sedimentary sequence overlies nonstratified crystalline rocks, in this case either older synvolcanic plutons or ancient crystalline basement rocks (e.g. Point Lake).

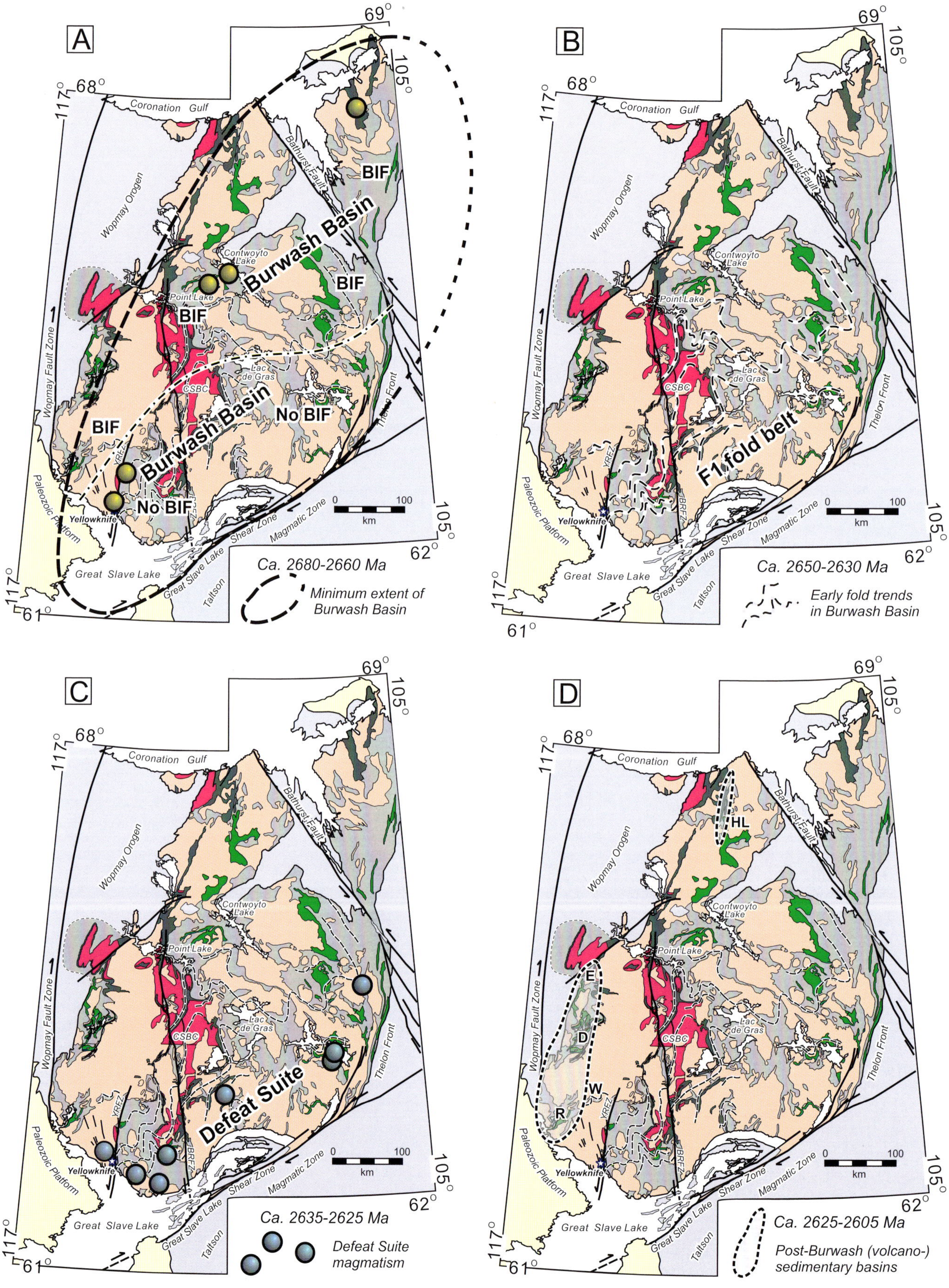
A
Coronation Gulf
Bathurst Fault
Wopmay Orogen
Wopmay Fault Zone
BIF
Burwash Basin
Contwoyto Lake
Point Lake
Lac de Gras
CSBC
No BIF
Thelon Front
YRFZ
Yellowknife
Paleozoic Platform
Great Slave Lake
Great Slave Lake Shear Zone
Magmatic Zone
Taltson
Ca. 2680-2660 Ma
Minimum extent of Burwash Basin
B
F1 fold belt
Ca. 2650-2630 Ma
Early fold trends in Burwash Basin
C
Defeat Suite
Ca. 2635-2625 Ma
Defeat Suite magmatism
D
HL
E
D
W
R
Ca. 2625-2605 Ma
Post-Burwash (volcano-) sedimentary basins
km
0
100
68°
69°
105°
117°
61°
62°

(Henderson, 1972; Jenner et al., 1981; Henderson, 1985; Yamishita and Creaser, 1999; Ferguson et al., 2005). The main axis of the basin and subsequent structural trends appear to have been northeast-southwest (Fig. 15), distinctly across the north-south isotopic boundaries that track the nature of deep basement. This interpretation is based on the following observations:

- Identical Burwash Formation turbidites extend from near the type area of Yellowknife (e.g. Henderson, 1972) to the northeastern Slave (Figs. 10, 15A).
- Banded iron formations in the turbidites are restricted to the northwest half of the craton (Fig. 15A) suggesting a northeast to southwest facies boundary or tectonic trend across the basin (see also Padgham, 1992). This argument is complicated however by the observation that some iron formations in the western part of the craton are hosted by a younger turbidite package, the Damoti Lake sequence (see Fig. 10E).
- Earliest folds in the turbidites, which formed between 2650 and 2630 Ma, have northeasterly trends after qualitative "unfolding" of younger superimposed fold generations (Fig. 15B) (e.g. Bleeker, 1996).
- The earliest plutonic suite that intrudes folded Burwash strata, the ca. 2630 Ma Defeat Suite, appears to form a northeast-southwest-trending magmatic belt across the southeastern half of the craton (Fig. 15C) (Davis and Bleeker, 1999).

With more and better U-Pb zircon ages, a tentative "volcanic line" of 2661 Ma felsic volcanic complexes, coeval with turbidite sedimentation, has begun to emerge (Fig. 15A; Bleeker and Davis, unpublished data). This volcanic line also trends northeast-southwest, from the Hope Bay belt to Yellowknife, and may represent the first recognition of a linear arc system.

Mineralization

The immature greywackes and mudstones of the Burwash Basin contain few primary mineral deposits other than banded iron formations (Fig. 6G). The latter occur intercalated in greywackes scattered across a broad swath in the northern part of the craton (Fig. 15A), from the Goose Lake and George Lake areas to the Point Lake area, and from there to the southwestern Slave. Many of the BIFs are highly magnetic. Although of scientific interest for the understanding of facies boundaries, basin evolution, and geochemistry, they are uneconomic in terms of their ferrous metal content.

Within and adjacent to the Back River Volcanic Complex, Jefferson et al. (1989) described three distinct iron formation horizons: 1) intercalated within the volcanic rocks of the Back River Complex itself; 2) at the upper contact of the Back River volcanics; and 3) an upper horizon well within the turbidites of the Burwash Formation. Lateral facies changes exhibited by the mineralogy of these iron formation units vary from oxide facies (magnetite±hematite and quartz), through silicate facies (chert-grunerite±stilpnomelane±chlorite), to carbonate facies (siderite ±quartz).

The principal type of economic mineralization within Burwash Formation metaturbidites is epigenetic gold mineralization hosted by the intercalated BIFs (Padgham, 1992). The most important example of this deposit type is the Lupin deposit on the southern shores of Contwoyto Lake (Bullis et al., 1991a,b; Geusebroek and Duke, 2004), which has been a significant gold producer from 1982 to 2003, yielding between 3 and 4 million ounces of Au (Normin, 2005). Other examples, such as George Lake (Fig. 16) and Goose Lake, occur throughout the northern Slave craton and may become economic with elevated gold prices and better access. The general model for these deposits is that the host BIFs served as chemical traps for gold-bearing, low-salinity, mixed H_2O-CO_2 fluids during metamorphism and deformation (Bullis et al., 1991b; see also Kerswill, 1993; Phillips, 1993). Destabilization of the Au-carrying sulphur complexes, due to interaction with reduced Fe-rich host rocks, led to alteration and gold deposition, either in veins or in fluid-altered and sulphidized zones of the iron formations. The structural timing of these epigenetic deposits is generally syn- to late-kinematic and syn- to late-metamorphic, i.e.,consistent with maximum fluid production deeper in the thickened structural-metamorphic pile. The most likely source for the fluids, and the gold, is metamorphic devolatilization of a voluminous, immature, sediment pile and its volcanic substrate (Phillips, 1993). Sporadic iron formations provided accidental traps to the migrating fluids, with discrete structures locally playing a role in increased focusing of fluid flow. Similar processes also led to gold-bearing quartz veins within metaturbidites (e.g. laminated veins along sheared bedding planes, saddle reefs) (Boyle, 1961, 1986), but without a large-scale focusing mechanism these occurrences and deposits tend to be of small size, although locally of high grade. Examples are the Ptarmigan and Discovery mines in proximity to Yellowknife (Fig. 4) (e.g. Brophy, 1987).

Ca. 2.65-2.63 Ga Closure of the Burwash Basin

Turbidite sedimentation in the Burwash Basin came to an end sometime before 2650 Ma, the age of the oldest recorded granitoid (feldspar porphyry) pluton intruding Burwash strata (Point Lake area, Mueller et al., 2001). Subsequent tectonic events record the closure and folding of the Burwash Basin (D_1) prior to 2630 Ma (see F_1 fold belt in Fig. 15B). The latter age constraint is provided by early plutons of the Defeat Suite (Davis and Bleeker, 1999), a distinct and possibly subduction-related magmatic suite across the southern (and southeastern) Slave craton (Fig. 15C).

FIGURE 15. Thematic maps illustrating key stratigraphic and structural aspects of the Slave craton through time. **(A)** Minimum extent of the ca. 2680 to 2660 Ma Burwash Basin, based on continuity and geochronological similarity of turbiditic greywackes across large parts of the craton. The dash-dot line separates areas with intercalated iron formations (northwest) from those lacking iron formations (southeast). Yellow spheres highlight localities with precisely dated 2661 Ma volcanism closely associated with turbidite sedimentation. The linear trend may reflect a 2661 Ma magmatic line in a general arc-like setting. **(B)** General trends of the F_1 fold belt in the Burwash Formation and correlatives, trending northeast-southwest across the craton. **(C)** Defeat Suite plutons dated between 2635 and 2625 Ma. This apparent trend of arc-like plutons parallels the F_1 fold belt. **(D)** Areas in the Slave craton with younger volcano-sedimentary packages that post-date deposition and folding of the Burwash Formation: the ca. 2620 Ma Damoti Lake (D) assemblage and the ca. 2612 to 2616 Ma High Lake (HL) assemblage. The Damoti Lake assemblage extends at least from Emile River (E) to Russell Lake (R), and possibly to Wheeler Lake (W). Its full extent is not known.

Closure of the highly extended, but largely ensialic back-arc basin allowed considerable shortening and mobility but with a structural style dominated, at least at high structural levels, by fairly systematic, mostly upright, northeast-southwest trending fold trains (e.g. Bleeker and Beaumont-Smith, 1995). At deeper levels, e.g., along the basement-cover interface, the fold trains must have detached allowing differential shortening of the basement and cover (e.g. Kusky, 1991). The folded Burwash strata do not represent an outboard accretionary prism however (cf. Kusky, 1991), which would require a trench setting rather than the more likely ensialic back-arc setting; hence, there is little evidence for a discrete "Contwoyto terrane" (Kusky, 1989) in the central Slave craton.

Interestingly, the northeast-southwest structural grain of the F_1 fold belt is also recognized in the lithospheric mantle (Grütter et al., 1999). Shallow subduction (either from the southeast or northwest?) may have emplaced distinct mantle slabs (Davis et al., 2003b). These processes terminated with docking of an outboard terrane (e.g. Fig. 13), either in the southeast or the northwest; however, this terrane is not preserved within the present limits of the Slave craton. Finally, crustal thickening led to uplift and erosional exhumation of folded Burwash strata and partial unroofing of Defeat Suite plutons. Detrital zircons of Defeat Suite age are recorded in younger sedimentary packages (e.g. Fig. 10F).

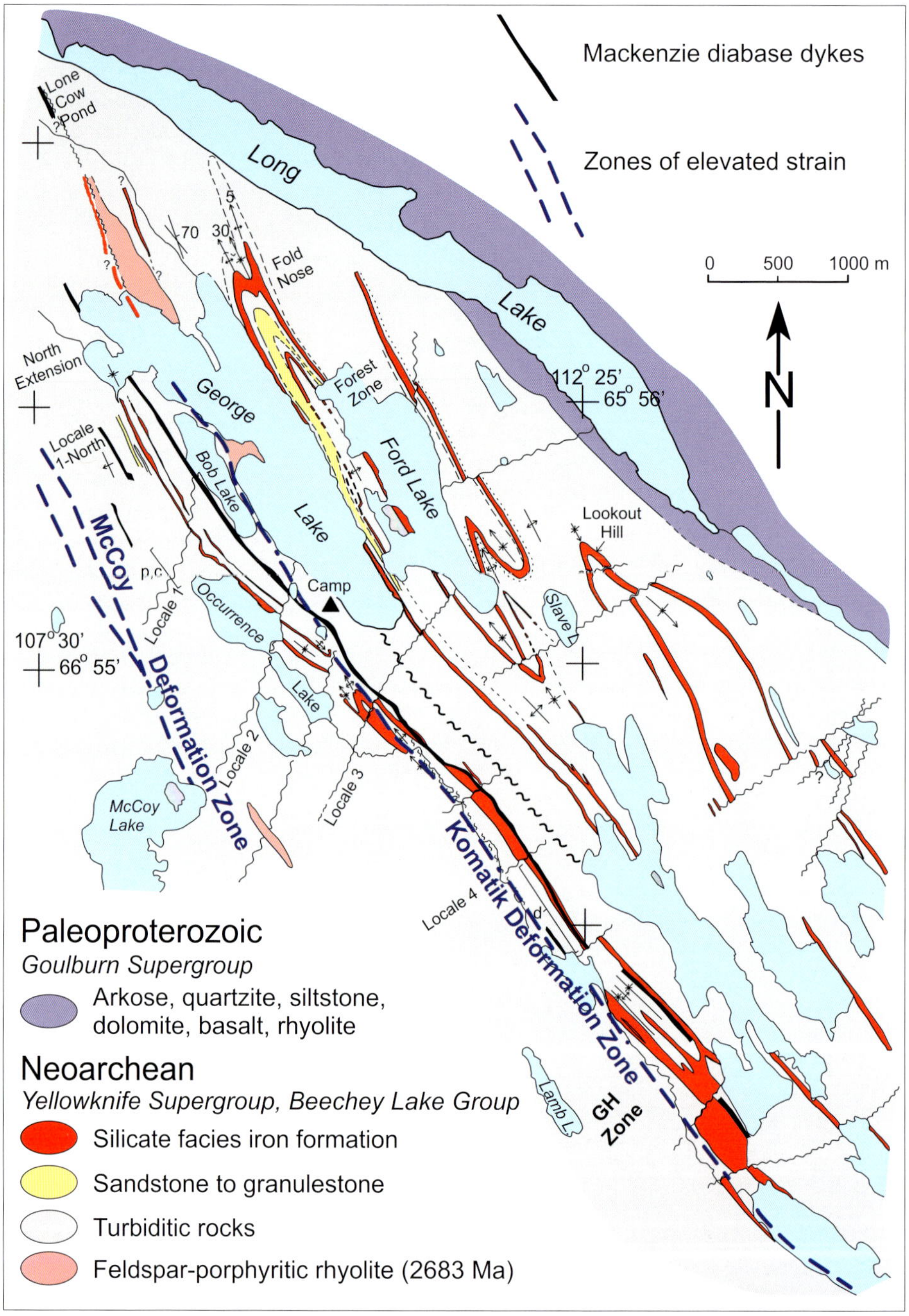

FIGURE 16. Geology of the George Lake area, northeastern Slave craton, where significant epigenetic gold mineralization is hosted by tightly folded and metamorphosed iron formation in Burwash-age metaturbidites (after Jefferson et al., 1992).

Mineralization

Folding and incipient crustal thickening (D_1), and the onset of regional metamorphism, together with Defeat Suite plutonism, must have initiated devolatilization reactions and metamorphic fluid flow. Although these events thus likely kick-started the development of epigenetic gold mineralization, they were followed and overprinted by much more intense metamorphic events ca. 20 to 30 million years later, during D_2 deformation.

Arc-generation and subduction processes almost certainly modified the mantle lithosphere below the Slave craton, possibly creating the starting conditions for what is now a thick diamondiferous mantle root (e.g. Davis et al., 2003b). Interestingly, trends of similar mantle domains, based on indicator mineral chemistry, appear to parallel the northeast-southwest trends of the Burwash Basin and D_1 folding (Grütter et al., 1999).

Post-2.63 Ga Turbidites

Along the western and northwestern margin of the craton, younger turbidites containing ca. 2630 Ma detrital zircons (Figs. 10E, F, and 15D) (Sircombe and Bleeker, unpublished SHRIMP data, Pehrsson and Villeneuve, 1999; Bennett et al., 2005; Bennett, 2006) record a migration of sedimentation and tectonic activity to the northwest. Deposition was coeval with uplift and erosional unroofing of Defeat plutons

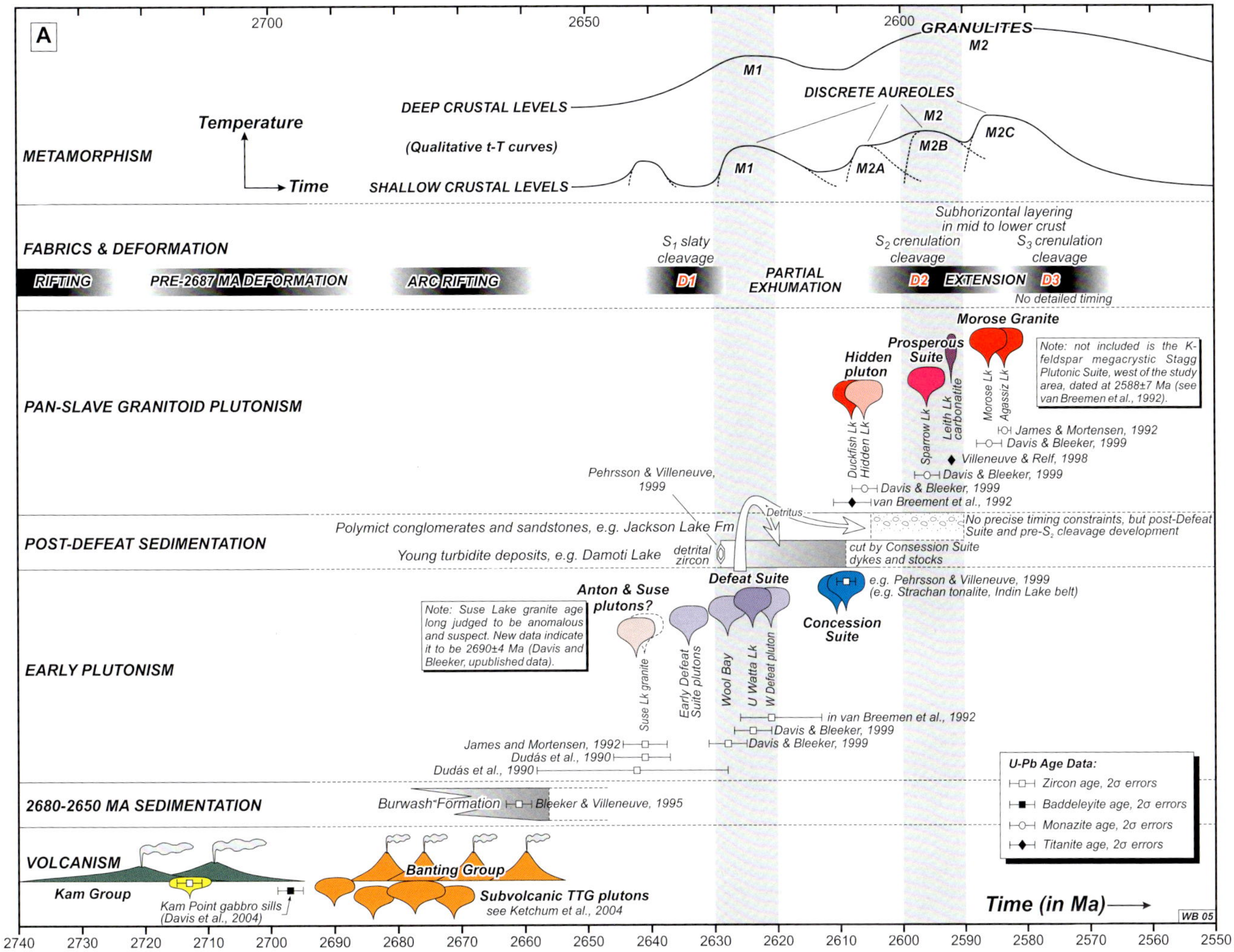

FIGURE 17. Time charts of stratigraphic and structural events in the Slave craton. **(A)** Detailed chart for key events in the Yellowknife Domain (updated from Davis and Bleeker, 1999). **(B)** Extended time chart for entire Slave craton (excluding pre-2900 Ma history of the Central Slave Basement Complex). Note breaks in horizontal scale (time) at two places. Typical detrital age spectra (1, 2a, b, 3, and 4) are shown for some of the main stratigraphic units: 1 = quartzites of the Central Slave Cover Group; 2 = turbiditic greywackes of Burwash Formation (s.s., 2a) and correlatives in the eastern Slave (Back River, 2b); the Damoti Lake turbidites (3); and synorogenic conglomerates (4). Known mafic dyke swarms are shown by narrow black bars (or grey, if poorly dated).

and tightly folded Burwash Formation strata. Shortly following their deposition, these younger turbidites were shortened and intruded by ca. 2616 to 2608 Ma tonalite-granodiorite plutons similar to the Concession Suite of the Contwoyto Lake area (Davis et al., 1994; Pehrsson and Villeneuve, 1999).

In the multiply folded, metamorphosed, and intermittently exposed terrain of the western Slave craton, it has proven difficult to distinguish these younger turbiditic greywackes from Burwash Basin turbidites. There is no sharply defined demarcation line that separates the two turbidite packages and recognition of the younger sequence relies heavily on the absence of Defeat Suite-age plutons and the presence of <2640 Ma detrital zircons. Preliminary work suggests that the younger turbidite sequence contains abundant intercalated iron formations, mostly of silicate facies, those of the Damoti Lake area representing one of the more significant examples. Many of the iron formations are "lean" (i.e., ironstones of low to moderate Fe content), comprising background turbiditic greywacke variably enriched in metamorphic garnet, other Fe-rich silicates, and/or disseminated Fe sulphides.

A distinct volcaniclastic sediment and greywacke package, associated with felsic volcanic rocks and subvolcanic intrusions, occurs along the tightly folded synclinal core of the High Lake greenstone belt of the northern Slave craton (see Figs. 10D, 15D) (Henderson et al., 2000). This package, dated at approximately 2616 to 2612 Ma (Henderson et al., 2000), is of significance in that it is one of the few examples of a preserved volcano-sedimentary carapace coeval with one of the major plutonic suites, i.e., the Concession Suite.

Mineralization

Types of mineralization within the younger (turbiditic) greywacke packages are similar to those in folded Burwash Basin strata. A principal example of epigenetic gold mineralization is that hosted by silicate facies iron formation in the Damoti Lake area. Similar iron formations occur all along the western margin of the Slave craton, from the Russell Lake area in the south to the Emile River area in the north (see Fig. 15D), and most have been moderately explored for gold and base metals. Scattered gold mineralization also occurs in numerous "lean" iron formations throughout the

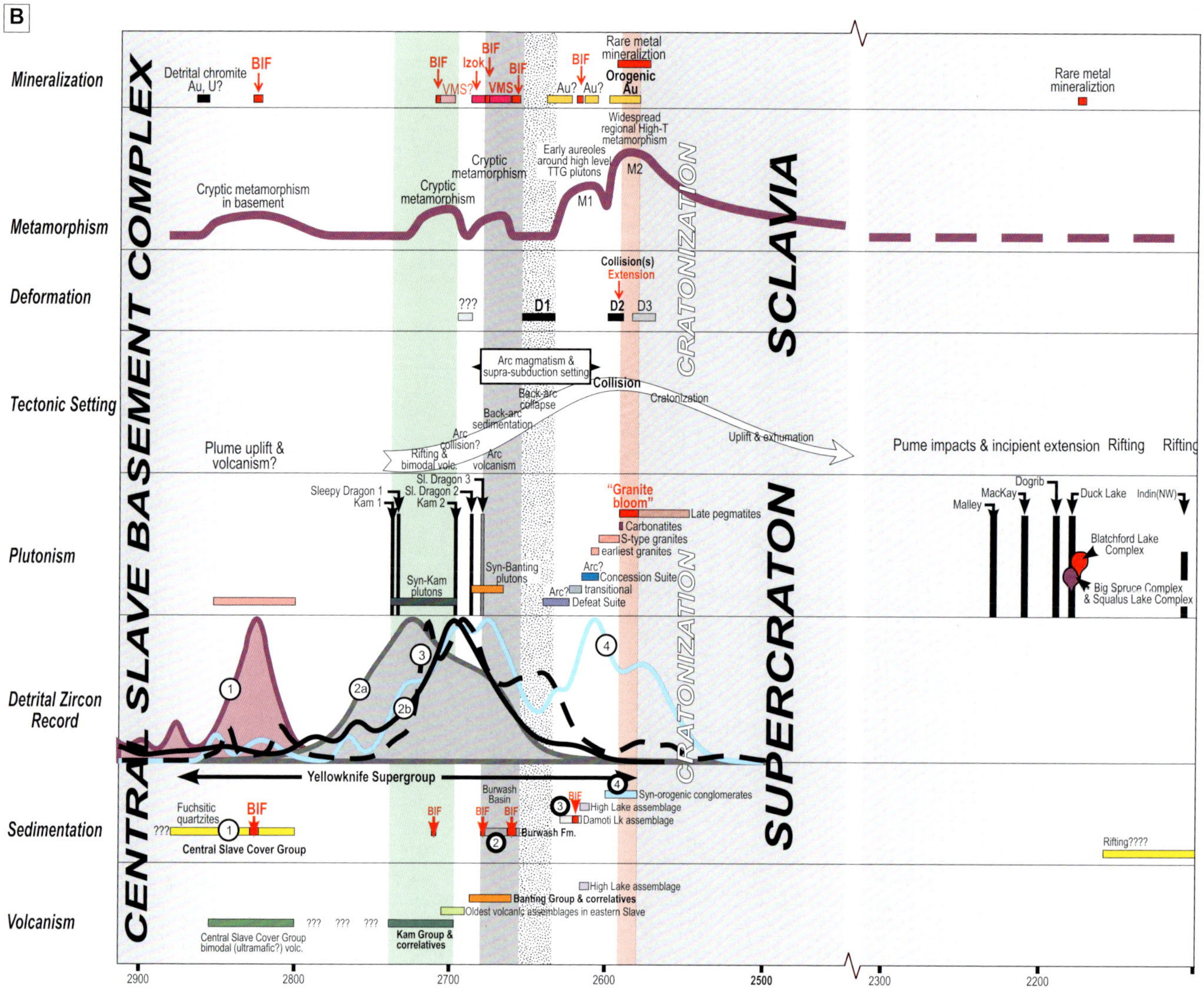

Figure 17 continued. (B) Extended time chart for entire Slave craton (excluding pre-2900 Ma history of the Central Slave Basement Complex). Note breaks in horizontal scale (time) at two places. Typical detrital age spectra (1, 2a, b, 3, and 4) are shown for some of the main stratigraphic units: 1 = quartzites of the Central Slave Cover Group; 2 = turbiditic greywackes of Burwash Formation (s.s., 2a) and correlatives in the eastern Slave (Back River, 2b); the Damoti Lake turbidites (3); and synorogenic conglomerates (4). Known mafic dyke swarms are shown by narrow black bars (or grey, if poorly dated).

western Slave (Padgham, 1992), e.g., the Wheeler and Germaine Lake areas ("W" in Fig. 15D). The stratigraphic status of the iron formations and the host turbidites in this particular area remains currently unresolved. Extensive field investigations have shown that an earlier reported 2612 Ma "tuff bed" (e.g. Isachsen and Bowring, 1994) is in fact another crosscutting porphyry dyke. Hence, the turbidites in this area are >2612 Ma and could be of Burwash age.

2.60-2.58 Ga, Final Orogenesis

Starting at ca. 2600 Ma, the entire craton was affected by cross-folding and significant further shortening (D_2), probably in response to final collision along a distant active margin of the growing supercraton Sclavia (Bleeker, 2003). Structural trends, such as axial planes of folds and the dominant cleavage resulting from this event, are generally north-south to northwest-southeast, but may refract strongly around more rigid plutons or other heterogeneities (e.g. Davis and Bleeker, 1999). Moderate overthickening of the crust led to high-T, low-P metamorphism (Thompson, 1989), widespread anatexis, the appearance of S-type granites, and a hot and weak lower crust. These processes culminated in ca. 2590 Ma extension and the craton-wide "granite bloom" (Kusky, 1993; Davis and Bleeker, 1999). The intrusion of carbonatites (Villeneuve and Relf, 1998) and involvement of other mantle-derived melts (e.g. sanukitoids, see Bennett, 2006) indicate a role for mantle processes (delamination?). Overall timing relationships are summarized in Figure 17A and B.

While peak temperatures were attained in the lower crust, large basement-cored domes were amplified by buoyancy-driven deformation (Fig. 3); lower crustal devolatilization reactions mobilized gold-bearing fluids; and synorogenic clastic basins formed and were immediately infolded into tight synclines (Bleeker, 2002). At least one of these synorogenic clastic basins, the Beaulieu Rapids Formation (Corcoran et al., 1999) may have formed as late as ca. 2580 Ma (Sircombe and Bleeker, unpublished SHRIMP data; see detrital zircon age spectra in Fig. 17B). Late strike-slip faults overprinted and truncated the synclinally infolded clastic

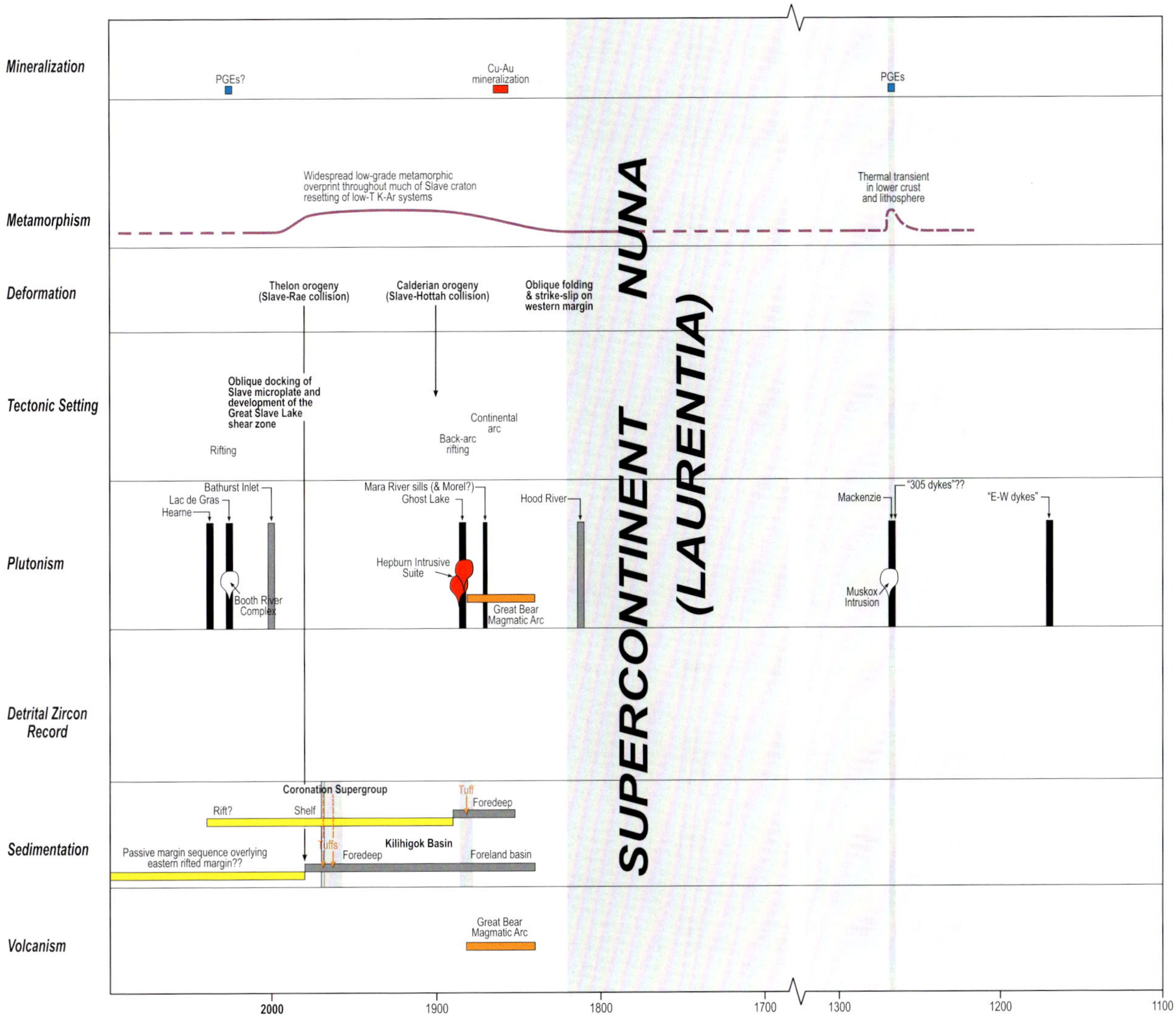

FIGURE 17B CONTINUED.

basins. Examples of such faults are the Yellowknife River Fault Zone in Yellowknife, and the Beaulieu River Fault Zone along the eastern flank of the Sleepy Dragon Complex (see Fig. 2). Over a period of ca. 70 million years, the lower crust cooled (Bethune et al., 1999), finally coupled with the mantle, and the Slave (within the larger context of Sclavia) became a craton.

Mineralization

Strong penetrative regional deformation, culminating between 2600 and 2590 Ma as constrained by synkinematic granite sheets (Davis and Bleeker, 1999), represents the most obvious deformation throughout much of the Slave craton. It must have driven moderate to significant crustal thickening and led to the main thermal peak of regional metamorphism in most areas. The D_2 deformation and associated metamorphism was the main driver for epigenetic gold mineralization throughout the Slave craton. In the Yellowknife greenstone belt, it led to formation of the approximately 15 million ounce Con-Giant Au deposit, along a complex system of mostly reverse shear zones. As is typical for this class of deposits, the Con-Giant system occurs mostly within moderate to strongly deformed basaltic rocks, in proximity to a regional stratigraphic break, the Yellowknife River Fault Zone (Figs. 2, 3). An asymmetric synclinal panel of synorogenic conglomerates (the Jackson Lake Formation) occurs along this fault zone. Similar relationships are observed in several other major Archean gold camps, most notably Timmins, Kirkland Lake, and Kalgoorlie. The critical control common to all these camps is localization of Au mineralization within significant bends of the regional fault zones; these bends were probably dilational during emplacement of the gold-bearing quartz veins, providing favoured pathways for focusing of upward fluid flow.

Although numerous other volcanic-hosted gold vein systems are known from the Slave craton, some of which saw brief production in the past, only one other major camp has emerged in recent years. This camp occurs in the Hope Bay belt, on the Coronation Gulf coast of the Slave craton, and consists of a string of deposits (Boston, Madrid, Doris; see Fig. 18) that are being readied for production. Elsewhere, overall potential for this class of deposits remains excellent, in spite of significant past exploration. Several greenstone belts throughout the Slave craton have very similar struc-

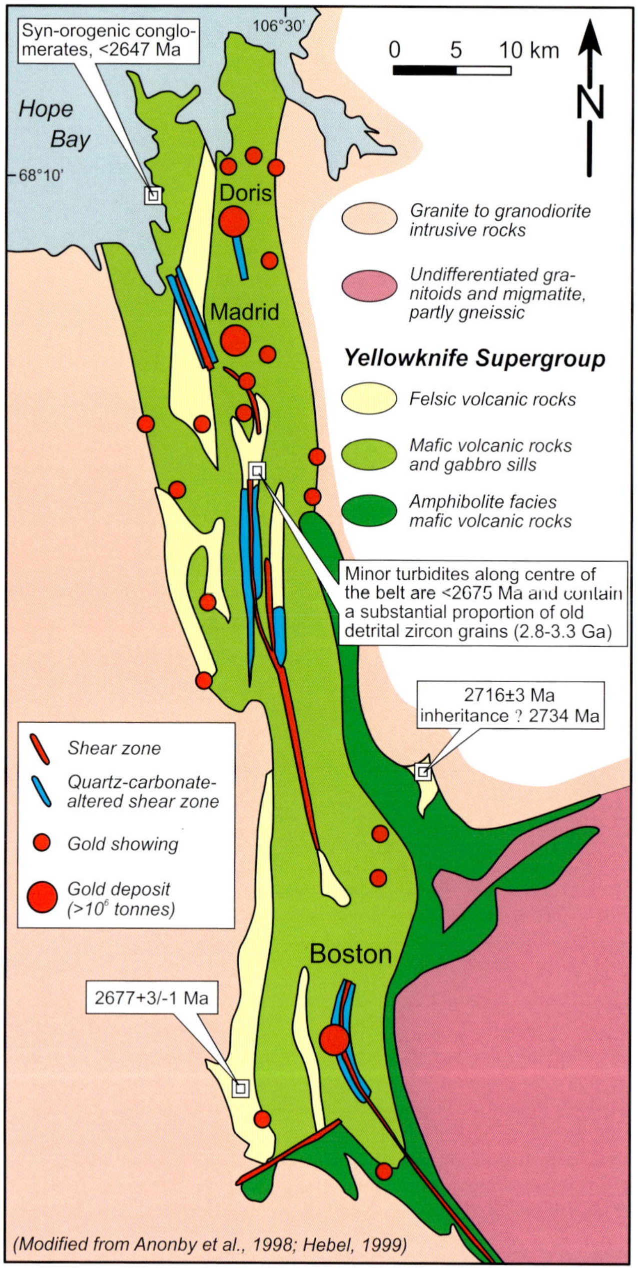

FIGURE 18. Simplified geology of the Hope Bay greenstone belt of the northeastern-most Slave craton, showing localities of gold mineralization (modified after Anonby et al., 1998; Hebel, 1999). The presence of a thick tholeiitic greenstone package with intercalated rhyolite, dated at 2716 ± 3 Ma, is reminiscent of the Yellowknife belt. The dated rhyolite contains inherited zircons that are 2734 Ma or older (Hebel, 1999), whereas minor turbidites along the centre of the belt contain a substantial fraction of Mesoarchean zircon grains. Collectively, these characteristics suggest an affinity with the Central Slave Basement Complex and its volcano-sedimentary cover (a rifted fragment thereof?), rather than a continuation of the juvenile greenstone belts of the eastern Slave craton.

tural-stratigraphic characteristics to that of the Yellowknife belt, including a thick, folded turbidite pile adjacent to a basaltic greenstone belt, a regional deformation zone, and a young, tightly infolded, synorogenic conglomerate package. The best examples are the Point Lake and Arcadia Bay areas. The Courageous Lake belt is equally interesting although it lacks a synorogenic conglomerate assemblage. This belt hosts the Slave craton's largest unexploited gold resource (Tundra-Fat deposit at 29.5 million tonnes containing 6.86 g/tonne gold).

Another class of mineral deposits related to final orogenesis is that of rare-element-enriched granitoids (Li, Be, Sn, Nb, Ta), in particular highly evolved anatectic granites and their late-stage pegmatites (e.g. Meintzer and Cerny, 1983; Meintzer et al., 1984; Meintzer and Wise, 1987). Tin (cassiterite) and Li (spodumene) were briefly mined from such pegmatites in the Yellowknife Domain, but several other pegmatite fields are known across the Slave craton (e.g. Tomascak and Cerny, 1992). From a global metallogenic point of view, these occurrences are of interest as they tend to be diagnostic for the presence of ancient, multiply recycled felsic crust.

Cratonization and Beyond

The youngest granite plutons of the Slave craton are ca. 2590 to 2580 Ma (James and Mortensen, 1992; van Breemen et al., 1992; Davis and Bleeker, 1999; Bennett et al., 2005; Ootes et al., 2005). Only some pegmatitic granites are known to be significantly younger, for instance a pegmatite in the Winter Lake belt dated by U-Pb monazite at 2550 ± 1 Ma (M. Villeneuve, pers. comm., 2006).

Between ca. 2595 and 2585 Ma, an enormous volume of granite was generated throughout much of the Slave craton. This "granite bloom", driven by moderate tectonic overthickening (D_1-D_2) and high intrinsic heat production, irreversibly transferred a significant fraction of heat-producing elements and lower crustal fluids (and Au) to the upper crust. In the lower crust, it must have involved large-scale migration of anatectic granitoid magmas, significant horizontal channel flow of partially molten rocks, development of horizontal layering, and flattening of the Moho discontinuity into a stable density configuration. Collectively, these processes allowed the lower crust to cool and stiffen over several tens of millions of years. U-Pb geochronology of lower crustal xenoliths shows that at depth high-grade metamorphic reactions and zircon growth continued to about 2510 Ma (Davis et al., 2003a). Finally, sufficient cooling allowed the crust to mechanically couple with the mantle (Bleeker, 2002). The end product was cratonic crust of high relative strength.

Following cratonization, there is an approximately 300 million year time span during which there are few recorded events within the Slave craton (Fig. 17B). Ca. 2.45 Ga magmatism, known from many other cratons around the world (e.g. Heaman, 1997), so far appears to be absent from the Slave craton.

At 2230 Ma, northeast-trending mafic dykes of the Malley swarm intruded the central Slave craton, providing the first evidence for Paleoproterozoic mantle-derived magmatism and attempted rifting events (LeCheminant and van Breemen, 1994). Subsequent to the Malley event, between 2200 and 2000 Ma, Slave crust was affected by as many as ten other mafic dyke swarms (see Ernst and Buchan, 2001, for a global data base, and 2004) and associated extension events before the craton finally broke out of the confines of its ancestral Sclavia supercraton. Details remain sketchy however. The eastern margin of the Slave craton, now involved in and overridden by the Thelon Orogen, almost certainly became established as a passive margin well before the western margin. The latter is referred to as the so-called Coronation margin and later became involved in Wopmay Orogen (e.g. Hoffman, 1980; Bowring and Grotzinger, 1992; Hildebrand and Bowring, 1999). Once liberated out of Sclavia, the Slave continental microplate likely experienced a drift phase as an independent craton, before being progressively incorporated into the growing Laurentian collage and the supercontinent of Nuna (Hoffman, 1988; Bleeker, 2003).

Paleoproterozoic amalgamation processes varied along the margins of the Slave craton. In the east, the Slave acted as a lower plate, being overridden by the west-vergent Thelon Orogen (Fig. 3). In the south, along the shores and islands of Great Slave Lake, deformation was mainly transpressional along an oblique suture that evolved into a major continental transform boundary (Hoffman, 1987; Hanmer et al., 1992). In the west, at least two arc terranes were involved (Hottah terrane and the younger Great Bear Magmatic Zone, Ghandi et al., 2001; Ghandi and van Breemen, 2005), followed by oblique folding and late-stage, dextral, strike-slip deformation along the Wopmay Fault Zone. Development of the Great Bear arc, between about 1880 and 1840 Ma, likely involved subduction of Paleoproterozoic lithosphere below the western Slave craton (e.g. Bostock, 1998; Heaman et al., 2002).

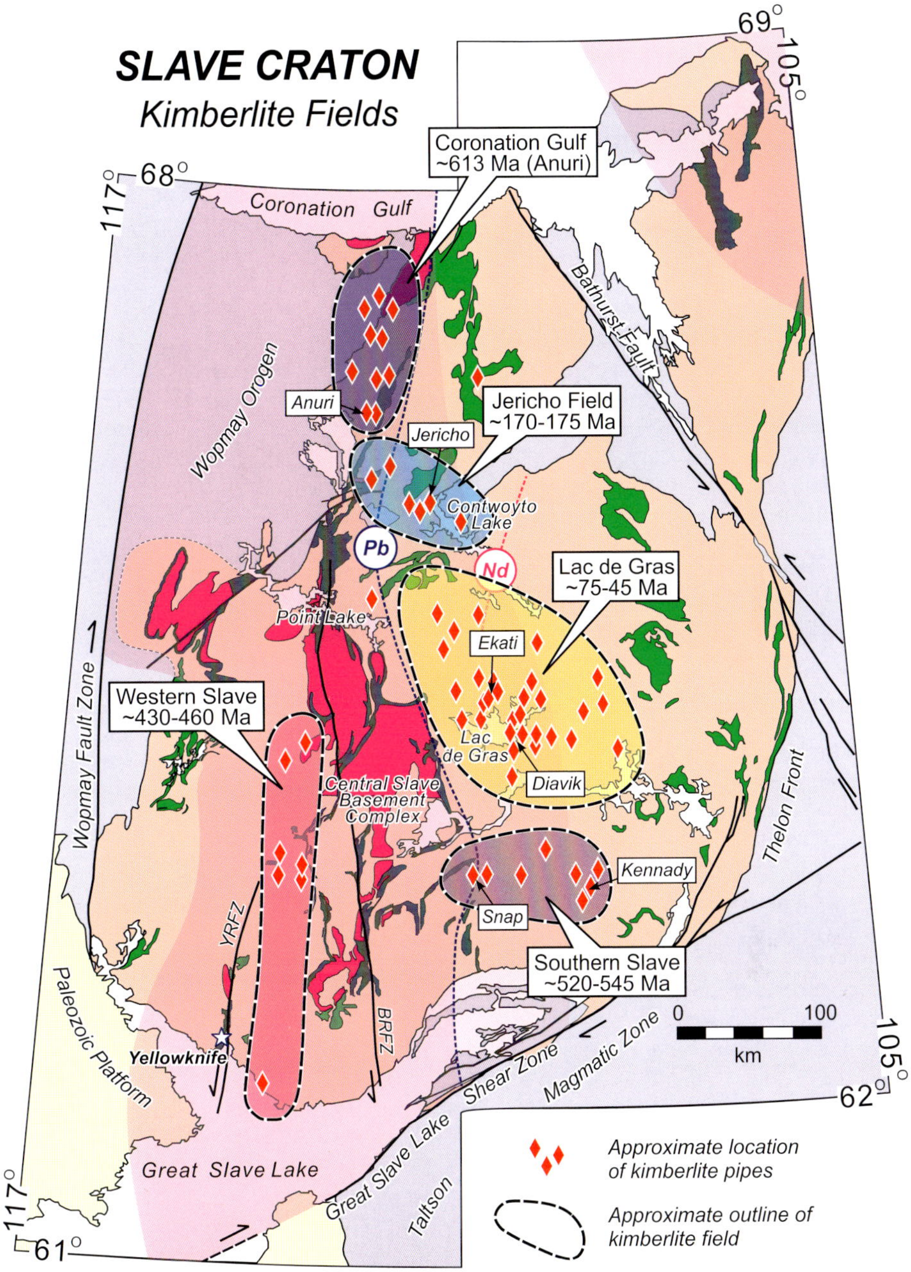

FIGURE 19. Kimberlite pipes and kimberlite fields of the Slave craton (Heaman et al., 2004; Lockhart et al., 2004). Nearly all kimberlites occur within the realm of the Central Slave Basement Complex or its buried eastern edge extending to the east of Lac de Gras (see Fig. 3). The Victoria Island Field, of Permian age (~256-286 Ma; off the map, towards the north), may also be associated with such basement as the Hope Bay block has many characteristics similar to that of the Yellowknife area. BRFZ = Beaulieu River Fault Zone, YRFZ = Yellowknife River Fault Zone.

Post-dating the assembly of Laurentia and Nuna, the Slave craton, particularly along its margins, became partially buried beneath intracontinental Proterozoic basins (e.g. Coppermine Homocline and Elu Basin, see Kerans et al., 1981; Hoffman and Hall, 1993). At ca. 1269 to 1267 Ma, the craton was partly uplifted and intruded by the giant Mackenzie dyke swarm, radiating from a plume centre west of Victoria Island (LeCheminant and Heaman, 1989, 1991; Baragar et al., 1996). Coeval flood basalts were erupted in basins now preserved as the Coppermine Homocline. This is the last major event affecting the core of the craton, although some younger mafic magmatic events did affect its edges (e.g. the 723 Ma Franklin event and the ca. 780 Ma Hottah sheets, Heaman et al., 1992; Harlan et al., 2003). Since that time, Slave crust has been "bobbing" gently up and down, with interior seas expanding and receding across the craton. Ordovician and Cretaceous-Tertiary sedimentary rocks and fossils are known as wall-rock fragments in some of the central Slave kimberlites (Nassichuk and McIntyre, 1995; Stasiuk and Nassichuk, 1995; Cookenboo et al., 1998; Stasiuk et al., 1999).

Despite the relative stability at the surface, melting events were triggered in the subcontinental mantle lithosphere and underlying asthenosphere, leaving their traces as clusters of kimberlites across the craton (Fig. 19). From the several hundreds of kimberlites now known across the craton, the following ages have been recorded: Cambrian, Siluro-Ordovician, Permian, Jurassic, Cretaceous, and finally Eocene (e.g. Heaman et al., 2003, 2004). The exact triggering events are still unclear but may relate to 1) mantle plumes or convective upwellings disturbing the stress and thermal state of the lithosphere (e.g. Sleep, 2003), or 2) rapidly changing far-field stress patterns within the North American plate due to events along its periphery, such as break-up and opening of the Iapetus Ocean, or changing subduction patterns along the Cordilleran margin (e.g. Snyder and Lockhart, 2005).

Mineralization

From the first attempted rifting events at ca. 2230 Ma (Malley dyke swarm) to final break-up of the supercraton Sclavia at 2.1 to 2.0 Ga, at least ten mafic and alkaline magmatic suites were emplaced within Slave crust, as dyke swarms and intrusive complexes. Several of these events are of interest for their known or potential mineral occurrences.

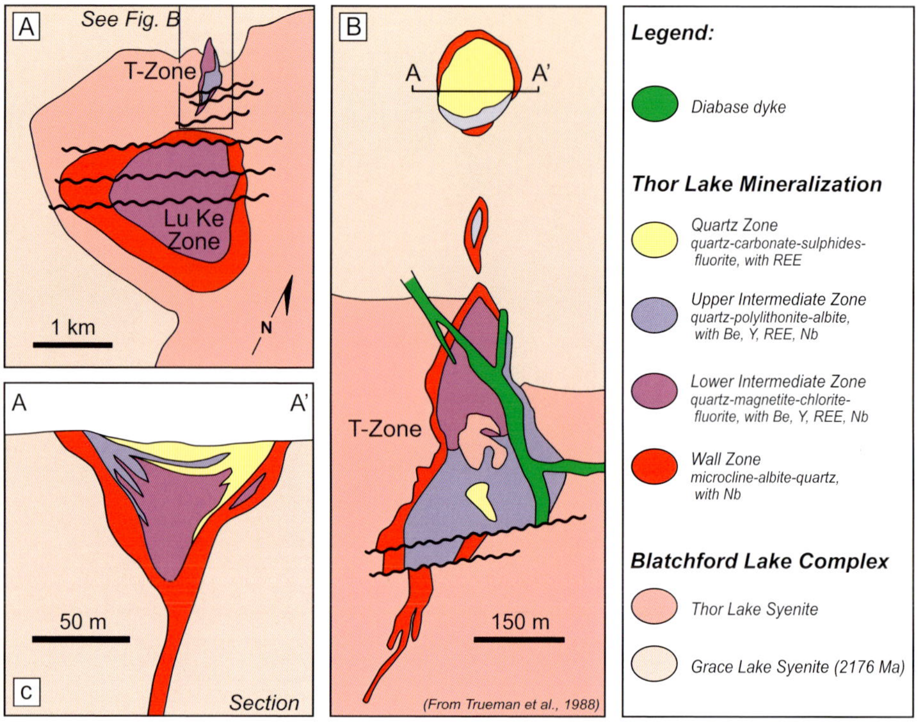

FIGURE 20. Geology of the Thor Lake rare element deposit (after Trueman et al., 1988). It occurs in the central syenite pluton of the Blatchford Lake Complex, an alkaline rift-related intrusive complex on the north shore of Great Slave Lake and a major component of the ca. 2180 Ma "Southwest Slave magmatic province" (see text).

A principal example is the ca. 2175 to 2185 Ma mafic and alkaline province within the southwestern part of the Slave craton (Fig. 4), comprising the Big Spruce (Cavell and Baadsgaard, 1986), Squalus (Stubley, 1996), and Blatchford (Davidson, 1978; Bowring et al., 1984; Sinclair et al., 1994) alkaline complexes, and the essentially contemporaneous Duck Lake mafic sill (2181 ± 2 Ma, Bleeker and Kamo, 2003) and Dogrib dyke swarm near Yellowknife (ca. 2180-2190 Ma, LeCheminant et al., 1997). All these events are proximal to either the southern or western margin of the Slave craton and are initial precursor events prior to final break-up[8].

The Blatchford Lake intrusive complex, on the north shore of Great Slave Lake, consists of a number of nested intrusions ranging from gabbro to granite and syenite. A central syenite body, the Thor Lake syenite, hosts rare metal mineralization (Be, Y, rare earths, Nb, Ta, Zr, and Ga) in hydrothermally altered breccias and veins (Fig. 20) (Trueman et al., 1988; see also Sinclair et al., 1994; Taylor and Pollard, 1996). Whereas the main granitic phases, and by inference the related syenite phases, have been dated at ca. 2176 Ma (Bowring et al., 1984; Sinclair et al., 1994), the age of the mineralization remains controversial and has been postulated to be significantly younger (ca. 2094 Ma, Bowring et al., 1984; and discussion in Sinclair et al., 1994). It seems counterintuitive, however, to dissociate in time (by ca. 80 million years) the high-temperature hydrothermal mineralization from the alkaline intrusions with which they are closely associated. In general terms, the Thor Lake deposits can be divided into a series of nested zones with a keel-shaped cross-section (Fig. 20): 1) a "Wall Zone", composed chiefly of quartz and feldspar, containing fluorite, sulphides, magnetite, carbonate, columbite, phenacite and several unidentified rare earth (REE)-bearing species; 2) a "Lower Intermediate Zone" that occurs predominately in the keels of the T-Zone deposits (Fig. 20), consisting of quartz, plagioclase, amphibole, biotite, muscovite, lepidolite, chlorite, carbonate, magnetite, with important accessory minerals consisting of fluorite, zircon, sulphides, phenacite, columbite and several REE minerals; 3) an "Upper Intermediate Zone" characterized by quartz, muscovite, albite, acmite and phenacite, with chlorite, fluorite, carbonates, magnetite, sulphides, and REE minerals as accessories; and 4) a vuggy "Quartz Zone" occupying the core of this complex (Trueman et al., 1988).

Another important example of rifting/extension related mafic magmatism is the large Booth River complex (Roscoe, 1985; Roscoe et al., 1987; Hulbert, 2005b) of the north-central Slave craton, outcropping from underneath the western extent of the Kilihigok Basin strata (see Fig. 4 for location). This complex has been dated at 2025 ± 2 Ma (Roscoe et al., 1987; Davis et al., 2004) and appears a magmatic focal point along the more extensive and slightly southward-fanning Lac de Gras dyke swarm, which is of similar age. The large, differentiated Booth River complex has attracted attention as a potential host for platinum group element (PGE) mineralization.

The much younger Muskox Intrusion (1267 Ma), on the northwestern margin of the craton and related to the large Mackenzie event (LeCheminant and Heaman, 1989), is

8. Based on these approximately coeval magmatic events, we tentatively define a "southwest Slave magmatic province", as shown in Figure 4.

another funnel-shaped layered intrusion that has attracted attention for reef-style PGE mineralization (Hulbert, 2005a).

The emplacement of diamondiferous kimberlite pipes represents the final metallogenic event for the Slave craton, with kimberlite ages spanning several discrete pulses throughout the Phanerozoic (Heaman et al., 2004). At Lac de Gras, in the central part of the craton, kimberlite pipes of Eocene age (ca. 55-50 Ma) form a dense cluster and now support two highly profitable diamond mines, Ekati and Diavik. Several other diamond mines are in various stages of development (e.g. Jericho, Kennady Lake). In just over a decade, diamonds have become the most profitable commodity within this ancient craton.

Summary

The Slave craton is a relatively small Archean craton with a geological knowledge base that is relatively mature. However, the following major questions remain:

1. What is the nature of the Hope Bay block in the northeast part of the craton? Does cryptic ancient basement reappear in this part of the craton? Is it perhaps a rifted fragment of the Central Slave Basement Complex?
2. What is the tectonic significance of pre-2687 Ma volcanic rocks in the eastern Slave? Do they form remnants of a ca. 2.7 Ga, exotic, intra-oceanic juvenile arc that collided with the extended Central Slave Basement Complex between 2697 and 2687 Ma? If so, where is the suture? Or do these volcanic rocks represent the oldest fill of narrow back-arc-like troughs formed by progressive rifting of the Central Slave Basement Complex in an overall arc setting?
3. What are the detailed trend and extent of particular magmatic (e.g. Defeat and Concession suites) and sedimentary belts (e.g. the post-Defeat turbidite basin along the western margin of the Slave)?
4. How did events inferred from the crustal geological record contribute to, or interfere with, formation of the subcontinental mantle lithosphere below the Slave craton?
5. How far does Slave mantle lithosphere extend below the Rae craton to the east?
6. What is the detailed break-up history for each of the Proterozoic margins of the Slave craton? In other words, how was the Slave crustal fragment liberated out of its ancestral supercraton Sclavia?
7. What is the detailed depositional record associated with rifting, thermal subsidence, and finally collision, along each of the margins of the Slave craton?
8. And, does Slave lithosphere extend northward all the way to the Innuitian front in the high Arctic (see Fig. 1)?

To many of these first-order questions, we currently have only rudimentary answers or mere guesses. More sophisticated answers will require more complete and more refined data sets. In particular, a greatly expanded geochronological database, both in quantity and precision, in conjunction with targeted fieldwork across the craton and its marginal belts, should quickly advance the state of knowledge.

In terms of mineral potential, much of the craton and all major greenstone belts have seen at least a first wave of exploration for major commodities. These investigations quickly led to the discovery of a number of large VMS deposits (e.g. Izok Lake), which await road access for economic production. Gold potential remains high, particularly in more remote greenstone belts that may not have seen adequate drill testing. In this respect, the Point Lake greenstone belt, and its extension further north, appears attractive as it has all the major attributes of a world-class gold camp.

Acknowledgements

We thank colleagues at the Geological Survey of Canada, particularly Bill Davis, John Kerswill, Dave Snyder, Sally Pehrsson, and Ken Buchan, for sharing their insights into the geology and metallogeny of the Slave craton. Kim Nguyen assisted with preparation of some of the figures. Thoughtful reviews by John Ketchum, Tim Kusky, and Jan Peter helped improve the manuscript.

References

Agashev, A.M., Pokhilenko, J.A., McDonald, J.A., Takazawa, E., Vavilov, M.A., Sobolev, N.V., and Watanabe, T., 2001, A unique kimberlite-carbonatite primary association in the Snap Lake dyke system, Slave Craton: Evidence from geochemical and isotopic studies: Slave-Kaapvaal Workshop, Merrickville, Ontario, Canada, September 5-9, Abstract.

Anonby, L., MacMaster, G., Kienlen, B., Ottenhof, P., Gebert, J., Doulos, D., McQuillan, F., and Bellwood, C., 1998, 1996/97 Geology, Geochemistry + Geophysics Report on the TOK 1-6 Concessions: BHP Minerals Canada Ltd. Company Report.

Aulbach, S., Griffin, W.L., Pearson, N.J., O'Reilly, S.Y., Kivi, K., and Doyle, B. J., 2004, Mantle formation and evolution, Slave Craton: constraints from HSE abundances and Re-Os isotope systematics of sulfide inclusions in mantle xenocrysts: Chemical Geology, v. 208, p. 61-88.

Baragar, W.R.A., and McGlynn, J.C., 1976, Early Archean basement in the Canadian Shield: A review of the evidence: Geological Survey of Canada, Paper 76-14.

Baragar, W.R. A., Ernst, R.E., Hulbert, L., and Peterson, T., 1996, Longitudinal petrochemical variation in the Mackenzie dyke swarm, northwestern Canadian Shield: Journal of Petrology, v. 37, p. 317-359.

Barrie, C.T., and Hannington, M.D., 1999, Classification of volcanic-associated massive sulfide deposits, *in* Barrie, C.T., and Hannington, M.D., eds., Volcanic-Associated Massive Sulfide Deposits: Processes and Examples in Modern and Ancient Settings: Reviews in Economic Geology, Society of Economic Geologists, v. 8, p. 2-10.

Beales, P., and Woodward, R., 2001, A Guide to Mineral Deposits of the Northwest Territories: The Department of Resources, Wildlife, and Economic Development, Government of the Northwest Territories, 182 p.

Bennett, V., 2006, A Multidisciplinary Investigation of the Formation, Growth and Evolution of Neoarchean Crust, Snare River Terrane: Southwestern Slave Province: Unpub. Ph.D. thesis, Memorial University of Newfoundland, St. John's, 510 p.

Bennett, V., Jackson, V.A., Rivers, T., Relf, C., Horan, P., and Tubrett, M., 2005, Geology and U-Pb geochronology of the Neoarchean Snare River terrane: Tracking evolving tectonic regimes and crustal growth mechanisms: Canadian Journal of Earth Sciences, v. 42, p. 895-934.

Bethune, K.M., Villeneuve, M.E., and Bleeker, W., 1999, Laser 40Ar/39Ar thermochronology of Archean rocks in Yellowknife Domain, southwestern Slave Province: insights into the cooling history of an Archean granite-greenstone terrane: Canadian Journal of Earth Sciences, v. 36, p. 1189-1206.

Bleeker, W., 1996, Thematic Structural Studies in the Slave Province, Northwest Territories: the Sleepy Dragon Complex: Geological Survey of Canada, Current Research 1996-C, 10 p.

——— 2001, The ca. 2680 Ma Raquette Lake Formation and Correlative Units Across the Slave Province, Northwest Territories: Evidence for a Craton-Scale Overlap Sequence: Geological Survey of Canada, Current Research, Paper 2001-C7, 26 p.

——— 2002, Archean tectonics: A review, with illustrations from the Slave craton, *in* Fowler, C.M.R., Ebinger, C. J., and Hawkesworth, C.J., eds.,

The Early Earth: Physical, Chemical and Biological Development: Geological Society, Special Publication No. 199, p. 151-181.

——— 2003, The late Archean record: A puzzle in ca. 35 pieces: Lithos, v. 71, p. 99-134.

——— 2004, Taking the pulse of planet Earth: A proposal for a new multi-disciplinary flagship project in Canadian solid earth sciences: Geoscience Canada, v. 31, p. 179-190.

Bleeker, W., and Beaumont-Smith, C., 1995, Thematic Structural Studies in the Slave Province: Preliminary Results and Implications for the Yellowknife Domain, Northwest Territories: Geological Survey of Canada, Current Research, Volume 1995-C, 10 p.

Bleeker, W., and Davis, W.J., 1999a, The 1991-1996 NATMAP Slave Province Project: Canadian Journal of Earth Sciences, v. 36, p. 1033-1238.

——— 1999b, The 1991-1996 NATMAP Slave Province Project: Introduction: Canadian Journal of Earth Sciences, v. 36, p. 1033-1042.

Bleeker, W., and Ernst, R., 2006, Short-lived mantle generated magmatic events and their dyke swarms: The key unlocking Earth's palaeogeographic record back to 2.6 Ga, *in* Hanski, E., Mertanen, S., Rämö, T., and Vuollo, J., eds., Dyke Swarms — Time Markers of Crustal Evolution: Balkema Publishers, Rotterdam, in press.

Bleeker, W., and Kamo, S., 2003, A precise age for the Duck Lake sill and its relevance for fitting the Slave in a global Archean context: 31st Yellowknife Geoscience Forum, November 19-21, Program & Abstracts of Talks & Posters, p. 7 8.

Bleeker, W., and Villeneuve, M., 1995, Structural studies along the Slave portion of the SNORCLE Transect, *in* Cook, F., and Erdmer, P.C., eds., Slave-NORthern Cordillera Lithospheric Evolution (SNORCLE), Report of 1995 Transect Meeting, April 8-9, University of Calgary, LITHOPROBE Report No. 44, p. 8-14.

Bleeker, W., Villeneuve, M., and Bethune, K., 1997, Thematic Structural Studies in the Slave Province, Northwest Territories: Contrasting Basement/Cover Relationships on Western and Southwestern Flanks of the Sleepy Dragon Complex: Geological Survey of Canada, Current Research 1997-C, 11 p.

Bleeker, W., Ketchum, J.W.F., and Davis, W.J., 1999a, The Central Slave Basement Complex, Part II: Age and tectonic significance of high-strain zones along the basement-cover contact: Canadian Journal of Earth Sciences, v. 36, p. 1111-1130.

Bleeker, W., Ketchum, J.W.F., Jackson, V.A., and Villeneuve, M.E., 1999b, The Central Slave Basement Complex, Part I: Its structural topology and autochthonous cover: Canadian Journal of Earth Sciences, v. 36, p. 1083-1109.

Bleeker, W., Stern, R., and Sircombe, K., 2000, Why the Slave Province, Northwest Territories, Got a Little Bigger: Geological Survey of Canada, Current Research 2000-C2, 9 p.

Bostock, M.G., 1998, Mantle stratigraphy and evolution of the Slave Province: Journal of Geophysical Research, v. 103, p. 21,183-21,200.

Bowring, S.A., and Grotzinger, J.P., 1992, Implications of new chronostratigraphy for tectonic evolution of Wopmay Orogen, northwest Canadian Shield: American Journal of Science, v. 292, p. 1-20.

Bowring, S.A., and Williams, I.S., 1999, Priscoan (4.00-4.03 Ga) orthogneisses from northwestern Canada: Contributions to Mineralogy and Petrology, v. 134, p. 3-16.

Bowring, S.A., Van Schmus, W.R., and Hoffman, P.F., 1984, U-Pb zircon ages from Athapuscow Aulacogen, East Arm of Great Slave Lake, N.W.T., Canada: Canadian Journal of Earth Sciences, v. 21, p. 1315-1324.

Bowring, S.A., Williams, I.S., and Compston, W., 1989, 3.96 Ga gneisses from the Slave Province, Northwest Territories, Canada: Geology, v. 17, p. 971-975.

Boyle, R. W., 1961, The Geology, Geochemistry and Origin of the Gold Deposits of the Yellowknife District: Geological Survey of Canada, Memoir 310, 193 p.

——— 1986, Gold deposits in turbidite sequences: their geology, geochemistry and history of the theories of their origin, *in* Keppie, J.D., Boyle, R.W., and Haynes, S.J., eds., Turbidite-Hosted Gold Deposits: Geological Association of Canada, Special Paper 32, p. 1-14.

Brophy, J.A., 1987, Tom and Ptarmigan mines, *in* Padgham, W.A., ed., Field Guide, Yellowknife Mining District: Geological Association of Canada, Mineral Deposit Division, Yellowknife, p. 195-198.

Bullis, H.R., Hureau, R.A., and Penner, B.D., 1991a, Distribution of gold and sulfides at Lupin, Northwest Territories: Economic Geology, v. 89, p. 1217-1227.

——— 1991b, Controls of gold and sulphide distribution at Lupin, N.W.T., Preliminary Reports on Geological Work in the NWT, Abstracts of Talks and Posters and Summaries of Work in Progress: Department of Indian Affairs and Northern Development, NWT Geology Division, Yellowknife, NWT, p. 18-19.

Carbno, G.B., and Canil, D., 2002, Mantle structure beneath the SW Slave craton, Canada: constraints from garnet geochemistry in the Drybones Bay kimberlite: Journal of Petrology, v. 43, p. 129-142.

Cavell, P.A., and Baadsgaard, H., 1986, Geochronology of the Big Spruce Lake alkaline intrusion: Canadian Journal of Earth Sciences, v. 23, p. 1-10.

Cerny, P., 1989, Exploration strategy and methods for pegmatite deposits of tantalum, *in* Cerny, P., and Saupe, P., eds., Lanthanides, Tantalum and Niobium: Springer Verlag, New York, p. 274-302.

Coffin, M.F., and Eldholm, O., 1994, Large igneous provinces: crustal structure, dimensions, and external consequences: Reviews of Geophysics, v. 32, p. 1-36.

——— 2001, Large igneous provinces: progenitors of some ophiolites? *in* Ernst, R. E., and Buchan, K. L., eds., Mantle plumes: Their Identification Through Time: Geological Society of America, Special Paper 352, p. 59-70.

Cook, F.A., van der Velden, A.J., Hall, K.W., and Roberts, B.J., 1999, Frozen subduction in Canada's Northwest Territories: Lithoprobe deep lithospheric reflection profiling of the western Canadian Shield: Tectonics, v. 18, p. 1-24.

Cookenboo, H.O., Orchard, M.J., and Daoud, D.K., 1998, Remnants of Paleozoic cover on the Archean Canadian Shield: Limestone xenoliths from kimberlite in the central Slave craton: Geology, v. 26, p. 391-394.

Corcoran, P.L., Mueller, W.U., and Padgham, W.A., 1999, Influence of tectonism and climate on lithofacies distribution and sandstone and conglomerate composition in the Archean Beaulieu Rapids Formation, Northwest Territories, Canada: Precambrian Research, v. 94, p. 175-204.

Cousens, B.L., 2000, Geochemistry of the Archean Kam Group, Yellowknife greenstone belt, Slave Province, Canada: Journal of Geology, v. 108, p. 181-197.

Cousens, B., Facey, K., and Falck, H., 2002, Geochemistry of the late Archean Banting Group, Yellowknife greenstone belt, Slave Province, Canada: simultaneous melting of upper mantle and juvenile mafic crust: Canadian Journal of Earth Sciences, v. 39, p. 1635-1656.

Covello, L., Roscoe, S.M., Donaldson, J.A., Roach, D., and Fyson, W.K., 1988, Archean Quartz Arenite and Ultramafic Rocks at Beniah Lake, Slave Structural Province, N.W.T.: Geological Survey of Canada, Current Research, Part C, Paper 88-1C, 10 p.

Creaser, R.A., Grütter, H., Carlson, J.A., and Crawford, B.B., 2004, Macrocrystal phologopite Rb-Sr dates from the Ekati property kimberlites, Slave Province, Canada: Evidence for multiple intrusive episodes in the Paleocene and Eocene: Lithos, v. 76, p. 399-414.

Davidson, A., 1978, The Blatchford Lake Intrusive Suite—An Aphebian Plutonic Complex in the Slave Province, Northwest Territories: Geological Survey of Canada, Paper 78-1A, 9 p.

Davis, W.J., and Bleeker, W., 1999, Timing of plutonism, deformation, and metamorphism in the Yellowknife Domain, Slave Province, Canada: Canadian Journal of Earth Sciences, v. 36, p. 1169-1187.

Davis, W.J., and Hegner, E., 1992, Neodymium isotopic evidence for the tectonic assembly of late Archean crust in the Slave Province, Northwest Canada: Contributions to Mineralogy and Petrology, v. 111, p. 493-504.

Davis, W.J., Fryers, B.J., and King, J.E., 1994, Geochemistry and evolution of late Archean plutonism and its significance to the tectonic development of the Slave craton: Precambrian Research, v. 67, p. 207-241.

Davis, W.J., Canil, D., MacKenzie, J.M., and Carbno, G.G., 2003a, Petrology and U-Pb geochronology of lower crustal xenoliths and the development of a craton, Slave Province, Canada: Lithos, v. 71, p. 541-573.

Davis, W.J., Jones, A.G., Bleeker, W., and Grütter, H., 2003b, Lithosphere development in the Slave craton: A linked crustal and mantle perspective: Lithos, v. 71, p. 575-589.

Davis, W.J., Bleeker, W., Hulbert, L., and Jackson, V., 2004, New geochronological results from the Slave Province Minerals and Geoscience Compilation and Synthesis Project, GSC Northern Resources Program: Yellowknife Geoscience Forum, Nov. 2004.

Dudás, F.O., Henderson, J.B., and Mortensen, J.K., 1990, U-Pb ages of zircons from the Anton Complex, southern Slave Province, Northwest Territories, Radiogenic Age and Isotopic Studies, Report 3: Geological Survey of Canada, Paper 89-2, p. 39-44.

Energy Mines and Resources Canada, 1989, 'NWT 48', *in* Canadian Mineral Deposits Not Being Mined in 1989:National Mineral Inventory, Mineral Policy Sector, Ottawa, Canada, Mineral Bulletin MR 223.

Ernst, R.E., and Buchan, K.L., 2001, Large mafic magmatic events through time and links to mantle-plume heads, *in* Ernst, R.E., and Buchan, K.L., eds., Mantle Plumes: Their Identification Through Time: Geological Society of America, Special Paper 352, p. 483-575.

——— 2004, Large Igneous Provinces (LIPs) in Canada and Adjacent Regions: 3 Ga to Present: Geoscience Canada, v. 31, p. 103-126.

Ernst, R.E., Buchan, K.L., and Campbell, I.H., 2005, Frontiers in Large Igneous Province research: Lithos, v. 79, p. 271-297.

Ferguson, M.E., Waldron, J.W.F., and Bleeker, W., 2005, The Archean deep-marine environment: turbidite architecture of the Burwash Formation, Slave Province, Northwest Territories: Canadian Journal of Earth Sciences, v. 42, p. 935-954.

Frith, R.A., and Fryer, B.J., 1985, Geochemistry and origin of the Regan Intrusive Suite and other granitoids in the northeastern Slave Province, northwest Canadian Shield: Canadian Journal of Earth Sciences, v. 22, p. 1048-1065.

Geusebroek, P.A., and Duke, N.A., 2004, An update on the geology of the Lupin Gold Mine, Nunavut, Canada: Exploration and Mining Geology, v. 13, p. 1-13.

Ghandi, S.S., and van Breemen, O., 2005, SHRIMP U-Pb geochronology of detrital zircons from the Treasure Lake Group: New evidence for Paleoproterozoic collisional tectonics in the southern Hottah Terrane, northwestern Canadian Shield: Canadian Journal of Earth Sciences, v. 42, p. 833-845.

Ghandi, S.S., Mortensen, J.K., Prasad, N., and van Breemen, O., 2001, Magmatic evolution of the southern Great Bear continental arc, northwestern Canadian Shield: geochronological constraints: Canadian Journal of Earth Sciences, v. 38, p. 767-785.

Gosselin, P., and Dubé, B., 2005, Gold Deposits of Canada: Distribution, Geological Parameters and Gold Content: Geological Survey of Canada, Open File Report 4896, 1 CD-ROM.

Graham, I., Burgess, J.L., Bryan, D., Ravenscroft, P. J., Thomas, E., Doyle, B.J., Hopkins, R., and Armstrong, K.A., 1999, Exploration history and geology of the Diavik kimberlites, Lac de Gras, Northwest Territories, Canada, *in* Gurney, J J., Gurney, J.L., Pascoe, M.D., and Richardson, S.H., eds., The J.B. Dawson Volume, Proceedings of the VIIth International Kimberlite Conference, Cape Town, p. 262-279.

Griffin, W.L., Doyle, B.J., Ryan, C.G., Pearson, N.J., O'Reilly, S., Davies, R., Kivi, K., van Achterberg, E., and Natapov, L., 1999, Layered mantle lithosphere in the Lac de Gras area, Slave craton: Composition, structure and origin: Journal of Petrology, v. 40, p. 705-727.

Grütter, H.S., Apter, D.B., and Kong, J., 1999, Crust-mantle coupling: evidence from mantle-derived xenocrystic garnets, *in* Gurney, J.J., Gurney, J.L., Pascoe, M.D., and Richardson, S.H., eds., The J.B. Dawson Volume: Proceedings of the VIIth International Kimberlite Conference, v. 1, p. 307-313.

Hanmer, S., Bowring, S., van Breemen, O., and Parrish, R., 1992, Great Slave Lake shear zone, NW Canada: Mylonitic record of Early Proterozoic continental convergence, collision and indentation: Journal of Structural Geology, v. 14, p. 757-773.

Harlan, S.S., Heaman, L., LeCheminant, A.N., and Premo, W.R., 2003, Gunbarrel mafic magmatic event: A key 780 Ma time marker for Rodinia plate reconstructions: Geology, v. 31, p. 1053-1056.

Heaman, L.M., 1997, Global mafic magmatism at 2.45 Ga: Remnants of an ancient large igneous province? Geology, v. 25, p. 299-302.

Heaman, L.M., LeCheminant, A.N., and Rainbird, R.H., 1992, Nature and timing of Franklin igneous events, Canada: Implications for a Late Proterozoic mantle plume and the break-up of Laurentia: Earth and Planetary Science Letters, v. 109, p. 117-131.

Heaman, L.M., Creaser, R.A., and Cookenboo, H.O., 2002, Extreme high-field-strength element enrichment in Jericho eclogite xenoliths: A cryptic record of Paleoproterozoic subduction, partial melting and metasomatism beneath the Slave craton, Canada: Geology, v. 30, p. 507-510.

Heaman, L.M., Kjarsgaard, B.A., and Creaser, R.A., 2003, The timing of kimberlite magmatism in North America: Implications for global kimberlite genesis and diamond exploration: Lithos, v. 71, p. 153-184.

——— 2004, The temporal evolution of North American kimberlites: Lithos, v. 76, p. 377-397.

Hebel, M.U., 1999, U-Pb Geochronology and Lithogeochemistry of the Hope Bay Greenstone Belt, Slave Structural Province, Northwest Territories, Canada: Unpub. M.Sc. thesis, The University of British Columbia, Vancouver, British Columbia, 96 p.

Helmstaedt, H., and Padgham, W.A., 1986, A new look at the stratigraphy of the Yellowknife Supergroup at Yellowknife, N.W.T.: Implications for the age of gold-bearing shear zones and Archean basin evolution: Canadian Journal of Earth Sciences, v. 23, p. 454-475.

Henderson, J.B., 1970, Stratigraphy of the Yellowknife Supergroup, Yellowknife Bay – Prosperous Lake Area, District of Mackenzie: Geological Survey of Canada, Paper 70-26, 12 p.

——— 1972, Sedimentology of Archean turbidites at Yellowknife, Northwest Territories: Canadian Journal of Earth Sciences, v. 9, p. 882-902.

——— 1981, Archean basin evolution in the Slave Province, Canada, *in* Kröner, A., ed., Plate Tectonics in the Precambrian. Developments in Precambrian Geology 4: Elsevier, Amsterdam, p. 213-235.

——— 1985, Geology of the Yellowknife-Hearne Lake Area, District of Mackenzie: A Segment Across an Archean Basin: Geological Survey of Canada, Memoir 414, 135 p.

Henderson, J.F., and Brown, I.C., 1966, Geology and Structure of the Yellowknife Greenstone Belt, District of Mackenzie: Geological Survey of Canada, Bulletin 141, 87 p.

Henderson, J.R., Henderson, M.N., Kerswill, J.A., and Dehls, J.F., 2000, Geology, High Lake Greenstone Belt, Nunavut: Geological Survey of Canada, "A" Series Map, 1945A, scale 1:100 000.

Hildebrand, R.S., and Bowring, S.A., 1999, Crustal recycling by slab failure: Geology, v. 27, p. 11-14.

Hoffman, P.F., 1980, Wopmay Orogen: A Wilson Cycle of Early Proterozoic age in the northwest of the Canadian Shield, *in* Strangway, D.W., ed., The Continental Crust and Its Mineral Deposits: Geological Association of Canada, Special Paper 20, p. 523-549.

——— 1987, Continental transform tectonics: Great Slave Lake shear zone (ca. 1.9 Ga), northwest Canada: Geology, v. 15, p. 785-788.

——— 1988, United plates of America, the birth of a craton: Early Proterozoic assembly and growth of Laurentia: Annual Reviews of Earth and Planetary Sciences, v. 16, p. 543-603.

——— 1989, Precambrian geology and tectonic history of North America, *in* Bally, A.W., and Palmer, A.R., eds., The Geology of North America - An Overview: Geological Society of America, Boulder, Colorado, p. 447-512.

Hoffman, P.F., and Hall, L., 1993, Geology, Slave craton and environs, District of Mackenzie, Northwest Territories: Geological Survey of Canada, Open File 2559, scale 1:1,000,000.

Hrabi, R.B., and Grant, J.W., 1999, Geology, Winter Lake Supracrustal Belt, Northwest Territories: Geological Survey of Canada, Open File 3676, 1:50 000.

Hrabi, R.B., Nelson, M.D., and Helmstaedt, H., 1995, Diverse Metavolcanic Sequences and Late Polymictic Conglomerate-Associated Metasedimentary Rocks in the Winter Lake Supracrustal Belt, Slave Province, Northwest Territories: Geological Survey of Canada, Current Research 1995-E, 12 p.

Hulbert, L.J., 2005a, Geology of the Muskox Intrusion and Associated Ni+Cu Occurrences: Geological Survey of Canada, Open File 4881, 1 CD-ROM.

——— 2005b, Geology of the Kilohigok Basin & Subjacent Booth River Igneous Complex and Mara River Igneous Sheets: Geological Survey of Canada, Open File 4880, 1 CD-ROM.

Iizuka, T., Horie, K., Komiya, T., Maruyama, S., Hirata, T., Hidaka, H., and Windley, B.F., 2006, 4.2 Ga zircon xenocryst in an Acasta gneiss from northwestern Canada: Evidence for early continental crust: Geology, v. 34, p. 245-248.

Isachsen, C.E., 1992, U-Pb Zircon Geochronology of the Yellowknife Volcanic Belt and Subjacent Rocks, N.W.T., Canada: Constraints on the Timing, Duration, and Mechanics of Greenstone Belt Formation: Unpub. Ph.D. thesis, Washington University, St. Louis, Missouri, 164 p.

Isachsen, C.E., and Bowring, S.A., 1994, Evolution of the Slave Craton: Geology, v. 22, p. 917-920.

——— 1997, The Bell Lake Group and Anton Complex. a basement-cover sequence beneath the Archean Yellowknife greenstone belt revealed and implicated in greenstone belt formation: Canadian Journal of Earth Sciences, v. 34, p. 169-189.

Isachsen, C.E., Bowring, S.A., and Padgham, W.A., 1991, U Pb geochronology of the Yellowknife volcanic belt, NWT, Canada: New constraints on the timing and duration of greenstone belt magmatism: Journal of Geology, v. 99, p. 55-67.

James, D.T., and Mortensen, J.K., 1992, An Archean metamorphic core complex in the southern Slave Province: Basement-cover structural relationships between the Sleepy Dragon Complex and the Yellowknife Supergroup: Canadian Journal of Earth Sciences, v. 29, p. 2133-2145.

Jefferson, C.W., Beaumon-Smith, C.J., and Lustwerk, R.L., 1989, Stratigraphy and Structural Setting of Iron-Formations and Gold in the Back River Area, District of Mackenzie, N.W.T.: Geological Survey of Canada, Current Research, Part C, 12 p.

Jefferson, C.W., Dufresne, M., Olson, R.A., and Rice, R., 1992, Stratigraphy and sedimentology of auriferous Archean iron formations in the vicinity of George Lake, eastern Slave Province, *in* Richardson, D.G., and Irving, M., eds., Project Summaries, Canada-Northwest Territories Mineral Development Subsidiary, Agreement 1987-1991: Geological Survey of Canada, Open File Report 2484, p. 199-203.

Jenner, G.A., Fryer, B.J., and McLennan, S.M., 1981, Geochemistry of the Archean Yellowknife Supergroup: Geochimica et Cosmochimica Acta, v. 45, p. 1111-1129.

Johnson, W., and Robinson, P., 1975, Geological Survey TB-CAD Group, Long Lac Mineral Exploration Limited: DIAND (Department of Indian and Northern Affairs), Mineral Assessment Report 080184.

Kasting, J.F., 1993, Earth's early atmosphere: Science, v. 259, p. 920-926.

Kerans, C., Ross, G.M., Donaldson, J.A., and Geldsetzer, H.J., 1981, Tectonism and depositional history of the Helikian Hornby Bay and Dismal Lakes Groups, District of Mackenzie, *in* Campbell, F.H.A., ed., Proterozoic Basins of Canada: Geological Survey of Canada, Paper 81-10, p. 157-182.

Kerswill, J.A., 1993, Models for iron-formation-hosted gold deposits, *in* Kirkham, R.V., Sinclair, W.D., Thorpe, R.I., and Duke, J.M., eds., Mineral Deposit Modeling: Geological Association of Canada, Special Paper 40, p. 171-200.

Ketchum, J.W.F., and Bleeker, W., 2000, New field and U-Pb data from the Central Slave Cover Group near Yellowknife and the Central Slave Basement Complex at Point Lake, *in* Cook, F., and Erdmer, P., eds., Slave-Northern Cordillera Lithospheric Experiment (SNORCLE) Transect Meeting, University of Calgary, Lithoprobe Report No. 72, p. 27-31.

——— 2001, Evolution of the Central Slave Basement Complex, Slave Craton, Canada: U-Pb constraints: 11th Annual V.M. Goldschmidt Conference, LPI Contribution, v. 1088, Abstract #3148.

Ketchum, J.W.F., Bleeker, W., and Stern, R.A., 2004, Evolution of an Archean basement complex and its autochthonous cover, southern Slave Province, Canada: Precambrian Research, v. 135, p. 149-176.

King, J., and Helmstaedt, H., 1997, The Slave Province, Northwest Territories, Canada, *in* de Wit, M.J., and Ahswal, L.D., eds., Greenstone Belts: Oxford Monograph on Geology and Geophysics, v. 35, p. 459-479.

Kopylova, M.G., and Russell, J.K., 2000, Chemical stratification of cratonic lithosphere: Constraints from the northern Slave craton, Canada: Earth and Planetary Science Letters, v. 181, p. 71-87.

Kusky, T.M., 1989, Accretion of the Archean Slave Province: Geology, v. 17, p. 63-67.

——— 1990, Evidence for Archean ocean opening and closing in the southern Slave Province: Tectonics, v. 9, p. 1533-1563.

——— 1991, Structural development of an Archean Orogen, western Point Lake, Northwest Territories: Tectonics, v. 10, p. 820-841.

——— 1993, Collapse of Archean orogens and the generation of late- to postkinematic granitoids: Geology, v. 21, p. 925-928.

Lambert, M.B., and van Staal, C.R., 1987, Archean Granite-Greenstone Boundary Relationships in the Beaulieu River Volcanic Belt, Slave Province, N.W.T.: Geological Survey of Canada, Current Research, Part A, Paper 87-1A, 14 p.

LeCheminant, A.N., and Heaman, L.M., 1989, Mackenzie igneous events, Canada: middle Proterozoic hotspot magmatism associated with ocean opening: Earth and Planetary Science Letters, v. 96, p. 38-48.

——— 1991, U-Pb ages for the 1.27 Ga Mackenzie igneous events, Canada: Support for a plume initiation model: Geological Association of Canada, Program with Abstracts, v. 16, p. A-73.

LeCheminant, A.N., and van Breemen, O., 1994, U-Pb ages of Proterozoic dyke swarms, Lac de Gras area, N.W.T.: Evidence for progressive break-up of an Archean supercontinent: Geological Association of Canada, Program with Abstracts, v. 19, p. 62.

LeCheminant, A., Buchan, K.L., and van Breemen, O., 1997, Paleoproterozoic continental break-up and reassembly: Evidence from 2.19 Ga diabase dyke swarms in the Slave and western Churchill provinces, Canada: Geological Association of Canada Annual Meeting, Ottawa '97, Abstract v. 22, p. A-86.

Lisle, T.E., and Seraphim, R.H., 1977, Report on the Thor Prospect of Highwood Resources Ltd.: DIAND (Department of Indian and Northern Affairs), Mineral Assessment Report 080720.

Lockhart, G., Grütter, H., and Carlson, J., 2004, Temporal, geomagnetic and related attributes of kimberlite magmatism at Ekati, Northwest Territories, Canada: Lithos, v. 77, p. 665-682.

Meintzer, R.E., and Cerny, P., 1983, Geological studies of rare-element pegmatites in the Yellowknife Basin of the NWT, *in* Mineral Industry Report, Northwest Territories: INAC, NAP, Geology Division, NWT, EGS 1983-2, p. 189-202.

Meintzer, R.E., and Wise, M.A., 1987, Granitoids and pegmatites of the Yellowknife area, N.W.T., *in* Padgham, W.A., ed., Field Guide, Yellowknife Mining District: Geological Association of Canada, Mineral Deposit Division, Yellowknife, p. 123-134.

Meintzer, R.E., Wise, M.A., and Cerny, P., 1984, Distribution and Structural Setting of Fertile Granites and Related Pegmatites in the Yellowknife Pegmatite Field, District of Mackenzie: Geological Survey of Canada, Current Research, Part A, Paper 84-1A, 9 p.

Miramar Mining Corporation, 2005, Company website, April 29, 2005, http://www.miramarmining.com/s/HopeBayProject.asp.

Mortensen, J.K., Thorpe, R.I., Padgham, W.A., King, J.E., and Davis, W.J., 1988, U-Pb zircon ages for felsic volcanism in Slave Province, N.W.T., *in* Radiogenic Age and Isotopic Studies: Report 2: Geological Survey of Canada, Paper 88-2, p. 85-95.

Mortensen, J.K., Henderson, J.B., Jackson, V.A., and Padgham, W.A., 1992a, U-Pb geochronology of Yellowknife Supergroup felsic volcanic rocks in the Russell Lake and Clan Lake areas, southwestern Slave Province, Northwest Territories, *in* Radiogenic Age and Isotopic Studies: Report 5: Geological Survey of Canada, Paper 91-2, p. 1-7.

Mortensen, J.K., Relf, C., Davis, W.J., and King, J.E., 1992b, U-Pb zircon ages from the Shallow Bay volcaniclastic belt, Contwoyto Lake area, Northwest Territories: Age constraints for Lupin-type iron-formation, *in* Radiogenic Age and Isotopic Studies: Report 5: Geological Survey of Canada, Paper 91-2, p. 9-15.

Mueller, W.U., Corcoran, P.L., Simard, R.L., and Mortensen, J.K., 2001, Continental arc-backarc evolution: the 2650-2710 Ma volcano-sedimentary sequences in the central Slave Province: Slave-NORthern Cordillera Lithospheric Evolution (SNORCLE) Transect and Cordilleran Tectonics Workshop Meeting, Pacific Geoscience Centre, Lithoprobe Report No. 79, February 22-25, p. 12-17.

Nassichuk, W.W., and McIntyre, D.J., 1995, Cretaceous and Tertiary Fossils Discovered in Kimberlites at Lac de Gras in the Slave Province, Northwest Territories: Geological Survey of Canada, Current Research 1995-B, 6 p.

Normin, 2005, The Northern Minerals Database 2005/04/29, C.S. Lord Northern Geoscience Centre, Yellowknife, Northwest Territories, http://www.nwtgeoscience.ca/normin/index.html.

Northern Miner, 1998, Kit expands gold resource at George Lake property: The Northern Miner, v. 84, p. 19.

Northrup, C.J., Isachsen, C., and Bowring, S.A., 1999, Tectonic evolution of the Point Lake greenstone belt and adjacent gneisses, central Slave craton, N.W.T., Canada: Canadian Journal of Earth Sciences, v. 36, p. 1043-1059.

Ootes, L., Buse, S., Davis, W.J., and Cousens, B., 2005, U-Pb Ages from the Northern Wecho River Area, Southwestern Slave Province, Northwest Territories: Constraints on Late Archean Plutonism and Metamorphism: Geological Survey of Canada, Current Research 2005-F2, 13 p.

Padgham, W.A., 1992, Mineral deposits in the Archean Slave Structural Province; Lithological and tectonic setting: Precambrian Research, v. 58, p. 1-24.

Padgham, W.A., and Fyson, W.K., 1992, The Slave Province: A distinct Archean craton: Canadian Journal of Earth Sciences, v. 29, p. 2072-2086.

Pehrsson, S.J., and Villeneuve, M.E., 1999, Deposition and imbrication of a 2670-2629 Ma supracrustal sequence in the Indin Lake area, southwestern Slave Province, Canada: Canadian Journal of Earth Sciences, v. 36, p. 1149-1168.

Phillips, G.N., 1993, Metamorphic fluids and gold: Mineralogical Magazine, v. 57, p. 365-374.

Roscoe, S.M., 1985, The Booth River Intrusive Suite, District of Mackenzie: Geological Survey of Canada, Current Research, Part A, Paper 85-1A, 4 p.

——— 1990, Quartzose Arenites and Possible Paleoplacers in Slave Structural Province, N.W.T.: Geological Survey of Canada, Current Research, Part C, Paper 90-1C, 8 p.

——— 1992, Archean paleoplacers, Slave Structural Province, NWT, *in* Richardson, D.G., and Irving, M., eds., Project Summaries, Canada-Northwest Territories Mineral Development Subsidiary Agreement: Geological Survey of Canada, Open File Report 2484, p. 169-174.

Roscoe, S.M., and Minter, W.E.L., 1993, Pyritic paleoplacer gold and uranium deposits, *in* Kirkham, R.V., Sinclair, W D., Thorpe, R.I., and Duke, J.M., eds., Mineral Deposit Modeling: Geological Association of Canada, Special Paper 40, p. 103-124.

Roscoe, S.M., Henderson, M.N., Hunt, P.A., and van Breemen, O., 1987, U-Pb zircon age of an alkaline granite body in the Booth River Intrusive Suite, N.W.T., *in* Radiogenic Age and Isotopic Studies: Report 1: Geological Survey of Canada, Paper 87-2, p. 95-100.

Roscoe, S.M., Stubley, M., and Roach, D., 1989, Archean Quartz Arenites and Pyritic Paleoplacers in the Beaulieu River Supracrustal Belt, Slave Structural Province, N.W.T.: Geological Survey of Canada, Current Research, Part C, Paper 89-1C, 16 p.

Ross, G M., Parrish, R.R., Villeneuve, M.E., and Bowring, S.A., 1991, Geophysics and geochronology of the crystalline basement of the Alberta basin, western Canada: Canadian Journal of Earth Sciences, v. 28, p. 512-522.

Sinclair, W.D., Hunt, P.A., and Birkett, T.C., 1994, U-Pb zircon and monazite ages of the Grace Lake Granite, Blatchford Lake Intrusive Suite, Slave Province, Northwest Territories, *in* Radiogenic Age and Isotopic Studies: Report 8: Geological Survey of Canada, Current Research 1994-F, p. 15-20.

Sircombe, K.N., Bleeker, W., and Stern, R.A., 2001, Detrital zircon geochronology and grain-size analysis of ~2800 Ma Mesoarchean proto-cratonic cover succession, Slave Province, Canada: Earth and Planetary Science Letters, v. 189, p. 207-220.

Sleep, N.H., 2003, Geodynamic implications of xenolith geotherms: Geochemistry Geophysics Geosystems, v. 4, no. 9, 1079, doi: 10.1029/2003GC000511.

Snyder, D.B., and Lockhart, G.D., 2005, Kimberlite trends in NW Canada: Journal of the Geological Society of London, v. 162, p. 737-740.

Stasiuk, L.D., and Nassichuk, W.W., 1995, Thermal History and Petrology of Wood and other Organic Inclusions in Kimberlite Pipes at Lac de Gras, Northwest Territories: Geological Survey of Canada, Current Research 1995-B, 10 p.

Stasiuk, L.D., Lockhart, G D., Nassichuk, W.W., and Carlson, J. A., 1999, Thermal maturity evaluation of dispersed organic matter inclusions from kimberlite pipes, Lac de Gras, Northwest Territories, Canada: International Journal of Coal Geology, v. 40, p. 1–25.

Stern, R.A., and Bleeker,W., 1998, Age of the world's oldest rocks refined using Canada's SHRIMP. The Acasta gneiss complex, Northwest Territories, Canada: Geoscience Canada, v. 25, p. 27-31.

Strand, P.D., 1996, Northwest Territories Current and Past Producing Mines Database, EGS 1996-11: C.S. Lord Northern Geoscience Centre, Yellowknife, Northwest Territories.

Stubley, M., 1996, Carp Lakes (85P)—Preliminary geological compilation: Indian and Northern Affairs Canada, Map EGS 1996-03, 1:250,000.

Taylor, R.P., and Pollard, P.J., 1996, Rare earth element mineralization in peralkaline systems: The T-zone REE-Y-Be deposit, Thor Lake, Northwest Territories, Canada, *in* Jones, A.P., Wall, F., and Williams, C.T., eds., Rare Earth Minerals: Chemistry, Origin and Ore Deposits: Chapman and Hall, p. 167-192.

Thompson, P.H., 1989, Moderate overthickening of thinned sialic crust and the origin of granitic magmatism and regional metamorphism in low-P – high-T terranes: Geology, v. 17, p. 520-523.

Thorpe, R.I., Cumming, G.L., and Mortensen, J.K., 1992, A significant Pb isotope boundary in the Slave Province and its probable relation to ancient basement in the western Slave Province, *in* Richardson, D.G., and Irving, M., comps., Project Summaries, Canada-Northwest Territories Mineral Development Agreement 1987-91: Geological Survey of Canada, Open File 2484, p. 179-184.

Tomascak, P., and Cerny, P., 1992, Granitic pegmatites in the Aylmer Lake basin, *in* Richardson, D.G., and Irving, M., comps., Project Summaries, Canada-Northwest Territories Mineral Development Subsidiary Agreement 1987-1991: Geological Survey of Canada, Open File 2484, p. 51-54.

Trueman, D.L., Pedersen, J.C., de St. Jorre, L., and Smith, D.G.W., 1988, The Thor Lake rare-metal deposits, Northwest Territories, *in* Taylor, R.P., and Strong, D.F., eds., Recent Advances in the Geology of Granite-Related Mineral Deposits: Canadian Institute of Mining and Metallurgy, Special Volume 39, p. 280-290.

van Breemen, O., Davis, W.J., and King, J.E., 1992, Temporal distribution of granitoid rocks in the Archean Slave Province, northwest Canadian Shield: Canadian Journal of Earth Sciences, v. 29, p. 2186-2199.

Villeneuve, M.E., and Henderson, J.B., 1998, U-Pb geochronology of Wijinnedi Lake area, Slave Province, District of Mackenzie, Northwest Territories, in Radiogenic age and isotopic studies, Report 11: Geological Survey of Canada, Current Research 1998-F, p. 99-106.

Villeneuve, M.E., and Relf, C., 1998, Tectonic setting of 2.6 Ga carbonatites in the Slave Province, northwest Canada: Journal of Petrology, v. 39, p. 1975-1986.

Williams, H., Hoffman, P.F., Lewry, J.F., Monger, J.W.H., and Rivers, T., 1991, Anatomy of North America: Thematic geologic portrayals of the continent: Tectonophysics, v. 187, p. 117-134.

Wolfden Resources Inc., 2005, Company website, High Lake property map and surrounding area, Figure no. 2, http://www.wolfdenresources.com.

Yamishita, K., and Creaser, R.A., 1999, Geochemical and Nd isotopic constraints for the origin of late Archean turbidites from the Yellowknife area, Northwest Territories, Canada: Geochimica et Cosmochimica Acta, v. 63, p. 2579-2598.

Tectonic Evolution and Metallogeny of the Southwestern Trans-Hudson Orogen

D. Corrigan, A.G. Galley, and S. Pehrsson

Geological Survey of Canada, 601 Booth Street, Ottawa, Ontario K1A 0E8
Corresponding author's email:dcorriga@nrcan.gc.ca

Abstract

The western Trans-Hudson Orogen records a complex, late Neoarchean to late Paleoproterozoic history of intraplate magmatism followed by rifting, formation of oceanic crust, arcs and back-arcs, intraoceanic and continental margin terrane accretion, ending with terminal collision between the juvenile internides and three Archean cratons. Distinct metallogenic associations are interpreted to reflect the evolving tectonostratigraphic, magmatic, structural, and thermal phases of the orogen, as a response to interaction between the different tectonic plates and crustal blocks. During the late Neoarchean, mafic plutonic rocks with locally high platinum group element (PGE) concentrations were emplaced in the Peter Lake Domain, along a proto-rifted margin, possibly in an intracontinental extensional environment. From about 2.07 to 1.90 Ga, a passive margin prism was deposited and unconformably overlain by a ca. 1.88 to 1.86 Ga foredeep basin along the rifted Hearne craton margin, leading to the formation of numerous sub-economic sedimented-hosted Cu, Cu-U, and Pb-Zn deposits. During the interval 1.92 to 1.88 Ga, volcanogenic massive sulphide (VMS) deposits (Cu-Zn ± Pb ± Ag ± Au) formed in paleotectonic environments that have been broadly linked to the various stages of oceanic arc evolution from nascent proto-arc to early rifted arc, to mature arc, to mature rifted evolved arc. This episode generated the world-class Flin Flon VMS mining camp as well as numerous other, less endowed but nonetheless prospective deposits. Relatively more modest sediment-hosted massive sulphide deposits also formed during that time period in back-arc or forearc basins. During the waning stages of primitive intraoceanic arc development, voluminous mafic to ultramafic flows and sills hosting the Thompson magmatic Ni-Cu-(PGE) deposit were emplaced along the Superior craton margin and its Paleoproterozoic cover sequence, potentially in an active, ensialic back-arc rift environment. Terminal collision between the Hearne, Sask, and Superior cratons triggered a final episode of orogenic Au mineralization during the interval 1.83 to 1.79 Ga.

Résumé

La partie occidentale de l'orogène trans-hudsonien témoigne d'une histoire complexe allant du Néo-archéen au Paléoprotérozoïque tardif, soulignée par un magmatisme intraplaque suivi d'extension crustale, de la formation de croûte et îles océaniques, puis de l'accrétion de terranes intraocéaniques et de marge continentale pour se terminer par des collisions terminales entre les internides juvéniles et trois cratons archéens. Des associations métallogéniques distinctes sont interprétées comme reflétant la successions des phases tectonostratigraphiques, magmatiques, structurales et thermales de l'orogène en réponse aux interactions entre les différentes plaques tectoniques et blocs crustaux. Pendant le Néo-archéen tardif, des roches plutoniques mafiques renfermant par endroits des concentrations élevées d'ÉGP ont été mises en place dans le domaine de Peter Lake, le long d'une proto-marge de divergence, peut-être dans un environnement d'extension intracontinentale. Entre environ 2,07 et 1,90 Ga avant le présent, un prisme de marge passive a été déposé et a été plus tard (il y a environ 1,88 à 1,86 Ga) recouvert en discordance par des dépôts dans une dépression péri-orogénique le long de la marge de divergence du craton de Hearne, ce qui a entraîné la formation de nombreux gisements subéconomiques de Cu, de Cu-U et de Pb-Zn dans des roches sédimentaires. Pendant l'intervalle de 1,92 à 1,88 Ga avant le présent, des gisements de sulfures massifs volcanogènes (SMV) (Cu-Zn ± Pb ± Ag ± Au) se sont formés dans des environnements paléotectoniques qui ont été généralement associés à divers stades d'évolution d'arcs océaniques, depuis celui de proto-arc naissant à celui d'arc de divergence précoce, puis à celui d'arc à maturité et enfin à celui d'arc de divergence à maturité évolué. C'est cet épisode qui a engendré les gisements de SMV de classe mondiale au camp minier de Flin Flon ainsi que de nombreux autres gisements moins riches mais néanmoins prometteurs. Des gisements relativement plus modestes de sulfures massifs dans des roches sédimentaires se sont formés pendant la même période dans des bassins d'arrière-arc ou d'avant-arc. Pendant les derniers stades d'évolution des arcs intra-océaniques, les volumineux filons-couches et coulées mafiques à ultramafiques renfermant les gisements magmatiques de Ni-Cu-(ÉGP) de Thompson et de Raglan ont été mis en place le long de la marge du craton du Supérieur et de sa séquence protérozoïque de couverture, peut-être dans un environnement de rift d'arrière-arc ensialique. Les collisions terminales entre les cratons de Hearne, de Sask et du Supérieur ont déclenché un ultime épisode de minéralisation orogénique aurifère pendant l'intervalle de 1,83 à 1,79 Ga avant le présent.

Introduction

From a global tectonic and metallogenic perspective, the Paleoproterozoic era (2.5 to 1.6 Ga) forms a unique period of Earth's evolution, highlighted by profound changes in continental plate configuration and tectonic processes (e.g. Griffin et al. 2003), ocean and atmospheric compositions (e.g. Anbar and Knoll, 2002), and the biosphere (e.g. Konhauser et al., 2002). By 2.5 Ga, Archean cratons had grown by tectonic and magmatic accretion into large, stable continental masses buoyed by thick, depleted lithospheric roots that could accommodate the deposition of laterally extensive passive margin sequences (e.g. Bédard, 2006; Percival, 2007). Wilson-Cycle tectonics also appeared in the Paleoproterozoic, resulting in the first documented obducted ophiolites and development of tectonostratigraphic sequences akin to those observed in the Phanerozoic (e.g.

Corrigan, D., Galley, A.G., Pehrsson, S., 2007, Tectonic evolution and metallogeny of the southwestern Trans-Hudson Orogen, *in* Goodfellow, W.D., ed., Mineral Deposits of Canada: A Synthesis of Major Deposit-Types, District Metallogeny, the Evolution of Geological Provinces, and Exploration Methods: Geological Association of Canada, Mineral Deposits Division, Special Publication No. 5, p. 881-902.

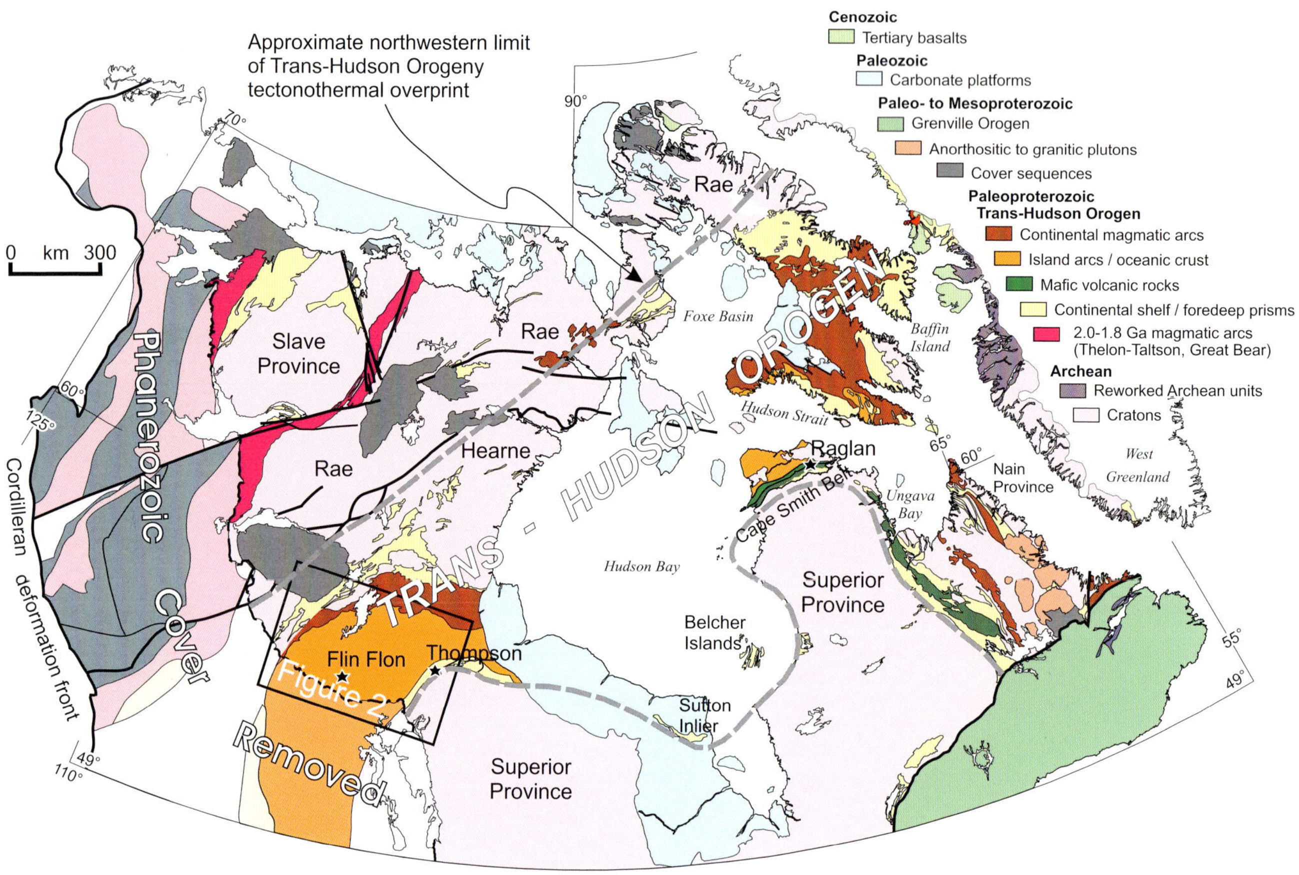

FIGURE 1. Schematic map of the Precambrian geology of Canada north of 49°N. Archean cratons are named. The predominantly buried Sask Craton is not shown. Gray dashed lines show the northwestern and southern extent of the ca. 1.80 Ga tectonothermal overprint of the Trans-Hudson Orogen. Stars indicate the location of major base metal mine camps in the Trans-Hudson Orogen.

Hoffman, 1990). The most significant period of crustal growth in the history of the North American Craton occurred by the merging of the Archean-age Slave, Rae, Hearne, Superior, North Atlantic, Sask, and Wyoming cratons during the interval 2.00 to 1.78 Ga and the accretion of juvenile crust in the intervening collisional suture zones.

The Trans-Hudson Orogen (Fig. 1) forms the largest Paleoproterozoic orogenic belt of Laurentia (Hoffman, 1988). It is the site of closure of the Manikewan Ocean (Stauffer, 1984), which resulted in the formation of juvenile crust and its eventual accretion to the Superior and Hearne Archean cratons as well as with a number of smaller intervening Archean to earliest Paleoproterozoic continental fragments, including the Sask Craton (Lewry et al., 1994; Hajnal et al., 1996). The Trans-Hudson Orogen is broadly coeval with the New Québec and Penokean orogens, which developed along the eastern and southern margins of the Superior Craton, respectively (Hoffman, 1988; Wardle and Hall, 2002). From a tectonostratigraphic perspective, the Trans-Hudson Orogen preserves a relatively complete record of its evolution, from early ca. 2.45 to 1.95 Ga rift to drift sedimentary assemblages deposited along Archean craton margins, to formation of ca. 2.0 to 1.88 Ga oceanic crust, oceanic islands and plateaus, oceanic and continental arcs, back-arc basins, as well as younger (1.88-1.83 Ga) foredeep and collisional basins (Ansdell, 2005). From a metallogenic perspective, the Trans-Hudson Orogen is particularly known for its world-class volcanogenic massive sulphide (VMS) deposits (e.g. Flin Flon, Ruttan) and magmatic Ni-Cu-PGE deposits hosted in mafic and ultramafic dykes and sills emplaced along the rifted Superior Craton margin (e.g. Thompson, Raglan) and in ocean arc or back-arc gabbronorites (e.g. Lynn Lake). On a lesser but still important scale, the Trans-Hudson Orogen contains important lode and shear-zone-hosted Au deposits (e.g. New Britannia, Seabee) and given its tectonostratigraphic and magmatic framework, has potential for Cu-Zn SEDEX (sedimentary exhalative) deposits, porphyry Cu-Au, and iron ore hosted in banded iron formation. In addition, there are diamonds in Cretaceous kimberlites that crosscut tectonically accreted Archean microcontinents that preserved thick lithospheric roots (e.g. Sask Craton).

This paper is an overview of the tectonic evolution of the southwestern Trans-Hudson Orogen and it aims at providing potential links between tectonostratigraphy and metallogeny. As is typically the case for overview papers, we cannot by any means cover every detail of the tectonic evolution and hopefully the references provided herein will guide the reader to the appropriate sources of information. The paper does not, however, provide an exhaustive list of all known deposits and mineral occurrences in the Trans-Hudson Orogen. For the latter, the reader should refer to detailed compilations and reports published by the Saskatchewan and Manitoba Geological Surveys, as well as the Geological

Survey of Canada (e.g. Ames, 1996; Beaumont-Smith and Böhm, 2003; Saskatchewan Geological Survey, 2003). This paper focuses on the orogen internides in Saskatchewan and Manitoba (Fig. 2), an area referred to as the "Reindeer zone of the Trans-Hudson Orogen" by Lewry and Collerson (1990). The latter more widely applied the term Trans-Hudson Orogen to the entire area affected by ca. 1.83 to 1.80 Ga "Hudsonian" tectonothermal overprint, which includes most of the Archean Hearne Province as well as a large part of the southeastern Rae Province, together with their Paleoproterozoic cover sequences (Fig. 1). This differs from the definition suggested by Hoffman (1990), who restricted the Trans-Hudson Orogen to that part of the Canadian Shield that includes Lewry and Collerson's (1990) "Reindeer zone", as well as the Sutton Inlier in northern Ontario, Belcher Islands in eastern Hudson Bay, and the Cape Smith Belt in northernmost Quebec.

From a metallogenic perspective, the southwestern Trans-Hudson Orogen preserves rocks from a wide range of tectonic and magmatic settings and hence, hosts a variety of mineral deposit types. Moreover, it has only been moderately thickened during its accretionary and collisional phase, perhaps due to the interfering effect of the Sask Craton on terminal Hearne-Superior collision (e.g. Saskatchewan Geological Survey, 2003) or to the dominantly transpressional nature of the main Superior Craton collision with the previously amalgamated Hearne/Sask Craton (e.g. Bleeker, 1990). In contrast with the remainder of the Trans-Hudson Orogen northeast of Manitoba, the internides (Reindeer zone) preserves wide areas that have been affected at most by upper greenschist to lower amphibolite metamorphic conditions, increasing the likelihood of preservation of mineral deposits originally emplaced at middle to upper crustal levels. At least four different metallotects can be defined, based on the geodynamic setting active during mineral deposition and their related deposit types (Fig. 2). The aim of this paper is to discuss the various metallogenic associations of the Trans-Hudson Orogen internides in the context of evolving tectonic, stratigraphic, and magmatic processes.

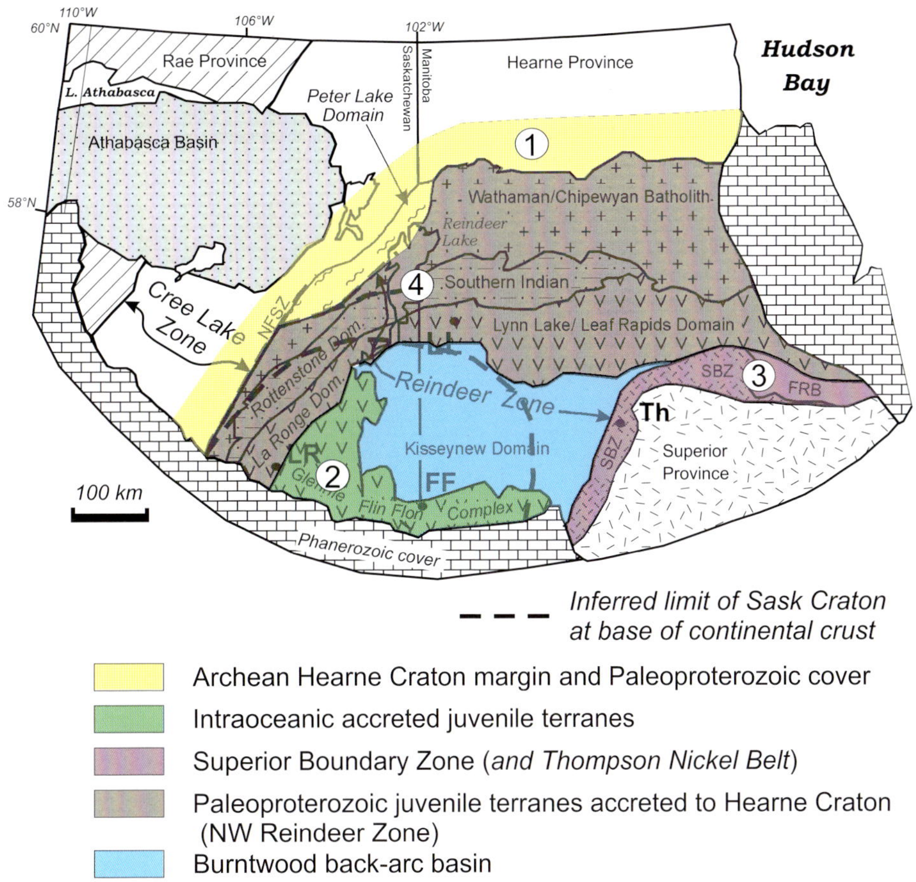

FIGURE 2. Schematic map showing the main domainal subdivisions of the southwestern Trans-Hudson Orogen and adjacent Archean cratons. The transparent colour overlay shows the five main tectonostratigraphic subdivisions (coloured) and four metallogenic associations (numbered 1 to 4 on diagram; see text for explanations). The Reindeer Zone (orogen internides) and Cree Lake Zone (reactivated Hearne Province and Paleoproterozoic cover rocks) are shown. Communities shown are FF = Flin Flon; LL = Lynn Lake; LR = La Ronge; Th = Thompson. Abbreviations are FRB = Fox River Belt; NFSZ = Needle Falls shear zone; SBZ =

Tectonic and Metallogenic Framework

The southwestern Trans-Hudson Orogen (Figs. 2, 3) forms a collage of tectonically accreted juvenile supracrustal belts, continental margin sedimentary rocks, an Archean- to earliest Paleoproterozoic-age microcontinent (Sask Craton), and reactivated rocks of the Archean Superior and Hearne forelands (Hoffman, 1988; Lewry and Collerson, 1990; Lewry et al., 1994; Lucas et al., 1996; Ansdell, 2005). On a very broad scale, at least five main crustal components are identified: i) the reactivated Archean-age Hearne and Superior Craton margins and their respective Paleoproterozoic cover sequences (Cree Lake zone and Superior boundary zone); ii) an intraoceanic accretionary complex (Glennie-Flin Flon complex), formed of ca. 1.92 to 1.88 Ga juvenile oceanic arcs, ocean plateaus, and associated marginal basins that were tectonically accreted in an intra-oceanic setting during the interval 1.88 to 1.87 Ga; iii) an accretionary orogen (northwestern Reindeer zone) formed of ca. 1.92 to 1.88 Ga juvenile volcanic arcs, back-arcs, and associated sedimentary basins that were thrust onto the Archean Hearne margin during the interval 1.92 to 1.87 Ga; iv) the ca. 1862 to 1848 Ma Wathaman-Chipewyan batholith, which consists of an Andean-type continental magmatic arc emplaced along the northwestern Reindeer zone; and v) a marginal basin and molasse subbasins (Kisseynew Domain) developed between the northwestern Reindeer zone and Glennie-Flin Flon complex during the interval 1.85 to 1.84 Ga. All of the above

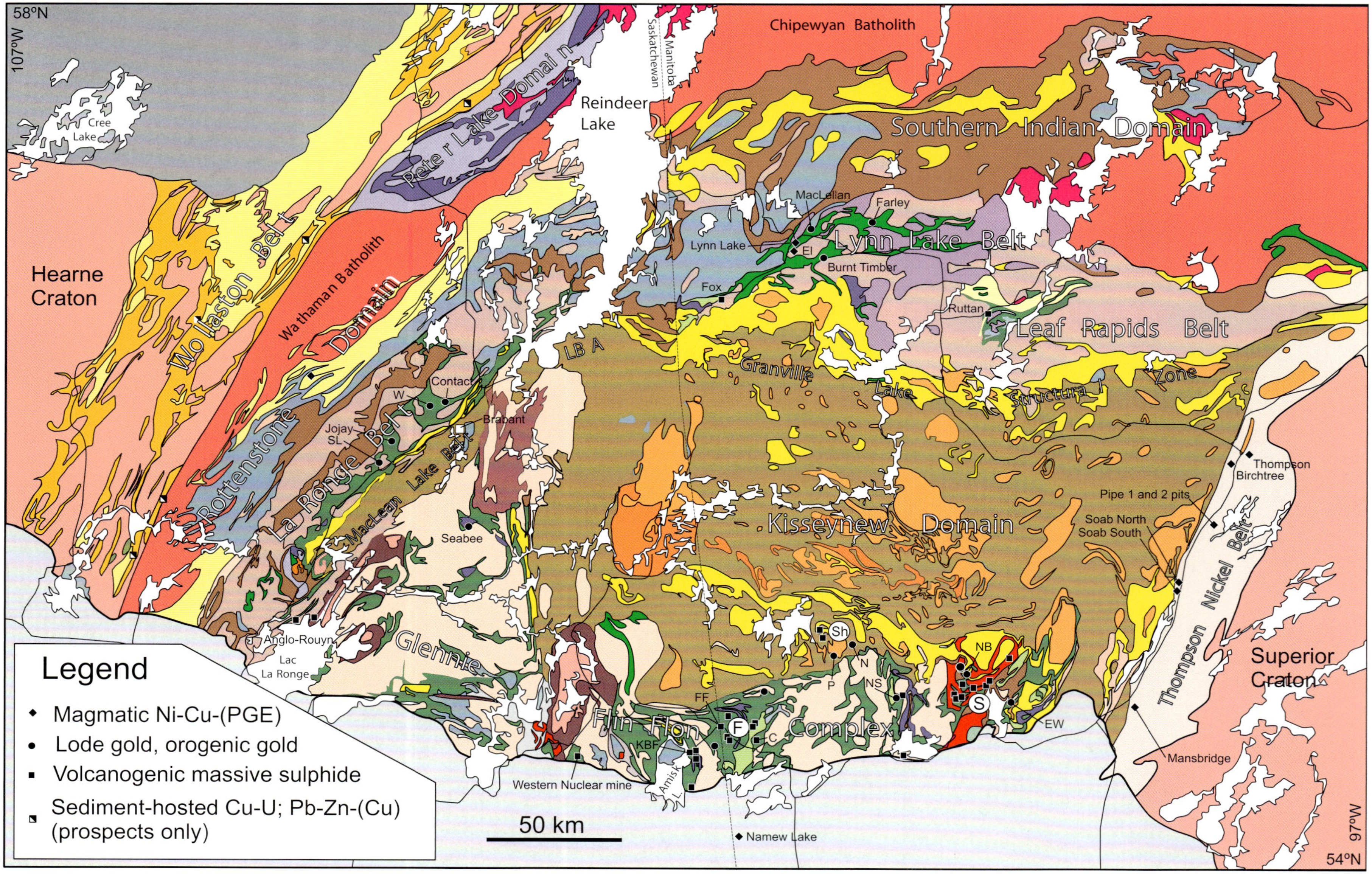

FIGURE 3. Simplified lithological map of the southwestern Trans-Hudson Orogen, showing distribution of past and active mines, except for sediment-hosted Cu-U, Pb-Zn-(Cu) in the Wollaston fold belt, which are occurrences and prospects only. Sources: Saskatchewan and Manitoba Geological Survey web-downloadable 1:250,000 scale bedrock geology compilations. Main mining camps in the Glennie-Flin Floncomplex are FF = Flin Flon; Sh = Sherridon; SL = Snow Lake. Abbreviations for gold and VMS mines in Glennie-Flin Flon complex are C = Centennial; EW = East Wekusko; FF = Flin Flon, Triple 7, Callinan; KBF = Konuto, Birch, Flexar; N = Nokomis; NB = New Britannia; NS = North Star; P = Puffy Lake. Abbreviations for gold mines in the La Ronge Belt: SL = Star Lake; W = Waddy Lake. Other abbreviations: LBA = Levesque Bay assemblage.

Figure 3 continued.

domains were penetratively deformed and metamorphosed at moderate to high temperatures and low to moderate pressures during the collisional stage of the Trans-Hudson Orogeny, where metamorphic conditions peaked during the interval 1.82 to 1.80 Ga (Lewry and Sibbald, 1980; Ansdell, 2005). The following sections provide a brief geological synthesis for each of the crustal domains outlined above and speculate about potential links between the various mineral deposits and tectonic evolution (Fig. 4).

Reactivated Archean Margins

The Trans-Hudson Orogen internides are flanked to the northwest (Cree Lake zone and Peter Lake Domain) and southeast (Superior boundary zone) by reactivated Archean-age rocks locally overlain by remnants of Paleoproterozoic cover sequences (Fig. 2). On the northwestern flank of the orogen, the uppermost amphibolite- to granulite-facies Cree Lake zone (Lewry and Sibbald, 1980) is composed of an interior zone formed predominantly of Archean age rocks with a strong Paleoproterozoic tectonothermal overprint and a marginal zone composed of more strongly reactivated and complexly interfolded Archean basement and Paleoproterozoic sedimentary cover (Fig. 3, Wollaston Domain). The interior zone (Mudjatik Domain) contains mostly homogeneous ca. 2.7 Ga granitoids (Chiarenzelli and Macdonald, 1986; Delaney et al., 1990) and layered granitic gneiss with minor metasedimentary screens (Macdonald and Thomas, 1983). In contrast, the Wollaston Domain comprises abundant Paleoproterozoic supracrustal rocks (Wollaston Group, Fig. 4A) infolded with ca. 2.7 to 2.5 Ga orthogneiss and granitoids, as well as isolated ca. 2.2 to 2.0 Ga felsic intrusions (Annesley et al., 1992, 1997). The Wollaston Group is interpreted as having been unconformably deposited on, and then complexly interfolded with, Archean-age basement rocks of the Hearne Province (Lewry and Collerson, 1990; Yeo, 1998; Tran et al., 2003). Structural trends in the Cree Lake zone are orogen-parallel, with dips ranging from subvertical to subhorizontal. Elongate granitoid basement domes are exposed along doubly plunging, northeast-trending, upright antiformal culminations (Macdonald and Thomas, 1983).

The Wollaston Group forms a thin to moderately thick cover on the Archean basement, and is best preserved in synformal keels produced by tight, orogen-parallel, doubly plunging folds. It is divided into two different successions separated by a major unconformity. The lowermost succession consists of coarse clastic rocks interbedded with mafic volcanic flows, which are overlain by quartzite, psammopelite, and pelite. Overall, the sequence corresponds to the "basal pelitic succession" referred to by Ray (1979), and interpreted as a rift-to-drift basin fill that was deposited on a passive margin developing on the southeastern Hearne Craton (Delaney, 1993; Coombe, 1994). A U-Pb age of 2075 ± 2 Ma (Ansdell et al., 2000) obtained on a synextensional quartz-feldspar porphyry rhyolite from the lower sequence is similar, within error, to ca. 2072 ± 3 Ma mafic magmatism along the northwest Superior Craton margin (Birthday Rapids gabbro dyke: Heaman and Corkery, 1996), suggesting that initial rifting along both craton margins was more or less contemporaneous, but does not imply any links between the two Archean cratons. The younger and thicker upper Wollaston succession, represented by the Daly Lake and Geikie River groups (Saskatchewan Geological Survey, 2003), consists of quartzite, psammopelite, pelite, and calc-silicate-bearing feldspathic arenite, with minor iron formation and marble. This sequence, which contains detrital zircons as young as ca. 1.88 Ga (Tran, 2001), is interpreted as a foreland basin deposited unconformably on the Lower Wollaston Group. The minimum age of the upper Wollaston Group is constrained at 1.85 Ga by crosscutting intrusions (Annesley et al., 1997). Paleoproterozoic high-grade metamorphic overprint in the Cree Lake zone is interpreted to have occurred during two distinct periods, during the intervals 2.3 to 2.0 Ga and 1.87 to 1.81 Ga (Bickford et al., 1994; Annesley et al., 1999; Orrell et al., 1999).

The Peter Lake Domain (Fig. 3) is a southern extension of the Hearne Province exposed between the Wathaman batholith (Rottenstone Domain) and Wollaston Domain. It is approximately 30 km wide by 150 km long and comprises a ca. 2.68 to 2.58 Ga orthogneiss component (basement complex), intruded by four temporally distinctive plutonic suites and overlain by the Campbell River Group (Maxeiner, 2006). The intrusive plutonic suites account for about one third of the domain and consist of ca. 2.58 to 2.56 Ga monzogranite, the 2.56 to 2.54 Ga Swan River complex, the 1.92 to 1.91 Ga Porter Bay complex, and isolated

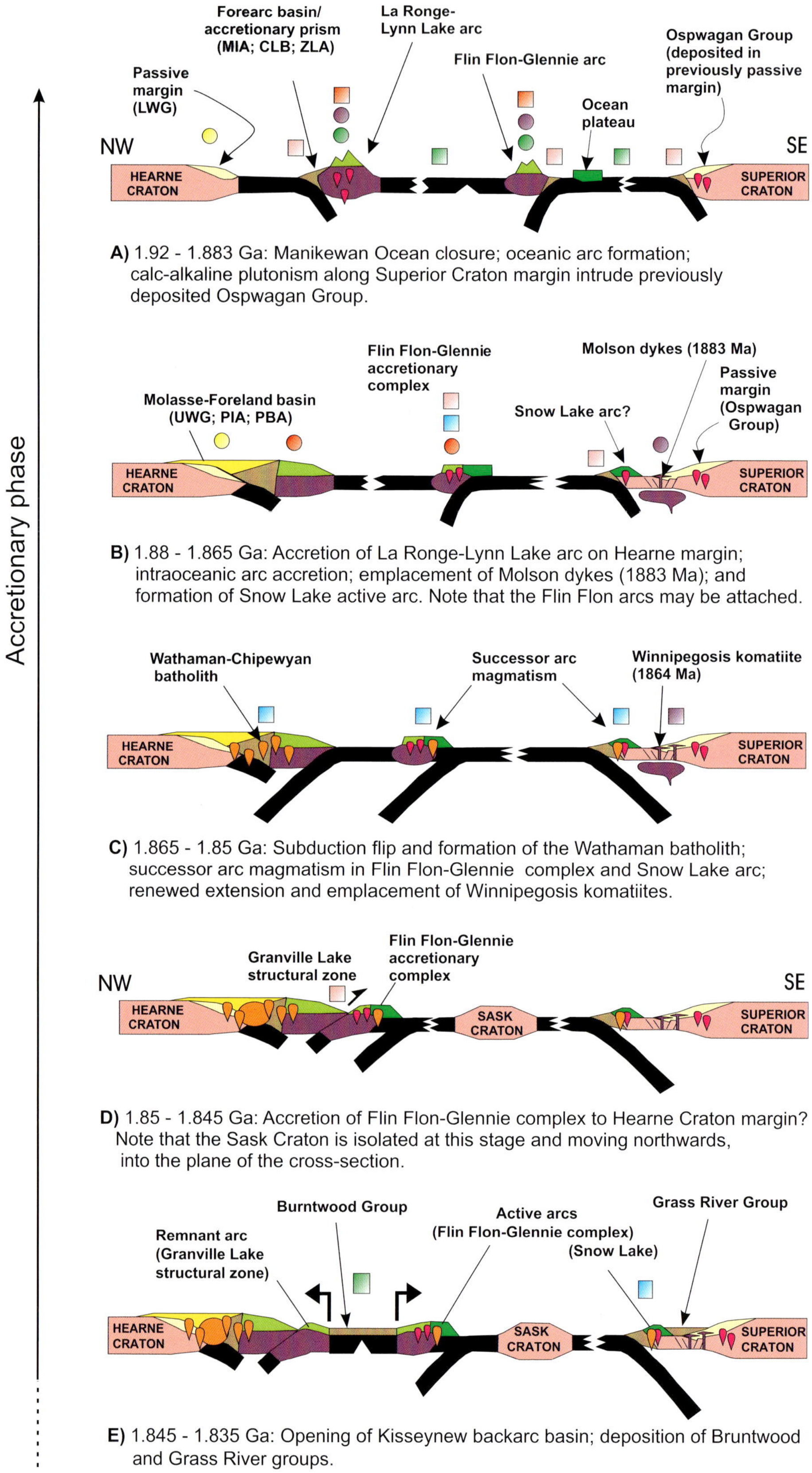

FIGURE 4. Series of schematic cross-sections representing the simplified tectonic evolution of the southwestern Trans-Hudson Orogen. Symbols represent known mineral deposits (circles) as well as potential for certain types of deposits (squares) corresponding to specific paleo-tectonic environments. See text for explanation. Abbreviations: CLB = Crew Lake Belt; MIA, Milton Island assemblage; PBA = Partridge Breast assemblage; PIA, Park Island assemblage; UWG = Upper Wollaston group; ZLA = Zed Lake assemblage.

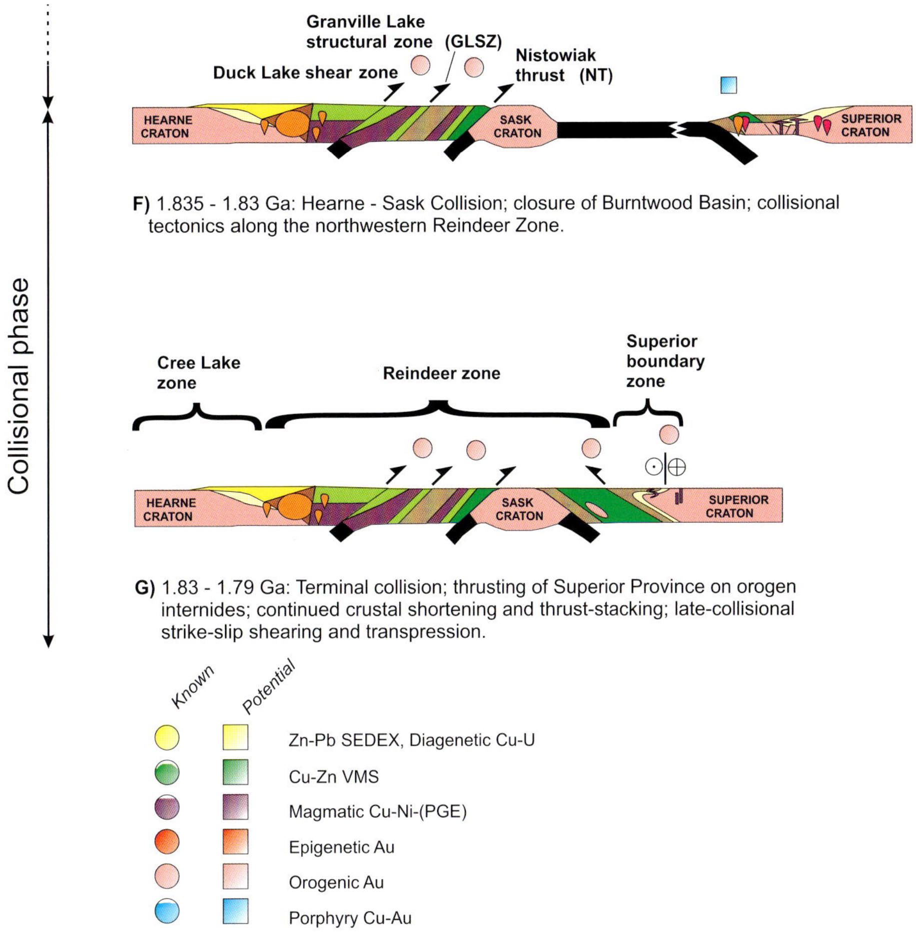

FIGURE 4 CONTINUED.

intrusions of the ca. 1.86-1.85 Ga Wathaman batholith (Rayner et al., 2005a,b; Maxeiner, 2006). The Swan River and Porter Bay complexes host the largest volume of mafic intrusive rock. The former includes gabbro, gabbro-norite, diorite, and granodiorite, with minor anorthosite and pyroxenite. Rocks of the Porter Bay complex are overall more intermediate in composition than the Swan River complex, but also contain layered and massive gabbro.

The Trans-Hudson Orogen is bound on the southeast by the Archean Superior Province (Fig. 1) which, in contrast with the Hearne Province, has been less extensively affected by Proterozoic deformation and metamorphism. The reactivated portion (Superior boundary zone) is a well defined feature whose exposed part is traceable from eastern and northeastern Manitoba, through northern Ontario and Quebec, to the Labrador Trough (Green et al., 1979). Along the Cape Smith Belt segment in northern Quebec, the Superior boundary zone hosts the Raglan mine (Lesher, 2007), the third largest Ni producer in Canada. Throughout its length, the Superior boundary zone consists of Archean rocks of the Superior Province that are unconformably overlain by Paleoproterozoic supracrustal rocks, with both basement and cover sequences deformed and metamorphosed as a result of ca. 1.83-1.78 Ga collision between the Superior Craton and Trans-Hudson Orogen internides (e.g. Bleeker, 1990; St-Onge et al., 1992). The Superior boundary zone in Manitoba is referred to as the Thompson Nickel Belt which consists of complexly deformed and metamorphosed 3.0 to 2.65 Ga rocks of the Superior Province (Weber, 1990), unconformably overlain by metasedimentary and metavolcanic rocks of the Ospwagan Group (Bleeker and Macek, 1996; Zwanzig, 2005). The deposition age of the lower Ospwagan Group is constrained by the youngest detrital zircons, which are ca. 2.53 Ga (Hamilton and Bleeker, 2002; Rayner et al., 2006) and the oldest crosscutting intrusions, dated at ca. 1.89 Ga (Zwanzig et al., 2003; Percival et al., 2004). Deposition may have been initiated as early as 2.09 to 2.07 Ga, which is inferred from the age of emplacement of the Cauchon and Birthday Rapids mafic dykes in Archean basement rocks (Heaman and Corkery, 1996). A single detrital zircon from the upper Ospwagan Group, dated at ca. 1974 Ma (Hamilton and Bleeker, 2002), suggests that at least the upper part of the stratigraphic sequence was deposited after that time. Subsequent to the deposition of the Ospwagan Group, the Thompson Nickel Belt underwent renewed rifting at ca. 1883 Ma, with emplacement of the Molson dyke swarm (Halls and Heaman, 1997; Hulbert et al., 2005) and contemporaneous continental arc plutonism along the western margin of the Superior Province, suggesting that both the dykes and plutons were emplaced in a continental back-arc environment (Percival et al., 2005) rather than in a simple plume-driven rift. Whether the active continental arc is now incorporated in the Snow Lake Belt (described below, see Fig. 4), or was completely removed by pre-collisional strike-slip shearing, is entirely conjectural. A second and apparently separate magmatic event along the Thompson Nickel Belt involved the eruption of the Winnipegosis komatiite at ca. 1864 Ma (Hulbert et al., 1994). This event is contemporaneous with, but geodynamically unrelated to, successor arc magmatism in the Glennie-Flin Flon Domain. Deposition of the coarse clastic to arkosic Grass River Group during the interval 1.845 to 1.835 Ga, contemporaneously with the Burntwood Group (Fig. 4E), preceded the 1.82-1.79 Ga collision between Superior Craton and Reindeer zone internides (e.g. White et al., 2002).

Magmatic Cu-Ni-(PGE) Deposits

Magmatic Ni-Cu-(PGE) deposits are known from both the Cree Lake zone and Superior boundary zone, but they differ significantly in tectonomagmatic nature and economic importance. The Thompson Nickel Belt is the second largest magmatic Cu-Ni-(PGE) district in Canada after Sudbury, with past production and current reserves estimated at more than 100 Mt at approximately 2.5% Ni (Layton-Matthews et al., 2007). To date, six mines have been in production and many prospects are scattered throughout the belt, particularly concentrated on the western margin of the belt. Nickel sulphide mineralization occurs in close association with ultramafic bodies that intrude metasedimentary gneiss of the Ospwagan Group (Peredery et al., 1982; Eckstrand and Hulbert, 2007). Recently, a U-Pb zircon age of 1880 ± 5 Ma was obtained on a mineralized ultramafic body in the Setting Lake area of the Thompson Nickel Belt (Hulbert et al., 2005), confirming earlier suggestions that the mineralized bodies were genetically related to the Molson dyke swarm (Heaman et al., 1986) and temporally correlative with the Raglan deposit in northern Quebec (Wodicka et al., 2002).

Along the northwestern margin of the Reindeer zone, the potential for magmatogenic Cu-Ni and platinum group element (PGE) occurrences has been known since the mid-1900s, and has been the object of ongoing exploration, although no economic deposit has been found to date (Saskatchewan Geological Survey, 2003). The Peter Lake Domain, in particular, hosts one of the largest mafic-ultramafic complexes in North America (Hulbert, 1988). Recent detailed mapping, however, suggests that the layered mafic-ultramafic complexes are associated with more intermediate magmatic compositions than originally thought and include monzodiorite, diorite and tonalite (Maxeiner, 2006, and references therein). Moreover, these collectively large magmatic complexes have been emplaced during three unrelated events spanning about 900 million years (Rayner et al., 2005a). Gabbros and norites with known PGE prospects are associated with the oldest, ca. 2.56 Ga mafic magmatic event represented by the Swan River complex (Maxeiner and Rayner, 2005). Although a reef model for magmatic Cu-Ni-PGE exploration has been favoured over the years and is still valid for exploration, there are indications that there might also be potential for other types of PGE enrichment, such as is found in Lac-des-Iles-type multiple intrusive sanukitoid suites (e.g. Sutcliffe et al., 1989), or hydrothermal enrichment (Hulbert, 1988). To date, exploration has yielded encouraging results (Maxeiner and Rayner, 2005) but occurrences investigated so far are subeconomic due to their limited size.

Sediment-Hosted Base Metal Deposits

The presence of sediment-hosted base metal mineralization in the south-central Wollaston Belt has been known since the early 1950s following exploration in the Foster Lakes area (Coombe, 1994). In general, mineralization occurs along the southeast margin of the belt (Fig. 3), likely reflecting a distinct phase in basin evolution (Delaney and Savage, 1998), and is in large part concentrated in the basal sedimentary succession and is therefore thought to be related to initial rifting and drifting of the Hearne continental margin, rather than to overlying syn-accretionary foredeep succession (Delaney, 1995a). Mineralization occurs in the form of oxide and silicate-facies iron formation, sandstone-hosted Cu and Cu-U, sandstone-hosted Pb-Zn deposits, as well as Zn- and Pb-rich pyritic mudstones interpreted as potential SEDEX-type deposits (Delaney, 1995a; Delaney and Savage, 1998). Copper and Cu-U deposits are mostly confined to the boundary between lower redbed conglomerate deposits and upper meta-arkose with interbedded calc-silicate members (Scott, 1973). The most significant showing of the Cu and Cu-U-type is the Thor prospect in the Duddridge Lake area, which consists of U-Cu mineralization within carbonate-bearing hematitic arkose (Bretzlaff, 1976). The latter attributed mineralization to epigenetic precipitation from circulating groundwaters in the local reducing environment of carbonate-rich arkoses. Potential sedimentary-exhalative Zn-Pb deposits of the Courtney Lake area are found in a thick rift-cover succession of graphitic mudstones, variably calcareous siliciclastic rocks, carbonates, iron formation and arkose overlying a basal rift-fill succession of conglomerate, and arkose and bimodal volcanics (Delaney and Savage, 1998). So far, the Wollaston Belt base metal deposits are subeconomic, with the largest consisting of the Janice Lake and George Lake occurrences (Fig. 3). The latter is estimated at 2.6 Mt grading 3.67% Zn and 0.53% Pb (Gracie and Lafrance, 1998). No similar deposits have been identified so far in the eastern extent of the Wollaston Belt in northern Manitoba. However, that part of the shield is much less well exposed, being extensively covered by glacial drift. Along the Superior Craton margin, the Paleoproterozoic age Ospwagan Group hosts abundant barren sulphidic clastic metasedimentary assemblages but has no known economic SEDEX or other type of mineral deposits.

Early Oceanic Crust (Manikewan Ocean)

Oceanic crust was produced in the Manikewan Ocean (Stauffer, 1984), which formed between the Hearne, Sask, and Superior cratons during the approximate interval of 2.07 to 1.92 Ga (Fig. 4A). This ocean basin likely extended northeastwards (present day coordinates), perhaps as far as the Ungava Peninsula in northern Quebec prior to its closure (e.g. Gibb and Walcott, 1971; Gibb, 1975). Remnants of this oceanic crust are found in the Flin Flon-Glennie complex (Lucas et al., 1994; Stern et al., 1995) and may potentially include the Watts Group of the Cape Smith Belt (Hoffman, 1988; St-Onge et al., 1992). In most localities where oceanic crust is preserved, it is not intact but thrust-sliced and structurally imbricated with ocean-island basalt (OIB), ocean-arc fragments, and associated sediments (e.g. Lucas et al., 1996). The only absolute age determination obtained thus far from a fragment of this oceanic crust is from a gabbro unit that is part of the Watts Group, which yielded a 1.99 Ga U-Pb zircon age (Parrish, 1989). Other smaller, thrust-imbricated crustal fragments with mid-ocean ridge basalt (MORB) and tholeiitic geochemical affinity occur in the La Ronge and Lynn Lake belts, but their tectonic settings are equivocal, due to limited preservation or uncertain affinity with surrounding rocks. Tectonized blocks of ultramafic rock exposed along the southern margin of the La Ronge Domain in Saskatchewan have been interpreted on the basis of geochemistry and field relationships as dismembered ophiolite (Maxeiner et al., 2005). Boudin trains of similar rock associations occur along the Granville Lake structural zone in Manitoba, and have also been interpreted as potential remnants or tectonic slivers of seafloor (Zwanzig, 2000). From a metallogenic perspective, the presence of rock assemblages of potential seafloor origin opens the question as to whether there is potential for Cu-Zn-rich VMS deposits associated with ocean spreading centres (e.g. Spooner, 1980; Höy,

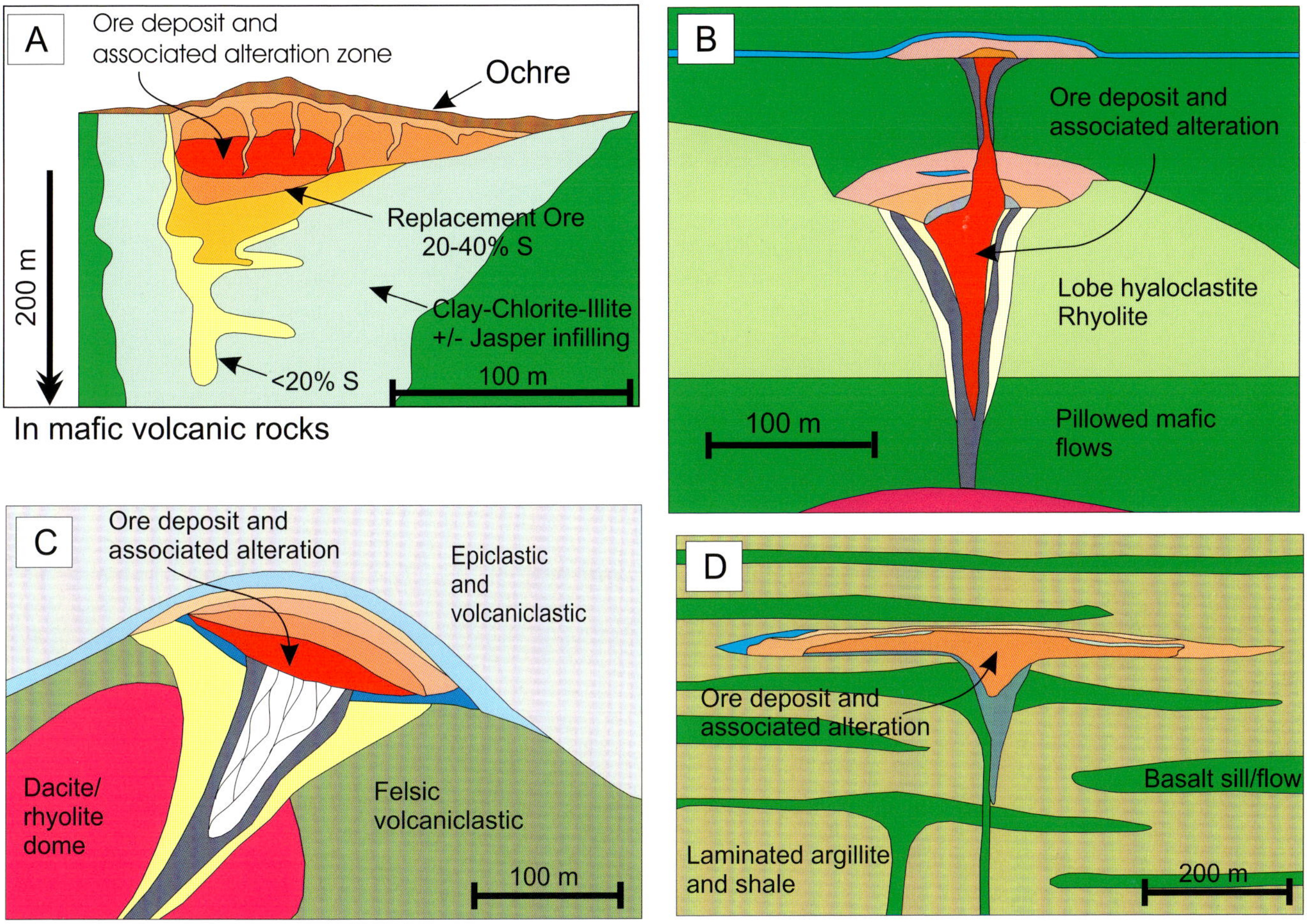

FIGURE 5. Set of four schematic cross-sections showing progressive arc evolution stages observed in various tectonostratigraphic units of the Flin Flon-Glennie complex, their relation to VMS deposits, and typical economic mineral associations. Examples linked to past and present producers are shown. **(A)** nascent/proto oceanic arc (e.g. Birch, Flexar, Konuto and Coronation deposits), **(B)** early rifted oceanic arc (e.g. Flin Flon, Callinan, Triple 7, Mandy, Schist, West Arm), **(C)** mature oceanic arc (Northstar, Don Jon, Pine Bay), and **(D)** mature rifted oceanic arc (White Lake, Cuprus).

1995; Galley et al., 2007). None have been formally identified thus far.

Intraoceanic Arcs and Related Supracrustal Rocks

The Glennie-Flin Flon complex (Ashton, 1999) represents remnants of ca. 1.92 to 1.88 Ga juvenile island arc, ocean floor, ocean plateaus, evolved arc, and associated sedimentary and plutonic rocks that developed during closure of the Manikewan Ocean (e.g. Stauffer, 1984, Syme and Bailes, 1993; Lucas et al., 1996; Whalen et al., 1999; Ansdell, 2005). From west to east it comprises the Snow Lake arc assemblage (Fig. 4B), the Amisk collage, the Hanson Lake block, and the Glennie Domain, and is interpreted as part of a single crustal entity (Glennie-Flin Flon complex) that formed at ca. 1.87 Ga as a result of intraoceanic accretion (Lewry and Collerson, 1990; Lucas et al., 1996). The various tectonostratigraphic assemblages forming the complex are fold-repeated and thrust-stacked, and tectonically overlie Archean to earliest Paleoproterozoic Sask Craton above a basal décollement that was active as early as about 1.84 Ga (Ashton et al., 2005). The lithotectonic evolution of the Glennie-Flin Flon complex is generally regarded as having involved four main development stages, consisting of i) 1.92 to 1.88 Ga formation of juvenile arcs, back-arc basins, and ocean plateaus (Fig. 4A); ii) 1.88 to 1.87 Ga intraoceanic accretion (Fig. 4B); iii) 1.87 to 1.84 Ga post-accretion development of successor arc intrusives and inter-arc basins; and iv) 1.84 to 1.83 Ga terminal collision stage (Lucas et al., 1996; Stern et al., 1999). In general, the pre-collisional arcs record an evolution from primitive arc tholeiites to evolved calc-alkaline arc rocks (Bailes and Galley, 1999; Maxeiner et al., 1999). Although the arcs were mostly juvenile, Nd isotopic data suggest variable input from Archean crust, perhaps recycled in sediments or as rifted crustal fragments locally incorporated in the early arc assemblages (Stern et al., 1995). In the Flin Flon Domain, older crust is actually preserved as thrust-imbricated slices interleaved with juvenile arc and back-arc crust (David and Syme, 1994; Lucas et al., 1996, Syme et al., 1999). To the east, the Snow Lake arc assemblage appears to contain an even greater proportion of Archean-derived crust, yielding evolved ε_{Nd} values of -0.4 to +2.4 in mafic rocks intruding the arc assemblages (Stern et al., 1995), as well as 2.65 to 2.82 Ga xenocrystic zircons in a 1.892 Ga rhyolitic breccia (David et al., 1996; Syme et al., 1999).

Volcanogenic Massive Sulphide Deposits of the Glennie--Flin Flon Complex

The Flin Flon-Glennie complex hosts the largest Paleoproterozoic VMS district in the world, with 24 deposits with aggregate tonnage of 143 Mt of sulphide in past and presently producing mines, and 64.3 Mt contained in a number of subeconomic or preproduction deposits (Syme et al., 1999; Galley et al., 2007). Most of the economic VMS deposits occur within two main fault-bounded crustal blocks, namely Flin Flon and Snow Lake, with other more isolated deposits occurring in the Hanson Lake Block and Glennie Domain (Fig. 3). Based on detailed mapping, tracer isotopes, and lithogeochemistry, Syme et al. (1999) concluded that all of the economic VMS deposits in the Flin Flon-Snow Lake portion of the Flin Flon-Glennie complex occur within oceanic arc tectono-stratigraphic assemblages, whereas only subeconomic Cu-Zn deposits occur in other paleo-environment settings. Most of the deposits can further be attributed to specific temporal evolution of the arc segments (Fig. 5), from nascent/proto oceanic arc environments to mature rifted oceanic arc (Galley, 1996; Bailes and Galley, 1999). In the Flin Flon Belt, the Konuto Lake mine is hosted in proto-arc-type mafic volcanic rocks of the Birch Lake arc assemblage, which was also host to the past-producing Coronation, Birch, and Flexar deposits (Saskatchewan Geological Survey, 2003). The Birch and Flexar deposits occur in chlorite schists derived from mafic metavolcanic rocks. The Coronation deposit occurs in andesitic flow and breccia, as well as mafic tuffs that were later intruded by granodiorite and gabbro. These deposits are characterized by chalcopyrite, pyrrhotite, pyrite, minor sphalerite, and local magnetite and arsenopyrite, with molybdenite and native Au present in some of the ore. The main deposits of the Flin Flon camp (Flin Flon, Triple 7, Callinan, Mandy, Schist, West Arm) occur within slightly more evolved, early rifted oceanic arc assemblages represented by quartz-phyric rhyolite units underlain by thick mafic volcaniclastic rocks (Bailes and Syme, 1989). Predominant minerals are pyrite, pyrrhotite, sphalerite, and chalcopyrite, and minor minerals include arsenopyrite, cubanite, galena, tetrahedrite-tennantite, sylvanite, tetradymite, altaite, magnetite, electrum, and Au. More mature oceanic arc assemblages host the Northstar, Don Jon, and Pine Bay Cu-Zn deposits. The latter are spatially related to a dacite-rhyolite dome that intrudes felsic volcaniclastic rocks.

The Snow Lake arc, located approximately 80 km east of Flin Flon, provides further examples of the variability of VMS deposit types in relation to arc tectono-magmatic evolution, with the main VMS deposits hosted in a 6 km section that records evolution from primitive to evolved to arc-rift sequences (Bailes and Galley, 1999). It hosts a number of present- and past-producing VMS deposits including Northstar, Snow Lake, and Chisel North.

To date, only a few economic deposits have been found outside of the Flin Flon-Snow Lake VMS district. They include the Anglo-Rouyn mine in the Glennie Domain, as well as the Western Nuclear mine and McIlvenna Bay deposit in the Hanson Lake Block. Mineralization at the Anglo-Rouyn mine is hosted in siliciclastic metasedimentary rocks interlayered with thick diopside-amphibole-epidote-bearing mafic gneiss that likely represents altered and metamorphosed mafic volcanic flows or sills. The entire area is highly tectonized and the ore zones forms en échelon lenses elongated parallel to the principal mineral lineation. Mineralization consists mostly of pyrite, pyrrhotite, and chalcopyrite, with minor magnetite, sphalerite, galena, and pentlandite. The mine, which operated between 1966 and 1972, produced 1.72 Mt grading 2.4% Cu, 1.2 g/t Au, and 5.5 g/t Ag (Saskatchewan Geological Survey, 2003). In terms of tectonic setting, the association of clastic sedimentary rocks with mafic sills and/or flows suggests either an oceanic back-arc setting or rifting of a mature oceanic arc (e.g. Fig. 4D). In the Hanson Lake Block, both the Western Nuclear and McIlvenna Bay deposits are hosted by predominantly felsic altered volcanic rocks with lesser mafic to intermediate flows, reminiscent of the early rifted to mature oceanic arc assemblages of the Flin Flon and Snow Lake camps. Mineralization at the Western Nuclear mine is characterized by numerous high-grade lenses of massive sulphides containing mostly sphalerite and galena, with minor arsenopyrite, pyrite, pyrrhotite, chalcopyrite, argentite, tetrahedrite, and Au. The mine was in operation from 1966 to 1969, during which time 147, 301 tonnes of ore grading 5.83% Pb, 9.99% Zn, and 0.51% Cu was produced (Saskatchewan Geological Survey, 2003). Located about 20 km southeast of the Western Nuclear mine, the McIlvenna Bay deposit is characterized by rounded to oval pyrite in a matrix of pyrite, sphalerite, and minor chalcopyrite, with local galena, marcasite, and magnetite, and consists of five subparallel mineralized lenses (Gracie and Tourigny, 2001). Three of the lenses have estimated reserves of 14.5 Mt grading 0.91% Cu, 6.08% Zn, 0.45% Pb, 0.45 g/t Au, and 24.4 g/t Ag. The two remaining lenses are relatively more Cu-rich and have reserves of 11.7 Mt ore grading 1.79% Cu, 0.48% Zn, 0.57 g/t Au, and 11.96 g/t Ag (Saskatchewan Geological Survey, 2003).

Magmatic Cu-Ni and PGE Deposits of the Glennie-Flin Flon Complex

Although less common than VMS deposits, magmatic Cu-Ni mineralization does occur in the Glennie-Flin Flon complex, with the only economic deposit being the Namew Lake mine, which occurs in the Flin Flon Belt beneath thin Paleozoic cover, approximately 40 km south of Flin Flon (Fig. 3). This deposit was mined from 1988 until 1993, yielding 2.5 Mt of ore grading 1.79% Ni and 0.63% Cu. It is hosted in an altered and deformed pyroxenite, interpreted on the basis of geochemistry as being formed from a komatiitic magma that incorporated sulphides by assimilation during its ascent through juvenile rocks of the Flin Flon Belt (Cumming and Krstic, 1991; Menard et al., 1996). Recent discoveries of PGE mineralization within a number of large mafic and mafic-ultramafic plutons and sills injected into supracrustal rocks, more specifically in the Bear Lake block near Flin Flon, has highlighted the potential for reef-, contact-, and Lac des Iles-type PGE deposits within the broader Flin Flon-Glennie Domain (Peck et al., 2000; Theyer, 2001; Theyer and Heine, 2003). These include the McBratney Lake PGE-Au occurrence (Olivo et al, 2002), as well as the Mikanagan Lake and Tartan Lake gabbroic complexes (Theyer and Heine, 2003). A U-Pb zircon age of 1840 ± 4 Ma reported for one of several gabbros that intrude the Bear Lake block (Bailes and Theyer, 2006) is within analytical error of a 1847 ± 6 Ma age reported for the Namew Lake deposit (Cumming and Krstic, 1991), suggesting that the magmatogenic Cu-Ni-(PGE) mineralization is not related to initial arc or early successor arc volcanism and intraoceanic accretion, but rather to late successor arc magmatism (see Fig. 4D, E).

Accreted Continental Margin Setting

La Ronge-Lynn Lake Domain

The La Ronge Domain (Saskatchewan) and the Lynn Lake/Leaf Rapids Domain (Manitoba) form a more or less continuous curvilinear belt approximately 50 km wide by 750 km long that is exposed from Lac La Ronge in north-central Saskatchewan to north-central Manitoba (Fig. 2). Traditionally, the La Ronge Domain has been separated into two major components, which consist of i) the 1905 to 1876 Ma (Bickford et al., 1987; Corrigan et al., 2001a) Central Metavolcanic Belt, interpreted as a composite volcanic, plutonic and sedimentary assemblage formed in an ensimatic island-arc setting transitional into an ensialic setting (Lewry, 1981; Watters and Pearce, 1987); and ii) the Crew Lake Belt, dominated by metasedimentary rocks of psammitic to pelitic composition that stratigraphically overlie rocks of the Central Metavolcanic Belt (Thomas, 1993). Following more recent fieldwork and analytical studies in the northeastern part of the La Ronge Domain, a more detailed subdivision of the tectonostratigraphic sequence has been established and is summarized below (e.g. Maxeiner et al., 2004). The Central Metavolcanic Belt has been divided into i) a lower succession of ultramafic tectonite, mafic to minor intermediate volcanic rocks, pelite, and exhalative horizons, as well as oxide- and silicate-facies iron formation (collectively named the Lawrence Point assemblage); and ii) an upper succession of felsic volcanic and pyroclastic rocks, interbedded with and laterally grading into a thick succession of volcaniclastic rocks (Reed Lake assemblage). The stratigraphic top of the Reed Lake assemblage conformably grades into the Milton Island assemblage, which consists of a thick, laterally continuous and homogeneous succession of graphite-bearing, mostly semi-pelitic to pelitic gneiss (Corrigan et al., 1998, 1999b). The Milton Island assemblage is tentatively correlated with the compositionally similar Crew Lake Belt to the southwest and the Zed Lake assemblage further east in Manitoba. Collectively, they are interpreted as remnants of a forearc basin formed on the northern flank of the La Ronge and Lynn Lake arcs, prior to the ca. 1.88 Ga collision with the Hearne Craton margin (Fig 4A, B).

The Lynn Lake Belt (Lewry et al., 1990) broadly correlates with the La Ronge Belt in the sense that both contain tectonically imbricated and complexly folded volcanic, plutonic, and sedimentary rocks of ca. 1.92 to 1.88 Ga age. However, there are some major differences as well. The Lynn Lake volcanic-sedimentary successions have been traditionally assigned to the Wasekwan Group (Bateman, 1945). Subsequently, Zwanzig et al. (1999) further subdivided the Wasekwan Group into a northern belt, predominantly consisting of mafic volcanic and volcaniclastic rocks of tholeiitic affinity, and a southern belt consisting of tholeiitic to calc-alkaline volcanic rocks with minor amounts of mid-oceanic ridge basalt (MORB). Rare felsic volcanic rocks from the northern belt have yielded U-Pb zircon ages ranging from 1915 +7/-6 Ma to 1886 ± 2 Ma (Baldwin et al., 1987; Beaumont-Smith and Böhm, 2002). This is significantly older than the southern belt, which has so far yielded U-Pb zircon ages of 1881 +3/-2 Ma, 1864 ±2 Ma, 1856 ±2 Ma, and ca.1842 Ma from felsic to intermediate volcanic rocks (Beaumont-Smith and Böhm, 2002, 2003). Recent Nd isotope data suggest that igneous rocks of the northern belt, which yield ε_{Nd} values of -0.2 to -0.3 and Ndt model ages of ca. 2.5 Ga, have been contaminated by Archean crust or sediments with a high proportion of Archean detritus (Beaumont-Smith and Böhm, 2003). In contrast, igneous rocks of the southern belt yield ε_{Nd} values of +4 to +5 and Ndt model ages <2.2 Ga, suggesting a juvenile origin (ibid.). Both the northern and parts of the southern volcanic belts are intruded by tonalitic plutons of the 1876 to 1871 Ma Pool Lake intrusive suite (Baldwin et al., 1987; Turek et al., 2000).

The Rusty Lake Belt occurs east of the Lynn Lake Belt and is separated from the latter by plutonic and metasedimentary rocks. It is subdivided into four fault-bounded blocks named the Northern, Ruttan, Karsakuwigamak, and Eastern blocks (Baldwin, 1987; Baldwin et al., 1987). Across the entire belt, metamorphic grade is typically represented by lower amphibolite facies and strain is variable, with primary textures and structures locally well preserved. The Northern block contains mostly polymictic conglomerate, siltstone, and metaturbidite, with minor pillowed basalt and sulphide-facies iron formation. The Ruttan block, which hosts the Ruttan Cu-Zn deposit, as well as the Eastern block comprise mafic flows and sills, minor dacitic and rhyolitic flows, volcaniclastic rocks, polymictic conglomerate, and metaturbidite (Powder Magazine Formation). Recent U-Pb geochronology in the Ruttan block has yielded an 1883 ± 2 Ma zircon age for a quartz-feldspar porphyry rhyolite, and a unimodal, 1880 to 1920 Ma age-range detrital source for the directly overlying Powder Magazine metaturbidite (Rayner and Corrigan, 2004). The dated rhyolite and metaturbidite are both intruded by a ca. 1882 ± 3 Ma diorite (unpubl. data, David Corrigan, 2007), suggesting that the Powder Magazine Formation was deposited during a relatively short time interval between approximately 1883 and 1882 Ma. The Karsakuwigamak block, interpreted as a caldera complex (Baldwin, 1987), consists of volcanic-derived conglomerate, sandstone, and siltstone overlain by an 8 km thick sequence of predominantly felsic volcanic flows and tuff (1878 ± 3 Ma; Baldwin et al., 1987), with minor mafic volcanic flows occurring near the base of the volcanic edifice.

Thus far, three temporally distinctive suites of plutonic rocks have been identified in the Lynn Lake Belt. An older suite, ranging in age from 1879 Ma to 1871 Ma, consists of mafic to felsic plutons that intrude both the northern and southern Lynn Lake Belt, as well as the Leaf Rapids Belt (Baldwin et al., 1987; Turek et al., 2000). This particular suite includes the 'El' gabbronorite, which hosts part of the Lynn Lake Ni-Cu deposit (Turek et al., 2000). A younger suite, represented mostly by K-feldspar megacrystic granite to granodiorite potentially related to the Wathaman/Chipewyan batholith, which is described in more detail below, was emplaced during the interval 1.85-1.86 Ga (Van Schmus and Schledewitz, 1986). A third suite consists of mafic to felsic plutons that were emplaced within a relatively short period during the interval 1831-1829 Ma (Turek et al.,

2000; Beaumont-Smith and Böhm, 2002; Rayner and Corrigan, 2004).

Rottenstone and Southern Indian Domains

The Rottenstone and Southern Indian domains structurally overly the La Ronge and Lynn Lake domains. Both are composed of variable proportions of metasedimentary and plutonic rocks, with minor screens of volcanic origin interpreted as higher metamorphic grade equivalents of rocks in the La Ronge and Lynn Lake belts (Lewry and Slimmon, 1985; Lewry et al., 1990). Based on U-Pb geochronology and unpublished Nd tracer isotope data, Corrigan et al. (1999b) interpreted a gneissic tonalite, granodiorite, and granite suite (Crowe Island complex) as the plutonic root of the Central Metavolcanic Belt. In the central Reindeer Lake area in Saskatchewan, rocks with potential back-arc affinity have been described in an area historically considered to be part of the Rottenstone Domain (Corrigan, 2000; Wright, 2001). Lithologies include mafic and felsic volcanic rocks, volcaniclastic rocks associated with siliciclastic and marly metasediments, as well as oxide-facies banded iron formation (Corrigan et al., 1999b). In the Clements Island area (informally referred to as the Clements Island Belt), the sedimentary and metavolcanic rocks are also associated with highly altered, cordierite-anthophyllite-bearing assemblages with small massive sulphide bands containing sphalerite and minor chalcopyrite (Pasquachi Bay showing), indicating potential for VMS or SEDEX deposits. A rhyolite from that belt yielded a U-Pb zircon age of 1905 +17/-5 Ma (Corrigan et al., 2001a), suggesting a temporal relationship with some of the older volcanic rocks of the La Ronge, Lynn Lake, and Flin Flon belts.

The Rottenstone Domain also contains a large proportion of the Milton Island assemblage, described in the *La Ronge-Lynn Lake Domain* section above, as well as the Park Island assemblage (Corrigan et al., 1998), which consists of fluvial to littoral facies siliciclastic metasedimentary rocks interpreted as a foredeep/foreland or molasse basin deposited on the previously accreted La Ronge arc assemblage (Fig. 4B). The stratigraphic equivalent of the Park Island assemblage in Manitoba is referred to as the Partridge Breast assemblage (Corrigan et al., 2001b). Both these units comprise detritus from the La Ronge and Lynn Lake supracrustal rocks and are intruded by the ca. 1.86 to 1.85 Ga Wathaman/Chipewyan batholith, suggesting deposition sometime during the interval 1.88 to 1.86 Ga. Most of the rocks within the Rottenstone and Southern Indian domains have juvenile Nd isotopic signatures (MacHattie, 2001; Corrigan et al., 2001b). However, the recent discovery of Archean gneiss in the southwestern Rottenstone Domain (Bickford et al., 2001) suggests further complexities within this crustal segment along strike.

Wathaman/Chipewyan Batholith

The Wathaman/Chipewyan batholith (Fig. 2) is a major intrusive complex that extends about 700 km from north-central Saskatchewan to northeastern Manitoba (McRitchie, 1977; Lewry et al., 1981; Fumerton et al., 1984; Meyer et al., 1992; MacHattie, 2001). It was emplaced during a relatively narrow time interval between ca. 1862 and 1850 Ma, in a continental arc setting (Fig. 4C). The main body of the batholith is mainly composed of K-feldspar megacrystic, biotite ± hornblende ± titanite-bearing granite, granodiorite, and quartz monzonite, with lesser amounts of diorite and gabbro (Corrigan et al., 1999a). On the basis of geochemistry and Nd tracer isotope data, MacHattie (2001) suggested that 1.86-1.85 Ga plutonic rocks in the La Ronge and Rottenstone domains may be related to the Wathaman/Chipewyan batholith, implying that these crustal blocks must have been already accreted to the Hearne margin prior to 1.86 Ga. The general state of strain in the batholith is quite variable and generally increases southwesterly, from extremely low in Manitoba to locally very high along its southwestern extent in Saskatchewan. In Saskatchewan, the batholith is bounded on its northwestern flank by the Needle Falls and Parker Lake shear zones (Macdonald and Thomas, 1983; Stauffer and Lewry, 1993), and on its southeastern margin by the Reilly Lake shear zone (Lafrance and Varga, 1996). It is important to note that the Needle Falls and Parker Lake shear zones merely rework the intrusive margin of the batholith and are not sutures (Corrigan et al., 1999a,b). The Wathaman/Chipewyan batholith, therefore, stitches the Archean Hearne margin with the La Ronge and Rottenstone domains, implying that the latter were accreted to the paleo-continental margin prior to ca. 1862 Ma. This has implications for the younger 1864 ± 4 Ma to ca. 1843 Ma calc-alkaline volcanic rocks of the Lynn Lake Belt, which may represent the earliest evidence for continental arc volcanism along the accreted margin (Beaumont-Smith and Böhm, 2003).

Mineral Deposits of the Northwestern Reindeer Zone

Volcanogenic Massive Sulphide Deposits

Although modest in comparison with the world-class Flin Flon Belt, the Leaf Rapids and Lynn Lake belts hosted past-producing Cu-Zn VMS deposits, including the Ruttan and Fox mines, respectively (Fig. 3). The 83 Mt (past production and proven reserves) Ruttan deposit, ranking in size amongst the top five percentile of all VMS deposits globally, is a relatively low-grade, bimodal-siliciclastic-type deposit (Barrie et al., 2005). It formed as a series of lenses that occur within a thick sequence of intermediate to felsic volcanic rocks overlying thick, arc tholeiitic basalts with the entire sequence overlain in turn by volcaniclastic and coarse to fine siliciclastic rocks (Ames, 1996; Ames and Taylor, 1996). The ore horizon is overlain by quartz- and feldspar-phyric felsic volcanic rocks and associated crystal ash tuff and tuff breccia, onto which a thick, fining-upwards proximal to distal volcaniclastic to siliciclastic sequence was emplaced (Powder Magazine Formation). The entire mine sequence is flanked to the southeast by the fault-bounded Karsakuwigamak block, which has been interpreted as a ca. 1878 Ma caldera complex dominated by felsic volcanic rocks (Baldwin et al., 1987). The association of a mafic to felsic volcanic edifice overlain by volcaniclastic rocks and flanked by a felsic caldera complex is reminiscent of the tectonic environment proposed for VMS mineralization in the Flin Flon District (Syme et al., 1999), and hints to a potentially similar paleotectonic environment for the Ruttan deposit. Mineralization at Ruttan consists of sphalerite, chalcopyrite, pyrite, and pyrrhotite, and is locally associated with anhydrite (Ames, 1996; Gale, 2003; Barrie et al., 2005).

Production exceeded 63 Mt of ore averaging 1.43% Zn, 1.22% Cu, 0.46 g/t Au, and 12.26% Ag.

The Fox mine, located in the Lynn Lake Belt approximately 45 km southwest of the town of Lynn Lake, occurs in highly altered mafic volcanic, volcaniclastic, and metasedimentary rocks of the Wasekwan Group, the main supracrustal unit of the Lynn Lake Belt. Both the Ruttan and Fox deposits are associated with intense alteration of host mafic to intermediate volcanic rocks, which produced cordierite-anthophyllite and cordierite-garnet-chlorite metamorphic assemblages associated with mineralization. A number of smaller Cu-Zn VMS occurrences have been identified in the La Ronge Belt, but none have proven economic thus far. However, their association with highly altered mafic volcanic and volcaniclastic rocks characterized by Mg and Al enrichment producing cordierite-anthophyllite ± garnet metamorphic assemblages, suggests a genesis similar to that of the Fox and Ruttan deposits. The former contained over 11 Mt grading 1.82% Cu and 1.78% Zn (McClenaghan et al., 2003).

Magmatic Ni-Cu-PGE Deposits

The accreted La Ronge, Rottenstone, and Lynn Lake belts are host to volumetrically small but locally important magmatic Cu-Ni-(PGE) occurrences and deposits that include the past-producing Lynn Lake and Rottenstone Lake mines (Fig. 3), as well as a number of subeconomic occurrences scattered across the northwestern Reindeer zone. Most of the Cu-Ni deposits occur in mafic and mafic-ultramafic plugs and sill-like bodies that are believed to be of subvolcanic setting and possible feeders to overlying tholeiitic volcanic sequences (Hulbert, 1995). The mineralized intrusions in this large metallotect consist of an early non-mineralized gabbroic phase injected by younger, more primitive norite-orthopyroxenite-peridotite bodies that are proximal to disseminated and massive Ni-Cu sulphide zones (Hulbert, 1995). However, other deposits, such as the one at Rottenstone Lake, occur in small harzburgitic plugs that intrude migmatitic clastic metasedimentary rocks (MacLachlan, 2003). The largest magmatic Cu-Ni deposits are located around the town of Lynn Lake, where the Lynn Lake and El plutons, mined between 1962 and 1983, yielded 20.2 Mt of ore grading 1.023% Ni and 0.535% Cu, making this the fourth largest producer in Canada after Sudbury, Thompson, and Raglan (Hulbert, 1988). The much smaller Rottenstone Lake deposit (Rottenstone Domain) was in operation from 1965 to 1968 and yielded 5,047 tons of concentrate grading 10.36% Ni, 5.74% Cu, 3.4 g/t Au, 18.2 g/t Ag, 11.3 g/t Pt, and 18.2 g/t Pd. Published reserves were estimated at approximately 55,000 tons at 2% Ni, 2% Cu, 7.5 g/t Pt, and 5.1 g/t Pd (Saskatchewan Geological Survey, 2003).

Potential for Other Magmatic-Related Deposits

The Wathaman/Chipewyan batholith and related satellite intrusions form a continental arc exposed along the northwestern flank of the Reindeer zone. This tectonomagmatic setting is normally favourable for epithermal Au as well as porphyry Cu-Au ± skarn (e.g. Sillitoe, 2000). However, in contrast to younger, less eroded regions from the Cordilleras, for example, the Wathaman/Chipewyan batholith represents a more deeply eroded equivalent, reducing the likelihood that high structural level intrusions and related hydrothermal systems are preserved. This might explain why no such deposit has been found so far in this part of the Reindeer zone. Recently, REE-rich carbonatite associated with an important zone of fenitization has been discovered in the Lynn Lake Belt (Mumin and Trott, 2003). However, further studies are necessary in order to fully understand the tectonic context of this new mineral deposit and its postulated link to iron oxide copper-gold (IOCG) or other types of hydrothermal deposits (see Corriveau et al., 2007).

Granville Lake Structural Zone

The variably strained, tectonized zone bounding the La Ronge-Lynn Lake-Leaf Rapids belts to the north and the predominantly turbiditic, ca. 1.85 to 1.84 Ga Burntwood Basin to the south, has been the focus of recent bedrock geological studies (e.g. Corrigan et al., 1998; Harper, 1998; Zwanzig, 2000; Maxeiner et al., 2004) and deep reflection seismic surveys (White et al., 2000; Corrigan et al., 2005). This tectonized zone has been recognized by different workers along its strike length (Fig. 3) and segments have been variably named "MacLean Lake Belt" (Lewry, 1984), "Levesque Bay assemblage" (Corrigan et al., 1998) and "Granville Lake structural zone" (Zwanzig, 2000). In the southern Reindeer Lake area, highly strained, structurally imbricated supracrustal rocks that include altered (carbonate alteration) mafic sills and flows, metapelite, sulphide-facies iron formation, as well as minor quartzite, calc-silicate gneiss (marl?) and marble (Levesque Bay assemblage: Corrigan et al., 1999b) have been interpreted as a distinctive supracrustal package, older than the Burntwood Group that forms most of the Kisseynew Domain to the south. The distinctive nature of the Levesque Bay assemblage has been confirmed by U-Pb zircon dating of two elongate tonalitic intrusions which yielded ages of ca. 1867 ± 3 Ma and 1897 +18/-15 Ma, respectively (Corrigan et al., 2001a), suggesting that at least part of this belt is older than 1897 Ma and therefore cannot be related to the Burntwood Group, as was originally inferred. Along strike to the southwest, this supracrustal succession is similar to, and sits at an equivalent tectonostratigraphic level as, the MacLean Lake Belt, which has been interpreted by Lewry (1984) as the thrust-imbricated remnant of a fore-arc basin. However, geochemical studies on mafic volcanic rocks of the MacLean Lake Belt and Levesque Bay assemblages yield mid-ocean ridge basalt (MORB) signatures, suggesting perhaps a back-arc setting (Harper and Hasanie, 1999; Demmans et al., 2002). Further east and more or less along strike in Manitoba, a similar, strongly imbricated supracrustal assemblage named the Granville Lake structural zone (GLSZ) contains mafic volcanic and volcaniclastic subunits that also yield MORB-like geochemical signatures (Zwanzig et al., 1999). At that locality, it corresponds to a series of well developed north-dipping, crustal-scale reflectors that are linked at depth to a 5 km jog in the Moho (White et al., 2000). At the orogen scale, this approximately 800 km linear feature, which includes the MacLean Lake Belt, the Levesque Bay assemblage, and the Granville Lake structural zone (Fig. 3), may represent a first-order tectonic break separating lower plate turbidites of the

Burntwood Group (Kisseynew Domain sensu stricto) from an upper plate imbricate tectonic stack that includes remnants of a back-arc basin, siliciclastic molassic sedimentary rocks (McLennan and Sickle meta-arkose) and rare ultramafic mega-boudins that include spinel lherzolite and pyroxenite (Corrigan et al., 1998; Zwanzig, 2000; Corrigan and Rayner, 2002). The Granville Lake structural zone may represent a ca. 1.85 Ga suture that records closure of an ocean basin that previously existed between the previously accreted La Ronge-Lynn Lake-Leaf Rapids arc terranes and the Flin Flon Belt (Zwanzig, 2000; Maxeiner et al., 2004; Corrigan et al., 2005) (see Fig. 4D).

From a metallogenic perspective, the postulated presence of a suture zone at the site of a former continental margin accretionary complex opens up a number of possibilities in terms of types of mineralization, including orogenic Au, VMS associated with oceanic crust, as well as Besshi-type VMS deposits. The Brabant Lake Cu-Zn-Ag-Au deposit (Fig. 3) occurs as elongate conformable lenses hosted predominantly in garnet-biotite paragneiss stratigraphically intercalated with altered mafic volcanic rocks of the MacLean Lake Belt. Further northeast along strike in the Levesque Bay assemblage, stratiform pyrrhotite and pyrite mineralization (barren sulphides) occurs in metamorphosed semi-pelitic rocks interlayered with altered amphibolite (Corrigan et al., 1998; Demmans et al., 2002), suggesting a similar metallogenic setting to the one reported for the Brabant Lake deposit. Even further along strike to the east of Reindeer Lake, Kaszycki (1989) reported one of the largest Cu-Zn-Au surficial sediment anomalies in the Granville Lake area in northern Manitoba. Assuming that the source of these anomalies is local and not the result of extensive down-ice surficial sediment dispersal, it would further indicate the potential for VMS and/or SEDEX deposits along the broad Granville Lake structural zone.

Back-Arc Basin

Kisseynew Domain

The Kisseynew Domain (Fig. 2) forms the youngest tectonostratigraphic entity in the Reindeer zone and consists, in large part, of migmatitic meta-turbidites of the Burntwood Group, interpreted to have been deposited in a ca. 1.85 to 1.84 Ga back-arc basin formed between the La Ronge/Lynn Lake and Flin Flon-Glennie arc complexes (Zwanzig, 1990; Ansdell et al., 1995) (see Fig. 4E). The Burntwood Group is unconformably and/or disconformably overlain along most of its perimeter by fluvial-alluvial sediments deposited between ca. 1845 and 1835 Ma, collectively interpreted as syncollisional molasse basins (Stauffer, 1990). These include the McLennan and Sickle arkoses on the northern flank, the Missi Group on the southern flank, and the Grass River Group on the eastern flank. Closure of the Burntwood back-arc basin may have occurred shortly after its deposition at ca. 1.84 Ga, perhaps as a result of the collision of the Sask craton with the Flin Flon-Glennie complex (Fig. 4F). This age coincides with the development of the Pelican thrust at the interface between the Sask Craton and overlying Glennie Domain (Ashton et al., 2005). Inversion of the Burntwood Basin is also reflected in the upward coarsening of sediments from turbidite to intraformational conglomerate, as reported in the northern part of the basin in Manitoba (Corrigan and Rayner, 2002). To date, there are no known VMS deposits associated with the Burntwood Group, nor has actual ocean floor been associated with the latter.

Archean Microcontinent

Sask Craton

Exposures of the predominantly buried Sask Craton (Lewry et al., 1994; Ansdell et al., 1995) have been identified in three, relatively small, basement windows in the Glennie and Flin Flon domains and, based on seismic images, this Archean fragment is interpreted to underlie a large part of the Reindeer zone at the Moho level (Hajnal et al., 2005). Where exposed, the Sask Craton footwall is separated from overlying juvenile Proterozoic rocks by high-metamorphic-grade mylonite zones with top-to-the southwest sense of shear (Ashton et al., 2005). Based on interpretations of regional aeromagnetic data and rare deep drillholes, a much larger segment of the Sask Craton is interpreted to occur south of the Glennie Domain, buried beneath Phanerozoic cover (Hajnal et al., 2005). Tracer isotope and U-Pb geochronological data obtained mainly from the limited surface exposures and drillholes suggest that the Sask Craton is predominantly formed of earliest Paleoproterozoic (2.40-2.50 Ga) felsic to mafic igneous rocks intruded in >3.1 Ga calc-alkaline orthogneiss (Chiarenzelli et al., 1998; Bickford et al., 2005; Rayner et al., 2005b). Although direct U-Pb age-dating of synthrusting structures indicates an initial collisional interaction between the Sask Craton and the Flin Flon-Glennie domains by 1826 Ma and perhaps as early as 1845 Ma (Ashton et al., 2005) (Fig. 4F), there is evidence in the form of ca. 2.4 to 2.5 Ga detrital zircons deposited during the interval 1.88 to 1.86 Ga in the northwestern Reindeer zone (Rayner and Corrigan, 2004) that crustal blocks or fragments of Sask-age crust might have been locally present in the 2.07 to 1.88 Ga Manikewan Ocean. What role the presence of the Sask Craton and similar-age crustal fragments might have had on arc nucleation and subsequent tectonostratigraphic evolution and metallogeny remains to be further tested. The Sask Craton hosts the Cretaceous age Fort-a-la-Corne diamondiferous kimberlites (Leckie et al., 1997).

Orogenic Lode Gold Deposits of the Reindeer Zone

Gold in the Reindeer zone is predominantly associated with juvenile arc assemblages and related metasedimentary rocks (metallotects '2' and '4' in Fig. 2). Gold deposits are either genetically associated with VMS deposits or occur as shear zone-hosted lode or orogenic Au. In general, Au is typically present in minor (<1 g/t) to high (>15 g/t) concentrations within the majority of the economic VMS deposits and is extracted as a by-product of base metal smelting and processing, together with Ag and other precious metals. Gold deposits of that type are locally important, as demonstrated by the approximately 3 million ounces extracted thus far from massive sulphide deposits of the Flin Flon camp (Saskatchewan Geological Survey, 2003). Some of the remobilized deposits are particularly well endowed, such as the 1.5 million ounce New Britannia past-producing mine in the Snow Lake Belt, which still ranks as the largest Paleoproterozoic-hosted Au deposit in Canada (Galley et al., 1986; Richardson and Ostry, 1996; Gagné et al., 2005).

However, by far the most common type of Au occurrence in the Reindeer zone consists of orogenic lode Au deposits (e.g. Ansdell and Kyser, 1992) that are particularly abundant in the La Ronge Belt, but occur as well in the Lynn Lake Belt and in the Glennie-Flin Flon complex. In the Lynn Lake Belt, Au deposits forming along the multi-stage reactivated Johnson shear zone (Beaumont-Smith and Böhm, 2003) fall into this category. The latter was formed in a previously accreted tectonic stack of volcanic rocks representing a variety of tectonic settings, which was affected by intense transpressional deformation during the collisional phase of the orogen during the interval 1.83 to 1.80 Ga (Fig. 4G). Gold mineralization in the La Ronge Belt is mostly structurally controlled and of quartz-vein type, but also includes other settings such as VMS-, epigenetic iron formation-, and sediment-hosted (Lafrance and Heaman, 2004). Host rocks for the structurally controlled examples include metavolcanic and metasedimentary rocks that were reactivated in syn-peak metamorphic high-temperature shear zones, but also include plutons of varied composition that intrude the supracrustal rocks. There are no mines that are presently active. Past-producers (locations shown in Fig. 3) include the Contact Lake mine (188,210 oz Au averaging 8 g/t), the Star Lake mine (76,947 oz Au averaging 14.7 g/t), the Jasper mine (82,697 oz Au averaging 18.8 g/t), the Jolu mine (203,751 oz Au averaging 14.4 g/t), and the Komis mine (26,885 oz at a grade of 0.22 oz/ton) (Saskatchewan Geological Survey, 2003).

In the Glennie Domain, the Seabee mine (Fig. 3) is presently the only operative Au mine in Saskatchewan. Gold ore at that locality is structurally controlled and occurs in arrays of en echelon quartz veins that were emplaced under ductile conditions in gabbroic rocks of the Laonil Lake intrusive complex (Tourigny, 2003). The quartz veins are isoclinally folded about steeply plunging axes. Gold occurs in visible native form as well as Au-bearing pyrite, chalcopyrite, and pyrrhotite. Other minerals present in the veins include tourmaline, barite, molybdenite, magnetite, actinolite-tremolite, sericite, sphalerite, and tellurides (Saskatchewan Geological Survey, 2003). The quartz veins and associated shear zones are spatially associated with pervasive potassic alteration, which is marked by replacement of hornblende by biotite in the host gabbro. The source of Au in this deposit was likely the mafic volcanic and/or minor greywacke country rocks into which the Laonil Lake intrusive complex was emplaced (Delaney, 1995b). Between 1991 and 2001, 500,000 oz of Au were produced from 2.1 Mt of ore. Reserves reported in 2002 were 1.6 Mt grading 7.9 g/t (0.245 oz/ton).

Summary and Discussion

The Trans-Hudson Orogen arguably forms one of the most extensive orogenic belts of Laurentia and hosts some of the richest and most varied Paleoproterozoic mineral deposits in the Canadian Shield. There are many factors that could have preferentially favoured preservation of economic mineral deposits in the Reindeer zone or southwestern part of the Trans-Hudson Orogen, but the most obvious is the relatively lower regional metamorphic grade preserved at surface, as well as the overall abundance of juvenile Paleoproterozoic crust in contrast to other parts of the orogen. There are many possible mechanisms that would have allowed such good preservation of supracrustal rocks, with the main ones possibly related to plate configuration and duration of convergence during orogenesis. For example, preservation of juvenile Proterozoic crust could have resulted from a 'shielding effect' of the buoyant Sask Craton (microcontinent) during the Superior-Hearne collision, or by the predominantly transpressional nature of the terminal collision between juvenile terranes of the Reindeer zone and the western edge of the Superior Craton, or a combination of both. Also, the duration of collision along the leading edge of the Superior Craton was relatively short lived (1.83-1.78 Ga), with most of the convergence back-stepping to the trailing edge of the Superior Province (Medicine Bow and Colorado orogens) from about 1.78 Ga onwards (Chamberlain, 1998; Sims and Stein, 2003). Whatever the actual cause(s), the result was the preservation of relatively large amounts of juvenile oceanic, arc-volcanic, and sedimentary rocks relative to older re-activated crust. This contrasts with the Ungava Peninsula indentor, for example, which accommodated a more orthogonal convergence, resulting in a greater amount of exhumation of juvenile supracrustal rocks, with the latter now mostly preserved in an orogen-parallel klippe overthrust on the Superior Craton margin (Lucas, 1989).

In the western Trans-Hudson Orogen, with some exceptions, each of the four broad metallogenic associations 1) mafic plutonic-hosted PGE and sediment-hosted base metal; 2) VMS and epigenetic Au; 3) magmatic Cu-Ni-(PGE); and 4) orogenic Au ± VMS ± magmatic Cu-Ni-(PGE)) can be linked to specific tectonic or geodynamic settings a) rifted continental margin; b) intraoceanic arc collage; c) ensialic back-arc; and d) accreted continental margin, respectively (Fig 2). In the Peter Lake Domain, early continental extension at ca. 2562 Ma produced widespread mantle-derived mafic magmatism, including layered gabbro and gabbronorite bodies, locally yielding high PGE values (Hulbert, 1988; Maxeiner and Rayner, 2005; Maxeiner, 2006). Much later, at ca. 2.07 Ga, actual rifting and drifting took place along the same approximate margin, leading to the deposition of numerous sub-economic sedimentary exhalative and sandstone-hosted Pb-Zn deposits. This lower sequence was unconformably overlain by a foredeep succession (Upper Wollaston Group) that contains numerous, although sub-economic, arkose- and conglomerate-hosted Cu and Cu-U deposits. The general stratigraphic sequence and type of mineralization observed in the lower Wollaston Group is similar to that described from other passive margin settings such as the Phanerozoic Selwyn Basin in the Yukon (Gordey, 1980; Abbott et al., 1987; Goodfellow, 2007). The lower Wollaston passive margin setting is also broadly similar in tectonic setting and age to the Black Angel Pb-Zn mine in west Greenland (Pedersen, 1981).

At the opposite margin of the Reindeer zone, along the Superior boundary zone, the Ospwagan Group does not appear to host any significant economic mineral deposits despite its similar paleo-tectonic setting to the Wollaston Group. However, remnants of this unit are relatively thin and poorly exposed (Scoates et al., 1977), limiting the potential discovery of deposits, if they were ever present. At least three Proterozoic mafic dyke 'events' occurred along that margin. The earliest, represented by the ca. 2.07 to 2.09 Ga

Birthday Rapids and Cauchon dykes (Heaman and Corkery, 1996), is synchronous with the Courtney Lake volcanics developed in the lower Wollaston Supergroup, suggesting that extension may have been more or less coeval along both the Hearne and Superior margins. The second event involved the emplacement of the Molson dykes at ca. 1883 Ma, with the latter emplaced both in the Archean basement and in the Proterozoic covers sequence, and correlated with the Thompson Nickel Belt magmatic Cu-Ni deposits (Hulbert et al., 2005). The third and final event involved emplacement of komatiitic to basaltic flows, dykes, and sills in the Bah Lake and Grass River groups, at about 1864 Ma (Zwanzig et al., 2003; Percival et al., 2005). At the orogen scale, is interesting to note that the Lynn Lake Ni deposits were hosted in relatively small, ca. 1871 Ma gabbro-norites that were penecontemporaneous with the Molson 'event', but were genetically unrelated to the latter, suggesting that although the Thompson Nickel Belt is a primary Cu-Ni metallotect, the accreted arc terranes (areas '2' and '4' in Fig. 2) within the orogen internides are also prospective.

The majority of economic VMS deposits in the Flin Flon Belt have been linked to oceanic arc complexes (Syme et al., 1999), with none thus far associated with ocean floor, ocean plateaus, or ocean island environments. However, the various deposits dispersed across the belt show a great variability with respect to temporal or geodynamic evolution of arc settings, from nascent arc, through early rifted arc, through mature and rifted oceanic arc (Fig. 4A). Most importantly, the very large deposits of the Flin Flon camp seem to be specifically related to the formation of caldera complexes in an early rifted oceanic arc, with a modern analogue potentially found in the Lau Basin (Syme et al., 1999) and a more ancient analogue in the Archean Noranda deposit in the Superior Province (e.g. Franklin et al., 1981). Outside of the Flin Flon Belt there exists a broader range of paleo-tectonic settings for VMS-type mineralization including some that are in many ways similar to those of the Flin Flon camp, including the past-producing Ruttan and Fox mines in the Lynn Lake and Leaf Rapids belts, respectively (Ames, 1996; Barrie et al., 2005). However, a number of smaller deposits that are mainly hosted by sediments with minor MORB-like mafic volcanic rocks suggest a more primitive oceanic crust environment typical of Besshi-type deposits (e.g. Fox, 1984). An example of this type of mineralization is found in the Brabant Lake deposit in the MacLean Lake Belt, as well as the past-producing Anglo Rouyn mine in the Glennie Domain. Extensive stratabound barren sulphide deposits observed in the Levesque Bay assemblage occur at a similar stratigraphic level as the Brabant Lake deposit and may represent more distal deposits related to the putative hydrothermal system.

In orogenic settings worldwide, orogenic Au or lode Au deposits are typically associated with greenschist-facies metamorphic conditions at convergent plate margins in accretionary and collisional orogens (Groves et al., 1998). Given the presence of greenschist to lowermost amphibolite-facies rock assemblages, the southwestern Trans-Hudson Orogen seems particularly prospective for that commodity. In the Reindeer zone of the Trans-Hudson Orogen, the two most common types of Au mineralization occur as either remobilized deposits associated with original VMS mineralization (Fig. 4B), or as structural or lode Au associated with earlier accretion or terminal collision (Fig. 4F, G). The relationship between Au, VMS deposits, and regional metamorphism has been the subject of numerous former and ongoing studies, with much of the evidence suggesting remobilization of Au immediately preceding or contemporaneous with regional metamorphism associated with the collisional phase of the Trans-Hudson Orogen (e.g. Froese and Moore, 1980; Galley et al., 1986).

Overall, the various styles of Au mineralization encountered in the Reindeer zone reflect the diversity of paleo-tectonic environments preserved in the orogen and the relatively widespread preservation of rocks that were originally formed at mid- to upper-crustal levels during accretion and subsequent collision. Given the abundance of Au showings, prospects, deposits, and past-producing mines in the La Ronge and Lynn Lake belts, it is interesting to note the paucity of reported mineralization in the Rusty Lake Belt, which shares many other lithological and tectonothermal similarities.

Apart from deposits of the Thompson Nickel Belt located along the Superior boundary zone, other Paleoproterozoic magmatic Ni-Cu-(PGE) deposits or occurrences are found elsewhere in the western Trans-Hudson Orogen, mainly within the accreted terranes of the Reindeer zone, but also in the reactivated Archean Hearne margin (Peter Lake Domain). Thus far, the largest exploited deposits outside of the Thompson Nickel Belt include the relatively small but rich deposits of the Lynn Lake Belt, with other smaller deposits and past-producing mines located in the Flin Flon (Namew Lake mine) and Rottenstone (Rottenstone mine) domains. Although the timing of emplacement of Thompson Nickel Belt and Lynn Lake intrusions that carry Ni-Cu-(PGE) mineralization is relatively similar (ca. 1880 Ma for the former (Hulbert et al., 2005) and ca. 1871 Ma for the latter (Turek et al., 2000)), the tectonic settings and magmatic associations are quite different. In the Thompson Nickel Belt, Ni-Cu-PGE mineralization is mainly hosted in metamorphosed ultramafic sills that intrude both the Superior Craton and Paleoproterozoic Ospwagan Group cover rocks, with the source of sulphur thought to have been derived in large part from the Pipe Formation (Macek and Bleeker, 1989). Current tectonomagmatic models for the generation of mafic-ultramafic dykes and sills of the Thompson Nickel Belt range from plume-generated to an ensialic back-arc setting (e.g. Percival et al., 2005, and references therein). In contrast, the Ni-Cu-(PGE) deposits of the Lynn Lake, La Ronge, and Flin Flon domains are exclusively related to subvolcanic mafic and ultramafic plutons interpreted as feeders to the overlying mafic volcanic sequences that were formed in an intraoceanic arc and successor arc settings (Hulbert, 1995; Theyer and Heine, 2003). The mostly PGE-rich, Ni-Cu-poor mineralization in the Peter Lake Domain formed in the ca. 2562 Ma Swan River complex (Maxeiner and Rayner, 2005). The latter intrudes mainly ca. 2.68 Ga orthogneiss and plutonic rocks, with very little sulphidic metasedimentary rocks. In terms of range of rock compositions and textures, the mineralized Peter Lake Domain intrusions can be compared with the Lac des Iles complex, which hosts the single operating PGE-only mine in Canada (Maxeiner, 2006).

Overall, the Reindeer zone of the Trans-Hudson Orogen represents a unique Proterozoic setting within Laurentia. This uniqueness is not necessarily defined by the types of deposits encountered, since most have characteristics of deposits described elsewhere in Archean to Phanerozoic examples, but by the diversity of mineral deposits and their relatively good state of preservation. As argued in the various sections above, the configuration of colliding Archean plates and microplates, as well as their relative vectors of displacement, was likely the most important factor influencing the preservation of juvenile Paleoproterozoic crust. Other parts of the Trans-Hudson Orogen that preserved relatively voluminous amounts of Paleoproterozoic cover rocks include the Belcher Island/Sutton re-entrant and the Labrador Trough (Clark and Wares, 2004). Conversely, parts of the Trans-Hudson Orogen that lie north of the northwest Superior and Ungava indentors were more profoundly exhumed, are generally of higher metamorphic grade, and have noticeably less Proterozoic juvenile supracrustal rocks preserved. Interestingly, what stands out in the Reindeer zone is the absence of some types of deposits, such as porphyry-type Cu-Mo±Au and skarn deposits, Cyprus-type or ophiolite-hosted VMS deposits, and metalliferous black shales. However, the present-day Reindeer zone likely represents a fraction of what was previously a much larger collage of juvenile rocks prior to terminal collision and is yet to be systematically explored for a wider range of mineral commodities and deposit types.

Acknowledegments

This synthesis has benefited tremendously from the past contributions of numerous researchers, as well as discussions with many colleagues. In particular, we are indebted to colleagues from the Saskatchewan and Manitoba surveys, industry, academia, and the Geological Survey of Canada for having generously shared data and ideas over the years. Comments from reviewers Kevin Ansdell and John Percival, as well as editor Wayne Goodfellow, greatly improved the clarity and contents of the manuscript. Kim Nguyen assisted in drafting Figure 3.

References

Abbott, J.G., Gordey, S.P., and Tempelman-Kluit, D.J., 1987, Setting of stratiform, sediment-hosted lead-zinc deposits in Yukon and northeastern British Columbia, *in* Morin, J.A., (ed.). Mineral Deposits of Northern Cordillera Symposium: Canadian Institute of Mining and Metallurgy, Special Volume 37, p. 1-18.

Ames, D.E., 1996, Stratigraphic and tectonic setting of the Paleoproterozoic Ruttan Cu-Zn VHMS deposit, Rusty Lake belt, Trans-Hudson Orogen, *in* Bonham-Carter, G.F., Galley, A.G., and Hall. G.E.M., eds., EXTECH I: A Multidisciplinary Approach to Massive Sulphide Research in the Rusty Lake-Snow Lake Greenstone Belts: Geological Survey of Canada, Bulletin 426, p. 15-43.

Ames, D.E., and Taylor, C., 1996, Geology of the West Anomaly orebody, Ruttan volcanic-hosted massive sulphide deposit, Proterozoic Rusty Lake belt, *in* Bonham-Carter, G.F., Galley, A.G., and Hall. G.E.M., eds., EXTECH I: A Multidisciplinary Approach to Massive Sulphide Research in the Rusty Lake-Snow Lake Greenstone Belts, Geological Survey of Canada, Bulletin 426, p. 45-76.

Anbar, A.D., and Knoll, A.H., 2002, Proterozoic ocean chemistry and evolution: A bioinorganic bridge?: Science, v. 297, p. 1137-1142.

Annesley, I.R., Madore, C., and Krogh, T.E., 1992, U-Pb zircon, titanite and monazite age from the Wollaston domain -A summary, *in* Summary of Investigations 1992: Saskatchewan Geological Survey, Miscellaneous Report 92-4, p. 61-65.

Annesley, I.R., Madore, C., Shi, R., and Krogh, T.E., 1997, U-Pb geochronology and thermotectonic events in the Wollaston Lake area, Wollaston Domain: A summary of 1994-1996 results, *in* Summary of investigations 1997: Saskatchewan Geological Survey, Miscellaneous Report 97-4, p. 162-173.

Annesley, I.R., Madore, C., Krogh, T.E., Kwok, Y.Y. and Kamo, S.L., 1999, New U-Pb zircon and monazite geochronological results for Archean and Paleoproterozoic basement to the southeastern part of the Athabasca Basin, Saskatchewan, *in* Summary of Investigations 1997: Saskatchewan Geological Survey, Miscellaneous Report 99-4.2, p. 90-99.

Ansdell, K.M., 2005, Tectonic evolution of the Manitoba-Saskatchewan segment of the Paleoproterozoic Trans-Hudson Orogen, Canada: Canadian Journal of Earth Sciences, v. 42, p. 741-759.

Ansdell, K.M., and Kyser, T.K., 1992, Mesothermal gold mineralization in a Proterozoic greenstone belt: Western Flin Flon Domain, Saskatchewan, Canada: Economic Geology, v. 87, p. 1496-1524.

Ansdell, K.M., Lucas, S.B., Connors, K., and Stern, R.A., 1995, Kisseynew metasedimentary gneiss belt, Trans-Hudson Orogen (Canada): Back-arc origin and collisional inversion: Geology, v. 23, p. 1039-1043.

Ansdell, K.M., MacNeil, A., Delaney, G.D., and Hamilton, M.A., 2000, Rifting and development of the Hearne passive margin: Age constraint from the Cook Lake area, Saskatchewan: Geological Association of Canada - Mineralogical Association of Canada, GeoCanada 2000, Program with Abstracts, v. 25.

Ashton, K.E., 1999, A proposed lithotectonic domain reclassification of the southeastern Reindeer Zone in Saskatchewan, *in* Summary of Investigations 1999, Volume 1: Saskatchewan Geological Survey, Saskatchewan Energy and Mines, Miscellaneous Report 99-4, p. 92-100.

Ashton, K.E., Lewry, J.F., Heaman, L.M., Hartlaub, R.P., Stauffer, M.R., and Tran, H.T., 2005, The Pelican Thrust Zone: Basal detachment between the Archean Sask Craton and Paleoproterozoic Flin Flon - Glennie complex, western Trans-Hudson Orogen: Canadian Journal of Earth Sciences, v. 42, p. 685-706.

Bailes, A.H., and Galley, A.G., 1999, Evolution of the Paleoproterozoic Snow Lake arc assemblage and geodynamic setting for associated volcanic-hosted massive sulphide deposits, Flin Flon Belt, Manitoba, Canada: Canadian Journal of Earth Sciences, v. 36, p. 1789-1805.

Bailes, A.H., and Syme, E.C., 1989, Geology of the Flin Flon - White Lake area: Manitoba Energy and Mines, Geological Report GR87-1.

Bailes, A.H., and Theyer, P., 2006, Wonderland gabbro: U-Pb age and geological interpretations, Manitoba (NTS 63K 12NE), *in* Report of Activities: Manitoba Geological Survey, Manitoba Science, Technology, Energy and Mines, p. 35-41.

Baldwin, D.A., 1987, Physical Volcanology of the Northwest Segment of the Karsakuwigamak Block, Proterozoic Rusty Lake Metavolcanic Belt, Northern Manitoba: Ph.D. thesis, University of Manitoba, Winnipeg, Manitoba, 438 p.

Baldwin, D.A., Syme, E.C., Zwanzig, H.V., Gordon, T.M., Hunt, P.A., and Stevens, R.D., 1987, U-Pb zircon ages from the Lynn Lake and Rusty Lake metavolcanic belts, Manitoba: Two ages of Proterozoic magmatism: Canadian Journal of Earth Sciences, v. 24, p. 1053-1063.

Barrie, C.T., Taylor, C., and Ames, D., 2005, Geology and metal contents of the Ruttan volcanogenic massive sulphide deposit, northern Manitoba, Canada: Mineralium Deposita, v. 39, p. 795-812.

Bateman, J.D., 1945, McVeigh Lake Area, Manitoba: Geological Survey of Canada, Paper 45-14, 34 p.

Beaumont-Smith, C.J., and Böhm, C.O., 2002, Structural analysis and geochronological studies in the Lynn Lake greenstone belt and its gold-bearing shear zones (NTS 64C10, 11, 12, 14, 15 and 16), *in* Report of Activities 2002: Manitoba Geological Survey, Manitoba Industry, Trade and Mines, p. 159-170.

——— 2003, Tectonic evolution and gold metallogeny of the Lynn lake greenstone belt, Manitoba (NTS 64C10, 11, 12, 14, 15 and 16), *in* Report of Activities 2003: Manitoba Geological Survey, Manitoba Industry, Trade and Mines, p. 39-49.

Bédard, J.H., 2006, A catalytic delamination-driven model for coupled genesis of Archean crust and sub-continental lithospheric mantle: Geochimica and Cosmochimica Acta, v. 70, p. 1188-1214.

Bickford, M.E., Van Schmus, W.R., Collerson, K.D., and Macdonald, R., 1987, U-Pb zircon geochronology project: New results and interpretations, *in* Summary of Investigations 1987: Saskatchewan Geological Survey, Miscellaneous Report 87-4, p. 17-22.

Bickford, M.E., Collerson, K.D., and Lewry, J.F., 1994, Crustal histories of the Hearne and Rae provinces, southwestern Canadian Shield, Saskatchewan: Constraints from geochronologic and isotopic data: Precambrian Research, v. 68, p. 1-21.

Bickford, M.E., Hamilton, M.A., Wortman, G.L., and Hill, B.M., 2001, Archean rocks in the southern Rottenstone Domain; significance for the evolution of the Trans-Hudson Orogen: Canadian Journal of Earth Sciences, v. 38, p. 1017-1025.

Bickford, M.E., Mock, T.D., Steinhart III, W.E., Collerson, K.D., and Lewry, J.F., 2005, Origin of the Archean Sask Craton and its extent within the Trans-Hudson Orogen: Evidence from Pb and Nd isotopic compositions of basement rocks and post-orogenic intrusions: Canadian Journal of Earth Sciences, v. 42, p. 659-684.

Bleeker, W., 1990, New structural-metamorphic constraints on Early Proterozoic oblique collision along the Thompson Nickel Belt, Manitoba, Canada, *in* Lewry, F., and Stauffer, M.R., eds., The Early Proterozoic Trans-Hudson Orogen of North America: Geological Association of Canada, Special Paper 37, p. 57-73.

Bleeker, W., and Maceck, J.J., 1996, Evolution of the Thompson Nickel Belt, Manitoba Setting of Ni-Cu Deposits in the Western Part of the Circum-Superior Boundary Zone: Geological Association of Canada - Mineralogical Association of Canada, Joint Annual Meeting, Field Trip Guidebook A-1, 44 p.

Bretzlaff, R.E., 1976, Genesis of Uranium Mineralization at Duddridge Lake, Saskatchewan: B.Sc. thesis, Carleton University, Ottawa, Ontario, 43 p.

Chamberlain, K.R., 1998, Medicine Bow Orogeny: Timing of deformation and model of crustal structure produced during continent - arc collision, ca. 1.78 Ga, southeastern Wyoming: Rocky Mountain Geology, v. 33, p. 259-277.

Chiarenzelli, J.R., and Macdonald, R., 1986, A U/ Pb zircon date for the Ennadai Group, *in* Summary of Investigations 1986: Saskatchewan Geological Survey, Miscellaneous Report 86-4, p. 112-113.

Chiarenzelli, J.R., Aspler, L.B., Villeneuve, M., and Lewry, J.F., 1998, Early Proterozoic evolution of the Saskatchewan craton and its allochthonous cover, Trans-Hudson Orogen: Journal of Geology, v. 106, p. 247-267.

Clark, T., and Wares, R., 2004, Synthèse Lithotectonique et Métallogénique de l'Orogène du Nouveau-Québec (Fosse du Labrador): Ministère des Ressources Naturelles, Québec, MM 2004-01, 180 p.

Coombe, W., 1994, Sediment-Hosted Base Metal Deposits of the Wollaston Domain, Northern Saskatchewan: Saskatchewan Energy and Mines, Report 213, 108 p.

Corrigan, D., 2000, Geology, central Reindeer Lake, Saskatchewan: Geological Survey of Canada, Open File 3750, 1 sheet.

Corrigan, D., and Rayner, N., 2002, Churchill River-Southern Indian Lake Targeted Geoscience Initiative (NTS 64B, 64C, 64G, 64H), Manitoba: Update and new findings, *in* Report of Activities 2002: Manitoba Industry, Trade and Mines, Manitoba Geological Survey, p. 144-158.

Corrigan D., Maxeiner, R., Bashforth, A., and Lucas, S., 1998, Preliminary report on the geology and tectonic history of the Trans-Hudson Orogen in the northwestern Reindeer Zone, Saskatchewan, *in* Current Research, Part C: Geological Survey of Canada, Paper 98-1C, p. 95-106.

Corrigan, D., MacHattie, T.G., and Chakungal, J., 1999a, The Wathaman batholith and its relation to the Peter Lake Domain: Insights from recent mapping along the Reindeer Lake transect, Trans-Hudson Orogen, *in* Summary of Investigations, 1999, Volume 2: Saskatchewan Geological Survey, Miscellaneous Report 99-4.2, p. 132-142.

Corrigan, D., Pehrsson, S.J., MacHattie, T.G., Piper, L., Wright, D., Lassen, B., and Chakungal, J., 1999b, Lithotectonic framework of the Trans-Hudson Orogen in the northwestern Reindeer Zone, Saskatchewan; an update from recent mapping along the Reindeer Lake transect, *in* Current Research, Part C: Geological Survey of Canada, Paper 99-1C, p. 169-178.

Corrigan, D., Maxeiner, R.O., and Harper, C., 2001a, Preliminary U-Pb results from the La Ronge-Lynn Lake Bridge project, *in* Summary of Investigations 2001, Volume 2: Saskatchewan Geological Survey, Miscellaneous Report 2001-4.2., p. 111-115.

Corrigan, D., Therriault, A. and Rayner, N., 2001b, Preliminary results from the Churchill River - Southern Indian Lake Targeted Geoscience Initiative, *in* Report of Activities, Manitoba Industry, Trade and Mines, Manitoba Geological Survey, p. 94-107.

Corrigan, D., Hajnal, Z., Németh, B, and Lucas, S.B., 2005, Tectonic framework of a Paleoproterozoic arc-continent to continent-continent collisional zone, Trans-Hudson Orogen, from geological and seismic reflection studies: Canadian Journal of Earth Sciences, v. 42, p. 421-434.

Corriveau, L., 2007, Iron oxide copper-gold deposits: A Canadian perspective, *in* Goodfellow, W.D., ed., Mineral Deposits of Canada: A Synthesis of Major Deposit-Types, District Metallogeny, the Evolution of Geological Provinces, and Exploration Methods: Geological Association of Canada, Mineral Deposits Division, Special Publication No. 5, p. 307-328.

Cumming, G.L., and Krstic, D., 1991, Geochronology at the Namew Lake Ni-Cu deposit, Flin Flon area, Manitoba, Canada: A Pb/Pb study of whole rocks and ore minerals: Canadian Journal of Earth Sciences, v. 28, p. 1328-1339.

David, J., and Syme, E.C., 1994, U-Pb geochronology of late Neoarchean tonalites in the Flin Flon Belt, Trans-Hudson Orogen: Surprise at surface: Canadian Journal of Earth Sciences, v. 31, p. 1785-1790.

David J., Bailes A.H., and Machado N., 1996, Evolution of the Snow Lake portion of the Palaeoproterozoic Flin Flon and Kisseynew belts, Trans-Hudson Orogen, Manitoba, Canada: Precambrian Research, v. 80, p. 107-124.

Delaney D.G., 1993, A re-examination of the context of U-Cu, Cu and U mineralization, Dudderidge Lake, Wollaston Domain, *in* Summary of Investigations 1993: Saskatchewan Geological Survey, Miscellaneous Report 93-4, p. 73-85.

——— 1995a, Investigations of sediment-hosted copper and copper-uranium mineralization, Wollaston Domain, *in* Richardson, D.G., ed., Investigations Completed by the Saskatchewan Geological Survey and the Geological Survey of Canada under the Geoscience Program of Canada-Saskatchewan Partnership Agreement on Mineral Development (1990-1993): Geological Survey of Canada, Open File 3119, p. 39-52.

Delaney, D.G., 1995b, Gold in the Glennie Domain, Trans-Hudson Orogen, *in* Richardson, D.G., ed., Investigations Completed by the Saskatchewan Geological Survey and the Geological Survey of Canada under the Geoscience Program of Canada-Saskatchewan Partnership Agreement on Mineral Development (1990-1993): Geological Survey of Canada, Open File 3119, p. 53-58.

Delaney, D.G., and Savage, D., 1998, Geological investigation of the context of quartzite-hosted Zn-Pb mineralization, Sito-Adams area, Wollaston Domain (parts of NTS 74A-4 and -5), *in* Summary of Investigations 1998: Saskatchewan Geological Survey, Miscellaneous Report 98-4, p. 29-35.

Delaney, D.G., Heaman, L.M., Kamo, S., Parrish, R.R., Slimmon, W.L., and Reilly, B.A., 1990, U-Pb sphene/zircon geochronological investigations, *in* Summary of Investigations 1990: Saskatchewan Geological Survey, Miscellaneous Report 90-4, p. 54-57.

Demmans, C.J., Maxeiner, R.O., Watters, B.R., and Coulson, I., 2002, The Brenton Island project. Geochemistry of amphibolites and tonalities of the northern Kisseynew Domain, Trans-Hudson Orogen, *in* Geological Association of Canada - Mineralogical Association of Canada, Joint Annual Meeting, Program with Abstracts, v. 27, p. 26-27.

Eckstrand, O.R., and Hulbert, L., 2007, Magmatic Nickel-Copper-Platinum Group Element Deposits, *in* Goodfellow, W.D., ed., Mineral Deposits of Canada: A Synthesis of Major Deposit-Types, District Metallogeny, the Evolution of Geological Provinces, and Exploration Methods: Geological Association of Canada, Mineral Deposits Division, Special Publication No. 5, p. 205-222.

Fox, J.S., 1984, Besshi-type volcanogenic sulphide deposits - a review: Canadian Institute of Mining and Metallurgy Bulletin, v. 77, p. 57-68.

Franklin, J.M., Lydon, J.W., and Sangster, D.M., 1981, Volcanic-associated massive sulfide deposits, *in* Skinner, B.J., ed., Economic Geology, 75th Anniversary Volume, p. 485-627.

Froese, E., and Moore, J.M., 1980, Metamorphism in the Snow Lake area, Manitoba: Geological Survey of Canada, Paper 78-27, 16 p.

Fumerton, S.L., Stauffer, M.R., and Lewry, J.F., 1984, The Wathaman Batholith: Largest known Precambrian pluton: Canadian Journal of Earth Sciences, v. 21, p. 1082-1097.

Gagné, S., Beaumont-Smith, C.J., Hynes, A., and Williams-Jones, A.E., 2005, Gold metallogenesis and tectonometamorphic history of selected deposits from the Snow Lake area and the southern flank of the Kisseynew Domain, west-central Manitoba (NTS 63J13, 64K10, 63K16 and 63N2), *in* Report of Activities 2005: Manitoba Geological Survey, Manitoba Industry, Economic Development and Mines, p. 20-27.

Gale, G., 2003, Vectoring volcanogenic sulphide deposits using rare earth elements and other pathfinder elements at the Ruttan mine, Manitoba (NTS 63B5), *in* Report of Activities, 2003, Manitoba Geological Survey, Manitoba Industry Trade and Mines, p. 54-73.

Galley, A.G., 1996, Paleoproterozoic volcanic-related massive sulphide deposits: tectonic and depositional environments: Notes from a GAC '96 Symposium: The Gangue, Geological Association of Canada, Mineral Deposits Division, v. 54, p. 10-15.

Galley, A.G., Ziehlke, D.V., Franklin, J.M., Ames, D.E., and Gordon, T.M., 1986, Gold mineralization in the Snow Lake - Wekusko Lake region, Manitoba, *in* Clark (ed.), A.L., ed., Gold in the Western Shield: Canadian Institute of Mining and Metallurgy, Special Volume 38, p. 379-398.

Galley, A.G., Hannington, M.D., and Jonasson, I.R., 2007, Volcanogenic massive sulphide deposits, *in* Goodfellow, W.D., ed., Mineral Deposits of Canada: A Synthesis of Major Deposit-Types, District Metallogeny, the Evolution of Geological Provinces, and Exploration Methods: Geological Association of Canada, Mineral Deposits Division, Special Publication No. 5, p. 141-161.

Gibb, R.A., 1975, Collisional tectonics in the Canadian Shield: Earth and Planetary Sciences Letters, v. 27, p. 378-382.

Gibb, R.A., and Walcott, R.I., 1971, A Precambrian suture in the Canadian Shield: Earth and Planetary Science Letters, v. 10, p. 417-422.

Goodfellow, W.D., 2007, Base metal metallogeny of the Selwyn Basin, Canada, *in* Goodfellow, W.D., ed., Mineral Deposits of Canada: A Synthesis of Major Deposit-Types, District Metallogeny, the Evolution of Geological Provinces, and Exploration Methods: Geological Association of Canada, Mineral Deposits Division, Special Publication No. 5, p. 553-579.

Gordey, S. P., 1980, Stratigraphic cross-section, Selwyn Basin to Mackenzie Platform, Nahanni map area, Yukon Territory and District of Mackenzie, *in* Current Research, Part A: Geological Survey of Canada, Paper 80-1A, p. 353-355.

Gracie, A.J., and Lafrance, B., 1998, Saskatchewan exploration and development highlights 1998: Saskatchewan Geological Survey, Miscellaneous Report 98-4, p. 1-16.

Gracie, A.J., and Tourigny, G., 2001, Saskatchewan exploration and development highlights 2001: Saskatchewan Geological Survey, Miscellaneous Report, 20 p.

Green, A.G., Cumming, G.L., and Cedarwell, D., 1979, Extension of the Superior-Churchill boundary zone into southern Canada: Canadian Journal of Earth Sciences, v. 16, p. 1691-1701.

Griffin, W.L., O'Reilley, S.Y., Abe, N., Aubach, S., Davies, R.M., Pearson, N.J., Doyle, B.J., and Kivi, K., 2003, The origin and evolution of Archean lithospheric mantle: Precambrian Research, v. 127, p. 19-41.

Groves, D.I., Goldfarb, M., Gebre-Mariam, M., Hagemann, S., and Robert, F., 1998, Orogenic gold deposits: A proposed classification in the context of their crustal distribution and relationship to other gold deposit types: Ore Geology Reviews, v. 13, p. 7-27.

Hajnal, Z., Lucas, S., White, D., Lewry, J., Bezdan, S., Stauffer, M.R., and Thomas, M.D., 1996, Seismic reflection images of high-angle faults and linked detachments in the Trans-Hudson Orogen: Tectonics, v. 15, p. 427-439.

Hajnal, Z., Lewry, J.F., White, D. Ashton, K., Clowes, R., Stauffer, M., Gyorfi, I., and Takacs, E., 2005, The Sask Craton and Hearne Province margin: Seismic reflection studies in the western Trans-Hudson Orogen: Canadian Journal of Earth Sciences, v. 42, p. 403-419.

Halls, H.C., and Heaman, L.M., 1997, New constraints on the Paleoproterozoic segment of the Superior Province apparent polar wander path from U-Pb dating of Molson Dykes, Manitoba: Geological Association of Canada - Mineralogical Association of Canada, Joint Annual Meeting, Program with Abstracts, v. 22, p. A61.

Hamilton, M.A., and Bleeker, W., 2002, SHRIMP U-Pb geochronology of the Ospwagan Group: Provenance and depositional age constraints of a Paleoproterozoic rift sequence, SE externides of the Trans-Hudson orogen: Geological Association of Canada - Mineralogical Association of Canada Annual Meeting, Program with Abstracts, v. 27, p. 44.

Harper, C.T., 1998, The La Ronge Domain - Glennie Domain transition: Street lake area (parts of NTS 64D-3 and -6),i Summary of Investigations 1998:Saskatchewan Geological Survey, Miscellaneous Report 98-4, p. 66-80.

Harper, C.T. and Hasanie, R., 1999, Geochemistry and tectonic implication of amphibolite gneisses of the MacLean Lake Belt, Waddy Lake - Southend region: LITHOPROBE Trans-Hudson Orogen Transect, 8th Transect Meeting, Saskatoon, February 25-26, 1999.

Heaman, L.M., and Corkery, T., 1996, U-Pb geochronology of the Split Lake Block, Manitoba: preliminary results, *in* Hajnal, Z., and Lewry, J., ed., Trans-Hudson Lithoprobe Transect: Lithoprobe Report, v. 55, p. 60-67.

Heaman, L.M., Machado, N., Krogh, T.E., and Weber, W., 1986, Presise U-Pb zircon ages for the Molson dyke swarm and the Fox River sill: Constraints for Early Proterozoic crustal evolution in northern Manitoba, Canada: Contributions to Mineralogy and Petrology, v. 94, p. 82-89.

Hoffman, P., 1988, United plates of America, the birth of a craton: Early Proterozoic assembly and growth of Laurentia: Annual Review of Earth and Planetary Science Letters, v. 16, p. 543-603.

——— 1990, Subdivision of the Churchill province and extent of the Trans-Hudson Orogen, *in* Lewry, J.F., and Stauffer, M.R., eds., The Early Proterozoic Trans-Hudson Orogen of North America: Geological Association of Canada, Special Paper 37, p. 15-39.

Höy, T., 1995, Cyprus massive sulphide Cu (Zn), *in* Lefebure, D.V., and Ray, G.E., eds., Selected British Columbia Mineral Deposit Profiles, Volume 1 - Metallics and Coal: British Columbia Ministry of Energy, Mines and Petroleum Resources, Open File 1995-20, p. 51-52.

Hulbert, L.J., 1988, Investigation of mafic and ultramafic rocks for nickel-copper and platinum group elements in northern Saskatchewan: preliminary findings, *in* Summary of Investigations 1988: Saskatchewan Geological Survey, Saskatchewan Energy and Mines, Miscellaneous Report 88-4, p. 152-154.

——— 1995, A geochemical investigation of mafic-ultramafic intrusions for metallogenic pathfinder elements in the La Ronge-Lynn Lake Greenstone Belt: An overview, *in* Richardson, D.G., ed., Investigations Completed by the Saskatchewan Geological Survey and the Geological Survey of Canada under the Geoscience Program of the Canada-Saskatchewan Partnership Agreement on Mineral Development (1990-1995): Geological Survey of Canada, Open File 3119, Saskatchewan Geological Survey, Open File Report 95-3, p. 139-146.

Hulbert, L. J., Stern, R., Kyser, T.K., Pearson, J., Lesher, M., and Grinenko, L., 1994, The Winnipegosis komatiite belt, central Manitoba: Manitoba Energy and Mines, Manitoba Mining and Minerals Convention 1994, Program and Abstracts, p. 21.

Hulbert, L.J., Hamilton, M.A., Horan, M.F., and Scoates, R.F.J., 2005, U-Pb zircon and Re-Os isotope geochronology of mineralized ultramafic intrusions and associated nickel ores from the Thompson Nickel Belt, Manitoba, Canada: Economic Geology, v. 100, p. 29-41.

Kaszycki, C.A. 1989. Surficial geology and drift prospecting, northwestern Manitoba, *in* Galley, A.G., comp., Investigations by the Geological Survey of Canada in Manitoba and Saskatchewan during the 1984-1989 Mineral Development Agreements: Geological Survey of Canada, Open File 2133, p. 131-154.

Konhauser, K.O., Hamade, T., Raiswell, R., Morris, R.C., Ferris, F.G., Southam, G., and Canfield, D.E., 2002, Could bacteria have formed the Precambrian banded iron formations? Geology v. 30, p. 1079-1082.

Lafrance, B., and Heaman, L., 2004, Structural controls on hypozonic orogenic gold mineralization in the La Ronge Domain, Trans-Hudson Orogen, Saskatchewan: Canadian Journal of Earth Sciences v. 41, p. 1453-1471.

Lafrance, B. and Varga, M., 1996, Structural studies of the Parker Lake Shear Zone and the Reilly Lake Shear Zone, Reindeer Lake, *in* Summary of Investigations 1996: Saskatchewan Geological Survey, Miscellaneous Report 96-4, p. 119-124.

Layton-Matthews, D.., Lesher, M., Burnham, O.M., Liwiang, J., Halden, N.M., Hulbert, L., and Peck, D.C., 2007, Magmatic Ni-Cu-(PGE) deposits of the Thompson Nickel Belt, *in* Goodfellow, W.D., ed., Mineral Deposits of Canada: A Synthesis of Major Deposit-Types, District Metallogeny, the Evolution of Geological Provinces, and Exploration Methods: Geological Association of Canada, Mineral Deposits Division, Special Publication No. 5, p. 409-432.

Leckie, D.A., Kjarsgaard, B.A., Bloch, J., McIntyre, D., McNeil, D., Stasiuk, L., and Heaman, L., 1997, Emplacement and reworking of Cretaceous, diamond-bearing, crater facies kimberlite in central Saskatchewan, Canada: Geological Society of America Bulletin, v. 109, p. 1000–1020.

Lesher, C.M., 2007, Ni-Cu-(PGE) Deposits in the Raglan Area, Cape Smith Belt, New Québec, *in* Goodfellow, W.D., ed., Mineral Deposits of Canada: A Synthesis of Major Deposit-Types, District Metallogeny, the Evolution of Geological Provinces, and Exploration Methods: Geological Association of Canada, Mineral Deposits Division, Special Publication No. 5, p. 351-386.

Lewry, J.F., 1981, Lower Proterozoic arc-microcontinent collisional tectonics in the western Churchill province: Nature, v. 294, p. 69-72.

——— 1984, Bedrock compilation, Lac la Ronge and Wapawekka areas (NTS 73P/73I), *in* Summary of Investigations 1984: Saskatchewan Geological Survey, Miscellaneous Report 84-4, p. 34-41.

Lewry, J.F., and Collerson, K.D., 1990, The Trans-Hudson Orogen: extent, subdivisions and problems, *in* Lewry, J.F., and Stauffer, M.R., eds., The Early Proterozoic Trans-Hudson Orogen of North America: Geological Association of Canada, Special Paper 37, p. 1-14.

Lewry, J.F., and Sibbald, T.I.I., 1980, Thermotectonic evolution of the Churchill Province in northern Saskatchewan: Tectonophysics, v. 68, p. 45-82.

Lewry, J.F., and Slimmon, W.L., 1985, Compilation bedrock geology, Lac La Ronge, NTS area 73P/73I: Saskatchewan Department of Energy and Mines, Report 225 (1:250,000 scale with marginal notes).

Lewry, J.F., Stauffer, M.R., and Fumerton, S., 1981, A Cordilleran-type batholithic belt in the Churchill Province in northern Saskatchewan: Precambrian Research, v. 14, p. 277-313.

Lewry, J.F., Thomas, D.J., Macdonald, R., and Chiarenzelli, J., 1990, Structural relations in accreted terranes of the Trans-Hudson Orogen, Saskatchewan: Telescoping in a collisional regime?, *in* Lewry, J.F., Stauffer, M.R., eds., The Early Proterozoic Trans-Hudson Orogen of North America:, Geological Association of Canada, Special Paper 37, p. 75-94.

Lewry, J.F., Hajnal, Z., Green, A., Lucas, S.B., White, D., Stauffer, M.R., Ashton, K.E., Weber, W., and Clowes, R.M., 1994, Structure of a Paleoproterozoic continent-continent collision zone; a Lithoprobe seismic reflection profile across the Trans-Hudson Orogen, Canada: Tectonophysics, v. 232, p. 143-160.

Lucas, S.B., 1989, Structural evolution of the Cape Smith thrust belt, and the role of out-of-sequence faulting in the thickening of mountain belts: Tectonics, v. 8, p. 655-676.

Lucas, S.B., White, D., Hajnal, Z., Lewry, J., Green, A., Clowes, R.M., Zwanzig, H., Ashton, K., Schledewitz, D., Stauffer, M., Norman, A., Williams, P.F., and Spence, G., 1994, Three-dimensional collisional structure of the Trans-Hudson Orogen, Canada: Tectonophysics, v. 232, p. 161-177.

Lucas, S.B., Stern, R.A., Syme, E.C., Reilly, B.A., and Thomas, D.J., 1996, Intraoceanic tectonics and the development of continental crust: 1.92-1.84 Ga evolution of the Flin Flon belt, Canada: Geological Society of America Bulletin, v. 108, p. 602-629.

Macdonald, R., and Thomas, M.W., 1983, Compilation bedrock geology, Reindeer Lake north, NTS area 64E: Saskatchewan Energy and Mines, Report 232 (1:250,000 scale map with marginal notes).

Maceck, J.J., and Bleeker, W., 1989, Thompson Nickel Belt Project: Pipe Pit mine, setting and Ospwagan Lakes, *in* Report of Field Activities 1989: Manitoba Energy and Mines, Geological Services, p. 73-87.

MacHattie, T.G., 2001, Petrogenesis of the Wathaman Batholith and La Ronge Domain Plutons in the Reindeer Lake Area, Trans-Hudson Orogen, Saskatchewan: M.Sc. Thesis, Memorial University, St. John's, Newfoundland.

MacLachlan, K., 2003, Preliminary investigation of the thermotectonic history of the central Rottenstone Domain, Hickson and Rottenstone lakes, Saskatchewan, *in* Summary of Investigations 2003: Saskatchewan Geological Survey, Saskatchewan Industry and Resources, Miscellaneous Report 2003-4.2, Paper A-5, 20 p., CD-ROM.

Maxeiner, R.O., 2006, Peter Lake Domain Project: 1:100 000-scale compilation of bedrock geology and platinum-group element potential maps, *in* Summary of Investigations 2006: Saskatchewan Geological Survey, Miscellaneous Report 2006-4.2, Paper A-9, 17 p., CD-ROM.

Maxeiner, R.O., and Rayner, N.M., 2005, PGE occurrences within layered intrusions of the 2.56 Ga Swan River Complex, Peter Lake Domain (NTS 64E), *in* Summary of Investigations 2005, Volume 2: Saskatchewan Geological Survey, Miscellaneous Report 2005-4.2, Paper A-5, 32 p., CD-ROM.

Maxeiner, R.O., Sibbald, T.I.I., Slimmon, W.L., Heaman, L.M., and Watters, B.R., 1999, Lithogeochemistry of volcano-plutonic assemblages of the southern Hanson Lake Block and Southeastern Glennie Domain, Trans-Hudson Orogen: Evidence for a single island arc complex: Canadian Journal of Earth Sciences, v. 36, p. 209-225.

Maxeiner, R.O., Harper, C., Corrigan, D., and MacDougall, D.G., 2004, La Ronge-Lynn Lake Bridge Project: Geology of the Southern Reindeer Lake area: Saskatchewan Industry and Resources, Open File Report 2003-1, 2 CD-ROM set and 3 maps.

Maxeiner, R.O., Corrigan, D., Harper, C.T., MacDougall, D.G., and Ansdell, K., 2005, Paleoproterozoic arc and ophiolitic rocks on the northwest-margin of the Trans-Hudson Orogen, Saskatchewan, Canada; their contribution to a revised tectonic framework for the orogen: Precambrian Research, v. 136, p. 67-106.

McClenaghan, S.H., Gale, G.H., and Lentz, D.R., 2003, Trace-element signatures of massive sulphides from the Fox mine, Lynn Lake (NTS 64C12), *in* Report of Activities 2003: Manitoba Geological Survey, Manitoba Industry, Economic Development and Mines, Paper GS-8, p. 51-53.

McRicthie, W.D., 1977, Reindeer Lake - Southern Indian Lake: Regional correlation, *in* Report of Field Activities, 1977: Manitoba Ministry of Research and Development, p. 13-18.

Menard, T., Lesher, C.M., Stowell, H.H., Price, D.P., Pickell, J.R., Onstott, T.C., and Hulbert, L., 1996, Geology, genesis, and metamorphic history of the Namew Lake Ni-Cu deposit, Manitoba: Economic Geology, v. 91, p. 1394-1413.

Meyer, M.T., Bickford, M.E., and Lewry, J.F., 1992, The Wathaman Batholith: An Early Proterozoic continental arc in the Trans-Hudson orogenic belt, Canada: Geological Society of America Bulletin, v. 104, p. 1073-1085.

Mumin, A.H. and Trott, M., 2003, Hydrothermal iron-sulphide copper-graphite mineralization in the northern Kisseynew Domain, Trans-Hudson Orogen, Manitoba (NTS 63O and 64B): evidence for deep-seated IOCG (Olympic Dam)-style metal deposition?, *in* Report of Activities 2003: Manitoba Geological Survey, Manitoba Industry, Economic Development and Mines, p. 79-85.

Olivo, G., Theyer, P., and Bursztyn, N., 2002, Platinum group element investigations in the Flin Flon greenstone belt: Petrography and mineralogy of the McBratney Lake PGE-Au occurrence (NTS 63K13), Manitoba, *in* Report of Activities 2002: Manitoba Geological Survey, Manitoba Industry, Trade and Mines, p. 94-99.

Orrell, S.E., Bickford, M.E. and Lewry, J.F., 1999, Crustal evolution and age of thermotectonic reworking in the western hinterland of the Trans-Hudson Orogen, northern Saskatchewan: Precambrian Research, v. 95, p. 187-223.

Parrish, R.R., 1989, U-Pb geochronology of the Cape Smith Belt and Sugluk Block, northern Quebec, *in* Lucas, S.B., Picard, C., and St-Onge, M.R., eds., Tectonic, Magmatic and Metallogenic Evolution of the Early Proterozoic Cape Smith Thrust Belt: Geoscience Canada, v. 16, p. 126-130.

Peck, D.C., Theyer, P., Hulbert, L., Xiong, J., Fedikow, M.A.F., and Cameron, H.D.M., 2000, Preliminary Exploration Database for Platinum-Group Elements in Manitoba: Manitoba Geological Survey, Manitoba Industry, Trade and Mines, Open File Report OF2000-5, 1 CD-ROM.

Pedersen, F.D., 1981, Polyphase deformation of the massive sulphide ore of the Black Angel mine, central West Greenland: Mineralium Deposita, v. 16, p. 157-176.

Percival, J.A., 2007, Geology and metallogeny of the Superior Province, Canada, *in* Goodfellow, W.D., ed., Mineral Deposits of Canada: A Synthesis of Major Deposit-Types, District Metallogeny, the Evolution of Geological Provinces, and Exploration Methods: Geological Association of Canada, Mineral Deposits Division, Special Publication No. 5, p. 903-928.

Percival, J.A., Whalen, J.B., and Rayner, N., 2004, Pikwitonei-Snow Lake, Manitoba transect, Trans-Hudson Orogen - Superior Margin Metallotect Project: Initial geological, isotopic, and U-Pb SHRIMP results, *in* Report of Activities 2004: Manitoba Geological Survey, Manitoba Industry, Economic Development and Mines, p. 120-134.

Percival, J.A., Whalen, J.B., and Rayner, N., 2005, Pikwitonei-Snow Lake, Manitoba transect (parts of NTS 63J, 63O and 63P), Trans-Hudson Orogen-Superior Margin Metallotect Project: New results and tectonic interpretation, *in* Report of Activities 2005: Manitoba Geological Survey, Manitoba Industry, Economic Development and Mines, p. 69-91.

Peredery, W.V., and Geological Staff, 1982, Geology and nickel sulphide deposits of the Thompson Belt, Manitoba, *in* Hutchinson, R.W.,

Spence, C.D., and Franklin, J.M., eds., Precambrian Sulphide Deposits: Geological Association of Canada, Special Paper 25, p. 165-209.

Ray, G.E., 1979, Reconnaissance bedrock geology, Wollaston Lake east (part of NTS area 64L), *in* Summary of Investigations 1979: Saskatchewan Geological Survey, Miscellaneous Report 79-10, p. 19-28 (plus preliminary map at 1: 100,000 scale).

Rayner, N., and Corrigan, D., 2004, Uranium-lead geochronological results from the Churchill River - Southern Indian Lake transect, northern Manitoba: Geological Survey of Canada, Current Research 2004-F1, 14 p.

Rayner, N.M., Maxeiner, R.O., and Corrigan, D., 2005a, Progress report on U-Pb geochronology results for the Peter Lake Domain Project, *in* Summary of Investigations 2005: Saskatchewan Geological Survey, Miscellaneous Report 2005-4.2, paper A-4, 12 p., CD-ROM.

Rayner, N., Stern, R.A., and Bickford, M.E., 2005b, Tectonic implications of new SHRIMP and TIMS U-Pb geochronology of rocks from the Sask Craton, Peter Lake Domain, and Hearne margin, Trans-Hudson Orogen, Saskatchewan: Canadian Journal of Earth Sciences, v. 42, p. 635-657.

Rayner, N.M., Zwanzig, H.V., and Percival, J.A., 2006, Detrital zircon provenance of the Pipe Formation, Ospwagan Group, Thompson Nickel Belt, Manitoba, NTS 63O8, *in* Report of Activities 2006: Manitoba Geological Survey, Manitoba Science, Technology, Energy and Mines, p. 116-124.

Richardson, D.J., and Ostry, G., 1996, Gold Deposits of Manitoba: Manitoba Energy and Mines, Geological Services, Economic Geology Report ER86-1, 114 p.

Saskatchewan Geological Survey, 2003, Geology, and Mineral and Petroleum Resources of Saskatchewan: Saskatchewan Industry and Resources, Miscellaneous Report 2003-7, 173 p.

Scoates, R.F.J., Macek, J.J., and Russell, J.K., 1977, Thompson nickel belt project, *in* Report of Field Activities 1977: Manitoba Mineral Resources Division, p. 47-53.

Scott, B.P., 1973, The Geology of the Pendelton Lake Area (West Half): Saskatchewan Department of Mineral Resources, Report 154, 58 p.

Sillitoe, R.H., 2000, Gold-rich porphyry deposits: Descriptive and genetic models and their role in exploration and discovery, in Hagman, S.G., and Brown, P.E. eds., Gold in 2000: Society of Economic Geologists, Reviews of Economic Geology, v. 13, p. 315-345.

Sims, P.K., and Stein, H.J., 2003, Tectonic evolution of the Proterozoic Colorado province, Southern Rocky Mountains: A summary and appraisal: Rocky Mountain Geology, v. 38, p. 183-204.

Spooner, E.T.C., 1980, Cu-pyrite mineralization and seawater convection in oceanic crust - the ophiolite ore deposits of Cyprus, *in* Strangway, D.W., ed., The Continental Crust and Its Mineral Deposits: Geological Association of Canada, Special Paper 20, p. 685-704.

St-Onge, M.R., Lucas, S.B., and Parrish, R.R., 1992, Terrane accretion in the internal zone of the Ungava orogen, northern Quebec. Part 1: Tectonostratigraphic assemblages and their tectonic implications: Canadian Journal of Earth Sciences, v. 29, p. 746-764.

Stauffer, M.R., 1984, Manikewan and early Proterozoic ocean in central Canada, its igneous history and orogenic closure: Precambrian Research, v. 25, p. 257-281.

——— 1990, The Missi Formation: An Aphebian molasse deposit in the Reindeer Lake Zone of the Trans-Hudson Orogen, Canada, *in* Lewry, J.F., and Stauffer, M.R., The Early Proterozoic Trans-Hudson Orogen of North America: Geological Association of Canada, Special Paper 37, p. 75-94.

Stauffer, M.R, and Lewry, J.F., 1993, Regional setting and kinematic features of the Needle Falls shear zone, Trans-Hudson Orogen: Canadian Journal of Earth Sciences, v. 30, p. 1338-1354.

Stern, R.A., Syme, E.C., and Lucas, S.B., 1995, Geochemistry of 1.9 Ga MORB- and OIB-like basalts from the Amisk collage, Flin Flon belt, Canada: Evidence for an intra-oceanic origin: Geochimica et Cosmochimica Acta, v. 59, p. 3131-3154.

Stern, R.A., Machado, N., Syme, E.C., Lucas, S.B., and David, J., 1999, Chronology of crustal growth and recycling in the Paleoproterozoic Amisk collage (Flin Flon Belt), Trans-Hudson Orogen, Canada: Canadian Journal of Earth Sciences, v. 36, p. 1807-1827.

Sutcliffe R.H., Sweeny J.M., and Edgar, A.D., 1989, The Lac des Iles complex, Ontario: Petrology and Platinum-group element mineralization in an Archean mafic intrusion: Canadian Journal of Earth Sciences, v. 26, p. 1408-1427.

Syme, E.C. and Bailes, A.H., 1993, Stratigraphic and tectonic setting of volcanogenic massive sulphide deposits, Flin Flon, Manitoba: Economic Geology, v. 88, p. 566-589.

Syme, E.C., Lucas, S.B., Bailes, A.H., and Stern, R.A., 1999, Contrasting arc and MORB-like assemblages in the Paleoproterozoic Flin Flon Belt, Manitoba, and the role of intra-arc extension in localizing volcanic-hosted massive sulphide deposits: Canadian Journal of Earth Sciences, v. 36, p. 1767-1788.

Theyer, P., 2001, Platinum-group element investigations in the Flin Flon greenstone belt: McBratney Lake occurrence and Josland Lake intrusion (parts of NTS 63K/9, /13), *in* Report of Activities 2001: Manitoba Geological Survey, Manitoba Industry, Trade and Mines, p. 33-39.

Theyer, P., and Heine, T.H., 2003, Platinum group element investigation in the Flin Flon greenstone belt: Mikanagan Lake and Tartan Lake gabbroic complexes, Manitoba (NTS 63K13): In Report of Activities 2003, Manitoba Industry, Trade and Mines, Manitoba Geological Survey, p. 25-28.

Thomas, D.J., 1993, Geology of the Star Lake-Otter Lake Portion of the Central Metavolcanic Belt, La Ronge Domain: Saskatchewan Geological Survey, Miscellaneous Report 236, 133 p.

Tourigny, G., 2003, Preliminary structural study of the gold-bearing shear zone system at the Seabee Mine, northern Saskatchewan, *in* Summary of Investigations 2003, Volume 2: Saskatchewan Geological Survey, Saskatchewan Industry and Resources, Miscellaneous Report 2003-4.2, Paper B-1, 11 p., CD-ROM.

Tran, H.T., 2001, Tectonic evolution of the Paleoproterozoic Wollaston Group in the Cree Lake Zone, northern Saskatchewan, Canada: Ph.D. thesis, University of Regina, Regina, Saskatchewan, 458 p.

Tran, H.T., Ansdell, K., Bethune, K., Watters, B., and Ashton, K., 2003, Nd isotope and geochemical constraints on the depositional setting of Paleoproterozoic metasedimentary rocks along the margin of the Archean Hearne Craton, Saskatchewan, Canada: Precambrian Research, v. 123, p. 1-28.

Turek, A., Woodhead, J., and Zwanzig, H.V., 2000, U-Pb age of the gabbro and other plutons at Lynn Lake (parts of NTS 64C), *in* Report of Activities 2000: Manitoba Geological Survey, Manitoba Industry, Trade and Mines, p. 97-104.

Van Schmus, W.R., and Schledewitz, D.C.P., 1986, U-Pb zircon geochronology of the Big Sand Lake area, northern Manitoba, *in* Report of Activities 1986: Manitoba Energy and Mines, Minerals Division, p. 207-210.

Wardle, R.J., and Hall, J., 2002, Proterozoic evolution of the northeastern Canadian Shield: Lithoprobe Eastern Canadian Shield Onshore-Offshore Transect (ECSOOT), introduction and summary: Canadian Journal of Earth Sciences, v. 39, p. 563-567.

Watters, B.R., and Pearce, J.A., 1987, Metavolcanic rocks of the La Ronge Domain in the Churchill Province, Saskatchewan geochemical evidence for a volcanic arc origin, *in* Pharaoh, T.C., Beckinsale, R.D. and Rickard, D., eds., Geochemistry and Mineralization of Proterozoic Volcanic Suites: Geological Society of London, Special Publication 33, p. 167-182.

Weber, W., 1990, The Churchill-Superior boundary zone, southeast margin of the Trans-Hudson orogen: A review, *in* Lewry, J.F., Stauffer, M.R., eds., The Early Proterozoic Trans-Hundson Orogen of North America: Geological Association of Canada, Special Paper 37, p. 41-55.

Whalen, J. B., Syme, E.C., and Stern, R.A., 1999, Geochemical and Nd isotopic evolution of Paleoproterozoic arc-type granitoids magmatism in the Flin Flon Belt, Trans-Hudson Orogen, Canada: Canadian Journal of Earth Sciences, v. 36, p. 227-250.

White, D.J., Zwanzig, H.V., and Hajnal, Z., 2000, Crustal suture preserved in the Paleoproterozoic Trans-Hudson Orogen, Canada: Geology, v. 29, p. 527-530.

White, D.J., Lucas, S.B., Bleeker, W., Hajnal, Z., Lewry, J.F., and Zwanzig, H.V., 2002, Suture-zone geometry along an irregular Paleoproterozoic margin: The Superior Boundary Zone, Manitoba, Canada: Geology, v. 30, p. 735-738.

Wodicka, N., Madore, L., Larbi, Y., and Vicker, P., 2002, Géochronologie U-Pb de filons-couches mafiques de la Ceinture de Cape Smith et de la Fosse du Labrador, *in* L'exploration Minérale au Québec: Notre Savoir, vos Découvertes: Séminaire d'information sur la recherche géologique, Programme et résumés 2002, Ministère des Ressources naturelles, Québec; DV 2002-10, 48 p.

Wright, D.M., 2001, Geology and Geochemistry of Variably Altered Metavolcanic Rocks of Southern Reindeer Lake, Saskatchewan,

Canada: M.Sc. thesis, University of Saskatchewan, Saskatoon, Saskatchewan, 153 p.

Yeo, G., 1998, A systems tract approach to the stratigraphy of paragneisses in the southeastern Wollaston Domain, *in* Summary of Investigations 1998: Saskatchewan Geological Survey, Saskatchewan Energy and Mines, Miscellaneous Report 98-4, p. 36-47.

Zwanzig, H., 1990, Kisseynew gneiss belt in Manitoba: Stratigraphy, structure, and tectonic evolution, *in* Lewry, J.F., and Stauffer, M.R., eds., The Early Proterozoic Trans-Hudson Orogen of North America: Geological Association of Canada, Special Paper 37, p. 95-120.

——— 2000, Geochemistry and tectonic framework of the Kisseynew Domain - Lynn Lake belt boundary (part of NTS 64P/13), *in* Report of Activities 2000: Manitoba Geological Survey, Manitoba Industry, Trades and Mines, p. 102-114.

——— 2005, Geochemistry, Sm-Nd isotope data and age constraints of the Bah Lake assemblage, Thompson Nickel Belt and Kisseynew Domain margin: Relation to Thompson-type ultramafic bodies and a tectonic model (NTS 63J, O and P), *in* Report of Activities 2005: Manitoba Geological Survey, Manitoba Industry, Economic Development and Mines, p. 40-53.

Zwanzig, H.V., Syme, E.C., and Gilbert, H.P., 1999, Updated trace element geochemistry of the ca. 1.9 Ga metavolcanic rocks in the Paleoproterozoic Lynn Lake Belt: Manitoba Industry, Trade and Mines, Geological Services, Open File Report 99-13, 46 p., plus accompanying map and diskette.

Zwanzig, H.V., Bohm, C., Protrel, A., and Machado, N., 2003, Field relations, U-Pb zircon ages and Nd model ages of granitoids intrusions along the Thompson Nickel Belt - Kisseynew Domain Boundary, Setting Lake area, Manitoba (NTS 63J15 and 63O2), *in* Report of Activities 2003: Manitoba Geological Survey, Manitoba Industry, Economic Development and Mines, p. 118-129.

Geology and Metallogeny of the Superior Province, Canada

John A. Percival

Geological Survey of Canada, 601Booth Street, Ottawa, Ontario K1A 0E8
Corresponding author's email: John.Percival@nrcan-rncan.gc.ca

Abstract

The Superior Province was assembled from continental fragments and intervening tracts of oceanic crust between 2.72 and 2.68 Ga. Continental blocks (Northern Superior, North Caribou, Winnipeg River, Marmion, Minnesota River Valley, Opatica, Goudalie) retain evidence for distinct evolution during the Paleo- and Mesoarchean and little evidence of continental breakup prior to amalgamation. Neoarchean volcano-plutonic rocks of arc affinity dominate oceanic and continent-margin settings, indicating widespread subduction prior to five collisional events between 2.72 and 2.68 Ga. The 3.0 Ga North Caribou superterrane collided with the Northern Superior superterrane between 2.72 and 2.71 Ga on the north and the Winnipeg River terrane to the south, trapping the English River flysch belt between 2.70 and 2.69 Ga. Docking of the juvenile western Wabigoon terrane occurred at a similar time, followed by collision of the Abitibi-Wawa terrane and syntectonic deposition of Quetico turbidites (2.698-2.690 Ga). The northeastern Superior Province east of Hudson and James bays is dominated by 2.78 to 2.69 Ga plutonic rocks emplaced into older crust (3.8-2.83 Ga). Recent discoveries highlight potential for iron-formation-hosted gold and magmatic Ni-PGE mineralization. Common magmatic and tectonic events are recognized in the northeastern and northwestern Superior, although domains and boundaries have not been correlated.

With the exception of Ni, Cr, Cu, PGE, and Ti deposits hosted by mafic-ultramafic intrusions (2.76-2.74 Ga), major mineral deposits of the Superior Province formed during the arc magmatic and collisional stages of orogenesis. Volcanogenic massive sulphide and rare gold deposits occur in 2.735 to 2.70 Ga calc-alkaline sequences, in back-arc or rifted arc settings. Orogenic lode-gold deposits formed along major and minor structures in greenstone belts and associated plutons during regional deformation between 2.71 and 2.68 Ga. Additional gold mineralization, sporadic rare metal pegmatites, and some hydrothermal deposits were formed between 2.68 and 2.60 Ga during widespread late hydrothermal activity. The occurrence of ancient (to 3.8 Ga) crust in the northern Superior Province may signal diamond potential but it remains uncertain whether cool lithosphere survived intense Neoarchean reworking.

Résumé

La Province du lac Supérieur s'est constituée par l'assemblage de fragments continentaux et d'étendues intercalaires de croûte océanique entre 2,72 et 2,68 Ga. Les blocs continentaux (Northern Superior, North Caribou, Winnipeg River, Marmion, Minnesota River Valley, Opatica et Goudalie) révèlent des signes d'une évolution distincte au Paléoarchéen et au Mésoarchéen et ne manifestent que peu d'indications d'un morcellement continental antérieur à leur assemblage. Les roches volcano-plutoniques du Néoarchéen présentant des affinités avec les roches d'arc dominent dans les cadres océanique et de marge continentale, ce qui indique une subduction généralisée avant les cinq épisodes de collision entre 2,72 et 2,68 Ga. Entre 2,72 et 2,71 Ga, le superterrane de North Caribou, âgé de 3,0 Ga, est entré en collision avec le superterrane de Northern Superior au nord et avec le terrane de Winnipeg River au sud, piégeant ainsi la ceinture de flysch d'English River entre 2,70 et 2,69 Ga. L'arrimage de la partie ouest du juvénile terrane de Wabigoon s'est produit à peu près à la même époque et a été suivi de la collision avec le terrane d'Abitibi-Wawa et du dépôt syntectonique des turbidites de Quetico (2,698-2,690 Ga). Des roches plutoniques datant de 2,78-2,69 Ga mises en place dans de la croûte plus ancienne (3,8-2,83 Ga) sont prédominantes dans la partie nord-est de la Province du lac Supérieur à l'est de la baie d'Hudson et de la baie James. Des découvertes récentes soulignent le potentiel de minéralisations aurifères dans des formations de fer ainsi que de gîtes de Ni-ÉGP dans des roches magmatiques. On a reconnu l'existence de nombreux événements magmatiques et tectoniques communs aux parties nord-est et nord-ouest de la Province du lac Supérieur, mais des corrélations n'ont pas encore été établies entre les domaines et les limites.

À l'exception des gîtes de Ni, Cr, Cu, ÉGP et Ti dans des intrusions mafiques-ultramafiques (2,76-2,74 Ga), les principaux gîtes minéraux de la Province du lac Supérieur se sont formés pendant les stades orogéniques d'arc magmatique et de collision. Il existe des gîtes de sulfures massifs volcanogènes et de rares gîtes d'or dans les séquences calco-alcalines datant de 2,735 à 2,70 Ga dans des cadres d'arrière-arc ou d'arc de divergence. Des gîtes d'or filoniens orogéniques se sont formés le long de structures majeures et mineures dans les ceintures de roches vertes et les plutons associés pendant la déformation régionale entre 2,71 et 2,68 Ga. D'autres minéralisations en or, de rares pegmatites métallifères sporadiques et certains gîtes hydrothermaux se sont formées de 2,68 à 2,60 Ga, pendant une période tardive d'activité hydrothermale répandue. L'occurrence de croûte ancienne (jusqu'à 3,8 Ga) dans la partie nord de la Province du lac Supérieur pourrait signaler un potentiel diamantifère, mais il est incertain que de la lithosphère froide ait pu survivre à l'intense remaniement du Néoarchéen.

Introduction

The Archean Superior Province represents the mining heartland of Canada, hosting major camps in the Abitibi district of Ontario-Quebec and the Red Lake region of western Ontario. In addition, the Paleoproterozoic Sudbury structure (1.85 Ga), located on the southeastern edge of the Superior Province, is one of the world's largest nickel producing regions. This report summarizes information on the geology and metallogeny of the Superior Province compiled in recent syntheses of Ontario (Fyon et al., 1992; Williams et al., 1992) and the Superior Province (Card, 1990; Card and Poulsen, 1998; Skulski and Villeneuve, 1999), augmented by recent work in the western (Corkery et al., 2000; Percival et

Percival, J.A., 2007, Geology and metallogeny of the Superior Province, Canada, *in* Goodfellow, W.D., ed., Mineral Deposits of Canada: A Synthesis of Major Deposit-Types, District Metallogeny, the Evolution of Geological Provinces, and Exploration Methods: Geological Association of Canada, Mineral Deposits Division, Special Publication No. 5, p. 903-928.

al., 2006b) and northeastern Superior Province (Labbé and Lacoste, 2004; Leclair et al., 2004a,b).

Major contributions to understanding the evolution of the Superior Province have come from U-Pb geochronological and geochemical studies over the past 30 years. Geochronology has provided an absolute time framework for the development of greenstone belt strata, the timing and duration of plutonic, structural, and metamorphic events, and serves as the basis for regional correlation. In addition to characterizing rock compositions, geochemical analyses have been used extensively in trace-element discrimination diagrams to infer geodynamic setting. This approach, which has been instrumental in identifying ore-forming environments, is based on the rationale that magma generation processes that impart diagnostic trace-element ratios are controlled by pressure, temperature, and partition coefficients, factors that were also valid in the past. Analyses of the Nd-Sm and Lu-Hf isotopic systems have been essential in defining the distribution of ancient continental and juvenile terranes, particularly within complexly deformed areas.

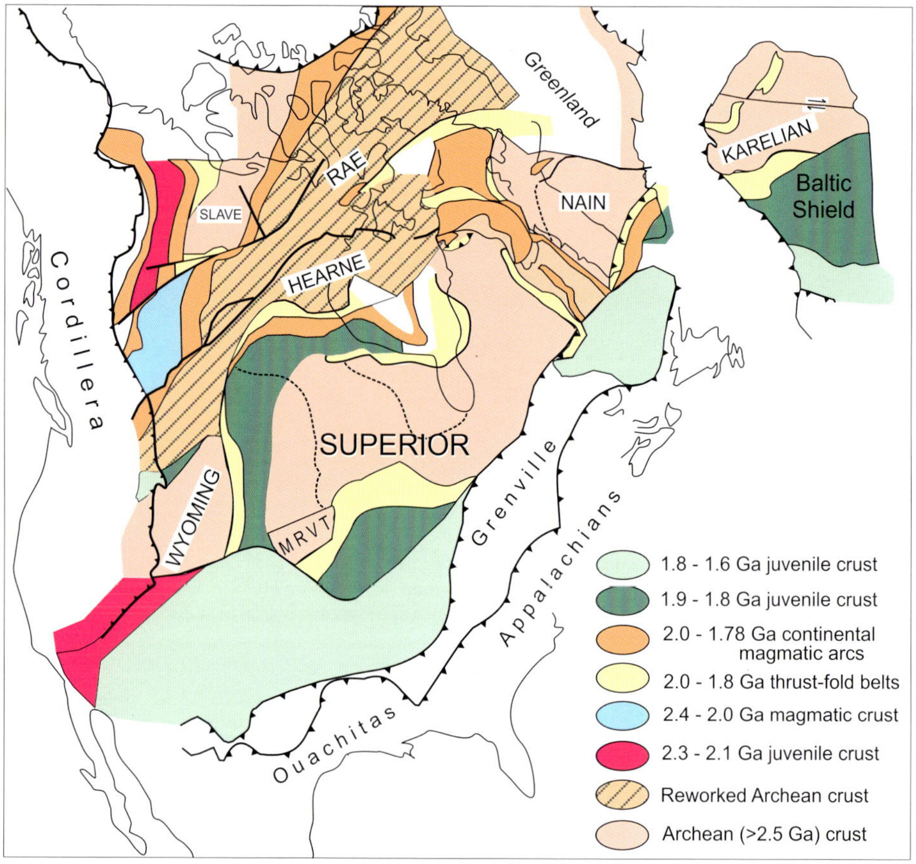

FIGURE 1. Tectonic map of North America, showing location of the Archean Superior Province at the core of the Canadian Shield (after Hoffman, 1989). Unlabelled white areas are knowledge gaps due to lack of exposure and geophysical coverage. Greenland and the Baltic shield are restored to their positions prior to opening of the Atlantic Ocean and Labrador Sea. Dashed lines on the Superior Province outline indicate the extent of Phanerozoic cover. MRVT: Minnesota River Valley terrane.

Superior Province

The Superior Province forms the core of the North American continent (Fig. 1) and is surrounded by provinces of Paleoproterozoic age on the west, north, and east, and Mesoproterozoic age (Grenville Province) on the southeast. Tectonic stability has prevailed since ca. 2.6 Ga in large parts of the Superior Province. Proterozoic and younger activity is limited to rifting of the margins, emplacement of numerous mafic dyke swarms (Buchan and Ernst, 2004), compressional reactivation, and large-scale rotation at ca. 1.9 Ga, and failed rifting at ca. 1.1 Ga. With the exception of the northwestern and northeastern Superior margins that were pervasively deformed and metamorphosed at 1.9 to 1.8 Ga, the craton has escaped ductile deformation.

Current views regard the Superior Province as a collage made up of small continental fragments of Mesoarchean age and Neoarchean oceanic plates with a complex history of aggregation between 2.72 and 2.68 Ga (Percival et al., 2004a, 2006b) and subsequent post-orogenic effects. Early interpretations of the Superior Province as an example of a purely autochthonous, vertical tectonic regime evolved by the early 1990s to accretionary terrane models in which most assemblage contacts were considered suspect rather than depositional. Recent work has succeeded in distinguishing sequences built upon continental basement (e.g. Red Lake greenstone belt, Sanborn-Barrie et al., 2001) from those of probable oceanic affinity (e.g. central Abitibi greenstone belt, Ayer et al., 2002), giving rise to improved models of tectonic evolution. Sedimentary rocks as old 2.48 Ga unconformably overlie Superior Province granites, indicating that most erosion had occurred prior to ca. 2.5 Ga.

The southern Superior Province (to latitude 52°N) is a major source of mineral wealth. Owing to its potential for base metals, gold, and other commodities, the Superior Province continues to attract mineral exploration in both established and frontier regions.

Geological Setting

A first-order feature of the Superior Province is its linear subprovinces of distinctive lithological and structural character, accentuated by subparallel boundary faults (e.g. Card and Ciesielski, 1986). Trends are generally east-west in the south, west-northwest in the northwest, and northwest in the northeastern Superior (Fig. 2). Recent work based on isotopic and zircon inheritance studies has provided a means of "seeing through" the latest structural and magmatic events, particularly in regions dominated by granitic rocks, revealing fundamental age domains across the Superior Province. Several domains of Mesoarchean age are recognized in spite of pervasive Neoarchean magmatism, metamorphism, and

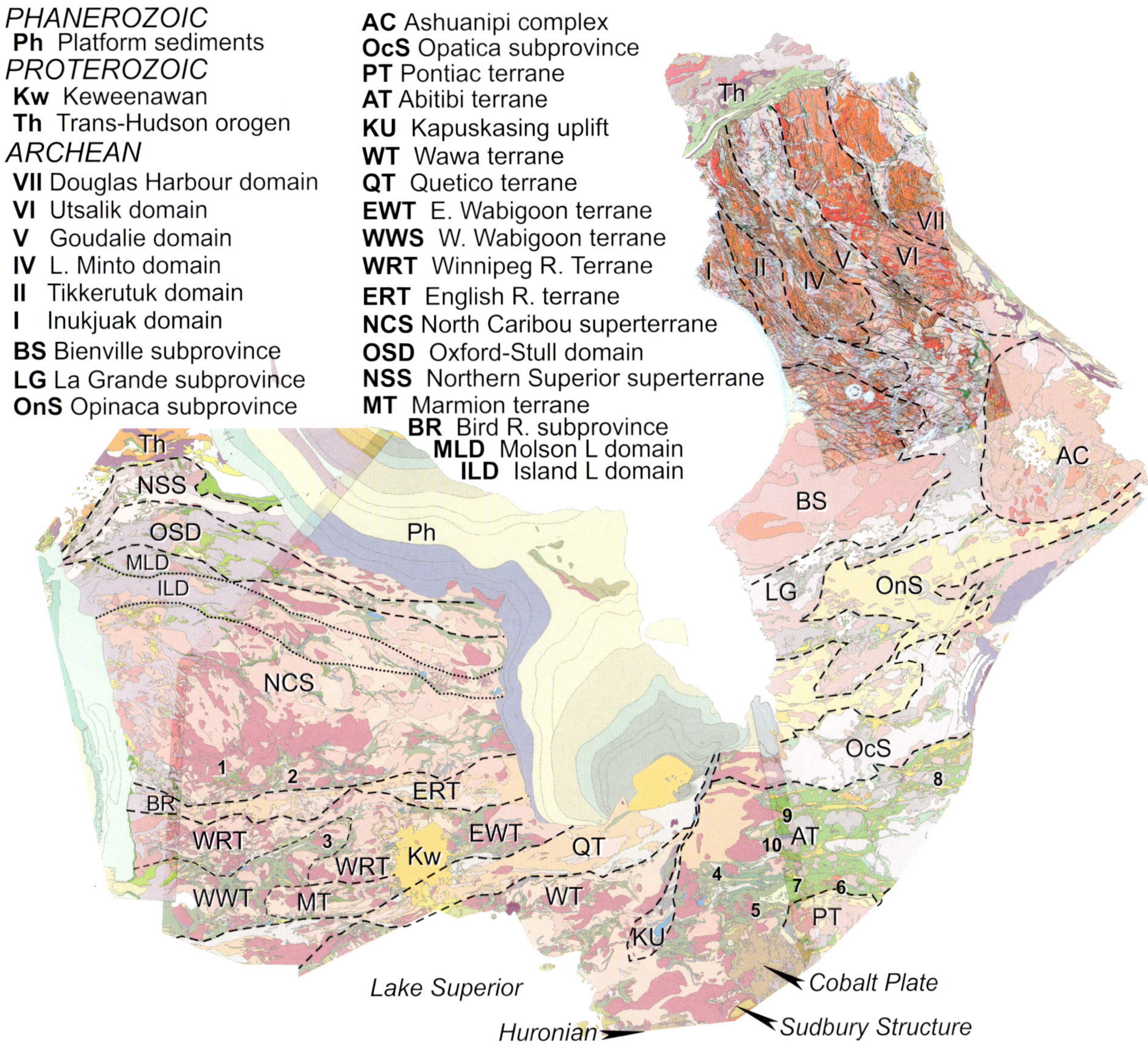

FIGURE 2. Mosaic map of the Superior Province showing major tectonic elements. Data sources: Manitoba (1965); Ontario (1992); Quebec (2002) and (Leclair, 2005). Major mineral districts: 1: Red Lake; 2: Confederation Lake; 3: Sturgeon Lake; 4: Timmins; 5: Kirkland Lake; 6: Cadillac; 7: Noranda; 8: Chibougamau; 9: Casa Berardi; 10: Normétal

deformation (Fig. 3). The term "terrane" is used in the sense of a geological domain with a distinct geological history prior to its amalgamation into the Superior Province during 2.72 to 2.68 Ga assembly events. "Superterranes" show evidence for internal amalgamation of terranes prior to the Neoarchean assembly. "Domains" are defined as distinct regions within a terrane or superterrane.

The oldest continental crust (up to 3.7 Ga) occurs in the Northern Superior superterrane (Skulski et al., 2000) and Inukjuak domain of the northeastern Superior province (David et al., 2003). A large region of ca. 3.0 Ga crust, the North Caribou superterrane (Thurston et al., 1991), which may extend into northern Quebec (Stott, 1997), has been interpreted as the continental nucleus to which other terranes were amalgamated during assembly of the Superior Province (cf. Goodwin, 1968; Thurston et al., 1991; Williams et al., 1992; Stott, 1997; Thurston, 2002). Farther south, the Winnipeg River (WR) and Marmion (MT) terranes are relatively small continental fragments dating back to 3.4 and 3.0 Ga, respectively (Beakhouse, 1991; Tomlinson et al., 2004). In the far south, the Minnesota River Valley terrane (MRVT), of unknown extent, contains remnants of crust as old as ca. 3.6 Ga (Goldich et al., 1984; Bickford et al., 2006; Schmitz et al., 2006). In central Quebec, the Opatica subprovince has Mesoarchean heritage, as do large parts of the northeastern Superior Province (Leclair et al., 2004a).

Domains of oceanic ancestry, identified by juvenile isotopic signatures and lack of Mesoarchean zircon inheritance, separate most of the continental fragments. These dominantly greenstone-granite terranes generally have long strike lengths and record geodynamic environments including oceanic floor, plateaux, island arc, and back-arc settings (e.g. Thurston, 1994). Examples include parts of the Oxford-Stull domain in the north, the western Wabigoon in the west, and the Wawa-Abitibi terrane in the southeastern Superior Province. Rocks formed in these environments host some of the largest massive sulphide deposits of the province, particularly in the Abitibi greenstone belt.

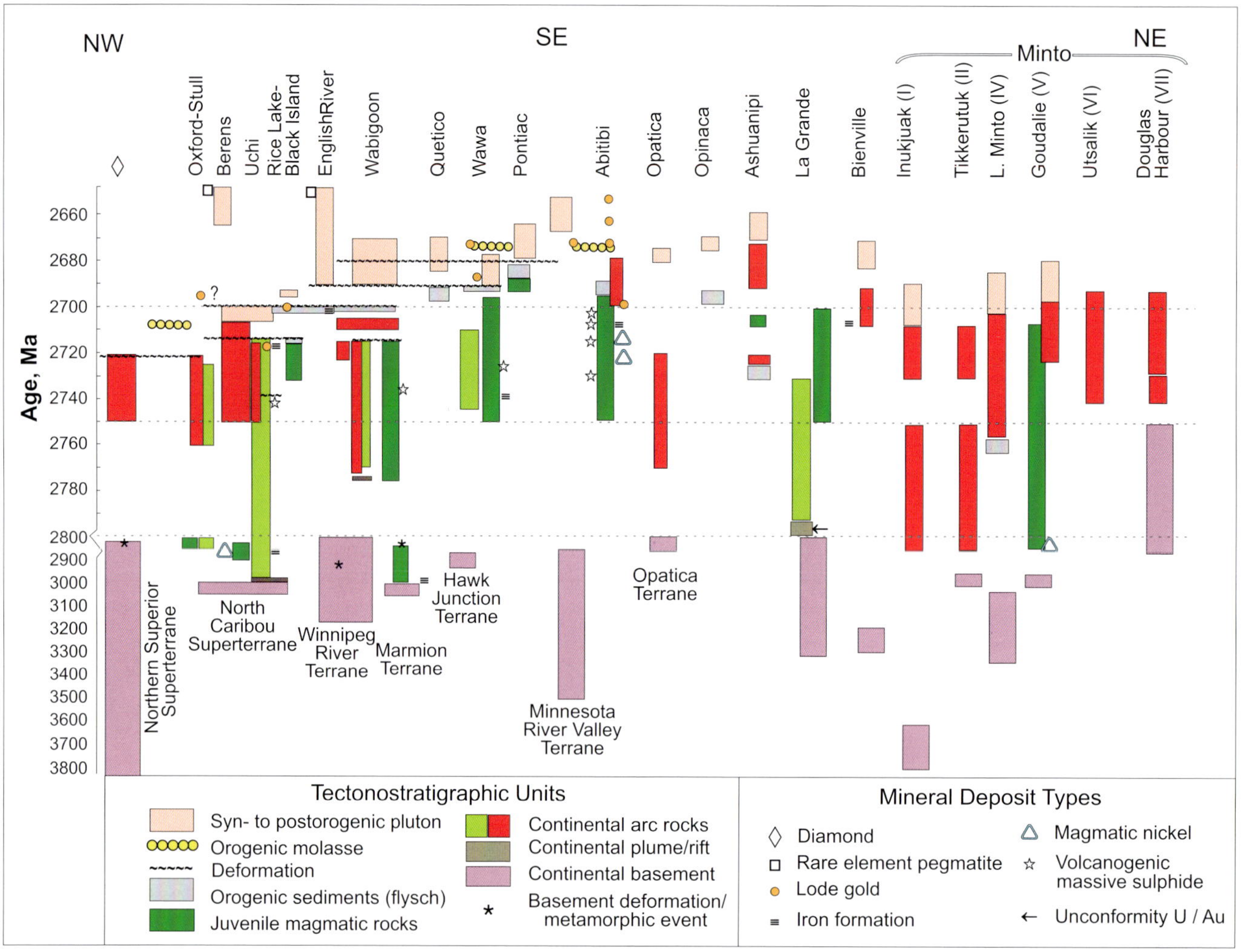

FIGURE 3. Time-stratigraphic correlation chart showing major tectonic elements of the Superior Province and ages of important mineral deposit types. Age data from Skulski and Villeneuve (1999) and more recent sources (see text for details).

Still younger features, the metasedimentary terranes (Breaks, 1991; Williams, 1991; Davis, 2002), separate some of the continental and oceanic domains. Extending across the entire province, these 50 to 100 km wide belts of metagreywacke, migmatite, and derived granite appear to represent thick synorogenic sequences that were deposited, deformed, and metamorphosed during collisional orogeny. Faults, developed in association with several distinct tectonic events, host a variety of orogenic gold deposits across the province.

Geophysical Setting

Early tomographic images of the Superior Province suggested lithosphere thickness of at least 250 km beneath the craton (Grand, 1987), and possibly as much as 350 km (van der Lee and Nolet, 1997). Analysis of shear-wave splitting indicates prominent east-west anisotropy in the lithosphere (Silver and Chan, 1988; Silver, 1996; Kay et al., 1999a,b), which has been attributed to mantle deformation during Archean tectonism. Musacchio et al. (2004) estimated upper mantle velocities in the 8.3 to 8.8 km.s^{-1} range for the western Superior Province, consistent with depleted harzburgite compositions, and lithosphere thickness is estimated at 300 km (Kendall et al., 2002). Beneath the western Superior Province, Kendall et al. (2002) distinguished a northern zone of isotropic upper mantle beneath the North Caribou superterrane that contrasts with a southern zone characterized by east-west anisotropy. These domains, which are separated by a subvertical high-velocity zone, extend to about 300 km depth (Sol et al., 2002). To the north beneath Paleoproterozoic crust of the Trans-Hudson Orogen, the lithosphere is thinner and notably more isotropic (Kendall et al., 2002). Domains resembling those defined seismically are also observed in the electrical conductivity structure. Craven et al. (2001) reported an essentially isotropic mantle beneath the North Caribou superterrane, in contrast to pronounced east-west anisotropy in the south. Subsequent analysis has modeled a steeply north-dipping, tabular, resistive zone separating the two domains, which corresponds approximately to the subvertical high-velocity zone. Both the seismic and electrical structures have been attributed to formation of the Superior craton through subduction-accretion processes (Craven et al., 2001; Kendall et al., 2002). Heat flow in the Superior Province averages 42 ± 8 mW.m^{-2} (Cheng et al., 2002; Rolandone et al., 2003) and the reduced heat flow (mantle component) is consistent with a thermal lithosphere at least 240 km thick (Jaupart et al., 1998), supporting the concept of a thick tectosphere.

Direct control on mantle lithosphere composition is generally acquired from xenoliths in kimberlite pipes. However, the known pipes are not ideally situated to provide comprehensive information. Pipes in the Attawapiskat area sample mantle close to the Superior margin, and indicate mainly lherzolitic compositions and a cool geotherm (Scully, 2000). The compositional range is similar to that from Kirkland Lake pipes (Vicker and Schulze, 1994; Schulze, 1996), where geothermobarometry suggests a warmer geotherm corresponding to surface heat flow of about 40 $mW.m^{-2}$. Taken together, observations of the Superior Province mantle indicate a cool refractory lithosphere typical of Archean cratons (Jordan, 1978). Individual mantle domains correspond to broad geological features (terranes) and anisotropies of seismic velocity and electrical conductivity are east-trending, coplanar to the dominant penetrative crustal structures.

Gravity and aeromagnetic trends generally correspond well with first-order crustal geological features. Metasedimentary belts define gravity and aeromagnetic lows, whereas greenstone belts form local gravity highs. Gravity data have been inverted to estimate crustal thickness, following removal of near-surface effects (e.g. Nitescu et al., 2003). This approach supports seismic observations of crust thinner, by as much as 9 km, beneath the English River terrane. Seismic reflection profiles in both western (White et al., 2003; Calvert et al., 2004) and eastern Superior (Calvert and Ludden, 1999) generally indicate gently north-dipping structures in southern low-grade regions, changing to southward dips in plutonic gneissic terranes farther north. The vergence patterns have been interpreted in terms of a doubly vergent orogen (White et al., 2003) or crustal-scale synclinorium (Hynes and Song, 2006). Reflectors locally extend beyond the Moho, where they have been interpreted as subduction scars (Calvert et al., 1995; White et al., 2003).

Western Superior Province

Northern Superior Superterrane

The Northern Superior superterrane (NSS) at the northern fringe of the Superior Province is dominated by granitic and gneissic rocks and has been identified based on fragmentary isotopic evidence from Ontario and Manitoba (Skulski et al., 1999). Supracrustal units in the Assean Lake complex of Manitoba include greywacke with detrital zircon ages up to 3.9 Ga, iron formation, and mafic to intermediate volcanic rocks (Böhm et al., 2000, 2003). The ancient rocks have been strongly reworked by granitoid magmatism at 3.2 to 3.1, 2.85 to 2.81, and 2.74 to 2.71 Ga (Fig. 3), representing evolution in a continental magmatic arc setting, followed by amphibolite-facies metamorphism at 2.68 and 2.61 Ga that may have resulted from collisions during tectonic assembly (Skulski et al., 2000; Böhm et al., 2003). The Northern Superior superterrane is bounded to the south by the North Kenyon fault, which juxtaposes it with the Oxford–Stull terrane and hosts local lode gold occurrences. Prominent ductile shear zones in the eastern Northern Superior superterrane may be Neoarchean or Paleoproterozoic in age. Exploration for diamondiferous kimberlite pipes has been promoted by identification of kimberlite indicator minerals in the Ontario-Manitoba border region (Stone, 2005) and general on-strike correlation with the Victor Pipe in the Attawapiskat area to the east. Correlation is uncertain between the Northern Superior superterrane and units of similar antiquity in the Inukjuak domain of northern Quebec.

Oxford-Stull Domain

The Oxford-Stull Lake domain, as defined by Thurston et al. (1991), contains some of the largest greenstone belts in the northwestern Superior Province, including the Knee Lake-Gods Lake and Stull Lake belts of Manitoba (Corkery and Skulski, 1998; Syme et al., 1999; Corkery et al., 2006). Mainly basaltic rocks of the Hayes River assemblage have been dated locally in the 2.83 Ga range, and volcanic rocks of the Oxford Lake assemblage fall between 2.729 and 2.719 Ga (Fig. 3; Corkery et al., 2000). The occurrence of coarse clastic sediments (Corkery et al., 2000) and alkaline volcanic rocks (Brooks et al., 1982) along some faults suggests that they were active during deposition of the Oxford Lake assemblage (Corkery et al., 1992). Mafic intrusions are common in some belts, such as at Big Trout Lake, and have some Ni -PGE potential. Plutons of tonalite, granodiorite, and granite underlie large parts of the terrane and yield ages between 2.83 and 2.69 Ga. Isotopic data suggest that the Oxford-Stull terrane evolved in an oceanic setting, possibly on the edge of thinned North Caribou crust (Parks et al., 2006), until ca. 2.71 Ga, when it was juxtaposed with the Northern Superior superterrane along the North Kenyon fault. It is bounded by the Stull-Wunnummin and Gods Lake Narrows shear zones in the south, which host numerous lode-gold occurrences.

North Caribou Superterrane

The North Caribou superterrane (Thurston et al., 1991) is the largest domain with Mesoarchean ancestry of the Superior Province. Basement consists of ca. 3.0 Ga juvenile plutonic and minor volcanic belts (Stevenson and Patchett, 1990; Stevenson, 1995; Corfu et al., 1998; Hollings et al., 1999; Henry et al., 2000), upon which were deposited early (2.98-2.85 Ga) rift-related and younger (2.85-2.71 Ga) arc sequences (Thurston and Chivers, 1990). It was severely reworked by continental arc magmatism at 2.75 to 2.70 Ga. The terrane has wide transitional margins in both the north and south.

In the north, the Oxford-Stull, Munro Lake, and Island Lake domains (Thurston et al., 1991) are inferred to have been formed on thinned crust of the North Caribou superterrane based on the presence of <3.0 model Nd ages of 2.9 to 2.71 Ga volcanic and plutonic rocks and the presence of <3.0 Ga detrital zircons in associated sedimentary units (Skulski et al., 2000). In the Ponask Lake area near the Ontario-Manitoba border, detrital zircons from a quartzite-komatiite sequence indicate a depositional age <2.865 Ga, inferred to date breakup at the northern North Caribou margin (Stone et al., 2004). The main phase of plutonism was followed by localized strain and shear-zone-hosted gold mineralization at ca. 2.685 Ga (Lin and Corfu, 2002), particularly in the Little Stull Lake area near the Ontario-Manitoba border (Jiang and Corkery, 1998; Lin and Jiang, 2001; Lin et al., 2006).

The central North Caribou superterrane is intruded by widespread tonalitic, dioritic, granodioritic, and granitic plutons that crystallized between 2.745 and 2.697 Ga at depths ranging from 18 to 10 km (0.6-0.3 GPa) (Corfu and Stone, 1998; Stone, 1998, 2000). Remnants of ca. 3.0 Ga tonalite and supracrustal rocks are sporadically preserved through the younger magmatism (Krogh et al., 1974; Corfu and Ayres, 1991; Henry et al., 1998; Whalen et al., 2003). Within the greenstone belts, thin packages of quartz arenite-carbonate-komatiite have variably been interpreted as platformal cover strata (Thurston and Chivers, 1990; Thurston et al., 1991) and plume-related rift deposits (Hollings and Kerrich, 1999; Percival et al., 2002a, 2006a). Evidence of early (>2.87 Ga) deformation is recorded in the North Caribou greenstone belt (Stott et al., 1989). The iron formation-hosted Musselwhite lode-gold deposit may have formed during development of structures associated with 2.87 Ga pluton emplacement, or during ca. 2.7 Ga events (Fyon et al., 1992). A few examples of porphyry-style Cu-Mo mineralization occur within the central North Caribou superterrane, including the ca. 2.70 Ga Setting Net Lake (Ayres et al., 1982) and Lang Lake occurrences (Findlay and Ayres, 1981).

The Uchi domain preserves a ca. 300 m.y. record of tectonostratigraphic evolution (Fig. 3) along the southern margin of the North Caribou superterrane (Stott and Corfu, 1991; Corfu and Stott, 1993b, 1996b; Hollings et al., 2000; Sanborn-Barrie et al., 2001, 2004). This region hosts some of the largest mineral deposits of the western Superior region, including the Red Lake gold camp. Aeromagnetic trends show the complex structural configuration of supracrustal rocks in a chain of greenstone belts separated by large lobes of plutonic material. The stratigraphic record preserved in the Rice Lake, Wallace Lake, Red Lake, Confederation Lake, Meen-Dempster, Pickle Lake, and Fort Hope greenstone belts reflects a history of rifting beginning ca. 2.99 Ga (Pirie, 1982; Corfu and Wallace, 1986; Corfu and Andrews, 1987; Tomlinson et al., 1998; Tomlinson et al., 2001; Percival et al., 2002a; Sasseville, 2002; Bailes et al., 2003; Sanborn-Barrie et al., 2004; Percival et al., 2006a; Sasseville et al., 2006), followed by a protracted period of continental arc magmatism at 2.94 to 2.91, 2.90 to 2.89, 2.85, and 2.75 to 2.72 Ga, punctuated by one or more unconformities (Stott, 1996; Beakhouse et al., 1999; Henry et al., 2000; Hollings et al., 2000; Rogers et al., 2000; Hollings, 2002; Rogers, 2002; Young, 2003; Sanborn-Barrie et al., 2004). The Rice Lake-Black Island belt contains juvenile rocks of probable back-arc affinity (Anderson, 2005; Bailes and Percival, 2005; Percival et al., 2006a). Metavolcanic rocks of the Uchi domain host both world-class gold deposits (Fig. 4) (Andrews et al., 1986; Brommecker, 1991; Poulsen et al., 1996; Dubé et al., 2004; Harris et al., 2006) and massive sulphide mineralization (Nunes and Thurston, 1980; Lesher et al., 1986; Rogers, 2002). Several deformation episodes are recognized within the greenstone belts, including pre-2.74, 2.73, 2.72, and 2.70 Ga events that have produced composite, steep, east-trending fabrics (Parker, 2000; Sanborn-Barrie et al., 2001, 2004; Dubé et al., 2003, 2004; Harris et al., 2006; Rogers, 2001; Young et al., 2006). Multiple ages of gold mineralization are indicated, with the main stage associated with D_2 deformation structures prior to 2.712 Ga and late-stage gold localization after 2.701 Ga (Corfu and Andrews, 1987; Dubé et al., 2004). Coarse clastic sedimentary rocks generally represent the youngest strata along the southern margin of the North Caribou superterrane (Devaney, 1999a,b). Where dated, these sequences contain detrital zircons as young as 2.703 Ga, and may be facies equivalents of the marine greywacke turbidites of the English River terrane to the south (e.g. Campbell, 1971; Stott, 1996).

Over 450 km of strike length, the east-trending Sydney Lake - Lake St. Joseph (SL-LSJ) fault separates rocks of the North Caribou margin to the north from metasedimentary schists and migmatitic rocks of the English River terrane to the south. The steeply dipping, 1 to 3 km wide, brittle-ductile fault zone is estimated to have accommodated about 30 km of right-lateral transcurrent displacement and 2.5 km of south-side-up movement (Stone, 1981). The Miniss River fault, which is cut and offset by the Sydney Lake - Lake St. Joseph fault, has an age of ca. 2.68 Ga (Bethune et al., 2000, 2006).

English River Terrane

The English River terrane (ERT) is thought to mark the suture between the North Caribou and Winnipeg River terranes (Corfu et al., 1995). Distinguished from adjacent regions by supracrustal rocks of metasedimentary origin, the English River terrane also displays high metamorphic grade, and a prominent east-west structural grain (Breaks, 1991). Based on the turbiditic nature of its chemically immature greywackes, the setting of the English River terrane has traditionally been considered a fore-arc basin (Langford and Morin, 1976) or accretionary prism (Breaks, 1991). Sedimentary facies vary from submarine fan on the northern margin, with associated banded iron formation (e.g., Griffith Mine), to deep-water wackes further south (Meyn and Palonen, 1980). Detrital zircon studies indicate that the sediments were deposited between 2.705 and 2.698 Ga, after arc activity in volcanic belts (Corfu et al., 1995; Davis, 1996a,b, 1998) and close to the time of collisional orogeny, thereby implying an origin as a synorogenic flysch basin. The small Melchett Lake greenstone belt in the central English River terrane comprises a juvenile, ca. 2.723 Ga calc-alkaline volcanic sequence (Corfu and Stott, 1993b, 1996b; Davis et al., 2000), possibly correlative with the Lake St. Joseph assemblage to the north. Metamorphic conditions in the English River terrane range from middle amphibolite facies near the margins, to low-pressure granulite facies (750-850°C at 0.6-0.7 Mpa) (Perkins and Chipera, 1985; Pan et al., 1999), coinciding with widespread generation of migmatite and diatexite at 2.691 Ga (Corfu et al., 1995). The main tectonothermal event was followed by a second thermal pulse at 2.669 Ga (op. cit. Pan et al., 1999), intrusion of ca. 2.65 Ga pegmatites (Corfu et al., 1995), and growth of hydrothermal minerals (Pan et al., 1999).

The dominant east-west structural grain of the terrane reflects upright to north-vergent F_2 folding of an earlier foliation, which is defined in many areas by migmatitic layering (Breaks, 1991; Hrabi and Cruden, 2001, 2006). The early foliation appears to be a composite fabric that includes primary layering and at least one set of early folds and axial planar foliation. It is particularly well expressed in the Lac Seul area (Sanborn-Barrie, 1988), where early, large-scale north-trending, west-vergent folds were delineated (Hynes, 1997).

Gravity (Nitescu et al., 2003), seismic reflection (White et al., 2003), and seismic refraction (Kay et al., 1999a) profiles

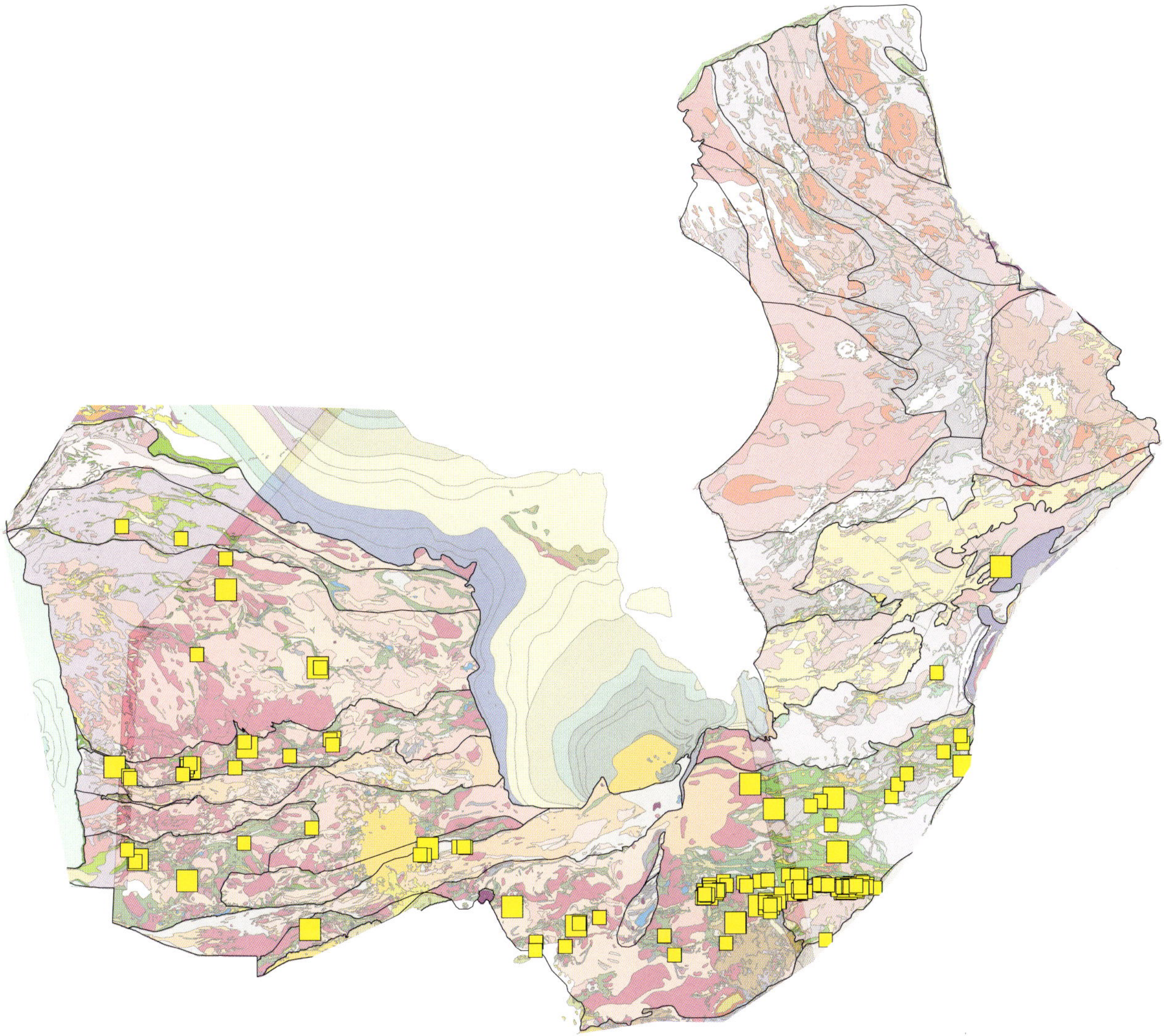

FIGURE 4. Gold deposits of the Superior Province (from Au deposit database, this volume). Symbol sizes are proportional to deposit size.

collectively indicate that the Moho beneath the combined English River – Winnipeg River terrane is elevated by about 8 km over that in adjacent terranes. A late to post-tectonic adjustment of this style could account for metamorphic pressure estimates up to 0.3 GPa higher within the English River terrane, as well as its slow cooling history (Corfu, 1996; Hanes and Archibald, 1999, 2001). The English River – Winnipeg River boundary separates dominantly metasedimentary rocks of the English River terrane from mainly metaplutonic rocks of the Winnipeg River terrane to the south. Between the two terranes lies the metavolcanic Bird River subprovince in eastern Manitoba, with mafic intrusion-hosted Cr, PGE deposits, and its narrow eastward extension, the Separation Lake greenstone belt (Breaks, 1991). These belts consist dominantly of largely juvenile mafic rocks with ages of ca. 2.78 to 2.73 Ga (Timmins et al., 1985). Depositional contacts have been inferred between English River clastic rocks and both volcanic strata of the Separation Lake belt (Hrabi and Cruden, 2006) and gneissic tonalitic basement to the east (Sanborn-Barrie, 1988). The Separation Lake greenstone belt appears to be in tectonic and intrusive contact with granitic rocks of the Winnipeg River terrane to the south (Blackburn and Young, 2000). The boundary zone is a focus for emplacement of ca. 2.646 Ga rare metal pegmatites (Larbi et al., 1999; Breaks et al., 2003; Smith et al., 2004), including the Tanco and Separation Rapids fields (Blackburn and Young, 2000).

Winnipeg River Terrane

The Winnipeg River terrane is a collective term used to describe the plutonic domain exposed north and east of the western Wabigoon terrane (Fig. 5). It consists of two main elements: 1) the Winnipeg River subprovince proper (Beakhouse, 1991), a more than 500 km long terrane composed of Mesoarchean metaplutonic rocks variably intruded by Neoarchean plutons; and 2) a largely Neoarchean plutonic domain, formerly referred to as the central Wabigoon granitoid complex (Percival et al., 2002b) and Wabigoon diapiric axis (Edwards and Sutcliffe, 1980; Thurston and Davis, 1985), that contains scattered remnants of Mesoarchean crust (Tomlinson and Percival, 2000; Whalen et al., 2002, 2004b; Tomlinson et al., 2004). With inheritance dating back to ca. 3.4 Ga (Henry et al., 2000; Tomlinson et al., 2003), the Winnipeg River terrane stands apart from the

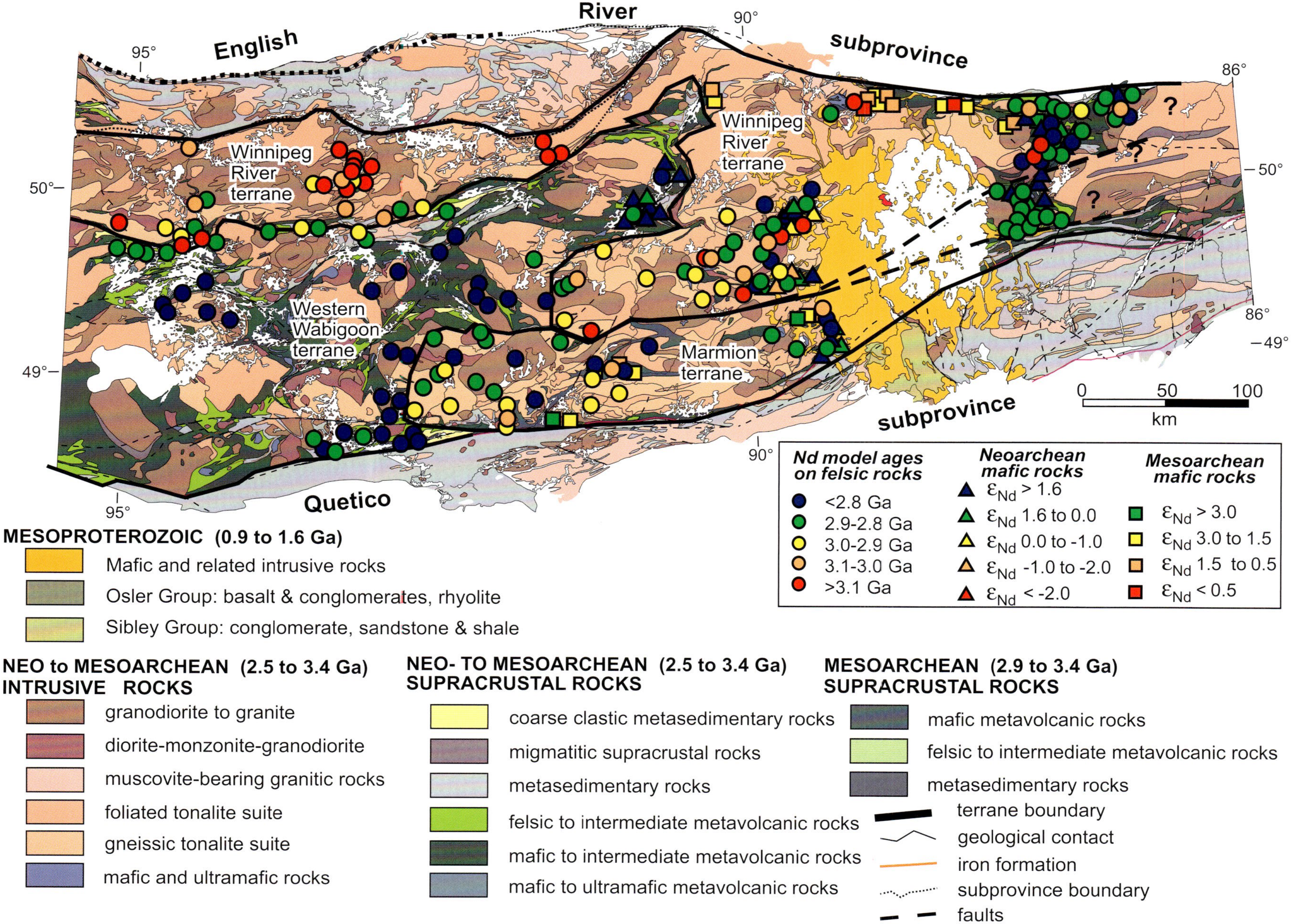

FIGURE 5. Distribution of terranes in the Wabigoon - Winnipeg River region, western Superior Province (modified after Tomlinson et al., 2004). Nd isotopic data used to establish terrane antiquity (see sources in Tomlinson et al., 2004) are shown as model ages for felsic rocks and as initial ε values for mafic rocks.

Northern Superior and North Caribou superterranes to the north (Fig. 3) and the Marmion domain to the south (described below). It also carries a distinct record of magmatic and structural events (Percival et al., 2004b; Melnyk et al., 2006), typically characterized by amphibolite- to granulite-facies metamorphism (Corfu, 1988).

The Mesoarchean history of the Winnipeg River terrane has remained cryptic due to extensive overprinting Neoarchean magmatism and deformation. Tonalitic rocks are the oldest units recognized, and include both 3.32 to 3.05 Ga gneissic (Corfu, 1988; Davis et al., 1988; Melnyk et al., 2006) and 3.04 Ga foliated varieties (Krogh et al., 1976). Similar isotopic signatures characterize younger (2.88, 2.84, and 2.83 Ga) tonalitic rocks, reflecting the antiquity of the basement (Beakhouse et al., 1988; Beakhouse and McNutt, 1991). Mafic volcanic belts older than ca. 3.0 Ga (Davis et al., 1988) and ca. 2.93 to 2.88 Ga volcanic rocks in the eastern Savant-Sturgeon greenstone belt (Sanborn-Barrie and Skulski, 1999; Sanborn-Barrie et al., 2002) are also considered part of the Winnipeg River terrane. Significant pulses of Neoarchean tonalite-granodiorite magmatism occurred at 2.715 to 2.705 Ga followed by emplacement of granites at ca. 2.70 to 2.69 Ga (Beakhouse et al., 1988; Corfu, 1988, 1996; Beakhouse, 1991; Cruden et al., 1997, 1998). A complex Neoarchean structural-metamorphic history began with deposition of metasedimentary rocks after 2.72 Ga (Melnyk et al., 2006). The supracrustal rocks and older gneisses were folded (D_3) between 2.717 and 2.713 Ga, prior to syntectonic injection of 2.713 to 2.707 Ga tonalite and granodiorite sheets accompanying D_4 horizontal extensional deformation (Melnyk et al., op. cit.). Upright folding during D_5 deformation took place after 2.705 Ga, and younger upright F_6 folds indicate a period of north-south compression associated with emplacement of 2.695 to 2.685 Ga granite and granodiorite (op. cit.). Late pegmatites and granites intruded during a dextral transpressive (D_7) regime (op. cit.).

The eastern Winnipeg River terrane hosts east-trending greenstone belts, including the Caribou Lake, Obonga, Garden Lake, and Heaven Lake belts, that have ages >3075 to 2703 Ma (Davis et al., 1988; Tomlinson et al., 2002, 2003). Dated granitoid units have ages in the range of 3075 to 2680 Ma (Davis et al., 1988; Whalen et al., 2002) and some of the oldest rocks have ε_{Nd} values of -1 to +1, suggesting derivation from even older crustal sources (Tomlinson et al., 2004). At least five generations of Neoarchean structures (D_1-D_5) have been recognized in complex tonalitic gneisses (Schwerdtner, 1992; Brown, 2002; Percival et al., 2003). Mineralization includes polymetallic occurrences in the Caribou Lake belt (Sutcliffe, 1988) and fractionated pegmatites along the English River terrane boundary (Breaks et al., 2003).

The Marmion terrane, formerly included as part of the south-central Wabigoon terrane, is now recognized to consist of 3.01 to 2.999 Ga tonalite (Davis and Jackson, 1988; Tomlinson et al., 2004), upon which the Steep Rock, known for its economic iron deposits, Finlayson Lake and Lumby Lake greenstone belts formed between 2.99 and 2.78 Ga (Stone et al., 2002; Tomlinson et al., 2003). Native silver occurs in a volcanic massive sulphide (VMS) environment within the Lumby Lake belt. Iron deposits in the Steep Rock belt have been upgraded through Cretaceous supergene enrichment (Machado, 1987).

A period of continental arc magmatism in the Winnipeg River terrane (2.72-2.70 Ga) (Corfu, 1988, 1996; Whalen et al., 2002; Melnyk et al., 2006) is attributed to north- and eastward subduction of oceanic rocks (Sanborn-Barrie and Skulski, 2006) followed by 2.708 to 2.701 Ga D_1 deformation. Post-2.704 Ga regional deformation (D_2) across the Wabigoon outlasted deposition of syncollisional coarse clastic sedimentary overlap sequences (i.e. Crowduck and Ament Bay assemblages) (Ayer and Davis, 1997; Fralick, 1997; Sanborn-Barrie and Skulski, 2006). Late faults with both strike-slip and dip-slip motion define the present terrane boundary (Gower and Clifford, 1981).

Wabigoon Terrane

The Wabigoon terrane has long been recognized as a composite terrane comprising volcanic-dominated domains with a central axis of variable-age plutonic rocks (Davis and Jackson, 1988; Percival et al., 2002b; Percival and Helmstaedt, 2004). It consists of distinct western and eastern segments.

Western Wabigoon Terrane

The western Wabigoon terrane is dominated by mafic volcanic rocks with large tonalitic plutons (Blackburn et al., 1991). Volcanic rocks range in composition from tholeiitic to calc-alkaline, and are interpreted to represent ocean floor or plateau and arc environments, respectively (Ayer and Davis, 1997; Ayer, 1998b; Ayer and Dostal, 2000; Wyman et al., 2000). Volcanic rocks were deposited between ca. 2.745 and 2.72 Ga (Corfu and Davis, 1992), with the exception of rare older rocks, such as the 2.775 Ga (Davis et al., 1988) Fourbay assemblage of oceanic plateau affinity (Sanborn-Barrie and Skulski, 1999) and minor younger (2.713-2.70 Ga) volcanic-sedimentary sequences. Plutonic rocks range from broadly synvolcanic batholiths composed of tonalite-diorite-gabbro (ca. 2.735-2.72 Ga) (Davis and Edwards, 1982; Corfu and Davis, 1992; Whalen et al., 2004b), to younger granodiorite batholiths and plutons (ca. 2.710 Ga) (Davis and Edwards, 1986; Sanborn-Barrie, 1988; Davis and Smith, 1991; Melnyk et al., 2006), monzodiorite plutons of sanukitoid affinity (ca. 2.698-2.690 Ga) (Stern and Hanson, 1991; Ayer, 1998a; Stevenson et al., 1999), and plutons and batholiths of monzogranite (2.69-2.66 Ga) (Schwerdtner et al., 1979; Sanborn-Barrie, 1988; Melnyk et al., 2000) (Fig. 3). Immature clastic metasedimentary sequences are preserved in narrow belts within volcanic sequences. They are commonly younger than the volcanic rocks, as illustrated by local unconformable relationships (Fralick, 1997) and geochronological constraints indicating deposition between ca. 2.711 and <2.702 Ga (Davis, 1996a,b, 1998; Fralick and Davis, 1999). Virtually all carry ancient (>3 Ga) detrital zircons indicating old components in source regions. At least two phases of deformation affected supracrustal rocks of the western Wabigoon terrane (Blackburn et al., 1991), with apparent diachroneity in the onset of deformation from ca. 2.709 Ga in the Lake of the Woods area (Davis and Smith, 1991; Ayer and Davis, 1997; Melnyk et al., 2006), to ca. 2.700 Ga in the Sioux Lookout - Savant area in the east (Sanborn-Barrie et al., 1998, 2002;

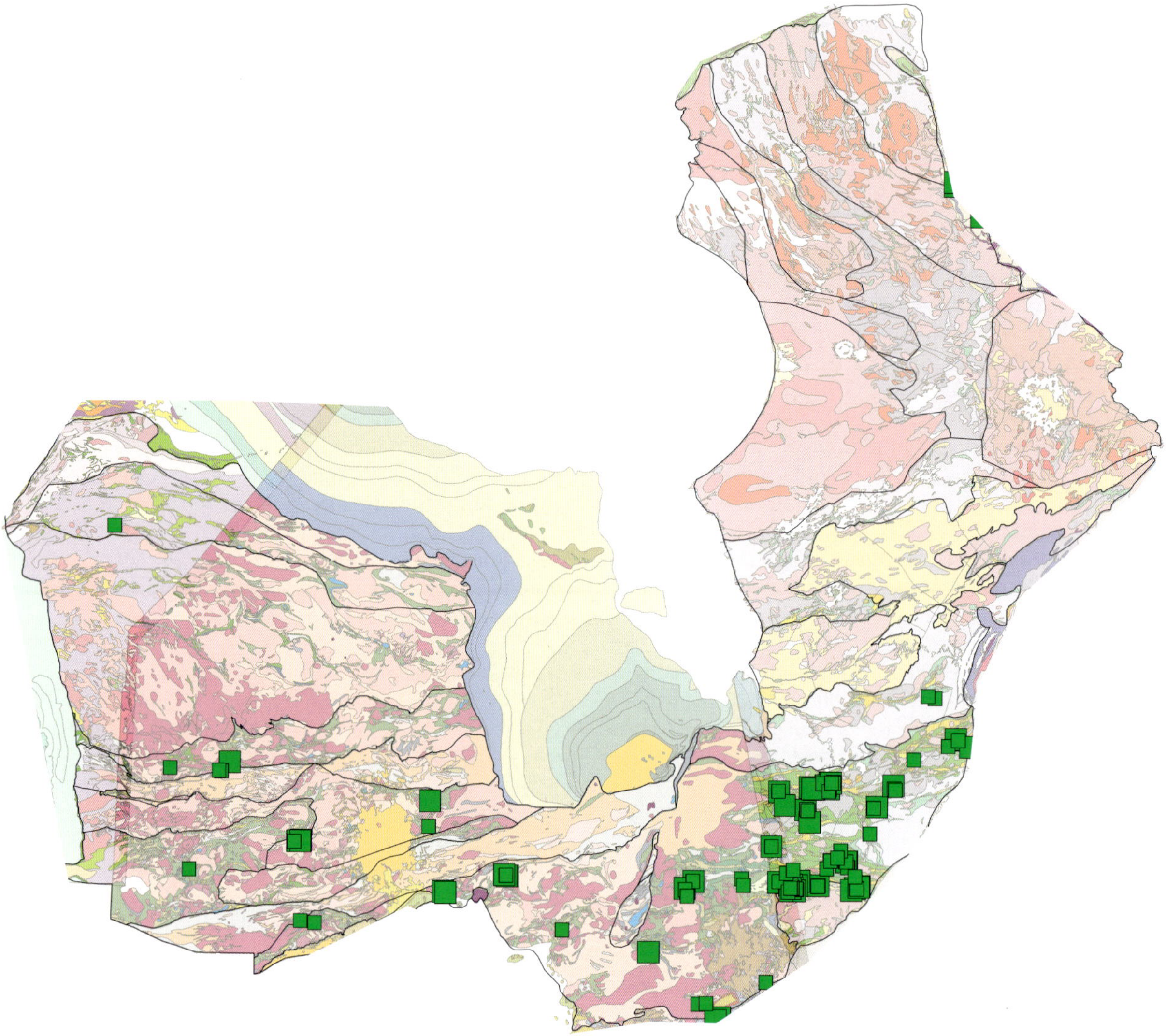

FIGURE 6. Volcanogenic massive sulphide deposits of the Superior Province (from VMS deposit database, this volume). Commodities include Cu, Zn, and Pb. Symbol sizes are proportional to deposit size.

Sanborn-Barrie and Skulski, 2006). These events involved at least local tectonic inversion, through thrust imbrication (Davis et al., 1988) and possible formation of nappe-like structures (Poulsen et al., 1980).

The Sturgeon - Savant greenstone belt, which hosted the Sturgeon Lake massive sulphide camp (Figs. 6, 7; Franklin et al., 1975; Galley et al., 2000), consists of several tectono-stratigraphic packages, including the Jutten assemblage of continental rift origin (Sanborn-Barrie et al., 2002), the ca. 2775 Ma Fourbay assemblage, and 2745 to 2735 Ma arc-like sequences, the Handy Lake and South Sturgeon assemblages (Davis et al., 1985; Sanborn-Barrie and Skulski, 1999; Sanborn-Barrie et al., 2002). The 2735 Ma Lewis Lake batholith (Whalen et al., 2004a) may have provided the heat source for seawater convection and massive sulphide mineralization (Galley et al., 2000). Younger (ca. 2718 Ma, Davis et al., 1988), high Fe, Ti basalt and minor dacite of the central Sturgeon assemblage represent a rifted arc sequence. Associated sedimentary rocks contain both arc (2745-2730 Ma) and continental (3.1-2.8 Ga) detritus (Skulski et al., 1998a). Two younger sedimentary sequences complete the stratigraphic record: 1) greywacke – iron formation (ca. 2.705 Ga) of the Warclub assemblage; and 2) sandstone and arkose (<2.698 Ga) of the synorogenic Ament Bay assemblage (Sanborn-Barrie et al., 2002). Two sets of ductile structures postdate <2.704 Ga rocks: 1) north-trending upright F_1 folds; and 2) east-trending upright D_2 folds and penetrative foliation. Pre-D_1 folds have been inferred locally (Sanborn-Barrie et al., 1998).

Eastern Wabigoon Terrane

The eastern Wabigoon terrane is a composite terrane with greenstone belts and intervening granitoid plutons that show variable Mesoarchean ancestry (Winnipeg River and Marmion) and Neoarchean oceanic affinity (Stott and Davis, 1999; Tomlinson et al., 2000; Stott et al., 2002). In the northwest part of the belt, the 3.0 to 2.92 Ga Toronto and Tashota assemblages may represent a continental margin sequence built on the Winnipeg River terrane based on their Nd isotopic character (Tomlinson et al., 2004). Calc-alkaline rocks of the 2.74 Ga Marshall assemblage have small massive sulphide deposits (Stott et al., 2002). The central part of the belt is dominated by rocks of oceanic affinity, including 2.78 to 2.738 Ga tholeiitic juvenile pillowed basalt and overlying 2.725 to 2.715 Ga calc-alkaline rocks (Stott et al., 2002). Parts of these assemblages contain widespread hydrothermal alteration and host small massive sulphide deposits (Fig. 6; op. cit.). Across the southern part of the eastern Wabigoon terrane, 2.78 to 2.74 Ga calc-alkaline volcanic rocks were possibly built on Marmion-age substrate (Tomlinson et al., 2004). Unconformably overlying clastic rocks (Albert-Gledhill and Conglomerate assemblages) were deposited after ca. 2.71 Ga. At least two deformation events affected the eastern Wabigoon domain: east-striking D_1 structures (<2.706 Ga) and east-striking, dextral transpressive D_2 shear zones (Stott et al., 2002). A 2.694 Ga pluton provides a lower limit on the age of D_2 deformation (Stott and Davis, 1999).

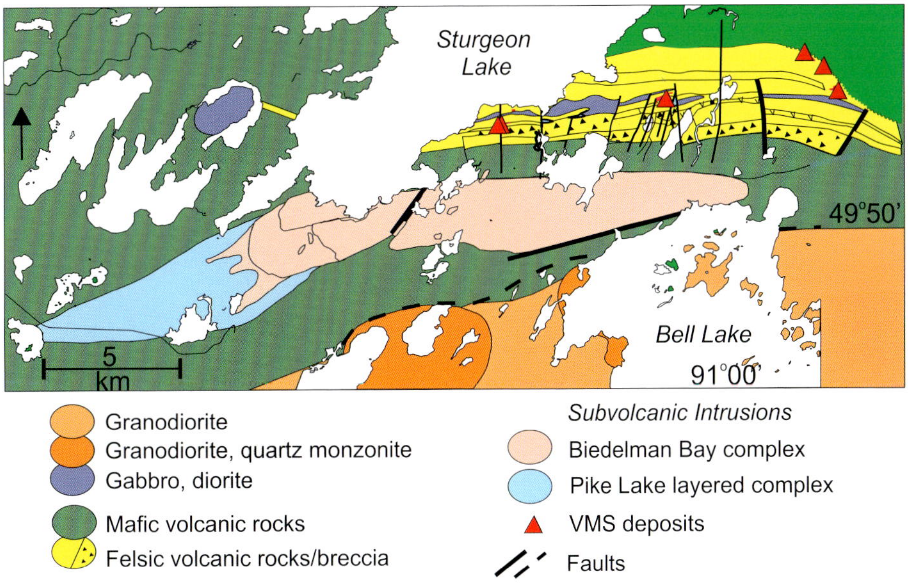

FIGURE 7. Simplified geological map of the Sturgeon Lake camp showing distribution of VMS deposits with respect to the Biedelman Bay subvolcanic intrusion (after Galley, 2003).

The Quetico - western Wabigoon boundary is well defined as the Seine River - Rainy Lake fault, in an area known for numerous gold showings and small polymetallic vein deposits (Poulsen, 1985). East of Lake Nipigon, the boundary with the eastern Wabigoon is an imbricate zone with an early history of structural telescoping (Devaney and Williams, 1989; Tomlinson et al., 1996) and the significant, structurally hosted Long Lac vein gold deposits (Lafrance et al., 2004). The Wabigoon-Quetico interface is also marked sporadically by <2.692 Ga coarse clastic rocks of the Seine assemblage (Fralick and Davis, 1999) that were deposited in transtensional basins (Blackburn et al., 1991) or delta fan environments (e.g. Fralick et al., 2006).

Quetico Terrane

The Quetico terrane consists dominantly of greywacke, derived migmatite, and granite. No stratigraphic sequence has been established within the steeply dipping, polydeformed and variably metamorphosed sedimentary succession; however, younging directions are dominantly to the north (Percival, 1989). Depositional age constraints indicate slightly older ages for the northern Quetico (2.698-2.696 Ga, Davis et al., 1990) than for the south (<2.692 Ga, Zaleski et al., 1999), consistent with accretionary prism geometry (Percival and Williams, 1989; Valli et al., 2004).

Several plutonic suites cut metasedimentary units, including early 2.696 Ga tonalite (Davis, 1996a). An early (D_1) deformation event predated emplacement of a chain of Alaskan-type mafic-ultramafic intrusions in the northern Quetico, which have some Ni-PGE potential (e.g. Pettigrew, 2004). They are associated with alkaline plutons including nepheline syenite and carbonatite with ages in the range 2.69 to 2.68 Ga (Lassen, 2004) and geochemical affinities with the Archean sanukitoid suite (cf. Stern et al., 1989; Stevenson et al., 1999; Lassen, 2004). Two subsequent deformation events (D_2, D_3) were followed by low-pressure, high-temperature metamorphism that reached upper-amphibolite and local granulite facies at ca. 2.67 to 2.65 Ga (Pan et al., 1994, 1998) in the central region and greenschist facies at the margins (Percival, 1989). Coeval, crust-derived granitic plutons and pegmatites, including ca. 2.67 Ga peraluminous granite and ca. 2.65 Ga biotite granite (e.g. Southwick, 1991) have sporadic rare-element mineralization (Breaks et al., 2003).

Tectonic models for the Quetico terrane have favoured forearc settings (e.g. Langford and Morin, 1976; Percival and Williams, 1989; Williams, 1991; Fralick et al., 2006). Depositional ages of ca. 2.698 to 2.690 Ga overlap those of late arc magmatism in the Wabigoon. The dominantly sanukitoid plutons of this age may have been triggered by slab breakoff (cf. Sajona et al., 2000).

The southern Quetico boundary separates metasedimentary rocks from the dominantly volcano-plutonic Wawa-Abitibi terrane to the south. Stratigraphic linkages between terranes are evident in some areas (Zaleski et al., 1999; Fralick et al., 2006), although at ca. 2.685 Ga dextral transpressive shear zones are common (Corfu and Stott, 1996a).

Wawa Terrane

Most workers accept a correlation between the Wawa and Abitibi terranes across the transverse Kapuskasing uplift. Within the Wawa terrane, volcanism appears to have begun with the 2.89 to 2.88 Ga Hawk assemblage. The 2.745 Ga Wawa and 2.72 Ga Greenwater and Manitouwadge assemblages indicate an oceanic setting. Volcanogenic massive sulphide deposits in the Wawa terrane have similar ages of 2.72 Ga (Figs. 3, 6; Corfu and Stott, 1986; Williams et al., 1991; Sage et al., 1996a,b). Polat et al. (1999) reported a variety of oceanic magma types from the Schreiber belt, and interpreted the belt as a tectonic mélange (Polat et al., 1998; Polat and Kerrich, 1999, 2001).

Relatively late-stage volcanism at ca. 2.695 Ga (Catfish assemblage) took place during D_1 thrusting. Subsequent calc-alkalic to alkalic magmatism (ca. 2.689 Ga, Corfu and

Stott, 1996a) and associated coarse clastic sedimentation (<2.689 Ga) was followed by emplacement of sanukitoid plutons (2.685-2.68 Ga) and dextral transpressive D_2 deformation.

Mineralization occurs in two main regions: the Michipicoten-Mishubishu belt in the Wawa area, and the Shebandowan-Schreiber belt to the west. The Michipicoten-Mishubishu belt contains mainly Fe and Au deposits with some Ni and vein Cu deposits. Iron deposits are in oxide-, sulphide- and carbonate-facies iron formations that lie stratigraphically above the 2.74 to 2.735 Ga Wawa assemblage. Gold deposits in this region occur in veins associated with shear zones within plutonic rocks of variable composition and age. The Shebandowan-Schreiber belt hosts important gold, iron, volcanic-hosted massive sulphide (e.g. Manitouwadge) (Peterson and Zaleski, 1999; Zaleski et al., 1999) and intrusion-hosted Ni deposits. The most significant is the Hemlo gold camp, a large disseminated deposit (Muir, 2003) in a strongly deformed, ca. 2.693 to 2.685 Ga volcano-sedimentary sequence (Davis and Lin, 2003). Gold was deposited during D_2 sinistral wrench deformation between 2.680 and 2.677 Ga (op. cit.), likely from fluids derived from granitoid rocks.

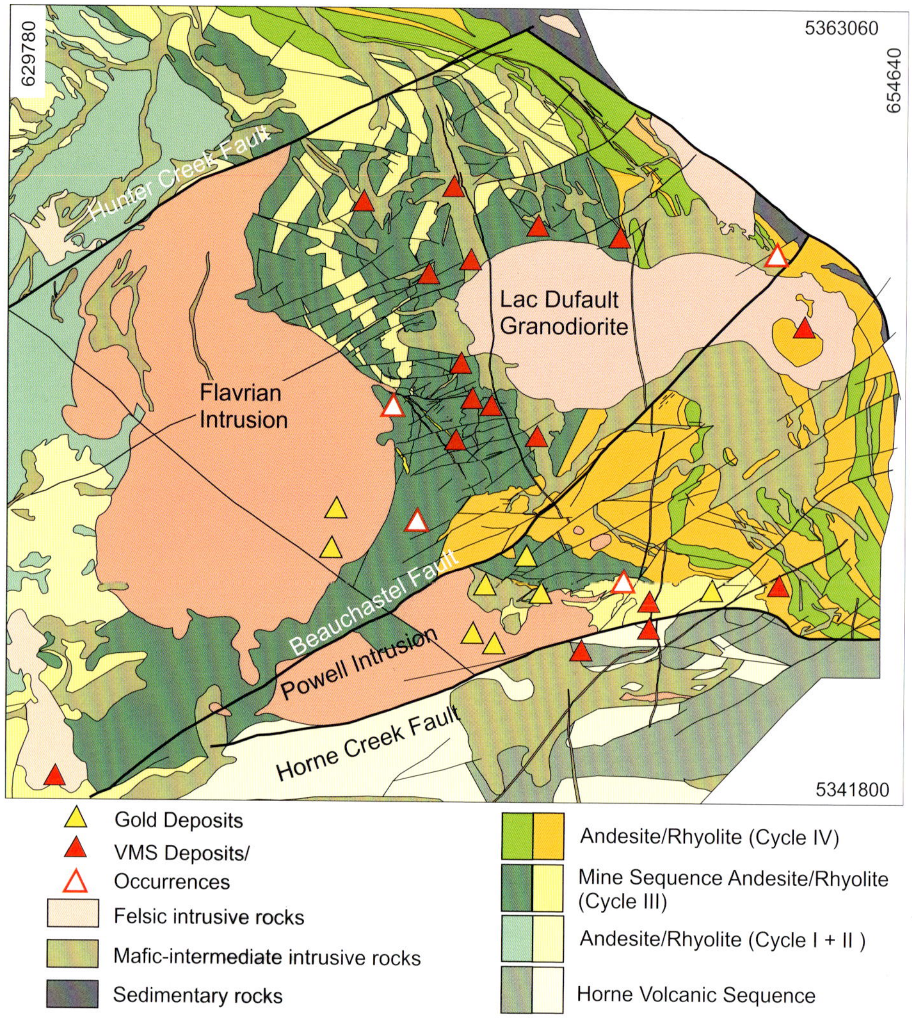

FIGURE 8. Simplified geological map of a cluster of VMS deposits in the Noranda camp. Fourteen deposits occur in bimodal volcanic rocks underlain by the Flavrian-Powell subvolcanic intrusion (modified after Galley et al., 2006).

The Great Lakes tectonic zone is the unexposed boundary between the Minnesota River Valley terrane and Wawa terrane, identified from aeromagnetic images (Sims and Day, 1993). It is inferred to dip northward based on the presence of isotopic inheritance in plutons of the Vermilion district of the southern Wawa-Abitibi terrane (Sims et al., 1997).

Kapuskasing Uplift

The Kapuskasing uplift represents a 500 km long fault-bounded structure that divides the Superior Province into eastern and western halves (Percival and West, 1994). Amphibolite- to granulite-facies tonalite gneiss, paragneiss, mafic gneiss, and anorthosite (2.765-2.66 Ga) represent mid- to lower crustal levels of the Abitibi-Wawa and Quetico terranes, exposed through east-directed thrusting and sinistral transcurrent motion during the Paleoproterozoic (Percival and McGrath, 1986; Percival and Peterman, 1994). Krogh (1993) noted the similarity between 2.66 to 2.62 Ga ages of high-grade metamorphism of Kapuskasing rocks and the timing of lode gold deposition in the Abitibi terrane (e.g. Zweng et al., 1993; Davis et al., 1994b), and proposed a genetic linkage. Brittle faults are cut by 1.89 and 1.1 Ga alkalic complexes, which have Nb, REE, and phosphate showings and prospects (Sage, 1991). Recent work suggests that the Kapuskasing structure accommodated a 20° counterclockwise rotation of the western Superior with respect to the east (Halls, 2004; Halls and Davis, 2004).

Eastern Superior Province

Abitibi Terrane

The Abitibi terrane hosts some of the richest mineral deposits of the Superior Province, including the giant Kidd Creek massive sulphide deposit (Hannington et al., 1999a) and the large gold camps of Ontario and Quebec (Figs. 2, 4; Robert and Poulsen, 1997; Poulsen et al., 2000). Views of the tectono-stratigraphic evolution of the Abitibi terrane have changed markedly from the allochthonous terrane concept introduced in the early 1990s (cf. Jackson and Fyon, 1991; Desrochers et al., 1993; Jackson et al., 1994), to a more traditional autochthonous stratigraphic framework supported by detailed and geochronological and volcanological studies (e.g. Heather, 1998; Ayer et al., 2002; Mueller and Mortensen, 2002). Stratigraphic complexities are explained in terms of evolution of oceanic geodynamic settings from plateau, to arc and rift environments (e.g. Thurston, 1994;

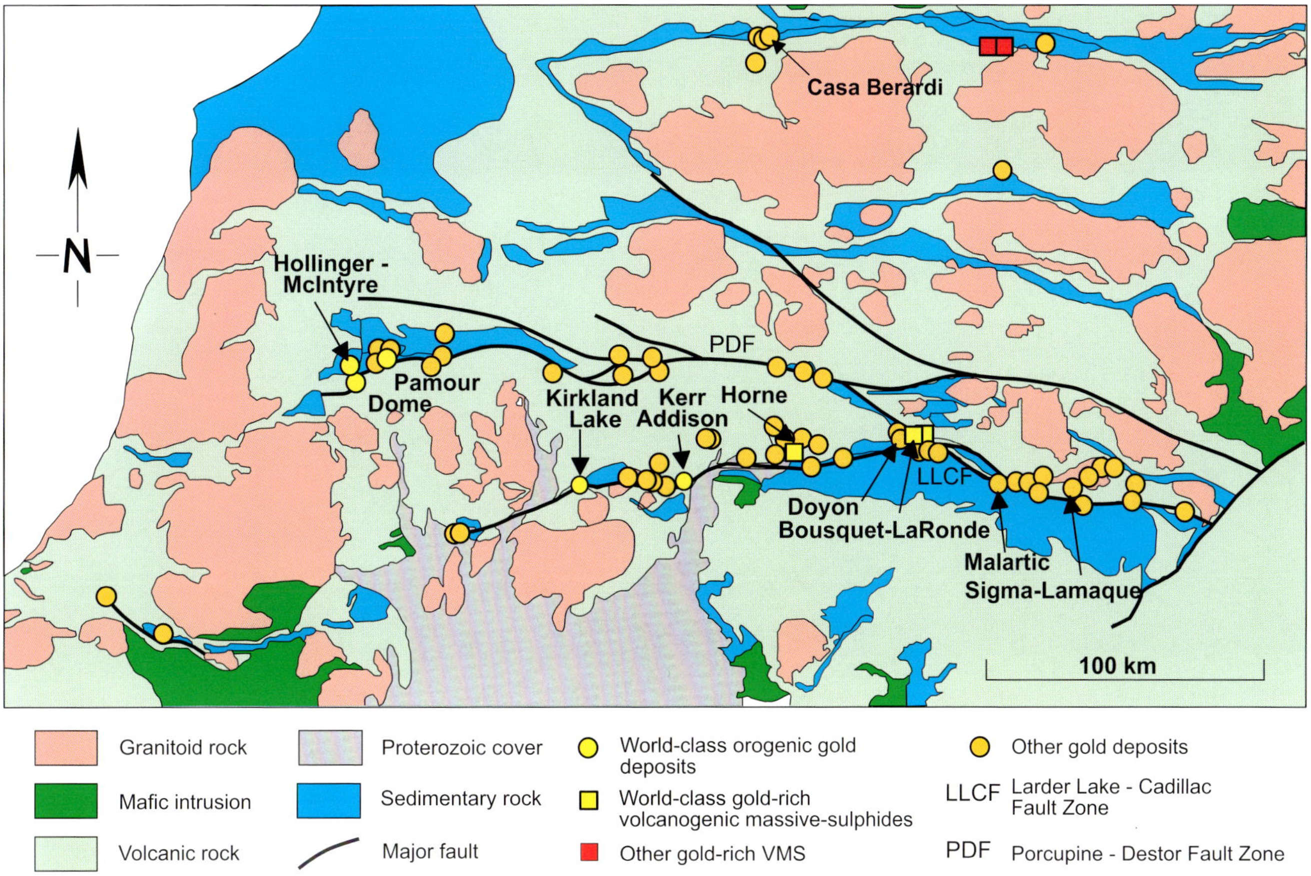

FIGURE 9. Simplified geological map of the Abitibi greenstone belt showing the distribution of major fault zones and gold deposits. Modified from Dubé and Gosselin (2006) and Poulsen et al. (2000).

Bédard and Ludden, 1997; Kerrich et al., 1999; Wyman et al., 1999, 2002).

The Abitibi terrane has traditionally been considered as northern, central, and southern regions with overlapping tectono-stratigraphic histories. In the northern Abitibi, volcanic assemblages are mainly 2.735 to 2.72 Ga (Ludden et al., 1986; Chown et al., 1992; Legault et al., 2002) and associated with layered intrusions, although units as old as 2.79 Ga are known (Rheaume et al., 2004). The central zone is dominated by plutonic rocks (Chown et al., 1992). Volcanic rocks of 2.71 to 2.695 Ga and their associated mineral deposits are abundant in the southern Abitibi belt (Fig. 8) (Dimroth et al., 1984; Daigneault et al., 2002; Dubé and Gosselin, 2006), although units as old as 2.75 Ga are present. The southern Abitibi has relatively young sedimentary-volcanic deposits including ca. 2.69 Ga greywackes of the Porcupine Group (Bleeker and Parrish, 1996) and 2.677 to 2.673 Ga conglomeratic and alkaline volcanic rocks of the Timiskaming Group (Davis, 2002).

The Matagami-Chibougamau camp of the northern Abitibi terrane is characterized mainly by Cu-Zn massive sulphide deposits, Cu-Zn vein deposits, and some lode-gold deposits. Volcanic strata hosting the sulphide mineralization range in age from ca. 2.73 to 2.715 Ga. Gold occurs in veins within shear zones and iron formations, or as disseminated mineralization associated with felsic intrusions (Figs. 4, 9; Card and Poulsen, 1998). In the central Abitibi, mineralization is restricted to the Normétal belt, and includes massive sulphides in 2.728 Ga volcanic rocks (Mortensen, 1993), as well as vein gold deposits. To the south, the Timmins-Val d'Or region hosts numerous gold deposits associated with the Porcupine-Destor deformation zone (Fig. 9), major Cu-Zn massive sulphide deposits in the Kidd assemblage (Hannington et al., 1999a; Wyman et al., 1999, 2002), komatiite and intrusion-related Ni deposits (Fig. 10), some pegmatitic deposits, and minor porphyry-type mineralization (Card and Poulsen, 1998; Jébrak and Doucet, 2002). The richly gold-mineralized Cadillac-Larder Lake break, which forms part of the southern boundary of the belt, is considered to represent a south-verging thrust carrying the Abitibi rocks over the Pontiac (Fig. 9) (Dimroth et al., 1984; Feng and Kerrich, 1991, 1992; Calvert et al., 1995; Calvert and Ludden, 1999; Ludden and Hynes, 2000; Poulsen et al., 2000; Davis, 2002). An alternative model views the fault zone as deep crustal material expelled on a fold detachment surface (Benn and Peschler, 2005).

Pontiac Terrane

Metasedimentary schist and paragneiss derived from turbiditic greywacke and minor conglomerate dominate the northern Pontiac terrane. Detrital zircons indicate depositional ages <2.685 Ga (Mortensen and Card, 1993; Davis, 2002), and tonalite, granodiorite, and granite plutons range in age from 2.682 to 2.66 Ga. The southern part of the belt includes the Baby-Belleterre greenstone belt, including volcanic rocks as young as 2.682 Ga (Mortensen and Card,

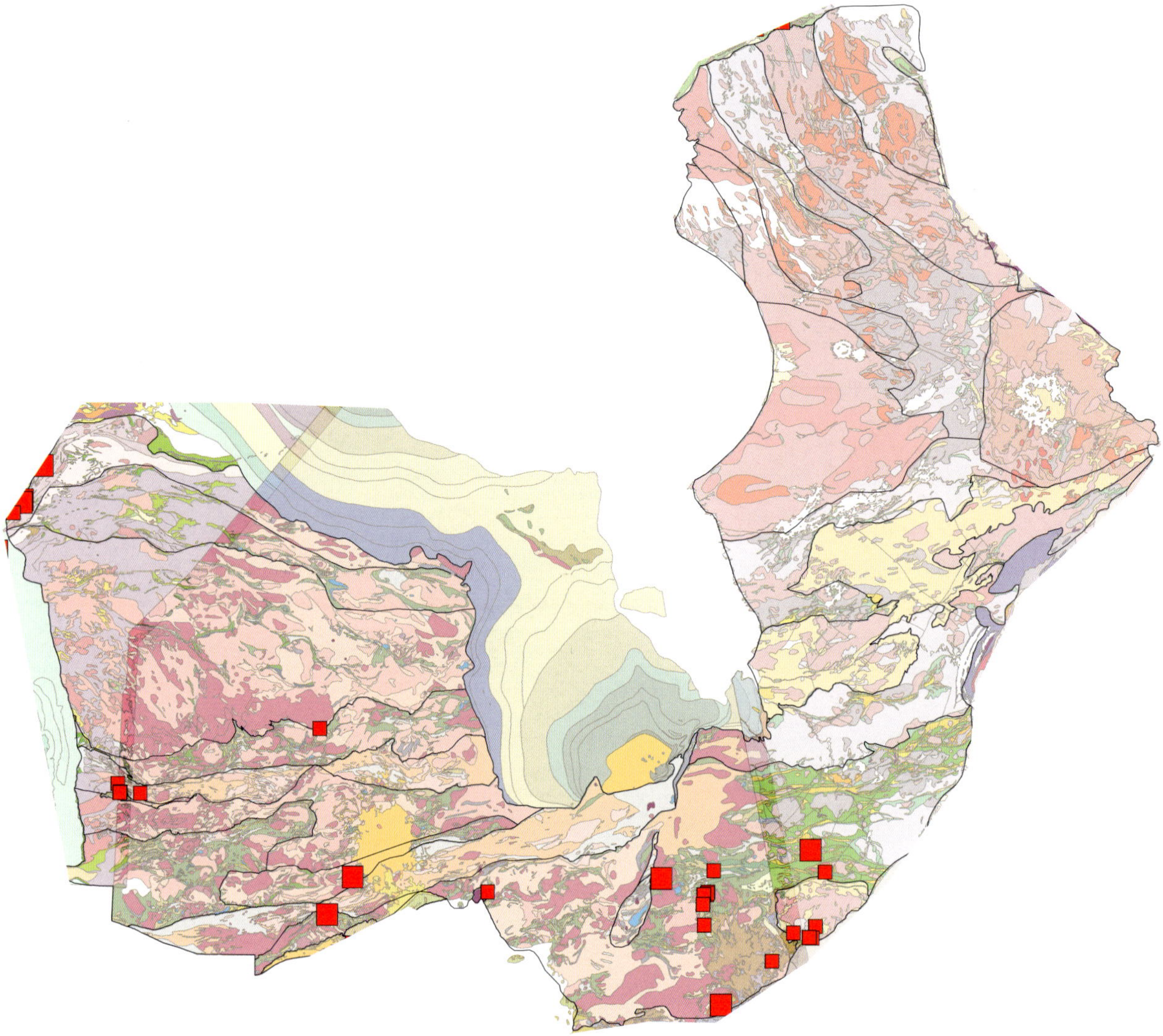

FIGURE 10. Nickel deposits of the Superior Province (from Ni deposit database, this volume). Deposits represent mainly intrusion-hosted Ni-Cu and Ni-Cu-PGE varieties. Symbol sizes are proportional to deposit size.

1993). The Pontiac terrane has been interpreted as a south-verging fold-thrust belt that was over-ridden by the southern Abitibi terrane (Benn et al., 1994; Calvert and Ludden, 1999; Davis, 2002).

The Benny-Belleterre mineral belt contains the quartz-vein hosted Belleterre gold deposit, as well as synvolcanic Ni-Cu sulphides (with some PGEs) in gabbroic sills and small plugs of dunite-gabbro.

Opatica Subprovince

This predominantly metaplutonic domain bounding the Abitibi terrane to the north consists of units ranging from 2.82 Ga tonalite, through 2.77 to 2.70 tonalite-granodiorite, to 2.68 Ga granite and pegmatite (Benn et al., 1992; Sawyer and Benn, 1993; Davis et al., 1994b). Polyphase deformation includes early, west-verging shear zones (<2.72 Ga), over-printed by 2.69 to 2.68 Ga south-vergent structures (Sawyer and Benn, 1993). The Frotet-Evans-Troilus greenstone belt has showings including Cu-Au veins, volcanogenic massive sulphide, iron formation, and intrusion-hosted Ni-Cu (Gosselin, 1996).

Opinaca Subprovince

Metagreywacke, derived migmatite, and granite characterize the Opinaca subprovince. Polydeformed schists occur at the belt margins, whereas the interior portions are metamorphosed to amphibolite and granulite facies. The northern boundary with the La Grande subprovince is a faulted stratigraphic contact (Goutier et al., 2002). The supracrustal rocks are cut by the 2.67 Ga Broadback River granite (Davis et al., 1994b). Mineralization in the Opinaca subprovince includes rare-metal occurrences within peraluminous granites and associated pegmatites (Goutier et al., 2002).

Ashuanipi Complex

The approximately 300 x 300 km Ashuanipi complex consists of high-grade metamorphic and plutonic rocks (Percival et al., 1992). Early sedimentary and volcanic rocks were intruded by a ca. 2.725 Ga tonalite-diorite suite (Percival et al., 2003), then deformed and metamorphosed to granulite facies and subsequently intruded by orthopyroxene-bearing diatexite (Percival, 1991), granite, granodiorite, and syenite plutons (2.65-2.62 Ga) and by A-type granites (2.57 Ga) (Leclair et al., 2004a).

Mineralization is restricted to small, high-grade gold showings in iron formation and skarns (Moritz and Chevé, 1992; Lapointe and Chown, 1993; Chevé and Brouillette, 1995). The occurrences are unusual in their setting in a granulite-facies terrane.

La Grande Subprovince

The La Grande subprovince consists of distinct sectors with variable tectonic setting. In the west, Mesoarchean basement (3.33-2.79 Ga) is unconformably overlain by the clastic Apple formation, which hosts uranium-gold occurrences (Roscoe and Donaldson, 1988), and 2.75 to 2.73 Ga volcanic strata (Goutier and Dion, 2004). Older strata are present in the Guyer-LG4 sector, including komatiites and related ca. 2.82 Ga sills, which contain Cu and massive sulphide mineralization. The Venus belt in the northeast contains Ni-Cu-PGE showings (Gosselin and Simard, 2001). Juvenile volcanic rocks (2.75-2.70 Ga) of the Eastmain sector are characterized by porphyry and other magmatic mineralization.

Bienville Subprovince

Plutonic rocks of the Bienville subprovince intrude the northern margin of the La Grande subprovince and mark a transition to dominantly granitic rocks to the north. The 200 km wide domain is underlain by foliated, gneissic, and massive granodiorite to granite with crystallization ages in the 2.73 to 2.68 Ga range (Skulski et al., 1998b; Percival et al., 2001). Xenocrystic zircons and Nd model ages indicate sources as old as 3.3 Ga.

Northeastern Superior Province (Minto Block)

This remote, far-north region of Quebec was under-explored prior to the early 1990s. Following the work of Percival et al. (1992, 1994, 2001), large parts were mapped for the first time at 1:250,000 (see synthesis in Leclair et al., 2004a). Metallogenic studies were carried out in parallel with regional mapping, and the major mineralization types identified (Labbé and Lacoste, 2004).

Domains of the Minto block were defined on the basis of lithology, and structural and aeromagnetic trends and intensity (Percival et al., 1992, 2001; Stern et al., 1994), and were refined by Leclair et al. (2004a) in light of regional map and isotopic constraints (Fig. 3). The Inukjuak domain in the west (Domain I of Leclair et al., 2004a) hosts the ancient (ca. 3.825 Ga) Nuvvuagittuq supracrustal belt (David et al., 2003) and ca. 3.65 Ga tonalitic gneisses, but is dominated by younger plutonic rocks (2.84-2.69 Ga). To the east, the Tikkerutuk domain (II) consists of pyroxene-bearing plutonic rocks,of 3.02 to 2.71 Ga age, that merge into the Bienville subprovince to the south, which is characterized by 2.71 to 2.69 Ga granite and granodiorite. Farther east, the Lake Minto domain (IV) contains metasedimentary and derived migmatitic rocks, in addition to 2.76 to 2.70 Ga pyroxene-bearing plutonic rocks (Percival and Mortensen, 2002). The Goudalie domain (V) forms a central spine consisting of relatively large volcano-sedimentary belts (2.88-2.71 Ga) with sparse 3.0 Ga tonalite and abundant pyroxene-bearing plutons (2.73-2.68 Ga). In the east-central Minto, the Utsalik domain (VI) consists of highly magnetic (Pilkington and Percival, 1999), 2.74 to 2.69 Ga pyroxene-bearing plutonic rocks with model ages in the 3.0 Ga range (Leclair et al., 2004a). To the northeast, Douglas Harbour domain (VII) consists of two large charnockitic massifs (2.74-2.73 Ga) within an older (2.88-2.75 Ga) tonalitic complex (Bédard, 2003; Bédard et al., 2003; Leclair et al., 2004a). The Diana complex in the northeast underwent pervasive ductile deformation during the Paleoproterozoic (Madore and Larbi, 2001). Percival and Skulski (2000) attributed the ca. 2.70 Ga deformation and high-grade metamorphism in the western Minto block to collisional processes, whereas Bédard et al. (2003) inferred magmatic and diapiric controls.

The northeastern Superior hosts a variety of mineral showings and small deposits that have been classified into syngenetic and epigenetic types (Labbé and Lacoste, 2004). Syngenetic mineralization, occurring mostly within supracrustal belts, includes Algoma-type iron formation, volcanogenic massive sulphide (Labbé and Lacoste, 2001), magmatic Ni-Cu (Labbé, 2005), mafic intrusion-hosted Fe-Ti-V, and porphyry U-Th and Mo (Labbé and Lacoste, 2004). Supracrustal rocks also host most epigenetic types, including disseminated polymetallic sulphide occurrences (Cu-Zn-Ag-Au, Labbé and Lacoste, 2001), gold in iron formation, vein quartz, and shear zones, sulphide Cu-Ni-Ag-Au in veins, rare earth elements in carbonated rocks, and minor U in veins.

Structural Style and Age of Deformation

Several first-order characteristics of the Superior Province point to the operation of plate tectonics during its assembly in the Neoarchean. Diagnostic features include 1) the presence of discrete continental and oceanic domains; 2) the existence of juvenile, mantle-derived magmatic rocks, including calc-alkaline basalt and sanukitoid-suite rocks, which suggest mantle metasomatism by fluids or melts in suprasubduction zone environments; 3) a southward-propagating array of five, temporally discrete, orogenic events over the 40 m.y. period between 2.72 and 2.68 Ga, which implies lateral accretion; 4) orogenic belts with length scales >1000 km, comparable to modern plate margin dimensions; 5) calc-alkaline granitoid batholiths with dimensions and compositions comparable to those of modern continental magmatic arcs, such as the South Patagonian batholith; 6) long strike-slip faults, indicating lateral movement (e.g. Sleep, 1992); and 7) gently dipping crustal panels and Moho offsets identified on teleseismic and magnetotelluric images in the Abitibi and Western Superior Lithoprobe transects. Collectively, these observations provide support for the concept that the Superior Province evolved at ca. 2.75 to 2.68 Ga through plate tectonic processes akin to those active today. It is important to note that this framework recognizes autochthonous development of stratigraphic sequences on continental margins or in oceanic regimes (i.e. within individual terranes, cf. Thurston, 2002). Thrust and nappe features have been identified locally through detailed structural and geochronological studies (e.g. Poulsen et al., 1980; Davis et al., 1988; Corfu and Ayres, 1991; Corfu and Stott, 1993a), although in general the structural style differs significantly from that of modern fold-thrust belts.

Steeply dipping foliation and steeply plunging folds are dominant in most domains of the Superior Province. In many regions these fabrics trend east-west or northwest and are the youngest penetrative structures in a polyphase chronology. Their nature has generally been attributed to transpressive strain late in the shortening history (e.g. Williams et al., 1992; Stott, 1997; Berclaz et al., 2004; Parmenter et al., 2006). Where evidence for earlier thrusting has been obtained, the structures responsible for stratigraphic repetition are generally not obvious. Accordingly, this style of deformation could be more common than is currently recognized. For example, early recumbent folds have been noted in the southern Wabigoon (Poulsen et al., 1980) and Quetico (Sawyer, 1983) terranes. Other early (D_1) structures appear to have formed in upright orientations. In the Red Lake and Sturgeon Lake belts, F_1 folds have steep plunges (Sanborn-Barrie et al., 1998, 2001; Sanborn-Barrie and Skulski, 1999, 2006). Similarly, the D_1 shear zone at Lake Winnipeg formed as a steep transcurrent structure (Percival et al., 2006a). The early (ca. 2.73 Ga) north-northwest-trending structures in the Red Lake and Confederation Lake belts are anomalous in light of their belt-scale extent and implication of east-west shortening. Similarly, the first folds in <2.704 Ga sedimentary rocks of the English River terrane are oriented north-northwest (Hynes, 1997). This recurring pattern could reflect structures defining the margins of early nappes or parautochthonous sheets, overprinted by subsequent folding and shortening (cf. Stott and Corfu, 1991).

Structural style varies widely on a regional scale. In the northwestern Superior, narrow sinuous belts of supracrustal rocks occupy synformal keels between plutonic massifs (Thurston et al., 1991), similar to the pattern of the eastern Minto block, where plutonic diapirs appear to dominate the structure (Bédard, 2003). Metasedimentary terranes have steep, linear features, including plutonic sheets. In the low-grade Abitibi terrane, the structure is dominated by open, upright folds, whereas structural patterns in the western Wabigoon terrane are controlled by large synvolcanic batholiths.

For the most part, the Superior Province has remained tectonically stable since ca. 2.6 Ga; however, ductile reworking occurred locally at the margins in response to ca. 1.9 to 1.8 Ga collisions. In the northwest, the Superior margin was affected by ca. 1.89 to 1.87 Ga continental arc magmatism and 1.82 to 1.79 Ga penetrative deformation related to collision of the Trans-Hudson Orogen (Percival et al., 2005). The Diana complex in the northeast has a similar ductile overprint (Madore and Larbi, 2001). Evidence of brittle Proterozoic reactivation is present locally in the form of new mineral growth along faults (Kamineni et al., 1992), including gold mineralization in parts of the Porcupine Destor fault (Fig. 9; Berger, 2001).

Metallogeny and Tectonics

Mineralization events are plotted with respect to stratigraphic age in Figure 3. It is evident that volcanogenic massive sulphide mineralization (Figs. 3, 6) generally formed between ca. 2.74 and 2.70 Ga, with rare older examples. The oldest (2.93-2.74 Ga) VMS deposits occur in the North Caribou superterrane and are characterized by distinct Ag, Pb-rich metal associations (Fyon et al., 1992; Parker, 1998). Progressively younger deposits formed further south, in the Uchi (e.g. 2.742 Ga South Bay deposit), Wabigoon (e.g. 2.735 Ga Sturgeon Lake deposits, Fig. 7), and Abitibi-Wawa terrane (e.g. 2.716-2.703 Ga Noranda (Figs. 6, 8), Kidd Creek, Horne, Laronde deposits), reflecting the north-to-south younging of arc magmatism. Examination of the setting of these VMS deposits has generally shown rifted arc or back-arc environments in continental or oceanic settings (e.g. Hannington et al., 1999a), associated with high-temperature magmatism including komatiites (Sproule et al., 2002) and FIII rhyolites (Hart et al., 2004). The link between a plume-arc setting and VMS mineralization appears particularly strong in the Abitibi terrane (Wyman et al., 1999, 2002). Gold-rich massive sulphide deposits such as the Horne and Laronde deposits represent a distinct class of VMS mineralization developed in oceanic settings including the Abitibi terrane (Hannington et al., 1999b; Dubé et al., 2006).

The Neoarchean is well known in Archean terranes globally for its massive sulphide "bonanza", and numerous authors have commented on potential underlying causes. Barley et al. (1998) proposed a Neoarchean global event (cf. Condie, 1998) in which plumes produced enriched oceanic crust that, when recycled in subduction zones, generated metal-rich derivative magmas. Similar circumstances of rifted oceanic arc settings surround the Paleoproterozoic VMS "bonanza" (Franklin et al., 1998). Formation of both the Neoarchean (2.73-2.70 Ga) and Paleoproterozoic (1.88-1.86 Ga) VMS deposits preceded collisional tectonism by 10 to 20 m.y., and it is plausible that impending collisions forced changes in plate stresses (subduction zone shallowing or roll-back), inducing extension within active arc systems that permitted access of deep-seated, metal-rich magmas.

Magmatic Cu-Ni-PGE deposits (Fig. 10) occur in a variety of settings within the Superior Province (Eckstrand and Hulbert, 2006). Komatiitic deposits of the Abitibi terrane resemble the Kambalda type. The remainder are intrusion-hosted, related to relatively young, mantle-derived gabbro (e.g Big Trout Lake, Lac des Iles), alkaline ultramafic (Pettigrew, 2004) and sanukitoid rocks (e.g. Entwine Lake).

At a broad scale, most lode-gold deposits (Fig. 4) are related to faults and shear zones, including terrane-bounding structures, although factors such as host rock type, alteration character, and structural chronology all play important roles (Dubé and Gosselin, 2006). These "orogenic gold" deposits formed late in the tectonic cycle, some in association with movement on major faults and others later, through fluid flow along dilatant zones (Robert and Poulsen, 1997; Poulsen et al., 2000; Dubé et al., 2004; Dubé and Gosselin, 2006). Where dated precisely, gold deposits reflect the general southward younging pattern of orogenies across the western Superior Province. For example, gold showings in the Stull Lake area in the north are associated with ca. 2.72 Ga faults (Lin et al., 2006), deposits in the Red Lake camp are ca. 2.713 Ga (Dubé et al., 2004), and deposits within the Abitibi terrane in the southeast are ca. 2.68 Ga or younger (Jemielita et al., 1990; Wong et al., 1991; Zweng et al., 1993). The timing of mineralization may be polyphase (Dubé et al., 2004), and coincides with late magmatic, metamorphic, and hydrothermal effects observed throughout the Superior Province (e.g. Krogh, 1993; Davis et al., 1994a; Moser et al., 1996; Percival and Skulski, 2000).

Fractionated pegmatites are common within metasedimentary terranes, at their margins, and along faults within volcanic-plutonic terranes. They carry rare-element mineralization at widespread locations across the Superior Province and generally represent the youngest Neoarchean event at ca. 2.66 to 2.64 Ga (Corkery et al., 1992; Smith et al., 2004). Some leucogranite plutons of crustal derivation have similar ages, implying a late, province-wide thermal anomaly that some authors have attributed to a widespread lithospheric delamination event. However, in light of the depleted lithosphere still present beneath the Superior Province (e.g. Kendall et al., 2002), including diamondiferous zones, the delamination hypothesis appears unlikely. Percival and Pysklywec (2004) suggested that the late hydrothermal overprint in the Superior Province is related to inversion of the mantle lithosphere cells, which led to cratonic stability.

Metallogenesis constitutes an integral part of Superior Province tectonic development, as illustrated by recurrent patterns of mineralization and associated events in several terranes. Volcanogenic massive sulphide deposits are generally associated with arc volcanic rocks and young systematically southward from ca. 2.74 Ga (Uchi) through 2.73 Ga (Wabigoon) to 2.72 to 2.71 Ga (Abitibi-Wawa). Gold deposits, formed during collisional orogenesis and subsequent transcurrent faulting, reflect progressive southward docking of terranes, from ca. 2.72 Ga in the northwest to 2.68 Ga in the south. Sanukitoid-suite rocks and their associated Cu-Ni-PGE mineralization generally follow collisional deformation within a few million years, again showing regional variation from 2.71 Ga in the north to 2.68 Ga in the south.

Paleoproterozoic and Mesoproterozoic Mafic Dykes and Related Rocks

Numerous Proterozoic dykes swarms transect the Superior Province, particularly in the south (Buchan and Ernst, 2004). The Matachewan swarm is volumetrically significant, making up almost 5% of bedrock area in northeastern Ontario. Some Ni-Cu mineralization is present in associated mafic sills within lower Huronian strata (e.g. James et al., 2002). Uranium mineralization in basal Huronian strata was derived from ca. 2.5 Ga erosion of the Superior Province to the north (Long, 2004). Significant mineralization is associated with the Paleoproterozoic Nipissing magmatic event, in form of vein silver deposits in sills of the Cobalt plate (Marshall and Watkinson, 2000). Mineralization related to the Keweenawan event (ca. 1.14-1.10 Ga) includes Cu, Ag in basalts and Ni-PGE showings in Logan sills in the Thunder Bay area (e.g. Hart, 2005). Alkaline intrusions of 1.9 to 1.1 Ga age host Ni-PGE mineralization, and carbonatites of similar age host numerous Nb-REE prospects (Sage, 1991). Apatite mined from the 1.88 Ga Cargill carbonatite complex was upgraded by a Cretaceous weathering event. Several kimberlite clusters have been identified within the Superior Province and include both Proterozoic (1.1 Ga, Sage, 2000) and Mesozoic examples (Heaman and Kjarsgaard, 2000). The 180 Ma Victor Pipe (Armstrong and Chatman, 2001) contains diamonds, as do 551 Ma occurrences in the Otish Mountains region (Moorhead et al., 2002).

Sudbury Structure

Situated on the southern edge of the Superior Province, the 1.85 Ga Sudbury Intrusive Complex represents one of the most richly mineralized bodies of the Canadian Shield. Current thinking regards the gabbro-noritic intrusion to have been generated in response to meteorite impact (Ames, 1999; Rousell et al., 2002, 2003; Therriault et al., 2002). Large Ni-PGE deposits, distributed around the perimeter of the intrusion, are localized primarily in the sublayer at the base of the differentiated sill (Therriault et al., 2002; Naldrett, 2003).

Knowledge Gaps in Superior Province

Most of the Superior Province has been mapped at scales of 1:250,000 or better. This regional information dates from the 1960s in parts of northern Ontario and Manitoba, and is very recent in northern Quebec. Mapping in regions with high mineral potential has been updated periodically. Remaining knowledge gaps are primarily the result of cover by Phanerozoic strata, glacial deposits, and water. In northern Ontario, recent interpretations using aeromagnetic data and drill core information have extended Superior Province units beneath the James Bay lowlands (Stott and Berdusco, 2000; Stott and Rayner, 2004). The James Bay and southern Hudson Bay region remains poorly understood as a result of cover and lack of aeromagnetic coverage. Improving our understanding of this area has several purposes: 1) resolving tectonic questions concerning the "big bend" in Superior Province structural trends, from east-west in the west and south, to north-south in the northeast (Fig. 2); 2) assessing the distribution of ancient rocks of the Northern Superior superterrane, with applications for diamond exploration; and 3) tracing northwest-trending, gold-mineralized faults and shear zones between northwestern Ontario and north-central Quebec.

Acknowledgements

Thanks are due to Beth Hillary for GIS support and to Ingrid Kjarsgaard for editorial persistence. Alain Leclair and Phil Thurston made relevant and helpful comments on an early version of the paper. Wayne Goodfellow is thanked for project coordination and management.

References

Ames, D.E., 1999, Geology and Regional Hydrothermal Alteration of the Crater-Fill, Onaping Formation; Association with Zn-Pb-Cu Mineralization, Sudbury Structure, Canada: Unpub. Ph.D. thesis, Carleton University, Ottawa, ON, 638 p.

Anderson, S.D., 2005, Preliminary results and economic significance of geological mapping in the Gem Lake area, southeast Rice Lake belt, Manitoba (NTS 52L11 and 14), *in* Report of Activities 2005: Manitoba Industry, Economic Development and Mines, Manitoba Geological Survey, p. 104-116.

Andrews, A.J., Hugon, H., Durocher, W.E., Corfu, F., and Lavigne, M.J., 1986, The anatomy of a gold-bearing greenstone belt; Red Lake, northwestern Ontario, Canada: Gold '86; An International Symposium on the Geology of Gold Deposits, p. 3-22.

Armstrong, J., and Chatman, J., 2001, Kimberlite indicator mineral database (KIDD): A public data compilation of till sample locations and kimberlite indicator mineral picking results, Slave Craton and environs, NWT and NU, Canada: Department of Indian and Northern Development, NWT Geology Division, EGS Open Report 2001-002, CD-ROM.

Ayer, J. A., 1998a, The mafic minerals of the Falcon Island ultrapotassic pluton, Lake of the Woods, Ontario; Progressive reduction during fractionation: The Canadian Mineralogist, v. 36, p. 49-66.

——— 1998b, Petrogenesis and Tectonic Evolution of the Lake of the Woods Greenstone Belt, Western Wabigoon Subprovince, Ontario, Canada: Unpub. Ph.D. thesis, University of Ottawa, Ottawa, ON, 213 p.

Ayer, J.A., and Davis, D.W., 1997, Neoarchean evolution of differing convergent margin assemblages in the Wabigoon Subprovince: geochemical and geochronological evidence from the Lake of the Woods greenstone belt, Superior Province, northwestern Ontario: Precambrian Research, v. 81, p. 155-178.

Ayer, J.A., and Dostal, J., 2000, Nd and Pb isotopes from the Lake of the Woods greenstone belt, northwestern Ontario; Implications for mantle evolution and the formation of crust in the southern Superior Province: Canadian Journal of Earth Sciences, v. 37, p. 1677-1689.

Ayer, J., Amelin, Y., Corfu, F., Kamo, S., Ketchum, J., Kwok, K., and Trowell, N., 2002, Evolution of the southern Abitibi greenstone belt based on U-Pb geochronology: autochthonous volcanic construction followed by plutonism, regional deformation and sedimentation: Precambrian Research, v. 115, p. 63-95.

Ayres, L.D., Averill, S.A., and Wolfe, WJ., 1982, An Archean molybdenite occurrence of possible porphyry type at Setting Net Lake, northwestern Ontario, Canada: Economic Geology, v. 77, p. 1105-1119.

Bailes, A., Percival, J.A., Corkery, M.T., McNicoll, V J., Tomlinson, K.Y., Rogers, N., Sasseville, C., Whalen, J.B., and Stone, D., 2003, Geology and tectonostratigraphic assemblages, West Uchi area: Geological Survey of Canada, Open File 1522, 1:250 000.

Bailes, A.H., and Percival, J.A., 2005, Geology of the Black Island area, Lake Winnipeg, Manitoba (parts of NTS 62P1, 7 and 8): Manitoba Geological Survey, Geoscientific Report GR2005-2, 33 p.

Barley, M.E., Krapez, B., Groves, D.I., and Kerrich, R., 1998, The late Archaean bonanza; metallogenic and environmental consequences of the interaction between mantle plumes, lithospheric tectonics and global cyclicity, *in* Percival, J.A., and Ludden, J.N., eds., Earth's Eolution Through the Precambrian: Precambrian Research, v. 91, p. 65-90.

Beakhouse, G.P., 1991, Winnipeg River subprovince, *in* Thurston, P.C., Williams, H.R., Sutcliffe, R.H., and Stott, G.M., eds., Geology of Ontario: Ontario Geological Survey, Special Volume 4, Part 1, p. 279-301.

Beakhouse, G.P., and McNutt, R.H., 1991, Contrasting types of late Archean plutonic rocks in northwestern Ontario; Implications for crustal evolution in the Superior Province: Precambrian Research, v. 49, p. 141-165.

Beakhouse, G.P., McNutt, R.H., and Krogh, T.E., 1988, Comparative Rb-Sr and U-Pb geochronology of late to post tectonic plutons in the Winnipeg River belt, northwestern Ontario, Canada: Chemical Geology, v. 72, p. 283-291.

Beakhouse, G.P., Heaman, L.M., and Creaser, R.A., 1999, Geochemical and U-Pb zircon geochronological constraints on the development of a Late Archean greenstone belt at Birch Lake, Superior Province, Canada: Precambrian Research, v. 97, p. 77-97.

Bédard, J.H., 2003, Evidence for regional-scale, pluton-driven, high-grade metamorphism in the Archaean Minto Block, northern Superior Province, Canada: Journal of Geology, v. 111, p. 183-205.

Bédard, J. H., Brouillette, P., Madore, L., and Berclaz, A., 2003, Archaean cratonization and deformation in the northern Superior Province, Canada; An evaluation of plate tectonic versus vertical tectonic models: Precambrian Research, v. 127, p. 61-87.

Bédard, L.P., and Ludden, J.N., 1997, Nd-isotope evolution of Archean plutonic rocks in the Opatica, Abitibi and Pontiac subprovinces, Quebec, Canada: Canadian Journal of Earth Sciences, v. 34, p. 286-299.

Benn, K., and Peschler, A.P., 2005, A detachment fold model for fault zones in the Late Archean Abitibi greenstone belt: Tectonophysics, v. 400, p. 85-104.

Benn, K., Sawyer, E.W., and Bouchez, J.L., 1992, Orogen parallel and transverse shearing in the Opatica belt, Quebec: Implications for the structure of the Abitibi belt: Canadian Journal of Earth Sciences, v. 29, p. 2429-2444.

Benn, K., Miles, W., Ghassemi, M.R., and Gillet, J., 1994, Crustal structure and kinematic framework of the northwestern Pontiac Subprovince, Quebec: An integrated structural and geophysical study: Canadian Journal of Earth Sciences, v. 31, p. 271-281.

Berclaz, A., Leclair, A., David, J., and Maurice, C., 2004, Structural evolution of the Northeastern Superior Province (NESP) over 1 billion years, with emphasis on greenstone belts and their relation with enclosing granitoids: Canadian Geophysical Union / American Geophysical Union / Society of Economic Geologists / Environmental and Engineering Geophysical Society Joint Assembly, EOS Transactions, Abstract, p. V31A-03.

Berger, B., 2001, Variations in styles of gold mineralization along the Porcupine-Destor deformation zone in Ontario: An exploration guide, *in* Summary of Field Work and other Activities: Ontario Geological Survey, Open File Report 6070, p. 9-1 to 9-13.

Bethune, K., Helmstaedt, H., and McNicoll, V.M., 2000, U-Pb geochronology bearing on the timing and nature of deformation along the Miniss River fault: Western Superior Transect Sixth Annual Workshop, p. 8-12.

——— 2006, Structural analysis of the Miniss River fault and allied faults bounding the central Wabigoon subprovince, western Superior Province: Canadian Journal of Earth Sciences, v. 43, p. 1031-1054.

Bickford, M.E., Wooden, J.L., and Bauer, R.L., 2006, SHRIMP study of zircons from Early Archean rocks in the Minnesota River Valley: Implications for the tectonic history of the Superior Province: Geological Society of America Bulletin, v. 118, p. 94-108.

Blackburn, C.E., and Young, J.B., 2000, Precambrian geology of the Separation Lake area, northwestern Ontario: Ontario Geological Survey, Open File Report 6001, 94 p.

Blackburn, C.E., John, G.W., Ayer, J., and Davis, D.W., 1991, Wabigoon Subprovince, *in* Thurston, P.C., Williams, H.R., Sutcliffe, R.H., and Stott, G.M., eds., Geology of Ontario: Ontario Geological Survey, Special Volume 4, Part 1, p. 303 -381.

Bleeker, W., and Parrish, R.R., 1996, Stratigraphy and U-Pb geochronology of Kidd Creek: Implications for the formation of giant massive sulphide deposits and the tectonic history of the Abitibi greenstone belt: Canadian Journal of Earth Sciences, v. 33, p. 1213-1231.

Böhm, C.O., Heaman, L.M., Creaser, R.A., and Corkery, M.T., 2000, Discovery of pre-3.5 Ga exotic crust at the northwestern Superior Province margin, Manitoba: Geology, v. 28, p. 75-78.

Böhm, C.O., Heaman, L.M., Stern, R.A., Corkery, M.T., and Creaser, R.A., 2003, Nature of Assean Lake ancient crust, Manitoba: A combined SHRIMP-ID-TIMS U-Pb geochronology and Sm-Nd isotope study: Precambrian Research, v. 126, p. 55-94.

Breaks, F.W., 1991, English River subprovince, *in* Thurston, P.C., Williams, H.R., Sutcliffe, R.H., and Stott, G.M., eds., Geology of Ontario: Ontario Geological Survey, Special Volume 4, Part 1, p. 239-277.

Breaks, F.W., Selway, J.B., and Tindle, A. G., 2003, Fertile peraluminous granites and related rare-element mineralization in pegmatites, Superior Province, northwest and northeast Ontario; Operation Treasure Hunt: Ontario Geological Survey, Open File Report 6099, 179 p.

Brommecker, R., 1991, The Structural Setting of Gold Occurrences in the S.E. Rice Lake Greenstone Belt, S.E. Manitoba: Unpub. M.Sc. thesis, Queen's University, Kingston, Ontario, 267 p.

Brooks, C., Ludden, J., Pigeon, Y., and Hubregtse, J.J.M.W., 1982, Volcanism of shoshonite to high-K andesite affinity in an Archean arc environment, Oxford Lake, Manitoba: Canadian Journal of Earth Sciences, v. 19, p. 55-67.

Brown, J.L., 2002, Neoarchean Evolution of the Western - Central Wabigoon Boundary Zone, Brightsand Forest area, Ontario: Unpub. M.Sc. thesis, University of Ottawa, Ottawa, Ontario, 240 p.

Buchan, K.L., and Ernst, R.E., 2004, Diabase dyke swarms and related units in Canada and adjacent regions [with accompanying notes]: Geological Survey of Canada, Map 2022A, 1:5,000,000.

Calvert, A., and Ludden, J.N., 1999, Archean continental assembly in the southeastern Superior Province of Canada: Tectonics, v. 18, p. 412-429.

Calvert, A.J., Sawyer, E.W., Davis, W.J., and Ludden, J.N., 1995, Archaean subduction inferred from seismic images of a mantle suture in the Superior Province: Nature, v. 375, p. 670-674.

Calvert, A.J., Cruden, A.R., and Hynes, A.J., 2004, Seismic evidence for preservation of the Archean Uchi granite-greenstone belt by crustal-scale extension: Tectonophysics, v. 388, p. 135-143.

Campbell, F.H.A., 1971, Stratigraphy and sedimentation of part of the Rice Lake group, Manitoba, *in* McRitchie, W.D., and Weber, W., eds., Geology and Geophysics of the Rice Lake Region, Southeastern Manitoba (Project Pioneer): Manitoba Department of Mines and Natural Resources, Mines Branch, Publication 71-1, p. 135-188.

Card, K.D., 1990, A review of the Superior Province of the Canadian shield, a product of Archean accretion: Precambrian Research, v. 48, p. 99-156.

Card, K.D., and Ciesielski, A., 1986, Subdivisions of the Superior Province of the Canadian Shield: Geoscience Canada, v. 13, p. 5-13.

Card, K.D., and Poulsen, K.H., 1998, Geology and mineral deposits of the Superior Province of the Canadian shield, *in* Lucas, S., and St-Onge, M.R., eds., Geology of the Precambrian Superior and Grenville Provinces and Precambrian Fossils in North America: Geological Survey of Canada, Geology of Canada 7, p. 13-194.

Cheng, L.Z., Mareschal, J.C., Jaupart, C., Rolandone, F., Gariepy, C., and Radigon, M., 2002, Simultaneous inversion of gravity and heat flow data; constraints on thermal regime, rheology and evolution of the Canadian Shield crust: Journal of Geodynamics, v. 34, p. 11-30.

Chevé, S., and Brouillette, P., 1995, Geologie et metallogenie de la partie nord-est de la Sous-Province d'Ashuanipi (Nouveau-Quebec), carte synthese, 23J, 23K, 23N et 23O.

Chown, E. H., Daigneault, R., Mueller, W., and Mortensesn, J.K., 1992, Tectonic evolution of the northern volcanic zone, Abitibi belt, Quebec: Canadian Journal of Earth Sciences, v. 29, p. 2211-2225.

Condie, K., 1998, Episodic continental growth and supercontinents; a mantle avalanche connection? Earth and Planetary Science Letters, v. 163, p. 97-108.

Corfu, F., 1988, Differential response of U-Pb systems in coexisting accessory minerals, Winnipeg River Subprovince, Canadian Shield: Implications for Archean growth and stabilization: Contributions to Mineralogy and Petrology, v. 98, p. 312-325.

——— 1996, Multistage zircon and titanite growth and inheritance in an Archean gneiss complex, Winnipeg River Subprovince, Ontario: Earth and Planetary Science Letters, v. 141, p. 175-186.

Corfu, F., and Andrews, A.J., 1987, Geochronological constraints on the timing of magmatism, deformation and gold mineralization in the Red Lake greenstone belt, northwestern Ontario: Canadian Journal of Earth Sciences, v. 24, p. 1302-1320.

Corfu, F., and Ayres, L.D., 1991, Unscrambling the stratigraphy of an Archaean greenstone belt: A U-Pb geochronological study of the Favourable Lake belt, northwestern Ontario: Precambrian Research, v. 50, p. 201-220.

Corfu, F., and Davis, D.W., 1992, A U-Pb geochronological framework for the western Superior Province, Onta, *in* Thurston, P. C., Williams, H. R., Sutcliffe, R. H., and Stott, G. M., eds., Geology of Ontario: Ontario Geological Survey, Special Volume 4, Pt. 2, p. 1335-1346.

Corfu, F., and Stone, D., 1998, Age structure and orogenic significance of the Berens River composite batholiths, western Superior Province: Canadian Journal of Earth Sciences, v. 35, p. 1089-1109.

Corfu, F., and Stott, G. M., 1986, U-Pb ages for late magmatism and regional deformation in the Shebandowan belt, Superior Province, Canada: Canadian Journal of Earth Sciences, v. 23, p. 1075-1082.

——— 1993a, U-Pb geochronology of the central Uchi Subprovince, Superior Province: Canadian Journal of Earth Sciences, v. 30, p. 1179-1196.

——— 1993b, Age and petrogenesis of two late Archean magmatic suites, northwestern Superior Province, Canada: zircon U-Pb and Lu-Hf isotopic relations: Journal of Petrology, v. 34, p. 817-838.

——— 1996a, Shebandowan greenstone belt, western Superior Province; U-Pb ages, tectonic implications and correlations: Geological Society of America Bulletin, v. 110, p. 1469-1484.

——— 1996b, Hf isotopic composition and age constraints on the evolution of the Archean central Uchi Subprovince, Ontario, Canada: Precambrian Research, v. 78, p. 53-63.

Corfu, F., and Wallace, H., 1986, U-Pb zircon ages for magmatism in the Red Lake greenstone belt, northwestern Ontario: Canadian Journal of Earth Sciences, v. 23, p. 27-42.

Corfu, F., Stott, G.M., and Breaks, F.W., 1995, U-Pb geochronology and evolution of the English River subprovince, an Archean low P - high T metasedimentary belt in the Superior Province: Tectonics, v. 14, p. 1220-1233.

Corfu, F., Davis, D.W., Stone, D., and Moore, M., 1998, Chronostratigraphic constraints on the genesis of Archean greenstone belts, northwestern Superior Province, Ontario, Canada: Precambrian Research, v. 92, p. 277-295.

Corkery, M. T., and Skulski, T., 1998, Geology of the Little Stull Lake Area (part of NTS 53K/10 and /7): Manitoba Energy and Mines, Geological Services, Report of Activities 1998, 8 p.

Corkery, M.T., Davis, D.W., and Lenton, P.G., 1992, Geochronological constraints on the development of the Cross Lake greenstone belt: Canadian Journal of Earth Sciences, v. 29, p. 2171-2185.

Corkery, M.T., Cameron, H.D. M., Lin, S., Skulski, T., Whalen, J.B., and Stern, R.A., 2000, Geological Investigations in the Knee Lake Belt (Parts of NTS 53L): Manitoba Industry, Trade and Mines, Manitoba Geological Survey, Report of Activities 2000, 7 p.

Corkery, M.T., Skulski, T., Stone, D., Syme, E.C., Bailes, A.H., Cameron, M.J., and Whalen, J.B., 2006, Geology and tectonostratigraphic assemblages, West Sachigo area, Manitoba: Ontario Geological Survey, Preliminary Map P.3463, 1:250 000.

Craven, J.A., Kurtz, R.D., Boerner, D.E., Skulski, T., Spratt, J., Ferguson, I.J., Wu, X., and Bailey, R.C., 2001, Conductivity of Western Superior Province Upper Mantle in Northwestern Ontario: Geological Survey of Canada, Current Research 2001-E6, 6 p.

Cruden, A.R., Davis, D.W., Menard, T., and Robin, P.-Y.R., 1997, Structural and geochronological relationships between the Winnipeg River and Wabigoon subprovinces: Implications for the terrane accretion model: Western Superior Transect Third Annual Workshop, p. 18-26.

Cruden, A.R., Davis, D.W., Melnyk, M.J., Robin, P.-Y.R., and Menard, T., 1998, Structural and geochronological observations at Kenora: Implications for the style and timing of deformation during the Kenoran orogeny, NW Ontario: Western Superior Transect Fourth Annual Workshop, p. 54-72.

Daigneault, R., Mueller, W.U., and Chown, E.H., 2002, Oblique Archean subduction: accretion and exhumation of an oceanic arc during dextral transpression, Southern Volcanic Zone, Abitibi Subprovince Canada: Precambrian Research, v. 115, p. 261-290.

David, J., Parent, M., Stevenson, R., Nadeau, P., and Godin, L., 2003, The Porpoise Cove supracrustal sequence, Inukjuak area: A unique example of Paleoarchean crust (ca. 3.8 Ga) in the Superior Province: Geological Association of Canada, Program with Abstracts, p. CD-ROM.

Davis, D. W., 1996a, Provenance and depositional age constraints on sedimentation in the western Superior transect area from U-Pb ages of zircons: Western Superior Transect Second Annual Workshop, p. 18-23.

——— 1996b, U-Pb age patterns for detrital zircons from the western Superior Province: Witnesses to orogeny: Geological Association of Canada, Program with Abstracts, p. A21.

——— 1998, Speculations on the formation and crustal structure of the Superior province from U-Pb geochronology: Western Superior Transect Fourth Annual Workshop, p. 21-28.

——— 2002, U-Pb geochronology of Archean metasedimentary rocks in the Pontiac and Abitibi subprovinces, Quebec, constraints on timing, provenance and regional tectonics: Precambrian Research, v. 115, p. 97-117.

Davis, D.W., and Edwards, G.R., 1982, Zircon U-Pb ages from the Kakagi Lake area, Wabigoon Subprovince, northwest Ontario: Canadian Journal of Earth Sciences, v. 19, p. 1235-1245.

——— 1986, Crustal evolution of Archean rocks in the Kakagi Lake area, Wabigoon Subprovince, Ontario, as interpreted from high precision U-Pb geochronology: Canadian Journal of Earth Sciences, v. 23, p. 182-192.

Davis, D.W., and Jackson, M., 1988, Geochronology of the Lumby Lake greenstone belt: a 3 Ga complex within the Wabigoon Subprovince, northwest Ontario: Geological Society of America Bulletin, v. 100, p. 818-824.

Davis, D.W., and Lin, S., 2003, Unraveling the geologic history of the Hemlo Archean gold deposit, Superior Province, Canada; a U-Pb geochronological study: Economic Geology, v. 98, p. 51-67.

Davis, D.W., and Smith, P.M., 1991, Archean gold mineralization in the Wabigoon Subprovince, a product of crustal accretion: Evidence from U-Pb geochronology in the Lake of the Woods area, Superior Province, Canada: Journal of Geology, v. 99, p. 337-353.

Davis, D.W., Krogh, T.E., Hinzer, J., and Nakamura, E., 1985, Zircon dating of polycyclic volcanism at Sturgeon Lake and implications for base metal mineralization: Economic Geology, v. 80, p. 1942-1952.

Davis, D.W., Sutcliffe, R.H., and Trowell, N.F., 1988, Geochronological constraints on the tectonic evolution of a late Archean greenstone belt, Wabigoon subprovince, northwest Ontario: Precambrian Research, v. 39, p. 171-191.

Davis, D.W., Pezzuto, F., and Ojakangas, R.W., 1990, The age and provenance of metasedimentary rocks in the Quetico subprovince, Ontario, from single zircon analyses: Implications for Archean sedimentation and tectonics in the Superior Province: Earth and Planetary Science Letters, v. 99, p. 195-205.

Davis, D.W., Schandl, E.S., and Wasteneys, H.A., 1994a, U-Pb dating of minerals in alteration halos of Superior Province massive sulphide deposits: Syngenesis versus metamorphism: Contributions to Mineralogy and Petrology, v. 115, p. 427-437.

Davis, D.W., Amelin, Y., Nowell, G.M., and Parrish, R.R., 2000, Constraints on crustal development in the Western Superior Lithoprobe Transect from Hf isotopes in zircons: Western Superior Transect Sixth Annual Workshop, p. 38-44.

Davis, W. J., Machado, N., Gariépy, C., Sawyer, E., and Benn, K., 1994b, U-Pb geochronology of the Opatica tonalite-gneiss belt and its relationship to the Abitibi greenstone belt, Superior Province, Quebec: Canadian Journal of Earth Sciences, v. 32, p. 113-127.

Desrochers, J.P., Hubert, C., Ludden, J.N., and Pilote, P., 1993, Accretion of oceanic plateau fragments in the Abitibi greenstone belt, Canada: Geology, v. 21, p. 451-454.

Devaney, J. R., 1999a, Late orogenic history of the Birch-Uchi Archean greenstone belt: Interpretation of clastic pull-apart basins marginal to a hypothesized indentor/ promontory: Western Superior Transect Sixth Annual Workshop, University of British Columbia, p. 55-65.

——— 1999b, Sedimentological and Volcanological Studies of the Southwest Birch-Uchi Greenstone Belt, Uchi Subprovince: Ontario Geological Survey, Summary of Field Work and Other Activities, Open File Report 6000, 12 p.

Devaney, J.R., and Williams, H.R., 1989, Evolution of an Archean subprovince boundary: a sedimentological and structural study of part of the Wabigoon-Quetico boundary in northern Ontario: Canadian Journal of Earth Sciences, v. 26, p. 1013-1026.

Dimroth, E., Rocheleau, M., and Mueller, W., 1984, Paleogeography, isostasy, and crustal evolution of the Archean Abitibi Belt: Comparison between the Rouyn-Noranda and Chibougamau-Chapais areas, *in* Guha, J., and Chown, E.H., eds., Chibougamau-Stratigraphy and Mineralization: Canadian Institute of Mining and Metallurgy, Special Volume 34, p. 73-91.

Dubé, B., and Gosselin, P., 2006, Greenstone-hosted quartz-carbonate vein deposits, *in* Goodfellow, W.D., ed., Mineral Deposits of Canada: A Synthesis of Major Deposit-Types, District Metallogeny, the Evolution of Geological Provinces, and Exploration Methods: Geological Association of Canada, Mineral Deposits Division, Special Publication 5, p. 49-73.

Dubé, B., Williamson, K., and Malo, M., 2003, Gold Mineralization within the Red Lake Mine Trend: Example from the Cochenour-Willans Mine Area, Red Lake, Ontario, with New Key Information from the Red Lake Mine and Potential Analogy with the Timmins Camp: Geological Survey of Canada, Current Research 2003-C21, 15 p.

Dubé, B., Williamson, K., McNicoll, V., Malo, M., Skulski, T., Twomey, T., and Sanborn-Barrie, M., 2004, Timing of gold mineralization at Red Lake, Northwestern Ontario, Canada: New constraints from U-Pb geochronology at the Goldcorp High-Grade Zone, Red Lake Mine and at the Madsen Mine: Economic Geology, v. 99, p. 1611-1641.

Dubé, B., Gosselin, P., Hannington, M., and Galley, A., 2006, Gold-rich volcanogenic massive sulphide deposits, *in* Goodfellow, W.D., ed., Mineral Deposits of Canada: A Synthesis of Major Deposit-Types, District Metallogeny, the Evolution of Geological Provinces, and Exploration Methods: Geological Association of Canada, Mineral Deposits Division, Special Publication 5, p. 75-94.

Eckstrand, O.R., and Hulbert, L., 2006, Magmatic nickel-copper-platinum group element deposits, *in* Goodfellow, W.D., ed., Mineral Deposits of Canada: A Synthesis of Major Deposit-Types, District Metallogeny, the Evolution of Geological Provinces, and Exploration Methods: Geological Association of Canada, Mineral Deposits Division, Special Publication 5, p. 205-222.

Edwards, G.R., and Sutcliffe, R.H., 1980, Archean granitoid terrains of the western Superior Province, Ontario: Geological Association of Canada, Program with Abstacts, p. 50.

Feng, R., and Kerrich, R., 1991, Single zircon age constraints on the tectonic juxtapositon of the Archean Abitibi greenstone belt and Pontiac Subprovince, Quebec, Canada: Geochimica et Cosmochimica Acta, v. 55, p. 3437-3441.

——— 1992, Geochemical evolution of granitoids from the Archean Abitibi southern volcanic zone and the Pontiac subprovince, Superior Province, Canada: Implications for tectonic history and source regions: Chemical Geology, v. 98, p. 23-70.

Findlay, D.J., and Ayres, L.D., 1981, Sheeted porphyry sill complexes and associated copper-molybdenum mineralization, Lang Lake, northwestern Ontario: Canadian Institute of Mining and Metallurgy Bulletin, v. 74, p. 52.

Fralick, P., 1997, Neoarchean evolution of the Wabigoon Subprovince: Evidence from the sedimentary record: Western Superior Transect Third Annual Workshop, University of British Columbia, p. 97-99.

Fralick, P., and Davis, D.W., 1999, The Seine-Couchiching problem revisited: sedimentology, geochronology and geochemistry of sedimentary units in the Rainy Lake and Sioux Lookout areas: Western Superior Transect, Fifth Annual Workshop, University of British Columbia, p. 66-75.

Fralick, P., Purdon, R.H., and Davis, D.W., 2006, Neoarchean trans-subprovince sediment transport in western Superior Province: Canadian Journal of Earth Sciences, v. 43, p. 1055-1070.

Franklin, J.M., Kasarda, J., and Poulsen, K.H., 1975, Petrology and chemistry of the alteration zone of the Mattabi massive sulfide deposit: Economic Geology, v. 70, p. 63-79.

Franklin, J.M., Gibson, H.L., and Galley, A.G., 1998, The Paleoproterozoic; a bonanza time for volcanogenic massive sulphide deposits, *in* Bottrill, T.J., and Sheahan, P., eds., The First Age of Giant Ore Formation; Stratigraphy, Tectonics and Mineralization in the Late Archean and Early Proterozoic: Annual Convention and Trade Show of the Prospectors and Developers Association of Canada, p. 115-127.

Fyon, A., Breaks, F.W., Heather, K.B., Jackson, S.L., Muir, T.L., Stott, G.M., and Thurston, P.C., 1992, Metallogeny of metallic mineral deposits in the Superior Province of Ontario, *in* Thurston, P.C., Williams, H.R., Sutcliffe, R.H., and Stott, G.M., eds., Geology of Ontario: Ontario Geological Survey, Special Volume 4, Part 2, p. 1091-1176.

Galley, A.G., 2003, Composite synvolcanic intrusions associated with Precambrian VMS-related hydrothermal systems: Mineralium Deposita, v. 38, p. 443-473.

Galley, A., van Breemen, O., and Franklin, J., 2000, The relationship between intrusion-hosted Cu-Mo mineralization and VMS deposits of the Archean Sturgeon Lake mining camp, northwestern Ontario: Economic Geology, v. 95, p. 1543-1550.

Galley, A., Hannington, M., and Jonasson, I., 2006, Volcanogenic massive sulphide deposits, *in* Goodfellow, W. D., ed., Mineral Resources of Canada: A Synthesis of Major Deposit-types, District Metallogeny, the Evolution of Geological Provinces, and Exploration Methods: Mineral Deposits Division, Geological Association of Canada, Special Publication 5, p. 141-161.

Géologie Québec, 2002, Carte géologique de la province de Québec: Ministère des Ressources, des Parcs et de la Faune, DV 2002-07, Digital release.

Goldich, S.S., Wooden, J.L., Ankenbauer, G.A., Levy, T.M., and Suda, R.U., 1984, Origin of the Morton Gneiss, southwestern Minnesota; Part I, Lithology, *in* Morey, G.B., and Hanson, G.N., eds., Selected Studies of Archean Gneisses and Lower Proterozoic Rocks, Southern Canadian Shield: Geological Society of America, Special Paper 182, p. 45-50.

Goodwin, A.M., 1968, Archean protocontinental growth and early crustal history of the Canadian shield: 23rd International Geological Congress, Prague, p. 69-89.

Gosselin, C., 1996, Synthèse géologique de la region do Frotet-Troilus: Ministère des Ressources Naturelles du Québec, ET-96-02, 21 p.

Gosselin, C., and Simard, M., 2001, Geology of the Lac Gayot Area (NTS 23M): Ministère des Ressources naturelles, Québec, RG 2000-03, 28 p.

Goutier, J., and Dion, C., 2004, Géologie et minéralisation de la Sous-province de La Grande, Baie-James: Exploration 2004, Ministere des Ressources naturelles, de la Faune et des Parcs.

Goutier, J., Dion, C., Oullette, M.C., Davis, D.W., David, J., and Parent, M., 2002, Géologie de la région du lac Guyer (33G/05, 33G/06 ET 33G/11): Gélogie Québec, RG 2001-15.

Gower, C.F., and Clifford, P.M., 1981, The structural geometry and geological history of Archean rocks at Kenora, northwestern Ontario; A proposed type area for the Kenoran Orogeny: Canadian Journal of Earth Sciences, v. 18, p. 1075-1091.

Grand, S.P., 1987, Tomographic inversion for shear velocity beneath the North American plate: Journal of Geophysical Research, v. 92, p. 14,065-14,090.

Halls, H.C., 2004, Proterozoic deformation of the Superior Province from paleomagnetic studies of dyke swarms: contributions to the origin of the Kapuskasing zone and Hudson Bay embayment: Lithoprobe Report 86, Posters with Abstracts, University of British Columbia, 1 p.

Halls, H.C., and Davis, D.W., 2004, Paleomagnetism and U-Pb geochronology of the 2.17 Ga Biscotasing dyke swarm, Ontario, Canada: Evidence for vertical-axis crustal rotation across the Kapuskasing zone: Canadian Journal of Earth Sciences, v. 41, p. 255-269.

Hanes, J.A., and Archibald, D.A., 1999, Post-orogenic tectonothermal history of the Archean western Superior Province of the Canadian Shield as determined by conventional and laser Ar-Ar data; An update, *in* Western Superior Transect Fifth Annual Workshop Lithoprobe Report #70, Lithoprobe Secretariat, University of British Columbia, p. 156-157.

——— 2001, Post-orogenic tectonothermal history of the Archean western Superior Province of the Canadian Shield as determined by conventional and laser Ar-Ar data: New data: 2001 Western Superior Transect Fifth Annual Workshop, University of British Columbia, p. 26-28.

Hannington, M.D., Barrie, C.T., and Bleeker, W., 1999a, The giant Kidd Creek volcanogenic massive sulfide deposit, western Abitibi Subprovince, Canada, *in* Hannington, M.D., and Barrie, C.T., eds., The Giant Kidd Creek Volcanogenic Massive Sulfide Deposit, Western Abitibi Subprovince, Canada: Economic Geology Monograph, 10, p. 1-30.

Hannington, M.D., Poulsen, K.H., Thompson, J.F.H., and Sillitoe, R.H., 1999b, Volcanogenic gold in the massive sulfide environment, *in* Barrie, C.T., and Hannington, M.D., eds., Volcanic-Associated Massive Sulfide Deposits: Processes and Examples in Modern and Ancient Settings: Reviews in Economic Geology, 8, p. 325-356.

Harris, J.R., Sanborn-Barrie, M., Panagapko, D.A., Skulski, T., and Parker, J.R., 2006, Gold prospectivity maps of the Red Lake greenstone belt: Application of GIS technology: Canadian Journal of Earth Sciences, v. 43, p. 865-893..

Hart, T.R., 2005, Project Unit 05-002. Lake Nipigon synthesis, *in* Summary of Field Work and Other Activities 2005: Ontario Geological Survey, Open File Report 6172, p. 12-1 to 12-6.

Hart, T.R., Gibson, H.L., and Lesher, C.M., 2004, Trace element geochemistry and petrogenesis of felsic volcanic rocks associated with volcanogenic massive Cu-Zn-Pb sulfide deposits: Economic Geology, v. 99, p. 1003-1013.

Heaman, L.M., and Kjarsgaard, B.A., 2000, Timing of eastern North American kimberlite magmatism; Continental extension of the Great Meteor Hotspot track? Earth and Planetary Science Letters, v. 178, p. 253-268.

Heather, K.B., 1998, New insights on the stratigraphy and structural geology of the southwestern Abitibi greenstone belt: Implications for the tectonic evolution and setting of mineral deposits in the Superior Province, The first age of giant ore formation: Stratigraphy, tectonics and mineralization in the late Archean and early Proterozoic: Prospectors and Developers Association of Canada, Annual Meeting, Toronto, p. 63-101.

Henry, P., Stevenson, R., and Gariepy, C., 1998, Late Archean mantle composition and crustal growth in the western Superior Province of Canada: Neodymium and lead isotopic evidence from the Wawa, Quetico, and Wabigoon subprovinces: Geochimica et Cosmochimica Acta, v. 62, p. 143-157.

Henry, P., Stevenson, R., Larbi, Y., and Gariepy, C., 2000, Nd isotopic evidence for Early to Late Archean (3.4-2.7 Ga) crustal growth in the Western Superior Province (Ontario, Canada): Tectonophysics, v. 322, p. 135-151.

Hoffman, P.F., 1989, Precambrian geology and tectonic history of North America, *in* Bally, A.W., and Palmer, A.R., eds., The Geology of North America - An Overview: Geological Society of America, Boulder, Colorado, A, p. 447-512.

Hollings, P., 2002, Archean Nb-enriched basalts in the northern Superior Province: Lithos, v. 64, p. 1-14.

Hollings, P., and Kerrich, R., 1999, Trace element systematics of ultramafic and mafic volcanic rocks from the 3 Ga North Caribou greenstone belt, northwestern Superior Province: Precambrian Research, v. 93, p. 257-279.

Hollings, P., Wyman, D.A., and Kerrich, R., 1999, Komatiite-basalt-rhyolite associations in northern Superior Province greenstone belts: Significance of plume-arc interaction in the generation of the protocontinental Superior Province: Lithos, v. 46, p. 137-161.

Hollings, P., Stott, G.M., and Wyman, D.A., 2000, Trace element geochemistry of the Meen-Dempster greenstone belt, Uchi subprovince, Superior Province, Canada: Back-arc development on the margins of an Archean protocontinent: Canadian Journal of Earth Sciences, v. 37, p. 1021-1038.

Hrabi, B., and Cruden, A.R., 2001, Three-dimensional geometry of the English River subprovince in the Separation Lake-Longlegged Lake area: Western Superior Transect Seventh Annual Workshop, University of British Columbia, p. 146-148.

——— 2006, Tectonic evolution of the English River Subprovince based on structural and geophysical studies in the Separation Rapids - Longlegged Lake area: Canadian Journal of Earth Sciences, v. 43, p. 947-966.

Hynes, A., 1997, Constraints on the tectonic evolution of the English River subprovince from studies in western Lac Seul: Western Superior Transect Third Annual Workshop, University of British Columbia, p. 37-45.

Hynes, A., and Song, Z., 2006, Variable unroofing in the western Superior Province - metamorphic evidence and possible origin: Canadian Journal of Earth Sciences, in press.

Jackson, S.L., and Fyon, J.A., 1991, The Western Abitibi Subprovince in Ontario, *in* Thurston, P.C., ed., Geology of Ontario: Ontario Geological Survey, Special Volume 4, Part 1, p. 405-482.

Jackson, S.L., Fyon, J.A., and Corfu, F., 1994, Review of Archean supracrustal assemblages of the southern Abitibi greenstone belt in Ontario, Canada: Products of microplate interaction within a large-scale plate-tectonic setting: Precambrian Research, v. 65, p. 183-205.

James, R.S., Easton, R.M., Peck, D.C., and Hrominchuck, J.L., 2002, The East Bull Lake intrusive suite: Remnants of an approximately 2.48 Ga large igneous and metallogenic province in the Sudbury area of the Canadian Shield: Economic Geology, v. 97, p. 1577-1606.

Jaupart, C., Mareschal, J.C., Guillou-Frottier, L., and Davale, A., 1998, Heat flow and thickness of the lithosphere in the Canadian Shield: Journal of Geophysical Research, v. 103, p. 15,269-15,286.

Jébrak, M., and Doucet, P., 2002, Geology and gold-molybdenum porphyry mineralisation of the Archean Taschereau-Launay plutons, Abitibi, Quebec: Precambrian Research, v. 115, p. 329-348.

Jemielita, R.A., Davis, D.W., and Krogh, T.E., 1990, U-Pb evidence for Abitibi gold mineralization postdating greenstone magmatism and metamorphism: Nature, v. 346, p. 831-834.

Jiang, D., and Corkery, M.T., 1998, A preliminary structural analysis of the Edmund Lake - Little Stull Lake area, northwestern Superior Province, Manitoba: Manitoba Energy and Mines, Geological Services, Report of Activities 1998, 7 p.

Jordan, T., 1978, Composition and development of the continental tectosphere: Nature, v. 274, p. 544-548.

Kamineni, D.C., Stone, D., and Peterman, Z.E., 1992, Early Proterozoic deformation in the western Superior Province, Canadian Shield: Geological Society of America Bulletin, v. 102, p. 1623-1634.

Kay, I., Musacchio, G., White, D., Asudeh, I., Roberts, B., Forsyth, D., Hajnal, Z., Koperwhats, B., and Farrell, D., 1999a, Imaging the Moho and Vp/Vs ratio in the western Superior Archean craton with wide-angle reflections: Geophysical Research Letters, v. 26, p. 2585-2588.

Kay, I., Sol, S., Kendall, J. M., Thomson, C., White, D., Asudeh, I., Roberts, B., and Francis, D., 1999b, Shear wave splitting observations in the Archean craton of western Superior: Geophysical Research Letters, v. 26, p. 2669-2672.

Kendall, J.M., Sol, S., Thomson, C.J., White, D.J., Asudeh, I., Snell, C.S., and Sutherland, F. H., 2002, Seismic heterogeneity and anisotropy in the western Superior Province, Canada: Insights into the evolution of an Archean craton, *in* Fowler, C.M.R., Ebinger, C.J., and Hawkesworth, C.J., eds., The Early Earth: Physical, Chemical, and Biological Development: Geological Society of London, Special Publications 199, p. 27-44.

Kerrich, R., Polat, A., Wyman, D.A., and Hollings, P., 1999, Trace element systematics of Mg- to Fe tholeiitic basalt suites of the Superior province: implications for Archean mantle reservoirs and greenstone belt genesis: Lithos, v. 46, p. 163-187.

Krogh, T.E., 1993, High precision U-Pb ages for granulite metamorphism and deformation in the Archean Kapuskasing structural zone, Ontario: Implications for the structure and development of the lower crust: Earth and Planetary Science Letters, v. 119, p. 1-18.

Krogh, T.E., Ermanovics, I.F., and Davis, G.L., 1974, Two episodes of metamorphism and deformation in the Archean rocks of the Canadian shield: Carnegie Institution of Washington, Geophysical Laboratory Yearbook, 3 p.

Labbé, J.-Y., 2005, Les elements du groupe du platine dans la partie nord-est de la Province du Supérieur: Ministère des Ressources naturelles du Québec, ET 2004-03, 35 p.

Labbé, J.-Y., and Lacoste, P., 2001, Environnements propices aux minérisations polymétalliques de type volcanogène dans le Grand-Nord québécois: Ministère des Ressources naturelles du Quebec, MB2001-07, 82 p.

——— 2004, Carte gitologique du nord-est de la Province du Supérieur, projet Grand-Nord: Ministère des Ressources naturelles, Faune et Parcs, Québec, Résumés des conferences et des photoprésentations, Québec Exploration 2004.

Lafrance, B., DeWolfe, J. C., and Stott, G.M., 2004, A structural reappraisal of the Beardmore-Geraldton Belt at the southern boundary of the Wabigoon Subprovince, Ontario, and implications for gold mineralization: Canadian Journal of Earth Sciences, v. 41, p. 217-235.

Langford, F.F., and Morin, J.A., 1976, The development of the Superior Province of northwestern Ontario by merging island arcs: American Journal of Science, v. 276, p. 1023-1034.

Lapointe, B., and Chown, E.H., 1993, Gold-bearing iron-formation in a granulite terrane of the Canadian Shield; a possible deep-level expression of an Archean gold-mineralizing system: Mineralium Deposita, v. 28, p. 191-197.

Larbi, Y., Stevenson, R., Breaks, F., Machado, N., and Gariépy, C., 1999, Age and isotopic compositions of late Archean leucogranites: implications for continental collision in the western Superior Province: Canadian Journal of Earth Sciences, v. 36, p. 495-510.

Lassen, B., 2004, Petrogenesis of the late Archean Quetico alkaline suite intrusions, western Superior Province, Canada: Unpub. Ph.D. thesis, University of Ottawa, Ottawa, ON, 240 p.

Leclair, A., 2005, Géologie du nord-est de la Province de Supérieur: Ministère des Ressources naturelles, Québec, Map DV 2004-04, 1:750,000.

Leclair, A., Berclaz, A., David, J., and Group, F.N.W., 2004a, The Northeastern Superior Province in Quebec's Far North: A Regional Synthesis, *in* Lithoprobe Report 86: Lithoprobe Secretariat, University of British Columbia.

Leclair, A., Berclaz, A., David, J., and Nord, G.d.T., 2004b, Synthèse géologique du nord-est de la Province du Supérieur (NEPS): Exploration 2004, Ministère des Ressources naturelles, Faunes et Parcs du Québec.

Legault, M., Gauthier, M., Jébrak, M., Davis, D.W., and Baillargeon, F., 2002, Evolution of the subaqueous to near-emergent Joutel volcanic complex, Northern Volcanic Zone, Abitibi Subprovince, Quebec, Canada: Precambrian Research, v. 115, p. 187-221.

Lesher, C.M., Goodwin, A.M., Campbell, I.H., and Gorton, M.P., 1986, Trace-element geochemistry of ore associated and barren, felsic metavolcanic rocks in the Superior Province, Canada: Canadian Journal of Earth Sciences, v. 23, p. 222-237.

Lin, S., and Corfu, F., 2002, Structural setting and geochronology of auriferous quartz veins at the High Rock Island gold deposit, northwestern Superior Province, Manitoba, Canada: Economic Geology, v. 97, p. 43-57.

Lin, S., and Jiang, D., 2001, Using along-strike variation in strain and kinematics to define the movement direction of curved transpressional shear zones: an example from northwestern Superior Province, Manitoba: Geology, v. 29, p. 767-770.

Lin, S., Corkery, M. T., Bailes, A.H., and Davis, D. W., 2006, Geology and geochronology of the Gods Lake Narrows area, Gods Lake greenstone belt, NW Superior Province: Canadian Journal of Earth Sciences, v. 43, p. 749-765.

Long, D.G.F., 2004, The tectonostratigraphic evolution of the Huronian basement and the subsequent basin fill: geological constraints on impact models of the Sudbury event: Precambrian Research, v. 129, p. 203-224.

Ludden, J., and Hynes, A., 2000, The Lithoprobe Abitibi-Grenville transect: Two billion years of crust formation and recycling in the Precambrian Shield of Canada: Canadian Journal of Earth Sciences, v. 37, p. 459-476.

Ludden, J.N., Hubert, C., and Gariépy, C., 1986, The tectonic evolution of the Abitibi greenstone belt of Canada: Geological Magazine, v. 123, p. 153-166.

Machado, A.D., 1987, On the origin and age of the Steep Rock buckshot: Chemical Geology, v. 60, p. 337-349.

Madore, L., and Larbi, Y., 2001, Geology of the Rivière Arnaud area (25D) and adjacent coastal areas (25C, 25E and 25F): Ministère des Ressources naturelles du Québec, RG2001-06, 33 p.

Manitoba, 1965, Geological map of Manitoba: Manitoba Department of Mines and Natural Resources, Map 65-1, 1:1 267 200.

Marshall, D., and Watkinson, D.H., 2000, The Cobalt mining district; silver sources, transport and deposition: Exploration and Mining Geology, v. 9, p. 81-90.

Melnyk, M.J., Cruden, A.R., and Davis, D.W., 2000, Structural geometry and deformational chronology of the Kenora gneisses: Western Superior Transect Sixth Annual Workshop, University of British Columbia, p. 82-89.

Melnyk, M.J., Cruden, A.R., Davis, D.W., and Stern, R.A., 2006, U-Pb ages of magmatism constraining regional deformation in the Winnipeg River subprovince and lake of the Woods greenstone belt: Evidence for Archean terrane accretion in the western Superior Province: Canadian Journal of Earth Sciences, v. 43, p. 967-993.

Meyn, H., and Palonen, P., 1980, Stratigraphy of an Archean submarine fan: Precambrian Research, v. 12, p. 257-285.

Moorhead, J., Girard, R., and Heaman, L., 2002, Caractérisation de kimberlites au Québec, *in* L'exploration minérale au Québec, séminaire d'information sur la recherche géologique, Programme et résumés: Ministère des Ressources naturelles, Québec, DV 2002-10, p. 36.

Moritz, R.P., and Chevé, S., 1992, Fluid-inclusion studies of high-grade metamorphic rocks of the Ashuanipi Complex, eastern Superior Province; constraints on the retrograde P-T path and implications for gold metallogeny: Canadian Journal of Earth Sciences, v. 29, p. 2309-2327.

Mortensen, J.K., 1993, U-Pb geochronology of the eastern Abitibi Subprovince; Part I Chibougamau-Matagami-Joutel region: Canadian Journal of Earth Sciences, v. 30, p. 11-28.

Mortensen, J.K., and Card, K.D., 1993, U-Pb age constraints for the magmatic and tectonic evolution of the Pontiac Subprovince, Quebec: Canadian Journal of Earth Sciences, v. 30, p. 1970-1980.

Moser, D., Heaman, L.M., and Hanes, J.A., 1996, Intracrustal extension of an Archean orogen revealed using single-grain U-Pb zircon geochronology: Tectonics, v. 15, p. 1093-1109.

Mueller, W.U., and Mortensen, J.K., 2002, Age constraints and characteristics of subaqueous volcanic construction, the Archean Hunter Mine Group, Abitibi greenstone belt: Precambrian Research, v. 115, p. 97-117.

Muir, T.L., 2003, Structural evolution of the Hemlo greenstone belt in the vicinity of the world-class Hemlo gold deposit: Canadian Journal of Earth Sciences, v. 40, p. 395-430.

Musacchio, G., White, D.J., Asudeh, I., and Thomson, C.J., 2004, Lithospheric structure and composition of the Archean western Superior Province from seismic refraction/wide-angle reflection and gravity modeling: Journal of Geophysical Research, v. 109, p. B03304.

Naldrett, A.J., 2003, From impact to riches; Evolution of geological understanding as seen at Sudbury, Canada: GSA Today, v. 13, p. 4-9.

Nitescu, B., Cruden, A.R., and Bailey, R.C., 2003, Topography of the crust-mantle interface under the western Superior craton from gravity data: Canadian Journal of Earth Sciences, v. 40, p. 1307-1320.

Nunes, P.D., and Thurston, P.C., 1980, Two hundred and twenty million years of Archean evolution: A zircon U-Pb age stratigraphic study of the Uchi-Confederation Lakes greenstone belt, northwestern Ontario: Canadian Journal of Earth Sciences, v. 17, p. 710-721.

Ontario, 1992, Bedrock geology of Ontario: Ontario Geological Survey, Maps 2541, 2542, 2543, 1:1 000 000.

Pan, Y., Fleet, M.E., and Williams, H.R., 1994, Granulite-facies metamorphism in the Quetico Subprovince, north of Manitouwadge, Ontario: Canadian Journal of Earth Sciences, v. 31, p. 1427-1439.

Pan, Y., Fleet, M.E., and Heaman, L.M., 1998, Thermo-tectonic evolution of an Archean accretionary complex: U-Pb geochronological constraints on granulites from the Quetico Subprovince, Ontario, Canada: Precambrian Research, v. 92, p. 117-128.

Pan, Y., Heaman, L.M., and Breaks, F.W., 1999, Thermo-tectonic evolution of the Umfreville-Conifer Lake granulite zone and the English River - Winnipeg River boundary zone: Preliminary U-Pb geochronological results: Western Superior Transect Fifth Annual Workshop, University of British Columbia, p. 104-108.

Parker, J., 1998, Regional metallogeny of the northwestern Superior Province Ontario: Volcanogenic massive sulphide (VMS) mineralization, *in* Summary of Field Work and other Activities: Ontario Geological Survey, Miscellaneous Paper 169, p. 143-155.

——— 2000, Gold mineralization and wall rock alteration in the Red Lake greenstone belt: A regional perspective: Ontario Geological Survey, Summary of Field Work and Other Activities, Open File Report 6032, p. 22-1 - 22-8.

Parks, J., Lin, S., Davis, D.W., and Corkery, M.T., 2006, Geochronological constrains on greenstone belt assembly in the Island Lake area, northwestern Superior Province: Canadian Journal of Earth Sciences, in press.

Parmenter, A.C., Lin, S., and Corkery, M.T., 2006, Structural evolution of the Cross Lake greenstone belt in the northwestern Superior Province, Manitoba: Implications for relationship between vertical and horizontal tectonism: Canadian Journal of Earth Sciences, in press.

Percival, J.A., 1989, A regional perspective of the Quetico metasedimentary belt, Superior Province, Canada: Canadian Journal of Earth Sciences, v. 26, p. 677-693.

——— 1991, Granulite-facies metamorphism and crustal magmatism in the Ashuanipi complex, Quebec-Labrador, Canada: Journal of Petrology, v. 32, p. 1261-1297.

Percival, J.A., and Helmstaedt, H., 2004, Insights on Archean continent - ocean assembly, western Superior Province, from new structural, geochemical and geochronological observations: Introduction and summary: Precambrian Research, v. 132, p. 209-212.

Percival, J.A., and McGrath, P.H., 1986, Deep crustal structure and tectonic history of the northern Kapuskasing uplift of Ontario: An integrated petrological-geophysical study: Tectonics, v. 5, p. 553-572.

Percival, J.A., and Mortensen, J.K., 2002, Water-deficient calc-alkaline magmatism in the northeastern Superior Province: significance of charnockitic magmatism: Journal of Petrology, v. 43, p. 1617-1650.

Percival, J.A., and Peterman, Z E., 1994, Rb-Sr biotite and whole-rock data from the Kapuskasing uplift and their bearing on the cooling and exhumation history: Canadian Journal of Earth Sciences, v. 31, p. 1172-1181.

Percival, J.A., and Pysklywec, R., 2004, Thermal-material linkage between Archean crust and mantle: American Geophysical Union, p. JA462.

Percival, J.A., and Skulski, T., 2000, Tectonothermal evolution of the northern Minto block, northeastern Superior Province, Canada: The Canadian Mineralogist, v. 38, p. 345-378.

Percival, J.A., and West, G.F., 1994, The Kapuskasing Uplift: A geological and geophysical synthesis: Canadian Journal of Earth Sciences, v. 31, p. 1256-1286.

Percival, J.A., and Williams, H.R., 1989, The Quetico accretionary complex, Superior Province, Canada: Geology, v. 17, p. 23-25.

Percival, J.A., Mortensen, J. K., Stern, R. A., and Card, K. D., 1992, Giant granulite terranes of northeastern Superior Province: The Ashuanipi complex and Minto block: Canadian Journal of Earth Sciences, v. 29, p. 2287-2308.

Percival, J.A., Stern, R.A., Mortensen, J.K., Card, K.D., and Bégin, N.J., 1994, Minto block Superior Province: Missing link in deciphering tectonic assembly of the craton at 2.7 Ga: Geology, v. 22, p. 839-842.

Percival, J.A., Stern, R.A., and Skulski, T., 2001, Crustal growth through successive arc magmatism: Reconnaissance U-Pb SHRIMP data from the northeastern Superior Province, Canada: Precambrian Research, v. 109, p. 203-238.

Percival, J.A., Bailes, A.H., and McNicoll, V., 2002a, Mesoarchean breakup, Neoarchean accretion in the western Superior craton, Lake Winnipeg Canada: Geological Association of Canada, Field Trip B3 Guidebook, 42 p.

Percival, J.A., Whalen, J.B., Tomlinson, K.Y., McNicoll, V., and Stott, G.M., 2002b, Geology and tectonostratigraphic assemblages, north-central Wabigoon Subprovince, Ontario: Geological Survey of Canada, Open File 4270, 1:250 000.

Percival, J.A., Stern, R.A., and Rayner, N., 2003, Archean adakites from the Ashuanipi complex, eastern Superior Province, Canada: Geochemistry, geochronology and tectonic significance: Contributions to Mineralogy and Petrology, v. 145, p. 265-280.

Percival, J.A., Bleeker, W., Cook, F.A., Rivers, T., Ross, G., and van Staal, C.R., 2004a, Panlithoprobe Workshop IV: Intra-orogen correlations and comparative orogenic anatomy: Geoscience Canada, v. 31, p. 23-39.

Percival, J.A., McNicoll, V., Brown, J.L., and Whalen, J.B., 2004b, Convergent margin tectonics, central Wabigoon subprovince, Superior Province, Canada: Precambrian Research, v. 132, p. 213-244.

Percival, J.A., Whalen, J.B., and Rayner, N., 2005, Pikwitonei-Snow Lake Manitoba transect (parts of NTS 63J, 63O and 63P), Trans-Hudson Orogen – Superior margin metallotect project: New results and tectonic interpretation, *in* Report of Activities 2005: Manitoba Industry, Economic Development and Mines, Manitoba Geological Survey, p. 69-91.

Percival, J.A., McNicoll, V., and Bailes, A.H., 2006a, Strike-slip juxtaposition of ca. 2.72 Ga juvenile arc and >2.98 Ga continent margin sequences and its implications for Archean terrane accretion, western Superior Province, Canada: Canadian Journal of Earth Sciences, v. 43, p. 895-927.

Percival, J.A., Sanborn-Barrie, M., Stott, G., Helmstaedt, H., and Skulski, T., 2006b, Tectonic evolution of the Western Superior Province from NATMAP and Lithoprobe studies: Canadian Journal of Earth Sciences, v. 43, p. 1085-1117.

Perkins, D., and Chipera, S., 1985, Garnet-orthopyroxene-plagioclase-quartz barometry: refinement and application to the English River subprovince and the Minnesota River valley: Contributions to Mineralogy and Petrology, v. 89, p. 40-48.

Peterson, V.L., and Zaleski, E., 1999, Structural history of the Manitouwadge greenstone belt and its volcanogenic Cu-Zn massive sulphide deposits, Wawa Subprovince, south-central Superior Province: Canadian Journal of Earth Sciences, v. 36, p. 605-625.

Pettigrew, N.T., 2004, Cu-Ni-PGE mineralization and petrogenesis of mafic-ultramafic intrusions in the western Quetico and Wabigoon subprovinces, northwestern Ontario, Canada: Unpub. M.Sc. thesis, University of Ottawa, Ottawa, ON, 153 p.

Pilkington, M., and Percival, J.A., 1999, Crustal magnetization and long-wavelength aeromagnetic anomalies of the Minto block, Quebec: Journal of Geophysical Research, v. 104, p. 7513-7526.

Pirie, J., 1982, Regional geological setting of gold deposits, eastern Red Lake area, northwestern Ontario, *in* Hodder, R.W., and Petruk, W., eds., Geology of Canadian Gold Deposits: Canadian Institute of Mining and Metallurgy, Special Volume 24, p. 171-183.

Polat, A., and Kerrich, R., 1999, Formation of an Archean tectonic melange in the Schreiber-Hemlo greenstone belt, Superior Province, Canada; implications for Archean subduction-accretion processes: Tectonics, v. 18, p. 733-755.

——— 2001, Magnesian andesites, Nb-enriched basalts-andesites, and adakites from late Archean 2.7 Ga Wawa greenstone belts, Superior Province, Canada: Implications for late Archean subduction zone petrogenetic processes: Contributions to Mineralogy and Petrology, v. 141, p. 36-52.

Polat, A., Kerrich, R., and Wyman, D., 1998, The late Archean Schreiber-Hemlo and White River-Dayohessarah greenstone belts, Superior Province; collages of oceanic plateaus, oceanic arcs, and subduction-accretion complexes: Tectonophysics, v. 289, p. 295-326.

Poulsen, K.H., 1985, Archean tectonics and mineralization at Rainy Lake, northwestern Ontario: Unpub. Ph.D. thesis, Queen's University, Kingston, ON, 341 p.

Poulsen, K.H., Borradaile, G.J., and Kehlenbeck, M.M., 1980, An inverted Archean succession at Rainy Lake, Ontario: Canadian Journal of Earth Sciences, v. 17, p. 1358-1369.

Poulsen, K.H., Weber, W., Brommecker, R., and Seneshen, D.N., 1996, Lithostratigraphic assembly and structural setting of gold mineralization in the eastern Rice Lake greenstone belt, Manitoba: Geological Association of Canada, Field Trip Guidebook A4, 106 p.

Poulsen, K.H., Robert, F., and Dubé, B., 2000, Geological classification of Canadian gold deposits: Geological Survey of Canada, Bulletin 540, 106 p.

Rheaume, P., Bandyayera, D., Fallara, F., Boudrias, G., and Cheng, L.Z., 2004, Géologie et métallogenie du secteur du lac aux Loutres, synthese métallogénique d'Urban-Barry (Phase 1 de 2): Gélogie Québec, Rapport Préliminaire RP 2004-5.

Robert, F., and Poulsen, K.H., 1997, World-class Archaean gold deposits in Canada: An overview: Australian Journal of Earth Sciences, v. 44, p. 329-351.

Rogers, N., 2001, Perliminary Report on the Stratigraphy and Structure of the Bee Lake Greenstone Belt, Superior Province, Northwestern Ontario: Geological Survey of Canada, Current Research, 2001-C17, 27 p.

——— 2002, Geology, Confederation Lake, Ontario: Geological Survey of Canada, Open File 4265, 1:50 000.

Rogers, N., McNicoll, V., van Staal, C.R., and Tomlinson, K.Y., 2000, Lithogeochemical studies in the Uchi-Confederation greenstone belt, northwestern Ontario: Implications for Archean tectonics: Geological Survey of Canada, Current Research 2000-C16, 11 p.

Rolandone, F., Mareschal, J.C., Jaupart, C., Gosselin, C., Bienfait, G., and Lapointe, R., 2003, Heat flow in the western superior province of the Canadian shield: Geophysical Research Letters, v. 30, p. 39-1 - 39-4.

Roscoe, S.M., and Donaldson, J.A., 1988, Uraniferous pyritic quartz pebble conglomerate and layered ultramafic intrusions in a sequence of quartzite, carbonate, iron formation and basalt of probable Archean age at Lac Sakami, Quebec: Geological Survey of Canada, Paper 88-1C, p. 117-121.

Rousell, D.H., Meyer, W., and Prevec, S.A., 2002, Bedrock geology and mineral deposits, *in* Rousell, D.H., and Jansons, K.J., eds., The Physical Environment of the City of Greater Sudbury: Ontario Geological Survey, Special Volume 5, p. 21-55.

Rousell, D.H., Federowich, J.S., and Dressler, B.O., 2003, Sudbury Breccia (Canada); A product of the 1850 Ma Sudbury event and host to footwall Cu-Ni-PGE deposits: Earth-Science Reviews, v. 60, p. 147-174.

Sage, R.P., 1991, Alkalic rock, carbonatite and kimberlite complexes of Ontario, Superior Province, *in* Thurston, P.C., Williams, H.R., Sutcliffe, R.H., and Stott, G.M., eds., Geology of Ontario: Ontario Geological Survey, Special Volume 4, Part 1, p. 683-709.

——— 2000, Kimberlites of the Lake Timiskaming Structural Zone: Supplement: Ontario Geological Survey, Open File Report 6018. 122 p.

Sage, R.P., Lightfoot, P.C., and Doherty, W., 1996a, Bimodal cyclical Archean basalts and rhyolites from the Michipicoten (Wawa) greenstone belt, Ontario; Geochemical evidence for magma contributions from the asthenospheric mantle and ancient continental lithosphere near the southern margin of the Superior Province: Precambrian Research, v. 76, p. 119-153.

——— 1996b, Geochemical characteristics of granitoid rocks from within the Archean Michipicoten greenstone belt, Wawa Subprovince, Superior Province, Canada; Implications for source regions and tectonic evolution: Precambrian Research, v. 76, p. 155-190.

Sajona, F.G., Maury, R.C., Pubellier, M., Leterrier, J., Bellon, H., and Cotton, J., 2000, Magmatic source enrichment by slab-derived melts in a young post-collision setting, central Mindanao, Philippines: Lithos, v. 54, p. 173-206.

Sanborn-Barrie, M., 1988, Geology of the tectonic boundary zone between the English River and Winnipeg River subprovinces, northwestern Ontario: Ontario Geological Survey, Summary of Field Work and Other Activities, Miscellaneous Paper 41, 10 p.

Sanborn-Barrie, M., and Skulski, T., 1999, 2.7 Ga tectonic assembly of continental margin and oceanic terranes in the Savant Lake-Sturgeon Lake greenstone belt, Ontario: Geological Survey of Canada, Current Research 1999-C, 12 p.

——— 2006, Structural evolution of the Savant Lake-Sturgeon Lake greenstone belt, western Superior Province: A record of continental arc - oceanic plateau collision at 2.7 Ga: Canadian Journal of Earth Sciences, v. 43, p. 995-1030.

Sanborn-Barrie, M., Skulski, T., and Whalen, J.B., 1998, Tectonostratigraphy of central Sturgeon Lake, Ontario: Deposition and deformation of submarine tholeiites and emergent calc-alkaline volcano-sedimentary sequences: Geological Survey of Canada, Current Research 1998-C, 12 p.

Sanborn-Barrie, M., Skulski, T., and Parker, J.R., 2001, Three hundred million years of tectonic history recorded by the Red Lake greenstone belt, Ontario: Geological Survey of Canada, Current Research 2001-C19, 19 p.

Sanborn-Barrie, M., Skulski, T., Percival, J.A., Whalen, J.B., Brown, J., and McNicoll, V., 2002, Geology and tectonostratigraphic assemblages, western Wabigoon Subprovince, Ontario: Geological Survey of Canada, Open File 4255, scale 1:250 000.

Sanborn-Barrie, M., Rogers, N., Skulski, T., Parker, J.R., McNicoll, V., and Devaney, J., 2004, Geology and tectonostratigraphic assemblages, east Uchi, Red Lake and Birch-Uchi belts, Ontario: Geological Survey of Canada, Open File 4256, scale 1:250,000.

Sasseville, C., 2002, Characteristics of Mesoarchean and Neoarchean Supracrustal Sequences at the Southern Margin of North Caribou Terrane in the Wallace Lake Greenstone Belt, Superior Province, Canada: Unpub. M.Sc. thesis, McGill University, Montreal, Quebec, 95 p.

Sasseville, C., Tomlinson, K. Y., Hynes, A., and McNicoll, V., 2006, Stratigraphy, structure, and geochronology of the 3.0-2.7 Ga Wallace Lake greenstone belt, Western Superior Province, SE Manitoba, Canada: Canadian Journal of Earth Sciences, v. 43, p. 929-945.

Sawyer, E.W., 1983, The structural history of a part of the Archaean Quetico metasedimentary belt, Superior Province, Canada: Precambrian Research, v. 22, p. 271-294.

Sawyer, E.W., and Benn, K., 1993, Structure of the high-grade Opatica Belt and adjacent lowgrade Abitibi Subprovince, Canada: An Archean mountain front: Journal of Structural Geology, v. 15, p. 1443-1458.

Schmitz, M.D., Bowring, S.A., Southwick, D.L., Boerboom, T.J., and Wirth, K.R., 2006, High-precision U-Pb geochronology in the Minnesota River Valley subprovince and its bearing on the Neoarchean to Paleoproterozoic evolution of the southern Superior Province: Geological Society of America Bulletin, v. 118, p. 82-93.

Schulze, D.J., 1996, Kimberlites in the vicinity of Kirkland Lake and Lake Timiskaming, Ontario and Quebec, *in* LeCheminant, A.N., Richardson, D.G., DiLabio, R.N.W., and Richardson, K.A., eds., Searching for Diamonds in Canada: Geological Survey of Canada, Open File 3228, p. 73-78.

Schwerdtner, W.M., 1992, Structural tests of diapir hypotheses in Archean crust of Ontario: Canadian Journal of Earth Sciences, v. 27, p. 387-402.

Schwerdtner, W.M., Stone, D., Osadetz, K., Morgan, J., and Stott, G.M., 1979, Granitoid complexes and the Archean tectonic record in the southern part of northwestern Ontario: Canadian Journal of Earth Sciences, v. 16, p. 1965-1977.

Scully, K.R., 2000, Ultramafic Xenoliths from the Attawapiskat Kimberlite Field, James Bay Lowlands, Ontario: A view into the sub-cratonic lithosphere of the western Superior Craton: Unpub. M.Sc. thesis, University of Toronto.

Silver, P.G., 1996, Seismic anisotropy beneath the continents: probing the depths of geology: Annual Review of Earth and Space Sciences, v. 24, p. 385-432.

Silver, P.G., and Chan, W.W., 1988, Implications for continental structure and evolution from seismic anisotropy: Nature, v. 335, p. 34-39.

Sims, P.K., and Day, W.C., 1993, The Great Lakes Tectonic Zone - revisited: United States Geological Survey, Bulletin 1904-S, 11 p.

Sims, P.K., Kotov, A.B., Neymark, L.A., and Peterman, Z.E., 1997, Nd isotopic evidence for middle and early Archean crust in the Wawa subprovince of Superior Province, Michigan, U.S.A.: Geological Association of Canada, Abstract Volume 23, p. A137.

Skulski, T., and Villeneuve, M., 1999, Geochronological compilation of the Superior Province, Manitoba, Ontario, Quebec: Geological Survey of Canada, Open File 3715, CD-ROM.

Skulski, T., Sanborn-Barrie, M., and Stern, R.A., 1998a, Did the Sturgeon Lake belt form near a continental margin? in Lithoprobe Report 65: Lithoprobe Secretariat, University of British Columbia, p.87-89.

Skulski, T., Stern, R. A., and Ciesielski, A., 1998b, Timing and sources of granitoid magmatism, Bienville subprovince, northern Quebec: Geological Association of Canada, Program with Abstracts, p. A174-A175.

Skulski, T., Percival, J.A. Whalen, J.B., and Stern, R.A., 1999, Archean crustal evolution in the northern Superior Province. in Tectonic and Magmatic Processes in Crustal Growth: A Pan-Lithoprobe Perspective: Lithoprobe Report 75, Lithoprobe Secretariat, University of British Columbia, p.128-129.

Skulski, T., Corkery, M.T., Stone, D., Whalen, J.B., and Stern, R.A., 2000, Geological and geochronological investigations in the Stull Lake - Edmund Lake greenstone belt and granitoid rocks of the northwestern Superior Province: Manitoba Industry, Trade and Mines, Manitoba Geological Survey, Report of Activities 2000, 12 p.

Sleep, N.H., 1992, Archean plate tectonics: what can be learned from continental geology? Canadian Journal of Earth Sciences, v. 29, p. 2066-2071.

Smith, S.R., Foster, G.L., Romer, R.L., Tindle, A.G., Kelley, S.P., Noble, S.R., Horstwood, M., and Breaks, F.W., 2004, U-Pb columbite-tantalite chronology of rare-element pegmatites using TIMS and Laser Ablation-Multi Collector-ICP-MS: Contributions to Mineralogy and Petrology, v. 147, p. 549-564.

Sol, S., Thomson, C.J., Kendall, J.-M., White, D., VanDecar, J.C., and Asudeh, I., 2002, Seismic tomographic images of the cratonic upper mantle beneath the Western Superior Province of the Canadian Shield - A remnant Archean slab? Physics of the Earth and Planetary Interiors, v. 134, p. 53-69.

Southwick, D L., 1991, On the genesis of Archean granite through two-stage melting of the Quetico accretionary prism at a transpressional plate boundary: Geological Society of America Bulletin, v. 103, p. 1385-1394.

Sproule, R.A., Lesher, C.M., Ayer, J.A., Thurston, P.C., and Herzberg, C.T., 2002, Spatial and temporal variations in the geochemistry of komatiites and komatiitic basalts in the Abitibi greenstone belt: Precambrian Research, v. 115, p. 153-186.

Stern, R.A., and Hanson, G.N., 1991, Archean high-Mg granodiorite: A derivative of light rare earth element-enriched monzodiorite of mantle origin: Journal of Petrology, v. 32, p. 201-238.

Stern, R.A., Shirey, S.B., and Hanson, G.N., 1989, Petrogenesis of mantle derived, LILE enriched Archean monzodiorite and trachyandesite (sanukitoids) in southwestern Superior Province: Canadian Journal of Earth Sciences, v. 26, p. 1688-1712.

Stern, R.A., Percival, J.A., and Mortensen, J.K., 1994, Geochemical evolution of the Minto block: a 2.7 Ga continental magmatic arc build on the Superior proto-craton: Precambrian Research, v. 65, p. 115-153.

Stevenson, R.K., 1995, Crust and mantle evolution in the late Archean: Evidence from a Sm-Nd isotopic study of the North Spirit Lake greenstone belt, northwestern Ontario: Geological Society of America Bulletin, v. 107, p. 1458-1467.

Stevenson, R.K., and Patchett, P.J., 1990, Implications for the evolution of continental crust from Hf isotope systematics of Archean detrital zircons: Geochimica et Cosmochimica Acta, v. 54, p. 1683-1697.

Stevenson, R., Henry, P., and Gariepy, C., 1999, Assimilation-fractional crystallization origin of Archean sanukitoid suites: Western Superior Province, Canada: Precambrian Research, v. 96, p. 83-99.

Stone, D., 1981, The Sydney Lake Fault Zone in Ontario and Manitoba, Canada: Unpub. Ph.D. thesis, University of Toronto, Ontario, 146 p.

——— 1998, Precambrian geology of the Berens River area, northwest Ontario: Ontario Geological Survey, Open File Report 5963, 115 p.

——— 2000, Temperature and pressure variations in suites of Archean felsic plutonic rocks, Berens River area, northwest Superior Province, Ontario, Canada: Canadian Mineralogist, v. 38, p. 455-470.

——— 2005, Geology of northern Superior area, Ontario: Ontario Geological Survey, Open File Report 6140, 94 p.

Stone, D., Tomlinson, K.Y., Davis, D.W., Fralick, P., Hall, J., Percival, J.A., and Pufahl, P., 2002, Geology and tectonostratigraphic assemblages, South-central Wabigoon Subprovince: Ontario Geological Survey, Preliminary Map P.3448, 1:250 000.

Stone, D., Corkery, M.T., Hallé, J., Ketchum, J., Lange, M., Skulski, T., and Whalen, J., 2004, Geology and tectonostratigraphic assemblages, eastern Sachigo Subprovince, Ontario and Manitoba: Ontario Geological Survey, Preliminary Map P.3462, 1:250 000.

Stott, G.M., 1996, The geology and tectonic history of the central Uchi Subprovince: Ontario Geological Survey Report 5952, 178 p.

——— 1997, The Superior Province, Canada, *in* de Wit, M.J., and Ashwal, L.D., eds., Greenstone Belts: Oxford Monograph on Geology and Geophysics 35, Clarendon, Oxford, p. 480-507.

Stott, G.M., and Berdusco, B.J., 2000, Precambrian features of the James Bay and Hudson Bay lowlands: Summary of Fieldwork and Other Activities 2000, Ontario Geological Survey, Open File Report 6032, p. 14-1 to 14-5.

Stott, G.M., and Corfu, F., 1991, Uchi subprovince, *in* Thurston, P.C., Williams, H.R., Sutcliffe, R.H., and Stott, G.M., eds., Geology of Ontario: Ontario Geological Survey Special Volume 4, Part 1, p. 145-238.

Stott, G.M., and Davis, D.W., 1999, Contributions to the tectonostratigraphic analysis of the Onaman-Tashota greenstone belt, eastern Wabigoon subprovince, *in* Harrap, R.M., and H. Helmstaedt, H., eds., Western Superior Transect Fifth Annual Workshop: Lithoprobe Report 70, Lithoprobe Secretariat, University of British Columbia, p. 122-124.

Stott, G.M., and Rayner, N., 2004. Discrimination of Archean domains in the Sachigo subprovince: in Summary of Field Work and Other Activities 2004: Ontario Geological Survey, Open File Report 6145, p. 10-1 to 10-8.

Stott, G.M., Corfu, F., Breaks, F.W., and Thurston, P.C., 1989, Multiple orogenesis in northwestern Superior Province: Geological Association of Canada, Abstracts 14, p. A56.

Stott, G.M., Davis, D.W., Parker, J.R., Straub, K.J., and Tomlinson, K.Y., 2002, Geology and tectonostratigraphic assemblages, eastern Wabigoon Subprovince, Ontario: Ontario Geological Survey Map P3449; Geological Survey of Canada Open File 4285, scale 1:250 000.

Sutcliffe, R.H., 1988, Geology of the Fletcher Lake area, District of Thunder Bay: Ontario Geological Survey, Report 251, 65 p.

Syme, E.C., Corkery, M.T., Bailes, A.H., Lin, S., Skulski, T., and Stern, R.A., 1999, Towards a new tectonostratigraphy for the Knee Lake greenstone belt, Sachigo subprovince, Manitoba, *in* Harrap, R.M., and H. Helmstaedt, H., eds., Western Superior Transect Fifth Annual Workshop: Lithoprobe Report 70, Lithoprobe Secretariat, University of British Columbia, p. 124-131.

Therriault, A.M., Fowler, A.D., and Grieve, R.A.F., 2002, The Sudbury igneous complex; a differentiated impact melt sheet: Economic Geology, v. 97, p. 1521-1540.

Thurston, P.C., 1994, Archean volcanic patterns, *in* Condie, K.C., ed., Archean Crustal Evolution: Developments in Precambrian Geology, 11, Elsevier, Amsterdam, p. 45-84.

——— 2002, Autochthonous development of Superior Province greenstone belts? Precambrian Research, v. 115, p. 11-36.

Thurston, P.C., and Chivers, K.M., 1990, Secular variation in greenstone sequence development emphasising Superior Province, Canada: Precambrian Research 46, p. 21-58.

Thurston, P.C., and Davis, D.W., 1985, The Wabigoon diapiric axis as a basement complex, *in* Summary of Field Work and Other Activities 1985: Ontario Geological Survey, Miscellaneous Paper 126, p. 138-141.

Thurston, P.C., Osmani, I.A., and Stone, D., 1991, Northwestern Superior Province: Review and terrane analysis, *in* Thurston, P.C., Williams, H.R., Sutcliffe, R.H., and Stott, G.M., eds., Geology of Ontario: Ontario Geological Survey Special Volume 4, Part 1, p. 81-144.

Timmins, E.A., Turek, A., Symons, D.T.A., and Smith, P.E., 1985, U-Pb zircon geochronology and paleomagnetism of the Bird River greenstone belt, Manitoba: Geological Association of Canada, Program with Abstracts 10, p. A62.

Tomlinson, K.Y., and Dickin, A.P., 2003, Geochemistry and neodymium isotopic character of granitoid rocks in the Lac Seul region of the Winnipeg River subprovince, northwestern Ontario, *in* Summary of Field Work and Other Activities, 2003: Ontario Geological Survey, Open File Report 6120, p. 13-1 to 13-8.

Tomlinson, K.Y., and Percival, J.A., 2000, Geochemistry and Nd isotopes of granitoid rocks in the Shikag-Garden lakes area, Ontario: Recycled Mesoarchean crust in the central Wabigoon Subprovince: Geological Survey of Canada, Current Research 2000-E12, 11 p. (online; http://www.nrcan.gc.ca/gsc/bookstore).

Tomlinson, K.Y., Hall, R.P., Hughes, D.J., and Thurston, P.C., 1996, Geochemistry and assemblage accretion of metavolcanic rocks in the Beardmore-Geraldton greenstone belt, Superior Province: Canadian Journal of Earth Sciences, v. 33, p. 1520-1533.

Tomlinson, K.Y., Stevenson, R.K., Hughes, D.J., Hall, R.P., Thurston, P.C., and Henry, P., 1998, The Red Lake greenstone belt, Superior Province: evidence of plume-related magmatism at 3 Ga and evidence of an older enriched source: Precambrian Research 89, p.59-76.

Tomlinson, K.Y., Stott, G.M., and Davis, D.W., 2000, Nd isotopes in the eastern Wabigoon subprovince: Implications for crustal recycling and correlations with the central Wabigoon, *in* Harrap, R.M., and H. Helmstaedt, H., eds., Western Superior Transect Sixth Annual Workshop: Lithoprobe Report 77, Lithoprobe Secretariat, University of British Columbia, p. 119-126.

Tomlinson, K.Y., Sasseville, C., and McNicoll, V., 2001, New U-Pg geochronology and structural interpretations from the Wallace Lake greenstone belt (North Caribou terrane): Implications for new regional correlations, *in* Harrap, R.M., and H. Helmstaedt, H., eds., Western Superior Transect Seventh Annual Workshop: Lithoprobe Report 80, Lithoprobe Secretariat, University of British Columbia, p. 8-9.

Tomlinson, K.Y., Davis, D.W., Percival, J.A., Hughes, D.J., and Thurston, P.C., 2002, Mafic to felsic magmatism and crustal recycling in the Obonga Lake greenstone belt, Western Superior Province: evidence from geochemistry, Nd isotopes and U-Pb geochronology: Precambrian Research, v. 114, p.295-325.

Tomlinson, K.Y., Davis, D.W., Stone, D., and Hart, T., 2003, New U-Pb and Nd isotopic evidence for crustal recycling and Archean terrane development in the south-central Wabigoon Subprovince, Canada: Contributions to Mineralogy and Petrology, v. 144, p. 684-702.

Tomlinson, K.Y., Stone, D., Stott, G.M., and Percival, J.A., 2004, Basement terranes and crustal recycling in the western Superior Province: Nd isotopic character of granitoid and felsic volcanic rocks in the Wabigoon subprovince, N. Ontario, Canada: Precambrian Research, v. 132, p. 245-274.

Valli, F., Guillot, S., and Hattori, K.H., 2004, Source and tectono-metamorphic evolution of mafic and pelitic metasedimentary rocks from the central Quetico metasedimentary belt, Archean Superior Province of Canada: Precambrian Research, v. 132, p. 155-177.

Van der Lee, S., and Nolet, G., 1997, Upper mantle S velocity structure of North America: Journal of Geophysical Research, v. 102, p. 22815-22838.

Vicker, P.A., and Schulze, D.J., 1994, Garnet peridotite xenoliths in Kirkland Lake kimberlite: Geological Association of Canada, Mineralogical Association of Canada, Program with Abstracts 19, p. 116.

Whalen, J.B., Percival, J.A., McNicoll, V., and Longstaffe, F.J., 2002, A mainly crustal origin for tonalitic granitoid rocks, Superior Province, Canada: Implications for late Archean tectonomagmatic processes: Journal of Petrology, v. 43, p. 1551-1570.

——— 2003, Intra-oceanic production of continental crust in a Th-depleted ca. 3.0 Ga arc complex, western Superior Province, Canada: Contributions to Mineralogy and Petrology, v. 156, p. 78-99.

——— 2004a, Geochemical and isotopic (Nd-O) evidence bearing on the origin of late- to post-orogenic high-K granitoid rocks in the Western Superior Province: Implications for late Archean tectonomagmatic processes: Precambrian Research, v. 132, p. 303-326.

Whalen, J.B., McNicoll, V., and Longstaffe, F.J., 2004b, Juvenile ca. 2.735-2.720 Ga high- and low-Al tonalitic plutons: implications for TTG and VMS petrogenesis, western Superior Province, Canada: Precambrian Research, v. 132, p. 275-301.

White, D.J., Musacchio, G., Helmstaedt, H.H., Harrap, R.M., Thurston, P.C., van der Velden, A., and Hall, K., 2003, Images of a lower-crustal oceanic slab: Direct evidence for tectonic accretion in the Archean western Superior province: Geology, v. 31, p. 997-1000.

Williams, H.R., 1991, Quetico subprovince, *in* Thurston, P.C., Williams, H.R., Sutcliffe, R.H., and Stott, G.M., eds., Geology of Ontario: Ontario Geological Survey, Special Volume 4, Part 1, p. 383-403.

Williams, H.R., Stott, G.M., Heather, K.B., Muir, T.L., and Sage, R.P., 1991, Wawa subprovince, *in* Thurston, P.C., Williams, H.R., Sutcliffe, R.H., and Stott, G.M., eds., Geology of Ontario: Ontario Geological Survey, Special Volume 4, Part 1, p. 485-539.

Williams, H.R., Stott, G.M., Thurston, P.C., Sutcliffe, R.H., Bennett, G., Easton, R.M., and Armstrong, D.K., 1992, Tectonic evolution of Ontario: Summary and synthesis, *in* Thurston, P.C., Williams, H.R., Sutcliffe, R.H., and Stott, G.M., eds., Geology of Ontario: Ontario Geological Survey, Special Volume 4, Part 1, p. 1255-1332.

Wong, L., Davis, D.W., Krogh, T.E., and Robert, F, 1991, U-Pb zircon and rutile chronology of Archean greenstone formation and gold mineralization in the Val d'Or region, Quebec: Earth and Planetary Science Letters, v. 104, p. 325-336.

Wyman, D., Ayer, J., and Devaney, J., 2000, Niobium-enriched basalts from the Wabigoon subprovince, Canada: Evidence for adakitic metasomatism above an Archean subduction zone: Earth and Planetary Science Letters, v. 179, p. 21-30.

Wyman, D.A., Bleeker, W., and Kerrich, R., 1999, A 2.7 Ga komatiite, low Ti tholeiite, arc tholeiite transition, and inferred proto-arc geodynamic setting of the Kidd Creek Deposit; Evidence from precise trace element data, *in* Hannington, M.D., and Barrie, C.T, eds., The Giant Kidd Creek Volcanogenic Massive Sulfide Eeposit, Western Abitibi Subprovince, Canada: Economic Geology, Monograph 10, p. 511-528.

Wyman, D.A., Kerrich, R., and Polat, A., 2002, Assembly of Archean cratonic mantle lithosphere and crust: Plume-arc interaction in the Abitibi-Wawa subduction-accretion complex: Precambrian Research, v. 115, p. 37-62.

Young, M.D., 2003, New structural, geochronological, and geochemical constraints on the tectonic assembly of the Archean Pickle Lake greenstone belt, Uchi subprovince, western Superior Province: Unpub. M.Sc. thesis: Queen's University, Kingston, Ontario, 182 p.

Young, M.D., McNicoll, V., Helmstaedt, H., Skulski, T., and Percival, J.A., 2006, Pickle Lake revisited: New structural, geochronological and geochemical constraints on greenstone belt assembly, western Superior Province, Canada: Canadian Journal of Earth Sciences, v. 43, p. 831-847.

Zaleski, E., van Breemen, O., and Peterson, V.L., 1999, Geological evolution of the Manitouwadge greenstone belt and Wawa-Quetico subprovince boundary, Superior Province, Ontario, constrained by U-Pb zircon dates of supracrustal and plutonic rocks: Canadian Journal of Earth Sciences, v. 36, p. 945-966.

Zweng, P.L., Mortensen, J.K., and Dalrymple, B., 1993, Thermochronology of the Camflo gold deposit, Malartic, Quebec: Implications for magmatic underplating and the formation of gold-bearing quartz veins: Economic Geology, v. 88, p. 1700-1721.

Large Igneous Provinces in Canada Through Time and their Metallogenic Potential

Richard E. Ernst

Ernst Geosciences, 43 Margrave Avenue, Ottawa, Ontario K1T 3Y2
and Department of Earth Sciences, Carleton University, Ottawa, Ontario K1S 5B6
Corresponding author's email: RichardErnst@ErnstGeosciences.com

Abstract

Large Igneous Provinces (LIPs) comprise flood basalts, major dyke swarms, sill provinces, and large layered mafic-ultramafic intrusions, and occur on average once per 20 million years back to 2600 Ma. There may also be Archean analogues. LIPs are key hosts for Ni-Cu-PGE, Cr, and Fe-Ti-V deposits. In some cases, there is also a spatial and temporal link with non-mafic intrusions such as carbonatites (with Au-Hg, REE-Ta-Nb deposits) and kimberlites (with diamonds). Since LIPs transport much thermal energy into the crust, they can drive hydrothermal systems that are capable of generating different types of mineral deposits. Basic information on more than 80 LIPs and interpreted LIP-fragments in Canada are presented, along with the generalized distribution of each. Our current understanding of LIPs through time is at a preliminary level, and much work, particularly high-precision geochronology, is needed to understand the full extent of LIPs through time and to assess their mineral potential.

Résumé

Les grandes provinces ignées (GPI) comprennent les basaltes de plateau, les essaims de dykes majeurs, les provinces de filons-couches et les grandes intrusions mafiques-ultramafiques stratifiées; les épisodes magmatiques qui les engendrent surviennent en moyenne à tous les 20 millions d'années depuis 2600 Ma. Des phénomènes analogues ont en outre pu se produire à l'Archéen. Les GPI sont les hôtes clés des gîtes de Ni-Cu-ÉGP, de Cr et de Fe-Ti-V. Dans certains cas, il existe en outre une relation spatiale et temporelle avec des intrusions non mafiques comme les carbonatites (qui peuvent renfermer des gîtes de Au-Hg, de ÉTR-Ta-Nb) et les kimberlites (qui peuvent contenir des diamants). Puisque les GPI déplacent de grandes quantités d'énergie thermique dans la croûte, elles peuvent entraîner la formation de systèmes hydrothermaux capables d'engendrer différents types de gîtes minéraux. On présente de l'information de base sur plus de 80 GPI et fragments interprétés de GPI au Canada avec une distribution généralisée pour chacune. Notre compréhension actuelle de l'évolution temporelle des GPI se situe encore au niveau préliminaire et des travaux considérables, en particulier de géochronologie de grande précision, seront nécessaires pour comprendre toute l'ampleur de l'évolution des GPI dans le temps et en évaluer le potentiel minéral.

Definition and Characteristics of Large Igneous Provinces

Earth history is punctuated by numerous periods during which large volumes of 'intraplate' mafic and silicic magma were emplaced in the crust (Fig. 1) by processes not associated with a 'normal' spreading ridge or subduction environments. The regions of the Earth that are affected by this intrusive and/or extrusive magmatism are termed Large Igneous Provinces (LIPs). LIPs consist of continental flood basalts, volcanic rifted margins, oceanic plateaus, and ocean-basin flood basalts (Coffin and Eldholm, 1994, 2001). Submarine ridges and seamount chains were part of the original definition (Coffin and Eldholm, 1994), but do not fit current definitions that emphasize a short duration of magmatic activity or sharp pulses of activity ("transient LIPs" of Coffin and Eldholm, 2001; Bryan and Ernst, 2006). Also important are the LIP plumbing systems of dyke swarms, sill provinces, and layered intrusions (Ernst and Buchan, 2001). Silicic magmatism is frequently associated with LIPs and in some cases they can be of high volume (e.g. Bryan et al., 2002, Dobretsov, 2005).

The size criterion for LIPs is controversial. We use the original criterion of an areal extent of at least 100,000 km^2 (Coffin and Eldholm, 1994, 2001; Ernst and Buchan, 2001; Ernst et al., 2005); note however that others, such as Courtillot and Renne (2003), prefer a more restrictive criterion of an areal extent of 1 million km^2 with a volume of 1 million km^3.

The expression of LIPs varies through Earth's history (Ernst and Buchan, 2001, Ernst et al., 2005). LIPs of Mesozoic and Cenozoic age are typically the best preserved and consist mainly of flood basalts. Those of Paleozoic and Proterozoic age are often more deeply eroded, and comprise flood basalt remnants and the exposed plumbing system represented by giant dyke swarms, sill provinces and layered intrusions. In the Archean, the most promising LIP candidates are greenstone belts containing tholeiite-komatiite sequences.

Many LIPs are linked to regional-scale uplift, continental rifting and breakup, and/or climactic/environmental crises (e.g. Courtillot et al., 1999, 2003; Pirajno, 2000; Campbell, 2001; Rainbird and Ernst, 2001; Şengör, 2001; Courtillot and Renne, 2003; Ernst and Buchan, 2003; Xu et al., 2004; Ernst et al., 2005), and to partial melting of an arriving mantle plume and the onset of continental breakup (e.g. White and Mackenzie, 1989; Campbell and Griffiths, 1990, Courtillot et al., 2003; Ernst et al., 2005). However, alternative non-plume origins are being forcefully argued and include back-arc rifting, delamination, edge convection, and bolide impact (e.g. Anderson, 2001; Foulger et al., 2005).

Since LIPs can be precisely dated, they provide key time markers in the stratigraphic record and therefore play an important role in unraveling the Earth's geological history. They are also an important target for magmatic Ni-Cu-PGE, Cr, Fe-Ti-V and other deposit-types (Naldrett, 1997, 1999; Pirajno, 2000; Schissel and Smail, 2001; Crocket, 2002;

Ernst, R.E., 2007, Large igneous provinces in Canada through time and their metallogenic potential, *in* Goodfellow, W.D., ed., Mineral Deposits of Canada: A Synthesis of Major Deposit-Types, District Metallogeny, the Evolution of Geological Provinces, and Exploration Methods: Geological Association of Canada, Mineral Deposits Division, Special Publication No. 5, p. 929-937.

Borisenko et al., 2006; Eckstrand and Hulbert, 2007). Table 1 profiles several of the most significant ore deposits that are associated with LIPs. The strongest association is with Ni-Cu-PGE deposits represented by Bushveld, Noril'sk, and Jinchuan intrusions with their world-class ore deposits (e.g. Naldrett, 1997). In addition, silicic LIPs (Bryan et al., 2002; Dobretsov, 2005) may be associated with Olympic Dam-type deposits (Cu-Au-Ag-U-REE) (e.g. Campbell, 1998; Corriveau, 2007). There is also a spatial association with kimberlites (diamonds) (e.g. Agashev et al., 2004; Kjarsgaard, 2007) and carbonatites (Au-Hg-REE-Ta-Nb deposits) (Bell, 2001). In addition to orthomagmatic ore deposits, the emplacement of LIPs produces a large thermal anomaly and convection in the crust. This large-scale fluid circulation engenders a variety of hydrothermal systems that form several types of hydrothermal ore deposits (Carlin-style, continental porphyry Mo, adularia-sericite epithermal systems, etc.; Pirajno, 2000).

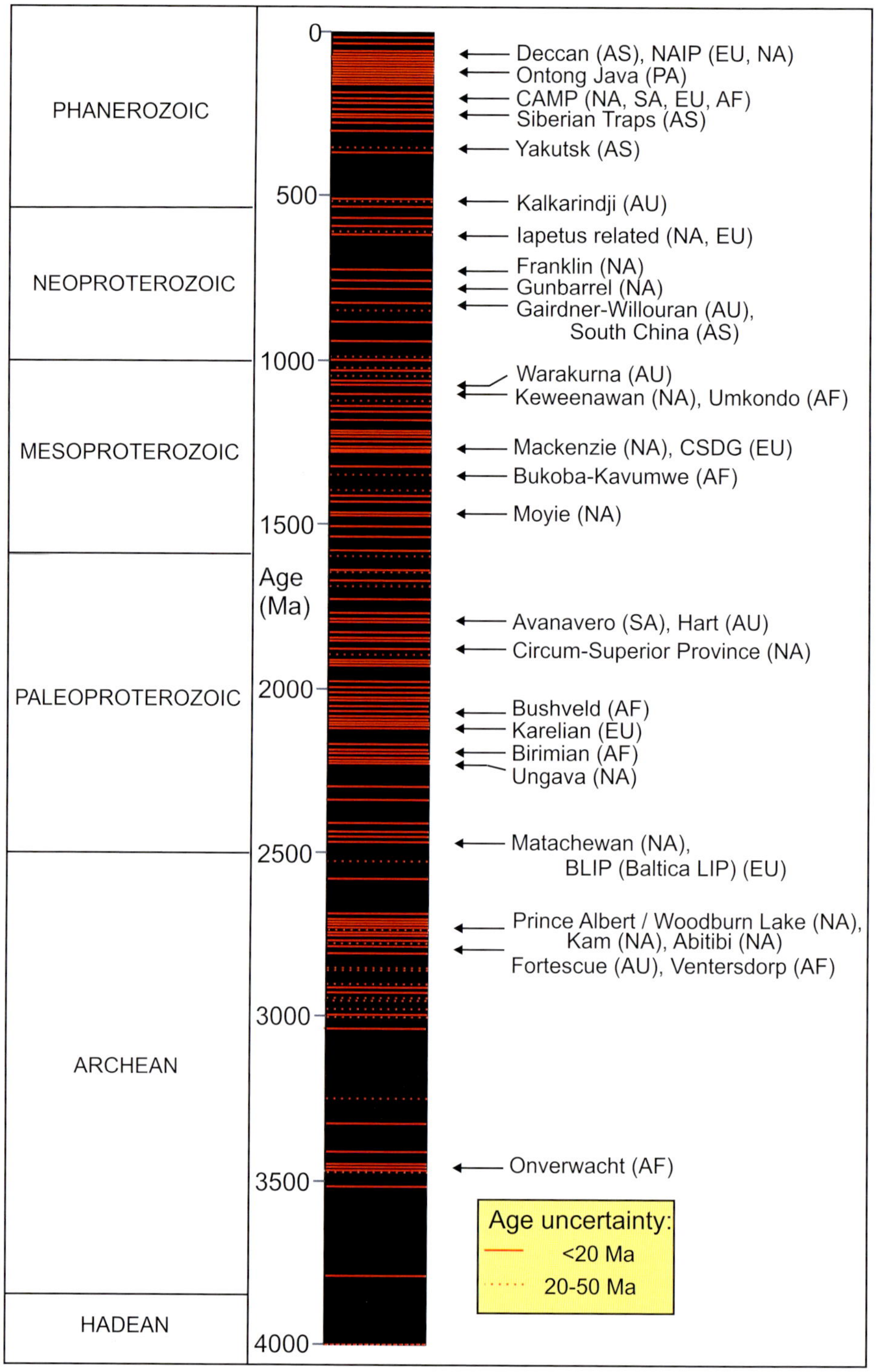

FIGURE 1. Global 'bar code' of large igneous provinces. AF = Africa, AN = Antarctica, AS = Asia, AU = Australia, EU = Europe, NA = North America, PA = Pacific Ocean, SA = South America.

The Role of the Large Igneous Provinces Concept in Exploration for Ni-Cu-PGE and Other Commodities

Applying the paradigm of LIPs to the search for commodities has several advantages that are discussed briefly below.

Large Igneous Provinces Represent a Systems Approach to Understanding Magmatic Events

LIPs consist of flood basalts (or their erosional remnants), dyke swarms (with radiating, linear, or arcuate geometries), sill provinces, layered mafic-ultramafic intrusions, and silicic magmatism (mainly associated with partial melting of lower crust). Underplated magma is also probably important, as are volcanic sequences associated with rifting. The LIP-system approach considers the pathway of the magma from mantle source areas, upward into a crustal plumbing system (dykes, sills, and layered intrusions) that distributes the magma (laterally and vertically) within the crust and also feeds surface lavas (e.g. Ernst and Buchan, 1997a,b). Understanding the magmatic pathways has ore deposit implications. Magma flowing through the plumbing system has an opportunity to interact with host rocks (e.g. with sulphur-rich and silica-rich sediments), which may cause the concentration and segregation of sulphides (e.g. Naldrett, 1997). In addition, extensive meteoric water and magmatic fluid circulation systems can develop in the host rocks adjacent to intrusions, particularly those feeders associated with extensive and prolonged magma flow. This circulation of hot fluids can lead to the formation of hydrothermal ore deposits associated with LIPs (e.g. Pirajno, 2000, 2004).

Table 1. Selected large igneous provinces with significant ore deposits.

Event	Region/Host Tectonic Terrane	Approx Size (Mkm^2)	Age	Commodity	Reference: E = event A = age C = commodity
Siberia	Siberia, Ural mtns, Central Asian Fold Belt	3.67	251 Ma	Norilsk (Ni-Cu-PGE)	(E, A, C) Dobretsov, 2005 Borisenko et al., 2006
Emeishan	China and Vietnam	0.66	258 Ma	Gabbroic intrusions Fe-Ti-V), ultramafic intrusions (Ni-Cu-PGE)	(E, A, C) Xu et al., 2004; Dobretsov, 2005 Borisenko et al., 2006
Yakutsk	Siberian craton (eastern margin)	0.96	360 Ma	Kimberlites (Diamonds)	(C) Agashev et al., 2004 (E, A) #44 in Ernst and Buchan, 2001
Guibei	South China Block	1.34	825 Ma	Jinchuan (Ni-Cu-PGE)	(E, A) Li, W.X. et al., 2003, 2005 (C) Li, X.H et al., 2005
Keweenawan	Mid-Continent Rift	0.42	1114-1085 Ma	Duluth (Cu-Ni-PGE)	(E, A) #157 in Buchan and Ernst, 2004 (C) Miller and Ripley, 1996
Mackenzie	Laurentia	2.71	1267 Ma	Muskox Intrusion (Cu-Ni-PGE)	(E, A) Baragar et al., 1996; (C) Hulbert, 2005
Gawler Range (felsic LIP)	Gawler Block, Australia	0.13	1600 Ma	Olympic Dam (Cu-Au-Ag-U-REE)	(E, A, C) Campbell et al., 1998
Sudbury Impact Event[1]		0.01	1850 Ma	Ni, Cu	(E, A, C) Naldrett, 2003
Circum-Superior 1800 Ma	Circum-Superior craton	0.1	1885-1865 Ma	Thompson (Ni), Raglan (Ni-Cu-PGE)	(E, A) www.largeigneousprovinces.org/ LOM.html (May, 2004) (C) Eckstrand and Hulbert, 2007
Bushveld-Molopo Farms	Kaapvaal craton	>0.09	2054 M	Bushveld (PGE), Phalaborwa carbonatite (Cu)	(E, A, C) www.largeigneousprovinces.org/ LOM.html (May 2005) (E) Eales and Cawthorn, 1996 (A for Phalaborwa) Fig. 4 in Hanson, 2003

1. Note that the Sudbury impact event is not strictly a large igneous province (it fails the size criterion), but is included herein as the type example of the production of significant volumes of mafic magmatism and associated Ni-Cu ore deposits by melting resulting from meteorite impact.

Feeding of Some Volcanics, Layered Intrusions, and Sills by Lateral Flow through Dykes from Distant Source Areas

It has been shown that giant dyke swarms with radiating geometry can transport magma laterally from source areas near the focal point of the radiating swarm, and emplace magma as sills or volcanics more than 1000 km away (e.g. Ernst and Buchan, 1997a,b). Thus an individual sill, volcanic unit, or layered intrusion may not originate from the immediately underlying mantle, but may have been transported laterally to its present location from a distal mantle source. This has implications for studies of whether a given mantle region has been previously or is subsequently tapped (e.g. 'second-stage' melt hypothesis for the origin of PGE-enriched magmas, Hamlyn and Keays, 1986). The phenomenon of long-distance lateral transport of magma is illustrated using a 2217 to 2210 Ma event from the eastern Superior and Southern provinces (Figs. 2, 3). The Nipissing sills in the Southern Province are postulated to originate via a radiating dyke swarm from a mantle source area 1300 km to the northeast (Buchan et al., 1998). Therefore, the mantle source for the Nipissing sills would have nothing to do with the mantle underneath the Southern Province that previously generated the 2500 to 2450 Ma Matachewan magmatic province [event #1a in Appendix 1 and Fig. 4]. Instead the relevant mantle would be located under the Labrador Trough (1300 km to the northeast).

Locating the Centre of a Mantle Plume Responsible for a Large Igneous Province

The plume centre region can be located using a variety of strategies including the centre of domal uplift, the focus of giant radiating dyke swarms, and the location of high-Mg magmas (e.g. Campbell et al., 1989; Ernst and Buchan, 2003). Formation of ore deposits is enhanced in conduits experiencing greater magma flow-through, and such conduits are more likely concentrated above the plume centre. Also, ultramafic magmas are better targets for mineralization and they also tend to be concentrated in the plume centre region.

Potential Linkage of Layered Intrusions and Dyke Swarms

The feeder system into or out of a layered intrusion is the loci of many Ni-Cu-PGE ore deposits (Naldrett, 1997). Therefore, it is important to investigate the link in LIPs between layered intrusions and dykes swarms, but in most LIPs the relationship is obscure. It is generally expected that layered intrusions spawn subswarms of dykes (e.g. Baragar et al., 1996), but it is also possible that layered intrusions can be spawned along a dyke swarm at widened portions of dykes (Ernst and Buchan, 1997b).

Dyke Swarms Emplaced into Cratonic Interiors have an Excellent Preservation Potential

Dyke swarms are generally better preserved than other parts of a LIP (e.g. volcanics, layered intrusions). These other parts may be concentrated in the cratonic margin and hence be highly deformed or reside in a cover sequence that is easily eroded (Halls, 1982; Bleeker and Ernst, 2006). Therefore, dyke swarms in the cratonic interior may be the only easily recognizable component of a given LIP to survive. However, once a LIP has been identified by its well preserved dyke swarm, then the rest of the LIP including layered intrusions (potentially hosting ore deposits) can be sought in more highly eroded and deformed regions along the cratonic margin.

Continental Reconstructions Allow Tracing of Metallogenic Belts between Blocks

A prominent example is offered by Bleeker and Ernst (2006) (see also Dahl et al., 2006; Fig. 3), where 2500 to 2450 Ma LIPs are used to reconstruct Baltica, Hearne, Wyoming, and Superior cratons. In this reconstruction, Karelia-Kola, Hearne, and Wyoming are placed against the southern Superior Craton (from east to west). This new reconstruction also makes some predictions about the correlation of younger LIPs between the cratonic blocks, ca. 2215 Ma magmatism in Superior and Karelia-Kola cratons, 2170 Ma magmatism in Superior and Wyoming cratons, and finally at 2100 to 2070 Ma in all four cratons. The latter event is inferred to mark the breakup of these four cratons.

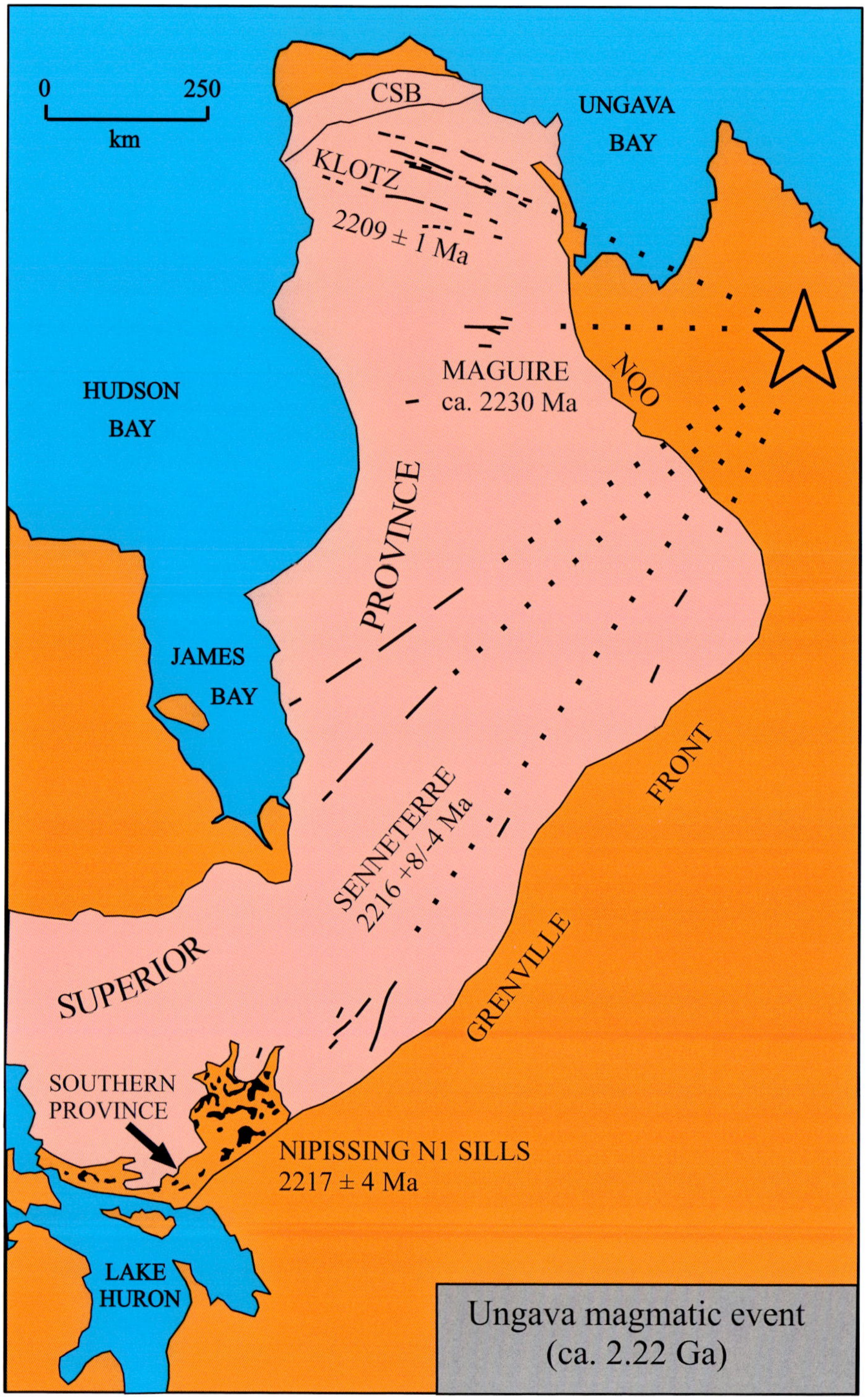

Figure 2. Nipissing-Ungava event (modified after Buchan et al., 1998, 2007). CSB is Cape Smith Belt, NQO is New Quebec Orogen. Star is interpreted plume centre.

Geochemical Analysis of Large Igneous Provinces to Assess Economic Potential

Recently, Zhang et al. (2006) reported the results of a comparative geochemical study of ten major LIPs in order to determine whether there are systematic differences between those LIPs with Ni and PGE potential and those lacking such potential. They concluded that LIPs that are fertile for Ni and PGE: 1) arrive from a deep-sourced mantle plume (FOZO [Focal Zone] source), 2) contain a significant percentage of high-MgO magmatism that is low in Al_2O_3 and Na_2O, but highly enriched in strongly incompatible elements, 3) have moderately high Os contents, and 4) an isotopic range between the Focal Zone (FOZO) and Enriched Mantle 1 (EM1). They conclude that interaction between plume-related magmas and moderately cratonic subcontinental lithospheric mantle (SCLM) is important in producing Ni-PGE fertility. In contrast, barren LIPs contain less high-MgO magmatism, and the isotopic signatures show interaction between isotopic reservoirs Focal Zone (FOZO) and Enriched Mantle 2 (EM2). Their conclusion for barren LIPs is that there may be a recycled component in the ascending plume, but limited interaction with subcontinental lithospheric mantle.

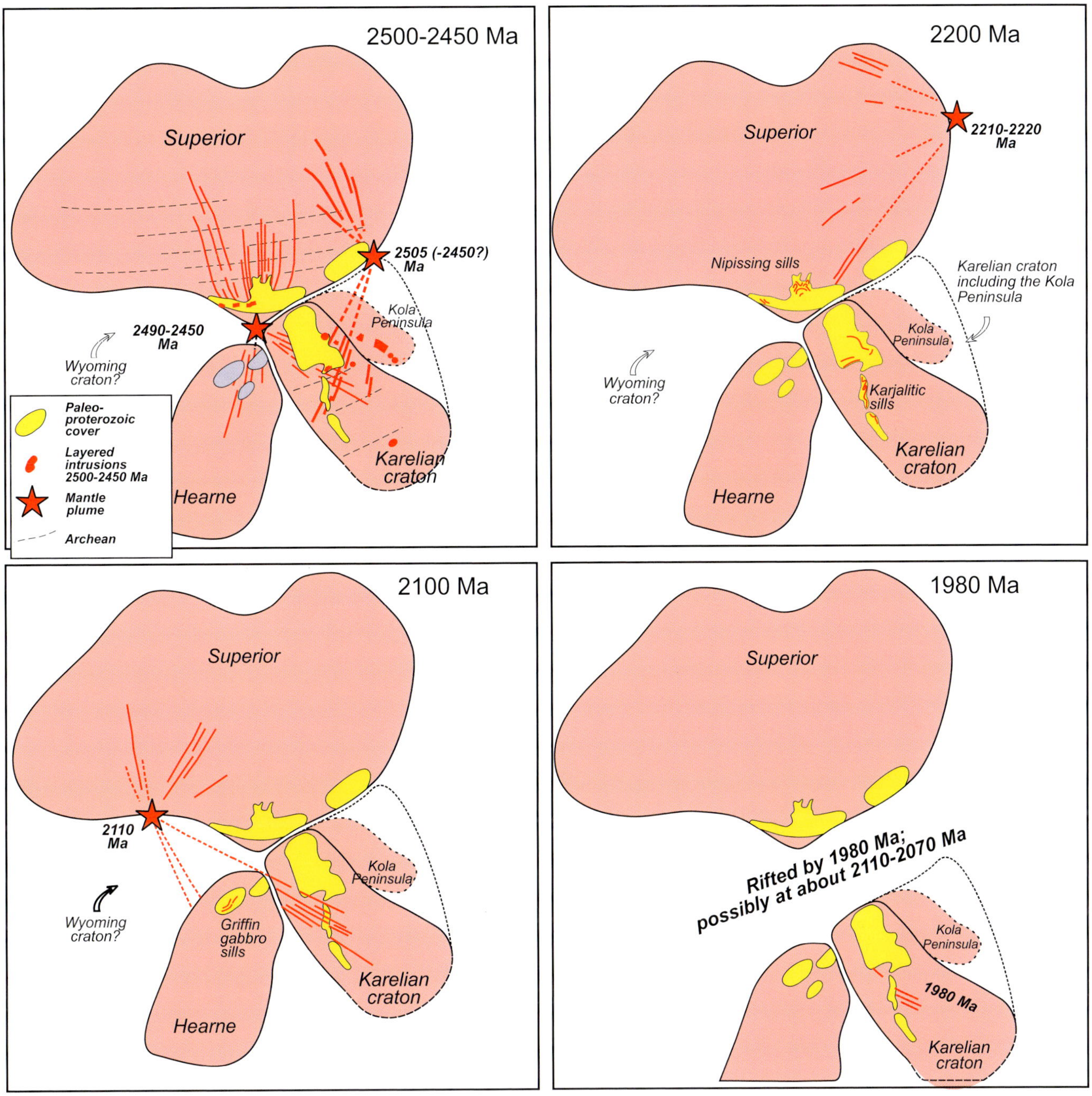

FIGURE 3. Reconstruction of Superior, Hearne, Baltic, and Wyoming cratons based on matching of magmatic events and tracing of coeval dyke swarms between the blocks. After Bleeker and Ernst (2006).

Canada's Record of Large Igneous Provinces and Interpreted Large Igneous Province Fragments

As outlined above, LIPs are potentially important tools for the exploration of a number of different ore deposit types. However, the LIP record is incompletely known. Important LIP compilations (mostly in the context of identifying plume-related events) have been attempted by Coffin and Eldholm (1994, 2001, for mostly the Cenozoic-Mesozoic record), Isley and Abbott (1999), and, for the Archean record, Arndt et al. (2001) and Tomlinson and Condie (2001). These and other sources were incorporated in the preparation of a detailed global listing by Ernst and Buchan (2001). Figure 4 shows the distribution of LIPs (and postulated LIP fragments) for Canada through time (updated after the more recent version of the database in Ernst and Buchan, 2004). Basic information on each event is provided in Appendices 1 and 2, and the PGE potential of some of these events is assessed in Hulbert (2002) and Ernst and Hulbert (2003).

Selected Canadian events that are significant from a PGE-Cu-Ni ore deposits' perspective are discussed below. The 2500 to 2450 Ma Matachewan event of the Superior Craton [event #1a in Appendix 1, Fig. 4] hosts the East Bull Lake and related layered intrusions of potential interest (James et al., 2002a,b). The 2217 to 2210 Ma Ungava event [event #2c in Appendix 1] fed the Nipissing sills of the southern Province, which have been historically important for Co-Ag-

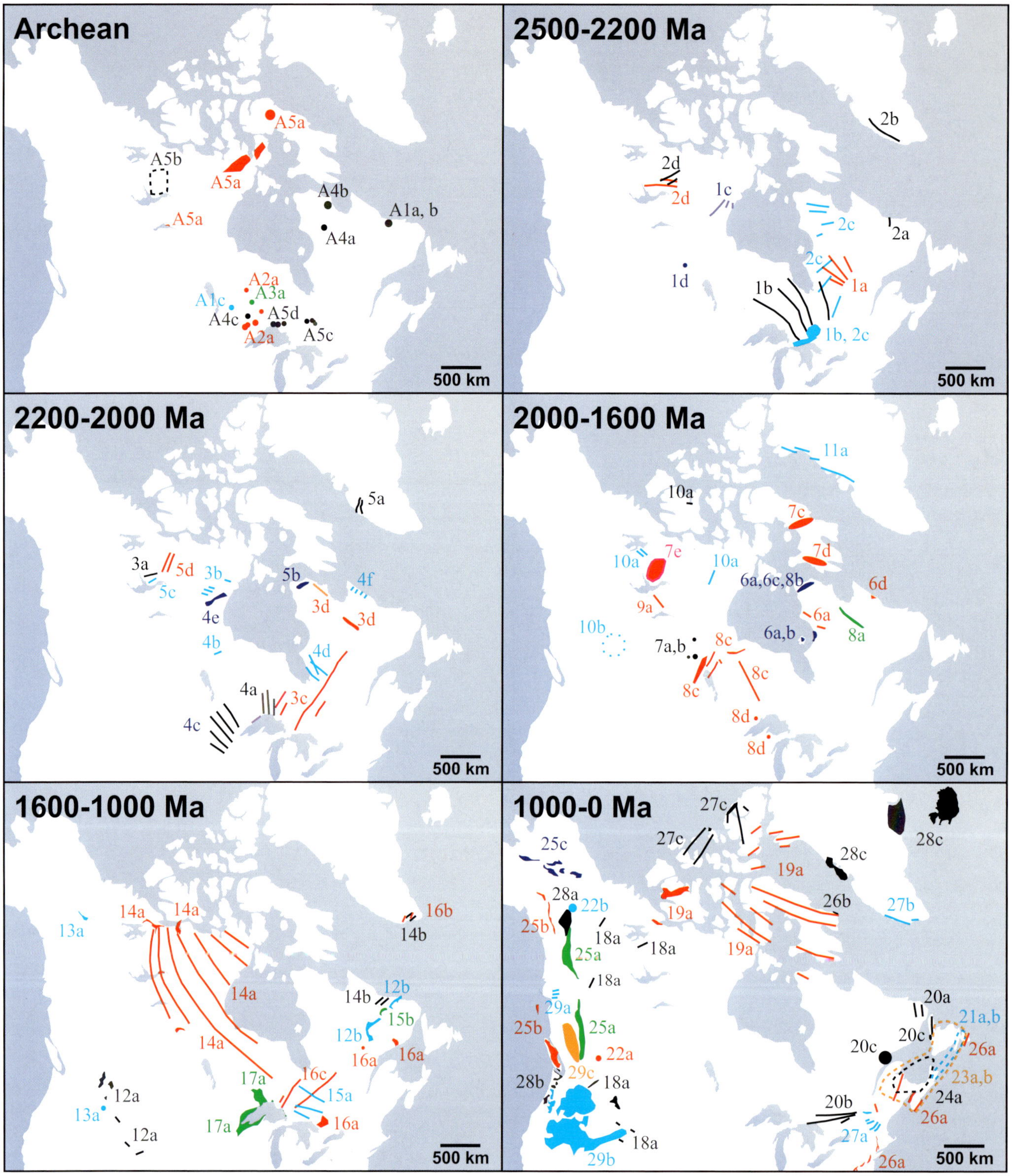

As mineralization (e.g. Marshall and Watkinson, 2000), and have PGE potential as well (e.g. James, et al., 2002b). A third significant event is the 1885 to 1865 Ma magmatism that circumscribes the Superior Craton [event #8 in Appendix 1, Fig. 4, see also Table 1], from the Labrador Trough, through the Cape Smith Belt, the Belcher Islands, Fox River Belt, Thompson Belt, Winipegosis komatiites, and also on the southern side of the Superior Craton in the Animkie Basin. Included in this belt are major Ni deposits of the Thompson and Raglan belts (Eckstrand and Hulbert, 2007; Layton-Matthews et al., 2007; Lesher 2007). The 1267 Ma Mackenzie event [event #14a] is the host for the well known and highly prospected Muskox layered intrusion (Hulbert, 2005, see also Table 1). The 1114 to 1085 Ma Keweenawan event [event #17a] is host to important Cu-Ni-PGE mineralization of the Duluth Complex (e.g. Nicholson et al., 1992;

FIGURE 4. Canadian record of large igneous provinces (updated from Ernst and Buchan, 2004). Numbers keyed to entries in Appendices 1 and 2. **Part A** (Archean): **A1a** Hunt River (3.11 Ga), **A1b** Florence Lake (2.99-2.98 Ga), **A1c** North Caribou (2.99 Ga), **A2a** Western Superior (2.93 Ga), **A3a** Pickle Crow (2.86 Ga), **A4a** Vizean (2.79 Ga), **A4b** Faribault-Thury Complex (ca. 2.88-2.71 Ga), **A4c** Fourbay sequence (2.78 Ga), **A5a** Prince Albert -Woodburn Group and possible correlatives (2.73 Ga), **A5b** Kam Group (2.73-2.70 Ga), **A5c** Abitibi 2.75-2.70 Ga, **A5d** Wawa (2.75-2.74 Ga). **Parts B-F** (Proterozoic-Phanerozoic): **1a** Mistassini (ca. 2.5 Ga), **1b** Matachewan (2.49-2.45 Ga), **1c** Kaminak (ca. 2.45 Ga), **1d** Mirond Lake (2.49 Ga),: **2a** Kikkertavak (2.235 Ga), **2b** BN-1 (2.21 Ga), **2c** Ungava (2.22-2.21 Ga), **2d** Malley- MacKay (2.23, 2.21 Ga), **3a** Dogrib (2.19 Ga), **3b** Tulemalu-MacQuoid (2.19 Ga), **3c** Biscotasing (2.17 Ga), **3d** Cramolet Lake- Payne River (2.17 Ga), **4a** Marathon (2.125-2.01 Ga), **4b** Cauchon Lake (2.09-2.07 Ga), **4c** Fort Frances (2.075 Ga), **4d** Lac Esprit (2.07 Ga), **4e** Griffin (2.11 Ga), **4f** Napaktok (? 2.12 Ga), **5a** Kangâmuit (2.05-2.04 Ga), **5b** Lower Povungnituk (2.04 or 1.96 Ga), **5c** Hearne (2.04 Ga), **5d** Lac de Gras (ca 2.025 Ga), **6a** Minto-Eskimo-Watts group (2.00 Ga), **6b** Flaherty-Haig (? 1.96 Ga), **6c** Upper Povungnituk (ca. 1.96 Ga), **6d** Mugford (1.95 Ga), **7a** Sandy Bay (1.90 Ga), **7b** Josland (1.88 Ga), **7c** Bravo Lake (1.90-1.88 Ga), **7d** Lake Harbour (ca. 1.87 Ga), **7e** Mara-Morel-Ghost (1.88-1.87 Ga), **8a** New Quebec Orogen-cycle 2 (1.88 Ga), **8b** Chukotat (1.89-1.87 Ga), **8c** Molson-Thompson-Winnipegosis-Pickle Crow (1.88- 1.86 Ga), **8d** Hemlock (1.88 Ga), **9a** Sparrow (1.83 Ga), **10a** Cleaver-Hadley Bay-MacRae Lake (1.75-1.74 Ga), **10b** Winagami (? 1.89-1.76 Ga), **11a** Melville Bugt (ca. 1.64 Ga), **12a** Moyie (1.47 Ga), **12b** Michael-Shabagamo (ca. 1.47 Ga), **13a** Hart River – Salmon River Arch (1.38 Ga), **14a** Mackenzie (1.27 Ga), **14b** Harp-Nain-Nutak-middle Gardar (1.28, 1.27 Ga), **15a** Sudbury [dykes] (1.24 Ga), **15b** Seal Lake - Mealy (1.24-1.25 Ga), **16a** Davy Group - Tshenukutish - Algonquin (1.18-1.16 Ga), **16b** Late Gardar (1.16 Ga), **16c** Abitibi [dykes] (1.14 Ga), **17a** Keweenawan (Mid-continent rift system) (1.11-1.09 Ga; main pulses: 1.11, 1.10 Ga), **18a** Gunbarrel (0.78 Ga), **19a** Franklin-Thule (0.72 Ga), **20a** Long Range (0.62 Ga), **20b** Grenville-Rideau (0.59 Ga), **20c** Sept-Îles – Catoctin (0.56 Ga), **21a** Harbour Main (ca. 0.62 Ga), **21b** Marystown (0.59-0.55 Ga), **22a** Hamill-Gog (0.57 Ga), **22b** Selwyn Basin (0.54-0.45 Ga), **23a** Middle Ordovician 'Overstep' Sequence (0.47-0.45 Ga), **23b** Late Ordovician-Silurian magmatism of Atlantic Canada (0.44-0.41 Ga), **24a** Magdalen (Maritimes) Basin (0.36-0.32 Ga), **25a** Cache Creek (Late Triassic and mid-Permian), **25b** Wrangellia (0.23 Ga), **25c** Ramparts Group (0.21 Ga), **26a** ENA (Eastern North America) portion of CAMP (Central Atlantic Magmatic Province) (0.20 Ga), **27a** New England-Québec (NEQ); Monteregian) (0.14-0.11 Ga), **27b** Trap (0.14 Ga), **27c** Sverdrup Basin Magmatic Province (part of High Arctic Large Igneous Province; HALIP) (0.13-0.09 Ga), **28a** Carmacks (0.07 Ga), **28b** Crescent (0.06-0.05 Ga), **28c** North Atlantic Igneous Province (NAIP) (0.06 Ga), **29a** Behm Canal (Tertiary 'Lamprophyre' Province) (ca. 0.023-0.005 Ga), **29b** Columbia River Basalt Group (mainly 0.017-0.0150 Ga), **29c** Chilcotin (Neogene-Paleogene; mainly Miocene).

Miller and Ripley, 1996). The 723 Ma Franklin event [event #19a] of northern Canada has been highly prospected for Ni-Cu-PGEs (Jefferson et al., 1994). The 230 Ma accreted oceanic plateau, Wrangellia [event 25b], has also been explored for Ni-Cu-PGE (e.g. Hulbert, 2002). The numerous additional LIP (and LIP fragment) events shown in Figure 4 and listed in Appendices 1 and 2 need to be systematically assessed for economic potential, using as a guide some of the strategies outlined in this paper.

Global Campaign to Study Large Igneous Provinces

As noted above, continental reconstructions are key to fully reconstructing LIP systems, and to evaluating their overall metallogenic potential. In this context, large mafic magmatic events offer the most effective and efficient approach for fulfilling the objective of obtaining continental reconstructions back to 2600 Ma (Bleeker and Ernst, 2006). The approach is two parts. Firstly, comparison of cratonic magmatic age-of-emplacement 'bar codes' can reveal which cratons were likely the nearest neighbours and over what time interval. Secondly, dyke swarms have simple geometry (often radiating and linear) and can therefore be used as piercing points between cratons to precisely constrain the reconstructions. However, to fulfill the potential of using the LIPs as a tool to reconstruct ancient continents back to 2600 Ma (and allow tracing of metallogenic belts between cratons), requires an improvement in the global LIP database. The most critical step is increased dating of the numerous undated but important magmatic units to better define the full extent of individual LIPs and identify new LIPs. For these and other reasons, a global LIP campaign focused around geochronology has been advocated (Bleeker, 2004; Ernst and Buchan, 2004; Ernst et al., 2005).

Conclusions

The temporal and spatial distribution of large igneous provinces provides an effective tool for understanding Earth's magmatic and tectonic processes, reconstructing paleogeography and ancient climates, and discovering world-class magmatic N-Cu-PGE deposits and other commodities. In the Canadian context, there are about 80 LIPs and interpreted (eroded and tectonic) fragments of LIPs. Apart from Sudbury (which is not strictly a LIP), the most prominent Ni-Cu-PGE deposits in Canada are linked with LIPs. Additional ore deposits for Ni-Cu-PGE, Fe-Ti-V and other commodities may be discovered by systematically applying the LIP paradigm, and by using the improving database of Canadian LIPs (Appendices 1 and 2; Figure 4).

Acknowledgements

I acknowledge and appreciate the collaborations over recent years with Wouter Bleeker, Ken Buchan, Larry Hulbert, and Franco Pirajno, which have contributed to my understanding of large igneous provinces and their relationships with ore deposits. Valuable reviews of the manuscript were provided by Wayne Goodfellow, Larry Hulbert, and Franco Pirajno. This work was partially funded by BHP Billiton mining company.

References

Agashev, A.M., Pokhilenko, N.P., Tolstov, A.V., Polyanichko, V.V., Mal'kovets, V.G., Sobolev, N.V., 2004, New age data on kimberlites from the Yakutian diamondiferous province: Doklady Earth Sciences, v. 399, p. 1142-1145.

Anderson, D.L., 2001, Top-down tectonics: Science, v. 293, p. 2016-2018.

Arndt, N., Bruzak, G., and Reischmann, T., 2001, The oldest continental and oceanic plateaus: Geochemistry of basalts and komatiites of the Pilbara Craton, Australia, *in* Ernst, R.E., and Buchan, K.L., eds., Mantle Plumes: Their Identification through Time: Geological Society of America, Special Paper 352, p. 359-387.

Baragar, W.R.A., Ernst, R.E., Hulbert, L., and Peterson, T., 1996, Longitudinal petrochemical variation in the Mackenzie dyke swarm, northwestern Canadian Shield: Journal of Petrology, v. 37, p. 317-359.

Bell, K., 2001, Carbonatites: Relationship to mantle-plume activity, *in* Ernst, R.E., and Buchan, K.L., eds., Mantle Plumes: Their Identification through Time: Geological Society of America, Special Paper 352, p. 267-290.

Bleeker, W., 2004, Taking the pulse of planet Earth: A proposal for a new multi-disciplinary flagship project in Canadian solid Earth sciences: Geoscience Canada, v. 31, p. 179-190.

Bleeker, W., and Ernst, R.E., 2006, Short-lived mantle generated magmatic events and their dyke swarms: The key unlocking Earth's paleogeographic record back to 2.6 Ga, *in* Hanski, E., Mertanen, S., Rämö, T., and Vuollo, J., eds., Dyke Swarms: Time Markers of Crustal Evolution: A.A. Balkema, Rotterdam, p. 3-26.

Borisenko, A.S., Sotnikov, V.I., Izokh, A.E., Polyakov, G.V., and Obolensky, A.A., 2006, Permo-Triassic mineralization in Asia and its relation to plume magmatism: Russian Geology and Geophysics, v. 47, p. 166-182.

Bryan, S.E., and Ernst, R.E. 2006, Proposed revision to large igneous province classification: http://www.mantleplumes.org/LIPClass2.html

Bryan, S.E., Riley, T.R., Jerram, D.A., Leat, P.T., and Stephens, C.J., 2002, Silicic volcanism: An under-valued component of large igneous provinces and volcanic rifted margins, *in* Menzies, M.A., Klemperer, S.L., Ebinger, C.J., and Baker J., eds., Magmatic Rifted Margins: Geological Society of America, Special Paper 362, p. 99-120.

Buchan, K.L., Mortensen, J.K., Card, K.D., and Percival, J.A., 1998, Paleomagnetism and U-Pb geochronology of diabase dyke swarms of Minto block, Superior Province, Quebec, Canada: Canadian Journal of Earth Sciences, v. 35, p. 1054-1069.

Buchan, K.L., Goutier, J., Hamilton, M.A., Ernst, R.E., and Matthews, W.A., 2007, Paleomagnetism, U-Pb geochronology and geochemistry of Lac Esprit and other dyke swarms, James Bay area, Quebec, and implications for Paleoproterozoic deformation of the Superior Province; Canadian Journal of Earth Sciences (in press).

Campbell, I.H., 1998, The mantle's chemical structure: Insights from the melting products of mantle plumes, *in* Jackson, I.N.S., ed., The Earth's Mantle: Composition, Structure and Evolution: Cambridge University Press, Cambridge, p. 259-310.

——— 2001, Identification of ancient mantle plumes, *in* Ernst, R.E., and Buchan, K.L., ed., Mantle Plumes: Their Identification through Time: Geological Society of America, Special Paper 352, p. 5-21.

Campbell, I.H., and Griffiths, R.W., 1990, Implications of mantle plume structure for the evolution of flood basalts: Earth and Planetary Science Letters, v. 99, p. 79-93.

Campbell, I.H., Griffiths, R.W., and Hill, R.I., 1989, Melting in an Archaean mantle plume: Heads it's basalts, tails it's komatiites: Nature, v. 339, p. 697-699.

Coffin, M.F., and Eldholm, O., 1994, Large igneous provinces: crustal structure, dimensions, and external consequences: Review of Geophysics, v. 32, p. 1-36.

——— 2001, Large igneous provinces: progenitors of some ophiolites?, *in* Ernst, R.E., and Buchan, K.L., eds., Mantle Plumes: Their Identification through Time. Geological Society of America, Special Paper 352, p. 59-70.

Corriveau, L., 2007, Iron oxide copper-gold deposits: A Canadian perspective, *in* Goodfellow, W.D., ed., Mineral Deposits of Canada: A Synthesis of Major Deposit-types, District Metallogeny, the Evolution of Geological Provinces, and Exploration Methods: Geological Association of Canada, Mineral Deposits Division, Special Publication 5, p. 307-328.

Courtillot, V.E., and Renne, P.R., 2003, On the ages of flood basalt events: C.R. Geoscience, v. 335, p. 113-140.

Courtillot, V., Jaupart, C., Manighetti, I., Tapponnier, P., and Besse, J., 1999, On causal links between flood basalts and continental breakup: Earth and Planetary Science Letters, v. 166, p. 177-195.

Courtillot, V., Davaille, A., Besse, J., and Stock, J., 2003, Three distinct types of hotspots in the Earth's mantle: Earth and Planetary Science Letters, v. 205, p. 295-308.

Crocket, J.H., 2002, Platinum-group element geochemistry of mafic and ultramafic rocks, in Cabri, L.J. (ed.) The Geology, Geochemistry, Mineralogy and Mineral Beneficiation of Platinum-Group Elements; Canadian Institute of Mining, Metallurgy and Petroleum, Special Volume 54, p. 177-210.

Dahl, P.S., Hamilton, M.A., Wooden, J.L., Foland, K.A., Frei, R., McCombs, J.A., and Holm, D.K. , 2006, 2480 Ma mafic magmatism in the northern Black Hills, South Dakota: A new link connecting the Wyoming and Superior cratons: Canadian Journal of Earth Sciences, v. 43, p. 1579-1600.

Dobretsov, N.L., 2005, 250 Ma Large Igneous Provinces of Asia: Siberian and Emeishan traps (plateau basalts) and associated granitoids: Russian Geology and Geophysics, v. 46, p. 870-890.

Eckstrand, O.R., and Hulbert, L., 2007, Magmatic nickel-copper-platinum group element deposits, *in* Goodfellow, W.D., ed., Mineral Deposits of Canada: A Synthesis of Major Deposit-types, District Metallogeny, the Evolution of Geological Provinces, and Exploration Methods: Geological Association of Canada, Mineral Deposits Division, Special Publication No. 5, p. 205-222.

Ernst, R.E., and Buchan, K.L., 1997a, Giant radiating dyke swarms: Their use in identifying pre-Mesozoic large igneous provinces and mantle plumes, *in* Mahoney, J.J. and Coffin, M.J., M.F., ed., Large Igneous Provinces: Continental, Oceanic and Planetary Flood Volcanism, American Geophysical Union, Geophysical Monograph 100, p. 297-333.

——— 1997b, Layered mafic intrusions: A model for their feeder systems and relationship with giant dyke swarms and mantle plume centres: South African Journal of Geology, v. 100, p. 319-334.

——— 2001, Large mafic magmatic events through time and links to mantle plume heads, *in* Ernst, R.E., and Buchan, K.L., ed., Mantle Plumes: Their Identification through Time: Geological Society of America, Special Paper 352, p. 483-575.

——— 2003, Recognizing mantle plumes in the geological record: Annual Review of Earth and Planetary Sciences, v. 31, p. 469-523.

——— 2004, Large igneous provinces (LIPs) in Canada and adjacent regions: 3 Ga to present: Geoscience Canada, v. 31, p. 103-126.

Ernst, R.E., and Hulbert, L.J., 2003, Background Pt-Pd levels in mafic Large Igneous Provinces (LIPs) in Canada: Geological Survey of Canada, Open File 4472.

Ernst, R.E., Buchan, K.L., and Campbell, I.H., 2005, Frontiers in Large Igneous Province research: Lithos, v. 79, p. 271-297.

Foulger, G.R., Natland, J.H., Presnall, D.C., and Anderson, D.L., 2005, Plates, Plumes, and Paradigms" Geological Society of America, Special Paper 388, p. 881.

Halls, H.C., 1982, The importance and potential of mafic dyke swarms in studies of geodynamic processes: Geoscience Canada, v. 9, p. 145-154.

Hamlyn, P.R., and Keays, R.R., 1986, Sulfur saturation and second-stage melts: Application to the Bushveld platinum metal deposits: Economic Geology, v. 81, p. 1431-1445.

Hulbert, L.J., 2002, Magmatic Platinum Group Element Environments in Canada: Present and Future Exploration Target Areas: Geological Association of Canada, Robinson Lecture 2002, Miscellaneous Paper #2, CD-ROM.

——— 2005, Geology of the Muskox Intrusion and Associated Ni+Cu Occurrences: Geological Survey of Canada, Open File 4881, 1 p.

Isley, A.E., and Abbott, D.H., 1999, Plume-related mafic volcanism and the deposition of banded iron formation: Journal of Geophysical Research, v. 104 (B7), p. 15461-15477.

James, D.T., Kamo, S., and Krogh, T., 2002a, Evolution of 3.1 and 3.0 Ga volcanic belts and a new thermotectonic model for the Hopedale Block, North Atlantic craton (Canada): Canadian Journal of Earth Sciences, v. 39, p. 687-710.

James, R.S., Easton, R.M., Peck, D.C., and Hrominchuk, J.L., 2002b, The East Bull Lake intrusive suite: Remnants of an approximately 2.48 Ga large igneous and metallogenic province in the Sudbury area of the Canadian Shield: Economic Geology, v. 97, p. 1577-1606.

Jefferson, C.W., Hulbert, L.J., Rainbird, R.H., Hall, G.E.M., Grégoire, D.C., and Grinenko, L.I., 1994, Mineral Resource Assessment of the Neoproterozoic Franklin Igneous Events of Arctic Canada: Comparison with the Permo-Triassic Noril'sk Talnakh Ni-Cu-PGE Deposits of Russia: Geological Survey of Canada, Open File 2789, 48 p.

Kjarsgaard, B.A., 2007, Kimberlite diamond deposits, *in* Goodfellow, W.D., ed., Mineral Deposits of Canada: A Synthesis of Major Deposit-types, District Metallogeny, the Evolution of Geological Provinces, and Exploration Methods: Geological Association of Canada, Mineral Deposits Division, Special Publication No. 5, p. 245-272.

Layton-Matthews, D., Lesher, C. M., Burnham, O. M., Liwanag, J., Haldend, N. M., Hulbert, L., and Peck, D. C., 2007, Magmatic Ni-Cu-(PGE) deposits of the Thompson Nickel Belt, *in* Goodfellow, W.D., ed., Mineral Deposits of Canada: A Synthesis of Major Deposit-types, District Metallogeny, the Evolution of Geological Provinces, and Exploration Methods: Geological Association of Canada, Mineral Deposits Division, Special Publication No. 5, p. 409-432.

Lesher, C.M., 2007, Ni-Cu-(PGE) Deposits in the Raglan area, Cape Smith Belt, New Quebec, *in* Goodfellow, W.D., ed., Mineral Deposits of Canada: A Synthesis of Major Deposit-types, District Metallogeny, the Evolution of Geological Provinces, and Exploration Methods: Geological Association of Canada, Mineral Deposits Division, Special Publication No. 5, p. 351-386.

Marshall, D., and Watkinson, D.H., 2000, The Cobalt mining district: Silver sources, transport and deposition: Exploration and Mining Geology, v. 9, p. 81-90.

Miller, J.D.J., and Ripley, E.M., 1996, Layered intrusions of the Duluth Complex, Minnesota, USA, *in* Cawthorn, R.G., ed., Layered Intrusions: Elsevier, Amsterdam, p. 257-301.

Naldrett, A.J., 1997, Key factors in the genesis of Norilsk, Sudbury, Jinchuan, Voiseys Bay and other world-class Ni-Cu-PGE deposits: Implications for exploration: Australian Journal of Earth Sciences v. 44, p. 283-315.

——— 1999, World-class Ni-Cu-PGE deposits: Key factors in their genesis; Mineralium Deposita, v. 34, p. 227-240.

Nicholson, S.W., Cannon, W.F., and Schulz, K.J., 1992, Metallogeny of the Midcontinent rift system of North America: Precambrian Research, v. 58, p. 355-386.

Pirajno, F., 2000, Ore Deposits and Mantle Plumes: Kluwer Academic, Dordrecht, 556 p.

——— 2004, Metallogeny in the Capricorn Orogen, Western Australia, the result of multiple ore-forming processes: Precambrian Research, v. 128, p. 411-439.

Rainbird, R., and Ernst, R.E., 2001, The sedimentary record of mantle-plume uplift, *in* Ernst, R.E., and Buchan. K.L., eds., Mantle Plumes: Their Identification through Time: Geological Society of America, Special Paper 352, p. 227-245.

Schissel, D., and Smail, R., 2001, Deep-mantle plumes and ore deposits, *in* Ernst, R.E., and Buchan, K.L., eds., Mantle Plumes: Their Identification through Time: Geological Society of America, Special Paper 352, p. 291-322.

Şengör, A.M.C., 2001, Elevation as indicator of mantle-plume activity, *in* Ernst, R.E., and Buchan, K.L., ed., Mantle Plumes: Their Identification through Time: Geological Society of America, Special Paper 352, p. 183-225.

Tomlinson, K.Y., and Condie, K.C., 2001, Archean mantle plumes: Evidence from greenstone belt geochemistry, *in* Ernst, R.E., and Buchan, K.L., eds., Mantle Plumes: Their Identification through Time: Geological Society of America, Special Paper 352, p. 341-357.

White, R.S., and McKenzie, D.P., 1989, Magmatism at rift zones: the generation of volcanic continental margins and flood basalts: Journal of Geophysical Research, v. 94, p. 7685-7729.

Xu, Y.-G., He, B., Chung, S-L., Menzies, M.A., and Frey, F.A., 2004, Geologic, geochemical, and geophysical consequences of plume involvement in the Emeishan flood-basalt province: Geology, v. 32, p. 917-920.

Zhang, M., Wang, K-L, O'Reilly, S.Y., Hronsky, J., and Griffin, W.L., 2006, Are all mantle plumes equal in nickel and PGE potential?: International Conference on Continental Volcanism, Guangzhou, China, Abstracts and Program, p. 178.

Overview of Geophysical Signatures Associated with Canadian Ore Deposits

K. Ford, P. Keating, and M.D. Thomas

Geological Survey of Canada, 601 Booth Street, Ottawa, Ontario K1A 0E8
Corresponding author's email:KFord@nrcan.gc.ca

Abstract

Canadian ore deposits have typical geophysical signatures, either at regional or detailed scales. The particular geophysical response of an ore deposit depends on the contrast between its physical properties and those of its host rock. The most important physical properties are density, magnetic susceptibility, electrical conductivity and chargeability, and radioactivity. Density and susceptibility contrasts have a local influence on the measurement of the Earth's gravity and magnetic fields. Electrical properties contrasts can be detected using natural fields, as in magnetotelluric methods, or by artificial electromagnetic fields. Radioactivity is a natural phenomenon and is related to the chemico-physical nature of the rocks and minerals. Some types of deposits do not have any direct geophysical response; the best example is gold mineralization. In that case, geophysics is used in an indirect fashion to map favourable structures that have a good geophysical response. Other types of deposits are directly detectable by geophysical techniques. For example, massive sulphide deposits generally produce significant electromagnetic, gravimetric, and magnetic responses.

Here, we first present the principal geophysical techniques currently in use in Canada. Only a brief and simplified description of each method is given since extensive descriptions are available in the geophysical literature. Some typical survey specifications are provided. This is followed by a description of the geophysical responses of the major mineral deposit types found in Canada. In each case, we discuss the physical properties that can give rise to a geophysical response and we present typical examples. Applicability and effectiveness of the techniques are also discussed.

Geophysics was instrumental in the discovery of many Canadian ore deposits. Although there are likely many shallow ore deposits still to be discovered, the next major discoveries are expected to be at greater depth. Hopefully, some current geophysical techniques can detect deposits at depths in excess of one kilometre.

Résumé

Les gîtes minéraux du Canada présentent des signatures géophysiques caractéristiques, que ce soit à l'échelle régionale ou locale. La réponse géophysique particulière d'un gîte dépend du contraste entre ses propriétés physiques et celles des roches encaissantes. Les plus importantes de ces propriétés sont la densité, la susceptibilité magnétique, la conductivité électrique et la chargeabilité ainsi que la radioactivité. Les contrastes de densité et de susceptibilité ont localement une influence sur les champs gravitationnel et magnétique de la Terre. Les contrastes des propriétés électriques peuvent être détectés en ayant recours aux champs naturels, comme le font les méthodes magnétotelluriques, à des champs électromagnétiques artificiels. La radioactivité est un phénomène naturel et est reliée à la nature physicochimique des roches et des minéraux. Certains types de gîtes ne présentent pas une réponse géophysique directe; le meilleur exemple est celui des minéralisations aurifères. Dans le cas de ces gîtes, la géophysique est appliquée de manière indirecte en permettant la cartographie de structures favorables qui présentent une bonne réponse géophysique. D'autres types de gîtes sont directement détectables par des méthodes géophysiques. Par exemple, les gîtes de sulfures massifs fournissent généralement d'importantes réponses électromagnétiques, gravimétriques et magnétiques.

Nous présentons ici les principales méthodes géophysiques actuellement utilisées au Canada. On ne fournit pour chaque méthode qu'une description brève et simplifiée puisque des descriptions élaborées sont disponibles dans la littérature géophysique. Des spécifications caractéristiques propres à certains types de levés sont offertes, suivies d'une description des réponses géophysiques des principaux types de gîtes au Canada. Dans chaque cas, nous discutons les propriétés physiques qui engendrent la réponse géophysique et présentons des exemples caractéristiques. L'applicabilité et l'efficacité des méthodes sont en outre discutées.

La géophysique a contribué à la découverte d'un grand nombre de gîtes minéraux du Canada. Bien qu'il reste vraisemblablement un grand nombre de gisements peu profonds à découvrir, les prochaines découvertes majeures se feront probablement à plus grande profondeur. Certaines des méthodes géophysiques courantes permettent heureusement la détection de gisements à des profondeurs excédant un kilomètre.

Introduction

This paper is a contribution to the Consolidating Canada's Geoscience Knowledge program, providing a synthesis of geophysical signatures that are associated with the nine major mineral deposit types addressed in this program. The detection of mineral deposits by geophysical methods is dependent primarily on a single factor, namely that the deposit displays physical or chemical attributes that differ significantly from those of the adjacent rock formations. Historically, the principal physical properties that have been the focus of geophysical exploration methods are density, magnetization (induced and remanent), conductivity, chargeability, radioactivity, and seismic velocity. The latter has generally been associated with the petroleum industry, though in recent years it has found increasing application in the search for orebodies.

Sometimes the desired deposit-type or mineralized target has a physical property (or properties) that permits direct discovery, for example Pb-Zn deposits (galena-sphalerite) have high densities that may be detected directly by a gravity survey. On the other hand, many base metal deposits are discovered by virtue of the physical properties of an associated non-economic mineral. A good example is a Cu-rich (chalcopyrite) volcanic massive sulphide (VMS) deposit targeted on the basis of a strong magnetic anomaly produced by genetically related pyrrhotite. Technically, this should be

Ford, K., Keating, P., and Thomas, M.D., 2007, Overview of geophysical signatures associated with Canadian ore deposits, *in* Goodfellow, W.D., ed., Mineral Deposits of Canada: A Synthesis of Major Deposit-Types, District Metallogeny, the Evolution of Geological Provinces, and Exploration Methods: Geological Association of Canada, Mineral Deposits Division, Special Publication No. 5, p. 939-970.

considered as indirect detection. However, because of the close spatial and genetic relationship between the pyrrhotitic orebody and the desired commodity (chalcopyrite), we consider this as a direct discovery, and would include in the same category the discovery of diamonds, which have a spatial and genetic association with their kimberlitic hosts. Again, signatures of physical and/or chemical properties of alteration assemblages associated with mineralizing processes, which may extend well beyond the economically mineralized target (e.g. K alteration), and which in principle could be categorized as an indirect target, are also classified as direct targets for the purpose of this report.

Geophysical surveys also often provide critical information in support of local or regional framework mapping and mineral exploration modeling, and this aspect is termed framework mapping.

In this report, a brief discussion of physical and chemical properties of ore, ore-associated minerals, and their hosts is presented, together with summary descriptions of the more commonly used geophysical methods that are based on these properties. The geophysical methods and their characteristic signatures are also illustrated in the context of several mineral deposit types.

Borehole geophysical logging is another aspect of geophysical exploration that employs a suite of methodologies to measure the aforementioned physical properties or related characteristics, and also commonly measures temperature. This subject is not discussed in this presentation.

Physical and Chemical Properties of Ore Minerals, Ore-Related Minerals and Common Lithologies

It has been noted that contrasts between the physical properties of ore minerals, together with ore-related minerals, and host rocks are critical to the successful application of geophysical methods. Chemical properties are included to cover properties measured by gamma-ray spectrometry surveys, which estimate abundances of K, Th, and U. Densities, magnetic susceptibilities, and electrical conductivities for several ore minerals, ore-associated minerals, and common rock types that may host mineral deposits are presented in Table 1. A complementary list of radioelement concentrations for several classes of rocks is provided in Table 2.

Principal Geophysical Exploration Methods

Geophysical methods have been applied in the search and delineation of orebodies in Canada since at least the beginning of the 20th century. Belland (1992), for example, reported use of a dip-needle magnetometer to map the Austin Brook iron deposit in the Bathurst Mining Camp, New Brunswick in 1903/04. Most of the first half of the 20th century witnessed the development and application of the magnetic, gravity, electrical, seismic, and radiation methods, all applied on the ground. Then, in the late 1940s, following World War II, magnetic and electromagnetic methods became airborne as a consequence of new technology developed during the war. Since that time, geophysical exploration has evolved tremendously in terms of instrumentation, acquisition, processing, global positioning, and analysis, benefiting from the parallel evolution of electronic computers. Examples of modern fixed-wing and rotary-wing aircraft

TABLE 1. Physical properties (density, magnetic susceptibility, electrical conductivity) of some common rock types, ore minerals, and ore-related minerals (from Thomas et al., 2000).

	Density g/cm^3			Magnetic SI x 10^{-3}			Conductivity mS/m		
Rock Type	Min.	Max.	Av.	Min.	Max.	Ave.	Min.	Max.	Av.
Sediments and Sedimentary Rocks									
Overburden			1.92						
Soil	1.2	2.4	1.92	0.01	1.26				
Clay	1.63	2.6	2.21				10	300	
Glaciolacustrine Clay						0.25	10	200	
Gravel	1.7	2.4	2				0.1	2	
Sand	1.7	2.3	2				0.1	2	
Glacial Till							0.5	20	
Saprolite (mafic volcanic rocks, schist)							50	500	
Saprolite (felsic volcanic rocks, granite, gneiss)							5	50	
Sandstone	1.61	2.76	2.35	0	20	0.4	1	20	
Shale	1.77	3.2	2.4	0.01	18	0.6	30	200	
Argillite							0.07	83.3	
Iron Formation							0.05	3300	
Limestone	1.93	2.9	2.55	0	3	0.3	0.01	1	
Dolomite	2.28	2.9	2.7	0	0.9	0.1	0.01	1	
Conglomerate							0.1	1	
Greywacke	2.6	2.7	2.65				0.09	0.24	
Coal			1.35			0.03	2	100	
Red Sediments			2.24	0.01	0.1				
Igneous Rocks									
Rhyolite	2.35	2.7	2.52	0.2	35				0.04
Andesite	2.4	2.8	2.61			160			
Granite	2.5	2.81	2.64	0	50	2.5			
Granodiorite	2.67	2.79	2.73						
Porphyry	2.6	2.89	2.74	0.3	200	60			
Quartz Porphyry				0	33	20	0.04	1.7	
Quartz Diorite	2.62	2.96	2.79						
Quartz Diorite, Dacite				38	191	83			
Diorite	2.72	2.99	2.85	0.6	120	85			
Diabase	2.5	3.2	2.91	1	160	55			0.03
Olivine Diabase						25			
Basalt	2.7	3.3	2.99	0.2	175	70			0.2
Gabbro	2.7	3.5	3.03	1	90	70			0.02
Hornblende Gabbro	2.98	3.18	3.08						
Peridotite	2.78	3.37	3.15	90	200	250			
Obsidian	2.2	2.4	2.3						
Pyroxenite	2.93	3.34	3.17			125			
Monzonite, Latite				33	135	85			
Acid Igneous Rocks	2.3	3.11	2.61	0	80	8			
Basic Igneous Rocks	2.09	3.17	2.79	0.5	97	25			
Mafic Volcanic Rocks							0.09	0.27	
Dacite	2.35	2.8	2.58						
Phonolite	2.45	2.71	2.59						
Trachyte	2.42	2.8	2.6	0	111	49			
Nepheline Syenite	2.53	2.7	2.61						
Syenite	2.6	2.95	2.77	0	111	49			
Anorthosite	2.64	2.94	2.78						
Norite	2.7	3.24	2.92						
Metamorphic Rocks									
Quartzite	2.5	2.7	2.6			4			
Schist	2.39	2.9	2.64	0.3	3	1.4			
Marble	2.6	2.9	2.75						
Serpentine	2.4	3.1	2.78	3	17				
Slate	2.7	2.9	2.79	0	35	6			
Gneiss	2.59	3	2.8	0.1	25				
Amphibolite	2.9	3.04	2.96			0.7			
Eclogite	3.2	3.54	3.37						
Granulite	2.52	2.73	2.65	3	30				
Phyllite	2.68	2.8	2.74			1.5			
Quartz Slate	2.63	2.91	2.77						
Chlorite Schist	2.75	2.98	2.87						
Skarn	2.95	3.15				2.5			1.25
Hornfels	2.9	3				0.31			0.05
Sulphide Minerals									
Chalcopyrite	4.1	4.3	4.2	0.02	0.4		1.11	6.67	
Galena	7.4	7.6	7.5			-0.03	1.11	1.47	
Pyrite	4.9	5.2	5	0.03	5.3		1.67	8.33	
Pyrrhotite	4.9	5.2	5		3200		6.25	5.00	
Sphalerite	3.5	4	3.75	-0.03	0.75		0.08	3.70	
Other									
Magnetite			5.04	1000	5700				1.92
Graphite	1.9	2.3	2.5	-0.08	0.2		1.01	3.57	

Compiled from Grant and West (1965), Keller and Frischknecht (1966), Carmichael (1982) Hunt et al. (1995), Palacky (1986), and Telford et al. (1990).

fitted with magnetometer, electromagnetic, and gamma-ray spectrometer installations are illustrated in Figure 1.

TABLE 2. Radioelement concentrations in different classes of rocks (after Killeen, 1979).

	Potassium (%)		Uranium (ppm)		Thorium (ppm)	
Rock Type	Mean	Range	Mean	Range	Mean	Range
Acid Extrusives	3.1	1.0-6.2	4.1	0.8-16.4	11.9	1.1-41.0
Acid Intrusives	3.4	0.1-7.6	4.5	0.1-30.0	25.7	0.1-253.1
Intermediate Extrusives	1.1	1.1-2.5	1.1	0.2-2.6	2.4	0.4-6.4
Intermediate Intrusives	2.1	0.1-6.2	3.2	0.1-23.4	12.2	0.4-106.0
Basic Extrusives	0.7	0.06-2.4	0.8	0.03-3.3	2.2	0.05-8.8
Basic Intrusives	0.8	0.01-2.6	0.8	0.01-5.7	2.3	0.03-15.0
Ultrabasic	0.3	0-0.8	0.3	0-1.6	1.4	0-7.5
Alkali Feldspathoidal Intermediate Extrusives	6.5	2.0-9.0	29.7	1.9-62.0	133.9	9.5-265.0
Alkali Feldspathoidal Intermediate Intrusives	4.2	1.0-9.9	55.8	0.3-720.0	132.6	0.4-880.0
Alkali Feldspathoidal Basic Extrusives	1.9	0.2-6.9	2.4	0.5-12.0	8.2	2.1-60.0
Alkali Feldspathoidal Basic Intrusives	1.8	0.3-4.8	2.3	0.4-5.4	8.4	2.8-19.6
Chemical Sedimentary Rocks	0.6	0.02-8.4	3.6	0.03-26.7	14.9	0.03-132.0
Carbonates	0.3	0.01-3.5	2	0.03-18.0	1.3	0.03-10.8
Detrital Sedimentary Rocks	1.5	0.01-9.7	4.8	0.01-80.0	12.4	0.2-362.0
Metamorphosed Igneous Rocks	2.5	0.1-6.1	4	0.1-148.5	14.8	0.1-104.2
Metamorphosed Sedimentary Rocks	2.1	0.01-5.3	3	0.1-53.4	12	0.1-91.4

Magnetic Method

The magnetic method is the oldest and most widely used geophysical exploration tool. An early discovery of a Canadian ore deposit by an aeromagnetic survey was that of the Marmora iron ore deposit in Ontario in 1949 (Reford, 1980). The effectiveness of the method depends mainly on the presence of magnetite in the rocks of the surveyed area. Another important magnetic mineral is pyrrhotite. The primary goal of magnetic surveys is direct detection of metallic orebodies through delineation of associated anomalies. Another objective is to determine trends, extents, and geometries of magnetic bodies in an area, and to interpret them in terms of geology. The technique is particularly effective in areas where few outcrops exist. Structural trends are faithfully reproduced in magnetic patterns, but assignment of rock type is ambiguous, since ranges of values of magnetic susceptibilities of different rock types may overlap (Table 1).

FIGURE 1. **(A)** Cessna 208B Caravan aircraft fitted with a magnetometer in the rear stinger. Image source: Sander Geophysics web site (www.sgl.com/pic_air.htm); **(B)** de Havilland Dash 7 aircraft fitted with a MEGATEM electromagnetic system (Fugro Airborne Surveys). Photo courtesy of Régis Dumont, Geological Survey of Canada; **(C)** Large volume (50.4 litres) airborne gamma ray spectrometer installation in a Cessna 208B Caravan aircraft. Photo courtesy of R.B.K. Shives, Geological Survey of Canada; **(D)** Eurocopter Astar 350B3 helicopter with a magnetometer in the forward mounted stinger and a gamma ray spectrometer system consisting of 33 litres of NaI detectors.

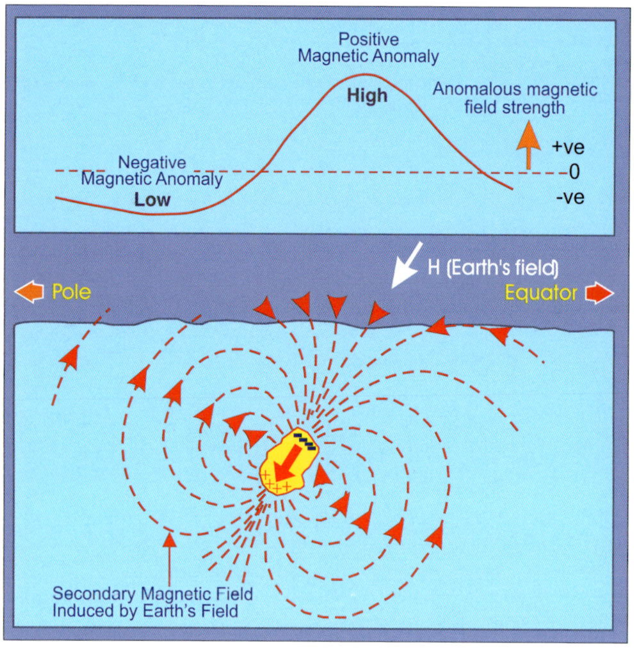

FIGURE 2. Induced magnetization in a geological body produces a local positive magnetic anomaly and a subsidiary negative anomaly.

Susceptibility may vary considerably, even within the same rock type.

In magnetic surveys, the intensity or strength of the Earth's total magnetic field is measured. The intensity of the total magnetic field over Canada ranges from about 52000 to more than 60000 nT. The total field includes contributions from the Earth's core and crust. A third component, originating from electrical currents in the upper atmosphere, is normally eliminated from survey data during processing. It is common practice to subtract the component of the field attributed to the Earth's core, described mathematically by the International Geomagnetic Reference Field, from the total field. The resultant field is termed the residual total magnetic field, and depending on location, values may be positive or negative. The Earth's magnetic field induces a secondary magnetic field in magnetic geological bodies, which locally enhances the Earth's field producing local positive culminations or anomalies (Fig. 2). Because of the dipolar nature of magnetization, negative anomalies may also be generated, though at high magnetic latitudes, such as within Canada, they are generally of low amplitude. The presence of reversed remanent magnetizations can, however, produce prominent negative anomalies.

Several derivatives of the residual total magnetic field provide value-added products that may contribute to the geological interpretation of magnetic data. They include maps of the first vertical derivative (= vertical gradient), 2nd vertical derivative, analytic signal, and tilt. The first vertical derivative (nT/m) is probably the most commonly used derived product. It may be derived mathematically from the total magnetic field, or alternatively it may be measured directly using a gradiometer, comprising two magnetometers separated vertically by 2 to 3 metres. Vertical gradient maps present a filtered picture of the magnetic field, emphasizing near-surface geological features. Gradient anomalies are narrower than corresponding total field features, hence magnetic anomalies produced by closely spaced geological units are better resolved by the vertical gradient. Vertical gradient maps are useful for mapping geological contacts, since theoretically the zero contour of the gradient coincides with contacts between contrasting magnetizations, provided the contacts are steep and the area is in high magnetic latitudes (Hood and Teskey, 1989).

Electrical Methods

Electrical methods, in common with the electromagnetic techniques, respond to the electrical conductivity of rocks and minerals, which may vary by 20 orders of magnitude (Grant and West, 1965). No other physical property varies that much. Native metals, such as copper and silver, are highly conductive, whereas minerals such as quartz are, for all practical purposes, nonconductive. Rock and mineral conductivity is a complex phenomenon. Current can be propagated in three different ways by electronic (= ohmic), electrolytic or dielectric conduction. Electronic conduction is effected by the presence of free electrons, and is the means by which current flows in metals. In electrolytic conduction, the current is carried by ions, and flows at a slower rate. In dielectric conduction, the current is known as the displacement current. In this case, there are very few or no current carriers and the electrons are slightly displaced relative to the atomic nuclei by an externally varying electric field. This very small separation between negative and positive charges is known as the dielectric polarization and produces the displacement current.

The electrical conductivity of rocks and mineral deposits is commonly measured in milliSiemens per metre (mS/m). Granite is essentially non-conductive, whereas the conductivity of shale ranges from 0.5 to 100 mS/m. Water content increases conductivity and may have a dramatic influence on its magnitude. Wet and dry tuffs, for example, have conductivities that differ by a factor of 100 (Telford et al., 1990). Different rock types have overlapping ranges of conductivity. The conductivities of massive sulphides may overlap those of other, nonmineralized materials such as graphite and clays. Conductive overburden, especially water-saturated clays, may generate electromagnetic anomalies that effectively mask the response of an underlying massive sulphide zone. Unequivocal identification of mineral deposits is, therefore, a difficult task. Where the conductivity of overburden is sufficiently uniform, the electromagnetic responses can be interpreted in terms of overburden thickness. Conductivity and resistivity are the inverse of each other, and both terms are commonly used.

Conductivities of common rocks and minerals are listed in Table 1. Massive sulphides, graphite, and salt water have high conductivities, exceeding 500 mS/m. Intermediate values, between 1 and 500 mS/m, are typical of sedimentary rocks, glacial sediments, weathered rock, alteration zones, and fresh water. Igneous and metamorphic rocks have low conductivities, less than 1 mS/m.

A number of electrical (and electromagnetic) techniques have been developed to take advantage of the high variability of rock and mineral conductivities. Electrical methods are applied on the ground, whereas electromagnetic methods may be employed on the ground or in the air using different airborne platforms. Some, like self-potential and the magnetotelluric method, use natural fields, while others such as DC resistivity or electromagnetic techniques use artificial sources. Here, discussion is restricted to the principal techniques used for mineral exploration.

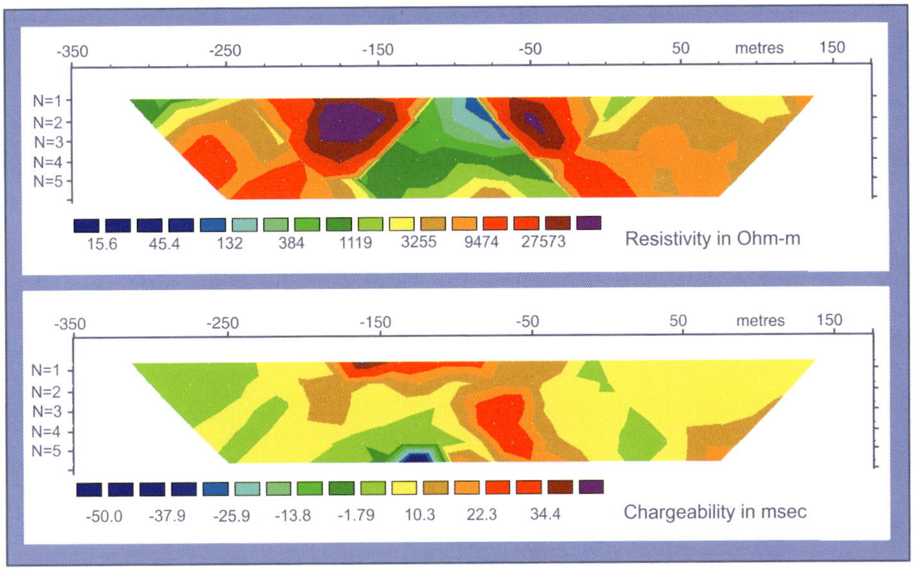

FIGURE 3. Pseudosections of resistivity and chargeability derived from a dipole-dipole resistivity-induced polarization survey. Electrode spacing varies from 1 to 5 m.

Electrical Resistivity Surveys

Measurements of resistivity, or rather apparent resistivity, represent one of the oldest geophysical survey techniques. Resistivity surveys are achieved by injecting current into the ground using two electrodes, and measuring the voltage at two other electrodes. Various electrode configurations may be used, but in all cases it is possible to compute the apparent resistivity of the subsurface at various depths, and the data can be used to generate a cross-section of the true resistivity (Fig. 3). The method is used both for the direct detection of orebodies, such as Mississippi Valley-type sulphide bodies for example, and to define the 3-D geometry of a target, such as a kimberlite pipe. Resistivity surveys are also used to map overburden thickness to help improve interpretation of ground gravity surveys. Conductivity, the inverse of resistivity, is commonly estimated from airborne electromagnetic data and used to map overburden.

Induced Polarization Surveys

Induced polarization (IP) has been used in mineral exploration since the mid-1950s. It is a rather complex phenomenon, but easy to measure. IP measures the chargeability of the ground, i.e. how well materials tend to retain electrical charges. Measurements are made either in the time-domain or the frequency-domain; their units are respectively milliseconds (msec) and percentage frequency effect (PFE). When a voltage applied between two electrodes is abruptly interrupted, the electrodes used to monitor the voltage do not register an instantaneous drop to zero, but rather record a fast initial decay followed by a slower decay. If the current is switched on again, the voltage will first increase at a very high rate and then build up slowly. This phenomenon is known as induced polarization. The technique is mostly concerned with measuring the electrical surface polarization of metallic minerals. This effect is induced by abrupt changes in electrical currents applied to the ground. Disseminated sulphides produce very good induced polarization responses. In theory, massive sulphides should have lower responses, but in practice they have very good responses. This is due to the mineralization halo generally surrounding massive sulphides. Clay minerals may also produce significant IP responses. More details can be found in various textbooks (e.g. Telford et al., 1990).

Induced polarization surveys are carried out along equally spaced lines perpendicular to the main geological strike. Two current electrodes are used to inject current into the ground, and two voltage electrodes are used to measure the decay voltage. Resistivity measurements are made concurrently. Various electrode layouts can be used (pole-dipole, dipole-dipole, etc.); varying the distance between the electrodes results in soundings to different depths, which may be used to map the variability of resistivity and chargeability with depth (Fig. 3). For dipole-dipole surveys, the distance between pairs of electrodes is kept constant and the separation between the voltage and current electrodes is increased. This distance is increased by integer multiples "n" of the distance between the voltage electrodes.

A new system, called Titan-24 developed by Quantec Geoscience, combines IP, resistivity, and magnetotelluric techniques, the latter using the Earth's natural field. This system can detect conductive deposits at a depth of 1.5 km.

Electromagnetic Methods

Electromagnetic (EM) techniques, both airborne and ground, are among the most commonly used methods in mineral exploration. They are capable of direct detection of conductive base-metal deposits, where large conductivity contrasts exist between the deposits and resistive host-rocks or thin overburden cover. The techniques have been highly successful in North America and Scandinavia. A multitude of other conductive sources, including swamps, shear and fracture zones, faults, and graphitic and barren metallic conductors, create a major source of ambiguity in the interpretation of EM anomalies.

Electromagnetic systems operate in either the frequency or the time domain. In either method, an EM field is transmitted, which on penetrating the ground and encountering conductive material generates a secondary field that may be measured by a receiver. The concept is illustrated for an airborne time-domain system in Figure 4. Different combinations and geometries of transmitters and receivers may be used.

In frequency-domain systems, a transmitter creates an alternating EM field. This primary field generates eddy cur-

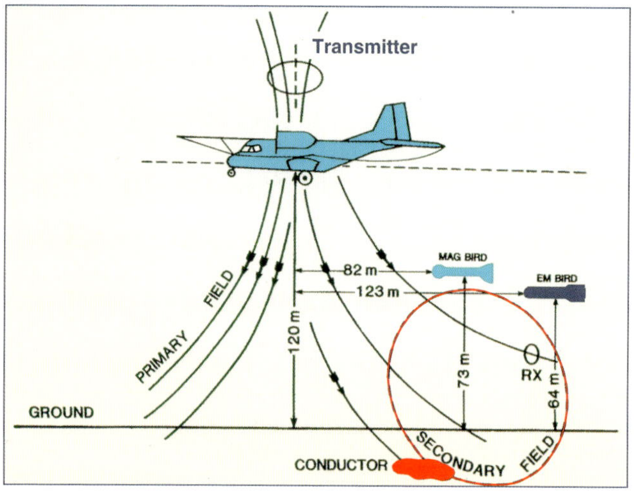

FIGURE 4. Schematic of one type of airborne electromagnetic system – in this case, a time-domain system (modified after Rowe et al., 1995).

rents in a conductive medium, which in turn produces a secondary EM field. This secondary field, detected by the receiver, is diagnostic of the electrical characteristics of the conductive medium excited by the primary field. In general, the secondary field is not in phase with the primary field. The EM receiver measures the in-phase and out-of-phase (quadrature) components of the secondary field, and the ratio of the secondary field to the primary field in parts per million (ppm). Time-domain electromagnetic systems transmit very short pulses. In general, these pulses last for a few milliseconds and are a few milliseconds away from each other. Pulses can have various shapes: half-sine, step, ramp, etc. For example, the MEGATEM airborne time-domain EM system often operates with a base frequency of 90 Hz, it uses half-sine pulses of 2 milliseconds, and pulses of opposite polarity are transmitted every 5.5 milliseconds. All such systems have a broad frequency range. The secondary field time-decay curve begins immediately at the end of the transmitter cut-off.

As EM responses are frequency dependent, modern airborne electromagnetic (AEM) systems use a wide range of frequencies to detect a large range of conductivities. A number of configurations for the transmitter and receiver coils are used to discriminate between horizontal and vertical conductors. The coplanar coil pair is more sensitive to horizontal conductors, while the coaxial coil pair is more sensitive to vertical conductors. The geometry and attitude (dip) of conductors also influence the shape of the anomalies. For a symmetric EM system, a vertical conductor produces a double peak anomaly when detected by the coplanar coil pair, and a single peak anomaly when measured by the coaxial coil pair (Fig. 5). It should be noted that fixed-wing time-domain electromagnetic systems are asymmetric because the receiver is towed behind and below the transmitter; this results in asymmetric responses that can also be interpreted in terms of dips (Fig. 6).

In the case of frequency-domain EM systems, a geologic formation that has high magnetic susceptibility and low conductivity will generate a large magnetic anomaly and a strong negative in-phase EM response that is opposite to the polarity of the response of a conductive body. In this situation, the AEM response is likely to reflect a shallower portion of the causative body than the magnetic response. It is also possible to determine the magnetic susceptibility of a rock and evaluate its magnetite content from AEM data (Fraser, 1973).

Airborne EM data can be converted into ground conductivities to produce a conductivity map. To obtain stable results, a number of simplifying assumptions or models must be employed. A homogeneous half-space model (Seigel and Pitcher, 1978), a model bounded by one plane surface (the upper part is infinitely resistive, e.g. air, while the bottom part has a finite resistivity and represents the Earth), assumes that the conductivity of the ground is uniform and that the Earth is flat and bottomless. This model is robust and generally provides realistic results. A second model is based on a single-layer earth, in which a homogeneous conductive layer of uniform thickness overlies a homogeneous half-space.

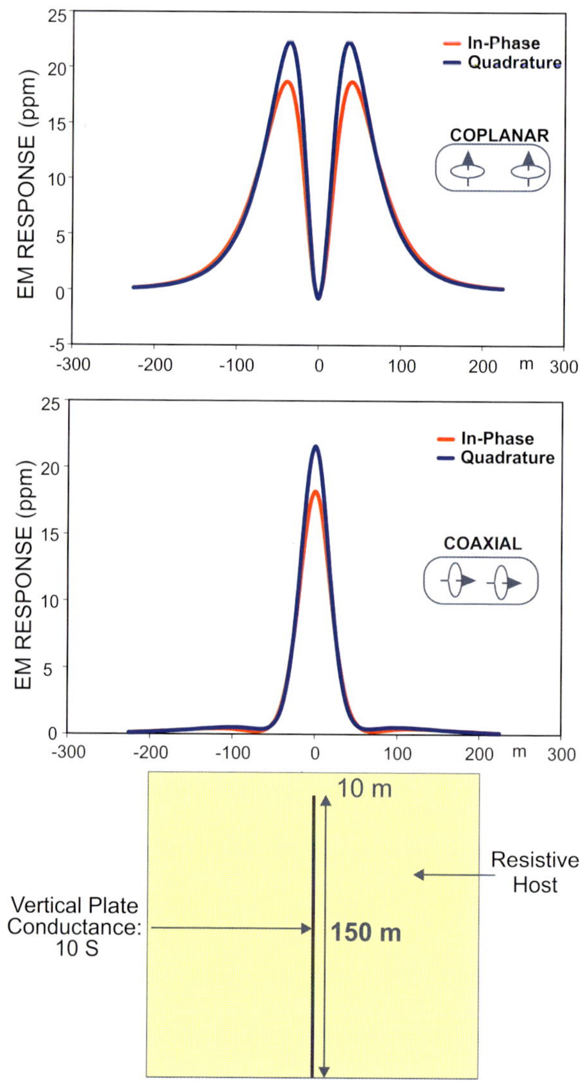

FIGURE 5. Electromagnetic in-phase and quadrature responses for a conductive thin vertical plate (conductance = 10 S; strike length = 300 m) hosted by a resistive (unresponsive) medium. Computation of the profiles assumes a transmitter frequency of 914 Hz and a survey elevation of 30 m above ground surface; the top of the plate is 10 m below surface. Transmitter-receiver coil separation is 7 m.

However, when the top layer is conductive and relatively thin, only the conductivity-thickness product can be determined. Conductance may be determined using a vertical half-plane model (Ghosh, 1972) in free space.

Apparent conductivity may be calculated for any measured frequency, different frequencies providing conductivity information for different parts of the crustal column. For example, in the Bathurst Mining Camp, conductivities calculated from mid-frequency (4433 Hz) coplanar EM data using a homogeneous half-space model best reflect the bedrock geology of the area. Conductivities calculated from low-frequency data (853 and 914 Hz) are generally associated with lithologies having high conductivity contrasts, whereas conductivities calculated from the high frequency EM data (32290 Hz) usually contain significant overburden responses.

The maximum depth at which a deposit can be detected by EM techniques depends on the size of the deposit, its conductivity, and the conductivity of the host rock and the overburden. Typically airborne time-domain EM can detect deposits/conductors at a depth of 250 m in the Abitibi mining belt, and as deep as 750 m in the Athabasca Basin. Ground time-domain EM techniques using large loops can detect conductive targets at depths in excess of 1 km.

Gamma-Ray Spectrometric Method

Early instruments measured only the total radioactivity from all sources, and were used primarily for uranium exploration. The first portable Geiger-Muller counters were built in 1932 at the University of British Columbia by G.M. Shrum and R. Smith and were tested on Quadra Island off the coast of British Columbia (Lang, 1953). The first practical portable counters were developed in the 1940s at the National Research Council (NRC) and used in 1944 by joint Geological Survey of Canada and NRC field crews on Great Bear Lake (Lang, 1953). The first scintillation counters were developed in 1949 and first used in ground radiation surveys of pitchblende deposits in the Lake Athabasca region (Brownell, 1950). In the 1960s, gamma-ray scintillation spectrometers were developed and calibrated to measure discrete windows within the spectrum of gamma-ray energies. This permitted determination of concentrations of individual radioelements.

Potassium (K), uranium (U), and thorium (Th) are the three most abundant, naturally occurring radioactive elements (Table 2). Their abundances (Table 2) and chemical properties provide insight into ore-related processes. For example, K is a major constituent of most rocks and is diagnostic of alteration associated with many mineral deposits; U and Th, usually present in trace amounts, are relatively mobile and immobile elements, respectively, under most surface conditions and are effective in the direct detection of mineralized rocks. Given the various physical and chemical characteristics of the radioactive daughter products in the ^{238}U decay series, disequilibrium can be a significant source of error. Disequilibrium occurs when one or more of the daughter products in the decay series is removed or added. In areas with extreme surficial weathering, disequilibrium is a significant concern. In Canada, due to recent glaciation, disequilibrium due to weathering is less of a concern.

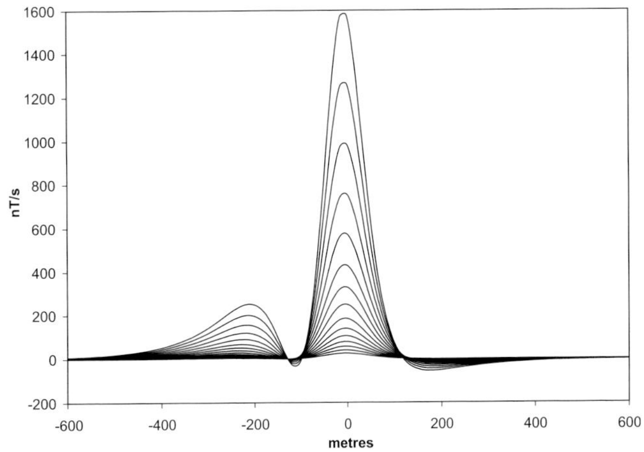

FIGURE 6. Response of the x-component (along line) of an airborne electromagnetic time-domain system (Megatem II). The conductor is a thin vertical plate having a conductance of 20 S, located at x = 0 and 120 m under the transmitter. The plane is flying from left to right. Note the presence of 2 peaks. The leading peak is an effect of the asymmetry of the system. The receiver is located in a bird 50 m under the transmitter and 130 m behind it.

Concentrations of uranium and thorium, when measured by gamma-ray spectrometry, are often reported as "equivalent uranium" (eU) and "equivalent thorium" (eTh) as a reminder that the measurement is based on the assumption that equilibrium exists. However, the Th decay series is almost always in equilibrium.

The effectiveness of gamma-ray spectrometry in geological mapping and mineral exploration depends on several factors: the extent to which measurable differences in the radioactive-element distributions relate to bedrock differences (normal lithological signatures); the extent to which these normal signatures are modified by mineralizing processes (K alteration); the extent to which the ore or associated alteration signatures are incorporated into surficial materials; and whether these surficial materials can be spatially related to bedrock sources. A schematic illustration of gamma-ray spectrometry anomalies associated with a porphyry Cu deposit is presented in Figure 7.

The attenuation of gamma rays by rock or soil prevents their emanation from depths greater than approximately 60 cm. This has profound implications for the interpretation of gamma-ray spectrometry surveys. Whereas magnetic, electromagnetic, gravimetric, or seismic sensors may detect features to depths of tens or hundreds of metres, often buried far below the mappable near-surface geology, gamma-ray data are interpretable in terms of the near-surface chemical composition of rocks and soils. A poor correlation between K, eU, and eTh distributions and other layers, such as magnetic or geological layers, reflects the inherently different sampling methods, and greater depth of penetration of the latter. In airborne surveys, a single measurement provides an estimate of the average surface concentration for an area of several thousand square metres. This single sample comprises variable proportions of exposed bedrock (fresh or weathered), overburden (wide variety possible – glacial tills, glacio-fluvial, colluvium, alluvium, loess), soils (clay, sand, loam, etc.) soil moisture, standing water (lakes, rivers, swamps, bogs), and vegetation. In almost all cases, the inhomogeneous nature of the surface (and therefore of each sam-

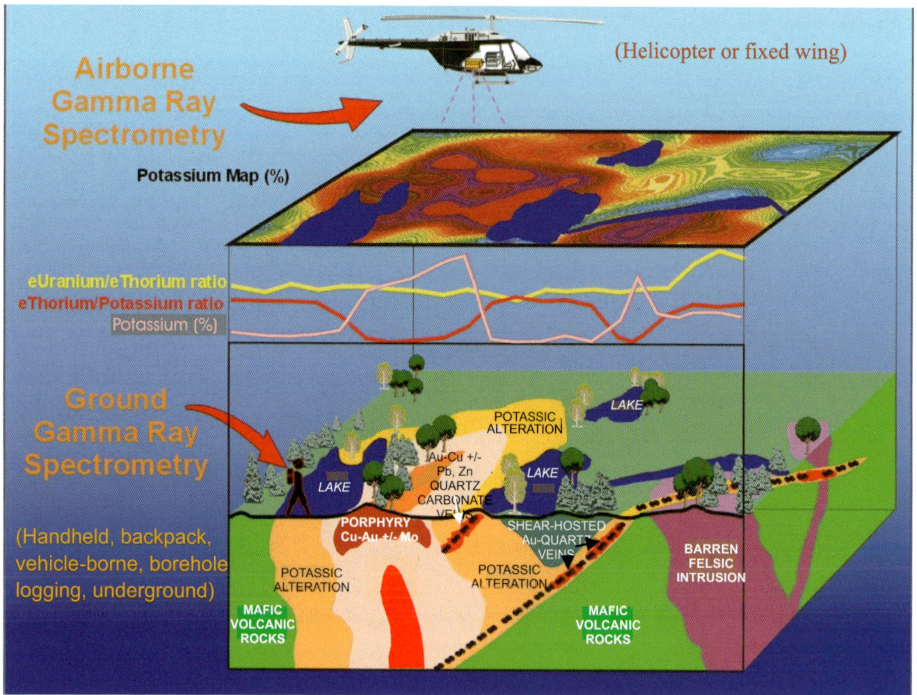

FIGURE 7. Schematic image of airborne and ground gamma-ray spectrometry anomalies associated with a porphyry Cu deposit. Figure courtesy of R.B.K Shives, Geological Survey of Canada.

ple) introduces variability such that the absolute value of K, eU, or eTh may be far less significant than the radioelement patterns. While K, eU, and eTh concentrations are considered to be the primary measured variables of a gamma-ray spectrometry survey, several derived products can provide important value-added information. These include total radioactivity (measured or derived), radioactive element ratios (eU/eTh, eU/K, and eTh/K), and the ternary radioelement map (Broome et al., 1987).

The effects of variations in soil moisture, amount of bedrock exposure, or source geometry can be minimized through the use of radioelement ratios. In mineralized systems, all three radioelements may be enriched or depleted relative to unaltered host rocks. However, one or more of the radioelements may be preferentially affected for valid geochemical reasons. In these cases, the ratios offer very sensitive alteration vectors, which may lead directly or indirectly to the mineralization, even where the individual radioelement patterns are ambiguous.

Gamma-ray spectrometry is particularly effective in granite- and gneiss-dominated terrains where radioactive element concentrations and contrasts are usually the highest (Table 2). However, depending on the flight-line spacing and radioactive element contrasts between different rock types, the technique may be effective in a variety of terrains. When airborne surveys employ both a magnetometer and gamma-ray spectrometer, the two techniques are often complementary, particularly in areas of complex geology.

Gravity Method

Gravity observations provide a measure of the Earth's gravity field, which is sensitive to variations in rock density. Local mass excesses or deficiencies produce, respectively, increases or decreases in the gravity field. These departures from the immediate background level are termed positive (Fig. 8) and negative anomalies, respectively. The unit of measurement used in geophysical studies is the milligal (mGal). In exploration for base metals, the gravity technique is commonly applied in follow-up investigations of magnetic, electrical, electromagnetic, or geochemical anomalies, and is particularly useful in assessing whether a conductivity anomaly is related to low-density graphite or a higher density sulphide deposit. It is also used as a primary exploration tool to detect the excess masses of base metal sulphide deposits. Gravity data may be used to estimate the size and tonnage of orebodies, and also contribute to exploration programs when used to map geology and structure that may favour the presence of ore deposits. Traditionally, gravity surveys have been carried out on the ground or on ice during winter. Surveys on water bodies have been accomplished on board ships, typically using specially designed marine gravity meters, and in some cases remotely controlled land gravity meters adapted for deployment on the ocean floor or a lake bottom (e.g. Goodacre et al., 1969). In recent years, significant improvements in the acquisition of airborne gravity data have made airborne surveys more attractive, in spite of the lower accuracy and resolution of the data. On the plus side, airborne surveys provide rapid coverage of large areas along flight-lines spaced much closer together than the spacing between available ground gravity observations that have been made as part of Canada's national gravity mapping program. Airborne surveys, therefore, provide a viable tool for upgrading gravity coverage in

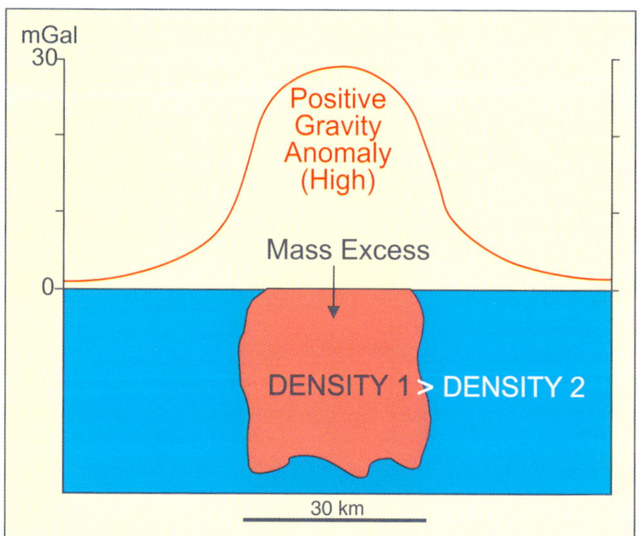

FIGURE 8. An excess mass in the crust locally enhances the gravity field producing a positive anomaly or gravity high.

most parts of Canada, and for establishing regional to semi-regional geological frameworks.

More promising for the direct detection of orebodies is airborne "tensor gravity". The method measures the gravity tensor, which includes nine tensor components corresponding to gravity gradients along three orthogonal directions, using a highly sensitive gravity gradiometer. One of the components of the tensor is the vertical gravity gradient. The tensor method has high resolution and can detect small targets, such as kimberlites, which characteristically have diameters of only a few hundred metres. At the present time, the method is expensive and the number of contractors capable of offering this service is limited. Lane (2004) presents a review and descriptions of the current state of the art of airborne systems.

Besides variations in crustal density, the gravity field is influenced by latitudinal position and changes in elevation. Therefore, several corrections are applied to observed gravity data to isolate variations related to crustal causes. A small correction is made to eliminate the effect of Earth tides. Two corrections are made to negate the elevation factor: the free-air correction (compensates for variation in the distance of the measurement point from the Earth's centre of mass), and the Bouguer correction (eliminates the gravity effect of the rock mass between the observation point and the datum). These corrections are applied relative to a vertical datum, commonly sea level. The difference between the corrected gravity observation and the theoretical value of gravity on the reference ellipsoid at the observation point is known as the Bouguer gravity anomaly, which is the gravity parameter most commonly displayed on gravity maps.

Airborne Survey Specifications and Instrumentation

Airborne surveys are normally flown in a direction perpendicular to the main geological strike of the survey area. Line spacing depends on the objectives of the survey: in regional surveys, it commonly varies between 400 m and 1 km, whereas in detailed surveys flown for mineral exploration it may be as little as 100 or 200 m. Flight elevation in fixed-wing airborne surveys with magnetic and electromagnetic sensors varies typically from 150 to 300 m for regional surveys, and is about 100 m for detailed surveys. The flight elevation in fixed-wing surveys that include gamma-ray spectrometry is normally 120 m, although it may vary between 100 and 150 m. In the case of helicopter-borne surveys, the flight elevation is generally lower, at around 60 m, and the magnetic and electromagnetic sensors are suspended by cables at elevations of 45 and 30 m above ground, respectively. For helicopter-borne surveys with gamma-ray spectrometry, the detectors are mounted in the helicopter and the nominal terrain clearance varies between 60 and 90 m, depending on local terrain conditions and the configuration of other survey equipment. Differential GPS navigation is used, and the estimated accuracy of the flight path is better than 10 m. A vertically mounted video camera is normally used for verification of the flight path.

Magnetometer System

Split-beam cesium vapour magnetometers, having a sensitivity of 0.005 nT, are commonly used for magnetic surveying. Magnetic data are recorded every 0.1 second. The magnetic data collected along the survey lines and the control lines are corrected for temporal (diurnal) variations in the magnetic field using ground-station magnetometer data. After editing the survey data, differences in magnetic values between traverse and control lines, established at intersections, are computer-analysed to obtain the levelling network. The magnetic data are then interpolated to a square grid having a dimension equal to approximately a quarter of the line spacing.

Electromagnetic System

Electromagnetic (EM) systems are either frequency-domain or time-domain systems. In Canada, frequency-domain surveys are flown with a helicopter. Time-domain systems can be operated in a helicopter or a fixed-wing aircraft. Line spacing usually ranges from about 100 to 250 m. As previously noted, when using a helicopter, the EM system is positioned at a height of approximately 30 m above ground. For fixed-wing systems, the plane and the transmitter fly at 120 m above ground, and the towed receiver is 50 m above ground.

Gamma-Ray Spectrometric System

A typical gamma-ray system includes a 256 channel spectrometer sampling data at 1 second intervals. Surveys flown with a fixed-wing aircraft usually consist of 50 litres of NaI(Tl) crystals in the main detector array with between 8 and 12 litres of NaI(Tl) crystals in the upward-looking detector array. Surveys flown using a helicopter will usually consist of 33 litres of NaI(Tl) crystals in the main detector array with 8 litres in the upward-looking array. After energy calibration of the spectra, counts from the main detector are recorded in five windows corresponding to thorium (2410-2810 keV), uranium (1660-1860 keV), potassium (1370-1570 keV), total radioactivity (400-2815 keV), and cosmic radiation (3000-6000 keV). Radiation in the upward-looking detector is recorded in a radon window (1660-1860 keV). Comprehensive descriptions of airborne and ground gamma-ray spectrometry surveying including fundamentals, instrumentation, calibration, data processing, and interpretation are covered by Grasty et al. (1991), Grasty and Minty (1995), Shives et al. (1995), Dickson and Scott (1997), Horsfall, (1997), Minty (1997), Minty et al. (1997), Wilford et al. (1997), and the International Atomic Energy Agency (2003) and references therein.

Mineral Deposit Types

Geophysical signatures are presented and discussed for nine major mineral deposit types: diamonds, lode Au, volcanogenic massive sulphides (VMS), sedimentary exhalative (SEDEX) base metals, Mississippi Valley-type (MVT) Pb-Zn deposits, porphyry Cu, unconformity-related U, Olympic Dam-type Cu-Au-Fe oxides, and magmatic Ni-Cu-PGE deposits. Comprehensive descriptions of these deposit types are included elsewhere in this volume (Corriveau, 2007; Dubé and Gosselin, 2007; Dubé et al., 2007; Eckstrand and Hulbert, 2007; Galley et al., 2007; Goodfellow and Lydon, 2007; Jefferson et al., 2007; Kjarsgaard, 2007; Paradis et al., 2007; Sinclair, 2007). The locations of deposits whose signatures are illustrated are plotted in Figure 9.

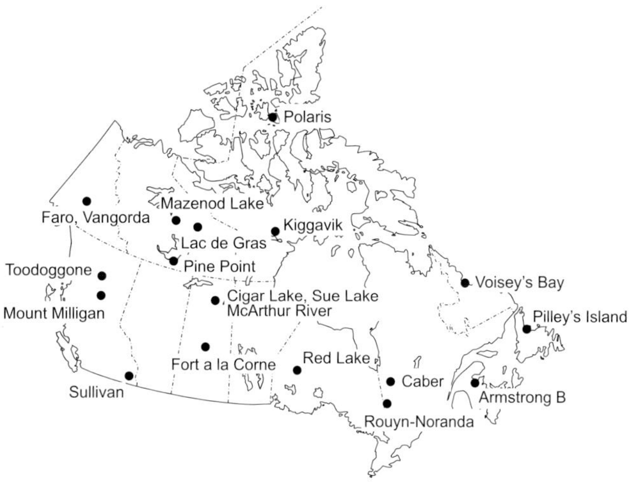

FIGURE 9. Locations of ore deposits for which geophysical signatures are illustrated.

Diamonds

Diamonds have been observed in a variety of geological environments, but typically only diamond-bearing kimberlites and lamproites, and paleoplacer and placer deposits derived from them are economic (Kjarsgaard, 1996, 2007; LeCheminant and Kjarsgaaard, 1996). Typically kimberlites occur as inverted cone-shaped diatremes or pipes. These are the principal exploration targets and will be the focus of much of the following discussion. Hoover and Campbell (1992) reported that kimberlites range in diameter from 100 to 5000 m, but are generally 400 to 1000 m in diameter, and have a depth extent of about 2000 m. In Canada, in the Slave Province on the BHP/Dia Met property, most of the kimberlites have diameters of less than 400 m (St. Pierre, 1999). In the Kirkland Lake area, kimberlites with maximum widths of less than 400 m have been reported by Brummer et al. (1992). Larger diameters have been documented in the James Bay Lowlands, ranging from 50 to 1500 m (Reed, 1993).

Many kimberlites are associated with distinct circular magnetic anomalies that are related to magnetite produced by deuteric alteration of olivine (Mitchell, 1986). Iron-rich ilmenite ($FeTiO_3$) has a magnetic susceptibility ranging from about 2 to 3800 x 10^{-3} SI (Hunt et al., 1995). Presumably the significant magnesium contents of ilmenites in kimberlites would result in lower susceptibilities, but the authors are not aware of any published values for such ilmenites. The primary magnetic mineral in the BHP/Dia Met kimberlites in the Slave Province is reported to be titanomagnetite (St. Pierre, 1999), which has a susceptibility of 130 to 620 x 10^{-3} SI (Hunt et al., 1995). Katsube and Kjarsgaard (1996) presented susceptibility values for Canadian kimberlites located in the Northwest Territories, Saskatchewan, and Ontario and indicated a range from about 1 to 100 x 10^{-3} SI. Clearly many kimberlites would produce positive magnetic anomalies, particularly where the host is sedimentary. In crystalline terrains, the sign of the anomaly would depend on the magnetic susceptibility of the host rocks.

A complication with respect to the expected sign of the magnetic anomaly is the presence of remanent magnetization, which may result in no magnetic anomaly or a negative anomaly (Hoover and Campbell, 1992), a situation observed with respect to kimberlites on the BHP/Dia Met property in the Slave Province, where significant components of remanent magnetization are evident (St. Pierre, 1999). Inspection of a detailed survey flown at a height of 120 m along lines spaced 250 m apart in the Lac de Gras area (Shives and Holman, 1995) reveals that about 20% of the kimberlite bodies produce no obvious magnetic response, whereas the responses of the remainder may be divided into approximately equal numbers of positive and negative responses. The variability in magnetic signatures in this area is illustrated in Figure 10A (e.g. the Leslie kimberlite has a positive signature, whereas the Grizzly pipe has a negative response). St. Pierre (1999) stated that the Grizzly pipe (Fig. 10B) has a strong negative remanent component. The Lac de Gras area is also covered by a regional survey flown at a height of 300 m along lines spaced 800 m apart, and in this case only 10% of the known kimberlites can be visually identified in the magnetic data. This shows the importance of flying high-resolution magnetic surveys when exploring for kimberlites. Such surveys are often flown at a line spacing of 100 m. Another factor influencing the signature is possible variability of susceptibility within a single pipe (Keating, 1996). Whether positive or negative, magnetic anomalies associated with kimberlites can be expected to be circular to oval-shaped, mimicking the cross-section shape of the kimberlite. Keating (1996) has noted that most aeromagnetic signatures of kimberlites in the Canadian Shield are circular, though ground surveys define more complex anomalies, which may be elongated and/or contain internal highs. An automatic method for detecting such circular anomalies related to kimberlites has been developed by Keating (1995).

Electromagnetic responses of kimberlites are generally associated with crater facies, which includes epiclastic rocks that may contain shales and mudstones, or with altered (weathered/metasomatized) near-surface parts of diatremes that contain secondary serpentine, calcite, dolomite, chlorite and clay minerals (Mitchell, 1986). The kimberlite is identified by the higher conductivity of the fine-grained sedimentary rocks and component clay minerals, which contrasts significantly with generally more resistive host rocks. Good examples of strong conductivity responses are observed in the Fort à la Corne area (Fig. 10D). St. Pierre (1999) reports that crater facies material in the BHP/Dia Met field has resistivities as low as 0.5 Ohm-m, which compares with typical resistivities of several thousand Ohm-m for hosting granites and gneisses. Because kimberlites in the Northwest Territories are generally located under lakes or swamps it may be difficult to determine if the observed EM response is due to the kimberlite itself, or to the increased thickness of the conductive overburden. However, some EM anomalies are clearly associated with the pipe itself; examples from the Lac de Gras area are presented by Smith et al. (1996). In the Kirkland Lake area, there are no EM anomalies associated with known kimberlites, nevertheless an increase of the conductivity due to a thicker conductive overburden is observed (Keating, 1995). Resistivities of Canadian kimberlites, published by Katsube and Kjarsgaard (1996), indicate a marked dichotomy, with values ranging from approximately 10 to 1000 Ohm-m, for crater and diatreme facies and from roughly 1000 to 100000 Ohm-m for hypabyssal facies.

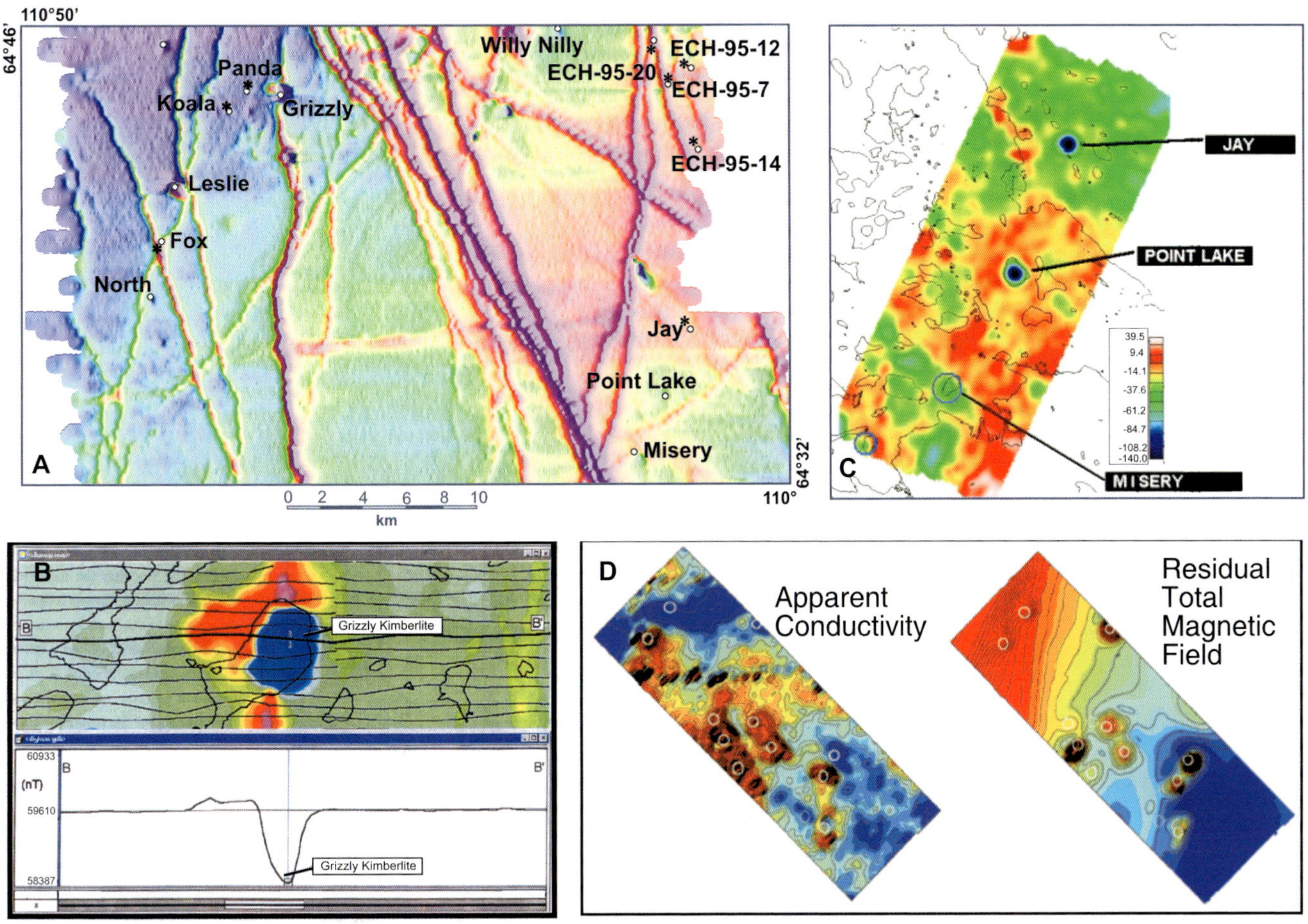

FIGURE 10. **(A)** Total magnetic field signatures of some kimberlite bodies in the Lac de Gras area. The flight elevation was 120 m and the line-spacing 250 m; **(B)** Negative total magnetic field anomaly over the Grizzly pipe, Lac de Gras area; flight elevation 30 m, flight-line spacing 125 m (adapted from St. Pierre, 1999); **(C)** Vertical gravity gradient from an airborne survey, Lac de Gras area (image from www.bhpbilliton.com); **(D)** Apparent conductivity and residual total magnetic field, Fort à la Corne area (image from www.fugroairborne.com). In all cases, the fields are represented by colour schemes that go from low values (blue shades) to high values (red shades).

Of significance to geophysical detection are the high densities of most minerals. Kjarsgaard's (1996) description of the mineralogical composition of kimberlites notes the inequigranular texture consequent on the presence of large macrocrysts and phenocrysts within a fine-grained matrix. Macrocrysts include the minerals olivine (density = 3.27-4.37 g/cm^3), Mg-ilmenite (4.7 g/cm^3), Ti-Cr-pyrope garnet (3.51 g/cm^3), clinopyroxene (3.2-3.3 g/cm^3), phlogopite (2.86 g/cm^3), enstatite (3.2-3.5 g/cm^3), and zircon (4.68 g/cm^3). Primary matrix minerals include olivine and a variety of minerals with densities that range between 2.2 g/cm^3 for primary serpentine and 4.7 g/cm^3 for ilmenite. The high densities of these minerals have the potential to create positive gravity anomalies, although this is tempered by the presence of the low density serpentine, which may have an overriding influence and produce negative anomalies. In fact, Hoover and Campbell (1992) reported that gravity anomalies associated with kimberlites are generally negative (of the order of 1 mGal amplitude), as a result of serpentinization and weathering of the constituent mafic rocks. In the Slave Province, St. Pierre (1999) noted mean densities of 1.9, 2.2, and 2.9 g/cm^3 for crater, diatreme, and hypabyssal facies, respectively, of several kimberlites. Contrasted against a mean background density of 2.7 g/cm^3 of hosting granites and gneisses, these values indicate the potential for both negative and positive gravity anomalies. However, since most kimberlites in the area comprise crater and diatreme facies, negative gravity anomalies prevail. In the Fort à la Corne area, kimberlites have a positive gravity signature because they are hosted in lighter sedimentary rocks (Cretaceous shales).

Airborne gravity gradiometry surveys have been successful in delineating kimberlites in the Lac de Gras area. Distinct signatures were imaged over the Jay and Point Lake bodies, but the Misery kimberlite produced little signal (Fig. 10C).

Individual kimberlites exhibit a wide range of radioactive element concentrations (Mitchell, 1986). Variations in K_2O content reflect petrographic divisions between micaeous K_2O-rich (0.09-5.04%) and mica-poor kimberlites (0.02-2.15%). Uranium and thorium concentrations range between 0.5 and 22.9 ppm (mean = 3.1 ppm) and between 2.8 and 920 ppm (mean=17 ppm), respectively.

Airborne gamma-ray spectrometry signatures of kimberlites are not well documented. Hoover and Campbell (1992) and Richardson (1996) report that radioactive element surveys have not been effective in the search for kimberlites. However, Hoover and Campbell (1992) observed that several papers had reported on the use of radioelement surveys

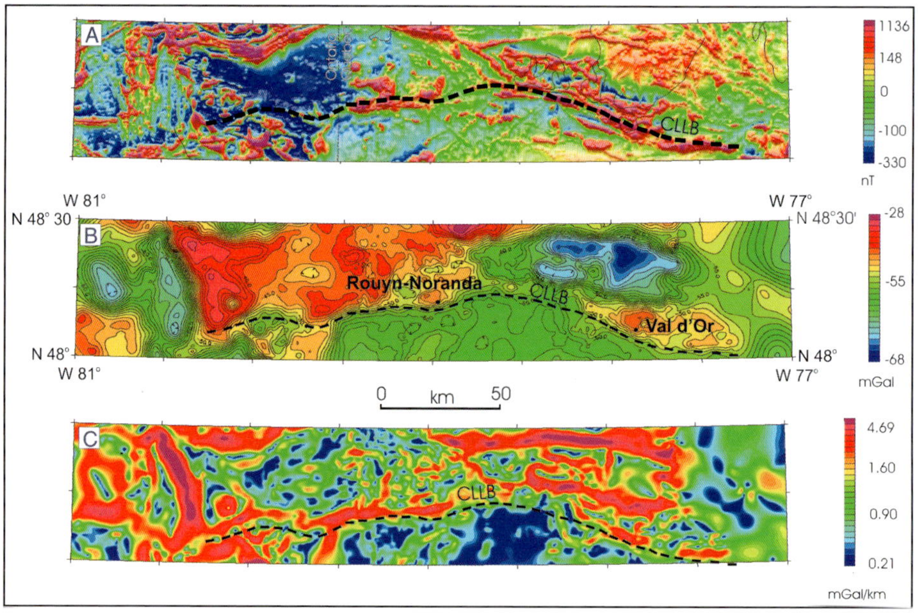

FIGURE 11. The Cadillac-Larder Lake Break (CLLB) superposed on **(A)** total magnetic field map; **(B)** Bouguer gravity anomaly map; and **C)** a map of the horizontal gradient of the Bouguer gravity anomaly

in Yakutia, Russia for discriminating between diamond-bearing basaltic kimberlites and barren micaceous kimberlites and carbonatites. An airborne gamma-ray spectrometry survey flown over the Lac de Gras area (Shives and Holman, 1995) did not detect discernable radioactive element signatures over any of the known kimberlites. This was to be expected given that most pipes are located under lakes and swamps that mask any radiometric responses. Nevertheless, in other kimberlite districts it might be expected that significant radioactive element signatures would be detectable with better exposure. For example, kimberlite intruded into sedimentary sequences with low radioactivity, such as limestone, and not covered by lakes or otherwise water-saturated, could have a detectable radioactive element signature. Mwenifumbo and Kjarsgaard (2002) report that borehole gamma-ray spectrometry measurements, conducted in conjunction with other borehole logging techniques, have been particularly useful in defining different phases of kimberlite pipes.

On the ground, pipe delineation can be done using standard geophysical techniques. Magnetic surveys are used to define the internal structure of the pipe, and eventually identify different intrusion phases (Macnae, 1995). Gravity surveys are used to determine the 3-D architecture of a pipe and investigate its depth extent. In the Fort à la Corne area, ground gravity was used to determine the depth extent and geometry of a pipe (Lehnert-Tiel et al., 1992). In this case, the gravity anomaly is positive because the kimberlite is located within less dense Cretaceous shales. Induced-polarization and resistivity surveys can be used to map the altered top part of the pipe (e.g. Macnae, 1979).

Seismic surveys are useful as a follow-up exploration tool for defining the geometry and internal structure/stratigraphy of kimberlites. Gendzwill and Matieshin (1996) provide a Canadian example of a seismic study conducted on the Smeaton kimberlite in the Fort à la Corne field, Saskatchewan. Vertical seismic reflection profiling conducted along 13 line-km permitted a refinement of subsurface information revealed by drillholes, yielding information on the shape and extent of the kimberlite and its stratigraphic relationships. It also supports a picture of a crater facies emplaced by several explosive episodes with continuing sedimentation between these events.

Lode Gold Deposits

A lode Au deposit is a hydrothermal deposit whose principal commodity is Au. Dubé and Gosselin (2007) provide a comprehensive synthesis of this deposit type and its various subcategories including 1) shear- and fault-zone-related deposits, principally greenstone-hosted quartz-carbonate vein deposits (orogenic, mesothermal, lode gold, shear-zone-related quartz-carbonate or gold-only deposits) associated with collisional tectonics, 2) intrusion-related deposits associated with felsic plutons of subaerial, oceanic, and continental setting, and 3) epithermal deposits (high- and low-sulphidation) associated with subaerial and shallow-marine environments (Lydon et al., 2004).

Shear- and fault-zone-related deposits are a major source of world gold production and represent about a quarter of Canada's production (Ash and Alldrick, 1996). Major Canadian examples include a number of deposits in the Timmins, Kirkland Lake, Red Lake, and Val D'or mining camps. Dubé and Gosselin (2007) describe these deposits as simple to complex networks of gold-bearing, laminated quartz-carbonate fault-fill veins. They are typically hosted by greenschist- to locally amphibolite-facies metamorphic rocks of predominantly mafic compositions. They occur near major compressional faults, but are usually localized along associated second-order faults and splays. Narrow zones (<1 m) of silicification, pyritization, and potassium metasomatism occur within broader zones (10s of metres) of carbonate alteration.

Low-sulphidation epithermal deposits are described by Panteleyev (1996a) as Au- and Ag-bearing quartz veins, stockworks, and breccias that exhibit open-space filling textures and are associated with volcanic-related hydrothermal and geothermal systems. High-sulphidation deposits occur as veins, vuggy breccias, and sulphide replacements associated with epizonal hydrothermal systems characterized by acid-leached, argillic, and siliceous alteration (Panteleyev, 1996b). Minerals relevant to geophysical detection include pyrite, sericite/illite, alunite, adularia, kaolinite, muscovite, magnetite, pyrrhotite, quartz, and carbonate minerals.

A variety of geophysical techniques are applicable to Au exploration. Specific methods will depend on lithological, mineralogical, and alteration characteristics of each deposit type. In general, for shear- and fault-zone (Au-quartz vein) deposits aeromagnetic data can provide valuable mapping

TABLE 3. Massive sulphide ore mineral density (g/cm^3) and magnetic susceptibility (10-3 SI).

Mineral	Density	Susceptibility
Barite	4.5	?
Chalcopyrite	4.2	0.4
Pyrite	5.02	5
Pyrrhotite	4.62	3200
Sphalerite	4	0.8
Galena	7.5	-0.03
Magnetite	5.18	5700
Hematite	5.26	40

Densities from Klein and Hurlbut (1985)
Susceptibilities from Hunt et al. (1995)

information by delineating lithologies, regional faults, and shear zones. An excellent example is the Cadillac Larder Lake Break (CLLB) in the Archean Abitibi belt, which extends for more than 200 km in an east-west direction. This major fault zone is characterized by the presence of numerous gold deposits. Although gold deposits have no geophysical response, the structures controlling gold mineralization may produce distinct geophysical signatures. Figure 11A, derived from the archives of the Geological Survey of Canada's Geophysical Data Repository, shows the residual total magnetic field in the area. The CLLB (black dotted line) is marked by a series of linear magnetic anomalies generally located on its north side. Figure 11B shows the Bouguer gravity anomaly over the same area. Data spacing is highly variable, between 500 m and over 10 km. Nevertheless, the CLLB is easily identifiable over most of its length. It is even easier to locate in Figure 11C, which displays the magnitude of the horizontal gradient of the Bouguer gravity anomaly used to map contacts between units of different densities.

At deposit-scale, magnetic lows can delineate areas of magnetite destruction associated with carbonate alteration. EM methods have also been used to map faults, veins, contacts, and alteration. IP methods and gamma-ray spectrometry may have local applications to map massive quartz veins (resistivity highs) and associated alteration (e.g. potassium highs (Shives et al., 1995)).

Klein and Bankey (1992) note that for epithermal styles of mineralization several types of geophysical signatures have been used to delineate favourable geology, structures, and alteration. These include regional-scale gravity lows that define favourable, thick, silicic volcanic sequences, or long-wavelength magnetic lows that are associated with alteration. Regional resistivity lows may be associated with weathered and altered volcanic sequences. Klein and Bankey (1992) state that while there are no geophysical signatures to directly detect epithermal vein mineralization, favourable deposit-scale structures and alteration can be measured. Gravity signatures are variable, ranging from highs associated with subvolcanic intrusions or structural highs within volcanic sequences to local gravity lows associated with zones of brecciation or fracturing. Alteration may result in local magnetic lows. Potassium highs are to be expected due to K alteration. However, naturally high K rock-types in the vicinity of mineralization may make identification of K enrichment associated with mineralization difficult. Low eTh/K ratios may be a more sensitive indicator of alteration (Shives et al., 1997). Resistivity lows may occur with associated sulphide mineralization, argillic alteration, and increased porosity related to open brecciation. However, resistivity highs will occur in zones of silicification or associated with intrusions or basement uplifts (Klein and Bankey, 1992). Associated pyritic alteration may cause high IP anomalies.

Some of these geophysical signatures are illustrated in Figure 12. In the Red Lake area, K enrichment is known to occur along the Madsen-Starrett-Olsen shear zone (Durocher, 1983) related to gold mineralization. Figure 12 shows results from a detailed, 250 metre line spaced, airborne gamma-ray spectrometry and total field magnetic survey (Hetu, 1991). The Madsen-Starrett-Olsen Zone shows a weak but distinct K anomaly that is unusual considering the underlying mafic volcanic lithologies. Higher amplitude K anomalies occur over granitic and felsic metavolcanics to the south and west. The K enrichment associated with the mineralized zones can be differentiated from the high K concentrations associated with the granitic and felsic metavolcanics by virtue of the fact that the mafic lithologies that host the mineralization have low Th concentrations. The associated K enrichment results in low eTh/K ratios. Similarly, in the Toodoggone area of British Columbia, a number of Au deposits and occurrences occur directly associated with or in close proximity to low eTh/K ratio anomalies.

Volcanogenic Massive Sulphide (VMS) Deposits

Volcanic massive sulphide (VMS) deposits form by discharge of hydrothermal solutions onto the seafloor, commonly near plate margins. A comprehensive synthesis of VMS deposits is provided by Galley et al. (2007). VMS deposits typically develop in the form of a concordant lens that is underlain by a discordant stockwork or stringer zone comprising vein-type sulphide mineralization located in a pipe of hydrothermally altered rock.

VMS deposits typically have density, magnetic, conductivity, and acoustic velocity properties that differ significantly from those of their host rocks. There is, therefore, enormous potential for direct detection of orebodies using geophysical methods that measure these properties. The most common sulphide mineral in VMS deposits is pyrite, which may be accompanied by subordinate pyrrhotite, chalcopyrite, sphalerite, and galena (Galley et al., 2007). Magnetite, hematite, and cassiterite are common nonsulphide metallic minerals, and the gangue mineral barite may also be present. Densities of these minerals range from 4.0 to 7.5 g/cm^3 (Table 3). Singularly or in combination, these minerals will have a large density contrast with respect to densities of host sedimentary and volcanic rocks; some typical values in the Bathurst Mining Camp range from 2.70 to 2.84 g/cm^3 (Thomas, 2003). Mean densities of massive to semimassive sulphides measured on drill core from several deposits in the Bathurst Mining Camp are naturally less than individual mineral densities, but nevertheless range from 3.58 to 4.42 g/cm^3 (Thomas, 2003). Most deposits in the Camp generate distinct gravity highs, the largest being associated with the Brunswick No. 6 deposit measured at approximately 4.0 mGal in amplitude prior to mining (Thomas, 2003).

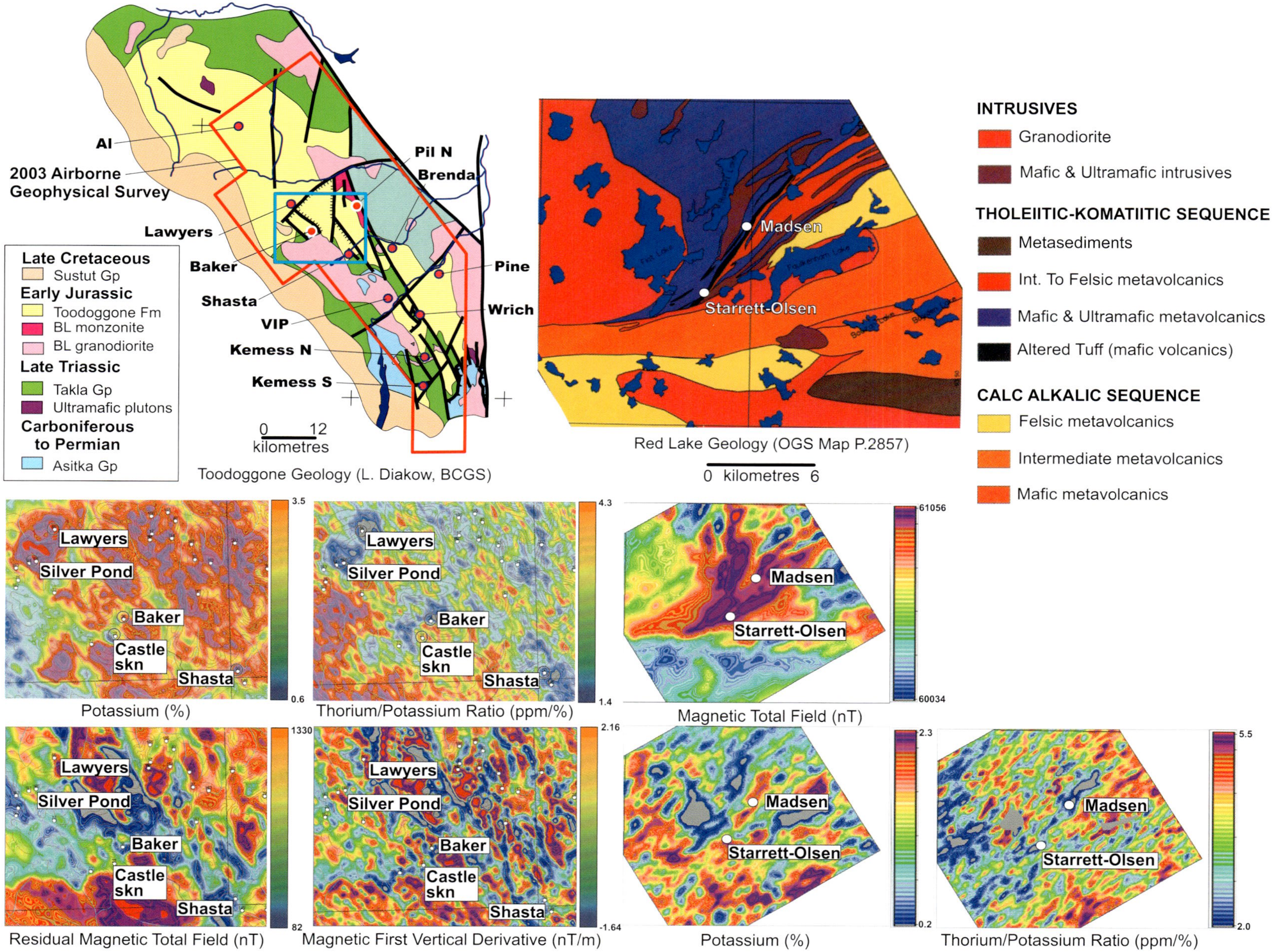

FIGURE 12. Airborne gamma-ray spectrometric and magnetic signatures over gold mineralization in the Toodoggone area, British Columbia (frcm Shives et al., 2004) and Madsen-Starrett-Olsen Zone, Red Lake area, Ontario (Hetu, 1991). In both examples, detailed structural information is provided by the aeromagnetic data, lithological information is provided by both the aeromagnetic and gamma-ray spectrometric data, and K enrichment due to alteration associated with mineralization is differentiated from high K lithologies by corresponding low Th/K ratios. White dots represent known mineralization.

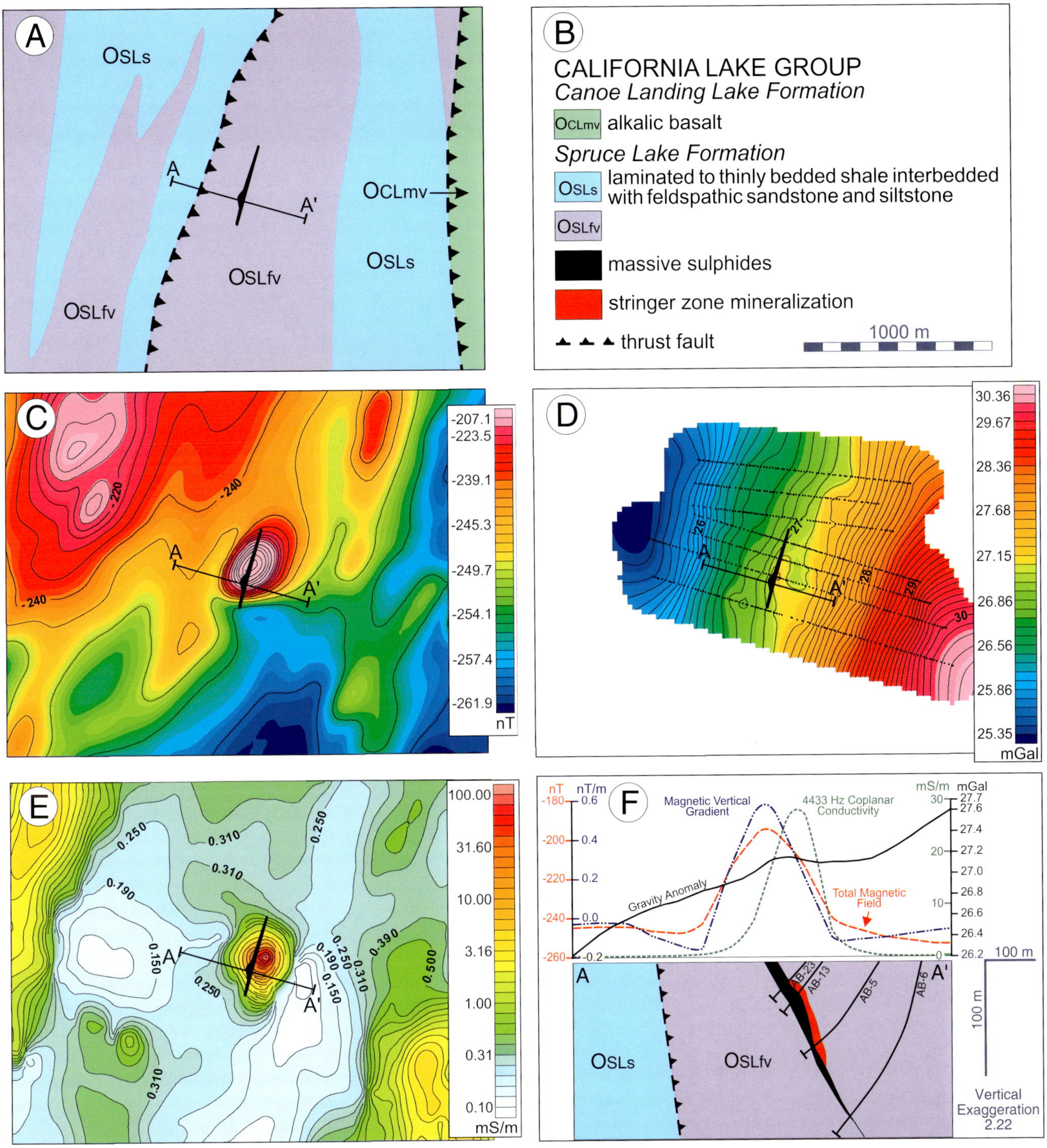

FIGURE 13. Geological and geophysical maps of the area containing the Armstrong B massive sulphide deposit, Bathurst mining camp, New Brunswick, together with geophysical profiles across the deposit. **(A)** Geological map (after van Staal, 1994); **(B)** Geological legend; **(C)** Map of magnetic field; **(D)** Map of gravity anomalies; **(E)** Map of conductivity (4433 Hz frequency; coplanar transmitter-receiver); and **(F)** Total magnetic field, vertical magnetic gradient, gravity anomaly, and conductivity (4433 Hz; coplanar) profiles crossing the Armstrong B deposit. The foregoing panels are based on images published by Thomas et al. (2000).

High magnetic susceptibilities of most sulphide minerals, except sphalerite and galena, ensure that prominent magnetic anomalies are also associated with VMS deposits. Bishop and Emerson (1999) note that sphalerite has no salient geophysical properties allowing its direct routine detection by geophysics. They note further that even though sphalerite occurs along with other detectable sulphides, detection of Zn-bearing deposits is very difficult. Pyrrhotite has the largest magnetic susceptibility (3200 x 10^{-3} SI) of the sulphide minerals and its presence is important to the detection of the sulphide body. Pyrite is by far the most common sulphide mineral in a sulphide lens (Galley et al., 2007) and has a susceptibility of 5 x 10^{-3} SI, which is considerably larger than susceptibilities (0.1 to 1.1 x 10^{-3} SI) of most host sedi-

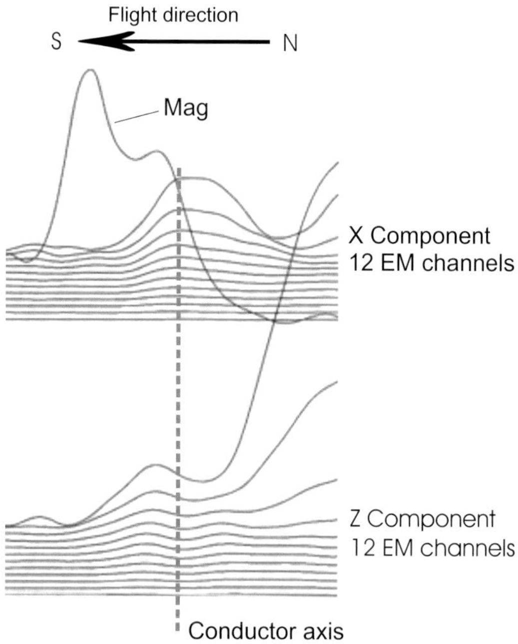

FIGURE 14. Electromagnetic (GEOTEM) profiles across the Caber volcanic massive sulphide deposit, 35 km west of Matagami, Quebec. Top channels are early time channels; the vertical scales are relative. There is an anomalous EM response on both the X and Z component. The higher amplitude, early time responses of the Z component are caused by a thickening of the overburden (~20 m) under the northern part of the profile. The profile is approximately 1 km long.

mentary and volcanic units in the Bathurst Mining Camp (Thomas, 1997). Thus, discernible positive signatures are expected over VMS deposits of significant size and should provide pointers to potentially economic mineralization because of the common association with chalcopyrite, sphalerite, or galena. Chalcopyrite and sphalerite have susceptibilities (Table 3) similar to those of sedimentary and volcanic hosts, and are unlikely to generate a strong magnetic signal. Galena, with a small negative susceptibility, also would have little influence on the magnetic field. Lydon (1984) observes that magnetite and hematite are two common nonsulphide metallic minerals occurring in sulphide lenses. For some VMS deposits, magnetite tends to be concentrated in the core of the stockwork and central, basal part of the overlying sulphide lens. The association of VMS deposits and magnetite, coupled with the large susceptibility of magnetite (5700 x 10^{-3} SI), provides a "remote" sensor for sulphide species via magnetic anomalies produced by magnetite. Hematite has a much smaller susceptibility (40 x 10^{-3} SI) than magnetite, but still has potential to produce sizable positive signals. Also of great importance in vectoring VMS deposits are temporally and spatially associated iron-formations, some of which contain magnetite (Peter et al., 2003), and produce prominent magnetic anomalies.

A comprehensive and illustrated description of gravity, magnetic, conductivity, and radiometric signatures for twenty sulphide deposits in the Bathurst Mining Camp, New Brunswick is presented by Thomas et al. (2000). Illustrations of geophysical signatures for the small Armstrong B VMS deposit are shown in Figure 13. The time-domain airborne EM signature (GEOTEM) for the Caber VMS deposit located west of Matagami, Quebec, obtained subsequent to discovery, is shown in Figure 14. This copper-zinc deposit, containing 1.3 MT at 1.3% Cu and 5.5% Zn, was discovered by BHP Minerals Canada in 1994. The deposit is located on the Key Tuffite horizon, along which are located the base metal VMS deposits of the Matagami Mining Camp. The strike length of the deposit is 200 to 250 m with a down-dip extent of 150 to 250 m. Bedrock is overlain by approximately 20 m of conductive overburden. Strong airborne and ground gamma-ray spectrometry signatures over K-altered mafic and felsic volcanic rocks associated with a massive sulphide deposit at Pilley's Island, Newfoundland (Shives et al., 1997) are shown in Figure 15.

SEDEX Base Metals Deposits

Sedimentary exhalative (SEDEX) sulphide deposits are found in sedimentary basins, usually in the form of conformable to semiconformable sheets or tabular lenses of stratiform sulphides (Lydon, 1995; Goodfellow and Lydon, 2007). Such bodies have typical aspect ratios of 20, maximum thicknesses of 5 to 20 m (Lydon, 1995), and may extend over a distance of more than 1 km (Goodfellow and Lydon, 2007). The principal ore minerals are sphalerite and galena. Chalcopyrite is sometimes concentrated in feeder zones of SEDEX deposits, but only rarely attains concentrations of economic interest. Pyrite is the most abundant sulphide, and pyrrhotite may be common.

The physical and chemical properties of SEDEX deposits make them amenable to detection by several geophysical techniques. In the Purcell Basin, southeastern British Columbia, sulphide mineralization associated with the Sullivan and smaller North Star and Stemwinder Pb-Zn-Ag deposits is reflected by strong finite conductors and positive magnetic anomalies (Fig. 16) (Lowe et al., 2000). Conductors coinciding with the western margin of the Sullivan deposit, where it is cut by the Sullivan fault, are attributed to an uneconomic massive pyrrhotite body underlying the western and shallowest part of the underground workings. A magnetic high associated with the deposit is also explained by this same body. Elevated K values and depleted eTh/K ratios, relative to those of unmineralized host rocks in the Sullivan-North Star corridor (Fig. 16), are associated with areas of hydrothermal sericitic alteration.

In the Anvil district, Yukon, a combination of magnetic, electromagnetic, and gravity methods detected the Vangorda, Faro, and Swim Lake deposits (Brock, 1973), with significant geophysical signatures being recorded over each deposit, though not necessarily for each method in every case. The Faro deposit was associated with a positive gravity anomaly more than 2.5 mGal in amplitude and a magnetic anomaly of about 300 nT amplitude (Figs. 17, 18). The Vangorda deposit also produced significant gravity and magnetic anomalies, having amplitudes of about 2 mGal and 800 nT, respectively (Fig. 19). Airborne electromagnetic surveys using a vertical coaxial transmitter-receiver configuration operating at 4000 Hz recorded strong in-phase responses from the Vangorda (up to 120 ppm) and Swim (190 ppm) deposits, but only weak to moderate responses over the Faro No. 1 (19 ppm) and Faro No. 2 (41 ppm) deposits (Brock, 1973).

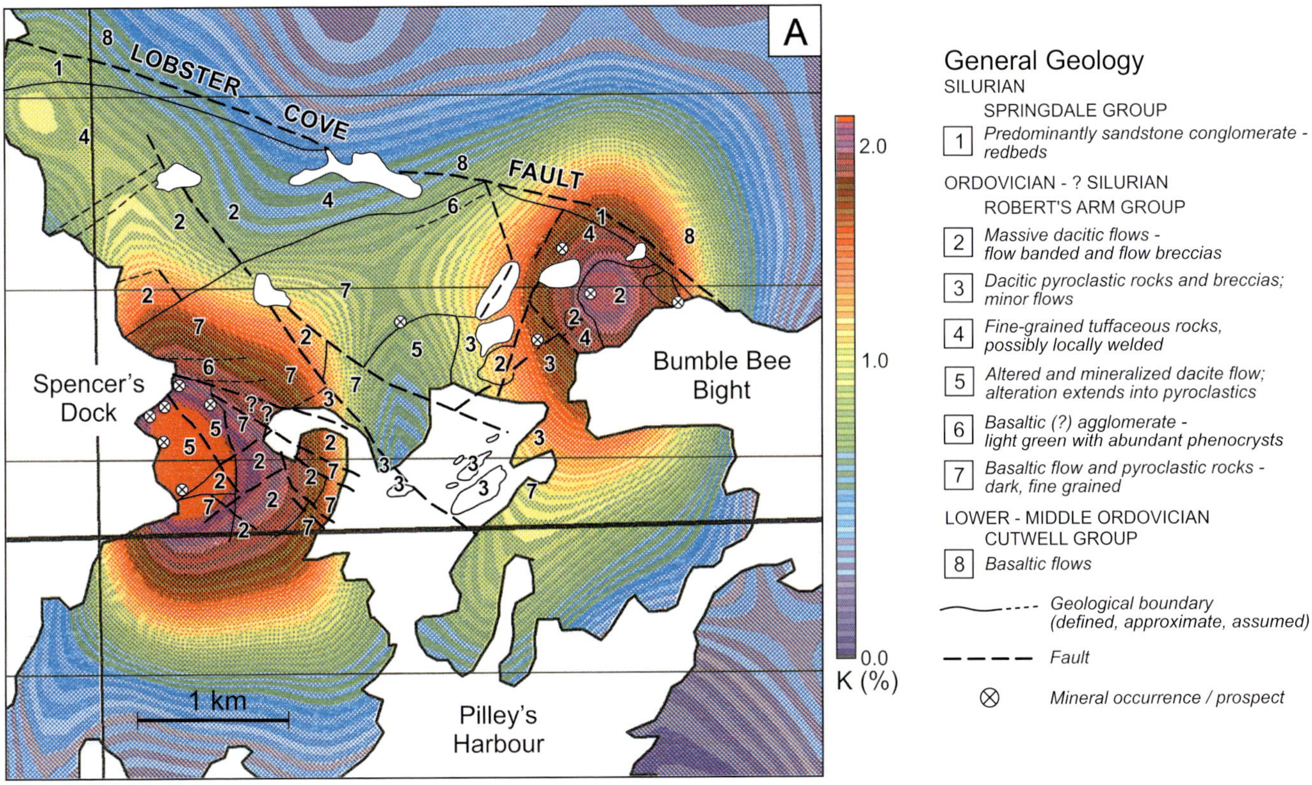

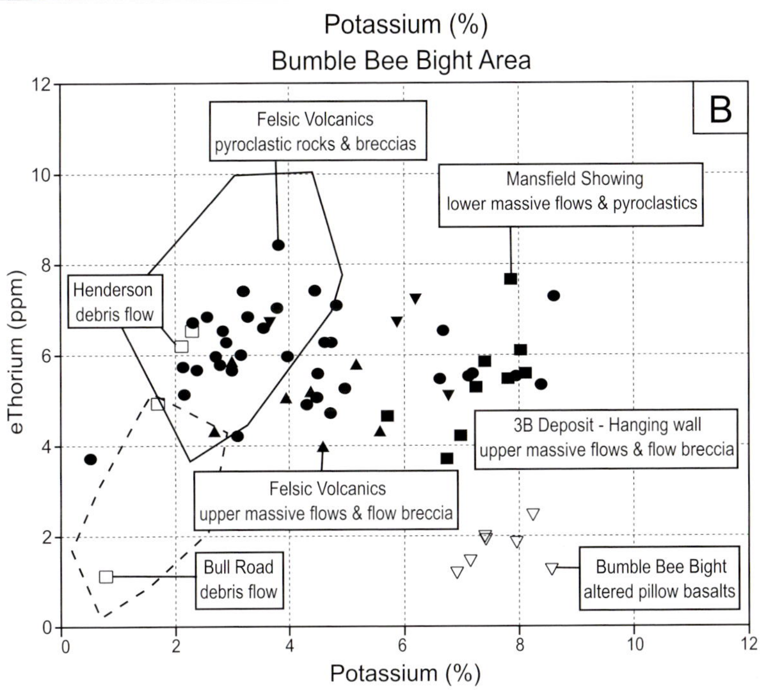

FIGURE 15. Airborne and ground gamma-ray spectrometry signatures over a K-altered volcanic sequence associated with volcanic massive sulphide mineralization, southern Pilley's Island, Newfoundland. **(A)** Potassium map derived from an airborne gamma-ray spectrometry survey flown with flight lines spaced 1000 m apart, and geology derived from Tuach et al. (1991); **(B)** Variations in eTh and K concentrations, as measured by ground gamma-ray spectrometry. Solid and dashed lines represent lithological signatures for unaltered felsic and mafic rocks, respectively, of the Roberts Arm Group.

Mississippi Valley-Type Lead-Zinc Deposits

Mississippi Valley-type (MVT) Pb-Zn deposits are typically stratabound, some are prismatic pipe-shaped bodies, hosted by limestone or dolomite in platform carbonate sequences (Sangster, 1995) and occur in clusters. Sphalerite and galena are the dominant ore minerals that characteristically occupy open spaces in carbonate breccias; replacement of host rocks is relatively rare. Most deposits or MVT districts occur below unconformities or nonconformities, related probably to minor uplift or warping, but few deposits have been affected by subsequent deformational events. Dimensions of the one hundred known orebodies in the Pine Point district vary from 60 to 2000 metres in length, 15 to 1000 metres in width, and 0.5 to 100 metres in thickness (Hannigan, 2007). Additional information on various aspects of MVT deposits may be found in Dewing et al. (2007), Hannigan (2007), Paradis and Nelson (2007), and Paradis et al. (2007).

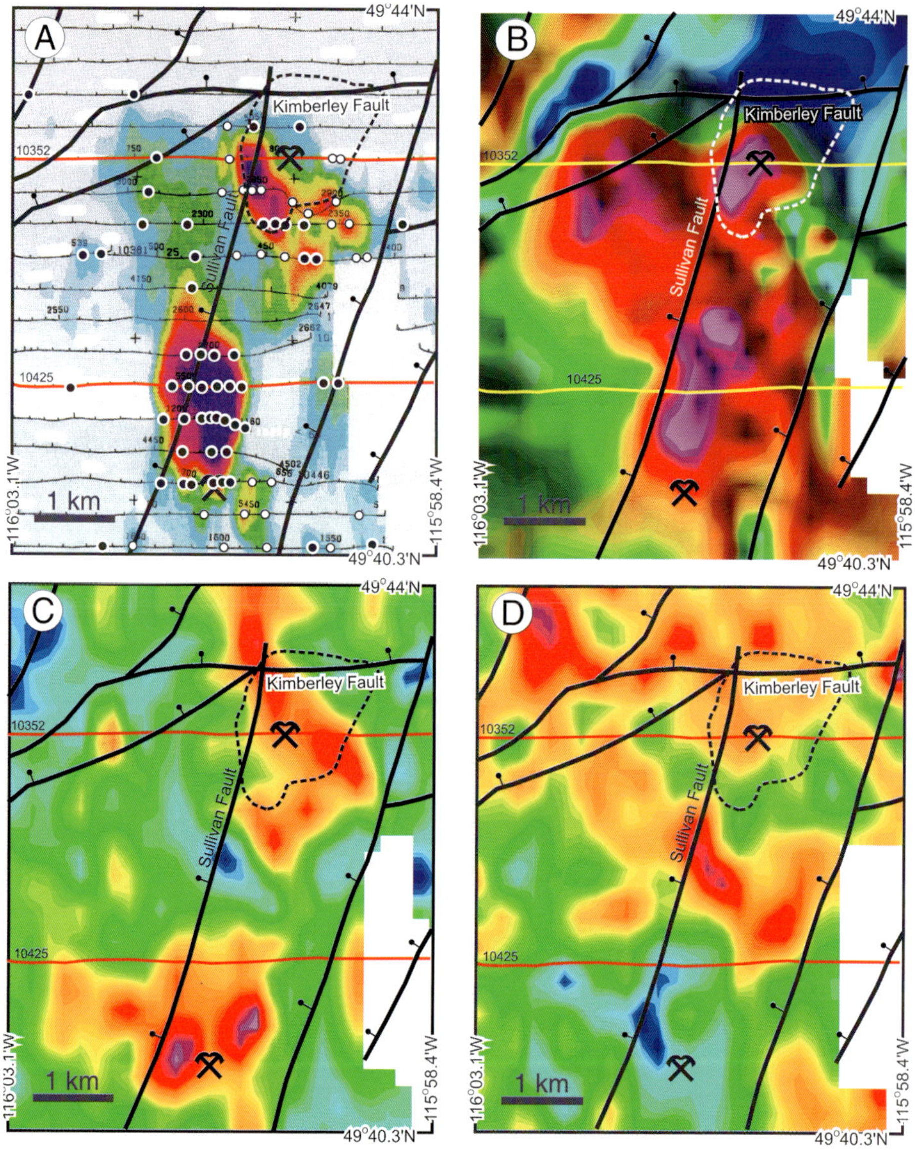

Figure 16. Geophysical images of the Sullivan - North Star area. Crossed-hammer symbols mark the centre of the Sullivan deposit in the north and the North Star deposit in the south. Dashed line shows the vertical projection of the economic limit of the Sullivan orebody. **(A)** EM (apparent conductivity estimated from 900 Hz coaxial data). Locations of calculated cultural and noncultural finite conductors are indicated by open and solid circles, respectively (conductivity values range from 0 to 5800 mS/m); **(B)** Magnetic anomalies (values range from -40 nT to +80 nT); **(C)** Potassium levels (values range from 0.86-2.83%); and **(D)** eTh/K ratios (values range from 3.09 to 6.99). In all cases, hot colours indicate high values and cool colours indicate low values. Information from Lowe et al. (2000).

The generally concentrated nature of the mineralized zones, high densities of sphalerite and galena, high conductivity of galena, and significant conductivities of commonly associated sulphides, such as pyrite, marcasite, and pyrrhotite, make the induced polarization (IP) and gravity methods the favoured approaches for exploration. At Pine Point, Seigel et al. (1968) attributed the discovery of the two Pyramid orebodies to time-domain IP surveys, with a subsequent gravity survey credited with focusing a drilling program and contributing to tonnage estimations. Distinct gravity (>0.8 mGal amplitude, Fig. 20) and chargeability (~21 ms, Fig. 21) anomalies were mapped over the tabular Pyramid No. 1 body. Turam electromagnetic data shows no discernable response, presumably due to the disconnected nature of marcasite common in the Pine Point deposit.

Lajoie and Klein (1979) also reported success with the IP and gravity methods in the Pine Point district, noting again the lack of electromagnetic responses, which was ascribed generally to calcite and dolomite gangue interrupting the conducting paths, though nonconductive and nonpolarizable sphalerite, the dominant mineral in the orebodies, plays a similar role. The IP method was considered the best exploration tool in this district, even for orebodies having small lateral dimensions, and the gravity method was regarded as an excellent complement to IP. The magnetic method may have some suc-

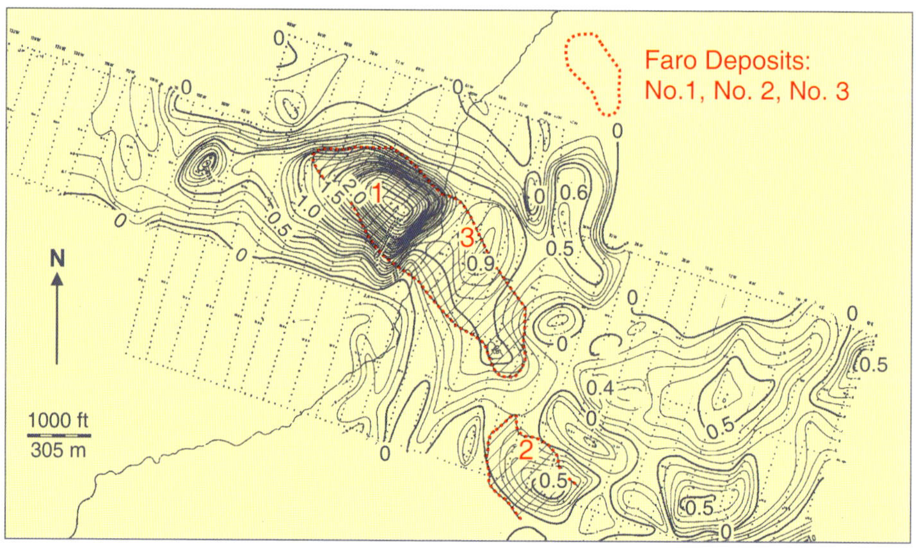

Figure 17. Residual Bouguer gravity anomaly map of the Faro SEDEX deposit, Yukon, which includes three separate orebodies, referred to as the No. 1, No. 2, and No. 3 deposits. This figure is a modified version of Figure 11 published by Brock (1973). Contour interval = 0.1 mGal.

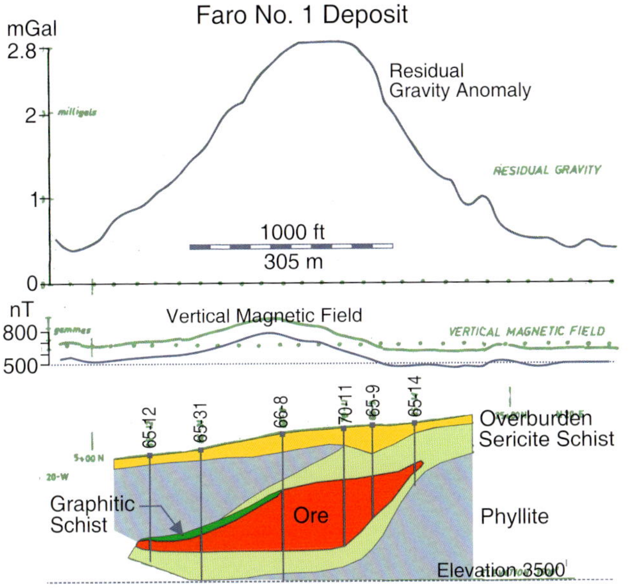

FIGURE 18. Bouguer gravity and vertical magnetic field profiles crossing the Faro No. 1 SEDEX deposit, Yukon (after Brock, 1973).

cess if the ore is associated with pyrrhotite, an example of a small, 20 nT amplitude magnetic anomaly conceding with a small uneconomic mineralized body is illustrated (Fig. 22). Lajoie and Klein (1979) describe results of seismic reflection experimentation in the region, designed to assess the method's capability as a tool for mapping, for finding ores directly and for outlining structures associated with mineralization. A successful result was the delineation of a large collapse structure within the hosting clastic reef complex. This structure may have influenced the development of an orebody located within it. In Nunavut, in the arctic islands, Hearst et al. (1994) reported on high-resolution seismic reflection profiling across the pyritic South Boundary Zone of the Nanisivik MVT deposit, concluding that sulphide mineralization could be detected to moderate depths of up to about 200 m if it had a minimum thickness of 5 to 10 m and

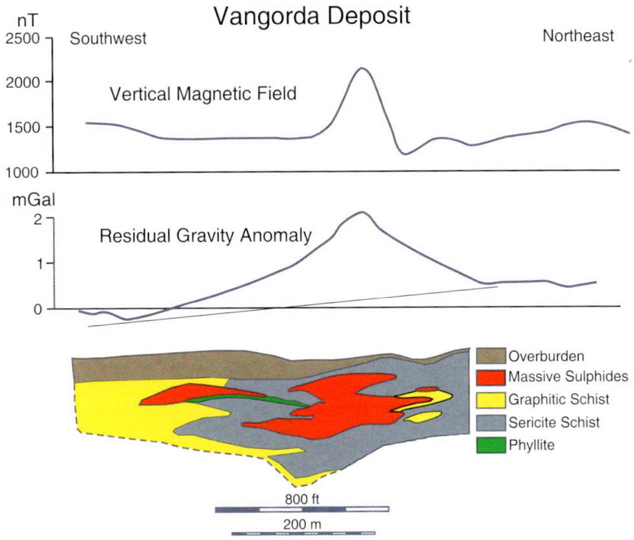

FIGURE 19. Bouguer gravity and vertical magnetic field profiles crossing the Vangorda SEDEX deposit, Yukon (after Brock, 1973).

a lateral extent on profile of at least 10 m. Also in the Arctic, the gravity method is credited with the discovery of the Polaris lead-zinc deposit, which produced an anomaly of about 0.9 mGal amplitude (Fig. 23).

Porphyry Copper Deposits

Porphyry deposits are a major source of production for Cu, Mo, and Re, and an important source for Au, Ag, and Sn. Sinclair, (2007) provides a comprehensive synthesis of porphyry deposits. Porphyry-style base and precious metal mineralization is spatially and genetically related to high level, epizonal and mesozonal felsic to intermediate porphyritic intrusions and adjacent host rocks. Mineralization may be in the form of stockwork quartz veins and veinlets, fractures, disseminations, and replacements containing pyrite, chalcopyrite, bornite, and magnetite. Deposit forms are quite variable and range in size from hundreds to thousands of metres laterally and with depth, and are commonly zoned

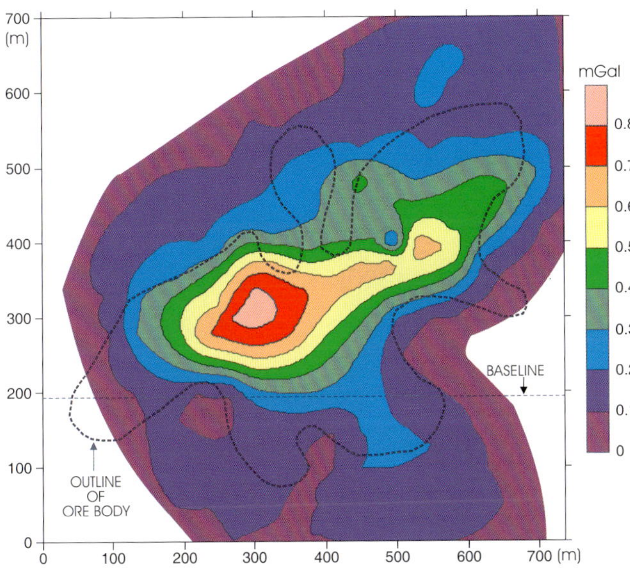

FIGURE 20. Bouguer gravity anomaly associated with the Pyramid No. 1 orebody, Pine Point District, Northwest Territories (after Seigel et al., 1968).

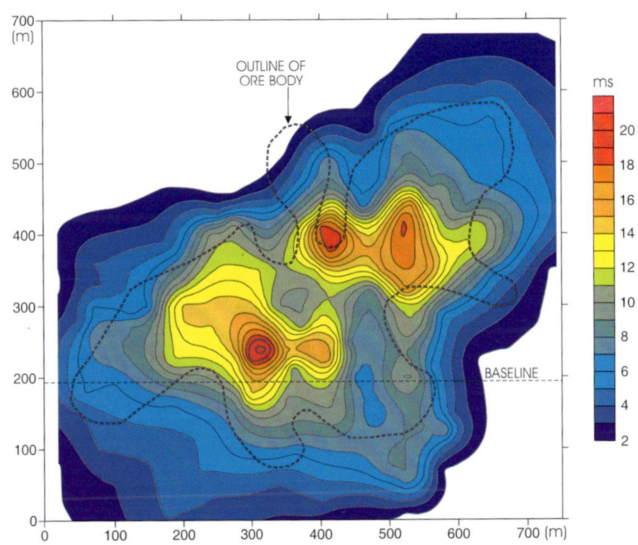

FIGURE 21. Chargeability anomaly associated with the Pyramid No. 1 orebody, Pine Point District, Northwest Territories (after Seigel et al., 1968).

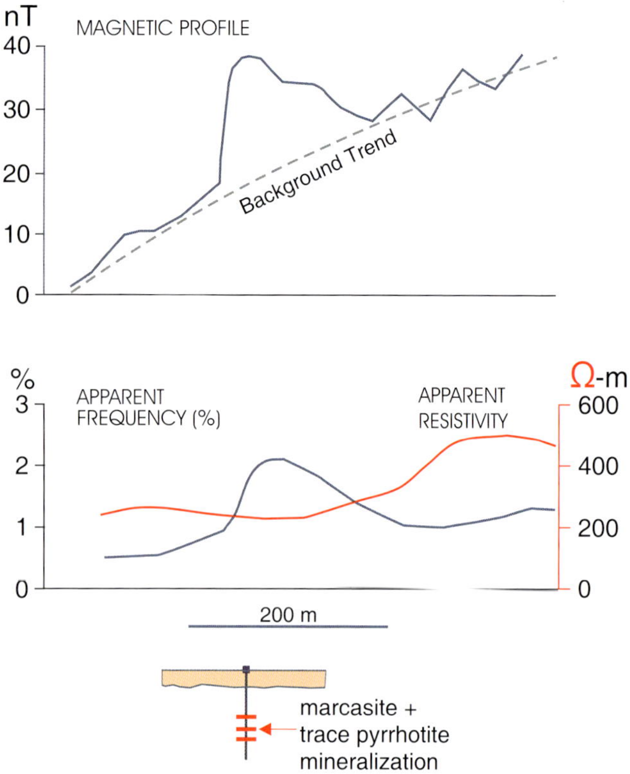

FIGURE 22. Magnetic, apparent resistivity and apparent frequency profiles crossing an uneconomic mineralized body, Pine Point District, Northwest Territories (after Lajoie and Klein, 1979).

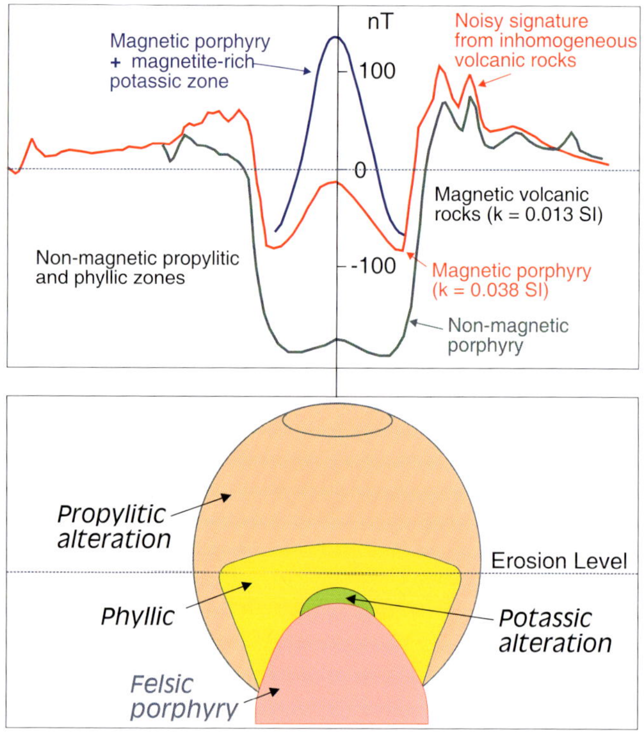

FIGURE 24. Magnetic signatures (assuming a vertical inducing field) associated with an idealized porphyry copper system based on figures in Clark et al. (1992) and Gunn and Dentith (1997).

with barren cores and generally concentric metal zones surrounded by barren pyritic haloes. Alteration associated with porphyry deposits is typically zoned from an inner potassic (biotite and/or K-feldspar) alteration zone, closely associated with mineralization, to a more extensive propylitic alteration zone consisting of quartz, chlorite, epidote, calcite, albite, and pyrite, which surrounds the inner potassic zone. Zones of phyllic and argillic alteration may occur between and overlap with the inner potassic and outer propylitic alteration zones. Minerals relevant to geophysical detection include magnetite, pyrite, chalcopyrite, biotite, K-feldspar, and sericite.

Broad, regional aeromagnetic anomalies commonly occur with genetically associated intrusive rocks, and provide a regional exploration target. Mineralization can be associated with magnetite-bearing rocks that would be delineated in an

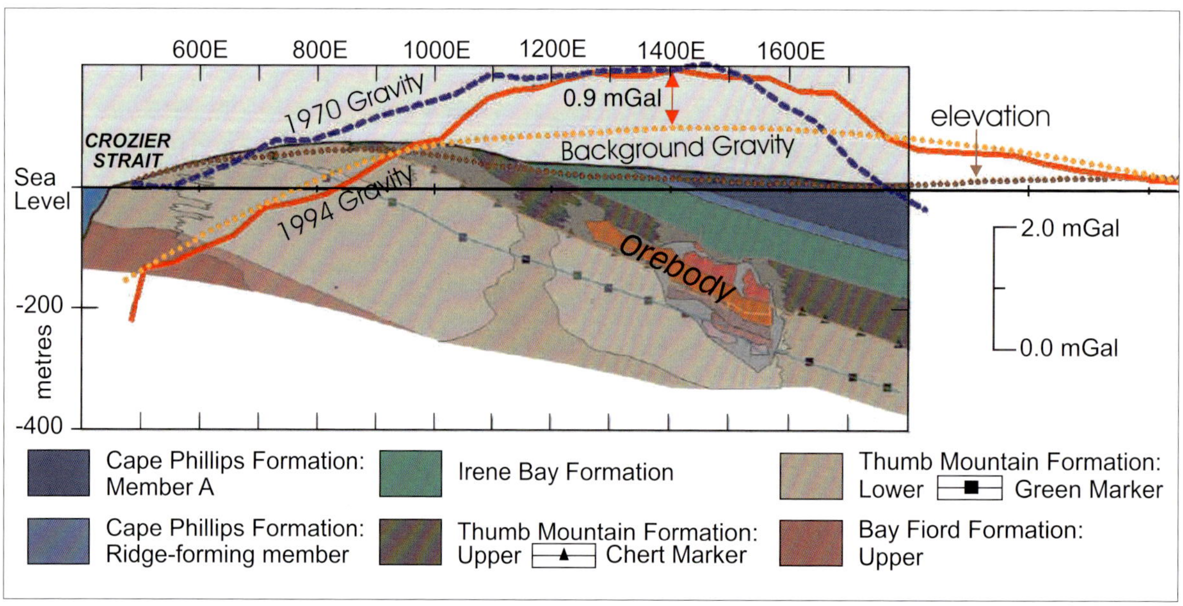

FIGURE 23. Gravity profiles along Line 2200N crossing the Polaris lead-zinc deposit (modified from figure provided by Bob Holroyd, Teck Cominco Ltd). The density used in the Bouguer correction is 2.67 g/cm^3.

aeromagnetic survey. Sillitoe (1979) described Au-rich porphyry Cu deposits from several localities worldwide, including examples from the intermontane zone of British Columbia, and noted that the Au is normally found in K-silicate alteration, which is associated with "an unusually high magnetite content", between 5 to 10% by volume. Mineralization, in some cases, is described as being associated with alteration/magnetite-rich zones on the order of 100s of metres in lateral extent; such zones would present compact and well defined magnetic targets. Hydrothermal alteration may destroy magnetite and may be manifested as a broad, smooth magnetic low. Gunn and Dentith (1997) note that such lows may be associated with zones of propylitic and phyllitic alteration within volcanic rocks capping porphyry intrusions. The volcanic rocks themselves, which are generally inhomogeneous, may produce erratic magnetic responses. Magnetic responses for an idealized porphyry Cu deposit, taken from Gunn and Dentith (1997) and based on Clark et al. (1992), are shown in Figure 24.

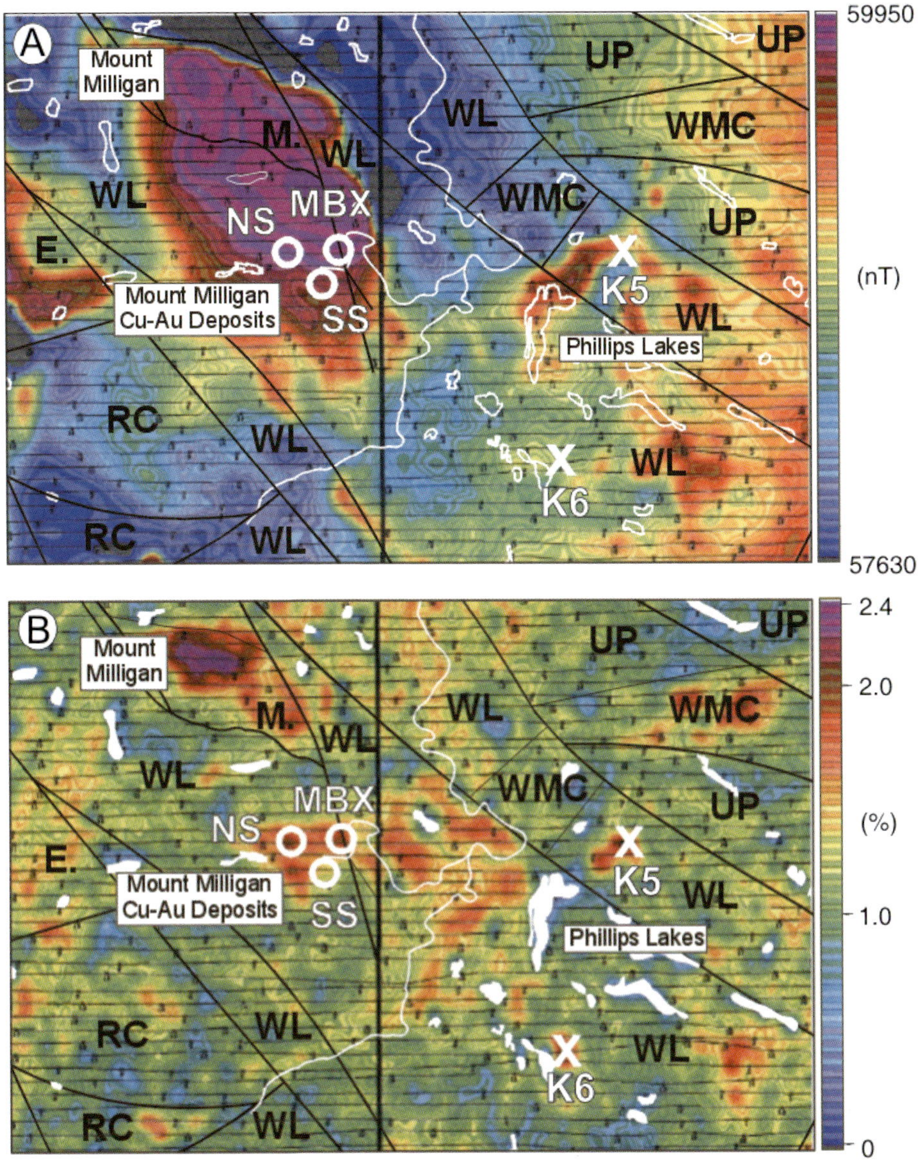

FIGURE 25. (A) Magnetic total field (nT) and (B) potassium (%) signatures in the area of the Mount Milligan porphyry copper deposits (MBX, NS, and SS), British Columbia. E - volcanic wacke; M - Mount Milligan Intrusive Complex, monzonite suite; RC - Rainbow Creek Formation; UP - Mesozoic sedimentary rocks; WL - Witch Lake Formation (Takla Group); WMC - Wolverine Metamorphic Complex.

Generally K and Th concentrations vary coincidently with protolith compositions, commonly increasing from mafic to felsic. Subsequent hydrothermal alteration associated with porphyry Cu deposits may disproportionably enrich K such that the ratio eTh/K produces a diagnostic low value. Airborne gamma-ray spectrometry surveys will map high K anomalies associated with hydrothermal alteration, but these anomalies may be difficult to distinguish from K anomalies associated with normal high-K rock-types. Low eTh/K ratio anomalies will distinguish K enrichment associated with biotite and K-feldspar alteration from that due to high-K lithologies (Shives et al., 1997). Induced polarization (IP) surveys will delineate pyritic alteration haloes that envelope cupriferous ore zones.

An example of aeromagnetic and airborne gamma-ray spectrometry signatures associated with a porphyry Cu deposit in the Canadian Cordillera is shown in Figure 25. The maps are compiled from airborne data collected along flight lines spaced 500 m apart (Geological Survey of Canada, 1991). At Mount Milligan, a broad magnetic high is associated with exposed and buried portions of the Mount Milligan Intrusive Complex (Fig. 25). High K concentrations are associated with bedrock exposures of the complex at Mount Milligan. To the south, discrete K anomalies are associated with the three main zones of the Mount Milligan deposit (MBX, NS, and SS) and provide better defined exploration targets relative to the regional magnetic signatures (Shives et al., 1997). Elsewhere in the survey area, despite extensive and thick overburden, other K anomalies (K5 and K6) are associated, respectively, with K-altered and mineralized andesitic volcanics (K5) and K-altered and mineralized intrusive boulders derived from an unexposed mineralized intrusive unit.

Unconformity-Related Uranium Deposits

Unconformity-related uranium deposits are the most significant high-grade, low-cost source of uranium in the world (Jefferson et al., 2007). In Canada, notable targets for exploration are the mid-Proterozoic sedimentary Athabasca and Thelon basins in the northwestern Canadian Shield. The

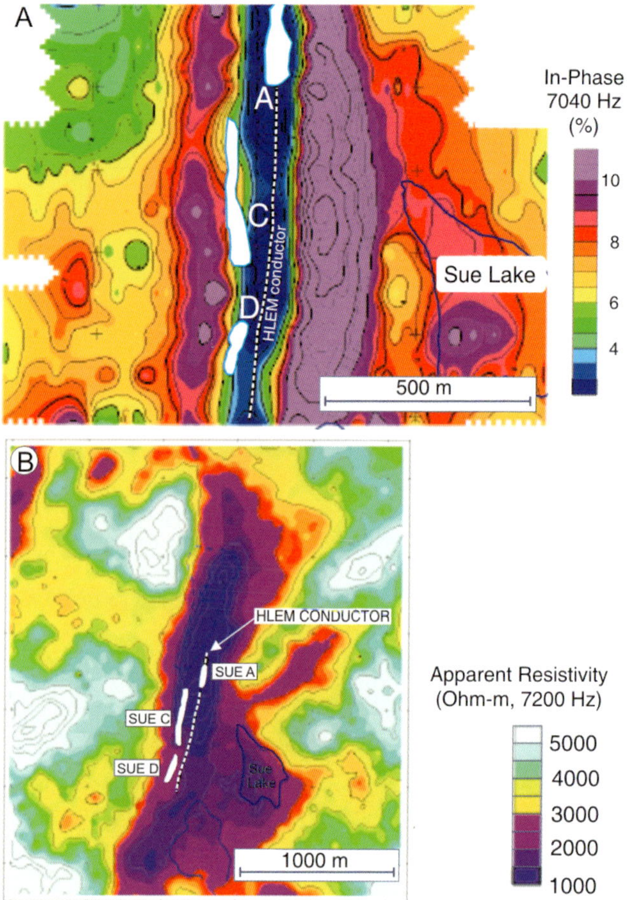

FIGURE 26. Electromagnetic responses over the Sue A, C, and D uranium deposits, Athabasca Basin. **(A)** MaxMin 1-10 horizontal loop EM in-phase response; **(B)** Dighem (Airborne) coplanar frequency domain apparent resistivity. Modified from figures published by Matthews et al. (1997).

deposits occur typically along or near unconformities between metamorphic basement rocks, commonly containing graphitic pelitic units, and overlying undeformed sedimentary successions consisting mainly of quartzose sandstone. Graphitic units provide a lithological control on mineralization (Ruzicka, 1995), acting as a reductant. Ruzicka (1995) recognized polymetallic (U-Ni-Co-As) deposits, occurring at the unconformity, and monometallic (U) varieties, generally positioned below the unconformity and rarely above. Faults and fractures intersecting the unconformity are key structures controlling localization of mineralization. Deposit geometry and orientation are characteristically controlled by these structures and the unconformity. Polymetallic orebodies are pod- or lense-shaped and aligned along the structure, whereas monometallic deposits occur as lenses in veins or thin veinlets in stockworks.

Apart from obvious radioactive properties and a high density (up to 9.7 g/cm^3), uraninite has no other physical property permitting direct detection by a geophysical technique. Airborne radiometric surveys conducted in 1967 provided the first clues to potential locations of mineralized targets in the Athabasca Basin, outlining many prospective anomalies (Schiller, 1979). Radioactive boulder trains were also of importance in many early discoveries in the Athabasca Basin (Matthews et al., 1997). Drilling of a radioactivity anomaly in an area containing many radioactive boulders in 1968 led to discovery of the Rabbit Lake deposit (Schiller, 1979). Airborne radiometric surveys combined with radioactive boulder studies laid the groundwork for the discovery of the Midwest deposit in 1979 (Scott, 1983). Ground gamma-ray spectrometry measurements also have a role in uranium exploration. For example, Shives et al. (2000) used them to map illitic alteration in sandstones of the Athabasca Basin.

The gravity method is a potential exploration tool where deposits are shallow, but is ineffective in deeper parts of the basin where deposits may be several hundreds of metres deep. Geophysical exploration is, therefore, essentially indirect and focussed on associated structures, alteration, and rock types. Recognition of the relationship between basement graphitic metapelites (conductive), steep faults, and mineralization has resulted in electromagnetic (EM) techniques, both airborne and ground, being the principal exploration methods that narrow the search for prospective zones. Examples of electromagnetic signatures over the cluster of Sue deposits are shown in Figure 26. A MaxMin horizontal loop electromagnetic (HLEM) in-phase response characterized by a linear north-south trending low flanked by two positive shoulders defines the position of a graphitic conductor associated with the deposits (Fig. 26A) (Matthews et al., 1997). This conductor coincides approximately with the axis of a parallel, coincident, linear, apparent resistivity low outlined by an airborne frequency domain survey (coplanar coil configuration) (Fig. 26B).

Electrical techniques, such as IP/resistivity surveys, have been successful in mapping alteration related to mineralization (McMullan et al., 1987; Matthews et al., 1997). At the Cigar Lake deposit, for example, results of time-domain EM surveys failed to record signatures that could be unequivocally tied to the alteration envelope surrounding the deposit, or to the clay-rich paleoregolith at the top of the basement, but resistivity techniques have been successful in mapping alteration (McMullan et al., 1987). An apparent resistivity section crossing the Cigar Lake deposit displays a triangular zone of relatively low apparent resistivities directly above the main pod of the deposit (Fig. 27). This indicates that mineralization-related hydrothermal alteration extends upwards

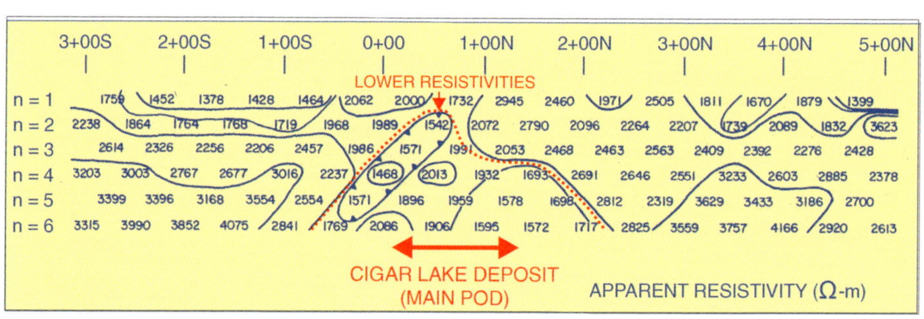

FIGURE 27. An apparent resistivity section crossing the Cigar Lake deposit, Athabasca Basin (after McMullan et al., 1987).

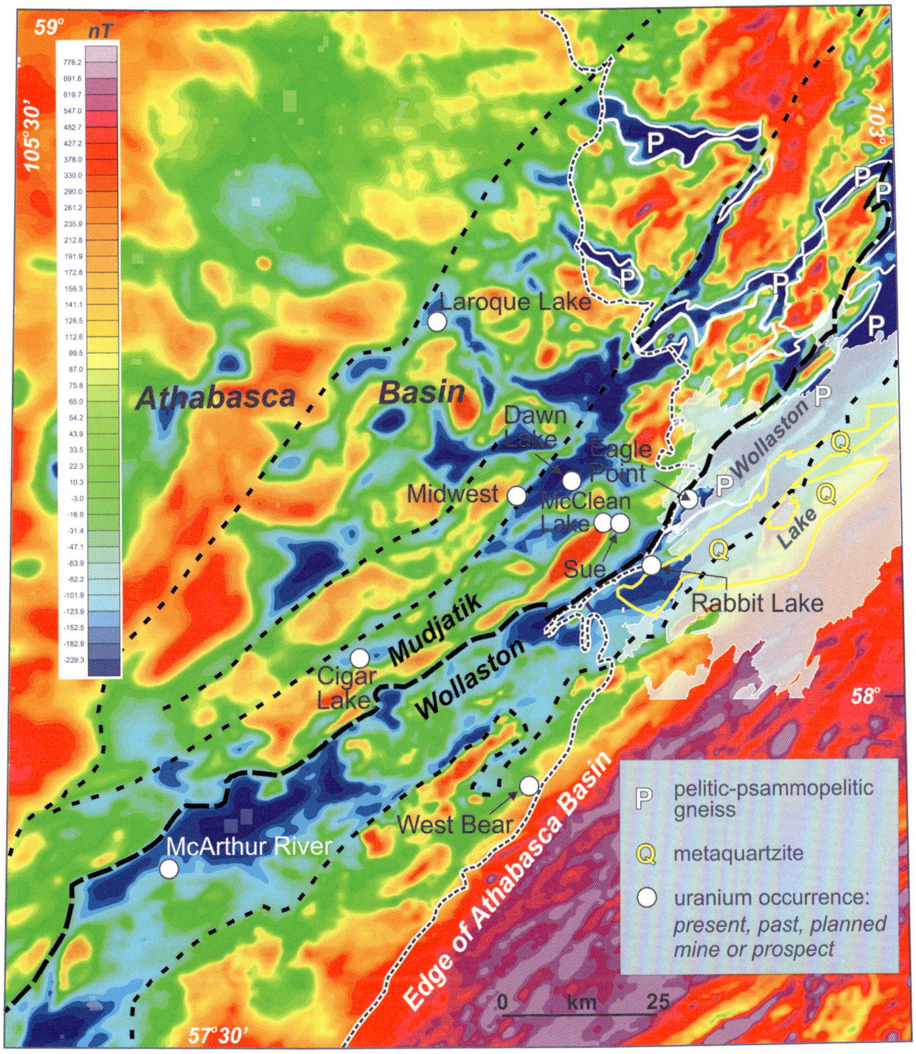

FIGURE 28. Aeromagnetic map (805 m line-spacing; 305 m flight-elevation) of the southeastern margin of the Athabasca Basin and adjacent Wollaston and Mudjatik domains of the Hearne structural province, Canadian Shield. Areas of low magnetic field (blue shades) outside the basin coincide with units of pelitic-psammopelitic gneiss and metaquartzite. Areas of low magnetic field within the basin are considered to signify the presence of similar units below the basin unconformity. These are desired targets for uranium exploration, as substantiated by the discovery of several deposits within or on the flanks of magnetic lows. Figure is based on a figure presented by Thomas and McHardy (in press).

into the Athabasca Basin for more than 400 m (McMullan et al., 1987). On a broader scale, aeromagnetic surveys have been used to map basement lithologies, effectively delineating pelitic metasedimentary rocks containing graphitic units associated with mineralization. These metasedimentary rocks coincide with magnetic lows (Fig. 28). Gravity surveys have been used to map and model faults and palaeotopographic highs on the basement, and to detect and delineate alteration halos characterized by silicification (increased density) or desilicification (decreased density), both within the basement and overlying sandstone sequences. An example of a negative gravity anomaly (~1 mGal amplitude) associated with an alteration zone within crystalline basement is observed on the southwest grid of the Kiggavik deposit near the margin of the Thelon Basin (Fig. 29). It is attributed mainly to quartz dissolution and filling of voids by water (Hasegawa et al., 1990). Seismic reflection surveys provide critical information on the location and geometry of structures, such as the basement unconformity and faults that offset it (Fig. 30) (White et al., in press). The seismic method has been effective also in outlining zones of silicification within basin sandstones, because of increases in seismic velocity related to silicification.

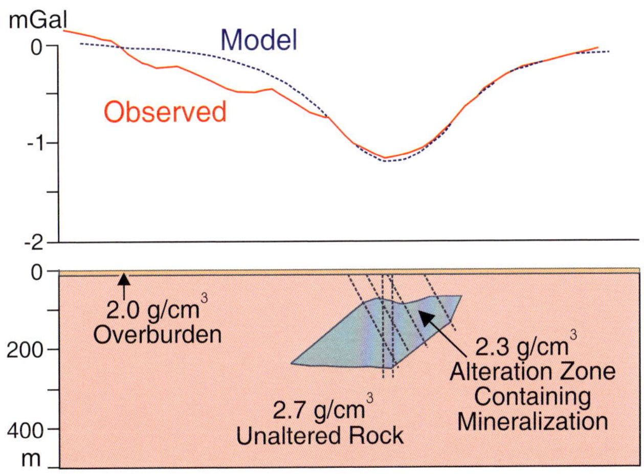

FIGURE 29. Gravity low related to uranium mineralization and alteration zone associated with the Kiggavik deposit (SW grid) within crystalline basement near the margin of the Thelon Basin (after Hasegawa et al., 1990).

Olympic Dam Type Iron Oxide-Copper-Gold Deposits

Iron oxide Cu-Au (IOCG) deposits represent an important source of Fe and are also a significant deposit type for Cu, U, and REE (Corriveau, 2007). While there are no producing mines in Canada, the most prospective targets for exploration are the Proterozoic granitic and gneissic terranes of the Canadian Shield, and parts of the Cordilleran and Appalachian orogens. Signifi-cant examples of IOCG-type deposits in Canada include the NICO and Sue Dianne deposits in the southern part of the Great Bear Magmatic Zone. Corriveau (2007) lists a number of other prospective Proterozoic IOCG districts in Canada, including the Paleoproterozoic Central Mineral Belt of Labrador and the Mesoproterozoic Wernecke Breccias in the Yukon Territory.

IOCG deposits exhibit a wide range of sulphide-deficient, low Ti, magnetite, and/or hematite orebodies of hydrothermal origin, often genetically associated with granitic to dioritic plutons with A-type geochemical affinities. They display a range of metal contents from

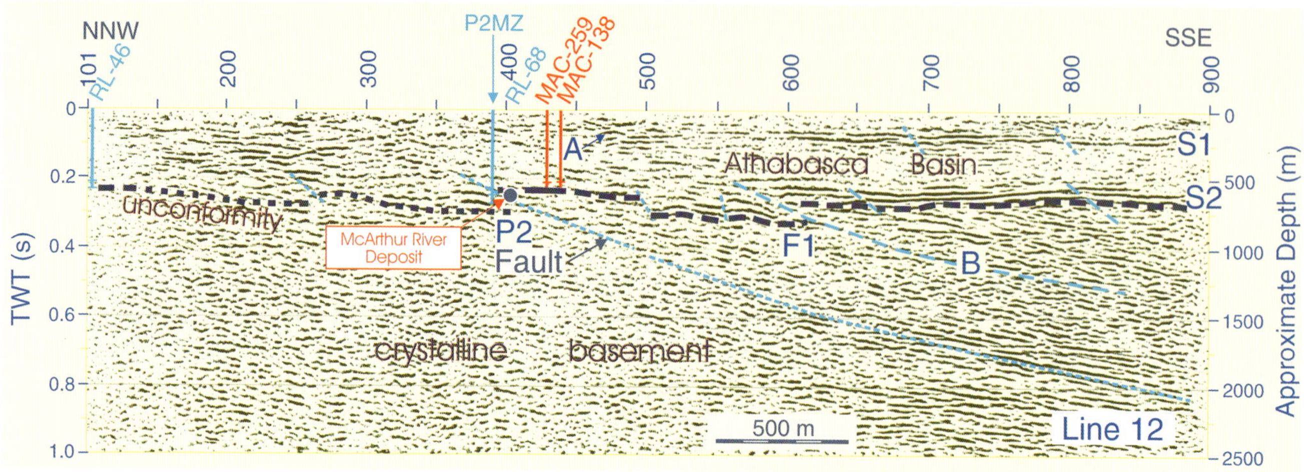

FIGURE 30. Reflection seismic section crossing the McArthur River uranium deposit, Athabasca Basin (modified from figure provided by D.J. White).

monometallic, Kiruna type (Fe ± P) to polymetallic Olympic Dam type (Fe ± Cu ± U ±Au ± REE) (Lefebure, 1995; Corriveau, 2007). These deposits occur typically as tabular and pipe-like breccia bodies, veins, and stratiform or discordant disseminations and massive lenses. Because these deposits occur in a wide variety of geological terranes, their host lithologies tend not to be diagnostic. However, the alteration assemblages can be diagnostic with a proximal potassic (sericite and K-feldspar) and Fe-oxide alteration superimposed on a generally more extensive calcic and sodic alteration. Minerals relevant to geophysical detection include Fe-oxide (hematite, martite, magnetite (low-Ti), specularite), Cu sulphides (chalcopyrite, bornite, chalcocite), uraninite, coffinite, sericite, and K-feldspar.

Smith (2002) presented a useful summary of geophysical signatures associated with ironoxide-Cu-Au deposits, and noted that the primary mineralogical characteristic is the abundance of magmatic Fe-oxide and the common relative lack of Fe sulphides. The abundance of mono- and polyphase iron oxide mineralization leads to significant regional and localized aeromagnetic anomalies. Potassic alteration often accompanies the introduction of Cu and Au, and U may also be present. Areas of K alteration may be broad and regionally extensive providing a large exploration target that will produce prominent airborne gamma-ray spectrometry anomalies. The strong magnetic and gamma-ray spectrometry signatures commonly associated with these deposits make combined airborne magnetic and gamma-ray spectrometry surveys an effective exploration tool. High eU and eU/eTh ratio anomalies directly measure U enrichment associated with mineralization, and high K and low eTh/K ratio anomalies are associated with sericitic and/or K-feldspar alteration that can be aerially extensive.

Geophysical signatures of the deposits themselves are however more complex. Whereas deposits are typically located in areas of significant magnetic relief, they do not always coincide with discrete magnetic anomalies. For example, a prominent magnetic anomaly coinciding with the Australian hematite-associated Olympic Dam deposit is apparently related to a deeper source (Smith, 2002). In the case of another Australian deposit, Ernest Henry, an apparently related magnetic response is similar to many other responses in the area that have no economic significance. This is a common characteristic because related Fe-oxide is generally more widely distributed than Cu and Au mineralization.

Smith (2002) reported that almost all known IOCG deposits produce a significant gravity response, observing that the Olympic Dam deposit produced a significant gravity signature; Ernest Henry also generated a distinct anomaly. Hematite associated with this type of deposit contributes to the gravity signature. As a result of the extensive nature of the disseminated mineralization and the electrical conductivity of breccia cores at some deposits, induced polarization and electromagnetic surveys may be locally applicable, and would complement regional magnetic and gamma-ray spectrometry surveys. Induced polarization and resistivity surveys have been widely used with notable success but will also respond to Fe-oxides and barren sulphides; electromagnetic and magnetotelluric methods have met with success in some cases (Smith, 2002).

A series of gamma-ray spectrometry images and a total magnetic field image compiled from data acquired in an airborne geophysical survey of the Mazenod Lake area, N.W.T., flown along lines spaced 500 m apart is illustrated in Figure 31 (Hetu et al., 1994). These images represent a subset of a larger dataset centred on polymetallic mineralization associated with a broad K anomaly that is coincident with a positive magnetic total field anomaly. At a regional scale, the low eTh/K ratio values distinguish the high K anomalies related to hydrothermal alteration from those associated with normal lithological variations. A gravity anomaly, amplitude 3 mGal, is coincident with the K and magnetic anomalies, and resistivity lows outline mineralized zones (Shives, et al., 1997). A geological cross-section with corresponding pseudo-magnetic and K profiles (from Gandhi et al., 1996) along line AB traversing the Mazenod Lake area Olympic Dam-type mineralization zone is displayed in Figure 32. Both K and magnetic field values are significantly enhanced over the zone.

Magmatic Nickel-Copper-Platinum Group Element Deposits

Eckstrand and Hulbert (2007) define magmatic nickel-copper-PGE deposits as a group of deposits with Ni, Cu, and platinum group elements (PGE) occurring as sulphides in a

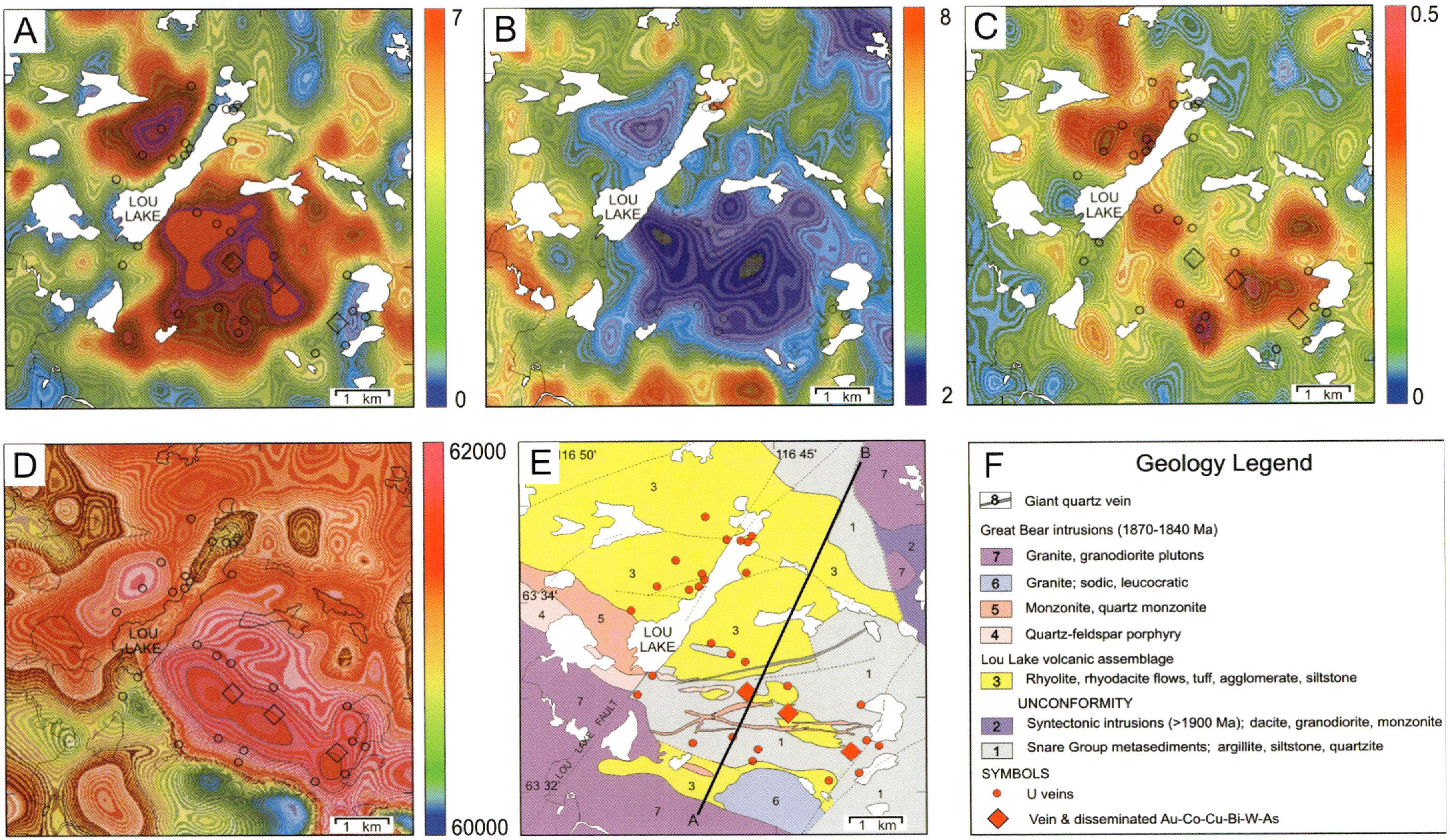

FIGURE 31. Potassium **(A)**, eThorium/Potassium **(B)**, eUranium/eThorium **(C)**; and Total magnetic field **(D)** images for an area near Mazenod Lake, Northwest Territories (Hetu et al., 1994). A geology map **(E)** and legend **(F)** (from Gandhi et al., 1996) are also illustrated.

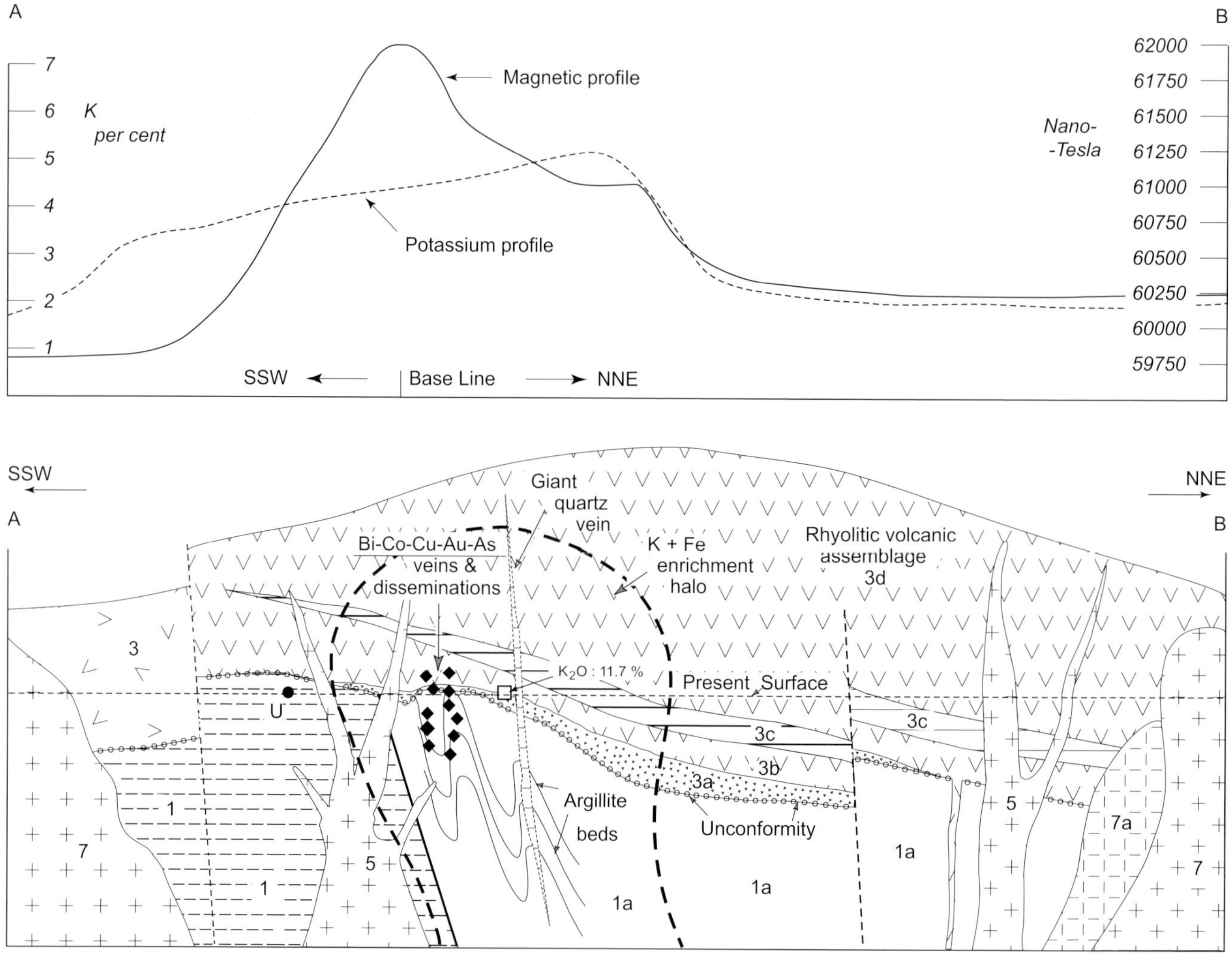

FIGURE 32. Total magnetic field and potassium profiles crossing the mineralized zone in the Mazenod Lake area, Northwest Territories. Legend for the geological cross-section: Unit 1a, quartzite, metasiltstone; Units 3a-3d, rhyolite and rhyodacite flows, ignimbrites, volcaniclastic rocks, and undivided rhyolitic rocks; Unit 5, monzonite and quartz monzonite; Unit 7a, dacite, subvolcanic porphyry (from Gandhi et al., 1996).

variety of mafic and ultramafic rocks. These deposits can be divided into two main types. The first type includes Ni-Cu deposits in which sulphide-rich ores are associated with differentiated mafic and/or ultramafic flows and sills. The second type is represented by deposits mined primarily for PGE's that are associated with sparsely dispersed sulphides in mafic/ultramafic layered intrusions.

In Ni-Cu deposits, Ni is usually the main economic commodity with Cu a co-product or by-product along with Co, PGE, and Au. The host rocks for these deposits include meteorite-impact mafic melts (Sudbury, Ontario), rift and continental flood basalt-associated mafic sills and dyke-like bodies (Noril'sk-Talnakh, Russia), komatiitic (magnesium-rich) volcanic flows and related sill-like intrusions (Thompson, Manitoba), and other mafic/ultramafic intrusions such as Voisey's Bay, Labrador (Eckstrand and Hulbert, 2007). PGE deposits can be divided into two main subtypes. These include reef-type or stratiform deposits hosted by well layered mafic/ultramafic intrusions (Merensky Reef, Bushveld Complex, South Africa) and magmatic breccia-type deposits hosted by stock-like or layered mafic/ultramafic intrusions (Lac des Iles, Ontario).

Both Ni-Cu-PGE and PGE deposits are associated with large mafic/ultramafic igneous complexes that can easily be detected, at a regional scale, by magnetic and gravity methods. This is the case, for example, for the Muskox, Thompson, and Raglan orebodies. Magnetics is therefore the main geophysical technique to map host rocks for this deposit type. The presence of pyrrhotite in the deposit may create a magnetic anomaly, although it can be lost in the strong magnetic anomalies of the associated intrusive rocks. In addition, gravity can be used to identify deposits at the detailed exploration scale because they have a high-density contrast with their host rocks.

The Ni-Cu-PGE sulphide deposits consist of highly conductive sulphides (pyrrhotite, pentlandite, and chalcopyrite) that produce large frequency domain EM anomalies. They sometimes can be difficult to detect with time-domain EM systems as their response is at the upper limit of the aperture of these systems, although they are identifiable from their slow decay. In the case of low-sulphide PGE deposits, high-resolution frequency-domain EM data are effective in identifying this type of mineralization, although they may generate only weak EM anomalies. However, high-resolution fre-

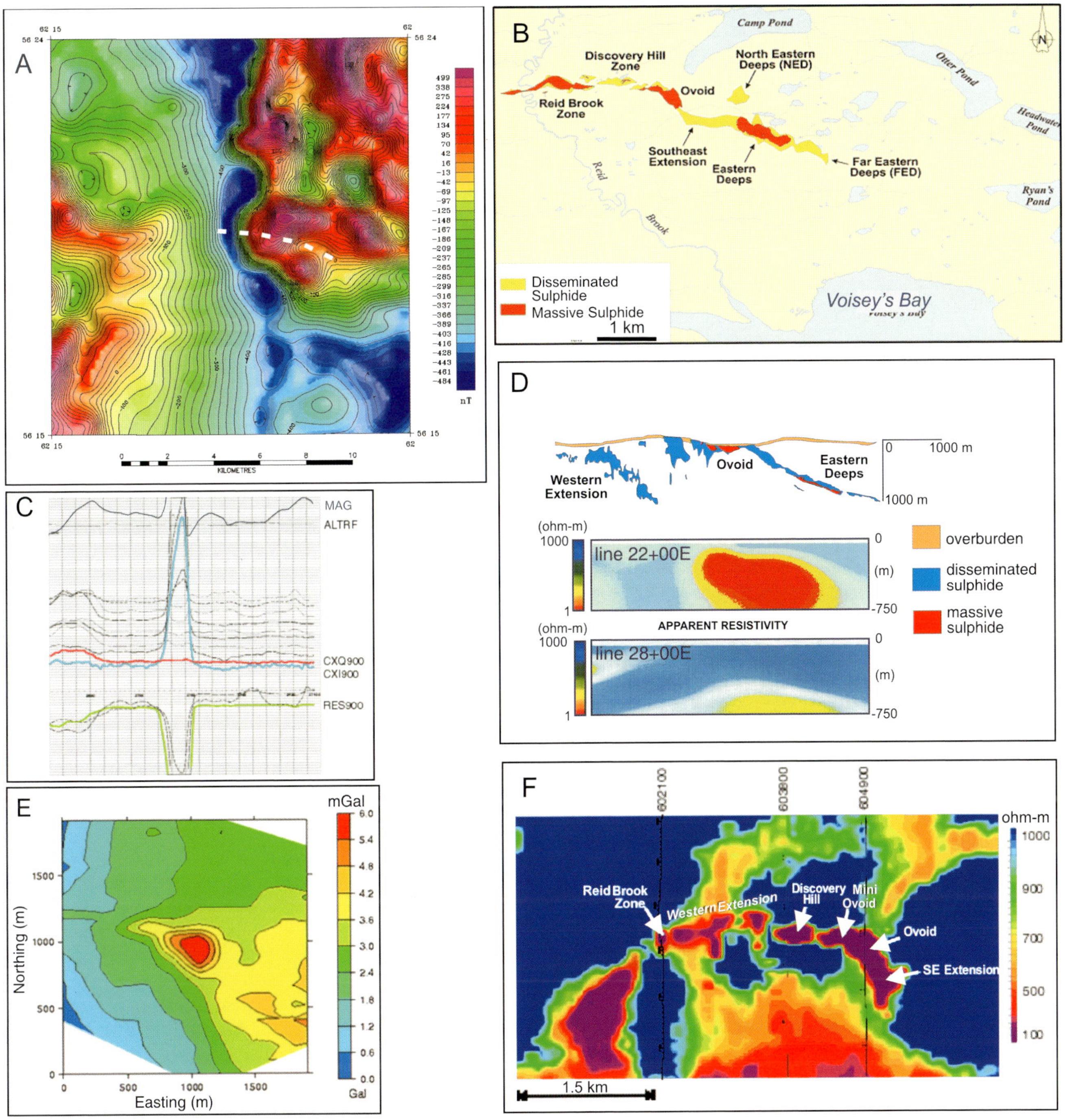

FIGURE 33. Geophysical signatures over nickel-copper-PGE orebodies, Voisey's Bay, Labrador. **(A)** Total magnetic field (white dashed line traces location of sulphide deposits); **(B)** Location and extent of the Voisey's Bay sulphide deposits (image from: www.gov.nf.ca); **(C)** strong EM response detected by a helicopter survey over the Ovoid Zone (image from: www.fugroairborne.com); **(D)** apparent resistivities calculated from an audio magnetotelluric survey over the Eastern Deeps massive sulphide body (after Balch, 1999); **(E)** Bouguer gravity anomaly, of approximately 4 mGal amplitude, over the Ovoid Zone (image from: http://www.geop.ubc.ca/ubcgif); **(F)** apparent resistivity at 900 Hz as defined by a Dighem HEM survey; the Voisey's Bay sulphide bodies are clearly outlined by an area of low resistivity (image from: www.fugroairborne.com).

quency-domain EM data can be corrected for the magnetic susceptibility of the mineralization and the host rocks and help highlight the response of low-sulphide PGE deposits (Huang and Fraser, 2000).

At the Voisey's Bay deposit, strong frequency-domain EM anomalies are observed over the Ovoid zone (Fig. 33). The apparent width of the deposit is 500 m, and the conductance is thousands of Siemens. Apparent resistivity is less than that of seawater. A small magnetic anomaly (not visible in the regional data shown above) is measured over the deposit. As a whole, there is no strong correlation between the total magnetic field and the known mineralization (Balch, 1999).

TABLE 4. Utility of geophysical methods in exploration for specific mineral deposit types.

Geophysical Method	Air or Ground	Application	Diamonds	Lode Gold	VMS Deposits	MVT Lead-Zinc Deposits	SEDEX Deposits	Porphyry Copper Deposits	Uranium Deposits	Olympic Dam-Type Deposits	Magmatic Ni-Cu-PGEs Deposits
MAGNETIC	Air	Geological Framework	High	High	High	High	High	High	High	High	High
		Direct Targeting	High	Ineffective	High	Ineffective	Moderate	High	Ineffective	High	Moderate
	Ground	Geological Framework	High	High	High	Moderate	High	High	High	High	Moderate
		Direct Targeting	High	Moderate	High	Ineffective	Moderate	High	Ineffective	High	Moderate
ELECTRO-MAGNETIC	Air	Geological Framework	Moderate	Moderate	Moderate	Ineffective	Moderate	Ineffective	High	Ineffective	High
		Direct Targeting	High	Moderate	High	Ineffective	Moderate	Ineffective	High	Moderate	High
	Ground	Geological Framework	High	Moderate	Moderate	Moderate	Moderate	Ineffective	High	Ineffective	High
		Direct Targeting	High	Ineffective	High	Moderate	Moderate	Ineffective	High	Moderate	High
ELECTRIC	Ground	Geological Framework	Ineffective	Ineffective	Moderate	Moderate	Ineffective	Ineffective	Ineffective	Moderate	High
		Direct Targeting	Moderate	Moderate	High	High	Ineffective	High	Moderate	Moderate	High
GRAVITY	Air	Geological Framework	Moderate	Moderate	Moderate	Moderate	Moderate	Moderate	Moderate	Moderate	High
		Direct Targeting	High	Ineffective	High	Moderate	Moderate	Ineffective	Ineffective	Moderate	High
	Ground	Geological Framework	High	Moderate	Moderate	Moderate	Moderate	Moderate	High	Moderate	High
		Direct Targeting	High	Ineffective	High	High	High	Moderate	Ineffective	Moderate	High
RADIOMETRIC	Air	Geological Framework	Moderate	High	High	Moderate	High	High	High	High	Moderate
		Direct Targeting	Ineffective	Moderate	Moderate	Ineffective	Ineffective	High	High	High	Ineffective
	Ground	Geological Framework	Ineffective	High	Moderate	Ineffective	Moderate	High	High	High	Moderate
		Direct Targeting	Ineffective	Moderate	Moderate	Ineffective	Ineffective	High	High	High	Ineffective
SEISMIC	Ground	Geological Framework	Moderate	Moderate	Moderate	Moderate	Moderate	Moderate	Moderate	Moderate	Moderate
		Direct Targeting	Moderate	Ineffective	Moderate	Moderate	Moderate	Moderate	Moderate	Moderate	Moderate

Qualitative Applicability Rating of Geophysical Method: High = Highly Effective; Moderate = Moderately Effective; Ineffective = Generally Ineffective

Conclusions

Geophysical techniques have made, and continue to make, a major contribution to exploration for a variety of mineral deposit types. This report provides a short overview of the more commonly used methods and illustrates the types of signatures that may be expected over specific deposit types. The suites of signatures are by no means comprehensive, and signatures may differ to varying degrees depending on local geology and mineralogy.

A qualitative assessment of the utility of the different geophysical techniques in the exploration for the different deposit types is attempted in Table 4 with respect to 1) direct detection and (2) framework mapping. In this table, direct detection also covers targets that are somewhat larger than the sought after orebody, such as a potassic alteration halo outlined by a radiometric survey. Framework mapping covers the aspect of geophysical mapping of lithology and structure, usually at a regional or semi-regional scale, but also including property-scale mapping. It is cautioned that Table 4 represents a qualitative evaluation of the efficacy of the techniques. As such it is rather generalized and it is acknowledged that examples may exist that might not be consistent with the ranking of a technique as presented.

Published documentation of geophysical attributes of Canadian ore deposits is limited and hence potentially valuable case histories that could benefit the explorationist are unavailable. In spite of this shortcoming, the existing documentation, together with numerous company web sites displaying geophysical images of properties and mineralization in various stages of exploration and evaluation, provide a reasonably complete spectrum of geophysical responses for several deposit types. Furthermore, the mineralogical and related rock property data for all types of ore deposits are sufficiently well known that a selection of suitable geophysical techniques can be readily made.

Even relatively new geophysical techniques, such as airborne scalar and full tensor gravimetry, deep EM methods that can investigate at depths of 1 or 2 km, tensor magnetics, etc., rely on the detection of physical property contrasts. The joint use of modern 3-D geophysical inversion, used to build physical property models, and 3-D geological models will certainly be essential for the discovery of future mineral deposits.

Acknowledgements

We thank Bob Holroyd and Boris Lum (Teck Cominco Limited) for the image of gravity profiles crossing the Polaris deposit, Nunavut, Sue Davis and Rob Shives (Geological Survey of Canada), respectively, for drafting services and discussions relating to gamma-ray spectrometry data, Wayne Goodfellow (Geological Survey of Canada) for editorial comment, and Bill Morris (McMaster University) for refereeing the manuscript.

References

Ash, C., and Alldrick, D., 1996, Au-quartz veins, *in* Lefebure, D.V., and Höy, T., eds., Selected British Columbia Mineral Deposit Profiles, Volume 2 – Metallic Deposits: British Columbia Ministry of Employment and Investment, Open File 1996-13, p. 53-56.

Balch, S.E., 1999, Ni-Cu sulphide deposits with examples from Voisey's Bay, *in* Lowe, C., Thomas, M.D., and Morris, W.A., eds., Geophysics in Mineral Exploration: Fundamentals and Case Histories: Geological Association of Canada, Short Course Notes, v. 14, p. 21-40.

Belland, M., 1992, The Birth of the Bathurst Mining Camp: A Development History of the Austin Brook Iron Mine and No. 6 Base-Metal Deposit: New Brunswick Department of Natural Resources and Energy, Popular Geology Paper 92-1, 56 p.

Bishop, J.R., and Emerson, D.W., 1999, Geophysical properties of zinc-bearing minerals: Australian Journal of Earth Sciences, v. 46, p. 311-328.

Brock, J.S., 1973, Geophysical exploration leading to the discovery of the Faro deposit: Canadian Mining and Metallurgical Bulletin, v. 66, p. 97-116.

Broome, J., Carson, J.M., Grant, J.A, and Ford, K.L., 1987, A Modified Ternary Radiolement Mapping Technique and Its Application to the South Coast of Newfoundland: Geological Survey of Canada, Paper 87-14, scale 1:500 000.

Brownell, G.M., 1950, Radiation surveys with a scintillation counter: Economic Geology, v. 45, p. 167-174.

Brummer, J.J., MacFadyen, D.A., and Pegg, C.C., 1992, Discovery of kimberlites in the Kirkland Lake area, Northern Ontario, Canada; Part II: Kimberlite discoveries, sampling, diamond content, ages and emplacement: Exploration Mining Geology, v. 1, p. 351-370.

Carmichael, R.S., 1982, Practical Handbook of Physical Properties of Rocks, Volume II: CRC Press, 340 p.

Clark, D.A., French, D.H., Lackie, M.A., and Schmidt, P.W., 1992, Magnetic petrology: Application of integrated rock magnetic and petrological techniques to geological interpretation of magnetic surveys: Exploration Geophysics, v. 23, p. 65-58.

Corriveau, L., 2007, Iron oxide copper-gold deposits: A Canadian perspective, *in* Goodfellow, W.D., ed., Mineral Deposits of Canada: A Synthesis of Major Deposit-Types, District Metallogeny, the Evolution of Geological Provinces, and Exploration Methods: Geological Association of Canada, Mineral Deposits Division, Special Publication No. 5, p. 307-328.

Dewing, K., Sharp, R.J., and Turner, E., 2007, Synopsis of the Polaris Zn-Pb district, Canadian Arctic Islands, Nunavut, *in* Goodfellow, W.D., ed., Mineral Deposits of Canada: A Synthesis of Major Deposit-Types, District Metallogeny, the Evolution of Geological Provinces, and Exploration Methods: Geological Association of Canada, Mineral Deposits Division, Special Publication No. 5, p. 655-672.

Dickson, B.L., and Scott, K.M., 1997, Interpretation of aerial gamma-ray surveys – adding the geochemical factors: AGSO Journal of Australian Geology and Geophysics, v. 17, p. 187-200.

Dubé, B., and Gosselin, P., 2007, Greenstone-hosted quartz-carbonate vein deposits, *in* Goodfellow, W.D., ed., Mineral Deposits of Canada: A Synthesis of Major Deposit-Types, District Metallogeny, the Evolution of Geological Provinces, and Exploration Methods: Geological Association of Canada, Mineral Deposits Division, Special Publication No. 5, p. 49-73.

Dubé, B., Gosselin, P., Hannington, M.D., and Galley, A., 2007, Gold-rich volcanogenic massive sulphide deposits, *in* Goodfellow, W.D., ed., Mineral Deposits of Canada: A Synthesis of Major Deposit-Types, District Metallogeny, the Evolution of Geological Provinces, and Exploration Methods: Geological Association of Canada, Mineral Deposits Division, Special Publication No. 5, p. 75-94.

Durocher, M.E., 1983, The nature of hydrothermal alteration associated with the Madsen-Starrett-Olsen deposits, Red lake area, *in* Colvine, A.C., ed., The Geology of Gold in Ontario: Ontario Geological Survey, Miscellaneous Paper 110, p. 123-140.

Eckstrand, O.R., and Hulbert, L., 2007, Magmatic nickel-copper-platinum group element deposits, *in* Goodfellow, W.D., ed., Mineral Deposits of Canada: A Synthesis of Major Deposit-Types, District Metallogeny, the Evolution of Geological Provinces, and Exploration Methods: Geological Association of Canada, Mineral Deposits Division, Special Publication No. 5, p. 205-222.

Fraser, D.C., 1973, Magnetite ore tonnage estimates from an aerial electromagnetic survey: Geoexploration, v. 11, p. 97-105.

Galley, A., Hannington, M., and Jonasson, I., 2007, Volcanogenic massive sulphide deposits, *in* Goodfellow, W.D., ed., Mineral Deposits of Canada: A Synthesis of Major Deposit-Types, District Metallogeny, the Evolution of Geological Provinces, and Exploration Methods: Geological Association of Canada, Mineral Deposits Division, Special Publication No. 5, p. 141-161.

Gandhi, S.S., Prasad, N., and Charbonneau, B.W., 1996, Geological and geophysical signatures of a large polymetallic exploration target at Lou Lake, southern Great Bear magmatic zone, Northwest Territories: Geological Survey of Canada, Current Research 1996-E, p. 147-158.

Gendzwill, D.J., and Matieshin, S.D., 1996, Seismic reflection survey of a kimberlite intrusion in the Fort à la Corne district, Saskatchewan, *in* Lecheminant, A.N., Richardson, D.G., DiLabio, R.N.W., and Richardson, K.A., eds., Searching for Diamonds in Canada: Geological Survey of Canada, Open File 3228, p. 251-253.

Geological Survey of Canada, 1991, Airborne geophysical survey, Mount Milligan area, British Columbia (NTS 93O/4W, N/1, N/2E): Open File 2535, scale 1:100 000.

Ghosh, M.K., 1972, Interpretation of airborne EM measurements based on thin sheet models: Research in Applied Geophysics, No. 4, University of Toronto.

Goodacre, A.K., Brule, B.G., and Cooper, R.V., 1969, Results of regional underwater gravity surveys in the Gulf of St. Lawrence: Department of Energy, Mines and Resources, Gravity Map Series of the Earth Physics Branch, No. 86, Ottawa.

Goodfellow, W.D., and Lydon, J., 2007, Sedimentary-exhalative (SEDEX) deposits, *in* Goodfellow, W.D., ed., Mineral Deposits of Canada: A Synthesis of Major Deposit-Types, District Metallogeny, the Evolution of Geological Provinces, and Exploration Methods: Geological Association of Canada, Mineral Deposits Division, Special Publication No. 5, p. 163-183.

Grant, F.S., and West, G.F., 1965, Interpretation Theory in Applied Geophysics: International Series in the Earth Sciences, McGraw-Hill Book Company, New York, 584 p.

Grasty, R.L., and Minty, B.R.S., 1995, A Guide to the Technical Specifications for Airborne Gamma-Ray Surveys: Australian Geological Survey Organization, Record 1995/60, 89 p.

Grasty, R.L., Mellander, H., and Parker, M., 1991, Airborne Gamma-Ray Spectrometer Surveying: International Atomic Energy Agency, Technical Report Series 323, Vienna, 97 p.

Gunn, P.J., and Dentith, M.C., 1997, Magnetic responses associated with mineral deposits: AGSO Journal of Australian Geology and Geophysics, v. 17, p. 145-158.

Hannigan, P.K., 2007, Metallogeny of the Pine Point Mississippi Valley-type zinc-lead district, southern Northwest Territories, *in* Goodfellow, W.D., ed., Mineral Deposits of Canada: A Synthesis of Major Deposit-Types, District Metallogeny, the Evolution of Geological Provinces, and Exploration Methods: Geological Association of Canada, Mineral Deposits Division, Special Publication No. 5, p. 609-632.

Hasegawa, K., Davidson, G.I., Wollenberg, P., and Iida, Y., 1990, Geophysical exploration for unconformity-related uranium deposits in the northeastern part of the Thelon Basin, Northwest Territories, Canada: Mining Geology, v. 40, p. 83-94.

Hearst, R.B., Morris, W.A., and Schieck, D.G., 1994, Reflection seismic profiling for massive sulphides in the high Arctic: 64th Annual International Meeting, Society of Exploration Geophysicists, p. 531-533.

Hetu, R.J., 1991, Airborne Geophysical Survey, Red Lake, Ontario: Geological Survey of Canada, Open File 2403, 41 p.

Hetu, R.J., Holman, P.B., Charbonneau, B.W., Prasad, N., and Gandhi, S.S., 1994, Multiparameter Airborne Geophysical Survey of the Mazenod Lake Area, Northwest Territories: Geological Survey of Canada, Open File 2806, 100 p.

Hood, P.J., and Teskey, D.J., 1989, Aeromagnetic gradiometer program of the Geological Survey of Canada: Geophysics, v. 54, p. 1012-1022.

Hoover, D.B., and Campbell, D.L., 1992, Geophysical models of diamond pipes, Cox and Singer model 12, *in* Hoover, D.B., Heran, W.D., and Hill, P.L., eds., The Geophysical Expression of Selected Mineral Deposits: United States Department of the Interior, Geological Survey, Open File Report 92-557, p. 85-88.

Horsfall, K.R., 1997, Airborne magnetic and gamma-ray data acquisition: AGSO Journal of Australian Geology and Geophysics, v. 17, p. 23-30.

Huang, H., and Fraser, D. C., 2000, Airborne resistivity and susceptibility mapping in magnetically polarizable areas. Geophysics, v. 65, p. 502-511.

Hunt, C.P., Moskowitz, B.M., and Banerjee, S.K., 1995, Magnetic properties of rocks and minerals, *in* Ahrens, T.J., ed., Rock Physics and Phase Relations: A Handbook of Physical Constants: American Geophysical Union Reference Shelf 3, American Geophysical Union, p. 189-204.

International Atomic Energy Agency, 2003, Guidelines for radioelement mapping using gamma ray spectrometry data. IAEA-TECDOC-1363, 173 p.

Jefferson, C.W., Thomas, D.J., Gandhi, S.S., Ramaekers, P., Delaney, G., Brisbin, D., Cutts, C., Quirt, D., Portella, P., and Olson, R.A., 2007, Unconformity-associated uranium deposits of the Athabasca Basin, Saskatchewan and Alberta, *in* Goodfellow, W. D., ed., Mineral Deposits of Canada: A Synthesis of Major Deposit-types, District Metallogeny, the Evolution of Geological Provinces, and Exploration Methods, Mineral Deposits Division, Geological Association of Canada, Special Publication 5, p. 273-305.

Katsube, T.J., and Kjarsgaard, B.A., 1996, Physical characteristics of Canadian kimberlites, *in* Lecheminant, A.N., Richardson, D.G., DiLabio, R.N.W., and Richardson, K.A., eds., Searching for Diamonds in Canada: Geological Survey of Canada, Open File 3228, p. 241-242.

Keating, P., 1995, A simple technique to identify magnetic anomalies due to kimberlite pipes: Mining and Exploration Geology, v. 4, p. 121-125.

——— 1996, Kimberlites and aeromagnetics, *in* Lecheminant, A.N., Richardson, D.G., DiLabio, R.N.W., and Richardson, K.A., eds., Searching for Diamonds in Canada: Geological Survey of Canada, Open File 3228, p. 233-236.

Keller, G.V., and Frischknecht, F.C., 1966, Electrical Methods in Geophysical Prospecting: International Series on Monographs on Electromagnetic Waves, Volume 10: Pergamon Press Inc., 519 p.

Killeen, P.G., 1979, Gamma-ray spectrometric methods in uranium exploration – application and interpretation, *in* Hood, P.J., ed., Geophysics and Geochemistry in Search for Metallic Ores: Geological Survey of Canada, Economic Geology Report 31, p. 163-230.

Kjarsgaard, B.A., 1996, Kimberlites, *in* Lecheminant, A.N., Richardson, D.G., DiLabio R.N.W., and Richardson, K.A., eds., Searching for Diamonds in Canada: Geological Survey of Canada, Open File 3228, p. 29-37.

——— 2007, Kimberlite diamond deposits, *in* Goodfellow, W.D., ed., Mineral Deposits of Canada: A Synthesis of Major Deposit-Types, District Metallogeny, the Evolution of Geological Provinces, and Exploration Methods: Geological Association of Canada, Mineral Deposits Division, Special Publication No. 5, p. 245-272.

Klein, C., and Hurlbut, C.S., 1985, Manual of Mineralogy (after J.D. Dana), 20th Edition: John Wiley and Sons, New York, 596 p.

Klein, D.P., and Bankey, V., 1992, Geophysical model of Creede, Comstock, Sado, Goldfield and related epithermal precious metal deposits, *in* Hoover, D.B., Heran, W.D., and Hill, P.L., eds., The Geophysical Expression of Selected Mineral Deposits: United States Department of the Interior, Geological Survey, Open File Report 92-557, p. 98-106.

Lajoie, J.L., and Klein, J., 1979, Geophysical exploration at the Pine Point Mines Ltd. zinc-lead property, Northwest Territories, Canada, *in* Hood, P.J., ed., Geophysics and Geochemistry in the Search for Metallic Ores: Proceedings of Exploration 1977, Geological Survey of Canada, Economic Geology Report 31, p. 652-664.

Lane, R.J.L., ed., 2004, Abstract from the ASEG-PESA Airborne Gravity 2004 Workshop: Geoscience Australia Record 2004/18, 141 p.

Lang, A.H., 1953, The development of portable geiger counters: Canadian Mining Journal, v. 74, p. 65-68.

Lecheminant, A.N., and Kjarsgaard, B.A., 1996, Part 1: Geology, petrology, and geotectonic controls – Introduction, *in* Lecheminant, A.N., Richardson, D.G., DiLabio, R.N.W., and Richardson, K.A., eds., Searching for Diamonds in Canada: Geological Survey of Canada, Open File 3228, p. 5-9.

Lefebure, D.V., 1995, Iron oxide breccias and veins P-Cu-Au-Ag-U, *in* Lefebure, D.V. and Ray, G.E., eds., Selected British Columbia Mineral Deposit Profiles, Volume 1 – Metallics and Coal: British Columbia Ministry of Energy of Employment and Investment, Open File 1995-20, p. 33-36.

Lehnert-Thiel, K., Loewer, R., Orr, R. G., and Robertshaw, P., 1992, Diamond-bearing kimberlites in Saskatchewan, Canada: The Fort à la Corne case history: Exploration and Mining Geology, v. 1, p. 391-403.

Lowe, C., Brown, D.A., Best, M.E., and Shives, R.B., 2000, High resolution geophysical survey of the Purcell Basin and Sullivan deposit: Implications for bedrock geology and mineral exploration, *in* Lydon, J.W., Slack, J.F., Höy, T., and Knapp, M.E., eds., The Geological Environment of the Sullivan Deposit, British Columbia: Geological Association of Canada, Mineral Deposits Division, Special Volume No. 1, p. 279-293.

Lydon, J.W., 1984, Volcanogenic massive sulphide deposits, Part 1: A descriptive model: Geoscience Canada, v. 11, p. 195-202.

——— 1995, Sedimentary exhalative sulphides (SEDEX), *in* Eckstrand, O.R., Sinclair, W.D., and Thorpe, R.I., eds., Geology of Canadian Mineral Deposit Types: Geological Survey of Canada, Geology of Canada, no. 8, p. 130-152 (also Geological Society of America, Geology of North America, v. P-1).

Lydon, J.W., Goodfellow, W.D., Dubé, B., Paradis, S., Sinclair, W.D., Corriveau, L., and Gosselin, P., 2004, A Preliminary Overview of Canada's Mineral Resources: Geological Survey of Canada, Open File 4668, 20 p.

Macnae, J.C., 1979, Kimberlites and exploration geophysics: Geophysics, v. 44, p. 1395-1416.

Macnae, J., 1995, Applications of geophysics for the detection and exploration of kimberlites and lamproites, *in* Griffin, W.L., ed., Diamond Exploration into the 21st Century: Journal of Geochemical Exploration, v. 53, p. 213-243.

Matthews, R., Koch, R., and Leppin, M., 1997, Paper 130: Advances in integrated exploration for unconformity uranium deposits in Western Canada, *in* Gubins, A.G., ed., Proceedings of Exploration 97: Fourth Decennial International Conference on Mineral Exploration: Geophysics and Geochemistry at the Millenium: p. 993-1024 (CD-ROM).

McMullan, S.R., Matthews, R.B., and Robertshaw, P., 1987, Exploration geophysics for Athabasca uranium deposits, *in* Garland, G.D., ed., Proceedings of Exploration '87: Third Decennial International Conference on Geophysical and Geochemical Exploration for Minerals and Groundwater: Ontario Geological Survey, Special Volume 3, p. 547-566.

Minty, B.R.S., 1997, Fundamentals of airborne gamma-ray spectrometry: AGSO Journal of Australian Geology and Geophysics, v. 17, p. 39-50.

Minty, B.R.S., Luyendyk, A.P.J., and Brodie, R.C., 1997, Calibration and data processing for airborne gamma-ray spectrometry: AGSO Journal of Australian Geology and Geophysics, v. 17, p. 51-62.

Mitchell, R.H., 1986, Kimberlites: Mineralogy, Geochemistry, and Petrography: Plenum Press, New York, 442 p.

Mwenifumbo, C.J., and Kjarsgaard, B., 2002, Kimberlite investigations using borehole geophysics: Abstracts of Saskatoon 2002 Geological Association of Canada-Mineralogical Association of Canada Convention.

Palacky, G., 1986, Airborne Resistivity Mapping: Geological Survey of Canada, Paper 86-2, 195 p.

Panteleyev, A., 1996a, Epithermal Au-Ag: Low Sulphidation, *in* Lefebure, D.V., and Höy, T., eds., Selected British Columbia Mineral Deposit Profiles, Volume 2 – Metallic Deposits: British Columbia Ministry of Employment and Investment, Open File 1996-13, p. 41-44.

——— 1996b, Epithermal Au-Ag-Cu: High Sulphidation, *in* Lefebure, D.V., and Höy, T., eds., Selected British Columbia Mineral Deposit Profiles, Volume 2 – Metallic Deposits: British Columbia Ministry of Employment and Investment, Open File 1996-13, p. 37-39.

Paradis, S., and Nelson, J.L., 2007, Metallogeny of the Robb Lake carbonate-hosted zinc-lead district, northeastern British Columbia, *in* Goodfellow, W.D., ed., Mineral Deposits of Canada: A Synthesis of Major Deposit-Types, District Metallogeny, the Evolution of Geological Provinces, and Exploration Methods: Geological Association of Canada, Mineral Deposits Division, Special Publication No. 5, p. 633-654.

Paradis, S., Hannigan, P., and Dewing, K., 2007, Mississippi Valley-type lead-zinc deposits, *in* Goodfellow, W.D., ed., Mineral Deposits of Canada: A Synthesis of Major Deposit-Types, District Metallogeny, the Evolution of Geological Provinces, and Exploration Methods: Geological Association of Canada, Mineral Deposits Division, Special Publication No. 5, p. 185-203.

Peter, J.M., Kjarsgaard, I.M., and Goodfellow, W.D., 2003. Hydrothermal sedimentary rocks of the Heath Steele belt, Bathurst mining camp, New Brunswick: Part 1. Mineralogy and mineral chemistry, *in* Goodfellow, W.D., McCutcheon, S.R., and Peter J.M., eds., Massive Sulfide

Deposits of the Bathurst Mining Camp, New Brunswick, and Northern Maine: Economic Geology, Monograph, 11, p. 361-390.

Reed, L.E., 1993, The application of geophysics to exploration for diamonds, *in* Dunne, K.P.E., and Grant, B., eds., Mid-Continent Diamonds: Geological Association of Canada – Mineralogical Association of Canada, Symposium Volume, p. 21-26.

Reford, M.S., 1980, Magnetic method: Geophysics, v. 45, p. 1640-1658.

Richardson, K.A., 1996, Part 3: Geophysical exploration and geographic information system (GIS) applications – Introduction, *in* Lecheminant, A.N., Richardson, D.G., DiLabio, R.N.W., and Richardson, K.A., eds., Searching for Diamonds in Canada: Geological Survey of Canada, Open File 3228, p. 225-228.

Rowe, J.D., Smith, R.S., and Beattie, D., 1995, Geotem/aeromagnetic definition of hydrocarbon alteration plumes in Nevada: Geoterrex, CGG Canada brochure, reprinted from CGG World, v. 26, April 1995.

Ruzicka, V., 1995, Unconformity-associated uranium, *in* Eckstrand, O.R., Sinclair, W.D., and Thorpe, R.I., eds., Geology of Canadian Mineral Deposit Types: Geological Survey of Canada, Geology of Canada, No. 8, p. 197-210 (also Geological Society of America, Geology of North America, v. P-1).

Sangster, D.F., 1995, Mississippi Valley-type lead zinc, *in* Eckstrand, O.R., Sinclair, W.D., and Thorpe, R.I., eds., Geology of Canadian Mineral Deposit Types: Geological Survey of Canada, Geology of Canada, No. 8, p. 253-261 (also Geological Society of America, Geology of North America, v. P-1).

Schiller, E.A., 1979, The history of uranium discoveries in northern Saskatchewan in the last decade, *in* Parslow, G.R., Uranium Exploration Techniques, 1978 November 16-17, Regina Symposium Proceedings: Saskatchewan Geological Society, Special Publication No. 4, p. 11-18.

Scott, F., 1983, Midwest Lake uranium discovery, *in* Cameron, E.M., ed., Uranium Exploration in Athabasca Basin, Saskatchewan, Canada: Geological Survey of Canada, Paper 82-11, p. 41-50.

Seigel, H.O., and Pitcher, D.H., 1978, Mapping Earth conductivities using a multifrequency airborne electromagnetic system: Geophysics, v. 43, p. 563-575.

Seigel, H.O., Hill, H.L., and Baird, J.G., 1968, Discovery case history of the Pyramid orebodies Pine Point, Northwest Territories, Canada: Geophysics, v. 33, p. 645-656.

Shives, R.B.K., and Holman, P.B., 1995, Digital Data for an Airborne Gamma-Ray Spectrometric, Magnetic, VLF-EM Survey, Paul Lake, Northwest Territories (parts of NTS 76D/9, 10): Geological Survey of Canada, Open File 3123.

Shives, R.B.K., Ford, K.L., and Charbonneau, B.W., 1995, Applications of Gamma-Ray Spectrometric/Magnetic/VLF-EM Surveys – Workshop Manual: Geological Survey of Canada, Open File 3061, 82 p.

Shives, R.B.K., Charbonneau, B.W., and Ford, K.L., 1997, The detection of potassic alteration by gamma-ray spectrometry - Recognition of alteration related to mineralization, *in* Gubins, A.G., ed., Proceedings of Exploration 97: Fourth Decennial International Conference on Mineral Exploration: Geophysics and Geochemistry at the Millenium, p. 741-752.

Shives, R.B.K., Wasyluik, K., and Zaluski, G., 2000, Detection of K-enrichment (illite) chimneys using ground gamma ray spectrometry, McArthur River area, northern Saskatchewan, *in* Summary of Investigations 2000, Volume 2: Saskatchewan Geological Survey, Saskatchewan Energy Mines, Miscellaneous Report 2000-4.2, p. 160-169.

Shives, R.B.K., Carson, J.M., Dumont, R., Ford, K.L., Holman, P.B., Diakow, L., 2004, Airborne multisensor geophysical survey, Toodoggone River Area, British Columbia (Parts of NTS 94D/15, E/2, 3, 5, 7, 10, 11): Geological Survey of Canada, Open File 4613, 1 CD-ROM.

Sillitoe, R.H., 1979, Some thoughts on gold-rich porphyry copper deposits: Mineralium Deposita, v.14, p. 161-174.

Sinclair, W.D., 2007, Porphyry deposits, *in* Goodfellow, W.D., ed., Mineral Deposits of Canada: A Synthesis of Major Deposit-Types, District Metallogeny, the Evolution of Geological Provinces, and Exploration Methods: Geological Association of Canada, Mineral Deposits Division, Special Publication No. 5, p. 223-243.

Smith, R.J., 2002, Geophysics of iron oxide copper-gold deposits, *in* Porter, T.M., ed., Hydrothermal Iron Oxide Copper-Gold and Related Deposits: A Global Perspective, Volume 2: PGC Publishing, Adelaide, p. 357-367.

Smith, R.S., Annan, A.P., Lemieux, J., and Pedersen, R.N., 1996, Application of a modified GEOTEM system to reconnaissance exploration for kimberlites in the Point Lake area, NWT, Canada: Geophysics, Society of Exploration Geophysists, v. 61, p. 82-92.

St. Pierre, M., 1999, Geophysical properties of the BHP/Dia Met kimberlites, NWT Canada, *in* Lowe, C., Thomas, M.D., and Morris, W.A., eds., Geophysics in Mineral Exploration: Fundamentals and Case Histories: Geological Association of Canada, Short Course Notes, v. 14, p. 63-72.

Telford, W.M., Geldart, L.P., and Sheriff, R.E., 1990, Applied Geophysics (Second Edition): Cambridge University Press, 770 p.

Thomas, M.D., 1997, Gravity and Magnetic Prospecting for Massive Sulphide Deposits: A Short Course Sponsored under the Bathurst Mining Camp Extech II Initiative: Geological Survey of Canada, Open File 3514, 66 p.

——— 2003, Gravity signatures of massive sulphide deposits, Bathurst Mining Camp, New Brunswick, Canada, *in* Goodfellow, W.D., McCutcheon, S.R., and Peter J.M., eds., Massive Sulfide Deposits of the Bathurst Mining Camp, New Brunswick, and Northern Maine: Economic Geology, Monograph, 11, p. 799-817.

Thomas, M.D., and McHardy, S., in press, Magnetic insights into basement geology in the area of McArthur River uranium deposit, Athabasca Basin, Saskatchewan, *in* Jefferson, C.W., and Delaney, G., eds., EXTECH IV: Geology and Uranium EXploration TECHnology of the Proterozoic Athabasca Basin, Saskatchewan, and Alberta: Geological Survey of Canada, Bulletin 588 (Saskatchewan Geological Society, Special Publication 17; Geological Association of Canada, Mineral Deposits Division, Special Publication 4).

Thomas, M.D., Walker, J.A., Keating, P., Shives, R., Kiss, F., and Goodfellow, W.D., 2000, Geophysical atlas of massive sulphide signatures, Bathurst Mining Camp, New Brunswick: Geological Survey of Canada, Open File 3887, New Brunswick Department of Natural Resources and Energy, Minerals and Energy Division, Open File 2000-4, 105 p.

Tuach, J., Hewton, R.S., and Cavey, G., 1991, Exploration targets for volcanogenic, base-metal sulphide deposits on Pilley's Island, Newfoundland, *in* Ore Horizons, Volume 1: Newfoundland and Labrador Department of Mines and Energy, Geological Surveys Branch, p. 89-98.

van Staal, C.R., 1994, Geology, Wildcat Brook, New Brunswick: Geological Survey of Canada, Map 1803A, scale 1:20 000.

White, D.J., Hajnal, Z., Roberts, B., Gyorfi, I., Reilkoff, B., Bellefleur, G., Mueller, C., Woelz, S., Mwenifumbo, J., Takacs, E., Schmitt, D.R., Brisbin, D., Jefferson, C.W., Koch, R., Powell, B., and Annesley, I.R., in press, Seismic methods for uranium exploration: An overview of EXTECH IV seismic studies at the McArthur River mining camp, Athabasca Basin, Saskatchewan, *in* Jefferson, C.W., and Delaney, G., eds., EXTECH IV: Geology and Uranium EXploration TECHnology of the Proterozoic Athabasca Basin, Saskatchewan and Alberta: Geological Survey of Canada, Bulletin 588 (Saskatchewan Geological Society, Special Publication 17; Geological Association of Canada, Mineral Deposits Division, Special Publication 4).

Wilford, J.R., Bierwirth, P.N., and Craig, M.A., 1997, Application of airborne gamma-ray spectrometry in soil/regolith mapping and applied geomorphology: AGSO Journal of Australian Geology and Geophysics, v. 17, p. 201-216.

Referenced Websites

http://www.bellgeo.com
http://www.bhpbilliton.com
http://www.fugroairborne.com
http://www.eos.ubc.ca/ubcgif
http://www.gov.nf.ca
http://www.quantecgeoscience.com
http://www.sgl.com

Web Sites of Interest

Geophysics Courses

http://geo.polymtl.ca, École Polytechnique de Montréal (Fr)
http://galitzin.mines.edu/INTROGP/index.jsp, Colorado School of Mines (En)
http://www.unites.uqam.ca/~sct/flux_de_chaleur/titre.html, UQAM (Fr)

http://www-ig.unil.ch/cours/, Université de Lausanne (En, Fr, Espagnol)
http://phineas.u-strasbg.fr/marquis/Enseignement/Public/Cours_Elmag/ EOST (Fr)
http://www.geop.ubc.ca/ubcgif/tutorials/index.html, UBC (En)
http://courses.geo.ucalgary.ca/, University of Calgary (En)
http://www.usask.ca/geology/, University of Saskatchewan (En)

Free Geophysics Books

http://sepwww.stanford.edu/sep/prof/index.html

Application of Seismic Methods to Mineral Exploration

Matthew Salisbury[1] and David Snyder[2]

1. Geological Survey of Canada, 1 Challenger Drive, Dartmouth, Nova Scotia B2Y 4A2
2. Geological Survey of Canada, 615 Booth Street, Ottawa, Ontario K1A 0E9
Corresponding author's email: msalisbu@nrcan-rncan.gc.ca

Abstract

Plots of compressional (Vp) and shear (Vs) wave velocity vs. density for rocks at elevated confining pressures show that velocities tend to increase with density along the well known Nafe-Drake curves for silicate rocks. Because of their high densities, however, many ores fall far to the right of the Nafe-Drake curves and display higher impedances than their common hosts, suggesting that it should be possible to detect and prospect for ores using high-resolution reflection techniques if the deposits meet the size, thickness, and presentation constraints required for reflection or diffraction. Experiments conducted by the Geological Survey of Canada, universities, and industry in hardrock mining camps across Canada over the past decade show that 2-D surveys are well suited to the determination of structure and the detection of orebodies, while 3-D surveys may be used for detection and delineation, and vertical seismic profiling (VSP) for delineation. Due to the small size of most deposits, the structural complexity of hard rock terranes and their low signal-to-noise ratios, the best results are obtained from carefully designed surveys using high frequency sources and customized processing sequences designed to identify both reflections and diffractors.

2-D and 3-D surveys have successfully detected and imaged large massive sulphide deposits such as the magmatic and volcanic massive sulphide (VMS) deposits in Sudbury and Bathurst and should also be useful for the detection of massive sedimentary exhalitive (SEDEX) and iron oxide copper gold (IOCG) deposits. Other types of deposits are more likely to be detected indirectly: lode gold and porphyry deposits by reflections from alteration haloes, unconformity Uranium deposits by haloes and basement offsets, and Mississippi Valley-type (MVT) deposits by white spots in otherwise reflective carbonates. Similarly, the high impedances between kimberlites and their hosts allow pipes to be delineated using VSP techniques.

Résumé

Des tracés des vitesses des ondes de compression (Vp) et de cisaillement (Vs) en fonction de la densité des roches soumises à des pressions de confinement élevées montrent que ces vitesses ont tendance à augmenter en fonction de la densité en suivant les courbes bien connues de Nafe-Drake pour les roches silicatées. Cependant, en raison de leur densité élevée, un grand nombre de minerais génèrent des courbes qui se situent loin à droite ce celles de Nafe-Drake et présentent des impédances plus élevées que les roches hôtes dans lesquelles on les trouve couramment, ce qui suggère qu'il pourrait être possible de les détecter à l'aide de méthodes de prospection par sismique-réflexion haute résolution si les gîtes sont conformes aux contraintes de taille, d'épaisseur et de présentation imposées par la sismique-réflexion ou la sismique-diffraction. Des expériences menées par la Commission géologique du Canada, des universités et l'industrie dans des camps miniers en roche dure d'un bout à l'autre du Canada au cours de la dernière décennie montrent que les levés 2D conviennent bien pour la détection et la détermination de la structure des corps minéralisés alors que les levés 3D peuvent être utilisés pour leur détection et leur délimitation et le profilage sismique vertical (PSV) pour leur délimitation. En raison de la petite taille de la plupart des gîtes, de la complexité structurale des terrains de roche dure et de leurs faibles rapports signal sur bruit, les meilleurs résultats sont obtenus dans le cadre de levés soigneusement conçus exploitant des sources haute fréquence et des séquences de traitement adaptées spécifiquement à l'identification des réflexions et des diffractions.

Des levés 2D et 3D ont permis de détecter avec succès et de produire des représentations de grands gîtes de sulfures massifs, comme les gisements magmatiques et les gisements de sulfures massifs volcanogènes à Sudbury et à Bathurst, et devraient également être utiles pour la détection de gîtes sedex et de gîtes d'oxydes de Fer-Cu-Au massifs. Il est plus vraisemblable que d'autres types de gîtes soient détectés de manière indirecte : les gîtes d'or primaires et les gîtes porphyriques d'après les réflexions de leurs auréoles d'altération, les gîtes d'uranium associés à des discordance d'après les auréoles et les décalages du socle et les gîtes de type Mississippi-Valley d'après des taches blanches dans des roches carbonatées par ailleurs réfléchissantes. De même, les impédances élevées entre les kimberlites et leurs roches hôtes permettent la délimitation des cheminées par des méthodes de PSV.

Overview

The mining industry has traditionally used geologic field mapping, electromagnetic and potential field techniques, and drilling to explore for new mineral deposits, but with new discoveries of large near-surface deposits becoming increasingly rare and the known reserves of most economic minerals in decline, it is clear that new deep exploration techniques are required to meet the future needs of industry and society. With gravity and magnetic methods unable to resolve targets beyond about 500 m, high-resolution seismic reflection techniques similar to those used by the petroleum industry, but modified for the hardrock environment, show the greatest potential for extending exploration to depths of 3 km, the current maximum depth of mining.

Reflectivity and Resolution: Basic Factors

Two basic factors govern whether or not a potential reflector can be detected and imaged by seismic reflection techniques: the difference in acoustic impedance between the deposit or horizon and its surroundings, and its geometry. The acoustic impedances (velocity-density products) of common rock types and ores are well known from laboratory

Salisbury, M., and Snyder, D., 2007, Application of seismic methods to mineral exploration, *in* Goodfellow, W.D., ed., Mineral Deposits of Canada: A Synthesis of Major Deposit-Types, District Metallogeny, the Evolution of Geological Provinces, and Exploration Methods: Geological Association of Canada, Mineral Deposits Division, Special Publication No. 5, p. 971-982.

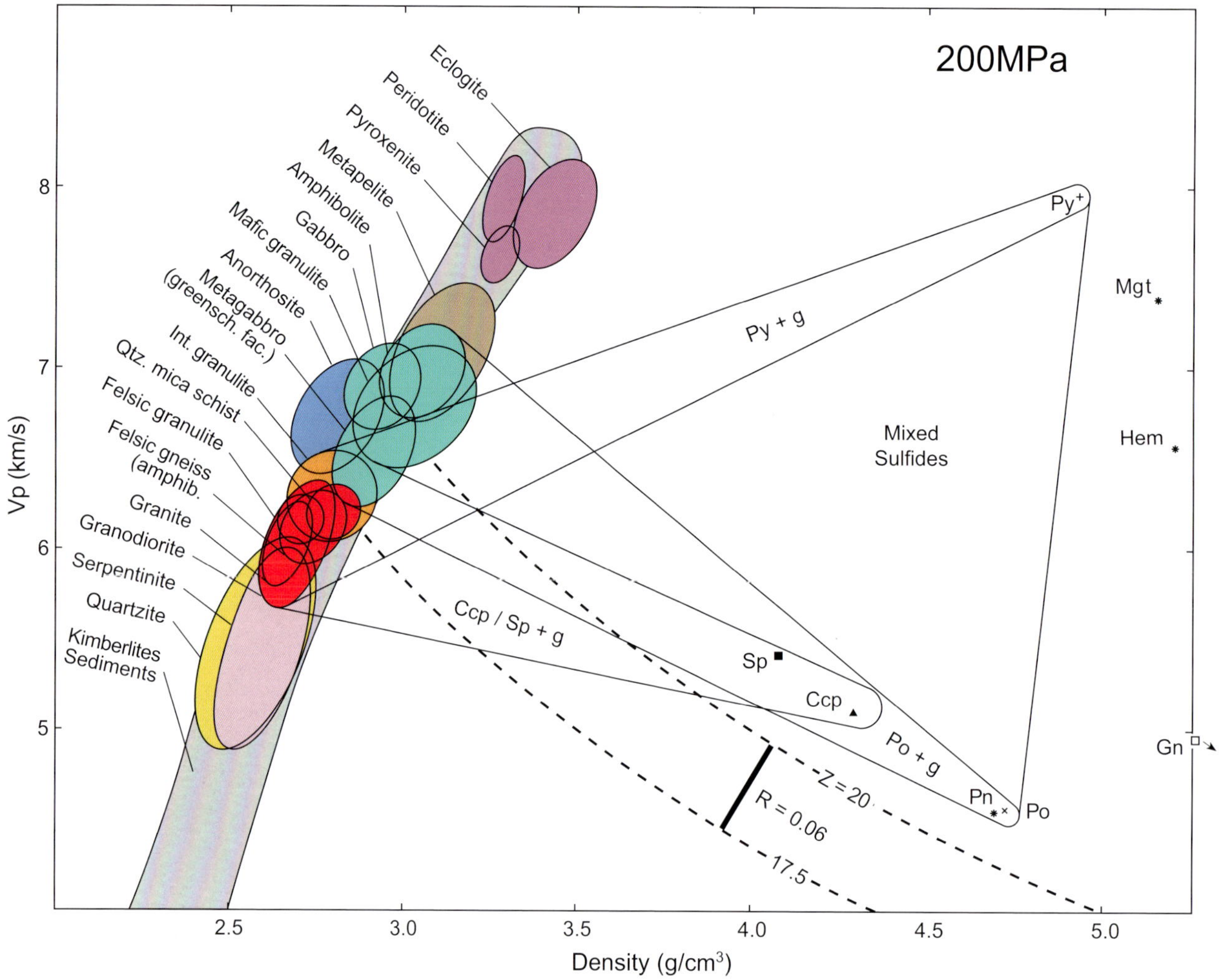

FIGURE 1. Lines of constant acoustic impedance (Z) superimposed on velocity-density fields and Nafe-Drake curve (grey) for common rocks at a standard confining pressure of 200 MPa (modified from Salisbury et al., 2003). Also shown are values for pyrite (Py), pentlandite (Pn), pyrrhotite (Po), chalcopyrite (Ccp), sphalerite (Sp), hematite (Hem), magnetite (Mgt), and fields for host rock-ore mixtures. A reflection coefficient (R) of 0.06 is sufficient to give a strong reflection. Galena (Gn, off scale) has a velocity of 3.7 km/s and a density of 7.5 g/cm^2.

and logging measurements (eg. Christensen, 1982; Salisbury and Iuliucci, 1999; Ji et al., 2002) and are summarized in Figure 1. As a rule of thumb, an impedance (Z) difference of 2.5 x 10^5 g/cm^{2}s between two rock types having velocities v_1 and v_2 and densities ρ_1 and ρ_2 will give a reflection coefficient,

$$R = \frac{v_2\rho_2 - v_1\rho_1}{v_2\rho_2 + v_1\rho_1} = \frac{Z_2 - Z_1}{Z_2 + Z_1}$$

of 0.06, which is sufficient to give a strong reflection if the geometry is appropriate (Salisbury et al., 1996, 2003). Thus contacts between mafic and felsic igneous rocks (Z~20 and 17.5, respectively) or between most sulphide ores ($Z \geq 22$) and most country rocks can be strongly reflective, whereas contacts between similar rocks, such as metagabbro and anorthosite, are not.

The second factor that governs whether or not a deposit can be resolved is its geometry, especially its size and depth of burial. The minimum thickness, t, that can be uniquely determined for a tabular deposit can be estimated from its tuning thickness,

$$t = v/(4f)$$

where v and f are the average velocity in the deposit and the dominant acoustic frequency used in surveying, respectively. This thickness, also called the Rayleigh limit or quarter wavelength criterion (Widess, 1973), is the thickness for which seismic waves reflecting off the top and bottom surfaces of the deposit constructively interfere. For thicker deposits, the two reflections separate in time, allowing the thickness to be uniquely determined. Thinner deposits can be detected as well, but the two reflection surfaces cannot be resolved and the combined reflection amplitude decreases to that of a single reflector.

Similarly, the smallest diameter deposit that can be resolved at any given depth, z, as a surface rather than a diffractor, is defined by the width of the first Fresnel zone,

$$d_F = (2zv/f)^{1/2}$$

where v is now the formation velocity. As in optics, the first Fresnel zone is the area of a planar reflector from which the reflected energy in the first quarter wavelength of a spherical wave front constructively interferes. Modeling results show that deposits as small as one wavelength across can still be detected as diffractors under ideal conditions but that smaller deposits will be undetectable due to attenuation (Berryhill,

1977). Thus in principle, a body which is 15 m thick by 60 m across could be detected as a point source at a depth of 1 km, assuming a formation velocity of 6 km/s and a dominant frequency of 100 Hz, and a body ≥350 m across could be resolved as a reflector. Deeper deposits have to be larger to be imaged, which is consistent with the economics of deep mining.

Given the resolution limits discussed above, seismic reflection can potentially be used for mineral exploration in three ways: 1) to identify and trace prospective structures and horizons at depth, much as the petroleum industry does today, 2) to detect and image new deposits directly, and 3) to delineate known deposits. Several different techniques have been developed by the petroleum industry over the past 70 years for shooting, acquiring, and processing seismic data, including 2-D, 3-D, and Vertical Seismic Profiling (VSP). These techniques will be briefly described below, along with the modifications that have been used to make them work in hardrock terranes. Refraction techniques will not be described because they currently lack the resolution needed for exploration.

Seismic Reflection Methods

2-D

If a sound wave from a seismic source at the surface reflects off a buried horizontal reflector, as in Figure 2a, the angle of incidence α will equal the angle of reflection and the signal will be received at a point that is equidistant from the common midpoint (CMP) at a time controlled by the velocity of sound and the total path length through the upper layer. Rays entering deeper layers will bend according to Snell's law at each horizon where the velocity changes. Thus in Figure 2a:

$$\frac{\textbf{Sin}\ \alpha_1}{\textbf{Sin}\ \beta_2} = \frac{v_1}{v_2}$$

where α_1 and β_2 are the angles of reflection and refraction at the interface and v_1 and v_2 are the acoustic velocities above and below.

While geophones are generally used as the receivers, a variety of sound sources can be used for seismic surveys on land, including impulsive sources such as Betsy guns and weight drop systems, dynamite in shallow boreholes, airguns in water-filled pits or fluid-filled pans attached to vehicles, and vibroseis trucks, which can be programmed to deliver a sweep-frequency waveform into the ground that rises (for example) from 10 to 100 Hz over a period of 12 seconds. As can be seen in Figure 2b, the sources and receivers are laid out at regular surveyed intervals along the seismic line, and each shot is monitored by a linear array of receivers since there is no way to predict *a priori* where the signal will intercept the surface if the reflectors are dipping or to predict the effects of refraction on deeper ray paths. The line is then shot by moving the shot point and the array forward in sync as the data is recorded until the line is completed. To improve the signal-to-noise ratio of the field records, multiple shots or multiple sweeps are usually conducted at each source point, a group of geophones is set out at each receiver position, and the recordings are summed. Depending on the length of the line, the spacing between receiver positions in the array, and the number of geophones in each group, it is not uncommon

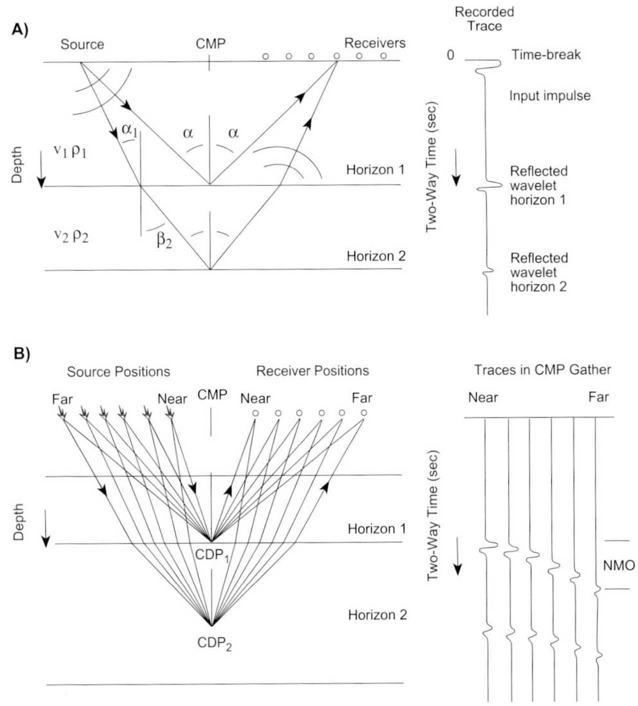

FIGURE 2. (A) Source-receiver geometry and recorded signal for simple 2-D reflection from buried horizontal reflectors (modified from Nelson, 1949). CMP is the common midpoint. α, β, v, and ρ are angle of incidence, angle of refraction, velocity, and density, respectively in layers 1 and 2. (B) Far- and near-source receiver positions for a typical CMP gather. NMO = normal moveout.

to have hundreds of geophone channels active at any one time.

After shooting, the geophone data is multiplexed and transmitted back to a central field computer where it is preprocessed and recorded for subsequent digital processing to remove noise and present the data as a geometrically correct plot of reflection amplitude versus time (depth) and distance along the seismic line. As can be seen in Table 1, conventional preprocessing and processing involves several steps (Yilmaz, 1987), but these are commonly varied to suit the problem being addressed. Preprocessing in the field involves demultiplexing and reformatting the data from the receiver array, environmental noise has to be edited out and amplitude adjustments, such as automatic gain control (AGC), need to be applied to correct for geometric spreading. In addition, the geometry of the survey (latitude, longitude, and elevation of shot and receiver positions) is established at this time and initial static corrections (time shifts) are applied to the data based on the geometry.

After preprocessing, the data is simplified by deconvolving all but the initial wavelet of the source signal from the received signal and all of the signals received from each common depth or midpoint (CMP) along the line are gathered as in Figure 2b. Velocities are then estimated as a function of time (velocity analysis) for selected CMP gathers by determining the velocity that best flattens the gather at each depth. This allows the static corrections to be improved by incorporating the effects of shallow low-velocity zones caused by weathering, which in turn, allows for reiterative improvement of the velocity analyses themselves. Once this is done, the signals are further improved by stacking the flattened gathers

TABLE 1. Conventional Processing Sequence (after Yilmaz, 1987).

1. Preprocessing
    - Demultiplex
    - Reformat
    - Edit
    - Geometric spreading correction
    - Set up field geometry
    - Application of field statics
2. Deconvolution
3. CMP sorting
4. Velocity analysis
5. Residual statics correction
6. Velocity re-analysis
7. NMO correction
    - Muting
    - Stacking
8. Time-variant band-pass filtering
8. Migration

(i.e. applying the normal moveout, or NMO correction) for each reflection point (bin) to improve the signal-to-noise ratio, the degree of improvement increasing with the square root of the number (fold) of shot-receiver pairs that are stacked for each CMP. Time-variant band pass filters are then applied to normalize the signal for attenuation effects and to examine the data at selected frequencies, and finally, the data is migrated to restore the geometry and the results adjusted for strike and crooked line effects. While the final objective is, of course, to present the reflection data as a function of depth, it should be emphasized that all of the processing steps outlined above are conducted in the time domain.

3-D

Two-dimensional surveys can be used for reconnaissance and to resolve simple structures at depth, but complicated structures causing out-of-plane reflections (sideswipe) can only be imaged using 3-D reflection techniques in which a 3-D volume of crust is ensonified and monitored using a planar, rather than an a linear, array of shots and receivers. In practice, this is accomplished by laying out thousands of geophones along parallel lines of receiver groups and then shooting to the entire array from each shot point along a series of orthogonal shot lines as in Figure 3. Although complicated by the fact that a typical 3-D survey contains orders of magnitude more data to process, the actual processing steps are fairly similar to those for 2-D surveys, although they use specialized 3-D migration algorithms. The end result, however, is a data cube that can be sliced to produce synthetic 2-D profiles in any arbitrary direction through the data, horizontal slices at arbitrary depths (time slices), horizon slices showing reflectivity variations in map plan for picked marker horizons, and 3-D tomographic images that can be viewed from any perspective.

Vertical Seismic Profiling (VSP), Borehole Seismic

2-D and 3-D seismic techniques are effective for mapping most structures at depth but less so for mapping steeply dipping structures (>60°) or to determine interval velocities or gradients where reflections are absent. These problems can be addressed, however, using borehole seismic techniques in which a seismometer is lowered to the bottom of a drillhole and then raised in stages as shots are fired to the seismometer and nearby reflectors from the surface (Fig. 4a). If the shot position is known and the borehole geometry has been

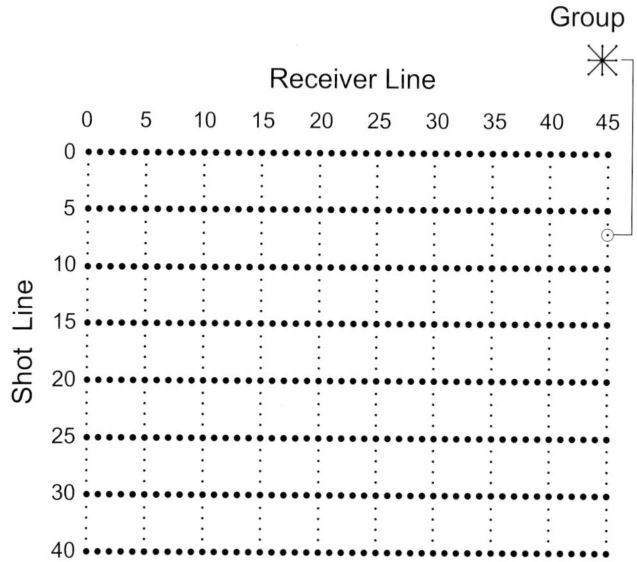

FIGURE 3. Shot (bold) and receiver (faint) positions for a typical 3-D reflection survey. Each receiver position is occupied by a group of geophones to increase signal-to-noise ratio. Fold increases inward.

determined from directional surveys, interval velocities can be calculated from the shot-receiver distances and direct arrival transit times (Fig. 4b), and the geometry of reflectors in the area can be determined using the common depth point (CDP) transform method (Dillon and Thomson, 1984). Directional ambiguities to reflectors can be removed through the use of three-component seismometers and more complex three-dimensional structures can be resolved through the use of linear and spatial arrays of shot points.

Seismic Exploration for Minerals – Special Considerations

While the seismic reflection methods used for petroleum and minerals exploration are similar, minerals exploration poses special problems that must be taken into account.

Most economic mineral deposits are found in hardrock, rather than sedimentary environments. Since the impedance contrasts and reflection coefficients between most common igneous and metamorphic rocks are smaller than those between sedimentary rocks (R~0.1 vs. 0.3), the signal-to-noise (S/N) ratio in minerals surveys will be low, making it more difficult to image structures in the country rock. Particular care must thus be taken during acquisition and processing to maximize the S/N ratio. This is typically done by using explosives in shallow boreholes filled with water or tamped with sand to ensure good source coupling to bedrock, by using cemented or clamped geophones whenever possible, by careful testing of different sources before conducting the survey itself, and by maximizing the fold (Eaton et al., 2003).

Velocities in igneous and metamorphic rocks are typically higher than sediments. Since wavelength varies with velocity for any given frequency, higher frequencies are required in hardrocks than sediments to achieve comparable resolution, which is another reason for using explosives.

Structures in hardrock terranes are often complex and steeply dipping, requiring high resolution (>100 Hz), high fold surveys to be resolved. Since reflections from dipping reflectors tend to arrive at the surface at unexpected loca-

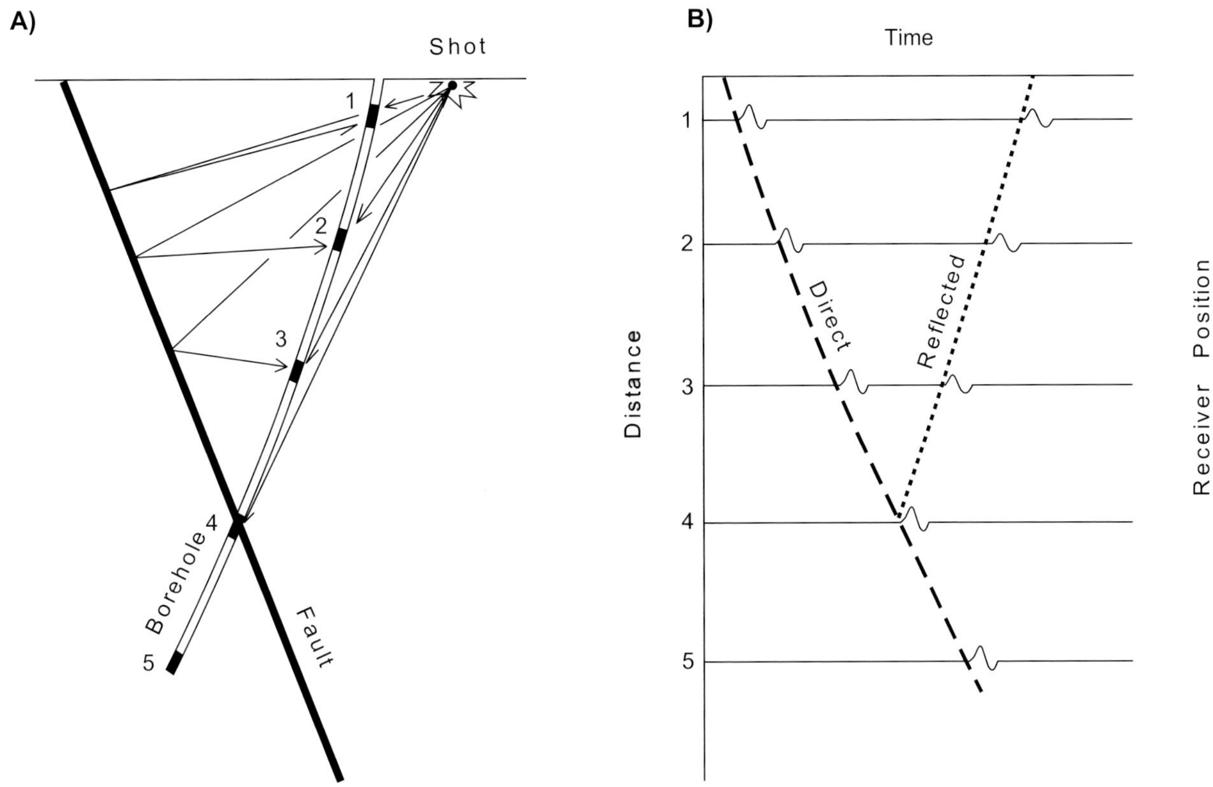

FIGURE 4. **(A)** Shot and receiver positions in a typical borehole or VSP seismic survey. **(B)** Distance-time plot showing transit times for direct and reflected arrivals from steeply dipping reflector.

tions, it is often necessary to model surveys in such terranes in advance in order to determine the optimum locations for receivers and to interpret the data.

Since even rich ore deposits are often fairly small (<1 km across), they commonly appear as diffractions rather than reflections on seismic reflection profiles. Processing of reflection data from mineral surveys thus often involves two processing streams, one to image structures such as folds or faults in country rock and one to locate, enhance, and trace diffractions to their sources using unmigrated data (Table 2). Prestack migration, a sophisticated processing technique in which the data is migrated without any prior stacking, is used in this second stream because conventional processing tends to treat small targets as noise. While prestack migration is a powerful and rewarding processing tool, it must be used with caution because it is both challenging and expensive.

Finally, since some types of ores only have small impedance contrasts with many common host rocks, it is often advisable to conduct laboratory measurements of the velocities and densities of the ores and host rocks in a potential survey area to determine whether reflections are even possible and the survey worth conducting.

While seismic reflection techniques can be used to explore for ore deposits at much greater depths than any other remote geophysical technique, they are expensive, with high-resolution 3-D surveys costing ~\$50K /km^2 for acquistion and processing (Adam et al., 2003), 2-D surveys ~\$6K /km, and a typical VSP survey to a depth of 500 m in a pre-existing borehole, ~\$30K total. Since 2-D surveys are best suited to structural reconnaissance, 3-D surveys to discovery and delineation, and VSP surveys to delineation, and the costs are significantly different, it is important to match the method to the intent of the survey and the type of deposit sought. As outlined below, this is best accomplished from knowledge of the geology and acoustic properties of both the country rocks and the deposits themselves.

Applications by Deposit Type

Diamonds

Although diamonds occasionally occur in placer deposits, their primary occurrence, both in Canada and worldwide, is

TABLE 2. Mineral Survey Processing Sequence (after Adam et al., 2003).

1. Preprocessing
Demultiplex
Set up field geometry
Edit
True amplitude recovery
2. Deconvolution
3. Band-pass filtering
4. Gain control
5. First break mute
6. Refraction statics corrections
7. Residual statics corrections
8a. Unmigrated Stack
Dip moveout (DMO) corrections
Velocity analysis
NMO correction
Stacking
Band-pass filter
8b. Migrated Stack
Prestack migration
Velocity analysis
Scaling
Band-pass filter

in kimberlite pipes, craters, and dykes in Archean terranes (Kjarsgaard, 1995). The pipes, which generally occur in swarms, are small (a few hundred metres across), carrot-shaped deposits with steep sides (75-85°). Feeder dykes, sills and blows are common at depth and relatively large crater facies deposits with shallow-dipping (0-35°) bedrock contacts may be present if erosion is shallow. Since they were emplaced as diatremes, usually during the Phanerozoic, the pipes often cut both PreCambrian basement and later sedimentary cover rocks.

Because diamonds are only trace constituents in even the richest kimberlites, their presence or absence cannot be ascertained using seismic methods. As can be seen in Figure 1, however, kimberlites often have much lower impedances than basement igneous and metamorphic rocks and most sedimentary rocks as well, which suggests that kimberlite contacts with the country rocks should make strong reflectors. This makes seismic methods potentially very useful for the delineation of kimberlite structures and for estimating their volumes, both important factors in diamond mining. The walls of pipes and dykes are usually steeply dipping, so they will be difficult to image, but shallow-dipping sill and crater facies contacts will be easily detectable using 2- and 3-D techniques. This suggests that it would be inappropriate to use seismic reflection to explore for pipes but that borehole techniques can be used to delineate their walls, as in the recent Victor kimberlite experiment (Bellefleur et al., 2005). 2- and 3-D methods can be used, however, to explore for kimberlite sills, to look for reflection terminations where sediments are cut by pipes, to delineate crater floors, and to look for whispering modes that are diagnostic of pipes (Urosevic and Evans, 1998, 2000). Alternatively, marine reflection techniques could be used to identify and map the tops of kimberlites in lake floors on the basis of their low reflectivity compared to shield rocks.

Lode-Gold Deposits

Most gold in Canada is mined from bedrock sources and those bedrock deposits that are mined largely, or exclusively for their gold content, are termed lode-gold deposits. While most occur in greenstone belts containing volcanic and sedimentary rocks of low to intermediate metamorphic grade, and there are many different kinds of these deposits, most belong to one of four groups, depending on differences in structure, composition, and origin:

- Epithermal gold deposits, in which gold and silver are found in fault-controlled veins and breccia in shallow hydrothemal deposits associated with silicic to intermediate volcanics and intrusives. Since the noble metals are only minor constituents of these deposits, their presence cannot be determined seismically. The most distinctive features of these deposits are the breccias that host the ores and the alteration haloes that surround them. Since the breccias and haloes will have low impedances due to their high chlorite, sericite, and clay contents, these deposits should produce strong reflections against their hosts, if contacts are sharp and not gradational. These structures would be detectable as diffractors using VSP and high-resolution, wide-angle 2- and 3-D seismic reflection techniques.
- Quartz-carbonate vein gold deposits, in which gold occurs in veins in obduction or collision-related faults and shear zones in deformed clastic sedimentary terranes or deformed volcanic/plutonic terranes containing arc and ophiolite assemblages. Vein deposits in clastic terranes lack aureoles, making them unattractive targets for direct detection and imaging. Seismic reflection can be used, however, to map promising structures such as fold axes or to track known deposits across fault offsets (Coflin et al., 1995). Since vein deposits in volcanic/plutonic terranes commonly occur in wide, schistose shear zones with carbonate alteration haloes and the country rocks display a wide range of impedances, the use of seismic reflection would be limited to mapping local and regional structure in support of conventional exploration.
- Iron-formation-hosted stratabound gold deposits, in which gold is found in intimate association with pyrrhotite or pyrite in strongly deformed iron formation. Whether the sulfides are conformable with the iron formation or concentrated in secondary fold axes, both iron formation and pyrite have high impedances, making these deposits excellent targets for seismic exploration and delineation.
- Disseminated and replacement gold deposits, in which gold is disseminated with other sulphides in structurally stratabound host-rock units composed of schists or carbonates. With the possible exception of the ores hosted by schists, which may have low impedances, these deposits have no distinguishing acoustic or structural characteristics to make them amenable to seismic prospecting.

Magmatic Ni-Cu-PGE Deposits

Magmatic sulphides, which have segregated from large intrusions or flows of mafic or ultramafic magma, are the principle source of Ni, Cu, and PGE minerals in Canada and the world. The sulphides are usually found as massive deposits that were concentrated by gravitational settling near the base of their magmatic hosts. In the case of the larger magma bodies, such as the Bushveld Complex and the Sudbury Complex (an impact structure) and intrusions associated with rifting and flood basalts, the magmas themselves have been differentiated, while the smaller komatiitic hosts remained undifferentiated.

As can be seen in Figure 1, seismic reflection techniques are well suited to mapping both the internal stratigraphy and structure of layered or stratified igneous complexes and their contacts with the country rock because ultramafics, mafics, and late differentiates all have sufficiently large impedance differences to reflect against each other and the ultramafics or mafics at the base will reflect against most country rocks, as will komatiites. This is clearly born out in the Bushveld Igneous Complex where 3-D surveys show cumulate stratigraphy and individual sulphide/platinoid and chromitite ore horizons with such clarity that they are routinely used for mine planning (Fig. 5; Duweke et al., 2002). Similarly, 2-D surveys conducted as part of an extended study of the Sudbury Basin (Figs. 6, 7) clearly show the granophyre/norite and footwall contacts as well as the deep structure of

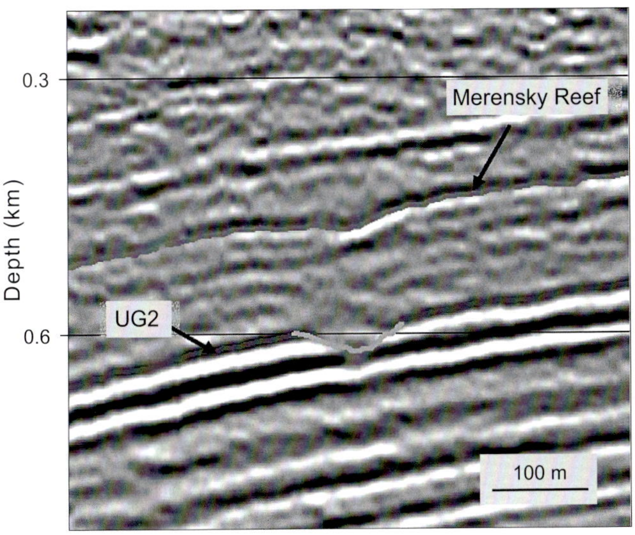

Figure 5. Section from 3-D survey in the Bushveld Complex, South Africa, showing sulphide/platinoid-bearing Merensky Reef and UG2 chromitite layer in gabbroic cumulates. Figure modified from Duweke et al., 2002.

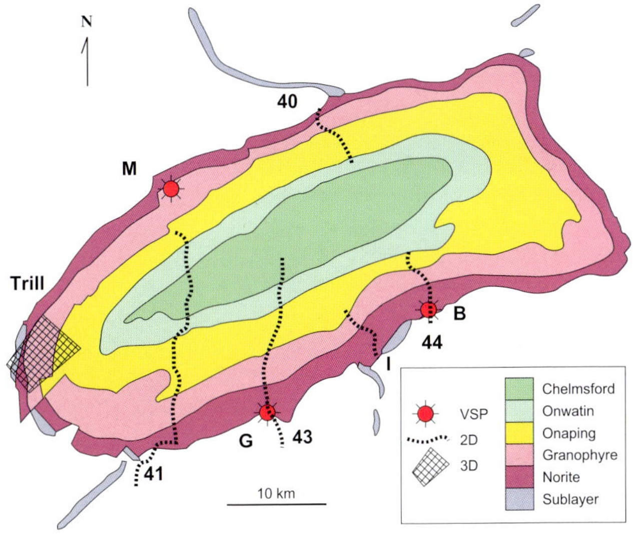

Figure 6. Locations of 2-D (lines 1, 40, 41, 43 and 44), 3-D (Trillabelle (Trill) deposit) and vertical seismic profiling (VSP) (McCreedy (M), Gertrude (G), and Blezard (B) deposits) surveys conducted by the Geological Survey of Canada, industry, and Lithoprobe in the Sudbury Basin.

the basin (Milkereit et al., 1992) and horizon slices though a 3-D cube at the Trillabelle deposit in Sudbury (Fig. 8) show clear images of the footwall (Milkereit et al., 1997). Since massive sulphides have considerably lower impedances than ultramafics, and markedly higher impedances than the sublayer norite or the country rock in Sudbury, large sulphide bodies imbedded along the footwall can be directly detected from their diffraction patterns in both 2-D and 3-D surveys (Figs. 8, 9; Milkereit et al., 1996). By inference, it should be possible to use 3-D techniques to explore for Ni-Cu-PGE deposits at the base or around the margins of any large mafic to ultramafic complex that picked up enough sulphur to precipitate sulphides.

Volcanic Massive Sulphide Deposits

Volcanic massive sulphide (VMS) deposits are squat to tabular, generally concordant deposits of massive sulphides formed by seafloor hydrothermal activity in oceanic, arc or back-arc environments. The deposits, which are typically found in sedimented grabens in mafic and felsic volcanics,

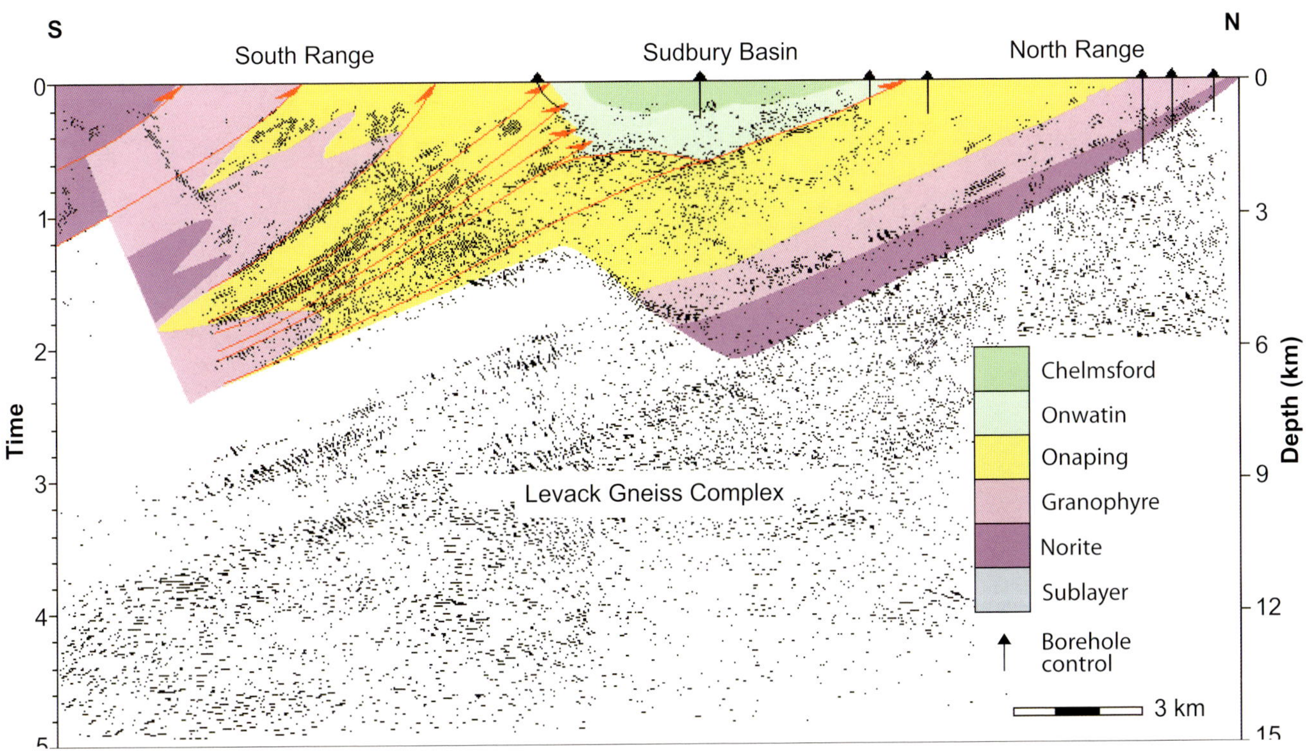

Figure 7. Lithoprobe line 41 showing reflectors associated with deep structure and lithologic contacts in the Sudbury Basin (modified from Milkereit et al., 1992). Location of seismic line 41 is shown in Figure 6.

are underlain by highly altered feeder or stringer zones and tend to be elongate and small to intermediate in size (<2 km). Although mined for Cu, Zn, and Pb, the deposits themselves are usually dominated by pyrite. As can be seen in Figure 1, massive sulphides, especially those rich in pyrite, have high acoustic impedances, making them excellent candidates for seismic exploration using high-resolution techniques. Indeed, in felsic settings, they are ideal candidates because they will be the *only* reflectors present and typically appear as P-wave, S-wave, and hybrid reflections (Bellefleur et al., 2004). Two-dimensional and VSP surveys have imaged such deposits in the Bathurst camp (Fig. 10) and the first deposit actually discovered using seismic reflection techniques was found from a 3-D exploration survey in the same camp (Fig. 11; Salisbury et al., 1997, 2003). Diffractions from massive sulphides and reflections from key stratigraphic horizons and structures have also been observed in 2-D and 3-D data from Matagami, Quebec (Adam et al., 1997, 2003; Calvert et al., 2003).

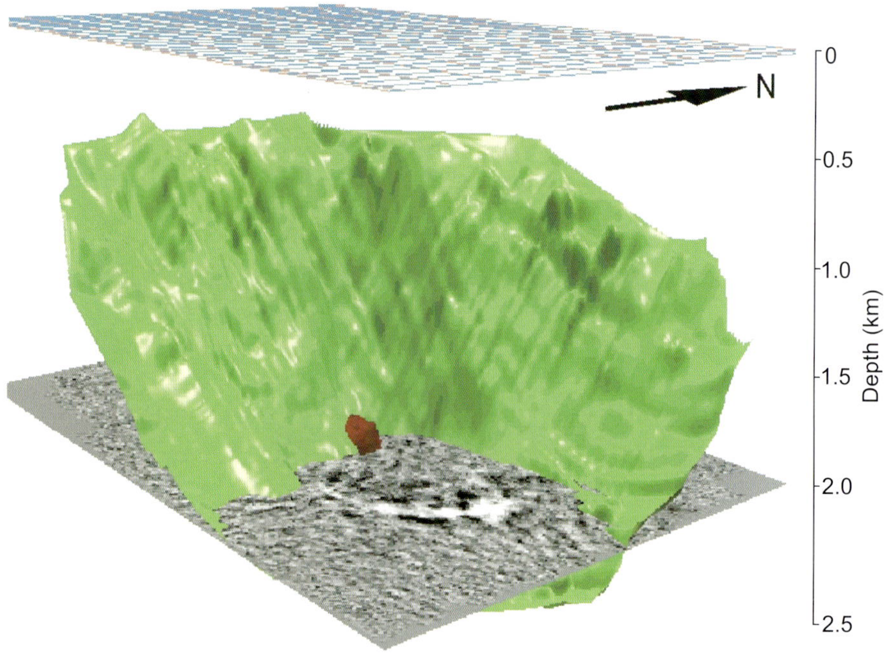

FIGURE 8. Reflection data from a 3-D cube in Sudbury showing the surface grid, horizon slice (green) corresponding to the footwall complex, and time slice at 2000 m depth. Note the bright diffraction ring in the time slice. Red spot represents known position of the Trillabelle deposit.

Sedimentary Exhalative Deposits

Sedimentary exhalative sulphide (SEDEX) deposits are small (<0.5 km) stratabound massive sulphide lenses formed by submarine hydrothemal venting of brines in the cover sequence of epicontinental and intracontinental rift basins during extension. As with VMS deposits, which are closely related, SEDEX deposits are mined for Zn and Pb, but pyrite and pyrrhotite are usually the predominant sulphides (Lydon, 1995). Since these ores will have considerably higher impedances than their sediment host rocks (Fig. 1), they should make strong reflectors. Seismic reflection surveys are well suited to mapping both the sediments and the fault structures

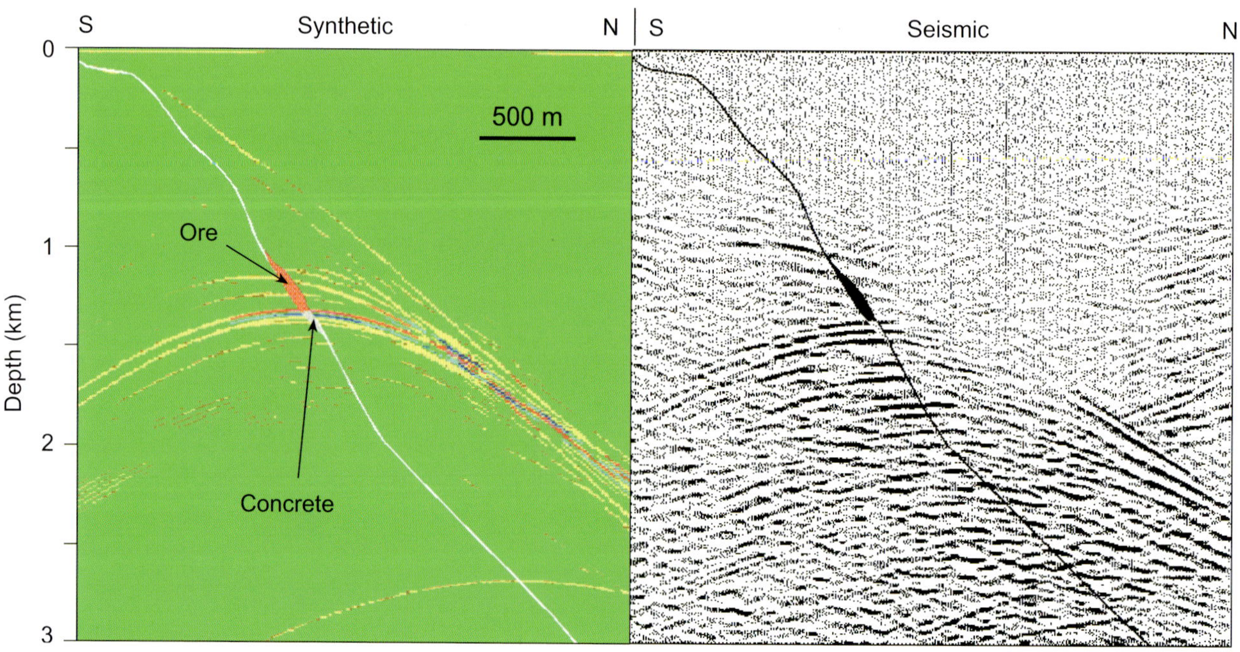

FIGURE 9. Synthetic (left) and observed (right) reflections from the massive sulphide deposit lying beneath line 43 along the footwall contact (line) near the Gertrude deposit, Sudbury Basin (G in Fig. 6). Note the diffractions at the top of the orebody, at the ore-concrete backfill contact and downdip from the deposit. Figure modified from Milkereit et al. (1996).

along which these deposits congregate, but the deposits themselves will tend to be diffractors, rather than reflectors, because of their small size. Like VMS deposits in felsic settings, SEDEX deposits should be excellent candidates for seismic exploration because they are hosted by low-impedance sediments, but modelling studies show that mafic dykes and sills in the survey area will compete for the interpreter's attention.

Mississippi Valley-Type Deposits

Mississippi Valley-Type deposits consist of swarms of small (≤100 m) orebodies in which Fe, Pb, and Zn sulphides fill the voids in thermokarst or solution breccias in carbonate platforms adjacent to large cratonic sedimentary basins. The deposits are typically stratiform at the district scale but individual bodies often cut vertically for short distances through the stratigraphy. These deposits will be difficult to detect and image seismically because they are small, the sulphides are diluted by breccia, and the host carbonates will have similar impedances to the ores (unless the carbonates are vuggy or the ores particularly rich in pyrite), making them acoustically transparent. Ironically, they might be detected as "white spots" in migrated high-resolution data because the breccias would show no stratigraphy, just as gas pockets obscure stratigraphy in marine sediments.

Porphyry Copper Deposits

Porphyry copper deposits are large (0.5 km across, 3 km deep) low grade deposits formed by hydrothermal activity and the crystallization of metals and volatiles from water-saturated, late-stage magmas at the roofs and peripheries of felsic to intermediate porphyritic intrusions in the root zones of andesitic volcanoes. The deposits are typically zoned both in terms of alteration and mineralization, with a potassic alteration zone characterized by biotite and K-spar near the intrusion, then a phyllic zone containing sericite and muscovite, an argillic zone containing kaolinite and clay, and an outer zone containing quartz, chlorite, and epidote further away. Copper mineralization is concentrated in the potassic and phyllic zones and is surrounded by an extensive, diagnostic pyritic halo. Gold and silver veins are found beyond the halo.

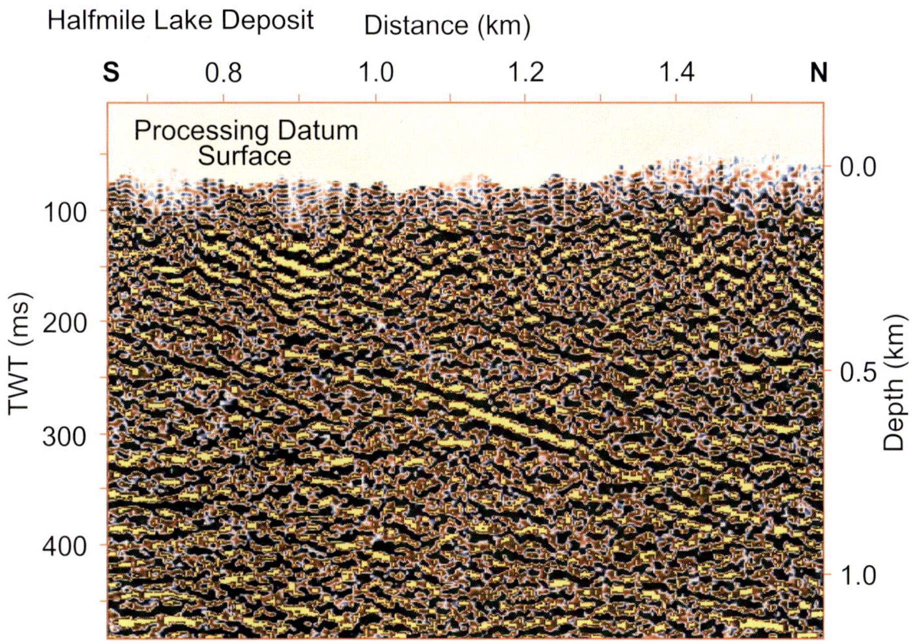

FIGURE 10. Two-dimensional multichannel seismic image of the Halfmile Lake deposit, Bathurst Camp. TWT = two-way travel time.

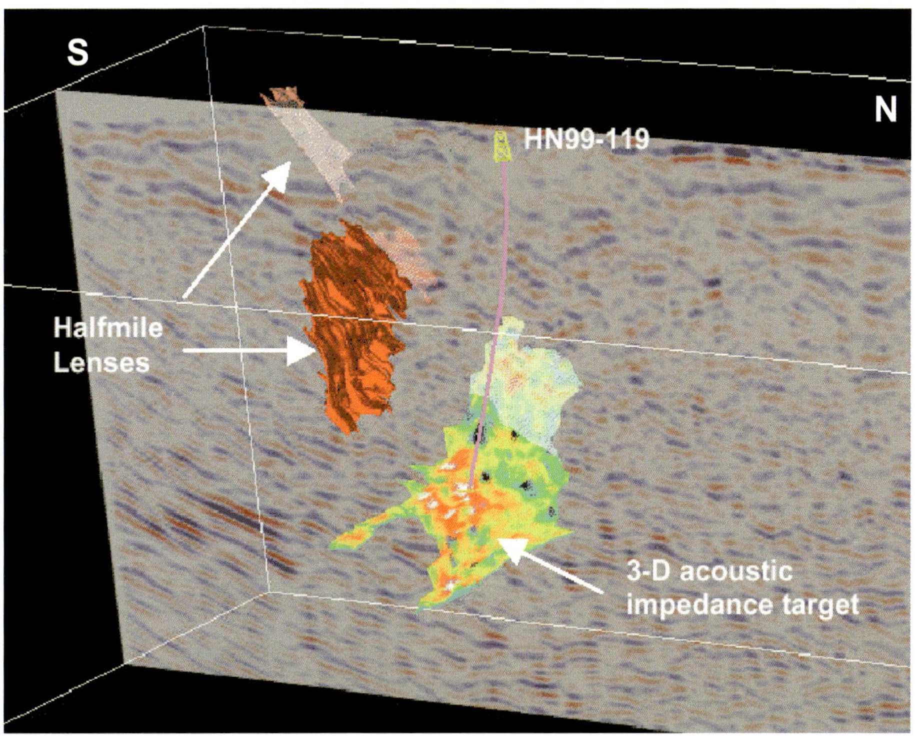

FIGURE 11. Cross-section through a 3-D seismic cube at the Halfmile Lake deposit, Bathurst Camp, showing the massive sulphide deposit discovered by seismic methods at 1300 m, plus earlier known deposits.

While the ore minerals in porphyry Cu deposits have significantly higher impedances than their felsic to intermediate hosts, the deposits themselves are unlikely to be detectable using seismic reflection methods because they are low grade. Thus the surveys that have been conducted over these

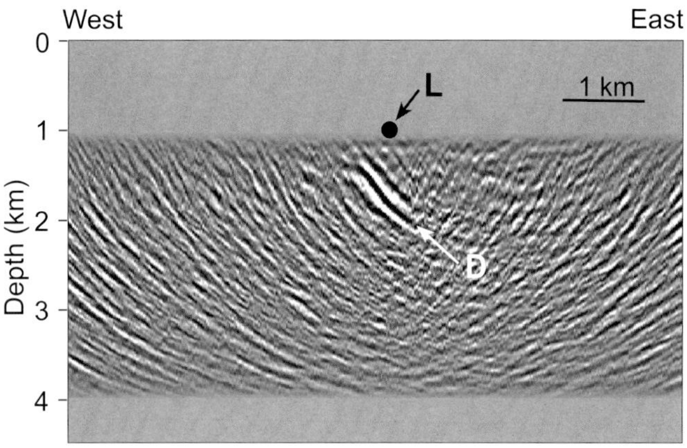

FIGURE 12. East-west section from a 3-D survey shot over the Louvicourt volcanic massive sulphide deposit, Val d'Or, Quebec. L shows the deposit location; reflection D is tentatively attributed to an alteration zone. Figure modified from Adam et al. (2004).

deposits to date have focussed on determining local and regional structure (Wright, 1981; Roy and Clowes, 2000). If the pyrite is sufficiently abundant in the pyritic haloes, however, the halo could be detected and used as an acoustic marker for the underlying Cu deposits. If the surrounding country rock is entirely felsic to intermediate, such haloes would be excellent targets because the rest of the complex would be acoustically transparent.

Unconformity Uranium Deposits

Unconformity U deposits are small (≤150m) pods and lenses of uranium oxide minerals (± various base metal sulphides, arsenides, and oxides) found at or near the basement unconformity in PreCambrian intracontinental sedimentary basins where the unconformity is cut by steep faults or fracture zones that focused return supergene circulation from basement, deep-seated hydrothermal fluids, or diagenetic-hydrothermal fluids from the basin fill. The unconformity is typically marked by a well developed, lateritic regolith and the deposits themselves are invariably surrounded by strong alteration haloes containing clays, chlorite, carbonates, and silica. Although high-grade uranium ores can have extremely high impedances (~30) because of their high densities, it is unlikely that the ore deposits themselves will be detectable using regional-scale seismic techniques because the deposits are small and the ores are generally dilute. High-resolution reflection methods can be used, however, to delineate these small tabular bodies for mining and their associated alteration haloes can likely be detected because of their low impedances and larger size. In addition, the faults associated with these deposits can be readily mapped as small basement offsets using 2- and 3-D seismic reflection techniques (Hajnal et al., 1997).

Iron Oxide Copper Gold Deposits

Iron oxide copper gold (IOCG) or Olympic Dam deposits, which are closely related to porphyry copper deposits, are small- to large-sized (<2 km), hematite-rich bodies associated with late-stage magmas from anorogenic granites or felsic plutons in continental arcs. The hematite, which occurs together with copper sulphides, uranium oxides, and gold and constitutes 30 to 70% of the rock, is found in the matrix of discordant felsic breccias cutting both extrusives and intrusives. Kiruna iron deposits, which occur in the same setting, are closely related to Olympic Dam deposits, but are hydrothermal in origin and occur as large (<4 km) tabular bodies composed largely of magnetite. Although both IOCG and Kiruna deposits occur in acoustically transparent, low-impedance settings, the deposits themselves are sharply defined and dominated by hematite and magnetite, which are very high-impedance minerals, potentially making both deposit types excellent targets for seismic exploration and delineation.

Knowledge Gaps

Because seismic reflection is still an emerging technology in hardrock terranes, it has only been used on an industrial scale to delineate structure in two deposit types (stratiform gold (Pretorius et al., 2003) and magmatic PGE deposits (Duweke et al., 2002)) and it has only been extensively tested for exploration with convincing results in VMS and magmatic Ni-Cu-PGE deposits. For most other deposit types, it has only been attempted on an experimental basis, or not at all, in part because of knowledge gaps.

Acoustic Properties of Metamorphic Aureoles

While massive sulphide deposits can be detected directly because of their high acoustic impedances, many other deposit types are too diffuse to be detected seismically. Several of these deposit types, however, are typically surrounded by or closely associated with metamorphic aureoles, which might serve as proxy seismic targets. Epithermal gold and unconformity uranium deposits, for example, are invariably associated with highly altered, low-grade metamorphics, and porphyry copper deposits are commonly surrounded by distinctive, concentric alteration haloes. Similarly, the basement surrounding the feeder dykes beneath VMS deposits is commonly strongly altered. Because these aureoles are generally composed of clays and phyllosilicates, they are likely to have low acoustic impedances. While a strong reflection has been tentatively attributed to such an alteration zone near the Louvicourt deposit at Val d'Or (Fig. 12; Adam et al., 2004), since the acoustic properties of aureoles have not been studied systematically either in the laboratory or by logging in any camp, it is not known whether their internal and external boundaries are acoustically sharp, and thus detectable, or gradational.

Magnetite-Hematite Systematics

While the velocities and densities of magnetite and hematite are known, the velocity-density systematics of their host rocks and alteration products have not been examined, making it difficult to model the acoustic behaviour of typical IOCG deposits. As for metamorphic aureoles, this can be remedied by systematic laboratory and logging studies in known deposits.

Shear Wave Velocities

Compressional wave velocities and densities are known for most ores and host rocks (with the exceptions noted above), but our knowledge of shear wave velocities is much more limited, making it difficult to impossible to use the full acoustic wavefield in seismic surveys. Until this is remedied by laboratory and logging studies, it will be difficult to move beyond the detection of orebodies and use seismic colour, amplitude versus offset (AVO) and attribute analysis for assay purposes.

Fault Zone Properties and Pathways

Since many deposit types are concentrated along fluid pathways in fault zones, it should be possible to narrow the search for ores or to search for them directly by imaging fault zones, provided the impedance contrast between the fault-zone material and the country rock, and the thickness of the fault zone itself, are sufficiently large to produce reflections. Although this could be a powerful aid to exploration, a major obstacle is that the fluid pathways in fault zones and the acoustic properties of fault-zone rocks are poorly understood because they have not been studied systematically. This could be easily remedied by conducting a series of acoustic property traverses across well known faults in major mining camps.

Properties and Geometries of Kimberlites

Finally, significant knowledge gaps exist involving the use of seismic methods to delineate kimberlite bodies. In particular, while there are strong suggestions that kimberlites should make strong reflectors, it is clear that the petrology and geometry of kimberlites is extremely variable. Systematic studies of the acoustic properties of kimberlites need to be conducted as a function of depth in order to model the seismic response of these bodies and determine the most efficient techniques for imaging them.

Acknowledgements

The research from which these conclusions have been drawn has been supported by INCO, Falconbridge, Noranda, Cominco, Rio Tinto, Kennecott Canada, DeBeers, NSERC, the GSCAtlantic, Mineral Resources and Continental Geoscience Divisions of the Geological Survey of Canada and CCGK Minerals Project X15.

References

Adam, E., Arnold, G., Beaudry, C., Matthews, L., Milkereit, B., Perron, G., and Pineault, R., 1997, Seismic exploration for VMS deposits, Matagami, Quebec, *in* Gubins, A.G., ed., Proceedings of Exploration 97: Fourth Decennial International Conference on Mineral Exploration, GEO/FX, p. 433-438.

Adam, E, Perron, G., Arnold, G., Matthews, L., and Milkereit, B., 2003, 3-D seismic imaging for VMS deposit exploration, Matagami, Quebec, *in* Eaton, D., Milkereit, B., and Salisbury, M., eds., Hardrock Seismic Exploration: Society of Exploration Geophysicists, Geophysical Development Series, v. 10, p. 229-246.

Adam, E., Milkereit, B., and Salmon, B., 2004, 3-D seismic exploration in the Val d'Or mining camp, Quebec: Society of Exploration Geophysicists International Exposition and 74th Annual Meeting, Denver, Colorado, October 10-15, Technical Program Expanded Abstracts, p. 1167-1170.

Bellefleur, G., Muller, C., Snyder, D., and Matthews, L., 2004, Downhole seismic imaging of a massive sulphide ore body with mode-converted waves, Halfmile Lake, New Brunswick: Geophysics, v. 69, p. 318-329.

Bellefleur, G., Matthews, L., Roberts, B., McMonnies, B., Salisbury, M., Snyder, D., Perron, G., and McGaughey, J., 2005, Downhole seismic imaging of the Victor kimberlite, James Bay lowlands, Ontario: A feasibility study, in Current Research: Geological Survey of Canada, Paper 2005-C1, p. 1-7.

Berryhill, J.R., 1977, Diffraction response for non-zero separation of source and receiver: Geophysics, v. 42, p. 1158-1176.

Calvert, A.J., Perron, G., and Li, Y., 2003, A comparison of 2-D seismic lines shot over the Ansil and Bell Allard mines in the Abitibi Greenstone belt, *in* Eaton, D., Milkereit, B., and Salisbury, M., eds., Hardrock Seismic Exploration: Society of Exploration Geophysicists, Geophysical Development Series, v. 10, p.164-177.

Christensen, N.I., 1982, Seismic velocities, *in* Carmichael, R.S., ed., Handbook of Physical Properties of Rocks, v. II: Boca Raton, CRC Press, p. 1-228.

Coflin, K., Salisbury, M., and Horne, R., 1995, High resolution multichannel reflection imaging of gold-bearing structures in the Meguma of Nova Scotia: Geological Assocation of Canada – Mineralogical Association of Canada, Program with Abstracts, v. 20, p. A18.

Duweke, W., Trickett, J.C., Tootal, K., and Slabbert, M., 2002, Three-dimensional reflection seismics as a tool to optimize mine design, planning and development in the Bushveld igneous complex: 64th EAGE Conference and Exhibition, Florence, Italy, May 27-30, Extended Abstrasts, p. D-20.

Dillon, P.B., and Thomson, R.C., 1984, Offset source VSP surveys and their image reconstruction: Geophysical Prospecting, v. 32, p. 790-811.

Eaton, D., Milkereit, B., and Salisbury, M., 2003, Hardrock seismic exploration; Mature technologies adapted to new exploration targets, *in* Eaton, D, Milkereit, B, and Salisbury, M., eds., Hardrock Seismic Exploration: Society of Exploration Geophysicists, Geophysical Development Series, v. 10, p. 1-6.

Hajnal, Z., Annesley, I.R., White, D., Matthews, R.B., Sopuck, V., Koch, R., Leppin, M., and Ahuja, S., 1997, Sedimentary-hosted mineral deposits: A high-resolution seismic survey in the Athabasca Basin, *in* Gubins, A.G., ed., Geophysics and Geochemistry at the Millenium: Proceedings of Exploration 97: Fourth Decennial International Conference on Mineral Exploration, GEO/FX, p. 421-432.

Ji, S., Wang, Q., and Xia, B., 2002, Handbook of Seismic Properties of Minerals, Rocks and Ores: Polytechnique International Press, 630 p.

Kjarsgaard, B., 1995, Kimberlite-hosted diamond, *in* Eckstrand, O.R., Sinclair, W.D., and Thorpe, R.I., eds., Geology of Canadian Mineral Deposit Types: Geological Survey of Canada, Geology of Canada No. 8, p. 560-568.

Lydon, J.W., 1995, Sedimentary exhalative sulphides, *in* Eckstrand, O.R., Sinclair, W.D., and Thorpe, R.I., eds., Geology of Canadian Mineral Deposit Types: Geological Survey of Canada, Geology of Canada No. 8, p.130-152.

Milkereit, B., Green, A., and the Sudbury Working Group, 1992, Deep geometry of the Sudbury Structure from seismic reflection profiling: Geology, v. 20, p. 807-811.

Milkereit, B., Eaton, D., Wu, J., Perron, G., Salisbury, M., Berrer, E.K., and Morrison, G., 1996, Seismic imaging of massive sulfide deposits: Part II. Reflection seismic profiling: Economic Geology, v. 91, p. 829-834.

Milkereit, B, Berrer, E.K., Watts, A., and Roberts, B., 1997, Development of 3-D seismic exploration technology for Ni-Cu deposits, Sudbury Basin, *in* Gubins, A.G., ed., Geophysics and Geochemistry at the Millenium: Proceedings of Exploration 97: Fourth Decennial International Conference on Mineral Exploration, GEO/FX, p. 439-448.

Nelson, H.R., 1949, New Technologies in Exploration Geophysics: Gulf Publishing, Houston, 281 p.

Pretorious, C.C., Muller, M.R., Larroque, M., and Wilkens, C., 2003, A review of 16 years of hardrock seismics on the Kaapvaal craton, *in* Eaton, D.W., Milkereit, B., and Salisbury, M., eds., Hardrock Seismic Exploration: Society of Exploration Geophysicists, Geophysical Development Series, v. 10, p. 247-268.

Roy, B., and Clowes, R., 2000, Seismic and potential-field imaging of the Guichon Creek batholith, British Columbia, Canada, to delineate structures hosting porphyry copper deposits: Geophysics, v. 65, p. 1418-1434.

Salisbury, M. and Iuliucci, R., 1999, Geological Survey of Canada/Dalhousie University Rock Properties Database, http://agc.bio.ns.ca/pubprod/rockproperties/index.html.

Salisbury, M., Milkereit, B., and Bleeker, W., 1996, Seismic imaging of massive sulfide deposits: Part 1. Rock Properties: Economic Geology, v., 91, p. 821-828.

Salisbury, M., Milkereit, B., Ascough, G.L., Adair, R., Schmitt, D., and Matthews, L., 1997, Physical properties and seismic imaging of massive sulphides, *in* Gubins, A.G., ed., Geophysics and Geochemistry at the Millenium: Proceedings of Exploration 97: Fourth Decennial

International Conference on Mineral Exploration, GEO/FX, p. 383-390.

Salisbury, M., Harvey, C., and Matthews, L., 2003, The acoustic properties of ores and host rocks in hardrock terranes, *in* Eaton, D., Milkereit, B., and Salisbury, M., eds., Hardrock Seismic Exploration: Society of Exploration Geophysicists, Geophysical Development Series, v. 10, p. 9-19.

Urosevic, M., and Evans, B., 1998, Seismic methods for the detection of kimberlite pipes: Exploration Geophysics, v. 29, p. 632-635.

Urosevic, M., and Evans, B., 2000, Surface and borehole seismic methods to delineate kimberlite pipes in Australia: The Leading Edge, v. 19, p. 756-758.

Widess, M., 1973, How thin is a thin bed? Geophysics, v. 38, p. 1176-1180.

Wright, P.M., 1981, Seismic methods in mineral exploration, *in* Skinner, B.J., ed., Economic Geology: 75th Anniversary Volume, 1905-1980: Economic Geology Publishing Company, New Haven, Connecticut, p. 863-870.

Yilmaz, O., 1987, Seismic Data Processing: Investigations in Geophysics, No. 2, Society of Exploration Geophysicists, Tulsa, Oklahoma, 516 p.

Indicator Mineral and Surficial Geochemical Exploration Methods for Kimberlite in Glaciated Terrain; Examples from Canada

M. B. McClenaghan and B. A. Kjarsgaard
601 Booth Street, Ottawa, Ontario K1A 0E8
Corresponding author's email: bmcclena@nrcan.gc.ca

Abstract

Diamond exploration in Canada focuses on kimberlite, the primary host rock for diamonds in this country. Kimberlite is a mineralogically and geochemically distinct point source that yields discrete dispersal trains in glacial sediments and dispersion fans in stream sediments. This paper provides a summary of selected diamond exploration techniques used in the glaciated terrain of Canada, focusing on indicator mineral methods. Kimberlite indicator minerals, in general, are able to survive long distance glacial transport. The relative abundance of specific minerals in till, and to a lesser extent in stream sediments, is a function of the mineralogy of kimberlite(s) in the source areas. Indicator mineral distributions observed at a regional scale define the net effect of glacial dispersal, often along different ice flow directions. Local scale distributions define individual dispersal trains. The 0.25 to 2.0 mm fraction of heavy mineral concentrates prepared from till and fluvial sediments is typically used for indicator mineral surveys, with the finer (0.25-0.50 mm) fraction often providing the best results. Till geochemistry is gaining popularity in diamond exploration because it is significantly cheaper than indicator mineral analysis and it can be performed quickly. Important kimberlite pathfinder elements include Mg, Ni, Cr, Co, Ca, Fe, Ti, Nb, Ta, REE, K, Rb, Sr, and Ba, but will depend on the composition of the country rocks versus the kimberlite. Soil selective leaches, soil gas, and hydrogeochemistry, still in the developmental stages for kimberlite exploration, may prove to be useful exploration techniques for application at a local scale.

Résumé

L'exploration à la recherche de diamants au Canada est centrée sur la kimberlite, la principale roche hôte des diamants au pays. La kimberlite est une source ponctuelle caractéristique des points de vue minéralogique et géochimique qui libère des traînées de dispersion distinctes dans les sédiments glaciaires et des cônes de dispersion dans les sédiments fluviaux. Dans la présente étude, on fournit un résumé de méthodes choisies d'exploration à la recherche de diamants utilisées dans les terrains anciennement englacés du Canada et en particulier des méthodes basées sur les minéraux indicateurs. En général, les minéraux indicateurs de kimberlite peuvent survivre au transport glaciaire sur de longues distances. L'abondance relative de minéraux spécifiques dans le till et, dans une moindre mesure, dans les sédiments fluviaux, dépend de la minéralogie de la (des) kimberlite(s) présente(s) dans les régions sources. La répartition des minéraux indicateurs observée à l'échelle régionale permet de définir l'effet net de la dispersion glaciaire, souvent dans différentes directions d'écoulement de la glace. La répartition à l'échelle locale permet de définir des traînées de dispersion individuelles. On utilise de manière caractéristique la fraction de 0,25 à 2,0 mm des concentrés de minéraux lourds extraits du till et des sédiments fluviaux dans le cadre des levés des minéraux indicateurs; la fraction plus fine (de 0,25 à 0,50 mm) fournit souvent les meilleurs résultats. Les méthodes géochimiques appliquées au till gagnent en popularité pour l'exploration à la recherche de diamants parce qu'elles sont beaucoup moins coûteuses et plus rapides à appliquer que l'analyse des minéraux indicateurs. Les éléments indicateurs importants pour la recherche de traînées de kimberlite sont Mg, Ni, Cr, Co, Ca, Fe, Ti, Nb, Ta, les terres rares, K, Rb, Sr et Ba, mais ceux à utiliser dépendront de la composition des roches encaissantes comparativement à celle de la kimberlite. Les méthodes fondées sur la lixiviation sélective des sols, sur les gaz présents dans les sols et sur l'hydrogéochimie en sont encore au stade de l'élaboration, mais pourraient s'avérer utiles pour l'exploration à l'échelle locale.

Introduction

Several major kimberlite fields in the world are located in glaciated terrain, including those in Yakutia, Archangel'sk, northern Finland, southwest Greenland, and Canada (Fig. 1). An understanding of the application of drift prospecting methods for glaciated terrain is essential for exploring for diamonds in these regions. In this terrain, glacial erosion is the principal means by which kimberlite has been dispersed, from a few tens of metres to tens of kilometres down-ice. Modern streams have also transported kimberlite debris, as boulders, cobbles, and discrete heavy mineral grains, up to several tens to hundreds of kilometres from source. As a result, boulder tracing, heavy mineral sampling using till or stream sediments and till geochemistry are the most common drift prospecting methods in glaciated terrain. Since the end of glaciation approximately 6000 to 10000 years ago, in situ weathering of kimberlite may have resulted in the geochemical transport of elements upwards from the kimberlites through overlying glacial sediments where they may be captured in surface soils and/or in groundwater. Thus, soil selective leaches, soil gas and hydrogeochemical methods, still in the developmental phase, may prove to be useful for kimberlite exploration in glaciated terrain.

In Canada, indicator mineral methods were first used to explore for kimberlites in the early 1960s and their continued application, in combination with geophysics, eventually led to the discovery of kimberlites fields in north and northeastern Ontario (Fig. 2) in the 1980s (Brummer et al., 1992a; Kjarsgaard and Levinson, 2002) and the highly diamondiferous kimberlites in the Lac de Gras field of the Northwest Territories (Fig. 2) in the early 1990s (Fipke et al., 1995a; Pell, 1997; J. Carlson et al., 1999). Since the discovery of the Lac de Gras kimberlites, many government agencies have undertaken regional diamond exploration surveys in Canada

McClenaghan, M.B., and Kjarsgaard, B.A., 2007, Indicator mineral and surficial geochemical exploration methods for kimberlite in glaciated terrain; Examples from Canada, *in* Goodfellow, W.D., ed., Mineral Deposits of Canada: A Synthesis of Major Deposit-Types, District Metallogeny, the Evolution of Geological Provinces, and Exploration Methods: Geological Association of Canada, Mineral Deposits Division, Special Publication No. 5, p. 983-1006.

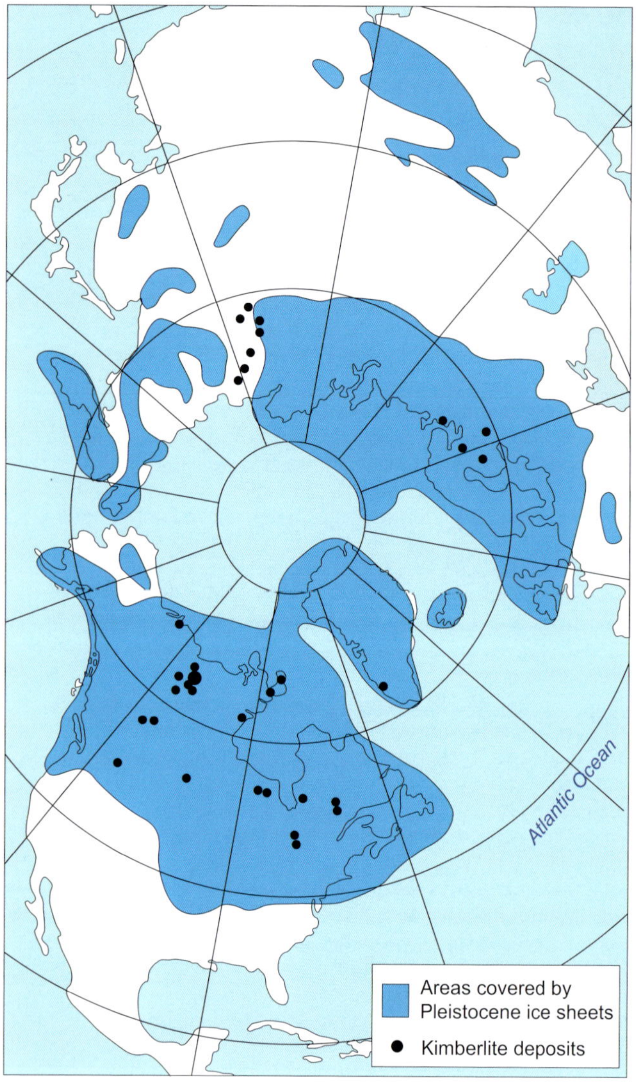

Figure 1. Areas of the northern hemisphere covered by Pleistocene ice sheets and location of major kimberlite deposits (modified from Eriksson, 1992).

(e.g. Dredge et al., 1995; Crabtree, 2003; Friske et al., 2003). Simultaneously, industry diamond exploration geologists conducted regional survey programs that led to the discovery of additional kimberlite fields in the Northwest Territories (Armstrong, 2001; Armstrong and Chatman, 2001), Nunavut, northern Quebec, and northern Alberta. A number of exploration case histories, describing the application of drift exploration methods in the discovery of kimberlites in glaciated terrain have been reported (e.g. Mckinlay et al., 1997; J. Carlson et al., 1999; S. Carlson et al., 1999; Doyle et al., 1999; Kong et al., 1999; Kirkley et al., 2003; Rikhotso et al., 2003). As well, case studies have been carried out around known kimberlites to document indicator mineral and surficial geochemical dispersal patterns in glacial sediments (e.g. McClenaghan et al., 1996, 1998, 1999a,b,c, 2002a,c, 2004; Fenton and Pawlowicz, 1997; Leckie et al., 1997; Eccles, 1998a,b; Kerr et al., 2000; McClenaghan et al., 2002a; Lehtonen et al., 2005) in order to refine existing and test new kimberlite exploration methodologies. In Canada, kimberlite indicator mineral methods are by far the most commonly used and successful of the surficial techniques that can be applied to diamond exploration in glaciated terrain. Thus, the majority of the published literature describing kimberlite-related exploration case histories, regional till or stream sediment surveys, and orientation studies in Canada deals with the use of indicator mineral methods. As a result, the emphasis of this paper is on indicator mineral methods, with numerous Canadian examples. Other surficial sediment techniques that may be applied to kimberlite exploration in glaciated terrain are also mentioned in the paper, although their applications are largely untested and/or undocumented in the literature.

Kimberlite

Key Attributes

Kimberlites are CO_2- and H_2O-rich ultramafic rocks of magmatic origin, derived from limited partial melting of the mantle at very high pressure (>10 GPa, or >300 km depth). The rocks have a distinctive inequigranular texture, and often appear hybrid in nature as they may contain mantle xenoliths, mantle xenocrysts, macrocrysts (a non-genetic term for crystals 1 to 20 mm in size, i.e. megacrysts and xenocrysts), crustal xenoliths, and euhedral to subhedral phenocrysts set in a groundmass matrix crystallized from the kimberlite magma. Macrocrysts include minerals derived from disaggregated mantle xenoliths, plus the kimberlite megacryst suite of minerals (Mg-ilmenite, Cr-poor Ti-pyrope garnet, Cr-diopside, phlogopite, enstatite, zircon, and olivine). Primary minerals include euhedral olivine and phlogopite phenocrysts and microphenocrysts in a fine-grained matrix of one or more of the following minerals: spinel, ilmenite, perovskite, monticellite, apatite, phlogopite-kinoshitalite mica, carbonate, olivine, and serpentine.

The basis for drift prospecting methods for kimberlites in glaciated terrain relies on two principles: 1) kimberlites contain a distinctive suite of resistant and abundant minerals known as kimberlite indicator minerals that includes Cr-pyrope garnet, pyrope-almandine garnet, Mg-ilmenite, Cr-

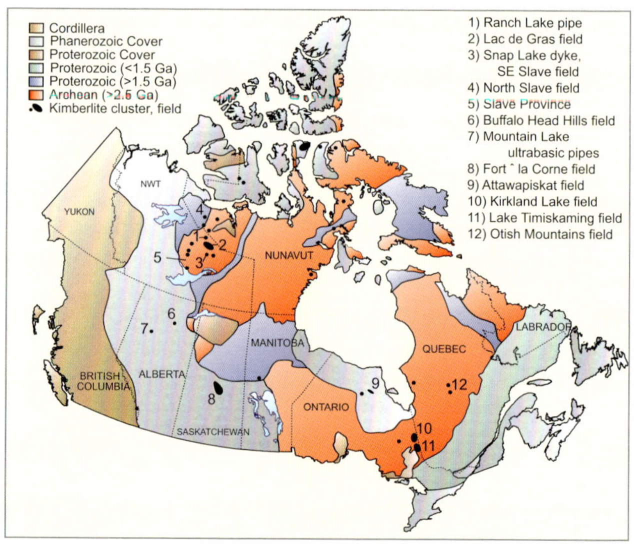

Figure 2. Kimberlite and selected ultrabasic locales in Canada, shown with a backdrop of the major geological terranes. Numbers refer to locations of kimberlite fields discussed in the text (modified from McClenaghan and Kjarsgaard, 2001).

FIGURE 3. Aerial photograph showing the glaciated landscape of the thin to moderately drift-covered bedrock in the Lac de Gras region and the location of the Point lake, Misery lake, Jaeger, and Jay kimberlites underlying glacially scoured lakes.

Weathering

In Canada, kimberlites range in age from ca. 1123 to 45 Ma (Heaman et al., 2004) and were subject to significant weathering prior to Quaternary glaciation. At the time of kimberlite emplacement, or shortly thereafter, macrocrysts and primary groundmass minerals are commonly replaced by serpentine and calcite. Furthermore, the highly porous nature of volcaniclastic kimberlite (crater and diatreme zone kimberlite) makes these rocks highly susceptible to post-emplacement alteration by weathering processes. Over time, this alteration leads to the development of a soft, friable, clay-rich regolith, often referred to as "yellow" or "blue" ground. Glacial erosion has removed varying amounts of the regolith and, in some cases, the more competent kimberlite below. For example, glacial erosion of the Peddie, McLean, and Seed kimberlites in the Lake Timiskaming kimberlite field in northeastern Ontario (Fig. 2) removed all regolith, leaving fresh glacially striated kimberlite beneath thick till. In contrast, glacial erosion of other kimberlites in the same field was less intense, leaving more than 6 metres of regolith at the top of some kimberlites (e.g. McClenaghan et al., 2004). Glacial erosion and entrainment of kimberlite debris was likely more robust where the preglacial regolith was thickest, thus the volume of regolith exposed to the glacier may have been a factor in determining the areal extent and abundance of kimberlite debris in glacial dispersal trains. Since kimberlite is commonly more susceptible to glacial erosion than the surrounding country rock, many kimberlites in glaciated terrain are found in bedrock depressions that are either in low, swampy ground under small lakes (Fig. 3) or covered by thick glacial sediments (Fig. 4). In the Buffalo

diopside, Cr-spinel, Mg-olivine, and enstatite and 2) kimberlites have a characteristic geochemical signature that is rich in both 'incompatible' elements Sr, Ba, LREE (La, Ce, Sm, Nd), Nb, Ta, Hf, Zr, P, and Ti and first order transition elements Mg, Ni, Cr, and Co of ultramafic affinity (Litinskii, 1961; Mitchell, 1986; Fipke et al., 1995b). These two distinctive features of kimberlite form the basis for indicator mineral and geochemical sampling of surficial sediments. Importantly, the distance of glacial dispersal down-ice for kimberlite indicator minerals and the kimberlite geochemical signature is related to the size of the kimberlite point source, the type of kimberlite being eroded (i.e, hypabyssal versus volcaniclastic) and the compositional contrast with the host rock. Note that kimberlites can be quite variable in size and may form thin dykes, sheets, and pipe-like bodies that can range from 50 to 1500 m in diameter.

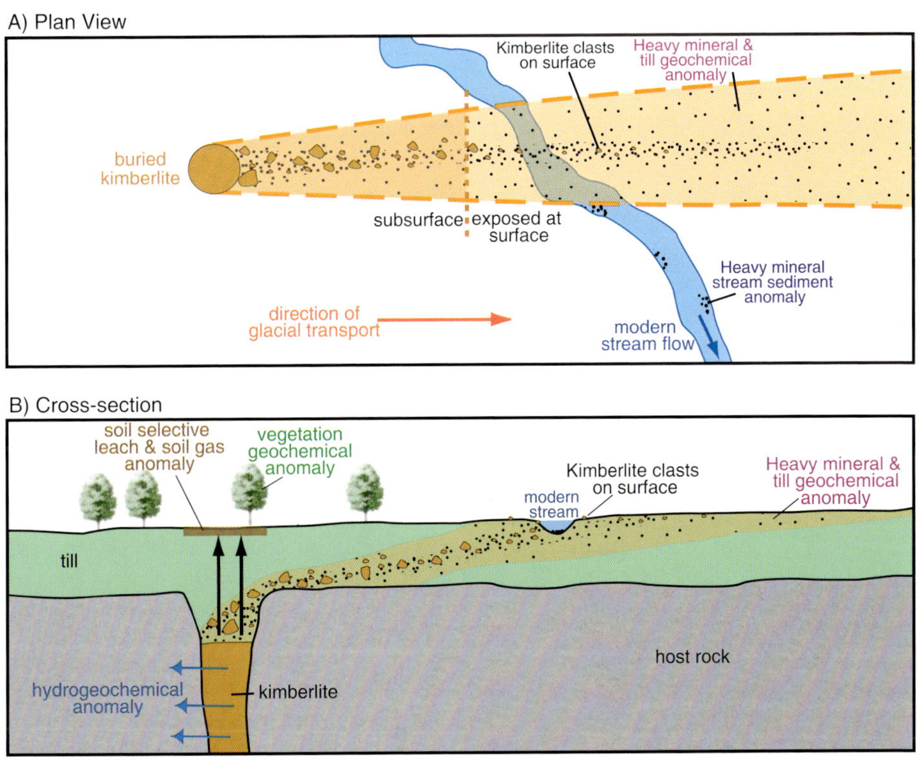

FIGURE 4. Schematic plan and cross-section views of clastic dispersal and chemical dispersion patterns in various media around a kimberlite in glaciated terrain.

FIGURE 5. Examples of typical colours and surface features of kimberlite indicator minerals: **(A)** purple and mauve Cr-pyrope, some retaining kelyphite (dark); **(B)** emerald green Cr-diopside; **(C)** opaque black Cr-spinel, some exhibiting slightly resorbed octahedral form; **(D)** black Mg-ilmenite, both single crystals and polycrystalline morphologies shown; **(E)** yellow Mg-olivine; **(F)** orange eclogitic pyrope-almandine garnets. Photos A to E from the Geological Survey of Canada, photo F from Mineral Services Canada.

Head Hills field, however, kimberlites are more competent than the surrounding Cretaceous sandstone and shale, and thus form topographic highs on the bedrock surface (Fenton et al., 2003; Skeleton et al., 2003).

Kimberlite Indicator Mineral Methods

The unique mineralogical signature of kimberlites enables the application of indicator mineral techniques commonly used for diamond exploration in southern Africa, Australia. and Russia to glaciated terrains of Canada and Fennoscandia. Several minerals that occur in kimberlite are useful indicators of the presence of this rock type (Gurney et al., 1993; Fipke et al., 1995b). These indicator minerals are visually (Fig. 5) and chemically distinct, mostly sand-sized (0.15-2.0 mm), sufficiently dense to be concentrated by gravity methods, and can typically (but not always) survive preglacial weathering, subsequent glacial transport, and postglacial weathering (Dummett et al., 1987; McCandless and Nash, 1996; Averill, 2001). Kimberlite indicator minerals include 1) xenocrysts derived from disaggregated peridotite and eclogite mantle xenoliths (olivine, enstatite, Cr-diopside, Cr-pyrope garnet, Cr-spinel, eclogitic pyrope-almandine garnet, omphacitic pyroxene, and diamond), 2) the associated megacryst suite of minerals (low-Cr Ti-pyrope, Mg-ilmenite, Cr-diopside, phlogopite, zircon and olivine), and 3) kimberlite-derived (i.e. phenocrystal) olivine and spinel. The physical characteristics of the kimberlite indicator minerals commonly used in glaciated terrain are summarized in Table 1. A subset of these indicator minerals (those derived from disaggregated peridotite and eclogite xenoliths) also can be used for a preliminary evaluation of the diamond potential of an area or a kimberlite body. The use of Cr-pyrope, pyrope-almandine garnet, Cr-diopside, Mg-ilmenite, and Cr-spinel in regional- and local-scale till and stream sediment sampling programs for kimberlite exploration is well established (J. Carlson et al., 1999; S. Carlson et al., 1999; Kong et al., 1999). However, few examples of glacial dispersal patterns for kimberlitic olivine have been published. In contrast to arid and tropical terrains, kimberlitic olivine easily survives glacial transport and it is visually distinct from olivine derived from other rock types (e.g. Afanase'ev et al., 1984; McClenaghan et al., 2004) and should be used along with the more common indicator minerals.

Indicator Mineral Chemistry

Various types of indicator minerals recovered by visual sorting from surficial media samples can be more precisely characterized by measuring the concentrations of major oxides and some trace elements by SEM-EDS or electron microprobe analysis (e.g. Sobolev, 1977; Fipke et al., 1989). Mineral analysis is undertaken in order to correctly differentiate indicator minerals from kimberlite versus those associated with other bedrock sources that can be visually similar. Simple identification of kimberlite-derived ilmenite, for example, utilizes a minimum threshold of 4 wt.% MgO (Fig. 6A). A more sophisticated crustal versus kimberlite ilmenite discrimination diagram using TiO_2 versus Cr_2O_3 (Fig. 6B) has recently been published by Wyatt et al. (2004).

TABLE 1. Summary of physical features of kimberlite indicator minerals (modified from Muggeridge, 1995). These features can be used to visually identify and distinguish kimberlite indicator minerals from other minerals in heavy mineral concentrates produced from kimberlites and surficial sediments.

Mineral	Composition[a]	Crystal system	Colour	Typical grain size (mm) normal (to rare)	Response to electromagnet (paramagnetism)	Specific gravity	Hardness (Moh)	Streak colour	Visible diagnostic features[b]	Main source rocks
Pyrope garnet	Mg Fe Ca Al silicate, *Cr, Ti*	Isometric	purple, crimson red, mauve	1-5 (to 100)	nonparamagnetic to very weakly paramagnetic	3.51	7.5		anhedral, kelyphite rim, characteristic colours, orange peel surface	Peridotite, kimberlite, lamprophyre *(certain basic volcanic rocks)*
Pyrope almandine garnet	Mg Fe Ca Al silicate, *Ti, Na*	Isometric	orange	1-5 (to 25)	nonparamagnetic to very weakly paramagnetic	3.7 to 3.8	7.5		anhedral, kelyphite rim, characteristic colours, orange peel surface	Eclogite
Mg-ilmenite (picroilmenite)	Mg Fe Ti oxide *Cr, Mn, Al*	Trigonal	black	1-20 (to 100)	strongly paramagnetic	4.5 to 5.0	5 to 6	black	anhedral, rounded or blocky, white porcellaneous leucoxene coating, perovskite overgrowth, glassy lustre on broken surfaces, conchoidal fracture	Kimberlite *(certain basic volcanic rocks)*
Cr-diopside	Ca Mg silicate, *Fe, Cr, Al, Na*	Monoclinic	emerald green	1-5 (to 50)	nonparamagnetic to very weakly paramagnetic	3.2 to 3.6	5 to 6		anhedral, blocky, cleavage in 2 directions, characteristic colour	Peridotite, kimberlite, *(certain ultrabasic rocks)*
Cr-spinel	Mg Cr Al Fe oxide, *Mn, Ti*	Isometric	black, reddish brown	0.5-1 (to 8)	moderately paramagnetic	4.3 to 4.57	5.5	dark brown	octahedral to irregular shape, reddish brown on edges of broken grains, glassy surface on broken grains	Peridotite, kimberlite, lamproite, carbonatite, various ultramafic plutonic and volcanic rocks
Forsteritic olivine	Mg silicate, *Fe, Ni, Mn*	Orthorhombic	pale yellow to green	2-10 (to 50)	nonparamagnetic to very weakly paramagnetic (non-kimberlitic grains with magnetite inclusions are strongly paramagnetic)	3.2 to 3.33	6 to 7		characteristic colour, irregular crystal apices, vermiform etching	Peridotite, kimberlite, lamproite, carbonatite, various ultramafic plutonic and volcanic rocks
Diamond	C native *(N, B)*	Isometric	colourless, pale colours (especially yellow and brown)	0.5-10 (to 30)	nonparamagnetic	3.52	10		adamantine lustre, crystal form, resorption features, step layering	Kimberlite, lamproite *(certain lamprophyres and high grade metamorphic rocks)*

a minor and trace elements listed in italics
b by naked eye, hand lens, or binocular microscope

The discrimination and classification of mantle-derived garnet versus crustal-derived garnet versus kimberlite megacryst garnet has been improved substantially since the early studies of Dawson and Stephens (1975), Sobolev (1971; 1977), Sobolev et al. (1973), Gurney and Switzer (1973), and Gurney (1984). Recently, Grütter et al. (2004) published a simple, easy to use garnet discrimination/classification scheme, based on a large dataset, using sound petrological principles. This classification allows for discrimination of: "G12" or wehrlite garnet; "G11" or high-Ti peridotite garnet; "G10" or harzburgite/dunite subcalcic garnet; "G9" or lherzolite Ca-saturated garnet; "G4" or pyroxenite – websterite – eclogite garnet; "G3" eclogite garnet, and "G1" megacryst garnet. Grütter et al. (2004) unfortunately retained the use of "G5" terminology in their updated garnet classification system. In the early 1990s in Canada, a number of sampling studies reported the occurrence of high iron "eclogitic G5" (nomenclature of Dawson and Stephens, 1975) garnets. Subsequently, Schulze (1995) suggested that many of these garnets are in fact of crustal origin. More recently, Fipke et al. (1995b) and Schulze (1997) have published a threshold of 22 wt.% FeO to separate mantle-derived garnets from Fe-rich crustal garnets.

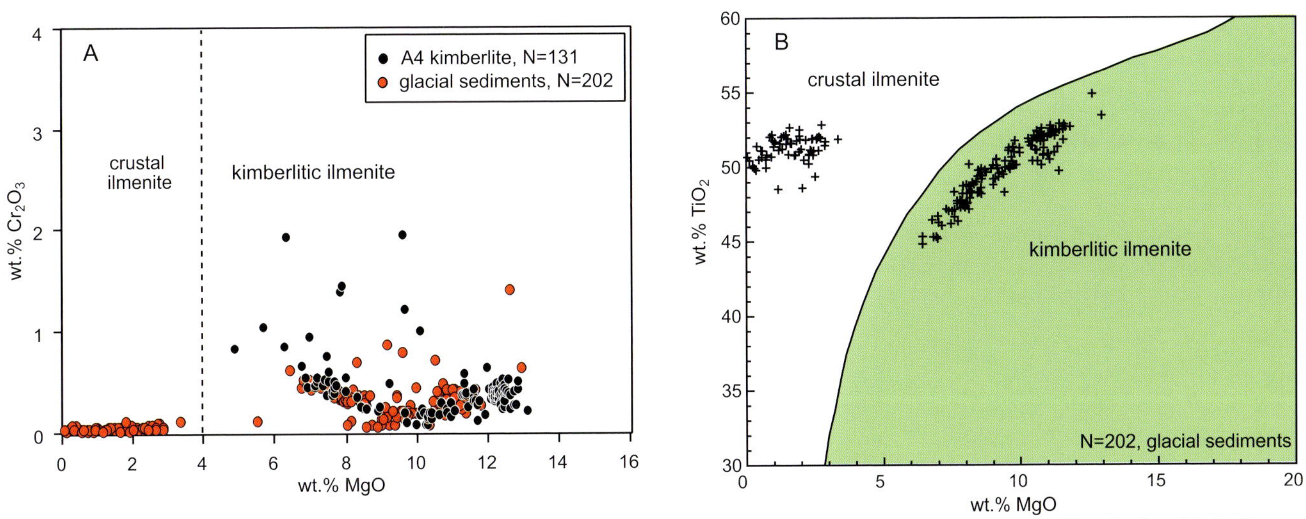

FIGURE 6. (A) Cr_2O_3 versus MgO bivariate plot for ilmenite from the A4 kimberlite, Kirkland Lake kimberlite field. Note that ilmenite from kimberlite contain >4 wt.% MgO and elevated Cr_2O_3; (B) TiO_2 versus Cr_2O_3 bivariate plot for ilmenite for the same data set (modified from McClenaghan et al., 1996). Curve separating kimberlitic from non-kimberlitic ilmenite is from Wyatt et al. (2004).

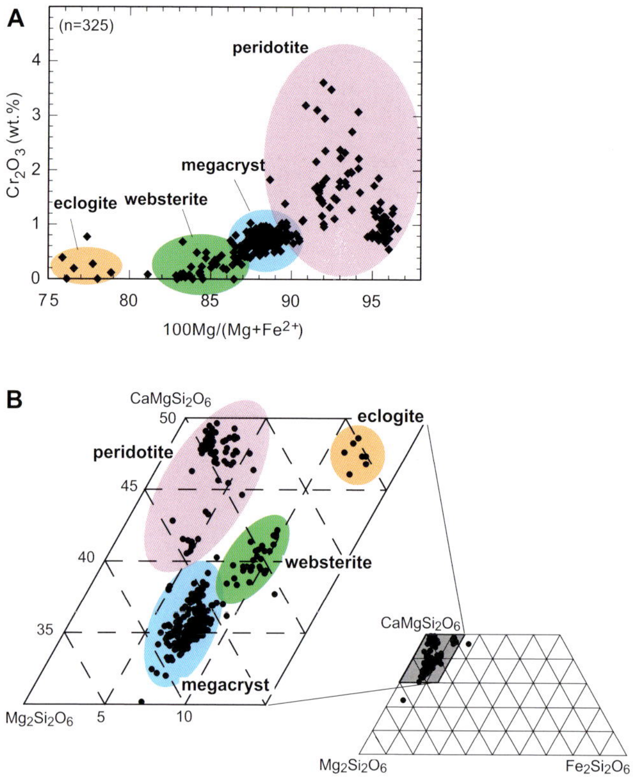

FIGURE 7. (A) Cr_2O_3 versus Mg# (100Mg/(Fe+Mg)) in diopside from kimberlite boulder SD-040, Lake Timiskaming kimberlite field; (B) projection of SD-040 kimberlite boulder clinopyroxene compositions into the pyroxene quadrilateral. The coloured ovals indicate the approximate compositional fields of diopside from eclogite, websterite, megacryst and peridotite lithologies. Data from McClenaghan et al. (2006b).

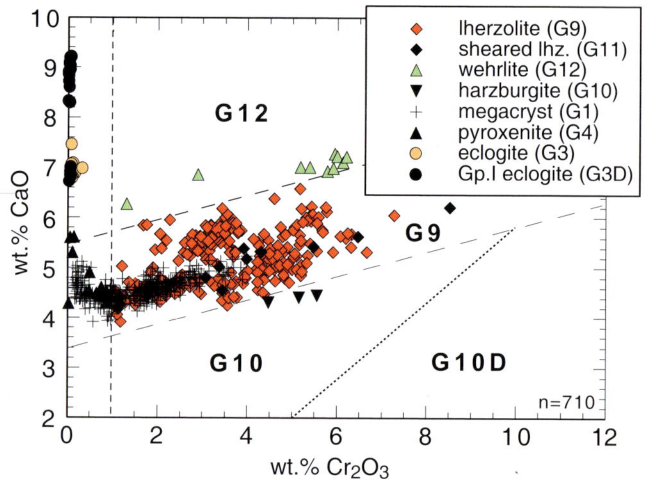

FIGURE 8. CaO versus Cr_2O_3 bivariate plot for garnet from the SD40 kimberlite boulder, Lake Timiskaming kimberlite field. Dashed diagonal lines between G9-G10 and G9-G12 garnets, dashed vertical line at 1 wt.% Cr_2O_3, and G10D-G10 discriminant line are from Grütter et al. (2004). Data from McClenaghan et al. (2006b).

Minerals that are more difficult to chemically classify as kimberlite-derived include olivine, spinel, and diopside. These minerals form phenocrysts in a wide variety of ultrabasic and basic rocks, and may also have primitive mineral chemistry, i.e. elevated Mg/(Mg+Fe) and Cr/(Cr+Al) ratios and high MgO and Cr_2O_3 concentrations. Kimberlite olivine phenocrysts are MgO-rich (forsteritic; Mg#>84) and contain variable, but high Ni concentrations (Mitchell, 1986; Fipke et al., 1995b) as do olivine from a variety of other rocks. However, olivine from disaggregated mantle peridotite typically has Mg#>91.5 with high Ni concentrations. Kimberlites and mantle xenoliths are the only rocks that contain very Cr-rich diopside with >1.5 wt.% Cr_2O_3, (Deer et al., 1982; Fipke et al., 1989, 1995b). But kimberlites also may contain diopside with <1.5% wt.% Cr_2O_3. Hence, Cr-diopside grains containing <1.5% wt.% Cr_2O_3 identified in surficial media require further discrimination to determine their bedrock source. A variety of discrimination plots are utilized (Boyd and Nixon, 1972; McCandless and Gurney, 1989; Nimis, 1998; Morris et al., 2002). Most use a combination of Mg# and Cr_2O_3 and Ca-Mg-Fe and Na-Al-Cr ternary relationships (e.g. Fig. 7).

Because diamond is a rare mineral in kimberlite (usually <<1.4 ppm), a subset of the kimberlite indicator minerals, termed "diamond indicators", is used to indicate the *potential* presence of diamond in these rocks. This subset is based on studies of silicate and oxide inclusions in diamond and minerals from diamond-bearing mantle xenoliths (e.g. Sobolev, 1971; Gurney and Switzer, 1973; Sobolev et al., 1973; Gurney, 1989; Fipke et al., 1995b). Specific diamond indicator minerals include subcalcic Cr-pyrope, "G10" garnet (harzburgite/dunite source rock), high Cr-Mg picrochromite (chromite-bearing harzburgite/dunite source rock), and high Na-Ti pyrope-almandine garnet (eclogite source rock). Many diamond inclusion peridotitic garnets have low-Ca, high-Cr compositions and these garnets are sought in diamond exploration (Gurney and Switzer, 1973; Sobolev et al., 1973; Fipke et al., 1989, 1995b). In the recent Grütter et al. (2004) classification system, discrimination of diamond subclass "G10" garnets is done on the CaO versus Cr_2O_3 bivariate plot (Fig. 8). In eclogitic garnet, elevated Na (and lesser Ti) contents are viewed as being diamond favorable (Fipke et al., 1989, 1995b; Schulze, 1997; Grütter and Quadling, 1999), and discrimination of diamond subclass eclogite "G3" garnet is still best accomplished by use of a TiO_2 versus Na_2O discriminant diagram (Fig. 9).

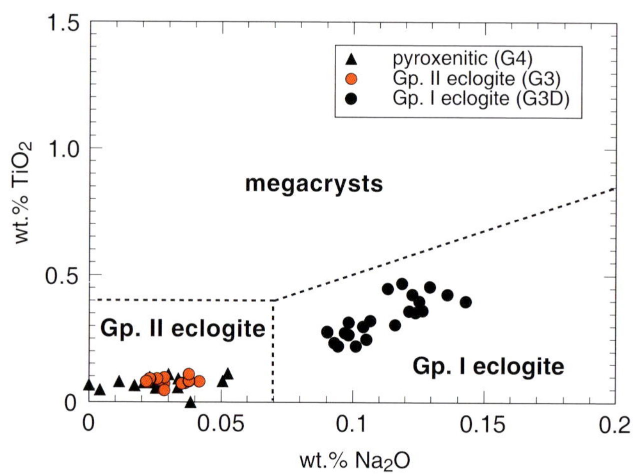

FIGURE 9. TiO_2 versus Na_2O bivariate plot for Cr-poor orange garnets from SD40 kimberlite boulder, Lake Timiskaming kimberlite field (fields from Schulze, 1997). Garnet nomenclature modified from Grütter et al. (2004). Data from McClenaghan et al. (2006b).

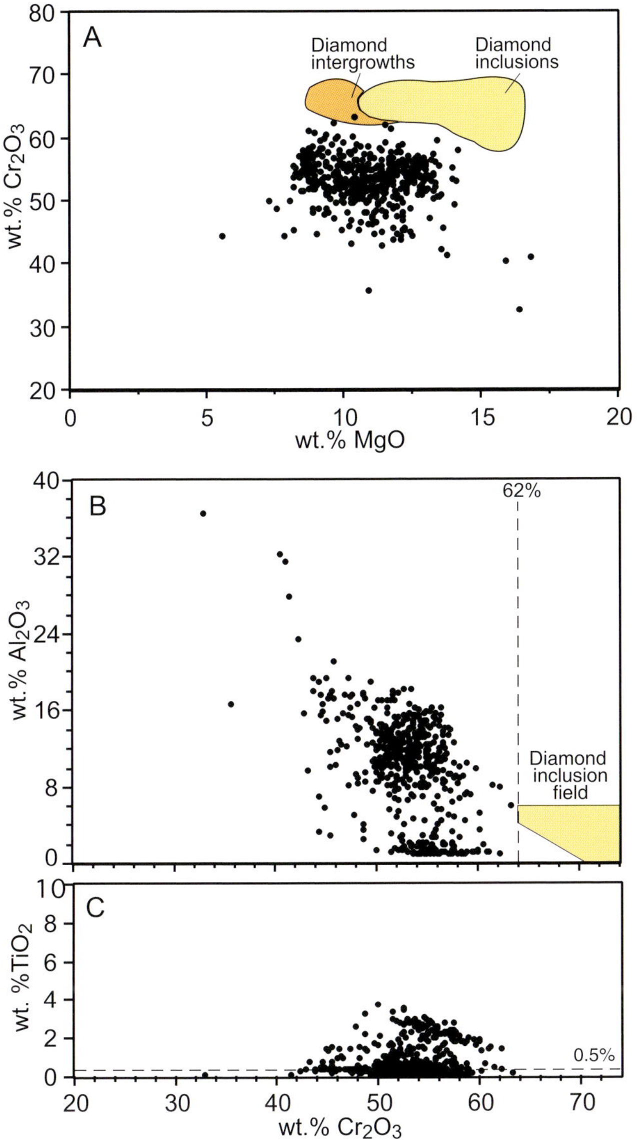

FIGURE 10. Plots for Cr-spinel from the Peddie kimberlite, Lake Timiskaming kimberlite field (N=505): **(A)** Cr_2O_3 versus MgO bivariate plot (diamond inclusion and intergrowth fields are from Fipke et al., 1995b); **(B)** Al_2O_3 versus Cr_2O_3 bivariate plot (diamond inclusion field is from Sobolev, 1977); **(C)** TiO_2 versus Cr_2O_3 bivariate plot (modified from McClenaghan et al., 1999a).

Cr-spinel with >61% wt.% Cr_2O_3, 10 to 16% wt.% MgO, <0.60 wt.% TiO_2, <8 wt.% Al_2O_3 and <6 wt.% Fe_2O_3 are typical of diamond inclusion spinel compositions (Sobolev, 1971, 1977; Gurney, 1984; Fipke et al., 1989, 1995b; Griffin et al., 1997; Grütter and Apter, 1998; Kjarsgaard, 1998) although kimberlites also contain spinel with a broader range of Cr_2O_3 and MgO (Fig. 10A). Sobolev (1971; 1977) provided further refinement of spinel discrimination by examination of Al_2O_3 and TiO_2 concentrations in conjunction with Cr_2O_3 contents (Fig. 10B,C). Note that a stoichiometric recalculation of total iron into FeO and Fe_2O_3 is also an important additional criterion.

Relative Abundance of Indicator Mineral Species

In nonglaciated terrains, chemical and physical degradation of indicator minerals changes their relative abundances in soils, wind-blown sediments, sheet wash, and stream sediments, thus the relative proportions of indicator minerals will vary with proximity to the kimberlite source (Mosig, 1980; Jennings, 1990). In Australia, for example, Cr-spinel is most resistant to weathering followed by, in decreasing order of resistance, Mg-ilmenite, Cr-pyrope, Cr-diopside, and olivine (Atkinson, 1989). The relative abundance of indicator mineral species in individual kimberlites varies significantly (e.g. Smirnov, 1959; Mitchell, 1986; Schulze, 1993a,b) and it is these variations in the kimberlite mineralogy that largely control the relative amounts of indicator mineral species in glacial and stream sediments (e.g. Kong et al., 1999). Decreases in the concentration of indicator minerals down-ice from kimberlites are primarily the result of dilution.

The relative abundance of indicator minerals may vary between kimberlite fields and thus it is important to recover all indicator mineral species instead of focusing on just one or two. For example, Cr-pyrope and Cr-diopside are the most abundant indicator minerals in kimberlite and till in the Lac de Gras kimberlite field in the central Slave region, while in the Kikerk Lake kimberlite cluster, in the north Slave region, Mg-ilmenite is the dominant indicator mineral (Fig. 11). The relative abundance of indicator minerals may also vary significantly between kimberlites within the same field. Where dispersal/dispersion trains from two kimberlites coalesce, the relative abundance of minerals may aid in distinguishing between the trains. The Diamond Lake kimberlite near Kirkland Lake, for example, contains >5000 grains Mg-ilmenite grains in a 10 kg sample (McClenaghan et al., 1998). In contrast, the C14 kimberlite, located 20 km to the northwest, is Mg-ilmenite poor and contains <10 grains per 10 kg (McClenaghan et al., 1999c). The large difference in relative abundance of Mg-ilmenite grains in these two kimberlites is reflected in the glacial sediments down-ice of each.

Indicator Mineral Size

Indicator mineral size in glacial sediments is controlled by mineral-forming processes in mantle rocks or kimberlite magma. Cr-pyrope in kimberlite is commonly 0.1 to 1.0 cm in diameter (Smirnov, 1959; Mitchell, 1986) and is highly susceptible to fracturing during decompression and cooling (Garvie and Robinson, 1984; McCandless, 1990). As a gla-

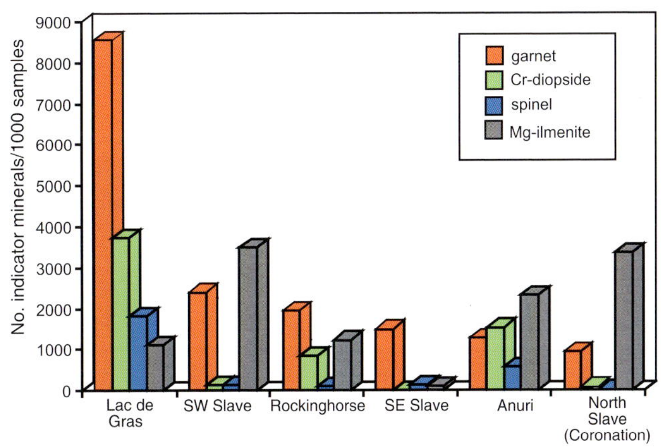

FIGURE 11. Comparison of the relative abundance of garnet, Cr-diopside, spinel, and Mg-ilmenite in till for six kimberlite fields in the Slave geological province, northern Canada (from J. Armstrong, unpub. data, 2003).

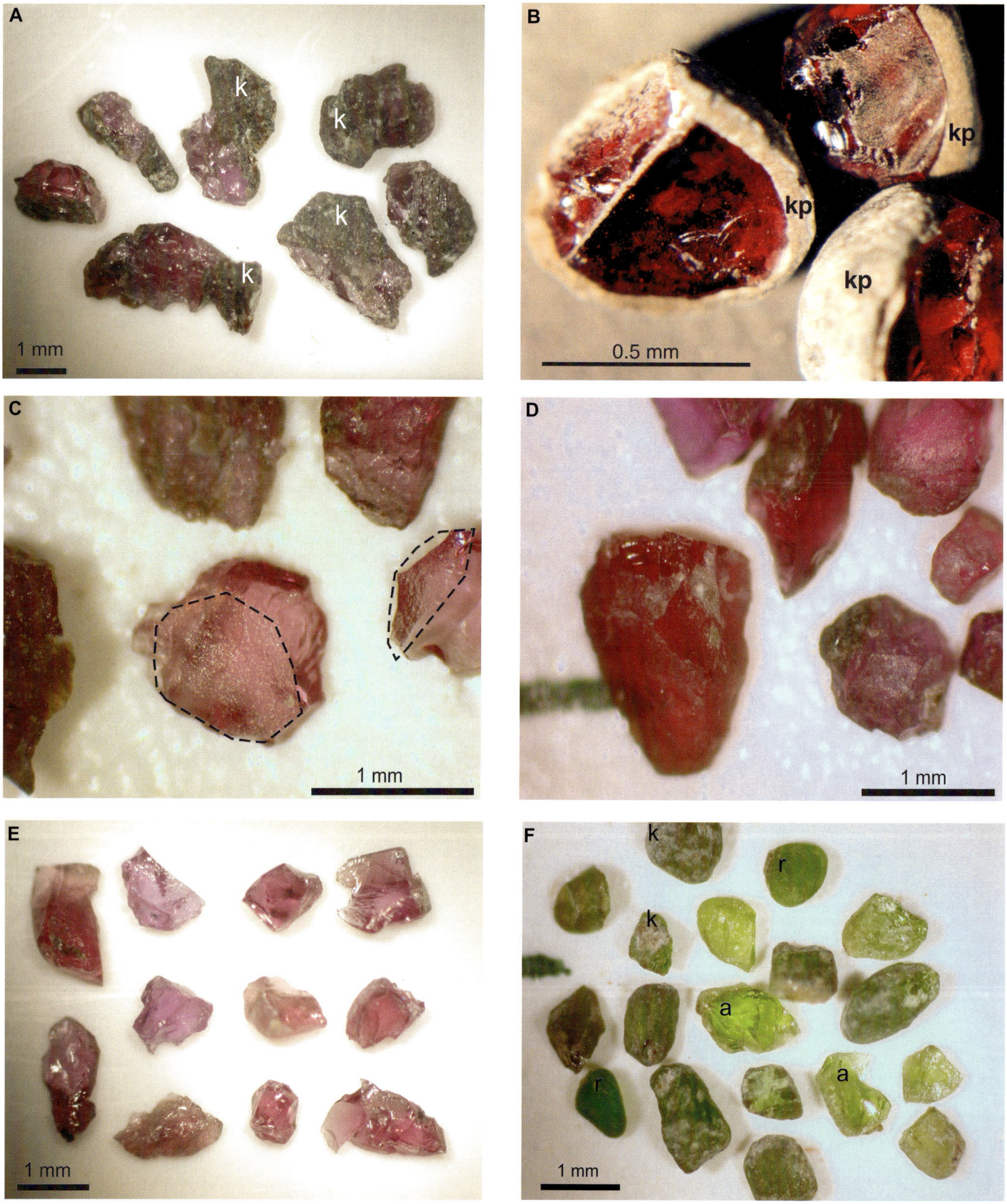

Figure 12. Colour photographs of indicator mineral grain surface textures from kimberlites: **(A)** Cr-pyrope with dark green-grey adhering kimberlite (k); **(B)** Cr-pyrope with dark green-grey kelyphite rims (kp); **(C)** Cr-pyrope displaying sub-kelyphite orange peel texture on surfaces, outlined by dashed line; **(D)** Cr-pyrope displaying sub-kelyphite frosted surfaces; **(E)** Cr-pyrope displaying fresh, angular broken surfaces; **(F)** Cr-diopside displaying rounded (r) to subrounded shape, adhering kimberlite (k), as well as broken angular surfaces (a).

FIGURE 12 CONTINUED. (G) chromite displaying octahedral to suboctahedral (resorbed) shapes.

cier entrains and transports kimberlite, Cr-pyrope preferentially breaks along the pre-existing fractures into angular grains that are dominantly sand sized (<2.0 mm). These grains appear to remain at this size regardless of glacial transport distance (Averill and McClenaghan, 1994).

Although indicator minerals occur in the fine sand (0.06-0.25 mm) fraction, grains in this size range are expensive and time consuming to recover due to their small size. Typically, indicator minerals are picked from the coarser sand (0.25-2.0 mm) sized material. Where extra grains are needed for indicator mineral chemistry studies, the fine sand fraction (0.125-0.25 mm) may also be processed and picked. Mg-ilmenite grains in sediments are commonly larger than other kimberlite indicator minerals because Mg-ilmenite occurs in kimberlite as a coarse mgeacryst phase and is intrinsically less susceptible to decompression fracturing. In sediment samples, especially coarse sorted gravel, it is not uncommon for Mg-ilmenite abundance in the 0.5 to 1.0 mm fraction to exceed abundance in the 0.25 to 0.5 mm fraction (S.A. Averill, pers. comm., 2006).

Indicator Mineral Shape and Surface Features

Surface features and textures on garnets (Table 1) may provide clues to a kimberlitic provenance, the type of transport medium (e.g. glacial versus fluvial), and the transport distance (Mosig, 1980; Afanase'ev et al., 1984; McCandless, 1990; Sobolev et al., 1993; Averill and McClenaghan, 1994; Golubev, 1995; McCandless and Nash, 1996; Garvie, 2003). Garnet may be partially covered with adhering kimberlite (Fig. 12A) or up to 2.0 mm of greenish black kelyphite reaction rims (Fig. 12B) composed mainly of pyroxene with spinel, phlogopite, or serpentine (Fediukova, 1977; Garvie and Robinson, 1984; Dredge et al., 1996). Kimberlite and kelyphite wear down or break off readily during abrasion (McCandless, 1990; McCandless and Nash, 1996) and therefore their presence/absence and thickness on individual grains may give some indication of the relative distance of glacial transport (Garvie, 2003). A finely pitted or microscopic "orange peel" textured surface underlies the kelyphite (Fig. 12C). This texture, formed on the surface of garnet, consists of minute triangular or rhombic pits that are impressions of radially arranged pyroxene crystals (Garvie and Robinson, 1984). Garnets in kimberlite may also exhibit frosted surfaces (Fig. 12D). Similar to kelyphite, the presence or absence of orange peel textures or frosted surfaces on garnet may give some indication of relative distance of glacial transport. As kimberlite or kelyphite is eroded from pyrope, the amount of orange peel or frosted surfaces exposed on grains increases (Mosig, 1980; Averill and McClenaghan, 1994; Garvie, 2003). Additional abrasion, however, will ultimately remove any sub-kelyphitic (orange peel or frosted) textures.

Garnets in kimberlite occur as spherical nodules to subangular fragments (Fig. 12E) that break apart along pre-existing fractures. It is these fractures that govern the size and shape of garnet grains in till down-ice as the garnet grains suffer very little wear, rounding, or breakdown as a result of glacial or glaciofluvial transport. Therefore, the roundness of garnet grains cannot be used to estimate distance of glacial transport. Similar preservation of original indicator mineral shape has been reported for indicator minerals that have undergone fluvial transport in Siberia (Afanase'ev et al., 1984). In contrast, indicator minerals transported by sheet wash and soil creep in Australia show pronounced wear and degradation over short transport distances (Mosig, 1980; Atkinson, 1989).

Similar to garnets, the shape and surface features of other indicator minerals in till may provide clues to relative transport distance in glaciated terrain. Cr-diopside grains may display adhering kimberlite and rounded to subrounded shape (Fig. 12F). Mg-ilmenite grains may exhibit adhering kimberlite, perovskite mantles, or delicate overgrowths of microcrystalline ilmenite. Chromite grains commonly exhibit varying degrees of resorbed octahedral crystal faces (Fig. 12G) and may display distorted crystal shapes (Yannan and Matsyuk, 1991; Lee et al., 2003).

Heavy Mineral Sampling and Analytical Methods

The main sediment types used for indicator mineral and geochemical surveys in glaciated terrain are kimberlite fragments (boulder, cobbles, pebbles), till, glaciofluvial sediments, beach sediments, and stream sediments. Kimberlite fragments are the most obvious and direct indicators of the presence of kimberlite. In some kimberlite fields, fragments have been deposited down-ice of kimberlites within and on top of till (e.g. Stewart et al., 1988; Papunen, 1995; Melnyk, 1997; J. Carlson et al., 1999) or within poorly sorted pebble to cobble gravel facies of eskers (Fig. 13) (e.g. Baker, 1982; Brummer et al., 1992a; McClenaghan et al., 2002b, 2006b; Kjarsgaard et al., 2004) and in modern fluvial sediments (Kong et al., 1999). Fragment distribution should be mapped systematically and sampled to determine indicator mineral chemistry and abundance.

Till is a first-cycle unsorted mixture of crushed rock and mineral fragments, from boulder to clay sized, transported by glaciers, blended with reworked sediments, and plastered onto the bed or released by melting at the base or surface of the ice (Klassen, 2001). Till is sampled to recover kimberlite indicator minerals and for geochemical analysis of its matrix (Fig. 4). Glaciofluvial sediments result from recycling of till by glacial meltwater, e.g. eskers, outwash, or beaches, and are sampled where sufficient sand-sized material is present. These sediments are useful at a regional-reconnaissance scale for detecting the presence of kimberlite (e.g. Lee, 1965; Golubev, 1995; McClenaghan, 2002) and are sampled to recover indicator minerals. Indicator minerals recovered from beach deposits formed from reworked esker deposits assisted in the initial discovery of the Lac de Gras kimberlites (Fipke et al., 1995a; Blusson, 1998). Glacial sediment sampling methods in permafrost and non-permafrost terrains are reviewed in Hirvas and Nenonen (1990), Kauranne et al. (1992), and McMartin and McClenaghan (2001).

Stream sediments, derived from erosion of bedrock and recycling of glacial sediments in modern day streams, are used at a reconnaissance scale to recover kimberlite indicator minerals (Fig. 4) in glaciated terrains having sufficient topographic relief (e.g. Dummett et al., 1987; Morris et al., 1994; Golubev, 1995; Morris et al., 1998; Steenfelt et al., 1999; Canil et al., 2005) or where surface sediments, such as peat or glaciolacustrine sediments, mask the underlying till (e.g. Brown et al., 1967; Wolfe et al., 1975; Kong et al., 1999; Crabtree, 2003; Friske et al., 2003). Because fluvial sediments may have undergone several cycles of transport from their bedrock source, it can be more difficult to trace indicator minerals found in them, as compared to indicator minerals in till that typically has a simpler transport history. Stream sediment sampling methods are reviewed by Maurice (1991), Ottensen and Theobald (1994), and Stendal and Theobald (1994).

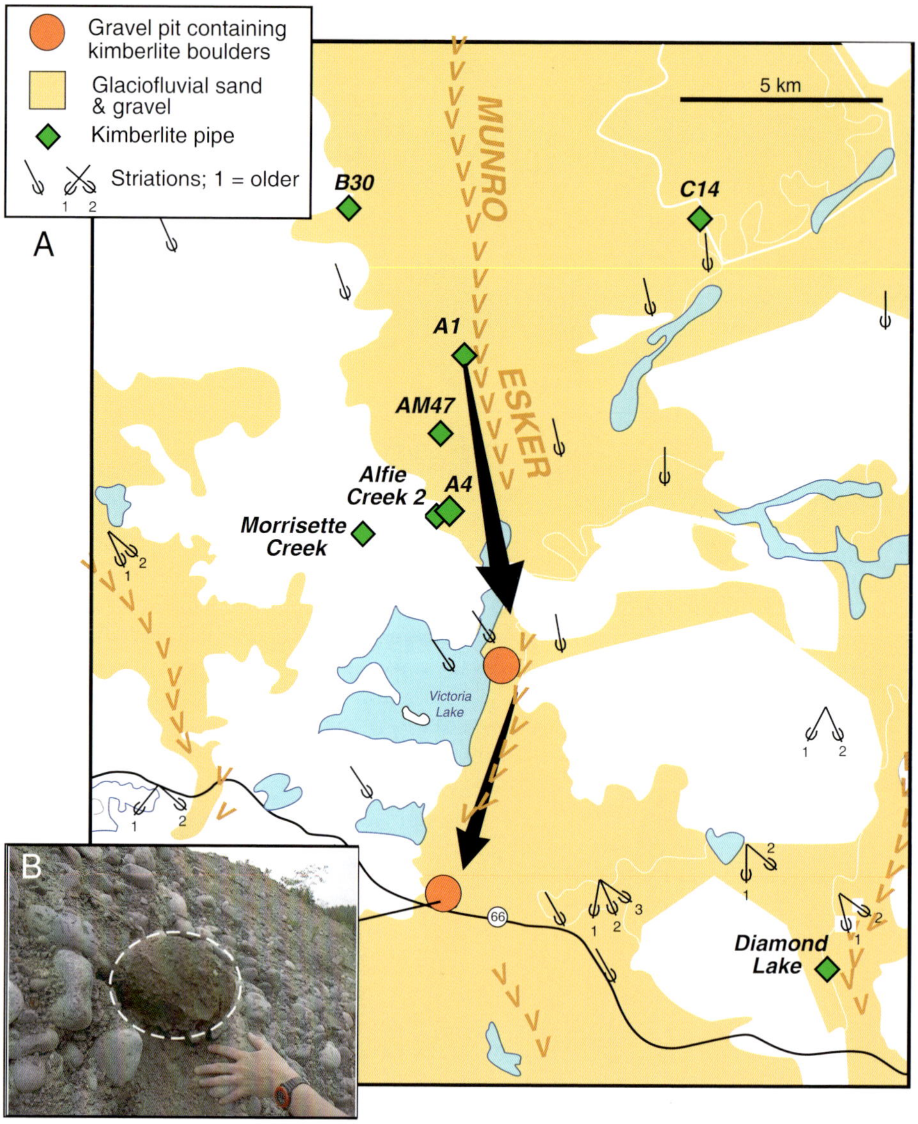

FIGURE 13. **(A)** Location of kimberlite boulders recovered from glaciofluvial sediments of the Munro Esker, relative to their source, the A1 kimberlite, which is 5 to 10 km upflow. Black arrows indicate approximate transport path of boulders in the esker (modified from Kjarsgaard et al., 2004); **(B)** highly weathered kimberlite boulder exposed in matrix-supported, poorly sorted, pebble- to small boulder-sized sediment in the Munro Esker.

For indicator mineral surveys, at least 10 to 20 kg (~5 to 10 litres) of sandy till, esker sediments, or stream sediments or 30 to 50 kg of clay-rich till are required to recover a sufficient weight of heavy minerals for visual sampling of indicator minerals (Averill, 1990, 2001; McClenaghan, 2005; Prior et al., 2005). An additional 1 kg (~0.5 litre) sample of till also may be collected for geochemical analysis of the till matrix. Indicator mineral concentrations should always be reported with respect to the weight of material sampled. To compare indicator mineral counts between till or stream sediment sampling programs, survey areas, or kimberlite fields, indicator counts should be standardized to a constant sample weight (Jennings, 1990). For example, McClenaghan et al. (1999a,b,c, 2004) standardized indicator mineral counts to a 10 kg sample weight in order to compare results from till samples down-ice of kimberlites in the Kirkland Lake and Lake Timiskaming kimberlite fields.

Sampling Strategies

Once a glaciated region is selected for diamond exploration, reconnaissance- to regional-scale surveys are conducted to identify specific areas for more focused exploration, and then follow-up of specific targets. In the following section, a number of examples are cited, however, these are not the only, or necessarily the optimum approach. For example, in Russia field panning of various types of sample media is an integral aspect of the exploration process. If present, fluvial, glaciofluvial (esker), or beach sediments are sampled initially for indicator minerals instead of till (e.g. Fipke et al., 1995b; Mckinlay et al., 1997; Blusson, 1998;

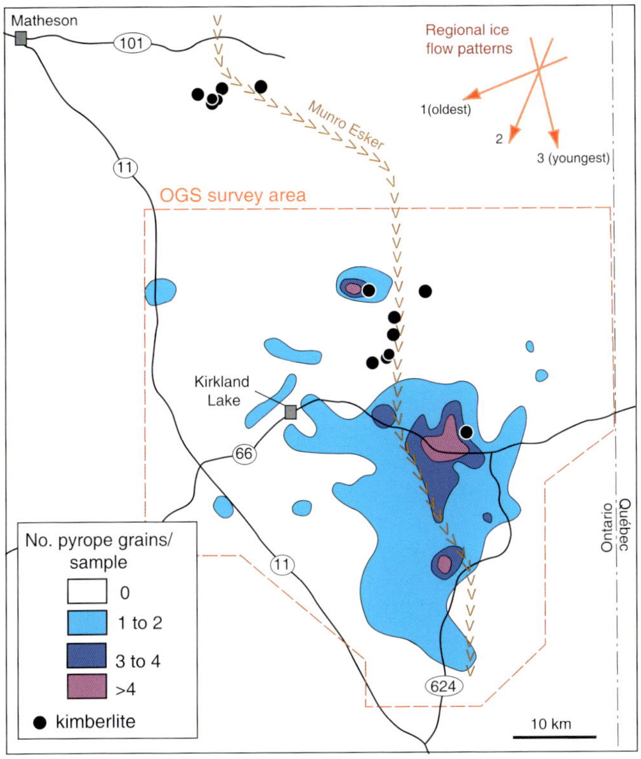

FIGURE 14. Distribution of pyrope in 5 to 8 kg till and glaciofluvial sediment samples collected by the Ontario Geological Survey as part of the KLIP regional geochemical survey (1979-1984) and the location of the subsequently discovered kimberlites (modified from Brummer et al., 1992a). Kimberlite locations are from Zalnieriunas and Sage (1995). Note pyrope abundances are for a ~1 g split of the total <2.0 mm heavy mineral concentrate, not from the entire 0.2 to 0.20 mm heavy mineral fraction.

Crabtree, 2003), because a relatively small number of samples represents the bedrock composition of a large area and sample collection is relatively rapid. Systematic follow-up is then undertaken by sampling till (e.g. Fipke et al., 1995a; Cookenboo et al., 1998). Further follow-up in areas of interest could include sampling till at a higher density and/or airborne geophysical surveys (e.g. Brummer et al., 1992a,b; Jennings, 1995; Mckinlay et al., 1997; J. Carlson et al., 1999; Graham et al., 1999). Follow-up till sampling typically consists of grid sampling along several lines perpendicular to ice-flow direction, located down-ice of indicator mineral or geophysical anomalies (e.g. McClenaghan et al., 2000b, c). In areas with thick till units, thick cover of glaciolacustrine sediments, or organic deposits overlying till (e.g. Kong et al., 1999; Crabtree, 2003), stream sediments are sampled above stream confluences with a sample density dictated by survey area size.

Heavy Mineral Sample Preparation

Heavy mineral concentrates are prepared from large till samples to concentrate kimberlite indicator minerals by removing components such as quartz, feldspar, most ferromagnesian minerals, carbonates, and rock fragments that may mask or dilute the indicator mineral signature (Shilts, 1975). The combinations of processing techniques used by exploration companies or government agencies for recovering indicator minerals are quite variable (e.g. Gregory and White, 1989; Davison, 1993; Towie and Seet, 1995; McClenaghan and Kjarsgaard, 2001). Glacial or stream sediment is disaggregated and the >2 mm fraction is removed for lithological analysis (pebble counts). Because of the large weight (10-50 kg) of <2 mm material to be processed, laboratories first preconcentrate the sample by density, grain size, or magnetic susceptibility before final density separation. Density preconcentration, which may include jig, shaking table, spiral, dense media separator (DMS), or pan, may be combined with the use of heavy liquid separation methods (e.g. using tetrabromoethane), prior to final concentration. The tabling method has the advantages of speed and relatively low cost (Gregory and White, 1989), and provides the opportunity to capture indicators of other commodities such as gold, base metals, and PGEs (McClenaghan, 2005). However, tabling methods are not as efficient as heavy liquid or dense media separator techniques (i.e. they are potentially prone to the loss of indicator minerals), which can be critical in reconnaissance-scale sampling when the recovery of every indicator mineral grain is crucial. Furthermore, modern dense media separator techniques are rapid and efficient, but only for particles >0.3 mm. Screening to recover the medium to very coarse sand-sized fraction, or paramagnetic separation to separate oxides and silicates, may be used instead of density preconcentration if recovery of gold grains is not a priority. Final density concentration is completed using heavy liquids such as methylene iodide (diluted with acetone), or a Magstream® separator using a specific gravity threshold of 3.1 to 3.2 g/cm^3 to ensure recovery of Cr-diopside, the least dense of the kimberlite indicator minerals (Table 1). The ferromagnetic fraction, composed almost exclusively of magnetite, is then removed using a hand magnet or roll separator, weighed and stored. The final product is a non-ferromagnetic heavy mineral concentrate consisting of grains between 0.25 and 2.0 mm, for picking of kimberlite indicator minerals. The ferromagnetic fraction is sometimes examined for Mg-ilmenite or kimberlite spinel.

Indicator Mineral Chemical Analysis

Visually selected indicator grains are counted and mounted on glass slides or in circular epoxy mounts. Mineral grains can be further screened at this stage by rapid qualitative analysis using an automated routine on a Scanning Electron Microscope (SEM) equipped with an EDS detector. Subsequently, mineral grains may undergo quantitative analysis by EDS on a SEM, or by EDS or WDS methods on an electron microprobe (EM) to determine their major and trace element composition. Mineral grains are then classified as being kimberlite-derived (or not) using the chemical criteria described above. Cr-pyrope, Cr-diopside, and olivine are the easiest minerals to identify visually and therefore the number of grains picked is usually equal to or only slightly more than the number confirmed by quantitative analysis. Mg-ilmenite may be visually similar to ilmenite from crustal rocks, as well as strongly resorbed Cr-spinel that has lost its octahedral form (Fig. 12G), thus, quantitative analysis may be required to confirm their identity (Wyatt et al., 2004). As a result, the number of Mg-ilmenite or chromite grains picked can be much greater than the number confirmed by quantitative analysis. Eclogite pyrope-almandine garnet identification must also be confirmed by quantitative analy-

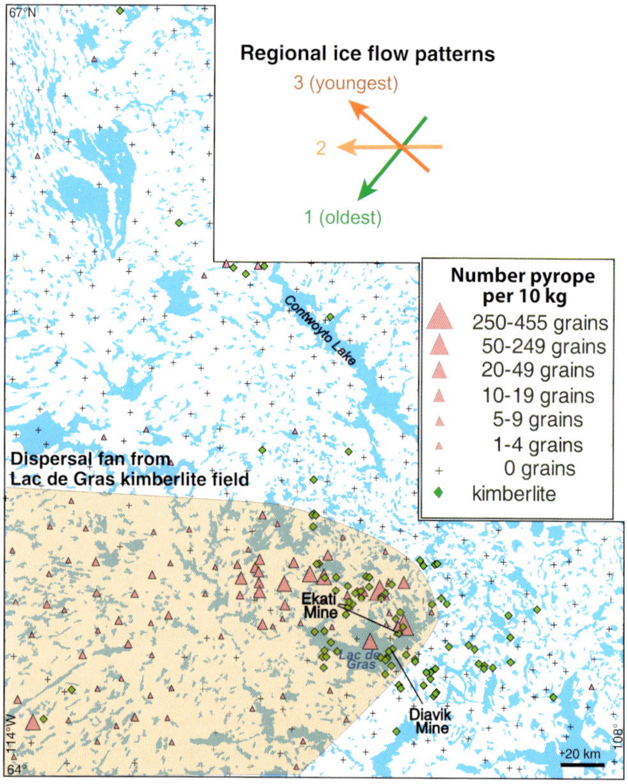

FIGURE 15. Abundance of pyrope in the 0.25 to 0.5 mm heavy mineral fraction of 10 kg surface till samples collected by the Geological Survey of Canada across the Lac de Gras region, northern Canada. Pyrope grains have been dispersed by three phases of ice flow to the southwest, west, and finally to the northeast. The highest concentrations are overlying and just west of the kimberlites centred on Lac de Gras (modified from Dredge et al., 1997; Kerr, 2000).

sis, as these garnets are similar in appearance to crustal-derived orange almandine or almandine-pyrope garnet.

Indicator Mineral Case Histories

The distribution of indicator minerals in till commonly defines broad fan-shaped patterns of dispersal from kimberlite fields that are many times larger than the fields themselves, and hence form large exploration targets. For example, the broad glacial dispersal fan (Fig. 14) that extends 30 km south (down-ice) from the Kirkland Lake kimberlite field (Fortescue et al., 1984; Brummer et al., 1992a,b) is the net result of three phases of ice flow to the southwest, south, and southeast. The glacial dispersal fan for the much larger Lac de Gras kimberlite field extends at least 180 km down-ice (west) reflecting three phases of ice flow (Dredge et al., 1995, 1997; Ward et al., 1995) across the field. At Lac de Gras, glacial dispersal from the field is best defined by high Cr-pyrope abundances in till (Fig. 15). Till in background areas contains no Cr-pyrope, whereas within the fan, Cr-pyrope counts vary between 1 and 455 grains. The lack of indicator minerals in till over the southeast part of the kimberlite field may be due to the presence of allochthonous till over this area, and/or thick till deposits that mask the underlying dispersal trains and/or the indicator-poor nature of these kimberlites. Indicator minerals in regional till and glaciofluvial sediment samples over large parts of the Canadian Prairie (Alberta, Saskatchewan, Manitoba) provide insight into the diamond potential of the region. Elevated Cr-pyrope concentrations in till form a broad, 300 km wide plume in western and central Saskatchewan (Fig. 16) derived from the Fort à la Corne and Candle Lake kimberlite fields. In contrast, elevated Cr-pyrope abun-

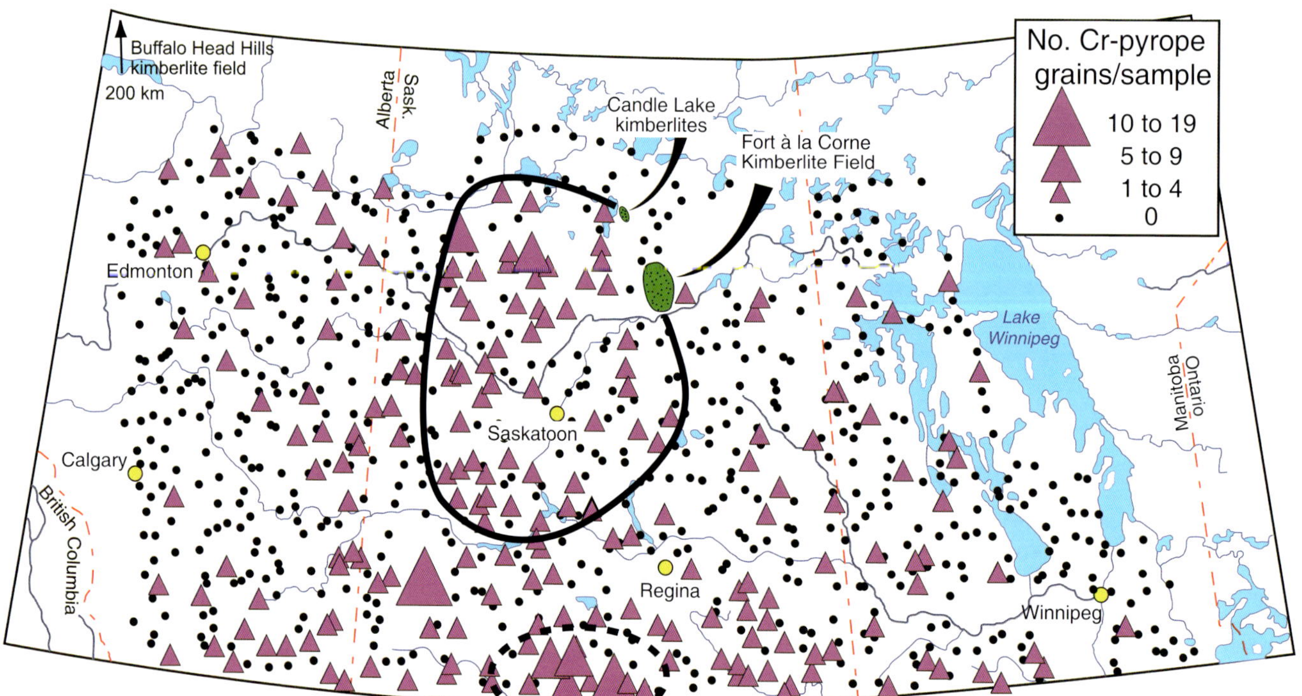

FIGURE 16. Abundance of Cr-pyrope in the 0.25 to 2.0 mm heavy mineral fraction of 25 kg surface till samples collected by the Geological Survey of Canada across the Canadian Prairie (Garrett and Thorleifson, 1996). The high concentrations of grains around Saskatoon, outlined by a solid black line, are derived from the Fort à la Corne kimberlite cluster and the high concentrations along southern border, outlined by dashed black line, are eroded from local Tertiary gravels.

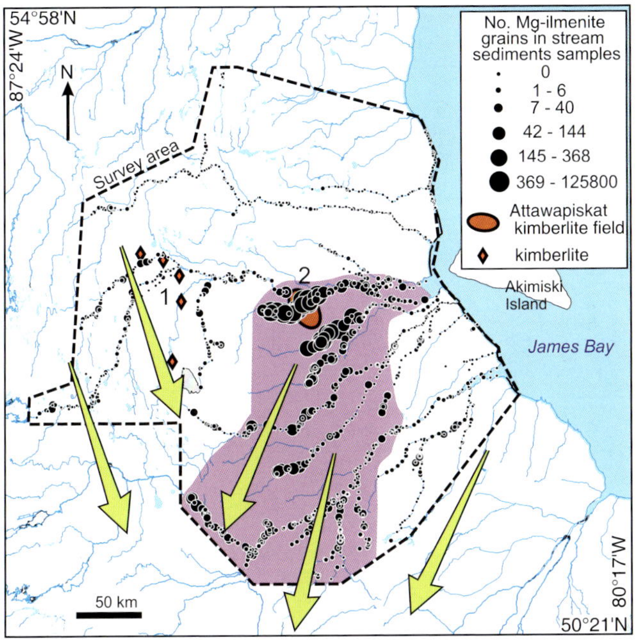

FIGURE 17. Abundance of Mg-ilmenite in the 0.25 to 2.0 mm heavy mineral fraction of 8 to 13 kg stream sediment samples around the Attawapiskat kimberlite field in northern Ontario (modified from Crabtree, 2003). Red diamonds (1) indicate location of the Kyle Lake kimberlites; red oval (2) indicates location of Jurassic age kimberlite cluster. Yellow arrows indicate the orientation of linear ice flow features (Prest et al., 1968; Crabtree, 2003). The dispersion fan, highlighted in purple, is the net effect of glacial transport from the Jurassic age kimberlite cluster to the southwest and subsequent postglacial fluvial transport to the northeast and east.

dances in southern Saskatchewan till are likely derived from local Tertiary gravels, which are suggested to contain fluvially derived material from rocks of the Montana Alkaline Province (Simpson, 1993; Kjarsgaard, 1995). In west-central Saskatchewan, a broad Cr-pyrope anomaly exists, but the source of the Cr-pyrope is unknown.

In the clay- and muskeg-covered terrain of the James Bay Lowland, indicator minerals in stream sediments clearly demarcate the presence of the Attawapiskat kimberlite field. At a reconnaissance scale, the signature of the field is best defined by Mg-ilmenite (Fig. 17), Cr-pyrope, and olivine, which are significantly more abundant in the stream sediments than Cr-diopside or spinel (Crabtree, 2003). The relative proportions of indicator minerals in the stream sediments reflect their relative abundance in the source Jurassic kimberlites to the north (Kong et al., 1999). The 300 km long dispersion fan of indicator minerals in stream sediments to the southwest of the field is the product of two modes of transport; glacial transport to the southwest and subsequent fluvial transport to the northeast and east in modern streams. In the thick (20-100 m), clay-rich drift covered terrain of the Buffalo Head Hills kimberlite field (Fig. 2), total counts of all indicator minerals in stream sediments indicate the presence of the kimberlites (Fig. 18). Indicator minerals have been glacially transported southwest and south from the kimberlites and then remobilized in small local streams flowing mainly to the east. The combined dispersal/dispersion of indicator minerals is detectable up to 30 km to the southwest

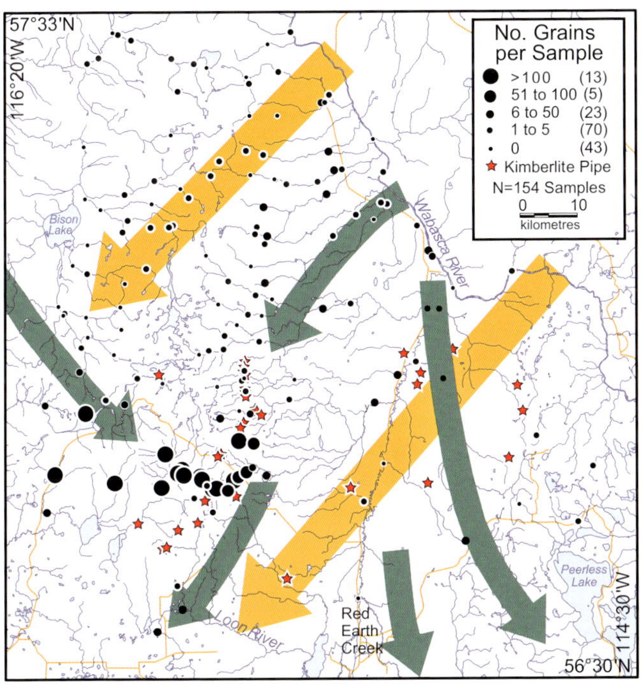

FIGURE 18. Regional ice-flow patterns and total kimberlite indicator mineral abundance in the 0.25 to 2.0 mm heavy mineral fraction of stream sediments normalized to 15 kg in the Buffalo Head Hills kimberlite field, northern Alberta. Yellow arrows indicate older ice flow directions during glacial maximum and green arrows indicate younger, late glacial ice-flow patterns. Red stars indicate the location of kimberlites (modified from Friske et al., 2003). Ice-flow arrows are modified from Paulen and McClenaghan (in press).

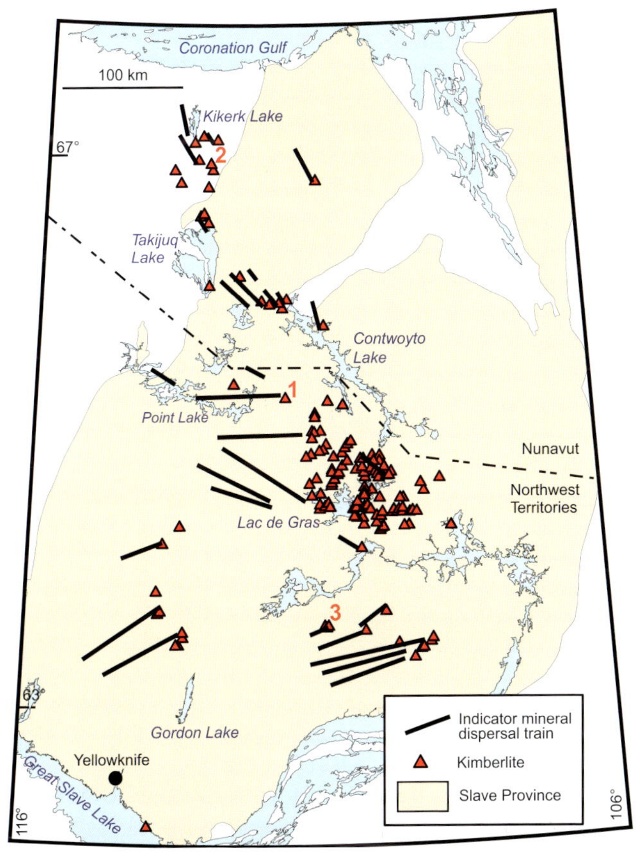

FIGURE 19. Summary of trends of the kimberlite indicator mineral glacial dispersal trains from individual kimberlites across the Slave Province (modified from Armstrong and Kjarsgaard, 2003). 1. Ranch Lake train (Fig. 20); 2. Potentilla train (Fig. 21); 3. Snap Lake train (Fig. 22).

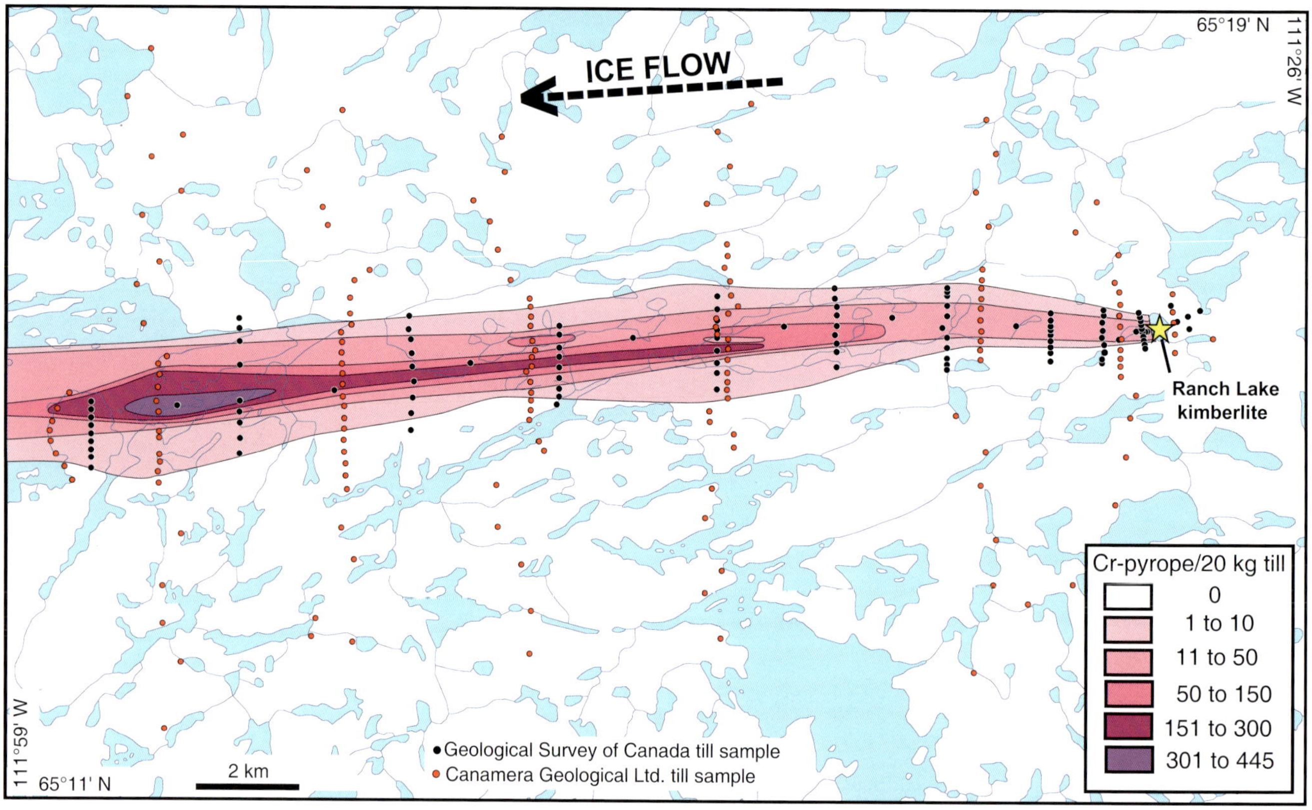

FIGURE 20. Ribbon-shaped dispersal train trending westward from the Ranch Lake kimberlite, Lac de Gras, as defined by Cr-pyrope concentrations in 20 kg till samples. The train was formed by a single phase of ice flow towards the west (modified from McClenaghan et al., 2002a).

and is dominated by olivine in the north and by a combination of Cr-pyrope, Mg-ilmenite, and chromite in the southwest. In this field, other minerals such as Cr-corundum and zircon may also be useful indicators as they have been reported in the kimberlites as well as in the local stream sediments (Friske et al., 2003; Hood and McCandless, 2004).

In contrast to the large dispersal fans derived from kimberlite fields, glacial dispersal of indicator minerals from individual kimberlites can result in the formation of ribbon- to fan-shaped dispersal trains (Fig. 4). In the Slave Province, numerous narrow ribbon-shaped dispersal trains have been identified (Armstrong and Kjarsgaard, 2003). These trains are sharp-edged linear features that often extend for tens of km down-ice and are interpreted to have formed by the last phase of ice flow across the region (e.g. Mckinlay et al., 1997; J. Carlson et al., 1999; Doyle et al., 1999; Graham et al., 1999; Armstrong and Kjarsgaard, 2003; Kirkley et al., 2003), which varies in direction within the Slave Province (Fig. 19). For example, the Ranch Lake train in the central part of the Slave

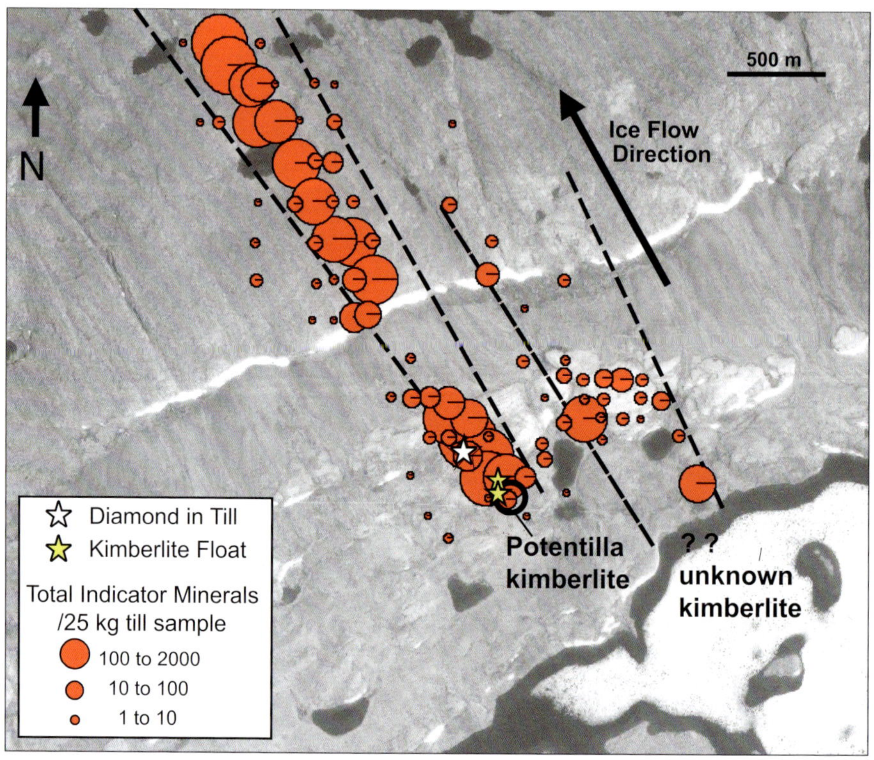

FIGURE 21. Ribbon-shaped dispersal train (outlined by dashed line) trending northwest from the Potentilla kimberlite in the North Slave kimberlite field of the north Slave Province. The train is defined by total indicator mineral abundance in the 0.4 to 1.25 mm fraction of 25 kg till samples, as well as the presence of kimberlite float and diamonds in till. A second train from an unknown kimberlite source is located to the east. The trains were produced by a single phase of ice flowing northwest (modified from Ashton Mining of Canada, unpublished data).

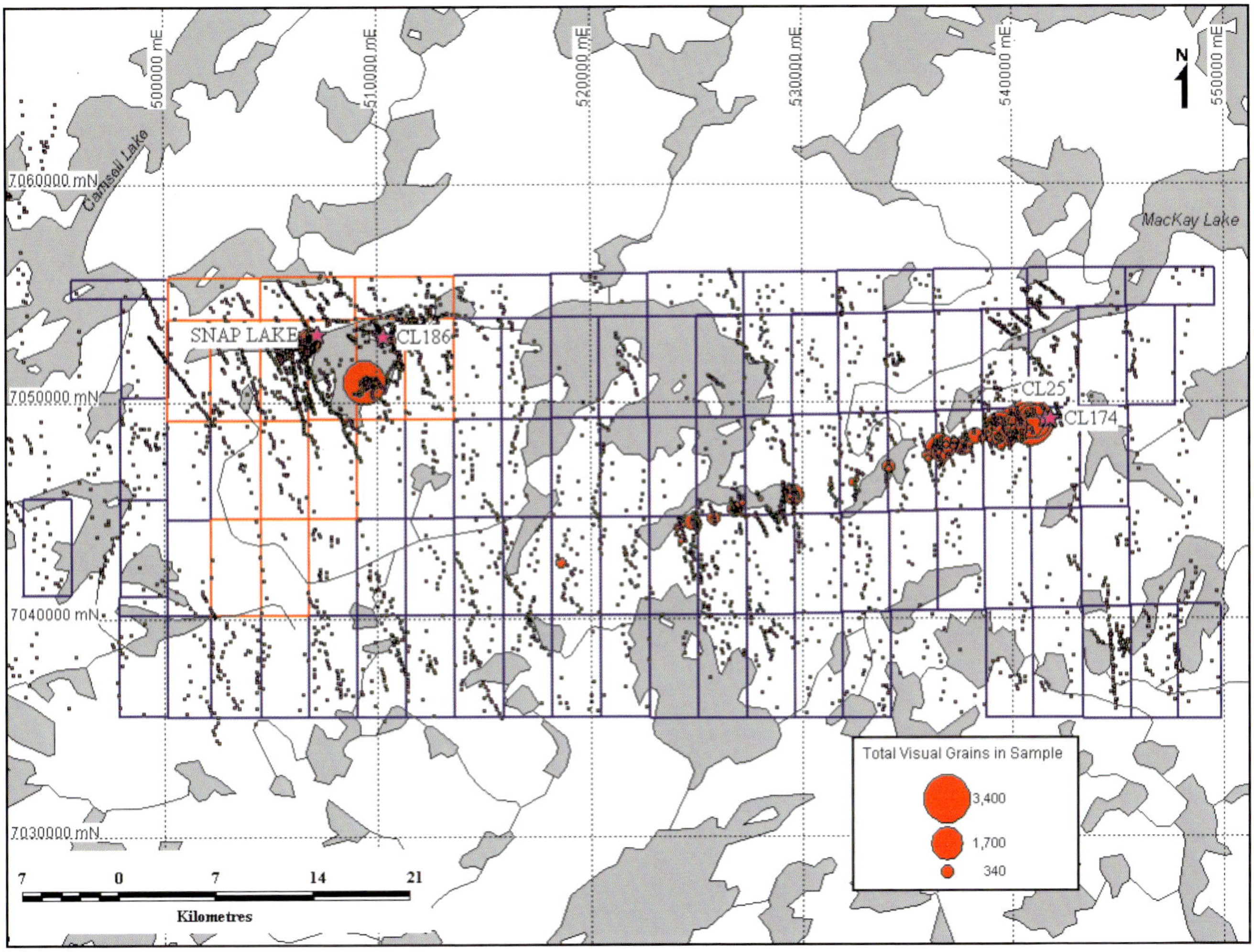

FIGURE 22. Short, wide dispersal train west of the Snap Lake kimberlite dyke and linear, ribbon-shaped dispersal trains trending west-southwest from the CL25 and CL174 kimberlites. Trains are defined by total indicator mineral counts in 25 to 50 kg till samples (Kirkley et al., 2003).

Province is a 70 km long, west-trending, ribbon-shaped train (Cookenboo, 1996) that is best defined by the abundance of Cr-pyrope (Fig. 20) and Cr-diopside, which are the dominant indicator minerals in the source kimberlite. Till samples within the train contain up to 545 Cr-diopside and 445 Cr-pyrope grains in 20 kg samples, while background samples do not contain any indicator minerals (McClenaghan et al., 2000b,c). In the Kikerk Lake kimberlite cluster, ribbon-shaped dispersal trains trend northwest (e.g. Fig. 21), while in the southeast part of the Slave Province, indicator mineral trains trend southwest and vary from long (>100 km) narrow, linear ribbons derived from kimberlite pipes to wide, short (<10 km) trains (Fig. 22) derived from the kimberlite sheet at Snap Lake (Armstrong and Kjarsgaard, 2003; Kirkley et al., 2003).

Fan-shaped trains are interpreted to have formed as a result of two or more phases of ice flow. A good example of fan-shaped dispersal is the south-southeastward-trending dispersal of indicator minerals from the Bucke and Gravel kimberlites in the Lake Timiskaming field (Fig. 23). The fan-shaped pattern was formed by ice flow to the south followed by flow to the southeast, producing an indicator mineral anomaly that widens with increasing distance down-ice. The fan extends at least 12 km down-ice and is best defined by Mg-ilmenite and Cr-pyrope, the most abundant indicator minerals in both kimberlites (Kjarsgaard et al., 2003). A second example, although not from Canada, has been documented by Lehtonen et al. (2005). They identified an indicator mineral fan extending down-ice from two kimberlites that was formed by two phases of ice flow to the east-southeast and southeast.

Geochemical Methods

Till Geochemistry

The use of till geochemistry in kimberlite exploration is increasing because of its versatility (regional- and local-scale surveys), initial low cost as compared to indicator mineral methods, and quick turnaround time. A kimberlite geochemical signature is detectable in till containing a minimum of hundreds of indicator minerals in a 10 kg sample (Fig. 24) (e.g. McClenaghan et al., 1996, 1999a, 2002a, 2004). Thus, this method is best used in conjunction with indicator mineral methods to identify kimberlite dispersal patterns (Gravel et al., 2005; Steenfelt et al., 2005) or for pre-screening/prioritizing bulk till samples prior to more costly and time consuming indicator mineral analysis. Because till geochemistry is already a well established method for precious and base metal exploration (McClenaghan et al., 2000a), its use in diamond exploration also allows for evaluation of a broad range of other commodities (e.g. gold and base metals). Till geochemistry also provides baseline environmental geochemical data for establishing pre-diamond mining soil conditions.

The application of till geochemistry to diamond exploration is based on the unusual major and trace element composition of kimberlites, thus elements such as Mg, Ni, Cr, Nb, Ta, Ti, Ba, K, Rb, Sr, P, and LREE should be the most useful kimberlite pathfinder elements (Atkinson, 1989; McClenaghan and Kjarsgaard, 2001). The presence of olivine in till will be reflected in elevated concentrations of Mg and Ni; high abundances of Cr-spinel, Cr-pyrope, and Cr-diopside will be reflected by high concentrations of Cr. Elevated values for Nb, Ta, and Ti reflect the presence of Mg- ilmenite and groundmass perovskite. Elevated Ba, K, and Rb are due to phlogopite; high levels of P and LREE are caused by apatite (and perovskite for LREE), and elevated Sr could reflect carbonate minerals and apatite. Because a set of pathfinder elements will depend on kimberlite composition as well as that of the surrounding bedrock, the diagnostic elements are region or location specific. For example, Cr, Ni, Ta, Nb, Sr, Ba, Co, Mg, Ca, K, Ti, P, and LREE serve a geochemical pathfinder elements in the Lake Timiskaming and Kirkland Lake kimberlite fields (McClenaghan et al., 1999a,b, 2004), whereas in the Lac de Gras area, the incompatible element-rich nature of the granitoid host rocks reduces the list of pathfinder elements to Ni, Cr, Ba, Co, Sr, K, Mg, and Mn (Wilkinson et al., 2001; McClenaghan et al., 2002c). Typically, element concentrations in the kimberlite that are ten times greater than that of the host rocks will produce down-ice geochemical anomalies in the till matrix. Till geochemical analyses should be carried out using a total digestion method, such as $LiBO_2$ fusion, Na_2O_2 fusion, or four-acid digestion and not with aqua regia in order to digest both oxide and silicate minerals.

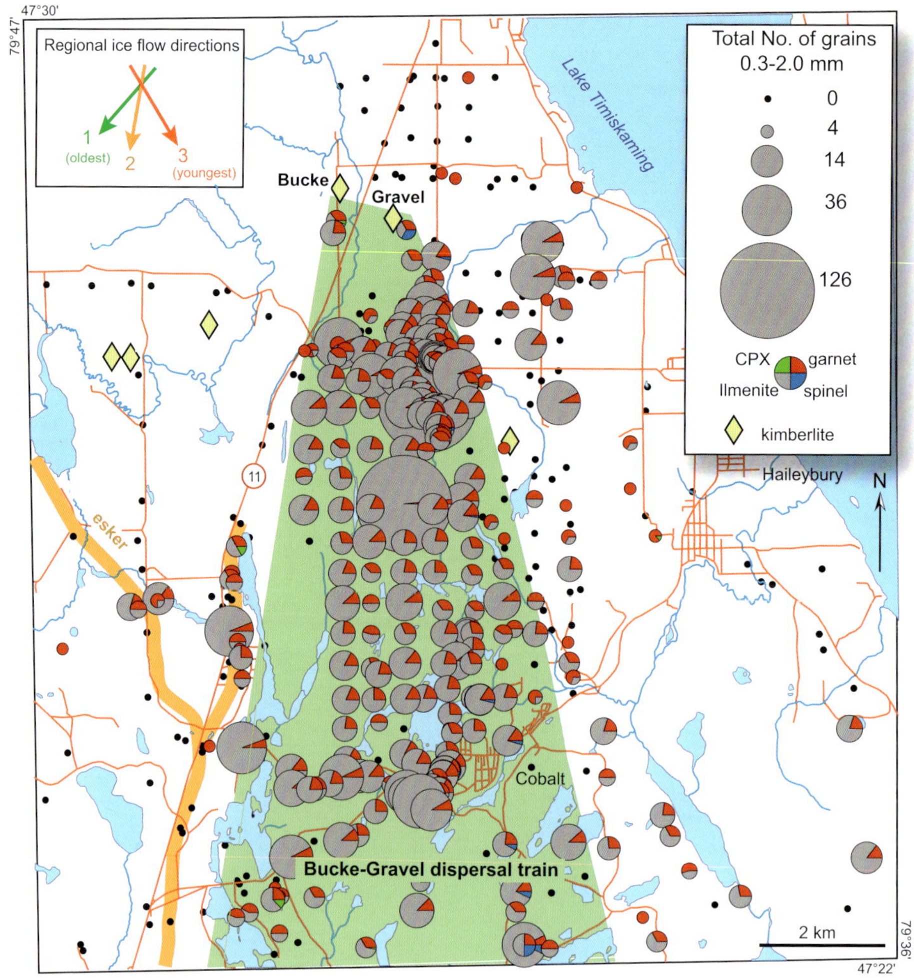

FIGURE 23. Fan-shaped dispersal train of indicator minerals (shaded green) trending south-southwest from the Bucke and Gravel kimberlites in the Lake Timiskaming kimberlite field of northeastern Ontario (unpublished data from De Beers, 2003). The train is defined by visually identified abundance of Mg-ilmenite and garnet in the 0.3 to 2.0 mm fraction of till and glaciofluvial sediments and its shape is the result of two phases of ice flow, southwest followed by southeast (modified from Kjarsgaard et al., 2003).

Case studies around known kimberlites in the Kirkland Lake and Lake Timiskaming fields indicate that the <0.063 mm fraction (silt + clay) provides a strong contrast between background and anomalous trace element concentrations related to kimberlite (Fig. 24). Anomalous elements in this fraction reflect silt-sized indicator minerals as well as clay-sized minerals from the fine-grained kimberlite matrix and its weathering products. Analyzing the -80 mesh (<0.177 mm) sand + silt + clay fraction of till is not recommended because it includes abundant sand-sized quartz and feldspar grains (Saarnisto, 1990), which will dilute and mask the geochemical signature of the kimberlite.

Soil Selective Leaches and Soil Gas

Recently, interest has been growing in the application of selective extraction geochemical techniques capable of detecting the presence of kimberlites buried beneath thick glacial sediments (Hamilton, 1998; Burt and Hamilton, 2004; McClenaghan et al., 2006a). These methods may prove to be useful for prioritizing geophysical targets (i.e. determining which targets are most likely to be kimberlites) prior to drilling in areas of thick (>10 m) glacial sediment cover where till is extremely thick, not accessible (i.e. clay covered), or not present. During kimberlite weathering, elements may be mobilized, transmitted vertically through the overlying glacial sediments, and deposited on mineral grains or organic matter in surface soils (Fig. 4). Suggested transport mechanisms may include some combination of diffusion, electrochemical, gaseous, or groundwater transport (e.g. Hamilton, 1998; Cameron et al., 2004). The mobile ions in the soil are loosely bound and readily extractable by weak, selective chemical digestions, such as ammonium acetate at pH 5 (AA5), Mobile Metal Ion® (MMI-D), or Enzyme Leach® and detectable using ICP-MS techniques. As of yet, only two case studies have been published on the application of this technique to kimberlite exploration in glaciated terrain, thus the suitability of this method to kimberlite exploration has not been fully evaluated.

In contrast to soil profiles in older tropical and arid terrains that have been developing for more than 100,000 years, kimberlitic geochemical signatures in Canada have had less than 10,000 years to potentially migrate upwards through thick glacial sediments to the surface and accumulate in soil. Thus, relative to other terrains, soil geochemical anomalies in glaciated terrain are expected to be weaker (i.e. lower concentrations, fewer elements), and sample collection from a specific depth range below the surface is important. For soil selective leach methods in glaciated terrain, either mineral soils (Ae, B, C horizons), or organic soils (humus, peat), are collected using soil horizon-based or depth-based protocols between 5 to 25 cm from surface. Importantly, only one type of sampling medium is collected, i.e. either mineral soil or organic soil. Kimberlite pathfinder elements will depend on the geochemical contrast between the kimberlite and its host rocks, as well as the thickness and nature of overlying glacial sediments (till, clay, sand, or organics), the groundwater regime, and the specific transport mechanism involved. Thus, some elements will be enriched in soil overlying kimberlites provided appropriate transport mechanisms were operative. Eccles (1998a) demonstrated that soil selective leaches using Enzyme Leach® were able to detect a geochemical signature in B horizon soil over the Mountain Lake ultrabasic diatreme hosted in Cretaceous sandstone. This case study, however, was conducted using soils developed on thin till enriched in clastically dispersed kimberlite debris. The kimberlite debris may have contributed to the selective leach signature. McClenaghan et al. (2006a) demonstrated that elevated concentrations of Ca, Sr, Na, Ni, Ba, Co, Mg, Mn, Pb, LREE, Ti, Cd, I, and U occur in mineral soils analyzed by ammonium acetate (pH 5) and MMI-D selective leaches over the 95-2 kimberlite in the Lake Timiskaming kimberlite field. This kimberlite, in contrast to the previous case study, is deeply buried by up to 60 m of mostly glaciofluvial sediments (Fig. 25), which contain no kimberlite debris from the 95-2 kimberlite. For many of the elements in soil over the 95-2 kimberlite, the response is skewed to the south part of the kimberlite and its southern margin. The skewed responses may reflect the presence of zones of enhanced groundwater flow (common at some kimberlite margins) or kimberlite-rich till overlying bedrock on the down-ice (south) margin of the kimberlite.

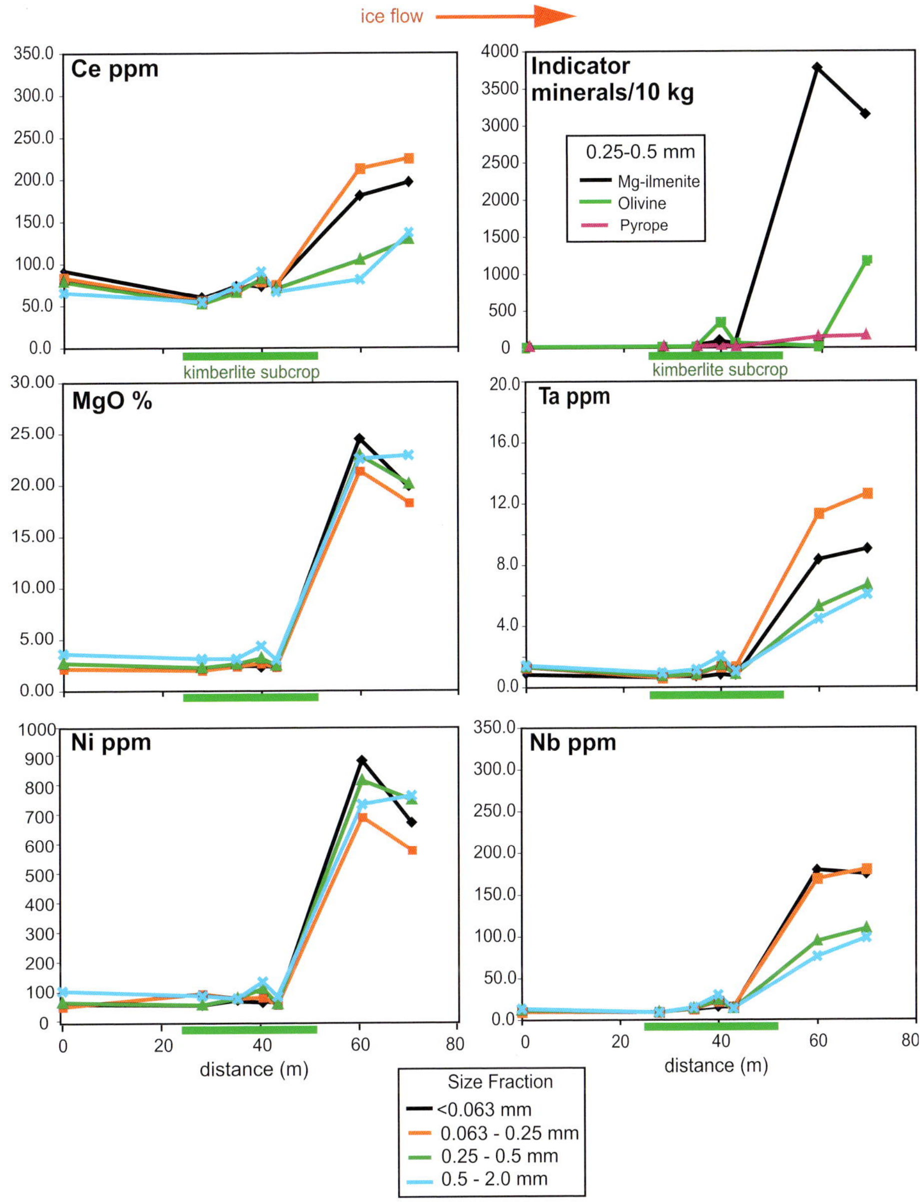

FIGURE 24. Indicator mineral abundance in the 0.25 to 0.5 mm fraction and concentrations of Ce, MgO, Ni, Ta, and Nb in four size fractions of till up-ice, overlying, and down-ice of the Triple B kimberlite, Lake Timiskaming field (modified from McClenaghan et al., 2004).

Gases in soil, which have migrated upwards from deeply buried kimberlites and/or been released in soils near surface through bacterial activity, may also indicate the presence of underlying kimberlite (Fig. 4). Primarily hydrocarbons, but also sulphur, halogens, and some other miscellaneous compounds are detected in soils using one of two laboratory methods, soil desorption pyrolysis (SDP®, Thiede et al.,

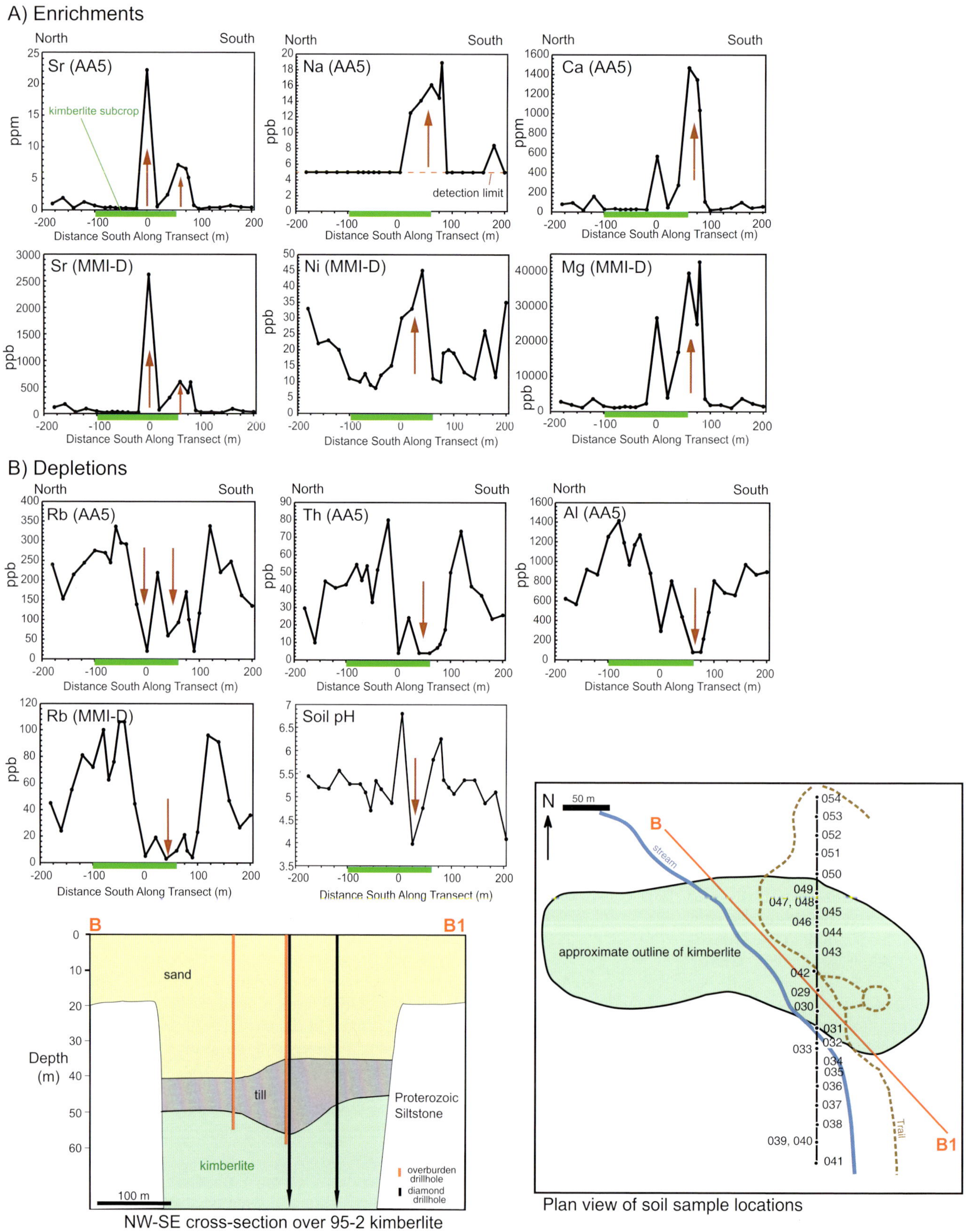

Figure 25. Element distribution patterns in Ae + B horizon soil developed on thick glaciofluvial sand overlying the 95-2 kimberlite, Lake Timiskaming field, determined by two selective leaches, ammonium acetate (pH 5) (AA5) and MMI-D®: A) elements enriched; and B) elements depleted in soils overlying the kimberlite. Soil sample locations and cross-section showing the glacial sediment thickness over the 95-2 kimberlite are also shown (modified from McClenaghan et al., 2006a).

2005) or soil gas hydrocarbons (SGH®). For example, a kimberlite in the Lac de Gras field was distinguished from a geophysically similar granitic intrusion by its elevated counts and sum of CFCs, sulphur gases, and inorganic compounds using the SDP® soil gas method (Fig. 26).

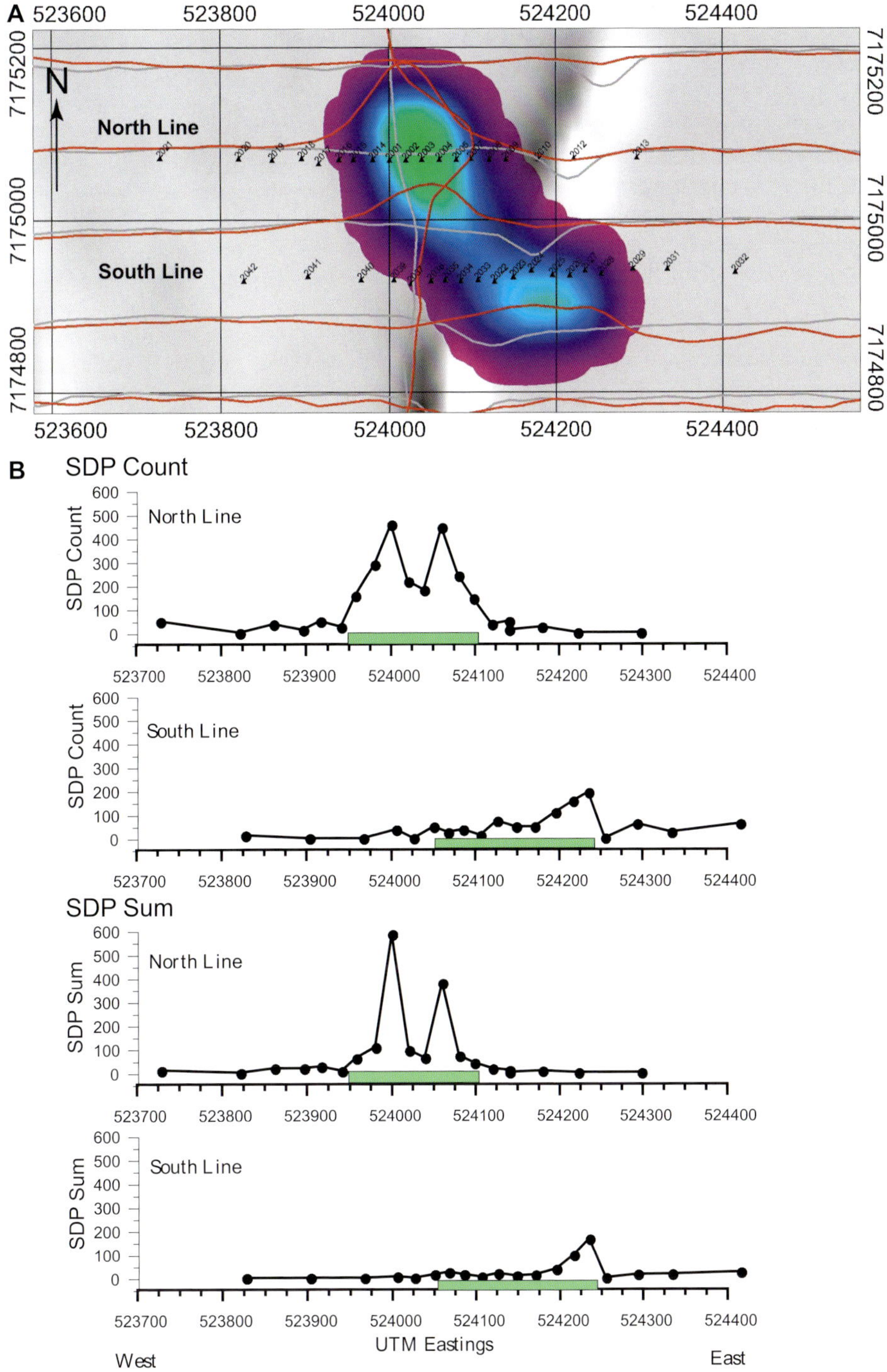

Figure 26. (A) Airborne EM response and the location of soil gas sample sites (north and south lines) across the Kodiak kimberlite, Lac de Gras field; and **(B)** SDP® soil gas count and sum values for a combination of CFCs, sulphur gases, and inorganic compounds in the soil samples collected along the two sample lines (BHP Billiton unpublished data, 2003).

Biogeochemistry

Similar to soil geochemical methods, biogeochemical methods are a local-scale exploration tool that may aid in prioritizing geophysical targets prior to drilling in areas of thick (>10 m) glacial sediment cover (Fig. 4). The few site-specific studies that have been conducted over some Siberian kimberlites in the glaciated terrain of northern Russia indicate the potential of biogeochemistry for kimberlite exploration (e.g. Kobets and Komarov, 1958; Litinskiy, 1964; Buks, 1965; Komogorova et al., 1986). More recently, biogeochemical studies over kimberlites in Canada (Dunn, 1993; McClenaghan and Dunn, 1995; Dunn and McClenaghan, 1996; Eccles, 1998b) support the Russian observations of anomalous geochemical (Ni, Rb, Sr, Cr, Co, Nb, Mg, P, and LREE) signatures in vegetation growing on glacial sediments over kimberlite. In general, biogeochemistry is probably not a cost-effective exploration tool.

Hydrogeochemistry

Groundwater interaction with kimberlite may produce aqueous geochemical anomalies (Fig. 4) typical of low-temperature serpentinization reactions, which are characterized by high pH (up to 12), very low concentrations of Mg, and very high concentrations of K (e.g. Sader et al., 2003, 2007). The trace element signatures of these waters will depend on the geochemical contrast between the kimberlite and its host rock. The identification of geochemical anomalies in groundwater and the unusual minerals that may precipitate from these groundwaters may be a potential exploration tool for the discovery of new kimberlites. Groundwater sampled from streams as part of regional stream sediment surveys or from seeps and springs as part of regional hydrogeochemical surveys will be the most useful media. In addition to the exploration benefits, groundwater geochemical data provide environmental baseline data on natural groundwater conditions prior to mining, as well as insights into the potential for alkaline mine drainage.

Summary and Conclusions

Kimberlite is a mineralogically and chemically distinct point source. In Canada, glacial dispersal of indicator minerals from kimberlite sources formed dispersal fan- and ribbon-shaped trains in till during the last glaciation.

Subsequently, the indicator minerals were redistributed in streams producing dispersion anomalies. A kimberlite's variable size (50-1500 m) and competence influences the size and extent of the dispersal/dispersion patterns, and whether they can be detected using kimberlite fragments, indicator minerals, or trace element geochemistry. Understanding ice flow history, unravelling the genetic history of different glacial (till, glaciofluvial) and post-glacial (fluvial, beach) sediments sampled, and identifying multiple till sheets in thick drift areas, such as the Prairies, northeastern Ontario, and the southern Slave, are essential to successful sampling, interpretation, and follow-up of anomalies. The ability to detect kimberlitic geochemical signatures in till, soils, vegetation, or groundwater will depend on the size of the kimberlite body and the geochemical contrast between it and its host rock, and the extent to which the geochemical signature has been diluted by non-kimberlitic material.

In glaciated terrain, kimberlites have been subjected primarily to mechanical erosion with the debris transported away from the source by glaciers and/or streams. The resultant dispersal/dispersion of indicator minerals resulted in trains and fans that are much larger exploration targets than the kimberlites themselves. Thus, indicator mineral methods using till or stream sediments (Fig. 4) combined with geophysical methods are best suited to kimberlite exploration in glaciated terrain. Indicator mineral dispersal trains may be the result of one or more phases of ice flow, can vary from a few tens of metres to in excess of 100 kilometres in length, and are variable in width. Dispersal is typically mapped using the abundance of indicator minerals in the 0.25 to 0.5 mm fraction of till or stream sediments, although in some cases the finer (0.15-0.25 mm) fraction may be useful. The total concentration of indicator minerals varies from single grains to tens of thousands of grains and will depend on a number of factors, including the initial content in the kimberlite source and the amount of dilution. However, not all kimberlites will have indicator mineral dispersal trains. Dispersal trains or fans may not be evident or detected where kimberlite debris was glacially transported down-ice beyond the range of sampling. Alternatively, the kimberlite may not have been subjected to glacial erosion (e.g. the kimberlite is buried by thick till sheets or other glacial sediments, or by younger sedimentary or volcanic cover sequences), or more rarely the kimberlite may not contain a sufficient volume of indicator minerals to produce a detectable train. Some dispersal trains may have gaps or be missing the head or some part of the tail. Not all kimberlites display an obvious geophysical expression, thus some indicator mineral trains may appear to have no source (e.g. Fig. 21). Where kimberlites occur close together, their trains may coalesce or overlap (e.g. Fig. 23). In this circumstance, indicator mineral chemistry and relative abundance of each indicator species will be helpful in identifying and distinguishing between multiple dispersal trains.

Local-scale orientation studies around known kimberlites can provide crucial information on the geochemical, mineralogical, and lithological signatures of till and fluvial sediments, on appropriate sampling methods, the size fractions to examine or analyze, and on local glacial stratigraphy and ice-flow patterns (Jennings, 1990). As a significant proportion of an exploration budget is spent on the initial sample collection, especially in remote areas, it is cost effective to explore for other commodities, not just diamonds. This cost effectiveness can be accommodated if indicator minerals of other commodities such as precious or base metals also are recovered from the same heavy mineral concentrates that are prepared for kimberlite indicator mineral picking (Averill, 2001; McClenaghan, 2005). The use of till geochemistry in combination with indicator mineral methods will further increase the cost effectiveness by expanding exploration to a broad range of commodities and providing baseline environmental data.

Where potential kimberlites have been identified by geophysical methods, soil selective leaches and soil gas, still in the testing phase, may be helpful at a local scale for prioritizing targets for drilling (Fig. 4). Hydrogeochemistry, when included as part of a regional stream survey, may provide additional insights into the presence of kimberlites nearby (Fig. 4). Soil, vegetation, and groundwater geochemical techniques may also provide baseline environmental data, which document natural conditions prior to mining. Many more case studies need to be published that document the application of these new geochemical methods and their limitations with respect to kimberlite exploration in glaciated terrain.

Acknowledgments

The authors thank the following individuals and companies for contributing information and figures: M. McCurdy, R. McNeil (GSC); J. Armstrong (Stornoway Diamond Corp., formerly of C.S. Lord Geoscience Centre); R. Paulen (Alberta Geological Survey); D. Crabtree (Ontario Geological Survey); W. Coker (BHP-Billiton); S. Windle (SDP Pty Ltd.); P. Sobie (MPH Consulting Ltd.); D. Boucher and B. Wyatt (De Beers Canada Exploration); T. Nowicki (Mineral Services Canada); B. Clements (Ashton Mining of Canada and their joint venture partners Stornoway Diamond Corp. and Caledonia Mining Corp). T. Barry and R. Knight (GSC) assisted with the preparation of the figures. T. Nowicki, S. Averill, and W. Goodfellow are thanked for their thorough and thoughtful reviews of the manuscript.

References

Afanase'ev, V.P., Varlamov, V.A., and Garanin, V.K., 1984, The abrasion of minerals in kimberlites in relation to the conditions and distance of their transportation: Geologiya i Geofizika, v. 25/10, p. 119-125.

Armstrong, J.P., 2001, A preliminary digital public data compilation of kimberlite indicator mineral chemistry (KIMC), Slave Craton and environs, NWT and NU, Canada: Department of Indian and Northern Development, NWT Geology Division, EGS Open Report 2001-001, CD-ROM.

Armstrong, J., and Chatman, J., 2001, Kimberlite indicator mineral database (KIDD): A public data compilation of till sample locations and kimberlite indicator mineral picking results, Slave Craton and environs, NWT and NU, Canada: Department of Indian and Northern Development, NWT Geology Division, EGS Open Report 2001-002, CD-ROM.

Armstrong, J.P., and Kjarsgaard, B.A., 2003, Geological setting of kimberlites in the Archean Slave Province: VIIIth International Kimberlite Conference, Slave Province and Northern Alberta Field Trip Guide Book, p. 31-38.

Atkinson, W.J., 1989, Diamond exploration philosophy, practice, and promises: A review, *in* Meyer, H.O. A., and Leonardos, O.H., eds., Proceedings of the Fifth International Kimberlite Conference, Brazil, v. 2, Diamonds: Characterization, Genesis and Exploration.: CPRM Special Publication 1B, p. 144 - 159, 1075-1107.

Averill, S.A., 1990, Drilling and sample processing methods for deep till geochemistry surveys: Making the right choices, *in* Application de la géologie du Quaternaire à l'exploration minérale: Association professionelle des géologues et des géophysiciens du Québec, p. 139-173.

——— 2001, The application of heavy indicator mineralogy in mineral exploration, *in* McClenaghan, M.B., Cook, S., and Bobrowsky, P.T., eds., Drift Exploration in Glaciated Terrain: Geological Society of London, Special Volume 185, p. 69-81.

Averill, S.A., and McClenaghan, M.B., 1994, Distribution and Character of Kimberlite Indicator Minerals in Glacial Sediments, C14 and Diamond Lake Kimberlite Pipes, Kirkland Lake, Ontario: Geological Survey of Canada, Open File 2819, 48 p.

Baker, C.L., 1982, Report on the sedimentology and provenance of sediments in eskers in the Kirkland Lake area and on finding of kimberlite float in Gauthier Township, *in* Wood, J., White, O.L., Barlow, R.B., and Colvine, A.C., eds., Summary of Field Work 1982: Ontario Geological Survey, Miscellaneous Publication 106, p. 125-127.

Blusson, S.L., 1998, Key steps to discovery of the Slave Craton diamond field, N.W.T., Canada, *in* Walton, G., and Jambor, J., eds., Pathways '98 Abstract Volume: Cordilleran Roundup, Vancouver, 1998, British Columbia and Yukon Chamber of Mines, p. 25.

Boyd, F.R., and Nixon, P.H., 1972, Ultramafic nodules from the Thaba Putsoa kimberlite pipe: Year Book - Carnegie Institution of Washington, v. 71, p. 362-373.

Brown, D.D., Bennett, G., and George, P.T., 1967, The source of alluvial kimberlite indicator minerals in the James Bay Lowland: Ontario Department of Mines, Miscellaneous Paper MP-7.

Brummer, J.J., MacFadyen, D.A., and Pegg, C.C., 1992a, Discovery of kimberlites in the Kirkland Lake area, northern Ontario, Canada. Part I: Early surveys and the surficial geology: Exploration and Mining Geology, v. 1, p. 339-350.

——— 1992b, Discovery of kimberlites in the Kirkland Lake area, northern Ontario, Canada. Part II: Kimberlite discoveries, sampling, diamond content, ages and emplacement: Exploration and Mining Geology, v. 1, p. 351-370.

Buks, I.I., 1965, The use of geobotanical method in the search for kimberlite tubes in the Yakutian polar region, *in* Chikishev, A.G., ed., Plant Indicators of Soils, Rocks and Subsurface Waters: Consultants Bureau, New York, p. 173-175.

Burt, A.K., and Hamilton, S.M., 2004, A comparison of selective leach signatures over kimberlites and other targets: Ontario Geological Survey, Open File Report 6142, 179 p.

Cameron, E.M., Hamilton, S.M., Leybourne, M.I., Hall, G.E.M., and McClenaghan, M.B., 2004, Finding deeply buried deposits using geochemistry: Geochemistry: Exploration, Environment, Analysis, v. 4, p. 7-32.

Canil, D., Mihalynuk, M., MacKenzie, J.M., Johnston, S.T., and Grant, B., 2005, Diamond in the Atlin-Nakina region, British Columbia: Insights from heavy minerals in stream sediments: Canadian Journal of Earth Sciences, v. 42, p. 2161-2171.

Carlson, J.A., Kirkley, M.B., Thomas, E.M., and Hillier, W.D., 1999, Recent Canadian kimberlite discoveries, *in* Gurney, J.J., Gurney, J.L., Pascoe, M.D., and Richardson, S.H., eds., The J.B. Dawson Volume, Proceedings of the VIIth International Kimberlite Conference, Cape Town: p. 81-89.

Carlson, S.M., Hillier, W.D., Hood, C.T., Pryde, R.P., and Skelton, D.N., 1999, The Buffalo Hills kimberlites: A newly discovered diamondiferous kimberlite province in north-central Alberta, Canada, *in* Gurney, J.J., Gurney, J.L., Pascoe, M.D., and Richardson, S.H., eds., The J.B. Dawson Volume, Proceedings of the VIIth International Kimberlite Conference, Cape Town: p. 109-116.

Cookenboo, H., 1996, Ranch Lake kimberlite in the central Slave craton; the mantle sample: The Gangue, Geological Association of Canada, Mineral Deposits Division, Report 52, 2 p.

Cookenboo, H.O., Dupuis, J., and Foulkes, J., 1998, Discovery and evaluation of the Jericho and 5034 kimberlite fields in the Slave Craton, northern Canada, *in* Walton, G., and Jambor, J., eds., Pathways '98 Abstract Volume: Cordilleran Roundup 1998. British Columbia and Yukon Chamber of Mines, p. 166-169.

Crabtree, D.C., 2003, Preliminary results from the James Bay Lowland indicator mineral sampling program: Ontario Geological Survey, Open File Report 6108, 115 p.

Davison, J.G., 1993, Diamond exploration samples: Laboratory processing, *in* Sheahan, P., and Chater, A., eds., Diamond: Exploration, Sampling and Evaluation: Prospectors and Developers Association of Canada, Short Course Proceedings, Toronto, p. 315-341.

Dawson, J.B., and Stephens, W.E., 1975, Statistical classification of garnets from kimberlite and associated xenoliths: Journal of Geology, v. 83, p. 589-607.

Deer, W.A., Howie, R.A., and Zussman, J., 1982, Rock Forming Minerals, Volume 1A, Orthosilicates, Second Edition: Longman Group Ltd., New York, USA, 579 p.

Doyle, B.J., Kivi, K., and Scott Smith, B.H., 1999, The Tli Kwi Cho (DO27 and DO18) diamondiferous kimberlite complex, Northwest Territories, Canada, *in* Gurney, J.J., Gurney, J.L., Pascoe, M.D., and Richardson, S.H., eds., The J.B. Dawson Volume, Proceedings of the VIIth International Kimberlite Conference, Cape Town: p. 194-204.

Dredge, L., Kjarsgaard, I., Ward, B., Kerr, D., and Stirling, J., 1995, Distribution and Chemistry of Kimberlite Indicator Minerals, Aylmer Lake map area, NWT (NTS 76C): Geological Survey of Canada, Open File 3080, 78 p.

Dredge, L.A., Ward, B.C., and Kerr, D.E., 1996, Morphology and kelyphite preservation on glacially transported pyrope grains, *in* LeCheminant, A.N., Richardson, D.G., DiLabio, R.N.W., and Richardson, K.A., eds., Searching for Diamonds in Canada: Geological Survey of Canada, Open File 3228, p. 197-204.

Dredge, L.A., Kerr, D.E., Kjarsgaard, I.M., Knight, R.D., and Ward, B.C., 1997, Kimberlite indicator minerals in till, central Slave Province, Northwest Territories: Geological Survey of Canada, Open File 3426, 1 poster.

Dummett, H.T., Fipke, C.E., and Blusson, S.L., 1987, Diamond exploration geochemistry in the North American Cordillera, *in* Elliot, I.L., and Smee, B.E., eds., Geoexpo/86 Exploration in the North American Cordillera: Association of Exploration Geochemists, p. 168-176.

Dunn, C.E., 1993, Diamondiferous kimberlite in Saskatchewan, Canada - A biogeochemical study: Journal of Geochemical Exploration, v. 47, p. 131-141.

Dunn, C.E., and McClenaghan, M B., 1996, Biogeochemical studies of kimberlites, *in* Le Cheminant, A.N., Richardson, D.G., DiLabio, R.N.W., and Richardson, K.A., eds., Searching for Diamonds in Canada: Geological Survey of Canada, Open File 3228, p. 219-223.

Eccles, D.R., 1998a, Enzyme leach-based soil geochemistry of the Mountain Lake Diatreme, Alberta: Alberta Geological Survey, Open File Report 1998-01, 30 p.

——— 1998b, Biogeochemical orientation survey of the Mountain Lake Diatreme, Alberta: Alberta Geological Survey, Open File Report 1998-06, 61 p.

Eriksson, K., 1992, Chapter 2: Glacigenic deposits, *in* Kauranne, L.K., Salminen, R., and Eriksson, K., eds., Handbook of Exploration Geochemistry, Volume 5, Regolith Exploration Geochemistry in Arctic and Temperate Terrains: Elsevier, Amsterdam, p. 13-40.

Fediukova, E., 1977, Kelyphite reaction rims in garnet peridotites: Acta Universitatis Carolinae-Geologica, v. 1977, p. 185-192.

Fenton, M.M., and Pawlowicz, J.G., 1997, Diamond Indicator Mineral Anomaly from Till Sample Site NAT95-134: Alberta Energy and Utilities Board, EUB/AGS Geo-note 1997-01, 7 p.

Fenton, M.M., Pawlowicz, J.G., Paulen, R.S., Prior, G.J., and Olson, R.A., 2003, Quaternary geology of northern Alberta: Implications for kimberlite exploration: VIIIth International Kimberlite Conference, Program with Abstracts, p. 125-126.

Fipke, C. E., Gurney, J. J., Moore, R. O., and Nassichuk, W. W., 1989, The development of advanced technology to distinguish between diamondiferous and barren diatremes: Geological Survey of Canada, Open File 2124, 90 p.

Fipke, C.E., Dummett, H.T., Moore, R.O., Carlson, J.A., Ashley, R.M., Gurney, J.J., and Kirkley, M.B., 1995a, History of the discovery of diamondiferous kimberlite in the Northwest Territories, Canada: Sixth International Kimberlite Conference, Novosibirsk, Siberia, Extended Abstract Volume, p. 158-160.

Fipke, C.E., Gurney, J.J., and Moore, R.O., 1995b, Diamond exploration techniques emphasizing indicator mineral geochemistry and Canadian examples: Geological Survey of Canada, Bulletin 423, 80 p.

Fortescue, J.A.C., Lourim, J., Gleeson, C.F., Jensen, L., and Baker, C.L., 1984, A synthesis and interpretation of basal till geochemical and mineralogical data obtained from the Kirkland Lake (KLIP) area (1979-1982), Part 1: Ontario Geological Survey, Open File Report 5506, 630 p.

Friske, P.W.B., Prior, G.J., McNeil, R.J., McCurdy, M.W., and Day, S.J.A., 2003, National Geochemical Reconnaissance (NGR) stream sediment and water survey in the Buffalo Head Hills area, northern Alberta (parts of NTS 84B, 84C, 84F and 84G), including analytical, mineralogical and kimberlite indicator mineral data from silts, heavy mineral concentrates and waters: Geological Survey of Canada, Open File 179, 1 CD-ROM.

Garrett, R.G., and Thorleifson, L.H., 1996, Kimberlite indicator mineral and soil geochemical reconnaissance of the Canadian Prairie region, *in* LeCheminant, A.N., Richardson, D.G., DiLabio, R.N.W., and Richardson, K.A., eds., Searching for Diamonds in Canada: Geological Survey of Canada, Open File 3228, p. 205-211.

Garvie, O.G., 2003, Surface textures found on kimberlite indicator minerals, *in* Indicator Mineral Methods in Mineral Exploration: Prospectors and Developers Association of Canada, Short Course Notes, p. 23.

Garvie, O.G., and Robinson, D.N., 1984, The formation of kelyphite and associated subkelyphitic and sculpted surfaces on pyrope from kimberlite, *in* Kornprobst, J., ed., Proceedings of the Third International Kimberlite Conference, Kimberlites I, Kimberlites and Related Rocks: Elsevier, Amsterdam, p. 371-382.

Golubev, Y K., 1995, Diamond exploration in glaciated terrain: A Russian perspective: Journal of Geochemical Exploration, v. 53, p. 265-275.

Graham, I., Burgess, J.L., Bryan, D., Ravenscroft, P.J., Thomas, E., Doyle, B.J., Hopkins, R., and Armstrong, K.A., 1999, Exploration history and geology of the Diavik kimberlites, Lac de Gras, Northwest Territories, Canada, *in* Gurney, J.J., Gurney, J.L., Pascoe, M.D., and Richardson, S.H., eds., The J.B. Dawson Volume, Proceedings of the VIIth International Kimberlite Conference, Cape Town: p. 262 279.

Gravel, J., Hrkac, R., and Richardson, P., 2005, Soil and lake sediment geochemistry in exploration for kimberlite: NWT, Canada: 22nd International Geochemical Exploration Symposium, Program and Abstracts, p. 54.

Gregory, G.P., and White, D.R., 1989, Collection and treatment of diamond exploration samples, *in* Ross, J., ed., Kimberlites and Related Rocks Volume 2 Their Crust/Mantle Setting, Diamonds and Diamond Exploration: Geological Society of Australia, Special Publication 14, p. 1123-1134.

Griffin, W.L., Fisher, N.I., Friedman, J.H., and Ryan, C.G., 1997, Statistical techniques for the classification of chromites in diamond exploration: Journal of Geochemical Exploration, v. 59, p. 233-249.

Grütter, H.S., and Apter, D.B., 1998, Kimberlite- and lamproite-borne chromite phenocrysts with "diamond-inclusion"–type chemistries: 7th International Kimberlite Conference, Cape Town, Extended Abstracts, p. 280-281.

Grütter, H.S., and Quadling, K.E., 1999, Can sodium in garnet be used to monitor eclogitic diamond potential? *in* Gurney, J.J., Gurney, J.L., Pascoe, M.D., and Richardson, S.H., eds., The J.B. Dawson Volume, Proceedings of the VIIth International Kimberlite Conference, Cape Town: p. 314-320.

Grütter, H.S., Gurney, J.J., Menzies, A.H., and Winter, F., 2004, An updated classification scheme for mantle-derived garnet, for use by diamond explorers: Lithos, v. 77, p. 841-857.

Gurney, J.J., 1984, A correlation between garnets and diamonds in kimberlites, *in* Kimberlite Occurrence and Origin: a basis for conceptual models in exploration: Geology Department and University Extension, University of Western Australia, Publication No. 8, p. 143-166.

——— 1989, Diamonds, *in* Ross, J., ed., Proceedings of the Fourth International Kimberlite Conference, v. 1. Kimberlites and Related Rocks: Their Composition, Occurrence, Origin and Emplacement: Geological Society of Australia, Special Publication 14, p. 935 -965.

Gurney, J.J., and Switzer, G., 1973, The discovery of garnets closely related to diamonds in the Finsch Pipe, South Africa: Contributions to Mineralogy and Petrology, v. 39, p. 103-116.

Gurney, J. ., Helmstaedt, H.H., and Moore, R.O., 1993, A review of the use and application of mantle mineral geochemistry in diamond exploration: Pure and Applied Chemistry, v. 65, p. 2423-2442.

Hamilton, S.M., 1998, Electrochemical mass-transport in overburden: A new model to account for the formation of selective leach geochemical anomalies in glacial terrain: Journal of Geochemical Exploration, v. 63, p. 155-172.

Heaman, L.M., Kjarsgaard, B.A., and Creaser, R.A., 2004, The temporal evolution of North American kimberlites: Lithos, v. 76, p. 377-397.

Hirvas, H., and Nenonen, K., 1990, Field methods for glacial indicator tracing, Chapter 12, *in* Kujansuu, R., and Saarnisto, M., eds., Glacial Indicator Tracing: A.A. Balkema, Rotterdam, p. 217-248.

Hood, C.T.S., and McCandless, T.E., 2004, Systematic variations in xenocryst mineral composition at the province scale, Buffalo Hills kimberlites, Alberta, Canada: Lithos, v. 77, p. 733-747.

Jennings, C.M.H., 1990, Exploration for diamondiferous kimberlites and lamproites, *in* Beck, L.S., and Harper, C.T., eds., Modern Exploration Techniques: Saskatchewan Geological Society, Special Publication 10, p. 139-148.

——— 1995, The exploration context of diamonds: Journal of Geochemical Exploration, v. 53, p. 113-124.

Kauranne, K., Salminen, R., and Eriksson, K., 1992, Handbook of Exploration Geochemistry, Volume 5. Regolith Exploration Geochemistry in Arctic and Temperate Terrains: Elsevier, Amsterdam, .549 p.

Kerr, D., 2000, Kimberlite indicator minerals in till, Lac de Gras region: Prospectors and Developers Convention, Toronto, Ontario, March 5-10, Poster.

Kerr, D.E., Smith, D., and Wilson, P., 2000, Anomalous kimberlite indicator mineral and gold grain abundances, Drybones Bay and Yellowknife area, Northwest Territories: Geological Survey of Canada, Open File D3861, 1 CD-ROM.

Kirkley, M., Mogg, T., and McBean, D., 2003, Snap Lake field trip guide, *in* Slave Province and Northern Alberta Field Trip Guide Book: VIIIth International Kimberlite Conference, p. 67-78.

Kjarsgaard, B.A., 1995, Research on kimberlites and applications of diamond exploration techniques in Saskatchewan, *in* Richardson, D.G., ed., Investigations Completed by the Saskatchewan Geological Survey and the Geological Survey of Canada Under the Geoscience Program of the Canada-Saskatchewan Partnership Agreement on Mineral Development (1990-1995): Geological Survey of Canada, Open File 3119, p. 167-181.

——— 1998, Compositional trends of spinel and mica in alkali minettes, southern Alberta, Canada: 7th International Kimberlite Conference, Cape Town, Extended Abstracts, p. 435–437.

Kjarsgaard, B.A., and Levinson, A.A., 2002, Diamonds in Canada: Gems and Gemology, v. 38, p. 208-238.

Kjarsgaard, B.A., McClenaghan, M B., Boucher, D.F., and Kivi, K., 2003, Kimberlites and ultrabasic rocks of the Wawa, Chapleau, Kirkland Lake and Lake Timiskaming area, *in* Northern Ontario Field Trip Guide Book: VIIIth International Kimberlite Conference, p. 1-38.

Kjarsgaard, I.M., McClenaghan, M.B., Kjarsgaard, B.A., and Heaman, L.M., 2004, Mineralogy of kimberlite boulders from eskers in the Kirkland Lake and Lake Timiskaming areas, northeastern Ontario, Canada: Lithos, v. 77, p. 705-731.

Klassen, R.A., 2001, A Quaternary geological perspective for geochemical exploration in glaciated terrain, *in* McClenaghan, M.B., Cook, S., and Bobrowsky, P.T., eds., Drift Exploration in Glaciated Terrain: Geological Society of London, Special Volume 185, p. 1-17.

Kobets, N.V., and Komarov, B.V., 1958, Some problems of methodology in prospecting for primary diamond deposits by aero methods: Bulletin of the Academy of Science, USSR, Geological Series 2, v. 80-86, (A.G.I. translation from Russian).

Komogorova, L.G., Stadnik, Y.V., and Federov, V.I., 1986, Phytogeochemical surveys within kimberlite bodies: Translated from: Fitogeokhimicheskiye issledovaniya v konturakh kimberlitovykh. tel. Doklady Akademii Nauk USSR, v. 29, p. 468-470.

Kong, J.M., Boucher, D.R., and Scott-Smith, B.H., 1999, Exploration and geology of the Attawapiskat kimberlites, James Bay Lowland, northern Ontario, Canada, *in* Gurney, J.J., Gurney, J.L., Pascoe, M.D., and Richardson, S.H., eds., The J.B. Dawson Volume, Proceedings of the VIIth International Kimberlite, Cape Town: p. 446-448.

Leckie, D.A., Kjarsgaard, B.A., Collins, M., Dufresne, M.D., Eccles, D.R., Heaman, L.M., Pierce, J., Stasiuk, L., and Sweet, R., 1997, Geology of a Late Cretaceous possible kimberlite at Mountain Lake, Alberta - chemistry, petrology, indicator minerals, aeromagnetic signature, age, stratigraphic position and setting: Geological Survey of Canada, Open File 3441, 202 p.

Lee, D.C., Maddren, J., and Griffin, B.J., 2003, The importance of chromite morphology in diamond exploration: VIIIth International Kimberlite Conference, Program with Abstracts, p. 133-134.

Lee, H.A., 1965, Investigation of eskers for mineral exploration: Geological Survey of Canada, Paper 65-14.

Lehtonen, M.L., Marmo, J.S., Nissinen, A.J., Johanson, B.S., and Pakkanen, L.K., 2005, Glacial dispersal studies using indicator minerals and till geochemistry around two eastern Finland kimberlites: Journal of Geochemical Exploration, v. 87, p. 19-43.

Litinskii, V.A., 1961, On the content of Ni, Cr, Ti, Nb and some other elements in kimberlites and the possibility of geochemical prospecting for kimberlite bodies: Geochemistry, v. 9, p. 813-821.

——— 1964, Application of metallometry and kappametry in prospecting for kimberlite bodies: International Geology Review, v. 6, p. 2027-2035.

Maurice, Y. T., 1991, The use of the heavy fraction of stream alluvium: application to low-energy glaciated environments in eastern Canada: Exploration Geochemistry Workshop, Geological Survey of Canada, Open File 2390, 5-1 to 5-24 p.

McCandless, T.E., 1990, Kimberlite xenocryst wear in high-energy fluvial systems: experimental studies: Journal of Geochemical Exploration, v. 37, p. 323-331.

McCandless, T.E., and Gurney, J.J., 1989, Sodium in garnet and potassium in clinopyroxene: criteria for classifying mantle eclogites, *in* Ross, J., ed., Kimberlites and Related Rocks: Geological Society of Australia, Special Publication 14, p. 827-832.

McCandless, T.E., and Nash, W.P., 1996, Detrital mantle indicator minerals in southwestern Wyoming, U.S.A.: Evaluation of mantle environment, igneous host, and diamond exploration significance: Exploration and Mining Geology, v. 5, p. 33-44.

McClenaghan, M. B., 2002, Indicator mineral and till geochemical methods for kimberlite exploration in glaciated terrain: Canadian Institute of Mining and Metallurgy and Petroleum Resources, v. 95, p. 79-84.

——— 2005, Indicator mineral methods in mineral exploration: Geochemistry: Exploration, Environment, Analysis, v. 5, p. 233-245.

McClenaghan, M.B., and Dunn, C.E., 1995, Biogeochemical Survey over Kimberlites in the Kirkland Lake Area, Northeastern Ontario: Geological Survey of Canada, Open File 3005, 69 p.

McClenaghan, M.B., and Kjarsgaard, B.A., 2001, Indicator mineral and geochemical methods for diamond exploration in glaciated terrain in Canada, *in* McClenaghan, M.B., Bobrowsky, P.T., Hall, G.E. M., and Cook, S.J., eds., Drift Exploration in Glaciated Terrain: Geological Society of London, Special Publication 185, p. 83-123.

McClenaghan, M.B., Kjarsgaard, I.M., Schulze, D.J., Stirling, J.A.R., Pringle, G., and Berger, B.R., 1996, Mineralogy and geochemistry of the B30 kimberlite and overlying glacial sediments, Kirkland Lake, Ontario: Geological Survey of Canada, Open File 3295, 245 p.

McClenaghan, M.B., Kjarsgaard, I.M., Schulze, D.J., Berger, B.R., Stirling, J.A.R., and Pringle, G., 1998, Mineralogy and Geochemistry of the Diamond Lake Kimberlite and Associated Esker Sediments, Kirkland Lake, Ontario: Geological Survey of Canada, Open File 3576, 200 p.

McClenaghan, M.B., Kjarsgaard, B.A., Kjarsgaard, I.M., Paulen, R.C., and Stirling, J.A.R., 1999a, Mineralogy and Geochemistry of the Peddie Kimberlite and Associated Glacial Sediments, Lake Timiskaming, Ontario: Geological Survey of Canada, Open File 3775, 190 p.

McClenaghan, M.B., Kjarsgaard, I.M., Kjarsgaard, B.A., Stirling, J.A.R., Pringle, G., and Berger, B.R., 1999b, Mineralogy and geochemistry of the A4 kimberlite and associated glacial sediments, Kirkland Lake, Ontario: Geological Survey of Canada, Open File 3769, 162 p.

McClenaghan, M.B., Kjarsgaard, I.M., Stirling, J.A.R., Pringle, G., Kjarsgaard, B.A., and Berger, B.R., 1999c, Mineralogy and Geochemistry of the C14 Kimberlite and Associated Glacial Sediments, Kirkland Lake, Ontario: Geological Survey of Canada, Open File 3719, 147 p.

McClenaghan, M.B., Thorleifson, L.H., and DiLabio, R.N.W., 2000a, Till geochemical and indicator mineral methods in mineral exploration: Ore Geology Reviews, v. 16, p. 145-166.

McClenaghan, M.B., Ward, B.C., Kjarsgaard, B.A., Kjarsgaard, I.M., Stirling, J.A.R., Dredge, L., and Kerr, D., 2000b, Indicator Mineral and Till Geochemical Dispersal Associated with the Ranch Lake Kimberlite, Lac de Gras Region, Northwest Territories: Geological Survey of Canada, Open File 3924, 198 p.

McClenaghan, M.B., Kjarsgaard, B.A., and Kjarsgaard, I.M., 2002a, Indicator mineral content and geochemistry of till around the Peddie kimberlite, Lake Timiskaming, Ontario: Geological Survey of Canada, Open File 4262, 132 p.

McClenaghan, M.B., Kjarsgaard, I.M., Kjarsgaard, B.A., and Heaman, L.M., 2002b, Mineralogy of kimberlite boulders from eskers in the Lake Timiskaming and Kirkland Lake areas, northeastern Ontario: Geological Survey of Canada, Open File 4361, 1 CD-ROM.

McClenaghan, M.B., Ward, B.C., Kjarsgaard, I.M., Kjarsgaard, B.A., Kerr, D.E., and Dredge, L.A., 2002c, Indicator mineral and till geochemical dispersal patterns associated with the Ranch Lake kimberlite, Lac de Gras region, NWT, Canada: Geochemistry: Exploration, Environment, Analysis, v. 2, p. 299-320.

McClenaghan, M.B., Kjarsgaard, I.M., and Kjarsgaard, B.A., 2004, Kimberlite indicator mineral chemistry and till geochemistry around the Seed and Triple B kimberlites, Lake Timiskaming, Ontario: Geological Survey of Canada, Open File 4822, 1 CD-ROM.

McClenaghan, M.B., Hamilton, S.M., Hall, G.E.M., and Burt, A.K., 2006a, Selective leach geochemistry of soils overlying the 95-2, B30, and A4 kimberlites, northeastern Ontario: Geological Survey of Canada, Open File Report 5069, 28 p.

McClenaghan, M.B., Kjarsgaard, I.M., Kjarsgaard, B.A., and Russell, H.A.J., 2006b, Indicator mineralogy of kimberlite boulders and sand samples from the Lac Baby and Sharp Lake eskers, Lake Timiskaming kimberlite field, western Quebec and northeastern Ontario: Geological Survey of Canada, Open File 5050, 1 CD-ROM.

Mckinlay, F.T., Williams, A.C., Kong, J., and Scott-Smith, B.H., 1997, An integrated exploration case history for diamonds, Hardy Lake Project, N.W.T., *in* Gubins, A.G., ed., Exploration '97, Proceedings of the Fourth Decennial International Conference on Mineral Exploration, p. 1029-1038.

McMartin, I., and McClenaghan, M.B., 2001, Till geochemistry and sampling techniques in glaciated Shield terrain, *in* McClenaghan, M.B., Bobrowsky, P.T., Hall, G.E.M., and Cook, S.J., eds., Drift Exploration in Glaciated Terrain: Geological Society of London, Special Volume 185, p. 83-123.

Melnyk, W., 1997, Snap Lake, NWT: Fourteenth Annual Cordilleran Geology and Exploration Roundup, Vancouver, Abstract Volume, p. 38.

Mitchell, R.H., 1986, Kimberlites: Mineralogy, Geochemistry and Petrology: Plenum Press, New York, 442 p.

Morris, T. F., Murray, C., and Crabtree, D. C., 1994, Results of overburden sampling for kimberlite heavy mineral indicators and gold grains, Michipicoten River- Wawa area, northern Ontario: Ontario Geological Survey, Open File Report 5908.

Morris, T.F., Crabtree, D.C., Sage, R.P., and Averill, S.A., 1998, Types, abundances and distribution of kimberlite indicator minerals in alluvial sediments, Wawa-Kinniwabi area, northeastern Ontario: implications for the presence of diamond-bearing kimberlite: Journal of Geochemical Exploration, v. 63, p. 217-235.

Morris, T.F., Sage, R.P., Ayer, J.A., and Crabtree, D.C., 2002, A study in clinopyroxene composition: implications for kimberlite exploration: Geochemistry: Exploration, Analysis, Environment, v. 2, p. 1-11.

Mosig, R.W., 1980, Morphology of indicator minerals, a guide to proximity of source, *in* Glover, J.E., and Grovers, D.E., eds., Kimberlites and Diamonds: Geology Department and Extension Service, University of Western Australia, Publication No. 5, p. 81-87.

Muggeridge, M.T., 1995, Pathfinder sampling techniques for locating primary sources of diamond; recovery of indicator minerals, diamonds and geochemical signatures: Journal of Geochemical Exploration, v. 53, p. 183-204.

Nimis, P., 1998, Evaluation of diamond potential from the composition of peridotitic chromian diopside: European Journal of Mineralogy, v. 10, p. 505-519.

Ottensen, R. T., and Theobald, P. K., 1994, Stream sediments in mineral exploration, *in* Hale, M., and Plant, J. A., eds., Drainage Geochemistry, Handbook of Exploration Geochemistry, 6, p. 147-184.

Papunen, H., 1995, Diamonds of northern Europe - A review, *in* Pasava, J., Kríbek, B., and Zák, K., eds., Mineral Deposits: From their Origin to their Environmental Impacts: A.A. Balkema, Rotterdam, p. 617-619.

Paulen, R.C., and McClenaghan, M.B., in press, Late Wisconsin ice flow history in the Buffalo Head Hills kimberlite field: Géographie Physique et Quaternaire.

Pell, J.A., 1997, Kimberlites in the Slave Craton, Northwest Territories, Canada: A preliminary review: Russian Geology and Geophysics, v. 38, p. 5-16.

Prest, V.K., Grant, D.R., and Rampton, V.N., 1968, Glacial map of Canada: Geological Survey of Canada, Map 1253A, scale 1:5,000,000.

Prior, G.J., Paulen, R.C., Pawlowicz, J.G., and Fenton, M.M., F., 2005, Kimberlite-indicator mineral till survey of the Sawn Lake area (NTS 84B/13), southern Buffalo Head Hills, Alberta: Alberta Energy and Utilities Board, EUB/AGS Geo-Note 2005-02, 103 p.

Rikhotso, C.T., Poniatowski, B.T., and Hetman, C.M., 2003, Overview of the exploration, evaluation an geology of the Gahcho Kué kimberlites, Northwest Territories, *in* Slave Province and Northern Alberta Field Trip Guide Book: VIIIth International Kimberlite Conference, p. 79-86.

Saarnisto, M., 1990, An outline of glacial indicator tracing, *in* Kujansuu, R., and Saarnisto, M., eds., Glacial Indicator Tracing: A.A. Balkema, Rotterdam, p. 1-13.

Sader, J.A., Leybourne, M.I., McClenaghan, M.B., and Hamilton, S.M., 2003, Geochemistry of groundwater from Jurassic kimberlites in the Kirkland Lake and Lake Timiskaming kimberlite fields, northeastern Ontario: Geological Survey of Canada, Open File 4515, 1 CD-ROM.

Sader, J.A., Leybourne, M.I., McClenaghan, M.B., and Hamilton, S.M., 2007, Low temperature serpentinization processes and kimberlite groundwater signatures in the Kirkland Lake and Lake Timiskaming kimberlite fields Ontario, Canada: Geochemistry: Exploration, Environment, Analysis, v. 7, p. 3-21.

Schulze, D.J., 1993a, Garnet xenocryst populations in North American kimberlites, *in* Sheahan P. and Chater, A., eds., Diamond: Exploration, Sampling and Evaluation: Prospectors and Developers Association of Canada, Short Course Proceedings, March 27, Toronto, Ontario, p. 359-377.

——— 1993b, An introduction to the recognition and significance of kimberlite indicator minerals, *in* Techniques in Exploration For Diamonds: Short Course Notes, May 13-14, Canadian Institute of Mining, Metallurgy and Petroleum Geology, Calgary, Alberta.

——— 1995, A guide to the recognition and significance of kimberlite indicator minerals, *in* Diamonds-Theory and Exploration: Geological Association of Canada, Short Course 20, p. 1-39.

——— 1997, The significance of eclogite and Cr-poor megacryst garnets in diamond exploration: Exploration and Mining Geology, v. 6, p. 349-366.

Shilts, W.W., 1975, Principles of geochemical exploration for sulphide deposits using shallow samples of glacial drift: Canadian Institute of Mining and Metallurgy Bulletin, v. 68, p. 73-80.

Simpson, M.A., 1993, Kimberlite indicator minerals in southwestern Saskatchewan: Saskatchewan Research Council, Publication No. R-1210-8-E91, 13 p.

Skeleton, D.N., Clements, B., McCandless, T.E., Hood, C., Aulbach, S., Davies, R., and Boyer, L.P., 2003, The Buffalo Head Hills kimberlite province, Alberta, *in* Slave Province and Northern Alberta Field Trip Guide Book: VIIIth International Kimberlite Conference, p. 11-19.

Smirnov, G.I., 1959, Mineralogy of Siberian kimberlites: International Geology Review, v. 1, p. 21-39.

Sobolev, N.V., 1971, On mineralogical criteria of a diamond potential of kimberlites: Geologika et Geophysika, v. 12, p. 70-78 (in Russian).

——— 1977, Deep seated inclusions in kimberlites and the problem of the composition of the upper mantle: American Geophysical Union, Washington, 279 p.

Sobolev, N.V., Lavrent'ev, Y.G., Pospelova, L.N., and Sobolev, E.V., 1973, Chrome-rich garnets from the kimberlites of Yakutia and their paragenesis: Contributions to Mineralogy and Petrology, v. 40, p. 39-52.

Sobolev, N.V., Pokhilenko, N.P., and Afanas'ev, V.P., 1993, Kimberlitic pyrope and chromite morphology and chemistry, as indicators of diamond grade in Yakutian and Arkhangelsk provinces, *in* Dunne, K.P.E., and Grant, B., eds., Mid-continent Diamonds: Geological Association of Canada, Symposium Volume, p. 63-69.

Steenfelt, A., Jensen, S.M., Larsen, L.M., and Stendal., H., 1999, Diamond exploration in southern West Greenland, *in* Cookenboo, H., ed., Diamond Exploration Methods and Case Studies: Association of Exploration Geochemists Short Course Notes, 19th International Geochemical Exploration Symposium, Vancouver, B.C., p. 76-84.

Steenfelt, A., Jensen, S.M., and Sand, K.K., 2005, Geochemical and mineralogical signature of kimberlite dykes in till and stream sediment data from southern west Greenland: 22nd International Geochemical Exploration Symposium, Association of Applied Geochemists, Program and Abstracts, p. 95.

Stendal, H., and Theobald, P.K., 1994, Heavy-mineral concentrates in geochemical exploration, *in* Hale, M., and Plant, J.A., eds., Drainage Geochemistry, Handbook of Exploration Geochemistry, v. 6, p. 185-225.

Stewart, R.A., Mayberry, S.W., and Pickerill, M.J., 1988, Composition of till in the vicinity of the Lake Ellen kimberlite and implications for the source of diamonds in glacial sediments of eastern Wisconsin, *in* MacDonald, D.R., and Mills, K.A., eds., Prospecting in Areas of Glaciated Terrain - 1988: Canadian Institute of Mining and Metallurgy, p. 103-120.

Thiede, D.S., Windle, S.J., Hall, G.E.M., McClenaghan, M.B., and Hamilton, S.M., 2005, SDP soil-gas geochemistry at Cross Lake, Ontario: 22nd International Geochemical Exploration Symposium, Association of Applied Geochemists, Program and Abstracts, p. 103-104.

Towie, N.J., and Seet, L.H., 1995, Diamond laboratory techniques: Journal of Geochemical Exploration, v. 53, p. 205-212.

Ward, B.C., Kjarsgaard, I.M., Dredge, L.A., Kerr, D.E., and Stirling, J.A.R., 1995, Distribution and chemistry of kimberlite indicator minerals, Lac de Gras map area, NWT (NTS 76D): Geological Survey of Canada, Open File 3079, 110 p.

Wilkinson, L., Harris, J.R., Kjarsgaard, B.A., McClenaghan, M.B., and DE.K., 2001, Influence of Till Thickness and Texture on Till Geochemistry in the Lac de Gras Area, Northwest Territories, with Applications for Regional Kimberlite Exploration: Geological Survey of Canada, Current Research 2001-C9, 17 p.

Wolfe, W.J., Lee, H.A., and Hicks, W.D., 1975, Heavy mineral indicators in alluvial and esker gravels of the Moose River Basin, James Bay Lowland: Ontario Geological Survey, Geoscience Report 126.

Wyatt, B.A., Baumgartner, M., Anckar, E., and Grütter, H., 2004, Compositional classification of kimberlitic and non-kimberlitic ilmenite: Lithos, v. 77, p. 819-840.

Yannan, B., and Matsyuk, S.S., 1991, Compositions of kimberlite spinels from Shangtung and Liaoning Provinces in China: Geochemistry International, v. 28, p. 89-96.

Zalnieriunas, R.V., and Sage, R.P., 1995, Known kimberlites of eastern Ontario, Ontario Geological Survey, Preliminary Map P3321.

Aqueous Geochemistry in Mineral Exploration

Matthew I. Leybourne

GNS Science, P.O. Box 30-368, Lower Hutt 5040, New Zealand

Corresponding author's email: m.leybourne@gns.cri.nz

Abstract

Ground and surface waters are important media for the geochemical exploration of many different styles of mineralization. Water-transported (hydromorphic) metals can develop anomalous concentrations and reveal hidden mineralization directly (dissolved phase) and by adsorption/precipitation reactions with suspended and bed-load stream sediments (labile fraction). Groundwater recharges to depth, resulting in greater likelihood of interacting with buried mineralization compared to surface geochemical methods. Advances in the understanding of ore formation processes, water-rock interaction, and metals transport/attenuation in the secondary environment are enhancing the efficacy of aqueous geochemical exploration. Overcoming the issue of false anomalies is aided by traditional and nontraditional isotopic techniques to more directly fingerprint metal sources, in particular through the use of Pb and S isotopes. Advances in analytical technologies should permit such isotopic analyses, traditionally not used by the exploration industry due to cost, to become as routine as elemental analyses are today.

This paper synthesizes the current thinking and state of the art of surficial geochemical methods that are useful to the mineral exploration industry. Examples of how ground and surface waters vector to mineralization are presented for a number of deposit types (volcanogenic massive sulphide, gold, porphyry copper, sedimentary exhalative, unconformity uranium, kimberlite diamond, and Cu-Ni-PGE). The most successful aqueous-phase indicators of mineralization are those that are ore-associated and mobile in solution.

Résumé

Les eaux souterraines et les eaux de surface constituent un important milieu pour l'exploration géochimique d'un grand nombre de styles de minéralisation différents. Les métaux transportés dans l'eau (hydromorphes) peuvent s'accumuler en concentrations anomales et révéler la présence de minéralisations cachées, soit directement (métaux dissous dans l'eau), soit indirectement par leur association (adsorption ou réaction de précipitation) aux sédiments en suspension ou aux sédiments de la charge de fond des cours d'eau (fraction labile). Les eaux souterraines sont alimentées en profondeur, ce qui fait qu'elles ont plus de chances d'interagir avec les minéralisations enfouies que ne le peuvent les méthodes géochimiques de surface. Les progrès accomplis dans la compréhension des processus minéralisateurs, des interactions entre l'eau et les roches ainsi que du transport ou de l'atténuation des métaux dans le milieu secondaire accentuent l'efficacité de l'exploration hydrogéochimique. Les méthodes isotopiques classiques et nouvelles, en particulier l'utilisation des isotopes de Pb et de S, facilitent l'élimination des fausses anomalies et permettent de caractériser plus directement les sources de métaux. Les progrès réalisés en matière de techniques d'analyse devraient permettre que l'emploi des analyses isotopiques de ce type, jusqu'ici peu répandu dans l'industrie de l'exploration pour des raisons de coût, devienne aussi courant que ne l'est aujourd'hui celui des analyses de la concentration des éléments chimiques.

Cette étude présente une synthèse des idées courantes et d'avant-garde en matière de techniques d'échantillonnage et d'analyse reliées aux méthodes géochimiques de surface, qui présentent un intérêt pratique pour l'industrie de l'exploration minière. Des exemples de la manière dont les eaux de surface et les eaux souterraines peuvent constituer des vecteurs pointant vers des concentrations minérales sont présentés pour un certain nombre de types de gîtes (gîtes de sulfures massifs volcanogènes, gîtes d'or, gîtes porphyriques de cuivre, gîtes de sulfures exhalatifs dans des roches sédimentaires, gîtes d'uranium associés à des discordances, gîtes de diamants dans des kimberlites et gîtes de Cu-Ni-ÉGP). Les meilleurs indicateurs de minéralisations dans l'eau sont ceux qui sont solubles et associés aux minerais.

Introduction

Aqueous geochemistry (including ground and surface water and stream sediments) has been applied successfully in mineral exploration for some time. However, many surface and near surface deposits have already been discovered. One of the challenges for mineral exploration, then, is to find new deposits, particularly in areas where either thick cover exists (Cameron et al., 2004; Reith et al., 2005) or where past exploration has already discovered shallow mineralization and exploration must extend to greater depth (Goodfellow et al., 2003). Although stream and lake sediment analyses have received the most attention over the last several decades, surface and ground water geochemistry has more recently received consideration (Cameron, 1977, 1978; Miller et al., 1982; Earle and Drever, 1983; Giblin and Snelling, 1983; Taisaev and Plyusnin, 1984; Boyle, 1988; Cidu et al., 1995; Mingqi et al., 1995; Giblin, 1994; Cameron et al., 1997, 2004; Leybourne et al., 1998, 2002; Leybourne et al., 2003; Phipps et al., 2004), in particular because of significant advances in analytical techniques and the widespread availability of inductively coupled plasma mass spectrometers (ICP-MS). Newer ICP-MS instruments allow routine multi-element analyses with low detection limits (sub-parts per trillion for many elements). Other advances have occurred in ore deposit models, which greatly influence exploration strategies, and in the area of partial (generally weak) extractions of soils and stream sediments, which in some cases have been shown to produce greater anomaly to background contrast than traditional strong-acid (e.g. aqua regia) digestions (Cameron et al., 2004). Indeed, the relative ease of multi-element analytical techniques has exacerbated the challenge of dealing with very large data sets and being able to extract the most relevant information for mineral exploration.

Because groundwaters penetrate deeply into the earth's crust, hydrogeochemistry offers the potential to explore into

Leybourne, M.I., 2007, Aqueous geochemistry in mineral exploration, *in* Goodfellow, W.D., ed., Mineral Deposits of Canada: A Synthesis of Major Deposit-Types, District Metallogeny, the Evolution of Geological Provinces, and Exploration Methods: Geological Association of Canada, Mineral Deposits Division, Special Publication No. 5, p.1007-1033.

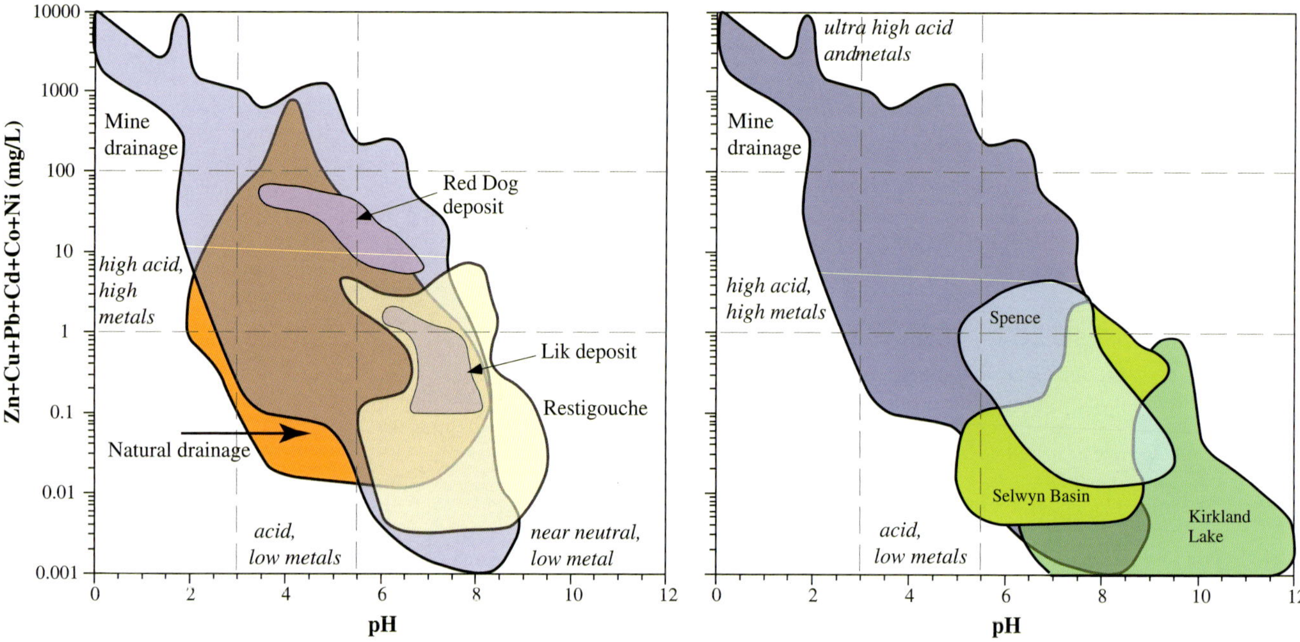

FIGURE 1. Ficklin diagram showing total dissolved metals (Zn, Cu, Pb, Co, Ni, Cd) for natural drainage waters from a variety of deposit types, as compiled by Plumlee et al. (1999). Also shown are fields for ground and surface waters from around the Restigouche (VMS) Deposit, Bathurst mining camp, Canada (Leybourne, 1998), the Spence (porphyry Cu) Deposit, Chile (Cameron and Leybourne, 2005; Leybourne and Cameron, 2006), the XY deposit and Nor occurrence (SEDEX), Selwyn Basin, Yukon (Goodfellow, 1983; Jonasson et al., 1987), the Lik and Red Dog deposits (SEDEX), Alaska (Kelley and Taylor, 1997) and kimberlites from the Kirkland Lake area (Sader et al., 2007). Diagram modified from (Plumlee et al., 1992; Plumlee et al., 1999). Fields are also shown for mine drainage waters for a number of deposit types and natural drainage waters from undisturbed deposits, as compiled by Plumlee at al. (1999). Note that there are a small number of waters in the Plumlee et al. (1999) compilation from Iron Mountain that extend to higher metal loads at negative pH values that are not shown here.

the third dimension and detect deeply buried mineralization, and therefore may prove more advantageous than other surficial geochemical methods. However, despite recent reviews (Giblin, 1994; Taufen, 1997) there are still relatively few detailed studies in the literature. Some of the advantages of using natural waters in mineral exploration include their chemical reactivity, physical mobility, relative ease and rapidity of sampling, greater uniformity compared to other sample media, and the fact that naturally low background values increase anomaly contrast (Giblin, 1994). In addition, because modern analytical methods typically require a sample in aqueous form, waters require less sample preparation than either sediments or rocks.

This paper synthesizes the current thinking and state of the art with respect to sampling, analysis, and anomaly type in aqueous geochemical methods that are useful to the mineral exploration industry. Although acid mine drainage studies have some utility in terms of understanding element associations between waters and ore, this review places more emphasis on studies around undisturbed mineralization because acid mine drainage represents conditions unlikely to be encountered during mineral exploration owing to greater surface area and generally much lower pH conditions associated with mine waste. However, some of the trends related to metal abundance and pH of ground and surface waters are similar between natural and disturbed systems, with the exception that natural systems do not attain such elevated metal loads and low pH values as some acid mine drainage systems (Fig. 1) (Plumlee et al., 1999). These systems are therefore characterized by the artificially accelerated oxidation of sulphide minerals.

Fundamentally, geochemical techniques for mineral exploration rely on the development of anomalies in the regional geochemical background. Critical to the devel-

TABLE 1. Major components of different deposit types.

Type of deposit	Major components	Minor components	Labile components[A]	Relatively immobile components[B]
VMS	Fe, S, Cu, Zn, Pb	Cd, Hg, Au, As, Sb, Ba, Bi, In	Fe, S, Zn, Cu, As, Cd, Hg, Sb	Pb, Bi, In, Au Ag, Ba
Porphyry Cu ± Mo	Cu, Mo, S		Cu, Mo, S, Fe, Se, As, Re	Ag, Au
SEDEX	Fe, S, Cu, Zn, Pb	Ag, Au, Ba, Cd	Fe, S, Zn, Cu, C	Pb, Ba, Au, Ag
Gold (vein)	Au, Ag	As, Sb, Se, Te, S, Hg	S, Se, As, Hg Te, Sb	Au, Ag
Ni-Cu-PGE	Ni, Cu, PGE	Cr, Co, S	Cu, S, PG	Co, Ni, Cr
Kimberlite (diamond)	Sr, Nb, Ba, Cr, Ni	LILE, HFSE, REE	Sr, LILE	Ba, HFSE, Nb Cr, Ni, REE
	U	Se, Mo, V, Cu, Pb	U, Se, Cu, Mo	U, Pb, V

A. Under oxidizing and near neutral conditions.

B. Under normal conditions; e.g., Ba is immobile in the presence of S as dissolved sulphate owing to insolubility of barite.

HFSE, high-field strength elements; LILE, large ion lithophile elements; REE, rare earth elements.

Table modified after McMartin and McClenaghan (2001).

TABLE 2. Summary of key aqueous geochemical pathfinders for different deposit types.

Type of deposit	Main pathfinders	Secondary pathfinders	Key analytical methods	Key publications	Notes
VMS	Zn	Low pH, Pb, SO_4,	ICP-MS – metals IC, ICP-OES - S	Cameron, 1978; Leybourne et al., 2003; Leybourne and Goodfellow, 2003	Sulphide-Pb sources typically isotopically distinct; Pb isotopes can fingerprint ore versus non-ore Pb
Porphyry Cu ± Mo	Distal – Se, Re, Mo, As Proximal - Cu	Pb, Zn	ICP-MS – metals IC, ICP-OES - S	Cameron et al., 2002; Cameron and Leybourne, 2005; Leybourne and Cameron, 2006, in press	S isotopes also useful as a complimentary vector
SEDEX	Zn	Ag, Au, Ba, Cd	ICP-MS – metals IC, ICP-OES – S TIMS, MC-ICP-MS – Pb isotopes	Goodfellow, 1983; Jonasson et al., 1987; Kelley and Taylor, 1997	Sulphide-Pb sources typically isotopically distinct; Pb isotopes can fingerprint ore versus non-ore Pb
Gold (vein)	Au	Se, As, Sb	Activated carbon preconcentration or BrCl (see text) ICP-MS, NAA	Giblin, 2001; Gray, 2001; Carey et al., 2003	
Cu-Ni-PGE	Ni, Cu, Pd	As, Cr, Co, S, PGE	ICP-MS – metals IC, ICP-OES - S	Hattori and Cameron, 2004	Pd mobility is enhanced under alkaline conditions relative to other Pathfinders
Kimberlite (diamond)	Low Mg, elevated K/Mg, pH » 10	Ni, Co, Cr, high Co/Mg and Ni/Mg	ICP-MS – metals IC, ICP-OES - S	Sader et al., 2003, in press	Also, formation of Mg hydroxides (brucite), silicates (serpentine) and carbonates (magnesite)
Unconformity uranium	Oxidizing – U, radon Reducing – Se, Mo	Se, Mo, As, V, Cu, Pb	ICP-MS – metals IC, ICP-OES – S TIMS, MC-ICP-MS – Pb isotopes	Deutscher et al., 1980; Langmuir and Chatham, 1980; Earle and Drever, 1983; Giblin and Snelling, 1983; Dickson and Giblin, 2006	Radiogenic $^{207}Pb/^{204}Pb$ and $^{206}Pb/^{204}Pb$ but non-radiogenic $^{208}Pb/^{204}Pb$ should prove useful

ICP-MS = inductively coupled plasma mass spectrometry; MC-ICP-MS = multi-collector ICP-MS; OES = optical emission spectroscopy; TIMS = thermal ionization mass spectrometry; IC = ion chromatograpy.

opment of effective exploration strategies is an understanding of the mechanisms of anomaly formation in different settings and for different sample media. There are two different types of anomalies that are pertinent to geochemical exploration. Soil and till geochemical anomalies are developed in areas where there has been mechanical transport of deposit material away from the site of mineralization (i.e. via glacial transport, solifluction, or mass wasting). In contrast, for ground or surface water, any anomalous geochemical signature is fundamentally the result of hydromorphic (water transported) dispersion. Stream sediments (both suspended and bed load) develop geochemical anomalies in two different ways. Analogous to soils, stream sediments may possess anomalous geochemistry as a result of mechanical dispersion of surface or near surface mineralization. In addition, loss of metals from solution results in mass transfer to the sediment phase, resulting in a hydromorphic component to the sediment geochemistry (Leybourne, 2001; Leybourne et al., 2003). Formation of Mn- and Fe-oxides and oxyhydroxides in particular are effective at scavenging trace metals from solution (Bau, 1999). If mineralization is not at or near surface, stream sediments can only develop a geochemical anomaly via hydromorphic dispersion. Thus, there is an important genetic link between stream sediment and stream water geochemical anomalies. However, the fundamental difference between stream sediment and stream water (and groundwater) as sample media is that the former represents a time-integrated signal of the mechanical with or without hydromorphic contributions, whereas the stream water represents a snapshot in time of the state of the geochemical system.

Recent advances in weak partial leaches have shown that geochemical anomalies occur in soils and till, in some cases over mineralization that is in excess of 100 m below surface. These anomalies have been postulated to form by a variety of processes, including capillary migration of ions, vapor transport, ground- and soil-water advection or diffusion, electrochemical transport (Hamilton, 1998), and seismic pumping (Cameron et al., 2002, 2004). Although there has been considerable debate as to which of these mechanisms is responsible for soil anomaly development, it is most likely that most of these mechanisms are viable, and that any one mechanism may be dominant depending on the type of mineralization, characteristics of the overburden, and climate. For example, recent work in the Atacama Desert of northern Chile has indicated that seismic pumping of groundwaters interacting with mineralization is a viable mechanism for producing significant anomalies in soils and gravels over porphyry Cu mineralization (Cameron et al., 2002, 2004). Clearly, some of the postulated mechanisms for these anomalies in soils will have limited applicability to aqueous geochemical exploration, but in some cases, as noted for porphyry Cu deposits in Chile, hydromorphic transport is fundamentally important in the development of geochemical anomalies and understanding the aqueous geochemistry of water-deposit interaction can greatly improve our ability to predict styles and composition of surficial anomalies.

Aqueous Geochemical Exploration for Specific Deposit Types

This section summarizes key considerations in the geochemical exploration of specific deposit types based on case studies using surface and ground waters. These examples are illustrative rather than comprehensive. In addition, the deposit types discussed are not exhaustive; however, the concepts are applicable to other styles of mineralization. Key geochemical features of the different deposit types discussed are presented in Table 1; key references and findings of

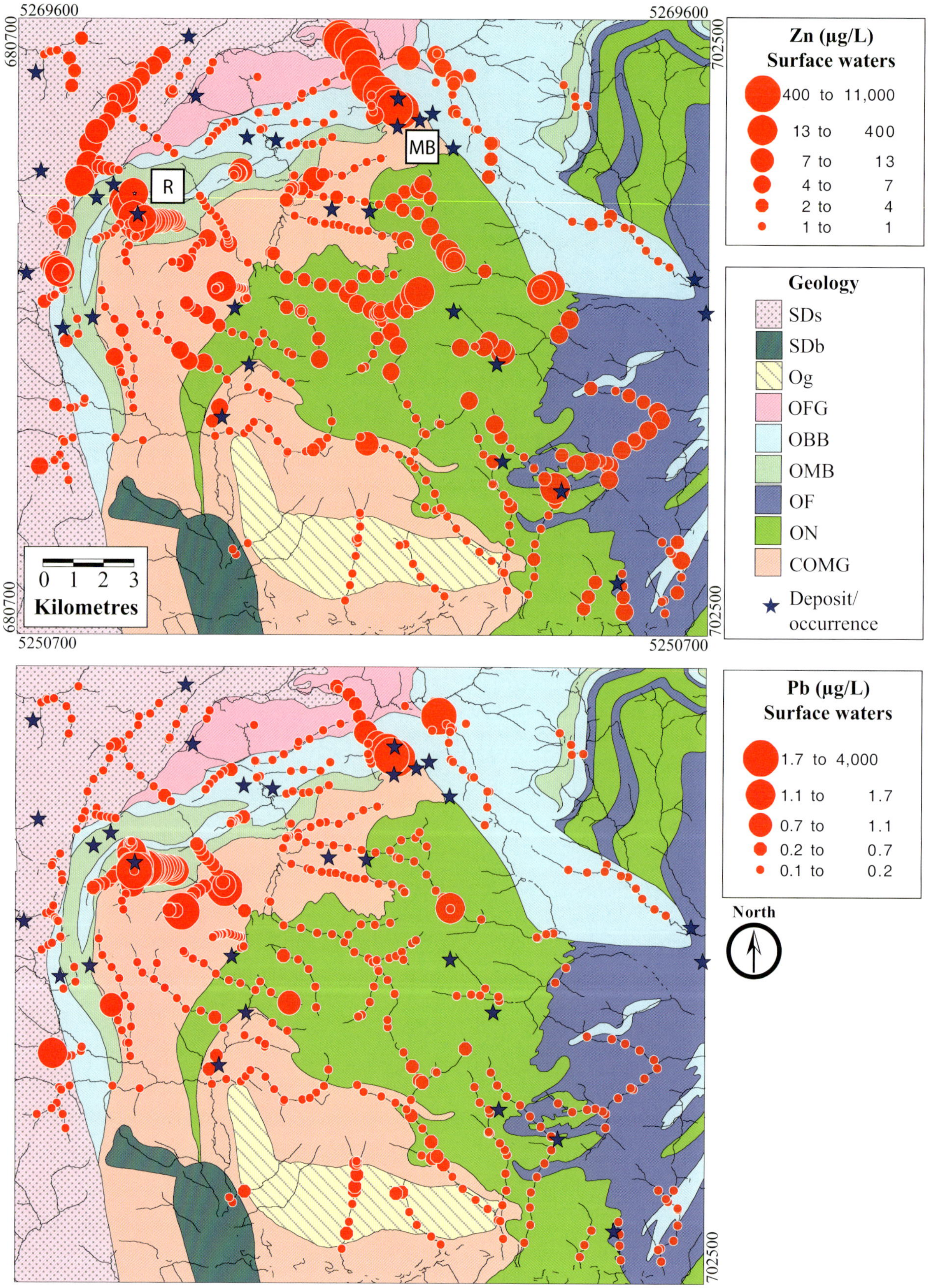

FIGURE 2. Proportional circle maps of Zn and Pb in surface waters of the Bathurst Mining Camp, New Brunswick, Canada. Also shown are deposits and showings/occurrences in the area of ~ 400 km^2. The two larger deposits are the Restigouche (R) and Murray Brook (MB) deposits, and the former was not mined at the time of sampling. It is only around these two deposits that Zn and Pb anomalies in surface waters are coincident (after Leybourne et al. (2003)). SDs, Silurian-Devonian sandstone/shale; SDb, gabbro/diabase; Og, Ordovician granite; OFG, Fournier Group; OBB, Boucher Brook Formation; OMB, Mount Brittain Formation; OF, Flat Landing Brook Formation; ON, Nepisiguit Falls Formation; COMG, Cambrian-Ordovician Miramichi Group.

aqueous studies around those deposit types are summarized in Table 2. Following these examples, the discussion section summarizes pertinent aspects of aqueous geochemical exploration in terms of what works and some important caveats.

Volcanogenic Massive Sulphide Deposits

Volcanogenic massive sulphide (VMS) deposits are variable in terms of host rock and sulphide compositions, but are characteristically dominated by metal sulphides, in particular by pyrite, pyrrhotite, sphalerite, galena, and chalcopyrite. Details of the geology and geochemistry of VMS mineralization are presented elsewhere in this volume (Dubé et al., 2007; Galley et al., 2007). These deposits are dominated by Fe, Zn, Pb, and Cu, with generally minor Cd, Hg, Au, As, Sb, Ba, Bi, and In (Table 1).

VMS deposits have received moderate attention in terms of studies related to aqueous geochemical exploration. Ground and surface waters were investigated for their efficacy in exploring for more deeply buried mineralization in the mature Bathurst mining camp, New Brunswick, Canada (Leybourne et al., 1998, 1999, 2002, 2003, 2006a; Leybourne, 2001; Leybourne and Goodfellow, 2003). These studies involved investigation of surface water and coincident stream sediment geochemistry over a moderately large scale (~ 400 km^2); samples were taken about every 500 m along all streams in the area (Fig. 2). This study area contains the previously (1989-1992) mined Murray Brook deposit (the gossan cap was mined for Au; in situ massive and supergene sulphides were left largely undisturbed), the smaller undisturbed (at the time of sampling; the deposit was subsequently mined) Restigouche deposit, and a number of other VMS occurrences (Fig. 2). At the Restigouche deposit, surface waters were collected at a denser spacing (Fig. 3) and groundwaters were collected from diamond drillholes within and distal from the deposit (Fig. 4).

For surface waters in the Bathurst mining camp, anomalies are most consistently developed for Zn, but around many occurrences, in addition to the Restigouche deposit, Pb, As, and Cu anomalies are also present. Zinc abundances in surface waters are low compared to other environments (only up to 90 µg/L; low compared to, for example, the Red Dog deposit, Fig. 1) but are clearly anomalous over the regional geochemical landscape (Leybourne et al., 2003) (Figs. 2, 3). Although there are no carbonate rocks in the study area (which is dominated by metamorphosed mafic and felsic volcanic rocks and metasediments derived largely from the felsic rocks), there are sufficient carbonate minerals (calcite, dolomite, and siderite) in veins cross-cutting mineralization and disseminated in the rock matrix to buffer the pH in most surface waters to near-neutral values, restricting the abundance of most metals in solution (e.g. Fig. 1). A result of these pH values is that metal dispersion in surface waters is generally restricted (typically <1 to 2 km), so that hydrogeochemical surveys require fairly tight (<500 m) sample spacing (Leybourne et al., 2003). A further result of this study was that anomalies in the surface waters mirrored those in the stream sediments and partial extractions of the sediments (Leybourne et al., 2003). To illustrate this point, Figure 5 shows detailed chemical profiles for stream waters and sediments in Charlotte Brook, a stream that crosses the Restigouche deposit close to its surface expression. Both water and sediment samples (and partial extraction) are clearly anomalous in Pb and Zn, with long dispersion trains.

Groundwaters proximal to the Restigouche deposit have elevated As, SO_4, and metal (Zn, Pb, Cu, Mo, Sb, Cd) contents that are significantly greater than local surface waters and groundwaters distal from mineralization (Leybourne and Goodfellow, 2003) (e.g. Zn, Pb; Figs. 4, 6). Metals will enter solutions as a result of oxidation of massive sulphide minerals in the deposit. In most cases, oxidation of sulphide minerals occurs where O_2-bearing waters interact with mineralization, although oxidation of sulphides can occur in the absence of dissolved oxygen (Hamilton, 1998). This control by dissolved O_2 is shown by Figure 6, which is a pseudo-cross-section of groundwater geochemical data (Pb, Zn) at the Restigouche deposit. Where wells intersect mineralization at greater depth (e.g. DDH B-80-7), metal contents are low in the waters recovered from those depths indicating that O_2-bearing waters do not penetrate to these depths. As with the surface waters, groundwaters around undisturbed VMS deposits in the Bathurst mining camp typically have near-neutral pH values, which generally restricts the mobility of base metal cations (Fig. 1A) owing to adsorption, primarily to Fe-oxyhydroxide minerals, and in rare cases due to precipitation of metal-sulphate minerals, e.g., anglesite [$PbSO_4$] is present in suspended sediments from the Restigouche deposit (Leybourne, 2001).

Based on these studies in the Bathurst mining camp and on other studies around VMS mineralization, it appears that the best and most widely dispersed indicator of this type of mineralization in both ground and surface waters is Zn. In many cases there are coincident anomalies of other ore-related elements/species, in particular Pb, Cu, As, Mo, Sb, and sulphate. However, some of these elements are relatively immobile under near-neutral conditions, so anomalies of these elements are not always present. Sulphate, although commonly anomalous down-flow of VMS deposits, is not in itself a reliable indicator of mineralization, given that it is a major anion in many waters. Similar results were found for groundwaters associated with the Myra Falls deposit, Vancouver Island, Canada (Phipps et al., 2004). The mountainous terrain and well developed fracture network has resulted in rapid recharge so that groundwaters interacting with sulphide ore have generally low salinity and are oxygenated, which should promote sulphide mineral oxidation. Zinc is the best indicator of massive sulphide mineralization, with a threshold value of 20 µg/L (Phipps et al., 2004). Zn anomalies at Myra Falls are commonly associated with more subtle anomalies in a suite of pathfinder elements associated with primary VMS mineralization and hydrothermal alteration, including Cu, Pb, Cd, Mn, As, Sb, Ba, and U (Phipps et al., 2004).

Porphyry Copper and Related Deposits

Porphyry Cu and porphyry Mo deposits are associated with intermediate and felsic intrusive rocks. Classic examples include the Chuquicamata, El Salvador, and Escondida deposits in Chile. Details of the geology and geochemistry of porphyry deposits are presented by Sillitoe (2000) and Sinclair (2007). Key geochemical characteristics include anomalous accumulation of As, Mo, Se, and Re in addition

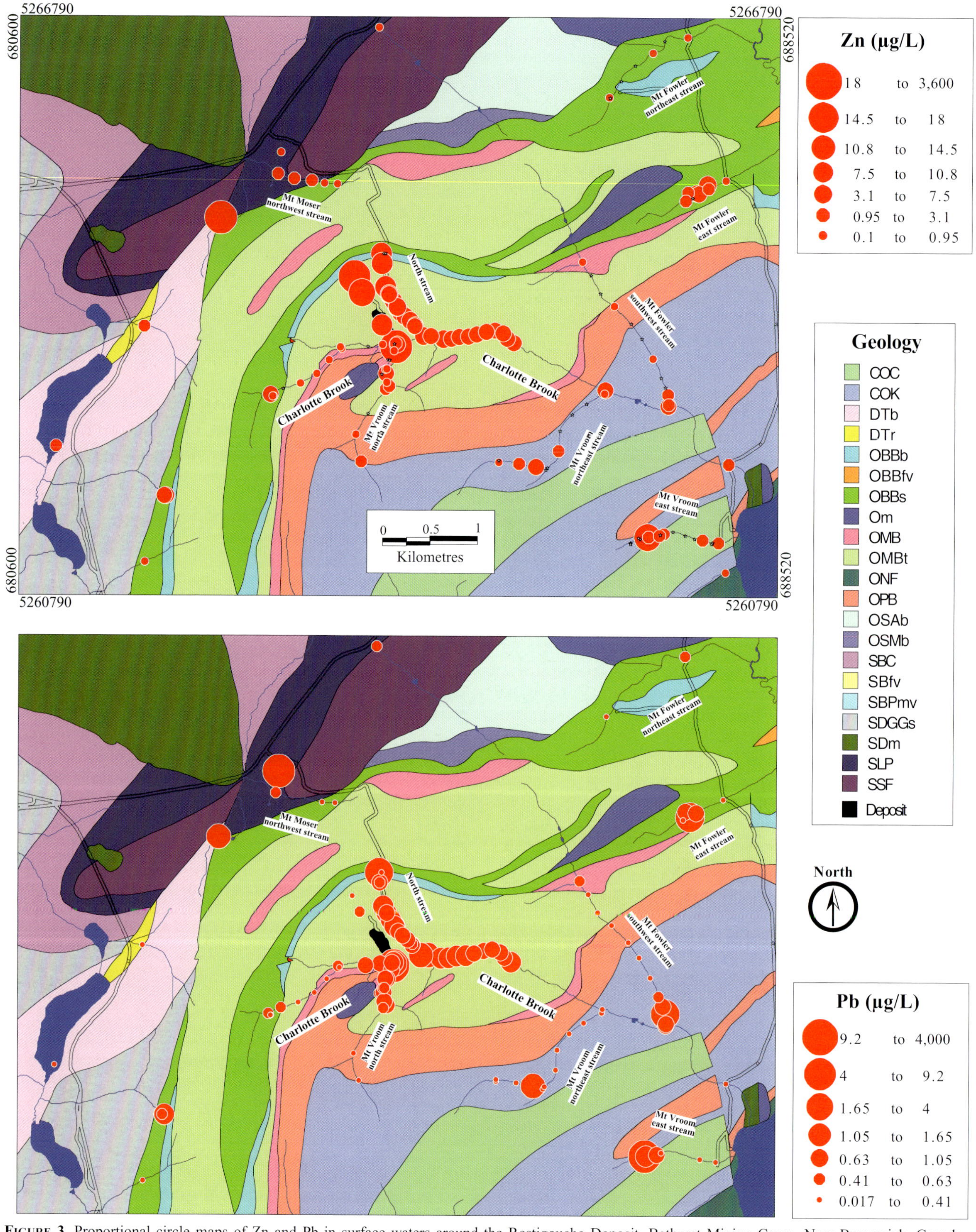

FIGURE 3. Proportional circle maps of Zn and Pb in surface waters around the Restigouche Deposit, Bathurst Mining Camp, New Brunswick, Canada. Modified from Leybourne (1998).

to Cu, and in many deposits Au, Pd, and Pt (Table 1). There have been a small number of studies of the aqueous geochemistry of waters associated with porphyry Cu and Mo deposits (Ficklin et al., 1981; Cameron et al., 2002; Cameron

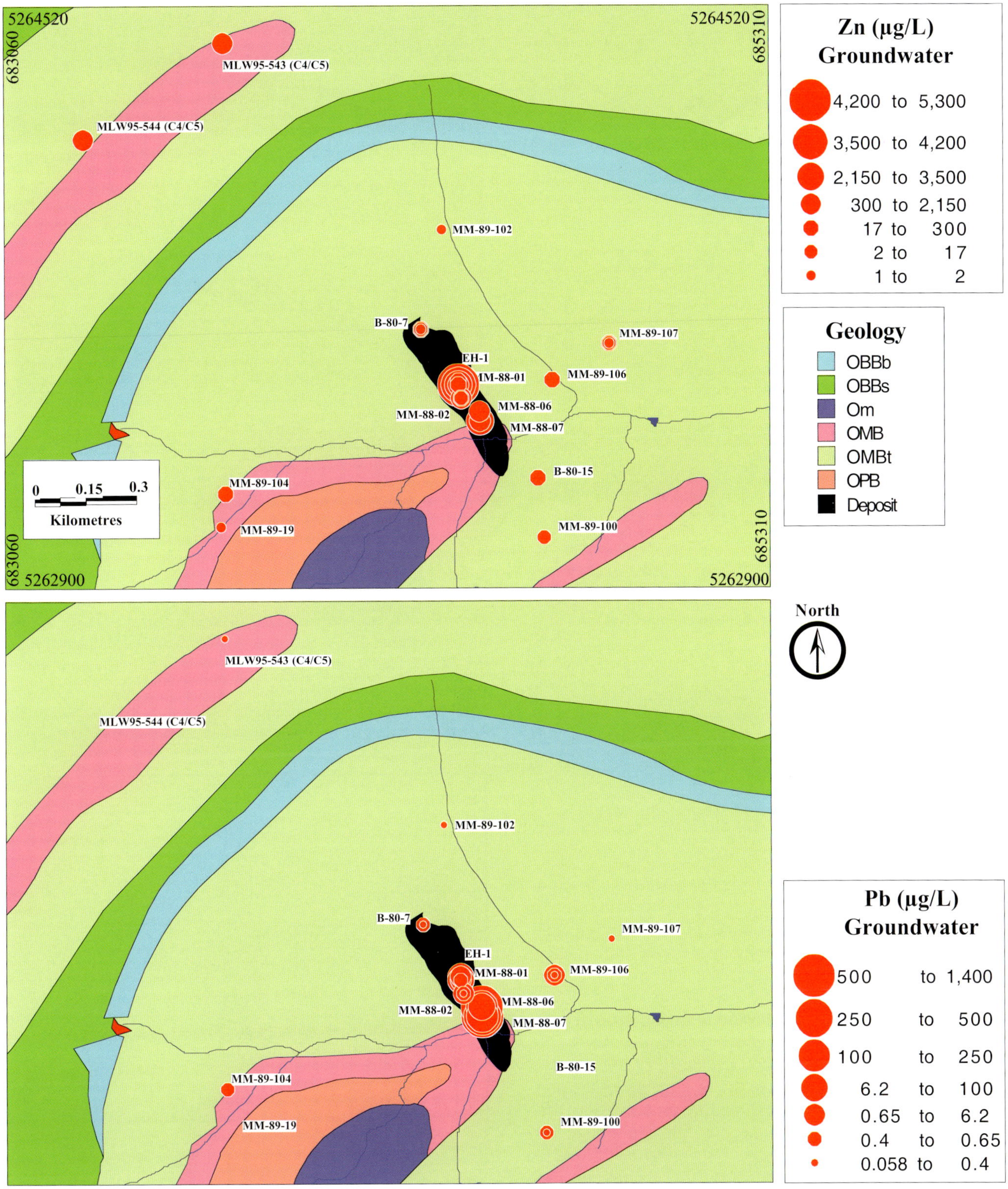

FIGURE 4. Proportional circle maps of Zn and Pb in groundwaters both within and around the Restigouche deposit, Bathurst Mining Camp, New Brunswick, Canada. Modified from Leybourne (1998).

and Leybourne, 2005; Leybourne and Cameron, 2006, in press). Probably the most comprehensive aqueous geochemical studies are those by Cameron and Leybourne (2005) around the Spence deposit in the Atacama Desert of northern Chile. At the Spence deposit, there appears to be two broad groups of waters, a relatively fresh regional groundwater end-member (with total dissolved solids typically <1000 mg/L) and a saline end-member (with total dissolved solids values up to and in excess of seawater) that has extensively interacted with Spence deposit porphyry Cu mineralization (Cameron et al., 2002; Cameron and Leybourne, 2005; Leybourne and Cameron, 2006) (Fig. 7). Up gradient of the

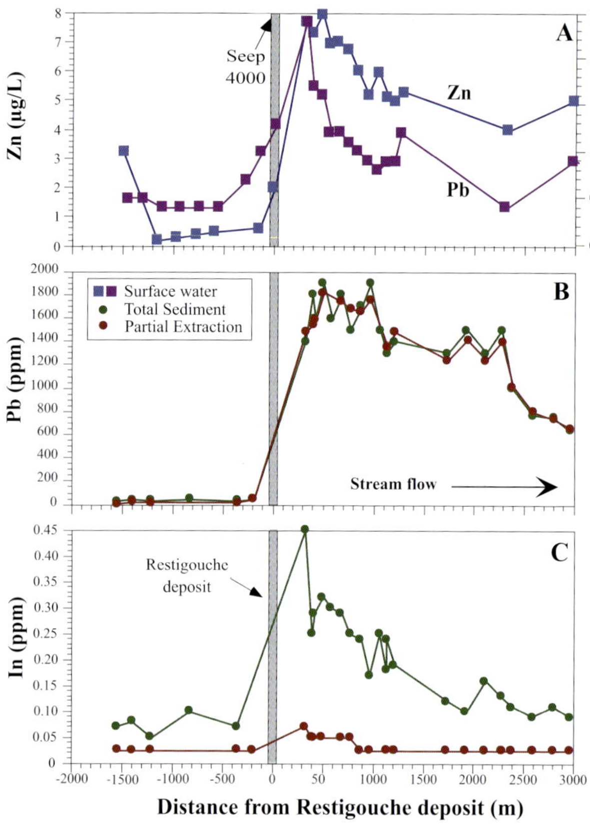

FIGURE 5. Downstream profiles of Zn and Pb **(A)** in stream waters, and Pb **(B)** and In **(C)** in stream sediments from Charlotte Brook draining the Restigouche deposit, Bathurst mining camp, Canada. Charlotte Brook flows across the deposit where it subcrops. Modified from Leybourne et al. (2003). Note the long dispersion for Zn in water and the atypically long dispersion for Pb, especially in the stream sediments. In this case, most of the Pb in the stream sediment is labile (little difference between total and partial extraction values). In contrast, indium is almost entirely transported mechanically, consistent with In being insoluble. For partial extraction analysis, sediments were treated with 0.25 M $NH_2OH{\bullet}HCl$ in 0.05 M HCl, using methods outlined in Hall et al. (1996b). See Leybourne et al. (2003) for details.

Spence deposit, groundwaters typically have low concentrations of porphyry Cu-associated elements, e.g., Cu, Mo, Se, Re, and As, and sulphate has isotopic compositions typical of salars and nonmineralized groundwaters in the Atacama Desert (e.g. Fig. 8 shows results for Cu and Se). Oxygen and hydrogen isotopes of these upflow waters are consistent with relatively recent recharge at elevations around 4000 m in the high Andes. Within and down gradient of the deposit, As, Se, Re, and Mo contents increase and remain elevated for at least 2 km from the deposit, owing to the fact that under the present pH and Eh conditions (near neutral, moderately oxidizing), these elements form oxyanions that are less readily adsorbed to oxyhydroxide or clay mineral surfaces than the metal cations (Cameron et al., 2002; Cameron and Leybourne, 2005; Leybourne and Cameron, 2006). In contrast, Cu is clearly anomalous in groundwaters in and around mineralization, but Cu decreases rapidly down gradient, owing to its adsorption to mineral surfaces (compare the results for Cu and Se within and downflow of mineralization, Fig. 8). Base metal contents in Spence deposit groundwaters are limited, as in other environments, by the generally near-neutral pH conditions (Fig. 1B). Thus, at the Spence

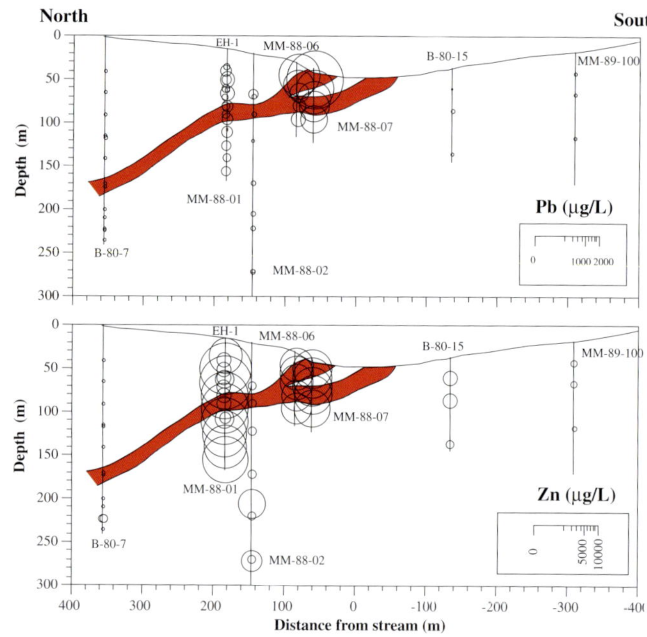

FIGURE 6. Proportional circle plots of Pb and Zn contents of groundwaters in boreholes projected onto a longitudinal cross-section of the Restigouche Deposit, Bathurst mining camp, Canada. Modified from Leybourne and Goodfellow (2003).

deposit, mineralization can be detected for long distances downflow of mineralization (at least several km) by anomalous contents of As, Se, Re, and Mo, with local definition of the target based on coincident anomalies in Cu abundance and S isotopes. Cameron and Leybourne (2005) also studied soils and waters around a large Cu (soil MMI) geochemical anomaly located in the Pampa del Tamarugal region of

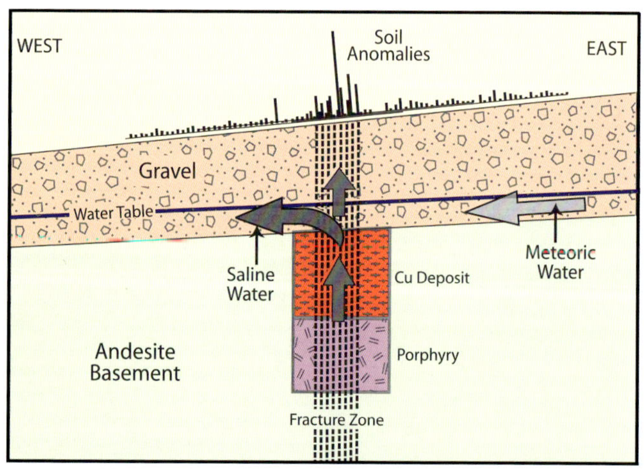

FIGURE 7. Conceptual model for the different groundwater-types and the formation of geochemical anomalies in groundwaters and soils at the Spence deposit. The porphyries and associated mineralization were introduced along a fault zone. Enhanced permeability in the zone permitted the upflow of saline formation fluids from the basement, which have then dispersed down-gradient to the west. Groundwaters of meteoric origin enter from the east and mix with the formation fluids. Reactivation of the fault has created a fracture zone in the overlying gravels. During earthquakes, saline formation fluids from the basement have been pumped to the surface. This has resulted in soil geochemical anomalies that reflect both the formation fluids (e.g. NaCl) and groundwater interaction with the porphyry copper deposit (e.g. As, Se, and Cu). Modified after Cameron and Leybourne (2005).

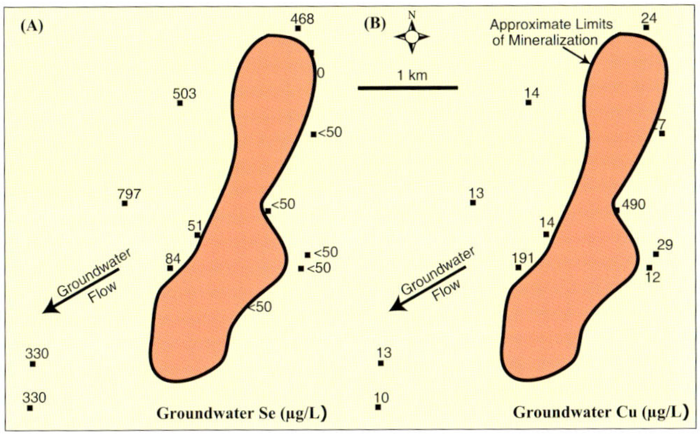

FIGURE 8. (A) Map showing the Se content of groundwater at the Spence Deposit, Atacama Desert, Chile. Note, waters on the upflow side of the deposit have Se abundances less than the detection limit of 50 µg/L. (B) Map showing Cu content of groundwater at the Spence Deposit. Note that elevated Cu is restricted to the immediate area of the deposit. After Cameron and Leybourne (2005).

northern Chile. Although soils here show elevated Cu, analyses of the waters and soils suggests that unlike at Spence, the Cu in the Tamarugal soil anomaly is the result of long periods of accumulation owing to evaporation of Cu-bearing waters from a distal source. S/Se ratios of these waters are similar to the regional waters upflow of the Spence deposit, and significantly elevated compared to waters that have clearly interacted with Spence mineralization (Fig. 9D). Note that Se/TDS ratios (where TDS is total dissolved solids) increase more with increasing Se compared to Na/TDS (Fig. 9C), indicating that Se is added in excess of increases in salinity, through interaction with mineralization (Leybourne et al., 2001).

Groundwaters were recovered from around the Casa Grande porphyry Cu-Mo system, Arizona (Ficklin et al., 1981). Groundwaters at Casa Grande appear to be anomalous in Mo, As, K, and SO_4 proximal to mineralization, although As anomalies are displaced from Mo and SO_4 anomalies, complicating interpretation in the absence of groundwater flow data. Groundwater Cu contents are not anomalous (Ficklin et al., 1981; Taufen, 1997). Waters around Casa Grande appear to have very similar characteristics in terms of metal association as waters from the Spence deposit. Similarly, Mo is mobile from porphyry Cu mineralization in the wetter climate of western Canada. Discharge of aqueous Mo into lakes via stream and groundwater flow has produced Mo anomalies in lake sediment in British Columbia (Cook, 2000).

In summary, porphyry Cu deposits are good candidates in terms of exploration using water chemistry; these deposits commonly contain anomalous concentrations of species that tend to be relatively mobile in neutral to slightly alkaline, oxidizing solutions, e.g., As, Se, Re, and Mo. As shown at the Spence deposit in Chile, these species form oxyanions and therefore migrate downflow for significant distances, providing a broad exploration target. Anomalies in these oxyanions can be refined through the use of S/Se ratios and S isotopes, to aid distinction from far-traveled saline fluids. Copper migration is typically restricted owing to adsorption to hydroxide surfaces, which also likely explains the lack of a Cu anomaly at Casa Grande.

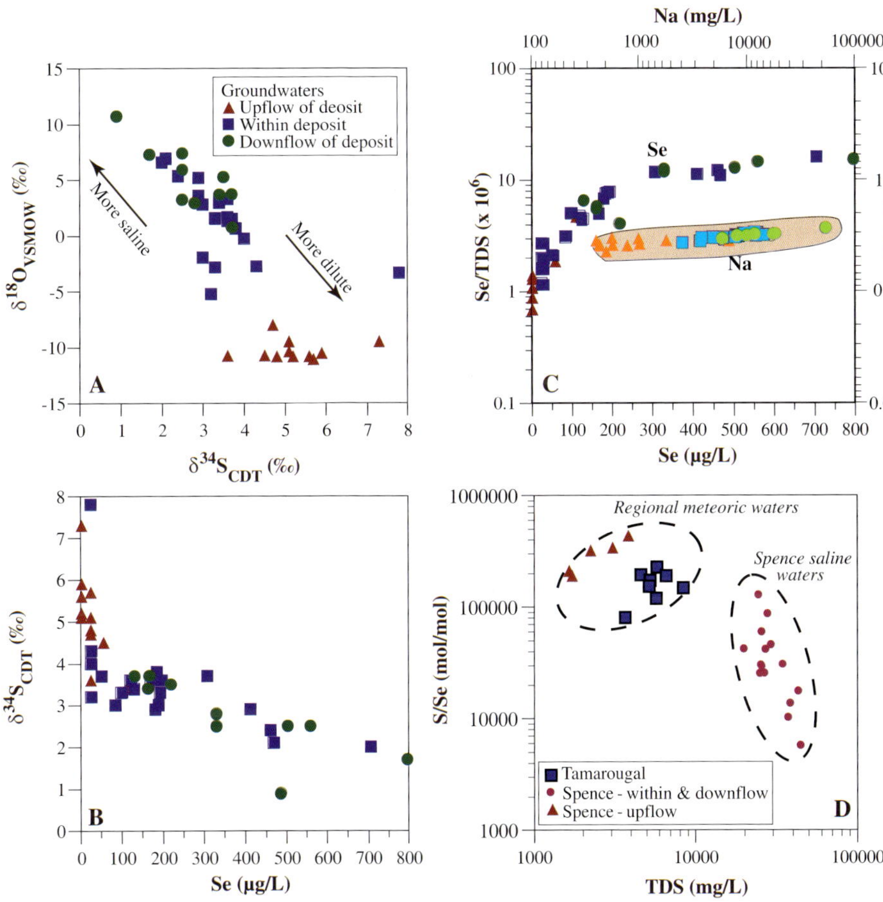

FIGURE 9. (A, B) Stable isotope ratios of S and O, and Se contents of groundwaters flowing into (east of deposit) and through the Spence deposit, Atacama Desert, northern Chile. These data show that groundwaters represent two broad groups, a more dilute upflow (east) groundwater with low Se contents, heavier $\delta^{34}S$ values and light $\delta^{18}O$ values (close to meteoric values for northern Chile), and a saline groundwater with $\delta^{34}S$ and Se contents largely controlled through interaction with porphyry copper mineralization. Modified from Leybourne and Cameron (2006). (C) Plot of Se/TDS versus Se (bottom and left axes) compared to Na/TDS (Total Dissolved Solids) versus Na (top and right axes). Note that Na, a conservative species, increases only slightly with increasing salinity, whereas Se/TDS increases to a greater extent, indicating that Se is added in excess of changes in salinity, from water-mineralization interaction. Modified from (Leybourne et al., 2001). (D) S/Se versus TDS for groundwaters from the Spence Deposit compared to those from the Pampa del Tamarugal. Despite a large soil geochemical anomaly at Tamarugal for Cu, the S/Se ratios in the waters strongly suggests that the source for the Cu is distal. Modified from Leybourne and Cameron (in press).

Sedimentary Exhalative Deposits

Sedimentary exhalative (SEDEX) deposits are characterized as tabular bodies of sphalerite and galena along with Fe-sulphides hosted in black shale and chert-bearing rocks, and are dominated by Zn, Pb, and Ag as commodities (Goodfellow and Lydon, 2007). Although SEDEX deposits are important as sources of base and precious metals, there are relatively few studies of waters associated with these deposits. Surface waters studies have been carried out around SEDEX deposits in the Selwyn Basin, Yukon (Goodfellow, 1983; Jonasson et al., 1987) and Alaska (Kelley and Taylor, 1997).

Kelley and Taylor (1997) reported the composition of surface waters flowing over three SEDEX deposits in Alaska: the Red Dog, Lik, and Drenchwater deposits. These deposits crop out at or near surface where crossed by streams. Waters downstream of these deposits are anomalous in Al, Fe, Mn, Cd, Co, Cu, Ni, Pb, and Zn (Fig. 1 shows the relationship between total metals and pH for the Lik and Red Dog deposits), although the longest dispersion (up to 1.7 km at Drenchwater and 14 km at Red Dog) are shown by Cd, Co, Fe, Ni, and Zn (Kelley and Taylor, 1997). Both of these deposits are also characterized by low pH waters (e.g. Red Dog, Fig. 1) proximal to and down-flow from mineralization, consistent with the lack of carbonate minerals in the host rocks (mainly black shales). In contrast, the Lik deposit lies within a section that contains platform carbonates in addition to carbonate veins that cross-cut mineralization (Kelley and Taylor, 1997). As a result, waters flowing downstream from the Lik deposit are near neutral and generally low in metals compared to the Red Dog and Drenchwater deposits (Fig. 1). However, one sample proximal to mineralization displayed anomalous Cd, Ni, Mn, and Zn; only Zn remained sufficiently elevated down-flow of the deposit to indicate the presence of mineralization (Zn ranged from 380 to 2000 μg/L, Kelley and Taylor, 1997). Thus, generation of low pH waters and elevated metal contents around these SEDEX deposits is dependent on local geological controls, specifically the buffering capacity of host rocks.

Similarly, Goodfellow (1983) studied surface waters flowing over the Howard's Pass (XY) and Nor SEDEX deposits in the Yukon, Canada. As with the Alaskan examples, where there are significant carbonates in the drainage system, metal contents are suppressed and Zn is the main (or only) pathfinder species (Goodfellow, 1983). Thus, at the XY deposit, only Zn shows a strong increase in abundance towards mineralization, whereas at the Nor occurrence stream waters were also anomalous in Pb and As, with stream sediments showing anomalous Cu, Hg, and Zn (Goodfellow, 1983). The XY deposit differs from the Nor occurrence in that at the XY deposit laminated sulphides are hosted by calcareous shales of the Road River Formation, whereas the Nor occurrence is hosted by pyritic and noncalcareous shales and cherts of the Lower Earn Group.

An important consideration for these examples from Alaska and the Yukon is that the mineralization is at or near surface and exposures are relatively fresh owing to the effects of glaciation (Jonasson et al., 1987; Miller, pers. comm. cited in Plumlee et al., 1999). Glaciation has thus resulted in enhanced weathering and, where buffering of solutions following water-sulphide interaction is limited, produces more elevated metal contents and longer dispersion than might be expected in unglaciated and intensely weathered terrain.

In summary, SEDEX deposits, like VMS deposits, can be expected to show elevated base metal contents in both ground and surface waters that have interacted with sulphides under oxiding conditions. Some SEDEX deposits have little or no carbonate in the immediate stratigraphy to mineralization, resulting in multi-element anomalies. Where mineralization is buffered, or more deeply buried with no surface erosion and clastic dispersion, Zn appears to be the dominant anomalous element.

Gold Deposits

The main impediment to Au exploration using hydrogeochemistry is the typically low solubility of Au in low-temperature waters (Vlassopoulos and Wood, 1990), because Au solubility is greatly dependent on the availability of suitable ligands. In most natural waters, Au solubility is controlled by the formation of $AuOH(H_2O)^0$ complexes, which limits Au solubility to a few parts per trillion (ppt) at best. Solubility is greatly enhanced in the presence of thiosulphate ($S_2O_3^{2-}$) and cyanide (CN^-) (Vlassopoulos and Wood, 1990). Bowell et al. (1993) investigated Au in soil pore waters and laterite profiles associated with Au mineralization at Ashanti, Ghana. Although Au was found to be mobile in soil pore waters owing to natural cyanide production (from decomposition of organic matter) and thiosulphate produced by weathering of sulphides, Au was mobile only on scale of a few centimetres (Bowell et al., 1993).

In Western Australia, Au mineralization is typically covered by thick regolith sequences. In this environment, groundwater geochemistry has proven to be successful in some cases, although interpretations of groundwater flow and modeling of water speciation is complicated by the presence of hypersaline waters (Carey et al., 2003). In a study around the St. Isles area, mesothermal Au mineralization is best reflected by elevated groundwater Au contents, which are as high as 52 ng/L, with speciation modeling suggesting that Au occurs primarily as $AuCl_2^-$ and/or AuI_2^- species in these saline waters (Carey et al., 2003). In this study, typical pathfinder elements for Au mineralization (such as Se, As, and Sb) showed no consistent variation with proximity to mineralization (Table 2). Similarly, Gray (2001) in a large study of groundwaters from the Yilgarn Craton of Western Australia found that groundwater Au was the best indicator of Au mineralization. Pathfinders such as As and Sb were only effective at near-neutral pH conditions; saline groundwaters in Western Australia are commonly acidic (acid-sulphate waters), resulting in adsorption of elements that form oxyanions (Gray, 2001). Various studies of groundwater Au contents in Australia indicate that Au is typically present at concentrations in excess of that predicted by equilibrium calculations (Giblin, 2001). Giblin (2001) suggested that this apparently enhanced Au solubility was either a consequence of Au-organic complexing not being taken into account in speciation calculations or to the propensity for Au to form fine colloids. In a study of Au contents of stream waters in the Bathurst mining camp, Canada, Leybourne et al. (2000a) used ultrafiltration (samples were unfiltered, filtered to

<0.45 μm, and <0.004 μm) to show that even in the presence of large amounts of cyanide, a strong Au-complexing agent, a maximum of 50% of the Au that passed the <0.45 μm filter passed the 0.004 μm filter and could be considered to be present in dissolved (non-colloidal) form.

In Nevada, groundwater Au contents are elevated (up to 4700 ng/L) proximal to mineralization under oxidizing conditions, whereas pathfinder species (As, Sb) are elevated in groundwaters where reducing conditions prevail (Grimes et al., 1995). Similarly, Cidu et al (1995) found for stream and spring waters in Sardinia, Italy, that Au was the best indicator of mineralization, although As and Sb were also typically anomalous close to mineralization. Sahoo and Pandalai (2000) studied groundwaters associated with Au mineralization in the Hutti-Maski schist belt of India. Although these authors apparently did not determine aqueous Au contents, they found that groundwater As and Sb contents delineated areas of Au mineralization (Sahoo and Pandalai, 2000).

Thiosulphate forms relatively strong complexes with Au at neutral to alkaline pH conditions (Webster, 1986). As Au mineralization is commonly associated with sulphide minerals, Au can be dispersed if ground and surface water conditions remain neutral (Benedetti and Boulègue, 1990, 1991). Indeed, the intermediary species formed during oxidation of sulphide to sulphate have been suggested as a prospecting tool in waters draining sulphide-bearing mineralization (Kondratenko, 1997).

In summary, the studies discussed above indicate that for ground and surface waters, aqueous Au contents provide the most consistent vectors to Au mineralization, although confidence in Au anomalies may be increased with coincident As and/or Sb anomalies. Water samples can be difficult to preserve for Au and Ag analyses, as these species tend to adsorb strongly to the walls of HDPE bottles acidified with nitric acid (Leybourne et al., 2000a). A separate aliquot can be acidified to 1% with BrCl; this keeps the Au, Ag, and Hg in solution, which can then be analyzed by ICP-MS to <2 ng/L. Alternatively, Au (and Ag) can be preconcentrated in the field using sachets of activated carbon, followed by analysis by NAA for detection limits of around 1 ng/L for Au and 1 μg/L for Ag (Giblin, 2001; Gray, 2001).

Magmatic Cu-Ni-PGE Deposits

The geological and geochemical characteristics of magmatic Cu-Ni-PGE deposits are reviewed by Eckstrand and Hulbert (2007) and Layton-Matthews et al. (2007). There has been little research into aqueous geochemical exploration for magmatic Ni-Cu-PGE mineralization. Some comments can be made, as there is good potential for water chemistry in this regard, in particular for those deposits that are sulphide-poor (especially in pyrrhotite) and therefore do not present a good geophysical target (Hattori and Cameron, 2004). The PGE are variably mobile in low-temperature aqueous systems with Pd the most mobile (Wood, 2002; Hattori and Cameron, 2004). Hattori and Cameron (2004) have suggested that Pd is mobile in surface waters around the Lac des Iles PGE deposit as hydroxide complexes (e.g. $Pd(OH)_2^0$, $Pd(OH)_3^-$, and $Pd(OH)_4^{2-}$); the low SO_4 and Cl contents of these waters precludes sulphate and chloride forming significant Pd complexes. As discussed above for porphyry copper systems, where metals and metalloids are complexed as neutral or negatively charged species, adsorption to clays and Fe and Mn oxides and oxyhydroxides is inhibited at neutral to slightly alkaline pH conditions (owing to surface charge considerations), consistent with the potential for significant aqueous dispersion of Pd (Hattori and Cameron, 2004). The Lac des Iles mineralization is sulphide poor (Hattori and Cameron, 2004). In contrast, many Ni-Cu-PGE deposits have abundant Ni and Fe sulphides. In sulphide-rich deposits, Pd (and other PGE) mobility in solution should be enhanced, in particular where pyrrhotite is abundant, as dissolution of Fe sulphides promotes more acidic conditions, which in turn will result in greater leaching of the PGE (Hattori and Cameron, 2004).

Given the mafic-ultramafic rock association of magmatic Ni-Cu-PGE deposits, typically water-rock reactions can be expected to result in relatively elevated pH values as a result of hydration of mafic Mg- and Fe-bearing minerals (Fritz et al., 1992; Stefánsson et al., 2001; Sader et al., 2007). In the case of PGE deposits with minor Fe-sulphides, pH values are likely to remain elevated (i.e. up to pH 10 or so), and Pd may prove to be the optimal pathfinder in solution, along with S and As, owing to their likely occurrence as oxyanions (Cameron and Hattori, 2003). Where significant Fe sulphides are present, as at most Cu-Ni-PGE deposits, pH values are likely to be anomalously low compared to waters more distal from mineralization, and the lower pH values will likely promote solution of Ni and Cu in addition to the PGE, so that multi-element anomalies should be more common (Table 2).

Kimberlite and Lamproite Diamonds

Diamonds are primarily extracted from ultramafic kimberlite and lamproite lithologies. Mineralogically these rocks are characterized by olivine, Cr-diopside, enstatite, Cr-poor Ti-pyrope garnet, Mg-ilmenite, and phlogopite (McClenaghan and Kjarsgaard, 2001). Kimberlites and lamproites are derived by small degrees of partial melting at significant depth (150-200 km) in the asthenospheric and/or subcontinental lithospheric mantle (Kjarsgaard, 2007). As such, they have distinctive geochemical signatures compared to most other mantle partial melts, with enrichment in the incompatible elements Sr, Ba, light REE (La, Ce, Pr, Nd, Sm), Nb, Ta, Hf, Zr, P, and Ti along with elevated Mg, Cr, Ni, and Co (McClenaghan and Kjarsgaard, 2001). Although it is a relatively simple task to geochemically distinguish these ultramafic rocks from their host rocks (typically felsic-dominated rocks of old shield areas), the distinction between diamond-bearing and barren kimberlite and lamproite is more problematic. Given this constraint, the use of aqueous geochemistry to explore for kimberlite in general and diamondiferous kimberlite more specifically, is therefore also geochemically challenging.

There have been almost no research studies specifically utilizing water chemistry for diamond/kimberlite exploration, although there have been a number of studies that have investigated the geochemical consequences of water-ultramafic rock interaction (Fritz et al., 1992; Allen and Seyfried, 2004; Palandri and Reed, 2004; Pokrovsky and Schott, 2004). New research in geochemical exploration for diamonds has involved groundwater geochemistry (Sader et

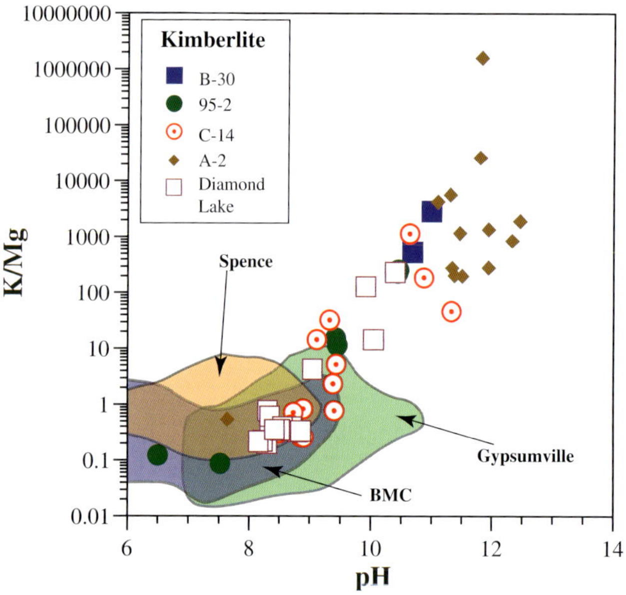

FIGURE 10. Plot of pH versus K/Mg ratio for groundwaters interacting with five different kimberlites from the Kirkland Lake area of northern Ontario, Canada. As increasing pH is due to hydrolysis reactions involving olivine and pyroxene, Mg is progressively buffered out of solution by brucite and magnesite precipitation (Sader et al.,2007). Shown for comparison are fields for groundwaters interacting with VMS mineralization in the Bathurst mining camp, New Brunswick, Canada (Leybourne and Goodfellow, 2003), groundwaters interacting with the Spence porphyry copper deposit, Atacama Desert, Chile (Cameron and Leybourne, 2005; Leybourne and Cameron, 2006), and from carbonate and crystalline rock aquifers around Gypsumville, MB, Canada (Leybourne, unpublished data).

al., 2003a,b, 2007). Where groundwater interacts with ultramafic rocks, hydration reactions of olivine and clinopyroxene result in elevated groundwater pH values, up to in excess of pH 12 (Fritz et al., 1992; Sader et al., 2007). These hydration reactions have the consequence of essentially buffering Mg out of solution in the formation of brucite [$Mg(OH)_2$] and/or serpentine minerals; groundwaters that have interacted extensively with kimberlites in northern Ontario are thus characterized as having elevated pH and highly atypical K/Mg ratios (Fig. 10). Further work is needed to better define trace element and/or isotopic anomalies in groundwaters that might provide vectors towards diamond-bearing ultramafic bodies, although elevated pH (Figs. 1, 10), anomalously low Mg and elevated K/Mg ratios in waters (Fig. 10) and the presence of key alteration minerals down-flow may assist exploration. Preliminary data (Sader and Leybourne, unpublished) suggests that some kimberlite waters are enriched in Co and Ni and have elevated Co/Mg and Ni/Mg ratios (Table 2).

Unconformity (Sandstone-Hosted) Uranium

Details related to the geology and geochemistry of unconformity uranium deposits is presented elsewhere in this volume (Jefferson et al., 2007). Exploration for uranium using groundwater geochemistry was investigated in the 1970s and early 1990s (e.g. Langmuir, 1978; Dyck, 1979; Langmuir and Chatham, 1980; Runnells and Lindberg, 1981; Earle and Drever, 1983; Giblin and Snelling, 1983) and has received renewed interest in the last few years (Singh et al., 2002; Pirlo and Giblin, 2004). Further insights into the relationships between water chemistry and U deposits can be gained from studies investigating the formation of uranium deposits and the long term stability of these deposits as proxies for the stability of buried nuclear waste materials (Yanase et al., 1995a,b; Arthur et al., 2006). Pirlo and Giblin (2004) provide a good summary of the history, strengths, and weaknesses of hydrogeochemistry with respect to uranium exploration. Previous studies of groundwater U exploration have focused on elemental determinations, U-series isotopes, He and He isotope anomalies, and the calculation of U-related mineral solubilities (Langmuir, 1978; Dyck, 1979; Deutscher et al., 1980; Langmuir and Chatham, 1980; Runnells and Lindberg, 1981; Earle and Drever, 1983; Giblin and Snelling, 1983; Toulhoat and Beaucaire, 1991; Singh et al., 2002; Pirlo and Giblin, 2004; Dickson and Giblin, 2006).

There are some significant challenges to the use of hydrogeochemistry to explore for U mineralization, in particular because of the strong redox control over U solubility and transport, and because U is a common trace element in a variety of rock types (Kotzer and Kyser, 1995; Dickson and Giblin, 2006). Despite these problems, many of the studies cited above suggest that the best aqueous indicator of U mineralization remains dissolved U contents (Table 2), assuming oxidizing conditions; other potential pathfinders are more problematic. In a large study of groundwaters in India (more than 600 groundwater samples), Singh et al. (2002) showed that U anomalies over known U-mineralization were coincident with anomalies in Na, K, Cl, SO_4, HCO_3, and TDS. However, none of these additional species are in any way diagnostic of U mineralization, limiting their utility. Similarly, Dickson and Giblin (2006) suggested that anomalous U contents were coincident with high Mg/total cations in groundwaters from the Pine Creek area, with REE, PO_4, F, and Cu in groundwaters from Olympic Dam, and with neutral pH, equilibrium with carbonate minerals and carbonaceous matter in groundwaters from sedimentary paleochannels in South Australia. Again, these coincident pathfinders are in no way diagnostic of U deposits. Finally, several studies have shown that calculation of mineral saturation indices (via geochemical software, see discussion) can provide vectors to U mineralization (e.g. Pirlo and Giblin, 2004; Dickson and Giblin, 2006); the issue remains that these calculated minerals (e.g. phengite) are not diagnostic. These calculations are useful additional sources of information for full data interpretation, but are unlikely to be successful in the absence of other pathfinders.

In summary, despite the fact that U is immobile under reducing conditions, in terms of elemental abundances, U appears to be the best indicator of unconformity-type U deposits. Clearly, exploration success will be enhanced if there are coincident anomalies in other elements (e.g. Se, V, Mo, and Cu; Table 2) that are typically enriched in U deposits. Molybdenum and Se may be especially useful in sulphide-poor reducing waters. There is some indication that waters are uncommonly Mg-rich and in equilibrium with phengite (Yanase et al., 1995a; Dickson and Giblin, 2006). One avenue that requires more research is the potential application of Pb and/or U-series isotopes in U exploration (Earle and Drever, 1983; Toulhoat and Beaucaire, 1991; Yanase et al., 1995b). Toulhoat and Beaucaire (1991) studied the Pb isotope geochemistry of groundwaters as an exploration tool for detecting buried U deposits. In this approach, groundwaters that have most interacted with U mineralization have the most radiogenic Pb isotopes with respect to $^{207}Pb/^{204}Pb$ and $^{206}Pb/^{204}Pb$ (from ^{235}U and ^{238}U decay, respectively) but not in $^{208}Pb/^{204}Pb$ (from ^{232}Th decay), owing to U deposits having anomalously high U/Pb but lower than typical Th/Pb.

Discussion: Aqueous Geochemistry in Mineral Exploration

Ground and Surface Water Sampling Techniques and Strategies

One of the limitations in using groundwater geochemistry in mineral exploration is the availability of suitable boreholes. In many areas of interest to mineral exploration, there are insufficient domestic or agricultural water wells for effective use of groundwater in an exploration program. Drilling is expensive and typically only occurs after a suitable exploration target has been defined. In Western Australia, thick regolith cover is an impediment to geochemical exploration of any kind and mineral exploration in this environment commonly involves drilling through cover to bedrock. The advantage of groundwater geochemistry is that groundwaters can provide a broader anomaly target than regolith geochemistry, with the result that the density of drillholes can potentially be reduced (Carey et al., 2003). In other environments, however, once drilling has commenced, groundwater composition may be useful in vectoring towards possible new targets and choosing locations for additional drillholes. Pauwels et al. (2002), for example, used some domestic wells in their surveys around two deposits in the Iberian Pyrite Belt, Spain, and were able to use groundwater geochemistry (anomalies in Cu, Zn, Pb, and Cd) to vector to known mineralization. Given the great expense of drilling exploration holes and the fact that groundwaters flow, there is only a small addition in cost to a drilling program to include an aqueous geochemical component, which may greatly enhance the value of the drilling, especially in the case of an apparently barren hole.

In areas where exploration holes or domestic wells are available, the sampling method requires some consideration. The principal sampling methods are bailer, down hole pump (e.g. double-valve or Grundfos), and straddle-packer systems (Figs. 11, 12). Flow-through bailers typically consist of a PVC tube (commonly around 1litre in volume) with one or two stop-valves that permit the flow of water through the tube during bailer descent in the well, because of the neutral buoyancy of the stop-valves. The valves are closed and the groundwater is trapped during retrieval of the bailer. The bailer method has the advantage that it is rapid, inexpensive, and easily learned for new users. The disadvantages of the bailer are 1) it is difficult to decontaminate the apparatus between samples and wells; 2) there is a fixed volume, which may not always be sufficient; and 3) the standing water column in a well is sampled, which may not represent the full variability of the well, in particular in crystalline rocks where groundwater flow is fracture dominated (Fig. 11). Another disadvantage of the bailer is that the method commonly results in disturbance of borehole walls and/or fracture zones, so that suspended sediment loads are typically greater than is the case for the packer (Leybourne, 2001).

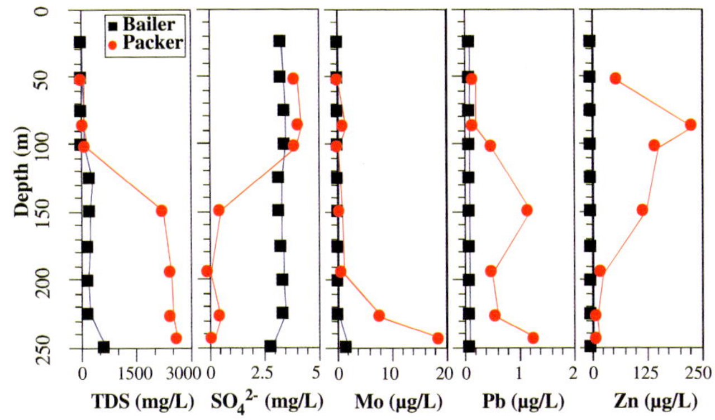

FIGURE 11. Down-hole variation plots of selected parameters for borehole MM-89-106 from the Restigouche deposit, Bathurst Mining Camp, New Brunswick, Canada. These plots illustrate the differences between samples collected using a bailer system and those using a straddle-packer system. In this case, the bailer samples indicate that there are significant chemical changes down hole, but do not reflect the full magnitude of those changes. Modified after Leybourne (1998).

Double-valve pumps operate in a similar fashion to bailers in that the pump head is lowered down the hole to the desired depth (typically at least 5 m below the water table) and the height of water column overlying the pump-head (i.e., the hydraulic head) results in groundwater flow into the chamber and up the PVC tubing to the static head level. Nitrogen gas or clean compressed air is then forced down one PVC tube into the sample head, resulting in return of groundwater to surface via another PVC tube. The advantage of the pump over the bailer is that successive samples can be taken until steady state conditions are reached for electrical conductivity (as a proxy for salinity), Eh (redox state), and pH. Once steady-state conditions are reached, the assumption is made that the system is sampling groundwaters that are representative of the borehole at that depth and the sample is taken (Leybourne et al., 2001; Leybourne and Cameron, 2006). This process not only allows greater confidence as to the degree to which the sample represents flow at the depth sampled, but also serves to decontaminate the sampling apparatus between boreholes. In some setups, the system can be pressurized prior to deployment in the well, so that the valves do not open until the pump reaches the depth at which the head of the overlying water column exceeds the pressure in the lines (Pauwels et al., 2002).

Bottomley et al. (1994) noted that sampling of depth-representative groundwaters in low-permeability fractured rock is difficult due to contamination by waters used in drilling the boreholes, mixing within open boreholes by groundwaters entering the hole from different fractures at different depths, and because fractures commonly have low hydraulic conductivities. These problems can be at least partly mitigated by the use of a straddle-packer system, which permits sampling of groundwaters from discrete intervals in a borehole (typically 1-3 m in length; Leybourne et al., 2002). The straddle-packer system is optimal for collecting samples that represent groundwater flow at the sampled depth, but suffers from high cost and time to collect samples. For exploration holes that are older than a few years, groundwater flow should have removed drilling fluids.

Surface waters are typically collected from the most actively flowing part of a stream. Samples are taken directly

FIGURE 12. **(A)** Sample chamber for a straddle-packer system, constructed at the Geological Survey of Canada. There is a unidirectional 1 psi cracking valve at the bottom of the sample chamber. **(B)** Upper sliding head packer. The packer is inflated when at the desired depth and seals the well. Note the white tubing at the base, which leads to the intake section. **(C)** Straddle-packer tubing spool and pressure apparatus. Constructed at the Geological Survey of Canada. All three photographs from the Restigouche deposit in the Bathurst Mining Camp, New Brunswick, Canada. **(D-F)** Down-hole pump apparatus. Note the tank of N_2 gas used to force sample back to surface. These photographs show sampling of groundwaters from kimberlites in the Kirkland Lake area of Northern Ontario, Canada.

from the stream into a sample bottle, or may be sampled using a peristaltic pump. The elemental budget of any stream can be parsed into the dissolved load, the suspended load, and the bed load.

Aqueous geochemists typically filter waters in situ, either to 0.45 or 0.20 μm. These filter pore sizes are operationally defined (Hall et al., 1996a), as they do not separate metals that are truly dissolved from those that are bound to fine colloids; many colloids are as small as <5 nm in size (e.g. Zanker et al., 2002). To separate truly dissolved aqueous species from those that may be associated with colloids requires ultrafiltration using either tangential flow or centrifuge techniques (Leybourne et al., 2000a; Zanker et al., 2002).

Most stream sediment studies only consider bed load geochemistry (Telmer et al., 2002). However, because metals undergo differential partitioning between the dissolved and particulate fraction (Leybourne, 2001), the suspended load is also a viable medium for geochemical exploration (Telmer et al., 2002). For example, Siegel (1990) investigated stream-suspended and bed load sediments in drainages associated with Mississippi Valley-type (MVT) Zn-Pb mineralization in Virginia, USA, and found that Zn formed the most consistent geochemical anomalies and that the suspended (>0.45 μm) sediment provided superior anomaly to background contrast compared to bed load sediments. One major disadvantage of using the suspended load is the difficulty of sample collection, especially for first- and second-order streams, which commonly have a very minor suspended component (Leybourne et al., 2003). For the study of suspended sediments over MVT mineralization, Siegel (1990) filtered approximately six litres of stream water per sample; this volume of filtration is impractical in large surveys (see below).

In geochemical exploration surveys where groundwaters are sampled, the groundwater suspended load may also be used; metals that are sparingly soluble in the dissolved load may be partitioned into the suspended sediment load and be more amenable to analysis, as was shown for groundwaters around undisturbed VMS deposits in the Bathurst mining camp (Leybourne, 2001). For example, suspended sediments in wells from the Bathurst mining camp commonly have Pb/Zn ratios that are greater than the aqueous phase and primary massive sulphide mineralization (Leybourne, 2001).

Depending on the nature of the survey, filtering may not be necessary. Cameron et al. (1997) showed that for Canadian northern lakes, pH and suspended sediment contents were sufficiently low that filtering and acidification were not required, with the caveat that analyses were carried out relatively rapidly after collection. Similarly, Leybourne and Cameron (2000a) showed that for groundwaters collected around the Spence porphyry Cu deposit in northern Chile, many of the elements of interest (e.g. S, Se, Re, As, and Mo) showed little or no difference between filtered (<0.45 μm) and unfiltered aliquots, although metal cations (e.g. Cu and Zn) showed large differences owing to adsorption to Fe, Mn, and Al oxyhydroxides.

Hydrology

For optimal interpretation of aqueous geochemical results in an exploration program, some understanding of the hydrology is required (Carey et al., 2003; Leybourne and Goodfellow, 2003; Gilliss et al., 2004). In order to develop either a surface geochemical anomaly or a distal groundwater geochemical anomaly of a deeply buried deposit, the hydrology must be conducive to return flow to the surface environment. In some cases, faults and/or fractures that originally focused fluid flow during ore formation may also serve as conduits to groundwater flow into and out of the deposit (e.g. Spence deposit, Fig. 7; Cameron and Leybourne, 2005). Furthermore, the Eh and pH conditions control, to a large extent, how ore minerals behave during water-rock interaction and the mobility and dispersion of ore-related metals/metalloids. Thus, if the Eh and pH conditions are not conducive to mineral dissolution/replacement and element transport, aqueous geochemical anomalies will not be formed. For example, Gilliss et al. (2004) investigated the aqueous geochemistry of groundwaters at the Tillex Cu-Zn deposit in Ontario, Canada and found that conditions proximal to the deposit are characterized by low Eh and near-neutral pH. Redox conditions at Tillex are such that dissolved sulphate is reduced to sulphide, restricting deposit sulphide mineral weathering and preventing the development of a significant aqueous geochemical plume (Gilliss et al., 2004).

Although some types of deposits are stratabound or controlled by faults and fractures, other deposits are channelized. This restriction of mineralization to channels has been noted in particular for some types of Cu-Ni-PGE and unconformity U deposits (Brand, 1999; Pirlo and Giblin, 2004). Channelization of mineralization is an important consideration for aqueous geochemical exploration because the orientation of the channels with respect to local hydrology will determine the extent to which mineralization is in contact with groundwaters. For example, at the Kambalda deposit, Ni-Cu-PGE mineralization is associated with komatiite flows, which were channelized by pre-existing topography. Mineralization at Kambalda forms linear features over 3 km long, but less than 300 m wide, and less than 5 m thick (Brand, 1999).

Streams are referred to as gaining or losing (i.e. influent and effluent) where shallow groundwater discharges into the stream or the stream recharges the shallow groundwater, respectively. Clearly, detection of buried mineralization will be more problematic where a losing stream crosses mineralization compared to a gaining stream. For example, the Restigouche deposit in the Bathurst mining camp crops out where Charlotte Brook crosses mineralization. Although stream waters show clear anomalies in ore-related elements across the deposit (e.g. Pb, Zn, Cu, and As; Fig. 5 shows Zn and Pb), sulphate contents are only slightly anomalous at best, consistent with Charlotte Brook being a losing reach proximal to mineralization at the time of sampling (Leybourne et al., 2003). In contrast, sediments are anomalous in Zn, Pb, and Cu for three kilometres downstream from the Restigouche deposit, indicating that Charlotte Brook is a gaining stream for at least part of the year (Fig. 5 shows water and sediment Pb contents).

Because of the temporal variability in surface water chemistry, stream water sampling is sensitive to timing of sampling with respect to perturbation in stream flow as a result of rainfall. In contrast, groundwaters typically integrate an annual to multi-annual signal, as revealed by the stable isotopes of oxygen and hydrogen (see Clark and Fritz, 1997). Surface waters primarily represent base flow (i.e. shallow groundwater discharge), combined with increases in discharge owing to rainfall or snowmelt. In areas with vegetation cover, the increase in stream discharge during a storm event may still be dominated by base flow (Caissie et al., 1996). In general, the stable isotopic composition (H, O) of precipitation varies significantly as a function of season, owing to the strong temperature dependence on the fractionation of these isotopes (Craig, 1961; Clark and Fritz, 1997). Thus, winter precipitation is typically depleted isotopically compared to summer precipitation.

Major Ions

Some studies have suggested that the major ion composition of ground and surface waters may provide useful vectors to mineralization, owing to the generally larger target provided by alteration halos around mineralization relative to the size of primary mineralization (e.g. Giblin, 1994; Taufen, 1997; Phipps et al., 2004;). Although the major ion chemistry of waters in a hydrogeochemical survey should be interpreted and considered in conjunction with the trace metal data, because of the complexity of water-rock reactions, cation exchange reactions, and the formation of secondary minerals, the major element composition of water should be used with great caution as an exploration guide in the absence of other (i.e. deposit-diagnostic) chemical data (Table 2). Some workers have suggested that soil Br and Cl anomalies develop over some deposit types as a result of electrochemical, gaseous, or advective transport (Clark, 1999; Hamilton, 1999). However, focusing on these anomalies is likely to cause more problems than it solves, simply because these elements are less likely to be diagnostic of mineralized zones. For example, work on groundwaters around the Spence deposit (Cameron et al., 2002; Cameron and Leybourne, 2005) suggests that in northern Chile saline groundwaters migrate to surface along fractures during seismic events. Because saline basinal fluids in northern Chile have elevated I, Br, Cl, and S contents, soils over fractures will be anomalous in I and Br regardless of whether or not the waters have interacted with porphyry Cu mineralization. In order to explore effectively using geochemical techniques, the explorationist needs to be able to identify mineralized rocks and be able to distinguish them from non-mineralized rocks of the same composition, therefor attention should be placed on what is geochemically distinct about the mineralization (Table 1).

Certainly, there are instances where there are clear major element changes that are diagnostic of a particular rock type or style of alteration, but these do not necessarily indicate mineralization. For example, Sader et al. (2007) showed that groundwaters interacting with kimberlite results in elevated pH and low Eh conditions, with a concomitant loss of Mg from solution and an increase in dissolved K contents (Fig. 10). However, these major ion changes occur regardless of whether the kimberlite is diamond bearing or not; any water-ultramafic rock interaction will likely produce similar changes in the major ion chemistry of groundwater. In contrast, Giblin (2001) suggested that waters interacting with ultramafic rocks will have Mg > Ca, Na, K.

Mineral Saturation and Speciation Calculations

Mineral saturation indices and metal speciation calculations are typically carried out using various software packages that include PHREEQC (Parkhurst, 1995), MINTEQA2 (Allison et al., 1991), and The Geochemists Workbench (Bethke, 1994). Some of the geochemical speciation/saturation computer programs are only available commercially, whereas other packages are available as free downloads via the web. Critical to the successful use of geochemical speciation and mineral saturation computer code is the realization that these programs should not be considered black boxes; some understanding of expected output greatly enhances the utility of these programs. In addition, output is only as good as the input data and the internal integrity of the thermodynamic database. In most cases, code is inherently based on equilibrium thermodynamics, so that kinetic (reaction rate) factors are not considered. Finally, much of the information of relevance to exploration is heavily dependent on the input values for pH and the redox state of the system; if these data are incomplete or poorly constrained, these calculations are not worth the effort.

Despite these caveats, calculating mineral saturation and element speciation can be instructive (Giblin, 1994; Leybourne, 1998; Gilliss et al., 2004; Pirlo and Giblin, 2004), for example, to better understand metal mobility or attenuation, and the occurrence of secondary mineralization, such as transported gossan (Boyle, 2003) or secondary copper mineralization. Sader et al. (2007) showed that reaction path modeling of water-kimberlite and water-gneiss interaction enhances understanding of likely meteoric water alteration assemblages, which can also be used as an exploration guide.

Metal Mobility and Solubility

The degree to which a water sample will be anomalous with respect to proximal or distal mineralization depends fundamentally on the solubility of the species that reflect that style of mineralization under the prevailing pH, Eh, and salinity conditions. Ideally, the geochemical explorationist would like any sample media to have anomalies in all species that define the style of mineralization. In reality, this is rarely, if ever, the case. However, although waters and sediments will not necessarily have anomalies in all the relevant elements, it is important to understand why this is the case and avoid the temptation to include elements that are not diagnostic of the style of mineralization (Tables 1, 2). Furthermore, care must be taken to interpret aqueous geochemical data with consideration of geochemical theory and analytical constraints. For example, although Ti is elevated in mafic and ultramafic igneous rocks compared to high-Si igneous and sedimentary lithologies, Ti is unlikely to be of much use in geochemical exploration because Ti is not particularly anomalous in mineralized mafic and ultramafic rocks compared to non-mineralized equivalents, Ti is particularly immobile in water, except in very rare cases, and Ti suffers from several potential spectral overlaps in an ICP plasma. Similarly, several studies have suggested that the rare earth elements (REE) may be useful in mineral exploration for non-REE deposits. In some cases this may be reasonable. For example, metal-rich hydrothermal fluids recovered from active vents at mid-ocean ridge crests show strong positive Eu anomalies, and these anomalies are also commonly observed in VMS deposits. In contrast, strong positive Eu anomalies are not common in other geological environments, so that the presence of these Eu anomalies could be used as an exploration guide (Volesky et al., 2000; Volesky, 2002; Leybourne and Cousens, 2005; Leybourne et al., 2000b, 2006b). Low-pH groundwaters proximal to VMS mineralization have also preserved the positive Eu anomalies (Leybourne et al., 2000b). Diamond-bearing kimberlites commonly have distinctive large ion lithophile elements (LILE) and high field strength elements (HFSE) (+ rare earth element) associations (McMartin and McClenaghan, 2001). In this case, aqueous anomalies in mobile species (K, Rb, Sr, and REE in some waters) may assist in differentiating between rock types, but may not be as useful in distinguishing mineralized from unmineralized rocks.

False anomalies may be produced in stream sediments owing to the accumulation of background metal concentrations by Fe- and Mn-oxyhydroxides. However, where stream sediment surveys reveal elevated Fe and Mn contents in association with anomalous ore-indicator metals, it would be wrong to assume that the results obviate the presence of mineralization. Firstly, Mn contents do increase towards the core of some hydrothermal systems, such as VMS (Large et al., 2001), so that high-Mn contents may also indicate proximal mineralization. Secondly, because the hydrology is so important in controlling the localization of hydromorphic anomalies, even relatively low-metal contents may indicate proximal mineralization. As discussed below, some isotopic systems can be used to more quantitatively assign a mineralization association where anomalous metal contents are associated with elevated Mn contents.

Pre-Cover, Cover, and Post-Cover History

Another issue is the pre-cover history of the rocks hosting mineralization. For example, regolith/saprolite development, and gossan formation can affect the type of anomaly developed in ground and surface waters. For example, at the Halfmile Lake deposit, Bathurst mining camp, mineralization crops out as a gossan (Boyle, 2003). As a result of differential mobility in gossan-forming groundwaters (i.e. Zn and Cu are more mobile than Pb), till overlying the deposit has relatively low Zn and Cu contents (Parkhill and Doiron, 2003). Thus, groundwaters interacting with this type of gossan will likely not have Zn and Cu anomalies. At Halfmile Lake, Pb contents are elevated in till down ice, and Pb isotopic ratios demonstrate that the Pb is associated with gossan mineralization over primary VMS mineralization (Hussein et al., 2003).

Stable and Radiogenic Isotopes

A variety of isotopic systems are traditionally not used in routine mineral exploration. Most of these are unlikely to become standard in mineral exploration, but discussion is warranted because there are cases where stable and/or radiogenic isotopic data may assist in data interpretation or the understanding of metal and water sources, and because advances in analytical methods are such that the cost of isotope analyses continues to decline and the simplicity of sample preparation increase. Isotopes of potential use in mineral exploration include the traditional stable isotopes (O, H, C, S), radiogenic isotopes (Pb, Sr), in addition to newer non-traditional isotope systems, such as Fe, Se, and Cu.

Analytical considerations include the ease of collecting, storing, and processing samples, and the cost of measuring element abundance. For example, traditionally Sr and Pb isotopes have been analyzed by thermal ionization mass spectrometry (TIMS). Measurements by TIMS are time consuming owing to the need for chemical separation of the species of interest. As such, application of these systems to mineral exploration has been limited, and most research studies have focused on Pb isotopes (Gulson, 1986; Gulson et al., 1988; Gulson et al., 1992; Bell and Franklin, 1993; Leybourne, 1998). However, recent advances in ICP-MS instrumentation, including dynamic reaction-cell ICP technologies (DRC-ICP-MS), high-resolution ICP-MS (HR), and multi-collector ICP-MS (MC), have resulted in the potential to more routinely and cost-effectively analyze geological samples for radiogenic and some light stable isotopes. For example, Doherty et al. (1997) used a standard quadrupole ICP-MS to measure Pb isotopes of till from the Restigouche deposit in the Bathurst mining camp. Although ICP-MS Pb isotope determinations are less precise than those by TIMS, for the purposes of detecting massive sulphide mineralization in the Bathurst mining camp and thus fingerprinting the Pb source, the ICP-MS had sufficient precision, with less sample preparation and analytical time compared to TIMS analyses on the same samples (Doherty et al., 1997). New multi-collector ICP-MS instruments are opening up new areas of research in non-traditional isotopes, some of which may find application in mineral exploration. These include the stable isotopes of Fe (Rouxel et al., 2004; Severmann et al., 2004), Cu (Ehrlich et al., 2004), and Se (Johnson and Bullen, 2004; Layton-Matthews et al., 2006).

Stable isotopes are a critical component in many aqueous geochemical studies because they provide important information regarding water sources, groundwater flow history, water-rock reactions, and groundwater evolution. For a good recent summary of the light stable isotopes in aqueous studies, see Clark and Fritz (1997). Stable and radiogenic isotopes have not found much application in exploration, owing to the cost and perceived difficulty in interpretation. Although stable isotope determinations are not overly expensive per isotope, especially compared to commercial rates for Pb and Sr by TIMS, adding a full suite (e.g. O, H, C, S) to an exploration program would significantly add to the cost of a survey, with additional resources required for sample collection, preparation, and data interpretation. However, although the light stable isotopes may not become routine, they can be a very useful addition to an orientation or more detailed study. For example, future successes in surficial geochemical exploration will rest with the ability of groundwaters to penetrate to depth, with subsequent return of pathfinder element-bearing fluids to surface. In that context, sampling of overland or shallow soil water flow following storm events will not be as useful as sampling base-flow conditions for stream water surveys, or deeper groundwater flow in the case of springs. In a study of stream, seep, and ground waters from two undisturbed VMS deposits in the Bathurst mining camp, Leybourne et al. (2006a) showed that surface waters were similar to shallow groundwaters based on the $\delta^{18}O$ and $\delta^{2}H$ compositions, indicating that surface water flow represented shallow groundwater discharge. In addition, saline groundwaters at the Restigouche deposit were found to be isotopically distinct from shallow groundwaters, consistent with the interpretation that the saline waters were paleowaters (Leybourne et al., 2006a). These saline waters were also enriched in Zn, Co, and Ni, but enhanced metal contents were a function of increased Cl complexing and, possibly, to deeper mineralization unrelated to VMS mineralization at the Restigouche deposit (Leybourne and Goodfellow, 2003). Similarly, O and H isotopes in groundwaters recovered around the undisturbed Spence porphyry Cu deposit in the hyper-arid Atacama Desert of northern Chile indicate that saline waters interacting with Cu mineralization (as shown by elevated Cu, As, Se, Re, Mo, and low $\delta^{34}S$; Fig. 9B) are paleowaters, possibly derived from sedimentary basin brines, and have mixed with regional groundwater flow, derived by recharge in the high Andes (Cameron and Leybourne, 2005; Leybourne and Cameron, 2006). Understanding the recharge history of groundwater and whether sampled waters are modern meteoric or paleowaters is important in terms of understanding local hydrology, in particular where groundwater composition is being used to vector to mineralization.

In contrast to the isotopes of O and H, which are unlikely to yield relationships directly linked to mineralization, sulphur isotopes can provide vectors to mineralization in addition to aiding interpretations of water sources and rock reactions. For example, Whitford et al. (1997) suggested that groundwater sulphur isotopic composition provides a vector to gold mineralization in Australia. Similarly, analysis of the sulphur isotopic composition of groundwaters at the Spence porphyry copper deposit in Chile show that groundwater $\delta^{34}S$ values decrease from values typical of regional waters and salars (Pueyo et al., 2001; Rech et al., 2003; Rissmann, 2003) to values closer to those typical of porphyry copper mineralization as waters flow into and through mineralization (Leybourne and Cameron, 2006). This increasing influence on groundwater sulphur isotopic composition is also evident in Se abundances (Fig. 9B; in addition to Re, Mo, and As, not shown). Recent work in the area that hosts the giant Broken Hill SEDEX deposit in Australia has also suggested that the S and O isotopic composition of dissolved sulphate can be used to vector to mineralization and provides a broader halo than groundwater base metal (Zn, Pb, Cu) contents (de Caritat and Kirste, 2005).

The generally low solubility of Pb in most waters means that ground and surface water Pb contents may not be sufficiently elevated proximal to mineralization to warrant consideration, in particular where pH values are buffered and/or mineralization is more deeply buried. However, because sulphide minerals do not incorporate U and Th into their structure, Pb derived from sulphide minerals will retain the isotopic composition of the ore-forming fluids, whereas Pb from surrounding U- and Th-rich host rocks will typically have more radiogenic Pb isotopic compositions. Therefore, Pb isotopes can "fingerprint" Pb sources, and indicate subsurface mineralization, even where aqueous Pb contents are low (Leybourne, 1998).

Partial and Sequential Extraction Techniques

A large number of schemes have been suggested in the literature, but essentially all reduce to the same basic idea; different extractions are designed to remove metals from different portions of the sediment (e.g. Chao, 1984; Hall et al., 1996b, 2003; McCarty et al., 1998; Gray et al., 1999). Most sequential extraction schemes have been designed for and applied to soils and till. However, a smaller number of studies have applied these partial leaches for stream sediments (Martin et al., 1984; Evans and Davies, 1994; Borovec, 1996; Leybourne et al., 2003). Most leaches are weak, designed to remove the more labile components in a sediment compared to more aggressive leaches such as aqua regia or total dissolution (e.g. fusion or $HF\text{-}HClO_4\text{-}HCl\text{-}HNO_3$ digestion).

Partial leaches have utility in geochemical exploration because they commonly result in an enhanced anomaly to background contrast. Furthermore, they provide insights into the speciation or binding sites of metals in soils and sediments. These insights are key to an improved understanding of the processes of metal migration in the secondary environment and the formation of geochemical anomalies (Hall et al., 1996b; Cameron et al., 2004). In addition, partial leaches can also aid in understanding metal bioavailability, metal sources, and the potential for metal remobilization with changes in surface conditions, especially pH (Hall et al., 1996b). Such changes resulting from a change in the pH conditions are especially important where mining of Fe-sulphide-rich deposits is being considered. In soils, loosely bound metals (weakly adsorbed, extracted with deionized water or weak acetate) may be attributable to a buried source of mineralization and/or to anthropogenic influences via atmospheric transport (Hall, 1998).

There are two areas in which these analytical schemes have application to aqueous geochemical exploration. Firstly, one of the principle reasons for using partial leaches is to try and capture the fraction of a till or soil elemental budget that has been transported in some way from buried mineralization to surface and added to the soil profile, either adsorbed to mineral surfaces or organic matter, precipitated/incorporated into pedogenic calcite or gypsum, or incorporated into oxyhydroxides of Fe and Mn. Although a variety of transport mechanisms have been proposed (Cameron et al., 2004), the processes by which these surface geochemical anomalies form is still not well understood. However, understanding groundwater geochemistry has greatly improved interpretations of metal transport processes in arid gravel-covered regions like Chile (Cameron et al., 2002) and glaciated terrains in Canada (Cameron et al., 2004). Secondly, although not all the proposed leaches are applicable to stream sediments, leaches that target the amorphous Mn- and Fe-oxyhydroxides can improve anomaly to background contrast, and also provide information regarding mechanical versus hydromorphic metal transport, which will help interpretations regarding vectoring to mineralization (e.g. compare the response of In and Pb in stream sediments from the Bathurst mining camp, Fig. 5).

Table 3. Advantages and disadvantages of surficial geochemical methods - EXTECH II, BMC, Canada.

Method	Advantages	Disadvantages	Reference
Surface waters	Ease of sampling; limited sample preparation (already in solution) for ICP analysis; instantaneous geochemical signature	Risk of contamination; potential of low background contrast; limited mobility of certain metals in near neutral pH waters	Leybourne, 1998; Leybourne et al., 2003
Groundwaters	Strong anomaly/background contrast, multiple samples can be taken at variable depths; hydrological information can be used to vector towards mineralization; potential for accessing deep mineralization	Expense of drilling holes, need to understand hydrology; depth specific samples difficult to collect except with expensive/time consuming packer system	Leybourne and Goodfellow, 2003
Stream sediments (total)	Ease of sampling; information about lithological variations	Post sample preparation (drying, sieving, dissolution); dilution of hydromorphic components by detrital components	Leybourne et al., 2003
Stream sediments (partial)	Separates hydromorphic from detrital contributions; generally greater dispersion than soils	Greater sample handling and preparation; more expensive than total analysis	Leybourne et al., 2003
Soil (total and sequential)	Deposit-related metals may be transported through thick cover (weak leaches)	Weak-leach soil anomalies may be difficult to interpret	Hall et al., 2003
Till	Glacial dispersion provides larger exploration target compared to other methods	Glacial movement may be complex; requires surface deposit; post collection sample preparation (drying, sieving, dissolution)	Parkhill and Doiron, 2003
Biogeochemistry	Tree roots may penetrate to tap more deeply buried mineralization; some plant species may preferentially take up ore-related metals	Optimally requires a single species of similar age over entire study area; metabolic requirement of plants for some metals may complicate interpretations	Leybourne et al., 1999
Gossans	Derived directly from original massive sulfide	Requires surface or near surface deposit; preservation of gossan during glaciation; distinguishing from gossan developed over non-sulphide Fe-rich rocks	Boyle, 2003

Recommendations for Sampling and Analytical Protocols of Waters in Mineral Exploration

From the above discussion and the examples presented, one can see that there are a large number of parameters that can be used to aid the detection of undiscovered mineralization and that there are a large number of potential issues and pitfalls that need to be avoided. However, it should be emphasized that despite these issues, aqueous geochemistry is a powerful tool because it is a relatively simple and if surveys are well designed, robust. In this section, recommendations are made regarding sampling and analysis to aid exploration geochemists in realizing this relative simplicity. Table 3 provides a summary of geochemical exploration techniques that were used in the Bathurst mining camp and indicates the strengths and weaknesses of the different approaches employed in that study (for all papers in this Bathurst mining camp study, see Goodfellow et al., 2003).

The collection and analysis of ground and surface waters in mineral exploration can be broken down into two different approaches, which, for the sake of this discussion, shall be termed the "routine exploration" and "exploration research" approaches. Giblin (2001) also provides useful guidelines for groundwater sampling in geochemical exploration. The routine approach is applicable in areas where some knowledge of the geochemical landscape and, ideally, the hydrology already exists. In areas where this knowledge base does not exist, or where techniques require refining, the research approach should be adopted in an orientation survey.

Routine Exploration

1. Groundwaters can be sampled rapidly with a bailer. If only one sample is to be taken from a well or drillhole, the sample should be taken at least 5 metres below water level. Surface waters are sampled directly from the actively flowing part of a stream. Ideally plastic gloves (cheap, disposable) should be worn during collection of water samples to ensure uniformity in sampling and prevent contamination. In any case, jewelry and watches should not be worn during sample collection.
2. If only one aliquot is taken, a 250 or 500 mL Nalgene HDPE bottle should be used. No pretreatment of Nalgene bottles is required, though the bottle and cap should be rinsed vigorously three times with sample water prior to filling the bottle (Cameron, 1978). For consistency, and assuming that samples will not be acidified, the bottle should be completely filled and stored in a cool place. If a separate aliquot is taken for anion and alkalinity determination, 125 or 60 mL bottles are sufficient.
3. Some parameters that are useful or critical to measure in the field are cheap and require minimal training of personnel. These field analyses should include, as a minimum, pH and conductivity. pH is very useful in subsequent data interpretation, but is also a useful guide in the field; large drops in pH values during a survey may indicate the presence of unbuffered sulphide oxidation, and sample density can be increased appropriately. Similarly, although the electrical conductivity of a water sample is only a rough measure of its salinity, sudden changes in salinity likely reflect additions of different, potentially deeper, groundwater. In addition, if all anions will not be measured, the conductivity can be used as a proxy for total dissolved solids. Meters for the

determination of pH and conductivity are relatively robust, inexpensive, and simple to calibrate.

4. In order to ensure internally consistent data that can be used with confidence by all members of the program, the following QA/QC procedures should be followed. Within every batch of twenty samples, a duplicate sample should be taken. If only small numbers of samples are being collected, a duplicate should be taken for every ten samples collected. Within every batch of twenty samples a bottle should be left empty, but given a sample number. Prior to submission to a laboratory, this bottle should be filled with a standard or, ideally, a certified reference material. Finally, several field blanks should be taken (bottle filled with deionized water) to test for any contamination in the bottles.
5. Because in a routine exploration survey samples do not need to be filtered or acidified, samples should be kept cold and analyzed as soon as feasible. For this kind of survey, ICP is the analytical method of choice, ideally a combination of ICP-MS and ICP-OES. The former permits the determination of essentially all of the trace species of interest in mineral exploration, with the exception that Au and Ag are not stable in solution in unpreserved bottles. ICP-OES permits the determination of major cations, metals that are present in high abundance (especially Fe, Mn, and Zn), and S (although note that not all commercial ICP-OES instruments operate at low enough wavelengths to measure S, around 185 nm). Although detailed interpretation of ground and surface waters requires full anion analyses, for a routine survey it may not be necessary to determine Cl or alkalinity (typically primarily carbonate alkalinity), whereas dissolved sulphate contents are useful as a secondary vector.
6. Cameron et al. (1997) showed that using a portable anode stripping voltammeter (ASV) to measure dissolved Zn was useful in guiding the exploration program; waters were analyzed immediately following each day's collection. Comparison of these field Zn analyses with subsequent ICP-MS analyses showed effectively no difference for Zn > 1 μg/L (Cameron et al., 1997).

Exploration Research

1. Although groundwaters can be collected using a bailer, for optimal results in an exploration research study a down-hole pump or, ideally, a straddle-packer system or equivalent should be used. As noted above, optimal results are obtained when samples can be collected from a specific depth, especially in fracture-flow dominated systems. In the case where down-hole pump or packer is used, sample return lines can be fitted to a flow-through cell to permit determination of field parameters (below) without exposing the groundwater to atmospheric oxygen. Surface water samples are collected in the same manner as above, using plastic gloves and taking the sample from the active part of the stream.
2. Field analyses should be more extensive in a more research-focused program. In addition to pH and conductivity, measurements should be made that try to capture the reduction-oxidation (redox) state of the aqueous system under study. There are a number of ways to measure redox, the most common being through the use of an Eh or ORP probe. Eh probes typically employ a Pt electrode. Unfortunately, Eh measurements performed this way are commonly only measure the redox state of a few species to which the platinum electrode responds. For example, Pt-Eh probes do not respond to the O_2-H_2O couple, so that there is typically little correlation between Eh measurements and the amount of dissolved oxygen. However, Pt-probes do respond well to the Fe^{3+}-Fe^{2+} couple, which given the common occurrence of Fe sulphides in many types of ore deposits, makes Eh measurements worthwhile. There are multiprobe meters that permit most or all of these measurements to be made simultaneously (Fig. 13). In addition, some are sufficiently narrow and can be fitted with long cables to permit in situ profiling of wells and/or boreholes. In situ profiling is useful because zones can be identified where changes in EC, pH, T, and Eh occur (Pauwels et al., 2002), information that can then be used to determine optimal sampling depths. However, these meters can cost several thousand dollars so care must be taken when deploying these down an exploration drillhole. Alternatively, these meters can also be fitted with flow-through cells, so that where groundwater samples are collected by pump or packer, pH, Eh, and DO can be measured without exposing the sample to the atmosphere. There are some other very simple tests of the redox state of a groundwater that can be measured in the field that are useful for full geochemical interpretation; these being measurements of dissolved ferrous (Fe^{2+}) iron and dissolved sulphide. These tests involve the use of a field spectrophotometer. The presence of dissolved ferrous Fe or sulphide is indicated by colour changes to a water sample the 1, 10 Phenanthroline method (Fe^{2+}) or the Methylene Blue method (HS^-) (e.g. Fig. 13).
3. For the research approach, a larger number of samples should be taken to allow the optimization of a more routine study in an area. Cameron and Leybourne (2000a,b, 2001, 2005) found that analyzing both an unfiltered and filtered (<0.45 μm; Fig. 13) aliquot was useful in demonstrating the different behaviour in species that behaved as cations (Cu, Fe) from those that were com-

FIGURE 13. **(A)** Measuring pH, T, conductivity, dissolved oxygen and Eh simultaneously with a multiprobe, Wichita River, Texas (photo courtesy of N. Miller). **(B)** Sterivex filters; the lower two filters were used to filter a groundwater with 0.75 mg/L Fe^{2+} and 0.21 mg/L HS- at the Kudz Ze Kayah deposit, Yukon, Canada (Layton-Matthews and Leybourne, unpublished data). **(C)** Example of a groundwater with significant Fe^{2+}, which develops a strong orange colour. Kudz Ze Kayah deposit, Yukon, Canada (Layton-Matthews and Leybourne, unpublished data). **(D)** Similarly, groundwaters with significant dissolved sulphide develop a blue colour, with depth of colour increasing with increasing HS^-. Kudz Ze Kayah deposit, Yukon, Canada (Layton-Matthews and Leybourne, unpublished data). **(E)** In situ measurement of pH, T, conductivity, dissolved oxygen, and Eh simultaneously with a multiprobe, Wichita River, Texas (photo courtesy of N. Miller). **(F)** For larger streams where it may be difficult to reach the active portion, a sample bottle can be fixed to the end of an extendable pole. Wichita River, Texas. **(G)** An example of precipitate that formed rapidly upon addition of $BaCl_2$ to a high-SO_4 water sample from the Wichita River, Texas (photo courtesy of N. Miller).

A
B
K2K03-1023
B
D
24019
25 mL
20 mL
10 mL
E
F
G
3-029 S

plexed as anions in northern Chile groundwaters (e.g. Se, As, Re, and Mo). Therefore, it is recommended that at least a filtered (to 0.2 or 0.45 μm) cation sample be taken, though an unfiltered aliquot should also be taken if resources permit. Use disposable cartridge filters (especially Millipore Sterivex-HV filters, catalog# SVHV010RS) and all-plastic syringes (no rubber stoppers, as these commonly contain significant Zn). A new filter and syringe should be used for each sample; filter the cation sample first and then the anion sample. The same filter and syringe can be used for both the cation and anion samples. Plastic gloves should be worn at all times during sampling; new gloves for every sample. Bottles should be rinsed with 20 to 30 mL of filtered water three times before filling the bottle.

4. The cation sample should be acidified to 1% by volume with 70% ultrapure nitric acid as soon as feasible. Because more detailed analyses will be performed, a 250 or 500 mL Nalgene bottle should be used. For the cation aliquot(s), sufficient headspace should be left in the bottle to permit addition of nitric acid. In contrast, an aliquot should also be taken for anion and alkalinity determination; this sample should be filtered, but otherwise is untreated, so the bottle should be completely filled. Additional samples should be taken based on the types of analyses to be performed. For sulphur isotope determination, $BaCl_2$ is typically added to a separate aliquot of sample. The $BaCl_2$ is highly soluble, whereas $BaSO_4$ is highly insoluble; if the sample has dissolved sulphate, it precipitates rapidly. The sample is allowed to sit for a few hours so that the $BaSO_4$ settles to the bottom and the remaining fluid is decanted, preserving the precipitate for isotopic analysis. This is also a useful field test in that the almost instantaneous formation of precipitate indicates the presence of dissolved sulphate and the amount of precipitate is obviously dependent on the concentration of sulphate. In a Au exploration program, an additional sample should be taken. Gold can either be preconcentrated in the field using sachets of activated carbon (Giblin, 2001; Gray, 2001) or by preservation to 1% with BrCl (Leybourne et al., 2000a).

5. As full a set of analyses should be performed as is practical. However, at a minimum, the analytical program should include determination of alkalinity by titration, anions (Cl, SO_4, PO4, Br, F, NO_3) by ion chromatography or liquid chromatography, major species by ICP-OES (e.g. S, Si, Fe, Mg, Ca, Na, K, B, Mn, Sr, Zn, and Cu), and trace elements by ICP-MS (Be, Li, Ti, Ag, Bi, Co, Cr, Cs, In, Fe, Mn, Pb, Rb, Sb, Tl, U, V, Al, Ba, Cd, Cu, Mo, As, Ni, Se, Zn, Sr, Re, Th, Y, and the REE). In many media, S and Cl are the dominant anions and their peaks on an ion chromatogram may mask those of less abundant species, or dilution may be required such that these minor peaks are too small to be resolved. In these cases, Br, I, and PO_4 can also be measured by ICP-MS, and F and NO_3 by ion specific electrode or calorimetric methods. Some species of interest are difficult to measure by quadrupole ICP-MS (e.g. Se, As in the presence of significant Cl) owing to mass overlaps by species that form in an ICP plasma. Where these species are required at low detection limits, hydride generation ICP-MS can be performed to improve detection limits and remove interferences (Hall and Pelchat, 1997a,b; Layton-Matthews et al., 2006).

Future Directions and Knowledge Gaps

1. The role of bacteria in controlling sulphide oxidation and silicate hydrolysis as well as metal mobility and attenuation is not well understood. In recent years, there has been an increasing appreciation that many reactions that were previously considered inorganic, are now recognized as least being facilitated, if not controlled, by biological processes. Recent studies have suggested that alteration of mid-ocean ridge basalt glass is bacterially mediated (Torsvik et al., 1998). For example, Reith et al. (2005) suggest that there is an association between Au mineralization and *Bacillus cereus* in soils in Australia. Supergene sphalerite mineralization was recently identified at the Mike deposit in Nevada (Bawden et al., 2003). Here, sulphur isotopes of framboidal sphalerite are very negative (as low as -70 ‰ $\delta^{34}S$, illustrating the control of sulphur-reducing bacteria on this unusual mineralization (>400,000 tonnes of Zn) (Bawden et al., 2003).
2. One of the important challenges for surficial geochemical exploration methods is to move beyond deposit-scale reconnaissance surveys (although, as is obvious from the preceding, more of these are needed) towards application of the methods in areas of prospective, but unknown, mineralization.
3. Greater emphasis needs to be placed on distinguishing real from false geochemical anomalies, in particular development of better models that relate redox and pH conditions of ground and surface waters with the element associations that can be expected given different styles of mineralization.
4. Increased use of non-traditional stable and radiogenic isotopes, both in terms of deposit characterization and in aqueous media for mineral exploration. For example, Cu and Se isotopes are relatively new systems being investigated. At the current state of understanding of these isotopes, we are still characterizing potential fractionations and cataloging values associated with different environments. However, in the future these systems may become well enough understood that they may become useful in mineral exploration. In addition, more research is needed on analytical methods, lowering the cost, sample preparation, and the speed of isotope analyses.
5. There is a continued need for better integration of real aqueous geochemical data and thermodynamic modeling programs, in particular the incorporation of kinetic thermodynamics and adsorption to Mn and clay minerals. In addition, the thermodynamic databases for saline waters (greater than seawater) are limited with respect to most of the trace metals of interest to exploration.
6. Accurate hydrological models are needed in a study area in order to connect surface aqueous geochemistry to groundwater flow and metal sources.
7. One of the great advantages of collecting ground and surface water samples during exploration is that the data

is also critical in establishing local and regional geochemical baselines should the program be successful in finding mineralization and the decision made to initiate mining.

Conclusions

The following recommendations can be made with respect to the use of hydrogeochemistry in mineral exploration:

1. In new areas or new styles of mineralization, orientation studies are important, in particular with a view to determining the elemental associations that are diagnostic of that style of mineralization. In these kinds of studies, emphasis should be placed on the characterization of groundwaters (via wells, springs), surface waters and partial leaches of stream (and groundwater suspended) sediments.
2. For groundwaters, determine the type of sampling apparatus to be used; routine exploration may only require a flow through bailer, whereas, more detailed studies will likely need some form of down-hole pump or straddle-packer system, particularly in fractured-rock media, in order to provide optimal understanding of the redox state of the groundwater system.
3. For surface waters, sampling should be carried out when base-flow conditions operate, i.e., in the dry season.
4. In stream sediment surveys, if resources are available, significant additional information on hydromorphic versus mechanical transport (and therefore potential depth to mineralization) can be gained by addition of partial leach schemes to bed load sediment samples.
5. If resources permit, isotopic analyses can prove invaluable. In particular, Pb isotopes can be used to fingerprint metal sources and S isotopes (in both SO_4^{2-} and HS^-, as appropriate) can provide vectors to mineralization. Analytical methods are rapidly advancing in terms of technology and more rapid and cost-effective techniques. It is likely that isotopic analyses, such as Pb and S, will become cheap and routine in a small number of years.

Acknowledgments

Wayne Goodfellow, Jan Peter, Dan Boyle, Dan Layton-Matthews, Eion Cameron, Beth McClenaghan, Stew Hamilton, Ian Jonasson, Gwendy Hall, Steve McCutcheon, Dave Lentz, and Tom Al are thanked for various discussions over the last few years on aqueous geochemistry, mineral exploration, and ore deposits. Dan Layton-Matthews, Will Doherty, Peter Belanger, Ian Clark, Wendy Abdi, Conrad Gregoire, Judy Vaive, and Gwendy Hall are thanked for their for help, advice, and support in various analytical labs over the last number of years. The geochemistry laboratory at the University of Texas at Dallas was able to operate with the help of Dan Layton-Matthews, various graduate students (especially Jamil Sader, Clinton Rissmann, and Randy Griffin) and support from University of Texas at Dallas Geosciences Department, National Science Foundation grant GEO-0224551, Canadian Mining Institute Research Organization (CAMIRO Deep Penetrating Geochemistry Phase II), BHP Ltd. (now BHP-Billiton Ltd), and the Geological Survey of Canada (TGI, Targeted Geoscience Initiative). Research in the Bathurst mining camp was funded by the Geological Survey of Canada EXTECH II (EXploration TECHnology) program. Finally, volume editor, Wayne Goodfellow, and reviewers, Eion Cameron and Paul Taufen, are thanked for critical comments that greatly improved the paper.

References

Allen, D.E., and Seyfried, W.E.J., 2004, Serpentinization and heat generation: Constraints from Lost City and Rainbow hydrothermal systems: Geochimica et Cosmochimica Acta, v. 68, p. 1347-1354.

Allison, J.D., Brown, D.S., and Novo-Gradec, K.J., 1991, MINTEQA2/PRODEFA2, a geochemical assessment model for environmental systems; Athens, GA, EPA/600/3-91/021, U.S. Environmental Protection Agency.

Arthur, R.C., Iwatsuki, T., Sasao, E., Metcalfe, R., Amano, K., and Ota, K., 2006, Geochemical constraints on the origin and stability of the Tono uranium deposit, Japan: Geochemistry: Exploration, Environment, Analysis, v. 6, p. 33-48.

Bau, M., 1999, Scavenging of dissolved yttrium and rare earths by precipitating iron oxyhydroxide: Experimental evidence for Ce oxidation, Y-Ho fractionation, and the lanthanide tetrad effect: Geochimica et Cosmochimica Acta, v. 63, p. 67-77.

Bawden, T.M., Einaudi, M.T., Bostick, B.C., Meibom, A., Wooden, J., Norby, J.W., Orobona, M.J.T., and Chamberlain, C.P., 2003, Extreme ^{34}S depletions in ZnS at the Mike gold deposit, Carlin Trend, Nevada: Evidence for bacteriogenic supergene sphalerite: Geology, v. 31, p. 913-916.

Bell, K., and Franklin, J.M., 1993, Application of lead isotopes to mineral exploration in glaciated areas: Geology, v. 21, p. 1143-1146.

Benedetti, M., and Boulègue, J., 1990, Transfer and deposition of gold in the Congo watershed: Earth and Planetary Science Letters, v. 100, p. 108-117.

——— 1991, Mechanism of gold transfer and deposition in a supergene environment: Geochimica et Cosmochimica Acta, v. 55, p. 1539-1547.

Bethke, C.M., 1994, The Geochemist's Workbench: University of Illinois.

Borovec, Z., 1996, Evaluation of the concentrations of trace elements in stream sediments by factor and cluster analysis and the sequential extraction procedure: Science of the Total Environment, v. 177, p. 237-250.

Bottomley, D.J., Gregoire, D.C., and Raven, K.G., 1994, Saline groundwaters and brines in the Canadian Shield: geochemical and isotopic evidence for a residual evaporite brine component: Geochimica et Cosmochimica Acta, v. 58, p. 1483-1498.

Bowell, R.J., Foster, R.P., and Gize, A.P., 1993, The mobility of gold in tropical rain forest soils: Economic Geology, v. 88, p. 999-1016.

Boyle, D.R., 1988, Application of groundwater geochemistry in mineral exploration, *in* Berry, M.J., ed., Current activities forum 1988, Program with abstracts: Geological Survey of Canada.

——— 2003, Preglacial weathering of massive sulfide deposits in the Bathurst mining camp: Economic geology, geochemistry, and exploration applications, *in* Goodfellow, W.D., McCutcheon, S.R., and Peter, J.M., eds., Massive Sulphide Deposits of the Bathurst Mining Camp, New Brunswick, and Northern Maine: Economic Geology, Monograph 11, p. 689-721.

Brand, N.W., 1999, Element ratios in nickel sulphide exploration: Vectoring towards ore environments: Journal of Geochemical Exploration, v. 67, p. 145-165.

Caissie, D., Pollock, T.L., and Cunjak, R.A., 1996, Variation in stream water chemistry and hydrograph separation in a small drainage basin: Journal of Hydrology, v. 178, p. 137-157.

Cameron, E.M., 1977, Geochemical dispersion in lake waters and sediments from massive sulphide mineralization, Agricola Lake, Northwest Territories: Journal of Geochemical Exploration, v. 7, p. 327-348.

——— 1978, Hydrogeochemical methods for base metal exploration in the northern Canadian Shield: Journal of Geochemical Exploration, v. 10, p. 219-243.

Cameron, E.M., and Hattori, K.H., 2003, Mobility of palladium in the surface environment: data from a regional lake sediment survey in northwestern Ontario: Geochemistry: Exploration, Environment, Analysis, v. 3, p. 299-311.

Cameron, E.M., and Leybourne, M.I., 2005, Relationship between groundwater chemistry and soil geochemical anomalies at the Spence copper porphyry deposit, Chile: Geochemistry: Exploration, Environment, Analysis, v. 5, p. 135-145.

Cameron, E.M., Marmont, C., and Hall, G.E.M., 1997, Water: A medium for exploration in northern terrains: Explore, v. 96, p. 1-6.

Cameron, E.M., Leybourne, M.I., and Kelley, D.L., 2002, Exploring for deeply covered mineral deposits: Formation of geochemical anomalies in northern Chile by earthquake-induced surface flooding of mineralized groundwaters: Geology, v. 30, p. 1007-1010.

Cameron, E.M., Hamilton, S.M., Leybourne, M.I., Hall, G.E.M., and McClenaghan, B., 2004, Finding deeply-buried deposits using geochemistry: Geochemistry: Exploration, Environment, Analysis, v. 4, p. 7-32.

Carey, M.L., McPhail, D.C., and Taufen, P.M., 2003, Groundwater flow in playa lake environments: Impact on gold and pathfinder element distributions in groundwaters surrounding mesothermal gold deposits, St.Ives area, Eastern Goldfields, Western Australia: Geochemistry: Exploration, Environment, Analysis, v. 3, p. 57-71.

Chao, T.T., 1984, Use of partial dissolution techniques in geochemical exploration: Journal of Geochemical Exploration, v. 20, p. 101-135.

Cidu, R., Fanfani, L., Shand, P., Edmunds, W.M., Dack, L.V.t., and Gijbels, R., 1995, Hydrogeochemical exploration for gold in the Osilo area, Sardinia, Italy: Applied Geochemistry, v. 10, p. 517-529.

Clark, I.D., and Fritz, P., 1997, Environmental Isotopes in Hydrogeology: Lewis Publishers, Boca Raton, Florida, 328 p.

Clark, J.R., 1999, Concepts and models for interpretation of enzyme leach data for mineral and petroleum exploration: Geoanalysis with emphasis on selective extractions: Association of Exploration Geochemistry Short Course, 19th International Geochemical Exploration symposium, April 10, 1999.

Cook, S.J., 2000, Distribution and dispersion of molybdenum in lake sediments adjacent to porphyry molybdenum mineralization, central British Columbia: Journal of Geochemical Exploration, v. 71, p. 13-50.

Craig, H., 1961, Isotopic variations in meteoric waters: Science, v. 133, p. 1702-1703.

de Caritat, P., and Kirste, D., 2005, Groundwater chemical and isotopic composition: a mineral exploration tool for areas of regolith cover - lessons from the Curnamona Province: CRC LEME-GSWA Mineral Exploration Seminar, Kalgoorlie, WA, February 5, 2005, p. 81-82.

Deutscher, R.L., Mann, A.W., and Giblin, A.M., 1980, Ground water geochemistry in the vicinity of the Jabiluka deposits, *in* Ferguson, J., and Goleby, A.B., eds., Uranium in the Pine Creek Geosyncline; Proceedings of the International Uranium Symposium on the Pine Creek Geosyncline., International Atomic Energy Agency, Vienna, p. 477-486.

Dickson, B., and Giblin, A., 2006, New horizons for groundwater geochemistry in uranium exploration: AESC 2006, Melbourne, Australia.

Doherty, W., Gregoire, D.C., Goodfellow, W.D., Firard, I., and Lightfoot, P.C., 1997, Determination of lead isotope ratios by ICP Mass Spectrometry: Application to base metal exploration in the Bathurst Mining Camp: Ottawa '97, Joint Annual Meeting of the Geological Association of Canada/Mineralogical Association of Canada, 1997, p. A-39.

Dubé, B., Gosselin, P. Mercier-Langevin, P., Hannington, M., and Galley, A., 2007, Gold-rich volcanogenic massive sulphide deposits, *in* Goodfellow, W.D., ed., Mineral Deposits of Canada: A Synthesis of Major Deposit-Types, District Metallogeny, the Evolution of Geological Provinces, and Exploration Methods: Geological Association of Canada, Mineral Deposits Division, Special Publication No. 5, p.75-94.

Dyck, W., 1979, Application of hydrogeochemistry to the search for uranium, *in* Hood, P.J., ed., Geophysics and Geochemistry in the Search for Metallic Ores: Proceedings of Exploration '77, an International Symposium, Ottawa, Canada, October 1977: Economic Geology, Report 31, p. 489-510.

Earle, S.A.M., and Drever, G.L., 1983, Hydrogeochemical exploration for uranium within the Athabasca Basin, northern Saskatchewan: Journal of Geochemical Exploration, v. 19, p. 57-73.

Eckstrand, O.R., and Hulbert, L., 2007, Magmatic Nickel-Copper-Platinum Group Element Deposits, *in* Goodfellow, W.D., ed., Mineral Deposits of Canada: A Synthesis of Major Deposit-Types, District Metallogeny, the Evolution of Geological Provinces, and Exploration Methods: Geological Association of Canada, Mineral Deposits Division, Special Publication No. 5, p. 205-222.

Ehrlich, S., Butler, I., Halicz, L., Rockard, D., Oldroyd, A., and Matthews, A., 2004, Experimental study of the copper isotope fractionation between aqueous Cu(II) and covellite, CuS: Chemical Geology, v. 209, p. 259-269.

Evans, D., and Davies, B.E., 1994, The influence of channel morphology on the chemical partitioning of Pb and Zn in contaminated river sediments: Applied Geochemistry, v. 9, p. 45-52.

Ficklin, W.H., Preston, D.J., Welsch, E.P., and Nowlan, G.A., 1981, Analytical Results for 78 Ground-Water Samples from the Casa Grande Vicinity, Arizona:United States Geological Survey, Open File Report 81-0960, 16 p.

Fritz, P., Clark, I.D., Fontes, J.-C., Whiticar, M.J., and Faber, E., 1992, Deuterium and ^{13}C evidence for low temperature production of hydrogen and methane in a highly alkaline groundwater environment on Oman, *in* Kharaka, Y., and Maest, A.S., eds., Proceedings of the 7th International Symposium on Water-Rock Interaction: Balkema, p. 793-796.

Galley, A.G., Hannington, M.D., and Jonasson, I.R., 2007, Volcanogenic massive sulphide deposits, *in* Goodfellow, W.D., ed., Mineral Deposits of Canada: A Synthesis of Major Deposit-Types, District Metallogeny, the Evolution of Geological Provinces, and Exploration Methods: Geological Association of Canada, Mineral Deposits Division, Special Publication No. 5, p. 141-161.

Giblin, A., 1994, Natural waters as sample media in drainage geochemistry, *in* Hale, M., and Plant, J.A., eds., Handbook of Exploration Geochemistry, Drainage Geochemistry, v. 6: Elsevier, p. 269-303.

——— 2001, Groundwaters: Geochemical Pathfinders to Concealed Ore Deposits, CSIRO Exploration and Mining, 70 p.

Giblin, A.M., and Snelling, A.A., 1983, Application of hydrogeochemistry to uranium exploration in the Pine Creek geosyncline, Northern Territory, Australia: Journal of Geochemical Exploration, v. 19, p. 33-55.

Gilliss, M.L., Al, T.A., Blowes, D.W., Hall, G.E.M., and MacLean, B., 2004, Geochemical dispersion in groundwater from a weathered Cu-Zn deposit in glaciated terrain: Geochemistry-Exploration Environment Analysis, v. 4, p. 291-305.

Goodfellow, W.D., 1983, Stream Sediment and Water Geochemistry of the Howard's Pass (XY) Zn-Pb Deposit and Nor Zn-Pb-Ba Occurrence, Selwyn Basin and Northwest Territories: Geological Survey of Canada, Open File Report 845, 12 p.

Goodfellow, W.D., and Lydon, J.W., 2007, Sedimentary-exhalative (SEDEX) deposits, *in* Goodfellow, W.D., ed., Mineral Deposits of Canada: A Synthesis of Major Deposit-Types, District Metallogeny, the Evolution of Geological Provinces, and Exploration Methods: Geological Association of Canada, Mineral Deposits Division, Special Publication No. 5, p. 163-183.

Goodfellow, W.D., McCutcheon, S.R., and Peter, J.M., eds., 2003, Massive Sulphide Deposits of the Bathurst Mining Camp, New Brunswick, and Northern Maine: Economic Geology, Monograph 11, 930 p.

Gray, D.J., 2001, Hydrogeochemistry in the Yilgarn Craton. Geochemistry: Exploration, Environment, Analysis, v. 1, p. 253-264.

Gray, D.J., Wildman, J.E., and Longman, G.D., 1999, Selective and partial extraction analyses of transported overburden for gold exploration in the Yilgarn Craton, Western Australia: Journal of Geochemical Exploration, v. 67, p. 51-66.

Grimes, D.J., Ficklin, W.H., Meier, A.L., and McHugh, J.B., 1995, Anomalous gold, antimony, arsenic, and tungsten in ground water and alluvium around disseminated gold deposits along the Getchell Trend, Humboldt County, Nevada: Journal of Geochemical Exploration, v. 52, p. 351-371.

Gulson, B.L., 1986, Lead Isotopes in Mineral Exploration: Elsevier, Amsterdam, 245 p.

Gulson, B.L., Large, R.R., and Porritt, P.M., 1988, Gold exploration using lead isotopes at Tennant Creek, Australia: Applied Geochemistry, v. 3, p. 243-254.

Gulson, B.L., Church, S.E., Mizon, K.J., and Meier, A.L., 1992, Lead isotopes in iron and manganese oxide coatings and their use as an exploration guide for concealed mineralization: Applied Geochemistry, v. 7, p. 495-511.

Hall, G.E.M., 1998, Analytical perspective on trace element species of interest in exploration: Journal of Geochemical Exploration, v. 61, p. 1-19.

Hall, G.E.M., and Pelchat, J.-C., 1997a, Analysis of geological materials for bismuth, antimony, selenium and tellurium by continuous flow hydride generation inductively coupled plasma mass spectrometry part 1.Mutual hydride interferences: Journal of Analytical Atomic Spectrometry, v. 12, p. 97-102.

——— 1997b, Analysis of geological materials for bismuth, antimony, selenium and tellurium by continuous flow hydride generation inductively coupled plasma mass spectrometry part 2.Methodology and results: Journal of Analytical Atomic Spectrometry, v. 12, p. 103-106.

Hall, G.E.M., Bonham-Carter, G.F., Horowitz, A.J., Lum, K., Lemieux, C., Quemerais, B., and Garbarino, J.R., 1996a, The effect of using different 0.45 µm filter membranes on 'dissolved' element concentrations in natural waters: Applied Geochemistry, v. 11, p. 243-249.

Hall, G.E.M., Vaive, J.E., Beer, R., and Hoashi, M., 1996b, Selective leaches revisited, with emphasis on the amorphous Fe oxyhydroxide phase extraction: Journal of Geochemical Exploration, v. 56, p. 59-78.

Hall, G.E.M., Parkhill, M.A., and Bonham-Carter, G.F., 2003, Conventional and selective leach geochemical exploration methods applied to humus and B horizon soil overlying the Restigouche VMS deposit, Bathurst Mining Camp, New Brunswick, *in* Goodfellow, W.D., McCutcheon, S.R., and Peter, J.M., eds., Massive Sulphide Deposits of the Bathurst Mining Camp, New Brunswick, and Northern Maine: Economic Geology, Monograph, 11, p. 763-782.

Hamilton, S.M., 1998, Electrochemical mass-transport in overburden: a new model to account for the formation of selective leach geochemical anomalies in glacial terrain: Journal of Geochemical Exploration, v. 63, p. 155-172.

——— 1999, Transport mechanisms responsible for selective leach anomalies, geoanalysis with emphasis on selective extractions: Association of Exploration Geochemists, Short Course.

Hattori, K.H., and Cameron, E.M., 2004, Using the high mobility of palladium in surface media in exploration for platinum group element deposits: evidence from the Lac des Iles region, northwestern Ontario: Economic Geology, v. 99, p. 157-171.

Jefferson, C.W., Thomas, D J., Gandhi, S.S., Ramaekers, P., Delaney, G., Brisban, D., Cutts, C., Portella, P., and Olson, R.A., 2007, Unconformity-associated uranium deposits of the Athabasca Basin, Saskatchewan and Alberta, *in* Goodfellow, W.D., ed., Mineral Deposits of Canada: A Synthesis of Major Deposit-Types, District Metallogeny, the Evolution of Geological Provinces, and Exploration Methods: Geological Association of Canada, Mineral Deposits Division, Special Publication No. 5, p. xx-xx.

Johnson, T.M., and Bullen, T.D., 2004, Mass-dependent fractionation of selenium and chromium isotopes in low-temperature environments, *in* Johnson, C.M., Beard, B.L., and Albarède, F., eds., Geochemistry of Non-Traditional Stable Isotopes: Mineralogical Society of America, Geochemical Society, Reviews in Mineralogy and Geochemistry, v. 55, p. 289-317.

Jonasson, I.R., Goodfellow, W.D., Walker, D.A., and Jackson, L.E., 1987, Secondary zinc mineralization in post-glacial sediments and talus at Howard's Pass, Yukon, Canada - mineralogy, geochemistry and genesis, in Balasubramaniam, K.S. ed., Weathering; Its Products and Deposits, Volume II: Products - Deposits – Geotechnics: Theophrastus Publishers, Greece, p. 523-577.

Kelley, K.D., and Taylor, C.D., 1997, Environmental geochemistry of shale-hosted Ag-Pb-Zn massive sulfide deposits in northwest Alaska: Natural background concentrations of metals in water from mineralized areas: Applied Geochemistry, v. 12, p. 397-409.

Kjarsgaard, B.A., 2007, Kimberlite diamond deposits, *in* Goodfellow, W.D., ed., Mineral Deposits of Canada: A Synthesis of Major Deposit-Types, District Metallogeny, the Evolution of Geological Provinces, and Exploration Methods: Geological Association of Canada, Mineral Deposits Division, Special Publication No. 5, p. 245-271.

Kondratenko, L.A., 1997, Geochemistry of sulfur oxoanions in waters of hypergenesis zone: evidence from sulfide deposits of Transbaikalia: Geochemistry International, v. 35, p. 904-912.

Kotzer, T.G., and Kyser, T.K., 1995, Petrogenesis of the Proterozoic Athabasca Basin, northern Saskatchewan, Canada, and its relation to diagenesis, hydrothermal uranium mineralization and paleohydrology: Chemical Geology, v. 120, p. 45-89.

Langmuir, D., 1978, Uranium solution-mineral equilibria at low temperatures with applications to sedimentary ore deposits: Geochimica et Cosmochimica Acta, v. 42, p. 547-569.

Langmuir, D., and Chatham, J.R., 1980, Groundwater prospecting for sandstone-type uranium deposits: A preliminary comparison of the merits of mineral-solution equilibria, and single-element tracer methods: Journal of Geochemical Exploration, v. 13, p. 201-219.

Large, R.R., Allen, R.L., Blake, M.D., and Herrmann, W., 2001, Hydrothermal alteration and volatile element halos for the Rosebery K Lens volcanic-hosted massive sulfide deposit, Western Tasmania: Economic Geology, v. 96, p. 1055-1072.

Layton-Matthews, D., Leybourne, M.I., Peter, J.M., and Scott, S.D., 2006, Determination of Selenium Isotopic Ratios by Continuous-Hydride-Generation Inductively-Coupled-Plasma Mass Spectrometry: Journal of Analytical Atomic Spectroscopy, v. 21, p. 41-49.

Layton-Matthews, D.., Lesher, M., Burnham, O.M., Liwiang, J., Halden, N.M., Hulbert, L., and Peck, D.C., 2007, Magmatic Ni-Cu-(PGE) deposits of the Thompson Nickel Belt, *in* Goodfellow, W.D., ed., Mineral Deposits of Canada: A Synthesis of Major Deposit-Types, District Metallogeny, the Evolution of Geological Provinces, and Exploration Methods: Geological Association of Canada, Mineral Deposits Division, Special Publication No. 5, p. 409-432.

Leybourne, M.I., 1998, Hydrochemistry of ground and Surface Waters Associated with Massive Sulphide Deposits, Bathurst Mining Camp, New Brunswick: Halfmile Lake and Restigouche deposits: Ph.D.thesis, University of Ottawa, 723 p.

——— 2001, Mineralogy and geochemistry of suspended sediments from groundwaters associated with undisturbed Zn-Pb massive sulphide deposits: Canadian Mineralogist, v. 39, p. 1597-1616.

Leybourne, M.I., and Cameron, E.M., 2000a, CAMIRO Deep-Penetrating Geochemistry, Phase II: Composition of Groundwaters at the Spence Deposit: Part A—Major Elements and Stable Isotopes, CAMIRO Deep-Penetrating Geochemistry, Phase II, 18 p.

——— 2000b, CAMIRO Deep-Penetrating Geochemistry, Phase II: Composition of Groundwaters at the Spence Deposit: Part B—Metals Associated with Supergene Cu Mineralization, CAMIRO Deep-Penetrating Geochemistry, Phase II, 14 p.

——— 2006, Composition of groundwaters associated with porphyry-Cu deposits, Atacama Desert, Chile: Elemental and isotopic constraints on water sources and water-rock reactions: Geochimica et Cosmochimica Acta, v. 70, p. 1616-1635.

——— in press, Composition of soils and groundwaters at Pampa del Tamarugal: anatomy of a false (groundwater transported) geochemical anomaly: Economic Geology.

Leybourne, M.I., and Cousens, B.L., 2005, Rare earth elements (REE) and Nd and Sr isotopes in groundwater and suspended sediments from the Bathurst Mining Camp, New Brunswick: water-rock reactions and elemental fractionation, *in* Johannesson, K.H., ed., Rare Earth Elements in Groundwater Flow Systems: Water Science and Technology Library, v. 51, Springer, p. 254-293.

Leybourne, M.I., and Goodfellow, W.D., 2003, Processes of metal solution and transport in ground and surface waters draining massive sulfide deposits, Bathurst Mining Camp, New Brunswick, *in* Goodfellow, W.D., McCutcheon, S.R., and Peter, J.M., eds., Massive Sulphide Deposits of the Bathurst Mining Camp, New Brunswick, and Northern Maine, Economic Geology, Monograph 11, p. 723-740.

Leybourne, M.I., Goodfellow, W.D., and Boyle, D.R., 1998, Hydrogeochemical, isotopic, and rare earth element evidence for contrasting water-rock interactions at two undisturbed Zn-Pb massive sulphide deposits, Bathurst Mining Camp, N.B., Canada: Journal of Geochemical Exploration, v. 64, p. 237-261.

Leybourne, M.I., Goodfellow, W.D., and Boyle, D.R., 1999, Stream water, stream sediment, and vegetation of the Restigouche - Murray Brook area, Bathurst Mining Camp, New Brunswick (parts of NTS 21O7, 21O8, 21O9 and 21O10): Geological Survey of Canada, Open-File D3763, 1 CD-ROM.

Leybourne, M.I., Goodfellow, W.D., Boyle, D.R., and Hall, G.E.M., 2000a, Form and distribution of anthropogenic gold mobilized in surface waters and sediments from a gossan tailings pile, Murray Brook massive sulfide deposit, New Brunswick, Canada: Applied Geochemistry, v. 15, p. 629-646.

Leybourne, M.I., Goodfellow, W.D., Boyle, D.R., and Hall, G.M., 2000b, Rapid development of negative Ce anomalies in surface waters and contrasting REE patterns in groundwaters associated with Zn-Pb massive sulphide deposits: Applied Geochemistry, v. 15, p. 695-723.

Leybourne, M.I., Cameron, E.M., and Miller, N., 2001, CAMIRO Deep-Penetrating Geochemistry, Phase II: Composition of Groundwaters at the Spence Deposit: Part C—Results of 2000 sampling, CAMIRO Deep-Penetrating Geochemistry, Phase II, 15 p.

Leybourne, M.I., Goodfellow, W.D., and Boyle, D.R., 2002, Sulfide oxidation and groundwater transport of base metals at the Halfmile Lake and Restigouche Zn-Pb massive sulfide deposits, Bathurst Mining Camp, New Brunswick: Geochemistry: Exploration, Environment, Analysis, v. 2, p. 37-44.

Leybourne, M.I., Boyle, D.R., and Goodfellow, W.D., 2003, Interpretation of stream water and sediment geochemistry in the Bathurst Mining Camp, New Brunswick: Applications to mineral exploration, *in* Goodfellow, W.D., McCutcheon, S.R., and Peter, J.M., eds., Massive Sulphide Deposits of the Bathurst Mining Camp, New Brunswick, and Northern Maine, Economic Geology, Monograph 11, p. 741-762.

Leybourne, M.I., Clark, I.D., and Goodfellow, W.D., 2006a, Stable isotope geochemistry of ground and surface waters associated with undisturbed massive sulfide deposits; constraints on origin and water-rock reactions: Chemical Geology, v. 231, p. 300-325.

Leybourne, M.I., Peter, J.M., Layton-Matthews, D., Volesky, J., and Boyle, D.R., 2006b, Mobility and fractionation of rare earth elements during supergene weathering and gossan formation and chemical modification of massive sulfide gossan: Geochimica et Cosmochimica Acta, v. 70, p. 1097-1112.

Martin, H., Wilhelm, E., Laville-Timsit, L., Lecomte, P., Sondag, F., and Warnant, P., 1984, Enhancement of stream-sediment geochemical anomalies in Belgium and France by selective extractions and mineral separations: Journal of Geochemical Exploration, v. 20, p. 179-203.

McCarty, D.K., Moore, J.N., and Marcus, W.A., 1998, Mineralogy and trace element association in an acid mine drainage iron oxide precipitate; comparison of selective extractions: Applied Geochemistry, v. 13, p. 165-176.

McClenaghan, M.B., and Kjarsgaard, B.A., 2001, Indicator mineral and geochemical methods for diamond exploration in glaciated terrain in Canada, *in* McClenaghan, M.B., Bobrowsky, P.T., Hall, G.E.M., and Cook, S.J., eds., Drift Exploration in Glaciated Terrain: Geological Society of London, Special Publication 185, p. 83-123.

McMartin, I., and McClenaghan, M.B., 2001, Till geochemistry and sampling techniques in glaciated shield terrain: a review, *in* McClenaghan, M.B., Bobrowsky, P.T., Hall, G.E.M., and Cook, S.J., eds., Drift Exploration in Glaciated Terrain: Geological Society of London, Special Publication 185, p. 19-43.

Miller, W.R., Ficklin, W.H., and Learned, R.E., 1982, Hydrogeochemical prospecting for porphyry copper deposits in the tropical-marine climate of Puerto Rico: Journal of Geochemical Exploration, v. 16, p. 217-233.

Mingqi, W., Tianxiang, R., Yinhan, L., and Shuxing, L., 1995, Stream water, an ideal medium for rapidly locating polymetallic mineralization in forest-swamp terrain: a case study in the Da Hinggan Mts., northern Heilongjiang, China: Journal of Geochemical Exploration, v. 55, p. 257-264.

Palandri, J.L., and Reed, M.H., 2004, Geochemical models of metasomatism in ultramafic systems: serpentinization, rodingitization, and sea floor carbonate chimney precipitation: Geochimica et Cosmochimica Acta, v. 68, p. 1115-1133.

Parkhill, M.A., and Doiron, A., 2003, Quaternary geology of the Bathurst Mining Camp and implications for base metal exploration using drift prospecting, *in* Goodfellow, W.D., McCutcheon, S.R., and Peter, J.M., eds., Massive Sulphide Deposits of the Bathurst Mining Camp, New Brunswick, and Northern Maine, Economic Geology, Monograph 11, p. 631-660.

Parkhurst, D.L., 1995, User's guide to PHREEQC—A computer program for speciation, reaction-path, advective-transport, and inverse geochemical calculations: United StatesGeological Survey, Lakewood, Colorado.

Pauwels, H., Tercier-Waeber, M.-L., Arenas, M., Castroviejo, R., Deschamps, Y., Lassin, A., Graziottin, F., and Elorza, F.-J., 2002, Chemical characteristics of groundwater around two massive sulphide deposits in an area of previous mining contamination, Iberian Pyrite Belt, Spain: Journal of Geochemical Exploration, v. 75, p. 17-41.

Phipps, G.C., Boyle, D.R., and Clark, I.D., 2004, Groundwater geochemistry and exploration methods: Myra Falls volcanogenic massive-sulphide deposits, Vancouver Island, British Columbia, Canada: Geochemistry: Exploration, Environment, Analysis, v. 4, p. 329-340.

Pirlo, M.C., and Giblin, A.M., 2004, Application of groundwater-mineral equilibrium calculations to geochemical exploration for sediment-hosted uranium: observations from the Frome Embayment, South Australia: Geochemistry: Exploration, Environment, Analysis, v. 4, p. 113-127.

Plumlee, G.S., Smith, K.S., Ficklin, W.H., and Briggs, P.H., 1992, Geological and geochemical controls on the composition of mine drainages in mineralized areas: 7th International Water-Rock Interaction Conference, Park City, Utah, July, 1992, 1992, p. 419-422.

Plumlee, G.S., Smith, K.S., Montour, M.R., Ficklin, W.H., and Mosier, E.L., 1999, Geologic controls on the composition of natural waters and mine waters, *in* Filipek, L.H., and Plumlee, G.S., eds., The Environmental Geochemistry of Mineral Deposits Part B: Case Studies and Research Topics: Society of Economic Geologists, Reviews in Economic Geology 6B, p. 373-432.

Pokrovsky, O.S., and Schott, J., 2004, Experimental study of brucite dissolution and precipitation in aqueous solutions: surface speciation and chemical affinity control: Geochimica et Cosmochimica Acta, v. 68, p. 31-45.

Pueyo, J.J., Chong, G., and Jensen, A., 2001, Neogene evaporites in desert volcanic environments: Atacama Desert, northern Chile: Sedimentology, v. 48, p. 1411-1431.

Rech, J.A., Quade, J., and Hart, W.S., 2003, Isotopic evidence for the source of Ca and S in soil gypsum, anhydrite and calcite in the Atacama Desert, Chile: Geochimica et Cosmochimica Acta, v. 67, p. 575-586.

Reith, F., McPhail, D.C., and Christy, A.G., 2005, Bacillus cereus, gold and associated elements in soil and other regolith samples from Tomakin Park Gold Mine in southeastern New South Wales, Australia: Journal of Geochemical Exploration, v. 85, p. 81-98.

Rissmann, C.F.W., 2003, Dilute Hydrothermal Fluids within the Groundwaters of the Monturaqui Basin, northern Chile: The Role of Volcano Socompa: M.Sc. thesis, University of Texas at Dallas, 95 p.

Rouxel, O., Fouquet, Y., and Ludden, J.N., 2004, Subsurface processes at the Lucky Strike hydrothermal field, Mid-Atlantic Ridge: Evidence from sulfur, selenium, and iron isotopes: Geochimica et Cosmochimica Acta, v. 68, p. 2295-2311.

Runnells, D.D., and Lindberg, R.D., 1981, Hydrogeochemical exploration models for sedimentary uranium deposits: use of the computer model WATEQFC: Journal of Geochemical Exploration, v. 15, p. 51-62.

Sader, J.A., Leybourne, M.I., McClenaghan, B.M., and Hamilton, S.M., 2003a, Geochemistry of groundwater from Jurassic kimberlites in the Kirkland Lake and Lake Timiskaming kimberlite fields, northeastern Ontario, Geological Survey of Canada, Open File, 4515, 1 CD-ROM.

Sader, J.A., Leybourne, M.I., McClenaghan, B.M., Hamilton, S.M., and Robertson, K., 2003b, Groundwater interaction with kimberlites – A geochemical investigation in northeastern Ontario: Explore, v. 118, p. 1-4.

Sader, J.A., Leybourne, M.I., McClenaghan, B.M., and Hamilton, S.M., 2007, Low temperature serpentinization processes and kimberlite groundwater signatures in the Kirkland Lake and Lake Timiskiming kimberlite fields Ontario, Canada: Implications for diamond exploration: Geochemistry: Exploration, Environment, Analysis, v. 7, p. 3-21.

Sahoo, N.R., and Pandalai, H.S., 2000, Secondary geochemical dispersion in the Precambrian auriferous Hutti-Maski schist belt, Raichur district, Karnataka, India.Part I: anomalies of As, Sb, Hg and Bi in soil and groundwater: Journal of Geochemical Exploration, v. 71, p. 269-289.

Severmann, S., Johnson, C.M., Beard, B.L., German, C.R., Edmonds, H.N., Chiba, H., and Green, D.R.H., 2004, The effect of plume processes on the Fe isotope composition of hydrothermally derived Fe in the deep ocean as inferred from the Rainbow vent site, Mid-Atlantic Ridge, 36°14'N: Earth and Planetary Science Letters, v. 225, p. 63-76.

Siegel, F.R., 1990, Exploration for Mississippi-Valley type stratabound Zn ores with stream suspensates and stream sediments, Virginia, U.S.A.: Journal of Geochemical Exploration, v. 38, p. 265-283.

Sillitoe, R.H., 2000, Gold-rich porphyry deposits: descriptive and genetic models and their role in exploration and discovery: Reviews in Economic Geology, v. 13, p. 315-344.

Sinclair, W.D., 2007, Porphyry deposits, *in* Goodfellow, W.D., ed., Mineral Deposits of Canada: A Synthesis of Major Deposit-Types, District Metallogeny, the Evolution of Geological Provinces, and Exploration Methods: Geological Association of Canada, Mineral Deposits Division, Special Publication No. 5, p. xx-xx.

Singh, R.V., Sinha, R.M., Bisht, B.S., and Banerjee, D.C., 2002, Hydrogeochemical exploration for unconformity-related uranium mineralization: Example from Palnadu sub-basin, Cuddapah Basin, Andhra Pradesh, India: Journal of Geochemical Exploration, v. 76, p. 71-92.

Stefánsson, A., Gíslason, S.R., and Arnórsson, S., 2001, Dissolution of primary minerals in natural waters II.Mineral saturation state: Chemical Geology, v. 172, p. 251-276.

Taisaev, T.T., and Plyusnin, A.M., 1984, Hydrogeochemical prospecting of gold in an alpine bald mountain zone: Journal of Geochemical Exploration, v. 21, p. 355-360.

Taufen, P.M., 1997, Ground waters and surface waters in exploration geochemical surveys, *in* Gubins, A.G., ed., Proceedings of Exploration 97: Fourth Decennial International Conference on Mineral Exploration: Toronto, Canada, Sept 14-18, 1997, p. 271-284.

Telmer, K., Pass, H.E., and Cook, S., 2002, Combining dissolved, suspended and bed load stream geochemistry to explore for volcanic massive sulfide deposits: Big Salmon Complex, northern British Columbia: Journal of Geochemical Exploration, v. 75, p. 107-121.

Torsvik, T., Furnes, H., Muehlenbachs, K., Thorseth, I.H., and Tumyr, O., 1998, Evidence for microbial activity at the glass-alteration interface in oceanic basalts: Earth and Planetary Science Letters, v. 162, p. 165-176.

Toulhoat, P., and Beaucaire, C., 1991, Comparison between lead isotopes, $^{234}U/^{238}U$ activity ratio and saturation index in hydrogeochemical exploration for concealed uranium deposits: Journal of Geochemical Exploration, v. 41, p. 181-196.

Vlassopoulos, D., and Wood, S.A., 1990, Gold speciation in natural waters: I. Solubility and hydrolysis reactions of gold in aqueous solution: Geochimica et Cosmochimica Acta, v. 54, p. 3-12.

Volesky, J.C., 2002, Massive Sulfide Deposits of the Wadi Bidah Mineral District, Saudi Arabia; Geologic Control of Mineralization, Remote Sensing and Mineral Exploration, Geochemical Exploration and Petrogenesis: Ph.D. thesis, University of Texas at Dallas, 115 p.

Volesky, J.C., Leybourne, M.I., and Boyle, D.R., 2000, Mobility of rare earth elements during weathering of massive sulfide deposits, Bathurst Mining Camp, New Brunswick, Canada: Geological Society of America Annual Meeting, Reno, Nevada, 2000.

Webster, J.G., 1986, The solubility of gold and silver in the system Au-Ag-SO_2-H_2O at 25°C and 1 atm: Geochimica et Cosmochimica Acta, v. 50, p. 1837-1845.

Whitford, D., Andrew, A., Carr, G., and Giblin, A., 1997, Multi-isotopic studies of groundwaters: exploration for concealed mineralization: Exploration and Mining Research News, v. 7, p. 5-7.

Wood, S.A., 2002, The aqueous geochemistry of the platinum group elements with applications to ore deposits: Canadian Institute of Mining and Metallurgy, Special Volume 54, p. 211-249.

Yanase, N., Payne, T.E., and Sekine, K., 1995a, Groundwater geochemistry in the Koongarra ore deposit, Australia (I): Implications for uranium migration: Geochemical Journal, v. 29, p. 1-29.

——— 1995b, Groundwater geochemistry in the Koongarra ore deposit, Australia (II): Activity ratios and migration mechanisms of uranium series radionuclides: Geochemical Journal, v. 29, p. 31-54.

Zanker, H., Moll, H., Richter, W., Brendler, V., Hennig, C., Reich, T., Kluge, A., and Huttig, G., 2002, The colloid chemistry of acid rock drainage solution from an abandoned Zn-Pb-Ag mine: Applied Geochemistry, v. 17, p. 633-648.

ALTERATION VECTORS TO METAMORPHOSED HYDROTHERMAL SYSTEMS IN GNEISSIC TERRANES

ANNE-LAURE BONNET[1] AND LOUISE CORRIVEAU[2]

1. INRS-Eau, Terre, Environnement, 490 de la Couronne, Quebec, Quebec G1K 9A9
2. Geological Survey of Canada, 490 de la Couronne, Quebec, Quebec G1K 9A9
Corresponding author's email: lcorrive@nrcan.gc.ca

Abstract

Alteration mapping is recognized as an effective tool for the discovery of hydrothermal ore deposits in both greenfield and brownfield exploration terranes. Similar to their non- or weakly-metamorphosed equivalents, metamorphosed hydrothermal alteration zones are also key elements in vectoring to ore. However, in gneiss terranes, the recognition of metamorphosed hydrothermal alteration is severely hampered by their resemblance to metasediments and metamorphosed paleosoils. This paper presents strategies and observation protocols to recognize, during fieldwork and with a reasonable degree of confidence, hydrothermal alteration zones that have been metamorphosed to upper amphibolite- and granulite-facies conditions. The geological and mineralogical criteria discussed are useful vectors to mineralization, especially among felsic gneisses. These criteria are adapted from strategies developed for volcanogenic massive sulphide deposits but can be applied to any deposit type where alteration zones are sufficiently large to be recognized during regional mapping, such as epithermal, porphyry, or iron oxide Cu-Au deposits. Once a hydrothermal system is suspected, more detailed geological mapping can reveal enough information to determine the intensity and the spatial distribution of the system, establish targets for systematic lithogeochemical and geophysical exploration, and test different metallogenic models, thereby increasing the probability of discovery. A proposed practical guide is included on the DVD at the back of this volume to further develop concepts presented in this paper. It includes some case examples and a photo atlas to illustrate rock types, outcrop aspects, mineral assemblages, and textures of alteration zones and their potential host volcanic rocks.

Résumé

La cartographie de zones d'altération est reconnue comme un outil efficace d'exploration pour la découverte de gîtes hydrothermaux en contextes d'exploration et de cartographie régionales ainsi qu'au sein de camps miniers. À l'instar de leurs équivalents peu ou faiblement métamorphisés, les zones d'altération hydrothermale métamorphisées représentent elles aussi des vecteurs de minéralisation. Toutefois, leur identification lors de la cartographie de terrains gneissiques est sévèrement entravée par leur ressemblance avec des roches métasédimentaires et des paléosols métamorphisés. Cet article présente des stratégies et des clés d'observation pour reconnaître, sur le terrain et avec un degré de confiance raisonnable, les zones d'altération hydrothermale métamorphisées aux conditions du faciès des amphibolites supérieur et du faciès des granulites. Les critères géologiques et minéralogiques discutés fournissent des vecteurs pour la découverte de minéralisations, particulièrement au sein de vastes étendues de gneiss felsiques. Ils s'appuient sur des stratégies développées pour les gîtes de sulfures massifs volcanogènes, mais sont aussi applicables à tout type de gîte qui s'accompagne de zones d'altération hydrothermale cartographiables à l'échelle régionale tels que les gîtes épithermaux, les gîtes porphyriques ou les gîtes d'oxydes de fer-Cu-Au. Une fois la présence d'un système hydrothermal supposée, une cartographie plus détaillée permet de déterminer l'intensité et la configuration du système, d'établir des cibles pour des programmes d'exploration lithogéochimique et géophysique systématiques, de mettre à l'essai différents modèles métallogéniques et, ce faisant, d'augmenter la probabilité de découverte. Un guide pratique a été ajouté sur le DVD joint au présent volume afin de développer plus en profondeur les concepts abordés dans cet article. Il comprend un atlas photographique et des exemples types afin de bien illustrer les types de roche, les aspects mégascopiques, les associations minérales et les textures des zones d'altération et de leurs roches hôtes volcaniques potentielles.

Introduction

Felsic volcano-plutonic environments represent first-order targets for mineral exploration of volcanogenic massive sulphide Cu-Au (VMS Cu-Au), epithermal, porphyry Cu (Mo-Au), and iron oxide Cu-Au (IOCG) deposits (e.g. Ohmoto, 1996; Corbett and Leach, 1998; Hedenquist et al., 2000; Sillitoe, 2000, 2003; Large et al., 2001). Once metamorphosed at high grade, these settings form large gneissic terranes, some hosting deposits with well documented metamorphosed alteration zones that provide effective vectors to ore (Table 1; Hodges and Manojlovic, 1993). Among those are the commonly recognized indicators for metamorphosed VMS deposits, the chloritic alteration zones expressed as cordierite-anthophyllite schists at amphibolite facies and the meta-exhalites (Tables 1, 2; Schreurs and Westra, 1985; Trägårdh, 1991; Bernier, 1992; Spry et al., 2000 and references therein).

Sericitic, argillic, and advanced argillic alteration zones are also key alteration types of many deposits, including VMS deposits (Tables 1, 2). However, as their gneissic derivatives commonly resemble metapelites or metamorphosed paleosoils (Bonnet et al., 2005), challenges remain concerning their recognition among gneissic terranes where mineralization has not been found or explored for (e.g. Allard, 1978; Froese, 1985, 1998). This can also be true within ore deposits themselves. For example, early papers on the Challenger gold deposit (e.g. Tomkins and Mavrogenes, 2002) interpreted the host to mineralization as metapelites, whereas more recent work is highlighting hydrothermal protoliths with a volcanic precursor and an epithermal origin (C. McFarlane, written comm., 2006; Table 1). Though the mineral potential of gneissic terranes and the importance of adapting alteration vectors for their successful exploration have been recognized for decades (Allard, 1978; Gauthier et

Bonnet, A-L., and Corriveau, L., 2007, Alteration vectors to metamorphosed hydrothermal systems in gneissic terranes, *in* Goodfellow, W.D., ed., Mineral Deposits of Canada: A Synthesis of Major Deposit-Types, District Metallogeny, the Evolution of Geological Provinces, and Exploration Methods: Geological Association of Canada, Mineral Deposits Division, Special Publication No. 5, p. 1035-1049.

TABLE 1. Examples of metamorphosed hydrothermal deposits and occurrences in gneiss terranes from mostly the Grenville and Superior provinces.

Name	Locality	Metamorphosed Alteration	Protolith	Precursor	Age (Ga)	Deposit Type	Metamorphic Facies	Geodynamic Context	References
Bondy Gneiss Complex	Grenville Province (Central Metasedimentary Belt, southwest Quebec)	Kfs, Bt-Grt, Crd-Opx, Sil-Grt, Sil-Qtz, or Bt-Opx-Mag gneiss; garnetite; albitite; magnetitite; calc-silicate rock; tourmalinite; Cu and Au mineralization	Potassic, sericitic, chloritic, argillic, advanced argillic, sodic and Fe-oxide alteration zones	Rhyolitic to basaltic volcanic rocks	1.4-1.3	Cu-Au VMS and/or Fe oxide hydrothermal system	Granulite	Volcanism and plutonism in a back-arc and island arc setting developed on a thin continental substrate of Laurentia	Corriveau et al., 1997; Blein et al., 2003, 2004
La Romaine Supracrustal Belt	Grenville Province (Natashquan domain, southeast Quebec)	Qtz-Crd gneiss, migmatized Grt-Bt-Sil gneiss, Qtz-Sil-Ms-Feox nodules and veins in quartzofeldspathic gneiss, ironstone, carbonate, and calc-silicate rocks, garnetite, diopsidite, epidosite, Cu mineralization	Sericitic, advanced argillic, silicic, Ca-Fe alteration zones	Felsic-intermediate pyroclastic deposits, mafic lava, mafic sill	1.5	High-sulphidation VMS-type hydrothermal system	Granulite	1.5 Ga volcanism and sedimentation within an intra continental arc rift (Pinwarian)	Corriveau and Bonnet, 2005; Bonnet et al., 2005
Challenger	Gawler craton, Australia	Aluminous gneiss quartz-rich gneiss	Sericitic alteration?	Volcanic	>2.5	Epithermal?	Granulite	Uncertain	C. McFarlane, written comm., 2006
Montauban	Grenville Province (Mauricie-Portneuf sector, central Québec)	Nodular Sil gneiss, Crd-Ath gneiss, Grt-Ghn quartzitic gneiss, Au-Ag mineralization	Chloritic, chlorite-sericite alteration zones, siliceous exhalite	Felsic and mafic volcanic rocks, sediments	1.4	VMS	Upper Amphibolite	Island arc (back-arc) volcanism and hydrothermal activity	Bernier and MacLean, 1993
Prieska	Namaqua Province (South Africa)	Ged gneiss, Qtz-perthite-Sil gneiss, Cal-Anh-Brt rock	Chloritic alteration, silicification, and sericitization	Dacitic tuff, sediments	1.3	VMS	Granulite	Back-arc magmatism	Theart et al., 1989
Ruostesuo	Vihanti-Pyhäsalmi district, central Finland	Crd-Oam-Grt-bearing gneiss	Chloritic alteration	Felsic volcanic rocks, basalt and basaltic andesite	1.9-2.0	VMS	Granulite	Rifting of a primitive oceanic island arc	Roberts et al., 2003
Coulon Belt	Superior Province (north-west Quebec)	Sil nodular gneiss; Crd-Ath and Bt gneiss	Chloritic and sericitic alteration zones	Felsic and mafic volcanic rocks; sedimentary rocks	?	VMS	Granulite		Archer et al., 2004; Huot et al., 2004
Geco	Superior Province (Quetico, Ontario)	Oam-Crd-Grt gneiss, Sil-Ms-Qtz schist, iron formation	Chloritic and sericitic alteration zones, sea-floor precipitate / near-seafloor replacement	Felsic volcanic rocks	2.7	VMS	Upper Amphibolite		Zaleski and Peterson, 1995

Abbreviations from Kretz (1983), except Ghn = gahnite, Oam = orthoamphibole, Feox = iron oxide.

al., 1985; Allard and Carpenter, 1988; Thompson et al., 1997; Spry et al., 2000), it is through the routine recognition of all gneissic alteration types that effective diversification of mineral exploration will be achieved from the known, greenstone belt-hosted, mining camps into the largely uncharted gneissic and granitic belts of shield terranes, the 'pinkstone belts'.

This paper presents a strategy for the recognition and charac-terization of metamorphosed hydrothermal alteration zones, in particular fossil sericitic, argillic, and advanced argillic alteration zones, that can point to potential VMS-type mineralization in gneissic paleo-volcanoplutonic belts (e.g. Boliden: Hallberg, 2001; Gossan Hill: Sharpe and Gemmell, 2002). The strategy is based on methodical use of geological and mineralogical criteria, and observation protocols to prognosticate potentially fertile area during fieldwork and from existing reports and maps. Once a hydrothermal system is identified in the field with a reasonable degree of confidence and mapped appropriately, exploration programs can follow their usual course. Detailed discussion and illustration of method and case examples are provided in the accompanying photo atlas and guide (Bonnet and Corriveau, 2007).

Table 2. Diagnostic mineralogy and geochemistry of greenschist- and granulite-grade metamorphosed alteration associated with volcanogenic massive suphide deposits.

Alteration type	Diagnostic minerals greenschist facies	Diagnostic minerals granulite facies	Diagnostic composition	Similar rocks (at granulite facies)
Advanced argillic	Kaolinite, Pyrophyllite, Andalusite, Corundum, Topaz	Sillimanite, Kyanite, Quartz	Al_2O_3, SiO_2	Laterite
Argillic	Sericite, Illite, Pyrophyllite	Sillimanite, Kyanite, Quartz, Biotite, Cordierite, Garnet	Al_2O_3, SiO_2, K_2O, Fe_2O_3, MgO	Pelite
Sericitic	Sericite, Illite, Quartz	Biotite, K-feldspar, Sillimanite, Kyanite, Quartz, Cordierite, Garnet	K_2O, Al_2O_3, Fe_2O_3, ±MgO, $\pm SiO_2$	Pelite
Chloritic	Chlorite, Quartz, Sericite	Cordierite, Orthopyroxene, Orthoamphibole, Phlogopite, Sillimanite, Kyanite	Fe_2O_3, MgO, $\pm Al_2O_3$, $\pm SiO_2$	Pelite
Carbonate Propylitic	Carbonate (Fe, Mg), Epidote, Chlorite, Sericite, Feldspar	Carbonate, Grossular Epidote, Hornblende, Diopside, Orthopyroxene	Fe_2O_3, CaO	Calc-silicate rock of sedimentary origin, marble or mafic rock

Gneissic Terranes as Exploration Targets

The identification of mineralized volcanic belts within gneissic terranes is a constant reminder that these settings include tectonically juxtaposed surface and near-surface environments that can be as prospective as any other ones that were never deeply buried during their evolution (Fig. 1A, Table 1; Zaleski and Peterson, 1995; Archer et al., 2004). However, there remains a lingering perception that they represent large crustal segments formed at and exhumed from great depth where large-scale hydrothermally derived ore deposits tend not to be formed (Fig. 1B) and where the likelihood of preserving high-level deposits such as epithermal, porphyry Cu, and high-sulphidation VMS deposits is considered poor due to rapid denudation (Sillitoe, 1973; Gustafson and Titley, 1978). That porphyry and epithermal deposits are being found in ancient terranes, even at granulite facies, is a reminder that tectonic burial may be so effective as to preserve high-level mineral deposits (Dubé et al., 1998; Kontak et al., 2002). A last but not least problem is that partial melting, remobilization, and obliteration of primary textures and structures associated with regional high-grade metamorphism and deformation are, in some cases, sufficiently significant to conceal the presence of entire volcanic belts and VMS-type hydrothermal systems within undifferentiated gneiss complexes (Blein et al., 2003; Corriveau and Bonnet, 2005). Initial lithostratigraphic geometry and primary textures and structures of volcanic belts can, however, be preserved in areas of low strain, where metamorphic events were transient or in

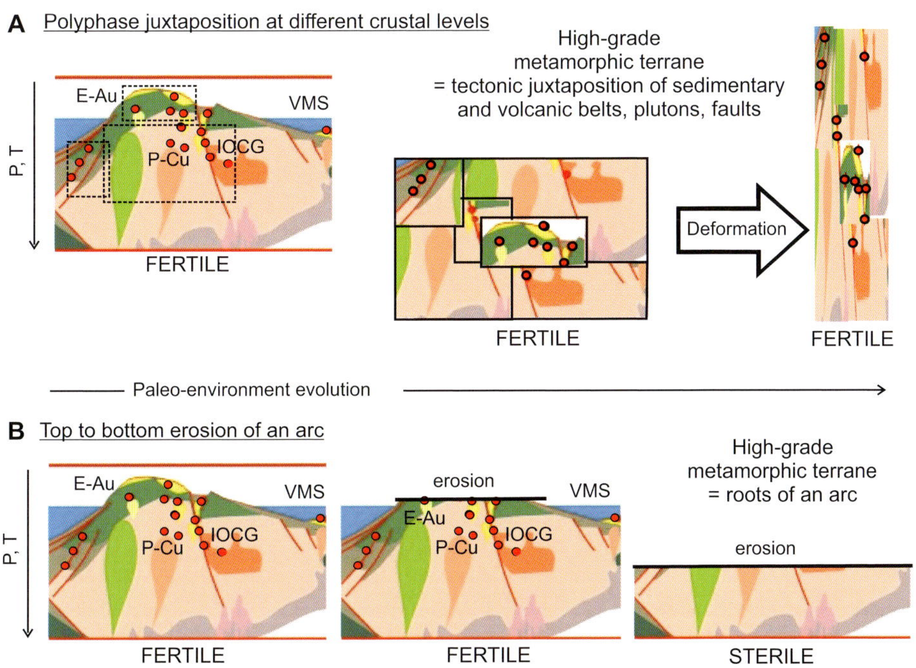

Figure 1. Schematic diagrams of a magmatic arc metamorphosed at granulite facies. **(A)** Settings that include fertile surface and near-surface volcanoplutonic belts that have been tectonically buried relatively intact though thoroughly metamorphosed and juxtaposed to other settings during orogenesis. Red dots locate potential settings for epithermal gold (E-Au), porphyry copper (P-Cu), volcanogenic massive sulphide (VMS Cu-Au), and iron oxide Cu-Au-U-REE (IOCG) deposits. **(B)** Top to bottom erosion of an arc and its impact on the preservation of different deposit types.

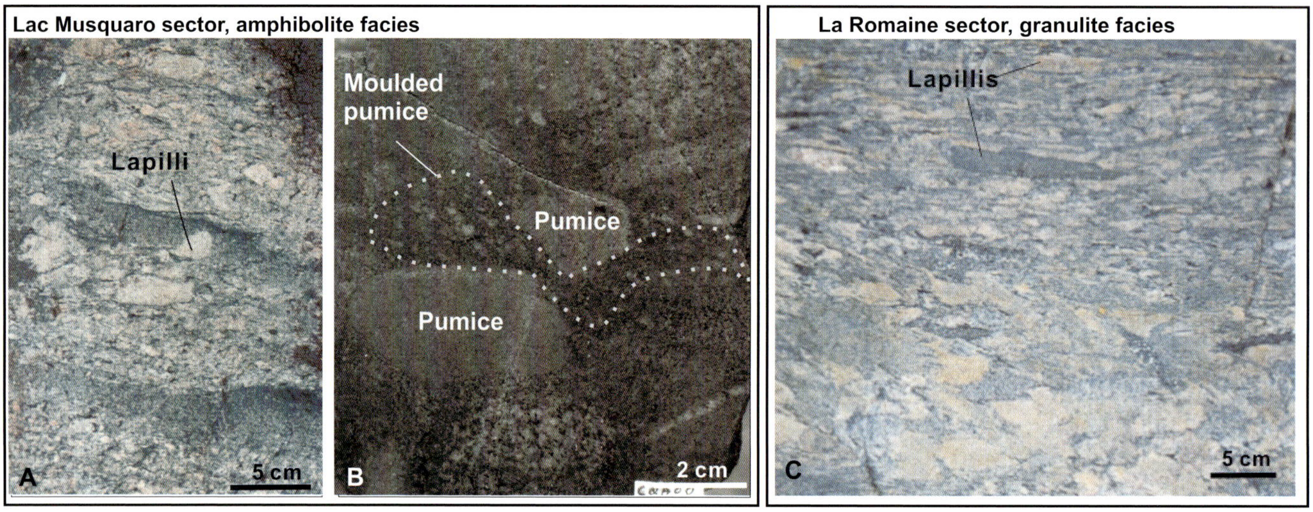

FIGURE 2. Volcaniclastic and pyroclastic rocks (1.5 Ga) at amphibolite and granulite facies from the Musquaro Lake and the La Romaine area in the eastern Grenville Province, Quebec (location in Fig. 1 of Corriveau et al., 2007). **(A)** Polygenic lapillistone. **(B)** Primary moulding of pumices of intermediate composition around more competent felsic lapilli; a texture diagnostic of a pyroclastic origin and proximal volcanic environment. **(C)** Polygenic lapillistone at granulite facies from La Romaine.

cases that competency was particularly uniform or high (e.g. Nadeau et al., 1999; Mueller and Corcoran, 2001; Gauthier et al., 2004). Such preservation can extend to very delicate textures, including those of quartz phenocryst (Stevens and Barron, 2002), phenoclast, lapilli (Fig. 2A), pumice (Fig. 2B), and scoria (Fig. 2C), not only at amphibolite facies but also at granulite facies (Figs. 2C, 3A; Corriveau and Bonnet, 2005). Where observed, relic primary textures permit the identification of metamorphosed volcanic environments during regional mapping and subsequent lithogeochemical interpretation of paleo-environments (Slagstad et al., 2004; Bonnet et al., 2005; LaFlèche et al., 2005). High-grade metamorphism and deformation may also be beneficial in terms of mineral potential by concentrating ore in structural traps and increasing grain size and purity (Gauthier, 1993; Marshall et al., 2000; Mavrogenes et al., 2001; Gauthier and Chartrand, 2005).

Influence of Metamorphism on Alteration Zones

Hydrothermal fluids involved in the genesis of VMS deposits are neutral to acid (pH 7 to 2), hot (T = 150 and 400ºC), and derived mostly from seawater (e.g. VMS) with or without a magmatic component (Lydon, 1988; Yang and Scott, 1996). As they circulate within the Earth's crust, hydrothermal fluids react with country rocks through mineral leaching and chemical precipitation and, due to their high fluid/rock ratios, may lead to the formation of significant alteration zones. These include chloritic (+Mg, +Fe, -Ca, -Na, -K), sericitic and argillic (+K, +Mg, ±Fe, -Ca, -Na), advanced argillic (+Al, +Si, -Fe, -Mg, -Na, -Ca, -K), and propylitic or carbonate (+Fe, +CO_2, +Mg, -Na) alteration. Through the chemical changes associated with hydrothermal leaching and precipitation, the composition of alteration zones and consequently their mineral assemblages commonly acquire a diagnostic character that differs from that of their unaltered precursor (Galley, 1995; Large et al., 2001). Following high-grade metamorphism and despite the associated obliteration of primary mineralogy and dehydration, rock compositions remain largely unchanged except in rare cases where anatectic melt remobilization is significant during partial melting (Sawyer, 2001) or where hydrothermal alteration took place during metamorphism through external influx of fluids. Consequently, in most cases, metamorphosed alteration zones preserve the imprint of the chemical changes that the original protolith has undergone during premetamorphic hydrothermal processes (Trägårdh, 1991; Hannington et al., 2003; Roberts et al., 2003; Blein et al., 2004; Stanton, 2004).

Rock composition and physical conditions, such as pressure and temperature (e.g. Rosenberg et al., 2000) constrain mineral assemblages, mineral composition, and modal mineral abundance of metamorphic rocks, including those forming metamorphosed hydrothermal alteration zones and those affected by synmetamorphic sulphidation and oxidation processes. As pressure and temperature change, mineral phases react with each other to form new mineral assemblages. Such mineralogical changes obey the phase rule; hence the pressure-temperature stability field of different metamorphic assemblages can be predicted, and experimentally reproduced and calibrated for the most common rock compositions and then graphically represented using petrogenetic grids and phase diagrams (AFM, A'CF, A'KF, etc; see Chapter 5 in Spear, 1993; Chapter 3.2 in Bonnet and Corriveau, 2007). The plotting of the composition of altered rocks and unaltered precursors in phase diagrams allows their respective stable mineral assemblages and modal mineral abundance to be determined. As grain size is enhanced by high-grade metamorphism, mineral assemblages and modal mineral abundance can be established routinely during fieldwork, thereby serving as a user-friendly tool to recognize different types of alteration during mapping. This is done by positioning the observed rock assemblage in a chemically appropriate chemographic phase diagram and visualizing potential chemical departure of a suspected hydrothermally altered rock from the composition of a 'normal' protolith (Fig. 3, Tables 2, 3). This exercise provides a means of identifying hydrothermal alteration in the field, a strategy that can be greatly facilitated where primary tex-

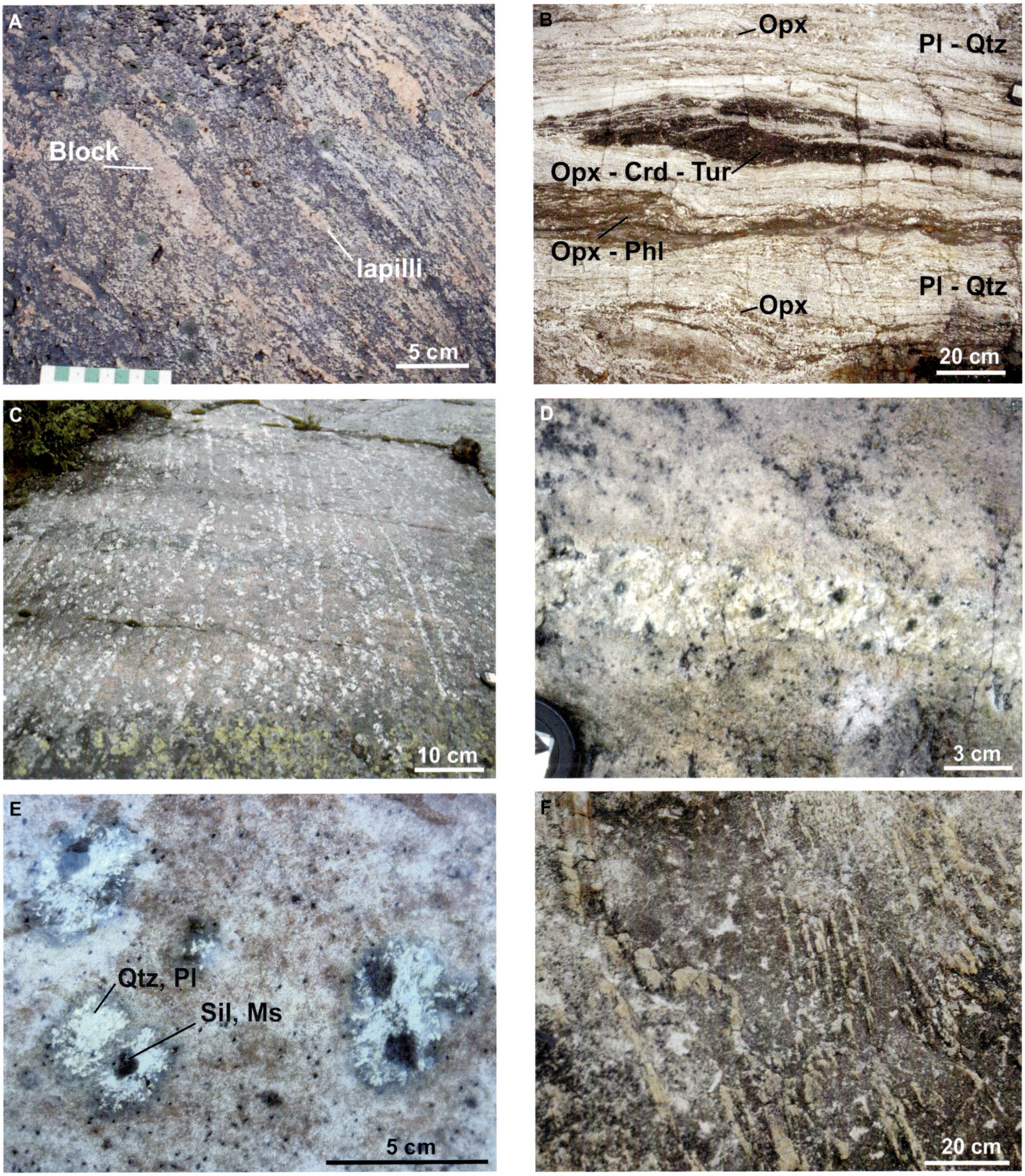

FIGURE 3. Photographs of metamorphosed argillic (**A**), chloritic (**B**), and sericitic (**C** to **F**) alteration types. (**A**) Granulite-facies biotite-garnet-sillimanite gneiss with preserved fragments interpreted as volcanic, observed in the alteration zones of the Bondy Gneiss Complex (location in Fig. 1 of Corriveau et al., 2007). Note that the volcanic precursor may have been similar to that of the lapillistone in Figure 2C. Relict textures are particularly difficult to find in the polydeformed Bondy Gneiss Complex (Harris et al., 2001). (**B**) Granulite-facies cordierite-orthopyroxene-bearing white gneiss, observed in the Bondy Gneiss Complex (Blein et al., 2004). (**C**) Network of nodular aluminous veins in a tuffaceous unit metamorphosed to upper amphibolite facies in the Musquaro Lake area. (**D**) Detail of a nodular aluminous vein seen in photo **C**. (**E**) Aluminous nodules with fibrolite-rich cores and feldspar-rich rims disseminated within a tuffaceous unit in the Musquaro Lake area. (**F**) Network of aluminous veins in a tuffaceous unit metamorphosed to granulite facies in the La Romaine area. Abbreviations: Crd=cordierite, Krn=kornerupine, Ms=muscovite, Opx=orthopyroxene, Pl=plagioclase, Qtz=quartz, Sil=sillimanite.

tures of the precursor rocks are also preserved, such as in Figure 3A, in contrast to the more current situation where they are not (Fig. 3B).

Above upper amphibolite facies, partial melting and leucosome formation become intrinsic parts of metamorphic reaction products (Fig. 4). Under these conditions, the mineral assemblage of the leucosomes is an important tracer of the metamorphic reactions that took place in a rock, information that in turns permits the in situ character of observed granitic veins versus those extraneous to the host rock to be determined with a reasonable degree of confidence. Appraisal of the extent of melting and melt extraction is important to distinguish altered protolith from restites derived from metasediment, as discussed below. It is also important during sampling for geochemistry. Of particular interest is mild to intense depletion of Na or K or enrichment of Al or Si during premetamorphic hydrothermal activity since these chemical changes increase the temperature at which significant melting may occur and reduce metamorphic remobilization of not only melt but also potentially of premetamorphic ore. The lowering of melting temperatures also increases the probability of preserving primary textures (e.g. Fig. 3A), which facilitates the study and lithogeochemical sampling of protoliths.

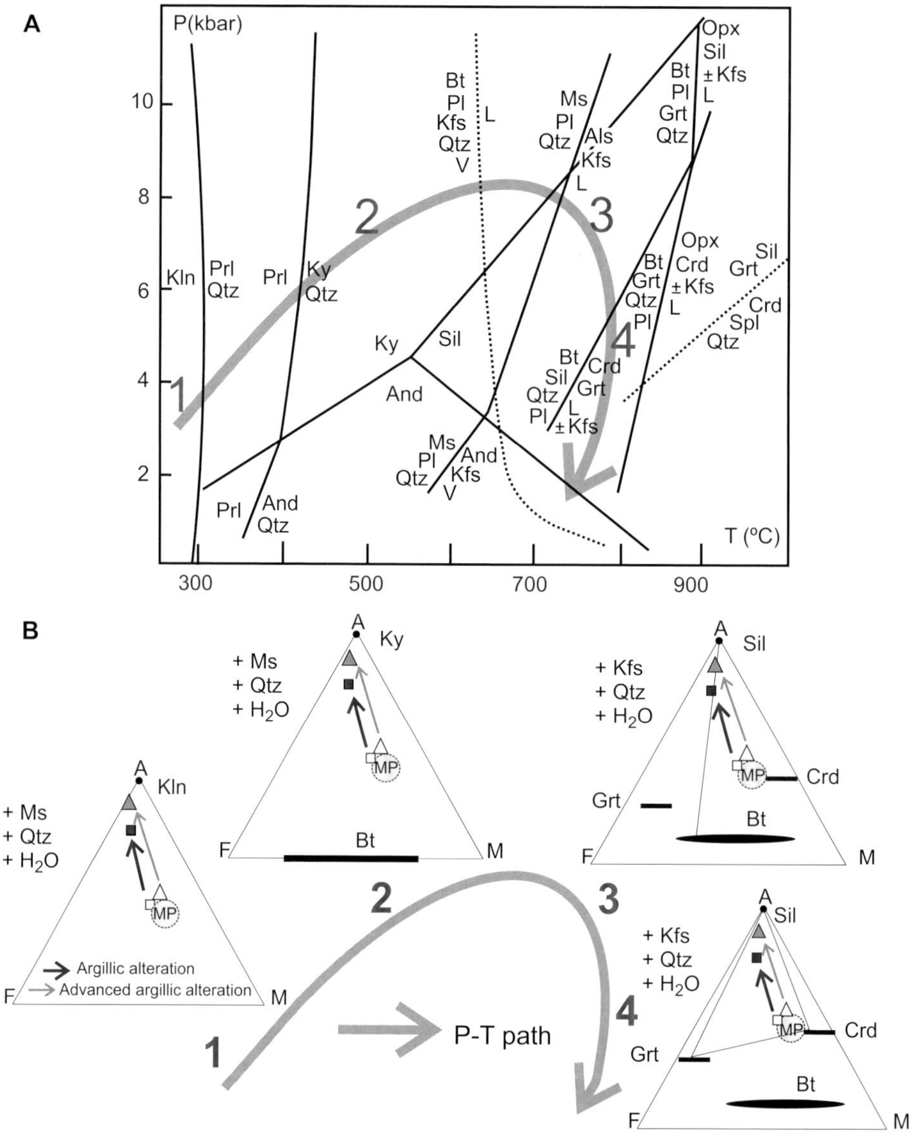

FIGURE 4. Petrogenetic grid for metapelite-like rocks showing the modelled pressure-temperature path for an argillic (dark grey square) and an advanced argillic (medium grey triangle) alteration types. The examples plotted here are from the Maronia porphyry Cu-Mo deposit (square; data of Melfos et al., 2002) and the Bondy Gneiss Complex (triangle; data of Blein et al., 2003, 2004). The composition and the mineralogy of these alteration types are represented in AFM diagrams (A = Al_2O_3-K_2O, F = FeO, M = MgO; molar proportion) at stages 1, 2, 3, and 4 of the pressure-temperature path (projection from muscovite for stages 1 and 2, and from K-feldspar for stages 3 and 4). In these AFM diagrams, the composition of the hydrothermally altered rock is compared with that of their precursor (empty square and triangle) and of metapelites (MP). The position of samples will only change from one diagram to the other if the projection is done from a different mineral as it is the composition that ultimately governs where samples plot on these diagrams. In contrast, the stability of minerals changes at each stage, governed by metamorphic reactions that are represented in the petrogenetic grid as calibrated by Spear and Cheney (1989) and Pattison et al. (2003). Mineral abbreviation is after Kretz (1983). L = liquid.

Advanced Argillic, Argillic, Chloritic, and Sericitic Alteration Zones

The principal oxides present in advanced argillic, chloritic, and sericitic alteration zones are SiO_2-Al_2O_3-FeO-MgO-K_2O-H_2O. In these alteration zones, CaO and Na_2O are very low in abundance and exert little influence on the overall mineral assemblage (beyond hindering the formation of Ca and Na minerals). As a result, the effects of metamorphism can be predicted using grids and phase diagrams designed for metapelites (aluminous metasedimentary rocks). These grids include those calibrated for the systems SiO_2-Al_2O_3-H_2O (ASH) and SiO_2-Al_2O_3-(FeO or MgO)-K_2O-H_2O (KFASH, KMASH, KFMASH; Fig. 3.1 in Bonnet and Corriveau, 2007). The ASH-type grids and relevant phase diagrams are used to model metamorphic evolution of advanced argillic alteration, which is characterized by chemical compositions that are low in Fe, Mg, and K. In contrast, the KFASH grids are useful to portray argillic and sericitic alteration that is characterized by low Mg but high Fe and K. Finally the KMASH and KFMASH systems may be used to better characterize metamorphosed chloritic alteration zones that are rich in Mg and, in some cases, Fe.

Modeling of argillic alteration mineralogical variations as a function of pressure and temperature is shown in Figure 4 using a particularly well documented argillic alteration of the outer margin of a porphyritic microgranite hosting the Maronia porphyry Cu-Mo deposit (Melfos et al., 2002). The

Table 3. Main indicator minerals of high-grade metamorphosed hydrothermal alteration.

Mineral	Potential indicator of
Orthoamphibole $(Mg,Fe)_7(SiO_{22})(OH,F)_2$ = Anthophyllite, Gedrite	Fe-Mg alteration
Orthopyroxene $(Mg,Fe)\ SiO_3$	
Cordierite $Al_3(Fe, Mg)_2(Si_5AlO_{18})$	Mg alteration
Talc $Mg_6(Si_8O_{20})(OH)_4$	
Magnetite $Fe_2^{3+}Fe^{2+}O_4$	Fe alteration
Hematite Fe_2O_3	
Biotite $K(Mg,Fe)_3AlSi_3O_{10}(OH)_2$	K alteration in particular in mafic rocks
Muscovite $K_2Al_4(Al_2Si_6O_{20})(OH,F)_4$	
Phlogopite $KMg_3(AlSi_3O_{10})(OH)_2$	K and Mg enrichment in aluminous gneiss
Phengite $K_2(Al,Mg)_4(Al_2Si_6O_{20})(OH,F)_4$	
Sillimanite Al_2SiO_5	Al influx or severe leaching of more mobile elements
Topaz $Al_2(SiO_4)(OH, F)_2$	
Spinel $(Mg, Fe, Zn)Al_2O_4$	
Garnet $(Fe, Mg, Mn)_3\ Al_2Si_3O_{12}$	Al enrichment through leaching of more mobile elements
Carbonates $Ca(Fe,Mg)(CO_3)_2$	Ca alteration
Epidote $Ca(Al,Fe)_3(OH)(SiO_4)_3$	
Diopside $CaMg(SiO_3)_2$	
Grossular $Ca_3Al_2Si_3O_{12}$	
Plagioclase Na-Ca(Si_3AlO_8)	Ca or Na alteration
Tourmaline $NaSi_6O_{10}(BO_3)_3(OH,F)_4$	B enrichment
Sulphides	S enrichment
Rutile TiO_2	Ti enrichment
Barite $BaSO_4$	Ba enrichment
Fluorite CaF_2	F enrichment
Apatite $Ca_5(PO_4)_3(F,Cl,OH)$	P enrichment
Monazite $(PO_4)(Ce, La, Di)$	P and REE influx or leaching of more mobile elements
Zircon $Zr(SiO_4)$	Zr influx or leaching of more mobile elements

altered rocks are metamorphosed to a low-grade mineral assemblage dominated by kaolinite, quartz, and in lesser amount sericite (Stage 1 in Fig. 4). These rocks are enriched in Si, Al, and K but depleted in Fe, Ca, Na, and Mg. The chemical composition of the alteration is distinct from that of its precursor and is significantly more aluminous than that of a pelite (Fig. 4). During prograde metamorphism, kaolinite breaks down to pyrophyllite and subsequently to kyanite, in both cases forming metamorphic quartz as a product (from Stage 1 to Stage 2 in Fig. 4) (Hemley et al., 1980). If the temperature increases further, sillimanite replaces kyanite between Stage 2 and Stage 3 (e.g. Powell et al., 1999). At upper-amphibolite-facies temperature conditions, muscovite in the presence of quartz is destabilized to form K-feldspar and sillimanite with a granitic leucosome (Stage 3 in Fig. 4). Available Fe and Mg in the system are largely deposited as biotite. Garnet or cordierite can also be stable depending on their composition, although they are unstable together under upper-amphibolite-facies conditions. The modal proportion of sillimanite in the hydrothermally altered rock will be particularly high, reflecting the aluminous composition of the protolith. In fact, the modal value of sillimanite will be higher than that of a metapelite as the Al contents is significantly higher than that of average composite shale (Gromet et al., 1984) and pelite (Fig. 4) (using the lever rule to calculate modal proportion of AFM phases, sillimanite/biotite would be in a ratio of 3 to 1 for the Maronia argillic alteration). Between Stage 3 and Stage 4, the biotite originally present reacts with sillimanite to form garnet and cordierite until one of the reactants disappears (Stage 4 in Fig. 4). A granitic melt is formed and will commonly host the garnet and cordierite produced. With sillimanite being in excess, biotite will be consumed and the stable mineral assemblage will consist of sillimanite, K-feldspar, quartz, garnet, and cordierite (Table 2).

Advanced argillic alteration types consist predominantly of Si and Al, although K, Fe, Mg, Na, and Ca are present in minor amounts. In low-grade metamorphic terranes, the dominant minerals are quartz, kaolinite, dickite, and alunite. During prograde metamorphism, the aluminous hydrothermal minerals are destabilized to form, according to the prevailing pressure-temperature conditions, pyrophyllite (during Stage 1, Fig. 4), andalusite (if at low pressure, Fig. 4), kyanite (along the pressure-temperature path of Fig. 4, reaching Stage 2), or sillimanite (Stage 3, Fig. 4). The aluminosilicates are very abundant compared to other AFM minerals, as the composition of the rock is particularly rich in Si and Al (Fig. 4). Biotite in Stage 1 to Stage 3 is in a proportion of 1 to 9 with respect to sillimanite. At granulite facies, the assemblages will be dominated by sillimanite and quartz (Table 2). In contrast to argillic alteration, garnet and cordierite would be absent or very minor in abundance, as the system does not contain enough Fe and Mg to form significant amounts of these minerals (Stage 4, Fig. 4); biotite would have completely broken down. Thus, at granulite facies, metamorphosed advanced argillic alteration forms very atypical quartz- and sillimanite-rich rocks, such as those observed at the granulite-facies Challenger gold deposit (Tomkins and Mavrogenes, 2002) and in the hydrothermal system of the Bondy Gneiss Complex (Blein et al., 2003, 2004). Topaz, allanite, and rutile can also be present in such peraluminous rocks and are associated with iron oxides and sulphides. They are not, however, a product of reactions in 'normal' pelite and consequently are not reported on Figure 4 used to conceptually model the evolution of aluminous alteration zones in this example.

In low-grade metamorphic terranes, chloritic alteration (Si+Al+Mg+Fe± K) forms a ferromagnesian schist dominated by chlorite, quartz, sericite, and minor amount of sulphides or tourmaline. The petrogenetic grids calibrated for the KFMAS system indicate that muscovite, phlogopite, and quartz appear above 425°C from the breakdown of chlorite and sericite (Wang et al., 1986; Trägårdh, 1991). Assemblages with cordierite+phlogopite, and subsequently cordierite+K-feldspar or cordierite+orthoamphibole (anthophyllite, gedrite)±garnet appear following the destabilization of chlorite, muscovite, and phlogopite (Schade et al., 1989; Trägårdh, 1991). If there is enough Al and Fe in the system, chlorite reacts with muscovite and garnet to form staurolite. Staurolite-bearing rocks are recognized in the Linda deposit (Zaleski et al., 1991) whereas rocks with cordierite+quartz+orthoamphibole+garnet+phlogopite are observed in the alteration halos surrounding VMS deposits in the mining district of Vihanti-Pyhäsalmi and Ruostesuo in Finland (Roberts et al., 2003) or with the Prieska deposit in South Africa (Theart et al., 1989). Rocks with anthophyllite+garnet+orthopyroxene-sulphide are associated also with the Coulon prospect in Quebec (Huot et al., 2004). These rocks are interpreted as chloritic alteration products of felsic or mafic volcanic rocks, metamorphosed at upper amphibolite-granulites facies. Finally at granulite facies, the orthoamphibole is destabilized completely and garnet-cordierite-orthopyroxene or cordierite-orthopyroxene assemblages are formed (Schreurs and Westra, 1985; Blein et al., 2004; Fig. 3B; Table 2, Plate X in Bonnet and Corriveau, 2007). Moore and Waters (1990), Bernier and MacLean (1993), and Hallberg (1994) suggest that cordierite-phlogopite and phlogopite-rich gneisses are diagnostic of magnesian alteration zones.

Sericitic alteration (Si, Al, K, and Fe or Mg) metamorphosed at low grade consists mainly of sericite, quartz, and sulphides. These minerals are destabilized during prograde metamorphism to form biotite, K-feldspar, aluminosilicates, and quartz at amphibolite and granulite facies (Table 2). Theart et al. (1989) illustrate that, at the Prieska deposit, aluminous gneisses with a quartz, K-feldspar, and sillimanite assemblage (containing 70-80% SiO_2) correspond to silicified and sericitized sediments. Gneisses with sillimanite nodules near the Montauban (Grenville Province), Coulon and Geco (Superior Province) deposits and prospects are interpreted as metamorphosed sericitic alteration zones (Bernier and MacLean, 1993; Zaleski and Peterson, 1995; Huot et al., 2004). Such nodular rocks are also found within the volcanic units of the La Romaine Supracrustal Belt and Musquaro extension of the Wakeham Group (Fig. 3C-F; Plate XI in Bonnet and Corriveau, 2007; Corriveau et al., 2007). The spatial association of such nodular units with mineralized zones at Montauban led Gauthier et al. (1985) to use these units as lithological vectors to volcanic-hosted Au mineralization. They are now being recognized at the Challenger deposit providing further evidence for significant premetamorphic hydrothermal activity at this Au deposit (C. McFarlane, written comm., 2006). Similar nodular units, formally interpreted as diatexites (Perreault and Heaman, 2003), may represent metamorphosed sericitic alteration and silicification, and provide a potential target for VMS exploration in the Brador's Bay supracrustal belt (eastern Grenville Province) (Corriveau et al., 2007; Plate X in Bonnet and Corriveau, 2007).

Carbonate and Propylitic Alteration Zones

Carbonate and propylitic alteration is composed of variable proportions of Ca, Al, Mg, Fe, and Si. The petrogenetic grids used to predict the mineral assemblages formed during their metamorphism are those calibrated for carbonate or mafic rocks (Fig. 3.1 in Bonnet and Corriveau, 2007). In low-grade metamorphic terranes, the dominant minerals are chlorite, various carbonates, epidote, plagioclase, and quartz. At amphibolite facies, amphibole (actinolite and at higher grade hornblende) and anorthite appear from the breakdown of chlorite and epidote, although epidote can remain stable under high pressure and temperature conditions in rocks with a very calcic composition (Spear, 1993). At the transition between upper amphibolite and granulite facies, hornblende and quartz are destabilized and a new mineral assemblage of orthopyroxene, clinopyroxene, and plagioclase appears. In Fe-rich systems, Fe-rich silicates, such as ferrous garnet and clinopyroxene, prevail (Spear, 1993) whereas in Ca-rich systems, Ca-rich silicates, such as grossular, diopside, and epidote, are dominant (Table 2). Chemical changes associated with carbonate or propylitic alteration are not very large, and the composition of these alteration types is commonly very similar to that of calc-silicate rocks of sedimentary or metasomatic origin (Pan and Fleet, 1992). Thus high-grade metamorphosed carbonate or propylitic alteration is difficult to identify in the field and is not commonly documented in the literature.

Limits Concerning the Use of the Petrogenetic Grids

Petrogenetic grids must be used carefully to interpret metamorphosed hydrothermal alteration zones, as their calibration does not take into account many of the trace and minor elements that occur in such rocks. Such elements can influence the stability of mineral assemblages if they occur in unusually high abundances relative to rocks used to calibrate the grids, or if they are contained in the reacting minerals such as chlorite and biotite (F, Ba), feldspar (Ba, Sr), staurolite and gahnite (Zn), and garnet (Mn) (e.g. Spry and Scott, 1986). Pelite generally contains very small quantities of Mn, P, S, B, F, Zr, Zn, Ba, Sr, and other trace elements (Gromet et al., 1984). These elements have therefore very little influence on the stability of the minerals that contain them and are generally not taken into account in the calibration of the petrogenetic grids. However, in aluminous rocks derived from altered protoliths, these elements can be relatively abundant. Zaleski et al. (1991) demonstrate that, because of incorporation of F in biotite and Zn in staurolite and gahnite, the assemblages kyanite-biotite and staurolite-anthophyllite-gahnite in the alteration zones of the Linda deposit (Manitoba) crystallized at a lower grade than those observed in aluminous rocks of sedimentary origin. These authors conclude that in this case, the use of the petrogenetic grids calibrated for the pelite is not suitable for rigorous interpretation of isograds and pressure-temperature metamorphic reaction conditions. On the other hand, the grids are useful in predicting the range of possible reactions and assemblages in the field, and where mineral assemblages are not those expected for normal metamorphosed rocks. Such discrepancies between observed and predicted metamorphic assemblages may indicate unusual compositions related to hydrothermal alteration.

Strategies, Criteria, and Diagnostics

Advanced argillic, sericitic, chloritic, carbonate, and siliceous alteration are useful exploration vectors for detecting VMS deposits. Once deformed and metamorphosed, such alteration is represented by aluminous, ferromagnesian, calc-silicate, or siliceous gneisses commonly with atypical mineral assemblages and modal mineral abundance with respect to sedimentary, volcanic, or plutonic protoliths. Their diagnostic geological traits are critical to protolith assessment (Fig. 5) in gneissic terranes as they provide criteria to distinguish them from metasediments or metamorphosed paleosoils with whom they share field exposure affinities. In under-explored areas, mineralization may not be reported or may appear insignificant; alteration mapping is required at this stage to locate mineralized areas. It also provides constraints on the intensity and the spatial distribution of hydrothermal systems, targets for systematic lithogeochemical and geophysical exploration, and supportive evidence for metallogenic model(s), all of which may contribute to the mineral resources assessment of the studied area and potentially the discovery of new deposits (e.g. Winston Lake deposit, Ontario; Thomas, 1991).

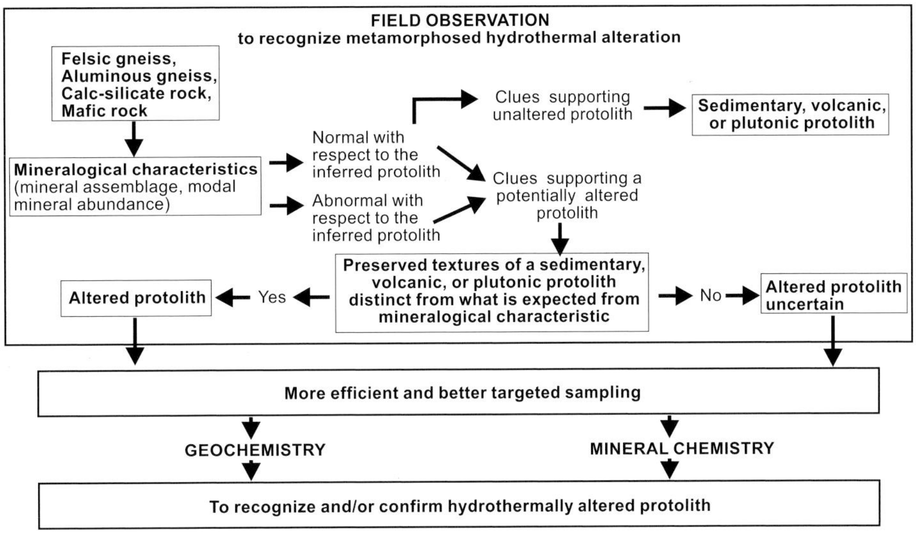

FIGURE 5. Field strategy for the recognition of metamorphosed hydrothermal alteration and by extension of hydrothermal systems possibly affiliated to mineral deposits.

Field Mineralogical, Textural, and Structural Analysis

A metamorphosed volcano-plutonic belt with VMS-type hydrothermal alteration zones will have the appearance of a felsic gneiss complex containing minor proportions of amphibolite as well as aluminous, ferromagnesian, calc-silicate, or siliceous 'metasedimentary' rocks with atypical textures, mineralogical characteristics, and modal abundances (Figs. 5, 6A-C). Diagnostic lithologies such as sillimanite-quartz rock, garnetite, and cordierite-orthoamphibole/orthopyroxene quartzofeldspathic or mafic gneiss (Fig. 3B) may be the first sign that hydrothermal alteration is present and that the geological context may be a near surface environment (Corriveau et al., 1997). In such cases, searching for volcanic textures among otherwise nondescript quartzofeldspathic gneisses may provide the first conclusive evidence of volcanic activity within a studied gneissic complex (Fig. 6A,B; Blein et al., 2003; Corriveau and Bonnet, 2005). Rock types that appear more 'normal' but have some unusual abundance of certain minerals may also be indicators of premetamorphic hydrothermal alteration (Table 3). However, different processes may lead to the formation of similar rocks and protolith assessment rarely provides a single solution (e.g. Reinhardt, 1987; Chapter 4 in Bonnet and Corriveau, 2007). For example, alteration processes that involve highly acidic hot fluids severely leach host rocks, leaving high-alumina assemblages (advanced argillic alteration). Such high-alumina assemblages may be formed also from low-temperature surface weathering, causing lateritization, or from partial melting of pelite with significant melt extraction. Consequently, peraluminous gneiss can have, as a protolith, a rock affected by an advanced argillic alteration, a laterite, or a restitic pelite (Gromet et al., 1984; Lemiere et al., 1986). Textures, such as aluminous nodules in quartzo-feldspathic gneiss (Fig. 3C-F) and fragments (lapillis or block) in aluminous gneiss (Figs. 3A, 6C) adjacent to lapillistone units with fragments of similar size and rock types (Fig. 6A,B), provide firmer evidence of hydrothermal alteration.

Metamorphosed hydrothermal alteration zones commonly consist of rocks that attract the interest of field geologists and tend to be located and discussed in geological reports or assessment files where they are interpreted as restites based on the unusually high modal abundance of aluminous or mafic minerals (e.g. the diatexites of Perreault and Heaman, 2003). Though such units do not become instant exploration targets, they warrant closer examination and a search for other key vectors to possible mineralization, such as meta-exhalite (coticule, garnetite, iron formation, tourmalinite, and quartz-gahnite rock; Spry et al., 2000).

Hydrothermally altered rocks that have a bulk composition, hence a mineral assemblage and a modal abundance, similar or fairly similar to that of sedimentary, volcanic, or plutonic units, (Fig. 7) may only be determined as hydrothermal in origin by lithogeochemistry (Fig. 7D). U-Pb Sensitive High Resolution Ion MicroProbe (SHRIMP) dating on zircon also has potential for constraining genetic models of ore formation in ancient high-grade metamorphosed terranes (Claoué-Long et al., 1990; Yeats et al., 1996; Williams, 1998; Hartmann et al., 2000). For example, such data can highlight the presence of a detrital zircon population (Fig. 7B), providing conclusive evidence of a sedimentary origin or of a single magmatic zircon population (Fig. 6D) that is consistent with a volcanic or plutonic origin.

Follow-up lithogeochemistry using alteration indices (Ishikawa et al., 1976; Hashigushi et al., 1983; Hodges and Manojlovic, 1993; Large et al., 2001; Piché and Jébrak, 2004) or mass-balances techniques (MacLean and Kranidiotis, 1987; MacLean, 1990; MacLean and Barrett, 1993; Leitch and Lentz, 1994) will help refine field interpretations. The NORMAT norm calculation of Piché and Jébrak (2004) is of particular use as it provides a measure of relative alteration intensity and characterization of alteration types through a series of alteration indices that are inde-

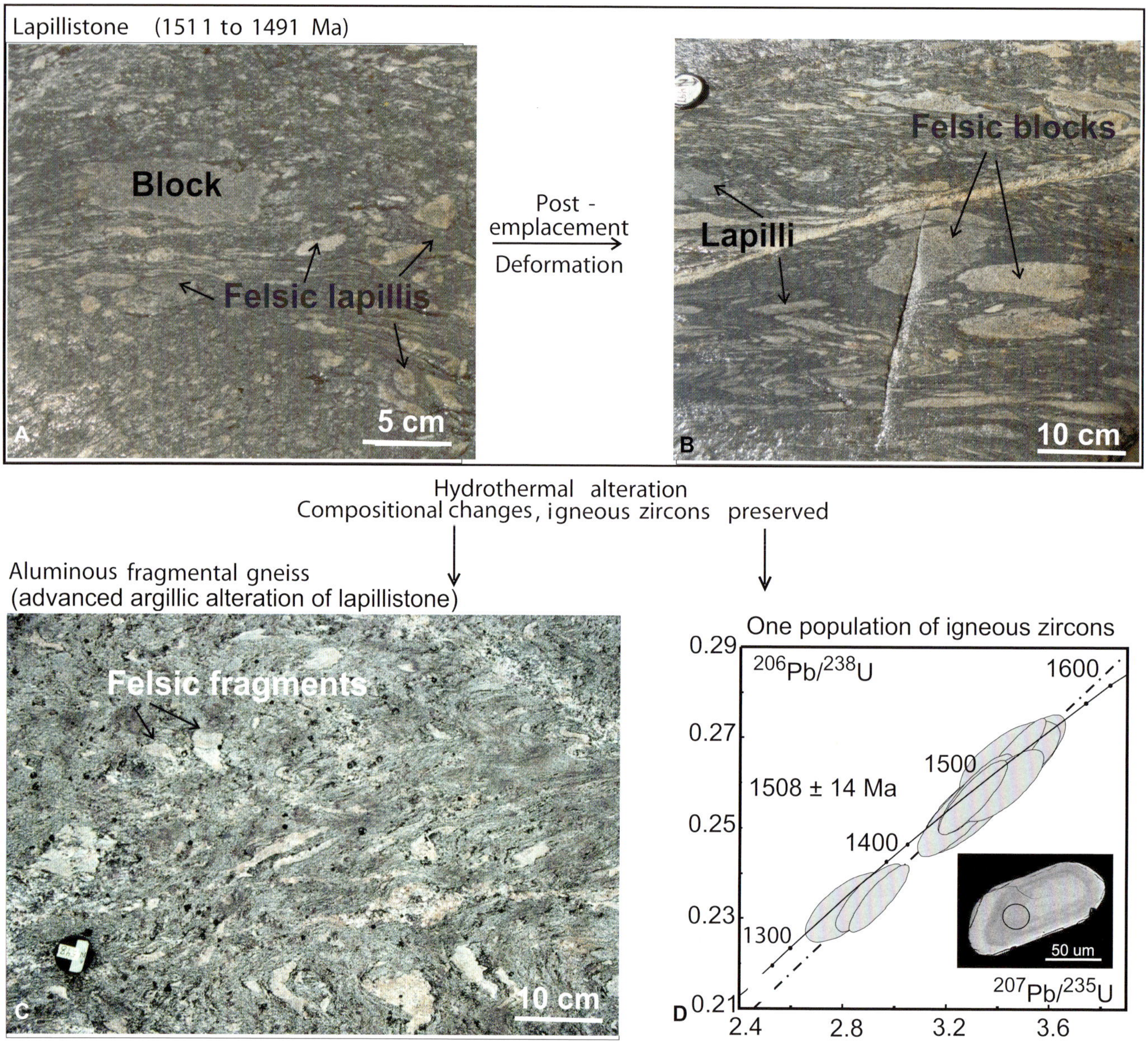

FIGURE 6. Field interpretation of fragmental aluminous gneiss in the granulite-facies La Romaine Supracrustal Belt and subsequent testing. **(A)** Precursor lapillistone with fabric interpreted as largely primarily based on volcanic textures. **(B)** Deformed equivalent in the same outcrop. **(C)** Adjacent garnet-sillimanite-biotite gneiss (aluminous unit) with fine-grained quartzo-feldspathic fragments of similar size, shape, and composition as those of (B). **(D)** SHRIMP zircon analyses from the garnet-sillimanite-biotite gneiss with well preserved volcanic fragments on the concordia diagram. The U-Pb crystallization age of 1508 ± 14 Ma (95%) is consistent with zircon U-Pb crystallization ages obtained on least altered lapillistone.

pendent of the composition of the precursor rock as long as the protolith was not peraluminous. These norm calculations are currently only available for greenschist-facies settings (Piché and Jébrak, 2006) but are being developed for higher grade rocks.

In order to identify residues of partial melting in a reliable way, it is important to empirically establish the degree of segregation and remobilization of anatectic melts, as well as the types of crystallization sites. For this purpose, it is important to (1) compare the mineralogical assemblages of the host gneisses and their leucosomes to know if they can be interpreted in terms of metamorphic reactions generating anatectic melts (Sawyer, 1999), and (2) document the morphology, structures, and textures of the leucosomes to estimate their degree of remobilization (Sawyer, 2001). If the mineralogical assemblages and the textures and structures of leucosomes are diagnostic of in situ leucosomes, the units that are rich in aluminous and ferromagnesian phases cannot be restites resulting from significant melt extraction (see Fig. 5 in Boggs and Corriveau, 2004) and are candidates for a hydrothermal alteration protolith, which can be tested by lithogeochemistry and U-Pb geochronology. In contrast, if textures are diagnostic of a significant degree of anatectic melting associated with severe melt extraction, then a restitic origin is a likely interpretation. However, special conditions are required for this to happen (Vigneresse et al., 1996) and it is fortunately more the exception than the rule, at least within the Canadian Precambrian Shield. Lithogeochemical protolith studies may lead to spurious results and may not be worth attempting for exploration purposes.

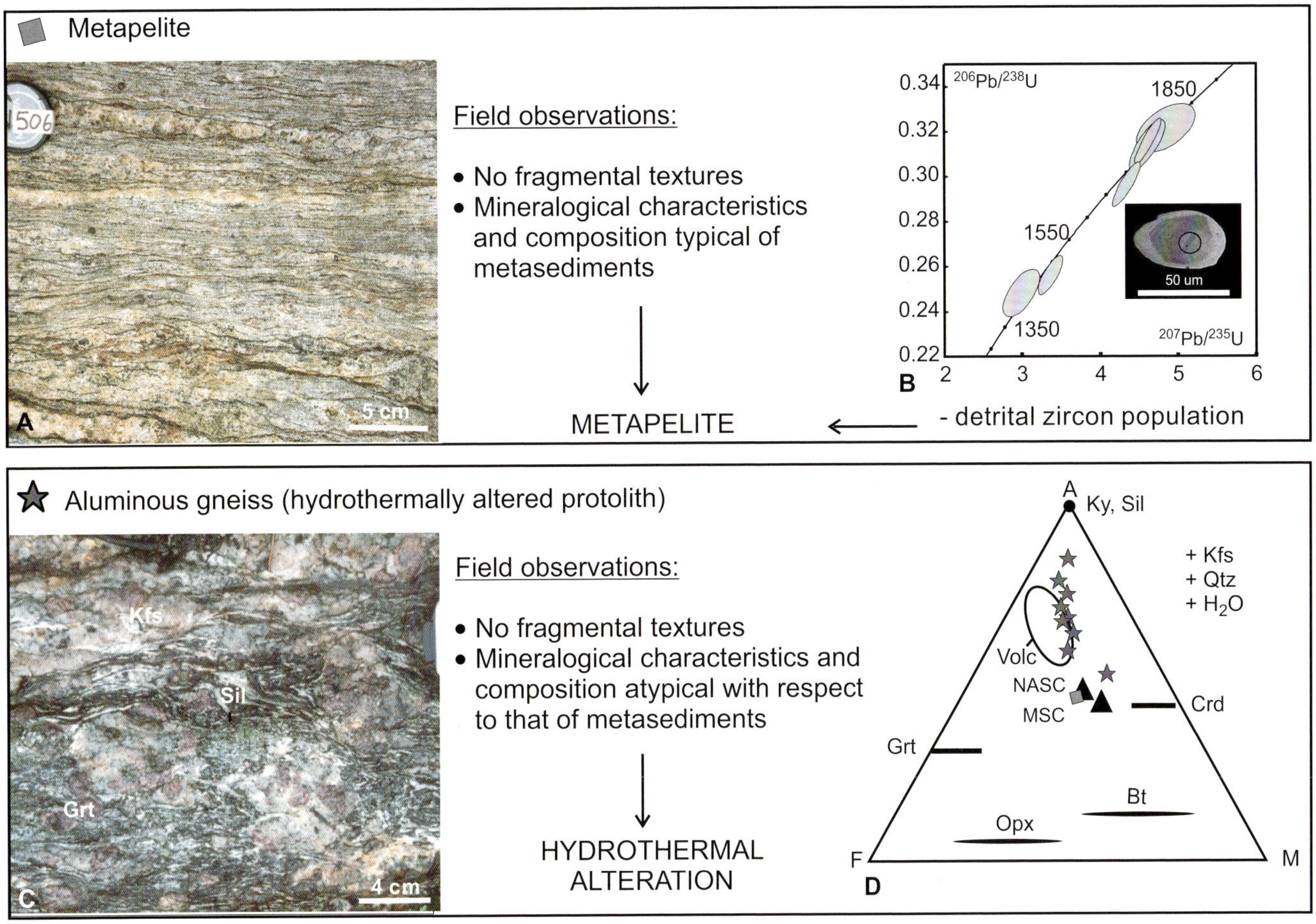

Figure 7. Procedures for recognizing the origin of metapelite-like aluminous gneisses in the granulite-facies La Romaine Supracrustal Belt and subsequent testing. **(A)** Garnet-sillimanite-biotite gneiss interpreted as a metapelite based on its mineralogy and textural characteristics.**(B)** SHRIMP zircon analyses from the metapelite on the concordia diagram. The gneiss has a detrital population of zircon, which confirms its sedimentary origin. **(C)** Garnet-sillimanite-biotite gneiss interpreted as a metamorphosed hydrothermal alteration based on its mineralogy and texture. **(D)** AFM diagrams (A = Al_2O_3-K_2O, F = FeO, M = MgO; molar proportion) showing the compositional differences between garnet-sillimanite-biotite gneiss of hydrothermal origin and metapelite (MSC and NASC values from Gromet et al., 1984).

Metamorphic Alteration Vectors

Alteration mapping based on visual estimates of rock-types, mineral modal abundance, and mineral assemblages are effective in qualitatively measuring the variations in major elements associated with the premetamorphic hydrothermal alteration and in establishing the chemical zonation with a reasonable degree of confidence. Such zonation reflects major element mobility during fluid-rock reaction in the upflow zone (e.g. Al, Fe, Mg, Ca, Na, and K) and represents geochemical indicators of the type and intensity of alteration (Barrett and MacLean, 1994; Galley, 1995; Large et al., 2001). These tools are also applicable to metamorphosed, weakly to intensely altered rocks whose precursor is known (Hodges and Manojlovic, 1993; Hallberg, 2001; Gale, 2003). Alteration vectors may be represented graphically using chemical diagrams or on AFM, A'CF, or A'KF phase diagrams, as discussed above, in order to visualize the whole-rock composition and the mineralogical assemblages of the altered rocks and their precursors. This allows (1) the chemical changes associated with the hydrothermal alteration to be determined by comparing the metamorphic characteristics of the altered rocks and their precursors, and by extrapolation, their compositions; (2) the establishment of the link between the metamorphic assemblages observed and the primary hydrothermal mineralogical assemblages; and (3) the development of metallogenic models (e.g. Roberts et al., 2003). The field models can then be refined with lithogeochemical studies (Large et al., 2001).

The main alteration vectors are represented in Figure 8A. In AFM diagrams, the vectors pointing towards the apices A, F, and M, respectively, model the enrichment of Al, Fe, and Mg, and characterize advanced argillic or siliceous alteration (Al-Si), argillic alteration (Al-Fe), and chloritic alteration (Mg ± Fe). The vector moving away from apex A characterizes sericitic- or potassic-alteration types as apex A takes into account Al and K (A = Al_2O_3 + K_2O in mole proportion; Spear, 1993). In this type of diagram, acidic or siliceous argillic alteration is localized close to apex A near the field of aluminosilicates (kaolinite, pyrophyllite, kyanite, and/or sillimanite, depending on the metamorphic facies); they are thus characterized by high contents of these minerals. Chloritic alteration is localized close to the apex M, thus close to the field of magnesian minerals such as chlorite, anthophyllite, gedrite, cordierite, phlogopite, and/or orthopyroxene (depending on the metamorphic facies). In the same manner, sericitic alteration is characterized by high content of sericite or, at higher metamorphic grade, of biotite or K-feldspar. Finally, iron-bearing alteration will be domi-

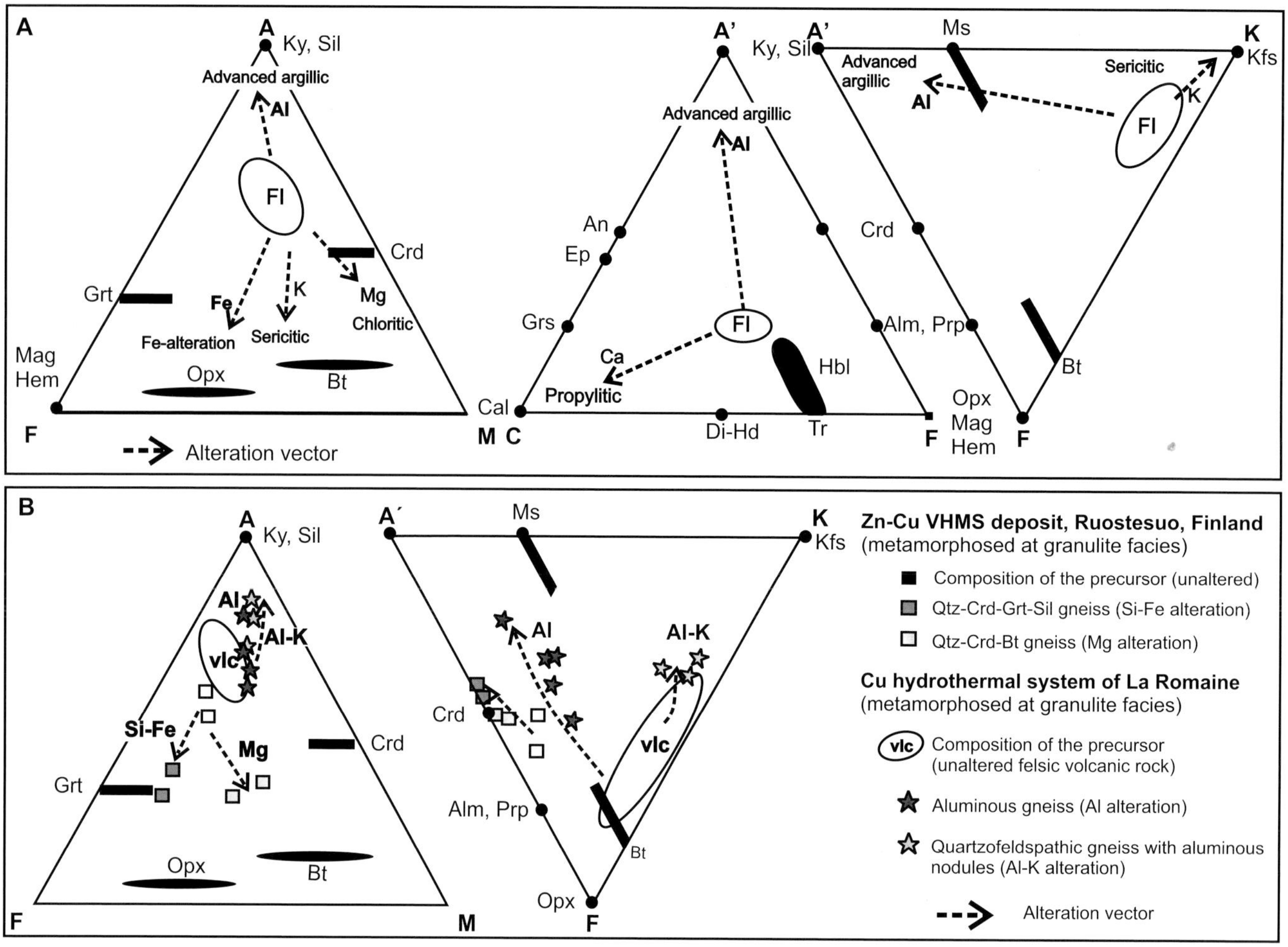

FIGURE 8. Alteration vectors and their control on modal composition and mineral paragenesis of metamorphosed hydrothermal alteration zones. **(A)** Ternary diagrams A'CF (A'=Al_2O_3+Fe_2O_3,-(K_2O+Na_2O), C=CaO, F=FeO+MnO+MgO; molar proportion) A'KF (A'=Al_2O_3+Fe_2O_3,-(K_2O+Na_2O+CaO), K=K_2O, F=FeO+MnO+MgO), and AFM (A=Al_2O_3-K_2O, F=FeO, M=MgO; molar proportion) showing geochemical changes of the main alteration zones associated to volcanogenic massive sulphide deposits. **(B)** Ternary diagrams A'KF and AFM showing the alteration vectors (dotted arrow) for Si-Fe- and Mg-alteration zones of the granulite-grade Ruostesuo Zn-Cu deposit (square; data from Roberts et al., 2003) and for advanced argillic (aluminous gneiss) and sericitic (gneiss with aluminous nodules and veins) alteration of felsic pyroclastic rocks from the La Romaine hydrothermal system (star; data from Bonnet et al., 2005).

nated by magnetite, hematite, iron sulphides, and, in certain cases, garnet. The A' KF and A' CF diagrams are particularly useful to represent carbonate (Ca) or propylitic alteration (Ca-Fe) localized close to the apex C as well as sericitic alteration (K-Al) localized close to the apex K. These alteration types are characterized by significant contents of calcic minerals (calcite, epidote, grossular, anorthite, and/or diopside), iron-bearing minerals (magnetite, hematite, garnet, and/or iron sulphides), and potassic minerals (sericite, biotite, and/or K-feldspar), depending on pressure-temperature conditions. In such diagrams, the FeO and MgO components are not differentiated and fall near apex F, hence they do not permit magnesian alteration to be differentiated from iron alteration.

To illustrate these points, the chemical changes recorded by the alteration zones of the granulite facies Ruostesuo Zn-Cu deposit are presented on ternary diagrams (Fig. 8B). This deposit comprises gneisses with quartz-cordierite-biotite and quartz-garnet-cordierite-sillimanite assemblages that respectively represent magnesian (enriched in Mg-Al) and iron (enriched in Fe-Al) alteration of a felsic volcanic precursor (Roberts et al., 2003). In an AFM diagram, these gneisses define vectors towards the F and M apices and in the diagram A' KF they project along a vector directed towards cordierite (between apices A' and F). Such a chemographic projection was also applied to rocks of the La Romaine hydrothermal system (Fig. 8B). This system includes metamorphosed pumaceous tuffs with aluminous veins and nodules, and aluminous gneisses of volcanic origin (Bonnet et al., 2005). The aluminous nodules or veins are composed mainly of quartz, muscovite, sillimanite, and iron oxides. These nodules are thus enriched in K, Al, and Fe, which differs from that of the least altered felsic volcanic rocks hosting them. The metatuffs with aluminous nodules or veins are thus typical of sericitic alteration and define vectors towards the apices A and K in the AFM and A' KF diagrams, respectively. The aluminous volcanic rocks are characterized by high contents of sillimanite indicating an Al-rich composition compared to that of their volcanic precursor. Such rocks are characteristic of an advanced argillic alteration (Bonnet et al., 2005). These gneisses are thus aligned towards apices A and A' in the AFM and A' KF diagrams respectively.

Conclusions

Volcano-plutonic belts are potentially fertile for hydrothermal ore deposits such as VMS deposits and represent potential targets for exploration. In high-grade metamorphic terranes alteration zones, where well developed, provide important vectors to ore. In felsic gneiss complexes where volcanic rocks are not even known, refinements to mapping strategies and protocols provide a means to recognize potential hydrothermal systems even though evidence for volcanic rocks and subvolcanic intrusions have been partially to completely obliterated during high-grade metamorphism. Metamorphism of hydrothermal alteration zones gives rise to diagnostic rock types, mineral assemblages, and modal abundances that can be used as vectors to mineralization. Widespread hydrothermal alteration that is chemically distinct from host rock lithologies can be recognized with a reasonable degree of confidence during regional mapping as long as it is not confused with metasedimentary rocks. The field strategies and tools proposed to evaluate gneissic protoliths and to recognize hydrothermal alteration zones are based on the following conditions:

1. Metamorphic characteristics of the gneiss (mineral assemblages and modal mineral abundance) reflecting the chemical composition of their protoliths and the physical conditions of the metamorphism;
2. Primary structures and textures that are locally preserved;
3. Field indicators diagnostic of hydrothermal activity, such as the presence of meta-exhalite, stockworks, or actual mineralization.

The geological criteria presented provide a method of identifying a fossil hydrothermal system with a reasonable degree of confidence using vectors to mineralization. These field criteria are not exact tools but represent permissive evidence for a hydrothermal origin of certain gneisses. They can also be applied to any deposit type associated to volcano-plutonic belts where alteration zones are sufficiently large enough to be recognized during regional mapping such as epithermal, porphyry, and iron oxide Cu-Au deposits. Once a hydrothermal system is identified, it is possible to target specific areas for more detailed mapping and exploration. These geological criteria are also useful in guiding the lithogeochemical sampling of key lithologies and in testing metallogenic models.

Acknowledgements

Funding for this work was provided by the DIVEX research group and the Geological Survey of Canada and is based on the knowledge gained in the eastern Grenville Province in the course of a Targeted Geoscience Initiative project of Natural Resources Canada in partnership with the Ministère des Ressources Naturelles et de la Faune, Québec (MNRF) and the Geological Survey of Newfoundland and Labrador. Financial support was also granted by l'Institut national de la recherche scientifique - Eau-Terre-Environnement (INRS-ETE) for the doctoral studies of A.-L. Bonnet under the co-supervision of M.R. LaFlèche (INRS-ETE) and Alain Tremblay (University of Quebec at Montreal). The authors thank P. Archer, M. Chapdelaine, and F. Huot for their contributions concerning the Coulon property when it belonged to the Virginia-Noranda consortium, C. Guilmette, S. Parson, O. Rabeau, G. Scherrer, and K. Williamson for preliminary research for this paper, L. Nadeau, W. Goodfellow, J. Lydon, O. van Breemen, and B. Dubé (Geological Survey of Canada), G. Beaudoin (Laval Unversity), M. Auclair, V. Bodycomb, M. Gauthier, and M. Jébrak (University of Quebec at Montreal), M.R. LaFlèche and M. Malo (INRS), P. Verpaelst, T. Clark, and S. Perreault (MRNF), S. Cadéron (U. Polytechnique), and the expert committee of DIVEX for their logistic and scientific contributions, and L. Dubé (INRS) for the digitalization of the photographs. Special thanks are expressed to G. Beaudoin and D. Lentz, journal reviewers, and W. Goodfellow, editor, for their scientific and editorial comments that significantly clarified the manuscript.

References

Allard, G.O., 1978, Pétrologie et potentiel économique du prolongement du sillon de roches vertes de Chibougamau dans la Province de Grenville: Ministère des Richesses naturelles, Québec, DPV-604.

Allard, G.O., and Carpenter, R.H., 1988, Mineralogical anomalies in metamorphosed terrains, a neglected but promising exploration tool: International Conference on the Geochemical Evolution of the Continental Crust, Pocos de Cladas, Brazil, p. 229-236.

Archer, P., Chapdelaine, M., and Huot, F., 2004, Découverte de sulfures massifs volcanogènes de haut-grade métamorphique dans la Ceinture de Coulon, Baie James: Québec Exploration 2004, Résumés des Conférences et des Photoprésentations, Québec, p. 13.

Barrett, T.J., and MacLean, W.H., 1994, Chemostratigraphy and hydrothermal alteration in exploration for VMS deposits in greenstones and younger volcanic rocks, *in* Lentz, D.R., ed., Alteration and Alteration Processes Associated with Ore-Forming Systems: Geological Association of Canada, Short Course Notes, v. 11, p. 433-467.

Bernier, L.R., 1992, Lithogeochemistry and Geothermobarometry of Mineralized Cordierite-Orthoamphibole and Related Rocks at Atik Lake, Manitoba, Némiscau and Montauban, Québec: Ph.D. thesis, McGill University, Montreal, Quebec, 372 p.

Bernier, L.R., and MacLean, W.H., 1993, Lithogeochemistry of a metamorphosed VMS alteration zone at Montauban, Grenville Province: Exploration and Mining Geology, v. 2, p. 367-386.

Blein, O., LaFlèche, M.R., and Corriveau, L., 2003, Geochemistry of the granulitic Bondy Gneiss Complex: a 1.4 Ga arc in the Central Metasedimentary Belt, Grenville Province, Canada: Precambrian Research, v. 120, p. 193-218.

Blein, O., Corriveau, L., and LaFlèche, M.R., 2004, Cordierite-orthopyroxene white gneiss: A key to unveiling pre-metamorphic hydrothermal activity in the Bondy gneiss complex, Grenville Province, Québec, *in* Tollo, R., Corriveau, L., McLelland, J., and Bartholomew, M., eds., Proterozoic Tectonic Evolution of the Grenville Orogen in eastern North America: Geological Society of America, Memoir 197, p. 19-33.

Boggs, K.J.E., and Corriveau, L., 2004, Granulite-facies P-T-t paths and the influence of retrograde cation diffusion during polyphase orogenesis, western Grenville Province, Québec, *in* Tollo, R., Corriveau, L., McLelland, J., and Bartholomew, M., eds., Proterozoic Tectonic Evolution of the Grenville Orogen in eastern North America: Geological Society of America, Memoir 197, p. 35-64.

Bonnet, A.-L., and Corriveau, L., 2007, Atlas et outils de reconnaissance de systèmes hydrothermaux métamorphisés dans les terrains gneissiques, *in* Goodfellow, W.D., ed., Mineral Deposits of Canada: A Synthesis of Major Deposit-types, District Metallogeny, the Evolution of Geological Provinces, and Exploration Methods, Special Publication 5, Mineral Deposits Division, Geological Association of Canada (DVD), 95 p.

Bonnet, A.-L., Corriveau, L., and LaFlèche, M.R., 2005, Chemical imprint of highly metamorphosed volcanic-hosted hydrothermal alterations in the La Romaine Supracrustal Belt, eastern Grenville Province, Quebec: Canadian Journal of Earth Sciences, v. 42, p. 1783-1814.

Claoué-Long, J.C., King, R.W., and Kerrich, R., 1990, Archean hydrothermal zircon in the Abitibi greenstone belt: Constraints on the timing of gold mineralization: Earth Planetary Science Letters, v. 89, p. 109-128.

Corbett, G.J., and Leach, T.M., 1998, Southwest Pacific Rim Gold-Copper Systems: Structure, Alteration, and Mineralization: Society of Economic Geologists, Special Publication No. 6, 236 p.

Corriveau, L., and Bonnet, A.-L., 2005, Pinwarian (1.5 Ga) volcanism and hydrothermal activity at the eastern margin of the Wakeham Group, Grenville Province, Quebec: Canadian Journal of Earth Sciences, v. 42, p. 1749-1782.

Corriveau, L., Blein, O., and LaFlèche, M., 1997, Progress Report on the Bondy Gneiss Complex and Its Cupriferous Hydrothermal System, Mont-Laurier Area: Ministère des Ressources naturelles, de la Faune et des Parcs, Québec, GM 54804, 133 p. and 1 map.

Corriveau, L., Perreault, S., and Davidson, A., 2007, Prospective metallogenic settings of the Grenville Province, *in* Goodfellow, W.D., ed., Mineral Deposits of Canada: A Synthesis of Major Deposit-types, District Metallogeny, the Evolution of Geological Provinces, and Exploration Methods: Geological Association of Canada, Mineral Deposits Division, Special Publication 5, p. xx-xx.

Dubé, B., Dunning, G., and Lauzière, K., 1998, Geology of the Hope Brook Mine, Newfoundland, Canada: A preserved late Proterozoic high-sulfidation epithermal gold deposit and its implications for exploration: Economic Geology, v. 93, p. 405-436.

Froese, E., 1985, Anthophyllite-bearing rocks in the Flin Flon-Sherridon area, Manitoba: Geological Survey of Canada, Paper 85-01B, p. 541-544

——— 1998, Metamorphism of hydrothermally altered rocks: Geological Survey of Canada, Paper 1998-E, p. 193 196.

Gale, G.H., 2003, Vectoring volcanogenic massive sulphide deposits using rare earth elements and other pathfinder elements at the Ruttan mine, Manitoba (NTS 63B5), *in* Report of Activities 2003: Manitoba Industry, Economic Development and Mines, Manitoba Geological Survey, p. 54-73.

Galley, A.G., 1995, Target vectoring using lithogeochemistry: Applications to the exploration for volcanic-hosted massive sulphide deposits: The Canadian Institute of Mining Bulletin, v. 88, p. 15-27.

Gauthier, M., 1993, L'aspect que prennent les amas sulfurés en terrain métamorphique: incidences sur la prospection minière: Bulletin de l'Association professionnelle des Géologues et des Géophysiciens du Québec, v. 10, p. 17-22.

Gauthier, M., and Chartrand, F., 2005, Metallogeny of the Grenville Province revisited: Canadian Journal of Earth Sciences, v. 42, p. 1719-1734.

Gauthier, M., Morin, G., and Marcoux, P., 1985, Minéralisations aurifères de la partie centrale de la Province de Grenville, Bouclier Canadien: Canadian Mining and Metallurgy Bulletin, v. 78, p. 60-69.

Gauthier, M., Corriveau, L., and Chouteau, M., 2004, Metamorphosed and Metamorphogenic Ore Deposits of the Central Metasedimentary Belt, southwestern Québec and southeastern Ontario, Grenville Province: Premières Journées De Launay, Post-symposium Field Trip Guidebook, 20 p.

Gromet, L.P., Dymek, R.F., Haskin, L.A., and Korotev, R.L., 1984, The "North American shale composite": Its compilation, major and trace element characteristics: Geochimica et Cosmochimica Acta, v. 48, p. 2469-2482.

Gustafson, L.B., and Titley, S.R, 1978, Preface, *in* Porphyry Deposits of the southwestern Pacific Islands and Australia: Economic Geology, v. 73, p. 597-599.

Hallberg, A., 1994, The Enåsen gold deposit, central Sweden. 1. A Palaeoproterozoic high-sulphidation epithermal gold mineralization: Mineralium Deposita, v. 29, p. 150-162.

——— 2001, Rock classification, magmatic affinity, and hydrothermal alteration at Boliden, Skellefte district, Sweden - a desk-top approach to whole rock geochemistry, *in* Weihed P., ed., Economic Geology Research: Sveriges Geologiska Undersökning, v. 833, p. 93-131.

Hannington, M.D., Kjarsgaard, I.M., Galley, A.G., and Taylor, B., 2003, Mineral-chemical studies of metamorphosed hydrothermal alteration in the Kristineberg volcanogenic massive sulphide district, Sweden: Mineralium Deposita, v. 38, p. 423-442.

Hartmann, L.A., Leite, J.A.D., Da Silva, L.C., Remus, M.V.D., McNaughton, N.J., Groves, D.I., Fletcher, I.R., Santos, J.O.S., and Vasconcellos, M.A.Z., 2000, Advances in SHRIMP geochronology and their impact on understanding the tectonic and metallogenic evolution of southern Brazil: Australian Journal of Earth Sciences, v. 47, p. 829-844.

Harris, L., Rivard, B., and Corriveau, L., 2001, Structure of the Lac Nominingue-Mont-Laurier region, Central Metasedimentary Belt, Québec Grenville Province: Canadian Journal of Earth Sciences, v. 38, p. 787-802.

Hashiguchi, H., Yamada, R., and Inoue, T., 1983, Practical application of low Na_2O anomalies in footwall acid lava for delimiting promising areas around the Kosaka and Fukazawa Kuroko deposits, Akita Prefectural, Japan, *in* Ohmoto, H. and Skinner, B.J., eds., The Kuroko Massive Sulfide Deposits: Economic Geology Monograph, v. 5, p. 387-394.

Hedenquist, J.W., Arribas, A., and Gonzalez-Urien, E., 2000, Exploration for epithermal gold deposits: Reviews in Economic Geology, v. 13, p. 245-277.

Hemley, J.J., Montoya, J.W., Marinenko, J.W., and Luce, R.W., 1980, Equilibria in the system Al_2O_3-SiO_2-H_2O and some general implications for alteration/mineralization processes: Economic Geology, v. 75, p. 210-228.

Hodges, D.J., and Manojlovic, P.M., 1993, Application of lithogeochemistry to exploration for deep VMS deposits in high-grade metamorphic rocks, Snow Lake, Manitoba: Journal of Geochemical Exploration, v. 48, p. 201-224.

Huot, F., Chapdelaine, M., and Archer, P., 2004, Technical Report and Recommendations, Reconnaissance Program, Coulon Project, Québec: Mines d'Or Virginia Inc, Internal Report, 34 p.

Ishikawa, Y., Sawaguchi, T., Iwaya, S., and Horiuchi, M., 1976, Delineation of prospecting targets for Kuroko deposits based on modes of volcanism of underlying dacite and alteration haloes: Mining Geology, v. 26, p. 105-117.

Kontak, D.J., DeWolfe, J., and Finck, P.W., 2002, The Coxheath plutonic-volcanic belt (NTS 11K/01): A linked porphyry-epithermal mineralized system of Precambrian age: Nova Scotia, Report of Activities, p. 69-87.

Kretz, R., 1983, Symbols for rock-forming minerals: American Mineralogist, v. 68, p. 277-279.

LaFlèche, M.R., Birkett, T., and Corriveau, L., 2005, Crustal development at the pre-Grenvillian Laurentian margin: a record from contrasting geochemistry of mafic and ultramafic orthogneisses in the Chochocouane River area, Quebec: Canadian Journal of Earth Sciences, v. 42, p. 1653-1675.

Large, R.R., Gemmel, J.B., and Paulick, H., 2001, The alteration box plot: a simple approach to understanding the relationship between alteration mineralogy and lithogeochemistry associated with volcanic-hosted massive sulphide deposits: Economic Geology, v. 96, p. 957-971.

Leitch, C.H.B., and Lentz, D.R., 1994, The Gresens approach to mass balance constraints of alteration systems: methods, pitfalls, examples, *in* Lentz, D.R., ed., Alteration and Alteration Processes Associated with Ore-Forming Systems: Geological Association of Canada, Short Course Notes, v. 11, p. 161-192.

Lemiere, B., Delfour, J., Moine, B., Piboule, M., Isnard, P., and Tegyey, M., 1986, Hydrothermal alteration and the formation of aluminous haloes around sulfide deposits. A model for laterites at Chizeuil (Morvan, France). Mineralium Deposita, v. 21, p. 147-155.

Lydon, J.W., 1988, Volcanogenic massive sulfide deposits. Part 2. Genetic model: Geoscience Canada, v. 15, p. 43-65.

MacLean, W.H., 1990, Mass changes calculations in altered rock series: Mineralium Deposita, v. 25, p. 44-49.

MacLean, W.H., and Barrett, T.J., 1993, Lithogeochemical techniques using immobile element: Journal of Exploration Geochemistry, v. 48, p. 109-133.

MacLean, W.H., and Kranidiotis, P., 1987, Immobile elements as monitors of mass transfer in hydrothermal alteration: Phelps Dodge massive sulphide deposit, Matagami, Quebec: Economic Geology, v. 82, p. 951-962.

Marshall, B., Vokes, F.M., and Larocque, A.C.L., 2000, Regional metamorphic remobilization: Upgrading and formation of ore deposits, *in* Spry, P.G., Marshall, B., and Vokes, F.M., eds., Metamorphic and Metamorphogenic Ore Deposits: Reviews in Economic Geology, v. 11, p. 19-38.

Mavrogenes, J.A., MacIntosh, I.W., and Ellis, D.J., 2001, Partial melting of the Broken Hill galena-sphalerite ore: Experimental studies in the system PbS-FeS-ZnS-(Ag2S): Economic Geology, v. 96, p. 205-210.

Melfos, V., Vavelidis, M., Christofides, G., and Seidel, E., 2002, Origin and evolution of the Tertiary Maronia porphyry copper-molybdenum deposit, Thrace, Greece: Mineralium Deposita, v. 37, p. 648-668.

Moore, J.M., and Waters, D.J., 1990, Geochemistry and origin of cordierite-orthoamphibole / orthopyroxene-phlogopite rocks from Namaqualand, South Africa: Chemical Geology, v. 85, p. 77-100.

Mueller, W.U., and Corcoran, P.L., 2001, Volcano-sedimentary processes operating on a marginal continental arc: The Archean Raquette Lake

Formation, Slave Province, Canada: Sedimentary Geology, v. 141-142, p. 169-204.

Nadeau, L., Brouillette, P., and Hébert, C., 1999, New observations on relict volcanic features in medium-grade gneiss of the Montauban Group, Grenville Province, Québec: Geological Survey of Canada, Paper 1999-1E, p. 149-160.

Ohmoto, H., 1996, Formation of volcanogenic massive sulfide deposits: the Kuroko perspective: Ore Geology Reviews, v. 10, p. 135-177.

Pan, Y., and Fleet, M.E., 1992, Calc-silicate alteration in the Hemlo gold deposit, Ontario: Mineral assemblages, P-T-X constraints and significance: Economic Geology, v. 87, p. 1101-1120.

Pattison, D.M., Chacko, T., Farquhar, J., and McFarlane, C.R.M., 2003, Temperatures of granulite-facies metamorphism: constraints from experimental phase equilibria and thermobarometry corrected for retrograde exchange: Journal of Petrology, v. 44, p. 867-900.

Perreault, S., and Heaman, L., 2003, Géologie et géochronologie de la Basse-Côte-Nord (entre Chevery et Blanc Sablon) dans la Province géologique de Grenville, *in* Brisebois, D. and Clark, T., eds., Géologie et Ressources Minérales de la Partie est de la Province de Grenville: Ministère des Ressources naturelles, de la Faune et des Parcs, Québec, DV 2002-03, p. 119-146.

Piché, M., and Jébrak, M., 2004, Normative minerals and alteration indices developed for mineral exploration: Journal of Geochemical Exploration, v. 82, p. 59-77.

——— 2006, Determination of alteration facies using the NORMAT normative mineral alteration index: Selbaie Cu-Zn deposit, northern Abitibi greenstone belt, Canada: Canadian Journal of Earth Sciences, v. 43, p. 1877-1885.

Powell, W.G., Pattison, D.R.M., and Johnston, P., 1999, Metamorphic history of the Hemlo gold deposit from Al2SiO5 mineral assemblages, with implications for the timing of mineralization: Canadian Journal of Earth Sciences, v. 36, p. 33-46.

Reinhardt, J., 1987, Cordierite-anthophyllite rocks from north-west Queensland, Australia: Metamorphosed magnesian pelites: Journal of Metamorphic Geology, v. 5, p. 451-472.

Roberts, M.D., Oliver, N.H.S., Fairclough, M.C., Höltta, P.S., and Lahtinen, R., 2003, Geochemical and oxygen isotope signature of sea-floor alteration associated with a polydeformed and highly metamorphosed massive sulfide deposit, Ruostesuo, Central Finland: Economic Geology, v. 98, p. 535-556.

Rosenberg, J.L., Spry, P.G., Jacobson, C.E., and Vokes, F.M., 2000, The effects of sulfidation and oxidation during metamorphism on the compositionally varied rocks adjacent to the Bleikvassli Zn-Pb-(Cu) deposit, Nordland, Norway: Mineralium Deposita, v. 35, p. 714-726.

Sawyer, E.W., 1999, Criteria for the recognition of partial melting: Physics and Chemistry of the Earth, v. 24, p. 269-279.

——— 2001, Melt segregation in the continental crust: distribution and movement of melt in anatectic rocks: Journal of Metamorphic Geology, v. 19, p. 291-309.

Schade, J., Cornell, D.H., and Theart, H.F.J., 1989, Rare earth element and isotopic evidence for the genesis of the Prieska massive sulphide deposit, South Africa: Economic Geology, v. 84, p. 49-63.

Schreurs, J., and Westra, L., 1985, Cordierite-orthopyroxene rocks: the granulite facies equivalents of the Orijarvi cordierite-anthophyllite rocks in West Uusimaa, southwest Finland: Lithos, v. 18, p. 215-228.

Sharpe, R., and Gemmell, J.B., 2002, The Archean Cu-Zn magnetite-rich Gossan Hill volcanic-hosted massive sulfide deposit, Western Australia: Genesis of a multistage hydrothermal system: Economic Geology, v. 97, p. 517-539.

Sillitoe, R.H., 1973, Tops and bottoms of porphyry copper deposits: Economic Geology, v. 68, p. 799-815.

——— 2000, Gold-rich porphyry deposits: Descriptive and genetic models and their role in exploration: Reviews in Economic Geology, v. 13, p. 315-345.

——— 2003, Iron oxide-copper-gold deposits: An Andean view: Mineralium Deposita, v. 38, p. 787-812.

Slagstad, T., Culshaw, N.G., Jamieson, R.A, and Ketchum, J.W.F., 2004, Early Mesoproterozoic tectonic history of the southwestern Grenville Province, Ontario: Constraints from geochemistry and geochronology of high-grade gneisses, *in* Tollo, R.P., Corriveau, L., McLelland, J., and Bartholomew, M., eds., Proterozoic Tectonic Evolution of the Grenville Orogen in North America: Geological Society of America, Memoir 197, p. 209-242.

Spear, F.S., 1993, Metamorphic Phase Equilibria and Pressure-Temperature-Time Paths: Mineralogical Society of America Monograph, 799 p.

Spear, F.S., and Cheney, J.T., 1989, A petrogenetic grid for pelitic schists in the system SiO_2-Al_2O_3-FeO-MgO-K_2O-H_2O: Contributions to Mineralogy and Petrology, v. 101, p. 149-164.

Spry, P.G., and Scott, S.D., 1986, Zincian spinel and staurolite as guide to ore in the Appalachians and Scandinavian Caledonides: The Canadian Mineralogist, v. 24, p. 147-163.

Spry, P.G., Peter, J.M., and Slack, J.F., 2000, Meta-exhalites as exploration guides to ore, *in* Spry, P.G., Marshall, B., and Vokes, F.M., eds., Metamorphic and Metamorphogenic Ore Deposits: Reviews in Economic Geology, v. 11, p.163-201.

Stanton, R.L., 2004, Exhalative metasediments - clues to the real nature of regional metamorphic processes: Mineralogy and Petrology, v. 80, p. 111-122.

Stevens, B.P.J., and Barron, L.M., 2002, Volcanic textures in the Palaeoproterozoic Hores gneiss, Broken Hill, Australia: Quarterly Notes, v. 113, p. 1-22.

Theart H.F.J., Cornell, D.H., and Schade, J., 1989, Geochemistry and metamorphism of the Prieska Zn-Cu deposit, South Africa: Economic Geology, v. 84, p. 34-48.

Thomas, D., 1991, Geology, Mineralogy and Hydrothermal Alteration at the Winston Lake Massive Sulphide Deposit, Marathon, Ontario: M.Sc. thesis, Queen's University, Kingston, Ontario, 231 p.

Thompson, P.H., Russell, I., Paul, D., Kerswill, J.A., and Froese, E., 1997, Regional geology and mineral potential of the Winter Lake-Lac de Gras area, central Slave Province, Northwest Territories: Geological Survey of Canada, Paper 1995-C, p. 107-119.

Tomkins, A.G., and Mavrogenes, J.A., 2002, Mobilization of gold as polymetallic melt during pelite anatexis at the Challenger deposit, South Australia: A metamorphosed Archean gold deposit: Economic Geology, v. 97, p. 1249-1271.

Trägårdh, J., 1991, Metamorphism of magnesium-altered felsic volcanic rocks from Bergslagen, central Sweden: A transition from Mg-chlorite-to cordierite-rich rocks: Ore Geology Reviews, v. 6, p. 485-497.

Vigneresse, J.L., Barbey, P., and Cuney, M., 1996, Rheological transitions during partial melting and crystallization with application to felsic magma segregation and transfer: Journal of Petrology, v. 37, p. 1579-1600.

Wang, G.-F., Banno, S., and Tackeushi, K., 1986, Reactions to define the biotite isograd in the Ryoke metamorphic belt, Kii Peninsula, Japan: Contributions to Mineralogy and Petrology, v. 93, p. 9-17.

Williams, I., 1998, U-Th-Pb geochronology by ion microprobe, *in* McKibben, M.A., Shanks, W.C., and Ridley, W.I., eds., Applications of Microanalytical Techniques to Understanding Mineralizing Processes: Reviews in Economic Geology, v. 7, p. 1-35.

Yang, K., and Scott, S.D., 1996, Possible contribution of a metal-rich magmatic fluid to a seafloor hydrothermal system: Nature, v. 383, p. 420-423.

Yeats, C.J., McNaughton, N.J., and Groves, D.I., 1996, SHRIMP U/Pb geochronological constraints on Archean volcanic-hosted massive sulfide and lode gold mineralization at Mount Gibson, Yilgarn Craton, western Australia: Economic Geology, v. 91, p. 1354-1371.

Zaleski, E., and Peterson, V.L., 1995, Depositional setting and deformation of massive sulfide deposits, iron-formation, and associated alteration in the Manitouwadge Greenstone Belt, Superior Province, Ontario: Economic Geology, v. 90, p. 2244-2261.

Zaleski, E., Froese, E., and Gordon, T.M., 1991, Metamorphic petrology of Fe-Zn-Mg-Al alteration at the Linda volcanogenic massive sulphide deposit, Snow Lake, Manitoba: The Canadian Mineralogist, v. 29, p. 995-1017.

3-D Geological Modelling Supporting Mineral Exploration

Eric de Kemp

Geological Survey of Canada, 615 Booth Street, Ottawa, Ontario, K1A 0E8
Corresponding author's email: edekemp@nrcan.gc.ca

Abstract

Mineral exploration is evolving into a more rigorous quantitative science. 3-D GIS provides support for this activity through an environment in which a rich and diverse set of exploration-related observations can be analyzed and interpreted. The core tools for developing and exploiting mine-, camp-, and regional-scale 3-D common earth models for the purpose of targeting new ore or specific geologic relationships are now here. It is now incumbent on the industry, with its wealth of knowledge of specific ore forming processes, its rich archive of 3-D data sets, and with a definite need to find the difficult and deeper ore, to capitalize on this new technology. Industry, government, and academia are encouraged to develop partnered 3-D interpretation teams focused on interpreting multi-scale 3-D maps in mineral-endowed belts. It is essential that these exercises are supported with appropriate human resources in order for 3-D modelling to achieve its expected goals of enhancing the mineral targeting process and ultimately increasing mineral wealth. New technologies are emerging for automating the model construction process, leaving the expert free to interpret and test exploration criteria. Advances in 3-D joint geologic and geophysical inversion are also contributing to a more rigorous interpretation workflow. When applied by focused multi-scale interpretation teams, this approach has great potential to make a significant impact on enhancing near-mine to greenfields mineral exploration activities.

Résumé

L'exploration minière évolue actuellement vers une science quantitative davantage rigoureuse. Les SIG 3-D fournissent une appui à cette activité en offrant un environnement comprenant un ensemble riche et diversifié d'observations associées à l'exploration qui peuvent être analysées et interprétées. Les outils fondamentaux pour l'élaboration et l'exploitation de modèles 3-D communs de la Terre aux échelles de la mine, du camp et de la région afin de cibler du nouveau minerai ou des relations géologiques spécifiques existent déjà. C'est désormais à l'industrie, avec les connaissances approfondies des processus de formation de minerais spécifiques dont elle dispose, avec ses riches archives de jeux de données 3-D et en raison de la nécessité de trouver des minerais plus difficiles à exploiter et plus profonds, qu'il revient de tirer avantage de cette nouvelle technologie. L'industrie, les gouvernements et le monde universitaire sont encouragés à constituer en partenariat des équipes d'interprétation 3-D axées sur l'interprétation de cartes 3-D à de multiples échelles des zones riches en minéraux. Il est essentiel que ces exercices soient appuyés par les ressources humaines qui conviennent pour que la modélisation 3-D fournisse les résultats attendus quant à l'amélioration du processus de ciblage de minéraux et ultimement quant à l'accroissement de la richesse minérale. On met actuellement au point de nouvelles technologies d'automatisation du processus d'élaboration de modèles laissant aux experts toute la latitude voulue pour l'interprétation et la mise à l'essai des critères d'exploration. Les progrès en inversion géologique et géophysique 3-D combinée contribuent à l'établissement d'un déroulement plus rigoureux du travail. Lorsqu'elle est appliquée par des équipes d'interprétation multi-échelles, cette approche offre de grandes possibilités d'amélioration des activités d'exploration minière, que ce soit à proximité des mines existantes ou en terrain vierge.

Introduction

One of the recent welcome trends in mineral exploration at the deposit scale is the focus on developing common spatial depositories for geologic data, and interpretations in what is called a 'common earth model' (McGaughey and Morrison, 2000). These 3-D knowledge cubes are becoming valuable assets for companies that want to make exploration decisions in a more rigorous and risk-managed environment. There is now a range of tools available to conduct 3-D data visualization, integration, modelling, and targeting with the aim of quantifying key relationships, and estimating where more ore might be located in 3-D space. In addition, perhaps a more exciting development is the possibility that we can start to make geologic linkages between orebodies at wider scales, beyond the mine head frames into more sparse data situations. In the rapidly evolving field of 3-D scientific computing, a few important developments have emerged and are being capitalized on by the industry to better test mineral exploration hypotheses and to communicate what is hidden in the ground to the expert and lay person alike.

As important as these technical developments are, what is most encouraging is the trend toward more strategic corporate 3-D data management plans and the adoption of a team-oriented interpretive approach. 3-D Geographical Information Systems (GIS) are emerging as the new environment in which all the domain experts, stratigraphers, structural geologists, geophysicists, geochemists, and resource geologists, along with the management team, can take better control of their data and enhance its value through specific interpretive and targeting exercises. Although there are a variety of technical and computational challenges, and as always there are limits to what can be done with each given data set, these developments nevertheless represent an exciting leap forward. Certainly, the major mineral exploration companies have moved far beyond the days when the only way to see their complete orebody was from a sliding glass panel display box in the foyer of the mine office. Now we can take advantage of complex 3-D spatial query tools that can express accurately the local exploration model, and analyze and validate those

de Kemp, E.A., 2007, 3-D geological modelling supporting mineral exploration, *in* Goodfellow, W.D., ed., Mineral Deposits of Canada: A Synthesis of Major Deposit Types, District Metallogeny, the Evolution of Geological Provinces, and Exploration Methods: Geological Association of Canada, Mineral Deposits Division, Special Publication No. 5, p. 1051-1061.

query results through drilling, 3-D mapping, or by running validation experiments.

Deposit-Scale 3-D Targeting

The core task of the mineral explorationist is to find new ore. It is thus essential that new targets for where this ore might be are generated, prioritized, and tested. How these targets are generated and the critical reasoning, or lack thereof, that supports a given suite of targets has provided rich fodder for many campfire tales. These days, however, we are facing the ever-increasing challenge to make reasoned targets in which exploration criteria are less direct, possibly with more uncertainty embedded in the supporting information and for more remote or deeper ore. This starts in the less risky areas proximal to existing mines and within the mine itself, extending known ore zones. These are the 'data-dense' exploration domains for which resource assessment geologists and mine explorationists work more closely as a team toward targeting and confirming new ore.

One of the first, and usually most daunting, tasks at the mineral deposit scale is the initial characterization, data conversion, and harmonization of relevant historical and thematic spatial information. 2-D GIS is now mature enough to handle most of these information sources whereby we can represent, query, and extract important spatial elements from proprietary and public spatial data bases. However, it is no easy task to pull together a historical data set in 3-D. The need for quality assessment, data mining, and rescue of key information is a task in itself. In older or abandoned mines, where interest has long waned, there may be reason for a sober second look at the information. These areas can have data from many different map projections, usually with several custom mine grids, each with its own vertical datum. Accurate topographic bases in which data were plotted may have not been available at the time, and the drilling program may have been tied to a prospectors baseline for which no accurate location can be established. For example, in a recently developed 3-D model of the Kidd Creek volcanogenic massive sulphide mine, at least three historical data sets from separate parts of the deposit needed to be transformed and leveled from their individual custom grids into a single UTM-based coordinate system. In smaller deposits, waves of historic activity can result in the complete separation of data sets from one another, such as between underground drilling logs and surface exploration logs, resulting in a serious loss of accuracy to the overall model. Drill core collected through decades of exploration, more often then not, have variable quality and a mixture of lithology codes that requires an expert to harmonize and place in an acceptable classification system (Vandermost and de Kemp, 2006). Most jurisdictions have no obligation to submit assessment information of such drill logs in a given format. Structural data is usually absent in what are clearly structurally controlled orebodies. Massive amounts of drilling in these mines are undertaken, usually for economic considerations with no or little oriented core to assess structural and stratigraphic continuity. Different standards for data collection, different data structures, different depth leveling, lost cadastral information, and the lack of corporate memory and expertise to resolve these problems represents one of the single most important challenges to 3-D modelling exercises. Due consideration to these issues is essential to move forward, and it is important that this be done before 3-D modelling can begin.

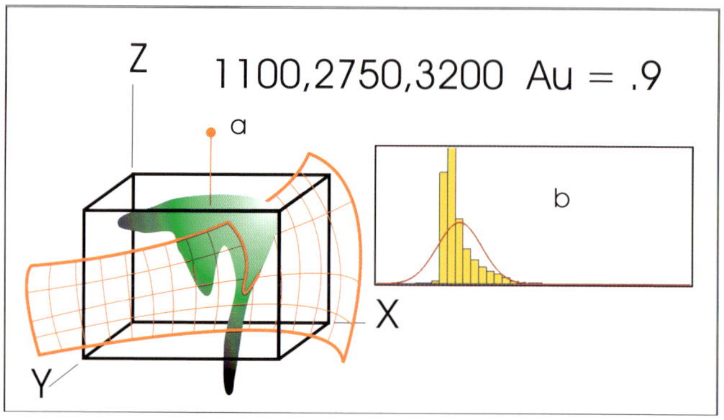

Figure 1. Two simple types of 3-D GIS queries used to examine orebody properties (shaded green region); **(a)** Gold value at a point along a drill core; **(b)** Gold distribution over a specific intersecting ore fault surface.

Mineral explorationists working at the mine and near-mine scale are dealing, on a day-to-day basis, with a host of 3-D geologic problems related to orebody definition. Because these orebodies and the host geologic features are 3-D objects, it is essential that exploration be given a 3-D treatment. In aid of this task, the explorationist now has available some simple spatial functions to investigate key spatial and attribute relationships that will impact exploration criteria for extending ore at the local scale. Functions for extracting the metal content values at specific locations from within an interpolated model (Fig. 1) or for the geometric selection of samples based on proximity measures (Fig. 2) in a true 3-D coordinate space, are used to do initial deposit-scale characterization.

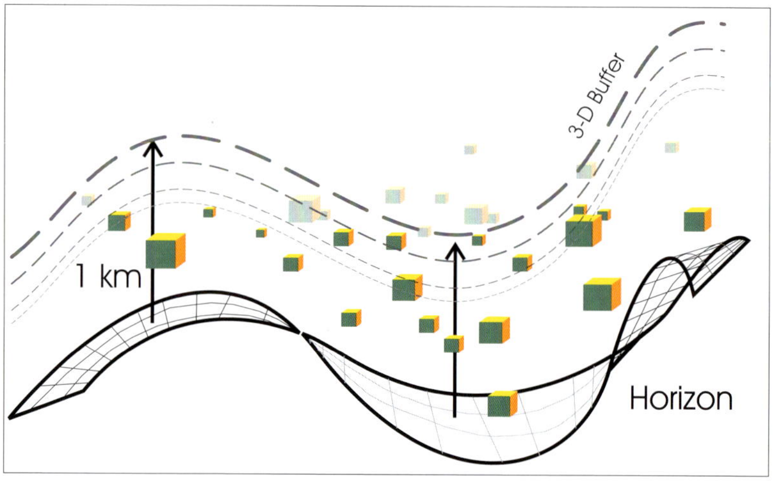

Figure 2. A 3-D buffer query used to select data points from drill core within 1 km vertical direction from a folded horizon.

This kind of analytical work also benefits from a rich set of formula processing tools in which specific calculations, weights, and reclassifications can be undertaken using geochemistry, rock property, and lithology codes. These functions are useful for preparing data for mineral and geochemical modelling using geostatistics or other forms of mathematical estimation (Fig. 3) that produce 3-D continuous models at the mine scale. The key enhancement for mineral exploration is that, with 3-D GIS, the underlying attribute data is linked to a 3-D visualization environment where properties and the relationships of those properties to each other can be interpreted in a more realistic geometric context. The 3-D visualization engine underlying each of these systems is moving toward providing more interpretive support to the explorationist through access to data analysis workflows, more intuitive 2-D and 3-D graphics manipulation, and drawing tools with rich 3-D cartographic quality symbolization sets. Although still in its infancy, a real 3-D mapping environment is emerging, and when used effectively can dramatically enhance the work of the explorationist.

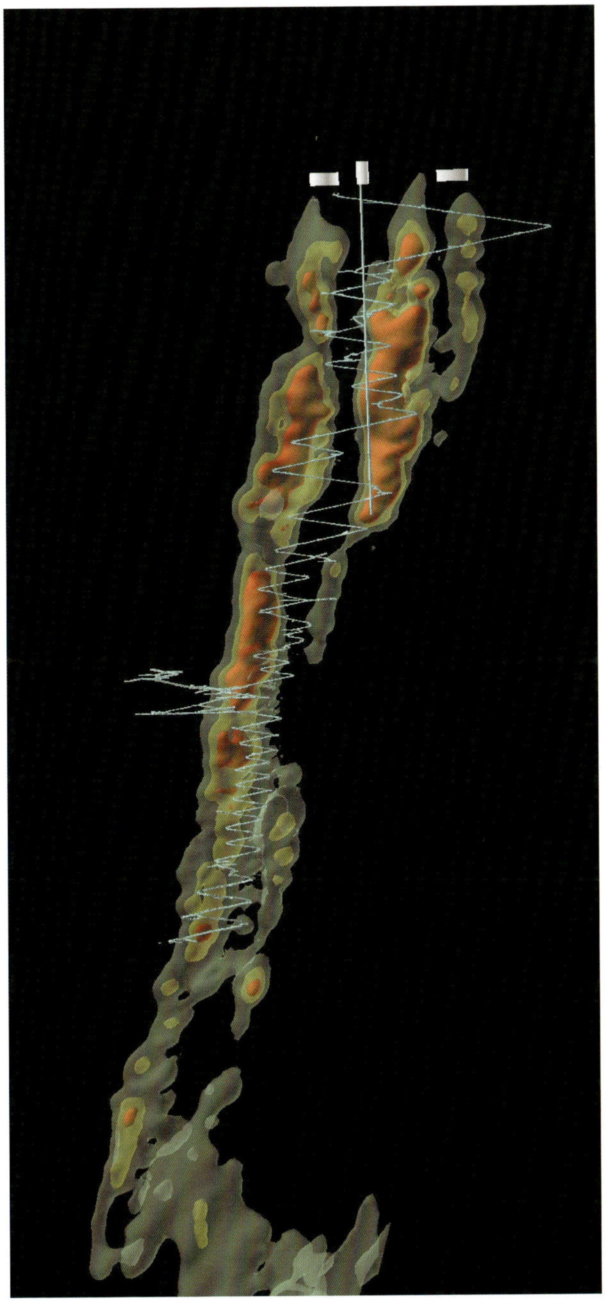

FIGURE 3. Geostatistical simulation of Cu ore showing the model results at 1% Cu (green halos), 2% Cu (yellow halos), and 3% Cu (red halos). Vertical view looking east, which represents a paleostratigraphic map view. Mine ramps and main shaft depicted in blue. Kidd Creek VMS mine Timmins, Canada. (Data provided courtesy of Xstrata Copper.)

A significant development in the mineral exploration community in recent years is the notion of a 'common earth model' borrowed essentially from the best practices of the oil and gas industry (Fig. 4). This is a new paradigm for which our knowledge of a given deposit is stored in a 3-D GIS so that it can be used by the explorationist. In essence, it is a mechanism for attributing the commonly shared 3-D space with all the relevant pieces of information that are needed to make a 3-D targeting exercise meaningful. It is a multi-parameter spatial model that provides the possibility of cross-discipline target validation and an easy means to examine spatial and attribute relationships.

For example, in a common earth model, it is a straightforward task to see if the results of a geophysical inversion match the distribution of physical rock properties collected down a borehole through the model. Determining the effectiveness of an alteration index for ore vectoring can be achieved with a common earth model through a simple cross-plot between the 3-D distance to ore parameter with the co-located values of the alteration index. This type of analysis can be refined so that our common earth model follows structural trends to allow distance calculations and the geostatistics used for estimating continuous geochemical model parameters to occur in the predeformation state (Fig. 5).

3-D spatial query and analysis functions are all data dependant. They do not magically create new insights into ore genesis or generate high-quality targets where the data are

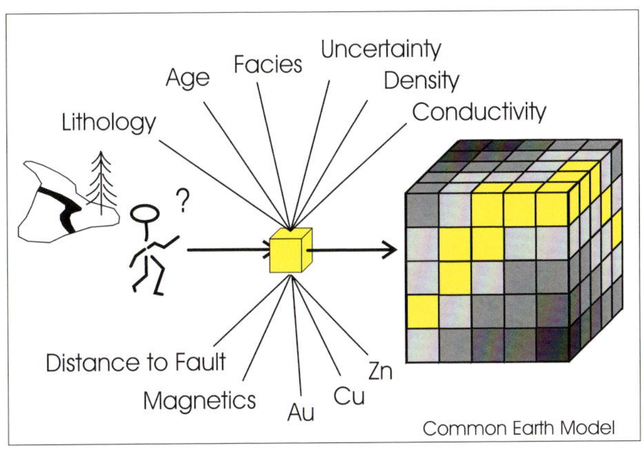

FIGURE 4. Common earth model concept, implemented as a regular partitioned multi-parameter voxel model. Implementations may vary depending on exploration requirements, with cell sizes varying in dimension, regularity, and size. Common earth models may also have a non-space partitioned implementation using an atomic-based vector data structure to model folded and faulted horizons. In that case, properties are mapped to internal vertices.

not sufficiently dense or reliable. Like all geotechnologies, they are simply an extension of the investigative toolbox for the explorationist, and need to be used appropriately and in a reasonable context. At the working level of a mine, mineral exploration targeting is now a fully digital 3-D remote predictive exercise, and is slowly developing a set of best practices or a protocol that explorationists can use (Figs. 6, 7). This protocol is now emerging with many deposit-scale examples, leading the way to a more rigorous and managed approach that uses 3-D GIS as its core technology (Apel, 2006; de Kemp, 2006; Sprague et al., 2006). The keys to victory in these cases are the development of targeting teams that define very clearly the following:

- Why they need to engage in a 3-D modelling exercise,
- What the target is that they are looking for, how it is characterized,
- What pieces of information are required to support specific exploration-oriented queries, and
- How they will judge the success of a targeting exercise.

With experience, as the time and care that are being spent on a 3-D integration effort increases, so should the practical benefits be realized as high-quality targets are identified.

In addition to the practice of quantitative targeting at the mine scale, there is a need for more advanced visualization to support an explorationist's natural ability to make inferences through 3-D visual inspection. The danger, however, in using realistic rendering tools with geoscience data is that the quality and distribution hardly ever justify the portrayal of an interpretation as a highly accurate model. One could ask the same question if the level of detail on 2-D interpreted maps is justified, given the distribution of outcrops and observation sites. The estimation of confidence maps or a quality distribution property is at least a start in embedding the uncertainty estimates into these models in order for the user to have some idea of how speculative or interpreted they are. Nevertheless, in some cases at the mine scale we do have quite extensive control to produce highly accurate models (Fig. 8). In that case, it is essential to have a visualization

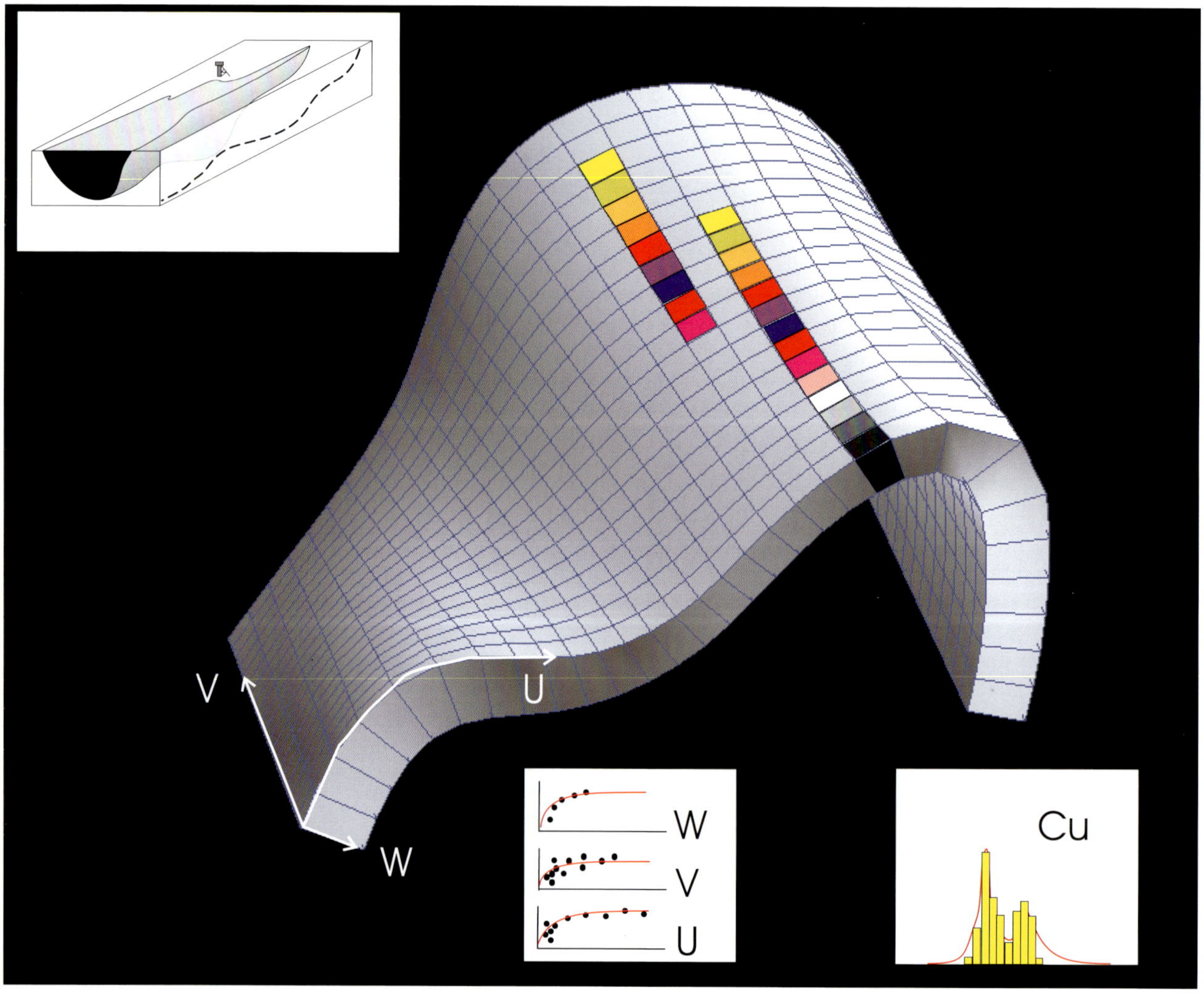

FIGURE 5. Irregular-spaced stratigraphic grid for modelling predeformation ore. Variogram analysis for kridging of 3-D data is undertaken in model space optimized for seeing the horizon in its primary state. Griding is required only along the horizon so more regional-scale models are possible if supported by the data.

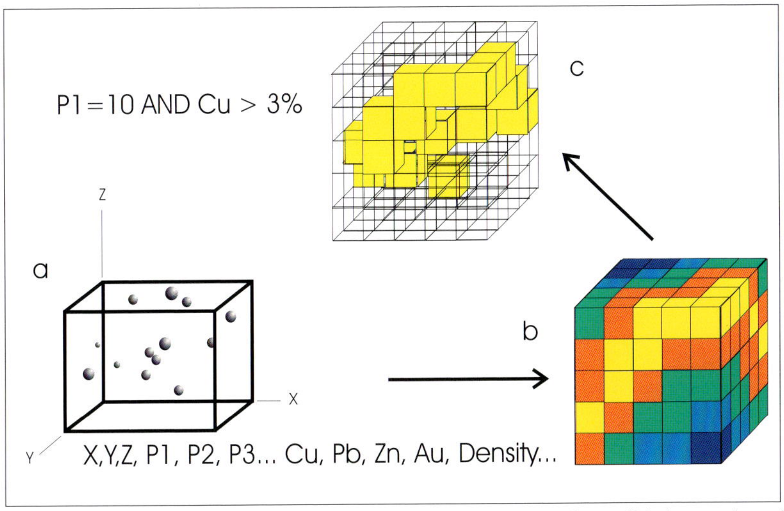

FIGURE 6. An example of a geospatial query, using a property P1, perhaps a lithology code, and a Cu abundance greater then 3 wt %. **(a)** The starting data set, which is a set of samples from drill core, surface, or drifts that support a model estimation; **(b)** A combination of all these estimated parameters in a common earth model; **(c)** A selected set, resulting from a query for which the criteria is true. The exploration target is depicted in yellow.

system that doesn't compromise the representation of key features that are essential to targeting (Caumon et al., 2005). This is especially true when dealing with thin marker horizons, complex folding and faulting, or the presence of linear and localized anomalies.

Regional Scale 'Greenfields' 3-D Interpretation

Most large deposits in the developed world that are exposed at or near the surface have almost certainly been found, and the ones that are detectable with the most common geophysical tools are generally not deep and not very big. This forces all of us in the exploration business to go back to being geologists, because orebodies are geological features. They occur in a geologic setting, within a specific stratigraphic, structural, and tectonic environment (Galley, 2003), and unless we develop a good understanding of the geologic controls and specific spatial distribution of those controls over wider areas, we will not be very successful at predicting deep ore. The task of making useful 3-D geologic models beyond the mine head frames and encompassing a whole mining camp or a region beyond it represents a big leap forward in helping reduce risk for brown and greenfields exploration projects (Fig. 9). Camp-scale, regional-scale, and even crustal-scale studies focused on broad-based resources assessments, tectonic synthesis, and basin analysis are being undertaken by governments in concert with academia and industry (Brun et al., 2001; Goleby et al., 2002; Ross et al., 2004; Le Carlier de Veslud et al., 2005). Several of these 3-D studies, along with the methodology used to develop regional- and camp-scale models (Fallara et al., 2006), have been successfully undertaken and have made a significant positive impact in generating more local exploration activity (Marquis et al., 2003; Fallara and Legault, 2004; Fallara et al., 2004a–c; Rabeau and Legault, 2006).

There is no exact definition of what a regional-scale model size would be, other then it is beyond a few mine head frames at the edge of what would be considered a camp-scale model, probably between 1:10,000 and 1:250,000 scale in extent. This scale also coincides with the 'greenfields' areas referred to by the exploration community. Greenfields is a fuzzy term used for areas in which we have minimal knowledge of the target mineral distribution and which are seen as the most high risk by industry, due to a perceived high cost/benefit ratio (Bowler, 2002). Beyond this we are dealing with large crustal-scale models that are not directly related to targeting or mineral compilation exercise, but nevertheless are important in defining critical metalotects or the geometry of controlling crustal features in a broad geological context. In the regional setting, 3-D data is sparse, either due to poor outcrop or lack of mapping. The data is also irregularly distributed and difficult to use because most of it is not in usable GIS format. Structural field observations that constrain local geometries of faults, horizons, and intrusion contacts and from which depth projections are based, are heterogeneous and of varying quality. Hence, the need to begin thinking about how we model not just the geology but the uncertainty associated with our interpretations (Chilès et al., 2004). There is, ultimately, a limit to what can be done in 3-D with complex geology and a limited data set. In many cases it may be necessary to acquire new truly 3-D data such as from a seismic survey, or to perform systematic field mapping to get a handle on local geometries. With this dramatic data density drop-off beyond the mine head frame, it is necessary to make our best-guess interpretations in the form of cross and longitudinal sections and down plunge projections. These can be done manually or with a drawing package and brought into a 3-D interpretive environment, or the 2-D map data can be imported and visualized directly in a 3-D system such as Gocad® or GeoModeller® and sections interpreted there. Much development is now underway to enhance this process so that an intuitive interpretive environment is available in which there is simultaneous linkage to interpreted objects in 2-D and 3-D.

Despite the challenge in making 3-D regional models due to poor data control, there is nevertheless tremendous potential for benefit (Fig. 10). If nothing else, the exercise will help in understanding the geology, define more clearly missing geological components, and focus field studies. Once the 3-D model is made, the benefits are similar to those realized from making 2-D regional maps. They potentially reduce the level of uncertainty of a given area by providing a first-order framework in which further geological analysis can proceed. They provide a regional 3-D geometric framework for tectonic modelling and mineral prospectivity studies, and can help target areas or blocks for detailed thematic studies. 3-D models, or 3-D maps as they are often referred to, quickly summarize complex geometry and property relationships to management or those not familiar with a given domain. This is not insignificant when funding is often directly dependant on a managers ability to sell project activities inside and outside an organization. Most importantly, 3-D regional models have the potential to link disparate geology from adjacent

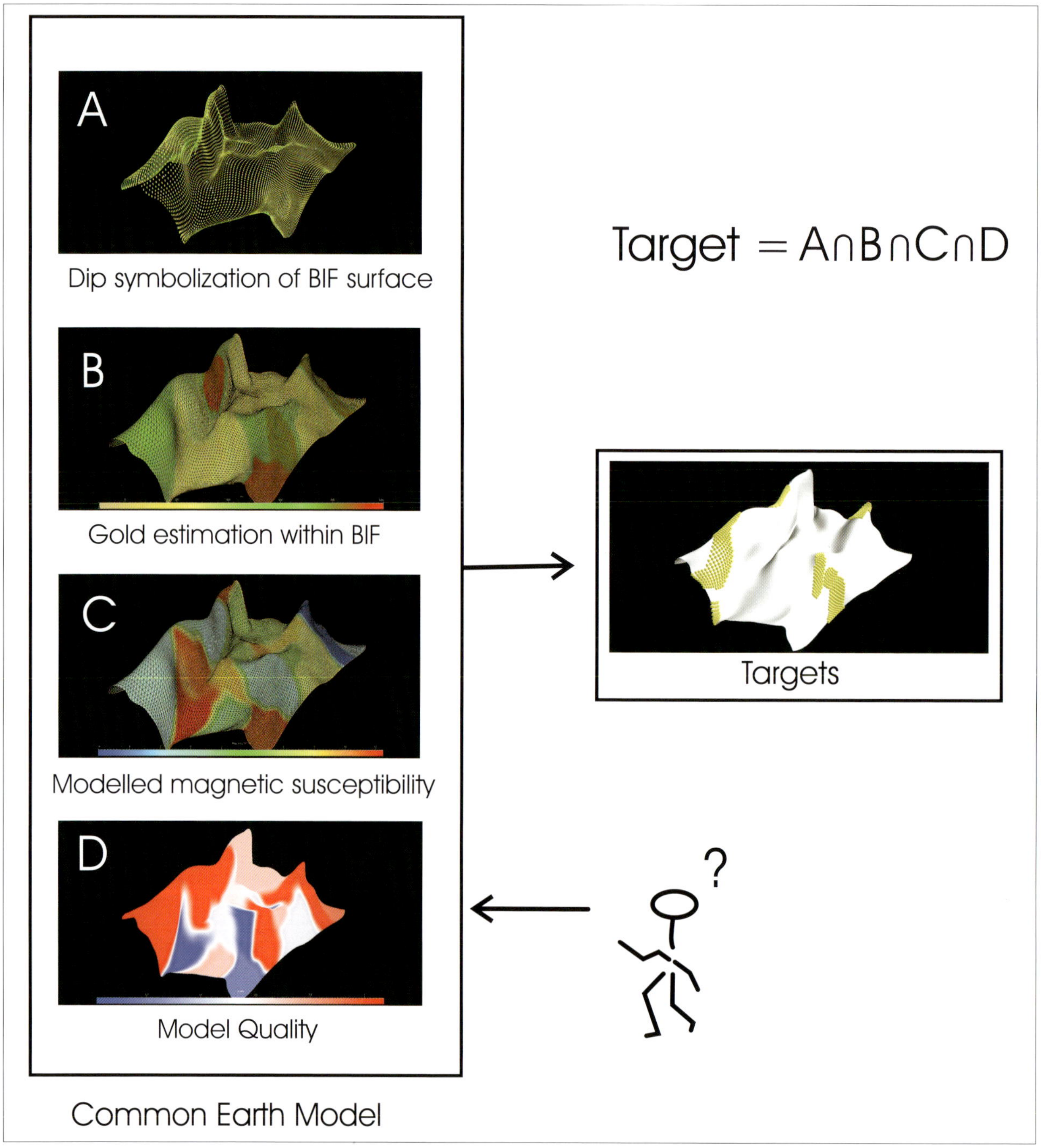

FIGURE 7. A common earth model implemented as a multi-parameterized surface. Each estimated thematic subset from A to D is given a specific value criteria cutoff which is then intersected to form the target result. **(A)** Banded iron formation (BIF) surface model. Exploration criteria could focus on surface geometry characteristics such as hinge zones, dips, ranges, etc. **(B)** Geochemical or assay values from proximal samples can be mapped to the BIF surface and the values interpolated as a surface property. **(C)** Geophysical inversion models could be sampled and mapped to the BIF surface. **(D)** A quality property could be developed based on metrics such as proximity to observations or confidence of historic sample sites. For this example, the target selection was based on near-hinge Au estimates greater then 600 ppm and low magnetic susceptibility (due to gruneritization) with a reasonable quality estimate of greater then 0.7.

fault blocks through more critical geometric analysis, ultimately making the greenfields areas more attractive.

Traditional exploration approaches stick to developing an exploration model from 2-D map-based targets of shallow anomalies, and focus energy as close as possible to the known deposits (Fig. 11). Going outside of the camp is often perceived as risky. With a 3-D interpreted model, the 3-D target is placed in the context of a 3-D geologic object. The exploration strategy places an emphasis on geologic understanding and targeting that looks for deeper ore, in conjunction with constraining the geometry of the geology (Fig. 12). With this methodology we will, at a minimum, gain a much better understanding of stratigraphic and structural relationships that

FIGURE 8. Complex volcanogenic massive sulphide orebody interpreted from dense drill core data. Note thin fingered segments of ore, disconnected ore pods, and folded geometries. Detailed geometries can be lost when gridding at a fixed scale unless cell size is extremely fine, in which case file size can become unmanageable. White contour depth interval is 10 meters; black contour interval is 50 meters.

link geology from familiar to unfamiliar ground, and perhaps even hit more prospective deep mineralization.

One of the great successes in the oil and gas industry is the scale integration of local observations with more regional interpretations. It is not uncommon for a key productive or source horizon to be the geological target that shapes the exploration strategy of a whole field (Srivastava et al., 2004 ; Albertão et al., 2005; Klinkby et al., 2005). This success has largely been due to the seismic industry being able to link and upscale local geological interpretations, from detailed electric borehole logs to regional 2-D and 3-D interpretations derived from seismic surveys. These 3-D models tend to get developed in more mature fields as production wanes and geological knowledge becomes more essential in targeting. Many of the petroleum software workflows are geared for this type of exercise, linking key constraints from observations to the 3-D interpretation and model construction process. Similar upscaling methods are needed for our domain in mining. For example, it should be a simple and smooth operation to link a rock property database from the boreholes of a region to constrain a 3-D geophysical inversion of a wider area or to constrain a regional fault network hierarchy to locally observed crosscutting relationships, while interpretations in the 3-D geologic model are made or adjusted accordingly. These operations are possible now, but they are neither simple nor smooth.

Besides the much needed improvements to software and workflows, there are more fundamental challenges we face in trying to upscale outcrop and borehole interpretations with other mine models and the regional picture. Common geological features are not always present between the various scales, and these features if present may not be easily imaged with remote methods provided by seismic, electro-magnetic, and potential fields methods. Steep structural orientations, high frequency folds, cryptic early structural histories, and poor physical rock property contrasts can all conspire to make seismic and other geophysical methods less then ideal for upscaling our geological interpretations (Woodall and Duncan, 1993). Harmonization of geological interpretations from historical boreholes is an ongoing challenge to upscaling, because various geologists through the years may have used a range of rock classification systems, may have lacked experience, or had a specialty which resulted in inappropriate classifications. This all points to the fact that, for the most part, exploration models are not very well scale integrated. In order for the benefits of 3-D modelling to be achieved this has to be improved. There needs to be a very strong linkage between the local-scale interpretations from outcrop, mine mapping, and borehole logs to camp- and regional-scale mapping activities. This is not easy when various personnel, and occasionally competing companies or agencies, are working on adjacent ground. However, by combining resources in a regional or greenfields 3-D model interpretation exercise, this can be mitigated. Once conversations begin between those working at the local mine site who do detailed facies, structural analysis, and geochemical work, and those trying to solve bigger regional-scale stratigraphic problems, such as fault network reconstruction and mineral potential analysis,

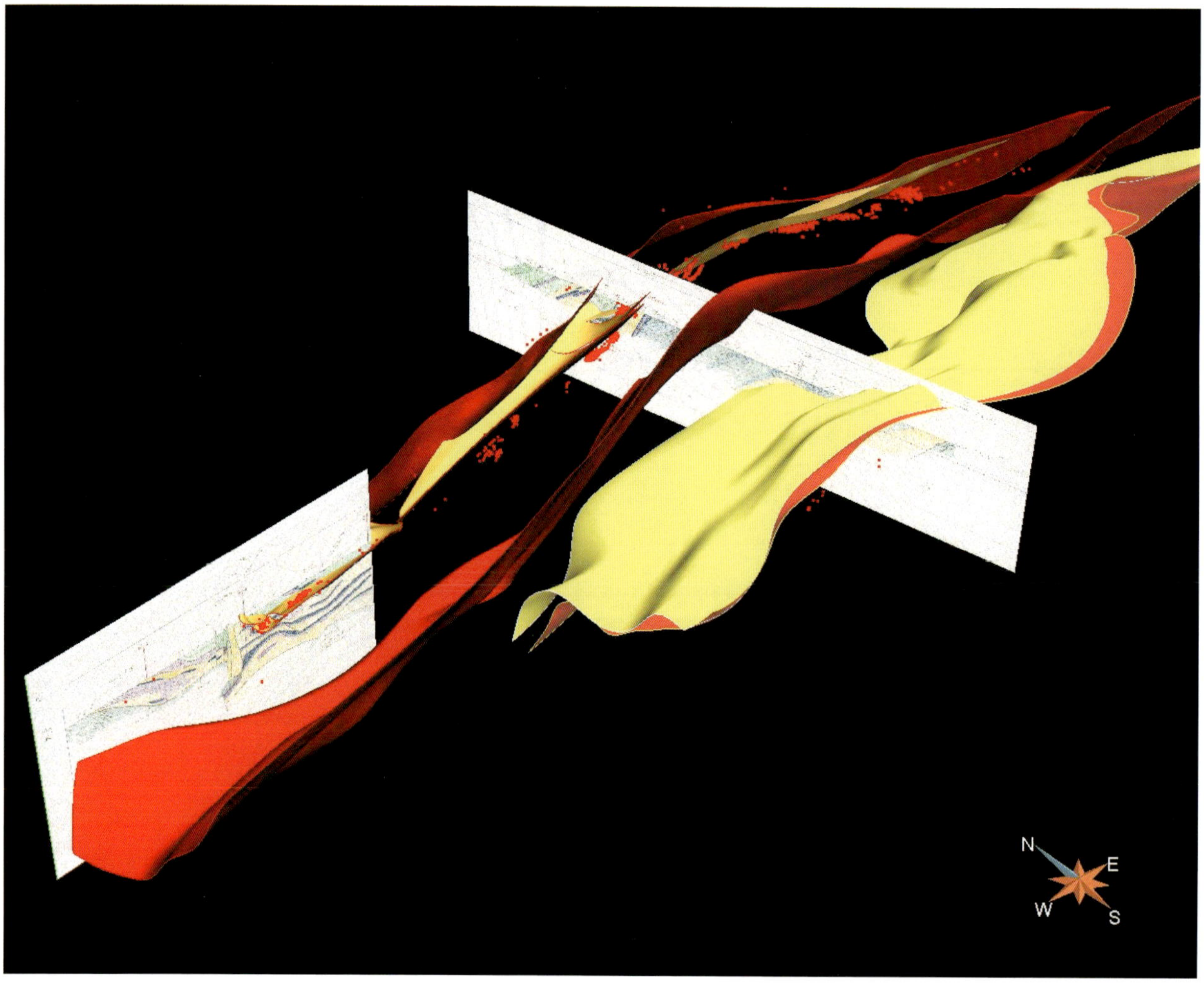

FIGURE 9. 3-D regional structural model, Kattiniq (Raglan Mine) region, Nunavik territory of northern Québec. Thrust faults (red surfaces) and horizon tops (yellow surfaces) developed with SPARSE© using GSC map data and interpreted cross sections. Model is roughly 80 km × 30 km and 8 km depth (de Kemp et al., 2007).

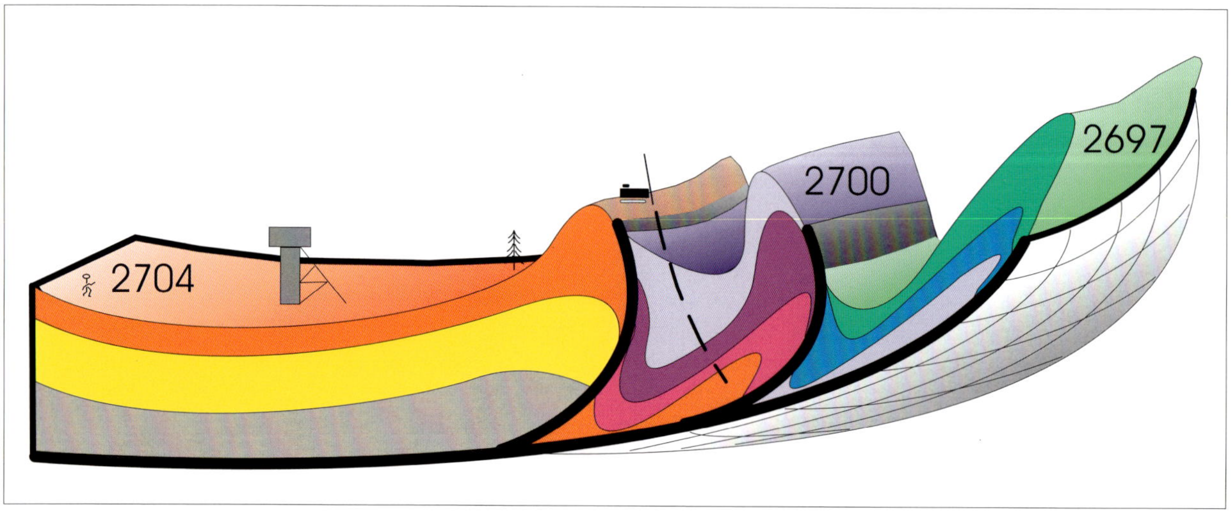

FIGURE 10. Simple structural and stratigraphic 3-D model or 3-D map that is used as a geometric framework into which properties can be embedded. Numbers indicate the absolute age of individual stratigraphic units. Example depicts deep drilling to confirm a favorable stratigraphic target. Geology visualization and query representation, (i.e., single property range, facies, or lithologies) when doing geochemical, geophysical, or mineral potential modelling, can be controlled separately with different fault blocks or a combination of fault blocks.

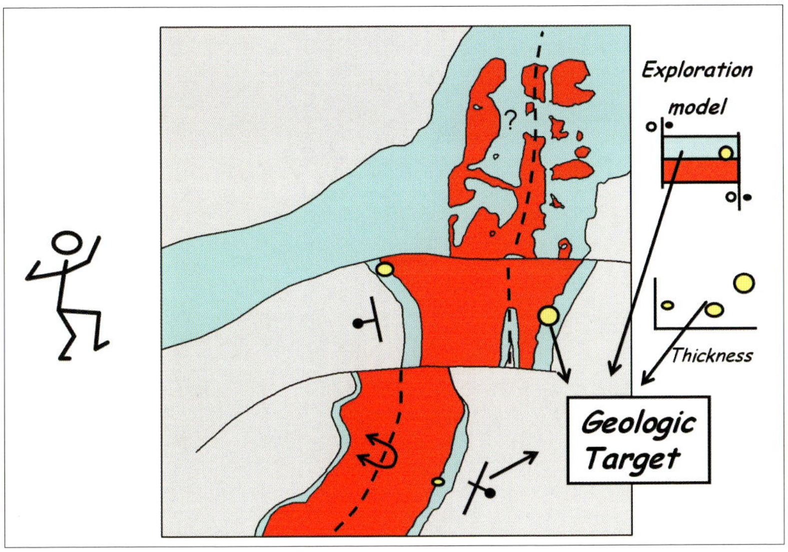

FIGURE 11. Traditional 2-D map-based approach to regional exploration targeting. All exploration criteria are combined into a set of relationships between geologic features and the significant occurrences or existing mines. Mineral occurrences or deposits depicted as yellow dots with size indicating relative mineral abundance. Without structural expertise, it is difficult to interpret the 2-D geologic target and its relationship to the occurrences. Example shown of polydeformed stratiform ore in the limbs of an anticline. Note fault block at top of map without known mineral occurrences qualifies as a 'greenfield' area.

exploration models will become truly scale integrated. The resultant models that emerge from these kinds of mixed-scale 3-D interpretation teams will tend to be more useful for reducing greenfields risk for explorationists, and even for deeper targeting of geological structural and stratigraphic controls or actual deposits themselves.

Future Trends

The notion that we can develop 3-D knowledge cubes or regions of 'glass earth' in which complex geology and its deeply hidden treasures can be revealed has captured the imagination of many. Perhaps more acutely, those (usually in upper-management) who see the instant perceptive appeal and educational benefit to stakeholders are easily enamoured of the idea that producing 3-D spinning models on web sites and DVDs is of great benefit to all. The reality is that this enthusiasm generally wanes, when there is somber consideration of the monumental effort and staff commitment required to qualify, integrate, and compile the working information needed as a starting point before attempting to make these models. Often there is a great disconnect between what people expect a model can do for them and what it will deliver in terms of specific relationship information or real exploration-oriented analysis possibilities. This is why it is essential to define up front what type of 3-D model will be made, who are the clients for such a model, and what resources in terms of technical staff familiar with 3-D methods and geologists with expert knowledge of the region, are available to develop it. This 3-D capacity development, as well as the research gap in developing new,

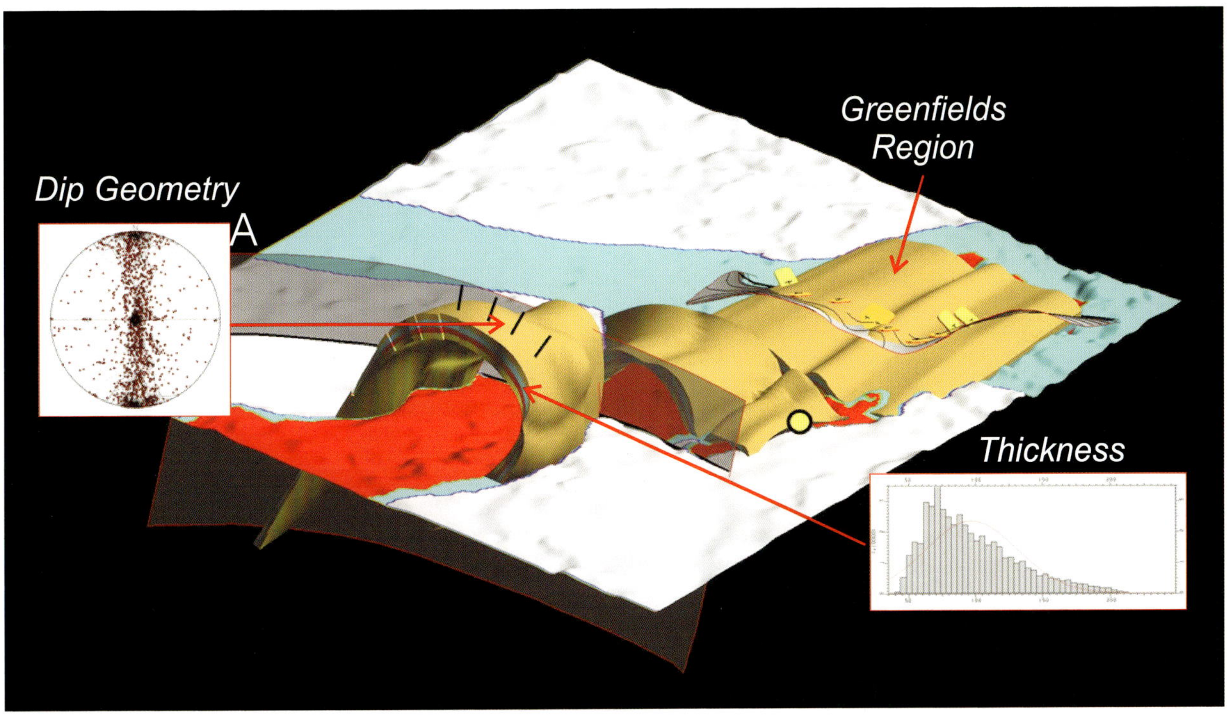

FIGURE 12. Example of an interpreted 3-D regional model from 2-D map data shown in Figure 11, presenting a more complete and intuitive geometric framework from which targeting can take place, perhaps even reducing risk in greenfields areas. Note structural style from the right, open near cylindrical folding, middle fault block characterized by a tighter doubly plunging antiformal structure, far left block bounds a dextrally offset antiform with overturned limb. Variability of thickness of target horizon depicted in histogram of the 3-D startigraphic model. Local mine site depicted as yellow dot.

more efficient methods for 3-D interpretation represents a much bigger challenge that the entire industry and academia face.

On the brighter side, there is a wide range of 3-D modelling research and development activity that will have direct impact on the mineral industry in the near future. At the mine scale, there exist new techniques for ore estimation and geochemical modelling (Journel and Kyriakidis, 2004), 3-D spatial targeting (Apel, 2006; Böhme and Apel, 2006), joint geophysical and geological inversion (Bosch et al., 2001, 2002), automating geologic reconstruction (Courrioux et al., 2001; Mallet, 2002; Guillen et al., 2004; Smirnoff et al., 2006, 2007; Frank et al., 2007), restoration (Dunbar and Cook, 2003; Moretti et al., 2006), and simulation of geological process (Jessell et al., 2001; Hobbs et al., 2007). The joint-modelling methods try to produce structural models that balance the results of several types of inversions using composite geological and geophysical fitness tests. All of these inversion methods individually produce non-unique solutions and are driven by underlying mathematical assumptions that are simplifications of the real world (Courrioux et al., 2001; Jessell, 2001; Putz et al., 2006). However, they may converge to a better solution space if they are used collectively. In this approach, a geological structural and stratigraphic model would be able to match the modelled rock property distribution from a geophysical inversion and visa versa (Brun et al., 2001; Guillen et al., 2004; Martelet et al., 2004). These are of course exciting developments for the mineral exploration community. However, it takes significant time and energy to move from early adoption of new technology by a few mavericks to putting in place a common best practice for our industry. Managers and policy makers are thus strongly encouraged to enhance any partnerships which try to speed up the delivery of these 3-D exploration technologies so that we can all take advantage of this opportunity.

Conclusions

New geotechnologies for the mineral exploration community are constantly emerging, but the one key technology that is demonstrating the ability to pull all these developments together under one roof is 3-D GIS. It is where geological modelling can be undertaken in a rigorous and quantitative way. With 3-D GIS, there is an opportunity to make data-driven or knowledge-driven predictions and interpretations that can be tested through cross-validation experiments or direct drilling of targets. It is an environment that, in the coming years, will see a richer set of tools and workflows for specific types of exploration tasks and deposit types. It is an environment that will likely follow the same trend as 2-D GIS has in the past, becoming more user friendly, more of an integration tool, and developing more relevant applications that are connected to the business case required for modern exploration.

Acknowledgements

The contributions of data and expertise from Xstrata Copper and Mira Geoscience is gratefully acknowledged. This research was supported in large part through funding from the Geological Survey of Canada, Targeted Geoscience Initiative (TGI-3), Deep Search project. Many thanks to S. Paradis whose comments greatly improved the paper. Earth Science Sector contribution number 20070072

References

Albertão, G.A., Grell, A.P., Badolato, D., and Dos Santos, L.R., 2005, 3D geological modeling in a turbidite system with complex stratigraphic-structural framework—An example from Campos Basin, Brazil [abs.]: Society of Petroleum Engineers, SPE 2005 Annual Technical Conference and Exhibition, Proceedings, p. 1015–1024.

Apel, M., 2006, Predict: A 3-D weights of evidence mineral potential modelling plug-in to Gocad: http://www.geo.tu-freiberg.de/~apelm/predict.htm.

Böhme, M., and Apel, M., 2006, Predictive 3-D mineral potential modelling of the Noranda district, Canada: Annual Gocad for Mining Users Meeting, Montréal, Oct. 4-6, 2006, Presentation.

Bosch, M., Guillen, A., and Ledru, P., 2001, Lithologic tomography: An application to geophysical data from the Cadomian belt of northern Brittany, France: Tectonophysics, v. 331 p. 197–227.

Bosch, M., Zamora, M., and Utama, W., 2002, Lithology discrimination from physical rock properties: Geophysics, v. 67, p. 573–581.

Bowler, J., 2002, Ministerial inquiry into Greenfields Exploration in Western Australia: Western Australia Department of Industry and Resources, 146 p.

Brun, J.-P., Guennoc, P., Truffert, C., and Vairon, J., 2001, Cadomian tectonics in northern Brittany: A contribution of 3-D crustal-scale modelling: Tectonophysics, v. 331 (1–2), p. 229–246.

Caumon, G., de Kemp, E.A., and Bosquet, F., 2005, Visualization of 3-D geological maps: An example using volumetric clipping with hardware [abs.]: International Association of Mathematical Geology, Annual Conference, Toronto, Canada, August 21–26, Proceedings, v. 1, p. 261–268.

Chilès, J.P., Aug, C., Guillen, A., and Lees, T., 2004, Modelling the geometry of geological units and its uncertainty in 3D from structural data: The potential field method: Australasian Institute of Mining and Metallurgy, Orebody Modelling and Strategic Mine Planning [abs.]: Uncertainty and Risk Management, Perth, WA, November 22–24, 2004, Proceedings, p. 313–320.

Courrioux, G., Nullans, S., Guillen, A., Boissonat, J. D., Repusseau, P., Renaud, X., and Thibaut, M., 2001, 3D volumetric modelling of Cadomian terranes (Northern Brittany, France): An automatic method using voronoi diagrams: Tectonophysics, v. 331, p. 186–196.

de Kemp, E.A., 2006, 3-D Interpretive Mapping: An Extension of GIS Technologies for the Geoscientist: Harris, J.R. ed., GIS for the Earth Sciences, Geological Association of Canada Special Publication, v. 44, p. 591–612.

de Kemp, E.A., Sanborn-Barrie, M., and St-Onge, M.R., 2007, 3-D model Kattiniq (Raglan Mine) region *in* St-Onge, M.R., Lamothe, D., Henderson, I., Ford, A., eds., Digital geoscience atlas of the Cape Smith Belt and adjacent domains, Ungava Peninsula, Quebec-Nunavut: Geological Survey of Canada, Open File, v. 5117 (DVD).

Dunbar, J.A., and Cook, R.W., 2003, Palinspastic reconstruction of structure maps: An automatied finite element approach with heterogeneous strain: Journal of Structural Geology, v. 26, p. 1021–1036.

Fallara, F., and Legault, M., 2004, Modele 3D Geo-integre de joutel, synthese metallogenique (phase 2/2): MRNFP, Report 3D 2004-03, CD-ROM.

Fallara, F., LaFrance, B., Cheng, L.Z., Boudrais, G., Cote, J., Bedard, N., Lei, Y., Dube, B., and Galley, A.G., 2004a, Modele 3D geo-integre de la Formation de Bousquet, zone volcanique sud de la sous-province de l'Abitibi, Québec: MRNFP, Report 3D, 2004-04, CD-ROM.

Fallara, F., Legault, M., Cheng, L.Z., Rabeau, O., and Goutier, J., 2004b, Modele 3D geo-integre d'un segment de la Faille de Porcupine-Destor, synthese metallogenique de duparquet (phase 1/2) MRNFP, Report 3D 2004-01, CD-ROM.

Fallara, F., Rheaume, P., Cheng, L.Z., Boudrias, G., and Bandyayera, D., 2004c, Modele 3D geo-integre du secteur du Lac aux Loutres, synthese metallogenique d'Urban-Barry (phase 1/2): MRNFP, Report 3D 2004-02, CD-ROM.

Fallara, F., Legault, M., and Rabeau, O., 2006, 3-D integrated geological modeling in the Abitibi subprovince (Québec, Canada): Techniques and applications: Exploration and Mining Geology, v. 15 (1–2), p. 27–42.

Frank, T., Tertois, A.-L., and Mallet, J.-L., 2007, 3D reconstruction of complex geological interfaces from irregularly distributed and noisy point data: Computers and Geosciences, in press.

Galley, A.G., 2003, Composite synvolcanic intrusions associated with Precambrian VMS-related hydrothermal systems: Mineralium Deposita, v. 38(4), p. 443-473.

Goleby, B.R., Korsch, R.J., Fomin, T., Bell, B., Nicoll, M.G., Drummond, B.J., and Owen, A.J., 2002, Preliminary 3-D geological model of the Kalgoorlie region, Yilgarn Craton, Western Australia, based on deep seismic-reflection and potential-field data: Australian Journal of Earth Sciences, v. 49(6), p. 917–933.

Guillen, A., Courrioux, G., Calcagno, P., Lane, R., Lees, T., and McInerney, P., 2004, Constrained gravity 3D litho-inversion applied to Broken Hill [abs.]: Australian Society of Exploration Geophysicists, 17th Geophysical Conference and Exhibition, Sydney, August 15–19, 2004, Proceedings, p. 6.

Hobbs, B., Regenauer-Lieb, K., and Ord, A., 2007, Thermodynamics of folding in the middle to lower crust: Geology, v. 35, p. 175–178.

Jessell, M., 2001, Three-dimensional geological modelling of potential-field data: Computers and Geosciences, v. 27, p. 455–465.

Jessell, M., Bons, P., Evans, L., Barr, T.D., and Stuewe, K., 2001, Elle; The numerical simulation of metamorphic and deformation microstructures: Computers and Geosciences, v. 27, p. 17–30.

Journel, A.G., and Kyriakidis, P.C., 2004, Evaluation of mineral reservers: A simulation approach: London, Oxford University Press, 216 p.

Klinkby, L., Kristensen, L., Nielsen, E.B., Zinck-Jørgensen, K., and Stemmerik, L., 2005, Mapping and characterization of thin chalk reservoirs using data integration: The Kraka field, Danish North Sea: Petroleum Geoscience, v. 11, p. 113–124.

Le Carlier de Veslud, C., Bourgeois, O., Diraison, M., and Ford, M., 2005, 3D stratigraphic and structural synthesis of the Dannemarie basin, Upper Rhine Graben: Bulletin de la Societe Geologique de France, v. 176, p. 433–442.

Mallet, J.L., 2002, Geomodeling, Oxford University Press, 599 p.

Marquis, R., Bois, D., and McGaughey, J., 2003, Quantitative geology using 3D common-earth modelling [abs.]: Prospectors and Developers Association of Canada, PDAC Annual Convention, Toronto, Canada, March 12, 2003, p. 5.

Martelet, G., Calcagno, P., Gumiaux, C., Truffert, C., Bitri, A., Gapais, D., and Brun, J.P., 2004, Integrated 3D geophysical and geological modelling of the Hercynian suture zone in the Champtoceaux area (south Brittany, France): Tectonophysics, v. 382, p. 117–128.

McGaughey, J., and Morrison, K., 2000, The common earth model: A revolution in exploration data integration [abs.]: Prospectors and Developers Association of Canada, Mining Millenium 2000, Toronto, March 5–10, 2000.

Moretti, I., Lepage, F., and Guiton, M., 2006, KINE3D: A new 3D restoration method based on a mixed approach linking geometry and geomechanics: Oil and Gas Science and Technology, v. 61, p. 277–289.

Putz, M., Stuwe, K., Jessell, M., and Calcagno, P., 2006, Three-dimensional model and late stage warping of the Plattengneis shear zone in the Eastern Alps: Tectonophysics, v. 412, p. 87–103.

Rabeau, O., and Legault, M., 2006, Étude métallogénique et modélisation 3D de la Faille de Cadillac dans le secteur de Rouyn-Noranda. MRNFP, RP 2006-03, 8 p.

Ross, M., Parent, M., and Lefebvre, R., 2004, 3D geologic framework models for regional hydrogeology and land-use management: A case study from a Quaternary basin of southwestern Quebec, Canada Hydrogeology Journal, v. 13, p. 690–707.

Smirnoff, A., Boisvert, E., and Paradis, S.J., 2006, Building 3D geological models with the support vector machine [abs.]: Geological Society of America, Annual Meeting, Philadelphia, 22–25 October 2006, p. 39-6.

——2007, Support vector machine for 3D modeling from sparse geological information of various origins: Computers and Geosciences, in press.

Sprague, K., de Kemp, E., Wong, W., McGaughey, J., Perron, G., and Barrie, T., 2006, Spatial targeting using queries in a 3-D GIS environment with application to mineral exploration: Computers and Geosciences, v. 32, p. 396–418.

Srivastava, A.K., Samanta, B.G., Singh, V., and Sen, G., 2004, Utilization of seismic attributes for reservoir mapping: A case study from the Cambay Basin, India: First Break, v. 22(4), p. 31–37.

Vandermost, A., and de Kemp, E.A., 2006, Increasing the interpretive potential and mineral prospectivity through 3-D data exploration, property characterization, and exploration model reconciliation [abs.]: Society of Economic Geologists, Annual Meeting, Keystone, CO, 14–16 May, 2006, p. 387–389.

Woodall, R., and Duncan, I.J., 1993, The third dimension: A geoscience challenge for the 21st century: Australian Institute of Mining and Metallurgy, Parkville, Victoria, Australia (AUS), 39–40 p.